Student, Teacher, and Parent One-Stop Internet Resources

Log on to physicspp.com

ONLINE STUDY TOOLS

- Problem of the Week
- Section Self-Check Quizzes
- Chapter Review Tests
- Standardized Test Practice
- Vocabulary PuzzleMaker
- Multilingual Science Glossary
- Concepts in Motion
- Personal Tutors

ONLINE RESEARCH

- Prescreened Web Links
- Internet Physics Labs
- Alternate CBL™ Lab Instructions
- In the News

ONLINE STUDENT EDITION

- Complete Student Edition available at physicspp.com

FOR TEACHERS

- Teacher Forum
- Teaching Today—Professional Development

SAFETY SYMBOLS

SAFETY SYMBOLS	HAZARD	EXAMPLES	PRECAUTION	REMEDY
DISPOSAL	Special disposal procedures need to be followed.	Certain chemicals, living organisms	Do not dispose of these materials in the sink or trash can.	Dispose of wastes as directed by your teacher.
BIOLOGICAL	Organisms or other biological materials that might be harmful to humans	Bacteria, fungi, blood, unpreserved tissues, plant materials	Avoid skin contact with these materials. Wear mask or gloves.	Notify your teacher if you suspect contact with material. Wash hands thoroughly.
EXTREME TEMPERATURE	Objects that can burn skin by being too cold or too hot	Boiling liquids, hot plates, dry ice, liquid nitrogen	Use proper protection when handling.	Go to your teacher for first aid.
SHARP OBJECT	Use of tools or glassware that can easily puncture or slice skin	Razor blades, pins, scalpels, pointed tools, dissecting probes, broken glass	Practice common-sense behavior and follow guidelines for use of the tool.	Go to your teacher for first aid.
FUME	Possible danger to respiratory tract from fumes	Ammonia, acetone, nail polish remover, heated sulfur, moth balls	Make sure there is good ventilation. Never smell fumes directly. Wear a mask.	Leave foul area and notify your teacher immediately.
ELECTRICAL	Possible danger from electrical shock or burn	Improper grounding, liquid spills, short circuits, exposed wires	Double-check setup with teacher. Check condition of wires and apparatus.	Do not attempt to fix electrical problems. Notify your teacher immediately.
IRRITANT	Substances that can irritate the skin or mucous membranes of the respiratory tract	Pollen, moth balls, steel wool, fiberglass, potassium permanganate	Wear dust mask and gloves. Practice extra care when handling these materials.	Go to your teacher for first aid.
CHEMICAL	Chemicals that can react with and destroy tissue and other materials	Bleaches such as hydrogen peroxide; acids such as sulfuric acid, hydrochloric acid; bases such as ammonia, sodium hydroxide	Wear goggles, gloves, and an apron.	Immediately flush the affected area with water and notify your teacher.
TOXIC	Substance may be poisonous if touched, inhaled, or swallowed.	Mercury, many metal compounds, iodine, poinsettia plant parts	Follow your teacher's instructions.	Always wash hands thoroughly after use. Go to your teacher for first aid.
FLAMMABLE	Flammable chemicals may be ignited by open flame, spark, or exposed heat.	Alcohol, kerosene, potassium permanganate	Avoid open flames and heat when using flammable chemicals.	Notify your teacher immediately. Use fire safety equipment if applicable.
OPEN FLAME	Open flame in use, may cause fire.	Hair, clothing, paper, synthetic materials	Tie back hair and loose clothing. Follow teacher's instruction on lighting and extinguishing flames.	Notify your teacher immediately. Use fire safety equipment if applicable.

Eye Safety
Proper eye protection should be worn at all times by anyone performing or observing science activities.

Clothing Protection
This symbol appears when substances could stain or burn clothing.

Radioactivity
This symbol appears when radioactive materials are used.

Handwashing
After the lab, wash hands with soap and water before removing goggles.

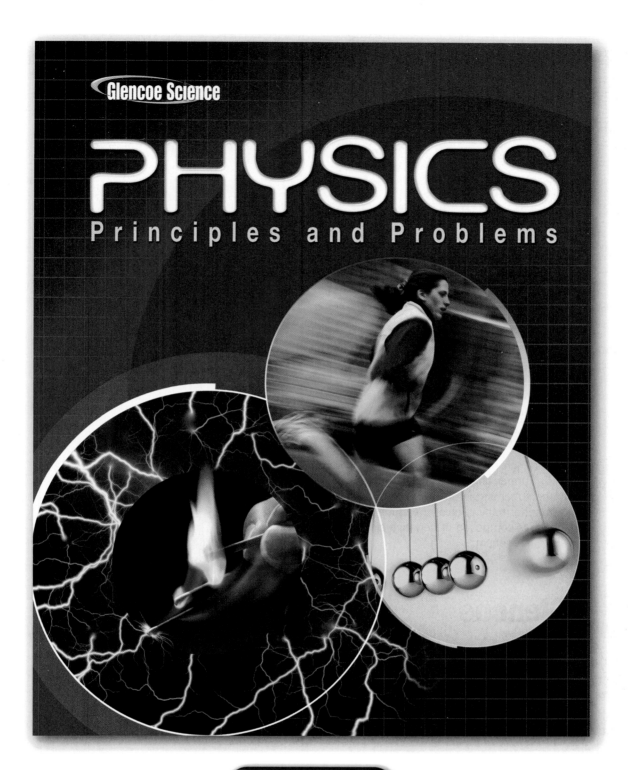

Glencoe Science

PHYSICS

Principles and Problems

AUTHORS

Zitzewitz ▪ Elliott ▪ Haase ▪ Harper ▪ Herzog
Nelson ▪ Nelson ▪ Schuler ▪ Zorn

Glencoe

A Glencoe Program

PHYSICS

Principles and Problems

Physics Online

Visit the Physics Web site

physicspp.com

You'll find: Applets, Animated Illustrations, Personal Tutors, Problem of the Week, Standardized Test Practice, Section Self-Check Quizzes, Chapter Review Tests, Online Student Edition, Web Links, Internet Physics Labs, Alternate CBL™ Lab Instructions, Vocabulary PuzzleMaker, In the News, Textbook Updates, Teacher Forum, Teaching Today—Professional Development **and much more!**

Cover Images Each cover image features a major concept taught in physics. The runner and the colliding spheres represent motion. In addition, the spheres demonstrate the conservation of momentum. Fire represents thermodynamics—the study of thermal energy—and lightning, which is composed of negative electric charges, represents electricity and magnetism.

The McGraw-Hill Companies

Glencoe

Send all inquiries to:
Glencoe/McGraw-Hill
8787 Orion Place
Columbus, OH 43240-4027

ISBN: 978-0-07-880721-3
MHID: 0-07-880721-2

Printed in the United States of America.

2 3 4 5 6 7 8 9 10 027 10 09 08

Paul W. Zitzewitz, lead author, is a professor of physics at the University of Michigan–Dearborn. He received his B.A. from Carleton College, and his M.A. and Ph.D. from Harvard University, all in physics. Dr. Zitzewitz has taught physics to undergraduates for 32 years, and is an active experimenter in the field of atomic physics with more than 50 research papers. He was named a Fellow of the American Physical Society for his contributions to physics and science education for high school and middle school teachers and students. He has been the president of the Michigan section of the American Association of Physics Teachers and chair of the American Physical Society's Forum on Education.

Todd George Elliott C.E.T., C.Tech., teaches in the Electrotechnology Department at Mohawk College of Applied Arts and Technology, Hamilton, Ontario, Canada. He received technology diplomas in electrical and electronics engineering technology from Niagara College. Todd has held various positions in the fields of semiconductor manufacturing, optical encoding, and electrical design. He is a pioneer in the field of distance education and is a developer of electrical/electronic technology courses, and works closely with major community colleges.

David G. Haase is an Alumni Distinguished Undergraduate Professor of Physics at North Carolina State University. He earned a B.A. in physics and mathematics at Rice University and an M.A. and a Ph.D. in physics at Duke University where he was a J.B. Duke Fellow. He has been an active researcher in experimental low temperature and nuclear physics. He teaches undergraduate and graduate physics courses and has worked many years in K-12 teacher training. He is the founding director of The Science House at NC State which annually serves over 3000 teachers and 20,000 students across North Carolina. He has co-authored over 100 papers in experimental physics and in science education. He is a Fellow of the American Physical Society. He received the Alexander Holladay Medal for Excellence, NC State University; the Pegram Medal for Physics Teaching Excellence; and was chosen 1990 Professor of the Year in the state of North Carolina by the Council for the Advancement and Support of Education (CASE).

Kathleen A. Harper is an instructional consultant with Faculty & TA Development and an instructor in physics at The Ohio State University. She received her M.A. in physics and B.S. in electrical engineering and applied physics from Case Western Reserve University, and her Ph.D. in physics from The Ohio State University. Her research interests include the teaching and learning of problem-solving skills and the development of alternative problem formats.

Michael R. Herzog consults for the New York State Education Department on physics curriculum and test development, after having taught physics for 27 years at Hilton Central High School. He holds a B.A. in physics from Amherst College and M.S. and M.A. degrees in engineering and education from the University of Rochester. He serves on the executive committee of the New York State Section of AAPT and is a founding member of the New York State Physics Mentors organization.

Jane Bray Nelson teaches at University High School in Orlando, Florida. She received her bachelor's degree from Florida State University, and her M.A. from Memphis State University. She is a National Board Certified Teacher in Adolescents and Young Adults–Science. She has received a Toyota TAPESTRY Award and a Tandy Scholars Award. In addition, she has received the Disney American Teacher Award in Science, the National Presidential Award for Science Teaching, the Florida High School Science Teacher Award, and been inducted into the National Teacher's Hall of Fame.

Jim Nelson teaches at University High School in Orlando, Florida. He received his bachelor's degree in physics from Lebanon Valley College and M.A.'s in secondary education from Temple University and in physics from Clarkson University. He has received the AAPT Distinguished Service Award, the AAPT Excellence in Pre-College Physics Teaching award, and the National Presidential Award for Science Teaching. Jim is the PI of the Physics Teaching Resource Agent program, and has served on the executive board of AAPT as high school representative and as president.

Charles A. Schuler is a writer of textbooks about electricity, electronics, industrial electronics, ISO 9000, and digital signal processing. He taught electronics and electrical engineering technology at California University of Pennsylvania for 30 years. He also developed a course for the Honors Program at California called "Scientific Inquiry." He received his B.S. from California University of Pennsylvania and his Ed.D. from Texas A&M University, where he was an NDEA Fellow.

Margaret K. Zorn is a science and mathematics writer from Yorktown, Virginia. She received an A.B. from the University of Georgia and an M.Sc. in physics from the University of Florida. Ms. Zorn previously taught algebra and calculus. She was a laboratory researcher in the field of detector development for experimental particle physics.

Contributing Writers

Contributing writers provided labs, standardized test practice, features, as well as problems for the additional problems appendix.

Christa Bedwin
Science Writer
Montreal, Canada

Thomas Bright
Physics Teacher
Concord High School
Concord, NC

David C. Haas
Science Writer
Granville, OH

Pat Herak
Science Teacher
Westerville City Schools
Westerville, OH

Mark Kinsler, Ph.D.
Science/Engineering Writer
Lancaster, OH

David Kinzer
Optical Engineer, Science Writer
Baraboo, WI

Craig Kramer
Physics Teacher
Bexley High School
Bexley, OH

Suzanne Lyons
Science Writer
Auburn, CA

Jack Minot
Physics Teacher
Bexley High School
Bexley, OH

Steven F. Moeckel
Science Writer
Troy, OH

David J. Olney
Science Writer
Mattapoisett, MA

Julie A.O. Smallfield
Science Writer
Spartanburg, SC

Amiee Wagner
Physics Teacher
Columbus State Community College
Columbus, OH

Teacher Advisory Board

The Teacher Advisory Board gave the editorial staff and design team feedback on the content and design of the 2005 edition of ***Physics: Principles and Problems.***

Kathleen M. Bartley
Physics Teacher
Westville High School
Westville, IN

Wayne Fisher, NBCT
Physics Teacher
Myers Park High School
Charlotte, NC

Stan Greenbaum
Physics Teacher
Gorton High School
Yonkers, NY

Stan Hutto, M.S.
Science Department Chair
Alamo Heights High School
San Antonio, TX

Martha S. Lai
Physics Teacher
Massey Hill Classical High School
Fayetteville, NC

Gregory MacDougall
Science Specialist
Central Savannah River Area
University of South Carolina–Aiken
Aiken, SC

Jane Bray Nelson
Physics Teacher
University High School
Orlando, FL

Jim Nelson
Physics Teacher
University High School
Orlando, FL

Safety Consultant

Kenneth Russell Roy, Ph.D.
Director of Science and Safety
Glastonbury Public Schools
Glastonbury, CT

Teacher Reviewers

Maria C. Aparicio
Physics Teacher
Spanish River High School
Boca Raton, FL

Daniel Barber
Physics Teacher
Klein Forest High School
Houston, TX

Tom Bartik
Department Chairman
Southside High School
Chocowinity, NC

Bob Beebe
Physics Teacher
Elbert High School
Elbert, CO

Patti R. Boles
Physics Teacher
East Rowan High School
Salisbury, NC

Julia Bridgewater
Physics Teacher
Ramona High School
Ramona, CA

Jim Broderick
Physics Teacher
Antelope Valley High School
Lancaster, CA

Hobart G. Cook
Physics Teacher
Cummings High School
Burlington, NC

Jason Craigo
Physics Teacher
Oberlin High School
Oberlin, OH

Gregory Cruz
Physics Teacher
Vanguard High School
Ocala, FL

Sue Cundiff
Physics Teacher
Gulf Breeze High School
Gulf Breeze, FL

Terry Elmer
Physics Teacher
Red Creek Central High School
Red Creek, NY

Hank Grizzle
Physics Teacher
Quemado High School
Quemado, NM

Shirley Hartnett
Physics Teacher
JC Birdlebough High School
Phoenix, NY

Mary S. Heltzel
Physics Teacher
Salina High School
Salina, OK

Tracy Hood
Physics Teacher
Plainfield High School
Plainfield, IN

Pam Hughes
Physics Teacher
Cherokee High School
Cherokee, IA

Kathy Jacquez
Physics Teacher
Fairmont Senior High School
Fairmont, WV

Wilma Jones
Physics Teacher
Taft Alternative Academy
Lawton, OK

Gene Kutscher
Science Chairman
Roslyn High School
Roslyn Heights, NY

Megan Lewis-Schroeder
Physics Teacher
Bellaire High School
Bellaire, MI

Mark Lutzenhiser
Physics Teacher
Sequim High School
Sequim, WA

Jill McLean
Physics Teacher
Centennial High School
Champaign, IL

Bradley E. Miller, Ph.D.
Physics Teacher
East Chapel Hill High School
Chapel Hill, NC

Don Rotsma
Physics Teacher
Galena High School
Reno, NV

David Shoemaker
Physics Teacher
Mechanicsburg Area High School
Mechanicsburg, PA

Consultants

Solomon Bililign, Ph.D.
Professor and Chair
Department of Physics
North Carolina A&T State University
Greensboro, NC

Juan R. Burciaga, Ph.D.
Visiting Professor
Department of Physics and Astronomy
Vassar College
Poughkeepsie, NY

Valentina French, Ph.D.
Associate Professor of Physics
Department of Physics
Indiana State University
Terre Haute, IN

Godfrey Gumbs, Ph.D.
Chianta-Stoll Professor of Physics
Department of Physics and Astronomy
Hunter College of the City University of New York
New York, NY

Ruth Howes, Ph.D.
Professor of Physics
Department of Physics
Marquette University
Milwaukee, WI

Lewis E. Johnson, Ph.D.
Assistant Professor
Department of Physics
Florida A&M University
Tallahassee, FL

Sally Koutsoliotas, Ph.D.
Associate Professor
Department of Physics
Bucknell University
Lewisburg, PA

Jun Qing Lu, Ph.D.
Assistant Professor
East Carolina University
Greenville, NC

William A. Mendoza, Ph.D.
Assistant Professor of Physics and Engineering
Department of Physics and Engineering
Jacksonville University
Jacksonville, FL

Jesús Pando, Ph.D.
Assistant Professor of Physics
Department of Physics
DePaul University
Chicago, IL

David W. Peakheart, Ph.D.
Lecturer
Department of Physics and Engineering
University of Central Oklahoma
Edmond, OK

Toni D. Sauncy, Ph.D.
Assistant Professor of Physics
Department of Physics
Angelo State University
San Angelo, TX

Sally Seidel, Ph.D.
Associate Professor of Physics
Department of Physics and Astronomy
University of New Mexico
Albuquerque, NM

Sudha Srinivas, Ph.D.
Associate Professor of Physics
Department of Physics
Central Michigan University
Mt. Pleasant, MI

Alma C. Zook, Ph.D.
Professor of Physics
Department of Physics and Astronomy
Pomona College
Claremont, CA

LAUNCH Lab

PHYSICS LAB

MINI LAB

Technology and Society

Future Technology

How it Works

eXtreme physics

Applying Math and Physics

PROBLEM-SOLVING Strategies

Connecting Math to Physics

APPLYING PHYSICS

Concepts In Motion

Interactive Figures Enhance and enrich your knowledge of physics concepts through animations of visuals.

Physics Online For additional animations, go to physicspp.com.

The online Personal Tutor presents teachers explaining major concepts in physics, step-by-step solutions to mathematical problems, and mathematical concepts.

Chapter 1 A Physics Toolkit

What You'll Learn

- You will use mathematical tools to measure and predict.
- You will apply accuracy and precision when measuring.
- You will display and evaluate data graphically.

Why It's Important

The measurement and mathematics tools presented here will help you to analyze data and make predictions.

Satellites Accurate and precise measurements are important when constructing and launching a satellite—errors are not easy to correct later. Satellites, such as the *Hubble Space Telescope* shown here, have revolutionized scientific research, as well as communications.

Think About This ▶

Physics research has led to many new technologies, such as satellite-based telescopes and communications. What are some other examples of tools developed from physics research in the last 50 years?

physicspp.com

Do all objects fall at the same rate?

Question

Mass is the quantity of matter an object has. Does mass affect the rate at which an object falls?

Procedure

Greek philosopher Aristotle drew conclusions about nature based on his observations. Italian Renaissance scientist Galileo conducted investigations to draw conclusions about nature. Both were interested in falling objects. Aristotle reasoned that heavier objects fall faster than lighter objects. Galileo investigated to determine if this is true.

1. Tape four pennies together in a stack.
2. Place the stack of pennies on your hand and place a single penny beside them.
3. **Observe** Which is heaviest and pushes down on your hand the most?
4. **Observe** Drop the single penny and the stack of pennies at the same time and observe the rate at which they fall.
5. **Record Data** Drop the two sets of pennies three times from a height of 2 m. Use a stopwatch to record the time it takes each to fall. Record your data.

Analysis

What do you conclude about mass and falling objects?

Critical Thinking Design an experiment to determine other factors that affect the rate of fall.

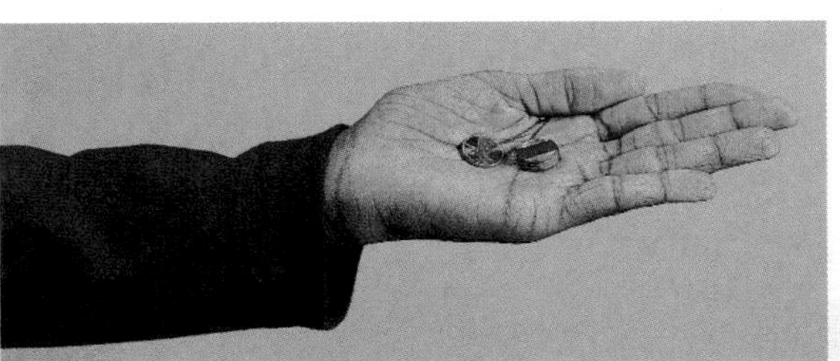

1.1 Mathematics and Physics

Objectives

- **Demonstrate** scientific methods.
- **Use** the metric system.
- **Evaluate** answers using dimensional analysis.
- **Perform** arithmetic operations using scientific notation.

Vocabulary

physics
dimensional analysis
significant digits
scientific method
hypothesis
scientific law
scientific theory

What do you think of when you see the word *physics?* Many people picture a chalkboard covered with formulas and mathematics: $E = mc^2$, $I = V/R$, $d = \frac{1}{2}at^2 + v_0t + d_0$. Perhaps you picture scientists in white lab coats, or well-known figures such as Marie Curie and Albert Einstein. Or, you might think of the many modern technologies created with physics, such as weather satellites, laptop computers, or lasers.

What is Physics?

Physics is a branch of science that involves the study of the physical world: energy, matter, and how they are related. Physicists investigate the motions of electrons and rockets, the energy in sound waves and electric circuits, the structure of the proton and of the universe. The goal of this course is to help you understand the physical world.

People who study physics go on to many different careers. Some become scientists at universities and colleges, at industries, or in research institutes. Others go into related fields, such as astronomy, engineering, computer science, teaching, or medicine. Still others use the problem-solving skills of physics to work in business, finance, or other very different disciplines.

Figure 1-1 Physicists use mathematics to represent many different phenomena—a trait sometimes spoofed in cartoons.

Mathematics in Physics

Physics uses mathematics as a powerful language. As illustrated in **Figure 1-1,** this use of mathematics often is spoofed in cartoons. In physics, equations are important tools for modeling observations and for making predictions. Physicists rely on theories and experiments with numerical results to support their conclusions. For example, think back to the Launch Lab. You can predict that if you drop a penny, it will fall. But how fast? Different models of falling objects give different answers to how the speed of the object changes, or on what the speed depends, or which objects will fall. By measuring how an object falls, you can compare the experimental data with the results predicted by different models. This tests the models, allowing you to pick the best one, or to develop a new model.

▶ EXAMPLE **Problem 1**

Electric Current The potential difference (V), or voltage, across a circuit equals the current (I) multiplied by the resistance (R) in the circuit. That is, V (volts) = I (amperes) × R (ohms). What is the resistance of a lightbulb that has a 0.75 amperes current when plugged into a 120-volt outlet?

1 Analyze the Problem

- Rewrite the equation.
- Substitute values.

Known:	**Unknown:**
I = 0.75 amperes	R = ?
V = 120 volts	

2 Solve for the Unknown

Rewrite the equation so the unknown is alone on the left.

$V = IR$

$IR = V$ — Reflexive property of equality

$R = \frac{V}{I}$ — Divide both sides by I.

$= \frac{120 \text{ volts}}{0.75 \text{ amperes}}$ — Substitute 120 volts for V, 0.75 amperes for I.

$= 160$ ohms — Resistance will be measured in ohms.

Math Handbook
Isolating a Variable
page 845

3 Evaluate the Answer

- **Are the units correct?** 1 volt = 1 ampere-ohm, so the answer in volts/ampere is in ohms, as expected.
- **Does the answer make sense?** 120 is divided by a number a little less than 1, so the answer should be a little more than 120.

PRACTICE Problems

- Additional Problems, Appendix B
- Solutions to Selected Problems, Appendix C

For each problem, give the rewritten equation you would use and the answer.

1. A lightbulb with a resistance of 50.0 ohms is used in a circuit with a 9.0-volt battery. What is the current through the bulb?
2. An object with uniform acceleration a, starting from rest, will reach a speed of v in time t according to the formula $v = at$. What is the acceleration of a bicyclist who accelerates from rest to 7.00 m/s in 4.00 s?
3. How long will it take a scooter accelerating at 0.400 m/s^2 to go from rest to a speed of 4.00 m/s?
4. The pressure on a surface is equal to the force divided by the area: $P = F/A$. A 53-kg woman exerts a force (weight) of 520 newtons. If the pressure exerted on the floor is 32,500 N/m^2, what is the area of the soles of her shoes?

Does it make sense? Sometimes you will work with unfamiliar units, as in Example Problem 1, and you will need to use estimation to check that your answer makes sense mathematically. At other times you can check that an answer matches your experience, as shown in **Figure 1-2.** When you work with falling objects, for example, check that the time you calculate an object will take to fall matches your experience—a copper ball dropping 5 m in 0.002 s, or in 17 s, doesn't make sense.

The Math Handbook in the back of this book contains many useful explanations and examples. Refer to it as needed.

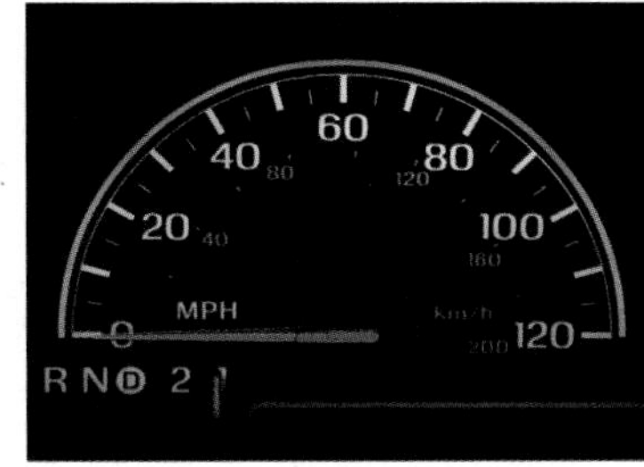

Figure 1-2 What is a reasonable range of values for the speed of an automobile?

SI Units

To communicate results, it is helpful to use units that everyone understands. The worldwide scientific community and most countries currently use an adaptation of the metric system to state measurements. The Système International d'Unités, or SI, uses seven base quantities, which are shown in **Table 1-1.** These base quantities were originally defined in terms of direct measurements. Other units, called derived units, are created by combining the base units in various ways. For example, energy is measured in joules, where 1 joule equals one kilogram-meter squared per second squared, or $1 \text{ J} = 1 \text{ kg·m}^2/\text{s}^2$. Electric charge is measured in coulombs, where $1 \text{ C} = 1 \text{ A·s}$.

Table 1-1
SI Base Units

Base Quantity	Base Unit	Symbol
Length	meter	m
Mass	kilogram	kg
Time	second	s
Temperature	kelvin	K
Amount of a substance	mole	mol
Electric current	ampere	A
Luminous intensity	candela	cd

■ **Figure 1-3** The standards for the kilogram and meter are shown. The International Prototype Meter originally was measured as the distance between two marks on a platinum-iridium bar, but as methods of measuring time became more precise than those for measuring length, the meter came to be defined as the distance traveled by light in a vacuum in 1/299 792 458 s.

Scientific institutions have been created to define and regulate measures. The SI system is regulated by the International Bureau of Weights and Measures in Sèvres, France. This bureau and the National Institute of Science and Technology (NIST) in Gaithersburg, Maryland keep the standards of length, time, and mass against which our metersticks, clocks, and balances are calibrated. Examples of two standards are shown in **Figure 1-3.** NIST works on many problems of measurement, including industrial and research applications.

You probably learned in math class that it is much easier to convert meters to kilometers than feet to miles. The ease of switching between units is another feature of the metric system. To convert between SI units, multiply or divide by the appropriate power of 10. Prefixes are used to change SI units by powers of 10, as shown in **Table 1-2.** You often will encounter these prefixes in daily life, as in, for example, milligrams, nanoseconds, and gigabytes.

Table 1-2

Prefixes Used with SI Units

Prefix	Symbol	Multiplier	Scientific Notation	Example
femto-	f	0.000000000000001	10^{-15}	femtosecond (fs)
pico-	p	0.000000000001	10^{-12}	picometer (pm)
nano-	n	0.000000001	10^{-9}	nanometer (nm)
micro-	μ	0.000001	10^{-6}	microgram (μg)
milli-	m	0.001	10^{-3}	milliamps (mA)
centi-	c	0.01	10^{-2}	centimeter (cm)
deci-	d	0.1	10^{-1}	deciliter (dL)
kilo-	k	1000	10^{3}	kilometer (km)
mega-	M	1,000,000	10^{6}	megagram (Mg)
giga-	G	1,000,000,000	10^{9}	gigameter (Gm)
tera-	T	1,000,000,000,000	10^{12}	terahertz (THz)

Dimensional Analysis

You can use units to check your work. You often will need to use different versions of a formula, or use a string of formulas, to solve a physics problem. To check that you have set up a problem correctly, write out the equation or set of equations you plan to use. Before performing calculations, check that the answer will be in the expected units, as shown in step 3 of Example Problem 1. For example, if you are finding a speed and you see that your answer will be measured in s/m or m/s^2, you know you have made an error in setting up the problem. This method of treating the units as algebraic quantities, which can be cancelled, is called **dimensional analysis.**

Math Handbook
Dimensional Analysis
page 847

Dimensional analysis also is used in choosing conversion factors. A conversion factor is a multiplier equal to 1. For example, because 1 kg = 1000 g, you can construct the following conversion factors:

$$1 = \frac{1 \text{ kg}}{1000 \text{ g}} \qquad 1 = \frac{1000 \text{ g}}{1 \text{ kg}}$$

Choose a conversion factor that will make the units cancel, leaving the answer in the correct units. For example, to convert 1.34 kg of iron ore to grams, do as shown below.

$$1.34\ \cancel{\text{kg}}\left(\frac{1000\ \text{g}}{1\ \cancel{\text{kg}}}\right) = 1340\ \text{g}$$

You also might need to do a series of conversions. To convert 43 km/h to m/s, do the following:

$$\left(\frac{43\ \cancel{\text{km}}}{1\ \cancel{\text{h}}}\right)\left(\frac{1000\ \text{m}}{1\ \cancel{\text{km}}}\right)\left(\frac{1\ \cancel{\text{h}}}{60\ \cancel{\text{min}}}\right)\left(\frac{1\ \cancel{\text{min}}}{60\ \text{s}}\right) = 12\ \text{m/s}$$

Personal Tutor For an online tutorial on dimensional analysis, visit physicspp.com.

PRACTICE Problems

- Additional Problems, Appendix B
- Solutions to Selected Problems, Appendix C

Use dimensional analysis to check your equation before multiplying.

5. How many megahertz is 750 kilohertz?

6. Convert 5021 centimeters to kilometers.

7. How many seconds are in a leap year?

8. Convert the speed 5.300 m/s to km/h.

Significant Digits

Suppose you use a meterstick to measure a pen, and you find that the end of the pen is just past 14.3 cm. This measurement has three valid digits: two you are sure of, and one you estimated. The valid digits in a measurement are called **significant digits.** The last digit given for any measurement is the uncertain digit. All nonzero digits in a measurement are significant.

Are all zeros significant? No. For example, in the measurement 0.0860 m, the first two zeros serve only to locate the decimal point and are not significant. The last zero, however, is the estimated digit and is significant. The measurement 172,000 m could have 3, 4, 5, or 6 significant digits. This ambiguity is one reason to use scientific notation: it is clear that the measurement 1.7200×10^5 m has five significant digits.

Math Handbook
Significant Digits
pages 833–836

Arithmetic with significant digits When you perform any arithmetic operation, it is important to remember that the result never can be more precise than the least-precise measurement.

To add or subtract measurements, first perform the operation, then round off the result to correspond to the least-precise value involved. For example, 3.86 m + 2.4 m = 6.3 m because the least-precise measure is to one-tenth of a meter.

To multiply or divide measurements, perform the calculation and then round to the same number of significant digits as the least-precise measurement. For example, 409.2 km/11.4 L = 35.9 km/L, because the least-precise measure has three significant digits.

Some calculators display several additional digits, as shown in **Figure 1-4,** while others round at different points. Be sure to record your answers with the correct number of digits. Note that significant digits are considered only when calculating with measurements. There is no uncertainty associated with counting (4 washers) or exact conversion factors (24 hours in 1 day).

Figure 1-4 This answer to 3.9 ÷ 7.2 should be rounded to two significant digits, 0.54.

PRACTICE Problems

- Additional Problems, Appendix B
- Solutions to Selected Problems, Appendix C

Solve the following problems.

9. a. 6.201 cm + 7.4 cm + 0.68 cm + 12.0 cm
  b. 1.6 km + 1.62 m + 1200 cm

10. a. 10.8 g − 8.264 g
  b. 4.75 m − 0.4168 m

11. a. 139 cm × 2.3 cm
  b. 3.2145 km × 4.23 km

12. a. 13.78 g ÷ 11.3 mL
  b. 18.21 g ÷ 4.4 cm^3

MINI LAB

Measuring Change

Collect five identical washers and a spring that will stretch measurably when one washer is suspended from it.

1. **Measure** the length of the spring with zero, one, two, and three washers suspended from it.
2. **Graph** the length of the spring versus the mass.
3. **Predict** the length of the spring with four and five washers.
4. **Test** your prediction.

Analyze and Conclude

5. **Describe** the shape of the graph. How did you use it to predict the two new lengths?

Scientific Methods

In physics class, you will make observations, do experiments, and create models or theories to try to explain your results or predict new answers, as shown in **Figure 1-5.** This is the essence of a **scientific method.** All scientists, including physicists, obtain data, make predictions, and create compelling explanations that quantitatively describe many different phenomena. The experiments and results must be reproducible; that is, other scientists must be able to recreate the experiment and obtain similar data. Written, oral, and mathematical communication skills are vital to every scientist.

A scientist often works with an idea that can be worded as a **hypothesis,** which is an educated guess about how variables are related. How can the hypothesis be tested? Scientists conduct experiments, take measurements, and identify what variables are important and how they are related. For example, you might find that the speed of sound depends on the medium through which sound travels, but not on the loudness of the sound. You can then predict the speed of sound in a new medium and test your results.

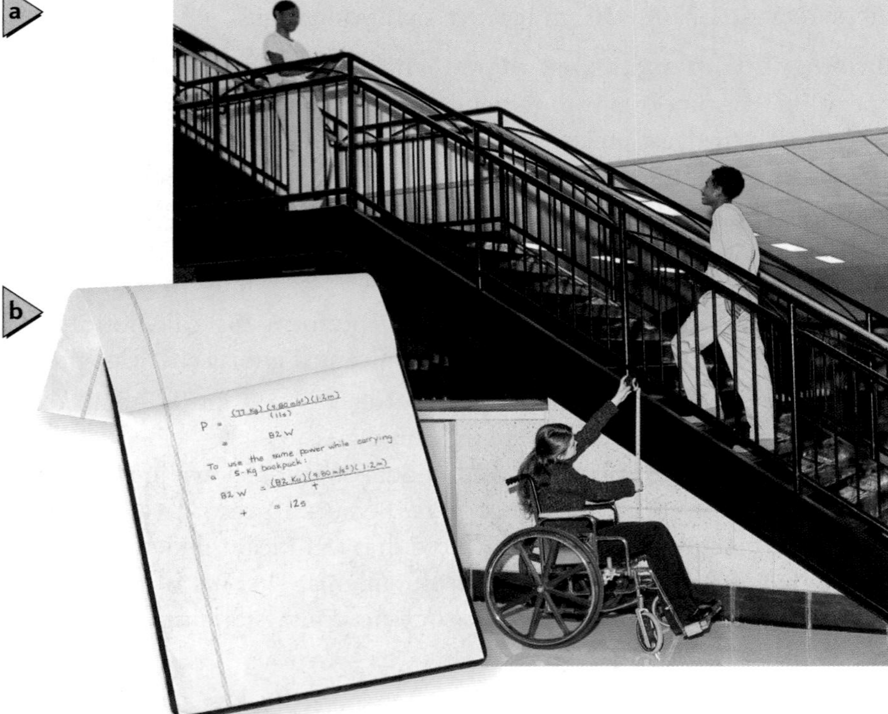

Figure 1-5 These students are conducting an experiment to determine how much power they produce climbing the stairs **(a).** They use their data to predict how long it would take an engine with the same power to lift a different load **(b).**

Figure 1-6 In the mid-1960s, Arno Penzias and Robert Wilson were trying to eliminate the constant background noise in an antenna to be used for radio astronomy. They tested systems, duct-taped seams, and cleared out pigeon manure, but the noise persisted. This noise is now understood to be the cosmic microwave background radiation, and is experimental support for the Big Bang theory.

Models, laws, and theories An idea, equation, structure, or system can model the phenomenon you are trying to explain. Scientific models are based on experimentation. Recall from chemistry class the different models of the atom that were in use over time—new models were developed to explain new observations and measurements.

If new data do not fit a model, both are re-examined. **Figure 1-6** shows a historical example. If a very well-established model is questioned, physicists might first look at the new data: can anyone reproduce the results? Were there other variables at work? If the new data are born out by subsequent experiments, the theories have to change to reflect the new findings. For example, in the nineteenth century it was believed that linear markings on Mars showed channels, as shown in **Figure 1-7a.** As telescopes improved, scientists realized that there were no such markings, as shown in **Figure 1-7b.** In recent times, again with better instruments, scientists have found features that suggest Mars once had running and standing water on its surface, as shown in **Figure 1-7c.** Each new discovery has raised new questions and areas for exploration.

A **scientific law** is a rule of nature that sums up related observations to describe a pattern in nature. For example, the law of conservation of charge states that in the various changes matter can undergo, the electric charge before and after stays the same. The law of reflection states that the angle of incidence for a light beam on a reflective surface equals the angle of reflection. Notice that the laws do not explain why these phenomena happen, they simply describe them.

Figure 1-7 Drawings of early telescope observations **(a)** showed channels on Mars; recent photos taken with improved telescopes do not **(b).** In this photo of Mars' surface from the *Mars Global Surveyor* spacecraft **(c),** these layered sedimentary rocks suggest that sedimentary deposits might have formed in standing water.

a
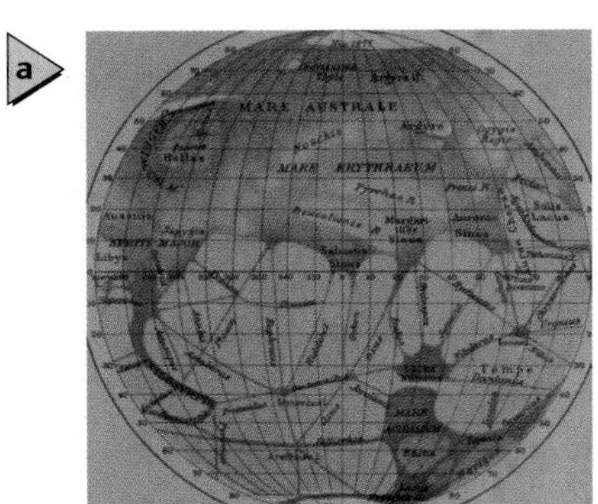

b

c
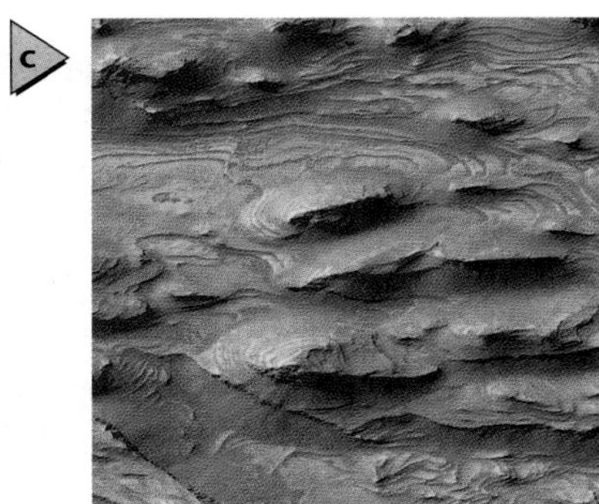

■ **Figure 1-8** Theories are changed and modified as new experiments provide insight and new observations are made. The theory of falling objects has undergone many revisions.

Greek philosophers proposed that objects fall because they seek their natural places. The more massive the object, the faster it falls.

Revision

Galileo showed that the speed at which an object falls depends on the amount of time it falls, not on its mass.

Revision

Galileo's statement is true, but Newton revised the reason why objects fall. Newton proposed that objects fall because the object and Earth are attracted by a force. Newton also stated that there is a force of attraction between any two objects with mass.

Revision

Galileo's and Newton's statements still hold true. However, Einstein suggested that the force of attraction between two objects is due to mass causing the space around it to curve.

A **scientific theory** is an explanation based on many observations supported by experimental results. Theories may serve as explanations for laws. A theory is the best available explanation of why things work as they do. For example, the theory of universal gravitation states that all the mass in the universe is attracted to other mass. Laws and theories may be revised or discarded over time, as shown in **Figure 1-8.** Notice that this use of the word *theory* is different from the common use, as in "I have a theory about why it takes longer to get to school on Fridays." In scientific use, only a very well-supported explanation is called a theory.

1.1 Section Review

13. **Math** Why are concepts in physics described with formulas?

14. **Magnetism** The force of a magnetic field on a charged, moving particle is given by $F = Bqv$, where F is the force in kg·m/s^2, q is the charge in A·s, and v is the speed in m/s. B is the strength of the magnetic field, measured in teslas, T. What is 1 tesla described in base units?

15. **Magnetism** A proton with charge 1.60×10^{-19} A·s is moving at 2.4×10^5 m/s through a magnetic field of 4.5 T. You want to find the force on the proton.
    - **a.** Substitute the values into the equation you will use. Are the units correct?
    - **b.** The values are written in scientific notation, $m\times10^n$. Calculate the 10^n part of the equation to estimate the size of the answer.
    - **c.** Calculate your answer. Check it against your estimate from part b.
    - **d.** Justify the number of significant digits in your answer.

16. **Magnetism** Rewrite $F = Bqv$ to find v in terms of F, q, and B.

17. **Critical Thinking** An accepted value for the acceleration due to gravity is 9.801 m/s^2. In an experiment with pendulums, you calculate that the value is 9.4 m/s^2. Should the accepted value be tossed out to accommodate your new finding? Explain.

Physics Online physicspp.com/self_check_quiz

1.2 Measurement

When you visit the doctor for a checkup, many measurements are taken: your height, weight, blood pressure, and heart rate. Even your vision is measured and assigned a number. Blood might be drawn so measurements can be made of lead or cholesterol levels. Measurements quantify our observations: a person's blood pressure isn't just "pretty good," it's 110/60, the low end of the good range.

What is a measurement? A **measurement** is a comparison between an unknown quantity and a standard. For example, if you measure the mass of a rolling cart used in an experiment, the unknown quantity is the mass of the cart and the standard is the gram, as defined by the balance or spring scale you use. In the Mini Lab in Section 1.1, the length of the spring was the unknown and the centimeter was the standard.

► **Objectives**

- **Distinguish** between accuracy and precision.
- **Determine** the precision of measured quantities.

► **Vocabulary**

measurement
precision
accuracy

Comparing Results

As you learned in Section 1.1, scientists share their results. Before new data are fully accepted, other scientists examine the experiment, looking for possible sources of error, and try to reproduce the results. Results often are reported with an uncertainty. A new measurement that is within the margin of uncertainty confirms the old measurement.

For example, archaeologists use radiocarbon dating to find the age of cave paintings, such as those from the Lascaux cave, in **Figure 1-9,** and the Chauvet cave. Radiocarbon dates are reported with an uncertainty. Three radiocarbon ages from a panel in the Chauvet cave are 30,940 ± 610 years, 30,790 ± 600 years, and 30,230 ± 530 years. While none of the measurements exactly match, the uncertainties in all three overlap, and the measurements confirm each other.

Figure 1-9 Drawings of animals from the Lascaux cave in France. By dating organic material in the cave, such as pigments and torch marks, scientists are able to suggest dates at which these cave paintings were made. Each date is reported with an uncertainty to show how precise the measurement is.

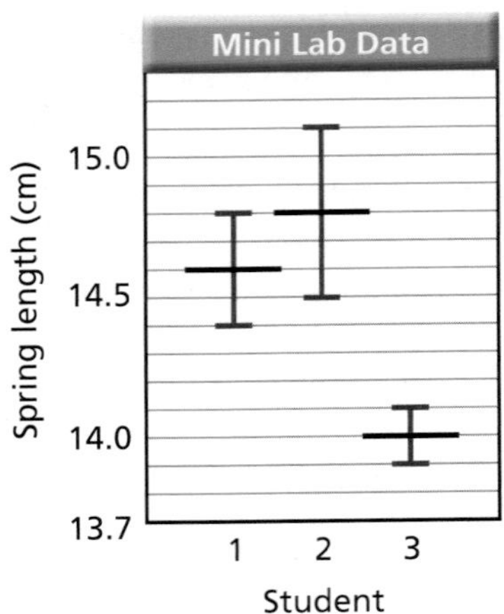

Figure 1-10 Three students took multiple measurements. Are the measurements in agreement? Is student 1's result reproducible?

Suppose three students performed the Mini Lab from Section 1.1 several times, starting with springs of the same length. With two washers on the spring, student 1 made repeated measurements, which ranged from 14.4 cm to 14.8 cm. The average of student 1's measurements was 14.6 cm, as shown in **Figure 1-10.** This result was reported as (14.6 ± 0.2) cm. Student 2 reported finding the spring's length to be (14.8 ± 0.3) cm. Student 3 reported a length of (14.0 ± 0.1) cm.

Could you conclude that the three measurements are in agreement? Is student 1's result reproducible? The results of students 1 and 2 overlap; that is, they have the lengths 14.5 cm to 14.8 cm in common. However, there is no overlap and, therefore, no agreement, between their results and the result of student 3.

Precision Versus Accuracy

Both precision and accuracy are characteristics of measured values. How precise and accurate are the measurements of the three students? The degree of exactness of a measurement is called its **precision.** Student 3's measurements are the most precise, within ± 0.1 cm. The measurements of the other two students are less precise because they have a larger uncertainty.

Precision depends on the instrument and technique used to make the measurement. Generally, the device that has the finest division on its scale produces the most precise measurement. The precision of a measurement is one-half the smallest division of the instrument. For example, the graduated cylinder in **Figure 1-11a** has divisions of 1 mL. You can measure an object to within 0.5 mL with this device. However, the smallest division on the beaker in **Figure 1-11b** is 50 mL. How precise were your measurements in the MiniLab?

The significant digits in an answer show its precision. A measure of 67.100 g is precise to the nearest thousandth of a gram. Recall from Section 1.1 the rules for performing operations with measurements given to different levels of precision. If you add 1.2 mL of acid to a beaker containing 2.4×10^2 mL of water, you cannot say you now have 2.412×10^2 mL of fluid, because the volume of water was not measured to the nearest tenth of a milliliter, but to 100 times that.

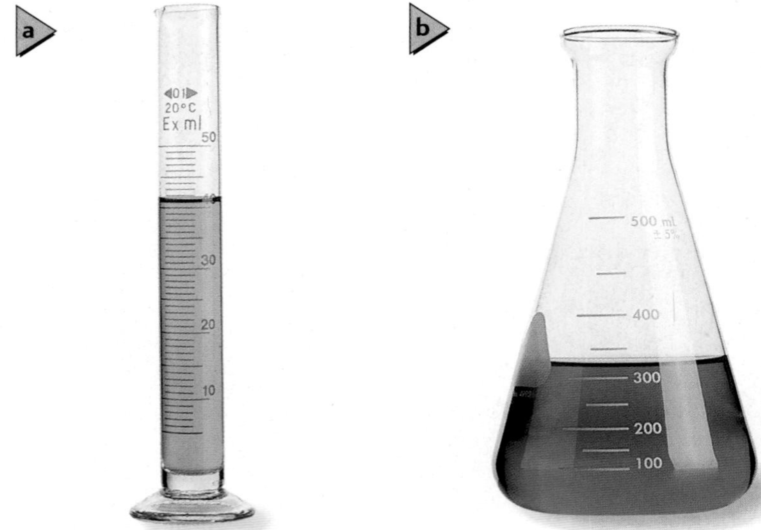

Figure 1-11 The graduated cylinder contains 41 ± 0.5 mL **(a).** The flask contains 325 mL ± 25 mL **(b).**

Concepts In Motion
Interactive Figure To see an animation on precision versus accuracy, visit physicspp.com.

Accuracy describes how well the results of a measurement agree with the "real" value; that is, the accepted value as measured by competent experimenters. If the length of the spring that the three students measured had been 14.8 cm, then student 2 would have been most accurate and student 3 least accurate. How accurate do you think your measurements in the Mini Lab on page 8 were? What might have led someone to make inaccurate measurements? How could you check the accuracy of measurements?

A common method for checking the accuracy of an instrument is called the two-point calibration. First, does the instrument read zero when it should? Second, does it give the correct reading when it is measuring an accepted standard, as shown in **Figure 1-12?** Regular checks for accuracy are performed on critical measuring instruments, such as the radiation output of the machines used to treat cancer.

Figure 1-12 Accuracy is checked by measuring a known value.

Techniques of Good Measurement

To assure accuracy and precision, instruments also have to be used correctly. Measurements have to be made carefully if they are to be as precise as the instrument allows. One common source of error comes from the angle at which an instrument is read. Scales should be read with one's eye directly above the measure, as shown in **Figure 1-13a.** If the scale is read from an angle, as shown in **Figure 1-13b,** you will get a different, and less accurate, value. The difference in the readings is caused by *parallax,* which is the apparent shift in the position of an object when it is viewed from different angles. To experiment with parallax, place your pen on a ruler and read the scale with your eye directly over the tip, then read the scale with your head shifted far to one side.

Figure 1-13 By positioning the scale head on **(a),** your results will be more accurate than if you read your measurements at an angle **(b).** How far did parallax shift the measurement in b?

a

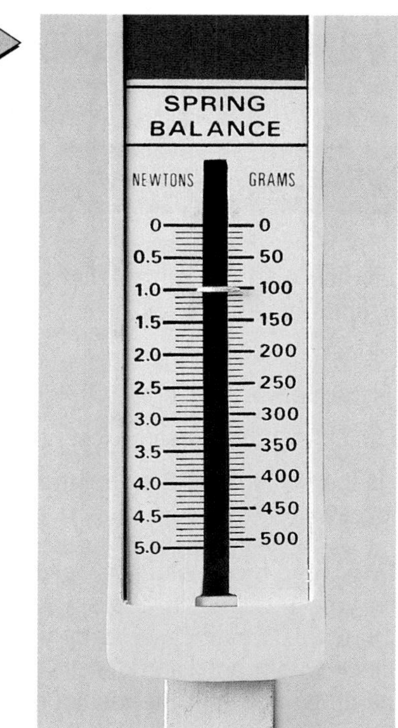

b

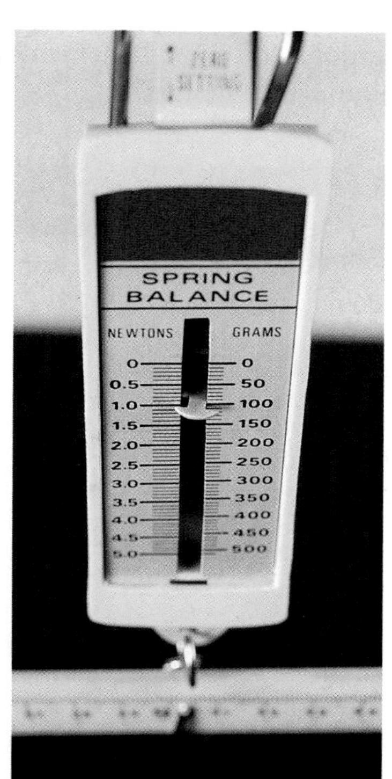

APPLYING PHYSICS

Distance to the Moon For over 25 years, scientists have been accurately measuring the distance to the Moon by shining lasers through telescopes. The laser beam reflects off reflectors placed on the surface of the Moon by Apollo astronauts. They have determined that the average distance between the centers of Earth and the Moon is 385,000 km, and it is known with an accuracy of better than one part in 10 billion. Using this laser technique, scientists also have discovered that the Moon is receding from Earth at about 3.8 cm/yr.

■ **Figure 1-14** A series of expeditions succeeded in placing a GPS receiver on top of Mount Everest. This improved the accuracy of the altitude measurement: Everest's peak is 8850 m, not 8848 m, above sea level.

The Global Positioning System, or GPS, offers an illustration of accuracy and precision in measurement. The GPS consists of 24 satellites with transmitters in orbit and numerous receivers on Earth. The satellites send signals with the time, measured by highly accurate atomic clocks. The receiver uses the information from at least four satellites to determine latitude, longitude, and elevation. (The clocks in the receivers are not as accurate as those on the satellites.)

Receivers have different levels of precision. A device in an automobile might give your position to within a few meters. Devices used by geophysicists, as in **Figure 1-14,** can measure movements of millimeters in Earth's crust.

The GPS was developed by the United States Department of Defense. It uses atomic clocks, developed to test Einstein's theories of relativity and gravity. The GPS eventually was made available for civilian use. GPS signals now are provided worldwide free of charge and are used in navigation on land, at sea, and in the air, for mapping and surveying, by telecommunications and satellite networks, and for scientific research into earthquakes and plate tectonics.

1.2 Section Review

18. Accuracy Some wooden rulers do not start with 0 at the edge, but have it set in a few millimeters. How could this improve the accuracy of the ruler?

19. Tools You find a micrometer (a tool used to measure objects to the nearest 0.01 mm) that has been badly bent. How would it compare to a new, high-quality meterstick in terms of its precision? Its accuracy?

20. Parallax Does parallax affect the precision of a measurement that you make? Explain.

21. Error Your friend tells you that his height is 182 cm. In your own words, explain the range of heights implied by this statement.

22. Precision A box has a length of 18.1 cm and a width of 19.2 cm, and it is 20.3 cm tall.

a. What is its volume?

b. How precise is the measure of length? Of volume?

c. How tall is a stack of 12 of these boxes?

d. How precise is the measure of the height of one box? of 12 boxes?

23. Critical Thinking Your friend states in a report that the average time required to circle a 1.5-mi track was 65.414 s. This was measured by timing 7 laps using a clock with a precision of 0.1 s. How much confidence do you have in the results of the report? Explain.

Physics Online physicspp.com/self_check_quiz

1.3 Graphing Data

A well-designed graph can convey information quickly and simply. Patterns that are not immediately evident in a list of numbers take shape when the data are graphed. In this section, you will develop graphing techniques that will enable you to display, analyze, and model data.

▸ **Objectives**

- **Graph** the relationship between independent and dependent variables.
- **Interpret** graphs.
- **Recognize** common relationships in graphs.

▸ **Vocabulary**

independent variable
dependent variable
line of best fit
linear relationship
quadratic relationship
inverse relationship

Identifying Variables

When you perform an experiment, it is important to change only one factor at a time. For example, **Table 1-3** gives the length of a spring with different masses attached, as measured in the Mini Lab. Only the mass varies; if different masses were hung from different types of springs, you wouldn't know how much of the difference between two data pairs was due to the different masses and how much to the different springs.

Table 1-3
Length of a Spring for Different Masses

Mass Attached to Spring (g)	Length of Spring (cm)
0	13.7
5	14.1
10	14.5
15	14.9
20	15.3
25	15.7
30	16.0
35	16.4

A *variable* is any factor that might affect the behavior of an experimental setup. The **independent variable** is the factor that is changed or manipulated during the experiment. In this experiment, the mass was the independent variable. The **dependent variable** is the factor that depends on the independent variable. In this experiment, the amount that the spring stretched depended on the mass. An experiment might look at how radioactivity varies with time, how friction changes with weight, or how the strength of a magnetic field depends on the distance from a magnet.

One way to analyze data is to make a line graph. This shows how the dependent variable changes with the independent variable. The data from Table 1-3 are graphed in black in **Figure 1-15.** The line in blue, drawn as close to all the data points as possible, is called a **line of best fit.** The line of best fit is a better model for predictions than any one point that helps determine the line. The problem-solving strategy on the next page gives detailed instructions for graphing data and sketching a line of best fit.

Figure 1-15 The independent variable, mass, is on the horizontal axis. The graph shows that the length of the spring increases as the mass suspended from the spring increases.

Concepts In Motion
Interactive Figure To see an animation on graphing data, visit physicspp.com.

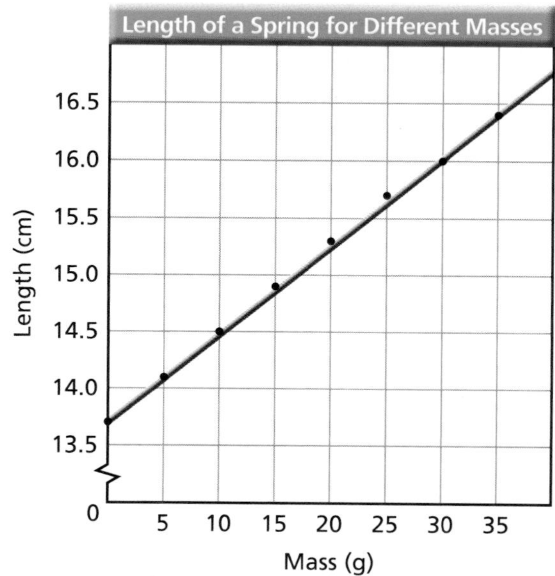

PROBLEM-SOLVING Strategies

Plotting Line Graphs

Connecting **Math to Physics**

Use the following steps to plot line graphs from data tables.

1. Identify the independent and dependent variables in your data. The independent variable is plotted on the horizontal axis, the *x*-axis. The dependent variable is plotted on the vertical axis, the *y*-axis.
2. Determine the range of the independent variable to be plotted.
3. Decide whether the origin (0, 0) is a valid data point.
4. Spread the data out as much as possible. Let each division on the graph paper stand for a convenient unit. This usually means units that are multiples of 2, 5, or 10.
5. Number and label the horizontal axis. The label should include the units, such as *Mass (grams).*
6. Repeat steps 2–5 for the dependent variable.
7. Plot the data points on the graph.
8. Draw the best-fit straight line or smooth curve that passes through as many data points as possible. This is sometimes called *eyeballing*. Do not use a series of straight line segments that connect the dots. The line that looks like the best fit to you may not be exactly the same as someone else's. There is a formal procedure, which many graphing calculators use, called the least-squares technique, that produces a unique best-fit line, but that is beyond the scope of this textbook.
9. Give the graph a title that clearly tells what the graph represents.

Math Handbook
Graphs of Relations
pages 848–852

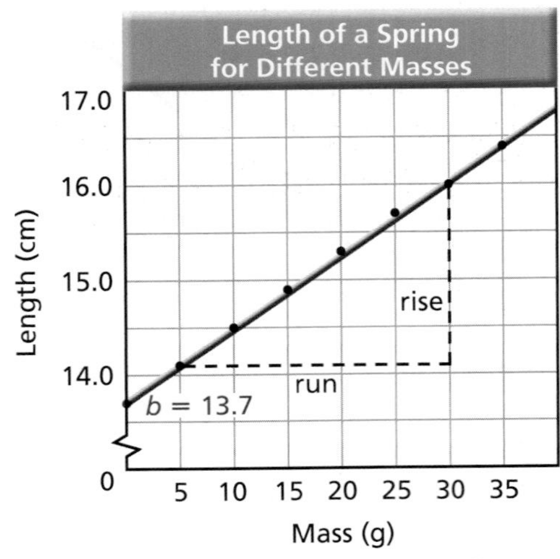

Figure 1-16 To find an equation of the line of best fit for a linear relationship, find the slope and *y*-intercept.

Linear Relationships

Scatter plots of data may take many different shapes, suggesting different relationships. (The line of best fit may be called a curve of best fit for nonlinear graphs.) Three of the most common relationships will be shown in this section. You probably are familiar with them from math class.

When the line of best fit is a straight line, as in Figure 1-15, the dependent variable varies linearly with the independent variable. There is a **linear relationship** between the two variables. The relationship can be written as an equation.

Linear Relationship Between Two Variables $y = mx + b$

Find the *y*-intercept, *b*, and the slope, *m*, as illustrated in **Figure 1-16.** Use points on the line—they may or may not be data points.

The slope is the ratio of the vertical change to the horizontal change. To find the slope, select two points, A and B, far apart on the line. The vertical change, or rise, Δy, is the difference between the vertical values of A and B. The horizontal change, or run, Δx, is the difference between the horizontal values of A and B.

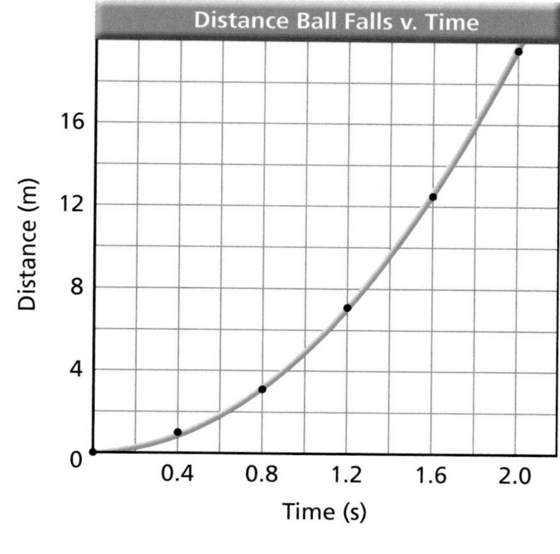

Figure 1-17 This graph indicates a quadratic, or parabolic, relationship.

Slope $m = \frac{rise}{run} = \frac{\Delta y}{\Delta x}$

The slope of a line is equal to the rise divided by the run, which also can be expressed as the change in y divided by the change in x.

$$\text{In Figure 1-16: } m = \frac{(16.0 \text{ cm} - 14.1 \text{ cm})}{(30 \text{ g} - 5 \text{ g})}$$
$$= 0.08 \text{ cm/g}$$

If y gets smaller as x gets larger, then $\Delta y/\Delta x$ is negative, and the line slopes downward.

The y-intercept, b, is the point at which the line crosses the y-axis, and it is the y-value when the value of x is zero. In this example, $b = 13.7$ cm. When $b = 0$, or $y = mx$, the quantity y is said to vary directly with x.

Nonlinear Relationships

Figure 1-17 shows the distance a brass ball falls versus time. Note that the graph is not a straight line, meaning the relationship is not linear. There are many types of nonlinear relationships in science. Two of the most common are the quadratic and inverse relationships. The graph in Figure 1-17 is a **quadratic relationship,** represented by the following equation.

Quadratic Relationship Between Two Variables
$y = ax^2 + bx + c$

A quadratic relationship exists when one variable depends on the square of another.

Math Handbook
Quadratic Graphs page 852
Quadratic Equations page 846

A computer program or graphing calculator easily can find the values of the constants a, b, and c in this equation. In this case, the equation is $d = 5t^2$. See the Math Handbook in the back of the book for more on making and using line graphs.

CHALLENGE PROBLEM

An object is suspended from spring 1, and the spring's elongation (the distance it stretches) is X_1. Then the same object is removed from the first spring and suspended from a second spring. The elongation of spring 2 is X_2. X_2 is greater than X_1.

1. On the same axes, sketch the graphs of the mass versus elongation for both springs.
2. Is the origin included in the graph? Why or why not?
3. Which slope is steeper?
4. At a given mass, $X_2 = 1.6\, X_1$. If $X_2 = 5.3$ cm, what is X_1?

Figure 1-18 This graph shows the inverse relationship between resistance and current. As resistance increases, current decreases.

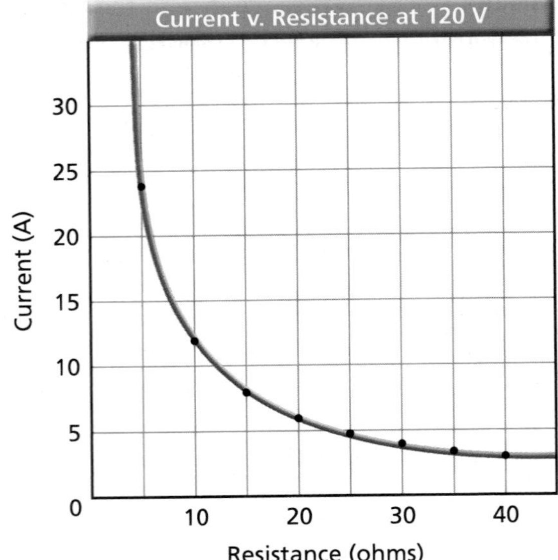

The graph in **Figure 1-18** shows how the current in an electric circuit varies as the resistance is increased. This is an example of an **inverse relationship,** represented by the following equation.

Inverse Relationship $y = \frac{a}{x}$

A hyperbola results when one variable depends on the inverse of the other.

The three relationships you have learned about are a sample of the simple relations you will most likely try to derive in this course. Many other mathematical models are used. Important examples include sinusoids, used to model cyclical phenomena, and exponential growth and decay, used to study radioactivity. Combinations of different mathematical models represent even more complex phenomena.

PRACTICE Problems

• Additional Problems, Appendix B
• Solutions to Selected Problems, Appendix C

24. The mass values of specified volumes of pure gold nuggets are given in **Table 1-4.**

a. Plot mass versus volume from the values given in the table and draw the curve that best fits all points.

b. Describe the resulting curve.

c. According to the graph, what type of relationship exists between the mass of the pure gold nuggets and their volume?

d. What is the value of the slope of this graph? Include the proper units.

e. Write the equation showing mass as a function of volume for gold.

f. Write a word interpretation for the slope of the line.

Table 1-4
Mass of Pure Gold Nuggets

Volume (cm^3)	Mass (g)
1.0	19.4
2.0	38.6
3.0	58.1
4.0	77.4
5.0	96.5

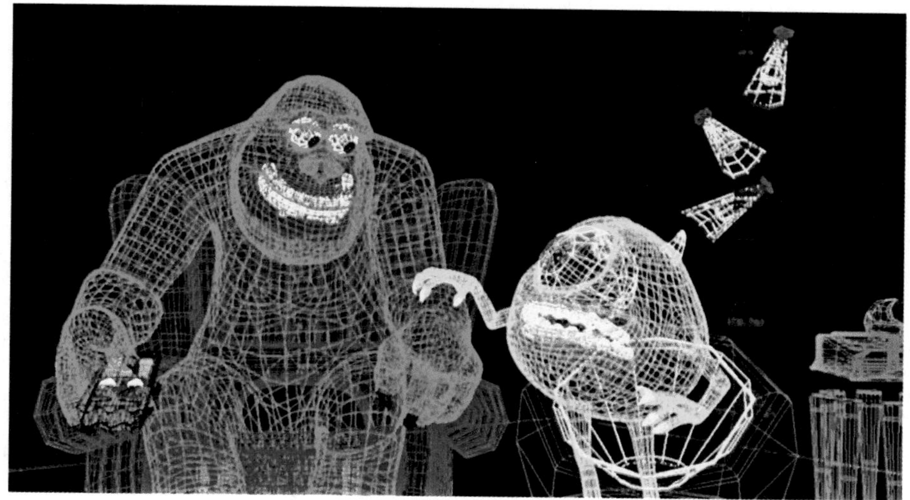

■ **Figure 1-19** Computer animators use mathematical models of the real world to create a convincing fictional world. They need to accurately portray how beings of different sizes move, how hair or clothing move with a character, and how light and shadows fall, among other physics topics.

Predicting Values

When scientists discover relations like the ones shown in the graphs in this section, they use them to make predictions. For example, the equation for the linear graph in **Figure 1-16** is as follows:

$$y = (0.08 \text{ cm/g})x + 13.7 \text{ cm}$$

Relations, either learned as formulas or developed from graphs, can be used to predict values you haven't measured directly. How far would the spring in Table 1-3 stretch with 49 g of mass?

$$y = (0.08 \text{ cm/g})(49 \text{ g}) + 13.7 \text{ cm}$$
$$= 18 \text{ cm}$$

It is important to decide how far you can extrapolate from the data you have. For example, 49 kg is a value far outside the ones measured, and the spring might break rather than stretch that far.

Physicists use models to accurately predict how systems will behave: what circumstances might lead to a solar flare, how changes to a circuit will change the performance of a device, or how electromagnetic fields will affect a medical instrument. People in all walks of life use models in many ways. One example is shown in **Figure 1-19.** With the tools you have learned in this chapter, you can answer questions and produce models for the physics questions you will encounter in the rest of this textbook.

1.3 Section Review

25. Make a Graph Graph the following data. Time is the independent variable.

Time (s)	0	5	10	15	20	25	30	35
Speed (m/s)	12	10	8	6	4	2	2	2

26. Interpret a Graph What would be the meaning of a nonzero *y*-intercept to a graph of total mass versus volume?

27. Predict Use the relation illustrated in Figure 1-16 to determine the mass required to stretch the spring 15 cm.

28. Predict Use the relation in Figure 1-18 to predict the current when the resistance is 16 ohms.

29. Critical Thinking In your own words, explain the meaning of a shallower line, or a smaller slope than the one in Figure 1-16, in the graph of stretch versus total mass for a different spring.

Physics Online physicspp.com/self_check_quiz

PHYSICS LAB • Internet

Exploring Objects in Motion

Physics is a science that is based upon experimental observations. Many of the basic principles used to describe and understand mechanical systems, such as objects in linear motion, can be applied later to describe more complex natural phenomena. How can you measure the speed of the vehicles in a video clip?

QUESTION

What types of measurements could be made to find the speed of a vehicle?

Objectives

- **Observe** the motion of the vehicles seen in the video.
- **Describe** the motion of the vehicles.
- **Collect and organize data** on the vehicle's motion.
- **Calculate** a vehicle's speed.

Safety Precautions

Materials

Internet access is required.
watch or other timer

Procedure

1. Visit physicspp.com/internet_lab to view the Chapter 1 lab video clip.
2. The video footage was taken in the midwestern United States at approximately noon. Along the right shoulder of the road are large, white, painted rectangles. These types of markings are used in many states for aerial observation of traffic. They are placed at 0.322-km (0.2-mi) intervals.
3. **Observe** What type of measurements might be taken? Prepare a data table, such as the one shown on the next page. Record your observations of the surroundings, other vehicles, and markings. On what color vehicle is the camera located, and what color is the pickup truck in the lane to the left?
4. **Measure and Estimate** View the video again and look for more details. Is the road smooth? In what direction are the vehicles heading? How long does it take each vehicle to travel two intervals marked by the white blocks? Record your observations and data.

Data Table			
Marker	Distance (km)	White Vehicle Time (s)	Gray Pickup Time (s)

Analyze

1. Summarize your qualitative observations.
2. Summarize your quantitative observations.
3. **Make and Use Graphs** Graph both sets of data on one pair of axes.
4. **Estimate** What is the speed of the vehicles in km/s and km/h?
5. **Predict** How far will each vehicle travel in 5 min?

Conclude and Apply

1. **Measure** What is the precision of the distance and time measurements?
2. **Measure** What is the precision of your speed measurement? On what does it depend?
3. **Use Variables, Constants, and Controls** Describe the independent and the dependent variables in this experiment.
4. **Compare and Contrast** Which vehicle's graph has a steeper slope? What is the slope equal to?
5. **Infer** What would a horizontal line mean on the graph? A line with a steeper slope?

Going Further

Speed is distance traveled divided by the amount of time to travel that distance. Explain how you could design your own experiment to measure speed in the classroom using remote-controlled cars. What would you use for markers? How precisely could you measure distance and time? Would the angle at which you measured the cars passing the markers affect the results? How much? How could you improve your measurements? What units make sense for speed? How far into the future could you predict the cars' positions? If possible, carry out the experiment and summarize your results.

Real-World Physics

When the speedometer is observed by a front-seat passenger, the driver, and a passenger in the rear driver's-side seat, readings of 90 km/h, 100 km/h, and 110 km/h, respectively, are observed. Explain the differences.

Share Your Data

Design an Experiment Visit physicspp.com/internet_lab to post your experiment for measuring speed in the classroom using remote-controlled cars. Include your list of materials, your procedure, and your predictions for the accuracy of your lab. If you actually perform your lab, post your data and results as well.

To find out more about measurement, visit the Web site: physicspp.com

Future Technology

Computer History and Growth

Each pixel of the animations or movies you watch, and each letter of the instant messages you send presents your computer with several hundred equations. Each equation must be solved in a few billionths of a second—if it takes a bit longer, you might complain that your computer is slow.

Early Computers The earliest computers could solve very complex arrays of equations, just as yours can, but it took them a lot longer to do so. There were several reasons for this. First, the mathematics of algorithms (problem-solving strategies) still was new. Computer scientists were only beginning to learn how to arrange a particular problem, such as the conversion of a picture into an easily-transmittable form, so that it could be solved by a machine.

UNIVAC 1, an early computer, filled an entire room.

Machine Size Second, the machines were physically large. Computers work by switching patterns of electric currents that represent binary numbers. A 16-bit machine works with binary numbers that are 16 bits long. If a 64-bit number must be dealt with, the machine must repeat the same operation four times. A 32-bit machine would have to repeat the operation only twice, thus making it that much faster. But a 32-bit machine is four times the size of a 16-bit machine; that is, it has four times as many wires and transistor switches, and even 8-bit machines were the size of the old UNIVAC shown above.

Moreover, current travels along wires at speeds no greater than about two-thirds the speed of light. This is a long time if the computer wires are 15 m long and must move information in less than 10^{-9} s.

Memory Third, electronic memories were extremely expensive. You may know that a larger memory lets your computer work faster. When one byte of memory required eight circuit boards, 1024 bytes (or 1 K) of memory was enormous. Because memory was so precious, computer programs had to be written with great cleverness. Astronants got to the Moon with 64 K of memory in *Apollo's* on-board computers.

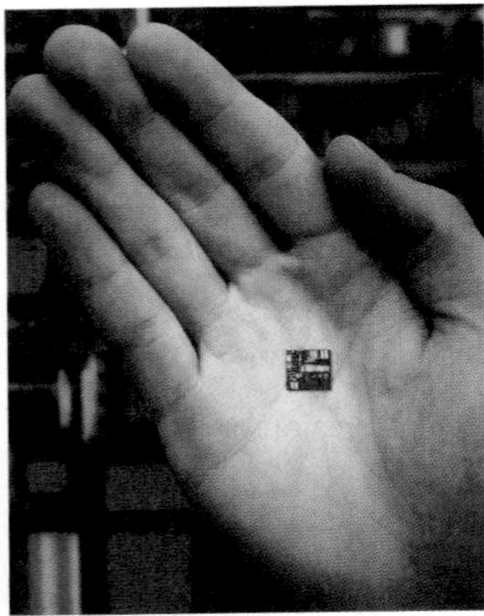

Processor chips used in today's computers are tiny compared to the old computer systems.

When Jack St. Clair Kilby and others invented the integrated circuit around 1960, the size and cost of computer circuitry dropped drastically. Physically smaller, and thus faster, machines could be built and very large memories became possible. Today, the transistors on a chip are now smaller than bacteria.

The cost and size of computers have dropped so much that your cell phone has far more computing power than most big office machines of the 1970s.

Going Further

1. **Research** A compression protocol makes a computer file smaller and less prone to transmission errors. Look up the terms *.jpg*, *.mp3*, *.mpeg*, and *.midi* and see how they apply to the activities you do on your computer.
2. **Calculate** Using the example here, how long does it take for a binary number to travel 15 m? How many such operations could there be each second?

Chapter 1

Study Guide

1.1 Mathematics and Physics

Vocabulary

- physics (p. 3)
- dimensional analysis (p. 6)
- significant digits (p. 7)
- scientific method (p. 8)
- hypothesis (p. 8)
- scientific law (p. 9)
- scientific theory (p. 10)

Key Concepts

- Physics is the study of matter and energy and their relationships.
- Dimensional analysis is used to check that an answer will be in the correct units.
- The result of any mathematical operation with measurements never can be more precise than the least-precise measurement involved in the operation.
- The scientific method is a systematic method of observing, experimenting, and analyzing to answer questions about the natural world.
- Scientific ideas change in response to new data.
- Scientific laws and theories are well-established descriptions and explanations of nature.

1.2 Measurement

Vocabulary

- measurement (p. 11)
- precision (p. 12)
- accuracy (p. 13)

Key Concepts

- New scientific findings must be reproducible; that is, others must be able to measure and find the same results.
- All measurements are subject to some uncertainty.
- Precision is the degree of exactness with which a quantity is measured. Scientific notation shows how precise a measurement is.
- Accuracy is the extent to which a measurement matches the true value.

1.3 Graphing Data

Vocabulary

- independent variable (p. 15)
- dependent variable (p. 15)
- line of best fit (p. 15)
- linear relationship (p. 16)
- quadratic relationship (p. 17)
- inverse relationship (p. 18)

Key Concepts

- Data are plotted in graphical form to show the relationship between two variables.
- The line that best passes through or near graphed data is called the line of best fit. It is used to describe the data and to predict where new data would lie on the graph.
- A graph in which data points lie on a straight line is a graph of a linear relationship. In the equation, m and b are constants.

$$y = mx + b$$

- The slope of a straight-line graph is the vertical change (rise) divided by the horizontal change (run) and often has a physical meaning.

$$m = \frac{rise}{run} = \frac{\Delta y}{\Delta x}$$

- The graph of a quadratic relationship is a parabolic curve. It is represented by the equation below. The constants a, b, and c can be found with a computer or a graphing calculator; simpler ones can be found using algebra.

$$y = ax^2 + bx + c$$

- The graph of an inverse relationship between x and y is a hyperbolic curve. It is represented by the equation below, where a is a constant.

$$y = \frac{a}{x}$$

Chapter 1 Assessment

Concept Mapping

30. Complete the following concept map using the following terms: *hypothesis, graph, mathematical model, dependent variable, measurement.*

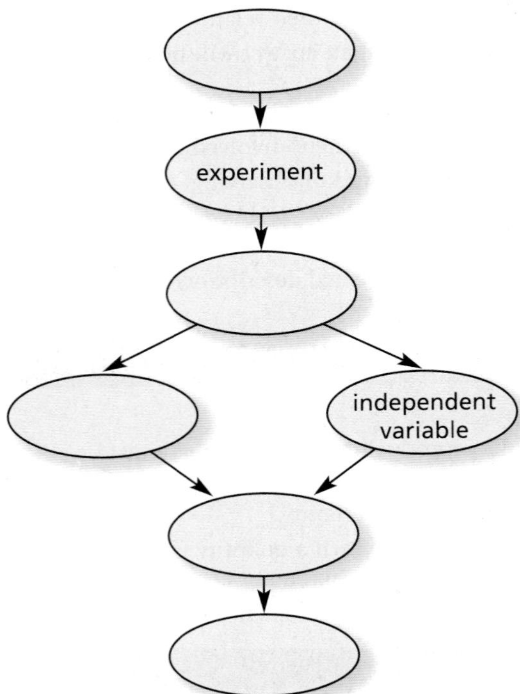

Mastering Concepts

31. Describe a scientific method. (1.1)

32. Why is mathematics important to science? (1.1)

33. What is the SI system? (1.1)

34. How are base units and derived units related? (1.1)

35. Suppose your lab partner recorded a measurement as 100 g. (1.1)
    - **a.** Why is it difficult to tell the number of significant digits in this measurement?
    - **b.** How can the number of significant digits in such a number be made clear?

36. Give the name for each of the following multiples of the meter. (1.1)

    a. $\frac{1}{100}$ m **b.** $\frac{1}{1000}$ m **c.** 1000 m

37. To convert 1.8 h to minutes, by what conversion factor should you multiply? (1.1)

38. Solve each problem. Give the correct number of significant digits in the answers. (1.1)
    - **a.** 4.667×10^4 g + 3.02×10^5 g
    - **b.** (1.70×10^2 J) ÷ (5.922×10^{-4} cm^3)

39. What determines the precision of a measurement? (1.2)

40. How does the last digit differ from the other digits in a measurement? (1.2)

41. A car's odometer measures the distance from home to school as 3.9 km. Using string on a map, you find the distance to be 4.2 km. Which answer do you think is more accurate? What does *accurate* mean? (1.2)

42. How do you find the slope of a linear graph? (1.3)

43. For a driver, the time between seeing a stoplight and stepping on the brakes is called reaction time. The distance traveled during this time is the reaction distance. Reaction distance for a given driver and vehicle depends linearly on speed. (1.3)
    - **a.** Would the graph of reaction distance versus speed have a positive or a negative slope?
    - **b.** A driver who is distracted has a longer reaction time than a driver who is not. Would the graph of reaction distance versus speed for a distracted driver have a larger or smaller slope than for a normal driver? Explain.

44. During a laboratory experiment, the temperature of the gas in a balloon is varied and the volume of the balloon is measured. Which quantity is the independent variable? Which quantity is the dependent variable? (1.3)

45. What type of relationship is shown in **Figure 1-20?** Give the general equation for this type of relation. (1.3)

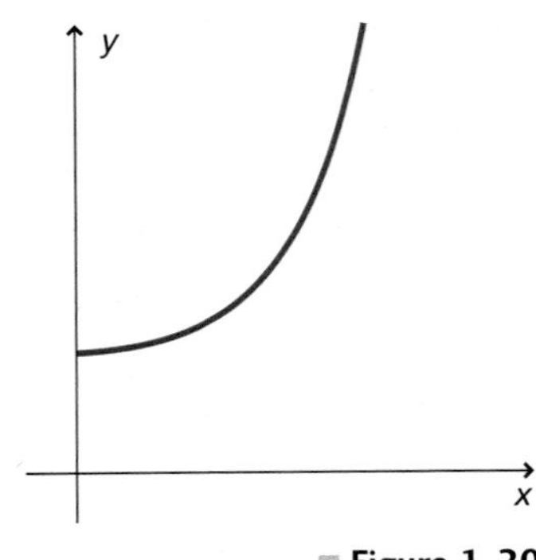

Figure 1-20

46. Given the equation $F = mv^2/R$, what relationship exists between each of the following? (1.3)
    - **a.** F and R
    - **b.** F and m
    - **c.** F and v

 For more problems, go to Additional Problems, Appendix B.

Applying Concepts

47. **Figure 1-21** gives the height above the ground of a ball that is thrown upward from the roof of a building, for the first 1.5 s of its trajectory. What is the ball's height at $t = 0$? Predict the ball's height at $t = 2$ s and at $t = 5$ s.

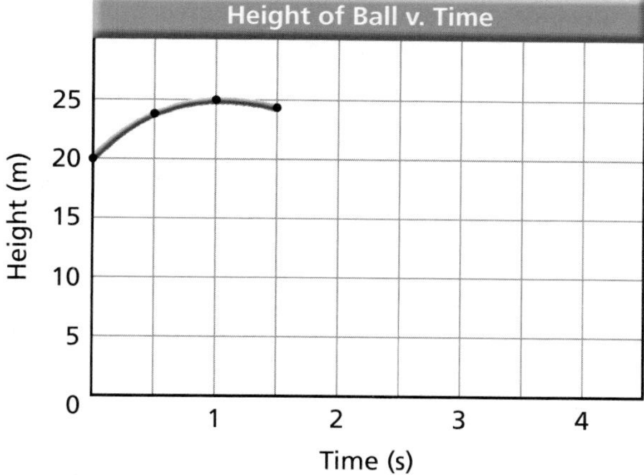

Figure 1-21

48. Is a scientific method one set of clearly defined steps? Support your answer.

49. Explain the difference between a scientific theory and a scientific law.

50. **Density** The density of a substance is its mass per unit volume.
a. Give a possible metric unit for density.
b. Is the unit for density a base unit or a derived unit?

51. What metric unit would you use to measure each of the following?
a. the width of your hand
b. the thickness of a book cover
c. the height of your classroom
d. the distance from your home to your classroom

52. **Size** Make a chart of sizes of objects. Lengths should range from less than 1 mm to several kilometers. Samples might include the size of a cell, the distance light travels in 1 s, and the height of a room.

53. **Time** Make a chart of time intervals. Sample intervals might include the time between heartbeats, the time between presidential elections, the average lifetime of a human, and the age of the United States. Find as many very short and very long examples as you can.

54. **Speed of Light** Two students measure the speed of light. One obtains $(3.001 \pm 0.001)\times10^8$ m/s; the other obtains $(2.999 \pm 0.006)\times10^8$ m/s.
a. Which is more precise?
b. Which is more accurate? (You can find the speed of light in the back of this textbook.)

55. You measure the dimensions of a desk as 132 cm, 83 cm, and 76 cm. The sum of these measures is 291 cm, while the product is 8.3×10^5 cm^3. Explain how the significant digits were determined in each case.

56. **Money** Suppose you receive \$5.00 at the beginning of a week and spend \$1.00 each day for lunch. You prepare a graph of the amount you have left at the end of each day for one week. Would the slope of this graph be positive, zero, or negative? Why?

57. Data are plotted on a graph, and the value on the y-axis is the same for each value of the independent variable. What is the slope? Why? How does y depend on x?

58. **Driving** The graph of braking distance versus car speed is part of a parabola. Thus, the equation is written $d = av^2 + bv + c$. The distance, d, has units in meters, and velocity, v, has units in meters/second. How could you find the units of a, b, and c? What would they be?

59. How long is the leaf in **Figure 1-22?** Include the uncertainty in your measurement.

Figure 1-22

60. The masses of two metal blocks are measured. Block A has a mass of 8.45 g and block B has a mass of 45.87 g.
a. How many significant digits are expressed in these measurements?
b. What is the total mass of block A plus block B?
c. What is the number of significant digits for the total mass?
d. Why is the number of significant digits different for the total mass and the individual masses?

61. **History** Aristotle said that the speed of a falling object varies inversely with the density of the medium through which it falls.
a. According to Aristotle, would a rock fall faster in water (density 1000 kg/m^3), or in air (density 1 kg/m^3)?
b. How fast would a rock fall in a vacuum? Based on this, why would Aristotle say that there could be no such thing as a vacuum?

62. Explain the difference between a hypothesis and a scientific theory.

63. Give an example of a scientific law.

64. What reason might the ancient Greeks have had not to question the hypothesis that heavier objects fall faster than lighter objects? *Hint: Did you ever question which falls faster?*

65. **Mars** Explain what observations led to changes in scientists' ideas about the surface of Mars.

66. A graduated cylinder is marked every mL. How precise a measurement can you make with this instrument?

Mastering Problems

1.1 Mathematics and Physics

67. Convert each of the following measurements to meters.
   a. 42.3 cm
   b. 6.2 pm
   c. 21 km
   d. 0.023 mm
   e. 214 μm
   f. 57 nm

68. Add or subtract as indicated.
   a. 5.80×10^9 s + 3.20×10^8 s
   b. 4.87×10^{-6} m − 1.93×10^{-6} m
   c. 3.14×10^{-5} kg + 9.36×10^{-5} kg
   d. 8.12×10^7 g − 6.20×10^6 g

69. Rank the following mass measurements from least to greatest: 11.6 mg, 1021 μg, 0.000006 kg, 0.31 mg.

70. State the number of significant digits in each of the following measurements.
   a. 0.00003 m
   b. 64.01 fm
   c. 80.001 m
   d. 0.720 μg
   e. 2.40×10^6 kg
   f. 6×10^8 kg
   g. 4.07×10^{16} m

71. Add or subtract as indicated.
   a. 16.2 m + 5.008 m + 13.48 m
   b. 5.006 m + 12.0077 m + 8.0084 m
   c. 78.05 cm^2 − 32.046 cm^2
   d. 15.07 kg − 12.0 kg

72. Multiply or divide as indicated.
   a. $(6.2\times10^{18}\ \text{m})(4.7\times10^{-10}\ \text{m})$
   b. $(5.6\times10^{-7}\ \text{m})/(2.8\times10^{-12}\ \text{s})$
   c. $(8.1\times10^{-4}\ \text{km})(1.6\times10^{-3}\ \text{km})$
   d. $(6.5\times10^{5}\ \text{kg})/(3.4\times10^{3}\ \text{m}^3)$

73. **Gravity** The force due to gravity is $F = mg$ where $g = 9.80\ \text{m/s}^2$.
   a. Find the force due to gravity on a 41.63-kg object.
   b. The force due to gravity on an object is 632 kg·m/s^2. What is its mass?

74. **Dimensional Analysis** Pressure is measured in pascals, where 1 Pa = 1 kg/m·s^2. Will the following expression give a pressure in the correct units?

$$\frac{(0.55\ \text{kg})(2.1\ \text{m/s})}{9.8\ \text{m/s}^2}$$

1.2 Measurement

75. A water tank has a mass of 3.64 kg when it is empty and a mass of 51.8 kg when it is filled to a certain level. What is the mass of the water in the tank?

76. The length of a room is 16.40 m, its width is 4.5 m, and its height is 3.26 m. What volume does the room enclose?

77. The sides of a quadrangular plot of land are 132.68 m, 48.3 m, 132.736 m, and 48.37 m. What is the perimeter of the plot?

78. How precise a measurement could you make with the scale shown in **Figure 1-23**?

Figure 1-23

79. Give the measure shown on the meter in **Figure 1-24** as precisely as you can. Include the uncertainty in your answer.

Figure 1-24

 For more problems, go to Additional Problems, Appendix B.

80. Estimate the height of the nearest door frame in centimeters. Then measure it. How accurate was your estimate? How precise was your estimate? How precise was your measurement? Why are the two precisions different?

81. Base Units Give six examples of quantities you might measure in a physics lab. Include the units you would use.

82. Temperature The temperature drops from 24°C to 10°C in 12 hours.

a. Find the average temperature change per hour.

b. Predict the temperature in 2 more hours if the trend continues.

c. Could you accurately predict the temperature in 24 hours?

1.3 Graphing Data

83. Figure 1-25 shows the masses of three substances for volumes between 0 and 60 cm^3.

a. What is the mass of 30 cm^3 of each substance?

b. If you had 100 g of each substance, what would be their volumes?

c. In one or two sentences, describe the meaning of the slopes of the lines in this graph.

d. What is the y-intercept of each line? What does it mean?

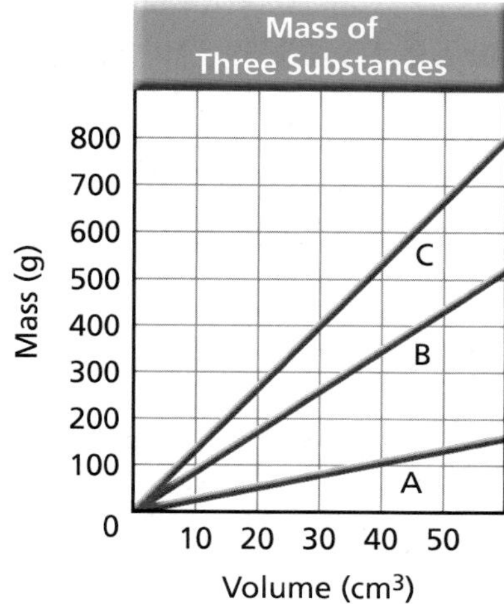

Figure 1-25

84. During a class demonstration, a physics instructor placed a mass on a horizontal table that was nearly frictionless. The instructor then applied various horizontal forces to the mass and measured the distance it traveled in 5 seconds for each force applied. The results of the experiment are shown in **Table 1-5.**

Table 1-5
Distance Traveled with Different Forces

Force (N)	Distance (cm)
5.0	24
10.0	49
15.0	75
20.0	99
25.0	120
30.0	145

a. Plot the values given in the table and draw the curve that best fits all points.

b. Describe the resulting curve.

c. Use the graph to write an equation relating the distance to the force.

d. What is the constant in the equation? Find its units.

e. Predict the distance traveled when a 22.0-N force is exerted on the object for 5 s.

85. The physics instructor from the previous problem changed the procedure. The mass was varied while the force was kept constant. Time and distance were measured, and the acceleration of each mass was calculated. The results of the experiment are shown in **Table 1-6.**

Table 1-6
Acceleration of Different Masses

Mass (kg)	Acceleration (m/s^2)
1.0	12.0
2.0	5.9
3.0	4.1
4.0	3.0
5.0	2.5
6.0	2.0

a. Plot the values given in the table and draw the curve that best fits all points.

b. Describe the resulting curve.

c. According to the graph, what is the relationship between mass and the acceleration produced by a constant force?

d. Write the equation relating acceleration to mass given by the data in the graph.

e. Find the units of the constant in the equation.

f. Predict the acceleration of an 8.0-kg mass.

86. During an experiment, a student measured the mass of 10.0 cm^3 of alcohol. The student then measured the mass of 20.0 cm^3 of alcohol. In this way, the data in **Table 1-7** were collected.

Table 1-7
The Mass Values of Specific Volumes of Alcohol

Volume (cm^3)	Mass (g)
10.0	7.9
20.0	15.8
30.0	23.7
40.0	31.6
50.0	39.6

a. Plot the values given in the table and draw the curve that best fits all the points.
b. Describe the resulting curve.
c. Use the graph to write an equation relating the volume to the mass of the alcohol.
d. Find the units of the slope of the graph. What is the name given to this quantity?
e. What is the mass of 32.5 cm^3 of alcohol?

Mixed Review

87. Arrange the following numbers from most precise to least precise

0.0034 m 45.6 m 1234 m

88. **Figure 1-26** shows the toroidal (doughnut-shaped) interior of the now-dismantled Tokamak Fusion Test Reactor. Explain why a width of 80 m would be an unreasonable value for the width of the toroid. What would be a reasonable value?

Figure 1-26

89. You are cracking a code and have discovered the following conversion factors: 1.23 longs = 23.0 mediums, and 74.5 mediums = 645 shorts. How many shorts are equal to one long?

90. You are given the following measurements of a rectangular bar: length = 2.347 m, thickness = 3.452 cm, height = 2.31 mm, mass = 1659 g. Determine the volume, in cubic meters, and density, in g/cm^3, of the beam. Express your results in proper form.

91. A drop of water contains 1.7×10^{21} molecules. If the water evaporated at the rate of one million molecules per second, how many years would it take for the drop to completely evaporate?

92. A 17.6-gram sample of metal is placed in a graduated cylinder containing 10.0 cm^3 of water. If the water level rises to 12.20 cm^3, what is the density of the metal?

Thinking Critically

93. **Apply Concepts** It has been said that fools can ask more questions than the wise can answer. In science, it is frequently the case that one wise person is needed to ask the right question rather than to answer it. Explain.

94. **Apply Concepts** Find the approximate mass of water in kilograms needed to fill a container that is 1.40 m long and 0.600 m wide to a depth of 34.0 cm. Report your result to one significant digit. (Use a reference source to find the density of water.)

95. **Analyze and Conclude** A container of gas with a pressure of 101 kPa has a volume of 324 cm^3 and a mass of 4.00 g. If the pressure is increased to 404 kPa, what is the density of the gas? Pressure and volume are inversely proportional.

96. **Design an Experiment** How high can you throw a ball? What variables might affect the answer to this question?

97. **Calculate** If the Sun suddenly ceased to shine, how long would it take Earth to become dark? (You will have to look up the speed of light in a vacuum and the distance from the Sun to Earth.) How long would it take the surface of Jupiter to become dark?

Writing in Physics

98. Research and describe a topic in the history of physics. Explain how ideas about the topic changed over time. Be sure to include the contributions of scientists and to evaluate the impact of their contributions on scientific thought and the world outside the laboratory.

99. Explain how improved precision in measuring time would have led to more accurate predictions about how an object falls.

 For more problems, go to Additional Problems, Appendix B.

Multiple Choice

1. Two laboratories use radiocarbon dating to measure the age of two wooden spear handles found in the same grave. Lab A finds an age of 2250 ± 40 years for the first object; lab B finds an age of 2215 ± 50 years for the second object. Which of the following is true?

- (A) Lab A's reading is more accurate than lab B's.
- (B) Lab A's reading is less accurate than lab B's.
- (C) Lab A's reading is more precise than lab B's.
- (D) Lab A's reading is less precise than lab B's.

2. Which of the following is equal to 86.2 cm?

- (A) 8.62 m
- (C) 8.62×10^{-4} km
- (B) 0.862 mm
- (D) 862 dm

3. Jario has a problem to do involving time, distance, and velocity, but he has forgotten the formula. The question asks him for a measurement in seconds, and the numbers that are given have units of m/s and km. What could Jario do to get the answer in seconds?

- (A) Multiply the km by the m/s, then multiply by 1000.
- (B) Divide the km by the m/s, then multiply by 1000.
- (C) Divide the km by the m/s, then divide by 1000.
- (D) Multiply the km by the m/s, then divide by 1000.

4. What is the slope of the graph?

- (A) 0.25 m/s^2
- (C) 2.5 m/s^2
- (B) 0.4 m/s^2
- (D) 4.0 m/s^2

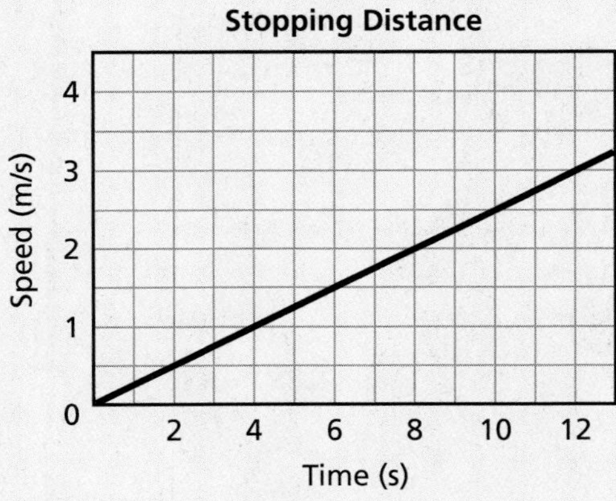

5. Which formula is equivalent to $D = \frac{m}{V}$?

- (A) $V = \frac{m}{D}$
- (C) $V = \frac{mD}{V}$
- (B) $V = Dm$
- (D) $V = \frac{D}{m}$

Extended Answer

6. You want to calculate an acceleration, in units of m/s^2, given a force, in N, and the mass, in g, on which the force acts. ($1\ N = 1\ kg{\cdot}m/s^2$)

- **a.** Rewrite the equation $F = ma$ so a is in terms of m and F.
- **b.** What conversion factor will you need to multiply by to convert grams to kilograms?
- **c.** A force of 2.7 N acts on a 350-g mass. Write the equation you will use, including the conversion factor, to find the acceleration.

7. Find an equation for a line of best fit for the data shown below.

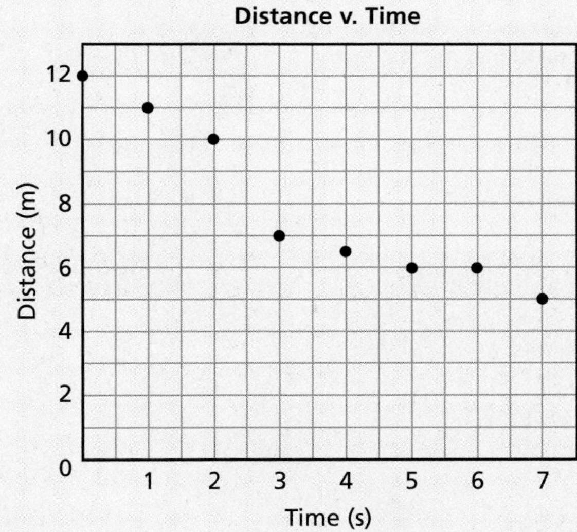

✓ Test-Taking TIP

Skip Around if You Can

You may want to skip over difficult questions and come back to them later, after you've answered the easier questions. This will guarantee more points toward your final score. In fact, other questions may help you answer the ones you skipped. Just be sure you fill in the correct ovals on your answer sheet.

Chapter 2

Representing Motion

What You'll Learn

- You will represent motion through the use of words, motion diagrams, and graphs.
- You will use the terms *position, distance, displacement,* and *time interval* in a scientific manner to describe motion.

Why It's Important

Without ways to describe and analyze motion, travel by plane, train, or bus would be chaotic at best. Times and speeds determine the winners of races as well as transportation schedules.

Running a Marathon As one runner passes another, the speed of the overtaking runner is greater than the speed of the other runner.

Think About This ▶
How can you represent the motion of two runners?

physicspp.com

Which car is faster?

Question

In a race between two toy cars, can you explain which car is faster?

Procedure

1. Obtain two toy cars, either friction cars or windup cars. Place the cars on your lab table or other surface recommended by your teacher.
2. Decide on a starting line for the race.
3. Release both cars from the same starting line at the same time. Note that if you are using windup cars, you will need to wind them up before you release them. Be sure to pull the cars back before release if they are friction cars.
4. **Observe** Watch the two cars closely as they move and determine which car is moving faster.
5. Repeat steps 1–3, but this time collect one type of data to support your conclusion about which car is faster.

Analysis

What data did you collect to show which car was moving faster? What other data could you collect to determine which car is faster?

Critical Thinking Write an operational definition of average speed.

2.1 Picturing Motion

In the previous chapter, you learned about the scientific processes that will be useful in your study of physics. In this chapter, you will begin to use these tools to analyze motion. In subsequent chapters, you will apply them to all kinds of movement using sketches, diagrams, graphs, and equations. These concepts will help you to determine how fast and how far an object will move, whether the object is speeding up or slowing down, and whether it is standing still or moving at a constant speed. Perceiving motion is instinctive—your eyes naturally pay more attention to moving objects than to stationary ones. Movement is all around you—from fast trains and speedy skiers to slow breezes and lazy clouds. Movements travel in many directions, such as the straight-line path of a bowling ball in a lane's gutter, the curved path of a tether ball, the spiral of a falling kite, and the swirls of water circling a drain.

► **Objectives**

- **Draw** motion diagrams to describe motion.
- **Develop** a particle model to represent a moving object.

► **Vocabulary**

motion diagram
particle model

All Kinds of Motion

Figure 2-1 An object in motion changes its position as it moves. In this photo, the camera was focused on the rider, so the blurry background indicates that the rider's position has changed.

What comes to your mind when you hear the word *motion?* A speeding automobile? A spinning ride at an amusement park? A baseball soaring over a fence for a home run? Or a child swinging back and forth in a regular rhythm? When an object is in motion, as shown in **Figure 2-1,** its position changes. Its position can change along the path of a straight line, a circle, an arc, or a back-and-forth vibration.

Some of the types of motion described above appear to be more complicated than others. When beginning a new area of study, it is generally a good idea to begin with what appears to be the least complicated situation, learn as much as possible about it, and then gradually add more complexity to that simple model. In the case of motion, you will begin your study with movement along a straight line.

Movement along a straight line Suppose that you are reading this textbook at home. At the beginning of Chapter 2, you glance over at your pet hamster and see that he is sitting in a corner of his cage. Sometime later, you look over again, and you see that he now is sitting by the food dish in the opposite corner of the cage. You can infer that he has moved from one place to another in the time in between your observations. Thus, a description of motion relates to place and time. You must be able to answer the questions of where and when an object is positioned to describe its motion. Next, you will look at some tools that are useful in determining when an object is at a particular place.

Motion Diagrams

Consider an example of straight-line motion: a runner is jogging along a straight path. One way of representing the motion of the runner is to create a series of images showing the positions of the runner at equal time intervals. This can be done by photographing the runner in motion to obtain a series of images.

Suppose you point a camera in a direction perpendicular to the direction of motion, and hold it still while the motion is occurring. Then you take a series of photographs of the runner at equal time intervals. **Figure 2-2** shows what a series of consecutive images for a runner might look like. Notice that the runner is in a different position in each image, but everything in the background remains in the same position. This indicates that, relative to the ground, only the runner is in motion. What is another way of representing the runner's motion?

Concepts In Motion
Interactive Figure To see an animation on motion diagrams versus particle motion, visit physicspp.com.

Figure 2-2 If you relate the position of the runner to the background in each image over equal time intervals, you will conclude that she is in motion.

Suppose that you stacked the images from Figure 2-2, one on top of the other. **Figure 2-3** shows what such a stacked image might look like. You will see more than one image of the moving runner, but only a single image of the motionless objects in the background. A series of images showing the positions of a moving object at equal time intervals is called a **motion diagram.**

Figure 2-3 Stacking a series of images taken at regular time intervals and combining them into one image creates a motion diagram for the runner for one portion of her run. Reducing the runner's motion to a series of single points results in a particle model of her motion.

The Particle Model

Keeping track of the motion of the runner is easier if you disregard the movement of the arms and legs, and instead concentrate on a single point at the center of her body. In effect, you can disregard the fact that she has some size and imagine that she is a very small object located precisely at that central point. A **particle model** is a simplified version of a motion diagram in which the object in motion is replaced by a series of single points. To use the particle model, the size of the object must be much less than the distance it moves. The internal motions of the object, such as the waving of the runner's arms are ignored in the particle model. In the photographic motion diagram, you could identify one central point on the runner, such as a dot centered at her waistline, and take measurements of the position of the dot. The bottom part of Figure 2-3 shows the particle model for the runner's motion. You can see that applying the particle model produces a simplified version of the motion diagram. In the next section, you will learn how to create and use a motion diagram that shows how far an object moved and how much time it took to move that far.

2.1 Section Review

1. **Motion Diagram of a Runner** Use the particle model to draw a motion diagram for a bike rider riding at a constant pace.

2. **Motion Diagram of a Bird** Use the particle model to draw a simplified motion diagram corresponding to the motion diagram in **Figure 2-4** for a flying bird. What point on the bird did you choose to represent it?

Figure 2-4

3. **Motion Diagram of a Car** Use the particle model to draw a simplified motion diagram corresponding to the motion diagram in **Figure 2-5** for a car coming to a stop at a stop sign. What point on the car did you use to represent it?

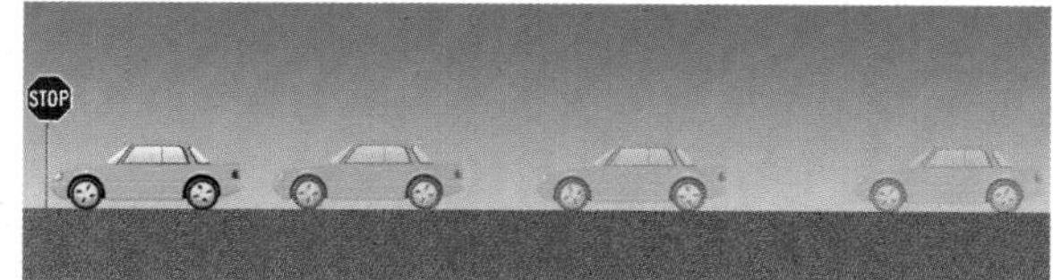

Figure 2-5

4. **Critical Thinking** Use the particle model to draw motion diagrams for two runners in a race, when the first runner crosses the finish line as the other runner is three-fourths of the way to the finish line.

2.2 Where and When?

▶ **Objectives**

- **Define** coordinate systems for motion problems.
- **Recognize** that the chosen coordinate system affects the sign of objects' positions.
- **Define** displacement.
- **Determine** a time interval.
- **Use** a motion diagram to answer questions about an object's position or displacement.

▶ **Vocabulary**

coordinate system
origin
position
distance
magnitude
vectors
scalars
resultant
time interval
displacement

Would it be possible to take measurements of distance and time from a motion diagram, such as the motion diagram of the runner? Before taking the photographs, you could place a meterstick or a measuring tape on the ground along the path of the runner. The measuring tape would tell you where the runner was in each image. A stopwatch or clock within the view of the camera could tell the time. But where should you place the end of the measuring tape? When should you start the stopwatch?

Coordinate Systems

When you decide where to place the measuring tape and when to start the stopwatch, you are defining a **coordinate system,** which tells you the location of the zero point of the variable you are studying and the direction in which the values of the variable increase. The **origin** is the point at which both variables have the value zero. In the example of the runner, the origin, represented by the zero end of the measuring tape, could be placed 6 m to the left of the tree. The motion is in a straight line; thus, your measuring tape should lie along that straight line. The straight line is an axis of the coordinate system. You probably would place the tape so that the meter scale increases to the right of the zero, but putting it in the opposite direction is equally correct. In **Figure 2-6a,** the origin of the coordinate system is on the left.

You can indicate how far away the runner is from the origin at a particular time on the simplified motion diagram by drawing an arrow from the origin to the point representing the runner, as shown in **Figure 2-6b.** This arrow represents the runner's **position,** the separation between an object and the origin. The length of the arrow indicates how far the object is from the origin, or the object's **distance** from the origin. The arrow points from the origin to the location of the moving object at a particular time.

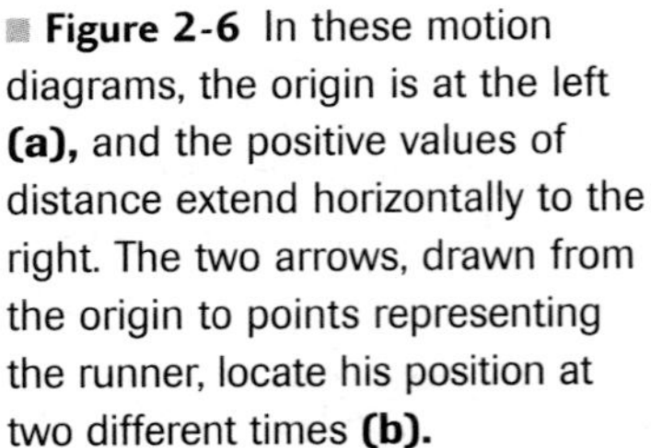

■ **Figure 2-6** In these motion diagrams, the origin is at the left **(a),** and the positive values of distance extend horizontally to the right. The two arrows, drawn from the origin to points representing the runner, locate his position at two different times **(b).**

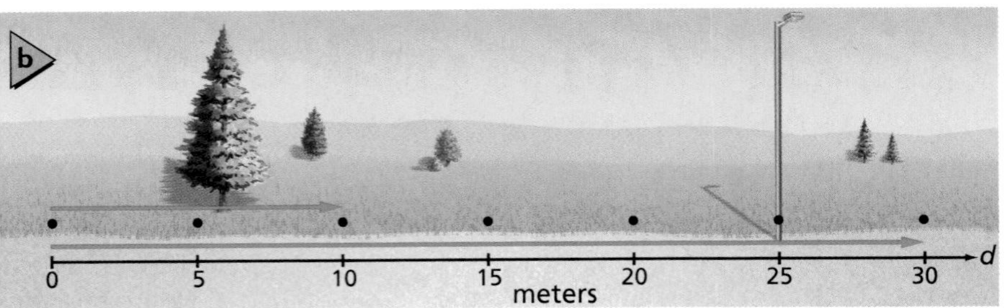

Is there such a thing as a negative position? Suppose you chose the coordinate system just described, placing the origin 4 m left of the tree with the *d*-axis extending in a positive direction to the right. A position 9 m to the left of the tree, 5 m left of the origin, would be a negative position, as shown in **Figure 2-7.** In the same way, you could discuss a time before the stopwatch was started.

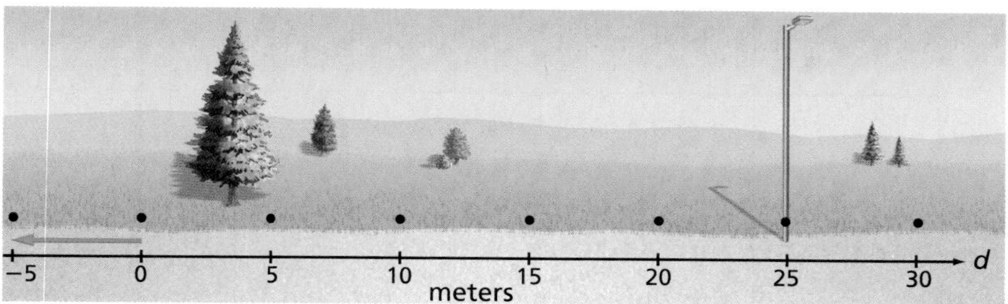

Figure 2-7 The arrow drawn on this motion diagram indicates a negative position.

Vectors and scalars Quantities that have both size, also called **magnitude,** and direction, are called **vectors,** and can be represented by arrows. Quantities that are just numbers without any direction, such as distance, time, or temperature, are called **scalars.** This textbook will use boldface letters to represent vector quantities and regular letters to represent scalars.

You already know how to add scalars; for example, $0.6 + 0.2 = 0.8$. How do you add vectors? Think about how you would solve the following problem. Your aunt asks you to get her some cold medicine at the store nearby. You walk 0.5 km east from your house to the store, buy the cold medicine, and then walk another 0.2 km east to your aunt's house. How far from the origin are you at the end of your trip? The answer, of course, is 0.5 km east + 0.2 km east = 0.7 km east. You also could solve this problem graphically, using the following method.

Using a ruler, measure and draw each vector. The length of a vector should be proportional to the magnitude of the quantity being represented, so you must decide on a scale for your drawing. For example, you might let 1 cm on paper represent 0.1 km. The important thing is to choose a scale that produces a diagram of reasonable size with a vector that is about 5–10 cm long. The vectors representing the two segments that made up your trip to your aunt's house are shown in **Figure 2-8,** drawn to a scale of 1 cm, which represents 0.1 km. The vector that represents the total of these two, shown here with a dotted line, is 7 cm long. According to the established scale, you were 0.7 km from the origin at the end of your trip. The vector that represents the sum of the other two vectors is called the **resultant.** The resultant always points from the tail of the first vector to the tip of the last vector.

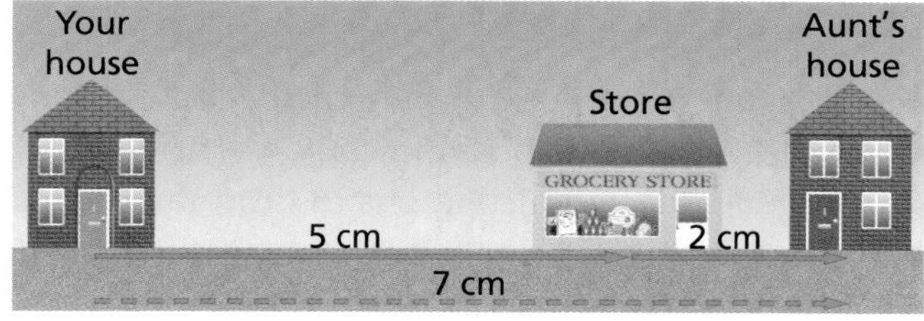

Figure 2-8 Add two vectors by placing them tip to tail. The resultant points from the tail of the first vector to the tip of the last vector.

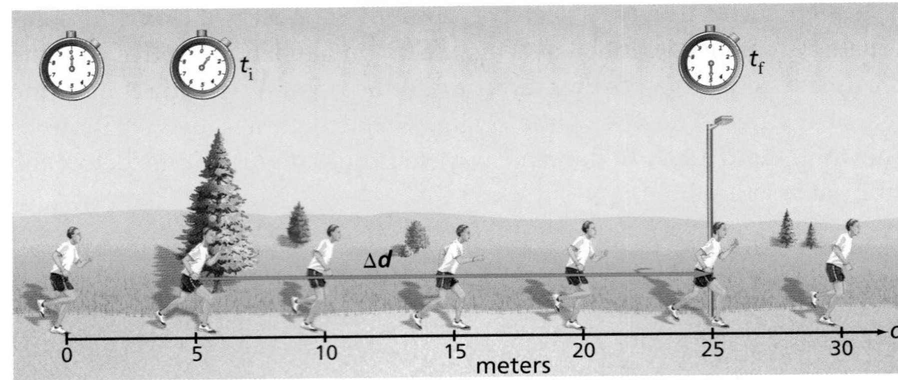

Figure 2-9 You can see that it took the runner 4.0 s to run from the tree to the lamppost. The initial position of the runner is used as a reference point. The vector from position 1 to position 2 indicates both the direction and amount of displacement during this time interval.

- Displacement vectors are shown in green.

Time Intervals and Displacements

When analyzing the runner's motion, you might want to know how long it took the runner to travel from the tree to the lamppost. This value is obtained by finding the difference in the stopwatch readings at each position. Assign the symbol t_i to the time when the runner was at the tree and the symbol t_f to the time when he was at the lamppost. The difference between two times is called a **time interval.** A common symbol for a time interval is Δt, where the Greek letter delta, Δ, is used to represent a change in a quantity. The time interval is defined mathematically as follows.

Time Interval $\Delta t = t_f - t_i$

The time interval is equal to the final time minus the initial time.

Although i and f are used to represent the initial and final times, they can be the initial and final times of any time interval you choose. In the example of the runner, the time it takes for him to go from the tree to the lamppost is $t_f - t_i = 5.0\ \text{s} - 1.0\ \text{s} = 4.0\ \text{s}$. How did the runner's position change when he ran from the tree to the lamppost, as shown in **Figure 2-9?** The symbol $\boldsymbol{d}$ may be used to represent position. In common speech, a position refers to a place; but in physics, a position is a vector with its tail at the origin of a coordinate system and its tip at the place.

Figure 2-9 shows $\Delta \boldsymbol{d}$, an arrow drawn from the runner's position at the tree to his position at the lamppost. This vector represents his change in position, or **displacement,** during the time interval between t_i and t_f. The length of the arrow represents the distance the runner moved, while the direction the arrow points indicates the direction of the displacement. Displacement is mathematically defined as follows.

Displacement $\Delta \boldsymbol{d} = \boldsymbol{d}_f - \boldsymbol{d}_i$

Displacement is equal to the final position minus the initial position.

Again, the initial and final positions are the beginning and end of any interval you choose. Also, while position can be considered a vector, it is common practice when doing calculations to drop the boldface, and use signs and magnitudes. This is because position usually is measured from the origin, and direction typically is included with the position indication.

Figure 2-10 Start with two vectors, ***A*** and ***B*** **(a).** To subtract vector ***B*** from vector ***A,*** first reverse vector ***B,*** then add them together to obtain the resultant, ***R*** **(b).**

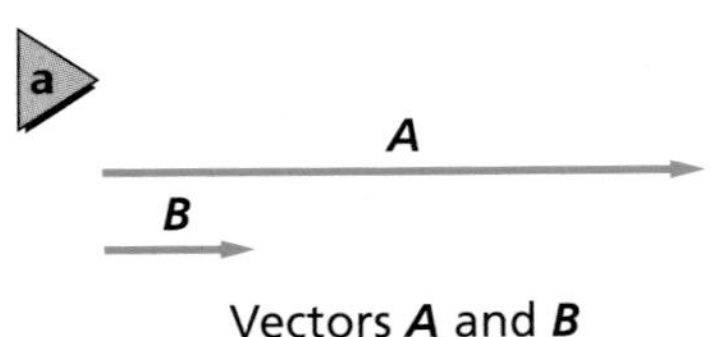

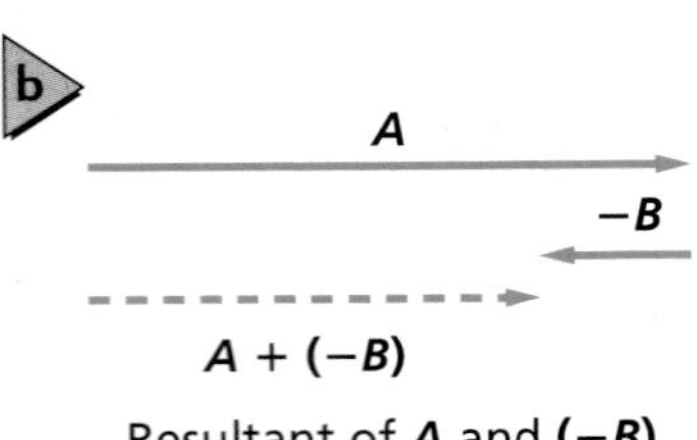

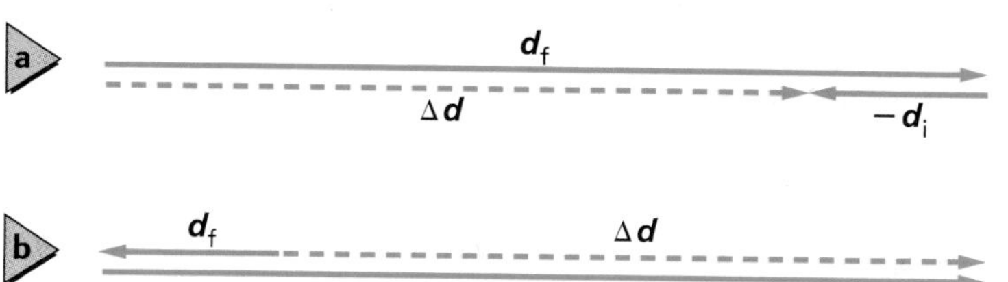

Figure 2-11 The displacement of the runner during the 4.0-s time interval is found by subtracting d_f from d_i. In **(a)** the origin is at the left, and in **(b)** it is at the right. Regardless of your choice of coordinate system, Δd is the same.

How do you subtract vectors? Reverse the subtracted vector and add. This is because $A - B = A + (-B)$. **Figure 2-10a** shows two vectors, A, 4 cm long pointing east, and B, 1 cm long also pointing east. **Figure 2-10b** shows $-B$, which is 1 cm long pointing west. Figure 2-10b shows the resultant of A and $-B$. It is 3 cm long pointing east.

To determine the length and direction of the displacement vector, $\Delta d = d_f - d_i$, draw $-d_i$, which is d_i reversed. Then draw d_f and copy $-d_i$ with its tail at d_f's tip. Add d_f and $-d_i$. In the example of the runner, his displacement is $d_f - d_i = 25.0\text{ m} - 5.0\text{ m} = 20.0\text{ m}$. He moved to the right of the tree. To completely describe an object's displacement, you must indicate the distance it traveled and the direction it moved. Thus, displacement, a vector, is not identical to distance, a scalar; it is distance and direction.

What would happen if you chose a different coordinate system; that is, if you measured the position of the runner from another location? Look at Figure 2-9, and suppose you change the right side of the d-axis to be zero. While the vectors drawn to represent each position change, the length and direction of the displacement vector does not, as shown in **Figures 2-11a** and **b.** The displacement, Δd, in the time interval from 1.0 s to 5.0 s does not change. Because displacement is the same in any coordinate system, you frequently will use displacement when studying the motion of an object. The displacement vector is always drawn with its flat end, or tail, at the earlier position, and its point, or tip, at the later position.

2.2 Section Review

5. Displacement The particle model for a car traveling on an interstate highway is shown below. The starting point is shown.

Here • • • • • • There

Make a copy of the particle model, and draw a vector to represent the displacement of the car from the starting time to the end of the third time interval.

6. Displacement The particle model for a boy walking to school is shown below.

Home • • • • • • • • • • School

Make a copy of the particle model, and draw vectors to represent the displacement between each pair of dots.

7. Position Two students compared the position vectors they each had drawn on a motion diagram to show the position of a moving object at the same time. They found that their vectors did not point in the same direction. Explain.

8. Critical Thinking A car travels straight along the street from the grocery store to the post office. To represent its motion you use a coordinate system with its origin at the grocery store and the direction the car is moving in as the positive direction. Your friend uses a coordinate system with its origin at the post office and the opposite direction as the positive direction. Would the two of you agree on the car's position? Displacement? Distance? The time interval the trip took? Explain.

2.3 Position-Time Graphs

► **Objectives**

- **Develop** position-time graphs for moving objects.
- **Use** a position-time graph to interpret an object's position or displacement.
- **Make** motion diagrams, pictorial representations, and position-time graphs that are equivalent representations describing an object's motion.

► **Vocabulary**

position-time graph
instantaneous position

When analyzing motion, particularly when it is more complex than the examples considered so far, it often is useful to represent the motion of an object in a variety of ways. As you have seen, a motion diagram contains useful information about an object's position at various times and can be helpful in determining the displacement of an object during time intervals. Graphs of the object's position and time also contain this information.

Review Figure 2-9, the motion diagram for the runner with a location to the left of the tree chosen as the origin. From this motion diagram, you can organize the times and corresponding positions of the runner, as in **Table 2-1.**

Using a Graph to Find Out Where and When

The data from **Table 2-1** can be presented by plotting the time data on a horizontal axis and the position data on a vertical axis, which is called a **position-time graph.** The graph of the runner's motion is shown in **Figure 2-12.** To draw this graph, first plot the runner's recorded positions. Then, draw a line that best fits the recorded points. Notice that this graph is not a picture of the path taken by the runner as he was moving—the graphed line is sloped, but the path that he ran was flat. The line represents the most likely positions of the runner at the times between the recorded data points. (Recall from Chapter 1 that this line often is referred to as a best-fit line.) This means that even though there is no data point to tell you exactly when the runner was 30.0 m beyond his starting point or where he was at t = 4.5 s, you can use the graph to estimate his position. The following example problem shows how. Note that before estimating the runner's position, the questions first are restated in the language of physics in terms of positions and times.

Concepts In MOtion
Interactive Figure To see an animation on position-time graphs, visit physicspp.com.

Table 2-1
Position v. Time

Time t (s)	Position d (m)
0.0	0.0
1.0	5.0
2.0	10.0
3.0	15.0
4.0	20.0
5.0	25.0
6.0	30.0

Figure 2-12 A position-time graph for the runner can be created by plotting his known position at each of several times. After these points are plotted, the line that best fits them is drawn. The best-fit line indicates the runner's most likely positions at the times between the data points.

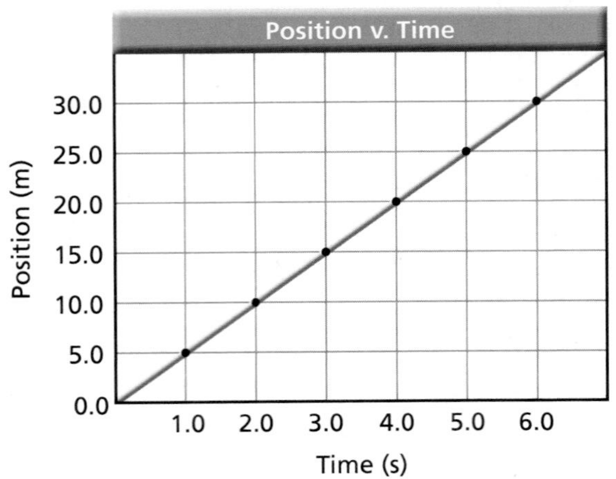

▶ EXAMPLE Problem 1

When did the runner whose motion is described in Figure 2-12 reach 30.0 m beyond the starting point? Where was he after 4.5 s?

1 Analyze the Problem

- Restate the questions.

  Question 1: At what time was the position of the object equal to 30.0 m?

  Question 2: What was the position of the object at 4.5 s?

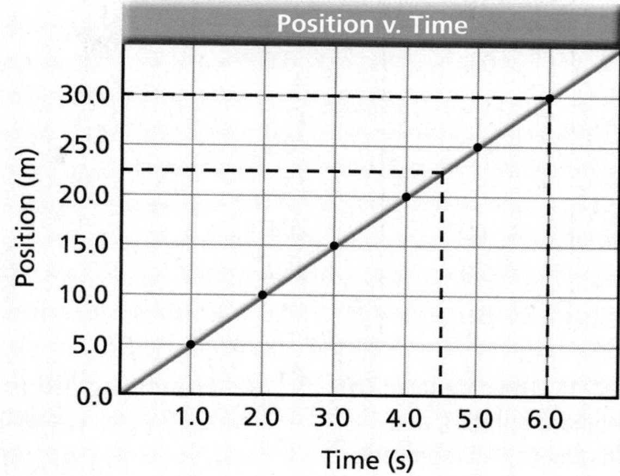

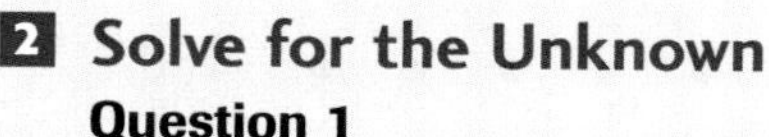

2 Solve for the Unknown

Question 1

Examine the graph to find the intersection of the best-fit line with a horizontal line at the 30.0-m mark. Next, find where a vertical line from that point crosses the time axis. The value of t there is 6.0 s.

Question 2

Find the intersection of the graph with a vertical line at 4.5 s (halfway between 4.0 s and 5.0 s on this graph). Next, find where a horizontal line from that point crosses the position axis. The value of d is approximately 22.5 m.

The two intersections are shown on the graph above.

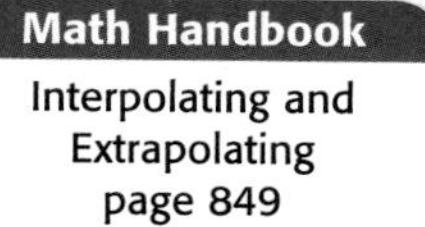

▶ PRACTICE Problems

• Additional Problems, Appendix B
• Solutions to Selected Problems, Appendix C

For problems 9–11, refer to ***Figure 2-13.***

9. Describe the motion of the car shown by the graph.

10. Draw a motion diagram that corresponds to the graph.

11. Answer the following questions about the car's motion. Assume that the positive d-direction is east and the negative d-direction is west.

a. When was the car 25.0 m east of the origin?

b. Where was the car at 1.0 s?

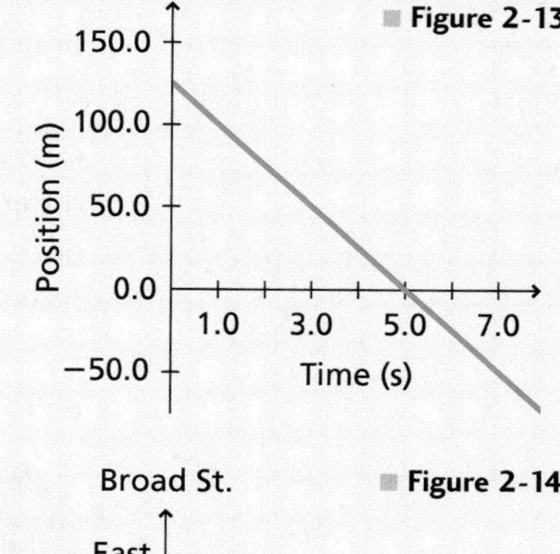

■ **Figure 2-13**

12. Describe, in words, the motion of the two pedestrians shown by the lines in **Figure 2-14.** Assume that the positive direction is east on Broad Street and the origin is the intersection of Broad and High Streets.

13. Odina walked down the hall at school from the cafeteria to the band room, a distance of 100.0 m. A class of physics students recorded and graphed her position every 2.0 s, noting that she moved 2.6 m every 2.0 s. When was Odina in the following positions?

a. 25.0 m from the cafeteria

b. 25.0 m from the band room

c. Create a graph showing Odina's motion.

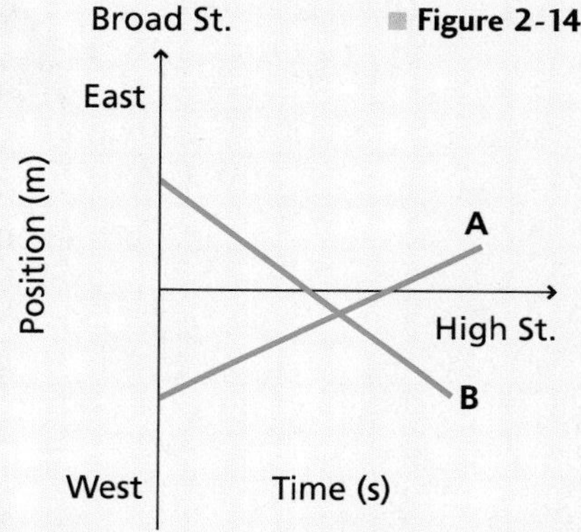

■ **Figure 2-14**

a

Table 2-1

Position v. Time

Time t (s)	Position d (m)
0.0	0.0
1.0	5.0
2.0	10.0
3.0	15.0
4.0	20.0
5.0	25.0
6.0	30.0

b

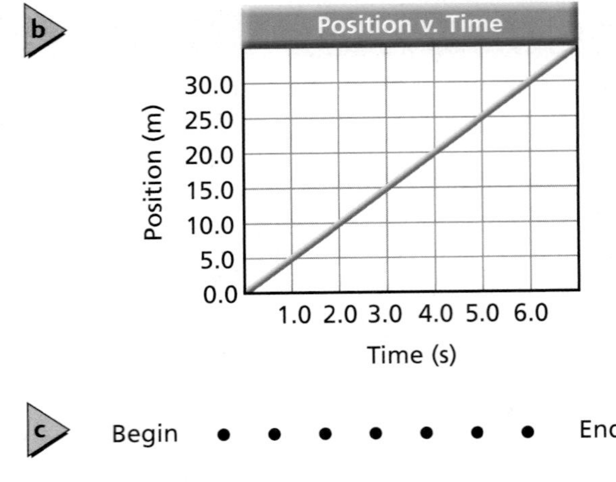

c Begin • • • • • • • End

Figure 2-15 The data table **(a)**, position-time graph **(b)**, and particle model **(c)** all represent the same moving object.

How long did the runner spend at any location? Each position has been linked to a time, but how long did that time last? You could say "an instant," but how long is that? If an instant lasts for any finite amount of time, then the runner would have stayed at the same position during that time, and he would not have been moving. However, as he was moving, an instant is not a finite period of time. This means that an instant of time lasts zero seconds. The symbol $\boldsymbol{d}$ represents the **instantaneous position** of the runner—the position at a particular instant.

Equivalent representations As shown in **Figure 2-15,** you now have several different ways to describe motion: words, pictures (or pictorial representations), motion diagrams, data tables, and position-time graphs. All of these representations are equivalent. That is, they can all contain the same information about the runner's motion. However, depending on what you want to find out about an object's motion, some of the representations will be more useful than others. In the pages that follow, you will get some practice constructing these equivalent representations and learning which ones are the easiest to use in solving different kinds of problems.

Considering the motion of multiple objects A position-time graph for two different runners in a race is shown in Example Problem 2. When and where does one runner pass the other? First, you need to restate this question in physics terms: At what time do the two objects have the same position? You can evaluate this question by identifying the point on the position-time graph at which the lines representing the two objects intersect.

CHALLENGE PROBLEM

Niram, Oliver, and Phil all enjoy exercising and often go to a path along the river for this purpose. Niram bicycles at a very consistent 40.25 km/h, Oliver runs south at a constant speed of 16.0 km/h, and Phil walks south at a brisk 6.5 km/h. Niram starts biking north at noon from the waterfalls. Oliver and Phil both start at 11:30 A.M. at the canoe dock, 20.0 km north of the falls.

1. Draw position-time graphs for each person.
2. At what time will the three exercise enthusiasts be within the smallest distance interval?
3. What is the length of that distance interval?

EXAMPLE Problem 2

When and where does runner B pass runner A?

1 Analyze the Problem

- Restate the question.
  At what time do A and B have the same position?

2 Solve for the Unknown

In the figure at right, examine the graph to find the intersection of the line representing the motion of A with the line representing the motion of B.

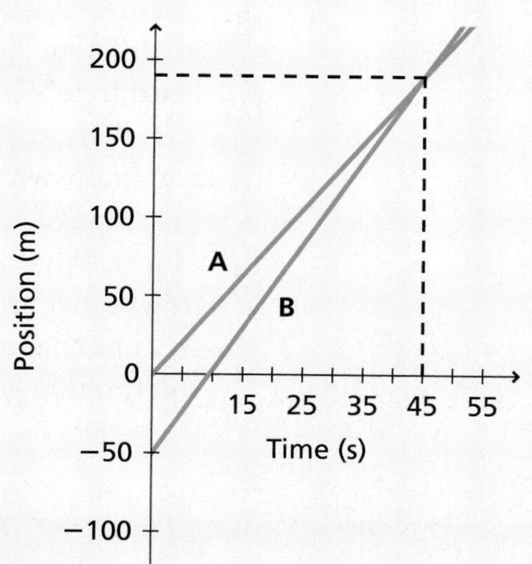

These lines intersect at 45 s and at about 190 m.

B passes A about 190 m beyond the origin, 45 s after A has passed the origin.

Math Handbook
Interpolating and Extrapolating
page 849

PRACTICE Problems

• Additional Problems, Appendix B
• Solutions to Selected Problems, Appendix C

For problems 14–17, refer to the figure in Example Problem 2.

14. What event occurred at $t = 0.0$ s?

15. Which runner was ahead at $t = 48.0$ s?

16. When runner A was at 0.0 m, where was runner B?

17. How far apart were runners A and B at $t = 20.0$ s?

18. Juanita goes for a walk. Sometime later, her friend Heather starts to walk after her. Their motions are represented by the position-time graphs in **Figure 2-16.**

a. How long had Juanita been walking when Heather started her walk?

b. Will Heather catch up to Juanita? How can you tell?

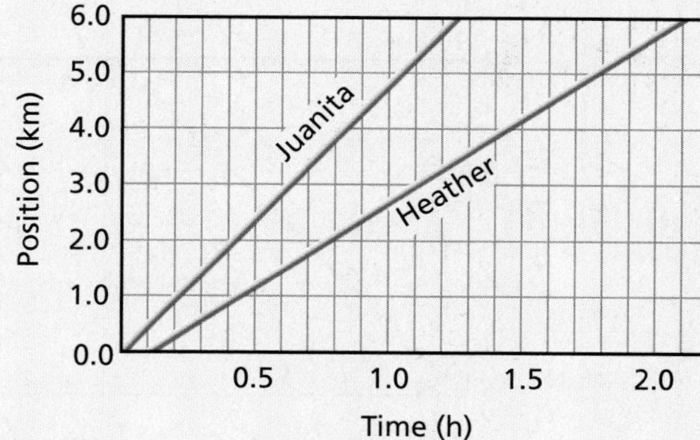

Figure 2-16

As you have seen, you can represent the motion of more than one object on a position-time graph. The intersection of two lines tells you when the two objects have the same position. Does this mean that they will collide? Not necessarily. For example, if the two objects are runners and if they are in different lanes, they will not collide. Later in this textbook, you will learn to represent motion in two dimensions.

Is there anything else that you can learn from a position-time graph? Do you know what the slope of a line means? In the next section, you will use the slope of a line on a position-time graph to determine the velocity of an object. What about the area under a plotted line? In Chapter 3, you will draw other graphs and learn to interpret the areas under the plotted lines. In later chapters you will continue to refine your skills with creating and interpreting graphs.

2.3 Section Review

19. Position-Time Graph From the particle model in **Figure 2-17** of a baby crawling across a kitchen floor, plot a position-time graph to represent his motion. The time interval between successive dots is 1 s.

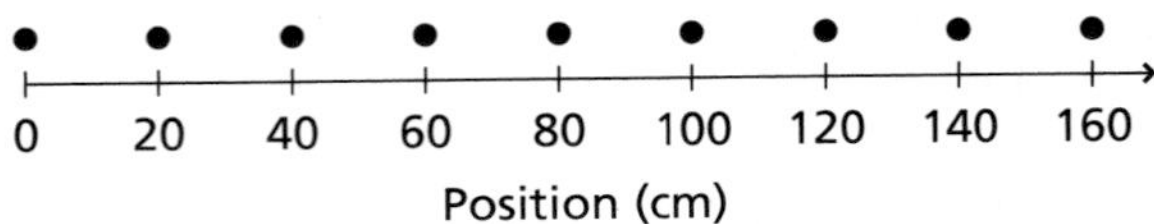

Figure 2-17

20. Motion Diagram Create a particle model from the position-time graph of a hockey puck gliding across a frozen pond in **Figure 2-18.**

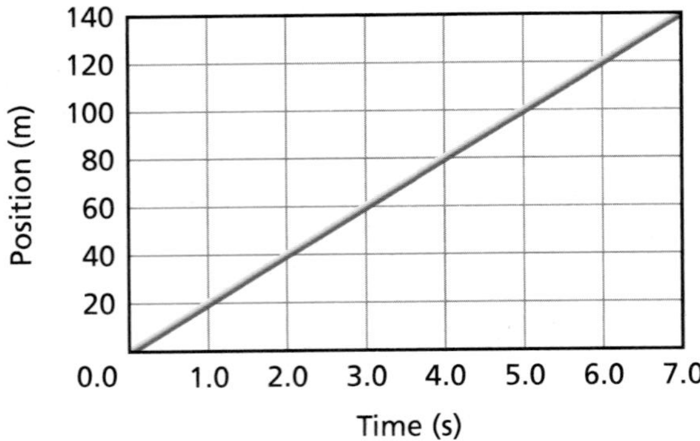

Figure 2-18

For problems 21–23, refer to Figure 2-18.

21. Time Use the position-time graph of the hockey puck to determine when it was 10.0 m beyond the origin.

22. Distance Use the position-time graph of the hockey puck to determine how far it moved between 0.0 s and 5.0 s.

23. Time Interval Use the position-time graph for the hockey puck to determine how much time it took for the puck to go from 40 m beyond the origin to 80 m beyond the origin.

24. Critical Thinking Look at the particle model and position-time graph shown in **Figure 2-19.** Do they describe the same motion? How do you know? Do not confuse the position coordinate system in the particle model with the horizontal axis in the position-time graph. The time intervals in the particle model are 2 s.

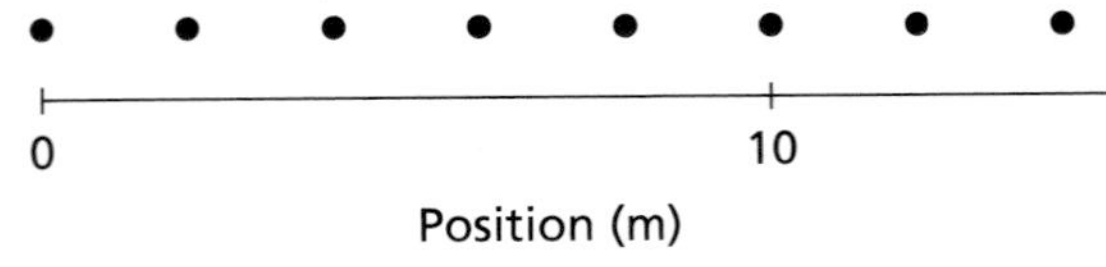

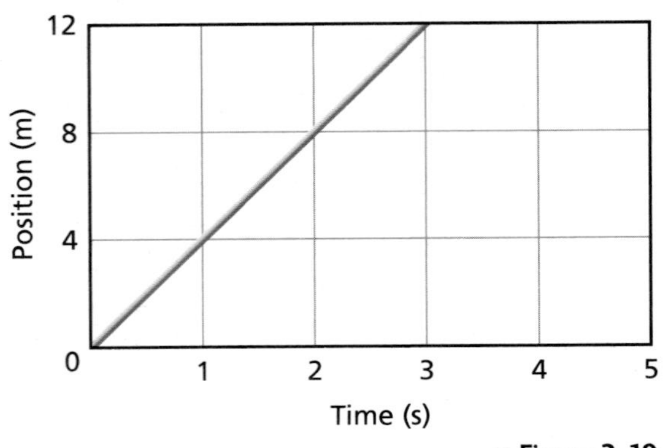

Figure 2-19

2.4 How Fast?

▶ Objectives

- **Define** velocity.
- **Differentiate** between speed and velocity.
- **Create** pictorial, physical, and mathematical models of motion problems.

▶ Vocabulary

average velocity
average speed
instantaneous velocity

You have learned how to use a motion diagram to show an object's movement. How can you measure how fast it is moving? With devices such as a meterstick and a stopwatch, you can measure position and time. Can this information be used to describe the rate of motion?

Velocity

Suppose you recorded two joggers on one motion diagram, as shown in **Figure 2-20a.** From one frame to the next you can see that the position of the jogger in red shorts changes more than that of the one wearing blue. In other words, for a fixed time interval, the displacement, $\Delta \boldsymbol{d}$, is greater for the jogger in red because she is moving faster. She covers a larger distance than the jogger in blue does in the same amount of time. Now, suppose that each jogger travels 100.0 m. The time interval, Δt, would be smaller for the jogger in red than for the one in blue.

Average velocity From the example of the joggers, you can see that both the displacement, $\Delta \boldsymbol{d}$, and time interval, Δt, might be needed to create the quantity that tells how fast an object is moving. How might they be combined? Compare the lines representing the red and blue joggers in the position-time graphs in **Figure 2-20b.** The slope of the red jogger's line is steeper than the slope of the blue jogger's line. A steeper slope indicates a greater change in displacement during each time interval.

Recall from Chapter 1 that to find the slope, you first choose two points on the line. Next, you subtract the vertical coordinate (d in this case) of the first point from the vertical coordinate of the second point to obtain the rise of the line. After that, you subtract the horizontal coordinate (t in this case) of the first point from the horizontal coordinate of the second point to obtain the run. Finally, you divide the rise by the run to obtain the slope. The slopes of the two lines shown in Figure 2-20b are found as follows:

$$\text{Red slope} = \frac{d_f - d_i}{t_f - t_i} = \frac{6.0\text{ m} - 2.0\text{ m}}{3.0\text{ s} - 1.0\text{ s}} = 2.0\text{ m/s}$$

$$\text{Blue slope} = \frac{d_f - d_i}{t_f - t_i} = \frac{3.0\text{ m} - 2.0\text{ m}}{3.0\text{ s} - 2.0\text{ s}} = 1.0\text{ m/s}$$

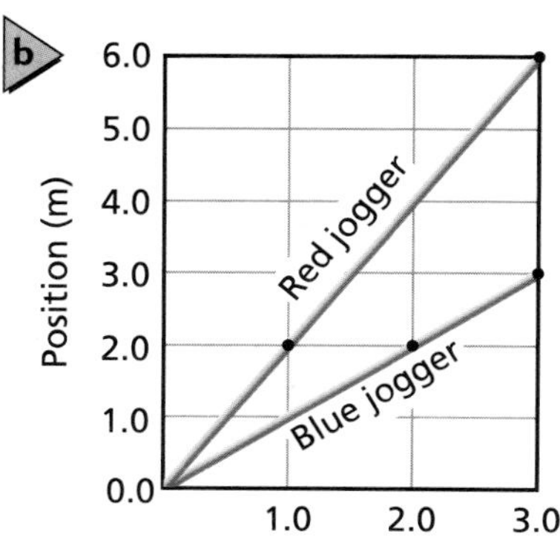

■ **Figure 2-20** The red jogger's displacement is greater than the displacement of the blue jogger in each time interval because the jogger in red is moving faster than the jogger in blue **(a).** The position-time graph represents the motion of the red and blue joggers. The points used to calculate the slope of each line are shown **(b).**

APPLYING PHYSICS

Speed Records The world record for the men's 100-m dash is 9.78 s, established in 2002 by Tim Montgomery. The world record for the women's 100-m dash is 10.65 s, established in 1998 by Marion Jones. These sprinters often are referred to as the world's fastest man and woman.

There are some important things to notice about this comparison. First, the slope of the faster runner is a greater number, so it is reasonable to assume that this number might be connected with the runner's speed. Second, look at the units of the slope, meters per second. In other words, the slope tells how many meters the runner moved in 1 s. These units are similar to miles per hour, which also measure speed. Looking at how the slope is calculated, you can see that slope is the change in position, divided by the time interval during which that change took place, or $(d_f - d_i) / (t_f - t_i)$, or $\Delta \boldsymbol{d}/\Delta t$. When $\Delta \boldsymbol{d}$ gets larger, the slope gets larger; when Δt gets larger, the slope gets smaller. This agrees with the interpretation above of the movements of the red and blue joggers. The slope of a position-time graph for an object is the object's **average velocity** and is represented by the ratio of the change of position to the time interval during which the change occurred.

Average Velocity $\bar{\boldsymbol{v}} \equiv \frac{\Delta \boldsymbol{d}}{\Delta t} = \frac{\boldsymbol{d}_f - \boldsymbol{d}_i}{t_f - t_i}$

Average velocity is defined as the change in position, divided by the time during which the change occurred.

The symbol $\equiv$ means that the left-hand side of the equation is defined by the right-hand side.

- Velocity vectors are **red.**
- Displacement vectors are **green.**

It is a common misconception to say that the slope of a position-time graph gives the speed of the object. Consider the slope of the position-time graph shown in **Figure 2-21.** The slope of this position-time graph is -5.0 m/s. As you can see the slope indicates both the magnitude and direction. Recall that average velocity is a quantity that has both magnitude and direction. The slope of a position-time graph indicates the average velocity of the object and not its speed. Take another look at Figure 2-21. The slope of the graph is -5.0 m/s and thus, the object has a velocity of -5.0 m/s. The object starts out at a positive position and moves toward the origin. It moves in the negative direction at a rate of 5.0 m/s.

Average speed The absolute value of the slope of a position-time graph tells you the **average speed** of the object; that is, how fast the object is moving. The sign of the slope tells you in what direction the object is moving. The combination of an object's average speed, $\bar{v}$, and the direction in which it is moving is the average velocity, $\bar{\boldsymbol{v}}$. Thus, for the object represented in Figure 2-21, the average velocity is -5.0 m/s, or 5.0 m/s in the negative direction. Its average speed is 5.0 m/s. Remember that if an object moves in the negative direction, then its displacement is negative. This means that the object's velocity always will have the same sign as the object's displacement.

As you consider other types of motion to analyze in future chapters, sometimes the velocity will be the important quantity to consider, while at other times, the speed will be the important quantity. Therefore, it is a good idea to make sure that you understand the differences between velocity and speed so that you will be sure to use the right one later.

Figure 2-21 The object whose motion is represented here is moving in the negative direction at a rate of 5.0 m/s.

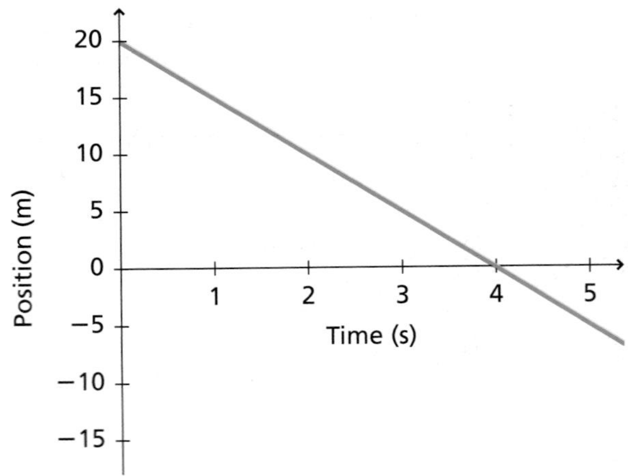

▶ EXAMPLE Problem 3

The graph at the right describes the motion of a student riding his skateboard along a smooth, pedestrian-free sidewalk. What is his average velocity? What is his average speed?

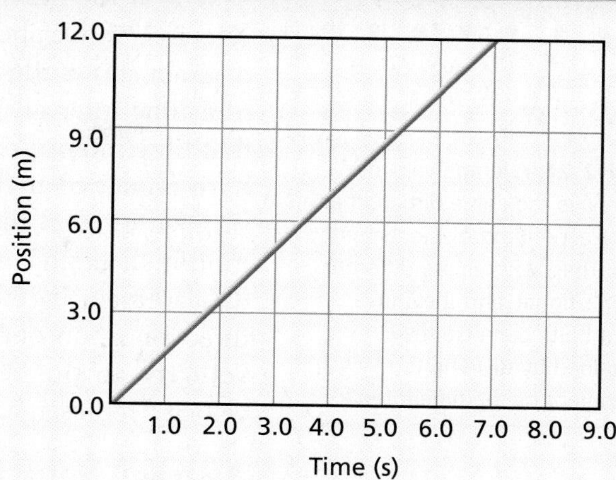

1 Analyze and Sketch the Problem

- Identify the coordinate system of the graph.

2 Solve for the Unknown

Unknown:

$\bar{\mathbf{v}} = ?$ $\quad \bar{v} = ?$

Find the average velocity using two points on the line.

$$\bar{v} = \frac{\Delta d}{\Delta t}$$ **Use magnitudes with signs indicating directions.**

$$= \frac{d_2 - d_1}{t_2 - t_1}$$

$$= \frac{12.0\text{ m} - 6.0\text{ m}}{8.0\text{ s} - 4.0\text{ s}}$$ **Substitute $d_2 = 12.0$ m, $d_1 = 6.0$ m, $t_2 = 8.0$ s, $t_1 = 4.0$ s.**

$\bar{\mathbf{v}} = 1.5$ m/s in the positive direction

The average speed, $\bar{v}$, is the absolute value of the average velocity, or 1.5 m/s.

Personal Tutor For an online tutorial on average velocity and average speed, visit physicspp.com.

3 Evaluate the Answer

- **Are the units correct?** m/s are the units for both velocity and speed.
- **Do the signs make sense?** The positive sign for the velocity agrees with the coordinate system. No direction is associated with speed.

▶ PRACTICE Problems

- Additional Problems, Appendix B
- Solutions to Selected Problems, Appendix C

25. The graph in **Figure 2-22** describes the motion of a cruise ship during its voyage through calm waters. The positive *d*-direction is defined to be south.

a. What is the ship's average speed?

b. What is its average velocity?

26. Describe, in words, the motion of the cruise ship in the previous problem.

27. The graph in **Figure 2-23** represents the motion of a bicycle. Determine the bicycle's average speed and average velocity, and describe its motion in words.

28. When Marilyn takes her pet dog for a walk, the dog walks at a very consistent pace of 0.55 m/s. Draw a motion diagram and position-time graph to represent Marilyn's dog walking the 19.8-m distance from in front of her house to the nearest fire hydrant.

Time (s)
1 2 3 4
0
−1
−2
Position (m)

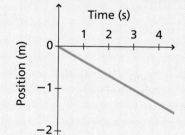

■ **Figure 2-22**

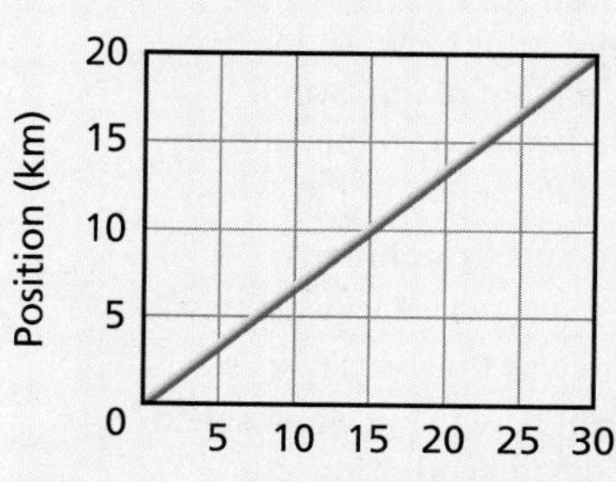

■ **Figure 2-23**

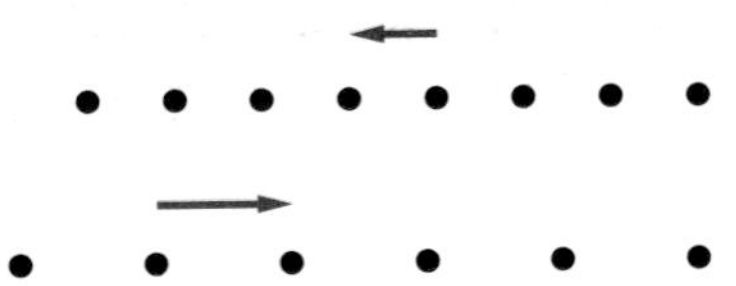

Figure 2-24 Average velocity vectors have the same direction as their corresponding displacement vectors. Their lengths are different, but proportional, and they have different units because they are obtained by dividing the displacement by the time interval.

Instantaneous Velocity

Why is the quantity $\Delta \boldsymbol{d}/\Delta t$ called average velocity? Why isn't it called velocity? Think about how a motion diagram is constructed. A motion diagram shows the position of a moving object at the beginning and end of a time interval. It does not, however, indicate what happened within that time interval. During the time interval, the speed of the object could have remained the same, increased, or decreased. The object may have stopped or even changed direction. All that can be determined from the motion diagram is an average velocity, which is found by dividing the total displacement by the time interval in which it took place. The speed and direction of an object at a particular instant is called the **instantaneous velocity.** In this textbook, the term *velocity* will refer to instantaneous velocity, represented by the symbol $\boldsymbol{v}$.

Average Velocity on Motion Diagrams

How can you show average velocity on a motion diagram? Although the average velocity is in the same direction as displacement, the two quantities are not measured using the same units. Nevertheless, they are proportional—when displacement is greater during a given time interval, so is average velocity. A motion diagram is not a precise graph of average velocity, but you can indicate the direction and magnitude of the average velocity on it. Imagine two cars driving down the road at different speeds. A video camera records their motions at the rate of one frame every second. Imagine that each car has a paintbrush attached to it that automatically descends and paints a line on the ground for half a second every second. The faster car would paint a longer line on the ground. The vectors you draw on a motion diagram to represent the velocity are like the lines made by the paintbrushes on the ground below the cars. Red is used to indicate velocity vectors on motion diagrams. **Figure 2-24** shows the particle models, complete with velocity vectors, for two cars: one moving to the right and the other moving more slowly to the left.

Using equations Any time you graph a straight line, you can find an equation to describe it. There will be many cases for which it will be more efficient to use such an equation instead of a graph to solve problems. Take another look at the graph in Figure 2-21 on page 44 for the object moving with a constant velocity of −5.0 m/s. Recall from Chapter 1 that any straight line can be represented by the formula: $y = mx + b$ where y is the quantity plotted on the vertical axis, m is the slope of the line, x is the quantity plotted on the horizontal axis, and b is the y-intercept of the line.

For the graph in Figure 2-21, the quantity plotted on the vertical axis is position, and the variable used to represent position is $\boldsymbol{d}$. The slope of the line is −5.0 m/s, which is the object's average velocity, $\bar{\boldsymbol{v}}$. The quantity plotted on the horizontal axis is time, t. The y-intercept is 20.0 m. What does this 20.0 m represent? By inspecting the graph and thinking about how the object moves, you can figure out that the object was at a position of 20.0 m when $t = 0.0$ s. This is called the initial position of the object, and is designated $\boldsymbol{d}_i$. **Table 2-2** summarizes how the general variables in the straight-line formula are changed to the specific variables you have been using to describe motion. It also shows the numerical values for the two constants in this equation.

MINI LAB

Instantaneous Velocity Vectors

1. Attach a 1-m-long string to your hooked mass.
2. Hold the string in one hand with the mass suspended.
3. Carefully pull the mass to one side and release it.
4. **Observe** the motion, the speed, and direction of the mass for several swings.
5. Stop the mass from swinging.
6. Draw a diagram showing instantaneous velocity vectors at the following points: top of the swing, midpoint between top and bottom, bottom of the swing, midpoint between bottom and top, and back to the top of the swing.

Analyze and Conclude

7. Where was the velocity greatest?
8. Where was the velocity least?
9. **Explain** how the average speed can be determined using your vector diagram.

Based on the information shown in Table 2-2, the equation $y = mx + b$ becomes $\boldsymbol{d} = \bar{\boldsymbol{v}}t + \boldsymbol{d}_i$, or, by inserting the values of the constants, $d = (-5.0 \text{ m/s})t + 20.0 \text{ m}$. This equation describes the motion that is represented in Figure 2-21. You can check this by plugging a value of t into the equation and seeing that you obtain the same value of $\boldsymbol{d}$ as when you read it directly from the graph. To conduct an extra check to be sure the equation makes sense, take a look at the units. You cannot set two items with different units equal to each other in an equation. In this equation, the left-hand side is a position, so its units are meters. The first term on the right-hand side multiplies meters per second times seconds, so its units are also meters. The last term is in meters, too, so the units on this equation are valid.

▶ Connecting **Math to Physics**

Table 2-2
Comparison of Straight Lines with Position-Time Graphs

General Variable	Specific Motion Variable	Value in Figure 2-21
y	$\boldsymbol{d}$	
m	$\bar{\boldsymbol{v}}$	−5.0 m/s
x	t	
b	$\boldsymbol{d}_i$	20.0 m

Equation of Motion for Average Velocity $\boldsymbol{d} = \bar{\boldsymbol{v}}t + \boldsymbol{d}_i$

An object's position is equal to the average velocity multiplied by time plus the initial position.

This equation gives you another way to represent the motion of an object. Note that once a coordinate system is chosen, the direction of $\boldsymbol{d}$ is specified by positive and negative values, and the boldface notation can be dispensed with, as in "d-axis." However, the boldface notation for velocity will be retained to avoid confusing it with speed.

Your toolbox of representations now includes words, motion diagrams, pictures, data tables, position-time graphs, and an equation. You should be able to use any one of these representations to generate at least the characteristics of the others. You will get more practice with this in the rest of this chapter and also in Chapter 3 as you apply these representations to help you solve problems.

2.4 Section Review

*For problems 29–31, refer to **Figure 2-25.***

29. Average Speed Rank the position-time graphs according to the average speed of the object, from greatest average speed to least average speed. Specifically indicate any ties.

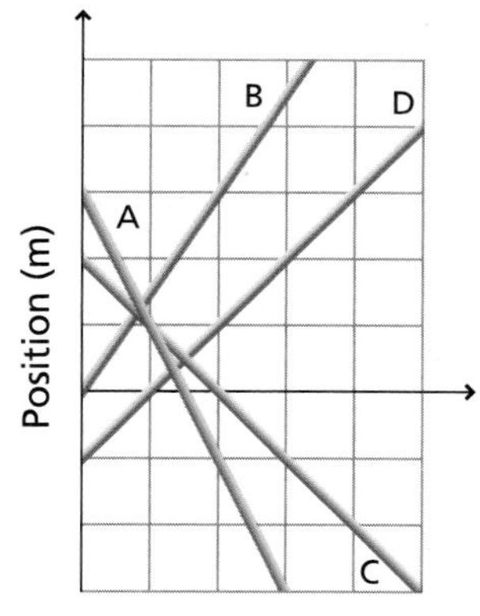

Figure 2-25

30. Average Velocity Rank the graphs according to average velocity, from greatest average velocity to least average velocity. Specifically indicate any ties.

31. Initial Position Rank the graphs according to the object's initial position, from most positive position to most negative position. Specifically indicate any ties. Would your ranking be different if you had been asked to do the ranking according to initial distance from the origin?

32. Average Speed and Average Velocity Explain how average speed and average velocity are related to each other.

33. Critical Thinking In solving a physics problem, why is it important to create pictorial and physical models before trying to solve an equation?

PHYSICS LAB

Creating Motion Diagrams

Alternate CBL instructions can be found on the Web site.
physicspp.com

In this activity you will construct motion diagrams for two toy cars. A motion diagram consists of a series of images showing the positions of a moving object at equal time intervals. Motion diagrams help describe the motion of an object. By looking at a motion diagram you can determine whether an object is speeding up, slowing down, or moving at constant speed.

QUESTION

How do the motion diagrams of a fast toy car and a slow toy car differ?

Objectives

- **Measure in SI** the location of a moving object.
- **Recognize spatial relationships** of moving objects.
- **Describe** the motion of a fast and slow object.

Safety Precautions

Materials

video camera
two toy windup cars
meterstick
foam board

Procedure

1. Mark a starting line on the lab table or the surface recommended by your teacher.
2. Place both toy cars at the starting line and release them at the same time. Be sure to wind them up before releasing them.
3. Observe both toy cars and determine which one is faster.
4. Place the slower toy car at the starting line.
5. Place a meterstick parallel to the path the toy car will take.
6. Have one of the members of your group get ready to operate the video camera.
7. Release the slower toy car from the starting line. Be sure to wind up the toy car before releasing it.
8. Use the video camera to record the slower toy car's motion parallel to the meterstick.
9. Set the recorder to play the tape frame by frame. Replay the video tape for 0.5 s, pressing the pause button every 0.1 s (3 frames).
10. Determine the toy car's position for each time interval by reading the meterstick on the video tape and record it in the data table.
11. Repeat steps 5–10 with the faster car.
12. Place a piece of foam board at an angle of approximately 30° to form a ramp.
13. Place the meterstick on the ramp so that it is parallel to the path the toy car will take.
14. Place the slower toy car at the top of the ramp and repeat steps 6–10.

Data Table 1	
Time (s)	Position of the Slower Toy Car (cm)
0.0	
0.1	
0.2	
0.3	
0.4	
0.5	

Data Table 2	
Time (s)	Position of the Faster Toy Car (cm)
0.0	
0.1	
0.2	
0.3	
0.4	
0.5	

Data Table 3	
Time (s)	Position of the Slower Toy Car on the Ramp (cm)
0.0	
0.1	
0.2	
0.3	
0.4	
0.5	

Analyze

1. Draw a motion diagram for the slower toy car using the data you collected.
2. Draw a motion diagram for the faster toy car using the data you collected.
3. Using the data you collected, draw a motion diagram for the slower toy car rolling down the ramp.

Conclude and Apply

How is the motion diagram of the faster toy car different from the motion diagram of the slower toy car?

Going Further

1. Draw a motion diagram for a car moving at a constant speed.
2. What appears to be the relationship between the distances between points in the motion diagram of a car moving at a constant speed?
3. Draw a motion diagram for a car that starts moving fast and then begins to slow down.
4. What happens to the distance between points in the motion diagram in the previous question as the car slows down?
5. Draw a motion diagram for a car that starts moving slowly and then begins to speed up.
6. What happens to the distance between points in the motion diagram in the previous question as the car speeds up?

Real-World Physics

Suppose a car screeches to a halt to avoid an accident. If that car has antilock brakes that pump on and off automatically every fraction of a second, what might the tread marks on the road look like? Include a drawing along with your explanation of what the pattern of tread marks on the road might look like.

To find out more about representing motion, visit the Web site: physicspp.com

eXtreme physics

Accurate Time

What time is it really? You might use a clock to find out what time it is at any moment. A clock is a device that counts regularly recurring events in order to measure time. Suppose the clock in your classroom reads 9:00. Your watch, however, reads 8:55, and your friend's watch reads 9:02. So what time is it, really? Which clock or watch is accurate?

Many automated processes are controlled by clocks. For example, an automated bell that signals the end of a class period is controlled by a clock. Thus, if you wanted to be on time for a class you would have to synchronize your watch to the one controlling the bell. Other processes, such as GPS navigation, space travel, internet synchronization, transportation, and communication, rely on clocks with extreme precision and accuracy. A reliable standard clock that can measure when exactly one second has elapsed is needed.

The Standard Cesium Clock Atomic clocks, such as cesium clocks, address this need. Atomic clocks measure the number of times the atom used in the clock switches its energy state. Such oscillations in an atom's energy occur very quickly and regularly. The National Institute of Standards and Technology (NIST) currently uses the oscillations of the cesium atom to determine the standard 1-s interval. One second is defined as the duration of 9,192,631,770 oscillations of the cesium atom.

The cesium atom has a single electron in its outermost energy level. This outer electron spins and behaves like a miniature magnet. The cesium nucleus also spins and acts like a miniature magnet. The nucleus and electron may spin in such a manner that their north magnetic poles are aligned. The nucleus and electron also may spin in a way that causes opposite poles to be aligned. If the poles are aligned, the cesium atom is in one energy state. If they are oppositely aligned, the atom is in another energy state. A microwave with a particular frequency can strike a cesium atom and cause the outside spinning electron to switch its magnetic pole orientation and change the atom's energy state. As a result, the atom emits light. This occurs at cesium's natural frequency of 9,192,631,770 cycles/s. This principle was used to design the cesium clock.

The cesium clock, *NIST-F1*, located at the NIST laboratories in Boulder, Colorado is among the most accurate clocks in the world.

How Does the Cesium Clock Work? The cesium clock consists of cesium atoms and a quartz crystal oscillator, which produces microwaves. When the oscillator's microwave signal precisely equals cesium's natural frequency, a large number of cesium atoms will change their energy state. Cesium's natural frequency is equal to 9,192,631,770 microwave cycles. Thus, there are 9,192,631,770 cesium energy level changes in 1 s. Cesium clocks are so accurate that a modern cesium clock is off by less than 1 s in 20 million years.

Going Further

1. **Research** What processes require the precise measurement of time?
2. **Analyze and Conclude** Why is the precise measurement of time essential to space navigation?

Chapter 2

Study Guide

2.1 Picturing Motion

Vocabulary
- motion diagram (p. 33)
- particle model (p. 33)

Key Concepts
- A motion diagram shows the position of an object at successive times.
- In the particle model, the object in the motion diagram is replaced by a series of single points.

2.2 Where and When?

Vocabulary
- coordinate system (p. 34)
- origin (p. 34)
- position (p. 34)
- distance (p. 34)
- magnitude (p. 35)
- vectors (p. 35)
- scalars (p. 35)
- resultant (p. 35)
- time interval (p. 36)
- displacement (p. 36)

Key Concepts
- You can define any coordinate system you wish in describing motion, but some are more useful than others.
- A time interval is the difference between two times.

$$\Delta t = t_f - t_i$$

- A vector drawn from the origin of the coordinate system to the object indicates the object's position.
- Change in position is displacement, which has both magnitude and direction.

$$\Delta \boldsymbol{d} = \boldsymbol{d}_f - \boldsymbol{d}_i$$

- The length of the displacement vector represents how far the object was displaced, and the vector points in the direction of the displacement.

2.3 Position-Time Graphs

Vocabulary
- position-time graph (p. 38)
- instantaneous position (p. 40)

Key Concepts
- Position-time graphs can be used to find the velocity and position of an object, as well as where and when two objects meet.
- Any motion can be described using words, motion diagrams, data tables, and graphs.

2.4 How Fast?

Vocabulary
- average velocity (p. 44)
- average speed (p. 44)
- instantaneous velocity (p. 46)

Key Concepts
- The slope of an object's position-time graph is the average velocity of the object's motion.

$$\bar{\boldsymbol{v}} = \frac{\Delta \boldsymbol{d}}{\Delta t} = \frac{\boldsymbol{d}_f - \boldsymbol{d}_i}{t_f - t_i}$$

- The average speed is the absolute value of the average velocity.
- An object's velocity is how fast it is moving and in what direction it is moving.
- An object's initial position, $\boldsymbol{d}_i$, its constant average velocity, $\bar{\boldsymbol{v}}$, its position, $\boldsymbol{d}$, and the time, t, since the object was at its initial position are related by a simple equation.

$$\boldsymbol{d} = \bar{\boldsymbol{v}}t + \boldsymbol{d}_i$$

Chapter 2

Assessment

Concept Mapping

34. Complete the concept map below using the following terms: *words, equivalent representations, position-time graph.*

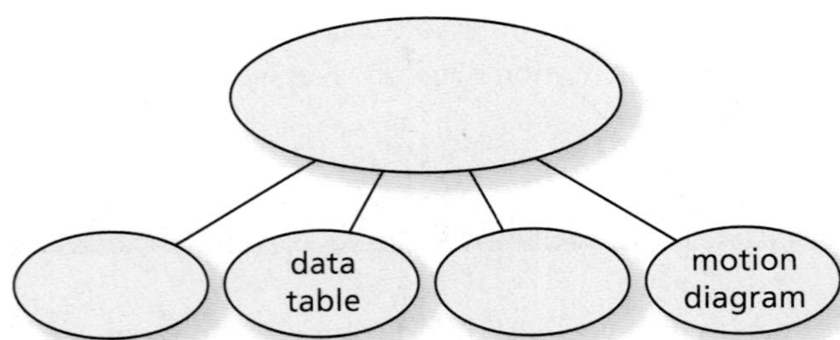

Mastering Concepts

35. What is the purpose of drawing a motion diagram? (2.1)

36. Under what circumstances is it legitimate to treat an object as a point particle? (2.1)

37. The following quantities describe location or its change: position, distance, and displacement. Briefly describe the differences among them. (2.2)

38. How can you use a clock to find a time interval? (2.2)

39. In-line Skating How can you use the position-time graphs for two in-line skaters to determine if and when one in-line skater will pass the other one? (2.3)

40. Walking Versus Running A walker and a runner leave your front door at the same time. They move in the same direction at different constant velocities. Describe the position-time graphs of each. (2.4)

41. What does the slope of a position-time graph measure? (2.4)

42. If you know the positions of a moving object at two points along its path, and you also know the time it took for the object to get from one point to the other, can you determine the particle's instantaneous velocity? Its average velocity? Explain. (2.4)

Applying Concepts

43. Test the following combinations and explain why each does not have the properties needed to describe the concept of velocity: $\Delta d + \Delta t$, $\Delta d - \Delta t$, $\Delta d \times \Delta t$, $\Delta t/\Delta d$.

44. Football When can a football be considered a point particle?

45. When can a football player be treated as a point particle?

46. Figure 2-26 is a graph of two people running.
 a. Describe the position of runner A relative to runner B at the y-intercept.
 b. Which runner is faster?
 c. What occurs at point P and beyond?

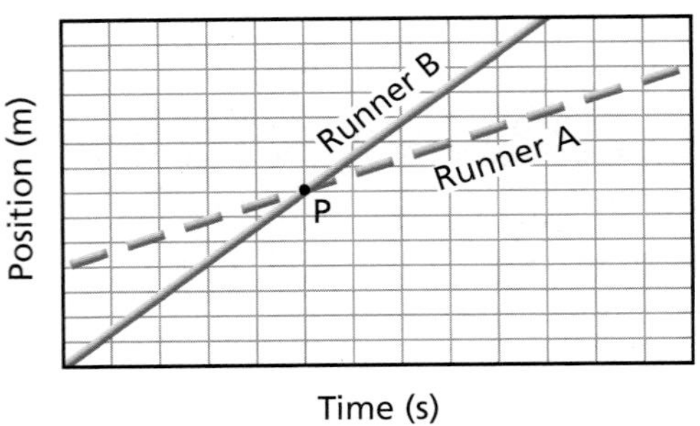

Figure 2-26

47. The position-time graph in **Figure 2-27** shows the motion of four cows walking from the pasture back to the barn. Rank the cows according to their average velocity, from slowest to fastest.

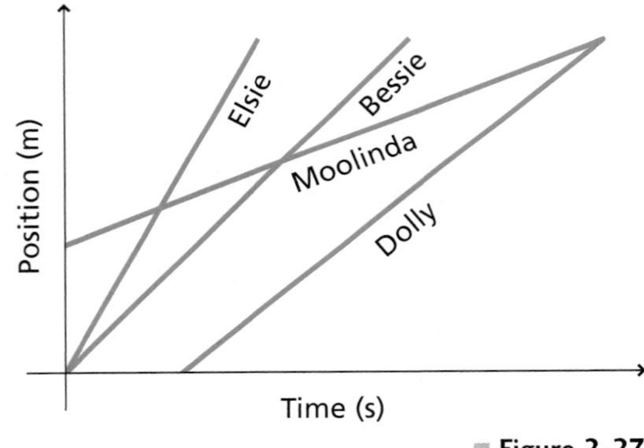

Figure 2-27

48. Figure 2-28 is a position-time graph for a rabbit running away from a dog.
 a. Describe how this graph would be different if the rabbit ran twice as fast.
 b. Describe how this graph would be different if the rabbit ran in the opposite direction.

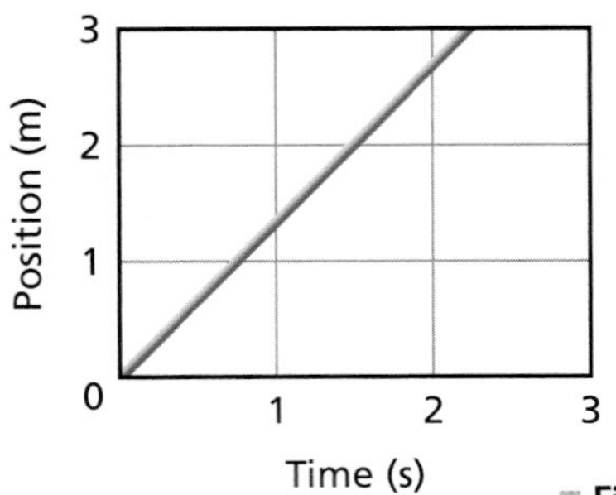

Figure 2-28

 For more problems, go to Additional Problems, Appendix B.

Mastering Problems

2.4 How Fast?

49. A bike travels at a constant speed of 4.0 m/s for 5.0 s. How far does it go?

50. Astronomy Light from the Sun reaches Earth in 8.3 min. The speed of light is 3.00×10^8 m/s. How far is Earth from the Sun?

51. A car is moving down a street at 55 km/h. A child suddenly runs into the street. If it takes the driver 0.75 s to react and apply the brakes, how many meters will the car have moved before it begins to slow down?

52. Nora jogs several times a week and always keeps track of how much time she runs each time she goes out. One day she forgets to take her stopwatch with her and wonders if there's a way she can still have some idea of her time. As she passes a particular bank, she remembers that it is 4.3 km from her house. She knows from her previous training that she has a consistent pace of 4.0 m/s. How long has Nora been jogging when she reaches the bank?

53. Driving You and a friend each drive 50.0 km. You travel at 90.0 km/h; your friend travels at 95.0 km/h. How long will your friend have to wait for you at the end of the trip?

Mixed Review

54. Cycling A cyclist maintains a constant velocity of +5.0 m/s. At time $t = 0.0$ s, the cyclist is +250 m from point A.

a. Plot a position-time graph of the cyclist's location from point A at 10.0-s intervals for 60.0 s.

b. What is the cyclist's position from point A at 60.0 s?

c. What is the displacement from the starting position at 60.0 s?

55. Figure 2-29 is a particle model for a chicken casually walking across the road. Time intervals are every 0.1 s. Draw the corresponding position-time graph and write the equation to describe the chicken's motion.

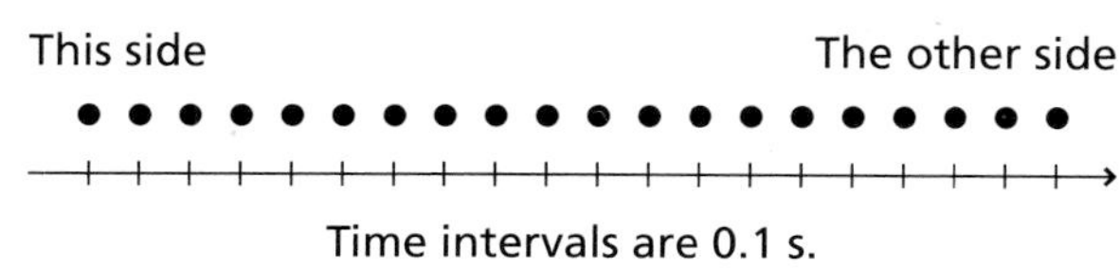

Figure 2-29

56. Figure 2-30 shows position-time graphs for Joszi and Heike paddling canoes in a local river.

a. At what time(s) are Joszi and Heike in the same place?

b. How much time does Joszi spend on the river before he passes Heike?

c. Where on the river does it appear that there might be a swift current?

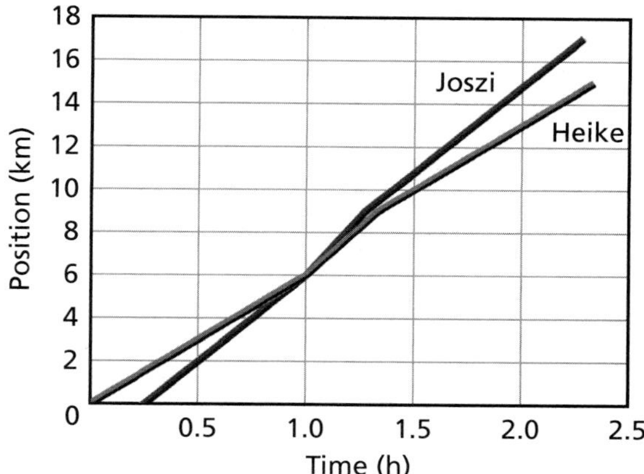

Figure 2-30

57. Driving Both car A and car B leave school when a stopwatch reads zero. Car A travels at a constant 75 km/h, and car B travels at a constant 85 km/h.

a. Draw a position-time graph showing the motion of both cars. How far are the two cars from school when the stopwatch reads 2.0 h? Calculate the distances and show them on your graph.

b. Both cars passed a gas station 120 km from the school. When did each car pass the gas station? Calculate the times and show them on your graph.

58. Draw a position-time graph for two cars traveling to the beach, which is 50 km from school. At noon, Car A leaves a store that is 10 km closer to the beach than the school is and moves at 40 km/h. Car B starts from school at 12:30 P.M. and moves at 100 km/h. When does each car get to the beach?

59. Two cars travel along a straight road. When a stopwatch reads $t = 0.00$ h, car A is at $d_A = 48.0$ km moving at a constant 36.0 km/h. Later, when the watch reads $t = 0.50$ h, car B is at $d_B = 0.00$ km moving at 48.0 km/h. Answer the following questions, first, graphically by creating a position-time graph, and second, algebraically by writing equations for the positions d_A and d_B as a function of the stopwatch time, t.

a. What will the watch read when car B passes car A?

b. At what position will car B pass car A?

c. When the cars pass, how long will it have been since car A was at the reference point?

60. **Figure 2-31** shows the position-time graph depicting Jim's movement up and down the aisle at a store. The origin is at one end of the aisle.
   a. Write a story describing Jim's movements at the store that would correspond to the motion represented by the graph.
   b. When does Jim have a position of 6.0 m?
   c. How much time passes between when Jim enters the aisle and when he gets to a position of 12.0 m? What is Jim's average velocity between 37.0 s and 46.0 s?

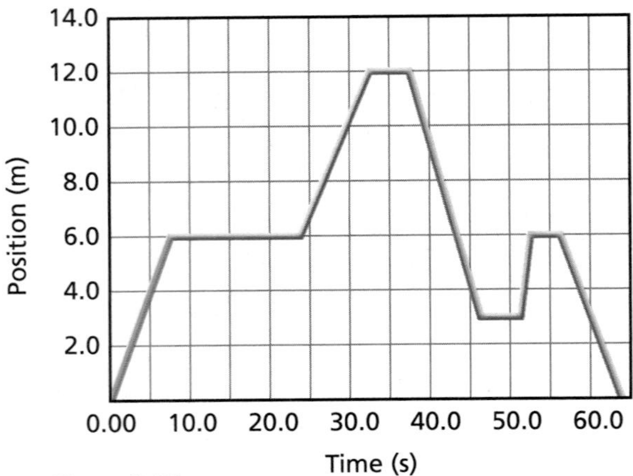

Figure 2-31

Thinking Critically

61. **Apply Calculators** Members of a physics class stood 25 m apart and used stopwatches to measure the time at which a car traveling on the highway passed each person. Their data are shown in **Table 2-3.**

Table 2-3	
Position v. Time	
Time (s)	**Position (m)**
0.0	0.0
1.3	25.0
2.7	50.0
3.6	75.0
5.1	100.0
5.9	125.0
7.0	150.0
8.6	175.0
10.3	200.0

Use a graphing calculator to fit a line to a position-time graph of the data and to plot this line. Be sure to set the display range of the graph so that all the data fit on it. Find the slope of the line. What was the speed of the car?

62. **Apply Concepts** You plan a car trip for which you want to average 90 km/h. You cover the first half of the distance at an average speed of only 48 km/h. What must your average speed be in the second half of the trip to meet your goal? Is this reasonable? Note that the velocities are based on half the distance, not half the time.

63. **Design an Experiment** Every time a particular red motorcycle is driven past your friend's home, his father becomes angry because he thinks the motorcycle is going too fast for the posted 25 mph (40 km/h) speed limit. Describe a simple experiment you could do to determine whether or not the motorcycle is speeding the next time it is driven past your friend's house.

64. **Interpret Graphs** Is it possible for an object's position-time graph to be a horizontal line? A vertical line? If you answer yes to either situation, describe the associated motion in words.

Writing in Physics

65. Physicists have determined that the speed of light is 3.00×10^8 m/s. How did they arrive at this number? Read about some of the series of experiments that were done to determine light's speed. Describe how the experimental techniques improved to make the results of the experiments more accurate.

66. Some species of animals have good endurance, while others have the ability to move very quickly, but for only a short amount of time. Use reference sources to find two examples of each quality and describe how it is helpful to that animal.

Cumulative Review

67. Convert each of the following time measurements to its equivalent in seconds. (Chapter 1)
   a. 58 ns
   b. 0.046 Gs
   c. 9270 ms
   d. 12.3 ks

68. State the number of significant digits in the following measurements. (Chapter 1)
   a. 3218 kg
   b. 60.080 kg
   c. 801 kg
   d. 0.000534 kg

69. Using a calculator, Chris obtained the following results. Rewrite the answer to each operation using the correct number of significant digits. (Chapter 1)
   a. 5.32 mm + 2.1 mm = 7.4200000 mm
   b. 13.597 m × 3.65 m = 49.62905 m^2
   c. 83.2 kg − 12.804 kg = 70.3960000 kg

 For more problems, go to Additional Problems, Appendix B.

Multiple Choice

1. Which of the following statements would be true about the particle model motion diagram for an airplane taking off from an airport?

   (A) The dots would form an evenly spaced pattern.

   (B) The dots would be far apart at the beginning, but get closer together as the plane accelerated.

   (C) The dots would be close together to start with, and get farther apart as the plane accelerated.

   (D) The dots would be close together to start, get farther apart, and become close together again as the airplane leveled off at cruising speed.

2. Which of the following statements about drawing vectors is false?

   (A) A vector diagram is needed to solve all physics problems properly.

   (B) The length of the vector should be proportional to the data.

   (C) Vectors can be added by measuring the length of each vector and then adding them together.

   (D) Vectors can be added in straight lines or in triangle formations.

Use this graph for problems 3–5.

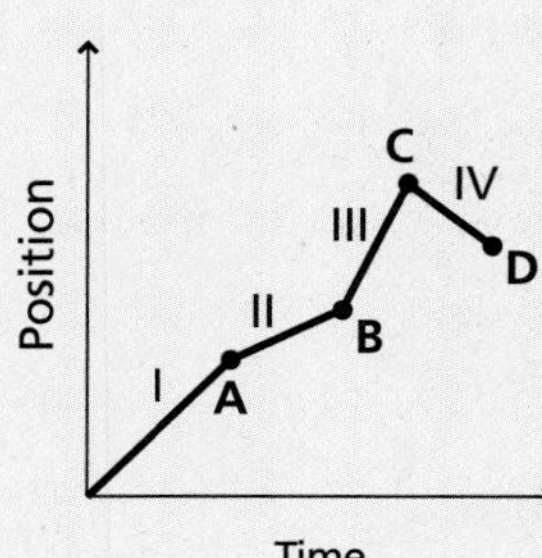

3. The graph shows the motion of a person on a bicycle. When does the person have the greatest velocity?

   (A) section I
   (B) section III
   (C) point D
   (D) point B

4. When is the person on the bicycle farthest away from the starting point?

   (A) point A
   (B) point B
   (C) point C
   (D) point D

5. Over what interval does the person on the bicycle travel the greatest distance?

   (A) section I
   (B) section II
   (C) section III
   (D) point IV

6. A squirrel descends an 8-m tree at a constant speed in 1.5 min. It remains still at the base of the tree for 2.3 min, and then walks toward an acorn on the ground for 0.7 min. A loud noise causes the squirrel to scamper back up the tree in 0.1 min to the exact position on the branch from which it started. Which of the following graphs would accurately represent the squirrel's vertical displacement from the base of the tree?

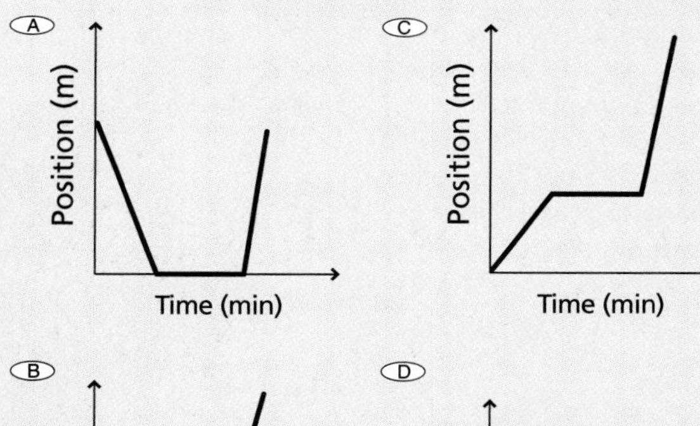

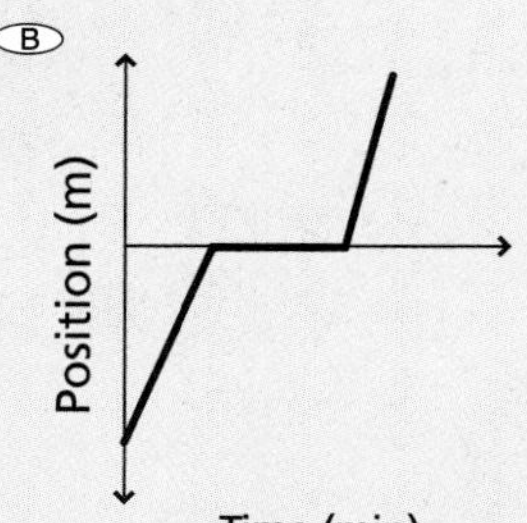

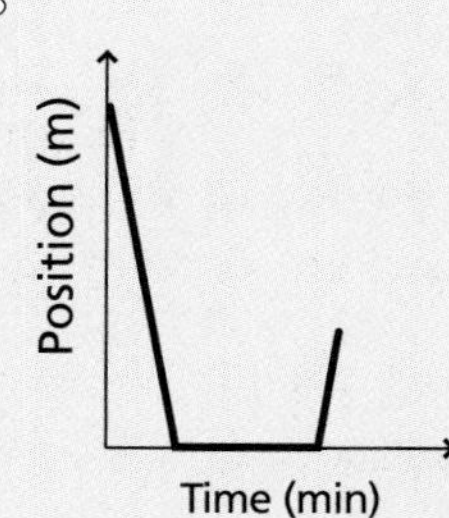

Extended Answer

7. Find a rat's total displacement at the exit if it takes the following path in a maze: start, 1.0 m north, 0.3 m east, 0.8 m south, 0.4 m east, finish.

✓ Test-Taking TIP

Stock up on Supplies

Bring all your test-taking tools: number two pencils, black and blue pens, erasers, correction fluid, a sharpener, a ruler, a calculator, and a protractor.

Chapter 3

Accelerated Motion

What You'll Learn

- You will develop descriptions of accelerated motion.
- You will use graphs and equations to solve problems involving moving objects.
- You will describe the motion of objects in free fall.

Why It's Important

Objects do not always move at constant velocities. Understanding accelerated motion will help you better decribe the motion of many objects.

Acceleration Cars, planes, subways, elevators, and other common forms of transportation often begin their journeys by speeding up quickly, and end by stopping rapidly.

Think About This ▶

The driver of a dragster on the starting line waits for the green light to signal the start of the race. At the signal, the driver will step on the gas pedal and try to speed up as quickly as possible. As the car speeds up, how will its position change?

physicspp.com

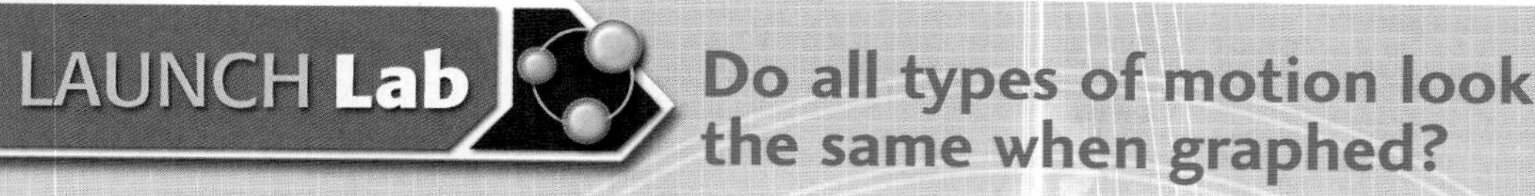

Do all types of motion look the same when graphed?

Question

How does a graph showing constant speed compare to a graph of a vehicle speeding up?

Procedure

1. Clamp a spark timer to the back edge of a lab table.
2. Cut a piece of timer tape approximately 50 cm in length, insert it into the timer, and tape it to vehicle 1.
3. Turn on the timer and release the vehicle. Label the tape with the vehicle number.
4. Raise one end of the lab table 8–10 cm by placing a couple of bricks under the back legs. ***CAUTION: Make sure the lab table remains stable.***
5. Repeat steps 2–4 with vehicle 2, but hold the vehicle in place next to the timer and release it after the timer has been turned on. Catch the vehicle before it falls.
6. **Construct and Organize Data** Mark the first dark dot where the timer began as *zero.* Measure the distance to each dot from the zero dot for 10 intervals and record your data.
7. **Make and Use Graphs** Make a graph of total distance versus interval number. Place data for both vehicles on the same plot and label each graph.

Analysis

Which vehicle moved with constant speed? Which one sped up? Explain how you determined this by looking at the timer tape.

Critical Thinking Describe the shape of each graph. How does the shape of the graph relate to the type of motion observed?

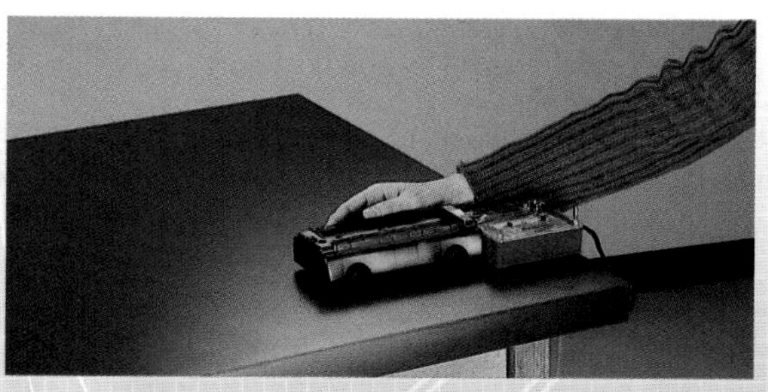

3.1 Acceleration

Uniform motion is one of the simplest kinds of motion. You learned in Chapter 2 that an object in uniform motion moves along a straight line with an unchanging velocity. From your own experiences, you know, however, that few objects move in this manner all of the time. In this chapter, you will expand your knowledge of motion by considering a slightly more complicated type of motion. You will be presented with situations in which the velocity of an object changes, while the object's motion is still along a straight line. Examples of objects and situations you will encounter in this chapter include automobiles that are speeding up, drivers applying brakes, falling objects, and objects thrown straight upward. In Chapter 6, you will continue to add to your knowledge of motion by analyzing some common types of motion that are not confined to a straight line. These include motion along a circular path and the motion of thrown objects, such as baseballs.

► **Objectives**

- **Define** acceleration.
- **Relate** velocity and acceleration to the motion of an object.
- **Create** velocity-time graphs.

► **Vocabulary**

velocity-time graph
acceleration
average acceleration
instantaneous acceleration

MINI LAB

A Steel Ball Race 

If two steel balls are released at the same instant, will the steel balls get closer or farther apart as they roll down a ramp?

1. Assemble an inclined ramp from a piece of U-channel or two metersticks taped together.

2. Measure 40 cm from the top of the ramp and place a mark there. Place another mark 80 cm from the top.

3. Predict whether the steel balls will get closer or farther apart as they roll down the ramp.

4. At the same time, release one steel ball from the top of the ramp and the other steel ball from the 40-cm mark.

5. Next, release one steel ball from the top of the ramp. As soon as it reaches the 40-cm mark, release the other steel ball from the top of the ramp.

Analyze and Conclude

6. Explain your observations in terms of velocities.

7. Do the steel balls have the same velocity as they roll down the ramp? Explain.

8. Do they have the same acceleration? Explain.

Changing Velocity

You can feel a difference between uniform and nonuniform motion. Uniform motion feels smooth. You could close your eyes and it would feel as though you were not moving at all. In contrast, when you move along a curve or up and down a roller coaster, you feel pushed or pulled.

Consider the motion diagrams shown in **Figure 3-1.** How would you describe the motion of the person in each case? In one diagram, the person is motionless. In another, she is moving at a constant speed. In a third, she is speeding up, and in a fourth, she is slowing down. How do you know which one is which? What information do the motion diagrams contain that could be used to make these distinctions?

The most important thing to notice in these motion diagrams is the distance between successive positions. You learned in Chapter 2 that motionless objects in the background of motion diagrams do not change positions. Therefore, because there is only one image of the person in **Figure 3-1a,** you can conclude that she is not moving; she is at rest. **Figure 3-1b** is like the constant-velocity motion diagrams in Chapter 2. The distances between images are the same, so the jogger is moving at a constant speed. The distance between successive positions changes in the two remaining diagrams. If the change in position gets larger, the jogger is speeding up, as shown in **Figure 3-1c.** If the change in position gets smaller, as in **Figure 3-1d,** the jogger is slowing down.

What does a particle-model motion diagram look like for an object with changing velocity? **Figure 3-2** shows the particle-model motion diagrams below the motion diagrams of the jogger speeding up and slowing down. There are two major indicators of the change in velocity in this form of the motion diagram. The change in the spacing of the dots and the differences in the lengths of the velocity vectors indicate the changes in velocity. If an object speeds up, each subsequent velocity vector is longer. If the object slows down, each vector is shorter than the previous one. Both types of motion diagrams give an idea of how an object's velocity is changing.

Velocity-Time Graphs

Just as it was useful to graph a changing position versus time, it also is useful to plot an object's velocity versus time, which is called a **velocity-time,** or ***v-t*** **graph. Table 3-1** on the next page shows the data for a car that starts at rest and speeds up along a straight stretch of road.

Figure 3-1 By noting the distance the jogger moves in equal time intervals, you can determine that the jogger is standing still **(a),** moving at a constant speed **(b),** speeding up **(c),** and slowing down **(d).**

Figure 3-2 The particle-model version of the motion diagram indicates the runner's changing velocity not only by the change in spacing of the position dots, but also by the change in length of the velocity vectors.

Concepts In Motion
Interactive Figure To see an animation on velocity-time graphs, visit physicspp.com.

The velocity-time graph obtained by plotting these data points is shown in **Figure 3-3.** The positive direction has been chosen to be the same as that of the motion of the car. Notice that this graph is a straight line, which means that the car was speeding up at a constant rate. The rate at which the car's velocity is changing can be found by calculating the slope of the velocity-time graph.

The graph shows that the slope is (10.0 m/s)/(2.00 s), or 5.00 m/s^2. This means that every second, the velocity of the car increased by 5.00 m/s. Consider a pair of data points that are separated by 1 s, such as 4.00 s and 5.00 s. At 4.00 s, the car was moving at a velocity of 20.0 m/s. At 5.00 s, the car was traveling at 25.0 m/s. Thus, the car's velocity increased by 5.00 m/s in 1.00 s. The rate at which an object's velocity changes is called the **acceleration** of the object. When the velocity of an object changes at a constant rate, it has a constant acceleration.

Average and Instantaneous Acceleration

The **average acceleration** of an object is the change in velocity during some measurable time interval divided by that time interval. Average acceleration is measured in m/s^2. The change in velocity at an instant of time is called **instantaneous acceleration.** The instantaneous acceleration of an object can be found by drawing a tangent line on the velocity-time graph at the point of time in which you are interested. The slope of this line is equal to the instantaneous acceleration. Most of the situations considered in this textbook involve motion with acceleration in which the average and instantaneous accelerations are equal.

Figure 3-3 The slope of a velocity-time graph is the acceleration of the object represented.

Table 3-1
Velocity v. Time

Time (s)	Velocity (m/s)
0.00	0.00
1.00	5.00
2.00	10.0
3.00	15.0
4.00	20.0
5.00	25.0

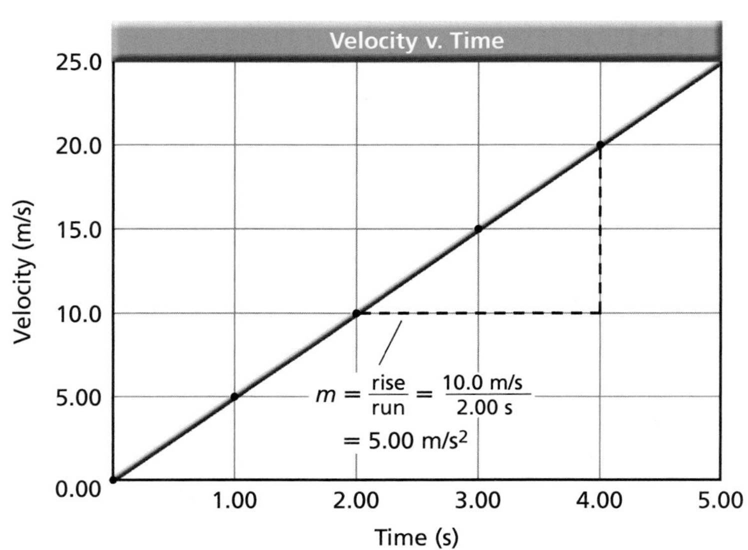

Figure 3-4 Looking at two consecutive velocity vectors and finding the difference between them yields the average acceleration vector for that time interval.

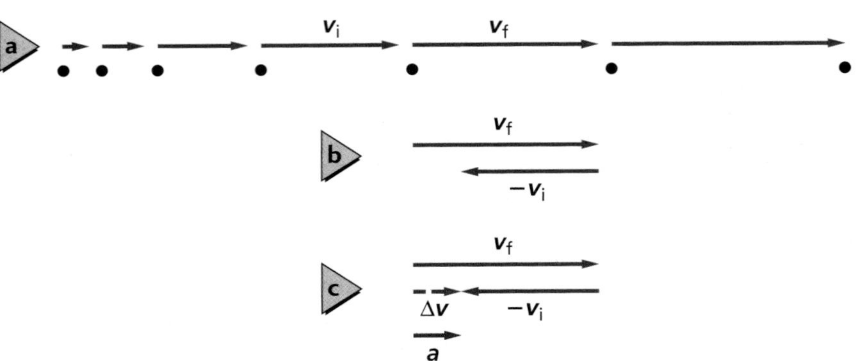

Color Convention

- Acceleration vectors are **violet.**
- Velocity vectors are **red.**
- Displacement vectors are **green.**

Displaying Acceleration on a Motion Diagram

For a motion diagram to give a full picture of an object's movement, it also should contain information about acceleration. This can be done by including average acceleration vectors. These vectors will indicate how the velocity is changing. To determine the length and direction of an average acceleration vector, subtract two consecutive velocity vectors, as shown in **Figures 3-4a** and **b.** That is, $\Delta \boldsymbol{v} = \boldsymbol{v}_f - \boldsymbol{v}_i = \boldsymbol{v}_f + (-\boldsymbol{v}_i)$. Then divide by the time interval, Δt. In Figures 3-4a and b, the time interval, Δt, is 1 s. This vector, $(\boldsymbol{v}_f - \boldsymbol{v}_i)/1$ s, shown in violet in **Figure 3-4c,** is the average acceleration during that time interval. The velocities $\boldsymbol{v}_i$ and $\boldsymbol{v}_f$ refer to the velocities at the beginning and end of a chosen time interval.

▶ EXAMPLE **Problem 1**

Velocity and Acceleration How would you describe the sprinter's velocity and acceleration as shown on the graph?

1 Analyze and Sketch the Problem

- From the graph, note that the sprinter's velocity starts at zero, increases rapidly for the first few seconds, and then, after reaching about 10.0 m/s, remains almost constant.

Known:	**Unknown:**
v = varies	a = ?

2 Solve for the Unknown

Draw a tangent to the curve at t = 1.0 s and t = 5.0 s.

Solve for acceleration at 1.0 s:

$$a = \frac{\text{rise}}{\text{run}}$$

The slope of the line at 1.0 s is equal to the acceleration at that time.

$$= \frac{11.0\text{ m/s} - 2.8\text{ m/s}}{2.4\text{ s} - 0.00\text{ s}}$$

$$= 3.4\text{ m/s}^2$$

Math Handbook
Slope
page 850

Solve for acceleration at 5.0 s:

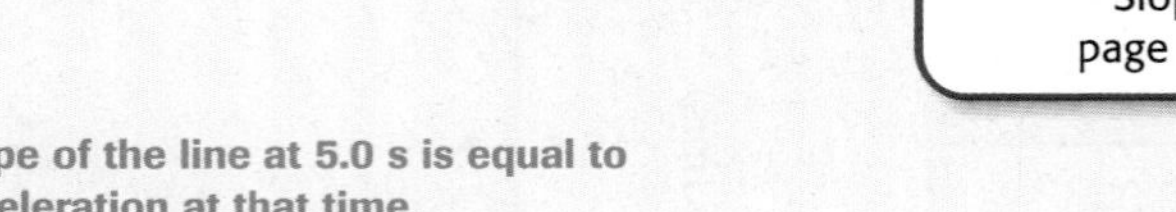

$$a = \frac{\text{rise}}{\text{run}}$$

The slope of the line at 5.0 s is equal to the acceleration at that time.

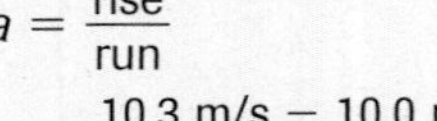

$$= \frac{10.3\text{ m/s} - 10.0\text{ m/s}}{10.0\text{ s} - 0.00\text{ s}}$$

$$= 0.030\text{ m/s}^2$$

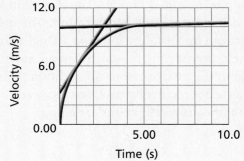

The acceleration is not constant because it changes from 3.4 m/s^2 to 0.03 m/s^2 at 5.0 s.
The acceleration is in the direction chosen to be positive because both values are positive.

3 Evaluate the Answer

- **Are the units correct?** Acceleration is measured in m/s^2.

• Additional Problems, Appendix B
• Solutions to Selected Problems, Appendix C

1. A dog runs into a room and sees a cat at the other end of the room. The dog instantly stops running but slides along the wood floor until he stops, by slowing down with a constant acceleration. Sketch a motion diagram for this situation, and use the velocity vectors to find the acceleration vector.
2. **Figure 3-5** is a *v-t* graph for Steven as he walks along the midway at the state fair. Sketch the corresponding motion diagram, complete with velocity vectors.
3. Refer to the *v-t* graph of the toy train in **Figure 3-6** to answer the following questions.
   a. When is the train's speed constant?
   b. During which time interval is the train's acceleration positive?
   c. When is the train's acceleration most negative?
4. Refer to Figure 3-6 to find the average acceleration of the train during the following time intervals.
   a. 0.0 s to 5.0 s **b.** 15.0 s to 20.0 s **c.** 0.0 s to 40.0 s
5. Plot a *v-t* graph representing the following motion. An elevator starts at rest from the ground floor of a three-story shopping mall. It accelerates upward for 2.0 s at a rate of 0.5 m/s^2, continues up at a constant velocity of 1.0 m/s for 12.0 s, and then experiences a constant downward acceleration of 0.25 m/s^2 for 4.0 s as it reaches the third floor.

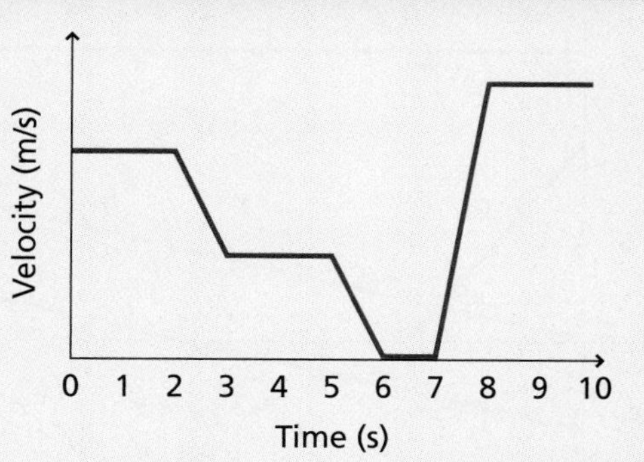

Figure 3-5

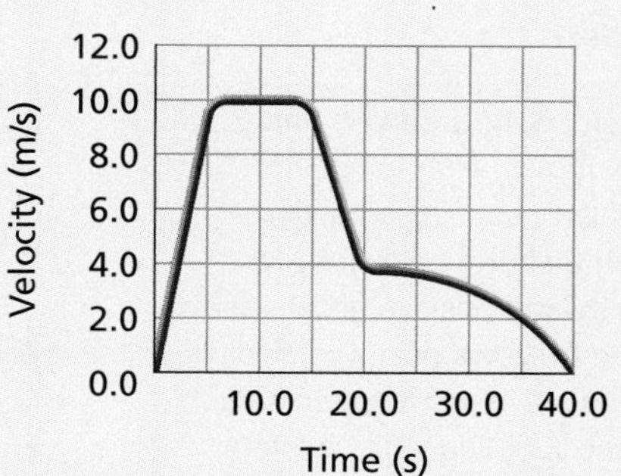

Figure 3-6

Positive and Negative Acceleration

Consider the four situations shown in **Figure 3-7a.** The first motion diagram shows an object moving in the positive direction and speeding up. The second motion diagram shows the object moving in the positive direction and slowing down. The third shows the object speeding up in the negative direction, and the fourth shows the object slowing down as it moves in the negative direction. **Figure 3-7b** shows the velocity vectors for the second time interval of each diagram, along with the corresponding acceleration vectors. Note Δt is equal to 1 s.

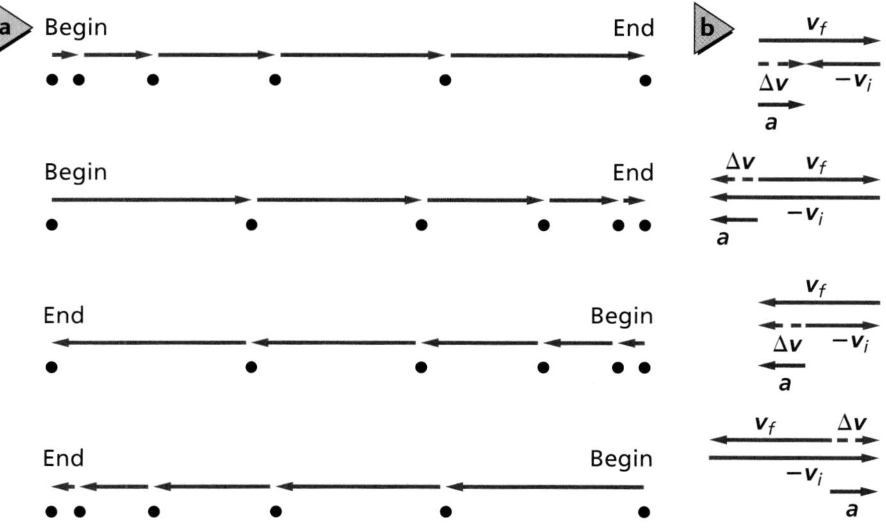

Figure 3-7 These four motion diagrams represent the four different possible ways to move along a straight line with constant acceleration **(a).** When the velocity vectors of the motion diagram and acceleration vectors point in the same direction, an object's speed increases. When they point in opposite directions, the object slows down **(b).**

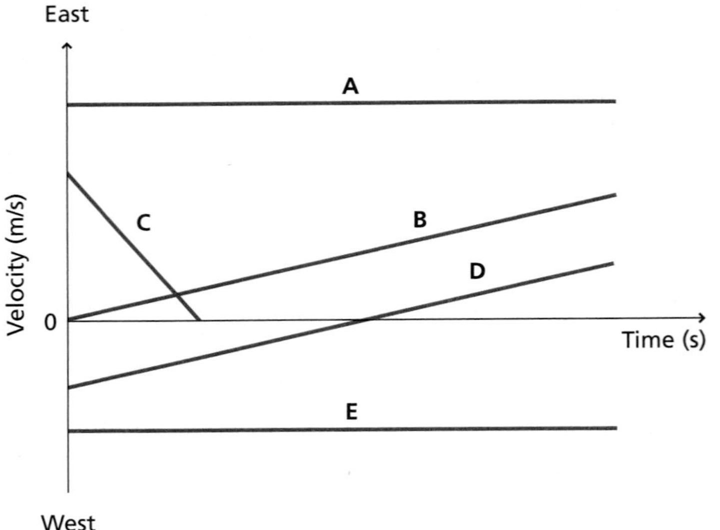

■ **Figure 3-8** Graphs A and E show motion with constant velocity in opposite directions. Graph B shows both positive velocity and positive acceleration. Graph C shows positive velocity and negative acceleration. Graph D shows motion with constant positive acceleration that slows down while velocity is negative and speeds up when velocity is positive.

In the first and third situations when the object is speeding up, the velocity and acceleration vectors point in the same direction in each case, as shown in Figure 3-7b. In the other two situations in which the acceleration vector is in the opposite direction from the velocity vectors, the object is slowing down. In other words, when the object's acceleration is in the same direction as its velocity, the object's speed increases. When they are in opposite directions, the speed decreases. Both the direction of an object's velocity and its direction of acceleration are needed to determine whether it is speeding up or slowing down. An object has a positive acceleration when the acceleration vector points in the positive direction and a negative acceleration, when the acceleration vector points in the negative direction. The sign of acceleration does not indicate whether the object is speeding up or slowing down.

Determining Acceleration from a *v-t* Graph

Velocity and acceleration information also is contained in velocity-time graphs. Graphs A, B, C, D, and E, shown in **Figure 3-8,** represent the motions of five different runners. Assume that the positive direction has been chosen to be east. The slopes of Graphs A and E are zero. Thus, the accelerations are zero. Both Graphs A and E show motion at a constant velocity—Graph A to the east and Graph E to the west. Graph B shows motion with a positive velocity. The slope of this graph indicates a constant, positive acceleration. You also can infer from Graph B that the speed increased because it shows positive velocity and acceleration. Graph C has a negative slope. Graph C shows motion that begins with a positive velocity, slows down, and then stops. This means that the acceleration and velocity are in opposite directions. The point at which Graphs C and B cross shows that the runners' velocities are equal at that point. It does not, however, give any information about the runners' positions.

Graph D indicates movement that starts out toward the west, slows down, and for an instant gets to zero velocity, and then moves east with increasing speed. The slope of Graph D is positive. Because the velocity and acceleration are in opposite directions, the speed decreases and equals zero at the time the graph crosses the axis. After that time, the velocity and acceleration are in the same direction and the speed increases.

Calculating acceleration How can you describe acceleration mathematically? The following equation expresses average acceleration as the slope of the velocity-time graph.

Average Acceleration $\bar{a} \equiv \frac{\Delta v}{\Delta t} = \frac{v_f - v_i}{t_f - t_i}$

Average acceleration is defined as the change in velocity, divided by the time it takes to make that change.

Suppose you run wind sprints back and forth across the gym. You first run at 4.0 m/s toward the wall. Then, 10.0 s later, you run at 4.0 m/s away from the wall. What is your average acceleration if the positive direction is toward the wall?

$$\bar{a} \equiv \frac{\Delta v}{\Delta t} = \frac{v_f - v_i}{t_f - t_i}$$

$$= \frac{(-4.0\ \text{m/s}) - (4.0\ \text{m/s})}{10.0\ \text{s}} = \frac{-8.0\ \text{m/s}}{10.0\ \text{s}} = -0.80\ \text{m/s}^2$$

The negative sign indicates that the direction of acceleration is away from the wall. The velocity changes when the direction of motion changes, because velocity includes the direction of motion. A change in velocity results in acceleration. Thus, acceleration also is associated with a change in the direction of motion.

▶ EXAMPLE Problem 2

Acceleration **Describe the motion of a ball as it rolls up a slanted driveway. The ball starts at 2.50 m/s, slows down for 5.00 s, stops for an instant, and then rolls back down at an increasing speed. The positive direction is chosen to be up the driveway, and the origin is at the place where the motion begins. What is the sign of the ball's acceleration as it rolls up the driveway? What is the magnitude of the ball's acceleration as it rolls up the driveway?**

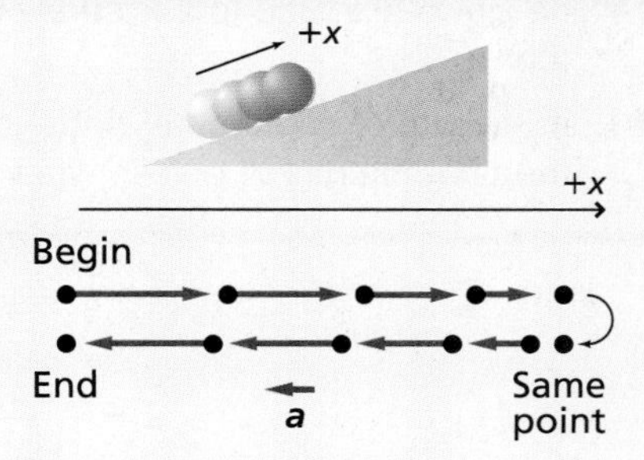

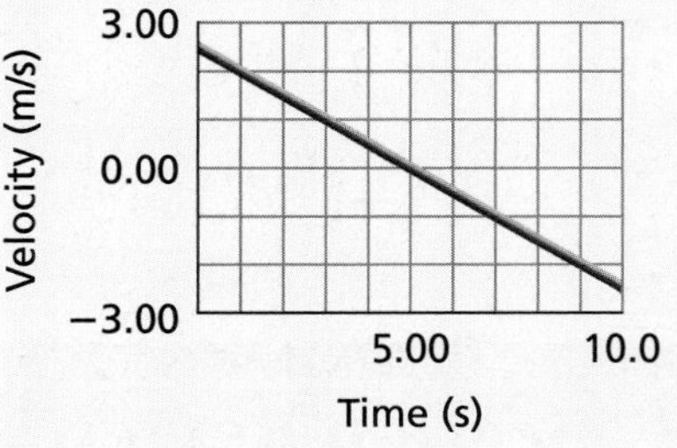

1 Analyze and Sketch the Problem

- Sketch the situation.
- Draw the coordinate system based on the motion diagram.

Known:	Unknown:
$v_i = +2.5$ m/s	$a = ?$
$v_f = 0.00$ m/s at $t = 5.00$ s	

2 Solve for the Unknown

Find the magnitude of the acceleration from the slope of the graph.

Solve for the change in velocity and the time taken to make that change.

$$\Delta v = v_f - v_i$$
$$= 0.00\ \text{m/s} - 2.50\ \text{m/s} \quad \text{Substitute } v_f = 0.00 \text{ m/s at } t_f = 5.00 \text{ s}, v_i = 2.50 \text{ m/s at } t_i = 0.00 \text{ s}$$
$$= -2.50\ \text{m/s}$$

$$\Delta t = t_f - t_i$$
$$= 5.00\ \text{s} - 0.00\ \text{s} \quad \text{Substitute } t_f = 5.00 \text{ s}, t_i = 0.00 \text{ s}$$
$$= 5.00\ \text{s}$$

Solve for the acceleration.

$$a = \frac{\Delta v}{\Delta t}$$
$$= \frac{-2.50\ \text{m/s}}{5.00\ \text{s}} \quad \text{Substitute } \Delta v = -2.50 \text{ m/s}, \Delta t = 5.00 \text{ s}$$
$$= -0.500\ \text{m/s}^2 \text{ or } 0.500\ \text{m/s}^2 \text{ down the driveway}$$

Math Handbook
Operations with Significant Digits pages 835–836

3 Evaluate the Answer

- **Are the units correct?** Acceleration is measured in m/s^2.
- **Do the directions make sense?** In the first 5.00 s, the direction of the acceleration is opposite to that of the velocity, and the ball slows down.

▶ PRACTICE Problems

- Additional Problems, Appendix B
- Solutions to Selected Problems, Appendix C

6. A race car's velocity increases from 4.0 m/s to 36 m/s over a 4.0-s time interval. What is its average acceleration?

7. The race car in the previous problem slows from 36 m/s to 15 m/s over 3.0 s. What is its average acceleration?

8. A car is coasting backwards downhill at a speed of 3.0 m/s when the driver gets the engine started. After 2.5 s, the car is moving uphill at 4.5 m/s. If uphill is chosen as the positive direction, what is the car's average acceleration?

9. A bus is moving at 25 m/s when the driver steps on the brakes and brings the bus to a stop in 3.0 s.

a. What is the average acceleration of the bus while braking?

b. If the bus took twice as long to stop, how would the acceleration compare with what you found in part **a?**

10. Rohith has been jogging to the bus stop for 2.0 min at 3.5 m/s when he looks at his watch and sees that he has plenty of time before the bus arrives. Over the next 10.0 s, he slows his pace to a leisurely 0.75 m/s. What was his average acceleration during this 10.0 s?

11. If the rate of continental drift were to abruptly slow from 1.0 cm/y to 0.5 cm/y over the time interval of a year, what would be the average acceleration?

There are several parallels between acceleration and velocity. Both are rates of change: acceleration is the time rate of change of velocity, and velocity is the time rate of change of position. Both acceleration and velocity have average and instantaneous forms. You will learn later in this chapter that the area under a velocity-time graph is equal to the object's displacement and that the area under an acceleration-time graph is equal to the object's velocity.

3.1 Section Review

12. Velocity-Time Graph What information can you obtain from a velocity-time graph?

13. Position-Time and Velocity-Time Graphs Two joggers run at a constant velocity of 7.5 m/s toward the east. At time $t = 0$, one is 15 m east of the origin and the other is 15 m west.

a. What would be the difference(s) in the position-time graphs of their motion?

b. What would be the difference(s) in their velocity-time graphs?

14. Velocity Explain how you would use a velocity-time graph to find the time at which an object had a specified velocity.

15. Velocity-Time Graph Sketch a velocity-time graph for a car that goes east at 25 m/s for 100 s, then west at 25 m/s for another 100 s.

16. Average Velocity and Average Acceleration A canoeist paddles upstream at 2 m/s and then turns around and floats downstream at 4 m/s. The turnaround time is 8 s.

a. What is the average velocity of the canoe?

b. What is the average acceleration of the canoe?

17. Critical Thinking A police officer clocked a driver going 32 km/h over the speed limit just as the driver passed a slower car. Both drivers were issued speeding tickets. The judge agreed with the officer that both were guilty. The judgement was issued based on the assumption that the cars must have been going the same speed because they were observed next to each other. Are the judge and the police officer correct? Explain with a sketch, a motion diagram, and a position-time graph.

Physics Online physicspp.com/self_check_quiz

3.2 Motion with Constant Acceleration

You have learned that the definition of average velocity can be algebraically rearranged to show the new position after a period of time, given the initial position and the average velocity. The definition of average acceleration can be manipulated similarly to show the new velocity after a period of time, given the initial velocity and the average acceleration.

► **Objectives**

- **Interpret** position-time graphs for motion with constant acceleration.
- **Determine** mathematical relationships among position, velocity, acceleration, and time.
- **Apply** graphical and mathematical relationships to solve problems related to constant acceleration.

Velocity with Average Acceleration

If you know an object's average acceleration during a time interval, you can use it to determine how much the velocity changed during that time. The definition of average acceleration,

$\bar{a} \equiv \frac{\Delta v}{\Delta t}$, can be rewritten as follows:

$$\Delta v = \bar{a}\Delta t$$

$$v_f - v_i = \bar{a}\Delta t$$

The equation for final velocity with average acceleration can be written as follows.

Final Velocity with Average Acceleration $v_f = v_i + \bar{a}\Delta t$

The final velocity is equal to the initial velocity plus the product of the average acceleration and time interval.

In cases in which the acceleration is constant, the average acceleration, $\bar{a}$, is the same as the instantaneous acceleration, a. This equation can be rearranged to find the time at which an object with constant acceleration has a given velocity. It also can be used to calculate the initial velocity of an object when both the velocity and the time at which it occurred are given.

PRACTICE Problems

• Additional Problems, Appendix B
• Solutions to Selected Problems, Appendix C

18. A golf ball rolls up a hill toward a miniature-golf hole. Assume that the direction toward the hole is positive.

a. If the golf ball starts with a speed of 2.0 m/s and slows at a constant rate of 0.50 m/s^2, what is its velocity after 2.0 s?

b. What is the golf ball's velocity if the constant acceleration continues for 6.0 s?

c. Describe the motion of the golf ball in words and with a motion diagram.

19. A bus that is traveling at 30.0 km/h speeds up at a constant rate of 3.5 m/s^2. What velocity does it reach 6.8 s later?

20. If a car accelerates from rest at a constant 5.5 m/s^2, how long will it take for the car to reach a velocity of 28 m/s?

21. A car slows from 22 m/s to 3.0 m/s at a constant rate of 2.1 m/s^2. How many seconds are required before the car is traveling at 3.0 m/s?

Table 3-2	
Position-Time Data for a Car	
Time (s)	**Position (m)**
0.00	0.00
1.00	2.50
2.00	10.0
3.00	22.5
4.00	40.0
5.00	62.5

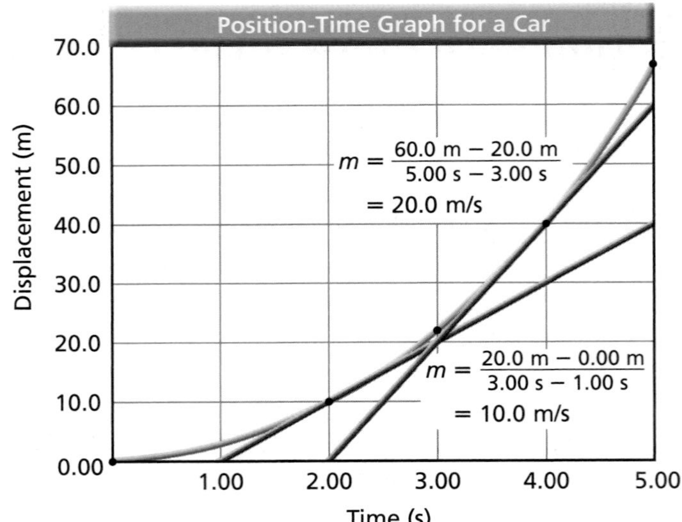

Figure 3-9 The slope of a position-time graph of a car moving with a constant acceleration gets steeper as time goes on.

Position with Constant Acceleration

You have learned that an object experiencing constant acceleration changes its velocity at a constant rate. How does the position of an object with constant acceleration change? The position data at different time intervals for a car with constant acceleration are shown in **Table 3-2.**

The data from Table 3-2 are graphed in **Figure 3-9.** The graph shows that the car's motion is not uniform: the displacements for equal time intervals on the graph get larger and larger. Notice that the slope of the line in Figure 3-9 gets steeper as time goes on. The slopes from the position-time graph can be used to create a velocity-time graph. Note that the slopes shown in Figure 3-9 are the same as the velocities graphed in **Figure 3-10a.**

A unique position-time graph cannot be created using a velocity-time graph because it does not contain any information about the object's position. However, the velocity-time graph does contain information about the object's displacement. Recall that for an object moving at a constant velocity, $v = \bar{v} = \Delta d/\Delta t$, so $\Delta d = v\Delta t$. On the graph in **Figure 3-10b,** v is the height of the plotted line above the t-axis, while Δt is the width of the shaded rectangle. The area of the rectangle, then, is $v\Delta t$, or Δd. Thus, the area under the v-t graph is equal to the object's displacement.

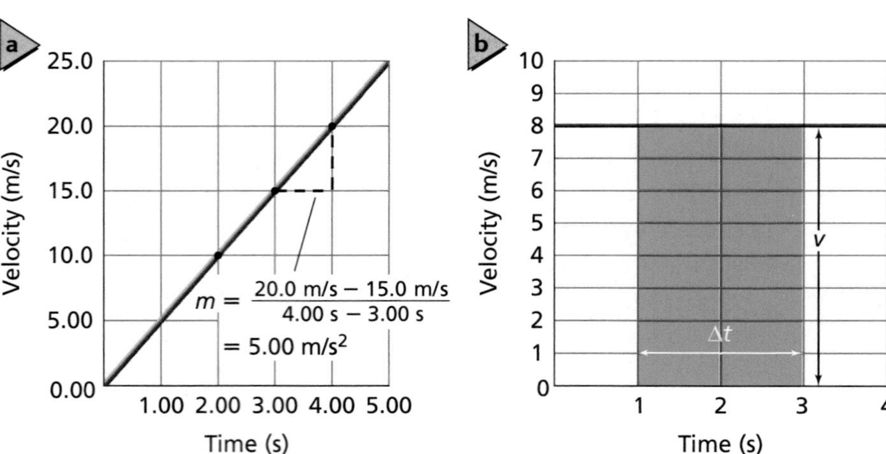

Figure 3-10 The slopes of the p-t graph in Figure 3-9 are the values of the corresponding v-t graph **(a).** For any v-t graph, the displacement during a given time interval is the area under the graph **(b).**

▶ EXAMPLE **Problem 3**

Finding the Displacement from a *v-t* Graph The *v-t* graph below shows the motion of an airplane. Find the displacement of the airplane at $\Delta t = 1.0$ s and at $\Delta t = 2.0$ s.

1 Analyze and Sketch the Problem

- The displacement is the area under the *v-t* graph.
- The time intervals begin at $t = 0.0$.

Known:	Unknown:
$v = +75$ m/s	$\Delta d = ?$
$\Delta t = 1.0$ s	
$\Delta t = 2.0$ s	

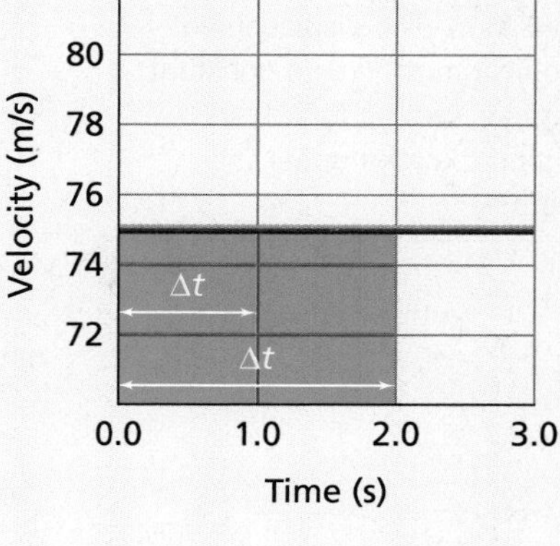

2 Solve for the Unknown

Solve for displacement during $\Delta t = 1.0$ s.

$$\Delta d = v\Delta t$$
$$= (+75\text{ m/s})(1.0\text{ s})$$ Substitute $v = +75$ m/s, $\Delta t = 1.0$ s
$$= +75\text{ m}$$

Solve for displacement during $\Delta t = 2.0$ s.

$$\Delta d = v\Delta t$$
$$= (+75\text{ m/s})(2.0\text{ s})$$ Substitute $v = +75$ m/s, $\Delta t = 2.0$ s
$$= +150\text{ m}$$

Math Handbook
Operations with Significant Digits
pages 835–836

3 Evaluate the Answer

- **Are the units correct?** Displacement is measured in meters.
- **Do the signs make sense?** The positive sign agrees with the graph.
- **Is the magnitude realistic?** Moving a distance equal to about one football field is reasonable for an airplane.

▶ PRACTICE **Problems**

• Additional Problems, Appendix B
• Solutions to Selected Problems, Appendix C

22. Use **Figure 3-11** to determine the velocity of an airplane that is speeding up at each of the following times.

a. 1.0 s **b.** 2.0 s **c.** 2.5 s

23. Use dimensional analysis to convert an airplane's speed of 75 m/s to km/h.

24. A position-time graph for a pony running in a field is shown in **Figure 3-12.** Draw the corresponding velocity-time graph using the same time scale.

25. A car is driven at a constant velocity of 25 m/s for 10.0 min. The car runs out of gas and the driver walks in the same direction at 1.5 m/s for 20.0 min to the nearest gas station. The driver takes 2.0 min to fill a gasoline can, then walks back to the car at 1.2 m/s and eventually drives home at 25 m/s in the direction opposite that of the original trip.

a. Draw a *v-t* graph using seconds as your time unit. Calculate the distance the driver walked to the gas station to find the time it took him to walk back to the car.

b. Draw a position-time graph for the situation using the areas under the velocity-time graph.

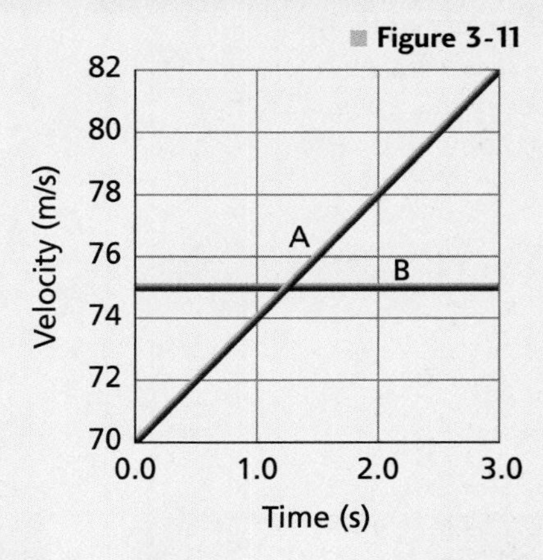

■ **Figure 3-11**

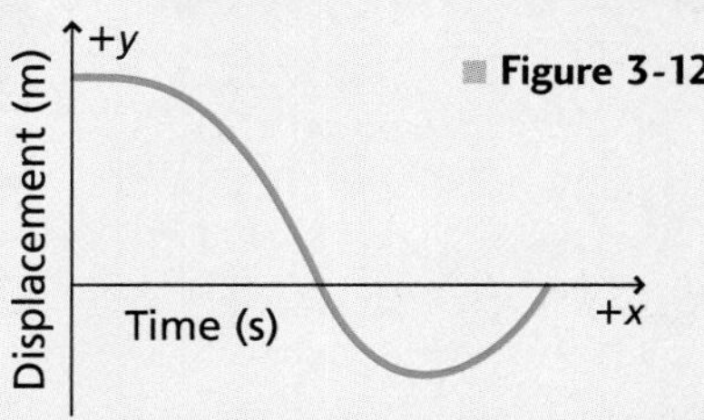

■ **Figure 3-12**

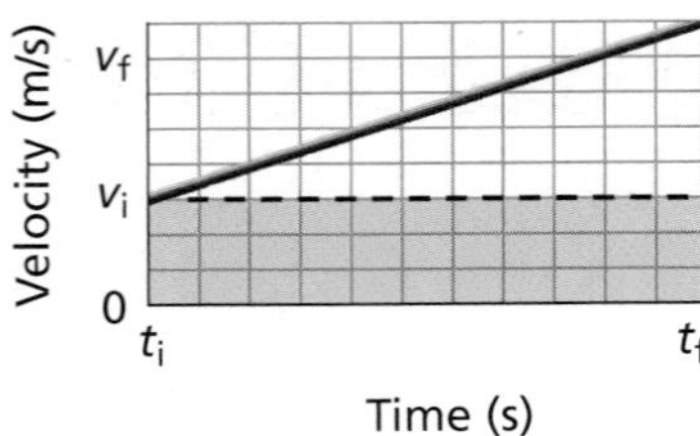

Figure 3-13 The displacement of an object moving with constant acceleration can be found by computing the area under the *v-t* graph.

The area under the *v-t* graph is equal to the object's displacement. Consider the *v-t* graph in **Figure 3-13** for an object moving with constant acceleration that started with an initial velocity of v_i. What is the object's displacement? The area under the graph can be calculated by dividing it into a rectangle and a triangle. The area of the rectangle can be found by $\Delta d_{\text{rectangle}} = v_i\Delta t$, and the area of the triangle can be found by $\Delta d_{\text{triangle}} = \frac{1}{2}\Delta v\Delta t$. Because average acceleration, $\bar{a}$, is equal to $\Delta v/\Delta t$, Δv can be rewritten as $\bar{a}\Delta t$. Substituting $\Delta v = \bar{a}\Delta t$ into the equation for the triangle's area yields $\Delta d_{\text{triangle}} = \frac{1}{2}(\bar{a}\Delta t)\Delta t$, or $\frac{1}{2}\bar{a}(\Delta t)^2$. Solving for the total area under the graph results in the following:

$$\Delta d = \Delta d_{\text{rectangle}} + \Delta d_{\text{triangle}} = v_i(\Delta t) + \frac{1}{2}\bar{a}(\Delta t)^2$$

When the initial or final position of the object is known, the equation can be written as follows:

$$d_f - d_i = v_i(\Delta t) + \frac{1}{2}\bar{a}(\Delta t)^2 \quad \text{or} \quad d_f = d_i + v_i(\Delta t) + \frac{1}{2}\bar{a}(\Delta t)^2$$

If the initial time is $t_i = 0$, the equation then becomes the following.

Position with Average Acceleration $d_f = d_i + v_it_f + \frac{1}{2}\bar{a}t_f^2$

An object's final position is equal to the sum of its initial position, the product of the initial velocity and the final time, and half the product of the acceleration and the square of the final time.

An Alternative Expression

Often, it is useful to relate position, velocity, and constant acceleration without including time. Rearrange the equation $v_f = v_i + \bar{a}t_f$ to solve for time: $t_f = \frac{v_f - v_i}{\bar{a}}$.

Rewriting $d_f = d_i + v_it_f + \frac{1}{2}\bar{a}t_f^2$ by substituting t_f yields the following:

$$d_f = d_i + v_i\frac{v_f - v_i}{\bar{a}} + \frac{1}{2}\bar{a}\left(\frac{v_f - v_i}{\bar{a}}\right)^2$$

This equation can be solved for the velocity, v_f, at any time, t_f.

Velocity with Constant Acceleration $v_f^2 = v_i^2 + 2\bar{a}(d_f - d_i)$

The square of the final velocity equals the sum of the square of the initial velocity and twice the product of the acceleration and the displacement since the initial time.

APPLYING PHYSICS

Drag Racing A dragster driver tries to obtain maximum acceleration over a 402-m (quarter-mile) course. The fastest time on record for the 402-m course is 4.480 s. The highest final speed on record is 147.63 m/s (330.23 mph).

The three equations for motion with constant acceleration are summarized in **Table 3-3.** Note that in a multi-step problem, it is useful to add additional subscripts to identify which step is under consideration.

Connecting **Math to Physics**

Table 3-3

Equations of Motion for Uniform Acceleration

Equation	Variables	Initial Conditions
$v_f = v_i + \bar{a}t_f$	$t_f, v_f, \bar{a}$	v_i
$d_f = d_i + v_it_f + \frac{1}{2}\bar{a}t_f^2$	$t_f, d_f, \bar{a}$	d_i, v_i
$v_f^2 = v_i^2 + 2\bar{a}(d_f - d_i)$	$d_f, v_f, \bar{a}$	d_i, v_i

▶ EXAMPLE Problem 4

Displacement An automobile starts at rest and speeds up at 3.5 m/s² after the traffic light turns green. How far will it have gone when it is traveling at 25 m/s?

1 Analyze and Sketch the Problem

- Sketch the situation.
- Establish coordinate axes.
- Draw a motion diagram.

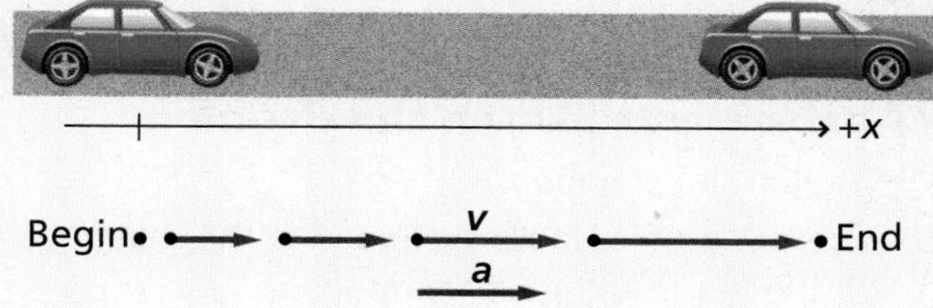

Begin• → → v → → •End
a →

Known:	Unknown:
$d_i = 0.00$ m	$d_f = ?$
$v_i = 0.00$ m/s	
$v_f = 25$ m/s	
$\bar{a} = a = 3.5$ m/s²	

2 Solve for the Unknown

Solve for d_f.

$$v_f^2 = v_i^2 + 2a(d_f - d_i)$$

$$d_f = d_i + \frac{v_f^2 - v_i^2}{2a}$$

$$= 0.00 \text{ m} + \frac{(25 \text{ m/s})^2 - (0.00 \text{ m/s})^2}{2(3.5 \text{ m/s}^2)}$$ Substitute $d_i = 0.00$ m, $v_f = 25$ m/s, $v_i = 0.00$ m/s

$$= 89 \text{ m}$$

Math Handbook
Order of Operations
page 843

3 Evaluate the Answer

- **Are the units correct?** Position is measured in meters.
- **Does the sign make sense?** The positive sign agrees with both the pictorial and physical models.
- **Is the magnitude realistic?** The displacement is almost the length of a football field. It seems large, but 25 m/s is fast (about 55 mph); therefore, the result is reasonable.

▶ PRACTICE Problems

Additional Problems, Appendix B

26. A skateboarder is moving at a constant velocity of 1.75 m/s when she starts up an incline that causes her to slow down with a constant acceleration of −0.20 m/s². How much time passes from when she begins to slow down until she begins to move back down the incline?

27. A race car travels on a racetrack at 44 m/s and slows at a constant rate to a velocity of 22 m/s over 11 s. How far does it move during this time?

28. A car accelerates at a constant rate from 15 m/s to 25 m/s while it travels a distance of 125 m. How long does it take to achieve this speed?

29. A bike rider pedals with constant acceleration to reach a velocity of 7.5 m/s over a time of 4.5 s. During the period of acceleration, the bike's displacement is 19 m. What was the initial velocity of the bike?

▶ EXAMPLE Problem 5

Two-Part Motion You are driving a car, traveling at a constant velocity of 25 m/s, when you see a child suddenly run onto the road. It takes 0.45 s for you to react and apply the brakes. As a result, the car slows with a steady acceleration of 8.5 m/s² and comes to a stop. What is the total distance that the car moves before it stops?

1 Analyze and Sketch the Problem

- Sketch the situation.
- Choose a coordinate system with the motion of the car in the positive direction.
- Draw the motion diagram and label v and a.

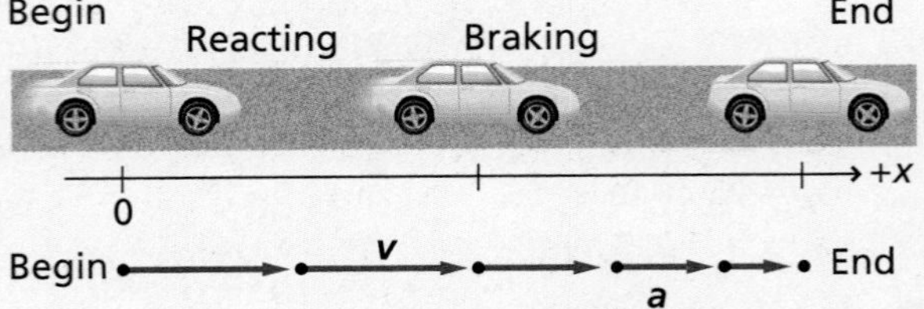

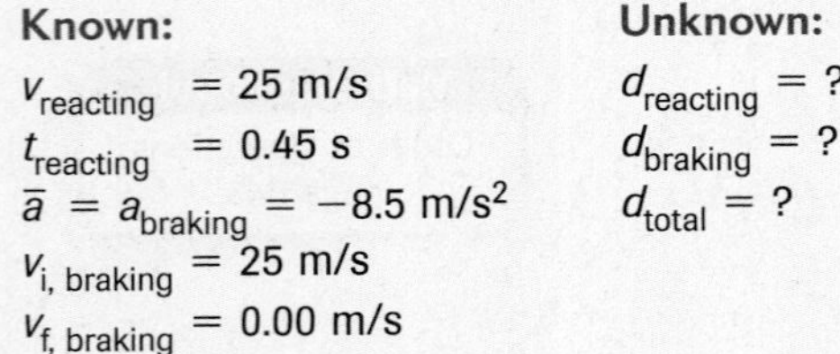

Known:	Unknown:
$v_{reacting} = 25$ m/s	$d_{reacting} = ?$
$t_{reacting} = 0.45$ s	$d_{braking} = ?$
$\bar{a} = a_{braking} = -8.5$ m/s²	$d_{total} = ?$
$v_{i,\ braking} = 25$ m/s	
$v_{f,\ braking} = 0.00$ m/s	

2 Solve for the Unknown

Reacting:

Solve for the distance the car travels at a constant speed.

$$d_{reacting} = v_{reacting}t_{reacting}$$

$$d_{reacting} = (25\ \text{m/s})(0.45\ \text{s}) = 11\ \text{m}$$

Substitute $v_{reacting} = 25$ m/s, $t_{reacting} = 0.45$ s

Braking:

Solve for the distance the car moves while braking.

$$v_{f,\ braking}{}^2 = v_{reacting}{}^2 + 2a_{braking}(d_{braking})$$

Solve for $d_{braking}$.

Math Handbook
Isolating a Variable
page 845

$$d_{braking} = \frac{v_{f,\ braking}{}^2 - v_{reacting}{}^2}{2a_{braking}}$$

$$= \frac{(0.00\ \text{m/s})^2 - (25\ \text{m/s})^2}{2(-8.5\ \text{m/s}^2)}$$

Substitute $v_{f,\ braking} = 0.00$ m/s, $v_{reacting} = 25$ m/s, $a_{braking} = -8.5$ m/s²

$$= 37\ \text{m}$$

The total distance traveled is the sum of the reaction distance and the braking distance.

Solve for d_{total}.

$$d_{total} = d_{reacting} + d_{braking}$$

$$= 11\ \text{m} + 37\ \text{m}$$

Substitute $d_{reacting} = 11$ m, $d_{braking} = 37$ m

$$= 48\ \text{m}$$

3 Evaluate the Answer

- **Are the units correct?** Distance is measured in meters.
- **Do the signs make sense?** Both $d_{reacting}$ and $d_{braking}$ are positive, as they should be.
- **Is the magnitude realistic?** The braking distance is small because the magnitude of the acceleration is large.

PRACTICE Problems

• Additional Problems, Appendix B
• Solutions to Selected Problems, Appendix C

30. A man runs at a velocity of 4.5 m/s for 15.0 min. When going up an increasingly steep hill, he slows down at a constant rate of 0.05 m/s^2 for 90.0 s and comes to a stop. How far did he run?

31. Sekazi is learning to ride a bike without training wheels. His father pushes him with a constant acceleration of 0.50 m/s^2 for 6.0 s, and then Sekazi continues at 3.0 m/s for another 6.0 s before falling. What is Sekazi's displacement? Solve this problem by constructing a velocity-time graph for Sekazi's motion and computing the area underneath the graphed line.

32. You start your bicycle ride at the top of a hill. You coast down the hill at a constant acceleration of 2.00 m/s^2. When you get to the bottom of the hill, you are moving at 18.0 m/s, and you pedal to maintain that speed. If you continue at this speed for 1.00 min, how far will you have gone from the time you left the hilltop?

33. Sunee is training for an upcoming 5.0-km race. She starts out her training run by moving at a constant pace of 4.3 m/s for 19 min. Then she accelerates at a constant rate until she crosses the finish line, 19.4 s later. What is her acceleration during the last portion of the training run?

You have learned several different tools that you can apply when solving problems dealing with motion in one dimension: motion diagrams, graphs, and equations. As you gain more experience, it will become easier to decide which tools are most appropriate in solving a given problem. In the following section, you will practice using these tools to investigate the motion of falling objects.

3.2 Section Review

34. Acceleration A woman driving at a speed of 23 m/s sees a deer on the road ahead and applies the brakes when she is 210 m from the deer. If the deer does not move and the car stops right before it hits the deer, what is the acceleration provided by the car's brakes?

35. Displacement If you were given initial and final velocities and the constant acceleration of an object, and you were asked to find the displacement, what equation would you use?

36. Distance An in-line skater first accelerates from 0.0 m/s to 5.0 m/s in 4.5 s, then continues at this constant speed for another 4.5 s. What is the total distance traveled by the in-line skater?

37. Final Velocity A plane travels a distance of 5.0×10^2 m while being accelerated uniformly from rest at the rate of 5.0 m/s^2. What final velocity does it attain?

38. Final Velocity An airplane accelerated uniformly from rest at the rate of 5.0 m/s^2 for 14 s. What final velocity did it attain?

39. Distance An airplane starts from rest and accelerates at a constant 3.00 m/s^2 for 30.0 s before leaving the ground.

a. How far did it move?

b. How fast was the airplane going when it took off?

40. Graphs A sprinter walks up to the starting blocks at a constant speed and positions herself for the start of the race. She waits until she hears the starting pistol go off, and then accelerates rapidly until she attains a constant velocity. She maintains this velocity until she crosses the finish line, and then she slows down to a walk, taking more time to slow down than she did to speed up at the beginning of the race. Sketch a velocity-time and a position-time graph to represent her motion. Draw them one above the other on the same time scale. Indicate on your *p-t* graph where the starting blocks and finish line are.

41. Critical Thinking Describe how you could calculate the acceleration of an automobile. Specify the measuring instruments and the procedures that you would use.

3.3 Free Fall

- **Objectives**
  - **Define** acceleration due to gravity.
  - **Solve** problems involving objects in free fall.
- **Vocabulary**
  free fall
  acceleration due to gravity

Drop a sheet of paper. Crumple it, and then drop it again. Drop a rock or a pebble. How do the three motions compare with each other? Do heavier objects fall faster than lighter ones? A light, spread-out object, such as a smooth sheet of paper or a feather, does not fall in the same manner as something more compact, such as a pebble. Why? As an object falls, it bumps into particles in the air. For an object such as a feather, these little collisions have a greater effect than they do on pebbles or rocks. To understand the behavior of falling objects, first consider the simplest case: an object such as a rock, for which the air does not have an appreciable effect on its motion. The term used to describe the motion of such objects is **free fall,** which is the motion of a body when air resistance is negligible and the action can be considered due to gravity alone.

Acceleration Due to Gravity

About 400 years ago, Galileo Galilei recognized that to make progress in the study of the motion of falling objects, the effects of the substance through which the object falls have to be ignored. At that time, Galileo had no means of taking position or velocity data for falling objects, so he rolled balls down inclined planes. By "diluting" gravity in this way, he could make careful measurements even with simple instruments.

Galileo concluded that, neglecting the effect of the air, all objects in free fall had the same acceleration. It didn't matter what they were made of, how much they weighed, what height they were dropped from, or whether they were dropped or thrown. The acceleration of falling objects, given a special symbol, g, is equal to 9.80 m/s^2. It is now known that there are small variations in g at different places on Earth, and that 9.80 m/s^2 is the average value.

The **acceleration due to gravity** is the acceleration of an object in free fall that results from the influence of Earth's gravity. Suppose you drop a rock. After 1 s, its velocity is 9.80 m/s downward, and 1 s after that, its velocity is 19.60 m/s downward. For each second that the rock is falling, its downward velocity increases by 9.80 m/s. Note that g is a positive number. When analyzing free fall, whether you treat the acceleration as positive or negative depends upon the coordinate system that you use. If your coordinate system defines upward to be the positive direction, then the acceleration due to gravity is equal to $-g$; if you decide that downward is the positive direction, then the acceleration due to gravity is $+g$.

A strobe photo of a dropped egg is shown in **Figure 3-14.** The time interval between the images is 0.06 s. The displacement between each pair of images increases, so the speed is increasing. If the upward direction is chosen as positive, then the velocity is becoming more and more negative.

Figure 3-14 An egg accelerates at 9.80 m/s^2 in free fall. If the upward direction is chosen as positive, then both the velocity and the acceleration of this egg in free fall are negative.

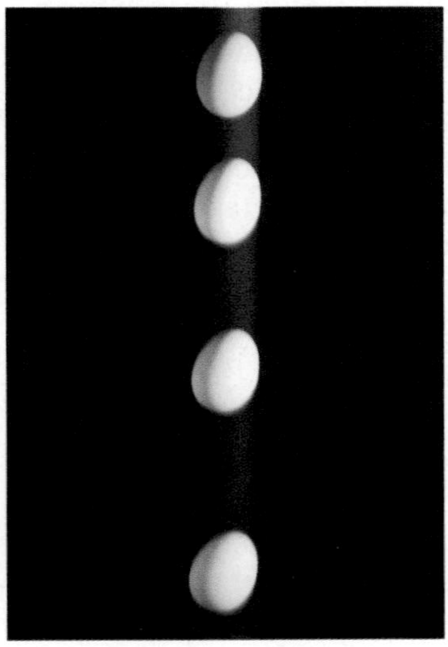

Ball thrown upward Instead of a dropped egg, could this photo also illustrate a ball thrown upward? If upward is chosen to be the positive direction, then the ball leaves the hand with a positive velocity of, for example, 20.0 m/s. The acceleration is downward, so a is negative. That is, $a = -g = -9.80$ m/s^2. Because the velocity and acceleration are in opposite directions, the speed of the ball decreases, which is in agreement with the strobe photo.

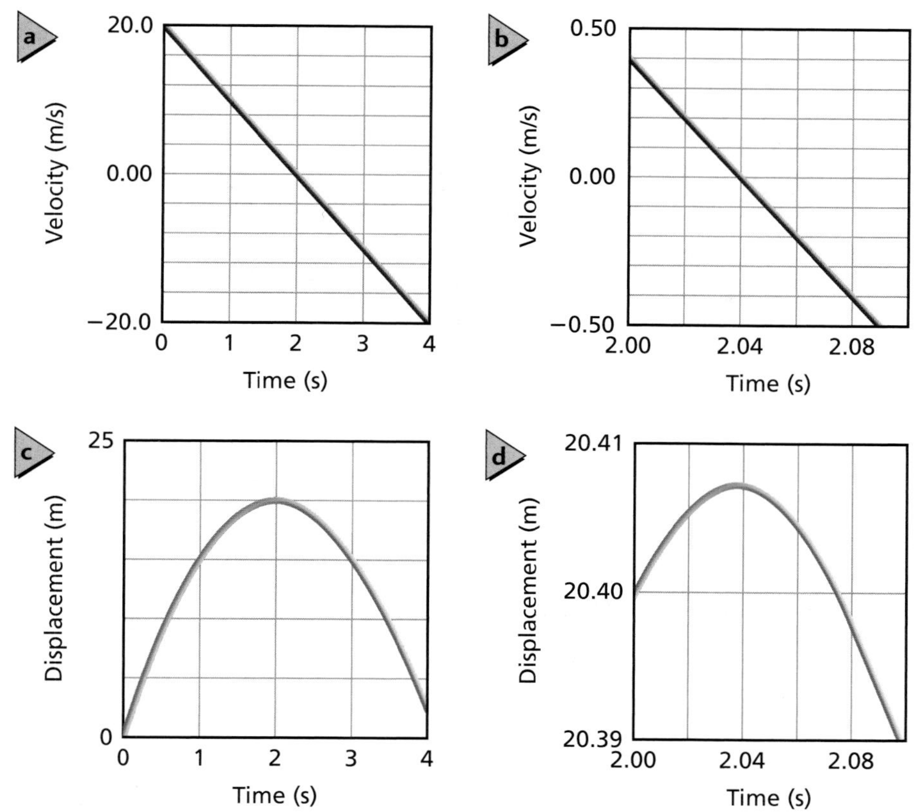

■ **Figure 3-15** In a coordinate system in which the upward direction is positive, the velocity of the thrown ball decreases until it becomes zero at 2.04 s. Then it increases in the negative direction as the ball falls **(a, b).** The *p-t* graphs show the height of the ball at corresponding time intervals **(c, d).**

Concepts In Motion

Interactive Figure To see an animation on acceleration due to gravity, physicspp.com.

After 1 s, the ball's velocity is reduced by 9.80 m/s, so it now is traveling at 10.2 m/s. After 2 s, the velocity is 0.4 m/s, and the ball still is moving upward. What happens during the next second? The ball's velocity is reduced by another 9.80 m/s, and is equal to −9.4 m/s. The ball now is moving downward. After 4 s, the velocity is −19.2 m/s, meaning that the ball is falling even faster. **Figure 3-15a** shows the velocity-time graph for the ball as it goes up and comes back down. At around 2 s, the velocity changes smoothly from positive to negative. **Figure 3-15b** shows a closer view of the *v-t* graph around that point. At an instant of time, near 2.04 s, the ball's velocity is zero. Look at the position-time graphs in **Figure 3-15c** and **d,** which show how the ball's height changes. How are the ball's position and velocity related? The ball reaches its maximum height at the instant of time when its velocity is zero.

At 2.04 s, the ball reaches its maximum height and its velocity is zero. What is the ball's acceleration at that point? The slope of the line in the *v-t* graphs in Figure 3-15a and 3-15b is constant at $-9.80\ m/s^2$.

Often, when people are asked about the acceleration of an object at the top of its flight, they do not take the time to fully analyze the situation, and respond that the acceleration at this point is zero. However, this is not the case. At the top of the flight, the ball's velocity is 0 m/s. What would happen if its acceleration were also zero? Then the ball's velocity would not be changing and would remain at 0 m/s. If this were the case, the ball would not gain any downward velocity and would simply hover in the air at the top of its flight. Because this is not the way objects tossed in the air behave on Earth, you know that the acceleration of an object at the top of its flight must not be zero. Further, because you know that the object will fall from that height, you know that the acceleration must be downward.

Free-fall rides Amusement parks use the concept of free fall to design rides that give the riders the sensation of free fall. These types of rides usually consist of three parts: the ride to the top, momentary suspension, and the plunge downward. Motors provide the force needed to move the cars to the top of the ride. When the cars are in free fall, the most massive rider and the least massive rider will have the same acceleration. Suppose the free-fall ride at an amusement park starts at rest and is in free fall for 1.5 s. What would be its velocity at the end of 1.5 s? Choose a coordinate system with a positive axis upward and the origin at the initial position of the car. Because the car starts at rest, v_i would be equal to 0.00 m/s. To calculate the final velocity, use the equation for velocity with constant acceleration.

$$\begin{aligned} v_f &= v_i + \bar{a}t_f \\ &= 0.00\ \text{m/s} + (-9.80\ \text{m/s}^2)(1.5\ \text{s}) \\ &= -15\ \text{m/s} \end{aligned}$$

How far does the car fall? Use the equation for displacement when time and constant acceleration are known.

$$\begin{aligned} d_f &= d_i + v_i t_f + \tfrac{1}{2}\bar{a}t_f^2 \\ &= 0.00\ \text{m} + (0.00\ \text{m/s})(1.5\ \text{s}) + \tfrac{1}{2}(-9.80\ \text{m/s}^2)(1.5\ \text{s})^2 \\ &= -11\ \text{m} \end{aligned}$$

PRACTICE Problems

• Additional Problems, Appendix B
• Solutions to Selected Problems, Appendix C

42. A construction worker accidentally drops a brick from a high scaffold.

a. What is the velocity of the brick after 4.0 s?

b. How far does the brick fall during this time?

43. Suppose for the previous problem you choose your coordinate system so that the opposite direction is positive.

a. What is the brick's velocity after 4.0 s?

b. How far does the brick fall during this time?

44. A student drops a ball from a window 3.5 m above the sidewalk. How fast is it moving when it hits the sidewalk?

45. A tennis ball is thrown straight up with an initial speed of 22.5 m/s. It is caught at the same distance above the ground.

a. How high does the ball rise?

b. How long does the ball remain in the air? *Hint: The time it takes the ball to rise equals the time it takes to fall.*

46. You decide to flip a coin to determine whether to do your physics or English homework first. The coin is flipped straight up.

a. If the coin reaches a high point of 0.25 m above where you released it, what was its initial speed?

b. If you catch it at the same height as you released it, how much time did it spend in the air?

CHALLENGE PROBLEM

You notice a water balloon fall past your classroom window. You estimate that it took the balloon about t seconds to fall the length of the window and that the window is about y meters high. Suppose the balloon started from rest. Approximately how high above the top of the window was it released? Your answer should be in terms of t, y, g, and numerical constants.

Remember to define the positive direction when establishing your coordinate system. As motion problems increase in complexity, it becomes increasingly important to keep all the signs consistent. This means that any displacement, velocity, or acceleration that is in the same direction as the one chosen to be positive will be positive. Thus, any displacement, velocity, or acceleration that is in the direction opposite to the one chosen to be positive should be indicated with a negative sign. Sometimes it might be appropriate to choose upward as positive. At other times, it might be easier to choose downward as positive. You can choose either direction you want, as long as you stay consistent with that convention throughout the solution of that particular problem. Suppose you solve one of the practice problems on the preceding page again, choosing the direction opposite to the one you previously designated as the positive direction for the coordinate system. You should arrive at the same answer, provided that you assigned signs to each of the quantities that were consistent with the coordinate system. It is important to be consistent with the coordinate system to avoid getting the signs mixed up.

3.3 Section Review

47. Maximum Height and Flight Time Acceleration due to gravity on Mars is about one-third that on Earth. Suppose you throw a ball upward with the same velocity on Mars as on Earth.

a. How would the ball's maximum height compare to that on Earth?

b. How would its flight time compare?

48. Velocity and Acceleration Suppose you throw a ball straight up into the air. Describe the changes in the velocity of the ball. Describe the changes in the acceleration of the ball.

49. Final Velocity Your sister drops your house keys down to you from the second floor window. If you catch them 4.3 m from where your sister dropped them, what is the velocity of the keys when you catch them?

50. Initial Velocity A student trying out for the football team kicks the football straight up in the air. The ball hits him on the way back down. If it took 3.0 s from the time when the student punted the ball until he gets hit by the ball, what was the football's initial velocity?

51. Maximum Height When the student in the previous problem kicked the football, approximately how high did the football travel?

52. Critical Thinking When a ball is thrown vertically upward, it continues upward until it reaches a certain position, and then it falls downward. At that highest point, its velocity is instantaneously zero. Is the ball accelerating at the highest point? Devise an experiment to prove or disprove your answer.

PHYSICS LAB • Internet

Alternate CBL instructions can be found on the Web site.
physicspp.com

Acceleration Due to Gravity

Small variations in the acceleration due to gravity, g, occur at different places on Earth. This is because g varies with distance from the center of Earth and is influenced by the subsurface geology. In addition, g varies with latitude due to Earth's rotation.

For motion with constant acceleration, the displacement is $d_f - d_i = v_i(t_f - t_i) + \frac{1}{2}a(t_f - t_i)^2$. If $d_i = 0$ and $t_i = 0$, then the displacement is $d_f = v_i t_f + \frac{1}{2}at_f^2$. Dividing both sides of the equation by t_f yields the following: $d_f/t_f = v_i + \frac{1}{2}at_f$. The slope of a graph of d_f/t_f versus t_f, is equal to $\frac{1}{2}a$. The initial velocity, v_i, is determined by the y-intercept. In this activity, you will be using a spark timer to collect free-fall data and use it to determine the acceleration due to gravity, g.

QUESTION

How does the value of *g* vary from place to place?

Objectives

- **Measure** free-fall data.
- **Make and use graphs** of velocity versus time.
- **Compare and contrast** values of g for different locations.

Safety Precautions

- **Keep clear of falling masses.**

Materials

spark timer
timer tape
1-kg mass
C-clamp
stack of newspapers
masking tape

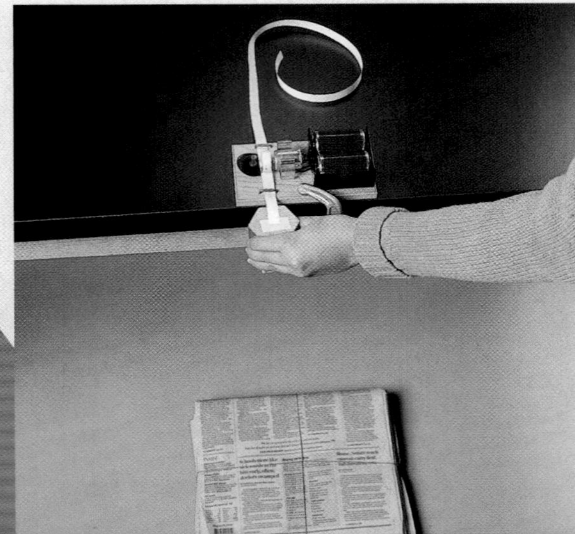

Procedure

1. Attach the spark timer to the edge of the lab table with the C-clamp.
2. If the timer needs to be calibrated, follow your teacher's instructions or those provided with the timer. Determine the period of the timer and record it in your data table.
3. Place the stack of newspapers on the floor, directly below the timer so that the mass, when released, will not damage the floor.
4. Cut a piece of timer tape approximately 70 cm in length and slide it into the spark timer.
5. Attach the timer tape to the 1-kg mass with a small piece of masking tape. Hold the mass next to the spark timer, over the edge of the table so that it is above the newspaper stack.
6. Turn on the spark timer and release the mass.
7. Inspect the timer tape to make sure that there are dots marked on it and that there are no gaps in the dot sequence. If your timer tape is defective, repeat steps 4–6 with another piece of timer tape.
8. Have each member of your group perform the experiment and collect his or her own data.
9. Choose a dot near the beginning of the timer tape, a few centimeters from the point where the timer began to record dots, and label it *0*. Label the dots after that *1, 2, 3, 4, 5,* etc. until you get near the end where the mass is no longer in free fall. If the dots stop, or the distance between them begins to get smaller, the mass is no longer in free fall.

Data Table	
Time period (#/s)	

Interval	Distance (cm)	Time (s)	Speed (cm/s)
1			
2			
3			
4			
5			
6			
7			
8			

10. Measure the total distance to each numbered dot from the zero dot, to the nearest millimeter and record it in your data table. Using the timer period, record the total time associated with each distance measurement and record it in your data table.

Analyze

1. **Use Numbers** Calculate the values for speed and record them in the data table.
2. **Make and Use Graphs** Draw a graph of speed versus time. Draw the best-fit straight line for your data.
3. Calculate the slope of the line. Convert your result to m/s^2.

Conclude and Apply

1. Recall that the slope is equal to $\frac{1}{2}a$. What is the acceleration due to gravity?
2. Find the relative error for your experimental value of g by comparing it to the accepted value.

   Relative error =

   $$\frac{\text{Accepted value} - \text{Experimental value}}{\text{Accepted value}} \times 100$$
3. What was the mass's velocity, v_i, when you began measuring distance and time?

Going Further

What is the advantage of measuring several centimeters away from the beginning of the timer tape rather than from the very first dot?

Real-World Physics

Why do designers of free-fall amusement-park rides design exit tracks that gradually curve toward the ground? Why is there a stretch of straight track?

Share Your Data

Communicate the average value of g to others. Go to physicspp.com/internet_lab and post the name of your school, city, state, elevation above sea level, and average value of g for your class. Obtain a map for your state and a map of the United States. Using the data posted on the Web site by other students, mark the values for g at the appropriate locations on the maps. Do you notice any variation in the acceleration due to gravity for different locations, regions and elevations?

To find out more about accelerated motion, visit the Web site: physicspp.com

Time Dilation at High Velocities

Can time pass differently in two reference frames? How can one of a pair of twins age more than the other?

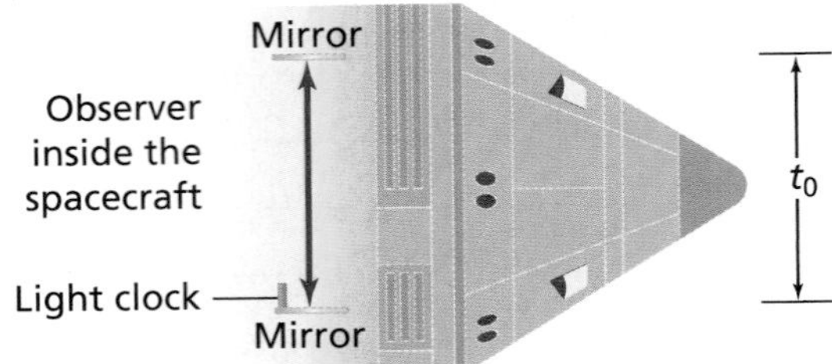

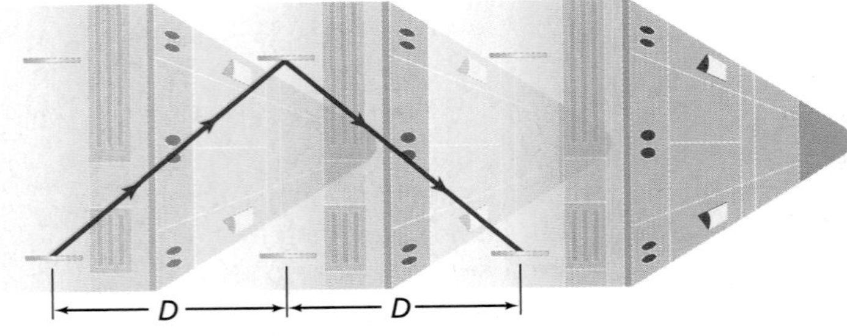

Light Clock Consider the following thought experiment using a light clock. A light clock is a vertical tube with a mirror at each end. A short pulse of light is introduced at one end and allowed to bounce back and forth within the tube. Time is measured by counting the number of bounces made by the pulse of light. The clock will be accurate because the speed of a pulse of light is always c, which is 3×10^8 m/s, regardless of the velocity of the light source or the observer.

Suppose this light clock is placed in a very fast spacecraft. When the spacecraft goes at slow speeds, the light beam bounces vertically in the tube. If the spacecraft is moving fast, the light beam still bounces vertically—at least as seen by the observer in the spacecraft.

A stationary observer on Earth, however, sees the pulse of light move diagonally because of the movement of the spacecraft. Thus, to the stationary observer, the light beam moves a greater distance. Distance = velocity × time, so if the distance traveled by the light beam increases, the product (velocity × time) also must increase.

Because the speed of the light pulse, c, is the same for any observer, time must be increasing for the stationary observer. That is, the stationary observer sees the moving clock ticking slower than the same clock on Earth.

Suppose the time per tick seen by the stationary observer on Earth is t_s, the time seen by the observer on the spacecraft is t_o, the length of the light clock is ct_o, the velocity of the spacecraft is v, and the speed of light is c. For every tick, the spacecraft moves vt_s and the light pulse moves ct_o. This leads to the following equation:

$$t_s = \frac{t_o}{\sqrt{1 - \left(\frac{v^2}{c^2}\right)}}$$

To the stationary observer, the closer v is to c, the slower the clock ticks. To the observer on the spacecraft, however, the clock keeps perfect time.

Time Dilation This phenomenon is called time dilation and it applies to every process associated with time aboard the spacecraft. For example, biological aging will proceed more slowly in the spacecraft than on Earth. So if the observer on the spacecraft is one of a pair of twins, he or she would age more slowly than the other twin on Earth. This is called the twin paradox. Time dilation has resulted in a lot of speculation about space travel. If spacecraft were able to travel at speeds close to the speed of light, trips to distant stars would take only a few years for the astronaut.

Going Further

1. **Calculate** Find the time dilation t_s/t_o for Earth's orbit about the Sun if v_{Earth} = 10,889 km/s.
2. **Calculate** Derive the equation for t_s above.
3. **Discuss** How is time dilation similar to or different from time travel?

Chapter 3

Study Guide

3.1 Acceleration

Vocabulary

- velocity-time graph (p. 58)
- acceleration (p.59)
- average acceleration (p. 59)
- instantaneous acceleration (p. 59)

Key Concepts

- A velocity-time graph can be used to find the velocity and acceleration of an object.
- The average acceleration of an object is the slope of its velocity-time graph.

$$\bar{a} \equiv \frac{\Delta v}{\Delta t} = \frac{v_f - v_i}{t_f - t_i}$$

- Average acceleration vectors on a motion diagram indicate the size and direction of the average acceleration during a time interval.
- When the acceleration and velocity are in the same direction, the object speeds up; when they are in opposite directions, the object slows down.
- Velocity-time graphs and motion diagrams can be used to determine the sign of an object's acceleration.

3.2 Motion with Constant Acceleration

Key Concepts

- If an object's average acceleration during a time interval is known, the change in velocity during that time can be found.

$$v_f = v_i + \bar{a}\Delta t$$

- The area under an object's velocity-time graph is its displacement.
- In motion with constant acceleration, there are relationships among the position, velocity, acceleration, and time.

$$d_f = d_i + v_i t_f + \frac{1}{2}\bar{a}t_f^2$$

- The velocity of an object with constant acceleration can be found using the following equation.

$$v_f^2 = v_i^2 + 2\bar{a}(d_f - d_i)$$

3.3 Free Fall

Vocabulary

- free fall (p. 72)
- acceleration due to gravity (p. 72)

Key Concepts

- The acceleration due to gravity on Earth, g, is 9.80 m/s^2 downward. The sign associated with g in equations depends upon the choice of the coordinate system.
- Equations for motion with constant acceleration can be used to solve problems involving objects in free fall.

Chapter 3

Assessment

Concept Mapping

53. Complete the following concept map using the following symbols or terms: *d, velocity, m/s²*, *v, m, acceleration.*

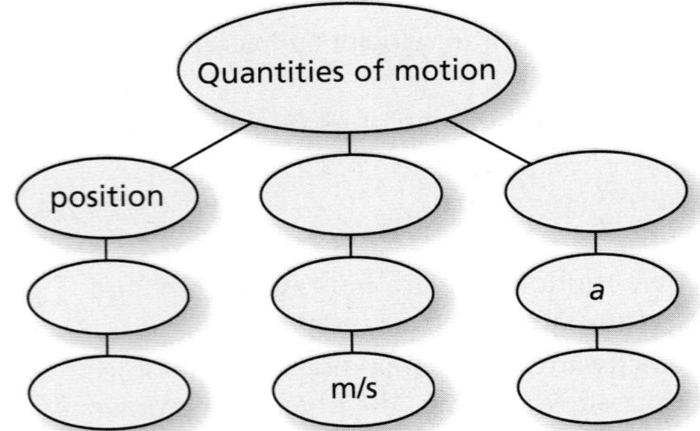

Mastering Concepts

54. How are velocity and acceleration related? (3.1)

55. Give an example of each of the following. (3.1)

a. an object that is slowing down, but has a positive acceleration

b. an object that is speeding up, but has a negative acceleration

56. **Figure 3-16** shows the velocity-time graph for an automobile on a test track. Describe how the velocity changes with time. (3.1)

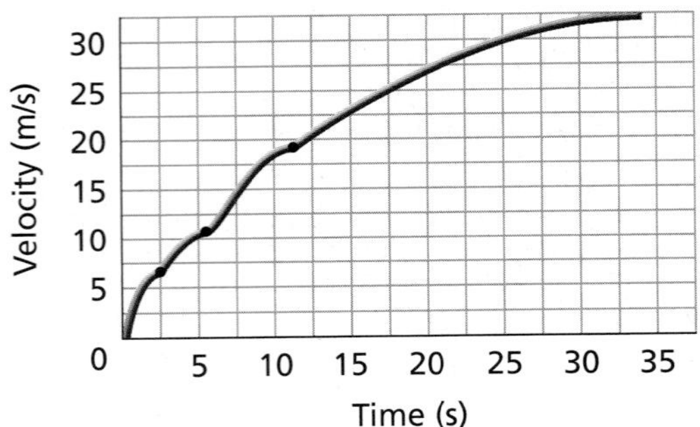

■ **Figure 3-16**

57. What does the slope of the tangent to the curve on a velocity-time graph measure? (3.1)

58. Can a car traveling on an interstate highway have a negative velocity and a positive acceleration at the same time? Explain. Can the car's velocity change signs while it is traveling with constant acceleration? Explain. (3.1)

59. Can the velocity of an object change when its acceleration is constant? If so, give an example. If not, explain. (3.1)

60. If an object's velocity-time graph is a straight line parallel to the *t*-axis, what can you conclude about the object's acceleration? (3.1)

61. What quantity is represented by the area under a velocity-time graph? (3.2)

62. Write a summary of the equations for position, velocity, and time for an object experiencing motion with uniform acceleration. (3.2)

63. Explain why an aluminum ball and a steel ball of similar size and shape, dropped from the same height, reach the ground at the same time. (3.3)

64. Give some examples of falling objects for which air resistance cannot be ignored. (3.3)

65. Give some examples of falling objects for which air resistance can be ignored. (3.3)

Applying Concepts

66. Does a car that is slowing down always have a negative acceleration? Explain.

67. **Croquet** A croquet ball, after being hit by a mallet, slows down and stops. Do the velocity and acceleration of the ball have the same signs?

68. If an object has zero acceleration, does it mean its velocity is zero? Give an example.

69. If an object has zero velocity at some instant, is its acceleration zero? Give an example.

70. If you were given a table of velocities of an object at various times, how would you find out whether the acceleration was constant?

71. The three notches in the graph in Figure 3-16 occur where the driver changed gears. Describe the changes in velocity and acceleration of the car while in first gear. Is the acceleration just before a gear change larger or smaller than the acceleration just after the change? Explain your answer.

72. Use the graph in Figure 3-16 and determine the time interval during which the acceleration is largest and the time interval during which the acceleration is smallest.

73. Explain how you would walk to produce each of the position-time graphs in **Figure 3-17.**

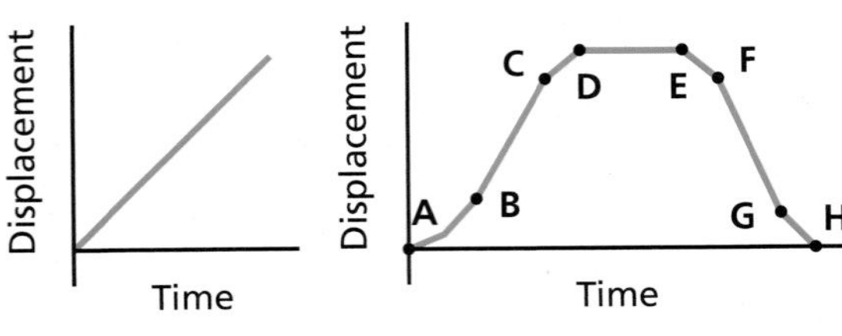

■ **Figure 3-17**

74. Draw a velocity-time graph for each of the graphs in **Figure 3-18.**

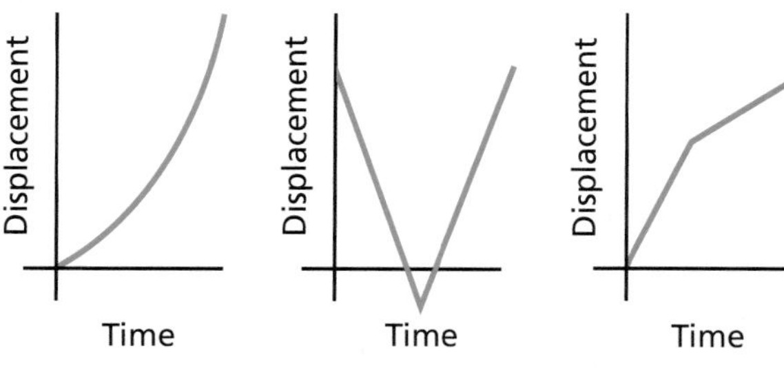

Figure 3-18

75. An object shot straight up rises for 7.0 s before it reaches its maximum height. A second object falling from rest takes 7.0 s to reach the ground. Compare the displacements of the two objects during this time interval.

76. **The Moon** The value of g on the Moon is one-sixth of its value on Earth.
   a. Would a ball that is dropped by an astronaut hit the surface of the Moon with a greater, equal, or lesser speed than that of a ball dropped from the same height to Earth?
   b. Would it take the ball more, less, or equal time to fall?

77. **Jupiter** The planet Jupiter has about three times the gravitational acceleration of Earth. Suppose a ball is thrown vertically upward with the same initial velocity on Earth and on Jupiter. Neglect the effects of Jupiter's atmospheric resistance and assume that gravity is the only force on the ball.
   a. How does the maximum height reached by the ball on Jupiter compare to the maximum height reached on Earth?
   b. If the ball on Jupiter were thrown with an initial velocity that is three times greater, how would this affect your answer to part **a?**

78. Rock A is dropped from a cliff and rock B is thrown upward from the same position.
   a. When they reach the ground at the bottom of the cliff, which rock has a greater velocity?
   b. Which has a greater acceleration?
   c. Which arrives first?

Mastering Problems

3.1 Acceleration

79. A car is driven for 2.0 h at 40.0 km/h, then for another 2.0 h at 60.0 km/h in the same direction.
   a. What is the car's average velocity?
   b. What is the car's average velocity if it is driven 1.0×10^2 km at each of the two speeds?

80. Find the uniform acceleration that causes a car's velocity to change from 32 m/s to 96 m/s in an 8.0-s period.

81. A car with a velocity of 22 m/s is accelerated uniformly at the rate of 1.6 m/s^2 for 6.8 s. What is its final velocity?

82. Refer to **Figure 3-19** to find the acceleration of the moving object at each of the following times.
   a. during the first 5.0 s of travel
   b. between 5.0 s and 10.0 s
   c. between 10.0 s and 15.0 s
   d. between 20.0 s and 25.0 s

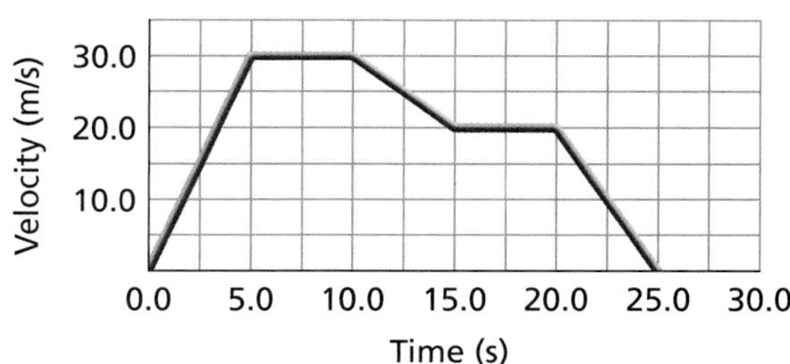

Figure 3-19

83. Plot a velocity-time graph using the information in **Table 3-4,** and answer the following questions.
   a. During what time interval is the object speeding up? Slowing down?
   b. At what time does the object reverse direction?
   c. How does the average acceleration of the object in the interval between 0.0 s and 2.0 s differ from the average acceleration in the interval between 7.0 s and 12.0 s?

Table 3-4
Velocity v. Time

Time (s)	Velocity (m/s)
0.00	4.00
1.00	8.00
2.00	12.0
3.00	14.0
4.00	16.0
5.00	16.0
6.00	14.0
7.00	12.0
8.00	8.00
9.00	4.00
10.0	0.00
11.0	−4.00
12.0	−8.00

84. Determine the final velocity of a proton that has an initial velocity of 2.35×10^5 m/s and then is accelerated uniformly in an electric field at the rate of -1.10×10^{12} m/s^2 for 1.50×10^{-7} s.

85. Sports Cars Marco is looking for a used sports car. He wants to buy the one with the greatest acceleration. Car A can go from 0 m/s to 17.9 m/s in 4.0 s; car B can accelerate from 0 m/s to 22.4 m/s in 3.5 s; and car C can go from 0 to 26.8 m/s in 6.0 s. Rank the three cars from greatest acceleration to least, specifically indicating any ties.

86. Supersonic Jet A supersonic jet flying at 145 m/s experiences uniform acceleration at the rate of 23.1 m/s^2 for 20.0 s.
a. What is its final velocity?
b. The speed of sound in air is 331 m/s. What is the plane's speed in terms of the speed of sound?

3.2 Motion with Constant Acceleration

87. Refer to Figure 3-19 to find the distance traveled during the following time intervals.
a. $t = 0.0$ s and $t = 5.0$ s
b. $t = 5.0$ s and $t = 10.0$ s
c. $t = 10.0$ s and $t = 15.0$ s
d. $t = 0.0$ s and $t = 25.0$ s

88. A dragster starting from rest accelerates at 49 m/s^2. How fast is it going when it has traveled 325 m?

89. A car moves at 12 m/s and coasts up a hill with a uniform acceleration of -1.6 m/s^2.
a. What is its displacement after 6.0 s?
b. What is its displacement after 9.0 s?

90. Race Car A race car can be slowed with a constant acceleration of -11 m/s^2.
a. If the car is going 55 m/s, how many meters will it travel before it stops?
b. How many meters will it take to stop a car going twice as fast?

91. A car is traveling 20.0 m/s when the driver sees a child standing on the road. She takes 0.80 s to react, then steps on the brakes and slows at 7.0 m/s^2. How far does the car go before it stops?

92. Airplane Determine the displacement of a plane that experiences uniform acceleration from 66 m/s to 88 m/s in 12 s.

93. How far does a plane fly in 15 s while its velocity is changing from 145 m/s to 75 m/s at a uniform rate of acceleration?

94. Police Car A speeding car is traveling at a constant speed of 30.0 m/s when it passes a stopped police car. The police car accelerates at 7.0 m/s^2. How fast will it be going when it catches up with the speeding car?

95. Road Barrier The driver of a car going 90.0 km/h suddenly sees the lights of a barrier 40.0 m ahead. It takes the driver 0.75 s to apply the brakes, and the average acceleration during braking is -10.0 m/s^2.
a. Determine whether the car hits the barrier.
b. What is the maximum speed at which the car could be moving and not hit the barrier 40.0 m ahead? Assume that the acceleration doesn't change.

3.3 Free Fall

96. A student drops a penny from the top of a tower and decides that she will establish a coordinate system in which the direction of the penny's motion is positive. What is the sign of the acceleration of the penny?

97. Suppose an astronaut drops a feather from 1.2 m above the surface of the Moon. If the acceleration due to gravity on the Moon is 1.62 m/s^2 downward, how long does it take the feather to hit the Moon's surface?

98. A stone that starts at rest is in free fall for 8.0 s.
a. Calculate the stone's velocity after 8.0 s.
b. What is the stone's displacement during this time?

99. A bag is dropped from a hovering helicopter. The bag has fallen for 2.0 s. What is the bag's velocity? How far has the bag fallen?

100. You throw a ball downward from a window at a speed of 2.0 m/s. How fast will it be moving when it hits the sidewalk 2.5 m below?

101. If you throw the ball in the previous problem up instead of down, how fast will it be moving when it hits the sidewalk?

102. Beanbag You throw a beanbag in the air and catch it 2.2 s later.
a. How high did it go?
b. What was its initial velocity?

Mixed Review

103. A spaceship far from any star or planet experiences uniform acceleration from 65.0 m/s to 162.0 m/s in 10.0 s. How far does it move?

104. Figure 3-20 is a strobe photo of a horizontally moving ball. What information about the photo would you need and what measurements would you make to estimate the acceleration?

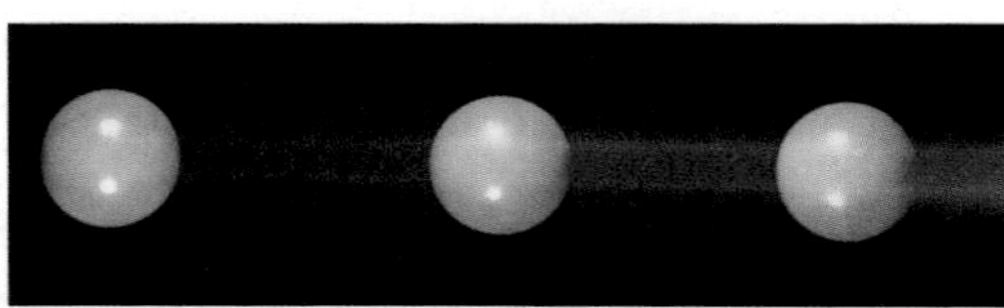

Figure 3-20

 For more problems, go to Additional Problems, Appendix B.

105. **Bicycle** A bicycle accelerates from 0.0 m/s to 4.0 m/s in 4.0 s. What distance does it travel?

106. A weather balloon is floating at a constant height above Earth when it releases a pack of instruments.
- **a.** If the pack hits the ground with a velocity of −73.5 m/s, how far did the pack fall?
- **b.** How long did it take for the pack to fall?

107. **Baseball** A baseball pitcher throws a fastball at a speed of 44 m/s. The acceleration occurs as the pitcher holds the ball in his hand and moves it through an almost straight-line distance of 3.5 m. Calculate the acceleration, assuming that it is constant and uniform. Compare this acceleration to the acceleration due to gravity.

108. The total distance a steel ball rolls down an incline at various times is given in **Table 3-5.**
- **a.** Draw a position-time graph of the motion of the ball. When setting up the axes, use five divisions for each 10 m of travel on the *d*-axis. Use five divisions for 1 s of time on the *t*-axis.
- **b.** Calculate the distance the ball has rolled at the end of 2.2 s.

Table 3-5

Distance v. Time

Time (s)	Distance (m)
0.0	0.0
1.0	2.0
2.0	8.0
3.0	18.0
4.0	32.0
5.0	50.0

109. Engineers are developing new types of guns that might someday be used to launch satellites as if they were bullets. One such gun can give a small object a velocity of 3.5 km/s while moving it through a distance of only 2.0 cm.
- **a.** What acceleration does the gun give this object?
- **b.** Over what time interval does the acceleration take place?

110. **Sleds** Rocket-powered sleds are used to test the responses of humans to acceleration. Starting from rest, one sled can reach a speed of 444 m/s in 1.80 s and can be brought to a stop again in 2.15 s.
- **a.** Calculate the acceleration of the sled when starting, and compare it to the magnitude of the acceleration due to gravity, 9.80 m/s^2.
- **b.** Find the acceleration of the sled as it is braking and compare it to the magnitude of the acceleration due to gravity.

111. The velocity of a car changes over an 8.0-s time period, as shown in **Table 3-6.**
- **a.** Plot the velocity-time graph of the motion.
- **b.** Determine the displacement of the car during the first 2.0 s.
- **c.** What displacement does the car have during the first 4.0 s?
- **d.** What is the displacement of the car during the entire 8.0 s?
- **e.** Find the slope of the line between t = 0.0 s and t = 4.0 s. What does this slope represent?
- **f.** Find the slope of the line between t = 5.0 s and t = 7.0 s. What does this slope indicate?

Table 3-6

Velocity v. Time

Time (s)	Velocity (m/s)
0.0	0.0
1.0	4.0
2.0	8.0
3.0	12.0
4.0	16.0
5.0	20.0
6.0	20.0
7.0	20.0
8.0	20.0

112. A truck is stopped at a stoplight. When the light turns green, the truck accelerates at 2.5 m/s^2. At the same instant, a car passes the truck going 15 m/s. Where and when does the truck catch up with the car?

113. **Safety Barriers** Highway safety engineers build soft barriers, such as the one shown in **Figure 3-21,** so that cars hitting them will slow down at a safe rate. A person wearing a safety belt can withstand an acceleration of -3.0×10^2 m/s^2. How thick should barriers be to safely stop a car that hits a barrier at 110 km/h?

Figure 3-21

114. Karate The position-time and velocity-time graphs of George's fist breaking a wooden board during karate practice are shown in **Figure 3-22.**

a. Use the velocity-time graph to describe the motion of George's fist during the first 10 ms.

b. Estimate the slope of the velocity-time graph to determine the acceleration of his fist when it suddenly stops.

c. Express the acceleration as a multiple of the gravitational acceleration, $g = 9.80$ m/s^2.

d. Determine the area under the velocity-time curve to find the displacement of the fist in the first 6 ms. Compare this with the position-time graph.

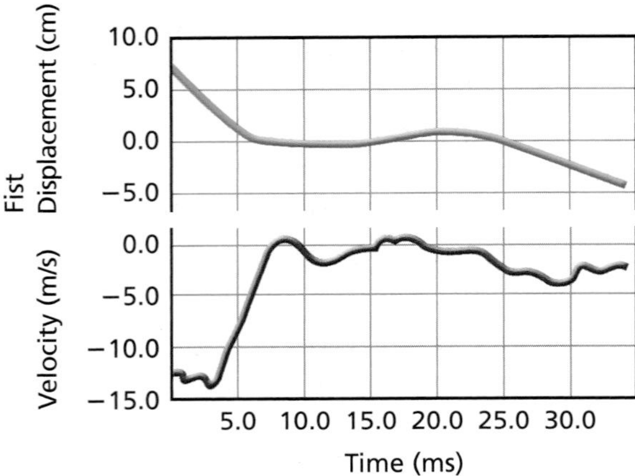

Figure 3-22

115. Cargo A helicopter is rising at 5.0 m/s when a bag of its cargo is dropped. The bag falls for 2.0 s.

a. What is the bag's velocity?

b. How far has the bag fallen?

c. How far below the helicopter is the bag?

Thinking Critically

116. Apply CBLs Design a lab to measure the distance an accelerated object moves over time. Use equal time intervals so that you can plot velocity over time as well as distance. A pulley at the edge of a table with a mass attached is a good way to achieve uniform acceleration. Suggested materials include a motion detector, CBL, lab cart, string, pulley, C-clamp, and masses. Generate distance-time and velocity-time graphs using different masses on the pulley. How does the change in mass affect your graphs?

117. Analyze and Conclude Which has the greater acceleration: a car that increases its speed from 50 km/h to 60 km/h, or a bike that goes from 0 km/h to 10 km/h in the same time? Explain.

118. Analyze and Conclude An express train, traveling at 36.0 m/s, is accidentally sidetracked onto a local train track. The express engineer spots a local train exactly 1.00×10^2 m ahead on the same track and traveling in the same direction. The local engineer is unaware of the situation. The express engineer jams on the brakes and slows the express train at a constant rate of 3.00 m/s^2. If the speed of the local train is 11.0 m/s, will the express train be able to stop in time, or will there be a collision? To solve this problem, take the position of the express train when the engineer first sights the local train as a point of origin. Next, keeping in mind that the local train has exactly a 1.00×10^2 m lead, calculate how far each train is from the origin at the end of the 12.0 s it would take the express train to stop (accelerate at -3.00 m/s^2 from 36 m/s to 0 m/s).

a. On the basis of your calculations, would you conclude that a collision will occur?

b. The calculations that you made do not allow for the possibility that a collision might take place before the end of the 12 s required for the express train to come to a halt. To check this, take the position of the express train when the engineer first sights the local train as the point of origin and calculate the position of each train at the end of each second after the sighting. Make a table showing the distance of each train from the origin at the end of each second. Plot these positions on the same graph and draw two lines. Use your graph to check your answer to part **a.**

Writing in Physics

119. Research and describe Galileo's contributions to physics.

120. Research the maximum acceleration a human body can withstand without blacking out. Discuss how this impacts the design of three common entertainment or transportation devices.

Cumulative Review

121. Solve the following problems. Express your answers in scientific notation. (Chapter 1)

a. 6.2×10^{-4} m + 5.7×10^{-3} m

b. 8.7×10^{8} km − 3.4×10^{7} m

c. $(9.21\times10^{-5}$ cm$)(1.83\times10^{8}$ cm$)$

d. $(2.63\times10^{-6}$ m$)/(4.08\times10^{6}$ s$)$

122. The equation below describes the motion of an object. Create the corresponding position-time graph and motion diagram. Then write a physics problem that could be solved using that equation. Be creative. $d = (35.0 \text{ m/s})\, t - 5.0$ m (Chapter 2)

 For more problems, go to Additional Problems, Appendix B.

Multiple Choice

Use the following information to answer the first two questions.

A ball rolls down a hill with a constant acceleration of 2.0 m/s^2. The ball starts at rest and travels for 4.0 s before it stops.

1. How far did the ball travel before it stopped?

   (A) 8.0 m (C) 16 m
   (B) 12 m (D) 20 m

2. What was the ball's velocity just before it stopped?

   (A) 2.0 m/s (C) 12 m/s
   (B) 8.0 m/s (D) 16 m/s

3. A driver of a car enters a new 110-km/h speed zone on the highway. The driver begins to accelerate immediately and reaches 110 km/h after driving 500 m. If the original speed was 80 km/h, what was the driver's rate of acceleration?

   (A) 0.44 m/s^2 (C) 8.4 m/s^2
   (B) 0.60 m/s^2 (D) 9.80 m/s^2

4. A flowerpot falls off the balcony of a penthouse suite 85 m above the street. How long does it take to hit the ground?

   (A) 4.2 s (C) 8.7 s
   (B) 8.3 s (D) 17 s

5. A rock climber's shoe loosens a rock, and her climbing buddy at the bottom of the cliff notices that the rock takes 3.20 s to fall to the ground. How high up the cliff is the rock climber?

   (A) 15.0 m (C) 50.0 m
   (B) 31.0 m (D) 1.00×10^2 m

6. A car traveling at 91.0 km/h approaches the turnoff for a restaurant 30.0 m ahead. If the driver slams on the brakes with an acceleration of -6.40 m/s^2, what will be her stopping distance?

   (A) 14.0 m (C) 50.0 m
   (B) 29.0 m (D) 100.0 m

7. What is the correct formula manipulation to find acceleration when using the equation $v_f^2 = v_i^2 + 2ad$?

   (A) $(v_f^2 - v_i^2)/d$ (C) $(v_f + v_i)^2/2d$
   (B) $(v_f^2 + v_i^2)/2d$ (D) $(v_f^2 - v_i^2)/2d$

8. The graph shows the motion of a farmer's truck. What is the truck's total displacement? Assume that north is the positive direction.

   (A) 150 m south (C) 300 m north
   (B) 125 m north (D) 600 m south

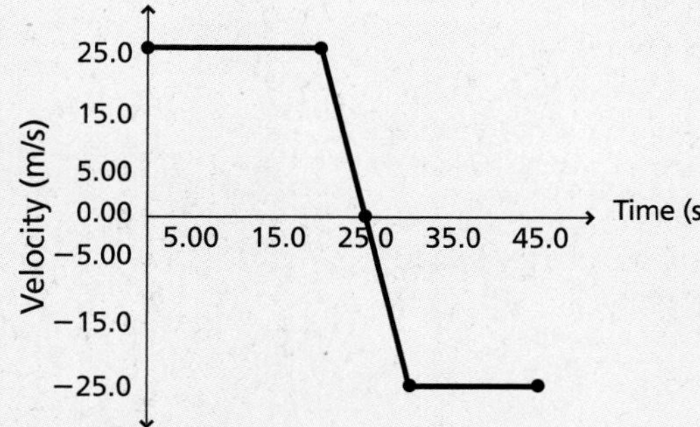

9. How can the instantaneous acceleration of an object with varying acceleration be calculated?

   (A) by calculating the slope of the tangent on a distance-time graph
   (B) by calculating the area under the graph on a distance-time graph
   (C) by calculating the area under the graph on a velocity-time graph
   (D) by calculating the slope of the tangent on a velocity-time graph

Extended Answer

10. Graph the following data, and then show calculations for acceleration and displacement after 12.0 s on the graph.

Time (s)	Velocity (m/s)
0.00	8.10
6.00	36.9
9.00	51.3
12.00	65.7

✓ Test-Taking TIP

Tables

If a test question involves a table, skim the table before reading the question. Read the title, column heads, and row heads. Then read the question and interpret the information in the table.

Chapter 4 Forces in One Dimension

What You'll Learn

- You will use Newton's laws to solve problems.
- You will determine the magnitude and direction of the net force that causes a change in an object's motion.
- You will classify forces according to the agents that cause them.

Why It's Important

Forces act on you and everything around you at all times.

Sports A soccer ball is headed by a player. Before play began, the ball was motionless. During play, the ball started, stopped, and changed directions many times.

Think About This ▶

What causes a soccer ball, or any other object, to stop, start, or change direction?

physicspp.com

LAUNCH Lab

Which force is stronger?

Question

What forces can act on an object that is suspended by a string?

Procedure

1. Tie a piece of heavy cord around the middle of a book. Tie one piece of lightweight string to the center of the cord on the top of the book. Tie another piece to the bottom.
2. While someone holds the end of the top lightweight string so that the book is suspended in the air, pull very slowly, but firmly, on the end of the bottom lightweight string. Record your observations. ***CAUTION: Keep feet clear of falling objects.***
3. Replace the broken string and repeat step 2, but this time pull very fast and very hard on the bottom string. Record your observations.

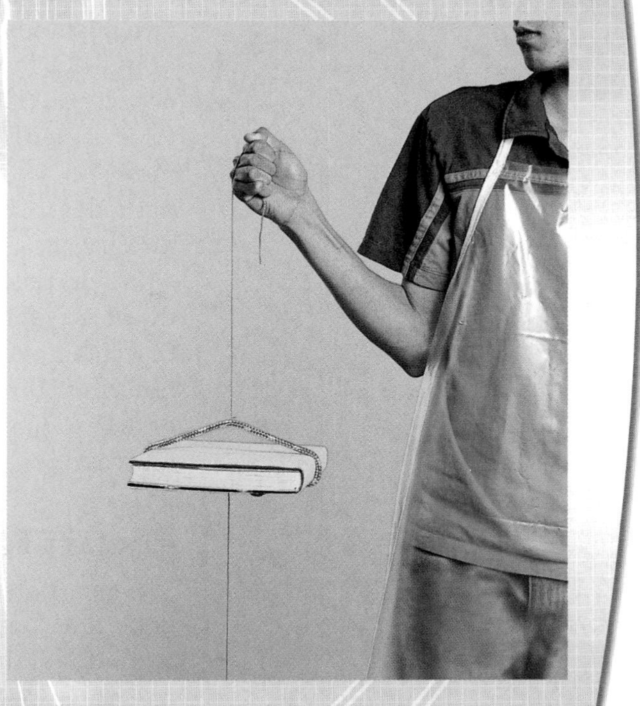

Analysis

Which string broke in step 2? Why? Which string broke in step 3? Why?

Critical Thinking Draw a diagram of the experimental set-up. Use arrows to show the forces acting on the book.

4.1 Force and Motion

Imagine that a train is speeding down a railroad track at 80 km/h when suddenly the engineer sees a truck stalled at a railroad crossing ahead. The engineer applies the brakes to try to stop the train before it crashes into the truck. Because the brakes cause an acceleration in the direction opposite the train's velocity, the train will slow down. Imagine that, in this case, the engineer is able to stop the train just before it crashes into the truck. But what if instead of moving at 80 km/h the train had been moving at 100 km/h? What would have to happen for the train to avoid hitting the truck? The acceleration provided by the train's brakes would have to be greater because the engineer still has the same distance in which to stop the train. Similarly, if the train was going 80 km/h but had been much closer to the truck when the engineer started to apply the brake, the acceleration also would need to be greater because the train would need to stop in less time.

▸ **Objectives**

- **Define** force.
- **Apply** Newton's second law to solve problems.
- **Explain** the meaning of Newton's first law.

▸ **Vocabulary**

force
free-body diagram
net force
Newton's second law
Newton's first law
inertia
equilibrium

Force and Motion

What happened to make the train slow down? A **force** is a push or pull exerted on an object. Forces can cause objects to speed up, slow down, or change direction as they move. When an engineer applies the brakes, the brakes exert a force on the wheels and cause the train to slow down. Based on the definitions of velocity and acceleration, this can be restated as follows: a force exerted on an object causes that object's velocity to change; that is, a force causes an acceleration.

Consider a textbook resting on a table. How can you cause it to move? Two possibilites are that you can push on it or you can pull on it. The push or pull is a force that you exert on the textbook. If you push harder on an object, you have a greater effect on its motion. The direction in which the force is exerted also matters—if you push the book to the right, the book moves in a different direction from the direction it would move if you pushed it to the left. The symbol $\boldsymbol{F}$ is a vector and represents the size and direction of a force, while F represents only the magnitude.

When considering how a force affects motion, it is important to identify the object of interest. This object is called the system. Everything around the object that exerts forces on it is called the external world. In the case of the book in **Figure 4-1,** the book is the system. Your hand and gravity are parts of the external world that can interact with the book by pushing or pulling on it and potentially causing its motion to change.

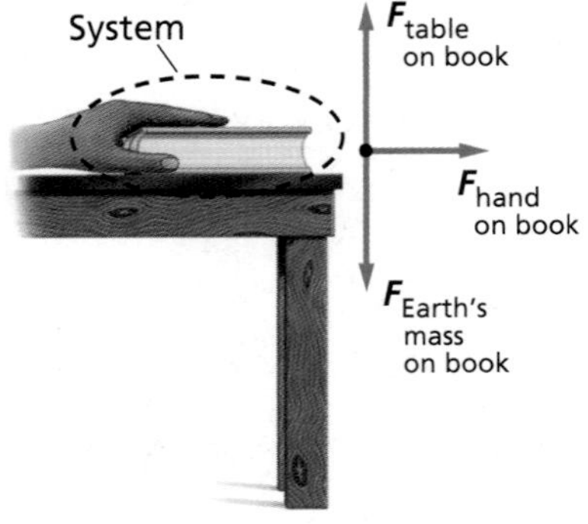

Figure 4-1 The book is the system. The table, the hand, and Earth's mass (through gravity) all exert forces on the book.

Contact Forces and Field Forces

Again, think about the different ways in which you could move a textbook. You could touch it directly and push or pull it, or you could tie a string around it and pull on the string. These are examples of contact forces. A contact force exists when an object from the external world touches a system and thereby exerts a force on it. If you are holding this physics textbook right now, your hands are exerting a contact force on it. If you place the book on a table, you are no longer exerting a force on the book. The table, however, is exerting a force because the table and the book are in contact.

There are other ways in which you could change the motion of the textbook. You could drop it, and as you learned in Chapter 3, it would accelerate as it falls to the ground. The gravitational force of Earth acting on the book causes this acceleration. This force affects the book whether or not Earth is actually touching it. This is an example of a field force. Field forces are exerted without contact. Can you think of other kinds of field forces? If you have ever experimented with magnets, you know that they exert forces without touching. You will investigate magnetism and other similar forces in more detail in future chapters. For now, the only field force that you need to consider is the gravitational force.

Forces result from interactions; thus, each force has a specific and identifiable cause called the agent. You should be able to name the agent exerting each force, as well as the system upon which the force is exerted. For example, when you push your textbook, your hand (the agent) exerts a force on the textbook (the system). If there are not both an agent and a system, a force does not exist. What about the gravitational force? If you allow your textbook to fall, the agent is the mass of Earth exerting a field force on the book.

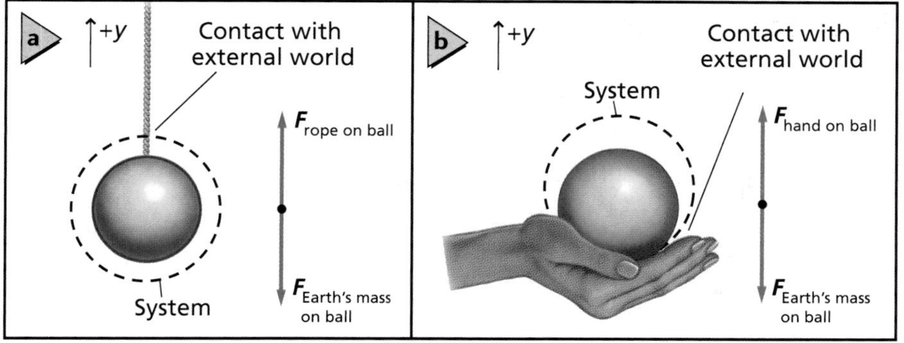

Figure 4-2 To make a physical model of the forces acting on an object, apply the particle model and draw an arrow to represent each force. Label each force, including its agent.

Concepts In Motion
Interactive Figure To see an animation on free-body diagrams, visit physicspp.com.

Free-body diagrams Just as pictorial models and motion diagrams are useful in solving problems about motion, similar representations will help you to analyze how forces affect motion. The first step in solving any problem is to create a pictorial model. For example, to represent the forces on a ball tied to a string or held in your hand, sketch the situations, as shown in **Figures 4-2a** and **4-2b.** Circle the system and identify every place where the system touches the external world. It is at these places that contact forces are exerted. Identify the contact forces. Then identify any field forces on the system. This gives you the pictorial model.

To make a physical representation of the forces acting on the ball in Figures 4-2a and 4-2b, apply the particle model and represent the object with a dot. Represent each force with a blue arrow that points in the direction that the force is applied. Try to make the length of each arrow proportional to the size of the force. Often, you will draw these diagrams before you know the magnitudes of all the forces. In such cases, make your best estimate. Always draw the force arrows pointing away from the particle, even when the force is a push. Make sure that you label each force. Use the symbol $\boldsymbol{F}$ with a subscript label to identify both the agent and the object on which the force is exerted. Finally, choose a direction to be positive and indicate this off to the side of your diagram. Usually, you select the positive direction to be in the direction of the greatest amount of force. This typically makes the problem easiest to solve by reducing the number of negative values in your calculations. This type of physical model, which represents the forces acting on a system, is called a **free-body diagram.**

PRACTICE Problems

• Additional Problems, Appendix B
• Solutions to Selected Problems, Appendix C

For each of the following situations, specify the system and draw a motion diagram and a free-body diagram. Label all forces with their agents, and indicate the direction of the acceleration and of the net force. Draw vectors of appropriate lengths.

1. A flowerpot falls freely from a windowsill. (Ignore any forces due to air resistance.)

2. A sky diver falls downward through the air at constant velocity. (The air exerts an upward force on the person.)

3. A cable pulls a crate at a constant speed across a horizontal surface. The surface provides a force that resists the crate's motion.

4. A rope lifts a bucket at a constant speed. (Ignore air resistance.)

5. A rope lowers a bucket at a constant speed. (Ignore air resistance.)

Force and Acceleration

How does an object move when one or more forces are exerted on it? One way to find out is by doing experiments. As before, begin by considering a simple situation. Once you fully understand that situation, then you can add more complications to it. In this case, begin with one controlled force exerted horizontally on an object. The horizontal direction is a good place to start because gravity does not act horizontally. Also, to reduce complications resulting from the object rubbing against the surface, do the experiments on a very smooth surface, such as ice or a very well-polished table, and use an object with wheels that spin easily. In other words, you are trying to reduce the resistance to motion in the situation.

To determine how force, acceleration, and velocity are related, you need to be able to exert a constant and controlled force on an object. How can you exert such a controlled force? A stretched rubber band exerts a pulling force; the farther you stretch it, the greater the force with which it pulls back. If you always stretch the rubber band the same amount, you always exert the same force. **Figure 4-3a** shows a rubber band, stretched a constant 1 cm, pulling a low-resistance cart. If you perform this experiment and determine the cart's velocity for some period of time, you could construct a graph like the one shown in **Figure 4-3b.** Does this graph look different from what you expected? What do you notice about the velocity? The constant increase in the velocity is a result of the constant acceleration the stretched rubber band gives the cart.

How does this acceleration depend upon the force? To find out, repeat the experiment, this time with the rubber band stretched to a constant 2 cm, and then repeat it again with the rubber band stretched longer and longer each time. For each experiment, plot a velocity-time graph like the one in Figure 4-3b, calculate the acceleration, and then plot the accelerations and forces for all the trials to make a force-acceleration graph, as shown in **Figure 4-4a.** What is the relationship between the force and acceleration? It's a linear relationship where the greater the force is, the greater the resulting acceleration. As you did in Chapters 2 and 3, you can apply the straight-line equation $y = mx + b$ to this graph.

Figure 4-3 Because the rubber band is stretched a constant amount, it applies a constant force on the cart, which is designed to be low-friction **(a).** The cart's motion can be graphed and shown to be a linear relationship **(b).**

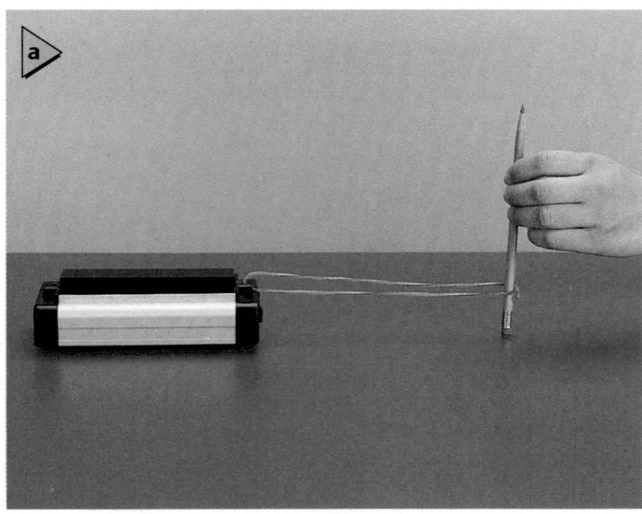

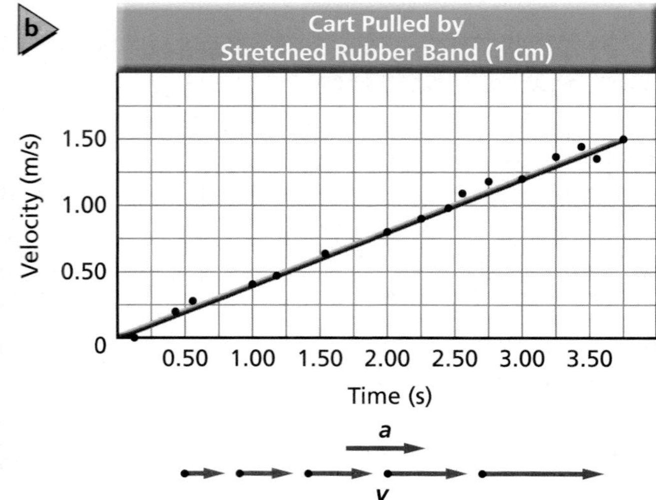

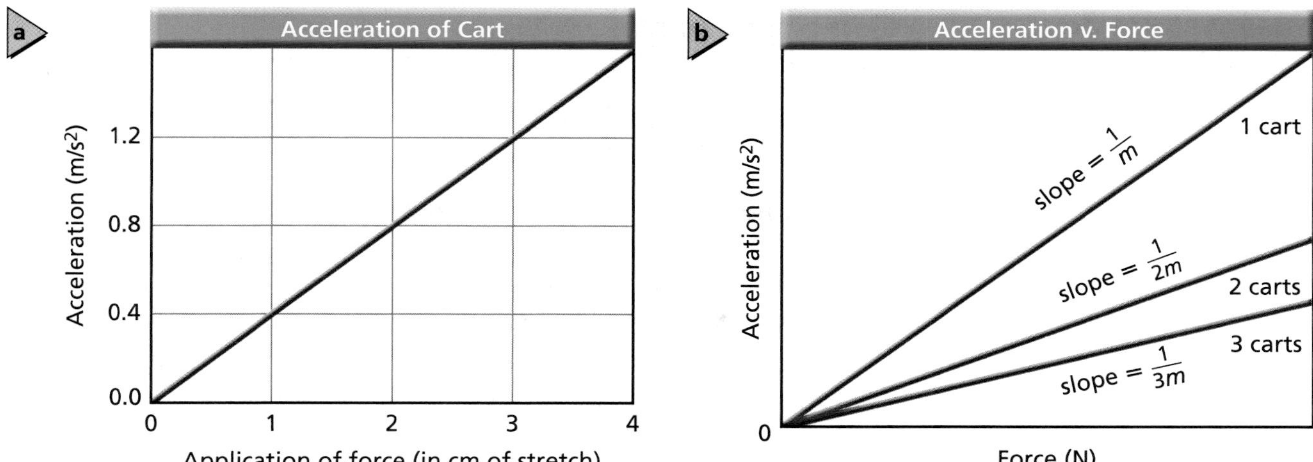

Figure 4-4 The graph shows that as the force increases, so does the acceleration **(a).** You can see that the slope of the force-acceleration graph depends upon the number of carts **(b).**

What is the physical meaning of this slope? Perhaps it describes something about the object that is accelerating. What happens if the object changes? Suppose that a second, identical cart is placed on top of the first, and then a third cart is added. The rubber band would be pulling two carts, and then three. A plot of the force versus acceleration for one, two, and three carts is shown in **Figure 4-4b.** The graph shows that if the same force is applied in each situation, the acceleration of two carts is $\frac{1}{2}$ the acceleration of one cart, and the acceleration of three carts is $\frac{1}{3}$ the acceleration of one cart. This means that as the number of carts is increased, a greater force is needed to produce the same acceleration. In this example, you would have to stretch the rubber band farther to get a greater amount of force. The slopes of the lines in Figure 4-4b depend upon the number of carts; that is, the slope depends on the total mass of the carts. If the slope, k in this case, is defined as the reciprocal of the mass ($k = 1/m$), then $a = F/m$, or $F = ma$.

What information is contained in the equation $a = F/m$? It tells you that a force applied to an object causes that object to experience a change in motion—the force causes the object to accelerate. It also tells you that for the same object, if you double the force, you will double the object's acceleration. Lastly, if you apply the same force to several different objects, the one with the most mass will have the smallest acceleration and the one with the least mass will have the greatest acceleration.

What are the proper units for measuring force? Because $F = ma$, one unit of force causes a 1-kg mass to accelerate at 1 m/s^2, so one force unit has the dimensions 1 $kg \cdot m/s^2$. The unit 1 $kg \cdot m/s^2$ is called the newton, represented by N. One newton of force applied to a 1-kg object will cause it to have an acceleration of 1 m/s^2. Do these units make sense? Think about a sky diver who is falling through the air. The properties affecting his motion are his mass and the acceleration due to the gravitational force, so these units do make sense. **Table 4-1** shows the magnitude of some common forces.

Table 4-1
Common Forces

Description	F (N)
Force of gravity on a coin (nickel)	0.05
Force of gravity on 1 lb (0.45 kg) of sugar	4.5
Force of gravity on a 150-lb (70-kg) person	686
Force of an accelerating car	3000
Force of a rocket motor	5,000,000

Combining Forces

What happens if you and a friend each push a table and exert 100 N of force on it? When you and your friend push together, you give the table a greater acceleration than when you push against each other. In fact, when you push together, you give the table twice the acceleration that it would have if just one of you applied 100 N of force. When you push on the table in opposite directions with the same amount of force, as in **Figure 4-5a,** the table does not move at all.

Figure 4-5b and **c** show free-body diagrams for these two situations. **Figure 4-5d** shows a third free-body diagram in which your friend pushes on the table twice as hard as you in the opposite direction. Below each free-body diagram is a vector representing the total result of the two forces. When the force vectors are in the same direction, they can be replaced by one vector with a length equal to their combined length. When the forces are in opposite directions, the resulting vector is the length of the difference between the two vectors. Another term for the vector sum of all the forces on an object is the **net force.**

You also can analyze the situation mathematically. Assume that you are pushing the table in the positive direction with a 100 N force in the above cases. In the first case, your friend is pushing with a negative force of 100 N. Adding them together gives a total force of 0 N, which means there is no acceleration. In the second case, your friend's force is 100 N, so the total force is 200 N in the positive direction and the table accelerates in the positive direction. In the third case, your friend's force is -200 N, so the total force is -100 N and the table will accelerate in the negative direction.

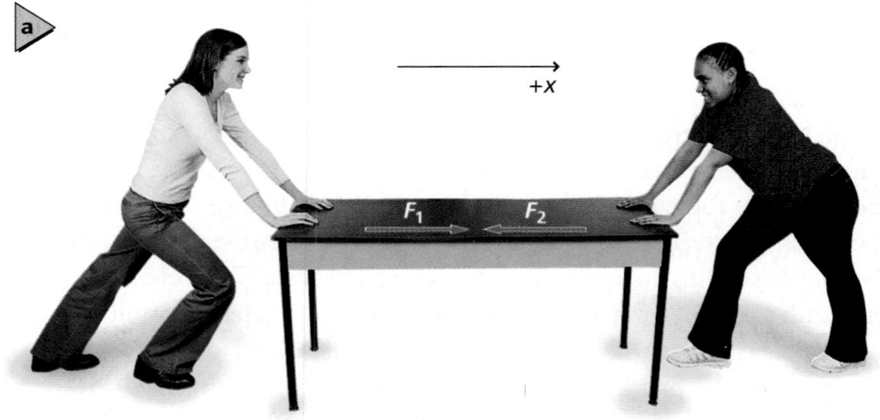

b	c	d
	F_1 = 100 N	
F_2 = 100 N F_1 = 100 N	F_2 = 100 N	F_2 = 200 N F_1 = 100 N
F_{net} = 0 N	F_{net} = 200 N	F_{net} = 100 N
Equal forces Opposite directions	Equal forces Same direction	Unequal forces Opposite directions

Figure 4-5 Pushing on the table with equal force in opposite directions **(a)** results in no net force on the table, as shown by the vector addition in the free-body diagram **(b).** However, there is a net force applied in **(c)** and **(d),** as shown by the free-body diagrams.

Newton's Second Law

You could conduct a series of experiments in which you and your friend vary the net force exerted on the table and measure the acceleration in each case. You would find that the acceleration of the table is proportional to the net force exerted on it and inversely proportional to its mass. In other words, if the net force of you and your friend acting on the table is 100 N, the table will experience the same acceleration as it would if only you were acting on it with a force of 100 N. Taking this into account, the mathematical relationship among force, mass, and acceleration can be rewritten in terms of the net force. The observation that the acceleration of an object is proportional to the net force and inversely proportional to the mass of the object being accelerated is **Newton's second law,** which is represented by the following equation.

Newton's Second Law $a = \frac{F_{net}}{m}$

The acceleration of an object is equal to the sum of the forces acting on the object, divided by the mass of the object.

Notice that Newton's second law can be rearranged to the form $F = ma$, which you learned about previously. If the table that you and your friend were pushing was 15.0 kg and the two of you each pushed with a force of 50.0 N in the same direction, what would be the acceleration of the table? To find out, calculate the net force, 50.0 N + 50.0 N = 100.0 N, and apply Newton's second law by dividing the net force of 100.0 N by the mass of the table, 15.0 kg, to get an acceleration of 6.67 m/s^2.

Here is a useful strategy for finding how the motion of an object depends on the forces exerted on it. First, identify all the forces acting on the object. Draw a free-body diagram showing the direction and relative strength of each force acting on the system. Then, add the forces to find the net force. Next, use Newton's second law to calculate the acceleration. Finally, if necessary, use kinematics to find the velocity or position of the object. When you learned about kinematics in Chapters 2 and 3, you studied the motion of objects without regard for the causes of motion. You now know that an unbalanced force, a net force, is the cause of a change in velocity (an acceleration).

PRACTICE Problems

- Additional Problems, Appendix B
- Solutions to Selected Problems, Appendix C

6. Two horizontal forces, 225 N and 165 N, are exerted on a canoe. If these forces are applied in the same direction, find the net horizontal force on the canoe.

7. If the same two forces as in the previous problem are exerted on the canoe in opposite directions, what is the net horizontal force on the canoe? Be sure to indicate the direction of the net force.

8. Three confused sleigh dogs are trying to pull a sled across the Alaskan snow. Alutia pulls east with a force of 35 N, Seward also pulls east but with a force of 42 N, and big Kodiak pulls west with a force of 53 N. What is the net force on the sled?

Newton's First Law

What is the motion of an object with no net force acting on it? A stationary object with no net force acting on it will stay at its position. Consider a moving object, such as a ball rolling on a surface. How long will the ball continue to roll? It will depend on the quality of the surface. If the ball is rolled on a thick carpet that offers much resistance, it will come to rest quickly. If it is rolled on a hard, smooth surface that offers little resistance, such as a bowling alley, the ball will roll for a long time with little change in velocity. Galileo did many experiments, and he concluded that in the ideal case of zero resistance, horizontal motion would never stop. Galileo was the first to recognize that the general principles of motion could be found by extrapolating experimental results to the ideal case, in which there is no resistance to slow down an object's motion.

In the absence of a net force, the motion (or lack of motion) of both the moving ball and the stationary object continues as it was. Newton recognized this and generalized Galileo's results in a single statement. This statement, "an object that is at rest will remain at rest, and an object that is moving will continue to move in a straight line with constant speed, if and only if the net force acting on that object is zero," is called **Newton's first law.**

Table 4-2			
Some Types of Forces			
Force	**Symbol**	**Definition**	**Direction**
Friction	$\boldsymbol{F}_{\text{f}}$	The contact force that acts to oppose sliding motion between surfaces	Parallel to the surface and opposite the direction of sliding
Normal	$\boldsymbol{F}_{\text{N}}$	The contact force exerted by a surface on an object	Perpendicular to and away from the surface
Spring	$\boldsymbol{F}_{\text{sp}}$	A restoring force; that is, the push or pull a spring exerts on an object	Opposite the displacement of the object at the end of the spring
Tension	$\boldsymbol{F}_{\text{T}}$	The pull exerted by a string, rope, or cable when attached to a body and pulled taut	Away from the object and parallel to the string, rope, or cable at the point of attachment
Thrust	$\boldsymbol{F}_{\text{thrust}}$	A general term for the forces that move objects such as rockets, planes, cars, and people	In the same direction as the acceleration of the object, barring any resistive forces
Weight	$\boldsymbol{F}_{\text{g}}$	A field force due to gravitational attraction between two objects, generally Earth and an object	Straight down toward the center of Earth

Inertia Newton's first law is sometimes called the law of inertia. Is inertia a force? No. **Inertia** is the tendency of an object to resist change. If an object is at rest, it tends to remain at rest. If it is moving at a constant velocity, it tends to continue moving at that velocity. Forces are results of interactions between two objects; they are not properties of single objects, so inertia cannot be a force. Remember that because velocity includes both the speed and direction of motion, a net force is required to change either the speed or direction of an object's motion.

Equilibrium According to Newton's first law, a net force is something that causes the velocity of an object to change. If the net force on an object is zero, then the object is in **equilibrium.** An object is in equilibrium if it is at rest or if it is moving at a constant velocity. Note that being at rest is simply a special case of the state of constant velocity, $v = 0$. Newton's first law identifies a net force as something that disturbs a state of equilibrium. Thus, if there is no net force acting on the object, then the object does not experience a change in speed or direction and is in equilibrium.

By understanding and applying Newton's first and second laws, you can often figure out something about the relative sizes of forces, even in situations in which you do not have numbers to work with. Before looking at an example of this, review **Table 4-2,** which lists some of the common types of forces. You will be dealing with many of these throughout your study of physics.

When analyzing forces and motion, it is important to keep in mind that the world is dominated by resistance. Newton's ideal, resistance-free world is not easy to visualize. If you analyze a situation and find that the result is different from a similar experience that you have had, ask yourself if this is because of the presence of resistance. In addition, many terms used in physics have everyday meanings that are different from those understood in physics. When talking or writing about physics issues, be careful to use these terms in their precise, scientific way.

APPLYING PHYSICS

Shuttle Engine Thrust The Space Shuttle Main Engines (SSMEs) each are rated to provide 1.6 million N of thrust. Powered by the combustion of hydrogen and oxygen, the SSMEs are throttled anywhere from 65 percent to 109 percent of their rated thrust.

4.1 Section Review

9. **Force** Identify each of the following as either **a, b,** or **c:** weight, mass, inertia, the push of a hand, thrust, friction, air resistance, spring force, and acceleration.
   a. a contact force
   b. a field force
   c. not a force

10. **Inertia** Can you feel the inertia of a pencil? Of a book? If you can, describe how.

11. **Free-Body Diagram** Draw a free-body diagram of a bag of sugar being lifted by your hand at a constant speed. Specifically identify the system. Label all forces with their agents and make the arrows the correct lengths.

12. **Direction of Velocity** If you push a book in the forward direction, does this mean its velocity has to be forward?

13. **Free-Body Diagram** Draw a free-body diagram of a water bucket being lifted by a rope at a decreasing speed. Specifically identify the system. Label all forces with their agents and make the arrows the correct lengths.

14. **Critical Thinking** A force of 1 N is the only force exerted on a block, and the acceleration of the block is measured. When the same force is the only force exerted on a second block, the acceleration is three times as large. What can you conclude about the masses of the two blocks?

Physics Online physicspp.com/self_check_quiz

4.2 Using Newton's Laws

▶ **Objectives**

- **Describe** how the weight and the mass of an object are related.
- **Differentiate** between actual weight and apparent weight.

▶ **Vocabulary**

apparent weight
weightlessness
drag force
terminal velocity

Newton's second law describes the connection between the cause of a change in an object's velocity and the resulting displacement. This law identifies the relationship between the net force exerted on an object and the object's acceleration.

Using Newton's Second Law

What is the weight force, $\mathbf{F}_g$, exerted on an object of mass m? Newton's second law can help answer this question. Consider the pictorial and physical models in **Figure 4-6,** which show a free-falling ball in midair. With what objects is it interacting? Because it is touching nothing and air resistance can be neglected, the only force acting on it is $\mathbf{F}_g$. You know from Chapter 3 that the ball's acceleration is $\mathbf{g}$. Newton's second law then becomes $\mathbf{F}_g = m\mathbf{g}$. Both the force and the acceleration are downward. The magnitude of an object's weight is equal to its mass times the acceleration it would have if it were falling freely. It is important to keep in mind that even when an object is not experiencing free-fall, the gravitational force of Earth is still acting on the object.

This result is true on Earth, as well as on any other planet, although the magnitude of g will be different on other planets. Because the value of g is much less on the Moon than on Earth, astronauts who landed on the Moon weighed much less while on the Moon, even though their mass had not changed.

■ **Figure 4-6** The net force on the ball is the weight force, $\mathbf{F}_g$.

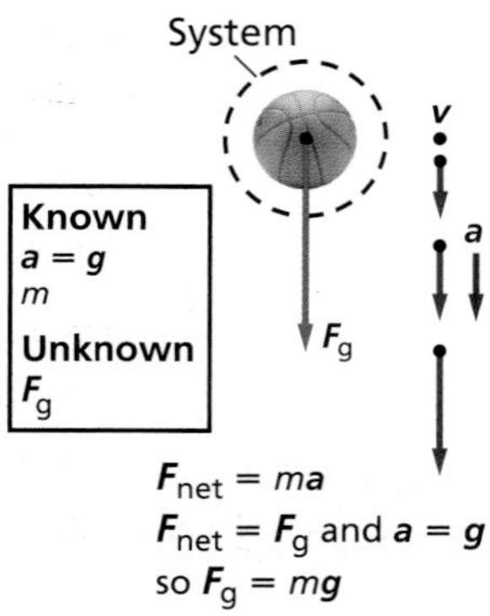

Scales A bathroom scale contains springs. When you stand on the scale, the scale exerts an upward force on you because you are in contact with it. Because you are not accelerating, the net force acting on you must be zero. Therefore, the spring force, F_{sp}, pushing up on you must be the same magnitude as your weight, F_g, pulling down on you, as shown in the pictorial and physical models in **Figure 4-7.** The reading on the scale is determined by the amount of force the springs inside it exert on you. A spring scale, therefore, measures weight, not mass. If you were on a different planet, the compression of the spring would be different, and consequently, the scale's reading would be different. Remember that the proper unit for expressing mass is kilograms and because weight is a force, the proper unit used to express weight is the newton.

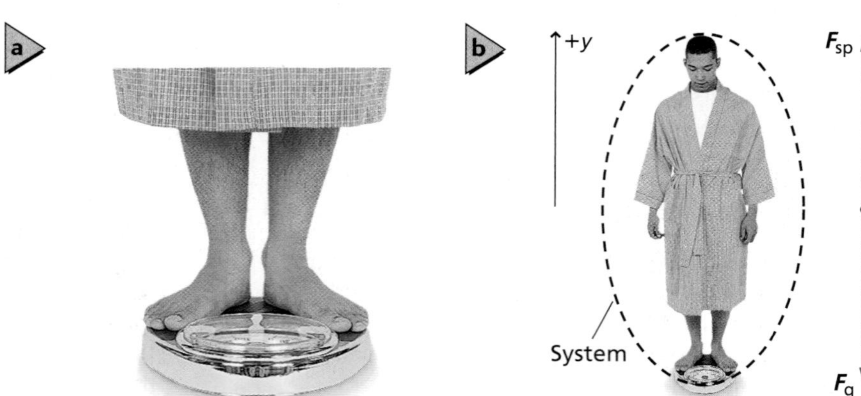

■ **Figure 4-7** The upward force of the spring in the scale is equal to your weight when you step on the bathroom scale **(a).** The free-body diagram in **(b)** shows that the system is in equilibrium because the force of the spring is equal to your weight.

▶ EXAMPLE Problem 1

Fighting Over a Pillow Anudja is holding a pillow, with a mass of 0.30 kg, when Sarah decides that she wants it and tries to pull it away from Anudja. If Sarah pulls horizontally on the pillow with a force of 10.0 N and Anudja pulls with a horizontal force of 11.0 N, what is the horizontal acceleration of the pillow?

1 Analyze and Sketch the Problem

- Sketch the situation.
- Identify the pillow as the system and the direction in which Anudja pulls as positive.
- Draw the free-body diagram. Label the forces.

Known:

$m = 0.30$ kg

$F_{\text{Anudja on pillow}} = 11.0$ N

$F_{\text{Sarah on pillow}} = 10.0$ N

Unknown:

$a = ?$

2 Solve for the Unknown

$$F_{\text{net}} = F_{\text{Anudja on pillow}} + (-F_{\text{Sarah on pillow}})$$

Use Newton's second law.

$$a = \frac{F_{\text{net}}}{m}$$

$$= \frac{F_{\text{Anudja on pillow}} + (-F_{\text{Sarah on pillow}})}{m}$$ Substitute $F_{\text{net}} = F_{\text{Anudja on pillow}} + (-F_{\text{Sarah on pillow}})$

$$= \frac{11.0\text{ N} - 10.0\text{ N}}{0.30\text{ kg}}$$ Substitute $F_{\text{Anudja on pillow}} = 11.0$ N, $F_{\text{Sarah on pillow}} = 10.0$ N, $m = 0.30$ kg

$$= 3.3\text{ m/s}^2$$

$\mathbf{a} = 3.3$ m/s^2 toward Anudja

Math Handbook
Operations with Significant Digits
pages 835–836

3 Evaluate the Answer

- **Are the units correct?** m/s^2 is the correct unit for acceleration.
- **Does the sign make sense?** The acceleration is in the positive direction, which is expected, because Anudja is pulling in the positive direction with a greater force than Sarah is pulling in the negative direction.
- **Is the magnitude realistic?** It is a reasonable acceleration for a light pillow.

▶ PRACTICE Problems

- Additional Problems, Appendix B
- Solutions to Selected Problems, Appendix C

15. You place a watermelon on a spring scale at the supermarket. If the mass of the watermelon is 4.0 kg, what is the reading on the scale?

16. Kamaria is learning how to ice-skate. She wants her mother to pull her along so that she has an acceleration of 0.80 m/s^2. If Kamaria's mass is 27.2 kg, with what force does her mother need to pull her? (Neglect any resistance between the ice and Kamaria's skates.)

17. Taru and Reiko simultaneously grab a 0.75-kg piece of rope and begin tugging on it in opposite directions. If Taru pulls with a force of 16.0 N and the rope accelerates away from her at 1.25 m/s^2, with what force is Reiko pulling?

18. In **Figure 4-8,** the block has a mass of 1.2 kg and the sphere has a mass of 3.0 kg. What are the readings on the two scales? (Neglect the masses of the scales.)

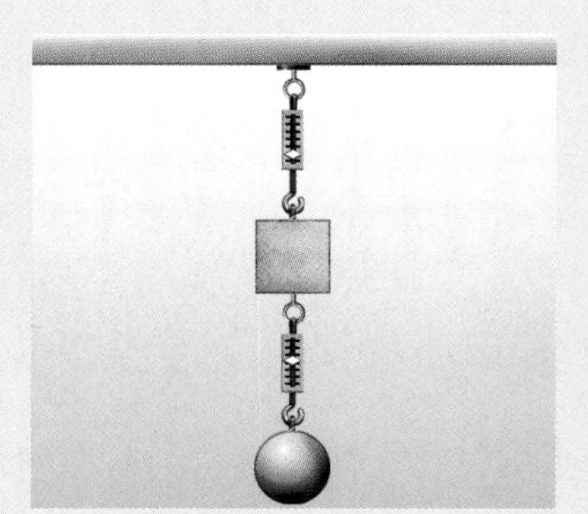

Figure 4-8

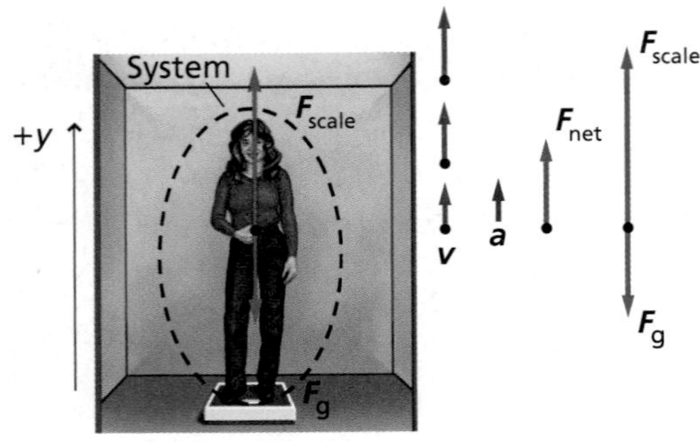

Figure 4-9 If you stand on a scale in an elevator accelerating upward, the scale must exert an upward force greater than the downward force of your weight.

Concepts In Motion
Interactive Figure To see an animation on apparent weight, visit physicspp.com.

Apparent weight What is weight? Because the weight force is defined as $\mathbf{F_g} = m\mathbf{g}$, $\mathbf{F_g}$ changes when $\mathbf{g}$ varies. On or near the surface of Earth, $\mathbf{g}$ is approximately constant, so an object's weight does not change appreciably as it moves around near Earth's surface. If a bathroom scale provides the only upward force on you, then it reads your weight. What would it read if you stood with one foot on the scale and one foot on the floor? What if a friend pushed down on your shoulders or up on your elbows? Then there would be other contact forces on you, and the scale would not read your weight.

What happens if you stand on a scale in an elevator? As long as the elevator is in equilibrium, the scale will read your weight. What would the scale read if the elevator accelerates upward? **Figure 4-9** shows the pictorial and physical models for this situation. You are the system, and upward is the positive direction. Because the acceleration of the system is upward, the upward force of the scale must be greater than the downward force of your weight. Therefore, the scale reading is greater than your weight. If you ride in an elevator like this, you would feel heavier because the floor would press harder on your feet. On the other hand, if the acceleration is downward, then you would feel lighter, and the scale would have a lower reading. The force exerted by the scale is called the apparent weight. An object's **apparent weight** is the force an object experiences as a result of all the forces acting on it, giving the object an acceleration.

Imagine that the cable holding the elevator breaks. What would the scale read then? The scale and you would both accelerate with $\mathbf{a} = -\mathbf{g}$. According to this formula, the scale would read zero and your apparent weight would be zero. That is, you would be weightless. However, **weightlessness** does not mean that an object's weight is actually zero; rather, it means that there are no contact forces pushing up on the object, and the object's apparent weight is zero.

PROBLEM-SOLVING Strategies

Force and Motion

When solving force and motion problems, use the following strategies.

1. Read the problem carefully, and sketch a pictorial model.
2. Circle the system and choose a coordinate system.
3. Determine which quantities are known and which are unknown.
4. Create a physical model by drawing a motion diagram showing the direction of the acceleration, and create a free-body diagram showing the net force.
5. Use Newton's laws to link acceleration and net force.
6. Rearrange the equation to solve for the unknown quantity.
7. Substitute known quantities with their units into the equation and solve.
8. Check your results to see if they are reasonable.

▶ EXAMPLE Problem 2

Real and Apparent Weight Your mass is 75.0 kg, and you are standing on a bathroom scale in an elevator. Starting from rest, the elevator accelerates upward at 2.00 m/s^2 for 2.00 s and then continues at a constant speed. Is the scale reading during acceleration greater than, equal to, or less than the scale reading when the elevator is at rest?

1 Analyze and Sketch the Problem

- Sketch the situation.
- Choose a coordinate system with the positive direction as upward.
- Draw the motion diagram. Label $\boldsymbol{v}$ and $\boldsymbol{a}$.
- Draw the free-body diagram. The net force is in the same direction as the acceleration, so the upward force is greater than the downward force.

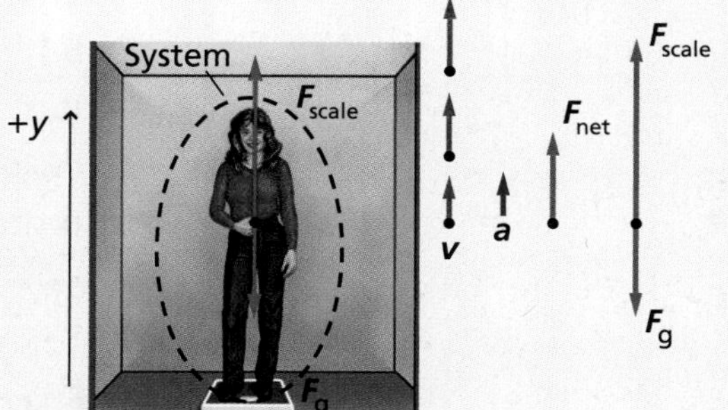

Known:	**Unknown:**
$m = 75.0$ kg	$F_{scale} = ?$
$a = 2.00$ m/s^2	
$t = 2.00$ s	
$g = 9.80$ N	

2 Solve for the Unknown

$F_{net} = ma$

$F_{net} = F_{scale} + (-F_g)$ — **F_g is negative because it is in the negative direction defined by the coordinate system.**

Solve for F_{scale}.

$F_{scale} = F_{net} + F_g$

Elevator at rest:

$$\begin{aligned} F_{scale} &= F_{net} + F_g && \text{The elevator is not accelerating. Thus, } F_{net} = 0.00\text{ N.} \\ &= F_g && \text{Substitute } F_{net} = 0.00\text{ N} \\ &= mg && \text{Substitute } F_g = mg \\ &= (75.0\text{ kg})(9.80\text{ m/s}^2) && \text{Substitute } m = 75.0\text{ kg}, g = 9.80\text{ m/s}^2 \\ &= 735\text{ N} \end{aligned}$$

Acceleration of the elevator:

$$\begin{aligned} F_{scale} &= F_{net} + F_g \\ &= ma + mg && \text{Substitute } F_{net} = ma, F_g = mg \\ &= m(a + g) && \text{Substitute } m = 75.0\text{ kg}, a = 2.00\text{ m/s}^2, g = 9.80\text{ m/s}^2 \\ &= (75.0\text{ kg})(2.00\text{ m/s}^2 + 9.80\text{ m/s}^2) \\ &= 885\text{ N} \end{aligned}$$

The scale reading when the elevator is accelerating (885 N) is larger than the scale reading at rest (735 N).

3 Evaluate the Answer

- **Are the units correct?** kg·m/s^2 is the force unit, N.
- **Does the sign make sense?** The positive sign agrees with the coordinate system.
- **Is the magnitude realistic?** F_{scale} is larger than it would be at rest when F_{scale} would be 735 N, so the magnitude is reasonable.

Personal Tutor For an online tutorial on real and apparent weight, visit physicspp.com.

PRACTICE Problems

- Additional Problems, Appendix B
- Solutions to Selected Problems, Appendix C

19. On Earth, a scale shows that you weigh 585 N.

a. What is your mass?

b. What would the scale read on the Moon ($g = 1.60$ m/s^2)?

20. Use the results from Example Problem 2 to answer questions about a scale in an elevator on Earth. What force would be exerted by the scale on a person in the following situations?

a. The elevator moves at constant speed.

b. It slows at 2.00 m/s^2 while moving upward.

c. It speeds up at 2.00 m/s^2 while moving downward.

d. It moves downward at constant speed.

e. It slows to a stop at a constant magnitude of acceleration.

Drag Force and Terminal Velocity

It is true that the particles in the air around an object exert forces on it. Air actually exerts a huge force, but in most cases, it exerts a balanced force on all sides, and therefore it has no net effect. Can you think of any experiences that help to prove that air exerts a force? When you stick a suction cup on a smooth wall or table, you remove air from the "inside" of it. The suction cup is difficult to remove because of the net force of the air on the "outside."

So far, you have neglected the force of air on an object moving through the air. In actuality, when an object moves through any fluid, such as air or water, the fluid exerts a drag force on the moving object in the direction opposite to its motion. A **drag force** is the force exerted by a fluid on the object moving through the fluid. This force is dependent on the motion of the object, the properties of the object, and the properties of the fluid that the object is moving through. For example, as the speed of the object increases, so does the magnitude of the drag force. The size and shape of the object also affects the drag force. The drag force is also affected by the properties of the fluid, such as its viscosity and temperature.

CHALLENGE PROBLEM

An air-track glider passes through a photoelectric gate at an initial speed of 0.25 m/s. As it passes through the gate, a constant force of 0.40 N is applied to the glider in the same direction as its motion. The glider has a mass of 0.50 kg.

1. What is the acceleration of the glider?

2. It takes the glider 1.3 s to pass through a second gate. What is the distance between the two gates?

3. The 0.40-N force is applied by means of a string attached to the glider. The other end of the string passes over a resistance-free pulley and is attached to a hanging mass, m. How big is m?

4. Derive an expression for the tension, T, in the string as a function of the mass, M, of the glider, the mass, m, of the hanging mass, and g.

If you drop a table-tennis ball, as in **Figure 4-10,** it has very little velocity at the start, and thus only a small drag force. The downward force of gravity is much stronger than the upward drag force, so there is a downward acceleration. As the ball's velocity increases, so does the drag force. Soon, the drag force equals the force of gravity. When this happens, there is no net force, and so there is no acceleration. The constant velocity that is reached when the drag force equals the force of gravity is called the **terminal velocity.**

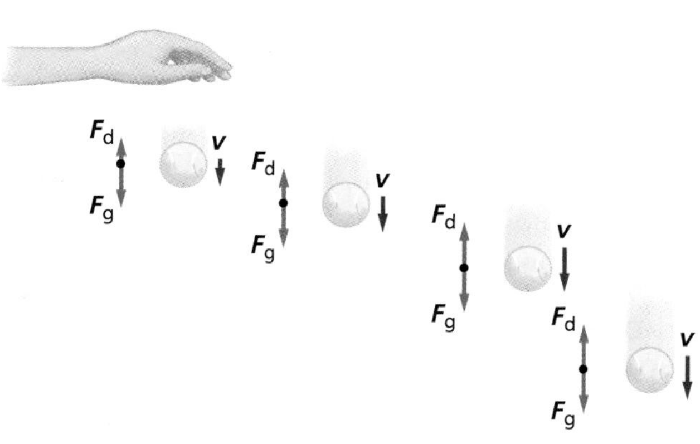

Figure 4-10 The drag force on an object increases as its velocity increases. When the drag force increases to the point that it equals the force of gravity, the object will no longer be accelerated.

When light objects with large surface areas are falling, the drag force has a substantial effect on their motion, and they quickly reach terminal velocity. Heavier, more-compact objects are not affected as much by the drag force. For example, the terminal velocity of a table-tennis ball in air is 9 m/s, that of a basketball is 20 m/s, and that of a baseball is 42 m/s. Competitive skiers increase their terminal velocities by decreasing the drag force on them. They hold their bodies in an egg shape and wear smooth clothing and streamlined helmets. Sky divers can increase or decrease their terminal velocity by changing their body orientation and shape. A horizontal, spread-eagle shape produces the slowest terminal velocity, about 60 m/s. Because a parachute changes the shape of the sky diver when it opens, a sky diver becomes part of a very large object with a correspondingly large drag force and a terminal velocity of about 5 m/s.

4.2 Section Review

21. Lunar Gravity Compare the force holding a 10.0-kg rock on Earth and on the Moon. The acceleration due to gravity on the Moon is 1.62 m/s^2.

22. Real and Apparent Weight You take a ride in a fast elevator to the top of a tall building and ride back down while standing on a bathroom scale. During which parts of the ride will your apparent and real weights be the same? During which parts will your apparent weight be less than your real weight? More than your real weight? Sketch free-body diagrams to support your answers.

23. Acceleration Tecle, with a mass of 65.0 kg, is standing by the boards at the side of an ice-skating rink. He pushes off the boards with a force of 9.0 N. What is his resulting acceleration?

24. Motion of an Elevator You are riding in an elevator holding a spring scale with a 1-kg mass suspended from it. You look at the scale and see that it reads 9.3 N. What, if anything, can you conclude about the elevator's motion at that time?

25. Mass Marcos is playing tug-of-war with his cat using a stuffed toy. At one instant during the game, Marcos pulls on the toy with a force of 22 N, the cat pulls in the opposite direction with a force of 19.5 N, and the toy experiences an acceleration of 6.25 m/s^2. What is the mass of the toy?

26. Acceleration A sky diver falls at a constant speed in the spread-eagle position. After he opens his parachute, is the sky diver accelerating? If so, in which direction? Explain your answer using Newton's laws.

27. Critical Thinking You have a job at a meat warehouse loading inventory onto trucks for shipment to grocery stores. Each truck has a weight limit of 10,000 N of cargo. You push each crate of meat along a low-resistance roller belt to a scale and weigh it before moving it onto the truck. However, right after you weigh a 1000-N crate, the scale breaks. Describe a way in which you could apply Newton's laws to figure out the approximate masses of the remaining crates.

4.3 Interaction Forces

► **Objectives**

- **Define** Newton's third law.
- **Explain** the tension in ropes and strings in terms of Newton's third law.
- **Define** the normal force.
- **Determine** the value of the normal force by applying Newton's second law.

► **Vocabulary**

interaction pair
Newton's third law
tension
normal force

You have learned that when an agent exerts a net force upon an object, the object undergoes acceleration. You know that this force can be either a field force or a contact force. But what causes the force? If you experiment with two magnets, you can feel each magnet pushing or pulling the other. Similarly, if you pull on a lever, you can feel the lever pulling back against you. Which is the agent and which is the object?

Identifying Interaction Forces

Imagine that you and a friend are each wearing in-line skates (with all the proper safety gear), and your friend is standing right in front of you, with his back to you. You push your friend so that he starts rolling forward. What happens to you? You move backwards. Why? Recall that a force is the result of an interaction between two objects. When you push your friend forward, you come into contact with him and exert a force that moves him forward. However, because he is also in contact with you, he also exerts a force on you, and this results in a change in your motion.

Forces always come in pairs. Consider you (Student A) as one system and your friend (Student B) as another. What horizontal forces act on each of the two systems? **Figure 4-11** shows the free-body diagram for the systems. Looking at this diagram, you can see that each system experiences a force exerted by the other. The two forces, $\boldsymbol{F}_{\text{A on B}}$ and $\boldsymbol{F}_{\text{B on A}}$, are the forces of interaction between the two of you. Notice the symmetry in the subscripts: A on B and B on A. What do you notice about the directions of these forces? What do you expect to be true about their relative magnitudes?

Figure 4-11 When you exert a force on your friend to push him forward, he exerts an equal and opposite force on you, which causes you to move backwards.

The forces $\boldsymbol{F}_{\text{A on B}}$ and $\boldsymbol{F}_{\text{B on A}}$ are an interaction pair. An **interaction pair** is two forces that are in opposite directions and have equal magnitude. Sometimes, this also is called an action-reaction pair of forces. This might suggest that one causes the other; however, this is not true. For example, the force of you pushing your friend doesn't cause your friend to exert a force on you. The two forces either exist together or not at all. They both result from the contact between the two of you.

Newton's Third Law

The force of you on your friend is equal in magnitude and opposite in direction to the force of your friend on you. This is summarized in **Newton's third law,** which states that all forces come in pairs. The two forces in a pair act on different objects and are equal in strength and opposite in direction.

Newton's Third Law $\boldsymbol{F}_{\text{A on B}} = -\boldsymbol{F}_{\text{B on A}}$

The force of A on B is equal in magnitude and opposite in direction of the force of B on A.

Consider the situation of you holding a book in your hand. Draw one free-body diagram each for you and for the book. Are there any interaction pairs? When identifying interaction pairs, keep in mind that they always will occur in two different free-body diagrams, and they always will have the symmetry of subscripts noted on the previous page. In this case, there is one interaction pair, $\boldsymbol{F}_{\text{book on hand}}$ and $\boldsymbol{F}_{\text{hand on book}}$. Notice also that each object has a weight. If the weight force is due to an interaction between the object and Earth's mass, then shouldn't each of these objects also exert a force on Earth? If this is the case, shouldn't Earth be accelerating?

Consider a soccer ball sitting on a table. The table, in turn, is sitting on Earth, as shown in **Figure 4-12.** First, analyze the forces acting on the ball. The table exerts an upward force on the ball, and the mass of Earth exerts a downward gravitational force on the ball. Even though these forces are in the opposite direction on the same object, they are not an interaction pair. They are simply two forces acting on the same object, not the interaction between two objects. Consider the ball and the table together. In addition to the upward force exerted by the table on the ball, the ball exerts a downward force on the table. This is one pair of forces. The ball and Earth also have an interaction pair. Thus, the interaction pairs related to the soccer ball are $\boldsymbol{F}_{\text{ball on table}} = -\boldsymbol{F}_{\text{table on ball}}$ and $\boldsymbol{F}_{\text{ball on Earth}} = -\boldsymbol{F}_{\text{Earth on ball}}$. It is important to keep in mind that an interaction pair must consist of two forces of equal magnitude pointing in opposite directions. These opposing forces must act on two different objects that can exert a force against each other.

The acceleration caused by the force of an object interacting with Earth is usually a very small number. Under most circumstances, the number is so small that for problems involving falling or stationary objects, Earth can be treated as part of the external world rather than as a second system. Consider Example Problem 3 using the following problem-solving strategies.

Figure 4-12 A soccer ball on a table on Earth is part of two interaction pairs—the interaction between the ball and table and the interaction between the ball and Earth. (Not to scale)

PROBLEM-SOLVING **Strategies**

Interaction Pairs

Use these strategies to solve problems in which there is an interaction between objects in two different systems.

1. Separate the system or systems from the external world.
2. Draw a pictorial model with coordinate systems for each system and a physical model that includes free-body diagrams for each system.
3. Connect interaction pairs by dashed lines.
4. To calculate your answer, use Newton's second law to relate the net force and acceleration for each system.
5. Use Newton's third law to equate the magnitudes of the interaction pairs and give the relative direction of each force.
6. Solve the problem and check the units, signs, and magnitudes for reasonableness.

MINI LAB

Tug-of-War Challenge

In a tug-of-war, predict how the force you exert on your end of the rope compares to the force your opponent exerts if you pull and your opponent just holds the rope.

1. Predict how the forces compare if the rope moves in your direction.

2. Test your prediction. ***CAUTION: Do not suddenly let go of the rope.***

Analyze and Conclude

3. Compare the force on your end of the rope to the force on your opponent's end of the rope. What happened when you started to move your opponent's direction?

▶ EXAMPLE Problem 3

Earth's Acceleration When a softball with a mass of 0.18 kg is dropped, its acceleration toward Earth is equal to g, the acceleration due to gravity. What is the force on Earth due to the ball, and what is Earth's resulting acceleration? Earth's mass is 6.0×10^{24} kg.

1 Analyze and Sketch the Problem

- Draw free-body diagrams for the two systems: the ball and Earth.
- Connect the interaction pair by a dashed line.

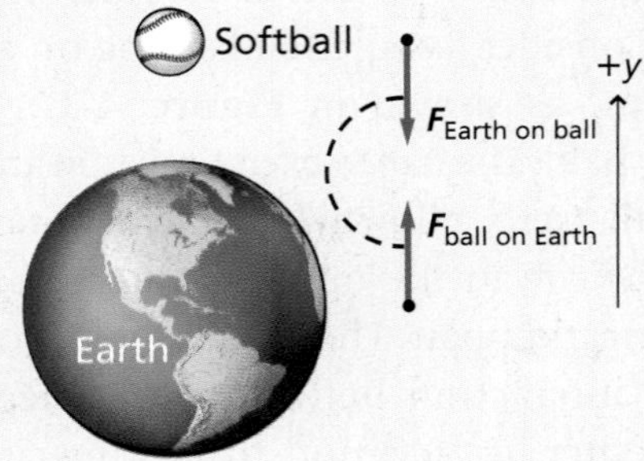

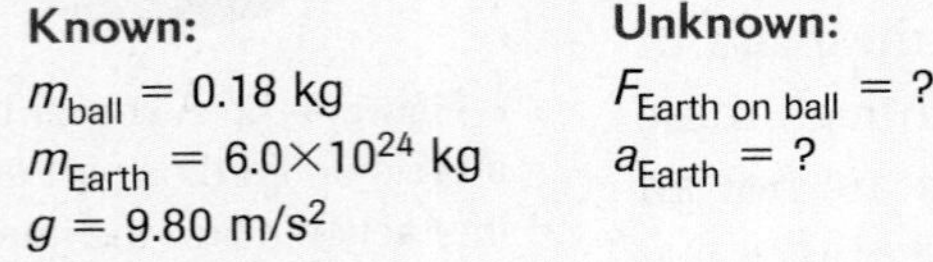

Known:	Unknown:
$m_{\text{ball}} = 0.18$ kg	$F_{\text{Earth on ball}} = ?$
$m_{\text{Earth}} = 6.0\times10^{24}$ kg	$a_{\text{Earth}} = ?$
$g = 9.80$ m/s^2	

2 Solve for the Unknown

Use Newton's second law to find the weight of the ball.

$$F_{\text{Earth on ball}} = m_{\text{ball}}a$$
$$= m_{\text{ball}}(-g) \qquad \textbf{Substitute } a = -g$$
$$= (0.18\text{ kg})(9.80\text{ m/s}^2) \qquad \textbf{Substitute } m_{\text{ball}} = 0.18\text{ kg}, g = 9.80\text{ m/s}^2$$
$$= -1.8\text{ N}$$

Use Newton's third law to find $F_{\text{ball on Earth}}$.

$$F_{\text{ball on Earth}} = -F_{\text{Earth on ball}}$$
$$= -(-1.8\text{ N}) \qquad \textbf{Substitute } F_{\text{Earth on ball}} = -1.8\text{ N}$$
$$= +1.8\text{ N}$$

Math Handbook
Operations with Scientific Notation
pages 842–843

Use Newton's second law to find a_{Earth}.

$$a_{\text{Earth}} = \frac{F_{\text{net}}}{m_{\text{Earth}}}$$
$$= \frac{1.8\text{ N}}{6.0\times10^{24}\text{ kg}} \qquad \textbf{Substitute } F_{\text{net}} = 1.8\text{ N}, m_{\text{Earth}} = 6.0\times10^{24}\text{ kg}$$
$$= 2.9\times10^{-25}\text{ m/s}^2 \text{ toward the softball}$$

3 Evaluate the Answer

- **Are the units correct?** Dimensional analysis verifies force in N and acceleration in m/s^2.
- **Do the signs make sense?** Force and acceleration should be positive.
- **Is the magnitude realistic?** Because of Earth's large mass, the acceleration should be small.

▶ PRACTICE Problems

• Additional Problems, Appendix B
• Solutions to Selected Problems, Appendix C

28. You lift a relatively light bowling ball with your hand, accelerating it upward. What are the forces on the ball? What forces does the ball exert? What objects are these forces exerted on?

29. A brick falls from a construction scaffold. Identify any forces acting on the brick. Also identify any forces that the brick exerts and the objects on which these forces are exerted. (Air resistance may be ignored.)

30. You toss a ball up in the air. Draw a free-body diagram for the ball while it is still moving upward. Identify any forces acting on the ball. Also identify any forces that the ball exerts and the objects on which these forces are exerted.

■ **Figure 4-13**

31. A suitcase sits on a stationary airport luggage cart, as in **Figure 4-13.** Draw a free-body diagram for each object and specifically indicate any interaction pairs between the two.

Forces of Ropes and Strings

Tension is simply a specific name for the force exerted by a string or rope. A simplification within this textbook is the assumption that all strings and ropes are massless. To understand tension in more detail, consider the situation in **Figure 4-14,** where a bucket hangs from a rope attached to the ceiling. The rope is about to break in the middle. If the rope breaks, the bucket will fall; thus, before it breaks, there must be forces holding the rope together. The force that the top part of the rope exerts on the bottom part is $\boldsymbol{F}_{\text{top on bottom}}$. Newton's third law states that this force must be part of an interaction pair. The other member of the pair is the force that the bottom part exerts on the top, $\boldsymbol{F}_{\text{bottom on top}}$. These forces, equal in magnitude but opposite in direction, also are shown in Figure 4-14.

Think about this situation in another way. Before the rope breaks, the bucket is in equilibrium. This means that the force of its weight downward must be equal in magnitude but opposite in direction to the tension in the rope upward. Similarly, if you look at the point in the rope just above the bucket, it also is in equilibrium. Therefore, the tension of the rope below it pulling down must be equal to the tension of the rope above it pulling up. You can move up the rope, considering any point in the rope, and see that the tension forces are pulling equally in both directions. Because the very bottom of the rope has a tension equal to the weight of the bucket, the tension everywhere in the rope is equal to the weight of the bucket. Thus, the tension in the rope is the weight of all objects below it. Because the rope is assumed to be massless, the tension everywhere in the rope is equal to the bucket's weight.

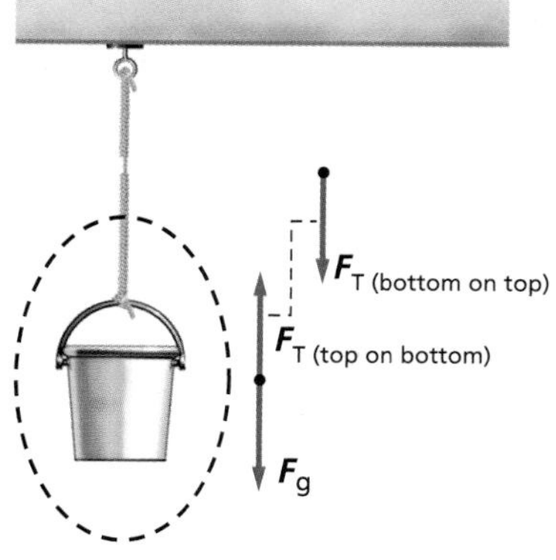

Figure 4-14 The tension in the rope is equal to the weight of all the objects hanging from it.

Tension forces also are at work in a tug-of-war, like the one shown in **Figure 4-15.** If team A, on the left, is exerting a force of 500 N and the rope does not move, then team B, on the right, also must be pulling with a force of 500 N. What is the tension in the rope in this case? If each team pulls with 500 N of force, is the tension 1000 N? To decide, think of the rope as divided into two halves. The left-hand end is not moving, so the net force on it is zero. Thus, $F_{\text{A on rope}} = F_{\text{right on left}} = 500$ N. Similarly, $F_{\text{B on rope}} = F_{\text{left on right}} = 500$ N. But the two tensions, $F_{\text{right on left}}$ and $F_{\text{left on right}}$, are an interaction pair, so they are equal and opposite. Thus, the tension in the rope equals the force with which each team pulls, or 500 N. To verify this, you could cut the rope in half, tie the ends to a spring scale, and ask the two teams each to pull with 500 N of force. You would see that the scale reads 500 N.

Figure 4-15 In a tug-of-war, the teams exert equal and opposite forces on each other via the tension in the rope.

▶ EXAMPLE Problem 4

Lifting a Bucket **A 50.0-kg bucket is being lifted by a rope. The rope will not break if the tension is 525 N or less. The bucket started at rest, and after being lifted 3.0 m, it is moving at 3.0 m/s. If the acceleration is constant, is the rope in danger of breaking?**

1 Analyze and Sketch the Problem

- Draw the situation and identify the forces on the system.
- Establish a coordinate system with the positive axis upward.
- Draw a motion diagram including v and a.
- Draw the free-body diagram, labeling the forces.

Known:		Unknown:
$m = 50.0$ kg	$v_f = 3.0$ m/s	$F_T = ?$
$v_i = 0.0$ m/s	$d = 3.0$ m	

+y
v
a
F_T (rope on bucket)
F_{net}
System
F_g (Earth's mass on bucket)

2 Solve for the Unknown

F_{net} is the sum of the positive force of the rope pulling up, F_T, and the negative weight force, $-F_g$, pulling down as defined by the coordinate system.

$$F_{net} = F_T + (-F_g)$$
$$F_T = F_{net} + F_g$$
$$= ma + mg \quad \text{Substitute } F_{net} = ma, F_g = mg$$
$$= m(a + g)$$

v_i, v_f, and d are known. Use this motion equation to solve for a.

$$v_f^2 = v_i^2 + 2ad$$
$$a = \frac{v_f^2 - v_i^2}{2d}$$
$$= \frac{v_f^2}{2d} \quad \text{Substitute } v_i = 0.0 \text{ m/s}^2$$

Math Handbook
Isolating a Variable
page 845

$$F_T = m(a + g)$$
$$= m\left(\frac{v_f^2}{2d} + g\right) \quad \text{Substitute } a = \frac{v_f^2}{2d}$$
$$= (50.0 \text{ kg})\left(\frac{(3.0 \text{ m/s})^2}{2(3.0 \text{ m})} + 9.80 \text{ m/s}^2\right) \quad \text{Substitute } m = 50.0 \text{ kg}, v_f = 3.0 \text{ m/s}, d = 3.0 \text{ m}, g = 9.80 \text{ m/s}^2$$
$$= 570 \text{ N}$$

The rope is in danger of breaking because the tension exceeds 525 N.

3 Evaluate the Answer

- **Are the units correct?** Dimensional analysis verifies kg·m/s^2, which is N.
- **Does the sign make sense?** The upward force should be positive.
- **Is the magnitude realistic?** The magnitude is a little larger than 490 N, which is the weight of the bucket. $F_g = mg = (50.0 \text{ kg})(9.80 \text{ m/s}^2) = 490$ N

▶ PRACTICE Problems

• Additional Problems, Appendix B
• Solutions to Selected Problems, Appendix C

32. You are helping to repair a roof by loading equipment into a bucket that workers hoist to the rooftop. If the rope is guaranteed not to break as long as the tension does not exceed 450 N and you fill the bucket until it has a mass of 42 kg, what is the greatest acceleration that the workers can give the bucket as they pull it to the roof?

33. Diego and Mika are trying to fix a tire on Diego's car, but they are having trouble getting the tire loose. When they pull together, Mika with a force of 23 N and Diego with a force of 31 N, they just barely get the tire to budge. What is the magnitude of the strength of the force between the tire and the wheel?

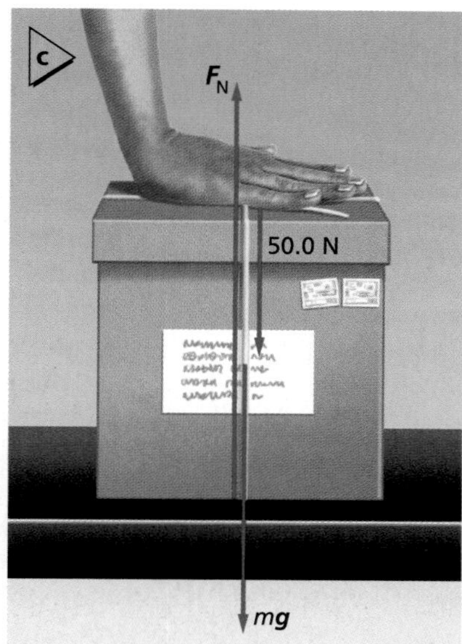

The Normal Force

Any time two objects are in contact, they each exert a force on each other. Think about a box sitting on a table. There is a downward force due to the gravitational attraction of Earth. There also is an upward force that the table exerts on the box. This force must exist, because the box is in equilibrium. The **normal force** is the perpendicular contact force exerted by a surface on another object.

The normal force always is perpendicular to the plane of contact between two objects, but is it always equal to the weight of an object as in **Figure 4-16a?** What if you tied a string to the box and pulled up on it a little bit, but not enough to move the box, as shown in **Figure 4-16b?** When you apply Newton's second law to the box, you see that $F_N + F_{\text{string on box}} - F_g = ma = 0$, which rearranges to $F_N = F_g - F_{\text{string on box}}$.

You can see that in this case, the normal force exerted by the table on the box is less than the box's weight, F_g. Similarly, if you pushed down on the box on the table as shown in **Figure 4-16c,** the normal force would be more than the box's weight. Finding the normal force will be important in the next chapter, when you begin dealing with resistance.

Figure 4-16 The normal force on an object is not always equal to its weight. In **(a)** the normal force is equal to the object's weight. In **(b)** the normal force is less than the object's weight. In **(c)** the normal force is greater than the object's weight.

4.3 Section Review

34. Force Hold a book motionless in your hand in the air. Identify each force and its interaction pair on the book.

35. Force Lower the book from problem 34 at increasing speed. Do any of the forces or their interaction-pair partners change? Explain.

36. Tension A block hangs from the ceiling by a massless rope. A second block is attached to the first block and hangs below it on another piece of massless rope. If each of the two blocks has a mass of 5.0 kg, what is the tension in each rope?

37. Tension If the bottom block in problem 36 has a mass of 3.0 kg and the tension in the top rope is 63.0 N, calculate the tension in the bottom rope and the mass of the top block.

38. Normal Force Poloma hands a 13-kg box to 61-kg Stephanie, who stands on a platform. What is the normal force exerted by the platform on Stephanie?

39. Critical Thinking A curtain prevents two tug-of-war teams from seeing each other. One team ties its end of the rope to a tree. If the other team pulls with a 500-N force, what is the tension? Explain.

Physics Online physicspp.com/self_check_quiz

PHYSICS LAB • Internet

Forces in an Elevator

Alternate CBL instructions can be found on the Web site.
physicspp.com

Have you ever been in a fast-moving elevator? Was the ride comfortable? How about an amusement ride that quickly moves upward or one that free-falls? What forces are acting on you during your ride? In this experiment, you will investigate the forces that affect you during vertical motion when gravity is involved with a bathroom scale. Many bathroom scales measure weight in pounds mass (lbm) or pounds force (lbf) rather than newtons. In the experiment, you will need to convert weights measured on common household bathroom scales to SI units.

QUESTION

What one-dimensional forces act on an object that is moving in a vertical direction in relation to the ground?

Objectives

- **Measure** Examine forces that act on objects that move vertically.
- **Compare and Contrast** Differentiate between actual weight and apparent weight.
- **Analyze and Conclude** Share and compare data of the acceleration of elevators.

Safety Precautions

- **Use caution when working around elevator doors.**
- **Do not interfere with normal elevator traffic.**
- **Watch that the mass on the spring scale does not fall and hit someone's feet or toes.**

Materials

elevator
bathroom scale
spring scale
mass

Procedure

1. Securely attach a mass to the hook on a spring scale. Record the force of the mass in the data table.
2. Accelerate the mass upward, then move it upward at a constant velocity, and then slow the mass down. Record the greatest amount of force on the scale, the amount of force at constant velocity, and the lowest scale reading.
3. Get your teacher's permission and proceed to an elevator on the ground floor. Before entering the elevator, measure your weight on a bathroom scale. Record this weight in the data table.

Data Table	
Force (step 1)	
Highest Reading (step 2)	
Reading at Constant Velocity (step 2)	
Lowest Reading (step 2)	
Your Weight (step 3)	
Highest Reading (step 4)	
Reading at Constant Velocity (step 5)	
Lowest Reading (step 6)	

4. Place the scale in the elevator. Step on the scale and record the mass at rest. Select the highest floor that the elevator goes up to. Once the elevator starts, during its upward acceleration, record the highest reading on the scale in the data table.
5. When the velocity of the elevator becomes constant, record the reading on the scale in the data table.
6. As the elevator starts to decelerate, watch for the lowest reading on the scale and record it in the data table.

Analyze

1. **Explain** In step 2, why did the mass appear to gain weight when being accelerated upward? Provide a mathematical equation to summarize this concept.
2. **Explain** Why did the mass appear to lose weight when being decelerated at the end of its movement during step 3? Provide a mathematical equation to summarize this concept.
3. **Measure in SI** Most bathroom scales read in pounds mass (lbm). Convert your reading in step 4 in pounds mass to kilograms. (1 kg = 2.21 lbm) (Note: skip this step if your scale measures in kilograms.)
4. **Measure in SI** Some bathroom scales read in pounds force (lbf). Convert all of the readings you made in steps 4–6 to newtons. (1 N = 0.225 lbf)
5. **Analyze** Calculate the acceleration of the elevator at the beginning of your elevator trip using the equation $F_{scale} = ma + mg$.
6. **Use Numbers** What is the acceleration of the elevator at the end of your trip?

Conclude and Apply

How can you develop an experiment to find the acceleration of an amusement park ride that either drops rapidly or climbs rapidly?

Going Further

How can a bathroom scale measure both pounds mass (lbm) and pounds force (lbf) at the same time?

Real-World Physics

Forces on pilots in high-performance jet airplanes are measured in *g*'s or *g*-force. What does it mean if a pilot is pulling 6 *g*'s in a power climb?

Share Your Data

Communicate You can visit physicspp.com/internet_lab to post the acceleration of your elevator and compare it to other elevators around the country, maybe even the world. Post a description of your elevator's ride so that a comparison of acceleration versus ride comfort can be evaluated.

To find out more about forces and acceleration, visit the Web site: physicspp.com

How it Works Bathroom Scale

The portable weighing scale was patented in 1896 by John H. Hunter. People used coin-operated scales, usually located in stores, to weigh themselves until the advent of the home bathroom scale in 1946. How does a bathroom scale work?

1 There are two long and two short levers that are attached to each other. Brackets in the lid of the scale sit on top of the levers to help evenly distribute your weight on the levers.

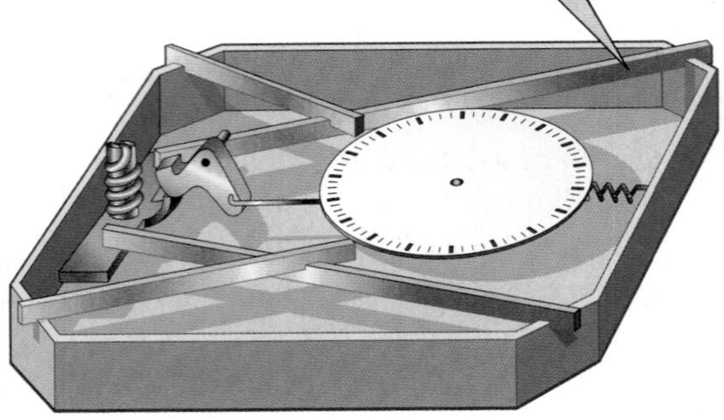

3 As the calibrating plate is pushed down by weight on the scale, the crank pivots. This, in turn, moves the rack and rotates the pinion. As a result, the dial on the scale rotates.

Dial

Main spring

Crank

Dial spring

Pinion

Calibrating plate

F_g

Rack

Lever

4 When the spring force, F_{sp}, from the main spring being stretched is equal to F_g, the crank, rack, and pinion no longer move, and your weight is shown on the dial.

2 The long levers rest on top of a calibrating plate that has the main spring attached to it. When you step on the scale, your weight, F_g, is exerted on the levers, which, in turn, exert a force on the calibrating plate and cause the main spring to stretch.

Thinking Critically

1. **Hypothesize** Most springs in bathroom scales cannot exert a force larger than 20 lbs (89 N). How is it possible that you don't break the scale every time you step on it? *(Hint: Think about exerting a large force near the pivot of a see-saw.)*
2. **Solve** If the largest reading on most scales is 240 lbs (1068 N) and the spring can exert a maximum of 20 lbs (89 N), what ratio does the lever use?

Chapter 4

Study Guide

4.1 Force and Motion

Vocabulary

- force (p. 88)
- free-body diagram (p. 89)
- net force (p. 92)
- Newton's second law (p. 93)
- Newton's first law (p. 94)
- inertia (p. 95)
- equilibrium (p. 95)

Key Concepts

- An object that experiences a push or a pull has a force exerted on it.
- Forces have both direction and magnitude.
- Forces may be divided into contact and field forces.
- In a free-body diagram, always draw the force vectors leading away from the object, even if the force is a push.
- The forces acting upon an object can be added using vector addition to find the net force.
- Newton's second law states that the acceleration of a system equals the net force acting on it, divided by its mass.

$$\boldsymbol{a} = \frac{\boldsymbol{F}_{\text{net}}}{m}$$

- Newton's first law states that an object that is at rest will remain at rest, and an object that is moving will continue to move in a straight line with constant speed, if and only if the net force acting on that object is zero.
- An object with no net force acting on it is in equilibrium.

4.2 Using Newton's Laws

Vocabulary

- apparent weight (p. 98)
- weightlessness (p. 98)
- drag force (p. 100)
- terminal velocity (p. 101)

Key Concepts

- The weight of an object depends upon the acceleration due to gravity and the mass of the object.
- An object's apparent weight is the force an object experiences as a result of the contact forces acting on it, giving the object an acceleration.
- An object with no apparent weight experiences weightlessness.
- The effect of drag on an object's motion is determined by the object's weight, size, and shape.
- If a falling object reaches a velocity such that the drag force is equal to the object's weight, it maintains that velocity, called the terminal velocity.

4.3 Interaction Forces

Vocabulary

- interaction pair (p. 102)
- Newton's third law (p. 102)
- tension (p. 105)
- normal force (p. 107)

Key Concepts

- All forces result from interactions between objects.
- Newton's third law states that the two forces that make up an interaction pair of forces are equal in magnitude, but opposite in direction and act on different objects.

$$\boldsymbol{F}_{\text{A on B}} = -\boldsymbol{F}_{\text{B on A}}$$

- In an interaction pair, $\boldsymbol{F}_{\text{A on B}}$ does not cause $\boldsymbol{F}_{\text{B on A}}$. The two forces either exist together or not at all.
- Tension is the specific name for the force exerted by a rope or string.
- The normal force is a support force resulting from the contact of two objects. It is always perpendicular to the plane of contact between the two objects.

Chapter 4 Assessment

Concept Mapping

40. Complete the following concept map using the following term and symbols: *normal*, F_T, F_g.

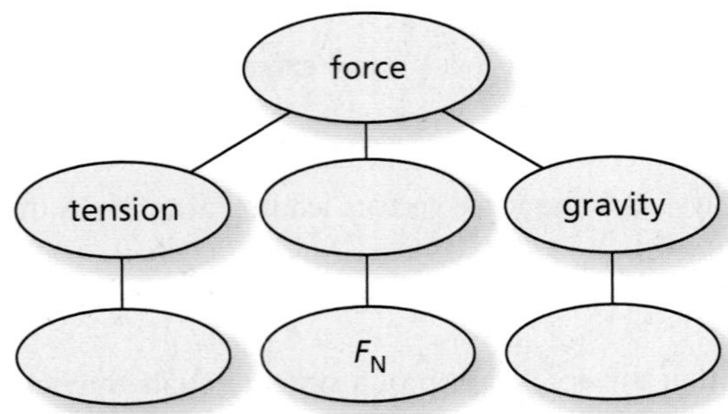

Mastering Concepts

41. A physics book is motionless on the top of a table. If you give it a hard push with your hand, it slides across the table and slowly comes to a stop. Use Newton's laws to answer the following questions. (4.1)

a. Why does the book remain motionless before the force of your hand is applied?

b. Why does the book begin to move when your hand pushes hard enough on it?

c. Under what conditions would the book remain in motion at a constant speed?

42. Cycling Why do you have to push harder on the pedals of a single-speed bicycle to start it moving than to keep it moving at a constant velocity? (4.1)

43. Suppose that the acceleration of an object is zero. Does this mean that there are no forces acting on it? Give an example supporting your answer. (4.2)

44. Basketball When a basketball player dribbles a ball, it falls to the floor and bounces up. Is a force required to make it bounce? Why? If a force is needed, what is the agent involved? (4.2)

45. Before a sky diver opens her parachute, she may be falling at a velocity higher than the terminal velocity that she will have after the parachute opens. (4.2)

a. Describe what happens to her velocity as she opens the parachute.

b. Describe the sky diver's velocity from when her parachute has been open for a time until she is about to land.

46. If your textbook is in equilibrium, what can you say about the forces acting on it? (4.2)

47. A rock is dropped from a bridge into a valley. Earth pulls on the rock and accelerates it downward. According to Newton's third law, the rock must also be pulling on Earth, yet Earth does not seem to accelerate. Explain. (4.3)

48. Ramon pushes on a bed that has been pushed against a wall, as in **Figure 4-17.** Draw a free-body diagram for the bed and identify all the forces acting on it. Make a separate list of all the forces that the bed applies to other objects. (4.3)

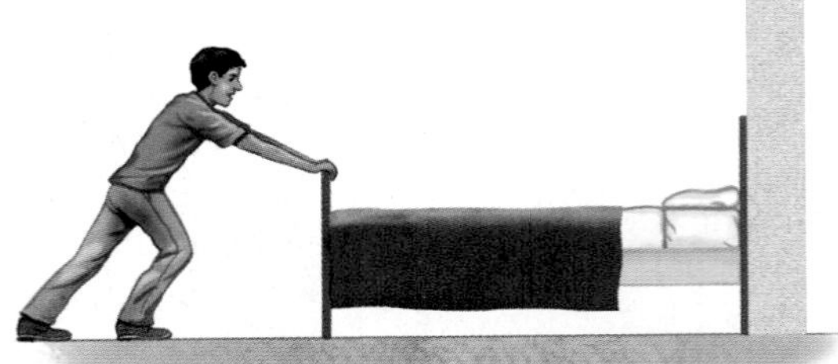

■ **Figure 4-17**

49. Figure 4-18 shows a block in three different situations. Rank them according to the magnitude of the normal force between the block and the surface, greatest to least. Specifically indicate any ties. (4.3)

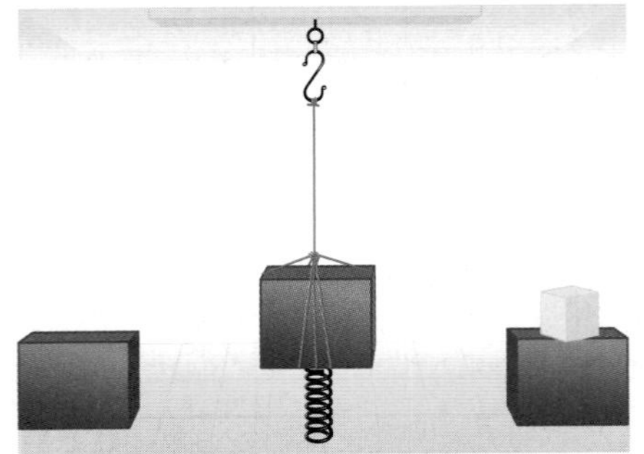

■ **Figure 4-18**

50. Explain why the tension in a massless rope is constant throughout it. (4.3)

51. A bird sits on top of a statue of Einstein. Draw free-body diagrams for the bird and the statue. Specifically indicate any interaction pairs between the two diagrams. (4.3)

52. Baseball A slugger swings his bat and hits a baseball pitched to him. Draw free-body diagrams for the baseball and the bat at the moment of contact. Specifically indicate any interaction pairs between the two diagrams. (4.3)

Applying Concepts

53. Whiplash If you are in a car that is struck from behind, you can receive a serious neck injury called whiplash.

a. Using Newton's laws, explain what happens to cause such an injury.

b. How does a headrest reduce whiplash?

 For more problems, go to Additional Problems, Appendix B.

54. Space Should astronauts choose pencils with hard or soft lead for making notes in space? Explain.

55. When you look at the label of the product in **Figure 4-19** to get an idea of how much the box contains, does it tell you its mass, weight, or both? Would you need to make any changes to this label to make it correct for consumption on the Moon?

Figure 4-19

56. From the top of a tall building, you drop two table-tennis balls, one filled with air and the other with water. Both experience air resistance as they fall. Which ball reaches terminal velocity first? Do both hit the ground at the same time?

57. It can be said that 1 kg is equivalent to 2.21 lb. What does this statement mean? What would be the proper way of making the comparison?

58. You toss a ball straight up into the air.
- **a.** Draw a free-body diagram for the ball at three points during its motion: on the way up, at the very top, and on the way down. Specifically identify the forces acting on the ball and their agents.
- **b.** What is the velocity of the ball at the very top of the motion?
- **c.** What is the acceleration of the ball at this same point?

Mastering Problems

4.1 Force and Motion

59. What is the net force acting on a 1.0-kg ball in free-fall?

60. Skating Joyce and Efua are skating. Joyce pushes Efua, whose mass is 40.0-kg, with a force of 5.0 N. What is Efua's resulting acceleration?

61. A car of mass 2300 kg slows down at a rate of 3.0 m/s^2 when approaching a stop sign. What is the magnitude of the net force causing it to slow down?

62. Breaking the Wishbone After Thanksgiving, Kevin and Gamal use the turkey's wishbone to make a wish. If Kevin pulls on it with a force 0.17 N larger than the force Gamal pulls with in the opposite direction, and the wishbone has a mass of 13 g, what is the wishbone's initial acceleration?

4.2 Using Newton's Laws

63. What is your weight in newtons?

64. Motorcycle Your new motorcycle weighs 2450 N. What is its mass in kilograms?

65. Three objects are dropped simultaneously from the top of a tall building: a shot put, an air-filled balloon, and a basketball.
- **a.** Rank the objects in the order in which they will reach terminal velocity, from first to last.
- **b.** Rank the objects according to the order in which they will reach the ground, from first to last.
- **c.** What is the relationship between your answers to parts a and b?

66. What is the weight in pounds of a 100.0-N wooden shipping case?

67. You place a 7.50-kg television on a spring scale. If the scale reads 78.4 N, what is the acceleration due to gravity at that location?

68. Drag Racing A 873-kg (1930-lb) dragster, starting from rest, attains a speed of 26.3 m/s (58.9 mph) in 0.59 s.
- **a.** Find the average acceleration of the dragster during this time interval.
- **b.** What is the magnitude of the average net force on the dragster during this time?
- **c.** Assume that the driver has a mass of 68 kg. What horizontal force does the seat exert on the driver?

69. Assume that a scale is in an elevator on Earth. What force would the scale exert on a 53-kg person standing on it during the following situations?
- **a.** The elevator moves up at a constant speed.
- **b.** It slows at 2.0 m/s^2 while moving upward.
- **c.** It speeds up at 2.0 m/s^2 while moving downward.
- **d.** It moves downward at a constant speed.
- **e.** It slows to a stop while moving downward with a constant acceleration.

70. A grocery sack can withstand a maximum of 230 N before it rips. Will a bag holding 15 kg of groceries that is lifted from the checkout counter at an acceleration of 7.0 m/s^2 hold?

71. A 0.50-kg guinea pig is lifted up from the ground. What is the smallest force needed to lift it? Describe its resulting motion.

72. **Astronomy** On the surface of Mercury, the gravitational acceleration is 0.38 times its value on Earth.
   a. What would a 6.0-kg mass weigh on Mercury?
   b. If the gravitational acceleration on the surface of Pluto is 0.08 times that of Mercury, what would a 7.0-kg mass weigh on Pluto?

73. A 65-kg diver jumps off of a 10.0-m tower.
   a. Find the diver's velocity when he hits the water.
   b. The diver comes to a stop 2.0 m below the surface. Find the net force exerted by the water.

74. **Car Racing** A race car has a mass of 710 kg. It starts from rest and travels 40.0 m in 3.0 s. The car is uniformly accelerated during the entire time. What net force is exerted on it?

4.3 Interaction Forces

75. A 6.0-kg block rests on top of a 7.0-kg block, which rests on a horizontal table.
   a. What is the force (magnitude and direction) exerted by the 7.0-kg block on the 6.0-kg block?
   b. What is the force (magnitude and direction) exerted by the 6.0-kg block on the 7.0-kg block?

76. **Rain** A raindrop, with mass 2.45 mg, falls to the ground. As it is falling, what magnitude of force does it exert on Earth?

77. A 90.0-kg man and a 55-kg man have a tug-of-war. The 90.0-kg man pulls on the rope such that the 55-kg man accelerates at 0.025 m/s^2. What force does the rope exert on the 90.0-kg man?

78. Male lions and human sprinters can both accelerate at about 10.0 m/s^2. If a typical lion weighs 170 kg and a typical sprinter weighs 75 kg, what is the difference in the force exerted on the ground during a race between these two species?

79. A 4500-kg helicopter accelerates upward at 2.0 m/s^2. What lift force is exerted by the air on the propellers?

80. Three blocks are stacked on top of one another, as in **Figure 4-20.** The top block has a mass of 4.6 kg, the middle one has a mass of 1.2 kg, and the bottom one has a mass of 3.7 kg. Identify and calculate any normal forces between the objects.

Figure 4-20

Mixed Review

81. The dragster in problem 68 completed a 402.3-m (0.2500-mi) run in 4.936 s. If the car had a constant acceleration, what was its acceleration and final velocity?

82. **Jet** A 2.75×10^6-N catapult jet plane is ready for takeoff. If the jet's engines supply a constant thrust of 6.35×10^6 N, how much runway will it need to reach its minimum takeoff speed of 285 km/h?

83. The dragster in problem 68 crossed the finish line going 126.6 m/s. Does the assumption of constant acceleration hold true? What other piece of evidence could you use to determine if the acceleration was constant?

84. Suppose a 65-kg boy and a 45-kg girl use a massless rope in a tug-of-war on an icy, resistance-free surface as in **Figure 4-21.** If the acceleration of the girl toward the boy is 3.0 m/s^2, find the magnitude of the acceleration of the boy toward the girl.

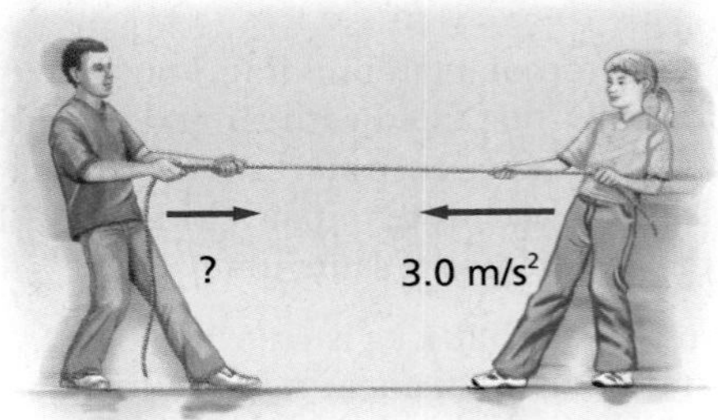

Figure 4-21

85. **Space Station** Pratish weighs 588 N and is weightless in a space station. If she pushes off the wall with a vertical acceleration of 3.00 m/s^2, determine the force exerted by the wall during her push off.

86. **Baseball** As a baseball is being caught, its speed goes from 30.0 m/s to 0.0 m/s in about 0.0050 s. The mass of the baseball is 0.145 kg.
   a. What is the baseball's acceleration?
   b. What are the magnitude and direction of the force acting on it?
   c. What are the magnitude and direction of the force acting on the player who caught it?

87. **Air Hockey** An air-hockey table works by pumping air through thousands of tiny holes in a table to support light pucks. This allows the pucks to move around on cushions of air with very little resistance. One of these pucks has a mass of 0.25 kg and is pushed along by a 12.0-N force for 9.0 s.
   a. What is the puck's acceleration?
   b. What is the puck's final velocity?

 For more problems, go to Additional Problems, Appendix B.

88. A student stands on a bathroom scale in an elevator at rest on the 64th floor of a building. The scale reads 836 N.
   a. As the elevator moves up, the scale reading increases to 936 N. Find the acceleration of the elevator.
   b. As the elevator approaches the 74th floor, the scale reading drops to 782 N. What is the acceleration of the elevator?
   c. Using your results from parts a and b, explain which change in velocity, starting or stopping, takes the longer time.

89. **Weather Balloon** The instruments attached to a weather balloon in **Figure 4-22** have a mass of 5.0 kg. The balloon is released and exerts an upward force of 98 N on the instruments.
   a. What is the acceleration of the balloon and instruments?
   b. After the balloon has accelerated for 10.0 s, the instruments are released. What is the velocity of the instruments at the moment of their release?
   c. What net force acts on the instruments after their release?
   d. When does the direction of the instruments' velocity first become downward?

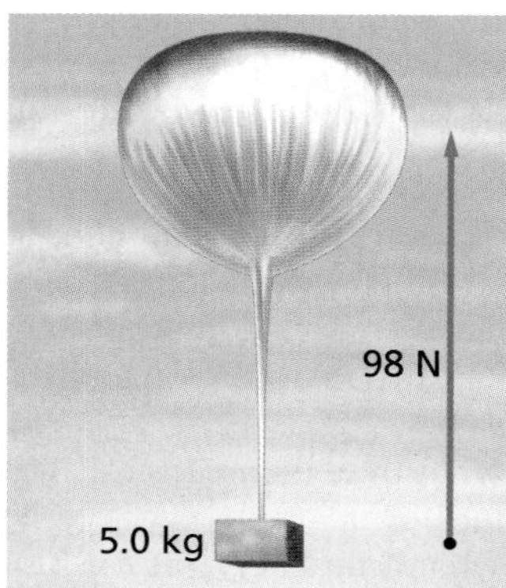

Figure 4-22

90. When a horizontal force of 4.5 N acts on a block on a resistance-free surface, it produces an acceleration of 2.5 m/s^2. Suppose a second 4.0-kg block is dropped onto the first. What is the magnitude of the acceleration of the combination if the same force continues to act? Assume that the second block does not slide on the first block.

91. Two blocks, masses 4.3 kg and 5.4 kg, are pushed across a frictionless surface by a horizontal force of 22.5 N, as shown in **Figure 4-23.**
   a. What is the acceleration of the blocks?
   b. What is the force of the 4.3-kg block on the 5.4-kg block?
   c. What is the force of the 5.4-kg block on the 4.3-kg block?

Figure 4-23

92. Two blocks, one of mass 5.0 kg and the other of mass 3.0 kg, are tied together with a massless rope as in **Figure 4-24.** This rope is strung over a massless, resistance-free pulley. The blocks are released from rest. Find the following.
   a. the tension in the rope
   b. the acceleration of the blocks

   Hint: you will need to solve two simultaneous equations.

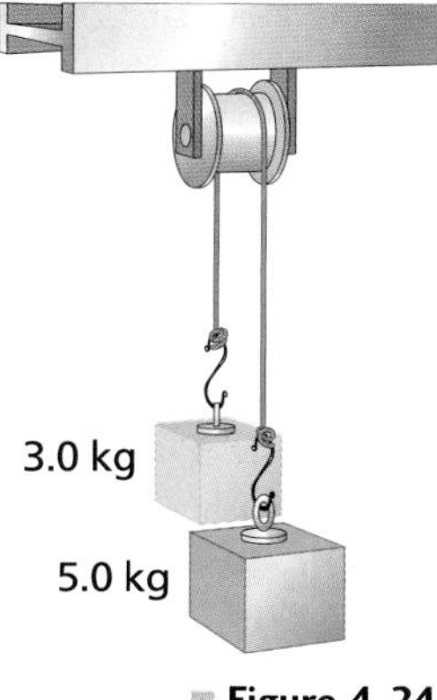

Figure 4-24

Thinking Critically

93. **Formulate Models** A 2.0-kg mass, m_A, and a 3.0-kg mass, m_B, are connected to a lightweight cord that passes over a frictionless pulley. The pulley only changes the direction of the force exerted by the rope. The hanging masses are free to move. Choose coordinate systems for the two masses with the positive direction being up for m_A and down for m_B.
   a. Create a pictorial model.
   b. Create a physical model with motion and free-body diagrams.
   c. What is the acceleration of the smaller mass?

94. **Use Models** Suppose that the masses in problem 93 are now 1.00 kg and 4.00 kg. Find the acceleration of the larger mass.

95. **Infer** The force exerted on a 0.145-kg baseball by a bat changes from 0.0 N to 1.0×10^4 N in 0.0010 s, then drops back to zero in the same amount of time. The baseball was going toward the bat at 25 m/s.
   a. Draw a graph of force versus time. What is the average force exerted on the ball by the bat?
   b. What is the acceleration of the ball?
   c. What is the final velocity of the ball, assuming that it reverses direction?

96. **Observe and Infer** Three blocks that are connected by massless strings are pulled along a frictionless surface by a horizontal force, as shown in **Figure 4-25.**
   a. What is the acceleration of each block?
   b. What are the tension forces in each of the strings?
   Hint: Draw a separate free-body diagram for each block.

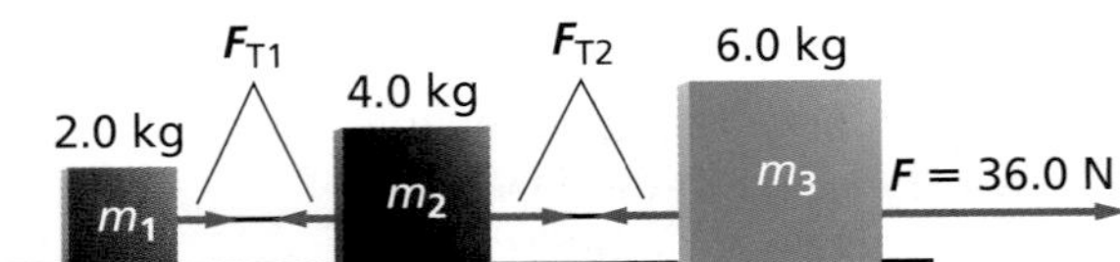

Figure 4-25

97. **Critique** Using the Example Problems in this chapter as models, write a solution to the following problem. A block of mass 3.46 kg is suspended from two vertical ropes attached to the ceiling. What is the tension in each rope?

98. **Think Critically** Because of your physics knowledge, you are serving as a scientific consultant for a new science-fiction TV series about space exploration. In episode 3, the heroine, Misty Moonglow, has been asked to be the first person to ride in a new interplanetary transport for use in our solar system. She wants to be sure that the transport actually takes her to the planet she is supposed to be going to, so she needs to take a testing device along with her to measure the force of gravity when she arrives. The script writers don't want her to just drop an object, because it will be hard to depict different accelerations of falling objects on TV. They think they'd like something involving a scale. It is your job to design a quick experiment Misty can conduct involving a scale to determine which planet in our solar system she has arrived on. Describe the experiment and include what the results would be for Pluto ($g = 0.30$ m/s^2), which is where she is supposed to go, and Mercury ($g = 3.70$ m/s^2), which is where she actually ends up.

99. **Apply Concepts** Develop a CBL lab, using a motion detector, that graphs the distance a free-falling object moves over equal intervals of time. Also graph velocity versus time. Compare and contrast your graphs. Using your velocity graph, determine the acceleration. Does it equal g?

Writing in Physics

100. Research Newton's contributions to physics and write a one-page summary. Do you think his three laws of motion were his greatest accomplishments? Explain why or why not.

101. Review, analyze, and critique Newton's first law. Can we prove this law? Explain. Be sure to consider the role of resistance.

102. Physicists classify all forces into four fundamental categories: gravitational, electromagnetic, strong nuclear, and weak nuclear. Investigate these four forces and describe the situations in which they are found.

Cumulative Review

103. **Cross-Country Skiing** Your friend is training for a cross-country skiing race, and you and some other friends have agreed to provide him with food and water along his training route. It is a bitterly cold day, so none of you wants to wait outside longer than you have to. Taro, whose house is the stop before yours, calls you at 8:25 A.M. to tell you that the skier just passed his house and is planning to move at an average speed of 8.0 km/h. If it is 5.2 km from Taro's house to yours, when should you expect the skier to pass your house? (Chapter 2)

104. **Figure 4-26** is a position-time graph of the motion of two cars on a road. (Chapter 3)
   a. At what time(s) does one car pass the other?
   b. Which car is moving faster at 7.0 s?
   c. At what time(s) do the cars have the same velocity?
   d. Over what time interval is car B speeding up all the time?
   e. Over what time interval is car B slowing down all the time?

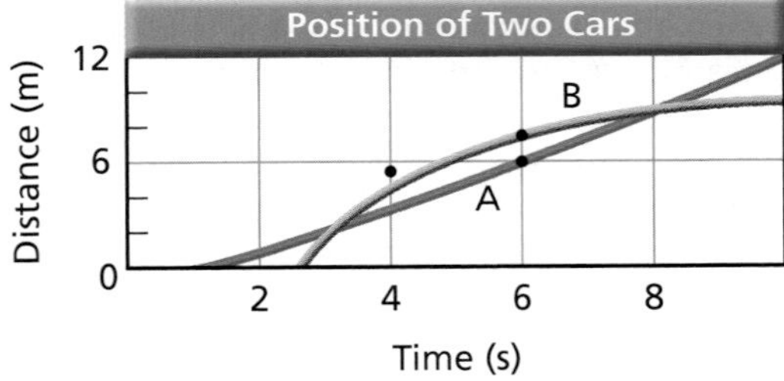

Figure 4-26

105. Refer to Figure 4-26 to find the instantaneous speed for the following: (Chapter 3)
   a. car B at 2.0 s
   b. car B at 9.0 s
   c. car A at 2.0 s

Multiple Choice

1. What is the acceleration of the car described by the graph below?

   (A) 0.20 m/s^2 (C) 1.0 m/s^2
   (B) 0.40 m/s^2 (D) 2.5 m/s^2

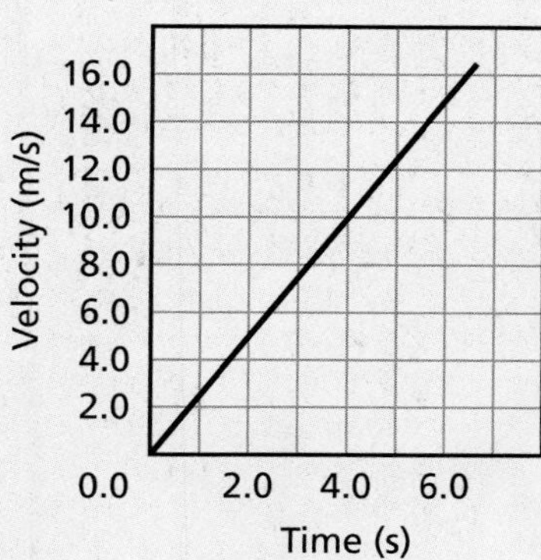

2. What distance will the car described by the above graph have traveled after 4.0 s?

   (A) 13 m (C) 80 m
   (B) 40 m (D) 90 m

3. If the car in the above graph maintains a constant acceleration, what will its velocity be after 10 s?

   (A) 10 km/h (C) 90 km/h
   (B) 25 km/h (D) 120 km/h

4. In a tug-of-war, 13 children, with an average mass of 30 kg each, pull westward on a rope with an average force of 150 N per child. Five parents, with an average mass of 60 kg each, pull eastward on the other end of the rope with an average force of 475 N per adult. Assuming that the whole mass accelerates together as a single entity, what is the acceleration of the system?

   (A) 0.62 m/s^2 E (C) 3.4 m/s^2 E
   (B) 2.8 m/s^2 W (D) 6.3 m/s^2 W

5. What is the weight of a 225-kg space probe on the Moon? The acceleration of gravity on the Moon is 1.62 m/s^2.

   (A) 139 N (C) 1.35×10^3 N
   (B) 364 N (D) 2.21×10^3 N

6. A 45-kg child sits on a 3.2-kg tire swing. What is the tension in the rope that hangs from a tree branch?

   (A) 310 N (C) 4.5×10^2 N
   (B) 4.4×10^2 N (D) 4.7×10^2 N

7. The tree branch in problem 6 sags and the child's feet rest on the ground. If the tension in the rope is reduced to 220 N, what is the value of the normal force being exerted on the child's feet?

   (A) 2.2×10^2 N (C) 4.3×10^2 N
   (B) 2.5×10^2 N (D) 6.9×10^2 N

8. According the graph below, what is the force being exerted on the 16-kg cart?

   (A) 4 N (C) 16 N
   (B) 8 N (D) 32 N

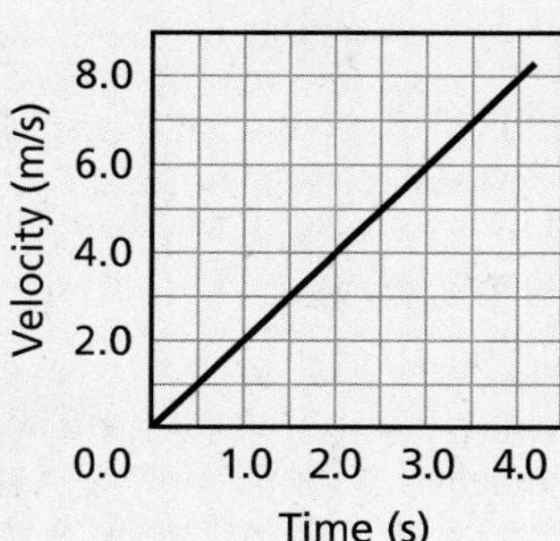

Extended Answer

9. Draw a free-body diagram of a dog sitting on a scale in an elevator. Using words and mathematical formulas, describe what happens to the apparent weight of the dog when: the elevator accelerates upward, the elevator travels at a constant speed downward, and the elevator falls freely downward.

✓Test-Taking TIP

Maximize Your Score

If possible, find out how your standardized test will be scored. In order to do your best, you need to know if there is a penalty for guessing, and if so, what the penalty is. If there is no random-guessing penalty at all, you should always fill in an answer, even if you have not read the question.

Chapter 5 Forces in Two Dimensions

What You'll Learn

- You will represent vector quantities both graphically and algebraically.
- You will use Newton's laws to analyze motion when friction is involved.
- You will use Newton's laws and your knowledge of vectors to analyze motion in two dimensions.

Why It's Important

Most objects experience forces in more than one dimension. A car being towed, for example, experiences upward and forward forces from the tow truck and the downward force of gravity.

Rock Climbing How do rock climbers keep from falling? This climber has more than one support point, and there are multiple forces acting on her in multiple directions.

Think About This ▶

A rock climber approaches a portion of the rock face that forces her to hang with her back to the ground. How will she use her equipment to apply the laws of physics in her favor and overcome this obstacle?

physicspp.com

Can 2 N + 2 N = 2 N?

Question

Under what conditions can two different forces equal one other force?

Procedure

1. **Measure** Use a spring scale to measure and record the weight of a 200-g object.
2. Obtain another spring scale, and attach one end of a 35-cm-long piece of string to the hooks on the bottom of each spring scale.
3. Tie one end of a 15-cm-long piece of string to the 200-g object. Loop the other end over the 35-cm-long piece of string and tie the end to the 200-g object. ***CAUTION: Avoid falling masses.***
4. Hold the spring scales parallel to each other so that the string between them forms a 120° angle. Move the string with the hanging object until both scales have the same reading. Record the readings on each scale.
5. **Collect and Organize Data** Slowly pull the string more and more horizontal while it is still supporting the 200-g object. Describe your observations.

Analysis

Does the sum of the forces measured by the two spring scales equal the weight of the hanging object? Is the sum greater than the weight? Less than the weight?

Critical Thinking Draw an equilateral triangle, with one side vertical, on a sheet of paper. If the two sides of the triangle are 2.0 N, explain the size of the third side. How is it possible that 2 N + 2 N = 2 N?

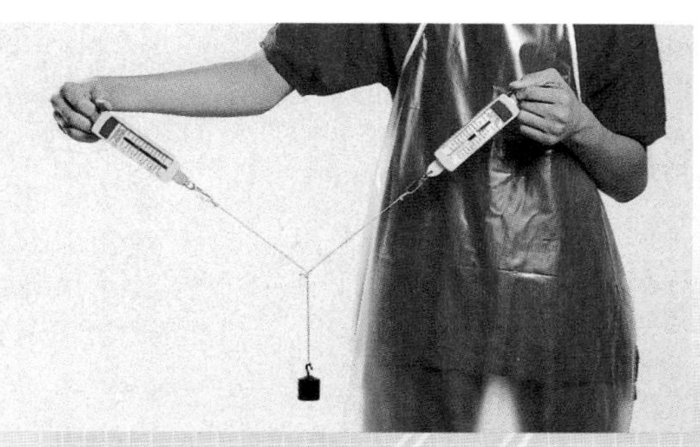

5.1 Vectors

How do rock climbers keep from falling in situations like the one shown on the preceding page? Notice that the climber has more than one support point and that there are multiple forces acting on her. She tightly grips crevices in the rock and has her feet planted on the rock face, so there are two contact forces acting on her. Gravity is pulling on her as well, so there are three total forces acting on the climber. One aspect of this situation that is different from the ones that you have studied in earlier chapters is that the forces exerted by the rock face on the climber are not horizontal or vertical forces. You know from previous chapters that you can pick your coordinate system and orient it in the way that is most useful to analyzing the situation. But what happens when the forces are not at right angles to each other? How can you set up a coordinate system and find for a net force when you are dealing with more than one dimension?

▶ **Objectives**

- **Evaluate** the sum of two or more vectors in two dimensions graphically.
- **Determine** the components of vectors.
- **Solve** for the sum of two or more vectors algebraically by adding the components of the vectors.

▶ **Vocabulary**

components
vector resolution

Vectors Revisited

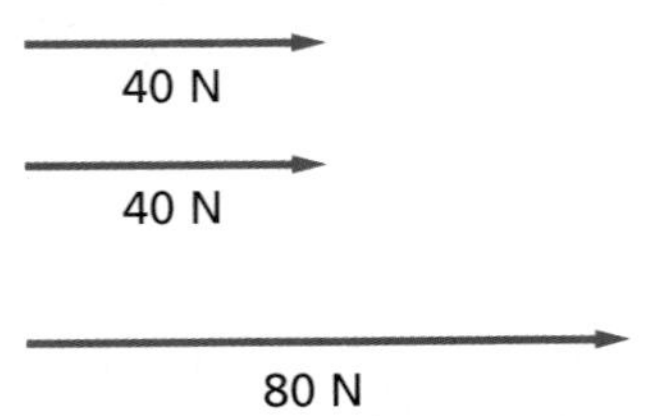

■ **Figure 5-1** The sum of the two 40-N forces is shown by the resultant vector below them.

Consider an example with force vectors. Recall the case in Chapter 4 in which you and a friend both pushed on a table together. Suppose that you each exerted 40 N of force to the right. **Figure 5-1** represents these vectors in a free-body diagram with the resultant vector, the net force, shown below it. The net force vector is 80 N, which is what you probably expected. But how was this net force vector obtained?

Vectors in Multiple Dimensions

The process for adding vectors works even when the vectors do not point along the same straight line. If you are solving one of these two-dimensional problems graphically, you will need to use a protractor, both to draw the vectors at the correct angles and also to measure the direction and magnitude of the resultant vector. You can add vectors by placing them tip-to-tail and then drawing the resultant of the vector by connecting the tail of the first vector to the tip of the second vector, as shown in **Figure 5-2**. **Figure 5-2a** shows the two forces in the free-body diagram. In **Figure 5-2b,** one of the vectors has been moved so that its tail is at the same place as the tip of the other vector. Notice that its length and direction have not changed. Because the length and direction are the only important characteristics of the vector, the vector is unchanged by this movement. This is always true: if you move a vector so that its length and direction are unchanged, the vector is unchanged. Now, as in **Figure 5-2c,** you can draw the resultant vector pointing from the tail of the first vector to the tip of the last vector and measure it to obtain its magnitude. Use a protractor to measure the direction of the resultant vector. Sometimes you will need to use trigonometry to determine the length or direction of resultant vectors. Remember that the length of the hypotenuse of a right triangle can be found by using the Pythagorean theorem. If you were adding together two vectors at right angles, vector ***A*** pointing north and vector ***B*** pointing east, you could use the Pythagorean theorem to find the magnitude of the resultant, *R*.

■ **Figure 5-2** Add vectors by placing them tip-to-tail and drawing the resultant from the tail of the first vector to the tip of the last vector.

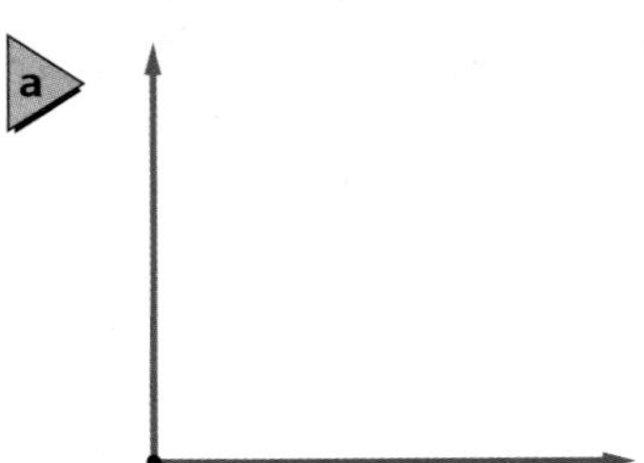

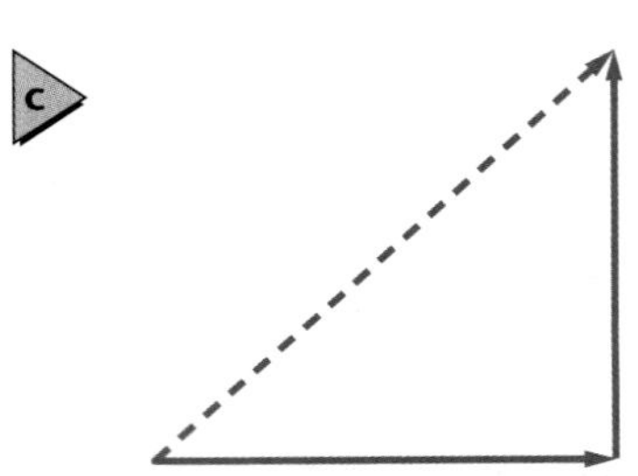

Pythagorean Theorem $R^2 = A^2 + B^2$

If vector *A* is at a right angle to vector *B*, then the sum of the squares of the magnitudes is equal to the square of the magnitude of the resultant vector.

If the two vectors to be added are at an angle other than 90°, then you can use the law of cosines or the law of sines.

Law of Cosines $R^2 = A^2 + B^2 - 2AB \cos \theta$

The square of the magnitude of the resultant vector is equal to the sum of the magnitudes of the squares of the two vectors, minus two times the product of the magnitudes of the vectors, multiplied by the cosine of the angle between them.

Law of Sines $\frac{R}{\sin \theta} = \frac{A}{\sin a} = \frac{B}{\sin b}$

The magnitude of the resultant, divided by the sine of the angle between two vectors, is equal to the magnitude of one of the vectors divided by the angle between that component vector and the resultant vector.

EXAMPLE Problem 1

Finding the Magnitude of the Sum of Two Vectors **Find the magnitude of the sum of a 15-km displacement and a 25-km displacement when the angle between them is 90° and when the angle between them is 135°.**

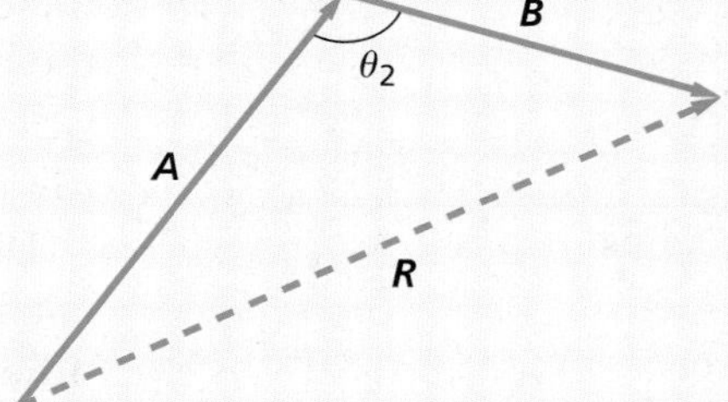

1 Analyze and Sketch the Problem

- Sketch the two displacement vectors, ***A*** and ***B***, and the angle between them.

Known:		Unknown:
$A = 25$ km	$\theta_1 = 90°$	$R = ?$
$B = 15$ km	$\theta_2 = 135°$	

2 Solve for the Unknown

When the angle is 90°, use the Pythagorean theorem to find the magnitude of the resultant vector.

$$R^2 = A^2 + B^2$$

$$R = \sqrt{A^2 + B^2}$$

$$= \sqrt{(25\ \text{km})^2 + (15\ \text{km})^2}$$ **Substitute $A = 25$ km, $B = 15$ km**

$$= 29\ \text{km}$$

Math Handbook
Square and Cube Roots
pages 839–840

When the angle does not equal 90°, use the law of cosines to find the magnitude of the resultant vector.

$$R^2 = A^2 + B^2 - 2AB(\cos \theta_2)$$

$$R = \sqrt{A^2 + B^2 - 2AB(\cos \theta_2)}$$

$$= \sqrt{(25\ \text{km})^2 + (15\ \text{km})^2 - 2(25\ \text{km})(15\ \text{km})(\cos 135°)}$$ **Substitute $A = 25$ km, $B = 15$ km, $\theta_2 = 135°$**

$$= 37\ \text{km}$$

3 Evaluate the Answer

- **Are the units correct?** Each answer is a length measured in kilometers.
- **Do the signs make sense?** The sums are positive.
- **Are the magnitudes realistic?** The magnitudes are in the same range as the two combined vectors, but longer. This is because each resultant is the side opposite an obtuse angle. The second answer is larger than the first, which agrees with the graphical representation.

PRACTICE Problems

- **Additional Problems, Appendix B**
- **Solutions to Selected Problems, Appendix C**

1. A car is driven 125.0 km due west, then 65.0 km due south. What is the magnitude of its displacement? Solve this problem both graphically and mathematically, and check your answers against each other.
2. Two shoppers walk from the door of the mall to their car, which is 250.0 m down a lane of cars, and then turn 90° to the right and walk an additional 60.0 m. What is the magnitude of the displacement of the shoppers' car from the mall door? Solve this problem both graphically and mathematically, and check your answers against each other.
3. A hiker walks 4.5 km in one direction, then makes a 45° turn to the right and walks another 6.4 km. What is the magnitude of her displacement?
4. An ant is crawling on the sidewalk. At one moment, it is moving south a distance of 5.0 mm. It then turns southwest and crawls 4.0 mm. What is the magnitude of the ant's displacement?

Components of Vectors

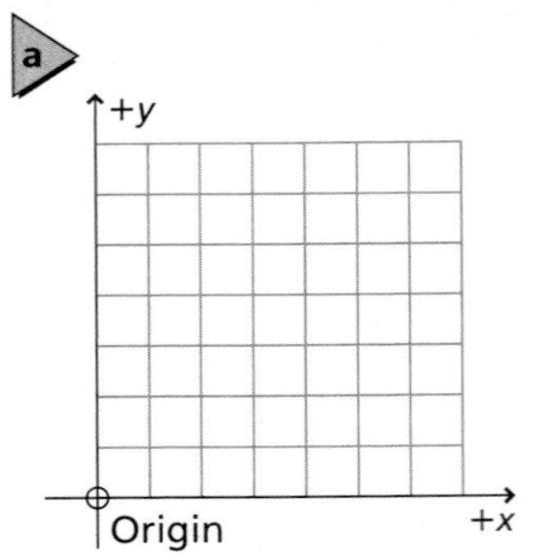

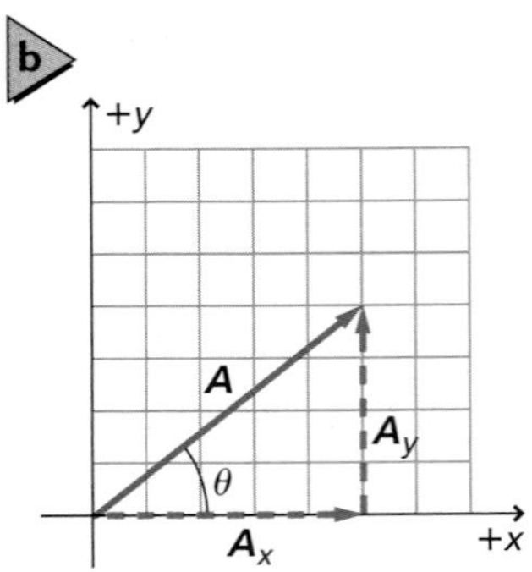

Figure 5-3 A coordinate system has an origin and two perpendicular axes **(a)**. The direction of a vector, ***A***, is measured counterclockwise from the *x*-axis **(b)**.

Choosing a coordinate system, such as the one in **Figure 5-3a,** is similar to laying a grid drawn on a sheet of transparent plastic on top of a vector problem. You have to choose where to put the center of the grid (the origin) and establish the directions in which the axes point. Notice that in the coordinate system shown in Figure 5-3a, the *x*-axis is drawn through the origin with an arrow pointing in the positive direction. The positive *y*-axis is located 90° counterclockwise from the positive *x*-axis and crosses the *x*-axis at the origin.

How do you choose the direction of the *x*-axis? There is never a single correct answer, but some choices make the problem easier to solve than others. When the motion you are describing is confined to the surface of Earth, it is often convenient to have the *x*-axis point east and the *y*-axis point north. When the motion involves an object moving through the air, the positive *x*-axis is often chosen to be horizontal and the positive *y*-axis vertical (upward). If the motion is on a hill, it's convenient to place the positive *x*-axis in the direction of the motion and the *y*-axis perpendicular to the *x*-axis.

Component vectors Defining a coordinate system allows you to describe a vector in a different way. Vector **A** shown in **Figure 5-3b,** for example, could be described as going 5 units in the positive *x*-direction and 4 units in the positive *y*-direction. You can represent this information in the form of two vectors like the ones labeled $\mathbf{A}_x$ and $\mathbf{A}_y$ in the diagram. Notice that $\mathbf{A}_x$ is parallel to the *x*-axis, and $\mathbf{A}_y$ is parallel to the *y*-axis. Further, you can see that if you add $\mathbf{A}_x$ and $\mathbf{A}_y$, the resultant is the original vector, **A**. A vector can be broken into its **components,** which are a vector parallel to the *x*-axis and another parallel to the *y*-axis. This can always be done and the following vector equation is always true.

$$\mathbf{A} = \mathbf{A}_x + \mathbf{A}_y$$

This process of breaking a vector into its components is sometimes called **vector resolution.** Notice that the original vector is the hypotenuse of a right triangle. This means that the magnitude of the original vector will always be larger than the magnitudes of either component vector.

Another reason for choosing a coordinate system is that the direction of any vector can be specified relative to those coordinates. The direction of a vector is defined as the angle that the vector makes with the *x*-axis, measured counterclockwise. In Figure 5-3b, the angle, θ, tells the direction of the vector, **A**. All algebraic calculations involve only the positive components of vectors, not the vectors themselves. In addition to measuring the lengths of the component vectors graphically, you can find the components by using trigonometry. The components are calculated using the equations below, where the angle, θ, is measured counterclockwise from the positive *x*-axis.

$$\cos\theta = \frac{\text{adjacent side}}{\text{hypotenuse}} = \frac{A_x}{A}; \text{ therefore, } A_x = A\cos\theta$$

$$\sin\theta = \frac{\text{opposite side}}{\text{hypotenuse}} = \frac{A_y}{A}; \text{ therefore, } A_y = A\sin\theta$$

When the angle that a vector makes with the *x*-axis is larger than 90°, the sign of one or more components is negative, as shown in **Figure 5-4.**

Figure 5-4 The sign of a component depends upon which of the four quadrants the component is in.

Second quadrant
$A_x < 0$
$A_y > 0$

+y

First quadrant
$A_x > 0$
$A_y > 0$

+x

$A_x < 0$
$A_y < 0$
Third quadrant

$A_x > 0$
$A_y < 0$
Fourth quadrant

Algebraic Addition of Vectors

You might be wondering why you need to resolve vectors into their components. The answer is that doing this often makes adding vectors together much easier mathematically. Two or more vectors (***A***, ***B***, ***C***, etc.) may be added by first resolving each vector into its *x*- and *y*-components. The *x*-components are added to form the *x*-component of the resultant: $R_x = A_x + B_x + C_x$. Similarly, the *y*-components are added to form the *y*-component of the resultant: $R_y = A_y + B_y + C_y$. This process is illustrated graphically in **Figure 5-5.** Because R_x and R_y are at a right angle (90°), the magnitude of the resultant vector can be calculated using the Pythagorean theorem, $R^2 = R_x^2 + R_y^2$. To find the angle or direction of the resultant, recall that the tangent of the angle that the vector makes with the *x*-axis is given by the following.

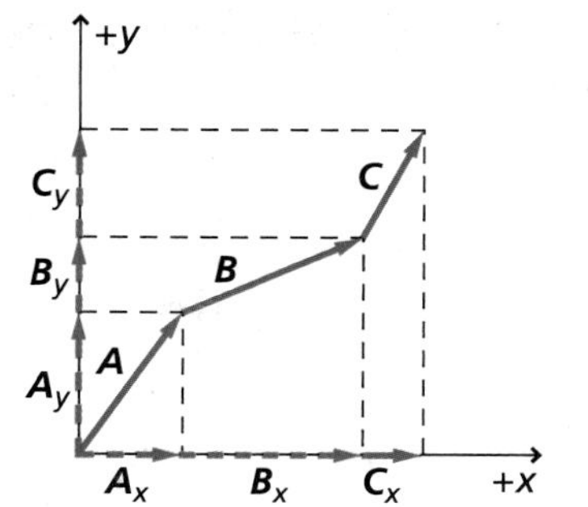

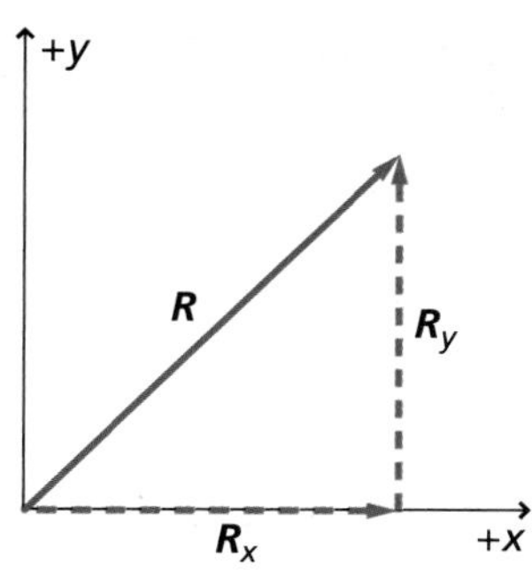

Figure 5-5 $\boldsymbol{R_x}$ is the sum of the *x*-components of ***A***, ***B***, and ***C***. $\boldsymbol{R_y}$ is the sum of the *y*-components. The vector sum of $\boldsymbol{R_x}$ and $\boldsymbol{R_y}$ is the vector sum of ***A***, ***B***, and ***C***.

Angle of the Resultant Vector $\theta = \tan^{-1}\left(\frac{R_y}{R_x}\right)$

The angle of the resultant vector is equal to the inverse tangent of the quotient of the *y*-component divided by the *x*-component of the resultant vector.

You can find the angle by using the $\tan^{-1}$ key on your calculator. Note that when $\tan\theta > 0$, most calculators give the angle between 0° and 90°, and when $\tan\theta < 0$, the angle is reported to be between 0° and −90°.

PROBLEM-SOLVING **Strategies**

Vector Addition

Use the following technique to solve problems for which you need to add or subtract vectors.

1. Choose a coordinate system.
2. Resolve the vectors into their *x*-components using $A_x = A\cos\theta$, and their *y*-components using $A_y = A\sin\theta$, where θ is the angle measured counterclockwise from the positive *x*-axis.
3. Add or subtract the component vectors in the *x*-direction.
4. Add or subtract the component vectors in the *y*-direction.
5. Use the Pythagorean theorem, $R = \sqrt{R_x^2 + R_y^2}$, to find the magnitude of the resultant vector.
6. Use $\theta = \tan^{-1}\left(\frac{R_y}{R_x}\right)$ to find the angle of the resultant vector.

Connecting **Math to Physics**

Math Review

$\sin\theta = \frac{\text{opposite side}}{\text{hypotenuse}} = \frac{R_y}{R}$

$\cos\theta = \frac{\text{adjacent side}}{\text{hypotenuse}} = \frac{R_x}{R}$

$\tan\theta = \frac{\text{opposite side}}{\text{adjacent side}} = \frac{R_y}{R_x}$

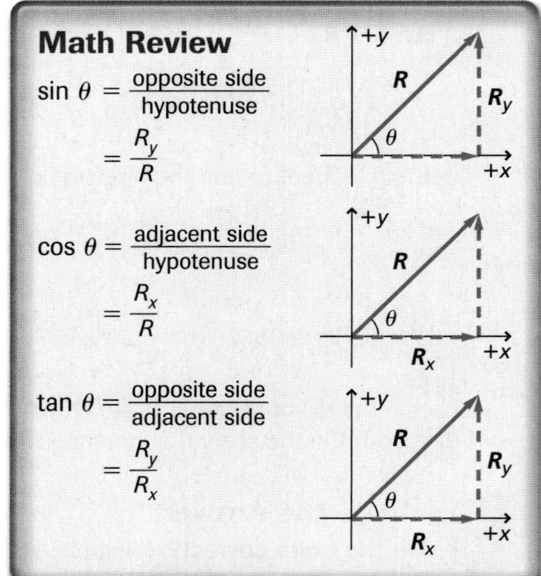

▶ EXAMPLE Problem 2

Finding Your Way Home A GPS receiver indicates that your home is 15.0 km and 40.0° north of west, but the only path through the woods leads directly north. If you follow the path 5.0 km before it opens into a field, how far, and in what direction, would you have to walk to reach your home?

1 Analyze and Sketch the Problem

- Draw the resultant vector, ***R***, from your original location to your home.
- Draw ***A***, the known vector, and draw ***B***, the unknown vector.

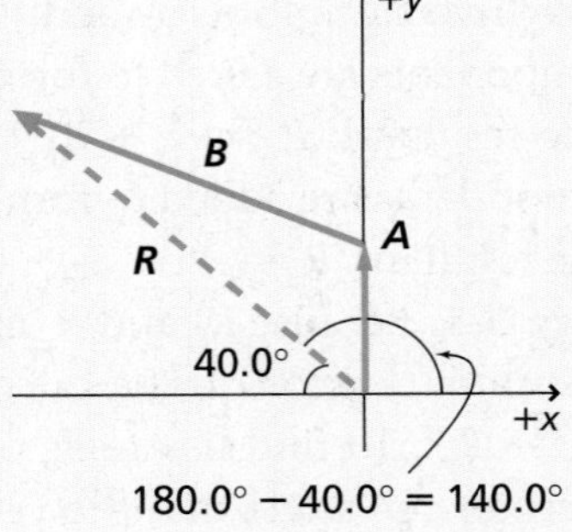

Known:
A = 5.0 km, due north
R = 15.0 km, 40.0° north of west
$\theta = 140.0°$

Unknown:
B = ?

2 Solve for the Unknown

Find the components of ***R***.

$$R_x = R\cos\theta$$
$$= (15.0\text{ km})\cos 140.0°$$ Substitute $R = 15.0$ km, $\theta = 140.0°$
$$= -11.5\text{ km}$$
$$R_y = R\sin\theta$$
$$= (15.0\text{ km})\sin 140.0°$$ Substitute $R = 15.0$ km, $\theta = 140.0°$
$$= 9.64\text{ km}$$

Because ***A*** is due north, $A_x = 0.0$ km and $A_y = 5.0$ km.

Math Handbook
Inverses of Sine, Cosine, and Tangent
page 856

Use the components of ***R*** and ***A*** to find the components of ***B***.

$$B_x = R_x - A_x$$
$$= -11.5\text{ km} - 0.0\text{ km}$$ Substitute $R_x = -11.5$ km, $A_x = 0.0$ km
$$= -11.5\text{ km}$$ The negative sign means that this component points west.
$$B_y = R_y - A_y$$
$$= 9.64\text{ km} - 5.0\text{ km}$$ Substitute $R_y = 9.64$ km, $A_y = 5.0$ km
$$= 4.6\text{ km}$$ This component points north.

Use the components of vector ***B*** to find the magnitude of vector ***B***.

$$B = \sqrt{B_x^2 + B_y^2}$$
$$= \sqrt{(-11.5\text{ km})^2 + (4.6\text{ km})^2}$$ Substitute $B_y = -4.6$ km, $B_x = -11.5$ km
$$= 12\text{ km}$$

Use the tangent to find the direction of vector ***B***.

$$\theta = \tan^{-1}\frac{B_y}{B_x}$$
$$= \tan^{-1}\frac{4.6\text{ km}}{-11.5\text{ km}}$$ Substitute $B_y = 4.6$ km, $B_x = -11.5$ km
$$= -22° \text{ or } 158°$$ Tangent of an angle is negative in quadrants II and IV, so two answers are possible.

Locate the tail of vector ***B*** at the origin of a coordinate system and draw the components B_x and B_y. The direction is in the second quadrant, at 158°, or 22° north of west. Thus, ***B*** = 12 km at 22° north of west.

3 Evaluate the Answer

- **Are the units correct?** Kilometers and degrees are correct.
- **Do the signs make sense?** They agree with the diagram.
- **Is the magnitude realistic?** The length of ***B*** should be longer than R_x because the angle between *A* and *B* is greater than 90°.

PRACTICE Problems

- Additional Problems, Appendix B
- Solutions to Selected Problems, Appendix C

Solve problems 5–10 algebraically. You may also choose to solve some of them graphically to check your answers.

5. Sudhir walks 0.40 km in a direction 60.0° west of north, then goes 0.50 km due west. What is his displacement?

6. Afua and Chrissy are going to sleep overnight in their tree house and are using some ropes to pull up a box containing their pillows and blankets, which have a total mass of 3.20 kg. The girls stand on different branches, as shown in **Figure 5-6,** and pull at the angles and with the forces indicated. Find the *x*- and *y*-components of the net force on the box. *Hint: Draw a free-body diagram so that you do not leave out a force.*

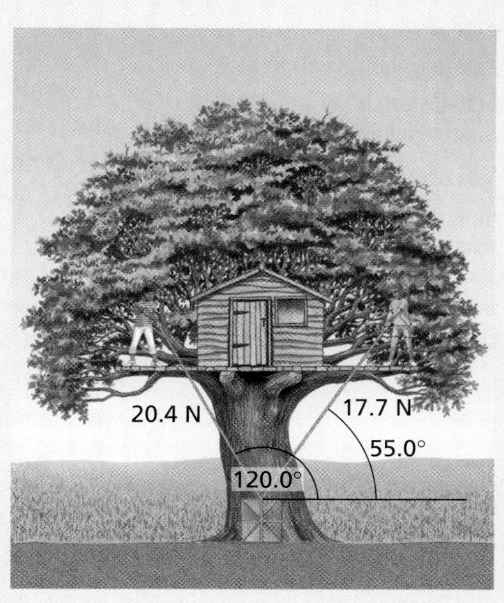

Figure 5-6
(Not to scale)

7. You first walk 8.0 km north from home, then walk east until your displacement from home is 10.0 km. How far east did you walk?

8. A child's swing is held up by two ropes tied to a tree branch that hangs 13.0° from the horizontal. If the tension in each rope is 2.28 N, what is the combined force (magnitude and direction) of the two ropes on the swing?

9. Could a vector ever be shorter than one of its components? Equal in length to one of its components? Explain.

10. In a coordinate system in which the *x*-axis is east, for what range of angles is the *x*-component positive? For what range is it negative?

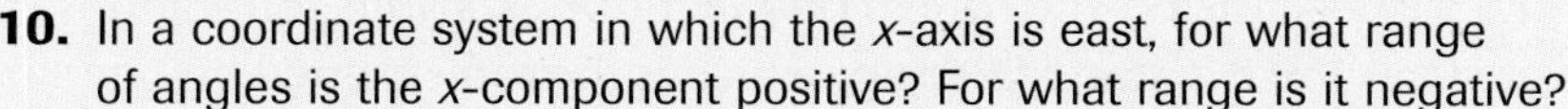

You will use these techniques to resolve vectors into their components throughout your study of physics. You will get more practice at it, particularly in the rest of this chapter and the next. Resolving vectors into components allows you to analyze complex systems of vectors without using graphical methods.

5.1 Section Review

11. Distance v. Displacement Is the distance that you walk equal to the magnitude of your displacement? Give an example that supports your conclusion.

12. Vector Difference Subtract vector ***K*** from vector ***L***, shown in **Figure 5-7.**

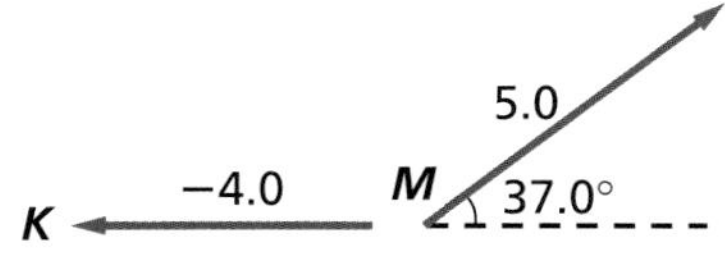

Figure 5-7

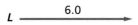

13. Components Find the components of vector ***M***, shown in Figure 5-7.

14. Vector Sum Find the sum of the three vectors shown in Figure 5-7.

15. Commutative Operations The order in which vectors are added does not matter. Mathematicians say that vector addition is commutative. Which ordinary arithmetic operations are commutative? Which are not?

16. Critical Thinking A box is moved through one displacement and then through a second displacement. The magnitudes of the two displacements are unequal. Could the displacements have directions such that the resultant displacement is zero? Suppose the box was moved through three displacements of unequal magnitude. Could the resultant displacement be zero? Support your conclusion with a diagram.

5.2 Friction

- **Objectives**
  - **Define** the friction force.
  - **Distinguish** between static and kinetic friction.
- **Vocabulary**
  - kinetic friction
  - static friction
  - coefficient of kinetic friction
  - coefficient of static friction

Push your hand across your desktop and feel the force called friction opposing the motion. Push your book across the desk. When you stop pushing, the book will continue moving for a little while, then it will slow down and stop. The frictional force acting on the book gave it an acceleration in the direction opposite to the one in which it was moving. So far, you have neglected friction in solving problems, but friction is all around you. You need it to both start and stop a bicycle and a car. If you have ever walked on ice, you understand the importance of friction.

Static and Kinetic Friction

There are two types of friction. Both always oppose motion. When you pushed your book across the desk, it experienced a type of friction that acts on moving objects. This force is known as **kinetic friction,** and it is exerted on one surface by another when the two surfaces rub against each other because one or both of them are moving.

To understand the other kind of friction, imagine trying to push a heavy couch across the floor. You give it a push, but it does not move. Because it does not move, Newton's laws tell you that there must be a second horizontal force acting on the couch, one that opposes your force and is equal in size. This force is **static friction,** which is the force exerted on one surface by another when there is no motion between the two surfaces. You might push harder and harder, as shown in **Figures 5-8a** and **5-8b,** but if the couch still does not move, the force of friction must be getting larger. This is because the static friction force acts in response to other forces. Finally, when you push hard enough, as shown in **Figure 5-8c,** the couch will begin to move. Evidently, there is a limit to how large the static friction force can be. Once your force is greater than this maximum static friction, the couch begins moving and kinetic friction begins to act on it instead of static friction.

A model for friction forces On what does a frictional force depend? The materials that the surfaces are made of play a role. For example, there is more friction between skis and concrete than there is between skis and snow. It may seem reasonable to think that the force of friction also might depend on either the surface area in contact or the speed of the motion, but experiments have shown that this is not true. The normal force between the two objects does matter, however. The harder one object is pushed against the other, the greater the force of friction that results.

Figure 5-8 There is a limit to the ability of the static friction force to match the applied force.

If you pull a block along a surface at a constant velocity, according to Newton's laws, the frictional force must be equal and opposite to the force with which you pull. You can pull a block of known mass along a table at a constant velocity and use a spring scale, as shown in **Figure 5-9,** to measure the force that you exert. You can then stack additional blocks on the block to increase the normal force and repeat the measurement.

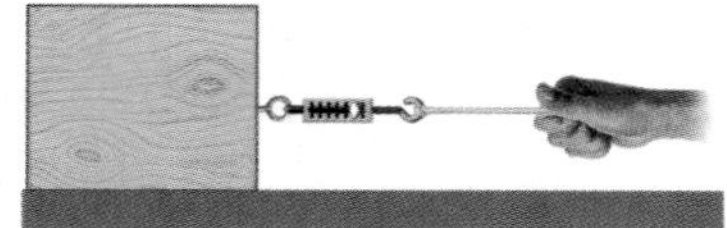

Figure 5-9 The spring scale pulls the block with a constant force.

Plotting the data will yield a graph like the one in **Figure 5-10.** There is a direct proportion between the kinetic friction force and the normal force. The different lines correspond to dragging the block along different surfaces. Note that the line corresponding to the sandpaper surface has a steeper slope than the line for the highly polished table. You would expect it to be much harder to pull the block along sandpaper than along a polished table, so the slope must be related to the magnitude of the resulting frictional force. The slope of this line, designated μ_k, is called the **coefficient of kinetic friction** between the two surfaces and relates the frictional force to the normal force, as shown below.

Kinetic Friction Force $F_{f,\ kinetic} = \mu_k F_N$

The kinetic friction force is equal to the product of the coefficient of the kinetic friction and the normal force.

The maximum static friction force is related to the normal force in a similar way as the kinetic friction force. Remember that the static friction force acts in response to a force trying to cause a stationary object to start moving. If there is no such force acting on an object, the static friction force is zero. If there is a force trying to cause motion, the static friction force will increase up to a maximum value before it is overcome and motion starts.

Static Friction Force $F_{f,\ static} \leq \mu_s F_N$

The static friction force is less than or equal to the product of the coefficient of the static friction and the normal force.

In the equation for the maximum static friction force, μ_s is the **coefficient of static friction** between the two surfaces, and $\mu_s F_N$ is the maximum static friction force that must be overcome before motion can begin. In Figure 5-8c, the static friction force is balanced the instant before the couch begins to move.

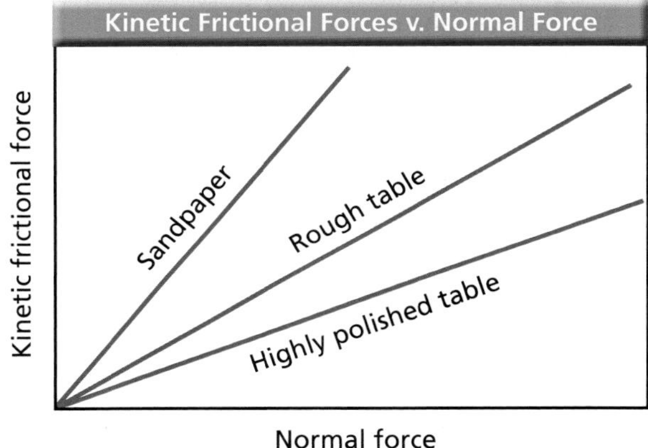

Figure 5-10 There is a linear relationship between the frictional force and the normal force.

▶ EXAMPLE Problem 3

Balanced Friction Forces You push a 25.0-kg wooden box across a wooden floor at a constant speed of 1.0 m/s. How much force do you exert on the box?

1 Analyze and Sketch the Problem

- Identify the forces and establish a coordinate system.
- Draw a motion diagram indicating constant v and $a = 0$.
- Draw the free-body diagram.

Known:	Unknown:
$m = 25.0$ kg	$F_p = ?$
$v = 1.0$ m/s	
$a = 0.0$ m/s^2	
$\mu_k = 0.20$ (Table 5-1)	

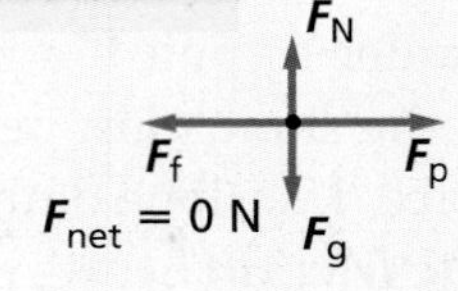

2 Solve for the Unknown

The normal force is in the y-direction, and there is no acceleration.

$$F_N = -F_g$$
$$= -mg \quad \text{Substitute } mg = F_g$$
$$= -(25.0 \text{ kg})(-9.80 \text{m/s}^2) \quad \text{Substitute } m = 25.0 \text{ kg}, g = -9.80 \text{ m/s}^2$$
$$= +245 \text{ N}$$

Math Handbook
Operations with Significant Digits
pages 835–836

The pushing force is in the x-direction; v is constant, thus there is no acceleration.

$$F_p = \mu_k mg$$
$$= (0.20)(25.0 \text{ kg})(9.80 \text{ m/s}^2) \quad \text{Substitute } \mu_k = 0.20, m = 25.0 \text{ kg}, g = 9.80 \text{ m/s}^2$$
$$= 49 \text{ N}$$
$$\mathbf{F}_p = +49 \text{ N, to the right}$$

3 Evaluate the Answer

- **Are the units correct?** Force is measured in kg·m/s^2 or N.
- **Does the sign make sense?** The positive sign agrees with the sketch.
- **Is the magnitude realistic?** The force is reasonable for moving a 25.0-kg box.

▶ PRACTICE Problems

- Additional Problems, Appendix B
- Solutions to Selected Problems, Appendix C

17. A girl exerts a 36-N horizontal force as she pulls a 52-N sled across a cement sidewalk at constant speed. What is the coefficient of kinetic friction between the sidewalk and the metal sled runners? Ignore air resistance.

18. You need to move a 105-kg sofa to a different location in the room. It takes a force of 102 N to start it moving. What is the coefficient of static friction between the sofa and the carpet?

19. Mr. Ames is dragging a box full of books from his office to his car. The box and books together have a combined weight of 134 N. If the coefficient of static friction between the pavement and the box is 0.55, how hard must Mr. Ames push the box in order to start it moving?

20. Suppose that the sled in problem 17 is resting on packed snow. The coefficient of kinetic friction is now only 0.12. If a person weighing 650 N sits on the sled, what force is needed to pull the sled across the snow at constant speed?

21. Suppose that a particular machine in a factory has two steel pieces that must rub against each other at a constant speed. Before either piece of steel has been treated to reduce friction, the force necessary to get them to perform properly is 5.8 N. After the pieces have been treated with oil, what will be the required force?

Table 5-1		
Typical Coefficients of Friction		
Surface	μ_s	μ_k
Rubber on dry concrete	0.80	0.65
Rubber on wet concrete	0.60	0.40
Wood on wood	0.50	0.20
Steel on steel (dry)	0.78	0.58
Steel on steel (with oil)	0.15	0.06

Note that the equations for the kinetic and maximum static friction forces involve only the magnitudes of the forces. The forces themselves, $\mathbf{F}_f$ and $\mathbf{F}_N$, are at right angles to each other. **Table 5-1** shows coefficients of friction between various surfaces. Although all the listed coefficients are less than 1.0, this does not mean that they must always be less than 1.0. For example, coefficients as large as 5.0 are experienced in drag racing.

▶ EXAMPLE **Problem 4**

Unbalanced Friction Forces If the force that you exert on the 25.0-kg box in Example Problem 3 is doubled, what is the resulting acceleration of the box?

1 Analyze and Sketch the Problem

- Draw a motion diagram showing v and a.
- Draw the free-body diagram with a doubled $\mathbf{F}_p$.

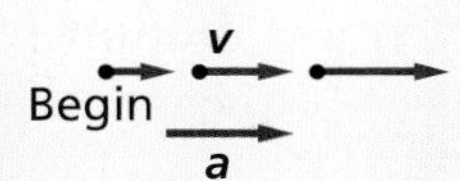

Known:

$m = 25.0$ kg $\quad \mu_k = 0.20$

$v = 1.0$ m/s $\quad F_p = 2(49\text{ N}) = 98$ N

Unknown:

$a = ?$

2 Solve for the Unknown

The normal force is in the y-direction, and there is no acceleration.

$$F_N = F_g$$
$$= mg$$

Substitute $F_g = mg$

In the x-direction there is an acceleration. So the forces must be unequal.

$$F_{net} = F_p - F_f$$
$$ma = F_p - F_f$$

Substitute $F_{net} = ma$

$$a = \frac{F_p - F_f}{m}$$

Math Handbook
Isolating a Variable
page 845

Find F_f and substitute it into the expression for a.

$$F_f = \mu_k F_N$$
$$= \mu_k mg$$

Substitute $F_N = mg$

$$a = \frac{F_p - \mu_k mg}{m}$$

Substitute $F_f = \mu_k mg$

$$= \frac{98\text{ N} - (0.20)(25.0\text{ kg})(9.80\text{ m/s}^2)}{25.0\text{ kg}}$$

Substitute $F_p = 98$ N, $m = 25.0$ kg, $\mu_k = 0.20$, $g = 9.80$ m/s^2

$$= 2.0\text{ m/s}^2$$

3 Evaluate the Answer

- **Are the units correct?** a is measured in m/s^2.
- **Does the sign make sense?** In this coordinate system, the sign should be positive.
- **Is the magnitude realistic?** If the force were cut in half, a would be zero.

PRACTICE Problems

- Additional Problems, Appendix B
- Solutions to Selected Problems, Appendix C

22. A 1.4-kg block slides across a rough surface such that it slows down with an acceleration of 1.25 m/s^2. What is the coefficient of kinetic friction between the block and the surface?

23. You help your mom move a 41-kg bookcase to a different place in the living room. If you push with a force of 65 N and the bookcase accelerates at 0.12 m/s^2, what is the coefficient of kinetic friction between the bookcase and the carpet?

24. A shuffleboard disk is accelerated to a speed of 5.8 m/s and released. If the coefficient of kinetic friction between the disk and the concrete court is 0.31, how far does the disk go before it comes to a stop? The courts are 15.8 m long.

25. Consider the force pushing the box in Example Problem 4. How long would it take for the velocity of the box to double to 2.0 m/s?

26. Ke Min is driving along on a rainy night at 23 m/s when he sees a tree branch lying across the road and slams on the brakes when the branch is 60.0 m in front of him. If the coefficient of kinetic friction between the car's locked tires and the road is 0.41, will the car stop before hitting the branch? The car has a mass of 2400 kg.

APPLYING PHYSICS

Causes of Friction All surfaces, even those that appear to be smooth, are rough at a microscopic level. If you look at a photograph of a graphite crystal magnified by a scanning tunneling microscope, the atomic level surface irregularities of the crystal are revealed. When two surfaces touch, the high points on each are in contact and temporarily bond. This is the origin of both static and kinetic friction. The details of this process are still unknown and are the subject of research in both physics and engineering.

Here are a few important things to remember when dealing with frictional situations. First, friction always acts in a direction opposite to the motion (or in the case of static friction, intended motion). Second, the magnitude of the force of friction depends on the magnitude of the normal force between the two rubbing surfaces; it does not necessarily depend on the weight of either object. Finally, multiplying the coefficient of static friction and the normal force gives you the maximum static friction force. Keep these things in mind as you review this section.

5.2 Section Review

27. Friction In this section, you learned about static and kinetic friction. How are these two types of friction similar? What are the differences between static and kinetic friction?

28. Friction At a wedding reception, you notice a small boy who looks like his mass is about 25 kg running part way across the dance floor, then sliding on his knees until he stops. If the kinetic coefficient of friction between the boy's pants and the floor is 0.15, what is the frictional force acting on him as he slides?

29. Velocity Derek is playing cards with his friends, and it is his turn to deal. A card has a mass of 2.3 g, and it slides 0.35 m along the table before it stops. If the coefficient of kinetic friction between the card and the table is 0.24, what was the initial speed of the card as it left Derek's hand?

30. Force The coefficient of static friction between a 40.0-kg picnic table and the ground below it is 0.43. What is the greatest horizontal force that could be exerted on the table while it remains stationary?

31. Acceleration Ryan is moving to a new apartment and puts a dresser in the back of his pickup truck. When the truck accelerates forward, what force accelerates the dresser? Under what circumstances could the dresser slide? In which direction?

32. Critical Thinking You push a 13-kg table in the cafeteria with a horizontal force of 20 N, but it does not move. You then push it with a horizontal force of 25 N, and it accelerates at 0.26 m/s^2. What, if anything, can you conclude about the coefficients of static and kinetic friction?

Physics Online physicspp.com/self_check_quiz

5.3 Force and Motion in Two Dimensions

► Objectives

- **Determine** the force that produces equilibrium when three forces act on an object.
- **Analyze** the motion of an object on an inclined plane with and without friction.

► Vocabulary

equilibrant

You have already worked with several situations dealing with forces in two dimensions. For example, when friction acts between two surfaces, you must take into account both the frictional force that is parallel to the surface and the normal force that is perpendicular to it. So far, you have considered only the motion along a level surface. Now you will use your skill in adding vectors to analyze situations in which the forces acting on an object are at angles other than 90°.

Equilibrium Revisited

Recall from Chapter 4 that when the net force on an object is zero, the object is in equilibrium. According to Newton's laws, the object will not accelerate because there is no net force acting on it; an object in equilibrium is motionless or moves with constant velocity. You have already analyzed several equilibrium situations in which two forces acted on an object. It is important to realize that equilibrium can occur no matter how many forces act on an object. As long as the resultant is zero, the net force is zero and the object is in equilibrium.

Figure 5-11a shows three forces exerted on a point object. What is the net force acting on the object? Remember that vectors may be moved if you do not change their direction (angle) or length. **Figure 5-11b** shows the addition of the three forces, ***A***, ***B***, and ***C***. Note that the three vectors form a closed triangle. There is no net force; thus, the sum is zero and the object is in equilibrium.

Suppose that two forces are exerted on an object and the sum is not zero. How could you find a third force that, when added to the other two, would add up to zero, and therefore cause the object to be in equilibrium? To find this force, first find the sum of the two forces already being exerted on the object. This single force that produces the same effect as the two individual forces added together is called the resultant force. The force that you need to find is one with the same magnitude as the resultant force, but in the opposite direction. A force that puts an object in equilibrium is called the **equilibrant. Figure 5-12** illustrates the procedure for finding this force for two vectors. Note that this general procedure works for any number of vectors.

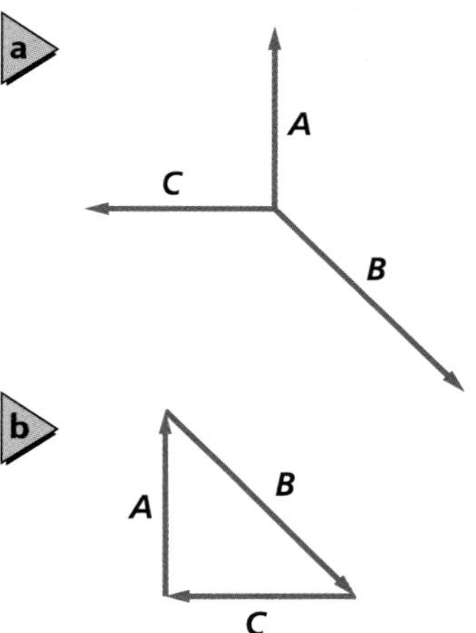

Figure 5-11 An object is in equilibrium when all the forces on it add up to zero.

Figure 5-12 The equilibrant is the same magnitude as the resultant, but opposite in direction.

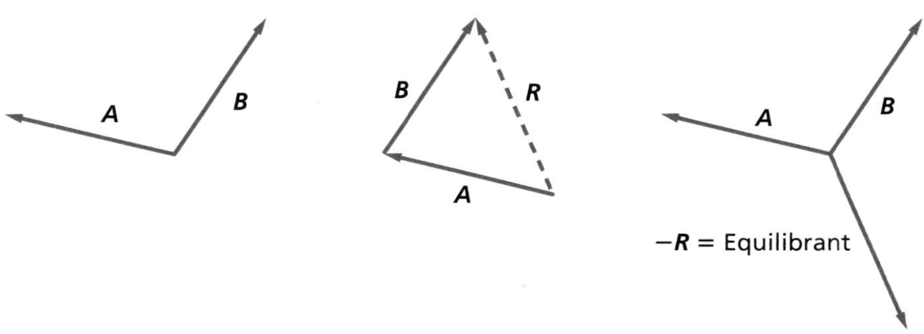

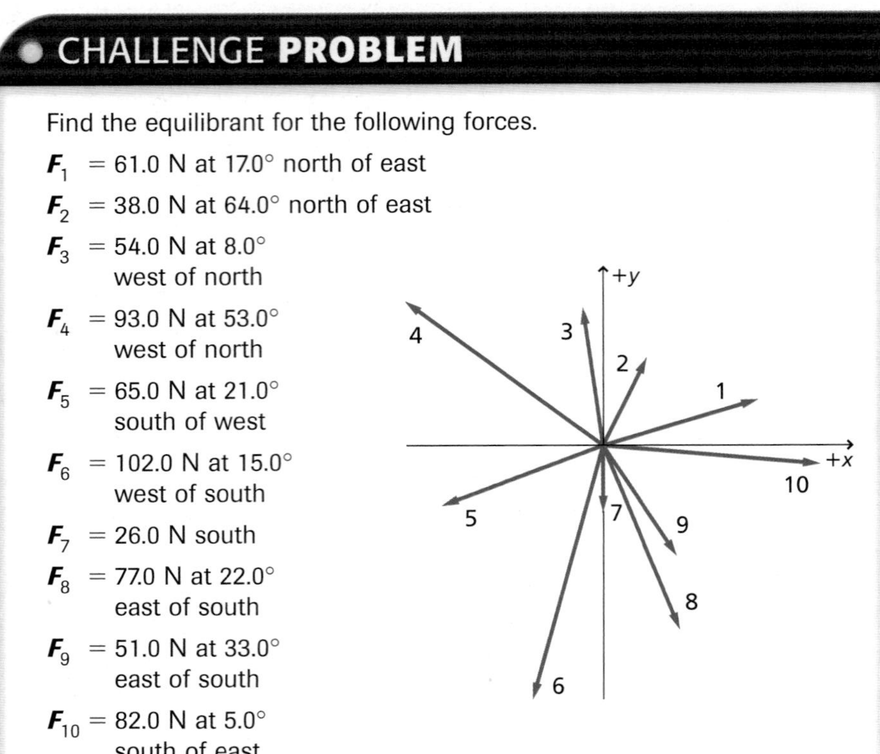

CHALLENGE PROBLEM

Find the equilibrant for the following forces.

$\boldsymbol{F}_1$ = 61.0 N at 17.0° north of east

$\boldsymbol{F}_2$ = 38.0 N at 64.0° north of east

$\boldsymbol{F}_3$ = 54.0 N at 8.0° west of north

$\boldsymbol{F}_4$ = 93.0 N at 53.0° west of north

$\boldsymbol{F}_5$ = 65.0 N at 21.0° south of west

$\boldsymbol{F}_6$ = 102.0 N at 15.0° west of south

$\boldsymbol{F}_7$ = 26.0 N south

$\boldsymbol{F}_8$ = 77.0 N at 22.0° east of south

$\boldsymbol{F}_9$ = 51.0 N at 33.0° east of south

$\boldsymbol{F}_{10}$ = 82.0 N at 5.0° south of east

Motion Along an Inclined Plane

You have applied Newton's laws to a variety of equilibrium situations, but only to motions that were either horizontal or vertical. How would you apply them in a situation like the one in **Figure 5-13a,** in which a skier glides down a slope?

Start by identifying the forces acting on the object, the skier, as shown in **Figure 5-13b** and sketching a free-body diagram. The gravitational force on the skier is in the downward direction toward the center of Earth. There is a normal force perpendicular to the hill, and the frictional forces opposing the skier's motion are parallel to the hill. The resulting free-body diagram is shown in **Figure 5-13c.** You can see that, other than the force of friction, only one force acts horizontally or vertically, and you know from experience that the acceleration of the skier will be along the slope. How do you find the net force that causes the skier to accelerate?

Concepts In Motion
Interactive Figure To see an animation on motion along an inclined plane, visit physicspp.com.

■ **Figure 5-13** A skier slides down a slope **(a).** Identify the forces that are acting upon the skier **(b)** and draw a free-body diagram describing those forces **(c).** It is important to draw the direction of the normal and the friction forces correctly in order to properly analyze these types of situations.

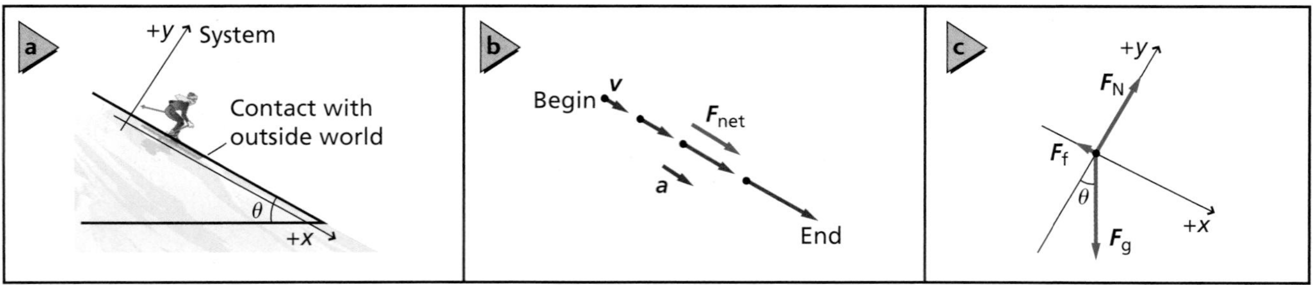

▶ EXAMPLE Problem 5

Components of Weight for an Object on an Incline A crate weighing 562 N is resting on a plane inclined 30.0° above the horizontal. Find the components of the weight forces that are parallel and perpendicular to the plane.

1 Analyze and Sketch the Problem

- Include a coordinate system with the positive *x*-axis pointing uphill.
- Draw the free-body diagram showing $\boldsymbol{F}_g$, the components $\boldsymbol{F}_{gx}$ and $\boldsymbol{F}_{gy}$, and the angle θ.

Known:	Unknown:
$F_g = 562$ N	$F_{gx} = ?$
$\theta = 30.0°$	$F_{gy} = ?$

2 Solve for the Unknown

$\boldsymbol{F}_{gx}$ and $\boldsymbol{F}_{gy}$ are negative because they point in directions opposite to the positive axes.

$$F_{gx} = -F_g(\sin\theta)$$
$$= -(562\text{ N})(\sin 30.0°) \quad \text{Substitute } F_g = 562,\ \theta = 30.0°$$
$$= -281\text{ N}$$

$$F_{gy} = -F_g(\cos\theta)$$
$$= -(562\text{ N})(\cos 30.0°) \quad \text{Substitute } F_g = 562,\ \theta = 30.0°$$
$$= -487\text{ N}$$

Personal Tutor For an online tutorial on components of weight, visit physicspp.com.

3 Evaluate the Answer

- **Are the units correct?** Force is measured in newtons.
- **Do the signs make sense?** The components point in directions opposite to the positive axes.
- **Are the magnitudes realistic?** The values are less than F_g.

▶ PRACTICE Problems

- Additional Problems, Appendix B
- Solutions to Selected Problems, Appendix C

33. An ant climbs at a steady speed up the side of its anthill, which is inclined 30.0° from the vertical. Sketch a free-body diagram for the ant.

34. Scott and Becca are moving a folding table out of the sunlight. A cup of lemonade, with a mass of 0.44 kg, is on the table. Scott lifts his end of the table before Becca does, and as a result, the table makes an angle of 15.0° with the horizontal. Find the components of the cup's weight that are parallel and perpendicular to the plane of the table.

35. Kohana, who has a mass of 50.0 kg, is at the dentist's office having her teeth cleaned, as shown in **Figure 5-14.** If the component of her weight perpendicular to the plane of the seat of the chair is 449 N, at what angle is the chair tilted?

36. Fernando, who has a mass of 43.0 kg, slides down the banister at his grandparents' house. If the banister makes an angle of 35.0° with the horizontal, what is the normal force between Fernando and the banister?

37. A suitcase is on an inclined plane. At what angle, relative to the vertical, will the component of the suitcase's weight parallel to the plane be equal to half the perpendicular component of its weight?

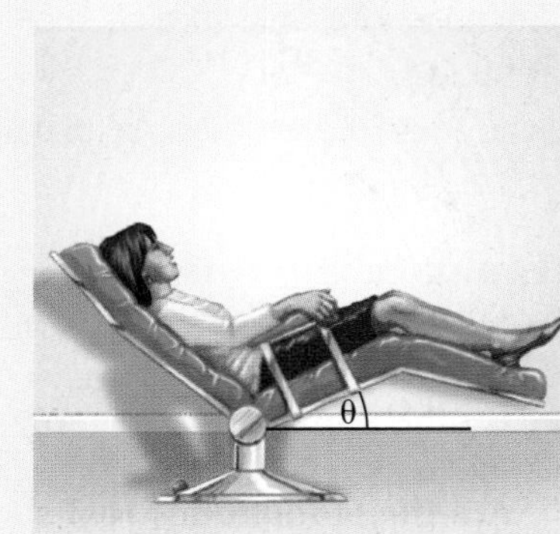

Figure 5-14

▶ EXAMPLE Problem 6

Skiing Downhill A 62-kg person on skis is going down a hill sloped at 37°. The coefficient of kinetic friction between the skis and the snow is 0.15. How fast is the skier going 5.0 s after starting from rest?

1 Analyze and Sketch the Problem

- Establish a coordinate system.
- Draw a free-body diagram showing the skier's velocity and direction of acceleration.
- Draw a motion diagram showing increasing v, and both a and F_{net} in the $+x$ direction, like the one shown in Figure 5-13.

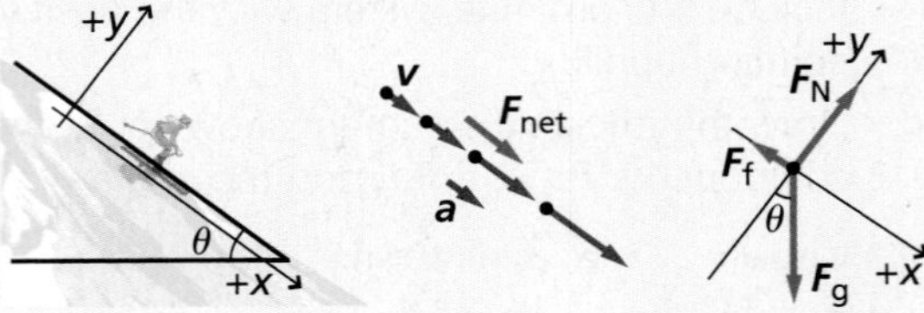

Known:	Unknown:
m = 62 kg	a = ?
θ = 37°	v_f = ?
μ_k = 0.15	
v_i = 0.0 m/s	
t = 5.0 s	

2 Solve for the Unknown

y-direction:

$F_{net,y} = ma_y$ — There is no acceleration in the y-direction, so $a_y = 0.0\ m/s^2$.

$= 0.0$ N

Solve for F_N.

$F_N - F_{gy} = F_{net,y}$ — F_{gy} is negative. It is in the negative direction as defined by the coordinate system.

$F_N = F_{gy}$ — Substitute $F_{net,y} = 0.0$ N and rearrange

$= mg(\cos\theta)$ — Substitute $F_{gy} = mg\cos\theta$

Math Handbook
Isolating a Variable
page 845

x-direction:

Solve for a.

$F_{net,x} = F_{gx} - F_f$ — F_f is negative because it is in the negative direction as defined by the coordinate system.

$ma_x = mg(\sin\theta) - \mu_k F_N$ — Substitute $F_{net,x} = ma$, $F_{gx} = mg\sin\theta$, $F_f = \mu_k F_N$

$= mg(\sin\theta) - \mu_k mg(\cos\theta)$ — Substitute $a = a_x$ because all the acceleration is in the x-direction; substitute $F_N = mg\cos\theta$

$a = g(\sin\theta - \mu_k\cos\theta)$

$= (9.80\ m/s^2)(\sin 37° - (0.15)\cos 37°)$ — Substitute $g = 9.80\ m/s^2$, $\theta = 37°$, $\mu_k = 0.15$

$= 4.7\ m/s^2$

Because v_i, a, and t are all known, use the following.

$v_f = v_i + at$

$= 0.0 + (4.7\ m/s^2)(5.0\ s)$ — Substitute $v_i = 0.0$ m/s, $a = 4.7\ m/s^2$, $t = 5.0$ s

$= 24$ m/s

3 Evaluate the Answer

- **Are the units correct?** Performing dimensional analysis on the units verifies that v_f is in m/s and a is in m/s^2.
- **Do the signs make sense?** Because v_f and a are both in the $+x$ direction, the signs do make sense.
- **Are the magnitudes realistic?** The velocity is fast, over 80 km/h (50 mph), but 37° is a steep incline, and the friction between the skis and the snow is not large.

PRACTICE Problems

- Additional Problems, Appendix B
- Solutions to Selected Problems, Appendix C

38. Consider the crate on the incline in Example Problem 5. Calculate the magnitude of the acceleration. After 4.00 s, how fast will the crate be moving?

39. If the skier in Example Problem 6 were on a 31° downhill slope, what would be the magnitude of the acceleration?

40. Stacie, who has a mass of 45 kg, starts down a slide that is inclined at an angle of 45° with the horizontal. If the coefficient of kinetic friction between Stacie's shorts and the slide is 0.25, what is her acceleration?

41. After the skier on the 37° hill in Example Problem 6 had been moving for 5.0 s, the friction of the snow suddenly increased and made the net force on the skier zero. What is the new coefficient of friction?

The most important decision in problems involving motion along a slope is what coordinate system to use. Because an object's acceleration is usually parallel to the slope, one axis, usually the *x*-axis, should be in that direction. The *y*-axis is perpendicular to the *x*-axis and perpendicular to the surface of the slope. With this coordinate system, you now have two forces, the normal and frictional forces, in the directions of the coordinate axes; however, the weight is not. This means that when an object is placed on an inclined plane, the magnitude of the normal force between the object and the plane will usually not be equal to the object's weight.

You will need to apply Newton's laws once in the *x*-direction and once in the *y*-direction. Because the weight does not point in either of these directions, you will need to break this vector into its *x*- and *y*-components before you can sum your forces in these two directions. Example Problem 5 and Example Problem 6 both showed this procedure.

MINI LAB

What's Your Angle?

Prop a board up so that it forms an inclined plane at a 45° angle. Hang a 500-g object from the spring scale.

1. **Measure** and record the weight of the object. Set the object on the bottom of the board and slowly pull it up the inclined plane at a constant speed.
2. **Observe and record** the reading on the spring scale.

Analyze and Conclude

3. **Calculate** the component of weight for the 500-g object that is parallel to the inclined plane.
4. **Compare** the spring-scale reading along the inclined plane with the component of weight parallel to the inclined plane.

5.3 Section Review

42. Forces One way to get a car unstuck is to tie one end of a strong rope to the car and the other end to a tree, then push the rope at its midpoint at right angles to the rope. Draw a free-body diagram and explain why even a small force on the rope can exert a large force on the car.

43. Mass A large scoreboard is suspended from the ceiling of a sports arena by 10 strong cables. Six of the cables make an angle of 8.0° with the vertical while the other four make an angle of 10.0°. If the tension in each cable is 1300.0 N, what is the scoreboard's mass?

44. Acceleration A 63-kg water skier is pulled up a 14.0° incline by a rope parallel to the incline with a tension of 512 N. The coefficient of kinetic friction is 0.27. What are the magnitude and direction of the skier's acceleration?

45. Equilibrium You are hanging a painting using two lengths of wire. The wires will break if the force is too great. Should you hang the painting as shown in **Figures 5-15a** or **5-15b?** Explain.

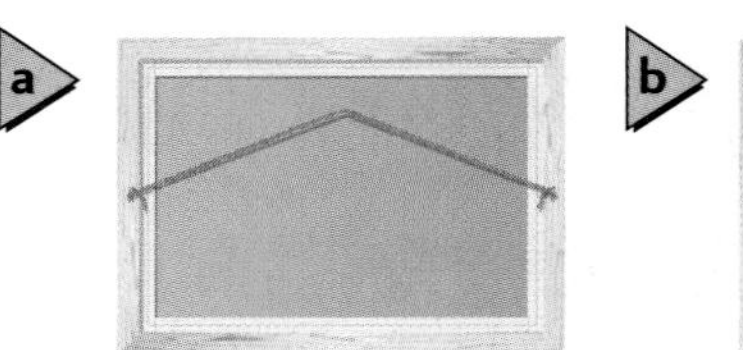

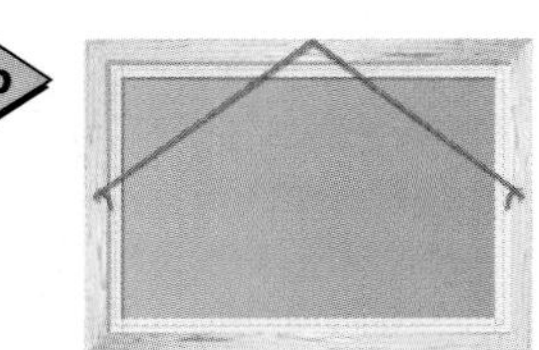

Figure 5-15

46. Critical Thinking Can the coefficient of friction ever have a value such that a skier would be able to slide uphill at a constant velocity? Explain why or why not. Assume there are no other forces acting on the skier.

PHYSICS LAB

The Coefficient of Friction

Alternate CBL instructions can be found on the Web site.
physicspp.com

Static and kinetic friction are forces that are a result of two surfaces in contact with each other. Static friction is the force that must be overcome to cause an object to begin moving, while kinetic friction occurs between two objects in motion relative to each other. The kinetic friction force, $F_{f,\,kinetic}$, is defined by $F_{f,\,kinetic} = \mu_k F_N$, where μ_k is the coefficient of kinetic friction and F_N is the normal force acting on the object. The maximum static frictional force, $F_{f,\,max\,static}$, is defined by $F_{f,\,static} = \mu_s F_N$ where μ_s is the coefficient of static friction and F_N is the normal force on the object. The maximum static frictional force that must be overcome before movement is able to begin is $\mu_s F_N$. If you apply a constant force to pull an object along a horizontal surface at a constant speed, then the frictional force opposing the motion is equal and opposite to the applied force, F_p. Therefore, $F_p = F_f$. The normal force is equal and opposite to the object's weight when the object is on a horizontal surface and the applied force is horizontal.

QUESTION

How can the coefficient of static and kinetic friction be determined for an object on a horizontal surface?

Objectives

- **Measure** the normal and frictional forces acting on an object starting in motion and already in motion.
- **Use numbers** to calculate μ_s and μ_k.
- **Compare and contrast** values of μ_s and μ_k.
- **Analyze** the kinetic friction results.
- **Estimate** the angle where sliding will begin for an object on an inclined plane.

Safety Precautions

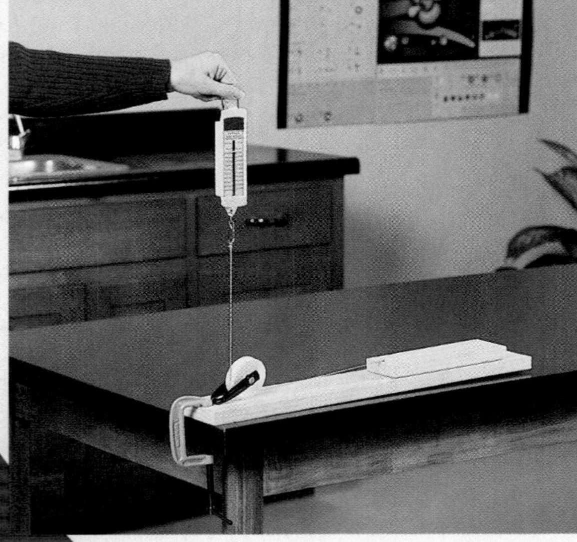

Materials

pulley
C-clamp
masking tape
wood surface
string (1 m)
spring scale, 0-5 N
wood block

Procedure

1. Check your spring scale to make sure that it reads zero when it is held vertically. If necessary, follow your teacher's instructions to zero it.
2. Attach the pulley to the edge of the table with a C-clamp.
3. Attach the string to the spring scale hook and the wood block.
4. Measure the weight of the block of wood, or other small object, and record the value as the normal force, F_N, in Data Tables 1, 2, and 3.
5. Unhook the string from the spring scale and run it through the pulley. Then reattach it to the spring scale.
6. Move the wood block as far away from the pulley as the string permits, while having it remain on the wood surface.
7. With the spring scale oriented vertically so that a right angle is formed between the wood block, the pulley, and the spring scale, slowly pull up on the spring scale. Observe the force that is necessary to cause the wood block to begin sliding. Record this value for the static frictional force in Data Table 1.

Material Table

Object material	
Surface material	

Data Table 1

F_N (N)	Static Friction Force, F_s (N)			
	Trial 1	Trial 2	Trial 3	Average

Data Table 2

F_N (N)	Kinetic Friction Force, F_f (N)			
	Trial 1	Trial 2	Trial 3	Average

Data Table 3

F_N(N)	F_s (N)	F_f (N)	μ_s	μ_k

Data Table 4 (Angle, θ, when sliding begins on an incline)

θ*	tan θ*

8. Repeat steps 6 and 7 for two additional trials.
9. Repeat steps 6 and 7. However, once the block begins sliding, pull just hard enough to keep it moving at a constant speed across the other horizontal surface. Record this force as the kinetic frictional force in Data Table 2.
10. Repeat step 9 for two additional trials.
11. Place the block on the end of the surface. Slowly raise one end of the surface to make an incline. Gently tap the block to cause it to move and overcome static friction. If the block stops, replace it at the top of the incline and repeat the procedure. Continue increasing the angle, θ, between the horizontal and the inclined surface, and tapping the block until it slides at a constant speed down the incline. Record the angle, θ, in Data Table 4.

Analyze

1. Average the data for the static frictional force, $F_{s,\ max}$, from the three trials and record the result in the last column of Data Table 1 and in Data Table 3.
2. Average the data for the kinetic frictional force, F_f, from the three trials and record the result in the last column of Data Table 2 and in Data Table 3.
3. Use the data in Data Table 3 to calculate the coefficient of static friction, μ_s, and record the value in Data Table 3.
4. Use the data in Data Table 3 to calculate the coefficient of kinetic friction, μ_k, and record the value in Data Table 3.
5. Calculate tan θ for your value in Data Table 4.

Conclude and Apply

1. **Compare and Contrast** Examine your values for μ_s and μ_k. Explain whether your results are reasonable or not.
2. **Use Models** Draw a free-body diagram showing the forces acting on the block if it is placed on an incline of angle θ. Make certain that you include the force due to friction in your diagram.
3. From your diagram, assuming that the angle, θ, is where sliding begins, what does tan θ represent?
4. Compare your value for tan θ (experimental), μ_s, and μ_k.

Going Further

Repeat the experiment with additional surfaces that have different characteristics.

Real-World Physics

If you were downhill skiing and wished to determine the coefficient of kinetic friction between your skis and the slope, how could you do this? Be specific about how you could find a solution to this problem.

To find out more about friction, visit the Web site: physicspp.com

Technology and Society

Roller Coasters

Why are roller coasters fun? A roller-coaster ride would be no fun at all if not for the forces acting on the coaster car and the rider. What forces do riders experience as they ride a roller coaster? The force of gravity acts on the rider and the coaster car in the downward direction. The seat of the car exerts a force on the rider in the opposite direction. When the coaster car makes a turn, the rider experiences a force in the opposite direction. Also, there are forces present due to the friction between the rider and the seat, the side of the car, and the safety bar.

The Force Factor

Designers of roller coasters take into account the magnitude of the forces exerted on the rider. They design the coaster in such a way that the forces thrill the rider without causing injury or too much discomfort.

Designers measure the amount of force exerted on the rider by calculating the force factor. The force factor is equal to the force exerted by the seat on the rider divided by the weight of the rider. Suppose the rider weighs about 68 kg. When the roller coaster is at the bottom of a hill, the rider may experience a force factor of 2. That means that at the bottom of the hill, the rider will feel as though he or she weighs twice as much, or in this case 136 kg. Conversely, at the top of a hill the force factor may be 0.5 and the rider will feel as though he or she weighs half his or her normal weight. Thus, designers create excitement by designing portions that change the rider's apparent weight.

The thrill of a roller-coaster ride is produced by the forces acting on the rider and the rider's reaction to visual cues.

The Thrill Factors Roller-coaster designers manipulate the way in which the body perceives the external world to create that "thrilling" sensation. For example, the roller coaster moves up the first hill very slowly, tricking the rider into thinking that the hill is higher than it is.

The organs of the inner ear sense the position of the head both when it is still and when it is moving. These organs help maintain balance by providing information to the brain. The brain then sends nerve impulses to the skeletal muscles to contract or relax to maintain balance. The constant change in position during a roller-coaster ride causes the organs of the inner ear to send conflicting messages to the brain. As a result, the skeletal muscles contract and relax throughout the ride.

You know that you are moving at high speeds because your eyes see the surroundings move past at high speed. So, designers make use of the surrounding landscape along with twists, turns, tunnels, and loops to give the rider plenty of visual cues. These visual cues, along with the messages from the inner ear, can result in disorientation and in some cases, nausea. To enthusiasts the disorientation is part of the thrill.

In order to attract visitors, amusement parks are constantly working on designing new rides that take the rider to new thrill levels. As roller-coaster technology improves, your most thrilling roller-coaster ride may be over the next hill.

Going Further

1. **Compare and Contrast** Compare and contrast your experience as a rider in the front of a roller coaster versus the back of it. Explain your answer in terms of the forces acting on you.
2. **Critical Thinking** While older roller coasters rely on chain systems to pull the coaster up the first hill, newer ones depend on hydraulic systems to do the same job. Research each of these two systems. What do you think are the advantages and disadvantages of using each system?

Chapter 5

Study Guide

5.1 Vectors

Vocabulary

- components (p. 122)
- vector resolution (p. 122)

Key Concepts

- When two vectors are at right angles, you can use the Pythagorean theorem to determine the magnitude of the resultant vector.

$$R^2 = A^2 + B^2$$

- The law of cosines and law of sines can be used to find the magnitude of the resultant of any two vectors.

$$R^2 = A^2 + B^2 - 2AB\cos\theta$$
$$\frac{R}{\sin\theta} = \frac{A}{\sin a} = \frac{B}{\sin b}$$

- The components of a vector are projections of the component vectors.

$$\cos\theta = \frac{\text{adjacent side}}{\text{hypotenuse}} = \frac{A_x}{A}\text{; therefore, } A_x = A\cos\theta$$
$$\sin\theta = \frac{\text{opposite side}}{\text{hypotenuse}} = \frac{A_y}{A}\text{; therefore, } A_y = A\sin\theta$$
$$\theta = \tan^{-1}\left(\frac{R_y}{R_x}\right)$$

- Vectors can be summed by separately adding the x- and y-components.

5.2 Friction

Vocabulary

- kinetic friction (p. 126)
- static friction (p. 126)
- coefficient of kinetic friction (p. 127)
- coefficient of static friction (p. 127)

Key Concepts

- A frictional force acts when two surfaces touch.
- The frictional force is proportional to the force pushing the surfaces together.
- The kinetic friction force is equal to the coefficient of kinetic friction times the normal force.

$$F_{\text{f, kinetic}} = \mu_k F_N$$

- The static friction force is less than or equal to the coefficient of static friction times the normal force.

$$F_{\text{f, static}} \leq \mu_k F_N$$

5.3 Force and Motion in Two Dimensions

Vocabulary

- equilibrant (p. 131)

Key Concepts

- The force that must be exerted on an object to cause it to be in equilibrium is called the equilibrant.
- The equilibrant is found by finding the net force on an object, then applying a force with the same magnitude but opposite direction.
- An object on an inclined plane has a component of the force of gravity in a direction parallel to the plane; the component can accelerate the object down the plane.

Chapter 5

Assessment

Concept Mapping

47. Complete the concept map below with the terms *sine, cosine,* or *tangent* to indicate whether each function is positive or negative in each quadrant. Some circles could remain blank, and others can have more than one term.

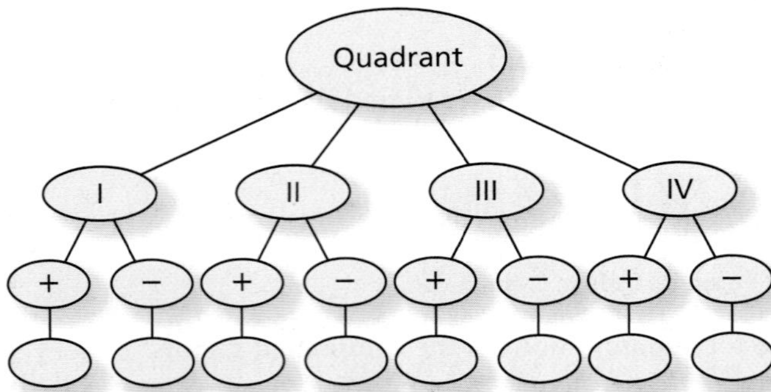

Mastering Concepts

48. How would you add two vectors graphically? (5.1)

49. Which of the following actions is permissible when you graphically add one vector to another: moving the vector, rotating the vector, or changing the vector's length? (5.1)

50. In your own words, write a clear definition of the resultant of two or more vectors. Do not explain how to find it; explain what it represents. (5.1)

51. How is the resultant displacement affected when two displacement vectors are added in a different order? (5.1)

52. Explain the method that you would use to subtract two vectors graphically. (5.1)

53. Explain the difference between **A** and *A*. (5.1)

54. The Pythagorean theorem usually is written $c^2 = a^2 + b^2$. If this relationship is used in vector addition, what do *a*, *b*, and *c* represent? (5.1)

55. When using a coordinate system, how is the angle or direction of a vector determined with respect to the axes of the coordinate system? (5.1)

56. What is the meaning of a coefficient of friction that is greater than 1.0? How would you measure it? (5.2)

57. **Cars** Using the model of friction described in this textbook, would the friction between a tire and the road be increased by a wide rather than a narrow tire? Explain. (5.2)

58. Describe a coordinate system that would be suitable for dealing with a problem in which a ball is thrown up into the air. (5.3)

59. If a coordinate system is set up such that the positive *x*-axis points in a direction 30° above the horizontal, what should be the angle between the *x*-axis and the *y*-axis? What should be the direction of the positive *y*-axis? (5.3)

60. Explain how you would set up a coordinate system for motion on a hill. (5.3)

61. If your textbook is in equilibrium, what can you say about the forces acting on it? (5.3)

62. Can an object that is in equilibrium be moving? Explain. (5.3)

63. What is the sum of three vectors that, when placed tip to tail, form a triangle? If these vectors represent forces on an object, what does this imply about the object? (5.3)

64. You are asked to analyze the motion of a book placed on a sloping table. (5.3)
    a. Describe the best coordinate system for analyzing the motion.
    b. How are the components of the weight of the book related to the angle of the table?

65. For a book on a sloping table, describe what happens to the component of the weight force parallel to the table and the force of friction on the book as you increase the angle that the table makes with the horizontal. (5.3)
    a. Which components of force(s) increase when the angle increases?
    b. Which components of force(s) decrease?

Applying Concepts

66. A vector that is 1 cm long represents a displacement of 5 km. How many kilometers are represented by a 3-cm vector drawn to the same scale?

67. **Mowing the Lawn** If you are pushing a lawn mower across the grass, as shown in **Figure 5-16,** can you increase the horizontal component of the force that you exert on the mower without increasing the magnitude of the force? Explain.

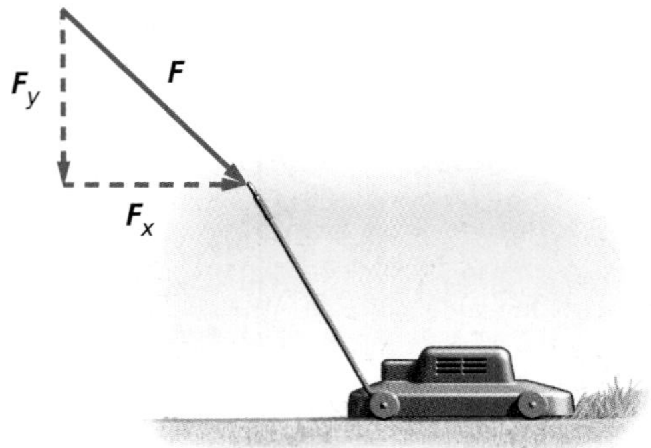

Figure 5-16

For more problems, go to Additional Problems, Appendix B.

68. A vector drawn 15 mm long represents a velocity of 30 m/s. How long should you draw a vector to represent a velocity of 20 m/s?

69. What is the largest possible displacement resulting from two displacements with magnitudes 3 m and 4 m? What is the smallest possible resultant? Draw sketches to demonstrate your answers.

70. How does the resultant displacement change as the angle between two vectors increases from 0° to 180°?

71. A and B are two sides of a right triangle, where $\tan \theta = A/B$.
   a. Which side of the triangle is longer if $\tan \theta$ is greater than 1.0?
   b. Which side is longer if $\tan \theta$ is less than 1.0?
   c. What does it mean if $\tan \theta$ is equal to 1.0?

72. **Traveling by Car** A car has a velocity of 50 km/h in a direction 60° north of east. A coordinate system with the positive x-axis pointing east and a positive y-axis pointing north is chosen. Which component of the velocity vector is larger, x or y?

73. Under what conditions can the Pythagorean theorem, rather than the law of cosines, be used to find the magnitude of a resultant vector?

74. A problem involves a car moving up a hill, so a coordinate system is chosen with the positive x-axis parallel to the surface of the hill. The problem also involves a stone that is dropped onto the car. Sketch the problem and show the components of the velocity vector of the stone.

75. **Pulling a Cart** According to legend, a horse learned Newton's laws. When the horse was told to pull a cart, it refused, saying that if it pulled the cart forward, according to Newton's third law, there would be an equal force backwards; thus, there would be balanced forces, and, according to Newton's second law, the cart would not accelerate. How would you reason with this horse?

76. **Tennis** When stretching a tennis net between two posts, it is relatively easy to pull one end of the net hard enough to remove most of the slack, but you need a winch to take the last bit of slack out of the net to make the top almost completely horizontal. Why is this true?

77. The weight of a book on an inclined plane can be resolved into two vector components, one along the plane, and the other perpendicular to it.
   a. At what angle are the components equal?
   b. At what angle is the parallel component equal to zero?
   c. At what angle is the parallel component equal to the weight?

78. **TV Towers** The transmitting tower of a TV station is held upright by guy wires that extend from the top of the tower to the ground. The force along the guy wires can be resolved into perpendicular and parallel components with respect to the ground. Which one is larger?

Mastering Problems

5.1 Vectors

79. **Cars** A car moves 65 km due east, then 45 km due west. What is its total displacement?

80. Find the horizontal and vertical components of the following vectors, as shown in **Figure 5-17.**
   a. E
   b. F
   c. A

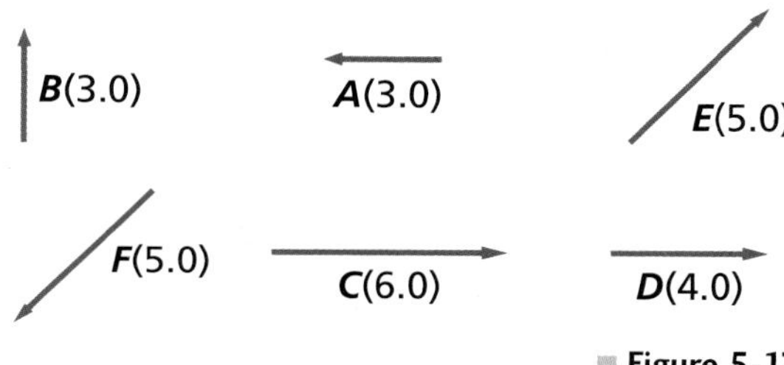

Figure 5-17

81. Graphically find the sum of the following pairs of vectors, whose characteristics are shown in Figure 5-17.
   a. D and A
   b. C and D
   c. C and A
   d. E and F

82. Graphically add the following sets of vectors, as shown in Figure 5-17.
   a. A, C, and D
   b. A, B, and E
   c. B, D, and F

83. You walk 30 m south and 30 m east. Find the magnitude and direction of the resultant displacement both graphically and algebraically.

84. **Hiking** A hiker's trip consists of three segments. Path A is 8.0 km long heading 60.0° north of east. Path B is 7.0 km long in a direction due east. Path C is 4.0 km long heading 315° counterclockwise from east.
   a. Graphically add the hiker's displacements in the order A, B, C.
   b. Graphically add the hiker's displacements in the order C, B, A.
   c. What can you conclude about the resulting displacements?

85. What is the net force acting on the ring in **Figure 5-18?**

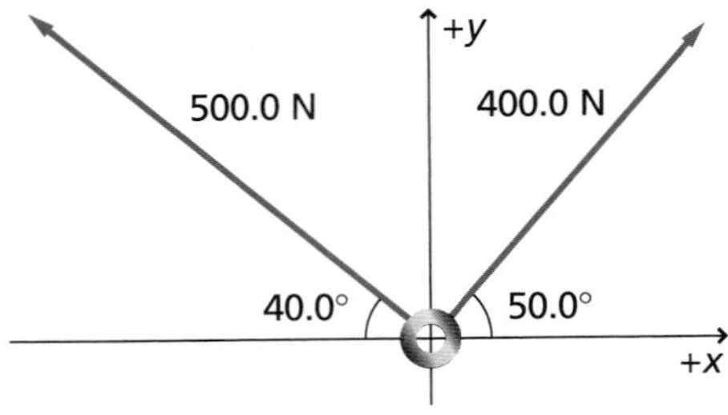

Figure 5-18

86. What is the net force acting on the ring in **Figure 5-19?**

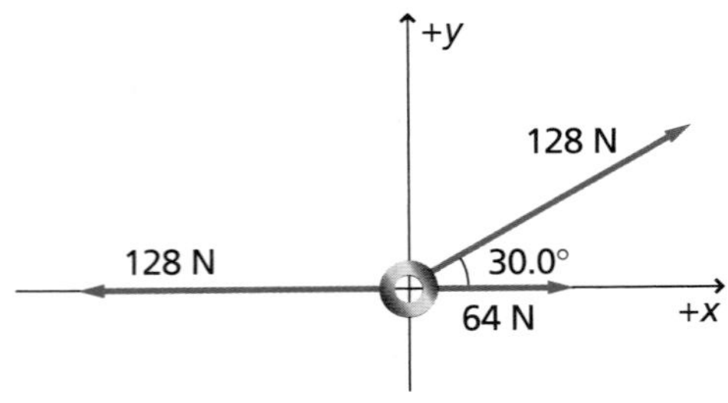

Figure 5-19

87. A Ship at Sea A ship at sea is due into a port 500.0 km due south in two days. However, a severe storm comes in and blows it 100.0 km due east from its original position. How far is the ship from its destination? In what direction must it travel to reach its destination?

88. Space Exploration A descent vehicle landing on Mars has a vertical velocity toward the surface of Mars of 5.5 m/s. At the same time, it has a horizontal velocity of 3.5 m/s.
a. At what speed does the vehicle move along its descent path?
b. At what angle with the vertical is this path?

89. Navigation Alfredo leaves camp and, using a compass, walks 4 km E, then 6 km S, 3 km E, 5 km N, 10 km W, 8 km N, and, finally, 3 km S. At the end of three days, he is lost. By drawing a diagram, compute how far Alfredo is from camp and which direction he should take to get back to camp.

5.2 Friction

90. If you use a horizontal force of 30.0 N to slide a 12.0-kg wooden crate across a floor at a constant velocity, what is the coefficient of kinetic friction between the crate and the floor?

91. A 225-kg crate is pushed horizontally with a force of 710 N. If the coefficient of friction is 0.20, calculate the acceleration of the crate.

92. A force of 40.0 N accelerates a 5.0-kg block at 6.0 m/s^2 along a horizontal surface.
a. How large is the frictional force?
b. What is the coefficient of friction?

93. Moving Appliances Your family just had a new refrigerator delivered. The delivery man has left and you realize that the refrigerator is not quite in the right position, so you plan to move it several centimeters. If the refrigerator has a mass of 180 kg, the coefficient of kinetic friction between the bottom of the refrigerator and the floor is 0.13, and the static coefficient of friction between these same surfaces is 0.21, how hard do you have to push horizontally to get the refrigerator to start moving?

94. Stopping at a Red Light You are driving a 2500.0-kg car at a constant speed of 14.0 m/s along a wet, but straight, level road. As you approach an intersection, the traffic light turns red. You slam on the brakes. The car's wheels lock, the tires begin skidding, and the car slides to a halt in a distance of 25.0 m. What is the coefficient of kinetic friction between your tires and the wet road?

5.3 Force and Motion in Two Dimensions

95. An object in equilibrium has three forces exerted on it. A 33.0-N force acts at 90.0° from the x-axis and a 44.0-N force acts at 60.0° from the x-axis. What are the magnitude and direction of the third force?

96. Five forces act on an object: (1) 60.0 N at 90.0°, (2) 40.0 N at 0.0°, (3) 80.0 N at 270.0°, (4) 40.0 N at 180.0°, and (5) 50.0 N at 60.0°. What are the magnitude and direction of a sixth force that would produce equilibrium?

97. Advertising Joe wishes to hang a sign weighing 7.50×10^2 N so that cable A, attached to the store, makes a 30.0° angle, as shown in **Figure 5-20.** Cable B is horizontal and attached to an adjoining building. What is the tension in cable B?

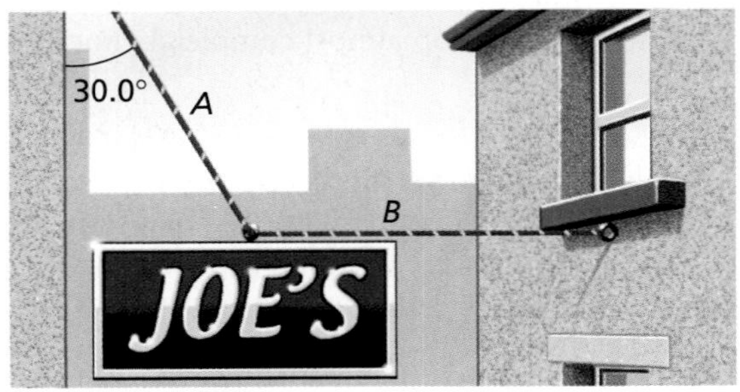

Figure 5-20

 For more problems, go to Additional Problems, Appendix B.

98. A street lamp weighs 150 N. It is supported by two wires that form an angle of 120.0° with each other. The tensions in the wires are equal.

a. What is the tension in each wire supporting the street lamp?

b. If the angle between the wires supporting the street lamp is reduced to 90.0°, what is the tension in each wire?

99. A 215-N box is placed on an inclined plane that makes a 35.0° angle with the horizontal. Find the component of the weight force parallel to the plane's surface.

100. **Emergency Room** You are shadowing a nurse in the emergency room of a local hospital. An orderly wheels in a patient who has been in a very serious accident and has had severe bleeding. The nurse quickly explains to you that in a case like this, the patient's bed will be tilted with the head downward to make sure the brain gets enough blood. She tells you that, for most patients, the largest angle that the bed can be tilted without the patient beginning to slide off is 32.0° from the horizontal.

a. On what factor or factors does this angle of tilting depend?

b. Find the coefficient of static friction between a typical patient and the bed's sheets.

101. Two blocks are connected by a string over a frictionless, massless pulley such that one is resting on an inclined plane and the other is hanging over the top edge of the plane, as shown in **Figure 5-21.** The hanging block has a mass of 16.0 kg, and the one on the plane has a mass of 8.0 kg. The coefficient of kinetic friction between the block and the inclined plane is 0.23. The blocks are released from rest.

a. What is the acceleration of the blocks?

b. What is the tension in the string connecting the blocks?

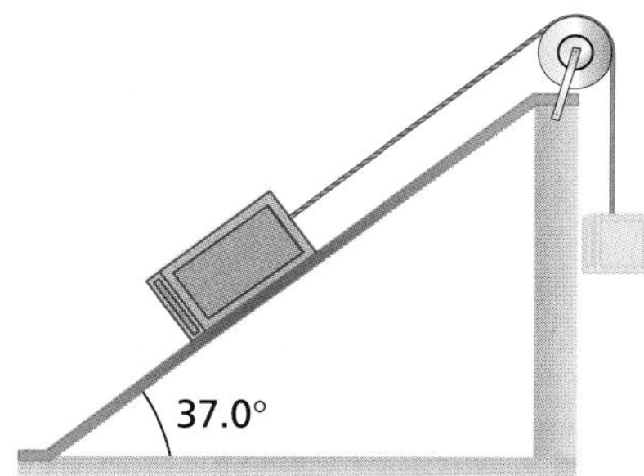

Figure 5-21

102. In **Figure 5-22,** a block of mass M is pushed with a force, F, such that the smaller block of mass m does not slide down the front of it. There is no friction between the larger block and the surface below it, but the coefficient of static friction between the two blocks is μ_s. Find an expression for F in terms of M, m, μ_s, and g.

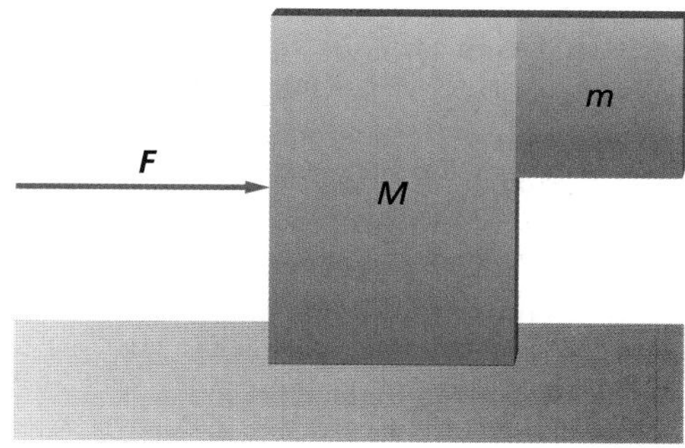

Figure 5-22

Mixed Review

103. The scale in **Figure 5-23** is being pulled on by three ropes. What net force does the scale read?

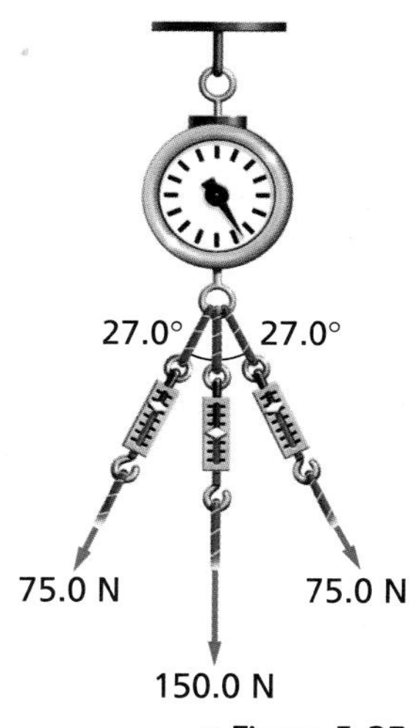

Figure 5-23

104. **Sledding** A sled with a mass of 50.0 kg is pulled along flat, snow-covered ground. The static friction coefficient is 0.30, and the kinetic friction coefficient is 0.10.

a. What does the sled weigh?

b. What force will be needed to start the sled moving?

c. What force is needed to keep the sled moving at a constant velocity?

d. Once moving, what total force must be applied to the sled to accelerate it at 3.0 m/s^2?

105. Mythology Sisyphus was a character in Greek mythology who was doomed in Hades to push a boulder to the top of a steep mountain. When he reached the top, the boulder would slide back down the mountain and he would have to start all over again. Assume that Sisyphus slides the boulder up the mountain without being able to roll it, even though in most versions of the myth, he rolled it.

a. If the coefficient of kinetic friction between the boulder and the mountainside is 0.40, the mass of the boulder is 20.0 kg, and the slope of the mountain is a constant 30.0°, what is the force that Sisyphus must exert on the boulder to move it up the mountain at a constant velocity?

b. If Sisyphus pushes the boulder at a velocity of 0.25 m/s and it takes him 8.0 h to reach the top of the mountain, what is the mythical mountain's vertical height?

106. Landscaping A tree is being transported on a flatbed trailer by a landscaper, as shown in **Figure 5-24.** If the base of the tree slides on the trailer, the tree will fall over and be damaged. If the coefficient of static friction between the tree and the trailer is 0.50, what is the minimum stopping distance of the truck, traveling at 55 km/h, if it is to accelerate uniformly and not have the tree slide forward and fall on the trailer?

Figure 5-24

Thinking Critically

107. Use Models Using the Example Problems in this chapter as models, write an example problem to solve the following problem. Include the following sections: Analyze and Sketch the Problem, Solve for the Unknown (with a complete strategy), and Evaluate the Answer. A driver of a 975-kg car traveling 25 m/s puts on the brakes. What is the shortest distance it will take for the car to stop? Assume that the road is concrete, the force of friction of the road on the tires is constant, and the tires do not slip.

108. Analyze and Conclude Margaret Mary, Doug, and Kako are at a local amusement park and see an attraction called the Giant Slide, which is simply a very long and high inclined plane. Visitors at the amusement park climb a long flight of steps to the top of the 27° inclined plane and are given canvas sacks. They sit on the sacks and slide down the 70-m-long plane. At the time when the three friends walk past the slide, a 135-kg man and a 20-kg boy are each at the top preparing to slide down. "I wonder how much less time it will take the man to slide down than it will take the boy," says Margaret Mary. "I think the boy will take less time," says Doug. "You're both wrong," says Kako. "They will reach the bottom at the same time."

a. Perform the appropriate analysis to determine who is correct.

b. If the man and the boy do not take the same amount of time to reach the bottom of the slide, calculate how many seconds of difference there will be between the two times.

Writing in Physics

109. Investigate some of the techniques used in industry to reduce the friction between various parts of machines. Describe two or three of these techniques and explain the physics of how they work.

110. Olympics In recent years, many Olympic athletes, such as sprinters, swimmers, skiers, and speed skaters, have used modified equipment to reduce the effects of friction and air or water drag. Research a piece of equipment used by one of these types of athletes and the way it has changed over the years. Explain how physics has impacted these changes.

Cumulative Review

111. Add or subtract as indicated and state the answer with the correct number of significant digits. (Chapter 1)

a. 85.26 g + 4.7 g

b. 1.07 km + 0.608 km

c. 186.4 kg − 57.83 kg

d. 60.08 s − 12.2 s

112. You ride your bike for 1.5 h at an average velocity of 10 km/h, then for 30 min at 15 km/h. What is your average velocity? (Chapter 3)

113. A 45-N force is exerted in the upward direction on a 2.0-kg briefcase. What is the acceleration of the briefcase? (Chapter 4)

 For more problems, go to Additional Problems, Appendix B.

Multiple Choice

1. Two tractors pull against a 1.00×10^3-kg log. If the angle of the tractors' chains in relation to each other is 18.0°, and each tractor pulls with a force of 8×10^2 N, what forces will they be able to exert?

   Ⓐ 250 N Ⓒ 1.58×10^3 N

   Ⓑ 1.52×10^3 N Ⓓ 1.60×10^3 N

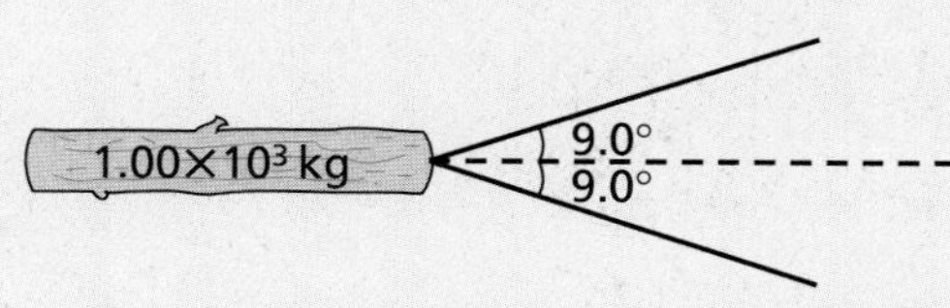

2. An airplane pilot tries to fly directly east with a velocity of 800.0 km/h. If a wind comes from the southwest at 80.0 km/h, what is the relative velocity of the airplane to the surface of Earth?

   Ⓐ 804 km/h, 5.7° N of E

   Ⓑ 858 km/h, 3.8° N of E

   Ⓒ 859 km/h, 4.0° N of E

   Ⓓ 880 km/h 45° N of E

3. For a winter fair, some students decide to build 30.0-kg wooden pull-carts on sled skids. If two 90.0-kg passengers get in, how much force will the puller have to exert to move a pull-cart? The coefficient of maximum static friction between the cart and the snow is 0.15.

   Ⓐ 1.8×10^2 N Ⓒ 2.1×10^3 N

   Ⓑ 3.1×10^2 N Ⓓ 1.4×10^4 N

4. It takes a minimum force of 280 N to move a 50.0-kg crate. What is the coefficient of maximum static friction between the crate and the floor?

   Ⓐ 0.18 Ⓒ 1.8

   Ⓑ 0.57 Ⓓ 5.6

5. What is the *y*-component of a 95.3-N force that is exerted at 57.1° to the horizontal?

   Ⓐ 51.8 N Ⓒ 114 N

   Ⓑ 80.0 N Ⓓ 175 N

6. A string exerts a force of 18 N on a box at an angle of 34° from the horizontal. What is the horizontal component of the force on the box?

   Ⓐ 10 N Ⓒ 21.7 N

   Ⓑ 15 N Ⓓ 32 N

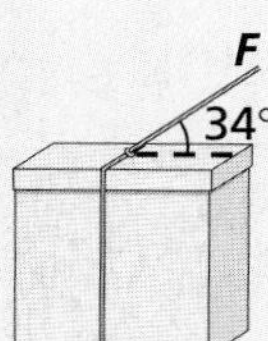

7. Sukey is riding her bicycle on a path when she comes around a corner and sees that a fallen tree is blocking the way 42 m ahead. If the coefficient of friction between her bicycle's tires and the gravel path is 0.36, and she is traveling at 50.0 km/h, how much stopping distance will she require? Sukey and her bicycle, together, have a mass of 95 kg.

   Ⓐ 3.00 m Ⓒ 8.12 m

   Ⓑ 4.00 m Ⓓ 27.3 m

Extended Answer

8. A man starts from a position 310 m north of his car and walks for 2.7 min in a westward direction at a constant velocity of 10 km/h. How far is he from his car when he stops?

9. Jeeves is tired of his 41.2-kg son sliding down the banister, so he decides to apply an extremely sticky paste that increases the coefficient of static friction to 0.72 to the top of the banister. What will be the magnitude of the static friction force on the boy if the banister is at an angle of 52.4° from the horizontal?

✓ Test-Taking TIP

Calculators Are Only Machines

If your test allows you to use a calculator, use it wisely. Figure out which numbers are relevant, and determine the best way to solve the problem before you start punching keys.

Chapter 6

Motion in Two Dimensions

What You'll Learn

- You will use Newton's laws and your knowledge of vectors to analyze motion in two dimensions.
- You will solve problems dealing with projectile and circular motion.
- You will solve relative-velocity problems.

Why It's Important

Almost all types of transportation and amusement-park attractions contain at least one element of projectile or circular motion or are affected by relative velocities.

Swinging Around
Before this ride starts to move, the seats hang straight down from their supports. When the ride speeds up, the seats swing out at an angle.

Think About This ▶
When the swings are moving around the circle at a constant speed, are they accelerating?

physicspp.com

LAUNCH Lab: How can the motion of a projectile be described?

Question

Can you describe a projectile's motion in both the horizontal and the vertical directions?

Procedure

1. With a marked grid in the background, videotape a ball that is launched with an initial velocity only in the horizontal direction.
2. **Make and Use Graphs** On a sheet of graph paper, draw the location of the ball every 0.1 s (3 frames).
3. Draw two motion diagrams: one for the ball's horizontal motion and one for its vertical motion.

Analysis

How does the vertical motion change as time passes? Does it increase, decrease, or stay the same? How does the horizontal motion change as time passes? Does it increase, decrease, or stay the same?

Critical Thinking Describe the motion of a horizontally launched projectile.

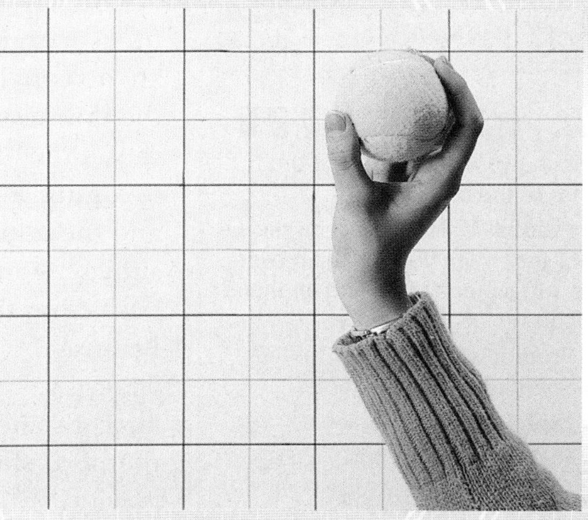

6.1 Projectile Motion

If you observed the movement of a golf ball being hit from a tee, a frog hopping, or a free throw being shot with a basketball, you would notice that all of these objects move through the air along similar paths, as do baseballs, arrows, and bullets. Each path is a curve that moves upward for a distance, and then, after a time, turns and moves downward for some distance. You may be familiar with this curve, called a parabola, from math class.

An object shot through the air is called a **projectile.** A projectile can be a football, a bullet, or a drop of water. After a projectile is launched, what forces are exerted on the projectile? You can draw a free-body diagram of a launched projectile and identify all the forces that are acting on it. No matter what the object is, after a projectile has been given an initial thrust, if you ignore air resistance, it moves through the air only under the force of gravity. The force of gravity is what causes the object to curve downward in a parabolic flight path. Its path through space is called its **trajectory.** If you know the force of the initial thrust on a projectile, you can calculate its trajectory.

► Objectives

- **Recognize** that the vertical and horizontal motions of a projectile are independent.
- **Relate** the height, time in the air, and initial vertical velocity of a projectile using its vertical motion, and then determine the range using the horizontal motion.
- **Explain** how the trajectory of a projectile depends upon the frame of reference from which it is observed.

► Vocabulary

projectile
trajectory

Independence of Motion in Two Dimensions

Think about two softball players warming up for a game, tossing a ball back and forth. What does the path of the ball through the air look like? It looks like a parabola, as you just learned. Imagine that you are standing directly behind one of the players and you are watching the softball as it is being tossed. What would the motion of the ball look like? You would see it go up and back down, just like any object that is tossed straight up in the air. If you were watching the softball from a hot-air balloon high above the field, what motion would you see then? You would see the ball move from one player to the other at a constant speed, just like any object that is given an initial horizontal velocity, such as a hockey puck sliding across ice. The motion of projectiles is a combination of these two motions.

Why do projectiles behave in this way? After a softball leaves a player's hand, what forces are exerted on the ball? If you ignore air resistance, there are no contact forces on the ball. There is only the field force of gravity in the downward direction. How does this affect the ball's motion? Gravity causes the ball to have a downward acceleration.

Figure 6-1 shows the trajectories of two softballs. One was dropped and the other was given an initial horizontal velocity of 2.0 m/s. What is similar about the two paths? Look at their vertical positions. During each flash from the strobe light, the heights of the two softballs are the same. Because the change in vertical position is the same for both, their average vertical velocities during each interval are also the same. The increasingly large distance traveled vertically by the softballs, from one time interval to the next, shows that they are accelerated downward due to the force of gravity. Notice that the horizontal motion of the launched ball does not affect its vertical motion. A projectile launched horizontally has no initial vertical velocity. Therefore, its vertical motion is like that of an object dropped from rest. The downward velocity increases regularly because of the acceleration due to gravity.

MINI LAB

Over the Edge

Obtain two balls, one twice the mass of the other.

1. **Predict** which ball will hit the floor first when you roll them over the surface of a table and let them roll off the edge.
2. **Predict** which ball will hit the floor furthest from the table.
3. **Explain** your predictions.
4. **Test** your predictions.

Analyze and Conclude

5. Does the mass of the ball affect its motion? Is mass a factor in any of the equations for projectile motion?

Concepts In Motion

Interactive Figure To see an animation on independence of motion in two directions, visit physicspp.com.

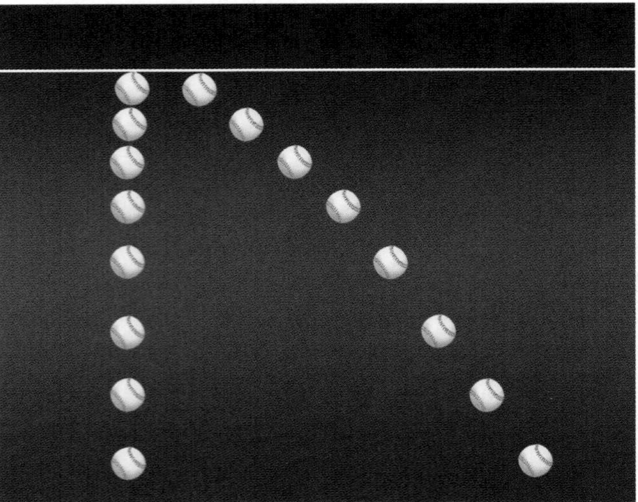

Figure 6-1 The ball on the right was given an initial horizontal velocity. The ball on the left was dropped at the same time from rest. Note that the vertical positions of the two objects are the same during each flash.

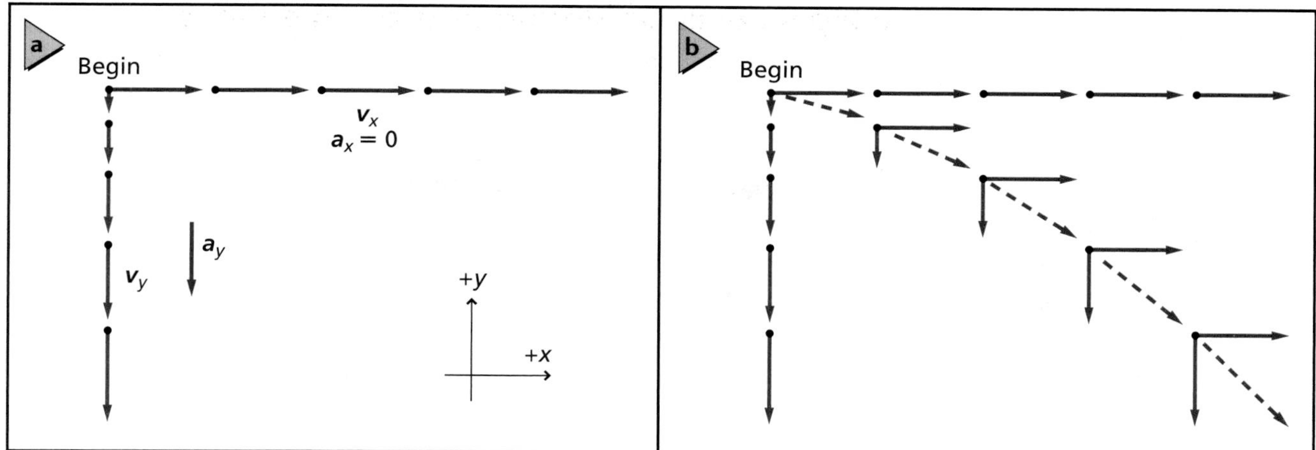

■ **Figure 6-2** A object's motion can be broken into its x- and y-components **(a).** When the horizontal and vertical components of the ball's velocity are combined **(b),** the resultant vectors are tangent to a parabola.

Separate motion diagrams for the horizontal and vertical motions are shown in **Figure 6-2a.** The vertical-motion diagram represents the motion of the dropped ball. The horizontal-motion diagram shows the constant velocity in the x-direction of the launched ball. This constant velocity in the horizontal direction is exactly what should be expected because there is no horizontal force acting on the ball.

In **Figure 6-2b,** the horizontal and vertical components are added to form the total velocity vector for the projectile. You can see how the combination of constant horizontal velocity and uniform vertical acceleration produces a trajectory that has a parabolic shape.

PROBLEM-SOLVING **Strategies**

Motion in Two Dimensions

Projectile motion in two dimensions can be determined by breaking the problem into two connected one-dimensional problems.

1. Divide the projectile motion into a vertical motion problem and a horizontal motion problem.
2. The vertical motion of a projectile is exactly that of an object dropped or thrown straight up or straight down. A gravitational force acts on the object and accelerates it by an amount, g. Review Section 3.3 on free fall to refresh your problem-solving skills for vertical motion.
3. Analyzing the horizontal motion of a projectile is the same as solving a constant velocity problem. No horizontal force acts on a projectile when drag due to air resistance is neglected. Consequently, there are no forces acting in the horizontal direction and therefore, no horizontal acceleration; $a_x = 0.0$ m/s. To solve, use the same methods that you learned in Section 2.4.
4. Vertical motion and horizontal motion are connected through the variable of time. The time from the launch of the projectile to the time it hits the target is the same for both vertical motion and horizontal motion. Therefore, solving for time in one of the dimensions, vertical or horizontal, automatically gives you time for the other dimension.

▶ PRACTICE Problems

• Additional Problems, Appendix B
• Solutions to Selected Problems, Appendix C

1. A stone is thrown horizontally at a speed of 5.0 m/s from the top of a cliff that is 78.4 m high.
   a. How long does it take the stone to reach the bottom of the cliff?
   b. How far from the base of the cliff does the stone hit the ground?
   c. What are the horizontal and vertical components of the stone's velocity just before it hits the ground?
2. Lucy and her friend are working at an assembly plant making wooden toy giraffes. At the end of the line, the giraffes go horizontally off the edge of the conveyor belt and fall into a box below. If the box is 0.6 m below the level of the conveyor belt and 0.4 m away from it, what must be the horizontal velocity of giraffes as they leave the conveyor belt?
3. You are visiting a friend from elementary school who now lives in a small town. One local amusement is the ice-cream parlor, where Stan, the short-order cook, slides his completed ice-cream sundaes down the counter at a constant speed of 2.0 m/s to the servers. (The counter is kept very well polished for this purpose.) If the servers catch the sundaes 7.0 cm from the edge of the counter, how far do they fall from the edge of the counter to the point at which the servers catch them?

Projectiles Launched at an Angle

When a projectile is launched at an angle, the initial velocity has a vertical component, as well as a horizontal component. If the object is launched upward, like a ball tossed straight up in the air, it rises with slowing speed, reaches the top of its path, and descends with increasing speed. **Figure 6-3a** shows the separate vertical- and horizontal-motion diagrams for the trajectory. In the coordinate system, the positive *x*-axis is horizontal and the positive *y*-axis is vertical. Note the symmetry. At each point in the vertical direction, the velocity of the object as it is moving upward has the same magnitude as when it is moving downward. The only difference is that the directions of the two velocities are opposite.

Figure 6-3b defines two quantities associated with the trajectory. One is the maximum height, which is the height of the projectile when the vertical velocity is zero and the projectile has only its horizontal-velocity component. The other quantity depicted is the range, *R*, which is the horizontal distance that the projectile travels. Not shown is the flight time, which is how much time the projectile is in the air. For football punts, flight time often is called hang time.

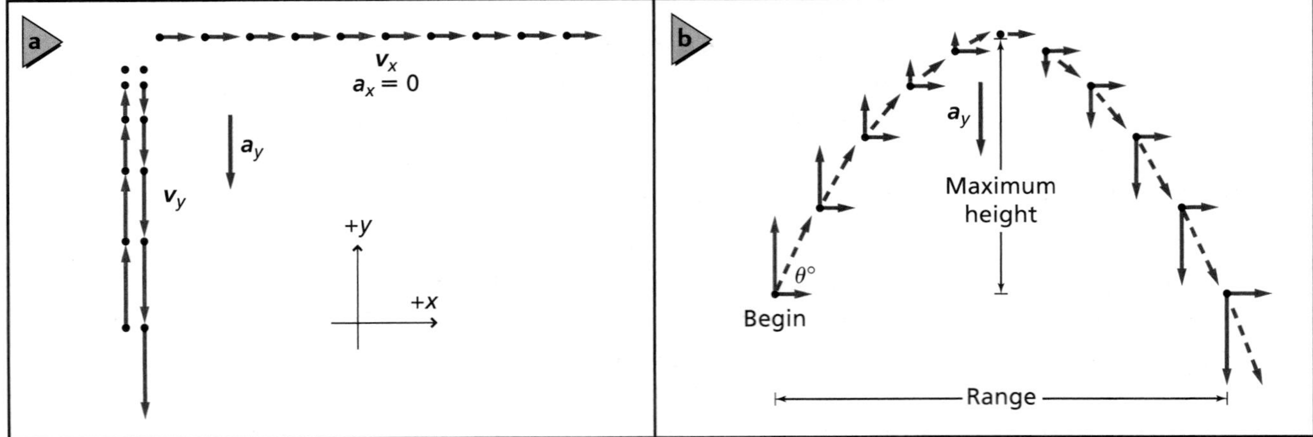

Figure 6-3 The vector sum of $\mathbf{v}_x$ and $\mathbf{v}_y$ at each position points in the direction of the flight.

▶ EXAMPLE Problem 1

The Flight of a Ball A ball is launched at 4.5 m/s at 66° above the horizontal. What are the maximum height and flight time of the ball?

1 Analyze and Sketch the Problem

- Establish a coordinate system with the initial position of the ball at the origin.
- Show the positions of the ball at the beginning, at the maximum height, and at the end of the flight.
- Draw a motion diagram showing ***v*, *a*,** and $\boldsymbol{F}_{net}$.

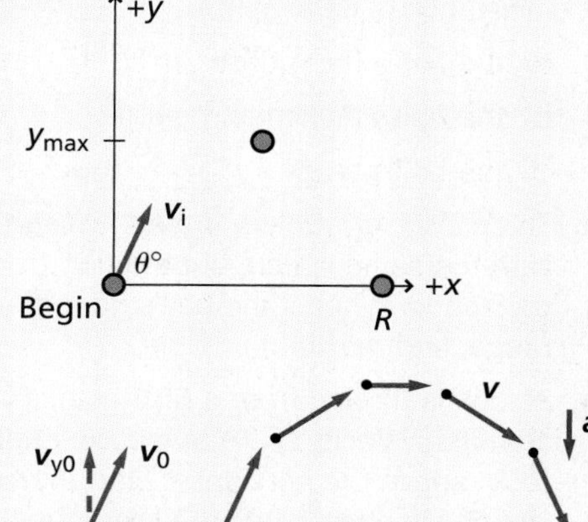

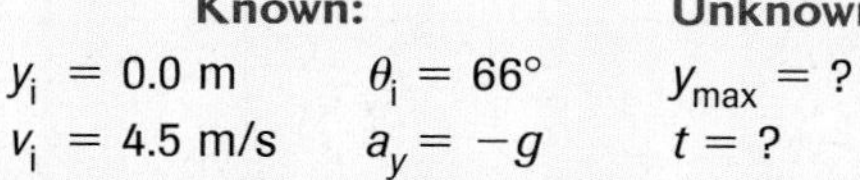

Known:		Unknown:
$y_i = 0.0$ m	$\theta_i = 66°$	$y_{max} = ?$
$v_i = 4.5$ m/s	$a_y = -g$	$t = ?$

2 Solve for the Unknown

Find the y-component of v_i.

$$v_{yi} = v_i(\sin\theta_i)$$
$$= (4.5\text{ m/s})(\sin 66°) \quad \text{Substitute } v_i = 4.5\text{ m/s}, \theta_i = 66°$$
$$= 4.1\text{ m/s}$$

Find an expression for time.

$$v_y = v_{yi} + a_y t$$
$$= v_{yi} - gt \quad \text{Substitute } a_y = -g$$
$$t = \frac{v_{yi} - v_y}{g} \quad \text{Solve for } t.$$

Physics Online
Personal Tutor For an animation on the flight of a ball, visit physicspp.com.

Solve for the maximum height.

$$y_{max} = y_i + v_{yi}t + \frac{1}{2}at^2$$
$$= y_i + v_{yi}\left(\frac{v_{yi} - v_y}{g}\right) + \frac{1}{2}(-g)\left(\frac{v_{yi} - v_y}{g}\right)^2 \quad \text{Substitute } t = \frac{v_{yi} - v_y}{g}, a = -g$$
$$= 0.0\text{ m} + (4.1\text{ m/s})\left(\frac{4.1\text{ m/s} - 0.0\text{ m/s}}{9.80\text{ m/s}^2}\right) + \frac{1}{2}(-9.80\text{ m/s}^2)\left(\frac{4.1\text{ m/s} - 0.0\text{ m/s}}{9.80\text{ m/s}^2}\right)^2$$
$$= 0.86\text{ m}$$

Substitute $y_i = 0.0$ m, $v_{yi} = 4.1$ m/s, $v_y = 0.0$ m/s at y_{max}, $g = 9.80\text{ m/s}^2$

Solve for the time to return to the launching height.

$$y_f = y_i + v_{yi}t + \frac{1}{2}at^2$$
$$0.0\text{ m} = 0.0\text{ m} + v_{yi}t - \frac{1}{2}gt^2 \quad \text{Substitute } y_f = 0.0\text{ m}, y_i = 0.0\text{ m}, a = -g$$
$$t = \frac{-v_{yi} \pm \sqrt{v_{yi}^2 - 4\left(-\frac{1}{2}g\right)(0.0\text{ m})}}{2\left(-\frac{1}{2}g\right)} \quad \text{Use the quadratic formula to solve for } t.$$
$$= \frac{-v_{yi} \pm v_{yi}}{-g}$$
$$= \frac{2v_{yi}}{g} \quad \text{0 is the time the ball left the launch, so use this solution.}$$
$$= \frac{(2)(4.1\text{ m/s})}{(9.80\text{ m/s}^2)} \quad \text{Substitute } v_{yi} = 4.1\text{ m/s}, g = 9.80\text{ m/s}^2$$
$$= 0.84\text{ s}$$

3 Evaluate the Answer

- **Are the units correct?** Dimensional analysis verifies that the units are correct.
- **Do the signs make sense?** All should be positive.
- **Are the magnitudes realistic?** 0.84 s is fast, but an initial velocity of 4.5 m/s makes this time reasonable.

PRACTICE Problems

- Additional Problems, Appendix B
- Solutions to Selected Problems, Appendix C

4. A player kicks a football from ground level with an initial velocity of 27.0 m/s, 30.0° above the horizontal, as shown in **Figure 6-4.** Find each of the following. Assume that air resistance is negligible.
   a. the ball's hang time
   b. the ball's maximum height
   c. the ball's range
5. The player in problem 4 then kicks the ball with the same speed, but at 60.0° from the horizontal. What is the ball's hang time, range, and maximum height?
6. A rock is thrown from a 50.0-m-high cliff with an initial velocity of 7.0 m/s at an angle of 53.0° above the horizontal. Find the velocity vector for when it hits the ground below.

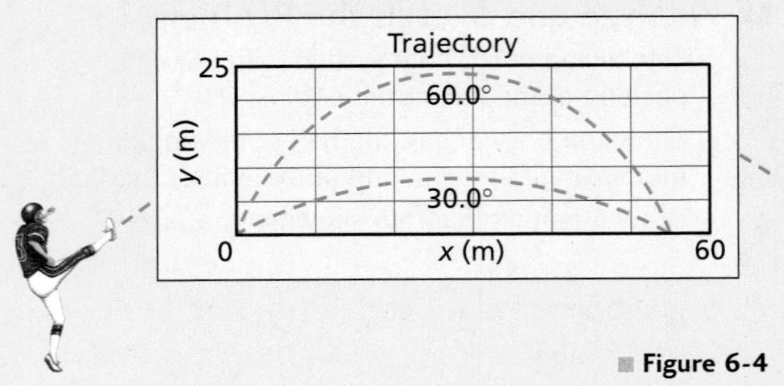

Figure 6-4

Trajectories Depend upon the Viewer

Suppose you toss a ball up and catch it while riding in a bus. To you, the ball would seem to go straight up and straight down. But what would an observer on the sidewalk see? The observer would see the ball leave your hand, rise up, and return to your hand, but because the bus would be moving, your hand also would be moving. The bus, your hand, and the ball would all have the same horizontal velocity. Thus, the trajectory of the ball would be similar to that of the ball in Example Problem 1.

Air resistance So far, air resistance has been ignored in the analysis of projectile motion. While the effects of air resistance are very small for some projectiles, for others, the effects are large and complex. For example, dimples on a golf ball reduce air resistance and maximize its range. In baseball, the spin of the ball creates forces that can deflect the ball. For now, just remember that the force due to air resistance does exist and it can be important.

6.1 Section Review

7. **Projectile Motion** Two baseballs are pitched horizontally from the same height, but at different speeds. The faster ball crosses home plate within the strike zone, but the slower ball is below the batter's knees. Why does the faster ball not fall as far as the slower one?

8. **Free-Body Diagram** An ice cube slides without friction across a table at a constant velocity. It slides off the table and lands on the floor. Draw free-body and motion diagrams of the ice cube at two points on the table and at two points in the air.

9. **Projectile Motion** A softball is tossed into the air at an angle of 50.0° with the vertical at an initial velocity of 11.0 m/s. What is its maximum height?

10. **Projectile Motion** A tennis ball is thrown out a window 28 m above the ground at an initial velocity of 15.0 m/s and 20.0° below the horizontal. How far does the ball move horizontally before it hits the ground?

11. **Critical Thinking** Suppose that an object is thrown with the same initial velocity and direction on Earth and on the Moon, where g is one-sixth that on Earth. How will the following quantities change?
    a. v_x
    b. the object's time of flight
    c. y_{max}
    d. R

physicspp.com/self_check_quiz

6.2 Circular Motion

Consider an object moving in a circle at a constant speed, such as a stone being whirled on the end of a string or a fixed horse on a merry-go-round. Are these objects accelerating? At first, you might think that they are not because their speeds do not change. However, remember that acceleration is the change in velocity, not just the change in speed. Because their direction is changing, the objects must be accelerating.

> **Objectives**
> - **Explain** why an object moving in a circle at a constant speed is accelerated.
> - **Describe** how centripetal acceleration depends upon the object's speed and the radius of the circle.
> - **Identify** the force that causes centripetal acceleration.
>
> **Vocabulary**
> uniform circular motion
> centripetal acceleration
> centripetal force

Describing Circular Motion

Uniform circular motion is the movement of an object or particle trajectory at a constant speed around a circle with a fixed radius. The position of an object in uniform circular motion, relative to the center of the circle, is given by the position vector $\boldsymbol{r}$, shown in **Figure 6-5a.** As the object moves around the circle, the length of the position vector does not change, but its direction does. To find the object's velocity, you need to find its displacement vector over a time interval. The change in position, or the object's displacement, is represented by $\Delta \boldsymbol{r}$. **Figure 6-5b** shows two position vectors: $\boldsymbol{r}_1$ at the beginning of a time interval, and $\boldsymbol{r}_2$ at the end of the time interval. Remember that a position vector is a displacement vector with its tail at the origin. In the vector diagram, $\boldsymbol{r}_1$ and $\boldsymbol{r}_2$ are subtracted to give the resultant $\Delta \boldsymbol{r}$, the displacement during the time interval. You know that a moving object's average velocity is $\Delta \boldsymbol{d}/\Delta t$, so for an object in circular motion, $\bar{\boldsymbol{v}} = \Delta \boldsymbol{r}/\Delta t$. The velocity vector has the same direction as the displacement, but a different length. You can see in **Figure 6-6a** that the velocity is at right angles to the position vector, which is tangent to its circular path. As the velocity vector moves around the circle, its direction changes but its length remains the same.

What is the direction of the object's acceleration? Figure 6-6a shows the velocity vectors $\boldsymbol{v}_1$ and $\boldsymbol{v}_2$ at the beginning and end of a time interval. The difference in the two vectors, $\Delta \boldsymbol{v}$, is found by subtracting the vectors, as shown in **Figure 6-6b.** The average acceleration, $\bar{\boldsymbol{a}} = \Delta \boldsymbol{v}/\Delta t$, is in the same direction as $\Delta \boldsymbol{v}$; that is, toward the center of the circle. Repeat this process for several other time intervals when the object is in different locations on the circle. As the object moves around the circle, the direction of the acceleration vector changes, but its length remains the same. Notice that the acceleration vector of an object in uniform circular motion always points in toward the center of the circle. For this reason, the acceleration of such an object is called center-seeking or **centripetal acceleration.**

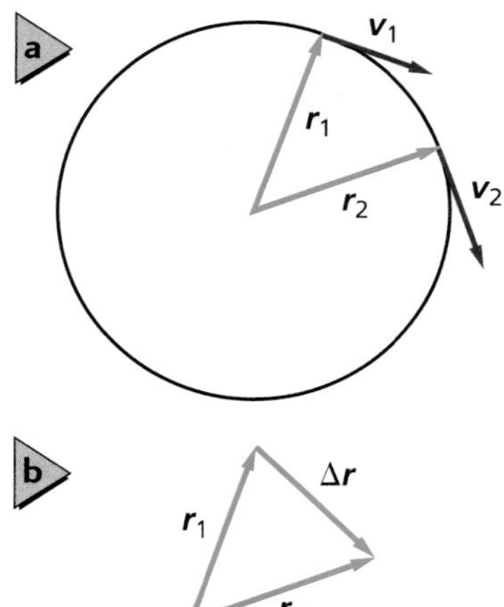

Figure 6-5 The displacement, Δr, of an object in circular motion, divided by the time interval in which the displacement occurs, is the object's average velocity during that time interval.

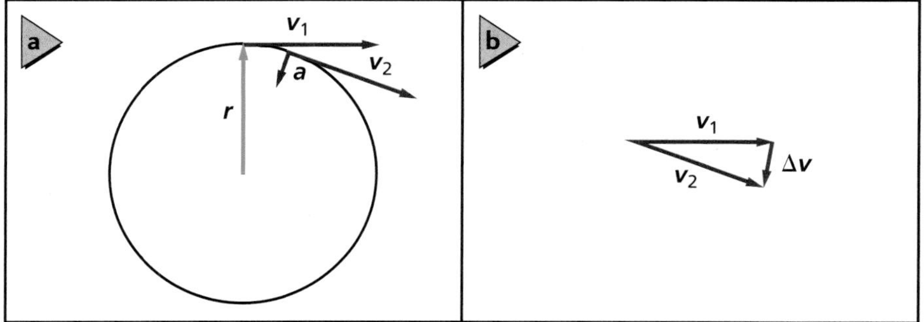

Figure 6-6 The direction of the change in velocity is toward the center of the circle, and so the acceleration vector also points to the center of the circle.

Centripetal Acceleration

What is the magnitude of an object's centripetal acceleration? Compare the triangle made from the position vectors in Figure 6-5b with the triangle made by the velocity vectors in Figure 6-6b. The angle between $\boldsymbol{r}_1$ and $\boldsymbol{r}_2$ is the same as that between $\boldsymbol{v}_1$ and $\boldsymbol{v}_2$. Therefore, the two triangles formed by subtracting the two sets of vectors are similar triangles, and the ratios of the lengths of two corresponding sides are equal. Thus, $\Delta r/r = \Delta v/v$. The equation is not changed if both sides are divided by Δt.

$$\frac{\Delta r}{r\Delta t} = \frac{\Delta v}{v\Delta t}$$

However, $v = \Delta r/\Delta t$ and $a = \Delta v/\Delta t$.

$$\frac{1}{r}\left(\frac{\Delta r}{\Delta t}\right) = \frac{1}{v}\left(\frac{\Delta v}{\Delta t}\right)$$

Substituting $v = \Delta r/\Delta t$ in the left-hand side and $a = \Delta v/\Delta t$ in the right-hand side gives the following equation.

$$\frac{v}{r} = \frac{a}{v}$$

Solve this equation for acceleration and give it the special symbol a_c, for centripetal acceleration.

Centripetal Acceleration $a_c = \frac{v^2}{r}$

Centripetal acceleration always points to the center of the circle. Its magnitude is equal to the square of the speed, divided by the radius of motion.

APPLYING PHYSICS

Space Elevators Scientists are considering the use of space elevators as a low-cost transportation system to space. A cable would be anchored to a station at Earth's equator, and the cable would extend almost 35,800 km from Earth's surface. The cable would be attached to a counterweight and would stay extended due to centripetal force. Special magnetically powered vehicles would then travel along the cable.

How can you measure the speed of an object moving in a circle? One way is to measure its period, T, the time needed for the object to make one complete revolution. During this time, the object travels a distance equal to the circumference of the circle, $2\pi r$. The object's speed, then, is represented by $v = 2\pi r/T$. If this expression is substituted for v in the equation for centripetal acceleration, the following equation is obtained.

$$a_c = \frac{\left(\frac{2\pi r}{T}\right)^2}{r} = \frac{4\pi^2 r}{T^2}$$

Because the acceleration of an object moving in a circle is always in the direction of the net force acting on it, there must be a net force toward the center of the circle. This force can be provided by any number of agents. For Earth circling the Sun, the force is the Sun's gravitational force on Earth, as you'll learn in Chapter 7. When a hammer thrower swings the hammer, as in **Figure 6-7,** the force is the tension in the chain attached to the massive ball. When an object moves in a circle, the net force toward the center of the circle is called the **centripetal force.** To accurately analyze centripetal acceleration situations, you must identify the agent of the force that causes the acceleration. Then you can apply Newton's second law for the component in the direction of the acceleration in the following way.

Figure 6-7 When the thrower lets go, the hammer initially moves in a straight line that is tangent to the point of release. Then it follows a trajectory like that of any object released into the air with an initial horizontal velocity.

Newton's Second Law for Circular Motion $F_{net} = ma_c$

The net centripetal force on an object moving in a circle is equal to the object's mass, times the centripetal acceleration.

When solving problems, you have found it useful to choose a coordinate system with one axis in the direction of the acceleration. For circular motion, the direction of the acceleration is always toward the center of the circle. Rather than labeling this axis *x* or *y*, call it *c*, for centripetal acceleration. The other axis is in the direction of the velocity, tangent to the circle. It is labeled *tang* for tangential. You will apply Newton's second law in these directions, just as you did in the two-dimensional problems in Chapter 5. Remember that centripetal force is just another name for the net force in the centripetal direction. It is the sum of all the real forces, those for which you can identify agents that act along the centripetal axis.

In the case of the hammer thrower in Figure 6-7, in what direction does the hammer fly when the chain is released? Once the contact force of the chain is gone, there is no force accelerating the hammer toward the center of the circle, so the hammer flies off in the direction of its velocity, which is tangent to the circle. Remember, if you cannot identify the agent of the force, then it does not exist.

▶ EXAMPLE **Problem 2**

Uniform Circular Motion **A 13-g rubber stopper is attached to a 0.93-m string. The stopper is swung in a horizontal circle, making one revolution in 1.18 s. Find the tension force exerted by the string on the stopper.**

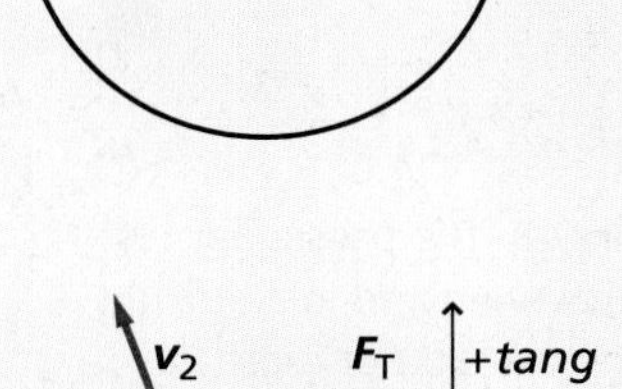

1 Analyze and Sketch the Problem

- Draw a free-body diagram for the swinging stopper.
- Include the radius and the direction of motion.
- Establish a coordinate system labeled *tang* and *c*. The directions of a and F_T are parallel to *c*.

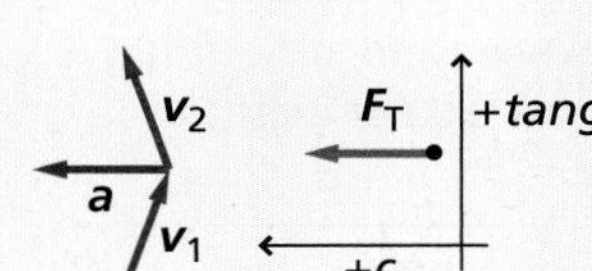

Known:	**Unknown:**
$m = 13$ g	$F_T = ?$
$r = 0.93$ m	
$T = 1.18$ s	

2 Solve for the Unknown

Find the centripetal acceleration.

$$a_c = \frac{4\pi^2 r}{T^2}$$

$$= \frac{4\pi^2(0.93\ \text{m})}{(1.18\ \text{s})^2} \qquad \text{Substitute } r = 0.93\ \text{m},\ T = 1.18\ \text{s}$$

$$= 26\ \text{m/s}^2$$

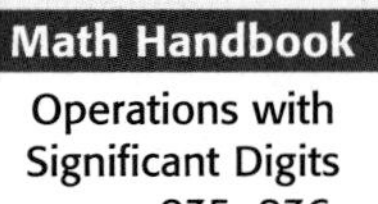

Math Handbook
Operations with Significant Digits pages 835–836

Use Newton's second law to find the tension in the string.

$$F_T = ma_c$$

$$= (0.013\ \text{kg})(26\ \text{m/s}^2) \qquad \text{Substitute } m = 0.013\ \text{kg},\ a_c = 26\ \text{m/s}^2$$

$$= 0.34\ \text{N}$$

3 Evaluate the Answer

- **Are the units correct?** Dimensional analysis verifies that a is in m/s² and F is in N.
- **Do the signs make sense?** The signs should all be positive.
- **Are the magnitudes realistic?** The force is almost three times the weight of the stopper, and the acceleration is almost three times that of gravity, which is reasonable for such a light object.

PRACTICE Problems

• Additional Problems, Appendix B
• Solutions to Selected Problems, Appendix C

12. A runner moving at a speed of 8.8 m/s rounds a bend with a radius of 25 m. What is the centripetal acceleration of the runner, and what agent exerts force on the runner?

13. A car racing on a flat track travels at 22 m/s around a curve with a 56-m radius. Find the car's centripetal acceleration. What minimum coefficient of static friction between the tires and road is necessary for the car to round the curve without slipping?

14. An airplane traveling at 201 m/s makes a turn. What is the smallest radius of the circular path (in km) that the pilot can make and keep the centripetal acceleration under 5.0 m/s^2?

15. A 45-kg merry-go-round worker stands on the ride's platform 6.3 m from the center. If her speed as she goes around the circle is 4.1 m/s, what is the force of friction necessary to keep her from falling off the platform?

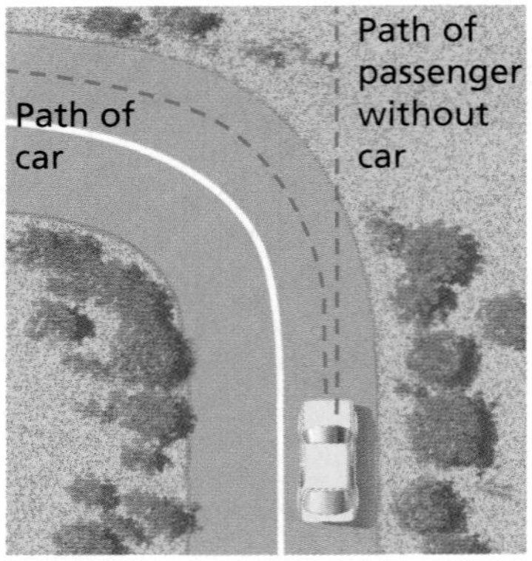

■ **Figure 6-8** The passenger would move forward in a straight line if the car did not exert an inward force.

A Nonexistent Force

If a car makes a sharp left turn, a passenger on the right side might be thrown against the right door. Is there an outward force on the passenger? Consider a similar situation. If a car in which you are riding stops suddenly, you will be thrown forward into your safety belt. Is there a forward force on you? No, because according to Newton's first law, you will continue moving with the same velocity unless there is a net force acting on you. The safety belt applies the force that accelerates you to a stop. **Figure 6-8** shows a car turning to the left as viewed from above. A passenger in the car would continue to move straight ahead if it were not for the force of the door acting in the direction of the acceleration; that is, toward the center of the circle. Thus, there is no outward force on the passenger. The so-called centrifugal, or outward force, is a fictitious, nonexistent force. Newton's laws are able to explain motion in both straight lines and circles.

6.2 Section Review

16. Uniform Circular Motion What is the direction of the force that acts on the clothes in the spin cycle of a washing machine? What exerts the force?

17. Free-Body Diagram You are sitting in the back-seat of a car going around a curve to the right. Sketch motion and free-body diagrams to answer the following questions.

a. What is the direction of your acceleration?

b. What is the direction of the net force that is acting on you?

c. What exerts this force?

18. Centripetal Force If a 40.0-g stone is whirled horizontally on the end of a 0.60-m string at a speed of 2.2 m/s, what is the tension in the string?

19. Centripetal Acceleration A newspaper article states that when turning a corner, a driver must be careful to balance the centripetal and centrifugal forces to keep from skidding. Write a letter to the editor that critiques this article.

20. Centripetal Force A bowling ball has a mass of 7.3 kg. If you move it around a circle with a radius of 0.75 m at a speed of 2.5 m/s, what force would you have to exert on it?

21. Critical Thinking Because of Earth's daily rotation, you always move with uniform circular motion. What is the agent that supplies the force that accelerates you? How does this motion affect your apparent weight?

physicspp.com/self_check_quiz

6.3 Relative Velocity

> ▶ **Objectives**
> - **Analyze** situations in which the coordinate system is moving.
> - **Solve** relative-velocity problems.

Suppose that you are in a school bus that is traveling at a velocity of 8 m/s in a positive direction. You walk with a velocity of 3 m/s toward the front of the bus. If a friend of yours is standing on the side of the road watching the bus with you on it go by, how fast would your friend say that you are moving? If the bus is traveling at 8 m/s, this means that the velocity of the bus is 8 m/s, as measured by your friend in a coordinate system fixed to the road. When you are standing still, your velocity relative to the road is also 8 m/s, but your velocity relative to the bus is zero. Walking at 1 m/s toward the front of the bus means that your velocity is measured relative to the bus. The problem can be rephrased as follows: Given the velocity of the bus relative to the road and your velocity relative to the bus, what is your velocity relative to the road?

A vector representation of this problem is shown in **Figure 6-9a.** After studying it, you will find that your velocity relative to the street is 9 m/s, the sum of 8 m/s and 1 m/s. Suppose that you now walk at the same speed toward the rear of the bus. What would be your velocity relative to the road? **Figure 6-9b** shows that because the two velocities are in opposite directions, the resultant velocity is 7 m/s, the difference between 8 m/s and 1 m/s. You can see that when the velocities are along the same line, simple addition or subtraction can be used to determine the relative velocity.

Take a closer look at how these results were obtained and see if you can find a mathematical rule to describe how velocities are combined in these relative-velocity situations. For the above situation, you can designate the velocity of the bus relative to the road as $\boldsymbol{v}_{b/r}$, your velocity relative to the bus as $\boldsymbol{v}_{y/b}$, and the velocity of you relative to the road as $\boldsymbol{v}_{y/r}$. To find the velocity of you relative to the road in both cases, you vectorially added the velocities of you relative to the bus and the bus relative to the road. Mathematically, this is represented as $\boldsymbol{v}_{y/b} + \boldsymbol{v}_{b/r} = \boldsymbol{v}_{y/r}$. The more general form of this equation is as follows.

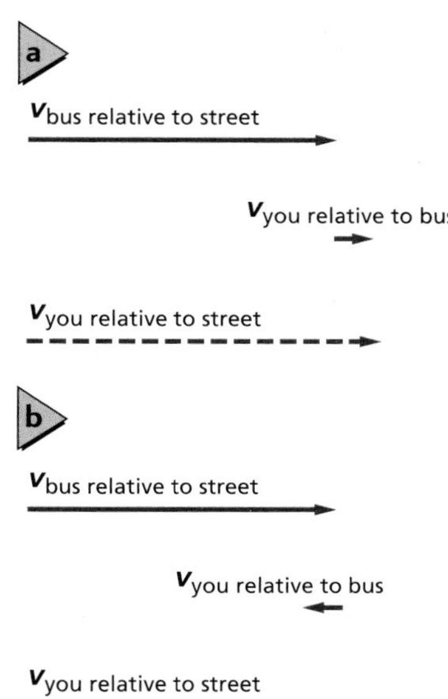

Figure 6-9 When a coordinate system is moving, two velocities are added if both motions are in the same direction and one is subtracted from the other if the motions are in opposite directions.

Relative Velocity $\boldsymbol{v}_{a/b} + \boldsymbol{v}_{b/c} = \boldsymbol{v}_{a/c}$

The relative velocity of object a to object c is the vector sum of object a's velocity relative to object b and object b's velocity relative to object c.

CHALLENGE **PROBLEM**

Phillipe whirls a stone of mass *m* on a rope in a perfect horizontal circle above his head such that the stone is at a height, *h*, above the ground. The circle has a radius of *r*, and the tension in the rope is *T*. Suddenly the rope breaks and the stone falls to the ground. The stone travels a horizontal distance, *s*, from the time the rope breaks until it impacts the ground. Find a mathematical expression for *s* in terms of *T*, *r*, *m*, and *h*. Does your expression change if Phillipe is walking 0.50 m/s relative to the ground?

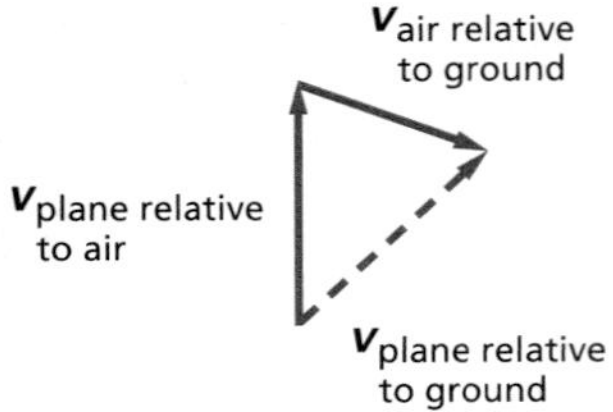

Figure 6-10 The plane's velocity relative to the ground can be obtained by vector addition.

This method for adding relative velocities also applies to motion in two dimensions. For example, airline pilots cannot expect to reach their destinations by simply aiming their planes along a compass direction. They must take into account the plane's speed relative to the air, which is given by their airspeed indicators, and their direction of flight relative to the air. They also must consider the velocity of the wind at the altitude they are flying relative to the ground. These two vectors must be combined, as shown in **Figure 6-10**, to obtain the velocity of the airplane relative to the ground. The resultant vector tells the pilot how fast and in what direction the plane must travel relative to the ground to reach its destination. A similar situation occurs for boats traveling on water with a flowing current.

▶ EXAMPLE Problem 3

Relative Velocity of a Marble **Ana and Sandra are riding on a ferry boat that is traveling east at a speed of 4.0 m/s. Sandra rolls a marble with a velocity of 0.75 m/s north, straight across the deck of the boat to Ana. What is the velocity of the marble relative to the water?**

1 Analyze and Sketch the Problem

- Establish a coordinate system.
- Draw vectors to represent the velocities of the boat relative to the water and the marble relative to the boat.

Known:

$v_{b/w} = 4.0$ m/s

$v_{m/b} = 0.75$ m/s

Unknown:

$v_{m/w} = ?$

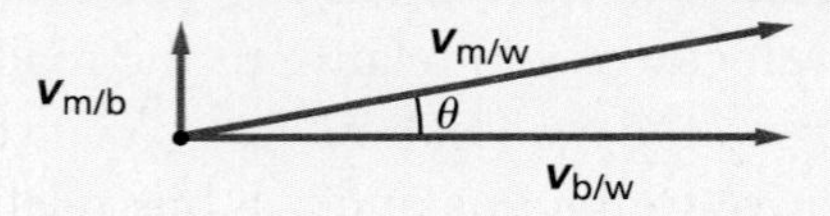

+y

2 Solve for the Unknown

Because the two velocities are at right angles, use the Pythagorean theorem.

$$v_{m/w}^2 = v_{b/w}^2 + v_{m/b}^2$$

$$v_{m/w} = \sqrt{v_{b/w}^2 + v_{m/b}^2}$$

$$= \sqrt{(4.0\ \text{m/s})^2 + (0.75\ \text{m/s})^2}$$ **Substitute $v_{b/w} = 4.0$ m/s, $v_{m/b} = 0.75$ m/s**

$$= 4.1\ \text{m/s}$$

Math Handbook
Inverses of Sine, Cosine, and Tangent
page 856

Find the angle of the marble's motion.

$$\theta = \tan^{-1}\left(\frac{v_{m/b}}{v_{b/w}}\right)$$

$$= \tan^{-1}\left(\frac{0.75\ \text{m/s}}{4.0\ \text{m/s}}\right)$$ **Substitute $v_{b/w} = 4.0$ m/s, $v_{m/b} = 0.75$ m/s**

$= 11°$ north of east

The marble is traveling 4.1 m/s at 11° north of east.

3 Evaluate the Answer

- **Are the units correct?** Dimensional analysis verifies that the velocity is in m/s.
- **Do the signs make sense?** The signs should all be positive.
- **Are the magnitudes realistic?** The resulting velocity is of the same order of magnitude as the velocities given in the problem.

PRACTICE Problems

- Additional Problems, Appendix B
- Solutions to Selected Problems, Appendix C

22. You are riding in a bus moving slowly through heavy traffic at 2.0 m/s. You hurry to the front of the bus at 4.0 m/s relative to the bus. What is your speed relative to the street?

23. Rafi is pulling a toy wagon through the neighborhood at a speed of 0.75 m/s. A caterpillar in the wagon is crawling toward the rear of the wagon at a rate of 2.0 cm/s. What is the caterpillar's velocity relative to the ground?

24. A boat is rowed directly upriver at a speed of 2.5 m/s relative to the water. Viewers on the shore see that the boat is moving at only 0.5 m/s relative to the shore. What is the speed of the river? Is it moving with or against the boat?

25. An airplane flies due north at 150 km/h relative to the air. There is a wind blowing at 75 km/h to the east relative to the ground. What is the plane's speed relative to the ground?

Another example of combined relative velocities is the navigation of migrating neotropical songbirds. In addition to knowing in which direction to fly, a bird must account for its speed relative to the air and its direction relative to the ground. If a bird tries to fly over the Gulf of Mexico into too strong a headwind, it will run out of energy before it reaches the other shore and will perish. Similarly, the bird must account for crosswinds or it will not reach its destination. You can add relative velocities even if they are at arbitrary angles by using the graphical methods that you learned in Chapter 5.

Biology Connection

Remember that the key to properly analyzing a two-dimensional relative-velocity situation is drawing the proper triangle to represent the three velocities. Once you have this triangle, you simply apply your knowledge of vector addition from Chapter 5. If the situation contains two velocities that are perpendicular to each other, you can find the third by applying the Pythagorean theorem; however, if the situation has no right angles, you will need to use one or both of the laws of sines and cosines.

6.3 Section Review

26. Relative Velocity A fishing boat with a maximum speed of 3 m/s relative to the water is in a river that is flowing at 2 m/s. What is the maximum speed the boat can obtain relative to the shore? The minimum speed? Give the direction of the boat, relative to the river's current, for the maximum speed and the minimum speed relative to the shore.

27. Relative Velocity of a Boat A powerboat heads due northwest at 13 m/s relative to the water across a river that flows due north at 5.0 m/s. What is the velocity (both magnitude and direction) of the motorboat relative to the shore?

28. Relative Velocity An airplane flies due south at 175 km/h relative to the air. There is a wind blowing at 85 km/h to the east relative to the ground. What are the plane's speed and direction relative to the ground?

29. A Plane's Relative Velocity An airplane flies due north at 235 km/h relative to the air. There is a wind blowing at 65 km/h to the northeast relative to the ground. What are the plane's speed and direction relative to the ground?

30. Relative Velocity An airplane has a speed of 285 km/h relative to the air. There is a wind blowing at 95 km/h at 30.0° north of east relative to Earth. In which direction should the plane head to land at an airport due north of its present location? What is the plane's speed relative to the ground?

31. Critical Thinking You are piloting a boat across a fast-moving river. You want to reach a pier directly opposite your starting point. Describe how you would navigate the boat in terms of the components of your velocity relative to the water.

PHYSICS LAB • Design Your Own

On Target

In this activity, you will analyze several factors that affect the motion of a projectile and use your understanding of these factors to predict the path of a projectile. Finally, you will design a projectile launcher and hit a target a known distance away.

QUESTION

What factors affect the path of a projectile?

Objectives

- **Formulate models** and then summarize the factors that affect the motion of a projectile.
- **Use models** to predict where a projectile will land.

Safety Precautions

Possible Materials

duct tape
plastic ware
rubber bands
paper clips
paper
masking tape
wood blocks
nails
hammer
PVC tubing
handsaw
scissors
coat hanger
chicken wire
wire cutter

Procedure

1. Brainstorm and list as many factors as you are able to think of that may affect the path of a projectile.
2. Create a design for your projectile launcher and decide what object will be your projectile shot by your launcher.
3. Taking the design of your launcher into account, determine which two factors are most likely to have a significant effect on the flight path of your projectile.
4. Check the design of your launcher and discuss your two factors with your teacher and make any necessary changes to your setup before continuing.
5. Create a method for determining what effect these two factors will have on the path of your projectile.
6. Have your teacher approve your method before collecting data.

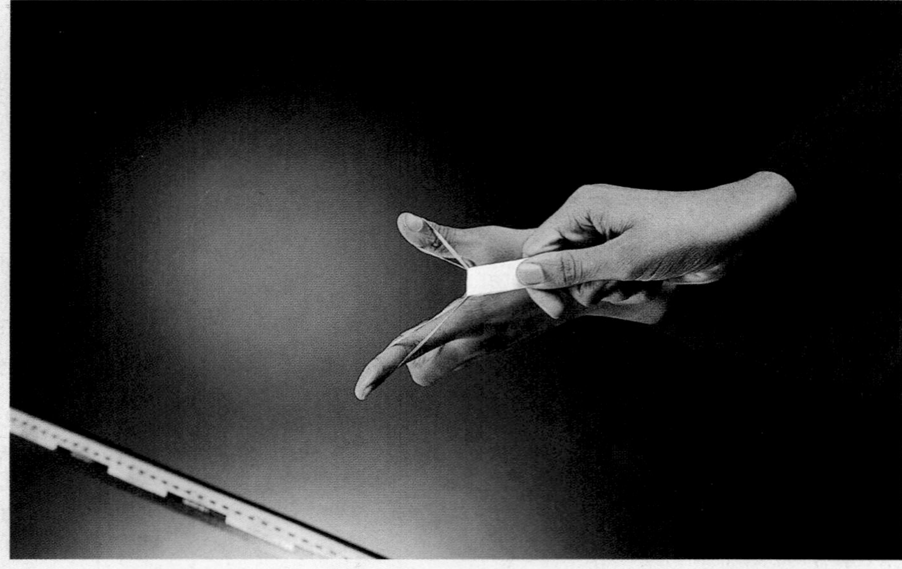

Data Table 1	
Launch Angle (deg)	**Distance Projectile Travels (cm)**

Data Table 2	
Distance Rubber Band Is Pulled Back (cm)	**Distance Projectile Travels (cm)**

Analyze

1. **Make and Use Graphs** Make graphs of your data to help you predict how to use your launcher to hit a target.
2. **Analyze** What are the relationships between each variable you have tested and the distance the projectile travels?

Conclude and Apply

1. What were the main factors influencing the path of the projectile?
2. Predict the conditions necessary to hit a target provided by your teacher.
3. **Explain** If you have a perfect plan and still miss the target on your first try, is there a problem with the variability of laws of physics? Explain.
4. Launch your projectile at the target. If you miss, make the necessary adjustments and try again.

Going Further

1. How might your data have varied if you did this experiment outside? Would there be any additional factors affecting the motion of your projectile?
2. How might the results of your experiment be different if the target was elevated above the height of the launcher?
3. How might your experiment differ if the launcher was elevated above the height of the target?

Real-World Physics

1. When a kicker attempts a field goal, do you think it is possible for him to miss because he kicked it too high? Explain.
2. If you wanted to hit a baseball as far as possible, what would be the best angle to hit the ball?

To find out more about projectile motion, visit the Web site: **physicspp.com**

Future Technology

Spinning Space Stations

There is a lot going on aboard the *International Space Station* (ISS). Scientists from different countries are conducting experiments and making observations. They have seen water drops form as floating spheres and have grown peas in space to test whether crops can be grown in weightlessness.

One goal of the ISS is to examine the effects on the human body when living in space for prolonged periods of time. If negative health effects can be identified, perhaps they can be prevented. This could give humans the option of living in space for long periods of time.

Harmful effects of weightlessness have been observed. On Earth, muscles have gravity to push and pull against. Muscles weaken from disuse if this resistance is removed. Bones can weaken for the same reason. Also, blood volume can decrease. On Earth, gravity pulls blood downward so it collects in the lower legs. In weightlessness, the blood can more easily collect in an astronaut's head. The brain senses the extra blood and sends a signal to make less of it.

Long-term life in space is hindered by the practical challenges of weightlessness as well. Imagine how daily life would change. Everything must be strapped or bolted down. You would have to be strapped down to a bed to sleep in one. Your life would be difficult in a space station unless the space station could be modified to simulate gravity. How could this be done?

The Rotating Space Station Have you ever been on a human centrifuge—a type of amusement park ride that uses centripetal force? Everyone stands against the walls of a big cylinder. Then the cylinder begins to rotate faster and faster until the riders are pressed against the walls. Because of the centripetal acceleration, the riders are held there so that even when the floor drops down they are held securely against the walls of the whirling container.

A space station could be designed that uses the effects of centripetal motion as a replacement for gravity. Imagine a space station in the form of a large ring. The space station and all the objects and occupants inside would float weightlessly inside. If the ring were made to spin, unattached objects would be held against the ring's outer edge because of the centripetal motion. If the space station spun at the right rate and if it had the right diameter, the centripetal motion would cause the occupants to experience a force of the same magnitude as gravity. The down direction in the space station would be what an observer outside the station would see as radially outward, away from the ring's center.

This is an artist's rendition of a rotating space station.

Centripetal acceleration is directly proportional to the distance from the center of a rotating object. A rotating space station could be built in the form of concentric rings, each ring experiencing a different gravity. The innermost rings would experience the smallest gravity, while outermost rings would experience the largest force. You could go from floating peacefully in a low-gravity ring to standing securely in the simulated Earth-gravity ring.

Going Further

1. **Research** What factors must engineers take into account in order to make a rotating space station that can simulate Earth's gravity?
2. **Apply** You are an astronaut aboard a rotating space station. You feel pulled by gravity against the floor. Explain what is really going on in terms of Newton's laws and centripetal force.
3. **Critical Thinking** What benefits does a rotating space station offer its occupants? What are the negative features?

Chapter 6

Study Guide

6.1 Projectile Motion

Vocabulary

- projectile (p. 147)
- trajectory (p. 147)

Key Concepts

- The vertical and horizontal motions of a projectile are independent.
- The vertical motion component of a projectile experiences a constant acceleration.
- When there is no air resistance, the horizontal motion component does not experience an acceleration and has constant velocity.
- Projectile problems are solved by first using the vertical motion to relate height, time in the air, and initial vertical velocity. Then the distance traveled horizontally is found.
- The range of a projectile depends upon the acceleration due to gravity and upon both components of the initial velocity.
- The curved flight path that is followed by a projectile is called a parabola.

6.2 Circular Motion

Vocabulary

- uniform circular motion (p. 153)
- centripetal acceleration (p. 153)
- centripetal force (p. 154)

Key Concepts

- An object moving in a circle at a constant speed accelerates toward the center of the circle, and therefore, it has centripetal acceleration.
- Centripetal acceleration depends directly on the square of the object's speed and inversely on the radius of the circle.

$$a_c = \frac{v^2}{r}$$

- The centripetal acceleration for an object traveling in a circle can also be expressed as a function of its period, T.

$$a_c = \frac{4\pi^2 r}{T^2}$$

- A net force must be exerted toward the circle's center to cause centripetal acceleration.

$$F_{net} = ma_c$$

- The velocity vector of an object with a centripetal acceleration is always tangent to the circular path.

6.3 Relative Velocity

Key Concepts

- Vector addition can be used to solve problems involving relative velocities.

$$\boldsymbol{v}_{a/b} + \boldsymbol{v}_{b/c} = \boldsymbol{v}_{a/c}$$

- The key to properly analyzing a two-dimensional relative-velocity problem is drawing the proper triangle to represent all three velocity vectors.

Chapter 6 Assessment

Concept Mapping

32. Use the following terms to complete the concept map below: *constant speed, horizontal part of projectile motion, constant acceleration, relative-velocity motion, uniform circular motion.*

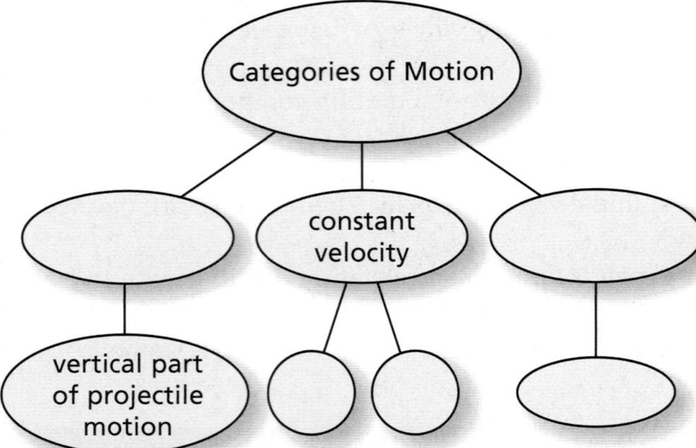

Mastering Concepts

33. Consider the trajectory of the cannonball shown in **Figure 6-11.** (6.1)
    a. Where is the magnitude of the vertical-velocity component largest?
    b. Where is the magnitude of the horizontal-velocity component largest?
    c. Where is the vertical-velocity smallest?
    d. Where is the magnitude of the acceleration smallest?

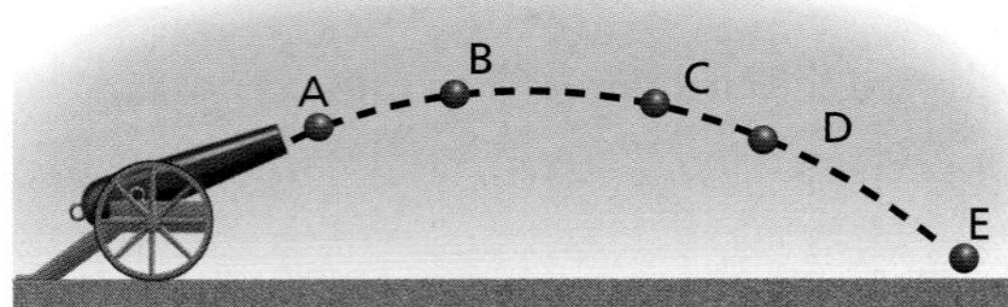

Figure 6-11

34. A student is playing with a radio-controlled race car on the balcony of a sixth-floor apartment. An accidental turn sends the car through the railing and over the edge of the balcony. Does the time it takes the car to fall depend upon the speed it had when it left the balcony? (6.1)

35. An airplane pilot flying at constant velocity and altitude drops a heavy crate. Ignoring air resistance, where will the plane be relative to the crate when the crate hits the ground? Draw the path of the crate as seen by an observer on the ground. (6.1)

36. Can you go around a curve with the following accelerations? Explain.
    a. zero acceleration
    b. constant acceleration (6.2)

37. To obtain uniform circular motion, how must the net force depend on the speed of the moving object? (6.2)

38. If you whirl a yo-yo about your head in a horizontal circle, in what direction must a force act on the yo-yo? What exerts the force? (6.2)

39. Why is it that a car traveling in the opposite direction as the car in which you are riding on the freeway often looks like it is moving faster than the speed limit? (6.3)

Applying Concepts

40. **Projectile Motion** Analyze how horizontal motion can be uniform while vertical motion is accelerated. How will projectile motion be affected when drag due to air resistance is taken into consideration?

41. **Baseball** A batter hits a pop-up straight up over home plate at an initial velocity of 20 m/s. The ball is caught by the catcher at the same height that it was hit. At what velocity does the ball land in the catcher's mitt? Neglect air resistance.

42. **Fastball** In baseball, a fastball takes about $\frac{1}{2}$ s to reach the plate. Assuming that such a pitch is thrown horizontally, compare the distance the ball falls in the first $\frac{1}{4}$ s with the distance it falls in the second $\frac{1}{4}$ s.

43. You throw a rock horizontally. In a second horizontal throw, you throw the rock harder and give it even more speed.
    a. How will the time it takes the rock to hit the ground be affected? Ignore air resistance.
    b. How will the increased speed affect the distance from where the rock left your hand to where the rock hits the ground?

44. **Field Biology** A zoologist standing on a cliff aims a tranquilizer gun at a monkey hanging from a distant tree branch. The barrel of the gun is horizontal. Just as the zoologist pulls the trigger, the monkey lets go and begins to fall. Will the dart hit the monkey? Ignore air resistance.

45. **Football** A quarterback throws a football at 24 m/s at a 45° angle. If it takes the ball 3.0 s to reach the top of its path and the ball is caught at the same height at which it is thrown, how long is it in the air? Ignore air resistance.

46. **Track and Field** You are working on improving your performance in the long jump and believe that the information in this chapter can help. Does the height that you reach make any difference to your jump? What influences the length of your jump?

 For more problems, go to Additional Problems, Appendix B.

47. Imagine that you are sitting in a car tossing a ball straight up into the air.
   a. If the car is moving at a constant velocity, will the ball land in front of, behind, or in your hand?
   b. If the car rounds a curve at a constant speed, where will the ball land?

48. You swing one yo-yo around your head in a horizontal circle. Then you swing another yo-yo with twice the mass of the first one, but you don't change the length of the string or the period. How do the tensions in the strings differ?

49. **Car Racing** The curves on a race track are banked to make it easier for cars to go around the curves at high speeds. Draw a free-body diagram of a car on a banked curve. From the motion diagram, find the direction of the acceleration.
   a. What exerts the force in the direction of the acceleration?
   b. Can you have such a force without friction?

50. **Driving on the Highway** Explain why it is that when you pass a car going in the same direction as you on the freeway, it takes a longer time than when you pass a car going in the opposite direction.

Mastering Problems

6.1 Projectile Motion

51. You accidentally throw your car keys horizontally at 8.0 m/s from a cliff 64-m high. How far from the base of the cliff should you look for the keys?

52. The toy car in **Figure 6-12** runs off the edge of a table that is 1.225-m high. The car lands 0.400 m from the base of the table.
   a. How long did it take the car to fall?
   b. How fast was the car going on the table?

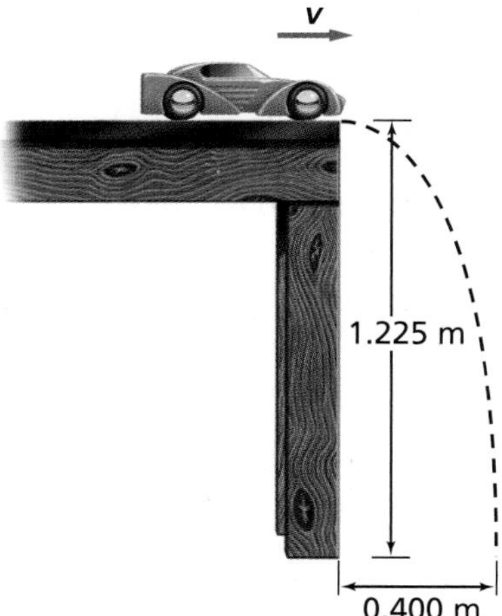

Figure 6-12

53. A dart player throws a dart horizontally at 12.4 m/s. The dart hits the board 0.32 m below the height from which it was thrown. How far away is the player from the board?

54. The two baseballs in **Figure 6-13** were hit with the same speed, 25 m/s. Draw separate graphs of y versus t and x versus t for each ball.

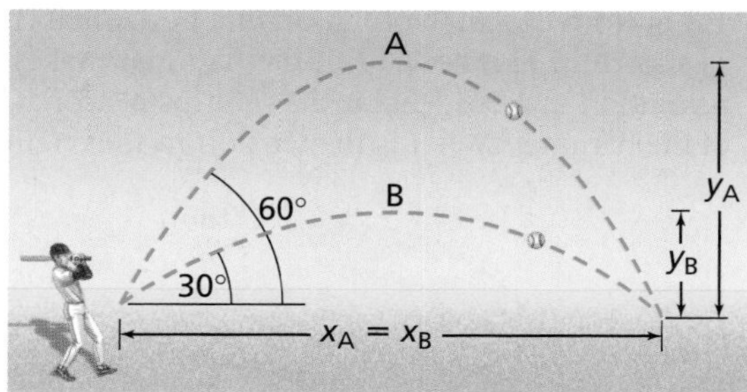

Figure 6-13

55. **Swimming** You took a running leap off a high-diving platform. You were running at 2.8 m/s and hit the water 2.6 s later. How high was the platform, and how far from the edge of the platform did you hit the water? Ignore air resistance.

56. **Archery** An arrow is shot at 30.0° above the horizontal. Its velocity is 49 m/s, and it hits the target.
   a. What is the maximum height the arrow will attain?
   b. The target is at the height from which the arrow was shot. How far away is it?

57. **Hitting a Home Run** A pitched ball is hit by a batter at a 45° angle and just clears the outfield fence, 98 m away. If the fence is at the same height as the pitch, find the velocity of the ball when it left the bat. Ignore air resistance.

58. **At-Sea Rescue** An airplane traveling 1001 m above the ocean at 125 km/h is going to drop a box of supplies to shipwrecked victims below.
   a. How many seconds before the plane is directly overhead should the box be dropped?
   b. What is the horizontal distance between the plane and the victims when the box is dropped?

59. **Diving** Divers in Acapulco dive from a cliff that is 61 m high. If the rocks below the cliff extend outward for 23 m, what is the minimum horizontal velocity a diver must have to clear the rocks?

60. **Jump Shot** A basketball player is trying to make a half-court jump shot and releases the ball at the height of the basket. Assuming that the ball is launched at 51.0°, 14.0 m from the basket, what speed must the player give the ball?

6.2 Circular Motion

61. Car Racing A 615-kg racing car completes one lap in 14.3 s around a circular track with a radius of 50.0 m. The car moves at a constant speed.
  a. What is the acceleration of the car?
  b. What force must the track exert on the tires to produce this acceleration?

62. Hammer Throw An athlete whirls a 7.00-kg hammer 1.8 m from the axis of rotation in a horizontal circle, as shown in **Figure 6-14.** If the hammer makes one revolution in 1.0 s, what is the centripetal acceleration of the hammer? What is the tension in the chain?

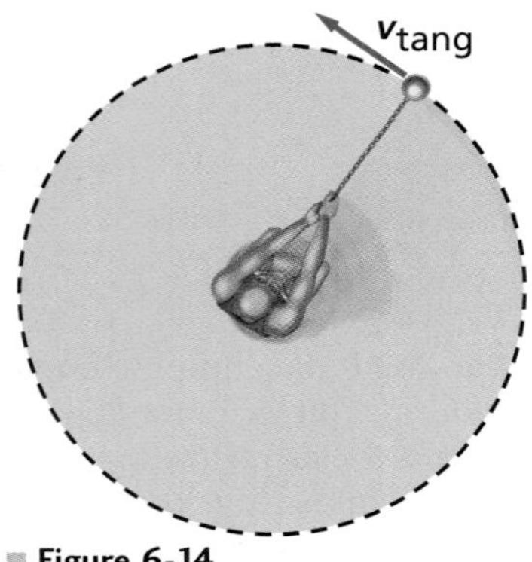

Figure 6-14

63. A coin is placed on a vinyl stereo record that is making $33\frac{1}{3}$ revolutions per minute on a turntable.
  a. In what direction is the acceleration of the coin?
  b. Find the magnitude of the acceleration when the coin is placed 5.0, 10.0, and 15.0 cm from the center of the record.
  c. What force accelerates the coin?
  d. At which of the three radii in part **b** would the coin be most likely to fly off the turntable? Why?

64. A rotating rod that is 15.3 cm long is spun with its axis through one end of the rod so that the other end of the rod has a speed of 2010 m/s (4500 mph).
  a. What is the centripetal acceleration of the end of the rod?
  b. If you were to attach a 1.0-g object to the end of the rod, what force would be needed to hold it on the rod?

65. Friction provides the force needed for a car to travel around a flat, circular race track. What is the maximum speed at which a car can safely travel if the radius of the track is 80.0 m and the coefficient of friction is 0.40?

66. A carnival clown rides a motorcycle down a ramp and around a vertical loop. If the loop has a radius of 18 m, what is the slowest speed the rider can have at the top of the loop to avoid falling? *Hint: At this slowest speed, the track exerts no force on the motorcycle at the top of the loop.*

67. A 75-kg pilot flies a plane in a loop as shown in **Figure 6-15.** At the top of the loop, when the plane is completely upside-down for an instant, the pilot hangs freely in the seat and does not push against the seat belt. The airspeed indicator reads 120 m/s. What is the radius of the plane's loop?

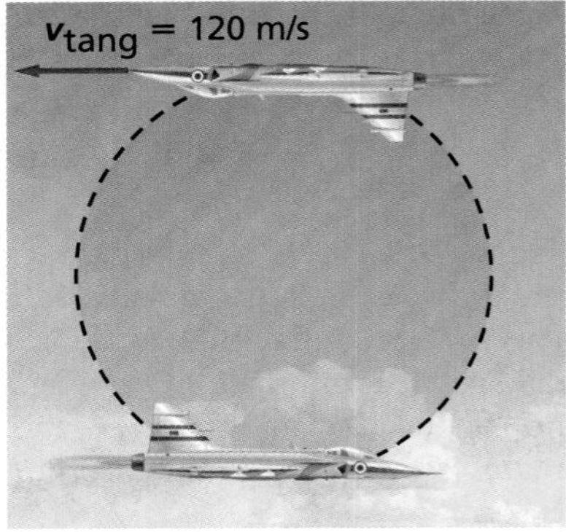

Figure 6-15

6.3 Relative Velocity

68. Navigating an Airplane An airplane flies at 200.0 km/h relative to the air. What is the velocity of the plane relative to the ground if it flies during the following wind conditions?
  a. a 50.0-km/h tailwind
  b. a 50.0-km/h headwind

69. Odina and LaToya are sitting by a river and decide to have a race. Odina will run down the shore to a dock, 1.5 km away, then turn around and run back. LaToya will also race to the dock and back, but she will row a boat in the river, which has a current of 2.0 m/s. If Odina's running speed is equal to LaToya's rowing speed in still water, which is 4.0 m/s, who will win the race? Assume that they both turn instantaneously.

70. Crossing a River You row a boat, such as the one in **Figure 6-16,** perpendicular to the shore of a river that flows at 3.0 m/s. The velocity of your boat is 4.0 m/s relative to the water.
  a. What is the velocity of your boat relative to the shore?
  b. What is the component of your velocity parallel to the shore? Perpendicular to it?

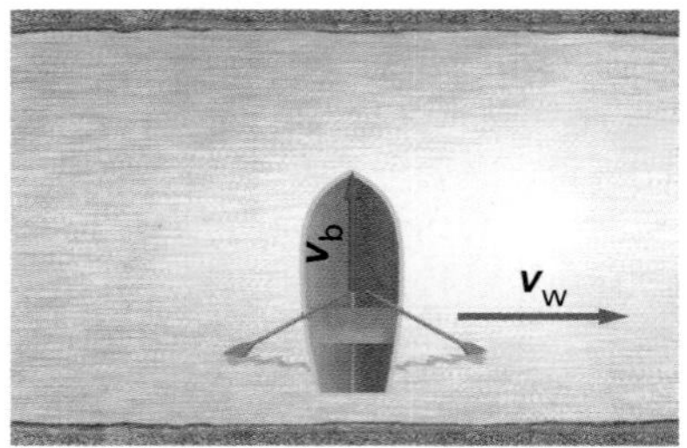

Figure 6-16

71. **Studying the Weather** A weather station releases a balloon to measure cloud conditions that rises at a constant 15 m/s relative to the air, but there is also a wind blowing at 6.5 m/s toward the west. What are the magnitude and direction of the velocity of the balloon?

72. **Boating** You are boating on a river that flows toward the east. Because of your knowledge of physics, you head your boat 53° west of north and have a velocity of 6.0 m/s due north relative to the shore.
   - **a.** What is the velocity of the current?
   - **b.** What is the speed of your boat relative to the water?

73. **Air Travel** You are piloting a small plane, and you want to reach an airport 450 km due south in 3.0 h. A wind is blowing from the west at 50.0 km/h. What heading and airspeed should you choose to reach your destination in time?

Mixed Review

74. Early skeptics of the idea of a rotating Earth said that the fast spin of Earth would throw people at the equator into space. The radius of Earth is about 6.38×10^3 km. Show why this idea is wrong by calculating the following.
   - **a.** the speed of a 97-kg person at the equator
   - **b.** the force needed to accelerate the person in the circle
   - **c.** the weight of the person
   - **d.** the normal force of Earth on the person, that is, the person's apparent weight

75. **Firing a Missile** An airplane, moving at 375 m/s relative to the ground, fires a missile forward at a speed of 782 m/s relative to the plane. What is the speed of the missile relative to the ground?

76. **Rocketry** A rocket in outer space that is moving at a speed of 1.25 km/s relative to an observer fires its motor. Hot gases are expelled out the back at 2.75 km/s relative to the rocket. What is the speed of the gases relative to the observer?

77. Two dogs, initially separated by 500.0 m, are running towards each other, each moving with a constant speed of 2.5 m/s. A dragonfly, moving with a constant speed of 3.0 m/s, flies from the nose of one dog to the other, then turns around instantaneously and flies back to the other dog. It continues to fly back and forth until the dogs run into each other. What distance does the dragonfly fly during this time?

78. A 1.13-kg ball is swung vertically from a 0.50-m cord in uniform circular motion at a speed of 2.4 m/s. What is the tension in the cord at the bottom of the ball's motion?

79. **Banked Roads** Curves on roads often are banked to help prevent cars from slipping off the road. If the posted speed limit for a particular curve of radius 36.0 m is 15.7 m/s (35 mph), at what angle should the road be banked so that cars will stay on a circular path even if there were no friction between the road and the tires? If the speed limit was increased to 20.1 m/s (45 mph), at what angle should the road be banked?

80. The 1.45-kg ball in **Figure 6-17** is suspended from a 0.80-m string and swung in a horizontal circle at a constant speed such that the string makes an angle of 14.0° with the vertical.
   - **a.** What is the tension in the string?
   - **b.** What is the speed of the ball?

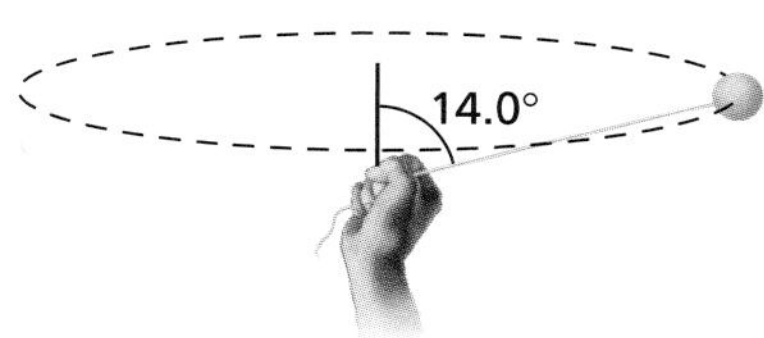

Figure 6-17

81. A baseball is hit directly in line with an outfielder at an angle of 35.0° above the horizontal with an initial velocity of 22.0 m/s. The outfielder starts running as soon as the ball is hit at a constant velocity of 2.5 m/s and barely catches the ball. Assuming that the ball is caught at the same height at which it was hit, what was the initial separation between the hitter and outfielder? *Hint: There are two possible answers.*

82. **A Jewel Heist** You are serving as a technical consultant for a locally produced cartoon. In one episode, two criminals, Shifty and Lefty, have stolen some jewels. Lefty has the jewels when the police start to chase him, and he runs to the top of a 60.0-m tall building in his attempt to escape. Meanwhile, Shifty runs to the convenient hot-air balloon 20.0 m from the base of the building and untethers it, so it begins to rise at a constant speed. Lefty tosses the bag of jewels horizontally with a speed of 7.3 m/s just as the balloon begins its ascent. What must the velocity of the balloon be for Shifty to easily catch the bag?

Thinking Critically

83. Apply Concepts Consider a roller-coaster loop like the one in **Figure 6-18.** Are the cars traveling through the loop in uniform circular motion? Explain.

Figure 6-18

84. Use Numbers A 3-point jump shot is released 2.2 m above the ground and 6.02 m from the basket. The basket is 3.05 m above the floor. For launch angles of 30.0° and 60.0°, find the speed the ball needs to be thrown to make the basket.

85. Analyze For which angle in problem 84 is it more important that the player get the speed right? To explore this question, vary the speed at each angle by 5 percent and find the change in the range of the attempted shot.

86. Apply Computers and Calculators A baseball player hits a belt-high (1.0 m) fastball down the left-field line. The ball is hit with an initial velocity of 42.0 m/s at 26°. The left-field wall is 96.0 m from home plate at the foul pole and is 14-m high. Write the equation for the height of the ball, y, as a function of its distance from home plate, x. Use a computer or graphing calculator to plot the path of the ball. Trace along the path to find how high above the ground the ball is when it is at the wall.

a. Is the hit a home run?

b. What is the minimum speed at which the ball could be hit and clear the wall?

c. If the initial velocity of the ball is 42.0 m/s, for what range of angles will the ball go over the wall?

87. Analyze Albert Einstein showed that the rule you learned for the addition of velocities does not work for objects moving near the speed of light. For example, if a rocket moving at velocity v_A releases a missile that has velocity v_B relative to the rocket, then the velocity of the missile relative to an observer that is at rest is given by $v = (v_A + v_B)/(1 + v_Av_B/c^2)$, where c is the speed of light, 3.00×10^8 m/s. This formula gives the correct values for objects moving at slow speeds as well. Suppose a rocket moving at 11 km/s shoots a laser beam out in front of it. What speed would an unmoving observer find for the laser light? Suppose that a rocket moves at a speed $c/2$, half the speed of light, and shoots a missile forward at a speed of $c/2$ relative to the rocket. How fast would the missile be moving relative to a fixed observer?

88. Analyze and Conclude A ball on a light string moves in a vertical circle. Analyze and describe the motion of this system. Be sure to consider the effects of gravity and tension. Is this system in uniform circular motion? Explain your answer.

Writing in Physics

89. Roller Coasters If you take a look at vertical loops on roller coasters, you will notice that most of them are not circular in shape. Research why this is so and explain the physics behind this decision by the coaster engineers.

90. Many amusement-park rides utilize centripetal acceleration to create thrills for the park's customers. Choose two rides other than roller coasters that involve circular motion and explain how the physics of circular motion creates the sensations for the riders.

Cumulative Review

91. Multiply or divide, as indicated, using significant digits correctly. (Chapter 1)

a. $(5\times10^8 \text{ m})(4.2\times10^7 \text{ m})$

b. $(1.67\times10^{-2} \text{ km})(8.5\times10^{-6} \text{ km})$

c. $(2.6\times10^4 \text{ kg})/(9.4\times10^3 \text{ m}^3)$

d. $(6.3\times10^{-1} \text{ m})/(3.8\times10^2 \text{ s})$

92. Plot the data in **Table 6-1** on a position-time graph. Find the average velocity in the time interval between 0.0 s and 5.0 s. (Chapter 3)

Table 6-1
Position v. Time

Clock Reading t (s)	Position d (m)
0.0	30
1.0	30
2.0	35
3.0	45
4.0	60
5.0	70

93. Carlos and his older brother Ricardo are at the grocery store. Carlos, with mass 17.0 kg, likes to hang on the front of the cart while Ricardo pushes it, even though both boys know this is not safe. Ricardo pushes the cart, with mass 12.4 kg, with his brother hanging on it such that they accelerate at a rate of 0.20 m/s^2. (Chapter 4)

a. With what force is Ricardo pushing?

b. What is the force the cart exerts on Carlos?

 For more problems, go to Additional Problems, Appendix B.

Multiple Choice

1. A 1.60-m-tall girl throws a football at an angle of 41.0° from the horizontal and at an initial velocity of 9.40 m/s. How far away from the girl will it land?

   (A) 4.55 m
   (B) 5.90 m
   (C) 8.90 m
   (D) 10.5 m

2. A dragonfly is sitting on a merry-go-round 2.8 m from the center. If the tangential velocity of the ride is 0.89 m/s, what is the centripetal acceleration of the dragonfly?

   (A) 0.11 m/s^2
   (B) 0.28 m/s^2
   (C) 0.32 m/s^2
   (D) 2.2 m/s^2

3. The centripetal force on a 0.82-kg object on the end of a 2.0-m massless string being swung in a horizontal circle is 4.0 N. What is the tangential velocity of the object?

   (A) 2.8 m/s^2
   (B) 3.1 m/s^2
   (C) 4.9 m/s^2
   (D) 9.8 m/s^2

4. A 1000-kg car enters an 80.0-m-radius curve at 20.0 m/s. What centripetal force must be supplied by friction so the car does not skid?

   (A) 5.0 N
   (B) 2.5×10^2 N
   (C) 5.0×10^3 N
   (D) 1.0×10^3 N

5. A jogger on a riverside path sees a rowing team coming toward him. If the jogger is moving at 10 km/h, and the boat is moving at 20 km/h, how quickly does the jogger approach the boat?

   (A) 3 m/s
   (B) 8 m/s
   (C) 40 m/s
   (D) 100 m/s

6. What is the maximum height obtained by a 125-g apple that is slung from a slingshot at an angle of 78° from the horizontal with an initial velocity of 18 m/s?

   (A) 0.70 m
   (B) 16 m
   (C) 32 m
   (D) 33 m

7. An orange is dropped at the same time a bullet is shot from a gun. Which of the following is true?

   (A) The acceleration due to gravity is greater for the orange because the orange is heavier.
   (B) Gravity acts less on the bullet than on the orange because the bullet is moving so fast.
   (C) The velocities will be the same.
   (D) The two objects will hit the ground at the same time.

Extended Answer

8. A colorfully feathered lead cannonball is shot horizontally out of a circus cannon 25 m/s from the high-wire platform on one side of a circus ring. If the high-wire platform is 52 m above the 80-m diameter ring, will the performers need to adjust their cannon (will the ball land inside the ring, or past it)?

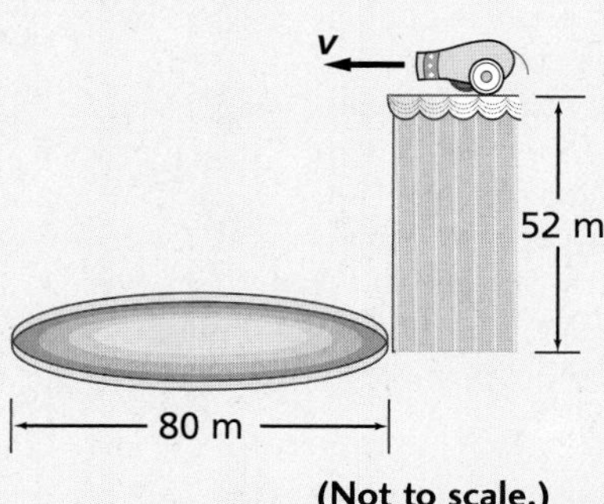

(Not to scale.)

9. A mythical warrior swings a 5.6-kg mace on the end of a magically massless 86-cm chain in a horizontal circle above his head. The mace makes one full revolution in 1.8 s. Find the tension in the magical chain.

✓ Test-Taking TIP

Practice Under Testlike Conditions

Answer all of the questions in the time provided without referring to your book. Did you complete the test? Could you have made better use of your time? What topics do you need to review?

Chapter 7

Gravitation

What You'll Learn

- You will learn the nature of gravitational force.
- You will relate Kepler's laws of planetary motion to Newton's laws of motion.
- You will describe the orbits of planets and satellites using the law of universal gravitation.

Why It's Important

Kepler's laws and the law of universal gravitation will help you understand the motion of planets and satellites.

Comets Comet Hale-Bopp was discovered by Alan Hale and Thomas Bopp in 1995. The comet entered the inner solar system in 1997 and was visible from Joshua Tree National Park in California, providing spectacular views of its white dust tail and blue ion tail.

Think About This ▶

Comets orbit the Sun just as planets and stars do. How can you describe the orbit of a comet such as Hale-Bopp?

physicspp.com

Can you model Mercury's motion?

Question

Do planets in our solar system have circular orbits or do they travel in some other path?

Procedure

1. Use the data table to plot the orbit of Mercury using the scale 10 cm = 1 AU. Note that one astronomical unit, AU, is Earth's distance from the Sun. 1 AU is equal to 1.5×10^8 km.
2. Calculate the distance in cm for each distance measured in AU.
3. Mark the center of your paper and draw a horizontal zero line and a vertical zero line going through it.
4. Place your protractor on the horizontal line and center it on the center point. Measure the degrees and place a mark.
5. Place a ruler connecting the center and the angle measurement. Mark the distance in centimeters for the corresponding angle. You will need to place the protractor on the vertical zero line for certain angle measurements.
6. Once you have marked all the data points, draw a line connecting them.

Analysis

Describe the shape of Mercury's orbit. Draw a line going through the Sun that represents the longest axis of the orbit, called the major axis.

Critical Thinking

How does the orbit of Mercury compare to the orbit of comet Hale-Bopp, shown on page 170?

Mercury's Orbit	
θ (°)	*d* (AU)
4	0.35
61	0.31
122	0.32
172	0.38
209	0.43
239	0.46
266	0.47
295	0.44
330	0.40
350	0.37

7.1 Planetary Motion and Gravitation

► **Objectives**
- **Relate** Kepler's laws to the law of universal gravitation.
- **Calculate** orbital speeds and periods.
- **Describe** the importance of Cavendish's experiment.

► **Vocabulary**

Kepler's first law
Kepler's second law
Kepler's third law
gravitational force
law of universal gravitation

Since ancient times, the Sun, Moon, planets, and stars had been assumed to revolve around Earth. Nicholas Copernicus, a Polish astronomer, noticed that the best available observations of the movements of planets and stars did not fully agree with the Earth-centered model. The results of his many years of work were published in 1543, when Copernicus was on his deathbed. His book showed that the motion of planets is much more easily understood by assuming that Earth and other planets revolve around the Sun.

Tycho Brahe was born a few years after the death of Copernicus. As a boy of 14 in Denmark, Brahe observed an eclipse of the Sun on August 21, 1560, and vowed to become an astronomer.

Brahe studied astronomy as he traveled throughout Europe for five years. He did not use telescopes. Instead, he used huge instruments that he designed and built in his own shop on the Danish island of Hven. He spent the next 20 years carefully recording the exact positions of the planets and stars. Brahe concluded that the Sun and the Moon orbit Earth and that all other planets orbit the Sun.

Figure 7-1 Among the huge astronomical instruments that Tycho Brahe constructed to use on Hven **(a)** were an astrolabe **(b)** and a sextant **(c).**

Kepler's Laws

Johannes Kepler, a 29-year-old German, became one of Brahe's assistants when he moved to Prague. Brahe trained his assistants to use instruments, such as those shown in **Figure 7-1.** Upon his death in 1601, Kepler inherited 30 years' worth of Brahe's observations. He studied Brahe's data and was convinced that geometry and mathematics could be used to explain the number, distance, and motion of the planets. Kepler believed that the Sun exerted a force on the planets and placed the Sun at the center of the system. After several years of careful analysis of Brahe's data on Mars, Kepler discovered the laws that describe the motion of every planet and satellite.

Kepler's first law states that the paths of the planets are ellipses, with the Sun at one focus. An ellipse has two foci, as shown in **Figure 7-2.** Like planets and stars, comets also orbit the Sun in elliptical orbits. Comets are divided into two groups—long-period comets and short-period comets—based on orbital periods, each of which is the time it takes the comet to complete one revolution. Long-period comets have orbital periods longer than 200 years, and short-period comets have orbital periods shorter than 200 years. Comet Hale-Bopp, with a period of 2400 years, is an example of a long-period comet. Comet Halley, with a period of 76 years, is an example of a short-period comet.

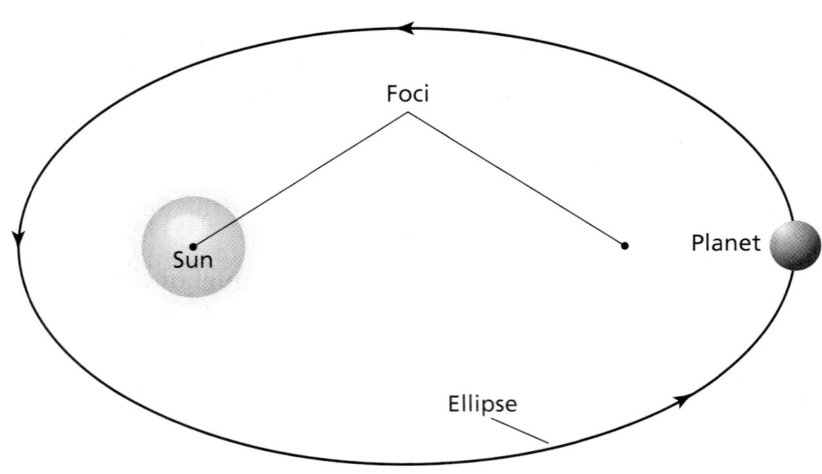

Figure 7-2 Planets orbit the Sun in elliptical orbits with the Sun at one focus. (Illustration not to scale)

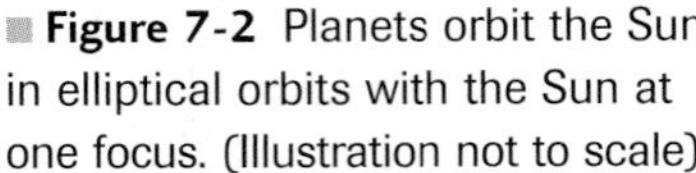

Interactive Figure To see an animation on Kepler's first law, visit physicspp.com.

Kepler found that the planets move faster when they are closer to the Sun and slower when they are farther away from the Sun. Thus, **Kepler's second law** states that an imaginary line from the Sun to a planet sweeps out equal areas in equal time intervals, as illustrated in **Figure 7-3.**

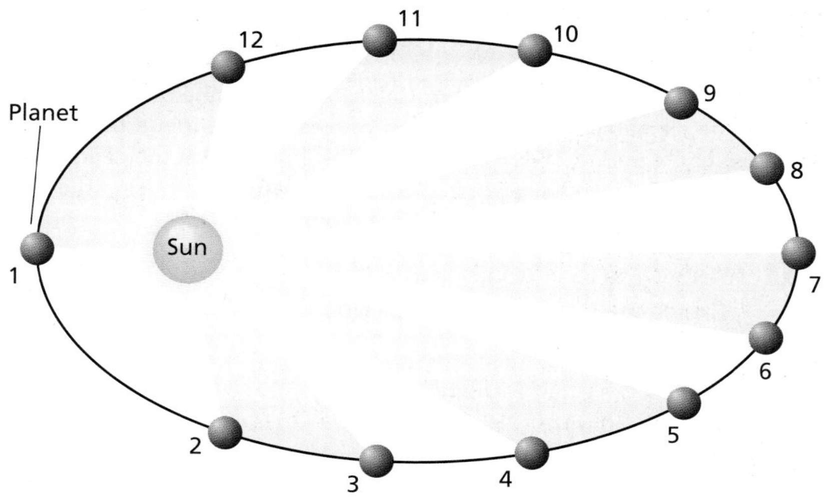

Figure 7-3 A planet moves fastest when it is close to the Sun and slowest when it is farther from the Sun. Equal areas are swept out in equal amounts of time. (Illustration not to scale)

Kepler also found that there is a mathematical relationship between periods of planets and their mean distances away from the Sun. **Kepler's third law** states that the square of the ratio of the periods of any two planets revolving about the Sun is equal to the cube of the ratio of their average distances from the Sun. Thus, if the periods of the planets are T_A and T_B, and their average distances from the Sun are r_A and r_B, Kepler's third law can be expressed as follows.

Kepler's Third Law $\left(\frac{T_A}{T_B}\right)^2 = \left(\frac{r_A}{r_B}\right)^3$

The squared quantity of the period of object A divided by the period of object B, is equal to the cubed quantity of object A's average distance from the Sun, divided by object B's average distance from the Sun.

Concepts In Motion
Interactive Figure To see an animation on Kepler's second law and third law, visit physicspp.com.

Note that the first two laws apply to each planet, moon, and satellite individually. The third law, however, relates the motion of several objects about a single body. For example, it can be used to compare the planets' distances from the Sun, shown in **Table 7-1,** to their periods about the Sun. It also can be used to compare distances and periods of the Moon and artificial satellites orbiting Earth.

Table 7-1
Solar System Data

Name	Average Radius (m)	Mass (kg)	Mean Distance From Sun (m)
Sun	6.96×10^{8}	1.99×10^{30}	–
Mercury	2.44×10^{6}	3.30×10^{23}	5.79×10^{10}
Venus	6.05×10^{6}	4.87×10^{24}	1.08×10^{11}
Earth	6.38×10^{6}	5.97×10^{24}	1.50×10^{11}
Mars	3.40×10^{6}	6.42×10^{23}	2.28×10^{11}
Jupiter	7.15×10^{7}	1.90×10^{27}	7.78×10^{11}
Saturn	6.03×10^{7}	5.69×10^{26}	1.43×10^{12}
Uranus	2.56×10^{7}	8.68×10^{25}	2.87×10^{12}
Neptune	2.48×10^{7}	1.02×10^{26}	4.50×10^{12}
Pluto	1.20×10^{6}	1.25×10^{22}	5.87×10^{12}

▶ EXAMPLE Problem 1

Callisto's Distance from Jupiter Galileo measured the orbital sizes of Jupiter's moons using the diameter of Jupiter as a unit of measure. He found that Io, the closest moon to Jupiter, had a period of 1.8 days and was 4.2 units from the center of Jupiter. Callisto, the fourth moon from Jupiter, had a period of 16.7 days. Using the same units that Galileo used, predict Callisto's distance from Jupiter.

1 Analyze and Sketch the Problem

- Sketch the orbits of Io and Callisto.
- Label the radii.

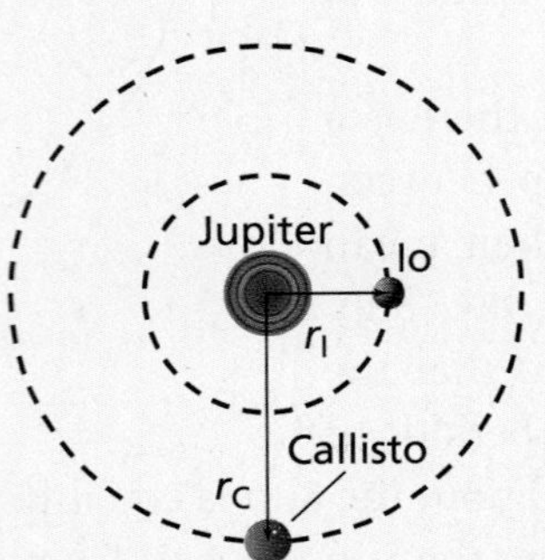

Known:	Unknown:
T_C = 16.7 days	r_C = ?
T_I = 1.8 days	
r_I = 4.2 units	

2 Solve for the Unknown

Solve Kepler's third law for r_C.

$$\left(\frac{T_C}{T_I}\right)^2 = \left(\frac{r_C}{r_I}\right)^3$$

$$r_C^3 = r_I^3\left(\frac{T_C}{T_I}\right)^2$$

$$r_C = \sqrt[3]{r_I^3\left(\frac{T_C}{T_I}\right)^2}$$

$$= \sqrt[3]{(4.2\text{ units})^3\left(\frac{16.7\text{ days}}{1.8\text{ days}}\right)^2}$$ **Substitute r_I = 4.2 units, T_C = 16.7 days, T_I = 1.8 days**

$$= \sqrt[3]{6.4\times10^3\text{ units}^3}$$

$$= 19\text{ units}$$

Math Handbook
Isolating a Variable
page 845

3 Evaluate the Answer

- **Are the units correct?** r_C should be in Galileo's units, like r_I.
- **Is the magnitude realistic?** The period is large, so the radius should be large.

▶ PRACTICE Problems

- Additional Problems, Appendix B
- Solutions to Selected Problems, Appendix C

1. If Ganymede, one of Jupiter's moons, has an orbital period of 7.15 days, how many units are there in its orbital radius? Use the information given in Example Problem 1.
2. An asteroid revolves around the Sun with a mean orbital radius twice that of Earth's. Predict the period of the asteroid in Earth years.
3. From Table 7-1, on page 173, you can find that, on average, Mars is 1.52 times as far from the Sun as Earth is. Predict the time required for Mars to orbit the Sun in Earth days.
4. The Moon has a period of 27.3 days and a mean distance of 3.90×10^5 km from the center of Earth.
   - **a.** Use Kepler's laws to find the period of a satellite in orbit 6.70×10^3 km from the center of Earth.
   - **b.** How far above Earth's surface is this satellite?
5. Using the data in the previous problem for the period and radius of revolution of the Moon, predict what the mean distance from Earth's center would be for an artificial satellite that has a period of exactly 1.00 day.

Newton's Law of Universal Gravitation

In 1666, 45 years after Kepler published his work, Newton began his studies of planetary motion. He found that the magnitude of the force, *F*, on a planet due to the Sun varies inversely with the square of the distance, *r*, between the centers of the planet and the Sun. That is, *F* is proportional to $1/r^2$. The force, ***F***, acts in the direction of the line connecting the centers of the two objects.

It is quoted that the sight of a falling apple made Newton wonder if the force that caused the apple to fall might extend to the Moon, or even beyond. He found that both the apple's and Moon's accelerations agreed with the $1/r^2$ relationship. According to his own third law, the force Earth exerts on the apple is exactly the same as the force the apple exerts on Earth. The force of attraction between two objects must be proportional to the objects' masses, and is known as the **gravitational force.**

Newton was confident that the same force of attraction would act between any two objects, anywhere in the universe. He proposed his **law of universal gravitation,** which states that objects attract other objects with a force that is proportional to the product of their masses and inversely proportional to the square of the distance between them. This can be represented by the following equation.

Law of Universal Gravitation $F = G\frac{m_1 m_2}{r^2}$

The gravitational force is equal to the universal gravitational constant, times the mass of object 1, times the mass of object 2, divided by the distance between the centers of the objects, squared.

According to Newton's equation, *F* is directly proportional to m_1 and m_2. Thus, if the mass of a planet near the Sun were doubled, the force of attraction would be doubled. Use the Connecting Math to Physics feature below to examine how changing one variable affects another. **Figure 7-4** illustrates the inverse square law graphically.

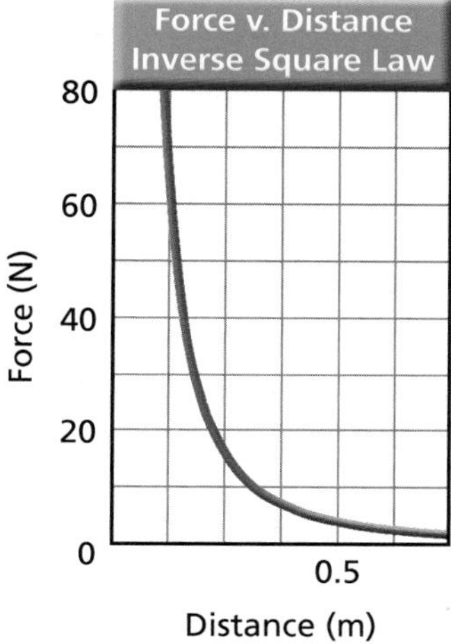

Figure 7-4 The change in gravitational force with distance follows the inverse square law.

Connecting Math to Physics

Direct and Inverse Relationships Newton's law of universal gravitation has both direct and inverse relationships.

$F \propto m_1 m_2$		$F \propto \frac{1}{r^2}$	
Change	**Result**	**Change**	**Result**
$2m_1 m_2$	$2F$	$2r$	$\frac{1}{4}F$
$3m_1 m_2$	$3F$	$3r$	$\frac{1}{9}F$
$2m_1\ 3m_2$	$6F$	$\frac{1}{2}r$	$4F$
$\frac{1}{2}\ m_1 m_2$	$\frac{1}{2}F$	$\frac{1}{3}r$	$9F$

Universal Gravitation and Kepler's Third Law

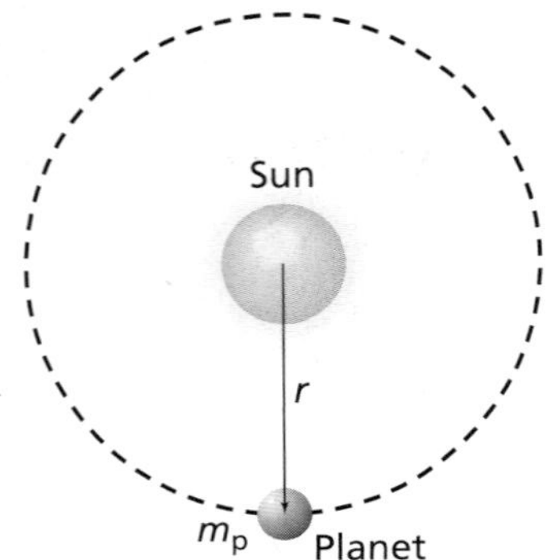

Figure 7-5 A planet with mass m_p and orbital radius r, orbits the Sun with mass m_S. (Illustration not to scale)

Newton stated his law of universal gravitation in terms that applied to the motion of planets about the Sun. This agreed with Kepler's third law and confirmed that Newton's law fit the best observations of the day.

Consider a planet orbiting the Sun, as shown in **Figure 7-5.** Newton's second law of motion, $F_{net} = ma$, can be written as $F_{net} = m_p a_c$, where F is the gravitational force, m_p is the mass of the planet, and a_c is the centripetal acceleration of the planet. For simplicity, assume circular orbits. Recall from your study of uniform circular motion in Chapter 6 that, for a circular orbit, $a_c = 4\pi^2 r/T^2$. This means that $F_{net} = m_p a_c$ may now be written as $F_{net} = m_p 4\pi^2 r/T^2$. In this equation, T is the time required for the planet to make one complete revolution about the Sun. If you set the right side of this equation equal to the right side of the law of universal gravitation, you arrive at the following result:

$$G\frac{m_S m_p}{r^2} = \frac{m_p 4\pi^2 r}{T^2}$$

$$T^2 = \left(\frac{4\pi^2}{Gm_S}\right)r^3$$

$$\text{Thus, } T = \sqrt{\left(\frac{4\pi^2}{Gm_S}\right)r^3}$$

The period of a planet orbiting the Sun can be expressed as follows.

Period of a Planet Orbiting the Sun $T = 2\pi\sqrt{\frac{r^3}{Gm_S}}$

The period of a planet orbiting the Sun is equal to 2π times the square root of the orbital radius cubed, divided by the product of the universal gravitational constant and the mass of the Sun.

Squaring both sides makes it apparent that this equation is Kepler's third law of planetary motion: the square of the period is proportional to the cube of the distance that separates the masses. The factor $4\pi^2/Gm_S$ depends on the mass of the Sun and the universal gravitational constant. Newton found that this derivation applied to elliptical orbits as well.

CHALLENGE PROBLEM

Astronomers have detected three planets that orbit the star Upsilon Andromedae. Planet B has an average orbital radius of 0.059 AU and a period of 4.6170 days. Planet C has an average orbital radius of 0.829 AU and a period of 241.5 days. Planet D has an average orbital radius of 2.53 AU and a period of 1284 days. (Distances are given in astronomical units (AU)—Earth's average distance from the Sun. The distance from Earth to the Sun is 1.00 AU.)

1. Do these planets obey Kepler's third law?
2. Find the mass of the star Upsilon Andromedae in units of the Sun's mass.

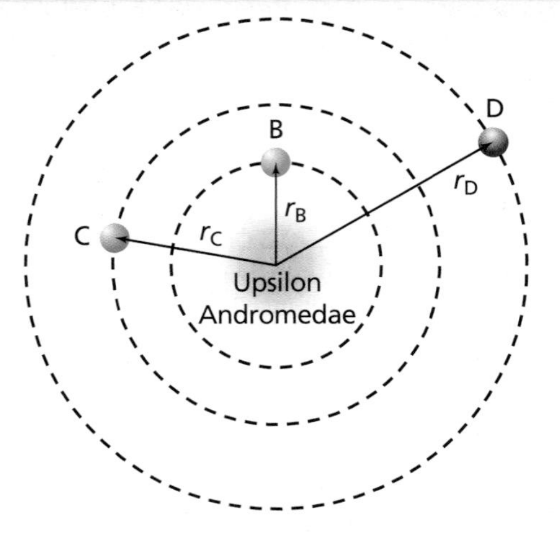

Measuring the Universal Gravitational Constant

How large is the constant, G? As you know, the force of gravitational attraction between two objects on Earth is relatively small. The slightest attraction, even between two massive bowling balls, is difficult to detect. In fact, it took 100 years from the time of Newton's work for scientists to develop an apparatus that was sensitive enough to measure the force of gravitational attraction.

Cavendish's experiment In 1798, Englishman Henry Cavendish used equipment similar to the apparatus shown in **Figure 7-6** to measure the gravitational force between two objects. The apparatus had a horizontal rod with two small lead spheres attached to each end. The rod was suspended at its midpoint by a thin wire so that it could rotate. Because the rod was suspended by a thin wire, the rod and spheres were very sensitive to horizontal forces. To measure G, Cavendish placed two large lead spheres in a fixed position, close to each of the two small spheres, as shown in **Figure 7-7.** The force of attraction between the large and the small spheres caused the rod to rotate. When the force required to twist the wire equaled the gravitational force between the spheres, the rod stopped rotating. By measuring the angle through which the rod turned, Cavendish was able to calculate the attractive force between the objects. The angle through which the rod turned is measured using the beam of light that is reflected from the mirror. He measured the masses of the spheres and the distance between their centers. Substituting these values for force, mass, and distance into Newton's law of universal gravitation, he found an experimental value for G: when m_1 and m_2 are measured in kilograms, r in meters, and F in newtons, then $G = 6.67\times10^{-11}\ \text{N}\cdot\text{m}^2/\text{kg}^2$.

Figure 7-6 Modern Cavendish balances are used to measure the gravitational forces between two objects.

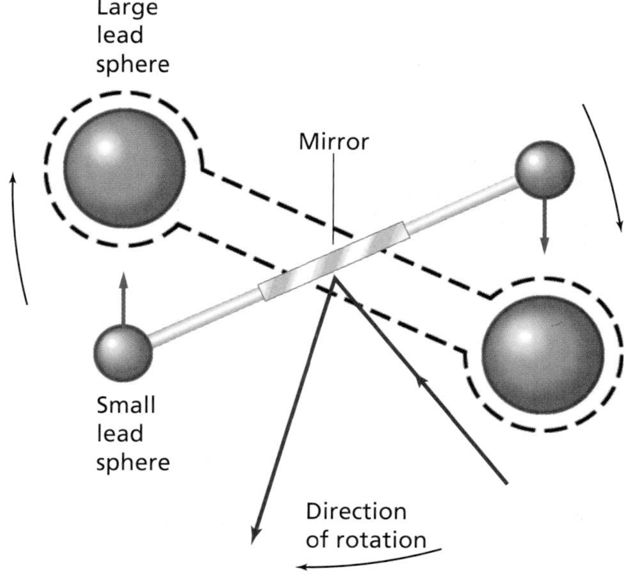

Figure 7-7 When the large lead spheres are placed near the small lead spheres, the gravitational attraction between the spheres causes the rod to rotate. The rotation is measured with the help of the reflected light ray.

Concepts In Motion
Interactive Figure To see an animation on Cavendish's experiment, visit physicspp.com.

The importance of *G* Cavendish's experiment often is called "weighing Earth," because his experiment helped determine Earth's mass. Once the value of G is known, not only the mass of Earth, but also the mass of the Sun can be determined. In addition, the gravitational force between any two objects can be calculated using Newton's law of universal gravitation. For example, the attractive gravitational force, F_g, between two bowling balls of mass 7.26 kg, with their centers separated by 0.30 m, can be calculated as follows:

$$F_g = \frac{(6.67\times10^{-11}\ \text{N}\cdot\text{m}^2/\text{kg}^2)(7.26\ \text{kg})(7.26\ \text{kg})}{(0.30\ \text{m})^2} = 3.9\times10^{-8}\ \text{N}$$

You know that on Earth's surface, the weight of an object of mass m is a measure of Earth's gravitational attraction: $F_g = mg$. If Earth's mass is represented by m_E and Earth's radius is represented by r_E, the following is true:

$$F_g = G\frac{m_E m}{r_E^{\,2}} = mg, \text{ and so } g = G\frac{m_E}{r_E^{\,2}}$$

This equation can be rearranged to solve for m_E.

$$m_E = \frac{g r_E^{\,2}}{G}$$

Using $r_E = 6.38\times10^6$ m, $g = 9.80$ m/s^2, and $G = 6.67\times10^{-11}$ N·m^2/kg^2, the following result is obtained for Earth's mass:

$$m_E = \frac{(9.80\ \text{m/s}^2)(6.38\times10^6\ \text{m})^2}{6.67\times10^{-11}\ \text{N}\cdot\text{m}^2/\text{kg}^2} = 5.98\times10^{24}\ \text{kg}$$

When you compare the mass of Earth to that of a bowling ball, you can see why the gravitational attraction between everyday objects is not easily observed. Cavendish's experiment determined the value of G, confirmed Newton's prediction that a gravitational force exists between two objects, and helped calculate the mass of Earth.

7.1 Section Review

6. **Neptune's Orbital Period** Neptune orbits the Sun with an orbital radius of 4.495×10^{12} m, which allows gases, such as methane, to condense and form an atmosphere, as shown in **Figure 7-8.** If the mass of the Sun is 1.99×10^{30} kg, calculate the period of Neptune's orbit.

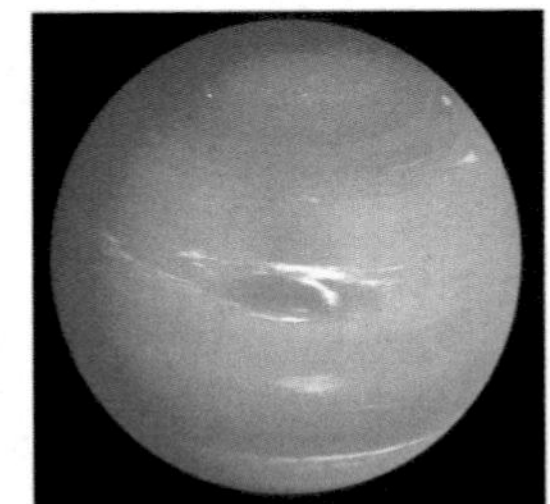

Figure 7-8

7. **Gravity** If Earth began to shrink, but its mass remained the same, what would happen to the value of g on Earth's surface?

8. **Gravitational Force** What is the gravitational force between two 15-kg packages that are 35 cm apart? What fraction is this of the weight of one package?

9. **Universal Gravitational Constant** Cavendish did his experiment using lead spheres. Suppose he had replaced the lead spheres with copper spheres of equal mass. Would his value of G be the same or different? Explain.

10. **Laws or Theories?** Kepler's three statements and Newton's equation for gravitational attraction are called "laws." Were they ever theories? Will they ever become theories?

11. **Critical Thinking** Picking up a rock requires less effort on the Moon than on Earth.
    - **a.** How will the weaker gravitational force on the Moon's surface affect the path of the rock if it is thrown horizontally?
    - **b.** If the thrower accidentally drops the rock on her toe, will it hurt more or less than it would on Earth? Explain.

Physics Online physicspp.com/self_check_quiz

7.2 Using the Law of Universal Gravitation

The planet Uranus was discovered in 1781. By 1830, it was clear that the law of gravitation didn't correctly predict its orbit. Two astronomers proposed that Uranus was being attracted by the Sun and by an undiscovered planet. They calculated the orbit of such a planet in 1845, and, one year later, astronomers at the Berlin Observatory found the planet now called Neptune. How do planets, such as Neptune, orbit the Sun?

▶ **Objectives**

- **Solve** orbital motion problems.
- **Relate** weightlessness to objects in free fall.
- **Describe** gravitational fields.
- **Compare** views on gravitation.

▶ **Vocabulary**

gravitational field
inertial mass
gravitational mass

Orbits of Planets and Satellites

Newton used a drawing similar to the one shown in **Figure 7-9** to illustrate a thought experiment on the motion of satellites. Imagine a cannon, perched high atop a mountain, firing a cannonball horizontally with a given horizontal speed. The cannonball is a projectile, and its motion has both vertical and horizontal components. Like all projectiles on Earth, it would follow a parabolic trajectory and fall back to the ground.

If the cannonball's horizontal speed were increased, it would travel farther across the surface of Earth, and still fall back to the ground. If an extremely powerful cannon were used, however, the cannonball would travel all the way around Earth, and keep going. It would fall toward Earth at the same rate that Earth's surface curves away. In other words, the curvature of the projectile would continue to just match the curvature of Earth, so that the cannonball would never get any closer or farther away from Earth's curved surface. The cannonball would, therefore, be in orbit.

Newton's thought experiment ignored air resistance. For the cannonball to be free of air resistance, the mountain on which the cannon is perched would have to be more than 150 km above Earth's surface. By way of comparison, the mountain would have to be much taller than the peak of Mount Everest, the world's tallest mountain, which is only 8.85 km in height. A cannonball launched from a mountain that is 150 km above Earth's surface would encounter little or no air resistance at an altitude of 150 km, because the mountain would be above most of the atmosphere. Thus, a cannonball or any object or satellite at or above this altitude could orbit Earth for a long time.

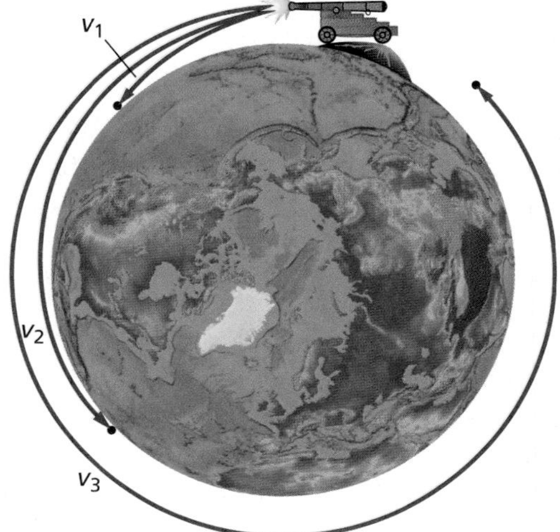

Figure 7-9 The horizontal speed v_1 is not high enough and the cannonball falls to the ground. With a higher speed, v_2, the cannonball travels farther. The cannonball travels all the way around Earth when the horizontal speed, v_3, is high enough.

Concepts In Motion
Interactive Figure To see an animation on the law of universal gravitation, visit physicspp.com.

APPLYING PHYSICS

Geosynchronous Orbit The *GOES-12* weather satellite orbits Earth once a day at an altitude of 35,785 km. The orbital speed of the satellite matches Earth's rate of rotation. Thus, to an observer on Earth, the satellite appears to remain above one spot on the equator. Satellite dishes on Earth can be directed to one point in the sky and not have to change position as the satellite orbits.

A satellite in an orbit that is always the same height above Earth moves in uniform circular motion. Recall that its centripetal acceleration is given by $a_c = v^2/r$. Newton's second law, $F_{net} = ma_c$, can thus be rewritten as $F_{net} = mv^2/r$. If Earth's mass is m_E, then this expression combined with Newton's law of universal gravitation produces the following equation:

$$G\frac{m_E m}{r^2} = \frac{mv^2}{r}$$

Solving for the speed of a satellite in circular orbit about Earth, v, yields the following.

Speed of a Satellite Orbiting Earth $v = \sqrt{\frac{Gm_E}{r}}$

The speed of a satellite orbiting Earth is equal to the square root of the universal gravitational constant times the mass of Earth, divided by the radius of the orbit.

Figure 7-10 *Landsat 7*, a remote sensing satellite, has a mass of about 2200 kg and orbits Earth at an altitude of about 705 km.

A satellite's orbital period A satellite's orbit around Earth is similar to a planet's orbit about the Sun. Recall that the period of a planet orbiting the Sun is expressed by the following equation:

$$T = 2\pi\sqrt{\frac{r^3}{Gm_S}}$$

Thus, the period for a satellite orbiting Earth is given by the following equation.

Period of a Satellite Orbiting Earth $T = 2\pi\sqrt{\frac{r^3}{Gm_E}}$

The period for a satellite orbiting Earth is equal to 2π times the square root of the radius of the orbit cubed, divided by the product of the universal gravitational constant and the mass of Earth.

The equations for the speed and period of a satellite can be used for any object in orbit about another. The mass of the central body will replace m_E in the equations, and r will be the distance between the centers of the orbiting body and the central body. If the mass of the central body is much greater than the mass of the orbiting body, then r is equal to the distance between the centers of the orbiting body and the central body. Orbital speed, v, and period, T, are independent of the mass of the satellite. Are there any factors that limit the mass of a satellite?

Geology Connection

A satellite's mass *Landsat 7*, shown in **Figure 7-10,** is an artificial satellite that provides images of Earth's continental surfaces. *Landsat* images have been used to create maps, study land use, and monitor resources and global changes. The *Landsat 7* system enables researchers to monitor small-scale processes, such as deforestation, on a global scale. Satellites, such as *Landsat 7*, are accelerated to the speeds necessary for them to achieve orbit by large rockets, such as shuttle-booster rockets. Because the acceleration of any mass must follow Newton's second law of motion, $F_{net} = ma$, more force is required to launch a more massive satellite into orbit. Thus, the mass of a satellite is limited by the capability of the rocket used to launch it.

▶ EXAMPLE Problem 2

Orbital Speed and Period Assume that a satellite orbits Earth 225 km above its surface. Given that the mass of Earth is 5.97×10^{24} kg and the radius of Earth is 6.38×10^{6} m, what are the satellite's orbital speed and period?

1 Analyze and Sketch the Problem

- Sketch the situation showing the height of the satellite's orbit.

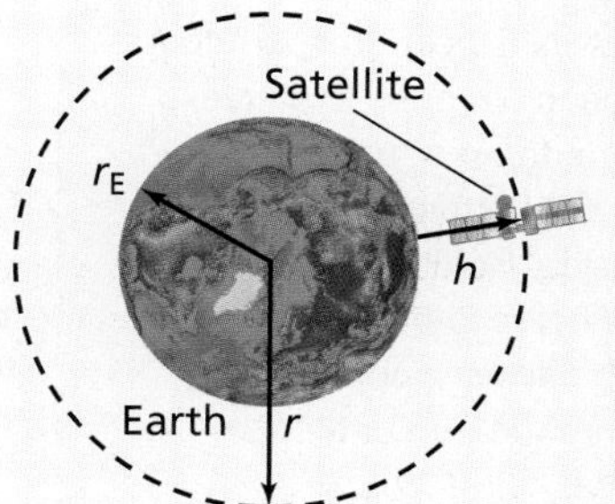

Known:	Unknown:
$h = 2.25\times10^{5}$ m	$v = ?$
$r_E = 6.38\times10^{6}$ m	$T = ?$
$m_E = 5.97\times10^{24}$ kg	
$G = 6.67\times10^{-11}\ \text{N·m}^2/\text{kg}^2$	

2 Solve for the Unknown

Determine the orbital radius by adding the height of the satellite's orbit to Earth's radius.

$$r = h + r_E$$
$$= 2.25\times10^{5}\ \text{m} + 6.38\times10^{6}\ \text{m} = 6.61\times10^{6}\ \text{m}$$

Substitute $h = 2.25\times10^{5}$ m, $r_E = 6.38\times10^{6}$ m

Solve for the speed.

$$v = \sqrt{\frac{Gm_E}{r}}$$
$$= \sqrt{\frac{(6.67\times10^{-11}\ \text{N·m}^2/\text{kg}^2)(5.97\times10^{24}\ \text{kg})}{6.61\times10^{6}\ \text{m}}}$$

Substitute $G = 6.67\times10^{-11}\ \text{N·m}^2/\text{kg}^2$, $m_E = 5.97\times10^{24}$ kg, $r = 6.61\times10^{6}$ m

$$= 7.76\times10^{3}\ \text{m/s}$$

Solve for the period.

$$T = 2\pi\sqrt{\frac{r^3}{Gm_E}}$$
$$= 2\pi\sqrt{\frac{(6.61\times10^{6}\ \text{m})^3}{(6.67\times10^{-11}\ \text{N·m}^2/\text{kg}^2)(5.97\times10^{24}\ \text{kg})}}$$

Substitute $r = 6.61\times10^{6}$ m, $G = 6.67\times10^{-11}\ \text{N·m}^2/\text{kg}^2$, $m_E = 5.97\times10^{24}$ kg

$$= 5.35\times10^{3}\ \text{s}$$

This is approximately 89 min, or 1.5 h.

Physics Online
Personal Tutor For an online tutorial on orbital speed and period, visit physicspp.com.

3 Evaluate the Answer

- **Are the units correct?** The unit for speed is m/s and the unit for period is s.

▶ PRACTICE Problems

• Additional Problems, Appendix B
• Solutions to Selected Problems, Appendix C

For the following problems, assume a circular orbit for all calculations.

12. Suppose that the satellite in Example Problem 2 is moved to an orbit that is 24 km larger in radius than its previous orbit. What would its speed be? Is this faster or slower than its previous speed?

13. Use Newton's thought experiment on the motion of satellites to solve the following.
  a. Calculate the speed that a satellite shot from a cannon must have to orbit Earth 150 km above its surface.
  b. How long, in seconds and minutes, would it take for the satellite to complete one orbit and return to the cannon?

14. Use the data for Mercury in Table 7-1 on page 173 to find the following.
  a. the speed of a satellite that is in orbit 260 km above Mercury's surface
  b. the period of the satellite

MINI LAB

Weightless Water

This activity is best done outdoors. Use a pencil to poke two holes through a foam or paper cup: one on the bottom and the other on the side. Hold your fingers over the two holes to block them as your lab partner pours colored water into the cup until it is two-thirds full.

1. Predict what will happen as the cup is allowed to fall.

2. Test your prediction: drop the cup and watch closely.

Analyze and Conclude

3. Describe your observations.

4. Explain your results.

Acceleration Due To Gravity

The acceleration of objects due to Earth's gravity can be found by using Newton's law of universal gravitation and his second law of motion. For a free-falling object, m, the following is true:

$$F = G\frac{m_E m}{r^2} = ma, \text{ so } a = G\frac{m_E}{r^2}$$

Because $a = g$ and $r = r_E$ on Earth's surface, the following equation can be written:

$$g = G\frac{m_E}{{r_E}^2}, \text{ thus, } m_E = \frac{g{r_E}^2}{G}$$

You found above that $a = G\frac{m_E}{r^2}$ for a free-falling object. Substituting the above expression for m_E yields the following:

$$a = G\frac{\left(\frac{g{r_E}^2}{G}\right)}{r^2}$$

$$a = g\left(\frac{r_E}{r}\right)^2$$

This shows that as you move farther from Earth's center, that is, as r becomes larger, the acceleration due to gravity is reduced according to this inverse square relationship. What happens to your weight, m_g, as you move farther and farther from Earth's center?

Weight and weightlessness You probably have seen photos similar to the one in **Figure 7-11** in which astronauts are on the space shuttle in an environment often called "zero-g" or "weightlessness." The shuttle orbits about 400 km above Earth's surface. At that distance, $g = 8.7$ m/s^2, only slightly less than on Earth's surface. Thus, Earth's gravitational force is certainly not zero in the shuttle. In fact, gravity causes the shuttle to orbit Earth. Why, then, do the astronauts appear to have no weight?

Remember that you sense weight when something, such as the floor or your chair, exerts a contact force on you. But if you, your chair, and the floor all are accelerating toward Earth together, then no contact forces are exerted on you. Thus, your apparent weight is zero and you experience weightlessness. Similarly, the astronauts experience weightlessness as the shuttle and everything in it falls freely toward Earth.

Figure 7-11 Astronaut Chiaki Mukai experiences weightlessness on board the space shuttle *Columbia*, as the shuttle and everything in it falls freely toward Earth.

The Gravitational Field

Recall from Chapter 6 that many common forces are contact forces. Friction is exerted where two objects touch, for example, when the floor and your chair or desk push on you. Gravity, however, is different. It acts on an apple falling from a tree and on the Moon in orbit. It even acts on you in midair as you jump up or skydive. In other words, gravity acts over a distance. It acts between objects that are not touching or that are not close together. Newton was puzzled by this concept. He wondered how the Sun could exert a force on planet Earth, which is hundreds of millions of kilometers away.

The answer to the puzzle arose from a study of magnetism. In the 19th century, Michael Faraday developed the concept of a field to explain how a magnet attracts objects. Later, the field concept was applied to gravity. Any object with mass is surrounded by a **gravitational field** in which another object experiences a force due to the interaction between its mass and the gravitational field, $\mathbf{g}$, at its location. This is expressed by the following equation.

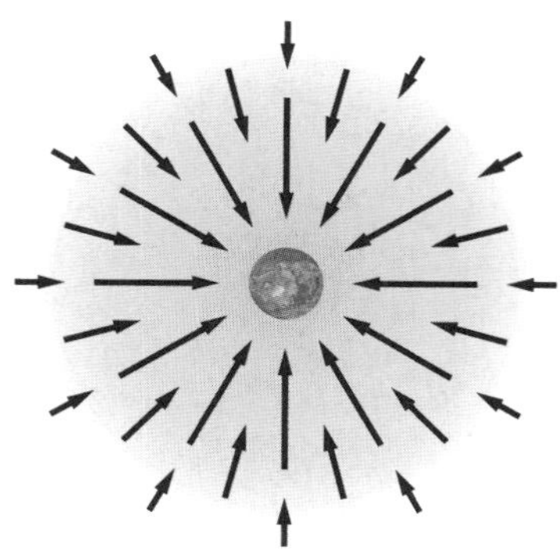

Figure 7-12 Vectors representing Earth's gravitational field all point toward Earth's center. The field is weaker farther from Earth.

Gravitational Field $g = \frac{Gm}{r^2}$

The gravitational field is equal to the universal gravitational constant times the object's mass, divided by the square of the distance from the object's center. The direction is toward the mass's center.

Suppose the gravitational field is created by the Sun. Then a planet of mass m has a force exerted on it that depends on its mass and the magnitude of the gravitational field at its location. That is, $\mathbf{F} = m\mathbf{g}$, toward the Sun. The force is caused by the interaction of the planet's mass with the gravitational field at its location, not with the Sun millions of kilometers away. To find the gravitational field caused by more than one object you would calculate both gravitational fields and add them as vectors.

The gravitational field can be measured by placing an object with a small mass, m, in the gravitational field and measuring the force, F, on it. The gravitational field can be calculated using $g = F/m$. The gravitational field is measured in N/kg, which is also equal to m/s^2.

On Earth's surface, the strength of the gravitational field is 9.80 N/kg, and its direction is toward Earth's center. The field can be represented by a vector of length g pointing toward the center of the object producing the field. You can picture the gravitational field of Earth as a collection of vectors surrounding Earth and pointing toward it, as shown in **Figure 7-12.** The strength of the field varies inversely with the square of the distance from the center of Earth. The gravitational field depends on Earth's mass, but not on the mass of the object experiencing it.

Two Kinds of Mass

Recall that when the concept of mass was discussed in Chapter 4, it was defined as the slope of a graph of force versus acceleration. That is, mass is equal to the ratio of the net force exerted on an object to its acceleration. This kind of mass, related to the inertia of an object, is called **inertial mass** and is represented by the following equation.

Inertial Mass $m_{\text{inertial}} = \frac{F_{\text{net}}}{a}$

Inertial mass is equal to the net force exerted on the object divided by the acceleration of the object.

The inertial mass of an object is measured by exerting a force on the object and measuring the object's acceleration using an inertial balance, such as the one shown in **Figure 7-13.** The more inertial mass an object has, the less it is affected by any force—the less acceleration it undergoes. Thus, the inertial mass of an object is a measure of the object's resistance to any type of force.

Figure 7-13 An inertial balance allows you to calculate the inertial mass of an object from the period (T) of the back-and-forth motion of the object. Calibration masses, such as the cylindrical ones shown here, are used to create a graph of T^2 versus the mass. The period of the unknown mass is then measured, and the inertial mass is determined from the calibration graph.

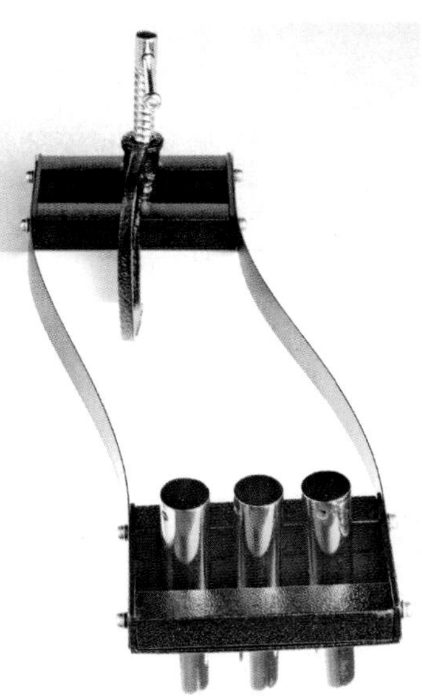

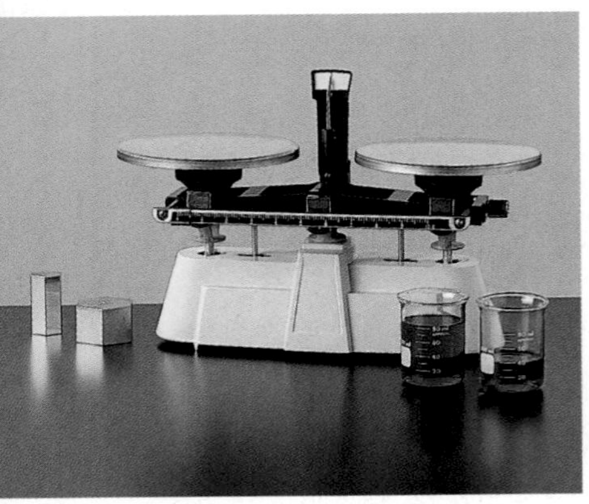

Figure 7-14 The platform balance shown here allows you to measure the force on a mass due to Earth's gravity.

Newton's law of universal gravitation, $F = Gm_1m_2/r^2$, also involves mass, but a different kind of mass. Mass as used in the law of universal gravitation determines the size of the gravitational force between two objects and is called **gravitational mass.** It can be measured using a simple balance, such as the one shown in **Figure 7-14.** If you measure the attractive force exerted on an object by another object of mass, m, at a distance, r, then you can define the gravitational mass in the following way.

Gravitational Mass $m_{\text{grav}} = \dfrac{r^2F_{\text{grav}}}{Gm}$

The gravitational mass of an object is equal to the distance between the objects squared, times the gravitational force, divided by the product of the universal gravitational constant, times the mass of the other object.

How different are these two kinds of mass? Suppose you have a watermelon in the trunk of your car. If you accelerate the car forward, the watermelon will roll backwards, relative to the trunk. This is a result of its inertial mass—its resistance to acceleration. Now, suppose your car climbs a steep hill at a constant speed. The watermelon will again roll backwards. But this time, it moves as a result of its gravitational mass. The watermelon is being attracted downward toward Earth. Newton made the claim that inertial mass and gravitational mass are equal in magnitude. This hypothesis is called the principle of equivalence. All experiments conducted so far have yielded data that support this principle. Albert Einstein also was intrigued by the principle of equivalence and made it a central point in his theory of gravity.

Einstein's Theory of Gravity

Newton's law of universal gravitation allows us to calculate the gravitational force that exists between two objects because of their masses. The concept of a gravitational field allows us to picture the way gravity acts on objects that are far away. Einstein proposed that gravity is not a force, but rather, an effect of space itself. According to Einstein, mass changes the space around it. Mass causes space to be curved, and other bodies are accelerated because of the way they follow this curved space.

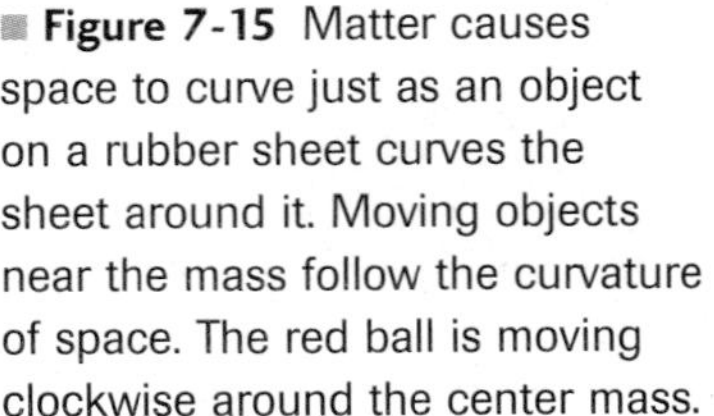

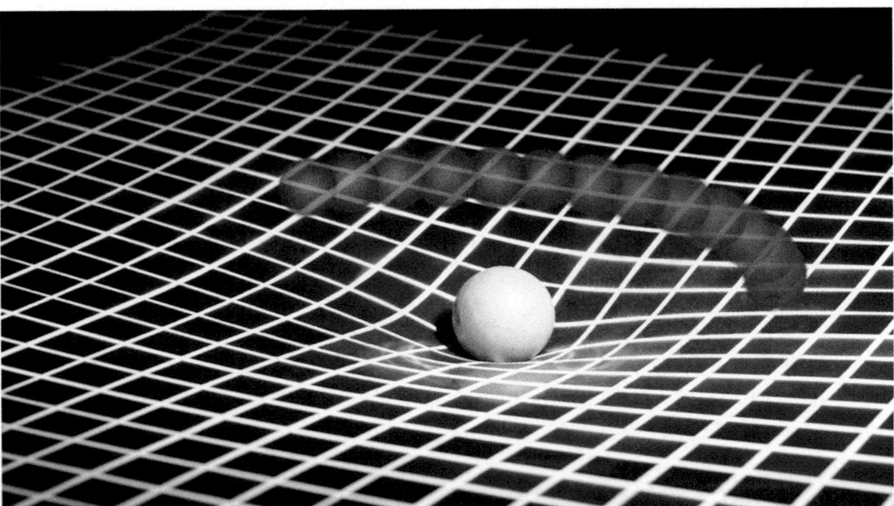

Figure 7-15 Matter causes space to curve just as an object on a rubber sheet curves the sheet around it. Moving objects near the mass follow the curvature of space. The red ball is moving clockwise around the center mass.

One way to picture how space is affected by mass is to compare space to a large, two-dimensional rubber sheet, as shown in **Figure 7-15.** The yellow ball on the sheet represents a massive object. It forms an indentation. A red ball rolling across the sheet simulates the motion of an object in space. If the red ball moves near the sagging region of the sheet, it will be accelerated. In the same way, Earth and the Sun are attracted to one another because of the way space is distorted by the two objects.

Einstein's theory, called the general theory of relativity, makes many predictions about how massive objects affect one another. In every test conducted to date, Einstein's theory has been shown to give the correct results.

Deflection of light Einstein's theory predicts the deflection or bending of light by massive objects. Light follows the curvature of space around the massive object and is deflected, as shown in **Figure 7-16.** In 1919, during an eclipse of the Sun, astronomers found that light from distant stars that passed near the Sun was deflected in agreement with Einstein's predictions.

Another result of general relativity is the effect on light from very massive objects. If an object is massive and dense enough, the light leaving it will be totally bent back to the object. No light ever escapes the object. Objects such as these, called black holes, have been identified as a result of their effects on nearby stars. The radiation produced when matter is pulled into black holes has also been helpful in their detection.

While Einstein's theory provides very accurate predictions of gravity's effects, it is still incomplete. It does not explain the origin of mass or how mass curves space. Physicists are working to understand the deeper meaning of gravity and the origin of mass itself.

Figure 7-16 The light from a distant star bends due to the Sun's gravitational field, thereby changing the apparent position of the star. (Illustration not to scale)

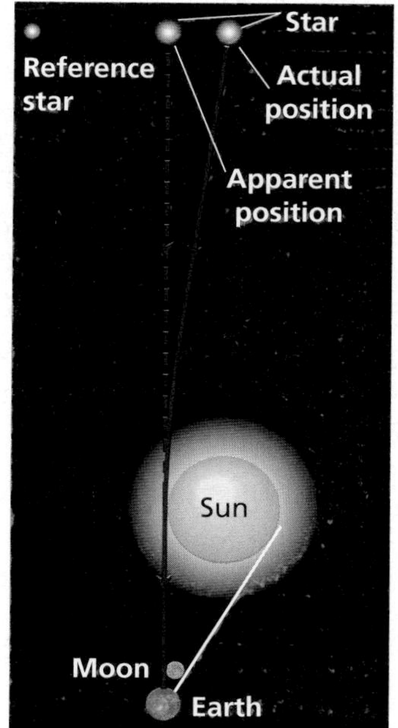

7.2 Section Review

15. Gravitational Fields The Moon is 3.9×10^5 km from Earth's center and 1.5×10^8 km from the Sun's center. The masses of Earth and the Sun are 6.0×10^{24} kg and 2.0×10^{30} kg, respectively.

a. Find the ratio of the gravitational fields due to Earth and the Sun at the center of the Moon.

b. When the Moon is in its third quarter phase, as shown in **Figure 7-17,** its direction from Earth is at right angles to the Sun's direction. What is the net gravitational field due to the Sun and Earth at the center of the Moon?

Figure 7-17 (Not to scale)

Moon

Sun

Earth

16. Gravitational Field The mass of the Moon is 7.3×10^{22} kg and its radius is 1785 km. What is the strength of the gravitational field on the surface of the Moon?

17. A Satellite's Mass When the first artificial satellite was launched into orbit by the former Soviet Union in 1957, U.S. president Dwight D. Eisenhower asked his scientific advisors to calculate the mass of the satellite. Would they have been able to make this calculation? Explain.

18. Orbital Period and Speed Two satellites are in circular orbits about Earth. One is 150 km above the surface, the other 160 km.

a. Which satellite has the larger orbital period?

b. Which one has the greater speed?

19. Theories and Laws Why is Einstein's description of gravity called a "theory," while Newton's is a "law?"

20. Weightlessness Chairs in an orbiting spacecraft are weightless. If you were on board such a spacecraft and you were barefoot, would you stub your toe if you kicked a chair? Explain.

21. Critical Thinking It is easier to launch a satellite from Earth into an orbit that circles eastward than it is to launch one that circles westward. Explain.

PHYSICS LAB

Modeling the Orbits of Planets and Satellites

In this experiment, you will analyze a model that will show how Kepler's first and second laws of motion apply to orbits of objects in space. Kepler's first law states that orbits of planets are ellipses, with the Sun at one focus. Kepler's second law states that an imaginary line from the Sun to a planet sweeps out equal areas in equal time intervals.

The shape of the elliptical orbit is defined by eccentricity, *e*, which is the ratio of the distance between the foci to the length of the major axis. When an object is at its farthest distance from the Sun along the major axis, it is at aphelion. When the object is at its closest distance from the Sun along the major axis, it is at perihelion.

QUESTION

What is the shape of the orbits of planets and satellites in the solar system?

Objectives

- **Formulate models** to infer the shape of orbits of planets and satellites.
- **Collect and organize data** for aphelion distances and perihelion distances of objects as they orbit the Sun.
- **Draw conclusions** about Kepler's first and second laws of motion.

Safety Precautions

- **Pins are sharp and can puncture the skin.**

Materials

piece of cardboard
sheet of blank, white paper
two push pins
metric ruler
sharp pencil or pen
four small pieces of tape
string (25 cm)

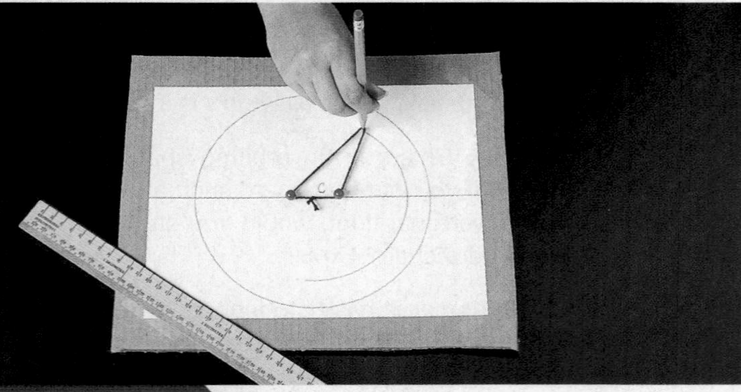

Procedure

1. Place a piece of paper on a piece of cardboard using tape at the four corners.
2. Draw a line across the center of the paper, along the length of the paper. This line represents the major axis.
3. Mark the center of the line and label it *C*.
4. Use the string to tie a loop, which, when stretched, has a length of 10 cm. For each object listed in the data table, calculate the distance between the foci, *d*, using the following equation:

$$d = \frac{2e(10.0\text{ cm})}{e + 1}$$

5. For the circle, place a pin at C. Put the loop of string over the pin and pull it tight with your pencil. Move the pencil in a circular fashion around the center, letting the string guide it.
6. For the next object, place one pin a distance of *d*/2 from C along the major axis.
7. Place a second pin a distance of *d*/2 on the opposite side of C. The two pins represent the foci. One focus is the location of the Sun.
8. Put the loop of string over both pins and pull it tight with your pencil. Move the pencil in a circular fashion, letting the string guide it.
9. Using the same paper, repeat steps 6-8 for each of the listed objects.
10. After all of the orbits are plotted, label each orbit with the name and eccentricity of the object plotted.

Data Table

Object	Eccentricity (e)	d (cm)	Measured A	Measured P	Experimental e	% Error
Circle	0					
Earth	0.017					
Pluto	0.25					
Comet	0.70					

Analyze

1. Measure the aphelion distance, A, by measuring the distance between one focus and the farthest point in the orbit along the major axis. Record your data in the data table.
2. Measure the perihelion distance, P, by measuring the closest distance between one focus and the closest point in the orbit along the major axis. Record the data in the data table.
3. Calculate the experimental eccentricity for each of the objects and record your data in the data table. Use the following equation:
$$e = \frac{A - P}{A + P}$$
4. **Error Analysis** Calculate the percent error for each object using the experimental eccentricities compared to the known eccentricities. Record your values in the data table.
5. **Analyze** Why is the shape of the orbit with $e = 0$ a circle?
6. **Compare** How does Earth's orbit compare to a circle?
7. **Observe** Which of the orbits truly looks elliptical?

Conclude and Apply

1. Does the orbit model you constructed obey Kepler's first law? Explain.
2. Kepler studied the orbit data of Mars ($e = 0.093$) and concluded that planets move about the Sun in elliptical orbits. What would Kepler have concluded if he had been on Mars and studied Earth's orbit?
3. Where does a planet travel fastest: at aphelion or perihelion? Why?
4. Kepler's second law helps to determine the ratio between Pluto's velocity at aphelion and perihelion (v_A/v_P). To determine this ratio, first calculate the area swept out by Pluto's orbit. This area is approximately equal to the area of a triangle: Area $= \frac{1}{2}$ (distance to the Sun) current velocity $\times$ time. If the area that the orbit sweeps out in a fixed amount of time, such as 30 days, is the same at aphelion and perihelion, this relationship can be written
$$\frac{1}{2}Pv_Pt = \frac{1}{2}Av_At$$
What is the ratio v_P/v_A for Pluto?
5. Pluto's minimum orbital velocity is 3.7 km/s. What are the values for v_P and v_A?

Going Further

1. You used rough approximations to look at Kepler's second law. Suggest an experiment to obtain precise results to confirm the second law.
2. Design an experiment to verify Kepler's third law.

Real-World Physics

Does a communications or weather satellite that is orbiting Earth follow Kepler's laws? Collect data to verify your answer.

To find out more about gravitation, visit the Web site: physicspp.com

Black Holes

What would happen if you were to travel to a black hole? Your body would be stretched, flattened, and eventually pulled apart. What is a black hole? What is known about black holes?

A black hole is one of the possible final stages in the evolution of a star. When fusion reactions stop in the core of a star that is at least 20 times more massive than the Sun, the core collapses forever, compacting matter into an increasingly smaller volume. The infinitely small, but infinitely dense, object that remains is called a singularity. The force of gravity is so immense in the region around the singularity that nothing, not even light, can escape it. This region is called a black hole.

Nothing Can Escape In 1917, German mathematician Karl Schwarzschild verified, mathematically, that black holes could exist. Schwarzschild used solutions to Einstein's theory of general relativity to describe the properties of black holes. He derived an expression for a radius, called the Schwarzschild radius, within which neither light nor matter escapes the force of gravity of the singularity. The Schwarzschild radius is represented by the following equation:

$$R_s = \frac{2GM}{c^2}$$

In this equation, G is Newton's universal gravitational constant, M is the mass of the black hole, and c is the speed of light. The edge of the sphere defined by the Schwarzschild radius is called the event horizon. At the event horizon, the escape velocity equals the speed of light. Because nothing travels faster than the speed of light, objects that cross the event horizon can never escape.

Indirect and Direct Evidence Black holes have three physical properties that can theoretically be measured—mass, angular momentum, and electric charge. A black hole's mass can be determined by the gravitational field it generates. Mass is calculated by using a modified form of Kepler's third law of planetary motion. Studies using NASA's *Rossi X-ray Timing Explorer* have shown that black holes spin just as stars and planets do. A black hole spins because it retains the angular momentum of the star that formed it. Even though a black hole's electric charge has not been measured, scientists hypothesize that a black hole may become charged when an excess of one type of electric charge falls into it. Super-heated gases in a black hole emit X rays, which can be detected by X-ray telescopes, such as the space-based *Chandra X-ray Observatory*.

Although not everything is known about black holes, there is direct and indirect evidence of their existence. Continued research and special missions will provide a better understanding of black holes.

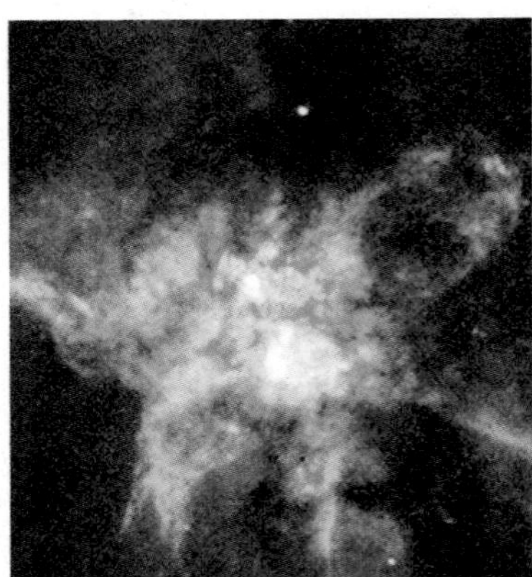

***Hubble* visible image of galaxy NGC 6240.**

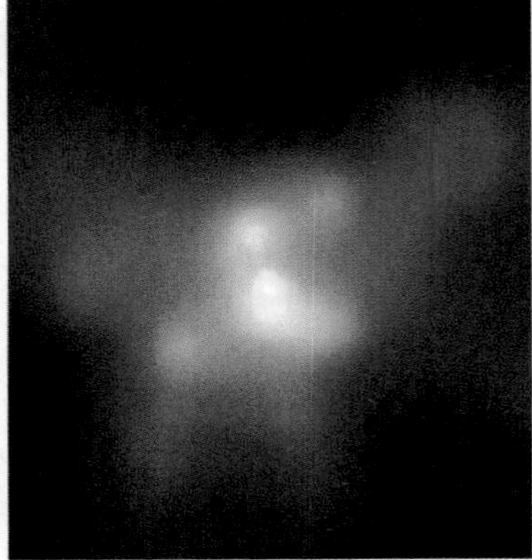

***Chandra* X-ray image of two black holes (blue) in NGC 6240.**

Going Further

Solve The escape velocity of an object leaving the event horizon can be represented by the following equation:

$$v = \sqrt{\frac{2GM}{R_s}}$$

In this equation, G is Newton's universal gravitational constant, M is the mass of the black hole, and R_s is the radius of the black hole. Show that the escape velocity equals the speed of light.

Chapter 7

Study Guide

7.1 Planetary Motion and Gravitation

Vocabulary

- Kepler's first law (p. 172)
- Kepler's second law (p. 173)
- Kepler's third law (p. 173)
- gravitational force (p. 175)
- law of universal gravitation (p. 175)

Key Concepts

- Kepler's first law states that planets move in elliptical orbits, with the Sun at one focus.
- Kepler's second law states that an imaginary line from the Sun to a planet sweeps out equal areas in equal times.
- Kepler's third law states that the square of the ratio of the periods of any two planets is equal to the cube of the ratio of their distances from the Sun.

$$\left(\frac{T_A}{T_B}\right)^2 = \left(\frac{r_A}{r_B}\right)^3$$

- Newton's law of universal gravitation states that the gravitational force between any two objects is directly proportional to the product of their masses and inversely proportional to the square of the distance between their centers. The force is attractive and along a line connecting their centers.

$$F = G\frac{m_1 m_2}{r^2}$$

- Newton's law of universal gravitation can be used to rewrite Kepler's third law to relate the radius and period of a planet to the mass of the Sun.

$$T^2 = \left(\frac{4\pi^2}{Gm_S}\right)r^3$$

7.2 Using the Law of Universal Gravitation

Vocabulary

- gravitational field (p. 183)
- inertial mass (p. 183)
- gravitational mass (p. 184)

Key Concepts

- The speed of an object in circular orbit is given by the following expression.

$$v = \sqrt{\frac{Gm_E}{r}}$$

- The period of a satellite in a circular orbit is given by the following expression.

$$T = 2\pi\sqrt{\frac{r^3}{Gm_E}}$$

- All objects have gravitational fields surrounding them.

$$g = \frac{Gm}{r^2}$$

- Gravitational mass and inertial mass are two essentially different concepts. The gravitational and inertial masses of an object, however, are numerically equal.

$$m_{inertial} = \frac{F_{net}}{a} \qquad m_{grav} = \frac{r^2 F_{grav}}{Gm}$$

- Einstein's general theory of relativity describes gravitational attraction as a property of space itself.

Concept Mapping

22. Create a concept map using these terms: *planets, stars, Newton's law of universal gravitation, Kepler's first law, Kepler's second law, Kepler's third law, Einstein's general theory of relativity.*

Mastering Concepts

23. In 1609, Galileo looked through his telescope at Jupiter and saw four moons. The name of one of the moons that he saw is Io. Restate Kepler's first law for Io and Jupiter. (7.1)

24. Earth moves more slowly in its orbit during summer in the northern hemisphere than it does during winter. Is it closer to the Sun in summer or in winter? (7.1)

25. Is the area swept out per unit of time by Earth moving around the Sun equal to the area swept out per unit of time by Mars moving around the Sun? (7.1)

26. Why did Newton think that a force must act on the Moon? (7.1)

27. How did Cavendish demonstrate that a gravitational force of attraction exists between two small objects? (7.1)

28. What happens to the gravitational force between two masses when the distance between the masses is doubled? (7.1)

29. According to Newton's version of Kepler's third law, how would the ratio T^2/r^3 change if the mass of the Sun were doubled? (7.1)

30. How do you answer the question, "What keeps a satellite up?" (7.2)

31. A satellite is orbiting Earth. On which of the following does its speed depend? (7.2)
    a. mass of the satellite
    b. distance from Earth
    c. mass of Earth

32. What provides the force that causes the centripetal acceleration of a satellite in orbit? (7.2)

33. During space flight, astronauts often refer to forces as multiples of the force of gravity on Earth's surface. What does a force of $5g$ mean to an astronaut? (7.2)

34. Newton assumed that a gravitational force acts directly between Earth and the Moon. How does Einstein's view of the attractive force between the two bodies differ from Newton's view? (7.2)

35. Show that the dimensions of g in the equation $g = F/m$ are in m/s^2. (7.2)

36. If Earth were twice as massive but remained the same size, what would happen to the value of g? (7.2)

Applying Concepts

37. **Golf Ball** The force of gravity acting on an object near Earth's surface is proportional to the mass of the object. **Figure 7-18** shows a tennis ball and golf ball in free fall. Why does a tennis ball not fall faster than a golf ball?

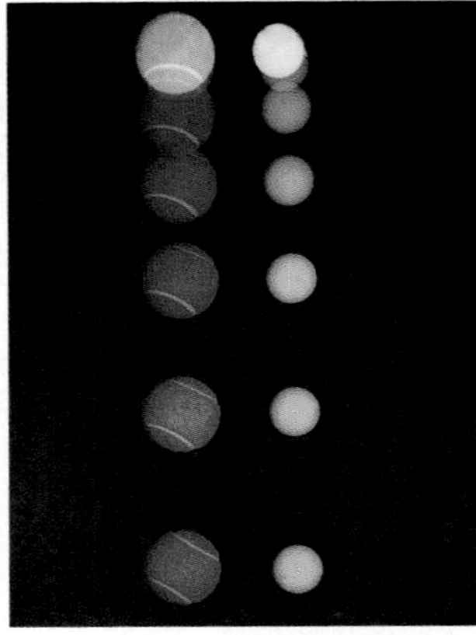

Figure 7-18

38. What information do you need to find the mass of Jupiter using Newton's version of Kepler's third law?

39. The mass of Pluto was not known until a satellite of the object was discovered. Why?

40. Decide whether each of the orbits shown in **Figure 7-19** is a possible orbit for a planet.

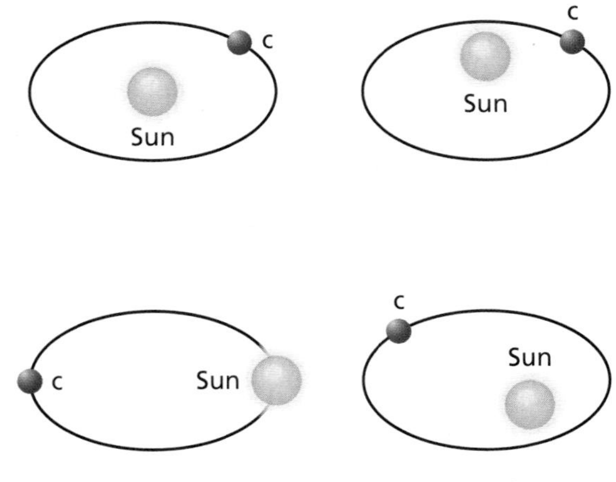

Figure 7-19

41. The Moon and Earth are attracted to each other by gravitational force. Does the more-massive Earth attract the Moon with a greater force than the Moon attracts Earth? Explain.

42. What would happen to the value of G if Earth were twice as massive, but remained the same size?

43. **Figure 7-20** shows a satellite orbiting Earth. Examine the equation $v = \sqrt{\frac{Gm_E}{r}}$, relating the speed of an orbiting satellite and its distance from the center of Earth. Does a satellite with a large or small orbital radius have the greater velocity?

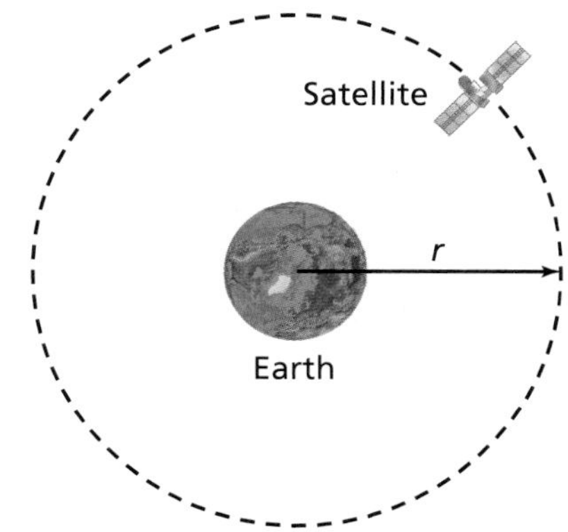

Figure 7-20 (Not to scale)

44. Space Shuttle If a space shuttle goes into a higher orbit, what happens to the shuttle's period?

45. Mars has about one-ninth the mass of Earth. **Figure 7-21** shows satellite M, which orbits Mars with the same orbital radius as satellite E, which orbits Earth. Which satellite has a smaller period?

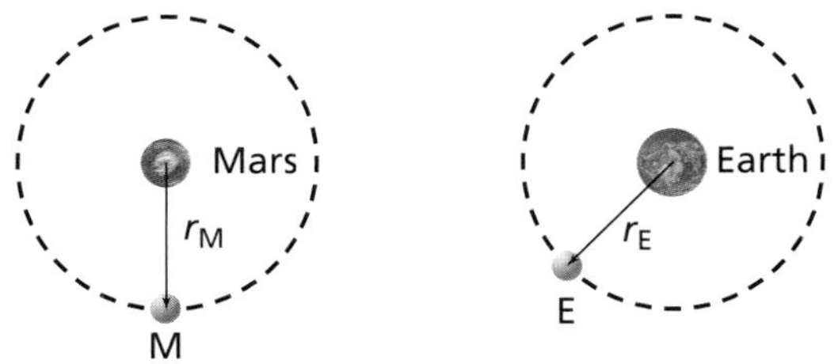

Figure 7-21 (Not to scale)

46. Jupiter has about 300 times the mass of Earth and about ten times Earth's radius. Estimate the size of *g* on the surface of Jupiter.

47. A satellite is one Earth radius above the surface of Earth. How does the acceleration due to gravity at that location compare to acceleration at the surface of Earth?

48. If a mass in Earth's gravitational field is doubled, what will happen to the force exerted by the field upon the mass?

49. Weight Suppose that yesterday your body had a mass of 50.0 kg. This morning you stepped on a scale and found that you had gained weight.

a. What happened, if anything, to your mass?

b. What happened, if anything, to the ratio of your weight to your mass?

50. As an astronaut in an orbiting space shuttle, how would you go about "dropping" an object down to Earth?

51. Weather Satellites The weather pictures that you see every day on TV come from a spacecraft in a stationary position relative to the surface of Earth, 35,700 km above Earth's equator. Explain how it can stay in exactly the same position day after day. What would happen if it were closer? Farther out? *Hint: Draw a pictorial model.*

Mastering Problems

7.1 Planetary Motion and Gravitation

52. Jupiter is 5.2 times farther from the Sun than Earth is. Find Jupiter's orbital period in Earth years.

53. **Figure 7-22** shows a Cavendish apparatus like the one used to find G. It has a large lead sphere that is 5.9 kg in mass and a small one with a mass of 0.047 kg. Their centers are separated by 0.055 m. Find the force of attraction between them.

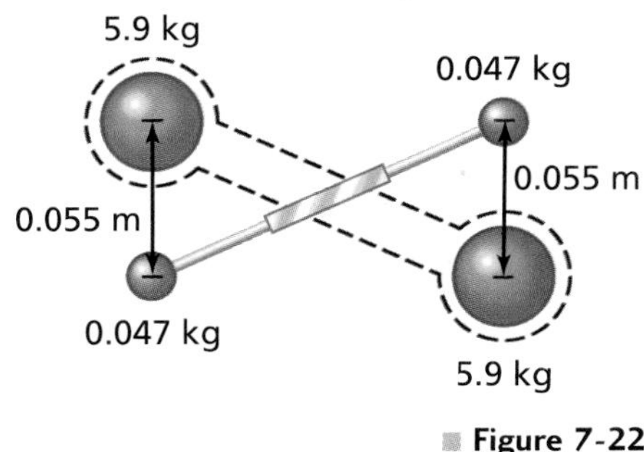

Figure 7-22

54. Use Table 7-1 on p. 173 to compute the gravitational force that the Sun exerts on Jupiter.

55. Tom has a mass of 70.0 kg and Sally has a mass of 50.0 kg. Tom and Sally are standing 20.0 m apart on the dance floor. Sally looks up and sees Tom. She feels an attraction. If the attraction is gravitational, find its size. Assume that both Tom and Sally can be replaced by spherical masses.

56. Two balls have their centers 2.0 m apart, as shown in **Figure 7-23.** One ball has a mass of 8.0 kg. The other has a mass of 6.0 kg. What is the gravitational force between them?

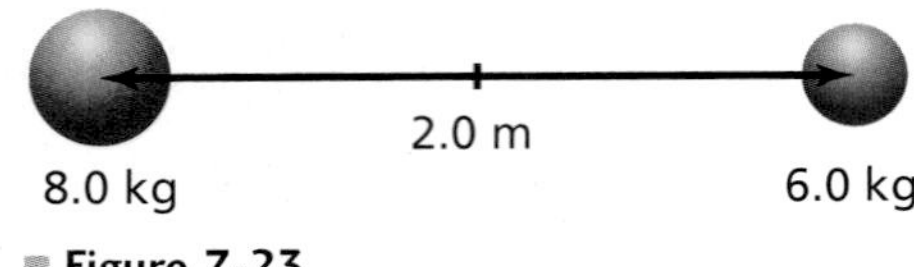

Figure 7-23

57. Two bowling balls each have a mass of 6.8 kg. They are located next to each other with their centers 21.8 cm apart. What gravitational force do they exert on each other?

58. Assume that you have a mass of 50.0 kg. Earth has a mass of 5.97×10^{24} kg and a radius of 6.38×10^{6} m.
   a. What is the force of gravitational attraction between you and Earth?
   b. What is your weight?

59. The gravitational force between two electrons that are 1.00 m apart is 5.54×10^{-71} N. Find the mass of an electron.

60. A 1.0-kg mass weighs 9.8 N on Earth's surface, and the radius of Earth is roughly 6.4×10^{6} m.
   a. Calculate the mass of Earth.
   b. Calculate the average density of Earth.

61. **Uranus** Uranus requires 84 years to circle the Sun. Find Uranus's orbital radius as a multiple of Earth's orbital radius.

62. **Venus** Venus has a period of revolution of 225 Earth days. Find the distance between the Sun and Venus as a multiple of Earth's orbital radius.

63. If a small planet, D, were located 8.0 times as far from the Sun as Earth is, how many years would it take the planet to orbit the Sun?

64. Two spheres are placed so that their centers are 2.6 m apart. The force between the two spheres is 2.75×10^{-12} N. What is the mass of each sphere if one sphere is twice the mass of the other sphere?

65. The Moon is 3.9×10^{5} km from Earth's center and 1.5×10^{8} km from the Sun's center. If the masses of the Moon, Earth, and the Sun are 7.3×10^{22} kg, 6.0×10^{24} kg, and 2.0×10^{30} kg, respectively, find the ratio of the gravitational forces exerted by Earth and the Sun on the Moon.

66. **Toy Boat** A force of 40.0 N is required to pull a 10.0-kg wooden toy boat at a constant velocity across a smooth glass surface on Earth. What force would be required to pull the same wooden toy boat across the same glass surface on the planet Jupiter?

67. Mimas, one of Saturn's moons, has an orbital radius of 1.87×10^{8} m and an orbital period of about 23.0 h. Use Newton's version of Kepler's third law to find Saturn's mass.

68. The Moon is 3.9×10^{8} m away from Earth and has a period of 27.33 days. Use Newton's version of Kepler's third law to find the mass of Earth. Compare this mass to the mass found in problem 60.

69. **Halley's Comet** Every 74 years, comet Halley is visible from Earth. Find the average distance of the comet from the Sun in astronomical units (AU).

70. Area is measured in m^2, so the rate at which area is swept out by a planet or satellite is measured in m^2/s.
   a. How quickly is an area swept out by Earth in its orbit about the Sun?
   b. How quickly is an area swept out by the Moon in its orbit about Earth? Use 3.9×10^{8} m as the average distance between Earth and the Moon, and 27.33 days as the period of the Moon.

7.2 Using the Law of Universal Gravitation

71. **Satellite** A geosynchronous satellite is one that appears to remain over one spot on Earth, as shown in **Figure 7-24.** Assume that a geosynchronous satellite has an orbital radius of 4.23×10^{7} m.
   a. Calculate its speed in orbit.
   b. Calculate its period.

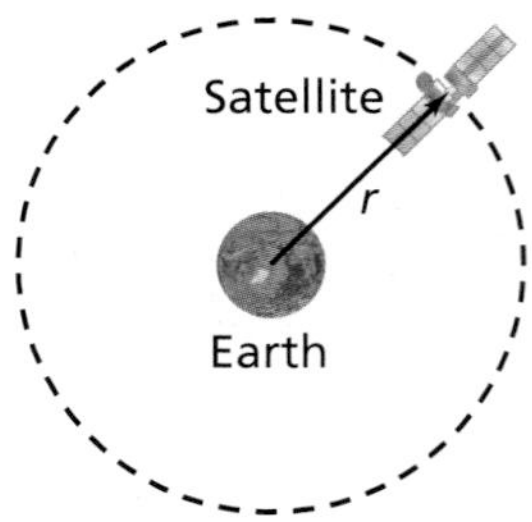

Figure 7-24 (Not to scale)

72. **Asteroid** The asteroid Ceres has a mass of 7×10^{20} kg and a radius of 500 km.
   a. What is g on the surface of Ceres?
   b. How much would a 90-kg astronaut weigh on Ceres?

73. **Book** A 1.25-kg book in space has a weight of 8.35 N. What is the value of the gravitational field at that location?

74. The Moon's mass is 7.34×10^{22} kg, and it is 3.8×10^{8} m away from Earth. Earth's mass is 5.97×10^{24} kg.
   a. Calculate the gravitational force of attraction between Earth and the Moon.
   b. Find Earth's gravitational field at the Moon.

 For more problems, go to Additional Problems, Appendix B.

75. Two 1.00-kg masses have their centers 1.00 m apart. What is the force of attraction between them?

76. The radius of Earth is about 6.38×10^3 km. A 7.20×10^3-N spacecraft travels away from Earth. What is the weight of the spacecraft at the following distances from Earth's surface?
a. 6.38×10^3 km
b. 1.28×10^4 km

77. Rocket How high does a rocket have to go above Earth's surface before its weight is half of what it is on Earth?

78. Two satellites of equal mass are put into orbit 30.0 m apart. The gravitational force between them is 2.0×10^{-7} N.
a. What is the mass of each satellite?
b. What is the initial acceleration given to each satellite by gravitational force?

79. Two large spheres are suspended close to each other. Their centers are 4.0 m apart, as shown in **Figure 7-25.** One sphere weighs 9.8×10^2 N. The other sphere has a weight of 1.96×10^2 N. What is the gravitational force between them?

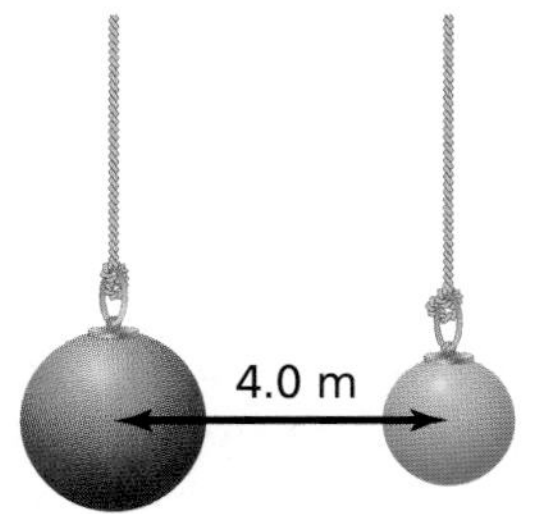

Figure 7-25

80. Suppose the centers of Earth and the Moon are 3.9×10^8 m apart, and the gravitational force between them is about 1.9×10^{20} N. What is the approximate mass of the Moon?

81. On the surface of the Moon, a 91.0-kg physics teacher weighs only 145.6 N. What is the value of the Moon's gravitational field at its surface?

82. The mass of an electron is 9.1×10^{-31} kg. The mass of a proton is 1.7×10^{-27} kg. An electron and a proton are about 0.59×10^{-10} m apart in a hydrogen atom. What gravitational force exists between the proton and the electron of a hydrogen atom?

83. Consider two spherical 8.0-kg objects that are 5.0 m apart.
a. What is the gravitational force between the two objects?
b. What is the gravitational force between them when they are 5.0×10^1 m apart?

84. If you weigh 637 N on Earth's surface, how much would you weigh on the planet Mars? Mars has a mass of 6.42×10^{23} kg and a radius of 3.40×10^6 m.

85. Using Newton's version of Kepler's third law and information from Table 7-1 on page 173, calculate the period of Earth's Moon if the orbital radius were twice the actual value of 3.9×10^8 m.

86. Find the value of g, acceleration due to gravity, in the following situations.
a. Earth's mass is triple its actual value, but its radius remains the same.
b. Earth's radius is tripled, but its mass remains the same.
c. Both the mass and radius of Earth are doubled.

87. Astronaut What would be the strength of Earth's gravitational field at a point where an 80.0-kg astronaut would experience a 25.0 percent reduction in weight?

Mixed Review

88. Use the information for Earth in Table 7-1 on page 173 to calculate the mass of the Sun, using Newton's version of Kepler's third law.

89. Earth's gravitational field is 7.83 N/kg at the altitude of the space shuttle. At this altitude, what is the size of the force of attraction between a student with a mass of 45.0 kg and Earth?

90. Use the data from Table 7-1 on page 173 to find the speed and period of a satellite that orbits Mars 175 km above its surface.

91. Satellite A satellite is placed in orbit, as shown in **Figure 7-26,** with a radius that is half the radius of the Moon's orbit. Find the period of the satellite in units of the period of the Moon.

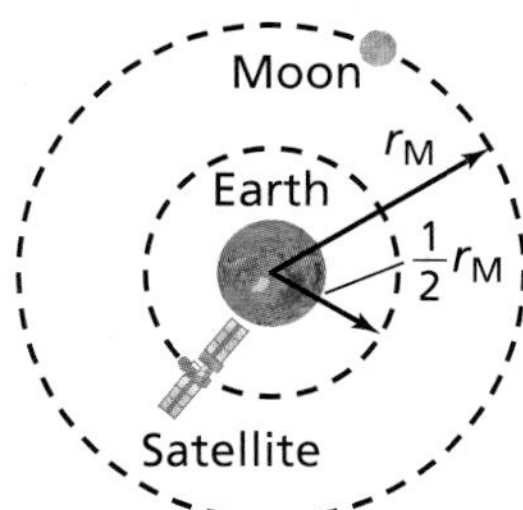

Figure 7-26

92. Cannonball The Moon's mass is 7.3×10^{22} kg and its radius is 1785 km. If Newton's thought experiment of firing a cannonball from a high mountain were attempted on the Moon, how fast would the cannonball have to be fired? How long would it take the cannonball to return to the cannon?

93. The period of the Moon is one month. Answer the following questions assuming that the mass of Earth is doubled.
  a. What would the period of the Moon be? Express your results in months.
  b. Where would a satellite with an orbital period of one month be located?
  c. How would the length of a year on Earth be affected?

94. How fast would a planet of Earth's mass and size have to spin so that an object at the equator would be weightless? Give the period of rotation of the planet in minutes.

95. Car Races Suppose that a Martian base has been established and car races are being considered. A flat, circular race track has been built for the race. If a car can achieve speeds of up to 12 m/s, what is the smallest radius of a track for which the coefficient of friction is 0.50?

96. *Apollo 11* On July 19, 1969, *Apollo 11's* revolution around the Moon was adjusted to an average orbit of 111 km. The radius of the Moon is 1785 km, and the mass of the Moon is 7.3×10^{22} kg.
  a. How many minutes did *Apollo 11* take to orbit the Moon once?
  b. At what velocity did *Apollo 11* orbit the Moon?

Thinking Critically

97. Analyze and Conclude Some people say that the tides on Earth are caused by the pull of the Moon. Is this statement true?
  a. Determine the forces that the Moon and the Sun exert on a mass, *m*, of water on Earth. Your answer will be in terms of *m* with units of N.
  b. Which celestial body, the Sun or the Moon, has a greater pull on the waters of Earth?
  c. Determine the difference in force exerted by the Moon on the water at the near surface and the water at the far surface (on the opposite side) of Earth, as illustrated in **Figure 7-27.** Again, your answer will be in terms of *m* with units of N.

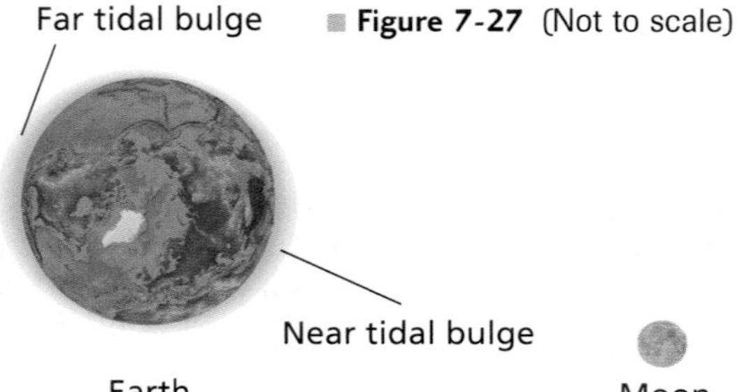

■ **Figure 7-27** (Not to scale)

  d. Determine the difference in force exerted by the Sun on water at the near surface and on water at the far surface (on the opposite side) of Earth.
  e. Which celestial body has a greater difference in pull from one side of Earth to the other?
  f. Why is the statement that the tides result from the pull of the Moon misleading? Make a correct statement to explain how the Moon causes tides on Earth.

98. Make and Use Graphs Use Newton's law of universal gravitation to find an equation where x is equal to an object's distance from Earth's center, and y is its acceleration due to gravity. Use a graphing calculator to graph this equation, using 6400–6600 km as the range for x and 9–10 m/s^2 as the range for y. The equation should be of the form $y = c(1/x^2)$. Trace along this graph and find y for the following locations.
  a. at sea level, 6400 km
  b. on top of Mt. Everest, 6410 km
  c. in a typical satellite orbit, 6500 km
  d. in a much higher orbit, 6600 km

Writing in Physics

99. Research and describe the historical development of the measurement of the distance between the Sun and Earth.

100. Explore the discovery of planets around other stars. What methods did the astronomers use? What measurements did they take? How did they use Kepler's third law?

Cumulative Review

101. Airplanes A jet airplane took off from Pittsburgh at 2:20 P.M. and landed in Washington, DC, at 3:15 P.M. on the same day. If the jet's average speed while in the air was 441.0 km/h, what is the distance between the cities? (Chapter 2)

102. Carolyn wants to know how much her brother Jared weighs. He agrees to stand on a scale for her, but only if they are riding in an elevator. If he steps on the scale while the elevator is accelerating upward at 1.75 m/s^2 and the scale reads 716 N, what is Jared's usual weight on Earth? (Chapter 4)

103. Potato Bug A 1.0-g potato bug is walking around the outer rim of an upside-down flying disk. If the disk has a diameter of 17.2 cm and the bug moves at a rate of 0.63 cm/s, what is the centripetal force acting on the bug? What agent provides this force? (Chapter 6)

 For more problems, go to Additional Problems, Appendix B.

Multiple Choice

1. Two satellites are in orbit around a planet. One satellite has an orbital radius of 8.0×10^6 m. The period of rotation for this satellite is 1.0×10^6 s. The other satellite has an orbital radius of 2.0×10^7 m. What is this satellite's period of rotation?

Ⓐ 5.0×10^5 s
Ⓑ 2.5×10^6 s
Ⓒ 4.0×10^6 s
Ⓓ 1.3×10^7 s

2. The illustration below shows a satellite in orbit around a small planet. The satellite's orbital radius is 6.7×10^4 km and its speed is 2.0×10^5 m/s. What is the mass of the planet around which the satellite orbits? ($G = 6.7\times10^{-11}$ N·m^2/kg^2)

Ⓐ 2.5×10^{18} kg
Ⓑ 4.0×10^{20} kg
Ⓒ 2.5×10^{23} kg
Ⓓ 4.0×10^{28} kg

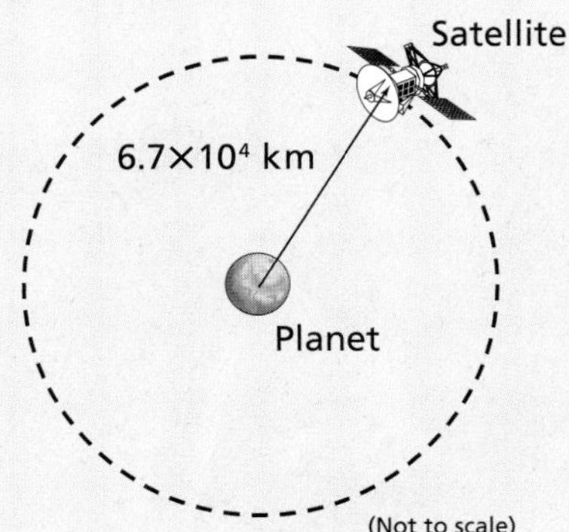

3. Two satellites are in orbit around the same planet. Satellite A has a mass of 1.5×10^2 kg, and satellite B has a mass of 4.5×10^3 kg. The mass of the planet is 6.6×10^{24} kg. Both satellites have the same orbital radius of 6.8×10^6 m. What is the difference in the orbital periods of the satellites?

Ⓐ no difference
Ⓑ 1.5×10^2 s
Ⓒ 2.2×10^2 s
Ⓓ 3.0×10^2 s

4. A moon revolves around a planet with a speed of 9.0×10^3 m/s. The distance from the moon to the center of the planet is 5.4×10^6 m. What is the orbital period of the moon?

Ⓐ $1.2\pi\times10^2$ s
Ⓑ $6.0\pi\times10^2$ s
Ⓒ $1.2\pi\times10^3$ s
Ⓓ $1.2\pi\times10^9$ s

5. A moon in orbit around a planet experiences a gravitational force not only from the planet, but also from the Sun. The illustration below shows a moon during a solar eclipse, when the planet, the moon, and the Sun are aligned. The moon has a mass of about 3.9×10^{21} kg. The mass of the planet is 2.4×10^{26} kg, and the mass of the Sun is 2.0×10^{30} kg. The distance from the moon to the center of the planet is 6.0×10^8 m, and the distance from the moon to the Sun is 1.5×10^{11} m. What is the ratio of the gravitational force on the moon due to the planet, compared to its gravitational force due to the Sun during the solar eclipse?

Ⓐ 0.5
Ⓑ 2.5
Ⓒ 5.0
Ⓓ 7.5

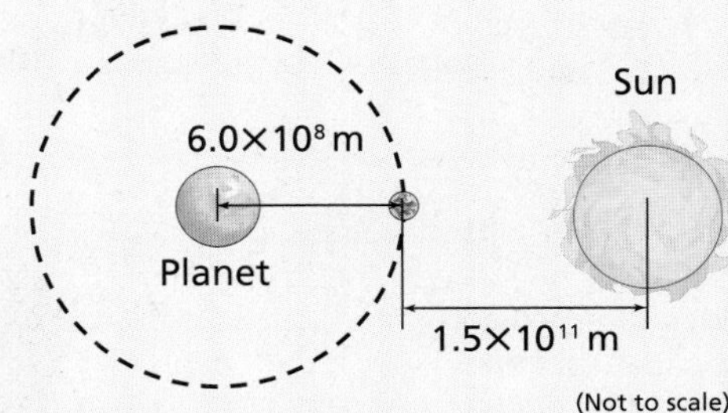

Extended Answer

6. Two satellites are in orbit around a planet. Satellite S_1 takes 20 days to orbit the planet at a distance of 2×10^5 km from the center of the planet. Satellite S_2 takes 160 days to orbit the planet. What is the distance of Satellite S_2 from the center of the planet?

✓ Test-Taking TIP

Plan Your Work and Work Your Plan

Plan your workload so that you do a little work each day, rather than a lot of work all at once. The key to retaining information is repeated review and practice. You will retain more if you study one hour a night for five days in a row instead of cramming the night before a test.

Chapter 8

Rotational Motion

What You'll Learn

- You will learn how to describe and measure rotational motion.
- You will learn how torque changes rotational velocity.
- You will explore factors that determine the stability of an object.
- You will learn the nature of centrifugal and Coriolis "forces."

Why It's Important

You encounter many rotating objects in everyday life, such as CDs, wheels, and amusement-park rides.

Spin Rides Amusement-park rides that spin are designed to thrill riders using the physics of rotational motion. The thrill is produced by a "force" that is present only when the ride spins.

Think About This ▶

Why do people who ride amusement-park rides that spin in circles, such as this one, experience such strong physical reactions?

physicspp.com

LAUNCH Lab How do different objects rotate as they roll?

Question
Do different objects of similar size and mass roll at the same rate on an incline?

Procedure

1. You will need a meterstick, a piece of foam board, a ball, a solid can, and a hollow can.
2. Position the foam board on a 20° incline.
3. Place the meterstick horizontally across the foam board, near the top of the incline, and hold it.
4. Place the ball, solid can, and hollow can against the meterstick. The solid can and hollow can should be placed sideways.
5. Simultaneously, release the three objects by lifting the meterstick.
6. As each object accelerates down the incline, due to gravity, observe the order in which each object reaches the bottom.
7. Repeat steps 2–5 two more times.

Analysis
List the objects in order from the greatest to the least acceleration.

Critical Thinking Which of the objects' properties may have contributed to their behavior? List the properties that were similar and those that were different for each object.

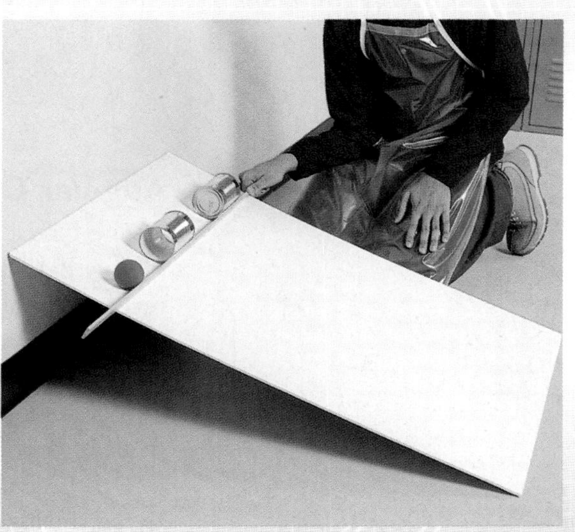

8.1 Describing Rotational Motion

You probably have observed a spinning object many times. How would you measure such an object's rotation? Find a circular object, such as a CD. Mark one point on the edge of the CD so that you can keep track of its position. Rotate the CD to the left (counterclockwise), and as you do so, watch the location of the mark. When the mark returns to its original position, the CD has made one complete revolution. How can you measure a fraction of one revolution? It can be measured in several different ways. A grad is $\frac{1}{400}$ of a revolution, whereas a degree is $\frac{1}{360}$ of a revolution. In mathematics and physics, yet another form of measurement is used to describe fractions of revolutions. In one revolution, a point on the edge travels a distance equal to 2π times the radius of the object. For this reason, the **radian** is defined as $\frac{1}{2\pi}$ of a revolution. In other words, one complete revolution is equal to 2π radians. A radian is abbreviated "rad."

▶ Objectives

- **Describe** angular displacement.
- **Calculate** angular velocity.
- **Calculate** angular acceleration.
- **Solve** problems involving rotational motion.

▶ Vocabulary

radian
angular displacement
angular velocity
angular acceleration

■ **Figure 8-1** The pie chart shows the radian measure of the most common angles, measured in the counterclockwise direction. Each angle is measured from $\theta = 0$.

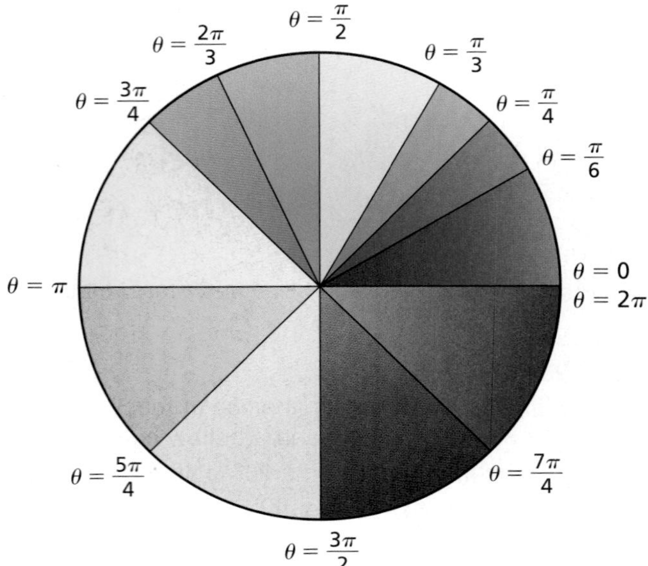

Angular Displacement

The Greek letter theta, θ, is used to represent the angle of revolution. **Figure 8-1** shows the angles in radians for several common fractions of a revolution. Note that counterclockwise rotation is designated as positive, while clockwise is negative. As an object rotates, the change in the angle is called **angular displacement.**

As you know, Earth makes one complete revolution, or 2π rad, in 24 h. In 12 h, its rotation is through π rad. Through what angle does Earth rotate in 6 h? Because 6 h is one-fourth of a day, Earth rotates through an angle of $\frac{\pi}{2}$ rad during that period. Earth's rotation as seen from the north pole is positive. Is it positive or negative when viewed from the south pole?

How far does a point on a rotating object move? You already found that a point on the edge of an object moves 2π times the radius in one revolution. In general, for rotation through an angle, θ, a point at a distance, r, from the center, as shown in **Figure 8-2,** moves a distance given by $d = r\theta$. If r is measured in meters, you might think that multiplying it by θ rad would result in d being measured in m·rad. However, this is not the case. Radians indicate the ratio between d and r. Thus, d is measured in m.

■ **Figure 8-2** The dashed line shows the path of the point on the CD as the CD rotates counterclockwise about its center.

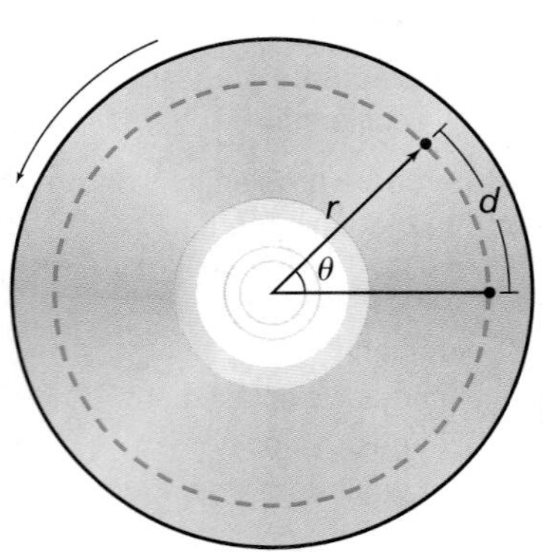

Angular Velocity

How fast does a CD spin? How do you determine its speed of rotation? Recall from Chapter 2 that velocity is displacement divided by the time taken to make the displacement. Likewise, the **angular velocity** of an object is angular displacement divided by the time taken to make the displacement. Thus, the angular velocity of an object is given by the following equation, where angular velocity is represented by the Greek letter omega, ω.

Angular Velocity of an Object $\omega = \frac{\Delta\theta}{\Delta t}$

The angular velocity is equal to the angular displacement divided by the time required to make the rotation.

Recall that if the velocity changes over a time interval, the average velocity is not equal to the instantaneous velocity at any given instant. Similarly, the angular velocity calculated in this way is actually the average angular velocity over a time interval, Δt. Instantaneous angular velocity is equal to the slope of a graph of angular position versus time.

Angular velocity is measured in rad/s. For Earth, $\omega_E = (2\pi \text{ rad})/(24.0 \text{ h})(3600 \text{ s/h}) = 7.27\times10^{-5}$ rad/s. In the same way that counterclockwise rotation produces positive angular displacement, it also results in positive angular velocity.

If an object's angular velocity is ω, then the linear velocity of a point a distance, r, from the axis of rotation is given by $v = r\omega$. The speed at which an object on Earth's equator moves as a result of Earth's rotation is given by $v = r\omega = (6.38\times10^6 \text{ m})(7.27\times10^{-5} \text{ rad/s}) = 464$ m/s. Earth is an example of a rotating, rigid body. Even though different points on Earth rotate different distances in each revolution, all points rotate through the same angle. All parts of a rigid body rotate at the same rate. The Sun, on the other hand, is not a rigid body. Different parts of the Sun rotate at different rates. Most objects that we will consider in this chapter are rigid bodies.

Astronomy Connection

Angular Acceleration

What if angular velocity is changing? For example, if a car were accelerated from 0.0 m/s to 25 m/s in 15 s, then the angular velocity of the wheels also would change from 0.0 rad/s to 78 rad/s in the same 15 s. The wheels would undergo **angular acceleration,** which is defined as the change in angular velocity divided by the time required to make the change. Angular acceleration, α, is represented by the following equation.

Angular Acceleration of an Object $\alpha = \frac{\Delta\omega}{\Delta t}$

Angular acceleration is equal to the change in angular velocity divided by the time required to make that change.

Angular acceleration is measured in rad/s^2. If the change in angular velocity is positive, then the angular acceleration also is positive. Angular acceleration defined in this way is also the average angular acceleration over the time interval Δt. One way to find the instantaneous angular acceleration is to find the slope of a graph of angular velocity as a function of time. The linear acceleration of a point at a distance, r, from the axis of an object with angular acceleration, α, is given by $a = r\alpha$. **Table 8-1** is a summary of linear and angular relationships.

Table 8-1			
Linear and Angular Measures			
Quantity	**Linear**	**Angular**	**Relationship**
Displacement	d (m)	θ (rad)	$d = r\theta$
Velocity	v (m/s)	ω (rad/s)	$v = r\omega$
Acceleration	a (m/s^2)	α (rad/s^2)	$a = r\alpha$

PRACTICE Problems

• Additional Problems, Appendix B
• Solutions to Selected Problems, Appendix C

1. What is the angular displacement of each of the following hands of a clock in 1 h? State your answer in three significant digits.
   a. the second hand
   b. the minute hand
   c. the hour hand
2. If a truck has a linear acceleration of 1.85 m/s^2 and the wheels have an angular acceleration of 5.23 rad/s^2, what is the diameter of the truck's wheels?
3. The truck in the previous problem is towing a trailer with wheels that have a diameter of 48 cm.
   a. How does the linear acceleration of the trailer compare with that of the truck?
   b. How do the angular accelerations of the wheels of the trailer and the wheels of the truck compare?
4. You want to replace the tires on your car with tires that have a larger diameter. After you change the tires, for trips at the same speed and over the same distance, how will the angular velocity and number of revolutions change?

Angular frequency A rotating object can make many revolutions in a given amount of time. For instance, a spinning wheel can go through several complete revolutions in 1 min. Thus, the number of complete revolutions made by the object in 1 s is called angular frequency. Angular frequency is $f = \omega/2\pi$. In the next section, you will explore the factors that cause the angular frequency to change.

8.1 Section Review

5. **Angular Displacement** A movie lasts 2 h. During that time, what is the angular displacement of each of the following?
   a. the hour hand
   b. the minute hand
6. **Angular Velocity** The Moon rotates once on its axis in 27.3 days. Its radius is 1.74×10^6 m.
   a. What is the period of the Moon's rotation in seconds?
   b. What is the frequency of the Moon's rotation in rad/s?
   c. What is the linear speed of a rock on the Moon's equator due only to the Moon's rotation?
   d. Compare this speed with the speed of a person on Earth's equator due to Earth's rotation.
7. **Angular Displacement** The ball in a computer mouse is 2.0 cm in diameter. If you move the mouse 12 cm, what is the angular displacement of the ball?
8. **Angular Displacement** Do all parts of the minute hand on a watch have the same angular displacement? Do they move the same linear distance? Explain.
9. **Angular Acceleration** In the spin cycle of a clothes washer, the drum turns at 635 rev/min. If the lid of the washer is opened, the motor is turned off. If the drum requires 8.0 s to slow to a stop, what is the angular acceleration of the drum?
10. **Critical Thinking** A CD-ROM has a spiral track that starts 2.7 cm from the center of the disk and ends 5.5 cm from the center. The disk drive must turn the disk so that the linear velocity of the track is a constant 1.4 m/s. Find the following.
    a. the angular velocity of the disk (in rad/s and rev/min) for the start of the track
    b. the disk's angular velocity at the end of the track
    c. the disk's angular acceleration if the disk is played for 76 min

Physics Online physicspp.com/self_check_quiz

8.2 Rotational Dynamics

How do you start the rotation of an object? That is, how do you change its angular velocity? Suppose you have a soup can that you want to spin. If you wrap a string around it and pull hard, you could make the can spin rapidly. Later in this chapter, you will learn why gravity, the force of Earth's mass on the can, acts on the center of the can. The force of the string, on the other hand, is exerted at the outer edge of the can, and at right angles to the line from the center of the can, to the point where the string leaves the can's surface.

You have learned that a force changes the velocity of a point object. In the case of a soup can, a force that is exerted in a very specific way changes the angular velocity of an extended object, which is an object that has a definite shape and size. Consider how you open a door: you exert a force. How can you exert the force to open the door most easily? To get the most effect from the least force, you exert the force as far from the axis of rotation as possible, as shown in **Figure 8-3.** In this case, the axis of rotation is an imaginary vertical line through the hinges. The doorknob is near the outer edge of the door. You exert the force on the doorknob at right angles to the door, away from the hinges. Thus, the magnitude of the force, the distance from the axis to the point where the force is exerted, and the direction of the force determine the change in angular velocity.

Lever arm For a given applied force, the change in angular velocity depends on the **lever arm,** which is the perpendicular distance from the axis of rotation to the point where the force is exerted. If the force is perpendicular to the radius of rotation, as it was with the soup can, then the lever arm is the distance from the axis, r. For the door, it is the distance from the hinges to the point where you exert the force, as illustrated in **Figure 8-4a,** on the next page. If the force is not perpendicular, the perpendicular component of the force must be found.

The force exerted by the string around the can is perpendicular to the radius. If a force is not exerted perpendicular to the radius, however, the lever arm is reduced. To find the lever arm, extend the line of the force until it forms a right angle with a line from the center of rotation. The distance between the intersection and the axis is the lever arm. Thus, using trigonometry, the lever arm, L, can be calculated by the equation $L = r \sin \theta$, as shown in **Figure 8-4b.** In this equation, r is the distance from the axis of rotation to the point where the force is exerted, and θ is the angle between the force and the radius from the axis of rotation to the point where the force is applied.

► Objectives

- **Describe** torque and the factors that determine it.
- **Calculate** net torque.
- **Calculate** the moment of inertia.

► Vocabulary

lever arm
torque
moment of inertia
Newton's second law for rotational motion

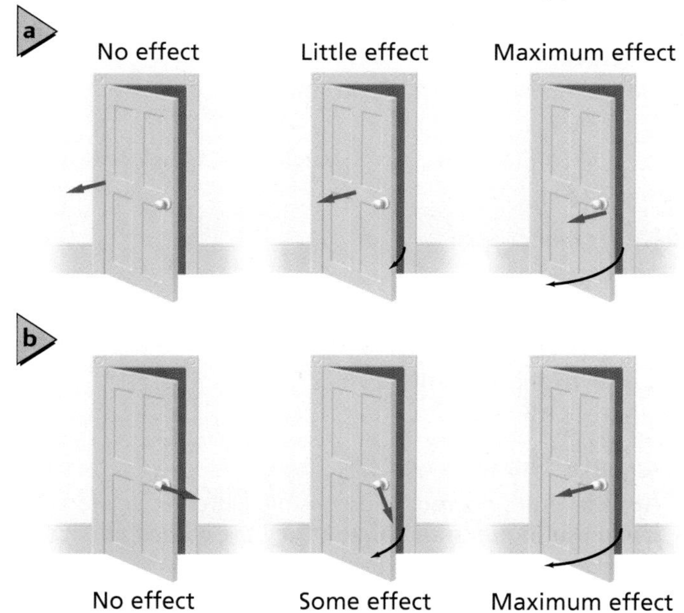

Figure 8-3 When opening a door that is free to rotate about its hinges, the greatest torque is produced when the force is applied farthest from the hinges **(a),** at an angle perpendicular to the door **(b).**

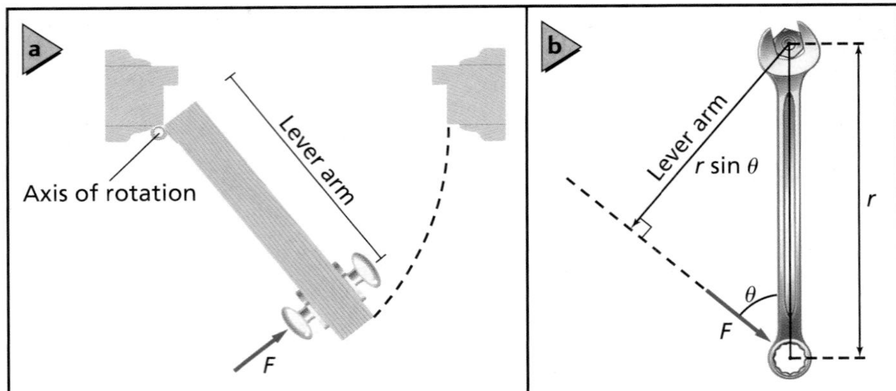

■ **Figure 8-4** The lever arm is along the width of the door, from the hinge to the point where the force is exerted **(a).** The lever arm is equal to $r \sin \theta$, when the angle, θ, between the force and the radius of rotation is not equal to 90° **(b).**

Torque is a measure of how effectively a force causes rotation. The magnitude of torque is the product of the force and the lever arm. Because force is measured in newtons, and distance is measured in meters, torque is measured in newton-meters (N·m). Torque is represented by the Greek letter tau, τ. The equation for torque is shown below.

Torque $\tau = Fr \sin \theta$

Torque is equal to the force times the lever arm.

▶ EXAMPLE Problem 1

Lever Arm A bolt on a car engine needs to be tightened with a torque of 35 N·m. You use a 25-cm-long wrench and pull on the end of the wrench at an angle of 60.0° from the perpendicular. How long is the lever arm, and how much force do you have to exert?

1 Analyze and Sketch the Problem

- Sketch the situation. Find the lever arm by extending the force vector backwards until a line that is perpendicular to it intersects the axis of rotation.

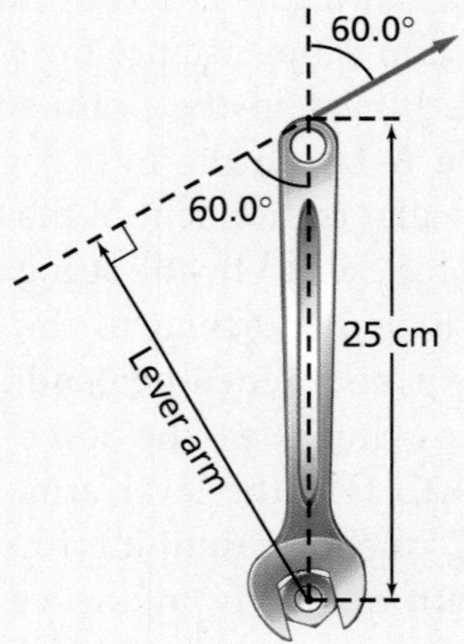

Known:		Unknown:
$r = 0.25$ m	$\tau = 35$ N·m	$L = ?$
$\theta = 60.0°$		$F = ?$

2 Solve for the Unknown

Solve for the length of the lever arm.

$$L = r \sin \theta$$
$$= (0.25 \text{ m})(\sin 60.0°) \quad \text{Substitute } r = 0.25 \text{ m}, \theta = 60.0°$$
$$= 0.22 \text{ m}$$

Solve for the force.

$$\tau = Fr \sin \theta$$
$$F = \frac{\tau}{r \sin \theta}$$
$$= \frac{35 \text{ N·m}}{(0.25 \text{ m})(\sin 60.0°)} \quad \text{Substitute } \tau = 35 \text{ N·m}, r = 0.25 \text{ m}, \theta = 60.0°$$
$$= 1.6\times10^2 \text{ N}$$

Math Handbook
Trigonometric Ratios
page 855

3 Evaluate the Answer

- **Are the units correct?** Force is measured in newtons.
- **Does the sign make sense?** Only the magnitude of the force needed to rotate the wrench clockwise is calculated.

▷ PRACTICE Problems

- Additional Problems, Appendix B
- Solutions to Selected Problems, Appendix C

11. Consider the wrench in Example Problem 1. What force is needed if it is applied to the wrench at a point perpendicular to the wrench?

12. If a torque of 55.0 N·m is required and the largest force that can be exerted by you is 135 N, what is the length of the lever arm that must be used?

13. You have a 0.234-m-long wrench. A job requires a torque of 32.4 N·m, and you can exert a force of 232 N. What is the smallest angle, with respect to the vertical, at which the force can be exerted?

14. You stand on the pedal of a bicycle. If you have a mass of 65 kg, the pedal makes an angle of 35° above the horizontal, and the pedal is 18 cm from the center of the chain ring, how much torque would you exert?

15. If the pedal in problem 14 is horizontal, how much torque would you exert? How much torque would you exert when the pedal is vertical?

Finding Net Torque

Try the following experiment. Get two pencils, some coins, and some transparent tape. Tape two identical coins to the ends of the pencil and balance it on the second pencil, as shown in **Figure 8-5.** Each coin exerts a torque that is equal to its weight, F_g, times the distance, r, from the balance point to the center of the coin, as follows:

$$\tau = F_g r$$

But the torques are equal and opposite in direction. Thus, the net torque is zero:

$$\tau_1 - \tau_2 = 0$$

or

$$F_{g1} r_1 - F_{g2} r_2 = 0$$

How can you make the pencil rotate? You could add a second coin on top of one of the two coins, thereby making the two forces different. You also could slide the balance point toward one end or the other of the pencil, thereby making the two distances different.

Figure 8-5 The torque exerted by the first coin, $F_{g1}r_1$, is equal and opposite in direction to the torque exerted by the second coin, $F_{g2}r_2$, when the pencil is balanced.

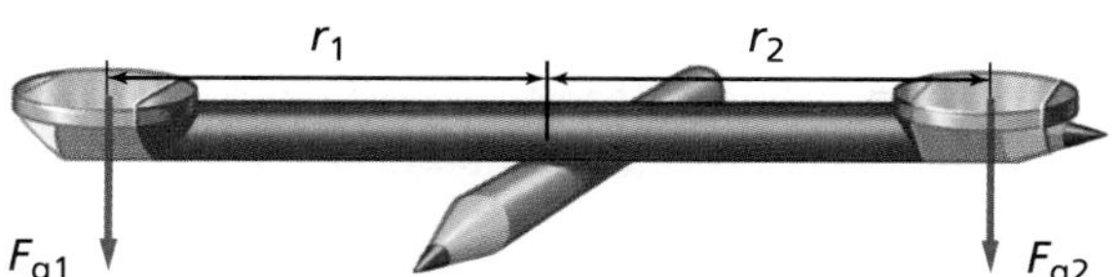

Concepts In Motion
Interactive Figure To see an animation on finding net torque, visit physicspp.com physicspp.com.

▶ EXAMPLE Problem 2

Balancing Torques Kariann (56 kg) and Aysha (43 kg) want to balance on a 1.75-m-long seesaw. Where should they place the pivot point?

1 Analyze and Sketch the Problem

- Sketch the situation.
- Draw and label the vectors.

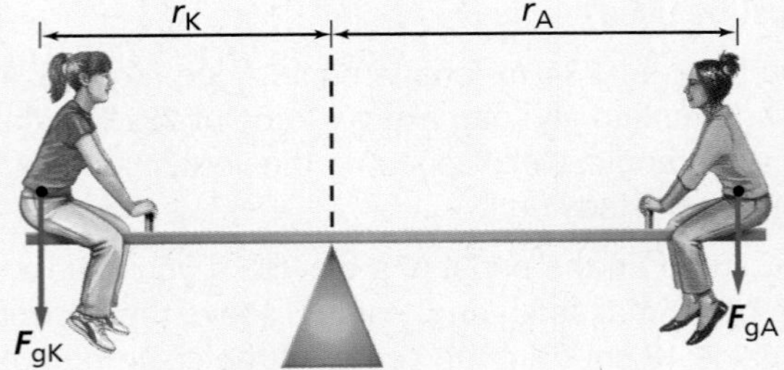

Known:	**Unknown:**
$m_K = 56$ kg	$r_K = ?$
$m_A = 43$ kg	$r_A = ?$
$r_K + r_A = 1.75$ m	

2 Solve for the Unknown

Find the two forces.

Kariann:

$$F_{gK} = m_K g$$
$$= (56\text{ kg})(9.80\text{ m/s}^2) \quad \textbf{Substitute } m_K = 56\text{ kg}, g = 9.80\text{ m/s}^2$$
$$= 5.5\times10^2\text{ N}$$

Aysha:

$$F_{gA} = m_A g$$
$$= (43\text{ kg})(9.80\text{ m/s}^2) \quad \textbf{Substitute } m_A = 43\text{ kg}, g = 9.80\text{ m/s}^2$$
$$= 4.2\times10^2\text{ N}$$

Define Kariann's distance in terms of the length of the seesaw and Aysha's distance.

$$r_K = 1.75\text{ m} - r_A$$

When there is no rotation, the sum of the torques is zero.

$$F_{gK}r_K = F_{gA}r_A$$
$$F_{gK}r_K - F_{gA}r_A = 0.0\text{ N·m}$$
$$F_{gK}(1.75\text{ m} - r_A) - F_{gA}r_A = 0.0\text{ N·m} \quad \textbf{Substitute } r_K = 1.75\text{ m} - r_A$$

Solve for r_A.

$$F_{gK}(1.75\text{ m}) - F_{gK}(r_A) - F_{gA}r_A = 0.0\text{ N·m}$$
$$F_{gK}r_A + F_{gA}r_A = F_{gK}(1.75\text{ m})$$
$$(F_{gK} + F_{gA})\, r_A = F_{gK}(1.75\text{ m})$$
$$r_A = \frac{F_{gK}(1.75\text{ m})}{(F_{gK} + F_{gA})}$$
$$= \frac{(5.5\times10^2\text{ N})(1.75\text{ m})}{(5.5\times10^2\text{ N} + 4.2\times10^2\text{ N})} \quad \textbf{Substitute } F_{gK} = 5.5\times10^2\text{ N}, F_{gA} = 4.2\times10^2\text{ N}$$
$$= 0.99\text{ m}$$

Math Handbook
Isolating a Variable
page 845

3 Evaluate the Answer

- **Are the units correct?** Distance is measured in meters.
- **Do the signs make sense?** Distances are positive.
- **Is the magnitude realistic?** Aysha is about 1 m from the center, so Kariann is about 0.75 m away from it. Because Kariann's weight is greater than Aysha's weight, the lever arm on Kariann's side should be shorter. Aysha is farther from the pivot, as expected.

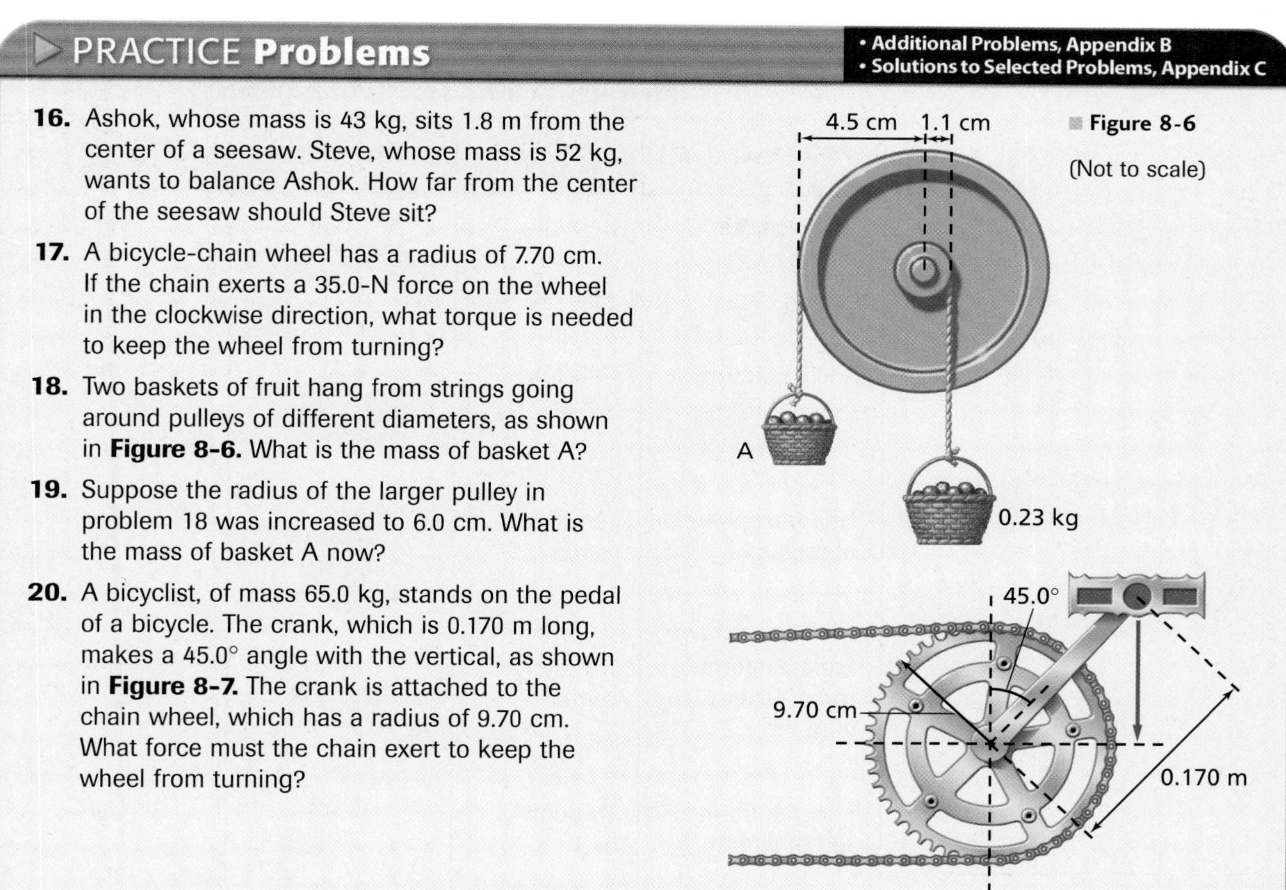

16. Ashok, whose mass is 43 kg, sits 1.8 m from the center of a seesaw. Steve, whose mass is 52 kg, wants to balance Ashok. How far from the center of the seesaw should Steve sit?

17. A bicycle-chain wheel has a radius of 7.70 cm. If the chain exerts a 35.0-N force on the wheel in the clockwise direction, what torque is needed to keep the wheel from turning?

18. Two baskets of fruit hang from strings going around pulleys of different diameters, as shown in **Figure 8-6.** What is the mass of basket A?

19. Suppose the radius of the larger pulley in problem 18 was increased to 6.0 cm. What is the mass of basket A now?

20. A bicyclist, of mass 65.0 kg, stands on the pedal of a bicycle. The crank, which is 0.170 m long, makes a 45.0° angle with the vertical, as shown in **Figure 8-7.** The crank is attached to the chain wheel, which has a radius of 9.70 cm. What force must the chain exert to keep the wheel from turning?

The Moment of Inertia

If you exert a force on a point mass, its acceleration will be inversely proportional to its mass. How does an extended object rotate when a torque is exerted on it? To observe firsthand, recover the pencil, the coins, and the transparent tape that you used earlier in this chapter. First, tape the coins at the ends of the pencil. Hold the pencil between your thumb and forefinger, and wiggle it back and forth. Take note of the forces that your thumb and forefinger exert. These forces create torques that change the angular velocity of the pencil and coins.

Now move the coins so that they are only 1 or 2 cm apart. Wiggle the pencil as before. Did the amount of torque and force need to be changed? The torque that was required was much less this time. Thus, the amount of mass is not the only factor that determines how much torque is needed to change angular velocity; the location of that mass also is relevant.

The resistance to rotation is called the **moment of inertia,** which is represented by the symbol I and has units of mass times the square of the distance. For a point object located at a distance, r, from the axis of rotation, the moment of inertia is given by the following equation.

Moment of Inertia of a Point Mass $I = mr^2$

The moment of inertia of a point mass is equal to the mass of the object times the square of the object's distance from the axis of rotation.

Table 8-2			
Moments of Inertia for Various Objects			
Object	**Location of Axis**	**Diagram**	**Moment of Inertia**
Thin hoop of radius r	Through central diameter	Axis; r	mr^2
Solid, uniform cylinder of radius r	Through center	Axis; r	$\frac{1}{2}mr^2$
Uniform sphere of radius r	Through center	Axis; r	$\frac{2}{5}mr^2$
Long, uniform rod of length l	Through center	Axis; l	$\frac{1}{12}ml^2$
Long, uniform rod of length l	Through end	Axis; l	$\frac{1}{3}ml^2$
Thin, rectangular plate of length l and width w	Through center	Axis; l; w	$\frac{1}{12}m(l^2 + w^2)$

■ **Figure 8-8** The moment of inertia of a book depends on the axis of rotation. The moment of inertia of the book in **(a)** is larger than the moment of inertia of the book in **(b)** because the average distance of the book's mass from the rotational axis is larger.

As you have seen, the moment of inertia for complex objects, such as the pencil and coins, depends on how far the coins are from the axis of rotation. A bicycle wheel, for example, has almost all of its mass in the rim and tire. Its moment of inertia is almost exactly equal to mr^2, where r is the radius of the wheel. For most objects, however, the mass is distributed continuously and so the moment of inertia is less than mr^2. For example, as shown in **Table 8-2,** for a solid cylinder of radius r, $I = \frac{1}{2}mr^2$, while for a solid sphere, $I = \frac{2}{5}mr^2$.

The moment of inertia also depends on the location of the rotational axis, as illustrated in **Figure 8-8.** To observe this first-hand, hold a book in the upright position, by placing your hands at the bottom of the book. Feel the torque needed to rock the book towards you, and then away from you. Now put your hands in the middle of the book and feel the torque needed to rock the book toward you and then away from you. Note that much less torque is needed when your hands are placed in the middle of the book because the average distance of the book's mass from the rotational axis is much less in this case.

▶ EXAMPLE Problem 3

Moment of Inertia A simplified model of a twirling baton is a thin rod with two round objects at each end. The length of the baton is 0.65 m, and the mass of each object is 0.30 kg. Find the moment of inertia of the baton if it is rotated about the midpoint between the round objects. What is the moment of inertia of the baton when it is rotated around one end? Which is greater? Neglect the mass of the rod.

1 Analyze and Sketch the Problem

- Sketch the situation. Show the baton with the two different axes of rotation and the distances from the axes of rotation to the masses.

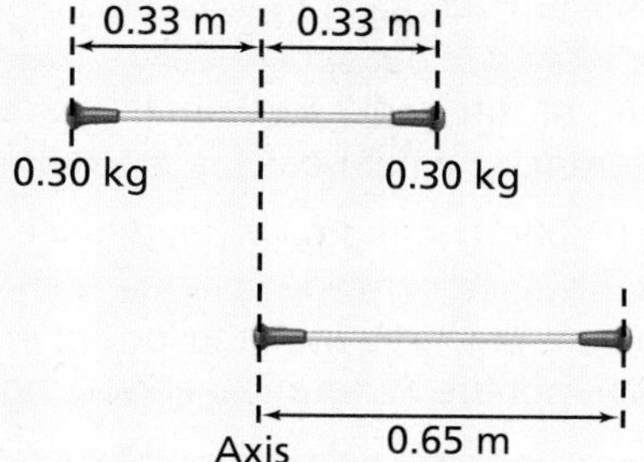

Known:	Unknown:
$m = 0.30$ kg	$I = ?$
$l = 0.65$ m	

2 Solve for the Unknown

Calculate the moment of inertia of each mass separately.

Rotating about the center of the rod:

$$r = \frac{1}{2}l$$

$$= \frac{1}{2}(0.65\text{ m})$$ **Substitute $l = 0.65$ m**

$$= 0.33\text{ m}$$

$$I_{\text{single mass}} = mr^2$$

$$= (0.30\text{ kg})(0.33\text{ m})^2$$ **Substitute $m = 0.30$ kg, $r = 0.33$ m**

$$= 0.033\text{ kg}\cdot\text{m}^2$$

Find the moment of inertia of the baton.

$$I = 2I_{\text{single mass}}$$

$$= 2(0.033\text{ kg}\cdot\text{m}^2)$$ **Substitute $I_{\text{single mass}} = 0.033$ kg·m²**

$$= 0.066\text{ kg}\cdot\text{m}^2$$

Math Handbook
Operations with Significant Digits
pages 835–836

Rotating about one end of the rod:

$$I_{\text{single mass}} = mr^2$$

$$= (0.30\text{ kg})(0.65\text{ m})^2$$ **Substitute $m = 0.30$ kg, $r = 0.65$ m**

$$= 0.13\text{ kg}\cdot\text{m}^2$$

Find the moment of inertia of the baton.

$$I = I_{\text{single mass}}$$

$$= 0.13\text{ kg}\cdot\text{m}^2$$

The moment of inertia is greater when the baton is swung around one end.

3 Evaluate the Answer

- **Are the units correct?** Moment of inertia is measured in kg·m^2.
- **Is the magnitude realistic?** Masses and distances are small, and so are the moments of inertia. Doubling the distance increases the moment of inertia by a factor of 4. Thus, doubling the distance overcomes having only one mass contributing.

PRACTICE Problems

Additional Problems, Appendix B

21. Two children of equal masses sit 0.3 m from the center of a seesaw. Assuming that their masses are much greater than that of the seesaw, by how much is the moment of inertia increased when they sit 0.6 m from the center?

22. Suppose there are two balls with equal diameters and masses. One is solid, and the other is hollow, with all its mass distributed at its surface. Are the moments of inertia of the balls equal? If not, which is greater?

23. **Figure 8-9** shows three massive spheres on a rod of very small mass. Consider the moment of inertia of the system, first when it is rotated about sphere A, and then when it is rotated about sphere C. Are the moments of inertia the same or different? Explain. If the moments of inertia are different, in which case is the moment of inertia greater?

24. Each sphere in the previous problem has a mass of 0.10 kg. The distance between spheres A and C is 0.20 m. Find the moment of inertia in the following instances: rotation about sphere A, rotation about sphere C.

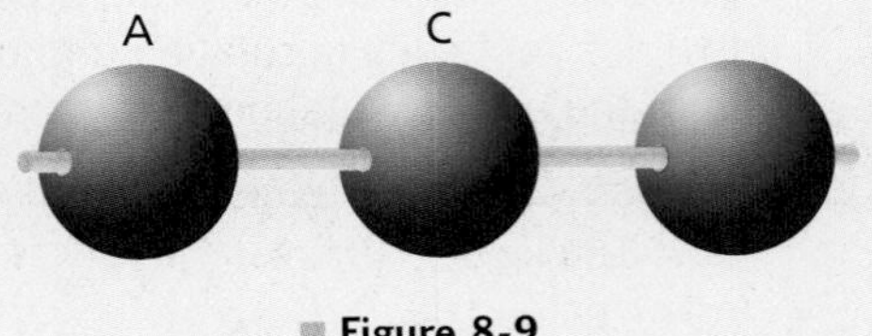

Figure 8-9

Newton's Second Law for Rotational Motion

Newton's second law for linear motion is expressed as $a = F_{net}/m$. If you rewrite this equation to represent rotational motion, acceleration is replaced by angular acceleration, α, force is replaced by net torque, τ_{net}, and mass is replaced by moment of inertia, I. Thus, **Newton's second law for rotational motion** states that angular acceleration is directly proportional to the net torque and inversely proportional to the moment of inertia. This law is expressed by the following equation.

Newton's Second Law for Rotational Motion $\alpha = \frac{\tau_{net}}{I}$

The angular acceleration of an object is equal to the net torque on the object, divided by the moment of inertia.

Recall the coins taped on the pencil. To change the direction of rotation of the pencil—to give it angular acceleration—you had to apply torque to the pencil. The greater the moment of inertia, the more torque needed to produce the same angular acceleration.

CHALLENGE PROBLEM

Rank the objects shown in the diagram according to their moments of inertia about the indicated axes. All spheres have equal masses and all separations are the same.

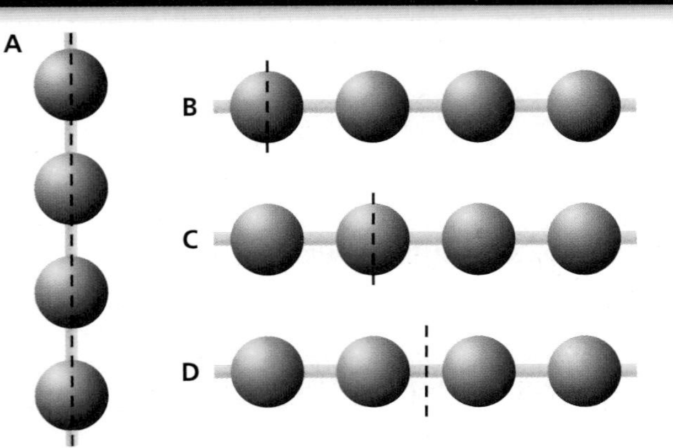

▶ EXAMPLE Problem 4

Torque A solid steel wheel has a mass of 15 kg and a diameter of 0.44 m. It starts at rest. You want to make it rotate at 8.0 rev/s in 15 s.

a. What torque must be applied to the wheel?

b. If you apply the torque by wrapping a strap around the outside of the wheel, how much force should you exert on the strap?

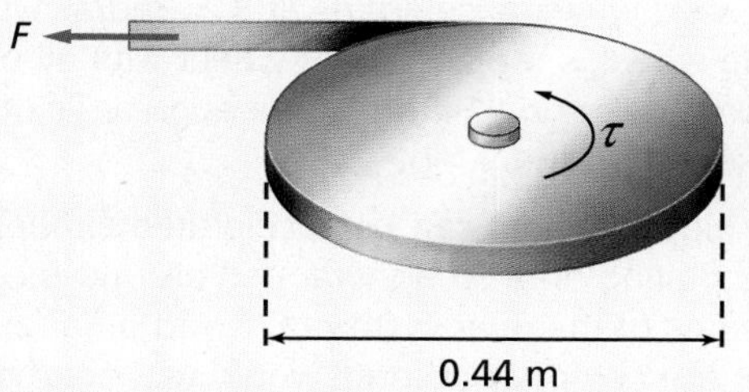

1 Analyze and Sketch the Problem

- Sketch the situation. The torque must be applied in a counterclockwise direction; force must be exerted as shown.

Known:	Unknown:
$m = 15$ kg	$\alpha = ?$
$r = \frac{1}{2}(0.44\text{ m}) = 0.22\text{ m}$	$I = ?$
$\omega_i = 0.0$ rad/s	$\tau = ?$
$\omega_f = 2\pi(8.0\text{ rev/s})$	$F = ?$
$t = 15$ s	

Physics Online

Personal Tutor For an online tutorial on torque, visit physicspp.com.

2 Solve for the Unknown

a. Solve for angular acceleration.

$$\alpha = \frac{\Delta\omega}{\Delta t}$$

$$= \frac{2\pi(8.0\text{ rev/s}) - (0.0\text{ rad/s})}{15\text{ s}}$$ Substitute $\omega_f = 2\pi(8.0\text{ rev/s})$, $\omega_i = 0.0\text{ rad/s}$

$$= 3.4\text{ rad/s}^2$$

Solve for the moment of inertia.

$$I = \frac{1}{2}mr^2$$

$$= \frac{1}{2}(15\text{ kg})(0.22\text{ m})^2$$ Substitute $m = 15$ kg, $r = 0.22$ m

$$= 0.36\text{ kg}\cdot\text{m}^2$$

Solve for torque.

$$\tau = I\alpha$$

$$= (0.36\text{ kg}\cdot\text{m}^2)(3.4\text{ rad/s}^2)$$ Substitute $I = 0.36\text{ kg}\cdot\text{m}^2$, $\alpha = 3.4\text{ rad/s}^2$

$$= 1.2\text{ kg}\cdot\text{m}^2/\text{s}^2$$

$$= 1.2\text{ N}\cdot\text{m}$$

Math Handbook
Operations with Significant Digits pages 835–836

b. Solve for force.

$$\tau = Fr$$

$$F = \frac{\tau}{r}$$

$$= \frac{1.2\text{ N}\cdot\text{m}}{0.22\text{ m}}$$ Substitute $\tau = 1.2\text{ N}\cdot\text{m}$, $r = 0.22$ m

$$= 5.5\text{ N}$$

3 Evaluate the Answer

- **Are the units correct?** Torque is measured in N·m and force is measured in N.
- **Is the magnitude realistic?** Despite its large mass, the small size of the wheel makes it relatively easy to spin.

PRACTICE Problems

- Additional Problems, Appendix B
- Solutions to Selected Problems, Appendix C

25. Consider the wheel in Example Problem 4. If the force on the strap were twice as great, what would be the speed of rotation of the wheel after 15 s?

26. A solid wheel accelerates at 3.25 rad/s^2 when a force of 4.5 N exerts a torque on it. If the wheel is replaced by a wheel with all of its mass on the rim, the moment of inertia is given by $I = mr^2$. If the same angular velocity were desired, what force would have to be exerted on the strap?

27. A bicycle wheel can be accelerated either by pulling on the chain that is on the gear or by pulling on a string wrapped around the tire. The wheel's radius is 0.38 m, while the radius of the gear is 0.14 m. If you obtained the needed acceleration with a force of 15 N on the chain, what force would you need to exert on the string?

28. The bicycle wheel in problem 27 is used with a smaller gear whose radius is 0.11 m. The wheel can be accelerated either by pulling on the chain that is on the gear or by pulling string that is wrapped around the tire. If you obtained the needed acceleration with a force of 15 N on the chain, what force would you need to exert on the string?

29. A disk with a moment of inertia of 0.26 kg·m^2 is attached to a smaller disk mounted on the same axle. The smaller disk has a diameter of 0.180 m and a mass of 2.5 kg. A strap is wrapped around the smaller disk, as shown in **Figure 8-10.** Find the force needed to give this system an angular acceleration of 2.57 rad/s^2.

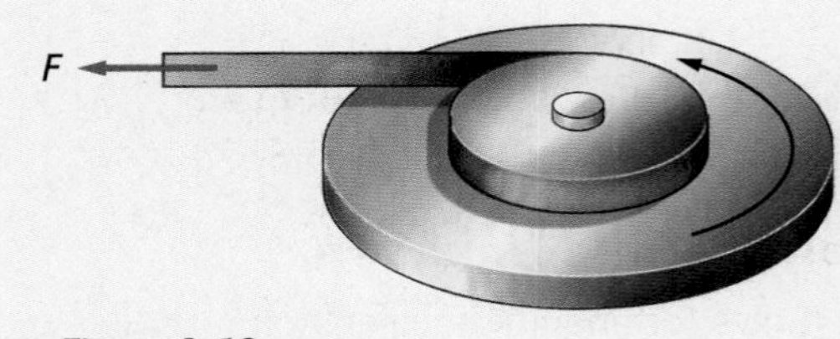

Figure 8-10

In summary, changes in the amount of torque applied to an object, or changes in the moment of inertia, affect the rate of rotation. In this section, you learned how Newton's second law of motion applies to rotational motion. In the next section, you will learn how to keep objects from rotating.

8.2 Section Review

30. Torque Vijesh enters a revolving door that is not moving. Explain where and how Vijesh should push to produce a torque with the least amount of force.

31. Lever Arm You try to open a door, but you are unable to push at a right angle to the door. So, you push the door at an angle of 55° from the perpendicular. How much harder would you have to push to open the door just as fast as if you were to push it at 90°?

32. Net Torque Two people are pulling on ropes wrapped around the edge of a large wheel. The wheel has a mass of 12 kg and a diameter of 2.4 m. One person pulls in a clockwise direction with a 43-N force, while the other pulls in a counterclockwise direction with a 67-N force. What is the net torque on the wheel?

33. Moment of Inertia Refer to Table 8-2 on page 206 and rank the moments of inertia from least to greatest of the following objects: a sphere, a wheel with almost all of its mass at the rim, and a solid disk. All have equal masses and diameters. Explain the advantage of using the one with the least moment of inertia.

34. Newton's Second Law for Rotational Motion A rope is wrapped around a pulley and pulled with a force of 13.0 N. The pulley's radius is 0.150 m. The pulley's rotational speed goes from 0.0 to 14.0 rev/min in 4.50 s. What is the moment of inertia of the pulley?

35. Critical Thinking A ball on an extremely low-friction, tilted surface, will slide downhill without rotating. If the surface is rough, however, the ball will roll. Explain why, using a free-body diagram.

8.3 Equilibrium

Why are some vehicles more likely than others to roll over when involved in an accident? What causes a vehicle to roll over? The answer lies in the design of the vehicle. In this section, you will learn some of the factors that cause an object to tip over.

▶ **Objectives**

- **Define** center of mass.
- **Explain** how the location of the center of mass affects the stability of an object.
- **Define** the conditions for equilibrium.
- **Describe** how rotating frames of reference give rise to apparent forces.

▶ **Vocabulary**

center of mass
centrifugal "force"
Coriolis "force"

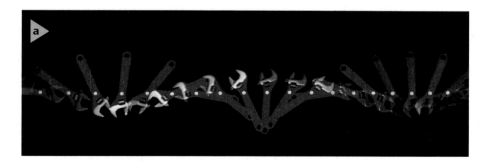

The Center of Mass

How does an object rotate around its center of mass? A wrench may spin about its handle or end-over-end. Does any single point on the wrench follow a straight path? **Figure 8-11a** shows the path of the wrench. You can see that there is a single point whose path traces a straight line, as if the wrench could be replaced by a point particle at that location. The **center of mass** of an object is the point on the object that moves in the same way that a point particle would move.

Locating the center of mass How can you locate the center of mass of an object? First, suspend the object from any point. When the object stops swinging, the center of mass is along the vertical line drawn from the suspension point as shown in **Figure 8-11b.** Draw the line. Then, suspend the object from another point. Again, the center of mass must be below this point. Draw a second vertical line. The center of mass is at the point where the two lines cross, as shown in **Figure 8-11c.** The wrench, racket, and all other freely rotating objects rotate about an axis that goes through their center of mass. Where is the center of mass of a person located?

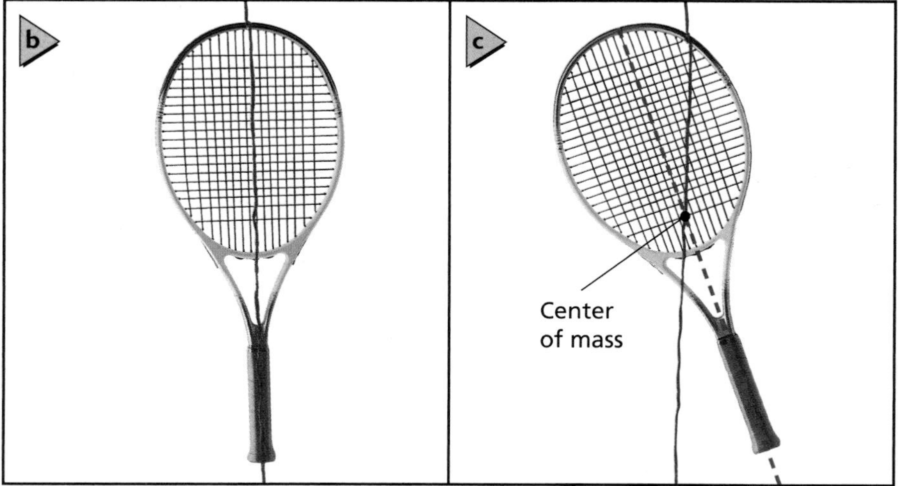

Figure 8-11 The path of the center of mass of a wrench is a straight line **(a).** The center of mass of an object such as a tennis racket can be found by first suspending it from any point **(b).** The point where the strings intersect is the location of the racket's center of mass **(c).**

Figure 8-12 The upward motion of the ballet dancer's head is less than the upward motion of the center of mass. Thus, the head and torso move in a nearly horizontal path. This creates an illusion of floating.

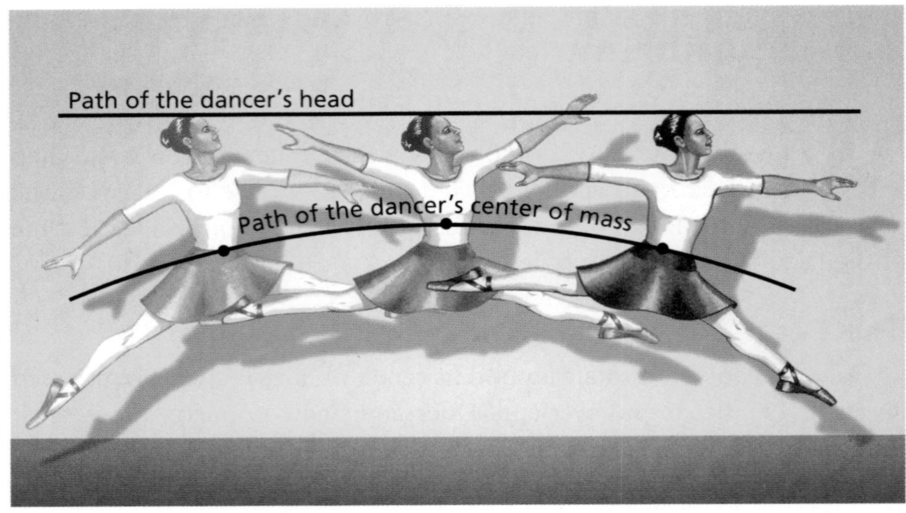

APPLYING PHYSICS

The Fosbury-Flop In high jumping, a technique called the Fosbury-Flop allows a high jumper to clear the bar when it is placed at the highest position. This is possible because the athlete's center of mass passes below the bar as he or she somersaults over the bar, with his or her back toward it.

The center of mass of a human body For a person who is standing with his or her arms hanging straight down, the center of mass is a few centimeters below the navel, midway between the front and back of the person's body. It is slightly higher in young children, because of their relatively larger heads. Because the human body is flexible, however, its center of mass is not fixed. If you raise your hands above your head, your center of mass rises 6 to 10 cm. A ballet dancer, for example, can appear to be floating on air by changing her center of mass in a leap. By raising her arms and legs while in the air, as shown in **Figure 8-12,** the dancer moves her center of mass closer to her head. The path of the center of mass is a parabola, so the dancer's head stays at almost the same height for a surprisingly long time.

Center of Mass and Stability

What factors determine whether a vehicle is stable or prone to roll over in an accident? To understand the problem, think about tipping over a box. A tall, narrow box, standing on end, tips more easily than a low, broad box. Why? To tip a box, as shown in **Figure 8-13,** you must rotate it about a corner. You pull at the top with a force, $\boldsymbol{F}$, applying a torque, τ_F. The weight of the box, acting on the center of mass, $\boldsymbol{F}_g$, applies an opposing torque, τ_w. When the center of mass is directly above the point of support, τ_w is zero. The only torque is the one applied by you. As the box rotates farther, its center of mass is no longer above its base of support, and both torques act in the same direction. At this point, the box tips over rapidly.

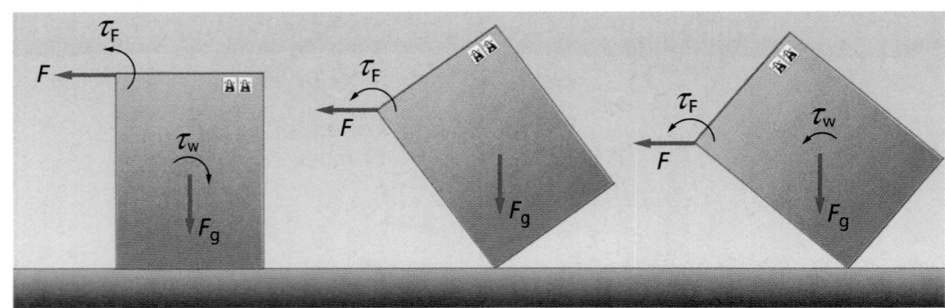

Figure 8-13 The bent arrows show the direction of the torque produced by the force exerted to tip over a box.

Stability An object is said to be stable if an external force is required to tip it. The box in Figure 8-13 is stable as long as the direction of the torque due to its weight, τ_w tends to keep it upright. This occurs as long as the box's center of mass lies above its base. To tip the box over, you must rotate its center of mass around the axis of rotation until it is no longer above the base of the box. To rotate the box, you must lift its center of mass. The broader the base, the more stable the object is. For this reason, if you are standing on a bus that is weaving through traffic and you want to avoid falling down, you need to stand with your feet spread apart.

Why do vehicles roll over? **Figure 8-14** shows two vehicles rolling over. Note that the one with the higher center of mass does not have to be tilted very far for its center of mass to be outside its base—its center of mass does not have to be raised as much as the other vehicle's. The lower the location of an object's center of mass, the greater its stability.

You are stable when you stand flat on your feet. When you stand on tiptoe, however, your center of mass moves forward directly above the balls of your feet, and you have very little stability. A small person can use torque, rather than force, to defend himself or herself against a stronger person. In judo, aikido, and other martial arts, the fighter uses torque to rotate the opponent into an unstable position, where the opponent's center of mass does not lie above his or her feet.

In summary, if the center of mass is outside the base of an object, it is unstable and will roll over without additional torque. If the center of mass is above the base of the object, it is stable. If the base of the object is very narrow and the center of mass is high, then the object is stable, but the slightest force will cause it to tip over.

MINI LAB

Spinning Tops

1. Cut out two cardboard disks of 10-cm and 15-cm diameter.
2. Use a pencil with an eraser that has rounded edges from use. If it is new, rub it on paper to round it.
3. Spin the pencil and try to make it stand on the eraser. Repeat several times and record your observations.
4. Carefully push the pencil through the center of the 10-cm disk.
5. Spin the pencil with the disk and try to make it stand on the eraser.
6. Move the disk to different points on the pencil. Spin and record your observations.
7. Repeat steps 4-6 with the 15-cm disk.

Analyze and Conclude

8. **Sequence** the three trials in order from least to most stable.
9. **Describe** the location of the pencil's center of mass.
10. **Analyze** the placement of the disk and its effect on stability.

Conditions for Equilibrium

If your pen is at rest, what is needed to keep it at rest? You could either hold it up or place it on a desk or some other surface. An upward force must be exerted on the pen to balance the downward force of gravity. You must also hold the pen so that it will not rotate. An object is said to be in static equilibrium if both its velocity and angular velocity are zero or constant. Thus, for an object to be in static equilibrium, it must meet two conditions. First, it must be in translational equilibrium; that is, the net force exerted on the object must be zero. Second, it must be in rotational equilibrium; that is, the net torque exerted on the object must be zero.

Concepts In MOtion

Interactive Figure To see an animation on stability, visit physicspp.com.

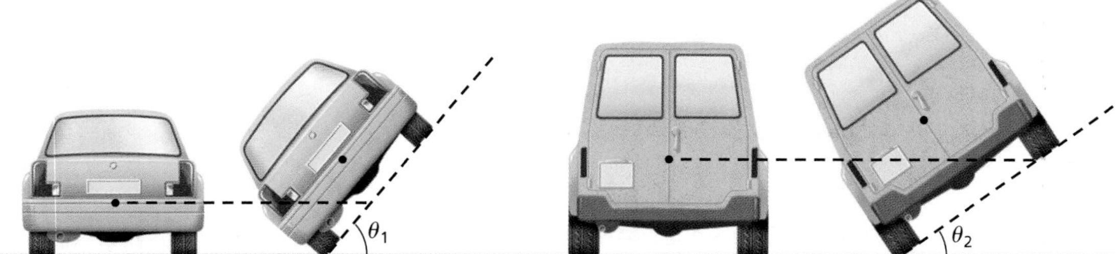

Figure 8-14 Larger vehicles have a higher center of mass than smaller ones. The higher the center of mass, the smaller the tilt needed to cause the vehicle's center of mass to move outside its base and cause the vehicle to roll over.

▶ EXAMPLE Problem 5

Static Equilibrium **A 5.8-kg ladder, 1.80 m long, rests on two sawhorses. Sawhorse A is 0.60 m from one end of the ladder, and sawhorse B is 0.15 m from the other end of the ladder. What force does each sawhorse exert on the ladder?**

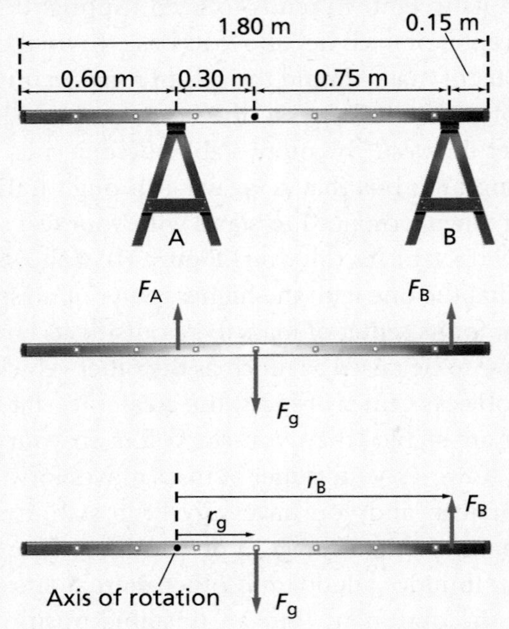

1 Analyze and Sketch the Problem

- Sketch the situation.
- Choose the axis of rotation at the point where F_A acts on the ladder. Thus, the torque due to F_A is zero.

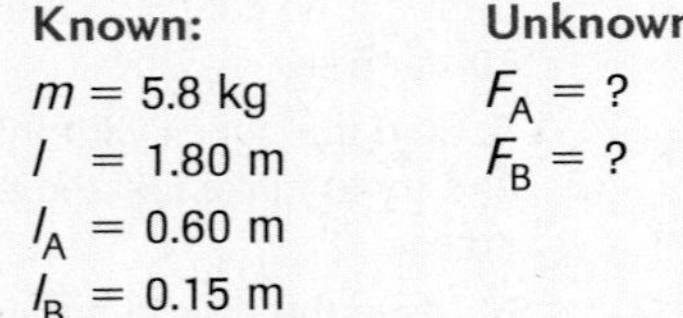

Known:	Unknown:
$m = 5.8$ kg	$F_A = ?$
$l = 1.80$ m	$F_B = ?$
$l_A = 0.60$ m	
$l_B = 0.15$ m	

2 Solve for the Unknown

For a ladder that has a constant density, the center of mass is at the center rung.

The net force is the sum of all forces on the ladder.

$$F_{net} = F_A + F_B + (-F_g)$$

$$0.0\text{ N} = F_A + F_B - F_g$$

The ladder is in translational equilibrium, so the net force exerted on it is zero.

Solve for F_A.

$$F_A = F_g - F_B$$

Find the torques due to F_g and F_B.

$$\tau_g = -r_g F_g$$ **τ_g is in the clockwise direction.**

$$\tau_B = +r_B F_B$$ **τ_B is in the counterclockwise direction.**

The net torque is the sum of all torques on the object.

$$\tau_{net} = \tau_B + \tau_g$$

$$0.0\text{ N·m} = \tau_B + \tau_g$$ **The ladder is in rotational equilibrium, so $\tau_{net} = 0.0$ N·m.**

$$\tau_B = -\tau_g$$

$$r_B F_B = r_g F_g$$ **Substitute $\tau_B = r_B F_B$, $\tau_g = -r_g F_g$**

Solve for F_B.

Math Handbook
Isolating a Variable
page 845

$$F_B = \frac{r_g F_g}{r_B}$$

$$= \frac{r_g mg}{r_B}$$ **Substitute $F_g = mg$**

Using the expression $F_A = F_g - F_B$, substitute in the expressions for F_B and F_g.

$$F_A = F_g - F_B$$

$$= F_g - \frac{r_g mg}{r_B}$$ **Substitute $F_B = \frac{r_g mg}{r_B}$**

$$= mg - \frac{r_g mg}{r_B}$$ **Substitute $F_g = mg$**

$$= mg\left(1 - \frac{r_g}{r_B}\right)$$

Solve for r_g.

$r_g = \frac{1}{2} - l_A$ — For a ladder, which has a constant density, the center of mass is at the center rung.

$= 0.90\text{ m} - 0.60\text{ m}$ — Substitute $\frac{l}{2} = 0.90\text{ m}$, $l_A = 0.60\text{ m}$

$= 0.30\text{ m}$

Solve for r_B.

$r_B = (0.90\text{ m} - l_B) + (0.90\text{ m} - l_A)$

$= (0.90\text{ m} - 0.15\text{ m}) + (0.90\text{ m} - 0.60\text{ m})$ — Substitute $l_B = 0.15\text{ m}$, $l_A = 0.60\text{ m}$

$= 0.75\text{ m} + 0.30\text{ m}$

$= 1.05\text{ m}$

Calculate F_B.

$F_B = \frac{r_g mg}{r_B}$

$= \frac{(0.30\text{ m})(5.8\text{ kg})(9.80\text{ m/s}^2)}{(1.05\text{ m})}$ — Substitute $r_g = 0.30\text{ m}$, $m = 5.8\text{ kg}$, $g = 9.80\text{ m/s}^2$, $r_B = 1.05\text{ m}$

$= 16\text{ N}$

Calculate F_A.

$F_A = mg\left(1 - \frac{r_g}{r_B}\right)$

$= \left(1 - \frac{(0.30\text{ m})}{(1.05\text{ m})}\right)(5.8\text{ kg})(9.80\text{ m/s}^2)$ — Substitute $r_g = 0.30\text{ m}$, $m = 5.8\text{ kg}$, $g = 9.80\text{ m/s}^2$, $r_B = 1.05\text{ m}$

$= 41\text{ N}$

3 Evaluate the Answer

- **Are the units correct?** Forces are measured in newtons.
- **Do the signs make sense?** Both forces are upward.
- **Is the magnitude realistic?** The forces add up to the weight of the ladder, and the force exerted by the sawhorse closer to the center of mass is greater, which is correct.

PRACTICE Problems

- Additional Problems, Appendix B
- Solutions to Selected Problems, Appendix C

36. What would be the forces exerted by the two sawhorses if the ladder in Example Problem 5 had a mass of 11.4 kg?

37. A 7.3-kg ladder, 1.92 m long, rests on two sawhorses, as shown in **Figure 8-15.** Sawhorse A, on the left, is located 0.30 m from the end, and sawhorse B, on the right, is located 0.45 m from the other end. Choose the axis of rotation to be the center of mass of the ladder.

a. What are the torques acting on the ladder?

b. Write the equation for rotational equilibrium.

c. Solve the equation for F_A in terms of F_g.

d. How would the forces exerted by the two sawhorses change if A were moved very close to, but not directly under, the center of mass?

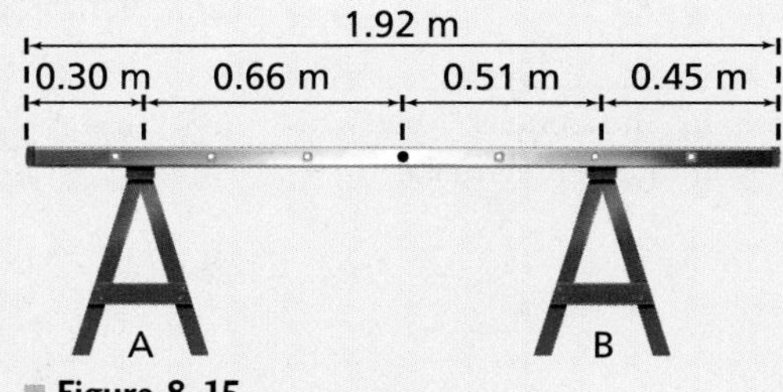

Figure 8-15

38. A 4.5-m-long wooden plank with a 24-kg mass is supported in two places. One support is directly under the center of the board, and the other is at one end. What are the forces exerted by the two supports?

39. A 85-kg diver walks to the end of a diving board. The board, which is 3.5 m long with a mass of 14 kg, is supported at the center of mass of the board and at one end. What are the forces on the two supports?

Rotating Frames of Reference

When you are on a on a rapidly spinning amusement-park ride, it feels like a strong force is pushing you to the outside. A pebble on the floor of the ride would accelerate outward without a horizontal force being exerted on it in the same direction. The pebble would not move in a straight line. In other words, Newton's laws would not apply. This is because rotating frames of reference are accelerated frames. Newton's laws are valid only in inertial or nonaccelerated frames.

Motion in a rotating reference frame is important to us because Earth rotates. The effects of the rotation of Earth are too small to be noticed in the classroom or lab, but they are significant influences on the motion of the atmosphere and therefore, on climate and weather.

Centrifugal "Force"

Suppose you fasten one end of a spring to the center of a rotating platform. An object lies on the platform and is attached to the other end of the spring. As the platform rotates, an observer on the platform sees the object stretch the spring. The observer might think that some force toward the outside of the platform is pulling on the object. This apparent force is called **centrifugal "force."** It is not a real force because there is no physical outward push on the object. Still, this "force" seems real, as anyone who has ever been on an amusement-park ride can attest.

As the platform rotates, an observer on the ground sees things differently. This observer sees the object moving in a circle. The object accelerates toward the center because of the force of the spring. As you know, the acceleration is centripetal acceleration and is given by $a_c = v^2/r$. It also can be written in terms of angular velocity, as $a_c = \omega^2 r$. Centripetal acceleration is proportional to the distance from the axis of rotation and depends on the square of the angular velocity. Thus, if you double the rotational frequency, the acceleration increases by a factor of 4.

The Coriolis "Force"

A second effect of rotation is shown in **Figure 8-16.** Suppose a person standing at the center of a rotating disk throws a ball toward the edge of the disk. Consider the horizontal motion of the ball as seen by two observers and ignore the vertical motion of the ball as it falls.

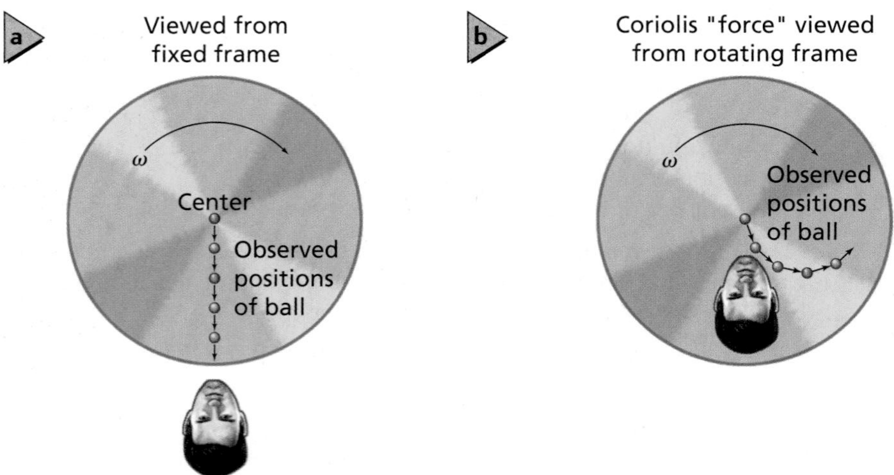

Figure 8-16 The Coriolis "force" exists only in rotating reference frames.

An observer standing outside the disk, as shown in **Figure 8-16a,** sees the ball travel in a straight line at a constant speed toward the edge of the disk. However, the other observer, who is stationed on the disk and rotating with it, as shown in **Figure 8-16b,** sees the ball follow a curved path at a constant speed. A force seems to be acting to deflect the ball. This apparent force is called the **Coriolis "force."** Like the centrifugal "force," the Coriolis "force" is not a real force. It seems to exist because we observe a deflection in horizontal motion when we are in a rotating frame of reference.

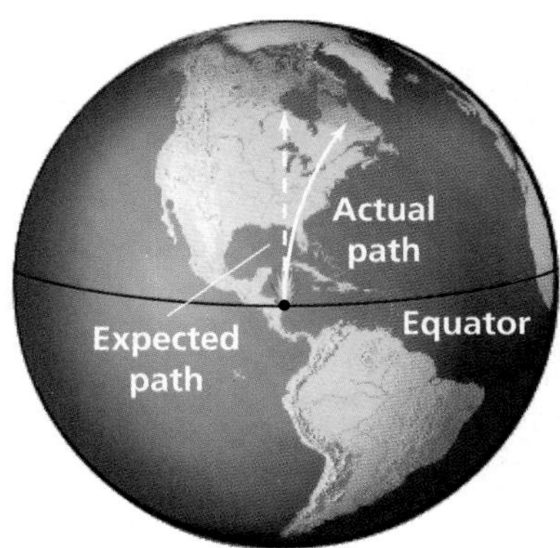

Figure 8-17 An observer on Earth sees the Coriolis "force" cause a projectile fired due north to deflect to the right of the intended target.

Coriolis "force" due to Earth Suppose a cannon is fired from a point on the equator toward a target due north of it. If the projectile were fired directly northward, it would also have an eastward velocity component because of the rotation of Earth. This eastward speed is greater at the equator than at any other latitude. Thus, as the projectile moves northward, it also moves eastward faster than points on Earth below it do. The result is that the projectile lands east of the target as shown in **Figure 8-17.** While an observer in space would see Earth's rotation, an observer on Earth could claim that the projectile missed the target because of the Coriolis "force" on the rocket. Note that for objects moving toward the equator, the direction of the apparent force is westward. A projectile will land west of the target when fired due south.

The direction of winds around high- and low-pressure areas results from the Coriolis "force." Winds flow from areas of high to low pressure. Because of the Coriolis "force" in the northern hemisphere, winds from the south go to the east of low-pressure areas. Winds from the north, however, end up west of low-pressure areas. Therefore, winds rotate counterclockwise around low-pressure areas in the northern hemisphere. In the southern hemisphere however, winds rotate clockwise around low-pressure areas.

Meteorology Connection

Most amusement-park rides thrill the riders because they are in accelerated reference frames while on the ride. The "forces" felt by roller-coaster riders at the tops and bottoms of hills, and when moving almost vertically downward, are mostly related to linear acceleration. On Ferris wheels, rotors, other circular rides, and on the curves of roller coasters, centrifugal "forces" provide most of the excitement.

8.3 Section Review

40. Center of Mass Can the center of mass of an object be located in an area where the object has no mass? Explain.

41. Stability of an Object Why is a modified vehicle with its body raised high on risers less stable than a similar vehicle with its body at normal height?

42. Conditions for Equilibrium Give an example of an object for each of the following conditions.

a. rotational equilibrium, but not translational equilibrium

b. translational equilibrium, but not rotational equilibrium

43. Center of Mass Where is the center of mass of a roll of masking tape?

44. Locating the Center of Mass Describe how you would find the center of mass of this textbook.

45. Rotating Frames of Reference A penny is placed on a rotating, old-fashioned record turntable. At the highest speed, the penny starts sliding outward. What are the forces acting on the penny?

46. Critical Thinking You have read about how the spin of Earth on its axis affects the winds. Predict the direction of the flow of surface ocean currents in the northern and southern hemispheres.

Physics Online physicspp.com/self_check_quiz

PHYSICSLAB

Translational and Rotational Equilibrium

Alternate CBL instructions can be found on the Web site.
physicspp.com

For maintenance on large buildings, scaffolding can be hung on the outside. In order for the scaffolding to support workers, it must be in translational and rotational equilibrium. If two or more forces act on the scaffolding, each can produce a rotation about either end. Scaffolding with uniform mass distribution acts as though all of the mass is concentrated at its center. In translational equilibrium the object is not accelerating; thus, the upward and downward forces are equal.

In order to achieve rotational equilibrium, the sum of all the clockwise torques must equal the sum of all the counterclockwise torques as measured from a pivot point. That is, the net torque must be zero. In this lab you will model scaffolding hung from two ropes using a meterstick and spring scales, and use numbers to measure the forces on the scaffolding.

QUESTION

What conditions are required for equilibrium when parallel forces act on an object?

Objectives

- **Collect and organize data** about the forces acting on the scaffolding.
- **Describe** clockwise and counterclockwise torque.
- **Compare and contrast** translational and rotational equilibrium.

Safety Precautions

- **Use care to avoid dropping masses.**

Materials

meterstick
two 0-5 N spring scales
two ring stands
two Buret clamps
500-g hooked mass
200-g hooked mass

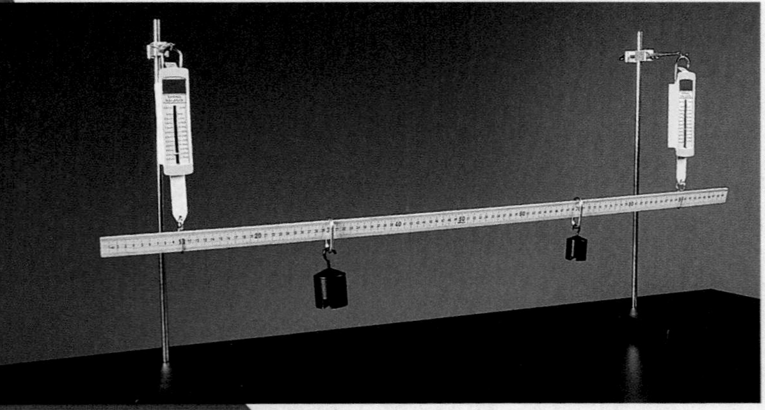

Procedure

The left spring scale will be considered a pivot point for the purposes of this lab. Therefore, the lever arm will be measured from this point.

1. Place the ring stands 80 cm apart.
2. Attach a Buret clamp to each of the ring stands.
3. Verify that the scales are set to zero before use. If the scales need to be adjusted, ask your teacher for assistance.
4. Hang a spring scale from each Buret clamp attached to a ring stand.
5. Hook the meterstick onto the spring scale in such a manner that the 10-cm mark is supported by one hook and the 90-cm mark is supported by the other hook.
6. Read each spring scale and record the force in Data Table 1.
7. Hang a 500-g mass on the meterstick at the 30-cm mark. This point should be 20-cm from the left scale.
8. Read each spring scale and record the force in Data Table 1.
9. Hang a 200-g mass on the meterstick at the 70-cm mark. This point should be 60 cm from the left scale.
10. Read each spring scale and record the force in Data Table 1.

Data Table 1

Object Added	Distance From Left Scale (m)	Left Scale Reading (N)	Right Scale Reading (N)
Meterstick	0.4		
500-g mass	0.2		
200-g mass	0.6		

Data Table 2

Object Added	τ_c	τ_{cc}	Lever Arm (m)	Force (N)
Meterstick				
500-g mass				
200-g mass				
Right scale				

Data Table 3

Object Added	τ_c (N·m)	τ_{cc} (N·m)
Meterstick		
500-g mass		
200-g mass		
Right scale		
$\Sigma\tau$		

Analyze

1. **Calculate** Find the mass of the meterstick.
2. **Calculate** Find the force, or weight, that results from each object and record it in Data Table 2. For the right scale, read the force it exerts and record it in Data Table 2.
3. Using the point where the left scale is attached as a pivot point, identify the forces located elsewhere that cause the scaffold to rotate clockwise or counterclockwise. Mark these in Data Table 2 with an *x*.
4. Record the lever arm distance of each force from the pivot point in Data Table 2.
5. **Use Numbers** Calculate the torque for each object by multiplying the force and lever arm distance. Record these values in Data Table 3.

Conclude and Apply

1. Is the system in translational equilibrium? How do you know?
2. Draw a free-body diagram of your system, showing all the forces.
3. Compare and contrast the sum of the clockwise torques, $\Sigma\tau_c$, and the counterclockwise torques, $\Sigma\tau_{cc}$.
4. What is the percent difference between $\Sigma\tau_c$ and $\Sigma\tau_{cc}$?

Going Further

Use additional masses at locations of your choice with your teacher's permission and record your data.

Real-World Physics

Research the safety requirements in your area for putting up, using, and dismantling scaffolding.

To find out more about rotational motion, visit the Web site: physicspp.com

Technology and Society

The Stability of Sport-Utility Vehicles

Why are sport-utility vehicles more flippable? Many believe that the large size of the sport-utility vehicle makes it more stable and secure. But, a sport-utility vehicle, as well as other tall vehicles such as vans, is much more likely to roll over than a car.

The Problem A sport-utility vehicle has a high center of mass which makes it more likely to topple. Another factor that affects rollover is the static stability factor, which is the ratio of the track width to the center of mass. Track width is defined as half the distance between the two front wheels. The higher the static stability factor, the more likely a vehicle will stay upright.

Many sport-utility vehicles have a center of mass 13 or 15 cm higher than passenger cars. Their track width, however, is about the same as that of passenger cars. Suppose the stability factor for a sport-utility vehicle is 1.06 and 1.43 for a car. Statistics show that in a single-vehicle crash, the sport-utility vehicle has a 37 percent chance of rolling over, while the car has a 10.6 percent chance of rolling over.

However, the static stability factor oversimplifies the issue. Weather and driver behavior are also contributers to rollover crashes. Vehicle factors, such as tires, suspension systems, inertial properties, and advanced handling systems all play a role as well.

It is true that most rollover crashes occur when a vehicle swerves off the road and hits a rut, soft soil, or other surface irregularity. This usually occurs when a driver is not paying proper attention or is speeding. Safe drivers greatly reduce their chances of being involved in a rollover accident by paying attention and driving at the correct speed. Still, weather and driver behavior being equal, the laws of physics indicate that sport-utility vehicles carry an increased risk.

What Is Being Done? Some models are being built with wider track widths or stronger roofs. Optional side-curtain air bags have sensors to keep the bags inflated for up to 6 s, rather than the usual fraction of a second. This will cushion passengers if the vehicle should flip several times.

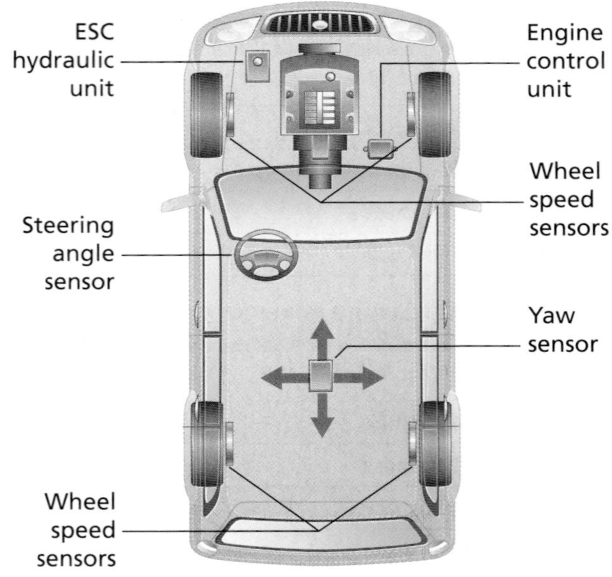

An ESC system processes information from the sensors and automatically applies the brakes to individual wheels when instability is detected.

A promising new technology called Electronic Stability Control (ESC) can be used to prevent rollover accidents. An ESC system has electronic sensors that detect when a vehicle begins to spin due to oversteering, and also when it begins to slide in a plowlike manner because of understeering. In these instances, an ESC system automatically applies the brakes at one or more wheels, thereby reorienting the vehicle in the right direction.

Safe driving is the key to preventing many automobile accidents. Knowledge of the physics behind rollover accidents and the factors that affect rollover accidents may help make you an informed, safe driver.

Going Further

1. **Hypothesize** In a multi-vehicle accident, sport-utility vehicles generally fare better than the passenger cars involved in the accident. Why is this so?
2. **Debate the Issue** ESC is a life-saving technology. Should it be mandatory in all sport-utility vehicles? Why or why not?

Chapter 8

Study Guide

8.1 Describing Rotational Motion

Vocabulary

- radian (p. 197)
- angular displacement (p. 198)
- angular velocity (p. 198)
- angular acceleration (p. 199)

Key Concepts

- Angular position and its changes are measured in radians. One complete revolution is 2π rad.
- Angular velocity is given by the following equation.

$$\omega = \frac{\Delta\theta}{\Delta t}$$

- Angular acceleration is given by the following equation.

$$\alpha = \frac{\Delta\omega}{\Delta t}$$

- For a rotating, rigid object, the angular displacement, velocity, and acceleration can be related to the linear displacement, velocity, and acceleration for any point on the object.

$$d = r\theta \qquad v = r\omega \qquad a = r\alpha$$

8.2 Rotational Dynamics

Vocabulary

- lever arm (p. 201)
- torque (p. 202)
- moment of inertia (p. 205)
- Newton's second law for rotational motion (p. 208)

Key Concepts

- When torque is exerted on an object, its angular velocity changes.
- Torque depends on the magnitude of the force, the distance from the axis of rotation at which it is applied, and the angle between the force and the radius from the axis of rotation to the point where the force is applied.

$$\tau = Fr \sin \theta$$

- The moment of inertia of an object depends on the way the object's mass is distributed about the rotational axis. For a point object:

$$I = mr^2$$

- Newton's second law for rotational motion states that angular acceleration is directly proportional to the net torque and inversely proportional to the moment of inertia.

$$\alpha = \frac{\tau_{net}}{I}$$

8.3 Equilibrium

Vocabulary

- center of mass (p. 211)
- centrifugal "force" (p. 216)
- Coriolis "force" (p. 217)

Key Concepts

- The center of mass of an object is the point on the object that moves in the same way that a point particle would move.
- An object is stable against rollover if its center of mass is above its base.
- An object is in equilibrium if there are no net forces exerted on it and if there are no net torques acting on it.
- Centrifugal "force" and the Coriolis "force" are two apparent forces that appear when a rotating object is analyzed from a coordinate system that rotates with it.

Concept Mapping

47. Complete the following concept map using the following terms: *angular acceleration, radius, tangential acceleration, centripetal acceleration.*

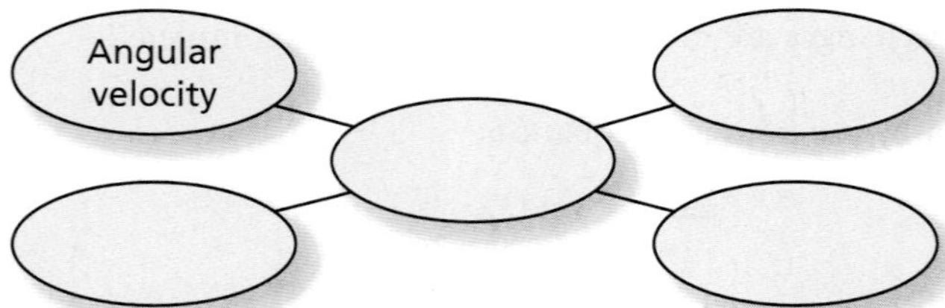

Mastering Concepts

48. A bicycle wheel rotates at a constant 25 rev/min. Is its angular velocity decreasing, increasing, or constant? (8.1)

49. A toy rotates at a constant 5 rev/min. Is its angular acceleration positive, negative, or zero? (8.1)

50. Do all parts of Earth rotate at the same rate? Explain. (8.1)

51. A unicycle wheel rotates at a constant 14 rev/min. Is the total acceleration of a point on the tire inward, outward, tangential, or zero? (8.1)

52. Think about some possible rotations of your textbook. Are the moments of inertia about these three axes the same or different? Explain. (8.2)

53. Torque is important when tightening bolts. Why is force not important? (8.2)

54. Rank the torques on the five doors shown in **Figure 8-18** from least to greatest. Note that the magnitude of all the forces is the same. (8.2)

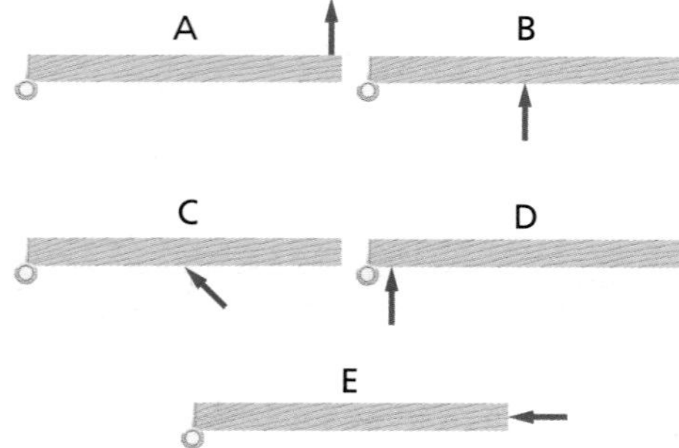

Figure 8-18

55. Explain how you can change an object's angular frequency. (8.2)

56. To balance a car's wheel, it is placed on a vertical shaft and weights are added to make the wheel horizontal. Why is this equivalent to moving the center of mass until it is at the center of the wheel? (8.3)

57. A stunt driver maneuvers a monster truck so that it is traveling on only two wheels. Where is the center of mass of the truck? (8.3)

58. Suppose you stand flat-footed, then you rise and balance on tiptoe. If you stand with your toes touching a wall, you cannot balance on tiptoe. Explain. (8.3)

59. Why does a gymnast appear to be floating on air when she raises her arms above her head in a leap? (8.3)

60. Why is a vehicle with wheels that have a large diameter more likely to roll over than a vehicle with wheels that have a smaller diameter? (8.3)

Applying Concepts

61. Two gears are in contact and rotating. One is larger than the other, as shown in **Figure 8-19.** Compare their angular velocities. Also compare the linear velocities of two teeth that are in contact.

Figure 8-19

62. Videotape When a videotape is rewound, why does it wind up fastest towards the end?

63. Spin Cycle What does a spin cycle of a washing machine do? Explain in terms of the forces on the clothes and water.

64. How can you experimentally find the moment of inertia of an object?

65. Bicycle Wheels Three bicycle wheels have masses that are distributed in three different ways: mostly at the rim, uniformly, and mostly at the hub. The wheels all have the same mass. If equal torques are applied to them, which one will have the greatest angular acceleration? Which one will have the least?

66. Bowling Ball When a bowling ball leaves a bowler's hand, it does not spin. After it has gone about half the length of the lane, however, it does spin. Explain how its rotation rate increased and why it does not continue to increase.

 For more problems, go to Additional Problems, Appendix B.

67. **Flat Tire** Suppose your car has a flat tire. You get out your tools and find a lug wrench to remove the nuts off the bolt studs. You find it impossible to turn the nuts. Your friend suggests ways you might produce enough torque to turn them. What three ways might your friend suggest?

68. **Tightrope Walkers** Tightrope walkers often carry long poles that sag so that the ends are lower than the center as shown in **Figure 8-20.** How does such a pole increase the tightrope walker's stability? *Hint: Consider both center of mass and moment of inertia.*

Figure 8-20

69. **Merry-Go-Round** While riding a merry-go-round, you toss a key to a friend standing on the ground. For your friend to be able to catch the key, should you toss it a second or two before you reach the spot where your friend is standing or wait until your friend is directly behind you? Explain.

70. Why can you ignore forces that act on the axis of rotation of an object in static equilibrium when determining the net torque?

71. In solving problems about static equilibrium, why is the axis of rotation often placed at a point where one or more forces are acting on the object?

Mastering Problems

8.1 Describing Rotational Motion

72. A wheel is rotated so that a point on the edge moves through 1.50 m. The radius of the wheel is 2.50 m, as shown in **Figure 8-21.** Through what angle (in radians) is the wheel rotated?

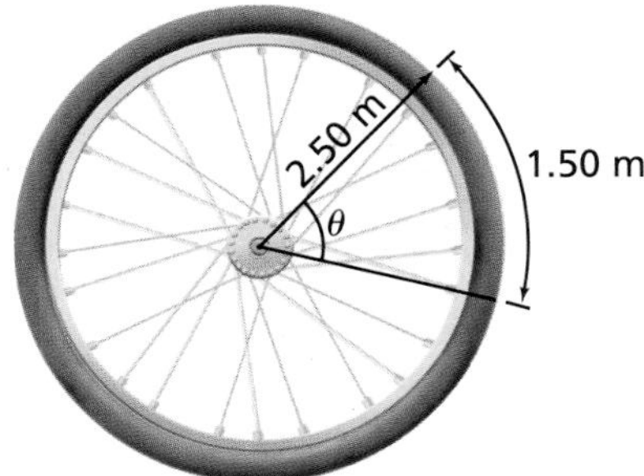

Figure 8-21

73. The outer edge of a truck tire that has a radius of 45 cm has a velocity of 23 m/s. What is the angular velocity of the tire in rad/s?

74. A steering wheel is rotated through 128°, as shown in **Figure 8-22.** Its radius is 22 cm. How far would a point on the steering wheel's edge move?

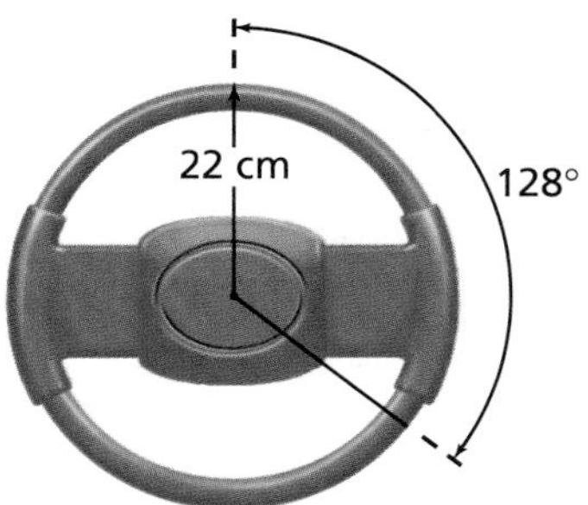

Figure 8-22

75. **Propeller** A propeller spins at 1880 rev/min.
  a. What is its angular velocity in rad/s?
  b. What is the angular displacement of the propeller in 2.50 s?

76. The propeller in the previous problem slows from 475 rev/min to 187 rev/min in 4.00 s. What is its angular acceleration?

77. An automobile wheel with a 9.00 cm radius, as shown in **Figure 8-23,** rotates at 2.50 rad/s. How fast does a point 7.00 cm from the center travel?

Figure 8-23

78. **Washing Machine** A washing machine's two spin cycles are 328 rev/min and 542 rev/min. The diameter of the drum is 0.43 m.
  a. What is the ratio of the centripetal accelerations for the fast and slow spin cycles? Recall that $a_c = \frac{v^2}{r}$ and $v = rw$
  b. What is the ratio of the linear velocity of an object at the surface of the drum for the fast and slow spin cycles?

79. Find the maximum centripetal acceleration in terms of g for the washing machine in problem 78.

80. A laboratory ultracentrifuge is designed to produce a centripetal acceleration of 0.35×10^6 *g* at a distance of 2.50 cm from the axis. What angular velocity in rev/min is required?

8.2 Rotational Dynamics

81. Wrench A bolt is to be tightened with a torque of 8.0 N·m. If you have a wrench that is 0.35 m long, what is the least amount of force you must exert?

82. What is the torque on a bolt produced by a 15-N force exerted perpendicular to a wrench that is 25 cm long, as shown in **Figure 8-24?**

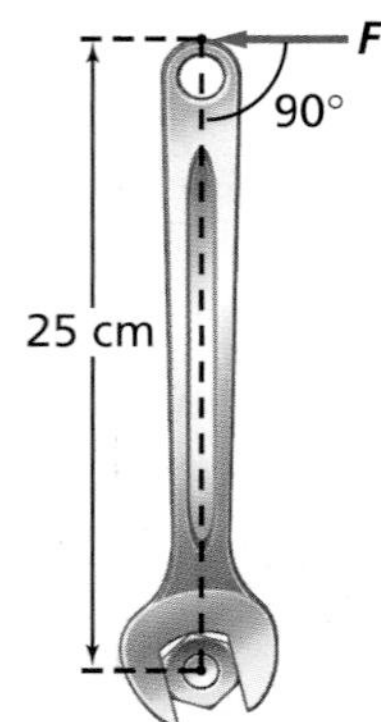

Figure 8-24

83. A toy consisting of two balls, each 0.45 kg, at the ends of a 0.46-m-long, thin, lightweight rod is shown in **Figure 8-25.** Find the moment of inertia of the toy. The moment of inertia is to be found about the center of the rod.

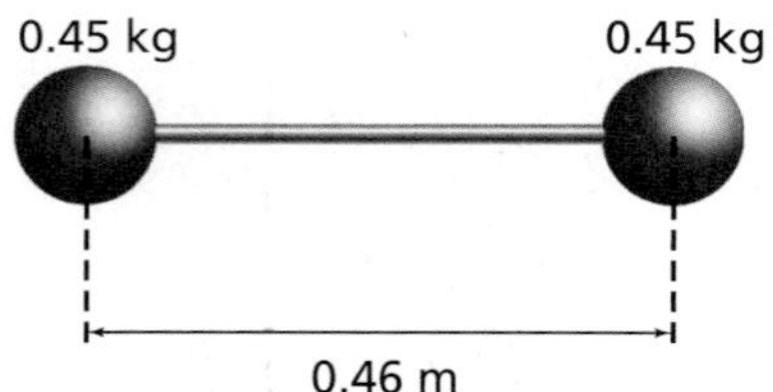

Figure 8-25

84. A bicycle wheel with a radius of 38 cm is given an angular acceleration of 2.67 rad/s^2 by applying a force of 0.35 N on the edge of the wheel. What is the wheel's moment of inertia?

85. Toy Top A toy top consists of a rod with a diameter of 8.0-mm and a disk of mass 0.0125 kg and a diameter of 3.5 cm. The moment of inertia of the rod can be neglected. The top is spun by wrapping a string around the rod and pulling it with a velocity that increases from zero to 3.0 m/s over 0.50 s.

a. What is the resulting angular velocity of the top?
b. What force was exerted on the string?

8.3 Equilibrium

86. A 12.5-kg board, 4.00 m long, is being held up on one end by Ahmed. He calls for help, and Judi responds.

a. What is the least force that Judi could exert to lift the board to the horizontal position? What part of the board should she lift to exert this force?
b. What is the greatest force that Judi could exert to lift the board to the horizontal position? What part of the board should she lift to exert this force?

87. Two people are holding up the ends of a 4.25-kg wooden board that is 1.75 m long. A 6.00-kg box sits on the board, 0.50 m from one end, as shown in **Figure 8-26.** What forces do the two people exert?

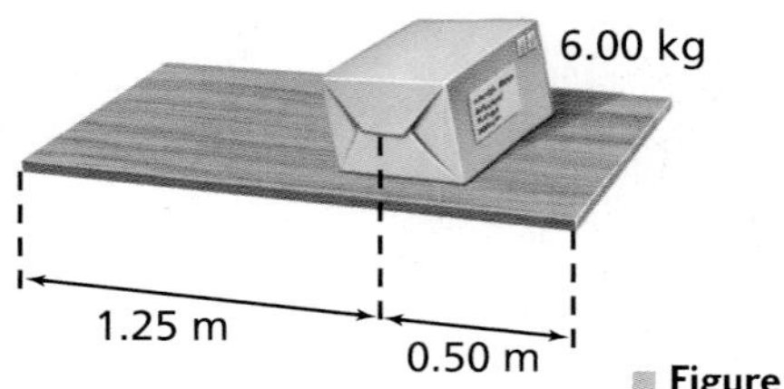

Figure 8-26

88. A car's specifications state that its weight distribution is 53 percent on the front tires and 47 percent on the rear tires. The wheel base is 2.46 m. Where is the car's center of mass?

Mixed Review

89. A wooden door of mass, *m*, and length, *l*, is held horizontally by Dan and Ajit. Dan suddenly drops his end.

a. What is the angular acceleration of the door just after Dan lets go?
b. Is the acceleration constant? Explain.

90. Topsoil Ten bags of topsoil, each weighing 175 N, are placed on a 2.43-m-long sheet of wood. They are stacked 0.50 m from one end of the sheet of wood, as shown in **Figure 8-27.** Two people lift the sheet of wood, one at each end. Ignoring the weight of the wood, how much force must each person exert?

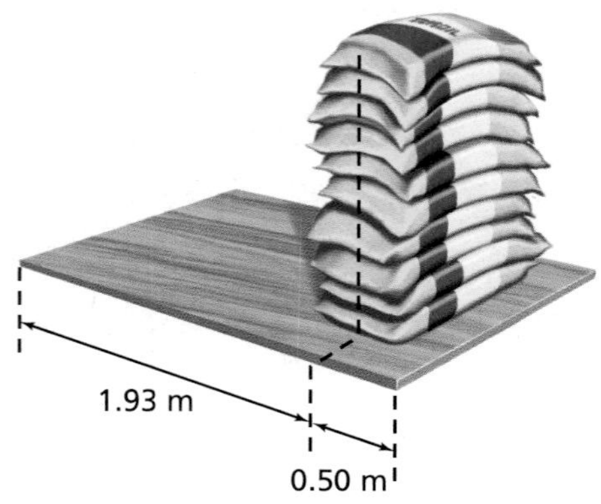

Figure 8-27

 For more problems, go to Additional Problems, Appendix B.

91. Basketball A basketball is rolled down the court. A regulation basketball has a diameter of 24.1 cm, a mass of 0.60 kg, and a moment of inertia of 5.8×10^{-3} kg·m^2. The basketball's initial velocity is 2.5 m/s.
  a. What is its initial angular velocity?
  b. The ball rolls a total of 12 m. How many revolutions does it make?
  c. What is its total angular displacement?

92. The basketball in the previous problem stops rolling after traveling 12 m.
  a. If its acceleration was constant, what was its angular acceleration?
  b. What torque was acting on it as it was slowing down?

93. A cylinder with a 50 cm diameter, as shown in **Figure 8-28,** is at rest on a surface. A rope is wrapped around the cylinder and pulled. The cylinder rolls without slipping.
  a. After the rope has been pulled a distance of 2.50 m at a constant speed, how far has the center of mass of the cylinder moved?
  b. If the rope was pulled a distance of 2.50 m in 1.25 s, how fast was the center of mass of the cylinder moving?
  c. What is the angular velocity of the cylinder?

Figure 8-28

94. Hard Drive A hard drive on a modern computer spins at 7200 rpm (revolutions per minute). If the drive is designed to start from rest and reach operating speed in 1.5 s, what is the angular acceleration of the disk?

95. Speedometers Most speedometers in automobiles measure the angular velocity of the transmission and convert it to speed. How will increasing the diameter of the tires affect the reading of the speedometer?

96. A box is dragged across the floor using a rope that is a distance h above the floor. The coefficient of friction is 0.35. The box is 0.50 m high and 0.25 m wide. Find the force that just tips the box.

97. The second hand on a watch is 12 mm long. What is the velocity of its tip?

98. Lumber You buy a 2.44-m-long piece of 10 cm × 10 cm lumber. Your friend buys a piece of the same size and cuts it into two lengths, each 1.22 m long, as shown in **Figure 8-29.** You each carry your lumber on your shoulders.
  a. Which load is easier to lift? Why?
  b. Both you and your friend apply a torque with your hands to keep the lumber from rotating. Which load is easier to keep from rotating? Why?

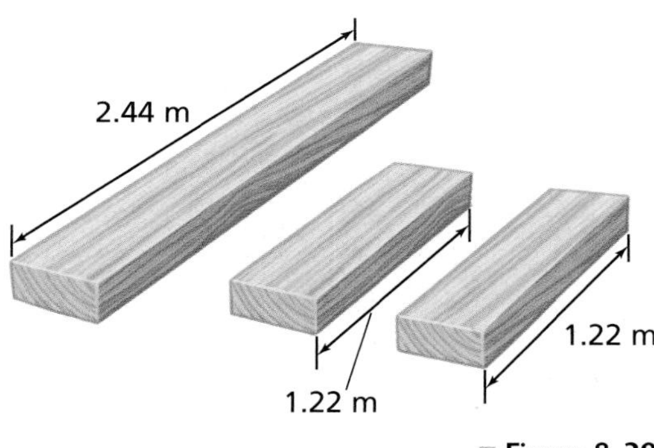

Figure 8-29

99. Surfboard Harris and Paul carry a surfboard that is 2.43 m long and weighs 143 N. Paul lifts one end with a force of 57 N.
  a. What force must Harris exert?
  b. What part of the board should Harris lift?

100. A steel beam that is 6.50 m long weighs 325 N. It rests on two supports, 3.00 m apart, with equal amounts of the beam extending from each end. Suki, who weighs 575 N, stands on the beam in the center and then walks toward one end. How close to the end can she come before the beam begins to tip?

Thinking Critically

101. Apply Concepts Consider a point on the edge of a rotating wheel.
  a. Under what conditions can the centripetal acceleration be zero?
  b. Under what conditions can the tangential (linear) acceleration be zero?
  c. Can the tangential acceleration be nonzero while the centripetal acceleration is zero? Explain.
  d. Can the centripetal acceleration be nonzero while the tangential acceleration is zero? Explain.

102. Apply Concepts When you apply the brakes in a car, the front end dips. Why?

103. Analyze and Conclude A banner is suspended from a horizontal, pivoted pole, as shown in **Figure 8-30.** The pole is 2.10 m long and weighs 175 N. The banner, which weighs 105 N, is suspended 1.80 m from the pivot point or axis of rotation. What is the tension in the cable supporting the pole?

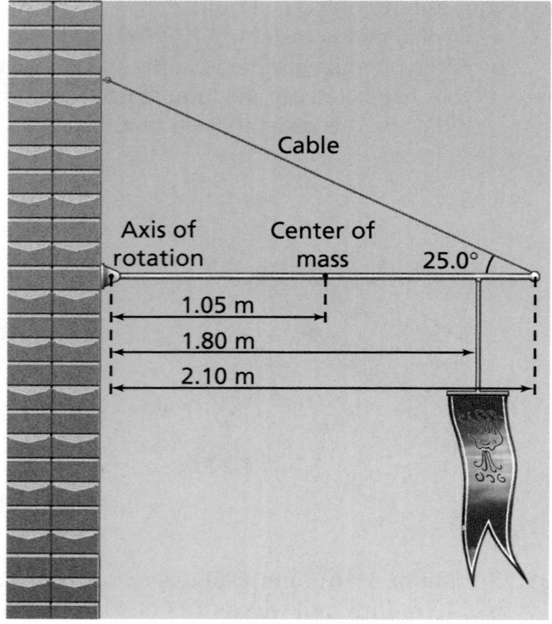

Figure 8-30

104. Analyze and Conclude A pivoted lamp pole is shown in **Figure 8-31.** The pole weighs 27 N, and the lamp weighs 64 N.

a. What is the torque caused by each force?

b. Determine the tension in the rope supporting the lamp pole.

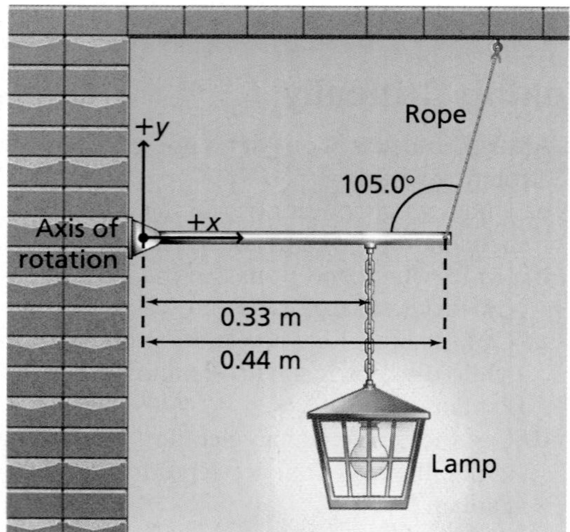

Figure 8-31

105. Analyze and Conclude Gerald and Evelyn carry the following objects up a flight of stairs: a large mirror, a dresser, and a television. Evelyn is at the front end, and Gerald is at the bottom end. Assume that both Evelyn and Gerald exert only upward forces.

a. Draw a free-body diagram showing Gerald and Evelyn exerting the same force on the mirror.

b. Draw a free-body diagram showing Gerald exerting more force on the bottom of the dresser.

c. Where would the center of mass of the television have to be so that Gerald carries all the weight?

Writing in Physics

106. Astronomers know that if a natural satellite is too close to a planet, it will be torn apart by tidal forces. The difference in the gravitational force on the part of the satellite nearest the planet and the part farthest from the planet is stronger than the forces holding the satellite together. Research the Roche limit and determine how close the Moon would have to orbit Earth to be at the Roche limit.

107. Automobile engines are rated by the torque that they produce. Research and explain why torque is an important quantity to measure.

Cumulative Review

108. Two blocks, one of mass 2.0 kg and the other of mass 3.0 kg, are tied together with a massless rope. This rope is strung over a massless, resistance-free pulley. The blocks are released from rest. Find the following. (Chapter 4)

a. the tension in the rope

b. the acceleration of the blocks

109. Eric sits on a see-saw. At what angle, relative to the vertical, will the component of his weight parallel to the plane be equal to one-third the perpendicular component of his weight? (Chapter 5)

110. The pilot of a plane wants to reach an airport 325 km due north in 2.75 hours. A wind is blowing from the west at 30.0 km/h. What heading and airspeed should be chosen to reach the destination on time? (Chapter 6)

111. A 60.0-kg speed skater with a velocity of 18.0 m/s comes into a curve of 20.0-m radius. How much friction must be exerted between the skates and ice to negotiate the curve? (Chapter 6)

 For more problems, go to Additional Problems, Appendix B.

Multiple Choice

1. The illustration below shows two boxes on opposite ends of a board that is 3.0 m long. The board is supported in the middle by a fulcrum. The box on the left has a mass, m_1, of 25 kg, and the box on the right has a mass, m_2, of 15 kg. How far should the fulcrum be positioned from the left side of the board in order to balance the masses horizontally?

   (A) 0.38 m (C) 1.1 m
   (B) 0.60 m (D) 1.9 m

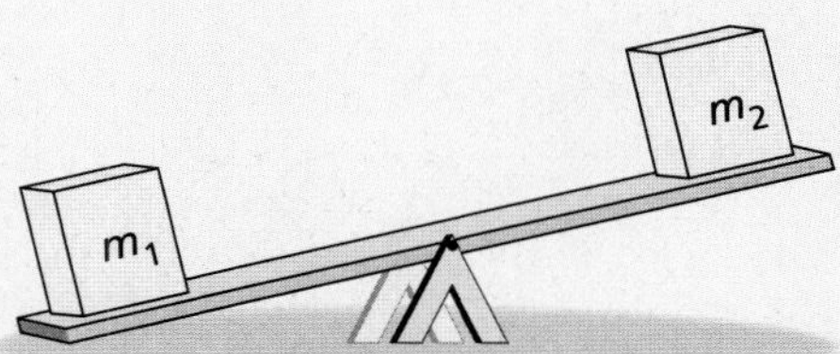

2. A force of 60 N is exerted on one end of a 1.0-m-long lever. The other end of the lever is attached to a rotating rod that is perpendicular to the lever. By pushing down on the end of the lever, you can rotate the rod. If the force on the lever is exerted at an angle of 30°, what torque is exerted on the lever? (sin 30° = 0.5; cos 30° = 0.87; tan 30° = 0.58)

   (A) 30 N (C) 60 N
   (B) 52 N (D) 69 N

3. A child attempts to use a wrench to remove a nut on a bicycle. Removing the nut requires a torque of 10 N·m. The maximum force the child is capable of exerting at a 90° angle is 50 N. What is the length of the wrench the child must use to remove the nut?

   (A) 0.1 m (C) 0.2 m
   (B) 0.15 m (D) 0.25 m

4. A car moves a distance of 420 m. Each tire on the car has a diameter of 42 cm. Which of the following shows how many revolutions each tire makes as they move that distance?

   (A) $\frac{5.0\times10^{1}}{\pi}$ rev (C) $\frac{1.5\times10^{2}}{\pi}$ rev
   (B) $\frac{1.0\times10^{2}}{\pi}$ rev (D) $\frac{1.0\times10^{3}}{\pi}$ rev

5. A thin hoop with a mass of 5.0 kg rotates about a perpendicular axis through its center. A force of 25 N is exerted tangentially to the hoop. If the hoop's radius is 2.0 m, what is its angular acceleration?

   (A) 1.3 rad/s (C) 5.0 rad/s
   (B) 2.5 rad/s (D) 6.3 rad/s

6. Two of the tires on a farmer's tractor have diameters of 1.5 m. If the farmer drives the tractor at a linear velocity of 3.0 m/s, what is the angular velocity of each tire?

   (A) 2.0 rad/s (C) 4.0 rad/s
   (B) 2.3 rad/s (D) 4.5 rad/s

Extended Answer

7. You use a 25-cm long wrench to remove the lug nuts on a car wheel, as shown in the illustration below. If you pull up on the end of the wrench with a force of 2.0×10^{2} N at an angle of 30°, what is the torque on the wrench? (sin 30° = 0.5, cos 30° = 0.87)

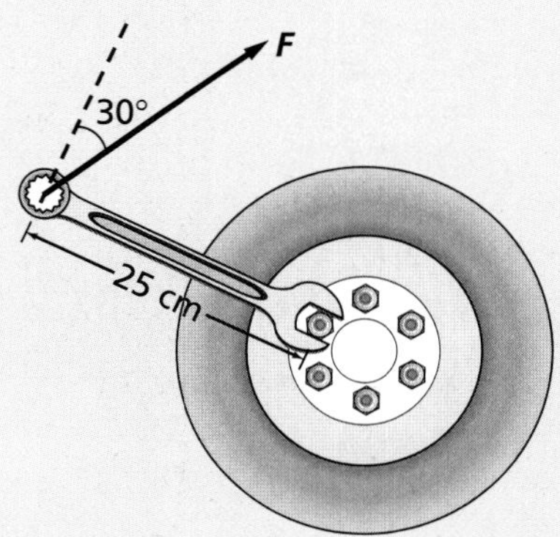

✓Test-Taking TIP

When Eliminating, Cross It Out

Consider each answer choice individually and cross out the ones you have eliminated. If you cannot write in the test booklet, use the scratch paper to list and cross off the answer choices. You will save time and stop yourself from choosing an answer you have mentally eliminated.

Chapter 9

Momentum and Its Conservation

What You'll Learn

- You will describe momentum and impulse and apply them to the interactions between objects.
- You will relate Newton's third law of motion to conservation of momentum.
- You will explore the momentum of rotating objects.

Why It's Important

Momentum is the key to success in many sporting events, including baseball, football, ice hockey, and tennis.

Baseball Every baseball player dreams of hitting a home run. When a player hits the ball, at the moment of collision, the ball and the bat are deformed by the collision. The resulting change in momentum determines the batter's success.

Think About This ▶

What is the force on a baseball bat when a home run is hit out of the park?

physicspp.com

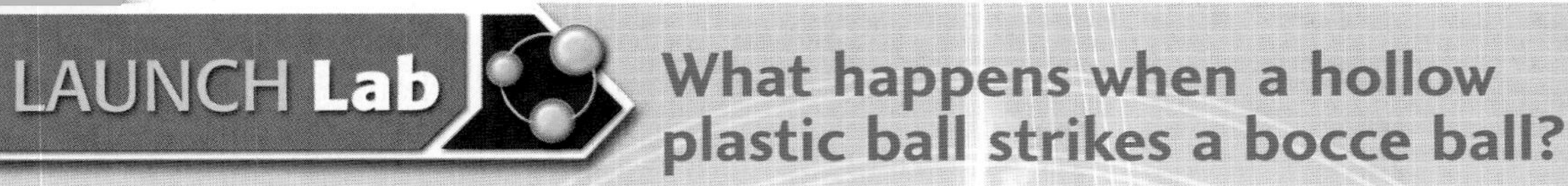

LAUNCH Lab — What happens when a hollow plastic ball strikes a bocce ball?

Question
What direction will a hollow plastic ball and a bocce ball move after a head-on collision?

Procedure

1. Roll a bocce ball and a hollow plastic ball toward each other on a smooth surface.
2. Observe the direction each one moves after the collision.
3. Repeat the experiment, this time keeping the bocce ball stationary, while rolling the hollow plastic ball toward it.
4. Observe the direction each one moves after the collision.
5. Repeat the experiment one more time, but keep the hollow plastic ball stationary, while rolling the bocce ball toward it.
6. Observe the direction each one moves after the collision.

Analysis
What factors affect how fast the balls move after the collision? What factors determine the direction each one moves after the collision?

Critical Thinking What factor(s) would cause the bocce ball to move backward after colliding with the hollow plastic ball?

9.1 Impulse and Momentum

It is always exciting to watch a baseball player hit a home run. The pitcher fires the baseball toward the plate. The batter swings at the baseball and the baseball recoils from the impact of the bat at high speed. Rather than concentrating on the force between the baseball and bat and their resulting accelerations, as in previous chapters, you will approach this collision in a different way in this chapter. The first step in analyzing this type of interaction is to describe what happens before, during, and after the collision between the baseball and bat. You can simplify the collision between the baseball and the bat by making the assumption that all motion is in the horizontal direction. Before the collision, the baseball moves toward the bat. During the collision, the baseball is squashed against the bat. After the collision, however, the baseball moves at a higher velocity away from the bat, and the bat continues in its path, but at a slower velocity.

▶ **Objectives**
- **Define** the momentum of an object.
- **Determine** the impulse given to an object.
- **Define** the angular momentum of an object.

▶ **Vocabulary**
impulse
momentum
impulse-momentum theorem
angular momentum
angular impulse-angular momentum theorem

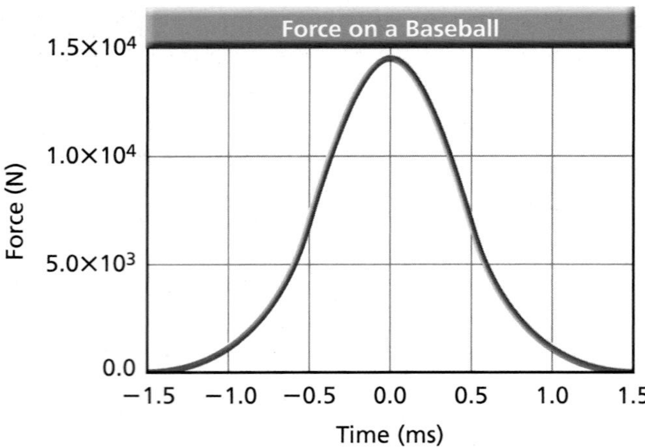

Figure 9-1 The force acting on a baseball increases, then rapidly decreases during a collision, as shown in this force-time graph.

Concepts In Motion
Interactive Figure To see an animation on impulse and momentum, visit physicspp.com.

Impulse and Momentum

How are the velocities of the ball, before and after the collision, related to the force acting on it? Newton's second law of motion describes how the velocity of an object is changed by a net force acting on it. The change in velocity of the ball must have been caused by the force exerted by the bat on the ball. The force changes over time, as shown in **Figure 9-1.** Just after contact is made, the ball is squeezed, and the force increases. After the force reaches its maximum, which is more than 10,000 times the weight of the ball, the ball recovers its shape and snaps away from the bat. The force rapidly returns to zero. This whole event takes place within about 3.0 ms. How can you calculate the change in velocity of the baseball?

Impulse Newton's second law of motion, $\boldsymbol{F} = m\boldsymbol{a}$, can be rewritten by using the definition of acceleration as the change in velocity divided by the time needed to make that change. It can be represented by the following equation:

$$\boldsymbol{F} = m\boldsymbol{a} = m\left(\frac{\Delta \boldsymbol{v}}{\Delta t}\right)$$

Multiplying both sides of the equation by the time interval, Δt, results in the following equation:

$$\boldsymbol{F}\Delta t = m\Delta \boldsymbol{v}$$

Impulse, or $\boldsymbol{F}\Delta t$, is the product of the average force on an object and the time interval over which it acts. Impulse is measured in newton-seconds. For instances in which the force varies with time, the magnitude of an impulse is found by determining the area under the curve of a force-time graph, such as the one shown in Figure 9-1.

The right side of the equation, $m\Delta \boldsymbol{v}$, involves the change in velocity: $\Delta \boldsymbol{v} = \boldsymbol{v}_f - \boldsymbol{v}_i$. Therefore, $m\Delta \boldsymbol{v} = m\boldsymbol{v}_f - m\boldsymbol{v}_i$. The product of the object's mass, m, and the object's velocity, $\boldsymbol{v}$, is defined as the **momentum** of the object. Momentum is measured in kg·m/s. An object's momentum, also known as linear momentum, is represented by the following equation.

Momentum $\boldsymbol{p} = m\boldsymbol{v}$

The momentum of an object is equal to the mass of the object times the object's velocity.

Recall the equation $\boldsymbol{F}\Delta t = m\Delta \boldsymbol{v} = m\boldsymbol{v}_f - m\boldsymbol{v}_i$. Because $m\boldsymbol{v}_f = \boldsymbol{p}_f$ and $m\boldsymbol{v}_i = \boldsymbol{p}_i$, this equation can be rewritten as follows: $\boldsymbol{F}\Delta t = m\Delta \boldsymbol{v} = \boldsymbol{p}_f - \boldsymbol{p}_i$. The right side of this equation, $\boldsymbol{p}_f - \boldsymbol{p}_i$, describes the change in momentum of an object. Thus, the impulse on an object is equal to the change in its momentum, which is called the **impulse-momentum theorem.** The impulse-momentum theorem is represented by the following equation.

Impulse-Momentum Theorem $\boldsymbol{F}\Delta t = \boldsymbol{p}_f - \boldsymbol{p}_i$

The impulse on an object is equal to the object's final momentum minus the object's initial momentum.

Color Convention

- Momentum and impulse vectors are orange.
- Force vectors are **blue.**
- Acceleration vectors are **violet.**
- Velocity vectors are **red.**
- Displacement vectors are green.

If the force on an object is constant, the impulse is the product of the force multiplied by the time interval over which it acts. Generally, the force is not constant, however, and the impulse is found by using an average force multiplied by the time interval over which it acts, or by finding the area under a force-time graph.

Because velocity is a vector, momentum also is a vector. Similarly, impulse is a vector because force is a vector. This means that signs will be important for motion in one dimension.

APPLYING PHYSICS

Running Shoes Running is hard on the feet. When a runner's foot strikes the ground, the force exerted by the ground on it is as much as four times the runner's weight. The cushioning in an athletic shoe is designed to reduce this force by lengthening the time interval over which the force is exerted.

Using the Impulse-Momentum Theorem

What is the change in momentum of a baseball? From the impulse-momentum theorem, you know that the change in momentum is equal to the impulse acting on it. The impulse on a baseball can be calculated by using a force-time graph. In Figure 9-1, the area under the curve is approximately 13.1 N·s. The direction of the impulse is in the direction of the force. Therefore, the change in momentum of the ball also is 13.1 N·s. Because 1 N·s is equal to 1 kg·m/s, the momentum gained by the ball is 13.1 kg·m/s in the direction of the force acting on it.

Assume that a batter hits a fastball. Before the collision of the ball and bat, the ball, with a mass of 0.145 kg, has a velocity of −38 m/s. Assume that the positive direction is toward the pitcher. Therefore, the baseball's momentum is $p_i = (0.145\ \text{kg})(-38\ \text{m/s}) = -5.5\ \text{kg·m/s}$.

What is the momentum of the ball after the collision? Solve the impulse-momentum theorem for the final momentum: $p_f = p_i + F\Delta t$. The ball's final momentum is the sum of the initial momentum and the impulse. Thus, the ball's final momentum is calculated as follows.

$$p_f = p_i + 13.1\ \text{kg·m/s}$$
$$= -5.5\ \text{kg·m/s} + 13.1\ \text{kg·m/s} = +7.6\ \text{kg·m/s}$$

What is the baseball's final velocity? Because $p_f = mv_f$, solving for $\boldsymbol{v}_f$ yields the following:

$$v_f = \frac{p_f}{m} = \frac{+7.6\ \text{kg·m/s}}{+0.145\ \text{kg·m/s}} = +52\ \text{m/s}$$

A speed of 52 m/s is fast enough to clear most outfield fences if the baseball is hit in the correct direction.

Using the Impulse-Momentum Theorem to Save Lives

A large change in momentum occurs only when there is a large impulse. A large impulse can result either from a large force acting over a short period of time or from a smaller force acting over a long period of time.

What happens to the driver when a crash suddenly stops a car? An impulse is needed to bring the driver's momentum to zero. According to the impulse-momentum equation, $\boldsymbol{F}\Delta t = \boldsymbol{p}_f - \boldsymbol{p}_i$. The final momentum, $\boldsymbol{p}_f$, is zero. The initial momentum, $\boldsymbol{p}_i$, is the same with or without an air bag. Thus, the impulse, $\boldsymbol{F}\Delta t$, also is the same. An air bag, such as the one shown in **Figure 9-2,** reduces the force by increasing the time interval during which it acts. It also exerts the force over a larger area of the person's body, thereby reducing the likelihood of injuries.

Figure 9-2 An air bag is inflated during a collision when the force due to the impact triggers the sensor. The chemicals in the air bag's inflation system react and produce a gas that rapidly inflates the air bag.

▶ EXAMPLE Problem 1

Average Force A 2200-kg vehicle traveling at 94 km/h (26 m/s) can be stopped in 21 s by gently applying the brakes. It can be stopped in 3.8 s if the driver slams on the brakes, or in 0.22 s if it hits a concrete wall. What average force is exerted on the vehicle in each of these stops?

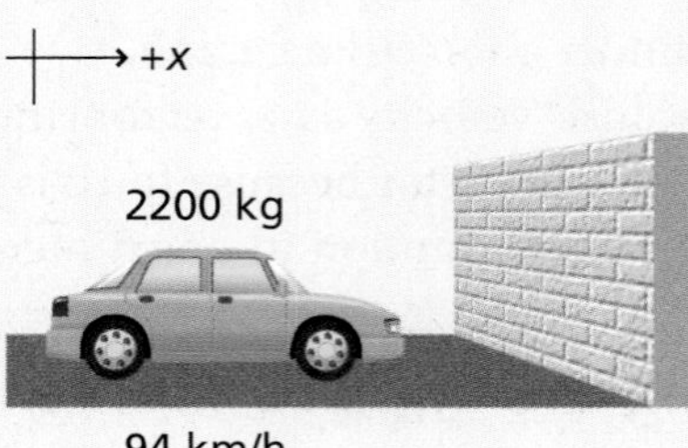

1 Analyze and Sketch the Problem

- Sketch the system.
- Include a coordinate axis and select the positive direction to be the direction of the velocity of the car.
- Draw a vector diagram for momentum and impulse.

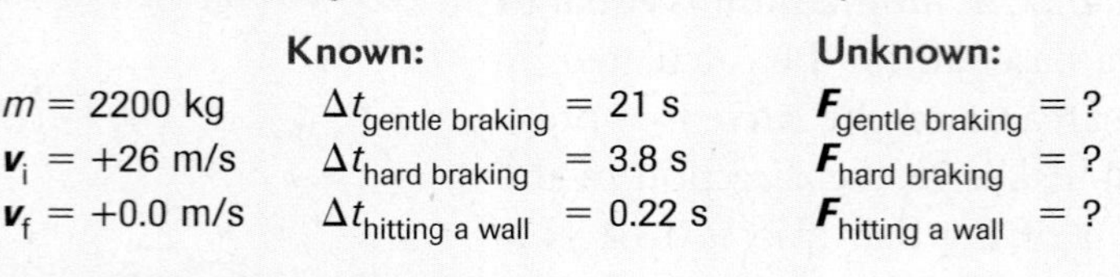

Known:		Unknown:
$m = 2200$ kg	$\Delta t_{\text{gentle braking}} = 21$ s	$\boldsymbol{F}_{\text{gentle braking}} = ?$
$\boldsymbol{v}_i = +26$ m/s	$\Delta t_{\text{hard braking}} = 3.8$ s	$\boldsymbol{F}_{\text{hard braking}} = ?$
$\boldsymbol{v}_f = +0.0$ m/s	$\Delta t_{\text{hitting a wall}} = 0.22$ s	$\boldsymbol{F}_{\text{hitting a wall}} = ?$

Vector diagram

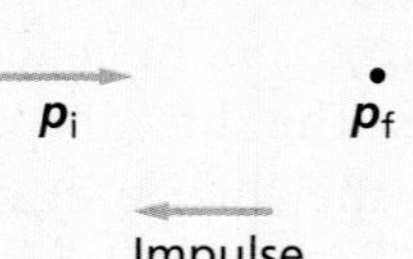

2 Solve for the Unknown

Determine the initial momentum, p_i.

$$p_i = mv_i$$
$$= (2200\ \text{kg})(+26\ \text{m/s})$$ **Substitute $m = 2200$ kg, $v_i = +26$ m/s**
$$= +5.7\times10^4\ \text{kg·m/s}$$

Determine the final momentum, p_f.

$$p_f = mv_f$$
$$= (2200\ \text{kg})(+0.0\ \text{m/s})$$ **Substitute $m = 2200$ kg, $v_f = +0.0$ m/s**
$$= +0.0\ \text{kg·m/s}$$

Apply the impulse-momentum theorem to obtain the force needed to stop the vehicle.

$$F\Delta t = p_f - p_i$$
$$F\Delta t = (+0.0\ \text{kg·m/s}) - (5.7\times10^4\ \text{kg·m/s})$$ **Substitute $p_f = 0.0$ kg·m/s, $p_i = 5.7\times10^4$ kg·m/s**
$$= -5.7\times10^4\ \text{kg·m/s}$$
$$F = \frac{-5.7\times10^4\ \text{kg·m/s}}{\Delta t}$$

$$F_{\text{gentle braking}} = \frac{-5.7\times10^4\ \text{kg·m/s}}{21\ \text{s}}$$ **Substitute $\Delta t_{\text{gentle braking}} = 21$ s**
$$= -2.7\times10^3\ \text{N}$$

$$F_{\text{hard braking}} = \frac{-5.7\times10^4\ \text{kg·m/s}}{3.8\ \text{s}}$$ **Substitute $\Delta t_{\text{hard braking}} = 3.8$ s**
$$= -1.5\times10^4\ \text{N}$$

$$F_{\text{hitting a wall}} = \frac{-5.7\times10^4\ \text{kg·m/s}}{0.22\ \text{s}}$$ **Substitute $\Delta t_{\text{hitting a wall}} = 0.22$ s**
$$= -2.6\times10^5\ \text{N}$$

Math Handbook
Operations with Significant Digits pages 835–836

3 Evaluate the Answer

- **Are the units correct?** Force is measured in newtons.
- **Does the direction make sense?** Force is exerted in the direction opposite to the velocity of the car and thus, is negative.
- **Is the magnitude realistic?** People weigh hundreds of newtons, so it is reasonable that the force needed to stop a car would be in the thousands of newtons. The impulse is the same for all three stops. Thus, as the stopping time is shortened by more than a factor of 10, the force is increased by more than a factor of 10.

PRACTICE Problems

- Additional Problems, Appendix B
- Solutions to Selected Problems, Appendix C

1. A compact car, with mass 725 kg, is moving at 115 km/h toward the east. Sketch the moving car.

a. Find the magnitude and direction of its momentum. Draw an arrow on your sketch showing the momentum.

b. A second car, with a mass of 2175 kg, has the same momentum. What is its velocity?

2. The driver of the compact car in the previous problem suddenly applies the brakes hard for 2.0 s. As a result, an average force of 5.0×10^{3} N is exerted on the car to slow it down.

a. What is the change in momentum; that is, the magnitude and direction of the impulse, on the car?

b. Complete the "before" and "after" sketches, and determine the momentum and the velocity of the car now.

3. A 7.0-kg bowling ball is rolling down the alley with a velocity of 2.0 m/s. For each impulse, shown in **Figures 9-3a** and **9-3b,** find the resulting speed and direction of motion of the bowling ball.

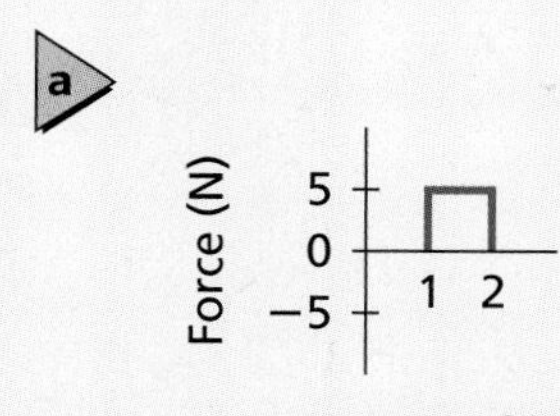

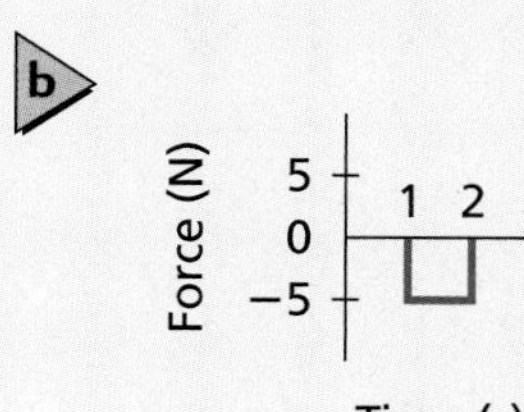

Figure 9-3

4. The driver accelerates a 240.0-kg snowmobile, which results in a force being exerted that speeds up the snowmobile from 6.00 m/s to 28.0 m/s over a time interval of 60.0 s.

a. Sketch the event, showing the initial and final situations.

b. What is the snowmobile's change in momentum? What is the impulse on the snowmobile?

c. What is the magnitude of the average force that is exerted on the snowmobile?

5. Suppose a 60.0-kg person was in the vehicle that hit the concrete wall in Example Problem 1. The velocity of the person equals that of the car both before and after the crash, and the velocity changes in 0.20 s. Sketch the problem.

a. What is the average force exerted on the person?

b. Some people think that they can stop their bodies from lurching forward in a vehicle that is suddenly braking by putting their hands on the dashboard. Find the mass of an object that has a weight equal to the force you just calculated. Could you lift such a mass? Are you strong enough to stop your body with your arms?

Angular Momentum

As you learned in Chapter 8, the angular velocity of a rotating object changes only if torque is applied to it. This is a statement of Newton's law for rotational motion, $\tau = I\Delta\omega/\Delta t$. This equation can be rearranged in the same way as Newton's second law of motion was, to produce $\tau\Delta t = I\Delta\omega$.

The left side of this equation, $\tau\Delta t$, is the angular impulse of the rotating object. The right side can be rewritten as $\Delta\omega = \omega_f - \omega_i$. The product of a rotating object's moment of inertia and angular velocity is called **angular momentum,** which is represented by the symbol L. The angular momentum of an object can be represented by the following equation.

Angular Momentum $L = I\omega$

The angular momentum of an object is equal to the product of the object's moment of inertia and the object's angular velocity.

Angular momentum is measured in kg·m^2/s. Just as the linear momentum of an object changes when an impulse acts on it, the angular momentum of an object changes when an angular impulse acts on it. Thus, the angular impulse on the object is equal to the change in the object's angular momentum, which is called the **angular impulse-angular momentum theorem.** The angular impulse-angular momentum theorem is represented by the following equation.

Angular Impulse-Angular Momentum Theorem $\tau\Delta t = L_f - L_i$

The angular impulse on an object is equal to the object's final angular momentum minus the object's initial angular momentum.

If there are no forces acting on an object, its linear momentum is constant. If there are no torques acting on an object, its angular momentum is also constant. Because an object's mass cannot be changed, if its momentum is constant, then its velocity is also constant. In the case of angular momentum, however, the object's angular velocity does not remain constant. This is because the moment of inertia depends on the object's mass and the way it is distributed about the axis of rotation or revolution. Thus, the angular velocity of an object can change even if no torques are acting on it.

Astronomy Connection

Consider, for example, a planet orbiting the Sun. The torque on the planet is zero because the gravitational force acts directly toward the Sun. Therefore, the planet's angular momentum is constant. When the distance between the planet and the Sun decreases, however, the planet's moment of inertia of revolution in orbit about the Sun also decreases. Thus, the planet's angular velocity increases and it moves faster. This is an explanation of Kepler's second law of planetary motion, based on Newton's laws of motion.

Figure 9-4 The diver's center of mass is in front of her feet as she gets ready to dive **(a).** As the diver changes her moment of inertia by moving her arms and legs to increase her angular momentum, the location of the center of mass changes, but the path of the center of mass remains a parabola **(b).**

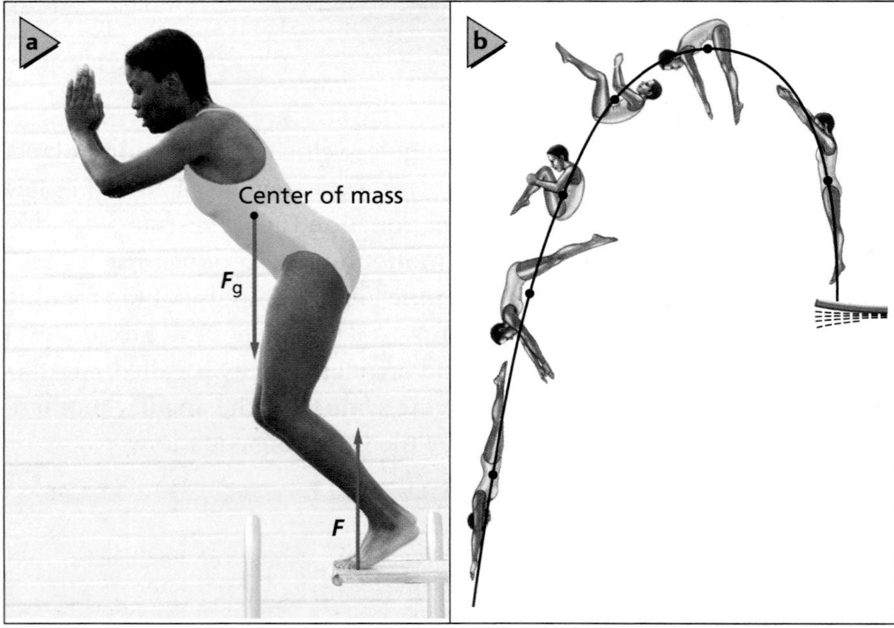

Consider the diver in **Figure 9-4.** How does she start rotating her body? She uses the diving board to apply an external torque to her body. Then, she moves her center of mass in front of her feet and uses the board to give a final upward push to her feet. This torque acts over time, Δt, and thus increases the angular momentum of the diver.

Before the diver reaches the water, she can change her angular velocity by changing her moment of inertia. She may go into a tuck position, grabbing her knees with her hands. By moving her mass closer to the axis of rotation, the diver decreases her moment of inertia and increases her angular velocity. When she nears the water, she stretches her body straight, thereby increasing the moment of inertia and reducing the angular velocity. As a result, she goes straight into the water.

An ice-skater uses a similar method to spin. To begin rotating on one foot, the ice-skater applies an external torque to her body by pushing a portion of the other skate into the ice, as shown in **Figure 9-5.** If she pushes on the ice in one direction, the ice will exert a force on her in the opposite direction. The force results in a torque if the force is exerted some distance away from the pivot point, and in a direction that is not toward it. The greatest torque for a given force will result if the push is perpendicular to the lever arm.

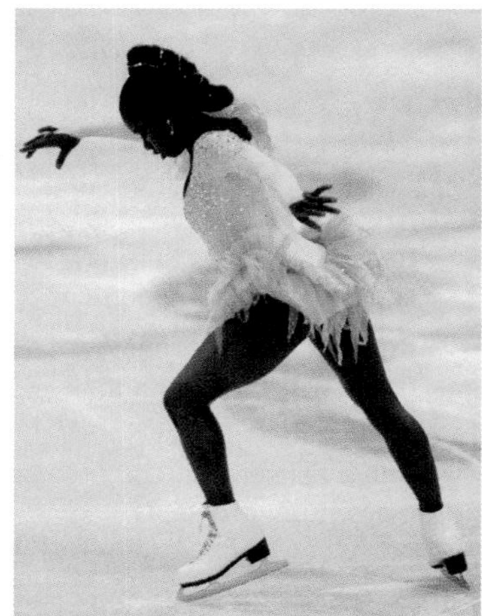

Figure 9-5 To spin on one foot, an ice-skater extends one leg and pushes on the ice. The ice exerts an equal and opposite force on her body and produces an external torque.

The ice-skater then can control her angular velocity by changing her moment of inertia. Both arms and one leg can be extended from the body to slow the rotation, or pulled in close to the axis of rotation to speed it up. To stop spinning, another torque must be exerted by using the second skate to create a way for the ice to exert the needed force.

9.1 Section Review

6. **Momentum** Is the momentum of a car traveling south different from that of the same car when it travels north at the same speed? Draw the momentum vectors to support your answer.

7. **Impulse and Momentum** When you jump from a height to the ground, you let your legs bend at the knees as your feet hit the floor. Explain why you do this in terms of the physics concepts introduced in this chapter.

8. **Momentum** Which has more momentum, a supertanker tied to a dock or a falling raindrop?

9. **Impulse and Momentum** A 0.174-kg softball is pitched horizontally at 26.0 m/s. The ball moves in the opposite direction at 38.0 m/s after it is hit by the bat.
   a. Draw arrows showing the ball's momentum before and after the bat hits it.
   b. What is the change in momentum of the ball?
   c. What is the impulse delivered by the bat?
   d. If the bat and softball are in contact for 0.80 ms, what is the average force that the bat exerts on the ball?

10. **Momentum** The speed of a basketball as it is dribbled is the same when the ball is going toward the floor as it is when the ball rises from the floor. Is the basketball's change in momentum equal to zero when it hits the floor? If not, in which direction is the change in momentum? Draw the basketball's momentum vectors before and after it hits the floor.

11. **Angular Momentum** An ice-skater spins with his arms outstretched. When he pulls his arms in and raises them above his head, he spins much faster than before. Did a torque act on the ice-skater? If not, how could his angular velocity have increased?

12. **Critical Thinking** An archer shoots arrows at a target. Some of the arrows stick in the target, while others bounce off. Assuming that the masses of the arrows and the velocities of the arrows are the same, which arrows produce a bigger impulse on the target? *Hint: Draw a diagram to show the momentum of the arrows before and after hitting the target for the two instances.*

9.2 Conservation of Momentum

▸ **Objectives**

- **Relate** Newton's third law to conservation of momentum.
- **Recognize** the conditions under which momentum is conserved.
- **Solve** conservation of momentum problems.

▸ **Vocabulary**

closed system
isolated system
law of conservation of momentum
law of conservation of angular momentum

In the first section of this chapter, you learned how a force applied during a time interval changes the momentum of a baseball. In the discussion of Newton's third law of motion, you learned that forces are the result of interactions between two objects. The force of a bat on a ball is accompanied by an equal and opposite force of the ball on the bat. Does the momentum of the bat, therefore, also change?

Two-Particle Collisions

The bat, the hand and arm of the batter, and the ground on which the batter is standing are all objects that interact when a batter hits the ball. Thus, the bat cannot be considered a single object. In contrast to this complex system, examine for a moment the much simpler system shown in **Figure 9-6,** the collision of two balls.

During the collision of the two balls, each one briefly exerts a force on the other. Despite the differences in sizes and velocities of the balls, the forces that they exert on each other are equal and opposite, according to Newton's third law of motion. These forces are represented by the following equation: $\boldsymbol{F}_{\text{D on C}} = -\boldsymbol{F}_{\text{C on D}}$

How do the impulses imparted by both balls compare? Because the time intervals over which the forces are exerted are the same, the impulses must be equal in magnitude but opposite in direction. How did the momenta of the balls change as a result of the collision?

According to the impulse-momentum theorem, the change in momentum is equal to the impulse. Compare the changes in the momenta of the two balls.

$$\text{For ball C: } \boldsymbol{p}_{\text{Cf}} - \boldsymbol{p}_{\text{Ci}} = \boldsymbol{F}_{\text{D on C}}\Delta t$$

$$\text{For ball D: } \boldsymbol{p}_{\text{Df}} - \boldsymbol{p}_{\text{Di}} = \boldsymbol{F}_{\text{C on D}}\Delta t$$

Because the time interval over which the forces were exerted is the same, the impulses are equal in magnitude, but opposite in direction. According to Newton's third law of motion, $-\boldsymbol{F}_{\text{C on D}} = \boldsymbol{F}_{\text{D on C}}$. Thus,

$$\boldsymbol{p}_{\text{Cf}} - \boldsymbol{p}_{\text{Ci}} = -(\boldsymbol{p}_{\text{Df}} - \boldsymbol{p}_{\text{Di}}), \text{ or } \boldsymbol{p}_{\text{Cf}} + \boldsymbol{p}_{\text{Df}} = \boldsymbol{p}_{\text{Ci}} + \boldsymbol{p}_{\text{Di}}.$$

This equation states that the sum of the momenta of the balls is the same before and after the collision. That is, the momentum gained by ball D is equal to the momentum lost by ball C. If the system is defined as the two balls, the momentum of the system is constant, and therefore, momentum is conserved for the system.

Before Collision (initial)

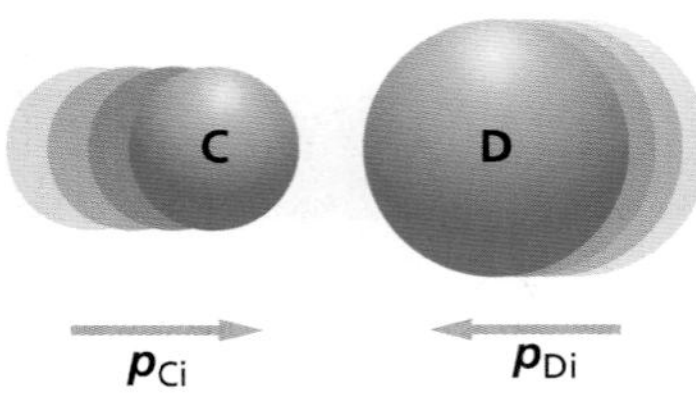

During Collision

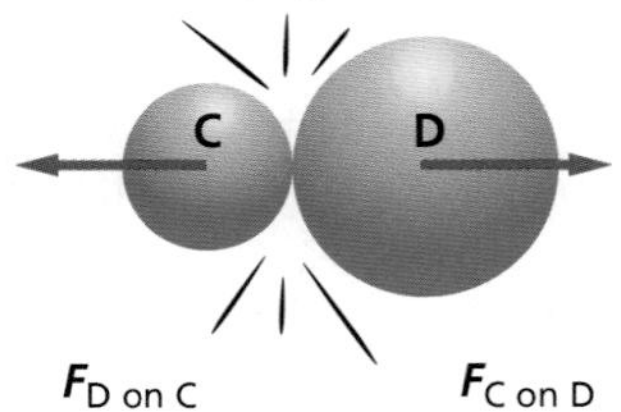

After Collision (final)

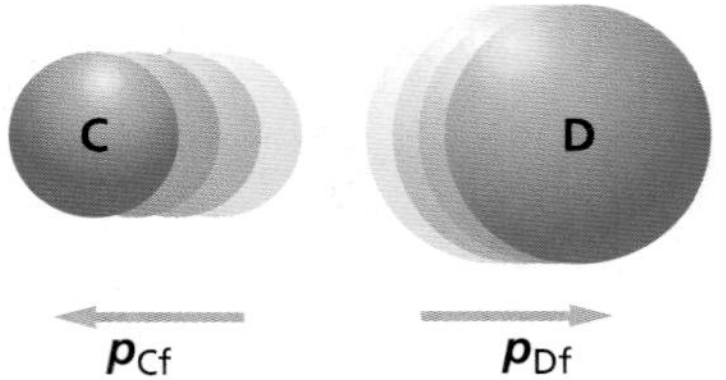

Figure 9-6 When two balls collide, they exert forces on each other that change their momenta.

Momentum in a Closed, Isolated System

Under what conditions is the momentum of the system of two balls conserved? The first and most obvious condition is that no balls are lost and no balls are gained. Such a system, which does not gain or lose mass, is said to be a **closed system.** The second condition required to conserve the momentum of a system is that the forces involved are internal forces; that is, there are no forces acting on the system by objects outside of it.

When the net external force on a closed system is zero, the system is described as an **isolated system.** No system on Earth can be said to be absolutely isolated, however, because there will always be some interactions between a system and its surroundings. Often, these interactions are small enough to be ignored when solving physics problems.

Systems can contain any number of objects, and the objects can stick together or come apart in a collision. Under these conditions, the **law of conservation of momentum** states that the momentum of any closed, isolated system does not change. This law will enable you to make a connection between conditions, before and after an interaction, without knowing any of the details of the interaction.

▶ EXAMPLE Problem 2

Speed A 1875-kg car going 23 m/s rear-ends a 1025-kg compact car going 17 m/s on ice in the same direction. The two cars stick together. How fast do the two cars move together immediately after the collision?

1 Analyze and Sketch the Problem

- Define the system.
- Establish a coordinate system.
- Sketch the situation showing the "before" and "after" states.
- Draw a vector diagram for the momentum.

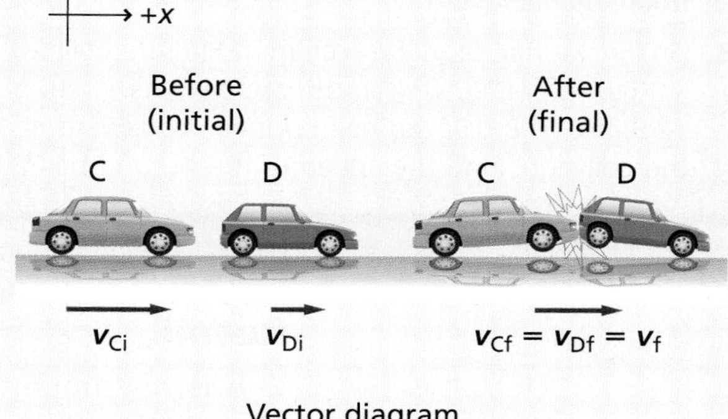

Known:

$m_C = 1875$ kg

$v_{Ci} = +23$ m/s

$m_D = 1025$ kg

$v_{Di} = +17$ m/s

Unknown:

$v_f = ?$

Vector diagram

p_{Ci} p_{Di} p_f

$p_i = p_{Ci} + p_{Di}$

2 Solve for the Unknown

Momentum is conserved because the ice makes the total external force on the cars nearly zero.

$$p_i = p_f$$

$$p_{Ci} + p_{Di} = p_{Cf} + p_{Df}$$

$$m_C v_{Ci} + m_D v_{Di} = m_C v_{Cf} + m_D v_{Df}$$

Because the two cars stick together, their velocities after the collision, denoted as v_f, are equal.

$$v_{Cf} = v_{Df} = v_f$$

$$m_C v_{Ci} + m_D v_{Di} = (m_C + m_D)v_f$$

Math Handbook
Order of Operations
page 843

Solve for v_f.

$$v_f = \frac{(m_C v_{Ci} + m_D v_{Di})}{(m_C + m_D)}$$

$$= \frac{(1875\text{ kg})(+23\text{ m/s}) + (1025\text{ kg})(+17\text{ m/s})}{(1875\text{ kg} + 1025\text{ kg})}$$

Substitute $m_C = 1875$ kg, $v_{Ci} = +23$ m/s, $m_D = 1025$ kg, $v_{Di} = +17$ m/s

$$= +21\text{ m/s}$$

3 Evaluate the Answer

- **Are the units correct?** Velocity is measured in m/s.
- **Does the direction make sense?** v_i and v_f are in the positive direction; therefore, v_f should be positive.
- **Is the magnitude realistic?** The magnitude of v_f is between the initial speeds of the two cars, but closer to the speed of the more massive one, so it is reasonable.

PRACTICE Problems

- Additional Problems, Appendix B
- Solutions to Selected Problems, Appendix C

13. Two freight cars, each with a mass of 3.0×10^5 kg, collide and stick together. One was initially moving at 2.2 m/s, and the other was at rest. What is their final speed?

14. A 0.105-kg hockey puck moving at 24 m/s is caught and held by a 75-kg goalie at rest. With what speed does the goalie slide on the ice?

15. A 35.0-g bullet strikes a 5.0-kg stationary piece of lumber and embeds itself in the wood. The piece of lumber and bullet fly off together at 8.6 m/s. What was the original speed of the bullet?

16. A 35.0-g bullet moving at 475 m/s strikes a 2.5-kg bag of flour that is on ice, at rest. The bullet passes through the bag, as shown in **Figure 9-7,** and exits it at 275 m/s. How fast is the bag moving when the bullet exits?

17. The bullet in the previous problem strikes a 2.5-kg steel ball that is at rest. The bullet bounces backward after its collision at a speed of 5.0 m/s. How fast is the ball moving when the bullet bounces backward?

18. A 0.50-kg ball that is traveling at 6.0 m/s collides head-on with a 1.00-kg ball moving in the opposite direction at a speed of 12.0 m/s. The 0.50-kg ball bounces backward at 14 m/s after the collision. Find the speed of the second ball after the collision.

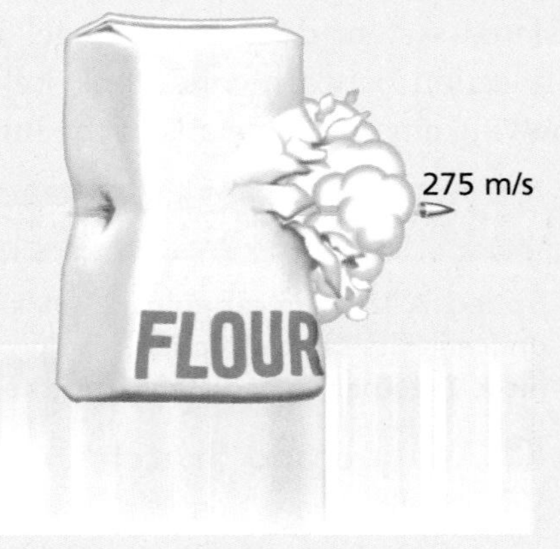

Figure 9-7

Recoil

It is very important to define a system carefully. The momentum of a baseball changes when the external force of a bat is exerted on it. The baseball, therefore, is not an isolated system. On the other hand, the total momentum of two colliding balls within an isolated system does not change because all forces are between the objects within the system.

Can you find the final velocities of the two in-line skaters in **Figure 9-8?** Assume that they are skating on a smooth surface with no external forces. They both start at rest, one behind the other.

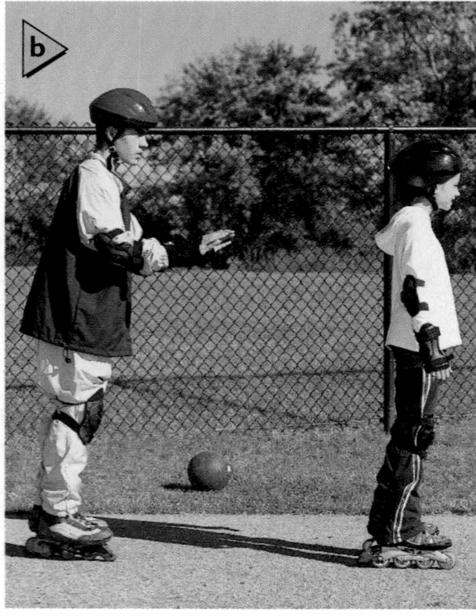

Figure 9-8 The internal forces exerted by Skater C, the boy, and Skater D, the girl, cannot change the total momentum of the system.

Skater C, the boy, gives skater D, the girl, a push. Now, both skaters are moving, making this situation similar to that of an explosion. Because the push was an internal force, you can use the law of conservation of momentum to find the skaters' relative velocities. The total momentum of the system was zero before the push. Therefore, it must be zero after the push.

Before		After
$\boldsymbol{p}_{Ci} + \boldsymbol{p}_{Di}$	=	$\boldsymbol{p}_{Cf} + \boldsymbol{p}_{Df}$
0	=	$\boldsymbol{p}_{Cf} + \boldsymbol{p}_{Df}$
$\boldsymbol{p}_{Cf}$	=	$-\boldsymbol{p}_{Df}$
$m_C\boldsymbol{v}_{Cf}$	=	$-m_D\boldsymbol{v}_{Df}$

The coordinate system was chosen so that the positive direction is to the left. The momenta of the skaters after the push are equal in magnitude but opposite in direction. The backward motion of skater C is an example of recoil. Are the skaters' velocities equal and opposite? The last equation shown above, for the velocity of skater C, can be rewritten as follows:

$$\boldsymbol{v}_{Cf} = \left(\frac{-m_D}{m_C}\right)\boldsymbol{v}_{Df}$$

The velocities depend on the skaters' relative masses. If skater C has a mass of 68.0 kg and skater D's mass is 45.4 kg, then the ratio of their velocities will be 68.0 : 45.4, or 1.50. The less massive skater moves at the greater velocity. Without more information about how hard skater C pushed skater D, however, you cannot find the velocity of each skater.

Propulsion in Space

How does a rocket in space change its velocity? The rocket carries both fuel and oxidizer. When the fuel and oxidizer combine in the rocket motor, the resulting hot gases leave the exhaust nozzle at high speed. If the rocket and chemicals are the system, then the system is a closed system. The forces that expel the gases are internal forces, so the system is also an isolated system. Thus, objects in space can accelerate by using the law of conservation of momentum and Newton's third law of motion.

A NASA space probe, called *Deep Space 1*, performed a flyby of an asteroid a few years ago. The most unusual of the 11 new technologies on board was an ion engine that exerts as much force as a sheet of paper resting on a person's hand. The ion engine shown in **Figure 9-9,** operates differently from a traditional rocket engine. In a traditional rocket engine, the products of the chemical reaction taking place in the combustion chamber are released at high speed from the rear. In the ion engine, however, xenon atoms are expelled at a speed of 30 km/s, producing a force of only 0.092 N. How can such a small force create a significant change in the momentum of the probe? Instead of operating for only a few minutes, as the traditional chemical rockets do, the ion engine can run continuously for days, weeks, or months. Therefore, the impulse delivered by the engine is large enough to increase the momentum of the 490-kg spacecraft until it reaches the speed needed to complete its mission.

Figure 9-9 The xenon atoms in the ion engine are ionized by bombarding them with electrons. Then, the positively charged xenon ions are accelerated to high speeds.

MINI LAB

Rebound Height

An object's momentum is the product of its mass and velocity.

1. Drop a large rubber ball from about 15 cm above a table.
2. Measure and record the ball's rebound height.
3. Repeat steps 1–2 with a small rubber ball.
4. Hold the small rubber ball on top of, and in contact with, the large rubber ball.
5. Release the two rubber balls from the same height, so that they fall together.
6. Measure the rebound heights of both rubber balls.

Analyze and Conclude

7. Describe the rebound height of each rubber ball dropped by itself.
8. Compare and contrast the rebound heights from number 7 with those from number 6.
9. Explain your observations.

▶ EXAMPLE Problem 3

Speed An astronaut at rest in space fires a thruster pistol that expels 35 g of hot gas at 875 m/s. The combined mass of the astronaut and pistol is 84 kg. How fast and in what direction is the astronaut moving after firing the pistol?

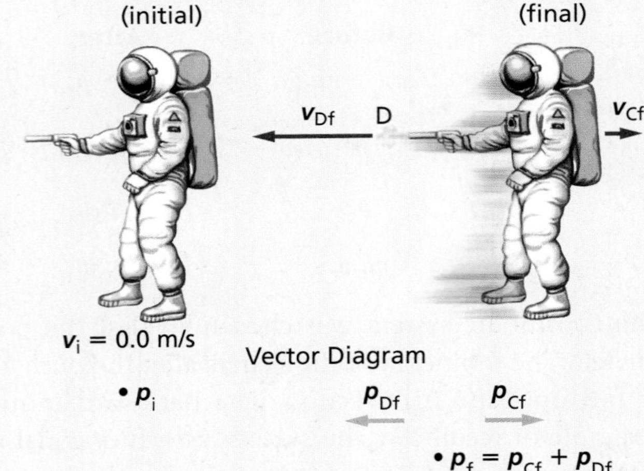

1 Analyze and Sketch the Problem

- Define the system.
- Establish a coordinate axis.
- Sketch the "before" and "after" conditions.
- Draw a vector diagram showing momenta.

Known:

$m_C = 84$ kg

$m_D = 0.035$ kg

$v_{Ci} = v_{Di} = +0.0$ m/s

$v_{Df} = -875$ m/s

Unknown:

$v_{Cf} = ?$

2 Solve for the Unknown

The system is the astronaut, the gun, and the chemicals that produce the gas.

$$p_i = p_{Ci} + p_{Di} = +0.0 \text{ kg·m/s}$$

Before the pistol is fired, all parts of the system are at rest; thus, the initial momentum is zero.

Use the law of conservation of momentum to find p_f.

$$p_i = p_f$$

$$+0.0 \text{ kg·m/s} = p_{Cf} + p_{Df}$$

$$p_{Cf} = -p_{Df}$$

The momentum of the astronaut is equal in magnitude, but opposite in direction to the momentum of the gas leaving the pistol.

Solve for the final velocity of the astronaut, v_{Cf}.

$$m_C v_{Cf} = -m_D v_{Df}$$

$$v_{Cf} = \left(\frac{-m_D v_{Df}}{m_C}\right)$$

$$= \frac{-(0.035 \text{ kg})(-875 \text{ m/s})}{84 \text{ kg}}$$

Substitute $m_D = 0.035$ kg, $v_{Df} = -875$ m/s, $m_C = 84$ kg

$$= +0.36 \text{ m/s}$$

Math Handbook
Isolating a Variable
page 845

3 Evaluate the Answer

- **Are the units correct?** The velocity is measured in m/s.
- **Does the direction make sense?** The velocity of the astronaut is in the opposite direction to that of the expelled gas.
- **Is the magnitude realistic?** The astronaut's mass is much larger than that of the gas, so the velocity of the astronaut is much less than that of the expelled gas.

▶ PRACTICE Problems

• Additional Problems, Appendix B
• Solutions to Selected Problems, Appendix C

19. A 4.00-kg model rocket is launched, expelling 50.0 g of burned fuel from its exhaust at a speed of 625 m/s. What is the velocity of the rocket after the fuel has burned? *Hint: Ignore the external forces of gravity and air resistance.*

20. A thread holds a 1.5-kg cart and a 4.5-kg cart together. After the thread is burned, a compressed spring pushes the carts apart, giving the 1.5-kg cart a speed of 27 cm/s to the left. What is the velocity of the 4.5-kg cart?

21. Carmen and Judi dock a canoe. 80.0-kg Carmen moves forward at 4.0 m/s as she leaves the canoe. At what speed and in what direction do the canoe and Judi move if their combined mass is 115 kg?

Two-Dimensional Collisions

Up until now, you have looked at momentum in only one dimension. The law of conservation of momentum holds for all closed systems with no external forces. It is valid regardless of the directions of the particles before or after they interact. But what happens in two or three dimensions? **Figure 9-10** shows the result of billiard ball C striking stationary billiard ball D. Consider the two billiard balls to be the system. The original momentum of the moving ball is $\boldsymbol{p}_{\text{Ci}}$ and the momentum of the stationary ball is zero. Therefore, the momentum of the system before the collision is equal to $\boldsymbol{p}_{\text{Ci}}$.

After the collision, both billiard balls are moving and have momenta. As long as the friction with the tabletop can be ignored, the system is closed and isolated. Thus, the law of conservation of momentum can be used. The initial momentum equals the vector sum of the final momenta, so $\boldsymbol{p}_{\text{Ci}} = \boldsymbol{p}_{\text{Cf}} + \boldsymbol{p}_{\text{Df}}$.

The equality of the momenta before and after the collision also means that the sum of the components of the vectors before and after the collision must be equal. Suppose the *x*-axis is defined to be in the direction of the initial momentum, then the *y*-component of the initial momentum is equal to zero. Therefore, the sum of the final *y*-components also must be zero:

$$\boldsymbol{p}_{\text{Cf}, y} + \boldsymbol{p}_{\text{Df}, y} = 0$$

The *y*-components are equal in magnitude but are in the opposite direction and, thus, have opposite signs. The sum of the horizontal components also is equal:

$$\boldsymbol{p}_{\text{Ci}} = \boldsymbol{p}_{\text{Cf}, x} + \boldsymbol{p}_{\text{Df}, x}$$

Figure 9-10 The law of conservation of momentum holds for all isolated, closed systems, regardless of the directions of objects before and after a collision.

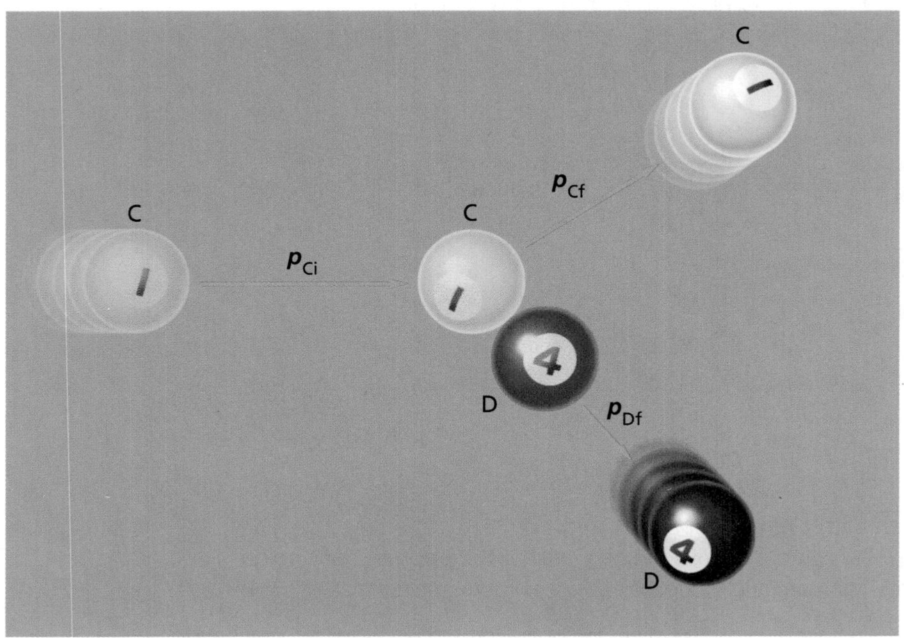

▶ EXAMPLE Problem 4

Speed A 1325-kg car, C, moving north at 27.0 m/s, collides with a 2165-kg car, D, moving east at 11.0 m/s. The two cars are stuck together. In what direction and with what speed do they move after the collision?

Before (initial)

After (final)

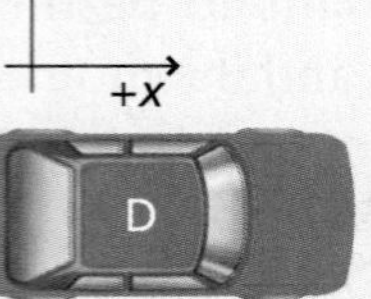

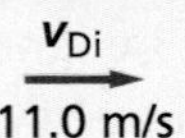

1 Analyze and Sketch the Problem

- Define the system.
- Sketch the "before" and "after" states.
- Establish the coordinate axis with the *y*-axis north and the *x*-axis east.
- Draw a momentum-vector diagram.

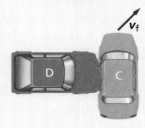

Known:	Unknown:
$m_C = 1325$ kg	$v_{f,x} = ?$
$m_D = 2165$ kg	$v_{f,y} = ?$
$v_{Ci,y} = 27.0$ m/s	$\theta = ?$
$v_{Di,x} = 11.0$ m/s	

Vector Diagram

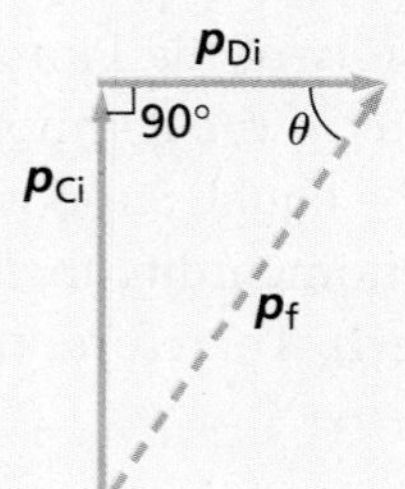

2 Solve for the Unknown

Determine the initial momenta of the cars and the momentum of the system.

$$p_{Ci} = m_C v_{Ci,y}$$
$$= (1325\ \text{kg})(27.0\ \text{m/s})$$ **Substitute $m_C = 1325$ kg, $v_{Ci,y} = 27.0$ m/s**
$$= 3.58\times10^4\ \text{kg·m/s (north)}$$

$$p_{Di} = m_D v_{Di,x}$$
$$= (2165\ \text{kg})(11.0\ \text{m/s})$$ **Substitute $m_D = 2165$ kg, $v_{Di,x} = 11.0$ m/s**
$$= 2.38\times10^4\ \text{kg·m/s (east)}$$

Use the law of conservation of momentum to find p_f.

$$p_{f,x} = p_{i,x} = 2.38\times10^4\ \text{kg·m/s}$$ **Substitute $p_{i,x} = p_{Di} = 2.38\times10^4$ kg·m/s**
$$p_{f,y} = p_{i,y} = 3.58\times10^4\ \text{kg·m/s}$$ **Substitute $p_{i,y} = p_{Ci} = 3.58\times10^4$ kg·m/s**

Use the diagram to set up equations for $p_{f,x}$ and $p_{f,y}$.

$$p_f = \sqrt{(p_{f,x})^2 + (p_{f,y})^2}$$
$$= \sqrt{(2.38\times10^4\ \text{kg·m/s})^2 + (3.58\times10^4\ \text{kg·m/s})^2}$$ **Substitute $p_{f,x} = 2.38\times10^4$ kg·m/s, $p_{f,y} = 3.58\times10^4$ kg·m/s**
$$= 4.30\times10^4\ \text{kg·m/s}$$

Solve for θ.

$$\theta = \tan^{-1}\left(\frac{p_{f,y}}{p_{f,x}}\right)$$
$$= \tan^{-1}\left(\frac{3.58\times10^4\ \text{kg·m/s}}{2.38\times10^4\ \text{kg·m/s}}\right)$$ **Substitute $p_{f,y} = 3.58\times10^4$ kg·m/s, $p_{f,x} = 2.38\times10^4$ kg·m/s**
$$= 56.4°$$

Determine the final speed.

$$v_f = \frac{p_f}{(m_C + m_D)}$$
$$= \frac{4.30\times10^4\ \text{kg·m/s}}{(1325\ \text{kg} + 2165\ \text{kg})}$$ **Substitute $p_f = 4.30\times10^4$ kg·m/s, $m_C = 1325$ kg, $m_D = 2165$ kg**
$$= 12.3\ \text{m/s}$$

Personal Tutor For an online tutorial on speed, visit physicspp.com.

3 Evaluate the Answer

- **Are the units correct?** The correct unit for speed is m/s.
- **Do the signs make sense?** Answers are both positive and at the appropriate angles.
- **Is the magnitude realistic?** The cars stick together, so v_f must be smaller than v_{Ci}.

▶ PRACTICE Problems

- Additional Problems, Appendix B
- Solutions to Selected Problems, Appendix C

22. A 925-kg car moving north at 20.1 m/s collides with a 1865-kg car moving west at 13.4 m/s. The two cars are stuck together. In what direction and at what speed do they move after the collision?

23. A 1383-kg car moving south at 11.2 m/s is struck by a 1732-kg car moving east at 31.3 m/s. The cars are stuck together. How fast and in what direction do they move immediately after the collision?

24. A stationary billiard ball, with a mass of 0.17 kg, is struck by an identical ball moving at 4.0 m/s. After the collision, the second ball moves 60.0° to the left of its original direction. The stationary ball moves 30.0° to the right of the moving ball's original direction. What is the velocity of each ball after the collision?

25. A 1345-kg car moving east at 15.7 m/s is struck by a 1923-kg car moving north. They are stuck together and move with an initial velocity of 14.5 m/s at $\theta = 63.5°$. Was the north-moving car exceeding the 20.1 m/s speed limit?

Conservation of Angular Momentum

Like linear momentum, angular momentum can be conserved. The **law of conservation of angular momentum** states that if no net external torque acts on an object, then its angular momentum does not change. This is represented by the following equation.

Law of Conservation of Angular Momentum $L_i = L_f$

An object's initial angular momentum is equal to its final angular momentum.

For example, Earth spins on its axis with no external torques. Its angular momentum is constant. Thus, Earth's angular momentum is conserved. As a result, the length of a day does not change. A spinning ice-skater also demonstrates conservation of angular momentum. **Figure 9-11a** shows an ice-skater spinning with his arms extended. When he pulls in his arms, as shown in **Figure 9-11b,** he begins spinning faster. Without an external torque, his angular momentum does not change; that is, $L = I\omega$ is constant. Thus, the ice-skater's increased angular velocity must be accompanied by a decreased moment of inertia. By pulling his arms close to his body, the ice-skater brings more mass closer to the axis of rotation, thereby decreasing the radius of rotation and decreasing his moment of inertia. You can calculate changes in angular velocity using the law of conservation of angular momentum.

$$L_i = L_f$$

$$\text{thus, } I_i\omega_i = I_f\omega_f$$

$$\frac{\omega_f}{\omega_i} = \frac{I_i}{I_f}$$

Because frequency is $f = \omega/2\pi$, the above equation can be rewritten as follows:

$$\frac{2\pi(f_f)}{2\pi(f_i)} = \frac{I_i}{I_f}$$

$$\text{thus, } \frac{f_f}{f_i} = \frac{I_i}{I_f}$$

■ **Figure 9-11** When the ice-skater's arms are extended, the moment of inertia increases and his angular velocity decreases **(a).** When his arms are closer to his body the moment of inertia decreases and results in an increased angular velocity **(b).**

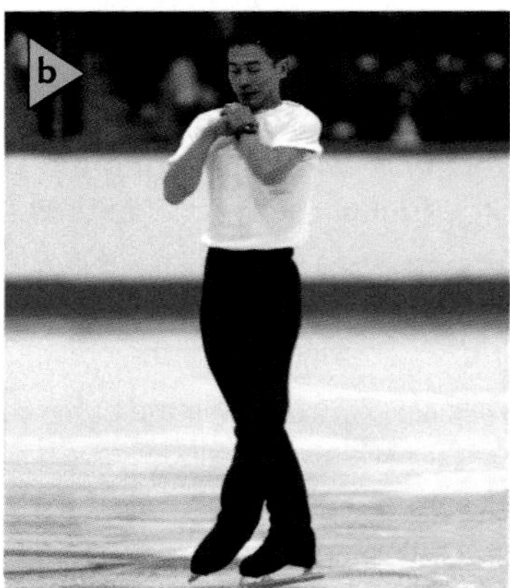

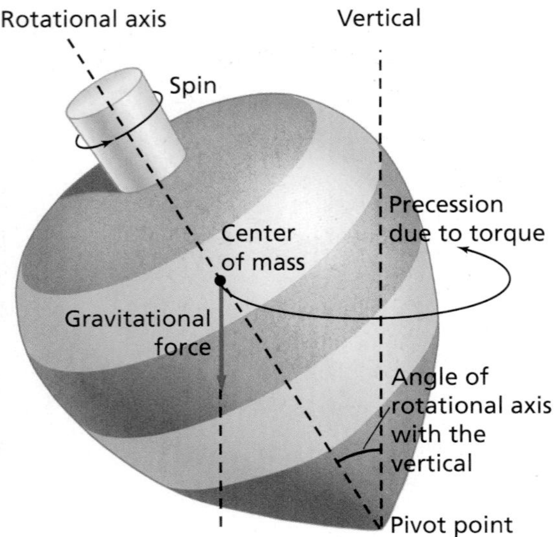

Figure 9-12 The upper end of the top precesses due to the torque acting on the top.

Notice that because f, ω, and I appear as ratios in these equations, any units may be used, as long as the same unit is used for both values of the quantity.

If a torque-free object starts with no angular momentum, it must continue to have no angular momentum. Thus, if part of an object rotates in one direction, another part must rotate in the opposite direction. For example, if you switch on a loosely held electric drill, the drill body will rotate in the direction opposite to the rotation of the motor and bit.

Consider a ball thrown at a weather vane. The ball, moving in a straight line, can start the vane rotating. Consider the ball and vane to be a system. With no external torques, angular momentum is conserved. The vane spins faster if the ball has a large mass, m, a large velocity, v, and hits at right angles as far as possible from the pivot of the vane. The angular momentum of a moving object, such as the ball, is given by $L = mvr$, where r is the perpendicular distance from the axis of rotation.

Tops and Gyroscopes

Because of the conservation of angular momentum, the direction of rotation of a spinning object can be changed only by applying a torque. If you played with a top as a child, you may have spun it by pulling the string wrapped around its axle. When a top is vertical, there is no torque on it, and the direction of its rotation does not change. If the top is tipped, as shown in **Figure 9-12,** a torque tries to rotate it downward. Rather than tipping over, however, the upper end of the top revolves, or precesses slowly about the vertical axis. Because Earth is not a perfect sphere, the Sun exerts a torque on it, causing it to precess. It takes about 26,000 years for Earth's rotational axis to go through one cycle of precession.

CHALLENGE **PROBLEM**

Your friend was driving her 1265-kg car north on Oak Street when she was hit by a 925-kg compact car going west on Maple Street. The cars stuck together and slid 23.1 m at 42° north of west. The speed limit on both streets is 22 m/s (50 mph). Assume that momentum was conserved during the collision and that acceleration was constant during the skid. The coefficient of kinetic friction between the tires and the pavement is 0.65.

1. Your friend claims that she wasn't speeding, but that the driver of other car was. How fast was your friend driving before the crash?
2. How fast was the other car moving before the crash? Can you support your friend's case in court?

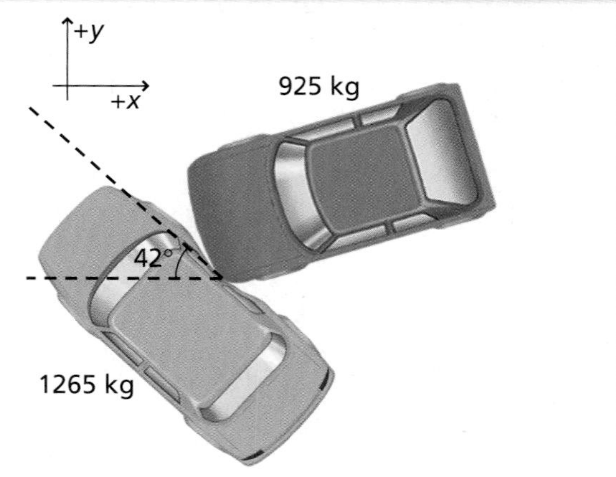

A gyroscope, such as the one shown in **Figure 9-13,** is a wheel or disk that spins rapidly around one axis while being free to rotate around one or two other axes. The direction of its large angular momentum can be changed only by applying an appropriate torque. Without such a torque, the direction of the axis of rotation does not change. Gyroscopes are used in airplanes, submarines, and spacecraft to keep an unchanging reference direction. Giant gyroscopes are used in cruise ships to reduce their motion in rough water. Gyroscopic compasses, unlike magnetic compasses, maintain direction even when they are not on a level surface.

Figure 9-13 Because the orientation of the spin axis of the gyroscope does not change even when it is moved, the gyroscope can be used to fix direction.

A football quarterback uses the gyroscope effect to make an accurate forward pass. As he throws, he spins, or spirals the ball. If the quarterback throws the ball in the direction of its spin axis of rotation, the ball keeps its pointed end forward, thereby reducing air resistance. Thus, the ball can be thrown far and accurately. If its spin direction is slightly off, the ball wobbles. If the ball is not spun, it tumbles end over end.

The flight of a plastic disk also is stabilized by spin. A well-spun plastic disk can fly many meters through the air without wobbling. You are able to perform tricks with a yo-yo because its fast rotational speed keeps it rotating in one plane.

9.2 Section Review

26. Angular Momentum The outer rim of a plastic disk is thick and heavy. Besides making it easier to catch, how does this affect the rotational properties of the plastic disk?

27. Speed A cart, weighing 24.5 N, is released from rest on a 1.00-m ramp, inclined at an angle of 30.0° as shown in **Figure 9-14.** The cart rolls down the incline and strikes a second cart weighing 36.8 N.

a. Calculate the speed of the first cart at the bottom of the incline.

b. If the two carts stick together, with what initial speed will they move along?

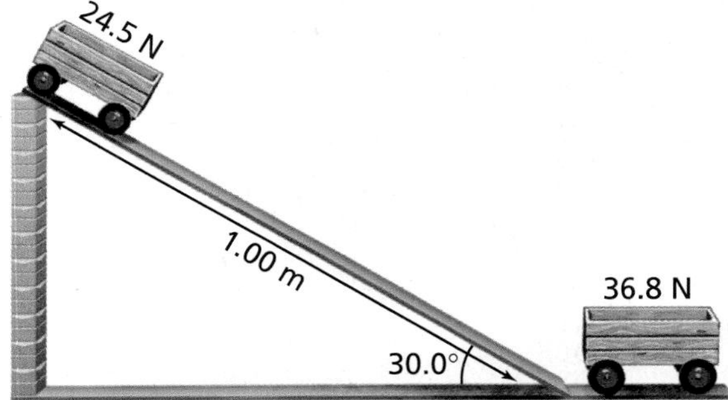

Figure 9-14

28. Conservation of Momentum During a tennis serve, the racket of a tennis player continues forward after it hits the ball. Is momentum conserved in the collision? Explain, making sure that you define the system.

29. Momentum A pole-vaulter runs toward the launch point with horizontal momentum. Where does the vertical momentum come from as the athlete vaults over the crossbar?

30. Initial Momentum During a soccer game, two players come from opposite directions and collide when trying to head the ball. They come to rest in midair and fall to the ground. Describe their initial momenta.

31. Critical Thinking You catch a heavy ball while you are standing on a skateboard, and then you roll backward. If you were standing on the ground, however, you would be able to avoid moving while catching the ball. Explain both situations using the law of conservation of momentum. Explain which system you use in each case.

PHYSICS LAB • Internet

Sticky Collisions

Alternate CBL instructions can be found on the Web site.
physicspp.com

In this activity, one moving cart will strike a stationary cart. During the collision, the two carts will stick together. You will measure mass and velocity, both before and after the collision. You then will calculate the momentum both before and after the collision.

QUESTION

How is the momentum of a system affected by a sticky collision?

Objectives

- **Describe** how momentum is transferred during a collision.
- **Calculate** the momenta involved.
- **Interpret data** from a collision.
- **Draw conclusions** that support the law of conservation of momentum.

Safety Precautions

Materials

Internet access required

Procedure

1. View Chapter 9 lab video clip 1 at physicspp.com/internet_lab to determine the mass of the carts.
2. Record the mass of each cart.
3. Watch video clip 2: Cart 1 strikes Cart 2.
4. In the video, three frames represent 0.1 s, and the main gridlines are separated by 10 cm. Record in the data table the distance Cart 1 travels in 0.1 s before the collision.
5. Observe the collision. Record in the data table the distance the Cart 1-Cart 2 system travels in 0.1 s after the collision.
6. Repeat steps 3–5 for video clip 3: Carts 1 and 3 strike Cart 2.
7. Repeat steps 3–5 for video clip 4: Cart 1 strikes Carts 2 and 3.
8. Repeat steps 3-5 for video clip 5: Carts 1 and 3 strike Carts 2 and 4.

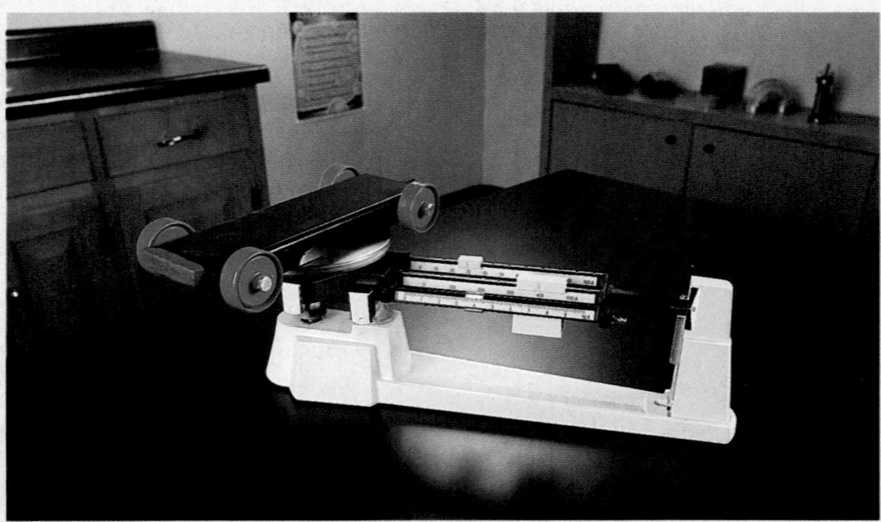

Data Tables

Cart	Mass (kg)
1	
2	
3	
4	

Time of Approach (s)	Distance Covered in Approach (cm)	Initial Velocity (cm/s)	Mass of Approaching Cart(s) (g)	Initial Momentum (g·cm/s)	Time of Departure (s)	Distance Covered in Departure (cm)	Final Velocity (cm/s)	Mass of Departing Cart(s) (g)	Final Momentum (g·cm/s)
0.1					0.1				
0.1					0.1				
0.1					0.1				
0.1					0.1				

Analyze

1. Calculate the initial and final velocities for each of the cart systems.
2. Calculate the initial and final momentum for each of the cart systems.
3. **Make and Use Graphs** Make a graph showing final momentum versus initial momentum for all the video clips.

Conclude and Apply

1. What is the relationship between the initial momentum and the final momentum of the cart systems in a sticky collision?
2. In theory, what should be the slope of the line in your graph?
3. The initial and final data numbers may not be the same due to the precision of the instruments, friction, and other variables. Is the initial momentum typically greater or less than the final momentum? Explain.

Going Further

1. Describe what the velocity and momentum data might look like if the carts did not stick together, but rather, bounced off of each other.
2. Design an experiment to test the impact of friction during the collision of the cart systems. Predict how the slope of the line in your graph will change with your experiment, and then try your experiment.

Real-World Physics

1. Suppose a linebacker collides with a stationary quarterback and they become entangled. What will happen to the velocity of the linebacker-quarterback system if momentum is conserved?
2. If a car rear-ends a stationary car so that the two cars become attached, what will happen to the velocity of the first car? The second car?

Share Your Data

Interpret Data Visit physicspp.com/internet_lab to post your findings from the experiment testing the impact of friction during the collisions of the cart systems. Examine your data, and graph final momentum versus initial momentum. Notice how close to or far off the slope is from 1.00.

To find out more about momentum, visit the Web site: physicspp.com

Future Technology

Solar Sailing

Nearly 400 years ago, Johannes Kepler observed that comet tails appeared to be blown by a solar breeze. He suggested that ships would be able to travel in space with sails designed to catch this breeze. Thus, the idea for solar sails was born.

How Does a Solar Sail Work? A solar sail is a spacecraft without an engine. A solar sail works like a giant fabric mirror that is free to move. Solar sails usually are made of 5-micron-thick aluminized polyester film or polyimide film with a 100-nm-thick aluminum layer deposited on one side to form the reflective surface.

Reflected sunlight, rather than rocket fuel, provides the force. Sunlight is made up of individual particles called photons. Photons have momentum, and when a photon bounces off a solar sail, it transfers its momentum to the sail, which propels the spacecraft along.

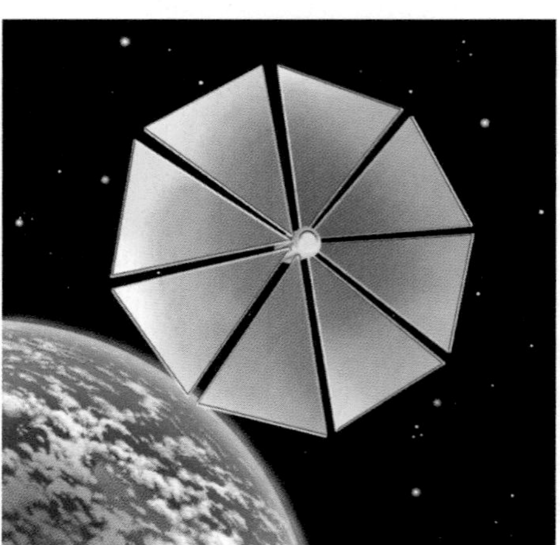

This artist's rendering shows *Cosmos 1*, the first solar sail launched into space.

The force of impacting photons is small in comparison to the force rocket fuel can supply. So, small sails experience only a small amount of force from sunlight, while larger sails experience a greater force. Thus, solar sails may be a kilometer or so across.

What speeds can a solar sail achieve? This depends on the momentum transferred to the sail by photons, as well as the sail's mass. To travel quickly through the solar system, a sail and the spacecraft should be lightweight.

Photons supplied by the Sun are constant. They impact the sail every second of every hour of every day during a space flight. The Sun's continuous supply of photons over time allows the sail to build up huge velocities and enables the spacecraft to travel great distances within a convenient time frame. Rockets require enormous amounts of fuel to move large masses, but solar sails only require photons from the Sun. Thus, solar sails may be a superior way to move large masses over great distances in outer space.

Future Journeys The *Cosmos 1* mission, a privately-funded international venture, launched the first solar-sail prototype. *Cosmos 1* was launched toward space on the tip of a submarine-launched rocket on June 21, 2005. The spacecraft looked like a flower with eight huge, solar-sail petals. Even though the goals were modest, *Cosmos 1* never had the chance to test its new technology. The first stage of the rocket never completed its scheduled burn, preventing *Cosmos 1* from entering orbit.

Solar sails are important, not only for travel, but also for creating new types of space and Earth weather monitoring stations. These stations would be able to provide greater coverage of Earth and more advanced warning of solar storms that cause problems to communication and electric power grids. It is hoped that in the next few decades, solar sails will be used as interplanetary shuttles because of their ability to travel great distances in convenient time frames. Vast distances could someday be traversed by vehicles that do not consume any fuel.

Going Further

1. **Research** how solar sails can help provide advanced warning of solar storms.
2. **Critical Thinking** A certain solar-sail model is predicted to take more time to reach Mars than a rocket-propelled spacecraft would, but less time to go to Pluto than a rocket-propelled spacecraft would. Explain why this is so.

Chapter 9 Study Guide

9.1 Impulse and Momentum

Vocabulary

- impulse (p. 230)
- momentum (p. 230)
- impulse-momentum theorem (p. 230)
- angular momentum (p. 233)
- angular impulse-angular momentum theorem (p. 234)

Key Concepts

- When doing a momentum problem, first examine the system before and after the event.
- The momentum of an object is the product of its mass and velocity and is a vector quantity.

$$\boldsymbol{p} = m\boldsymbol{v}$$

- The impulse on an object is the average net force exerted on the object multiplied by the time interval over which the force acts.

$$\text{Impluse} = \boldsymbol{F}\Delta t$$

- The impulse on an object is equal to the change in momentum of the object.

$$\boldsymbol{F}\Delta t = \boldsymbol{p}_{\mathrm{f}} - \boldsymbol{p}_{\mathrm{i}}$$

- The angular momentum of a rotating object is the product of its moment of inertia and its angular velocity.

$$L = I\omega$$

- The angular impulse-angular momentum theorem states that the angular impulse on an object is equal to the change in the object's angular momentum.

$$\tau\Delta t = L_{\mathrm{f}} - L_{\mathrm{i}}$$

9.2 Conservation of Momentum

Vocabulary

- closed system (p. 236)
- isolated system (p. 237)
- law of conservation of momentum (p. 237)
- law of conservation of angular momentum (p. 243)

Key Concepts

- According to Newton's third law of motion and the law of conservation of momentum, the forces exerted by colliding objects on each other are equal in magnitude and opposite in direction.
- Momentum is conserved in a closed, isolated system.

$$\boldsymbol{p}_{\mathrm{f}} = \boldsymbol{p}_{\mathrm{i}}$$

- The law of conservation of momentum can be used to explain the propulsion of rockets.
- Vector analysis is used to solve momentum-conservation problems in two dimensions.
- The law of conservation of angular momentum states that if there are no external torques acting on a system, then the angular momentum is conserved.

$$L_{\mathrm{i}} = L_{\mathrm{f}}$$

- Because angular momentum is conserved, the direction of rotation of a spinning object can be changed only by applying a torque.

Chapter 9

Assessment

Concept Mapping

32. Complete the following concept map using the following terms: *mass, momentum, average force, time over which the force is exerted.*

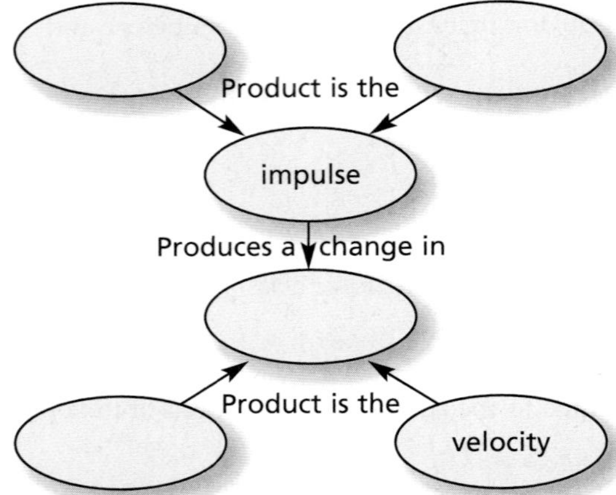

Mastering Concepts

33. Can a bullet have the same momentum as a truck? Explain. (9.1)

34. A pitcher throws a curve ball to the catcher. Assume that the speed of the ball doesn't change in flight. (9.1)

a. Which player exerts the larger impulse on the ball?

b. Which player exerts the larger force on the ball?

35. Newton's second law of motion states that if no net force is exerted on a system, no acceleration is possible. Does it follow that no change in momentum can occur? (9.1)

36. Why are cars made with bumpers that can be pushed in during a crash? (9.1)

37. An ice-skater is doing a spin. (9.1)

a. How can the skater's angular momentum be changed?

b. How can the skater's angular velocity be changed without changing the angular momentum?

38. What is meant by "an isolated system?" (9.2)

39. A spacecraft in outer space increases its velocity by firing its rockets. How can hot gases escaping from its rocket engine change the velocity of the craft when there is nothing in space for the gases to push against? (9.2)

40. A cue ball travels across a pool table and collides with the stationary eight ball. The two balls have equal masses. After the collision, the cue ball is at rest. What must be true regarding the speed of the eight ball? (9.2)

41. Consider a ball falling toward Earth. (9.2)

a. Why is the momentum of the ball not conserved?

b. In what system that includes the falling ball is the momentum conserved?

42. A falling basketball hits the floor. Just before it hits, the momentum is in the downward direction, and after it hits the floor, the momentum is in the upward direction. (9.2)

a. Why isn't the momentum of the basketball conserved even though the bounce is a collision?

b. In what system is the momentum conserved?

43. Only an external force can change the momentum of a system. Explain how the internal force of a car's brakes brings the car to a stop. (9.2)

44. Children's playgrounds often have circular-motion rides. How could a child change the angular momentum of such a ride as it is turning? (9.2)

Applying Concepts

45. Explain the concept of impulse using physical ideas rather than mathematics.

46. Is it possible for an object to obtain a larger impulse from a smaller force than it does from a larger force? Explain.

47. Foul Ball You are sitting at a baseball game when a foul ball comes in your direction. You prepare to catch it bare-handed. To catch it safely, should you move your hands toward the ball, hold them still, or move them in the same direction as the moving ball? Explain.

48. A 0.11-g bullet leaves a pistol at 323 m/s, while a similar bullet leaves a rifle at 396 m/s. Explain the difference in exit speeds of the two bullets, assuming that the forces exerted on the bullets by the expanding gases have the same magnitude.

49. An object initially at rest experiences the impulses described by the graph in **Figure 9-15.** Describe the object's motion after impulses A, B, and C.

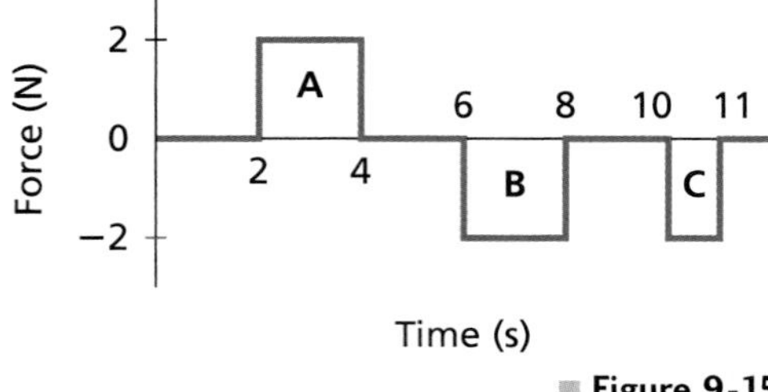

Figure 9-15

50. During a space walk, the tether connecting an astronaut to the spaceship breaks. Using a gas pistol, the astronaut manages to get back to the ship. Use the language of the impulse-momentum theorem and a diagram to explain why this method was effective.

51. Tennis Ball As a tennis ball bounces off a wall, its momentum is reversed. Explain this action in terms of the law of conservation of momentum. Define the system and draw a diagram as a part of your explanation.

52. Imagine that you command spaceship *Zeldon,* which is moving through interplanetary space at high speed. How could you slow your ship by applying the law of conservation of momentum?

53. Two trucks that appear to be identical collide on an icy road. One was originally at rest. The trucks are stuck together and move at more than half the original speed of the moving truck. What can you conclude about the contents of the two trucks?

54. Explain, in terms of impulse and momentum, why it is advisable to place the butt of a rifle against your shoulder when first learning to shoot.

55. Bullets Two bullets of equal mass are shot at equal speeds at blocks of wood on a smooth ice rink. One bullet, made of rubber, bounces off of the wood. The other bullet, made of aluminum, burrows into the wood. In which case does the block of wood move faster? Explain.

Mastering Problems

9.1 Impulse and Momentum

56. Golf Rocío strikes a 0.058-kg golf ball with a force of 272 N and gives it a velocity of 62.0 m/s. How long was Rocío's club in contact with the ball?

57. A 0.145-kg baseball is pitched at 42 m/s. The batter hits it horizontally to the pitcher at 58 m/s.

a. Find the change in momentum of the ball.

b. If the ball and bat are in contact for 4.6×10^{-4} s, what is the average force during contact?

58. Bowling A force of 186 N acts on a 7.3-kg bowling ball for 0.40 s. What is the bowling ball's change in momentum? What is its change in velocity?

59. A 5500-kg freight truck accelerates from 4.2 m/s to 7.8 m/s in 15.0 s by the application of a constant force.

a. What change in momentum occurs?

b. How large of a force is exerted?

60. In a ballistics test at the police department, Officer Rios fires a 6.0-g bullet at 350 m/s into a container that stops it in 1.8 ms. What is the average force that stops the bullet?

61. Volleyball A 0.24-kg volleyball approaches Tina with a velocity of 3.8 m/s. Tina bumps the ball, giving it a speed of 2.4 m/s but in the opposite direction. What average force did she apply if the interaction time between her hands and the ball was 0.025 s?

62. Hockey A hockey player makes a slap shot, exerting a constant force of 30.0 N on the hockey puck for 0.16 s. What is the magnitude of the impulse given to the puck?

63. Skateboarding Your brother's mass is 35.6 kg, and he has a 1.3-kg skateboard. What is the combined momentum of your brother and his skateboard if they are moving at 9.50 m/s?

64. A hockey puck has a mass of 0.115 kg and is at rest. A hockey player makes a shot, exerting a constant force of 30.0 N on the puck for 0.16 s. With what speed does it head toward the goal?

65. Before a collision, a 25-kg object was moving at +12 m/s. Find the impulse that acted on the object if, after the collision, it moved at the following velocities.

a. +8.0 m/s

b. −8.0 m/s

66. A 0.150-kg ball, moving in the positive direction at 12 m/s, is acted on by the impulse shown in the graph in **Figure 9-16.** What is the ball's speed at 4.0 s?

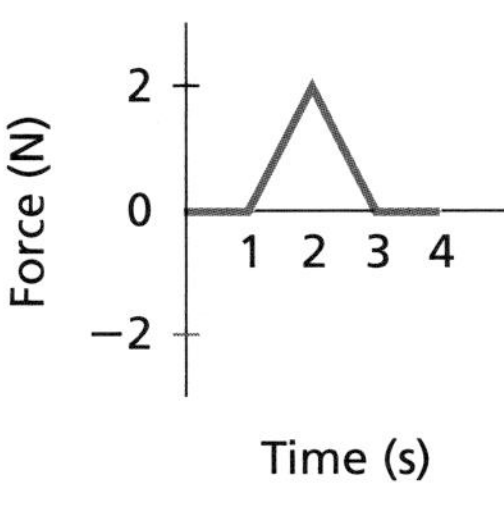

Figure 9-16

67. Baseball A 0.145-kg baseball is moving at 35 m/s when it is caught by a player.

a. Find the change in momentum of the ball.

b. If the ball is caught with the mitt held in a stationary position so that the ball stops in 0.050 s, what is the average force exerted on the ball?

c. If, instead, the mitt is moving backward so that the ball takes 0.500 s to stop, what is the average force exerted by the mitt on the ball?

68. Hockey A hockey puck has a mass of 0.115 kg and strikes the pole of the net at 37 m/s. It bounces off in the opposite direction at 25 m/s, as shown in **Figure 9-17.**

a. What is the impulse on the puck?

b. If the collision takes 5.0×10^{-4} s, what is the average force on the puck?

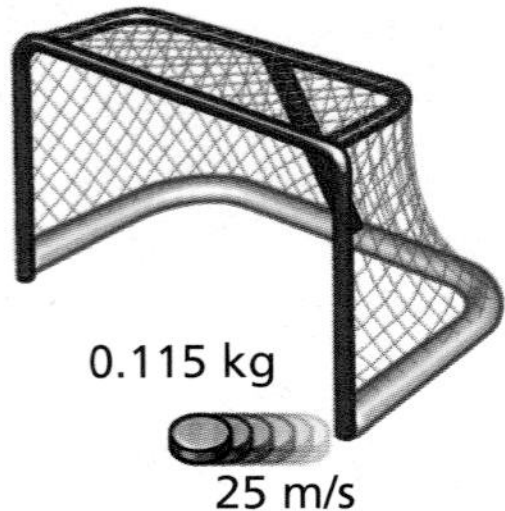

Figure 9-17

69. A nitrogen molecule with a mass of 4.7×10^{-26} kg, moving at 550 m/s, strikes the wall of a container and bounces back at the same speed.

a. What is the impulse the molecule delivers to the wall?

b. If there are 1.5×10^{23} collisions each second, what is the average force on the wall?

70. Rockets Small rockets are used to make tiny adjustments in the speeds of satellites. One such rocket has a thrust of 35 N. If it is fired to change the velocity of a 72,000-kg spacecraft by 63 cm/s, how long should it be fired?

71. An animal rescue plane flying due east at 36.0 m/s drops a bale of hay from an altitude of 60.0 m, as shown in **Figure 9-18.** If the bale of hay weighs 175 N, what is the momentum of the bale the moment before it strikes the ground? Give both magnitude and direction.

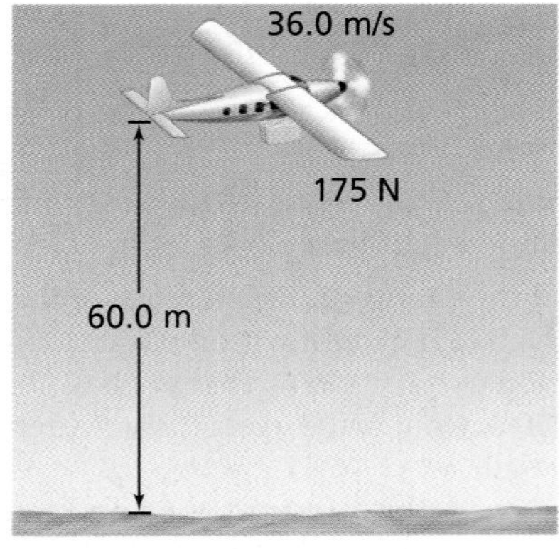

Figure 9-18

72. Accident A car moving at 10.0 m/s crashes into a barrier and stops in 0.050 s. There is a 20.0-kg child in the car. Assume that the child's velocity is changed by the same amount as that of the car, and in the same time period.

a. What is the impulse needed to stop the child?

b. What is the average force on the child?

c. What is the approximate mass of an object whose weight equals the force in part b?

d. Could you lift such a weight with your arm?

e. Why is it advisable to use a proper restraining seat rather than hold a child on your lap?

9.2 Conservation of Momentum

73. Football A 95-kg fullback, running at 8.2 m/s, collides in midair with a 128-kg defensive tackle moving in the opposite direction. Both players end up with zero speed.

a. Identify the "before" and "after" situations and draw a diagram of both.

b. What was the fullback's momentum before the collision?

c. What was the change in the fullback's momentum?

d. What was the change in the defensive tackle's momentum?

e. What was the defensive tackle's original momentum?

f. How fast was the defensive tackle moving originally?

74. Marble C, with mass 5.0 g, moves at a speed of 20.0 cm/s. It collides with a second marble, D, with mass 10.0 g, moving at 10.0 cm/s in the same direction. After the collision, marble C continues with a speed of 8.0 cm/s in the same direction.

a. Sketch the situation and identify the system. Identify the "before" and "after" situations and set up a coordinate system.

b. Calculate the marbles' momenta before the collision.

c. Calculate the momentum of marble C after the collision.

d. Calculate the momentum of marble D after the collision.

e. What is the speed of marble D after the collision?

75. Two lab carts are pushed together with a spring mechanism compressed between them. Upon release, the 5.0-kg cart repels one way with a velocity of 0.12 m/s, while the 2.0-kg cart goes in the opposite direction. What is the velocity of the 2.0-kg cart?

 For more problems, go to Additional Problems, Appendix B.

76. A 50.0-g projectile is launched with a horizontal velocity of 647 m/s from a 4.65-kg launcher moving in the same direction at 2.00 m/s. What is the launcher's velocity after the launch?

77. A 12.0-g rubber bullet travels at a velocity of 150 m/s, hits a stationary 8.5-kg concrete block resting on a frictionless surface, and ricochets in the opposite direction with a velocity of -1.0×10^2 m/s, as shown in **Figure 9-19.** How fast will the concrete block be moving?

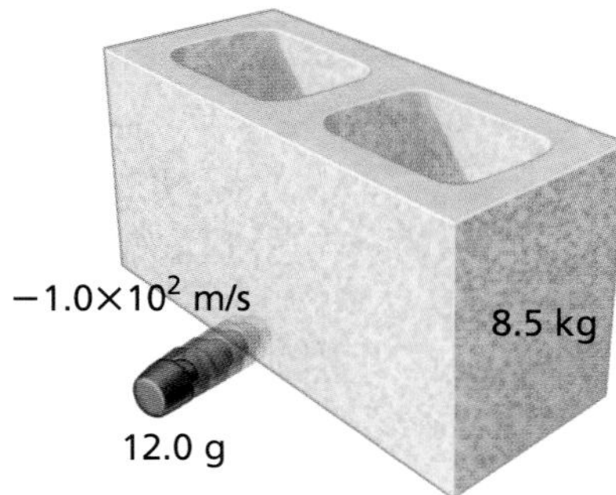

Figure 9-19

78. Skateboarding Kofi, with mass 42.00 kg, is riding a skateboard with a mass of 2.00 kg and traveling at 1.20 m/s. Kofi jumps off and the skateboard stops dead in its tracks. In what direction and with what velocity did he jump?

79. Billiards A cue ball, with mass 0.16 kg, rolling at 4.0 m/s, hits a stationary eight ball of similar mass. If the cue ball travels 45° above its original path and the eight ball travels 45° below the horizontal, as shown in **Figure 9-20,** what is the velocity of each ball after the collision?

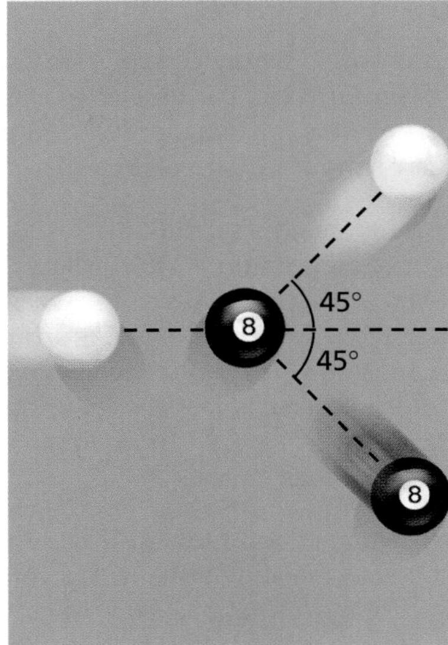

Figure 9-20

80. A 2575-kg van runs into the back of an 825-kg compact car at rest. They move off together at 8.5 m/s. Assuming that the friction with the road is negligible, calculate the initial speed of the van.

81. In-line Skating Diego and Keshia are on in-line skates and stand face-to-face, then push each other away with their hands. Diego has a mass of 90.0 kg and Keshia has a mass of 60.0 kg.

a. Sketch the event, identifying the "before" and "after" situations, and set up a coordinate axis.

b. Find the ratio of the skaters' velocities just after their hands lose contact.

c. Which skater has the greater speed?

d. Which skater pushed harder?

82. A 0.200-kg plastic ball moves with a velocity of 0.30 m/s. It collides with a second plastic ball of mass 0.100 kg, which is moving along the same line at a speed of 0.10 m/s. After the collision, both balls continue moving in the same, original direction. The speed of the 0.100-kg ball is 0.26 m/s. What is the new velocity of the 0.200-kg ball?

Mixed Review

83. A constant force of 6.00 N acts on a 3.00-kg object for 10.0 s. What are the changes in the object's momentum and velocity?

84. The velocity of a 625-kg car is changed from 10.0 m/s to 44.0 m/s in 68.0 s by an external, constant force.

a. What is the resulting change in momentum of the car?

b. What is the magnitude of the force?

85. Dragster An 845-kg dragster accelerates on a race track from rest to 100.0 km/h in 0.90 s.

a. What is the change in momentum of the dragster?

b. What is the average force exerted on the dragster?

c. What exerts that force?

86. Ice Hockey A 0.115-kg hockey puck, moving at 35.0 m/s, strikes a 0.365-kg jacket that is thrown onto the ice by a fan of a certain hockey team. The puck and jacket slide off together. Find their velocity.

87. A 50.0-kg woman, riding on a 10.0-kg cart, is moving east at 5.0 m/s. The woman jumps off the front of the cart and lands on the ground at 7.0 m/s eastward, relative to the ground.

a. Sketch the "before" and "after" situations and assign a coordinate axis to them.

b. Find the cart's velocity after the woman jumps off.

88. Gymnastics Figure 9-21 shows a gymnast performing a routine. First, she does giant swings on the high bar, holding her body straight and pivoting around her hands. Then, she lets go of the high bar and grabs her knees with her hands in the tuck position. Finally, she straightens up and lands on her feet.

a. In the second and final parts of the gymnast's routine, around what axis does she spin?

b. Rank in order, from greatest to least, her moments of inertia for the three positions.

c. Rank in order, from greatest to least, her angular velocities in the three positions.

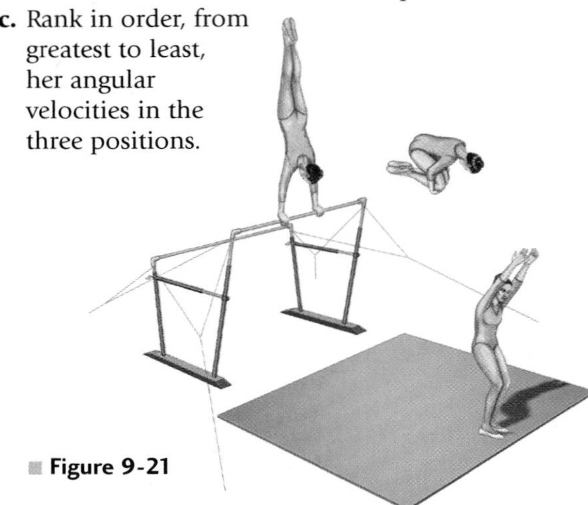

Figure 9-21

89. A 60.0-kg male dancer leaps 0.32 m high.

a. With what momentum does he reach the ground?

b. What impulse is needed to stop the dancer?

c. As the dancer lands, his knees bend, lengthening the stopping time to 0.050 s. Find the average force exerted on the dancer's body.

d. Compare the stopping force with his weight.

Thinking Critically

90. Apply Concepts A 92-kg fullback, running at 5.0 m/s, attempts to dive directly across the goal line for a touchdown. Just as he reaches the line, he is met head-on in midair by two 75-kg linebackers, both moving in the direction opposite the fullback. One is moving at 2.0 m/s and the other at 4.0 m/s. They all become entangled as one mass.

a. Sketch the event, identifying the "before" and "after" situations.

b. What is the velocity of the football players after the collision?

c. Does the fullback score a touchdown?

91. Analyze and Conclude A student, holding a bicycle wheel with its axis vertical, sits on a stool that can rotate without friction. She uses her hand to get the wheel spinning. Would you expect the student and stool to turn? If so, in which direction? Explain.

92. Analyze and Conclude Two balls during a collision are shown in **Figure 9-22,** which is drawn to scale. The balls enter from the left, collide, and then bounce away. The heavier ball, at the bottom of the diagram, has a mass of 0.600 kg, and the other has a mass of 0.400 kg. Using a vector diagram, determine whether momentum is conserved in this collision. Explain any difference in the momentum of the system before and after the collision.

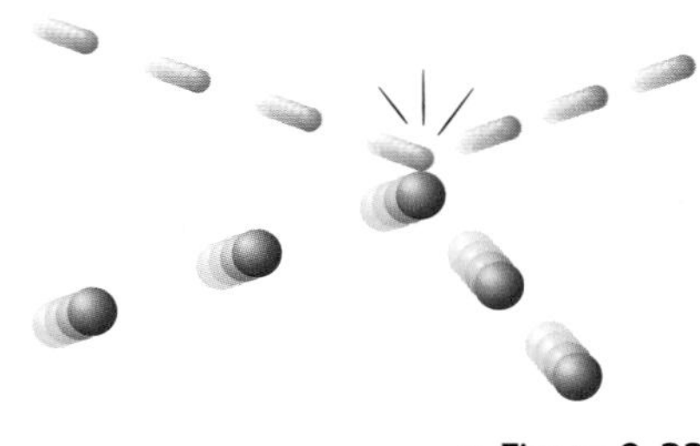

Figure 9-22

Writing in Physics

93. How can highway barriers be designed to be more effective in saving people's lives? Research this issue and describe how impulse and change in momentum can be used to analyze barrier designs.

94. While air bags save many lives, they also have caused injuries and even death. Research the arguments and responses of automobile makers to this statement. Determine whether the problems involve impulse and momentum or other issues.

Cumulative Review

95. A 0.72-kg ball is swung vertically from a 0.60-m string in uniform circular motion at a speed of 3.3 m/s. What is the tension in the cord at the top of the ball's motion? (Chapter 6)

96. You wish to launch a satellite that will remain above the same spot on Earth's surface. This means the satellite must have a period of exactly one day. Calculate the radius of the circular orbit this satellite must have. *Hint: The Moon also circles Earth and both the Moon and the satellite will obey Kepler's third law. The Moon is 3.9×10^8 m from Earth and its period is 27.33 days.* (Chapter 7)

97. A rope is wrapped around a drum that is 0.600 m in diameter. A machine pulls with a constant 40.0 N force for a total of 2.00 s. In that time, 5.00 m of rope is unwound. Find α, ω at 2.00 s, and I. (Chapter 8)

 For more problems, go to Additional Problems, Appendix B.

Multiple Choice

1. When a star that is much larger than the Sun nears the end of its lifetime, it begins to collapse, but continues to rotate. Which of the following describes the conditions of the collapsing star's moment of inertia (I), angular momentum (L), and angular velocity (ω)?

   Ⓐ I increases, L stays constant, ω decreases
   Ⓑ I decreases, L stays constant, ω increases
   Ⓒ I increases, L increases, ω increases
   Ⓓ I increases, L increases, ω stays constant

2. A 40.0-kg ice-skater glides with a speed of 2.0 m/s toward a 10.0-kg sled at rest on the ice. The ice-skater reaches the sled and holds on to it. The ice-skater and the sled then continue sliding in the same direction in which the ice-skater was originally skating. What is the speed of the ice-skater and the sled after they collide?

   Ⓐ 0.4 m/s   Ⓒ 1.6 m/s
   Ⓑ 0.8 m/s   Ⓓ 3.2 m/s

3. A bicyclist applies the brakes and slows the motion of the wheels. The angular momentum of each wheel then decreases from 7.0 kg·m^2/s to 3.5 kg·m^2/s over a period of 5.0 s. What is the angular impulse on each wheel?

   Ⓐ −0.7 kg·m^2/s
   Ⓑ −1.4 kg·m^2/s
   Ⓒ −2.1 kg·m^2/s
   Ⓓ −3.5 kg·m^2/s

4. A 45.0-kg ice-skater stands at rest on the ice. A friend tosses the skater a 5.0-kg ball. The skater and the ball then move backwards across the ice with a speed of 0.50 m/s. What was the speed of the ball at the moment just before the skater caught it?

   Ⓐ 2.5 m/s   Ⓒ 4.0 m/s
   Ⓑ 3.0 m/s   Ⓓ 5.0 m/s

5. What is the difference in momentum between a 50.0-kg runner moving at a speed of 3.00 m/s and a 3.00×10^3-kg truck moving at a speed of only 1.00 m/s?

   Ⓐ 1275 kg·m/s   Ⓒ 2850 kg·m/s
   Ⓑ 2550 kg·m/s   Ⓓ 2950 kg·m/s

6. When the large gear in the diagram rotates, it turns the small gear in the opposite direction at the same linear speed. The larger gear has twice the radius and four times the mass of the smaller gear. What is the angular momentum of the larger gear as a function of the angular momentum of the smaller gear? *Hint: The moment of inertia for a disk is $\frac{1}{2}mr^2$, where m is mass and r is the radius of the disk.*

   Ⓐ $-2L_{small}$ Ⓒ $-8L_{small}$
   Ⓑ $-4L_{small}$ Ⓓ $-16L_{small}$

7. A force of 16 N exerted against a rock with an impulse of 0.8 kg·m/s causes the rock to fly off the ground with a speed of 4.0 m/s. What is the mass of the rock?

   Ⓐ 0.2 kg
   Ⓑ 0.8 kg
   Ⓒ 1.6 kg
   Ⓓ 4.0 kg

Extended Answer

8. A 12.0-kg rock falls to the ground. What is the impulse on the rock if its velocity at the moment it strikes the ground is 20.0 m/s?

✓ Test-Taking TIP

If It Looks Too Good To Be True

Beware of answer choices in multiple-choice questions that seem ready-made and obvious. Remember that only one answer choice for each question is correct. The rest are made up by test-makers to distract you. This means that they might look very appealing. Check each answer choice carefully before making your final selection.

Chapter 10

Energy, Work, and Simple Machines

What You'll Learn

- You will recognize that work and power describe how the external world changes the energy of a system.
- You will relate force to work and explain how machines ease the load.

Why It's Important

Simple machines and the compound machines formed from them make many everyday tasks easier to perform.

Mountain Bikes A multispeed mountain bicycle with shock absorbers allows you to match the ability of your body to exert forces, to do work, and to deliver power climbing steep hills, traversing flat terrain at high speeds, and safely descending hills.

Think About This ▶

How does a multispeed mountain bicycle enable a cyclist to ride over any kind of terrain with the least effort?

physicspp.com

What factors affect energy?

Question

What factors affect the energy of falling objects and their ability to do work?

Procedure

1. Place about 2 cm of fine sand in the bottom of a pie plate or baking pan.
2. Obtain a variety of metal balls or glass marbles of different sizes.
3. Hold a meterstick vertically in one hand, with one end just touching the surface of the sand. With the other hand, drop one of the balls into the sand. Record the height from which you dropped the ball.
4. Carefully remove the ball from the sand, so as not to disturb the impact crater it made. Measure the depth of the crater and how far sand was thrown from the crater.
5. Record the mass of the ball.
6. Smooth out the sand in the pie plate and perform steps 3–5 with different sizes of balls and drop them from varying heights. Be sure to drop different sizes of balls from the same height, as well as the same ball from different heights.

Analysis

Compare your data for the different craters. Is there an overall trend to your data? Explain.

Critical Thinking As the balls are dropped into the sand, they do work on the sand. Energy can be defined as the ability of an object to do work on itself or its surroundings. Relate the trend(s) you found in this lab to the energy of the balls. How can the energy of a ball be increased?

10.1 Energy and Work

In Chapter 9, you learned about the conservation of momentum. You learned that you could examine the state of a system before and after an impulse acted on it without knowing the details about the impulse. The law of conservation of momentum was especially useful when considering collisions, during which forces sometimes changed dramatically. Recall the discussion in Chapter 9 of the two skaters who push each other away. While momentum is conserved in this situation, the skaters continue to move after pushing each other away; whereas before the collision, they were at rest. When two cars crash into each other, momentum is conserved. Unlike the skaters, however, the cars, which were moving prior to the collision, became stationary after the crash. The collision probably resulted in a lot of twisted metal and broken glass. In these types of situations, some other quantity must have been changed as a result of the force acting on each system.

► **Objectives**
- **Describe** the relationship between work and energy.
- **Calculate** work.
- **Calculate** the power used.

► **Vocabulary**

work
energy
kinetic energy
work-energy theorem
joule
power
watt

Work and Energy

Recall that change in momentum is the result of an impulse, which is the product of the average force exerted on an object and the time of the interaction. Consider a force exerted on an object while the object moves a certain distance. Because there is a net force, the object will be accelerated, $a = F/m$, and its velocity will increase. Examine Table 3-3 in Chapter 3, on page 68, which lists equations describing the relationships among position, velocity, and time for motion under constant acceleration. Consider the equation involving acceleration, velocity, and distance: $2ad = v_f^2 - v_i^2$. If you use Newton's second law to replace a with F/m and multiply both sides by $m/2$, you obtain $Fd = \frac{1}{2}mv_f^2 - \frac{1}{2}mv_i^2$.

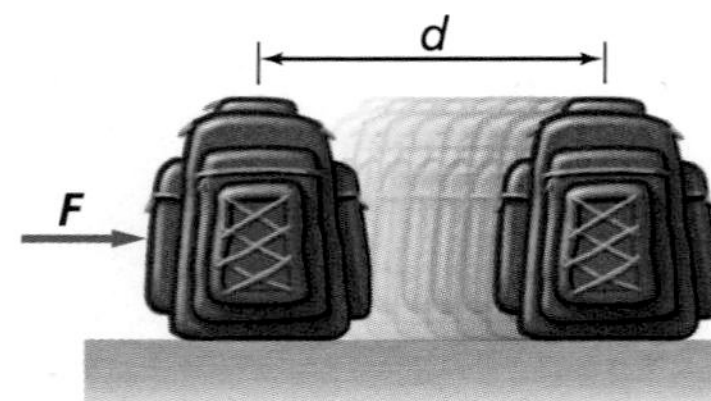

■ **Figure 10-1** Work is done when a constant force, *F*, is exerted on the backpack in the direction of motion and the backpack moves a distance, *d*.

Work The left side of the equation describes something that was done to the system by the external world (the environment). A force, F, was exerted on an object while the object moved a distance, d, as shown in **Figure 10-1**. If F is a constant force, exerted in the direction in which the object is moving, then **work,** W, is the product of the force and the object's displacement.

Work $W = Fd$

Work is equal to a constant force exerted on an object in the direction of motion, times the object's displacement.

You probably have used the word *work* in many other ways. For example, a computer might work well, learning physics can be hard work, and you might work at an after-school job. To physicists, however, work has a very precise meaning.

Recall that $Fd = \frac{1}{2}mv_f^2 - \frac{1}{2}mv_i^2$. Rewriting the equation $W = Fd$ results in $W = \frac{1}{2}mv_f^2 - \frac{1}{2}mv_i^2$. The right side of the equation involves the object's mass and its velocities after and before the force was exerted. The quantity $\frac{1}{2}mv_i^2$ describes a property of the system.

Kinetic energy What property of a system does $\frac{1}{2}mv_i^2$ describe? A massive, fast-moving vehicle can do damage to objects around it, and a baseball hit at high speed can rise high into the air. That is, an object with this property can produce a change in itself or the world around it. This property, the ability of an object to produce a change in itself or the world around it, is called **energy.** The fast-moving vehicle and the baseball possess energy that is associated with their motion. This energy resulting from motion is called **kinetic energy** and is represented by the symbol KE.

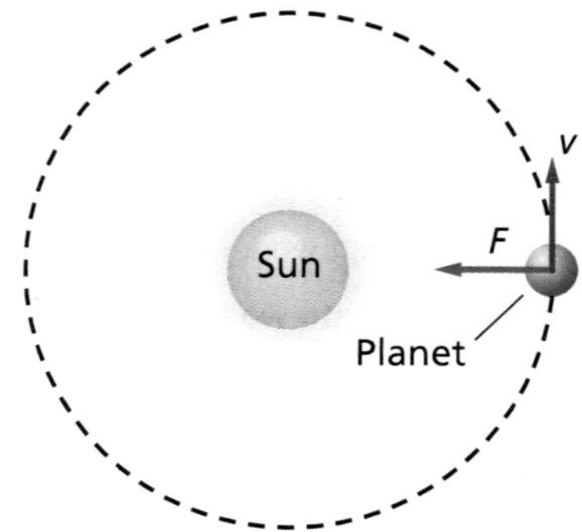

■ **Figure 10-2** If a planet is in a circular orbit, then the force is perpendicular to the direction of motion. Consequently, the gravitational force does no work on the planet.

Kinetic Energy $KE = \frac{1}{2}mv^2$

The kinetic energy of an object is equal to $\frac{1}{2}$ times the mass of the object multiplied by the speed of the object squared.

Substituting KE into the equation $W = \frac{1}{2}mv_f^2 - \frac{1}{2}mv_i^2$ results in $W = KE_f - KE_i$. The right side is the difference, or change, in kinetic energy. The **work-energy theorem** states that when work is done on an object, the result is a change in kinetic energy. The work-energy theorem can be represented by the following equation.

Work-Energy Theorem $W = \Delta KE$

Work is equal to the change in kinetic energy.

The relationship between work done and the change in energy that results was established by nineteenth-century physicist James Prescott Joule. To honor his work, a unit of energy is called a **joule** (J). For example, if a 2-kg object moves at 1 m/s, it has a kinetic energy of 1 $kg{\cdot}m^2/s^2$, or 1 J.

Recall that a system is the object of interest and the external world is everything else. For example, one system might be a box in a warehouse and the external world might consist of yourself, Earth's mass, and anything else external to the box. Through the process of doing work, energy can move between the external world and the system.

Notice that the direction of energy transfer can go both ways. If the external world does work on a system, then W is positive and the energy of the system increases. If, however, a system does work on the external world, then W is negative and the energy of the system decreases. In summary, work is the transfer of energy by mechanical means.

Calculating Work

The first equation used to calculate work is $W = Fd$. This equation, however, holds only for constant forces exerted in the direction of motion. What happens if the force is exerted perpendicular to the direction of motion? An everyday example of this is the motion of a planet around the Sun, as shown in **Figure 10-2**. If the orbit is circular, then the force is always perpendicular to the direction of motion. Recall from Chapter 6 that a perpendicular force does not change the speed of an object, only its direction. Consequently, the speed of the planet doesn't change. Therefore, its kinetic energy also is constant. Using the equation $W = \Delta KE$, you can see that when KE is constant, $\Delta KE = 0$ and thus, $W = 0$. This means that if F and d are at right angles, then $W = 0$.

Because the work done on an object equals the change in energy, work also is measured in joules. One joule of work is done when a force of 1 N acts on an object over a displacement of 1 m. An apple weighs about 1 N. Thus, when you lift an apple a distance of 1 m, you do 1 J of work on it.

Constant force exerted at an angle You've learned that a force exerted in the direction of motion does an amount of work given by $W = Fd$. A force exerted perpendicular to the motion does no work. What work does a force exerted at an angle do? For example, what work does the person pushing the car in **Figure 10-3a** do? You know that any force can be replaced by its components. If the coordinate system shown in **Figure 10-3b** is used, the 125-N force, $\mathbf{F}$, exerted in the direction of the person's arm, has two components. The magnitude of the horizontal component, F_x, is related to the magnitude of the force, F, by a cosine function: $\cos 25.0° = F_x/F$. By solving for F_x, you obtain $F_x = F \cos 25.0° = (125\text{ N})(\cos 25.0°) = 113\text{ N}$. Using the same method, the vertical component $F_y = -F \sin 25.0° = -(125\text{ N})(\sin 25.0°) = -52.8\text{ N}$, where the negative sign shows that the force is downward. Because the displacement is in the x direction, only the x-component does work. The y-component does no work.

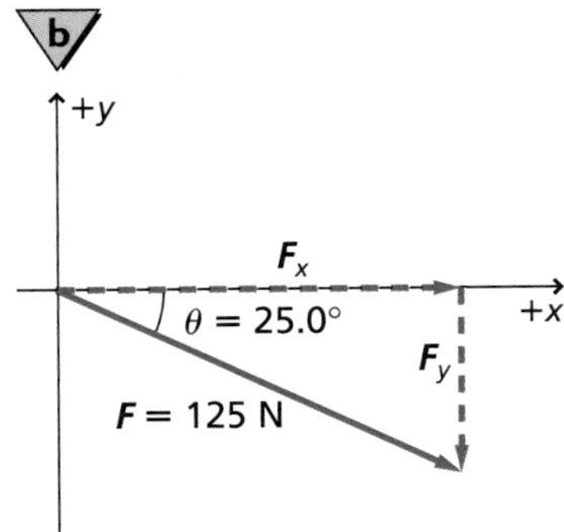

Figure 10-3 If a force is applied to a car at an angle, the net force doing the work is the component that acts in the direction of the displacement.

Concepts In Motion

Interactive Figure To see an animation on work by a force acting at an angle, visit physicspp.com.

The work you do when you exert a force on an object, at an angle to the direction of motion, is equal to the component of the force in the direction of the displacement, multiplied by the distance moved. The magnitude of the component force acting in the direction of displacement is found by multiplying the magnitude of force, F, by the cosine of the angle between F and the direction of the displacement: $F_x = F \cos \theta$. Thus, the work done is represented by the following equation.

Work (Angle Between Force and Displacement) $W = Fd \cos \theta$

Work is equal to the product of force and displacement, times the cosine of the angle between the force and the direction of the displacement.

Other agents exert forces on the pushed car as well. Which of these agents do work? Earth's gravity acts downward, the ground exerts a normal force upward, and friction exerts a horizontal force opposite the direction of motion. The upward and downward forces are perpendicular to the direction of motion and do no work. For these forces, $\theta = 90°$, which makes $\cos \theta = 0$, and thus, $W = 0$.

The work done by friction acts in the direction opposite that of motion—at an angle of 180°. Because $\cos 180° = -1$, the work done by friction is negative. Negative work done by a force exerted by something in the external world reduces the kinetic energy of the system. If the person in Figure 10-3a were to stop pushing, the car would quickly stop moving—its energy of motion would be reduced. Positive work done by a force increases the energy, while negative work decreases it. Use the problem-solving strategies below when you solve problems related to work.

PROBLEM-SOLVING Strategies

Work

When solving work-related problems, use the following strategies.

1. Sketch the system and show the force that is doing the work.
2. Draw the force and displacement vectors of the system.
3. Find the angle, θ, between each force and displacement.
4. Calculate the work done by each force using $W = Fd \cos \theta$.
5. Calculate the net work done. Check the sign of the work using the direction of energy transfer. If the energy of the system has increased, the work done by that force is positive. If the energy has decreased, then the work done by that force is negative.

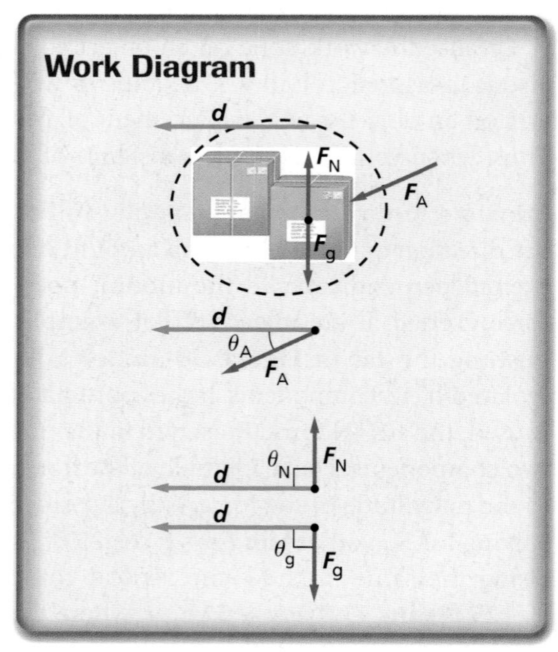

▶ EXAMPLE Problem 1

Work and Energy A 105-g hockey puck is sliding across the ice. A player exerts a constant 4.50-N force over a distance of 0.150 m. How much work does the player do on the puck? What is the change in the puck's energy?

1 Analyze and Sketch the Problem

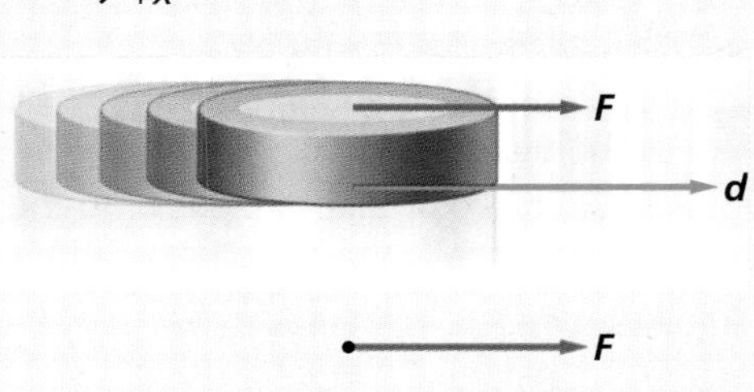

- Sketch the situation showing initial conditions.
- Establish a coordinate system with $+x$ to the right.
- Draw a vector diagram.

Known:	Unknown:
$m = 105$ g	$W = ?$
$F = 4.50$ N	$\Delta KE = ?$
$d = 0.150$ m	

2 Solve for the Unknown

Use the equation for work when a constant force is exerted in the same direction as the object's displacement.

$$W = Fd$$
$$= (4.50\ \text{N})(0.150\ \text{m}) \quad \text{Substitute } F = 4.50\ \text{N}, d = 0.150\ \text{m}$$
$$= 0.675\ \text{N·m}$$
$$= 0.675\ \text{J} \quad 1\ \text{J} = 1\ \text{N·m}$$

Math Handbook
Operations with Significant Digits
pages 835–836

Use the work-energy theorem to determine the change in energy of the system.

$$W = \Delta KE$$
$$\Delta KE = 0.675\ \text{J} \quad \text{Substitute } W = 0.675\ \text{J}$$

3 Evaluate the Answer

- **Are the units correct?** Work is measured in joules.
- **Does the sign make sense?** The player (external world) does work on the puck (the system). So the sign of work should be positive.

▶ PRACTICE Problems

• Additional Problems, Appendix B
• Solutions to Selected Problems, Appendix C

1. Refer to Example Problem 1 to solve the following problem.
  a. If the hockey player exerted twice as much force, 9.00 N, on the puck, how would the puck's change in kinetic energy be affected?
  b. If the player exerted a 9.00-N force, but the stick was in contact with the puck for only half the distance, 0.075 m, what would be the change in kinetic energy?

2. Together, two students exert a force of 825 N in pushing a car a distance of 35 m.
  a. How much work do the students do on the car?
  b. If the force was doubled, how much work would they do pushing the car the same distance?

3. A rock climber wears a 7.5-kg backpack while scaling a cliff. After 30.0 min, the climber is 8.2 m above the starting point.
  a. How much work does the climber do on the backpack?
  b. If the climber weighs 645 N, how much work does she do lifting herself and the backpack?
  c. What is the change in the climber's energy?

▶ EXAMPLE Problem 2

Force and Displacement at an Angle A sailor pulls a boat a distance of 30.0 m along a dock using a rope that makes a 25.0° angle with the horizontal. How much work does the sailor do on the boat if he exerts a force of 255 N on the rope?

1 Analyze and Sketch the Problem

- Establish coordinate axes.
- Sketch the situation showing the boat with initial conditions.
- Draw a vector diagram showing the force and its component in the direction of the displacement.

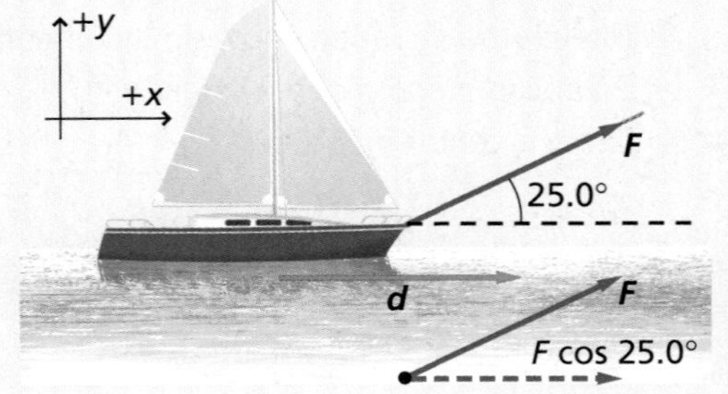

Known:	Unknown:
$F = 255$ N	$W = ?$
$d = 30.0$ m	
$\theta = 25.0°$	

2 Solve for the Unknown

Use the equation for work done when there is an angle between the force and displacement.

$$W = Fd \cos \theta$$
$$= (255\ \text{N})(30.0\ \text{m})(\cos 25.0°) \quad \text{Substitute } F = 255\ \text{N},\ d = 30.0\ \text{m},\ \theta = 25.0°$$
$$= 6.93 \times 10^3\ \text{J}$$

3 Evaluate the Answer

- **Are the units correct?** Work is measured in joules.
- **Does the sign make sense?** The sailor does work on the boat, which agrees with a positive sign for work.

Math Handbook
Trigonometric Ratios page 855

▶ PRACTICE Problems

• Additional Problems, Appendix B
• Solutions to Selected Problems, Appendix C

4. If the sailor in Example Problem 2 pulled with the same force, and along the same distance, but at an angle of 50.0°, how much work would he do?

5. Two people lift a heavy box a distance of 15 m. They use ropes, each of which makes an angle of 15° with the vertical. Each person exerts a force of 225 N. How much work do they do?

6. An airplane passenger carries a 215-N suitcase up the stairs, a displacement of 4.20 m vertically, and 4.60 m horizontally.

a. How much work does the passenger do?

b. The same passenger carries the same suitcase back down the same set of stairs. How much work does the passenger do now?

7. A rope is used to pull a metal box a distance of 15.0 m across the floor. The rope is held at an angle of 46.0° with the floor, and a force of 628 N is applied to the rope. How much work does the force on the rope do?

8. A bicycle rider pushes a bicycle that has a mass of 13 kg up a steep hill. The incline is 25° and the road is 275 m long, as shown in **Figure 10-4.** The rider pushes the bike parallel to the road with a force of 25 N.

a. How much work does the rider do on the bike?

b. How much work is done by the force of gravity on the bike?

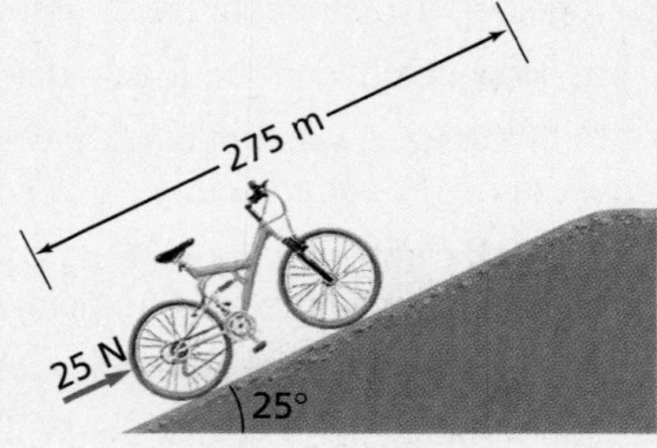

Figure 10-4 (Not to scale)

Finding work done when forces change A graph of force versus displacement lets you determine the work done by a force. This graphical method can be used to solve problems in which the force is changing. **Figure 10-5a** shows the work done by a constant force of 20.0 N that is exerted to lift an object a distance of 1.50 m. The work done by this constant force is represented by $W = Fd = (20.0\text{ N})(1.50\text{ m}) = 30.0\text{ J}$. The shaded area under the graph is equal to (20.0 N)(1.50 m), or 30.0 J. The area under a force-displacement graph is equal to the work done by that force, even if the force changes. **Figure 10-5b** shows the force exerted by a spring, which varies linearly from 0.0 to 20.0 N as it is compressed 1.50 m. The work done by the force that compressed the spring is the area under the graph, which is the area of a triangle, $\frac{1}{2}$(base)(altitude), or $W = \frac{1}{2}(20.0\text{ N})(1.50\text{ m}) = 15.0\text{ J}$.

Work done by many forces Newton's second law of motion relates the net force on an object to its acceleration. In the same way, the work-energy theorem relates the net work done on a system to its energy change. If several forces are exerted on a system, calculate the work done by each force, and then add the results.

■ **Figure 10-5** Work can be obtained graphically by finding the area under a force-displacement graph.

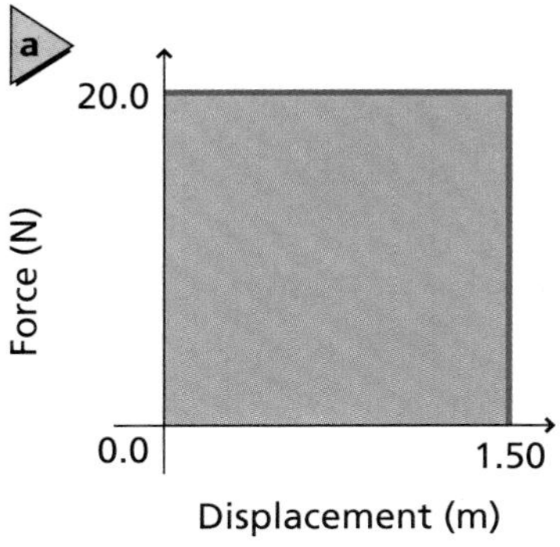

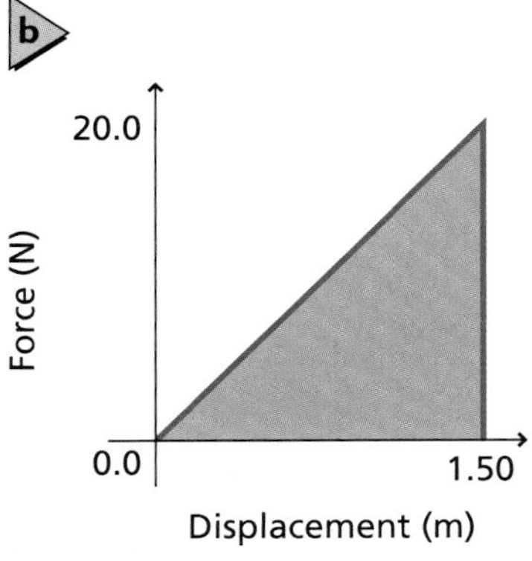

Power

Until now, none of the discussions of work has mentioned the time it takes to move an object. The work done by a person lifting a box of books is the same whether the box is lifted onto a shelf in 2 s or each book is lifted separately so that it takes 20 min to put them all on the shelf. Although the work done is the same, the rate at which it is done is different. **Power** is the work done, divided by the time taken to do the work. In other words, power is the rate at which the external force changes the energy of the system. It is represented by the following equation.

Power $P = \frac{W}{t}$

Power is equal to the work done, divided by the time taken to do the work.

Consider the three students in **Figure 10-6.** The girl hurrying up the stairs is more powerful than the boy who is walking up the stairs. Even though the same work is accomplished by both, the girl accomplishes it in less time and thus develops more power. In the case of the two students walking up the stairs, both accomplish work in the same amount of time.

Power is measured in watts (W). One **watt** is 1 J of energy transferred in 1 s. A watt is a relatively small unit of power. For example, a glass of water weighs about 2 N. If you lift it 0.5 m to your mouth, you do 1 J of work. If you lift the glass in 1 s, you are doing work at the rate of 1 W. Because a watt is such a small unit, power often is measured in kilowatts (kW). One kilowatt is equal to 1000 W.

■ **Figure 10-6** These students are doing work at different rates while climbing the stairs.

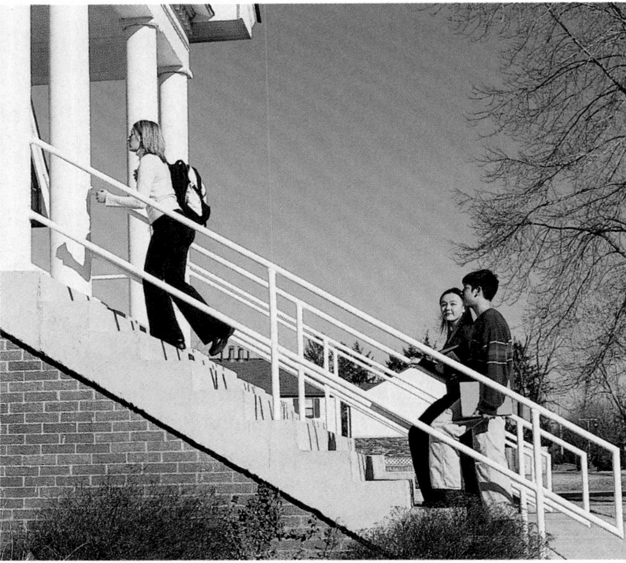

▶ EXAMPLE Problem 3

Power An electric motor lifts an elevator 9.00 m in 15.0 s by exerting an upward force of 1.20×10^4 N. What power does the motor produce in kW?

1 Analyze and Sketch the Problem

- Sketch the situation showing the elevator with initial conditions.
- Establish a coordinate system with up as positive.
- Draw a vector diagram for the force and displacement.

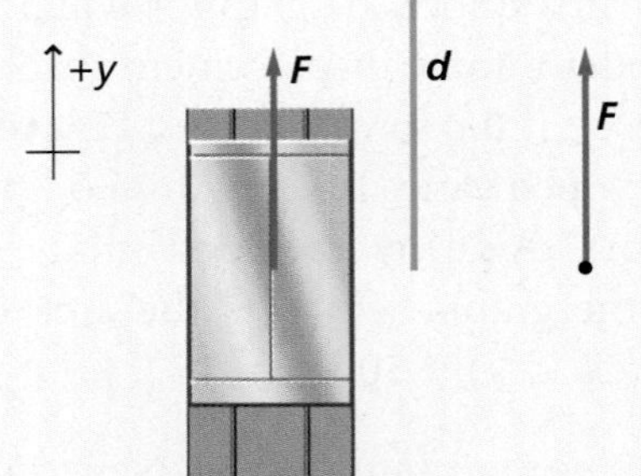

Known:	**Unknown:**
$d = 9.00$ m	$P = ?$
$t = 15.0$ s	
$F = 1.20\times10^4$ N	

2 Solve for the Unknown

Solve for power.

$$P = \frac{W}{t}$$

$$= \frac{Fd}{t} \quad \text{Substitute } W = Fd$$

$$= \frac{(1.20\times10^4\ \text{N})(9.00\ \text{m})}{(15.0\ \text{s})} \quad \text{Substitute } F = 1.20\times10^4\ \text{N},\ d = 9.00\ \text{m},\ t = 15.0\ \text{s}$$

$$= 7.20\ \text{kW}$$

Math Handbook
Operations with Scientific Notation
pages 842–843

3 Evaluate the Answer

- **Are the units correct?** Power is measured in J/s.
- **Does the sign make sense?** The positive sign agrees with the upward direction of the force.

▶ PRACTICE Problems

• Additional Problems, Appendix B
• Solutions to Selected Problems, Appendix C

9. A box that weighs 575 N is lifted a distance of 20.0 m straight up by a cable attached to a motor. The job is done in 10.0 s. What power is developed by the motor in W and kW?

10. You push a wheelbarrow a distance of 60.0 m at a constant speed for 25.0 s, by exerting a 145-N force horizontally.

a. What power do you develop?

b. If you move the wheelbarrow twice as fast, how much power is developed?

11. What power does a pump develop to lift 35 L of water per minute from a depth of 110 m? (1 L of water has a mass of 1.00 kg.)

12. An electric motor develops 65 kW of power as it lifts a loaded elevator 17.5 m in 35 s. How much force does the motor exert?

13. A winch designed to be mounted on a truck, as shown in **Figure 10-7,** is advertised as being able to exert a 6.8×10^3-N force and to develop a power of 0.30 kW. How long would it take the truck and the winch to pull an object 15 m?

14. Your car has stalled and you need to push it. You notice as the car gets going that you need less and less force to keep it going. Suppose that for the first 15 m, your force decreased at a constant rate from 210.0 N to 40.0 N. How much work did you do on the car? Draw a force-displacement graph to represent the work done during this period.

Figure 10-7

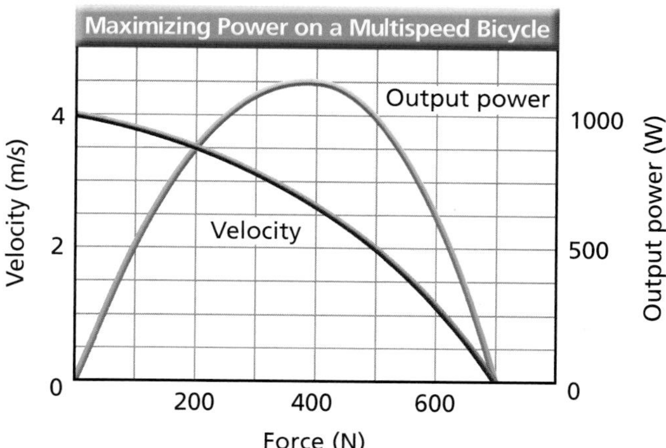

Figure 10-8 When riding a multispeed bicycle, if the muscles in your body exert a force of 400 N and the speed is 2.6 m/s, the power output is over 1000 W.

You may have noticed in Example Problem 3 that when the force and displacement are in the same direction, $P = Fd/t$. However, because the ratio d/t is the speed, power also can be calculated using $P = Fv$.

When you are riding a multispeed bicycle, how do you choose the correct gear? You want to get your body to deliver the largest amount of power. By considering the equation $P = Fv$ you can see that either zero force or zero speed results in no power delivered. The muscles cannot exert extremely large forces, nor can they move very fast. Thus, some combination of moderate force and moderate speed will produce the largest amount of power. **Figure 10-8** shows that in this particular situation, the maximum power output is over 1000 W when the force is about 400 N and speed is about 2.6 m/s. All engines—not just humans—have these limitations. Simple machines often are designed to match the force and speed that the engine can deliver to the needs of the job. You will learn more about simple machines in the next section.

APPLYING PHYSICS

Tour de France A bicyclist in the Tour de France rides at about 8.94 m/s for more than 6 h a day. The power output of the racer is about 1 kW. One-fourth of that power goes into moving the bike against the resistance of the air, gears, and tires. Three-fourths of the power is used to cool the racer's body.

10.1 Section Review

15. **Work** Murimi pushes a 20-kg mass 10 m across a floor with a horizontal force of 80 N. Calculate the amount of work done by Murimi.

16. **Work** A mover loads a 185-kg refrigerator into a moving van by pushing it up a 10.0-m, friction-free ramp at an angle of inclination of 11.0°. How much work is done by the mover?

17. **Work and Power** Does the work required to lift a book to a high shelf depend on how fast you raise it? Does the power required to lift the book depend on how fast you raise it? Explain.

18. **Power** An elevator lifts a total mass of 1.1×10^3 kg a distance of 40.0 m in 12.5 s. How much power does the elevator generate?

19. **Work** A 0.180-kg ball falls 2.5 m. How much work does the force of gravity do on the ball?

20. **Mass** A forklift raises a box 1.2 m and does 7.0 kJ of work on it. What is the mass of the box?

21. **Work** You and a friend each carry identical boxes from the first floor of a building to a room located on the second floor, farther down the hall. You choose to carry the box first up the stairs, and then down the hall to the room. Your friend carries it down the hall on the first floor, then up a different stairwell to the second floor. Who does more work?

22. **Work and Kinetic Energy** If the work done on an object doubles its kinetic energy, does it double its velocity? If not, by what ratio does it change the velocity?

23. **Critical Thinking** Explain how to find the change in energy of a system if three agents exert forces on the system at once.

Physics Online physicspp.com/self_check_quiz

10.2 Machines

▸ **Objectives**

- **Demonstrate** a knowledge of the usefulness of simple machines.
- **Differentiate** between ideal and real machines in terms of efficiency.
- **Analyze** compound machines in terms of combinations of simple machines.
- **Calculate** efficiencies for simple and compound machines.

▸ **Vocabulary**

machine
effort force
resistance force
mechanical advantage
ideal mechanical advantage
efficiency
compound machine

Everyone uses machines every day. Some are simple tools, such as bottle openers and screwdrivers, while others are complex, such as bicycles and automobiles. Machines, whether powered by engines or people, make tasks easier. A **machine** eases the load by changing either the magnitude or the direction of a force to match the force to the capability of the machine or the person.

Benefits of Machines

Consider the bottle opener in **Figure 10-9.** When you use the opener, you lift the handle, thereby doing work on the opener. The opener lifts the cap, doing work on it. The work that you do is called the input work, W_i. The work that the machine does is called the output work, W_o.

Recall that work is the transfer of energy by mechanical means. You put work into a machine, such as the bottle opener. That is, you transfer energy to the opener. The opener, in turn, does work on the cap, thereby transferring energy to it. The opener is not a source of energy, and therefore, the cap cannot receive more energy than the amount of energy that you put into the opener. Thus, the output work can never be greater than the input work. The machine simply aids in the transfer of energy from you to the bottle cap.

Mechanical advantage The force exerted by a person on a machine is called the **effort force,** F_e. The force exerted by the machine is called the **resistance force,** F_r. As shown in Figure 10-9a, F_e is the upward force exerted by the person using the bottle opener and F_r is the upward force exerted by the bottle opener. The ratio of resistance force to effort force, F_r/F_e, is called the **mechanical advantage,** MA, of the machine.

Mechanical Advantage $$MA = \frac{F_r}{F_e}$$

The mechanical advantage of a machine is equal to the resistance force divided by the effort force.

■ **Figure 10-9** A bottle opener is an example of a simple machine. It makes opening a bottle easier, but it does not lessen the work required to do so.

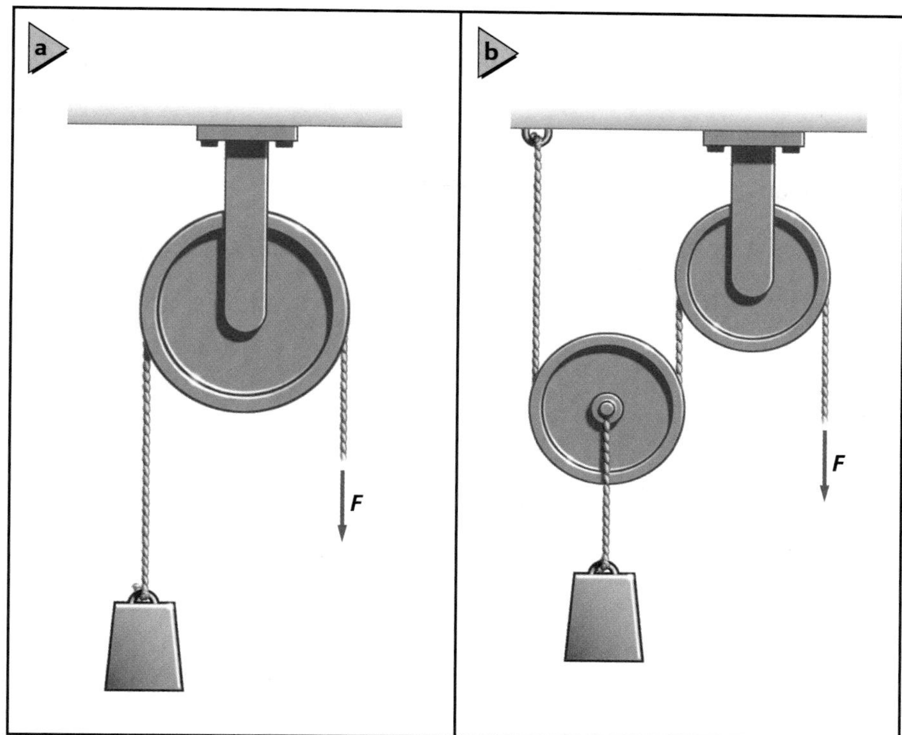

Figure 10-10 A fixed pulley has a mechanical advantage equal to 1 **(a).** A pulley system with a movable pulley has a mechanical advantage greater than 1 **(b).**

In a fixed pulley, such as the one shown in **Figure 10-10a,** the forces, F_e and F_r, are equal, and consequently *MA* is 1. What is the advantage of this machine? The fixed pulley is useful, not because the effort force is lessened, but because the direction of the effort force is changed. Many machines, such as the bottle opener shown in Figure 10-9 and the pulley system shown in **Figure 10-10b,** have a mechanical advantage greater than 1. When the mechanical advantage is greater than 1, the machine increases the force applied by a person.

You can write the mechanical advantage of a machine in another way using the definition of work. The input work is the product of the effort force that a person exerts, F_e, and the distance his or her hand moved, d_e. In the same way, the output work is the product of the resistance force, F_r, and the displacement of the load, d_r. A machine can increase force, but it cannot increase energy. An ideal machine transfers all the energy, so the output work equals the input work: $W_o = W_i$ or $F_r d_r = F_e d_e$.

This equation can be rewritten $F_r/F_e = d_e/d_r$. Recall that mechanical advantage is given by $MA = F_r/F_e$. Therefore, for an ideal machine, **ideal mechanical advantage,** *IMA*, is equal to the displacement of the effort force, divided by the displacement of the load. The ideal mechanical advantage can be represented by the following equation.

Ideal Mechanical Advantage $IMA = \frac{d_e}{d_r}$

The ideal mechanical advantage of an ideal machine is equal to the displacement of the effort force, divided by the displacement of the load.

Note that you measure the distances moved to calculate the ideal mechanical advantage, but you measure the forces exerted to find the actual mechanical advantage.

Efficiency In a real machine, not all of the input work is available as output work. Energy removed from the system means that there is less output work from the machine. Consequently, the machine is less efficient at accomplishing the task. The **efficiency** of a machine, *e*, is defined as the ratio of output work to input work.

Personal Tutor For an online tutorial on percentage, visit physicspp.com.

Efficiency $e = \frac{W_o}{W_i} \times 100$

The efficiency of a machine (in %) is equal to the output work, divided by the input work, multiplied by 100.

An ideal machine has equal output and input work, $W_o/W_i = 1$, and its efficiency is 100 percent. All real machines have efficiencies of less than 100 percent.

Efficiency can be expressed in terms of the mechanical advantage and ideal mechanical advantage. Efficiency, $e = W_o/W_i$, can be rewritten as follows:

$$\frac{W_o}{W_i} = \frac{F_r d_r}{F_e d_e}$$

Because $MA = F_r/F_e$ and $IMA = d_e/d_r$, the following expression can be written for efficiency.

Efficiency $e = \frac{MA}{IMA} \times 100$

The efficiency of a machine (in %) is equal to its mechanical advantage, divided by the ideal mechanical advantage, multiplied by 100.

A machine's design determines its ideal mechanical advantage. An efficient machine has an *MA* almost equal to its *IMA*. A less-efficient machine has a small *MA* relative to its *IMA*. To obtain the same resistance force, a greater force must be exerted in a machine of lower efficiency than in a machine of higher efficiency.

CHALLENGE PROBLEM

An electric pump pulls water at a rate of 0.25 m^3/s from a well that is 25 m deep. The water leaves the pump at a speed of 8.5 m/s.

1. What power is needed to lift the water to the surface?
2. What power is needed to increase the pump's kinetic energy?
3. If the pump's efficiency is 80 percent, how much power must be delivered to the pump?

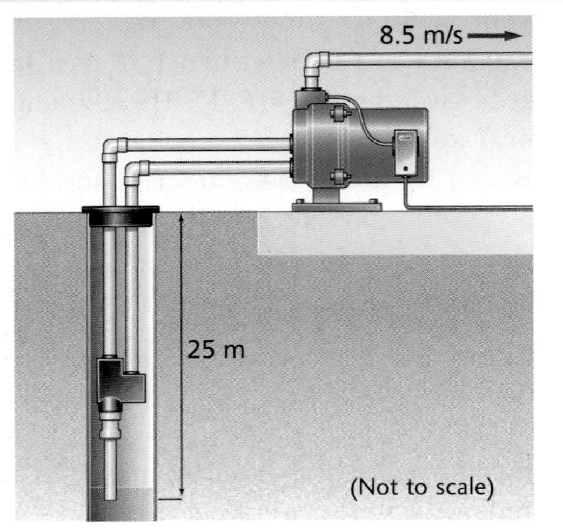

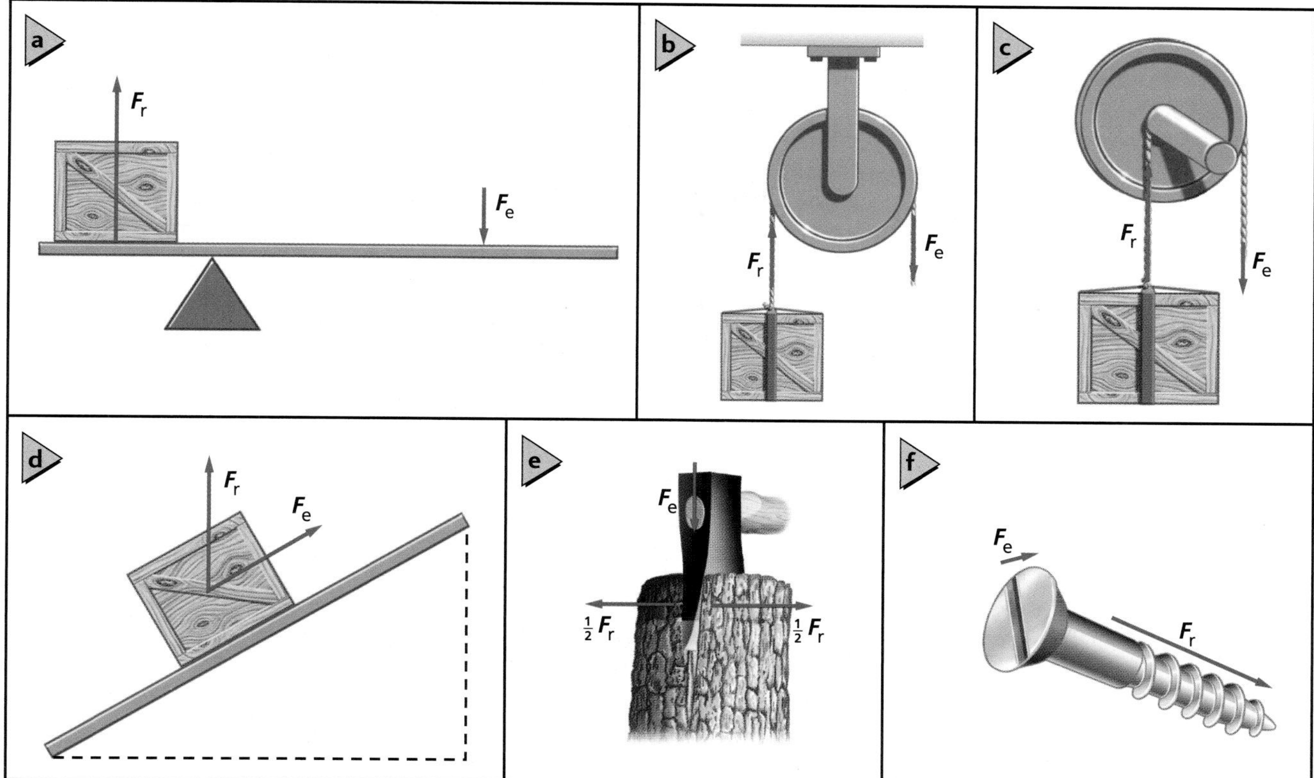

■ **Figure 10-11** Simple machines include the lever **(a),** pulley **(b),** wheel and axle **(c),** inclined plane **(d),** wedge **(e),** and screw **(f).**

Compound Machines

Most machines, no matter how complex, are combinations of one or more of the six simple machines: the lever, pulley, wheel and axle, inclined plane, wedge, and screw. These machines are shown in **Figure 10-11.**

The *IMA* of all the machines shown in Figure 10-11 is the ratio of distances moved. For machines, such as the lever and the wheel and axle, this ratio can be replaced by the ratio of the distance between the place where the force is applied and the pivot point. A common version of the wheel and axle is a steering wheel, such as the one shown in **Figure 10-12.** The *IMA* is the ratio of the radii of the wheel and axle.

A machine consisting of two or more simple machines linked in such a way that the resistance force of one machine becomes the effort force of the second is called a **compound machine.**

■ **Figure 10-12** The *IMA* for the steering wheel is r_e/r_r.

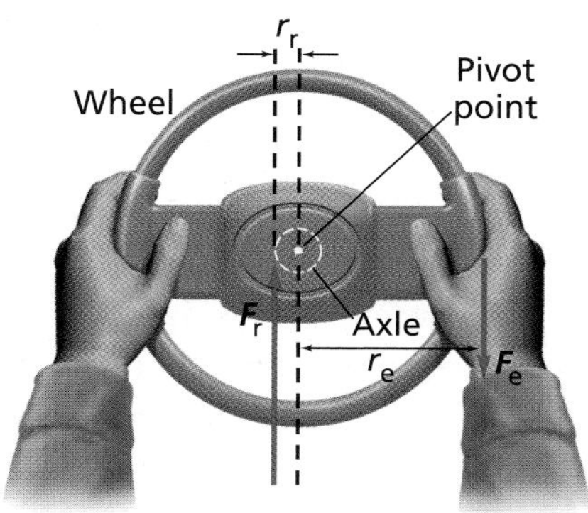

Figure 10-13 A series of simple machines combine to transmit the force that the rider exerts on the pedal to the road.

In a bicycle, the pedal and front gear act like a wheel and axle. The effort force is the force that the rider exerts on the pedal, $F_{rider\ on\ pedal}$. The resistance is the force that the front gear exerts on the chain, $F_{gear\ on\ chain}$, as shown in **Figure 10-13.** The chain exerts an effort force on the rear gear, $F_{chain\ on\ gear}$, equal to the force exerted on the chain. This gear and the rear wheel act like another wheel and axle. The resistance force is the force that the wheel exerts on the road, $F_{wheel\ on\ road}$. According to Newton's third law, the ground exerts an equal forward force on the wheel, which accelerates the bicycle forward.

The *MA* of a compound machine is the product of the *MA*s of the simple machines from which it is made. For example, in the case of the bicycle illustrated in Figure 10-13, the following is true.

$$MA = MA_{machine\ 1} \times MA_{machine\ 2}$$

$$MA = \left(\frac{F_{gear\ on\ chain}}{F_{rider\ on\ pedal}}\right)\left(\frac{F_{wheel\ on\ road}}{F_{chain\ on\ gear}}\right) = \frac{F_{wheel\ on\ road}}{F_{rider\ on\ pedal}}$$

The *IMA* of each wheel-and-axle machine is the ratio of the distances moved.

For the pedal gear, $IMA = \frac{\text{pedal radius}}{\text{front gear radius}}$

For the rear wheel, $IMA = \frac{\text{rear gear radius}}{\text{wheel radius}}$

For the bicycle, then,

$$IMA = \left(\frac{\text{pedal radius}}{\text{front gear radius}}\right)\left(\frac{\text{rear gear radius}}{\text{wheel radius}}\right)$$

$$= \left(\frac{\text{rear gear radius}}{\text{front gear radius}}\right)\left(\frac{\text{pedal radius}}{\text{wheel radius}}\right)$$

Because both gears use the same chain and have teeth of the same size, you can count the number of teeth to find the *IMA*, as follows.

$$IMA = \left(\frac{\text{teeth on rear gear}}{\text{teeth on front gear}}\right)\left(\frac{\text{pedal arm length}}{\text{wheel radius}}\right)$$

Shifting gears on a bicycle is a way of adjusting the ratio of gear radii to obtain the desired *IMA*. You know that if the pedal of a bicycle is at the top or bottom of its circle, no matter how much downward force you exert, the pedal will not turn. The force of your foot is most effective when the force is exerted perpendicular to the arm of the pedal; that is, when the torque is largest. Whenever a force on a pedal is specified, assume that it is applied perpendicular to the arm.

MINI LAB

Wheel and Axle

The gear mechanism on your bicycle multiplies the distance that you travel. What does it do to the force?

1. Mount a wheel and axle system on a sturdy support rod.

2. Wrap a 1-m-long piece of string clockwise around the axle.

3. Wrap another piece of 1-m-long string counterclockwise around the large diameter wheel.

4. Hang a 500-g mass from the end of the string on the larger wheel. ***CAUTION: Avoid dropping the mass.***

5. Pull the string from the axle down so that the mass is lifted by about 10 cm.

Analyze and Conclude

6. What did you notice about the force on the string in your hand?

7. What did you notice about the distance that your hand needed to move to lift the mass? Explain the results in terms of the work done on both strings.

▶ EXAMPLE Problem 4

Mechanical Advantage You examine the rear wheel on your bicycle. It has a radius of 35.6 cm and has a gear with a radius of 4.00 cm. When the chain is pulled with a force of 155 N, the wheel rim moves 14.0 cm. The efficiency of this part of the bicycle is 95.0 percent.

a. What is the *IMA* of the wheel and gear?

b. What is the *MA* of the wheel and gear?

c. What is the resistance force?

d. How far was the chain pulled to move the rim 14.0 cm?

1 Analyze and Sketch the Problem

- Sketch the wheel and axle.
- Sketch the force vectors.

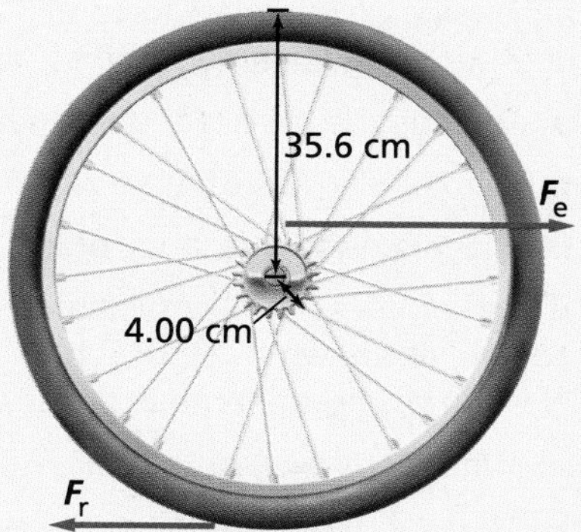

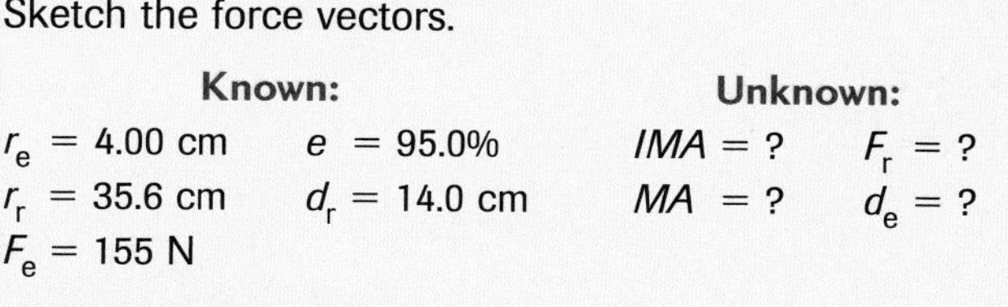

Known:		Unknown:	
r_e = 4.00 cm	e = 95.0%	IMA = ?	F_r = ?
r_r = 35.6 cm	d_r = 14.0 cm	MA = ?	d_e = ?
F_e = 155 N			

2 Solve for the Unknown

a. Solve for *IMA*.

$$IMA = \frac{r_e}{r_r}$$

For a wheel-and-axle machine, *IMA* is equal to the ratio of radii.

$$= \frac{4.00\text{ cm}}{35.6\text{ cm}}$$

Substitute r_e = 4.00 cm, r_r = 35.6 cm

$$= 0.112$$

b. Solve for *MA*.

$$e = \frac{MA}{IMA} \times 100$$

$$MA = \left(\frac{e}{100}\right) \times IMA$$

$$= \left(\frac{95.0}{100}\right) \times 0.112$$

Substitute e = 95.0%, IMA = 0.112

$$= 0.106$$

c. Solve for force.

$$MA = \frac{F_r}{F_e}$$

$$F_r = (MA)(F_e)$$

$$= (0.106)(155\text{ N})$$

Substitute MA = 0.106, F_e = 155 N

$$= 16.4\text{ N}$$

d. Solve for distance.

Math Handbook
Isolating a Variable
page 845

$$IMA = \frac{d_e}{d_r}$$

$$d_e = (IMA)(d_r)$$

$$= (0.112)(14.0\text{ cm})$$

Substitute IMA = 0.112, d_r = 14.0 cm

$$= 1.57\text{ cm}$$

3 Evaluate the Answer

- **Are the units correct?** Force is measured in newtons and distance in centimeters.
- **Is the magnitude realistic?** *IMA* is low for a bicycle because a greater F_e is traded for a greater d_r. *MA* is always smaller than *IMA*. Because *MA* is low, F_r also will be low. The small distance the axle moves results in a large distance covered by the wheel. Thus, d_e should be very small.

PRACTICE Problems

- Additional Problems, Appendix B
- Solutions to Selected Problems, Appendix C

24. If the gear radius in the bicycle in Example Problem 4 is doubled, while the force exerted on the chain and the distance the wheel rim moves remain the same, what quantities change, and by how much?

25. A sledgehammer is used to drive a wedge into a log to split it. When the wedge is driven 0.20 m into the log, the log is separated a distance of 5.0 cm. A force of 1.7×10^4 N is needed to split the log, and the sledgehammer exerts a force of 1.1×10^4 N.

a. What is the *IMA* of the wedge?

b. What is the *MA* of the wedge?

c. Calculate the efficiency of the wedge as a machine.

26. A worker uses a pulley system to raise a 24.0-kg carton 16.5 m, as shown in **Figure 10-14.** A force of 129 N is exerted, and the rope is pulled 33.0 m.

a. What is the *MA* of the pulley system?

b. What is the efficiency of the system?

27. You exert a force of 225 N on a lever to raise a 1.25×10^3-N rock a distance of 13 cm. If the efficiency of the lever is 88.7 percent, how far did you move your end of the lever?

28. A winch has a crank with a 45-cm radius. A rope is wrapped around a drum with a 7.5-cm radius. One revolution of the crank turns the drum one revolution.

a. What is the ideal mechanical advantage of this machine?

b. If, due to friction, the machine is only 75 percent efficient, how much force would have to be exerted on the handle of the crank to exert 750 N of force on the rope?

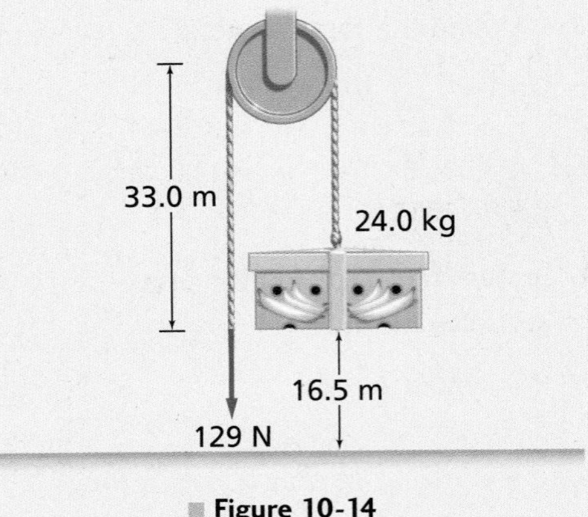

Figure 10-14

Multi-gear bicycle On a multi-gear bicycle, the rider can change the *MA* of the machine by choosing the size of one or both gears. When accelerating or climbing a hill, the rider increases the ideal mechanical advantage to increase the force that the wheel exerts on the road. To increase the *IMA*, the rider needs to make the rear gear radius large compared to the front gear radius (refer to the *IMA* equation on page 270). For the same force exerted by the rider, a larger force is exerted by the wheel on the road. However, the rider must rotate the pedals through more turns for each revolution of the wheel.

On the other hand, less force is needed to ride the bicycle at high speed on a level road. The rider needs to choose a gear that has a small rear gear and a large front gear that will result in a smaller *IMA*. Thus, for the same force exerted by the rider, a smaller force is exerted by the wheel on the road. However, in return, the rider does not have to move the pedals as far for each revolution of the wheel.

An automobile transmission works in the same way. To accelerate a car from rest, large forces are needed and the transmission increases the *IMA*. At high speeds, however, the transmission reduces the *IMA* because smaller forces are needed. Even though the speedometer shows a high speed, the tachometer indicates the engine's low angular speed.

The Human Walking Machine

Movement of the human body is explained by the same principles of force and work that describe all motion. Simple machines, in the form of levers, give humans the ability to walk and run. The lever systems of the human body are complex. However each system has the following four basic parts.

1. a rigid bar (bone)
2. a source of force (muscle contraction)
3. a fulcrum or pivot (movable joints between bones)
4. a resistance (the weight of the body or an object being lifted or moved)

Figure 10-15 shows the parts of the lever system in a human leg. Lever systems of the body are not very efficient, and mechanical advantages are low. This is why walking and jogging require energy (burn calories) and help people lose weight.

When a person walks, the hip acts as a fulcrum and moves through the arc of a circle, centered on the foot. The center of mass of the body moves as a resistance around the fulcrum in the same arc. The length of the radius of the circle is the length of the lever formed by the bones of the leg. Athletes in walking races increase their velocity by swinging their hips upward to increase this radius. A tall person's body has lever systems with less mechanical advantage than a short person's does. Although tall people usually can walk faster than short people can, a tall person must apply a greater force to move the longer lever formed by the leg bones. How would a tall person do in a walking race? What are the factors that affect a tall person's performance? Walking races are usually 20 or 50 km long. Because of the inefficiency of their lever systems and the length of a walking race, very tall people rarely have the stamina to win.

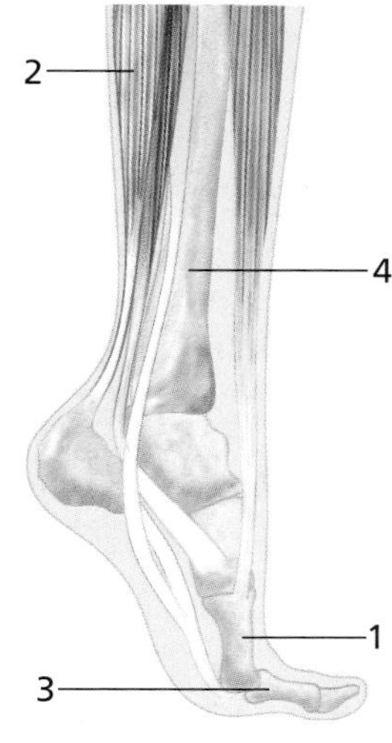

Figure 10-15 The human walking machine.

10.2 Section Review

29. Simple Machines Classify the tools below as a lever, a wheel and axle, an inclined plane, a wedge, or a pulley.

a. screwdriver
b. pliers
c. chisel
d. nail puller

30. *IMA* A worker is testing a multiple pulley system to estimate the heaviest object that he could lift. The largest downward force he could exert is equal to his weight, 875 N. When the worker moves the rope 1.5 m, the object moves 0.25 m. What is the heaviest object that he could lift?

31. Compound Machines A winch has a crank on a 45-cm arm that turns a drum with a 7.5-cm radius through a set of gears. It takes three revolutions of the crank to rotate the drum through one revolution. What is the *IMA* of this compound machine?

32. Efficiency Suppose you increase the efficiency of a simple machine. Do the *MA* and *IMA* increase, decrease, or remain the same?

33. Critical Thinking The mechanical advantage of a multi-gear bicycle is changed by moving the chain to a suitable rear gear.

a. To start out, you must accelerate the bicycle, so you want to have the bicycle exert the greatest possible force. Should you choose a small or large gear?

b. As you reach your traveling speed, you want to rotate the pedals as few times as possible. Should you choose a small or large gear?

c. Many bicycles also let you choose the size of the front gear. If you want even more force to accelerate while climbing a hill, would you move to a larger or smaller front gear?

PHYSICS LAB •

Stair Climbing and Power

Can you estimate the power you develop as you climb a flight of stairs? Climbing stairs requires energy. As the weight of the body moves through a distance, work is done. Power is a measure of the rate at which work is done. In this activity you will try to maximize the power you develop by applying a vertical force up a flight of stairs over a period of time.

QUESTION

What can you do to increase the power you develop as you climb a flight of stairs?

Objectives

- **Predict** the factors that affect power.
- **Calculate** the power developed.
- **Define** power operationally.
- **Interpret** force, distance, work, time and power data.
- **Make and use graphs** of work versus time, power versus force, and power versus time.

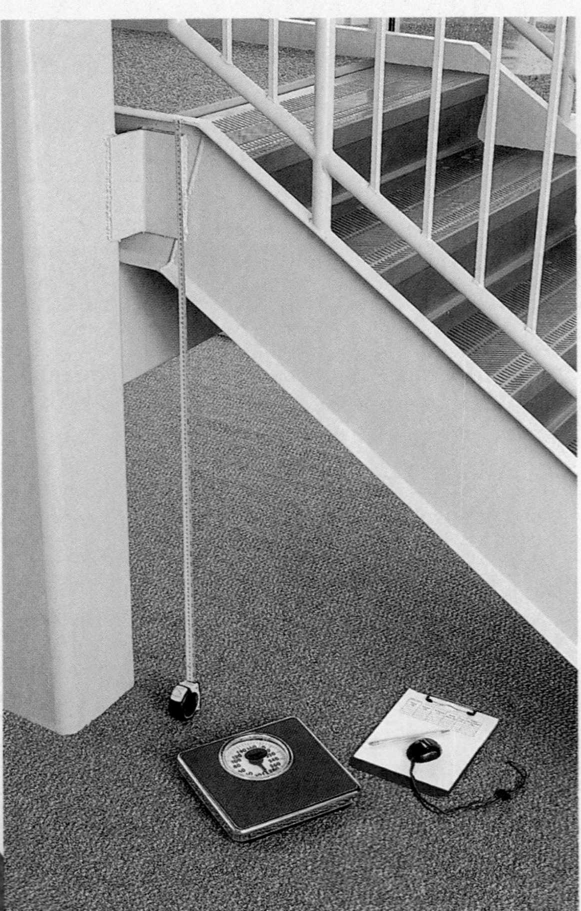

Safety Precautions

- **Avoid wearing loose clothing.**

Materials

meterstick (or tape measure)
stopwatch
bathroom scale

Procedure

1. Measure and record the mass of each person in your group using a bathroom scale. If the scale does not have kilogram units, convert the weight in pounds to kilograms. Recall that 2.2 lbs = 1 kg.
2. Measure the distance from the floor to the top of the flight of stairs you will climb. Record it in the data table.
3. Have each person in your group climb the flight of stairs in a manner that he or she thinks will maximize the power developed.
4. Use your stopwatch to measure the time it takes each person to perform this task. Record your data in the data table.

Data Table

Mass (kg)	Weight (N)	Distance (m)	Work Done (J)	Time (s)	Power Generated (W)

Analyze

1. **Calculate** Find each person's weight in newtons and record it in the data table.
2. Calculate the work done by each person.
3. Calculate the power developed by each person in your group as he or she climbs the flight of stairs.
4. **Make and Use Graphs** Use the data you calculated to draw a graph of work versus time and draw the best-fit line.
5. Draw a graph of power versus work and draw the best-fit line.
6. Draw a graph of power versus time and draw the best-fit line.

Conclude and Apply

1. Did each person in your group have the same power rating? Why or why not?
2. Which graph(s) showed a definite relationship between the two variables?
3. Explain why this relationship exists.
4. Write an operational definition of power.

Going Further

1. What three things can be done to increase the power you develop while climbing the flight of stairs?
2. Why were the fastest climbers not necessarily the ones who developed the most power?
3. Why were the members of your group with more mass not necessarily the ones who developed the most power?
4. Compare and contrast your data with those of other groups in your class.

Real-World Physics

1. Research a household appliance that has a power rating equal to or less than the power you developed by climbing the stairs.
2. Suppose an electric power company in your area charges $0.06/kWh. If you charged the same amount for the power you develop climbing stairs, how much money would you earn by climbing stairs for 1 h?
3. If you were designing a stair climbing machine for the local health club, what information would you need to collect? You decide that you will design a stair climbing machine with the ability to calculate the power developed. What information would you have the machine collect in order to let the climber know how much power he or she developed?

Physics Online

To find out more about energy, work, and simple machines, visit the Web site: physicspp.com

How it Works Bicycle Gear Shifters

In a multispeed bicycle with two or three front gears and from five to eight rear gears, front and rear derailleurs (shifters) are employed to position the chain. Changing the combination of front and rear gears varies the *IMA* of the system. A larger *IMA* reduces effort in climbing hills. A lower *IMA* allows for greater speed on level ground, but more effort is required.

$$IMA = \frac{\text{number of teeth on rear gear}}{\text{number of teeth on front gear}}$$

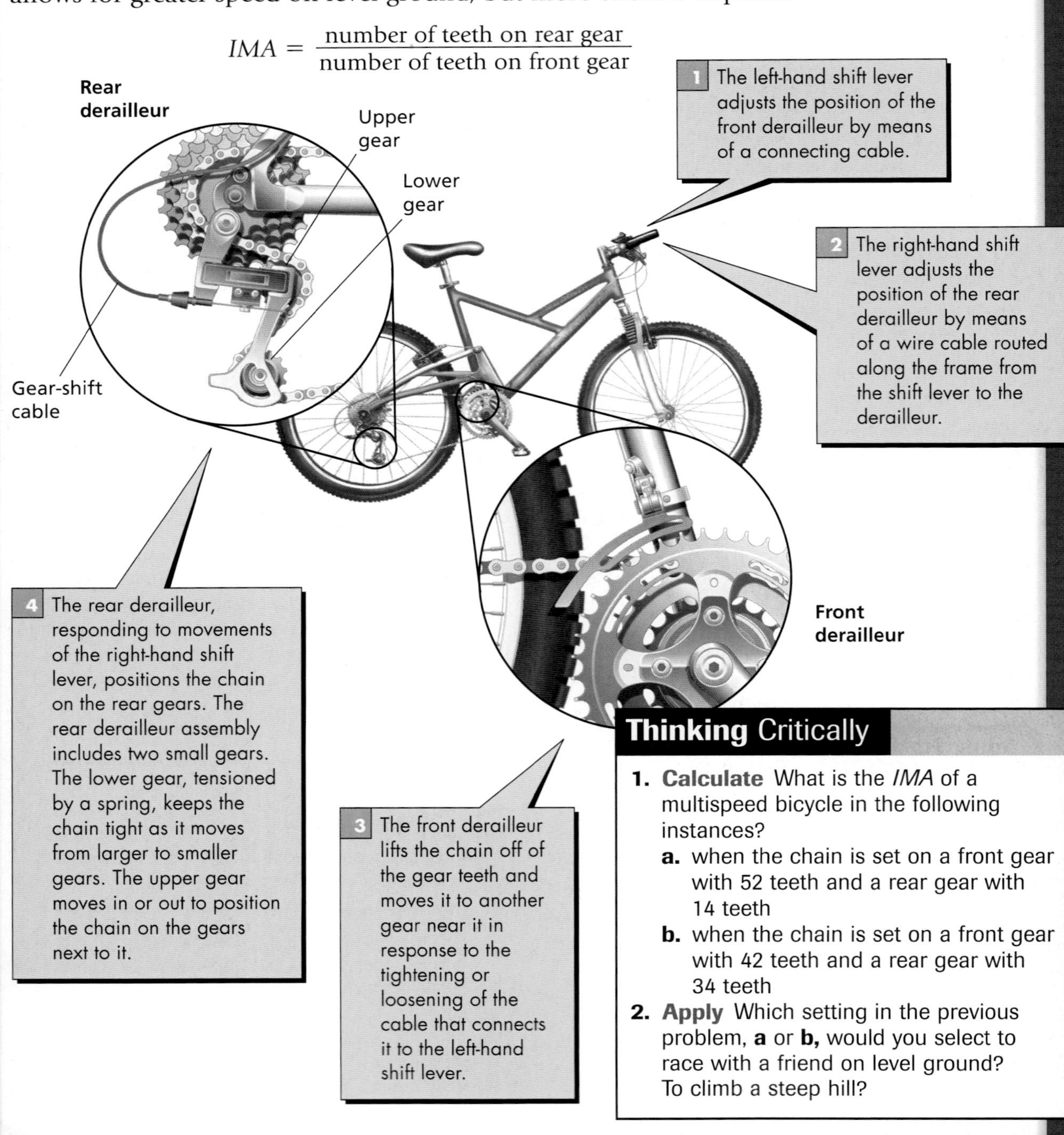

Rear derailleur
Upper gear
Lower gear
Gear-shift cable
Front derailleur
1 The left-hand shift lever adjusts the position of the front derailleur by means of a connecting cable.
2 The right-hand shift lever adjusts the position of the rear derailleur by means of a wire cable routed along the frame from the shift lever to the derailleur.
3 The front derailleur lifts the chain off of the gear teeth and moves it to another gear near it in response to the tightening or loosening of the cable that connects it to the left-hand shift lever.
4 The rear derailleur, responding to movements of the right-hand shift lever, positions the chain on the rear gears. The rear derailleur assembly includes two small gears. The lower gear, tensioned by a spring, keeps the chain tight as it moves from larger to smaller gears. The upper gear moves in or out to position the chain on the gears next to it.

Thinking Critically

1. **Calculate** What is the *IMA* of a multispeed bicycle in the following instances?
   a. when the chain is set on a front gear with 52 teeth and a rear gear with 14 teeth
   b. when the chain is set on a front gear with 42 teeth and a rear gear with 34 teeth
2. **Apply** Which setting in the previous problem, **a** or **b,** would you select to race with a friend on level ground? To climb a steep hill?

Chapter 10

Study Guide

10.1 Energy and Work

Vocabulary

- work (p. 258)
- energy (p. 258)
- kinetic energy (p. 258)
- work-energy theorem (p. 258)
- joule (p. 259)
- power (p. 263)
- watt (p. 263)

Key Concepts

- Work is the transfer of energy by mechanical means.

$$W = Fd$$

- A moving object has kinetic energy.

$$KE = \frac{1}{2}mv^2$$

- The work done on a system is equal to the change in energy of the system.

$$W = \Delta KE$$

- Work is the product of the force exerted on an object and the distance the object moves in the direction of the force.

$$W = Fd \cos \theta$$

- The work done can be determined by calculating the area under a force-displacement graph.
- Power is the rate of doing work, that is the rate at which energy is transferred.

$$P = \frac{W}{t}$$

10.2 Machines

Vocabulary

- machine (p. 266)
- effort force (p. 266)
- resistance force (p. 266)
- mechanical advantage (p. 266)
- ideal mechanical advantage (p. 267)
- efficiency (p. 268)
- compound machine (p. 269)

Key Concepts

- Machines, whether powered by engines or humans, do not change the amount of work done, but they do make the task easier.
- A machine eases the load, either by changing the magnitude or the direction of the force exerted to do work.
- The mechanical advantage, *MA*, is the ratio of resistance force to effort force.

$$MA = \frac{F_r}{F_e}$$

- The ideal mechanical advantage, *IMA*, is the ratio of the distances moved.

$$IMA = \frac{d_e}{d_r}$$

- The efficiency of a machine is the ratio of output work to input work.

$$e = \frac{W_o}{W_i} \times 100$$

- In all real machines, *MA* is less than *IMA*.
- The efficiency of a machine can by found from the real and ideal mechanical advantages.

$$e = \frac{MA}{IMA} \times 100$$

Chapter 10 Assessment

Concept Mapping

34. Create a concept map using the following terms: *force, displacement, direction of motion, work, change in kinetic energy.*

Mastering Concepts

35. In what units is work measured? (10.1)

36. Suppose a satellite revolves around Earth in a circular orbit. Does Earth's gravity do any work on the satellite? (10.1)

37. An object slides at constant speed on a frictionless surface. What forces act on the object? What work is done by each force? (10.1)

38. Define *work* and *power.* (10.1)

39. What is a watt equivalent to in terms of kilograms, meters, and seconds? (10.1)

40. Is it possible to get more work out of a machine than you put into it? (10.2)

41. Explain how the pedals of a bicycle are a simple machine. (10.2)

Applying Concepts

42. Which requires more work, carrying a 420-N backpack up a 200-m-high hill or carrying a 210-N backpack up a 400-m-high hill? Why?

43. Lifting You slowly lift a box of books from the floor and put it on a table. Earth's gravity exerts a force, magnitude mg, downward, and you exert a force, magnitude mg, upward. The two forces have equal magnitudes and opposite directions. It appears that no work is done, but you know that you did work. Explain what work was done.

44. You have an after-school job carrying cartons of new copy paper up a flight of stairs, and then carrying recycled paper back down the stairs. The mass of the paper does not change. Your physics teacher says that you do not work all day, so you should not be paid. In what sense is the physics teacher correct? What arrangement of payments might you make to ensure that you are properly compensated?

45. You carry the cartons of copy paper down the stairs, and then along a 15-m-long hallway. Are you working now? Explain.

46. Climbing Stairs Two people of the same mass climb the same flight of stairs. The first person climbs the stairs in 25 s; the second person does so in 35 s.

a. Which person does more work? Explain your answer.

b. Which person produces more power? Explain your answer.

47. Show that power delivered can be written as $P = Fv \cos \theta$.

48. How can you increase the ideal mechanical advantage of a machine?

49. Wedge How can you increase the mechanical advantage of a wedge without changing its ideal mechanical advantage?

50. Orbits Explain why a planet orbiting the Sun does not violate the work-energy theorem.

51. Claw Hammer A claw hammer is used to pull a nail from a piece of wood, as shown in **Figure 10-16.** Where should you place your hand on the handle and where should the nail be located in the claw to make the effort force as small as possible?

Figure 10-16

Mastering Problems

10.1 Energy and Work

52. The third floor of a house is 8 m above street level. How much work is needed to move a 150-kg refrigerator to the third floor?

53. Haloke does 176 J of work lifting himself 0.300 m. What is Haloke's mass?

54. Football After scoring a touchdown, an 84.0-kg wide receiver celebrates by leaping 1.20 m off the ground. How much work was done by the wide receiver in the celebration?

55. Tug-of-War During a tug-of-war, team A does 2.20×10^5 J of work in pulling team B 8.00 m. What force was team A exerting?

56. To keep a car traveling at a constant velocity, a 551-N force is needed to balance frictional forces. How much work is done against friction by the car as it travels from Columbus to Cincinnati, a distance of 161 km?

 For more problems, go to Additional Problems, Appendix B.

57. Cycling A cyclist exerts a force of 15.0 N as he rides a bike 251 m in 30.0 s. How much power does the cyclist develop?

58. A student librarian lifts a 2.2-kg book from the floor to a height of 1.25 m. He carries the book 8.0 m to the stacks and places the book on a shelf that is 0.35 m above the floor. How much work does he do on the book?

59. A force of 300.0 N is used to push a 145-kg mass 30.0 m horizontally in 3.00 s.
a. Calculate the work done on the mass.
b. Calculate the power developed.

60. Wagon A wagon is pulled by a force of 38.0 N exerted on the handle at an angle of 42.0° with the horizontal. If the wagon is pulled in a circle of radius 25.0 m, how much work is done?

61. Lawn Mower Shani is pushing a lawn mower with a force of 88.0 N along a handle that makes an angle of 41.0° with the horizontal. How much work is done by Shani in moving the lawn mower 1.2 km to mow the yard?

62. A 17.0-kg crate is to be pulled a distance of 20.0 m, requiring 1210 J of work to be done. If the job is done by attaching a rope and pulling with a force of 75.0 N, at what angle is the rope held?

63. Lawn Tractor A 120-kg lawn tractor, shown in **Figure 10-17,** goes up a 21° incline that is 12.0 m long in 2.5 s. Calculate the power that is developed by the tractor.

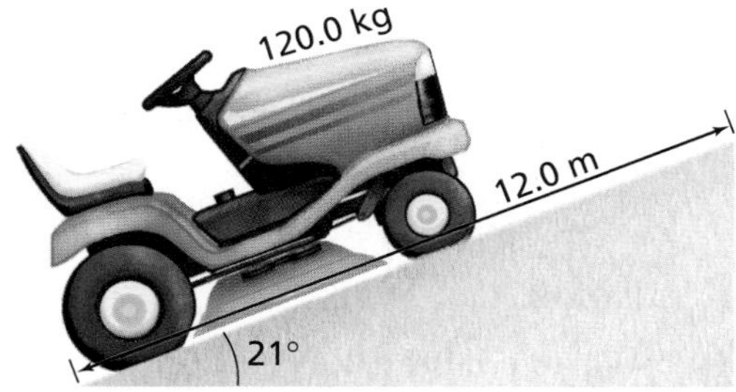

Figure 10-17

64. You slide a crate up a ramp at an angle of 30.0° by exerting a 225-N force parallel to the ramp. The crate moves at a constant speed. The coefficient of friction is 0.28. How much work did you do on the crate as it was raised a vertical distance of 1.15 m?

65. Piano A 4.2×10^{3}-N piano is to be slid up a 3.5-m frictionless plank at a constant speed. The plank makes an angle of 30.0° with the horizontal. Calculate the work done by the person sliding the piano up the plank.

66. Sled Diego pulls a 4.5-kg sled across level snow with a force of 225 N on a rope that is 35.0° above the horizontal, as shown in **Figure 10-18.** If the sled moves a distance of 65.3 m, how much work does Diego do?

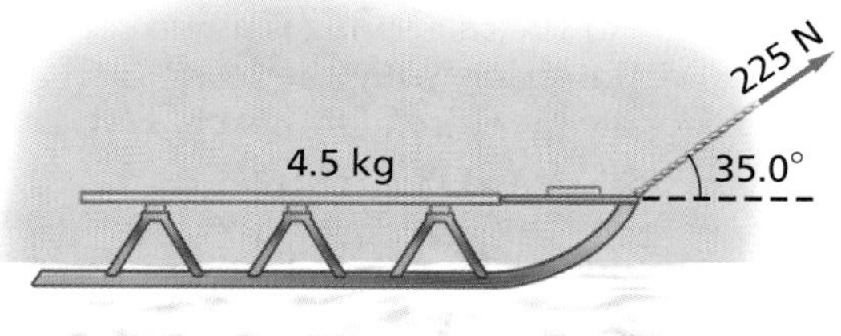

Figure 10-18

67. Escalator Sau-Lan has a mass of 52 kg. She rides up the escalator at Ocean Park in Hong Kong. This is the world's longest escalator, with a length of 227 m and an average inclination of 31°. How much work does the escalator do on Sau-Lan?

68. Lawn Roller A lawn roller is pushed across a lawn by a force of 115 N along the direction of the handle, which is 22.5° above the horizontal. If 64.6 W of power is developed for 90.0 s, what distance is the roller pushed?

69. John pushes a crate across the floor of a factory with a horizontal force. The roughness of the floor changes, and John must exert a force of 20 N for 5 m, then 35 N for 12 m, and then 10 N for 8 m.
a. Draw a graph of force as a function of distance.
b. Find the work John does pushing the crate.

70. Maricruz slides a 60.0-kg crate up an inclined ramp that is 2.0-m long and attached to a platform 1.0 m above floor level, as shown in **Figure 10-19.** A 400.0-N force, parallel to the ramp, is needed to slide the crate up the ramp at a constant speed.
a. How much work does Maricruz do in sliding the crate up the ramp?
b. How much work would be done if Maricruz simply lifted the crate straight up from the floor to the platform?

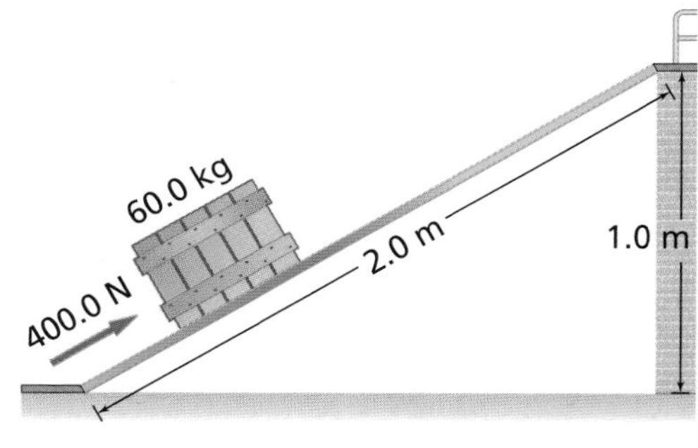

Figure 10-19

71. **Boat Engine** An engine moves a boat through the water at a constant speed of 15 m/s. The engine must exert a force of 6.0 kN to balance the force that the water exerts against the hull. What power does the engine develop?

72. In **Figure 10-20,** the magnitude of the force necessary to stretch a spring is plotted against the distance the spring is stretched.
   a. Calculate the slope of the graph, k, and show that $F = kd$, where $k = 25$ N/m.
   b. Find the amount of work done in stretching the spring from 0.00 m to 0.20 m by calculating the area under the graph from 0.00 m to 0.20 m.
   c. Show that the answer to part **b** can be calculated using the formula $W = \frac{1}{2}kd^2$, where W is the work, $k = 25$ N/m (the slope of the graph), and d is the distance the spring is stretched (0.20 m).

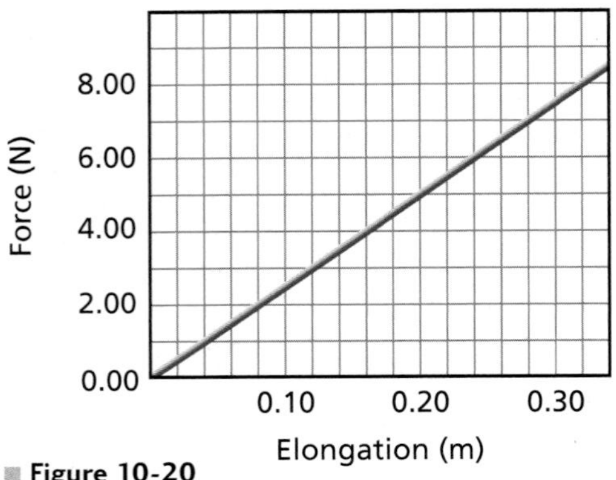

Figure 10-20

73. Use the graph in Figure 10-20 to find the work needed to stretch the spring from 0.12 m to 0.28 m.

74. A worker pushes a crate weighing 93 N up an inclined plane. The worker pushes the crate horizontally, parallel to the ground, as illustrated in **Figure 10-21.**
   a. The worker exerts a force of 85 N. How much work does he do?
   b. How much work is done by gravity? (Be careful with the signs you use.)
   c. The coefficient of friction is $\mu = 0.20$. How much work is done by friction? (Be careful with the signs you use.)

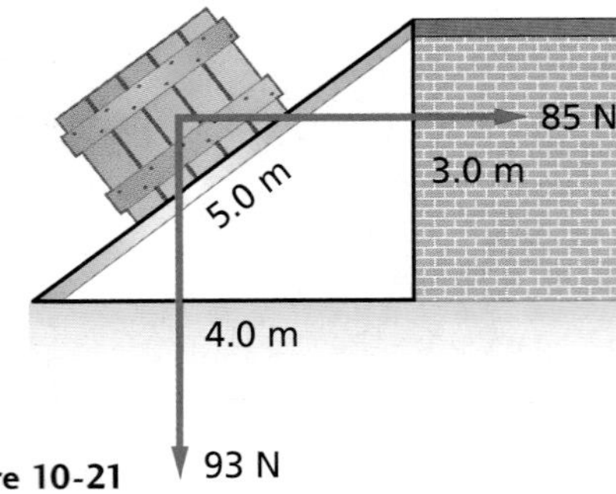

Figure 10-21

75. **Oil Pump** In 35.0 s, a pump delivers 0.550 m^3 of oil into barrels on a platform 25.0 m above the intake pipe. The oil's density is 0.820 g/cm^3.
   a. Calculate the work done by the pump.
   b. Calculate the power produced by the pump.

76. **Conveyor Belt** A 12.0-m-long conveyor belt, inclined at 30.0°, is used to transport bundles of newspapers from the mail room up to the cargo bay to be loaded onto delivery trucks. Each newspaper has a mass of 1.0 kg, and there are 25 newspapers per bundle. Determine the power that the conveyor develops if it delivers 15 bundles per minute.

77. A car is driven at a constant speed of 76 km/h down a road. The car's engine delivers 48 kW of power. Calculate the average force that is resisting the motion of the car.

78. The graph in **Figure 10-22** shows the force and displacement of an object being pulled.
   a. Calculate the work done to pull the object 7.0 m.
   b. Calculate the power that would be developed if the work was done in 2.0 s.

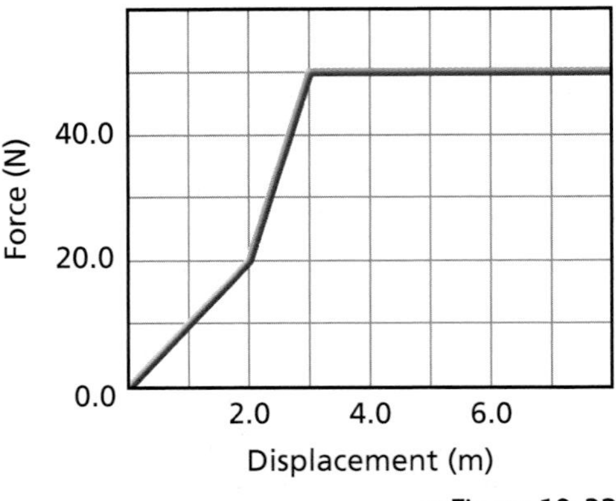

Figure 10-22

10.2 Machines

79. **Piano** Takeshi raises a 1200-N piano a distance of 5.00 m using a set of pulleys. He pulls in 20.0 m of rope.
   a. How much effort force would Takeshi apply if this were an ideal machine?
   b. What force is used to balance the friction force if the actual effort is 340 N?
   c. What is the output work?
   d. What is the input work?
   e. What is the mechanical advantage?

80. **Lever** Because there is very little friction, the lever is an extremely efficient simple machine. Using a 90.0-percent-efficient lever, what input work is required to lift an 18.0-kg mass through a distance of 0.50 m?

 For more problems, go to Additional Problems, Appendix B.

81. A pulley system lifts a 1345-N weight a distance of 0.975 m. Paul pulls the rope a distance of 3.90 m, exerting a force of 375 N.
a. What is the ideal mechanical advantage of the system?
b. What is the mechanical advantage?
c. How efficient is the system?

82. A force of 1.4 N is exerted through a distance of 40.0 cm on a rope in a pulley system to lift a 0.50-kg mass 10.0 cm. Calculate the following.
a. the *MA*
b. the *IMA*
c. the efficiency

83. A student exerts a force of 250 N on a lever, through a distance of 1.6 m, as he lifts a 150-kg crate. If the efficiency of the lever is 90.0 percent, how far is the crate lifted?

84. What work is required to lift a 215-kg mass a distance of 5.65 m, using a machine that is 72.5 percent efficient?

85. The ramp in **Figure 10-23** is 18 m long and 4.5 m high.
a. What force, parallel to the ramp (F_A), is required to slide a 25-kg box at constant speed to the top of the ramp if friction is disregarded?
b. What is the *IMA* of the ramp?
c. What are the real *MA* and the efficiency of the ramp if a parallel force of 75 N is actually required?

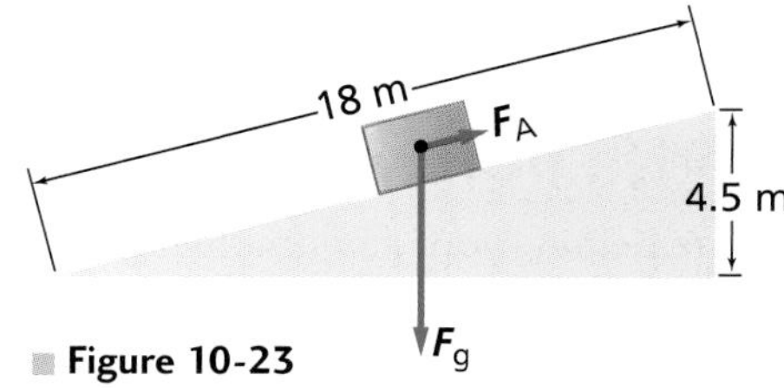

Figure 10-23

86. Bicycle Luisa pedals a bicycle with a gear radius of 5.00 cm and a wheel radius of 38.6 cm, as shown in **Figure 10-24.** If the wheel revolves once, what is the length of the chain that was used?

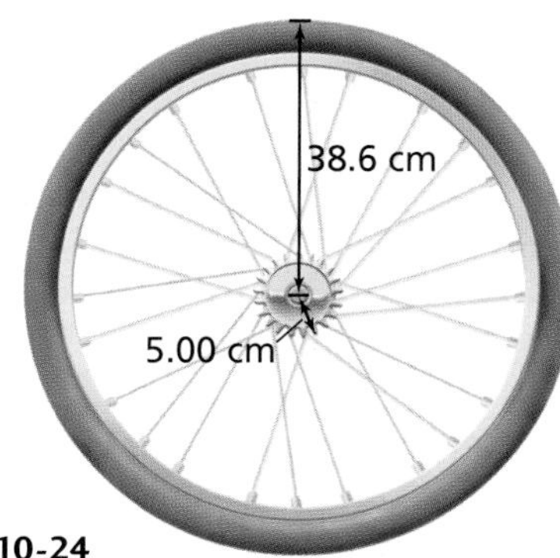

Figure 10-24

87. Crane A motor with an efficiency of 88 percent operates a crane with an efficiency of 42 percent. If the power supplied to the motor is 5.5 kW, with what constant speed does the crane lift a 410-kg crate of machine parts?

88. A compound machine is constructed by attaching a lever to a pulley system. Consider an ideal compound machine consisting of a lever with an *IMA* of 3.0 and a pulley system with an *IMA* of 2.0.
a. Show that the IMA of this compound machine is 6.0.
b. If the compound machine is 60.0 percent efficient, how much effort must be applied to the lever to lift a 540-N box?
c. If you move the effort side of the lever 12.0 cm, how far is the box lifted?

Mixed Review

89. Ramps Isra has to get a piano onto a 2.0-m-high platform. She can use a 3.0-m-long frictionless ramp or a 4.0-m-long frictionless ramp. Which ramp should Isra use if she wants to do the least amount of work?

90. Brutus, a champion weightlifter, raises 240 kg of weights a distance of 2.35 m.
a. How much work is done by Brutus lifting the weights?
b. How much work is done by Brutus holding the weights above his head?
c. How much work is done by Brutus lowering them back to the ground?
d. Does Brutus do work if he lets go of the weights and they fall back to the ground?
e. If Brutus completes the lift in 2.5 s, how much power is developed?

91. A horizontal force of 805 N is needed to drag a crate across a horizontal floor with a constant speed. You drag the crate using a rope held at an angle of 32°.
a. What force do you exert on the rope?
b. How much work do you do on the crate if you move it 22 m?
c. If you complete the job in 8.0 s, what power is developed?

92. Dolly and Ramp A mover's dolly is used to transport a refrigerator up a ramp into a house. The refrigerator has a mass of 115 kg. The ramp is 2.10 m long and rises 0.850 m. The mover pulls the dolly with a force of 496 N up the ramp. The dolly and ramp constitute a machine.
a. What work does the mover do?
b. What is the work done on the refrigerator by the machine?
c. What is the efficiency of the machine?

93. Sally does 11.4 kJ of work dragging a wooden crate 25.0 m across a floor at a constant speed. The rope makes an angle of 48.0° with the horizontal.
   a. How much force does the rope exert on the crate?
   b. What is the force of friction acting on the crate?
   c. What work is done by the floor through the force of friction between the floor and the crate?

94. **Sledding** An 845-N sled is pulled a distance of 185 m. The task requires 1.20×10^4 J of work and is done by pulling on a rope with a force of 125 N. At what angle is the rope held?

95. An electric winch pulls an 875-N crate up a 15° incline at 0.25 m/s. The coefficient of friction between the crate and incline is 0.45.
   a. What power does the winch develop?
   b. If the winch is 85 percent efficient, what is the electrical power that must be delivered to the winch?

Thinking Critically

96. **Analyze and Conclude** You work at a store, carrying boxes to a storage loft that is 12 m above the ground. You have 30 boxes with a total mass of 150 kg that must be moved as quickly as possible, so you consider carrying more than one up at a time. If you try to move too many at once, you know that you will go very slowly, resting often. If you carry only one box at a time, most of the energy will go into raising your own body. The power (in watts) that your body can develop over a long time depends on the mass that you carry, as shown in **Figure 10-25.** This is an example of a power curve that applies to machines as well as to people. Find the number of boxes to carry on each trip that would minimize the time required. What time would you spend doing the job? Ignore the time needed to go back down the stairs and to lift and lower each box.

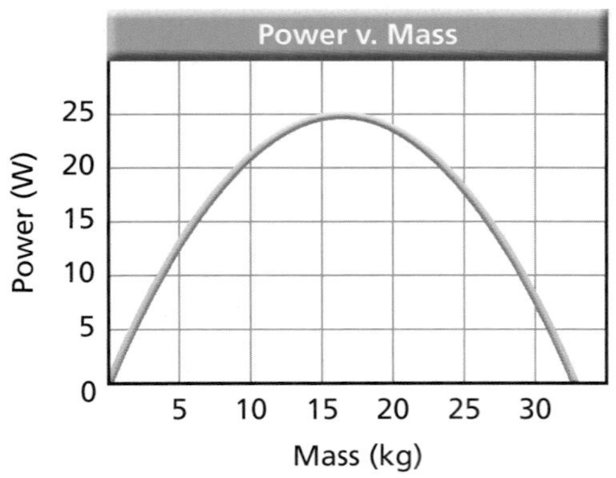

Figure 10-25

97. **Apply Concepts** A sprinter of mass 75 kg runs the 50.0-m dash in 8.50 s. Assume that the sprinter's acceleration is constant throughout the race.
   a. What is the average power of the sprinter over the 50.0 m?
   b. What is the maximum power generated by the sprinter?
   c. Make a quantitative graph of power versus time for the entire race.

98. **Apply Concepts** The sprinter in the previous problem runs the 50.0-m dash in the same time, 8.50 s. However, this time the sprinter accelerates in the first second and runs the rest of the race at a constant velocity.
   a. Calculate the average power produced for that first second.
   b. What is the maximum power that the sprinter now generates?

Writing in Physics

99. Just as a bicycle is a compound machine, so is an automobile. Find the efficiencies of the component parts of the power train (engine, transmission, wheels, and tires). Explore possible improvements in each of these efficiencies.

100. The terms *force, work, power,* and *energy* often mean the same thing in everyday use. Obtain examples from advertisements, print media, radio, and television that illustrate meanings for these terms that differ from those used in physics.

Cumulative Review

101. You are helping your grandmother with some gardening and have filled a garbage can with weeds and soil. Now you have to move the garbage can across the yard and realize it is so heavy that you will need to push it, rather than lift it. If the can has a mass of 24 kg, the coefficient of kinetic friction between the can's bottom and the muddy grass is 0.27, and the coefficient of static friction between those same surfaces is 0.35, how hard do you have to push horizontally to get the can to just start moving? (Chapter 5)

102. **Baseball** If a major league pitcher throws a fastball horizontally at a speed of 40.3 m/s (90 mph) and it travels 18.4 m (60 ft, 6 in), how far has it dropped by the time it crosses home plate? (Chapter 6)

103. People sometimes say that the Moon stays in its orbit because the "centrifugal force just balances the centripetal force, giving no net force." Explain why this idea is wrong. (Chapter 8)

 For more problems, go to Additional Problems, Appendix B.

Multiple Choice

1. A pulley system consists of two fixed pulleys and two movable pulleys that lift a load that has a weight of 300 N. If the effort force used to lift the load is 100 N, what is the mechanical advantage of the system?

   (A) $\frac{1}{3}$ (C) 3

   (B) $\frac{3}{4}$ (D) 6

2. The box in the diagram is being pushed up the ramp with a force of 100.0 N. If the height of the ramp is 3.0 m, what is the work done on the box? (sin 30° = 0.50, cos 30° = 0.87, tan 30° = 0.58)

   (A) 150 J (C) 450 J

   (B) 260 J (D) 600 J

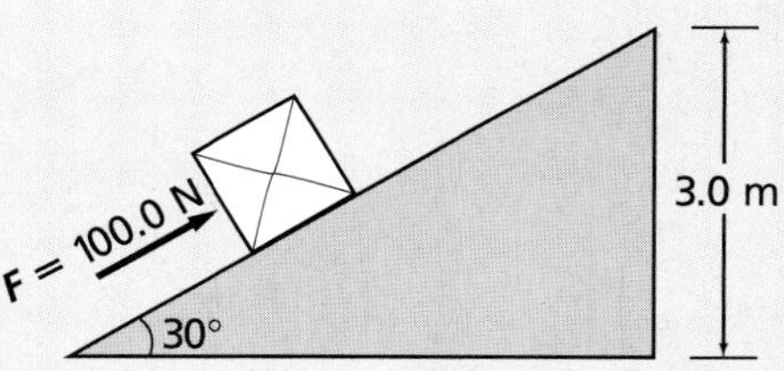

3. A compound machine used to raise heavy boxes consists of a ramp and a pulley. The efficiency of pulling a 100-kg box up the ramp is 50%. If the efficiency of the pulley is 90%, what is the overall efficiency of the compound machine?

   (A) 40% (C) 50%

   (B) 45% (D) 70%

4. A skater with a mass of 50.0 kg slides across an icy pond with negligible friction. As he approaches a friend, both he and his friend hold out their hands, and the friend exerts a force in the direction opposite to the skater's movement, which slows the skater's speed from 2.0 m/s^2 to 1.0 m/s^2. What is the change in the skater's kinetic energy?

   (A) 25 J (C) 100 J

   (B) 75 J (D) 150 J

5. A 20.0-N block is attached to the end of a rope, and the rope is looped around a pulley system. If you pull the opposite end of the rope a distance of 2.00 m, the pulley system raises the block a distance of 0.40 m. What is the pulley system's ideal mechanical advantage?

   (A) 2.5 (C) 5.0

   (B) 4.0 (D) 10.0

6. Two people carry identical 40.0-N boxes up a ramp. The ramp is 2.00 m long and rests on a platform that is 1.00 m high. One person walks up the ramp in 2.00 s, and the other person walks up the ramp in 4.00 s. What is the difference in power the two people use to carry the boxes up the ramp?

   (A) 5.00 W (C) 20.0 W

   (B) 10.0 W (D) 40.0 W

7. A 4-N soccer ball sits motionless on a field. A player's foot exerts a force of 5 N on the ball for a distance of 0.1 m, and the ball rolls a distance of 10 m. How much kinetic energy does the ball gain from the player?

   (A) 0.5 J (C) 9 J

   (B) 0.9 J (D) 50 J

Extended Answer

8. The diagram shows a box being pulled by a rope with a force of 200.0 N along a horizontal surface. The angle the rope makes with the horizontal is 45°. Calculate the work done on the box and the power required to pull it a distance of 5.0 m in 10.0 s. (sin 45° = cos 45° = 0.7)

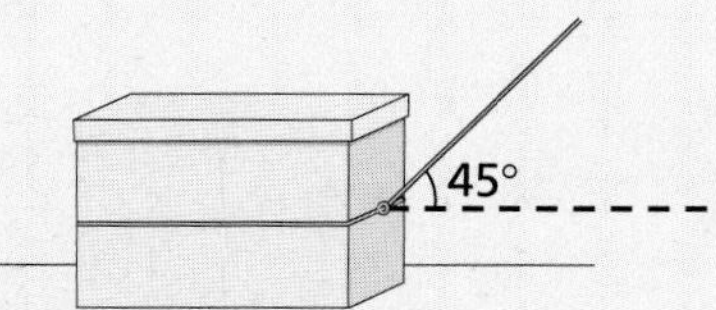

✓ Test-Taking TIP

Beat the Clock and then Go Back

As you take a practice test, pace yourself to finish each section just a few minutes early so you can go back and check over your work.

Chapter 11

Energy and Its Conservation

What You'll Learn

- You will learn that energy is a property of an object that can change the object's position, motion, or its environment.
- You will learn that energy changes from one form to another, and that the total amount of energy in a closed system remains constant.

Why It's Important

Energy turns the wheels of our world. People buy and sell energy to operate electric appliances, automobiles, and factories.

Skiing The height of the ski jump determines the energy the skier has at the bottom of the ramp before jumping into the air and flying many meters down the slope. The distance that the ski jumper travels depends on his or her use of physical principles such as air resistance, balance, and energy.

Think About This ▶

How does the height of the ski ramp affect the distance that the skier can jump?

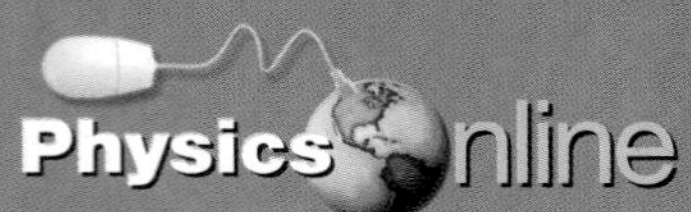

physicspp.com

LAUNCH Lab How can you analyze the energy of a bouncing basketball?

Question

What is the relationship between the height a basketball is dropped from and the height it reaches when it bounces back?

Procedure

1. Place a meterstick against a wall. Choose an initial height from which to drop a basketball. Record the height in the data table.
2. Drop the ball and record how high the ball bounced.
3. Repeat steps 1 and 2 by dropping the basketball from three other heights.
4. **Make and Use Graphs** Construct a graph of bounce height (y) versus drop height (x). Find the best-fit line.

Analysis

Use the graph to find how high a basketball would bounce if it were dropped from a height of 10.0 m.

When the ball is lifted and ready to drop, it possesses energy. What are the factors that influence this energy?

Critical Thinking Why doesn't the ball bounce back to the height from which it was dropped?

11.1 The Many Forms of Energy

The word *energy* is used in many different ways in everyday speech. Some fruit-and-cereal bars are advertised as energy sources. Athletes use energy in sports. Companies that supply your home with electricity, natural gas, or heating fuel are called energy companies.

Scientists and engineers use the term *energy* much more precisely. As you learned in the last chapter, work causes a change in the energy of a system. That is, work transfers energy between a system and the external world.

In this chapter, you will explore how objects can have energy in a variety of ways. Energy is like ice cream—it comes in different varieties. You can have vanilla, chocolate, or peach ice cream. They are different varieties, but they are all ice cream and serve the same purpose. However, unlike ice cream, energy can be changed from one variety to another. In this chapter, you will learn how energy is transformed from one variety (or form) to another and how to keep track of the changes.

▶ **Objectives**

- **Use** a model to relate work and energy.
- **Calculate** kinetic energy.
- **Determine** the gravitational potential energy of a system.
- **Identify** how elastic potential energy is stored.

▶ **Vocabulary**

rotational kinetic energy
gravitational potential energy
reference level
elastic potential energy

A Model of the Work-Energy Theorem

In the last chapter, you were introduced to the work-energy theorem. You learned that when work is done on a system, the energy of that system increases. On the other hand, if the system does work, then the energy of the system decreases. These are abstract ideas, but keeping track of energy is much like keeping track of your spending money.

If you have a job, the amount of money that you have increases every time you are paid. This process can be represented with a bar graph, as shown in **Figure 11-1a.** The orange bar represents how much money you had to start with, and the blue bar represents the amount that you were paid. The green bar is the total amount that you possess after the payment. An accountant would say that your cash flow was positive. What happens when you spend the money that you have? The total amount of money that you have decreases. As shown in **Figure 11-1b,** the bar that represents the amount of money that you had before you bought that new CD is higher than the bar that represents the amount of money remaining after your shopping trip. The difference is the cost of the CD. Cash flow is shown as a bar below the axis because it represents money going out, which can be shown as a negative number. Energy is similar to your spending money. The amount of money that you have changes only when you earn more or spend it. Similarly, energy can be stored, and when energy is spent, it affects the motion of a system.

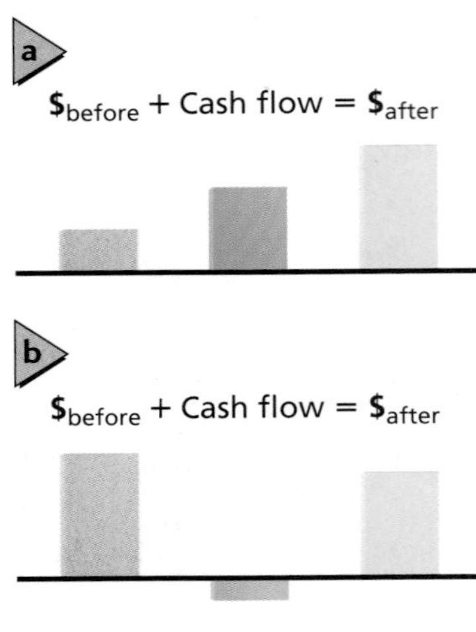

Figure 11-1 When you earn money, the amount of cash that you have increases **(a).** When you spend money, the amount of cash that you have decreases **(b).**

Throwing a ball Gaining and losing energy also can be illustrated by throwing and catching a ball. In Chapter 10, you learned that when you exert a constant force, F, on an object through a distance, d, in the direction of the force, you do an amount of work, represented by $W = Fd$. The work is positive because the force and motion are in the same direction, and the energy of the object increases by an amount equal to W. Suppose the object is a ball, and you exert a force to throw the ball. As a result of the force you apply, the ball gains kinetic energy. This process is shown in **Figure 11-2a.** You can again use a bar graph to represent the process. This time, the height of the bar represents the amount of work, or energy, measured in joules. The kinetic energy after the work is done is equal to the sum of the initial kinetic energy plus the work done on the ball.

Figure 11-2 The kinetic energy after throwing or catching a ball is equal to the kinetic energy before plus the input work.

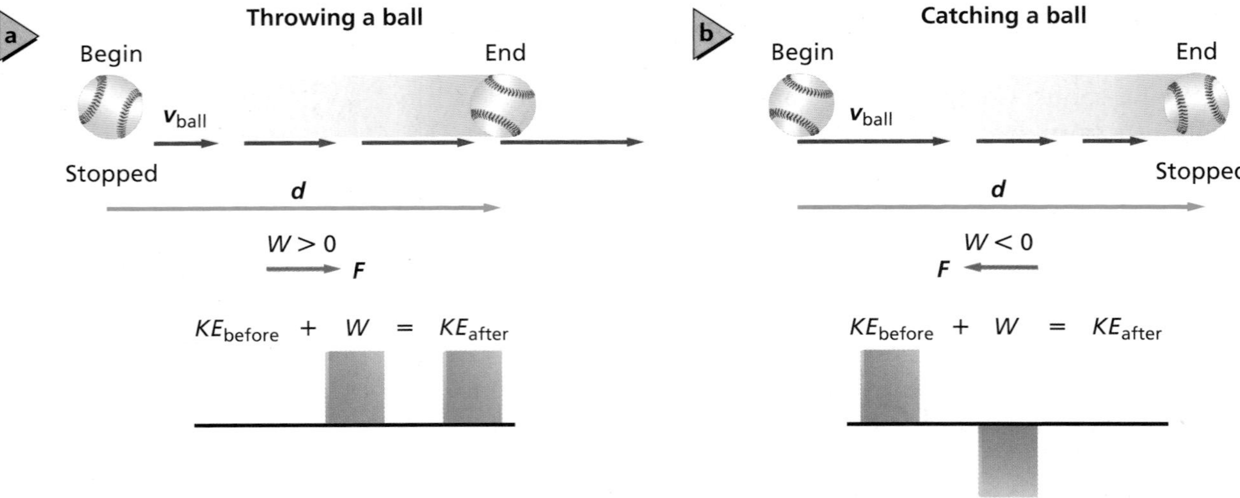

Catching a ball What happens when you catch a ball? Before hitting your hands or glove, the ball is moving, so it has kineticc energy. In catching it, you exert a force on the ball in the direction opposite to its motion. Therefore, you do negative work on it, causing it to stop. Now that the ball is not moving, it has no kinetic energy. This process and the bar graph that represents it are shown in **Figure 11-2b.** Kinetic energy is always positive, so the initial kinetic energy of the ball is positive. The work done on the ball is negative and the final kinetic energy is zero. Again, the kinetic energy after the ball has stopped is equal to the sum of the initial kinetic energy plus the work done on the ball.

Kinetic Energy

Recall that kinetic energy, $KE = \frac{1}{2}mv^2$, where m is the mass of the object and v is the magnitude of its velocity. The kinetic energy is proportional to the object's mass. A 7.26-kg shot put thrown through the air has much more kinetic energy than a 0.148-kg baseball with the same velocity, because the shot put has a greater mass. The kinetic energy of an object is also proportional to the square of the object's velocity. A car speeding at 20 m/s has four times the kinetic energy of the same car moving at 10 m/s. Kinetic energy also can be due to rotational motion. If you spin a toy top in one spot, does it have kinetic energy? You might say that it does not because the top is not moving anywhere. However, to make the top rotate, someone had to do work on it. Therefore, the top has **rotational kinetic energy.** This is one of the several varieties of energy. Rotational kinetic energy can be calculated using $KE_{rot} = \frac{1}{2}I\omega^2$, where I is the object's moment of inertia and ω is the object's angular velocity.

The diver, shown in **Figure 11-3a,** does work as she pushes off of the diving board. This work produces both linear and rotational kinetic energies. When the diver's center of mass moves as she leaps, linear kinetic energy is produced. When she rotates about her center of mass, as shown in **Figure 11-3b,** rotational kinetic energy is produced. Because she is moving toward the water and rotating at the same time while in the tuck position, she has both linear and rotational kinetic energy. When she slices into the water, as shown in **Figure 11-3c,** she has linear kinetic energy.

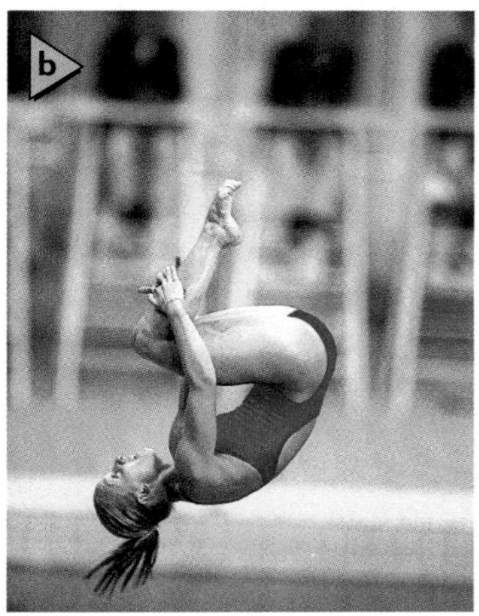

Figure 11-3 The diver does work as she pushes off of the diving board **(a).** This work produces rotational kinetic energy as she rotates about her center of mass **(b)** and she has linear kinetic energy when she slices into the water **(c).**

PRACTICE Problems

• Additional Problems, Appendix B
• Solutions to Selected Problems, Appendix C

1. A skater with a mass of 52.0 kg moving at 2.5 m/s glides to a stop over a distance of 24.0 m. How much work did the friction of the ice do to bring the skater to a stop? How much work would the skater have to do to speed up to 2.5 m/s again?
2. An 875.0-kg compact car speeds up from 22.0 m/s to 44.0 m/s while passing another car. What are its initial and final energies, and how much work is done on the car to increase its speed?
3. A comet with a mass of 7.85×10^{11} kg strikes Earth at a speed of 25.0 km/s. Find the kinetic energy of the comet in joules, and compare the work that is done by Earth in stopping the comet to the 4.2×10^{15} J of energy that was released by the largest nuclear weapon ever built.

Figure 11-4 Money in the form of bills, quarters, and pennies are different forms of the same thing.

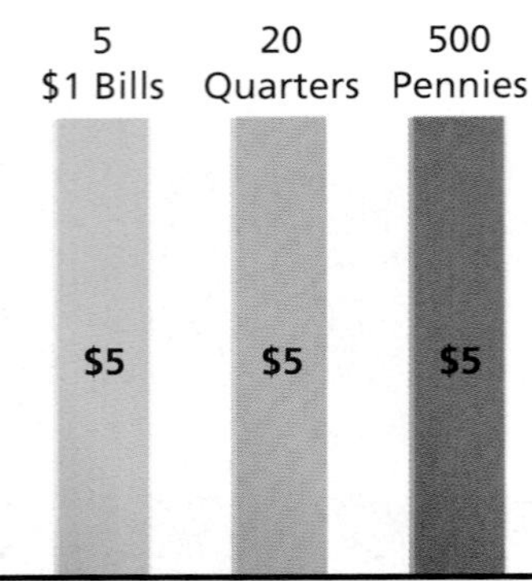

$\$5 = \$1.00 \times 5$
$\$5 = \$0.25 \times 20$
$\$5 = \$0.01 \times 500$

Stored Energy

Imagine a group of boulders high on a hill. These boulders have been lifted up by geological processes against the force of gravity; thus, they have stored energy. In a rock slide, the boulders are shaken loose. They fall and pick up speed as their stored energy is converted to kinetic energy.

In the same way, a small, spring-loaded toy, such as a jack-in-the-box, has stored energy, but the energy is stored in a compressed spring. While both of these examples represent energy stored by mechanical means, there are many other means of storing energy. Automobiles, for example, carry their energy stored in the form of chemical energy in the gasoline tank. Energy is made useful or causes motion when it changes from one form to another.

How does the money model that was discussed earlier illustrate the transformation of energy from one form to another? Money, too, can come in different forms. You can have one five-dollar bill, 20 quarters, or 500 pennies. In all of these cases, you still have five dollars. The height of the bar graph in **Figure 11-4** represents the amount of money in each form. In the same way, you can use a bar graph to represent the amount of energy in various forms that a system has.

Gravitational Potential Energy

Look at the oranges being juggled in **Figure 11-5.** If you consider the system to be only one orange, then it has several external forces acting on it. The force of the juggler's hand does work, giving the orange its original kinetic energy. After the orange leaves the juggler's hand, only the force of gravity acts on it. How much work does gravity do on the orange as its height changes?

Figure 11-5 Kinetic and potential energy are constantly being exchanged when juggling.

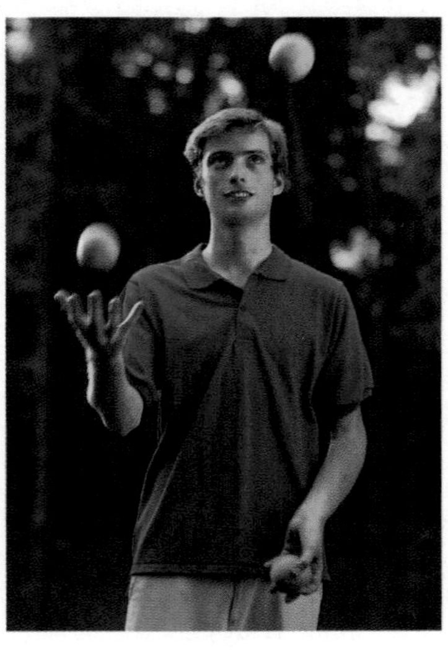

Work done by gravity Let *h* represent the orange's height measured from the juggler's hand. On the way up, its displacement is upward, but the force on the orange, F_g, is downward, so the work done by gravity is negative: $W_g = -mgh$. On the way back down, the force and displacement are in the same direction, so the work done by gravity is positive: $W_g = mgh$. Thus, while the orange is moving upward, gravity does negative work, slowing the orange to a stop. On the way back down, gravity does positive work, increasing the orange's speed and thereby increasing its kinetic energy. The orange recovers all of the kinetic energy it originally had when it returns to the height at which it left the juggler's hand. It is as if the orange's kinetic energy is stored in another form as the orange rises and is transformed back to kinetic energy as the orange falls.

Consider a system that consists of an object plus Earth. The gravitational attraction between the object and Earth is a force that always does work on the object as it moves. If the object moves away from Earth, energy is stored in the system as a result of the gravitational force between the object and Earth. This stored energy is called **gravitational potential energy** and is represented by the symbol *PE*. The height to which the object has risen is determined by using a **reference level,** the position where *PE* is defined to be zero. For an object with mass, *m*, that has risen to a height, *h*, above the reference level, gravitational potential energy is represented by the following equation.

Gravitational Potential Energy $PE = mgh$

The gravitational potential energy of an object is equal to the product of its mass, the acceleration due to gravity, and the distance from the reference level.

In the equation for gravitational potential energy, g is the acceleration due to gravity. Gravitational potential energy, like kinetic energy, is measured in joules.

Kinetic energy and potential energy of a system Consider the energy of a system consisting of an orange used by the juggler plus Earth. The energy in the system exists in two forms: kinetic energy and gravitational potential energy. At the beginning of the orange's flight, all the energy is in the form of kinetic energy, as shown in **Figure 11-6a.** On the way up, as the orange slows down, energy changes from kinetic energy to potential energy. At the highest point of the orange's flight, the velocity is zero. Thus, all the energy is in the form of gravitational potential energy. On the way back down, potential energy changes back into kinetic energy. The sum of kinetic energy and potential energy is constant at all times because no work is done on the system by any external forces.

Reference levels In Figure 11-6a, the reference level is the juggler's hand. That is, the height of the orange is measured from the juggler's hand. Thus, at the juggler's hand, $h = 0$ m and $PE = 0$ J. You can set the reference level at any height that is convenient for solving a given problem.

Suppose the reference level is set at the highest point of the orange's flight. Then, $h = 0$ m and the system's $PE = 0$ J at that point, as illustrated in **Figure 11-6b.** The potential energy of the system is negative at the beginning of the orange's flight, zero at the highest point, and negative at the end of the orange's flight. If you were to calculate the total energy of the system represented in Figure 11-6a, it would be different from the total energy of the system represented in Figure 11-6b. This is because the reference levels are different in each case. However, the total energy of the system in each situation would be constant at all times during the flight of the orange. Only changes in energy determine the motion of a system.

APPLYING PHYSICS

Potential Energy of an Atom It is interesting to consider the relative sizes of potential energy per atom. For instance, a carbon atom has a mass of about 2×10^{-26} kg. If you lift it 1 m above the ground, its gravitational potential energy is about 2×10^{-25} J. The electrostatic energy that holds the electrons on the atom has a value of about 10^{-19} J, and the nuclear potential energy that holds the nucleus together is greater than 10^{-12} J. The nuclear potential energy is at least a million million times greater than the gravitational potential energy.

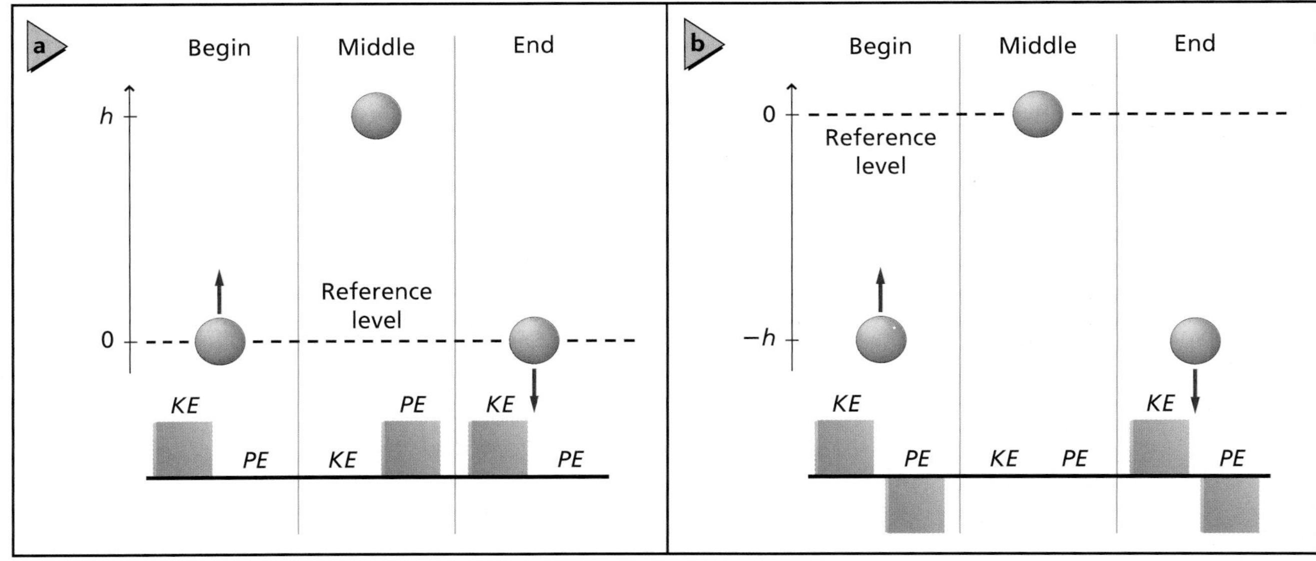

Figure 11-6 The energy of an orange is converted from one form to another in various stages of its flight **(a).** Note that the choice of a reference level is arbitrary, but that the total energy remains constant **(b).**

▶ EXAMPLE Problem 1

Gravitational Potential Energy You lift a 7.30-kg bowling ball from the storage rack and hold it up to your shoulder. The storage rack is 0.610 m above the floor and your shoulder is 1.12 m above the floor.

a. When the bowling ball is at your shoulder, what is the bowling ball's gravitational potential energy relative to the floor?

b. When the bowling ball is at your shoulder, what is its gravitational potential energy relative to the storage rack?

c. How much work was done by gravity as you lifted the ball from the rack to shoulder level?

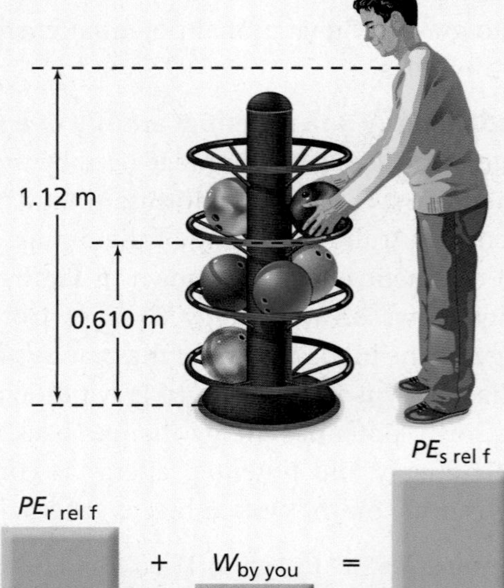

1 Analyze and Sketch the Problem

- Sketch the situation.
- Choose a reference level.
- Draw a bar graph showing the gravitational potential energy with the floor as the reference level.

Known:	**Unknown:**
$m = 7.30$ kg	$PE_{s\ rel\ f} = ?$
$h_r = 0.610$ m (relative to the floor)	$PE_{s\ rel\ r} = ?$
$h_s = 1.12$ m (relative to the floor)	
$g = 9.80$ m/s^2	

2 Solve for the Unknown

a. Set the reference level to be at the floor.

Solve for the potential energy of the ball at shoulder level.

$PE_{s\ rel\ f} = mgh_s$

$= (7.30\text{ kg})(9.80\text{ m/s}^2)(1.12\text{ m})$ **Substitute $m = 7.30$ kg, $g = 9.80$ m/s^2, $h = 1.12$ m**

$= 80.1\text{ J}$

b. Set the reference level to be at the rack height.

Solve for the height of your shoulder relative to the rack.

$h = h_s - h_r$

Solve for the potential energy of the ball.

$PE_{s\ rel\ r} = mgh$

$= mg(h_s - h_r)$ **Substitute $h = h_s - h_r$**

$= (7.30\text{ kg})(9.80\text{ m/s}^2)(1.12\text{ m} - 0.610\text{ m})$ **Substitute $m = 7.3$ kg, $g = 9.80$ m/s^2, $h_s = 1.12$ m, $h_r = 0.610$ m**

$= 36.5\text{ J}$ **This also is equal to the work done by you.**

Math Handbook
Order of Operations
page 843

c. The work done by gravity is the weight of the ball times the distance the ball was lifted.

$W = Fd$

$= -(mg)h$ **Because the weight opposes the motion of lifting, the work is negative.**

$= -(mg)(h_s - h_r)$

$= -(7.30\text{ kg})(9.80\text{ m/s}^2)(1.12\text{ m} - 0.610\text{ m})$ **Substitute $m = 7.30$ kg, $g = 9.80$ m/s^2, $h_s = 1.12$ m, $h_r = 0.610$ m**

$= -36.5\text{ J}$

3 Evaluate the Answer

- **Are the units correct?** The potential energy and work are both measured in joules.
- **Is the magnitude realistic?** The ball should have a greater potential energy relative to the floor than relative to the rack, because the ball's distance above the reference level is greater.

PRACTICE Problems

- Additional Problems, Appendix B
- Solutions to Selected Problems, Appendix C

4. In Example Problem 1, what is the potential energy of the bowling ball relative to the rack when it is on the floor?
5. If you slowly lower a 20.0-kg bag of sand 1.20 m from the trunk of a car to the driveway, how much work do you do?
6. A boy lifts a 2.2-kg book from his desk, which is 0.80 m high, to a bookshelf that is 2.10 m high. What is the potential energy of the book relative to the desk?
7. If a 1.8-kg brick falls to the ground from a chimney that is 6.7 m high, what is the change in its potential energy?
8. A warehouse worker picks up a 10.1-kg box from the floor and sets it on a long, 1.1-m-high table. He slides the box 5.0 m along the table and then lowers it back to the floor. What were the changes in the energy of the box, and how did the total energy of the box change? (Ignore friction.)

Elastic Potential Energy

When the string on the bow shown in **Figure 11-7** is pulled, work is done on the bow, storing energy in it. Thus, the energy of the system increases. Identify the system as the bow, the arrow, and Earth. When the string and arrow are released, energy is changed into kinetic energy. The stored energy in the pulled string is called **elastic potential energy,** which is often stored in rubber balls, rubber bands, slingshots, and trampolines.

Energy also can be stored in the bending of an object. When stiff metal or bamboo poles were used in pole-vaulting, the poles did not bend easily. Little work was done on the poles, and consequently, the poles did not store much potential energy. Since flexible fiberglass poles were introduced, however, record pole-vaulting heights have soared.

Figure 11-7 Elastic potential energy is stored in the string of this bow. Before the string is released, the energy is all potential **(a).** As the string is released, the energy is transferred to the arrow as kinetic energy **(b).**

Figure 11-8 When a pole-vaulter jumps, elastic potential energy is changed into kinetic energy and gravitational potential energy.

A pole-vaulter runs with a flexible pole and plants its end into the socket in the ground. When the pole-vaulter bends the pole, as shown in **Figure 11-8**, some of the pole-vaulter's kinetic energy is converted to elastic potential energy. When the pole straightens, the elastic potential energy is converted to gravitational potential energy and kinetic energy as the pole-vaulter is lifted as high as 6 m above the ground. Unlike stiff metal poles or bamboo poles, fiberglass poles have an increased capacity for storing elastic potential energy. Thus, pole-vaulters are able to clear bars that are set very high.

Mass Albert Einstein recognized yet another form of potential energy: mass itself. He said that mass, by its very nature, is energy. This energy, E_0, is called rest energy and is represented by the following famous formula.

Rest Energy $E_0 = mc^2$

The rest energy of an object is equal to the object's mass times the speed of light squared.

According to this formula, stretching a spring or bending a vaulting pole causes the spring or pole to gain mass. In these cases, the change in mass is too small to be detected. When forces within the nucleus of an atom are involved, however, the energy released into other forms, such as kinetic energy, by changes in mass can be quite large.

11.1 Section Review

9. Elastic Potential Energy You get a spring-loaded toy pistol ready to fire by compressing the spring. The elastic potential energy of the spring pushes the rubber dart out of the pistol. You use the toy pistol to shoot the dart straight up. Draw bar graphs that describe the forms of energy present in the following instances.

a. The dart is pushed into the gun barrel, thereby compressing the spring.

b. The spring expands and the dart leaves the gun barrel after the trigger is pulled.

c. The dart reaches the top of its flight.

10. Potential Energy A 25.0-kg shell is shot from a cannon at Earth's surface. The reference level is Earth's surface. What is the gravitational potential energy of the system when the shell is at 425 m? What is the change in potential energy when the shell falls to a height of 225 m?

11. Rotational Kinetic Energy Suppose some children push a merry-go-round so that it turns twice as fast as it did before they pushed it. What are the relative changes in angular momentum and rotational kinetic energy?

12. Work-Energy Theorem How can you apply the work-energy theorem to lifting a bowling ball from a storage rack to your shoulder?

13. Potential Energy A 90.0-kg rock climber first climbs 45.0 m up to the top of a quarry, then descends 85.0 m from the top to the bottom of the quarry. If the initial height is the reference level, find the potential energy of the system (the climber and Earth) at the top and at the bottom. Draw bar graphs for both situations.

14. Critical Thinking Karl uses an air hose to exert a constant horizontal force on a puck, which is on a frictionless air table. He keeps the hose aimed at the puck, thereby creating a constant force as the puck moves a fixed distance.

a. Explain what happens in terms of work and energy. Draw bar graphs.

b. Suppose Karl uses a different puck with half the mass of the first one. All other conditions remain the same. How will the kinetic energy and work differ from those in the first situation?

c. Explain what happened in parts **a** and **b** in terms of impulse and momentum.

Physics Online physicspp.com/self_check_quiz

11.2 Conservation of Energy

Consider a ball near the surface of Earth. The sum of gravitational potential energy and kinetic energy in that system is constant. As the height of the ball changes, energy is converted from kinetic energy to potential energy, but the total amount of energy stays the same.

▶ **Objectives**

- **Solve** problems using the law of conservation of energy.
- **Analyze** collisions to find the change in kinetic energy.

▶ **Vocabulary**

law of conservation of energy
mechanical energy
thermal energy
elastic collision
inelastic collision

Conservation of Energy

In our everyday world, it may not seem as if energy is conserved. A hockey puck eventually loses its kinetic energy and stops moving, even on smooth ice. A pendulum stops swinging after some time. The money model can again be used to illustrate what is happening in these cases.

Suppose you have a total of \$50 in cash. One day, you count your money and discover that you are \$3 short. Would you assume that the money just disappeared? You probably would try to remember whether you spent it, and you might even search for it. In other words, rather than giving up on the conservation of money, you would try to think of different places where it might have gone.

Law of conservation of energy Scientists do the same thing as you would if you could not account for a sum of money. Whenever they observe energy leaving a system, they look for new forms into which the energy could have been transferred. This is because the total amount of energy in a system remains constant as long as the system is closed and isolated from external forces. The **law of conservation of energy** states that in a closed, isolated system, energy can neither be created nor destroyed; rather, energy is conserved. Under these conditions, energy changes from one form to another while the total energy of the system remains constant.

Conservation of mechanical energy The sum of the kinetic energy and gravitational potential energy of a system is called **mechanical energy, E.** In any given system, if no other forms of energy are present, mechanical energy is represented by the following equation.

Mechanical Energy of a System $E = KE + PE$

The mechanical energy of a system is equal to the sum of the kinetic energy and potential energy if no other forms of energy are present.

Imagine a system consisting of a 10.0-N ball and Earth, as shown in **Figure 11-9.** Suppose the ball is released from 2.00 m above the ground, which you set to be the reference level. Because the ball is not yet moving, it has no kinetic energy. Its potential energy is represented by the following equation:

$$PE = mgh = (10.0\ \text{N})(2.00\ \text{m}) = 20.0\ \text{J}$$

The ball's total mechanical energy, therefore, is 20.0 J. As the ball falls, it loses potential energy and gains kinetic energy. When the ball is 1.00 m above Earth's surface: $PE = mgh = (10.0\ \text{N})(1.00\ \text{m}) = 10.0\ \text{J}$.

Figure 11-9 A decrease in potential energy is equal to the increase in kinetic energy.

Concepts In Motion
Interactive Figure To see an animation on conservation of mechanical energy, visit physicspp.com.

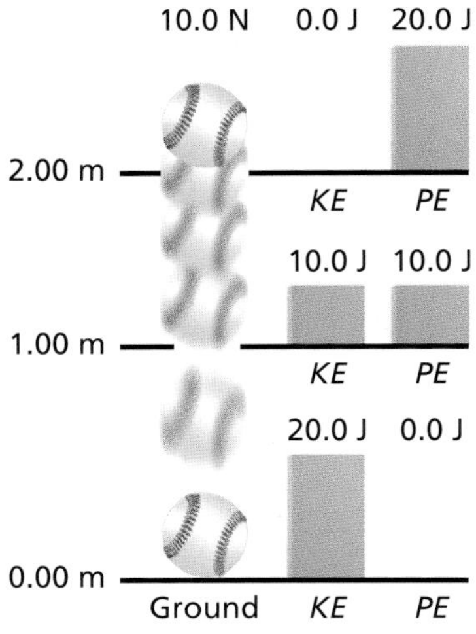

10.0 N
2.00 m — KE 0.0 J, PE 20.0 J
1.00 m — KE 10.0 J, PE 10.0 J
0.00 m — KE 20.0 J, PE 0.0 J
Ground

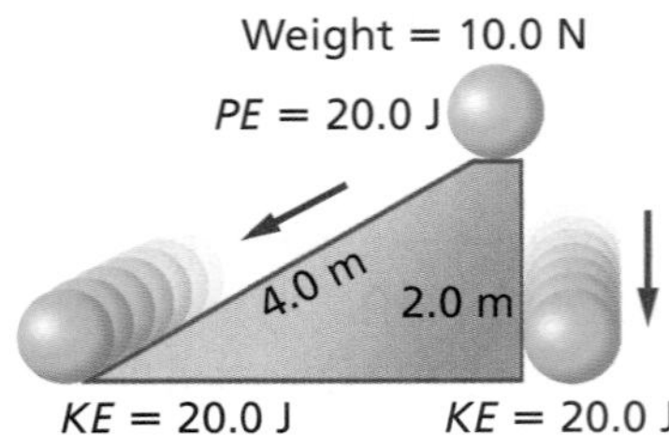

Figure 11-10 The path that an object follows in reaching the ground does not affect the final kinetic energy of the object.

What is the ball's kinetic energy when it is at a height of 1.00 m? The system consisting of the ball and Earth is closed and isolated because no external forces are acting upon it. Hence, the total energy of the system, E, remains constant at 20.0 J.

$$E = KE + PE, \text{ so } KE = E - PE$$

$$KE = 20.0\text{ J} - 10.0\text{ J} = 10.0\text{ J}$$

When the ball reaches ground level, its potential energy is zero, and its kinetic energy is 20.0 J. The equation that describes conservation of mechanical energy can be written as follows.

Conservation of Mechanical Energy

$$KE_{before} + PE_{before} = KE_{after} + PE_{after}$$

When mechanical energy is conserved, the sum of the kinetic energy and potential energy present in the system before the event is equal to the sum of the kinetic energy and potential energy in the system after the event.

Figure 11-11 For the simple harmonic motion of a pendulum bob **(a),** the mechanical energy—the sum of the potential and kinetic energies—is a constant **(b).**

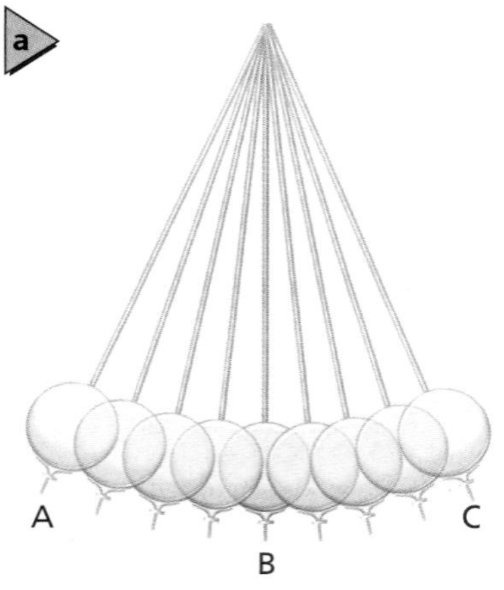

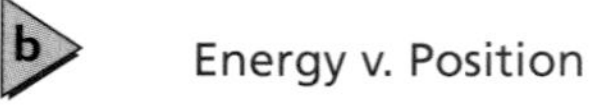

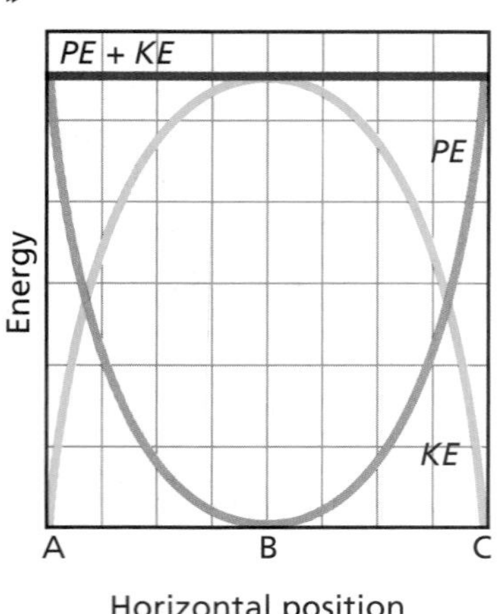

What happens if the ball does not fall down, but rolls down a ramp, as shown in **Figure 11-10?** If there is no friction, there are no external forces acting on the system. Thus, the system remains closed and isolated. The ball still moves down a vertical distance of 2.00 m, so its loss of potential energy is 20.0 J. Therefore, it gains 20.0 J of kinetic energy. As long as there is no friction, the path that the ball takes does not matter.

Roller coasters In the case of a roller coaster that is nearly at rest at the top of the first hill, the total mechanical energy in the system is the coaster's gravitational potential energy at that point. Suppose some other hill along the track were higher than the first one. The roller coaster would not be able to climb the higher hill because the energy required to do so would be greater than the total mechanical energy of the system.

Skiing Suppose you ski down a steep slope. When you begin from rest at the top of the slope, your total mechanical energy is simply your gravitational potential energy. Once you start skiing downhill, your gravitational potential energy is converted to kinetic energy. As you ski down the slope, your speed increases as more of your potential energy is converted to kinetic energy. In ski jumping, the height of the ramp determines the amount of energy that the jumper has to convert into kinetic energy at the beginning of his or her flight.

Pendulums The simple oscillation of a pendulum also demonstrates conservation of energy. The system is the pendulum bob and Earth. Usually, the reference level is chosen to be the height of the bob at the lowest point, when it is at rest. If an external force pulls the bob to one side, the force does work that gives the system mechanical energy. At the instant the bob is released, all the energy is in the form of potential energy, but as the bob swings downward, the energy is converted to kinetic energy. **Figure 11-11** shows a graph of the changing potential and kinetic energies of a pendulum. When the bob is at the lowest point, its gravitational potential energy is zero, and its kinetic energy is equal to the total mechanical energy in the system. Note that the total mechanical energy of the system is constant if we assume that there is no friction. You will learn more about pendulums in Chapter 14.

Loss of mechanical energy The oscillations of a pendulum eventually come to a stop, a bouncing ball comes to rest, and the heights of roller-coaster hills get lower and lower. Where does the mechanical energy in such systems go? Any object moving through the air experiences the forces of air resistance. In a roller coaster, there are frictional forces between the wheels and the tracks.

When a ball bounces off of a surface, all of the elastic potential energy that is stored in the deformed ball is not converted back into kinetic energy after the bounce. Some of the energy is converted into thermal energy and sound energy. As in the cases of the pendulum and the roller coaster, some of the original mechanical energy in the system is converted into another form of energy within members of the system or transmitted to energy outside the system, as in air resistance. Usually, this new energy causes the temperature of objects to rise slightly. You will learn more about this form of energy, called **thermal energy,** in Chapter 12. The following strategies will be helpful to you when solving problems related to conservation of energy.

PROBLEM-SOLVING **Strategies**

Conservation of Energy

When solving problems related to the conservation of energy, use the following strategies.

1. Carefully identify the system. Make sure it is closed. In a closed system, no objects enter or leave the system.
2. Identify the forms of energy in the system.
3. Identify the initial and final states of the system.
4. Is the system isolated?
   a. If there are no external forces acting on the system, then the system is isolated and the total energy of the system is constant.
   $$E_{before} = E_{after}$$
   b. If there are external forces, then the following is true.
   $$E_{before} + W = E_{after}$$
5. If mechanical energy is conserved, decide on the reference level for potential energy. Draw bar graphs showing initial and final energy like the bar graphs shown to the right.

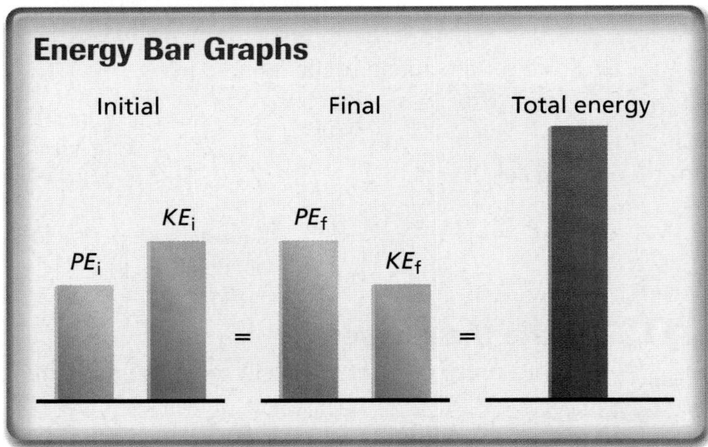

▶ EXAMPLE Problem 2

Conservation of Mechanical Energy During a hurricane, a large tree limb, with a mass of 22.0 kg and a height of 13.3 m above the ground, falls on a roof that is 6.0 m above the ground.

a. Ignoring air resistance, find the kinetic energy of the limb when it reaches the roof.

b. What is the speed of the limb when it reaches the roof?

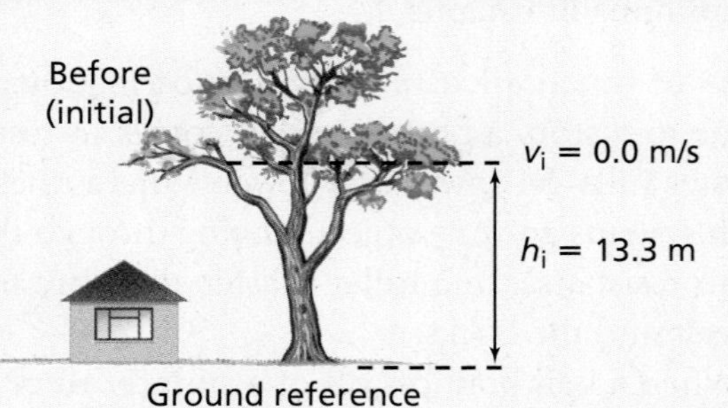

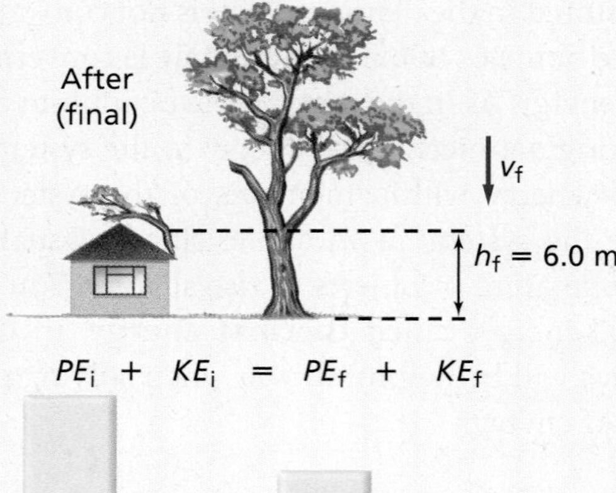

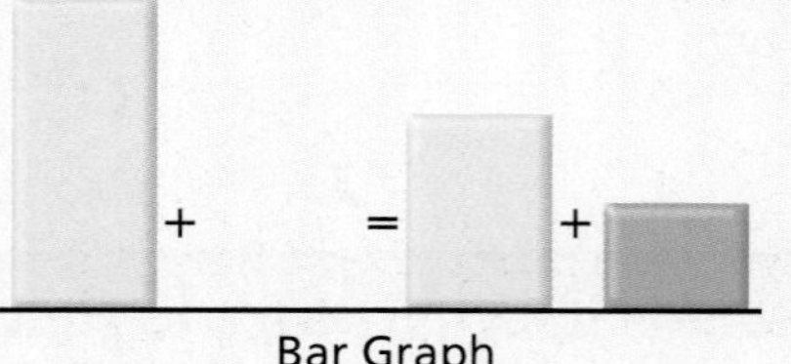

1 Analyze and Sketch the Problem

- Sketch the initial and final conditions.
- Choose a reference level.
- Draw a bar graph.

Known:

$m = 22.0$ kg $\quad g = 9.80$ m/s^2

$h_{limb} = 13.3$ m $\quad v_i = 0.0$ m/s

$h_{roof} = 6.0$ m $\quad KE_i = 0.0$ J

Unknown:

$PE_i = ?$ $\quad KE_f = ?$

$PE_f = ?$ $\quad v_f = ?$

2 Solve for the Unknown

a. Set the reference level as the height of the roof.

Solve for the initial height of the limb relative to the roof.

$$h = h_{limb} - h_{roof}$$
$$= 13.3 \text{ m} - 6.0 \text{ m} \qquad \text{Substitute } h_{limb} = 13.3 \text{ m}, h_{roof} = 6.0 \text{ m}$$
$$= 7.3 \text{ m}$$

Solve for the initial potential energy of the limb.

$$PE_i = mgh$$
$$= (22.0 \text{ kg})(9.80 \text{ m/s}^2)(7.3 \text{ m}) \qquad \text{Substitute } m = 22.0 \text{ kg}, g = 9.80 \text{ m/s}^2, h = 7.3 \text{ m}$$
$$= 1.6\times10^3 \text{ J}$$

Identify the initial kinetic energy of the limb.

$$KE_i = 0.0 \text{ J} \qquad \text{The tree limb is initially at rest.}$$

The kinetic energy of the limb when it reaches the roof is equal to its initial potential energy because energy is conserved.

$$KE_f = PE_i \qquad PE_f = 0.0 \text{ J because } h = 0.0 \text{ m at the reference level.}$$
$$= 1.6\times10^3 \text{ J}$$

b. Solve for the speed of the limb.

$$KE_f = \frac{1}{2}mv_f^2$$
$$v_f = \sqrt{\frac{2KE_f}{m}}$$
$$= \sqrt{\frac{2(1.6\times10^3 \text{ J})}{22.0 \text{ kg}}} \qquad \text{Substitute } KE_f = 1.6\times10^3 \text{ J}, m = 22.0 \text{ kg}$$
$$= 12 \text{ m/s}$$

Physics Online

Personal Tutor For an online tutorial on square and cube roots, visit physicspp.com.

3 Evaluate the Answer

- **Are the units correct?** Velocity is measured in m/s and energy is measured in kg·m^2/s^2 = J.
- **Do the signs make sense?** *KE* and the magnitude of velocity are always positive.

PRACTICE Problems

- Additional Problems, Appendix B
- Solutions to Selected Problems, Appendix C

15. A bike rider approaches a hill at a speed of 8.5 m/s. The combined mass of the bike and the rider is 85.0 kg. Choose a suitable system. Find the initial kinetic energy of the system. The rider coasts up the hill. Assuming there is no friction, at what height will the bike come to rest?

16. Suppose that the bike rider in problem 15 pedaled up the hill and never came to a stop. In what system is energy conserved? From what form of energy did the bike gain mechanical energy?

17. A skier starts from rest at the top of a 45.0-m-high hill, skis down a 30° incline into a valley, and continues up a 40.0-m-high hill. The heights of both hills are measured from the valley floor. Assume that you can neglect friction and the effect of the ski poles. How fast is the skier moving at the bottom of the valley? What is the skier's speed at the top of the next hill? Do the angles of the hills affect your answers?

18. In a belly-flop diving contest, the winner is the diver who makes the biggest splash upon hitting the water. The size of the splash depends not only on the diver's style, but also on the amount of kinetic energy that the diver has. Consider a contest in which each diver jumps from a 3.00-m platform. One diver has a mass of 136 kg and simply steps off the platform. Another diver has a mass of 102 kg and leaps upward from the platform. How high would the second diver have to leap to make a competitive splash?

Analyzing Collisions

A collision between two objects, whether the objects are automobiles, hockey players, or subatomic particles, is one of the most common situations analyzed in physics. Because the details of a collision can be very complex during the collision itself, the strategy is to find the motion of the objects just before and just after the collision. What conservation laws can be used to analyze such a system? If the system is isolated, then momentum and energy are conserved. However, the potential energy or thermal energy in the system may decrease, remain the same, or increase. Therefore, you cannot predict whether or not kinetic energy is conserved. **Figure 11-12** and **Figure 11-13** on the next page show three different kinds of collisions. In case 1, the momentum of the system before and after the collision is represented by the following:

$$p_i = p_{Ci} + p_{Di} = (1.00\ \text{kg})(1.00\ \text{m/s}) + (1.00\ \text{kg})(0.00\ \text{m/s}) = 1.00\ \text{kg·m/s}$$

$$p_f = p_{Cf} + p_{Df} = (1.00\ \text{kg})(-0.20\ \text{m/s}) + (1.00\ \text{kg})(1.20\ \text{m/s}) = 1.00\ \text{kg·m/s}$$

Thus, in case 1, the momentum is conserved. Look again at Figure 11-13 and verify for yourself that momentum is conserved in cases 2 and 3.

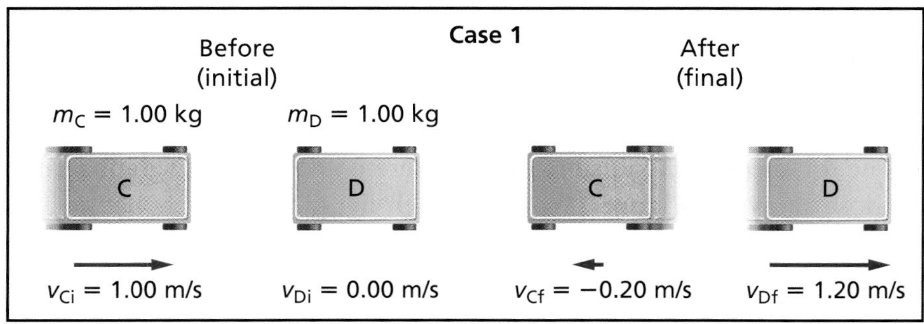

Figure 11-12 Two moving objects can have different types of collisions. Case 1: the two objects move apart in opposite directions.

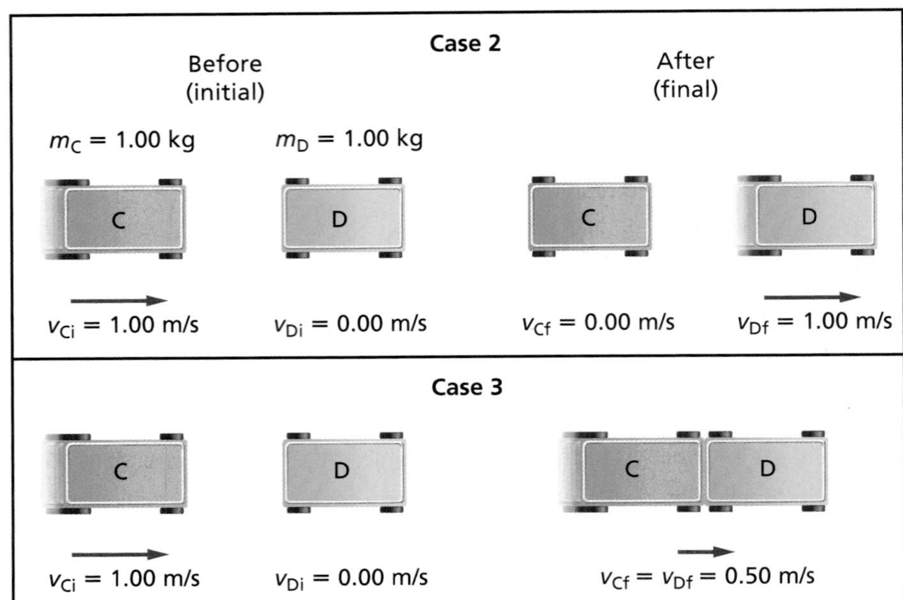

Figure 11-13 Case 2: the moving object comes to rest and the stationary object begins to move. Case 3: the two objects are stuck together and move as one.

Next, consider the kinetic energy of the system in each of these cases. For case 1 the kinetic energy of the system before and after the collision is represented by the following equations:

$$KE_{Ci} + KE_{Di} = \frac{1}{2}(1.00 \text{ kg})(1.00 \text{ m/s})^2 + \frac{1}{2}(1.00 \text{ kg})(0.00 \text{ m/s})^2$$
$$= 0.50 \text{ J}$$

$$KE_{Cf} + KE_{Df} = \frac{1}{2}(1.00 \text{ kg})(-0.20 \text{ m/s})^2 + \frac{1}{2}(1.00 \text{ kg})(1.20 \text{ m/s})^2$$
$$= 0.74 \text{ J}$$

In case 1, the kinetic energy of the system increased. If energy in the system is conserved, then one or more of the other forms of energy must have decreased. Perhaps when the two carts collided, a compressed spring was released, adding kinetic energy to the system. This kind of collision is sometimes called a superelastic or explosive collision.

After the collision in case 2, the kinetic energy is equal to:

$$KE_{Cf} + KE_{Df} = (1.0 \text{ kg})(0.00 \text{ m/s})^2 + \frac{1}{2}(1.0 \text{ kg})(1.0 \text{ m/s})^2 = 0.50 \text{ J}$$

Kinetic energy remained the same after the collision. This type of collision, in which the kinetic energy does not change, is called an **elastic collision.** Collisions between hard, elastic objects, such as those made of steel, glass, or hard plastic, often are called nearly elastic collisions.

After the collision in case 3, the kinetic energy is equal to:

$$KE_{Cf} + KE_{Df} = \frac{1}{2}(1.00 \text{ kg})(0.50 \text{ m/s})^2 + \frac{1}{2}(1.00 \text{ kg})(0.50 \text{ m/s})^2 = 0.25 \text{ J}$$

Kinetic energy decreased and some of it was converted to thermal energy. This kind of collision, in which kinetic energy decreases, is called an **inelastic collision.** Objects made of soft, sticky material, such as clay, act in this way.

The three kinds of collisions can be represented using bar graphs, such as those shown in **Figure 11-14.** Although the kinetic energy before and after the collisions can be calculated, only the change in other forms of energy can be found. In automobile collisions, kinetic energy is transferred into other forms of energy, such as heat and sound.

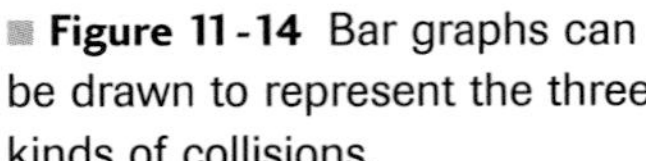

Figure 11-14 Bar graphs can be drawn to represent the three kinds of collisions.

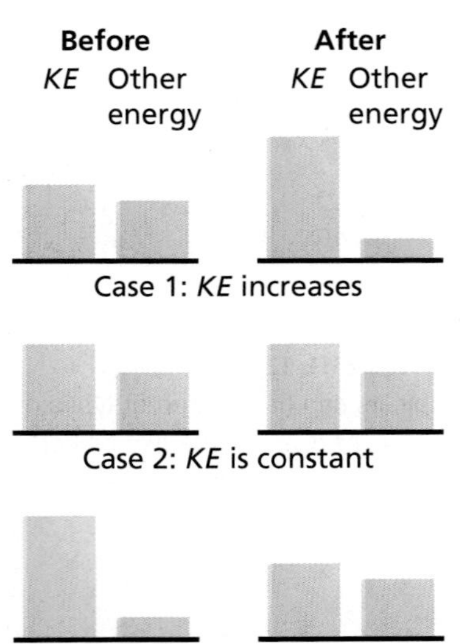

▶ EXAMPLE Problem 3

Kinetic Energy In an accident on a slippery road, a compact car with a mass of 575 kg moving at 15.0 m/s smashes into the rear end of a car with mass 1575 kg moving at 5.00 m/s in the same direction.

a. What is the final velocity if the wrecked cars lock together?

b. How much kinetic energy was lost in the collision?

c. What fraction of the original kinetic energy was lost?

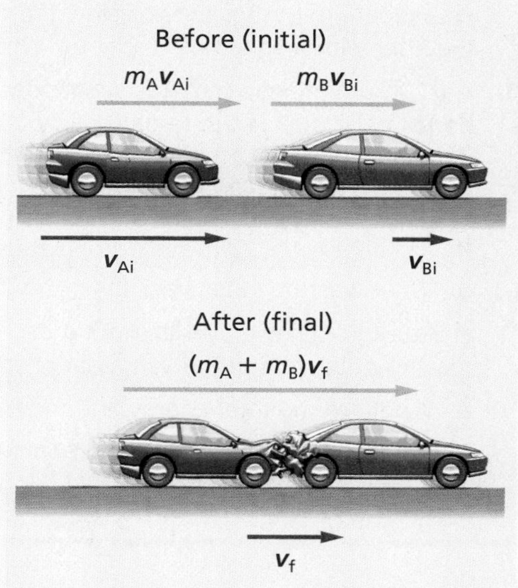

1 Analyze and Sketch the Problem

- Sketch the initial and final conditions.
- Sketch the momentum diagram.

Known:

$m_A = 575$ kg $m_B = 1575$ kg
$v_{Ai} = 15.0$ m/s $v_{Bi} = 5.00$ m/s
$v_{Af} = v_{Bf} = v_f$

Unknown:

$v_f = ?$ $\Delta KE = KE_f - KE_i = ?$
Fraction of KE_i lost, $\Delta KE/KE_i = ?$

2 Solve for the Unknown

a. Use the conservation of momentum equation to find the final velocity.

$$p_{Ai} + p_{Bi} = p_{Af} + p_{Bf}$$
$$m_A v_{Ai} + m_B v_{Bi} = (m_A + m_B)v_f$$
$$v_f = \frac{(m_A v_{Ai} + m_B v_{Bi})}{(m_A + m_B)}$$
$$= \frac{(575 \text{ kg})(15.0 \text{ m/s}) + (1575 \text{ kg})(5.00 \text{ m/s})}{(575 \text{ kg} + 1575 \text{ kg})}$$
Substitute $m_A = 575$ kg, $v_{Ai} = 15.0$ m/s, $m_B = 1575$ kg, $v_{Bi} = 5.00$ m/s

$= 7.67$ m/s, in the direction of the motion before the collision

Math Handbook
Isolating a Variable
page 845

b. To determine the change in kinetic energy of the system, KE_f and KE_i are needed.

$$KE_f = \frac{1}{2}mv^2$$
$$= \frac{1}{2}(m_A + m_B)v_f^2$$
Substitute $m = m_A + m_B$
$$= \frac{1}{2}(575 \text{ kg} + 1575 \text{ kg})(7.67 \text{ m/s})^2$$
Substitute $m_A = 575$ kg, $m_B = 1575$ kg, $v_f = 7.67$ m/s
$$= 6.32\times10^4 \text{ J}$$

$$KE_i = KE_{Ai} + KE_{Bi}$$
$$= \frac{1}{2}m_A v_{Ai}^2 + \frac{1}{2}m_B v_{Bi}^2$$
Substitute $KE_{Ai} = \frac{1}{2}m_A v_{Ai}^2$, $KE_{Bi} = \frac{1}{2}m_B v_{Bi}^2$
$$= \frac{1}{2}(575 \text{ kg})(15.0 \text{ m/s})^2 + \frac{1}{2}(1575 \text{ kg})(5.00 \text{ m/s})^2$$
Substitute $m_A = 575$ kg, $m_B = 1575$ kg, $v_{Ai} = 15.0$ m/s, $v_{Bi} = 5.00$ m/s
$$= 8.44\times10^4 \text{ J}$$

Solve for the change in kinetic energy of the system.

$$\Delta KE = KE_f - KE_i$$
$$= 6.32\times10^4 \text{ J} - 8.44\times10^4 \text{ J}$$
Substitute $KE_f = 6.32\times10^4$ J, $KE_i = 8.44\times10^4$ J
$$= -2.12\times10^4 \text{ J}$$

c. Calculate the fraction of the original kinetic energy that is lost.

$$\frac{\Delta KE}{KE_i} = \frac{-2.12\times10^4 \text{ J}}{8.44\times10^4 \text{ J}}$$
Substitute $\Delta KE = -2.11\times10^4$ J, $KE_i = 8.44\times10^4$ J

$= -0.251 = 25.1\%$ of the original kinetic energy in the system was lost.

3 Evaluate the Answer

- **Are the units correct?** Velocity is measured in m/s; energy is measured in J.
- **Does the sign make sense?** Velocity is positive, consistent with the original velocities.

PRACTICE Problems

- Additional Problems, Appendix B
- Solutions to Selected Problems, Appendix C

19. An 8.00-g bullet is fired horizontally into a 9.00-kg block of wood on an air table and is embedded in it. After the collision, the block and bullet slide along the frictionless surface together with a speed of 10.0 cm/s. What was the initial speed of the bullet?

20. A 0.73-kg magnetic target is suspended on a string. A 0.025-kg magnetic dart, shot horizontally, strikes the target head-on. The dart and the target together, acting like a pendulum, swing 12.0 cm above the initial level before instantaneously coming to rest.

a. Sketch the situation and choose a system.

b. Decide what is conserved in each part and explain your decision.

c. What was the initial velocity of the dart?

21. A 91.0-kg hockey player is skating on ice at 5.50 m/s. Another hockey player of equal mass, moving at 8.1 m/s in the same direction, hits him from behind. They slide off together.

a. What are the total energy and momentum in the system before the collision?

b. What is the velocity of the two hockey players after the collision?

c. How much energy was lost in the collision?

In collisions, you can see how momentum and energy are really very different. Momentum is almost always conserved in a collision. Energy is conserved only in elastic collisions. Momentum is what makes objects stop. A 10.0-kg object moving at 5.00 m/s will stop a 20.0-kg object moving at 2.50 m/s if they have a head-on collision. However, in this case, the smaller object has much more kinetic energy. The kinetic energy of the smaller object is $KE = \frac{1}{2}(10.0\text{ kg})(5.0\text{ m/s})^2 = 125\text{ J}$. The kinetic energy of the larger object is $KE = \frac{1}{2}(20.0\text{ kg})(2.50\text{ m/s})^2 = 62.5\text{ J}$. Based on the work-energy theorem, you can conclude that it takes more work to make the 10.0-kg object move at 5.00 m/s than it does to move the 20.0-kg object at 2.50 m/s. It sometimes is said that in automobile collisions, the momentum stops the cars but it is the energy in the collision that causes the damage.

It also is possible to have a collision in which nothing collides. If two lab carts sit motionless on a table, connected by a compressed spring, their total momentum is zero. If the spring is released, the carts will be forced to move away from each other. The potential energy of the spring will be transformed into the kinetic energy of the carts. The carts will still move away from each other so that their total momentum is zero.

CHALLENGE PROBLEM

A bullet of mass m, moving at speed v_1, goes through a motionless wooden block and exits with speed v_2. After the collision, the block, which has mass m_B, is moving.

1. What is the final speed, v_B, of the block?

2. How much energy was lost to the bullet?

3. How much energy was lost to friction inside the block?

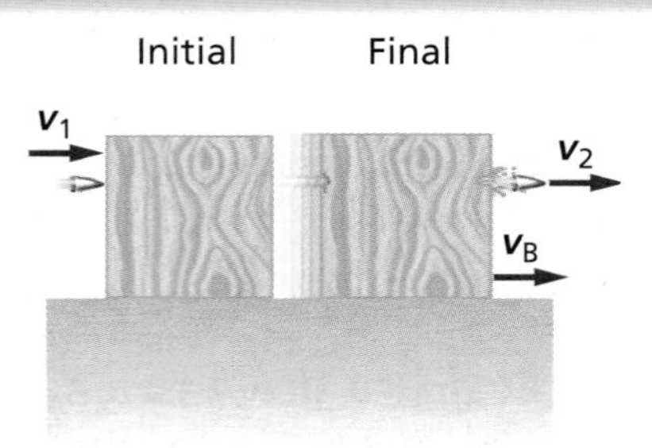

It is useful to remember two simple examples of collisions. One is the elastic collision between two objects of equal mass, such as when a cue ball with velocity, $\boldsymbol{v}$, hits a motionless billiard ball head-on. In this case, after the collision, the cue ball is motionless and the other ball rolls off at velocity, $\boldsymbol{v}$. It is easy to prove that both momentum and energy are conserved in this collision.

The other simple example is to consider a skater of mass m, with velocity $\boldsymbol{v}$, running into another skater of equal mass who happens to be standing motionless on the ice. If they hold on to each other after the collision, they will slide off at a velocity of $\frac{1}{2}\boldsymbol{v}$ because of the conservation of momentum. The final kinetic energy of the pair would be equal to $KE = \frac{1}{2}(2m)(\frac{1}{2}v)^2 = \frac{1}{4}mv^2$, which is half the initial kinetic energy. This is because the collision was inelastic.

You have investigated examples in which the conservation of energy, and sometimes the conservation of momentum, can be used to calculate the motions of a system of objects. These systems would be too complicated to comprehend using only Newton's second law of motion. The understanding of the forms of energy and how energy flows from one form to another is one of the most useful concepts in science. The term *energy conservation* appears in everything from scientific papers to electric appliance commercials. Scientists use the concept of energy to explore topics much more complicated than colliding billiard balls.

MINI LAB

Energy Exchange

1. Select different-sized steel balls and determine their masses.

2. Stand a spring-loaded laboratory cart on end with the spring mechanism pointing upward.

3. Place a ball on top of the spring mechanism and press down until the ball is touching the cart.

4. Quickly release the ball so that the spring shoots it upward. ***CAUTION: Stay clear of the ball when launching.***

5. Repeat the process several times, and measure the average height.

6. Estimate how high the other sizes of steel balls will rise.

Analyze and Conclude

7. Classify the balls by height attained. What can you conclude?

11.2 Section Review

22. Closed Systems Is Earth a closed, isolated system? Support your answer.

23. Energy A child jumps on a trampoline. Draw bar graphs to show the forms of energy present in the following situations.

a. The child is at the highest point.

b. The child is at the lowest point.

24. Kinetic Energy Suppose a glob of chewing gum and a small, rubber ball collide head-on in midair and then rebound apart. Would you expect kinetic energy to be conserved? If not, what happens to the energy?

25. Kinetic Energy In table tennis, a very light but hard ball is hit with a hard rubber or wooden paddle. In tennis, a much softer ball is hit with a racket. Why are the two sets of equipment designed in this way? Can you think of other ball-paddle pairs in sports? How are they designed?

26. Potential Energy A rubber ball is dropped from a height of 8.0 m onto a hard concrete floor. It hits the floor and bounces repeatedly. Each time it hits the floor, it loses $\frac{1}{5}$ of its total energy. How many times will it bounce before it bounces back up to a height of only about 4 m?

27. Energy As shown in **Figure 11-15,** a 36.0-kg child slides down a playground slide that is 2.5 m high. At the bottom of the slide, she is moving at 3.0 m/s. How much energy was lost as she slid down the slide?

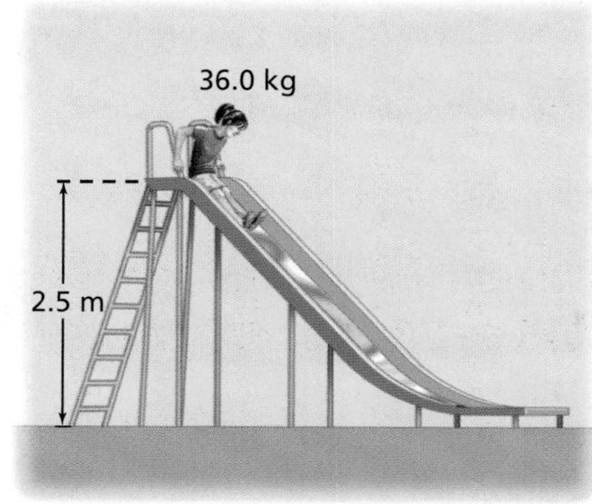

Figure 11-15

28. Critical Thinking A ball drops 20 m. When it has fallen half the distance, or 10 m, half of its energy is potential and half is kinetic. When the ball has fallen for half the amount of time it takes to fall, will more, less, or exactly half of its energy be potential energy?

PHYSICS LAB

Conservation of Energy

Alternate CBL instructions can be found on the Web site.
physicspp.com

There are many examples of situations where energy is conserved. One such example is a rock falling from a given height. If the rock starts at rest, at the moment the rock is dropped, it only has potential energy. As it falls, its potential energy decreases as its height decreases, but its kinetic energy increases. The sum of potential energy and kinetic energy remains constant if friction is neglected. When the rock is about to hit the ground, all of its potential energy has been converted to kinetic energy. In this experiment, you will model a falling object and calculate its speed as it hits the ground.

QUESTION

How does the transfer of an object's potential energy to kinetic energy demonstrate conservation of energy?

Objectives

- **Calculate** the speed of a falling object as it hits the ground by using a model.
- **Interpret data** to find the relationship between potential energy and kinetic energy of a falling object.

Safety Precautions

Figure 1

Figure 2

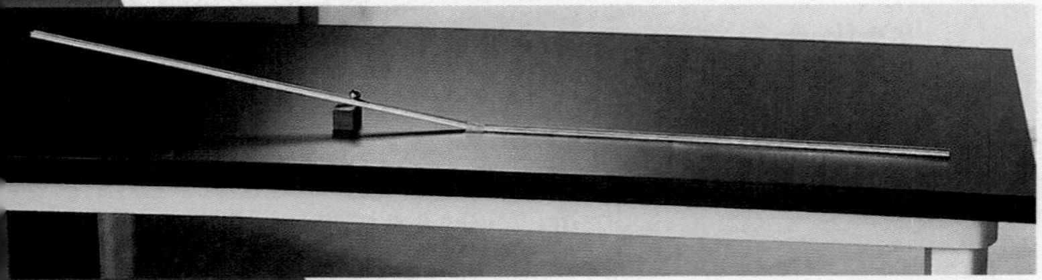

Figure 3

Materials

grooved track (two sections)
marble or steel ball
stopwatch
block of wood
electronic balance
metric ruler
graphing calculator

Procedure

1. Place the two sections of grooved track together, as shown in **Figure 1.** Raise one end of the track and place the block under it, about 5 cm from the raised end. Make sure the ball can roll smoothly across the junction of the two tracks.
2. Record the length of the level portion of the track in the data table. Place a ball on the track directly above the point supported by the block. Release the ball. Start the stopwatch when the ball reaches the level section of track. Stop timing when the ball reaches the end of the level portion of the track. Record the time required for the ball to travel that distance in the data table.
3. Move the support block so that it is under the midsection of the inclined track, as shown in **Figure 2.** Place the ball on the track just above the point supported by the block. Release the ball and measure the time needed for the ball to roll the length of the level portion of the track and record it in the data table. Notice that even though the incline is steeper, the ball is released from the same height as in step 2.
4. Calculate the speed of the ball on the level portion of the track in steps 2 and 3. Move the support block to a point about three-quarters down the length of the inclined track, as shown in **Figure 3.**

Data Table			
Release Height (m)	**Distance (m)**	**Time (s)**	**Speed (m/s)**
0.05			
0.05			
0.05			
0.01			
0.02			
0.03			

5. Predict the amount of time the ball will take to travel the length of the level portion of the track. Record your prediction. Test your prediction.
6. Place the support block at the midpoint of the inclined track (Figure 2). Measure a point on the inclined portion of the track that is 1.0 cm above the level portion of the track. Be sure to measure 1.0 cm above the level portion, and not 1.0 cm above the table.
7. Release the ball from this point and measure the time required for the ball to travel on the level portion of the track and record it in the data table.
8. Use a ruler to measure a point that is 2.0 cm above the level track. Release the ball from this point and measure the time required for the ball to travel on the level portion of the track. Record the time in the data table.
9. Repeat step 8 for 3.0 cm, 4.0 cm, 5.0 cm, 6.0 cm, 7.0 cm, and 8.0 cm. Record the times.

Analyze

1. **Infer** What effect did changing the slope of the inclined plane in steps 2–6 have on the speed of the ball on the level portion of the track?
2. **Analyze** Perform a power law regression for this graph using your graphing calculator. Record the equation of this function. Graph this by inputting the equation into Y=. Draw a sketch of the graph.
3. Using the data from step 9 for the release point of 8.0 cm, find the potential energy of the ball before it was released. Use an electronic balance to find the mass of the ball. Note that height must be in m, and mass in kg.
4. Using the speed data from step 9 for the release point of 8.0 cm calculate the kinetic energy of the ball on the level portion of the track. Remember, speed must be in m/s and mass in kg.

Conclude and Apply

1. Solve for speed, y, in terms of height, x. Begin by setting $PE_i = KE_f$.
2. How does the equation found in the previous question relate to the power law regression calculated earlier?
3. Suppose you want the ball to roll twice as fast on the level part of the track as it did when you released it from the 2-cm mark. Using the power law regression performed earlier, calculate the height from which you should release the ball.
4. Explain how this experiment only models dropping a ball and finding its kinetic energy just as it hits the ground.
5. **Compare and Contrast** Compare the potential energy of the ball before it is released (step 8) to the kinetic energy of the ball on the level track (step 9). Explain why they are the same or why they are different.
6. **Draw Conclusions** Does this experiment demonstrate conservation of energy? Explain.

Going Further

What are potential sources of error in the experiment, and how can they be reduced?

Real-World Physics

How does your favorite roller coaster demonstrate the conservation of energy by the transfer of potential energy to kinetic energy?

To find out more about energy, visit the Web site: **physicspp.com**

Technology and Society

Running Smarter

The Physics of Running Shoes Today's running shoes are high-tech marvels. They enhance performance and protect your body by acting as shock absorbers. How do running shoes help you win a race? They reduce your energy consumption, as well as allow you to use energy more efficiently. Good running shoes must be flexible enough to bend with your feet as you run, support your feet, and hold them in place. They must be lightweight and provide traction to prevent slipping.

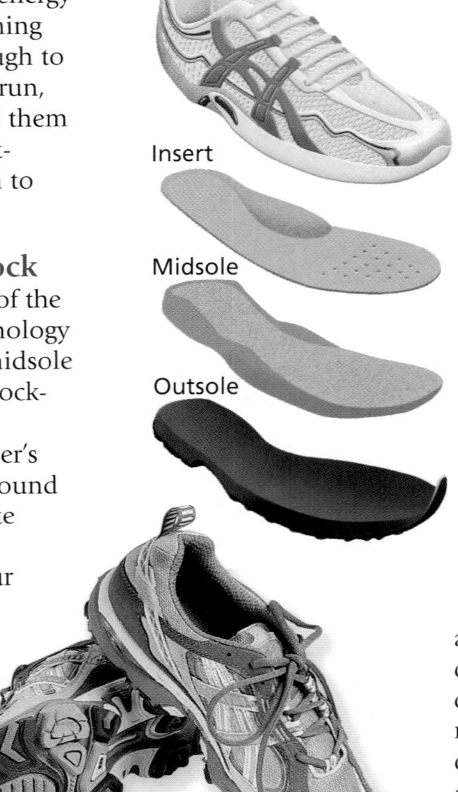

Running Shoes as Shock Absorbers

Today, much of the focus of running shoe technology centers on the cushioned midsole that plays a key role as a shock-absorber and performance enhancer. Each time a runner's foot hits the ground, the ground exerts an equal and opposite force on the runner's foot. This force can be nearly four times a runner's weight, causing aches and pains, shin splints, and damage to knees and ankles over long distances.

Cushioning is used in running shoes to decrease the force absorbed by the runner. As a runner's foot hits the ground and comes to a stop, its momentum changes. The change in momentum is $\Delta p = F\Delta t$, where F is the force on that object and Δt is the time during which the force acts. The cushioning causes the change of momentum to occur over an extended time and reduces the force of the foot on the ground. The decreased force reduces the damage to the runner's body.

Running Shoes Boost Performance

A shoe's cushioning system also affects energy consumption. The bones, muscles, ligaments, and tendons of the foot and leg are a natural cushioning system. But operating this system requires the body to use stored energy to contract muscles. So if a shoe can be worn that assists a runner's natural cushioning system, the runner does not expend as much of his or her own stored energy. The energy the runner saved can be spent to run farther or faster.

The cushioned midsole uses the law of conservation of energy to return as much of the energy to the runner as possible. The runner's kinetic energy transforms to elastic potential energy, plus heat, when the runner's foot hits the running surface. If the runner can reduce the amount of energy that is lost as heat, the runner's elastic potential energy can be converted back to useful kinetic energy.

Bouncy, springy, elastic materials that resist crushing over time commonly are used to create the cushioned midsole. Options now range from silicon gel pads to complex fluid-filled systems and even springs to give a runner extra energy efficiency.

Going Further

1. **Use Scientific Explanations** Use physics to explain why manufacturers put cushioned midsoles in running shoes.
2. **Analyze** Which surface would provide more cushioning when running: a grassy field or a concrete sidewalk? Explain why that surface provides better cushioning.
3. **Research** Some people prefer to run barefoot, even in marathon races. Why might this be so?

Chapter 11

Study Guide

11.1 The Many Forms of Energy

Vocabulary

- rotational kinetic energy (p. 287)
- gravitational potential energy (p. 288)
- reference level (p. 288)
- elastic potential energy (p. 291)

Key Concepts

- The kinetic energy of an object is proportional to its mass and the square of its velocity.
- The rotational kinetic energy of an object is proportional to the object's moment of inertia and the square of its angular velocity.
- When Earth is included in a system, the work done by gravity is replaced by gravitational potential energy.
- The gravitational potential energy of an object depends on the object's weight and its distance from Earth's surface.

$$PE = mgh$$

- The reference level is the position where the gravitational potential energy is defined to be zero.
- Elastic potential energy may be stored in an object as a result of its change in shape.
- Albert Einstein recognized that mass itself has potential energy. This energy is called rest energy.

$$E_0 = mc^2$$

11.2 Conservation of Energy

Vocabulary

- law of conservation of energy (p. 293)
- mechanical energy (p. 293)
- thermal energy (p. 295)
- elastic collision (p. 298)
- inelastic collision (p. 298)

Key Concepts

- The sum of kinetic and potential energy is called mechanical energy.

$$E = KE + PE$$

- If no objects enter or leave a system, the system is considered to be a closed system.
- If there are no external forces acting on a system, the system is considered to be an isolated system.
- The total energy of a closed, isolated system is constant. Within the system, energy can change form, but the total amount of energy does not change. Thus, energy is conserved.

$$KE_{before} + PE_{before} = KE_{after} + PE_{after}$$

- The type of collision in which the kinetic energy after the collision is less than the kinetic energy before the collision is called an inelastic collision.
- The type of collision in which the kinetic energy before and after the collision is the same is called an elastic collision.
- Momentum is conserved in collisions if the external force is zero. The mechanical energy may be unchanged or decreased by the collision, depending on whether the collision is elastic or inelastic.

Chapter 11 Assessment

Concept Mapping

29. Complete the concept map using the following terms: *gravitational potential energy, elastic potential energy, kinetic energy.*

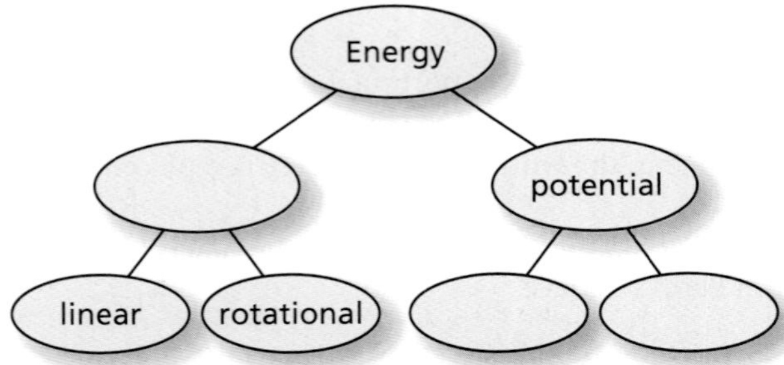

Mastering Concepts

Unless otherwise directed, assume that air resistance is negligible.

30. Explain how work and a change in energy are related. (11.1)

31. What form of energy does a wound-up watch spring have? What form of energy does a functioning mechanical watch have? When a watch runs down, what has happened to the energy? (11.1)

32. Explain how energy change and force are related. (11.1)

33. A ball is dropped from the top of a building. You choose the top of the building to be the reference level, while your friend chooses the bottom. Explain whether the energy calculated using these two reference levels is the same or different for the following situations. (11.1)
a. the ball's potential energy at any point
b. the change in the ball's potential energy as a result of the fall
c. the kinetic energy of the ball at any point

34. Can the kinetic energy of a baseball ever be negative? (11.1)

35. Can the gravitational potential energy of a baseball ever be negative? Explain without using a formula. (11.1)

36. If a sprinter's velocity increases to three times the original velocity, by what factor does the kinetic energy increase? (11.1)

37. What energy transformations take place when an athlete is pole-vaulting? (11.2)

38. The sport of pole-vaulting was drastically changed when the stiff, wooden poles were replaced by flexible, fiberglass poles. Explain why. (11.2)

39. You throw a clay ball at a hockey puck on ice. The smashed clay ball and the hockey puck stick together and move slowly. (11.2)
a. Is momentum conserved in the collision? Explain.
b. Is kinetic energy conserved? Explain.

40. Draw energy bar graphs for the following processes. (11.2)
a. An ice cube, initially at rest, slides down a frictionless slope.
b. An ice cube, initially moving, slides up a frictionless slope and instantaneously comes to rest.

41. Describe the transformations from kinetic energy to potential energy and vice versa for a roller-coaster ride. (11.2)

42. Describe how the kinetic energy and elastic potential energy are lost in a bouncing rubber ball. Describe what happens to the motion of the ball. (11.2)

Applying Concepts

43. The driver of a speeding car applies the brakes and the car comes to a stop. The system includes the car but not the road. Apply the work-energy theorem to the following situations.
a. The car's wheels do not skid.
b. The brakes lock and the car's wheels skid.

44. A compact car and a trailer truck are both traveling at the same velocity. Did the car engine or the truck engine do more work in accelerating its vehicle?

45. **Catapults** Medieval warriors used catapults to assault castles. Some catapults worked by using a tightly wound rope to turn the catapult arm. What forms of energy are involved in catapulting a rock to the castle wall?

46. Two cars collide and come to a complete stop. Where did all of their energy go?

47. During a process, positive work is done on a system, and the potential energy decreases. Can you determine anything about the change in kinetic energy of the system? Explain.

48. During a process, positive work is done on a system, and the potential energy increases. Can you tell whether the kinetic energy increased, decreased, or remained the same? Explain.

49. **Skating** Two skaters of unequal mass have the same speed and are moving in the same direction. If the ice exerts the same frictional force on each skater, how will the stopping distances of their bodies compare?

 For more problems, go to Additional Problems, Appendix B.

50. You swing a 55-g mass on the end of a 0.75-m string around your head in a nearly horizontal circle at constant speed, as shown in **Figure 11-16.**
- **a.** How much work is done on the mass by the tension of the string in one revolution?
- **b.** Is your answer to part **a** in agreement with the work-energy theorem? Explain.

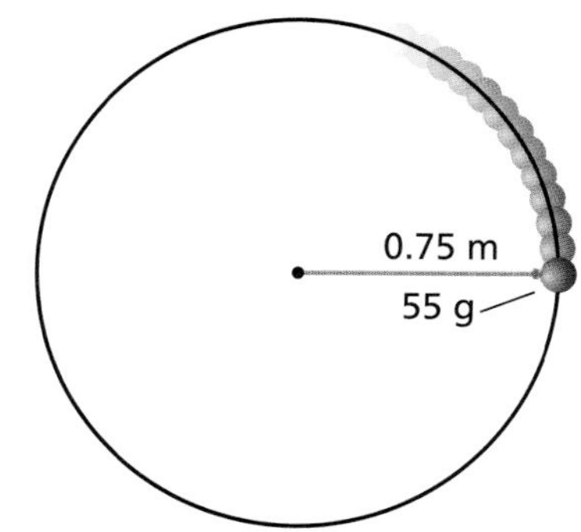

Figure 11-16

51. Give specific examples that illustrate the following processes.
- **a.** Work is done on a system, thereby increasing kinetic energy with no change in potential energy.
- **b.** Potential energy is changed to kinetic energy with no work done on the system.
- **c.** Work is done on a system, increasing potential energy with no change in kinetic energy.
- **d.** Kinetic energy is reduced, but potential energy is unchanged. Work is done by the system.

52. **Roller Coaster** You have been hired to make a roller coaster more exciting. The owners want the speed at the bottom of the first hill doubled. How much higher must the first hill be built?

53. Two identical balls are thrown from the top of a cliff, each with the same speed. One is thrown straight up, the other straight down. How do the kinetic energies and speeds of the balls compare as they strike the ground?

Mastering Problems

Unless otherwise directed, assume that air resistance is negligible.

11.1 The Many Forms of Energy

54. A 1600-kg car travels at a speed of 12.5 m/s. What is its kinetic energy?

55. A racing car has a mass of 1525 kg. What is its kinetic energy if it has a speed of 108 km/h?

56. Shawn and his bike have a combined mass of 45.0 kg. Shawn rides his bike 1.80 km in 10.0 min at a constant velocity. What is Shawn's kinetic energy?

57. Tony has a mass of 45 kg and is moving with a speed of 10.0 m/s.
- **a.** Find Tony's kinetic energy.
- **b.** Tony's speed changes to 5.0 m/s. Now what is his kinetic energy?
- **c.** What is the ratio of the kinetic energies in parts **a** and **b**? Explain.

58. Katia and Angela each have a mass of 45 kg, and they are moving together with a speed of 10.0 m/s.
- **a.** What is their combined kinetic energy?
- **b.** What is the ratio of their combined mass to Katia's mass?
- **c.** What is the ratio of their combined kinetic energy to Katia's kinetic energy? Explain.

59. **Train** In the 1950s, an experimental train, which had a mass of 2.50×10^4 kg, was powered across a level track by a jet engine that produced a thrust of 5.00×10^5 N for a distance of 509 m.
- **a.** Find the work done on the train.
- **b.** Find the change in kinetic energy.
- **c.** Find the final kinetic energy of the train if it started from rest.
- **d.** Find the final speed of the train if there had been no friction.

60. **Car Brakes** A 14,700-N car is traveling at 25 m/s. The brakes are applied suddenly, and the car slides to a stop, as shown in **Figure 11-17.** The average braking force between the tires and the road is 7100 N. How far will the car slide once the brakes are applied?

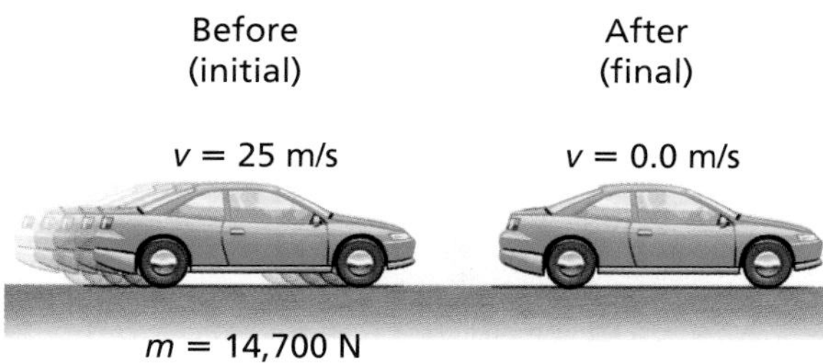

Figure 11-17

61. A 15.0-kg cart is moving with a velocity of 7.50 m/s down a level hallway. A constant force of 10.0 N acts on the cart, and its velocity becomes 3.20 m/s.
- **a.** What is the change in kinetic energy of the cart?
- **b.** How much work was done on the cart?
- **c.** How far did the cart move while the force acted?

62. How much potential energy does DeAnna, with a mass of 60.0 kg, gain when she climbs a gymnasium rope a distance of 3.5 m?

63. **Bowling** A 6.4-kg bowling ball is lifted 2.1 m into a storage rack. Calculate the increase in the ball's potential energy.

64. Mary weighs 505 N. She walks down a flight of stairs to a level 5.50 m below her starting point. What is the change in Mary's potential energy?

65. **Weightlifting** A weightlifter raises a 180-kg barbell to a height of 1.95 m. What is the increase in the potential energy of the barbell?

66. A 10.0-kg test rocket is fired vertically from Cape Canaveral. Its fuel gives it a kinetic energy of 1960 J by the time the rocket engine burns all of the fuel. What additional height will the rocket rise?

67. Antwan raised a 12.0-N physics book from a table 75 cm above the floor to a shelf 2.15 m above the floor. What was the change in the potential energy of the system?

68. A hallway display of energy is constructed in which several people pull on a rope that lifts a block 1.00 m. The display indicates that 1.00 J of work is done. What is the mass of the block?

69. **Tennis** It is not uncommon during the serve of a professional tennis player for the racket to exert an average force of 150.0 N on the ball. If the ball has a mass of 0.060 kg and is in contact with the strings of the racket, as shown in **Figure 11-18,** for 0.030 s, what is the kinetic energy of the ball as it leaves the racket? Assume that the ball starts from rest.

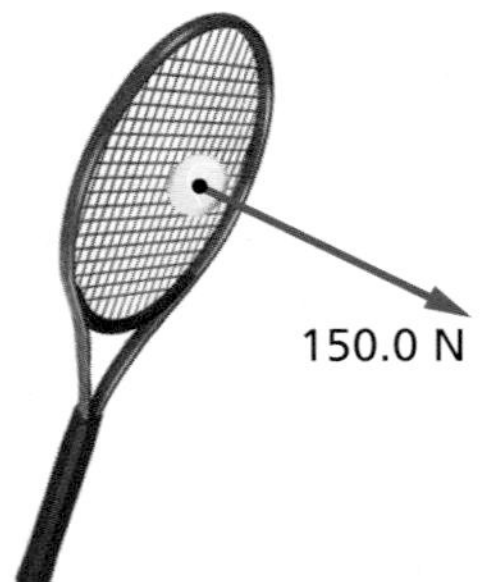

Figure 11-18

70. Pam, wearing a rocket pack, stands on frictionless ice. She has a mass of 45 kg. The rocket supplies a constant force for 22.0 m, and Pam acquires a speed of 62.0 m/s.
  a. What is Pam's final kinetic energy?
  b. What is the magnitude of the force?

71. **Collision** A 2.00×10^3-kg car has a speed of 12.0 m/s. The car then hits a tree. The tree doesn't move, and the car comes to rest, as shown in **Figure 11-19.**
  a. Find the change in kinetic energy of the car.
  b. Find the amount of work done as the front of the car crashes into the tree.
  c. Find the size of the force that pushed in the front of the car by 50.0 cm.

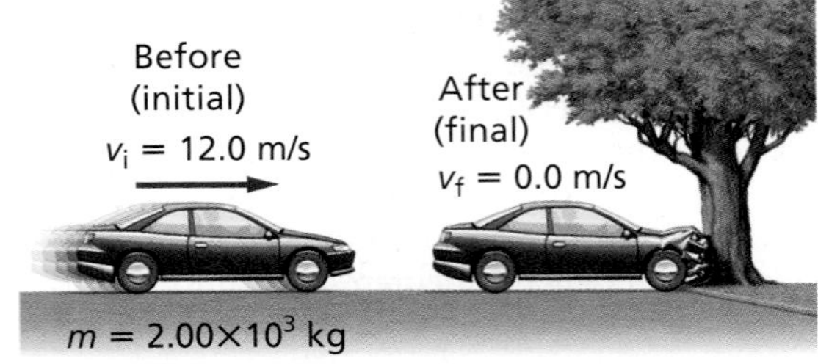

Figure 11-19

72. A constant net force of 410 N is applied upward to a stone that weighs 32 N. The upward force is applied through a distance of 2.0 m, and the stone is then released. To what height, from the point of release, will the stone rise?

11.2 Conservation of Energy

73. A 98.0-N sack of grain is hoisted to a storage room 50.0 m above the ground floor of a grain elevator.
  a. How much work was done?
  b. What is the increase in potential energy of the sack of grain at this height?
  c. The rope being used to lift the sack of grain breaks just as the sack reaches the storage room. What kinetic energy does the sack have just before it strikes the ground floor?

74. A 20-kg rock is on the edge of a 100-m cliff, as shown in **Figure 11-20.**
  a. What potential energy does the rock possess relative to the base of the cliff?
  b. The rock falls from the cliff. What is its kinetic energy just before it strikes the ground?
  c. What speed does the rock have as it strikes the ground?

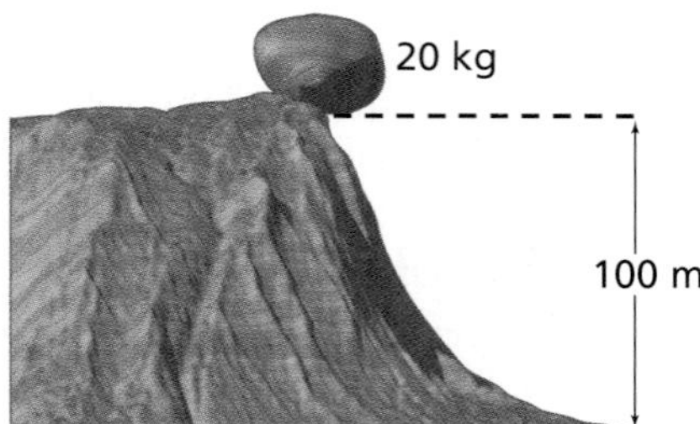

Figure 11-20

75. **Archery** An archer puts a 0.30-kg arrow to the bowstring. An average force of 201 N is exerted to draw the string back 1.3 m.
  a. Assuming that all the energy goes into the arrow, with what speed does the arrow leave the bow?
  b. If the arrow is shot straight up, how high does it rise?

 For more problems, go to Additional Problems, Appendix B.

76. A 2.0-kg rock that is initially at rest loses 407 J of potential energy while falling to the ground. Calculate the kinetic energy that the rock gains while falling. What is the rock's speed just before it strikes the ground?

77. A physics book of unknown mass is dropped 4.50 m. What speed does the book have just before it hits the ground?

78. **Railroad Car** A railroad car with a mass of 5.0×10^5 kg collides with a stationary railroad car of equal mass. After the collision, the two cars lock together and move off at 4.0 m/s, as shown in **Figure 11-21.**
   a. Before the collision, the first railroad car was moving at 8.0 m/s. What was its momentum?
   b. What was the total momentum of the two cars after the collision?
   c. What were the kinetic energies of the two cars before and after the collision?
   d. Account for the loss of kinetic energy.

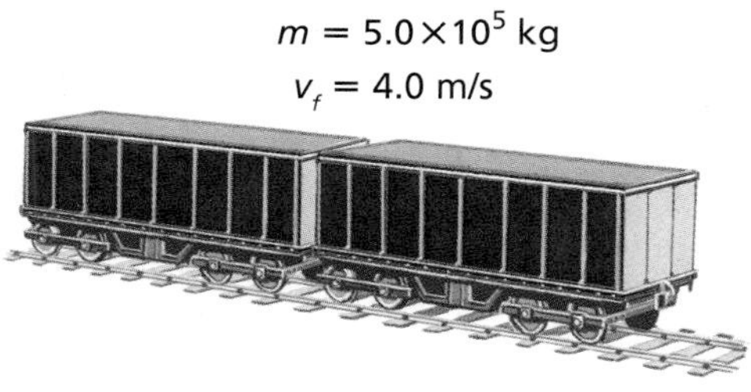

Figure 11-21

79. From what height would a compact car have to be dropped to have the same kinetic energy that it has when being driven at 1.00×10^2 km/h?

80. Kelli weighs 420 N, and she is sitting on a playground swing that hangs 0.40 m above the ground. Her mom pulls the swing back and releases it when the seat is 1.00 m above the ground.
   a. How fast is Kelli moving when the swing passes through its lowest position?
   b. If Kelli moves through the lowest point at 2.0 m/s, how much work was done on the swing by friction?

81. Hakeem throws a 10.0-g ball straight down from a height of 2.0 m. The ball strikes the floor at a speed of 7.5 m/s. What was the initial speed of the ball?

82. **Slide** Lorena's mass is 28 kg. She climbs the 4.8-m ladder of a slide and reaches a velocity of 3.2 m/s at the bottom of the slide. How much work was done by friction on Lorena?

83. A person weighing 635 N climbs up a ladder to a height of 5.0 m. Use the person and Earth as the system.
   a. Draw energy bar graphs of the system before the person starts to climb the ladder and after the person stops at the top. Has the mechanical energy changed? If so, by how much?
   b. Where did this energy come from?

Mixed Review

84. Suppose a chimpanzee swings through the jungle on vines. If it swings from a tree on a 13-m-long vine that starts at an angle of 45°, what is the chimp's velocity when it reaches the ground?

85. An 0.80-kg cart rolls down a frictionless hill of height 0.32 m. At the bottom of the hill, the cart rolls on a flat surface, which exerts a frictional force of 2.0 N on the cart. How far does the cart roll on the flat surface before it comes to a stop?

86. **High Jump** The world record for the men's high jump is about 2.45 m. To reach that height, what is the minimum amount of work that a 73.0-kg jumper must exert in pushing off the ground?

87. A stuntwoman finds that she can safely break her fall from a one-story building by landing in a box filled to a 1-m depth with foam peanuts. In her next movie, the script calls for her to jump from a five-story building. How deep a box of foam peanuts should she prepare?

88. **Football** A 110-kg football linebacker has a head-on collision with a 150-kg defensive end. After they collide, they come to a complete stop. Before the collision, which player had the greater momentum and which player had the greater kinetic energy?

89. A 2.0-kg lab cart and a 1.0-kg lab cart are held together by a compressed spring. The lab carts move at 2.1 m/s in one direction. The spring suddenly becomes uncompressed and pushes the two lab carts apart. The 2-kg lab cart comes to a stop, and the 1.0-kg lab cart moves ahead. How much energy did the spring add to the lab carts?

90. A 55.0-kg scientist roping through the top of a tree in the jungle sees a lion about to attack a tiny antelope. She quickly swings down from her 12.0-m-high perch and grabs the antelope (21.0 kg) as she swings. They barely swing back up to a tree limb out of reach of the lion. How high is this tree limb?

91. An 0.80-kg cart rolls down a 30.0° hill from a vertical height of 0.50 m as shown in **Figure 11-22.** The distance that the cart must roll to the bottom of the hill is 0.50 m/sin 30.0° = 1.0 m. The surface of the hill exerts a frictional force of 5.0 N on the cart. Does the cart roll to the bottom of the hill?

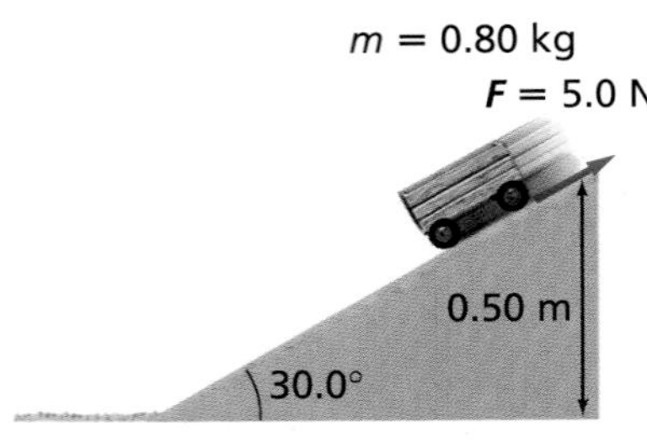

Figure 11-22

92. Object A, sliding on a frictionless surface at 3.2 m/s, hits a 2.0-kg object, B, which is motionless. The collision of A and B is completely elastic. After the collision, A and B move away from each other at equal and opposite speeds. What is the mass of object A?

93. Hockey A 90.0-kg hockey player moving at 5.0 m/s collides head-on with a 110-kg hockey player moving at 3.0 m/s in the opposite direction. After the collision, they move off together at 1.0 m/s. How much energy was lost in the collision?

Thinking Critically

94. Apply Concepts A golf ball with a mass of 0.046 kg rests on a tee. It is struck by a golf club with an effective mass of 0.220 kg and a speed of 44 m/s. Assuming that the collision is elastic, find the speed of the ball when it leaves the tee.

95. Apply Concepts A fly hitting the windshield of a moving pickup truck is an example of a collision in which the mass of one of the objects is many times larger than the other. On the other hand, the collision of two billiard balls is one in which the masses of both objects are the same. How is energy transferred in these collisions? Consider an elastic collision in which billiard ball m_1 has velocity $\boldsymbol{v}_1$ and ball m_2 is motionless.

a. If $m_1 = m_2$, what fraction of the initial energy is transferred to m_2?

b. If $m_1 >> m_2$, what fraction of the initial energy is transferred to m_2?

c. In a nuclear reactor, neutrons must be slowed down by causing them to collide with atoms. (A neutron is about as massive as a proton.) Would hydrogen, carbon, or iron atoms be more desirable to use for this purpose?

96. Analyze and Conclude In a perfectly elastic collision, both momentum and mechanical energy are conserved. Two balls, with masses m_A and m_B, are moving toward each other with speeds v_A and v_B, respectively. Solve the appropriate equations to find the speeds of the two balls after the collision.

97. Analyze and Conclude A 25-g ball is fired with an initial speed of v_1 toward a 125-g ball that is hanging motionless from a 1.25-m string. The balls have a perfectly elastic collision. As a result, the 125-g ball swings out until the string makes an angle of 37.0° with the vertical. What is v_1?

Writing in Physics

98. All energy comes from the Sun. In what forms has this solar energy come to us to allow us to live and to operate our society? Research the ways that the Sun's energy is turned into a form that we can use. After we use the Sun's energy, where does it go? Explain.

99. All forms of energy can be classified as either kinetic or potential energy. How would you describe nuclear, electric, chemical, biological, solar, and light energy, and why? For each of these types of energy, research what objects are moving and how energy is stored in those objects.

Cumulative Reveiw

100. A satellite is placed in a circular orbit with a radius of 1.0×10^7 m and a period of 9.9×10^3 s. Calculate the mass of Earth. *Hint: Gravity is the net force on such a satellite. Scientists have actually measured the mass of Earth this way.* (Chapter 7)

101. A 5.00-g bullet is fired with a velocity of 100.0 m/s toward a 10.00-kg stationary solid block resting on a frictionless surface. (Chapter 9)

a. What is the change in momentum of the bullet if it is embedded in the block?

b. What is the change in momentum of the bullet if it ricochets in the opposite direction with a speed of 99 m/s?

c. In which case does the block end up with a greater speed?

102. An automobile jack must exert a lifting force of at least 15 kN. (Chapter 10)

a. If you want to limit the effort force to 0.10 kN, what mechanical advantage is needed?

b. If the jack is 75% efficient, over what distance must the effort force be exerted in order to raise the auto 33 cm?

 For more problems, go to Additional Problems, Appendix B.

Multiple Choice

1. A bicyclist increases her speed from 4.0 m/s to 6.0 m/s. The combined mass of the bicyclist and the bicycle is 55 kg. How much work did the bicyclist do in increasing her speed?

   (A) 11 J (C) 55 J
   (B) 28 J (D) 550 J

2. The illustration below shows a ball swinging freely in a plane. The mass of the ball is 4.0 kg. Ignoring friction, what is the maximum speed of the ball as it swings back and forth?

   (A) 0.14 m/s (C) 7.0 m/s
   (B) 21 m/s (D) 49 m/s

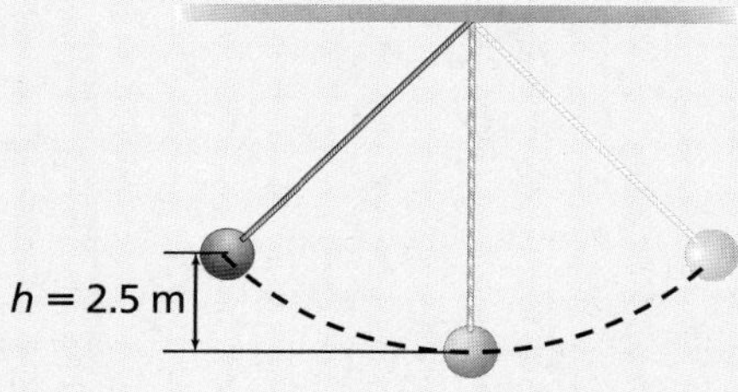

3. You lift a 4.5-kg box from the floor and place it on a shelf that is 1.5 m above the ground. How much energy did you use in lifting the box?

   (A) 9.0 J (C) 11 J
   (B) 49 J (D) 66 J

4. You drop a 6.0×10^{-2}-kg ball from a height of 1.0 m above a hard, flat surface. The ball strikes the surface and loses 0.14 J of its energy. It then bounces back upward. How much kinetic energy does the ball have just after it bounces off the flat surface?

   (A) 0.20 J (C) 0.45 J
   (B) 0.59 J (D) 0.73 J

5. You move a 2.5-kg book from a shelf that is 1.2 m above the ground to a shelf that is 2.6 m above the ground. What is the change in the book's potential energy?

   (A) 1.4 J (C) 3.5 J
   (B) 25 J (D) 34 J

6. A ball of mass m rolls along a flat surface with a speed of v_1. It strikes a padded wall and bounces back in the opposite direction. The energy of the ball after striking the wall is half its initial energy. Ignoring friction, which of the following expressions gives the ball's new speed as a function of its initial speed?

   (A) $\frac{1}{2}v_1$ (C) $\sqrt{2}(v_1)$
   (B) $\frac{\sqrt{2}}{2}(v_1)$ (D) $2v_1$

7. The illustration below shows a box on a curved, frictionless track. The box starts with zero velocity at the top of the track. It then slides from the top of the track to the horizontal part at the ground. Its velocity just at the moment it reaches the ground is 14 m/s. What is the height, h, from the ground to the top of the track?

   (A) 7 m (C) 10 m
   (B) 14 m (D) 20 m

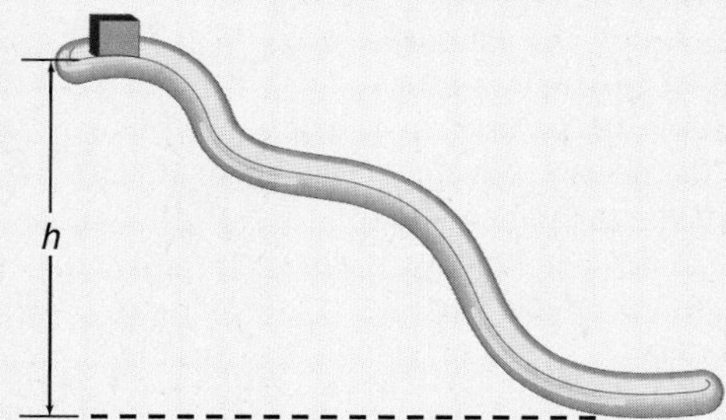

Extended Answer

8. A box sits on a platform supported by a compressed spring. The box has a mass of 1.0 kg. When the spring is released, it gives 4.9 J of energy to the box, and the box flies upward. What will be the maximum height above the platform reached by the box before it begins

✓ Test-Taking TIP

Use the Process of Elimination

On any multiple-choice test, there are two ways to find the correct answer to each question. Either you can choose the right answer immediately or you can eliminate the answers that you know are wrong.

Chapter 12

Thermal Energy

What You'll Learn

- You will learn how temperature relates to the potential and kinetic energies of atoms and molecules.
- You will distinguish heat from work.
- You will calculate heat transfer and the absorption of thermal energy.

Why It's Important

Thermal energy is vital for living creatures, chemical reactions, and the working of engines.

Solar Energy A strategy used to produce electric power from sunlight concentrates the light with many mirrors onto one collector that becomes very hot. The energy collected at a high temperature is then used to drive an engine, which turns an electric generator.

Think About This ▶
What forms of energy does light from the Sun take in the process of converting solar energy into useful work through an engine?

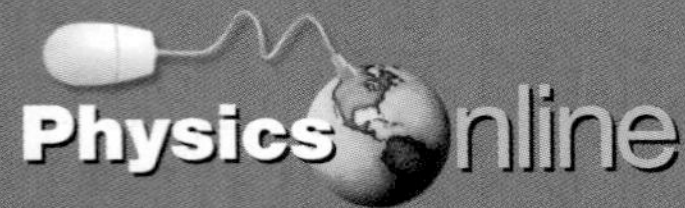

physicspp.com

What happens when you provide thermal energy by holding a glass of water?

Question
What happens to the temperature of water when you hold a glass of water in your hand?

Procedure

1. You will need to use a 250-mL beaker and 150 mL of water.
2. Fill the beaker with the 150 mL of water.
3. Record the initial temperature of the water by holding a thermometer in the water in the beaker. Note that the bulb end of the thermometer must not touch the bottom or sides of the beaker, nor should it touch a table or your hands.
4. Remove the thermometer and hold the beaker of water for 2 min by cupping it with both hands, as shown in the figure.
5. Have your lab partner record the final temperature of the water by placing the thermometer in the beaker. Be sure that the bulb end of the thermometer is not touching the bottom or sides of the beaker.

Analysis
Calculate the change in temperature of the water. If you had more water in the beaker, would it affect the change in temperature?
Critical Thinking Explain what caused the water temperature to change.

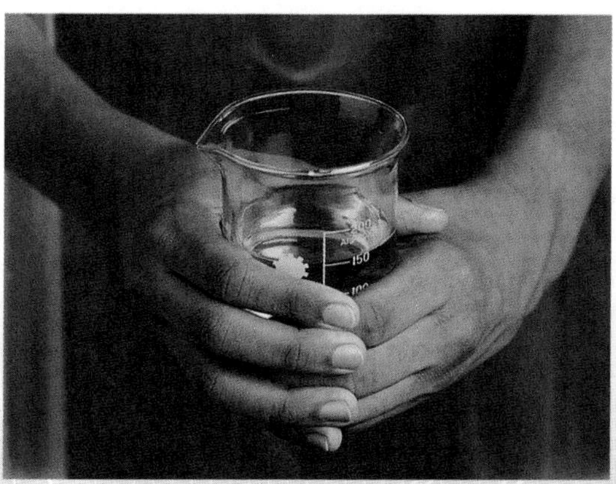

12.1 Temperature and Thermal Energy

The study of heat transformations into other forms of energy, called thermodynamics, began with the eighteenth-century engineers who built the first steam engines. These steam engines were used to power trains, factories, and water pumps for coal mines, and thus they contributed greatly to the Industrial Revolution in Europe and in the United States. In learning to design more efficient engines, the engineers developed new concepts about how heat is related to useful work. Although the study of thermodynamics began in the eighteenth century, it was not until around 1900 that the concepts of thermodynamics were linked to the motions of atoms and molecules in solids, liquids, and gases.

Today, the concepts of thermodynamics are widely used in various applications that involve heat and temperature. Engineers use the laws of thermodynamics to continually develop higher performance refrigerators, automobile engines, aircraft engines, and numerous other machines.

► **Objectives**

- **Describe** thermal energy and compare it to potential and kinetic energies.
- **Distinguish** temperature from thermal energy.
- **Define** specific heat and **calculate** heat transfer.

► **Vocabulary**

conduction
thermal equilibrium
heat
convection
radiation
specific heat

Thermal Energy

You already have studied how objects collide and trade kinetic energies. For example, the many molecules present in a gas have linear and rotational kinetic energies. The molecules also may have potential energy in their vibrations and bending. The gas molecules collide with each other and with the walls of their container, transferring energy among each other in the process. There are numerous molecules moving freely in a gas, resulting in many collisions. Therefore, it is convenient to discuss the total energy of the molecules and the average energy per molecule. The total energy of the molecules is called thermal energy, and the average energy per molecule is related to the temperature of the gas.

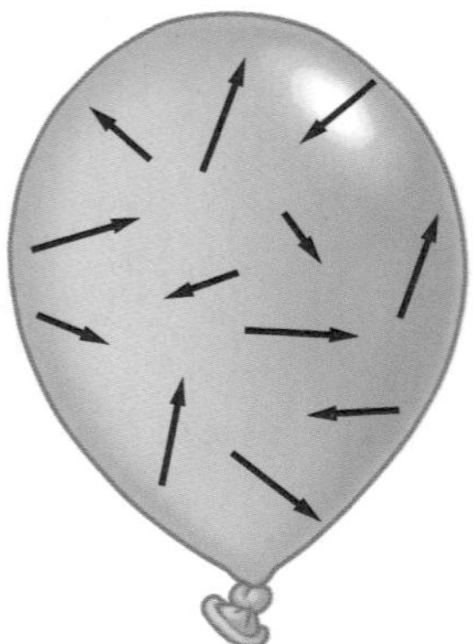

Figure 12-1 Helium atoms in a balloon collide with the rubber wall and cause the balloon to expand.

Hot objects What makes an object hot? When you fill up a balloon with helium, the rubber in the balloon is stretched by the repeated pounding from helium atoms. Each of the billions of helium atoms in the balloon collides with the rubber wall, bounces back, and hits the other side of the balloon, as shown in **Figure 12-1.** If you put a balloon in sunlight, you might notice that the balloon gets slightly larger. The energy from the Sun makes each of the gas atoms move faster and bounce off the rubber walls of the balloon more often. Each atomic collision with the balloon wall puts a greater force on the balloon and stretches the rubber. Thus, the balloon expands.

On the other hand, if you refrigerate a balloon, you will find that it shrinks slightly. Lowering the temperature slows the movement of the helium atoms. Hence, their collisions do not transfer enough momentum to stretch the balloon quite as much. Even though the balloon contains the same number of atoms, the balloon shrinks.

Solids The atoms in solids also have kinetic energy, but they are unable to move freely as gas atoms do. One way to illustrate the molecular structure of a solid is to picture a number of atoms that are connected to each other by springs. Because of the springs, the atoms bounce back and forth, with some bouncing more than others. Each atom has some kinetic energy and some potential energy from the springs that are attached to it. If a solid has N number of atoms, then the total thermal energy in the solid is equal to the average kinetic and potential energy per atom times N.

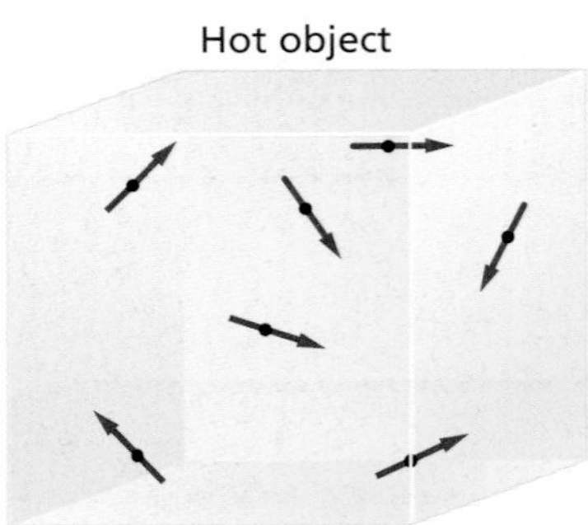

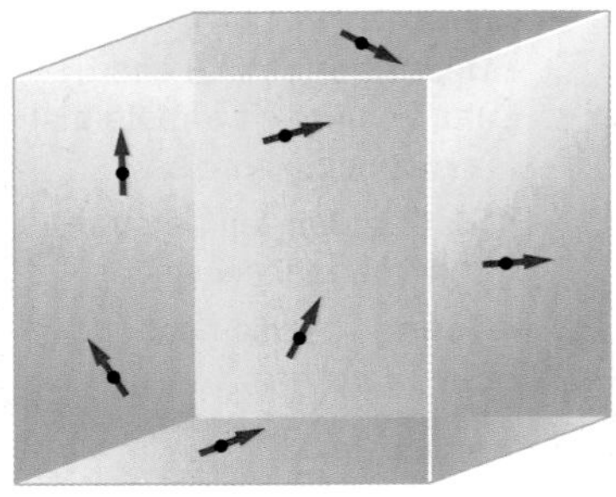

Figure 12-2 Particles in a hot object have greater kinetic and potential energies than particles in a cold object do.

Thermal Energy and Temperature

According to the previous discussion of gases and solids, a hot object has more thermal energy than a similar cold object, as shown in **Figure 12-2.** This means that, as a whole, the particles in a hot object have greater thermal energy than the particles in a cold object. This does not mean that all the particles in an object have exactly the same amount of energy; they have a wide range of energies. However, the average energy of the particles in a hot object is higher than the average energy of the particles in a cold object. To understand this, consider the heights of students in a twelfth-grade class. Although the students' heights vary, you can calculate the average height of the students in the class. This average is likely to be greater than the average height of students in a ninth-grade class, even though some ninth-grade students may be taller than some twelfth-grade students.

Temperature Temperature depends only on the average kinetic energy of the particles in the object. Because temperature depends on average kinetic energy, it does not depend on the number of atoms in an object. To understand this, consider two blocks of steel. The first block has a mass of 1 kg, and the second block has a mass of 2 kg. If the 1-kg block is at the same temperature as the 2-kg block, the average kinetic energy of the particles in each block is the same. However, the 2-kg block has twice the mass of the 1-kg block. Hence, the 2-kg block has twice the amount of particles as the 1-kg block. Thus, the total amount of kinetic energy of the particles in the 2-kg block is twice that of the 1-kg mass. Total kinetic energy is divided by the total number of particles in an object to calculate its average kinetic energy. Therefore, the thermal energy in an object is proportional to the number of particles in it. Temperature, however, is not dependent on the number of particles in an object.

Before Thermal Equilibrium

Hot object (A) Cold object (B)

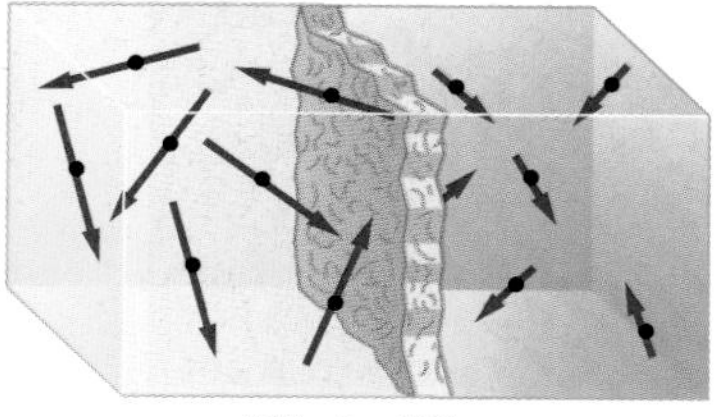

$KE_A > KE_B$

After Thermal Equilibrium

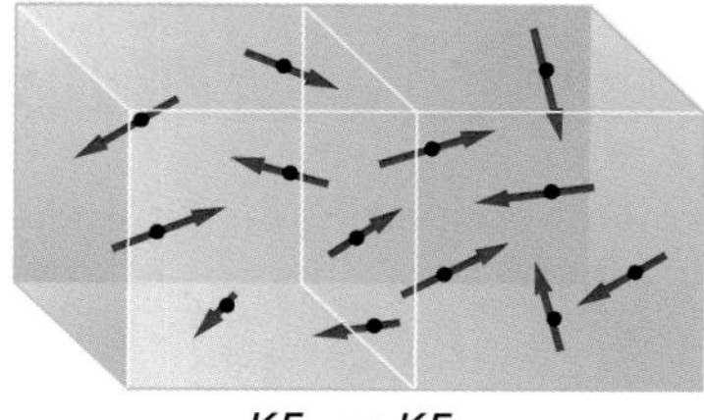

$KE_A = KE_B$

Figure 12-3 Thermal energy is transferred from a hot object to a cold object. When thermal equilibrium is reached, the transfer of energy between objects is equal.

Equilibrium and Thermometry

How do you measure your body temperature? For example, if you suspect that you have a fever, you might place a thermometer in your mouth and wait for a few minutes before checking the thermometer for your temperature reading. The microscopic process involved in measuring temperature involves collisions and energy transfers between the thermometer and your body. Your body is hot compared to the thermometer, which means that the particles in your body have greater thermal energy and are moving faster than the particles in the thermometer. When the cold glass tube of the thermometer touches your skin, which is warmer than the glass, the faster-moving particles in your skin collide with the slower-moving particles in the glass. Energy is then transferred from your skin to the glass particles by the process of **conduction,** which is the transfer of kinetic energy when particles collide. The thermal energy of the particles that make up the thermometer increases, while at the same time, the thermal energy of the particles in your skin decreases.

Thermal equilibrium As the particles in the glass gain more energy, they begin to give some of their energy back to the particles in your body. At some point, the rate of transfer of energy between the glass and your body becomes equal, and your body and the thermometer are then at the same temperature. At this point, your body and the thermometer are said to have reached **thermal equilibrium,** the state in which the rate of energy flow between two objects is equal and the objects are at the same temperature, as shown in **Figure 12-3.**

The operation of a thermometer depends on some property, such as volume, which changes with temperature. Many household thermometers contain colored alcohol that expands when heated and rises in a narrow tube. The hotter the thermometer, the more the alcohol expands and the higher it rises in the tube. In liquid-crystal thermometers, such as the one shown in **Figure 12-4,** a set of different kinds of liquid crystals is used. Each crystal's molecules rearrange at a specific temperature, which causes the color of the crystal to change and indicates the temperature by color. Medical thermometers and the thermometers that monitor automobile engines use very small, temperature-sensitive electronic circuits to take rapid measurements.

Figure 12-4 Thermometers use a change in physical properties to measure temperature. A liquid-crystal thermometer changes color with a temperature change.

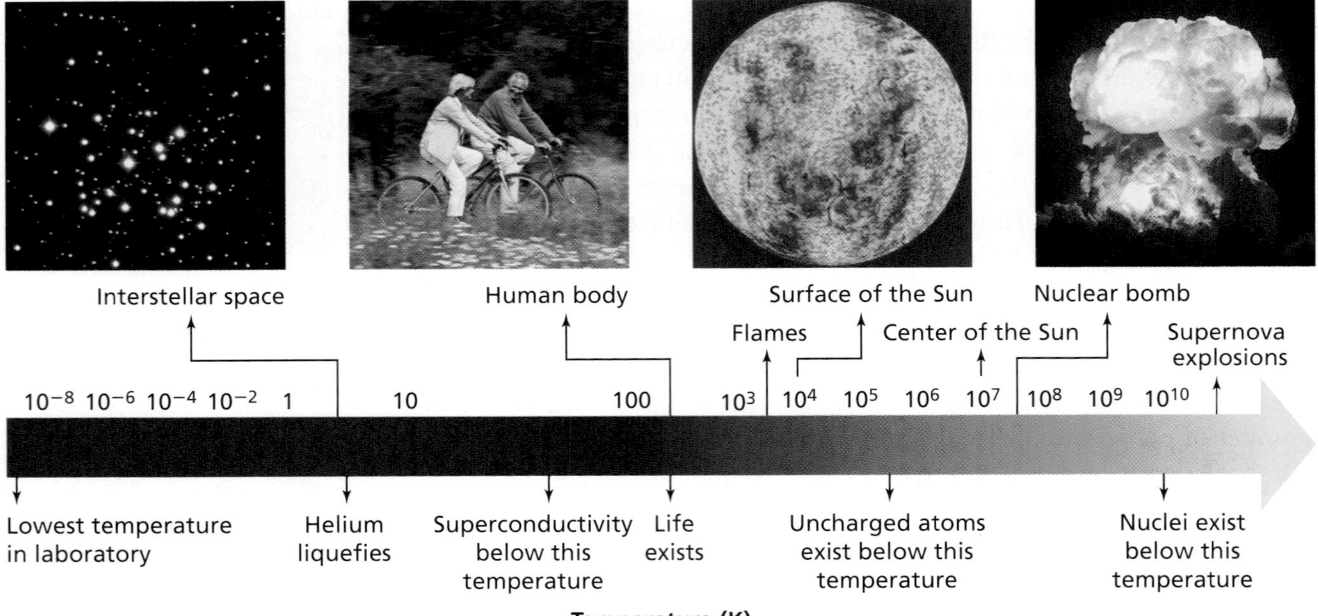

Figure 12-5 There is an extremely wide range of temperatures throughout the universe. Note that the scale has been expanded in areas of particular interest.

Temperature Scales: Celsius and Kelvin

Over the years, scientists developed temperature scales so that they could compare their measurements with those of other scientists. A scale based on the properties of water was devised in 1741 by Swedish astronomer and physicist Anders Celsius. On this scale, now called the Celsius scale, the freezing point of pure water is defined to be 0°C. The boiling point of pure water at sea level is defined to be 100°C.

Temperature limits The wide range of temperatures present in the universe is shown in **Figure 12-5.** Temperatures do not appear to have an upper limit. The interior of the Sun is at least 1.5×10^{7}°C. Temperatures do, however, have a lower limit. Generally, materials contract as they cool. If an ideal gas, such as the helium in a balloon is cooled, it contracts in such a way that it occupies a volume that is only the size of the helium atoms at −273.15°C. At this temperature, all the thermal energy that can be removed has been removed from the gas. It is impossible to reduce the temperature any further. Therefore, there can be no temperature lower than −273.15°C, which is called absolute zero.

The Celsius scale is useful for day-to-day measurements of temperature. It is not conducive for working on science and engineering problems, however, because it has negative temperatures. Negative temperatures suggest a molecule could have negative kinetic energy, which is not possible because kinetic energy is always positive. The solution to this issue is to use a temperature scale based on absolute zero.

The zero point of the Kelvin scale is defined to be absolute zero. On the Kelvin scale, the freezing point of water (0°C) is about 273 K and the boiling point of water is about 373 K. Each interval on this scale, called a kelvin, is equal to 1°C. Thus, $T_C + 273 = T_K$. **Figure 12-6** shows representative temperatures on the three most-common scales: Fahrenheit, Celsius, and Kelvin.

Very cold temperatures are reached by liquefying gases. Helium liquefies at 4.2 K, or −269°C. Even colder temperatures can be reached by making use of special properties of solids, helium isotopes, and atoms and lasers.

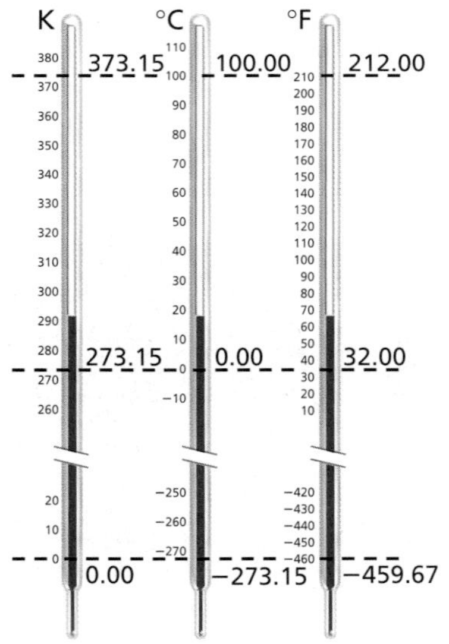

Figure 12-6 The three most-common temperature scales are Kelvin, Celsius, and Fahrenheit.

PRACTICE Problems

Additional Problems, Appendix B

1. Convert the following Kelvin temperatures to Celsius temperatures.
   - **a.** 115 K
   - **b.** 172 K
   - **c.** 125 K
   - **d.** 402 K
   - **e.** 425 K
   - **f.** 212 K
2. Find the Celsius and Kelvin temperatures for the following.
   - **a.** room temperature
   - **b.** a typical refrigerator
   - **c.** a hot summer day in North Carolina
   - **d.** a winter night in Minnesota

Heat and the Flow of Thermal Energy

When two objects come in contact with each other, they transfer energy. This energy that is transferred between the objects is called **heat.** Heat is described as the energy that always flows from the hotter object to the cooler object. Left to itself heat never flows from a colder object to a hotter object. The symbol Q is used to represent an amount of heat, which, like other forms of energy, is measured in joules. If Q has a negative value, heat has left the object; if Q has a positive value, heat has been absorbed by the object.

Conduction If you place one end of a metal rod in a flame, the hot gas particles in the flame conduct heat to the rod. The other end of the rod also becomes warm within a short period of time. Heat is conducted because the particles in the rod are in direct contact with each other.

Convection Thermal energy transfer can occur even if the particles in an object are not in direct contact with each other. Have you ever looked into a pot of water just about to boil? The water at the bottom of the pot is heated by conduction and rises to the top, while the colder water at the top sinks to the bottom. Heat flows between the rising hot water and the descending cold water. This motion of fluid in a liquid or gas caused by temperature differences is called **convection.** Atmospheric turbulence is caused by convection of gases in the atmosphere. Thunderstorms are excellent examples of large-scale atmospheric convection. Ocean currents that cause changes in weather patterns also result from convection.

Radiation The third method of thermal transfer, unlike the first two, does not depend on the presence of matter. The Sun warms Earth from over 150 million km away via **radiation,** which is the transfer of energy by electromagnetic waves. These waves carry the energy from the hot Sun through the vacuum of space to the much cooler Earth.

Meteorology Connection

APPLYING PHYSICS

Steam Heating In a steam heating system of a building, water is turned into steam in a boiler located in a maintenance area or the basement. The steam then flows through insulated pipes to each room in the building. In the radiator, the steam is condensed as liquid water and then flows back through pipes to the boiler to be revaporized. The hot steam physically carries the heat from the boiler, and that energy is released when the steam condenses in the radiator. Some disadvantages of steam heating are that it requires expensive boilers and pipes must contain steam under pressure.

Specific Heat

Some objects are easier to heat than others. On a bright summer day, the Sun warms the sand on a beach and the ocean water. However, the sand on the beach gets quite hot, while the ocean water stays relatively cool. When heat flows into an object, its thermal energy and temperature increase. The amount of the increase in temperature depends on the size of the object and on the material from which the object is made.

Table 12-1			
Specific Heat of Common Substances			
Material	**Specific Heat (J/kg·K)**	**Material**	**Specific Heat (J/kg·K)**
Aluminum	897	Lead	130
Brass	376	Methanol	2450
Carbon	710	Silver	235
Copper	385	Steam	2020
Glass	840	Water	4180
Ice	2060	Zinc	388
Iron	450		

The **specific heat** of a material is the amount of energy that must be added to the material to raise the temperature of a unit mass by one temperature unit. In SI units, specific heat, represented by C, is measured in J/kg·K. **Table 12-1** provides values of specific heat for some common substances. For example, 897 J must be added to 1 kg of aluminum to raise its temperature by 1 K. The specific heat of aluminum is therefore 897 J/kg·K.

The heat gained or lost by an object as its temperature changes depends on the mass, the change in temperature, and the specific heat of the substance. By using the following equation, you can calculate the amount of heat, Q, that must be transferred to change the temperature of an object.

Heat Transfer $Q = mC\Delta T = mC(T_f - T_i)$

Heat transfer is equal to the mass of an object times the specific heat of the object times the difference between the final and initial temperatures.

Liquid water has a high specific heat compared to the other substance in Table 12-1. When the temperature of 10.0 kg of water is increased by 5.0 K, the heat absorbed is $Q = (10.0\ \text{kg})(4180\ \text{J/kg·K})(5.0\ \text{K}) = 2.1\times10^5$ J. Remember that the temperature interval for kelvins is the same as that for Celsius degrees. For this reason, you can calculate ΔT in kelvins or in degrees Celsius.

▶ EXAMPLE **Problem 1**

Heat Transfer **A 5.10-kg cast-iron skillet is heated on the stove from 295 K to 450 K. How much heat had to be transferred to the iron?**

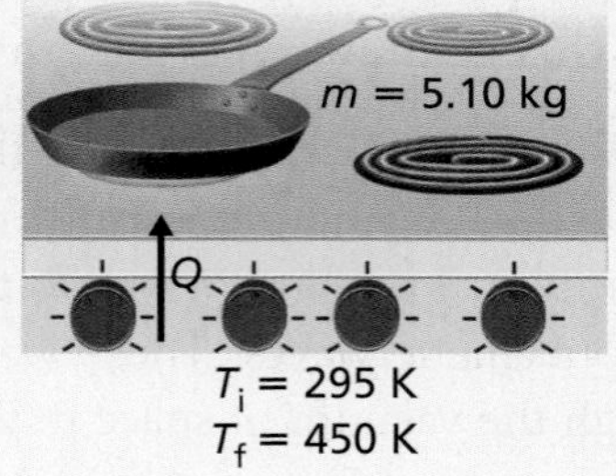

1 Analyze and Sketch the Problem

- Sketch the flow of heat into the skillet from the stove top.

Known:		Unknown:
$m = 5.10$ kg	$C = 450$ J/kg·K	$Q = ?$
$T_i = 295$ K	$T_f = 450$ K	

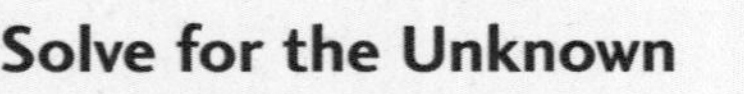

2 Solve for the Unknown

$$Q = mC(T_f - T_i)$$
$$= (5.10\ \text{kg})(450\ \text{J/kg·K})(450\ \text{K} - 295\ \text{K})$$ Substitute $m = 5.10$ kg, $C = 450$ J/kg·K, $T_f = 450$ K, $T_i = 295$ K
$$= 3.6\times10^5\ \text{J}$$

3 Evaluate the Answer

- **Are the units correct?** Heat is measured in J.
- **Does the sign make sense?** Temperature increased, so Q is positive.

Math Handbook
Order of Operations page 843

PRACTICE Problems

• Additional Problems, Appendix B
• Solutions to Selected Problems, Appendix C

3. When you turn on the hot water to wash dishes, the water pipes have to heat up. How much heat is absorbed by a copper water pipe with a mass of 2.3 kg when its temperature is raised from 20.0°C to 80.0°C?

4. The cooling system of a car engine contains 20.0 L of water (1 L of water has a mass of 1 kg).

a. What is the change in the temperature of the water if the engine operates until 836.0 kJ of heat is added?

b. Suppose that it is winter, and the car's cooling system is filled with methanol. The density of methanol is 0.80 g/cm^3. What would be the increase in temperature of the methanol if it absorbed 836.0 kJ of heat?

c. Which is the better coolant, water or methanol? Explain.

5. Electric power companies sell electricity by the kWh, where 1 kWh = 3.6×10^6 J. Suppose that it costs $0.15 per kWh to run an electric water heater in your neighborhood. How much does it cost to heat 75 kg of water from 15°C to 43°C to fill a bathtub?

Calorimetry: Measuring Specific Heat

A simple calorimeter, such as the one shown in **Figure 12-7,** is a device used to measure changes in thermal energy. A calorimeter is carefully insulated so that heat transfer to the external world is kept to a minimum. A measured mass of a substance that has been heated to a high temperature is placed in the calorimeter. The calorimeter also contains a known mass of cold water at a measured temperature. The heat released by the substance is transferred to the cooler water. The change in thermal energy of the substance is calculated using the resulting increase in the water temperature. More sophisticated types of calorimeters are used to measure chemical reactions and the energy content of various foods.

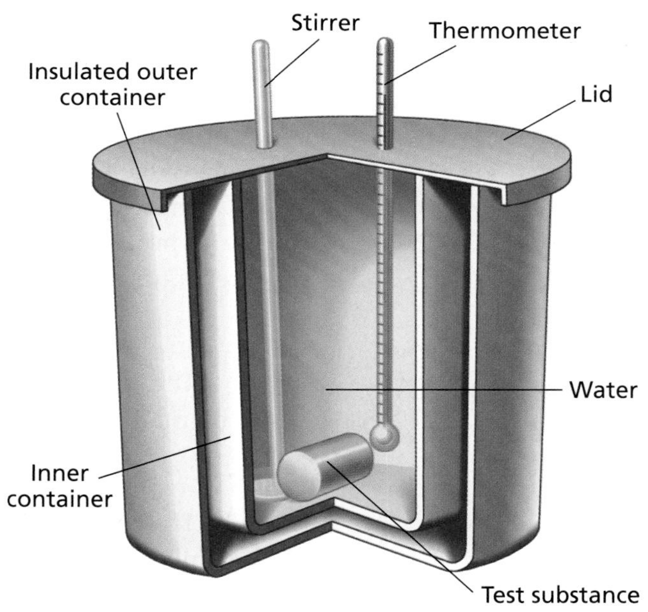

Figure 12-7 A calorimeter provides an isolated, closed system in which to measure energy transfer.

The operation of a calorimeter depends on the conservation of energy in an isolated, closed system. Energy can neither enter nor leave this system. As a result, if the energy of one part of the system increases, the energy of another part of the system must decrease by the same amount. Consider a system composed of two blocks of metal, block A and block B, shown in **Figure 12-8a** on the next page. The total energy of the system is constant, as represented by the following equation.

Conservation of Energy $E_A + E_B = \text{constant}$

In an isolated, closed system, the thermal energy of object A plus the thermal energy of object B is constant.

Figure 12-8 A system is composed of two model blocks at different temperatures that initially are separated **(a).** When the blocks are brought together, heat flows from the hotter block to the colder block **(b).** Total energy remains constant.

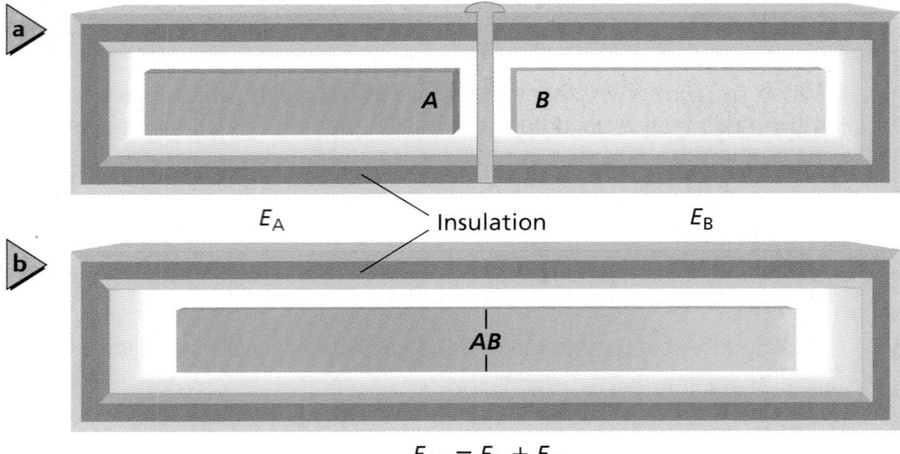

Suppose that the two blocks initially are separated but can be placed in contact with each other. If the thermal energy of block A changes by an amount ΔE_A, then the change in thermal energy of block B, ΔE_B, must be related by the equation, $\Delta E_A + \Delta E_B = 0$. Thus, $\Delta E_A = -\Delta E_B$. The change in energy of one block is positive, while the change in energy of the other block is negative. For the block whose thermal energy change is positive, the temperature of the block rises. For the block whose thermal energy change is negative, the temperature falls.

Assume that the initial temperatures of the two blocks are different. When the blocks are brought together, heat flows from the hotter block to the colder block, as shown in **Figure 12-8b.** The heat flow continues until the blocks are in thermal equilibrium, which is when the blocks have the same temperature.

In an isolated, closed system, the change in thermal energy is equal to the heat transferred because no work is done. Therefore, the change in energy for each block can be expressed by the following equation:

$$\Delta E = Q = mC\Delta T$$

The increase in thermal energy of block A is equal to the decrease in thermal energy of block B. Thus, the following relationship is true:

$$m_A C_A \Delta T_A + m_B C_B \Delta T_B = 0$$

The change in temperature is the difference between the final and initial temperatures; that is, $\Delta T = T_f - T_i$. If the temperature of a block increases, $T_f > T_i$, and ΔT is positive. If the temperature of the block decreases, $T_f < T_i$, and ΔT is negative. The final temperatures of the two blocks are equal. The following is the equation for the transfer of energy:

$$m_A C_A (T_f - T_A) + m_B C_B (T_f - T_B) = 0$$

To solve for T_f, expand the equation.

$$m_A C_A T_f - m_A C_A T_A + m_B C_B T_f - m_B C_B T_B = 0$$

$$T_f(m_A C_A + m_B C_B) = m_A C_A T_A + m_B C_B T_B$$

$$T_f = \frac{m_A C_A T_A + m_B C_B T_B}{m_A C_A + m_B C_B}$$

▶ EXAMPLE Problem 2

Transferring Heat in a Calorimeter A calorimeter contains 0.50 kg of water at 15°C. A 0.040-kg block of zinc at 115°C is placed in the water. What is the final temperature of the system?

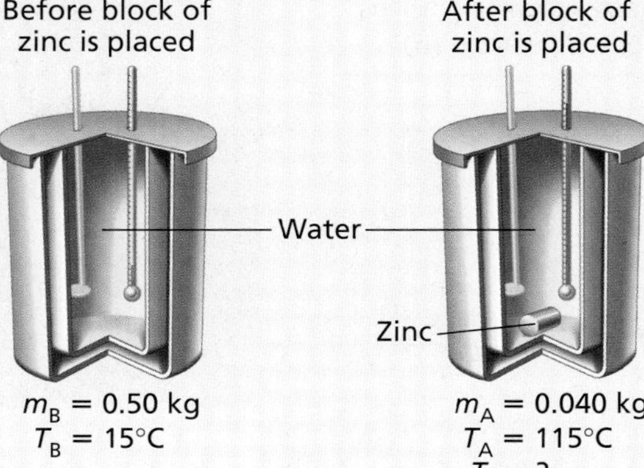

1 Analyze and Sketch the Problem

- Let zinc be sample A and water be sample B.
- Sketch the transfer of heat from the hotter zinc to the cooler water.

Known:

$m_A = 0.040$ kg
$C_A = 388$ J/kg·°C
$T_A = 115$°C
$m_B = 0.50$ kg
$C_B = 4180$ J/kg·°C
$T_B = 15.0$°C

Unknown:

$T_f = ?$

Math Handbook
Operations with Significant Digits
pages 835–836

2 Solve for the Unknown

Determine the final temperature using the following equation.

$$T_f = \frac{m_A C_A T_A + m_B C_B T_B}{m_A C_A + m_B C_B}$$

$$= \frac{(0.040\text{ kg})(388\text{ J/kg·°C})(115\text{°C}) + (0.50\text{ kg})(4180\text{ J/kg·°C})(15.0\text{°C})}{(0.040\text{ kg})(388\text{ J/kg·°C}) + (0.50\text{ kg})(4180\text{ J/kg·°C})}$$

Substitute $m_A = 0.040$ kg, $C_A = 388$ J/kg·°C, $T_A = 115$°C, $m_B = 0.50$ kg, $C_B = 4180$ J/kg·°C, $T_B = 15$°C

$$= 16\text{°C}$$

3 Evaluate the Answer

- **Are the units correct?** Temperature is measured in Celsius.
- **Is the magnitude realistic?** The answer is between the initial temperatures of the two samples, as is expected when using a calorimeter.

▶ PRACTICE Problems

- Additional Problems, Appendix B
- Solutions to Selected Problems, Appendix C

6. A 2.00×10^2-g sample of water at 80.0°C is mixed with 2.00×10^2 g of water at 10.0°C. Assume that there is no heat loss to the surroundings. What is the final temperature of the mixture?

7. A 4.00×10^2-g sample of methanol at 16.0°C is mixed with 4.00×10^2 g of water at 85.0°C. Assume that there is no heat loss to the surroundings. What is the final temperature of the mixture?

8. Three metal fishing weights, each with a mass of 1.00×10^2 g and at a temperature of 100.0°C, are placed in 1.00×10^2 g of water at 35.0°C. The final temperature of the mixture is 45.0°C. What is the specific heat of the metal in the weights?

9. A 1.00×10^2-g aluminum block at 100.0°C is placed in 1.00×10^2 g of water at 10.0°C. The final temperature of the mixture is 25.0°C. What is the specific heat of the aluminum?

Figure 12-9 A lizard regulates its body temperature by hiding under a rock when the atmosphere is hot **(a)** and sunbathing when the atmosphere gets cold **(b)**.

Biology Connection

Animals can be divided into two groups based on their body temperatures. Most are cold-blooded animals whose body temperatures depend on the environment. The others are warm-blooded animals whose body temperatures are controlled internally. That is, a warm-blooded animal's body temperature remains stable regardless of the temperature of the environment. In contrast, when the temperature of the environment is high, the body temperature of a cold-blooded animal also becomes high. A cold-blooded animal, such as the lizard shown in **Figure 12-9,** regulates this heat flow by hiding under a rock or crevice, thereby reducing its body temperature. Humans are warm-blooded and have a body temperature of about 37°C. To regulate its body temperature, a warm-blooded animal increases or decreases the level of its metabolic processes. Thus, a warm-blooded animal may hibernate in winter and reduce its body temperature to approach the freezing point of water.

12.1 Section Review

10. Temperature Make the following conversions.

- **a.** 5°C to kelvins
- **b.** 34 K to degrees Celsius
- **c.** 212°C to kelvins
- **d.** 316 K to degrees Celsius

11. Conversions Convert the following Celsius temperatures to Kelvin temperatures.

- **a.** 28°C
- **b.** 154°C
- **c.** 568°C
- **d.** −55°C
- **e.** −184°C

12. Thermal Energy Could the thermal energy of a bowl of hot water equal that of a bowl of cold water? Explain your answer.

13. Heat Flow On a dinner plate, a baked potato always stays hot longer than any other food. Why?

14. Heat The hard tile floor of a bathroom always feels cold to bare feet even though the rest of the room is warm. Is the floor colder than the rest of the room?

15. Specific Heat If you take a plastic spoon out of a cup of hot cocoa and put it in your mouth, you are not likely to burn your tongue. However, you could very easily burn your tongue if you put the hot cocoa in your mouth. Why?

16. Heat Chefs often use cooking pans made of thick aluminum. Why is thick aluminum better than thin aluminum for cooking?

17. Heat and Food It takes much longer to bake a whole potato than to cook french fries. Why?

18. Critical Thinking As water heats in a pot on a stove, the water might produce some mist above its surface right before the water begins to roll. What is happening, and where is the coolest part of the water in the pot?

Physics Online physicspp.com/self_check_quiz

12.2 Changes of State and the Laws of Thermodynamics

Eighteenth-century steam-engine builders used heat to turn liquid water into steam. The steam pushed a piston to turn the engine, and then the steam was cooled and condensed into a liquid again. Adding heat to the liquid water changed not only its temperature, but also its structure. You will learn that changing state means changing form as well as changing the way in which atoms store thermal energy.

Changes of State

The three most common states of matter are solids, liquids, and gases. As the temperature of a solid is raised, it usually changes to a liquid. At even higher temperatures, it becomes a gas. How can these changes be explained? Consider a material in a solid state. When the thermal energy of the solid is increased, the motion of the particles also increases, as does the temperature.

Figure 12-10 diagrams the changes of state as thermal energy is added to 1.0 g of water starting at 243 K (ice) and continuing until it reaches 473 K (steam). Between points A and B, the ice is warmed to 273 K. At some point, the added thermal energy causes the particles to move rapidly enough that their motion overcomes the forces holding the particles together in a fixed location. The particles are still touching each other, but they have more freedom of movement. Eventually, the particles become free enough to slide past each other.

Melting point At this point, the substance has changed from a solid to a liquid. The temperature at which this change occurs is the melting point of the substance. When a substance is melting, all of the added thermal energy goes to overcome the forces holding the particles together in the solid state. None of the added thermal energy increases the kinetic energy of the particles. This can be observed between points B and C in Figure 12-10, where the added thermal energy melts the ice at a constant 273 K. Because the kinetic energy of the particles does not increase, the temperature does not increase between points B and C.

Boiling point Once a solid is completely melted, there are no more forces holding the particles in the solid state. Adding more thermal energy again increases the motion of the particles, and the temperature of the liquid rises. In the diagram, this process occurs between points C and D. As the temperature increases further, some particles in the liquid acquire enough energy to break free from the other particles. At a specific temperature, known as the boiling point, further addition of energy causes the substance to undergo another change of state. All the added thermal energy converts the substance from the liquid state to the gaseous state.

▸ **Objectives**

- **Define** heats of fusion and vaporization.
- **State** the first and second laws of thermodynamics.
- **Distinguish** between heat and work.
- **Define** entropy.

▸ **Vocabulary**

heat of fusion
heat of vaporization
first law of thermodynamics
heat engine
entropy
second law of thermodynamics

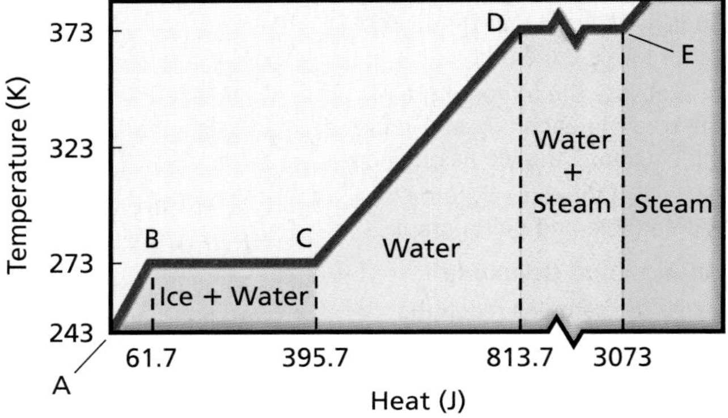

Figure 12-10 A plot of temperature versus heat added when 1.0 g of ice is converted to steam. Note that the scale is broken between points D and E.

Table 12-2		
Heats of Fusion and Vaporization of Common Substances		
Material	**Heat of Fusion H_f (J/kg)**	**Heat of Vaporization H_v (J/kg)**
Copper	2.05×10^5	5.07×10^6
Mercury	1.15×10^4	2.72×10^5
Gold	6.30×10^4	1.64×10^6
Methanol	1.09×10^5	8.78×10^5
Iron	2.66×10^5	6.29×10^6
Silver	1.04×10^5	2.36×10^6
Lead	2.04×10^4	8.64×10^5
Water (ice)	3.34×10^5	2.26×10^6

As in melting, the temperature does not rise while a liquid boils. In Figure 12-10, this transition is represented between points D and E. After the material is entirely converted to gas, any added thermal energy again increases the motion of the particles, and the temperature rises. Above point E, steam is heated to temperatures greater than 373 K.

Heat of fusion The amount of energy needed to melt 1 kg of a substance is called the **heat of fusion** of that substance. For example, the heat of fusion of ice is 3.34×10^5 J/kg. If 1 kg of ice at its melting point, 273 K, absorbs 3.34×10^5 J, the ice becomes 1 kg of water at the same temperature, 273 K. The added energy causes a change in state but not in temperature. The horizontal distance in Figure 12-10 from point B to point C represents the heat of fusion.

Heat of vaporization At normal atmospheric pressure, water boils at 373 K. The thermal energy needed to vaporize 1 kg of a liquid is called the **heat of vaporization.** For water, the heat of vaporization is 2.26×10^6 J/kg. The distance from point D to point E in Figure 12-10 represents the heat of vaporization. Every material has a characteristic heat of vaporization.

Between points A and B, there is a definite slope to the line as the temperature is raised. This slope represents the specific heat of the ice. The slope between points C and D represents the specific heat of water, and the slope above point E represents the specific heat of steam. Note that the slope for water is less than those of both ice and steam. This is because water has a greater specific heat than does ice or steam. The heat, Q, required to melt a solid of mass m is given by the following equation.

Heat Required to Melt a Solid $Q = mH_f$

The heat required to melt a solid is equal to the mass of the solid times the heat of fusion of the solid.

Similarly, the heat, Q, required to vaporize a mass, m, of liquid is given by the following equation.

Heat Required to Vaporize a Liquid $Q = mH_v$

The heat required to vaporize a liquid is equal to the mass of the liquid times the heat of vaporization of the liquid.

When a liquid freezes, an amount of heat, $Q = -mH_f$, must be removed from the liquid to turn it into a solid. The negative sign indicates that the heat is transferred from the sample to the external world. In the same way, when a vapor condenses to a liquid, an amount of heat, $Q = -mH_v$, must be removed from the vapor. The values of some heats of fusion, H_f, and heats of vaporization, H_v, are shown in **Table 12-2.**

MINI LAB

Melting

1. Label two foam cups *A* and *B*.
2. Measure and pour 75 mL of room-temperature water into each cup. Wipe up any spilled liquid.
3. Add an ice cube to cup A, and add ice water to cup B until the water levels are equal.
4. Measure the temperature of the water in each cup at 1-min intervals until the ice has melted.
5. Record the temperatures in a data table and plot a graph.

Analyze and Conclude

6. Do the samples reach the same final temperature? Why?

▶ EXAMPLE Problem 3

Heat Suppose that you are camping in the mountains. You need to melt 1.50 kg of snow at 0.0°C and heat it to 70.0°C to make hot cocoa. How much heat will be needed?

1 Analyze and Sketch the Problem

- Sketch the relationship between heat and water in its solid and liquid states.
- Sketch the transfer of heat as the temperature of the water increases.

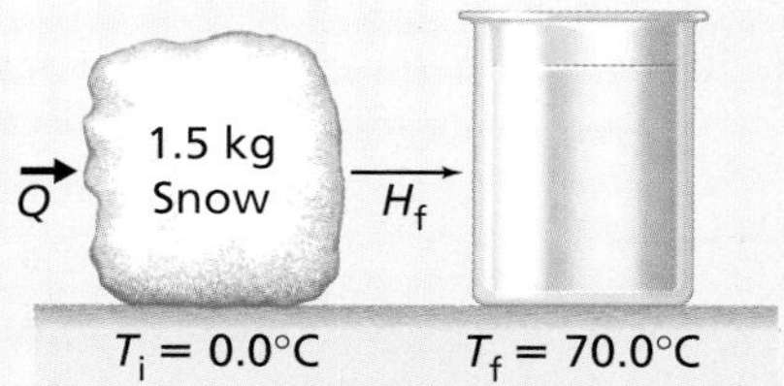

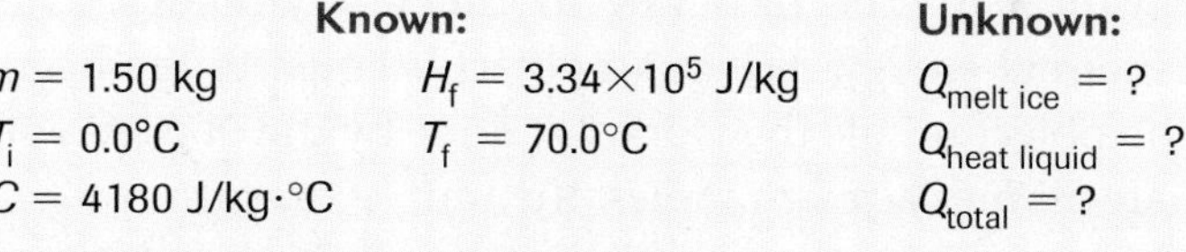

Known:		Unknown:
$m = 1.50$ kg	$H_f = 3.34\times10^5$ J/kg	$Q_{melt\ ice} = ?$
$T_i = 0.0°C$	$T_f = 70.0°C$	$Q_{heat\ liquid} = ?$
$C = 4180$ J/kg·°C		$Q_{total} = ?$

Physics Online

Personal Tutor For an online tutorial on heat, visit physicspp.com.

2 Solve for the Unknown

Calculate the heat needed to melt ice.

$$\begin{aligned} Q_{melt\ ice} &= mH_f \\ &= (1.50\text{ kg})(3.34\times10^5\text{ J/kg}) \\ &= 5.01\times10^5\text{ J} \\ &= 5.01\times10^2\text{ kJ} \end{aligned}$$

Substitute $m = 1.50$ kg, $H_f = 3.34\times10^5$ J/kg

Calculate the temperature change.

$$\begin{aligned} \Delta T &= T_f - T_i \\ &= 70.0°C - 0.0°C \\ &= 70.0°C \end{aligned}$$

Substitute $T_f = 70.0°C$, $T_i = 0.0°C$

Calculate the heat needed to raise the water temperature.

$$\begin{aligned} Q_{heat\ liquid} &= mC\Delta T \\ &= (1.50\text{ kg})(4180\text{ J/kg·°C})(70.0°C) \\ &= 4.39\times10^5\text{ J} \\ &= 4.39\times10^2\text{ kJ} \end{aligned}$$

Substitute $m = 1.50$ kg, $C = 4180$ J/kg·°C, $\Delta T = 70.0°C$

Calculate the total amount of heat needed.

$$\begin{aligned} Q_{total} &= Q_{melt\ ice} + Q_{heat\ liquid} \\ &= 5.01\times10^2\text{ kJ} + 4.39\times10^2\text{ kJ} \\ &= 9.40\times10^2\text{ kJ} \end{aligned}$$

Substitute $Q_{melt\ ice} = 5.01\times10^2$ kJ, $Q_{heat\ liquid} = 4.39\times10^2$ kJ

3 Evaluate the Answer

- **Are the units correct?** Energy units are in joules.
- **Does the sign make sense?** Q is positive when heat is absorbed.
- **Is the magnitude realistic?** The amount of heat needed to melt the ice is greater than the amount of heat needed to increase the water temperature by 70.0°C. It takes more energy to overcome the forces holding the particles in the solid state than to raise the temperature of water.

▶ PRACTICE Problems

- Additional Problems, Appendix B
- Solutions to Selected Problems, Appendix C

19. How much heat is absorbed by 1.00×10^2 g of ice at −20.0°C to become water at 0.0°C?

20. A 2.00×10^2-g sample of water at 60.0°C is heated to steam at 140.0°C. How much heat is absorbed?

21. How much heat is needed to change 3.00×10^2 g of ice at −30.0°C to steam at 130.0°C?

The First Law of Thermodynamics

Before thermal energy was linked to the motion of atoms, the study of heat and temperature was considered to be a separate science. The first law developed for this science was a statement about what thermal energy is and where it can go. As you know, you can heat a nail by holding it over a flame or by pounding it with a hammer. That is, you can increase the nail's thermal energy by adding heat or by doing work on it. We do not normally think that the nail does work on the hammer. However, the work done by the nail on the hammer is equal to the negative of the work done by the hammer on the nail. The **first law of thermodynamics** states that the change in thermal energy, ΔU, of an object is equal to the heat, Q, that is added to the object minus the work, W, done by the object. Note that ΔU, Q, and W are all measured in joules, the unit of energy.

The First Law of Thermodynamics $\Delta U = Q - W$

The change in thermal energy of an object is equal to the heat added to the object minus the work done by the object.

Thermodynamics also involves the study of the changes in thermal properties of matter. The first law of thermodynamics is merely a restatement of the law of conservation of energy, which states that energy is neither created nor destroyed, but can be changed into other forms.

Another example of changing the amount of thermal energy in a system is a hand pump used to inflate a bicycle tire. As a person pumps, the air and the hand pump become warm. The mechanical energy in the moving piston is converted into thermal energy of the gas. Similarly, other forms of energy, such as light, sound, and electric energy, can be changed into thermal energy. For example, a toaster converts electric energy into heat when it toasts bread, and the Sun warms Earth with light from a distance of over 150 million km away.

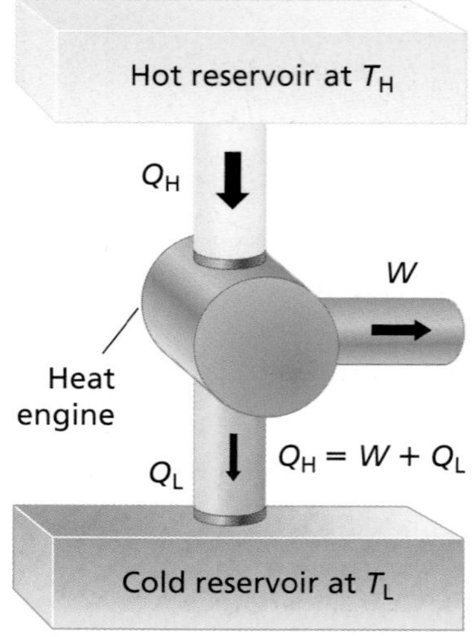

Figure 12-11 A heat engine transforms heat at high temperature into mechanical energy and low-temperature waste heat.

Heat engines The warmth that you experience when you rub your hands together is a result of the conversion of mechanical energy into thermal energy. The conversion of mechanical energy into thermal energy occurs easily. However, the reverse process, the conversion of thermal energy into mechanical energy, is more difficult. A device that is able to continuously convert thermal energy to mechanical energy is called a **heat engine.**

A heat engine requires a high-temperature source from which thermal energy can be removed; a low-temperature receptacle, called a sink, into which thermal energy can be delivered; and a way to convert the thermal energy into work. A diagram of a heat engine is shown in **Figure 12-11.** An automobile internal-combustion engine, such as the one shown in **Figure 12-12,** is one example of a heat engine. In the engine, a mixture of air and gasoline vapor is ignited and produces a high-temperature flame. Input heat, Q_H, flows from the flame to the air in the cylinder. The hot air expands and pushes on a piston, thereby changing thermal energy into mechanical energy. To obtain continuous mechanical energy, the engine must be returned to its starting condition. The heated air is expelled and replaced by new air, and the piston is returned to the top of the cylinder.

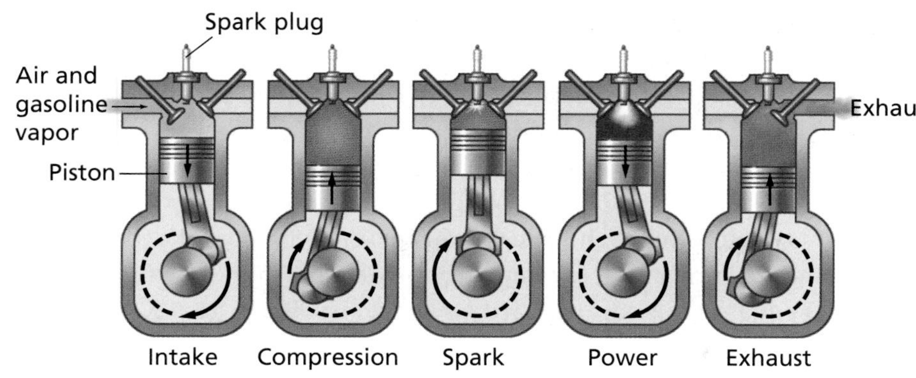

Figure 12-12 The heat produced by burning gasoline causes the gases that are produced to expand and to exert force and do work on the piston.

Concepts In Motion
Interactive Figure To see an animation on a heat engine, visit physicspp.com.

The entire cycle is repeated many times each minute. The thermal energy from the burning of gasoline is converted into mechanical energy, which eventually results in the movement of the car.

Not all of the thermal energy from the high-temperature flame in an automobile engine is converted into mechanical energy. When the automobile engine is functioning, the exhaust gases and the engine parts become hot. As the exhaust comes in contact with outside air and transfers heat to it, the temperature of the outside air is raised. In addition, heat from the engine is transferred to a radiator. Outside air passes through the radiator and the air temperature is raised.

All of this energy, Q_L, transferred out of the automobile engine is called waste heat, that is, heat that has not been converted into work. When the engine is working continuously, the internal energy of the engine does not change, or $\Delta U = 0 = Q - W$. The net heat going into the engine is $Q = Q_H - Q_L$. Thus, the work done by the engine is $W = Q_H - Q_L$. In an automobile engine, the thermal energy in the flame produces the mechanical energy and the waste heat that is expelled. All heat engines generate waste heat, and therefore no engine can ever convert all of the energy into useful motion or work.

Efficiency Engineers and car salespeople often talk about the fuel efficiency of automobile engines. They are referring to the amount of the input heat, Q_H, that is turned into useful work, W. The actual efficiency of an engine is given by the ratio W/Q_H. The efficiency could equal 100 percent only if all of the input heat were turned into work by the engine. Because there is always waste heat, even the most efficient engines fall short of 100-percent efficiency.

In solar collectors, heat is collected at high temperatures and used to drive engines. The Sun's energy is transmitted as electromagnetic waves and increases the internal energy of the solar collectors. This energy is then transmitted as heat to the engine, where it is turned into useful work and waste heat.

Refrigerators Heat flows spontaneously from a warm object to a cold object. However, it is possible to remove thermal energy from a colder object and add it to a warmer object if work is done. A refrigerator is a common example of a device that accomplishes this transfer with the use of mechanical work. Electric energy runs a motor that does work on a gas and compresses it.

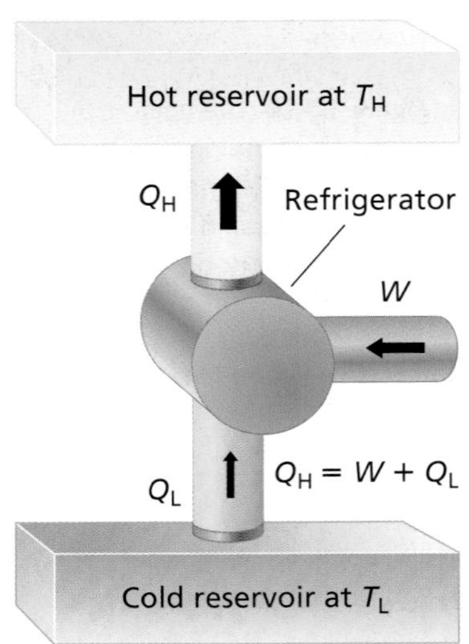

Figure 12-13 A refrigerator absorbs heat, Q_L, from the cold reservoir and gives off heat, Q_H, to the hot reservoir. Work, W, is done on the refrigerator.

The gas draws heat from the interior of the refrigerator, passes from the compressor through the condenser coils on the outside of the refrigerator, and cools into a liquid. Thermal energy is transferred into the air in the room. The liquid reenters the interior, vaporizes, and absorbs thermal energy from its surroundings. The gas returns to the compressor and the process is repeated. The overall change in the thermal energy of the gas is zero. Thus, according to the first law of thermodynamics, the sum of the heat removed from the refrigerator's contents and the work done by the motor is equal to the heat expelled, as shown in **Figure 12-13.**

Heat pumps A heat pump is a refrigerator that can be run in two directions. In the summer, the pump removes heat from a house and thus cools the house. In the winter, heat is removed from the cold outside air and transferred into the warmer house. In both cases, mechanical energy is required to transfer heat from a cold object to a warmer one.

PRACTICE Problems

• Additional Problems, Appendix B
• Solutions to Selected Problems, Appendix C

22. A gas balloon absorbs 75 J of heat. The balloon expands but stays at the same temperature. How much work did the balloon do in expanding?

23. A drill bores a small hole in a 0.40-kg block of aluminum and heats the aluminum by 5.0°C. How much work did the drill do in boring the hole?

24. How many times would you have to drop a 0.50-kg bag of lead shot from a height of 1.5 m to heat the shot by 1.0°C?

25. When you stir a cup of tea, you do about 0.050 J of work each time you circle the spoon in the cup. How many times would you have to stir the spoon to heat a 0.15-kg cup of tea by 2.0°C?

26. How can the first law of thermodynamics be used to explain how to reduce the temperature of an object?

The Second Law of Thermodynamics

Many processes that are consistent with the first law of thermodynamics have never been observed to occur spontaneously. Three such processes are presented in **Figure 12-14.** For example, the first law of thermodynamics does not prohibit heat flowing from a cold object to a hot object. However, when hot objects have been placed in contact with cold objects, the hot objects have never been observed to become hotter. Similarly, the cold objects have never been observed to become colder.

Entropy If heat engines completely converted thermal energy into mechanical energy with no waste heat, then the first law of thermodynamics would be obeyed. However, waste heat is always generated, and randomly distributed particles of a gas are not observed to spontaneously arrange themselves in specific ordered patterns. In the nineteenth century, French engineer Sadi Carnot studied the ability of engines to convert thermal energy into mechanical energy. He developed a logical proof that even an ideal engine would generate some waste heat. Carnot's result is best described in terms of a quantity called **entropy,** which is a measure of the disorder in a system.

Consistent with the First Law of Thermodynamics but do not occur spontaneously

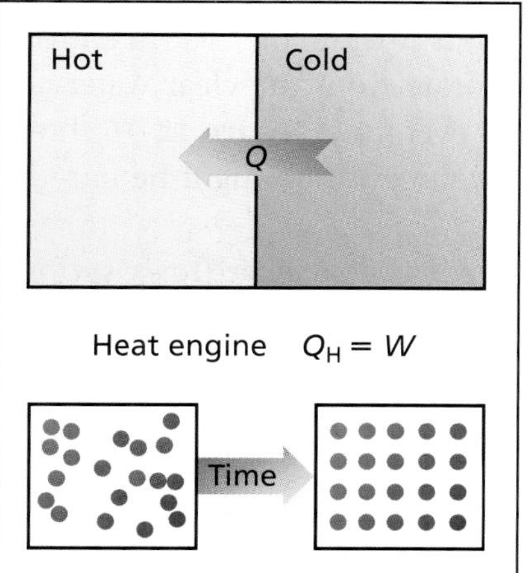

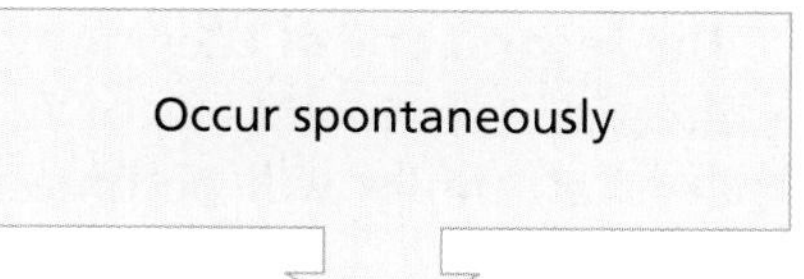

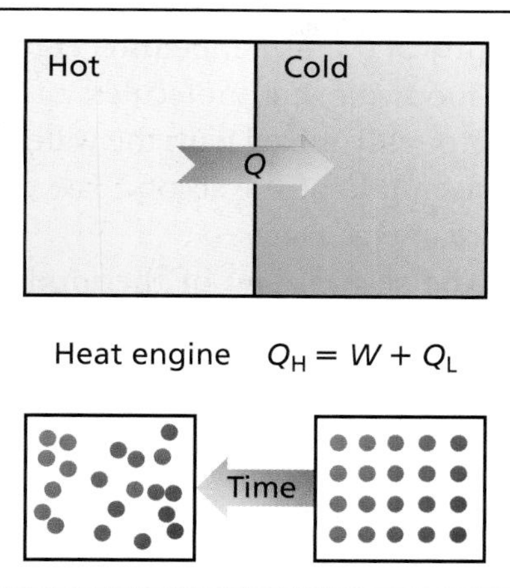

Figure 12-14 Many processes that do not violate the first law of thermodynamics do not occur spontaneously. The spontaneous processes obey both the first and second law of thermodynamics.

When a baseball is dropped and falls due to gravity, it possesses potential and kinetic energies that can be recovered to do work. However, when the baseball falls through the air, it collides with many air molecules that absorb some of its energy. This causes air molecules to move in random directions and at random speeds. The energy absorbed from the baseball causes more disorder among the molecules. The greater the range of speeds exhibited by the molecules, the greater the disorder, which in turn increases the entropy. It is highly unlikely that the molecules that have been dispersed in all directions will come back together, give their energies back to the baseball, and cause it to rise.

Entropy, like thermal energy, is contained in an object. If heat is added to an object, entropy is increased. If heat is removed from an object, entropy is decreased. If an object does work with no change in temperature, the entropy does not change, as long as friction is ignored. The change in entropy, ΔS, is expressed by the following equation, in which entropy has units of J/K and the temperature is measured in kelvins.

Change in Entropy $\Delta S = \frac{Q}{T}$

The change in entropy of an object is equal to the heat added to the object divided by the temperature of the object in kelvins.

CHALLENGE **PROBLEM**

Entropy has some interesting properties. Compare the following situations. Explain how and why these changes in entropy are different.

1. Heating 1.0 kg of water from 273 K to 274 K.
2. Heating 1.0 kg of water from 353 K to 354 K.
3. Completely melting 1.0 kg of ice at 273 K.
4. Heating 1.0 kg of lead from 273 K to 274 K.

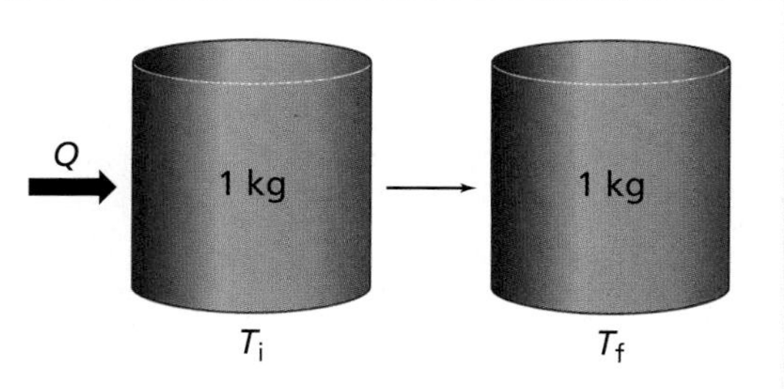

■ **Figure 12-15** The spontaneous mixing of the food coloring and water is an example of the second law of thermodynamics.

The **second law of thermodynamics** states that natural processes go in a direction that maintains or increases the total entropy of the universe. That is, all things will become more and more disordered unless some action is taken to keep them ordered. The increase in entropy and the second law of thermodynamics can be thought of as statements of the probability of events happening. **Figure 12-15** illustrates an increase in entropy as food-coloring molecules, originally separate from the clear water, are thoroughly mixed with the water molecules over time. **Figure 12-16** shows an example of the second law of thermodynamics that might be familiar to many teenagers.

The second law of thermodynamics predicts that heat flows spontaneously only from a hot object to a cold object. Consider a hot iron bar and a cold cup of water. On the average, the particles in the iron will be moving very fast, whereas the particles in the water will be moving slowly. When the bar is plunged into the water and thermal equilibrium is eventually reached, the average kinetic energy of the particles in the iron and the water will be the same. More particles now have an increased random motion than was true for the initial state. This final state is less ordered than the initial state. The fast particles are no longer confined solely to the iron, and the slower particles are no longer confined only to the water; all speeds are evenly distributed. The entropy of the final state is greater than that of the initial state.

Violations of the second law We take for granted many daily events that occur spontaneously, or naturally, in one direction. We would be shocked, however, if the reverse of the same events occurred spontaneously. You are not surprised when a metal spoon, heated at one end, soon becomes uniformly hot. Consider your reaction, however, if a spoon lying on a table suddenly, on its own, became red hot at one end and icy cold at the other. If you dive into a swimming pool, you take for granted that you push the water molecules away as you enter the water. However, you would be amazed if you were swimming in the pool and all the water molecules spontaneously threw you up onto the diving board. Neither of these imagined reverse processes would violate the first law of thermodynamics. They are simply examples of the countless events that do not occur because their processes would violate the second law of thermodynamics.

■ **Figure 12-16** If no work is done on a system, entropy spontaneously reaches a maximum.

The second law of thermodynamics and the increase in entropy also give new meaning to what has been commonly called the *energy crisis.* The energy crisis refers to the continued use of limited resources of fossil fuels, such as natural gas and petroleum. When you use a resource, such as natural gas to heat your home, you do not use up the energy in the gas. As the gas ignites, the internal chemical energy contained in the molecules of the gas is converted into thermal energy of the flame. The thermal energy of the flame is then transferred to thermal energy in the air of your home. Even if this warm air leaks to the outside, the energy is not lost. Energy has not been used up. The entropy, however, has increased.

The chemical structure of natural gas is very ordered. As you have learned, when a substance becomes warmer, the average kinetic energy of the particles in the substance increases. In contrast, the random motion of warmed air is very disordered. While it is mathematically possible for the original chemical order to be reestablished, the probability of this occurring is essentially zero. For this reason, entropy often is used as a measure of the unavailability of useful energy. The energy in the warmed air in a home is not as available to do mechanical work or to transfer heat to other objects as the original gas molecules were. The lack of usable energy is actually a surplus of entropy.

12.2 Section Review

27. Heat of Vaporization Old-fashioned heating systems sent steam into radiators in each room of a house. In the radiators, the steam condensed back to water. Analyze this process and explain how it heated a room.

28. Heat of Vaporization How much heat is needed to change 50.0 g of water at 80.0°C to steam at 110.0°C?

29. Heat of Vaporization The specific heat of mercury is 140 J/kg·°C. Its heat of vaporization is 3.06×10^5 J/kg. How much energy is needed to heat 1.0 kg of mercury metal from 10.0°C to its boiling point and vaporize it completely? The boiling point of mercury is 357°C.

30. Mechanical Energy and Thermal Energy James Joule carefully measured the difference in temperature of water at the top and bottom of a waterfall. Why did he expect a difference?

31. Mechanical Energy and Thermal Energy A man uses a 320-kg hammer moving at 5.0 m/s to smash a 3.0-kg block of lead against a 450-kg rock. When he measured the temperature he found that it had increased by 5.0°C. Explain how this happened.

32. Mechanical Energy and Thermal Energy Water flows over a fall that is 125.0 m high, as shown in **Figure 12-17.** If the potential energy of the water is all converted to thermal energy, calculate the temperature difference between the water at the top and the bottom of the fall.

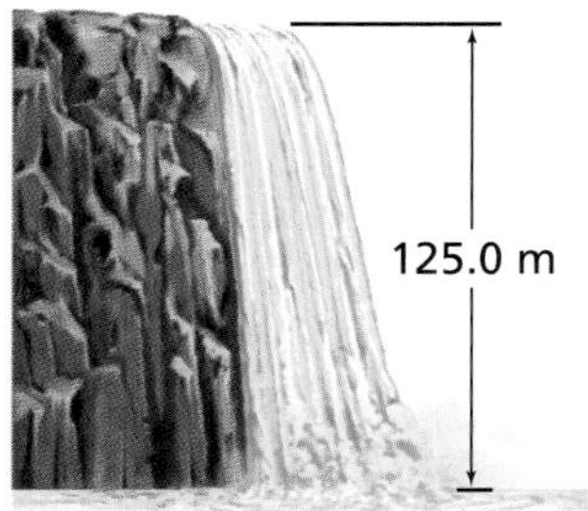

Figure 12-17

33. Entropy Evaluate why heating a home with natural gas results in an increased amount of disorder.

34. Critical Thinking A new deck of cards has all the suits (clubs, diamonds, hearts, and spades) in order, and the cards are ordered by number within the suits. If you shuffle the cards many times, are you likely to return the cards to their original order? Explain. Of what physical law is this an example?

PHYSICS LAB

Heating and Cooling

Alternate CBL instructions can be found on the Web site.
physicspp.com

When a beaker of water is set on a hot plate and the hot plate is turned on, heat is transferred. It first is transferred to the beaker and then to the water at the bottom of the beaker by conduction. The water then transfers heat from the bottom to the top by moving hot water to the top through convection. Once the heat source is removed or shut off, the water radiates thermal energy until it reaches room temperature. How quickly the water heats up is a function of the amount of heat added, the mass of the water, and the specific heat of water.

QUESTION

How does the constant supply of thermal energy affect the temperature of water?

Objectives

- **Measure,** in SI, temperature and mass.
- **Make and use graphs** to help describe the change in temperature of water as it heats up and cools down.
- **Explain** any similarities and differences in these two changes.

Safety Precautions

- **Be careful when using a hot plate. It can burn the skin.**

Materials

hot plate (or Bunsen burner)
250-mL ovenproof glass beaker
50–200 g of water
two thermometers (non-mercury)
stopwatch (or timer)

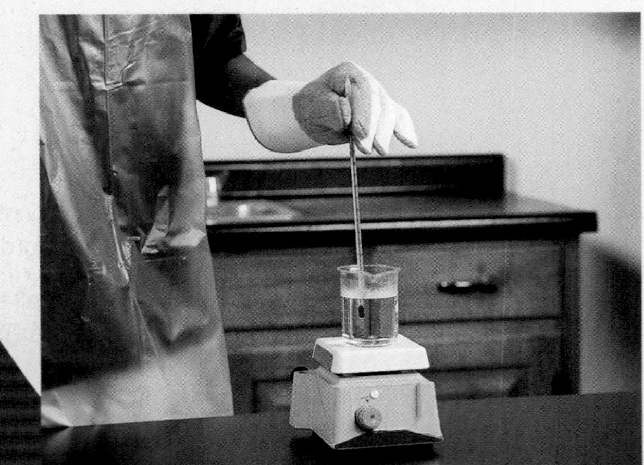

Procedure

1. Set the hot plate to the highest setting, or as recommended by your teacher. Allow a few minutes for the plate to heat up.
2. Measure the mass of the empty beaker.
3. Pour 150 mL of water into the beaker and measure the combined mass of the water and the beaker.
4. Calculate and record the mass of the water in the beaker.
5. Create a data and observations table.
6. Record the initial temperature of the water and the air in the classroom. Note that the bulb end of the thermometers must not touch the bottom or sides of the beaker, nor should it touch a table or your hands.
7. Place the beaker on the hot plate and record the temperature every minute for 5 min.
8. Carefully remove the beaker from the hot plate and record the temperature every minute for the next 10 min.
9. At the end of 10 min, record the temperature of the air.
10. Turn off the hot plate.
11. When finished, allow the equipment to cool and dispose of the water as instructed by your teacher.

Data Table	
Mass of water	
Initial air temperature	
Final air temperature	
Change in air temperature	

Time (min)	Temperature (°C)	Heating or Cooling

Analyze

1. Calculate the change in air temperature to determine if air temperature may be an extraneous variable.
2. Make a scatter-plot graph of temperature (vertical axis) versus time (horizontal axis). Use a computer or a calculator to construct the graph, if possible.
3. **Calculate** What was the change in water temperature as the water heated up?
4. **Calculate** What was the drop in water temperature when the heat source was removed?
5. Calculate the average slope for the temperature increase by dividing change in temperature by the amount of time the water was heating up.
6. Calculate the average slope for the temperature decrease by dividing change in temperature by the amount of time the heat source was removed.

Conclude and Apply

1. **Summarize** What was the change in water temperature when a heat source was applied?
2. **Summarize** What was the change in water temperature once the heat source was removed?
3. What would happen to the water temperature after the next 10 min? Would it continue cooling down forever?
4. Did the water appear to heat up or cool down quicker? Why do you think this is so? *Hint: Examine the slopes you calculated.*
5. **Hypothesize** Where did the thermal energy in the water go once the water began to cool down? Support your hypothesis.

Going Further

1. Does placing your thermometer at the top of the water in your beaker result in different readings than if it is placed at the bottom of the beaker? Explain.
2. Hypothesize what the temperature changes might look like if you had the following amounts of water in the beaker: 50 mL, 250 mL.
3. Suppose you insulated the beaker you were using. How would the beaker's ability to heat up and cool down be affected?

Real-World Physics

1. Suppose you were to use vegetable oil in the beaker instead of water. Hypothesize what the temperature changes might look like if you were to follow the same steps and perform the experiment.
2. If you were to take soup at room temperature and cook it in a microwave oven for 3 min, would the soup return to room temperature in 3 min? Explain your answer.

To find out more about thermal energy, visit the Web site: physicspp.com

HOW it Works The Heat Pump

Heat pumps, also called reversible air conditioners, were invented in the 1940s. They are used to heat and cool homes and hotel rooms. Heat pumps change from heaters to air conditioners by reversing the flow of refrigerant through the system.

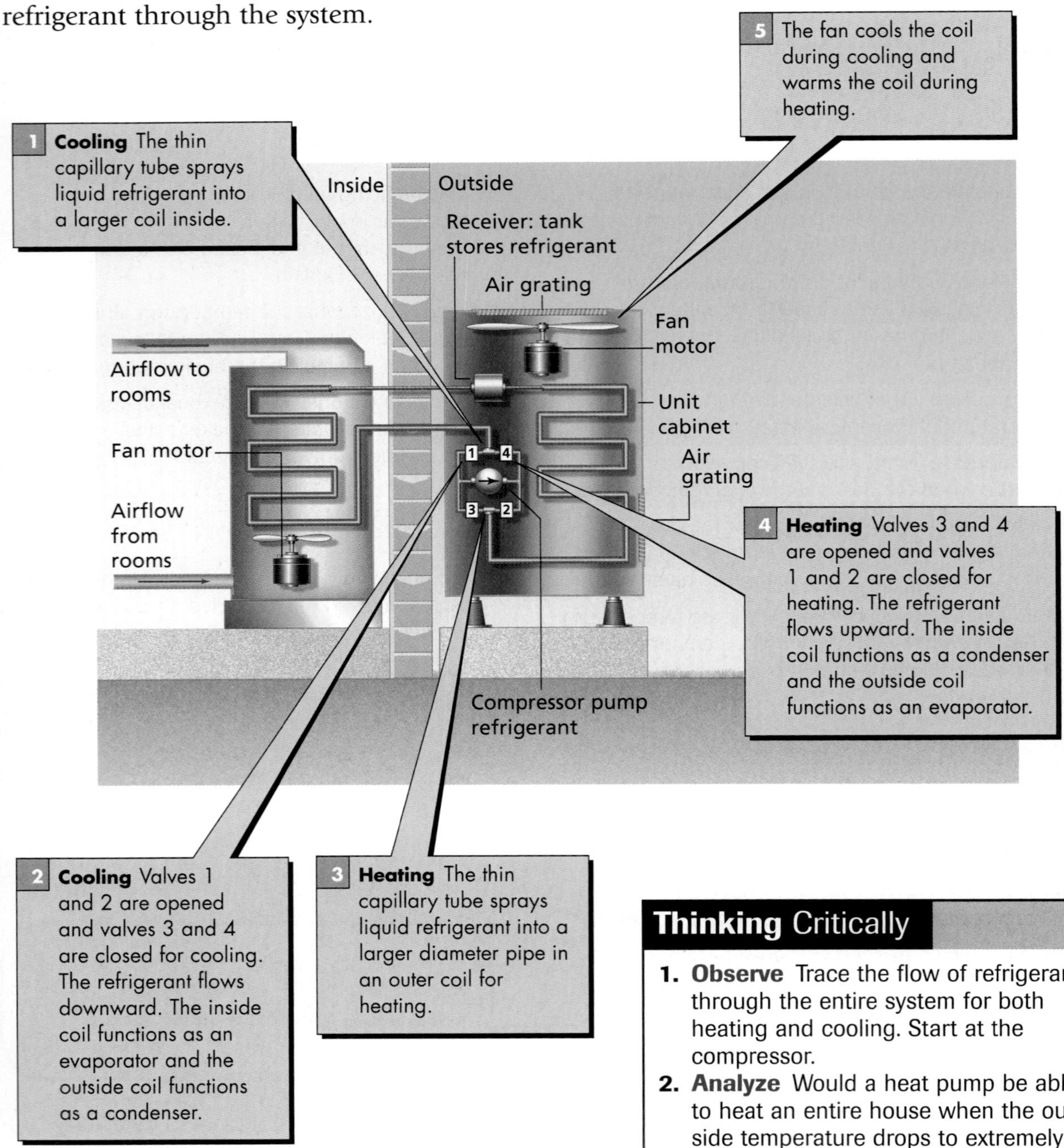

1 Cooling The thin capillary tube sprays liquid refrigerant into a larger coil inside.

2 Cooling Valves 1 and 2 are opened and valves 3 and 4 are closed for cooling. The refrigerant flows downward. The inside coil functions as an evaporator and the outside coil functions as a condenser.

3 Heating The thin capillary tube sprays liquid refrigerant into a larger diameter pipe in an outer coil for heating.

4 Heating Valves 3 and 4 are opened and valves 1 and 2 are closed for heating. The refrigerant flows upward. The inside coil functions as a condenser and the outside coil functions as an evaporator.

5 The fan cools the coil during cooling and warms the coil during heating.

Inside
Outside
Receiver: tank stores refrigerant
Air grating
Fan motor
Airflow to rooms
Unit cabinet
Fan motor
Air grating
Airflow from rooms
Compressor pump refrigerant

Thinking Critically

1. **Observe** Trace the flow of refrigerant through the entire system for both heating and cooling. Start at the compressor.
2. **Analyze** Would a heat pump be able to heat an entire house when the outside temperature drops to extremely cold levels?

Chapter 12

Study Guide

12.1 Temperature and Thermal Energy

Vocabulary

- conduction (p. 315)
- thermal equilibrium (p. 315)
- heat (p. 317)
- convection (p. 317)
- radiation (p. 317)
- specific heat (p. 318)

Key Concepts

- The temperature of a gas is proportional to the average kinetic energy of its particles.
- Thermal energy is a measure of the internal motion of an object's particles.
- A thermometer reaches thermal equilibrium with the object that it comes in contact with, and then a temperature-dependent property of the thermometer indicates the temperature.
- The Celsius and Kelvin temperature scales are used in scientific work. The magnitude of 1 K is equal to the magnitude of 1°C.
- At absolute zero, no more thermal energy can be removed from a substance.
- Heat is energy transferred because of a difference in temperature.

$$Q = mC\Delta T = mC(T_f - T_i)$$

- Specific heat is the quantity of heat required to raise the temperature of 1 kg of a substance by 1 K.
- In a closed, isolated system, heat may flow and change the thermal energy of parts of the system, but the total energy of the system is constant.

$$E_A + E_B = \text{constant}$$

12.2 Changes of State and the Laws of Thermodynamics

Vocabulary

- heat of fusion (p. 324)
- heat of vaporization (p. 324)
- first law of thermodynamics (p. 326)
- heat engine (p. 326)
- entropy (p. 328)
- second law of thermodynamics (p. 330)

Key Concepts

- The heat of fusion is the quantity of heat needed to change 1 kg of a substance from its solid to liquid state at its melting point.

$$Q = mH_f$$

- The heat of vaporization is the quantity of heat needed to change 1 kg of a substance from its liquid to gaseous state at its boiling point.

$$Q = mH_v$$

- Heat transferred during a change of state does not change the temperature of a substance.
- The change in energy of an object is the sum of the heat added to it minus the work done by the object.

$$\Delta U = Q - W$$

- A heat engine continuously converts thermal energy to mechanical energy.
- A heat pump and a refrigerator use mechanical energy to transfer heat from a region of lower temperature to one of higher temperature.
- Entropy is a measure of the disorder of a system.
- The change in entropy of an object is defined to be the heat added to the object divided by the temperature of the object.

$$\Delta S = \frac{Q}{T}$$

Chapter 12 Assessment

Concept Mapping

35. Complete the following concept map using the following terms: *heat, work, internal energy.*

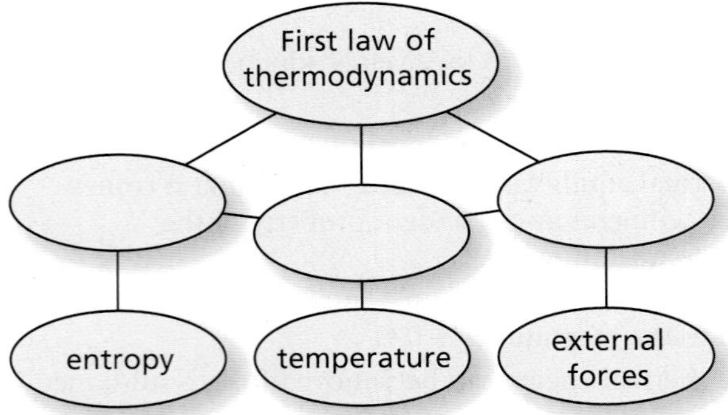

Mastering Concepts

36. Explain the differences among the mechanical energy of a ball, its thermal energy, and its temperature. (12.1)

37. Can temperature be assigned to a vacuum? Explain. (12.1)

38. Do all of the molecules or atoms in a liquid have the same speed? (12.1)

39. Is your body a good judge of temperature? On a cold winter day, a metal doorknob feels much colder to your hand than a wooden door does. Explain why this is true. (12.1)

40. When heat flows from a warmer object in contact with a colder object, do the two have the same temperature changes? (12.1)

41. Can you add thermal energy to an object without increasing its temperature? Explain. (12.2)

42. When wax freezes, does it absorb or release energy? (12.2)

43. Explain why water in a canteen that is surrounded by dry air stays cooler if it has a canvas cover that is kept wet. (12.2)

44. Which process occurs at the coils of a running air conditioner inside a house, vaporization or condensation? Explain. (12.2)

Applying Concepts

45. **Cooking** Sally is cooking pasta in a pot of boiling water. Will the pasta cook faster if the water is boiling vigorously or if it is boiling gently?

46. Which liquid would an ice cube cool faster, water or methanol? Explain.

47. Equal masses of aluminum and lead are heated to the same temperature. The pieces of metal are placed on a block of ice. Which metal melts more ice? Explain.

48. Why do easily vaporized liquids, such as acetone and methanol, feel cool to the skin?

49. Explain why fruit growers spray their trees with water when frost is expected to protect the fruit from freezing.

50. Two blocks of lead have the same temperature. Block A has twice the mass of block B. They are dropped into identical cups of water of equal temperatures. Will the two cups of water have equal temperatures after equilibrium is achieved? Explain.

51. **Windows** Often, architects design most of the windows of a house on the north side. How does putting windows on the south side affect the heating and cooling of the house?

Mastering Problems

12.1 Temperature and Thermal Energy

52. How much heat is needed to raise the temperature of 50.0 g of water from 4.5°C to 83.0°C?

53. A 5.00×10^2-g block of metal absorbs 5016 J of heat when its temperature changes from 20.0°C to 30.0°C. Calculate the specific heat of the metal.

54. **Coffee Cup** A 4.00×10^2-g glass coffee cup is 20.0°C at room temperature. It is then plunged into hot dishwater at a temperature of 80.0°C, as shown in **Figure 12-18.** If the temperature of the cup reaches that of the dishwater, how much heat does the cup absorb? Assume that the mass of the dishwater is large enough so that its temperature does not change appreciably.

Figure 12-18

55. A 1.00×10^2-g mass of tungsten at 100.0°C is placed in 2.00×10^2 g of water at 20.0°C. The mixture reaches equilibrium at 21.2°C. Calculate the specific heat of tungsten.

 For more problems, go to Additional Problems, Appendix B.

56. A 6.0×10^2-g sample of water at 90.0°C is mixed with 4.00×10^2 g of water at 22.0°C. Assume that there is no heat loss to the surroundings. What is the final temperature of the mixture?

57. A 10.0-kg piece of zinc at 71.0°C is placed in a container of water, as shown in **Figure 12-19.** The water has a mass of 20.0 kg and a temperature of 10.0°C before the zinc is added. What is the final temperature of the water and the zinc?

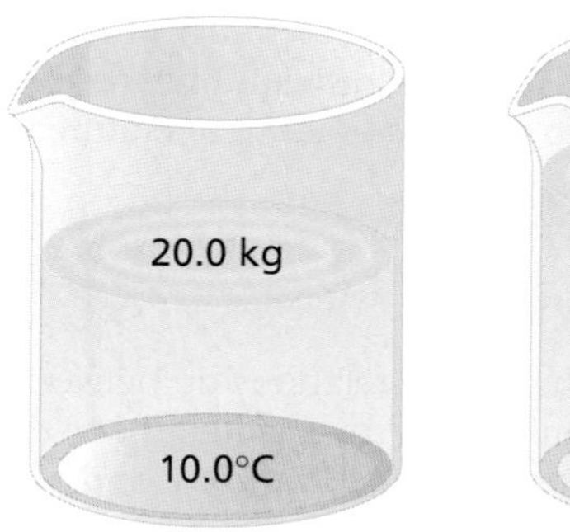

Figure 12-19

58. The kinetic energy of a compact car moving at 100 km/h is 2.9×10^5 J. To get a feeling for the amount of energy needed to heat water, what volume of water (in liters) would 2.9×10^5 J of energy warm from room temperature (20.0°C) to boiling (100.0°C)?

59. Water Heater A 3.0×10^2-W electric immersion heater is used to heat a cup of water, as shown in **Figure 12-20.** The cup is made of glass, and its mass is 3.00×10^2 g. It contains 250 g of water at 15°C. How much time is needed to bring the water to the boiling point? Assume that the temperature of the cup is the same as the temperature of the water at all times and that no heat is lost to the air.

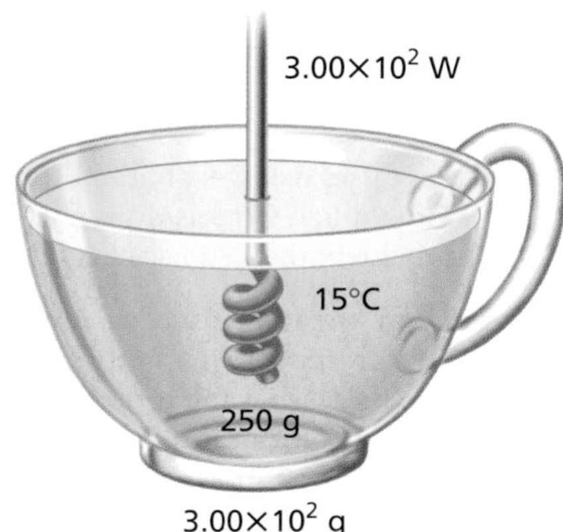

Figure 12-20

60. Car Engine A 2.50×10^2-kg cast-iron car engine contains water as a coolant. Suppose that the engine's temperature is 35.0°C when it is shut off, and the air temperature is 10.0°C. The heat given off by the engine and water in it as they cool to air temperature is 4.40×10^6 J. What mass of water is used to cool the engine?

12.2 Changes of State and the Laws of Thermodynamics

61. Years ago, a block of ice with a mass of about 20.0 kg was used daily in a home icebox. The temperature of the ice was 0.0°C when it was delivered. As it melted, how much heat did the block of ice absorb?

62. A 40.0-g sample of chloroform is condensed from a vapor at 61.6°C to a liquid at 61.6°C. It liberates 9870 J of heat. What is the heat of vaporization of chloroform?

63. A 750-kg car moving at 23 m/s brakes to a stop. The brakes contain about 15 kg of iron, which absorbs the energy. What is the increase in temperature of the brakes?

64. How much heat is added to 10.0 g of ice at −20.0°C to convert it to steam at 120.0°C?

65. A 4.2-g lead bullet moving at 275 m/s strikes a steel plate and comes to a stop. If all its kinetic energy is converted to thermal energy and none leaves the bullet, what is its temperature change?

66. Soft Drink A soft drink from Australia is labeled "Low-Joule Cola." The label says "100 mL yields 1.7 kJ." The can contains 375 mL of cola. Chandra drinks the cola and then wants to offset this input of food energy by climbing stairs. How high would Chandra have to climb if she has a mass of 65.0 kg?

Mixed Review

67. What is the efficiency of an engine that produces 2200 J/s while burning enough gasoline to produce 5300 J/s? How much waste heat does the engine produce per second?

68. Stamping Press A metal stamping machine in a factory does 2100 J of work each time it stamps out a piece of metal. Each stamped piece is then dipped in a 32.0-kg vat of water for cooling. By how many degrees does the vat heat up each time a piece of stamped metal is dipped into it?

69. A 1500-kg automobile comes to a stop from 25 m/s. All of the energy of the automobile is deposited in the brakes. Assuming that the brakes are about 45 kg of aluminum, what would be the change in temperature of the brakes?

70. Iced Tea To make iced tea, you start by brewing the tea with hot water. Then you add ice. If you start with 1.0 L of 90°C tea, what is the minimum amount of ice needed to cool it to 0°C? Would it be better to let the tea cool to room temperature before adding the ice?

71. A block of copper at 100.0°C comes in contact with a block of aluminum at 20.0°C, as shown in **Figure 12-21.** The final temperature of the blocks is 60.0°C. What are the relative masses of the blocks?

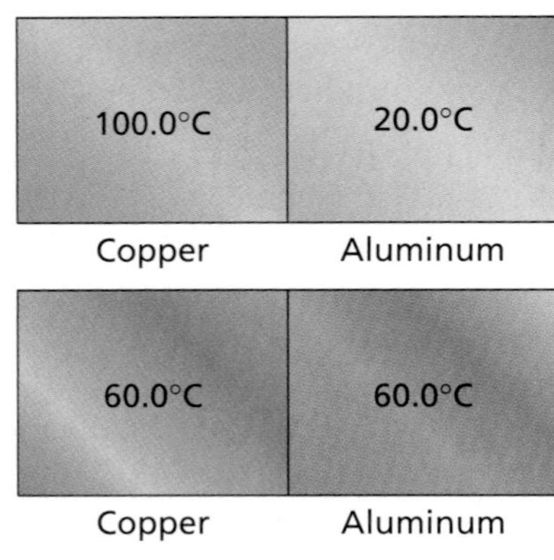

Figure 12-21

72. A 0.35-kg block of copper sliding on the floor hits an identical block moving at the same speed from the opposite direction. The two blocks come to a stop together after the collision. Their temperatures increase by 0.20°C as a result of the collision. What was their speed before the collision?

73. A 2.2-kg block of ice slides across a rough floor. Its initial velocity is 2.5 m/s and its final velocity is 0.50 m/s. How much of the ice block melted as a result of the work done by friction?

Thinking Critically

74. Analyze and Conclude A certain heat engine removes 50.0 J of thermal energy from a hot reservoir at temperature $T_H = 545$ K and expels 40.0 J of heat to a colder reservoir at temperature $T_L = 325$ K. In the process, it also transfers entropy from one reservoir to the other.

a. How does the operation of the engine change the total entropy of the reservoirs?

b. What would be the total entropy change in the reservoirs if $T_L = 205$ K?

75. Analyze and Conclude During a game, the metabolism of basketball players often increases by as much as 30.0 W. How much perspiration must a player vaporize per hour to dissipate this extra thermal energy?

76. Analyze and Conclude Chemists use calorimeters to measure the heat produced by chemical reactions. For instance, a chemist dissolves 1.0×10^{22} molecules of a powdered substance into a calorimeter containing 0.50 kg of water. The molecules break up and release their binding energy to the water. The water temperature increases by 2.3°C. What is the binding energy per molecule for this substance?

77. Apply Concepts All of the energy on Earth comes from the Sun. The surface temperature of the Sun is approximately 10^4 K. What would be the effect on our world if the Sun's surface temperature were 10^3 K?

Writing in Physics

78. Our understanding of the relationship between heat and energy was influenced by a soldier named Benjamin Thompson, Count Rumford, and a brewer named James Prescott Joule. Both relied on experimental results to develop their ideas. Investigate what experiments they did and evaluate whether or not it is fair that the unit of energy is called the joule and not the thompson.

79. Water has an unusually large specific heat and large heats of fusion and vaporization. Our weather and ecosystems depend upon water in all three states. How would our world be different if water's thermodynamic properties were like other materials, such as methanol?

Cumulative Review

80. A rope is wound around a drum with a radius of 0.250 m and a moment of inertia of 2.25 kg·m^2. The rope is connected to a 4.00-kg block. (Chapter 8)

a. Find the linear acceleration of the block.

b. Find the angular acceleration of the drum.

c. Find the tension, F_T, in the rope.

d. Find the angular velocity of the drum after the block has fallen 5.00 m.

81. A weight lifter raises a 180-kg barbell to a height of 1.95 m. How much work is done by the weight lifter in lifting the barbell? (Chapter 10)

82. In a Greek myth, the man Sisyphus is condemned by the gods to forever roll an enormous rock up a hill. Each time he reaches the top, the rock rolls back down to the bottom. If the rock has a mass of 215 kg, the hill is 33 m in height, and Sisyphus can produce an average power of 0.2 kW, how many times in 1 h can he roll the rock up the hill? (Chapter 11)

 For more problems, go to Additional Problems, Appendix B.

Multiple Choice

1. Which of the following temperature conversions is incorrect?

   Ⓐ $-273°C = 0\ K$ Ⓒ $298\ K = 571°C$
   Ⓑ $273°C = 546\ K$ Ⓓ $88\ K = -185°C$

2. What are the units of entropy?

   Ⓐ J/K Ⓒ J
   Ⓑ K/J Ⓓ kJ

3. Which of the following statements about thermal equilibrium is false?

   Ⓐ When two objects are at equilibrium, heat radiation between the objects continues to occur.
   Ⓑ Thermal equilibrium is used to create energy in a heat engine.
   Ⓒ The principle of thermal equilibrium is used for calorimetry calculations.
   Ⓓ When two objects are not at equilibrium, heat will flow from the hotter object to the cooler object.

4. How much heat is required to heat 87 g of methanol ice at 14 K to vapor at 340 K? (melting point = −97.6°C, boiling point = 64.6°C)

   Ⓐ 17 kJ Ⓒ 1.4×10^2 kJ
   Ⓑ 69 kJ Ⓓ 1.5×10^2 kJ

5. Which statement is true about energy, entropy, and changes of state?

   Ⓐ Freezing ice increases in energy as it gains molecular order as a solid.
   Ⓑ The higher the specific heat capacity of a substance, the higher its melting point will be.
   Ⓒ States of matter with increased kinetic energy have higher entropy.
   Ⓓ Energy and entropy cannot increase at the same time.

6. How much heat is needed to warm 363 mL of water in a baby bottle from 24°C to 38°C?

   Ⓐ 21 kJ Ⓒ 121 kJ
   Ⓑ 36 kJ Ⓓ 820 kJ

7. Why is there always some waste heat in a heat engine?

   Ⓐ Heat cannot flow from a cold object to a hot object.
   Ⓑ Friction slows the engine down.
   Ⓒ The entropy increases at each stage.
   Ⓓ The heat pump uses energy.

8. How much heat is absorbed from the surroundings when 81 g of 0.0°C ice in a beaker melts and warms to 10°C?

   Ⓐ 0.34 kJ Ⓒ 30 kJ
   Ⓑ 27 kJ Ⓓ 190 kJ

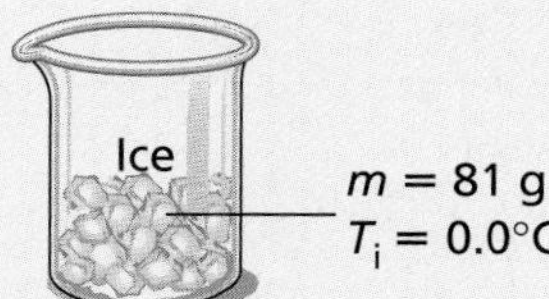

9. You do 0.050 J of work on the coffee in your cup each time you stir it. What would be the increase in entropy in 125 mL of coffee at 65°C when you stir it 85 times?

   Ⓐ 0.013 J/K Ⓒ 0.095 J/K
   Ⓑ 0.050 J Ⓓ 4.2 J

Extended Answer

10. What is the difference in heat required to melt 454 g of ice at 0.00°C, and to turn 454 g of water at 100.0°C into steam? Is the amount of this difference greater or less than the amount of energy required to heat the 454 g of water from 0.00°C to 100.0°C?

✓ Test-Taking TIP

Your Mistakes Can Teach You

The mistakes you make before the test are helpful because they show you areas in which you need more work. When calculating the heat needed to melt and warm a substance, remember to calculate the heat needed for melting as well as the heat needed for raising the temperature of the substance.

Chapter 13 States of Matter

What You'll Learn

- You will explain the expansion and contraction of matter caused by changes in temperature.
- You will apply Pascal's, Archimedes', and Bernoulli's principles in everyday situations.

Why It's Important

Fluids and the forces that they exert enable us to swim and dive, balloons to float, and planes to fly. Thermal expansion affects the designs of buildings, roads, bridges, and machines.

Submarines A nuclear submarine is designed to maneuver at all levels in the ocean. It must withstand great differences in pressure and temperature as it moves deeper under water.

Think About This ▶

How is the submarine able to float at the surface of the ocean and to dive far beneath it?

physicspp.com

Does it float or sink?

Question

How can you measure the buoyancy of objects?

Procedure

1. Obtain a small vial (with a cap or a seal) and a 500-mL graduated cylinder. Attach a rubber band to the vial in order to suspend it from a spring scale.
2. Use the spring scale to find the weight of the vial. Then, use the graduated cylinder to find the volume of water displaced by the sealed vial when it is floating. Record both of these figures. Immediately wipe up any spilled liquid.
3. Place one nickel into the vial and close the lid. Repeat the procedures in step 2, recording the weight of the vial and the nickel, as well as the volume of water displaced. Also, record whether the vial floats or sinks.
4. Repeat steps 2 and 3, each time adding one nickel until the vial no longer floats. When the vial sinks, use the spring scale to find the apparent weight of the vial. Be sure that the vial is not touching the graduated cylinder when it is suspended under water.

Analysis

Use the information you recorded to calculate the density of the vial-and-nickel system in each of your trials. Also, calculate the mass of the water displaced by the system in each trial. How does density appear to be related to floating?

Critical Thinking How does the mass of the vial-nickel system appear to be related to the mass of the water displaced by the system? Does this relationship hold regardless of whether the system is floating?

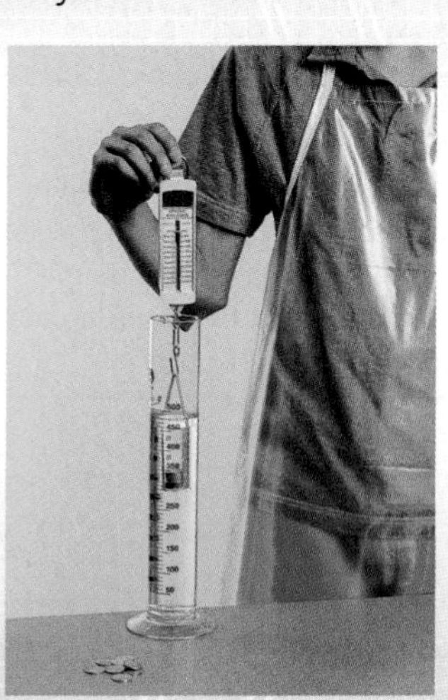

13.1 Properties of Fluids

Water and air are probably two of the most common substances in the everyday lives of people. We feel their effects when we drink, when we bathe, and literally with every breath we take. In your everyday experience, it might not seem that water and air have a great deal in common. If you think further about them, however, you will recognize that they have common properties. Both water and air flow, and unlike solids, neither one of them has a definite shape. Gases and liquids are two states of matter in which atoms and molecules have great freedom to move.

In this chapter, you will explore states of matter. Beginning with gases and liquids, you will learn about the principles that explain how matter responds to changes in temperature and pressure, how hydraulic systems can multiply forces, and how huge metallic ships can float on water. You also will investigate the properties of solids, discovering how they expand and contract, why some solids are elastic, and why some solids seem to straddle the line between solid and liquid.

▸ **Objectives**

- **Describe** how fluids create pressure.
- **Calculate** the pressure, volume, and number of moles of a gas.
- **Compare** gases and plasma.

▸ **Vocabulary**

fluids
pressure
pascal
combined gas law
ideal gas law
thermal expansion
plasma

Pressure

Figure 13-1 The ice cubes, which are solids, have definite shapes. However, the liquid water, a fluid, takes the shape of its container. What fluid is filling the space above the water?

Suppose that you put an ice cube in an empty glass. The ice cube has a certain mass and shape, and neither of these quantities depends on the size or shape of the glass. What happens, however, when the ice melts? Its mass remains the same, but its shape changes. The water flows to take the shape of its container and forms a definite, flat, upper surface, as in **Figure 13-1.** If you boiled the water, it would change into a gas in the form of water vapor, and it also would flow and expand to fill the room. However, the water vapor would not have any definite surface. Both liquids and gases are **fluids,** which are materials that flow and have no definite shape of their own. For now, you can assume that you are dealing with ideal fluids, whose particles take up no space and have no intermolecular attractive forces.

Pressure in fluids You have applied the law of conservation of energy to solid objects. Can this law also be applied to fluids? Work and energy can be defined if we introduce the concept of **pressure,** which is the force on a surface, divided by the area of the surface. Since pressure is force exerted over a surface, anything that exerts pressure is capable of producing change and doing work.

Pressure $P = \frac{F}{A}$

Pressure equals force divided by surface area.

Pressure (P) is a scalar quantity. In the SI system, the unit of pressure is the **pascal** (Pa), which is 1 N/m^2. Because the pascal is a small unit, the kilopascal (kPa), equal to 1000 Pa, is more commonly used. The force, F, on a surface is assumed to be perpendicular to the surface area, A. **Figure 13-2** illustrates the relationships between force, area, and pressure. **Table 13-1** shows how pressures vary in different situations.

Figure 13-2 The astronaut and the landing module both exert pressure on the lunar surface. If the lunar module had a mass of approximately 7300 kg and rested on four pads that were each 91 cm in diameter, what pressure did it exert on the Moon's surface? How could you estimate the pressure exerted by the astronaut?

Solids, liquids, and pressure Imagine that you are standing on the surface of a frozen lake. The forces that your feet exert on the ice are spread over the area of your shoes, resulting in pressure on the ice. Ice is a solid that is made up of vibrating water molecules, and the forces that hold the water molecules in place cause the ice to exert upward forces on your feet that equal your weight. If the ice melted, most of the bonds between the water molecules would be weakened. Although the molecules would continue to vibrate and remain close to each other, they also would slide past one another, and you would break through the surface. The moving water molecules would continue to exert forces on your body.

Gas particles and pressure The pressure exerted by a gas can be understood by applying the kinetic-molecular theory of gases. The kinetic-molecular theory explains the properties of an ideal gas. In reality, the particles of a gas take up space and have intermolecular attractive forces, but an ideal gas is an accurate model of a real gas under most conditions. According to the kinetic-molecular theory, the particles in a gas are in random motion at high speeds and undergoing elastic collisions with each other. When a gas particle hits a container's surface, it rebounds, which changes its momentum. The impulses exerted by many of these collisions result in gas pressure on the surface.

Atmospheric pressure On every square centimeter of Earth's surface at sea level, the atmospheric gas exerts a force of approximately 10 N, about the weight of a 1-kg object. The pressure of Earth's atmosphere on your body is so well balanced by your body's outward forces that you seldom notice it. You probably become aware of this pressure only when your ears pop as the result of pressure changes, as when you ride an elevator in a tall building or fly in an airplane. Atmospheric pressure is about 10 N per 1 cm^2 (10^{-4} m^2), which is about 1.0×10^5 N/m^2, or 100 kPa. Other planets in our solar system also have atmospheres. The pressure exerted by these atmospheres, however, varies widely. For example, the pressure at the surface of Venus is about 92 times the pressure at the surface of Earth, while the pressure at the surface of Mars is less than 1 percent of Earth's.

Table 13-1
Some Typical Pressures

Location	Pressure (Pa)
The center of the Sun	3×10^{16}
The center of Earth	4×10^{11}
The deepest ocean trench	1.1×10^{8}
Standard atmosphere	1.01325×10^{5}
Blood pressure	1.6×10^{4}
Air pressure on top of Mt. Everest	3×10^{4}
The best vacuum	1×10^{-13}

▶ EXAMPLE Problem 1

Calculating Pressure **A child weighs 364 N and sits on a three-legged stool, which weighs 41 N. The bottoms of the stool's legs touch the ground over a total area of 19.3 cm^2.**

a. **What is the average pressure that the child and stool exert on the ground?**

b. **How does the pressure change when the child leans over so that only two legs of the stool touch the floor?**

1 Analyze and Sketch the Problem

- Sketch the child and the stool, labeling the total force that they exert on the ground.
- List the variables, including the force that the child and stool exert on the ground and the areas for parts **a** and **b.**

Known:

$F_{g\ child} = 364$ N

$F_{g\ stool} = 41$ N

$F_{g\ total} = F_{g\ child} + F_{g\ stool} = 364\ N + 41\ N = 405$ N

$A_A = 19.3\ cm^2$

$A_B = \frac{2}{3} \times 19.3\ cm^2 = 12.9\ cm^2$

Unknown:

$P_A = ?$

$P_B = ?$

> **Math Handbook**
> Dimensional Calculations
> pages 846–847

2 Solve for the Unknown

Find each pressure.

$P = F/A$

a. $$P_A = \left(\frac{405\ N}{19.3\ cm^2}\right)\left(\frac{(100\ cm)^2}{(1\ m)^2}\right) = 2.10\times10^2\ kPa$$

Substitute $F = F_{g\ total} = 405$ N, $A = A_A = 19.3$ cm^2

b. $$P_B = \left(\frac{405\ N}{12.9\ cm^2}\right)\left(\frac{(100\ cm)^2}{(1\ m)^2}\right) = 3.14\times10^2\ kPa$$

Substitute $F = F_{g\ total} = 405$ N, $A = A_B = 12.9$ cm^2

3 Evaluate the Answer

- **Are the units correct?** The units for pressure should be Pa, and 1 N/m^2 = 1 Pa.

▷ PRACTICE Problems

• Additional Problems, Appendix B
• Solutions to Selected Problems, Appendix C

1. The atmospheric pressure at sea level is about 1.0×10^5 Pa. What is the force at sea level that air exerts on the top of a desk that is 152 cm long and 76 cm wide?
2. A car tire makes contact with the ground on a rectangular area of 12 cm by 18 cm. If the car's mass is 925 kg, what pressure does the car exert on the ground as it rests on all four tires?
3. A lead brick, 5.0 cm × 10.0 cm × 20.0 cm, rests on the ground on its smallest face. Lead has a density of 11.8 g/cm^3. What pressure does the brick exert on the ground?
4. In a tornado, the pressure can be 15 percent below normal atmospheric pressure. Suppose that a tornado occurred outside a door that is 195 cm high and 91 cm wide. What net force would be exerted on the door by a sudden 15 percent drop in normal atmospheric pressure? In what direction would the force be exerted?
5. In industrial buildings, large pieces of equipment must be placed on wide steel plates that spread the weight of the equipment over larger areas. If an engineer plans to install a 454-kg device on a floor that is rated to withstand additional pressure of 5.0×10^4 Pa, how large should the steel support plate be?

The Gas Laws

As scientists first studied gases and pressure, they began to notice some interesting relationships. The first relationship to emerge was named Boyle's law, after seventeenth-century chemist and physicist Robert Boyle. Boyle's law states that for a fixed sample of gas at constant temperature, the volume of the gas varies inversely with the pressure. Because the product of inversely related variables is a constant, Boyle's law can be written PV = constant, or $P_1V_1 = P_2V_2$. The subscripts that you see in the gas laws will help you keep track of different variables, such as pressure and volume, as they change throughout a problem. These variables can be rearranged to solve for an unknown pressure or volume. As shown in **Figure 13-3,** the relationship between the pressure and the volume of a gas is critical to the sport of scuba diving.

■ **Figure 13-3** The gas in the tank on the diver's back is at high pressure. This pressure is reduced by the regulator so that the pressure of the gas the diver breathes is equal to the water pressure. In the photo, you can see bubbles coming from the regulator.

A second relationship was discovered about 100 years after Boyle's work by Jacques Charles. When Charles cooled a gas, the volume shrank by $\frac{1}{273}$ of its original volume for every degree cooled, which is a linear relationship. At the time, Charles could not cool gases to the extremely low temperatures achieved in modern laboratories. In order to see what lower limits might be possible, he extended, or extrapolated, the graph of his data to these temperatures. This extrapolation suggested that if the temperature were reduced to −273°C, a gas would have zero volume. The temperature at which a gas would have zero volume is now called absolute zero, which is represented by the zero of the Kelvin temperature scale.

These experiments indicated that under constant pressure, the volume of a sample of gas varies directly with its Kelvin temperature, a result that is now called Charles's law. Charles's law can be written V/T = constant, or $V_1/T_1 = V_2/T_2$.

Combining Boyle's law and Charles's law relates the pressure, temperature, and volume of a fixed amount of ideal gas, which leads to the equation called the **combined gas law.**

Combined Gas Law $\frac{P_1V_1}{T_1} = \frac{P_2V_2}{T_2} = \text{constant}$

For a fixed amount of an ideal gas, the pressure times the volume, divided by the Kelvin temperature equals a constant.

As shown in **Figure 13-4,** the combined gas law reduces to Boyle's law under conditions of constant temperature and to Charles's law under conditions of constant pressure.

The ideal gas law You can use the kinetic-molecular theory to discover how the constant in the combined gas law depends on the number of particles, N. Suppose that the volume and temperature of an ideal gas are held constant. If the number of particles increases, the number of collisions that the particles make with the container will increase, thereby increasing the pressure. Removing particles decreases the number of collisions, and thus, decreases the pressure. You can conclude that the constant in the combined gas law equation is proportional to N.

$$\frac{PV}{T} = kN$$

The constant, k, is called Boltzmann's constant, and its value is 1.38×10^{-23} Pa·m^3/K. Of course, N, the number of particles, is a very large number. Instead of using N, scientists often use a unit called a mole. One mole (abbreviated mol and represented in equations by n) is similar to one dozen, except that instead of representing 12 items, one mole represents 6.022×10^{23} particles. This number is called Avogadro's number, after Italian scientist Amedeo Avogadro.

Avogadro's number is numerically equal to the number of particles in a sample of matter whose mass equals the molar mass of the substance. You can use this relationship to convert between mass and n, the number of moles present. Using moles instead of the number of particles changes Boltzmann's constant. This new constant is abbreviated R, and it has the value 8.31 Pa·m^3/mol·K. Rearranging, you can write the **ideal gas law** in its most familiar form.

Ideal Gas Law $PV = nRT$

For an ideal gas, the pressure times the volume is equal to the number of moles multiplied by the constant R and the Kelvin temperature.

Note that with the given value of R, volume must be expressed in m^3, temperature in K, and pressure in Pa. In practice, the ideal gas law predicts the behavior of gases remarkably well, except under conditions of high pressures or low temperatures.

MINI LAB

Pressure

How much pressure do you exert when standing on one foot? Have a partner trace your foot, and then use the outline to estimate its area.

1. **Determine** your weight in newtons and the area of the outline in m^2.
2. **Calculate** the pressure.
3. **Compare and contrast** the pressure you exert on the ground with the pressure exerted by various objects. For example, you could weigh a brick and determine the pressure it exerts when resting on different faces.

Analyze and Conclude

4. How do shoes with high heels affect the pressure that a person exerts on the ground?

Figure 13-4 You can use the combined gas law to derive both Boyle's and Charles's laws. What happens if you hold volume constant?

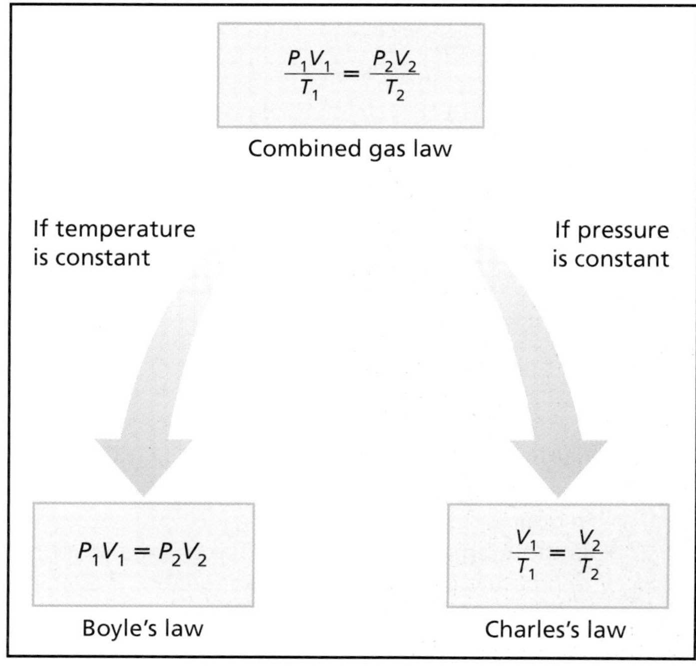

▶ EXAMPLE Problem 2

Gas Laws **A 20.0-L sample of argon gas at 273 K is at atmospheric pressure, 101.3 kPa. The temperature is lowered to 120 K, and the pressure is increased to 145 kPa.**

a. What is the new volume of the argon sample?

b. Find the number of moles of argon atoms in the argon sample.

c. Find the mass of the argon sample. The molar mass, *M*, of argon is 39.9 g/mol.

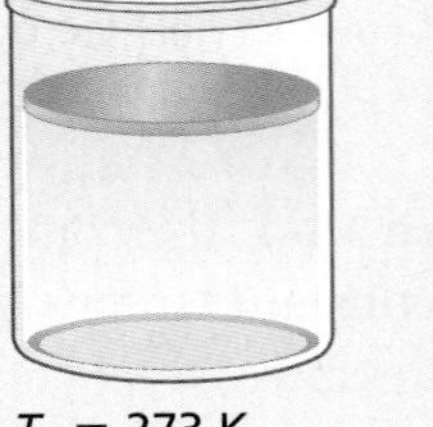

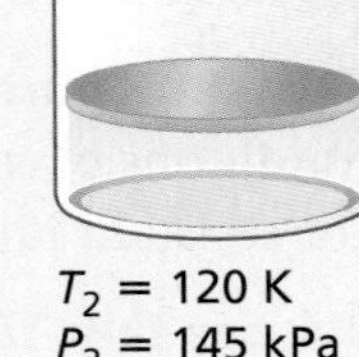

1 Analyze and Sketch the Problem

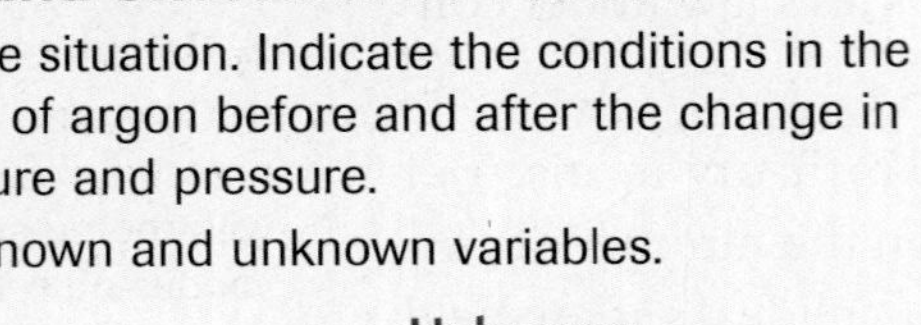

- Sketch the situation. Indicate the conditions in the container of argon before and after the change in temperature and pressure.
- List the known and unknown variables.

Known:	Unknown:
$V_1 = 20.0$ L	$V_2 = ?$
$P_1 = 101.3$ kPa	moles of argon = ?
$T_1 = 273$ K	mass of argon sample = ?
$P_2 = 145$ kPa	
$T_2 = 120$ K	
$R = 8.31\ \text{Pa·m}^3/\text{mol·K}$	
$M_{argon} = 39.9$ g/mol	

Personal Tutor For an online tutorial on gas laws, visit physicspp.com.

2 Solve for the Unknown

a. Use the combined gas law and solve for V_2.

$$\frac{P_1V_1}{T_1} = \frac{P_2V_2}{T_2}$$

$$V_2 = \frac{P_1V_1T_2}{P_2T_1}$$

$$= \frac{(101.3\text{ kPa})(20.0\text{ L})(120\text{ K})}{(145\text{ kPa})(273\text{ K})}$$

Substitute $P_1 = 101.3$ kPa, $P_2 = 145$ kPa, $V_1 = 20.0$ L, $T_1 = 273$ K, $T_2 = 120$ K

$$= 6.1\text{ L}$$

b. Use the ideal gas law and solve for n.

$$PV = nRT$$

$$n = \frac{PV}{RT}$$

$$= \frac{(101.3\times10^3\text{ Pa})(0.0200\text{ m}^3)}{(8.31\text{ Pa·m}^3/\text{mol·K})(273\text{ K})}$$

Substitute $P = 101.3\times10^3$ Pa, $V = 0.0200$ m³, $R = 8.31$ m³/mol·K, $T = 273$ K

$$= 0.893\text{ mol}$$

c. Use the molar mass to convert from moles of argon in the sample to mass of the sample.

$$m = Mn$$

$$m_{\text{argon sample}} = (39.9\text{ g/mol})(0.893\text{ mol})$$

Substitute $M = 39.9$ g/mol, $n = 0.893$ mol

$$= 35.6\text{ g}$$

3 Evaluate the Answer

- **Are the units correct?** The volume, V_2, is in liters, and the mass of the sample is in grams.
- **Is the magnitude realistic?** The change in volume is consistent with an increase in pressure and decrease in temperature. The calculated mass of the argon sample is reasonable.

PRACTICE Problems

- Additional Problems, Appendix B
- Solutions to Selected Problems, Appendix C

6. A tank of helium gas used to inflate toy balloons is at a pressure of 15.5×10^6 Pa and a temperature of 293 K. The tank's volume is 0.020 m^3. How large a balloon would it fill at 1.00 atmosphere and 323 K?
7. What is the mass of the helium gas in the previous problem? The molar mass of helium gas is 4.00 g/mol.
8. A tank containing 200.0 L of hydrogen gas at 0.0°C is kept at 156 kPa. The temperature is raised to 95°C, and the volume is decreased to 175 L. What is the new pressure of the gas?
9. The average molar mass of the components of air (mainly diatomic oxygen gas and diatomic nitrogen gas) is about 29 g/mol. What is the volume of 1.0 kg of air at atmospheric pressure and 20.0°C?

Thermal Expansion

As you applied the combined gas law, you discovered how gases expand as their temperatures increase. When heated, all forms of matter—solids, liquids, and gases—generally become less dense and expand to fill more space. This property, known as **thermal expansion,** has many useful applications, such as circulating air in a room. When the air near the floor of a room is warmed, gravity pulls the denser, colder air near the ceiling down, which pushes the warmer air upward. This circulation of air within a room is called a convection current. **Figure 13-5** shows convection currents starting as the hot air above the flames rises. You also can see convection currents in a pot of hot, but not boiling, water on a stove. When the pot is heated from the bottom, the colder and denser water sinks to the bottom where it is warmed and then pushed up by the continuous flow of cooler water from the top.

This thermal expansion occurs in most liquids. A good model for all liquids does not exist, but it is useful to think of a liquid as a finely ground solid. Groups of two, three, or more particles move together as if they were tiny pieces of a solid. When a liquid is heated, particle motion causes these groups to expand in the same way that particles in a solid are pushed apart. The spaces between groups increase. As a result, the whole liquid expands. With an equal change in temperature, liquids expand considerably more than solids, but not as much as gases.

Why ice floats Because matter expands as it is heated, you might predict that ice would be more dense than water, and therefore, it should sink. However, when water is heated from 0°C to 4°C, instead of expanding, it contracts as the forces between particles increase and the ice crystals collapse. These forces between water molecules are strong, and the crystals that make up ice have a very open structure. Even when ice melts, tiny crystals remain. These remaining crystals are melting, and the volume of the water decreases until the temperature reaches 4°C. However, once the temperature of water moves above 4°C, its volume increases because of greater molecular motion. The practical result is that water is most dense at 4°C and ice floats. This fact is very important to our lives and environment. If ice sank, lakes would freeze from the bottom each winter and many would never melt completely in the summer.

Figure 13-5 This image was made by a special technique that enables you to see different densities in the air. Convection currents are set up as warmer, less dense air rises and cooler, denser air sinks.

■ **Figure 13-6** The colorful lighting effects in neon signs are caused by luminous plasmas formed in the glass tubing.

Plasma

If you heat a solid, it melts to form a liquid. Further heating results in a gas. What happens if you increase the temperature still further? Collisions between the particles become violent enough to tear the electrons off the atoms, thereby producing positively charged ions. The gaslike state of negatively charged electrons and positively charged ions is called **plasma.** Plasma is considered to be another fluid state of matter.

The plasma state may seem to be uncommon; however, most of the matter in the universe is plasma. Stars consist mostly of plasma at extremely high temperatures. Much of the matter between stars and galaxies consists of energetic hydrogen that has no electrons. This hydrogen is in the plasma state. The primary difference between gas and plasma is that plasma can conduct electricity, whereas gas cannot. Lightning bolts are in the plasma state. Neon signs, such as the one shown in **Figure 13-6,** fluorescent bulbs, and sodium vapor lamps all contain glowing plasma.

13.1 Section Review

10. Pressure and Force Suppose that you have two boxes. One is 20 cm × 20 cm × 20 cm. The other is 20 cm × 20 cm × 40 cm.

a. How does the pressure of the air on the outside of the two boxes compare?

b. How does the magnitude of the total force of the air on the two boxes compare?

11. Meteorology A weather balloon used by meteorologists is made of a flexible bag that allows the gas inside to freely expand. If a weather balloon containing 25.0 m^3 of helium gas is released from sea level, what is the volume of gas when the balloon reaches a height of 2100 m, where the pressure is 0.82×10^5 Pa? Assume that the temperature is unchanged.

12. Gas Compression In a certain internal-combustion engine, 0.0021 m^3 of air at atmospheric pressure and 303 K is rapidly compressed to a pressure of 20.1×10^5 Pa and a volume of 0.0003 m^3. What is the final temperature of the compressed gas?

13. Density and Temperature Starting at 0°C, how will the density of water change if it is heated to 4°C? To 8°C?

14. The Standard Molar Volume What is the volume of 1.00 mol of a gas at atmospheric pressure and a temperature of 273 K?

15. The Air in a Refrigerator How many moles of air are in a refrigerator with a volume of 0.635 m^3 at a temperature of 2.00°C? If the average molar mass of air is 29 g/mol, what is the mass of the air in the refrigerator?

16. Critical Thinking Compared to the particles that make up carbon dioxide gas, the particles that make up helium gas are very small. What can you conclude about the number of particles in a 2.0-L sample of carbon dioxide gas compared to the number of particles in a 2.0-L sample of helium gas if both samples are at the same temperature and pressure?

Physics Online physicspp.com/self_check_quiz

13.2 Forces Within Liquids

The liquids considered thus far have been ideal liquids, in which the particles are totally free to slide past one another. The unexpected behavior of water between 0°C and 4°C, however, illustrates that in real fluids, particles exert electromagnetic forces of attraction, called **cohesive forces,** on each other. These and other forces affect the behavior of fluids.

▶ **Objectives**
- **Explain** how cohesive forces cause surface tension.
- **Explain** how adhesive forces cause capillary action.
- **Discuss** evaporative cooling and the role of condensation in cloud formation.

▶ **Vocabulary**

cohesive forces
adhesive forces

Cohesive Forces

Have you ever noticed that dewdrops on spiderwebs and falling drops of oil are nearly spherical? What happens when rain falls on a freshly washed and waxed car? The water drops bead up into rounded shapes, as shown in the spiderweb in **Figure 13-7.** All of these phenomena are examples of surface tension, which is the tendency of the surface of a liquid to contract to the smallest possible area. Surface tension is a result of the cohesive forces among the particles of a liquid.

Notice that beneath the surface of the liquid shown in **Figure 13-8a** on the next page, each particle of the liquid is attracted equally in all directions by neighboring particles, and even to the particles of the wall of the container. As a result, no net force acts on any of the particles beneath the surface. At the surface, however, the particles are attracted downward and to the sides, but not upward. There is a net downward force, which acts on the top layers and causes the surface layer to be slightly compressed. The surface layer acts like a tightly stretched rubber sheet or a film that is strong enough to support the weight of very light objects, such as the water strider in **Figure 13-8b** on the next page. The surface tension of water also can support a steel paper clip, even though the density of steel is nine times greater than that of water. Try it!

Figure 13-7 Rainwater beads up on a spider's web because water drops have surface tension.

Why does surface tension produce spherical drops? The force pulling the surface particles into a liquid causes the surface to become as small as possible, and the shape that has the least surface for a given volume is a sphere. The higher the surface tension of the liquid, the more resistant the liquid is to having its surface broken. For example, liquid mercury has much stronger cohesive forces than water does. Thus, liquid mercury forms spherical drops, even when it is placed on a smooth surface. On the other hand, liquids such as alcohol and ether have weaker cohesive forces. A drop of either of these liquids flattens out on a smooth surface.

Viscosity In nonideal fluids, the cohesive forces and collisions between fluid molecules cause internal friction that slows the fluid flow and dissipates mechanical energy. The measure of this internal friction is called the viscosity of the liquid. Water is not very viscous, but motor oil is very viscous. As a result of its viscosity, motor oil flows slowly over the parts of an engine to coat the metal and reduce rubbing. Lava, molten rock that flows from a volcano or vent in Earth's surface, is one of the most viscous fluids. There are several types of lava, and the viscosity of each type varies with composition and temperature.

Geology Connection

■ **Figure 13-8** Molecules in the interior of a liquid are attracted in all directions **(a).** A water strider can walk on water because molecules at the surface have a net inward attraction that results in surface tension **(b).**

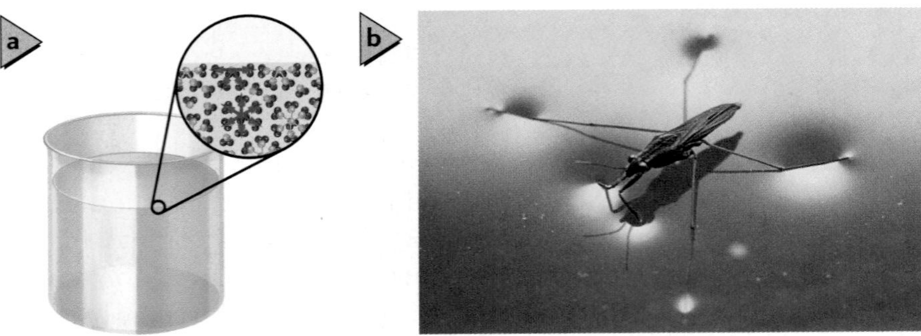

Adhesive Forces

Similar to cohesive forces, **adhesive forces** are electromagnetic attractive forces that act between particles of different substances. If a glass tube with a small inner diameter is placed in water, the water rises inside the tube. The water rises because the adhesive forces between glass and water molecules are stronger than the cohesive forces between water molecules. This phenomenon is called capillary action. The water continues to rise until the weight of the water that is lifted balances the total adhesive force between the glass and water molecules. If the radius of the tube increases, the volume and the weight of the water will increase proportionally faster than the surface area of the tube. Thus, water is lifted higher in a narrow tube than in a wider one. Capillary action causes molten wax to rise in a candle's wick and water to move up through the soil and into the roots of plants.

When a glass tube is placed in a beaker of water, the surface of the water climbs the outside of the tube, as shown in **Figure 13-9a.** The adhesive forces between the glass molecules and water molecules are greater than the cohesive forces between the water molecules. In contrast, the cohesive forces between mercury molecules are greater than the adhesive forces between the mercury and glass molecules, so the liquid does not climb the tube. These forces also cause the center of the mercury's surface to depress, as shown in **Figure 13-9b.**

APPLYING PHYSICS

▶ **Plants** Cohesive forces in liquids actually allow them to be stretched just like rubber bands. This stretching is difficult to achieve in the laboratory, but it is common in plants. The strength of the cohesive forces in water keeps the water from breaking and bubbling as it goes through the plant tissue to the leaves. If not for these forces, trees could not grow higher than about 10 m. ◀

Evaporation and Condensation

Why does a puddle of water disappear on a hot, dry day? As you learned in Chapter 12, the particles in a liquid are moving at random speeds. If a fast-moving particle can break through the surface layer, it will escape from the liquid. Because there is a net downward cohesive force at the surface, however, only the most energetic particles escape. This escape of particles is called evaporation.

■ **Figure 13-9** Water climbs the outside wall of this glass tube **(a),** while the mercury is depressed by the rod **(b).** The forces of attraction between mercury atoms are stronger than any adhesive forces between the mercury and the glass.

Evaporative cooling Evaporation has a cooling effect. On a hot day, your body perspires, and the evaporation of your sweat cools you down. In a puddle of water, evaporation causes the remaining liquid to cool down. Each time a particle with higher-than-average kinetic energy escapes from the water, the average kinetic energy of the remaining particles decreases. As you learned in Chapter 12, a decrease in average kinetic energy is a decrease in temperature. You can test this cooling effect by pouring a small amount of rubbing alcohol in the palm of your hand. Alcohol molecules evaporate easily because they have weak cohesive forces. As the molecules evaporate, the cooling effect is quite noticeable. A liquid that evaporates quickly is called a volatile liquid.

Have you ever wondered why humid days feel warmer than dry days at the same temperature? On a day that is humid, the water vapor content of the air is high. Because there are already many water molecules in the air, the water molecules in perspiration are less likely to evaporate from the skin. Evaporation is the body's primary cooling mechanism, so the body is not able to cool itself as effectively on a humid day.

Particles of liquid that have evaporated into the air can also return to the liquid phase if the kinetic energy or temperature decreases, a process called condensation. What happens if you bring a cold glass into a hot, humid area? The outside of the glass soon becomes coated with condensed water. Water molecules moving randomly in the air surrounding the glass strike the cold surface, and if they lose enough energy, the cohesive forces become strong enough to prevent their escape.

The air above any body of water, as shown in **Figure 13-10,** contains evaporated water vapor, which is water in the form of gas. If the temperature is reduced, the water vapor condenses around tiny dust particles in the air and produces droplets only 0.01 mm in diameter. A cloud of these droplets is called fog. Fog often forms when moist air is chilled by the cold ground. Fog also can form in your home. When a carbonated drink is opened, the sudden decrease in pressure causes the temperature of the gas in the container to drop, which condenses the water vapor dissolved in that gas.

Figure 13-10 Warm, moist, surface air rises until it reaches a height where the temperature is at the point at which water vapor condenses and forms clouds.

13.2 Section Review

17. Evaporation and Cooling In the past, when a baby had a high fever, the doctor might have suggested gently sponging off the baby with rubbing alcohol. Why would this help?

18. Surface Tension A paper clip, which has a density greater than that of water, can be made to stay on the surface of water. What procedures must you follow for this to happen? Explain.

19. Language and Physics The English language includes the terms *adhesive tape* and *working as a cohesive group.* In these terms, are *adhesive* and *cohesive* being used in the same context as their meanings in physics?

20. Adhesion and Cohesion In terms of adhesion and cohesion, explain why alcohol clings to the surface of a glass rod but mercury does not.

21. Floating How can you tell that the paper clip in problem 18 was not floating?

22. Critical Thinking On a hot, humid day, Beth sat on the patio with a glass of cold water. The outside of the glass was coated with water. Her younger sister, Jo, suggested that the water had leaked through the glass from the inside to the outside. Suggest an experiment that Beth could do to show Jo where the water came from.

Physics Online physicspp.com/self_check_quiz

13.3 Fluids at Rest and in Motion

▶ **Objectives**

- **Relate** Pascal's principle to simple machines and occurrences.
- **Apply** Archimedes' principle to buoyancy.
- **Apply** Bernoulli's principle to airflow.

▶ **Vocabulary**

Pascal's principle
buoyant force
Archimedes' principle
Bernoulli's principle
streamlines

You have learned how fluids exert pressure, the force per unit area. You also know that the pressure exerted by fluids changes; for example, atmospheric pressure drops as you climb a mountain. In this section, you will learn about the forces exerted by resting and moving fluids.

Fluids at Rest

If you have ever dived deep into a swimming pool or lake, you know that your body, especially your ears, is sensitive to changes in pressure. You may have noticed that the pressure you felt on your ears did not depend on whether your head was upright or tilted, but that if you swam deeper, the pressure increased.

Pascal's principle Blaise Pascal, a French physician, noted that the pressure in a fluid depends upon the depth of the fluid and has nothing to do with the shape of the fluid's container. He also discovered that any change in pressure applied at any point on a confined fluid is transmitted undiminished throughout the fluid, a fact that is now known as **Pascal's principle.** Every time you squeeze a tube of toothpaste, you demonstrate Pascal's principle. The pressure that your fingers exert at the bottom of the tube is transmitted through the toothpaste and forces the paste out at the top. Likewise, if you squeeze one end of a helium balloon, the other end of the balloon expands.

When fluids are used in machines to multiply forces, Pascal's principle is being applied. In a common hydraulic system, a fluid is confined to two connecting chambers, as shown in **Figure 13-11.** Each chamber has a piston that is free to move, and the pistons have different surface areas. If a force, F_1, is exerted on the first piston with a surface area of A_1, the pressure, P_1, exerted on the fluid can be determined by using the following equation.

$$P_1 = \frac{F_1}{A_1}$$

This equation is simply the definition of pressure: pressure equals the force per unit area. The pressure exerted by the fluid on the second piston, with a surface area A_2, can also be determined.

$$P_2 = \frac{F_2}{A_2}$$

According to Pascal's principle, pressure is transmitted without change throughout a fluid, so pressure P_2 is equal in value to P_1. You can determine the force exerted by the second piston by using $F_1/A_1 = F_2/A_2$ and solving for F_2. This force is shown by the following equation.

Force Exerted by a Hydraulic Lift $F_2 = \frac{F_1 A_2}{A_1}$

The force exerted by the second piston is equal to the force exerted by the first piston multiplied by the ratio of the area of the second piston to the area of the first piston.

■ **Figure 13-11** The pressure exerted by the force of the small piston is transmitted throughout the fluid and results in a multiplied force on the larger piston.

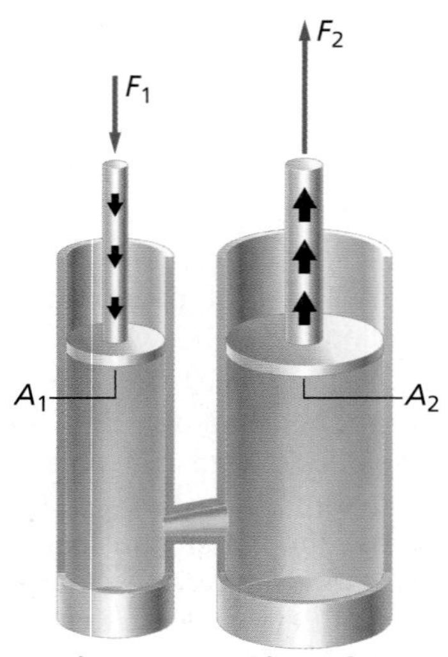

PRACTICE Problems

- Additional Problems, Appendix B
- Solutions to Selected Problems, Appendix C

23. Dentists' chairs are examples of hydraulic-lift systems. If a chair weighs 1600 N and rests on a piston with a cross-sectional area of 1440 cm^2, what force must be applied to the smaller piston, with a cross-sectional area of 72 cm^2, to lift the chair?

24. A mechanic exerts a force of 55 N on a 0.015 m^2 hydraulic piston to lift a small automobile. The piston that the automobile sits on has an area of 2.4 m^2. What is the weight of the automobile?

25. By multiplying a force, a hydraulic system serves the same purpose as a lever or seesaw. If a 400-N child standing on one piston is balanced by a 1100-N adult standing on another piston, what is the ratio of the areas of their pistons?

26. In a machine shop, a hydraulic lift is used to raise heavy equipment for repairs. The system has a small piston with a cross-sectional area of 7.0×10^{-2} m^2 and a large piston with a cross-sectional area of 2.1×10^{-1} m^2. An engine weighing 2.7×10^{3} N rests on the large piston.

a. What force must be applied to the small piston to lift the engine?

b. If the engine rises 0.20 m, how far does the smaller piston move?

Swimming Under Pressure

When you are swimming, you feel the pressure of the water increase as you dive deeper. This pressure is actually a result of gravity; it is related to the weight of the water above you. The deeper you go, the more water there is above you, and the greater the pressure. The pressure of the water is equal to the weight, F_g, of the column of water above you divided by the column's cross-sectional area, A. Even though gravity pulls only in the downward direction, the fluid transmits the pressure in all directions: up, down, and to the sides. You can find the pressure of the water by applying the following equation.

$$P = \frac{F_g}{A}$$

The weight of the column of water is $F_g = mg$, and the mass is equal to the density, ρ, of the water times its volume, $m = \rho V$. You also know that the volume of the water is the area of the base of the column times its height, $V = Ah$. Therefore, $F_g = \rho Ahg$. Substituting ρAhg for F_g in the equation for water pressure gives $P = F_g/A = \rho Ahg/A$. Divide A from the numerator and denominator to arrive at the simplest form of the equation for the pressure exerted by a column of water on a submerged body.

Pressure of Water on a Body $P = \rho hg$

The pressure that a column of water exerts on a body is equal to the density of water times the height of the column times the acceleration due to gravity.

This formula works for all fluids, not just water. The pressure of a fluid on a body depends on the density of the fluid, its depth, and g. If there were water on the Moon, the pressure of the water at any depth would be one-sixth as great as on Earth. As illustrated in **Figure 13-12,** submersibles, both crewed and robotic, have explored the deepest ocean trenches and encountered pressures in excess of 1000 times standard air pressure.

Figure 13-12 In 1960, the *Trieste,* a crewed submersible, descended to the bottom of the Marianas Trench, a depth of over 10,500 m. The crewed submersible *Alvin,* shown below, can safely dive to a depth of 4500 m.

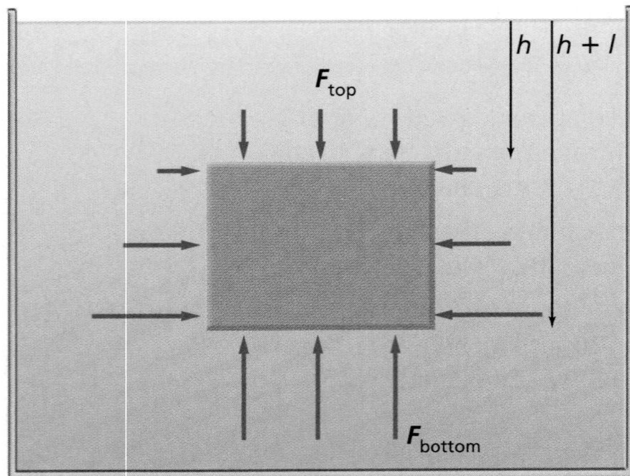

Figure 13-13 A fluid exerts a greater upward force on the bottom of an immersed object than the downward force on the top of the object. The net upward force is called the buoyant force.

Concepts In Motion

Interactive Figure To see an animation on buoyancy, visit physicspp.com.

Buoyancy What produces the upward force that allows you to swim? The increase in pressure with increasing depth creates an upward force called the **buoyant force.** By comparing the buoyant force on an object with its weight, you can predict whether the object will sink or float.

Suppose that a box is immersed in water. It has a height of l and its top and bottom each have a surface area of A. Its volume, then, is $V = lA$. Water pressure exerts forces on all sides, as shown in **Figure 13-13.** Will the box sink or float? As you know, the pressure on the box depends on its depth, h. To find out whether the box will float in water, you will need to determine how the pressure on the top of the box compares with the pressure from below the box. Compare these two equations:

$$F_{\text{top}} = P_{\text{top}} A = \rho h g A$$

$$F_{\text{bottom}} = P_{\text{bottom}} A = \rho(l + h) g A$$

On the four vertical sides, the forces are equal in all directions, so there is no net horizontal force. The upward force on the bottom is larger than the downward force on the top, so there is a net upward force. The buoyant force can now be determined.

$$\begin{aligned} F_{\text{buoyant}} &= F_{\text{bottom}} - F_{\text{top}} \\ &= \rho(l + h)gA - \rho hgA \\ &= \rho lgA = \rho Vg \end{aligned}$$

These calculations show the net upward force to be proportional to the volume of the box. This volume equals the volume of the fluid displaced, or pushed out of the way, by the box. Therefore, the magnitude of the buoyant force, ρVg, equals the weight of the fluid displaced by the object.

Buoyant Force $F_{\text{buoyant}} = \rho_{\text{fluid}} Vg$

The buoyant force on an object is equal to the weight of the fluid displaced by the object, which is equal to the density of the fluid in which the object is immersed multiplied by the object's volume and the acceleration due to gravity.

This relationship was discovered in the third century B.C. by Greek scientist Archimedes. **Archimedes' principle** states that an object immersed in a fluid has an upward force on it that is equal to the weight of the fluid displaced by the object. The force does not depend on the weight of the object, only on the weight of the displaced fluid.

Sink or float? If you want to know whether an object sinks or floats, you have to take into account all of the forces acting on the object. The buoyant force pushes up, but the weight of the object pulls it down. The difference between the buoyant force and the object's weight determines whether an object sinks or floats.

Suppose that you submerge three objects in a tank filled with water ($\rho_{\text{water}} = 1.00\times10^3$ kg/m^3). Each of the objects has a volume of 100 cm^3, or 1.00×10^{-4} m^3. The first object is a steel block with a mass of 0.90 kg.

The second is an aluminum soda can with a mass of 0.10 kg. The third is an ice cube with a mass of 0.090 kg. How will each item move when it is immersed in water? The upward force on all three objects, as shown in **Figure 13-14,** is the same, because all displace the same weight of water. This buoyant force can be calculated as follows.

$$F_{buoyant} = \rho_{water}Vg$$
$$= (1.00\times10^3\ kg/m^3)(1.00\times10^{-4}\ m^3)(9.80\ m/s^2)$$
$$= 0.980\ N$$

The weight of the block of steel is 8.8 N, much greater than the buoyant force. There is a net downward force, so the block will sink to the bottom of the tank. The net downward force, its apparent weight, is less than its real weight. All objects in a liquid, even those that sink, have an apparent weight that is less than when the object is in air. The apparent weight can be expressed by the equation $F_{apparent} = F_g - F_{buoyant}$. For the block of steel, the apparent weight is 8.8 N − 0.98 N, or 7.8 N.

The weight of the soda can is 0.98 N, the same as the weight of the water displaced. There is, therefore, no net force, and the can will remain wherever it is placed in the water. It has neutral buoyancy. Objects with neutral buoyancy are described as being weightless; their apparent weight is zero. This property is similar to that experienced by astronauts in orbit, which is why astronaut training sometimes takes place in swimming pools.

The weight of the ice cube is 0.88 N, less than the buoyant force, so there is a net upward force, and the ice cube will rise. At the surface, the net upward force will lift part of the ice cube out of the water. As a result, less water will be displaced, and the upward force will be reduced. The ice cube will float with enough volume in the water so that the weight of water displaced equals the weight of the ice cube. An object will float if its density is less than the density of the fluid in which it is immersed.

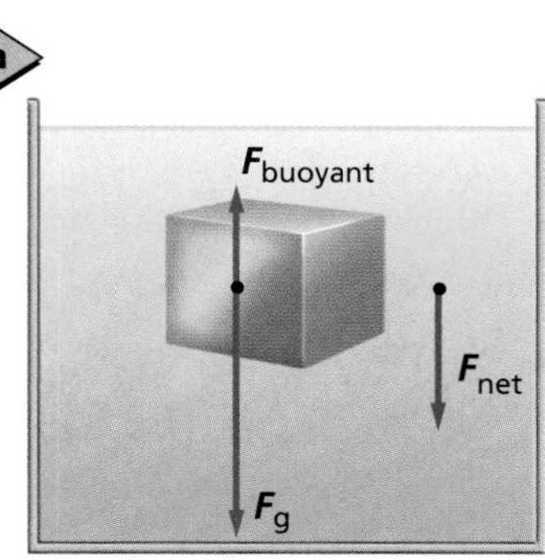

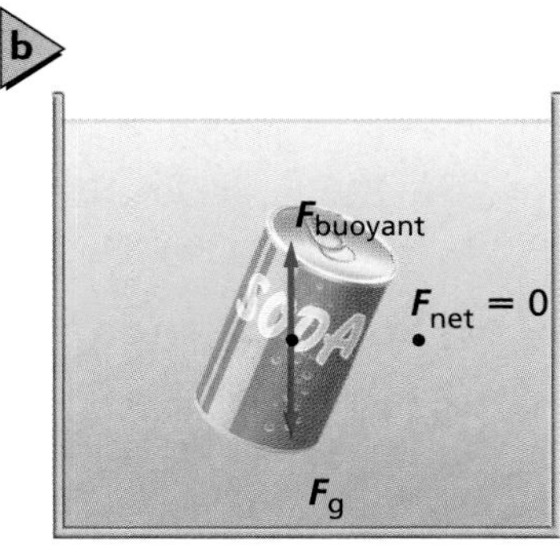

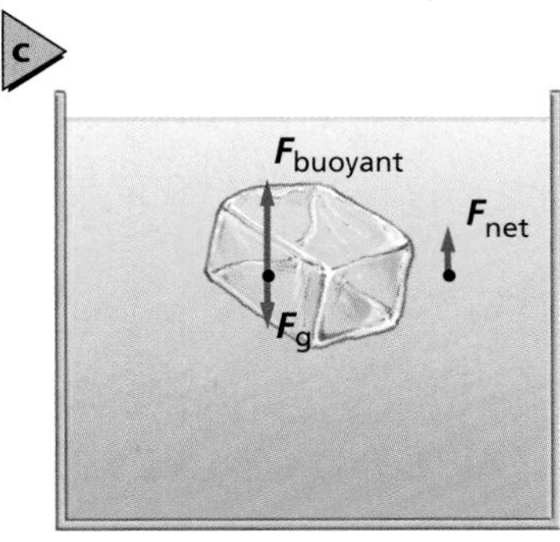

Figure 13-14 A block of steel **(a)**, an aluminum can of soda **(b)**, and an ice cube **(c)** all have the same volume, displace the same amount of water, and experience the same buoyant force. However, because their weights are different, the net forces on the three objects are also different.

Ships Archimedes' principle explains why ships can be made of steel and still float; if the hull is hollow and large enough so that the average density of the ship is less than the density of water, the ship will float. You may have noticed that a ship loaded with cargo rides lower in the water than a ship with an empty cargo hold. You can demonstrate this effect by fashioning a small boat out of folded aluminum foil. The boat should float easily, and it will ride lower in the water if you add a cargo of paper clips. If the foil is crumpled into a tight ball, the boat will sink because of its increased density. Similarly, the continents of Earth float upon a denser material below the surface. The drifting motion of these continental plates is responsible for the present shapes and locations of the continents.

Geology Connection

Other examples of Archimedes' principle in action include submarines and fishes. Submarines take advantage of Archimedes' principle as water is pumped into or out of a number of different chambers to change the submarine's average density, causing it to rise or sink. Fishes that have swim bladders also use Archimedes' principle to control their depths. Such a fish can expand or contract its swim bladder, just like you can puff up your cheeks. To move upward in the water, the fish expands its swim bladder to displace more water and increase the buoyant force. The fish moves downward by contracting the volume of its swim bladder.

▶ EXAMPLE Problem 3

Archimedes' Principle A cubic decimeter, 1.00×10^{-3} m^3, of a granite building block is submerged in water. The density of granite is 2.70×10^3 kg/m^3.

a. What is the magnitude of the buoyant force acting on the block?

b. What is the apparent weight of the block?

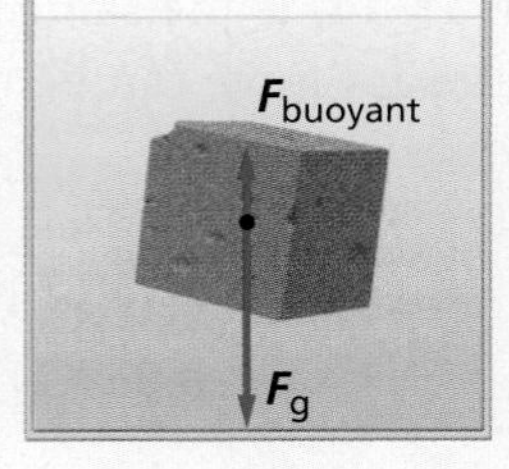

1 Analyze and Sketch the Problem

- Sketch the cubic decimeter of granite immersed in water.
- Show the upward buoyant force and the downward force due to gravity acting on the granite.

Known:	Unknown:
$V = 1.00\times10^{-3}$ m^3	$F_{buoyant} = ?$
$\rho_{granite} = 2.70\times10^3$ kg/m^3	$F_{apparent} = ?$
$\rho_{water} = 1.00\times10^3$ kg/m^3	

Math Handbook
Operations with Scientific Notation
pages 842–843

2 Solve for the Unknown

a. Calculate the buoyant force on the granite block.

$$F_{buoyant} = \rho_{water}Vg$$
$$= (1.00\times10^3\ \text{kg/m}^3)(1.00\times10^{-3}\ \text{m}^3)(9.80\ \text{m/s}^2)$$
$$= 9.80\ \text{N}$$

Substitute $\rho_{water} = 1.00\times10^3$ kg/m^3, $V = 1.00\times10^{-3}$ m^3, $g = 9.80$ m/s^2

b. Calculate the granite's weight and then find its apparent weight.

$$F_g = mg = \rho_{granite}Vg$$
$$= (2.70\times10^3\ \text{kg/m}^3)(1.00\times10^{-3}\ \text{m}^3)(9.80\ \text{m/s}^2)$$
$$= 26.5\ \text{N}$$

Substitute $\rho_{granite} = 2.70\times10^3$ kg/m^3, $V = 1.00\times10^{-3}$ m^3, $g = 9.80$ m/s^2

$$F_{apparent} = F_g - F_{buoyant}$$
$$= 26.5\ \text{N} - 9.80\ \text{N} = 16.7\ \text{N}$$

Substitute $F_g = 26.5$ N, $F_{buoyant} = 9.80$ N

3 Evaluate the Answer

- **Are the units correct?** The forces and apparent weight are in newtons, as expected.
- **Is the magnitude realistic?** The buoyant force is about one-third the weight of the granite, a sensible answer because the density of water is about one-third that of granite.

▷ PRACTICE Problems

• Additional Problems, Appendix B
• Solutions to Selected Problems, Appendix C

27. Common brick is about 1.8 times denser than water. What is the apparent weight of a 0.20 m^3 block of bricks under water?

28. A girl is floating in a freshwater lake with her head just above the water. If she weighs 610 N, what is the volume of the submerged part of her body?

29. What is the tension in a wire supporting a 1250-N camera submerged in water? The volume of the camera is 16.5×10^{-3} m^3.

30. Plastic foam is about 0.10 times as dense as water. What weight of bricks could you stack on a 1.0 m × 1.0 m × 0.10 m slab of foam so that the slab of foam floats in water and is barely submerged, leaving the bricks dry?

31. Canoes often have plastic foam blocks mounted under the seats for flotation in case the canoe fills with water. What is the approximate minimum volume of foam needed for flotation for a 480-N canoe?

Fluids in Motion: Bernoulli's Principle

Try the experiment shown in **Figure 13-15.** Hold a strip of notebook paper just under your lower lip. Then blow hard across the top surface. Why does the strip of paper rise? The blowing of the air has decreased the air pressure above the paper. Because this pressure decreases, the pressure in the still air below the paper pushes the paper upward. The relationship between the velocity and pressure exerted by a moving fluid is named for Swiss scientist Daniel Bernoulli. **Bernoulli's principle** states that as the velocity of a fluid increases, the pressure exerted by that fluid decreases. This principle is a statement of work and energy conservation as applied to fluids.

■ **Figure 13-15** Blowing across the surface of a sheet of paper demonstrates Bernoulli's principle.

One instance in which fluid velocity can increase is when it flows through a constriction. The nozzles on some garden hoses can be opened or narrowed so that the velocity of the water spray can be changed. You may have seen the water in a stream speed up as it passed through narrowed sections of the stream bed. As the nozzle of the hose and the stream channel become wider or narrower, the velocity of the fluid changes to maintain the overall flow of water. In addition to streams and hoses, the pressure of blood in our circulatory systems depends partly on Bernoulli's principle. Treatments of heart disease involve removing obstructions in the arteries and veins and preventing clots in the blood.

Consider a horizontal pipe completely filled with a smoothly flowing ideal fluid. If a certain mass of the fluid enters one end of the pipe, then an equal mass must come out the other end. Now consider a section of pipe with a cross section that becomes narrower, as shown in **Figure 13-16a.** To keep the same mass of fluid moving through the narrow section in a fixed amount of time, the velocity of the fluid must increase. As the fluid's velocity increases, so does its kinetic energy. This means that net work has been done on the swifter fluid. This net work comes from the difference between the work that was done to move the mass of fluid into the pipe and the work that was done by the fluid pushing the same mass out of the pipe. The work is proportional to the force on the fluid, which, in turn, depends on the pressure. If the net work is positive, the pressure at the input end of the section, where the velocity is lower, must be larger than the pressure at the output end, where the velocity is higher.

■ **Figure 13-16** Pressure P_1 is greater than P_2 because v_1 is less than v_2 **(a).** In a carburetor, low pressure in the narrow part of the tube draws fuel into the air flow **(b).**

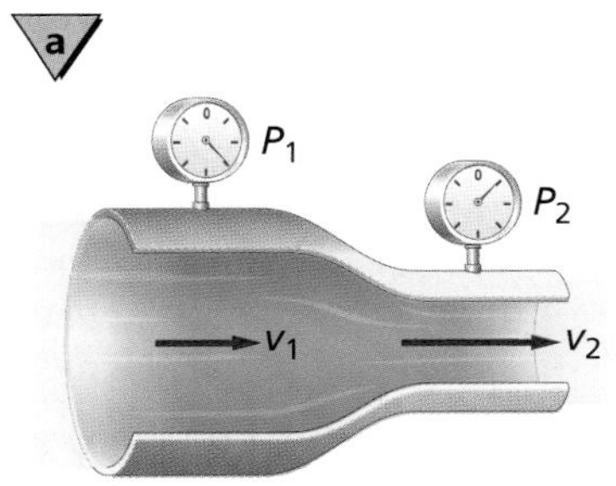

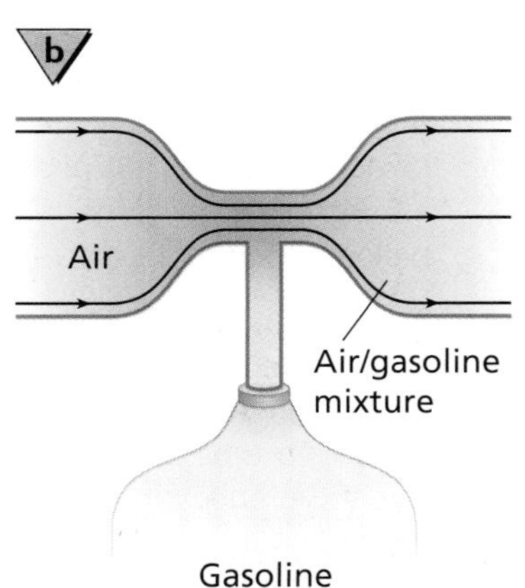

Applications of Bernoulli's principle There are many common applications of Bernoulli's principle, such as paint sprayers and perfume bottles. The simple atomizer on a perfume bottle works by blowing air across the top of a tube sunk into perfume, which creates lower pressure at the top of the tube than in the bottle. As a result, perfume is forced into the air flow. A gasoline engine's carburetor, which is where air and gas are mixed, is another common application of Bernoulli's principle. Part of the carburetor is a tube with a constriction, as shown in the diagram in **Figure 13-16b.** The pressure on the gasoline in the fuel supply is the same as the pressure in the thicker part of the tube. Air flowing through the narrow section of the tube, which is attached to the fuel supply, is at a lower pressure, so fuel is forced into the air flow. By regulating the flow of air in the tube, the amount of fuel mixed into the air can be varied. Newer cars tend to have fuel injectors rather than carburetors, but carburetors are common in older cars and the motors of small gasoline-powered machines, such as lawn mowers.

■ **Figure 13-17** The smooth streamlines show the air flowing above a car that is being tested in a wind tunnel.

Streamlines Automobile and aircraft manufacturers spend a great deal of time and money testing new designs in wind tunnels to ensure the greatest efficiency of movement through air. The flow of fluids around objects is represented by **streamlines,** as shown in **Figure 13-17.** Objects require less energy to move through a smooth streamlined flow.

Streamlines can best be illustrated by a simple demonstration. Imagine carefully squeezing tiny drops of food coloring into a smoothly flowing fluid. If the colored lines that form stay thin and well defined, the flow is said to be streamlined. Notice that if the flow narrows, the streamlines move closer together. Closely spaced streamlines indicate greater velocity and therefore reduced pressure. If streamlines swirl and become diffused, the flow of the fluid is said to be turbulent. Bernoulli's principle does not apply to turbulent flow.

13.3 Section Review

32. Floating and Sinking Does a full soda pop can float or sink in water? Try it. Does it matter whether or not the drink is diet? All soda pop cans contain the same volume of liquid, 354 mL, and displace the same volume of water. What is the difference between a can that sinks and one that floats?

33. Floating and Density A fishing bobber made of cork floats with one-tenth of its volume below the water's surface. What is the density of cork?

34. Floating in Air A helium balloon rises because of the buoyant force of the air lifting it. The density of helium is 0.18 kg/m^3, and the density of air is 1.3 kg/m^3. How large a volume would a helium balloon need to lift a 10-N lead brick?

35. Transmission of Pressure A toy rocket launcher is designed so that a child stomps on a rubber cylinder, which increases the air pressure in a launching tube and pushes a foam rocket into the sky. If the child stomps with a force of 150 N on a 2.5×10^{-3} m^2 area piston, what is the additional force transmitted to the 4.0×10^{-4} m^2 launch tube?

36. Pressure and Force An automobile weighing 2.3×10^4 N is lifted by a hydraulic cylinder with an area of 0.15 m^2.

a. What is the pressure in the hydraulic cylinder?

b. The pressure in the lifting cylinder is produced by pushing on a 0.0082 m^2 cylinder. What force must be exerted on this small cylinder to lift the automobile?

37. Displacement Which of the following displaces more water when it is placed in an aquarium?

a. A 1.0-kg block of aluminum or a 1.0-kg block of lead?

b. A 10-cm^3 block of aluminum or a 10-cm^3 block of lead?

38. Critical Thinking As you discovered in Practice Problem 4, a tornado passing over a house sometimes makes the house explode from the inside out. How might Bernoulli's principle explain this phenomenon? What could be done to reduce the danger of a door or window exploding outward?

Physics Online physicspp.com/self_check_quiz

13.4 Solids

How do solids and liquids differ? Solids are stiff, they can be cut in pieces, and they retain their shapes. You can push on solids. Liquids flow and if you push your finger on water, your finger will move through it. However, if you have ever watched butter warm and begin to lose its shape, you may have wondered if the line between solids and liquids is always distinct.

► **Objectives**

- **Relate** the properties of solids to their structures.
- **Explain** why solids expand and contract when the temperature changes.
- **Calculate** the expansion of solids.
- **Explain** the importance of thermal expansion.

► **Vocabulary**

crystal lattice
amorphous solid
coefficient of linear expansion
coefficient of volume expansion

Solid Bodies

Under certain conditions, solids and liquids are not easily distinguished. As bottle glass is heated through the molten state, the change from solid to liquid is so gradual that it is difficult to tell which is which. Some solids, such as crystalline quartz, are made of particles that are lined up in orderly patterns. Other solids, such as glass, are made of jumbled arrangements of particles, just like a liquid. As shown in **Figure 13-18,** quartz and fused quartz (also called quartz glass) are the same chemically, but their physical properties are quite different.

When the temperature of a liquid is lowered, the average kinetic energy of the particles decreases. As the particles slow down, the cohesive forces have more effect, and for many solids, the particles become frozen into a fixed pattern called a **crystal lattice,** shown in **Figure 13-19** on the next page. Although the cohesive forces hold the particles in place, the particles in a crystalline solid do not stop moving completely. Rather, they vibrate around their fixed positions. In other materials, such as butter and glass, the particles do not form a fixed crystalline pattern. Such a substance, which has no regular crystal structure but does have a definite volume and shape, is called an **amorphous solid.** Amorphous solids also are classified as viscous, or slowly flowing, liquids.

Pressure and freezing As a liquid becomes a solid, its particles usually fit more closely together than in the liquid state, making solids more dense than liquids. As you have learned, however, water is an exception because it is most dense at 4°C. Water is also an exception to another general rule. For most liquids, an increase in the pressure on the surface of the liquid increases its freezing point. Because water expands as it freezes, an increase in pressure forces the molecules closer together and opposes the freezing. Therefore, higher pressure lowers the freezing point of water very slightly.

a

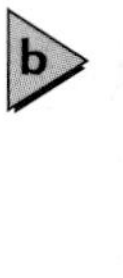

b

Figure 13-18 In crystalline quartz, the particles are arranged in an orderly pattern **(a).** A crystalline solid melts at a specific temperature. Fused quartz is the same chemical as crystalline quartz, but the particles are jumbled randomly in the solid. When fused quartz melts, its properties change slowly over a range of temperatures, allowing it to be worked in a fashion similar to everyday glass **(b).**

Figure 13-19 Ice, the solid form of water, has a larger volume than an equal mass of its liquid form **(a).** The crystalline structure of ice is in the form of a lattice **(b).**

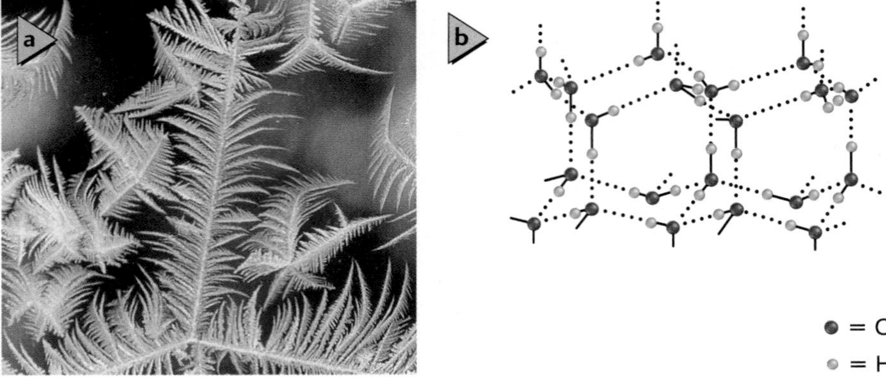

It has been hypothesized that the drop in water's freezing point caused by the pressure of an ice-skater's blades produces a thin film of liquid between the ice and the blades. Calculations of the pressure caused by even the sharpest blade show that the ice is still too cold to melt. More recent measurements, however, have shown that the friction between the blade and the ice generates enough thermal energy to melt the ice and create a thin layer of water. This explanation is supported by measurements of the spray of ice particles, which are considerably warmer than the ice itself. The same process of melting occurs during snow skiing.

Elasticity of solids External forces applied to a solid object may twist or bend it out of shape. The ability of a solid object to return to its original form when the external forces are removed is called the elasticity of the solid. If too much deformation occurs, the object will not return to its original shape because its elastic limit has been exceeded. Elasticity depends on the electromagnetic forces that hold the particles of a substance together. Malleability and ductility are two properties that depend on the structure and elasticity of a substance. Because gold can be flattened and shaped into thin sheets, it is said to be malleable. Copper is a ductile metal because it can be pulled into thin strands of wire.

Figure 13-20 The extreme temperatures of a summer day caused these railroad tracks to buckle.

Thermal Expansion of Solids

It is standard practice for engineers to design small gaps, called expansion joints, into concrete-and-steel highway bridges to allow for the expansion of parts in the heat of summer. Objects expand only a small amount when they are heated, but that small amount could be several centimeters in a 100-m-long bridge. If expansion gaps were not present, the bridge could buckle or parts of it could break. High temperatures can also damage railroad tracks that are laid without expansion joints, as shown in **Figure 13-20.** Some materials, such as the ovenproof glass that is used for cooking and laboratory experiments, are designed to have the least possible thermal expansion. Large telescope mirrors are made of a ceramic material that is designed to undergo essentially no thermal expansion.

To understand the expansion of heated solids, picture a solid as a collection of particles connected by springs that represent the attractive forces between the particles. When the particles get too close, the springs push them apart. When a solid is heated, the kinetic energy of the particles increases, and they vibrate rapidly and move farther apart, weakening the attractive forces between the particles. As a result, when the particles vibrate more violently with increased temperature, their average separation increases and the solid expands.

The change in length of a solid is proportional to the change in temperature, as shown in **Figure 13-21.** A solid will expand in length twice as much when its temperature is increased by 20°C than when it is increased by 10°C. The expansion also is proportional to its length. A 2-m bar will expand twice as much as a 1-m bar with the same change in temperature. The length, L_2, of a solid at temperature T_2 can be found with the following equation, where L_1 is the length at temperature T_1 and alpha, α, is the **coefficient of linear expansion.**

$$L_2 = L_1 + \alpha L_1(T_2 - T_1)$$

Using simple algebra, you can solve for α.

$$L_2 - L_1 = \alpha L_1(T_2 - T_1)$$

$$\Delta L = \alpha L_1 \Delta T$$

Coefficient of Linear Expansion $\alpha = \frac{\Delta L}{L_1 \Delta T}$

The coefficient of linear expansion is equal to the change in length, divided by the original length and the change in temperature.

The unit for the coefficient of linear expansion is 1/°C, or $(°C)^{-1}$. Since solids expand in three directions, the **coefficient of volume expansion,** β, is about three times the coefficient of linear expansion.

Coefficient of Volume Expansion $\beta = \frac{\Delta V}{V_1 \Delta T}$

The coefficient of volume expansion is equal to the change in volume divided by the original volume and the change in temperature.

Again, the unit for β is 1/°C, or $(°C)^{-1}$. The two coefficients of thermal expansion for a variety of materials are given in **Table 13-2.**

Table 13-2

Coefficients of Thermal Expansion at 20°C

Material	Coefficient of Linear Expansion, $\alpha(°C)^{-1}$	Coefficient of Volume Expansion, $\beta(°C)^{-1}$
Solids		
Aluminum	25×10^{-6}	75×10^{-6}
Glass (soft)	9×10^{-6}	27×10^{-6}
Glass (ovenproof)	3×10^{-6}	9×10^{-6}
Concrete	12×10^{-6}	36×10^{-6}
Copper	16×10^{-6}	48×10^{-6}
Liquids		
Methanol		1200×10^{-6}
Gasoline		950×10^{-6}
Water		210×10^{-6}

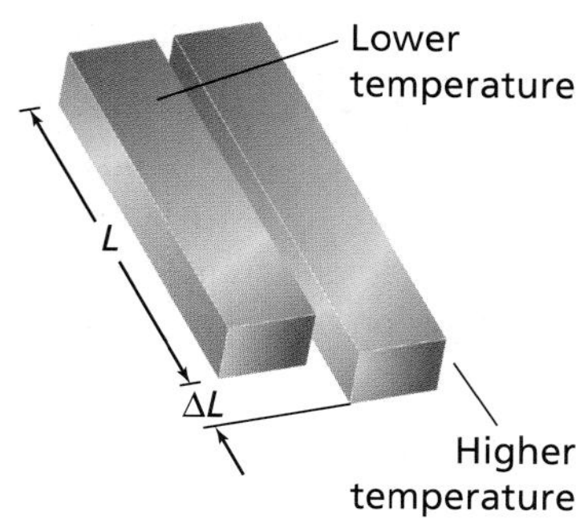

Figure 13-21 The change in length of a material is proportional to the original length and the change in temperature.

▶ EXAMPLE Problem 4

Linear Expansion A metal bar is 1.60 m long at room temperature, 21°C. The bar is put into an oven and heated to a temperature of 84°C. It is then measured and found to be 1.7 mm longer. What is the coefficient of linear expansion of this material?

1 Analyze and Sketch the Problem

- Sketch the bar, which is 1.7 mm longer at 84°C than at 21°C.
- Identify the initial length of the bar, L_1, and the change in length, ΔL.

Known:	Unknown:
$L_1 = 1.60$ m	$\alpha = ?$
$\Delta L = 1.7\times10^{-3}$ m	
$T_1 = 21°C$	
$T_2 = 84°C$	

2 Solve for the Unknown

Calculate the coefficient of linear expansion using the known length, change in length, and change in temperature.

$$\alpha = \frac{\Delta L}{L_1 \Delta T}$$

$$= \frac{1.7\times10^{-3}\ \text{m}}{(1.60\ \text{m})(84°\text{C} - 21°\text{C})}$$

Substitute $\Delta L = 1.7\times10^{-3}$ m, $L_1 = 1.60$ m, $\Delta T = (T_2 - T_1) = 84°C - 21°C$

$$= 1.7\times10^{-5}\ °\text{C}^{-1}$$

Math Handbook
Operations with Significant Digits
pages 835–836

3 Evaluate the Answer

- **Are the units correct?** The units are correctly expressed in $°C^{-1}$.
- **Is the magnitude realistic?** The magnitude of the coefficient is close to the accepted value for copper.

▶ PRACTICE Problems

• Additional Problems, Appendix B
• Solutions to Selected Problems, Appendix C

39. A piece of aluminum house siding is 3.66 m long on a cold winter day of −28°C. How much longer is it on a very hot summer day at 39°C?

40. A piece of steel is 11.5 cm long at 22°C. It is heated to 1221°C, close to its melting temperature. How long is it?

41. A 400-mL glass beaker at room temperature is filled to the brim with cold water at 4.4°C. When the water warms up to 30.0°C, how much water will spill from the beaker?

42. A tank truck takes on a load of 45,725 L of gasoline in Houston, where the temperature is 28.0°C. The truck delivers its load in Minneapolis, where the temperature is −12.0°C.

a. How many liters of gasoline does the truck deliver?

b. What happened to the gasoline?

43. A hole with a diameter of 0.85 cm is drilled into a steel plate. At 30.0°C, the hole exactly accommodates an aluminum rod of the same diameter. What is the spacing between the plate and the rod when they are cooled to 0.0°C?

44. A steel ruler is marked in millimeters so that the ruler is absolutely correct at 30.0°C. By what percentage would the ruler be incorrect at −30.0°C?

CHALLENGE PROBLEM

You need to make a 1.00-m-long bar that expands with temperature in the same way as a 1.00-m-long bar of copper would. As shown in the figure at the right, your bar must be made from a bar of iron and a bar of aluminum attached end to end. How long should each of them be?

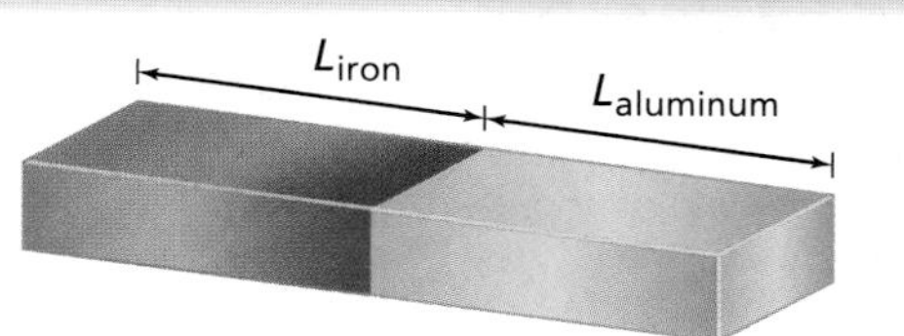

Applications of thermal expansion Different materials expand at different rates, as indicated by the different coefficients of expansion given in Table 13-2. Engineers must consider these different expansion rates when designing structures. Steel bars are often used to reinforce concrete, and therefore the steel and concrete must have the same expansion coefficient. Otherwise, the structure could crack on a hot day. Similarly, a dentist must use filling materials that expand and contract at the same rate as tooth enamel.

Different rates of expansion have useful applications. For example, engineers have taken advantage of these differences to construct a useful device called a bimetallic strip, which is used in thermostats. A bimetallic strip consists of two strips of different metals welded or riveted together. Usually, one strip is brass and the other is iron. When heated, brass expands more than iron does. Thus, when the bimetallic strip of brass and iron is heated, the brass part of the strip becomes longer than the iron part. The bimetallic strip bends with the brass on the outside of the curve. If the bimetallic strip is cooled, it bends in the opposite direction. The brass is then on the inside of the curve.

In a home thermostat, shown in **Figure 13-22,** the bimetallic strip is installed so that it bends toward an electric contact as the room cools. When the room cools below the setting on the thermostat, the bimetallic strip bends enough to make electric contact with the switch, which turns on the heater. As the room warms, the bimetallic strip bends in the other direction. When the room's temperature reaches the setting on the thermostat, the electric circuit is broken and the heater switches off.

Figure 13-22 In this thermostat, a coiled bimetallic strip controls the flow of mercury for opening and closing electrical switches.

13.4 Section Review

45. Relative Thermal Contraction On a hot day, you are installing an aluminum screen door in a concrete door frame. You want the door to fit well on a cold winter day. Should you make the door fit tightly in the frame or leave extra room?

46. States of Matter Why could candle wax be considered a solid? Why might it also be considered a viscous liquid?

47. Thermal Expansion Can you heat a piece of copper enough to double its length?

48. States of Matter Does Table 13-2 provide a way to distinguish between solids and liquids?

49. Solids and Liquids A solid can be defined as a material that can be bent and will resist bending. Explain how these properties relate to the binding of atoms in a solid, but do not apply to a liquid.

50. Critical Thinking The iron ring in **Figure 13-23** was made by cutting a small piece from a solid ring. If the ring in the figure is heated, will the gap become wider or narrower? Explain your answer.

Figure 13-23

PHYSICS LAB

Evaporative Cooling

Alternate CBL instructions can be found on the Web site.
physicspp.com

If you have ever spilled a small amount of rubbing alcohol on your skin, you probably noticed how cool it felt. You have learned that this coolness is caused by evaporation. In this experiment, you will test the rates at which different types of alcohol evaporate. An alcohol is a substance that has a hydroxyl functional group (–OH) attached to a carbon or a chain of carbons. From your observations of evaporative cooling, you will infer the relative strength of the cohesive forces in the tested alcohols.

QUESTION

How do the rates of evaporation compare for different alcohols?

Objectives

- **Collect and organize data** for the evaporation of alcohols.
- **Compare and contrast** the rates of evaporation for various alcohols.
- **Analyze** why some alcohols evaporate faster than others.
- **Infer** the relationship between cohesive forces and the rate of evaporation.

Safety Precautions

- **The chemicals used in this experiment are flammable and poisonous. Do not inhale the fumes from these chemicals. Do not have any open flame near these chemicals. Use in a well-ventilated room or fume hood.**
- **Avoid contact with the chemicals on your skin or clothing. Notify your teacher immediately if an accident or spill occurs.**
- **Wash your hands after the lab is over.**

Materials

methanol (methyl alcohol)
ethanol (ethyl alcohol)
2-propanol (isopropyl alcohol)
masking tape (two pieces)
thermometer (non-mercury)
filter paper (three pieces, 2.5 cm × 2.5 cm)
small rubber bands

Procedure

1. Wrap the thermometer with a square piece of filter paper fastened by a small rubber band. To do this, first slip the rubber band onto the thermometer. Then, wrap the paper around the thermometer and roll the rubber band over the wrapped paper. The paper should fit snugly over the thermometer's end.
2. Obtain a small beaker of methanol. Place the paper-covered end of the thermometer in the container of methanol. Do not let the container fall over. Keep the thermometer in the container for 1 min.

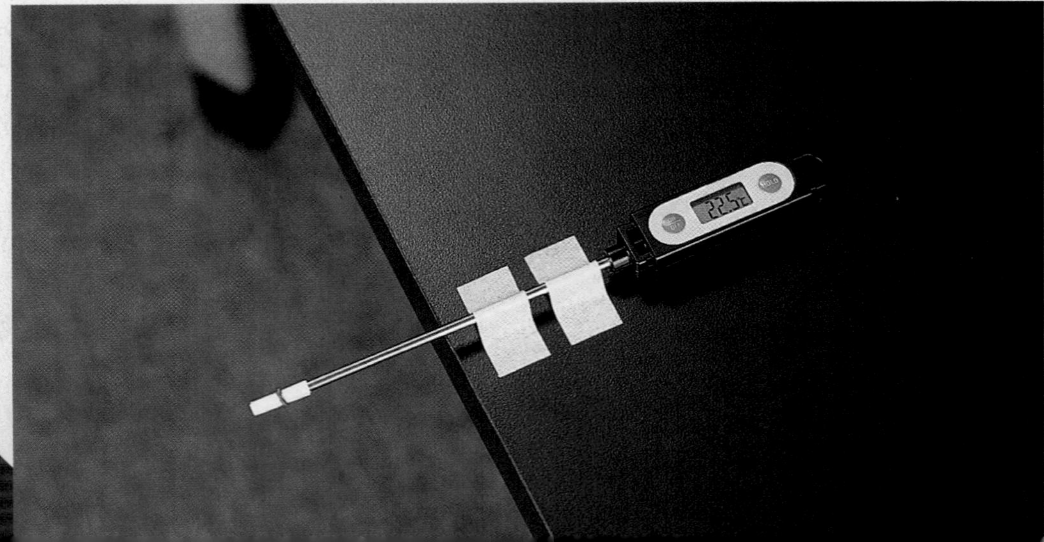

Data Table			
Liquid	**T_2 (°C)**	**T_1 (°C)**	**ΔT(°C)**
Methyl alcohol			
Ethyl alcohol			
Isopropyl alcohol			

3. After 1 min has elapsed, record the temperature reading on the thermometer in the data table under T_1. This is the initial temperature of the methanol.
4. Remove the thermometer from the methanol. Place the thermometer over the edge of a table top so that the thermometer's tip extends about 5 cm beyond the edge of the table. Use the masking tape to anchor the thermometer in place.
5. Observe the temperature during the experiment. After 4 min have elapsed, observe and record the temperature in the data table in the column marked T_2.
6. Roll the rubber band up the thermometer and dispose of the filter paper as directed by your teacher.
7. Repeat steps 1–6, but use ethanol as the liquid. Record your results in the data table.
8. Repeat steps 1–6, but use isopropyl alcohol as the liquid. Record your results in the data table.

Analyze

1. **Interpret Data** Did the thermometer show a temperature increase or decrease for your trials? Why?
2. Calculate ΔT for each of your liquids by finding the difference between the ending temperatures and the initial temperatures of the liquids ($T_2 - T_1$).
3. Using the chemical formulas for methanol (CH_3OH), ethanol (C_2H_5OH), and isopropyl alcohol (C_3H_7OH), determine the molar mass of each of the liquids you tested. You will need to refer to the periodic table to determine the molar masses.
4. **Infer** What can the ΔT for each trial tell you about the rates of evaporation of the alcohols?
5. **Think Critically** Why was paper used on the thermometer instead of using only the thermometer?

Conclude and Apply

1. Using the rates of evaporation of the alcohols you studied, how can you determine which alcohol had the strongest cohesive forces?
2. Which alcohol had the weakest cohesive forces?
3. What general trend did you find between the change in temperature (ΔT) and the molar mass of an alcohol?
4. **Hypothesize** Would a fan blowing in the lab room change the room's air temperature? Would it change the ΔT that you observed? Explain.

Going Further

Predict the size of ΔT for 1-butanol, which has the formula C_4H_9OH, relative to the alcohols that you tested.

Real-World Physics

The National Weather Service began using a new windchill index in 2001. The old chart was based on data derived from water-freezing experiments done in Antarctica in the 1940s. Explain how windchill relates to evaporative cooling, why this phenomenon is important in cold weather, and how the new chart improves upon the old chart.

To find out more about states of matter, visit the Web site: **physicspp.com**

eXtreme physics

A Strange Matter

You now are familiar with the four most common states of matter: solid, liquid, gas, and plasma. But did you know that there is a fifth state of matter? Meet the Bose-Einstein Condensate (BEC).

What is a Bose-Einstein Condensate?

The origins of the BEC are in the 1920s in Satyendra Nath Bose's studies of the quantum rules governing photon energies. Einstein applied Bose's equations to atoms. The equations showed that if the temperature of certain atoms were low enough, most of the atoms would be in the same quantum level. In other words, at extremely low temperatures, atoms that had occupied different energy states suddenly fall into the lowest possible energy state. At these temperatures, which are not found in nature, but are created in the lab through some ingenious technology, the atoms of a BEC cannot be distinguished. Even their positions are identical.

How is a BEC created?

The first BEC was created in 1995 by Eric Cornell and Carl Wieman of Boulder, Colorado. To make their BEC, Cornell and Wieman used rubidium atoms. They had to figure out how to cool these atoms to a lower temperature than had ever been achieved.

You might be surprised to learn that one important step in reaching the necessary temperatures was to use lasers to cool the rubidium atoms. Lasers can burn through metal, but they also can cool a sample of atoms if they are tuned so that their photons bounce off the atoms. In this case, the photons will carry off some of the atoms' energy, lowering the temperature of the sample. But the laser will not cool the sample unless it is precisely tuned.

When the laser is tuned to the proper frequency, the result is a sample of very cold atoms known as "optical molasses." How can the optical molasses be contained? Thermal contact between the optical molasses and a material container surely would warm the chilled atoms because any container would have a higher temperature than the atoms. So the container of choice for the BEC is a nonmaterial one—a container formed by combining lasers with a magnetic field.

While optical molasses is cold (about 1/10,000 K), it is not cold enough to form a BEC. Scientists use evaporative cooling to make the final step to the required temperatures. In evaporative cooling, the optical molasses is contained in a stronger magnetic container that allows the highest energy atoms to escape. The atoms with the lowest possible energy are left behind. These are the atoms that suddenly condense to form a BEC. The images below show the final cooling of a sample of atoms to form a BEC (the central lump) at fifty-billionths of a kelvin.

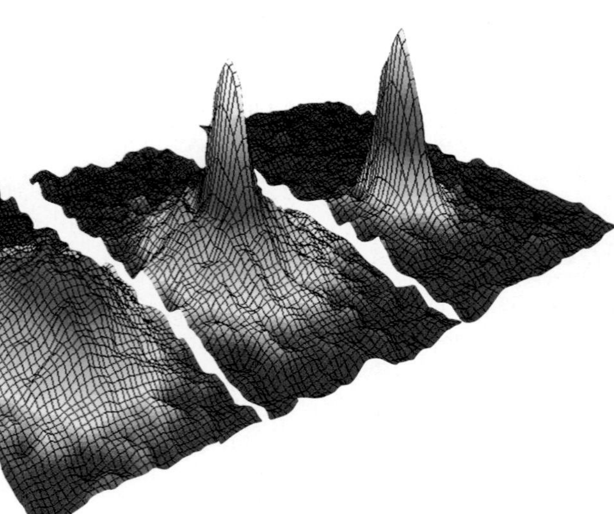

In these three images, the central peak forms as atoms condense to a BEC.

Going Further

1. **Evaluate** What sorts of difficulties might scientists have to overcome in finding applications for the BEC?
2. **Compare and Contrast** Is the evaporative cooling process that creates the BEC the same process that helps keep you cool on a hot day? Explain.

Chapter 13

Study Guide

13.1 Properties of Fluids

Vocabulary

- fluids (p. 342)
- pressure (p. 342)
- pascal (p. 342)
- combined gas law (p. 345)
- ideal gas law (p. 345)
- thermal expansion (p. 347)
- plasma (p. 348)

Key Concepts

- Matter in the fluid state flows and has no definite shape of its own.
- Pressure is the force divided by the area on which it is exerted.

$$P = \frac{F}{A}$$

- The combined gas law can be used to calculate changes in the volume, temperature, and pressure of an ideal gas.

$$\frac{P_1V_1}{T_1} = \frac{P_2V_2}{T_2}$$

- The ideal gas law can be written as follows.

$$PV = nRT$$

13.2 Forces Within Liquids

Vocabulary

- cohesive forces (p. 349)
- adhesive forces (p. 350)

Key Concepts

- Cohesive forces are the attractive forces that like particles exert on one another. Surface tension and viscosity both result from cohesive forces.
- Adhesive forces are the attractive forces that particles of different substances exert on one another. Capillary action results from adhesive forces.

13.3 Fluids at Rest and in Motion

Vocabulary

- Pascal's principle (p. 352)
- buoyant force (p. 354)
- Archimedes' principle (p. 354)
- Bernoulli's principle (p. 357)
- streamlines (p. 358)

Key Concepts

- According to Pascal's principle, an applied pressure change is transmitted undiminished throughout a fluid.

$$F_2 = \frac{F_1A_2}{A_1}$$

- Pressure at any depth is proportional to the fluid's weight above that depth.

$$P = \rho hg$$

- According to Archimedes' principle, the buoyant force equals the weight of the fluid displaced by an object.

$$F_{buoyant} = \rho_{fluid}Vg$$

- Bernoulli's principle states that the pressure exerted by a fluid decreases as its velocity increases.

13.4 Solids

Vocabulary

- crystal lattice (p. 359)
- amorphous solid (p. 359)
- coefficient of linear expansion (p. 361)
- coefficient of volume expansion (p. 361)

Key Concepts

- A crystalline solid has a regular pattern of particles, and an amorphous solid has an irregular pattern of particles.
- The thermal expansion of a solid is proportional to the temperature change and original size, and it depends on the material.

$$\alpha = \frac{\Delta L}{L_1\Delta T} \qquad \beta = \frac{\Delta V}{V_1\Delta T}$$

Chapter 13 Assessment

Concept Mapping

51. Complete the concept map below using the following terms: *density, viscosity, elasticity, pressure.* A term may be used more than once.

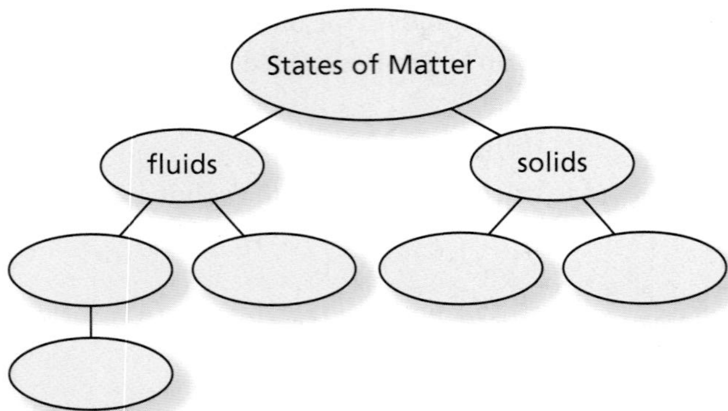

Mastering Concepts

52. How are force and pressure different? (13.1)

53. A gas is placed in a sealed container, and some liquid is placed in a container of the same size. The gas and liquid both have definite volume. How do they differ? (13.1)

54. In what way are gases and plasmas similar? In what way are they different? (13.1)

55. The Sun is made of plasma. How is this plasma different from the plasmas on Earth? (13.1)

56. **Lakes** A frozen lake melts in the spring. What effect does this have on the temperature of the air above the lake? (13.2)

57. **Hiking** Canteens used by hikers often are covered with canvas bags. If you wet the canvas bag covering a canteen, the water in the canteen will be cooled. Explain. (13.2)

58. What do the equilibrium tubes in **Figure 13-24** tell you about the pressure exerted by a liquid? (13.3)

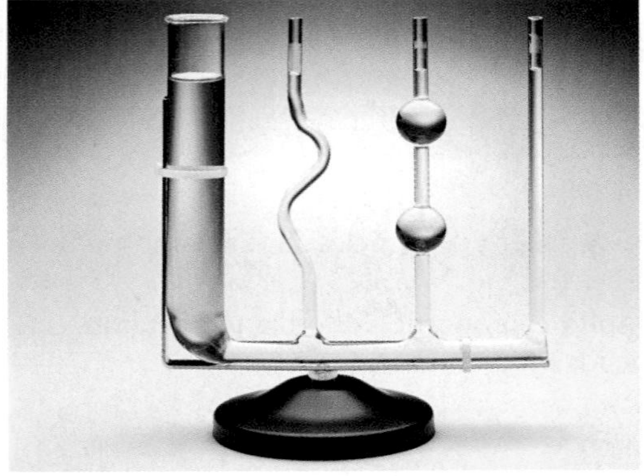

■ **Figure 13-24**

59. According to Pascal's principle, what happens to the pressure at the top of a container if the pressure at the bottom is increased? (13.3)

60. How does the water pressure 1 m below the surface of a small pond compare with the water pressure the same distance below the surface of a lake? (13.3)

61. Does Archimedes' principle apply to an object inside a flask that is inside a spaceship in orbit? (13.3)

62. A stream of water goes through a garden hose into a nozzle. As the water speeds up, what happens to the water pressure? (13.3)

63. How does the arrangement of atoms in a crystalline substance differ from that in an amorphous substance? (13.4)

64. Does the coefficient of linear expansion depend on the unit of length used? Explain. (13.4)

Applying Concepts

65. A rectangular box with its largest surface resting on a table is rotated so that its smallest surface is now on the table. Has the pressure on the table increased, decreased, or remained the same?

66. Show that a pascal is equivalent to a $kg/m{\cdot}s^2$.

67. **Shipping Cargo** Compared to an identical empty ship, would a ship filled with table-tennis balls sink deeper into the water or rise in the water? Explain.

68. Drops of mercury, water, ethanol, and acetone are placed on a smooth, flat surface, as shown in **Figure 13-25.** From this figure, what can you conclude about the cohesive forces in these liquids?

■ **Figure 13-25**

69. How deep would a water container have to be to have the same pressure at the bottom as that found at the bottom of a 10.0-cm deep beaker of mercury, which is 13.55 times as dense as water?

70. Alcohol evaporates more quickly than water does at the same temperature. What does this observation allow you to conclude about the properties of the particles in the two liquids?

71. Suppose you use a hole punch to make a circular hole in aluminum foil. If you heat the foil, will the size of the hole decrease or increase? Explain.

72. Equal volumes of water are heated in two narrow tubes that are identical, except that tube A is made of soft glass and tube B is made of ovenproof glass. As the temperature increases, the water level rises higher in tube B than in tube A. Give a possible explanation.

73. A platinum wire easily can be sealed in a glass tube, but a copper wire does not form a tight seal with the glass. Explain.

74. Five objects with the following densities are put into a tank of water.

a. 0.85 g/cm^3
b. 0.95 g/cm^3
c. 1.05 g/cm^3
d. 1.15 g/cm^3
e. 1.25 g/cm^3

The density of water is 1.00 g/cm^3. The diagram in **Figure 13-26** shows six possible positions of these objects. Select a position, from 1 to 6, for each of the five objects. Not all positions need to be selected.

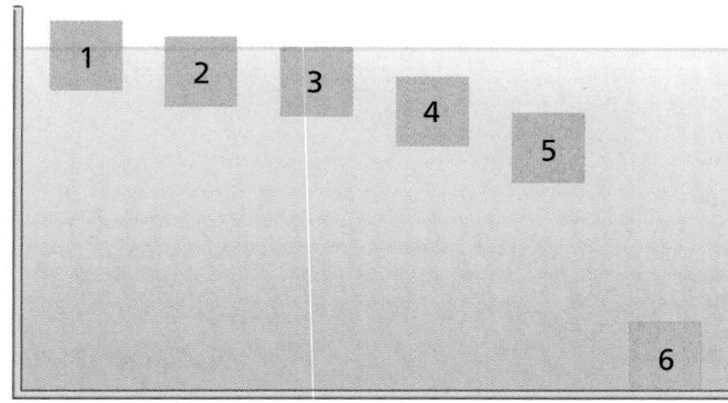

Figure 13-26

Mastering Problems

13.1 Properties of Fluids

75. Textbooks A 0.85-kg physics book with dimensions of 24.0 cm × 20.0 cm is at rest on a table.

a. What force does the book apply to the table?
b. What pressure does the book apply?

76. A 75-kg solid cylinder that is 2.5 m long and has an end radius of 7.0 cm stands on one end. How much pressure does it exert?

77. What is the total downward force of the atmosphere on the top of your head right now? Assume that the top of your head has an area of about 0.025 m^2.

78. Soft Drinks Sodas are made fizzy by the carbon dioxide (CO_2) dissolved in the liquid. An amount of carbon dioxide equal to about 8.0 L of carbon dioxide gas at atmospheric pressure and 300.0 K can be dissolved in a 2-L bottle of soda. The molar mass of CO_2 is 44 g/mol.

a. How many moles of carbon dioxide are in the 2-L bottle? (1 L = 0.001 m^3)
b. What is the mass of the carbon dioxide in the 2-L bottle of soda?

79. As shown in **Figure 13-27,** a constant-pressure thermometer is made with a cylinder containing a piston that can move freely inside the cylinder. The pressure and the amount of gas enclosed in the cylinder are kept constant. As the temperature increases or decreases, the piston moves up or down in the cylinder. At 0°C, the height of the piston is 20 cm. What is the height of the piston at 100°C?

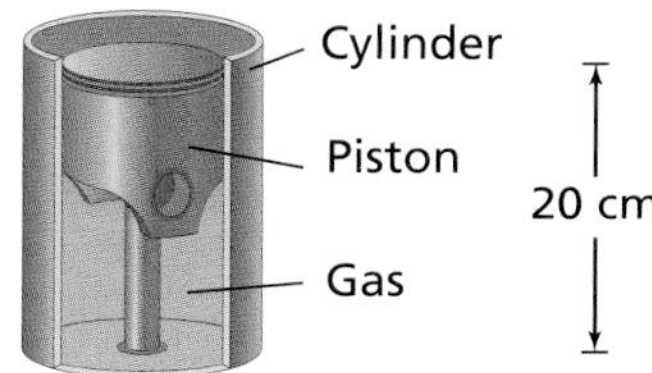

Figure 13-27

80. A piston with an area of 0.015 m^2 encloses a constant amount of gas in a cylinder with a volume of 0.23 m^3. The initial pressure of the gas is 1.5×10^5 Pa. A 150-kg mass is then placed on the piston, and the piston moves downward to a new position, as shown in **Figure 13-28.** If the temperature is constant, what is the new volume of the gas in the cylinder?

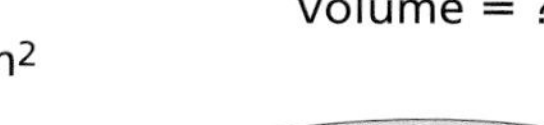

Volume = 0.23 m^3
Piston area = 0.015 m^2

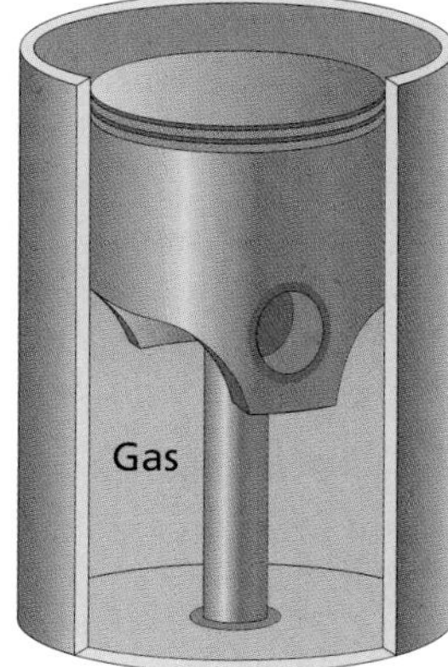

Figure 13-28

81. **Automobiles** A certain automobile tire is specified to be used at a gauge pressure of 30.0 psi, or 30.0 pounds per square inch. (One pound per square inch equals 6.90×10^3 Pa.) The term *gauge pressure* means the pressure above atmospheric pressure. Thus, the actual pressure in the tire is 1.01×10^5 Pa + (30.0 psi)(6.90×10^3 Pa/psi) = 3.08×10^5 Pa. As the car is driven, the tire's temperature increases, and the volume and pressure increase. Suppose you filled a car's tire to a volume of 0.55 m^3 at a temperature of 280 K. The initial pressure was 30.0 psi, but during the drive, the tire's temperature increased to 310 K and the tire's volume increased to 0.58 m^3.
   a. What is the new pressure in the tire?
   b. What is the new gauge pressure?

13.3 Fluids at Rest and in Motion

82. **Reservoirs** A reservoir behind a dam is 17-m deep. What is the pressure of the water at the following locations?
   a. the base of the dam
   b. 4.0 m from the top of the dam

83. A test tube standing vertically in a test-tube rack contains 2.5 cm of oil ($\rho = 0.81$ g/cm^3) and 6.5 cm of water. What is the pressure exerted by the two liquids on the bottom of the test tube?

84. **Antiques** An antique yellow metal statuette of a bird is suspended from a spring scale. The scale reads 11.81 N when the statuette is suspended in air, and it reads 11.19 N when the statuette is completely submerged in water.
   a. Find the volume of the statuette.
   b. Is the bird made of gold ($\rho = 19.3\times10^3$ kg/m^3) or gold-plated aluminum ($\rho = 2.7\times10^3$ kg/m^3)?

85. During an ecology experiment, an aquarium half-filled with water is placed on a scale. The scale shows a weight of 195 N.
   a. A rock weighing 8 N is added to the aquarium. If the rock sinks to the bottom of the aquarium, what will the scale read?
   b. The rock is removed from the aquarium, and the amount of water is adjusted until the scale again reads 195 N. A fish weighing 2 N is added to the aquarium. What is the scale reading with the fish in the aquarium?

86. What is the size of the buoyant force on a 26.0-N ball that is floating in fresh water?

87. What is the apparent weight of a rock submerged in water if the rock weighs 45 N in air and has a volume of 2.1×10^{-3} m^3?

88. What is the maximum weight that a balloon filled with 1.00 m^3 of helium can lift in air? Assume that the density of air is 1.20 kg/m^3 and that of helium is 0.177 kg/m^3. Neglect the mass of the balloon.

89. If a rock weighs 54 N in air and has an apparent weight of 46 N when submerged in a liquid with a density twice that of water, what will be its apparent weight when it is submerged in water?

90. **Oceanography** As shown in **Figure 13-29,** a large buoy used to support an oceanographic research instrument is made of a cylindrical, hollow iron tank. The tank is 2.1 m in height and 0.33 m in diameter. The total mass of the buoy and the research instrument is about 120 kg. The buoy must float so that one end is above the water to support a radio transmitter. Assuming that the mass of the buoy is evenly distributed, how much of the buoy will be above the waterline when it is floating?

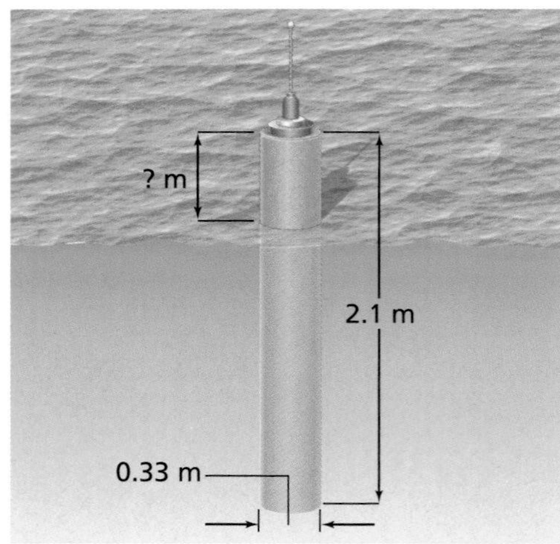

Figure 13-29

13.4 Solids

91. A bar of an unknown metal has a length of 0.975 m at 45°C and a length of 0.972 m at 23°C. What is its coefficient of linear expansion?

92. An inventor constructs a thermometer from an aluminum bar that is 0.500 m in length at 273 K. He measures the temperature by measuring the length of the aluminum bar. If the inventor wants to measure a 1.0-K change in temperature, how precisely must he measure the length of the bar?

93. **Bridges** How much longer will a 300-m steel bridge be on a 30°C day in August than on a −10°C night in January?

94. What is the change in length of a 2.00-m copper pipe if its temperature is raised from 23°C to 978°C?

 For more problems, go to Additional Problems, Appendix B.

95. What is the change in volume of a 1.0-m^3 concrete block if its temperature is raised 45°C?

96. Bridges Bridge builders often use rivets that are larger than the rivet hole to make the joint tighter. The rivet is cooled before it is put into the hole. Suppose that a builder drills a hole 1.2230 cm in diameter for a steel rivet 1.2250 cm in diameter. To what temperature must the rivet be cooled if it is to fit into the rivet hole, which is at 20.0°C?

97. A steel tank filled with methanol is 2.000 m in diameter and 5.000 m in height. It is completely filled at 10.0°C. If the temperature rises to 40.0°C, how much methanol (in liters) will flow out of the tank, given that both the tank and the methanol will expand?

98. An aluminum sphere is heated from 11°C to 580°C. If the volume of the sphere is 1.78 cm^3 at 11°C, what is the increase in volume of the sphere at 580°C?

99. The volume of a copper sphere is 2.56 cm^3 after being heated from 12°C to 984°C. What was the volume of the copper sphere at 12°C?

100. A square of iron plate that is 0.3300 m on each side is heated from 0°C to 95°C.

a. What is the change in the length of the sides of the square?

b. What is the relative change in area of the square?

101. An aluminum cube with a volume of 0.350 m^3 at 350.0 K is cooled to 270.0 K.

a. What is its volume at 270.0 K?

b. What is the length of a side of the cube at 270.0 K?

102. Industry A machinist builds a rectangular mechanical part for a special refrigerator system from two rectangular pieces of steel and two rectangular pieces of aluminum. At 293 K, the part is a perfect square, but at 170K, the part becomes warped, as shown in **Figure 13-30.** Which parts were made of steel and which were made of aluminum?

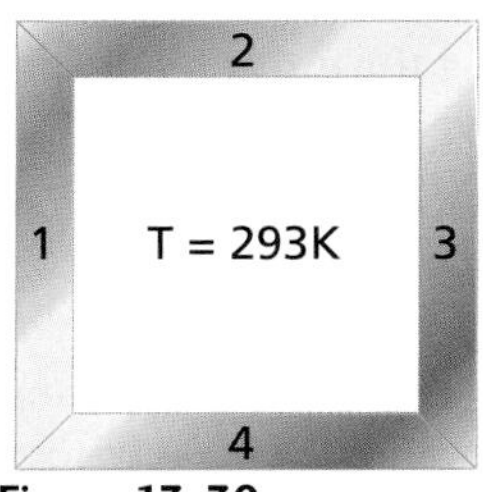

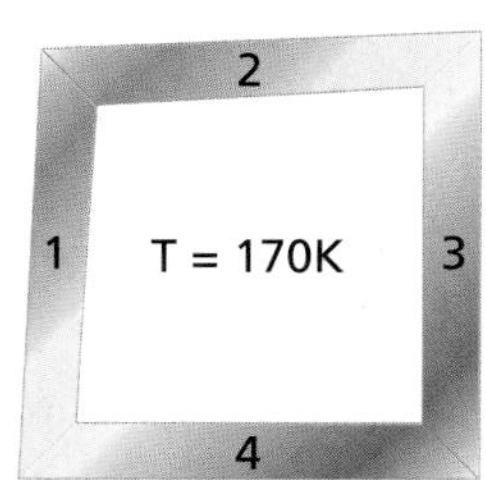

Figure 13-30

Mixed Review

103. What is the pressure on the hull of a submarine at a depth of 65 m?

104. Scuba Diving A scuba diver swimming at a depth of 5.0 m under water exhales a 4.2×10^{-6} m^3 bubble of air. What is the volume of that bubble just before it reaches the surface of the water?

105. An 18-N bowling ball floats with about half of the ball submerged.

a. What is the diameter of the bowling ball?

b. What would be the approximate apparent weight of a 36-N bowling ball?

106. An aluminum bar is floating in a bowl of mercury. When the temperature is increased, does the aluminum float higher or sink deeper into the mercury?

107. There is 100.0 mL of water in an 800.0-mL soft-glass beaker at 15.0°C. How much will the water level have dropped or risen when the beaker and water are heated to 50.0°C?

108. Auto Maintenance A hydraulic jack used to lift cars for repairs is called a three-ton jack. The large piston is 22 mm in diameter, and the small one is 6.3 mm in diameter. Assume that a force of 3 tons is 3.0×10^4 N.

a. What force must be exerted on the small piston to lift a 3-ton weight?

b. Most jacks use a lever to reduce the force needed on the small piston. If the resistance arm is 3.0 cm, how long must the effort arm of an ideal lever be to reduce the force to 100.0 N?

109. Ballooning A hot-air balloon contains a fixed volume of gas. When the gas is heated, it expands and pushes some gas out at the lower, open end. As a result, the mass of the gas in the balloon is reduced. Why would the air in a balloon have to be hotter to lift the same number of people above Vail, Colorado, which has an altitude of 2400 m, than above the tidewater flats of Virginia, which have an altitude of 6 m?

110. The Living World Some plants and animals are able to live in conditions of extreme pressure.

a. What is the pressure exerted by the water on the skin of a fish or worm that lives near the bottom of the Puerto Rico Trench, 8600 m below the surface of the Atlantic Ocean? Use 1030 kg/m^3 for the density of seawater.

b. What would be the density of air at that pressure, relative to its density above the surface of the ocean?

Thinking Critically

111. Apply Concepts You are washing dishes in the sink. A serving bowl has been floating in the sink. You fill the bowl with water from the sink, and it sinks to the bottom. Did the water level in the sink go up or down when the bowl was submerged?

112. Apply Concepts Persons confined to bed are less likely to develop bedsores if they use a waterbed rather than an ordinary mattress. Explain.

113. Analyze and Conclude One method of measuring the percentage of body fat is based on the fact that fatty tissue is less dense than muscle tissue. How can a person's average density be assessed with a scale and a swimming pool? What measurements does a physician need to record to find a person's average percentage of body fat?

114. Analyze and Conclude A downward force of 700 N is required to fully submerge a plastic foam sphere, as shown in **Figure 13-31.** The density of the foam is 95 kg/m^3.

a. What percentage of the sphere would be submerged if the sphere were released to float freely?

b. What is the weight of the sphere in air?

c. What is the volume of the sphere?

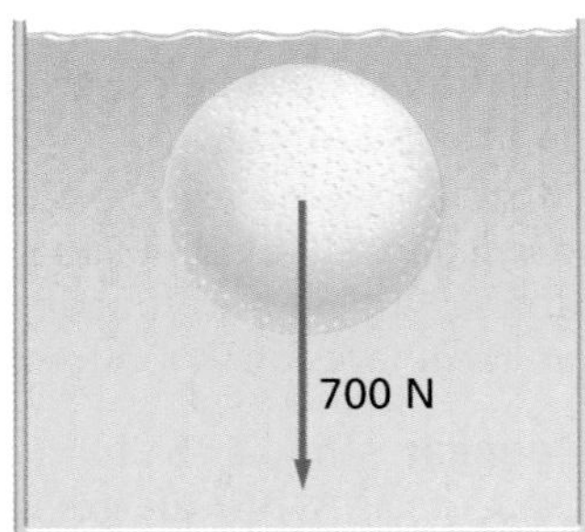

Figure 13-31

115. Apply Concepts Tropical fish for aquariums are often transported home from pet shops in transparent plastic bags filled mostly with water. If you placed a fish in its unopened transport bag in a home aquarium, which of the cases in **Figure 13-32** best represents what would happen? Explain your reasoning.

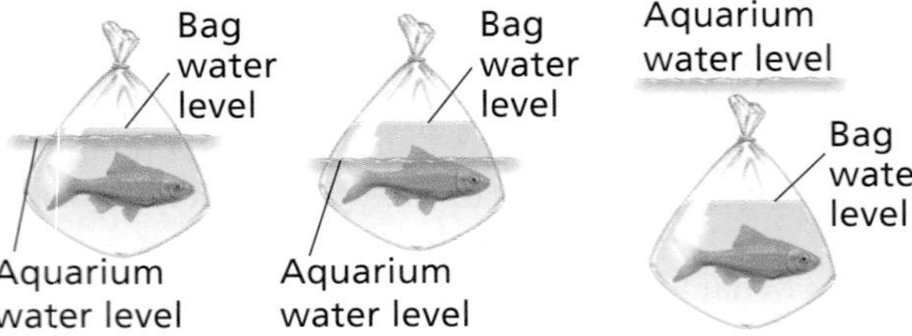

Figure 13-32

Writing in Physics

116. Some solid materials expand when they are cooled. Water between 4° and 0°C is the most common example, but rubber bands also expand in length when cooled. Research what causes this expansion.

117. Research Joseph Louis Gay-Lussac and his contributions to the gas laws. How did Gay-Lussac's work contribute to the discovery of the formula for water?

Cumulative Review

118. Two blocks are connected by a string over a frictionless, massless pulley such that one is resting on an inclined plane and the other is hanging over the top edge of the plane, as shown in **Figure 13-33.** The hanging block has a mass of 3.0 kg and the block on the plane has a mass of 2.0 kg. The coefficient of kinetic friction between the block and the inclined plane is 0.19. Answer the following questions assuming the blocks are released from rest. (Chapter 5)

a. What is the acceleration of the blocks?

b. What is the tension in the string connecting the blocks?

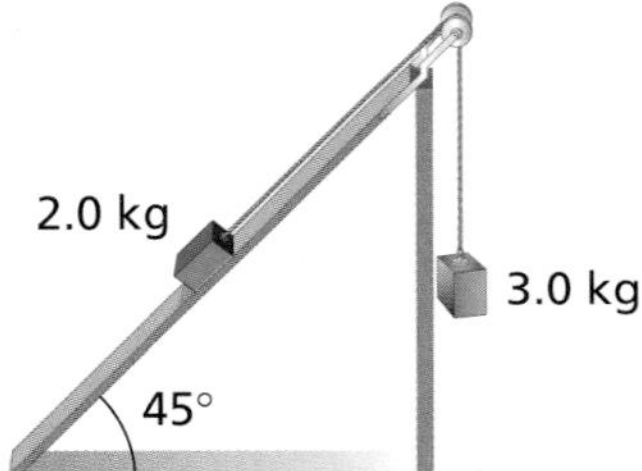

Figure 13-33

119. A compact car with a mass of 875 kg, moving south at 15 m/s, is struck by a full-sized car with a mass of 1584 kg, moving east at 12 m/s. The two cars stick together, and momentum is conserved. (Chapter 9)

a. Sketch the situation, assigning coordinate axes and identifying "before" and "after."

b. Find the direction and speed of the wreck immediately after the collision, remembering that momentum is a vector quantity.

c. The wreck skids along the ground and comes to a stop. The coefficient of kinetic friction while the wreck is skidding is 0.55. Assume that the acceleration is constant. How far does the wreck skid after impact?

120. A 188-W motor will lift a load at the rate (speed) of 6.50 cm/s. How great a load can the motor lift at this rate? (Chapter 10)

 For more problems, go to Additional Problems, Appendix B.

Multiple Choice

1. Gas with a volume of 10.0 L is trapped in an expandable cylinder. If the pressure is tripled and the temperature is increased by 80.0 percent (as measured on the Kelvin scale), what will be the new volume of the gas?

   Ⓐ 2.70 L Ⓒ 16.7 L
   Ⓑ 6.00 L Ⓓ 54.0 L

2. Nitrogen gas at standard atmospheric pressure, 101.3 kPa, has a volume of 0.080 m^3. If there are 3.6 mol of the gas, what is the temperature?

   Ⓐ 0.27 K Ⓒ 0.27°C
   Ⓑ 270 K Ⓓ 270°C

3. As diagrammed below, an operator applies a force of 200.0 N to the first piston of a hydraulic lift, which has an area of 5.4 cm^2. What is the pressure applied to the hydraulic fluid?

   Ⓐ 3.7×10^1 Pa Ⓒ 3.7×10^3 Pa
   Ⓑ 2.0×10^3 Pa Ⓓ 3.7×10^5 Pa

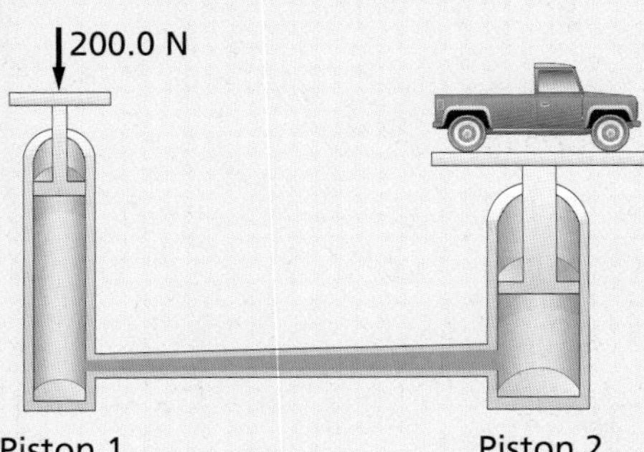

4. If the second piston in the lift diagrammed above exerts a force of 41,000 N, what is the area of the second piston?

   Ⓐ 0.0049 m^2 Ⓒ 0.11 m^2
   Ⓑ 0.026 m^2 Ⓓ 11 m^2

5. The density of cocobolo wood from Costa Rica is 1.10 g/cm^3. What is the apparent weight of a figurine that displaces 786 mL when submerged in a freshwater lake?

   Ⓐ 0.770 N Ⓒ 7.70 N
   Ⓑ 0.865 N Ⓓ 8.47 N

6. What is the buoyant force on a 17-kg object that displaces 85 L of water?

   Ⓐ 1.7×10^2 N Ⓒ 1.7×10^5 N
   Ⓑ 8.3×10^2 N Ⓓ 8.3×10^5 N

7. Which one of the following items does not contain matter in the plasma state?

   Ⓐ neon lighting
   Ⓑ stars
   Ⓒ lightning
   Ⓓ incandescent lighting

8. What is the mass of 365 mL of carbon dioxide gas at 3.0 atm pressure (1 atm = 101.3 kPa) and 24°C? The molar mass for carbon dioxide is 44.0 g/mol.

   Ⓐ 0.045 g Ⓒ 45 g
   Ⓑ 2.0 g Ⓓ 2.0 kg

Extended Answer

9. A balloon has a volume of 125 mL of air at standard atmospheric pressure, 101.3 kPa. If the balloon is anchored 1.27 m under the surface of a swimming pool, as illustrated in the diagram below, what is the new volume of the balloon?

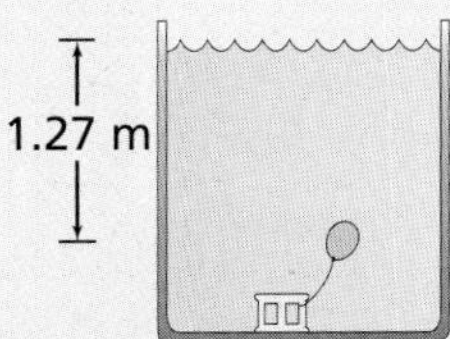

✓ Test-Taking TIP

Work Weak Muscles; Maintain Strong Ones

If you're preparing for a standardized test that covers many topics, it's sometimes difficult to focus on all the topics that require your attention. Focus most of your energy on your weaker areas and review your stronger topics frequently.

Chapter 14

Vibrations and Waves

What You'll Learn

- You will examine vibrational motion and learn how it relates to waves.
- You will determine how waves transfer energy.
- You will describe wave behavior and discuss its practical significance.

Why It's Important

Knowledge of the behavior of vibrations and waves is essential to the understanding of resonance and how safe buildings and bridges are built, as well as how communications through radio and television are achieved.

"Galloping Gertie" Shortly after it was opened to traffic, the Tacoma Narrows Bridge near Tacoma, Washington, began to vibrate whenever the wind blew (see inset). One day, the oscillations became so large that the bridge broke apart and collapsed into the water below.

Think About This ►

How could a light wind cause the bridge in the inset photo to vibrate with such large waves that it eventually collapsed?

physicspp.com

How do waves behave in a coiled spring?

Question

How do pulses that are sent down a coiled spring behave when the other end of the spring is stationary?

Procedure

1. Stretch out a coiled spiral spring, but do not overstretch it. One person should hold one end still, while the other person generates a sideways pulse in the spring. Observe the pulse while it travels along the spring and when it hits the held end. Record your observations.
2. Repeat step 1 with a larger pulse. Record your observations.
3. Generate a different pulse by compressing the spring at one end and letting go. Record your observations.
4. Generate a third type of pulse by twisting one end of the spring and then releasing it. Record your observations.

Analysis

What happens to the pulses as they travel through the spring? What happens as they hit the end of the spring? How did the pulse in step 1 compare to that generated in step 2?

Critical Thinking What are some properties that seem to control how a pulse moves through the spring?

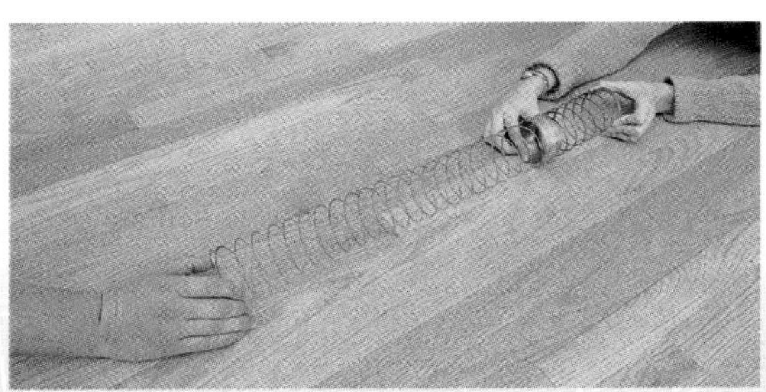

14.1 Periodic Motion

Objectives

- **Describe** the force in an elastic spring.
- **Determine** the energy stored in an elastic spring.
- **Compare** simple harmonic motion and the motion of a pendulum.

Vocabulary

periodic motion
simple harmonic motion
period
amplitude
Hooke's law
pendulum
resonance

You've probably seen a clock pendulum swing back and forth. You would have noticed that every swing followed the same path, and each trip back and forth took the same amount of time. This action is an example of vibrational motion. Other examples include a metal block bobbing up and down on a spring and a vibrating guitar string. These motions, which all repeat in a regular cycle, are examples of **periodic motion.**

In each example, the object has one position at which the net force on it is zero. At that position, the object is in equilibrium. Whenever the object is pulled away from its equilibrium position, the net force on the system becomes nonzero and pulls the object back toward equilibrium. If the force that restores the object to its equilibrium position is directly proportional to the displacement of the object, the motion that results is called **simple harmonic motion.**

Two quantities describe simple harmonic motion. The **period,** T, is the time needed for an object to repeat one complete cycle of the motion, and the **amplitude** of the motion is the maximum distance that the object moves from equilibrium.

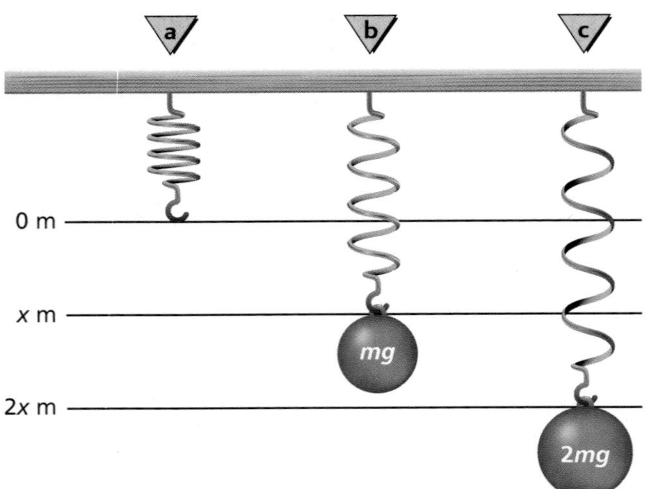

■ **Figure 14-1** The force exerted by a spring is directly proportional to the distance the spring is stretched.

The Mass on a Spring

How does a spring react to a force that is applied to it? **Figure 14-1a** shows a spring hanging from a support with nothing attached to it. The spring does not stretch because no external force is exerted on it. **Figure 14-1b** shows the same spring with an object of weight mg hanging from it. The spring has stretched by distance x so that the upward force it exerts balances the downward force of gravity acting on the object. **Figure 14-1c** shows the same spring stretched twice as far, $2x$, to support twice the weight, $2mg$, hanging from it. **Hooke's law** states that the force exerted by a spring is directly proportional to the amount that the spring is stretched. A spring that acts in this way is said to obey Hooke's law, which can be expressed by the following equation.

Hooke's Law $F = -kx$

The force exerted by a spring is equal to the spring constant times the distance the spring is compressed or stretched from its equilibrium position.

In this equation, k is the spring constant, which depends on the stiffness and other properties of the spring, and x is the distance that the spring is stretched from its equilibrium position. Not all springs obey Hooke's law, but many do. Those that do are called elastic.

Potential energy When a force is applied to stretch a spring, such as by hanging an object on its end, there is a direct linear relationship between the exerted force and the displacement, as shown by the graph in **Figure 14-2.** The slope of the graph is equal to the spring constant, given in units of newtons per meter. The area under the curve represents the work done to stretch the spring, and therefore equals the elastic potential energy that is stored in the spring as a result of that work. The base of the triangle is x, and the height is the force, which, according to the equation for Hooke's law, is equal to kx, so the potential energy in the spring is given by the following equation.

Potential Energy in a Spring $PE_{sp} = \frac{1}{2}kx^2$

The potential energy in a spring is equal to one-half times the product of the spring constant and the square of the displacement.

■ **Figure 14-2** The spring constant of a spring can be determined from the graph of force versus displacement of the spring.

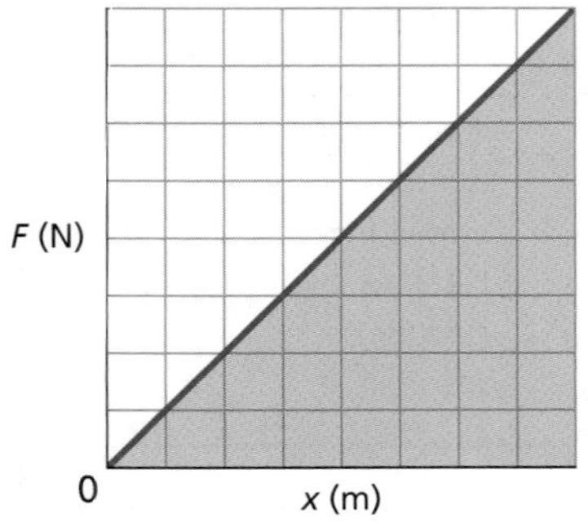

The units of the area, and thus, of the potential energy, are newton · meters, or joules.

How does the net force depend upon position? When an object hangs on a spring, the spring stretches until its upward force, $\mathbf{F}_{sp}$, balances the object's weight, $\mathbf{F}_g$, as shown in **Figure 14-3a.** The block is then in its equilibrium position. If you pull the object down, as in **Figure 14-3b,** the spring force increases, until it balances the forces exerted by your hand and gravity. When you let go of the object, it accelerates in the upward direction, as in **Figure 14-3c.** However, as the stretch of the spring is reduced, the

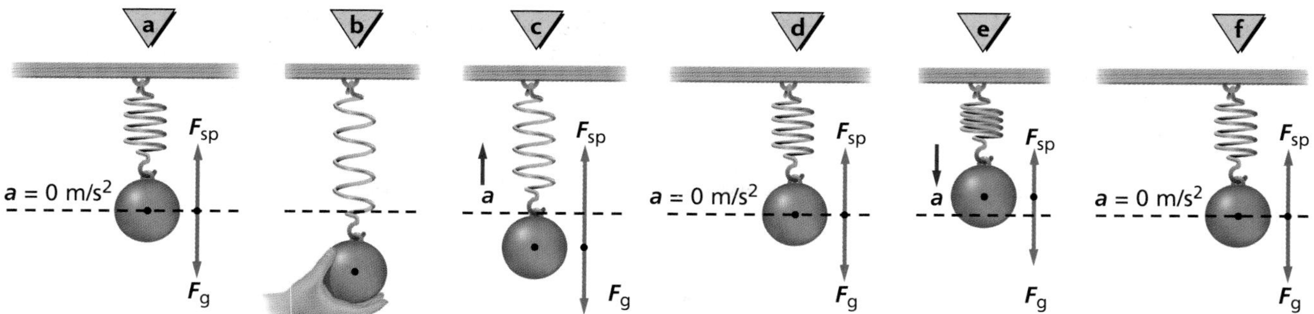

upward force decreases. In **Figure 14-3d,** the upward force of the spring and the object's weight are equal—there is no acceleration. Because there is no net force, the object continues its upward velocity, moving above the equilibrium position. In **Figure 14-3e,** the net force is in the direction opposite the displacement of the object and is directly proportional to the displacement, so the object moves with a simple harmonic motion. The object returns to the equilibrium position, as in **Figure 14-3f.**

Figure 14-3 Simple harmonic motion is demonstrated by the vibrations of an object hanging on a spring.

▶ EXAMPLE Problem 1

The Spring Constant and the Energy in a Spring A spring stretches by 18 cm when a bag of potatoes weighing 56 N is suspended from its end.

a. Determine the spring constant.

b. How much elastic potential energy is stored in the spring when it is stretched this far?

1 Analyze and Sketch the Problem

- Sketch the situation.
- Show and label the distance that the spring has stretched and its equilibrium position.

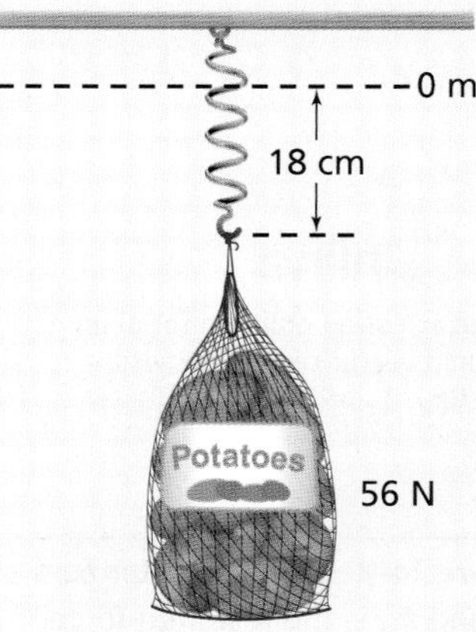

Known:	Unknown:
$x = 18$ cm	$k = ?$
$F = 56$ N	$PE_{sp} = ?$

2 Solve for the Unknown

a. Use $F = -kx$ and solve for k.

$$k = \frac{F}{x}$$

The minus sign can be dropped because it just means that the force is restoring.

$$= \frac{56\text{ N}}{0.18\text{ m}}$$

Substitute $F = 56$ N, $x = 0.18$ m

$$= 310\text{ N/m}$$

b.

$$PE_{sp} = \frac{1}{2}kx^2$$

$$= \frac{1}{2}(310\text{ N/m})(0.18\text{ m})^2$$

Substitute $k = 310$ N/m, $x = 0.18$ m

$$= 5.0\text{ J}$$

Math Handbook

Operations with Significant Digits pages 835–836

3 Evaluate the Answer

- **Are the units correct?** N/m are the correct units for the spring constant. (N/m)(m²) = N·m = J, which is the correct unit for energy.
- **Is the magnitude realistic?** The spring constant is consistent with a scale used, for example, to weigh groceries. The energy of 5.0 J is equal to the value obtained from $W = Fx = mgh$, when the average force of 28 N is applied.

PRACTICE Problems

- Additional Problems, Appendix B
- Solutions to Selected Problems, Appendix C

1. How much force is necessary to stretch a spring 0.25 m when the spring constant is 95 N/m?
2. A spring has a spring constant of 56 N/m. How far will it stretch when a block weighing 18 N is hung from its end?
3. What is the spring constant of a spring that stretches 12 cm when an object weighing 24 N is hung from it?
4. A spring with a spring constant of 144 N/m is compressed by a distance of 16.5 cm. How much elastic potential energy is stored in the spring?
5. A spring has a spring constant of 256 N/m. How far must it be stretched to give it an elastic potential energy of 48 J?

When the external force holding the object is released, as in Figure 14-3c, the net force and the acceleration are at their maximum, and the velocity is zero. As the object passes through the equilibrium point, Figure 14-3d, the net force is zero, and so is the acceleration. Does the object stop? No, it would take a net downward force to slow the object, and that will not exist until the object rises above the equilibrium position. When the object comes to the highest position in its oscillation, the net force and the acceleration are again at their maximum, and the velocity is zero. The object moves down through the equilibrium position to its starting point and continues to move in this vibratory manner. The period of oscillation, T, depends upon the mass of the object and the strength of the spring.

Automobiles Elastic potential energy is an important part of the design and building of today's automobiles. Every year, new models of cars are tested to see how well they withstand damage when they crash into barricades at low speeds. A car's ability to retain its integrity depends upon how much of the kinetic energy it had before the crash can be converted into the elastic potential energy of the frame after the crash. Many bumpers are modified springs that store energy as a car hits a barrier in a slow-speed collision. After the car stops and the spring is compressed, the spring returns to its equilibrium position, and the car recoils from the barrier.

Personal Tutor For an online tutorial on pendulums and vector resolution, visit physicspp.com.

Pendulums

Simple harmonic motion also can be demonstrated by the swing of a pendulum. A simple **pendulum** consists of a massive object, called the bob, suspended by a string or light rod of length l. After the bob is pulled to one side and released, it swings back and forth, as shown in **Figure 14-4.** The string or rod exerts a tension force, F_T, and gravity exerts a force, F_g, on the bob. The vector sum of the two forces produces the net force, shown at three positions in Figure 14-4. At the left and right positions shown in Figure 14-4, the net force and acceleration are maximum, and the velocity is zero. At the middle position in Figure 14-4, the net force and acceleration are zero, and the velocity is maximum. You can see that the net force is a restoring force; that is, it is opposite the direction of the displacement of the bob and is trying to restore the bob to its equilibrium position.

Figure 14-4 F_{net}, the vector sum of F_t and F_g, is the restoring force for the pendulum.

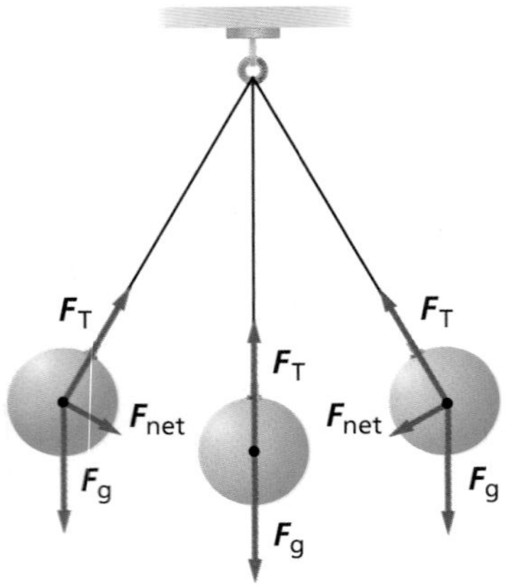

For small angles (less than about 15°) the restoring force is proportional to the displacement, so the movement is simple harmonic motion. The period of a pendulum is given by the following equation.

Period of a Pendulum $T = 2\pi\sqrt{\frac{l}{g}}$

The period of a pendulum is equal to two pi times the square root of the length of the pendulum divided by the acceleration due to gravity.

Notice that the period depends only upon the length of the pendulum and the acceleration due to gravity, not on the mass of the bob or the amplitude of oscillation. One application of the pendulum is to measure g, which can vary slightly at different locations on Earth.

EXAMPLE Problem 2

Finding g Using a Pendulum **A pendulum with a length of 36.9 cm has a period of 1.22 s. What is the acceleration due to gravity at the pendulum's location?**

1 Analyze and Sketch the Problem

- Sketch the situation.
- Label the length of the pendulum.

Known:	Unknown:
$l = 36.9$ cm	$g = ?$
$T = 1.22$ s	

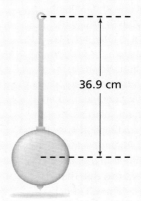

36.9 cm

2 Solve for the Unknown

$$T = 2\pi\sqrt{\frac{l}{g}}$$

Solve for g.

$$g = \frac{(2\pi)^2 l}{T^2}$$

$$= \frac{4\pi^2(0.369\text{ m})}{(1.22\text{ s})^2} \quad \text{Substitute } l = 0.369\text{ m}, T = 1.22\text{ s}$$

$$= 9.78\text{ m/s}^2$$

Math Handbook
Isolating a Variable
page 845

3 Evaluate the Answer

- **Are the units correct?** m/s^2 are the correct units for acceleration.
- **Is the magnitude realistic?** The calculated value of g is quite close to the standard value of g, 9.80 m/s^2. This pendulum could be at a high elevation above sea level.

PRACTICE Problems

• Additional Problems, Appendix B
• Solutions to Selected Problems, Appendix C

6. What is the period on Earth of a pendulum with a length of 1.0 m?
7. How long must a pendulum be on the Moon, where $g = 1.6\text{ m/s}^2$, to have a period of 2.0 s?
8. On a planet with an unknown value of g, the period of a 0.75-m-long pendulum is 1.8 s. What is g for this planet?

CHALLENGE PROBLEM

A car of mass m rests at the top of a hill of height h before rolling without friction into a crash barrier located at the bottom of the hill. The crash barrier contains a spring with a spring constant, k, which is designed to bring the car to rest with minimum damage.

1. Determine, in terms of m, h, k, and g, the maximum distance, x, that the spring will be compressed when the car hits it.
2. If the car rolls down a hill that is twice as high, how much farther will the spring be compressed?
3. What will happen after the car has been brought to rest?

APPLYING PHYSICS

Foucault Pendulum A Foucault pendulum has a long wire, about 16 m in length, with a heavy weight of about 109 kg attached to one end. According to Newton's first law of motion, a swinging pendulum will keep swinging in the same direction unless it is pushed or pulled in another direction. However, because Earth rotates every 24 h underneath the pendulum, to an observer it would seem as though the pendulum's direction of swing has changed. To demonstrate this, pegs are arranged in a circle on the floor beneath so that the swinging pendulum will knock them down as the floor rotates. At the north pole, this apparent rotation would be 15°/h.

Resonance

To get a playground swing going, you "pump" it by leaning back and pulling the chains at the same point in each swing, or a friend gives you repeated pushes at just the right times. **Resonance** occurs when small forces are applied at regular intervals to a vibrating or oscillating object and the amplitude of the vibration increases. The time interval between applications of the force is equal to the period of oscillation. Other familiar examples of resonance include rocking a car to free it from a snowbank and jumping rhythmically on a trampoline or a diving board. The large-amplitude oscillations caused by resonance can create stresses. Audiences in theater balconies, for example, sometimes damage the structures by jumping up and down with a period equal to the natural oscillation period of the balcony.

Resonance is a special form of simple harmonic motion in which the additions of small amounts of force at specific times in the motion of an object cause a larger and larger displacement. Resonance from wind, combined with the design of the bridge supports, may have caused the original Tacoma Narrows Bridge to collapse.

14.1 Section Review

9. **Hooke's Law** Two springs look alike but have different spring constants. How could you determine which one has the greater spring constant?

10. **Hooke's Law** Objects of various weights are hung from a rubber band that is suspended from a hook. The weights of the objects are plotted on a graph against the stretch of the rubber band. How can you tell from the graph whether or not the rubber band obeys Hooke's law?

11. **Pendulum** How must the length of a pendulum be changed to double its period? How must the length be changed to halve the period?

12. **Energy of a Spring** What is the difference between the energy stored in a spring that is stretched 0.40 m and the energy stored in the same spring when it is stretched 0.20 m?

13. **Resonance** If a car's wheel is out of balance, the car will shake strongly at a specific speed, but not when it is moving faster or slower than that speed. Explain.

14. **Critical Thinking** How is uniform circular motion similar to simple harmonic motion? How are they different?

Physics Online physicspp.com/self_check_quiz

14.2 Wave Properties

Objectives

- **Identify** how waves transfer energy without transferring matter.
- **Contrast** transverse and longitudinal waves.
- **Relate** wave speed, wavelength, and frequency.

Vocabulary

wave
wave pulse
periodic wave
transverse wave
longitudinal wave
surface wave
trough
crest
wavelength
frequency

Both particles and waves carry energy, but there is an important difference in how they do this. Think of a ball as a particle. If you toss the ball to a friend, the ball moves from you to your friend and carries energy. However, if you and your friend hold the ends of a rope and you give your end a quick shake, the rope remains in your hand. Even though no matter is transferred, the rope still carries energy through the wave that you created. A **wave** is a disturbance that carries energy through matter or space.

You have learned how Newton's laws of motion and principles of conservation of energy govern the behavior of particles. These laws and principles also govern the motion of waves. There are many kinds of waves that transmit energy, including the waves you cannot see.

Mechanical Waves

Water waves, sound waves, and the waves that travel down a rope or spring are types of mechanical waves. Mechanical waves require a medium, such as water, air, ropes, or a spring. Because many other waves cannot be directly observed, mechanical waves can serve as models.

Transverse waves The two disturbances shown in **Figure 14-5a** are called wave pulses. A **wave pulse** is a single bump or disturbance that travels through a medium. If the wave moves up and down at the same rate, a **periodic wave** is generated. Notice in Figure 14-5a that the rope is disturbed in the vertical direction, but the pulse travels horizontally. A wave with this type of motion is called a transverse wave. A **transverse wave** is one that vibrates perpendicular to the direction of the wave's motion.

Longitudinal waves In a coiled-spring toy, you can create a wave pulse in a different way. If you squeeze together several turns of the coiled-spring toy and then suddenly release them, pulses of closely-spaced turns will move away in both directions, as shown in **Figure 14-5b.** This is called a **longitudinal wave.** The disturbance is in the same direction as, or parallel to, the direction of the wave's motion. Sound waves are longitudinal waves. Fluids usually transmit only longitudinal waves.

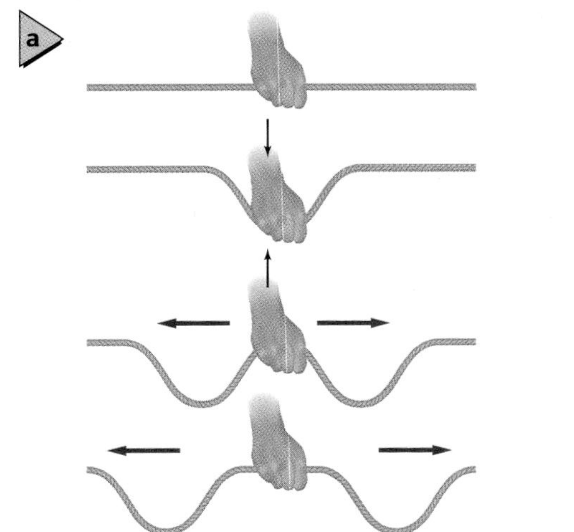

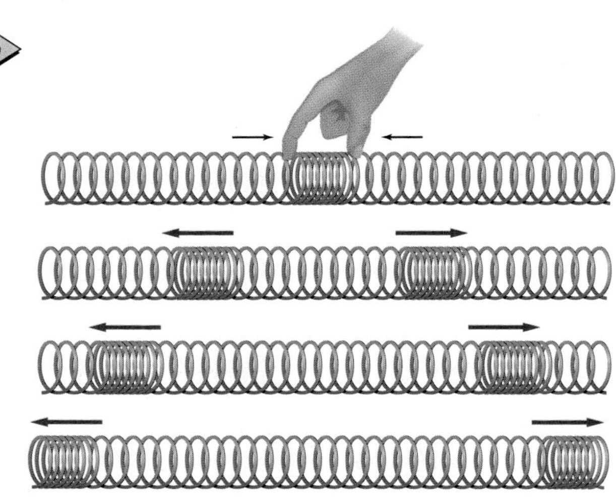

Figure 14-5 A quick shake of a rope sends out transverse wave pulses in both directions **(a).** The squeeze and release of a coiled-spring toy sends out longitudinal wave pulses in both directions **(b).**

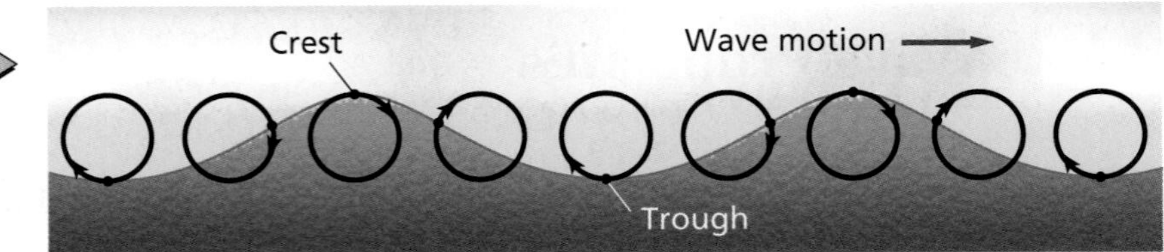

Figure 14-6 Surface waves have properties of both transverse and longitudinal waves **(a).** The paths of the individual particles are circular **(b).**

Surface waves Waves that are deep in a lake or ocean are longitudinal; at the surface of the water, however, the particles move in a direction that is both parallel and perpendicular to the direction of wave motion, as shown in **Figure 14-6.** Each of the waves is a **surface wave,** which has characteristics of both transverse and longitudinal waves. The energy of water waves usually comes from distant storms, whose energy initially came from the heating of Earth by solar energy. This energy, in turn, was carried to Earth by transverse electromagnetic waves from the Sun.

Measuring a Wave

There are many ways to describe or measure a wave. Some characteristics depend on how the wave is produced, whereas others depend on the medium through which the wave travels.

Speed How fast does a wave move? The speed of the pulse shown in **Figure 14-7** can be found in the same way as the speed of a moving car is determined. First, measure the displacement of the wave peak, Δd, then divide this by the time interval, Δt, to find the speed, given by $v = \Delta d/\Delta t$. The speed of a periodic wave can be found in the same way. For most mechanical waves, both transverse and longitudinal, the speed depends only on the medium through which the waves move.

Amplitude How does the pulse generated by gently shaking a rope differ from the pulse produced by a violent shake? The difference is similar to the difference between a ripple in a pond and an ocean breaker: they have different amplitudes. You have learned that the amplitude of a wave is the maximum displacement of the wave from its position of rest, or equilibrium. Two similar waves having different amplitudes are shown in **Figure 14-8.**

A wave's amplitude depends on how it is generated, but not on its speed. More work must be done to generate a wave with a greater amplitude. For example, strong winds produce larger water waves than those formed by gentle breezes. Waves with greater amplitudes transfer more energy.

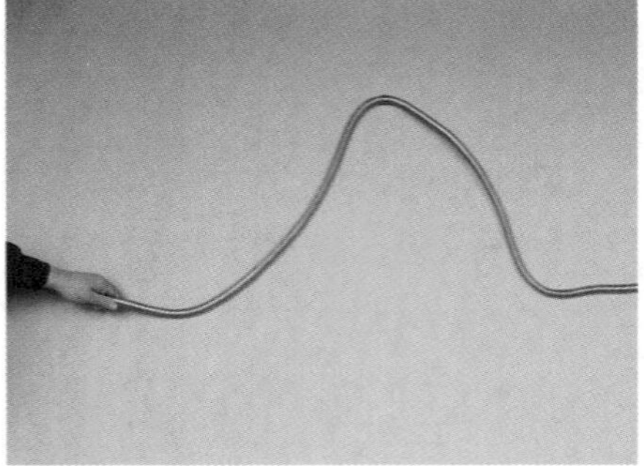

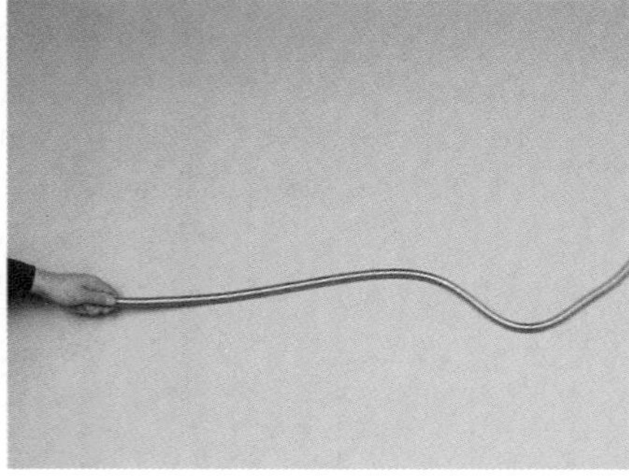

Figure 14-7 These two photographs were taken 0.20 s apart. During that time, the crest moved 0.80 m. The velocity of the wave is 4.0 m/s.

Whereas a small wave might move sand on a beach a few centimeters, a giant wave can uproot and move a tree. For waves that move at the same speed, the rate at which energy is transferred is proportional to the square of the amplitude. Thus, doubling the amplitude of a wave increases the amount of energy it transfers each second by a factor of 4.

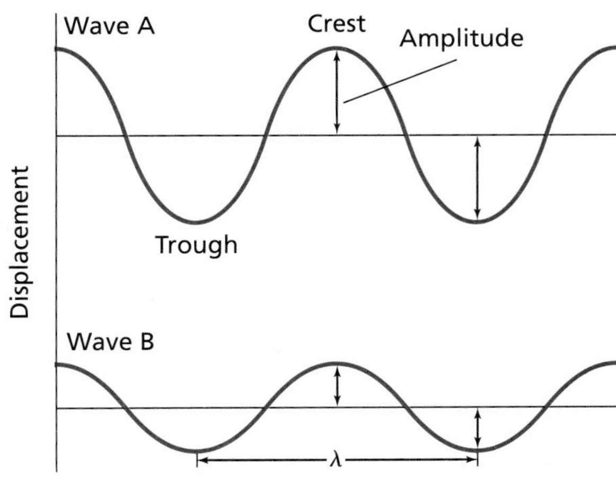

Figure 14-8 The amplitude of wave A is larger than that of wave B.

Wavelength Rather than focusing on one point on a wave, imagine taking a snapshot of the wave so that you can see the whole wave at one instant in time. Figure 14-8 shows each low point, called a **trough,** and each high point, called a **crest,** of a wave. The shortest distance between points where the wave pattern repeats itself is called the **wavelength.** Crests are spaced by one wavelength. Each trough also is one wavelength from the next. The Greek letter lambda, λ, represents wavelength.

Phase Any two points on a wave that are one or more whole wavelengths apart are in phase. Particles in the medium are said to be in phase with one another when they have the same displacement from equilibrium and the same velocity. Particles in the medium with opposite displacements and velocities are 180° out of phase. A crest and a trough, for example, are 180° out of phase with each other. Two particles in a wave can be anywhere from 0° to 180° out of phase with one another.

Period and frequency Although wave speed and amplitude can describe both pulses and periodic waves, period, T, and frequency, f, apply only to periodic waves. You have learned that the period of a simple harmonic oscillator, such as a pendulum, is the time it takes for the motion of the oscillator to complete one cycle. Such an oscillator is usually the source, or cause, of a periodic wave. The period of a wave is equal to the period of the source. In **Figures 14-9a** through **14-9d,** the period, T, equals 0.04 s, which is the time it takes the source to complete one cycle. The same time is taken by P, a point on the rope, to return to its initial phase.

Figure 14-9 One end of a string, with a piece of tape at point P, is attached to a blade vibrating 25 times per second. Note the change in position of point P over time.

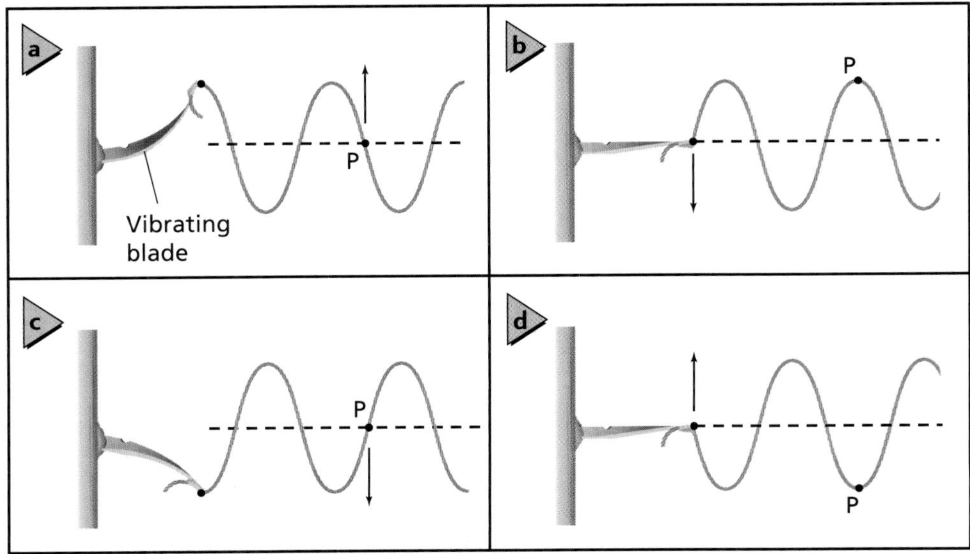

The **frequency** of a wave, f, is the number of complete oscillations it makes each second. Frequency is measured in hertz. One hertz (Hz) is one oscillation per second. The frequency and period of a wave are related by the following equation.

Frequency of a Wave $f = \frac{1}{T}$

The frequency of a wave is equal to the reciprocal of the period.

Both the period and the frequency of a wave depend only on its source. They do not depend on the wave's speed or the medium.

Although you can directly measure a wavelength, the wavelength depends on both the frequency of the oscillator and the speed of the wave. In the time interval of one period, a wave moves one wavelength. Therefore, the wavelength of a wave is the speed multiplied by the period, $\lambda = vT$. Because the frequency is usually more easily found than the period, this equation is most often written in the following way.

Wavelength $\lambda = \frac{v}{f}$

The wavelength of a wave is equal to the velocity divided by the frequency.

Picturing waves If you took a snapshot of a transverse wave on a spring, it might look like one of the waves shown in Figure 14-8. This snapshot could be placed on a graph grid to show more information about the wave, as in **Figure 14-10a.** Similarly, if you record the motion of a single particle, such as point P in Figure 14-9, that motion can be plotted on a displacement-versus-time graph, as in **Figure 14-10b.** The period is found using the time axis of the graph. Longitudinal waves can also be depicted by graphs, where the y-axis could represent pressure, for example.

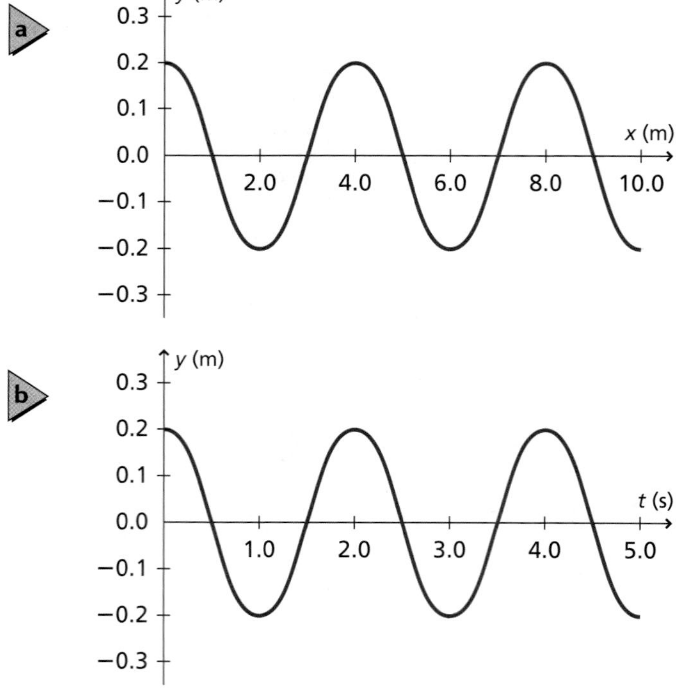

Figure 14-10 Waves can be represented by graphs. The wavelength of this wave is 4.0 m **(a).** The period is 2.0 s **(b).** The amplitude, or displacement, is 0.2 m in both graphs. If these graphs represent the same wave, what is its speed?

▶ EXAMPLE Problem 3

Characteristics of a Wave A sound wave has a frequency of 192 Hz and travels the length of a football field, 91.4 m, in 0.271 s.

a. What is the speed of the wave?

b. What is the wavelength of the wave?

c. What is the period of the wave?

d. If the frequency was changed to 442 Hz, what would be the new wavelength and period?

1 Analyze and Sketch the Problem

- Draw a model of the football field.
- Diagram a velocity vector.

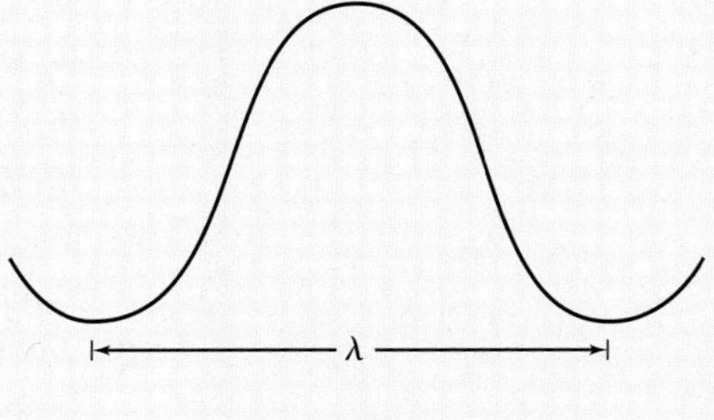

Known:	Unknown:
$f = 192$ Hz	$v = ?$
$d = 91.4$ m	$\lambda = ?$
$t = 0.271$ s	$T = ?$

2 Solve for the Unknown

a. Solve for v.

$$v = \frac{d}{t}$$

$$= \frac{91.4 \text{ m}}{0.271 \text{ s}}$$ Substitute $d = 91.4$ m, $t = 0.271$ s

$$= 337 \text{ m/s}$$

b. Solve for λ.

$$\lambda = \frac{v}{f}$$

$$= \frac{337 \text{ m/s}}{192 \text{ Hz}}$$ Substitute $v = 337$ m/s, $f = 192$ Hz

$$= 1.76 \text{ m}$$

c. Solve for T.

$$T = \frac{1}{f}$$

$$= \frac{1}{192 \text{ Hz}}$$ Substitute $f = 192$ Hz

$$= 0.00521 \text{ s}$$

d. $$\lambda = \frac{v}{f}$$

$$= \frac{337 \text{ m/s}}{442 \text{ Hz}}$$ Substitute $v = 337$ m/s, $f = 442$ Hz

$$= 0.762 \text{ m}$$

$$T = \frac{1}{f}$$

$$= \frac{1}{442 \text{ Hz}}$$ Substitute $f = 442$ Hz

$$= 0.00226 \text{ s}$$

Math Handbook
Operations with Significant Digits pages 835–836

3 Evaluate the Answer

- **Are the units correct?** Hz has the units s^{-1}, so (m/s)/Hz = (m/s)·s = m, which is correct.
- **Are the magnitudes realistic?** A typical sound wave travels approximately 343 m/s, so 337 m/s is reasonable. The frequencies and periods are reasonable for sound waves. 442 Hz is close to a 440-Hz A above middle-C on a piano.

PRACTICE Problems

• Additional Problems, Appendix B
• Solutions to Selected Problems, Appendix C

15. A sound wave produced by a clock chime is heard 515 m away 1.50 s later.

a. What is the speed of sound of the clock's chime in air?

b. The sound wave has a frequency of 436 Hz. What is the period of the wave?

c. What is the wave's wavelength?

16. A hiker shouts toward a vertical cliff 465 m away. The echo is heard 2.75 s later.

a. What is the speed of sound of the hiker's voice in air?

b. The wavelength of the sound is 0.750 m. What is its frequency?

c. What is the period of the wave?

17. If you want to increase the wavelength of waves in a rope, should you shake it at a higher or lower frequency?

18. What is the speed of a periodic wave disturbance that has a frequency of 3.50 Hz and a wavelength of 0.700 m?

19. The speed of a transverse wave in a string is 15.0 m/s. If a source produces a disturbance that has a frequency of 6.00 Hz, what is its wavelength?

20. Five pulses are generated every 0.100 s in a tank of water. What is the speed of propagation of the wave if the wavelength of the surface wave is 1.20 cm?

21. A periodic longitudinal wave that has a frequency of 20.0 Hz travels along a coil spring. If the distance between successive compressions is 0.600 m, what is the speed of the wave?

You probably have been intuitively aware that waves carry energy that can do work. You may have seen the massive damage done by the huge storm surge of a hurricane or the slower erosion of cliffs and beaches done by small, everyday waves. It is important to remember that while the amplitude of a mechanical wave determines the amount of energy it carries, only the medium determines the wave's speed.

14.2 Section Review

22. Speed in Different Media If you pull on one end of a coiled-spring toy, does the pulse reach the other end instantaneously? What happens if you pull on a rope? What happens if you hit the end of a metal rod? Compare and contrast the pulses traveling through these three materials.

23. Wave Characteristics You are creating transverse waves in a rope by shaking your hand from side to side. Without changing the distance that your hand moves, you begin to shake it faster and faster. What happens to the amplitude, wavelength, frequency, period, and velocity of the wave?

24. Waves Moving Energy Suppose that you and your lab partner are asked to demonstrate that a transverse wave transports energy without transferring matter. How could you do it?

25. Longitudinal Waves Describe longitudinal waves. What types of media transmit longitudinal waves?

26. Critical Thinking If a raindrop falls into a pool, it creates waves with small amplitudes. If a swimmer jumps into a pool, waves with large amplitudes are produced. Why doesn't the heavy rain in a thunderstorm produce large waves?

14.3 Wave Behavior

When a wave encounters the boundary of the medium in which it is traveling, it often reflects back into the medium. In other instances, some or all of the wave passes through the boundary into another medium, often changing direction at the boundary. In addition, many properties of wave behavior result from the fact that two or more waves can exist in the same medium at the same time—quite unlike particles.

▶ **Objectives**

- **Relate** a wave's speed to the medium in which the wave travels.
- **Describe** how waves are reflected and refracted at boundaries between media.
- **Apply** the principle of superposition to the phenomenon of interference.

▶ **Vocabulary**

incident wave
reflected wave
principle of superposition
interference
node
antinode
standing wave
wave front
ray
normal
law of reflection
refraction

Waves at Boundaries

Recall from Section 14.2 that the speed of a mechanical wave depends only on the properties of the medium it passes through, not on the wave's amplitude or frequency. For water waves, the depth of the water affects wave speed. For sound waves in air, the temperature affects wave speed. For waves on a spring, the speed depends upon the spring's tension and mass per unit length.

Examine what happens when a wave moves across a boundary from one medium into another, as in two springs of different thicknesses joined end-to-end. **Figure 14-11** shows a wave pulse moving from a large spring into a smaller one. The wave that strikes the boundary is called the **incident wave.** One pulse from the larger spring continues in the smaller spring, but at the specific speed of waves traveling through the smaller spring. Note that this transmitted wave pulse remains upward.

Some of the energy of the incident wave's pulse is reflected backward into the larger spring. This returning wave is called the **reflected wave.** Whether or not the reflected wave is upright or inverted depends on the characteristics of the two springs. For example, if the waves in the smaller spring have a higher speed because the spring is heavier or stiffer, then the reflected wave will be inverted.

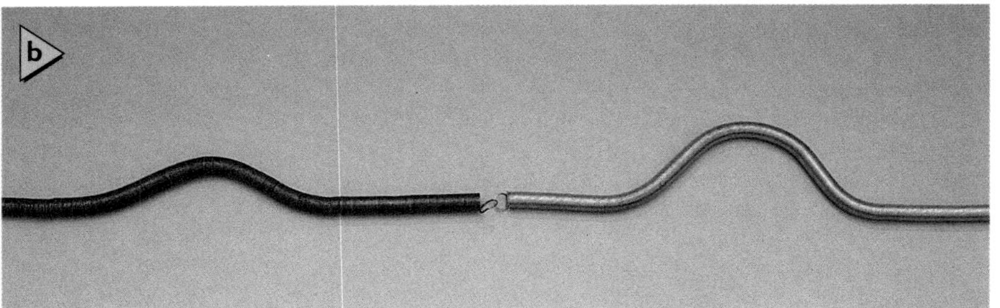

■ **Figure 14-11** The junction of the two springs is a boundary between two media. A pulse reaching the boundary **(a)** is partially reflected and partially transmitted **(b)**.

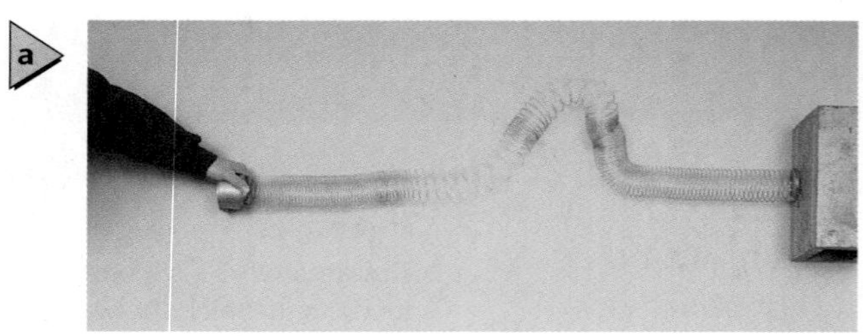

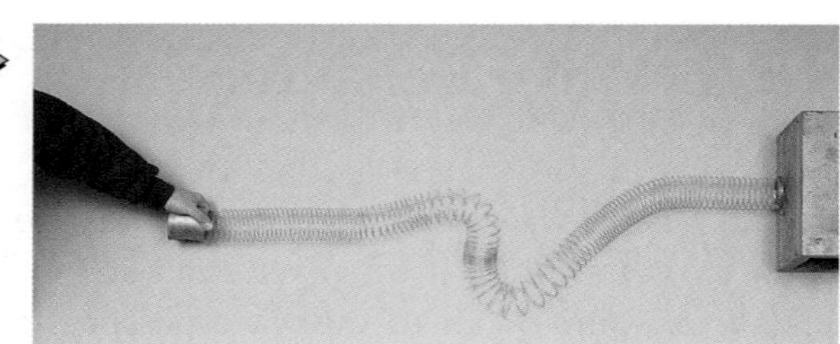

Figure 14-12 A pulse approaches a rigid wall **(a)** and is reflected back **(b).** Note that the amplitude of the reflected pulse is nearly equal to the amplitude of the incident pulse, but it is inverted.

What happens if the boundary is a wall rather than another spring? When a wave pulse is sent down a spring connected to a rigid wall, the energy transmitted is reflected back from the wall, as shown in **Figure 14-12.** The wall is the boundary of a new medium through which the wave attempts to pass. Instead of passing through, the pulse is reflected from the wall with almost exactly the same amplitude as the pulse of the incident wave. Thus, almost all the wave's energy is reflected back. Very little energy is transmitted into the wall. Also note that the pulse is inverted. If the spring were attached to a loose ring around a pole, a free-moving boundary, the wave would not be inverted.

Superposition of Waves

Suppose a pulse traveling down a spring meets a reflected pulse coming back. In this case, two waves exist in the same place in the medium at the same time. Each wave affects the medium independently. The **principle of superposition** states that the displacement of a medium caused by two or more waves is the algebraic sum of the displacements caused by the individual waves. In other words, two or more waves can combine to form a new wave. If the waves move in opposite directions, they can cancel or form a new wave of lesser or greater amplitude. The result of the superposition of two or more waves is called **interference.**

Figure 14-13 When two equal pulses meet, there is a point, called the node (N), where the medium remains undisturbed **(a).** Constructive interference results in maximum interference at the antinode (A) **(b).** If the opposite pulses have unequal amplitudes, cancellation is incomplete **(c).**

Concepts In Motion
Interactive Figure To see an animation on wave interference, visit physicspp.com.

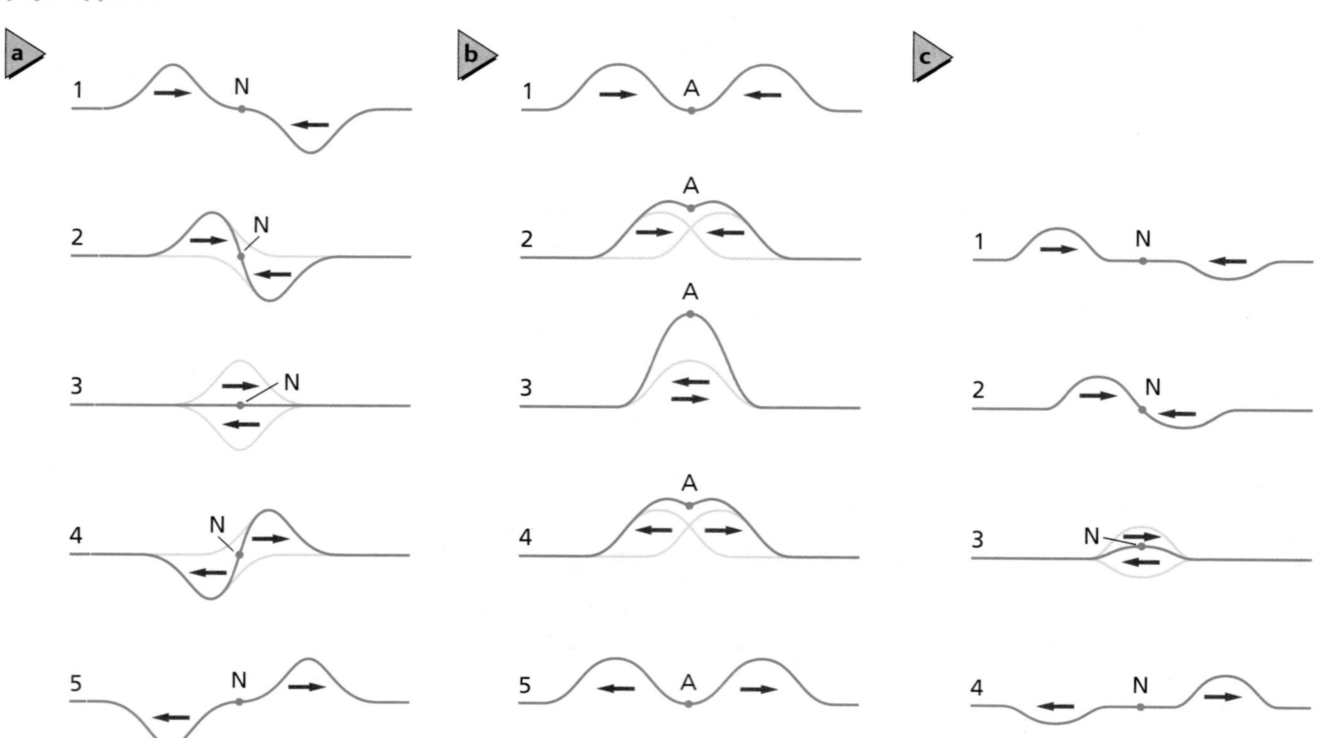

Wave interference Wave interference can be either constructive or destructive. When two pulses with equal but opposite amplitudes meet, the displacement of the medium at each point in the overlap region is reduced. The superposition of waves with equal but opposite amplitudes causes destructive interference, as shown in **Figure 14-13a.** When the pulses meet and are in the same location, the displacement is zero. Point N, which does not move at all, is called a **node.** The pulses continue to move and eventually resume their original form.

Constructive interference occurs when wave displacements are in the same direction. The result is a wave that has an amplitude greater than those of any of the individual waves. **Figure 14-13b** shows the constructive interference of two equal pulses. A larger pulse appears at point A when the two waves meet. Point A has the largest displacement and is called the **antinode.** The two pulses pass through each other without changing their shapes or sizes. If the pulses have unequal amplitudes, the resultant pulse at the overlap is the algebraic sum of the two pulses, as shown in **Figure 14-13c.**

Standing waves You can apply the concept of superimposed waves to the control of the formation of large amplitude waves. If you attach one end of a rope or coiled spring to a fixed point, such as a doorknob, and then start to vibrate the other end, the wave leaves your hand, is reflected at the fixed end, is inverted, and returns to your hand. When it reaches your hand, the reflected wave is inverted and travels back down the rope. Thus, when the wave leaves your hand the second time, its displacement is in the same direction as it was when it left your hand the first time.

What if you want to increase the amplitude of the wave that you create? Suppose you adjust the motion of your hand so that the period of vibration equals the time needed for the wave to make one round-trip from your hand to the door and back. Then, the displacement given by your hand to the rope each time will add to the displacement of the reflected wave. As a result, the oscillation of the rope in one segment will be much greater than the motion of your hand. You would expect this based on your knowledge of constructive interference. This large-amplitude oscillation is an example of mechanical resonance. The nodes are at the ends of the rope and an antinode is in the middle, as shown in **Figure 14-14a.** Thus, the wave appears to be standing still and is called a **standing wave.** You should note, however, that the standing wave is the interference of the two traveling waves moving in opposite directions. If you double the frequency of vibration, you can produce one more node and one more antinode in the rope. Then it appears to vibrate in two segments. Further increases in frequency produce even more nodes and antinodes, as shown in **Figures 14-14b** and **c.**

MINI LAB

Wave Interaction

With a coiled-spring toy, you can create pressure waves, as well as transverse waves of various amplitudes, speeds, and orientations.

1. **Design an experiment** to test what happens when waves from different directions meet.
2. Perform your experiment and record your observations.

Analyze and Conclude

3. Does the speed of either wave change?
4. Do the waves bounce off each other, or do they pass through each other?

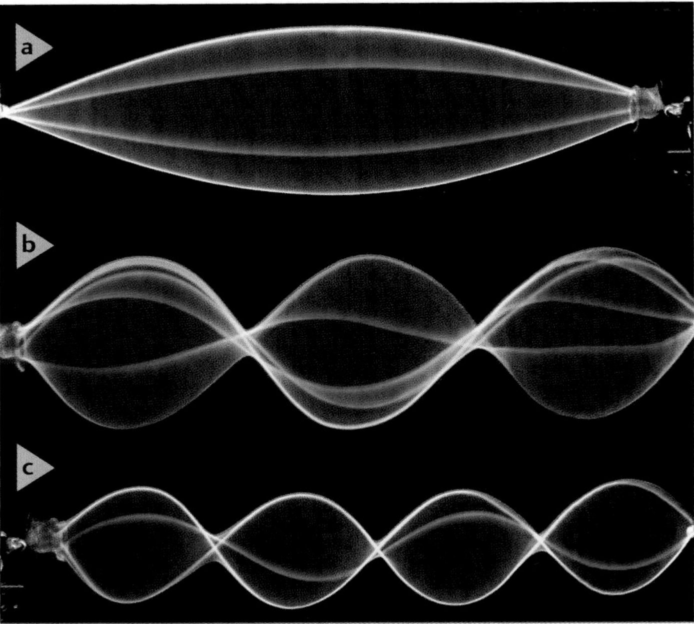

Figure 14-14 Interference produces standing waves in a rope. As the frequency is increased, as shown from top to bottom, the number of nodes and antinodes increases.

Figure 14-15 Circular waves spread outward from their source **(a).** The wave can be represented by circles drawn at their crests **(b).** Notice that the rays are perpendicular to the wave fronts.

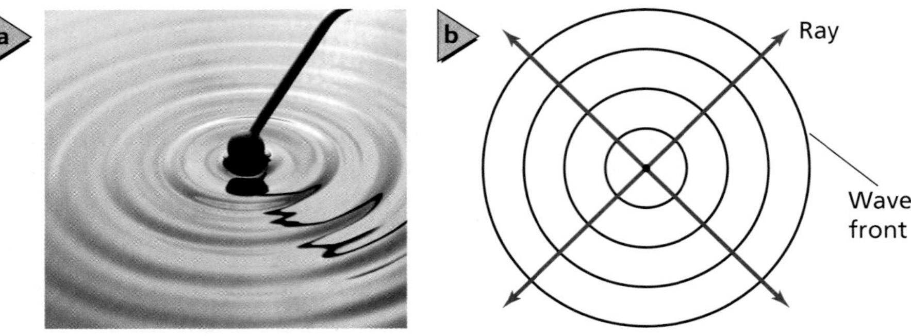

Waves in Two Dimensions

You have studied waves on a rope and on a spring reflecting from rigid supports, where the amplitude of the waves is forced to be zero by destructive interference. These mechanical waves move in only one dimension. However, waves on the surface of water move in two dimensions, and sound waves and electromagnetic waves will later be shown to move in three dimensions. How can two-dimensional waves be demonstrated?

Picturing waves in two dimensions When you throw a small stone into a calm pool of water, you see the circular crests and troughs of the resulting waves spreading out in all directions. You can sketch those waves by drawing circles to represent the wave crests. If you dip your finger into water with a constant frequency, the resulting sketch would be a series of concentric circles, called wave fronts, centered on your finger. A **wave front** is a line that represents the crest of a wave in two dimensions, and it can be used to show waves of any shape, including circular waves and straight waves. **Figure 14-15a** shows circular waves in water, and **Figure 14-15b** shows the wave fronts that represent those water waves. Wave fronts drawn to scale show the wavelengths of the waves, but not their amplitudes.

Whatever their shape, two-dimensional waves always travel in a direction that is perpendicular to their wave fronts. That direction can be represented by a **ray,** which is a line drawn at a right angle to the crest of the wave. When all you want to show is the direction in which a wave is traveling, it is convenient to draw rays instead of wave fronts.

Reflection of waves in two dimensions A ripple tank can be used to show the properties of two-dimensional waves. A ripple tank contains a thin layer of water. Vibrating boards produce wave pulses, or, in the case of

Figure 14-16 A wave pulse in a ripple tank is reflected by a barrier **(a).** The ray diagram models the wave in time sequence as it approaches the barrier and is then reflected to the right **(b).**

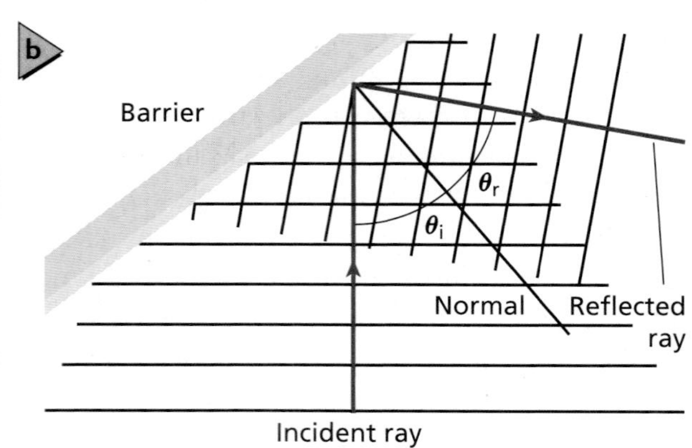

Figure 14-16a, traveling waves of water with constant frequency. A lamp above the tank produces shadows below the tank that show the locations of the crests of the waves. The wave pulse travels toward a rigid barrier that reflects the wave: the incident wave moves upward, and the reflected wave moves to the right.

The direction of wave motion can be modeled by a ray diagram. **Figure 14-16b** shows the ray diagram for the waves in the ripple tank. The ray representing the incident wave is the arrow pointing upward. The ray representing the reflected wave points to the right.

The direction of the barrier also is shown by a line, which is drawn at a right angle, or perpendicular, to the barrier, called the **normal.** The angle between the incident ray and the normal is called the angle of incidence. The angle between the normal and the reflected ray is called the angle of reflection. The **law of reflection** states that the angle of incidence is equal to the angle of reflection.

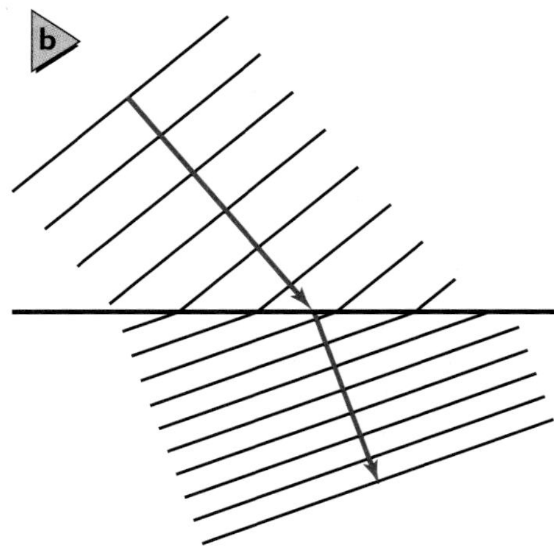

Refraction of waves in two dimensions A ripple tank also can be used to model the behavior of waves as they move from one medium into another. **Figure 14-17a** shows a glass plate placed in a ripple tank. The water above the plate is shallower than the water in the rest of the tank and acts like a different medium. As the waves move from deep to shallow water, their speed decreases, and the direction of the waves changes. Because the waves in the shallow water are generated by the waves in the deep water, their frequency is not changed. Based on the equation $\lambda = v/f$, the decrease in the speed of the waves means that the wavelength is shorter in the shallower water. The change in the direction of waves at the boundary between two different media is known as **refraction. Figure 14-17b** shows a wave front and ray model of refraction. When you study the reflection and refraction of light in Chapter 17, you will learn the law of refraction, called Snell's law.

You may not be aware that echoes are caused by the reflection of sound off hard surfaces, such as the walls of a large warehouse or a distant cliff face. Refraction is partly responsible for rainbows. When white light passes through a raindrop, refraction separates the light into its individual colors.

Figure 14-17 As the water waves move over a shallower region of the ripple tank where a glass plate is placed, they slow down and their wavelength decreases **(a).** Refraction can be represented by a diagram of wave fronts and rays **(b).**

14.3 Section Review

27. **Waves at Boundaries** Which of the following wave characteristics remain unchanged when a wave crosses a boundary into a different medium: frequency, amplitude, wavelength, velocity, and/or direction?

28. **Refraction of Waves** Notice in Figure 14-17a how the wave changes direction as it passes from one medium to another. Can two-dimensional waves cross a boundary between two media without changing direction? Explain.

29. **Standing Waves** In a standing wave on a string fixed at both ends, how is the number of nodes related to the number of antinodes?

30. **Critical Thinking** As another way to understand wave reflection, cover the right-hand side of each drawing in Figure 14-13a with a piece of paper. The edge of the paper should be at point N, the node. Now, concentrate on the resultant wave, shown in darker blue. Note that it acts like a wave reflected from a boundary. Is the boundary a rigid wall, or is it open-ended? Repeat this exercise for Figure 14-13b.

PHYSICS LAB • Design Your Own

Pendulum Vibrations

A pendulum can provide a simple model for the investigation of wave properties. In this experiment, you will design a procedure to use the pendulum to examine amplitude, period, and frequency of a wave. You also will determine the acceleration due to gravity from the period and length of the pendulum.

QUESTION

How can a pendulum demonstrate the properties of waves?

Objectives

- **Determine** what variables affect a pendulum's period.
- **Investigate** the frequency and period amplitude of a pendulum.
- **Measure** *g*, the acceleration due to gravity, using the period and length of a pendulum.

Safety Precautions

Possible Materials

string (1.5 m)
three sinkers
paper clip
ring stand with ring
stopwatch

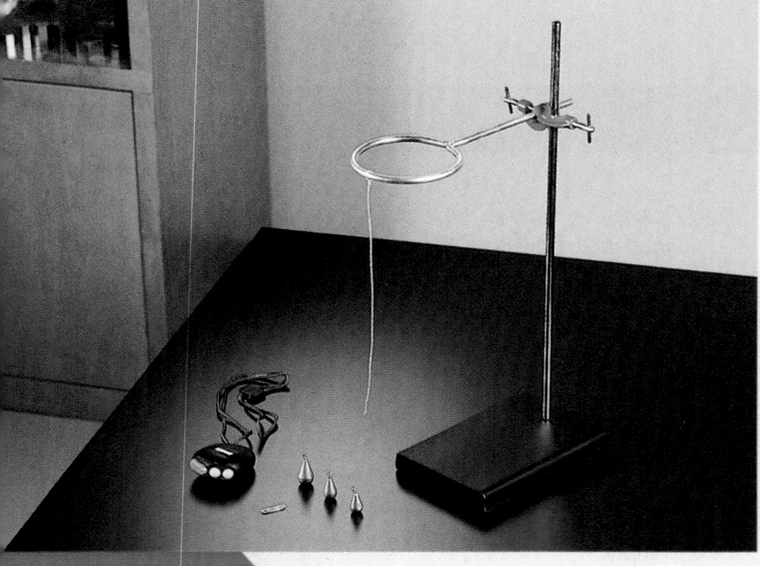

Procedure

1. Design a pendulum using a ring stand, a string with a paper clip, and a sinker attached to the paper clip. Be sure to check with your teacher and have your design approved before you proceed with the lab.
2. For this investigation, the length of the pendulum is the length of the string plus half the length of the bob. The amplitude is how far the bob is pulled from its equilibrium point. The frequency is the cycles/s of the bob. The period is the time for the bob to travel back and forth (one cycle). When collecting data for the period, find the time it takes to make ten cycles, and then calculate the period in s/cycles. When finding frequency, count how many cycles occur in 10 s, and then convert your value to cycles/s.
3. Design a procedure that keeps the mass of the bob and the amplitude constant, but varies the length. Determine the frequency and period of the pendulum. Record your results in the data table. Use several trials at several lengths to collect your data.
4. Design a procedure that keeps length and amplitude constant, but varies the mass of the bob. Determine the frequency and period of the pendulum. Record your results in the data table. Use several trials to collect your data.
5. Design a procedure that keeps length and mass of the bob constant, but varies the amplitude of the pendulum. Determine the frequency and period of the pendulum. Record your results in the data table. Use several trials to collect your data.
6. Design a procedure using the pendulum to calculate *g*, the acceleration due to gravity, using the equation $T = 2\pi\sqrt{\ell/g}$. T is the period, and ℓ is the length of the pendulum string. Remember to use several trials to collect your data.

Data Table 1

This data table format can be used for steps 2–5.

	Trial 1	Trial 2	Trial 3	Average	Period (s/cycle)	Frequency (cycles/s)
Length 1					——	
Length 2						——
Length 3						
Mass 1					——	
Mass 2						——
Mass 3						
Amplitude 1					——	
Amplitude 2						——
Amplitude 3						

Data Table 2

This data table format can be used for step 6, finding *g*.

	Trial 1	Trial 2	Trial 3	Average	Period (s/cycle)	Length of String (m)
Length 1						
Length 2						
Length 3						

Analyze

1. **Summarize** What is the relationship between the pendulum's amplitude and its period?
2. **Summarize** What is the relationship between the pendulum's bob mass and its period?
3. **Compare and Contrast** How are the period and length of a pendulum related?
4. **Analyze** Calculate g from your data in step 6.
5. **Error Analysis** What is the percent error of your experimental g value? What are some possible reasons for the difference between your experimental value of g and the accepted value of g?

Conclude and Apply

1. **Infer** What variable(s) affects a pendulum's period?
2. **Analyze** Why is it better to run three or more trials to obtain the frequency and period of each pendulum?
3. **Compare** How is the motion of a pendulum like that of a wave?
4. **Analyze and Conclude** When does the pendulum bob have the greatest kinetic energy?
5. **Analyze and Conclude** When does the pendulum bob have the greatest potential energy?

Going Further

Suppose you had a very long pendulum. What other observations could be made, over the period of a day, of this pendulum's motion?

Real-World Physics

Pendulums are used to drive some types of clocks. Using the observations from your experiments, what design problems are there in using your pendulum as a time-keeping instrument?

To find out more about the behavior of waves, visit the Web site: physicspp.com

Technology and Society

Earthquake Protection

An earthquake is the equivalent of a violent explosion somewhere beneath the surface of Earth. The mechanical waves that radiate from an earthquake are both transverse and longitudinal. Transverse waves shake a structure horizontally, while longitudinal waves cause vertical shaking. Earthquakes cannot be predicted or prevented, so we must construct our buildings to withstand them.

As our knowledge of earthquakes increases, existing buildings must be retrofitted to withstand newly discovered types of earthquake-related failures.

Reducing Damage Most bridges and parking ramps were built by stacking steel-reinforced concrete sections atop one another. Gravity keeps them in place. These structures are immensely strong under normal conditions, but they can be shaken apart by a strong earthquake. New construction codes dictate that their parts must be bonded together by heavy steel straps.

Earthquake damage to buildings also can be reduced by allowing a small amount of controlled movement between the building frame and its foundation. To minimize vertical shaking of a building, springs are inserted into the vertical members of the framework. These springs are made of a strong rubber compound compressed within heavy structural steel cylinders. Sideways shaking is diminished by placing sliding supports beneath the building columns. These allow the structure to remain stationary if the ground beneath it moves sideways.

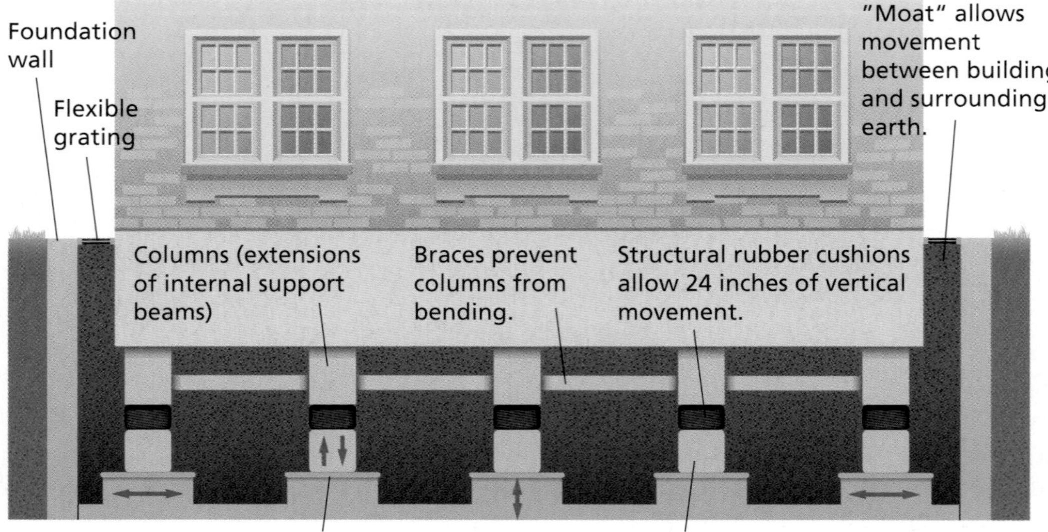
Foundation wall
Flexible grating
"Moat" allows movement between building and surrounding earth.
Columns (extensions of internal support beams)
Braces prevent columns from bending.
Structural rubber cushions allow 24 inches of vertical movement.
Special pads support the building, yet allow sliding if earth moves horizontally.
Column ends can slide 22 inches in any direction on smooth support pads.

New building designs reduce damage by earthquakes.

Long structures, like tunnels and bridges, must be constructed to survive vertical or horizontal shearing fractures of the earth beneath. The Bay Area Rapid Transit tunnel that runs beneath San Francisco Bay has flexible couplings for stability should the bay floor buckle.

Going Further

1. **Research** What is the framework of your school made of and how were the foundations built?
2. **Observe** Find a brick building that has a crack in one of its walls. See if you can tell why the crack formed and why it took the path that it did. What might this have to do with earthquakes?

Chapter 14

Study Guide

14.1 Periodic Motion

Vocabulary

- periodic motion (p. 375)
- simple harmonic motion (p. 375)
- period (p. 375)
- amplitude (p. 375)
- Hooke's law (p. 376)
- pendulum (p. 378)
- resonance (p. 380)

Key Concepts

- Periodic motion is any motion that repeats in a regular cycle.
- Simple harmonic motion results when the restoring force on an object is directly proportional to the object's displacement from equilibrium. Such a force obeys Hooke's law.

$$F = -kx$$

- The elastic potential energy stored in a spring that obeys Hooke's law is expressed by the following equation.

$$PE_{sp} = \frac{1}{2}kx^2$$

- The period of a pendulum can be found with the following equation.

$$T = 2\pi\sqrt{\frac{l}{g}}$$

14.2 Wave Properties

Vocabulary

- wave (p. 381)
- wave pulse (p. 381)
- periodic wave (p. 381)
- transverse wave (p. 381)
- longitudinal wave (p. 381)
- surface wave (p. 382)
- trough (p. 383)
- crest (p. 383)
- wavelength (p. 383)
- frequency (p. 384)

Key Concepts

- Waves transfer energy without transferring matter.
- In transverse waves, the displacement of the medium is perpendicular to the direction of wave motion. In longitudinal waves, the displacement is parallel to the direction of wave motion.
- Frequency is the number of cycles per second and is related to period by:

$$f = \frac{1}{T}$$

- The wavelength of a continuous wave can be found by using the following equation.

$$\lambda = \frac{v}{f}$$

14.3 Wave Behavior

Vocabulary

- incident wave (p. 387)
- reflected wave (p. 387)
- principle of superposition (p. 388)
- interference (p. 388)
- node (p. 389)
- antinode (p. 389)
- standing wave (p. 389)
- wave front (p. 390)
- ray (p. 390)
- normal (p. 391)
- law of reflection (p. 391)
- refraction (p. 391)

Key Concepts

- When a wave crosses a boundary between two media, it is partially transmitted and partially reflected.
- The principle of superposition states that the displacement of a medium resulting from two or more waves is the algebraic sum of the displacements of the individual waves.
- Interference occurs when two or more waves move through a medium at the same time.
- When two-dimensional waves are reflected from boundaries, the angles of incidence and reflection are equal.
- The change in direction of waves at the boundary between two different media is called refraction.

Chapter 14 Assessment

Concept Mapping

31. Complete the concept map using the following terms and symbols: *amplitude, frequency, v, λ, T.*

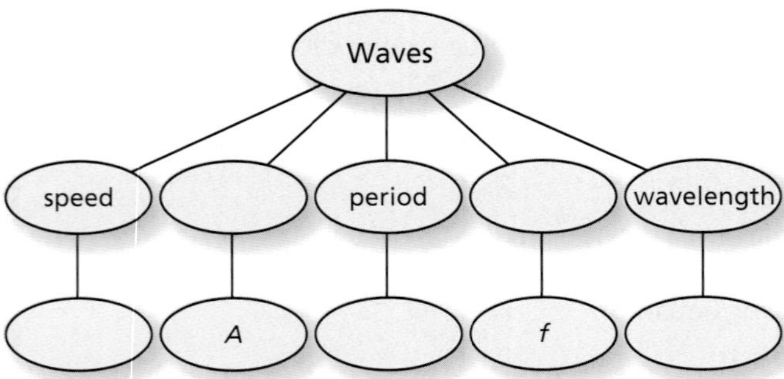

Mastering Concepts

32. What is periodic motion? Give three examples of periodic motion. (14.1)

33. What is the difference between frequency and period? How are they related? (14.1)

34. What is simple harmonic motion? Give an example of simple harmonic motion. (14.1)

35. If a spring obeys Hooke's law, how does it behave? (14.1)

36. How can the spring constant of a spring be determined from a graph of force versus displacement? (14.1)

37. How can the potential energy in a spring be determined from the graph of force versus displacement? (14.1)

38. Does the period of a pendulum depend on the mass of the bob? The length of the string? Upon what else does the period depend? (14.1)

39. What conditions are necessary for resonance to occur? (14.1)

40. How many general methods of energy transfer are there? Give two examples of each. (14.2)

41. What is the primary difference between a mechanical wave and an electromagnetic wave? (14.2)

42. What are the differences among transverse, longitudinal, and surface waves? (14.2)

43. Waves are sent along a spring of fixed length. (14.2)

a. Can the speed of the waves in the spring be changed? Explain.

b. Can the frequency of a wave in the spring be changed? Explain.

44. What is the wavelength of a wave? (14.2)

45. Suppose you send a pulse along a rope. How does the position of a point on the rope before the pulse arrives compare to the point's position after the pulse has passed? (14.2)

46. What is the difference between a wave pulse and a periodic wave? (14.2)

47. Describe the difference between wave frequency and wave velocity. (14.2)

48. Suppose you produce a transverse wave by shaking one end of a spring from side to side. How does the frequency of your hand compare with the frequency of the wave? (14.2)

49. When are points on a wave in phase with each other? When are they out of phase? Give an example of each. (14.2)

50. What is the amplitude of a wave and what does it represent? (14.2)

51. Describe the relationship between the amplitude of a wave and the energy it carries. (14.2)

52. When a wave reaches the boundary of a new medium, what happens to it? (14.3)

53. When a wave crosses a boundary between a thin and a thick rope, as shown in **Figure 14-18,** its wavelength and speed change, but its frequency does not. Explain why the frequency is constant. (14.3)

Figure 14-18

54. How does a spring pulse reflected from a rigid wall differ from the incident pulse? (14.3)

55. Describe interference. Is interference a property of only some types of waves or all types of waves? (14.3)

56. What happens to a spring at the nodes of a standing wave? (14.3)

57. **Violins** A metal plate is held fixed in the center and sprinkled with sugar. With a violin bow, the plate is stroked along one edge and made to vibrate. The sugar begins to collect in certain areas and move away from others. Describe these regions in terms of standing waves. (14.3)

58. If a string is vibrating in four parts, there are points where it can be touched without disturbing its motion. Explain. How many of these points exist? (14.3)

59. Wave fronts pass at an angle from one medium into a second medium, where they travel with a different speed. Describe two changes in the wave fronts. What does not change? (14.3)

 For more problems, go to Additional Problems, Appendix B.

Applying Concepts

60. A ball bounces up and down on the end of a spring. Describe the energy changes that take place during one complete cycle. Does the total mechanical energy change?

61. Can a pendulum clock be used in the orbiting *International Space Station?* Explain.

62. Suppose you hold a 1-m metal bar in your hand and hit its end with a hammer, first, in a direction parallel to its length, and second, in a direction at right angles to its length. Describe the waves produced in the two cases.

63. Suppose you repeatedly dip your finger into a sink full of water to make circular waves. What happens to the wavelength as you move your finger faster?

64. What happens to the period of a wave as the frequency increases?

65. What happens to the wavelength of a wave as the frequency increases?

66. Suppose you make a single pulse on a stretched spring. How much energy is required to make a pulse with twice the amplitude?

67. You can make water slosh back and forth in a shallow pan only if you shake the pan with the correct frequency. Explain.

68. In each of the four waves in **Figure 14-19,** the pulse on the left is the original pulse moving toward the right. The center pulse is a reflected pulse; the pulse on the right is a transmitted pulse. Describe the rigidity of the boundaries at A, B, C, and D.

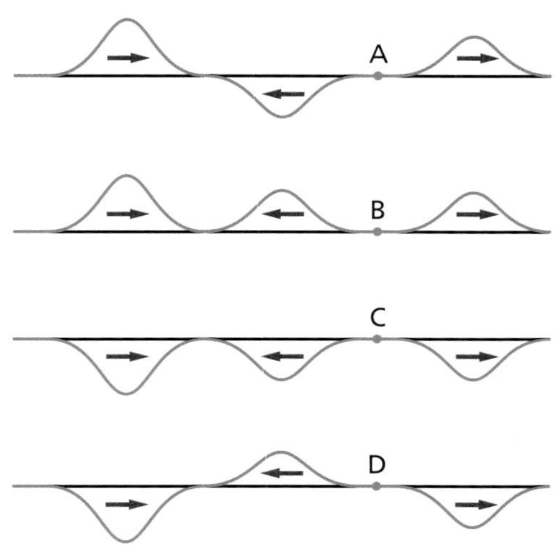

Figure 14-19

Mastering Problems

14.1 Periodic Motion

69. A spring stretches by 0.12 m when some apples weighing 3.2 N are suspended from it, as shown in **Figure 14-20.** What is the spring constant of the spring?

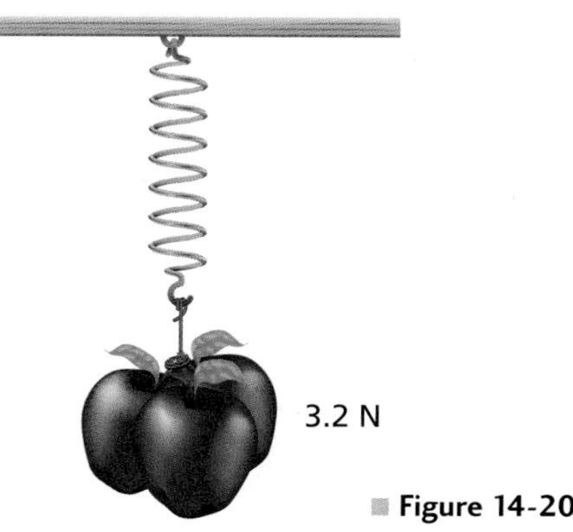

Figure 14-20

70. Car Shocks Each of the coil springs of a car has a spring constant of 25,000 N/m. How much is each spring compressed if it supports one-fourth of the car's 12,000-N weight?

71. How much potential energy is stored in a spring with a spring constant of 27 N/m if it is stretched by 16 cm?

72. Rocket Launcher A toy rocket-launcher contains a spring with a spring constant of 35 N/m. How far must the spring be compressed to store 1.5 J of energy?

73. Force-versus-length data for a spring are plotted on the graph in **Figure 14-21.**

a. What is the spring constant of the spring?

b. What is the energy stored in the spring when it is stretched to a length of 0.50 m?

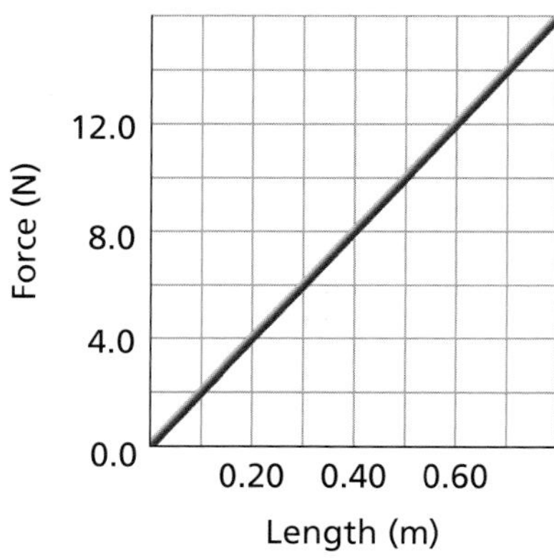

Figure 14-21

74. How long must a pendulum be to have a period of 2.3 s on the Moon, where $g = 1.6\ \text{m/s}^2$?

14.2 Wave Properties

75. **Building Motion** The Sears Tower in Chicago, shown in **Figure 14-22,** sways back and forth in the wind with a frequency of about 0.12 Hz. What is its period of vibration?

Figure 14-22

76. **Ocean Waves** An ocean wave has a length of 12.0 m. A wave passes a fixed location every 3.0 s. What is the speed of the wave?

77. Water waves in a shallow dish are 6.0-cm long. At one point, the water moves up and down at a rate of 4.8 oscillations/s.
   a. What is the speed of the water waves?
   b. What is the period of the water waves?

78. Water waves in a lake travel 3.4 m in 1.8 s. The period of oscillation is 1.1 s.
   a. What is the speed of the water waves?
   b. What is their wavelength?

79. **Sonar** A sonar signal of frequency 1.00×10^6 Hz has a wavelength of 1.50 mm in water.
   a. What is the speed of the signal in water?
   b. What is its period in water?
   c. What is its period in air?

80. A sound wave of wavelength 0.60 m and a velocity of 330 m/s is produced for 0.50 s.
   a. What is the frequency of the wave?
   b. How many complete waves are emitted in this time interval?
   c. After 0.50 s, how far is the front of the wave from the source of the sound?

81. The speed of sound in water is 1498 m/s. A sonar signal is sent straight down from a ship at a point just below the water surface, and 1.80 s later, the reflected signal is detected. How deep is the water?

82. Pepe and Alfredo are resting on an offshore raft after a swim. They estimate that 3.0 m separates a trough and an adjacent crest of each surface wave on the lake. They count 12 crests that pass by the raft in 20.0 s. Calculate how fast the waves are moving.

83. **Earthquakes** The velocity of the transverse waves produced by an earthquake is 8.9 km/s, and that of the longitudinal waves is 5.1 km/s. A seismograph records the arrival of the transverse waves 68 s before the arrival of the longitudinal waves. How far away is the earthquake?

14.3 Wave Behavior

84. Sketch the result for each of the three cases shown in **Figure 14-23,** when the centers of the two approaching wave pulses lie on the dashed line so that the pulses exactly overlap.

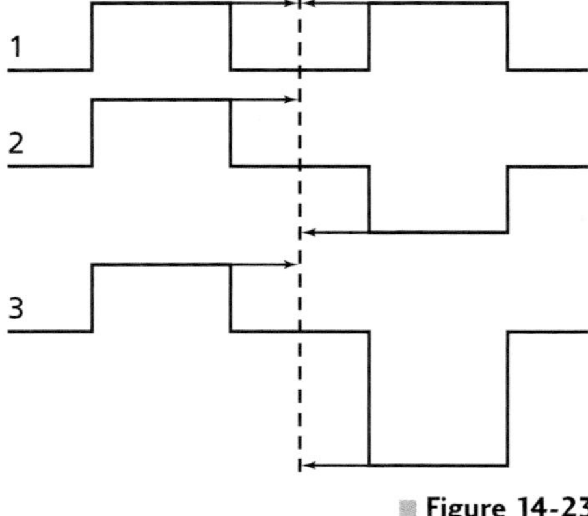

Figure 14-23

85. If you slosh the water in a bathtub at the correct frequency, the water rises first at one end and then at the other. Suppose you can make a standing wave in a 150-cm-long tub with a frequency of 0.30 Hz. What is the velocity of the water wave?

86. **Guitars** The wave speed in a guitar string is 265 m/s. The length of the string is 63 cm. You pluck the center of the string by pulling it up and letting go. Pulses move in both directions and are reflected off the ends of the string.
   a. How long does it take for the pulse to move to the string end and return to the center?
   b. When the pulses return, is the string above or below its resting location?
   c. If you plucked the string 15 cm from one end of the string, where would the two pulses meet?

 For more problems, go to Additional Problems, Appendix B.

87. Sketch the result for each of the four cases shown in **Figure 14-24,** when the centers of each of the two wave pulses lie on the dashed line so that the pulses exactly overlap.

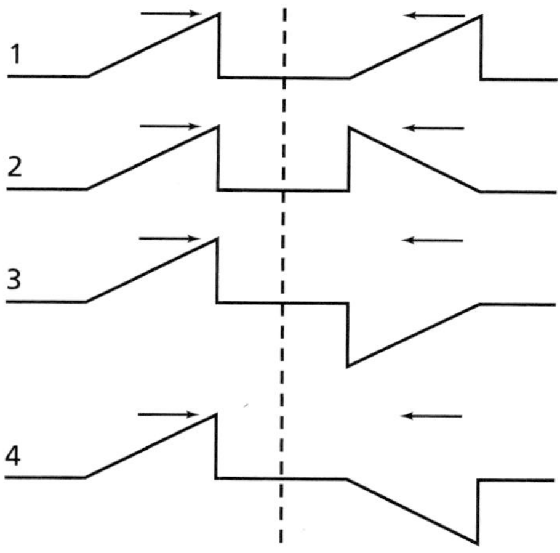

Figure 14-24

Mixed Review

88. What is the period of a pendulum with a length of 1.4 m?

89. The frequency of yellow light is 5.1×10^{14} Hz. Find the wavelength of yellow light. The speed of light is 3.00×10^{8} m/s.

90. Radio Wave AM-radio signals are broadcast at frequencies between 550 kHz (kilohertz) and 1600 kHz and travel 3.0×10^{8} m/s.

a. What is the range of wavelengths for these signals?

b. FM frequencies range between 88 MHz (megahertz) and 108 MHz and travel at the same speed. What is the range of FM wavelengths?

91. You are floating just offshore at the beach. Even though the waves are steadily moving in toward the beach, you don't move any closer to the beach.

a. What type of wave are you experiencing as you float in the water?

b. Explain why the energy in the wave does not move you closer to shore.

c. In the course of 15 s you count ten waves that pass you. What is the period of the waves?

d. What is the frequency of the waves?

e. You estimate that the wave crests are 3 m apart. What is the velocity of the waves?

f. After returning to the beach, you learn that the waves are moving at 1.8 m/s. What is the actual wavelength of the waves?

92. Bungee Jumper A high-altitude bungee jumper jumps from a hot-air balloon using a 540-m-bungee cord. When the jump is complete and the jumper is just suspended from the cord, it is stretched 1710 m. What is the spring constant of the bungee cord if the jumper has a mass of 68 kg?

93. The time needed for a water wave to change from the equilibrium level to the crest is 0.18 s.

a. What fraction of a wavelength is this?

b. What is the period of the wave?

c. What is the frequency of the wave?

94. When a 225-g mass is hung from a spring, the spring stretches 9.4 cm. The spring and mass then are pulled 8.0 cm from this new equilibrium position and released. Find the spring constant of the spring and the maximum speed of the mass.

95. Amusement Ride You notice that your favorite amusement-park ride seems bigger. The ride consists of a carriage that is attached to a structure so it swings like a pendulum. You remember that the carriage used to swing from one position to another and back again eight times in exactly 1 min. Now it only swings six times in 1 min. Give your answers to the following questions to two significant digits.

a. What was the original period of the ride?

b. What is the new period of the ride?

c. What is the new frequency?

d. How much longer is the arm supporting the carriage on the larger ride?

e. If the park owners wanted to double the period of the ride, what percentage increase would need to be made to the length of the pendulum?

96. Clocks The speed at which a grandfather clock runs is controlled by a swinging pendulum.

a. If you find that the clock loses time each day, what adjustment would you need to make to the pendulum so it will keep better time?

b. If the pendulum currently is 15.0 cm, by how much would you need to change the length to make the period lessen by 0.0400 s?

97. Bridge Swinging In the summer over the New River in West Virginia, several teens swing from bridges with ropes, then drop into the river after a few swings back and forth.

a. If Pam is using a 10.0-m length of rope, how long will it take her to reach the peak of her swing at the other end of the bridge?

b. If Mike has a mass that is 20 kg more than Pam, how would you expect the period of his swing to differ from Pam's?

c. At what point in the swing is *KE* at a maximum?

d. At what point in the swing is *PE* at a maximum?

e. At what point in the swing is *KE* at a minimum?

f. At what point in the swing is *PE* at a minimum?

98. You have a mechanical fish scale that is made with a spring that compresses when weight is added to a hook attached below the scale. Unfortunately, the calibrations have completely worn off of the scale. However, you have one known mass of 500.0 g that displaces the spring 2.0 cm.
   a. What is the spring constant for the spring?
   b. If a fish displaces the spring 4.5 cm, what is the mass of the fish?

99. **Car Springs** When you add a 45-kg load to the trunk of a new small car, the two rear springs compress an additional 1.0 cm.
   a. What is the spring constant for each of the springs?
   b. How much additional potential energy is stored in each of the car springs after loading the trunk?

100. The velocity of a wave on a string depends on how tightly the string is stretched, and on the mass per unit length of the string. If F_T is the tension in the string, and μ is the mass/unit length, then the velocity, v, can be determined by the following equation.

$$v = \sqrt{\frac{F_T}{\mu}}$$

A piece of string 5.30-m long has a mass of 15.0 g. What must the tension in the string be to make the wavelength of a 125-Hz wave 120.0 cm?

Thinking Critically

101. **Analyze and Conclude** A 20-N force is required to stretch a spring by 0.5 m.
   a. What is the spring constant?
   b. How much energy is stored in the spring?
   c. Why isn't the work done to stretch the spring equal to the force times the distance, or 10 J?

102. **Make and Use Graphs** Several weights were suspended from a spring, and the resulting extensions of the spring were measured. **Table 14-1** shows the collected data.

Table 14-1
Weights on a Spring

Force, F (N)	Extension, x (m)
2.5	0.12
5.0	0.26
7.5	0.35
10.0	0.50
12.5	0.60
15.0	0.71

   a. Make a graph of the force applied to the spring versus the spring length. Plot the force on the y-axis.
   b. Determine the spring constant from the graph.
   c. Using the graph, find the elastic potential energy stored in the spring when it is stretched to 0.50 m.

103. **Apply Concepts** Gravel roads often develop regularly spaced ridges that are perpendicular to the road, as shown in **Figure 14-25.** This effect, called washboarding, occurs because most cars travel at about the same speed and the springs that connect the wheels to the cars oscillate at about the same frequency. If the ridges on a road are 1.5 m apart and cars travel on it at about 5 m/s, what is the frequency of the springs' oscillation?

■ **Figure 14-25**

Writing in Physics

104. **Research** Christiaan Huygens' work on waves and the controversy between him and Newton over the nature of light. Compare and contrast their explanations of such phenomena as reflection and refraction. Whose model would you choose as the best explanation? Explain why.

Cumulative Review

105. A 1400-kg drag racer automobile can complete a one-quarter mile (402 m) course in 9.8 s. The final speed of the automobile is 250 mi/h (112 m/s). (Chapter 11)
   a. What is the kinetic energy of the automobile?
   b. What is the minimum amount of work that was done by its engine? Why can't you calculate the total amount of work done?
   c. What was the average acceleration of the automobile?

106. How much water would a steam engine have to evaporate in 1 s to produce 1 kW of power? Assume that the engine is 20 percent efficient. (Chapter 12)

 For more problems, go to Additional Problems, Appendix B.

Stan

Multiple Choice

1. What is the value of the spring constant of a spring with a potential energy of 8.67 J when it's stretched 247 mm?

   (A) 70.2 N/m (C) 142 N/m
   (B) 71.1 N/m (D) 284 N/m

2. What is the force acting on a spring with a spring constant of 275 N/m that is stretched 14.3 cm?

   (A) 2.81 N (C) 39.3 N
   (B) 19.2 N (D) 3.93×10^{30} N

3. A mass stretches a spring as it hangs from the spring. What is the spring constant?

   (A) 0.25 N/m (C) 26 N/m
   (B) 0.35 N/m (D) 3.5×10^{2} N/m

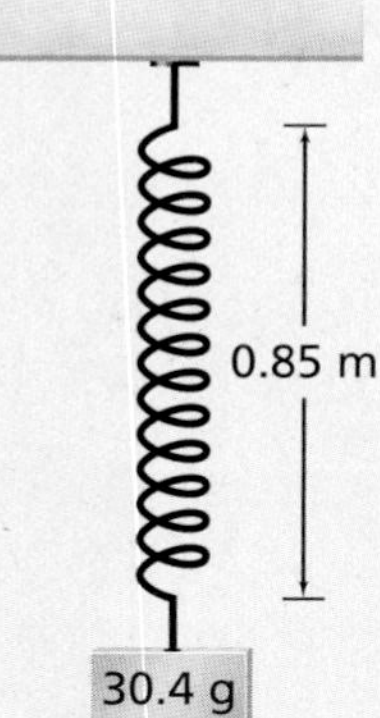

4. A spring with a spring constant of 350 N/m pulls a door closed. How much work is done as the spring pulls the door at a constant velocity from an 85.0-cm stretch to a 5.0-cm stretch?

   (A) 112 N·m (C) 224 N·m
   (B) 130 J (D) 1.12×10^{3} J

5. What is the correct rearrangement of the formula for the period of a pendulum to find the length of the pendulum?

   (A) $l = \frac{4\pi^2 g}{T^2}$ (C) $l = \frac{T^2 g}{(2\pi)^2}$
   (B) $l = \frac{gT}{4\pi^2}$ (D) $l = \frac{Tg}{2\pi}$

6. What is the frequency of a wave with a period of 3 s?

   (A) 0.3 Hz (C) $\frac{\pi}{3}$ Hz
   (B) $\frac{3}{c}$ Hz (D) 3 Hz

7. Which option describes a standing wave?

	Waves	Direction	Medium
(A)	Identical	Same	Same
(B)	Nonidentical	Opposite	Different
(C)	Identical	Opposite	Same
(D)	Nonidentical	Same	Different

8. A 1.2-m wave travels 11.2 m to a wall and back again in 4 s. What is the frequency of the wave?

   (A) 0.2 Hz (C) 5 Hz
   (B) 2 Hz (D) 9 Hz

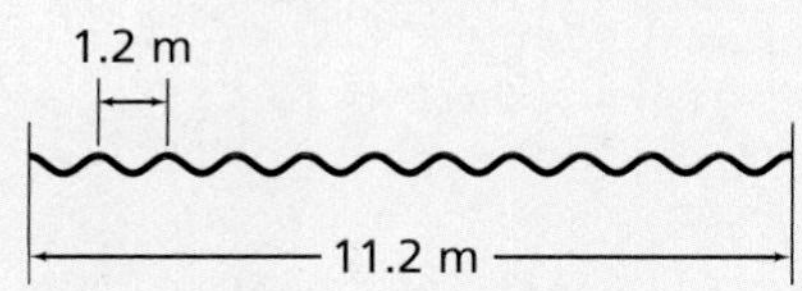

9. What is the length of a pendulum that has a period of 4.89 s?

   (A) 5.94 m (C) 24.0 m
   (B) 11.9 m (D) 37.3 m

Extended Answer

10. Use dimensional analysis of the equation $kx = mg$ to derive the units of k.

✓ Test-Taking TIP

Practice, Practice, Practice

Practice to improve your performance on standardized tests. Don't compare yourself to anyone else.

Chapter 15 Sound

What You'll Learn

- You will describe sound in terms of wave properties and behavior.
- You will examine some of the sources of sound.
- You will explain properties that differentiate between music and noise.

Why It's Important

Sound is an important means of communication and, in the form of music, cultural expression.

Musical Groups A small musical group might contain two or three instruments, while a marching band can contain 100 or more. The instruments in these groups form sounds in different ways, but they can create exciting music when they are played together.

Think About This ▶

How do the instruments in a musical group create the sounds that you hear? Why do various instruments sound different even when they play the same note?

physicspp.com

How can glasses produce musical notes?

Question

How can you use glasses to produce different musical notes, and how do glasses with stems compare to those without stems?

Procedure

1. Select a stemmed glass with a thin rim.
2. **Prepare** Carefully inspect the top edge of the glass for sharp edges. Notify your teacher if you observe any sharp edges. Be sure to repeat this inspection every time you select a different glass.
3. Place the glass on the table in front of you. Firmly hold the base with one hand. Wet your finger and slowly rub it around the top edge of the glass. ***CAUTION: Glass is fragile. Handle carefully.***
4. Record your observations. Increase or decrease the speed a little. What happens?
5. Select a stemmed glass that is larger or smaller than the first glass. Repeat steps 2-4.
6. Select a glass without a stem and repeat steps 2–4.

Analysis

Summarize your observations. Which glasses—stemmed, not stemmed, or both—were able to produce ringing tones? What factors affected the tones produced?

Critical Thinking

Propose a method for producing different notes from the same glass. Test your proposed method. Suggest a test to further investigate the properties of glasses that can produce ringing tones.

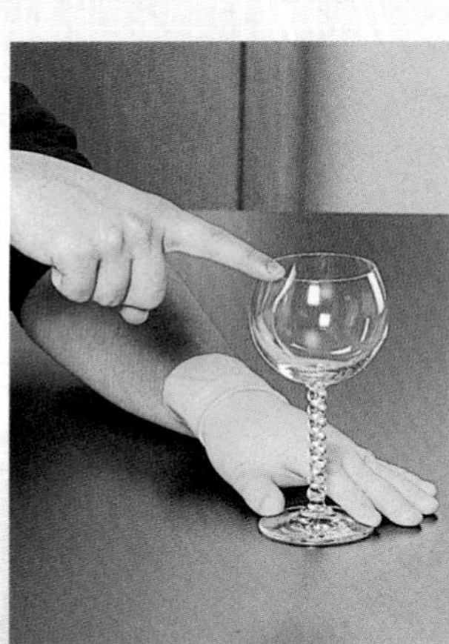

15.1 Properties and Detection of Sound

Objectives

- **Demonstrate** the properties that sound shares with other waves.
- **Relate** the physical properties of sound waves to our perception of sound.
- **Identify** some applications of the Doppler effect.

Vocabulary

sound wave
pitch
loudness
sound level
decibel
Doppler effect

Sound is an important part of existence for many living things. Animals can use sound to hunt, attract mates, and warn of the approach of predators. In humans, the sound of a siren can heighten our awareness of our surroundings, while the sound of music can soothe and relax us. From your everyday experiences, you already are familiar with several of the characteristics of sound, including volume, tone, and pitch. Without thinking about it, you can use these, and other characteristics, to categorize many of the sounds that you hear; for example, some sound patterns are characteristic of speech, while others are characteristic of a musical group. In this chapter, you will study the physical principles of sound, which is a type of wave.

In Chapter 14, you learned how to describe waves in terms of speed, frequency, wavelength, and amplitude. You also discovered how waves interact with each other and with matter. Knowing that sound is a type of wave allows you to describe some of its properties and interactions. First, however, there is a question that you need to answer: exactly what type of wave is sound?

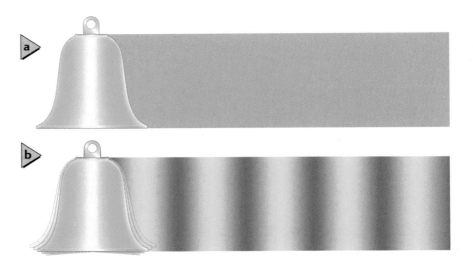

■ **Figure 15-1** Before the bell is struck, the air around it is a region of average pressure **(a).** Once the bell is struck, however, the vibrating edge creates regions of high and low pressure. The dark areas represent regions of higher pressure; the light areas represent regions of lower pressure **(b).** For simplicity, the diagram shows the regions moving in one direction; in reality, the waves move out from the bell in all directions.

Sound Waves

Put your fingers against your throat as you hum or speak. Can you feel the vibrations? Have you ever put your hand on the loudspeaker of a boom box? **Figure 15-1** shows a vibrating bell that also can represent your vocal cords, a loudspeaker, or any other sound source. As it moves back and forth, the edge of the bell strikes the particles in the air. When the edge moves forward, air particles are driven forward; that is, the air particles bounce off the bell with a greater velocity. When the edge moves backward, air particles bounce off the bell with a lower velocity.

The result of these velocity changes is that the forward motion of the bell produces a region where the air pressure is slightly higher than average. The backward motion produces slightly below-average pressure. Collisions among the air particles cause the pressure variations to move away from the bell in all directions. If you were to focus at one spot, you would see the value of the air pressure rise and fall, not unlike the behavior of a pendulum. In this way, the pressure variations are transmitted through matter.

Describing sound A pressure variation that is transmitted through matter is a **sound wave.** Sound waves move through air because a vibrating source produces regular variations, or oscillations, in air pressure. The air particles collide, transmitting the pressure variations away from the source of the sound. The pressure of the air oscillates about the mean air pressure, as shown in **Figure 15-2.** The frequency of the wave is the number of oscillations in pressure each second. The wavelength is the distance between successive regions of high or low pressure. Because the motion of the particles in air is parallel to the direction of the wave's motion, sound is a longitudinal wave.

The speed of sound in air depends on the temperature, with the speed increasing by about 0.6 m/s for each 1°C increase in air temperature. At room temperature (20°C), sound moves through air at sea level at a speed of 343 m/s. Sound also travels through solids and liquids. In general, the speed of sound is greater in

■ **Figure 15-2** A coiled spring models the compressions and rarefactions of a sound wave **(a).** The pressure of the air rises and falls as the sound wave propagates through the atmosphere **(b).** You can use a sine curve alone to model changes in pressure. Note that the positions of *x*, *y*, and *z* show that the wave, not matter, moves forward **(c).**

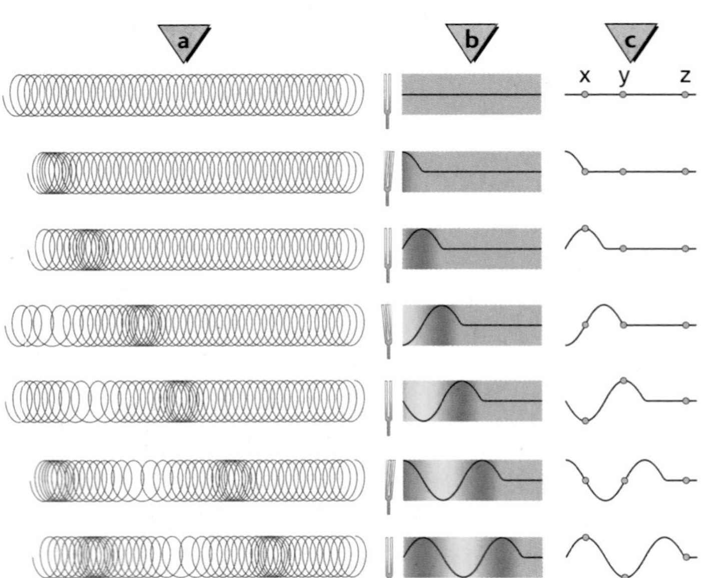

solids and liquids than in gases. **Table 15-1** lists the speeds of sound waves in various media. Sound cannot travel in a vacuum because there are no particles to collide.

Sound waves share the general properties of other waves. For example, they reflect off hard objects, such as the walls of a room. Reflected sound waves are called echoes. The time required for an echo to return to the source of the sound can be used to find the distance between the source and the reflective object. This principle is used by bats, by some cameras, and by ships that employ sonar. Two sound waves can interfere, causing dead spots at nodes where little sound can be heard. As you learned in Chapter 14, the frequency and wavelength of a wave are related to the speed of the wave by the equation $\lambda = v/f$.

Table 15-1
Speed of Sound in Various Media

Medium	m/s
Air (0°)	331
Air (20°)	343
Helium (0°)	972
Water (25°)	1493
Seawater (25°)	1533
Copper (25°)	3560
Iron (25°)	5130

PRACTICE Problems

• Additional Problems, Appendix B
• Solutions to Selected Problems, Appendix C

1. Find the wavelength in air at 20°C of an 18-Hz sound wave, which is one of the lowest frequencies that is detectable by the human ear.
2. What is the wavelength of an 18-Hz sound wave in seawater at 25°C?
3. Find the frequency of a sound wave moving through iron at 25°C with a wavelength of 1.25 m.
4. If you shout across a canyon and hear the echo 0.80 s later, how wide is the canyon?
5. A 2280-Hz sound wave has a wavelength of 0.655 m in an unknown medium. Identify the medium.

Detection of Pressure Waves

Sound detectors convert sound energy—the kinetic energy of the vibrating air particles—into another form of energy. A common detector is a microphone, which converts sound waves into electrical energy. A microphone consists of a thin disk that vibrates in response to sound waves and produces an electrical signal. You will learn about this transformation process in Chapter 25, during your study of electricity and magnetism.

The human ear As shown in **Figure 15-3,** the human ear is a detector that receives pressure waves and converts them to electrical impulses. Sound waves entering the auditory canal cause vibrations of the tympanic membrane. Three tiny bones then transfer these vibrations to fluid in the cochlea. Tiny hairs lining the spiral-shaped cochlea detect certain frequencies in the vibrating fluid. These hairs stimulate nerve cells, which send impulses to the brain and produce the sensation of sound.

The ear detects sound waves over a wide range of frequencies and is sensitive to an enormous range of amplitudes. In addition, human hearing can distinguish many different qualities of sound. Knowledge of both physics and biology is required to understand the complexities of the ear. The interpretation of sounds by the brain is even more complex, and it is not totally understood.

Figure 15-3 The human ear is a complex sense organ that translates sound vibrations into nerve impulses that are sent to the brain for interpretation. The malleus, incus, and stapes are the three bones of the middle ear that sometimes are referred to as the hammer, anvil, and stirrup.

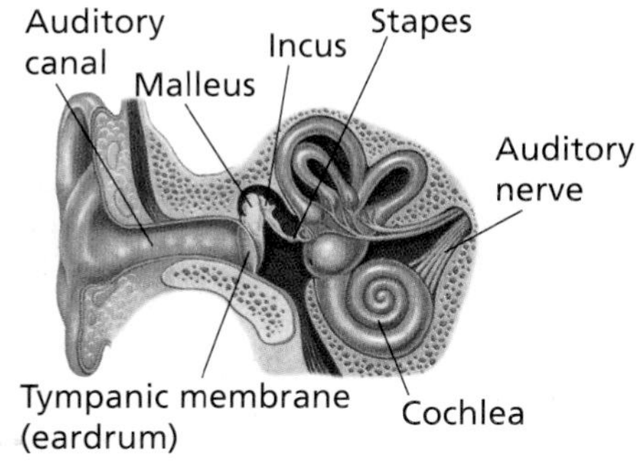

Perceiving Sound

Pitch Marin Mersenne and Galileo first determined that the **pitch** we hear depends on the frequency of vibration. Pitch can be given a name on the musical scale. For instance, the middle C note has a frequency of 262 Hz. The ear is not equally sensitive to all frequencies. Most people cannot hear sounds with frequencies below 20 Hz or above 16,000 Hz. Older people are less sensitive to frequencies above 10,000 Hz than are young people. By age 70, most people cannot hear sounds with frequencies above 8000 Hz. This loss affects the ability to understand speech.

Loudness Frequency and wavelength are two physical characteristics of sound waves. Another physical characteristic of sound waves is amplitude. Amplitude is the measure of the variation in pressure along a wave. In humans, sound is detected by the ear and interpreted by the brain. The **loudness** of a sound, as perceived by our sense of hearing, depends primarily on the amplitude of the pressure wave.

The human ear is extremely sensitive to pressure variations in sound waves, which is the amplitude of the wave. Recall from Chapter 13 that 1 atm of pressure equals 1.01×10^{5} Pa. The ear can detect pressure-wave amplitudes of less than one-billionth of an atmosphere, or 2×10^{-5} Pa. At the other end of the audible range, pressure variations of approximately 20 Pa or greater cause pain. It is important to remember that the ear detects only pressure variations at certain frequencies. Driving over a mountain pass changes the pressure on your ears by thousands of pascals, but this change does not take place at audible frequencies.

Because humans can detect a wide range in pressure variations, these amplitudes are measured on a logarithmic scale called the **sound level.** The unit of measurement for sound level is the **decibel** (dB). The sound level depends on the ratio of the pressure variation of a given sound wave to the pressure variation in the most faintly heard sound, 2×10^{-5} Pa. Such an amplitude has a sound level of 0 dB. A sound with a pressure amplitude ten times larger (2×10^{-4} Pa) is 20 dB. A pressure amplitude ten times larger than this is 40 dB. Most people perceive a 10-dB increase in sound level as about twice as loud as the original level. **Figure 15-4** shows the sound level for a variety of sounds. In addition to pressure variations, power and intensity of sound waves can be described by decibel scales.

Exposure to loud sounds, in the form of noise or music, has been shown to cause the ear to lose its sensitivity, especially to high frequencies. The longer a person is exposed to loud sounds, the greater the effect. A person can recover from short-term exposure in a period of hours, but the effects

Figure 15-4 This decibel scale shows the sound levels of some familiar sounds.

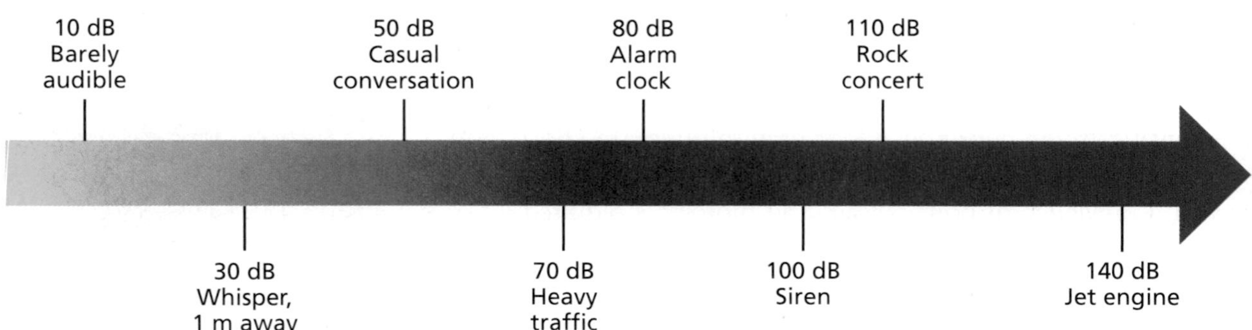

of long-term exposure can last for days or weeks. Long exposure to 100-dB or greater sound levels can produce permanent damage. Many rock musicians have suffered serious hearing loss, some as much as 40 percent. Hearing loss also can result from loud music being transmitted to stereo headphones from personal radios and CD players. In some cases, the listeners are unaware of just how high the sound levels really are. Cotton earplugs reduce the sound level only by about 10 dB. Special ear inserts can provide a 25-dB reduction. Specifically designed earmuffs and inserts as shown in **Figure 15-5,** can reduce the sound level by up to 45 dB.

Loudness, as perceived by the human ear, is not directly proportional to the pressure variations in a sound wave. The ear's sensitivity depends on both pitch and amplitude. Also, perception of pure tones is different from perception of a mixture of tones.

Figure 15-5 Continuous exposure to loud sounds can cause serious hearing loss. In many occupations, workers, such as this flight controller, must wear ear protection.

The Doppler Effect

Have you ever noticed that the pitch of an ambulance, fire, or police siren changed as the vehicle sped past you? The pitch was higher when the vehicle was moving toward you, then it dropped to a lower pitch as the source moved away. This frequency shift is called the **Doppler effect** and is shown in **Figure 15-6.** The sound source, S, is moving to the right with a speed of v_s. The waves that it emits spread in circles centered on the source at the time it produced the waves. As the source moves toward the sound detector, Observer A in **Figure 15-6a,** more waves are crowded into the space between them. The wavelength is shortened to λ_A. Because the speed of sound is not changed, more crests reach the ear per second, which means that the frequency of the detected sound increases. When the source is moving away from the detector, Observer B in Figure 15-6a, the wavelength is lengthened to λ_B and the detected frequency is lower. **Figure 15-6b** illustrates the Doppler effect for a moving source of sound on water waves in a ripple tank.

A Doppler shift also occurs if the detector is moving and the source is stationary. In this case, the Doppler shift results from the relative velocity of the sound waves and the detector. As the detector approaches the stationary source, the relative velocity is larger, resulting in an increase in the wave crests reaching the detector each second. As the detector recedes from the source, the relative velocity is smaller, resulting in a decrease in the wave crests reaching the detector each second.

Figure 15-6 As a sound-producing source moves toward an observer, the wavelength is shortened to λ_A; the wavelength is λ_B for waves produced by a source moving away from an observer **(a).** A moving wave-producing source illustrates the Doppler effect in a ripple tank **(b).**

Concepts In Motion
Interactive Figure To see an animation on the Doppler effect, visit physicspp.com.

a

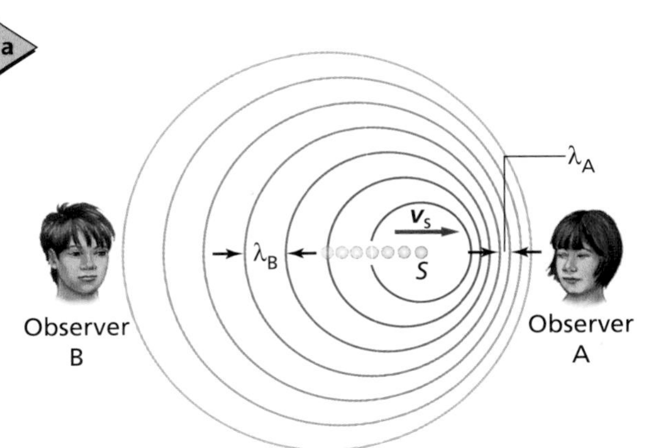

b

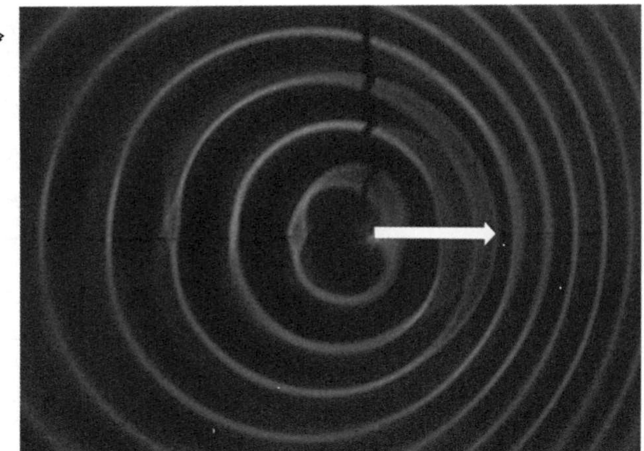

For both a moving source and a moving observer, the frequency that the observer hears can be calculated using the equation below.

Doppler Effect $f_d = f_s\left(\frac{v - v_d}{v - v_s}\right)$

The frequency perceived by a detector is equal to the velocity of the detector relative to the velocity of the wave, divided by the velocity of the source relative to the velocity of the wave, multiplied by the wave's frequency.

In the Doppler effect equation, v is the velocity of the sound wave, v_d is the velocity of the detector, v_s is the velocity of the sound's source, f_s is the frequency of the wave emitted by the source, and f_d is the frequency received by the detector. This equation applies when the source is moving, when the observer is moving, and when both are moving.

As you solve problems using the above equation, be sure to define the coordinate system so that the positive direction is from the source to the detector. The sound waves will be approaching the detector from the source, so the velocity of sound is always positive. Try drawing diagrams to confirm that the term $(v - v_d)/(v - v_s)$ behaves as you would predict based on what you have learned about the Doppler effect. Notice that for a source moving toward the detector (positive direction, which results in a smaller denominator compared to a stationary source) and for a detector moving toward the source (negative direction and increased numerator compared to a stationary detector), the detected frequency, f_d, increases. Similarily, if the source moves away from the detector or if the detector moves away from the source, then f_d decreases. Read the Connecting Math to Physics feature below to see how the Doppler effect equation reduces when the source or observer is stationary.

Personal Tutor For an online tutorial on reducing equations, visit physicspp.com.

▶ Connecting **Math to Physics**

Reducing Equations When an element in a complex equation is equal to zero, the equation might reduce to a form that is easier to use.

Stationary detector, source in motion: $v_d = 0$	Stationary source, detector in motion: $v_s = 0$
$f_d = f_s\left(\frac{v - v_d}{v - v_s}\right)$	$f_d = f_s\left(\frac{v - v_d}{v - v_s}\right)$
$= f_s\left(\frac{v}{v - v_s}\right)$	$= f_s\left(\frac{v - v_d}{v}\right)$
$= f_s\left(\frac{\frac{v}{v}}{\frac{v}{v} - \frac{v_s}{v}}\right)$	$= f_s\left(\frac{\frac{v}{v} - \frac{v_d}{v}}{\frac{v}{v}}\right)$
$= f_s\left(\frac{1}{1 - \frac{v_s}{v}}\right)$	$= f_s\left(\frac{1 - \frac{v_d}{v}}{1}\right)$
	$= f_s\left(1 - \frac{v_d}{v}\right)$

▶ EXAMPLE Problem 1

The Doppler Effect A trumpet player sounds C above middle C (524 Hz) while traveling in a convertible at 24.6 m/s. If the car is coming toward you, what frequency would you hear? Assume that the temperature is 20°C.

1 Analyze and Sketch the Problem

- Sketch the situation.
- Establish a coordinate axis. Make sure that the positive direction is from the source to the detector.
- Show the velocities of the source and detector.

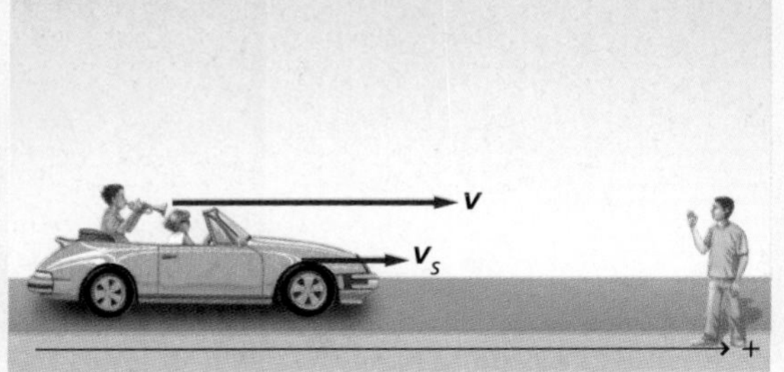

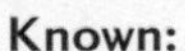

Known:	**Unknown:**
$v = +343$ m/s	$f_d = ?$
$v_s = +24.6$ m/s	
$v_d = 0$ m/s	
$f_s = 524$ Hz	

2 Solve for the Unknown

Use $f_d = f_s\left(\frac{v - v_d}{v - v_s}\right)$ with $v_d = 0$ m/s.

Math Handbook
Fractions
page 837

$$f_d = f_s\left(\frac{1}{1 - \frac{v_s}{v}}\right)$$

$$= 524\text{ Hz}\left(\frac{1}{1 - \frac{24.6\text{ m/s}}{343\text{ m/s}}}\right)$$ Substitute $v = +343$ m/s, $v_s = +24.6$ m/s, and $f_s = 524$ Hz

$$= 564\text{ Hz}$$

3 Evaluate the Answer

- **Are the units correct?** Frequency is measured in hertz.
- **Is the magnitude realistic?** The source is moving toward you, so the frequency should be increased.

▶ PRACTICE Problems

• Additional Problems, Appendix B
• Solutions to Selected Problems, Appendix C

6. Repeat Example Problem 1, but with the car moving away from you. What frequency would you hear?
7. You are in an auto traveling at 25.0 m/s toward a pole-mounted warning siren. If the siren's frequency is 365 Hz, what frequency do you hear? Use 343 m/s as the speed of sound.
8. You are in an auto traveling at 55 mph (24.6 m/s). A second auto is moving toward you at the same speed. Its horn is sounding at 475 Hz. What frequency do you hear? Use 343 m/s as the speed of sound.
9. A submarine is moving toward another submarine at 9.20 m/s. It emits a 3.50-MHz ultrasound. What frequency would the second sub, at rest, detect? The speed of sound in water is 1482 m/s.
10. A sound source plays middle C (262 Hz). How fast would the source have to go to raise the pitch to C sharp (271 Hz)? Use 343 m/s as the speed of sound.

Figure 15-7 In a process called echolocation, bats use the Doppler effect to locate prey.

The Doppler effect occurs in all wave motion, both mechanical and electromagnetic. It has many applications. Radar detectors use the Doppler effect to measure the speed of baseballs and automobiles. Astronomers observe light from distant galaxies and use the Doppler effect to measure their speeds and infer their distances. Physicians can detect the speed of the moving heart wall in a fetus by means of the Doppler effect in ultrasound. Bats use the Doppler effect to detect and catch flying insects. When an insect is flying faster than a bat, the reflected frequency is lower, but when the bat is catching up to the insect, as in **Figure 15-7,** the reflected frequency is higher. Not only do bats use sound waves to navigate and locate their prey, but they often must do so in the presence of other bats. This means they must discriminate their own calls and reflections against a background of many other sounds of many frequencies. Scientists continue to study bats and their amazing abilities to use sound waves.

Biology Connection

15.1 Section Review

11. **Graph** The eardrum moves back and forth in response to the pressure variations of a sound wave. Sketch a graph of the displacement of the eardrum versus time for two cycles of a 1.0-kHz tone and for two cycles of a 2.0-kHz tone.

12. **Effect of Medium** List two sound characteristics that are affected by the medium through which the sound passes and two characteristics that are not affected.

13. **Sound Properties** What physical characteristic of a sound wave should be changed to change the pitch of the sound? To change the loudness?

14. **Decibel Scale** How much greater is the sound pressure level of a typical rock band's music (110 dB) than a normal conversation (50 dB)?

15. **Early Detection** In the nineteenth century, people put their ears to a railroad track to get an early warning of an approaching train. Why did this work?

16. **Bats** A bat emits short pulses of high-frequency sound and detects the echoes.
    a. In what way would the echoes from large and small insects compare if they were the same distance from the bat?
    b. In what way would the echo from an insect flying toward the bat differ from that of an insect flying away from the bat?

17. **Critical Thinking** Can a trooper using a radar detector at the side of the road determine the speed of a car at the instant the car passes the trooper? Explain.

Physics Online physicspp.com/self_check_quiz

15.2 The Physics of Music

In the middle of the nineteenth century, German physicist Hermann Helmholtz studied sound production in musical instruments and the human voice. In the twentieth century, scientists and engineers developed electronic equipment that permits not only a detailed study of sound, but also the creation of electronic musical instruments and recording devices that allow us to listen to music whenever and wherever we wish.

▸ **Objectives**

- **Describe** the origin of sound.
- **Demonstrate** an understanding of resonance, especially as applied to air columns and strings.
- **Explain** why there are variations in sound among instruments and among voices.

▸ **Vocabulary**

closed-pipe resonator
open-pipe resonator
fundamental
harmonics
dissonance
consonance
beat

Sources of Sound

Sound is produced by a vibrating object. The vibrations of the object create particle motions that cause pressure oscillations in the air. A loudspeaker has a cone that is made to vibrate by electrical currents. The surface of the cone creates the sound waves that travel to your ear and allow you to hear music. Musical instruments such as gongs, cymbals, and drums are other examples of vibrating surfaces that are sources of sound.

The human voice is produced by vibrations of the vocal cords, which are two membranes located in the throat. Air from the lungs rushing through the throat starts the vocal cords vibrating. The frequency of vibration is controlled by the muscular tension placed on the vocal cords.

In brass instruments, such as the trumpet and tuba, the lips of the performer vibrate, as shown in **Figure 15-8a.** Reed instruments, such as the clarinet and saxophone, have a thin wooden strip, or reed, that vibrates as a result of air blown across it, as shown in **Figure 15-8b.** In flutes and organ pipes, air is forced across an opening in a pipe. Air moving past the opening sets the column of air in the instrument into vibration.

In stringed instruments, such as the piano, guitar, and violin, wires or strings are set into vibration. In the piano, the wires are struck; in the guitar, they are plucked; and in the violin, the friction of the bow causes the strings to vibrate. Often, the strings are attached to a sounding board that vibrates with the strings. The vibrations of the sounding board cause the pressure oscillations in the air that we hear as sound. Electric guitars use electronic devices to detect and amplify the vibrations of the guitar strings.

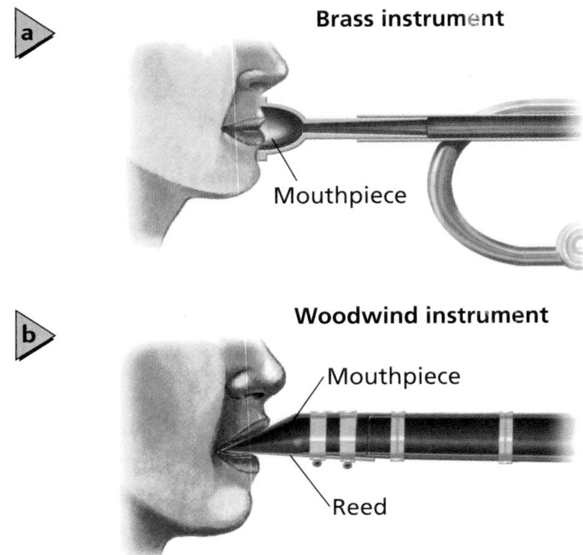

Figure 15-8 The shapes of the mouthpieces of a brass instrument **(a)** and a reed instrument **(b)** help determine the characteristics of the sound each instrument produces.

Resonance in Air Columns

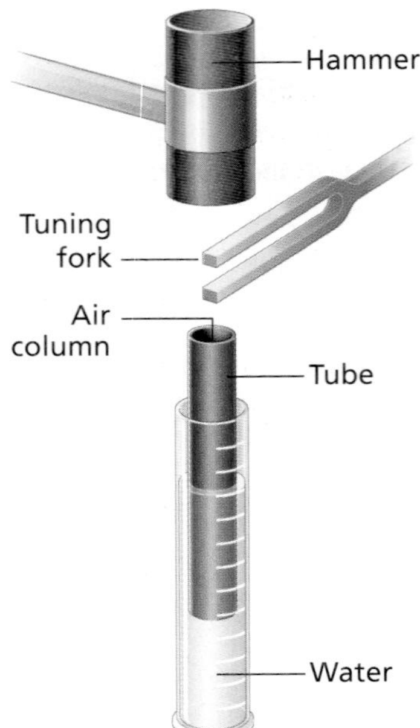

■ **Figure 15-9** Raising or lowering the tube changes the length of the air column. When the column is in resonance with the tuning fork, the sound is loudest.

If you have ever used just the mouthpiece of a brass or reed instrument, you know that the vibration of your lips or the reed alone does not make a sound with any particular pitch. The long tube that makes up the instrument must be attached if music is to result. When the instrument is played, the air within this tube vibrates at the same frequency, or in resonance, with a particular vibration of the lips or reed. Remember that resonance increases the amplitude of a vibration by repeatedly applying a small external force at the same natural frequency. The length of the air column determines the frequencies of the vibrating air that will be set into resonance. For many instruments, such as flutes, saxophones, and trombones, changing the length of the column of vibrating air varies the pitch of the instrument. The mouthpiece simply creates a mixture of different frequencies, and the resonating air column acts on a particular set of frequencies to amplify a single note, turning noise into music.

A tuning fork above a hollow tube can provide resonance in an air column, as shown in **Figure 15-9.** The tube is placed in water so that the bottom end of the tube is below the water surface. A resonating tube with one end closed to air is called a **closed-pipe resonator.** The length of the air column is changed by adjusting the height of the tube above the water. If the tuning fork is struck with a rubber hammer and the length of the air column is varied as the tube is lifted up and down in the water, the sound alternately becomes louder and softer. The sound is loud when the air column is in resonance with the tuning fork. A resonating air column intensifies the sound of the tuning fork.

Standing pressure wave How does resonance occur? The vibrating tuning fork produces a sound wave. This wave of alternate high- and low-pressure variations moves down the air column. When the wave hits the water surface, it is reflected back up to the tuning fork, as indicated in **Figure 15-10a.** If the reflected high-pressure wave reaches the tuning fork at the same moment that the fork produces another high-pressure wave, then the emitted and returning waves reinforce each other. This reinforcement of waves produces a standing wave, and resonance is achieved.

An **open-pipe resonator** is a resonating tube with both ends open that also will resonate with a sound source. In this case, the sound wave does not reflect off a closed end, but rather off an open end. The pressure of the reflected wave is inverted; for example, if a high-pressure wave strikes the open end, a low-pressure wave will rebound, as shown in **Figure 15-10b.**

■ **Figure 15-10** A tube placed in water is a closed-pipe resonator. In closed pipes, high pressure waves reflect as high pressure **(a).** In open pipes, the reflected waves are inverted **(b).**

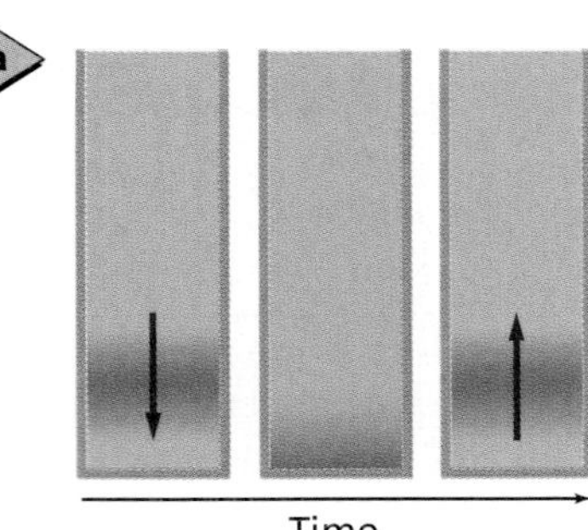

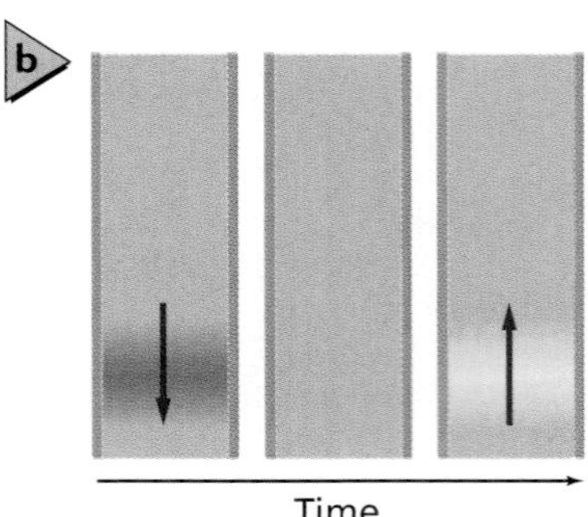

Resonance lengths A standing sound wave in a pipe can be represented by a sine wave, as shown in **Figure 15-11.** Sine waves can represent either the air pressure or the displacement of the air particles. You can see that standing waves have nodes and antinodes. In the pressure graphs, the nodes are regions of mean atmospheric pressure, and at the antinodes, the pressure oscillates between its maximum and minimum values.

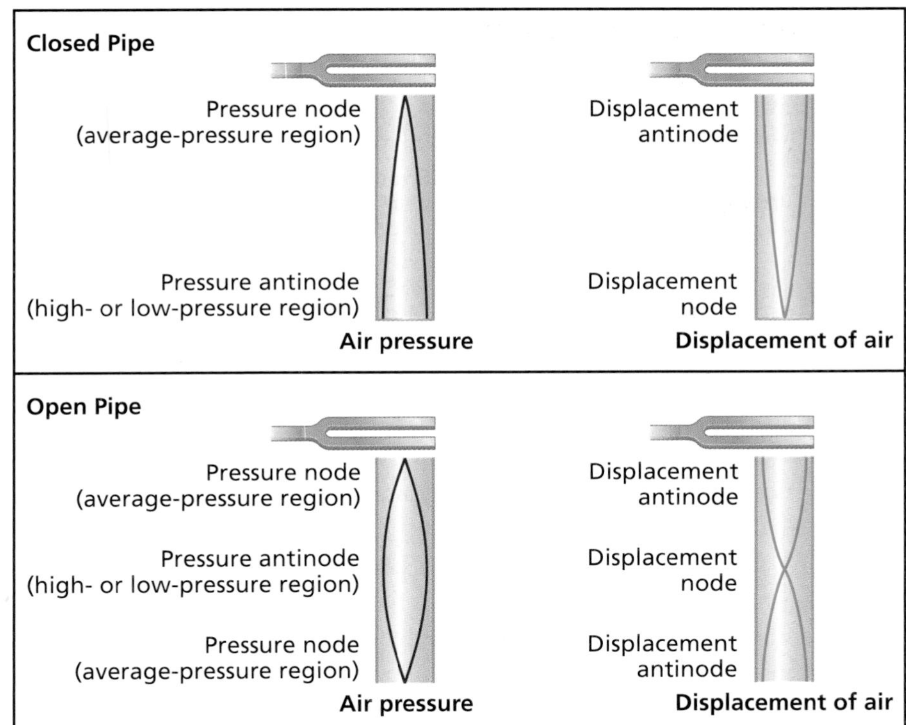

Figure 15-11 Sine waves represent standing waves in pipes.

Concepts In Motion
Interactive Figure To see an animation of resonance in closed and open pipes, visit physicspp.com.

In the case of the displacement graph, the antinodes are regions of high displacement and the nodes are regions of low displacement. In both cases, two antinodes (or two nodes) are separated by one-half wavelength.

Resonance frequencies in a closed pipe The shortest column of air that can have an antinode at the closed end and a node at the open end is one-fourth of a wavelength long, as shown in **Figure 15-12.** As the frequency is increased, additional resonance lengths are found at half-wavelength intervals. Thus, columns of length $\lambda/4$, $3\lambda/4$, $5\lambda/4$, $7\lambda/4$, and so on will all be in resonance with a tuning fork.

In practice, the first resonance length is slightly longer than one-fourth of a wavelength. This is because the pressure variations do not drop to zero exactly at the open end of the pipe. Actually, the node is approximately 0.4 pipe diameters beyond the end. Additional resonance lengths, however, are spaced by exactly one-half of a wavelength. Measurements of the spacing between resonances can be used to find the velocity of sound in air, as shown in the next Example Problem.

APPLYING PHYSICS

Hearing and Frequency
The human auditory canal acts as a closed-pipe resonator that increases the ear's sensitivity for frequencies between 2000 and 5000 Hz, but the full range of frequencies that people hear extends from 20 to 20,000 Hz. A dog's hearing extends to frequencies as high as 45,000 Hz, and a cat's extends to frequencies as high as 100,000 Hz.

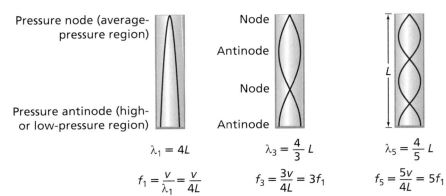

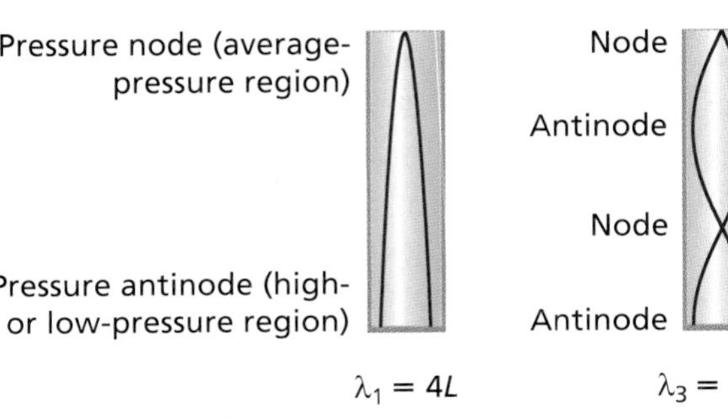

$\lambda_1 = 4L$

$f_1 = \frac{v}{\lambda_1} = \frac{v}{4L}$

Node
Antinode
Node
Antinode

$\lambda_3 = \frac{4}{3}L$

$f_3 = \frac{3v}{4L} = 3f_1$

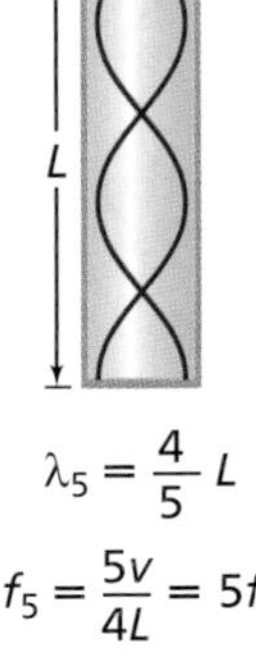

$\lambda_5 = \frac{4}{5}L$

$f_5 = \frac{5v}{4L} = 5f_1$

Figure 15-12 A closed pipe resonates when its length is an odd number of quarter wavelengths.

Figure 15-13 An open pipe resonates when its length is an even number of quarter wavelengths.

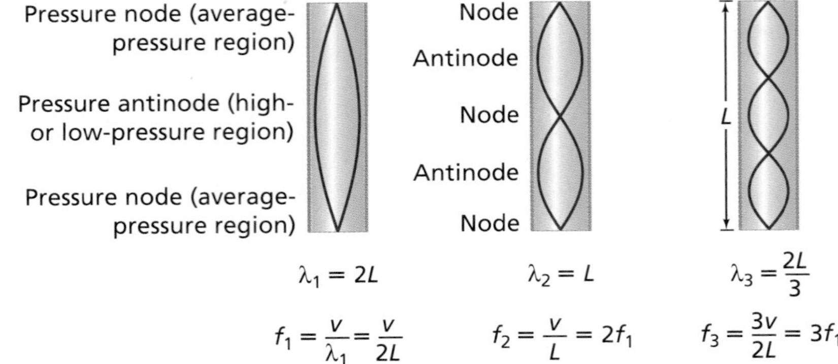

Resonance frequencies in an open pipe The shortest column of air that can have nodes at both ends is one-half of a wavelength long, as shown in **Figure 15-13.** As the frequency is increased, additional resonance lengths are found at half-wavelength intervals. Thus, columns of length $\lambda/2$, λ, $3\lambda/2$, 2λ, and so on will be in resonance with a tuning fork.

If open and closed pipes of the same length are used as resonators, the wavelength of the resonant sound for the open pipe will be half as long as that for the closed pipe. Therefore, the frequency will be twice as high for the open pipe as for the closed pipe. For both pipes, resonance lengths are spaced by half-wavelength intervals.

Hearing resonance Musical instruments use resonance to increase the loudness of particular notes. Open-pipe resonators include flutes and saxophones. Clarinets and the hanging pipes under marimbas and xylophones are examples of closed-pipe resonators. If you shout into a long tunnel, the booming sound you hear is the tunnel acting as a resonator. The seashell in **Figure 15-14** acts as a closed-pipe resonator.

Resonance on Strings

Although the waveforms on vibrating strings vary in shape, depending upon how they are produced, such as by plucking, bowing, or striking, they have many characteristics in common with standing waves on springs and ropes, which you studied in Chapter 14. A string on an instrument is clamped at both ends, and therefore, the string must have a node at each end when it vibrates. In **Figure 15-15,** you can see that the first mode of vibration has an antinode at the center and is one-half of a wavelength long. The next resonance occurs when one wavelength fits on the string, and additional standing waves arise when the string length is $3\lambda/2$, 2λ, $5\lambda/2$, and so on. As with an open pipe, the resonant frequencies are whole-number multiples of the lowest frequency.

The speed of a wave on a string depends on the tension of the string, as well as its mass per unit length. This makes it possible to tune a stringed instrument by changing the tension of its strings. The tighter the string, the faster the wave moves along it, and therefore, the higher the frequency of its standing waves.

Figure 15-14 A seashell acts as a closed-pipe resonator to amplify certain frequencies from the background noise.

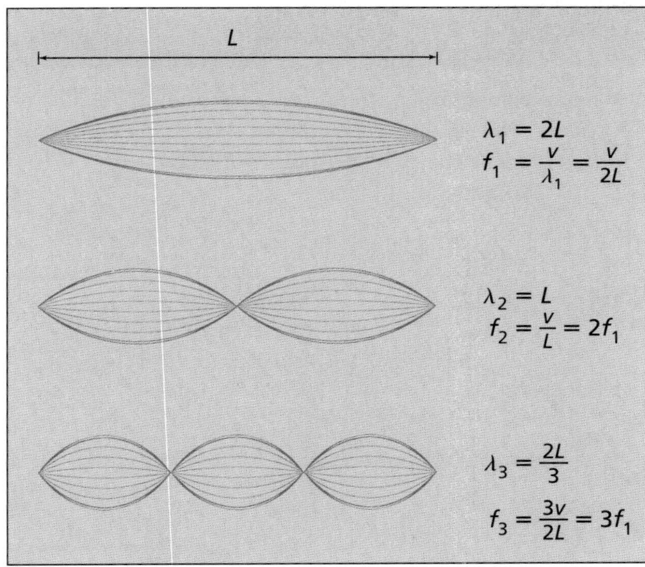

L
$\lambda_1 = 2L$
$f_1 = \frac{v}{\lambda_1} = \frac{v}{2L}$
$\lambda_2 = L$
$f_2 = \frac{v}{L} = 2f_1$
$\lambda_3 = \frac{2L}{3}$
$f_3 = \frac{3v}{2L} = 3f_1$

Figure 15-15 A string resonates with standing waves when its length is a whole number of half wavelengths.

Because strings are so small in cross-sectional area, they move very little air when they vibrate. This makes it necessary to attach them to a sounding board, which transfers their vibrations to the air and produces a stronger sound wave. Unlike the strings themselves, the sounding board should not resonate at any single frequency. Its purpose is to convey the vibrations of all the strings to the air, and therefore it should vibrate well at all frequencies produced by the instrument. Because of the complicated interactions among the strings, the sounding board, and the air, the design and construction of stringed instruments are complex processes, considered by many to be as much an art as a science.

Sound Quality

A tuning fork produces a soft and uninteresting sound. This is because its tines vibrate like simple harmonic oscillators and produce the simple sine wave shown in **Figure 15-16a.** Sounds made by the human voice and musical instruments are much more complex, like the wave in **Figure 15-16b.** Both waves have the same frequency, or pitch, but they sound very different. The complex wave is produced by using the principle of superposition to add waves of many frequencies. The shape of the wave depends on the relative amplitudes of these frequencies. In musical terms, the difference between the two waves is called timbre, tone color, or tone quality.

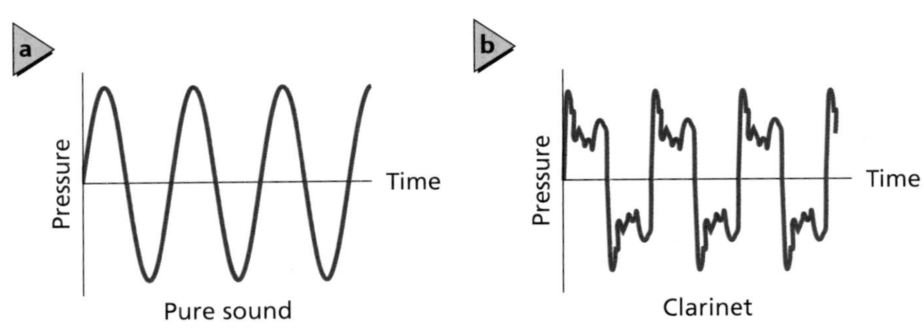

Figure 15-16 A graph of pure sound versus time **(a)** and a graph of clarinet sound waves versus time **(b)** are shown.

▶ EXAMPLE Problem 2

Finding the Speed of Sound Using Resonance When a tuning fork with a frequency of 392 Hz is used with a closed-pipe resonator, the loudest sound is heard when the column is 21.0 cm and 65.3 cm long. What is the speed of sound in this case? Is the temperature warmer or cooler than normal room temperature, which is 20°C? Explain your answer.

1 Analyze and Sketch the Problem

- Sketch the closed-pipe resonator.
- Mark the resonance lengths.

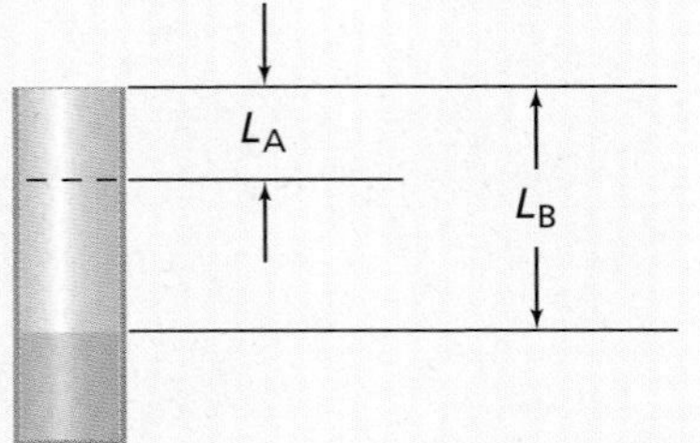

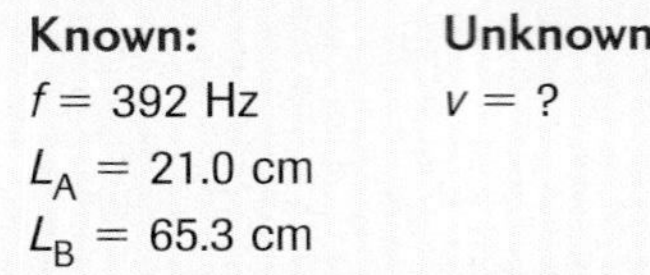

Known:	Unknown:
$f = 392$ Hz	$v = ?$
$L_A = 21.0$ cm	
$L_B = 65.3$ cm	

2 Solve for the Unknown

Solve for the length of the wave using the length-wavelength relationship for a closed pipe.

Math Handbook
Order of Operations page 843

$$L_B - L_A = \frac{1}{2}\lambda$$

$$\lambda = 2(L_B - L_A)$$ **Rearrange the equation for λ.**

$$= 2(0.653\text{ m} - 0.210\text{ m})$$ **Substitute $L_B = 0.653$ m, $L_A = 0.210$ m**

$$= 0.886\text{ m}$$

Use $\lambda = \frac{v}{f}$.

$$v = f\lambda$$ **Rearrange the equation for v.**

$$= (392\text{ Hz})(0.886\text{ m})$$ **Substitute $f = 392$ Hz, $\lambda = 0.886$ m**

$$= 347\text{ m/s}$$

The speed is slightly greater than the speed of sound at 20°C, indicating that the temperature is slightly higher than normal room temperature.

3 Evaluate the Answer

- **Are the units correct?** (Hz)(m) = (1/s)(m) = m/s. The answer's units are correct.
- **Is the magnitude realistic?** The speed is slightly greater than 343 m/s, which is the speed of sound at 20°C.

▶ PRACTICE Problems

- Additional Problems, Appendix B
- Solutions to Selected Problems, Appendix C

18. A 440-Hz tuning fork is held above a closed pipe. Find the spacing between the resonances when the air temperature is 20°C.

19. A 440-Hz tuning fork is used with a resonating column to determine the velocity of sound in helium gas. If the spacings between resonances are 110 cm, what is the velocity of sound in helium gas?

20. The frequency of a tuning fork is unknown. A student uses an air column at 27°C and finds resonances spaced by 20.2 cm. What is the frequency of the tuning fork? Use the speed calculated in Example Problem 2 for the speed of sound in air at 27°C.

21. A bugle can be thought of as an open pipe. If a bugle were straightened out, it would be 2.65-m long.

a. If the speed of sound is 343 m/s, find the lowest frequency that is resonant for a bugle (ignoring end corrections).

b. Find the next two resonant frequencies for the bugle.

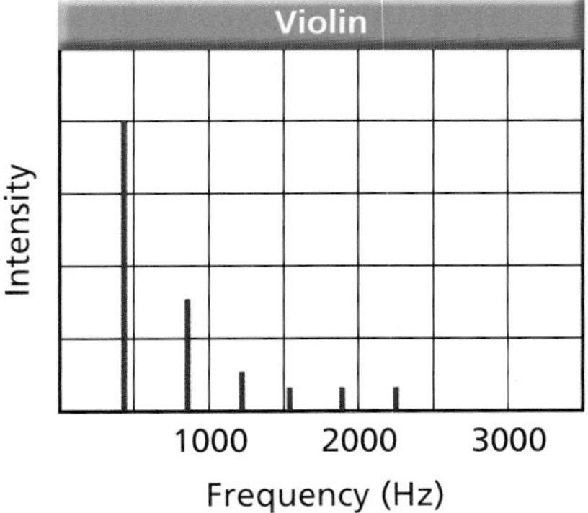

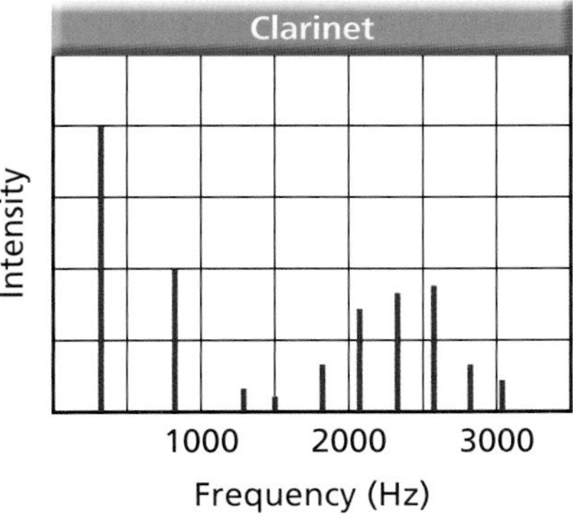

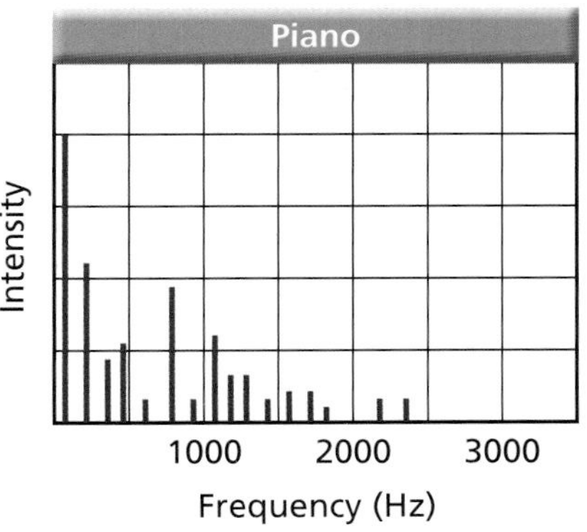

Figure 15-17 A violin, a clarinet, and a piano produce characteristic sound spectra. Each spectrum is unique, as is the timbre of the instrument.

The sound spectrum: fundamental and harmonics The complex sound wave in Figure 15-16b was made by a clarinet. Why does the clarinet produce such a sound wave? The air column in a clarinet acts as a closed pipe. Look back at Figure 15-12, which shows three resonant frequencies for a closed pipe. Because the clarinet acts as a closed pipe, for a clarinet of length L the lowest frequency, f_1, that will be resonant is $v/4L$. This lowest frequency is called the **fundamental.** A closed pipe also will resonate at $3f_1$, $5f_1$, and so on. These higher frequencies, which are odd-number multiples of the fundamental frequency, are called **harmonics.** It is the addition of these harmonics that gives a clarinet its distinctive timbre.

Some instruments, such as an oboe, act as open-pipe resonators. Their fundamental frequency, which is also the first harmonic, is $f_1 = v/2L$ with subsequent harmonics at $2f_1$, $3f_1$, $4f_1$, and so on. Different combinations and amplitudes of these harmonics give each instrument its own unique timbre. A graph of the amplitude of a wave versus its frequency is called a sound spectrum. The spectra of three instruments are shown in **Figure 15-17.**

CHALLENGE **PROBLEM**

1. Determine the tension, F_T, in a violin string of mass m and length L that will play the fundamental note at the same frequency as a closed pipe also of length L. Express your answer in terms of m, L, and the speed of sound in air, v. The equation for the speed of a wave on a string is $u = \sqrt{\frac{F_T}{\mu}}$, where F_T is the tension in the string and μ is the mass per unit length of the string.
2. What is the tension in a string of mass 1.0 g and 40.0 cm long that plays the same note as a closed pipe of the same length?

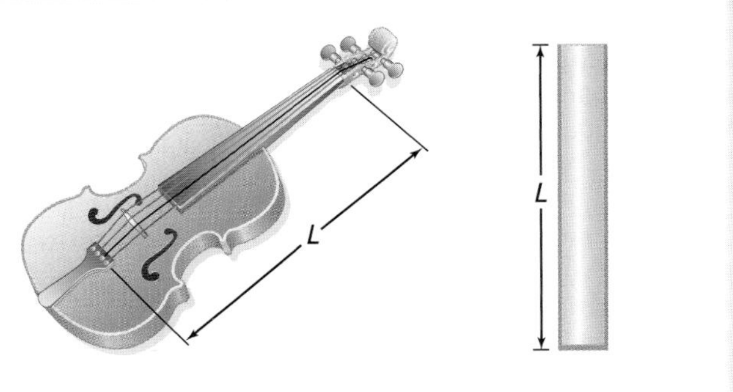

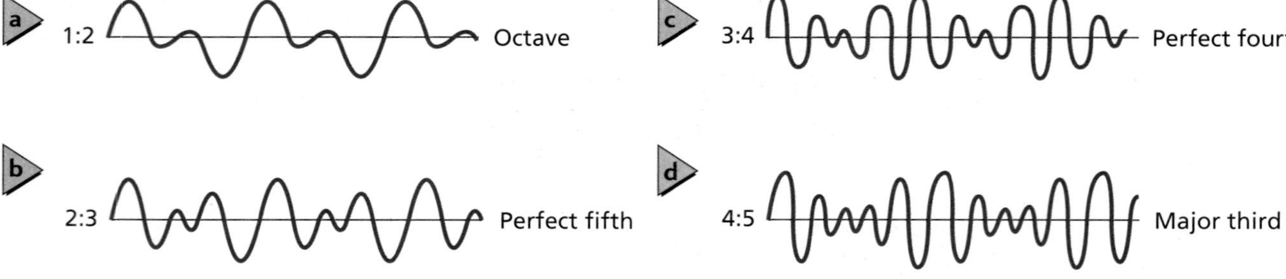

Figure 15-18 These time graphs show the superposition of two waves having the ratios of 1:2, 2:3, 3:4, and 4:5.

Consonance and dissonance When sounds that have two different pitches are played at the same time, the resulting sound can be either pleasant or jarring. In musical terms, several pitches played together are called a chord. An unpleasant set of pitches is called **dissonance.** If the combination is pleasant, the sounds are said to be in **consonance.**

What makes a sound pleasant to listen to? Different cultures have different definitions, but most Western cultures accept the definitions of Pythagoras, who lived in ancient Greece. Pythagoras experimented by plucking two strings at the same time. He noted that pleasing sounds resulted when the strings had lengths in small, whole-number ratios, for example 1:2, 2:3, or 3:4. This means that their pitches (frequencies) will also have small, whole-number ratios.

Musical intervals Two notes with frequencies related by the ratio 1:2 are said to differ by an octave. For example, if a note has a frequency of 440 Hz, a note that is one octave higher has a frequency of 880 Hz. The fundamental and its harmonics are related by octaves; the first harmonic is one octave higher than the fundamental, the second is two octaves higher, and so on. The sum of the fundamental and the first harmonic is shown in **Figure 15-18a.** It is the ratio of two frequencies, not the size of the interval between them, that determines the musical interval.

In other musical intervals, two pitches may be close together. For example, the ratio of frequencies for a "major third" is 4:5. A typical major third is made up of the notes C and E. The note C has a frequency of 262 Hz, so E has a frequency of (5/4)(262 Hz) = 327 Hz. In the same way, notes in a "fourth" (C and F) have a frequency ratio of 3:4, and those in a "fifth" (C and G) have a ratio of 2:3. Graphs of these pleasant sounds are shown in Figure 15-18. More than two notes sounded together also can produce consonance. The three notes called do, mi, and sol make a major chord. For at least 2500 years, this has been recognized as the sweetest of the three-note chords; it has the frequency ratio of 4:5:6.

MINI LAB

Sounds Good

Sometimes it can be pretty tough to tell just by looking whether an instrument will act as an open-pipe resonator or as a closed-pipe resonator. Ask a musician who plays a wind instrument to bring it to class.

1. **Measure** the length of the instrument.
2. Have the musician play the lowest possible note on the instrument.
3. **Determine** the frequency of the note using a frequency generator.

Analyze and Conclude

4. **Draw Conclusions** Did the tested instrument behave most like a closed-pipe resonator or most like an open-pipe resonator? Was the frequency the fundamental or one of the subsequent harmonics?
5. **Determine** the frequencies of the notes that would form an octave, a perfect fifth, a perfect fourth, and a major third with your observed note.

Beats

You have seen that consonance is defined in terms of the ratio of frequencies. When the ratio becomes nearly 1:1, the frequencies become very close. Two frequencies that are nearly identical interfere to produce high and low sound levels, as illustrated in **Figure 15-19.** This oscillation of wave amplitude is called a **beat.** The frequency of a beat is the magnitude of difference between the frequencies of the two waves, $f_{\text{beat}} = |f_A - f_B|$. When the difference is less than 7 Hz, the ear detects this as a pulsation of loudness. Musical instruments often are tuned by sounding one against another and adjusting the frequency of one until the beat disappears.

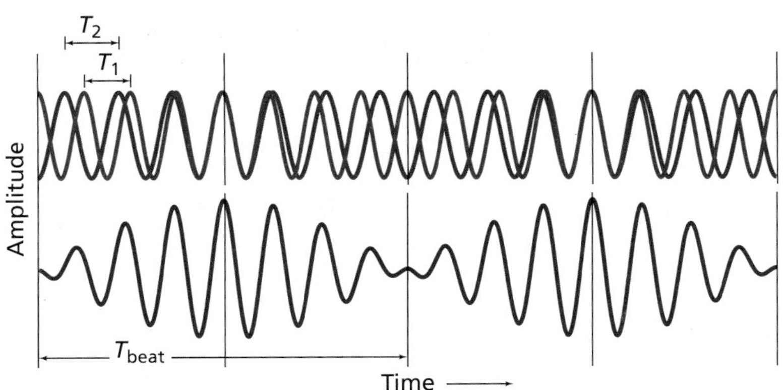

■ **Figure 15-19** Beats occur as a result of the superposition of two sound waves of slightly different frequencies.

Sound Reproduction and Noise

How often do you listen to music produced directly by a human voice or musical instrument? Most of the time, the music has been recorded and played through electronic systems. To reproduce the sound faithfully, the system must accommodate all frequencies equally. A good stereo system keeps the amplitudes of all frequencies between 20 and 20,000 Hz the same to within 3 dB.

A telephone system, on the other hand, needs only to transmit the information in spoken language. Frequencies between 300 and 3000 Hz are sufficient. Reducing the number of frequencies present helps reduce the noise. A noise wave is shown in **Figure 15-20.** Many frequencies are present with approximately the same amplitude. While noise is not helpful in a telephone system, some people claim that listening to noise has a calming effect. For this reason, some dentists use noise to help their patients relax.

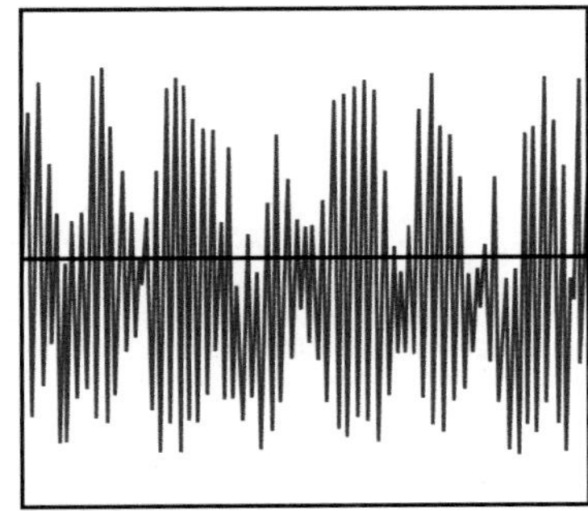

■ **Figure 15-20** Noise is composed of several frequencies and involves random changes in frequency and amplitude.

15.2 Section Review

22. **Origins of Sound** What is the vibrating object that produces sounds in each of the following?
    a. a human voice
    b. a clarinet
    c. a tuba
    d. a violin

23. **Resonance in Air Columns** Why is the tube from which a tuba is made much longer than that of a cornet?

24. **Resonance in Open Tubes** How must the length of an open tube compare to the wavelength of the sound to produce the strongest resonance?

25. **Resonance on Strings** A violin sounds a note of F sharp, with a pitch of 370 Hz. What are the frequencies of the next three harmonics produced with this note?

26. **Resonance in Closed Pipes** One closed organ pipe has a length of 2.40 m.
    a. What is the frequency of the note played by this pipe?
    b. When a second pipe is played at the same time, a 1.40-Hz beat note is heard. By how much is the second pipe too long?

27. **Timbre** Why do various instruments sound different even when they play the same note?

28. **Beats** A tuning fork produces three beats per second with a second, 392-Hz tuning fork. What is the frequency of the first tuning fork?

29. **Critical Thinking** Strike a tuning fork with a rubber hammer and hold it at arm's length. Then press its handle against a desk, a door, a filing cabinet, and other objects. What do you hear? Why?

PHYSICS LAB

Speed of Sound

Alternate CBL instructions can be found on the Web site.
physicspp.com

If a vibrating tuning fork is held above a closed pipe of the proper length, the air in the pipe will vibrate at the same frequency, f, as the tuning fork. By placing a glass tube in a large, water-filled graduated cylinder, the length of the glass tube can be changed by raising or lowering it in the water. The shortest column of air that will resonate occurs when the tube is one-fourth of a wavelength long. This resonance will produce the loudest sound, and the wavelength at this resonance is described by $\lambda = 4L$, where L is the length from the water to the open end of the pipe. In this lab, you will determine L, calculate λ, and calculate the speed of sound.

QUESTION

How can you use a closed-pipe resonator to determine the speed of sound?

Objectives

- **Collect and organize data** to obtain resonant points in a closed pipe.
- **Measure** the length of a closed-pipe resonator.
- **Analyze** the data to determine the speed of sound.

Safety Precautions

- **Immediately wipe up any spilled liquids.**
- **Glass is fragile. Handle with care.**

Materials

three tuning forks of known frequencies
graduated cylinder (1000-mL)
water
tuning fork mallet
metric ruler
thermometer (non-mercury)
glass tube (approximately 40 cm in length and 3.5 cm in diameter)

Procedure

1. Put on your safety goggles. Fill the graduated cylinder nearly to the top with water.
2. Measure the room air temperature and record it in Data Table 1.
3. Select a tuning fork and record its frequency in Data Tables 2 and 3.
4. Measure and record the diameter of the glass tube in Data Table 2.
5. Carefully place the glass tube into the water-filled graduated cylinder.
6. Hold the tuning fork by the base. Swiftly strike it on the side with the tuning fork mallet. Do not strike the tuning fork on the laboratory table or other hard surface.
7. Hold the vibrating fork over the open end of the glass tube and slowly raise the tube and the fork until the loudest sound is heard. Once this point is located, move the tube up and down slightly to determine the exact point of resonance. Measure the distance from the water to the top of the glass tube and record this distance in Data Table 2.
8. Repeat steps 3, 6, and 7 for two additional tuning forks and record your results as trials 2 and 3. The three tuning forks that you test should resonate at three different frequencies.
9. Empty the water from the graduated cylinder.

Data Table 1

Trial	Temperature (°C)	Accepted Speed of Sound (m/s)	Experimental Speed of Sound (m/s)
1			
2			
3			

Data Table 2

Trial	Tuning Fork Frequency (Hz)	Diameter (m)	Length of Tube Above Water (m)	Calculated Wavelength (m)
1				
2				
3				

Data Table 3

Trial	Tuning Fork Frequency (Hz)	Accepted Speed of Sound (m/s)	Corrected Calculated Wavelength (m)	Corrected Experimental Speed of Sound (m/s)
1				
2				
3				

Analyze

1. Calculate the accepted speed of sound using the relationship $v = 331 \text{ m/s} + 0.60T$, where v is the speed of sound at temperature T, and T is the air temperature in degrees Celsius. Record this as the accepted speed of sound in Data Tables 1 and 3 for all the trials.
2. Since the first resonant point is located when the tube is one-fourth of a wavelength above the water, use the measured length of the tube to determine the calculated wavelength for each trial. Record the calculated wavelengths in Data Table 2.
3. Multiply the values in Data Table 2 of wavelength and frequency to determine the experimental speed of sound and record this in Data Table 1 for each of the trials.
4. **Error Analysis** For each trial in Data Table 1, determine the relative error between the experimental and accepted speed of sound.

   Relative error =
   $$\frac{|\text{Accepted value} - \text{Experimental value}|}{\text{Accepted value}} \times 100$$
5. **Critique** To improve the accuracy of your calculations, the tube diameter must be taken into consideration. The following relationship provides a more accurate calculation of wavelength: $\lambda = 4(L + 0.4d)$, where λ is the wavelength, L is the length of the tube above the water, and d is the inside diameter of the tube. Using the values in Data Table 1 for length and diameter, recalculate λ and record it in Data Table 3 as the corrected wavelength. Calculate the corrected experimental speed of sound by multiplying the tuning fork frequency and corrected wavelength and record the new value for the corrected experimental speed of sound in Data Table 3.
6. **Error Analysis** For each trial in Data Table 3, determine the relative error between the corrected experimental speed and the accepted speed of sound. Use the same formula that you used in step 4, above.

Conclude and Apply

1. **Infer** In general, the first resonant point occurs when the tube length = $\lambda/4$. What are the next two lengths where resonance will occur?
2. **Think Critically** If you had a longer tube, would it be possible to locate another position where resonance occurs? Explain your answer.

Going Further

Which result produced the more accurate speed of sound?

Real-World Physics

Explain the relationship between the size of organ pipes and their resonant frequencies.

To find out more about the properties of sound waves, visit the Web site: physicspp.com

eXtreme physics

Sound Waves in the Sun

The study of wave oscillations in the Sun is called helioseismology. Naturally occurring sound waves (p waves), gravity waves, and surface gravity waves all occur in the Sun. All of these waves are composed of oscillating particles, but different forces cause the oscillations.

For sound waves, pressure differences cause the particles to oscillate. In the Sun, sound waves travel through the convective zone, which is just under the surface, or photosphere. The sound waves do not travel in a straight line, as shown in the image.

Ringing like a Bell The sound waves in the Sun cause the surface of the Sun to vibrate in the radial direction, much like a ringing bell vibrates. When a bell is rung, a clapper hits the bell in one place and standing waves are created. The surface of the Sun does have standing waves, but they are not caused by one large event. Instead, scientists hypothesize that many smaller disruptions in the convective zone start most of the sound waves in the Sun. Just like boiling water in a pot can be noisy, bubbles that are larger than the state of Texas form on the surface of the Sun and start sound waves.

Unlike a pot of boiling water, the sound coming from the Sun is much too low for us to hear. The A above middle C on the piano has a period of 0.00227 s ($f = 440$ Hz). The middle mode of oscillation of the waves in the Sun has a period of 5 min ($f = 0.003$ Hz).

Because we cannot hear the sound waves from the Sun, scientists measure the motion of the surface of the Sun to learn about its sound waves. Because a sound wave takes 2 h to travel from one side of the Sun to the other, the Sun must be observed for long time periods. This necessity makes observations from Earth difficult because the Sun is not visible during the night. In 1995, the *Solar and Heliospheric Observatory* (SOHO) was launched by NASA. This satellite orbits Earth such that it always can observe the Sun.

The motion of the surface of the Sun is measured by observing Doppler shifts in sunlight. The measured vibrations are a complicated pattern that equals the sum of all of the standing waves present in the Sun. Just like a ringing bell, many overtones are present in the Sun. Through careful analysis, the individual standing waves in the Sun and their intensities can be calculated.

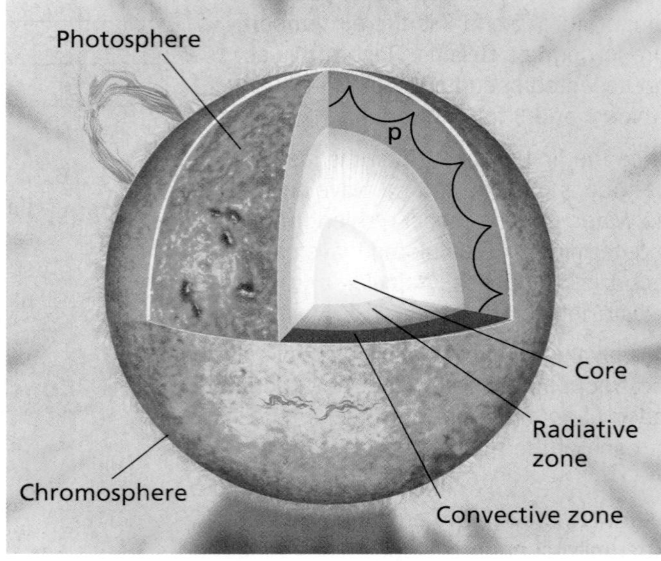

Sound waves (p waves) travel through the Sun's convective zone.

Results Because composition, temperature, and density affect the propagation of sound waves, the Sun's wave oscillations provide information about its interior. SOHO results have given insight into the rotation rate of the Sun as a function of latitude and depth, as well as the density and temperature of the Sun. These results are compared to theoretical calculations to improve our understanding of the Sun.

Going Further

1. **Hypothesize** How do scientists separate the surface motion due to sound waves from the motion due to the rotation of the Sun?
2. **Critical Thinking** Would sound waves in another star, similar to the Sun but different in size, have the same wavelength as sound waves in the Sun?

Chapter 15

Study Guide

15.1 Properties and Detection of Sound

Vocabulary

- sound wave (p. 404)
- pitch (p. 406)
- loudness (p. 406)
- sound level (p. 406)
- decibel (p. 406)
- Doppler effect (p. 407)

Key Concepts

- Sound is a pressure variation transmitted through matter as a longitudinal wave.
- A sound wave has frequency, wavelength, speed, and amplitude. Sound waves reflect and interfere.
- The speed of sound in air at room temperature (20°C) is 343 m/s. The speed increases roughly 0.6 m/s with each 1°C increase in temperature.
- Sound detectors convert the energy carried by a sound wave into another form of energy. The human ear is a highly efficient and sensitive detector of sound waves.
- The frequency of a sound wave is heard as its pitch.
- The pressure amplitude of a sound wave can be measured in decibels (dB).
- The loudness of sound as perceived by the ear and brain depends mainly on its amplitude.
- The Doppler effect is the change in frequency of sound caused by the motion of either the source or the detector. It can be calculated with the following equation.

$$f_d = f_s\left(\frac{v - v_d}{v - v_s}\right)$$

15.2 The Physics of Music

Vocabulary

- closed-pipe resonator (p. 412)
- open-pipe resonator (p. 412)
- fundamental (p. 417)
- harmonics (p. 417)
- dissonance (p. 418)
- consonance (p. 418)
- beat (p. 418)

Key Concepts

- Sound is produced by a vibrating object in a material medium.
- Most sounds are complex waves that are composed of more than one frequency.
- An air column can resonate with a sound source, thereby increasing the amplitude of its resonant frequency.
- A closed pipe resonates when its length is $\lambda/4$, $3\lambda/4$, $5\lambda/4$, and so on. Its resonant frequencies are odd-numbered multiples of the fundamental.
- An open pipe resonates when its length is $\lambda/2$, $2\lambda/2$, $3\lambda/2$, and so on. Its resonant frequencies are whole-number multiples of the fundamental.
- A clamped string has a node at each end and resonates when its length is $\lambda/2$, $2\lambda/2$, $3\lambda/2$, and so on, just as with an open pipe. The string's resonant frequencies are also whole-number multiples of the fundamental.
- The frequencies and intensities of the complex waves produced by a musical instrument determine the timbre that is characteristic of that instrument.
- The fundamental frequency and harmonics can be described in terms of resonance.
- Notes on a musical scale differ in frequency by small, whole-number ratios. An octave has a frequency ratio of 1:2.
- Two waves with almost the same frequency interfere to produce beats.

Chapter 15 Assessment

Concept Mapping

30. Complete the concept map below using the following terms: *amplitude, perception, pitch, speed.*

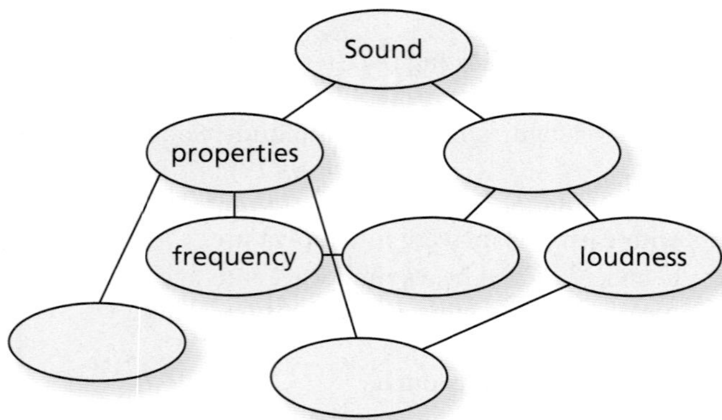

Mastering Concepts

31. What are the physical characteristics of sound waves? (15.1)

32. When timing the 100-m run, officials at the finish line are instructed to start their stopwatches at the sight of smoke from the starter's pistol and not at the sound of its firing. Explain. What would happen to the times for the runners if the timing started when sound was heard? (15.1)

33. Name two types of perception of sound and the physical characteristics of sound waves that correspond to them. (15.1)

34. Does the Doppler shift occur for only some types of waves or for all types of waves? (15.1)

35. Sound waves with frequencies higher than can be heard by humans, called ultrasound, can be transmitted through the human body. How could ultrasound be used to measure the speed of blood flowing in veins or arteries? Explain how the waves change to make this measurement possible. (15.1)

36. What is necessary for the production and transmission of sound? (15.2)

37. **Singing** How can a certain note sung by an opera singer cause a crystal glass to shatter? (15.2)

38. **Marching** In the military, as marching soldiers approach a bridge, the command "route step" is given. The soldiers then walk out-of-step with each other as they cross the bridge. Explain. (15.2)

39. **Musical Instruments** Why don't most musical instruments sound like tuning forks? (15.2)

40. **Musical Instruments** What property distinguishes notes played on both a trumpet and a clarinet if they have the same pitch and loudness? (15.2)

41. **Trombones** Explain how the slide of a trombone, shown in **Figure 15-21,** changes the pitch of the sound in terms of a trombone being a resonance tube. (15.2)

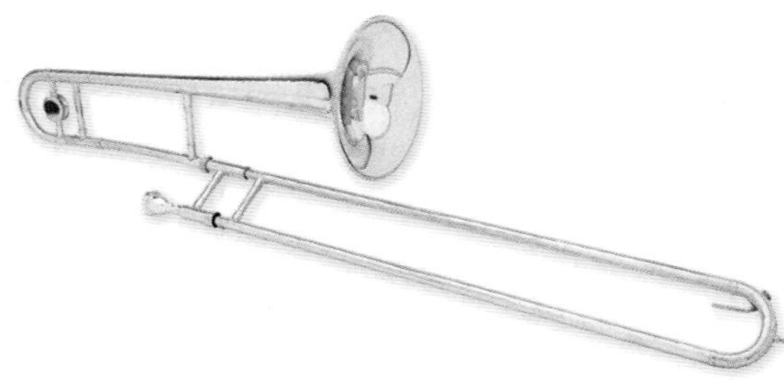

Figure 15-21

Applying Concepts

42. **Estimation** To estimate the distance in kilometers between you and a lightning flash, count the seconds between the flash and the thunder and divide by 3. Explain how this rule works. Devise a similar rule for miles.

43. The speed of sound increases by about 0.6 m/s for each degree Celsius when the air temperature rises. For a given sound, as the temperature increases, what happens to the following?

a. the frequency

b. the wavelength

44. **Movies** In a science-fiction movie, a satellite blows up. The crew of a nearby ship immediately hears and sees the explosion. If you had been hired as an advisor, what two physics errors would you have noticed and corrected?

45. **The Redshift** Astronomers have observed that the light coming from distant galaxies appears redder than light coming from nearer galaxies. With the help of **Figure 15-22,** which shows the visible spectrum, explain why astronomers conclude that distant galaxies are moving away from Earth.

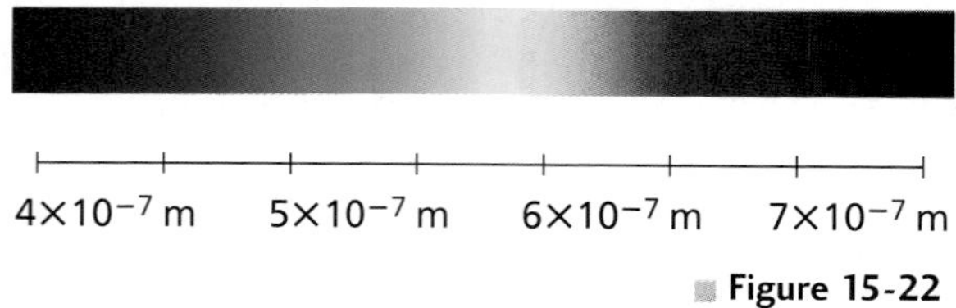

Figure 15-22

46. Does a sound of 40 dB have a factor of 100 (10^2) times greater pressure variation than the threshold of hearing, or a factor of 40 times greater?

 For more problems, go to Additional Problems, Appendix B.

47. If the pitch of sound is increased, what are the changes in the following?
a. the frequency
b. the wavelength
c. the wave velocity
d. the amplitude of the wave

48. The speed of sound increases with temperature. Would the pitch of a closed pipe increase or decrease when the temperature of the air rises? Assume that the length of the pipe does not change.

49. **Marching Bands** Two flutists are tuning up. If the conductor hears the beat frequency increasing, are the two flute frequencies getting closer together or farther apart?

50. **Musical Instruments** A covered organ pipe plays a certain note. If the cover is removed to make it an open pipe, is the pitch increased or decreased?

51. **Stringed Instruments** On a harp, **Figure 15-23a,** long strings produce low notes and short strings produce high notes. On a guitar, **Figure 15-23b,** the strings are all the same length. How can they produce notes of different pitches?

Figure 15-23

Mastering Problems

15.1 Properties and Detection of Sound

52. You hear the sound of the firing of a distant cannon 5.0 s after seeing the flash. How far are you from the cannon?

53. If you shout across a canyon and hear an echo 3.0 s later, how wide is the canyon?

54. A sound wave has a frequency of 4700 Hz and travels along a steel rod. If the distance between compressions, or regions of high pressure, is 1.1 m, what is the speed of the wave?

55. **Bats** The sound emitted by bats has a wavelength of 3.5 mm. What is the sound's frequency in air?

56. **Photography** As shown in **Figure 15-24,** some cameras determine the distance to the subject by sending out a sound wave and measuring the time needed for the echo to return to the camera. How long would it take the sound wave to return to such a camera if the subject were 3.00 m away?

Figure 15-24

57. Sound with a frequency of 261.6 Hz travels through water at 25°C. Find the sound's wavelength in water. Do not confuse sound waves moving through water with surface waves moving through water.

58. If the wavelength of a 4.40×10^{2}-Hz sound in freshwater is 3.30 m, what is the speed of sound in freshwater?

59. Sound with a frequency of 442 Hz travels through an iron beam. Find the wavelength of the sound in iron.

60. **Aircraft** Adam, an airport employee, is working near a jet plane taking off. He experiences a sound level of 150 dB.
a. If Adam wears ear protectors that reduce the sound level to that of a typical rock concert, what decrease in dB is provided?
b. If Adam then hears something that sounds like a barely audible whisper, what will a person not wearing the ear protectors hear?

61. **Rock Music** A rock band plays at an 80-dB sound level. How many times greater is the sound pressure from another rock band playing at each of the following sound levels?
a. 100 dB
b. 120 dB

62. A coiled-spring toy is shaken at a frequency of 4.0 Hz such that standing waves are observed with a wavelength of 0.50 m. What is the speed of propagation of the wave?

63. A baseball fan on a warm summer day (30°C) sits in the bleachers 152 m away from home plate.
a. What is the speed of sound in air at 30°C?
b. How long after seeing the ball hit the bat does the fan hear the crack of the bat?

64. On a day when the temperature is 15°C, a person stands some distance, d, as shown in **Figure 15-25,** from a cliff and claps his hands. The echo returns in 2.0 s. How far away is the cliff?

Figure 15-25 (Not to scale)

65. **Medical Imaging** Ultrasound with a frequency of 4.25 MHz can be used to produce images of the human body. If the speed of sound in the body is the same as in salt water, 1.50 km/s, what is the length of a 4.25-MHz pressure wave in the body?

66. **Sonar** A ship surveying the ocean bottom sends sonar waves straight down into the seawater from the surface. As illustrated in **Figure 15-26,** the first reflection, off of the mud at the sea floor, is received 1.74 s after it was sent. The second reflection, from the bedrock beneath the mud, returns after 2.36 s. The seawater is at a temperature of 25°C, and the speed of sound in mud is 1875 m/s.
   a. How deep is the water?
   b. How thick is the mud?

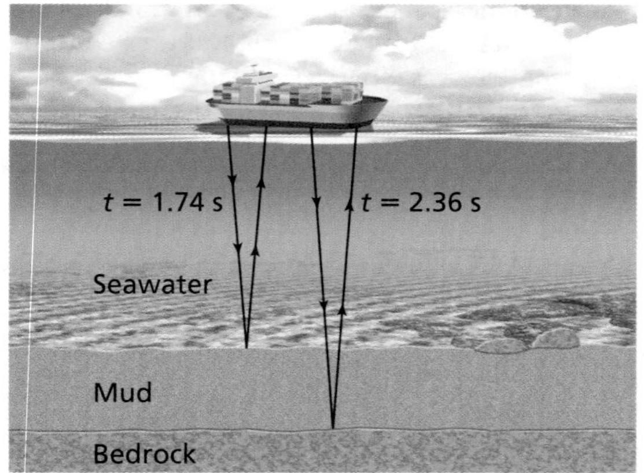

Figure 15-26 (Not to scale)

67. Determine the variation in sound pressure of a conversation being held at a sound level of 60 dB.

68. A fire truck is moving at 35 m/s, and a car in front of the truck is moving in the same direction at 15 m/s. If a 327-Hz siren blares from the truck, what frequency is heard by the driver of the car?

69. A train moving toward a sound detector at 31.0 m/s blows a 305-Hz whistle. What frequency is detected on each of the following?
   a. a stationary train
   b. a train moving toward the first train at 21.0 m/s

70. The train in the previous problem is moving away from the detector. What frequency is now detected on each of the following?
   a. a stationary train
   b. a train moving away from the first train at a speed of 21.0 m/s

15.2 The Physics of Music

71. A vertical tube with a tap at the base is filled with water, and a tuning fork vibrates over its mouth. As the water level is lowered in the tube, resonance is heard when the water level has dropped 17 cm, and again after 49 cm of distance exists from the water to the top of the tube. What is the frequency of the tuning fork?

72. **Human Hearing** The auditory canal leading to the eardrum is a closed pipe that is 3.0 cm long. Find the approximate value (ignoring end correction) of the lowest resonance frequency.

73. If you hold a 1.2-m aluminum rod in the center and hit one end with a hammer, it will oscillate like an open pipe. Antinodes of pressure correspond to nodes of molecular motion, so there is a pressure antinode in the center of the bar. The speed of sound in aluminum is 5150 m/s. What would be the bar's lowest frequency of oscillation?

74. One tuning fork has a 445-Hz pitch. When a second fork is struck, beat notes occur with a frequency of 3 Hz. What are the two possible frequencies of the second fork?

75. **Flutes** A flute acts as an open pipe. If a flute sounds a note with a 370-Hz pitch, what are the frequencies of the second, third, and fourth harmonics of this pitch?

76. **Clarinets** A clarinet sounds the same note, with a pitch of 370 Hz, as in the previous problem. The clarinet, however, acts as a closed pipe. What are the frequencies of the lowest three harmonics produced by this instrument?

77. **String Instruments** A guitar string is 65.0 cm long and is tuned to produce a lowest frequency of 196 Hz.
   a. What is the speed of the wave on the string?
   b. What are the next two higher resonant frequencies for this string?

 For more problems, go to Additional Problems, Appendix B.

78. **Musical Instruments** The lowest note on an organ is 16.4 Hz.
   a. What is the shortest open organ pipe that will resonate at this frequency?
   b. What is the pitch if the same organ pipe is closed?

79. **Musical Instruments** Two instruments are playing musical A (440.0 Hz). A beat note with a frequency of 2.5 Hz is heard. Assuming that one instrument is playing the correct pitch, what is the frequency of the pitch played by the second instrument?

80. A flexible, corrugated, plastic tube, shown in **Figure 15-27,** is 0.85 m long. When it is swung around, it creates a tone that is the lowest pitch for an open pipe of this length. What is the frequency?

Figure 15-27

81. The tube from the previous problem is swung faster, producing a higher pitch. What is the new frequency?

82. During normal conversation, the amplitude of a pressure wave is 0.020 Pa.
   a. If the area of an eardrum is 0.52 cm^2, what is the force on the eardrum?
   b. The mechanical advantage of the three bones in the middle ear is 1.5. If the force in part a is transmitted undiminished to the bones, what force do the bones exert on the oval window, the membrane to which the third bone is attached?
   c. The area of the oval window is 0.026 cm^2. What is the pressure increase transmitted to the liquid in the cochlea?

83. **Musical Instruments** One open organ pipe has a length of 836 mm. A second open pipe should have a pitch that is one major third higher. How long should the second pipe be?

84. As shown in **Figure 15-28,** a music box contains a set of steel fingers clamped at one end and plucked on the other end by pins on a rotating drum. What is the speed of a wave on a finger that is 2.4 cm long and plays a note of 1760 Hz?

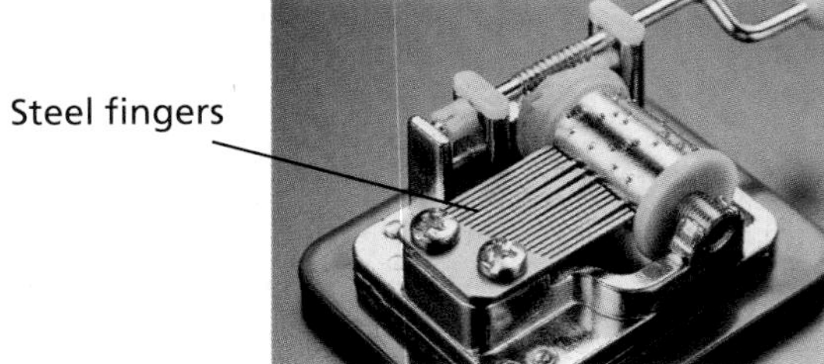

Figure 15-28

Mixed Review

85. An open organ pipe is 1.65 m long. What fundamental frequency note will it produce if it is played in helium at 0°C?

86. If you drop a stone into a well that is 122.5 m deep, as illustrated in **Figure 15-29,** how soon after you drop the stone will you hear it hit the bottom of the well?

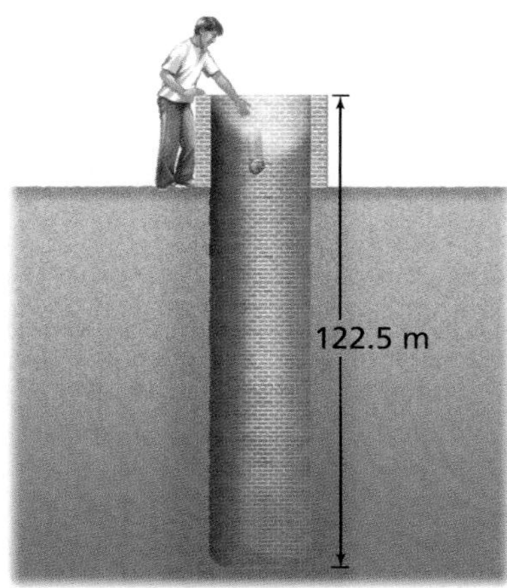

Figure 15-29 (Not to scale)

87. A bird on a newly discovered planet flies toward a surprised astronaut at a speed of 19.5 m/s while singing at a pitch of 945 Hz. The astronaut hears a tone of 985 Hz. What is the speed of sound in the atmosphere of this planet?

88. In North America, one of the hottest outdoor temperatures ever recorded is 57°C and one of the coldest is −62°C. What are the speeds of sound at those two temperatures?

89. A ship's sonar uses a frequency of 22.5 kHz. The speed of sound in seawater is 1533 m/s. What is the frequency received on the ship that was reflected from a whale traveling at 4.15 m/s away from the ship? Assume that the ship is at rest.

90. When a wet finger is rubbed around the rim of a glass, a loud tone of frequency 2100 Hz is produced. If the glass has a diameter of 6.2 cm and the vibration contains one wavelength around its rim, what is the speed of the wave in the glass?

91. **History of Science** In 1845, Dutch scientist Christoph Buys-Ballot developed a test of the Doppler effect. He had a trumpet player sound an A note at 440 Hz while riding on a flatcar pulled by a locomotive. At the same time, a stationary trumpeter played the same note. Buys-Ballot heard 3.0 beats per second. How fast was the train moving toward him?

92. You try to repeat Buys-Ballot's experiment from the previous problem. You plan to have a trumpet played in a rapidly moving car. Rather than listening for beat notes, however, you want to have the car move fast enough so that the moving trumpet sounds one major third above a stationary trumpet.
a. How fast would the car have to move?
b. Should you try the experiment? Explain.

93. Guitar Strings The equation for the speed of a wave on a string is $v = \sqrt{\frac{F_T}{\mu}}$, where F_T is the tension in the string and μ is the mass per unit length of the string. A guitar string has a mass of 3.2 g and is 65 cm long. What must be the tension in the string to produce a note whose fundamental frequency is 147 Hz?

94. A train speeding toward a tunnel at 37.5 m/s sounds its horn at 327 Hz. The sound bounces off the tunnel mouth. What is the frequency of the reflected sound heard on the train? *Hint: Solve the problem in two parts. First, assume that the tunnel is a stationary observer and find the frequency. Then, assume that the tunnel is a stationary source and find the frequency measured on the train.*

Thinking Critically

95. Make and Use Graphs The wavelengths of the sound waves produced by a set of tuning forks with given frequencies are shown in **Table 15-2** below.
a. Plot a graph of the wavelength versus the frequency (controlled variable). What type of relationship does the graph show?
b. Plot a graph of the wavelength versus the inverse of the frequency ($1/f$). What kind of graph is this? Determine the speed of sound from this graph.

Table 15-2
Tuning Forks

Frequency (Hz)	Wavelength (m)
131	2.62
147	2.33
165	2.08
196	1.75
220	1.56
247	1.39

96. Make Graphs Suppose that the frequency of a car horn is 300 Hz when it is stationary. What would the graph of the frequency versus time look like as the car approached and then moved past you? Complete a rough sketch.

97. Analyze and Conclude Describe how you could use a stopwatch to estimate the speed of sound if you were near the green on a 200-m golf hole as another group of golfers hit their tee shots. Would your estimate of the speed of sound be too large or too small?

98. Apply Concepts A light wave coming from a point on the left edge of the Sun is found by astronomers to have a slightly higher frequency than light from the right side. What do these measurements tell you about the Sun's motion?

99. Design an Experiment Design an experiment that could test the formula for the speed of a wave on a string. Explain what measurements you would make, how you would make them, and how you would use them to test the formula.

Writing in Physics

100. Research the construction of a musical instrument, such as a violin or French horn. What factors must be considered besides the length of the strings or tube? What is the difference between a quality instrument and a cheaper one? How are they tested for tone quality?

101. Research the use of the Doppler effect in the study of astronomy. What is its role in the big bang theory? How is it used to detect planets around other stars? To study the motions of galaxies?

Cumulative Review

102. Ball A, rolling west at 3.0 m/s, has a mass of 1.0 kg. Ball B has a mass of 2.0 kg and is stationary. After colliding with ball B, ball A moves south at 2.0 m/s. (Chapter 9)
a. Sketch the system, showing the velocities and momenta before and after the collision.
b. Calculate the momentum and velocity of ball B after the collision.

103. Chris carries a 10-N carton of milk along a level hall to the kitchen, a distance of 3.5 m. How much work does Chris do? (Chapter 10)

104. A movie stunt person jumps from a five-story building (22 m high) onto a large pillow at ground level. The pillow cushions her fall so that she feels a deceleration of no more than 3.0 m/s^2. If she weighs 480 N, how much energy does the pillow have to absorb? How much force does the pillow exert on her? (Chapter 11)

 For more problems, go to Additional Problems, Appendix B.

Multiple Choice

1. How does sound travel from its source to your ear?
   - (A) by changes in air pressure
   - (B) by vibrations in wires or strings
   - (C) by electromagnetic waves
   - (D) by infrared waves

2. Paulo is listening to classical music in the speakers installed in his swimming pool. A note with a frequency of 327 Hz reaches his ears while he is under water. What is the wavelength of the sound that reaches Paulo's ears? Use 1493 m/s for the speed of sound in water.
   - (A) 2.19 nm
   - (B) 4.88×10^{-5} m
   - (C) 2.19×10^{-1} m
   - (D) 4.57 m

3. The sound from a trumpet travels at 351 m/s in air. If the frequency of the note is 298 Hz, what is the wavelength of the sound wave?
   - (A) 9.93×10^{-4} m
   - (B) 0.849 m
   - (C) 1.18 m
   - (D) 1.05×10^{5} m

4. The horn of a car attracts the attention of a stationary observer. If the car is approaching the observer at 60.0 km/h and the horn has a frequency of 512 Hz, what is the frequency of the sound perceived by the observer? Use 343 m/s for the speed of sound in air.
   - (A) 488 Hz
   - (B) 512 Hz
   - (C) 538 Hz
   - (D) 600 Hz

5. As shown in the diagram below, a car is receding at 72 km/h from a stationary siren. If the siren is wailing at 657 Hz, what is the frequency of the sound perceived by the driver? Use 343 m/s for the speed of sound.
   - (A) 543 Hz
   - (B) 620 Hz
   - (C) 647 Hz
   - (D) 698 Hz

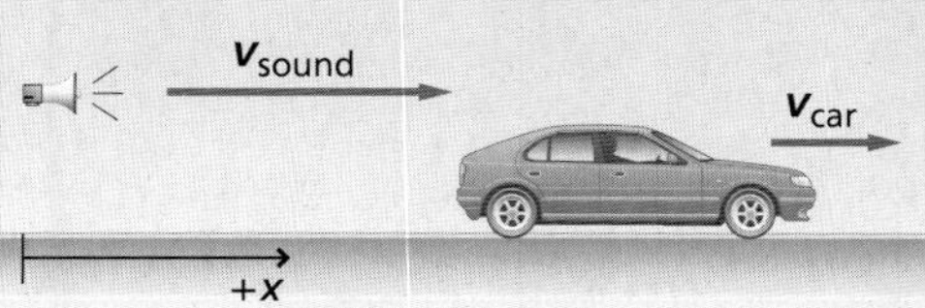

6. Reba hears 20 beats in 5.0 s when she plays two notes on her piano. She is certain that one note has a frequency of 262 Hz. What are the possible frequencies of the second note?
   - (A) 242 Hz or 282 Hz
   - (B) 258 Hz or 266 Hz
   - (C) 260 Hz or 264 Hz
   - (D) 270 Hz or 278 Hz

7. Which of the following pairs of instruments have resonant frequencies at each whole-number multiple of the lowest frequency?
   - (A) a clamped string and a closed pipe
   - (B) a clamped string and an open pipe
   - (C) an open pipe and a closed pipe
   - (D) an open pipe and a reed instrument

Extended Answer

8. The figure below shows the first resonance length of a closed air column. If the frequency of the sound is 488 Hz, what is the speed of the sound?

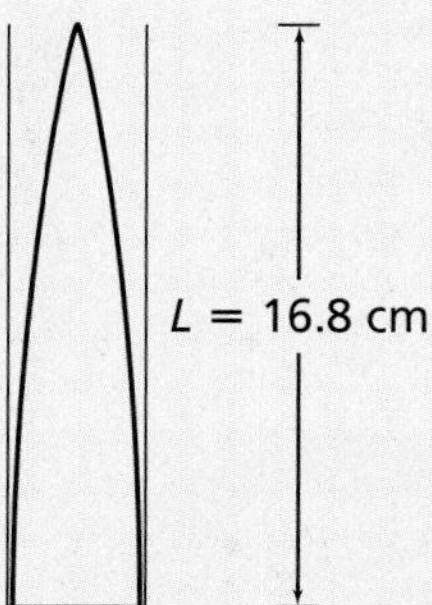

✓ Test-Taking TIP

Write It Down

Most tests ask you a large number of questions in a small amount of time. Write down your work whenever possible. Do math on paper, not in your head. Underline and reread important facts in passages and diagrams—do not try to memorize them.

Chapter 16 Fundamentals of Light

What You'll Learn

- You will understand sources of light and how light illuminates the universe around us.
- You will be able to describe the wave nature of light and some phenomena that reveal this nature.

Why It's Important

Light is a primary source of information about how the universe behaves. We all use information such as color, brightness, and shadow every day to interpret the events occurring around us.

Balloon Race You can tell the difference between the competing balloons because of the different colors visible in the sunlight. You can distinguish the balloons from the backgrounds because of color differences in the grass and sky.

Think About This ▶

What causes these differences in color? How are these colors related?

physicspp.com

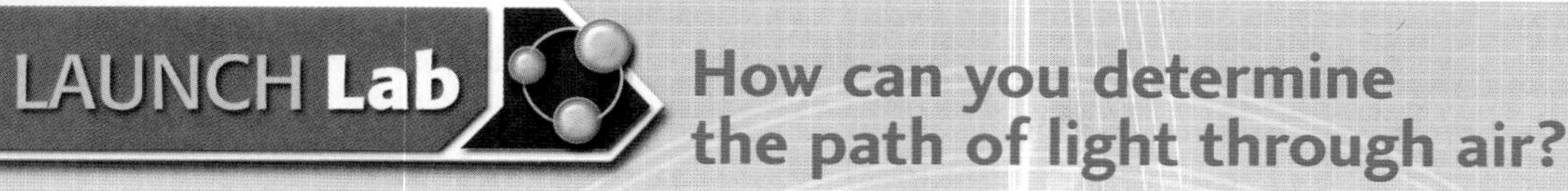

LAUNCH Lab How can you determine the path of light through air?

Question

What path does light take as it travels through the air?

Procedure

1. Punch a hole with a pushpin in the center of an index card.
2. Using clay, stand the index card so that its longer edge is on the table top.
3. Turn on a lamp and have one lab partner hold the lamp so that the lightbulb shines through the hole in the card. ***CAUTION: Lamp can get hot over time.***
4. Hold a mirror on the opposite side of the index card so that light coming through the hole strikes the mirror. Darken the room.
5. Angle the mirror so that it reflects the beam of light onto the back of the card. ***CAUTION: Be careful not to reflect the light beam into someone's eyes.***
6. Write down your observations.

Analysis

Describe the image of the reflected light beam that you see on the index card. Describe the path that the light beam takes.

Critical Thinking Can you see the light beam in the air? Why or why not?

16.1 Illumination

Light and sound are two methods by which you can receive information. Of the two, light seems to provide the greater variety of information. The human eye can detect tiny changes in the size, position, brightness, and color of an object. Our eyes usually can distinguish shadows from solid objects and sometimes distinguish reflections of objects from the objects themselves. In this section, you will learn where light comes from and how it illuminates the universe around you.

One of the first things that you ever discovered about light, although you may not have been conscious of your discovery, is that it travels in a straight line. How do you know this? When a narrow beam of light, such as that of a flashlight or sunlight streaming through a small window, is made visible by dust particles in the air, you see the path of the light as a straight line. When your body blocks sunlight, you see your outline in a shadow. Also, whenever you locate an object with your eyes and walk toward it, you most likely walk in a straight path. These things are possible only because light travels in straight lines. Based on this knowledge of how light travels, models have been developed that describe how light works.

► **Objectives**

- **Develop** the ray model of light.
- **Predict** the effect of distance on light's illumination.
- **Solve** problems involving the speed of light.

► **Vocabulary**

ray model of light
luminous source
illuminated source
opaque
transparent
translucent
luminous flux
illuminance

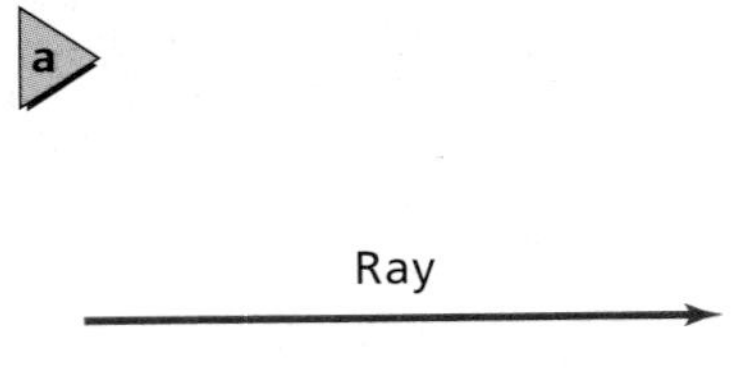

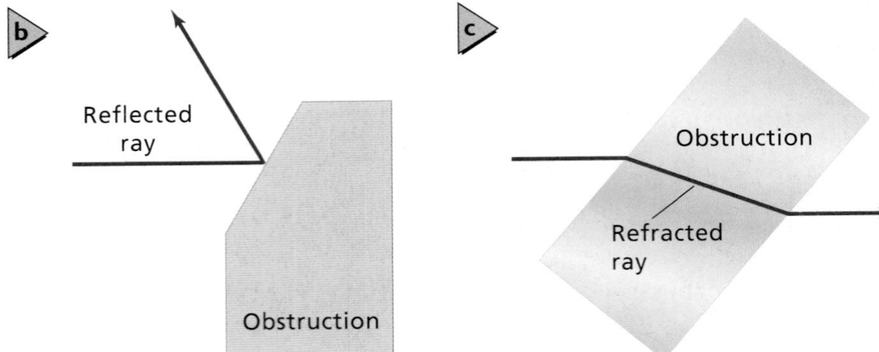

Figure 16-1 A ray is a straight line that represents the linear path of a narrow beam of light **(a).** A light ray can change direction if it is reflected **(b)** or refracted **(c).**

Ray Model of Light

Isaac Newton, whose laws of motion you studied in Chapter 6, believed that light is a stream of fast-moving, unimaginably tiny particles, which he called corpuscles. However, his model could not explain all of the properties of light. Experiments showed that light also behaves like a wave. In the **ray model of light,** light is represented as a ray that travels in a straight path, the direction of which can be changed only by placing an obstruction in the path, as shown in **Figure 16-1.** The ray model of light was introduced as a way to study how light interacts with matter, regardless of whether light is a particle or a wave. This study of light is called ray optics or geometric optics.

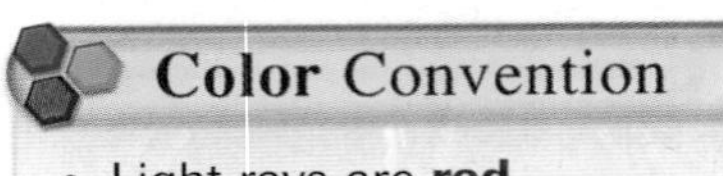

- Light rays are **red.**

Sources of light Rays of light come from sources of light. Our major source of light is the Sun. Other natural sources of light include flames, sparks, and even fireflies. In the past 100 years, humans have been able to produce several other kinds of light sources. Incandescent bulbs, fluorescent lamps, television screens, lasers, and tiny, light-emitting diodes (LEDs) are each a result of humans using electricity to produce light.

What is the difference between sunlight and moonlight? Sunlight, of course, is much, much brighter. There also is an important fundamental difference between the two. The Sun is a **luminous source,** an object that emits light. In contrast, the Moon is an **illuminated source,** an object that becomes visible as a result of the light reflecting off it, as shown in **Figure 16-2.** An incandescent lamp, such as a common lightbulb, is luminous because electrical energy heats a thin tungsten wire in the bulb and causes it to glow. An incandescent source emits light as a result of its high temperature. A bicycle reflector, on the other hand, works as an illuminated source. It is designed to become highly visible when it is illuminated by luminous automobile headlights.

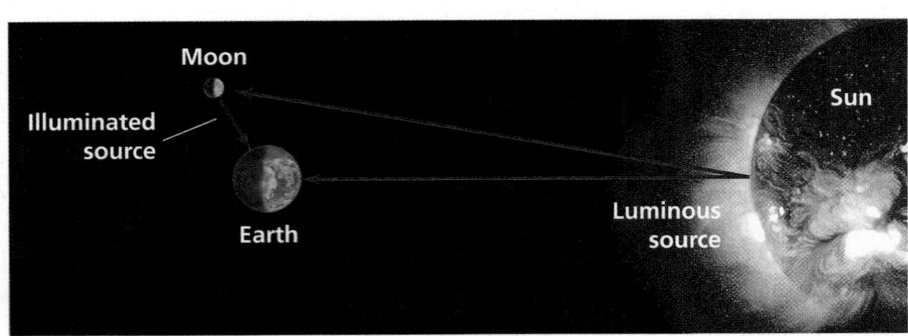

Figure 16-2 The Sun acts as a luminous source to Earth and the Moon. The Moon acts as an illuminated source to Earth. (Illustration not to scale)

Figure 16-3 The transparent glass allows objects to be seen through it **(a).** The translucent lamp shade allows light to pass through, although the lightbulb source itself is not visible **(b).** The opaque tarp covers the statue, preventing the statue from being seen **(c).**

Illuminated sources are visible to you because light is reflecting off or transmitting (passing) through the object to your eyes. Media, such as brick, that do not transmit light, but reflect some light, are **opaque** media. Media that transmit light, such as air and glass, are **transparent** media. Media that transmit light, but do not permit objects to be seen clearly through them, are **translucent** media. Lamp shades and frosted lightbulbs are examples of objects that are made of translucent media. All three types of media are illustrated in **Figure 16-3.** Transparent or translucent media not only transmit light, but they also can reflect a fraction of the light. For example, you often can see your reflection in a glass window.

Quantity of light The rate at which light energy is emitted from a luminous source is called the **luminous flux,** P. The unit of luminous flux is the lumen (lm). A typical 100-W incandescent lightbulb emits approximately 1750 lm. You can think of the luminous flux as a measure of the rate at which light rays come out of a luminous source. Imagine placing a lightbulb at the center of a 1-m-radius sphere, as shown in **Figure 16-4.** The lightbulb emits light in almost all directions. The 1750 lm of luminous flux characterizes all of the light that strikes the inside surface of the sphere in a given unit of time. Even if the sphere was 2 m in radius, the luminous flux of the lightbulb would be the same as for the 1-m-radius sphere, because the total number of light rays does not increase.

Once you know the quantity of light being emitted by a luminous source, you can determine the amount of illumination that the luminous source provides to an object, such as a book. The illumination of a surface, or the rate at which light strikes the surface, is called the **illuminance,** E. You can think of this as a measure of the number of light rays that strike a surface. Illuminance is measured in lux, lx, which is equivalent to lumens per square meter, lm/m^2.

Consider the setup shown in Figure 16-4. What is the illuminance of the sphere's inside surface? The equation for the surface area of a sphere is $4\pi r^2$, so the surface area of this sphere is $4\pi(1.00\text{ m})^2 = 4\pi\text{ m}^2$. The luminous flux striking each square meter of the sphere is $1750\text{ lm}/(4\pi\text{ m}^2) = 139\text{ lx}$. At a distance of 1.00 m from the bulb, 139 lm strikes each square meter. The illuminance of the inside of the sphere is 139 lx.

Figure 16-4 Luminous flux is the rate at which light is emitted from a luminous source, whereas illuminance is the rate at which light falls on a surface.

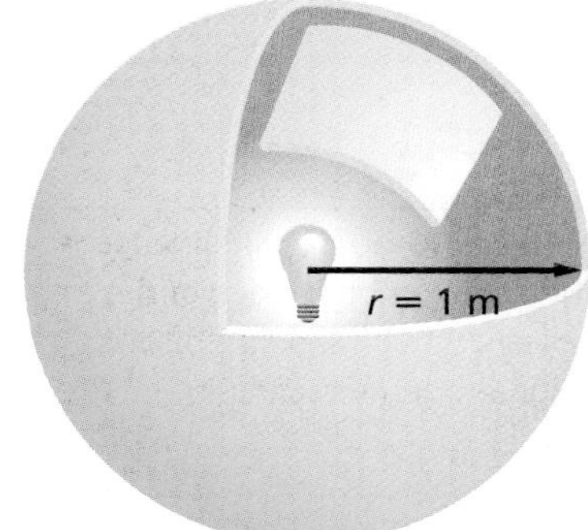

■ **Figure 16-5** The illuminance, *E*, produced by a point source of light varies inversely as the square of the distance from the light source.

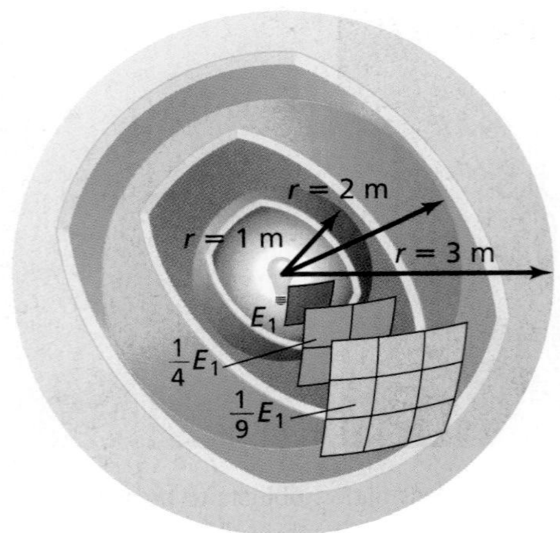

An inverse-square relationship What would happen if the sphere surrounding the lamp were larger? If the sphere had a radius of 2.00 m, the luminous flux still would total 1750 lm, but the area of the sphere would be $4\pi(2.00\text{ m})^2 = 16.0\pi\text{ m}^2$, four times larger than the 1.00-m sphere, as shown in **Figure 16-5.** The illuminance of the inside of the 2.00-m sphere is $1750\text{ lm}/(16.0\pi\text{ m}^2) = 34.8\text{ lx}$, so 34.8 lm strikes each square meter.

The illuminance on the inside surface of the 2.00-m sphere is one-fourth the illuminance on the inside of the 1.00-m sphere. In the same way, the inside of a sphere with a 3.00-m radius has an illuminance only $(1/3)^2$, or 1/9, as large as the 1.00-m sphere. Figure 16-5 shows that the illuminance produced by a point source is proportional to $1/r^2$, an inverse-square relationship. As the light rays spread out in straight lines in all directions from a point source, the number of light rays available to illuminate a unit of area decreases as the square of the distance from the point source.

Luminous intensity Some luminous sources are specified in candela, cd. A candela is not a measure of luminous flux, but of luminous intensity. The luminous intensity of a point source is the luminous flux that falls on 1 m^2 of the inside of a 1-m-radius sphere. Thus, luminous intensity is luminous flux divided by 4π. A bulb with 1750 lm of flux has an intensity of $1750\text{ lm}/4\pi = 139\text{ cd}$.

In **Figure 16-6,** the lightbulb is twice as far away from the screen as the candle. For the lightbulb to provide the same illuminance on the lightbulb side of the screen as the candle does on the candle side of the screen, the lightbulb would have to be four times brighter than the candle, and, therefore, the luminous intensity of the lightbulb would have to be four times the luminous intensity of the candle.

■ **Figure 16-6** The illuminance is the same on both sides of the screen, though the lightbulb is brighter than the candle.

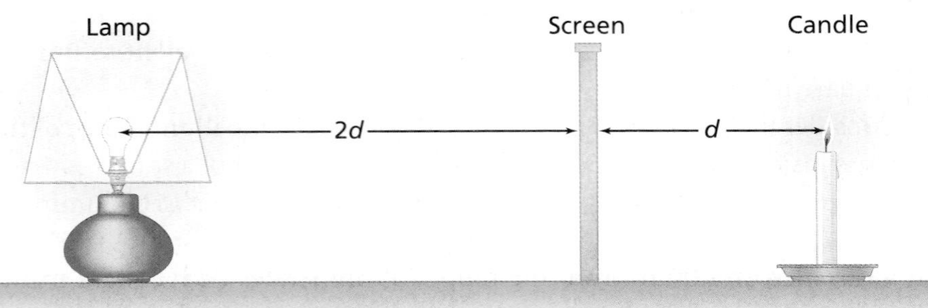

How to Illuminate a Surface

How would you increase the illuminance of your desktop? You could use a brighter bulb, which would increase the luminous flux, or you could move the light source closer to the surface of your desk, thereby decreasing the distance between the light source and the surface it is illuminating. To make the problem easier, you can use the simplification that the light source is a point source. Thus, the illuminance and distance will follow the inverse-square relationship. The problem is further simplified if you assume that light from the source strikes perpendicular to the surface of the desk. Using these simplifications, the illuminance caused by a point light source is represented by the following equation.

Point Source Illuminance $E = \frac{P}{4\pi r^2}$

If an object is illuminated by a point source of light, then the illuminance at the object is equal to the luminous flux of the light source, divided by the surface area of the sphere, whose radius is equal to the distance the object is from the light source.

Remember that the luminous flux of the light source is spreading out spherically in all directions, so only a fraction of the luminous flux is available to illuminate the desk. Use of this equation is valid only if the light from the luminous source strikes perpendicular to the surface it is illuminating. It is also only valid for luminous sources that are small enough or far enough away to be considered point sources. Thus, the equation does not give accurate values of illuminance for long, fluorescent lamps or incandescent lightbulbs that are close to the surfaces that they illuminate.

APPLYING PHYSICS

Illuminated Minds When deciding how to achieve the correct illuminance on students' desktops, architects must consider the luminous flux of the lights as well as the distance of the lights above the desktops. In addition, the efficiencies of the light sources are an important economic factor.

Connecting Math to Physics

Direct and Inverse Relationships The illuminance provided by a source of light has both a direct and an inverse relationship.

Math	Physics
$y = \frac{x}{az^2}$	$E = \frac{P}{4\pi r^2}$
If z is constant, then y is directly proportional to x. • When x increases, y increases. • When x decreases, y decreases.	If r is constant, then E is directly proportional to P. • When P increases, E increases. • When P decreases, E decreases.
If x is constant, then y is inversely proportional to z^2. • When z^2 increases, y decreases. • When z^2 decreases, y increases.	If P is constant, then E is inversely proportional to r^2. • When r^2 increases, E decreases. • When r^2 decreases, E increases.

▶ EXAMPLE Problem 1

Illumination of a Surface What is the illuminance at on your desktop if it is lighted by a 1750-lm lamp that is 2.50 m above your desk?

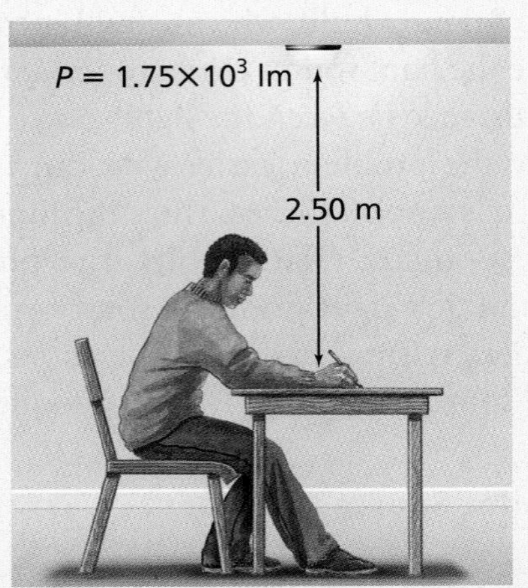

1 Analyze and Sketch the Problem

- Assume that the lightbulb is the point source.
- Diagram the position of the bulb and desktop. Label *P* and *r*.

Known:	Unknown:
$P = 1.75\times10^3$ lm	$E = ?$
$r = 2.50$ m	

2 Solve for the Unknown

The surface is perpendicular to the direction in which the light ray is traveling, so you can use the point source illuminance equation.

$$E = \frac{P}{4\pi r^2}$$

$$= \frac{1.75\times10^3\ \text{lm}}{4\pi(2.50\ \text{m})^2} \quad \textbf{Substitue } P = 1.75\times10^3 \textbf{ lm, } r = 2.50 \textbf{ m}$$

$$= 22.3\ \text{lm/m}^2$$

$$= 22.3\ \text{lx}$$

Physics Online

Personal Tutor For an online tutorial on the illumination of a surface, visit physicspp.com.

3 Evaluate the Answer

- **Are the units correct?** The units of luminance are $\text{lm/m}^2 = \text{lx}$, which the answer agrees with.
- **Do the signs make sense?** All quantities are positive, as they should be.
- **Is the magnitude realistic?** The illuminance is less than the luminous flux, which it should be at this distance.

▶ PRACTICE Problems

- Additional Problems, Appendix B
- Solutions to Selected Problems, Appendix C

1. A lamp is moved from 30 cm to 90 cm above the pages of a book. Compare the illumination on the book before and after the lamp is moved.
2. What is the illumination on a surface that is 3.0 m below a 150-W incandescent lamp that emits a luminous flux of 2275 lm?
3. Draw a graph of the illuminance produced by a 150-W incandescent lamp between 0.50 m and 5.0 m.
4. A 64-cd point source of light is 3.0 m above the surface of a desk. What is the illumination on the desk's surface in lux?
5. A public school law requires a minimum illuminance of 160 lx at the surface of each student's desk. An architect's specifications call for classroom lights to be located 2.0 m above the desks. What is the minimum luminous flux that the lights must produce?
6. A screen is placed between two lamps so that they illuminate the screen equally, as shown in **Figure 16-7.** The first lamp emits a luminous flux of 1445 lm and is 2.5 m from the screen. What is the distance of the second lamp from the screen if the luminous flux is 2375 lm?

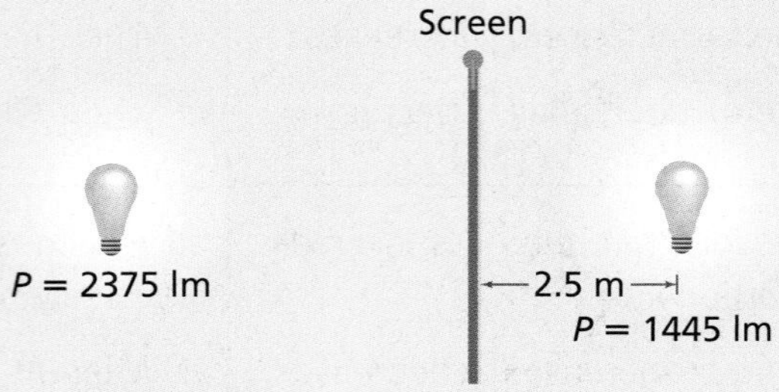

Figure 16-7 (Not to scale)

Engineers who design lighting systems must understand how the light will be used. If an even illumination is needed to prevent dark areas, the common practice is to evenly space normal lights over the area to be illuminated, as was most likely done with the lights in your classroom. Because such light sources do not produce truly even light, however, engineers also design special light sources that control the spread of the light, such that they produce even illuminations over large surface areas. Much work has been done in this field with automobile headlights.

The Speed of Light

For light to travel from a source to an object to be illuminated, it must travel across some distance. According to classical mechanics, if you can measure the distance and the time it takes to travel that distance, you can calculate a speed. Before the seventeenth century, most people believed that light traveled instantaneously. Galileo was the first to hypothesize that light has a finite speed, and to suggest a method of measuring its speed using distance and time. His method, however, was not precise enough, and he was forced to conclude that the speed of light is too fast to be measured over a distance of a few kilometers.

Danish astronomer Ole Roemer was the first to determine that light does travel with a measurable speed. Between 1668 and 1674, Roemer made 70 measurements of the 1.8-day orbital period of Io, one of Jupiter's moons. He recorded the times when Io emerged from Jupiter's shadow, as shown in **Figure 16-8.** He made his measurements as part of a project to improve maps by calculating the longitude of locations on Earth. This is an early example of the needs of technology driving scientific advances.

After making many measurements, Roemer was able to predict when the next eclipse of Io would occur. He compared his predictions with the actual measured times and found that Io's orbital period increased on average by about 13 s per orbit when Earth was moving away from Jupiter and decreased on average by about 13 s per orbit when Earth was approaching Jupiter. Roemer believed that Jupiter's moons were just as regular in their orbits as Earth's moon; thus, he wondered what might cause this discrepancy in the measurement of Io's orbital period.

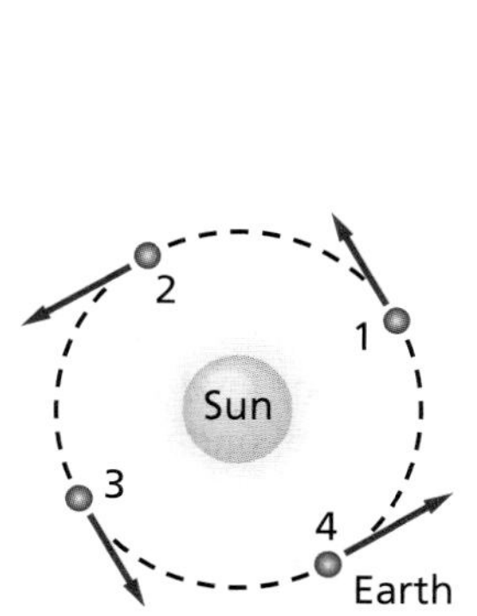

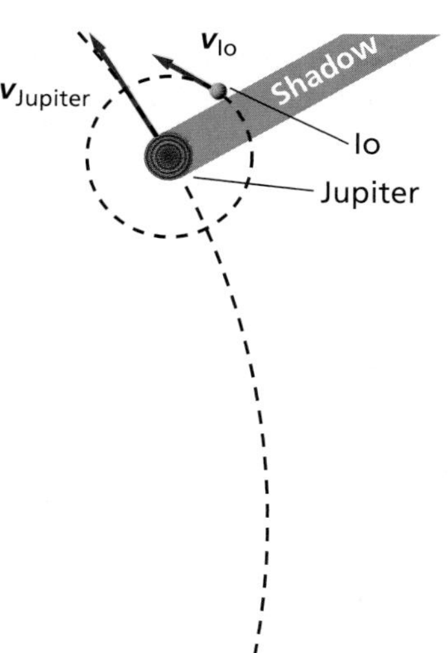

Figure 16-8 Roemer measured the time between eclipses as Io emerged from Jupiter's shadow. During successive eclipses, Io's measured orbital period became increasingly smaller or larger depending on whether Earth was moving toward (from position 3 to 1) or away from (from position 1 to 3) Jupiter. (Illustration not to scale)

Measurements of the speed of light Roemer concluded that as Earth moved away from Jupiter, the light from each new appearance of Io took longer to travel to Earth because of the increasing distance to Earth. Likewise, as Earth approached Jupiter, Io's orbital period seemed to decrease. Roemer noted that during the 182.5 days that it took Earth to travel from position 1 to position 3, as shown in Figure 16-8, there were (185.2 days)(1 Io eclipse/1.8 days) = 103 Io eclipses. Thus, for light to travel the diameter of Earth's orbit, he calculated that it takes (103 eclipses)(13 s/eclipse) = 1.3×10^3 s, or 22 min.

Using the presently known value of the diameter of Earth's orbit (2.9×10^{11} m), Roemer's value of 22 min gives a value for the speed of light of 2.9×10^{11} m/((22 min)(60 s/min)) = 2.2×10^8 m/s. Today, the speed of light is known to be closer to 3.0×10^8 m/s. Thus, light takes 16.5 min, not 22 min, to cross Earth's orbit. Nevertheless, Roemer had successfully proved that light moves at a finite speed.

Although many measurements of the speed of light have been made, the most notable were performed by American physicist Albert A. Michelson. Between 1880 and the 1920s, he developed Earth-based techniques to measure the speed of light. In 1926, Michelson measured the time required for light to make a round-trip between two California mountains 35 km apart. Michelson used a set of rotating mirrors to measure such small time intervals. Michelson's best result was $(2.997996 \pm 0.00004)\times10^8$ m/s. For this work, he became the first American to receive a Nobel prize in science.

The speed of light in a vacuum is a very important and universal value; thus it has its own special symbol, c. Based on the wave nature of light, which you will study in the next section, the International Committee on Weights and Measurements has measured and defined the speed of light in a vacuum to be $c = 299{,}792{,}458$ m/s. For many calculations, the value $c = 3.00\times10^8$ m/s is precise enough. At this speed, light travels 9.46×10^{12} km in a year. This amount of distance is called a light-year.

16.1 Section Review

7. **Use of Material Light Properties** Why might you choose a window shade that is translucent? Opaque?

8. **Illuminance** Does one lightbulb provide more illuminance than two identical lightbulbs at twice the distance? Explain.

9. **Luminous Intensity** Two lamps illuminate a screen equally—lamp A at 5.0 m, lamp B at 3.0 m. If lamp A is rated 75 cd, what is lamp B rated?

10. **Distance of a Light Source** Suppose that a lightbulb illuminating your desk provides only half the illuminance that it should. If it is currently 1.0 m away, how far should it be to provide the correct illuminance?

11. **Distance of Light Travel** How far does light travel in the time it takes sound to travel 1 cm in air at 20°C?

12. **Distance of Light Travel** The distance to the Moon can be found with the help of mirrors left on the Moon by astronauts. A pulse of light is sent to the Moon and returns to Earth in 2.562 s. Using the measured value of the speed of light to the same precision, calculate the distance from Earth to the Moon.

13. **Critical Thinking** Use the correct time taken for light to cross Earth's orbit, 16.5 min, and the diameter of Earth's orbit, 2.98×10^{11} m, to calculate the speed of light using Roemer's method. Does this method appear to be accurate? Why or why not?

Physics Online physicspp.com/self_check_quiz

16.2 The Wave Nature of Light

You probably have heard that light is composed of waves, but what evidence do you have that this is so? Suppose that you walk by the open door of the band-rehearsal room at school. You hear the music as you walk toward the rehearsal-room door long before you can see the band members through the door. Sound seems to have reached you by bending around the edge of the door, whereas the light that enables you to see the band members has traveled only in a straight line. If light is composed of waves, why doesn't light seem to act in the same way as sound does? In fact, light does act in the same way; however, the effect is much less obvious with light than with sound.

► **Objectives**

- **Describe** how diffraction demonstrates that light is a wave.
- **Predict** the effect of combining colors of light and mixing pigments.
- **Explain** phenomena such as polarization and the Doppler effect.

► **Vocabulary**

diffraction
primary color
secondary color
complementary color
primary pigment
secondary pigment
polarization
Malus's law

Diffraction and the Wave Model of Light

In 1665, Italian scientist Francesco Maria Grimaldi observed that the edges of shadows are not perfectly sharp. He introduced a narrow beam of light into a dark room and held a rod in front of the light such that it cast a shadow on a white surface. The shadow cast by the rod on the white surface was wider than the shadow should have been if light traveled in a straight line past the edges of the rod. Grimaldi also noted that the shadow was bordered by colored bands. Grimaldi recognized this phenomenon as **diffraction,** which is the bending of light around a barrier.

In 1678, Dutch scientist Christiaan Huygens argued in favor of a wave model to explain diffraction. According to Huygens' principle, all the points of a wave front of light can be thought of as new sources of smaller waves. These wavelets expand in every direction and are in step with one another. A flat, or plane, wave front of light consists of an infinite number of point sources in a line. As this wave front passes by an edge, the edge cuts the wave front such that each circular wavelet generated by each Huygens' point will propagate as a circular wave in the region where the original wave front was bent, as shown in **Figure 16-9.** This is diffraction.

Figure 16-9 According to Huygens' principle, the crest of each wave can be thought of as a series of point sources. Each point source creates a circular wavelet. All the wavelets combine to make a flat wave front, except at the edge where circular wavelets of the Huygens' points move away from the wave front.

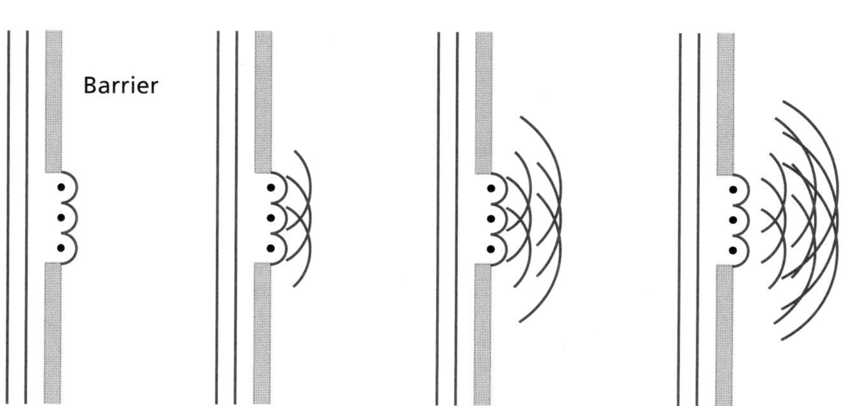

Concepts In Motion
Interactive Figure To see an animation on diffraction, visit physicspp.com.

Color

In 1666, possibly prompted by Grimaldi's publication of his diffraction results, Newton performed experiments on the colors produced when a narrow beam of sunlight passed through a glass prism, as shown in **Figure 16-10.** Newton called the ordered arrangement of colors a spectrum. Using his corpuscle model of light, he believed that particles of light were interacting with some unevenness in the glass to produce the spectrum.

Figure 16-10 White light, when passed through a prism, is separated into a spectrum of colors.

To test this assumption, Newton allowed the spectrum from one prism to fall on a second prism. If the spectrum was caused by irregularities in the glass, he reasoned that the second prism would increase the spread in colors. Instead, the second prism reversed the spreading of colors and recombined them to form white light. After more experiments, Newton concluded that white light is composed of colors, and that a property of the glass other than unevenness caused the light to separate into colors.

Based on the work of Grimaldi, Huygens, and others, we know that light has wave properties and that each color of light is associated with a wavelength. Visible light falls within the range of wavelengths from about 400 nm (4.00×10^{-7} m) to 700 nm (7.00×10^{-7} m), as shown in **Figure 16-11.** The longest wavelengths are seen as red light. As wavelength decreases, the color changes to orange, yellow, green, blue, indigo, and finally, violet.

Red (7.00×10^{-7} m) Violet (4.00×10^{-7} m)

Figure 16-11 The spectrum of visible light ranges from the long, red wavelength to the short, violet wavelength.

As white light crosses the boundary from air into glass and back into air in Figure 16-10, its wave nature causes each different color of light to be bent, or refracted, at a different angle. This unequal bending of the different colors causes the white light to be spread into a spectrum. This reveals that different wavelengths of light interact in different but predictable ways with matter.

Color by addition of light White light can be formed from colored light in a variety of ways. For example, when the correct intensities of red, green, and blue light are projected onto a white screen, as in **Figure 16-12,** the region where these three colors overlap on the screen will appear to be white. Thus, red, green, and blue light form white light when they are combined. This is called the additive color process, which is used in color-television tubes. A color-television tube contains tiny, dotlike sources of red, green, and blue light. When all three colors of light have the correct intensities, the screen appears to be white. For this reason, red, green, and blue are each called a **primary color.** The primary colors can be mixed in pairs to form three additional colors, as shown in Figure 16-12. Red and green light together produce yellow light, blue and green light produce cyan, and red and blue light produce magenta. The colors yellow, cyan, and magenta are each called a **secondary color,** because each is a combination of two primary colors.

Figure 16-12 Different combinations of blue, green, and red light can produce yellow, cyan, magenta, or white light.

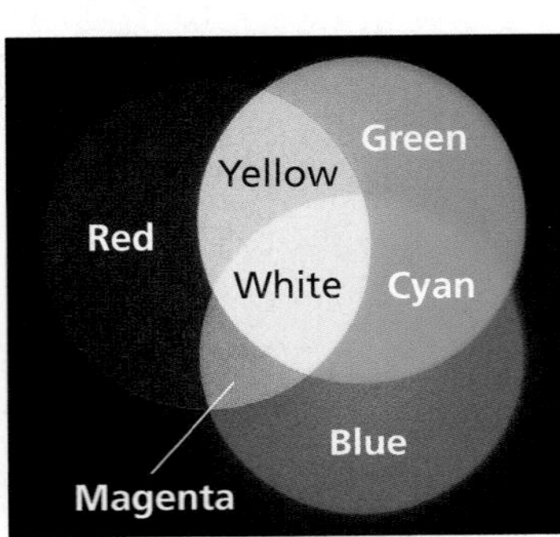

As shown in Figure 16-12, yellow light can be made from red light and green light. If yellow light and blue light are projected onto a white screen with the correct intensities, the surface will appear to be white. **Complementary colors** are two colors of light that can be combined to make white light. Thus, yellow is a complementary color of blue, and vice versa, because the two colors of light combine to make white light. In the same way, cyan and red are complementary colors, as are magenta and green. Yellowish laundry can be whitened with a bluing agent added to detergent.

Color by subtraction of light As you learned in the first section of this chapter, objects can reflect and transmit light. They also can absorb light. Not only does the color of an object depend on the wavelengths present in the light that illuminates the object, but an object's color also depends on what wavelengths are absorbed by the object and what wavelengths are reflected. The natural existence or artificial placement of dyes in the material of an object, or pigments on its surface, give the object color.

A dye is a molecule that absorbs certain wavelengths of light and transmits or reflects others. When light is absorbed, its energy is taken into the object that it strikes and is turned into other forms of energy. A red shirt is red because the dyes in it reflect red light to our eyes. When white light falls on the red object shown in **Figure 16-13,** the dye molecules in the object absorb the blue and green light and reflect the red light. When only blue light falls on the red object, very little light is reflected and the object appears to be almost black.

The difference between a dye and a pigment is that pigments usually are made of crushed minerals, rather than plant or insect extracts. Pigment particles can be seen with a microscope. A pigment that absorbs only one primary color and reflects two from white light is called a **primary pigment.** Yellow pigment absorbs blue light and reflects red and green light. Yellow, cyan, and magenta are the colors of primary pigments. A pigment that absorbs two primary colors and reflects one color is called a **secondary pigment.** The colors of secondary pigments are red (which absorbs green and blue light), green (which absorbs red and blue light), and blue (which absorbs red and green light). Note that the primary pigment colors are the secondary colors. In the same way, the secondary pigment colors are the primary colors.

MINI LAB

Color by Temperature

Some artists refer to red and orange as hot colors and green and blue as cool colors. Do colors really relate to temperature in this way?

1. Obtain a glass prism from your teacher.

2. Obtain a lamp with a dimmer switch from your teacher. Turn on the lamp and turn off the room light. Set the dimmer to minimum brightness of the lamp.

3. Slowly increase the brightness of the lamp. ***CAUTION: Lamp can get hot and burn skin.***

4. Observe the color of light produced by the prism and how it relates to the warmth of the lightbulb on your hand.

Analyze and Conclude

5. What colors appeared first when the light was dim?

6. What colors were the last to appear as you brightened the light?

7. How do these colors relate to the temperature of the filament?

Figure 16-13 The dyes in the dice selectively absorb and reflect various wavelengths of light. The dice are illuminated by white light **(a),** red light **(b),** and blue light **(c).**

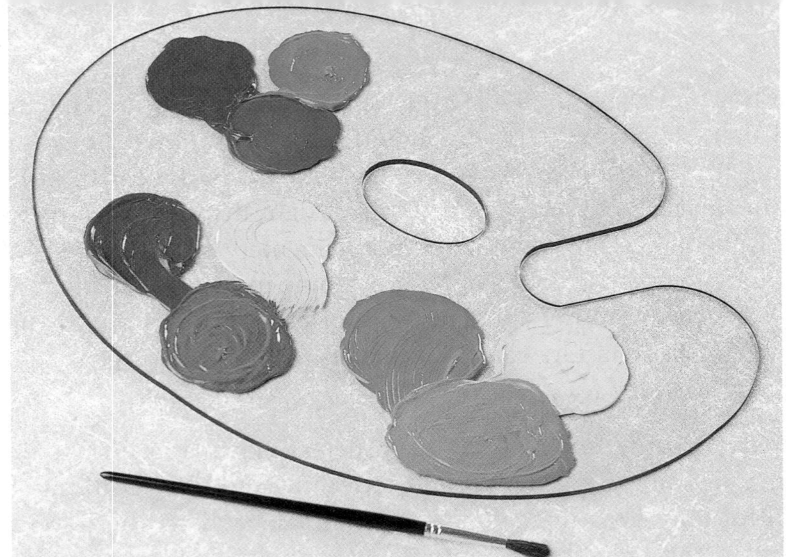

■ **Figure 16-14** The primary pigments are magenta, cyan, and yellow. Mixing these in pairs produces the secondary pigments: red, green, and blue.

Concepts In Motion
Interactive Figure To see an animation on primary colors of light, visit physicspp.com.

The primary and secondary pigments are shown in **Figure 16-14.** When the primary pigments yellow and cyan are mixed, the yellow absorbs blue light and the cyan absorbs red light. Thus, Figure 16-14 shows yellow and cyan combining to make green pigment. When yellow pigment is mixed with the secondary pigment, blue, which absorbs green and red light, all of the primary colors are absorbed, so the result is black. Thus, yellow and blue are complementary pigments. Cyan and red, as well as magenta and green, are also complementary pigments.

A color printer uses yellow, magenta, and cyan dots of pigment to make a color image on paper. Often, pigments that are used are finely ground compounds, such as titanium(IV) oxide (white), chromium(III) oxide (green), and cadmium sulfide (yellow). Pigments mix to form suspensions rather than solutions. Their chemical form is not changed in a mixture, so they still absorb and reflect the same wavelengths.

Chemistry Connection

Results in color You can now begin to understand the colors that you see in the photo at the beginning of this chapter. The plants on the hillside look green because of the chlorophyll in them. One type of chlorophyll absorbs red light and the other absorbs blue light, but they both reflect green light. The energy in the red and blue light that is absorbed is used by the plants for photosynthesis, which is the process by which green plants make food.

Biology Connection

In the same photo, the sky is bluish. Violet and blue light are scattered (repeatedly reflected) much more by molecules in the air than are other wavelengths of light. Green and red light are not scattered much by the air, which is why the Sun looks yellow or orange, as shown in **Figure 16-15.** However, violet and blue light from the Sun are scattered in all directions, illuminating the sky in a bluish hue.

■ **Figure 16-15** The Sun can appear to be a shade of yellow or orange because of the scattering of violet and blue light.

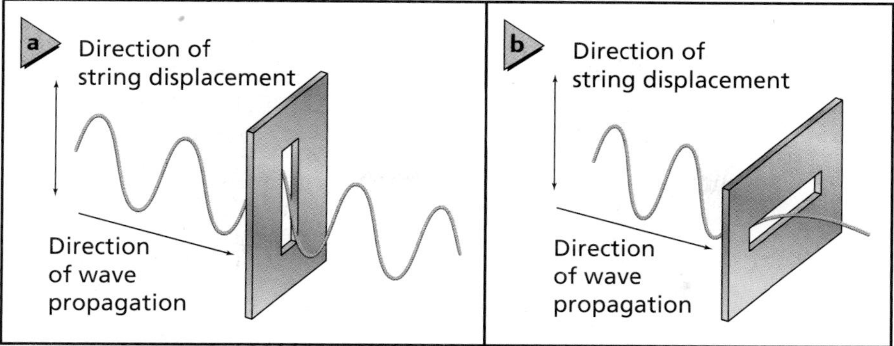

Figure 16-16 In this rope model of light, light is a single wave oriented in relation to the vertical plane and thus passes through a vertical polarizer **(a).** It cannot pass through a horizontal polarizer **(b).**

Polarization of Light

Have you ever looked at light reflected off a road through polarizing sunglasses? As you rotate the glasses, the road first appears to be dark, then light, and then dark again. Light from a lamp, however, changes very little as the glasses are rotated. Why is there a difference? Normal lamplight is not polarized. However, the light that is coming from the road is reflected and has become polarized. **Polarization** is the production of light in a single plane of oscillation.

Polarization by filtering Polarization can be understood by considering a rope model of light waves, as shown in **Figure 16-16.** The transverse mechanical waves in the rope represent transverse light waves. The slot represents what is referred to as the polarizing axis of a polarizing medium. When the rope waves are parallel to the slot, they pass through. When they are perpendicular to the slot, the waves are blocked. Polarizing media contain long molecules in which electrons can oscillate, or move back and forth, all in the same direction. As light travels past the molecules, the electrons can absorb light waves that oscillate in the same direction as the electrons. This process allows light waves vibrating in one direction to pass through, while the waves vibrating in the other direction are absorbed. The direction of a polarizing medium perpendicular to the long molecules is called the polarizing axis. Only waves oscillating parallel to that axis can pass through.

Ordinary light actually contains waves vibrating in every direction perpendicular to its direction of travel. If a polarizing medium is placed in a beam of ordinary light, only the components of the waves in the same direction as the polarizing axis can pass through. On average, half of the total light amplitude passes through, thereby reducing the intensity of the light by half. The polarizing medium produces light that is polarized. Such a medium is called a polarizing filter.

Polarization by reflection When you look through a polarizing filter at the light reflected by a sheet of glass and rotate the filter, you will see the light brighten and dim. The light is partially polarized along the plane of the glass when it is reflected. That is, the reflected ray contains a great deal of light vibrating parallel to the surface of the glass. The polarization of light reflected by roads is the reason why polarizing sunglasses reduce glare. The fact that the intensity of light reflected off a road varies as polarizing sunglasses are rotated suggests that the reflected light is partially polarized. Photographers can use polarizing filters over camera lenses to block reflected light, as shown in **Figure 16-17.**

Figure 16-17 This photo of a music store, taken without a polarizing filter, contains the glare of light off of the surface of the window **(a).** This photo of the same scene was taken with a polarizing filter **(b).**

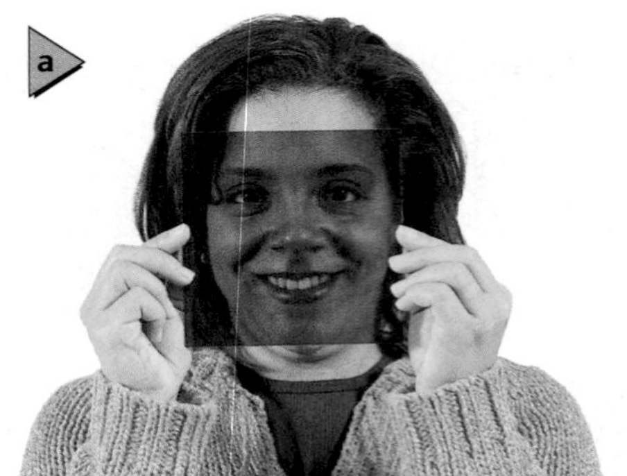

Figure 16-18 When two polarizing filters are arranged with their polarizing axes in parallel, a maximum amount of light passes through **(a).** When two polarizing filters are arranged with perpendicular axes, no light passes through **(b).**

Polarization analysis Suppose that you produce polarized light with a polarizing filter. What would happen if you place a second polarizing filter in the path of the polarized light? If the polarizing axis of the second filter is parallel to that of the first, the light will pass through, as shown in **Figure 16-18a.** If the polarizing axis of the second filter is perpendicular to that of the first, no light will pass through, as shown in **Figure 16-18b.**

The law that explains the reduction of light intensity as it passes through a second polarizing filter is called **Malus's law.** If the light intensity after the first polarizing filter is I_1, then a second polarizing filter, with its polarizing axis at an angle, θ, relative to the polarizing axis of the first, will result in a light intensity, I_2, that is equal to or less than I_1.

Malus's Law $I_2 = I_1\cos^2\theta$

The intensity of light coming out of a second polarizing filter is equal to the intensity of polarized light coming out of a first polarizing filter multiplied by the cosine, squared, of the angle between the polarizing axes of the two filters.

Using Malus's law, you can compare the light intensity coming out of the second polarizing filter to the light intensity coming out of the first polarizing filter, and thereby determine the orientation of the polarizing axis of the first filter relative to the second filter. A polarizing filter that uses Malus's law to accomplish this is called an analyzer. Analyzers can be used to determine the polarization of light coming from any source.

CHALLENGE PROBLEM

You place an analyzer filter between the two cross-polarized filters, such that its polarizing axis is not parallel to either of the two filters, as shown in the figure to the right.

1. You observe that some light passes through filter 2, though no light passed through filter 2 previous to inserting the analyzer filter. Why does this happen?
2. The analyzer filter is placed at an angle of θ relative to the polarizing axes of filter 1. Derive an equation for the intensity of light coming out of filter 2 compared to the intensity of light coming out of filter 1.

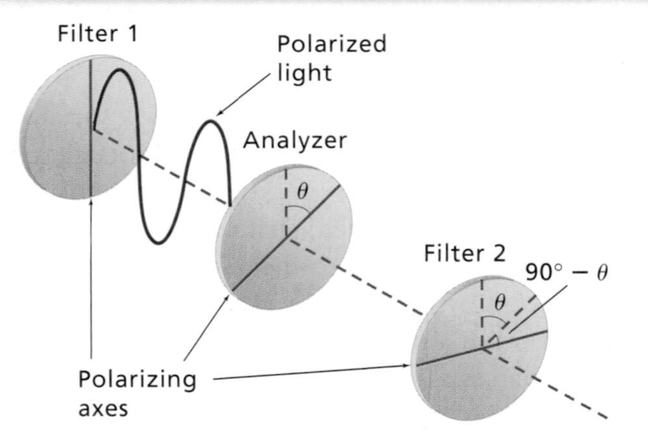

The Speed of a Light Wave

As you learned in Chapter 14, the wavelength, λ, of a wave is a function of its speed in the medium in which it is traveling and its constant frequency, f. Because light has wave properties, the same mathematical models used to describe waves in general can be used to describe light. For light of a given frequency traveling through a vacuum, wavelength is a function of the speed of light, c, which can be written as $\lambda_0 = c/f$. The development of the laser in the 1960s provided new methods of measuring the speed of light. The frequency of light can be counted with extreme precision using lasers and the time standard provided by atomic clocks. Measurements of wavelengths of light, however, are much less precise.

Different colors of light have different frequencies and wavelengths, but in a vacuum, they all travel at c. Because all wavelengths of light travel at the same speed in a vacuum, when you know the frequency of a light wave in a vacuum, you can calculate its wavelength, and vice versa. Thus, using precise measurements of light frequency and light speed, you can calculate a precise value of light wavelength.

Relative motion and light What happens if a source of light is traveling toward you or you are moving toward the light source? You learned in Chapter 15 that the frequency of a sound heard by the listener changes if either the source or the listener of the sound is moving. The same is true for light. However, when you consider the velocities of a source of sound and the observer, you are really considering each one's velocity relative to the medium through which the sound travels.

Because light waves are not vibrations of the particles of a mechanical medium, unlike sound waves, the Doppler effect of light can involve only the velocities of the source and the observer relative to each other. The magnitude of the difference between the velocities of the source and observer is called the relative speed. Remember that the only factors in the Doppler effect are the velocity components along the axis between the source and observer, as shown in **Figure 16-19.**

Figure 16-19 The observer and the light source have different velocities **(a).** The magnitude of the vector subtraction of the velocity components along the axis between the source of light and the observer of the light is referred to as the relative speed along the axis between the source and observer, v **(b).** (Illustration not to scale)

Concepts In Motion
Interactive Figure To see an animation on relative motion and light, visit physicspp.com.

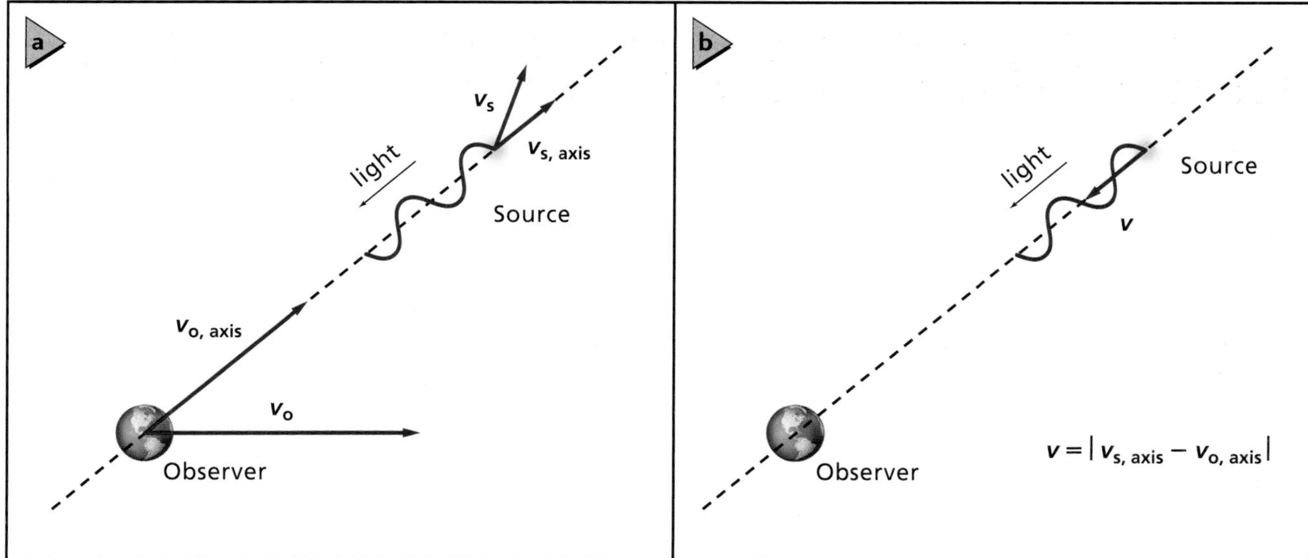

The Doppler effect To study the Doppler effect for light, the problem can be simplified by considering axial relative speeds that are much less than the speed of light ($v << c$). This simplification is used to develop the equation for the obseved light frequency, f_{obs}, which is the frequency of light as seen by an observer.

Observed Light Frequency $f_{obs} = f\left(1 \pm \frac{v}{c}\right)$

The observed frequency of light from a source is equal to the actual frequency of the light generated by the source, times the quantity 1 plus the relative speed along the axis between the source and the observer if they are moving toward each other, or 1 minus the relative speed if they are moving away from each other.

Because most observations of the Doppler effect for light have been made in the context of astronomy, the equation for the Doppler effect for light generally is written in terms of wavelength rather than frequency. Using the relationship $\lambda = c/f$ and the $v << c$ simplification, the following equation can be used for the Doppler shift, $\Delta\lambda$, which is the difference between the observed wavelength of light and the actual wavelength.

Doppler Shift $(\lambda_{obs} - \lambda) = \Delta\lambda = \pm\frac{v}{c}\lambda$

The difference between the observed wavelength of light and the actual wavelength of light generated by a source is equal to the actual wavelength of light generated by the source, times the relative speed of the source and observer, divided by the speed of light. This quantity is positive if they are moving away from each other or negative if they are moving toward each other.

A positive change in wavelength means that the light is red-shifted. This occurs when the relative velocity of the source is in a direction away from the observer. A negative change in wavelength means that the light is blue-shifted. This occurs when the relative velocity of the source is in a direction toward the observer. When the wavelength is red-shifted (lengthens), the observed frequency is lower as a result of the inverse relationship between the two variables, because the speed of light remains constant. When the wavelength is blue-shifted, the observed frequency is higher.

Researchers can determine how astronomical objects, such as galaxies, are moving relative to Earth by observing the Doppler shift of light. This is done by observing the spectrum of light coming from stars in the galaxy using a spectrometer, as shown in **Figure 16-20.** Elements that are present in the stars of galaxies emit specific wavelengths in the lab. A spectrometer is able to measure the Doppler shift of these wavelengths.

Figure 16-20 Three emission lines of hydrogen are visibly redshifted in the spectrum of quasar 3C 273, as indicated by the taglines outside the spectra. Their wavelengths are shifted approximately 16% of what they are in a laboratory setting.

PRACTICE Problems

Additional Problems, Appendix B

14. What is the frequency of oxygen's spectral line if its wavelength is 513 nm?

15. A hydrogen atom in a galaxy moving with a speed of 6.55×10^{6} m/s away from Earth emits light with a frequency of 6.16×10^{14} Hz. What frequency of light from that hydrogen atom would be observed by an astronomer on Earth?

16. A hydrogen atom in a galaxy moving with a speed of 6.55×10^{16} m/s away from Earth emits light with a wavelength of 4.86×10^{-7} m. What wavelength would be observed on Earth from that hydrogen atom?

17. An astronomer is looking at the spectrum of a galaxy and finds that it has an oxygen spectral line of 525 nm, while the laboratory value is measured at 513 nm. Calculate how fast the galaxy would be moving relative to Earth. Explain whether the galaxy is moving toward or away from Earth and how you know.

In 1929, Edwin Hubble suggested that the universe is expanding. Hubble reached his conclusion of the expanding universe by analyzing emission spectra from many galaxies. Hubble noticed that the spectral lines of familiar elements were at longer wavelengths than expected. The lines were shifted toward the red end of the spectrum. No matter what area of the skies he observed, the galaxies were sending red-shifted light to Earth. What do you think caused the spectral lines to be red-shifted? Hubble concluded that all galaxies are moving away from Earth.

Astronomy Connection

You have learned that some characteristics of light can be explained with a simple ray model of light, whereas others require a wave model of light. In Chapters 17 and 18, you will use both of these models to study how light interacts with mirrors and lenses. In Chapter 19, you will learn about other aspects of light that can be understood only through the use of the wave model of light.

16.2 Section Review

18. Addition of Light Colors What color of light must be combined with blue light to obtain white light?

19. Combination of Pigments What primary pigment colors must be mixed to produce red? Explain how red results using color subtraction for pigment colors.

20. Light and Pigment Interaction What color will a yellow banana appear to be when illuminated by each of the following?

a. white light

b. green and red light

c. blue light

21. Wave Properties of Light The speed of red light is slower in air and water than in a vacuum. The frequency, however, does not change when red light enters water. Does the wavelength change? If so, how?

22. Polarization Describe a simple experiment that you could do to determine whether sunglasses in a store are polarizing.

23. Critical Thinking Astronomers have determined that Andromeda, a neighboring galaxy to our own galaxy, the Milky Way, is moving toward the Milky Way. Explain how they determined this. Can you think of a possible reason why Andromeda is moving toward our galaxy?

Physics Online physicspp.com/self_check_quiz

PHYSICS LAB

Polarization of Light

Alternate CBL instructions can be found on the Web site.
physicspp.com

A light source that produces transverse light waves that are all in the same fixed plane is said to be polarized in that plane. A polarizing filter can be used to find light sources that produce polarized light. Some media can rotate the plane of polarization of light as it transmits the light. Such media are said to be optically active. In this activity, you will investigate these concepts of polarized light.

QUESTION

What types of luminous and illuminated light sources produce polarized light?

Objectives

- **Experiment** with various sources of light and polarizing filters.
- **Describe** the results of your experiment.
- **Recognize** possible uses of polarizing filters in everyday life.

Safety Precautions

- **Minimize the length of time you look directly at bright light sources.**
- **Do not do this lab with laser light sources.**
- **Do not look at the Sun, even if you are using polarizing filters.**
- **Light sources can get hot and burn skin.**

Materials

two polarizing filter sheets
incandescent light source
fluorescent light source
pieces of white and black paper
calculator with a liquid crystal display
clear, plastic protractor
mirror

Procedure

1. Take a polarizing filter and look at an incandescent light source. Rotate the filter. Write your observations in the data table.
2. Use a polarizing filter to look at a fluorescent light source. Rotate the filter. Write your observations in the data table.
3. Use a polarizing filter to observe light reflected off the surface of a mirror at approximately a 45° angle. Rotate the filter. Record your observations in the data table.

Data Table	
Light Source	Observations
1	
2	
3	
4	
5	
6	
7	
8	

4. Use a polarizing filter to observe light reflected off a white piece of paper at approximately a 45° angle. Rotate the filter. Record your observations in the data table.
5. Use a polarizing filter to observe light reflected off a piece of black paper at approximately a 45° angle. Rotate the filter. Record your observations in the data table.
6. Use a polarizing filter to observe a liquid crystal display on a calculator. Rotate the filter. Write your observations in the data table.
7. Place one polarizing filter on top of the other filter. Look at an incandescent light source through this set of the filters. Rotate one of the filters with respect to the other. Make a complete rotation. Record your observations in the data table.
8. Place a clear, plastic protractor between the two polarizing filters. Look at an incandescent light source with this. Do a complete rotation of one of the filters. Position the two filters the same way that produced no light in step 7. Record your observations in the data table.

Analyze

1. **Interpret Data** Does incandescent light produce polarized light? How do you know?
2. **Interpret Data** Does fluorescent light produce polarized light? How do you know?
3. **Interpret Data** Does reflection from a mirrored surface produce polarized light? How do you know?
4. **Compare and Contrast** How does reflected light from white paper compare to reflected light from black paper in terms of polarized light? Why are they different?
5. **Interpret Data** Is the light from liquid crystal displays polarized? How do you know?

Conclude and Apply

1. **Analyze and Conclude** How can two polarizing filters be used to prevent any light from passing through them?
2. **Analyze and Conclude** Why can the clear, plastic protractor between the polarizing filters be seen even though nothing else can be seen through the polarizing filters?
3. **Draw Conclusions** In general, what types of situations produce polarized light?

Going Further

1. On a sunny day, look at the polarization of blue sky near and far from the Sun using a polarizing filter. ***CAUTION: Do not look directly at the Sun.*** What characteristics of polarized light do you observe?
2. Is reflected light from clouds polarized? Make an observation to confirm your answer.

Real-World Physics

1. Why are high quality sunglasses made with polarizing lenses?
2. Why are polarizing sunglasses a better option than tinted sunglasses when driving a car?

To find out more about light, visit the Web site: **physicspp.com**

Technology and Society

Advances In Lighting

History has recorded the use of oil, candles, and gas to provide illumination in the dark hours of the night. However, there has always been inherent danger with the use of open flames to provide light. The invention of electric lighting in the nineteenth century provided brighter light and improved safety to the public.

The original form of electric light, which is still in common use, is the incandescent bulb. A tungsten filament is heated by electricity until it glows white. The tungsten does not burn, but it vaporizes, which eventually breaks the filament. Because the light is a result of heating the tungsten, this is not very efficient. Recent pursuits in electric lighting have produced longer-lasting, lower-heat light sources.

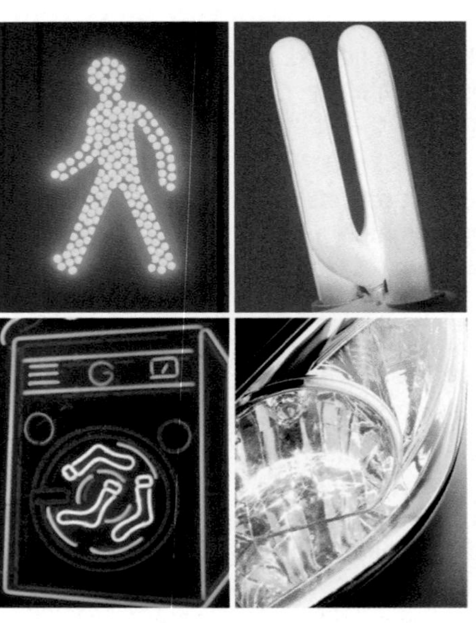

Clockwise from the upper left, the photos show LEDs, a fluorescent light, a halogen light, and gas-discharge lamps in the form of neon lights.

Quartz-Halogen Lamps To prevent a filament from breaking, the bulb can be made very small and filled with bromine or iodine gas. Tungsten ions from the filament combine with the gas molecules in the cooler parts of the lamp to make a compound, which circulates through the lamp and recombines with the filament. The light is very white and bright, but it also is very hot. An ordinary glass bulb would melt, so fused quartz, which has a very high melting point, is used.

Gas-Discharge Lamps This type of lamp is made of a glass tube with a wire electrode sealed into each end. All of the air is extracted and replaced by a very small amount of a specially chosen gas. A high voltage is applied across the electrodes. The electricity ionizes, or strips, some electrons from the gas atoms. An ionized gas is a good conductor, so electric current flows through it, causing the gas to glow brightly.

The use of a gas-discharge lamp depends upon the type of gas: neon for advertising, xenon for searchlights and camera flashes, and sodium vapor for streetlights. Each type of gas produces a different color, but the construction of each lamp is very similar.

Fluorescent Lamps The glow produced by mercury vapor is almost invisible because most of its spectrum is in the ultraviolet region, which is not visible. A fluorescent lamp is made by painting the inside of a mercury-discharge lamp with phosphor, a chemical that glows brightly when ultraviolet light strikes it. Fluorescent lights can be made in any color by changing the mixture of red, green, and blue phosphors. They have a long life and are economical to use, because they produce little heat and a great deal of light.

Light-Emitting Diodes The light of the future may be the white light-emitting diode, or LED. The LED produces white light by illuminating a tiny phosphor screen inside the LED with blue light. LEDs are bright enough to read by and produce almost no heat as they operate. They are so efficient that a car battery could power the lamps in a home for days without being recharged.

Going Further

1. **Observe** Novelty stores sell many devices that use lights. Examine some of them to see what types of lamp technology are used.
2. **Research** Find out about the inner construction, characteristic color, and typical uses of a few types of gas-discharge lamps.

Chapter 16

Study Guide

16.1 Illumination

Vocabulary
- ray model of light (p. 432)
- luminous source (p. 432)
- illuminated source (p. 432)
- opaque (p. 433)
- transparent (p. 433)
- translucent (p. 433)
- luminous flux (p. 433)
- illuminance (p. 433)

Key Concepts
- Light travels in a straight line through any uniform medium.
- Materials can be characterized as being transparent, translucent, or opaque, depending on the amount of light that they reflect, transmit, or absorb.
- The luminous flux of a light source is the rate at which light is emitted. It is measured in lumens (lm).
- Illuminance is the luminous flux per unit area. It is measured in lux (lx), or lumens per square meter (lm/m^2).
- For a point source, illuminance follows an inverse-square relationship with distance and a direct relationship with luminous flux.

$$E = \frac{P}{4\pi r^2}$$

- In a vacuum, light has a constant speed of $c = 3.00\times10^8$ m/s.

16.2 The Wave Nature of Light

Vocabulary
- diffraction (p. 439)
- primary color (p. 440)
- secondary color (p. 440)
- complementary colors (p. 441)
- primary pigment (p. 441)
- secondary pigment (p. 441)
- polarization (p. 443)
- Malus's law (p. 444)

Key Concepts
- Light can have wavelengths between 400 and 700 nm.
- White light is a combination of the spectrum of colors, each color having a different wavelength.
- Combining the primary colors, red, blue, and green, forms white light. Combinations of two primary colors form the secondary colors, yellow, cyan, and magenta.
- The primary pigments, cyan, magenta, and yellow, are used in combinations of two to produce the secondary pigments, red, blue, and green.
- Polarized light consists of waves oscillating in the same plane.
- When two polarizing filters are used to polarize light, the intensity of the light coming out of the last filter is dependent on the angle between the polarizing axes of the two filters.

$$I_2 = I_1\cos^2\theta$$

- Light waves traveling through a vacuum can be characterized in terms of frequency, wavelength, and the speed of light.

$$\lambda_0 = \frac{c}{f}$$

- Light waves are Doppler shifted based upon the relative speed along the axis of the observer and the source of light.

$$f_{obs} = f\left(1 \pm \frac{v}{c}\right)$$

$$\Delta\lambda = (\lambda_{obs} - \lambda) = \pm\frac{v}{c}\lambda$$

Chapter 16 Assessment

Concept Mapping

24. Complete the following concept map using the following terms: *wave, c, Doppler effect, polarization.*

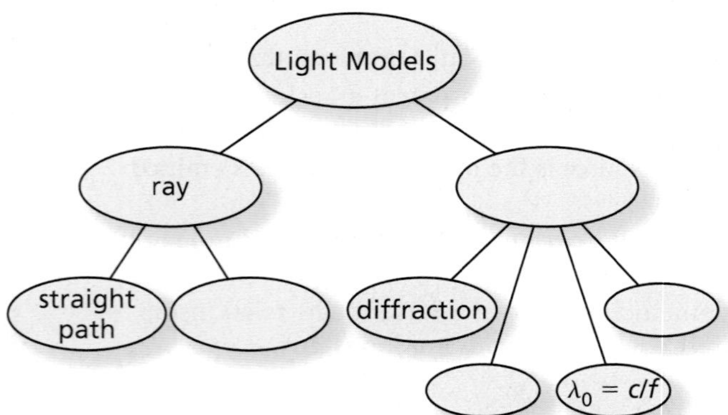

Mastering Concepts

25. Sound does not travel through a vacuum. How do we know that light does? (16.1)

26. Distinguish between a luminous source and an illuminated source. (16.1)

27. Look carefully at an ordinary, frosted, incandescent bulb. Is it a luminous or an illuminated source? (16.1)

28. Explain how you can see ordinary, nonluminous classroom objects. (16.1)

29. Distinguish among transparent, translucent, and opaque objects. (16.1)

30. To what is the illumination of a surface by a light source directly proportional? To what is it inversely proportional? (16.1)

31. What did Galileo assume about the speed of light? (16.1)

32. Why is the diffraction of sound waves more familiar in everyday experience than is the diffraction of light waves? (16.2)

33. What color of light has the shortest wavelength? (16.2)

34. What is the range of the wavelengths of light, from shortest to longest? (16.2)

35. Of what colors does white light consist? (16.2)

36. Why does an object appear to be black? (16.2)

37. Can longitudinal waves be polarized? Explain. (16.2)

38. If a distant galaxy were to emit a spectral line in the green region of the light spectrum, would the observed wavelength on Earth shift toward red light or toward blue light? Explain. (16.2)

39. What happens to the wavelength of light as the frequency increases? (16.2)

Applying Concepts

40. A point source of light is 2.0 m from screen A and 4.0 m from screen B, as shown in **Figure 16-21.** How does the illuminance at screen B compare with the illuminance at screen A?

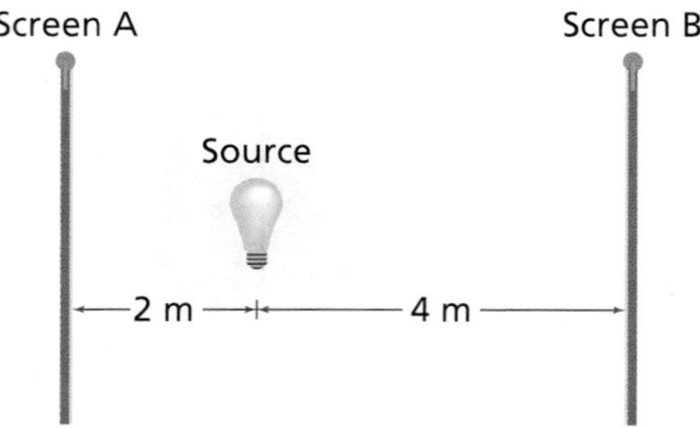

Figure 16-21

41. Reading Lamp You have a small reading lamp that is 35 cm from the pages of a book. You decide to double the distance.
   a. Is the illuminance at the book the same?
   b. If not, how much more or less is it?

42. Why are the insides of binoculars and cameras painted black?

43. Eye Sensitivity The eye is most sensitive to yellow-green light. Its sensitivity to red and blue light is less than 10 percent as great. Based on this knowledge, what color would you recommend that fire trucks and ambulances be painted? Why?

44. Streetlight Color Some very efficient streetlights contain sodium vapor under high pressure. They produce light that is mainly yellow with some red. Should a community that has these lights buy dark blue police cars? Why or why not?

*Refer to **Figure 16-22** for problems 45 and 46.*

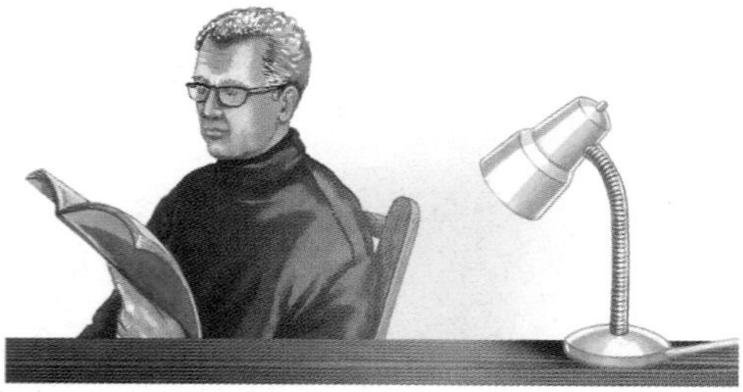

Figure 16-22

45. What happens to the illuminance at a book as the lamp is moved farther away from the book?

46. What happens to the luminous intensity of the lamp as it is moved farther away from the book?

 For more problems, go to Additional Problems, Appendix B.

47. **Polarized Pictures** Photographers often put polarizing filters over camera lenses to make clouds in the sky more visible. The clouds remain white, while the sky looks darker. Explain this based on your knowledge of polarized light.

48. An apple is red because it reflects red light and absorbs blue and green light.
    a. Why does red cellophane look red in reflected light?
    b. Why does red cellophane make a white lightbulb look red when you hold the cellophane between your eye and the lightbulb?
    c. What happens to the blue and green light?

49. You put a piece of red cellophane over one flashlight and a piece of green cellophane over another. You shine the light beams on a white wall. What color will you see where the two flashlight beams overlap?

50. You now put both the red and green cellophane pieces over one of the flashlights in Problem 49. If you shine the flashlight beam on a white wall, what color will you see? Explain.

51. If you have yellow, cyan, and magenta pigments, how can you make a blue pigment? Explain.

52. **Traffic Violation** Suppose that you are a traffic officer and you stop a driver for going through a red light. Further suppose that the driver draws a picture for you (**Figure 16-23**) and explains that the light looked green because of the Doppler effect when he went through it. Explain to him using the Doppler shift equation just how fast he would have had to be going for the red light (λ = 645 nm), to appear green (λ = 545 nm). *Hint: For the purpose of this problem, assume that the Doppler shift equation is valid at this speed.*

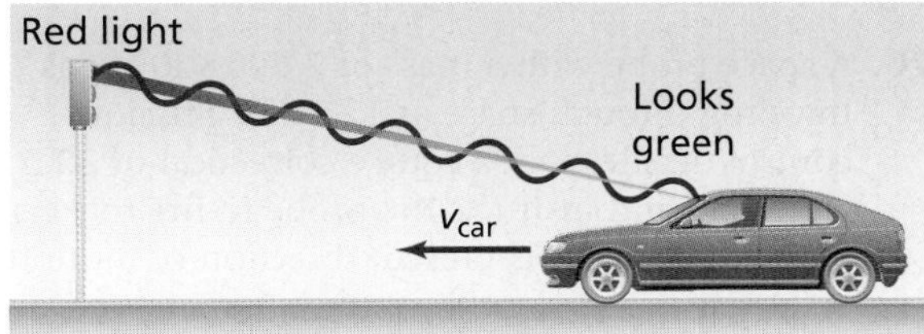

Figure 16-23

Mastering Problems

16.1 Illumination

53. Find the illumination 4.0 m below a 405-lm lamp.

54. Light takes 1.28 s to travel from the Moon to Earth. What is the distance between them?

55. A three-way bulb uses 50, 100, or 150 W of electric power to deliver 665, 1620, or 2285 lm in its three settings. The bulb is placed 80 cm above a sheet of paper. If an illumination of at least 175 lx is needed on the paper, what is the minimum setting that should be used?

56. **Earth's Speed** Ole Roemer found that the average increased delay in the disappearance of Io from one orbit around Jupiter to the next is 13 s.
    a. How far does light travel in 13 s?
    b. Each orbit of Io takes 42.5 h. Earth travels the distance calculated in part **a** in 42.5 h. Find the speed of Earth in km/s.
    c. Check to make sure that your answer for part **b** is reasonable. Calculate Earth's speed in orbit using the orbital radius, 1.5×10^{8} km, and the period, 1.0 yr.

57. A student wants to compare the luminous flux of a lightbulb with that of a 1750-lm lamp. The lightbulb and the lamp illuminate a sheet of paper equally. The 1750-lm lamp is 1.25 m away from the sheet of paper; the lightbulb is 1.08 m away. What is the lightbulb's luminous flux?

58. Suppose that you wanted to measure the speed of light by putting a mirror on a distant mountain, setting off a camera flash, and measuring the time it takes the flash to reflect off the mirror and return to you, as shown in **Figure 16-24.** Without instruments, a person can detect a time interval of about 0.10 s. How many kilometers away would the mirror have to be? Compare this distance with that of some known distances.

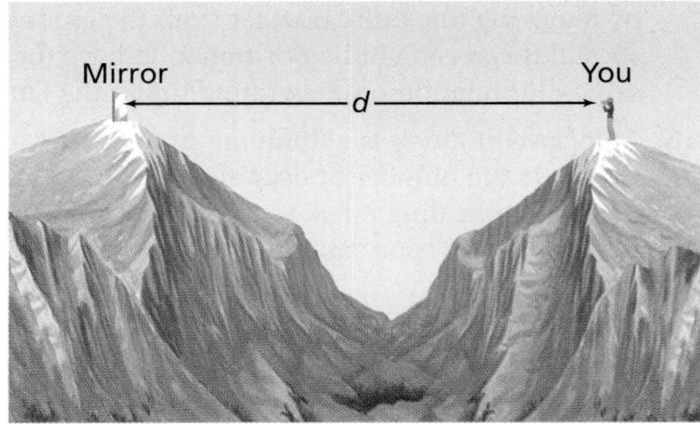

Figure 16-24

16.2 The Wave Nature of Light

59. Convert 700 nm, the wavelength of red light, to meters.

60. **Galactic Motion** How fast is a galaxy moving relative to Earth if a hydrogen spectral line of 486 nm is red-shifted to 491 nm?

61. Suppose that you are facing due east at sunrise. Sunlight is reflected off the surface of a lake, as shown in **Figure 16-25.** Is the reflected light polarized? If so, in what direction?

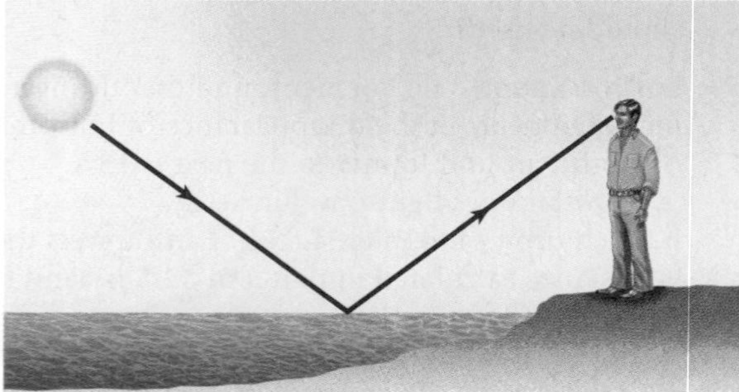

Figure 16-25

62. **Polarizing Sunglasses** In which direction should the transmission axis of polarizing sunglasses be oriented to cut the glare from the surface of a road: vertically or horizontally? Explain.

63. **Galactic Motion** A hydrogen spectral line that is known to be 434 nm is red-shifted by 6.50 percent in light coming from a distant galaxy. How fast is the galaxy moving away from Earth?

64. For any spectral line, what would be an unrealistic value of the apparent wavelength for a galaxy moving away from Earth? Why?

Mixed Review

65. **Streetlight Illumination** A streetlight contains two identical bulbs that are 3.3 m above the ground. If the community wants to save electrical energy by removing one bulb, how far from the ground should the streetlight be positioned to have the same illumination on the ground under the lamp?

66. An octave in music is a doubling of frequency. Compare the number of octaves that correspond to the human hearing range to the number of octaves in the human vision range.

67. A 10.0-cd point-source lamp and a 60.0-cd point-source lamp cast equal intensities on a wall. If the 10.0-cd lamp is 6.0 m from the wall, how far from the wall is the 60.0-cd lamp?

68. **Thunder and Lightning** Explain why it takes 5 s to hear thunder when lightning is 1.6 km away.

69. **Solar Rotation** Because the Sun rotates on its axis, one edge of the Sun moves toward Earth and the other moves away. The Sun rotates approximately once every 25 days, and the diameter of the Sun is 1.4×10^{9} m. Hydrogen on the Sun emits light of frequency 6.16×10^{14} Hz from the two sides of the Sun. What changes in wavelength are observed?

Thinking Critically

70. **Research** Why did Galileo's method for measuring the speed of light not work?

71. **Make and Use Graphs** A 110-cd light source is 1.0 m from a screen. Determine the illumination on the screen originally and for every meter of increasing distance up to 7.0 m. Graph the data.
    a. What is the shape of the graph?
    b. What is the relationship between illuminance and distance shown by the graph?

72. **Analyze and Conclude** If you were to drive at sunset in a city filled with buildings that have glass-covered walls, the setting Sun reflected off the building's walls might temporarily blind you. Would polarizing glasses solve this problem?

Writing in Physics

73. Write an essay describing the history of human understanding of the speed of light. Include significant individuals and the contribution that each individual made.

74. Look up information on the SI unit *candela, cd,* and explain in your own words the standard that is used to set the value of 1 cd.

Cumulative Review

75. A 2.0-kg object is attached to a 1.5-m long string and swung in a vertical circle at a constant speed of 12 m/s. (Chapter 7)
    a. What is the tension in the string when the object is at the bottom of its path?
    b. What is the tension in the string when the object is at the top of its path?

76. A space probe with a mass of 7.600×10^{3} kg is traveling through space at 125 m/s. Mission control decides that a course correction of 30.0° is needed and instructs the probe to fire rockets perpendicular to its present direction of motion. If the gas expelled by the rockets has a speed of 3.200 km/s, what mass of gas should be released? (Chapter 9)

77. When a 60.0-cm-long guitar string is plucked in the middle, it plays a note of frequency 440 Hz. What is the speed of the waves on the string? (Chapter 14)

78. What is the wavelength of a sound wave with a frequency of 17,000 Hz in water at 25°C? (Chapter 15)

 For more problems, go to Additional Problems, Appendix B.

Multiple Choice

1. In 1987, a supernova was observed in a neighboring galaxy. Scientists believed the galaxy was 1.66×10^{21} m away. How many years prior to the observation did the supernova explosion actually occur?

   (A) 5.53×10^{3} yr
   (B) 1.75×10^{5} yr
   (C) 5.53×10^{12} yr
   (D) 1.74×10^{20} yr

2. A galaxy is moving away at 5.8×10^{6} m/s. Its light appears to observers to have a frequency of 5.6×10^{14} Hz. What is the emitted frequency of the light?

   (A) 1.1×10^{13} Hz
   (B) 5.5×10^{14} Hz
   (C) 5.7×10^{14} Hz
   (D) 6.2×10^{14} Hz

3. Which of the following light color combinations is incorrect?

   (A) Red plus green produces yellow.
   (B) Red plus yellow produces magenta.
   (C) Blue plus green produces cyan.
   (D) Blue plus yellow produces white.

4. The illuminance of direct sunlight on Earth is about 1×10^{5} lx. A light on a stage has an intensity in a certain direction of 5×10^{6} cd. At what distance from the stage does a member of the audience experience an illuminance equal to that of sunlight?

   (A) 1.4×10^{-1} m
   (B) 7 m
   (C) 10 m
   (D) 5×10^{1} m

5. What is meant by the phrase *color by subtraction of light?*

   (A) Adding green, red, and blue light produces white light.
   (B) Exciting phosphors with electrons in a television produces color.
   (C) Paint color is changed by subtracting certain colors, such as producing blue paint from green by removing yellow.
   (D) The color that an object appears to be is a result of the material absorbing specific light wavelengths and reflecting the rest.

6. The illuminance due to a 60.0-W lightbulb at 3.0 m is 9.35 lx. What is the total luminous flux of the bulb?

   (A) 8.3×10^{-2} lm
   (B) 7.4×10^{-1} lm
   (C) 1.2×10^{2} lm
   (D) 1.1×10^{3} lm

7. Light from the Sun takes about 8.0 min to reach Earth. How far away is the Sun?

   (A) 2.4×10^{9} m
   (B) 1.4×10^{10} m
   (C) 1.4×10^{8} km
   (D) 2.4×10^{9} km

8. What is the frequency of 404 nm of light in a vacuum?

   (A) 2.48×10^{-3} Hz
   (B) 7.43×10^{5} Hz
   (C) 2.48×10^{6} Hz
   (D) 7.43×10^{14} Hz

Extended Answer

9. A celestial object is known to contain an element that emits light at a wavelength of 525 nm. The observed spectral line for this element is at 473 nm. Is the object approaching or receding, and at what speed?

10. Nonpolarized light of intensity I_o is incident on a polarizing filter, and the emerging light strikes a second polarizing filter, as shown in the figure. What is the light intensity emerging from the second polarizing filter?

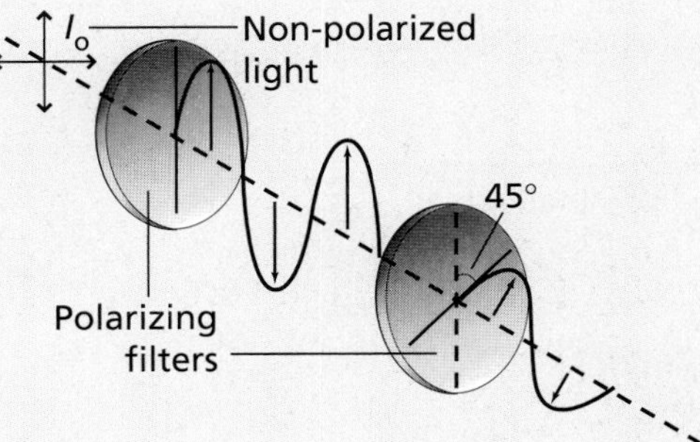

✓ Test-Taking TIP

Ask Questions

When you have a question about what will be on a test, the way a test is scored, the time limits placed on each section, or anything else, ask the instructor or the person giving the test.

Chapter 17 Reflection and Mirrors

What You'll Learn

- You will learn how light reflects off different surfaces.
- You will learn about the different types of mirrors and their uses.
- You will use ray tracing and mathematical models to describe images formed by mirrors.

Why It's Important

How light reflects off a surface into your eyes determines the reflection that you see. When you look down at the surface of a lake, you see an upright reflection of yourself.

Mountain Scene When you look across a lake, you might see a scene like the one in this photo. The image of the trees and mountains in the lake appears to you to be upside-down.

Think About This ▶

Why would the image you see of yourself in the lake be upright, while the image of the mountain is upside-down?

physicspp.com

LAUNCH Lab: How is an image shown on a screen?

Question

What types of mirrors are able to reflect an image onto a screen?

Procedure

1. Obtain an index card, a plane mirror, a concave mirror, a convex mirror, and a flashlight from your teacher.
2. Turn off the room lights and stand near the window.
3. Hold the index card in one hand. Hold the flat, plane mirror in the other hand.
4. Reflect the light coming through the window onto the index card. ***CAUTION: Do not look directly at the Sun or at the reflection of the Sun in a mirror.*** Slowly move the index card closer to and then farther away from the mirror and try to make a clear image of objects outside the window.
5. If you can project a clear image, this is called a real image. If you only see a fuzzy light on the index card then no real image is formed. Record your observations.
6. Repeat steps 3–5 with the concave and convex mirror.
7. Perform step 4 for each mirror with a flashlight and observe the reflection on the index card.

Analysis

Which mirror(s) produced a real image?

What are some things you notice about the image(s) you see?

Critical Thinking Based upon your observation of the flashlight images, propose an explanation of how a real image is formed.

17.1 Reflection from Plane Mirrors

▸ **Objectives**

- **Explain** the law of reflection.
- **Distinguish** between specular and diffuse reflection.
- **Locate** the images formed by plane mirrors.

▸ **Vocabulary**

specular reflection
diffuse reflection
plane mirror
object
image
virtual image

Undoubtedly, as long as there have been humans, they have seen their faces reflected in the quiet water of lakes and ponds. When you look at the surface of a body of water, however, you don't always see a clear reflection. Sometimes, the wind causes ripples in the water, and passing boats create waves. Disturbances on the surface of the water prevent the light from reflecting in a manner such that a clear reflection is visible.

Almost 4000 years ago, Egyptians understood that reflection requires smooth surfaces. They used polished metal mirrors to view their images. Sharp, well-defined, reflected images were not possible until 1857, however, when Jean Foucault, a French scientist, developed a method of coating glass with silver. Modern mirrors are produced using ever-increasing precision. They are made with greater reflecting ability by the evaporation of aluminum or silver onto highly polished glass. The quality of reflecting surfaces is even more important in applications such as lasers and telescopes. More than ever before, clear reflections in modern, optical instruments require smooth surfaces.

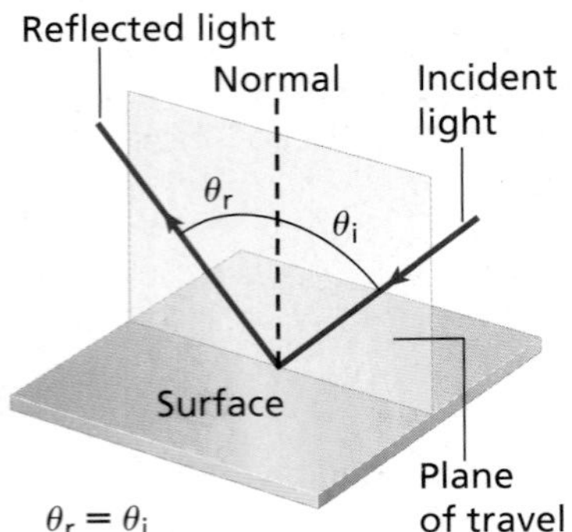

Figure 17-1 The incident ray and the reflected ray are in the same plane of travel.

Concepts In Motion
Interactive Figure To see an animation on the law of reflection, visit physicspp.com.

The Law of Reflection

What happens to the light that is striking this book? When you hold the book up to the light, you will see that no light passes through it. Recall from Chapter 16 that an object like this is called opaque. Part of the light is absorbed, and part is reflected. The absorbed light spreads out as thermal energy. The behavior of the reflected light depends on the type of surface and the angle at which the light strikes the surface.

Recall from Chapter 14 that when a wave traveling in two dimensions encounters a barrier, the angle of incidence is equal to the angle of reflection of the wave. This same two-dimensional reflection relationship applies to light waves. Consider what happens when you bounce-pass a basketball. The ball bounces in a straight line, as viewed from above, to the other player. Light reflects in the same way as a basketball. **Figure 17-1** shows a ray of light striking a reflecting surface. The normal is an imaginary line that is perpendicular to a surface at the location where light strikes the surface. The reflected ray, the incident ray, and the normal to the surface always will be in the same plane. Although the light is traveling in three dimensions, the reflection of the light is planar (two-dimensional). The planar and angle relationships are known together as the law of reflection.

Law of Reflection $\theta_r = \theta_i$

The angle that a reflected ray makes as measured from the normal to a reflective surface equals the angle that the incident ray makes as measured from the same normal.

Color Convention

- Light rays and wave fronts are **red.**
- Mirrors are light blue.

This law can be explained in terms of the wave model of light. **Figure 17-2** shows a wave front of light approaching a reflective surface. As each point along the wave front reaches the surface, it reflects off at the same angle as the preceding point. Because all points are traveling at the same speed, they all travel the same total distance in the same time. Thus, the wave front as a whole leaves the surface at an angle equal to its incident angle. Note that the wavelength of the light does not affect this process. Red, green, and blue light all follow this law.

Figure 17-2 A wave front of light approaches a reflective surface. Point P on the wave front strikes the surface first **(a).** Point Q strikes the surface after point P reflects at an angle equal to the incident angle **(b).** The process continues with all points reflecting off at angles equal to their incident angles, resulting in a reflected wave front **(c).**

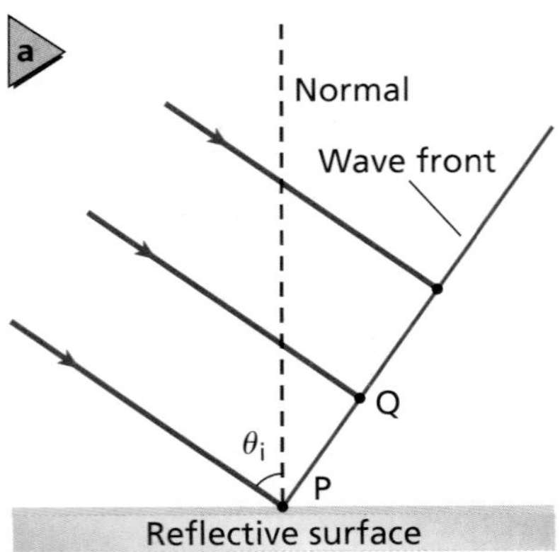

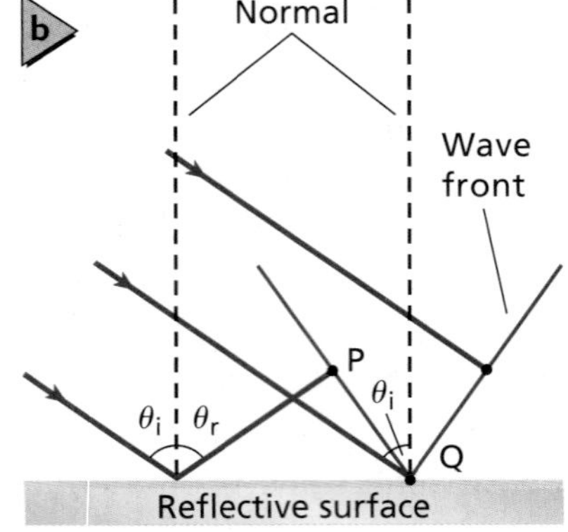

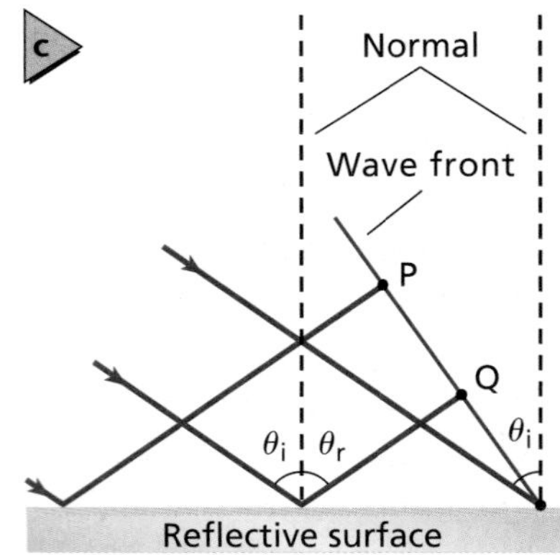

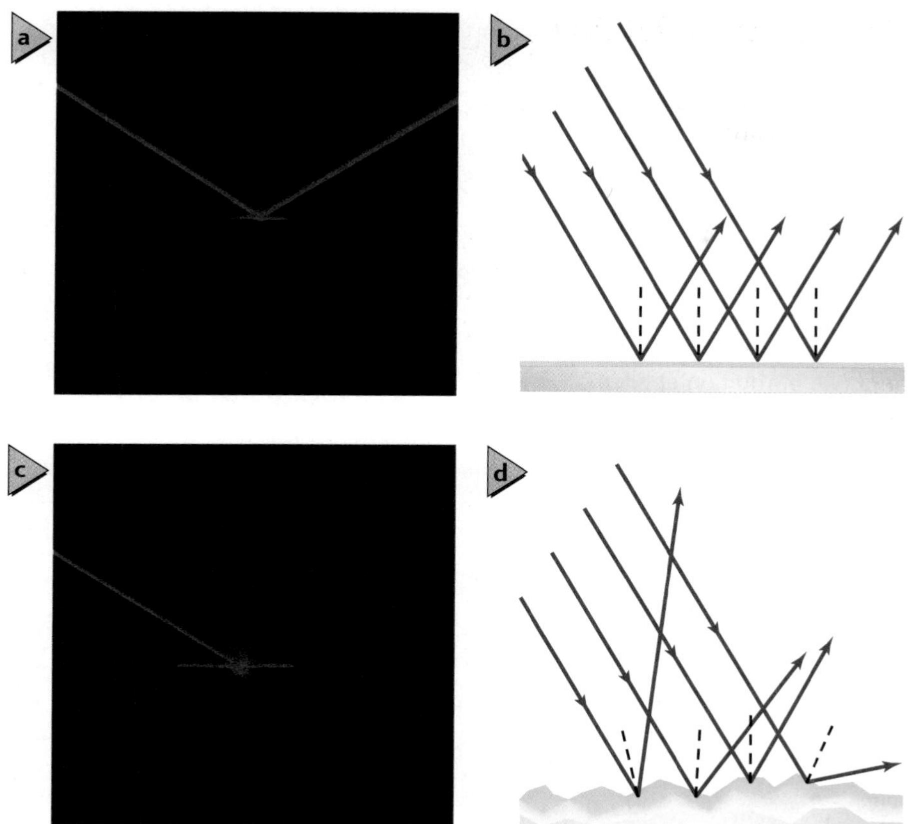

Figure 17-3 When a beam of light strikes a mirrored surface **(a),** the parallel rays in the beam reflect in parallel and maintain the light as a beam **(b).** When the light beam strikes a rough surface **(c),** the parallel rays that make up the beam are reflected from different microscopic surfaces, thereby diffusing the beam **(d).**

Smooth and rough surfaces Consider the beam of light shown in **Figure 17-3a.** All of the rays in this beam of light reflect off the surface parallel to one another, as shown in **Figure 17-3b.** This occurs only if the reflecting surface is not rough on the scale of the wavelength of light. Such a surface is considered to be smooth relative to the light. A smooth surface, such as a mirror, causes **specular reflection,** in which parallel light rays are reflected in parallel.

What happens when light strikes a surface that appears to be smooth, but actually is rough on the scale of the wavelength of light, such as the page of this textbook or a white wall? Is light reflected? How could you demonstrate this? **Figure 17-3c** shows a beam of light reflecting off a sheet of paper, which has a rough surface. All of the light rays that make up the beam are parallel before striking the surface, but the reflected rays are not parallel, as shown in **Figure 17-3d.** This is **diffuse reflection,** the scattering of light off a rough surface.

The law of reflection applies to both smooth and rough surfaces. For a rough surface, the angle that each incident ray makes with the normal equals the angle that its reflected ray makes with the normal. However, on a microscopic scale, the normals to the surface locations where the rays strike are not parallel. Thus, the reflected rays cannot be parallel. The rough surface prevents them from being parallel. In this case, a reflected beam cannot be seen because the reflected rays are scattered in different directions. With specular reflection, as with a mirror, you can see your face. But no matter how much light is reflected off a wall or a sheet of paper, you will never be able to use them as mirrors.

▶ EXAMPLE Problem 1

Changing the Angle of Incidence A light ray strikes a plane mirror at an angle of 52.0° to the normal. The mirror then rotates 35.0° around the point where the beam strikes the mirror so that the angle of incidence of the light ray decreases. The axis of rotation is perpendicular to the plane of the incident and the reflected rays. What is the angle of rotation of the reflected ray?

$\theta_{i,\ initial} = 52.0°$

Initial normal

$\theta_{r,\ initial} = 52.0°$

1 Analyze and Sketch the Problem

- Sketch the situation before the rotation of the mirror.
- Draw another sketch with the angle of rotation applied to the mirror.

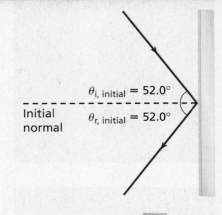

Known:	**Unknown:**
$\theta_{i,\ initial} = 52.0°$	$\Delta\theta_r = ?$
$\Delta\theta_{mirror} = 35.0°$	

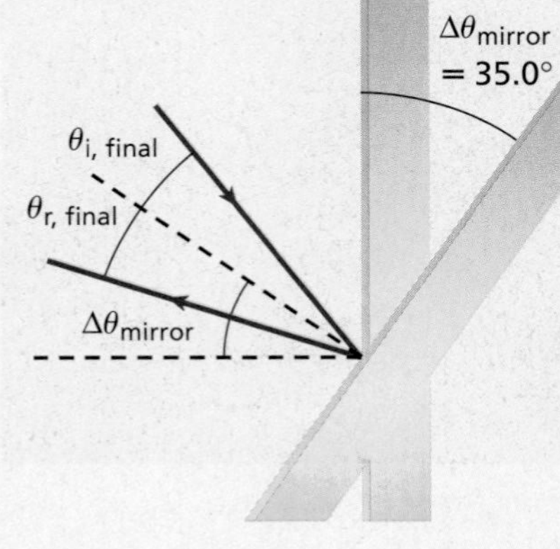

2 Solve for the Unknown

For the angle of incidence to reduce, rotate clockwise.

$$\theta_{i,\ final} = \theta_{i,\ initial} - \Delta\theta_{mirror}$$
$$= 52.0° - 35.0° \qquad \textbf{Substitute } \theta_{i,\ initial} = 52.0°, \Delta\theta_{mirror} = 35.0°$$
$$= 17.0° \text{ clockwise from the new normal}$$

Apply the law of reflection.

$$\theta_{r,\ final} = \theta_{i,\ final}$$
$$= 17.0° \text{ counterclockwise from the new normal} \qquad \textbf{Substitute } \theta_{i,\ final} = 17.0°$$

Using the two sketches, determine the angle through which the reflected ray has rotated.

$$\Delta\theta_r = 52.0° + 35.0° - 17.0°$$
$$= 70.0° \text{ clockwise from the original angle}$$

Math Handbook
Operations with Significant Digits
pages 835–836

3 Evaluate the Answer

- **Is the magnitude realistic?** Comparing the final sketch with the initial sketch shows that the angle the light ray makes with the normal decreases as the mirror rotates clockwise toward the light ray. It makes sense, then, that the reflected ray also rotates clockwise.

▶ PRACTICE Problems

- Additional Problems, Appendix B
- Solutions to Selected Problems, Appendix C

1. Explain why the reflection of light off ground glass changes from diffuse to specular if you spill water on it.
2. If the angle of incidence of a ray of light is 42°, what is each of the following?
   a. the angle of reflection
   b. the angle the incident ray makes with the mirror
   c. the angle between the incident ray and the reflected ray
3. If a light ray reflects off a plane mirror at an angle of 35° to the normal, what was the angle of incidence of the ray?
4. Light from a laser strikes a plane mirror at an angle of 38° to the normal. If the laser is moved so that the angle of incidence increases by 13°, what is the new angle of reflection?
5. Two plane mirrors are positioned at right angles to one another. A ray of light strikes one mirror at an angle of 30° to the normal. It then reflects toward the second mirror. What is the angle of reflection of the light ray off the second mirror?

Objects and Plane-Mirror Images

If you looked at yourself in a mirror this morning you saw your reflection in a plane mirror. A **plane mirror** is a flat, smooth surface from which light is reflected by specular reflection. To understand reflection from a mirror, you must consider the object of the reflection and the type of image that is formed. In Chapter 16, the word *object* was used to refer to sources of light. In describing mirrors, the word *object* is used in the same way, but with a more specific application. An **object** is a source of light rays that are to be reflected by a mirrored surface. An object can be a luminous source, such as a lightbulb, or an illuminated source, such as a girl, as shown in **Figure 17-4.** Most objects that you will work with in this chapter are a source of light that diverges, or spreads out from the source.

Figure 17-4 The lightbulb is a luminous source that produces diverging light by shining in all directions. The girl is an illuminated source that produces diverging light by the diffused reflection from her body of light that comes from the lightbulb.

Consider a single point on the bird in **Figure 17-5.** Light reflects diffusely from the crest of the bird, the object point. What happens to the light? Some of the light travels from the bird to the mirror and reflects. What does the boy see? Some of the reflected light from the bird hits his eyes. Because his brain processes this information as if the light has traveled a straight path, it seems to the boy as if the light had followed the dashed lines. The light seems to have come from a point behind the mirror, the image point.

The boy in Figure 17-5 sees rays of light that come from many points on the bird. The combination of the image points produced by reflected light rays forms the **image** of the bird. It is a **virtual image,** which is a type of image formed by diverging light rays. A virtual image is always on the opposite side of the mirror from the object. Images of real objects produced by plane mirrors are always virtual images.

Properties of Plane-Mirror Images

Looking at yourself in a mirror, you can see that your image appears to be the same distance behind the mirror as you are in front of the mirror. How could you test this? Place a ruler between you and the mirror. Where does the image touch the ruler? You also see that your image is oriented as you are, and it matches your size. This is where the expression *mirror image* originates. If you move toward the mirror, your image moves toward the mirror. If you move away, your image also moves away.

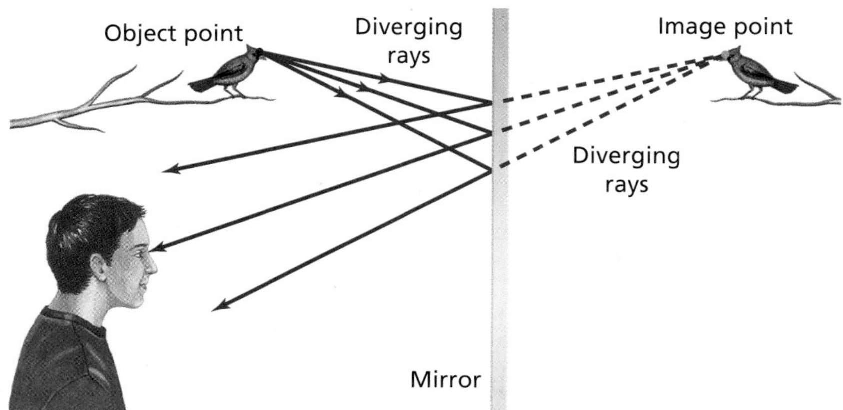

Figure 17-5 The reflected rays that enter the eye appear to originate at a point behind the mirror.

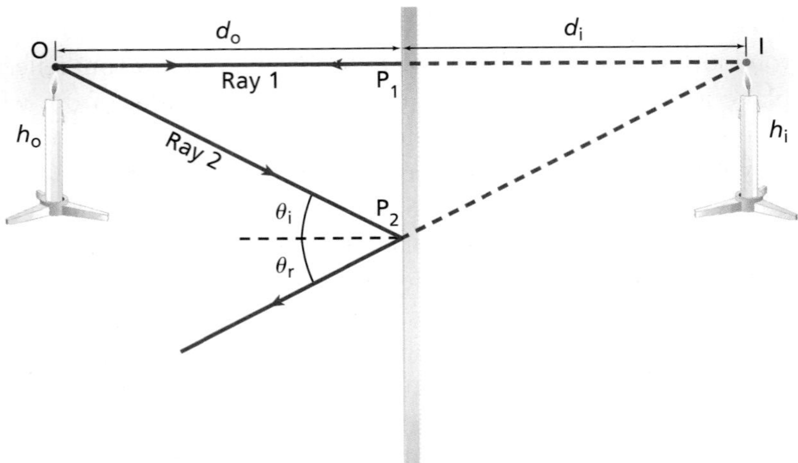

Figure 17-6 Light rays (two are shown) leave a point on the object. Some strike the mirror and are reflected into the eye. Sight lines, drawn as dashed lines, extend from the location on the mirror where the rays are reflected back to where they converge. The image is located where the sight lines converge: $d_i = -d_o$.

MINI LAB

Virtual Image Position

Suppose you are looking at your image in a plane mirror. Can you measure the location of the image?

1. Obtain a camera from your teacher with a focusing ring that has distances marked on it.

2. Stand 1.0 m from a mirror and focus on the edge of the mirror. Check the reading on the focusing ring. It should be 1.0 m.

3. Measure the position of the image by focusing the camera on your image. Check the reading on the focusing ring.

Analyze and Conclude

4. How far is the image behind the mirror?

5. Why is the camera able to focus on a virtual image that is behind the mirror even though there is no real object behind the mirror?

Image position and height The geometric model of **Figure 17-6** demonstrates why the distances are the same. Two rays from point O at the tip of the candle strike the mirror at point P_1 and point P_2, respectively. Both rays are reflected according to the law of reflection. The reflected rays are extended behind the mirror as sight lines, converging at point I, which is the image of point O. Ray 1, which strikes the mirror at an angle of incidence of 0°, is reflected back on itself, so the sight line is at 90° to the mirror, just as ray 1. Ray 2 is also reflected at an angle equal to the angle of incidence, so the sight line is at the same angle to the mirror as ray 2.

This geometric model reveals that line segments $\overline{OP_1}$ and $\overline{IP_1}$ are corresponding sides of two congruent triangles, OP_1P_2 and IP_1P_2. The position of the object with respect to the mirror, or the object position, d_o, has a length equal to the length of $\overline{OP_1}$. The apparent position of the image with respect to the mirror, or the image position, d_i, has a length equal to the length of $\overline{IP_1}$. Using the convention that image position is negative to indicate that the image is virtual, the following is true.

Plane-Mirror Image Position $d_i = -d_o$

With a plane mirror, the image position is equal to the negative of the object position. The negative sign indicates that the image is virtual.

You can draw rays from the object to the mirror to determine the size of the image. The sight lines of two rays originating from the bottom of the candle in Figure 17-6 will converge at the bottom of the image. Using the law of reflection and congruent-triangle geometry, the following is true of the object height, h_o, and image height, h_i, and any other dimension of the object and image.

Plane-Mirror Image Height $h_i = h_o$

With a plane mirror, image height is equal to object height.

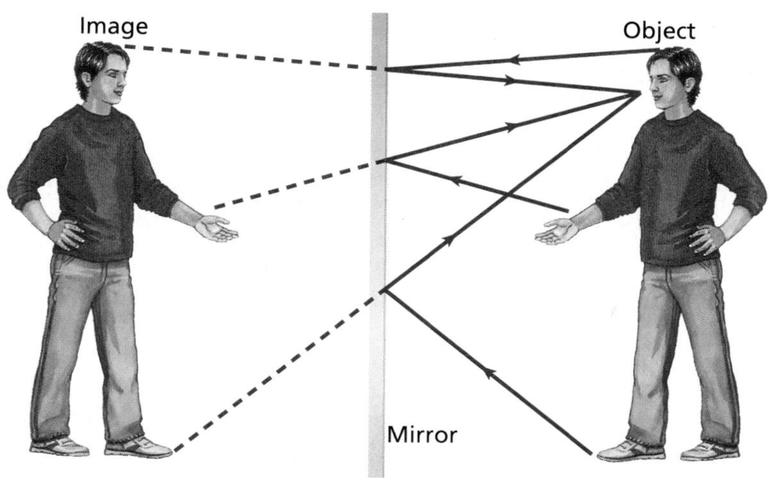

Figure 17-7 The image formed in a plane mirror is the same size as the object and is the same distance behind the mirror as the object is in front. However, when the boy blinks his right eye, the left eye of his image blinks.

Image orientation A plane mirror produces an image with the same orientation as the object. If you are standing on your feet, a plane mirror produces an image of you standing on your feet. If you are doing a headstand, the mirror shows you doing a headstand. However, there is a difference between you and the appearance of your image in a mirror. Follow the sight lines in **Figure 17-7.** The ray that diverges from the right hand of the boy converges at what appears to be the left hand of his image. Left and right appear to be reversed by a plane mirror. Why, then, are top and bottom not also reversed? This does not happen because a plane mirror does not really reverse left and right. The mirror in Figure 17-7 only reverses the boy's image such that it is facing in the opposite direction as the boy, or, in other words, it produces a front-to-back reversal.

Consider the image of the mountain in the photo at the beginning of the chapter. In this case, the image of the mountain can be described as upside down, but the image is actually a front-to-back reversal of your view of the actual mountain. Because the mirror (the lake surface) is horizontal, rather than vertical, your perspective, or angle of view, makes the image look upside down. Turn your book 90° counterclockwise and look at Figure 17-7 again. The actual boy is now facing down, and his image is facing up, upside down relative to the actual boy, just like the image of the mountain. The only thing that has changed is your perspective.

17.1 Section Review

6. **Reflection** A light ray strikes a flat, smooth, reflecting surface at an angle of 80° to the normal. What is the angle that the reflected ray makes with the surface of the mirror?

7. **Law of Reflection** Explain how the law of reflection applies to diffuse reflection.

8. **Reflecting Surfaces** Categorize each of the following as a specular or a diffuse reflecting surface: paper, polished metal, window glass, rough metal, plastic milk jug, smooth water surface, and ground glass.

9. **Image Properties** A 50-cm-tall dog stands 3 m from a plane mirror and looks at its image. What is the image position, height, and type?

10. **Image Diagram** A car is following another car down a straight road. The first car has a rear window tilted 45°. Draw a ray diagram showing the position of the Sun that would cause sunlight to reflect into the eyes of the driver of the second car.

11. **Critical Thinking** Explain how diffuse reflection of light off an object enables you to see an object from any angle.

Physics Online physicspp.com/self_check_quiz

17.2 Curved Mirrors

► **Objectives**

- **Explain** how concave and convex mirrors form images.
- **Describe** properties and uses of spherical mirrors.
- **Determine** the locations and sizes of spherical mirror images.

► **Vocabulary**

concave mirror
principal axis
focal point
focal length
real image
spherical aberration
magnification
convex mirror

If you look at the surface of a shiny spoon, you will notice that your reflection is different from what you see in a plane mirror. The spoon acts as a curved mirror, with one side curved inward and the other curved outward. The properties of curved mirrors and the images that they form depend on the shape of the mirror and the object's position.

Concave Mirrors

The inside surface of a shiny spoon, the side that holds food, acts as a concave mirror. A **concave mirror** has a reflective surface, the edges of which curve toward the observer. Properties of a concave mirror depend on how much it is curved. **Figure 17-8** shows how a spherical concave mirror works. In a spherical concave mirror, the mirror is shaped as if it were a section of a hollow sphere with an inner reflective surface. The mirror has the same geometric center, C, and radius of curvature, r, as a sphere of radius, r. The line that includes line segment CM is the **principal axis,** which is the straight line perpendicular to the surface of the mirror that divides the mirror in half. Point M is the center of the mirror where the principal axis intersects the mirror.

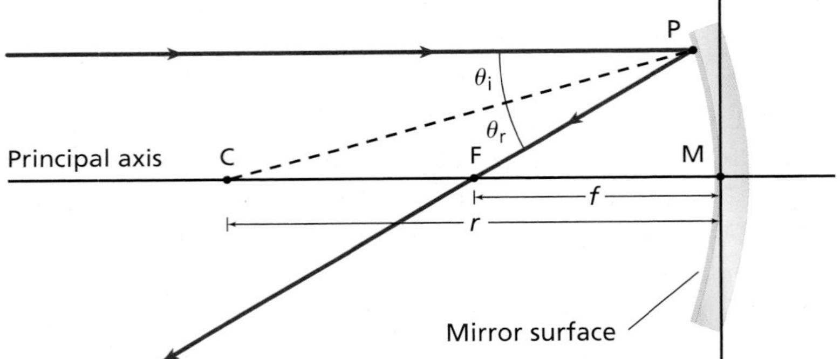

Figure 17-8 The focal point of a spherical concave mirror is located halfway between the center of curvature and the mirror surface. Rays entering parallel to the principal axis are reflected to converge at the focal point, F.

When you point the principal axis of a concave mirror toward the Sun, all the rays are reflected through a single point. You can locate this point by moving a sheet of paper toward and away from the mirror until the smallest and sharpest spot of sunlight is focused on the paper. This spot is called the **focal point** of the mirror, the point where incident light rays that are parallel to the principal axis converge after reflecting from the mirror. The Sun is a source of parallel light rays because it is very far away. All of the light that comes directly from the Sun must follow almost parallel paths to Earth, just as all of the arrows shot by an archer must follow almost parallel paths to hit within the circle of a bull's-eye.

When a ray strikes a mirror, it is reflected according to the law of reflection. Figure 17-8 shows that a ray parallel to the principal axis is reflected and crosses the principal axis at point F, the focal point. F is at the halfway point between M and C. The **focal length,** f, is the position of the focal point with respect to the mirror along the principal axis and can be expressed as $f = r/2$. The focal length is positive for a concave mirror.

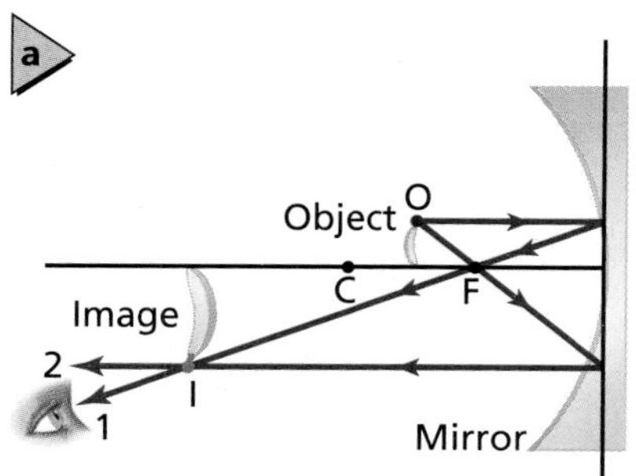

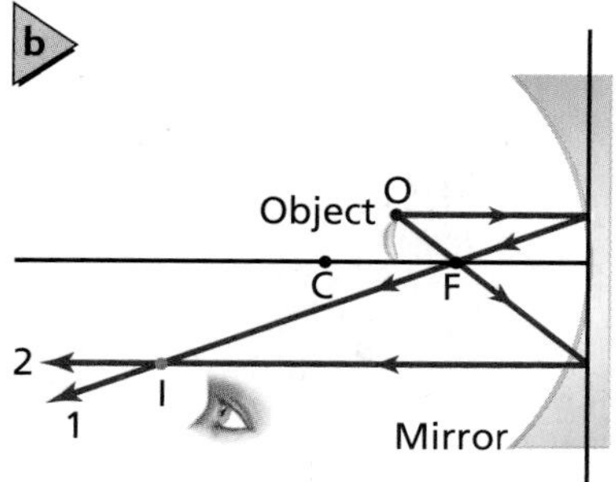

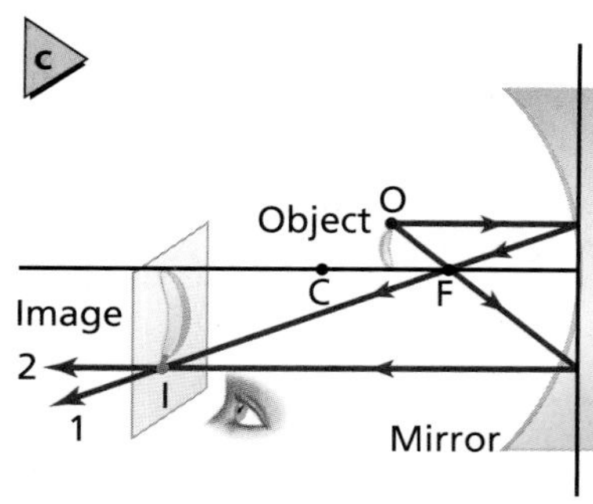

Figure 17-9 The real image, as seen by the unaided eye **(a).** The unaided eye cannot see the real image if it is not in a location to catch the rays that form the image **(b).** The real image as seen on a white opaque screen **(c).**

Graphical Method of Finding the Image

You have already drawn rays to follow the path of light that reflects off plane mirrors. This method is even more useful when applied to curved mirrors. Not only can the location of the image vary, but so can the orientation and size of the image. You can use a ray diagram to determine properties of an image formed by a curved mirror. **Figure 17-9** shows the formation of a **real image,** an image that is formed by the converging of light rays. The image is inverted and larger than the object. The rays actually converge at the point where the image is located. The point of intersection, I′, of the two reflected rays determines the position of the image. You can see the image floating in space if you place your eye so that the rays that form the image fall on your eye, as in **Figure 17-9a.** As **Figure 17-9b** shows, however, your eye must be oriented so as to see the rays coming from the image location. You cannot look at the image from behind. If you were to place a movie screen at this point, the image would appear on the screen, as shown in **Figure 17-9c.** You cannot do this with virtual images.

To more easily understand how ray tracing works with curved mirrors, you can use simple, one-dimensional objects, such as the arrow shown in **Figure 17-10a.** A spherical concave mirror produces an inverted real image if the object position, d_o, is greater than twice the focal length, f. The object is then beyond the center of curvature, C. If the object is placed between the center of curvature and the focal point, F, as shown in **Figure 17-10b,** the image is again real and inverted. However, the size of the image is now greater than the size of the object.

Concepts In Motion
Interactive Figure To see an animation on the graphical method of finding the image, visit physicspp.com.

Figure 17-10 When the object is farther from the mirror than C, the image is a real image that is inverted and smaller compared to the object **(a).** When the object is located between C and F, the real image is inverted, larger than the object, and located beyond C **(b).**

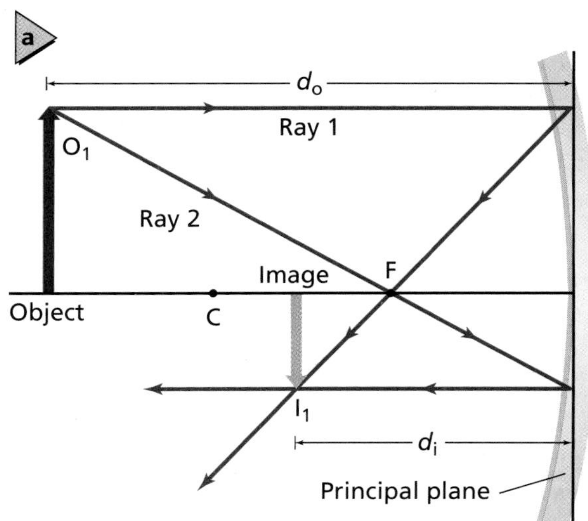

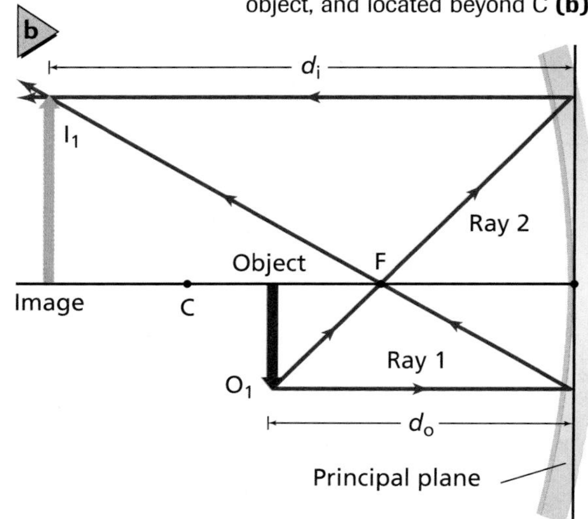

Figure 17-11 A Gregorian telescope produces a real image that is upright.

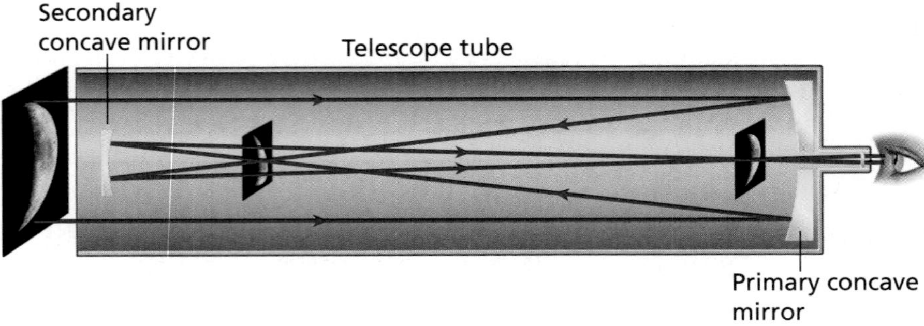

Astronomy Connection

How can the inverted real image created by a concave mirror be turned right-side up? In 1663, Scottish astronomer James Gregory developed the Gregorian telescope, shown in **Figure 17-11,** to resolve this problem. It is composed of a large concave mirror and a small concave mirror arranged such that the smaller mirror is outside of the focal point of the larger mirror. Parallel rays of light from distant objects strike the larger mirror and reflect toward the smaller mirror. The rays then reflect off the smaller mirror and form a real image that is oriented exactly as the object is.

PROBLEM-SOLVING **Strategies**

Using Ray Tracing to Locate Images Formed by Spherical Mirrors

Use the following strategies for spherical-mirror problems. Refer to Figure 17-10.

1. Using lined or graph paper, draw the principal axis of the mirror as a horizontal line from the left side to the right side of your paper, leaving six blank lines above and six blank lines below.
2. Place a point and a label on the principal axis the object, C, and F, as follows.
   a. If the mirror is a concave mirror and the object is beyond C, away from the mirror, place the mirror at the right side of the page, place the object at the left side of the page, and place C and F to scale.
   b. If the mirror is a concave mirror and the object is between C and F, place the mirror at the right side of the page, place C at the center of the paper, F halfway between the mirror and C, and the object to scale.
   c. For any other situation, place the mirror in the center of the page. Place the object or F (whichever is the greatest distance from the mirror) at the left side of the page, and place the other to scale.
3. To represent the mirror, draw a vertical line at the mirror point that extends the full 12 lines of space. This is the principal plane.
4. Draw the object as an arrow and label its top O_1. For concave mirrors, objects inside of C should not be higher than three lines high. For all other situations, the objects should be six lines high. The scale for the height of the object will be different from the scale along the principal axis.
5. Draw ray 1, the parallel ray. It is parallel to the principal axis and reflects off the principal plane and passes through F.
6. Draw ray 2, the focus ray. It passes through F, reflects off the principal plane, and is reflected parallel to the principal axis.
7. The image is located where rays 1 and 2 (or their sight lines) cross after reflection. Label the point I_1. The image is an arrow perpendicular from the principal axis to I_1.

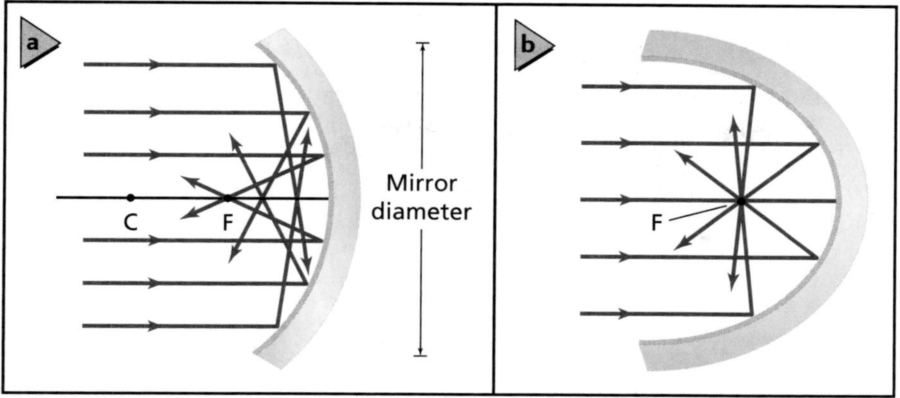

Figure 17-12 A concave spherical mirror reflects some rays, such that they converge at points other than the focus **(a).** A parabolic mirror focuses all parallel rays at a point **(b).**

Real image defects in concave mirrors In tracing rays, you have reflected the rays from the principal plane, which is a vertical line representing the mirror. In reality, rays are reflected off the mirror itself, as shown in **Figure 17-12a.** Notice that only parallel rays that are close to the principal axis, or paraxial rays, are reflected through the focal point. Other rays converge at points closer to the mirror. The image formed by parallel rays reflecting off a spherical mirror with a large mirror diameter and a small radius of curvature is a disk, not a point. This effect, called **spherical aberration,** makes an image look fuzzy, not sharp.

A mirror ground to the shape of a parabola, as in **Figure 17-12b,** suffers no spherical aberration. Because of the cost of manufacturing large, perfectly parabolic mirrors, many of the newest telescopes use spherical mirrors and smaller, specially-designed secondary mirrors or lenses to correct for spherical aberration. Also, spherical aberration is reduced as the ratio of the mirror's diameter, shown in Figure 17-12a, to its radius of curvature is reduced. Thus, lower-cost spherical mirrors can be used in lower-precision applications.

APPLYING PHYSICS

***Hubble* Trouble** In 1990, NASA launched the *Hubble Space Telescope* into orbit around Earth. *Hubble* was expected to provide clear images without atmospheric distortions. However, soon after it was deployed, *Hubble* was found to have a spherical aberration. In 1993, corrective optics, called COSTAR, were installed on *Hubble* to enable it to produce clear images.

Mathematical Method of Locating the Image

The spherical mirror model can be used to develop a simple equation for spherical mirrors. You must use the paraxial ray approximation, which states that only rays that are close to and almost parallel with the principal axis are used to form an image. Using this, in combination with the law of reflection, leads to the mirror equation, relating the focal length, f, object position, d_o, and image position, d_i, of a spherical mirror.

Mirror Equation $\frac{1}{f} = \frac{1}{d_i} + \frac{1}{d_o}$

The reciprocal of the focal length of a spherical mirror is equal to the sum of the reciprocals of the image position and the object position.

When using this equation to solve problems, it is important to remember that it is only approximately correct. It does not predict spherical aberration, because it uses the paraxial ray approximation. In reality, light coming from an object toward a mirror is diverging, so not all of the light is close to or parallel to the axis. When the mirror diameter is small relative to the radius of curvature to minimize spherical aberration, this equation predicts image properties more precisely.

▶ Connecting Math to Physics

Adding and Subtracting Fractions When using the mirror equation, you first use math to move the fraction that contains the quantity you are seeking to the left-hand side of the equation and everything else to the right. Then you combine the two fractions on the right-hand side using a common denominator that results from multiplying the denominators.

Math	Physics
$\frac{1}{x} = \frac{1}{y} + \frac{1}{z}$	$\frac{1}{f} = \frac{1}{d_i} + \frac{1}{d_o}$
$\frac{1}{y} = \frac{1}{x} - \frac{1}{z}$	$\frac{1}{d_i} = \frac{1}{f} - \frac{1}{d_o}$
$\frac{1}{y} = \left(\frac{1}{x}\right)\left(\frac{z}{z}\right) - \left(\frac{1}{z}\right)\left(\frac{x}{x}\right)$	$\frac{1}{d_i} = \left(\frac{1}{f}\right)\left(\frac{d_o}{d_o}\right) - \left(\frac{1}{d_o}\right)\left(\frac{f}{f}\right)$
$\frac{1}{y} = \frac{z - x}{xz}$	$\frac{1}{d_i} = \frac{d_o - f}{fd_o}$
$y = \frac{xz}{z - x}$	$d_i = \frac{fd_o}{d_o - f}$

Using this approach, the following relationships can be derived for image position, object position, and focal length:

$$d_i = \frac{fd_o}{d_o - f} \qquad d_o = \frac{fd_i}{d_i - f} \qquad f = \frac{d_i d_o}{d_o + d_i}$$

Magnification Another property of a spherical mirror is **magnification,** m, which is how much larger or smaller the image is relative to the object. In practice, this is a simple ratio of the image height to the object height. Using similar-triangle geometry, this ratio can be written in terms of image positon and object position.

Magnification $m \equiv \frac{h_i}{h_o} = \frac{-d_i}{d_o}$

The magnification of an object by a spherical mirror, defined as the image height divided by the object height, is equal to the negative of the image position, divided by the object position.

Image position is positive for a real image when using the preceding equations. Thus, the magnification is negative, which means that the image is inverted compared to the object. If the object is beyond point C, the absolute value of the magnification for the real image is less than 1. This means that the image is smaller than the object. If the object is placed between point C and point F, the absolute value of the magnification for the real image is greater than 1. Thus, the image is larger than the object.

▶ EXAMPLE Problem 2

Real Image Formation by a Concave Mirror A concave mirror has a radius of 20.0 cm. A 2.0-cm-tall object is 30.0 cm from the mirror. What is the image position and image height?

1 Analyze and Sketch the Problem

- Draw a diagram with the object and the mirror.
- Draw two principal rays to locate the image in the diagram.

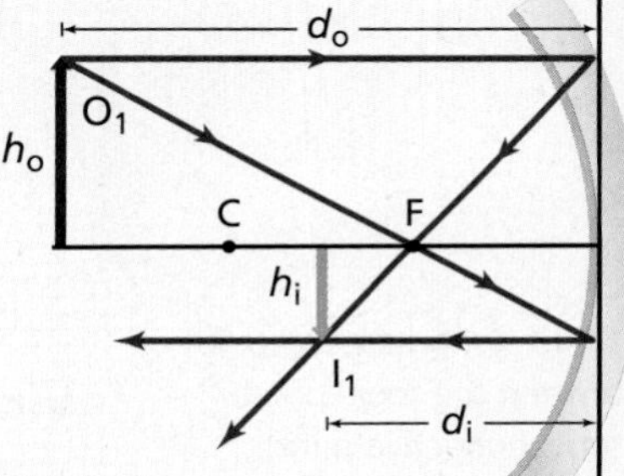

Known:	Unknown:
$h_o = 2.0$ cm	$d_i = ?$
$d_o = 30.0$ cm	$h_i = ?$
$r = 20.0$ cm	

2 Solve for the Unknown

Focal length is half the radius of curvature.

$$f = \frac{r}{2}$$

$$= \frac{20.0\text{ cm}}{2} \quad \text{Substitute } r = 20.0\text{ cm}$$

$$= 10.0\text{ cm}$$

Use the mirror equation and solve for image position.

$$\frac{1}{f} = \frac{1}{d_i} + \frac{1}{d_o}$$

$$d_i = \frac{fd_o}{d_o - f}$$

$$= \frac{(10.0\text{ cm})(30.0\text{ cm})}{30.0\text{ cm} - 10.0\text{ cm}} \quad \text{Substitute } f = 10.0\text{ cm}, d_o = 30.0\text{ cm}$$

$$= 15.0\text{ cm (real image, in front of the mirror)}$$

Math Handbook
Fractions
page 837

Use the magnification equation and solve for image height.

$$m \equiv \frac{h_i}{h_o} = \frac{-d_i}{d_o}$$

$$h_i = \frac{-d_i h_o}{d_o}$$

$$= \frac{-(15.0\text{ cm})(2.0\text{ cm})}{30.0\text{ cm}} \quad \text{Substitute } d_i = 15.0\text{ cm}, h_o = 2.0\text{ cm}, d_o = 30.0\text{ cm}$$

$$= -1.0\text{ cm (inverted, smaller image)}$$

3 Evaluate the Answer

- **Are the units correct?** All positions are in centimeters.
- **Do the signs make sense?** Positive position and negative height agree with the drawing.

▶ PRACTICE Problems

Additional Problems, Appendix B

12. Use a ray diagram, drawn to scale, to solve Example Problem 2.

13. An object is 36.0 cm in front of a concave mirror with a 16.0-cm focal length. Determine the image position.

14. A 3.0-cm-tall object is 20.0 cm from a 16.0-cm-radius concave mirror. Determine the image position and image height.

15. A concave mirror has a 7.0-cm focal length. A 2.4-cm-tall object is 16.0 cm from the mirror. Determine the image height.

16. An object is near a concave mirror of 10.0-cm focal length. The image is 3.0 cm tall, inverted, and 16.0 cm from the mirror. What are the object position and object height?

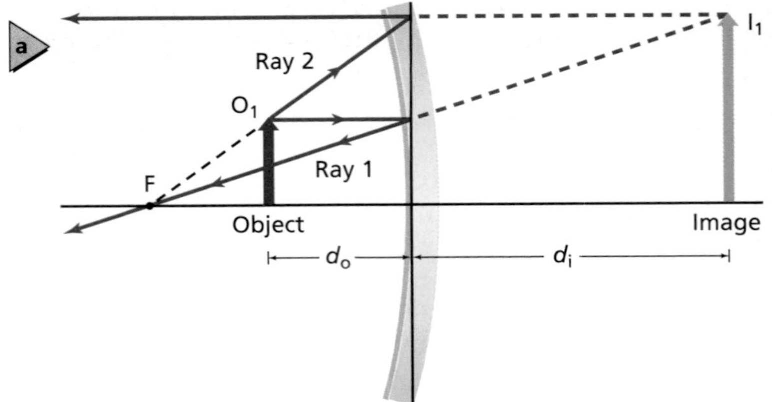

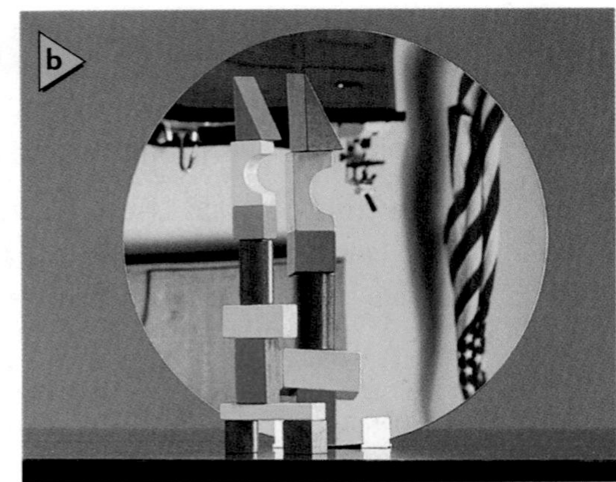

Figure 17-13 When an object is located between the focal point and a spherical concave mirror, a virtual image that is upright and larger compared to the object is formed behind the mirror **(a),** as shown with the stack of blocks **(b).** What else do you see in this picture?

Virtual Images with Concave Mirrors

You have seen that as an object approaches the focal point, F, of a concave mirror, the image moves farther away from the mirror. If the object is at the focal point, all reflected rays are parallel. They never meet, therefore, and the image is said to be at infinity, so the object could never be seen. What happens if the object is moved even closer to the mirror?

What do you see when you move your face close to a concave mirror? The image of your face is right-side up and behind the mirror. A concave mirror produces a virtual image if the object is located between the mirror and the focal point, as shown in the ray diagram in **Figure 17-13a.** Again, two rays are drawn to locate the image of a point on an object. As before, ray 1 is drawn parallel to the principal axis and reflected through the focal point. Ray 2 is drawn as a line from the point on the object to the mirror, along a line defined by the focal point and the point on the object. At the mirror, ray 2 is reflected parallel to the principal axis. Note that ray 1 and ray 2 diverge as they leave the mirror, so there cannot be a real image. However, sight lines extended behind the mirror converge, showing that the virtual image forms behind the mirror.

When you use the mirror equation to solve problems involving concave mirrors for which an object is between the mirror and the focal point, you will find that the image position is negative. The magnification equation gives a positive magnification greater than 1, which means that the image is upright and larger compared to the object, like the image in **Figure 17-13b.**

CHALLENGE **PROBLEM**

An object of height h_o is located at d_o relative to a concave mirror with focal length f.

1. Draw and label a ray diagram showing the focal length and location of the object if the image is located twice as far from the mirror as the object. Prove your answer mathematically. Calculate the focal length as a function of object position for this placement.
2. Draw and label a ray diagram showing the location of the object if the image is located twice as far from the mirror as the focal point. Prove your answer mathematically. Calculate the image height as a function of the object height for this placement.
3. Where should the object be located so that no image is formed?

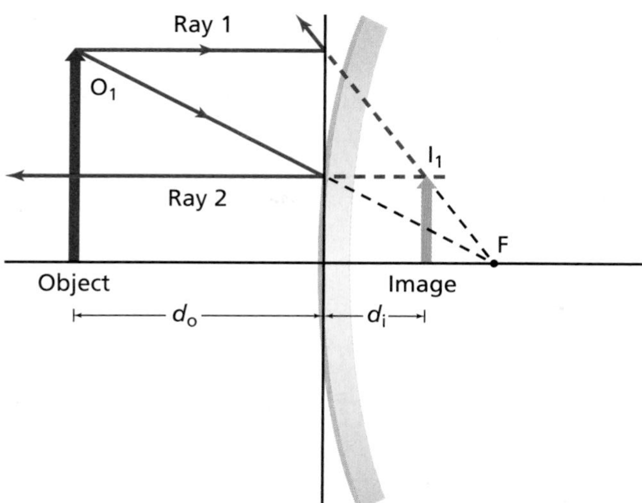

Figure 17-14 A convex mirror always produces virtual images that are upright and smaller compared to the object.

Convex Mirrors

In the first part of this chapter, you learned that the inner surface of a shiny spoon acts as a concave mirror. If you turn the spoon around, the outer surface acts as a **convex mirror,** a reflective surface with edges that curve away from the observer. What do you see when you look at the back of a spoon? You see an upright, but smaller image of yourself.

Properties of a spherical convex mirror are shown in **Figure 17-14.** Rays reflected from a convex mirror always diverge. Thus, convex mirrors form virtual images. Points F and C are behind the mirror. In the mirror equation, f and d_i are negative numbers because they are both behind the mirror.

The ray diagram in Figure 17-14 represents how an image is formed by a spherical convex mirror. The figure uses two rays, but remember that there are an infinite number of rays. Ray 1 approaches the mirror parallel to the principal axis. The reflected ray is drawn along a sight line from F through the point where ray 1 strikes the mirror. Ray 2 approaches the mirror on a path that, if extended behind the mirror, would pass through F. The reflected part of ray 2 and its sight line are parallel to the principal axis. The two reflected rays diverge, and the sight lines intersect behind the mirror at the location of the image. An image produced by a single convex mirror is a virtual image that is upright and smaller compared to the object.

The magnification equation is useful for determining the apparent dimensions of an object as seen in a spherical convex mirror. If you know the diameter of an object, you can multiply by the magnification fraction to see how the diameter changes. You will find that the diameter is smaller, as are all other dimensions. This is why the objects appear to be farther away than they actually are for convex mirrors.

Field of view It may seem that convex mirrors would have little use because the images that they form are smaller than the objects. However, this property of convex mirrors does have practical uses. By forming smaller images, convex mirrors enlarge the area, or field of view, that an observer sees, as shown in **Figure 17-15.** Also, the center of this field of view is visible from any angle of an observer off the principal axis of the mirror; thus, the field of view is visible from a wide perspective. For this reason, convex mirrors often are used in cars as passenger-side rearview mirrors.

Figure 17-15 Convex mirrors produce images that are smaller than the objects. This increases the field of view for observers.

▶ EXAMPLE Problem 3

Image in a Security Mirror A convex security mirror in a warehouse has a −0.50-m focal length. A 2.0-m-tall forklift is 5.0 m from the mirror. What are the image position and image height?

1 Analyze and Sketch the Problem

- Draw a diagram with the mirror and the object.
- Draw two principal rays to locate the image in the diagram.

Known:	Unknown:
$h_o = 2.0$ m	$d_i = ?$
$d_o = 5.0$ m	$h_i = ?$
$f = -0.50$ m	

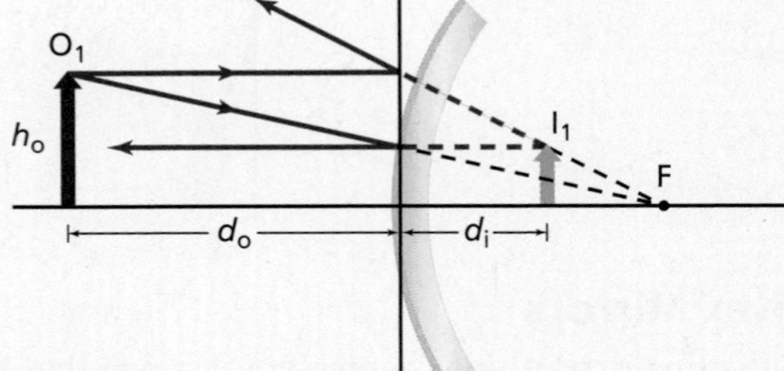

2 Solve for the Unknown

Use the mirror equation and solve for image position.

$$\frac{1}{f} = \frac{1}{d_i} + \frac{1}{d_o}$$

$$d_i = \frac{fd_o}{d_o - f}$$

$$= \frac{(-0.50\text{ m})(5.0\text{ m})}{5.0\text{ m} - 0.50\text{ m}}$$ **Substitute $f = -0.50$ m, $d_o = 5.0$ m**

$$= -0.45\text{ m (virtual image, behind the mirror)}$$

Math Handbook
Isolating a Variable
page 845

Use the magnification equation and solve for image height.

$$m \equiv \frac{h_i}{h_o} = \frac{-d_i}{d_o}$$

$$h_i = \frac{-d_i h_o}{d_o}$$

$$= \frac{-(-0.45\text{ m})(2.0\text{ m})}{(5.0\text{ m})}$$ **Substitute $d_i = -0.45$ m, $h_o = 2.0$ m, $d_o = 5.0$ m**

$$= 0.18\text{ m (upright, smaller image)}$$

3 Evaluate the Answer

- **Are the units correct?** All positions are in meters.
- **Do the signs make sense?** A negative position indicates a virtual image; a positive height indicates an image that is upright. These agree with the diagram.

▶ PRACTICE Problems

• Additional Problems, Appendix B
• Solutions to Selected Problems, Appendix C

17. An object is located 20.0 cm in front of a convex mirror with a −15.0-cm focal length. Find the image position using both a scale diagram and the mirror equation.
18. A convex mirror has a focal length of −13.0 cm. A lightbulb with a diameter of 6.0 cm is placed 60.0 cm from the mirror. What is the lightbulb's image position and diameter?
19. A convex mirror is needed to produce an image that is three-fourths the size of an object and located 24 cm behind the mirror. What focal length should be specified?
20. A 7.6-cm-diameter ball is located 22.0 cm from a convex mirror with a radius of curvature of 60.0 cm. What are the ball's image position and diameter?
21. A 1.8-m-tall girl stands 2.4 m from a store's security mirror. Her image appears to be 0.36 m tall. What is the focal length of the mirror?

Table 17-1					
Single-Mirror System Properties					
Mirror Type	f	d_o	d_i	m	**Image**
Plane	N/A	$d_o > 0$	$\|d_i\| = d_o$ (negative)	Same size	Virtual
Concave	+	$d_o > r$	$r > d_i > f$	Reduced, inverted	Real
		$r > d_o > f$	$d_i > r$	Enlarged, inverted	Real
		$f > d_o > 0$	$\|d_i\| > d_o$ (negative)	Enlarged	Virtual
Convex	–	$d_o > 0$	$\|f\| > \|d_i\| > 0$ (negative)	Reduced	Virtual

Mirror Comparison

How do the various types of mirrors compare? **Table 17-1** compares the properties of single-mirror systems with objects that are located on the principal axis of the mirror. Virtual images are always behind the mirror, which means that the image position is negative. When the absolute value of a magnification is between zero and one, the image is smaller than the object. A negative magnification means the image is inverted relative to the object. Notice that the single plane mirror and convex mirror produce only virtual images, whereas the concave mirror can produce real images or virtual images. Plane mirrors give simple reflections, and convex mirrors expand the field of view. A concave mirror acts as a magnifier when an object is within the focal length of the mirror.

17.2 Section Review

22. Image Properties If you know the focal length of a concave mirror, where should you place an object so that its image is upright and larger compared to the object? Will this produce a real or virtual image?

23. Magnification An object is placed 20.0 cm in front of a concave mirror with a focal length of 9.0 cm. What is the magnification of the image?

24. Object Position The placement of an object in front of a concave mirror with a focal length of 12.0 cm forms a real image that is 22.3 cm from the mirror. What is the object position?

25. Image Position and Height A 3.0-cm-tall object is placed 22.0 cm in front of a concave mirror having a focal length of 12.0 cm. Find the image position and height by drawing a ray diagram to scale. Verify your answer using the mirror and magnification equations.

26. Ray Diagram A 4.0-cm-tall object is located 14.0 cm from a convex mirror with a focal length of −12.0 cm. Draw a scale ray diagram showing the image position and height. Verify your answer using the mirror and magnification equations.

27. Radius of Curvature A 6.0-cm-tall object is placed 16.4 cm from a convex mirror. If the image of the object is 2.8 cm tall, what is the radius of curvature of the mirror?

28. Focal Length A convex mirror is used to produce an image that is two-thirds the size of an object and located 12 cm behind the mirror. What is the focal length of the mirror?

29. Critical Thinking Would spherical aberration be less for a mirror whose height, compared to its radius of curvature, is small or large? Explain.

PHYSICS LAB

Concave Mirror Images

Alternate CBL instructions can be found on the Web site.
physicspp.com

A concave mirror reflects light rays that arrive parallel to the principal axis through the focal point. Depending on the object position, different types of images can be formed. Real images can be projected onto a screen while virtual images cannot. In this experiment you will investigate how changing the object position affects the image location and type.

QUESTION

What are the conditions needed to produce real and virtual images using a concave mirror?

Objectives

- **Collect and organize data** of object and image positions.
- **Observe** real and virtual images.
- **Summarize** conditions for production of real and virtual images with a concave mirror.

Safety Precautions

- **Do not look at the reflection of the Sun in a mirror or use a concave mirror to focus sunlight.**

Materials

concave mirror
flashlight
screen support
mirror holder
two metersticks
four meterstick supports
screen
lamp with a 15-W lightbulb

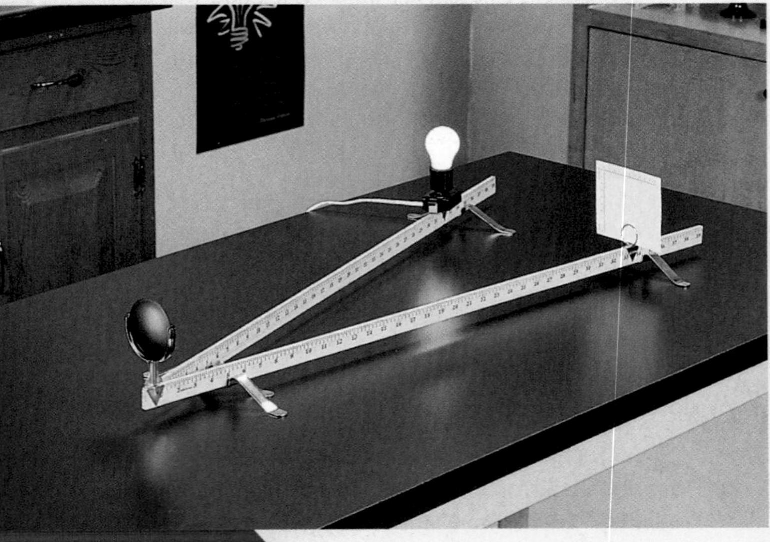

Procedure

1. Determine the focal length of your concave mirror by using the following procedure. ***CAUTION: Do not use the Sun to perform this procedure.*** Reflect light from a flashlight onto a screen and slowly move the screen closer or farther away from the mirror until a sharp, bright image is visible. Measure the distance between the screen and the mirror along the principal axis. Record this value as the actual focal length of the mirror, f.
2. On the lab table, set up two metersticks on supports in a V orientation. Place the zero measurement ends at the apex of the two metersticks.
3. Place the mirror in a mirror holder and place it at the apex of the two metersticks.
4. Using the lamp as the object of the reflection, place it on one meterstick at the opposite end from the apex. Place the screen, supported by a screen support, on the other meterstick at the opposite end from the apex.
5. Turn the room lights off.
6. Turn on the lamp. ***CAUTION: Do not touch the hot lightbulb.*** Measure object position, d_o, and record this as Trial 1. Measure the object height, h_o, and record it as Trial 1. This is measured as the actual height of the lightbulb, or glowing filament if the bulb is clear.
7. Adjust the mirror or metersticks, as necessary, such that the reflected light shines on the screen. Slowly move the screen back and forth along the meterstick until a sharp image is seen. Measure image position, d_i, and the image height, h_i, and record these as Trial 1.

Data Table

Trial	d_o (cm)	d_i (cm)	h_o (cm)	h_i (cm)
1				
2				
3				
4				
5				

Calculation Table

Trial	$\frac{1}{d_o}$ (cm^{-1})	$\frac{1}{d_i}$ (cm^{-1})	$\frac{1}{d_o}+\frac{1}{d_i}$ (cm^{-1})	f_{calc} (cm)	% error
1					
2					
3					
4					
5					

8. Move the lamp closer to the mirror so that d_o is twice the focal length, f. Record this as Trial 2. Move the screen until an image is obtained on the screen. Measure d_i and h_i, and record these as Trial 2.
9. Move the lamp closer to the mirror so that d_o is a few centimeters larger than f. Record this as Trial 3. Move the screen until an image is obtained on the screen. Measure d_i and h_i, and record these as Trial 3.
10. Move the lamp so that d_o is equal to f. Record this as Trial 4 data. Move the screen back and forth and try to obtain an image. What do you observe?
11. Move the lamp so that d_o is less than f by a few centimeters. Record this as Trial 5. Move the screen back and forth and try to obtain an image. What do you observe?

Analyze

1. **Use Numbers** Calculate $1/d_o$ and $1/d_i$ and enter the values in the calculation table.
2. **Use Numbers** Calculate the sum of $1/d_o$ and $1/d_i$ and enter the values in the calculation table. Calculate the reciprocal of this number and enter it in the calculation table as f_{calc}.
3. **Error Analysis** Compare the experimental focal length, f_{calc}, with f, the accepted focal length, by finding the percent error.

$$\text{percent error} = \frac{|f - f_{calc}|}{f} \times 100$$

Conclude and Apply

1. **Classify** What type of image was observed in each of the trials?
2. **Analyze** What conditions cause real images to be formed?
3. **Analyze** What conditions cause virtual images to be formed?

Going Further

1. What are the conditions needed for the image to be larger than the object?
2. Review the methods used for data collection. Identify sources of error and what might be done to improve accuracy.

Real-World Physics

What advantage would there be in using a telescope with a concave mirror?

To find out more about reflection, visit the Web site: **physicspp.com**

Adaptive Optical Systems

Objects in space are difficult to observe from Earth because they twinkle. Our moving, unevenly-heated atmosphere refracts their light in a chaotic manner. It's like trying to look at a small object through the bottom of an empty, clear, glass jar while rotating the jar.

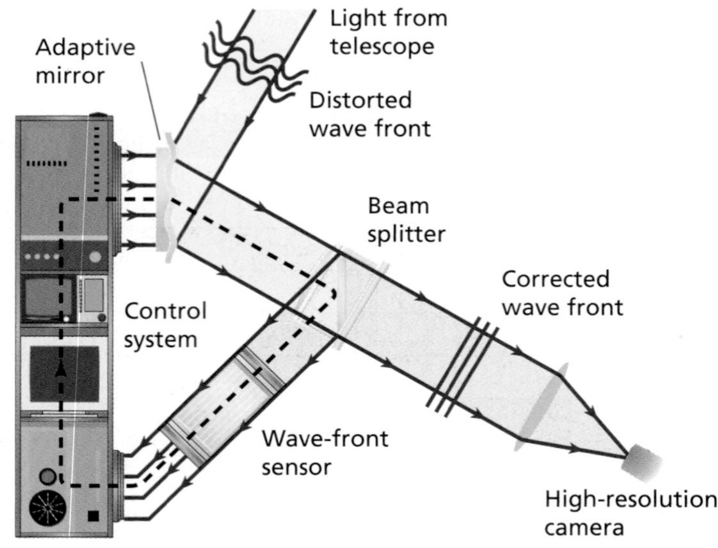

AOS compensates for distortion when viewing Titan, Saturn's largest moon.

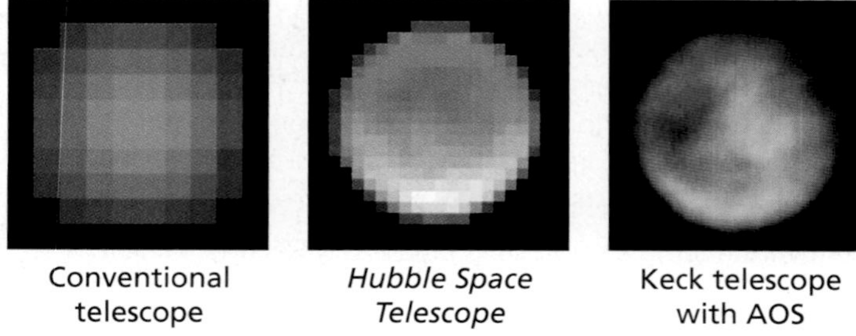

Conventional telescope | *Hubble Space Telescope* | Keck telescope with AOS

Flexible Adaptive Mirror

An adaptive optical system (AOS) continuously compensates for the distortion of the atmosphere, removing the twinkle from star images to allow astronomers to view and photograph steady images of the most distant objects in the visible universe.

An AOS directs the magnified image of the stars from the telescope onto a flexible adaptive mirror made of thin glass. This mirror is stretched across 20–30 movable pistons that can poke or pull the surface of the mirror into many complicated shapes. Each piston is driven by a fast, computer-controlled motor. When the mirror surface is shaped into just the right pattern at just the right time, it will compensate for the convective movement of the atmosphere between the telescope and the star, and reflect a clear image to the observer or camera.

Wave-Front Sensor To detect the atmospheric distortion at each instant of time, a wave-front sensor looks at a single star through the telescope. This device has an array of tiny lenses (lenslets) in several rows. Each lenslet forms an image of the star on a sensitive screen behind it. The position of each image can be read by the computer.

If each image is not directly behind its lenslet, then the computer knows that the star's light waves are being distorted by the atmosphere. A star is a distant point source of light, so it should produce plane waves. Distorted images of a star are non-planar light waves, and these uneven waves cause the images of the star behind some of the lenslets to be displaced.

The computer looks at this error and calculates how the adaptive mirror should be wrinkled to bring each of the lenslet images back into place. The star image reflected to the observer then will be correct, and a clear image of all other objects (like galaxies and planets) in the vicinity also will be seen clearly. The adaptive mirror is re-shaped about 1000 times per second.

Going Further

1. **Research** What is done if there is not a suitable star for the wave-front sensor to analyze in a region of space under observation?
2. **Apply** Research how adaptive optics will be used in the future to correct vision.

Chapter 17

Study Guide

17.1 Reflection from Plane Mirrors

Vocabulary

- specular reflection (p. 459)
- diffuse reflection (p. 459)
- plane mirror (p. 461)
- object (p. 461)
- image (p. 461)
- virtual image (p. 461)

Key Concepts

- According to the law of reflection, the angle that an incident ray makes with the normal equals the angle that the reflected ray makes with the normal.

$$\theta_r = \theta_i$$

- The law of reflection applies to both smooth and rough surfaces. A rough surface, however, has normals that go in lots of different directions, which means that parallel incident rays will not be reflected in parallel.
- A smooth surface produces specular reflection. A rough surface produces diffuse reflection.
- Specular reflection results in the formation of images that appear to be behind plane mirrors.
- An image produced by a plane mirror is always virtual, is the same size as the object, has the same orientation, and is the same distance from the mirror as the object.

$$d_i = -d_o \qquad h_i = h_o$$

17.2 Curved Mirrors

Vocabulary

- concave mirror (p. 464)
- principal axis (p. 464)
- focal point (p. 464)
- focal length (p. 464)
- real image (p. 465)
- spherical aberration (p. 467)
- magnification (p. 468)
- convex mirror (p. 471)

Key Concepts

- You can locate the image created by a spherical mirror by drawing two rays from a point on the object to the mirror. The intersection of the two reflected rays is the image of the object point.
- The mirror equation gives the relationship among image position, object position, and focal length of a spherical mirror.

$$\frac{1}{f} = \frac{1}{d_i} + \frac{1}{d_o}$$

- The magnification of a mirror image is given by equations relating either the positions or the heights of the image and the object.

$$m \equiv \frac{h_i}{h_o} = \frac{-d_i}{d_o}$$

- A single concave mirror produces a real image that is inverted when the object position is greater than the focal length.
- A single concave mirror produces a virtual image that is upright when the object position is less than the focal length.
- A single convex mirror always produces a virtual image that is upright and smaller compared to the object.
- By forming smaller images, convex mirrors make images seem farther away and produce a wide field of view.
- Mirrors can be used in combinations to produce images of any size, orientation, and location desired. The most common use of combinations of mirrors is as telescopes.

Concept Mapping

30. Complete the following concept map using the following terms: *convex, upright, inverted, real, virtual.*

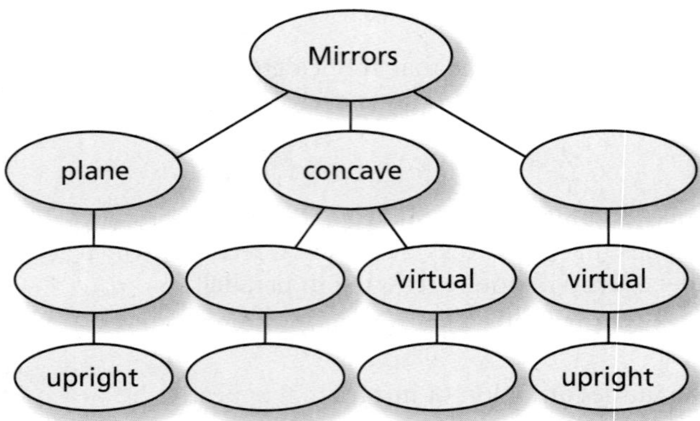

Mastering Concepts

31. How does specular reflection differ from diffuse reflection? (17.1)

32. What is meant by the phrase "normal to the surface"? (17.1)

33. Where is the image produced by a plane mirror located? (17.1)

34. Describe the properties of a plane mirror. (17.1)

35. A student believes that very sensitive photographic film can detect a virtual image. The student puts photographic film at the location of a virtual image. Does this attempt succeed? Explain. (17.1)

36. How can you prove to someone that an image is a real image? (17.1)

37. An object produces a virtual image in a concave mirror. Where is the object located? (17.2)

38. What is the defect that all concave spherical mirrors have and what causes it? (17.2)

39. What is the equation relating the focal point, object position, and image position? (17.2)

40. What is the relationship between the center of curvature and the focal length of a concave mirror? (17.2)

41. If you know the image position and object position relative to a curved mirror, how can you determine the mirror's magnification? (17.2)

42. Why are convex mirrors used as rearview mirrors? (17.2)

43. Why is it impossible for a convex mirror to form a real image? (17.2)

Applying Concepts

44. Wet Road A dry road is more of a diffuse reflector than a wet road. Based on **Figure 17-16,** explain why a wet road appears blacker to a driver than a dry road does.

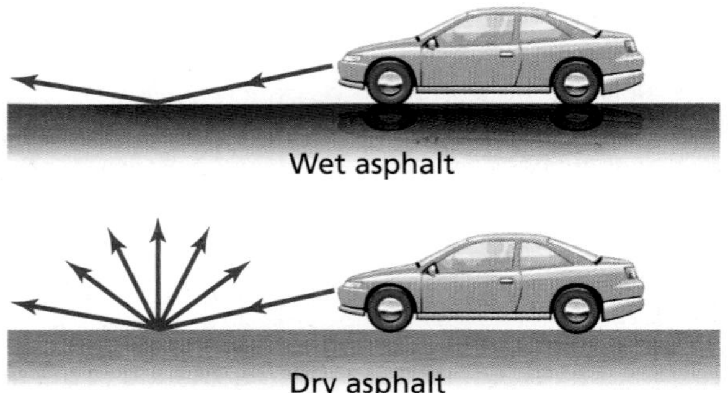

Figure 17-16

Richard Megna/Fundamental Photographs

45. Book Pages Why is it desirable that the pages of a book be rough rather than smooth and glossy?

46. Locate and describe the physical properties of the image produced by a concave mirror when the object is located at the center of curvature.

47. An object is located beyond the center of curvature of a spherical concave mirror. Locate and describe the physical properties of the image.

48. Telescope You have to order a large concave mirror for a telescope that produces high-quality images. Should you order a spherical mirror or a parabolic mirror? Explain.

49. Describe the properties of the image seen in the single convex mirror in **Figure 17-17.**

Figure 17-17

 For more problems, go to Additional Problems, Appendix B.

50. List all the possible arrangements in which you could use a spherical mirror, either concave or convex, to form a real image.

51. List all possible arrangements in which you could use a spherical mirror, either concave or convex, to form an image that is smaller compared to the object.

52. Rearview Mirrors The outside rearview mirrors of cars often carry the warning "Objects in the mirror are closer than they appear." What kind of mirrors are these and what advantage do they have?

Mastering Problems

17.1 Reflection from Plane Mirrors

53. A ray of light strikes a mirror at an angle of 38° to the normal. What is the angle that the reflected angle makes with the normal?

54. A ray of light strikes a mirror at an angle of 53° to the normal.
  a. What is the angle of reflection?
  b. What is the angle between the incident ray and the reflected ray?

55. A ray of light incident upon a mirror makes an angle of 36° with the mirror. What is the angle between the incident ray and the reflected ray?

56. Picture in a Mirror Penny wishes to take a picture of her image in a plane mirror, as shown in **Figure 17-18.** If the camera is 1.2 m in front of the mirror, at what distance should the camera lens be focused?

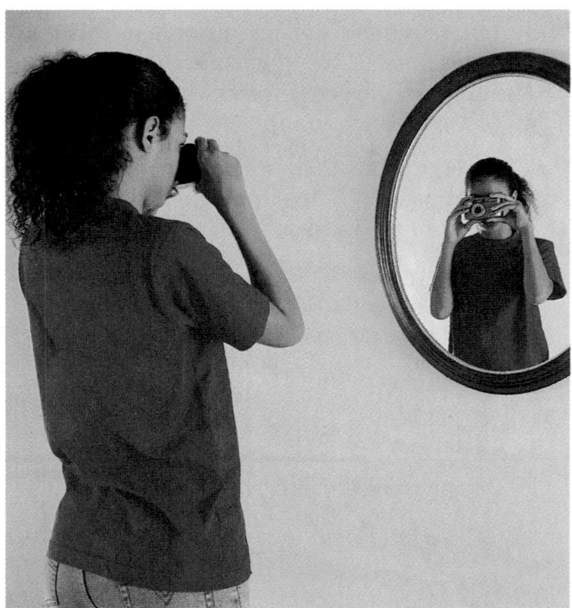

Figure 17-18

57. Two adjacent plane mirrors form a right angle, as shown in **Figure 17-19.** A light ray is incident upon one of the mirrors at an angle of 30° to the normal.
  a. What is the angle at which the light ray is reflected from the other mirror?
  b. A retroreflector is a device that reflects incoming light rays back in a direction opposite to that of the incident rays. Draw a diagram showing the angle of incidence on the first mirror for which the mirror system acts as a retroreflector.

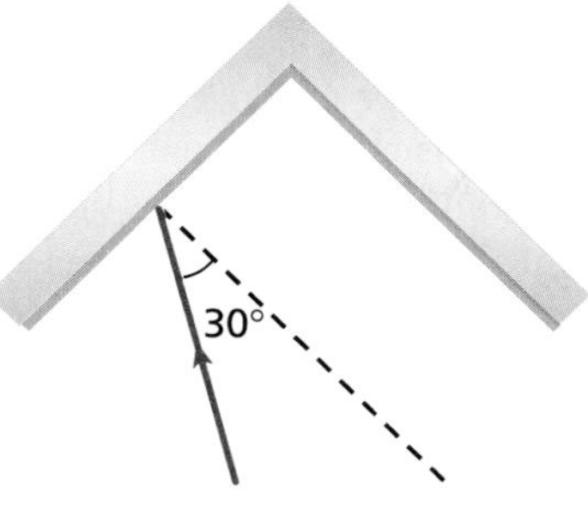

Figure 17-19

58. Draw a ray diagram of a plane mirror to show that if you want to see yourself from your feet to the top of your head, the mirror must be at least half your height.

59. Two plane mirrors are connected at their sides so that they form a 45° angle between them. A light ray strikes one mirror at an angle of 30° to the normal and then reflects off the second mirror. Calculate the angle of reflection of the light ray off the second mirror.

60. A ray of light strikes a mirror at an angle of 60° to the normal. The mirror is then rotated 18° clockwise, as shown in **Figure 17-20.** What is the angle that the reflected ray makes with the mirror?

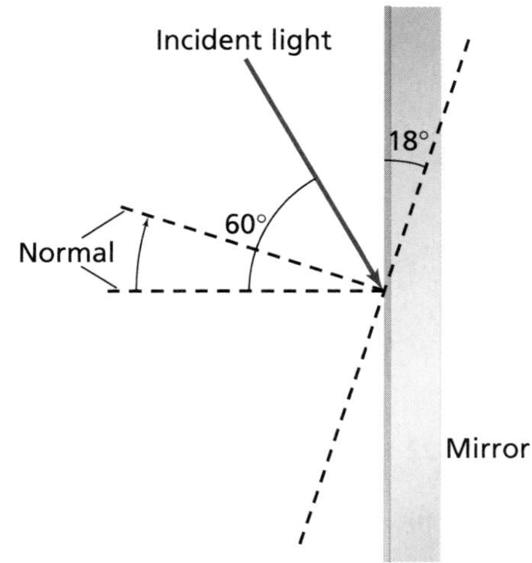

Figure 17-20

17.2 Curved Mirrors

61. A concave mirror has a focal length of 10.0 cm. What is its radius of curvature?

62. An object located 18 cm from a convex mirror produces a virtual image 9 cm from the mirror. What is the magnification of the image?

63. **Fun House** A boy is standing near a convex mirror in a fun house at a fair. He notices that his image appears to be 0.60 m tall. If the magnification of the mirror is $\frac{1}{3}$, what is the boy's height?

64. Describe the image produced by the object in **Figure 17-21** as real or virtual, inverted or upright, and smaller or larger than the object.

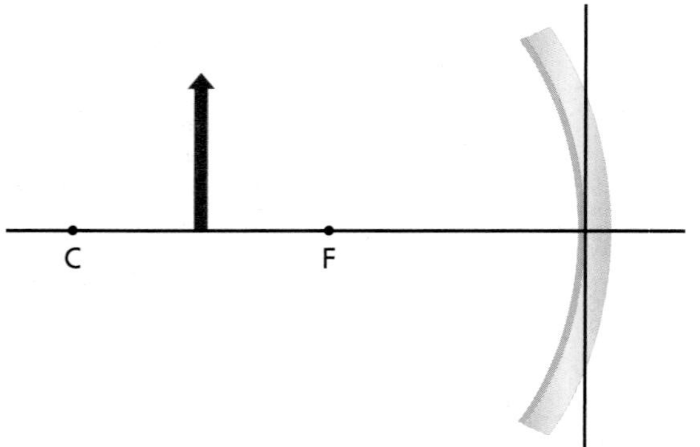

Figure 17-21

65. **Star Image** Light from a star is collected by a concave mirror. How far from the mirror is the image of the star if the radius of curvature is 150 cm?

66. Find the image position and height for the object shown in **Figure 17-22.**

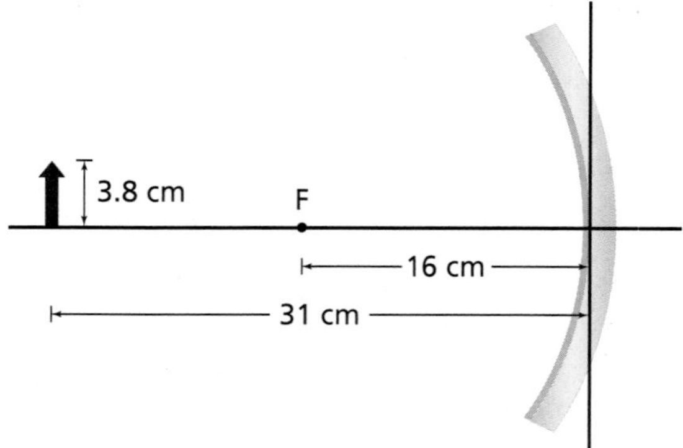

Figure 17-22

67. **Rearview Mirror** How far does the image of a car appear behind a convex mirror, with a focal length of −6.0 m, when the car is 10.0 m from the mirror?

68. An object is 30.0 cm from a concave mirror of 15.0 cm focal length. The object is 1.8 cm tall. Use the mirror equation to find the image position. What is the image height?

69. **Dental Mirror** A dentist uses a small mirror with a radius of 40 mm to locate a cavity in a patient's tooth. If the mirror is concave and is held 16 mm from the tooth, what is the magnification of the image?

70. A 3.0-cm-tall object is 22.4 cm from a concave mirror. If the mirror has a radius of curvature of 34.0 cm, what are the image position and height?

71. **Jeweler's Mirror** A jeweler inspects a watch with a diameter of 3.0 cm by placing it 8.0 cm in front of a concave mirror of 12.0-cm focal length.
a. Where will the image of the watch appear?
b. What will be the diameter of the image?

72. Sunlight falls on a concave mirror and forms an image that is 3.0 cm from the mirror. An object that is 24 mm tall is placed 12.0 cm from the mirror.
a. Sketch the ray diagram to show the location of the image.
b. Use the mirror equation to calculate the image position.
c. How tall is the image?

73. Shiny spheres that are placed on pedestals on a lawn are convex mirrors. One such sphere has a diameter of 40.0 cm. A 12-cm-tall robin sits in a tree that is 1.5 m from the sphere. Where is the image of the robin and how tall is the image?

Mixed Review

74. A light ray strikes a plane mirror at an angle of 28° to the normal. If the light source is moved so that the angle of incidence increases by 34°, what is the new angle of reflection?

75. Copy **Figure 17-23** on a sheet of paper. Draw rays on the diagram to determine the height and location of the image.

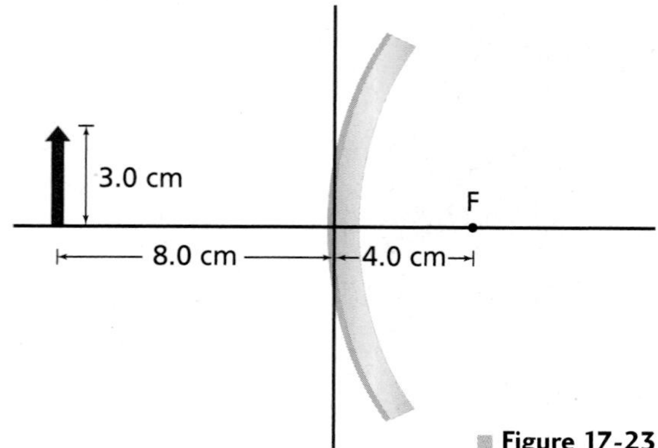

Figure 17-23

 For more problems, go to Additional Problems, Appendix B.

76. An object is located 4.4 cm in front of a concave mirror with a 24.0-cm radius. Locate the image using the mirror equation.

77. A concave mirror has a radius of curvature of 26.0 cm. An object that is 2.4 cm tall is placed 30.0 cm from the mirror.
a. Where is the image position?
b. What is the image height?

78. What is the radius of curvature of a concave mirror that magnifies an object by a factor of +3.2 when the object is placed 20.0 cm from the mirror?

79. A convex mirror is needed to produce an image one-half the size of an object and located 36 cm behind the mirror. What focal length should the mirror have?

80. Surveillance Mirror A convenience store uses a surveillance mirror to monitor the store's aisles. Each mirror has a radius of curvature of 3.8 m.
a. What is the image position of a customer who stands 6.5 m in front of the mirror?
b. What is the image height of a customer who is 1.7 m tall?

81. Inspection Mirror A production-line inspector wants a mirror that produces an image that is upright with a magnification of 7.5 when it is located 14.0 mm from a machine part.
a. What kind of mirror would do this job?
b. What is its radius of curvature?

82. The object in **Figure 17-24** moves from position 1 to position 2. Copy the diagram onto a sheet of paper. Draw rays showing how the image changes.

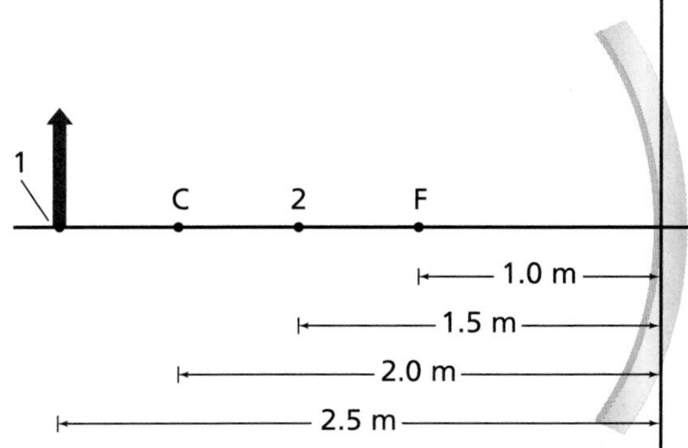

Figure 17-24

83. A ball is positioned 22 cm in front of a spherical mirror and forms a virtual image. If the spherical mirror is replaced with a plane mirror, the image appears 12 cm closer to the mirror. What kind of spherical mirror was used?

84. A 1.6-m-tall girl stands 3.2 m from a convex mirror. What is the focal length of the mirror if her image appears to be 0.28 m tall?

85. Magic Trick A magician uses a concave mirror with a focal length of 8.0 m to make a 3.0-m-tall hidden object, located 18.0 m from the mirror, appear as a real image that is seen by his audience. Draw a scale ray diagram to find the height and location of the image.

86. A 4.0-cm-tall object is placed 12.0 cm from a convex mirror. If the image of the object is 2.0 cm tall, and the image is located at −6.0 cm, what is the focal length of the mirror? Draw a ray diagram to answer the question. Use the mirror equation and the magnification equation to verify your answer.

Thinking Critically

87. Apply Concepts The ball in **Figure 17-25** slowly rolls toward the concave mirror on the right. Describe how the size of the ball's image changes as it rolls along.

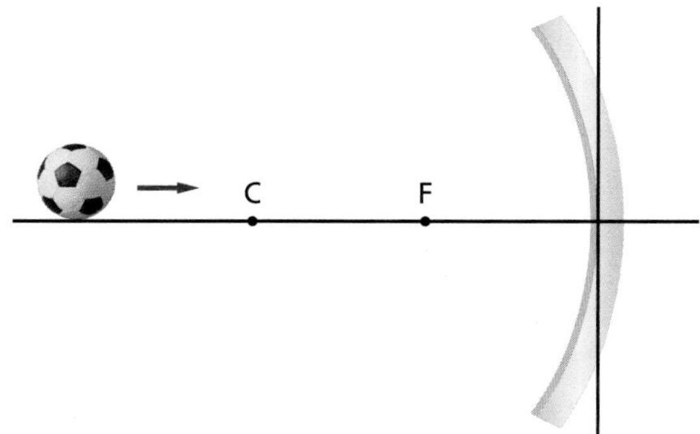

Figure 17-25

88. Analyze and Conclude The object in **Figure 17-26** is located 22 cm from a concave mirror. What is the focal length of the mirror?

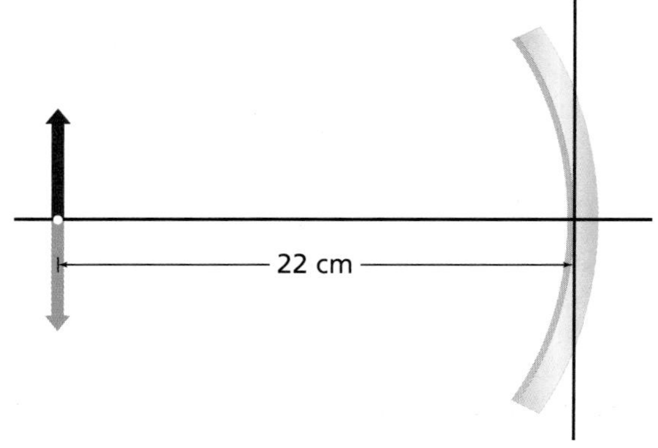

Figure 17-26

89. **Use Equations** Show that as the radius of curvature of a concave mirror increases to infinity, the mirror equation reduces to the relationship between the object position and the image position for a plane mirror.

90. **Analyze and Conclude** An object is located 6.0 cm from a plane mirror. If the plane mirror is replaced with a concave mirror, the resulting image is 8.0 cm farther behind the mirror. Assuming that the object is located between the focal point and the concave mirror, what is the focal length of the concave mirror?

91. **Analyze and Conclude** The layout of the two-mirror system shown in Figure 17-11 is that of a Gregorian telescope. For this question, the larger concave mirror has a radius of curvature of 1.0 m, and the smaller mirror is located 0.75 m away. Why is the secondary mirror concave?

92. **Analyze and Conclude** An optical arrangement used in some telescopes is the Cassegrain focus, shown in **Figure 17-27.** This telescope uses a convex secondary mirror that is positioned between the primary mirror and the focal point of the primary mirror.

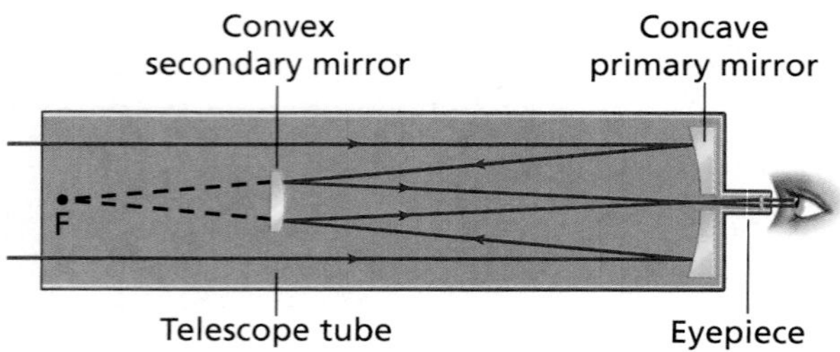

Figure 17-27

a. A single convex mirror produces only virtual images. Explain how the convex mirror in this telescope functions within the system of mirrors to produce real images.

b. Are the images produced by the Cassegrain focus upright or inverted? How does this relate to the number of times that the light crosses?

Writing in Physics

93. Research a method used for grinding, polishing, and testing mirrors used in reflecting telescopes. You may report either on methods used by amateur astronomers who make their own telescope optics, or on a method used by a project at a national laboratory. Prepare a one-page report describing the method, and present it to the class.

94. Mirrors reflect light because of their metallic coating. Research and write a summary of one of the following:

a. the different types of coatings used and the advantages and disadvantages of each

b. the precision optical polishing of aluminum to such a degree of smoothness that no glass is needed in the process of making a mirror

Cumulative Review

95. A child runs down the school hallway and then slides on the newly waxed floor. He was running at 4.7 m/s before he started sliding and he slid 6.2 m before stopping. What was the coefficient of friction of the waxed floor? (Chapter 11)

96. A 1.0 g piece of copper falls from a height of 1.0×10^3 m from an airplane to the ground. Because of air resistance it reaches the ground moving at a velocity of 70.0 m/s. Assuming that half of the energy lost by the piece was distributed as thermal energy to the copper, how much did it heat during the fall? (Chapter 12)

97. It is possible to lift a person who is sitting on a pillow made from a large sealed plastic garbage bag by blowing air into the bag through a soda straw. Suppose that the cross-sectional area of the person sitting on the bag is 0.25 m^2 and the person's weight is 600 N. The soda straw has a cross-sectional area of 2×10^{-5} m^2. With what pressure must you blow into the straw to lift the person that is sitting on the sealed garbage bag? (Chapter 13)

98. What would be the period of a 2.0-m-long pendulum on the Moon's surface? The Moon's mass is 7.34×10^{22} kg, and its radius is 1.74×10^6 m. What is the period of this pendulum on Earth? (Chapter 14)

99. **Organ pipes** An organ builder must design a pipe organ that will fit into a small space. (Chapter 15)

a. Should he design the instrument to have open pipes or closed pipes? Explain.

b. Will an organ constructed with open pipes sound the same as one constructed with closed pipes? Explain.

100. Filters are added to flashlights so that one shines red light and the other shines green light. The beams are crossed. Explain in terms of waves why the light from both flashlights is yellow where the beams cross, but revert back to their original colors beyond the intersection point. (Chapter 16)

 For more problems, go to Additional Problems, Appendix B.

Multiple Choice

1. Where is the object located if the image that is produced by a concave mirror is smaller than the object?
   - (A) at the mirror's focal point
   - (B) between the mirror and the focal point
   - (C) between the focal point and center of curvature
   - (D) past the center of curvature

2. What is the focal length of a concave mirror that magnifies, by a factor of +3.2, an object that is placed 30 cm from the mirror?
   - (A) 23 cm
   - (B) 32 cm
   - (C) 44 cm
   - (D) 46 cm

3. An object is placed 21 cm in front of a concave mirror with a focal length of 14 cm. What is the image position?
   - (A) −42 cm
   - (B) −8.4 cm
   - (C) 8.4 cm
   - (D) 42 cm

4. The light rays in the illustration below do not properly focus at the focal point. This problem occurs with ________.
   - (A) all spherical mirrors
   - (B) all parabolic mirrors
   - (C) only defective spherical mirrors
   - (D) only defective parabolic mirrors

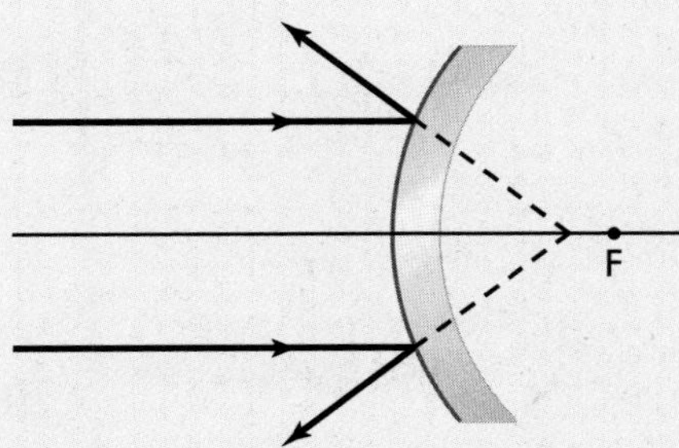

5. A ray of light strikes a plane mirror at an angle of 23° to the normal. What is the angle between the reflected ray and the mirror?
   - (A) 23°
   - (B) 46°
   - (C) 67°
   - (D) 134°

6. A concave mirror produces an inverted image that is 8.5 cm tall, located 34.5 cm in front of the mirror. If the focal point of the mirror is 24.0 cm, then what is the height of the object that is reflected?
   - (A) 2.3 cm
   - (B) 3.5 cm
   - (C) 14 cm
   - (D) 19 cm

7. A concave mirror with a focal length of 16.0 cm produces an image located 38.6 cm from the mirror. What is the distance of the object from the front of the mirror?
   - (A) 2.4 cm
   - (B) 11.3 cm
   - (C) 22.6 cm
   - (D) 27.3 cm

8. A convex mirror is used to produce an image that is three-fourths the size of an object and located 8.4 cm behind the mirror. What is the focal length of the mirror?
   - (A) −34 cm
   - (B) −11 cm
   - (C) −6.3 cm
   - (D) −4.8 cm

9. A cup sits 17 cm from a concave mirror. The image of the book appears 34 cm in front of the mirror. What are the magnification and orientation of the cup's image?
   - (A) 0.5, inverted
   - (B) 0.5, upright
   - (C) 2.0, inverted
   - (D) 2.0, upright

Extended Answer

10. A 5.0-cm-tall object is located 20.0 cm from a convex mirror with a focal length of −14.0 cm. Draw a scale-ray diagram showing the image height.

✓ Test-Taking TIP

Your Answers Are Better Than the Test's

When you know how to solve a problem, solve it before looking at the answer choices. Often, more than one answer choice will look good, so do the calculations first, and arm yourself with your answer before looking.

Chapter 18 Refraction and Lenses

What You'll Learn

- You will learn how light changes direction and speed when it travels through different materials.
- You will compare properties of lenses and the images that they form.
- You will learn about different applications of lenses, including how lenses in your eyes enable you to see.

Why It's Important

Some light travels in a straight path from objects to your eyes. Some light is reflected before it reaches you. Other light follows a path that appears to be bent.

Wavy Trees If you swim underwater, you will notice that things underwater look normal, but objects above the surface of the water appear to be distorted by the waves on the surface.

Think About This ▶
What causes the images of the trees to be wavy?

physicspp.com

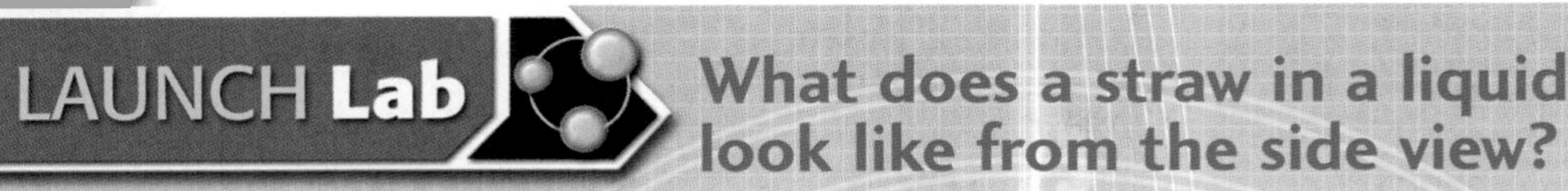

What does a straw in a liquid look like from the side view?

Question
Does a straw look different when observed through water, oil, and corn syrup?

Procedure

1. Fill one 400-mL beaker with water.
2. Fill a second 400-mL beaker halfway with corn syrup and the rest with water (pour slowly as to avoid mixing the two liquids).
3. Fill a third 400-mL beaker halfway with water and the rest with cooking oil (pour slowly as to avoid mixing the two liquids).
4. Place a straw gently in each beaker and lean it on the spout.
5. Observe each straw through the side of the beaker as you slowly turn the beaker.
6. Make a data table to record descriptions of the straws' appearance in each solution.

Analysis
In which containers does the straw appear to be broken? Are all amounts of break the same? When does the straw not appear to be broken? Explain.

Critical Thinking Form a hypothesis as to when a solid object appears to be broken and when it does not. Be sure to include an explanation of the amount of break.

18.1 Refraction of Light

Looking at the surface of a swimming pool on a summer day, you can see sunlight reflecting off the water. You can see objects that are in the pool because some of the sunlight travels into the water and reflects off the objects. When you look closely at objects in the water, however, you will notice that they look distorted. For example, things beneath the surface look closer than normal, the feet of a person standing still in the pool appear to move back and forth, and lines along the bottom of the pool seem to sway with the movement of the water. These effects occur because light changes direction as it passes from water to air.

As you learned in Chapter 16, the path of light is bent as it crosses the boundary between two media due to refraction. The amount of refraction depends on properties of the two media and on the angle at which the light strikes the boundary. As waves travel along the surface of the water, the boundary between the air and water moves up and down, and tilts back and forth. The path of light leaving the water shifts as the boundary moves, causing objects under the surface to appear to waver.

▶ **Objectives**

- **Solve** problems involving refraction.
- **Explain** total internal reflection.
- **Explain** some optical effects caused by refraction.

▶ **Vocabulary**

index of refraction
Snell's law of refraction
critical angle
total internal reflection
dispersion

Figure 18-1 Light bends toward the normal as it moves from air to glass and bends away from the normal as it moves from glass to air **(a).** The bending of light makes objects appear to be shifted from their actual locations **(b).**

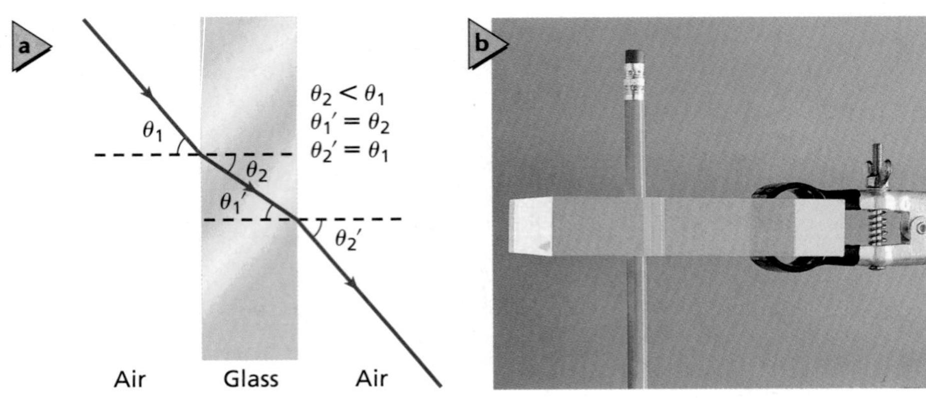

Snell's Law of Refraction

What happens when you shine a narrow beam of light at the surface of a piece of glass? As you can see in **Figure 18-1,** it bends as it crosses the boundary from air to glass. The bending of light, called refraction, was first studied by René Descartes and Willebrord Snell around the time of Kepler and Galileo.

To discuss the results of Descartes and Snell, you have to define two angles. The angle of incidence, θ_1, is the angle at which the light ray strikes the surface. It is measured from the normal to the surface. The angle of refraction, θ_2, is the angle at which the transmitted light leaves the surface. It also is measured with respect to the normal. In 1621, Snell found that when light passed from air into a transparent substance, the sines of the angles were related by the equation $\sin \theta_1 / \sin \theta_2 = n$. Here n is a constant that depends on the substance, not on the angles, and is called the **index of refraction.** The indices of refraction for some substances are listed in **Table 18-1.** The relationship found by Snell is valid when light goes across a boundary between any two materials. This more general equation is known as **Snell's law of refraction.**

Color Convention

- Refracting media and lenses are light blue.

Snell's Law of Refraction $n_1 \sin \theta_1 = n_2 \sin \theta_2$

The product of the index of refraction of the first medium and the sine of the angle of incidence is equal to the product of the index of refraction of the second medium and the sine of the angle of refraction.

Table 18-1

Indices of Refraction for Yellow Light (λ = 589 nm in vacuum)

Medium	n
Vacuum	1.00
Air	1.0003
Water	1.33
Ethanol	1.36
Crown glass	1.52
Quartz	1.54
Flint glass	1.62
Diamond	2.42

Figure 18-1 shows how Snell's law applies when light travels through a piece of glass with parallel surfaces, such as a windowpane. The light is refracted both when it enters the glass and again when it leaves the glass. When light goes from air to glass it moves from material with a lower index of refraction to one with a higher index of refraction. That is, $n_1 < n_2$. To keep the two sides of the equation equal, one must have $\sin \theta_1 > \sin \theta_2$. The light beam is bent toward the normal to the surface.

When light travels from glass to air it moves from material having a higher index of refraction to one with a lower index. In this case, $n_1 > n_2$. To keep the two sides of the equation equal one must have $\sin \theta_1' < \sin \theta_2'$. That is, the light is bent away from the normal. Note that the direction of the ray when it leaves the glass is the same as it was before it struck the glass, but it is shifted from its original position.

▶ EXAMPLE Problem 1

Angle of Refraction **A light beam in air hits a sheet of crown glass at an angle of 30.0°. At what angle is the light beam refracted?**

1 Analyze and Sketch the Problem

- Make a sketch of the air and crown glass boundary.
- Draw a ray diagram.

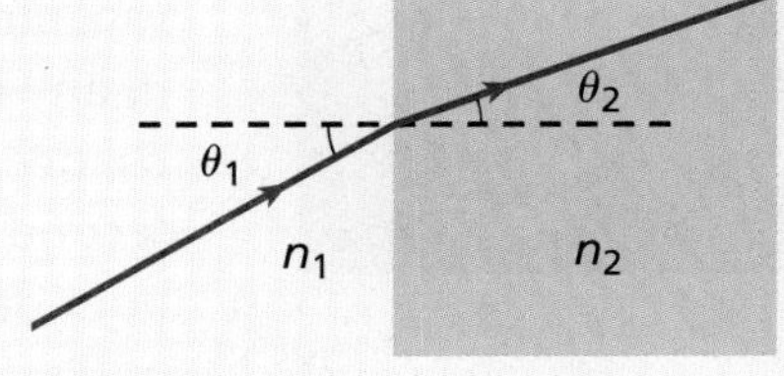

Known:

$\theta_1 = 30.0°$

$n_1 = 1.00$

$n_2 = 1.52$

Unknown:

$\theta_2 = ?$

2 Solve for the Unknown

Use Snell's law to solve for the sine of the angle of refraction.

$$n_1 \sin \theta_1 = n_2 \sin \theta_2$$

$$\sin \theta_2 = \left(\frac{n_1}{n_2}\right) \sin \theta_1$$

$$\theta_2 = \sin^{-1}\left(\left(\frac{n_1}{n_2}\right) \sin \theta_1\right)$$

$$= \sin^{-1}\left(\left(\frac{1.00}{1.52}\right) \sin 30.0°\right)$$

Substitute $n_1 = 1.00$, $n_2 = 1.52$, $\theta_1 = 30.0°$

$$= 19.2°$$

3 Evaluate the Answer

- **Are the units correct?** Angles are expressed in degrees.
- **Is the magnitude realistic?** The index of refraction, n_2, is greater than the index of refraction, n_1. Therefore, the angle of refraction, θ_2, must be less than the angle of incidence, θ_1.

Physics Online

Personal Tutor For an online tutorial on the angle of refraction, visit physicspp.com.

▶ PRACTICE Problems

• Additional Problems, Appendix B
• Solutions to Selected Problems, Appendix C

1. A laser beam in air is incident upon ethanol at an angle of incidence of 37.0°. What is the angle of refraction?
2. Light in air is incident upon a piece of crown glass at an angle of incidence of 45.0°. What is the angle of refraction?
3. Light passes from air into water at 30.0° to the normal. Find the angle of refraction.
4. Light in air is incident upon a diamond facet at 45.0°. What is the angle of refraction?
5. A block of unknown material is submerged in water. Light in the water is incident on the block at an angle of incidence of 31°. The angle of refraction of the light in the block is 27°. What is the index of refraction of the material of the block?

Refraction is responsible for the Moon appearing red during a lunar eclipse. A lunar eclipse occurs when Earth blocks sunlight from the Moon. As a result, you might expect the Moon to be completely dark. Instead, light refracts through Earth's atmosphere and bends around Earth toward the Moon. Recall that Earth's atmosphere scatters most of the blue and green light. Thus, mostly red light illuminates the Moon. Because the Moon reflects most colors of light equally well, it reflects the red light back to Earth, and therefore the Moon appears to be red.

Astronomy Connection

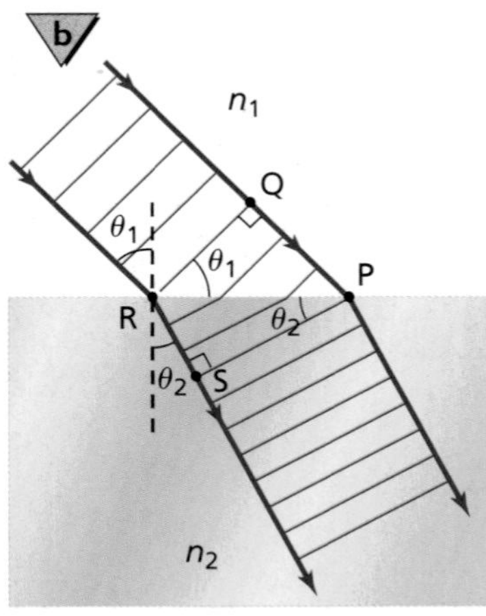

Figure 18-2 Light moves from air to glass to air again **(a).** Light slows down and bends toward the normal when it enters a region of a higher index of refraction **(b).**

Wave Model of Refraction

The wave model of light was developed almost 200 years after Snell published his research. An understanding that light interacts with atoms when traveling through a medium, such that it moves more slowly than in a vacuum, was achieved three hundred years after Snell's work. The wave relationship that you learned in Chapter 16 for light traveling through a vacuum, $\lambda_0 = c/f$, can be rewritten as $\lambda = v/f$, where v is the speed of light in any medium and λ is the wavelength. The frequency of light, f, does not change when it crosses a boundary. That is, the number of oscillations per second that arrive at a boundary is the same as the number that leave the boundary and transmit through the refracting medium. So, the wavelength of light, λ, must decrease when light slows down. Wavelength in a medium is shorter than wavelength in a vacuum.

What happens when light travels from a region with a high speed into one with a low speed, as shown in **Figure 18-2a?** The diagram in **Figure 18-2b** shows a beam of light as being made up of a series of parallel, straight wave fronts. Each wave front represents the crest of a wave and is perpendicular to the direction of the beam. The beam strikes the surface at an angle, θ_1. Consider the triangle PQR. Because the wave fronts are perpendicular to the direction of the beam, $\angle$PQR is a right angle and $\angle$QRP is equal to θ_1. Therefore, sin θ_1 is equal to the distance between P and Q divided by the distance between P and R.

$$\sin \theta_1 = \frac{\overline{PQ}}{\overline{PR}}$$

The angle of refraction, θ_2, can be related in a similar way to the triangle PSR. In this case

$$\sin \theta_2 = \frac{\overline{RS}}{\overline{PR}}$$

By taking the ratio of the sines of the two angles, $\overline{PR}$ is canceled, leaving the following equation:

$$\frac{\sin \theta_2}{\sin \theta_1} = \frac{\overline{RS}}{\overline{PQ}}$$

Figure 18-2b is drawn such that the distance between P and Q is equal to the length of three wavelengths of light in medium 1, or $\overline{PQ} = 3\lambda_1$. In a similar way, $\overline{RS} = 3\lambda_2$. Substituting these two values into the previous equation and canceling the common factor of 3 provides an equation that relates the angles of incidence and refraction with the wavelength of the light in each medium.

$$\frac{\sin \theta_2}{\sin \theta_1} = \frac{3\lambda_2}{3\lambda_1} = \frac{\lambda_2}{\lambda_1}$$

Using $\lambda = v/f$ in the above equation and canceling the common factor of f, the equation is rewritten as follows:

$$\frac{\sin \theta_2}{\sin \theta_1} = \frac{v_2}{v_1}$$

Snell's law also can be written as a ratio of the sines of the angles of incidence and refraction.

$$\frac{\sin \theta_2}{\sin \theta_1} = \frac{n_1}{n_2}$$

Index of refraction Using the transitive property of equality, the previous two equations lead to the following equation:

$$\frac{n_1}{n_2} = \frac{v_2}{v_1}$$

In a vacuum, $n = 1$ and $v = c$. If either medium is a vacuum, then the equation is simplified to an equation that relates the index of refraction to the speed of light in a medium.

Index of Refraction $n = \frac{c}{v}$

The index of refraction of a medium is equal to the speed of light in a vacuum divided by the speed of light in the medium.

This definition of the index of refraction can be used to find the wavelength of light in a medium compared to the wavelength the light would have in a vacuum. In a medium with an index of refraction n the speed of light is given by $v = c/n$. The wavelength of the light in a vacuum is $\lambda_0 = c/f$. Solve for frequency, and substitute $f = c/\lambda_0$ and $v = c/n$ into $\lambda = v/f$. $\lambda = (c/n)/(c/\lambda_0) = \lambda_0/n$, and thus the wavelength of light in a medium is smaller than the wavelength in a vacuum.

Total Internal Reflection

The angle of refraction is larger than the angle of incidence when light passes into a medium of a lower index of refraction, as shown in **Figure 18-3a.** This leads to an interesting phenomenon. As the angle of incidence increases, the angle of refraction increases. At a certain angle of incidence known as the **critical angle,** θ_c, the refracted light ray lies along the boundary of the two media, as shown in **Figure 18-3b.**

Recall from Chapter 16 that when light strikes a transparent boundary, even though much of the light is transmitted, some is reflected. **Total internal reflection** occurs when light traveling from a region of a higher index of refraction to a region of a lower index of refraction strikes the boundary at an angle greater than the critical angle such that all light reflects back into the region of the higher index of refraction, as shown in **Figure 18-3c.** To construct an equation for the critical angle of any boundary, you can use Snell's law and substitute $\theta_1 = \theta_c$ and $\theta_2 = 90.0°$.

Critical Angle for Total Internal Reflection $\sin \theta_c = \frac{n_2}{n_1}$

The sine of the critical angle is equal to the index of refraction of the refracting medium divided by the index of refraction of the incident medium.

Total internal reflection causes some curious effects. Suppose that you are looking up at the surface from underwater in a calm pool. You might see an upside-down reflection of another nearby object that also is underwater or a reflection of the bottom of the pool itself. The surface of the water acts like a mirror. Likewise, when you are standing on the side of a pool, it is possible for things below the surface of the water to not be visible to you. When a swimmer is underwater, near the surface, and on the opposite side of the pool from you, you might not see him or her. This is because the light from his or her body is reflected.

Figure 18-3 Ray A is partially refracted and partially reflected **(a).** Ray B is refracted along the boundary of the medium and forms the critical angle **(b).** An angle of incidence greater than the critical angle results in the total internal reflection of Ray C, which follows the law of reflection **(c).**

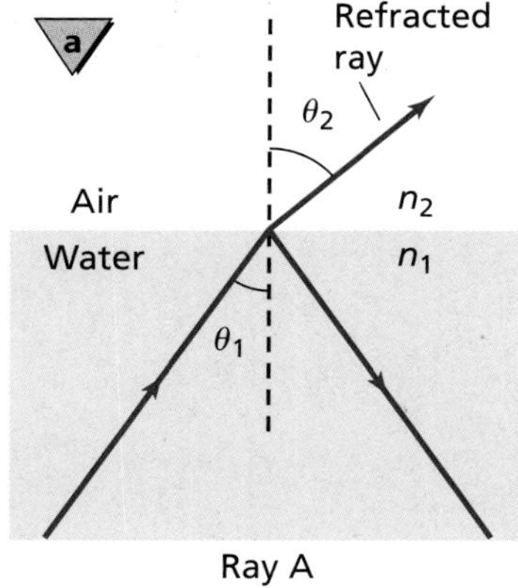

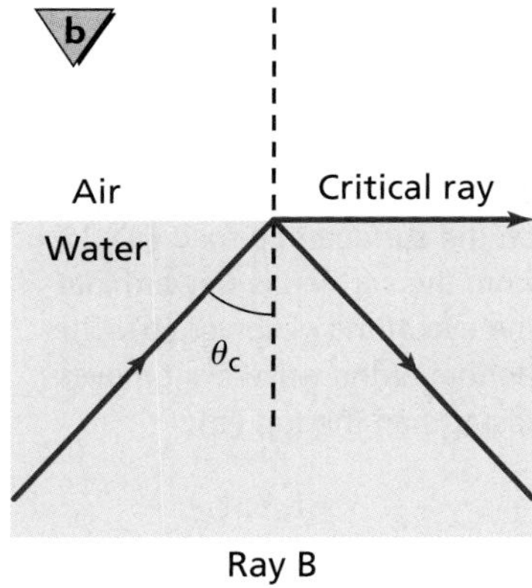

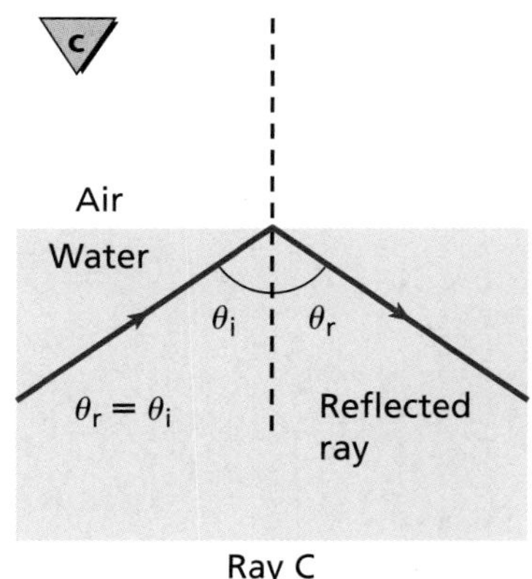

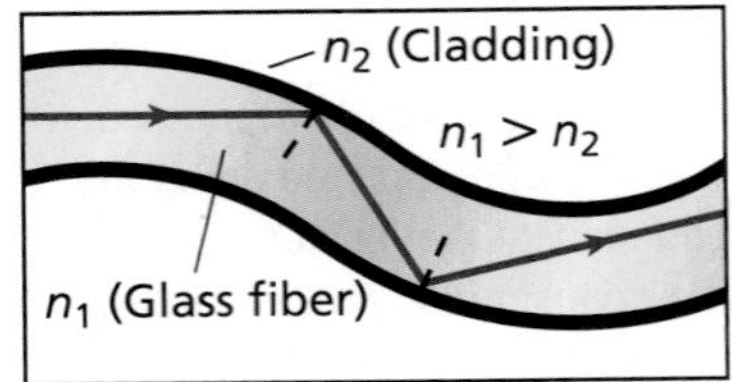

Figure 18-4 Light impulses from a source enter one end of the optical fiber. Each time the light strikes the surface, the angle of incidence is larger than the critical angle, and, therefore, the light is kept within the fiber.

Optical fibers are an important technical application of total internal reflection. As shown in **Figure 18-4,** the light traveling through the transparent fiber always hits the internal boundary of the optical fiber at an angle greater than the critical angle, so all of the light is reflected and none of the light is transmitted through the boundary. Thus, the light maintains its intensity over the distance of the fiber.

Mirages

On a hot summer day, you sometimes can see the mirage effect shown in **Figure 18-5a.** As you drive down a road, you see what appears to be the reflection of an oncoming car in a pool of water. The pool, however, disappears as you approach it. The mirage is the result of the Sun heating the road. The hot road heats the air above it and produces a thermal layering of air that causes light traveling toward the road to gradually bend upward. This makes the light appear to be coming from a reflection in a pool, as shown in **Figure 18-5b.**

Figure 18-5c shows how this occurs. As light from a distant object travels downward toward the road, the index of refraction of the air decreases as the air gets hotter, but the temperature change is gradual. Recall from Chapter 16 that light wave fronts are comprised of Huygens' wavelets. In the case of a mirage, the Huygens' wavelets closer to the ground travel faster than those higher up, causing the wave fronts to gradually turn upward. A similar phenomenon, called a superior mirage, occurs when a reflection of a distant boat appears above the boat. The water keeps the air that is closer to its surface cooler.

Figure 18-5 A mirage is seen on the surface of a road **(a).** Light from the car bends upward into the eye of the observer **(b).** The bottom of the wave front moves faster than the top **(c).**

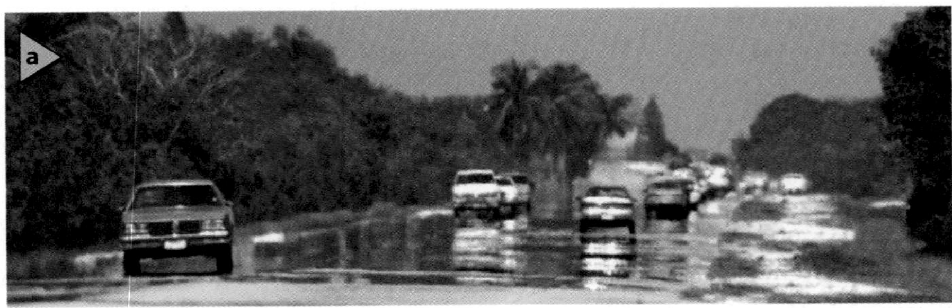

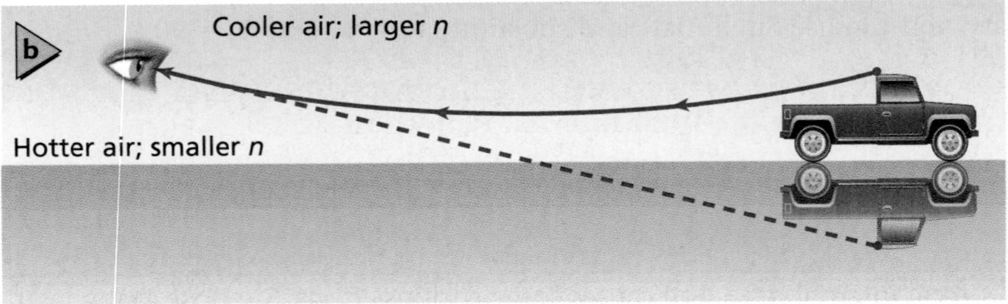

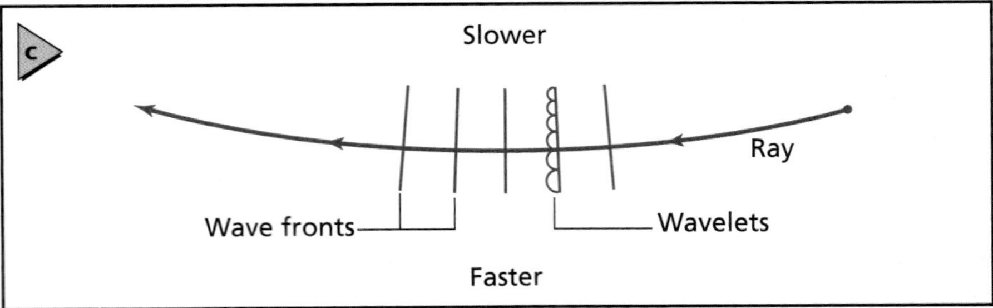

Dispersion of Light

The speed of light in a medium is determined by interactions between the light and the atoms that make up the medium. Recall from Chapters 12 and 13 that temperature and pressure are related to the energy of particles on the atomic level. The speed of light, and therefore, the index of refraction for a gaseous medium, can change slightly with temperature. In addition, the speed of light and the index of refraction vary for different wavelengths of light in the same liquid or solid medium.

You learned in Chapter 16 that white light separates into a spectrum of colors when it passes through a glass prism, as shown in **Figure 18-6a.** This phenomenon is called **dispersion.** If you look carefully at the light that passes through a prism, you will notice that violet is refracted more than red, as shown in **Figure 18-6b.** This occurs because the speed of violet light through glass is less than the speed of red light through glass. Violet light has a higher frequency than red light, which causes it to interact differently with the atoms of the glass. This results in glass having a slightly higher index of refraction for violet light than it has for red light.

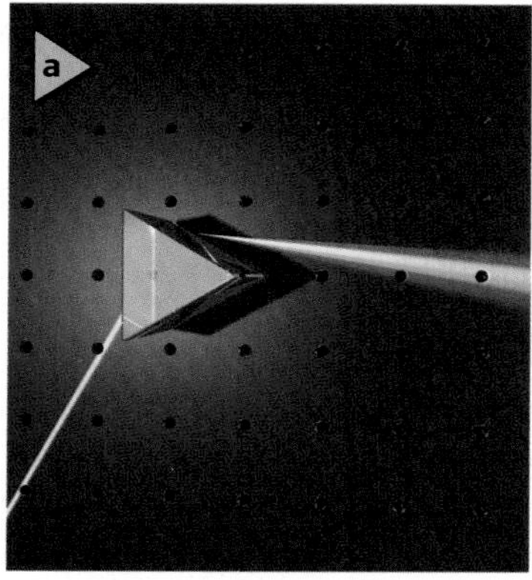

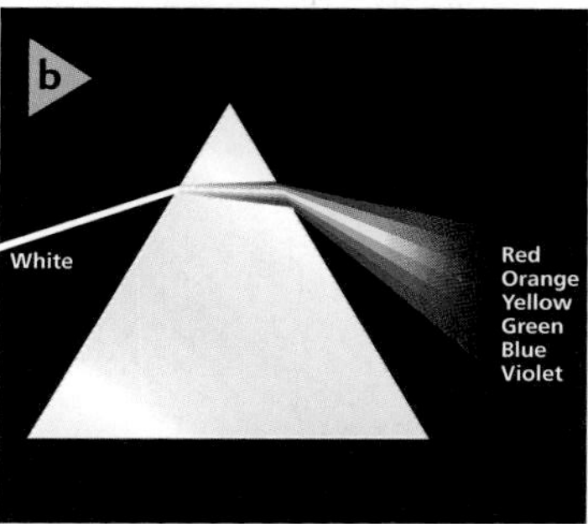

Figure 18-6 White light directed through a prism is dispersed into bands of different colors **(a).** Different colors of light bend different amounts when they enter a medium **(b).**

Rainbows A prism is not the only means of dispersing light. A rainbow is a spectrum formed when sunlight is dispersed by water droplets in the atmosphere. Sunlight that falls on a water droplet is refracted. Because of dispersion, each color is refracted at a slightly different angle, as shown in **Figure 18-7a.** At the back surface of the droplet, some of the light undergoes internal reflection. On the way out of the droplet, the light once again is refracted and dispersed.

Although each droplet produces a complete spectrum, an observer positioned between the Sun and the rain will see only a certain wavelength of light from each droplet. The wavelength depends on the relative positions of the Sun, the droplet, and the observer, as shown in **Figure 18-7b.** Because there are many droplets in the sky, a complete spectrum is visible. The droplets reflecting red light make an angle of 42° in relation to the direction of the Sun's rays; the droplets reflecting blue light make an angle of 40°.

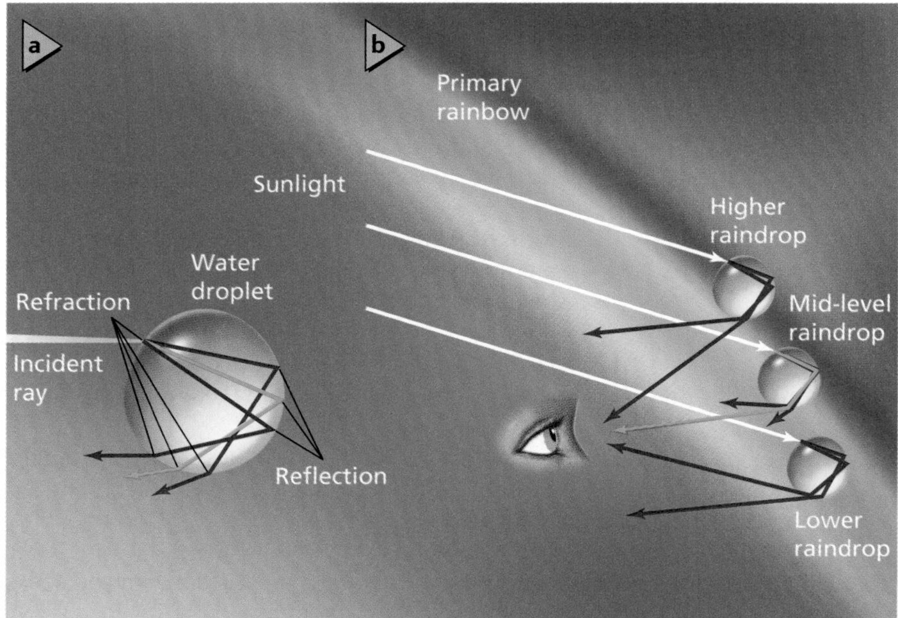

Figure 18-7 Rainbows form because white light is dispersed as it enters, reflects at the inside boundary, and exits the raindrops **(a).** Because of dispersion, only one color from each raindrop reaches an observer **(b).** (Illustration not to scale)

■ **Figure 18-8** A mist across your view allows for light comprising the entire spectrum of colors to reach your eyes in the form of a rainbow. Reflection from the raindrops sometimes enables you to see a second rainbow with the colors reversed.

Sometimes, you can see a faint second-order rainbow like the one shown in **Figure 18-8.** The second rainbow is outside of the first, is fainter, and has the order of the colors reversed. Light rays that are reflected twice inside water droplets produce this effect. Very rarely, a third rainbow is visible outside the second. What is your prediction about how many times light is reflected in the water droplets and the order of appearance of the colors for the third rainbow?

18.1 Section Review

6. Index of Refraction You notice that when a light ray enters a certain liquid from water, it is bent toward the normal, but when it enters the same liquid from crown glass, it is bent away from the normal. What can you conclude about the liquid's index of refraction?

7. Index of Refraction A ray of light in air has an angle of incidence of 30.0° on a block of unknown material and an angle of refraction of 20.0°. What is the index of refraction of the material?

8. Speed of Light Could an index of refraction ever be less than 1? What would this imply about the speed of light in that medium?

9. Speed of Light What is the speed of light in chloroform ($n = 1.51$)?

10. Total Internal Reflection If you were to use quartz and crown glass to make an optical fiber, which would you use for the cladding layer? Why?

11. Angle of Refraction A beam of light passes from water into polyethylene with $n = 1.50$. If $\theta_i = 57.5°$, what is the angle of refraction in the polyethylene?

12. Critical Angle Is there a critical angle for light traveling from glass to water? From water to glass?

13. Dispersion Why can you see the image of the Sun just above the horizon when the Sun itself has already set?

14. Critical Thinking In what direction can you see a rainbow on a rainy late afternoon? Explain.

18.2 Convex and Concave Lenses

The refraction of light in nature that forms rainbows and red lunar eclipses is beautiful, but refraction also is useful. In 1303, French physician Bernard of Gordon wrote of the use of lenses to correct eyesight. Around 1610, Galileo used two lenses to make a telescope, with which he discovered the moons of Jupiter. Since Galileo's time, lenses have been used in many instruments, such as microscopes and cameras. Lenses are probably the most useful of all optical devices.

▶ **Objectives**

- **Describe** how real and virtual images are formed by single convex and concave lenses.
- **Locate** images formed by lenses using ray tracing and equations.
- **Explain** how chromatic aberration can be reduced.

▶ **Vocabulary**

lens
convex lens
concave lens
thin lens equation
chromatic aberration
achromatic lens

Types of Lenses

A **lens** is a piece of transparent material, such as glass or plastic, that is used to focus light and form an image. Each of a lens's two faces might be either curved or flat. The lens in **Figure 18-9a** is called a **convex lens** because it is thicker at the center than at the edges. A convex lens often is called a converging lens because when surrounded by material with a lower index of refraction it refracts parallel light rays so that the rays meet at a point. The lens in **Figure 18-9b** is called a **concave lens** because it is thinner in the middle than at the edges. A concave lens often is called a diverging lens because when surrounded by material with a lower index of refraction rays passing through it spread out.

When light passes through a lens, refraction occurs at the two lens surfaces. Using Snell's law and geometry, you can predict the paths of rays passing through lenses. To simplify such problems, assume that all refraction occurs on a plane, called the principal plane, that passes through the center of the lens. This approximation, called the thin lens model, applies to all the lenses that you will learn about in this chapter section.

Lens equations The problems that you will solve involve spherical thin lenses, lenses that have faces with the same curvature as a sphere. Based on the thin lens model, as well as the other simplifications used in solving problems for spherical mirrors, equations have been developed that look exactly like the equations for spherical mirrors. The **thin lens equation** relates the focal length of a spherical thin lens to the object position and the image position.

Thin Lens Equation $\frac{1}{f} = \frac{1}{d_i} + \frac{1}{d_o}$

The inverse of the focal length of a spherical lens is equal to the sum of the inverses of the image position and the object position.

The magnification equation for spherical mirrors used in Chapter 17 also can be used for spherical thin lenses. It is used to determine the height and orientation of the image formed by a spherical thin lens.

Magnification $m \equiv \frac{h_i}{h_o} = \frac{-d_i}{d_o}$

The magnification of an object by a spherical lens, defined as the image height divided by the object height, is equal to the negative of the image position divided by the object position.

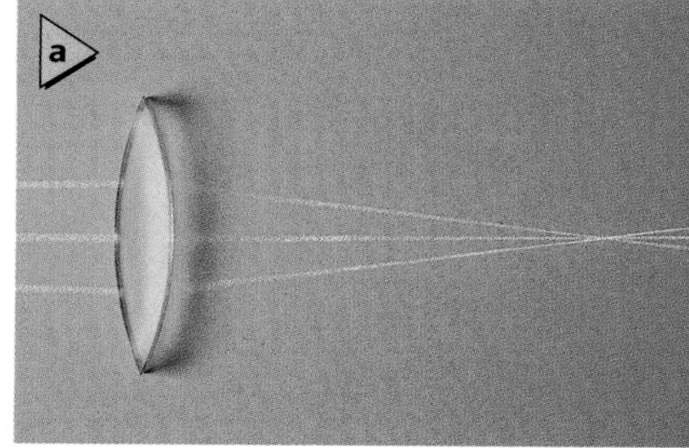

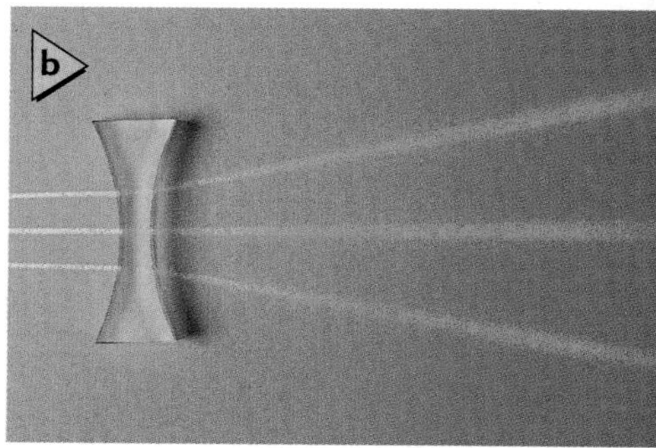

Figure 18-9 A convex lens causes rays of light to converge **(a).** A concave lens causes rays of light to diverge **(b).**

Table 18-2					
Properties of a Single Spherical Lens System					
Lens Type	**f**	**d_o**	**d_i**	**m**	**Image**
Convex	+	$d_o > 2f$	$2f > d_i > f$	Reduced Inverted	Real
		$2f > d_o > f$	$d_i > 2f$	Enlarged Inverted	Real
		$f > d_o > 0$	$\|d_i\| > d_o$ (negative)	Enlarged	Virtual
Concave	−	$d_o > 0$	$\|f\| > \|d_i\| > 0$ (negative)	Reduced	Virtual

Using the equations for lenses It is important that you use the proper sign conventions when using these equations. **Table 18-2** shows a comparison of the image position, magnification, and type of image formed by single convex and concave lenses when an object is placed at various object positions, d_o, relative to the lens. Notice the similarity of this table to Table 17-1 for mirrors. As with mirrors, the distance from the principal plane of a lens to its focal point is the focal length, f. The focal length depends upon the shape of the lens and the index of refraction of the lens material. Focal lengths and image positions can be negative.

For lenses, virtual images are always on the same side of the lens as the object, which means that the image position is negative. When the absolute value of a magnification is between zero and one, the image is smaller than the object. Magnifications with absolute values greater than one represent images that are larger than the objects. A negative magnification means the image is inverted compared to the object. Notice that a concave lens produces only virtual images, whereas a convex lens can produce real images or virtual images.

Convex Lenses and Real Images

As shown in **Figure 18-10a,** paper can be ignited by producing a real image of the Sun on the paper. Recall from Chapter 17 that the rays of the Sun are almost exactly parallel when they reach Earth. After being refracted by the lens, the rays converge at the focal point, F, of the lens. **Figure 18-10b** shows two focal points, one on each side of the lens. You could turn the lens around, and it will work the same.

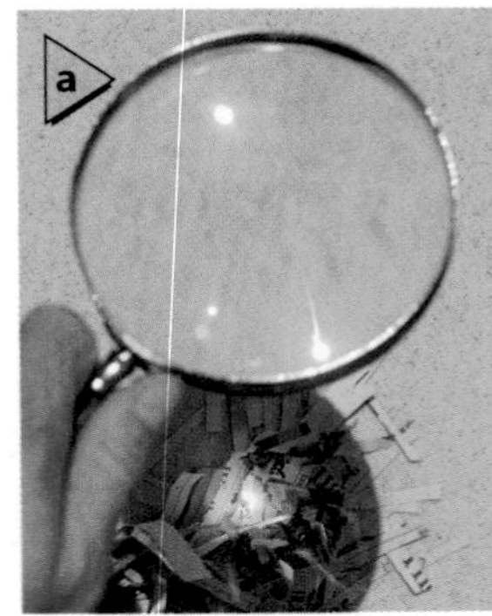

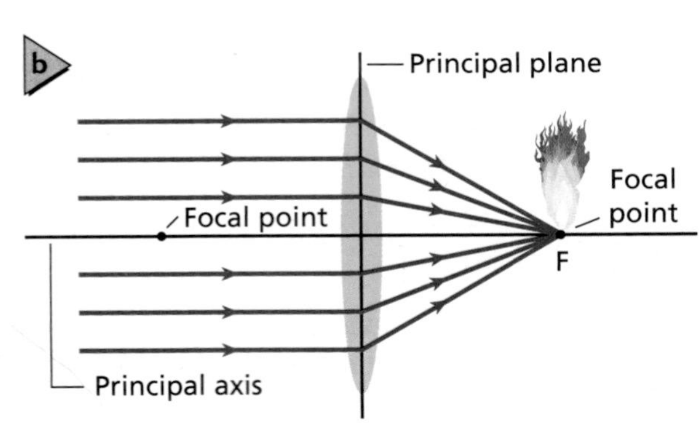

Figure 18-10 A converging lens can be used to ignite paper **(a).** Light entering parallel to the principal axis converges at the focal point of the lens, concentrating solar energy **(b).**

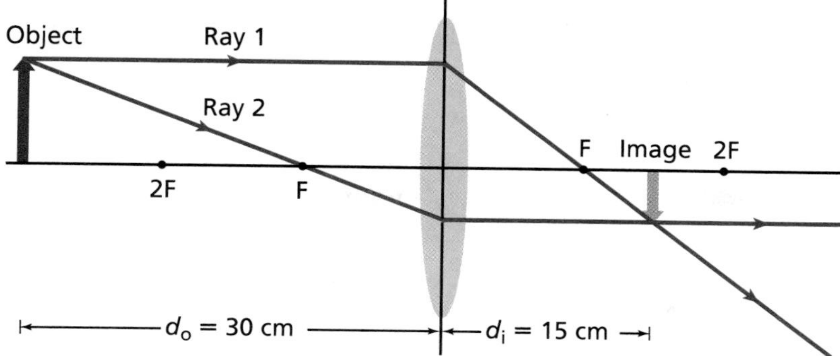

■ **Figure 18-11** When an object is placed at a distance greater than twice the focal length of the lens, a real image is produced that is inverted and smaller compared to the object. If the object is placed at the location of the image, you could locate the new image by tracing the same rays in the opposite direction.

Concepts In Motion
Interactive Figure To see an animation on ray diagrams, visit physicspp.com.

Ray diagrams In **Figure 18-11,** rays are traced from an object located far from a convex lens. For the purpose of locating the image, you only need to use two rays. Ray 1 is parallel to the principal axis. It refracts and passes through F on the other side of the lens. Ray 2 passes through F on its way to the lens. After refraction, its path is parallel to the principal axis. The two rays intersect at a point beyond F and locate the image. Rays selected from other points on the object converge at corresponding points to form the complete image. Note that this is a real image that is inverted and smaller compared to the object.

You can use Figure 18-11 to locate the image of an object that is closer to the lens than the object in the figure. If a refracted ray is reversed in direction, it will follow its original path in the reverse direction. This means that the image and object may be interchanged by changing the direction of the rays. Imagine that the path of light through the lens in Figure 18-11 is reversed and the object is at a distance of 15 cm from the right side of the lens. The new image, located 30 cm from the left side of the lens, is a real image that is inverted and larger compared to the object.

If the object is placed at twice the focal length from the lens at the point 2F, as shown in **Figure 18-12,** the image also is found at 2F. Because of symmetry, the image and object have the same size. Thus, you can conclude that if an object is more than twice the focal length from the lens, the image is smaller than the object. If the object is between F and 2F, then the image is larger than the object.

■ **Figure 18-12** When an object is placed at a distance equal to twice the focal length from the lens, the image is the same size as the object.

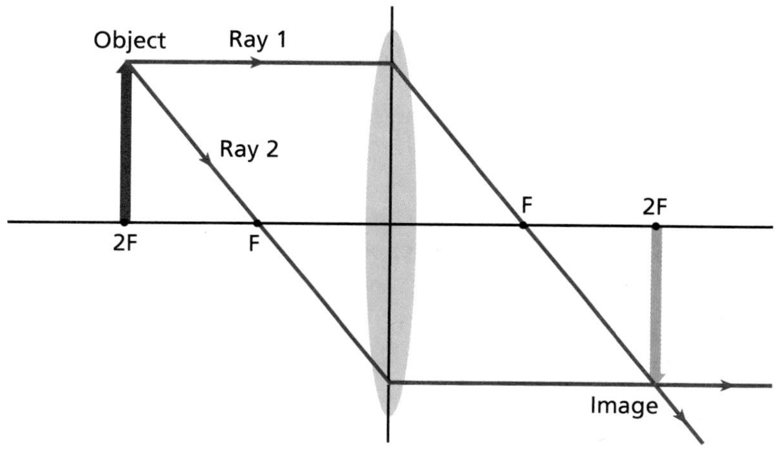

MINI LAB

Lens Masking Effects

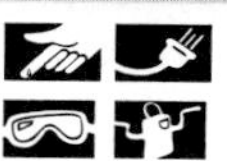

What happens when you mask, or cover, part of a lens? Does this cause only part of a real image to be formed by the lens?

1. Stick the edge of a convex lens into a ball of clay and place the lens on a tabletop. ***CAUTION: Lenses have sharp edges. Handle carefully.***

2. Use a small lamp on one side and a screen on the other side to get a sharp image of the lamp's lightbulb. ***CAUTION: Lamps get hot and can burn skin.***

3. Predict what will happen to the image if you place your hand over the top half of the lens. This is called masking.

4. Observe the effects of masking more of the lens and masking less of the lens.

Analyze and Conclude

5. How much of the lens is needed for a complete image?

6. What is the effect of masking the lens?

EXAMPLE Problem 2

An Image Formed by a Convex Lens An object is placed 32.0 cm from a convex lens that has a focal length of 8.0 cm.

a. Where is the image?

b. If the object is 3.0 cm high, how tall is the image?

c. What is the orientation of the image?

1 Analyze and Sketch the Problem

- Sketch the situation, locating the object and the lens.
- Draw the two principal rays.

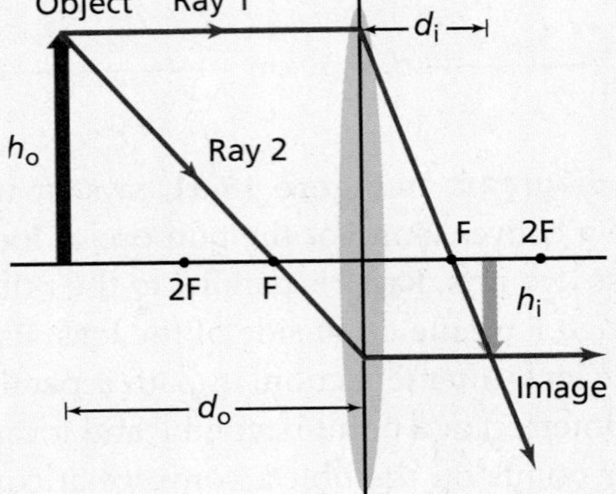

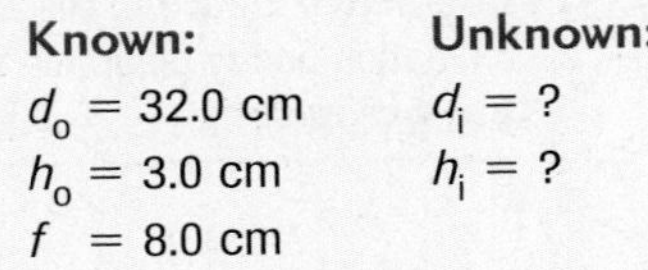

Known:	Unknown:
$d_o = 32.0$ cm	$d_i = ?$
$h_o = 3.0$ cm	$h_i = ?$
$f = 8.0$ cm	

2 Solve for the Unknown

a. Use the thin lens equation to determine d_i.

$$\frac{1}{f} = \frac{1}{d_i} + \frac{1}{d_o}$$

$$d_i = \frac{fd_o}{d_o - f}$$

$$= \frac{(8.0\text{ cm})(32.0\text{ cm})}{32.0\text{ cm} - 8.0\text{ cm}} \qquad \textbf{Substitute } f = 8.0 \textbf{ cm, } d_o = 32.0 \textbf{ cm}$$

$= 11$ cm (11 cm away from the lens on the side opposite the object)

b. Use the magnification equation and solve for image height.

$$m \equiv \frac{h_i}{h_o} = \frac{-d_i}{d_o}$$

$$h_i = \frac{-d_i h_o}{d_o}$$

$$= \frac{-(11\text{ cm})(3.0\text{ cm})}{32.0\text{ cm}} \qquad \textbf{Substitute } d_i = 11 \textbf{ cm, } h_o = 3.0 \textbf{ cm, } d_o = 32.0 \textbf{ cm}$$

$= -1.0$ cm (1.0 cm tall)

c. The negative sign in part **b** means that the image is inverted.

Math Handbook
Operations with Significant Digits
pages 835–836

3 Evaluate the Answer

- **Are the units correct?** All are in centimeters.
- **Do the signs make sense?** Image position is positive (real image) and image height is negative (inverted compared to the object), which make sense for a convex lens.

PRACTICE Problems

- Additional Problems, Appendix B
- Solutions to Selected Problems, Appendix C

15. A 2.25-cm-tall object is 8.5 cm to the left of a convex lens of 5.5-cm focal length. Find the image position and height.

16. An object near a convex lens produces a 1.8-cm-tall real image that is 10.4 cm from the lens and inverted. If the focal length of the lens is 6.8 cm, what are the object position and height?

17. An object is placed to the left of a convex lens with a 25-mm focal length so that its image is the same size as the object. What are the image and object positions?

18. Use a scale ray diagram to find the image position of an object that is 30 cm to the left of a convex lens with a 10-cm focal length.

19. Calculate the image position and height of a 2.0-cm-tall object located 25 cm from a convex lens with a focal length of 5.0 cm. What is the orientation of the image?

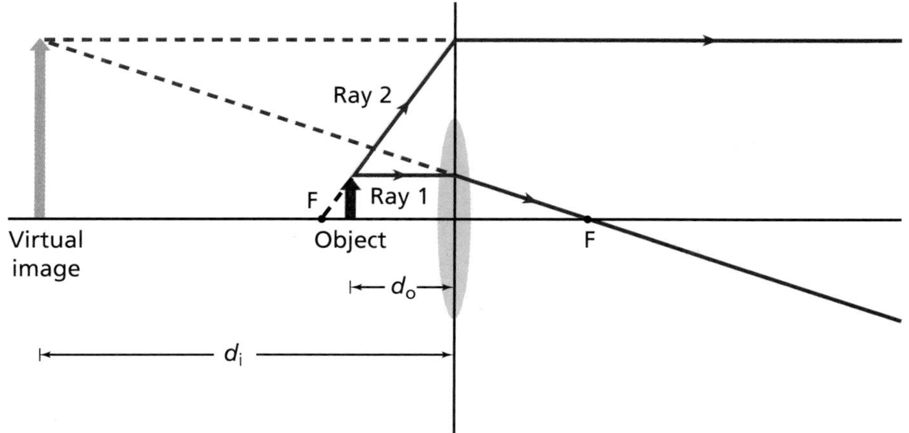

■ **Figure 18-13** The two principal rays show that a convex lens forms a virtual image that is upright and larger compared to the object when the object is located between the lens and the focal point. Because the principal rays are simply part of a model to help locate an image, they do not have to pass through the picture of the lens in a diagram. In reality, the image is formed only by the light that passes through the actual lens.

Convex Lenses and Virtual Images

When an object is placed at the focal point of a convex lens, the refracted rays will emerge in a parallel beam and no image will be seen. When the object is brought closer to the lens, the rays will diverge on the opposite side of the lens, and the rays will appear to an observer to come from a spot on the same side of the lens as the object. This is a virtual image that is upright and larger compared to the object.

Figure 18-13 shows how a convex lens forms a virtual image. The object is located between F and the lens. Ray 1, as usual, approaches the lens parallel to the principal axis and is refracted through the focal point, F. Ray 2 travels from the tip of the object, in the direction it would have if it had started at F on the object side of the lens. The dashed line from F to the object shows you how to draw ray 2. Ray 2 leaves the lens parallel to the principal axis. Rays 1 and 2 diverge as they leave the lens. Thus, no real image is possible. Drawing sight lines for the two rays back to their apparent intersection locates the virtual image. It is on the same side of the lens as the object, and it is upright and larger compared to the object. Note that the actual image is formed by light that passes through the lens, but you can still determine the location of the image by drawing rays that do not have to pass through the lens.

PRACTICE Problems

• Additional Problems, Appendix B
• Solutions to Selected Problems, Appendix C

20. A newspaper is held 6.0 cm from a convex lens of 20.0-cm focal length. Find the image position of the newsprint image.

21. A magnifying glass has a focal length of 12.0 cm. A coin, 2.0 cm in diameter, is placed 3.4 cm from the lens. Locate the image of the coin. What is the diameter of the image?

22. A convex lens with a focal length of 22.0 cm is used to view a 15.0-cm-long pencil located 10.0 cm away. Find the height and orientation of the image.

23. A stamp collector wants to magnify a stamp by 4.0 when the stamp is 3.5 cm from the lens. What focal length is needed for the lens?

24. A magnifier with a focal length of 30 cm is used to view a 1-cm-tall object. Use ray tracing to determine the location and size of the image when the magnifier is positioned 10 cm from the object.

■ **Figure 18-14** Concave lenses produce only virtual images that are upright and smaller compared to their objects.

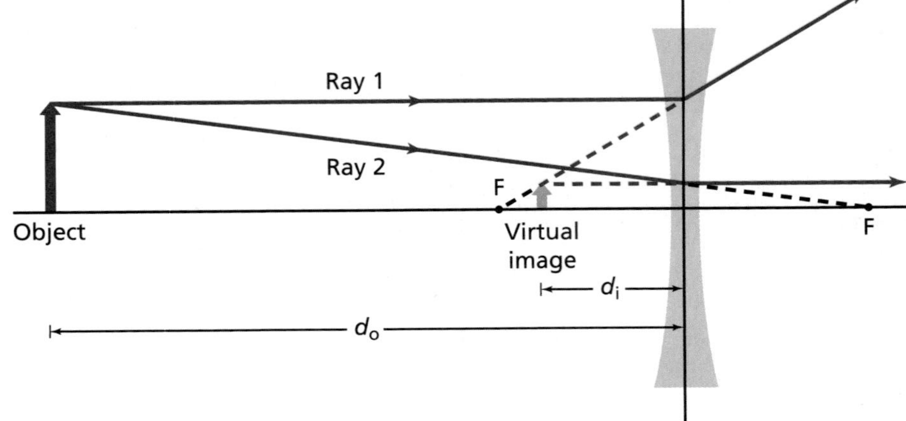

Concave Lenses

A concave lens causes all rays to diverge. **Figure 18-14** shows how such a lens forms a virtual image. Ray 1 approaches the lens parallel to the principal axis. It leaves the lens along a line that extends back through the focal point. Ray 2 approaches the lens as if it is going to pass through the focal point on the opposite side, and leaves the lens parallel to the principal axis. The sight lines of rays 1 and 2 intersect on the same side of the lens as the object. Because the rays diverge, they produce a virtual image. The image is located at the point from where the two rays apparently diverge. The image also is upright and smaller compared to the object. This is true no matter how far from the lens the object is located. The focal length of a concave lens is negative.

When solving problems for concave lenses using the thin lens equation, you should remember that the sign convention for focal length is different from that of convex lenses. If the focal point for a concave lens is 24 cm from the lens, you should use the value $f = -24$ cm in the thin lens equation. All images for a concave lens are virtual. Thus, if an image distance is given as 20 cm from the lens, then you should use $d_i = -20$ cm. The object position always will be positive.

Defects of Spherical Lenses

Throughout this section, you have studied lenses that produce perfect images at specific positions. In reality, spherical lenses, just like spherical mirrors, have intrinsic defects that cause problems with the focus and color of images. Spherical lenses exhibit an aberration associated with their spherical design, just as mirrors do. In addition, the dispersion of light through a spherical lens causes an aberration that mirrors do not exhibit.

Spherical aberration The model that you have used for drawing rays through spherical lenses suggests that all parallel rays focus at the same position. However, this is only an approximation. In reality, parallel rays that pass through the edges of a spherical lens focus at positions different from those of parallel rays that pass through the center. This inability of a spherical lens to focus all parallel rays to a single point is called spherical aberration. Making lens surfaces aspherical, such as in cameras, eliminates spherical aberration. In high-precision instruments, many lenses, often five or more, are used to form sharp, well-defined images.

Chromatic aberration Lenses have a second defect that mirrors do not have. A lens is like a prism, so different wavelengths of light are refracted at slightly different angles, as you can see in **Figure 18-15a.** Thus, the light that passes through a lens, especially near the edges, is slightly dispersed. An object viewed through a lens appears to be ringed with color. This effect is called **chromatic aberration.** The term *chromatic* comes from the Greek word *chromo,* which means "color."

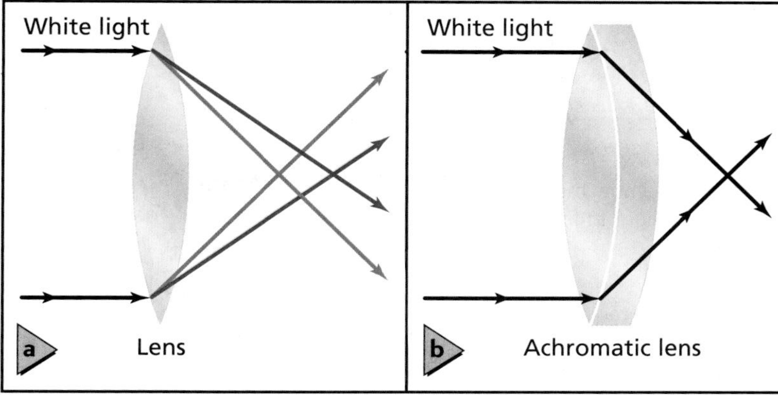

■ **Figure 18-15** All simple lenses have chromatic aberration, in which light of different wavelengths is focused at different points **(a).** An achromatic lens is a combination of lenses, which minimizes the chromatic defect **(b).**

Chromatic aberration is always present when a single lens is used. However, this defect can be greatly reduced by an **achromatic lens,** which is a system of two or more lenses, such as a convex lens with a concave lens, that have different indices of refraction. Such a combination of lenses is shown in **Figure 18-15b.** Both lenses in the figure disperse light, but the dispersion caused by the convex lens is almost canceled by the dispersion caused by the concave lens. The index of refraction of the convex lens is chosen so that the combination of lenses still converges the light.

18.2 Section Review

25. Magnification Magnifying glasses normally are used to produce images that are larger than the related objects, but they also can produce images that are smaller than the related objects. Explain.

26. Image Position and Height A 3.0-cm-tall object is located 2.0 cm from a convex lens having a focal length of 6.0 cm. Draw a ray diagram to determine the location and size of the image. Use the thin lens equation and the magnification equation to verify your answer.

27. Types of Lenses The cross sections of four different thin lenses are shown in **Figure 18-16.**

a. Which of these lenses, if any, are convex, or converging, lenses?

b. Which of these lenses, if any, are concave, or diverging, lenses?

■ **Figure 18-16**

28. Chromatic Aberration All simple lenses have chromatic aberration. Explain, then, why you do not see this effect when you look through a microscope.

29. Chromatic Aberration You shine white light through a convex lens onto a screen and adjust the distance of the screen from the lens to focus the red light. Which direction should you move the screen to focus the blue light?

30. Critical Thinking An air lens constructed of two watch glasses is placed in a tank of water. Copy **Figure 18-17** and draw the effect of this lens on parallel light rays incident on the lens.

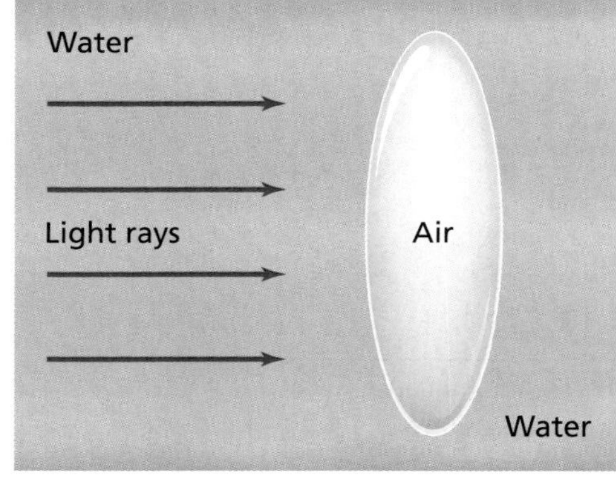

■ **Figure 18-17**

18.3 Applications of Lenses

► **Objectives**

- **Describe** how the eye focuses light to form an image.
- **Explain** nearsightedness and farsightedness and how eyeglass lenses correct these defects.
- **Describe** the optical systems in some common optical instruments.

► **Vocabulary**

nearsightedness
farsightedness

The properties that you have learned for the refraction of light through lenses are used in almost every optical instrument. In many cases, a combination of lenses and mirrors is used to produce clear images of small or faraway objects. Telescopes, binoculars, cameras, microscopes, and even your eyes contain lenses.

Lenses in Eyes

The eye is a remarkable optical device. As shown in **Figure 18-18,** the eye is a fluid-filled, almost spherical vessel. Light that is emitted or reflected off an object travels into the eye through the cornea. The light then passes through the lens and focuses onto the retina that is at the back of the eye. Specialized cells on the retina absorb this light and send information about the image along the optic nerve to the brain.

Focusing images Because of its name, you might assume that the lens of an eye is responsible for focusing light onto the retina. In fact, light entering the eye is primarily focused by the cornea because the air-cornea surface has the greatest difference in indices of refraction. The lens is responsible for the fine focus that allows you to clearly see both distant and nearby objects. Using a process called accommodation, muscles surrounding the lens can contract or relax, thereby changing the shape of the lens. This, in turn, changes the focal length of the eye. When the muscles are relaxed, the image of distant objects is focused on the retina. When the muscles contract, the focal length is shortened, and this allows images of closer objects to be focused on the retina.

Biology Connection

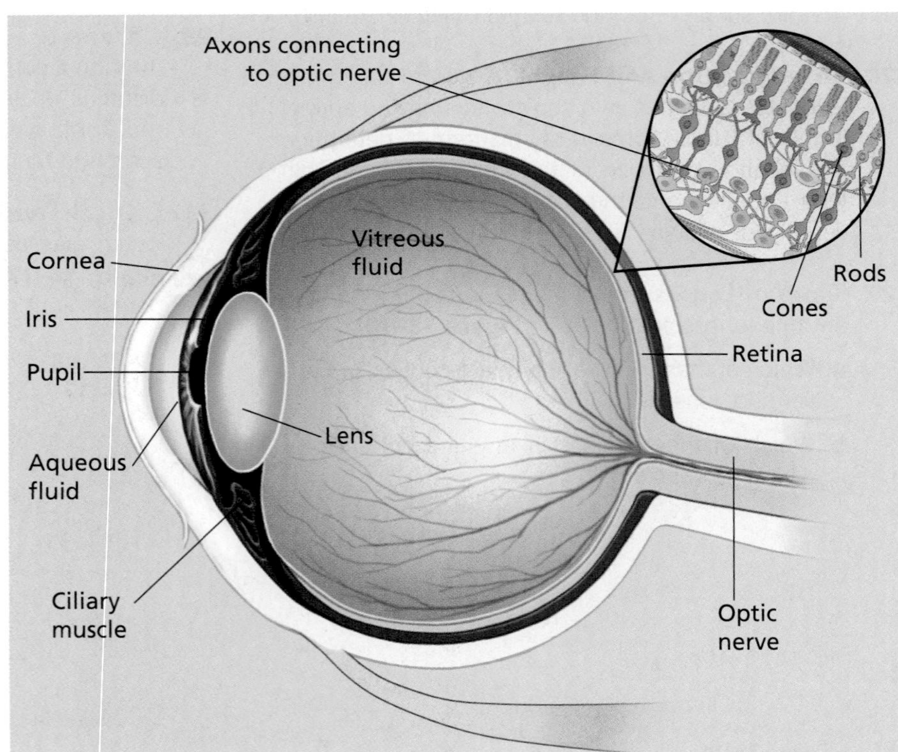

Figure 18-18 The human eye is complex and has many components that must work together.

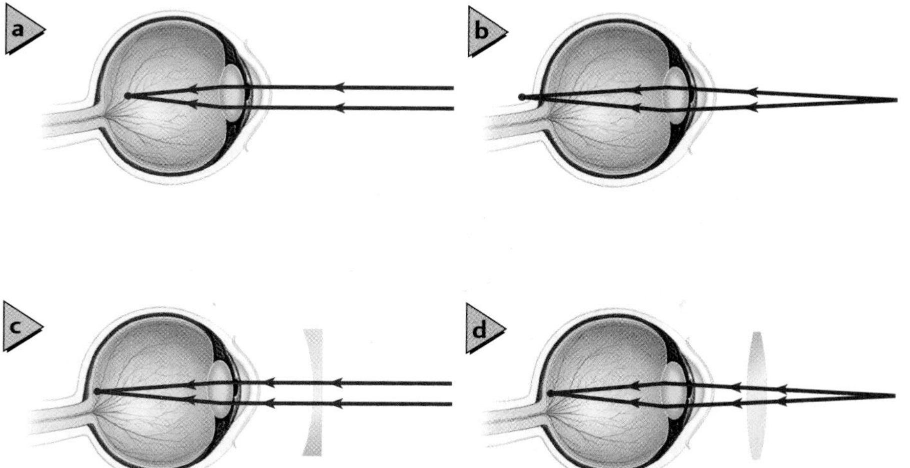

Figure 18-19 A nearsighted person cannot see distant objects clearly because images are focused in front of the retina **(a).** A concave lens corrects this defect **(c).** A farsighted person cannot see close objects clearly because images are focused behind the retina **(b).** A convex lens corrects this defect **(d).**

Nearsightedness and farsightedness The eyes of many people do not focus sharp images on the retina. Instead, images are focused either in front of the retina or behind it. External lenses, in the form of eyeglasses or contact lenses, are needed to adjust the focal length and move images to the retina. **Figure 18-19a** shows the condition of **nearsightedness,** or myopia, whereby the focal length of the eye is too short to focus light on the retina. Images are formed in front of the retina. As shown in **Figure 18-19b,** concave lenses correct this by diverging light, thereby increasing images' distances from the lens, and forming images on the retina.

You also can see in **Figure 18-19c** that **farsightedness,** or hyperopia, is the condition in which the focal length of the eye is too long. Images are therefore formed past the retina. A similar result is caused by the increasing rigidity of the lenses in the eyes of people who are more than about 45 years old. Their muscles cannot shorten the focal length enough to focus images of close objects on the retina. For either defect, convex lenses produce virtual images farther from the eye than the associated objects, as shown in **Figure 18-19d.** The images then become the objects for the eye lens and can be focused on the retina, thereby correcting the defect.

APPLYING PHYSICS

Contacts Contact lenses produce the same results as eyeglasses do. These small, thin lenses are placed directly on the corneas. A thin layer of tears between the cornea and lens keeps the lens in place. Most of the refraction occurs at the air-lens surface, where the difference in indices of refraction is greatest.

CHALLENGE PROBLEM

As light enters the eye, it first encounters the air/cornea interface. Consider a ray of light that strikes the interface between the air and a person's cornea at an angle of 30.0° to the normal. The index of refraction of the cornea is approximately 1.4.

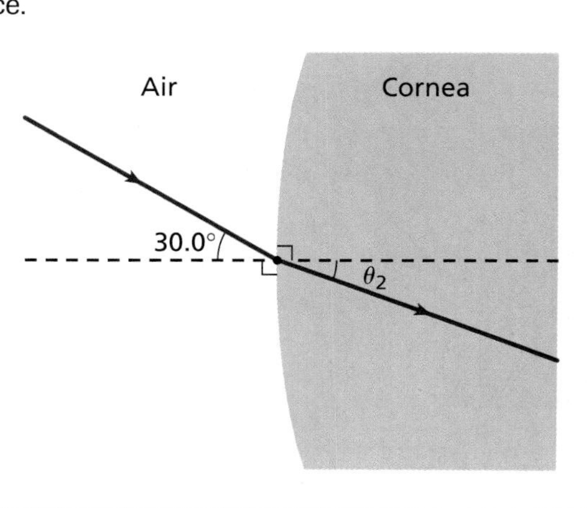

1. Use Snell's law to calculate the angle of refraction.
2. What would the angle of refraction be if the person was swimming underwater?
3. Is the refraction greater in air or in water? Does this mean that objects under water seem closer or more distant than they would in air?
4. If you want the angle of refraction for the light ray in water to be the same as it is for air, what should the new angle of incidence be?

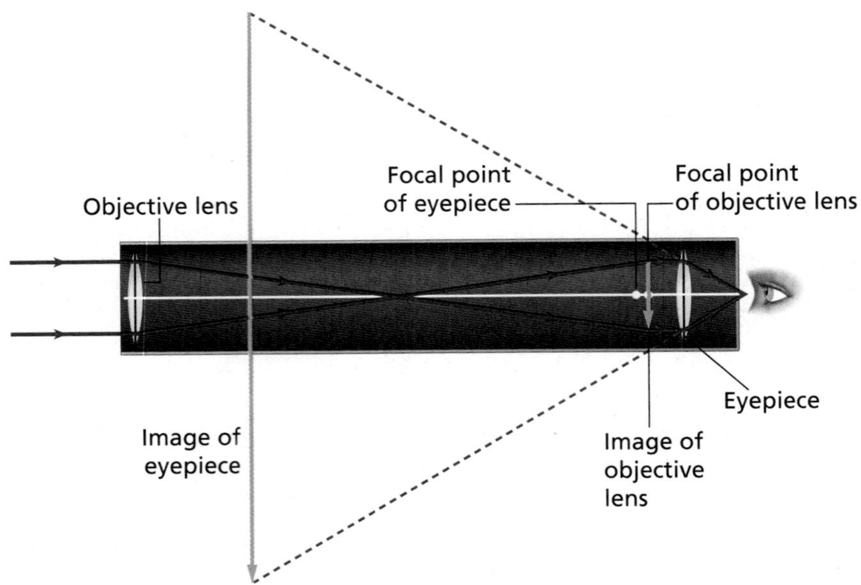

■ **Figure 18-20** An astronomical refracting telescope creates a virtual image that is inverted compared to the object. (Illustration not to scale)

Refracting Telescopes

An astronomical refracting telescope uses lenses to magnify distant objects. **Figure 18-20** shows the optical system for a Keplerian telescope. Light from stars and other astronomical objects is so far away that the rays can be considered parallel. The parallel rays of light enter the objective convex lens and are focused as a real image at the focal point of the objective lens. The image is inverted compared to the object. This image then becomes the object for the convex lens of the eyepiece. Notice that the eyepiece lens is positioned so that the focal point of the objective lens is between the eyepiece lens and its focal point. This means that a virtual image is produced that is upright and larger than the first image. However, because the first image was already inverted, the final image is still inverted. For viewing astronomical objects, an image that is inverted is acceptable.

In a telescope, the convex lens of the eyepiece is almost always an achromatic lens. Recall that an achromatic lens is a combination of lenses that function as one lens. The combination of lenses eliminates the peripheral colors, or chromatic aberration, that can form on images.

■ **Figure 18-21** Binoculars are like two side-by-side refracting telescopes.

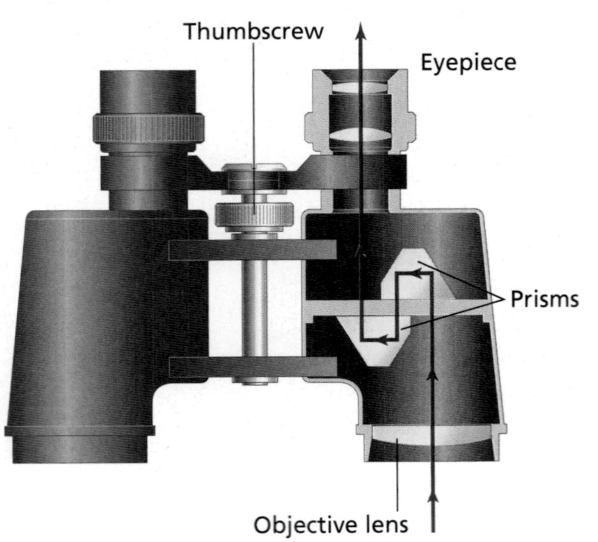

Binoculars

Binoculars, like telescopes, produce magnified images of faraway objects. **Figure 18-21** shows a typical binocular design. Each side of the binoculars is like a small telescope: light enters a convex objective lens, which inverts the image. The light then travels through two prisms that use total internal reflection to invert the image again, so that the viewer sees an image that is upright compared to the object. The prisms also extend the path along which the light travels and direct it toward the eyepiece of the binoculars. Just as the separation of your two eyes gives you a sense of three dimensions and depth, the prisms allow a greater separation of the objective lenses, thereby improving the three-dimensional view of a distant object.

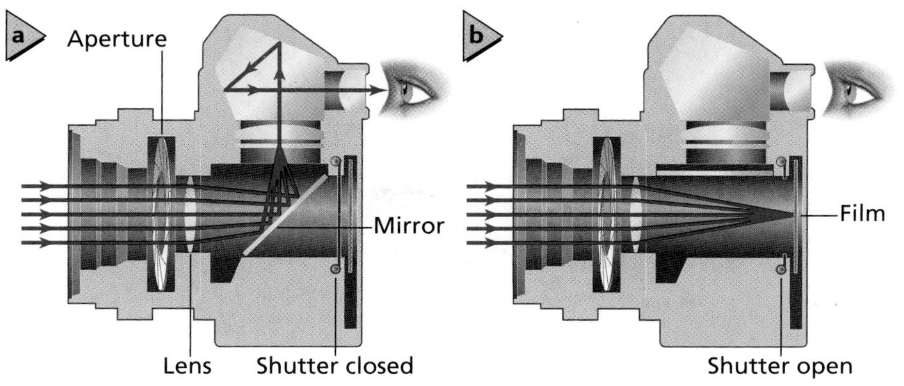

■ **Figure 18-22** The single-lens reflex camera shown here can divert the image formed by the lens through a prism for viewing **(a)** or directly to the film **(b).**

Cameras

Figure 18-22a shows the optical system used in a single-lens reflex camera. As light enters the camera, it passes through an achromatic lens. This lens system refracts the light much like a single convex lens would, forming an image that is inverted on the reflex mirror. The image is reflected upward to a prism that inverts and redirects the light to the viewfinder. When the person holding the camera takes a photograph, he or she presses the shutter-release button, which briefly raises the mirror, as shown in **Figure 18-22b.** The light, instead of being diverted upward to the prism, then travels along a straight path to form an image on the film.

■ **Figure 18-23** The objective lens and the eyepiece in this simple microscope produce an image that is inverted and larger compared to the object.

Concepts In Motion
Interactive Figure To see an animation on how a simple microscope works, visit physicspp.com.

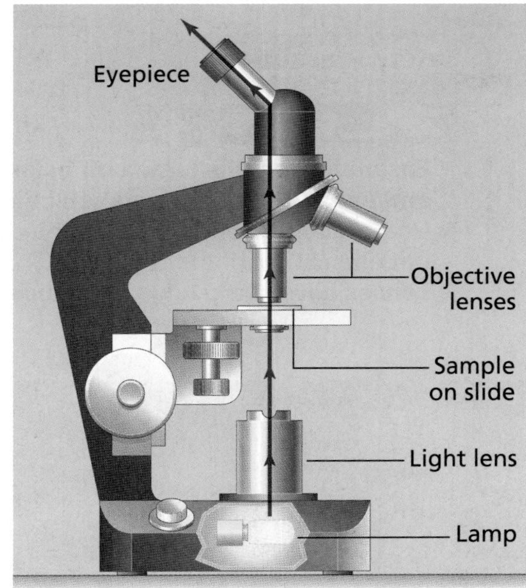

Microscopes

Like a telescope, a microscope has both an objective convex lens and a convex eyepiece. However, microscopes are used to view small objects. **Figure 18-23** shows the optical system used in a simple compound microscope. The object is located between one and two focal lengths from the objective lens. A real image is produced that is inverted and larger than the object. As with a telescope, this image then becomes the object for the eyepiece. This image is between the eyepiece and its focal point. A virtual image is produced that is upright and larger than the image of the objective lens. Thus, the viewer sees an image that is inverted and greatly larger than the original object.

18.3 Section Review

31. Refraction Explain why the cornea is the primary focusing element in the eye.

32. Lens Types Which type of lens, convex or concave, should a nearsighted person use? Which type should a farsighted person use?

33. Focal Length Suppose your camera is focused on a person who is 2 m away. You now want to focus it on a tree that is farther away. Should you move the lens closer to the film or farther away?

34. Image Why is the image that you observe in a refracting telescope inverted?

35. Prisms What are three benefits of having prisms in binoculars?

36. Critical Thinking When you use the highest magnification on a microscope, the image is much darker than it is at lower magnifications. What are some possible reasons for the darker image? What could you do to obtain a brighter image?

Physics Online physicspp.com/self_check_quiz

PHYSICS LAB

Convex Lenses and Focal Length

The thin lens equation states that the inverse of the focal length is equal to the sum of the inverses of the image position from the lens and the object position from the lens.

QUESTION

How is the image position with a thin convex lens related to the object position and the focal length?

Objectives

- **Make and use graphs** to describe the relationship between the image position with a thin convex lens and the object position.
- **Use models** to show that no matter the image position, the focal length is a constant.

Safety Precautions

- **Ensure the lamp is turned off before plugging and unplugging it from the electrical outlet.**
- **Use caution when handling lamps. They get hot and can burn the skin.**
- **Lenses have sharp edges. Handle carefully.**

Materials

25-W straight-line filament bulb
lamp base
thin convex lens
meterstick
lens holder
index card

Procedure

1. Place a meterstick on your lab table so that it is balancing on the thin side and the metric numbers are right side up.
2. Place a convex lens in a lens holder and set it on the meterstick on or between the 10-cm and 40-cm marks on the meterstick. (Distances will vary depending on the focal length of the lens used.)
3. Turn on the lamp and set it next to the meterstick so that the center of the lightbulb is even with the 0-cm end of the meterstick.
4. Hold an index card so that the lens is between the lamp and the index card.
5. Move the index card back and forth until an upside-down image of the lightbulb is as sharp as possible.
6. Record the distance of the lightbulb from the lens (d_o) and the distance of the image from the lens (d_i).
7. Move the lens to another spot between 10 cm and 40 cm and repeat steps 5 and 6. (Distances will vary depending on the focal length of the lens used.)
8. Repeat step 7 three more times.

Data Table		
Trial	d_o (cm)	d_i (cm)
1		
2		
3		
4		
5		

Calculation Table				
Trial	$\frac{1}{d_o}$ (cm^{-1})	$\frac{1}{d_i}$ (cm^{-1})	$\frac{1}{d_o} + \frac{1}{d_i}$ (cm^{-1})	f (cm)
1				
2				
3				
4				
5				

Analyze

1. **Make and Use Graphs** Make a scatter-plot graph of the image position (vertical axis) versus the object position (horizontal axis). Use a computer or calculator to construct the graph if possible.
2. **Use Numbers** Calculate $1/d_o$ and $1/d_i$ and enter the values in the calculation table.
3. **Use Numbers** Calculate the sum of $1/d_o$ and $1/d_i$ and enter the values in the calculation table. Calculate the reciprocal of this number and enter it in the calculation table as f.

Conclude and Apply

1. **Interpret Data** Looking at the graph, describe the relationship between d_o and d_i.
2. **Interpret Data** Find out the actual focal length of the lens from your teacher. How accurate are your calculations of f?
3. **Interpret Data** Compare the results of your focal length calculations of the five trials. Are your results similar?
4. **Lab Techniques** Why do you suppose you were instructed not to hold your lens closer than 10 cm or farther than 40 cm?

Going Further

1. Which measurement is more precise: d_o or d_i? Why do you think so?
2. What can you do to make either (or both) measurement(s) more accurate?

Real-World Physics

1. If you were to take a picture with a camera, first of a distant scene, then of an object less than a meter away, how should the distance between the lens and film be changed?
2. What are two ways in which the image projected onto your retina differs from the object you look at? (Remember the lens in your eye is also convex.)

To find out more about lenses and refraction, visit the Web site: physicspp.com

eXtreme physics

Gravitational Lenses

In 1979, astronomers at the Jodrell Bank Observatory in Great Britain discovered two quasars that were separated by only 7 arc seconds (seven 0.36th's of a degree). Measurements showed they should have been 500,000 light years apart. The two quasars seemed to fluctuate in brightness and in rhythm with each other. The most amazing thing, however, was that the quasars had identical spectra. There appeared to be two different objects, but the two objects were the same.

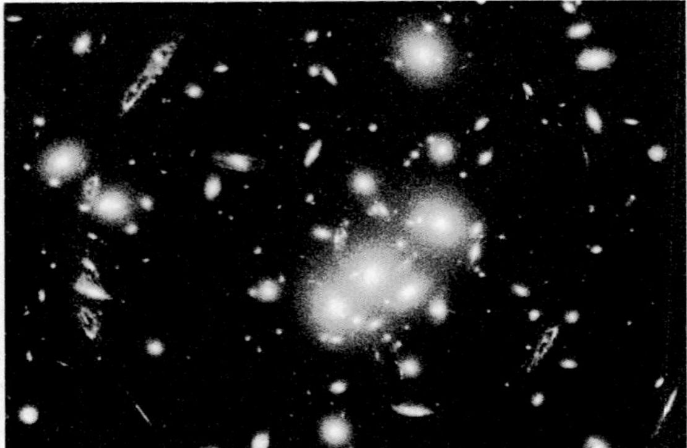

The blue shapes are multiple images of the same galaxy produced by gravitational lensing from galaxy cluster 0024+1654 in the center of the photo.

Further work by astronomers around the world confirmed that there was just a single quasar, and that its light was being distorted by a cluster of galaxies dominated by a massive elliptical galaxy lying in the line of sight between the quasar and Earth. The astronomers realized that they were seeing two images of one quasar. The galaxy acted like an imperfect convex lens, focusing the deflected light in such a way that two images were formed from one object. Why would they think that the light was bent?

Gravity and Light The astronomers remembered the work of Albert Einstein and his theory of relativity. Einstein proposed that light would be bent by the gravitational fields of massive objects. In the classical theory of space, known as Euclidean space, light travels in a straight line. According to Einstein, light bends when it comes near a massive object. In 1919, comparison of starlight before and during a solar eclipse proved Einstein's theory to be true.

In 1936, Einstein proposed the phenomenon of the gravitational lens. Because light can be bent by the gravitational fields of massive objects, virtual images of rings should be seen by observers on Earth when a massive object is between Earth and the object being observed. Einstein never observed such a phenomenon, but his theory of relativity supported the possible existence of gravitational lenses.

The illustration shows how light from a distant galaxy is bent around a galaxy cluster before reaching Earth.

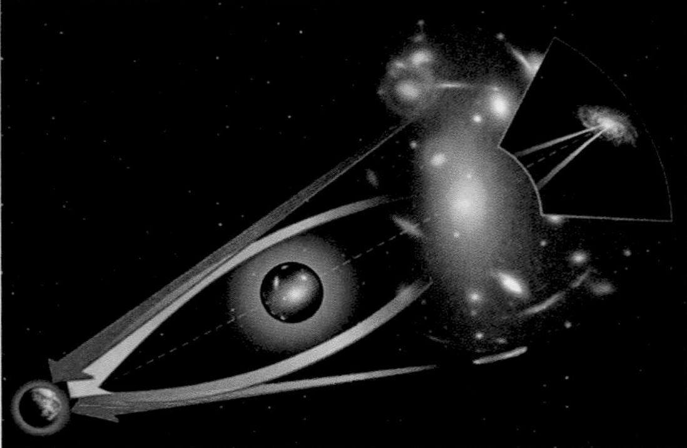

The Evidence As often occurs in science, once someone discovers something for the first time, many more supporting discoveries are made soon after. Since Einstein's proposals, and the discovery in 1979 of the double-image quasar, many more gravitational lenses have been observed. Both Einstein's rings and multiple images have been observed. Einstein's rings result when the gravitational lens and the light from the object are in near-perfect alignment. Multiple images are formed when the gravitational lens and the light from the object are not in perfect alignment. Over 50 gravitational lenses have been discovered.

Going Further

1. **Infer** Why was the discovery of gravitational lenses important?
2. **Compare and Contrast** How are gravitational lenses similar to convex lenses? How are they different?

Chapter 18

Study Guide

18.1 Refraction of Light

Vocabulary

- index of refraction (p. 486)
- Snell's law of refraction (p. 486)
- critical angle (p. 489)
- total internal reflection (p. 489)
- dispersion (p. 491)

Key Concepts

- The path of travel of light bends when it passes from a medium with an index of refraction, n_1, into a medium with a different index of refraction, n_2.

$$n_1 \sin \theta_1 = n_2 \sin \theta_2$$

- The ratio of the speed of light in a vacuum, c, to the speed of light in any medium, v, is the index of refraction, n, of the medium.

$$n = \frac{c}{v}$$

- When light traveling through a medium hits a boundary of a medium with a smaller index of refraction, if the angle of incidence exceeds the critical angle, θ_c, the light will be reflected back into the original medium by total internal reflection.

$$\sin \theta_c = \frac{n_2}{n_1}$$

18.2 Convex and Concave Lenses

Vocabulary

- lens (p. 493)
- convex lens (p. 493)
- concave lens (p. 493)
- thin lens equation (p. 493)
- chromatic aberration (p. 499)
- achromatic lens (p. 499)

Key Concepts

- The focal length, f; the object position, d_o; and the image position, d_i, for a lens are related by the thin lens equation.

$$\frac{1}{f} = \frac{1}{d_i} + \frac{1}{d_o}$$

- The magnification, m, of an image by a lens is defined in the same way as the magnification of an image by a mirror.

$$m \equiv \frac{h_i}{h_o} = \frac{-d_i}{d_o}$$

- A single convex lens produces a real image that is inverted when the object position is greater than the focal length. The image is reduced or enlarged, depending on the object position.
- A single convex lens produces a virtual image that is upright and larger than the object when the object is located between the lens and the focal point.
- A single concave lens always produces a virtual image that is upright and smaller than the object.
- All simple lenses have chromatic aberration. All lenses made with spherical surfaces have spherical aberration.

18.3 Applications of Lenses

Vocabulary

- nearsightedness (p. 501)
- farsightedness (p. 501)

Key Concepts

- Differences in indices of refraction between air and the cornea are primarily responsible for focusing light in the eye.
- Optical instruments use combinations of lenses to obtain clear images of small or distant objects.

Physics Online physicspp.com/vocabulary_puzzlemaker

Assessment

Concept Mapping

37. Complete the following concept map using the following terms: *inverted, larger, smaller, virtual.*

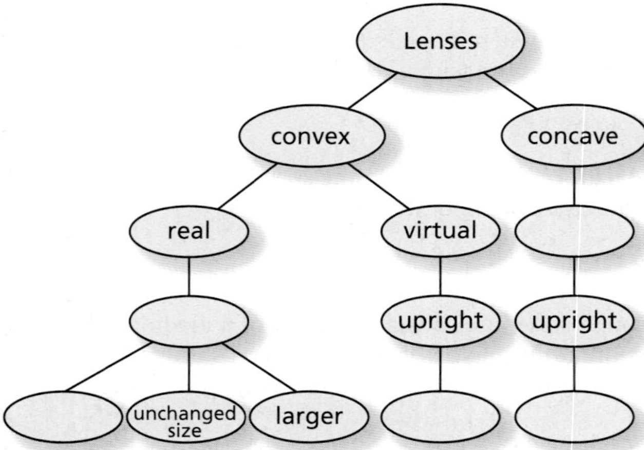

Mastering Concepts

38. How does the angle of incidence compare with the angle of refraction when a light ray passes from air into glass at a nonzero angle? (18.1)

39. How does the angle of incidence compare with the angle of refraction when a light ray leaves glass and enters air at a nonzero angle? (18.1)

40. Regarding refraction, what is the critical angle? (18.1)

41. Although the light coming from the Sun is refracted while passing through Earth's atmosphere, the light is not separated into its spectrum. What does this indicate about the speeds of different colors of light traveling through air? (18.1)

42. Explain why the Moon looks red during a lunar eclipse. (18.1)

43. How do the shapes of convex and concave lenses differ? (18.2)

44. Locate and describe the physical properties of the image produced by a convex lens when an object is placed some distance beyond 2F. (18.2)

45. What factor, other than the curvature of the surfaces of a lens, determines the location of the focal point of the lens? (18.2)

46. To project an image from a movie projector onto a screen, the film is placed between F and 2F of a converging lens. This arrangement produces an image that is inverted. Why does the filmed scene appear to be upright when the film is viewed? (18.2)

47. Describe why precision optical instruments use achromatic lenses. (18.2)

48. Describe how the eye focuses light. (18.3)

49. What is the condition in which the focal length of the eye is too short to focus light on the retina? (18.3)

50. What type of image is produced by the objective lens in a refracting telescope? (18.3)

51. The prisms in binoculars increase the distance between the objective lenses. Why is this useful? (18.3)

52. What is the purpose of a camera's reflex mirror? (18.3)

Applying Concepts

53. Which substance, A or B, in **Figure 18-24** has a larger index of refraction? Explain.

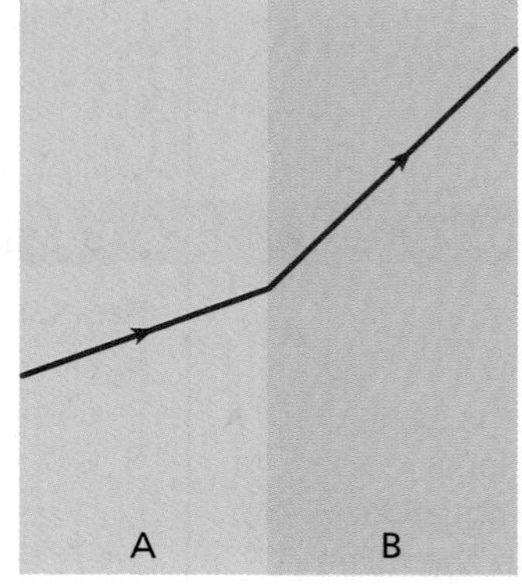

Figure 18-24

54. A light ray strikes the boundary between two transparent media. What is the angle of incidence for which there is no refraction?

55. How does the speed of light change as the index of refraction increases?

56. How does the size of the critical angle change as the index of refraction increases?

57. Which pair of media, air and water or air and glass, has the smaller critical angle?

58. **Cracked Windshield** If you crack the windshield of your car, you will see a silvery line along the crack. The glass has separated at the crack, and there is air in the crack. The silvery line indicates that light is reflecting off the crack. Draw a ray diagram to explain why this occurs. What phenomenon does this illustrate?

59. **Legendary Mirage** According to legend, Eric the Red sailed from Iceland and discovered Greenland after he had seen the island in a mirage. Describe how the mirage might have occurred.

 For more problems, go to Additional Problems, Appendix B.

60. A prism bends violet light more than it bends red light. Explain.

61. **Rainbows** Why would you never see a rainbow in the southern sky if you were in the northern hemisphere? In which direction should you look to see rainbows if you are in the southern hemisphere?

62. Suppose that Figure 18-14 is redrawn with a lens of the same focal length but a larger diameter. Explain why the location of the image does not change. Would the image be affected in any way?

63. A swimmer uses a magnifying glass to observe a small object on the bottom of a swimming pool. She discovers that the magnifying glass does not magnify the object very well. Explain why the magnifying glass is not functioning as it would in air.

64. Why is there chromatic aberration for light that goes through a lens but not for light that reflects from a mirror?

65. When subjected to bright sunlight, the pupils of your eyes are smaller than when they are subjected to dimmer light. Explain why your eyes can focus better in bright light.

66. **Binoculars** The objective lenses in binoculars form real images that are upright compared to their objects. Where are the images located relative to the eyepiece lenses?

Mastering Problems

18.1 Refraction of Light

67. A ray of light travels from air into a liquid, as shown in **Figure 18-25.** The ray is incident upon the liquid at an angle of 30.0°. The angle of refraction is 22.0°.

a. Using Snell's law, calculate the index of refraction of the liquid.

b. Compare the calculated index of refraction to those in Table 18-1. What might the liquid be?

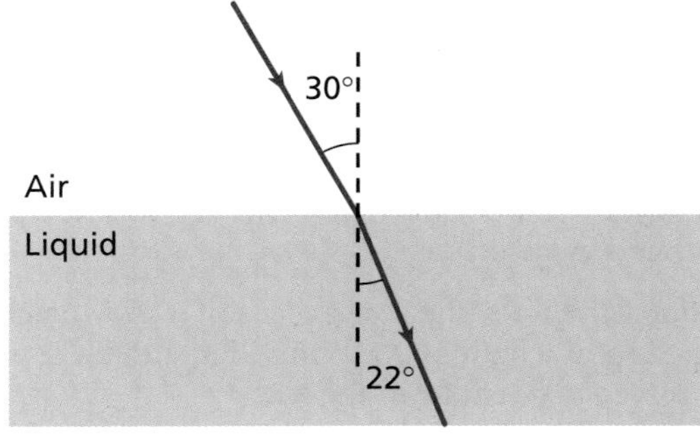

Figure 18-25

68. Light travels from flint glass into ethanol. The angle of refraction in the ethanol is 25.0°. What is the angle of incidence in the glass?

69. A beam of light strikes the flat, glass side of a water-filled aquarium at an angle of 40.0° to the normal. For glass, $n = 1.50$.

a. At what angle does the beam enter the glass?

b. At what angle does the beam enter the water?

70. Refer to Table 18-1. Use the index of refraction of diamond to calculate the speed of light in diamond.

71. Refer to Table 18-1. Find the critical angle for a diamond in air.

72. **Aquarium Tank** A thick sheet of plastic, $n = 1.500$, is used as the side of an aquarium tank. Light reflected from a fish in the water has an angle of incidence of 35.0°. At what angle does the light enter the air?

73. **Swimming-Pool Lights** A light source is located 2.0 m below the surface of a swimming pool and 1.5 m from one edge of the pool, as shown in **Figure 18-26.** The pool is filled to the top with water.

a. At what angle does the light reaching the edge of the pool leave the water?

b. Does this cause the light viewed from this angle to appear deeper or shallower than it actually is?

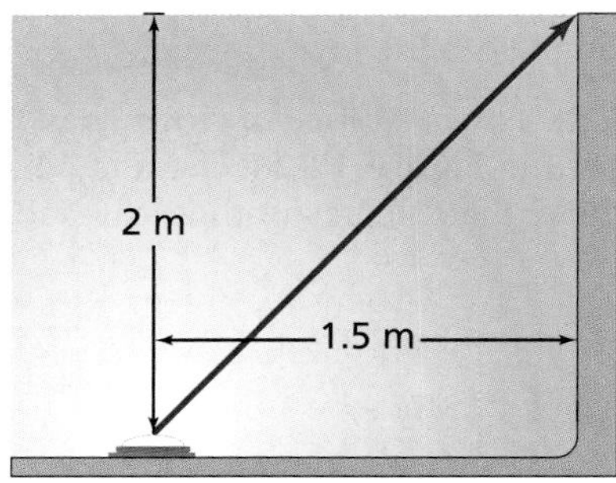

Figure 18-26 (Not to scale)

74. A diamond's index of refraction for red light, 656 nm, is 2.410, while that for blue light, 434 nm, is 2.450. Suppose that white light is incident on the diamond at 30.0°. Find the angles of refraction for red and blue light.

75. The index of refraction of crown glass is 1.53 for violet light, and it is 1.51 for red light.

a. What is the speed of violet light in crown glass?

b. What is the speed of red light in crown glass?

76. The critical angle for a special glass in air is 41.0°. What is the critical angle if the glass is immersed in water?

77. A ray of light in a tank of water has an angle of incidence of 55.0°. What is the angle of refraction in air?

78. The ray of light shown in **Figure 18-27** is incident upon a 60°-60°-60° glass prism, $n = 1.5$.
   a. Using Snell's law of refraction, determine the angle, θ_2, to the nearest degree.
   b. Using elementary geometry, determine the value of θ_1'.
   c. Determine θ_2'.

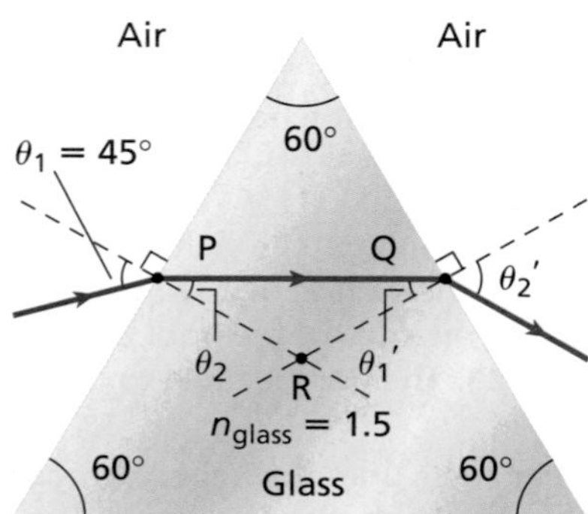

Figure 18-27

79. The speed of light in a clear plastic is 1.90×10^8 m/s. A ray of light strikes the plastic at an angle of 22.0°. At what angle is the ray refracted?

80. A light ray enters a block of crown glass, as illustrated in **Figure 18-28.** Use a ray diagram to trace the path of the ray until it leaves the glass.

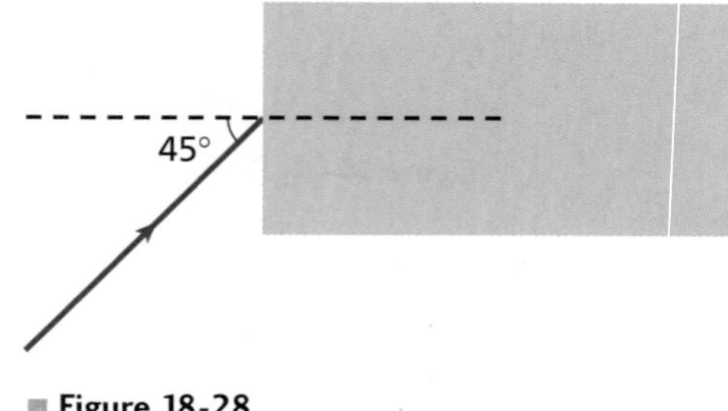

Figure 18-28

18.2 Convex and Concave Lenses

81. The focal length of a convex lens is 17 cm. A candle is placed 34 cm in front of the lens. Make a ray diagram to locate the image.

82. A converging lens has a focal length of 25.5 cm. If it is placed 72.5 cm from an object, at what distance from the lens will the image be?

83. If an object is 10.0 cm from a converging lens that has a focal length of 5.00 cm, how far from the lens will the image be?

84. A convex lens is needed to produce an image that is 0.75 times the size of the object and located 24 cm from the lens on the other side. What focal length should be specified?

85. An object is located 14.0 cm from a convex lens that has a focal length of 6.0 cm. The object is 2.4 cm high.
   a. Draw a ray diagram to determine the location, size, and orientation of the image.
   b. Solve the problem mathematically.

86. A 3.0-cm-tall object is placed 22 cm in front of a converging lens. A real image is formed 11 cm from the lens. What is the size of the image?

87. A 3.0-cm-tall object is placed 15.0 cm in front of a converging lens. A real image is formed 10.0 cm from the lens.
   a. What is the focal length of the lens?
   b. If the original lens is replaced with a lens having twice the focal length, what are the image position, size, and orientation?

88. A diverging lens has a focal length of 15.0 cm. An object placed near it forms a 2.0-cm-high image at a distance of 5.0 cm from the lens.
   a. What are the object position and object height?
   b. The diverging lens is now replaced by a converging lens with the same focal length. What are the image position, height, and orientation? Is it a virtual image or a real image?

18.3 Applications of Lenses

89. **Camera Lenses** Camera lenses are described in terms of their focal length. A 50.0-mm lens has a focal length of 50.0 mm.
   a. A camera with a 50.0-mm lens is focused on an object 3.0 m away. What is the image position?
   b. A 1000.0-mm lens is focused on an object 125 m away. What is the image position?

90. **Eyeglasses** To clearly read a book 25 cm away, a farsighted girl needs the image to be 45 cm from her eyes. What focal length is needed for the lenses in her eyeglasses?

91. **Copy Machine** The convex lens of a copy machine has a focal length of 25.0 cm. A letter to be copied is placed 40.0 cm from the lens.
   a. How far from the lens is the copy paper?
   b. How much larger will the copy be?

 For more problems, go to Additional Problems, Appendix B.

92. Camera A camera lens with a focal length of 35 mm is used to photograph a distant object. How far from the lens is the real image of the object? Explain.

93. Microscope A slide of an onion cell is placed 12 mm from the objective lens of a microscope. The focal length of the objective lens is 10.0 mm.

a. How far from the lens is the image formed?

b. What is the magnification of this image?

c. The real image formed is located 10.0 mm beneath the eyepiece lens. If the focal length of the eyepiece is 20.0 mm, where does the final image appear?

d. What is the final magnification of this compound system?

94. Telescope The optical system of a toy refracting telescope consists of a converging objective lens with a focal length of 20.0 cm, located 25.0 cm from a converging eyepiece lens with a focal length of 4.05 cm. The telescope is used to view a 10.0-cm-high object, located 425 cm from the objective lens.

a. What are the image position, height, and orientation as formed by the objective lens? Is this a real or virtual image?

b. The objective lens image becomes the object for the eyepiece lens. What are the image position, height, and orientation that a person sees when looking into the telescope? Is this a real or virtual image?

c. What is the magnification of the telescope?

Mixed Review

95. A block of glass has a critical angle of 45.0°. What is its index of refraction?

96. Find the speed of light in antimony trioxide if it has an index of refraction of 2.35.

97. A 3.0-cm-tall object is placed 20 cm in front of a converging lens. A real image is formed 10 cm from the lens. What is the focal length of the lens?

98. Derive $n = \sin\theta_1/\sin\theta_2$ from the general form of Snell's law of refraction, $n_1 \sin\theta_1 = n_2 \sin\theta_2$. State any assumptions and restrictions.

99. Astronomy How many more minutes would it take light from the Sun to reach Earth if the space between them were filled with water rather than a vacuum? The Sun is 1.5×10^8 km from Earth.

100. What is the focal length of the lenses in your eyes when you read a book that is 35.0 cm from them? The distance from each lens to the retina is 0.19 mm.

101. Apparent Depth Sunlight reflects diffusively off the bottom of an aquarium. **Figure 18-29** shows two of the many light rays that would reflect diffusively from a point off the bottom of the tank and travel to the surface. The light rays refract into the air as shown. The red dashed line extending back from the refracted light ray is a sight line that intersects with the vertical ray at the location where an observer would see the image of the bottom of the tank.

a. Compute the direction that the refracted ray will travel above the surface of the water.

b. At what depth does the bottom of the tank appear to be if you look into the water? Divide this apparent depth into the true depth and compare it to the index of refraction.

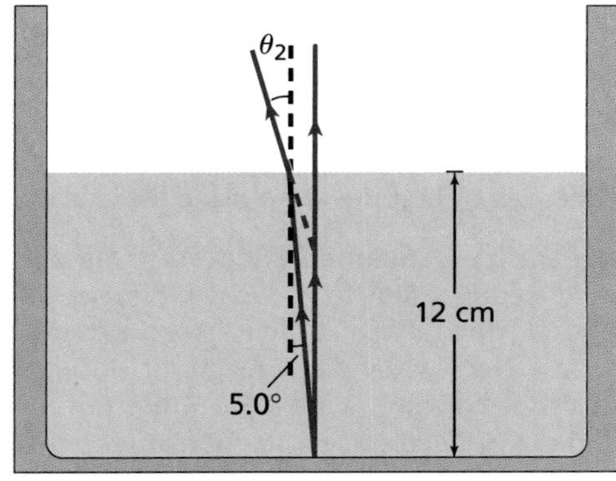

Figure 18-29

102. It is impossible to see through adjacent sides of a square block of glass with an index of refraction of 1.5. The side adjacent to the side that an observer is looking through acts as a mirror. **Figure 18-30** shows the limiting case for the adjacent side to not act like a mirror. Use your knowledge of geometry and critical angles to show that this ray configuration is not achievable when $n_{glass} = 1.5$.

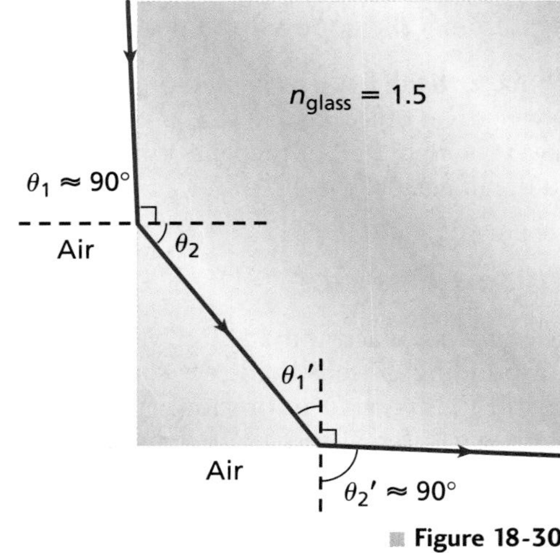

Figure 18-30

103. **Bank Teller Window** A 25-mm-thick sheet of plastic, $n = 1.5$, is used in a bank teller's window. A ray of light strikes the sheet at an angle of 45°. The ray leaves the sheet at 45°, but at a different location. Use a ray diagram to find the distance between the ray that leaves and the one that would have left if the plastic were not there.

Thinking Critically

104. **Recognize Spatial Relationships** White light traveling through air ($n = 1.0003$) enters a slab of glass, incident at exactly 45°. For dense flint glass, $n = 1.7708$ for blue light ($\lambda = 435.8$ nm) and $n = 1.7273$ for red light ($\lambda = 643.8$ nm). What is the angular dispersion of the red and blue light?

105. **Compare and Contrast** Find the critical angle for ice ($n = 1.31$). In a very cold world, would fiber-optic cables made of ice or those made of glass do a better job of keeping light inside the cable? Explain.

106. **Recognize Cause and Effect** Your lab partner used a convex lens to produce an image with $d_i = 25$ cm and $h_i = 4.0$ cm. You are examining a concave lens with a focal length of -15 cm. You place the concave lens between the convex lens and the original image, 10 cm from the image. To your surprise, you see a real image on the wall that is larger than the object. You are told that the image from the convex lens is now the object for the concave lens, and because it is on the opposite side of the concave lens, it is a virtual object. Use these hints to find the new image position and image height and to predict whether the concave lens changed the orientation of the original image.

107. **Define Operationally** Name and describe the effect that causes the rainbow-colored fringe commonly seen at the edges of a spot of white light from a slide or overhead projector.

108. **Think Critically** A lens is used to project the image of an object onto a screen. Suppose that you cover the right half of the lens. What will happen to the image?

Writing in Physics

109. The process of accommodation, whereby muscles surrounding the lens in the eye contract or relax to enable the eye to focus on close or distant objects, varies for different species. Investigate this effect for different animals. Prepare a report for the class showing how this fine focusing is accomplished for different eye mechanisms.

110. Investigate the lens system used in an optical instrument such as an overhead projector or a particular camera or telescope. Prepare a graphics display for the class explaining how the instrument forms images.

Cumulative Review

111. If you drop a 2.0 kg bag of lead shot from a height of 1.5 m, you could assume that half of the potential energy will be converted into thermal energy in the lead. The other half would go to thermal energy in the floor. How many times would you have to drop the bag to heat it by 10°C? (Chapter 12)

112. A blacksmith puts an iron hoop or tire on the outer rim of a wooden carriage wheel by heating the hoop so that it expands to a diameter greater than the wooden wheel. When the hoop cools, it contracts to hold the rim in place. If a blacksmith has a wooden wheel with a 1.0000-m diameter and wants to put a rim with a 0.9950-m diameter on the wheel, what is the minimum temperature change the iron must experience? ($\alpha_{iron} = 12\times10^{-6}$/°C) (Chapter 13)

113. A car sounds its horn as it approaches a pedestrian in a crosswalk. What does the pedestrian hear as the car brakes to allow him to cross the street? (Chapter 15)

114. Suppose that you could stand on the surface of the Sun and weigh yourself. Also suppose that you could measure the illuminance on your hand from the Sun's visible spectrum produced at that position. Next, imagine yourself traveling to a position 1000 times farther away from the center of the Sun as you were when standing on its surface. (Chapter 16)
   a. How would the force of gravity on you from the Sun at the new position compare to what it was at the surface?
   b. How would the illuminance on your hand from the Sun at the new position compare to what it was when you were standing on its surface? (For simplicity, assume that the Sun is a point source at both positions.)
   c. Compare the effect of distance upon the gravitational force and illuminance.

115. **Beautician's Mirror** The nose of a customer who is trying some face powder is 3.00-cm high and is located 6.00 cm in front of a concave mirror having a 14.0-cm focal length. Find the image position and height of the customer's nose by means of the following. (Chapter 17)
   a. a ray diagram drawn to scale
   b. the mirror and magnification equations

 For more problems, go to Additional Problems, Appendix B.

Multiple Choice

1. A flashlight beam is directed at a swimming pool in the dark at an angle of 46° with respect to the normal to the surface of the water. What is the angle of refraction of the beam in the water? (The refractive index for water is 1.33.)

Ⓐ 18° Ⓒ 33°
Ⓑ 30° Ⓓ 44°

2. The speed of light in diamond is 1.24×10^{8} m/s. What is the index of refraction of diamond?

Ⓐ 0.0422 Ⓒ 1.24
Ⓑ 0.413 Ⓓ 2.42

3. Which one of the items below is not involved in the formation of rainbows?

Ⓐ diffraction Ⓒ reflection
Ⓑ dispersion Ⓓ refraction

4. George's picture is being taken by Cami, as shown in the figure, using a camera which has a convex lens with a focal length of 0.0470 m. Determine George's image position.

Ⓐ 1.86 cm Ⓒ 4.82 cm
Ⓑ 4.70 cm Ⓓ 20.7 cm

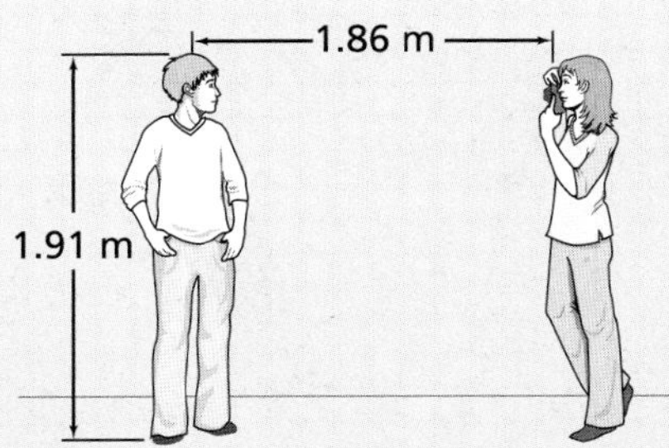

5. What is the magnification of an object that is 4.15 m in front of a camera that has an image position of 5.0 cm?

Ⓐ −0.83 Ⓒ 0.83
Ⓑ −0.012 Ⓓ 1.2

6. Which one of the items below is not involved in the formation of mirages?

Ⓐ heating of air near the ground
Ⓑ Huygens' wavelets
Ⓒ reflection
Ⓓ refraction

7. What is the image position for the situation shown in the figure?

Ⓐ −6.00 m Ⓒ 0.167 m
Ⓑ −1.20 m Ⓓ 0.833 m

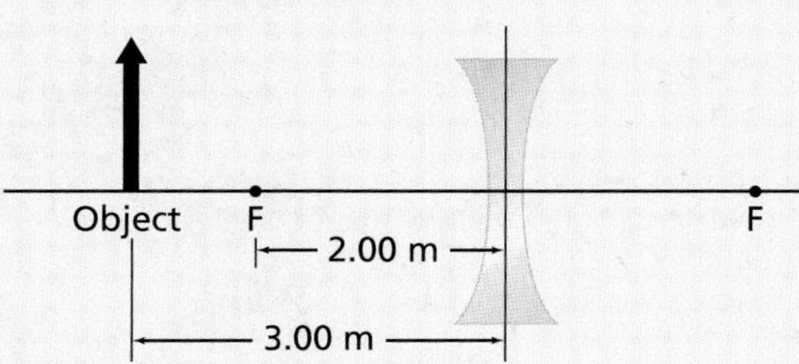

8. What is the critical angle for total internal reflection when light travels from glass (n = 1.52) to water (n = 1.33)?

Ⓐ 29.0° Ⓒ 48.8°
Ⓑ 41.2° Ⓓ 61.0°

9. What happens to the image formed by a convex lens when half of the lens is covered?

Ⓐ half of the image disappears
Ⓑ the image dims
Ⓒ the image gets blurry
Ⓓ the image inverts

Extended Answer

10. The critical angle for total internal reflection at a diamond-air boundary is 24.4°. What is the angle of refraction in the air if light is incident on the boundary at an angle of 20.0°?

11. An object that is 6.98 cm from a lens produces an image that is 2.95 cm from the lens on the same side of the lens. Determine the type of lens that is producing the image and explain how you know.

✓ Test-Taking TIP

Use as Much Time as You Can

You will not get extra points for finishing a test early. Work slowly and carefully to prevent careless errors that can occur when you are hurrying to finish.

Chapter 19 Interference and Diffraction

What You'll Learn

- You will learn how interference and diffraction patterns demonstrate that light behaves like a wave.
- You will learn how interference and diffraction patterns occur in nature and how they are used.

Why It's Important

Interference and diffraction can be seen all around you. Compact discs demonstrate diffraction, bubbles show interference, and the wings of a *Morpho* butterfly show both.

Bubble Solution Bubble solution in a container is transparent. However, if you suspend the solution in a grid of plastic, swirls of color can be seen. These colors are not caused by pigments or dyes in the soap, but by an effect of the wave nature of light.

Think About This ▶
How does bubble solution produce a rainbow of colors?

physicspp.com

LAUNCH Lab — Why does a compact disc reflect a rainbow of light?

Question

How is light affected when it reflects off a compact disc?

Procedure

1. Obtain a compact disc (CD or DVD), a light projector, and color light filters from your teacher.
2. Lay the compact disc on a table with the reflective surface facing up.
3. Place a color filter on the light projector.
4. Turn the light on and shine it on the compact disc, causing the light to reflect off it onto a white surface. ***CAUTION: Do not look directly into the projector light.***
5. Record your observations of the light on the screen.
6. Turn the light off and change to a different color filter.
7. Repeat steps 4–5 with the new color filter.
8. Repeat steps 4–5 with white light.

Analysis

Does the color of light affect the pattern? How is the reflection of white light different from that of single-color light?

Critical Thinking

Look closely at your observations of white light reflecting from the discs. Suggest possible sources for the bands of color.

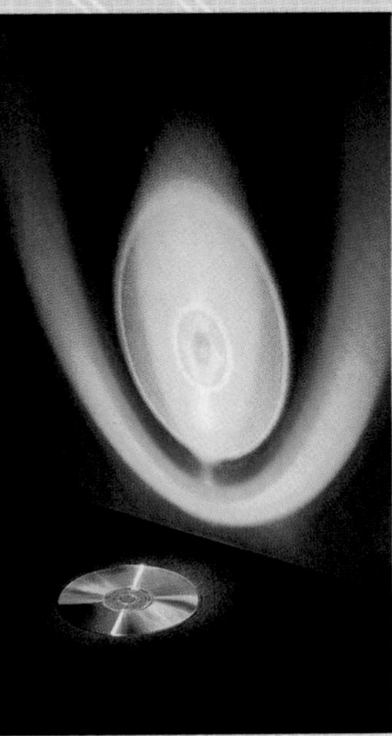

19.1 Interference

In Chapter 16, you learned that light sometimes acts like a wave. Light can be diffracted when it passes by an edge, just like water waves and sound waves can. In Chapters 17 and 18, you learned that reflection and refraction can be explained when light is modeled as a wave. What led scientists to believe that light has wave properties? They discovered that light could be made to interfere, which you will learn about in this section.

When you look at objects that are illuminated by a white light source such as a nearby lightbulb, you are seeing **incoherent light,** which is light with unsynchronized wave fronts. The effect of incoherence in waves can be seen in the example of heavy rain falling on a swimming pool. The surface of the water is choppy and does not have any regular pattern of wave fronts or standing waves. Because light waves have such a high frequency, incoherent light does not appear choppy to you. Instead, as light from an incoherent white light source illuminates an object, you see the superposition of the incoherent light waves as an even, white light.

▶ Objectives

- **Explain** how light falling on two slits produces an interference pattern.
- **Calculate** light wavelengths from interference patterns.
- **Apply** modeling techniques to thin-film interference.

▶ Vocabulary

incoherent light
coherent light
interference fringes
monochromatic light
thin-film interference

Figure 19-1 Smooth wave fronts of light are created by point sources **(a)** and by lasers **(b).**

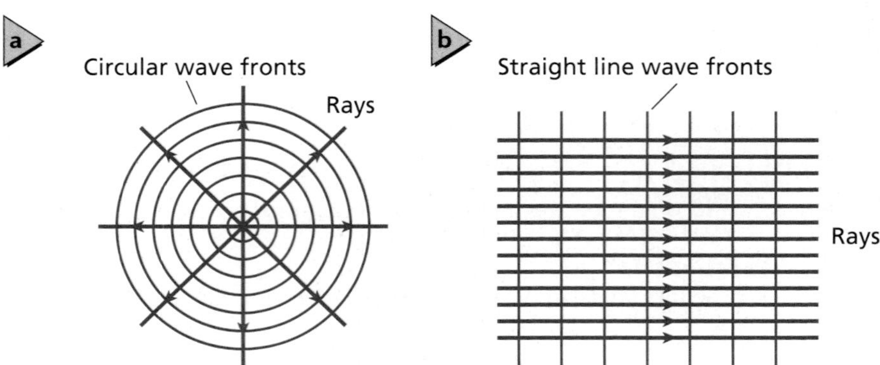

Interference of Coherent Light

The opposite of incoherent light is **coherent light,** which is light from two or more sources that add together in superposition to produce smooth wave fronts. A smooth wave front can be created by a point source, as shown in **Figure 19-1a.** A smooth wave front also can be created by multiple point sources when all point sources are synchronized, such as with a laser, as shown in **Figure 19-1b.** Only the superposition of light waves from coherent light sources can produce the interference phenomena that you will examine in this section.

English physician Thomas Young proved that light has wave properties when he produced an interference pattern by shining light from a single coherent source through two slits. Young directed coherent light at two closely spaced, narrow slits in a barrier. When the overlapping light from the two slits fell on an observing screen, the overlap did not produce even illumination, but instead created a pattern of bright and dark bands that Young called **interference fringes.** He explained that these bands must be the result of constructive and destructive interference of light waves from the two slits in the barrier.

Figure 19-2 These are double-slit interference patterns for blue light **(a),** for red light **(b),** and for white light **(c).**

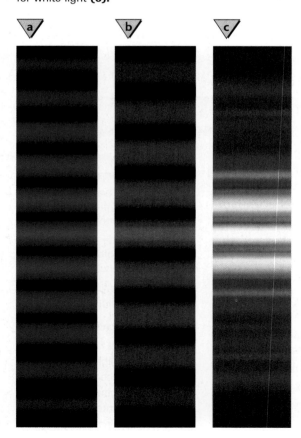

In a double-slit interference experiment that uses **monochromatic light,** which is light of only one wavelength, constructive interference produces a bright central band of the given color on the screen, as well as other bright bands of near-equal spacing and near-equal width on either side, as shown in **Figures 19-2a** and **19-2b.** The intensity of the bright bands decreases the farther the band is from the central band, as you can easily see in Figure 19-2a. Between the bright bands are dark areas where destructive interference occurs. The positions of the constructive and destructive interference bands depend on the wavelength of the light. When white light is used in a double-slit experiment, however, interference causes the appearance of colored spectra instead of bright and dark bands, as shown in **Figure 19-2c.** All wavelengths interfere constructively in the central bright band, and thus that band is white. The positions of the other colored bands result from the overlap of the interference fringes that occur where wavelengths of each separate color interfere constructively.

Figure 19-3 The coherent source that is created by a narrow single slit produces coherent, nearly cylindrical waves that travel to the two slits in the second barrier. Two coherent, nearly cylindrical waves leave the double slit.

Double-slit interference To create coherent light from incoherent light, Young placed a light barrier with a narrow slit in front of a monochromatic light source. Because the width of the slit is very small, only a coherent portion of the light passes through and is diffracted by the slit, producing nearly cylindrical diffracted wave fronts, as shown in **Figure 19-3.** Because a cylinder is symmetrical, the two portions of the wave front arriving at the second barrier with two slits will be in phase. The two slits at the second barrier then produce coherent, nearly cylindrical wave fronts that can then interfere, as shown in Figure 19-3. Depending on their phase relationship, the two waves can undergo constructive or destructive interference, as shown in **Figure 19-4.**

Figure 19-4 A pair of in-phase waves is created at the two slits. At some locations, the waves might undergo constructive interference to create a bright band **(a),** or destructive interference to create a dark band **(b).**

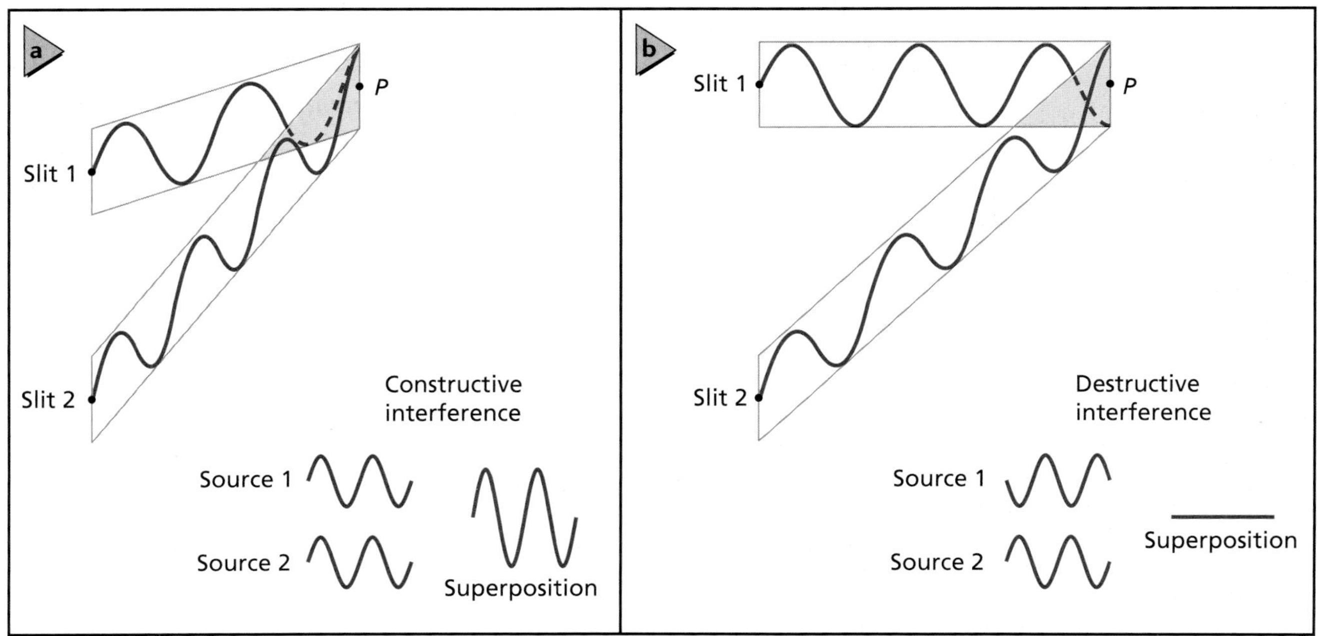

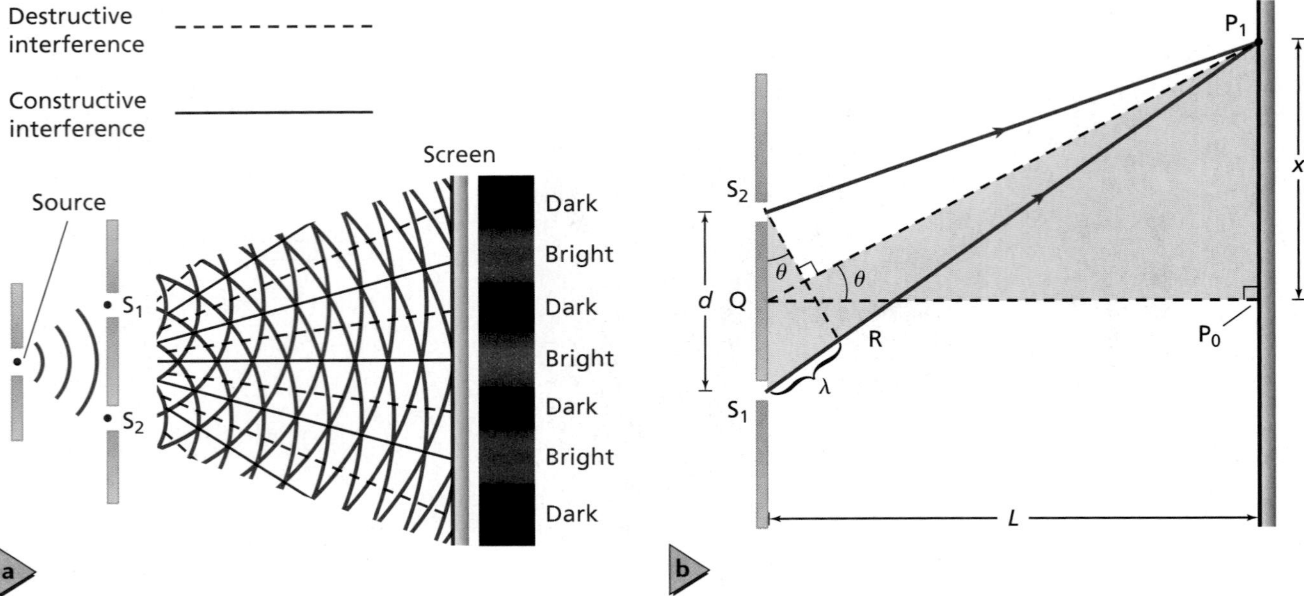

■ **Figure 19-5** The interference of monochromatic light that passes through the double slit produces bright and dark bands on a screen **(a).** This diagram **(b)** represents an analysis of the first bright band. The distance from the slits to the screen, *L*, is about 10^5 times longer than the separation, *d*, between the two slits. (Illustrations not to scale)

Concepts In Motion

Interactive Figure To see an animation on measuring the wavelength of light, visit physicspp.com.

Measuring the wavelength of light A top view of nearly cylindrical wave fronts and Young's double slit experiment are shown in **Figure 19-5a.** The wave fronts interfere constructively and destructively to form a pattern of light and dark bands. A typical diagram that is used to analyze Young's experiment is shown in **Figure 19-5b.** Light that reaches point P_0 travels the same distance from each slit. Because the waves are in phase, they interfere constructively on the screen to create the central bright band at P_0. There is also constructive interference at the first bright band, P_1, on either side of the central band, because line segment P_1S_1 is one wavelength, λ, longer than the line segment P_1S_2. Thus, the waves arrive at P_1 in phase.

There are two triangles shaded in the figure. The larger triangle is a right triangle, so $\tan \theta = x/L$. In the smaller triangle RS_1S_2, the side S_1R is the length difference of the two light paths, which is one wavelength. There are now two simplifications that make the problem easier to solve.

1. If L is much larger than d, then line segments S_1P_1 and S_2P_1 are nearly parallel to each other and to line segment QP_1, and triangle RS_1S_2 is very nearly a right triangle. Thus, $\sin \theta \approx \lambda/d$.
2. If the angle θ is small, then $\sin \theta$ is very nearly equal to $\tan \theta$.

With the above simplifications, the relationships $\tan \theta = x/L$, $\sin \theta \approx \lambda/d$, and $\sin \theta \approx \tan \theta$ combine to form the equation $x/L = \lambda/d$. Solving for λ gives the following.

Wavelength from Double-Slit Experiment $\lambda = \frac{xd}{L}$

The wavelength of light, as measured by a double slit, is equal to the distance on the screen from the central bright band to the first bright band, multiplied by the distance between the slits, divided by the distance to the screen.

Constructive interference from two slits occurs at locations, x_m, on either side of the central bright band, which are determined using the equation $m\lambda = x_m d/L$, where $m = 0, 1, 2$, etc., as limited by the small angle simplification. The central bright band occurs at $m = 0$. Frequently, the band given by $m = 1$ is called the first-order band, and so on.

▶ EXAMPLE Problem 1

Wavelength of Light **A double-slit experiment is performed to measure the wavelength of red light. The slits are 0.0190 mm apart. A screen is placed 0.600 m away, and the first-order bright band is found to be 21.1 mm from the central bright band. What is the wavelength of the red light?**

1 Analyze and Sketch the Problem

- Sketch the experiment, showing the slits and the screen.
- Draw the interference pattern with bands in appropriate locations.

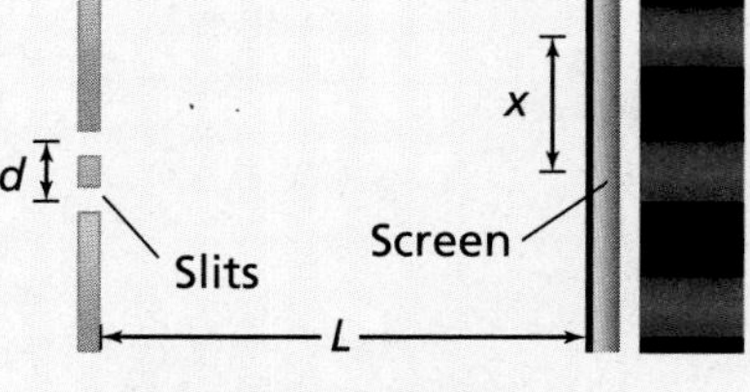

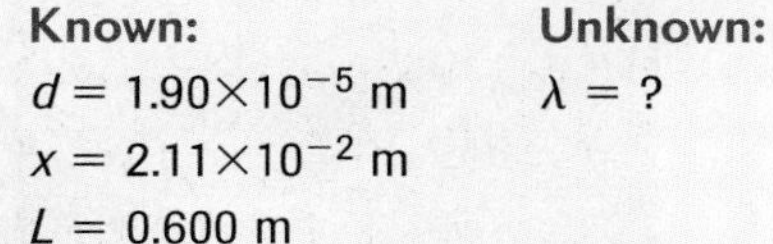

Known:	Unknown:
$d = 1.90\times10^{-5}$ m	$\lambda = ?$
$x = 2.11\times10^{-2}$ m	
$L = 0.600$ m	

Math Handbook
Operations with Scientific Notation
pages 842–843

2 Solve for the Unknown

$$\lambda = \frac{xd}{L}$$

$$= \frac{(2.11\times10^{-2}\text{ m})(1.90\times10^{-5}\text{ m})}{(0.600\text{ m})}$$ **Substitute $x = 2.11\times10^{-2}$ m, $d = 1.90\times10^{-5}$ m, $L = 0.600$ m**

$$= 6.68\times10^{-7}\text{ nm} = 668\text{ nm}$$

3 Evaluate the Answer

- **Are the units correct?** The answer is in units of length, which is correct for wavelength.
- **Is the magnitude realistic?** The wavelength of red light is near 700 nm, and that of blue is near 400 nm. Thus, the answer is reasonable for red light.

▶ PRACTICE Problems

• Additional Problems, Appendix B
• Solutions to Selected Problems, Appendix C

1. Violet light falls on two slits separated by 1.90×10^{-5} m. A first-order bright band appears 13.2 mm from the central bright band on a screen 0.600 m from the slits. What is λ?
2. Yellow-orange light from a sodium lamp of wavelength 596 nm is aimed at two slits that are separated by 1.90×10^{-5} m. What is the distance from the central band to the first-order yellow band if the screen is 0.600 m from the slits?
3. In a double-slit experiment, physics students use a laser with $\lambda = 632.8$ nm. A student places the screen 1.000 m from the slits and finds the first-order bright band 65.5 mm from the central line. What is the slit separation?
4. Yellow-orange light with a wavelength of 596 nm passes through two slits that are separated by 2.25×10^{-5} m and makes an interference pattern on a screen. If the distance from the central line to the first-order yellow band is 2.00×10^{-2} m, how far is the screen from the slits?

Young presented his findings in 1803, but was ridiculed by the scientific community. His conclusions did not begin to gain acceptance until 1820 after Jean Fresnel proposed a mathematical solution for the wave nature of light in a competition. One of the judges, Siméon Denis Poisson, showed that, if Fresnel was correct, a shadow of a circular object illuminated with coherent light would have a bright spot at the center of the shadow. Another judge, Jean Arago, proved this experimentally. Before this, both Poisson and Arago were skeptics of the wave nature of light.

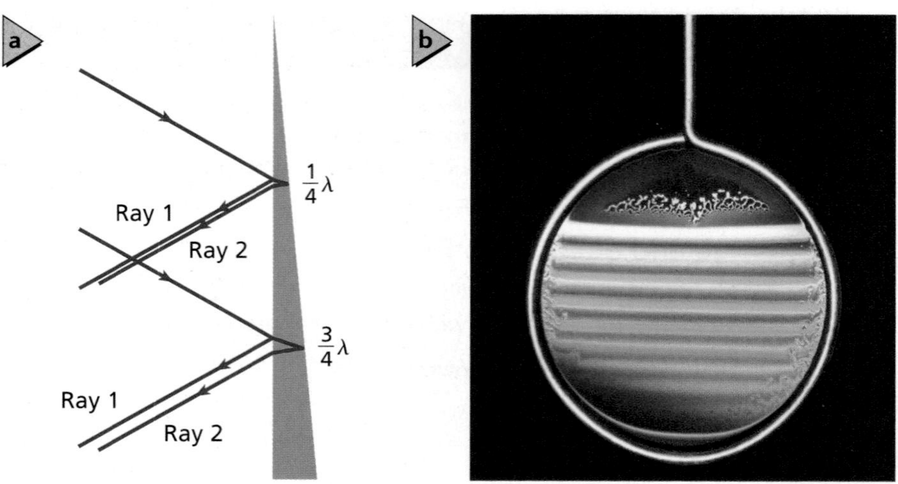

Figure 19-6 Each wavelength is reinforced where the soap film thickness is $\lambda/4$, $3\lambda/4$, $5\lambda/4$ **(a).** Because each color has a different wavelength, a series of color bands is reflected from the soap film **(b).**

Thin-Film Interference

Have you ever seen a spectrum of colors produced by a soap bubble or by the oily film on a water puddle in a parking lot? These colors were not the result of separation of white light by a prism or of absorption of colors in a pigment. The spectrum of colors was a result of the constructive and destructive interference of light waves due to reflection in a thin film, a phenomenon called **thin-film interference.**

If a soap film is held vertically, as in **Figure 19-6,** its weight makes it thicker at the bottom than at the top. The thickness varies gradually from top to bottom. When a light wave strikes the film, it is partially reflected, as shown by ray 1, and partially transmitted. The reflected and transmitted waves have the same frequency as the original. The transmitted wave travels through the film to the back surface, where, again, part is reflected, as shown by ray 2. This act of splitting each light wave from an incoherent source into a matched pair of waves means that the reflected light from a thin film is coherent.

APPLYING PHYSICS

Nonreflective Eyeglasses A thin film can be placed on the lenses of eyeglasses to keep them from reflecting wavelengths of light that are highly visible to the human eye. This prevents the glare of reflected light.

Color reinforcement How is the reflection of one color enhanced? This happens when the two reflected waves are in phase for a given wavelength. If the thickness of the soap film in Figure 19-6 is one-fourth of the wavelength of the wave in the film, $\lambda/4$, then the round-trip path length in the film is $\lambda/2$. In this case, it would appear that ray 2 would return to the front surface one-half wavelength out of phase with ray 1, and that the two waves would cancel each other based on the superposition principle. However, when a transverse wave is reflected from a medium with a slower wave speed, the wave is inverted. With light, this happens at a medium with a larger index of refraction. As a result, ray 1 is inverted on reflection; whereas ray 2 is reflected from a medium with a smaller index of refraction (air) and is not inverted. Thus, ray 1 and ray 2 are in phase.

If the film thickness, d, satisfies the requirement, $d = \lambda/4$, then the color of light with that wavelength will be most strongly reflected. Note that because the wavelength of light in the film is shorter than the wavelength in air, $d = \lambda_{film}/4$, or, in terms of the wavelength in air, $d = \lambda_{vacuum}/4n_{film}$. The two waves reinforce each other as they leave the film. Light with other wavelengths undergoes destructive interference.

As you know, different colors of light have different wavelengths. For a film of varying thickness, such as the one shown in Figure 19-6, the wavelength requirement will be met at different thicknesses for different colors. The result is a rainbow of color. Where the film is too thin to produce constructive interference for any wavelength of color, the film appears to be black. Notice in Figure 19-6b that the spectrum repeats. When the thickness of the film is $3\lambda/4$, the round-trip distance is $3\lambda/2$, and constructive interference occurs again. Any thickness equal to $1\lambda/4$, $3\lambda/4$, $5\lambda/4$, and so on, satisfies the conditions for constructive interference for a given wavelength.

Applications of thin-film interference The example of a film of soapy water in air involves constructive interference with one of two waves inverted upon reflection. In the chapter opener example of bubble solution, as the thickness of the film changes, the wavelength undergoing constructive interference changes. This creates a shifting spectrum of color on the surface of the film soap when it is under white light. In other examples of thin-film interference, neither wave or both waves might be inverted. You can develop a solution for any problem involving thin-film interference using the following strategies.

PROBLEM-SOLVING Strategies

Connecting **Math to Physics**

Thin-Film Interference

When solving thin-film interference problems, construct an equation that is specific to the problem by using the following strategies.

1. Make a sketch of the thin film and the two coherent waves. For simplicity, draw the waves as rays.
2. Read the problem. Is the reflected light enhanced or reduced? When it is enhanced, the two reflected waves undergo constructive interference. When the reflected light is reduced, the waves undergo destructive interference.
3. Are either or both waves inverted on reflection? If the index of refraction changes from a lower to a higher value, then the wave is inverted. If it changes from a higher to a lower value, there is no inversion.
4. Find the extra distance that the second wave must travel through the thin film to create the needed interference.
   a. If you need constructive interference and one wave is inverted OR you need destructive interference and either both waves or none are inverted, then the difference in distance is an odd number of half wavelengths: $(m + 1/2)\lambda_{film}$, where $m = 0, 1, 2$, etc.
   b. If you need constructive interference and either both waves or none are inverted OR you need destructive interference and one wave is inverted, then the difference is an integer number of wavelengths: $m\lambda_{film}$, where $m = 1, 2, 3$, etc.
5. Set the extra distance traveled by the second ray to twice the film thickness, $2d$.
6. Recall from Chapter 18 that $\lambda_{film} = \lambda_{vacuum}/n_{film}$.

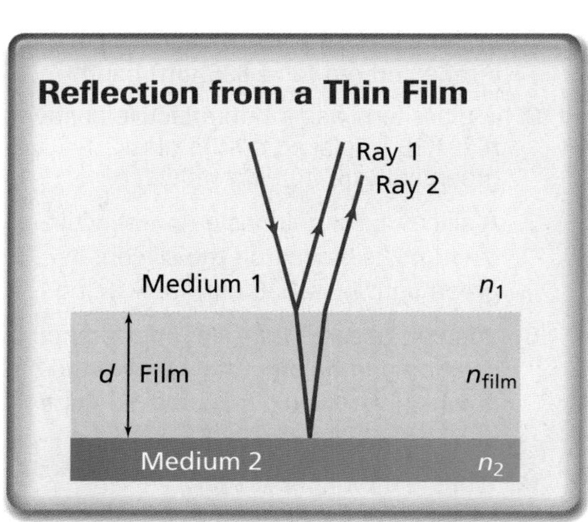

▶ EXAMPLE Problem 2

Oil and Water **You observe colored rings on a puddle and conclude that there must be an oil slick on the water. You look directly down on the puddle and see a yellow–green (λ = 555 nm) region. If the refractive index of oil is 1.45 and that of water is 1.33, what is the minimum thickness of oil that could cause this color?**

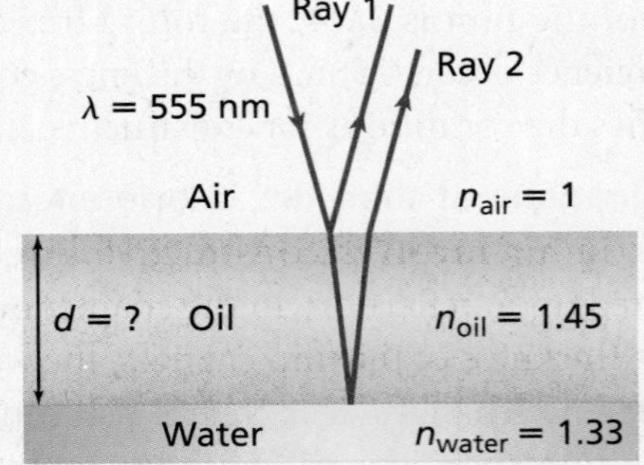

1 Analyze and Sketch the Problem

- Sketch the thin film and layers above and below it.
- Draw rays showing reflection off the top of the film as well as the bottom.

Known:	Unknown:
$n_{water} = 1.33$	$d = ?$
$n_{oil} = 1.45$	
$\lambda = 555$ nm	

2 Solve for the Unknown

Because $n_{oil} > n_{air}$, there is a phase inversion on the first reflection. Because $n_{water} < n_{oil}$, there is no phase inversion on the second reflection. Thus, there is one wave inversion. The wavelength in oil is less than it is in air.

Follow the problem-solving strategy to construct the equation.

$$2d = \left(m + \frac{1}{2}\right)\frac{\lambda}{n_{oil}}$$

Because you want the minimum thickness, $m = 0$.

$$d = \frac{\lambda}{4n_{oil}} \quad \textbf{Substitute } m = 0$$

$$= \frac{555\ \text{nm}}{(4)(1.45)} \quad \textbf{Substitute } \lambda = 555\ \textbf{nm}, n_{oil} = 1.45$$

$$= 95.7\ \text{nm}$$

Math Handbook
Operations with Significant Digits
pages 835–836

3 Evaluate the Answer

- **Are the units correct?** The answer is in nm, which is correct for thickness.
- **Is the magnitude realistic?** The minimum thickness is smaller than one wavelength, which is what it should be.

▶ PRACTICE Problems

• Additional Problems, Appendix B
• Solutions to Selected Problems, Appendix C

5. In the situation in Example Problem 2, what would be the thinnest film that would create a reflected red (λ = 635 nm) band?

6. A glass lens has a nonreflective coating placed on it. If a film of magnesium fluoride, $n = 1.38$, is placed on the glass, $n = 1.52$, how thick should the layer be to keep yellow-green light from being reflected?

7. A silicon solar cell has a nonreflective coating placed on it. If a film of sodium monoxide, $n = 1.45$, is placed on the silicon, $n = 3.5$, how thick should the layer be to keep yellow-green light (λ = 555 nm) from being reflected?

8. You can observe thin-film interference by dipping a bubble wand into some bubble solution and holding the wand in the air. What is the thickness of the thinnest soap film at which you would see a black stripe if the light illuminating the film has a wavelength of 521 nm? Use $n = 1.33$.

9. What is the thinnest soap film ($n = 1.33$) for which light of wavelength 521 nm will constructively interfere with itself?

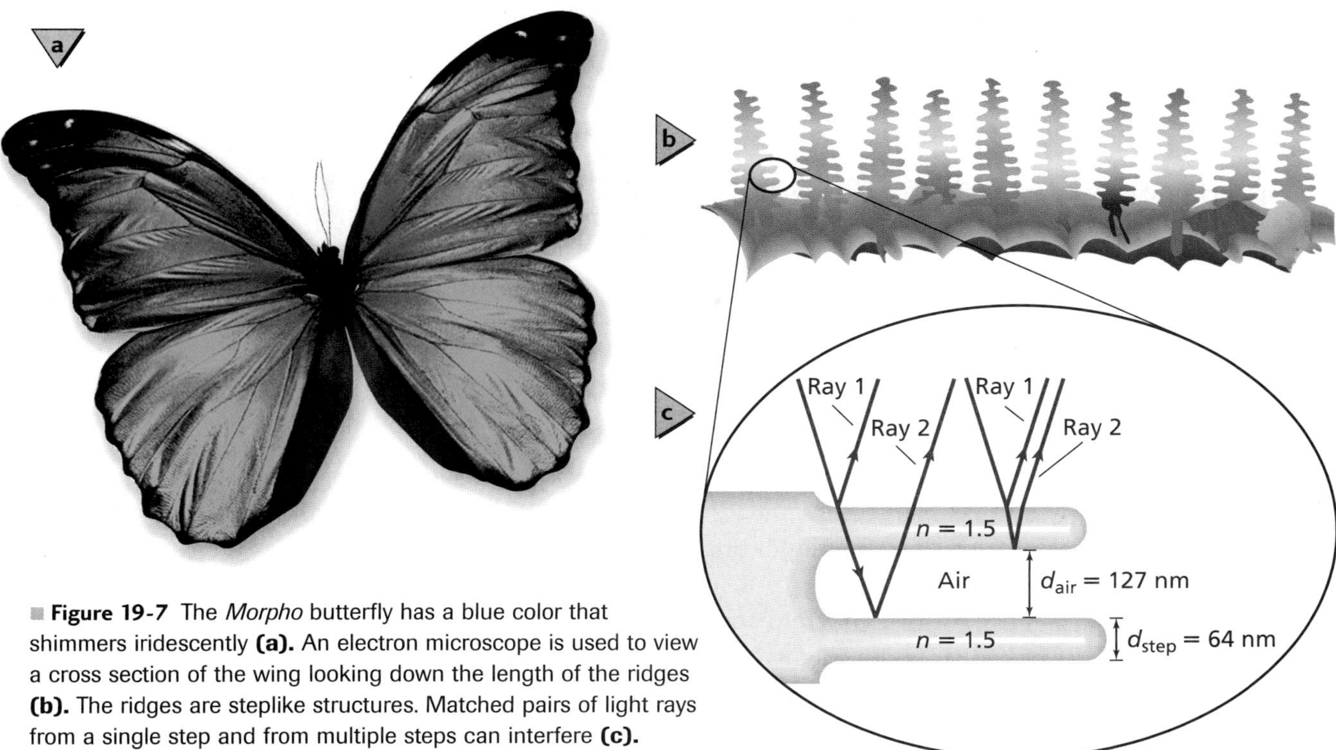

■ **Figure 19-7** The *Morpho* butterfly has a blue color that shimmers iridescently **(a).** An electron microscope is used to view a cross section of the wing looking down the length of the ridges **(b).** The ridges are steplike structures. Matched pairs of light rays from a single step and from multiple steps can interfere **(c).**

Thin-film interference also occurs naturally on the wings of the *Morpho* butterfly, shown in **Figure 19-7a.** The shimmering blue of the butterfly is caused by ridges that project up from the ground scales of the wings, as shown in **Figure 19-7b.** Light is reflected from and refracted through a series of steplike structures, as diagrammed in **Figure 19-7c,** forming a blue interference pattern that appears to shimmer to those who see the butterfly.

19.1 Section Review

10. Film Thickness Lucien is blowing bubbles and holds the bubble wand up so that a soap film is suspended vertically in the air. What is the second thinnest width of the soap film at which he could expect to see a bright stripe if the light illuminating the film has a wavelength of 575 nm? Assume the soap solution has an index of refraction of 1.33.

11. Bright and Dark Patterns Two very narrow slits are cut close to each other in a large piece of cardboard. They are illuminated by monochromatic red light. A sheet of white paper is placed far from the slits, and a pattern of bright and dark bands is seen on the paper. Describe how a wave behaves when it encounters a slit, and explain why some regions are bright while others are dark.

12. Interference Patterns Sketch the pattern described in problem 11.

13. Interference Patterns Sketch what happens to the pattern in problem 11 when the red light is replaced by blue light.

14. Film Thickness A plastic reflecting film ($n = 1.83$) is placed on an auto glass window ($n = 1.52$).

a. What is the thinnest film that will reflect yellow-green light?

b. Unfortunately, a film this thin cannot be manufactured. What is the next-thinnest film that will produce the same effect?

15. Critical Thinking The equation for wavelength from a double-slit experiment uses the simplification that θ is small so that $\sin\theta \approx \tan\theta$. Up to what angle is this a good approximation when your data has two significant figures? Would the maximum angle for a valid approximation increase or decrease as you increase the precision of your angle measurement?

19.2 Diffraction

Objectives

- **Explain** how diffraction gratings form diffraction patterns.
- **Describe** how diffraction gratings are used in grating spectrometers.
- **Discuss** how diffraction limits the ability to distinguish two closely spaced objects with a lens.

Vocabulary

diffraction pattern
diffraction grating
Rayleigh criterion

In Chapter 16, you learned that smooth wave fronts of light spread when they are diffracted around an edge. Diffraction was explained using Huygens' principle that a smooth wave front is made up of many small point-source wavelets. The cutting of coherent light on two edges spaced closely together produces a **diffraction pattern,** which is a pattern on a screen of constructive and destructive interference of Huygens' wavelets.

Single-Slit Diffraction

When coherent, blue light passes through a single, small opening that is larger than the wavelength of the light, the light is diffracted by both edges, and a series of bright and dark bands appears on a distant screen, as shown in **Figure 19-8.** Instead of the nearly equally spaced bands produced by two coherent sources in Young's double-slit experiment, this pattern has a wide, bright central band with dimmer, narrower bands on either side. When using red light instead of blue, the width of the bright central band increases. With white light, the pattern is a mixture of patterns of all the colors of the spectrum.

Figure 19-8 Notice the wide central band and the narrower bands on either side. A single-slit diffraction pattern for red light would have a wider central band than blue light as long as the same size slit is used for both colors.

To see how Huygens' wavelets produce the diffraction pattern, imagine a slit of width w as being divided into an even number of Huygens' points, as shown in **Figure 19-9.** Each Huygens' point acts as a point source of Huygens' wavelets. Divide the slit into two equal parts and choose one source from each part so that the pair is separated by a distance $w/2$. This pair of sources produces coherent, cylindrical waves that will interfere.

For any Huygens' wavelet produced in the top half, there will be another Huygens' wavelet in the bottom half, a distance $w/2$ away, that it will interfere with destructively to create a dark band on the screen. All similar pairings of Huygens' wavelets interfere destructively at dark bands. Conversely, a bright band on the screen is where pairings of Huygens' wavelets interfere constructively. In the dim regions between bright and dark bands, partial destructive interference occurs.

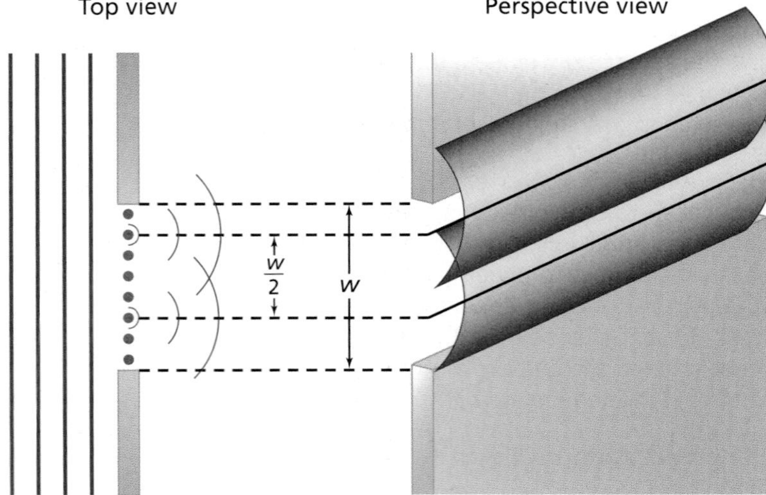

Figure 19-9 A slit of width w is divided into pairs of lines that form Huygens' wavelets, each pair separated by $w/2$.

Concepts In Motion
Interactive Figure To see an animation on single-slit diffraction, visit physicspp.com.

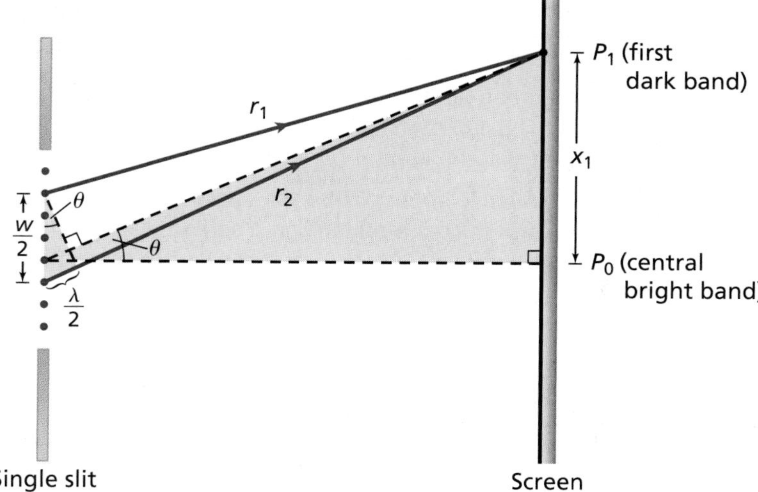

Figure 19-10 This diagram represents an analysis of the first dark band. The distance to the screen, *L*, is much larger than the slit width, *w*. (Illustration not to scale)

Diffraction pattern When the single slit is illuminated, a central bright band appears at location P_0 on the screen, as shown in **Figure 19-10.** The first dark band is at position P_1. At this location, the path lengths r_1 and r_2 of the two Huygens' wavelets differ by one-half wavelength, thereby producing a dark band by destructive interference. This model is mathematically similar to that of double-slit interference. A comparison of a single-slit diffraction pattern with a double-slit interference pattern using slits of the same width reveals that all of the bright interference fringes of the double-slit interference pattern fit within the width of the central bright band of the single-slit diffraction pattern. Double-slit interference results from the interference of the single-slit diffraction patterns of the two slits.

An equation now can be developed for the diffraction pattern produced by a single slit using the same simplifications that were used for double-slit interference, assuming that the distance to the screen is much larger than w. The separation distance between the sources of the two interfering waves is now $w/2$. To find the distance measured on the screen to the first dark band, x_1, note that the path length difference is now $\lambda/2$ because at the dark band there is destructive interference. As a result $x_1/L = \lambda/w$.

You can see from Figure 19-10 that it might be difficult to measure the distance to the first dark band from the center of the central bright band. A better method of determining x_1 is to measure the width of the central bright band, $2x_1$. The following equation gives the width of the central bright band from single-slit diffraction.

Width of Bright Band in Single-Slit Diffraction $2x_1 = \frac{2\lambda L}{w}$

The width of the central bright band is equal to the product of twice the wavelength times the distance to the screen, divided by the width of the slit.

Canceling the 2's out of the above equation gives you the distance from the center of the central bright band to where the first dark band occurs. The location of additional dark bands can be found where the path lengths differ by $3\lambda/2$, $5\lambda/2$, and so on. This can be expressed as $x_m = m\lambda L/w$, where $m = 1, 2, 3$, etc., as limited by the small angle simplication. When $m = 1$, this equation provides the position of the first-order dark band. The second-order dark band occurs at $m = 2$, and so forth.

PRACTICE Problems

- Additional Problems, Appendix B
- Solutions to Selected Problems, Appendix C

16. Monochromatic green light of wavelength 546 nm falls on a single slit with a width of 0.095 mm. The slit is located 75 cm from a screen. How wide will the central bright band be?

17. Yellow light with a wavelength of 589 nm passes through a slit of width 0.110 mm and makes a pattern on a screen. If the width of the central bright band is 2.60×10^{-2} m, how far is it from the slits to the screen?

18. Light from a He-Ne laser ($\lambda = 632.8$ nm) falls on a slit of unknown width. A pattern is formed on a screen 1.15 m away, on which the central bright band is 15 mm wide. How wide is the slit?

19. Yellow light falls on a single slit 0.0295 mm wide. On a screen that is 60.0 cm away, the central bright band is 24.0 mm wide. What is the wavelength of the light?

20. White light falls on a single slit that is 0.050 mm wide. A screen is placed 1.00 m away. A student first puts a blue-violet filter ($\lambda = 441$ nm) over the slit, then a red filter ($\lambda = 622$ nm). The student measures the width of the central bright band.

a. Which filter produced the wider band?

b. Calculate the width of the central bright band for each of the two filters.

Single-slit diffraction patterns make the wave nature of light noticeable when the slits are 10 to 100 times the wavelength of the light. Larger openings, however, cast sharp shadows, as Isaac Newton first observed. While the single-slit pattern depends on the wavelength of light, it is only when a large number of slits are put together that diffraction provides a useful tool for measuring wavelength.

CHALLENGE PROBLEM

You have several unknown substances and wish to use a single-slit diffraction apparatus to determine what each one is. You decide to place a sample of an unknown substance in the region between the slit and the screen and use the data that you obtain to determine the identity of each substance by calculating its index of refraction.

1. Come up with a general formula for the index of refraction of an unknown substance in terms of the wavelength of the light, λ_{vacuum}, the width of the slit, w, the distance from the slit to the screen, L, and the distance between the central bright band and the first dark band, x_1.
2. If the source you used had a wavelength of 634 nm, the slit width was 0.10 mm, the distance from the slit to the screen was 1.15 m, and you immersed the apparatus in water ($n_{substance} = 1.33$), then what would you expect the width of the center band to be?

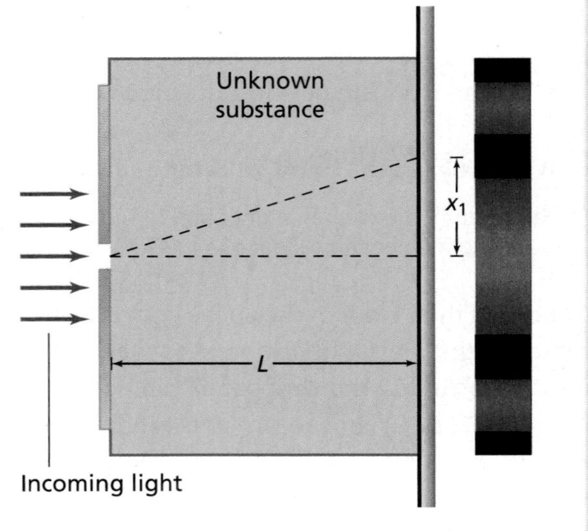

Diffraction Gratings

■ **Figure 19-11** Diffraction gratings are used to create diffraction patterns for the analysis of light sources.

Although double-slit interference and single-slit diffraction depend on the wavelength of light, diffraction gratings, such as those shown in **Figure 19-11,** are used to make precision measurements of wavelength. A **diffraction grating** is a device made up of many single slits that diffract light and form a diffraction pattern that is an overlap of single-slit diffraction patterns. Diffraction gratings can have as many as 10,000 slits per centimeter. That is, the spacing between the slits can be as small as 10^{-6} m, or 1000 nm.

One type of diffraction grating is called a transmission grating. A transmission grating can be made by scratching very fine lines with a diamond point on glass that transmits light. The spaces between the scratched lines act like slits. A less expensive type of grating is a replica grating. A replica grating is made by pressing a thin plastic sheet onto a glass grating. When the plastic is pulled away, it contains an accurate imprint of the scratches. Jewelry made from replica gratings, shown in **Figure 19-12a,** produces a spectrum.

Reflection gratings are made by inscribing fine lines on metallic or reflective glass surfaces. The color spectra produced when white light reflects off the surface of a CD or DVD is the result of a reflection grating, as shown in **Figure 19-12b.** If you were to shine monochromatic light on a DVD, the reflected light would produce a diffraction pattern on a screen. Transmission and reflection gratings produce similar diffraction patterns, which can be analyzed in the same manner.

Holographic diffraction gratings produce the brightest spectra. They are made by using a laser and mirrors to create an interference pattern consisting of parallel bright and dark lines. The pattern is projected on a piece of metal that is coated with a light-sensitive material. The light produces a chemical reaction that hardens the material. The metal is then placed in acid, which attacks the metal wherever it was not protected by the hardened material. The result is a series of hills and valleys in the metal identical to the original interference pattern. The metal can be used as a reflection grating or a replica transmission grating can be made from it. Because of the sinusoidal shape of the hills and valleys, the diffraction patterns are exceptionally bright.

■ **Figure 19-12** A transmission grating spectrum is created by jewelry made with replica gratings **(a).** Compact discs act as reflection gratings, creating a spectrum diffraction pattern when they are placed under white light **(b).**

Figure 19-13 A spectroscope is used to measure the wavelengths of light emitted by a light source.

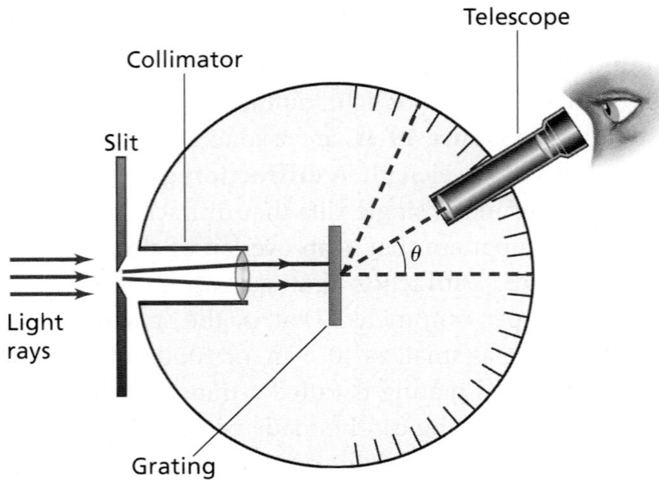

Measuring wavelength An instrument used to measure light wavelengths using a diffraction grating is called a grating spectroscope, as shown in the diagram in **Figure 19-13.** The source to be analyzed emits light that is directed to a slit. The light from the slit passes through a diffraction grating. The grating produces a diffraction pattern that is viewed through a telescope.

The diffraction pattern produced by a diffraction grating has narrow, equally spaced, bright lines, as shown in **Figure 19-14.** The larger the number of slits per unit length of the grating, the narrower the lines in the diffraction pattern. As a result, the distance between the bright lines can be measured much more precisely with a grating spectroscope than with a double slit.

Earlier in this chapter, you found that the interference pattern produced by a double slit could be used to calculate wavelength. An equation for the diffraction grating can be developed in the same way as for the double slit. However, with a diffraction grating, θ could be large, so the small angle simplification does not apply. Wavelength can be found by measuring the angle, θ, between the central bright line and the first-order bright line.

Wavelength from a Diffraction Grating $\lambda = d \sin \theta$

The wavelength of light is equal to the slit separation distance times the sine of the angle at which the first-order bright line occurs.

Constructive interference from a diffraction grating occurs at angles on either side of the central bright line given by the equation $m\lambda = d\sin \theta$, where $m = 0, 1, 2$, etc. The central bright line occurs at $m = 0$.

Figure 19-14 A grating was used to produce diffraction patterns for red light **(a)** and white light **(b).**

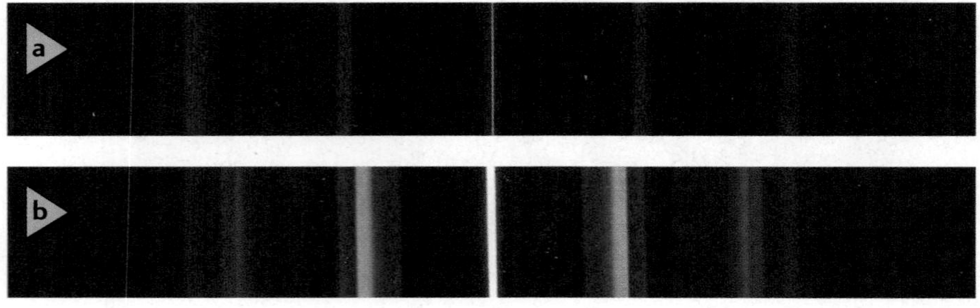

▶ EXAMPLE Problem 3

Using a DVD as a Diffraction Grating A student noticed the beautiful spectrum reflected off a rented DVD. She directed a beam from her teacher's green laser pointer at the DVD and found three bright spots reflected on the wall. The label on the pointer indicated that the wavelength was 532 nm. The student found that the spacing between the spots was 1.29 m on the wall, which was 1.25 m away. What is the spacing between the rows on the DVD?

1 Analyze and Sketch the Problem

- Sketch the experiment, showing the DVD as a grating and the spots on the wall.
- Identify and label the knowns.

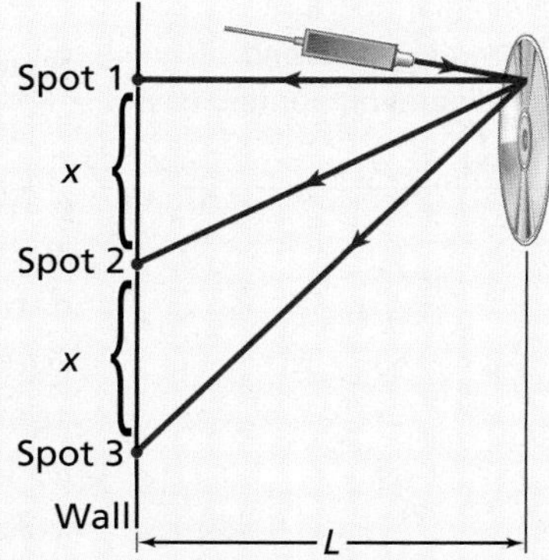

Known:	**Unknown:**
$x = 1.29$ m	$d = ?$
$L = 1.25$ m	
$\lambda = 532$ nm	

2 Solve for the Unknown

Find the angle between the central bright spot and one next to it using $\tan\theta = x/L$.

$$\theta = \tan^{-1}\left(\frac{x}{L}\right)$$

$$= \tan^{-1}\left(\frac{1.29\text{ m}}{1.25\text{ m}}\right) \quad \text{Substitute } x = 1.29\text{ m}, L = 1.25\text{ m}$$

$$= 45.9°$$

Use the diffraction grating wavelength and solve for d.

$$\lambda = d\sin\theta$$

$$d = \frac{\lambda}{\sin\theta}$$

$$= \frac{532\times10^{-9}\text{ m}}{\sin 45.9°} \quad \text{Substitute } \lambda = 532\times10^{-9}\text{ m}, \theta = 45.9°$$

$$= 7.41\times10^{-7}\text{ m}$$

Personal Tutor For an online tutorial on diffraction grating, visit physicspp.com.

3 Evaluate the Answer

- **Are the units correct?** The answer is in m, which is correct for separation.
- **Is the magnitude realistic?** With x and L almost the same size, d is close to λ.

▶ PRACTICE Problems

- Additional Problems, Appendix B
- Solutions to Selected Problems, Appendix C

21. White light shines through a grating onto a screen. Describe the pattern that is produced.

22. If blue light of wavelength 434 nm shines on a diffraction grating and the spacing of the resulting lines on a screen that is 1.05 m away is 0.55 m, what is the spacing between the slits in the grating?

23. A diffraction grating with slits separated by 8.60×10^{-7} m is illuminated by violet light with a wavelength of 421 nm. If the screen is 80.0 cm from the grating, what is the separation of the lines in the diffraction pattern?

24. Blue light shines on the DVD in Example Problem 3. If the dots produced on a wall that is 0.65 m away are separated by 58.0 cm, what is the wavelength of the light?

25. Light of wavelength 632 nm passes through a diffraction grating and creates a pattern on a screen that is 0.55 m away. If the first bright band is 5.6 cm from the central bright band, how many slits per centimeter does the grating have?

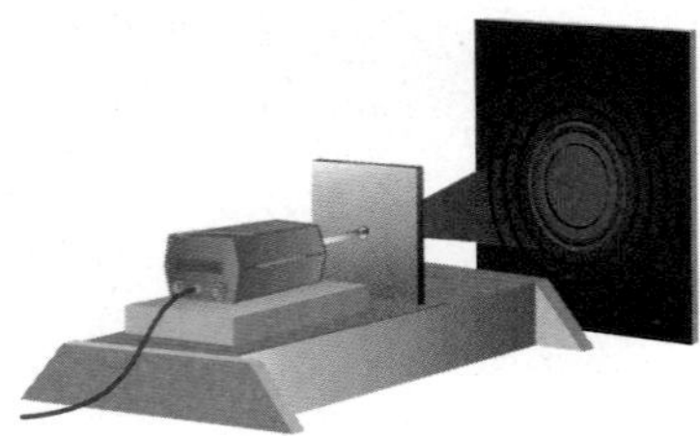

Figure 19-15 The diffraction pattern of a circular aperture produces alternating dark and bright rings. (Illustration not to scale)

In thin-film interference, the interference pattern is visible only within a narrow angle of view straight over the film. This would be the case for the *Morpho* butterfly's blue, shimmering interference pattern, if not for the layer of glass-like scales on top of the layer of ground scales. This glass-like scale layer acts as a diffraction grating and causes the blue, shimmering interference pattern to be spread to a diffraction pattern with a wider angle of view. Scientists believe that this makes the *Morpho* butterfly more visible to potential mates.

Resolving Power of Lenses

The circular lens of a telescope, a microscope, and even your eye acts as a hole, called an aperture, through which light is allowed to pass. An aperture diffracts the light, just as a single slit does. Alternating bright and dark rings occur with a circular aperture, as shown in **Figure 19-15.** The equation for an aperture is similar to that for a single slit. However, an aperture has a circular edge rather than the two edges of a slit, so slit width, w, is replaced by aperture diameter, D, and an extra geometric factor of 1.22 enters the equation, resulting in $x_1 = 1.22\lambda L/D$.

When light from a distant star is viewed through the aperture of a telescope, the image is spread out due to diffraction. If two stars are close enough together, the images may be blurred together, as shown in **Figure 19-16.** In 1879, Lord Rayleigh, a British physicist, mathematician, and Nobel prize winner, established a criterion for determining whether there is one or two stars in such an image. The **Rayleigh criterion** states that if the center of the bright spot of one star's image falls on the first dark ring of the second, the two images are at the limit of resolution. That is, a viewer will be able to tell that there are two stars rather than only one.

Astronomy Connection

If two images are at the limit of resolution, how far apart are the objects? Using the Rayleigh criterion, the centers of the bright spots of the two images are a distance of x_1 apart. Figure 19-16 shows that similar triangles can be used to find that $x_{obj}/L_{obj} = x_1/L$. Combining this with $x_1 = 1.22\lambda L/D$ to eliminate x_1/L, and solving for the separation distance between objects, x_{obj}, the following equation can be derived.

Rayleigh Criterion $$x_{obj} = \frac{1.22\lambda L_{obj}}{D}$$

The separation distance between objects that are at the limit of resolution is equal to 1.22, times the wavelength of light, times the distance from the circular aperture to the objects, divided by the diameter of the circular aperture.

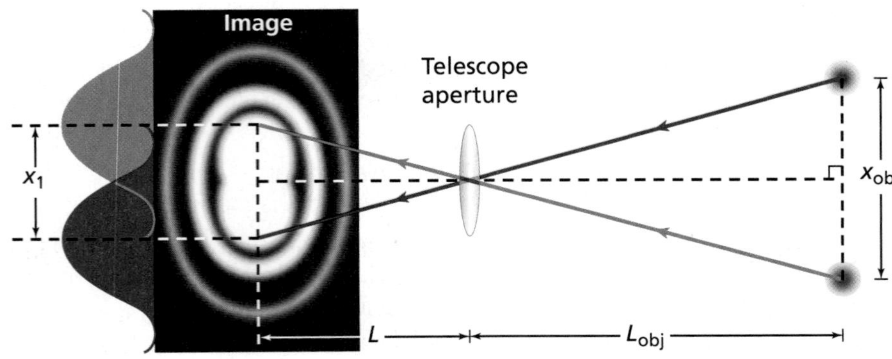

Figure 19-16 Similar-triangle geometry allows you to calculate the actual separation distance of objects. The blue and red colors are used only for the purpose of illustration. (Illustration not to scale)

Diffraction in the eye In bright light the eye's pupil is about 3 mm in diameter. The eye is most sensitive to yellow-green light where $\lambda = 550$ nm. So the Rayleigh criterion applied to the eye gives $x_{obj} = 2\times10^{-4} L_{obj}$. The distance between the pupil and retina is about 2 cm, so two barely resolved point sources would be separated by about 4 μm on the retina. The spacing between the light detectors, the cones, in the most sensitive part of the eye, the fovea, is about 2 μm. Thus, in the ideal case, the three adjacent cones would record light, dark, and light. It would seem that the eye is ideally constructed. If cones were closer together, they would see details of the diffraction pattern, not of the sources. If cones were farther apart, they would not be able to resolve all possible detail.

Applying the Rayleigh criterion to find the ability of the eye to separate two distance sources shows that the eye could separate two automobile headlamps (1.5 m apart) at a distance of 7 km. In practice, however, the eye is not limited by diffraction. Imperfections in the lens and the liquid that fills the eye reduce the eye's resolution to about five times that set by the Rayleigh criterion. Most people use their eyes for purposes other than resolving point sources. For example, the eye seems to have a built-in ability to detect straight edges.

Many telescope manufacturers advertise that their instruments are diffraction limited. That is, they claim that their telescopes can separate two point sources at the Rayleigh criterion. To reach this limit they must grind the mirrors and lenses to an accuracy of one-tenth of a wavelength, or about 55 nm. The larger the diameter of the mirror, the greater the resolution of the telescope. Unfortunately, light from planets or stars must go through Earth's atmosphere. The same variations in the atmosphere that cause stars to twinkle keep telescopes from reaching the diffraction limit. Because the *Hubble Space Telescope* is above Earth's atmosphere, the resolution of its images is much better than those of larger telescopes on Earth's surface.

MINI LAB

Retinal Projection Screen

Did you know that you can use the retina of your eyeball as a screen? ***CAUTION: Do not do the following with a laser or the Sun.***

1. Plug in and turn on an incandescent lamp with a straight filament. Stand about 2 m from the lamp.

2. Hold a diffraction grating in front of your eye so the color spectra are oriented horizontally.

3. Observe the color spectra patterns and draw your observations using colored pencils.

Analyze and Conclude

4. Which color is closest to the central bright line (the light filament)? Which is farthest?

5. How many spectra are you able to see on either side of the light?

6. Interpret Data Are your observations consistent with the equation for the wavelength from a diffraction grating?

19.2 Section Review

26. Distance Between First-Order Dark Bands Monochromatic green light of wavelength 546 nm falls on a single slit of width 0.080 mm. The slit is located 68.0 cm from a screen. What is the separation of the first dark bands on each side of the central bright band?

27. Diffraction Patterns Many narrow slits are close to each other and equally spaced in a large piece of cardboard. They are illuminated by monochromatic red light. A sheet of white paper is placed far from the slits, and a pattern of bright and dark bands is visible on the paper. Sketch the pattern that would be seen on the screen.

28. Line Spacing You shine a red laser light through one diffraction grating and form a pattern of red dots on a screen. Then you substitute a second diffraction grating for the first one, forming a different pattern. The dots produced by the first grating are spread out more than those produced by the second. Which grating has more lines per millimeter?

29. Rayleigh Criterion The brightest star in the winter sky in the northern hemisphere is Sirius. In reality, Sirius is a system of two stars that orbit each other. If the *Hubble Space Telescope* (diameter 2.4 m) is pointed at the Sirius system, which is 8.44 light-years from Earth, what is the minimum separation there would need to be between the stars in order for the telescope to be able to resolve them? Assume that the average light coming from the stars has a wavelength of 550 nm.

30. Critical Thinking You are shown a spectrometer, but do not know whether it produces its spectrum with a prism or a grating. By looking at a white-light spectrum, how could you tell?

PHYSICS LAB • Design Your Own

Double-Slit Interference of Light

Alternate CBL instructions can be found on the Web site.
physicspp.com

Light sometimes behaves as a wave. As coherent light strikes a pair of slits that are close together, the light passing through the slits will create a pattern of constructive and destructive interference on a screen. In this investigation you will develop a procedure and measure the wavelength of a monochromatic light source using two slits.

QUESTION

How can a double-slit interference pattern of light be used to measure the light's wavelength?

Objectives

- **Observe** a double-slit interference pattern of monochromatic light.
- **Calculate** the wavelength of light using a double-slit interference pattern.

Safety Precautions

- **Use laser protective eyewear approved by ANSI.**
- **Never look directly into the light of a laser.**

Possible Materials

laser pointer or laser to be tested
double-slit plate
laser pointer or laser of known wavelength
clothes pin to hold a laser pointer
clay ball to hold the double-slit plate
meterstick

Procedure

1. Determine which equation applies to double-slit interference.
2. Use a double slit of known slit-separation distance, *d*, or develop a method to determine *d*.
3. Sketch how light passes through a double slit to help you determine how *x* and *L* can be measured.
4. Using your sketch from step 3 and the list of possible materials provided in this lab, design the lab setup and write a procedure for performing the experiment.
5. Determine the values of *m* that would be invalid for the equation.
6. ***CAUTION: Looking directly into laser light could damage your eyes.***
7. Be sure to check with your teacher and have approval before you implement your design.
8. Perform your experiment. Write your data in a data table similar to the one on the next page.

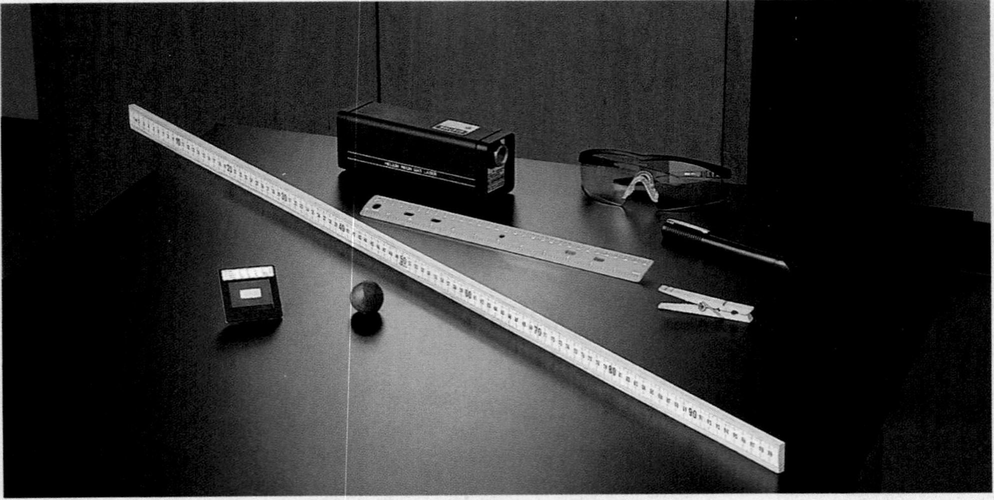

Data Table						
Source	Color	Accepted λ (m)	d (m)	m	x (m)	L (m)
				1		
				2		
				3		
				4		
				5		

Analyze

1. Adjust the distance of your slits from the screen. Is there a distance that allows you to collect the most data with the best precision?
2. Calculate the wavelength, λ, of your light source using m and measurements of x, d, and L.
3. **Error Analysis** Compare your calculated wavelength to the accepted value by determining the percentage of error.

Conclude and Apply

1. **Conclude** Did your procedure enable you to use a double-slit interference pattern to measure the wavelength of light? Explain.
2. **Estimate** what results you would get if you used a plate with a smaller slit separation distance, d, and performed the experiment exactly the same.
3. **Infer** How would your observations change if you used green light, but used the same double-slit plate and screen distance? What would you observe?

Going Further

1. **Use a Scientific Explanation** Describe why the double-slit interference pattern dims, brightens, and dims again as distance from the center of the pattern increases.
2. **Error Analysis** Describe several things you could do in the future to reduce systematic error in your experiment.
3. **Evaluate** Examine the measuring equipment you used and determine which equipment limited you the most on the precision of your calculations and which equipment gave you more precision than you needed, if any.
4. **Lab Techniques** What might be done to an experimental setup to use white light from a normal lightbulb to produce a double-slit interference pattern?

Real-World Physics

1. When white light shines through slits in a screen door, why is a pattern not visible in the shadow on a wall?
2. Would things look different if all of the light that illuminated your world was coherent? Explain.

To find out more about interference patterns, visit the Web site: physicspp.com

HOW it WORKS Holography

Holography is a form of photography that produces a three-dimensional image. Dennis Gabor made the first hologram in 1947, but holography was impractical until the invention of the gas laser in 1960. Holograms are used on credit cards to help prevent counterfeiting, and they may one day be used for ultra-high-density data storage. How is a hologram made?

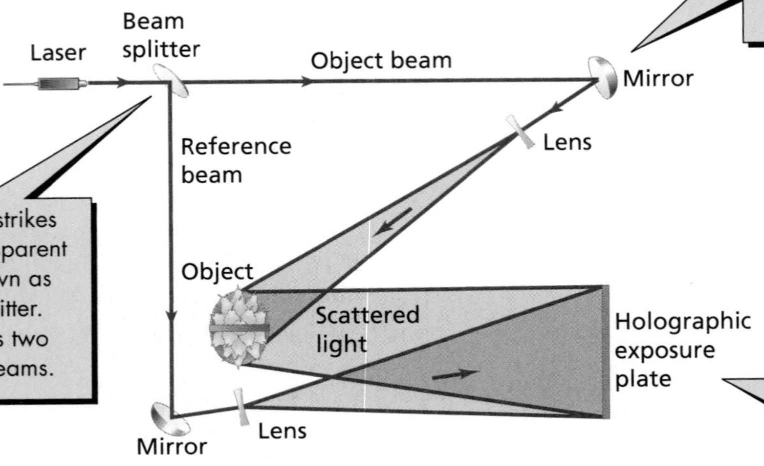

1 Laser light strikes a semitransparent mirror known as a beam splitter. This creates two coherent beams.

2 The reference and object beams are directed by mirrors, and are made to diverge by lenses.

3 Light scattered by reflection off the object, in this case a basket of pears, interferes with the reference beam. The interference pattern that is formed by the two beams is recorded on the holographic exposure plate.

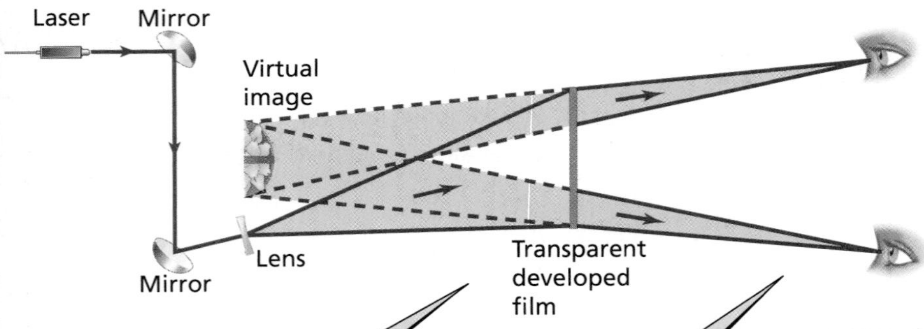

4 When a transparent film of the developed plate is placed in a diverging laser beam, light passing through the film creates a three-dimensional virtual image of the original object with rainbowlike bands of color.

5 A person sees the image as if viewing the original object through a window. Moving his or her head changes the perspective.

Thinking Critically

1. **Infer** A hologram records a complex pattern of constructive and destructive interference fringes. Why do you suppose a vibration-isolated surface is needed for good results?
2. **Use Scientific Explanations** Identify and explain where the following wave properties occur in the diagrams: reflection, refraction, and interference.

Chapter 19

Study Guide

19.1 Interference

Vocabulary

- incoherent light (p. 515)
- coherent light (p. 516)
- interference fringes (p. 516)
- monochromatic light (p. 516)
- thin-film interference (p. 520)

Key Concepts

- Incoherent light illuminates an object evenly, just as a lightbulb illuminates your desk.
- Only the superposition of light waves from coherent light sources can produce an interference pattern.
- Interference demonstrates that light has wave properties.
- Light passing through two closely spaced, narrow slits produces a pattern on a screen of dark and light bands called interference fringes.
- Interference patterns can be used to measure the wavelength of light.

$$\lambda = \frac{xd}{L}$$

- Interference patterns can result from the creation of coherent light at the refractive boundary of a thin film.

19.2 Diffraction

Vocabulary

- diffraction pattern (p. 524)
- diffraction grating (p. 527)
- Rayleigh criterion (p. 530)

Key Concepts

- Light passing through a narrow slit is diffracted, or spread out from a straight-line path, and produces a diffraction pattern on a screen.
- The diffraction pattern from a single slit has a bright central band that has a width equal to the distance between the first dark bands on either side of the bright central band.

$$2x_1 = \frac{2\lambda L}{w}$$

- Diffraction gratings consist of large numbers of slits that are very close together and produce narrow lines that result from an overlap of the single-slit diffraction patterns of all of the slits in the grating.
- Diffraction gratings can be used to measure the wavelength of light precisely or to separate light composed of different wavelengths.

$$\lambda = d \sin \theta$$

- Diffraction limits the ability of an aperture to distinguish two closely spaced objects.

$$x_{obj} = \frac{1.22\lambda L_{obj}}{D}$$

- If the central bright spot of one image falls on the first dark ring of the second image, the images are at the limit of resolution.

Chapter 19 Assessment

Concept Mapping

31. Monochromatic light of wavelength λ illuminates two slits in a Young's double-slit experiment setup that are separated by a distance, d. A pattern is projected onto a screen a distance, L, away from the slits. Complete the following concept map using λ, L, and d to indicate how you could vary them to produce the indicated change in the spacing between adjacent bright bands, x.

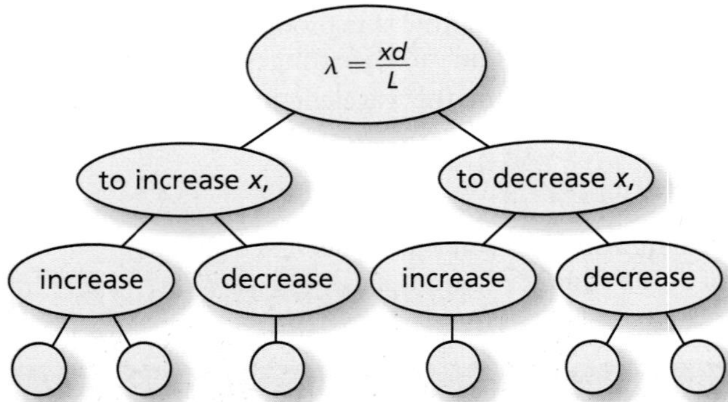

Mastering Concepts

32. Why is it important that monochromatic light was used to make the interference pattern in Young's interference experiment? (19.1)

33. Explain why the position of the central bright band of a double-slit interference pattern cannot be used to determine the wavelength of the light waves. (19.1)

34. Describe how you could use light of a known wavelength to find the distance between two slits. (19.1)

35. Describe in your own words what happens in thin-film interference when a dark band is produced by light shining on a soap film suspended in air. Make sure you include in your explanation how the wavelength of the light and thickness of the film are related. (19.1)

36. White light shines through a diffraction grating. Are the resulting red lines spaced more closely or farther apart than the resulting violet lines? Why? (19.2)

37. Why do diffraction gratings have large numbers of slits? Why are these slits so close together? (19.2)

38. Why would a telescope with a small diameter not be able to resolve the images of two closely spaced stars? (19.2)

39. For a given diffraction grating, which color of visible light produces a bright line closest to the central bright band? (19.2)

Applying Concepts

40. For each of the following examples, indicate whether the color is produced by thin-film interference, refraction, or the presence of pigments.
a. soap bubbles
b. rose petals
c. oil films
d. a rainbow

41. How can you tell whether a pattern is produced by a single slit or a double slit?

42. Describe the changes in a single-slit diffraction pattern as the width of the slit is decreased.

43. **Science Fair** At a science fair, one exhibition is a very large soap film that has a fairly consistent width. It is illuminated by a light with a wavelength of 432 nm, and nearly the entire surface appears to be a lovely shade of purple. What would you see in the following situations?
a. the film thickness was doubled
b. the film thickness was increased by half a wavelength of the illuminating light
c. the film thickness was decreased by one quarter of a wavelength of the illuminating light

44. What are the differences in the characteristics of the diffraction patterns formed by diffraction gratings containing 10^4 lines/cm and 10^5 lines/cm?

45. **Laser Pointer Challenge** You have two laser pointers, a red one and a green one. Your friends Mark and Carlos disagree about which has the longer wavelength. Mark insists that red light has a longer wavelength, while Carlos is sure that green has the longer wavelength. You have a diffraction grating handy. Describe what demonstration you would do with this equipment and how you would explain the results to Carlos and Mark to settle their disagreement.

46. **Optical Microscope** Why is blue light used for illumination in an optical microscope?

Mastering Problems

19.1 Interference

47. Light falls on a pair of slits 19.0 μm apart and 80.0 cm from a screen, as shown in **Figure 19-17.** The first-order bright band is 1.90 cm from the central bright band. What is the wavelength of the light?

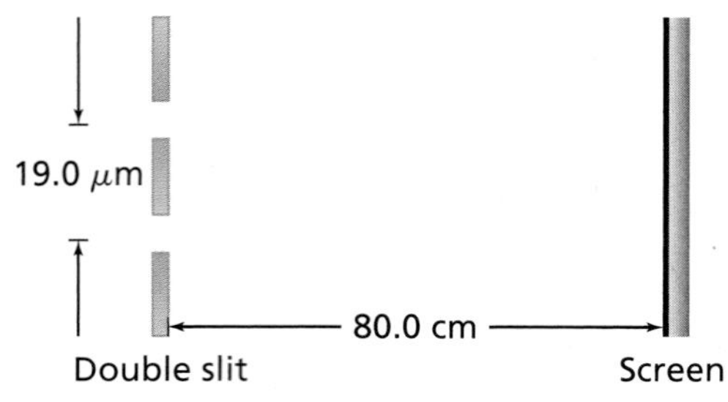

Figure 19-17 (Not to scale)

 For more problems, go to Additional Problems, Appendix B.

48. **Oil Slick** After a short spring shower, Tom and Ann take their dog for a walk and notice a thin film of oil ($n = 1.45$) on a puddle of water, producing different colors. What is the minimum thickness of a place where the oil creates constructive interference for light with a wavelength of 545 nm?

49. Light of wavelength 542 nm falls on a double slit. First-order bright bands appear 4.00 cm from the central bright band. The screen is 1.20 m from the slits. How far apart are the slits?

50. **Insulation Film** Winter is approaching and Alejandro is helping to cover the windows in his home with thin sheets of clear plastic ($n = 1.81$) to keep the drafts out. After the plastic is taped up around the windows such that there is air between the plastic and the glass panes, the plastic is heated with a hair dryer to shrink-wrap the window. The thickness of the plastic is altered during this process. Alejandro notices a place on the plastic where there is a blue stripe of color. He realizes that this is created by thin-film interference. What are three possible thicknesses of the portion of the plastic where the blue stripe is produced if the wavelength of the light is 4.40×10^{2} nm?

51. Samir shines a red laser pointer through three different double-slit setups. In setup A, the slits are separated by 0.150 mm and the screen is 0.60 m away from the slits. In setup B, the slits are separated by 0.175 mm and the screen is 0.80 m away. Setup C has the slits separated by 0.150 mm and the screen a distance of 0.80 m away. Rank the three setups according to the separation between the central bright band and the first-order bright band, from least to most separation. Specifically indicate any ties.

19.2 Diffraction

52. Monochromatic light passes through a single slit with a width of 0.010 cm and falls on a screen 100 cm away, as shown in **Figure 19-18.** If the width of the central band is 1.20 cm, what is the wavelength of the light?

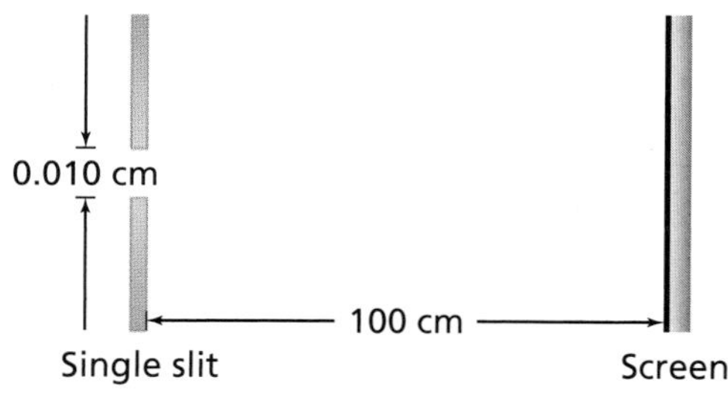

Figure 19-18 (Not to scale)

53. A good diffraction grating has 2.5×10^{3} lines per cm. What is the distance between two lines in the grating?

54. Light with a wavelength of 4.5×10^{-5} cm passes through a single slit and falls on a screen 100 cm away. If the slit is 0.015 cm wide, what is the distance from the center of the pattern to the first dark band?

55. ***Hubble Space Telescope*** Suppose the *Hubble Space Telescope,* 2.4 m in diameter, is in orbit 1.0×10^{5} m above Earth and is turned to view Earth, as shown in **Figure 19-19.** If you ignore the effect of the atmosphere, how large an object can the telescope resolve? Use $\lambda = 5.1\times10^{-7}$ m.

Figure 19-19

56. Monochromatic light with a wavelength of 425 nm passes through a single slit and falls on a screen 75 cm away. If the central bright band is 0.60 cm wide, what is the width of the slit?

57. **Kaleidoscope** Jennifer is playing with a kaleidoscope from which the mirrors have been removed. The eyehole at the end is 7.0 mm in diameter. If she can just distinguish two bluish-purple specks on the other end of the kaleidoscope separated by 40 μm, what is the length of the kaleidoscope? Use $\lambda = 650$ nm and assume that the resolution is diffraction limited through the eyehole.

58. **Spectroscope** A spectroscope uses a grating with 12,000 lines/cm. Find the angles at which red light, 632 nm, and blue light, 421 nm, have first-order bright lines.

Mixed Review

59. Record Marie uses an old $33\frac{1}{3}$ rpm record as a diffraction grating. She shines a laser, $\lambda = 632.8$ nm, on the record, as shown in **Figure 19-20.** On a screen 4.0 m from the record, a series of red dots 21 mm apart are visible.

a. How many ridges are there in a centimeter along the radius of the record?

b. Marie checks her results by noting that the ridges represent a song that lasts 4.01 minutes and takes up 16 mm on the record. How many ridges should there be in a centimeter?

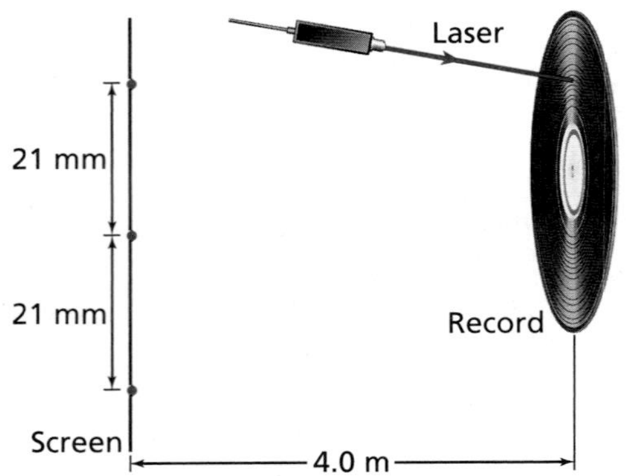

Figure 19-20 (Not to scale)

60. An anti-reflective coating, $n = 1.2$, is applied to a lens. If the thickness of the coating is 125 nm, what is (are) the color(s) of light for which complete destructive interference will occur? *Hint: Assume the lens is made out of glass.*

61. Camera When a camera with a 50-mm lens is set at f/8, its aperture has an opening 6.25 mm in diameter.

a. For light with $\lambda = 550$ nm, what is the resolution of the lens? The film is 50.0 mm from the lens.

b. The owner of a camera needs to decide which film to buy for it. The expensive one, called fine-grained film, has 200 grains/mm. The less costly, coarse-grained film has only 50 grains/mm. If the owner wants a grain to be no smaller than the width of the central bright spot calculated in part **a**, which film should he purchase?

Thinking Critically

62. Apply Concepts Yellow light falls on a diffraction grating. On a screen behind the grating, you see three spots: one at zero degrees, where there is no diffraction, and one each at $+30°$ and $-30°$. You now add a blue light of equal intensity that is in the same direction as the yellow light. What pattern of spots will you now see on the screen?

63. Apply Concepts Blue light of wavelength λ passes through a single slit of width w. A diffraction pattern appears on a screen. If you now replace the blue light with a green light of wavelength 1.5λ, to what width should you change the slit to get the original pattern back?

64. Analyze and Conclude At night, the pupil of a human eye has an aperture diameter of 8.0 mm. The diameter is smaller in daylight. An automobile's headlights are separated by 1.8 m.

a. Based upon Rayleigh's criterion, how far away can the human eye distinguish the two headlights at night? *Hint: Assume a wavelength of 525 nm.*

b. Can you actually see a car's headlights at the distance calculated in part **a**? Does diffraction limit your eyes' sensing ability? Hypothesize as to what might be the limiting factors.

Writing in Physics

65. Research and describe Thomas Young's contributions to physics. Evaluate the impact of his research on the scientific thought about the nature of light.

66. Research and interpret the role of diffraction in medicine and astronomy. Describe at least two applications in each field.

Cumulative Review

67. How much work must be done to push a 0.5-m^3 block of wood to the bottom of a 4-m-deep swimming pool? The density of wood is 500 kg/m^3. (Chapter 13)

68. What are the wavelengths of microwaves in an oven if their frequency is 2.4 GHz? (Chapter 14)

69. Sound wave crests that are emitted by an airplane are 1.00 m apart in front of the plane, and 2.00 m apart behind the plane. (Chapter 15)

a. What is the wavelength of the sound in still air?

b. If the speed of sound is 330 m/s, what is the frequency of the source?

c. What is the speed of the airplane?

70. A concave mirror has a 48.0-cm radius. A 2.0-cm-tall object is placed 12.0 cm from the mirror. Calculate the image position and image height. (Chapter 17)

71. The focal length of a convex lens is 21.0 cm. A 2.00-cm-tall candle is located 7.50 cm from the lens. Use the thin-lens equation to calculate the image position and image height. (Chapter 18)

 For more problems, go to Additional Problems, Appendix B.

Multiple Choice

1. What is the best possible explanation for why the colors of a thin film, such as a soap bubble or oil on water, appear to change and move as you watch?

   (A) because convective heat waves in the air next to the thin film distort the light

   (B) because the film thickness at any given location changes over time

   (C) because the wavelengths in sunlight vary over time

   (D) because your vision varies slightly over time

2. Light at 410 nm shines through a slit and falls on a flat screen as shown in the figure below. The width of the slit is 3.8×10^{-6} m. What is the width of the central bright band?

   (A) 0.024 m (C) 0.048 m

   (B) 0.031 m (D) 0.063 m

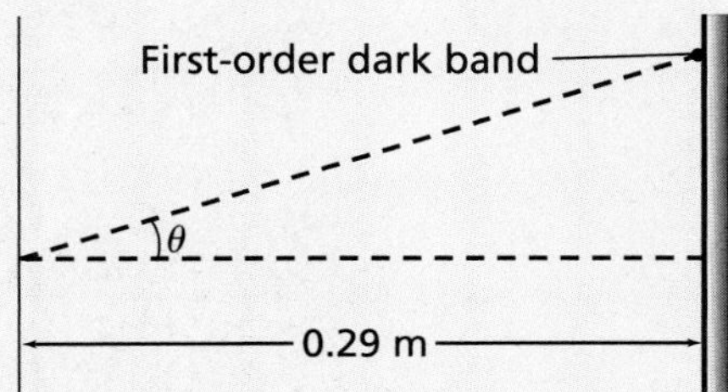

3. For the situation in problem 2, what is the angle, θ, of the first dark band?

   (A) 3.1° (C) 12.4°

   (B) 6.2° (D) 17°

4. Two stars 6.2×10^{4} light-years from Earth are 3.1 light-years apart. What is the smallest diameter telescope that could resolve them using 610 nm light?

   (A) 5.0×10^{-5} m (C) 1.5×10^{-2} m

   (B) 6.1×10^{-5} m (D) 1.5×10^{7} m

5. A grating has slits that are 0.055 mm apart. What is the angle of the first-order bright line for light with a wavelength of 650 nm?

   (A) 0.012° (C) 1.0°

   (B) 0.68° (D) 11°

6. A laser beam at 638 nm illuminates two narrow slits. The third-order band of the resulting pattern is 7.5 cm from the center bright band. The screen is 2.475 m from the slits. How far apart are the slits?

   (A) 5.8×10^{-8} m (C) 2.1×10^{-5} m

   (B) 6.3×10^{-7} m (D) 6.3×10^{-5} m

7. A flat screen is placed 4.200 m from a pair of slits that are illuminated by a beam of monochromatic light. On the screen, the separation between the central bright band and the second-order bright band is 0.082 m. The distance between the slits is 5.3×10^{-5} m. Determine the wavelength of the light.

   (A) 2.6×10^{-7} m (C) 6.2×10^{-7} m

   (B) 5.2×10^{-7} m (D) 1.0×10^{-6} m

8. A clown is blowing soap bubbles and you notice that the color of one region of a particularly large bubble matches the color of his nose. If the bubble is reflecting 6.5×10^{-7} m red light waves, and the index of refraction of the soap film is 1.41, what is the minimum thickness of the soap bubble at the location where it is reflecting red?

   (A) 1.2×10^{-7} m (C) 9.2×10^{-7} m

   (B) 3.5×10^{-7} m (D) 1.9×10^{-6} m

Extended Answer

9. A diffraction grating that has 6000 slits per cm produces a diffraction pattern that has a first-order bright line at 20° from the central bright line. What is the wavelength of the light?

✓ Test-Taking TIP

Don't Be Afraid To Ask For Help

If you are practicing for a test and you are having difficulty understanding why you got a question wrong or you are having difficulty even arriving at an answer, ask someone for help. It is important to ask for help before a test because you cannot ask for help during a test.

Chapter 20

Static Electricity

What You'll Learn

- You will observe the behavior of electric charges and analyze how these charges interact with matter.
- You will examine the forces that act between electric charges.

Why It's Important

Static electricity enables the operation of devices such as printers and copiers, but it has harmful effects on electronic components and in the form of lightning.

Lightning The tiny spark that you experience when you touch a doorknob and the dazzling display of lightning in a storm are both examples of the discharge of static electricity. The charging processes and the means of discharging are vastly different in scale, but they are similar in their fundamental nature.

Think About This ▶

What causes charge to build up in a thundercloud, and how does it discharge in the form of a spectacular lightning bolt?

physicspp.com

Which forces act over a distance?

Question
What happens when a plastic ruler is rubbed with wool and then brought near a pile of paper scraps?

Procedure

1. Place 15-20 scraps of paper from a hole punch on the table.
2. Take a plastic ruler and rub it with a piece of wool.
3. Bring the ruler close to the pieces of paper. Observe the effect the ruler has on the scraps of paper.

Analysis

What happens to the pieces of paper when the ruler is brought close to them? What happens to the pieces of paper that come in contact with the ruler? Did you observe any unexpected results when the ruler was brought close to the paper scraps? If so, describe these results.

Critical Thinking

What forces are acting on the pieces of paper before the ruler is brought close to them? What can you infer about the forces on the paper after the ruler is brought near?

Based on your answers to the previous questions, form a hypothesis that explains the effect the ruler has on the scraps of paper.

20.1 Electric Charge

▶ **Objectives**

- **Demonstrate** that charged objects exert forces, both attractive and repulsive.
- **Recognize** that charging is the separation, not the creation, of electric charges.
- **Describe** the differences between conductors and insulators.

▶ **Vocabulary**

electrostatics
neutral
insulator
conductor

You may have had the experience of rubbing your shoes on a carpet to create a spark when you touched someone. In 1752, Benjamin Franklin set off a flurry of research in the field of electricity when his famous kite experiment showed that lightning is similar to the sparks caused by friction. In his experiment, Franklin flew a kite with a key attached to the string. As a thunderstorm approached, the loose threads of the kite string began to stand up and repel one another, and when Franklin brought his knuckle close to the key, he experienced a spark. Electric effects produced in this way are called static electricity.

In this chapter, you will investigate **electrostatics,** the study of electric charges that can be collected and held in one place. The effects of electrostatics are observable over a vast scale, from huge displays of lightning to the submicroscopic world of atoms and molecules. Current electricity, which is produced by batteries and generators, will be explored in later chapters.

Charged Objects

Have you ever noticed the way that your hair is attracted to the comb when you comb your hair on a dry day or the way that your hair stands on end after it is rubbed with a balloon? Perhaps you also have found that socks sometimes stick together when you take them out of a clothes dryer. If so, you will recognize the attraction of the bits of paper to a plastic ruler demonstrated by the Launch Lab and shown in **Figure 20-1.** You might have noticed the way the paper pieces jumped up to the ruler as you worked through the Launch Lab. There must be a new, relatively strong force causing this upward acceleration because it is larger than the downward acceleration caused by the gravitational force of Earth.

There are other differences between this new force and gravity. Paper is attracted to a plastic ruler only after the ruler has been rubbed; if you wait a while, the attractive property of the ruler disappears. Gravity, on the other hand, does not require rubbing and does not disappear. The ancient Greeks noticed effects similar to that of the ruler when they rubbed amber. The Greek word for amber is *elektron,* and today this attractive property is called electric. An object that exhibits electric interaction after rubbing is said to be charged.

Like charges You can explore electric interactions with simple objects, such as transparent tape. Fold over about 5 mm of the end of a strip of tape for a handle, and then stick the remaining 8- to 12-cm-long part of the tape strip on a dry, smooth surface, such as your desktop. Stick a second, similar piece of tape next to the first. Quickly pull both strips off the desk and bring them near each other. A new property causes the strips to repel each other: they are electrically charged. Because they were prepared in the same way, they must have the same type of charge. Thus, you have demonstrated that two objects with the same type of charge repel each other.

■ **Figure 20-1** Rubbing a plastic ruler with wool produces a new force of attraction between the ruler and bits of paper. When the ruler is brought close to bits of paper, the attractive electric force accelerates the paper bits upward against the force of gravity.

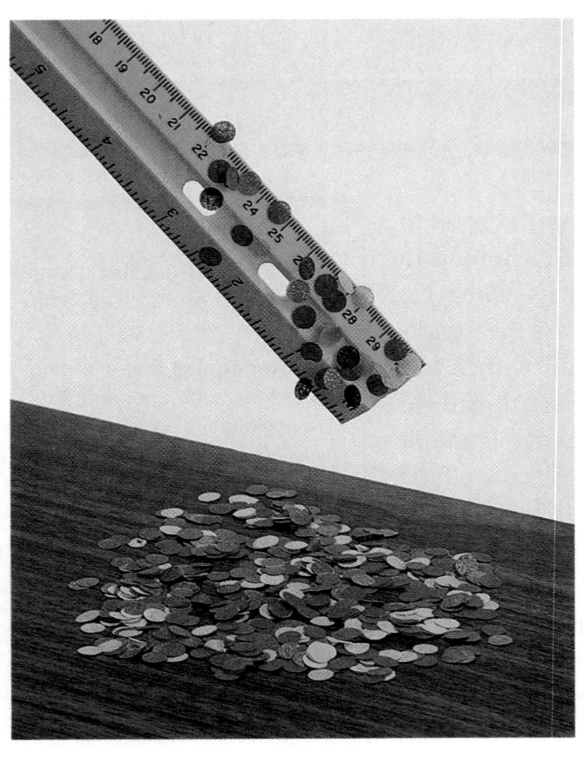

You can learn more about this charge by doing some simple experiments. You may have found that the tape is attracted to your hand. Are both sides attracted, or just one? If you wait a while, especially in humid weather, you will find that the electric charge disappears. You can restore it by again sticking the tape to the desk and pulling it off. You also can remove its charge by gently rubbing your fingers down both sides of the tape.

Opposite charges Now, stick one strip of tape on the desk and place the second strip on top of the first. As shown in **Figure 20-2a,** use the handle of the bottom strip of tape to pull the two off the desk together. Rub them with your fingers until they are no longer attracted to your hand. You now have removed all the electric charge. With one hand on the handle of one strip and the other on the handle of the second strip, quickly pull the two strips apart. You will find that they are now both charged. They once again are attracted to your hands. Do they still repel each other? No, they now attract each other. They are charged, but they are no longer charged alike. They have opposite charges and therefore attract each other.

Is tape the only object that you can charge? Once again, stick one strip of tape to the desk and the second strip on top. Label the bottom strip *B* and the top strip *T*. Pull the pair off together. Discharge them, then pull them apart. Stick the handle end of each strip to the edge of a table, the bottom of a lamp shade, or some similar object. The two should hang down a short distance apart. Finally, rub a comb or pen on your clothing and bring it near one strip of tape and then the other. You will find that one strip will be attracted to the comb, while the other will be repelled by it, as shown in **Figure 20-2b.** You now can explore the interactions of charged objects with the strips of tape.

Figure 20-2 Strips of tape can be given opposite charges **(a)** and then be used to demonstrate the interactions of like and opposite charges **(b).**

Experimenting with charge Try to charge other objects, such as glasses and plastic bags. Rub them with different materials, such as silk, wool, and plastic wrap. If the air is dry, scuff your shoes on carpet and bring your finger near the strips of tape. To test silk or wool, slip a plastic bag over your hand before holding the cloth. After rubbing, take your hand out of the bag and bring both the bag and cloth near the strips of tape.

Most charged objects will attract one strip and repel the other. You will never find an object that repels both strips of tape, although you might find some that attract both. For example, your finger will attract both strips. You will explore this effect later in this chapter.

Types of charge From your experiments, you can make a list of objects labeled *B*, for bottom, which have the same charge as the tape stuck on the desk. Another list can be made of objects labeled *T*, which have the same charge as the top strip of tape. There are only two lists, because there are only two types of charge. Benjamin Franklin called them positive and negative charges. Using Franklin's convention, when hard rubber and plastic are rubbed, they become negatively charged. When materials such as glass and wool are rubbed, they become positively charged.

Color Convention

- Positive charges are shown in **red.**
- Negative charges are shown in **blue.**

Just as you showed that an uncharged pair of tape strips became oppositely charged, you probably were able to show that if you rubbed plastic with wool, the plastic became negatively charged and the wool positively charged. The two kinds of charges were not created alone, but in pairs. These experiments suggest that matter normally contains both charges, positive and negative. Contact in some way separates the two. To explore this further, you must consider the microscopic picture of matter.

A Microscopic View of Charge

Electric charges exist within atoms. In 1897, J.J. Thomson discovered that all materials contain light, negatively charged particles that he called electrons. Between 1909 and 1911, Ernest Rutherford, a student of Thomson from New Zealand, discovered that the atom has a massive, positively charged nucleus. When the positive charge of the nucleus equals the negative charge of the surrounding electrons, then the atom is **neutral.**

Figure 20-3 When wool is used to charge a rubber rod, electrons are removed from the wool atoms and cling to the rubber atoms. In this way, both objects become charged.

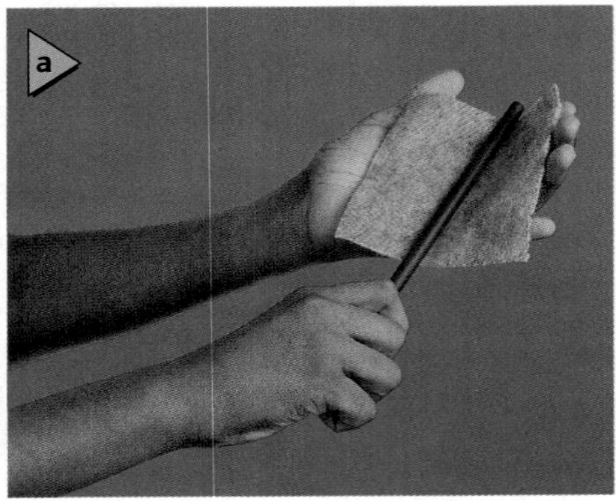

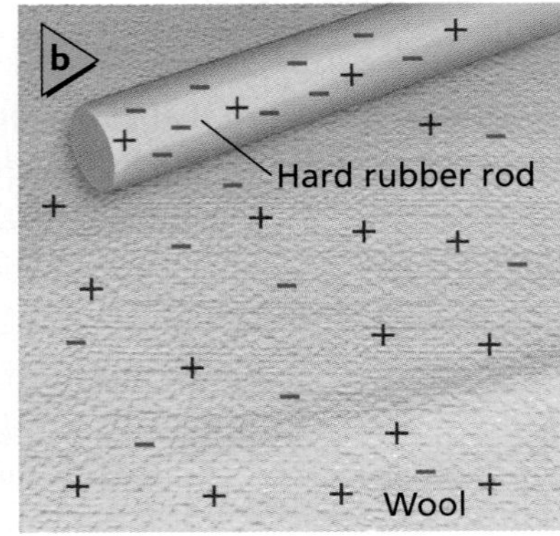

With the addition of energy, the outer electrons can be removed from atoms. An atom missing electrons has an overall positive charge, and consequently, any matter made of these electron-deficient atoms is positively charged. The freed electrons can remain unattached or become attached to other atoms, resulting in negatively charged particles. From a microscopic viewpoint, acquiring charge is a process of transferring electrons.

Separation of charge If two neutral objects are rubbed together, each can become charged. For instance, when rubber and wool are rubbed together, electrons from atoms on the wool are transferred to the rubber, as shown in **Figure 20-3.** The extra electrons on the rubber result in a net negative charge. The electrons missing from the wool result in a net positive charge. The combined total charge of the two objects remains the same. Charge is conserved, which is one way of saying that individual charges never are created or destroyed. All that happens is that the positive and negative charges are separated through a transfer of electrons.

Complex processes that affect the tires of a moving car or truck can cause the tires to become charged. Processes inside a thundercloud can cause the cloud bottom to become negatively charged and the cloud top to become positively charged. In both these cases, charge is not created, but separated.

APPLYING **PHYSICS**

Conductor or Insulator? It might be tempting to classify an element as solely a conductor or solely an insulator, but the classification can change depending on the form the element takes. For example, carbon in the form of diamond is an insulator, but carbon in the form of graphite can conduct charge. This is because the carbon atoms in diamonds are tightly bonded to four other carbons, while the carbon atoms in graphite form three stronger bonds and a fourth, weaker bond that allows electrons a limited amount of movement. As a result, graphite is a much better conductor than diamond, even though both are simply carbon atoms.

Conductors and Insulators

Hold a plastic rod or comb at its midpoint and rub only one end. You will find that only the rubbed end becomes charged. In other words, the charges that you transferred to the plastic stayed where they were put; they did not move. A material through which a charge will not move easily is called an electric **insulator.** The strips of tape that you charged earlier in this chapter acted in this way. Glass, dry wood, most plastics, cloth, and dry air are all good insulators.

Suppose that you support a metal rod on an insulator so that it is isolated, or completely surrounded by insulators. If you then touch the charged comb to one end of the metal rod, you will find that the charge spreads very quickly over the entire rod. A material that allows charges to move about easily is called an electric **conductor.** Electrons carry, or conduct, electric charge through the metal. Metals are good conductors because at least one electron on each atom of the metal can be removed easily. These electrons act as if they no longer belong to any one atom, but to the metal as a whole; consequently, they move freely throughout the piece of metal. **Figure 20-4** contrasts how charges behave when they are

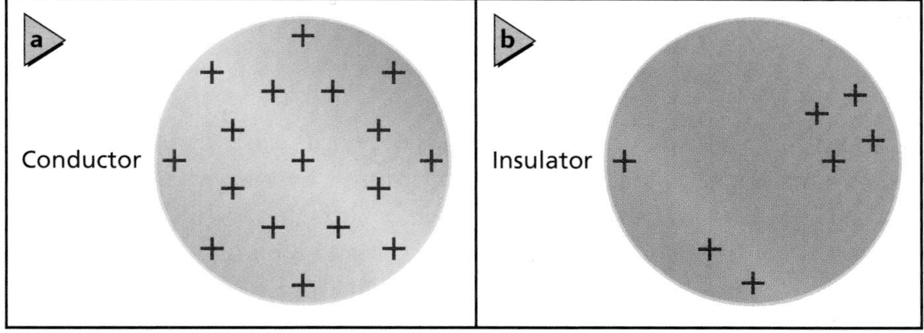

Figure 20-4 Charges placed on a conductor will spread over the entire surface **(a).** Charges placed on an insulator will remain where they are placed **(b).**

placed on a conductor with how they behave on an insulator. Copper and aluminum are both excellent conductors and are used commercially to carry electricity. Plasma, a highly ionized gas, and graphite also are good conductors of electric charge.

When air becomes a conductor Air is an insulator; however, under certain conditions, charges move through air as if it were a conductor. The spark that jumps between your finger and a doorknob after you have rubbed your feet on a carpet discharges you. In other words, you have become neutral because the excess charges have left you. Similarly, lightning discharges a thundercloud. In both of these cases, air became a conductor for a brief moment. Recall that conductors must have charges that are free to move. For a spark or lightning to occur, freely moving charged particles must be formed in the normally neutral air. In the case of lightning, excess charges in the cloud and on the ground are great enough to remove electrons from the molecules in the air. The electrons and positively or negatively charged atoms form a plasma, which is a conductor. The discharge of Earth and the thundercloud by means of this conductor forms a luminous arc called lightning. In the case of your finger and the doorknob, the discharge is called a spark.

20.1 Section Review

1. **Charged Objects** After a comb is rubbed on a wool sweater, it is able to pick up small pieces of paper. Why does the comb lose that ability after a few minutes?

2. **Types of Charge** In the experiments described earlier in this section, how could you find out which strip of tape, B or T, is positively charged?

3. **Types of Charge** A pith ball is a small sphere made of a light material, such as plastic foam, often coated with a layer of graphite or aluminum paint. How could you determine whether a pith ball that is suspended from an insulating thread is neutral, is charged positively, or is charged negatively?

4. **Charge Separation** A rubber rod can be charged negatively when it is rubbed with wool. What happens to the charge of the wool? Why?

5. **Conservation of Charge** An apple contains trillions of charged particles. Why don't two apples repel each other when they are brought together?

6. **Charging a Conductor** Suppose you hang a long metal rod from silk threads so that the rod is isolated. You then touch a charged glass rod to one end of the metal rod. Describe the charges on the metal rod.

7. **Charging by Friction** You can charge a rubber rod negatively by rubbing it with wool. What happens when you rub a copper rod with wool?

8. **Critical Thinking** It once was proposed that electric charge is a type of fluid that flows from objects with an excess of the fluid to objects with a deficit. Why is the current two-charge model better than the single-fluid model?

Physics Online physicspp.com/self_check_quiz

20.2 Electric Force

► **Objectives**

- **Summarize** the relationships between electric forces, charges, and distance.
- **Explain** how to charge objects by conduction and induction.
- **Develop** a model of how charged objects can attract a neutral object.
- **Apply** Coulomb's law to problems in one and two dimensions.

► **Vocabulary**

electroscope
charging by conduction
charging by induction
grounding
Coulomb's law
coulomb
elementary charge

Electric forces must be strong because they can easily produce accelerations larger than the acceleration caused by gravity. You also have learned that they can be either repulsive or attractive, while gravitational forces always are attractive. Over the years, many scientists made attempts to measure electric forces. Daniel Bernoulli, best known for his work with fluids, made some crude measurements in 1760. In the 1770s, Henry Cavendish showed that electric forces must obey an inverse square force law, but being extremely shy, he did not publish his work. His manuscripts were discovered over a century later, after all his work had been duplicated by others.

Forces on Charged Bodies

The forces that you observed on tape strips also can be demonstrated by suspending a negatively charged, hard rubber rod so that it turns easily, as shown in **Figure 20-5.** If you bring another negatively charged rod near the suspended rod, the suspended rod will turn away. The negative charges on the rods repel each other. It is not necessary for the rods to make contact. The force, called the electric force, acts at a distance. If a positively charged glass rod is suspended and a similarly charged glass rod is brought close, the two positively charged rods also will repel each other. If a negatively charged rod is brought near a positively charged rod, however, the two will attract each other, and the suspended rod will turn toward the oppositely charged rod. The results of your tape experiments and these actions of charged rods can be summarized in the following way:

- There are two kinds of electric charges: positive and negative.
- Charges exert forces on other charges at a distance.
- The force is stronger when the charges are closer together.
- Like charges repel; opposite charges attract.

Neither a strip of tape nor a large rod that is hanging in open air is a very sensitive or convenient way of determining charge. Instead, a device called an electroscope is used. An **electroscope** consists of a metal knob connected by a metal stem to two thin, lightweight pieces of metal foil, called leaves. **Figure 20-6** shows a neutral electroscope. Note that the leaves hang loosely and are enclosed to eliminate stray air currents.

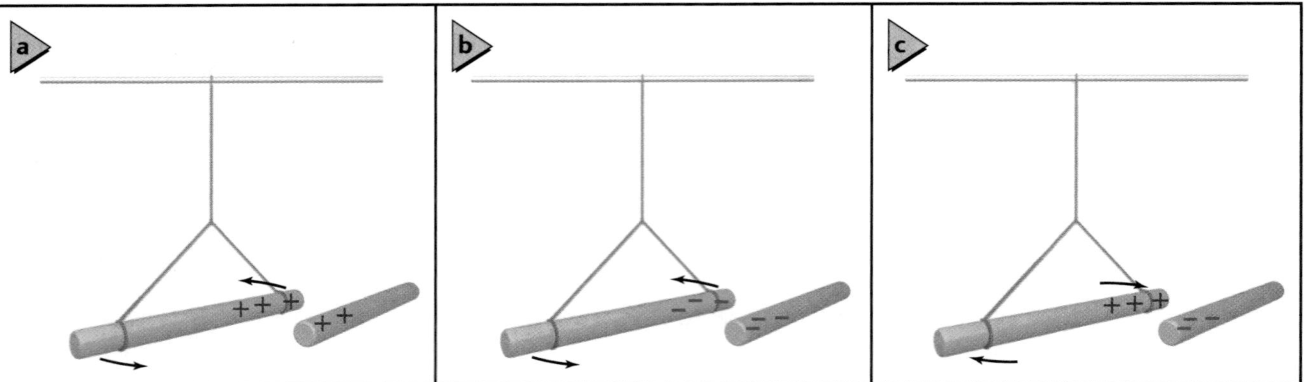

■ **Figure 20-5** A charged rod, when brought close to another charged and suspended rod, will attract or repel the suspended rod.

Charging by conduction When a negatively charged rod is touched to the knob of an electroscope, electrons are added to the knob. These charges spread over all the metal surfaces. As shown in **Figure 20-7a,** the two leaves are charged negatively and repel each other; therefore, they spread apart. The electroscope has been given a net charge. Charging a neutral body by touching it with a charged body is called **charging by conduction.** The leaves also will spread apart if the electroscope is charged positively. How, then, can you find out whether the electroscope is charged positively or negatively? The type of charge can be determined by observing the leaves when a rod of known charge is brought close to the knob. The leaves will spread farther apart if the rod and the electroscope have the same charge, as shown in **Figure 20-7b.** The leaves will fall slightly if the electroscope's charge is opposite that of the rod, as in **Figure 20-7c.**

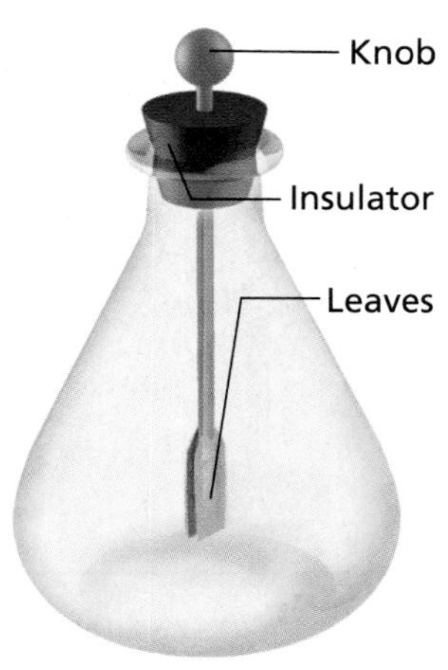

Figure 20-6 An electroscope is a device used for detecting charges. In a neutral electroscope, the leaves hang loosely, almost touching one another.

Separation of charge on neutral objects Earlier in this chapter, when you brought your finger near either charged strip of tape, the tape was attracted to your finger. Your finger, however, was neutral—it had equal amounts of positive and negative charge. You know that in conductors, charges can move easily, and that in the case of sparks, electric forces can change insulators into conductors. Given this information, you can develop a plausible model for the force that your finger exerted on the strips of tape.

Suppose you move your finger, or any uncharged object, close to a positively charged object. The negative charges in your finger will be attracted to the positively charged object, and the positive charges in your finger will be repelled. Your finger will remain neutral, but the positive and negative charges will be separated. The electric force is stronger for charges that are closer together; therefore, the separation results in an attractive force between your finger and the charged object. The force that a charged ruler exerts on neutral pieces of paper is the result of the same process, the separation of charges.

The negative charges at the bottom of thunderclouds also can cause charge separation in Earth. Positive charges in the ground are attracted to Earth's surface under the cloud. The forces of the charges in the cloud and those on Earth's surface can break molecules into positively and negatively charged particles. These charged particles are free to move, and they establish a conducting path from the ground to the cloud. The lightning that you observe occurs when a bolt travels at speeds on the order of 500,000 km/h along the conducting path and discharges the cloud.

Figure 20-7 A negatively charged electroscope will have its leaves spread apart **(a).** A negatively charged rod pushes electrons down to the leaves, causing them to spread farther apart **(b).** A positively charged rod attracts some of the electrons, causing the leaves to spread apart less **(c).**

Concepts In Motion
Interactive Figure To see an animation of change by induction, visit physicspp.com.

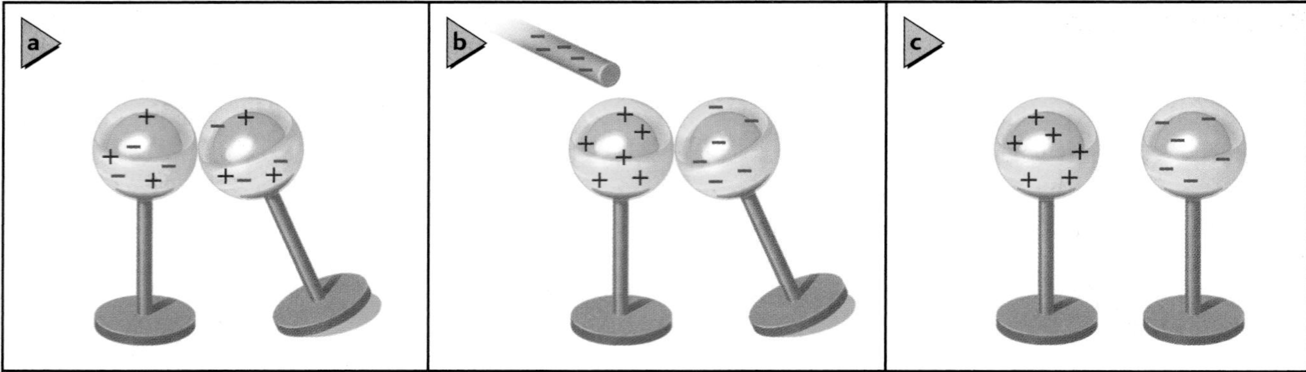

Figure 20-8 One method of charging by induction begins with neutral spheres that are touching **(a).** A charged rod is brought near them **(b),** then the spheres are separated and the charged rod is removed **(c).** The charges on the separated spheres are equal in magnitude, but opposite in sign.

Charging by induction Suppose that two identical, insulated metal spheres are touching, as shown in **Figure 20-8a.** When a negatively charged rod is brought close to one, as in **Figure 20-8b,** electrons from the first sphere will be forced onto the sphere farther from the rod and will make it negatively charged. The closer sphere is now positively charged. If the spheres are separated while the rod is nearby, each sphere will have a charge, and the charges will be equal but opposite, as shown in **Figure 20-8c.** This process of charging an object without touching it is called **charging by induction.**

A single object can be charged by induction through **grounding,** which is the process of connecting a body to Earth to eliminate excess charge. Earth is a very large sphere, and it can absorb great amounts of charge without becoming noticeably charged itself. If a charged body is touched to Earth, almost any amount of charge can flow to Earth.

If a negatively charged rod is brought close to the knob of an electroscope, as in **Figure 20-9a,** electrons are repelled onto the leaves. If the knob is then grounded on the side opposite the charged rod, electrons will be pushed from the electroscope into the ground until the leaves are neutral, as in **Figure 20-9b.** Removing the ground before the rod leaves the electroscope with a deficit of electrons, and it will be positively charged, as in **Figure 20-9c.** Grounding also can be used as a source of electrons. If a positive rod is brought near the knob of a grounded electroscope, electrons will be attracted from the ground, and the electroscope will obtain a negative charge. When this process is employed, the charge induced on the electroscope is opposite that of the object used to charge it. Because the rod never touches the electroscope, its charge is not transferred, and it can be used many times to charge objects by induction.

Figure 20-9 A negatively charged rod induces a separation of charges in an electroscope **(a).** The electroscope is grounded, and negative charges are pushed from the electroscope to the ground **(b).** The ground is removed before the rod, and the electroscope is left with a positive charge **(c).**

Coulomb's Law

You have seen that a force acts between two or more charged objects. In your experiments with tape, you found that the force depends on distance. The closer you brought the charged comb to the tape, the stronger the force was. You also found that the more you charged the comb, the stronger the force was. How can you vary the quantity of charge in a controlled way? This problem was solved in 1785 by French physicist Charles Coulomb. The type of apparatus used by Coulomb is shown in **Figure 20-10.** An insulating rod with small conducting spheres, A and A′, at each end was suspended by a thin wire. A similar sphere, B, was placed in contact with sphere A. When they were touched with a charged object, the charge spread evenly over the two spheres. Because they were the same size, they received equal amounts of charge. The symbol for charge is q. Therefore, the amount of charge on the spheres can be represented by the notation q_A and q_B.

Force depends on distance Coulomb found how the force between the two charged spheres depended on the distance. First, he carefully measured the amount of force needed to twist the suspending wire through a given angle. He then placed equal charges on spheres A and B and varied the distance, r, between them. The force moved A, which twisted the suspending wire. By measuring the deflection of A, Coulomb could calculate the force of repulsion. He showed that the force, F, varied inversely with the square of the distance between the centers of the spheres.

$$F \propto \frac{1}{r^2}$$

Force depends on charge To investigate the way in which the force depended on the amount of charge, Coulomb had to change the charges on the spheres in a measured way. He first charged spheres A and B equally, as before. Then he selected an uncharged sphere, C, of the same size as sphere B. When C was placed in contact with B, the spheres shared the charge that had been on B alone. Because the two were the same size, B then had only half of its original charge. Therefore, the charge on B was only one-half the charge on A. After Coulomb adjusted the position of B so that the distance, r, between A and B was the same as before, he found that the force between A and B was half of its former value. That is, he found that the force varied directly with the charge of the bodies.

$$F \propto q_A q_B$$

After many similar measurements, Coulomb summarized the results in a law now known as **Coulomb's law:** the magnitude of the force between charge q_A and charge q_B, separated by a distance r, is proportional to the magnitude of the charges and inversely proportional to the square of the distance between them.

$$F \propto \frac{q_A q_B}{r^2}$$

The unit of charge: the coulomb The amount of charge that an object has is difficult to measure directly. Coulomb's experiments, however, showed that the quantity of charge could be related to force. Thus, Coulomb could define a standard quantity of charge in terms of the amount of force that it produces. The SI standard unit of charge is called the **coulomb** (C).

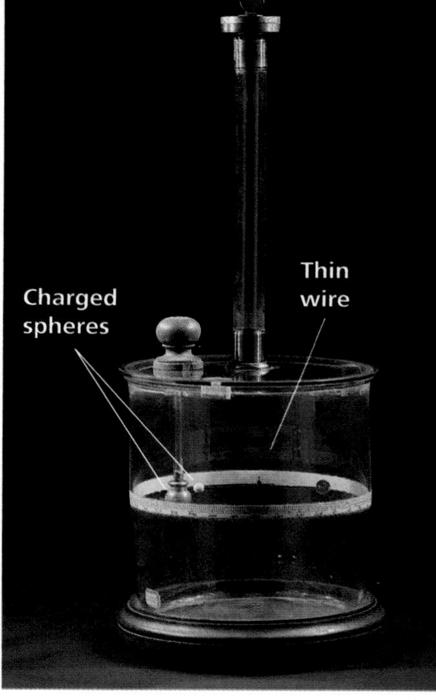

Figure 20-10 Coulomb used a similar type of apparatus to measure the force between two spheres, A and B. He observed the deflection of A while varying the distance between A and B.

MINI LAB

Investigating Induction and Conduction

Use a balloon and an electroscope to investigate charging by induction and charging by conduction.

1. **Predict** what will happen if you charge a balloon by rubbing it with wool and bring it near a neutral electroscope.
2. **Predict** what will happen if you touch the balloon to the electroscope.
3. **Test** your predictions.

Analyze and Conclude

4. **Describe** your results.
5. **Explain** the movements of the leaves in each step of the experiment. Include diagrams.
6. **Describe** the results if the wool had been used to charge the electroscope.

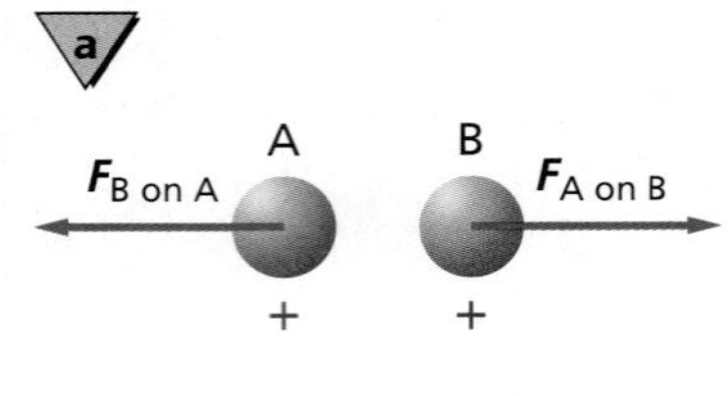

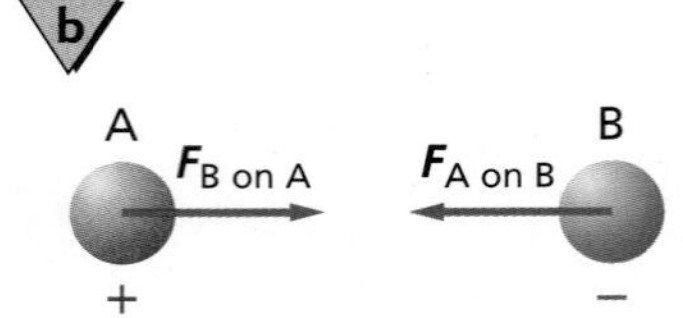

Figure 20-11 The rule for determining the direction of force is: like charges repel; unlike charges attract.

One coulomb is the charge of 6.24×10^{18} electrons or protons. A typical lightning bolt can carry 5 C to 25 C of charge. The charge on a single electron is 1.60×10^{-19} C. The magnitude of the charge of an electron is called the **elementary charge.** Even small pieces of matter, such as coins, contain up to 10^6 C of negative charge. This enormous amount of negative charge produces almost no external effects because it is balanced by an equal amount of positive charge. If the charge is unbalanced, even as small a charge as 10^{-9} C can result in large forces.

According to Coulomb's law, the magnitude of the force on charge q_A caused by charge q_B a distance r away can be written as follows.

Coulomb's Law $F = K\frac{q_A q_B}{r^2}$

The force between two charges is equal to Coulomb's constant, times the product of the two charges, divided by the square of the distance between them.

When the charges are measured in coulombs, the distance in meters, and the force in newtons, the constant, K, is 9.0×10^9 N·m²/C².

The Coulomb's law equation gives the magnitude of the force that charge q_A exerts on q_B and also the force that q_B exerts on q_A. These two forces are equal in magnitude but opposite in direction. You can observe this example of Newton's third law of motion in action when you bring two strips of tape with like charges together. Each exerts forces on the other. If you bring a charged comb near either strip of tape, the strip, with its small mass, moves readily. The acceleration of the comb and you is, of course, much less because of the much greater mass.

The electric force, like all other forces, is a vector quantity. Force vectors need both a magnitude and a direction. However, the Coulomb's law equation above gives only the magnitude of the force. To determine the direction, you need to draw a diagram and interpret charge relations carefully. If two positively charged objects, A and B, are brought near, the forces they exert on each other are repulsive, as shown in **Figure 20-11a.** If, instead, B is negatively charged, the forces are attractive, as shown in **Figure 20-11b.**

PROBLEM-SOLVING **Strategies**

Electric Force Problems

Use these steps to find the magnitude and direction of the force between charges.

1. Sketch the system showing all distances and angles to scale.
2. Diagram the vectors of the system.
3. Use Coulomb's law to find the magnitude of the force.
4. Use your diagram along with trigonometric relations to find the direction of the force.
5. Perform all algebraic operations on both the numbers and the units. Make sure that the units match the variables in question.
6. Consider the magnitude of your answer. Is it reasonable?

▶ EXAMPLE Problem 1

Coulomb's Law in Two Dimensions **Sphere A, with a charge of $+6.0\ \mu C$, is located near another charged sphere, B. Sphere B has a charge of $-3.0\ \mu C$ and is located 4.0 cm to the right of A.**

a. What is the force of sphere B on sphere A?

b. A third sphere, C, with a $+1.5$-μC charge, is added to the configuration. If it is located 3.0 cm directly beneath A, what is the new net force on sphere A?

1 Analyze and Sketch the Problem

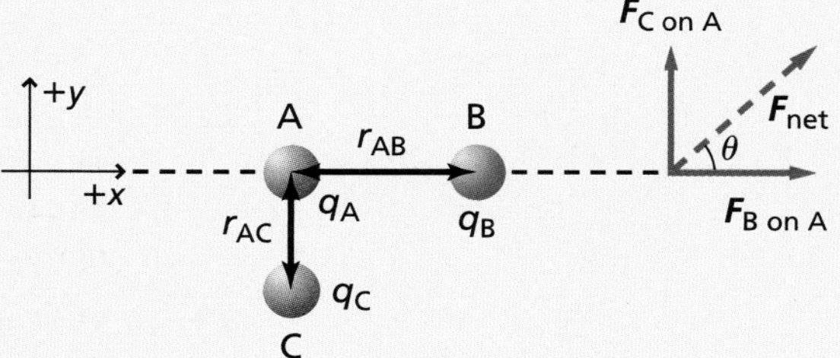

- Establish coordinate axes and sketch the spheres.
- Show and label the distances between the spheres.
- Diagram and label the force vectors.

Known:		Unknown:
$q_A = +6.0\ \mu C$	$r_{AB} = 4.0$ cm	$F_{B\ on\ A} = ?$
$q_B = -3.0\ \mu C$	$r_{AC} = 3.0$ cm	$F_{C\ on\ A} = ?$
$q_C = +1.5\ \mu C$		$F_{net} = ?$

Personal Tutor For an online tutorial on sine, cosine, and tangent, visit physicspp.com.

2 Solve for the Unknown

a. Find the force of sphere B on sphere A.

$$F_{B\ on\ A} = K\frac{q_A q_B}{r_{AB}^2}$$

$$= (9.0\times10^9\ N\cdot m^2/C^2)\frac{(6.0\times10^{-6}\ C)(3.0\times10^{-6}\ C)}{(4.0\times10^{-2}\ m)^2}$$

Substitute $q_A = 6.0\ \mu C$, $q_B = 3.0\ \mu C$, $r_{AB} = 4.0$ cm

$$= 1.0\times10^2\ N$$

Because spheres A and B have unlike charges, the force of B on A is to the right.

b. Find the force of sphere C on sphere A.

$$F_{C\ on\ A} = K\frac{q_A q_C}{r_{AC}^2}$$

$$= (9.0\times10^9\ N\cdot m^2/C^2)\frac{(6.0\times10^{-6}\ C)(1.5\times10^{-6}\ C)}{(3.0\times10^{-2}\ m)^2}$$

Substitute $q_A = 6.0\ \mu C$, $q_C = 1.5\ \mu C$, $r_{AC} = 3.0$ cm

$$= 9.0\times10^1\ N$$

Spheres A and C have like charges, which repel. The force of C on A is upward.

Find the vector sum of $\boldsymbol{F}_{B\ on\ A}$ and $\boldsymbol{F}_{C\ on\ A}$ to find $\boldsymbol{F}_{net}$ on sphere A.

$$F_{net} = \sqrt{F_{B\ on\ A}^2 + F_{C\ on\ A}^2}$$

$$= \sqrt{(1.0\times10^2\ N)^2 + (9.0\times10^1\ N)^2}$$

Substitute $F_{B\ on\ A} = 1.0\times10^2$ N, $F_{C\ on\ A} = 9.0\times10^1$ N

$$= 130\ N$$

$$\tan\theta = \frac{F_{C\ on\ A}}{F_{B\ on\ A}}$$

$$\theta = \tan^{-1}\left(\frac{F_{C\ on\ A}}{F_{B\ on\ A}}\right)$$

$$= \tan^{-1}\left(\frac{9.0\times10^1\ N}{1.0\times10^2\ N}\right)$$

Substitute $F_{C\ on\ A} = 9.0\times10^1$ N, $F_{B\ on\ A} = 1.0\times10^2$ N

$$= 42°$$

$$\boldsymbol{F}_{net} = 130\ N,\ 42° \text{ above the } x\text{-axis}$$

3 Evaluate the Answer

- **Are the units correct?** $(N\cdot m^2/C^2)(C)(C)/m^2 = N$. The units work out to be newtons.
- **Does the direction make sense?** Like charges repel; unlike charges attract.
- **Is the magnitude realistic?** The magnitude of the net force is in agreement with the magnitudes of the component forces.

PRACTICE Problems

- Additional Problems, Appendix B
- Solutions to Selected Problems, Appendix C

9. A negative charge of -2.0×10^{-4} C and a positive charge of 8.0×10^{-4} C are separated by 0.30 m. What is the force between the two charges?
10. A negative charge of -6.0×10^{-6} C exerts an attractive force of 65 N on a second charge that is 0.050 m away. What is the magnitude of the second charge?
11. The charge on B in Example Problem 1 is replaced by a charge of $+3.00\ \mu$C. Diagram the new situation and find the net force on A.
12. Sphere A is located at the origin and has a charge of $+2.0\times10^{-6}$ C. Sphere B is located at +0.60 m on the *x*-axis and has a charge of -3.6×10^{-6} C. Sphere C is located at +0.80 m on the *x*-axis and has a charge of $+4.0\times10^{-6}$ C. Determine the net force on sphere A.
13. Determine the net force on sphere B in the previous problem.

As you use the Coulomb's law equation, keep in mind that Coulomb's law is valid only for point charges or uniform spherical charge distributions. That is, a charged sphere may be treated as if all the charge were located at its center if the charge is spread evenly across its entire surface or throughout its volume. If a sphere is a conductor and another charge is brought near it, the charges on the sphere will be attracted or repelled, and the charge no longer will act as if it were at the sphere's center. Therefore, it is important to consider how large and how far apart two charged spheres are before applying Coulomb's law. The problems in this textbook assume that charged spheres are small enough and far enough apart to be considered point charges unless otherwise noted. When shapes such as long wires or flat plates are considered, Coulomb's law must be modified to account for the nonpoint charge distributions.

Application of Electrostatic Forces

There are many applications of electric forces on particles. For example, these forces can collect soot in smokestacks, thereby reducing air pollution, as shown in **Figure 20-12.** Tiny paint droplets, charged by

CHALLENGE PROBLEM

As shown in the figure on the right, two spheres of equal mass, *m*, and equal positive charge, *q*, are a distance, *r*, apart.

1. Derive an expression for the charge, *q*, that must be on each sphere so that the spheres are in equilibrium; that is, so that the attractive and repulsive forces between them are balanced.
2. If the distance between the spheres is doubled, how will that affect the expression for the value of *q* that you determined in the previous problem? Explain.
3. If the mass of each sphere is 1.50 kg, determine the charge on each sphere needed to maintain the equilibrium.

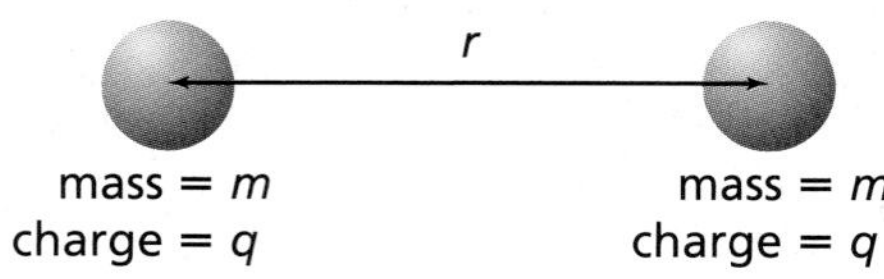

Figure 20-12 The fly ash being released by these smokestacks is a by-product of burning coal. Static-electricity precipitators can be used to reduce fly ash emissions.

induction, can be used to paint automobiles and other objects very uniformly. Photocopy machines use static electricity to place black toner on a page so that a precise reproduction of the original document is made. In other instances, applications are concerned with the control of static charge. For example, static charge can ruin film if it attracts dust, and electronic equipment can be damaged by the discharge of static charge. In these cases, applications are designed to avoid the buildup of static charge and to safely eliminate any charge that does build up.

20.2 Section Review

14. Force and Charge How are electric force and charge related? Describe the force when the charges are like charges and the force when the charges are opposite charges.

15. Force and Distance How are electric force and distance related? How would the force change if the distance between two charges were tripled?

16. Electroscopes When an electroscope is charged, the leaves rise to a certain angle and remain at that angle. Why do they not rise farther?

17. Charging an Electroscope Explain how to charge an electroscope positively using

a. a positive rod.

b. a negative rod.

18. Attraction of Neutral Objects What two properties explain why a neutral object is attracted to both positively and negatively charged objects?

19. Charging by Induction In an electroscope being charged by induction, what happens when the charging rod is moved away before the ground is removed from the knob?

20. Electric Forces Two charged spheres are held a distance, r, apart. One sphere has a charge of $+3\mu C$, and the other sphere has a charge of $+9\mu C$. Compare the force of the $+3\mu C$ sphere on the $+9\mu C$ sphere with the force of the $+9\mu C$ sphere on the $+3\mu C$ sphere.

21. Critical Thinking Suppose that you are testing Coulomb's law using a small, positively charged plastic sphere and a large, positively charged metal sphere. According to Coulomb's law, the force depends on $1/r^2$, where r is the distance between the centers of the spheres. As the spheres get close together, the force is smaller than expected from Coulomb's law. Explain.

Physics Online physicspp.com/self_check_quiz

PHYSICS LAB • Design Your Own

Charged Objects

In this chapter, you observed and studied phenomena that result from the separation of electric charges. You learned that hard rubber and plastic tend to become negatively charged when they are rubbed, while glass and wool tend to become positively charged. But what happens if two objects that tend to become negatively charged are rubbed together? Will electrons be transferred? If so, which material will gain electrons, and which will lose them? In this physics lab, you will design a procedure to further your investigations of positive and negative charges.

QUESTION

How can you test materials for their ability to hold positive and negative charges?

Objectives

- **Observe** that different materials tend to become positively or negatively charged.
- **Compare and contrast** the ability of materials to acquire and hold positive and negative charges.
- **Interpret data** to order a list of materials from strongest tendency to be negatively charged to strongest tendency to be positively charged.

Safety Precautions

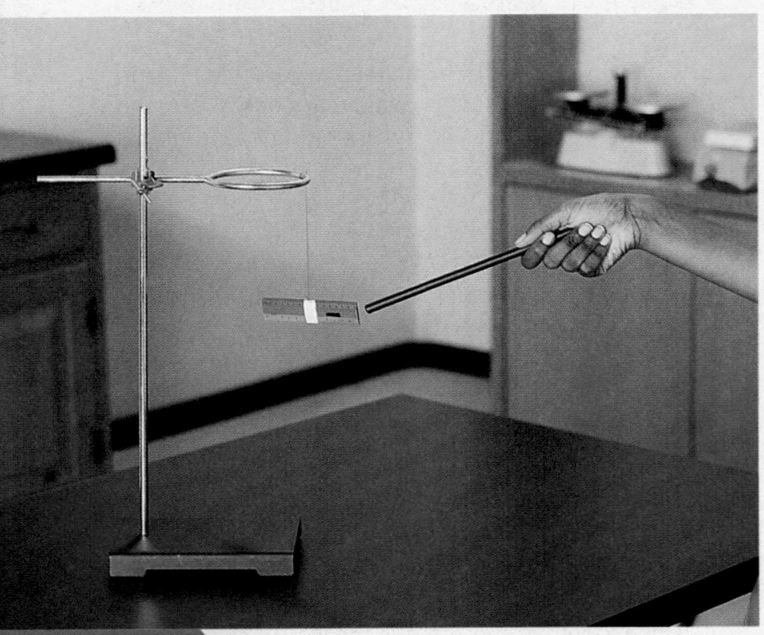

Materials

15-cm plastic ruler
thread
ring stand with ring
masking tape
materials to be charged, such as rubber rods, plastic rods, glass rods, PVC pipe, copper pipe, steel pipe, pencils, pens, wool, silk, plastic wrap, plastic sandwich bags, waxed paper, and aluminum foil

Procedure

1. Use the lab photo as a guide to suspend a 15-cm plastic ruler. It is advisable to wash the ruler in soapy water, then rinse and dry it thoroughly before each use, especially if it is a humid day. The thread should be attached at the midpoint of the ruler with two or three wraps of masking tape between the thread and ruler.
2. Use the following situations as a reference for types of charges a material can have: 1) a plastic ruler rubbed with wool gives the plastic ruler an excess negative charge and the wool an excess positive charge, and 2) a plastic ruler rubbed with plastic wrap gives the plastic ruler an excess positive charge and the plastic wrap an excess negative charge.

Data Table

Material 1	Material 2	Charge on Ruler (+, −, 0)	Observation of Ruler's Movements	Charge on Material 1 (+, −, 0)	Charge on Material 2 (+, −, 0)

3. Design a procedure to test which objects tend to become negatively charged and which tend to become positively charged. Try various combinations of materials and record your observations in the data table.
4. Develop a test to see if an object is neutral. Remember that a charged ruler may be attracted to a neutral object if it induces a separation of charge in the neutral object.
5. Be sure to check with your teacher and have your procedure approved before you proceed with your lab.

Analyze

1. **Observe and Infer** As you brought charged materials together, could you detect a force between the charged materials? Describe this force.
2. **Formulate Models** Make a drawing of the charge distribution on the two materials for one of your trials. Use this drawing to explain why the materials acted the way they did during your experiments with them.
3. **Draw Conclusions** Which materials hold an excess charge? Which materials do not hold a charge very well?
4. **Draw Conclusions** Which materials tend to become negatively charged? Which tend to become positively charged?
5. **Interpret Data** Use your data table to list the relative tendencies of materials to be positively or negatively charged.

Conclude and Apply

1. Explain what is meant by the phrases *excess charge* and *charge imbalance* when referring to static electricity.
2. Does excess charge remain on a material or does it dissipate over time?
3. Could you complete this physics lab using a metal rod in place of the suspended plastic ruler? Explain.
4. Clear plastic wrap seals containers of food. Why does plastic wrap cling to itself after it is pulled from its container?

Going Further

Review the information in your textbook about electroscopes. Redesign the lab using an electroscope, rather than a suspended ruler, to test for the type of charge on an object.

Real-World Physics

Trucks often have a rubber strap or a chain that drags along the road. Why are they used?

To find out more about static charge, visit the Web site: physicspp.com

Future Technology

Spacecraft and Static Electricity

Most objects on Earth do not build up substantial static-electric charges because a layer of moisture clings to surfaces, allowing charges to migrate to or from the ground. As you learned in this chapter, Earth can absorb almost any amount of charge. However, there is no moisture in space, and Earth is far away. Charged particles ejected from the Sun, or in the ionosphere, strike and cling to spacecraft, charging their surfaces to thousands of volts.

Plasma and Charging

In Chapter 13, you learned that plasma consists of free electrons and positive ions. Orbiting spacecraft are surrounded by a thin cloud of this plasma. The electrons in plasma can move far more easily than more massive positive ions. Thus, spacecraft surfaces tend to attract electrons and develop a negative charge. This negative charge eventually attracts some heavy positive ions, which strike the spacecraft and can damage its surface.

On the *International Space Station,* an additional difficulty stems from the array of solar panels that convert energy from the Sun into electricity. When the arrays are powering the space station, the voltage on the surface of the craft tends to be close to the voltage of the solar array. As a result, it is possible that an electric arc could form between the space station and the plasma that surrounds it.

Plasma Contactor Unit (PCU)

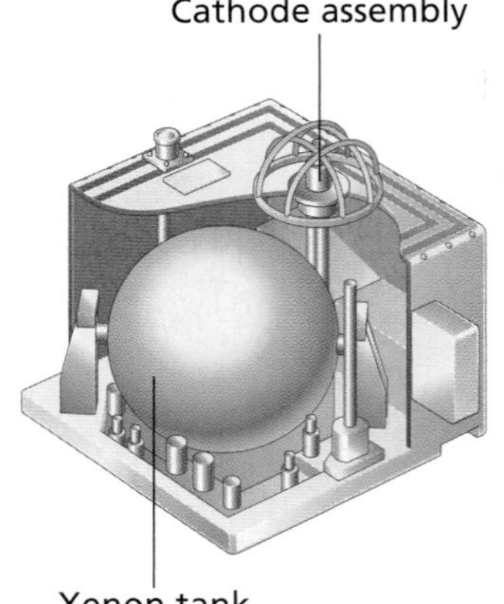

This is a cutaway drawing of the PCU.

Consequences of an Arc Arcs are extremely hot and carry a great deal of current. They can prematurely ignite retro-rockets or explosive bolts and interfere with the operation of the spacecraft's electronic equipment. The solar panels are particularly susceptible to arc damage. In addition to damage to the spacecraft's components, there is a remote chance that the buildup of charge might endanger astronauts on space walks.

To discharge the potential difference and protect craft and crew, the space station's skin must be connected by a conductor, called a plasma contactor, to the plasma cloud surrounding it. The connection begins on board the station, where a stream of xenon gas from a tank in the Plasma Contactor Unit (PCU) is ionized by an electric current. This ionization takes place in the cathode assembly. The ionized xenon, now in the plasma state, passes out of the craft through the cathode assembly. It is this stream of conductive plasma that connects the craft to the surrounding plasma cloud, thereby reducing the potential difference to safe levels.

Future Applications

Future spacecraft might integrate the plasma contactor into the propulsion system. For example, the Variable Specific Impulse Magnetoplasma Rocket (VASIMR) could use the plasma exhaust that it produces to provide an electric connection between the spacecraft and the surrounding plasma. Scientists think that this type of rocket could be used in the future to travel between planets.

Going Further

1. **Apply** What is the purpose of a plasma contactor? How is it similar to using your finger to ground an electroscope?
2. **Research** How could scientists assess the charge on the surface of the *International Space Station?*

Chapter 20 Study Guide

20.1 Electric Charge

Vocabulary

- electrostatics (p. 541)
- neutral (p. 543)
- insulator (p. 544)
- conductor (p. 544)

Key Concepts

- There are two kinds of electric charge, positive and negative. Interactions of these charges explain the attraction and repulsion that you observed in the strips of tape.
- Electric charge is not created or destroyed; it is conserved. Charging is the separation, not creation, of electric charges.
- Objects can be charged by the transfer of electrons. An area with excess electrons has a net negative charge; an area with a deficit of electrons has a net positive charge.
- Charges added to one part of an insulator remain on that part. Insulators include glass, dry wood, plastics, and dry air.
- Charges added to a conductor quickly spread over the surface of the object. In general, examples of conductors include graphite, metals, and matter in the plasma state.
- Under certain conditions, charges can move through a substance that is ordinarily an insulator. Lightning moving through air is one example.

20.2 Electric Force

Vocabulary

- electroscope (p. 546)
- charging by conduction (p. 547)
- charging by induction (p. 548)
- grounding (p. 548)
- Coulomb's law (p. 549)
- coulomb (p. 549)
- elementary charge (p. 550)

Key Concepts

- When an electroscope is charged, electric forces cause its thin metal leaves to spread apart.
- An object can be charged by conduction by touching it with a charged object.
- A charged object will induce a separation of charges within a neutral conductor. This process will result in an attractive force between the charged object and the neutral conductor.
- To charge a conductor by induction, a charged object is first brought near it, causing a separation of charges. Then, the conductor to be charged is separated, trapping opposite charges on the two halves.
- Grounding is the removal of excess charge by touching an object to Earth. Grounding can be used in the process of charging an electroscope by induction.
- Coulomb's law states that the force between two charged particles varies directly with the product of their charges and inversely with the square of the distance between them.

  $$F = K\frac{q_A q_B}{r^2}$$

  To determine the direction of the force, remember the following rule: like charges repel; unlike charges attract.
- The SI unit of charge is the coulomb. One coulomb (C) is the magnitude of the charge of 6.24×10^{18} electrons or protons. The elementary charge, the charge of a proton or electron, is 1.60×10^{-19} C.

Chapter 20 Assessment

Concept Mapping

22. Complete the concept map below using the following terms: *conduction, distance, elementary charge.*

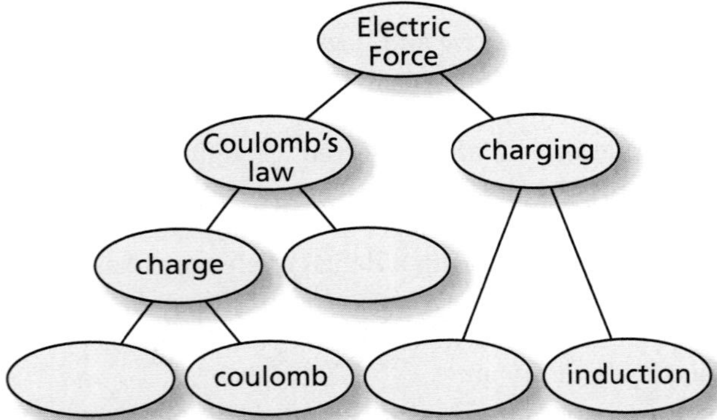

Mastering Concepts

23. If you comb your hair on a dry day, the comb can become positively charged. Can your hair remain neutral? Explain. (20.1)

24. List some insulators and conductors. (20.1)

25. What property makes metal a good conductor and rubber a good insulator? (20.1)

26. Laundry Why do socks taken from a clothes dryer sometimes cling to other clothes? (20.2)

27. Compact Discs If you wipe a compact disc with a clean cloth, why does the CD then attract dust? (20.2)

28. Coins The combined charge of all electrons in a nickel is hundreds of thousands of coulombs. Does this imply anything about the net charge on the coin? Explain. (20.2)

29. How does the distance between two charges impact the force between them? If the distance is decreased while the charges remain the same, what happens to the force? (20.2)

30. Explain how to charge a conductor negatively if you have only a positively charged rod. (20.2)

Applying Concepts

31. How does the charge of an electron differ from the charge of a proton? How are they similar?

32. Using a charged rod and an electroscope, how can you find whether or not an object is a conductor?

33. A charged rod is brought near a pile of tiny plastic spheres. Some of the spheres are attracted to the rod, but as soon as they touch the rod, they are flung off in different directions. Explain.

34. Lightning Lightning usually occurs when a negative charge in a cloud is transported to Earth. If Earth is neutral, what provides the attractive force that pulls the electrons toward Earth?

35. Explain what happens to the leaves of a positively charged electroscope when rods with the following charges are brought close to, but not touching, the electroscope.
  a. positive
  b. negative

36. As shown in **Figure 20-13,** Coulomb's law and Newton's law of universal gravitation appear to be similar. In what ways are the electric and gravitational forces similar? How are they different?

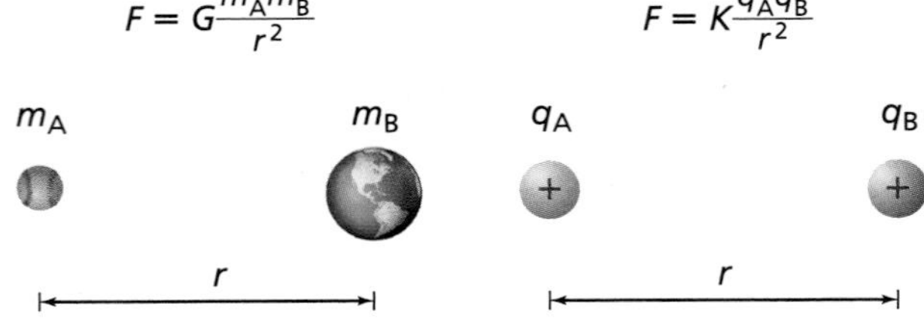

Figure 20-13 (Not to scale)

37. The constant, K, in Coulomb's equation is much larger than the constant, G, in the universal gravitation equation. Of what significance is this?

38. The text describes Coulomb's method for charging two spheres, A and B, so that the charge on B was exactly half the charge on A. Suggest a way that Coulomb could have placed a charge on sphere B that was exactly one-third the charge on sphere A.

39. Coulomb measured the deflection of sphere A when spheres A and B had equal charges and were a distance, r, apart. He then made the charge on B one-third the charge on A. How far apart would the two spheres then have had to be for A to have had the same deflection that it had before?

40. Two charged bodies exert a force of 0.145 N on each other. If they are moved so that they are one-fourth as far apart, what force is exerted?

41. Electric forces between charges are enormous in comparison to gravitational forces. Yet, we normally do not sense electric forces between us and our surroundings, while we do sense gravitational interactions with Earth. Explain.

 For more problems, go to Additional Problems, Appendix B.

Mastering Problems

20.2 Electric Force

42. Two charges, q_A and q_B, are separated by a distance, r, and exert a force, F, on each other. Analyze Coulomb's law and identify what new force would exist under the following conditions.
a. q_A is doubled
b. q_A and q_B are cut in half
c. r is tripled
d. r is cut in half
e. q_A is tripled and r is doubled

43. Lightning A strong lightning bolt transfers about 25 C to Earth. How many electrons are transferred?

44. Atoms Two electrons in an atom are separated by 1.5×10^{-10} m, the typical size of an atom. What is the electric force between them?

45. A positive and a negative charge, each of magnitude 2.5×10^{-5} C, are separated by a distance of 15 cm. Find the force on each of the particles.

46. A force of 2.4×10^{2} N exists between a positive charge of 8.0×10^{-5} C and a positive charge of 3.0×10^{-5} C. What distance separates the charges?

47. Two identical positive charges exert a repulsive force of 6.4×10^{-9} N when separated by a distance of 3.8×10^{-10} m. Calculate the charge of each.

48. A positive charge of 3.0 μC is pulled on by two negative charges. As shown in **Figure 20-14,** one negative charge, -2.0 μC, is 0.050 m to the west, and the other, -4.0 μC, is 0.030 m to the east. What total force is exerted on the positive charge?

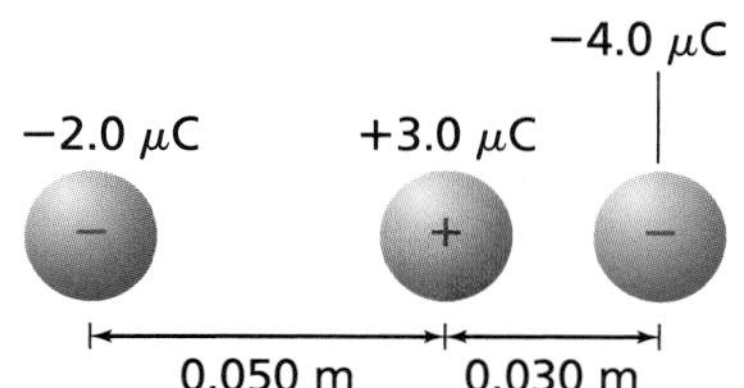

Figure 20-14

49. Figure 20-15 shows two positively charged spheres, one with three times the charge of the other. The spheres are 16 cm apart, and the force between them is 0.28 N. What are the charges on the two spheres?

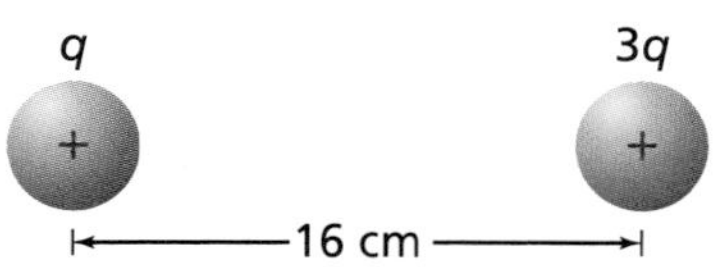

Figure 20-15

50. Charge in a Coin How many coulombs of charge are on the electrons in a nickel? Use the following method to find the answer.
a. Find the number of atoms in a nickel. A nickel has a mass of about 5 g. A nickel is 75 percent Cu and 25 percent Ni, so each mole of the coin's atoms will have a mass of about 62 g.
b. Find the number of electrons in the coin. On average, each atom has 28.75 electrons.
c. Find the coulombs on the electrons.

51. Three particles are placed in a line. The left particle has a charge of -55 μC, the middle one has a charge of $+45$ μC, and the right one has a charge of -78 μC. The middle particle is 72 cm from each of the others, as shown in **Figure 20-16.**
a. Find the net force on the middle particle.
b. Find the net force on the right particle.

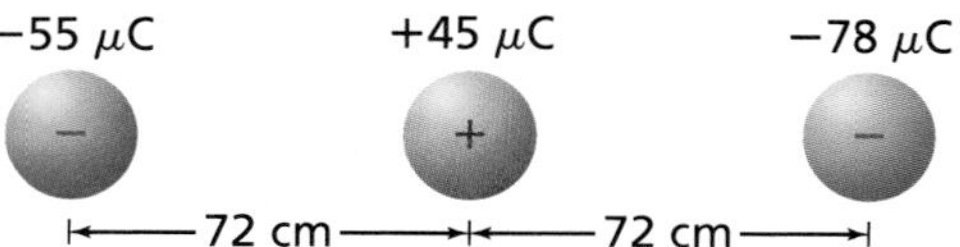

Figure 20-16

Mixed Review

52. A small metal sphere with charge 1.2×10^{-5} C is touched to an identical neutral sphere and then placed 0.15 m from the second sphere. What is the electric force between the two spheres?

53. Atoms What is the electric force between an electron and a proton placed 5.3×10^{-11} m apart, the approximate radius of a hydrogen atom?

54. A small sphere of charge 2.4 μC experiences a force of 0.36 N when a second sphere of unknown charge is placed 5.5 cm from it. What is the charge of the second sphere?

55. Two identically charged spheres placed 12 cm apart have an electric force of 0.28 N between them. What is the charge of each sphere?

56. In an experiment using Coulomb's apparatus, a sphere with a charge of 3.6×10^{-8} C is 1.4 cm from a second sphere of unknown charge. The force between the spheres is 2.7×10^{-2} N. What is the charge of the second sphere?

57. The force between a proton and an electron is 3.5×10^{-10} N. What is the distance between these two particles?

Thinking Critically

58. Apply Concepts Calculate the ratio of the electric force to the gravitational force between the electron and the proton in a hydrogen atom.

59. Analyze and Conclude Sphere A, with a charge of +64 μC, is positioned at the origin. A second sphere, B, with a charge of −16 μC, is placed at +1.00 m on the x-axis.

a. Where must a third sphere, C, of charge +12 μC be placed so there is no net force on it?

b. If the third sphere had a charge of +6 μC, where should it be placed?

c. If the third sphere had a charge of −12 μC, where should it be placed?

60. Three charged spheres are located at the positions shown in **Figure 20-17.** Find the total force on sphere B.

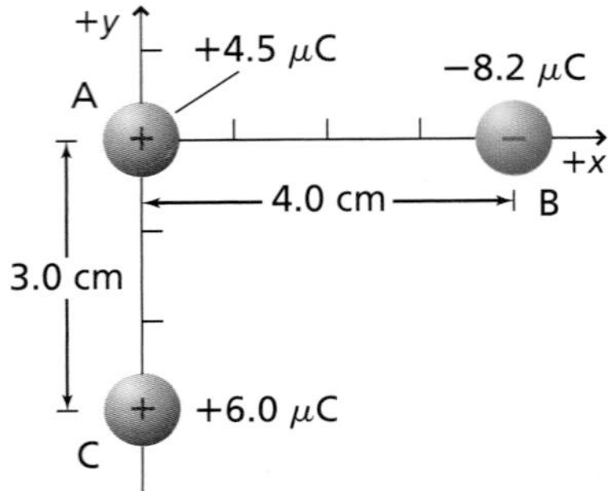

Figure 20-17

61. The two pith balls in **Figure 20-18** each have a mass of 1.0 g and an equal charge. One pith ball is suspended by an insulating thread. The other is brought to 3.0 cm from the suspended ball. The suspended ball is now hanging with the thread forming an angle of 30.0° with the vertical. The ball is in equilibrium with F_E, F_g, and F_T. Calculate each of the following.

a. F_g on the suspended ball

b. F_E

c. the charge on the balls

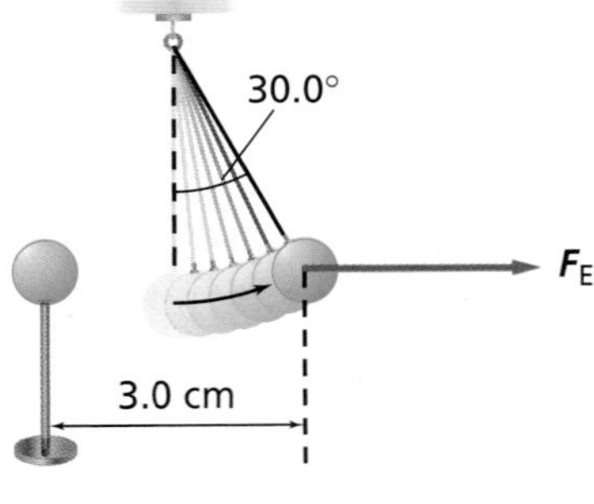

Figure 20-18

62. Two charges, q_A and q_B, are at rest near a positive test charge, q_T, of 7.2 μC. The first charge, q_A, is a positive charge of 3.6 μC located 2.5 cm away from q_T at 35°; q_B is a negative charge of −6.6 μC located 6.8 cm away at 125°.

a. Determine the magnitude of each of the forces acting on q_T.

b. Sketch a force diagram.

c. Graphically determine the resultant force acting on q_T.

Writing in Physics

63. History of Science Research several devices that were used in the seventeenth and eighteenth centuries to study static electricity. Examples that you might consider include the Leyden jar and the Wimshurst machine. Discuss how they were constructed and how they worked.

64. In Chapter 13, you learned that forces exist between water molecules that cause water to be denser as a liquid between 0°C and 4°C than as a solid at 0°C. These forces are electrostatic in nature. Research electrostatic intermolecular forces, such as van der Waals forces and dipole-dipole forces, and describe their effects on matter.

Cumulative Review

65. Explain how a pendulum can be used to determine the acceleration of gravity. (Chapter 14)

66. A submarine that is moving 12.0 m/s sends a sonar ping of frequency 1.50×10^3 Hz toward a seamount that is directly in front of the submarine. It receives the echo 1.800 s later. (Chapter 15)

a. How far is the submarine from the seamount?

b. What is the frequency of the sonar wave that strikes the seamount?

c. What is the frequency of the echo received by the submarine?

67. Security Mirror A security mirror is used to produce an image that is three-fourths the size of an object and is located 12.0 cm behind the mirror. What is the focal length of the mirror? (Chapter 17)

68. A 2.00-cm-tall object is located 20.0 cm away from a diverging lens with a focal length of 24.0 cm. What are the image position, height, and orientation? Is this a real or a virtual image? (Chapter 18)

69. Spectrometer A spectrometer contains a grating of 11,500 slits/cm. Find the angle at which light of wavelength 527 nm has a first-order bright band. (Chapter 19)

 For more problems, go to Additional Problems, Appendix B.

Stan

Multiple Choice

1. How many electrons have been removed from a positively charged electroscope if it has a net charge of 7.5×10^{-11} C?

   Ⓐ 7.5×10^{-11} electrons

   Ⓑ 2.1×10^{-9} electrons

   Ⓒ 1.2×10^{8} electrons

   Ⓓ 4.7×10^{8} electrons

2. The force exerted on a particle with a charge of 5.0×10^{-9} C by a second particle that is 4 cm away is 8.4×10^{-5} N. What is the charge of the second particle?

   Ⓐ 4.2×10^{-13} C   Ⓒ 3.0×10^{-9} C

   Ⓑ 2.0×10^{-9} C   Ⓓ 6.0×10^{-5} C

3. Three charges, A, B, and C, are located in a line, as shown below. What is the net force on charge B?

   Ⓐ 78 N toward A   Ⓒ 130 N toward A

   Ⓑ 78 N toward C   Ⓓ 210 N toward C

   $+8.5\times10^{-6}$ C $+3.1\times10^{-6}$ C $+6.4\times10^{-6}$ C

   A   B   C

   4.2 cm   2.9 cm

4. What is the charge on an electroscope that has an excess of 4.8×10^{10} electrons?

   Ⓐ 3.3×10^{-30} C   Ⓒ 7.7×10^{-9} C

   Ⓑ 4.8×10^{-10} C   Ⓓ 4.8×10^{10} C

5. Two charged bodies exert a force of 86 N on each other. If they are moved so that they are six times farther apart, what is the new force that they will exert on each other?

   Ⓐ 2.4 N   Ⓒ 86 N

   Ⓑ 14 N   Ⓓ 5.2×10^{2} N

6. Two equally charged bodies exert a force of 90 N on each other. If one of the bodies is exchanged for a body of the same size, but three times as much charge, what is the new force that they will exert on each other?

   Ⓐ 10 N   Ⓒ 2.7×10^{2} N

   Ⓑ 30 N   Ⓓ 8.1×10^{2} N

7. An alpha particle has a mass of 6.68×10^{-27} kg and a charge of 3.2×10^{-19} C. What is the ratio of the electrostatic force to the gravitational force between two alpha particles?

   Ⓐ 1   Ⓒ 2.3×10^{15}

   Ⓑ 4.8×10^{7} Ⓓ 3.1×10^{35}

8. Charging a neutral body by touching it with a charged body is called charging by ______.

   Ⓐ conduction   Ⓒ grounding

   Ⓑ induction   Ⓓ discharging

9. Macy rubs a balloon with wool, giving the balloon a charge of -8.9×10^{-14} C. What is the force between the balloon and a metal sphere that is charged to 25 C and is 2 km away?

   Ⓐ 8.9×10^{-15} N   Ⓒ 2.2×10^{-12} N

   Ⓑ 5.0×10^{-9} N   Ⓓ 5.6×10^{4} N

Extended Answer

10. According to the diagram, what is the net force exerted by charges A and B on charge C? In your answer, include a diagram showing the force vectors $\mathbf{F}_{\text{A on C}}$, $\mathbf{F}_{\text{B on C}}$, and $\mathbf{F}_{\text{net}}$.

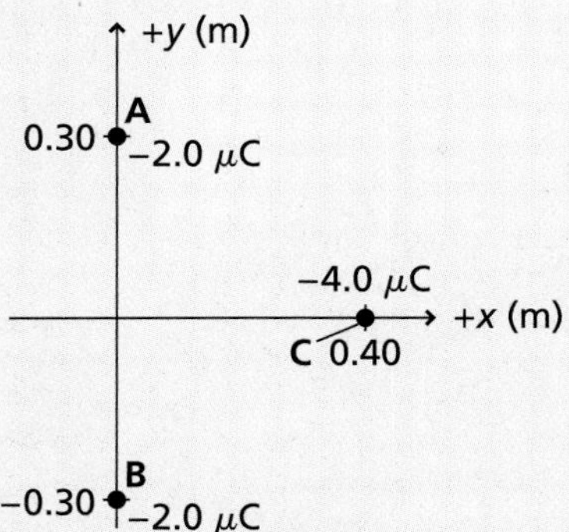

✓ Test-Taking TIP

Slow Down

Check to make sure you are answering the question that each problem is posing. Read the questions and answer choices very carefully. Remember that doing most of the problems and getting them right is always preferable to doing all of the problems and getting a lot of them wrong.

Chapter 21

Electric Fields

What You'll Learn

- You will relate electric fields to electric forces and distinguish between them.
- You will relate electric potential difference to work and energy.
- You will describe how charges are distributed on conductors.
- You will explain how capacitors store electric charges.

Why It's Important

Electricity is an essential form of energy for modern societies.

High-Energy Discharge A high-voltage generator produces the glow you see inside these discharge spheres.

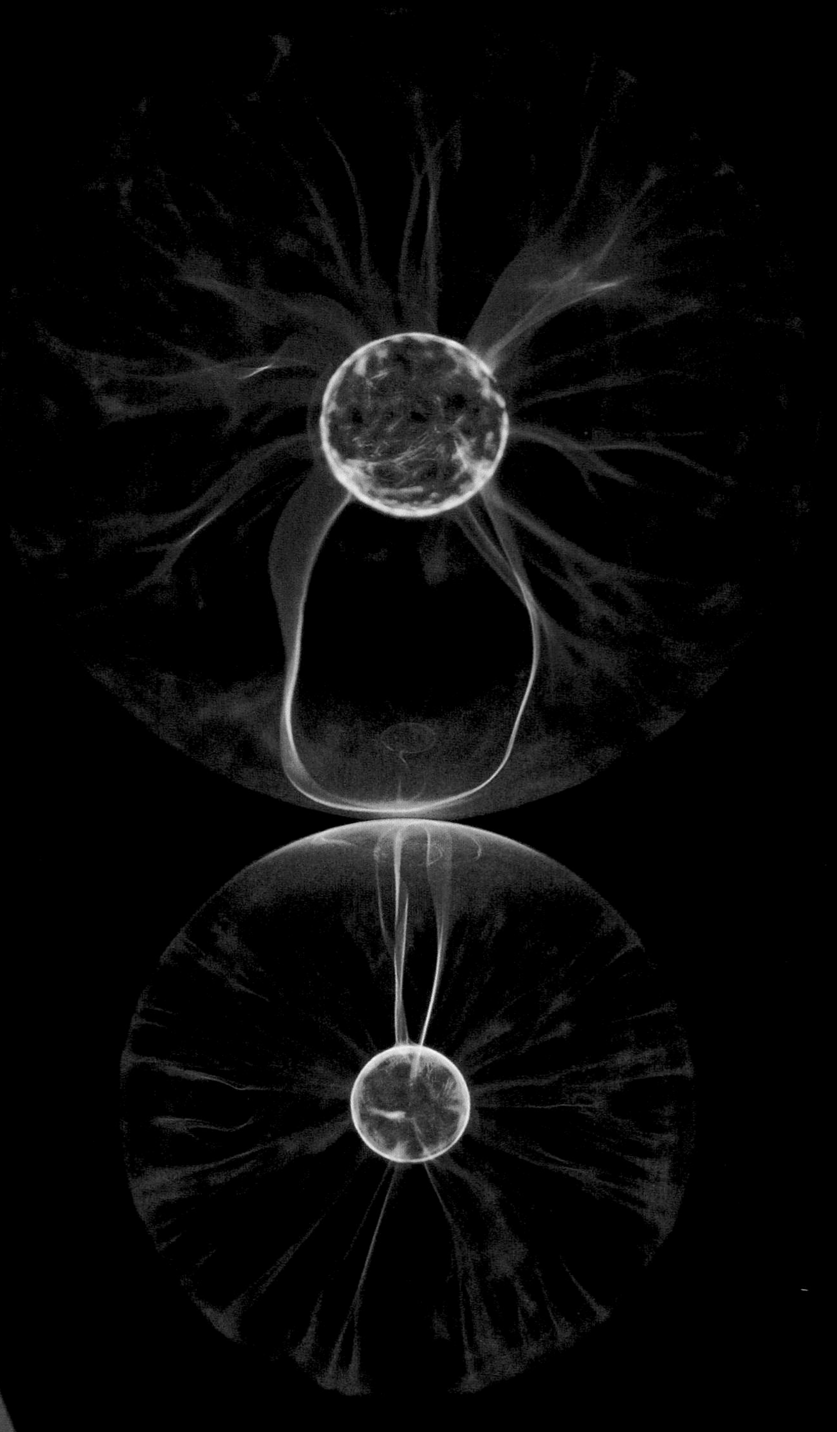

Think About This ▶

Why doesn't an ordinary lightbulb glow in the same way as these discharge spheres connected to a high-voltage generator?

How do charged objects interact at a distance?

Question
How is a charged object affected by interaction with other charged objects at a distance?

Procedure

1. Inflate and tie off two balloons. Attach a $\frac{1}{2}$-m length of string to each balloon.
2. Rub one balloon back and forth on your shirt six to eight times, causing it to become charged. Hang it from a cabinet, table, or other support by the string with a piece of tape.
3. Rub the second balloon the same way and then suspend it from its string.
4. **Observe** Slowly bring the second balloon toward the suspended one. How do the balloons behave? Tape the second balloon so it hangs by its string next to the first balloon.
5. **Observe** Bring your hand toward the charged balloons. What happens?

Analysis
What did you observe as the two balloons were brought near each other? What happened as your hand was brought near the balloons?

Critical Thinking With what two objects have you previously observed similar behaviors of action at a distance?

21.1 Creating and Measuring Electric Fields

Electric force, like gravitational force, which you studied in Chapter 8, varies inversely as the square of the distance between two point objects. Both forces can act from great distances. How can a force be exerted across what seems to be empty space? Michael Faraday suggested that because an electrically charged object, A, creates a force on another charged object, B, anywhere in space, object A must somehow change the properties of space. Object B somehow senses the change in space and experiences a force due to the properties of the space at its location. We call the changed property of space an **electric field.** An electric field means that the interaction is not between two distant objects, but between an object and the field at its location.

The forces exerted by electric fields can do work, transferring energy from the field to another charged object. This energy is something you use on a daily basis, whether you plug an appliance into an electric outlet or use a battery-powered, portable device. In this chapter, you will learn more about electric fields, forces, and electric energy.

▶ **Objectives**
- **Define** an electric field.
- **Solve** problems relating to charge, electric fields, and forces.
- **Diagram** electric field lines.

▶ **Vocabulary**
electric field
electric field line

The Electric Field

How can you measure an electric field? Place a small charged object at some location. If there is an electric force on it, then there is an electric field at that point. The charge on the object that is used to test the field, called the test charge, must be small enough that it doesn't affect other charges.

Color Convention

- Electric field lines are **indigo.**
- Positive charges are **red.**
- Negative charges are **blue.**

Consider **Figure 21-1,** which illustrates a charged object with a charge of q. Suppose you place the positive test charge at some point, A, and measure a force, $\boldsymbol{F}$. According to Coulomb's law, the force is directly proportional to the strength of the test charge, q'. That is, if the charge is doubled, so is the force. Therefore, the ratio of the force to the charge is a constant. If you divide the force, $\boldsymbol{F}$, by the test charge, q', you obtain a vector quantity, $\boldsymbol{F}/q'$. This quantity does not depend on the test charge, only on the force, $\boldsymbol{F}$, and the location of point A. The electric field at point A, the location of q', is represented by the following equation.

Electric Field Strength $\boldsymbol{E} = \frac{\boldsymbol{F}_{\text{on } q'}}{q'}$

The strength of an electric field is equal to the force on a positive test charge divided by the strength of the test charge.

The direction of an electric field is the direction of the force on a positive test charge. The magnitude of the electric field strength is measured in newtons per coulomb, N/C.

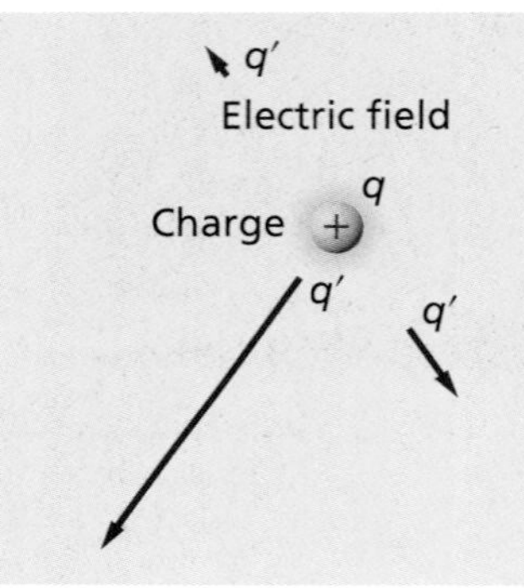

Figure 21-1 Arrows can be used to represent the magnitude and direction of the electric field about an electric charge at various locations.

A picture of an electric field can be made by using arrows to represent the field vectors at various locations, as shown in Figure 21-1. The length of the arrow is used to show the strength of the field. The direction of the arrow shows the field direction. To find the field from two charges, the fields from the individual charges are added vectorially. A test charge can be used to map out the field resulting from any collection of charges. Typical electric field strengths produced by charge collections are shown in **Table 21-1.**

An electric field should be measured only by a very small test charge. This is because the test charge also exerts a force on q. It is important that the force exerted by the test charge does not cause charge to be redistributed on a conductor, thereby causing q to move to another location and thus, changing the force on q' as well as the electric field strength being measured. A test charge always should be small enough so that its effect on q is negligible.

Table 21-1

Approximate Values of Typical Electric Fields

Field	Value (N/C)
Near a charged, hard-rubber rod	1×10^{3}
In a television picture tube	1×10^{5}
Needed to create a spark in air	3×10^{6}
At an electron's orbit in a hydrogen atom	5×10^{11}

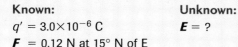

▶ EXAMPLE Problem 1

Electric Field Strength An electric field is measured using a positive test charge of 3.0×10^{-6} C. This test charge experiences a force of 0.12 N at an angle of 15° north of east. What are the magnitude and direction of the electric field strength at the location of the test charge?

1 Analyze and Sketch the Problem

- Draw and label the test charge, q'.
- Show and label the coordinate system centered on the test charge.
- Diagram and label the force vector at 15° north of east.

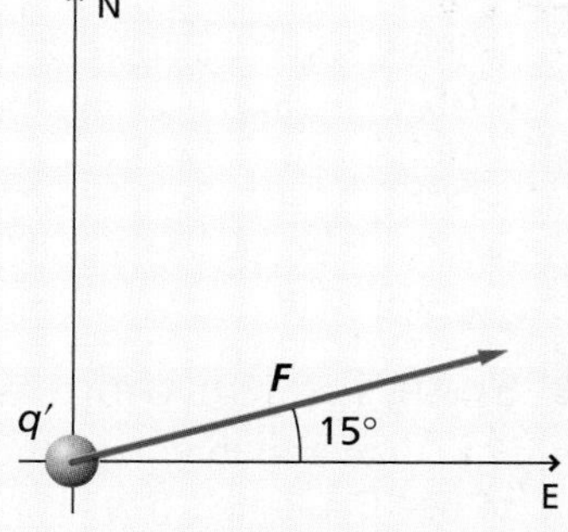

Known:	**Unknown:**
$q' = 3.0\times10^{-6}$ C	$\mathbf{E} = ?$
$\mathbf{F} = 0.12$ N at 15° N of E	

2 Solve for the Unknown

$$E = \frac{F}{q'}$$

$$= \frac{0.12\ \text{N}}{3.0\times10^{-6}\ \text{C}}$$ Substitute $F = 0.12$ N, $q' = 3.0\times10^{-6}$ C

$$= 4.0\times10^{4}\ \text{N/C}$$

The force on the test charge and the electric field are in the same direction.

$\mathbf{E} = 4.0\times10^{4}$ N/C at 15° N of E

Math Handbook
Operations with Significant Digits
pages 835–836

3 Evaluate the Answer

- **Are the units correct?** Electric field strength is correctly measured in N/C.
- **Does the direction make sense?** The field direction is in the direction of the force because the test charge is positive.
- **Is the magnitude realistic?** This field strength is consistent with the values listed in Table 21-1.

▶ PRACTICE Problems

- Additional Problems, Appendix B
- Solutions to Selected Problems, Appendix C

1. A positive test charge of 5.0×10^{-6} C is in an electric field that exerts a force of 2.0×10^{-4} N on it. What is the magnitude of the electric field at the location of the test charge?

2. A negative charge of 2.0×10^{-8} C experiences a force of 0.060 N to the right in an electric field. What are the field's magnitude and direction at that location?

3. A positive charge of 3.0×10^{-7} C is located in a field of 27 N/C directed toward the south. What is the force acting on the charge?

4. A pith ball weighing 2.1×10^{-3} N is placed in a downward electric field of 6.5×10^{4} N/C. What charge (magnitude and sign) must be placed on the pith ball so that the electric force acting on it will suspend it against the force of gravity?

5. You are probing the electric field of a charge of unknown magnitude and sign. You first map the field with a 1.0×10^{-6}-C test charge, then repeat your work with a 2.0×10^{-6}-C test charge.

- **a.** Would you measure the same forces at the same place with the two test charges? Explain.
- **b.** Would you find the same field strengths? Explain.

▶ EXAMPLE Problem 2

Electric Field Strength What is the electric field strength at a point that is 0.30 m to the right of a small sphere with a charge of -4.0×10^{-6} C?

1 Analyze and Sketch the Problem

- Draw and label the sphere and its charge, q, and the test charge, q'.
- Show and label the distance between the charges.
- Diagram and label the force vector acting on q'.

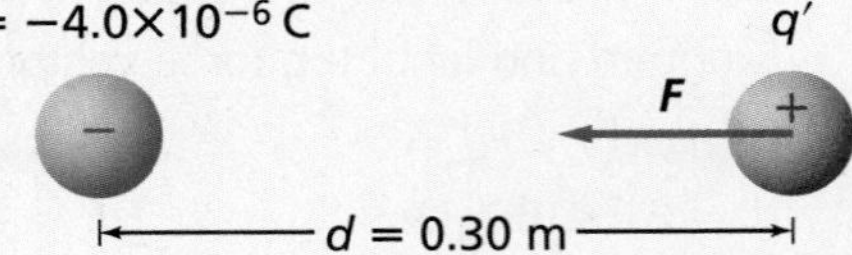

Known:	Unknown:
$q = -4.0\times10^{-6}$ C	$\boldsymbol{E} = ?$
$d = 0.30$ m	

2 Solve for the Unknown

The force and the magnitude of the test charge are unknown, so use Coulomb's law in combination with the electric field strength.

Math Handbook
Operations with Scientific Notation pages 842–843

$$E = \frac{F}{q'}$$

$$= K\frac{qq'}{d^2q'} \quad \text{Substitute } F = K\frac{qq'}{d^2}$$

$$= K\frac{q}{d^2}$$

$$= \left(9.0\times10^{9}\ \text{N}\cdot\text{m}^2/\text{C}^2\right)\frac{(-4.0\times10^{-6}\ \text{C})}{(0.30\ \text{m})^2} \quad \text{Substitute } K = 9.0\times10^{9}\ \text{N}\cdot\text{m}^2/\text{C}^2,\ q = -4.0\times10^{-6}\ \text{C},\ d = 0.30\ \text{m}$$

$$= -4.0\times10^{5}\ \text{N/C}$$

$\boldsymbol{E} = 4.0\times10^{5}$ N/C toward the sphere, or to the left

3 Evaluate the Answer

- **Are the units correct?** $(\text{N}\cdot\text{m}^2/\text{C}^2)(\text{C})/\text{m}^2 = \text{N/C}$. The units work out to be N/C, which is correct for electric field strength.
- **Does the direction make sense?** The negative sign indicates that the positive test charge is attracted toward the negative point charge.
- **Is the magnitude realistic?** This field strength is consistent with the values listed in Table 21-1.

▶ PRACTICE Problems

- Additional Problems, Appendix B
- Solutions to Selected Problems, Appendix C

6. What is the magnitude of the electric field strength at a position that is 1.2 m from a point charge of 4.2×10^{-6} C?

7. What is the magnitude of the electric field strength at a distance twice as far from the point charge in problem 6?

8. What is the electric field at a position that is 1.6 m east of a point charge of $+7.2\times10^{-6}$ C?

9. The electric field that is 0.25 m from a small sphere is 450 N/C toward the sphere. What is the charge on the sphere?

10. How far from a point charge of $+2.4\times10^{-6}$ C must a test charge be placed to measure a field of 360 N/C?

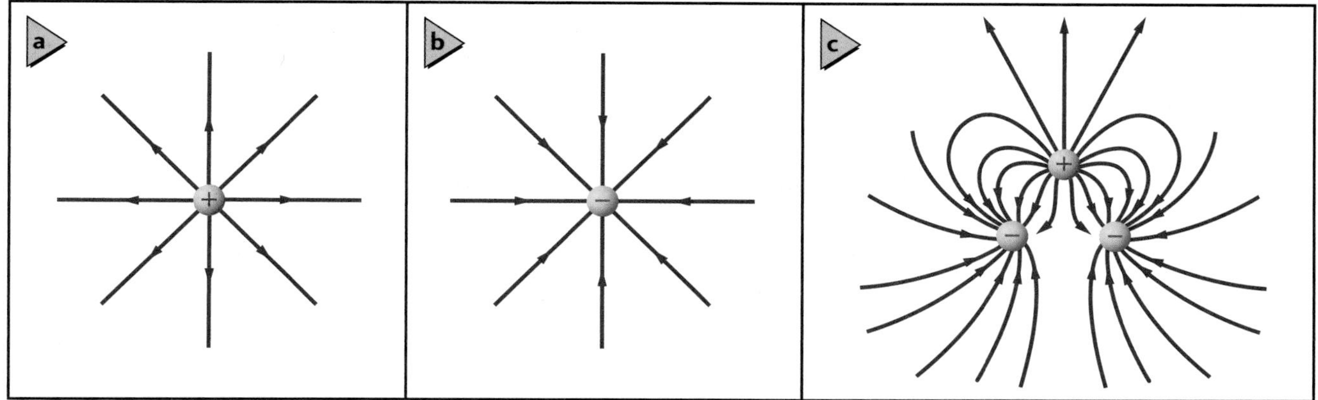

Figure 21-2 Lines of force are drawn perpendicularly away from a positively charged object **(a)** and perpendicularly into a negatively charged object **(b).** Electric field lines are shown between like charged and oppositely charged objects **(c).**

So far, you have measured an electric field at a single point. Now, imagine moving the test charge to another location. Measure the force on it again and calculate the electric field. Repeat this process again and again until you assign every location in space a measurement of the vector quantity of the electric field strength associated with it. The field is present even if there is no test charge to measure it. Any charge placed in an electric field experiences a force on it resulting from the electric field at that location. The strength of the force depends on the magnitude of the field, $\boldsymbol{E}$, and the magnitude of the charge, q. Thus, $\boldsymbol{F} = \boldsymbol{E}q$. The direction of the force depends on the direction of the field and the sign of the charge.

Picturing the Electric Field

A picture of an electric field is shown in **Figure 21-2.** Each of the lines used to represent the actual field in the space around a charge is called an **electric field line.** The direction of the field at any point is the tangent drawn to a field line at that point. The strength of the electric field is indicated by the spacing between the lines. The field is strong where the lines are close together. It is weaker where the lines are spaced farther apart. Although only two-dimensional models can be shown here, remember that electric fields exist in three dimensions.

The direction of the force on a positive test charge near another positive charge is away from the other charge. Thus, the field lines extend radially outward like the spokes of a wheel, as shown in **Figure 21-2a.** Near a negative charge, the direction of the force on the positive test charge is toward the negative charge, so the field lines point radially inward, as shown in **Figure 21-2b.** When there are two or more charges, the field is the vector sum of the fields resulting from the individual charges. The field lines become curved and the pattern is more complex, as shown in **Figure 21-2c.** Note that field lines always leave a positive charge and enter a negative charge, and that they never cross each other.

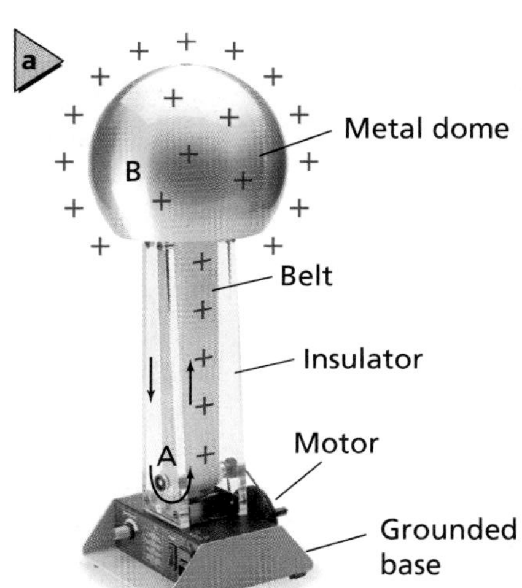

Figure 21-3 In the Van de Graaff generator **(a),** charge is transferred onto a moving belt at A, and from the belt to the metal dome at B. An electric motor does the work needed to increase the electric potential energy. When a person touches a Van de Graaff generator, the results can be dramatic **(b).**

Figure 21-4 Lines of force between unlike charges **(a, c)** and between like charges **(b, d)** describe the behavior of a positively charged object in a field. The top figures are computer tracings of electric field lines.

Concepts In Motion
Interactive Figure To see an animation on electric fields and force on a charge, visit physicspp.com.

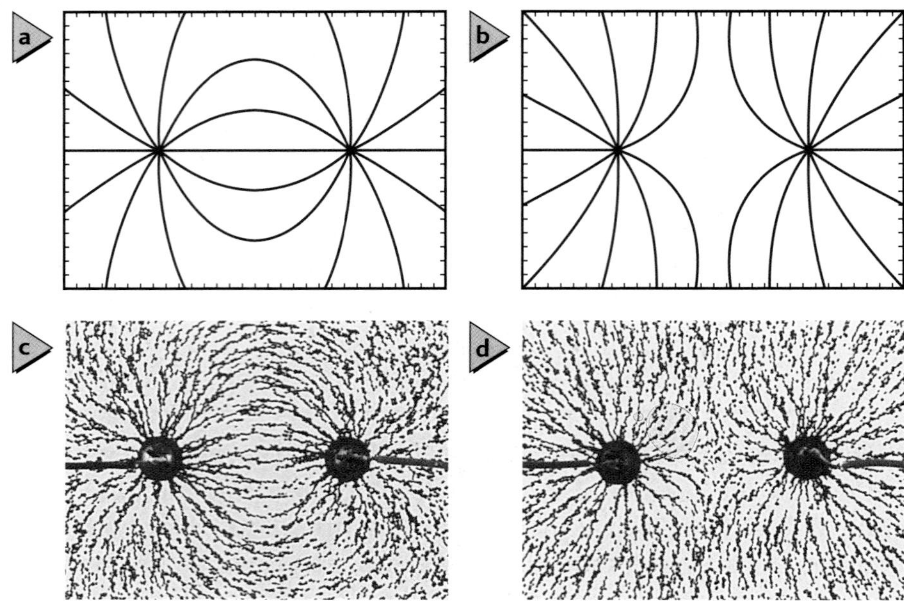

Robert Van de Graaff devised the high-voltage electrostatic generator in the 1930s. Van de Graaff's machine, shown in **Figure 21-3a** on the previous page, is a device that transfers large amounts of charge from one part of the machine to a metal terminal at the top of the device. Charge is transferred onto a moving belt at the base of the generator, position A, and is transferred off the belt at the metal dome at the top, position B. An electric motor does the work needed to increase the electric potential energy. A person touching the terminal of a Van de Graaff machine is charged electrically. The charges on the person's hairs repel each other, causing the hairs to follow the field lines, as shown in **Figure 21-3b.**

Another method of visualizing field lines is to use grass seed in an insulating liquid, such as mineral oil. The electric forces cause a separation of charge in each long, thin grass seed. The seeds then turn so that they line up along the direction of the electric field. The seeds thus form a pattern of the electric field lines, as in **Figure 21-4.** Field lines do not really exist. They are simply a means of providing a model of an electric field. Electric fields, on the other hand, do exist. Although they provide a method of calculating the force on a charged body, they do not explain why charged bodies exert forces on each other.

21.1 Section Review

11. Measuring Electric Fields Suppose you are asked to measure the electric field in space. How do you detect the field at a point? How do you determine the magnitude of the field? How do you choose the magnitude of the test charge? What do you do next?

12. Field Strength and Direction A positive test charge of magnitude 2.40×10^{-8} C experiences a force of 1.50×10^{-3} N toward the east. What is the electric field at the position of the test charge?

13. Field Lines In Figure 21-4, can you tell which charges are positive and which are negative? What would you add to complete the field lines?

14. Field Versus Force How does the electric field, $\boldsymbol{E}$, at the test charge differ from the force, $\boldsymbol{F}$, on it?

15. Critical Thinking Suppose the top charge in Figure 21-2c is a test charge measuring the field resulting from the two negative charges. Is it small enough to produce an accurate measure? Explain.

Physics Online physicspp.com/self_check_quiz

21.2 Applications of Electric Fields

As you have learned, the concept of energy is extremely useful in mechanics. The law of conservation of energy allows us to solve motion problems without knowing the forces in detail. The same is true in the study of electrical interactions. The work performed moving a charged particle in an electric field can result in the particle's gaining potential, or kinetic energy, or both. Because this chapter investigates charges at rest, only changes in potential energy will be discussed.

▶ **Objectives**

- **Define** electric potential difference.
- **Calculate** potential difference from the work required to move a charge.
- **Describe** how charges are distributed on solid and hollow conductors.
- **Solve** problems pertaining to capacitance.

▶ **Vocabulary**

electric potential difference
volt
equipotential
capacitor
capacitance

Energy and Electric Potential

Recall the change in gravitational potential energy of a ball when it is lifted, as shown in **Figure 21-5.** Both the gravitational force, $\mathbf{F}$, and the gravitational field, $\mathbf{g} = \mathbf{F}/m$, point toward Earth. If you lift a ball against the force of gravity, you do work on it, thereby increasing its potential energy.

The situation is similar with two unlike charges: they attract each other, and so you must do work to pull one charge away from the other. When you do the work, you transfer energy to the charge where that energy is stored as potential energy. The larger the test charge, the greater the increase in its potential energy, ΔPE.

Although the force on the test charge depends on its magnitude, q', the electric field it experiences does not. The electric field, $\mathbf{E} = \mathbf{F}/q'$, is the force per unit charge. In a similar way, the **electric potential difference,** ΔV, is defined as the work done moving a positive test charge between two points in an electric field divided by the magnitude of the test charge.

Electric Potential Difference $\Delta V = \frac{W_{\text{on } q'}}{q'}$

The difference in electrical potential is the ratio of the work needed to move a charge to the strength of that charge.

Electric potential difference is measured in joules per coulomb. One joule per coulomb is called a **volt** (J/C = V).

Consider the situation shown in **Figure 21-6** on the next page. The negative charge creates an electric field toward itself. Suppose you place a small positive test charge in the field at position A. It will experience a force in the direction of the field. If you now move the test charge away from the negative charge to position B, as in **Figure 21-6a,** you will have to exert a force, $\mathbf{F}$, on the charge. Because the force that you exert is in the same direction as the displacement, the work that you do on the test charge is positive. Therefore, there also will be a positive change in the electric potential difference. The change in potential difference does not depend on the magnitude of the test charge. It depends only on the field and the displacement.

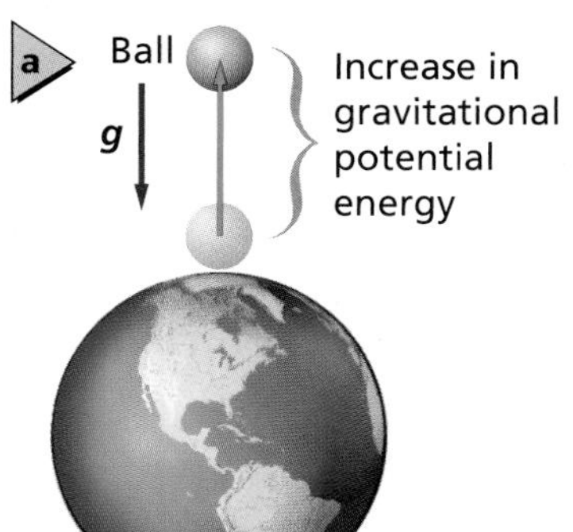

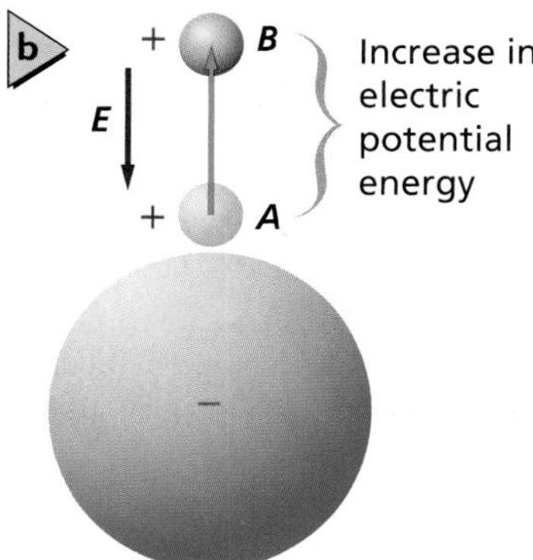

■ **Figure 21-5** Work is needed to move an object against the force of gravity **(a)** and against the electric force **(b).** In both cases, the potential energy of the object is increased.

Figure 21-6 Electric potential difference is determined by measuring the work per unit charge. If you move unlike charges apart, you increase the electric potential difference **(a).** If you move unlike charges closer together, you reduce the electric potential difference **(b).**

Concepts In Motion
Interactive Figure To see an animation on electric potential difference, visit physicspp.com.

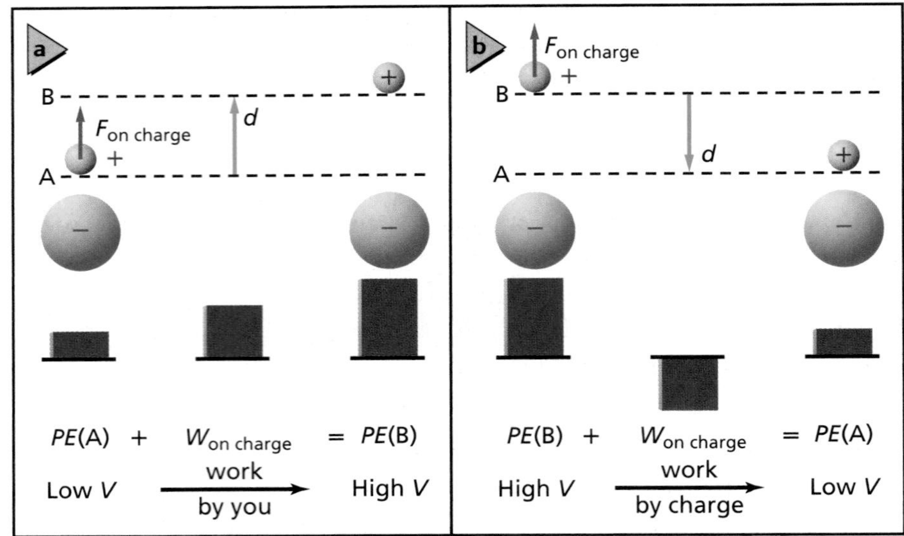

APPLYING PHYSICS

Static Electricity Modern electronic devices, such as personal computers, contain components that are easily damaged by static electric discharges. To prevent damage to these sensitive components during repair, a technician will wear a conductive strap around his or her wrist. The other end of this strap is clipped to a grounded piece of metal. The strap conducts charge away from the technician and eliminates any possible potential difference with the grounded equipment.

Suppose you now move the test charge back to position A from position B, as in **Figure 21-6b.** The force that you exert is now in the direction opposite the displacement, so the work that you do is negative. The electric potential difference is also negative. In fact, it is equal and opposite to the potential difference for the move from position A to position B. The electric potential difference does not depend on the path used to go from one position to another. It does depend on the two positions.

Is there always an electric potential difference between the two positions? Suppose you move the test charge in a circle around the negative charge. The force that the electric field exerts on the test charge is always perpendicular to the direction in which you moved it, so you do no work. Therefore, the electric potential difference is zero. Whenever the electric potential difference between two or more positions is zero, those positions are said to be at **equipotential.**

Only differences in potential energy can be measured. The same is true of electric potential; thus, only differences in electric potential are important. The electric potential difference from point A to point B is defined as $\Delta V = V_B - V_A$. Electric potential differences are measured with a voltmeter. Sometimes, the electric potential difference is simply called the voltage. Do not confuse electric potential difference, ΔV, with the unit for volts, V.

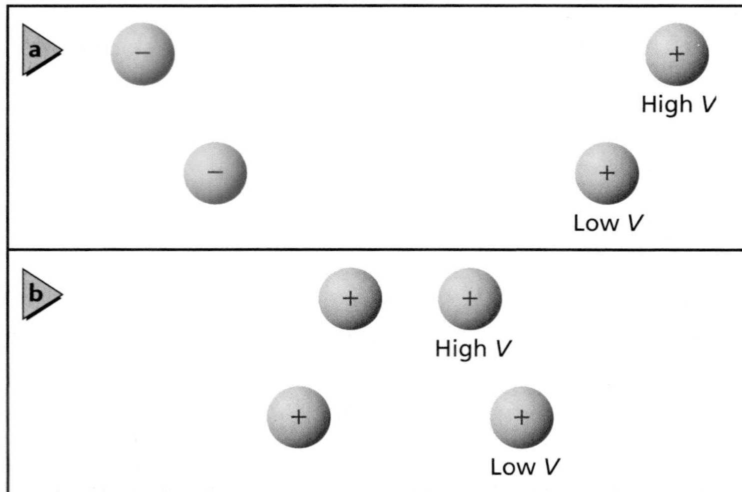

Figure 21-7 Electric potential is smaller when two unlike charges are closer together **(a)** and larger when two like charges are closer together **(b).**

You have seen that electric potential difference increases as a positive test charge is separated from a negative charge. What happens when a positive test charge is separated from a positive charge? There is a repulsive force between these two charges. Potential energy decreases as the two charges are moved farther apart. Therefore, the electric potential is smaller at points farther from the positive charge, as shown in **Figure 21-7.**

As you learned in Chapter 11, the potential energy of a system can be defined as zero at any reference point. In the same way, the electric potential of any point can be defined as zero. No matter what reference point is chosen, the value of the electric potential difference from point A to point B always will be the same.

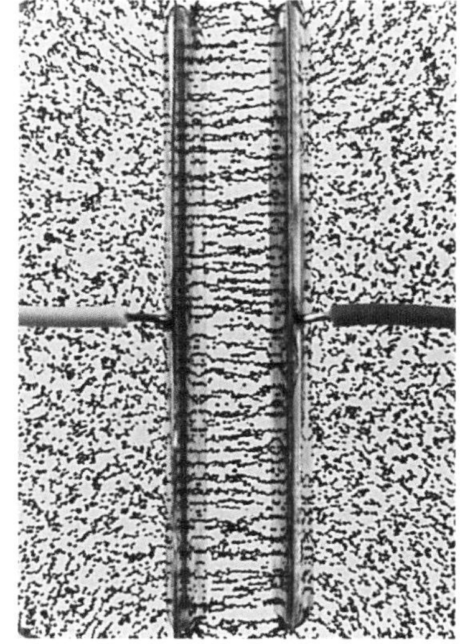

Figure 21-8 A representation of an electric field between parallel plates is shown.

The Electric Potential in a Uniform Field

A uniform electric force and field can be made by placing two large, flat, conducting plates parallel to each other. One is charged positively and the other is charged negatively. The electric field between the plates is constant, except at the edges of the plates, and its direction is from the positive to the negative plate. The pattern formed by the grass seeds pictured in **Figure 21-8** represents the electric field between parallel plates.

If a positive test charge, q', is moved a distance, d, in the direction opposite the electric field direction, the work done is found by the relationship $W_{\text{on } q'} = Fd$. Thus, the electric potential difference, the work done per unit charge, is $\Delta V = Fd/q' = (F/q')d$. Now, the electric field intensity is the force per unit charge, $E = F/q'$. Therefore, the electric potential difference, ΔV, between two points a distance, d, apart in a uniform field, E, is represented by the following equation.

Electric Potential Difference in a Uniform Field $\Delta V = Ed$

The electrical potential difference in a uniform field is equal to the product of electric field intensity and the distance moved by a charge.

The electric potential increases in the direction opposite the electric field direction. That is, the electric potential is higher near the positively charged plate. By dimensional analysis, the product of the units of E and d is (N/C)(m). This is equivalent to one J/C, which is the definition of 1 V.

PRACTICE Problems

• Additional Problems, Appendix B
• Solutions to Selected Problems, Appendix C

16. The electric field intensity between two large, charged, parallel metal plates is 6000 N/C. The plates are 0.05 m apart. What is the electric potential difference between them?

17. A voltmeter reads 400 V across two charged, parallel plates that are 0.020 m apart. What is the electric field between them?

18. What electric potential difference is applied to two metal plates that are 0.200 m apart if the electric field between them is 2.50×10^3 N/C?

19. When a potential difference of 125 V is applied to two parallel plates, the field between them is 4.25×10^3 N/C. How far apart are the plates?

20. A potential difference of 275 V is applied to two parallel plates that are 0.35 cm apart. What is the electric field between the plates?

▶ EXAMPLE Problem 3

Work Required to Move a Proton Between Charged Parallel Plates Two charged parallel plates are 1.5 cm apart. The magnitude of the electric field between the plates is 1800 N/C.

a. What is the electric potential difference between the plates?

b. What work is required to move a proton from the negative plate to the positive plate?

1 Analyze and Sketch the Problem

- Draw the plates separated by 1.5 cm.
- Label one plate with positive charges and the other with negative charges.
- Draw uniformly spaced electric field lines from the positive plate to the negative plate.
- Indicate the electric field strength between the plates.
- Place a proton in the electric field.

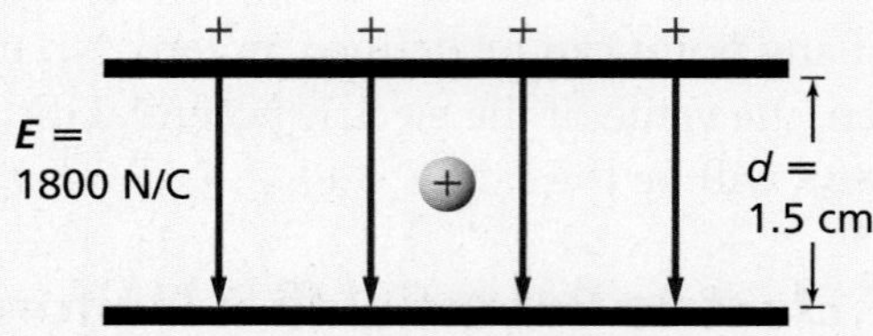

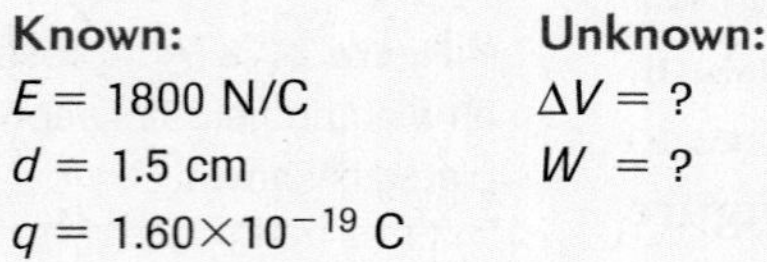

Known:	**Unknown:**
$E = 1800$ N/C	$\Delta V = ?$
$d = 1.5$ cm	$W = ?$
$q = 1.60\times10^{-19}$ C	

Math Handbook
Operations with Scientific Notation
pages 842–843

2 Solve for the Unknown

$$\Delta V = Ed$$
$$= (1800\text{ N/C})(0.015\text{ m}) \quad \textbf{Substitute } E = 1800\text{ N/C}, d = 0.015\text{ m}$$
$$= 27\text{ V}$$

$$\Delta V = \frac{W}{q}$$
$$W = q\Delta V$$
$$= (1.60\times10^{-19}\text{ C})(27\text{ V}) \quad \textbf{Substitute } q = 1.60\times10^{-19}\text{ C}, \Delta V = 27\text{ V}$$
$$= 4.3\times10^{-18}\text{ J}$$

3 Evaluate the Answer

- **Are the units correct?** (N/C)(m) = N·m/C = J/C = V. The units work out to be volts. C·V = C(J/C) = J, the unit for work.
- **Does the sign make sense?** Positive work must be done to move a positive charge toward a positive plate.
- **Is the magnitude realistic?** With such a small charge moved through a potential difference of a few volts, the work performed will be small.

▶ PRACTICE Problems

- Additional Problems, Appendix B
- Solutions to Selected Problems, Appendix C

21. What work is done when 3.0 C is moved through an electric potential difference of 1.5 V?

22. A 12-V car battery can store 1.44×10^{6} C when it is fully charged. How much work can be done by this battery before it needs recharging?

23. An electron in a television picture tube passes through a potential difference of 18,000 V. How much work is done on the electron as it passes through that potential difference?

24. If the potential difference in problem 18 is between two parallel plates that are 2.4 cm apart, what is the magnitude of the electric field between them?

25. The electric field in a particle-accelerator machine is 4.5×10^{5} N/C. How much work is done to move a proton 25 cm through that field?

Millikan's Oil-Drop Experiment

One important application of the uniform electric field between two parallel plates is the measurement of the charge of an electron. This first was determined by American physicist Robert A. Millikan in 1909. **Figure 21-9** shows the method used by Millikan to measure the charge carried by a single electron. First, fine oil drops were sprayed from an atomizer into the air. These drops were charged by friction with the atomizer as they were sprayed. Gravity acting on the drops caused them to fall, and a few of them entered the hole in the top plate of the apparatus. An electric potential difference then was placed across the two plates. The resulting electric field between the plates exerted a force on the charged drops. When the top plate was made positive enough, the electric force caused negatively charged drops to rise. The electric potential difference between the plates was adjusted to suspend a charged drop between the plates. At this point, the downward force of Earth's gravitational field and the upward force of the electric field were equal in magnitude.

The magnitude of the electric field, E, was determined from the electric potential difference between the plates. A second measurement had to be made to find the weight of the drop using the relationship mg, which was too tiny to measure by ordinary methods. To make this measurement, a drop first was suspended. Then, the electric field was turned off, and the rate of the fall of the drop was measured. Because of friction with the air molecules, the oil drop quickly reached terminal velocity, which was related to the mass of the drop by a complex equation. Using the measured terminal velocity to calculate mg and knowing E, the charge, q, could be calculated.

Charge on an electron Millikan found that there was a great deal of variation in the charges of the drops. When he used X rays to ionize the air and add or remove electrons from the drops, he noted, however, that the changes in the charge on the drops were always a multiple of 1.60×10^{-19} C. The changes were caused by one or more electrons being added to or removed from the drops. Millikan concluded that the smallest change in charge that could occur was the amount of charge of one electron. Therefore, Millikan proposed that each electron always has the same charge, 1.60×10^{-19} C. Millikan's experiment showed that charge is quantized. This means that an object can have only a charge with a magnitude that is some integral multiple of the charge of an electron.

MINI LAB

Electric Fields 

Tie a pith ball on the end of a 20-cm nylon thread and tie the other end to a plastic straw. Holding the straw horizontally, notice that the ball hangs straight down. Now rub a piece of wool on a 30 cm $\times$ 30 cm square of plastic foam to charge both objects. Stand the foam vertically. Hold the straw and touch the pith ball to the wool.

1. Predict what will happen when the ball is close to the foam.

2. Test your prediction by slowly bringing the hanging ball toward the charged plastic foam.

3. Predict the ball's behavior at different locations around the foam, and test your prediction.

4. Observe the angle of the thread as you move the pith ball to different regions around the foam.

Analyze and Conclude

5. Explain, in terms of the electric field, why the ball swings toward the charged plastic.

6. Compare the angle of the thread at various points around the foam. Why did it change?

7. Infer what the angle of the thread indicates about the strength and the direction of the electric field.

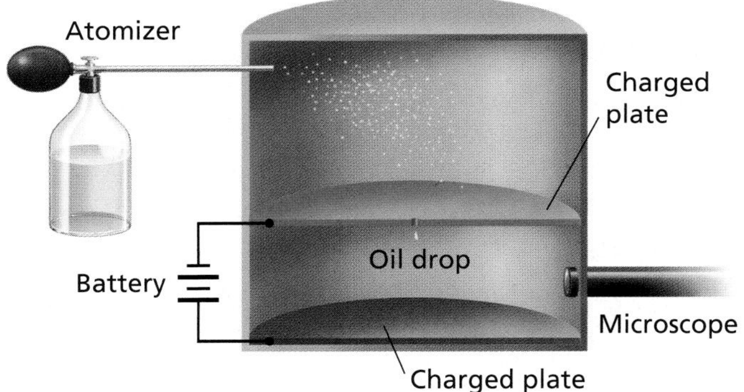

Figure 21-9 This illustration shows a cross-sectional view of the apparatus that Millikan used to determine the charge on an electron.

▶ EXAMPLE Problem 4

Finding the Charge on an Oil Drop In a Millikan oil-drop experiment, a drop has been found to weigh 2.4×10^{-14} N. The parallel plates are separated by a distance of 1.2 cm. When the potential difference between the plates is 450 V, the drop is suspended, motionless.

a. What is the charge on the oil drop?

b. If the upper plate is positive, how many excess electrons are on the oil drop?

1 Analyze and Sketch the Problem

- Draw the plates with the oil drop suspended between them.
- Draw and label vectors representing the forces.
- Indicate the potential difference and the distance between the plates.

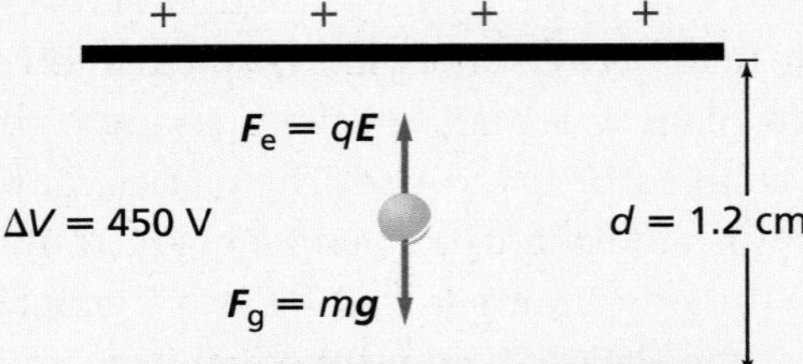

Known:	Unknown:
$\Delta V = 450$ V	charge on drop, $q = ?$
$F_g = 2.4\times10^{-14}$ N	number of electrons, $n = ?$
$d = 1.2$ cm	

2 Solve for the Unknown

To be suspended, the electric force and gravitational force must be balanced.

$$F_e = F_g$$

$$qE = F_g \qquad \text{Substitute } F_e = qE$$

$$\frac{q\Delta V}{d} = F_g \qquad \text{Substitute } E = \frac{\Delta V}{d}$$

Personal Tutor For an online tutorial on isolating variables, visit physicspp.com.

Solve for q.

$$q = \frac{F_g d}{\Delta V}$$

$$= \frac{(2.4\times10^{-14}\ \text{N})(0.012\ \text{m})}{450\ \text{V}} \qquad \text{Substitute } F_g = 2.4\times10^{-14}\ \text{N}, d = 0.012\ \text{m}, \Delta V = 450\ \text{V}$$

$$= 6.4\times10^{-19}\ \text{C}$$

Solve for the number of electrons on the drop.

$$n = \frac{q}{e}$$

$$= \frac{6.4\times10^{-19}\ \text{C}}{1.6\times10^{-19}\ \text{C}} \qquad \text{Substitute } q = 6.4\times10^{-19}\ \text{C}, e = 1.6\times10^{-19}\ \text{C}$$

$$= 4$$

3 Evaluate the Answer

- **Are the units correct?** N·m/V = J/(J/C) = C, the unit for charge.
- **Is the magnitude realistic?** This is a small whole number of elementary charges.

▶ PRACTICE Problems

• Additional Problems, Appendix B
• Solutions to Selected Problems, Appendix C

26. A drop is falling in a Millikan oil-drop apparatus with no electric field. What forces are acting on the oil drop, regardless of its acceleration? If the drop is falling at a constant velocity, describe the forces acting on it.

27. An oil drop weighs 1.9×10^{-15} N. It is suspended in an electric field of 6.0×10^{3} N/C. What is the charge on the drop? How many excess electrons does it carry?

28. An oil drop carries one excess electron and weighs 6.4×10^{-15} N. What electric field strength is required to suspend the drop so it is motionless?

29. A positively charged oil drop weighing 1.2×10^{-14} N is suspended between parallel plates separated by 0.64 cm. The potential difference between the plates is 240 V. What is the charge on the drop? How many electrons is the drop missing?

Sharing of Charge

All systems come to equilibrium when the energy of the system is at a minimum. For example, if a ball is placed on a hill, it finally will come to rest in a valley where its gravitational potential energy is smallest. This also would be the location where its gravitational potential has been reduced by the largest amount. This same principle explains what happens when an insulated, positively charged metal sphere, such as the one shown in **Figure 21-10,** touches a second, uncharged sphere.

The excess charges on sphere A repel each other, so when the neutral sphere, B, touches sphere A, there is a net force on the charges on A toward B. Suppose that you were to physically move the charges, individually, from A to B. When you move the first charge, the other charges on A would push it toward B, so, to control its speed, you would have to exert a force in the opposite direction. Therefore, you do negative work on it, and the electric potential difference from A to B is negative. When the next few charges are moved, they feel a small repulsive force from the charges already on B, but there is still a net positive force in that direction. At some point, the force pushing a charge off A will equal the repulsive force from the charges on B, and the electric potential difference is zero. After this point of equilibrium, work would have to be done to move the next charge to B, so this would not happen by itself and would require an increase in the energy of the system. However, if you did continue to move charges, the electric potential difference from A to B would then be positive. Thus, you can see that charges would move from A to B without external forces until there is no electric potential difference between the two spheres.

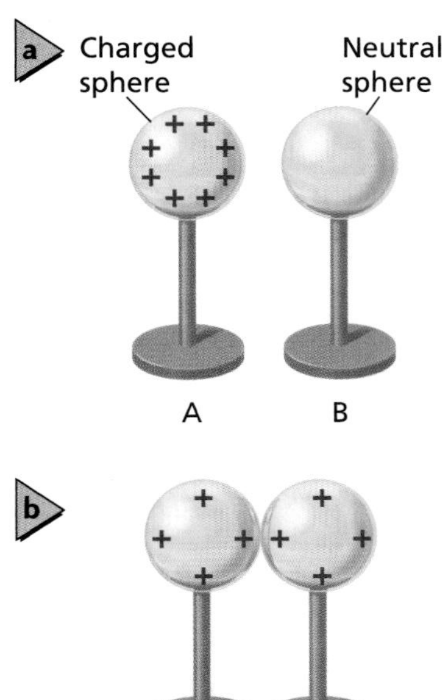

Figure 21-10 A charged sphere shares charge equally with a neutral sphere of equal size when they are placed in contact with each other.

Different sizes of spheres Suppose that the two spheres have different sizes, as in **Figure 21-11.** Although the total numbers of charges on the two spheres are the same, the larger sphere has a larger surface area, so the charges can spread farther apart, and the repulsive force between them is reduced. Thus, if the two spheres now are touched together, there will be a net force that will move charges from the smaller to the larger sphere. Again, the charges will move to the sphere with the lower electric potential until there is no electric potential difference between the two spheres. In this case, the larger sphere will have a larger charge when equilibrium is reached.

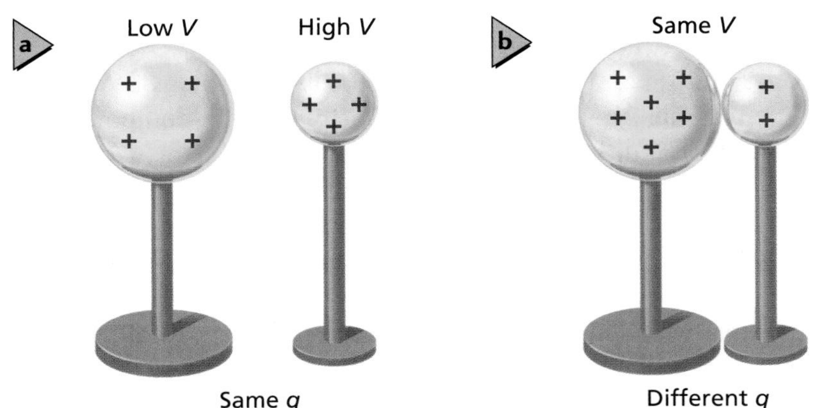

Figure 21-11 Charges are transferred from a sphere with high potential to a sphere with lower potential when they touch. The charges move to create no potential difference.

Figure 21-12 The ground wire on a fuel truck prevents ignition of the gasoline vapors.

The same principle explains how charges move on the individual spheres, or on any conductor. They distribute themselves so that the net force on each charge is zero. With no force, there is no electric field along the surface of the conductor. Thus, there is no electric potential difference anywhere on the surface. The surface of a conductor is, therefore, an equipotential surface.

If a charged body is grounded by touching Earth, almost any amount of charge can flow to Earth until the electric potential difference between that body and Earth is reduced to zero. Gasoline trucks, for example, can become charged by friction. If the charge on a gasoline truck were to jump to Earth through gasoline vapor, it could cause an explosion. To prevent this, a metal wire on the truck safely conducts the charge to the ground, as shown in **Figure 21-12.** Similarly, if a computer is not grounded, an electric potential difference between the computer and Earth can occur. If a person then touches the computer, charges could flow through the computer to the person and damage the equipment or hurt the person.

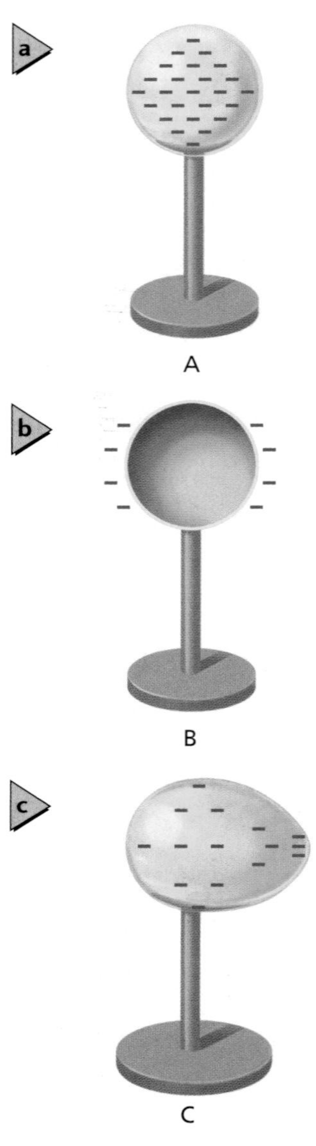

Figure 21-13 On a conducting sphere, **(a),** the charge is evenly distributed around the surface. The charges on the hollow sphere, **(b),** are entirely on the outer surface. In irregular shapes, **(c),** the charges will be closest together at sharp points.

Electric Fields Near Conductors

The charges on a conductor are spread as far apart as they can be to make the energy of the system as low as possible. The result is that all charges are on the surface of a solid conductor. If the conductor is hollow, excess charges will move to the outer surface. If a closed metal container is charged, there will be no charges on the inside surfaces of the container. In this way, a closed metal container shields the inside from electric fields. For example, people inside a car are protected from the electric fields generated by lightning. Likewise, on an open coffee can, there will be very few charges inside and none near the bottom. Even if the inner surface of an object is pitted or bumpy, giving it a larger surface area than the outer surface, the charge still will be entirely on the outside.

On the outside of a conductor, however, the electric field often is not zero. Even though the surface of a conductor is at an equipotential, the electric field around the outside of it depends on the shape of the conductor, as well as on the electric potential difference between it and Earth. The charges are closer together at sharp points of a conductor, as indicated in **Figure 21-13.** Therefore, the field lines are closer together and the field is stronger. This field can become so strong that when electrons are knocked off of atoms by passing cosmic rays, the electrons and resulting ions are accelerated by the field, causing them to strike other atoms, resulting in more ionization of atoms. This chain reaction is what results in the pink glow, such as that seen inside a gas-discharge sphere. If the field is strong enough, when the particles hit other molecules they will produce a stream of ions and electrons that form a plasma, which is a conductor. The result is a spark, or, in extreme cases, lightning. To reduce discharges and sparking, conductors that are highly charged or that operate at high potentials are made smooth in shape to reduce the electric fields.

In contrast, a lightning rod is pointed so that the electric field will be strong near the end of the rod. As the field accelerates electrons and ions, they form the start of a conducting path from the rod to the clouds. As a result of the rod's sharply pointed shape, charges in the clouds spark to the rod, rather than to a chimney or other high point on a house or other building. From the rod, a conductor takes the charges safely to the ground.

Lightning usually requires a potential difference of millions of volts between Earth and the clouds. Even a small gas-discharge tube operates at several thousand volts. Household wiring, on the other hand, does not normally carry a high enough potential difference to cause such discharges.

Storing Charges: The Capacitor

When you lift a book, you increase its gravitational potential energy. This can be interpreted as storing energy in a gravitational field. In a similar way, you can store energy in an electric field. In 1746, Dutch physician and physicist Pieter Van Musschenbroek invented a small device that could store a large electric charge. In honor of the city in which he worked, it was called a Leyden jar. Benjamin Franklin used a Leyden jar to store the charge from lightning and in many other experiments. A version of the Leyden jar is still in use today in electric equipment. This new device for storing a charge has a new form, is much smaller in size, and is called a **capacitor.**

As charge is added to an object, the electric potential difference between that object and Earth increases. For a given shape and size of an object, the ratio of charge stored to electric potential difference, $q/\Delta V$, is a constant called the **capacitance,** C. For a small sphere far from the ground, even a small amount of added charge will increase the electric potential difference. Thus, C is small. A larger sphere can hold more charge for the same increase in electric potential difference, and its capacitance is larger.

Capacitors are designed to have specific capacitances. All capacitors are made up of two conductors that are separated by an insulator. The two conductors have equal and opposite charges. Capacitors are used today in electric circuits to store charge. Commercial capacitors, such as those shown in **Figure 21-14,** typically contain strips of aluminum foil separated by thin plastic that are tightly rolled up to conserve space.

The capacitance of a capacitor is independent of the charge on it, and can be measured by first placing charge q on one plate and charge $-q$ on the other, and then measuring the electric potential difference, ΔV, that results. The capacitance is found by using the following equation, and is measured in farads, F.

Capacitance $C = \frac{q}{\Delta V}$

Capacitance is the ratio of charge on one plate to potential difference.

Figure 21-14 Various types of capacitors are pictured below.

The farad as a unit of measure One farad, F, named after Michael Faraday, is one coulomb per volt, C/V. Just as 1 C is a large amount of charge, 1 F is also a fairly large capacitance. Most capacitors used in modern electronics have capacitances between 10 picofarads (10×10^{-12} F) and 500 microfarads (500×10^{-6} F). However, memory capacitors that are used to prevent loss of memory in some computers can have capacitance from 0.5 F to 1.0 F. Note that if the charge is increased, the electric potential difference also increases. The capacitance depends only on the construction of the capacitor, not on the charge, q.

▶ EXAMPLE Problem 5

Finding Capacitance A sphere has an electric potential difference between it and Earth of 40.0 V when it has been charged to 2.4×10^{-6} C. What is its capacitance?

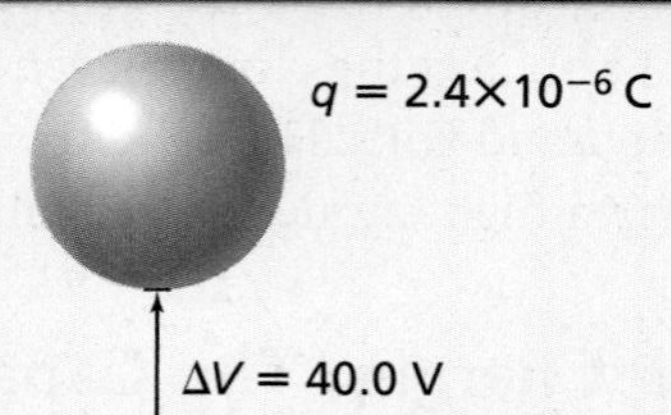

1 Analyze and Sketch the Problem

- Draw a sphere above Earth and label the charge and potential difference.

Known:	Unknown:
$\Delta V = 40.0$ V	$C = ?$
$q = 2.4\times10^{-6}$ C	

2 Solve for the Unknown

$$C = \frac{q}{\Delta V}$$

$$= \frac{2.4\times10^{-6}\text{ C}}{40.0\text{ V}}$$ Substitute $\Delta V = 40.0$ V, $q = 2.4\times10^{-6}$ C

$$= 6.0\times10^{-8}\text{ F}$$

$$= 0.060\ \mu\text{F}$$

Math Handbook
Operations with Significant Digits
pages 835–836

3 Evaluate the Answer

- **Are the units correct?** C/V = F. The units are farads.
- **Is the magnitude realistic?** A small capacitance would store a small charge at a low voltage.

▶ PRACTICE Problems

• Additional Problems, Appendix B
• Solutions to Selected Problems, Appendix C

30. A 27-μF capacitor has an electric potential difference of 45 V across it. What is the charge on the capacitor?

31. Both a 3.3-μF and a 6.8-μF capacitor are connected across a 24-V electric potential difference. Which capacitor has a greater charge? What is it?

32. The same two capacitors as in problem 31 are each charged to 3.5×10^{-4} C. Which has the larger electric potential difference across it? What is it?

33. A 2.2-μF capacitor first is charged so that the electric potential difference is 6.0 V. How much additional charge is needed to increase the electric potential difference to 15.0 V?

34. When a charge of 2.5×10^{-5} C is added to a capacitor, the potential difference increases from 12.0 V to 14.5 V. What is the capacitance of the capacitor?

CHALLENGE PROBLEM

The plates of a capacitor attract each other because they carry opposite charges. A capacitor consisting of two parallel plates that are separated by a distance, d, has capacitance, C.

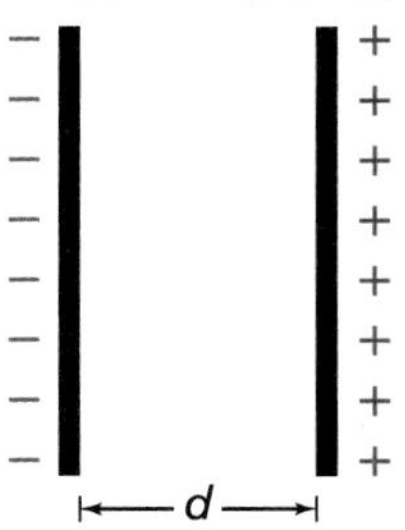

1. Derive an expression for the force between the two plates when the capacitor has charge, q.
2. What charge must be stored on a 22-μF capacitor to have a force of 2.0 N between the plates if they are separated by 1.5 mm?

Varieties of capacitors Capacitors have many shapes and sizes, as shown in Figure 21-14. Some are large enough to fill whole rooms and can store enough charge to create artificial lightning or power giant lasers that release thousands of joules of energy in a few billionths of a second. Capacitors in television sets can store enough charge at several hundred volts to be very dangerous if they are touched. These capacitors can remain charged for hours after the televisions have been turned off. This is why you should not open the case of a television or a computer monitor even if it is unplugged.

The capacitance of a capacitor is controlled by varying the surface area of the two conductors, or plates, within a capacitor, by the distance between the plates, and by the nature of the insulating material. Capacitors are named for the type of insulator, or dielectric, used to separate the plates, and include ceramic, mica, polyester, paper, and air. Higher capacitance is obtained by increasing the surface area and decreasing the separation of the plates. Certain dielectrics have the ability to effectively offset some of the charge on the plates and allow more charge to be stored.

21.2 Section Review

35. Potential Difference What is the difference between electric potential energy and electric potential difference?

36. Electric Field and Potential Difference Show that a volt per meter is the same as a newton per coulomb.

37. Millikan Experiment When the charge on an oil drop suspended in a Millikan apparatus is changed, the drop begins to fall. How should the potential difference on the plates be changed to bring the drop back into balance?

38. Charge and Potential Difference In problem 37, if changing the potential difference has no effect on the falling drop, what does this tell you about the new charge on the drop?

39. Capacitance How much charge is stored on a 0.47-μF capacitor when a potential difference of 12 V is applied to it?

40. Charge Sharing If a large, positively charged, conducting sphere is touched by a small, negatively charged, conducting sphere, what can be said about the following?

a. the potentials of the two spheres

b. the charges on the two spheres

41. Critical Thinking Referring back to Figure 21-3a, explain how charge continues to build up on the metal dome of a Van de Graaff generator. In particular, why isn't charge repelled back onto the belt at point B?

PHYSICS LAB

Charging of Capacitors

Alternate CBL instructions can be found on the Web site.
physicspp.com

A capacitor is an electric device that is made from two conductors, or plates, that are separated by an insulator. It is designed to have a specific capacitance. The capacitance depends on the physical characteristics and geometric arrangement of the conductors and the insulator. In the circuit schematic, the capacitor appears to create an open circuit, even when the switch is in the closed position. However, because capacitors store charge, when the switch is closed, charge from the battery will move to the capacitor. The equal, but opposite charges on the two plates within the capacitor establishes a potential difference, or voltage. As charge is added to the capacitor, the electric potential difference increases. In this laboratory activity you will examine the charging of several different capacitors.

QUESTION

How do the charging times of different capacitors vary with capacitance?

Objectives

- **Collect and organize** data on the rate of charge of different capacitors.
- **Compare and contrast** the rate of charging for different capacitances.
- **Make and use graphs** of potential difference versus time for several capacitors.

Safety Precautions

Materials

9-V battery
9-V battery clip
hook-up wires
switch
voltmeter
47-kΩ resistor
stopwatch
capacitors: 1000 μF, 500 μF, 240 μF

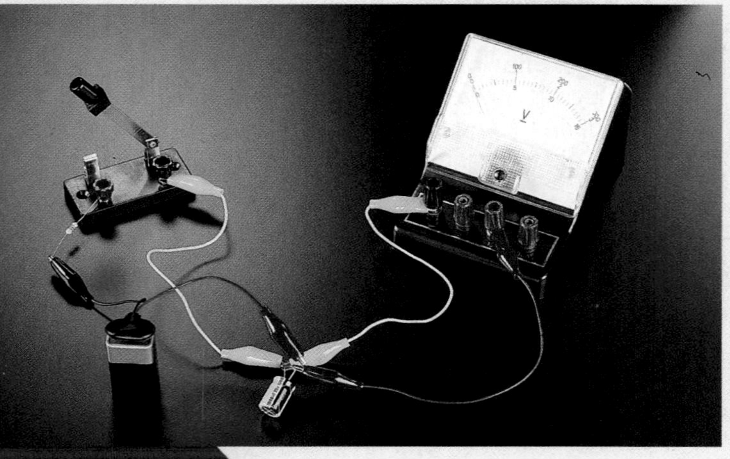

Procedure

1. Before you begin, leave the switch open (off). Do not attach the battery at this time. ***CAUTION: Be careful to avoid a short circuit, especially by permitting the leads from the battery clip to touch each other.*** Connect the circuit, as illustrated. Do this by connecting either end of the resistor to one side of the switch. The resistor is used to reduce the charging of the capacitor to a measurable rate. Connect the other end of the resistor to the negative side of the 9-V battery clip. Inspect your 1000-μF capacitor to determine whether either end is marked with a negative sign, or an arrow with negative signs on it, that points to the lead that is to be connected to the negative side of the battery. Connect this negative lead to the other side of the switch. Attach the unconnected (positive) lead of the capacitor to the positive lead from the battery clip.
2. Connect the positive terminal of the voltmeter to the positive side of the capacitor and the negative terminal to the negative side of the capacitor. Compare your circuit to the photo to verify your connections. Attach the battery after your teacher has inspected the circuit.
3. Prepare a data table having columns for time and potential difference on each of the three different capacitors.
4. One person should watch the time and another should record potential difference at the designated times. Close the switch and measure the voltage at 5-s intervals. Open the switch after you have collected data.

Data Table

Time (s)	Voltage (V) across 1000 μF	Voltage (V) across 500 μF	Voltage (V) across 240 μF	Time (s)	Voltage (V) across 1000 μF	Voltage (V) across 500 μF	Voltage (V) across 240 μF
0				55			
5				60			
10				65			
15				70			
20				75			
25				80			
30				85			
35				90			
40				95			
45				100			
50				105			

5. When you have completed the trial, take a short piece of wire and place it across both ends of the capacitor. This will cause the capacitor to discharge.
6. Replace the 1000-μF capacitor with a 500-μF capacitor. Repeat steps 4–5 and enter data into the appropriate columns of your data table for the 500-μF capacitor.
7. Replace the 500-μF capacitor with a 240-μF capacitor. Repeat steps 4–5 and enter data into the appropriate column of your data table for this last capacitor.

Analyze

1. **Observe and Infer** Does each capacitor charge to 9 V? Propose an explanation for the observed behavior.
2. **Make and Use Graphs** Prepare a graph that plots the time horizontally and the potential difference vertically. Make a separate labeled line for each capacitor.

Conclude and Apply

1. **Interpret Data** Does the voltage on the capacitor immediately jump to the battery's potential difference (9-V)? Explain the reason for the observed behavior.
2. **Infer** Does the larger capacitor require a longer time to become fully charged? Explain why or why not.

Going Further

1. The time for a capacitor to charge to the voltage of the battery depends upon its capacitance and the opposition to the flow of charge in the circuit. In this lab, the opposition to the flow of charge was controlled by the 47-kΩ resistor that was placed in the circuit. In circuits with a capacitor and resistance, such as in this activity, the time in seconds to charge the capacitor to 63.3 percent of the applied voltage is equal to the product of the capacitor and resistance. This is called the time constant. Therefore, $T = RC$, where T is in seconds, R is in ohms, and C is in microfarads. Calculate the time constant for each of the capacitors with the 47-kΩ resistor.
2. Compare your time constants to the values from your graph.

Real-World Physics

Explain Small, disposable, flash cameras, as well as regular electronic flash units, require time before the flash is ready to be used. A capacitor stores the energy for the flash. Explain what might be going on during the time you must wait to take your next picture.

To find out more about electric fields, visit the Web site: physicspp.com

HOW it Works Lightning Rods

Lightning can be very destructive because it creates huge currents in materials that are poor conductors and generates a great deal of heat. In addition to protecting a structure by dissipating some of the charge before lightning strikes, lightning rods are excellent conductors that provide a safe path for the current. Benjamin Franklin is credited with inventing the lightning rod in the 1750s.

3 Positive charges spark out from the lightning rod, meeting the step leader. The conducting path is complete and current neutralizes the separation of charges. Even if the strike does not hit the lightning rod directly, the massive current still can leap to the rod, which is the path of least resistance to the ground.

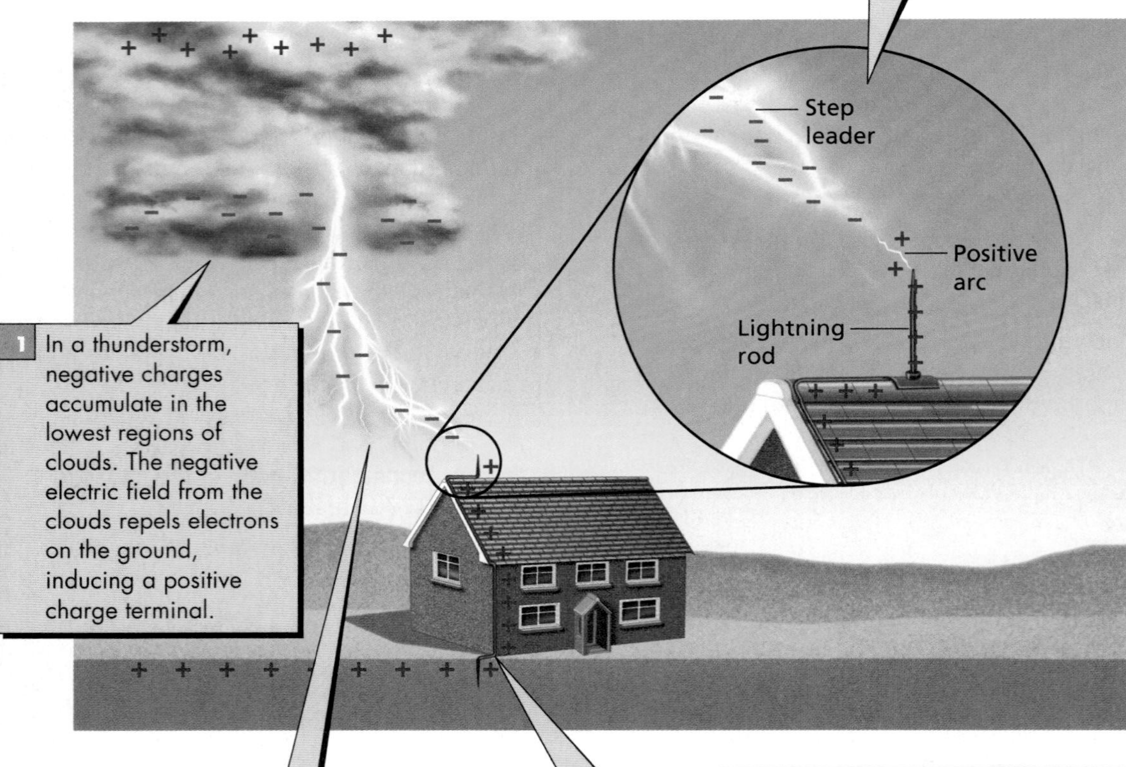

1 In a thunderstorm, negative charges accumulate in the lowest regions of clouds. The negative electric field from the clouds repels electrons on the ground, inducing a positive charge terminal.

2 The strong electric field accelerates electrons and ions, causing a chain reaction in the air, forming plasma. The ionized air is a conductor, and it branches out from the cloud forming what are called step leaders.

4 The current travels safely through the conductor to the ground terminal.

Thinking Critically

1. **Hypothesize** Along what path would the current travel if a house without a lightning rod were struck by lightning?
2. **Evaluate** Should the resistance between the ground terminal and Earth be high or low?
3. **Infer** What are the dangers of an incorrectly installed lightning rod system?

Chapter 21

Study Guide

21.1 Creating and Measuring Electric Fields

Vocabulary

- electric field (p. 563)
- electric field line (p. 567)

Key Concepts

- An electric field exists around any charged object. The field produces forces on other charged objects.
- The electric field is the force per unit charge.

$$\mathbf{E} = \frac{\mathbf{F}}{q'}$$

- The direction of the electric field is the direction of the force on a tiny, positive test charge.
- Electric field lines provide a picture of the electric field. They are directed away from positive charges and toward negative charges. They never cross, and their density is related to the strength of the field.

21.2 Applications of Electric Fields

Vocabulary

- electric potential difference (p. 569)
- volt (p. 569)
- equipotential (p. 570)
- capacitor (p. 577)
- capacitance (p. 577)

Key Concepts

- Electric potential difference is the change in potential energy per unit charge in an electric field.

$$\Delta V = \frac{W}{q'}$$

- Electric potential differences are measured in volts.
- The electric field between two parallel plates is uniform between the plates, except near the edges. In a uniform field, the potential difference is related to the field strength by the following.

$$\Delta V = Ed$$

- Robert Millikan's experiments showed that electric charge is quantized.
- Robert Millikan also showed that the negative charge carried by an electron is 1.60×10^{-19} C.
- Charges will move in a conductor until the electric potential is the same everywhere on the conductor.
- Grounding makes the potential difference between an object and Earth equal to zero.
- Grounding can prevent sparks resulting from a neutral object making contact with objects that have built-up charge on them.
- Electric fields are strongest near sharply pointed conductors.
- Capacitance is the ratio of the charge on an object to its electric potential difference.

$$C = \frac{q}{\Delta V}$$

- Capacitance is independent of the charge on an object and the electric potential difference across it.
- Capacitors are used to store charge.

Chapter 21

Assessment

Concept Mapping

42. Complete the concept map below using the following terms: *capacitance, field strength, J/C, work.*

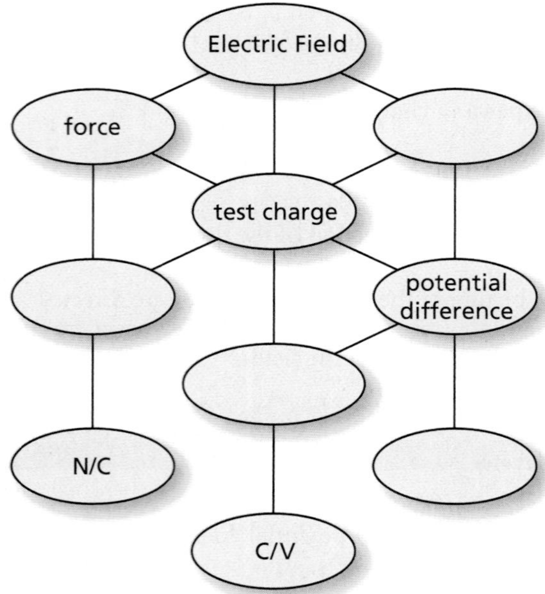

Mastering Concepts

43. What are the two properties that a test charge must have? (21.1)

44. How is the direction of an electric field defined? (21.1)

45. What are electric field lines? (21.1)

46. How is the strength of an electric field indicated with electric field lines? (21.1)

47. Draw some of the electric field lines between each of the following. (21.1)
    a. two like charges of equal magnitude
    b. two unlike charges of equal magnitude
    c. a positive charge and a negative charge having twice the magnitude of the positive charge
    d. two oppositely charged parallel plates

48. In **Figure 21-15,** where do the electric field lines leave the positive charge end? (21.1)

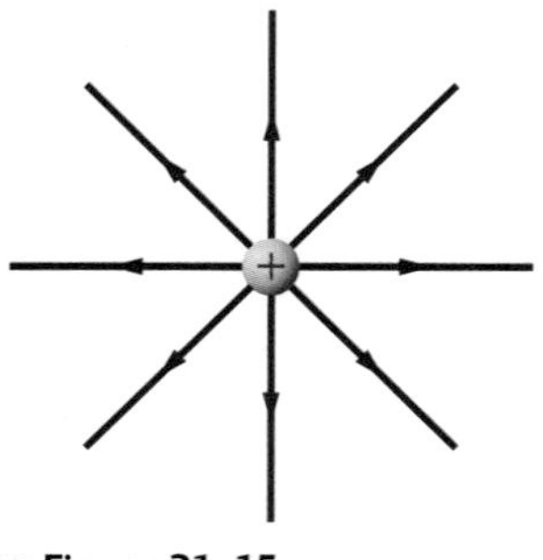

Figure 21-15

49. What SI unit is used to measure electric potential energy? What SI unit is used to measure electric potential difference? (21.2)

50. Define *volt* in terms of the change in potential energy of a charge moving in an electric field. (21.2)

51. Why does a charged object lose its charge when it is touched to the ground? (21.2)

52. A charged rubber rod that is placed on a table maintains its charge for some time. Why is the charged rod not discharged immediately? (21.2)

53. A metal box is charged. Compare the concentration of charge at the corners of the box to the charge concentration on the sides of the box. (21.2)

54. **Computers** Delicate parts in electronic equipment, such as those pictured in **Figure 21-16,** are contained within a metal box inside a plastic case. Why? (21.2)

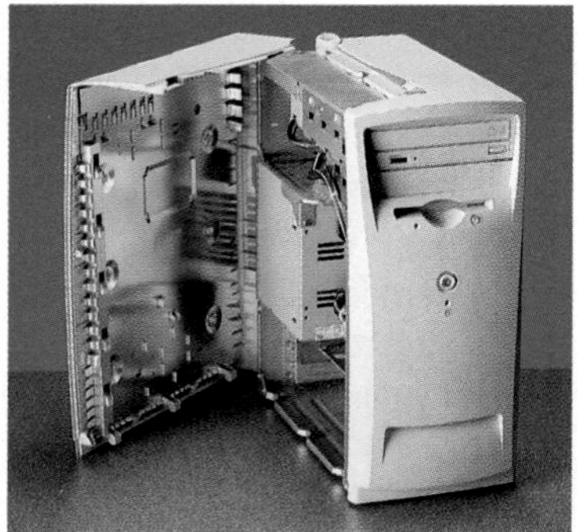

Figure 21-16

Applying Concepts

55. What happens to the strength of an electric field when the charge on the test charge is halved?

56. Does it require more energy or less energy to move a constant positive charge through an increasing electric field?

57. What will happen to the electric potential energy of a charged particle in an electric field when the particle is released and free to move?

58. **Figure 21-17** shows three spheres with charges of equal magnitude, with their signs as shown. Spheres y and z are held in place, but sphere x is free to move. Initially, sphere x is equidistant from spheres y and z. Choose the path that sphere x will begin to follow. Assume that no other forces are acting on the spheres.

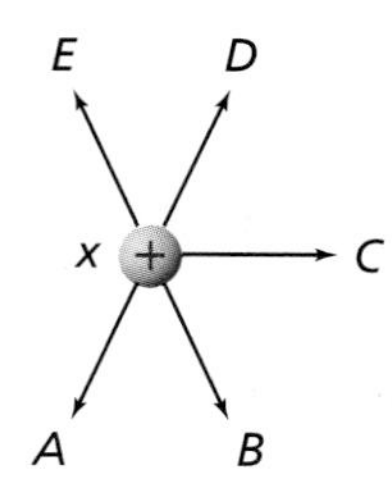

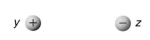

Figure 21-17

 For more problems, go to Additional Problems, Appendix B.

59. What is the unit of electric potential difference in terms of m, kg, s, and C?

60. What do the electric field lines look like when the electric field has the same strength at all points in a region?

61. **Millikan Oil-Drop Experiment** When doing a Millikan oil-drop experiment, it is best to work with drops that have small charges. Therefore, when the electric field is turned on, should you try to find drops that are moving rapidly or slowly? Explain.

62. Two oil drops are held motionless in a Millikan oil-drop experiment.
   a. Can you be sure that the charges are the same?
   b. The ratios of which two properties of the oil drops have to be equal?

63. José and Sue are standing on an insulating platform and holding hands when they are given a charge, as in **Figure 21-18.** José is larger than Sue. Who has the larger amount of charge, or do they both have the same amount?

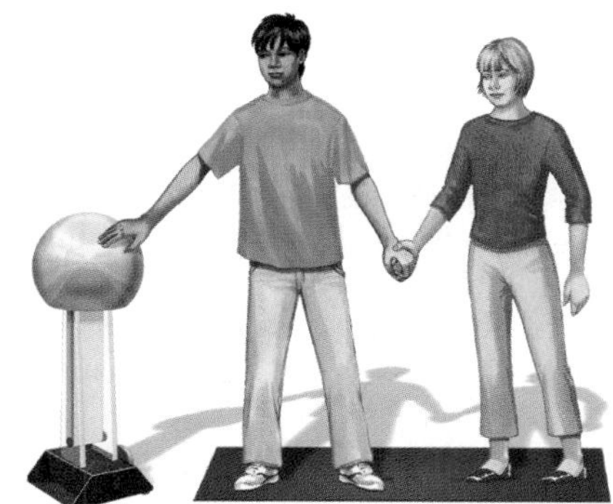

Figure 21-18

64. Which has a larger capacitance, an aluminum sphere with a 1-cm diameter or one with a 10-cm diameter?

65. How can you store different amounts of charge in a capacitor?

Mastering Problems

21.1 Creating and Measuring Electric Fields

The charge of an electron is -1.60×10^{-19} C.

66. What charge exists on a test charge that experiences a force of 1.4×10^{-8} N at a point where the electric field intensity is 5.0×10^{-4} N/C?

67. A positive charge of 1.0×10^{-5} C, shown in **Figure 21-19,** experiences a force of 0.30 N when it is located at a certain point. What is the electric field intensity at that point?

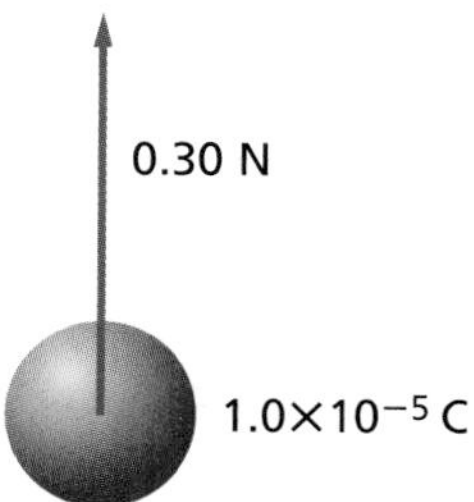

Figure 21-19

68. A test charge experiences a force of 0.30 N on it when it is placed in an electric field intensity of 4.5×10^{5} N/C. What is the magnitude of the charge?

69. The electric field in the atmosphere is about 150 N/C downward.
   a. What is the direction of the force on a negatively charged particle?
   b. Find the electric force on an electron with charge -1.6×10^{-19} C.
   c. Compare the force in part **b** with the force of gravity on the same electron (mass $= 9.1\times10^{-31}$ kg).

70. Carefully sketch each of the following.
   a. the electric field produced by a $+1.0$-μC charge
   b. the electric field resulting from a $+2.0$-μC charge (Make the number of field lines proportional to the change in charge.)

71. A positive test charge of 6.0×10^{-6} C is placed in an electric field of 50.0-N/C intensity, as in **Figure 21-20.** What is the strength of the force exerted on the test charge?

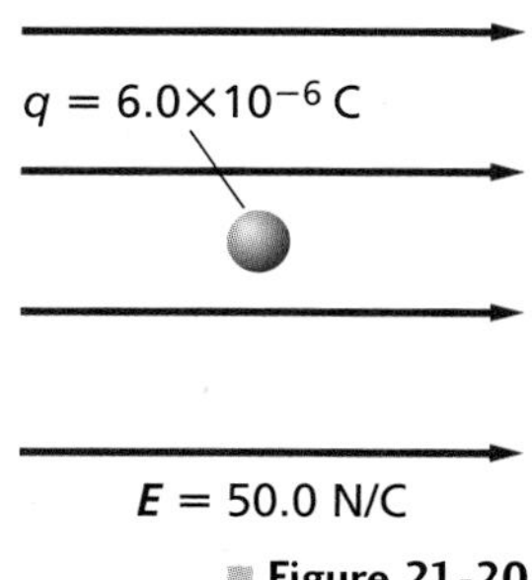

Figure 21-20

72. Charges X, Y, and Z all are equidistant from each other. X has a $+1.0$-μC charge, Y has a $+2.0$-μC charge, and Z has a small negative charge.
   a. Draw an arrow representing the force on charge Z.
   b. Charge Z now has a small positive charge on it. Draw an arrow representing the force on it.

73. In a television picture tube, electrons are accelerated by an electric field having a value of 1.00×10^{5} N/C.
a. Find the force on an electron.
b. If the field is constant, find the acceleration of the electron (mass = 9.11×10^{-31} kg).

74. What is the electric field strength 20.0 cm from a point charge of 8.0×10^{-7} C?

75. The nucleus of a lead atom has a charge of 82 protons.
a. What are the direction and magnitude of the electric field at 1.0×10^{-10} m from the nucleus?
b. What are the direction and magnitude of the force exerted on an electron located at this distance?

21.2 Applications of Electric Fields

76. If 120 J of work is performed to move 2.4 C of charge from the positive plate to the negative plate shown in **Figure 21-21,** what potential difference exists between the plates?

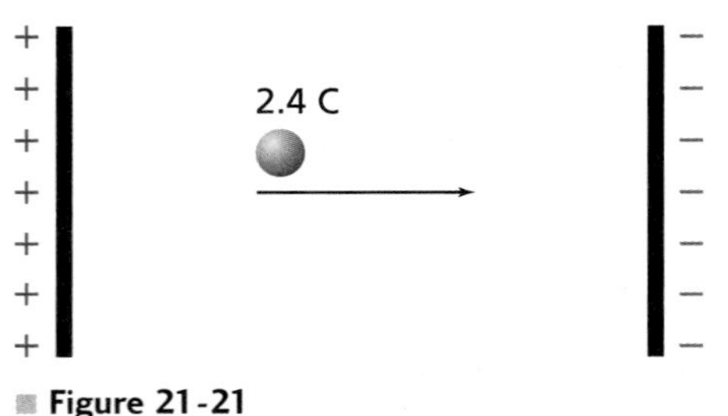

Figure 21-21

77. How much work is done to transfer 0.15 C of charge through an electric potential difference of 9.0 V?

78. An electron is moved through an electric potential difference of 450 V. How much work is done on the electron?

79. A 12-V battery does 1200 J of work transferring charge. How much charge is transferred?

80. The electric field intensity between two charged plates is 1.5×10^{3} N/C. The plates are 0.060 m apart. What is the electric potential difference, in volts, between the plates?

81. A voltmeter indicates that the electric potential difference between two plates is 70.0 V. The plates are 0.020 m apart. What electric field intensity exists between them?

82. A capacitor that is connected to a 45.0-V source contains 90.0 μC of charge. What is the capacitor's capacitance?

83. What electric potential difference exists across a 5.4-μF capacitor that has a charge of 8.1×10^{-4} C?

84. The oil drop shown in **Figure 21-22** is negatively charged and weighs 4.5×10^{-15} N. The drop is suspended in an electric field intensity of 5.6×10^{3} N/C.
a. What is the charge on the drop?
b. How many excess electrons does it carry?

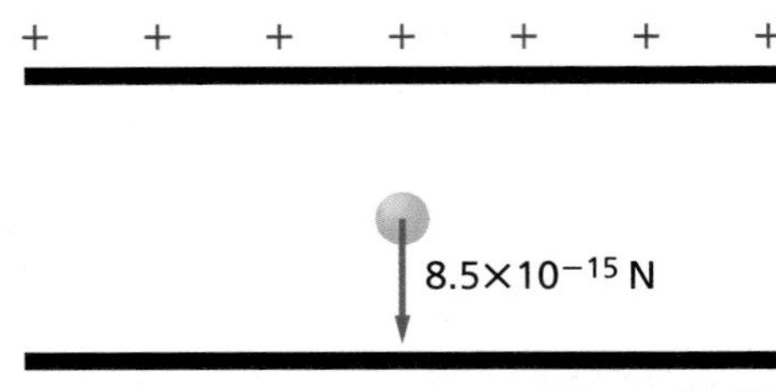

Figure 21-22

85. What is the charge on a 15.0-pF capacitor when it is connected across a 45.0-V source?

86. A force of 0.065 N is required to move a charge of 37 μC a distance of 25 cm in a uniform electric field, as in **Figure 21-23.** What is the size of the electric potential difference between the two points?

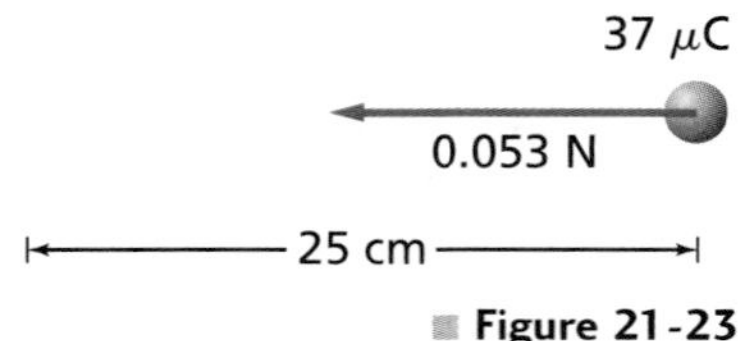

Figure 21-23

87. **Photoflash** The energy stored in a capacitor with capacitance C, and an electric potential difference, ΔV, is represented by $W = \frac{1}{2}C\Delta V^{2}$. One application of this is in the electronic photoflash of a strobe light, like the one in **Figure 21-24.** In such a unit, a capacitor of 10.0 μF is charged to 3.0×10^{2} V. Find the energy stored.

Figure 21-24

88. Suppose it took 25 s to charge the capacitor in problem 87.
a. Find the average power required to charge the capacitor in this time.
b. When this capacitor is discharged through the strobe lamp, it transfers all its energy in 1.0×10^{-4} s. Find the power delivered to the lamp.
c. How is such a large amount of power possible?

89. **Lasers** Lasers are used to try to produce controlled fusion reactions. These lasers require brief pulses of energy that are stored in large rooms filled with capacitors. One such room has a capacitance of 61×10^{-3} F charged to a potential difference of 10.0 kV.
   a. Given that $W = \frac{1}{2}C\Delta V^2$, find the energy stored in the capacitors.
   b. The capacitors are discharged in 10 ns (1.0×10^{-8} s). What power is produced?
   c. If the capacitors are charged by a generator with a power capacity of 1.0 kW, how many seconds will be required to charge the capacitors?

Mixed Review

90. How much work does it take to move 0.25 μC between two parallel plates that are 0.40 cm apart if the field between the plates is 6400 N/C?

91. How much charge is stored on a 0.22-μF parallel plate capacitor if the plates are 1.2 cm apart and the electric field between them is 2400 N/C?

92. Two identical small spheres, 25 cm apart, carry equal but opposite charges of 0.060 μC, as in **Figure 21-25.** If the potential difference between them is 300 V, what is the capacitance of the system?

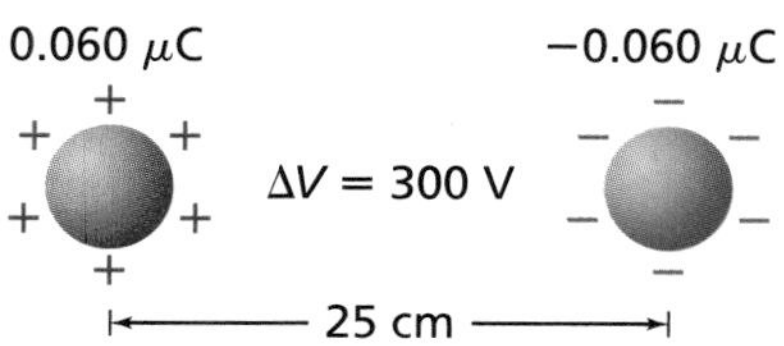

Figure 21-25

93. The plates of a 0.047 μF capacitor are 0.25 cm apart and are charged to a potential difference of 120 V. How much charge is stored on the capacitor?

94. What is the strength of the electric field between the plates of the capacitor in Problem 93 above?

95. An electron is placed between the plates of the capacitor in Problem 93 above, as in **Figure 21-26.** What force is exerted on that electron?

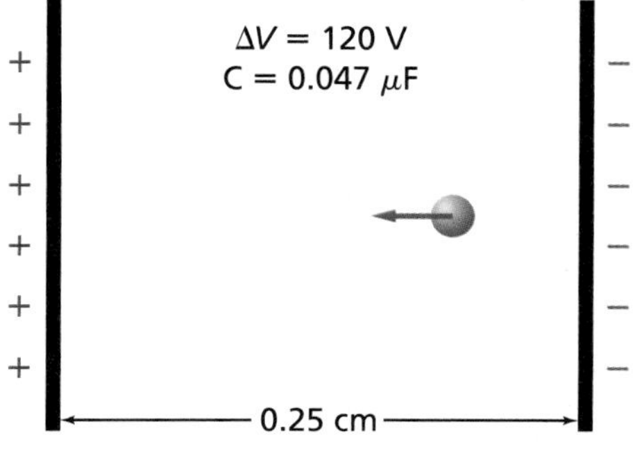

Figure 21-26

96. How much work would it take to move an additional 0.010 μC between the plates at 120 V in Problem 93?

97. The graph in **Figure 21-27** represents the charge stored in a capacitor as the charging potential increases. What does the slope of the line represent?

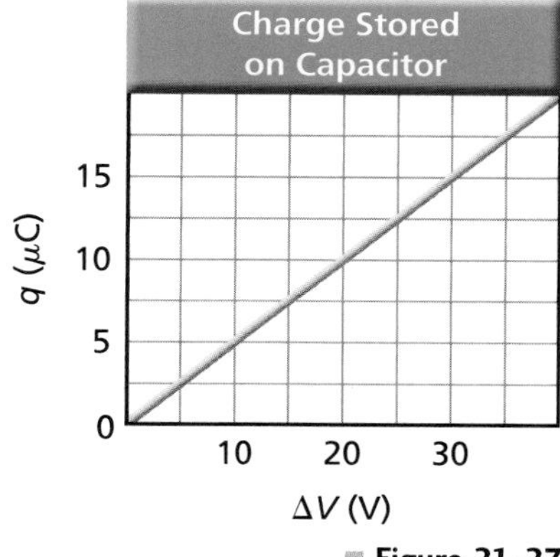

Figure 21-27

98. What is the capacitance of the capacitor represented by Figure 21-27?

99. What does the area under the graph line in Figure 21-27 represent?

100. How much work is required to charge the capacitor in problem 98 to a potential difference of 25 V?

101. The work found in Problem 100 above is not equal to $q\Delta V$. Why not?

102. Graph the electric field strength near a positive point charge as a function of distance from it.

103. Where is the field of a point charge equal to zero?

104. What is the electric field strength at a distance of zero meters from a point charge? Is there such a thing as a true point charge?

Thinking Critically

105. **Apply Concepts** Although a lightning rod is designed to carry charge safely to the ground, its primary purpose is to prevent lightning from striking in the first place. How does it do that?

106. **Analyze and Conclude** In an early set of experiments in 1911, Millikan observed that the following measured charges could appear on a single oil drop. What value of elementary charge can be deduced from these data?
   a. 6.563×10^{-19} C
   b. 8.204×10^{-19} C
   c. 11.50×10^{-19} C
   d. 13.13×10^{-19} C
   e. 16.48×10^{-19} C
   f. 18.08×10^{-19} C
   g. 19.71×10^{-19} C
   h. 22.89×10^{-19} C
   i. 26.13×10^{-19} C

107. **Analyze and Conclude** Two small spheres, A and B, lie on the x-axis, as in **Figure 21-28.** Sphere A has a charge of $+3.00\times10^{-6}$ C. Sphere B is 0.800 m to the right of sphere A and has a charge of -5.00×10^{-6} C. Find the magnitude and direction of the electric field strength at a point above the x-axis that would form the apex of an equilateral triangle with spheres A and B.

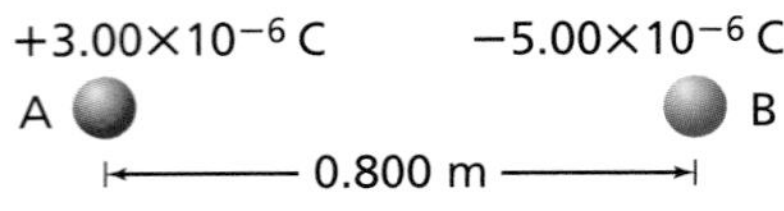

Figure 21-28

108. **Analyze and Conclude** In an ink-jet printer, drops of ink are given a certain amount of charge before they move between two large, parallel plates. The purpose of the plates is to deflect the charges so that they are stopped by a gutter and do not reach the paper. This is shown in **Figure 21-29.** The plates are 1.5-cm long and have an electric field of $E = 1.2\times10^{6}$ N/C between them. Drops with a mass $m = 0.10$ ng, and a charge $q = 1.0\times10^{-16}$ C, are moving horizontally at a speed, $v = 15$ m/s, parallel to the plates. What is the vertical displacement of the drops when they leave the plates? To answer this question, complete the following steps.

a. What is the vertical force on the drops?
b. What is their vertical acceleration?
c. How long are they between the plates?
d. How far are they displaced?

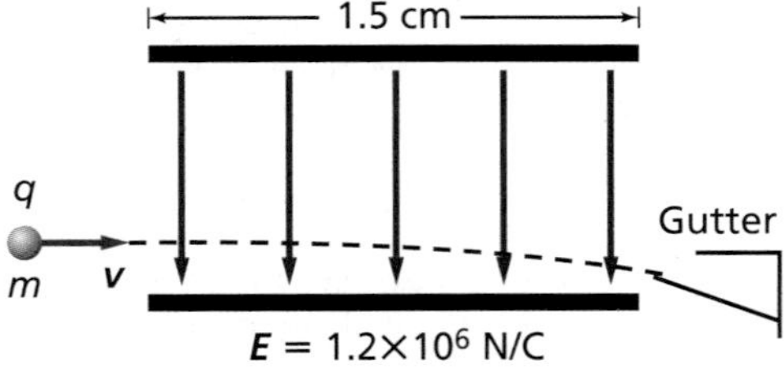

Figure 21-29

109. **Apply Concepts** Suppose the Moon had a net negative charge equal to $-q$, and Earth had a net positive charge equal to $+10q$. What value of q would yield the same magnitude of force that you now attribute to gravity?

Writing in Physics

110. Choose the name of an electric unit, such as coulomb, volt, or farad, and research the life and work of the scientist for whom it was named. Write a brief essay on this person and include a discussion of the work that justified the honor of having a unit named for him.

Cumulative Review

111. Michelson measured the speed of light by sending a beam of light to a mirror on a mountain 35 km away. (Chapter 16)

a. How long does it take light to travel the distance to the mountain and back?
b. Assume that Michelson used a rotating octagon with a mirror on each face of the octagon. Also assume that the light reflects from one mirror, travels to the other mountain, reflects off of a fixed mirror on that mountain, and returns to the rotating mirrors. If the rotating mirror has advanced so that when the light returns, it reflects off of the next mirror in the rotation, how fast is the mirror rotating?
c. If each mirror has a mass of 1.0×10^{1} g and rotates in a circle with an average radius of 1.0×10^{1} cm, what is the approximate centripetal force needed to hold the mirror while is it rotating?

112. **Mountain Scene** You can see an image of a distant mountain in a smooth lake just as you can see a mountain biker next to the lake because light from each strikes the surface of the lake at about the same angle of incidence and is reflected to your eyes. If the lake is about 100 m in diameter, the reflection of the top of the mountain is about in the middle of the lake, the mountain is about 50 km away from the lake, and you are about 2 m tall, then approximately how high above the lake does the top of the mountain reach? (Chapter 17)

113. A converging lens has a focal length of 38.0 cm. If it is placed 60.0 cm from an object, at what distance from the lens will the image be? (Chapter 18)

114. A force, F, is measured between two charges, Q and q, separated by a distance, r. What would the new force be for each of the following? (Chapter 20)

a. r is tripled
b. Q is tripled
c. both r and Q are tripled
d. both r and Q are doubled
e. all three, r, Q, and q, are tripled

 For more problems, go to Additional Problems, Appendix B.

Multiple Choice

1. Why is an electric field measured only by a small test charge?

   Ⓐ so the charge doesn't disturb the field

   Ⓑ because small charges have small momentum

   Ⓒ so its size doesn't nudge the charge to be measured aside

   Ⓓ because an electron always is used as the test charge and electrons are small

2. A force of 14 N exists on charge q, which is 2.1×10^{-9} C. What is the magnitude of the electric field?

   Ⓐ 0.15×10^{-9} N/C Ⓒ 29×10^{-9} N/C

   Ⓑ 6.7×10^{-9} N/C Ⓓ 6.7×10^{9} N/C

3. A positive test charge of 8.7 μC experiences a force of 8.1×10^{-6} N at an angle of 24° N of E. What are the magnitude and direction of the electric field strength at the location of the test charge?

   Ⓐ 7.0×10^{-8} N/C, 24° N of E

   Ⓑ 1.7×10^{-6} N/C, 24° S of W

   Ⓒ 1.1×10^{-3} N/C, 24° W of S

   Ⓓ 9.3×10^{-1} N/C, 24° N of E

4. What is the potential difference between two plates that are 18 cm apart with a field of 4.8×10^{3} N/C?

   Ⓐ 27 V Ⓒ 0.86 kV

   Ⓑ 86 V Ⓓ 27 kV

5. How much work is done on a proton to move it from the negative plate to a positive plate 4.3 cm away if the field is 125 N/C?

   Ⓐ 5.5×10^{-23} J Ⓒ 1.1×10^{-16} J

   Ⓑ 8.6×10^{-19} J Ⓓ 5.4 J

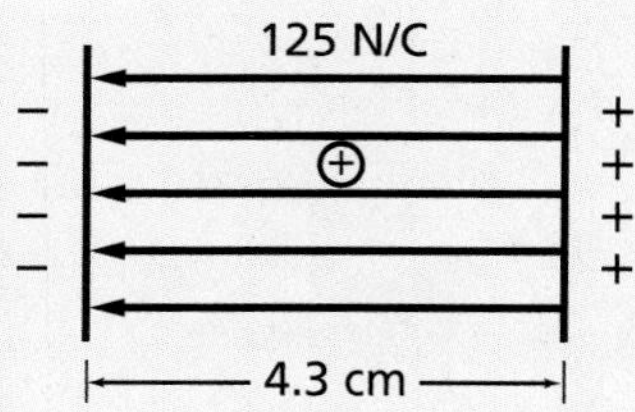

6. How was the magnitude of the field in Millikan's oil-drop experiment determined?

   Ⓐ using a measurable electromagnet

   Ⓑ from the electric potential between the plates

   Ⓒ from the magnitude of the charge

   Ⓓ by an electrometer

7. In an oil drop experiment, a drop with a weight of 1.9×10^{-14} N was suspended motionless when the potential difference between the plates that were 63 mm apart was 0.78 kV. What was the charge on the drop?

   Ⓐ -1.5×10^{-18} C Ⓒ -1.2×10^{-15} C

   Ⓑ -3.9×10^{-16} C Ⓓ -9.3×10^{-13} C

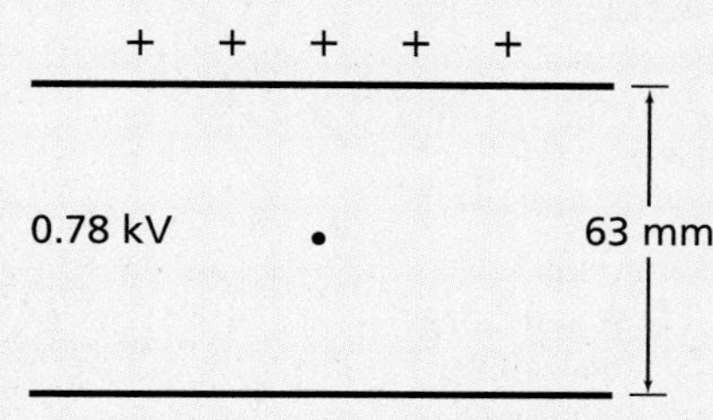

8. A capacitor has a capacitance of 0.093 μF. If the charge on the capacitor is 58 μC, what is the electrical potential difference?

   Ⓐ 5.4×10^{-12} V Ⓒ 6.2×10^{2} V

   Ⓑ 1.6×10^{-6} V Ⓓ 5.4×10^{3} V

Extended Answer

9. Assume 18 extra electrons are on an oil drop. Calculate the charge of the oil drop, and calculate the potential difference needed to suspend it if it has a weight of 6.12×10^{-14} N and the plates are 14.1 mm apart.

✓Test-Taking TIP

Use the Buddy System

Study in a group. A small study group works well because it allows you to draw from a broader base of skills and content knowledge. Keep your group small, question each other, and stay on target.

Chapter 22 Current Electricity

What You'll Learn

- You will explain energy transfer in circuits.
- You will solve problems involving current, potential difference, and resistance.
- You will diagram simple electric circuits.

Why It's Important

The electric tools and appliances that you use are based upon the ability of electric circuits to transfer energy resulting from potential difference, and thus, perform work.

Power Transmission Lines Transmission lines crisscross our country to transfer energy to where it is needed. This transfer is accomplished at high potential differences, often as high as 500,000 V.

Think About This ►
Transmission line voltages are too high to use safely in homes and businesses. Why are such high voltages used in transmission lines?

physicspp.com

Can you get a lightbulb to light?

Question

Given a wire, a battery, and a lightbulb, can you get the bulb to light?

Procedure

1. Obtain a lightbulb, a wire, and a battery. Try to find as many ways as possible to get the lightbulb to light. ***Caution: Wire is sharp and can cut skin. Wire can also get hot if connected across the battery.***
2. Diagram two ways in which you are able to get the lightbulb to work. Be sure to label the battery, the wire, and the bulb.
3. Diagram at least three ways in which you are not able to get the bulb to light.

Analysis

How did you know if electric current was flowing? What do your diagrams of the lit bulb have in common? What do your diagrams of the unlit bulb have in common? From your observations, what conditions seem to be necessary in order for the bulb to light?

Critical Thinking What causes electricity to flow through the bulb?

22.1 Current and Circuits

As you learned in Chapter 11, flowing water at the top of a waterfall has both potential and kinetic energy. However, the large amount of natural potential and kinetic energy available from resources such as Niagara Falls are of little use to people or manufacturers who are 100 km away, unless that energy can be transported efficiently. Electric energy provides the means to transfer large quantities of energy great distances with little loss. This transfer usually is done at high potential differences through power lines, such as those shown in the photo on the left. Once this energy reaches the consumer, it can easily be converted into another form or combination of forms, including sound, light, thermal energy, and motion.

Because electric energy can so easily be changed into other forms, it has become indispensable in our daily lives. Even quick glances around you will likely generate ample examples of the conversion of electric energy. Inside, lights to help you read at night, microwaves and electric ranges to cook food, computers, and stereos all rely on electricity for power. Outside, street lamps, store signs, advertisements, and the starters in cars all use flowing electric charges. In this chapter, you will learn how potential differences, resistance, and current are related. You also will learn about electric power and energy transfer.

▶ **Objectives**

- **Describe** conditions that create current in an electric circuit.
- **Explain** Ohm's law.
- **Design** closed circuits.
- **Differentiate** between power and energy in an electric circuit.

▶ **Vocabulary**

electric current
conventional current
battery
electric circuit
ampere
resistance
resistor
parallel connection
series connection

Producing Electric Current

In Chapter 21, you learned that when two conducting spheres touch, charges flow from the sphere at a higher potential to the one at a lower potential. The flow continues until there is no potential difference between the two spheres.

A flow of charged particles is an **electric current.** In **Figure 22-1a,** two conductors, A and B, are connected by a wire conductor, C. Charges flow from the higher potential difference of B to A through C. This flow of positive charge is called **conventional current.** The flow stops when the potential difference between A, B, and C is zero. You could maintain the electric potential difference between B and A by pumping charged particles from A back to B, as illustrated in **Figure 22-1b.** Since the pump increases the electric potential energy of the charges, it requires an external energy source to run. This energy could come from a variety of sources. One familiar source, a voltaic or galvanic cell (a common dry cell), converts chemical energy to electric energy. Several galvanic cells connected together are called a **battery.** A second source of electric energy—a photovoltaic cell, or solar cell—changes light energy into electric energy.

Figure 22-1 Conventional current is defined as positive charges flowing from the positive plate to the negative plate **(a).** A generator pumps the positive charges back to the positive plate and maintains the current **(b).** In most metals, negatively-charged electrons actually flow from the negative to the positive plate, creating the appearance of positive charges that are moving in the opposite direction.

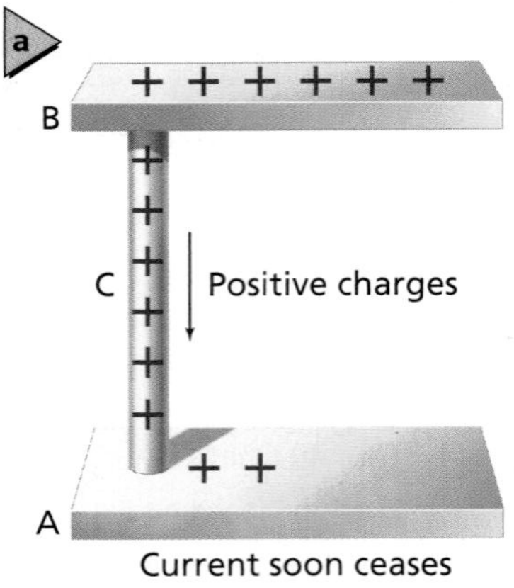

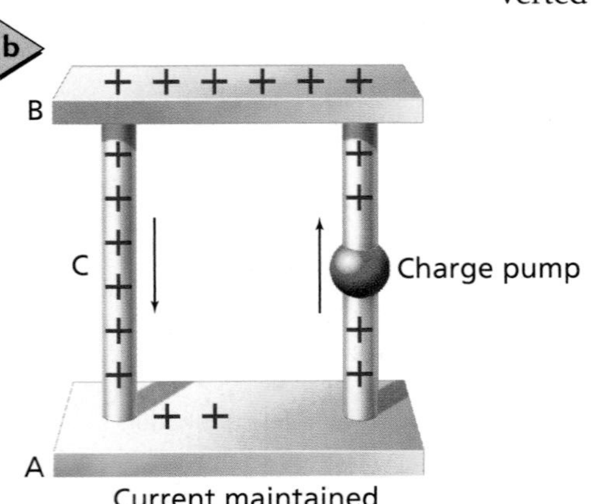

Electric Circuits

The charges in Figure 22-1b move around a closed loop, cycling from the pump to B, through C, to A and back to the pump. Any closed loop or conducting path allowing electric charges to flow is called an **electric circuit.** A circuit includes a charge pump, which increases the potential energy of the charges flowing from A to B, and a device that reduces the potential energy of the charges flowing from B to A. The potential energy lost by the charges, qV, moving through the device is usually converted into some other form of energy. For example, electric energy is converted to kinetic energy by a motor, to light energy by a lamp, and to thermal energy by a heater.

A charge pump creates the flow of charged particles that make up a current. Consider a generator driven by a waterwheel, such as the one pictured in **Figure 22-2a.** The water falls and rotates the waterwheel and generator. Thus, the kinetic energy of the water is converted to electric energy by the generator. The generator, like the charge pump, increases the electric potential difference, V. Energy in the amount qV is needed to increase the potential difference of the charges. This energy comes from the change in energy of the water. Not all of the water's kinetic energy, however, is converted to electric energy, as shown in **Figure 22-2b.**

If the generator attached to the waterwheel is connected to a motor, the charges in the wire flow into the motor. The flow of charges continues through the circuit back to the generator. The motor converts electric energy to kinetic energy.

Conservation of charge Charges cannot be created or destroyed, but they can be separated. Thus, the total amount of charge—the number of negative electrons and positive ions—in the circuit does not change. If one coulomb flows through the generator in 1 s, then one coulomb also will flow through the motor in 1 s. Thus, charge is a conserved quantity. Energy also is conserved. The change in electric energy, ΔE, equals qV. Because q is conserved,

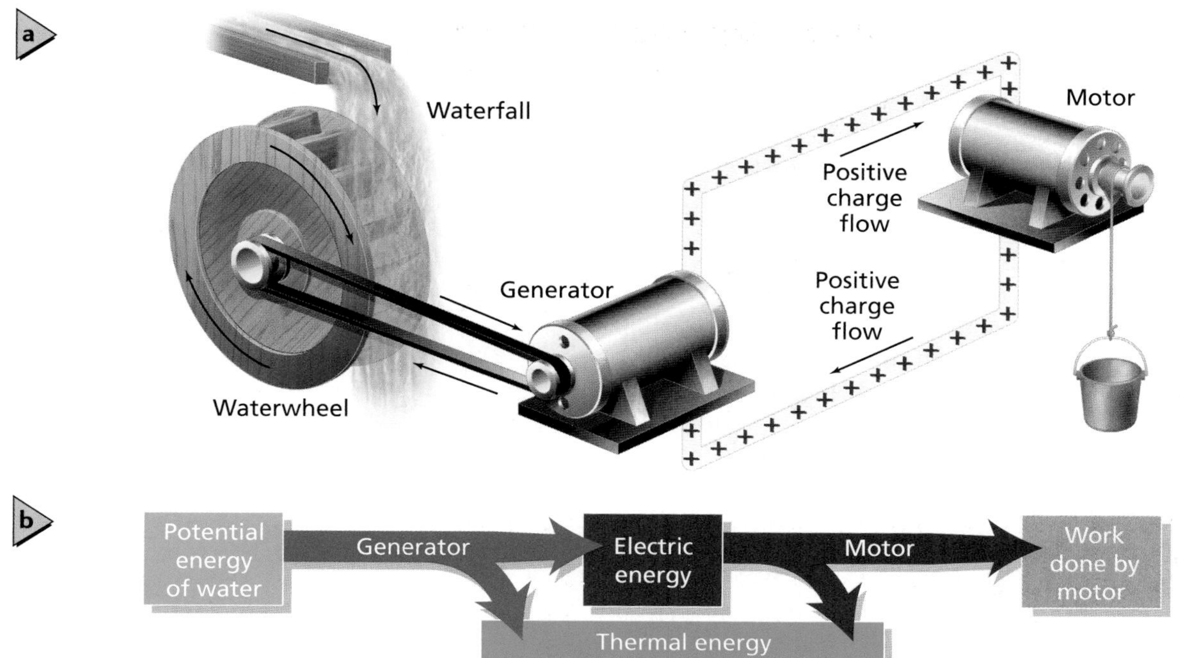

the net change in potential energy of the charges going completely around the circuit must be zero. The increase in potential difference produced by the generator equals the decrease in potential difference across the motor.

If the potential difference between two wires is 120 V, the waterwheel and the generator must do 120 J of work on each coulomb of charge that is delivered. Every coulomb of charge moving through the motor delivers 120 J of energy to the motor.

Figure 22-2 The potential energy of the waterfall is eventually converted into work done on the bucket **(a).** The production and use of electric current is not 100 percent efficient. Some thermal energy is produced by the splashing water, friction, and electric resistance **(b).**

Concepts In Motion

Interactive Figure To see an animation on current and circuits, visit physicspp.com.

Rates of Charge Flow and Energy Transfer

Power, which is defined in watts, W, measures the rate at which energy is transferred. If a generator transfers 1 J of kinetic energy to electric energy each second, it is transferring energy at the rate of 1 J/s, or 1 W. The energy carried by an electric current depends on the charge transferred, q, and the potential difference across which it moves, V. Thus, $E = qV$. Recall from Chapter 20 that the unit for the quantity of electric charge is the coulomb. The rate of flow of electric charge, q/t, called electric current, is measured in coulombs per second. Electric current is represented by I, so $I = q/t$. A flow of 1 C/s is called an **ampere,** A.

The energy carried by an electric current is related to the voltage, $E = qV$. Since current, $I = q/t$, is the rate of charge flow, the power, $P = E/t$, of an electric device can be determined by multiplying voltage and current. To derive the familiar form of the equation for the power delivered to an electric device, you can use $P = E/t$ and substitute $E = qV$ and $q = It$.

Power $P = IV$

Power is equal to the current times the potential difference.

If the current through the motor in Figure 22-2a is 3.0 A and the potential difference is 120 V, the power in the motor is calculated using the expression $P = (3.0\ \text{C/s})(120\ \text{J/C}) = 360\ \text{J/s}$, which is 360 W.

▶ EXAMPLE Problem 1

Electric Power and Energy A 6.0-V battery delivers a 0.50-A current to an electric motor connected across its terminals.

a. What power is delivered to the motor?

b. If the motor runs for 5.0 min, how much electric energy is delivered?

1 Analyze and Sketch the Problem

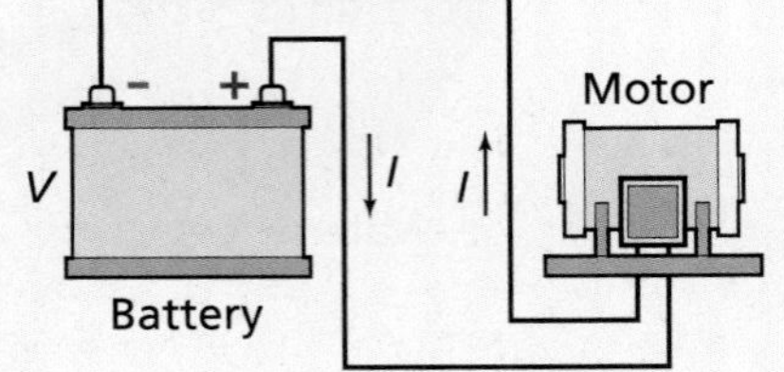

- Draw a circuit showing the positive terminal of a battery connected to a motor and the return wire from the motor connected to the negative terminal of the battery.
- Show the direction of conventional current.

Known:	Unknown:
$V = 6.0$ V	$P = ?$
$I = 0.50$ A	$E = ?$
$t = 5.0$ min	

2 Solve for the Unknown

a. Use $P = IV$ to find the power.

$$P = IV$$

$$P = (0.50\text{ A})(6.0\text{ V})$$ Substitue $I = 0.50$ A, $V = 6.0$ V

$$= 3.0\text{ W}$$

b. In Chapter 10, you learned that $P = E/t$. Solve for E to find the energy.

$$E = Pt$$

$$= (3.0\text{ W})(5.0\text{ min})$$ Substitute $P = 3.0$ W, $t = 5.0$ min

$$= (3.0\text{ J/s})(5.0\text{ min})\left(\frac{60\text{ s}}{1\text{ min}}\right)$$

$$= 9.0\times10^{2}\text{ J}$$

Math Handbook
Significant Digits
page 834

3 Evaluate the Answer

- **Are the units correct?** Power is measured in watts, and energy is measured in joules.
- **Is the magnitude realistic?** With relatively low voltage and current, a few watts of power is reasonable.

▶ PRACTICE Problems

• Additional Problems, Appendix B
• Solutions to Selected Problems, Appendix C

1. The current through a lightbulb connected across the terminals of a 125-V outlet is 0.50 A. At what rate does the bulb convert electric energy to light? (Assume 100 percent efficiency.)

2. A car battery causes a current of 2.0 A through a lamp and produces 12 V across it. What is the power used by the lamp?

3. What is the current through a 75-W lightbulb that is connected to a 125-V outlet?

4. The current through the starter motor of a car is 210 A. If the battery maintains 12 V across the motor, how much electric energy is delivered to the starter in 10.0 s?

5. A flashlight bulb is rated at 0.90 W. If the lightbulb drops 3.0 V, how much current goes through it?

Table 22-1		
Changing Resistance		
Factor	**How resistance changes**	**Example**
Length	Resistance increases as length increases.	L_1, L_2; $R_{L1} > R_{L2}$
Cross-sectional area	Resistance increases as cross-sectional area decreases.	A_1, A_2; $R_{A1} > R_{A2}$
Temperature	Resistance increases as temperature increases.	T_1, T_2; $R_{T1} > R_{T2}$
Material	Keeping length, cross-sectional area, and temperature constant, resistance varies with the material used.	*R* increases (upward): Platinum, Iron, Aluminum, Gold, Copper, Silver

Resistance and Ohm's Law

George Ohm (1787-1854) studied the relationship between current and potential difference. Ohm's Law states that current is directly proportional to the potential difference. Suppose two conductors have a potential difference between them. If they are connected with a copper rod, a large current is created. On the other hand, putting a glass rod between them creates almost no current. The property determining how much current will flow is called **resistance. Table 22-1** lists some of the factors that impact resistance. Resistance is measured by placing a potential difference across a conductor and dividing the voltage by the current. The resistance, *R*, is defined as the ratio of electric potential difference, *V*, to the current, *I*.

Resistance $R = \frac{V}{I}$

Resistance is equal to potential voltage divided by current.

The resistance of the conductor, *R*, is measured in ohms. One ohm (1 Ω) is the resistance permitting an electric charge of 1 A to flow when a potential difference of 1 V is applied across the resistance. A simple circuit relating resistance, current, and voltage is shown in **Figure 22-3.** The circuit is completed by a connection to an ammeter, which is a device that measures current.

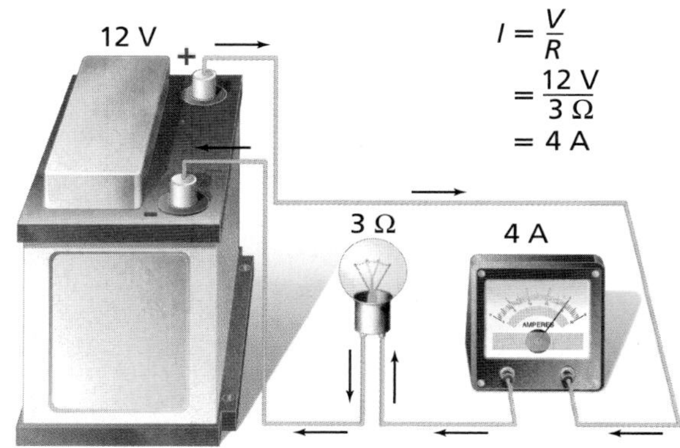

Figure 22-3 One ohm, Ω, is defined as 1 V/A. In a circuit with a 3-Ω resistance and a 12-V battery, there is a 4-A current.

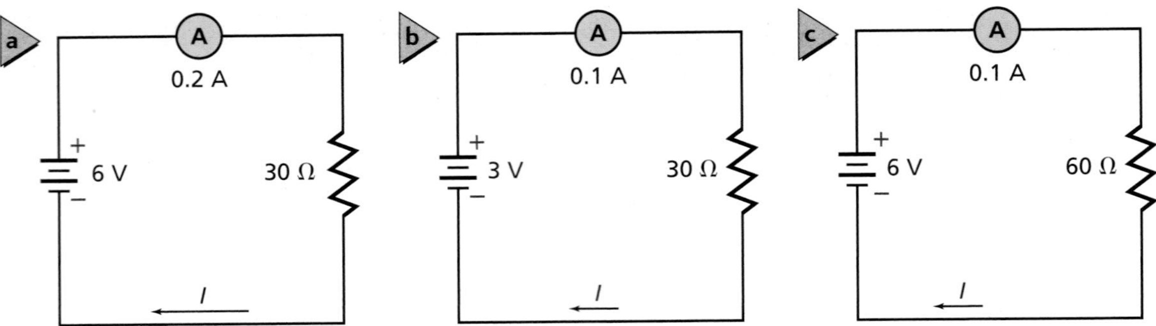

Figure 22-4 The current through a simple circuit **(a)** can be regulated by removing some of the dry cells **(b)** or by increasing the resistance of the circuit **(c)**.

The unit for resistance is named for German scientist Georg Simon Ohm, who found that the ratio of potential difference to current is constant for a given conductor. The resistance for most conductors does not vary as the magnitude or direction of the potential applied to it changes. A device having constant resistance independent of the potential difference obeys Ohm's law.

Most metallic conductors obey Ohm's law, at least over a limited range of voltages. Many important devices, however, do not. A radio and a pocket calculator contain many devices, such as transistors and diodes, that do not obey Ohm's law. Even a lightbulb has resistance that depends on its temperature and does not obey Ohm's law.

Wires used to connect electric devices have low resistance. A 1-m length of a typical wire used in physics labs has a resistance of about 0.03 Ω. Wires used in home wiring offer as little as 0.004 Ω of resistance for each meter of length. Because wires have so little resistance, there is almost no potential drop across them. To produce greater potential drops, a large resistance concentrated into a small volume is necessary. A **resistor** is a device designed to have a specific resistance. Resistors may be made of graphite, semiconductors, or wires that are long and thin.

There are two ways to control the current in a circuit. Because $I = V/R$, I can be changed by varying V, R, or both. **Figure 22-4a** shows a simple circuit. When V is 6 V and R is 30 Ω, the current is 0.2 A. How could the current be reduced to 0.1 A? According to Ohm's law, the greater the voltage placed across a resistor, the larger the current passing through it. If the current through a resistor is cut in half, the potential difference also is cut

Figure 22-5 A potentiometer can be used to change current in an electric circuit.

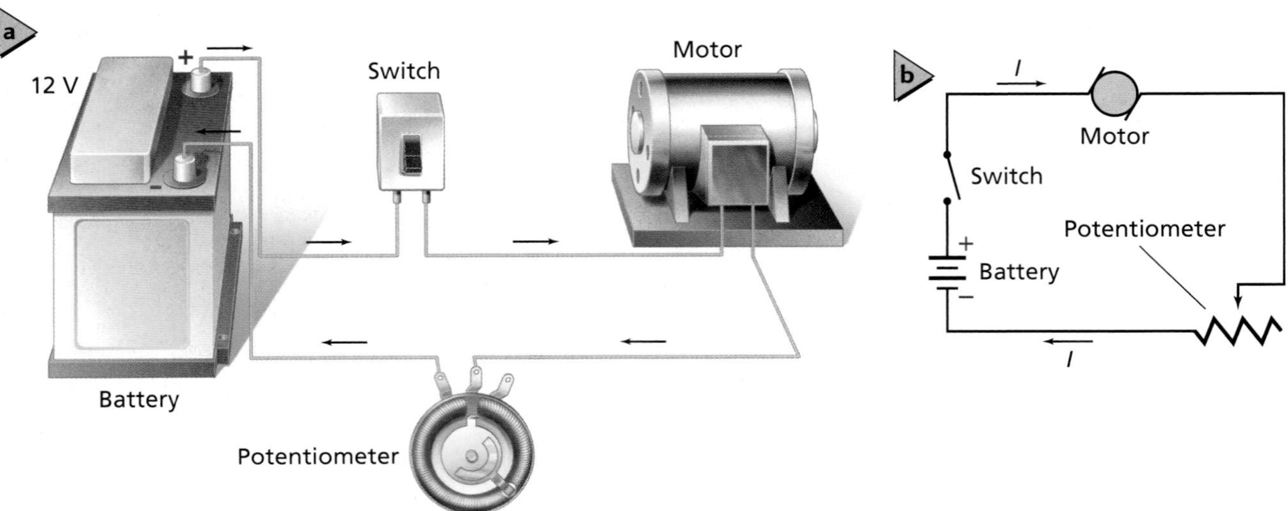

in half. In **Figure 22-4b,** the voltage applied across the resistor is reduced from 6 V to 3 V to reduce the current to 0.1 A. A second way to reduce the current to 0.1 A is to replace the 30-Ω resistor with a 60-Ω resistor, as shown in **Figure 22-4c.**

Resistors often are used to control the current in circuits or parts of circuits. Sometimes, a smooth, continuous variation of the current is desired. For example, the speed control on some electric motors allows continuous, rather than step-by-step, changes in the rotation of the motor. To achieve this kind of control, a variable resistor, called a potentiometer, is used. A circuit containing a potentiometer is shown in **Figure 22-5.** Some variable resistors consist of a coil of resistance wire and a sliding contact point. Moving the contact point to various positions along the coil varies the amount of wire in the circuit. As more wire is placed in the circuit, the resistance of the circuit increases; thus, the current changes in accordance with the equation $I = V/R$. In this way, the speed of a motor can be adjusted from fast, with little wire in the circuit, to slow, with a lot of wire in the circuit. Other examples of using variable resistors to adjust the levels of electrical energy can be found on the front of a TV: the volume, brightness, contrast, tone, and hue controls are all variable resistors.

APPLYING PHYSICS

Resistance The resistance of an operating 100-W lightbulb is about 140 Ω. When the lightbulb is turned off and at room temperature, its resistance is only about 10 Ω. This is because of the great difference between room temperature and the lightbulb's operating temperature.

The human body The human body acts as a variable resistor. When dry, skin's resistance is high enough to keep currents that are produced by small and moderate voltages low. If skin becomes wet, however, its resistance is lower, and the electric current can rise to dangerous levels. A current as low as 1 mA can be felt as a mild shock, while currents of 15 mA can cause loss of muscle control and currents of 100 mA can cause death.

Biology Connection

Diagramming Circuits

A simple circuit can be described in words. It can also be depicted by photographs or artists' drawings of the parts. Most frequently, however, an electric circuit is drawn using standard symbols for the circuit elements. Such a diagram is called a circuit schematic. Some of the symbols used in circuit schematics are shown in **Figure 22-6.**

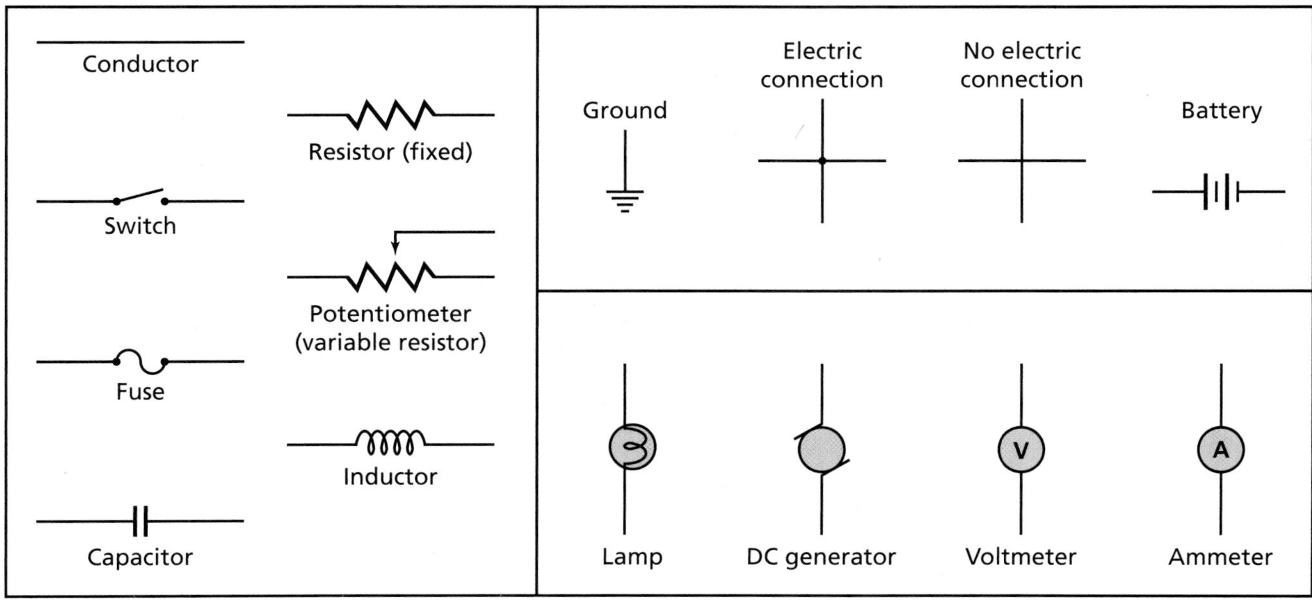

Figure 22-6 These symbols commonly are used to diagram electric circuits.

▶ EXAMPLE Problem 2

Current Through a Resistor A 30.0-V battery is connected to a 10.0-Ω resistor. What is the current in the circuit?

1 Analyze and Sketch the Problem

- Draw a circuit containing a battery, an ammeter, and a resistor.
- Show the direction of the conventional current.

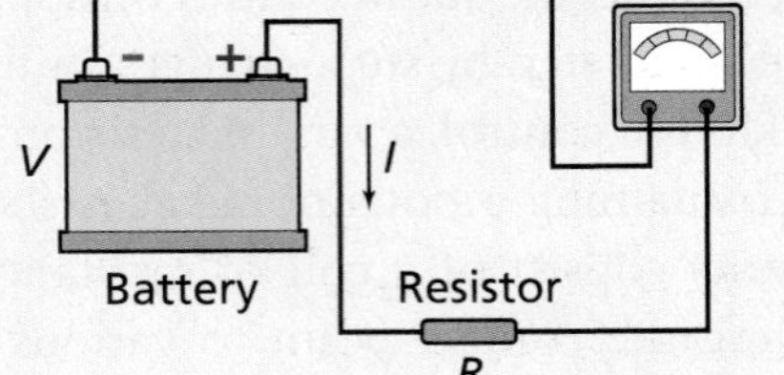

Known:	Unknown:
$V = 30.0$ V	$I = ?$
$R = 10.0\ \Omega$	

2 Solve for the Unknown

Use $I = V/R$ to determine the current.

$$I = \frac{V}{R}$$

$$= \frac{30.0\text{ V}}{10.0\ \Omega} \quad \text{Substitute } V = 30.0\text{ V}, R = 10.0\ \Omega$$

$$= 3.00\text{ A}$$

Math Handbook
Operations with Significant Digits
pages 835–836

3 Evaluate the Answer

- **Are the units correct?** Current is measured in amperes.
- **Is the magnitude realistic?** There is a fairly large voltage and a small resistance, so a current of 3.00 A is reasonable.

▶ PRACTICE Problems

• Additional Problems, Appendix B
• Solutions to Selected Problems, Appendix C

For all problems, assume that the battery voltage and lamp resistances are constant, no matter what current is present.

6. An automobile panel lamp with a resistance of 33 Ω is placed across a 12-V battery. What is the current through the circuit?

7. A motor with an operating resistance of 32 Ω is connected to a voltage source. The current in the circuit is 3.8 A. What is the voltage of the source?

8. A sensor uses 2.0×10^{-4} A of current when it is operated by a 3.0-V battery. What is the resistance of the sensor circuit?

9. A lamp draws a current of 0.50 A when it is connected to a 120-V source.

- **a.** What is the resistance of the lamp?
- **b.** What is the power consumption of the lamp?

10. A 75-W lamp is connected to 125 V.

- **a.** What is the current through the lamp?
- **b.** What is the resistance of the lamp?

11. A resistor is added to the lamp in the previous problem to reduce the current to half of its original value.

- **a.** What is the potential difference across the lamp?
- **b.** How much resistance was added to the circuit?
- **c.** How much power is now dissipated in the lamp?

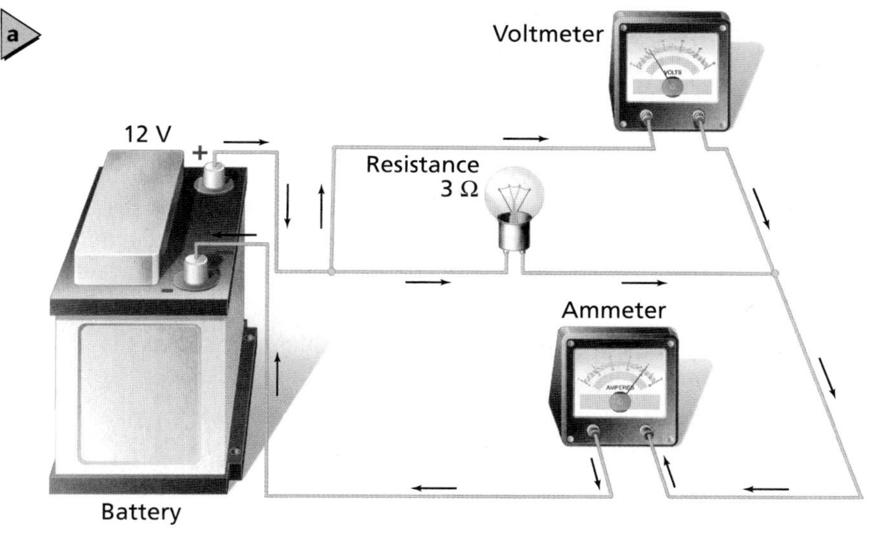

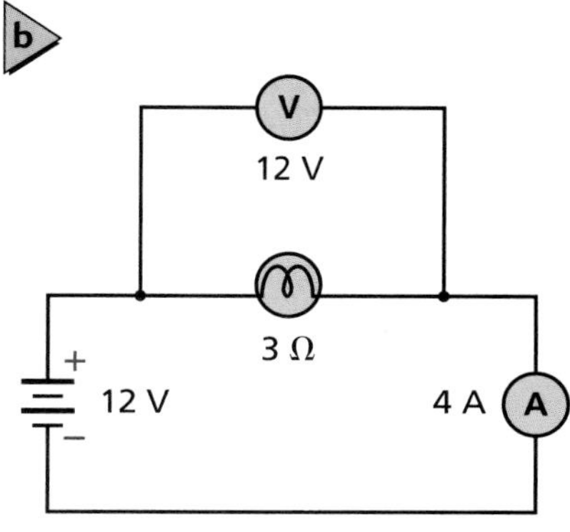

■ **Figure 22-7** A simple electric circuit is represented pictorially **(a)** and schematically **(b).**

An artist's drawing and a schematic of the same circuit are shown in **Figures 22-7a** and **22-7b.** Notice in both the drawing and the schematic that the electric charge is shown flowing out of the positive terminal of the battery. To draw schematic diagrams, use the problem-solving strategy below, and always set up a conventional current.

You learned that an ammeter measures current and a voltmeter measures potential differences. Each instrument has two terminals, usually labeled + and −. A voltmeter measures the potential difference across any component of a circuit. When connecting the voltmeter in a circuit, always connect the + terminal to the end of the circuit component that is closer to the positive terminal of the battery, and connect the − terminal to the other side of the component.

PROBLEM-SOLVING **Strategies**

Drawing Schematic Diagrams

Follow these steps when drawing schematic diagrams.

1. Draw the symbol for the battery or other source of electric energy, such as a generator, on the left side of the page. Put the positive terminal on top.
2. Draw a wire coming out of the positive terminal. When you reach a resistor or other device, draw the symbol for it.
3. If you reach a point where there are two current paths, such as at a voltmeter, draw a ——┴—— in the diagram. Follow one path until the two current paths join again. Then draw the second path.
4. Follow the current path until you reach the negative terminal of the battery.
5. Check your work to make sure that you have included all parts and that there are complete paths for the current to follow.

MINI LAB

Current Affairs

Do you think that current diminishes as it passes through different elements in the circuit? As a scientist, you can test this question.

1. Draw a circuit that includes a power supply and two miniature lamps.

2. Draw the circuit again and include an ammeter to measure the current between the power supply and the lamps.

3. In a third diagram, show the ammeter at a position to measure the current between the lamps.

Analyze and Conclude

4. Predict if the current between the lamps will be more than, less than, or the same as the current before the lamps. Explain.

5. Test your prediction by building the circuits. ***CAUTION: Wire is sharp and can cut skin.***

PRACTICE Problems

- Additional Problems, Appendix B
- Solutions to Selected Problems, Appendix C

12. Draw a circuit diagram to include a 60.0-V battery, an ammeter, and a resistance of 12.5 Ω in series. Indicate the ammeter reading and the direction of the current.

13. Draw a series-circuit diagram showing a 4.5-V battery, a resistor, and an ammeter that reads 85 mA. Determine the resistance and label the resistor. Choose a direction for the conventional current and indicate the positive terminal of the battery.

14. Add a voltmeter to measure the potential difference across the resistors in problems 12 and 13 and repeat the problems.

15. Draw a circuit using a battery, a lamp, a potentiometer to adjust the lamp's brightness, and an on-off switch.

16. Repeat the previous problem, adding an ammeter and a voltmeter across the lamp.

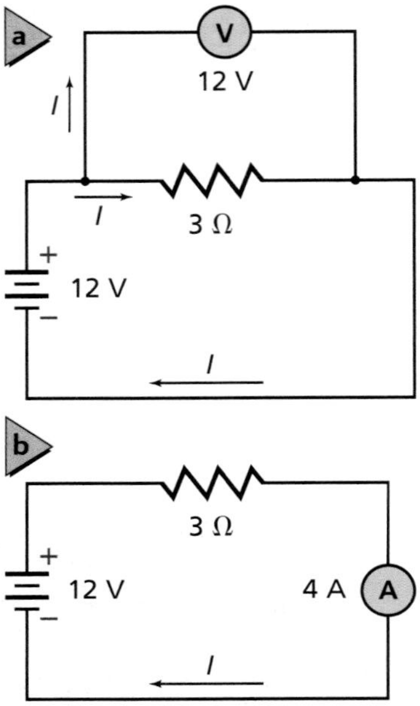

Figure 22-8 These schematics show a parallel **(a)** and a series circuit **(b).**

When a voltmeter is connected across another component, it is called a **parallel connection** because the circuit component and the voltmeter are aligned parallel to each other in the circuit, as diagrammed in **Figure 22-8a.** Any time the current has two or more paths to follow, the connection is labeled *parallel.* The potential difference across the voltmeter is equal to the potential difference across the circuit element. Always associate the words *voltage across* with a parallel connection.

An ammeter measures the current through a circuit component. The same current going through the component must go through the ammeter, so there can be only one current path. A connection with only one current path, called a **series connection,** is shown in **Figure 22-8b.** To add an ammeter to a circuit, the wire connected to the circuit component must be removed and connected to the ammeter instead. Then, another wire is connected from the second terminal of the ammeter to the circuit component. In a series connection, there can be only a single path through the connection. Always associate the words *current through* with a series connection.

22.1 Section Review

17. Schematic Draw a schematic diagram of a circuit that contains a battery and a lightbulb. Make sure the lightbulb will light in this circuit.

18. Resistance Joe states that because $R = V/I$, if he increases the voltage, the resistance will increase. Is Joe correct? Explain.

19. Resistance You want to measure the resistance of a long piece of wire. Show how you would construct a circuit with a battery, a voltmeter, an ammeter, and the wire to be tested to make the measurement. Specify what you would measure and how you would compute the resistance.

20. Power A circuit has 12 Ω of resistance and is connected to a 12-V battery. Determine the change in power if the resistance decreases to 9.0 Ω.

21. Energy A circuit converts 2.2×10^3 J of energy when it is operated for 3.0 min. Determine the amount of energy it will convert when it is operated for 1 h.

22. Critical Thinking We say that power is "dissipated" in a resistor. To dissipate is to use, to waste, or to squander. What is "used" when charge flows through a resistor?

22.2 Using Electric Energy

Many familiar household appliances convert electric energy to some other form, such as light, kinetic energy, sound, or thermal energy. When you turn on one of these appliances, you complete a circuit and begin converting electric energy. In this section, you will learn to determine the rate of energy conversion and the amount that is converted.

► **Objectives**

- **Explain** how electric energy is converted into thermal energy.
- **Explore** ways to deliver electric energy to consumers near and far.
- **Define** kilowatt-hour.

► **Vocabulary**

superconductor
kilowatt-hour

Energy Transfer in Electric Circuits

Energy that is supplied to a circuit can be used in many different ways. A motor converts electric energy to mechanical energy, and a lamp changes electric energy into light. Unfortunately, not all of the energy delivered to a motor or a lamp ends up in a useful form. Lightbulbs, especially incandescent lightbulbs, become hot. Motors are often far too hot to touch. In each case, some of the electric energy is converted into thermal energy. You will now examine some devices that are designed to convert as much energy as possible into thermal energy.

Heating a resistor Current moving through a resistor causes it to heat up because flowing electrons bump into the atoms in the resistor. These collisions increase the atoms' kinetic energy and, thus, the temperature of the resistor. A space heater, a hot plate, and the heating element in a hair dryer all are designed to convert electric energy into thermal energy. These and other household appliances, such as those pictured in **Figure 22-9,** act like resistors when they are in a circuit. When charge, q, moves through a resistor, its potential difference is reduced by an amount, V. As you have learned, the energy change is represented by qV. In practical use, the rate at which energy is changed—the power, $P = E/t$—is more important. Earlier, you learned that current is the rate at which charge flows, $I = q/t$, and that power dissipated in a resistor is represented by $P = IV$. For a resistor, $V = IR$. Thus, if you know I and R, you can substitute $V = IR$ into the equation for electric power to obtain the following.

Power $P = I^2R$

Power is equal to current squared times resistance.

Thus, the power dissipated in a resistor is proportional both to the square of the current passing through it and to the resistance. If you know V and R, but not I, you can substitute $I = V/R$ into $P = IV$ to obtain the following equation.

Power $P = \frac{V^2}{R}$

Power is equal to the potential difference squared divided by the resistance.

Figure 22-9 These appliances are designed to change electric energy into thermal energy.

The power is the rate at which energy is converted from one form to another. Energy is changed from electric to thermal energy, and the temperature of the resistor rises. If the resistor is an immersion heater or burner on an electric stovetop, for example, heat flows into cold water fast enough to bring the water to the boiling point in a few minutes.

If power continues to be dissipated at a uniform rate, then after time t, the energy converted to thermal energy will be $E = Pt$. Because $P = I^2R$ and $P = V^2/R$, the total energy to be converted to thermal energy can be written in the following ways.

Thermal Energy

$$E = Pt$$
$$E = I^2Rt$$
$$E = \left(\frac{V^2}{R}\right)t$$

Thermal energy is equal to the power dissipated multiplied by the time. It is also equal to the current squared multiplied by resistance and time as well as the voltage squared divided by resistance multiplied by time.

▶ EXAMPLE Problem 3

Electric Heat **A heater has a resistance of 10.0 Ω. It operates on 120.0 V.**

a. **What is the power dissipated by the heater?**

b. **What thermal energy is supplied by the heater in 10.0 s?**

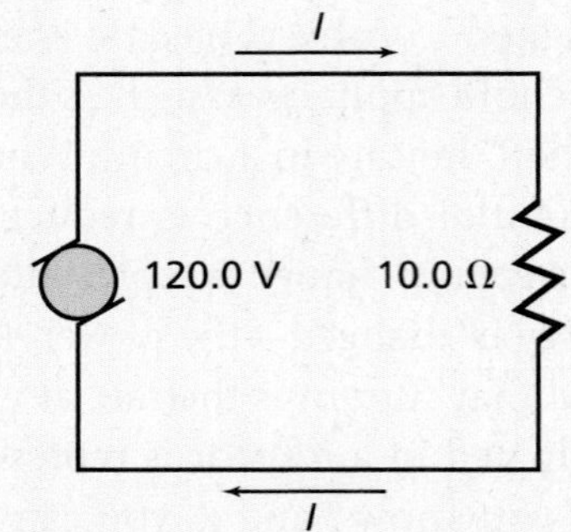

1 Analyze and Sketch the Problem

- Sketch the situation.
- Label the known circuit components, which are a 120.0-V potential difference source and a 10.0-Ω resistor.

Known:	Unknown:
$R = 10.0\ \Omega$	$P = ?$
$V = 120.0\ \text{V}$	$E = ?$
$t = 10.0\ \text{s}$	

2 Solve for the Unknown

a. Because R and V are known, use $P = V^2/R$.

$$P = \frac{(120.0\ \text{V})^2}{10.0\ \Omega}$$ Substitute $V = 120.0$ V, $R = 10.0\ \Omega$

$$= 1.44\ \text{kW}$$

b. Solve for the energy.

$$E = Pt$$
$$= (1.44\ \text{kW})(10.0\ \text{s})$$ Substitute $P = 1.44$ kW, $t = 10.0$ s
$$= 14.4\ \text{kJ}$$

Personal Tutor For an online tutorial on electric heat, visit physicspp.com.

3 Evaluate the Answer

- **Are the units correct?** Power is measured in watts, and energy is measured in joules.
- **Are the magnitudes realistic?** For power, $10^2 \times 10^2 \times 10^{-1} = 10^3$, so kilowatts is reasonable. For energy, $10^3 \times 10^1 = 10^4$, so an order of magnitude of 10,000 joules is reasonable.

▶ PRACTICE Problems

• Additional Problems, Appendix B
• Solutions to Selected Problems, Appendix C

23. A 15-Ω electric heater operates on a 120-V outlet.
  a. What is the current through the heater?
  b. How much energy is used by the heater in 30.0 s?
  c. How much thermal energy is liberated in this time?

24. A 39-Ω resistor is connected across a 45-V battery.
  a. What is the current in the circuit?
  b. How much energy is used by the resistor in 5.0 min?

25. A 100.0-W lightbulb is 22 percent efficient. This means that 22 percent of the electric energy is converted to light energy.
  a. How many joules does the lightbulb convert into light each minute it is in operation?
  b. How many joules of thermal energy does the lightbulb produce each minute?

26. The resistance of an electric stove element at operating temperature is 11 Ω.
  a. If 220 V are applied across it, what is the current through the stove element?
  b. How much energy does the element convert to thermal energy in 30.0 s?
  c. The element is used to heat a kettle containing 1.20 kg of water. Assume that 65 percent of the heat is absorbed by the water. What is the water's increase in temperature during the 30.0 s?

27. A 120-V water heater takes 2.2 h to heat a given volume of water to a certain temperature. How long would a 240-V unit operating with the same current take to accomplish the same task?

Superconductors A **superconductor** is a material with zero resistance. There is no restriction of current in superconductors, so there is no potential difference, V, across them. Because the power that is dissipated in a conductor is given by the product IV, a superconductor can conduct electricity without loss of energy. At present, almost all superconductors must be kept at temperatures below 100 K. The practical uses of superconductors include MRI magnets and in synchrotrons, which use huge amounts of current and can be kept at temperatures close to 0 K.

■ **Figure 22-10** In the year 2000, energy produced by Itaipú Dam met 24 percent of Brazil's electric energy needs and 95 percent of Paraguay's.

Transmission of Electric Energy

Hydroelectric facilities, such as the one at Itaipú Dam, shown in **Figure 22-10,** are capable of producing a great deal of energy. This hydroelectric energy often must be transmitted over long distances to reach homes and industries. How can the transmission occur with as little loss to thermal energy as possible?

Thermal energy is produced at a rate represented by $P = I^2R$. Electrical engineers call this unwanted thermal energy the joule heating loss, or I^2R loss. To reduce this loss, either the current, I, or the resistance, R, must be reduced.

All wires have some resistance, even though their resistance is small. The large wire used to carry electric current into a home has a resistance of 0.20 Ω for 1 km.

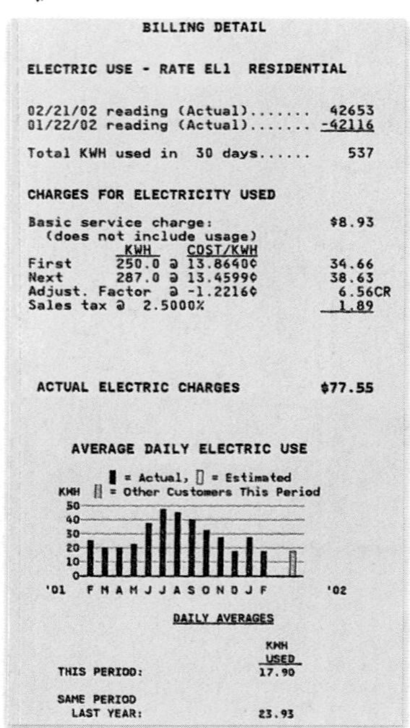

BILLING DETAIL

ELECTRIC USE - RATE EL1 RESIDENTIAL

02/21/02 reading (Actual)....... 42653
01/22/02 reading (Actual)....... -42116

Total KWH used in 30 days...... 537

CHARGES FOR ELECTRICITY USED

	KWH	COST/KWH	
Basic service charge: (does not include usage)			$8.93
First	250.0	@ 13.8640¢	34.66
Next	287.0	@ 13.4599¢	38.63
Adjust. Factor		@ -1.2216¢	6.56CR
Sales tax @ 2.5000%			1.89

ACTUAL ELECTRIC CHARGES $77.55

AVERAGE DAILY ELECTRIC USE

▌ = Actual, ▯ = Estimated
KWH ▯ = Other Customers This Period
50 40 30 20 10 0
'01 F M A M J J A S O N D J F '02

DAILY AVERAGES

	KWH USED
THIS PERIOD:	17.90
SAME PERIOD LAST YEAR:	23.93

Figure 22-11 Watt-hour meters measure the amount of electric energy used by a consumer **(a).** Meter readings then are used in calculating the cost of energy **(b).**

Suppose that a farmhouse were connected directly to a power plant 3.5 km away. The resistance in the wires needed to carry a current in a circuit to the home and back to the plant is represented by the following equation: $R = 2(3.5\text{ km})(0.20\ \Omega/\text{km}) = 1.4\ \Omega$. An electric stove might cause a 41-A current through the wires. The power dissipated in the wires is represented by the following relationships: $P = I^2R = (41\text{ A})^2\ (1.4\ \Omega) = 2400\text{ W}$.

All of this power is converted to thermal energy and, therefore, is wasted. This loss could be minimized by reducing the resistance. Cables of high conductivity and large diameter (and therefore low resistance) are available, but such cables are expensive and heavy. Because the loss of energy is also proportional to the square of the current in the conductors, it is even more important to keep the current in the transmission lines low.

How can the current in the transmission lines be kept low? The electric energy per second (power) transferred over a long-distance transmission line is determined by the relationship $P = IV$. The current is reduced without the power being reduced by an increase in the voltage. Some long-distance lines use voltages of more than 500,000 V. The resulting lower current reduces the I^2R loss in the lines by keeping the I^2 factor low. Long-distance transmission lines always operate at voltages much higher than household voltages in order to reduce I^2R loss. The output voltage from the generating plant is reduced upon arrival at electric substations to 2400 V, and again to 240 V or 120 V before being used in homes.

The Kilowatt-Hour

While electric companies often are called power companies, they actually provide energy rather than power. Power is the rate at which energy is delivered. When consumers pay their home electric bills, an example of which is shown in **Figure 22-11,** they pay for electric energy, not power.

The amount of electric energy used by a device is its rate of energy consumption, in joules per second (W) times the number of seconds that the device is operated. Joules per second times seconds, (J/s)s, equals the total amount of joules of energy. The joule, also defined as a watt-second, is a relatively small amount of energy, too small for commercial sales use. For this reason, electric companies measure energy sales in a unit of a

CHALLENGE PROBLEM

Use the figure to the right to help you answer the questions below.

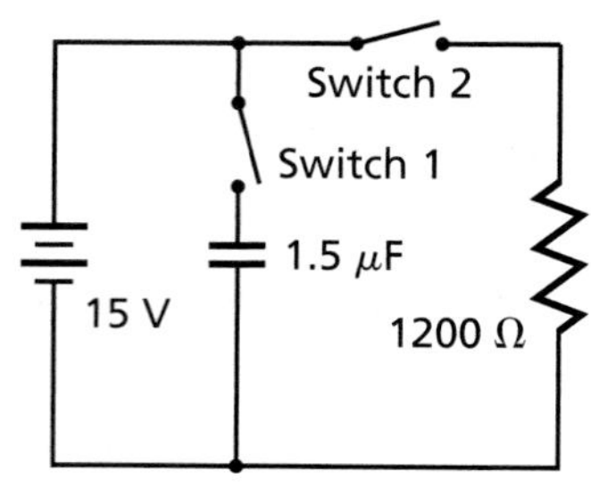

1. Initially, the capacitor is uncharged. Switch 1 is closed, and Switch 2 remains open. What is the voltage across the capacitor?
2. Switch 1 is now opened, and Switch 2 remains open. What is the voltage across the capacitor? Why?
3. Next, Switch 2 is closed, while Switch 1 remains open. What is the voltage across the capacitor and the current through the resistor immediately after Switch 2 is closed?
4. As time goes on, what happens to the voltage across the capacitor and the current through the resistor?

large number of joules called a kilowatt-hour, kWh. A **kilowatt-hour** is equal to 1000 watts delivered continuously for 3600 s (1 h), or 3.6×10^6 J. Not many household devices other than hot-water heaters, stoves, clothes dryers, microwave ovens, heaters, and hair dryers require more than 1000 W of power. Ten 100-W lightbulbs operating all at once use only 1 kWh of energy when they are left on for one full hour.

PRACTICE Problems

- Additional Problems, Appendix B
- Solutions to Selected Problems, Appendix C

28. An electric space heater draws 15.0 A from a 120-V source. It is operated, on the average, for 5.0 h each day.

a. How much power does the heater use?

b. How much energy in kWh does it consume in 30 days?

c. At $0.12 per kWh, how much does it cost to operate the heater for 30 days?

29. A digital clock has a resistance of 12,000 Ω and is plugged into a 115-V outlet.

a. How much current does it draw?

b. How much power does it use?

c. If the owner of the clock pays $0.12 per kWh, how much does it cost to operate the clock for 30 days?

30. An automotive battery can deliver 55 A at 12 V for 1.0 h and requires 1.3 times as much energy for recharge due to its less-than-perfect efficiency. How long will it take to charge the battery using a current of 7.5 A? Assume that the charging voltage is the same as the discharging voltage.

31. Rework the previous problem by assuming that the battery requires the application of 14 V when it is recharging.

You have learned several ways in which power companies solve the problems involved in transmitting electric current over great distances. You also have learned how power companies calculate electric bills and how to predict the cost of running various appliances in the home. The distribution of electric energy to all corners of Earth is one of the greatest engineering feats of the twentieth century.

22.2 Section Review

32. Energy A car engine drives a generator, which produces and stores electric charge in the car's battery. The headlamps use the electric charge stored in the car battery. List the forms of energy in these three operations.

33. Resistance A hair dryer operating from 120 V has two settings, hot and warm. In which setting is the resistance likely to be smaller? Why?

34. Power Determine the power change in a circuit if the applied voltage is decreased by one-half.

35. Efficiency Evaluate the impact of research to improve power transmission lines on society and the environment.

36. Voltage Why would an electric range and an electric hot-water heater be connected to a 240-V circuit rather than a 120-V circuit?

37. Critical Thinking When demand for electric power is high, power companies sometimes reduce the voltage, thereby producing a "brown-out." What is being saved?

PHYSICS LAB •

Voltage, Current, and Resistance

In this chapter, you studied the relationships between voltage, current, and resistance in simple circuits. Voltage is the potential difference that pushes current through a circuit, while resistance determines how much current will flow if a potential difference exists. In this activity, you will collect data and make graphs in order to investigate the mathematical relationships between voltage and current and between resistance and current.

QUESTION

What are the relationships between voltage and current and resistance and current?

Objectives

- **Measure** current in SI.
- **Describe** the relationship between the resistance of a circuit and the total current flowing through a circuit.
- **Describe** the relationship between voltage and the total current flowing through a circuit.
- **Make and use graphs** to show the relationships between current and resistance and between current and voltage.

Safety Precautions

- ***CAUTION: Resistors and circuits may become hot.***
- ***CAUTION: Wires are sharp and can cut skin.***

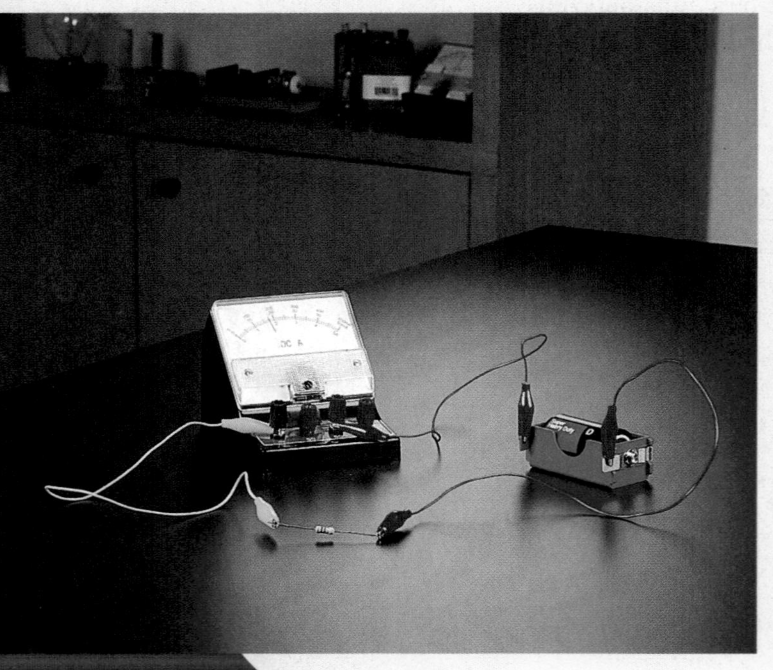

Materials

four 1.5-V D batteries
four D-battery holders
one 10-kΩ resistor
one 500-μA ammeter
five wires with alligator clips
one 20-kΩ resistor
one 30-kΩ resistor
one 40-kΩ resistor

Procedure

Part A

1. Place the D battery in the D-battery holder.
2. Create a circuit containing the D battery, 10-kΩ resistor, and 500-μA ammeter.
3. Record the values for resistance and current in Data Table 1. For resistance, use the value of the resistor. For current, read and record the value given by the ammeter.
4. Replace the 10-kΩ resistor with a 20-kΩ resistor.
5. Record the resistance and the current in Data Table 1.
6. Repeat steps 4–5, but replace the 20-kΩ resistor with a 30-kΩ resistor.
7. Repeat steps 4–5, but replace the 30-kΩ resistor with a 40-kΩ resistor.

Part B

8. Recreate the circuit that you made in step 2. Verify the current in the circuit and record the values for voltage and current in Data Table 2.
9. Add a second 1.5-V D battery to the setup and record the values for voltage and current in Data Table 2. When you are using more than one battery, record the sum of the batteries' voltages as the voltage in Data Table 2.
10. Repeat step 9 with three 1.5-V D batteries.
11. Repeat step 9 with four 1.5-V D batteries.

Data Table 1		
Voltage (V)	Resistance (kΩ)	Current (μA)
1.5		
1.5		
1.5		
1.5		

Data Table 2		
Voltage (V)	Resistance (kΩ)	Current (μA)
	10	
	10	
	10	
	10	

Analyze

1. **Make and Use Graphs** Graph the current versus the resistance. Place resistance on the *x*-axis and current on the *y*-axis.
2. **Make and Use Graphs** Graph the current versus the voltage. Place voltage on the *x*-axis and current on the *y*-axis.
3. **Error Analysis** Other than the values of the resistors, what factors could have affected the current in Part A? How might the effect of these factors be reduced?
4. **Error Analysis** Other than the added batteries, what factors could have affected the current in Part B? How might the effect of these factors be reduced?

Conclude and Apply

1. Looking at the first graph that you made, describe the relationship between resistance and current?
2. Why do you suppose this relationship between resistance and current exists?
3. Looking at the second graph that you made, how would you describe the relationship between voltage and current?
4. Why do you suppose this relationship between voltage and current exists?

Going Further

1. What would be the current in a circuit with a voltage of 3.0 V and a resistance of 20 kΩ? How did you determine this?
2. Could you derive a formula from your lab data to explain the relationship among voltage, current, and resistance? *Hint: Look at the graph of current versus voltage. Assume it is a straight line that goes through the origin.*
3. How well does your data match this formula? Explain.

Real-World Physics

1. Identify some common appliances that use 240 V rather than 120 V.
2. Why do the appliances that you identified require 240 V? What would be the consequences for running such an appliance on a 120-V circuit?

To find out more about current electricity, visit the Web site: physicspp.com

Technology and Society

Hybrid Cars

Meet the hybrid car. It is fuel-efficient, comfortable, safe, quiet, clean, and it accelerates well. Hybrid sales are growing and are expected to exceed 350,000 vehicles in 2008.

Why are they called hybrids? A vehicle is called a hybrid if it uses two or more sources of energy. For example, diesel-electric locomotives are hybrids. But the term *hybrid vehicle* usually refers to a car that uses gas and electricity.

Conventional cars have large engines that enable them to accelerate quickly and to drive up steep hills. But the engine's size makes it inefficient. In a hybrid, a lighter, more efficient gas engine meets most driving needs. When extra energy is needed, it is supplied by electricity from rechargeable batteries.

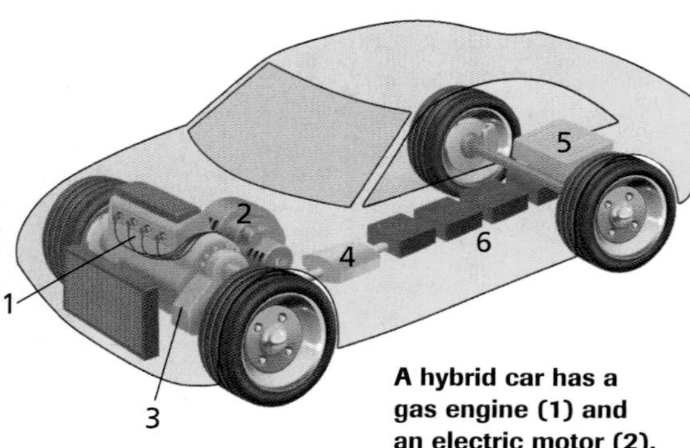

A hybrid car has a gas engine (1) and an electric motor (2).

How do hybrids work? The illustration above shows one type of hybrid, called a parallel hybrid. The small internal combustion engine (1) powers the car during most driving situations. The gas engine and electric motor (2) are connected to the wheels by the same transmission (3). Computerized electronics (4) decide when to use the electric motor, when to use the engine, and when to use both.

This type of hybrid has no external power source besides the gas in the fuel tank (5). Unlike an electric car, you don't need to plug the hybrid into an electric outlet to recharge the batteries (6). Rather, the batteries are recharged by a process called regenerative braking, as shown in the schematic diagram. In conventional vehicles, the brakes apply friction to the wheels, converting a vehicle's kinetic energy into heat. However, a hybrid's electric motor can act as a generator. When the electric motor slows the car, kinetic energy is converted to electric energy, which then recharges the batteries.

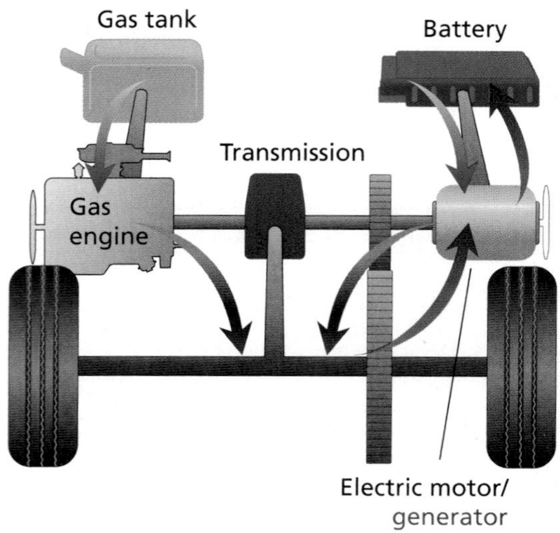

In regenerative breaking, energy from the moving car recharges the batteries.

Can hybrids benefit society? Hybrid cars improve gas mileage and reduce tailpipe emissions. Improved gas mileage saves on the cost of operating the car. Tailpipe emissions include carbon dioxide and carbon monoxide, as well as various hydrocarbons and nitrogen oxides. These emissions can contribute to certain problems, such as smog. Because hybrids improve gas mileage and reduce tailpipe emissions, many people feel that these cars are one viable way to help protect air quality and conserve fuel resources.

Going Further

1. **Analyze and Conclude** What is regenerative braking?
2. **Predict** Will increased sales of hybrids benefit society? Support your answer.

Chapter 22

Study Guide

22.1 Current and Circuits

Vocabulary

- electric current (p. 592)
- conventional current (p. 592)
- battery (p. 592)
- electric circuit (p. 592)
- ampere (p. 593)
- resistance (p. 595)
- resistor (p. 596)
- parallel connection (p. 600)
- series connection (p. 600)

Key Concepts

- Conventional current is defined as current in the direction in which a positive charge would move.
- Generators convert mechanical energy to electric energy.
- A circuit converts electric energy to heat, light, or some other useful output.
- As charge moves through a circuit, resistors cause a drop in potential energy.
- An ampere is equal to one coulomb per second (1 C/s).
- Power can be found by multiplying voltage times current.

$$P = IV$$

- The resistance of a device is given by the ratio of the device's voltage to its current.

$$R = \frac{V}{I}$$

- Ohm's law states that the ratio of potential difference to current is a constant for a given conductor. Any resistance that does not change with temperature, voltage, or the direction of charge flow obeys Ohm's law.
- Circuit current can be controlled by changing voltage, resistance, or both.

22.2 Using Electric Energy

Vocabulary

- superconductor (p. 603)
- kilowatt-hour (p. 605)

Key Concepts

- The power in a circuit is equal to the square of the current times the resistance, or to the voltage squared divided by the resistance.

$$P = I^2R \text{ or } P = \frac{V^2}{R}$$

- If power is dissipated at a uniform rate, the thermal energy converted equals power multiplied by time. Power also can be represented by I^2R and V^2/R to give the last two equations.

$$\begin{aligned} E &= Pt \\ &= I^2Rt \\ &= \left(\frac{V^2}{R}\right)t \end{aligned}$$

- Superconductors are materials with zero resistance. At present, the practical uses of superconductors are limited.
- Unwanted thermal energy produced in the transmission of electric energy is called the joule heating loss, or I^2R loss. The best way to minimize the joule heating loss is to keep the current in the transmission wires low. Transmitting at higher voltages enables current to be reduced without power being reduced.
- The kilowatt-hour, kWh, is an energy unit. It is equal to 3.6×10^6 J.

Assessment

Concept Mapping

38. Complete the concept map using the following terms: *watt, current, resistance.*

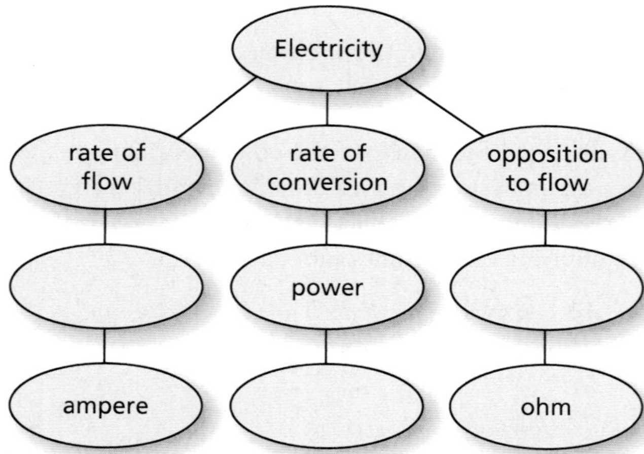

Mastering Concepts

39. Define the unit of electric current in terms of fundamental MKS units. (22.1)

40. How should a voltmeter be connected in **Figure 22-12** to measure the motor's voltage? (22.1)

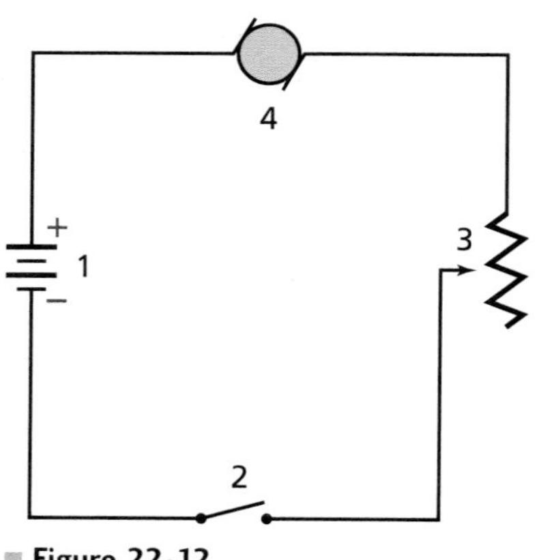

Figure 22-12

41. How should an ammeter be connected in Figure 22-12 to measure the motor's current? (22.1)

42. What is the direction of the conventional motor current in Figure 22-12? (22.1)

43. Refer to Figure 22-12 to answer the following questions. (22.1)

a. Which device converts electric energy to mechanical energy?
b. Which device converts chemical energy to electric energy?
c. Which device turns the circuit on and off?
d. Which device provides a way to adjust speed?

44. Describe the energy conversions that occur in each of the following devices. (22.1)

a. an incandescent lightbulb
b. a clothes dryer
c. a digital clock radio

45. Which wire conducts electricity with the least resistance: one with a large cross-sectional diameter or one with a small cross-sectional diameter? (22.1)

46. A simple circuit consists of a resistor, a battery, and connecting wires. (22.1)

a. Draw a circuit schematic of this simple circuit.
b. How must an ammeter be connected in a circuit for the current to be correctly read?
c. How must a voltmeter be connected to a resistor for the potential difference across it to be read?

47. Why do lightbulbs burn out more frequently just as they are switched on rather than while they are operating? (22.2)

48. If a battery is short-circuited by a heavy copper wire being connected from one terminal to the other, the temperature of the copper wire rises. Why does this happen? (22.2)

49. What electric quantities must be kept small to transmit electric energy economically over long distances? (22.2)

50. Define the unit of power in terms of fundamental MKS units. (22.2)

Applying Concepts

51. Batteries When a battery is connected to a complete circuit, charges flow in the circuit almost instantaneously. Explain.

52. Explain why a cow experiences a mild shock when it touches an electric fence.

53. Power Lines Why can birds perch on high-voltage lines without being injured?

54. Describe two ways to increase the current in a circuit.

55. Lightbulbs Two lightbulbs work on a 120-V circuit. One is 50 W and the other is 100 W. Which bulb has a higher resistance? Explain.

56. If the voltage across a circuit is kept constant and the resistance is doubled, what effect does this have on the circuit's current?

57. What is the effect on the current in a circuit if both the voltage and the resistance are doubled? Explain.

 For more problems, go to Additional Problems, Appendix B.

58. Ohm's Law Sue finds a device that looks like a resistor. When she connects it to a 1.5-V battery, she measures only 45×10^{-6} A, but when she uses a 3.0-V battery, she measures 25×10^{-3} A. Does the device obey Ohm's law?

59. If the ammeter in Figure 22-4a on page 596 were moved to the bottom of the diagram, would the ammeter have the same reading? Explain.

60. Two wires can be placed across the terminals of a 6.0-V battery. One has a high resistance, and the other has a low resistance. Which wire will produce thermal energy at a faster rate? Why?

Mastering Problems

22.1 Current and Circuits

61. A motor is connected to a 12-V battery, as shown in **Figure 22-13.**

a. How much power is delivered to the motor?

b. How much energy is converted if the motor runs for 15 min?

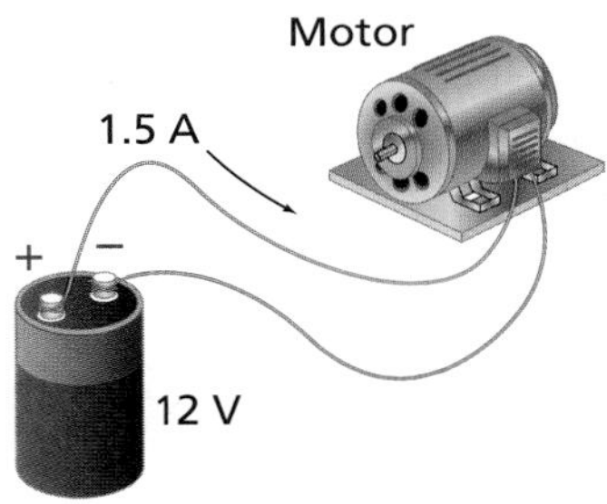

Figure 22-13

62. Refer to **Figure 22-14** to answer the following questions.

a. What should the ammeter reading be?

b. What should the voltmeter reading be?

c. How much power is delivered to the resistor?

d. How much energy is delivered to the resistor per hour?

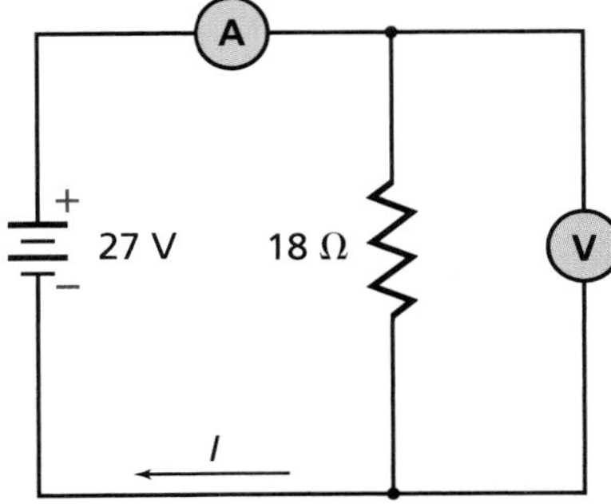

Figure 22-14

63. Refer to **Figure 22-15** to answer the following questions.

a. What should the ammeter reading be?

b. What should the voltmeter reading be?

c. How much power is delivered to the resistor?

d. How much energy is delivered to the resistor per hour?

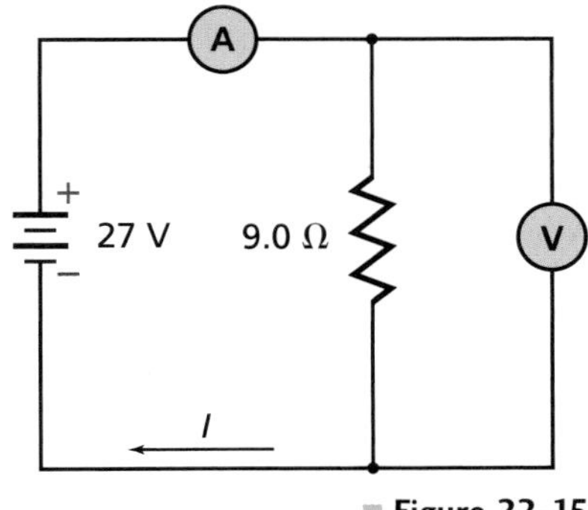

Figure 22-15

64. Refer to **Figure 22-16** to answer the following questions.

a. What should the ammeter reading be?

b. What should the voltmeter reading be?

c. How much power is delivered to the resistor?

d. How much energy is delivered to the resistor per hour?

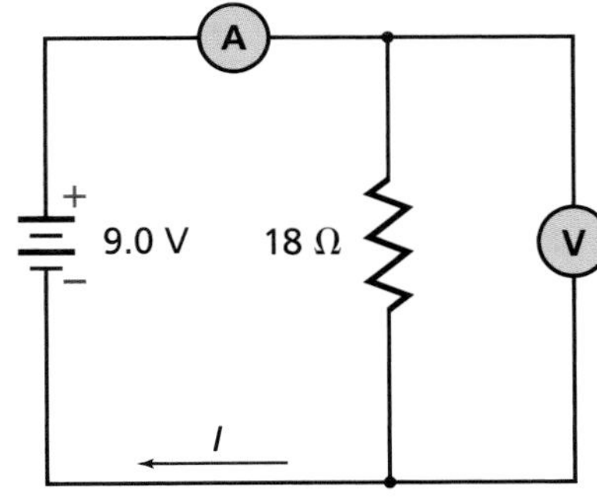

Figure 22-16

65. Toasters The current through a toaster that is connected to a 120-V source is 8.0 A. What power is dissipated by the toaster?

66. Lightbulbs A current of 1.2 A is measured through a lightbulb when it is connected across a 120-V source. What power is dissipated by the bulb?

67. A lamp draws 0.50 A from a 120-V generator.

a. How much power is delivered?

b. How much energy is converted in 5.0 min?

68. A 12-V automobile battery is connected to an electric starter motor. The current through the motor is 210 A.

a. How many joules of energy does the battery deliver to the motor each second?

b. What power, in watts, does the motor use?

69. **Dryers** A 4200-W clothes dryer is connected to a 220-V circuit. How much current does the dryer draw?

70. **Flashlights** A flashlight bulb is connected across a 3.0-V potential difference. The current through the bulb is 1.5 A.
   a. What is the power rating of the bulb?
   b. How much electric energy does the bulb convert in 11 min?

71. **Batteries** A resistor of 60.0 Ω has a current of 0.40 A through it when it is connected to the terminals of a battery. What is the voltage of the battery?

72. What voltage is applied to a 4.0-Ω resistor if the current is 1.5 A?

73. What voltage is placed across a motor with a 15-Ω operating resistance if there is 8.0 A of current?

74. A voltage of 75 V is placed across a 15-Ω resistor. What is the current through the resistor?

75. Some students connected a length of nichrome wire to a variable power supply to produce between 0.00 V and 10.00 V across the wire. They then measured the current through the wire for several voltages. The students recorded the data for the voltages used and the currents measured, as shown in **Table 22-2.**
   a. For each measurement, calculate the resistance.
   b. Graph *I* versus *V*.
   c. Does the nichrome wire obey Ohm's law? If not, for all the voltages, specify the voltage range for which Ohm's law holds.

Table 22-2

Voltage, *V* (volts)	Current, *I* (amps)	Resistance, *R* = *V*/*I* (amps)
2.00	0.0140	________
4.00	0.0270	________
6.00	0.0400	________
8.00	0.0520	________
10.00	0.0630	________
−2.00	−0.0140	________
−4.00	−0.0280	________
−6.00	−0.0390	________
−8.00	−0.0510	________
−10.00	−0.0620	________

76. Draw a series circuit diagram to include a 16-Ω resistor, a battery, and an ammeter that reads 1.75 A. Indicate the positive terminal and the voltage of the battery, the positive terminal of the ammeter, and the direction of conventional current.

77. A lamp draws a 66-mA current when connected to a 6.0-V battery. When a 9.0-V battery is used, the lamp draws 75 mA.
   a. Does the lamp obey Ohm's law?
   b. How much power does the lamp dissipate when it is connected to the 6.0-V battery?
   c. How much power does it dissipate at 9.0 V?

78. **Lightbulbs** How much energy does a 60.0-W lightbulb use in half an hour? If the lightbulb converts 12 percent of electric energy to light energy, how much thermal energy does it generate during the half hour?

79. The current through a lamp connected across 120 V is 0.40 A when the lamp is on.
   a. What is the lamp's resistance when it is on?
   b. When the lamp is cold, its resistance is 1/5 as great as it is when the lamp is hot. What is the lamp's cold resistance?
   c. What is the current through the lamp as it is turned on if it is connected to a potential difference of 120 V?

80. The graph in **Figure 22-17** shows the current through a device called a silicon diode.
   a. A potential difference of +0.70 V is placed across the diode. What is the resistance of the diode?
   b. What is the diode's resistance when a +0.60-V potential difference is used?
   c. Does the diode obey Ohm's law?

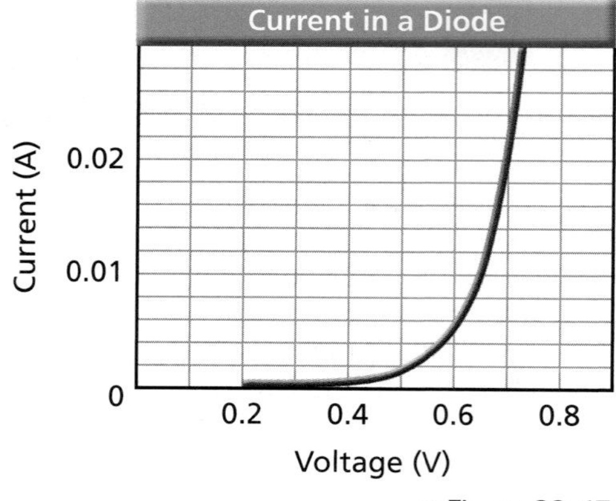

Figure 22-17

81. Draw a schematic diagram to show a circuit including a 90-V battery, an ammeter, and a resistance of 45 Ω connected in series. What is the ammeter reading? Draw arrows showing the direction of conventional current.

22.2 Using Electric Energy

82. **Batteries** A 9.0-V battery costs $3.00 and will deliver 0.0250 A for 26.0 h before it must be replaced. Calculate the cost per kWh.

83. What is the maximum current allowed in a 5.0-W, 220-Ω resistor?

84. A 110-V electric iron draws 3.0 A of current. How much thermal energy is developed in an hour?

85. For the circuit shown in **Figure 22-18,** the maximum safe power is 5.0×10^{1} W. Use the figure to find the following:
 a. the maximum safe current
 b. the maximum safe voltage

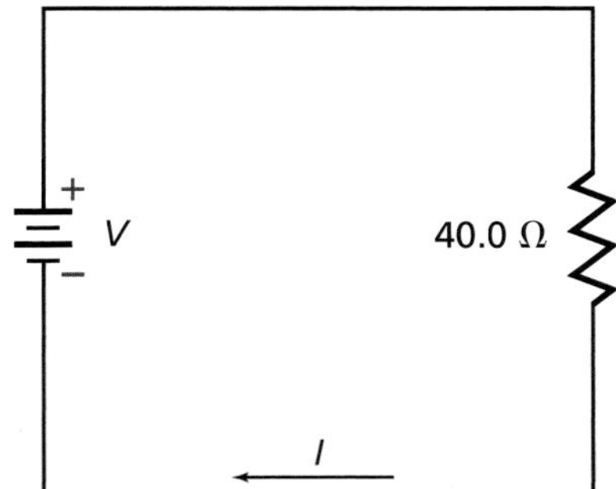

Figure 22-18

86. **Utilities Figure 22-19** represents an electric furnace. Calculate the monthly (30-day) heating bill if electricity costs $0.10 per kWh and the thermostat is on one-fourth of the time.

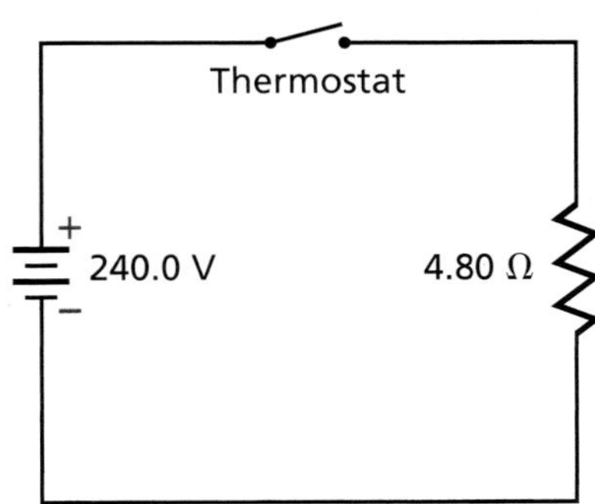

Figure 22-19

87. **Appliances** A window air conditioner is estimated to have a cost of operation of $50 per 30 days. This is based on the assumption that the air conditioner will run half of the time and that electricity costs $0.090 per kWh. Determine how much current the air conditioner will take from a 120-V outlet.

88. **Radios** A transistor radio operates by means of a 9.0-V battery that supplies it with a 50.0-mA current.
 a. If the cost of the battery is $2.49 and it lasts for 300.0 h, what is the cost per kWh to operate the radio in this manner?
 b. The same radio, by means of a converter, is plugged into a household circuit by a homeowner who pays $0.12 per kWh. What does it now cost to operate the radio for 300.0 h?

Mixed Review

89. If a person has $5, how long could he or she play a 200 W stereo if electricity costs $0.15 per kWh?

90. A current of 1.2 A is measured through a 50.0-Ω resistor for 5.0 min. How much heat is generated by the resistor?

91. A 6.0-Ω resistor is connected to a 15-V battery.
 a. What is the current in the circuit?
 b. How much thermal energy is produced in 10.0 min?

92. **Lightbulbs** An incandescent lightbulb with a resistance of 10.0 Ω when it is not lit and a resistance of 40.0 Ω when it is lit has 120 V placed across it.
 a. What is the current draw when the bulb is lit?
 b. What is the current draw at the instant the bulb is turned on?
 c. When does the lightbulb use the most power?

93. A 12-V electric motor's speed is controlled by a potentiometer. At the motor's slowest setting, it uses 0.02 A. At its highest setting, the motor uses 1.2 A. What is the range of the potentiometer?

94. An electric motor operates a pump that irrigates a farmer's crop by pumping 1.0×10^{4} L of water a vertical distance of 8.0 m into a field each hour. The motor has an operating resistance of 22.0 Ω and is connected across a 110-V source.
 a. What current does the motor draw?
 b. How efficient is the motor?

95. A heating coil has a resistance of 4.0 Ω and operates on 120 V.
 a. What is the current in the coil while it is operating?
 b. What energy is supplied to the coil in 5.0 min?
 c. If the coil is immersed in an insulated container holding 20.0 kg of water, what will be the increase in the temperature of the water? Assume 100 percent of the heat is absorbed by the water.
 d. At $0.08 per kWh, how much does it cost to operate the heating coil 30 min per day for 30 days?

96. **Appliances** An electric heater is rated at 500 W.
 a. How much energy is delivered to the heater in half an hour?
 b. The heater is being used to heat a room containing 50 kg of air. If the specific heat of air is 1.10 kJ/kg·°C, and 50 percent of the thermal energy heats the air in the room, what is the change in air temperature in half an hour?
 c. At $0.08 per kWh, how much does it cost to run the heater 6.0 h per day for 30 days?

Thinking Critically

97. **Formulate Models** How much energy is stored in a capacitor? The energy needed to increase the potential difference of a charge, q, is represented by $E = qV$. But in a capacitor, $V = q/C$. Thus, as charge is added, the potential difference increases. As more charge is added, however, it takes more energy to add the additional charge. Consider a 1.0-F "supercap" used as an energy storage device in a personal computer. Plot a graph of V as the capacitor is charged by adding 5.0 C to it. What is the voltage across the capacitor? The area under the curve is the energy stored in the capacitor. Find the energy in joules. Is it equal to the total charge times the final potential difference? Explain.

98. **Apply Concepts** A microwave oven operates at 120 V and requires 12 A of current. Its electric efficiency (converting AC to microwave radiation) is 75 percent, and its conversion efficiency from microwave radiation to heating water is also 75 percent.
    a. Draw a block power diagram similar to the energy diagram shown in Figure 22-2b on page 593. Label the function of each block according to total joules per second.
    b. Derive an equation for the rate of temperature increase ($\Delta T/s$) from the information presented in Chapter 12. Solve for the rate of temperature rise given the rate of energy input, the mass, and the specific heat of a substance.
    c. Use your equation to solve for the rate of temperature rise in degrees Celsius per second when using this oven to heat 250 g of water above room temperature.
    d. Review your calculations carefully for the units used and discuss why your answer is in the correct form.
    e. Discuss, in general terms, different ways in which you could increase the efficiency of microwave heating.
    f. Discuss, in efficiency terms, why microwave ovens are not useful for heating everything.
    g. Discuss, in general terms, why it is not a good idea to run microwave ovens when they are empty.

99. **Analyze and Conclude** A salesclerk in an appliance store states that microwave ovens are the most electrically efficient means of heating objects.
    a. Formulate an argument to refute the clerk's claim. *Hint: Think about heating a specific object.*
    b. Formulate an argument to support the clerk's claim. *Hint: Think about heating a specific object.*
    c. Formulate a diplomatic reply to the clerk.

100. **Apply Concepts** The sizes of 10-Ω resistors range from a pinhead to a soup can. Explain.

101. **Make and Use Graphs** The diode graph shown in Figure 22-17 on page 612 is more useful than a similar graph for a resistor that obeys Ohm's law. Explain.

102. **Make and Use Graphs** Based on what you have learned in this chapter, identify and prepare two parabolic graphs.

Writing in Physics

103. There are three kinds of equations encountered in science: (1) definitions, (2) laws, and (3) derivations. Examples of these are: (1) an ampere is equal to one coulomb per second, (2) force is equal to mass times acceleration, (3) power is equal to voltage squared divided by resistance. Write a one-page explanation of where "resistance is equal to voltage divided by current" fits. Before you begin to write, first research the three categories given above.

104. In Chapter 13, you learned that matter expands when it is heated. Research the relationship between thermal expansion and high-voltage transmission lines.

Cumulative Review

105. A person burns energy at the rate of about 8.4×10^6 J per day. How much does she increase the entropy of the universe in that day? How does this compare to the entropy increase caused by melting 20 kg of ice? (Chapter 12)

106. When you go up the elevator of a tall building, your ears might pop because of the rapid change in pressure. What is the pressure change caused by riding in an elevator up a 30-story building (150 m)? The density of air is about 1.3 kg/m^3 at sea level. (Chapter 13)

107. What is the wavelength in air of a 17-kHz sound wave, which is at the upper end of the frequency range of human hearing? (Chapter 15)

108. Light of wavelength 478 nm falls on a double slit. First-order bright bands appear 3.00 mm from the central bright band. The screen is 0.91 m from the slits. How far apart are the slits? (Chapter 19)

109. A charge of $+3.0 \times 10^{-6}$ C is 2.0 m from a second charge of $+6.0 \times 10^{-5}$ C. What is the magnitude of the force between them? (Chapter 20)

 For more problems, go to Additional Problems, Appendix B.

Multiple Choice

1. A 100-W lightbulb is connected to a 120-V electric line. What is the current that the lightbulb draws?

   (A) 0.8 A (C) 1.2 A
   (B) 1 A (D) 2 A

2. A 5.0-Ω resistor is connected to a 9.0-V battery. How much thermal energy is produced in 7.5 min?

   (A) 1.2×10^{2} J (C) 3.0×10^{3} J
   (B) 1.3×10^{3} J (D) 7.3×10^{3} J

3. The current in the flashlight shown below is 0.50 A, and the voltage is the sum of the voltages of the individual batteries. What is the power delivered to the bulb of the flashlight?

   (A) 0.11 W (C) 2.3 W
   (B) 1.1 W (D) 4.5 W

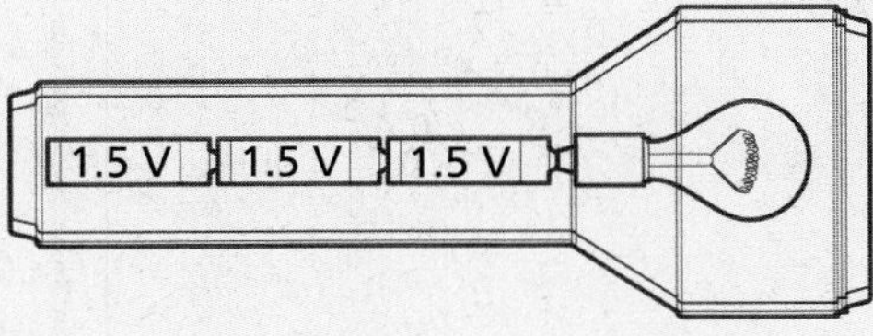

4. If the flashlight in the illustration above is left on for 3.0 min, how much electric energy is delivered to the bulb?

   (A) 6.9 J (C) 2.0×10^{2} J
   (B) 14 J (D) 4.1×10^{2} J

5. A current of 2.0 A flows through a circuit containing a motor with a resistance of 12 Ω. How much energy is converted if the motor runs for one minute?

   (A) 4.8×10^{1} J (C) 2.9×10^{3} J
   (B) 2.0×10^{1} J (D) 1.7×10^{5} J

6. What is the effect on the current in a simple circuit if both the voltage and the resistance are reduced by half?

   (A) divided by 2 (C) multiplied by 2
   (B) no change (D) multiplied by 4

7. A 50.0-Ω resistance causes a current of 5.00 mA to flow through a circuit connected to a battery. What is the power in the circuit?

   (A) 1.00×10^{-2} W (C) 1.25×10^{-3} W
   (B) 1.00×10^{-3} W (D) 2.50×10^{-3} W

8. How much electric energy is delivered to a 60.0-W lightbulb if the bulb is left on for 2.5 hours?

   (A) 4.2×10^{-2} J (C) 1.5×10^{2} J
   (B) 2.4×10^{1} J (D) 5.4×10^{5} J

Extended Answer

9. The diagram below shows a simple circuit containing a DC generator and a resistor. The table shows the resistances of several small electric devices. If the resistor in the diagram represents a hair dryer, what is the current in the circuit? How much energy does the hair dryer use if it runs for 2.5 min?

Device	Resistance (Ω)
Hair dryer	8.5 Ω
Heater	10.0 Ω
Small motor	12.0 Ω

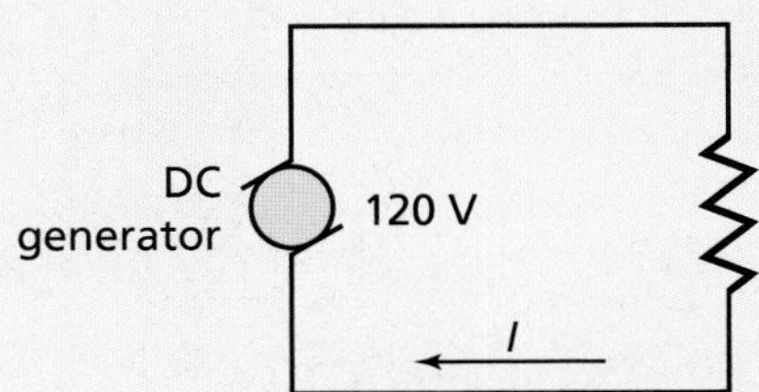

✓ Test-Taking TIP

More Than One Graphic

If a test question has more than one table, graph, diagram, or drawing with it, use them all. If you answer based on just one graphic, you probably will miss an important piece of information.

Chapter 23 Series and Parallel Circuits

What You'll Learn

- You will distinguish among series circuits, parallel circuits, and series-parallel combinations, and solve problems involving them.
- You will explain the function of fuses, circuit breakers, and ground-fault interrupters, and describe how ammeters and voltmeters are used in circuits.

Why It's Important

Electric circuits are the basis of every electric device, from electric lights to microwave ovens to computers. Learning how circuits work will help you understand how countless electric devices function.

Electric Load Centers Electric load centers form the link between the utility company and the circuits in a building. Each circuit breaker protects an individual circuit, which has the various loads connected in parallel.

Think About This ▶

Why are the building loads connected in parallel? How are the circuit breakers connected?

physicspp.com

How do fuses protect electric circuits?

Question

How does a fuse prevent an electric circuit from drawing too much current and creating a safety hazard?

Procedure

1. Connect the negative terminal of a 9-V battery to one terminal of a flashlight-bulb socket using a copper wire. ***CAUTION: Wire ends may be sharp and could cause cuts.***
2. Connect the other terminal of the bulb socket to a single strand of steel wool using copper wire. Make sure the strand of steel wool is suspended over a small glass container.
3. Connect the other end of the single strand of steel wool to a switch using another piece of copper wire. Make sure the switch is open (turned off).
4. Connect the other terminal of the switch to the positive terminal of a power supply or a battery.
5. **Hypothesize** Predict what will happen when the switch is closed (turned on).
6. **Observe** Close the switch and make observations of the strand of steel wool.
7. Repeat steps 1–6 using a thicker strand of steel wool, or twist several strands together to form a single, thicker strand.

Analysis

Explain how the thickness of a wire is related to how fast the wire will overheat and break apart. Why have circuit breakers replaced fuses in the electric circuit boxes of new homes?

Critical Thinking Why is it important to replace a burned-out fuse in a house or car electric circuit with one that has the correct rating?

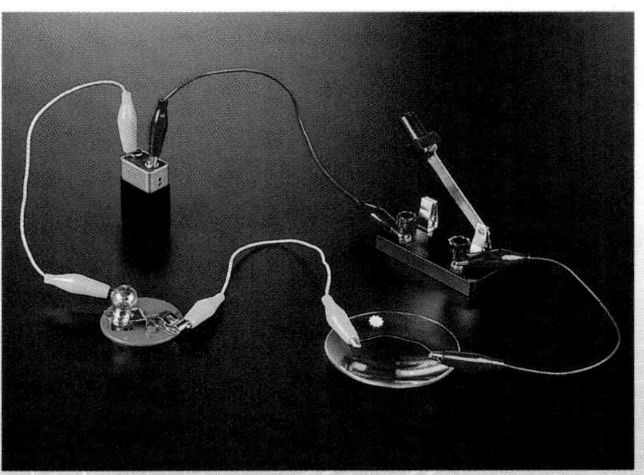

23.1 Simple Circuits

Although the connection may not immediately be clear to you, a mountain river can be used to model an electric circuit. From its source high in the mountains, the river flows downhill to the plains below. No matter which path the river takes, its change in elevation, from the mountaintop to the plain, is the same. Some rivers flow downhill in a single stream. Other rivers may split into two or more smaller streams as they flow over a waterfall or through a series of rapids. In this case, part of the river follows one path, while other parts of the river follow different paths. No matter how many paths the river takes, however, the total amount of water flowing down the mountain remains unchanged. In other words, the amount of water flowing downhill is not affected by the path it takes.

▶ **Objectives**
- **Describe** series and parallel circuits.
- **Calculate** currents, voltage drops, and equivalent resistances in series and parallel circuits.

▶ **Vocabulary**

series circuit
equivalent resistance
voltage divider
parallel circuit

■ **Figure 23-1** No matter what path a river or a stream takes down a mountain, the amount of water and the drop in elevation are the same.

How does the river shown in **Figure 23-1** model an electric circuit? The distance that the river drops is similar to the potential difference in a circuit. The amount of water flowing in the river is similar to current in a circuit. Narrow rapids create resistance and are similar to resistors in a circuit. What part of a river is similar to a battery or a generator in an electric circuit? The energy source needed to raise water to the top of the mountain is the Sun. Solar energy evaporates water from lakes and seas leading to the formation of clouds that release rain or snow that falls on the mountaintops. Continue to think about the mountain river model as you read about the current in electric circuits.

Series Circuits

Three students are connecting two identical lamps to a battery, as illustrated in **Figure 23-2.** Before they make the final connection to the battery, their teacher asks them to predict the brightnesses of the two lamps.

Each student knows that the brightness of a lamp depends on the current through it. The first student predicts that only the lamp close to the positive (+) terminal of the battery will light because all the current will be used up as thermal and light energy. The second student predicts that only part of the current will be used up, and the second lamp will glow, but less brightly than the first. The third student predicts that the lamps will be of equal brightness because current is a flow of charge and the charge leaving the first lamp has nowhere else to go in the circuit except through the second lamp. The third student reasons that because the current will be the same in each lamp, the brightness also will be the same. How do you predict the lights will behave?

If you consider the mountain river model for this circuit, you will realize that the third student is correct. Recall from Chapter 20 that charge cannot be created or destroyed. Because the charge in the circuit has only one path to follow and cannot be destroyed, the same amount of charge must leave a circuit as enters the circuit. This means that the current is the same everywhere in the circuit. If you connect three ammeters in the circuit, as shown in **Figure 23-3,** they all will show the same current. A circuit such as this, in which all current travels through each device, is called a **series circuit.**

If the current is the same throughout the circuit, what is used by the lamp to produce the thermal and light energy? Recall that power, the rate at which electric energy is converted, is represented by $P = IV$. Thus, if there is a potential difference, or voltage drop, across the lamp, then electric energy is being converted into another form. The resistance of the lamp is defined as $R = V/I$. Thus, the potential difference, also called the voltage drop, is $V = IR$.

■ **Figure 23-2** What is your prediction about the brightnesses of the two lightbulbs after the circuit is connected?

Concepts In Motion
Interactive Figure To see an animation on electric circuits, visit physicspp.com.

Current and resistance in a series circuit From the river model, you know that the sum of the drops in height is equal to the total drop from the top of the mountain to sea level. In an electric circuit, the increase in voltage provided by the generator or other energy source, V_{source}, is equal to the sum of voltage drops across lamps A and B, and is represented by the following equation:

$$V_{source} = V_A + V_B$$

To find the potential drop across a resistor, multiply the current in the circuit by the resistance of the individual resistor. Because the current through the lamps is the same, $V_A = IR_A$ and $V_B = IR_B$. Therefore, $V_{source} = IR_A + IR_B$, or $V_{source} = I(R_A + R_B)$. The current through the circuit is represented by the following equation:

$$I = \frac{V_{source}}{R_A + R_B}$$

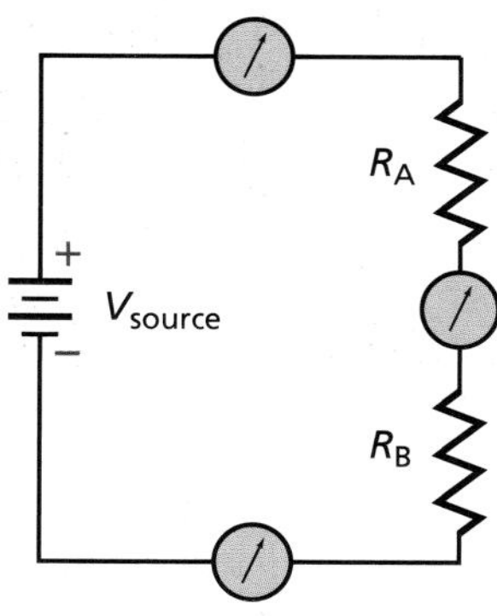

Figure 23-3 The ammeters show that the current is the same everywhere in a series circuit.

The same idea can be extended to any number of resistances in series, not just two. The same current would exist in the circuit with a single resistor, R, that has a resistance equal to the sum of the resistances of the two lamps. Such a resistance is called the **equivalent resistance** of the circuit. For resistors in series, the equivalent resistance is the sum of all the individual resistances, as expressed by the following equation.

Equivalent Resistance for Resistors in Series $R = R_A + R_B + \ldots$

The equivalent resistance of resistors in series equals the sum of the individual resistances of the resistors.

Notice that the equivalent resistance is greater than that of any individual resistor. Therefore, if the battery voltage does not change, adding more devices in series always decreases the current. To find the current through a series circuit, first calculate the equivalent resistance and then use the following equation.

Current $I = \frac{V_{source}}{R}$

Current in a series circuit is equal to the potential difference of the source divided by the equivalent resistance.

PRACTICE Problems

• Additional Problems, Appendix B
• Solutions to Selected Problems, Appendix C

1. Three 20-Ω resistors are connected in series across a 120-V generator. What is the equivalent resistance of the circuit? What is the current in the circuit?

2. A 10-Ω, 15-Ω, and 5-Ω resistor are connected in a series circuit with a 90-V battery. What is the equivalent resistance of the circuit? What is the current in the circuit?

3. A 9-V battery is in a circuit with three resistors connected in series.

- **a.** If the resistance of one of the resistors increases, how will the equivalent resistance change?
- **b.** What will happen to the current?
- **c.** Will there be any change in the battery voltage?

4. A string of holiday lights has ten bulbs with equal resistances connected in series. When the string of lights is connected to a 120-V outlet, the current through the bulbs is 0.06 A.

- **a.** What is the equivalent resistance of the circuit?
- **b.** What is the resistance of each bulb?

5. Calculate the voltage drops across the three resistors in problem 2, and verify that their sum equals the voltage of the battery.

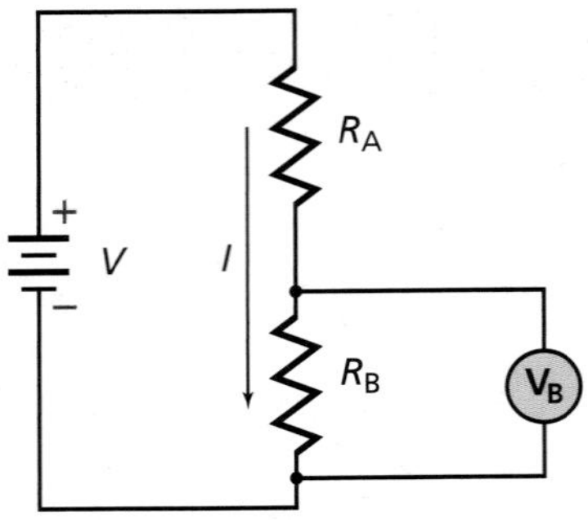

■ **Figure 23-4** In this voltage-divider circuit, the values of R_A and R_B are chosen such that the voltage drop across R_B is the desired voltage.

Voltage drops in a series circuit As current moves through any circuit, the net change in potential must be zero. This is because the circuit's electric energy source, the battery or generator, raises the potential an amount equal to the potential drop produced when the current passes through the resistors. Therefore, the net change is zero.

An important application of series resistors is a circuit called a voltage divider. A **voltage divider** is a series circuit used to produce a voltage source of desired magnitude from a higher-voltage battery. For example, suppose you have a 9-V battery but need a 5-V potential source. Consider the circuit shown in **Figure 23-4.** Two resistors, R_A and R_B, are connected in series across a battery of magnitude V. The equivalent resistance of the circuit is $R = R_A + R_B$. The current is represented by the following equation:

$$I = \frac{V}{R}$$
$$= \frac{V}{R_A + R_B}$$

The desired voltage, 5 V, is the voltage drop, V_B, across resistor R_B: $V_B = IR_B$. Into this equation, the earlier equation for current is substituted.

$$V_B = IR_B$$
$$= \left(\frac{V}{R_A + R_B}\right)R_B$$
$$= \frac{VR_B}{R_A + R_B}$$

Voltage dividers often are used with sensors, such as photoresistors. The resistance of a photoresistor depends upon the amount of light that strikes it. Photoresistors are made of semiconductors, such as silicon, selenium, or cadmium sulfide. A typical photoresistor can have a resistance of 400 Ω when light is striking it compared with a resistance of 400,000 Ω when the photoresistor is in the dark. The voltage output of a voltage divider that uses a photoresistor depends upon the amount of light striking the photoresistor sensor. This circuit can be used as a light meter, such as the one shown in **Figure 23-5.** In this device, an electronic circuit detects the potential difference and converts it to a measurement of illuminance that can be read on the digital display. The amplified voltmeter reading will drop as illuminance increases.

■ **Figure 23-5** The voltage output of this voltage divider depends upon the amount of light striking the photoresistor sensor **(a).** Light meters used in photography make use of a voltage divider **(b).**

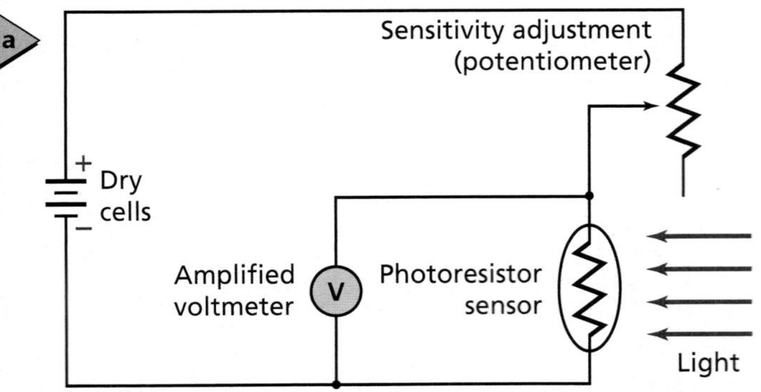

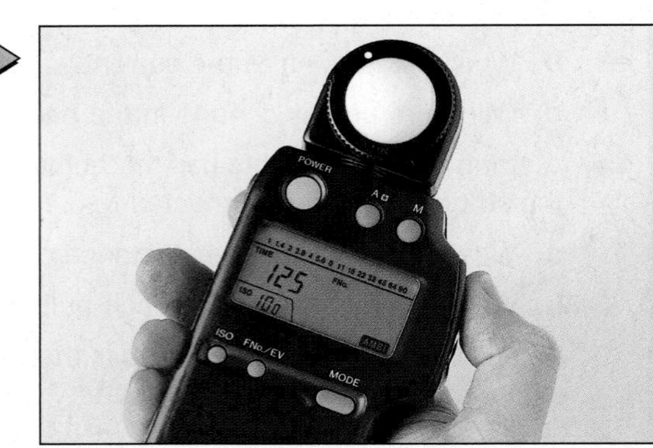

▶ EXAMPLE Problem 1

Voltage Drops in a Series Circuit Two resistors, 47.0 Ω and 82.0 Ω, are connected in series across a 45.0-V battery.

a. What is the current in the circuit?

b. What is the voltage drop across each resistor?

c. If the 47.0-Ω resistor is replaced by a 39.0-Ω resistor, will the current increase, decrease, or remain the same?

d. What is the new voltage drop across the 82.0-Ω resistor?

1 Analyze and Sketch the Problem

- Draw a schematic of the circuit.

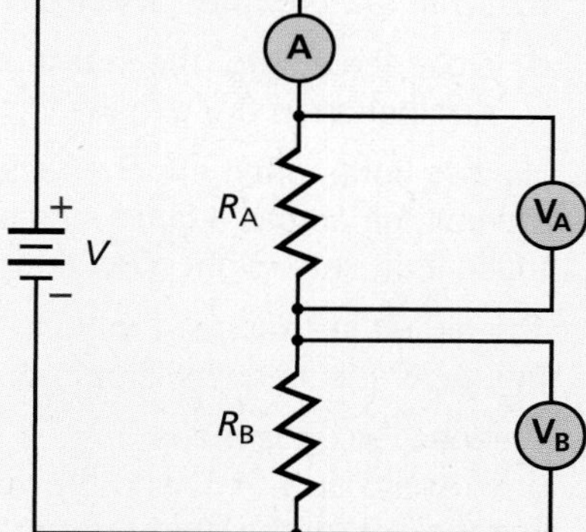

Known:	Unknown:
$V_{source} = 45.0$ V	$I = ?$
$R_A = 47.0$ Ω	$V_A = ?$
$R_B = 82.0$ Ω	$V_B = ?$

2 Solve for the Unknown

a. To determine the current, first find the equivalent resistance.

$$I = \frac{V_{source}}{R} \text{ and } R = R_A + R_B$$

$$= \frac{V_{source}}{R_A + R_B}$$ **Substitute $R = R_A + R_B$**

$$= \frac{45.0\text{ V}}{47.0\ \Omega + 82.0\ \Omega}$$ **Substitute $V_{source} = 45.0$ V, $R_A = 47.0$ Ω, $R_B = 82.0$ Ω**

$$= 0.349\text{ A}$$

b. Use $V = IR$ for each resistor.

$$V_A = IR_A$$

$$= (0.349\text{ A})(47.0\ \Omega)$$ **Substitute $I = 0.349$ A, $R_A = 47.0$ Ω**

$$= 16.4\text{ V}$$

$$V_B = IR_B$$

$$= (0.349\text{ A})(82.0\ \Omega)$$ **Substitute $I = 0.349$ A, $R_B = 82.0$ Ω**

$$= 28.6\text{ V}$$

Math Handbook
Operations with Significant Digits
pages 835–836

c. Calculate current, this time using 39.0 Ω as R_A.

$$I = \frac{V_{source}}{R_A + R_B}$$

$$= \frac{45.0\text{ V}}{39.0\ \Omega + 82.0\ \Omega}$$ **Substitute $V_{source} = 45.0$ V, $R_A = 39.0$ Ω, $R_B = 82.0$ Ω**

$$= 0.372\text{ A}$$

The current will increase.

d. Determine the new voltage drop in R_B.

$$V_B = IR_B$$

$$= (0.372\text{ A})(82.0\ \Omega)$$ **Substitute $I = 0.372$ A, $R_B = 82.0$ Ω**

$$= 30.5\text{ V}$$

3 Evaluate the Answer

- **Are the units correct?** Current is A = V/Ω; voltage is V = A·Ω.
- **Is the magnitude realistic?** For current, if $R > V$, $I < 1$. The voltage drop across any one resistor must be less than the voltage of the circuit. Both values of V_B are less than V_{source}, which is 45 V.

▶ PRACTICE Problems

- Additional Problems, Appendix B
- Solutions to Selected Problems, Appendix C

6. The circuit shown in Example Problem 1 is producing these symptoms: the ammeter reads 0 A, V_A reads 0 V, and V_B reads 45 V. What has happened?

7. Suppose the circuit shown in Example Problem 1 has these values: $R_A = 255\ \Omega$, $R_B = 292\ \Omega$, and $V_A = 17.0$ V. No other information is available.

a. What is the current in the circuit?

b. What is the battery voltage?

c. What are the total power dissipation and the individual power dissipations?

d. Does the sum of the individual power dissipations in the circuit equal the total power dissipation in the circuit? Explain.

8. Holiday lights often are connected in series and use special lamps that short out when the voltage across a lamp increases to the line voltage. Explain why. Also explain why these light sets might blow their fuses after many bulbs have failed.

9. The circuit in Example Problem 1 has unequal resistors. Explain why the resistor with the lower resistance will operate at a lower temperature.

10. A series circuit is made up of a 12.0-V battery and three resistors. The voltage across one resistor is 1.21 V, and the voltage across another resistor is 3.33 V. What is the voltage across the third resistor?

▶ EXAMPLE Problem 2

Voltage Divider A 9.0-V battery and two resistors, 390 Ω and 470 Ω, are connected as a voltage divider. What is the voltage across the 470-Ω resistor?

1 Analyze and Sketch the Problem

- Draw the battery and resistors in a series circuit.

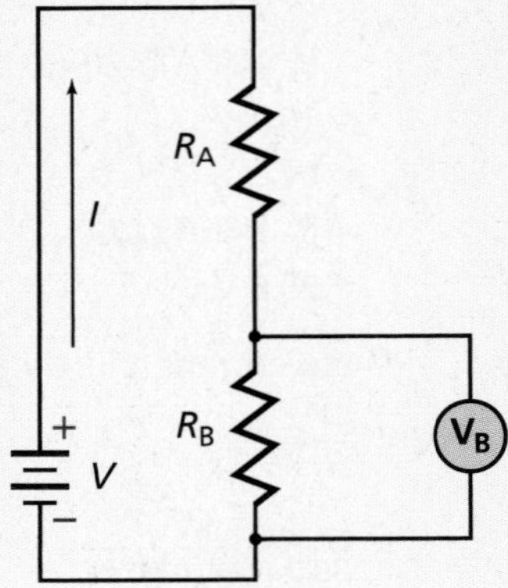

Known:	**Unknown:**
$V_{source} = 9.0$ V	$V_B = ?$
$R_A = 390\ \Omega$	
$R_B = 470\ \Omega$	

2 Solve for the Unknown

$$R = R_A + R_B$$

$$I = \frac{V_{source}}{R}$$

$$= \frac{V_{source}}{R_A + R_B} \qquad \text{Substitute } R = R_A + R_B$$

$$V_B = IR_B$$

$$= \frac{V_{source}R_B}{R_A + R_B} \qquad \text{Substitute } I = \frac{V_{source}}{R_A + R_B}$$

$$= \frac{(9.0\ \text{V})(470\ \Omega)}{390\ \Omega + 470\ \Omega} \qquad \text{Substitute } V_{source} = 9.0\ \text{V}, R_A = 390\ \Omega, R_B = 470\ \Omega$$

$$= 4.9\ \text{V}$$

Math Handbook
Order of Operations
page 843

3 Evaluate the Answer

- **Are the units correct?** The voltage is V = VΩ/Ω. The ohms cancel, leaving volts.
- **Is the magnitude realistic?** The voltage drop is less than the battery voltage. Because 470 Ω is more than half of the equivalent resistance, the voltage drop is more than half of the battery voltage.

PRACTICE Problems

- Additional Problems, Appendix B
- Solutions to Selected Problems, Appendix C

11. A 22-Ω resistor and a 33-Ω resistor are connected in series and placed across a 120-V potential difference.

a. What is the equivalent resistance of the circuit?

b. What is the current in the circuit?

c. What is the voltage drop across each resistor?

d. What is the voltage drop across the two resistors together?

12. Three resistors of 3.3 kΩ, 4.7 kΩ, and 3.9 kΩ are connected in series across a 12-V battery.

a. What is the equivalent resistance?

b. What is the current through the resistors?

c. What is the voltage drop across each resistor?

d. Find the total voltage drop across the three resistors.

13. A student makes a voltage divider from a 45-V battery, a 475-kΩ resistor, and a 235-kΩ resistor. The output is measured across the smaller resistor. What is the voltage?

14. Select a resistor to be used as part of a voltage divider along with a 1.2-kΩ resistor. The drop across the 1.2-kΩ resistor is to be 2.2 V when the supply is 12 V.

Parallel Circuits

Look at the circuit shown in **Figure 23-6.** How many current paths are there? The current from the generator can go through any of the three resistors. A circuit in which there are several current paths is called a **parallel circuit.** The three resistors are connected in parallel; both ends of the three paths are connected together. In the mountain river model, such a circuit is illustrated by three paths for the water over a waterfall. Some paths might have a large flow of water, while others might have a small flow. The sum of the flows, however, is equal to the total flow of water over the falls. In addition, regardless of which channel the water flows through, the drop in height is the same. Similarly, in a parallel electric circuit, the total current is the sum of the currents through each path, and the potential difference across each path is the same.

Figure 23-6 The parallel paths for current in this diagram are analogous to multiple paths that a river might take down a mountain.

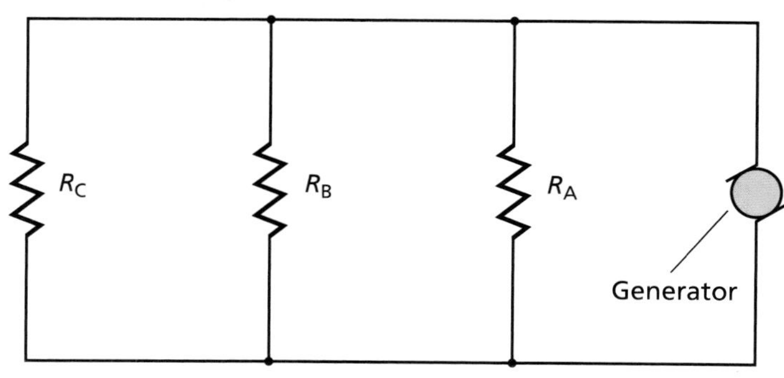

MINI LAB

Parallel Resistance

Hook up a power supply, a resistor, and an ammeter in a series circuit.

1. Predict what will happen to the current in the circuit when a second, identical resistor is added in parallel to the first.

2. Test your prediction.

3. Predict the new currents when the circuit contains three and four identical resistors in parallel.

4. Test your prediction.

Analyze and Conclude

5. Make a data table to show your results.

6. Explain your results. *(Hint: Include the idea of resistance.)*

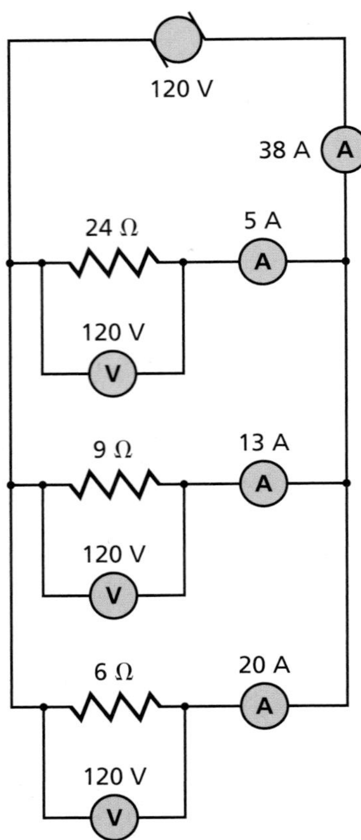

Figure 23-7 In a parallel circuit, the total current is equal to the sum of the currents in the individual paths.

What is the current through each resistor in a parallel electric circuit? It depends upon the individual resistances. For example, in **Figure 23-7,** the potential difference across each resistor is 120 V. The current through a resistor is given by $I = V/R$, so you can calculate the current through the 24-Ω resistor as $I = (120\ \text{V})/(24\ \Omega) = 5.0\ \text{A}$ and then calculate the currents through the other two resistors. The total current through the generator is the sum of the currents through the three paths, in this case, 38 A.

What would happen if the 6-Ω resistor were removed from the circuit? Would the current through the 24-Ω resistor change? That current depends only upon the potential difference across it and its resistance; because neither has changed, the current also is unchanged. The same is true for the current through the 9-Ω resistor. The branches of a parallel circuit are independent of each other. The total current through the generator, however, would change. The sum of the currents in the branches would be 18 A if the 6-Ω resistor were removed.

Resistance in a parallel circuit How can you find the equivalent resistance of a parallel circuit? In Figure 23-7, the total current through the generator is 38 A. Thus, the value of a single resistor that results in a 38-A current when a 120-V potential difference is placed across it can easily be calculated by using the following equation:

$$R = \frac{V}{I} = \frac{120\ \text{V}}{38\ \text{A}} = 3.2\ \Omega$$

Notice that this resistance is smaller than that of any of the three resistors in parallel. Placing two or more resistors in parallel always decreases the equivalent resistance of a circuit. The resistance decreases because each new resistor provides an additional path for current, thereby increasing the total current while the potential difference remains unchanged.

To calculate the equivalent resistance of a parallel circuit, first note that the total current is the sum of the currents through the branches. If I_A, I_B, and I_C are the currents through the branches and I is the total current, then $I = I_A + I_B + I_C$. The potential difference across each resistor is the same, so the current through each resistor, for example, R_A, can be found from $I_A = V/R_A$. Therefore, the equation for the sum of the currents is as follows:

$$\frac{V}{R} = \frac{V}{R_A} + \frac{V}{R_B} + \frac{V}{R_C}$$

Dividing both sides of the equation by V provides an equation for the equivalent resistance of the three parallel resistors.

Equivalent Resistance for Resistors in Parallel

$$\frac{1}{R} = \frac{1}{R_A} + \frac{1}{R_B} + \frac{1}{R_C} \ldots$$

The reciprocal of the equivalent resistance is equal to the sum of the reciprocals of the individual resistances.

This equation can be used for any number of resistors in parallel.

APPLYING PHYSICS

Testing Resistance Ohmmeters, which are used to measure resistance, work by passing a known voltage across a resistor and measuring the current. The resistance is then displayed. Some ohmmeters use potentials of less than 1 V to avoid damaging delicate electronic components, whereas others may use hundreds of volts to check the integrity of insulating materials.

▶ EXAMPLE Problem 3

Equivalent Resistance and Current in a Parallel Circuit Three resistors, 60.0 Ω, 30.0 Ω, and 20.0 Ω, are connected in parallel across a 90.0-V battery.

a. Find the current through each branch of the circuit.

b. Find the equivalent resistance of the circuit.

c. Find the current through the battery.

1 Analyze and Sketch the Problem

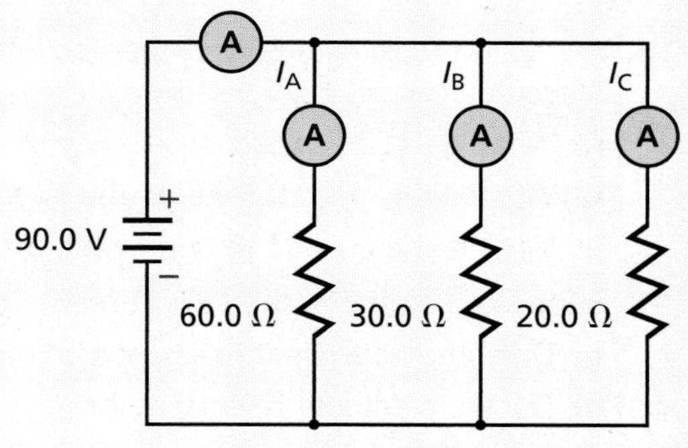

- Draw a schematic of the circuit.
- Include ammeters to show where you would measure each of the currents.

Known:	Unknown:	
$R_A = 60.0\ \Omega$	$I_A = ?$	$I = ?$
$R_B = 30.0\ \Omega$	$I_B = ?$	$R = ?$
$R_C = 20.0\ \Omega$	$I_C = ?$	
$V = 90.0$ V		

2 Solve for the Unknown

a. Because the voltage across each resistor is the same, use $I = \frac{V}{R}$ for each branch.

$$I_A = \frac{V}{R_A} = \frac{90.0\ \text{V}}{60.0\ \Omega} = 1.50\ \text{A}$$

Substitute $V = 90.0$ V, $R_A = 60.0\ \Omega$

$$I_B = \frac{V}{R_B} = \frac{90.0\ \text{V}}{30.0\ \Omega} = 3.00\ \text{A}$$

Substitute $V = 90.0$ V, $R_B = 30.0\ \Omega$

Math Handbook
Fractions page 837

$$I_C = \frac{V}{R_C} = \frac{90.0\ \text{V}}{20.0\ \Omega} = 4.50\ \text{A}$$

Substitute $V = 90.0$ V, $R_C = 20.0\ \Omega$

b. Use the equivalent resistance equation for parallel circuits.

$$\frac{1}{R} = \frac{1}{R_A} + \frac{1}{R_B} + \frac{1}{R_C} = \frac{1}{60.0\ \Omega} + \frac{1}{30.0\ \Omega} + \frac{1}{20.0\ \Omega} = \frac{1}{10.0\ \Omega}$$

Substitute $R_A = 60.0\ \Omega$, $R_B = 30.0\ \Omega$, $R_C = 20.0\ \Omega$

$$R = 10.0\ \Omega$$

c. Use $I = \frac{V}{R}$ to find the total current.

$$I = \frac{V}{R} = \frac{90.0\ \text{V}}{10.0\ \Omega} = 9.00\ \text{A}$$

Substitute $V = 90.0$ V, $R = 10.0\ \Omega$

Personal Tutor For an online tutorial on resistance and current in a parallel circuit, visit physicspp.com.

3 Evaluate the Answer

- **Are the units correct?** Current is measured in amps; resistance is measured in ohms.
- **Is the magnitude realistic?** The equivalent resistance is less than any single resistor. The current for the circuit, I, equals the sum of the current found for each resistor, $I_A + I_B + I_C$.

PRACTICE Problems

- Additional Problems, Appendix B
- Solutions to Selected Problems, Appendix C

15. Three 15.0-Ω resistors are connected in parallel and placed across a 30.0-V battery.

a. What is the equivalent resistance of the parallel circuit?

b. What is the current through the entire circuit?

c. What is the current through each branch of the circuit?

16. A 120.0-Ω resistor, a 60.0-Ω resistor, and a 40.0-Ω resistor are connected in parallel and placed across a 12.0-V battery.

a. What is the equivalent resistance of the parallel circuit?

b. What is the current through the entire circuit?

c. What is the current through each branch of the circuit?

17. Suppose that one of the 15.0-Ω resistors in problem 15 is replaced by a 10.0-Ω resistor.

a. Does the equivalent resistance change? If so, how?

b. Does the amount of current through the entire circuit change? If so, in what way?

c. Does the amount of current through the other 15.0-Ω resistors change? If so, how?

18. A 150-Ω branch in a circuit must be reduced to 93 Ω. A resistor will be added to this branch of the circuit to make this change. What value of resistance should be used and how must the resistor be connected?

19. A 12-Ω, 2-W resistor is connected in parallel with a 6.0-Ω, 4-W resistor. Which will become hotter if the voltage across them keeps increasing?

Series and parallel connections differ in how they affect a lighting cirucuit. Imagine a 60-W and a 100-W bulb are used in a lighting circuit. Recall that the brightness of a lightbulb is proportional to the power it dissipates, and that $P = I^2R$. When the bulbs are connected in parallel, each is connected across 120 V and the 100-W bulb glows more brightly. When connected in series, the current through each bulb is the same. Because the resistance of the 60-W bulb is greater than that of the 100-W bulb, the higher-resistance 60-W bulb dissipates more power and glows more brightly.

23.1 Section Review

20. Circuit Types Compare and contrast the voltages and the currents in series and parallel circuits.

21. Total Current A parallel circuit has four branch currents: 120 mA, 250 mA, 380 mA, and 2.1 A. How much current is supplied by the source?

22. Total Current A series circuit has four resistors. The current through one resistor is 810 mA. How much current is supplied by the source?

23. Circuits A switch is connected in series with a 75-W bulb to a source of 120 V.

a. What is the potential difference across the switch when it is closed (turned on)?

b. What is the potential difference across the switch if another 75-W bulb is added in series?

24. Critical Thinking The circuit in **Figure 23-8** has four identical resistors. Suppose that a wire is added to connect points A and B. Answer the following questions, and explain your reasoning.

a. What is the current through the wire?

b. What happens to the current through each resistor?

c. What happens to the current drawn from the battery?

d. What happens to the potential difference across each resistor?

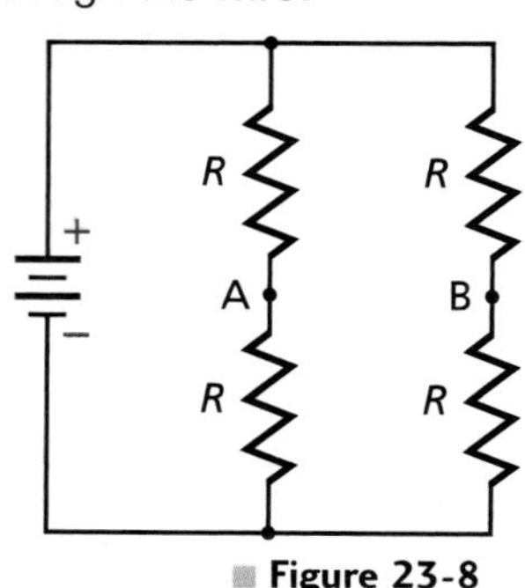

Figure 23-8

Physics Online physicspp.com/self_check_quiz

23.2 Applications of Circuits

You have learned about some of the elements of household wiring circuits. It is important to understand the requirements and limitations of these systems. Above all, you need to be aware of the safety measures that must be followed to prevent accidents and injuries.

► **Objectives**

- **Explain** how fuses, circuit breakers, and ground-fault interrupters protect household wiring.
- **Analyze** and solve problems involving combined series-parallel circuits.
- **Explain** how voltmeters and ammeters are used in circuits.

► **Vocabulary**

short circuit
fuse
circuit breaker
ground-fault interrupter
combination series-parallel circuit
ammeter
voltmeter

Safety Devices

In an electric circuit, fuses and circuit breakers act as safety devices. They prevent circuit overloads that can occur when too many appliances are turned on at the same time or when a short circuit occurs in one appliance. A **short circuit** occurs when a circuit with a very low resistance is formed. The low resistance causes the current to be very large. When appliances are connected in parallel, each additional appliance placed in operation reduces the equivalent resistance in the circuit and increases the current through the wires. This additional current might produce enough thermal energy to melt the wiring's insulation, cause a short circuit, or even begin a fire.

A **fuse** is a short piece of metal that melts when too large a current passes through it. The thickness of the metal used in the fuse is determined by the amount of current that the circuit is designed to handle safely. If a large, unsafe current passes through the circuit, the fuse melts and breaks the circuit. A **circuit breaker,** shown in **Figure 23-9,** is an automatic switch that opens when the current reaches a threshold value. If there is a current greater than the rated (threshold) value in the circuit, the circuit becomes overloaded. The circuit breaker opens and stops the current.

Current follows a single path from the power source through an electrical appliance and back to the source. Sometimes, a faulty appliance or an accidental drop of the appliance into water might create another current pathway. If this pathway flows through the user, serious injury could result. A current as small as 5 mA flowing through a person could result in electrocution. A **ground-fault interrupter** in an electric outlet prevents such injuries because it contains an electronic circuit that detects small differences in current caused by an extra current path and opens the circuit. Electric codes for buildings often require ground-fault interrupters to be used in bathroom, kitchen, and exterior outlets.

Figure 23-9 When too much current flows through the bimetallic strip, the heat that is generated causes the strip to bend and release the latch. The handle moves to the off position, causing the switch to open and break the circuit.

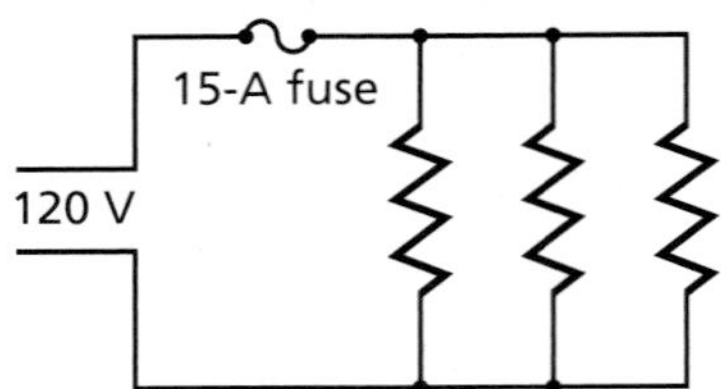

■ **Figure 23-10** The parallel wiring arrangement used in homes allows the simultaneous use of more than one appliance. However, if too many appliances are used at once, the fuse could melt.

Household applications **Figure 23-10** diagrams a parallel circuit used in the wiring of homes, and also shows some common appliances that would be connected in parallel. The current in any one circuit does not depend upon the current in the other circuits. Suppose that a 240-W television is plugged into a 120-V outlet. The current is represented by $I = P/V$. For the television, $I = (240\ \text{W})/(120\ \text{V}) = 2.0\ \text{A}$. When a 720-W curling iron is plugged into the outlet, its current draw is $I = (720\ \text{W})/(120\ \text{V}) = 6.0\ \text{A}$. Finally, a 1440-W hair dryer is plugged into the same outlet. The current through the hair dryer is $I = (1440\ \text{W})/(120\ \text{V}) = 12\ \text{A}$. The resistance of each appliance can be calculated using the equation $R = V/I$. The equivalent resistance of the three appliances is as follows.

$$\frac{1}{R} = \frac{1}{60\ \Omega} + \frac{1}{20\ \Omega} + \frac{1}{10\ \Omega}$$

$$= \frac{1}{6\ \Omega}$$

$$R = 6\ \Omega$$

A fuse is connected in series with the power source so that the entire current passes through it. The current through the fuse is calculated using the equivalent resistance.

$$I = \frac{V}{R} = \frac{120\ \text{V}}{6\ \Omega} = 20\ \text{A}$$

If the fuse in the circuit is rated as 15 A, the 20-A current would exceed the rating and cause the fuse to melt, or "blow," and cut off current.

Fuses and circuit breakers also protect against the large currents created by a short circuit. Without a fuse or a circuit breaker, the current caused by a short circuit easily could start a fire. For example, a short circuit could occur if the insulation on a lamp cord became old and brittle. The two wires in the cord might accidentally touch, resulting in a resistance in the wire of only 0.010 Ω. This resistance results in a huge current.

$$I = \frac{V}{R} = \frac{120\ \text{V}}{0.010\ \Omega} = 12{,}000\ \text{A}$$

Such a current would cause a fuse or a circuit breaker to open the circuit, thereby preventing the wires from becoming hot enough to start a fire.

• CHALLENGE **PROBLEM**

When the galvanometer, a device used to measure very small currents or voltages, in this circuit measures zero, the circuit is said to be balanced.

1. Your lab partner states that the only way to balance this circuit is to make all the resistors equal. Will this balance the circuit? Is there more than one way to balance this circuit? Explain.

2. Derive a general equation for a balanced circuit using the given labels. *Hint: Treat the circuit as a voltage divider.*

3. Which of the resistors can be replaced with a variable resistor and then used to balance the circuit?

4. Which of the resistors can be replaced with a variable resistor and then used as a sensitivity control? Why would this be necessary? How would it be used in practice?

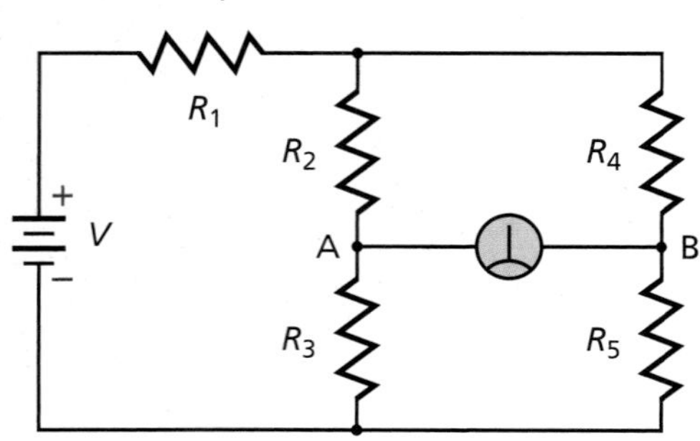

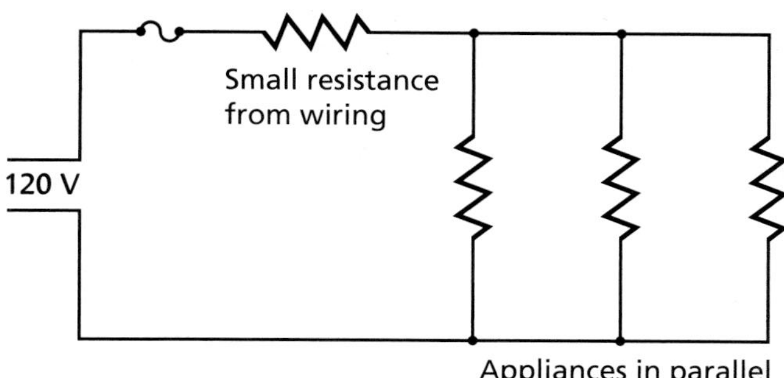

Figure 23-11 The small resistance inherent in wires is in series with the parallel resistances of household appliances.

Combined Series-Parallel Circuits

Have you ever noticed the light in your bathroom or bedroom dim when you turned on a hair dryer? The light and the hair dryer are connected in parallel across 120 V. Because of the parallel connection, the current through the light should not have changed when you turned on the hair dryer. Yet the light did dim, so the current must have changed. The dimming occurred because the house wiring had a small resistance. As shown in **Figure 23-11,** this resistance was in series with the parallel circuit. Such a circuit, which includes series and parallel branches, is called a **combination series-parallel circuit.** The following are strategies for analyzing such circuits.

PROBLEM-SOLVING **Strategies**

Series-Parallel Circuits

When analyzing a combination series-parallel circuit, use the following steps to break down the problem.

1. Draw a schematic diagram of the circuit.
2. Find any parallel resistors. Resistors in parallel have separate current paths. They must have the same potential differences across them. Calculate the single equivalent resistance of a resistor that can replace them. Draw a new schematic using that resistor.
3. Are any resistors (including the equivalent resistor) now in series? Resistors in series have one and only one current path through them. Calculate a single new equivalent resistance that can replace them. Draw a new schematic diagram using that resistor.
4. Repeat steps 2 and 3 until you can reduce the circuit to a single resistor. Find the total circuit current. Then go backwards through the circuits to find the currents through and the voltages across individual resistors.

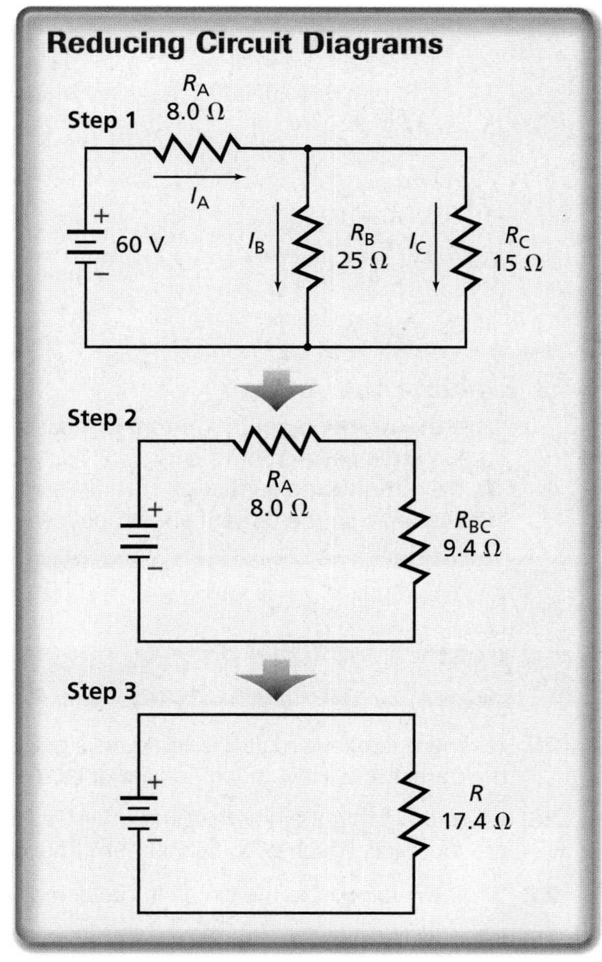

▶ EXAMPLE Problem 4

Series-Parallel Circuit A hair dryer with a resistance of 12.0 Ω and a lamp with a resistance of 125 Ω are connected in parallel to a 125-V source through a 1.50-Ω resistor in series. Find the current through the lamp when the hair dryer is on.

1 Analyze and Sketch the Problem

- Draw the series-parallel circuit including the hair dryer and lamp.
- Replace R_A and R_B with a single equivalent resistance, R_p.

Known:		Unknown:	
$R_A = 125\ \Omega$	$R_C = 1.50\ \Omega$	$I = ?$	$I_A = ?$
$R_B = 12.0\ \Omega$	$V_{source} = 125\ V$	$R = ?$	$R_p = ?$

2 Solve for the Unknown

Find the equivalent resistance for the parallel circuit, then find the equivalent resistance for the entire circuit, and then calculate the current.

$$\frac{1}{R_p} = \frac{1}{R_A} + \frac{1}{R_B} = \frac{1}{125\ \Omega} + \frac{1}{12.0\ \Omega}$$ **Substitute $R_A = 125\ \Omega$, $R_B = 12.0\ \Omega$**

$$R_p = 10.9\ \Omega$$

$$R = R_C + R_p = 1.50\ \Omega + 11.0\ \Omega = 12.4\ \Omega$$ **Substitute $R_C = 1.50\ \Omega$, $R_p = 10.9\ \Omega$**

$$I = \frac{V_{source}}{R} = \frac{125\ V}{12.4\ \Omega} = 10.1\ A$$ **Substitute $V_{source} = 125\ V$, $R = 12.4\ \Omega$**

$$V_C = IR_C = (10.1\ A)(1.50\ \Omega) = 15.2\ V$$ **Substitute $I = 10.1\ A$, $R_C = 1.50\ \Omega$**

$$V_A = V_{source} - V_C = 125\ V - 15.2\ V = 1.10\times10^2\ V$$ **Substitute $V_{source} = 125\ V$, $V_C = 15.2\ V$**

$$I_A = \frac{V_A}{R_A} = \frac{1.10\times10^2\ V}{125\ \Omega} = 0.880\ A$$ **Substitute $V_A = 1.10\times10^2\ V$, $R_A = 125\ \Omega$**

Math Handbook
Operations with Significant Digits
pages 835–836

3 Evaluate the Answer

- **Are the units correct?** Current is measured in amps, and potential drops are measured in volts.
- **Is the magnitude realistic?** The resistance is greater than the voltage, so the current should be less than 1 A.

▶ PRACTICE Problems

• Additional Problems, Appendix B
• Solutions to Selected Problems, Appendix C

25. A series-parallel circuit has three resistors: one dissipates 2.0 W, the second 3.0 W, and the third 1.5 W. How much current does the circuit require from a 12-V battery?

26. There are 11 lights in series, and they are in series with two lights in parallel. If the 13 lights are identical, which of them will burn brightest?

27. What will happen to the circuit in problem 26 if one of the parallel lights burns out?

28. What will happen to the circuit in problem 26 if one of the parallel lights shorts out?

Ammeters and Voltmeters

An **ammeter** is a device that is used to measure the current in any branch or part of a circuit. If, for example, you wanted to measure the current through a resistor, you would place an ammeter in series with the resistor. This would require opening the current path and inserting an ammeter. Ideally, the use of an ammeter should not change the current in the circuit. Because the current would decrease if the ammeter increased the resistance in the circuit, the resistance of an ammeter is designed to be as low as possible. **Figure 23-12a** shows an ammeter as a meter placed in parallel with a 0.01-Ω resistor. Because the resistance of the ammeter is much less than that of the resistors, the current decrease is negligible.

Another instrument, called a **voltmeter,** is used to measure the voltage drop across a portion of a circuit. To measure the potential drop across a resistor, a voltmeter is connected in parallel with the resistor. Voltmeters are designed to have a very high resistance so as to cause the smallest possible change in currents and voltages in the circuit. Consider the circuit shown in **Figure 23-12b.** A voltmeter is shown as a meter in series with a 10-kΩ resistor. When the voltmeter is connected in parallel with R_B, the equivalent resistance of the combination is smaller than R_B alone. Thus, the total resistance of the circuit decreases, and the current increases. The value of R_A has not changed, but the current through it has increased, thereby increasing the potential drop across it. The battery, however, holds the potential drop across R_A and R_B constant. Thus, the potential drop across R_B must decrease. The result of connecting a voltmeter across a resistor is to lower the potential drop across it. The higher the resistance of the voltmeter, the smaller the voltage change. Practical meters have resistances of 10 MΩ.

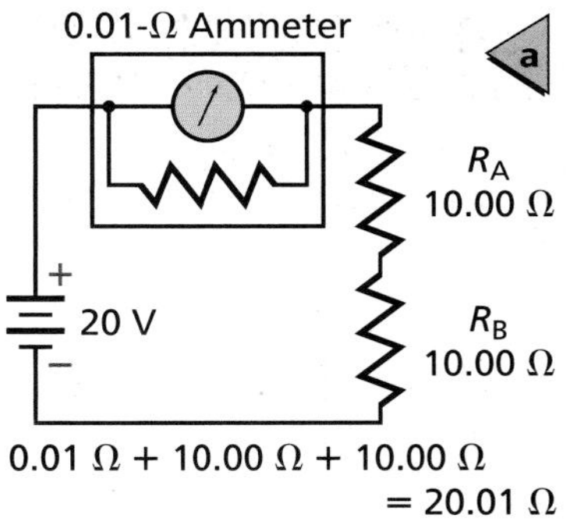

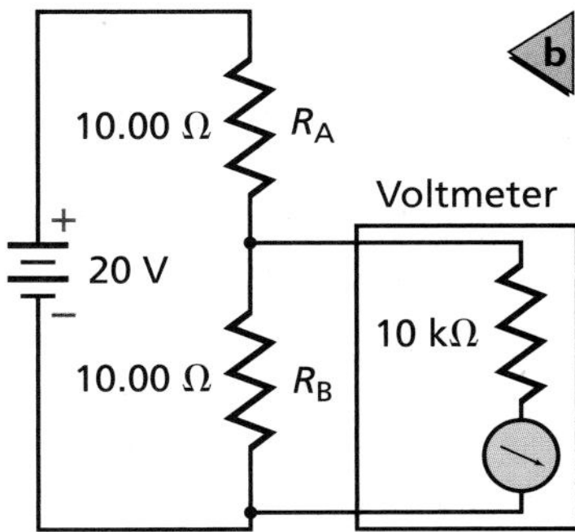

Figure 23-12 An ammeter is connected in series with two resistors **(a).** The small resistance of the ammeter slightly alters the current in the circuit. A voltmeter is connected in parallel with a resistor **(b).** The high resistance of the voltmeter results in a negligible change in the circuit current and voltage.

23.2 Section Review

*Refer to **Figure 23-13** for questions 29–33, and 35. The bulbs in the circuit are identical.*

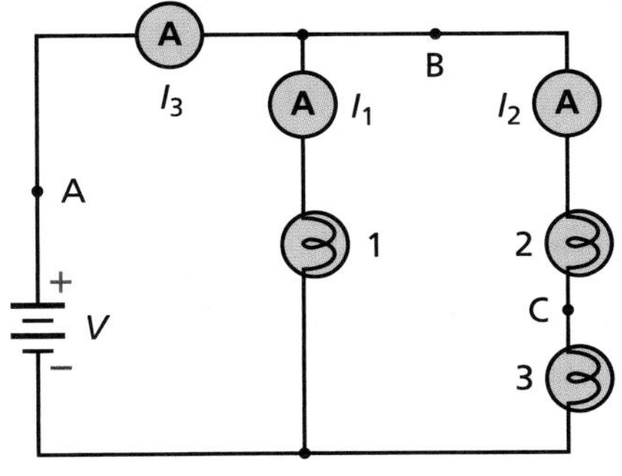

Figure 23-13

29. Brightness How do the bulb brightnesses compare?

30. Current If I_3 measures 1.7 A and I_1 measures 1.1 A, how much current is flowing in bulb 2?

31. Circuits in Series The wire at point C is broken and a small resistor is inserted in series with bulbs 2 and 3. What happens to the brightnesses of the two bulbs? Explain.

32. Battery Voltage A voltmeter connected across bulb 2 measures 3.8 V, and a voltmeter connected across bulb 3 measures 4.2 V. What is the battery voltage?

33. Circuits Using the information from problem 32, determine if bulbs 2 and 3 are identical.

34. Circuit Protection Describe three common safety devices associated with household wiring.

35. Critical Thinking Is there a way to make the three bulbs in Figure 23-13 burn with equal intensity without using any additional resistors? Explain.

PHYSICS LAB

Series and Parallel Circuits

Alternate CBL instructions can be found on the Web site.
physicspp.com

In every circuit there is a relationship among current, potential difference, and resistance in electric circuits. In this experiment, you will investigate how the relationship of current, potential difference, and resistance in series circuits compares to that in parallel circuits.

QUESTION

How do relationships among current, potential difference, and resistance compare in series and parallel circuits?

Objectives

- **Describe** the relationship among current, potential difference, and resistance in a series circuit.
- **Summarize** the relationship among current, potential difference, and resistance in a parallel circuit.
- **Collect data** for current and potential difference using electric meters.
- **Calculate** resistance in a lightbulb from current and potential-difference data.

Safety Precautions

- **Hazard from electric shock is minimal because of the low currents used in this experiment. This experiment should not be carried out using current from an AC circuit, as this current is deadly.**
- **Handle wire ends with care as they may be sharp and could cause cuts.**

Materials

low-voltage power supply
two light sockets
two small lightbulbs
ammeter or multimeter (0–500-mA scale)
voltmeter or multimeter (0–30-V scale)
about ten copper wires with alligator clips

Procedure

1. Wire two lightbulb sockets in series with an ammeter and a low-voltage power supply. Observe the correct polarity when wiring the ammeter.
2. Screw the lightbulbs into the sockets. Turn on the power supply. Adjust the power control so that the bulbs are dimly lit.
3. Unscrew one of the bulbs. Record your observations in the data table.
4. Screw in the bulb again and find the potential difference across both sets of bulbs by placing the positive probe of the voltmeter on the positive end of the circuit and the negative probe on the negative end of the circuit. Record your data in the data table.
5. Find the potential difference across each individual lightbulb by placing the positive probe of the voltmeter on the positive end of a bulb and the negative probe on the negative end of the bulb. Record your data in the data table. Repeat for the other bulb in series.
6. Place the ammeter at various locations in the series circuit. Record these currents in the data table.
7. Wire the two lightbulb sockets in parallel with the low-voltage power supply and in series with an ammeter.

Data Table	
Step	**Observations**
3	
4	
5	
6	
8	
9	
10	
11	

8. Screw the lightbulbs into the sockets. Turn on the power supply. Adjust the power control so that the bulbs are dimly lit. Record the current shown on the ammeter in the data table.
9. Check the potential difference across the entire circuit and across each lightbulb. Record the values in the data table.
10. Place the voltmeter probes across one of the lightbulbs. Now unscrew one of the lightbulbs. Record your observations of both lightbulbs, and record the current and potential difference read by the meters in the data table.
11. Return the lightbulb you removed in the previous step to its socket. Now unscrew the other lightbulb. Record your observations of both lightbulbs, and record the current and potential difference read by the meters in the data table.

Analyze

1. Calculate the resistance of the pair of lightbulbs in the series circuit.
2. Calculate the resistance of each lightbulb in the series circuit.
3. How does the resistance of the pair of lightbulbs compare to the individual resistance of each lightbulb?
4. How does the potential difference across the individual lightbulbs compare to the potential difference across the pair of lightbulbs in the series circuit?
5. Calculate the resistance of each of the lightbulbs while they are in the parallel circuit. How does this compare to the resistance calculated for the bulbs in the series circuit?

Conclude and Apply

1. Summarize the relationship among current, potential difference, and resistance in a series circuit.
2. Summarize the relationship between current and potential difference in a parallel circuit.

Going Further

Repeat the experiment using lightbulbs of different voltage ratings (for example: 1.5 V, 3.0 V, and 6.0 V).

Real-World Physics

1. The lightbulbs in most homes all are rated for 120 V, no matter how many bulbs there are. How is the ability to use any number of same-voltage bulbs affected by the way in which the bulbs are wired (series or parallel)?
2. Why do lights in a home dim when a large appliance, such as an air conditioner, is turned on?

To find out more about series and parallel circuits, visit the Web site: physicspp.com

HOW it Works Ground Fault Circuit Interrupters (GFCI)

A ground fault occurs when electricity takes an incorrect path to ground, such as through a person's body. Charles Dalziel, an engineering professor at the University of California, was an expert on the effects of electric shock. When he realized that ground faults were the cause of many electrocutions, he invented a device to prevent such accidents. How does a ground fault circuit interrupter (GFCI) work?

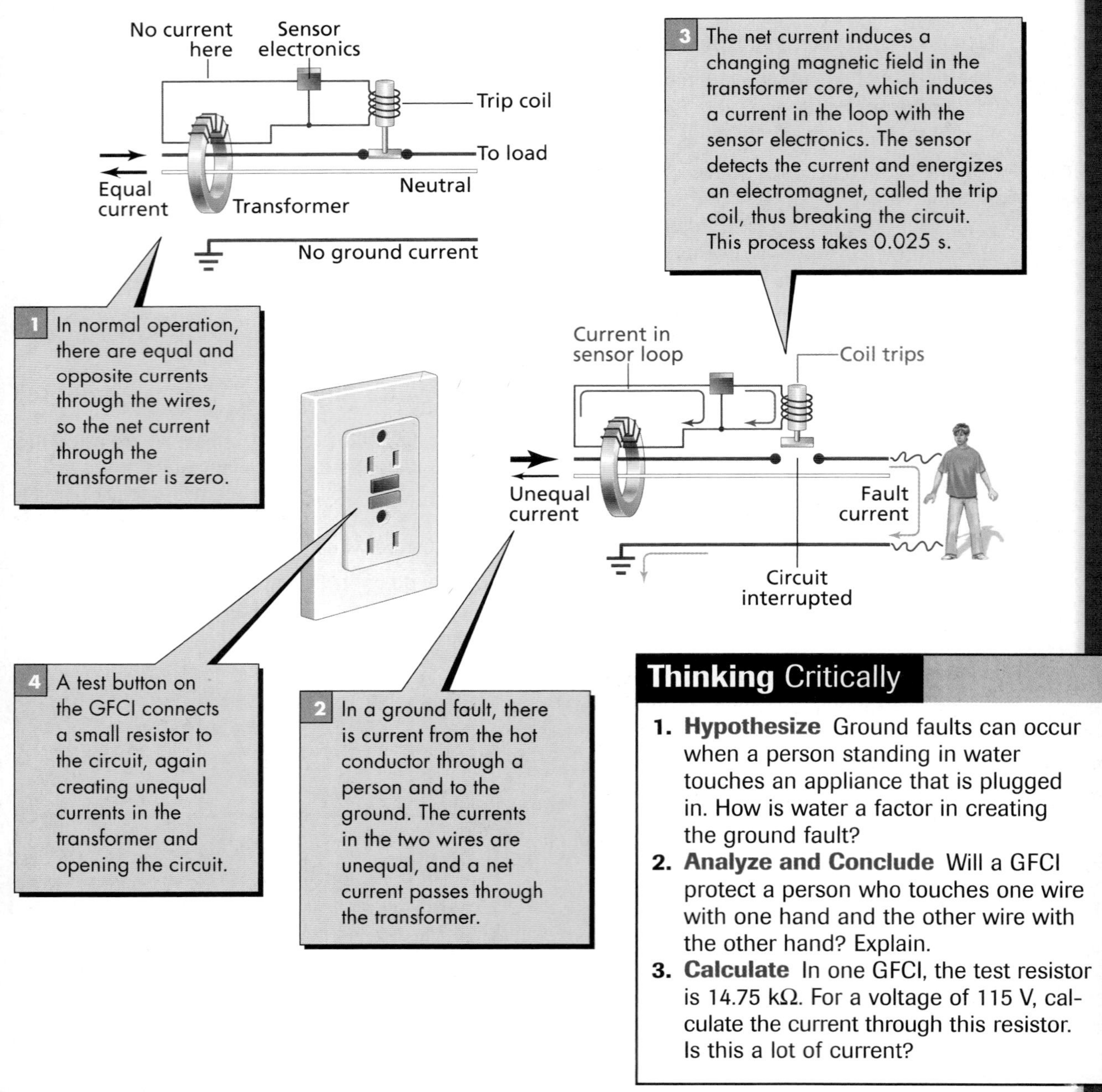

No current here
Sensor electronics
Trip coil
To load
Equal current
Transformer
Neutral
No ground current
1 In normal operation, there are equal and opposite currents through the wires, so the net current through the transformer is zero.
2 In a ground fault, there is current from the hot conductor through a person and to the ground. The currents in the two wires are unequal, and a net current passes through the transformer.
3 The net current induces a changing magnetic field in the transformer core, which induces a current in the loop with the sensor electronics. The sensor detects the current and energizes an electromagnet, called the trip coil, thus breaking the circuit. This process takes 0.025 s.
4 A test button on the GFCI connects a small resistor to the circuit, again creating unequal currents in the transformer and opening the circuit.
Current in sensor loop
Coil trips
Unequal current
Fault current
Circuit interrupted

Thinking Critically

1. **Hypothesize** Ground faults can occur when a person standing in water touches an appliance that is plugged in. How is water a factor in creating the ground fault?
2. **Analyze and Conclude** Will a GFCI protect a person who touches one wire with one hand and the other wire with the other hand? Explain.
3. **Calculate** In one GFCI, the test resistor is 14.75 kΩ. For a voltage of 115 V, calculate the current through this resistor. Is this a lot of current?

Chapter 23

Study Guide

23.1 Simple Circuits

Vocabulary

- series circuit (p. 618)
- equivalent resistance (p. 619)
- voltage divider (p. 620)
- parallel circuit (p. 623)

Key Concepts

- The current is the same everywhere in a simple series circuit.
- The equivalent resistance of a series circuit is the sum of the resistances of its parts.

$$R = R_A + R_B + \ldots$$

- The current in a series circuit is equal to the potential difference divided by the equivalent resistance.

$$I = \frac{V_{source}}{R}$$

- The sum of the voltage drops across resistors that are in series is equal to the potential difference applied across the combination.
- A voltage divider is a series circuit used to produce a voltage source of desired magnitude from a higher-voltage battery.
- The voltage drops across all branches of a parallel circuit are the same.
- In a parallel circuit, the total current is equal to the sum of the currents in the branches.
- The reciprocal of the equivalent resistance of parallel resistors is equal to the sum of the reciprocals of the individual resistances.

$$\frac{1}{R} = \frac{1}{R_A} + \frac{1}{R_B} + \frac{1}{R_C} + \ldots$$

- If any branch of a parallel circuit is opened, there is no current in that branch. The current in the other branches is unchanged.

23.2 Applications of Circuits

Vocabulary

- short circuit (p. 627)
- fuse (p. 627)
- circuit breaker (p. 627)
- ground-fault interrupter (p. 627)
- combination series-parallel circuit (p. 629)
- ammeter (p. 631)
- voltmeter (p. 631)

Key Concepts

- A fuse or circuit breaker, placed in series with appliances, creates an open circuit when dangerously high currents flow.
- A complex circuit is a combination of series and parallel branches. Any parallel branch first is reduced to a single equivalent resistance. Then, any resistors in series are replaced by a single resistance.
- An ammeter is used to measure the current in a branch or part of a circuit. An ammeter always has a low resistance and is connected in series.
- A voltmeter measures the potential difference (voltage) across any part or combination of parts of a circuit. A voltmeter always has a high resistance and is connected in parallel with the part of the circuit being measured.

Chapter 23 Assessment

Concept Mapping

36. Complete the concept map using the following terms: *series circuit, $R = R_1 + R_2 + R_3$, constant current, parallel circuit, constant potential.*

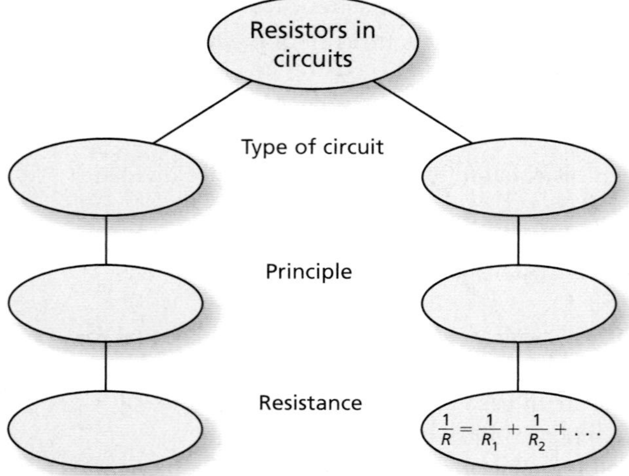

Mastering Concepts

37. Why is it frustrating when one bulb burns out on a string of holiday tree lights connected in series? (23.1)

38. Why does the equivalent resistance decrease as more resistors are added to a parallel circuit? (23.1)

39. Several resistors with different values are connected in parallel. How do the values of the individual resistors compare with the equivalent resistance? (23.1)

40. Why is household wiring constructed in parallel instead of in series? (23.1)

41. Why is there a difference in equivalent resistance between three 60-Ω resistors connected in series and three 60-Ω resistors connected in parallel? (23.1)

42. Compare the amount of current entering a junction in a parallel circuit with that leaving the junction. (A junction is a point where three or more conductors are joined.) (23.1)

43. Explain how a fuse functions to protect an electric circuit. (23.2)

44. What is a short circuit? Why is a short circuit dangerous? (23.2)

45. Why is an ammeter designed to have a very low resistance? (23.2)

46. Why is a voltmeter designed to have a very high resistance? (23.2)

47. How does the way in which an ammeter is connected in a circuit differ from the way in which a voltmeter is connected? (23.2)

Applying Concepts

48. What happens to the current in the other two lamps if one lamp in a three-lamp series circuit burns out?

49. Suppose the resistor, R_A, in the voltage divider in Figure 23-4 is made to be a variable resistor. What happens to the voltage output, V_B, of the voltage divider if the resistance of the variable resistor is increased?

50. Circuit A contains three 60-Ω resistors in series. Circuit B contains three 60-Ω resistors in parallel. How does the current in the second 60-Ω resistor of each circuit change if a switch cuts off the current to the first 60-Ω resistor?

51. What happens to the current in the other two lamps if one lamp in a three-lamp parallel circuit burns out?

52. An engineer needs a 10-Ω resistor and a 15-Ω resistor, but there are only 30-Ω resistors in stock. Must new resistors be purchased? Explain.

53. If you have a 6-V battery and many 1.5-V bulbs, how could you connect them so that they light but do not have more than 1.5 V across each bulb?

54. Two lamps have different resistances, one larger than the other.
   a. If the lamps are connected in parallel, which is brighter (dissipates more power)?
   b. When the lamps are connected in series, which lamp is brighter?

55. For each of the following, write the form of circuit that applies: series or parallel.
   a. The current is the same everywhere throughout the entire circuit.
   b. The total resistance is equal to the sum of the individual resistances.
   c. The voltage drop across each resistor in the circuit is the same.
   d. The voltage drop in the circuit is proportional to the resistance.
   e. Adding a resistor to the circuit decreases the total resistance.
   f. Adding a resistor to the circuit increases the total resistance.
   g. If the current through one resistor in the circuit goes to zero, there is no current in the entire circuit.
   h. If the current through one resistor in the circuit goes to zero, the current through all other resistors remains the same.
   i. This form is suitable for house wiring.

56. **Household Fuses** Why is it dangerous to replace the 15-A fuse used to protect a household circuit with a fuse that is rated at 30 A?

 For more problems, go to Additional Problems, Appendix B.

Mastering Problems

23.1 Simple Circuits

57. Ammeter 1 in **Figure 23-14** reads 0.20 A.
  a. What should ammeter 2 indicate?
  b. What should ammeter 3 indicate?

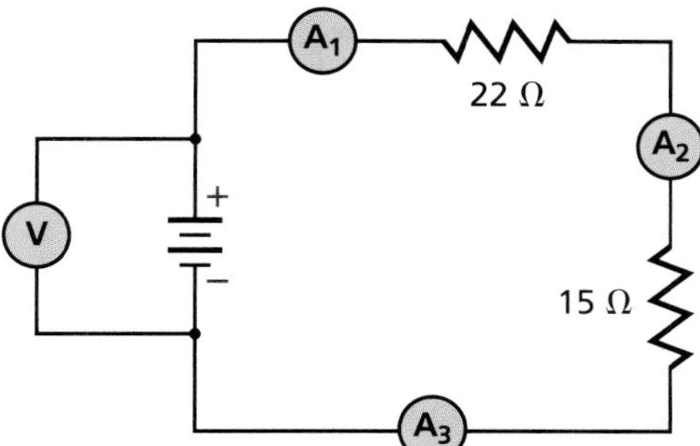

Figure 23-14

58. Calculate the equivalent resistance of these series-connected resistors: 680 Ω, 1.1 kΩ, and 10 kΩ.

59. Calculate the equivalent resistance of these parallel-connected resistors: 680 Ω, 1.1 kΩ, and 10.2 kΩ.

60. A series circuit has two voltage drops: 5.50 V and 6.90 V. What is the supply voltage?

61. A parallel circuit has two branch currents: 3.45 A and 1.00 A. What is the current in the energy source?

62. Ammeter 1 in Figure 23-14 reads 0.20 A.
  a. What is the total resistance of the circuit?
  b. What is the battery voltage?
  c. How much power is delivered to the 22-Ω resistor?
  d. How much power is supplied by the battery?

63. Ammeter 2 in Figure 23-14 reads 0.50 A.
  a. Find the voltage across the 22-Ω resistor.
  b. Find the voltage across the 15-Ω resistor.
  c. What is the battery voltage?

64. A 22-Ω lamp and a 4.5-Ω lamp are connected in series and placed across a potential difference of 45 V as shown in **Figure 23-15.**
  a. What is the equivalent resistance of the circuit?
  b. What is the current in the circuit?
  c. What is the voltage drop across each lamp?
  d. What is the power dissipated in each lamp?

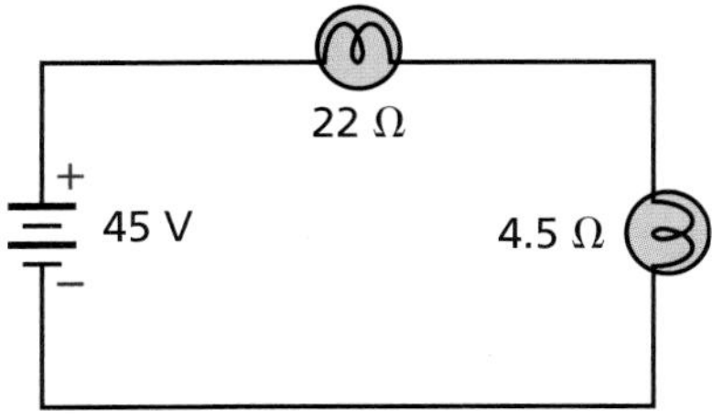

Figure 23-15

65. Refer to **Figure 23-16** to answer the following questions.
  a. What should the ammeter read?
  b. What should voltmeter 1 read?
  c. What should voltmeter 2 read?
  d. How much energy is supplied by the battery per minute?

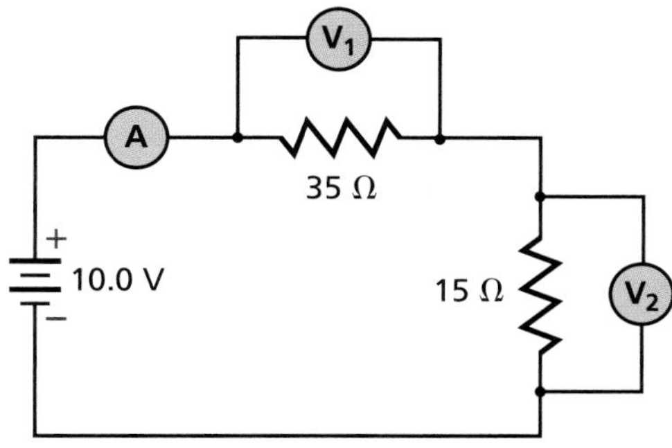

Figure 23-16

66. For **Figure 23-17,** the voltmeter reads 70.0 V.
  a. Which resistor is the hottest?
  b. Which resistor is the coolest?
  c. What will the ammeter read?
  d. What is the power supplied by the battery?

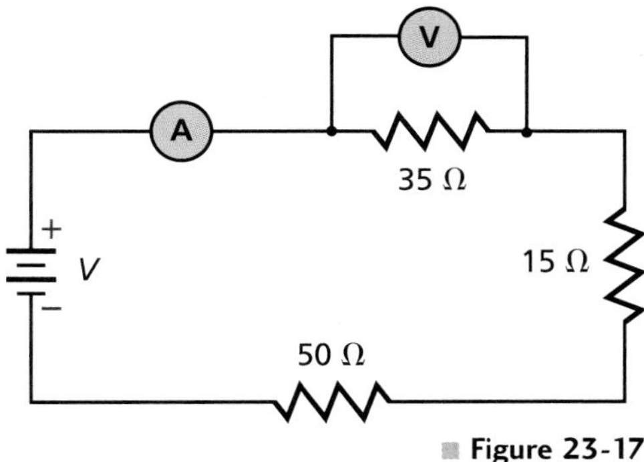

Figure 23-17

67. For **Figure 23-18,** the battery develops 110 V.
  a. Which resistor is the hottest?
  b. Which resistor is the coolest?
  c. What will ammeter 1 read?
  d. What will ammeter 2 read?
  e. What will ammeter 3 read?
  f. What will ammeter 4 read?

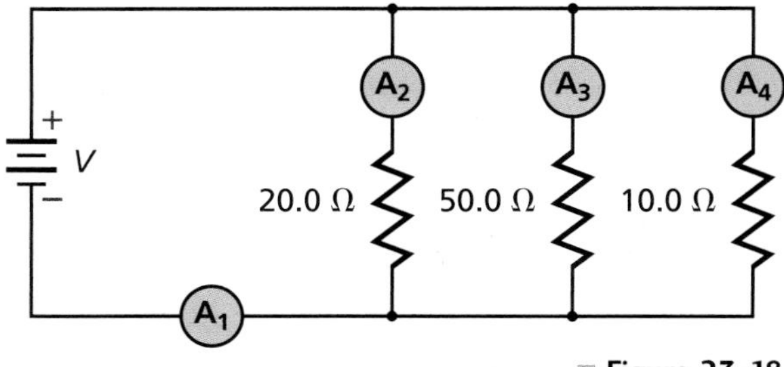

Figure 23-18

68. For Figure 23-18, ammeter 3 reads 0.40 A.
a. What is the battery voltage?
b. What will ammeter 1 read?
c. What will ammeter 2 read?
d. What will ammeter 4 read?

69. What is the direction of the conventional current in the 50.0-Ω resistor in Figure 23-18?

70. The load across a battery consists of two resistors, with values of 15 Ω and 47 Ω, connected in series.
a. What is the total resistance of the load?
b. What is the voltage of the battery if the current in the circuit is 97 mA?

71. Holiday Lights A string of 18 identical holiday tree lights is connected in series to a 120-V source. The string dissipates 64 W.
a. What is the equivalent resistance of the light string?
b. What is the resistance of a single light?
c. What power is dissipated by each light?

72. One of the lights in problem 71 burns out. The light shorts out the bulb filament when it burns out. This drops the resistance of the lamp to zero.
a. What is the resistance of the light string now?
b. Find the power dissipated by the string.
c. Did the power increase or decrease when the bulb burned out?

73. A 16.0-Ω and a 20.0-Ω resistor are connected in parallel. A difference in potential of 40.0 V is applied to the combination.
a. Compute the equivalent resistance of the parallel circuit.
b. What is the total current in the circuit?
c. What is the current in the 16.0-Ω resistor?

74. Amy needs 5.0 V for an integrated-circuit experiment. She uses a 6.0-V battery and two resistors to make a voltage divider. One resistor is 330 Ω. She decides to make the other resistor smaller. What value should it have?

75. Pete is designing a voltage divider using a 12-V battery and a 82-Ω resistor as R_B. What resistor should be used as R_A if the output voltage across R_B is to be 4.0 V?

76. Television A typical television dissipates 275 W when it is plugged into a 120-V outlet.
a. Find the resistance of the television.
b. The television and 2.5-Ω wires connecting the outlet to the fuse form a series circuit that works like a voltage divider. Find the voltage drop across the television.
c. A 12-Ω hair dryer is plugged into the same outlet. Find the equivalent resistance of the two appliances.
d. Find the voltage drop across the television and the hair dryer.

23.2 Applications of Circuits

77. Refer to **Figure 23-19** and assume that all the resistors are 30.0 Ω. Find the equivalent resistance.

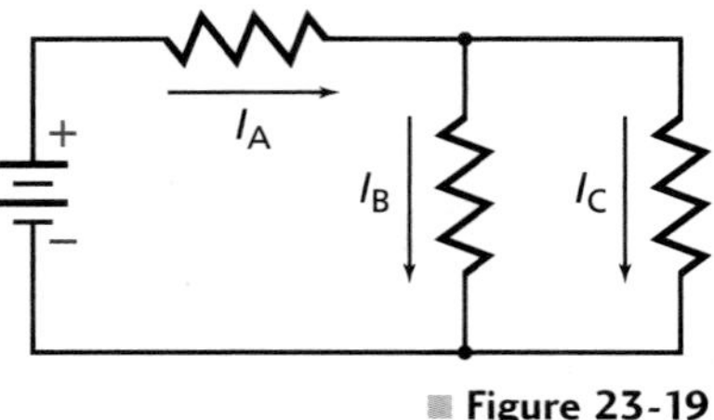

Figure 23-19

78. Refer to Figure 23-19 and assume that each resistor dissipates 120 mW. Find the total dissipation.

79. Refer to Figure 23-19 and assume that I_A = 13 mA and I_B = 1.7 mA. Find I_C.

80. Refer to Figure 23-19 and assume that I_B = 13 mA and I_C = 1.7 mA. Find I_A.

81. Refer to **Figure 23-20** to answer the following questions.
a. Determine the total resistance.
b. Determine the current through the 25-Ω resistor.
c. Which resistor is the hottest? Coolest?

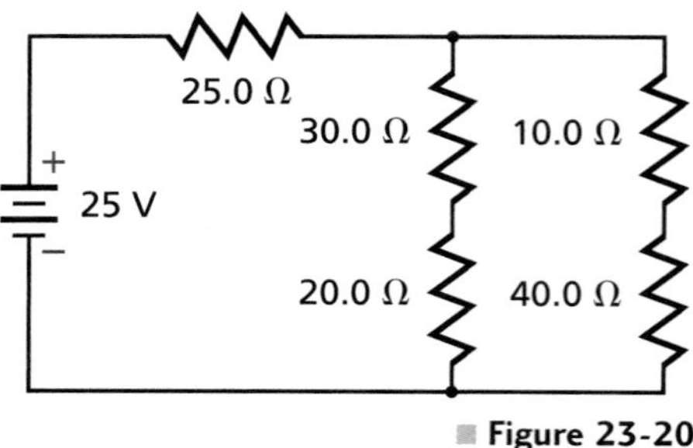

Figure 23-20

82. A circuit contains six 60-W lamps with a resistance of 240-Ω each and a 10.0-Ω heater connected in parallel. The voltage across the circuit is 120 V. Find the current in the circuit for the following situations.
a. Four lamps are turned on.
b. All of the lamps are turned on.
c. Six lamps and the heater are operating.

83. If the circuit in problem 82 has a 12-A fuse, will the fuse melt if all the lamps and the heater are on?

84. During a laboratory exercise, you are supplied with a battery of potential difference *V*, two heating elements of low resistance that can be placed in water, an ammeter of very small resistance, a voltmeter of extremely high resistance, wires of negligible resistance, a beaker that is well insulated and has negligible heat capacity, and 0.10 kg of water at 25°C. By means of a diagram and standard symbols, show how these components should be connected to heat the water as rapidly as possible.

 For more problems, go to Additional Problems, Appendix B.

85. If the voltmeter used in problem 84 holds steady at 45 V and the ammeter reading holds steady at 5.0 A, estimate the time in seconds required to completely vaporize the water in the beaker. Use 4.2 kJ/kg·°C as the specific heat of water and 2.3×10^6 J/kg as the heat of vaporization of water.

86. **Home Circuit** A typical home circuit is shown in **Figure 23-21.** The wires to the kitchen lamp each have a resistance of 0.25 Ω. The lamp has a resistance of 0.24 kΩ. Although the circuit is parallel, the lead lines are in series with each of the components of the circuit.

a. Compute the equivalent resistance of the circuit consisting of just the lamp and the lead lines to and from the lamp.

b. Find the current to the lamp.

c. Find the power dissipated in the lamp.

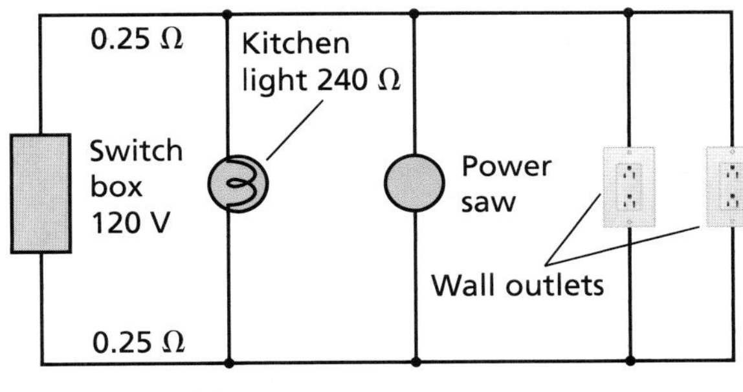

Figure 23-21

Mixed Review

87. A series circuit has two voltage drops: 3.50 V and 4.90 V. What is the supply voltage?

88. A parallel circuit has two branch currents: 1.45 A and 1.00 A. What is the current in the energy source?

89. A series-parallel circuit has three resistors, dissipating 5.50 W, 6.90 W, and 1.05 W, respectively. What is the supply power?

90. Determine the maximum safe power in each of three 150-Ω, 5-W resistors connected in series.

91. Determine the maximum safe power in each of three 92-Ω, 5-W resistors connected in parallel.

92. A voltage divider consists of two 47-kΩ resistors connected across a 12-V battery. Determine the measured output for the following.

a. an ideal voltmeter

b. a voltmeter with a resistance of 85 kΩ

c. a voltmeter with a resistance of 10×10^6 Ω

93. Determine the maximum safe voltage that can be applied across the three series resistors in **Figure 23-22** if all three are rated at 5.0 W.

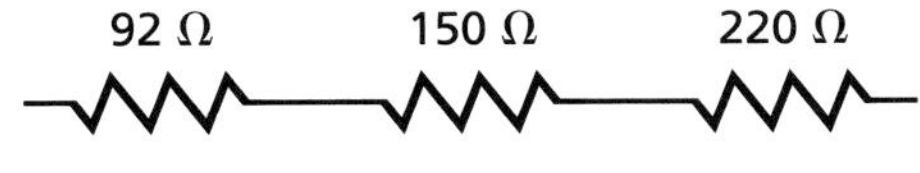

Figure 23-22

94. Determine the maximum safe total power for the circuit in problem 93.

95. Determine the maximum safe voltage that can be applied across three parallel resistors of 92 Ω, 150 Ω, and 220 Ω, as shown in **Figure 23-23,** if all three are rated at 5.0 W.

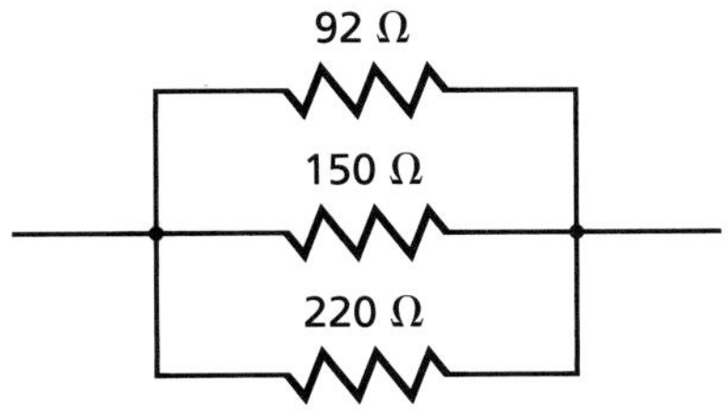

Figure 23-23

Thinking Critically

96. **Apply Mathematics** Derive equations for the resistance of two equal-value resistors in parallel, three equal-value resistors in parallel, and N equal-value resistors in parallel.

97. **Apply Concepts** Three-way lamps, of the type in **Figure 23-24,** having a rating of 50 W, 100 W, and 150 W, are common. Draw four partial schematic diagrams that show the lamp filaments and the switch positions for each brightness level, as well as the off position. (You do not need to show the energy source.) Label each diagram.

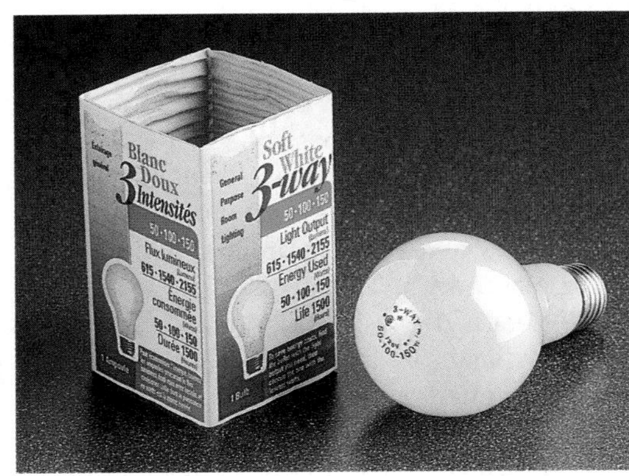

Figure 23-24

98. **Apply Concepts** Design a circuit that will light one dozen 12-V bulbs, all to the correct (same) intensity, from a 48-V battery.
   a. Design A requires that should one bulb burn out, all other bulbs continue to produce light.
   b. Design B requires that should one bulb burn out, those bulbs that continue working must produce the correct intensity.
   c. Design C requires that should one bulb burn out, one other bulb also will go out.
   d. Design D requires that should one bulb burn out, either two others will go out or no others will go out.

99. **Apply Concepts** A battery consists of an ideal source of potential difference in series with a small resistance. The electric energy of the battery is produced by chemical reactions that occur in the battery. However, these reactions also result in a small resistance that, unfortunately, cannot be completely eliminated. A flashlight contains two batteries in series, as shown in **Figure 23-25.** Each has a potential difference of 1.50 V and an internal resistance of 0.200 Ω. The bulb has a resistance of 22.0 Ω.
   a. What is the current through the bulb?
   b. How much power does the bulb dissipate?
   c. How much greater would the power be if the batteries had no internal resistance?

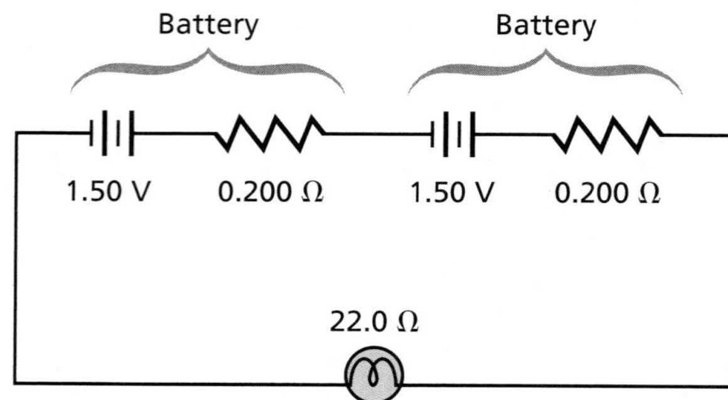

Figure 23-25

100. **Apply Concepts** An ohmmeter is made by connecting a 6.0-V battery in series with an adjustable resistor and an ideal ammeter. The ammeter deflects full-scale with a current of 1.0 mA. The two leads are touched together and the resistance is adjusted so that 1.0 mA flows.
   a. What is the resistance of the adjustable resistor?
   b. The leads are now connected to an unknown resistance. What resistance would produce a current of half-scale, 0.50 mA? Quarter-scale, 0.25 mA? Three-quarters-scale, 0.75 mA?
   c. Is the ohmmeter scale linear? Explain.

Writing in Physics

101. Research Gustav Kirchhoff and his laws. Write a one-page summary of how they apply to the three types of circuits presented in this chapter.

Cumulative Review

102. **Airplane** An airplane flying through still air produces sound waves. The wave fronts in front of the plane are spaced 0.50 m apart and those behind the plane are spaced 1.50 m apart. The speed of sound is 340 m/s. (Chapter 15)
   a. What would be the wavelength of the sound waves if the airplane were not moving?
   b. What is the frequency of the sound waves produced by the airplane?
   c. What is the speed of the airplane?
   d. What is the frequency detected by an observer located directly in front of the airplane?
   e. What is the frequency detected by an observer located directly behind the airplane?

103. An object is located 12.6 cm from a convex mirror with a focal length of −18.0 cm. What is the location of the object's image? (Chapter 17)

104. The speed of light in a special piece of glass is 1.75×10^{8} m/s. What is its index of refraction? (Chapter 18)

105. **Monocle** An antireflective coating with an index of refraction of 1.4 is applied to a monocle with an index of refraction of 1.52. If the thickness of the coating is 75 nm, what is/are the wavelength(s) of light for which complete destructive interference will occur? (Chapter 19)

106. Two charges of 2.0×10^{-5} C and 8.0×10^{-6} C experience a force between them of 9.0 N. How far apart are the two charges? (Chapter 20)

107. A field strength, $\boldsymbol{E}$, is measured a distance, d, from a point charge, Q. What would happen to the magnitude of $\boldsymbol{E}$ in the following situations? (Chapter 21)
   a. d is tripled
   b. Q is tripled
   c. both d and Q are tripled
   d. the test charge q' is tripled
   e. all three, d, Q, and q', are tripled

108. The current flow in a 12-V circuit drops from 0.55 A to 0.44 A. Calculate the change in resistance. (Chapter 22)

 For more problems, go to Additional Problems, Appendix B.

Multiple Choice

Use the following circuit diagram to answer questions 1–3.

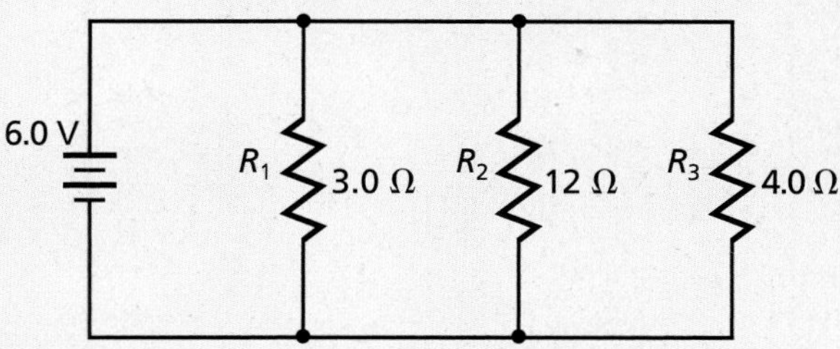

1. What is the equivalent resistance of the circuit?

   (A) $\frac{1}{19}$ Ω (C) 1.5 Ω
   (B) 1.0 Ω (D) 19 Ω

2. What is the current in the circuit?

   (A) 0.32 A (C) 1.2 A
   (B) 0.80 A (D) 4.0 A

3. How much current is in R_3?

   (A) 0.32 A (C) 2.0 A
   (B) 1.5 A (D) 4.0 A

4. What would a voltmeter placed across R_2 read?

   (A) 0.32 V (C) 3.8 V
   (B) 1.5 V (D) 6.0 V

Use the following circuit diagram to answer questions 5 and 6.

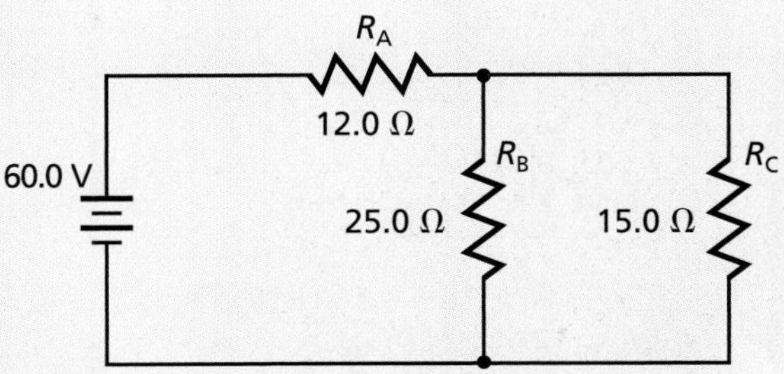

5. What is the equivalent resistance of the circuit?

   (A) 8.42 Ω (C) 21.4 Ω
   (B) 10.7 Ω (D) 52.0 Ω

6. What is the current in the circuit?

   (A) 1.15 A (C) 2.80 A
   (B) 2.35 A (D) 5.61 A

7. Nina connects eight 12-Ω lamps in series. What is the total resistance of the circuit?

   (A) 0.67 Ω (C) 12 Ω
   (B) 1.5 Ω (D) 96 Ω

8. Which statement is true?

   (A) The resistance of a typical ammeter is very high.
   (B) The resistance of a typical voltmeter is very low.
   (C) Ammeters have zero resistance.
   (D) A voltmeter causes a small change in current.

Extended Answer

9. Chris is throwing a tailgate party before a nighttime football game. To light the tailgate party, he connects 15 large outdoor lamps to his 12.0-V car battery. Once connected, the lamps do not glow. An ammeter shows that the current through the lamps is 0.350 A. If the lamps require a 0.500-A current in order to work, how many lamps must Chris remove from the circuit?

10. A series circuit has an 8.0-V battery and four resistors, $R_1 = 4.0$ Ω, $R_2 = 8.0$ Ω, $R_3 = 13.0$ Ω, and $R_4 = 15.0$ Ω. Calculate the current and the power in the circuit.

✓ Test-Taking TIP

Take a Break

If you have the opportunity to take a break or get up from your desk during a test, take it. Getting up and moving around will give you extra energy and help you clear your mind. During the break, think about something other than the test so you'll be able to begin again with a fresh start.

Chapter 24 Magnetic Fields

What You'll Learn

- You will assign forces of attraction or repulsion between magnetic poles.
- You will relate magnetism to electric charge and electricity.
- You will describe how electromagnetism can be harnessed for practical applications.

Why It's Important

Magnetism is the basis for many technologies. Information on the hard drive of a computer is stored as a magnetic pattern.

Atom Smashers An accelerator tube, such as the one pictured, is surrounded by superconducting magnets. There is no magnetic field at the center of the tube where high-energy particles travel. If the particles stray from the center, they receive a magnetic push to keep them there.

Think About This ▶

How do forces applied by magnets cause particles to accelerate? Can any particle be accelerated?

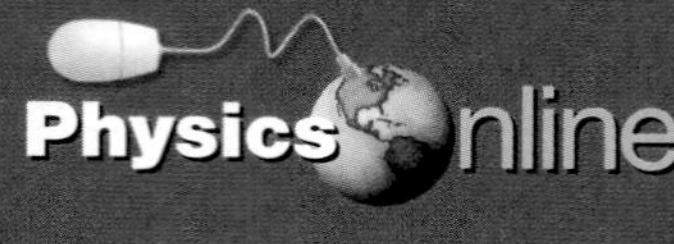

physicspp.com

In which direction do magnetic fields act?

Question

What would be the direction of force on a magnetized object in a magnetic field?

Procedure

1. Place a bar magnet horizontally in front of you so that the north pole faces left.
2. Place a second bar magnet horizontally next to, and 5.0 cm away from the first (you should be able to place the compass between the magnets). The north pole also should be facing the left.
3. Draw your setup on a sheet of paper. Be sure to label the poles.
4. Place a compass by the two magnets. Draw the direction the arrow is pointing.
5. Continue to move the compass to other positions, each time drawing the direction it points until you have drawn 15–20 arrows.
6. Repeat steps 3–5, this time with the two north poles facing each other.

Analysis

What did the red end of the compass needle typically point toward? Away from? Why might some of the arrows not point to either location stated in question 1?

Critical Thinking What you have diagrammed with your arrows is called a magnetic field. Recall what a gravitational field and an electric field are, and define *magnetic field.*

24.1 Magnets: Permanent and Temporary

The existence of magnets and magnetic fields has been known for more than 2000 years. Chinese sailors employed magnets as navigational compasses approximately 900 years ago. Throughout the world, early scientists studied magnetic rocks, called lodestones. Today, magnets play an increasingly important role in our everyday lives. Electric generators, simple electric motors, television sets, cathode-ray displays, tape recorders, and computer hard drives all depend on the magnetic effects of electric currents.

If you have ever used a compass or picked up tacks or paper clips with a magnet, you have observed some effects of magnetism. You even might have made an electromagnet by winding wire around a nail and connecting it to a battery. The properties of magnets become most obvious when you experiment with two of them. To enhance your study of magnetism, you can experiment with magnets, such as those shown in **Figure 24-1** on the next page.

▶ **Objectives**

- **Describe** the properties of magnets and the origin of magnetism in materials.
- **Compare and contrast** various magnetic fields.

▶ **Vocabulary**

polarized
magnetic fields
magnetic flux
first right-hand rule
solenoid
electromagnet
second right-hand rule
domain

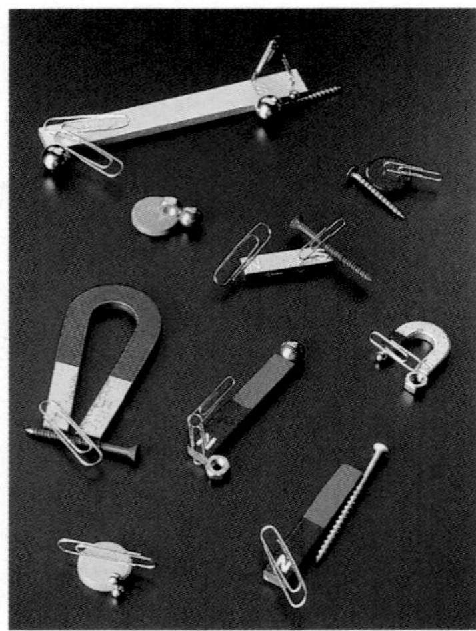

Figure 24-1 Common magnets are available in most hardware stores.

General Properties of Magnets

Suspend a magnet from a thread, as in **Figure 24-2a.** If you use a bar magnet, you might have to tie a yoke around it to keep it horizontal. When the magnet comes to rest, is it lined up in any particular direction? Now rotate the magnet so that it points in a different direction. When you release the magnet, does it come to rest in the same direction? If so, in which direction does it point?

You should have found that the magnet lined up in a north-south direction. Mark the end that points to the north with the letter *N* for reference. From this simple experiment, you can conclude that a magnet is **polarized;** that is, it has two distinct and opposite ends. One of the poles is the north-seeking pole; the other is the south-seeking pole. A compass is nothing more than a small magnet, mounted so that it is free to turn.

Suspend another magnet to determine the north end, and mark it as you did with the first magnet. While one magnet is suspended, observe the interaction of the two magnets by bringing the other magnet near, as in **Figure 24-2b.** What happens as you bring the two ends that were pointing north, the north poles, toward each other? Now try it with the south poles. Lastly, what happens as you bring opposite poles (the north pole of one magnet and the south pole of the other magnet) toward each other?

You should have observed that the two north poles repelled each other, as did the two south poles. However, the north pole of one magnet should have attracted the south pole of the other magnet. Like poles repel; unlike poles attract. Magnets always have two opposite magnetic poles. If you break a magnet in half, you create two smaller magnets, and each will have two poles. Scientists have tried to break magnets into separate north and south poles, called monopoles, but no one has succeeded, not even on the microscopic level.

Knowing that magnets always orient themselves in a north-south direction, it may occur to you that Earth itself is a giant magnet. Because opposite poles attract and the north pole of a compass magnet points north, the south pole of the Earth-magnet must be near Earth's geographic north pole.

a

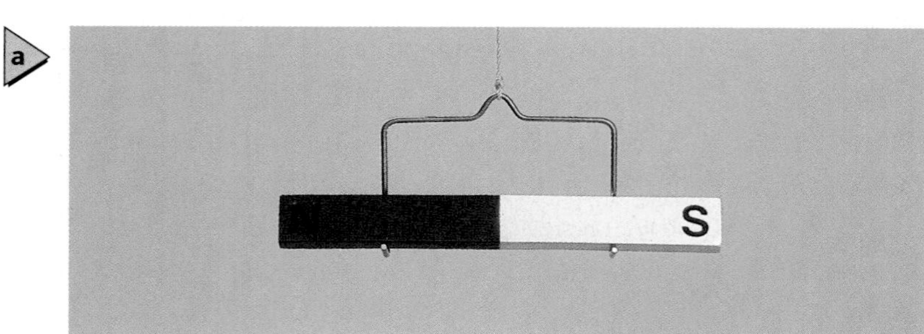

b

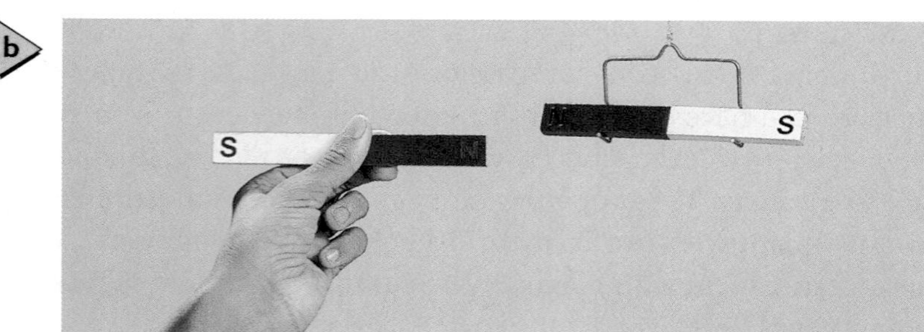

Figure 24-2 If you suspend a magnet by a thread, it will align itself with magnetic properties in Earth **(a).** The magnet's north pole will point north. If you then move the north pole of a second magnet toward the north pole of the suspended magnet, the suspended magnet will move away **(b).**

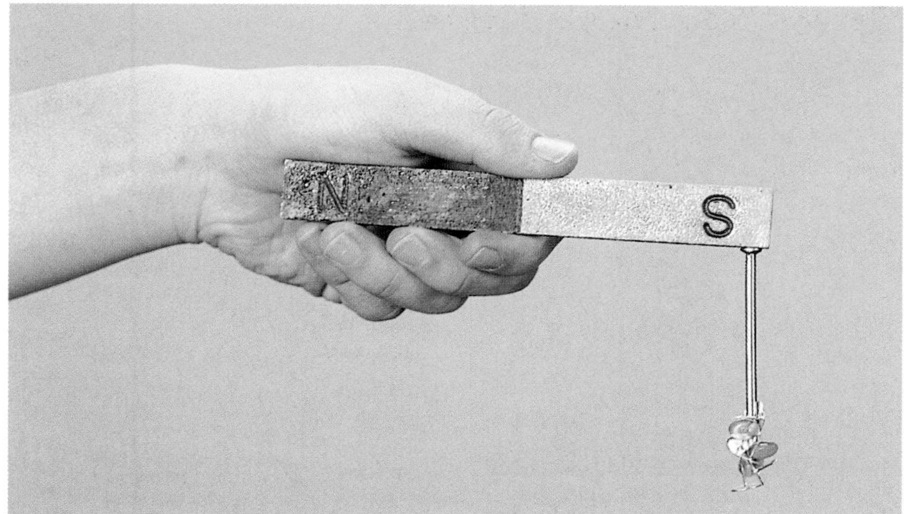

Figure 24-3 A common nail is attracted to a magnet. In the process, the nail itself becomes magnetized. Here you can see when a magnet is touching the nail, the nail is able to attract other metal objects. However, if you separate the magnet from the nail, some of the objects will drop off because the nail will have lost some of its magnetism.

How do magnets affect other materials? As you probably discovered as a child, magnets attract things besides other magnets, such as nails, tacks, paper clips, and many other metal objects. Unlike the interaction between two magnets, however, either end of a magnet will attract either end of a piece of metal. How can you explain this behavior? First, you can touch a magnet to a nail and then touch the nail to smaller metal pieces. The nail itself becomes a magnet, as shown in **Figure 24-3.** The magnet causes the nail to become polarized. The direction of polarization of the nail depends on the polarization of the magnet. If you pull away the magnet, the nail loses some of its magnetization and will no longer exhibit as much attraction for other metal objects.

If you repeat the experiment shown in Figure 24-3 with a piece of soft iron (iron with a low carbon content) in place of a nail, you will notice that the iron loses all of its attraction for the other metal objects when the magnet is pulled away. This is because soft iron is a temporary magnet. A nail has other material in it to make it harder and allows it to retain some of its magnetism when a permanent magnet is pulled away.

Permanent magnets The magnetism of permanent magnets is produced in the same way in which you created the magnetism of the nail. Because of the microscopic structure of the magnet material, the induced magnetism becomes permanent. Many permanent magnets are made of an iron alloy called ALNICO V, that contains a mix of **al**uminum, **ni**ckel, and **co**balt. A variety of rare earth elements, such as neodymium and gadolinium, produce permanent magnets that are extremely strong for their size.

Magnetic Fields Around Permanent Magnets

When you experimented with two magnets, you noticed that the forces between magnets, both attraction and repulsion, occur not only when the magnets touch each other, but also when they are held apart. In the same way that long-range electric and gravitational forces can be described by electric and gravitational fields, magnetic forces can be described by the existence of fields around magnets. These **magnetic fields** are vector quantities that exist in a region in space where a magnetic force occurs.

Figure 24-4 The magnetic field of a bar magnet shows up clearly in three dimensions when the magnet is suspended in glycerol with iron filings **(a).** It is, however, easier to set up a magnet on a sheet of paper covered with iron filings to see the pattern in two dimensions **(b).**

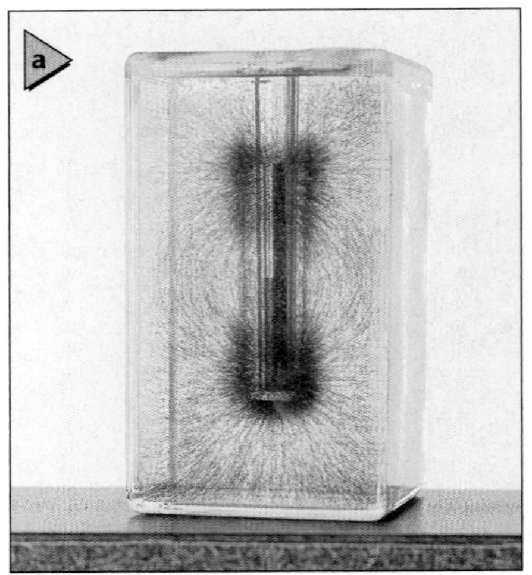

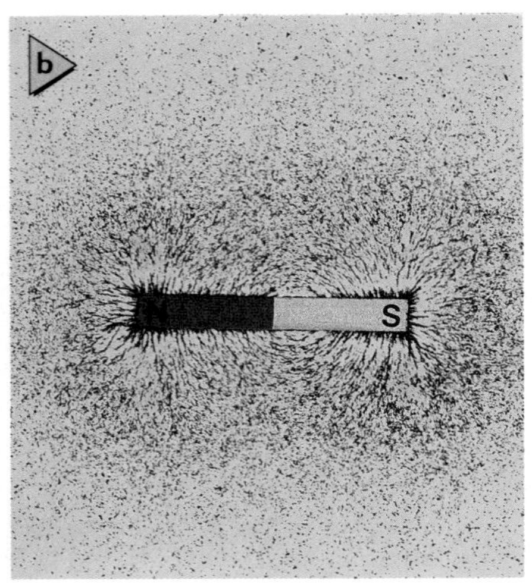

The presence of a magnetic field around a magnet can be shown using iron filings. Each long, thin, iron filing becomes a small magnet by induction. Just like a tiny compass needle, the iron filing rotates until it is parallel to the magnetic field. **Figure 24-4a** shows filings in a glycerol solution surrounding a bar magnet. The three-dimensional shape of the field is visible. In **Figure 24-4b,** the filings make up a two-dimensional plot of the field, which can help you visualize magnetic field lines. Filings also can show how the field can be distorted by an object.

Magnetic field lines Note that magnetic field lines, like electric field lines, are imaginary. They are used to help us visualize a field, and they also provide a measure of the strength of the magnetic field. The number of magnetic field lines passing through a surface is called the **magnetic flux.** The flux per unit area is proportional to the strength of the magnetic field. As you can see in Figure 24-4, the magnetic flux is most concentrated at the poles; thus, this is where the magnetic field strength is the greatest.

The direction of a magnetic field line is defined as the direction in which the north pole of a compass points when it is placed in the magnetic field. Outside the magnet, the field lines emerge from the magnet at its north pole and enter the magnet at its south pole, as illustrated in **Figure 24-5.** What happens inside the magnet? There are no isolated poles on which field lines can start or stop, so magnetic field lines always travel inside the magnet from the south pole to the north pole to form closed loops.

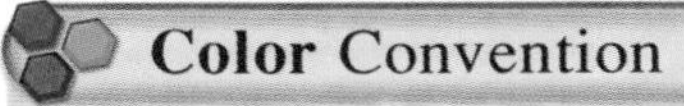

- Positive charges are **red.**
- Negative charges are **blue.**
- Electric field lines are **indigo.**
- Magnetic field lines are **green.**

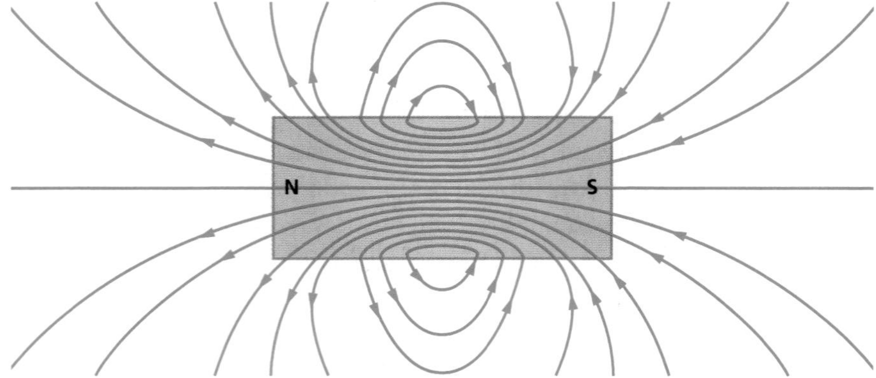

Figure 24-5 Magnetic field lines can be visualized as closed loops leaving the north pole of a magnet and entering the south pole of the same magnet.

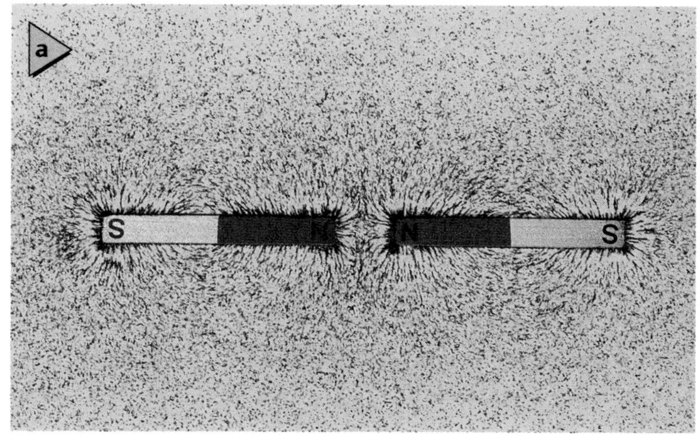

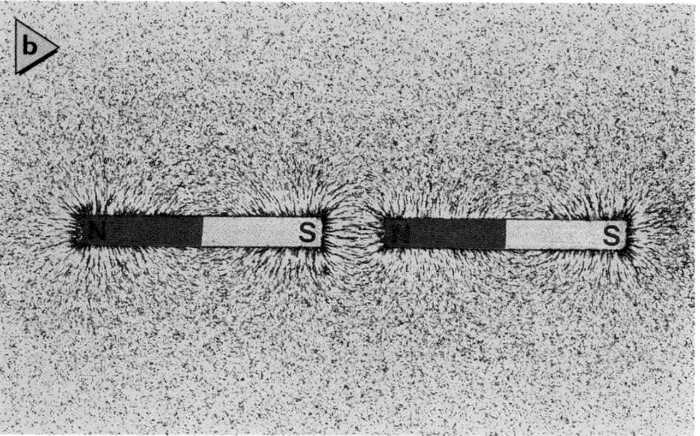

■ **Figure 24-6** The magnetic field lines indicated by iron filings on paper clearly show that like poles repel **(a)** and unlike poles attract **(b).** The iron filings do not form continuous lines between like poles. Between a north and a south pole, however, the iron filings show that field lines run directly between the two magnets.

What kinds of magnetic fields are produced by pairs of bar magnets? You can visualize these fields by placing two magnets on a sheet of paper, and then sprinkling the paper with iron filings. **Figure 24-6a** shows the field lines between two like poles. In contrast, two unlike poles (north and south) placed close together produce the pattern shown in **Figure 24-6b.** The filings show that the field lines between two unlike poles run directly from one magnet to the other.

Forces on objects in magnetic fields Magnetic fields exert forces on other magnets. The field produced by the north pole of one magnet pushes the north pole of a second magnet away in the direction of the field line. The force exerted by the same field on the south pole of the second magnet is attractive in a direction opposite the field lines. The second magnet attempts to line up with the field, just like a compass needle.

When a sample made of iron, cobalt, or nickel is placed in the magnetic field of a permanent magnet, the field lines become concentrated within the sample. Lines leaving the north pole of the magnet enter one end of the sample, pass through it, and leave the other end. Thus, the end of the sample closest to the magnet's north pole becomes the sample's south pole, and the sample is attracted to the magnet.

Concepts In Motion

Interactive Figure To see an animation on magnestism, visit physicspp.com.

PRACTICE Problems

• Additional Problems, Appendix B
• Solutions to Selected Problems, Appendix C

1. If you hold a bar magnet in each hand and bring your hands close together, will the force be attractive or repulsive if the magnets are held in the following ways?

a. the two north poles are brought close together

b. a north pole and a south pole are brought together

2. **Figure 24-7** shows five disk magnets floating above each other. The north pole of the top-most disk faces up. Which poles are on the top side of each of the other magnets?

3. A magnet attracts a nail, which, in turn, attracts many small tacks, as shown in Figure 24-3 on page 645. If the north pole of the permanent magnet is the left end, as shown, which end of the nail is the south pole?

4. Why do magnetic compasses sometimes give false readings?

■ **Figure 24-7**

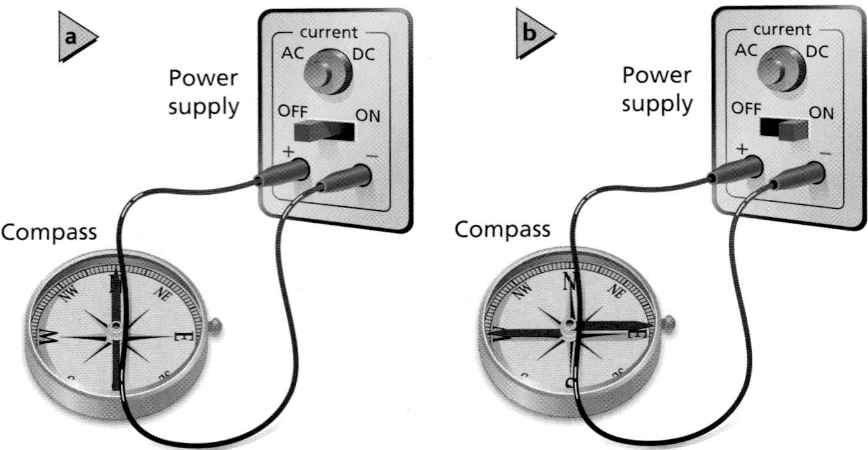

■ **Figure 24-8** Using an apparatus similar to the one shown **(a),** Oersted was able to demonstrate a connection between magnetism and electricity by applying current to the wire **(b).**

■ **Figure 24-9** The magnetic field produced by the current in a wire through a cardboard disk shows up as concentric circles of iron filings around the wire.

Electromagnetism

In 1820, Danish physicist Hans Christian Oersted was experimenting with electric currents in wires. Oersted laid a wire across the top of a small compass and connected the ends of the wire to complete an electrical circuit, as shown in **Figure 24-8a.** He had expected the needle to point toward the wire or in the same direction as the current in the wire. Instead, he was amazed to see that the needle rotated until it pointed perpendicular to the wire, as shown in **Figure 24-8b.** The forces on the compass magnet's poles were perpendicular to the direction of current in the wire. Oersted also found that when there was no current in the wire, no magnetic forces existed.

If a compass needle turns when placed near a wire carrying an electric current, it must be the result of a magnetic field created by the current. You easily can show the magnetic field around a current-carrying wire by placing a wire vertically through a horizontal piece of cardboard on which iron filings are sprinkled. When there is current through the wire, the filings will form a pattern of concentric circles, around the wire, as shown in **Figure 24-9.**

The circular lines indicate that magnetic field lines around current-carrying wires form closed loops in the same way that field lines about permanent magnets form closed loops. The strength of the magnetic field around a long, straight wire is proportional to the current in the wire. The strength of the field also varies inversely with the distance from the wire. A compass shows the direction of the field lines. If you reverse the direction of the current, the compass needle also reverses its direction, as shown in **Figure 24-10a.**

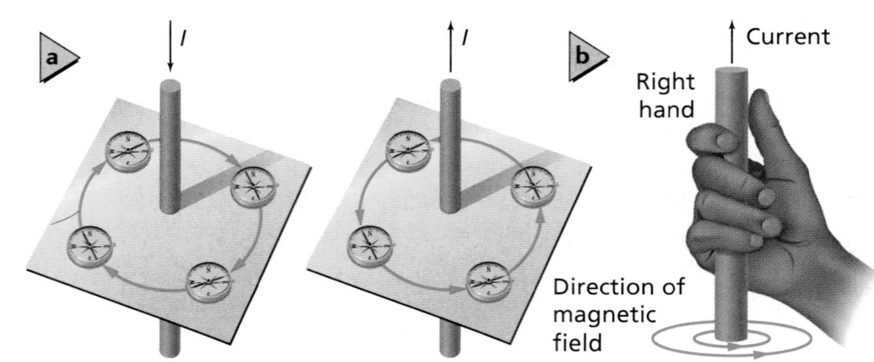

■ **Figure 24-10** The magnetic field produced by current in a straight-wire conductor reverses when the current in the wire is reversed **(a).** The first right-hand rule for a straight, current-carrying wire shows the direction of the magnetic field **(b).**

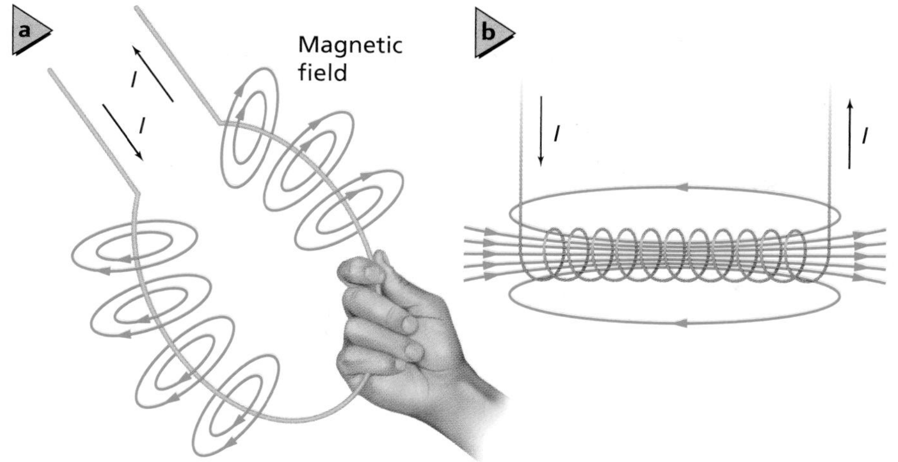

Figure 24-11 The magnetic field around a circular loop of current-carrying wire can be modeled with the aid of the first right-hand rule **(a).** A current in a solenoid creates a magnetic field with the field from each coil adding to all the others **(b).**

Concepts In Motion
Interactive Figure To see an animation on the first and second right-hand rules, visit physicspp.com.

The **first right-hand rule** is a method you can use to determine the direction of a magnetic field relative to the direction of conventional current. Imagine holding a length of insulated wire with your right hand. Keep your thumb pointed in the direction of the conventional (positive) current. The fingers of your hand circle the wire and point in the direction of the magnetic field, as illustrated in **Figure 24-10b.**

Magnetic field near a coil An electric current in a single circular loop of wire forms a magnetic field all around the loop. Applying the right-hand rule to any part of the wire loop, it can be shown that the direction of the field inside the loop is always the same. In **Figure 24-11a,** the field is always up, or out of the page. Outside the loop, it is always down, or into the page. When a wire is looped several times to form a coil and a current is allowed to flow through the coil, the field around all the loops is in the same direction, as shown in **Figure 24-11b.** A long coil of wire consisting of many loops is called a **solenoid.** The field from each loop in a solenoid adds to the fields of the other loops and creates a greater total field strength.

When there is an electric current in a coil of wire, the coil has a field similar to a permanent magnet. When this current-carrying coil is brought close to a suspended bar magnet, one end of the coil repels the north pole of the magnet. Thus, the current-carrying coil has a north and a south pole and is itself a magnet. This type of magnet, which is created when current flows through a wire coil, is called an **electromagnet.** The strength of the field is proportional to the current in the coil. The magnetic field produced by each loop is the same. Because these fields are in the same direction, increasing the number of loops increases the strength of the magnetic field.

The strength of an electromagnet also can be increased by placing an iron rod or core inside the coil. The core supports the magnetic field better than air does. It increases the magnetic field because the field of the solenoid creates a temporary magnetic field in the core, just as a nearby permanent magnet does when brought near a metal object.

The **second right-hand rule** is a method you can use to determine the direction of the field produced by an electromagnet relative to the flow of conventional current. Imagine holding an insulated coil with your right hand. If you then curl your fingers around the loops in the direction of the conventional (positive) current, as in **Figure 24-12,** your thumb will point toward the north pole of the electromagnet.

APPLYING PHYSICS

Electromagnets Cranes for moving iron and steel in industrial settings frequently use electromagnets. One such magnet, which operates at 230 V and draws 156 A, can lift over 11,300 kg!

Figure 24-12 The second right-hand rule can be used to determine the polarity of an electromagnet.

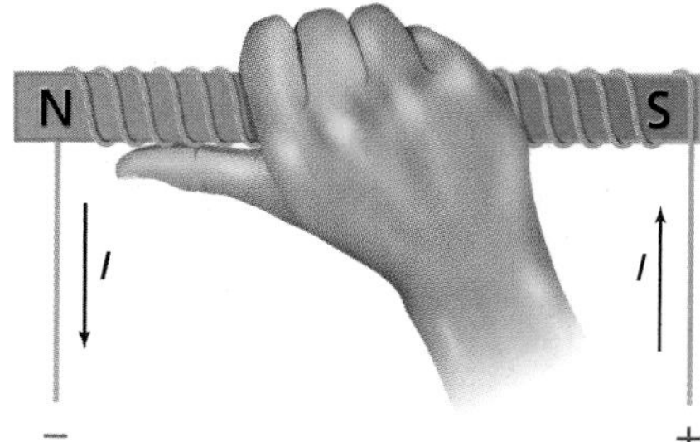

PRACTICE Problems

- Additional Problems, Appendix B
- Solutions to Selected Problems, Appendix C

5. A long, straight, current-carrying wire runs from north to south.

a. A compass needle placed above the wire points with its north pole toward the east. In what direction is the current flowing?

b. If a compass is put underneath the wire, in which direction will the compass needle point?

6. How does the strength of a magnetic field, 1 cm from a current-carrying wire, compare with each of the following?

a. the strength of the field that is 2 cm from the wire

b. the strength of the field that is 3 cm from the wire

7. A student makes a magnet by winding wire around a nail and connecting it to a battery, as shown in **Figure 24-13.** Which end of the nail, the pointed end or the head, will be the north pole?

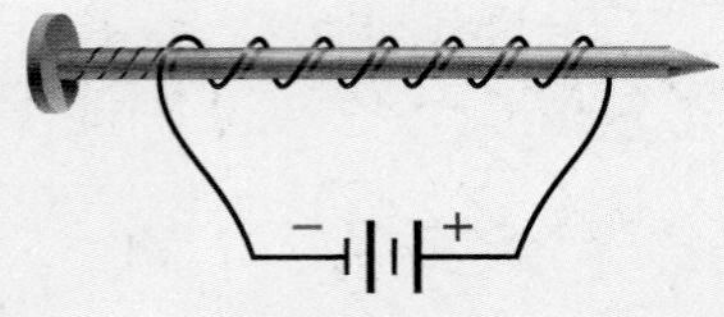

Figure 24-13

8. You have a spool of wire, a glass rod, an iron rod, and an aluminum rod. Which rod should you use to make an electromagnet to pick up steel objects? Explain.

9. The electromagnet in problem 8 works well, but you decide that you would like to make its strength adjustable by using a potentiometer as a variable resistor. Is this possible? Explain.

A Microscopic Picture of Magnetic Materials

Recall that when you put a piece of iron, nickel, or cobalt next to a magnet, the element also becomes magnetic, and it develops north and south poles. The magnetism, however, is only temporary. The creation of this temporary polarity depends on the direction of the external field. When you take away the external field, the element loses its magnetism. The three elements—iron, nickel, and cobalt—behave like electromagnets in many ways. They have a property called ferromagnetism.

In the early nineteenth century, French scientist André-Marie Ampère knew that the magnetic effects of an electromagnet are the result of electric current through its loops. He proposed a theory of magnetism in iron to explain this behavior. Ampère reasoned that the effects of a bar magnet must result from tiny loops of current within the bar.

Magnetic domains Although the details of Ampère's reasoning were wrong, his basic idea was correct. Each electron in an atom acts like a tiny electromagnet. When the magnetic fields of the electrons in a group of neighboring atoms are all aligned in the same direction, the group is called a **domain.** Although they may contain 10^{20} individual atoms, domains are still very small—usually from 10 to 1000 microns. Thus, even a small sample of iron contains a huge number of domains.

When a piece of iron is not in a magnetic field, the domains point in random directions, and their magnetic fields cancel one another out. If, however, a piece of iron is placed in a magnetic field, the domains tend to align with the external field, as shown in **Figure 24-14.** In the case of a temporary magnet, after the external field is removed, the domains return to their random arrangement. In a permanent magnet, the iron has been alloyed with other substances to keep the domains aligned after the external magnetic field is removed.

MINI LAB

3-D Magnetic Fields

Tie a string to the middle of a nail so that the nail will hang horizontally. Put a small piece of tape around the string where it wraps around the nail so that the string will not slip. Insert the nail into a coil and apply a voltage to the coil. Turn off the power and remove the nail from the coil. Now hold the string to suspend the nail.

1. Predict how the nail will behave in the presence of a permanent magnet.

2. Test your prediction.

Analyze and Conclude

3. Explain what evidence you have that the nail became magnetized.

4. Make a 3-D drawing that shows the magnetic field around the magnet.

Recording media Electromagnets make up the recording heads of audiocassette and videotape recorders. Recorders create electrical signals that represent the sounds or pictures being recorded. The electric signals produce currents in the recording head that create magnetic fields. When magnetic recording tape, which has many tiny bits of magnetic material bonded to thin plastic, passes over the recording head, the domains of the bits are aligned by the magnetic fields of the head. The directions of the domains' alignments depend on the direction of the current in the head and become a magnetic record of the sounds or pictures being recorded. The magnetic material on the tape allows the domains to keep their alignments until a strong enough magnetic field is applied to change them again. On a playback of the tape, the signal, produced by currents generated as the head passes over the magnetic particles, goes to an amplifier and a pair of loudspeakers or earphones. When a previously recorded tape is used to record new sounds, an erase head produces a rapidly alternating magnetic field that randomizes the directions of the domains on the tape.

A magnetic history of the Earth Rocks containing iron have recorded the history of the varying directions of Earth's magnetic field. Rocks on the seafloor were produced when molten rock poured out of cracks in the bottom of the oceans. As they cooled, the rocks were magnetized in the direction of Earth's field at the time. As a result of seafloor spreading, the rocks farther from the cracks are older than those near the cracks. Scientists who first examined seafloor rocks were surprised to find that the direction of the magnetization in different rocks varied. They concluded from their data that the north and south magnetic poles of Earth have exchanged places many times in Earth's history. The origin of Earth's magnetic field is not well understood. How this field might reverse direction is even more of a mystery.

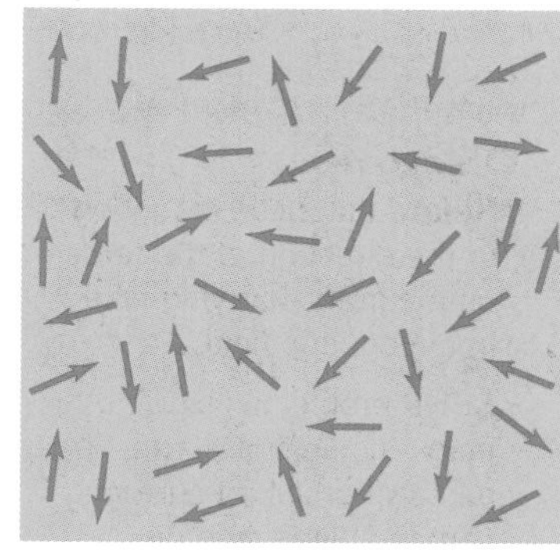

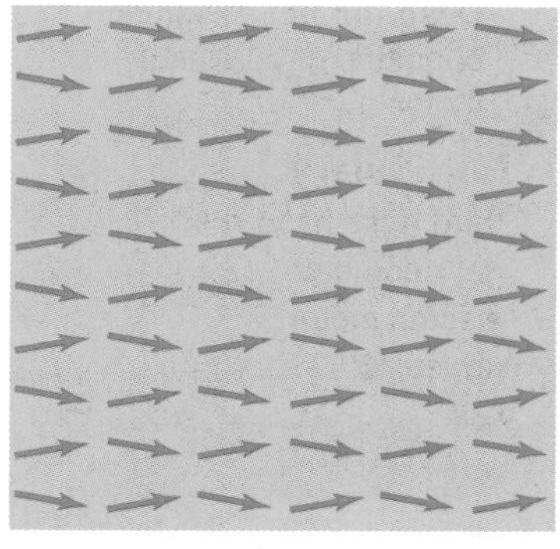

Figure 24-14 A piece of iron **(a)** becomes a magnet only when its domains align **(b)**.

24.1 Section Review

10. Magnetic Fields Is a magnetic field real, or is it just a means of scientific modeling?

11. Magnetic Forces Identify some magnetic forces around you. How could you demonstrate the effects of those forces?

12. Magnetic Fields A current-carrying wire is passed through a card on which iron filings are sprinkled. The filings show the magnetic field around the wire. A second wire is close to and parallel to the first wire. There is an identical current in the second wire. If the two currents are in the same direction, how will the first magnetic field be affected? How will it be affected if the two currents are in opposite directions?

13. Direction of a Magnetic Field Describe the right-hand rule used to determine the direction of a magnetic field around a straight, current-carrying wire.

14. Electromagnets A glass sheet is placed over an active electromagnet, and iron filings sprinkled on the sheet create a pattern on it. If this experiment is repeated with the polarity of the power supply reversed, what observable differences will result? Explain.

15. Critical Thinking Imagine a toy containing two parallel, horizontal metal rods, one above the other. The top rod is free to move up and down.

a. The top rod floats above the lower one. If the top rod's direction is reversed, however, it falls down onto the lower rod. Explain why the rods could behave in this way.

b. Assume that the top rod was lost and replaced with another one. In this case, the top rod falls on top of the bottom rod no matter what its orientation is. What type of replacement rod must have been used?

Physics Online physicspp.com/self_check_quiz

24.2 Forces Caused by Magnetic Fields

▶ **Objectives**

- **Relate** magnetic induction to the direction of the force on a current-carrying wire in a magnetic field.
- **Solve** problems involving magnetic field strength and the forces on current-carrying wires, and on moving, charged particles in magnetic fields.
- **Describe** the design and operation of an electric motor.

▶ **Vocabulary**

third right-hand rule
galvanometer
electric motor
armature

As you learned in the previous section, while Ampère was studying the behaviors of magnets, he noted that an electric current produces a magnetic field similar to that of a permanent magnet. Because a magnetic field exerts forces on permanent magnets, Ampère hypothesized that there is also a force on a current-carrying wire when it is placed in a magnetic field.

Forces on Currents in Magnetic Fields

The force on a wire in a magnetic field can be demonstrated using the arrangement shown in **Figure 24-15.** A battery produces current in a wire directly between two bar magnets. Recall that the direction of the magnetic field between two magnets is from the north pole of one magnet to the south pole of the other magnet. When there is a current in the wire, a force is exerted on the wire. Depending on the direction of the current, the force on the wire either pushes it down, as shown in **Figure 24-15a,** or pulls it up, as shown in **Figure 24-15b.** Michael Faraday discovered that the force on the wire is at right angles to both the direction of the magnetic field and the direction of the current.

Determining the force's direction Faraday's description of the force on a current-carrying wire does not completely describe the direction because the force can be upward or downward. The direction of the force on a current-carrying wire in a magnetic field can be found by using **the third right-hand rule.** This technique is illustrated in **Figure 24-16.** The magnetic field is represented by the symbol ***B***, and its direction is represented by a series of arrows. To use the third right-hand rule, point the fingers of your right hand in the direction of the magnetic field, and point your thumb in the direction of the conventional (positive) current in the wire. The palm of your hand will be facing in the direction of the force acting on the wire. When drawing a directional arrow that is into or out of the page, direction is indicated with crosses and dots, respectively. Think of the crosses as the tail feathers of the arrow, and the dots as the arrowhead.

Soon after Oersted announced his discovery that the direction of the magnetic field in a wire is perpendicular to the flow of electric current in the wire, Ampère was able to demonstrate the forces that current-carrying wires exert on each other. **Figure 24-17a** shows the direction of the magnetic field around each of the current-carrying wires, which is determined by the first right-hand rule. By applying the third right-hand rule to either wire, you can show why the wires attract each other. **Figure 24-17b** demonstrates the opposite situation. When currents are in opposite directions, the wires have a repulsive force between them.

Figure 24-15 Current-carrying wires experience forces when they are placed in magnetic fields. In this case the force can be down **(a),** or up **(b),** depending on the direction of the current.

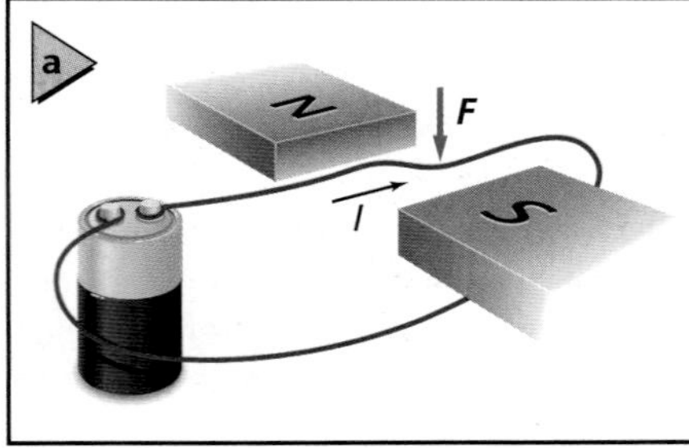

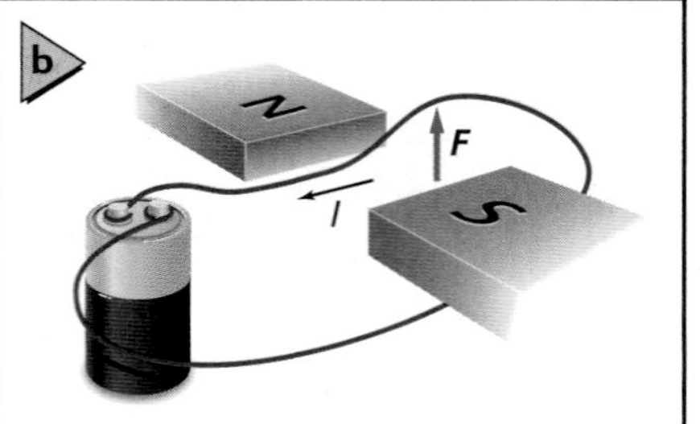

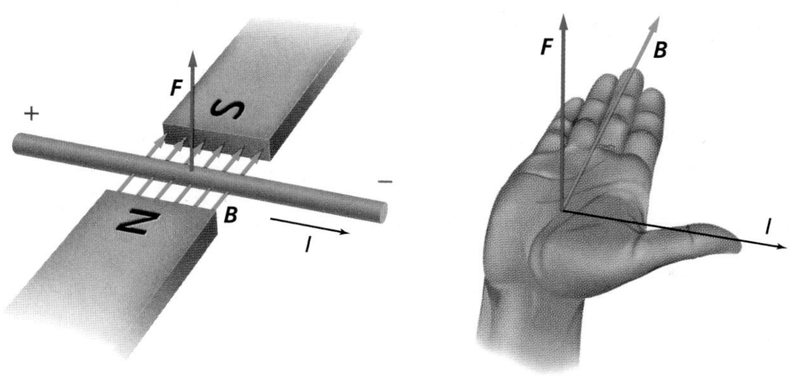

Figure 24-16 The third right-hand rule can be used to determine the direction of force when the current and magnetic field are known.

Concepts In Motion
Interactive Figure To see an animation on the third right-hand rule, visit physicspp.com.

Force on a wire resulting from a magnetic field It is possible to determine the force of magnetism exerted on a current-carrying wire passing through a magnetic field at right angles to the wire. Experiments show that the magnitude of the force, F, on the wire, is proportional to the strength of the field, B, the current, I, in the wire, and the length, L, of the wire in the magnetic field. The relationship of these four factors is as follows:

Force on a Current-Carrying Wire in a Magnetic Field $F = ILB$

The force on a current-carrying wire in a magnetic field is equal to the product of the current, the length of the wire, and the magnetic field strength.

The strength of a magnetic field, B, is measured in teslas, T. 1 T is equivalent to 1 N/A·m.

Note that if the wire is not perpendicular to the magnetic field, a factor of $\sin \theta$ is introduced in the above equation, resulting in $F = ILB \sin \theta$. As the wire becomes parallel to the magnetic field, the angle θ becomes zero, and the force is reduced to zero. When $\theta = 90°$, the equation is again $F = ILB$.

Loudspeakers

One use of the force on a current-carrying wire in a magnetic field is in a loudspeaker. A loudspeaker changes electric energy to sound energy using a coil of fine wire mounted on a paper cone and placed in a magnetic field. The amplifier driving the loudspeaker sends a current through the coil. The current changes direction between 20 and 20,000 times each second, depending on the pitch of the tone it represents. A force exerted on the coil, because it is in a magnetic field, pushes the coil either into or out of the field, depending on the direction of the current. The motion of the coil causes the cone to vibrate, thereby creating sound waves in the air.

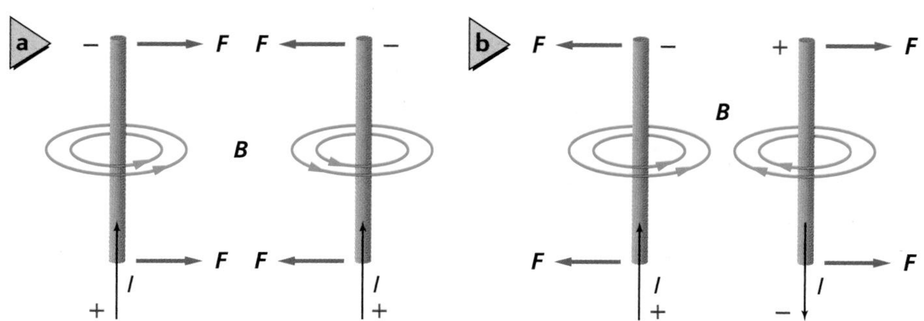

Figure 24-17 Two current-carrying conductors are attracted when the currents are in the same direction **(a),** and are repelled when the currents are in opposite directions **(b).**

▶ EXAMPLE Problem 1

Calculate the Strength of a Magnetic Field A straight wire carrying a 5.0-A current is in a uniform magnetic field oriented at right angles to the wire. When 0.10 m of the wire is in the field, the force on the wire is 0.20 N. What is the strength of the magnetic field, B?

1 Analyze and Sketch the Problem

- Sketch the wire and show the direction of the current with an arrow, the magnetic field lines labeled B, and the force on the wire, F.
- Determine the direction of the force using the third right-hand rule. The field, wire, and force are all at right angles.

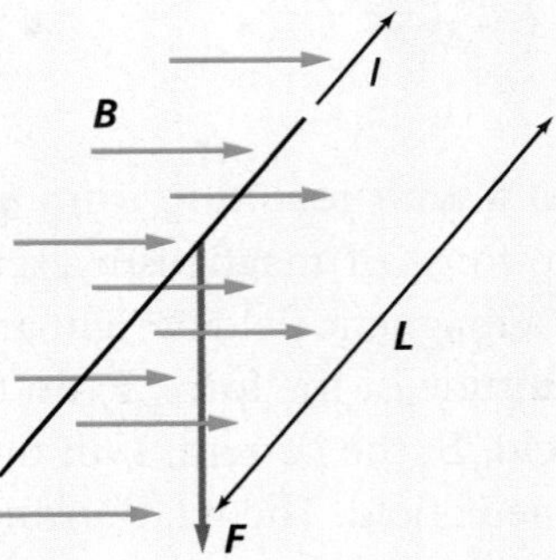

Known:	Unknown:
$I = 5.0$ A	$B = ?$
$L = 0.10$ m	
$F = 0.20$ N	

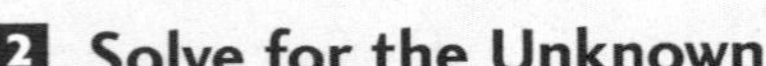

2 Solve for the Unknown

B is uniform and because B and I are perpendicular to each other, $F = ILB$.

$$F = ILB$$

Solve for B.

$$B = \frac{F}{IL}$$

$$= \frac{0.20\text{ N}}{(5.0\text{ A})(0.10\text{ m})} \quad \text{Substitute } F = 0.20\text{ N}, I = 5.0\text{ A}, L = 0.10\text{ m}$$

$$= 0.40\text{ N/A·m} = 0.40\text{ T}$$

B is 0.40 T from left to right and perpendicular to I and F.

Math Handbook
Operations with Significant Digits
pages 835–836

3 Evaluate the Answer

- **Are the units correct?** The answer is in teslas, the correct unit for a magnetic field.
- **Is the magnitude realistic?** The current and the length make the magnetic field fairly large, which is realistic.

▶ PRACTICE Problems

- Additional Problems, Appendix B
- Solutions to Selected Problems, Appendix C

16. What is the name of the rule used to predict the direction of force on a current-carrying wire at right angles to a magnetic field? Identify what must be known to use this rule.

17. A wire that is 0.50 m long and carrying a current of 8.0 A is at right angles to a 0.40-T magnetic field. How strong is the force that acts on the wire?

18. A wire that is 75 cm long, carrying a current of 6.0 A, is at right angles to a uniform magnetic field. The magnitude of the force acting on the wire is 0.60 N. What is the strength of the magnetic field?

19. A 40.0-cm-long copper wire carries a current of 6.0 A and weighs 0.35 N. A certain magnetic field is strong enough to balance the force of gravity on the wire. What is the strength of the magnetic field?

20. How much current will be required to produce a force of 0.38 N on a 10.0 cm length of wire at right angles to a 0.49-T field?

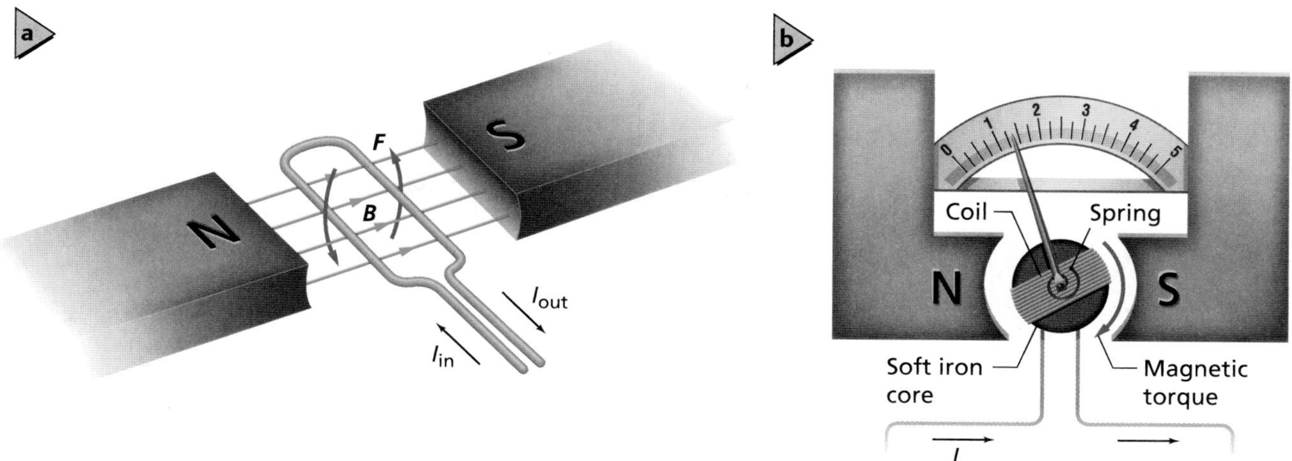

■ **Figure 24-18** If a wire loop is placed in a magnetic field when there is a current, the loop will rotate **(a).** The coil in a galvanometer rotates in proportion to the magnitude of the current **(b).**

Galvanometers

The forces exerted on a loop of wire in a magnetic field can be used to measure current. If a small loop of current-carrying wire is placed in the strong magnetic field of a permanent magnet, as in **Figure 24-18a,** it is possible to measure very small currents. The current passing through the loop goes in one end of the loop and out the other end. Applying the third right-hand rule to each side of the loop, note that one side of the loop is forced down, while the other side of the loop is forced up. The resulting torque rotates the loop, and the magnitude of the torque acting on the loop is proportional to the magnitude of the current. This principle is used in a galvanometer. A **galvanometer** is a device used to measure very small currents, and therefore, it can be used as a voltmeter or an ammeter.

A small spring in the galvanometer exerts a torque that opposes the torque that results from the flow of current through the wire loop; thus, the amount of rotation is proportional to the current. The meter is calibrated by finding out how much the coil turns when a known current is sent through it, as shown in **Figure 24-18b.** The galvanometer can then be used to measure unknown currents.

Many galvanometers produce full-scale deflections with as little as 50 μA (50×10^{-6} A) of current. The resistance of the coil of wire in a sensitive galvanometer is about 1000 Ω. To measure larger currents, a galvanometer can be converted into an ammeter by placing a resistor with resistance smaller than the galvanometer in parallel with the meter, as shown in **Figure 24-19a.** Most of the current, I_s, passes through the resistor, called the shunt, because the current is inversely proportional to resistance; whereas only a few microamps, I_m, flow through the galvanometer. The resistance of the shunt is chosen according to the desired deflection scale.

A galvanometer also can be connected as a voltmeter. To make a voltmeter, a resistor, called the multiplier, is placed in series with the meter, as shown in **Figure 24-19b.** The galvanometer measures the current through the multiplier. The current is represented by $I = V/R$, where V is the voltage across the voltmeter and R is the effective resistance of the galvanometer and the multiplier resistor. Now suppose you want the needle of a voltmeter to move across the entire scale when 10 V is placed across it. The resistor is chosen so that at 10 V, the meter is deflected full-scale by the current through the meter and the resistor.

■ **Figure 24-19** A galvanometer is connected for use as an ammeter **(a),** and a galvanometer is connected for use as a voltmeter **(b).**

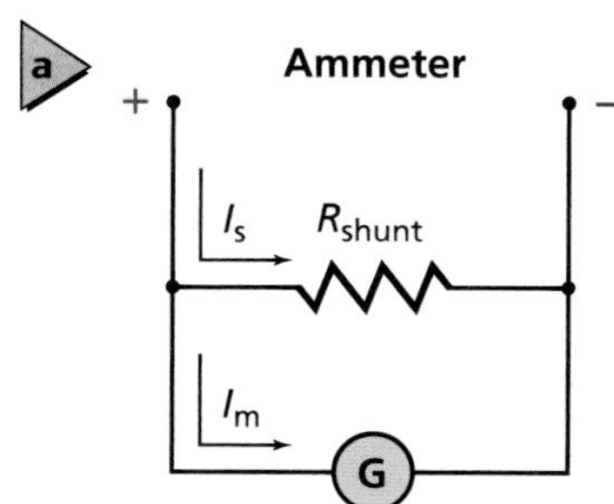

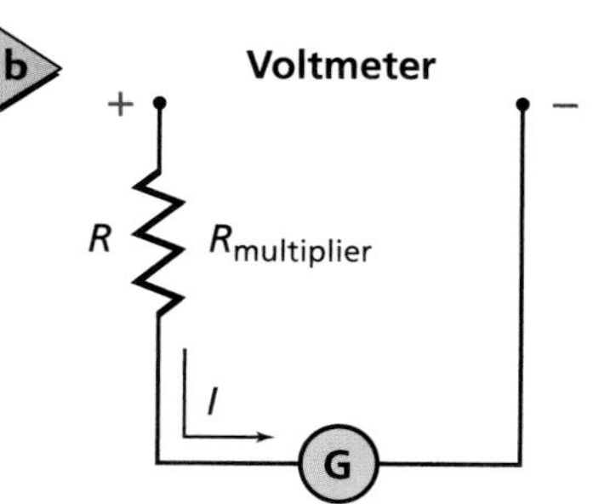

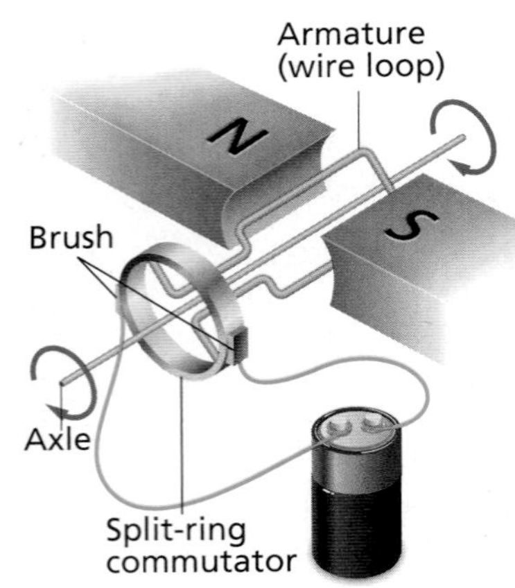

Figure 24-20 In an electric motor, split-ring commutators allow the current in the wire loops to change direction and thus enable the loops in the motor to rotate 360°.

Concepts In Motion
Interactive Figure To see an animation on electric motors, visit physicspp.com.

Electric motors You have seen how the simple loop of wire used in a galvanometer cannot rotate more than 180°. The forces push the right side of the loop up and the left side of the loop down until the loop reaches the vertical position. The loop will not continue to turn because the forces are still up and down, now parallel to the loop, and can cause no further rotation.

How can you allow the loop to continue to rotate? The current through the loop must reverse direction just as the loop reaches its vertical position. This reversal allows the loop to continue rotating, as illustrated in **Figure 24-20.** To reverse current direction, an electric connection is made between contacts, called brushes, and a ring that is split into two halves, called a split-ring commutator. Brushes, which are usually pieces of graphite, make contact with the commutator and allow current to flow into the loop. As the loop rotates, so does the commutator. The split ring is arranged so that each half of the commutator changes brushes just as the loop reaches the vertical position. Changing brushes reverse the current in the loop. As a result, the direction of the force on each side of the loop is reversed, and the loop continues to rotate. This process repeats at each half-turn, causing the loop to spin in the magnetic field. The result is an **electric motor,** which is an apparatus that converts electric energy into rotational kinetic energy.

Although only one loop is indicated in Figure 24-20, in an electric motor, the wire coil, called the **armature,** is made of many loops mounted on a shaft or axle. The total force acting on the armature is proportional to $nILB$, where n is the total number of turns on the armature, B is the strength of the magnetic field, I is the current, and L is the length of wire in each turn that moves through the magnetic field. The magnetic field is produced either by permanent magnets or by an electromagnet, called a field coil. The torque on the armature, and, as a result, the speed of the motor, is controlled by varying the current through the motor.

CHALLENGE PROBLEM

The figure shows two identical motors with a common shaft. For simplicity, the commutators are not shown. Each armature coil consists of 48 turns of wire with rectangular dimensions of 17 cm wide by 35 cm deep. The armature resistance is 12 Ω. The red wire travels to the left (along half the width) and then back to the rear of the motor (along the depth). The magnetic field is 0.21 T. The diameter of the pulley is 7.2 cm. A rope fixed to the pulley and the floor prevents the motor shaft from turning.

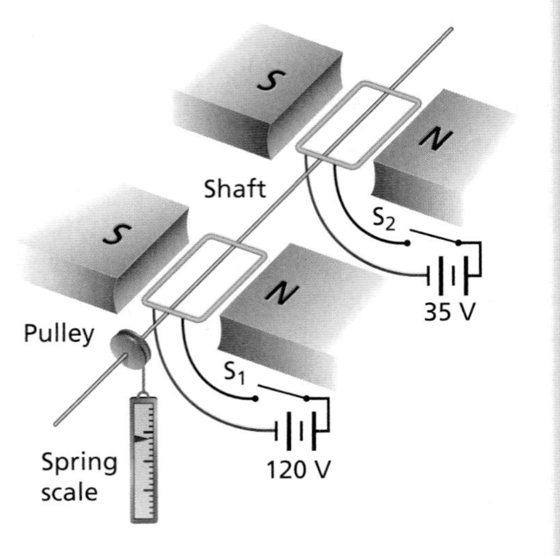

1. Given $F = ILB$, derive an equation for the torque on the armature for the position shown.
2. With S_1 closed and S_2 open, determine the torque on the shaft and the force on the spring scale.
3. With both switches closed, determine the torque on the shaft and the force on the spring scale.
4. What happens to torque if the armature is in a different position?

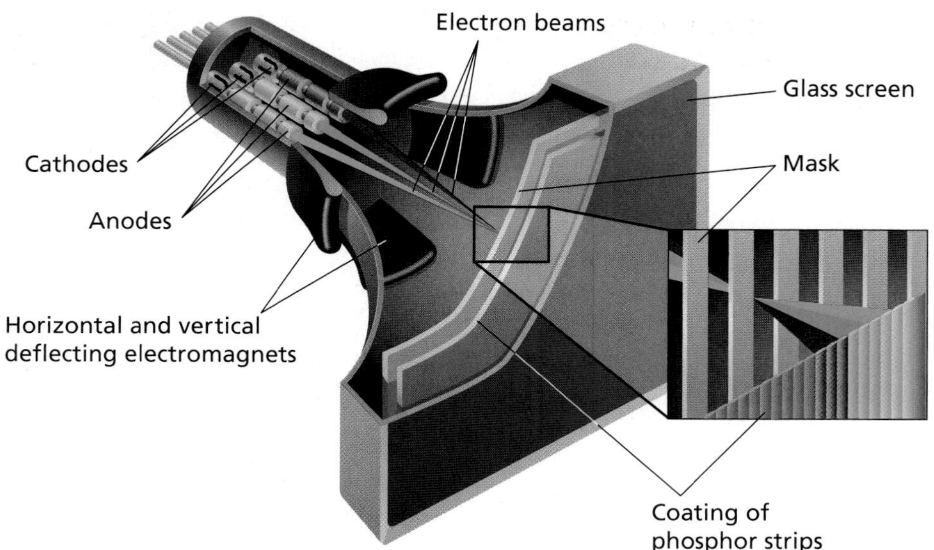

Figure 24-21 Pairs of magnets deflect the electron beam vertically and horizontally to form pictures for viewing.

The Force on a Single Charged Particle

Charged particles do not have to be confined to a wire. They also can move in a vacuum where the air particles have been removed to prevent collisions. A picture tube, also called a cathode-ray tube, in a computer monitor or television set uses electrons deflected by magnetic fields to form the pictures on the screen, as illustrated in **Figure 24-21.** Electric fields pull electrons off atoms in the negative electrode, or cathode. Other electric fields gather, accelerate, and focus the electrons into a narrow beam. Magnetic fields control the motion of the beam back-and-forth and up-and-down across the screen. The screen is coated with a phosphor that glows when it is struck by the electrons, thereby producing the picture.

The force produced by a magnetic field on a single electron depends on the velocity of the electron, the strength of the field, and the angle between directions of the velocity and the field. Consider a single electron moving in a wire of length L. The electron is moving perpendicular to the magnetic field. The current, I, is equal to the charge per unit time entering the wire, $I = q/t$. In this case, q is the charge of the electron and t is the time it takes to move the distance, L. The time required for a particle with speed v to travel distance L is found by using the equation of motion, $d = vt$, or, in this case, $t = L/v$. As a result, the equation for the current, $I = q/t$, can be replaced by $I = qv/L$. Therefore, the force on a single electron moving perpendicular to a magnetic field of strength B can be found.

Force of a Magnetic Field on a Charged, Moving Particle $F = qvB$

The force on a particle moving in a magnetic field is equal to the charge of the particle, its velocity, and the field strength.

The particle's charge is measured in coulombs, C, its velocity in meters per second, m/s, and the strength of the magnetic field in teslas, T.

The direction of the force is perpendicular to both the velocity of the particle and the magnetic field. The direction given by the third right-hand rule is for positively charged particles. For electrons, the force is in the opposite direction.

▶ EXAMPLE Problem 2

Force on a Charged Particle in a Magnetic Field **A beam of electrons travels at 3.0×10^6 m/s through a uniform magnetic field of 4.0×10^{-2} T at right angles to the field. How strong is the force acting on each electron?**

1 Analyze and Sketch the Problem

- Draw the beam of electrons and its direction of motion; the magnetic field of lines, labeled B; and the force on the electron beam, F. Remember that the force is opposite the force given by the third right-hand rule because of the electron's negative charge.

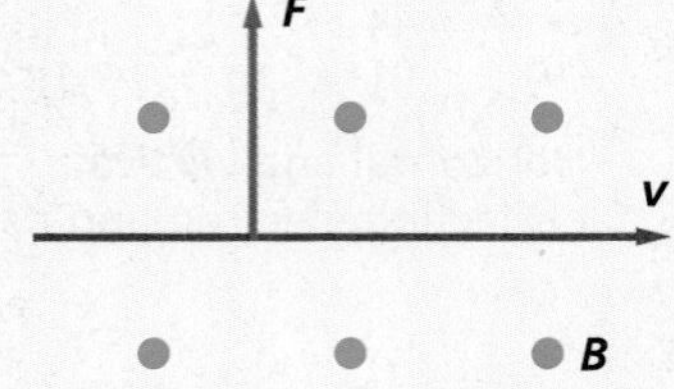

Known:	Unknown:
$v = 3.0\times10^6$ m/s	$F = ?$
$B = 4.0\times10^{-2}$ T	
$q = -1.60\times10^{-19}$ C	

2 Solve for the Unknown

$$F = qvB$$
$$= (-1.60\times10^{-19}\text{ C})(3.0\times10^6\text{ m/s})(4.0\times10^{-2}\text{ T})$$
$$= -1.9\times10^{-14}\text{ N}$$

Substitute $q = -1.60\times10^{-19}$ C, $v = 3.0\times10^6$ m/s, $B = 4.0\times10^{-2}$ T

3 Evaluate the Answer

- **Are the units correct?** T = N/(A·m), and A = C/s; so T = N·s/(C·m). Thus, (T·C·m)/s = N, the unit for force.
- **Does the direction make sense?** Use the third right-hand rule to verify that the directions of the forces are correct, recalling that the force on the electron is opposite the force given by the third right-hand rule.
- **Is the magnitude realistic?** Forces on electrons and protons are always small fractions of a newton.

Personal Tutor For an online tutorial on force on a charged particle, visit physicspp.com.

▶ PRACTICE Problems

- Additional Problems, Appendix B
- Solutions to Selected Problems, Appendix C

21. In what direction does the thumb point when using the third right-hand rule for an electron moving at right angles to a magnetic field?

22. An electron passes through a magnetic field at right angles to the field at a velocity of 4.0×10^6 m/s. The strength of the magnetic field is 0.50 T. What is the magnitude of the force acting on the electron?

23. A stream of doubly ionized particles (missing two electrons, and thus, carrying a net charge of two elementary charges) moves at a velocity of 3.0×10^4 m/s perpendicular to a magnetic field of 9.0×10^{-2} T. What is the magnitude of the force acting on each ion?

24. Triply ionized particles in a beam carry a net positive charge of three elementary charge units. The beam enters a magnetic field of 4.0×10^{-2} T. The particles have a speed of 9.0×10^6 m/s. What is the magnitude of the force acting on each particle?

25. Doubly ionized helium atoms (alpha particles) are traveling at right angles to a magnetic field at a speed of 4.0×10^4 m/s. The field strength is 5.0×10^{-2} T. What force acts on each particle?

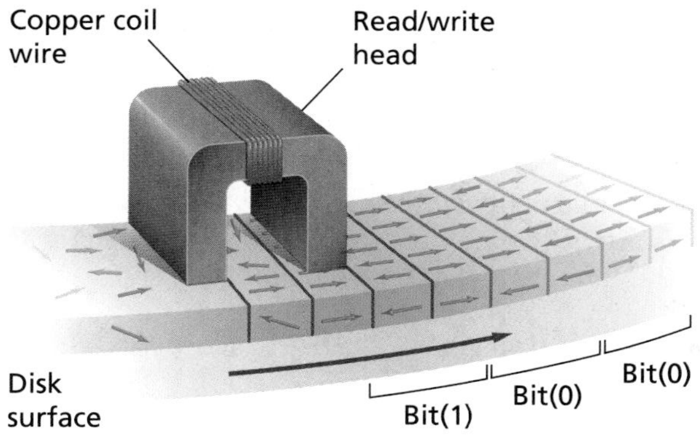

■ **Figure 24-22** Information is written to a computer disk by changing the magnetic field in a read/write head as the media passes beneath it. This causes magnetic particles in the media to align themselves in a pattern that represents the stored information.

Storing Information with Magnetic Media

Data and software commands for computers are processed digitally in bits. Each bit is identified as either a 0 or a 1. How are these bits stored? The surface of a computer storage disk is covered with an even distribution of magnetic particles within a film. The direction of the particles' domains changes in response to a magnetic field. During recording onto the disk, current is routed to the disk drive's read/write head, which is an electromagnet composed of a wire-wrapped iron core. The current through the wire induces a magnetic field in the core.

When the read/write head passes over the spinning storage disk, as in **Figure 24-22,** the domains of atoms in the magnetic film line up in bands. The orientation of the domains depends on the direction of the current.

Two bands code for one bit of information. Two bands magnetized with the poles oriented in the same direction represent 0. Two bands represent 1 with poles oriented in opposite directions. The recording current always reverses when the read/write head begins recording the next data bit.

To retrieve data, no current is sent to the read/write head. Rather, the magnetized bands in the disk induce current in the coil as the disk spins beneath the head. Changes in the direction of the induced current are sensed by the computer and interpreted as 0's and 1's.

24.2 Section Review

26. Magnetic Forces Imagine that a current-carrying wire is perpendicular to Earth's magnetic field and runs east-west. If the current is east, in which direction is the force on the wire?

27. Deflection A beam of electrons in a cathode-ray tube approaches the deflecting magnets. The north pole is at the top of the tube; the south pole is on the bottom. If you are looking at the tube from the direction of the phosphor screen, in which direction are the electrons deflected?

28. Galvanometers Compare the diagram of a galvanometer in Figure 24-18 on page 655 with the electric motor in Figure 24-20 on page 656. How is the galvanometer similar to an electric motor? How are they different?

29. Motors When the plane of the coil in a motor is perpendicular to the magnetic field, the forces do not exert a torque on the coil. Does this mean that the coil does not rotate? Explain.

30. Resistance A galvanometer requires 180 μA for full-scale deflection. What is the total resistance of the meter and the multiplier resistor for a 5.0-V full-scale deflection?

31. Critical Thinking How do you know that the forces on parallel current-carrying wires are a result of magnetic attraction between wires, and not a result of electrostatics? *Hint: Consider what the charges are like when the force is attractive. Then consider what the forces are when three wires carry currents in the same direction.*

PHYSICS LAB • Design Your Own

Creating an Electromagnet

An electromagnet uses the magnetic field generated by a current to magnetize a piece of metal. In this activity, you will construct an electromagnet and test one variable that you think might affect the strength of it.

QUESTION

What is one variable that determines the strength of an electromagnet?

Objectives

- **Hypothesize** which variables might affect the strength of an electromagnet.
- **Observe** the effects on an electromagnet's strength.
- **Collect and organize data** comparing the chosen variable and magnet strength.
- **Make and use graphs** to help identify a relationship between a controlling variable and a responding variable.
- **Analyze and conclude** what the effect is of the chosen variable on magnet strength.

Safety Precautions

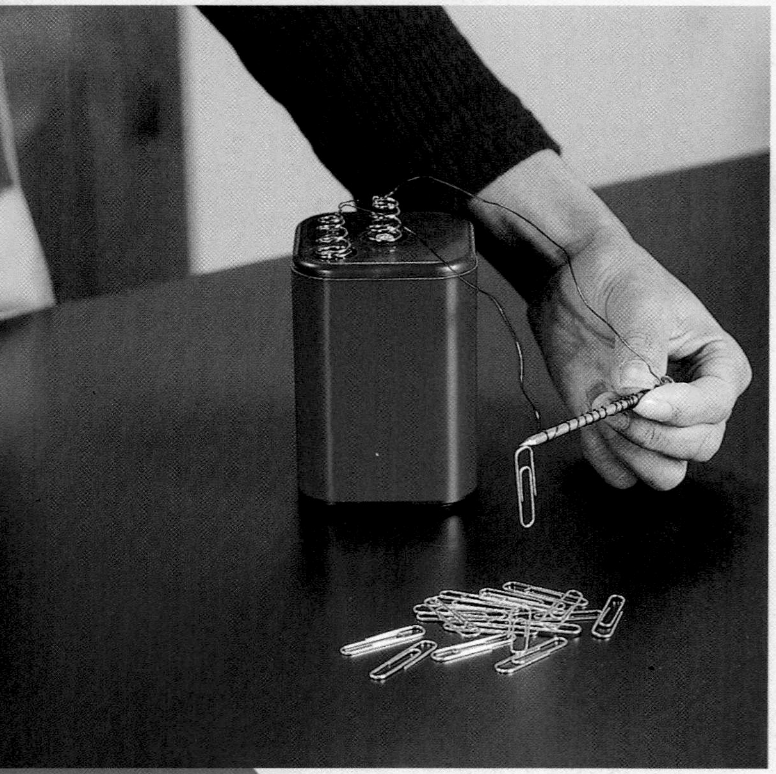

Possible Materials

large paper clips
small paper clips
steel BBs
wire
steel nail
6-V lantern batteries
9-V batteries
DC power source

Procedure

1. List the materials you will use to make your electromagnet.
2. List all the possible variables you think could affect the strength of an electromagnet.
3. Choose the one variable you will vary to determine whether it does, in fact, affect the strength of an electromagnet.
4. Determine a method to detect the strength of the magnetic field produced by the electromagnet.
5. Have your teacher approve your lists before continuing.
6. Write a brief procedure for your experiment. Be sure to include all the values for the variables you will be keeping constant.
7. Create a data table like the one on the following page that displays the two quantities you will measure.
8. Build your electromagnet by using a nail and a length of wire. Wrap the wire around the nail. Be sure to leave several inches from both ends of the wire sticking out from your coil to allow attachment to the power source. ***CAUTION: The end of the nail or wire may be sharp. Exercise care when handling these materials to avoid being cut or scraped.***

Data Table	
Number of ______________	Number of ______________

9. Have your teacher inspect your magnet before continuing.
10. Perform your experiment and record your data. ***CAUTION: If you are using BBs in your experiment, avoid possible injury by immediately picking up any BBs that should happen to fall to the floor.***

Analyze

1. **Make and Use Graphs** Create a graph showing the relationship between your two variables.
2. What were the variables that you attempted to control in this experiment? Were there any you were unable to control?
3. If you evaluated the strength of the electromagnet by the amount of material it could pick up, how did you try to control any error from the magnet attracting only whole numbers of objects?

Conclude and Apply

1. What is the relationship between your chosen variable and the strength of a magnet?
2. What variables did other students in your class find that also affected the strength of an electromagnet?
3. Were there any variables, by any group, that were found not to affect the strength of the electromagnet?

Going Further

1. Compare the various variables students found that affected magnet strength. Did any of the variables appear to greatly increase strength without much change in the independent variable? If so, which ones?
2. If you wanted to increase magnet strength, which method seems the most cost effective? Explain.
3. If you need to easily vary the strength of an electromagnet, how would you suggest that be done?

Real-World Physics

1. If you needed to create a stronger electromagnet for use in a small space, such as inside a laptop computer, what method would you use to increase the electromagnetic strength, given the size constraints?
2. Some buildings have electromagnets to hold fire doors open when the building is occupied. These magnets are mounted to the wall, like a door stop, behind the door. Thinking about the actions a fire alarm system would need to perform to control a fire, what is the advantage of using a system like this to hold the doors? How might a system like this be an advantage, or a disadvantage, in the event of a natural disaster?
3. Some electric bells work by having an arm strike the side of a metal dome-shaped bell. How might an electromagnet be used to make this bell work? How might the bell be wired to allow the arm to strike repeatedly (continual ringing) until the power supply is removed?

To find out more about magnetic fields, visit the Web site: physicspp.com

The Hall Effect

Something as simple as magnetic fields deflecting charged particles has led to a revolution in how we measure or detect the movement of things, such as bicycle wheels and automotive crankshafts. It all starts when current passes through a wide, flat conductor, in the presence of a magnetic field.

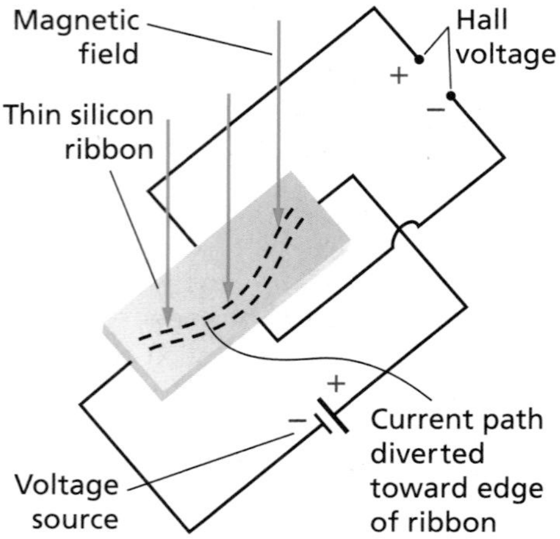

A magnetic field forces more electrons to the edge of a thin metal strip. This creates the Hall voltage.

The magnetic lines of force are perpendicular to the ribbon's broad surface. This makes the flowing electrons crowd into one side of the ribbon. Because there are more electrons on one edge of the ribbon than on the other, a voltage, called the *Hall voltage,* is generated across the width of the ribbon. The magnitude of the Hall voltage is dependent upon the strength of the magnetic field.

E.H. Hall discovered this effect in 1879. Its industrial and scientific significance were discovered only recently because the Hall voltage is small in ribbons of conventional metals. Now, very thin layers of semiconducting silicon yield substantial Hall voltages.

The Hall effect can be used to explore conduction in different types of materials. The sign of the Hall-effect voltage gives the sign of the moving charge, and the magnitude of the voltage tells us about the density and velocity of the charge. Such experiments have shown that in copper and most other metals, electrons carry the charge, but in zinc it is the positive charges that move.

A Useful Sensor Engineers have developed the Hall-effect sensor. These tiny black plastic devices contain a thin film of silicon with wires connected, as shown in the diagram. The Hall voltage wires are connected to a tiny amplifier so that other instruments can detect it.

If a permanent magnet is moved near a Hall-effect sensor, the voltage from the amplifier will increase. Thus, the sensor can be used to detect the proximity of the magnet.

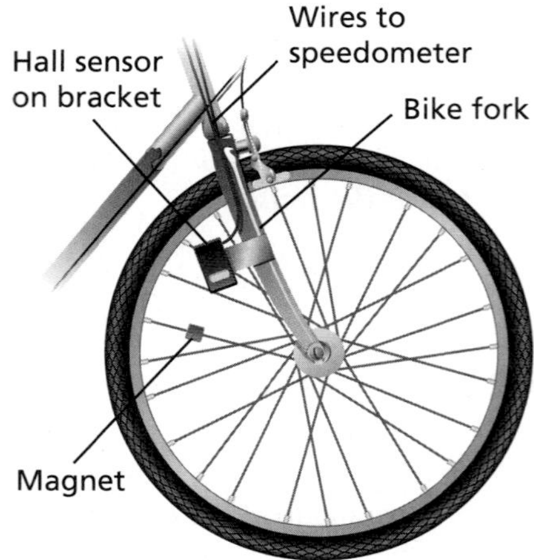

Bicycle speedometers use a Hall-effect sensor to display the speed at which a bicycle is moving.

Everyday Applications Bicycle speedometers use a permanent magnet attached to the front wheel. Each revolution of the wheel brings the magnet close to a Hall-effect sensor. The resulting pulses are counted and timed. Hall-effect sensors also are used to time the spark in automobile engines. When a magnet mounted on the crankshaft or distributor rotor moves near a sensor, a voltage pulse is produced, and the ignition system instantly fires the spark plug.

Going Further

1. **Analyze** Why are the Hall-voltage electrodes positioned directly across from each other? What if they weren't?
2. **Critical Thinking** Might a strong magnetic field applied across a conducting ribbon change the resistance of that ribbon as a result of the Hall effect? Consider what you learned about the cross-sectional areas of wires.

Chapter 24 Study Guide

24.1 Magnets: Permanent and Temporary

Vocabulary

- polarized (p. 644)
- magnetic field (p. 645)
- magnetic flux (p. 646)
- first right-hand rule (p. 649)
- solenoid (p. 649)
- electromagnet (p. 649)
- second right-hand rule (p. 649)
- domain (p. 650)

Key Concepts

- Like magnetic poles repel; unlike magnetic poles attract.
- Magnetic fields exit from the north pole of a magnet and enter its south pole.
- Magnetic field lines always form closed loops.
- A magnetic field exists around any carrying-current wire.
- A coil of wire carrying a current has a magnetic field. The field about the coil is like the field about a permanent magnet.

24.2 Forces Caused by Magnetic Fields

Vocabulary

- third right-hand rule (p. 652)
- galvanometer (p. 655)
- electric motor (p. 656)
- armature (p. 656)

Key Concepts

- The strength of a magnetic field is measured in teslas.
- When a current-carrying wire is placed in a magnetic field, there exists a force on the wire, perpendicular to both the field and the wire.
- The force on a current-carrying wire in a magnetic field is proportional to the current flow, the length of the wire, and the field strength.

$$F = ILB$$

- A galvanometer consists of a loop of wire in a magnetic field, and is used to measure small currents. When current is passed through the loop, a force on the wire loop results in a deflection of the loop.
- A galvanometer can be used as an ammeter by adding a shunt resistor in parallel with the galvanometer.
- A galvanometer can be used as a voltmeter by adding a multiplier resistor in series with the galvanometer.
- A loudspeaker functions by varying the current through a coil that is placed in a magnetic field. The coil is attached to a paper cone that moves when the coil moves. As the current varies, the cone vibrates, thereby producing sound.
- An electric motor consists of a coil of wire placed in a magnetic field. When there is a current in the coil, the coil rotates as a result of the force on the wire in the magnetic field. Complete 360° rotation is achieved by using a commutator to switch the direction of the current in the coil as the coil rotates.
- The force that a magnetic field exerts on a charged particle depends on three factors: the charge of the particle, the velocity of the particle, and the strength of the field. The direction of the force is perpendicular to both the field and the particle's velocity.

$$F = qvB$$

- Computer monitors and television screens function by using magnets to focus and direct particles on phosphor screens. When particles strike the screen, light is emitted, and produces images on the screen.

Chapter 24 Assessment

Concept Mapping

32. Complete the following concept map using the following: *right-hand rule, F = qvB,* and *F = ILB.*

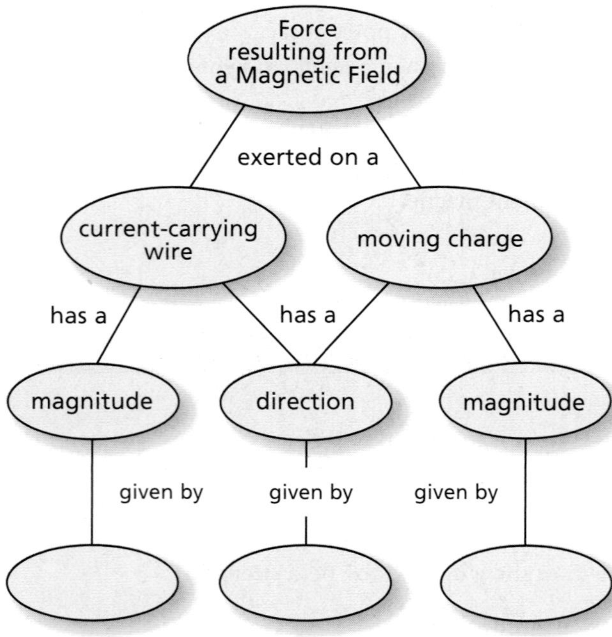

Mastering Concepts

33. State the rule for magnetic attraction and repulsion. (24.1)

34. Describe how a temporary magnet differs from a permanent magnet. (24.1)

35. Name the three most important common magnetic elements. (24.1)

36. Draw a small bar magnet and show the magnetic field lines as they appear around the magnet. Use arrows to show the direction of the field lines. (24.1)

37. Draw the magnetic field between two like magnetic poles and then between two unlike magnetic poles. Show the directions of the fields. (24.1)

38. If you broke a magnet in two, would you have isolated north and south poles? Explain. (24.1)

39. Describe how to use the first right-hand rule to determine the direction of a magnetic field around a straight current-carrying wire. (24.1)

40. If a current-carrying wire is bent into a loop, why is the magnetic field inside the loop stronger than the magnetic field outside? (24.1)

41. Describe how to use the second right-hand rule to determine the polarity of an electromagnet. (24.1)

42. Each electron in a piece of iron is like a tiny magnet. The iron, however, may not be a magnet. Explain. (24.1)

43. Why will dropping or heating a magnet weaken it? (24.1)

44. Describe how to use the third right-hand rule to determine the direction of force on a current-carrying wire placed in a magnetic field. (24.2)

45. A strong current suddenly is switched on in a wire. No force acts on the wire, however. Can you conclude that there is no magnetic field at the location of the wire? Explain. (24.2)

46. What kind of meter is created when a shunt is added to a galvanometer? (24.2)

Applying Concepts

47. A small bar magnet is hidden in a fixed position inside a tennis ball. Describe an experiment that you could do to find the location of the north pole and the south pole of the magnet.

48. A piece of metal is attracted to one pole of a large magnet. Describe how you could tell whether the metal is a temporary magnet or a permanent magnet.

49. Is the magnetic force that Earth exerts on a compass needle less than, equal to, or greater than the force that the compass needle exerts on Earth? Explain.

50. Compass Suppose you are lost in the woods but have a compass with you. Unfortunately, the red paint marking the north pole of the compass needle has worn off. You have a flashlight with a battery and a length of wire. How could you identify the north pole of the compass?

51. A magnet can attract a piece of iron that is not a permanent magnet. A charged rubber rod can attract an uncharged insulator. Describe the different microscopic processes producing these similar phenomena.

52. A current-carrying wire runs across a laboratory bench. Describe at least two ways in which you could find the direction of the current.

53. In which direction, in relation to a magnetic field, would you run a current-carrying wire so that the force on it, resulting from the field, is minimized, or even made to be zero?

 For more problems, go to Additional Problems, Appendix B.

54. Two wires carry equal currents and run parallel to each other.
  a. If the two currents are in opposite directions, where will the magnetic field from the two wires be larger than the field from either wire alone?
  b. Where will the magnetic field from both be exactly twice as large as from one wire?
  c. If the two currents are in the same direction, where will the magnetic field be exactly zero?

55. How is the range of a voltmeter changed when the resistor's resistance is increased?

56. A magnetic field can exert a force on a charged particle. Can the field change the particle's kinetic energy? Explain.

57. A beam of protons is moving from the back to the front of a room. It is deflected upward by a magnetic field. What is the direction of the field causing the deflection?

58. Earth's magnetic field lines are shown in **Figure 24-23.** At what location, poles or equator, is the magnetic field strength greatest? Explain.

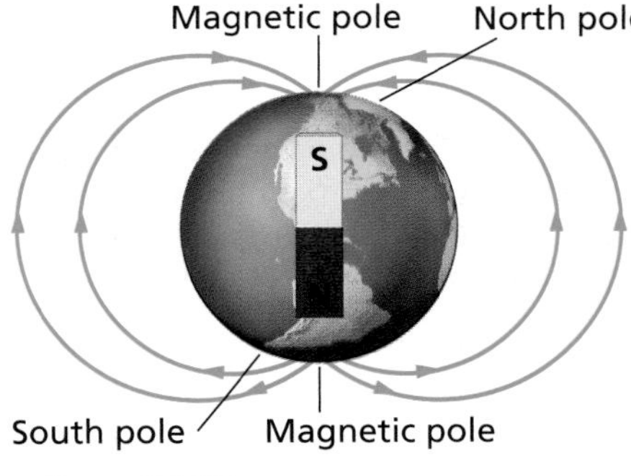

Figure 24-23

Mastering Problems

24.1 Magnets: Permanent and Temporary

59. As the magnet below in **Figure 24-24** moves toward the suspended magnet, what will the magnet suspended by the string do?

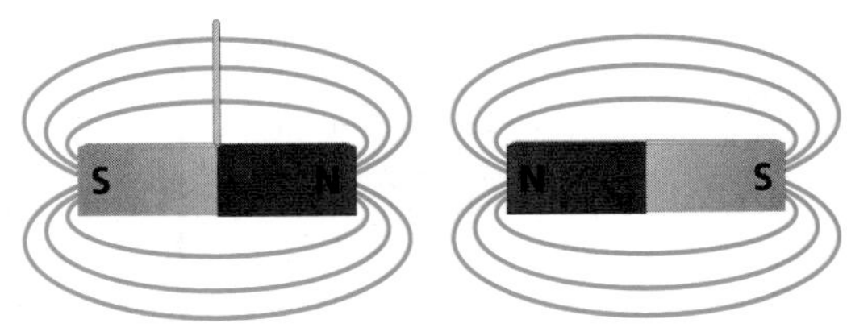

Figure 24-24

60. As the magnet in **Figure 24-25** moves toward the suspended magnet, what will the magnet that is suspended by the string do?

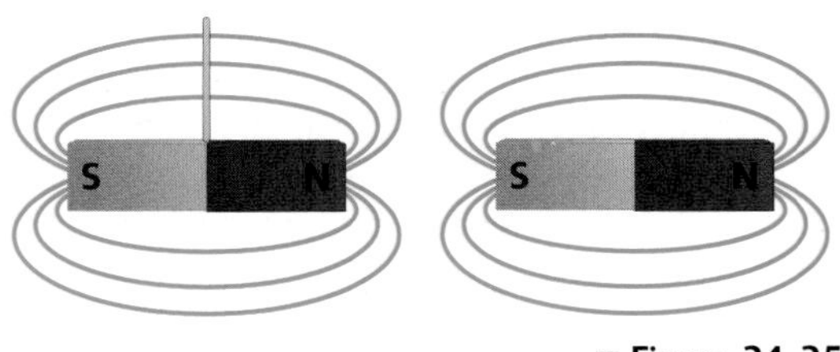

Figure 24-25

61. Refer to **Figure 24-26** to answer the following questions.
  a. Where are the poles?
  b. Where is the north pole?
  c. Where is the south pole?

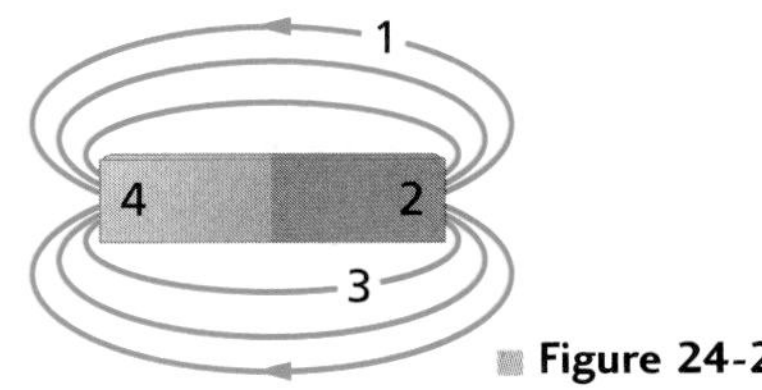

Figure 24-26

62. **Figure 24-27** shows the response of a compass in two different positions near a magnet. Where is the south pole of the magnet located?

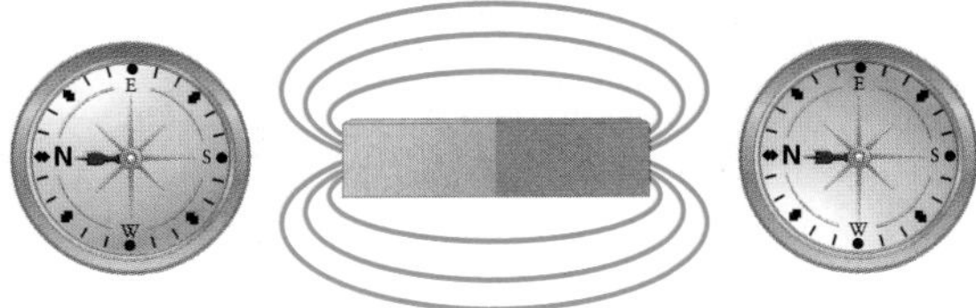

Figure 24-27

63. A wire that is 1.50 m long and carrying a current of 10.0 A is at right angles to a uniform magnetic field. The force acting on the wire is 0.60 N. What is the strength of the magnetic field?

64. A conventional current flows through a wire, as shown in **Figure 24-28.** Copy the wire segment and sketch the magnetic field that the current generates.

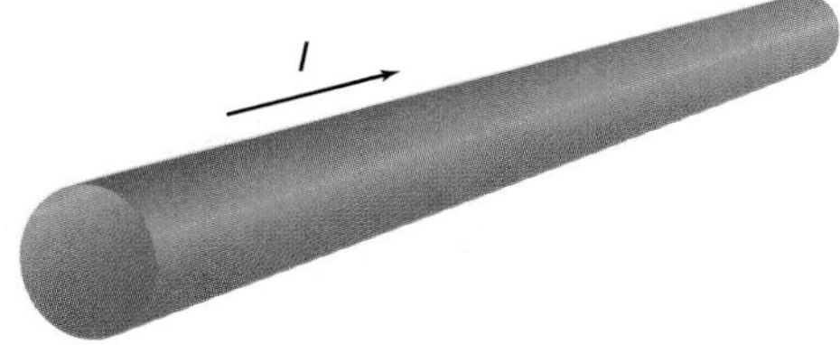

Figure 24-28

65. The current is coming straight out of the page in **Figure 24-29.** Copy the figure and sketch the magnetic field that the current generates.

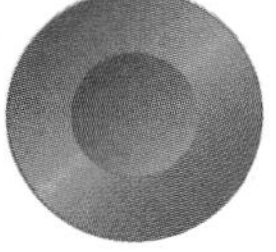

Figure 24-29

66. **Figure 24-30** shows the end view of an electromagnet with current flowing through it.
  a. What is the direction of the magnetic field inside the loops?
  b. What is the direction of the magnetic field outside the loops?

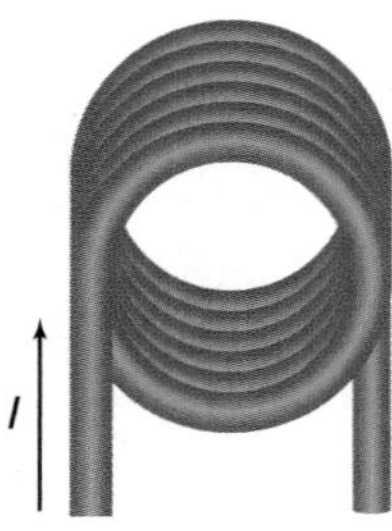

Figure 24-30

67. **Ceramic Magnets** The repulsive force between two ceramic magnets was measured and found to depend on distance, as given in **Table 24-1.**
  a. Plot the force as a function of distance.
  b. Does this force follow an inverse square law?

Table 24-1

Separation, d (cm)	Force, F (N)
1.0	3.93
1.2	0.40
1.4	0.13
1.6	0.057
1.8	0.030
2.0	0.018
2.2	0.011
2.4	0.0076
2.6	0.0053
2.8	0.0038
3.0	0.0028

24.2 Forces Caused by Magnetic Fields

68. The arrangement shown in **Figure 24-31** is used to convert a galvanometer to what type of device?

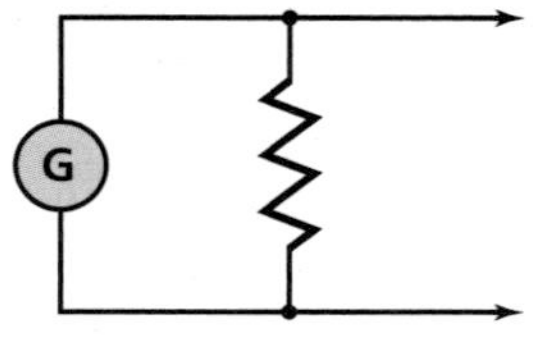

Figure 24-31

69. What is the resistor shown in Figure 24-31 called?

70. The arrangement shown in **Figure 24-32** is used to convert a galvanometer to what type of device?

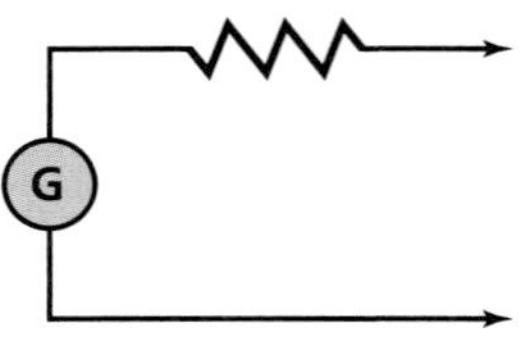

Figure 24-32

71. What is the resistor shown in Figure 24-32 called?

72. A current-carrying wire is placed between the poles of a magnet, as shown in **Figure 24-33.** What is the direction of the force on the wire?

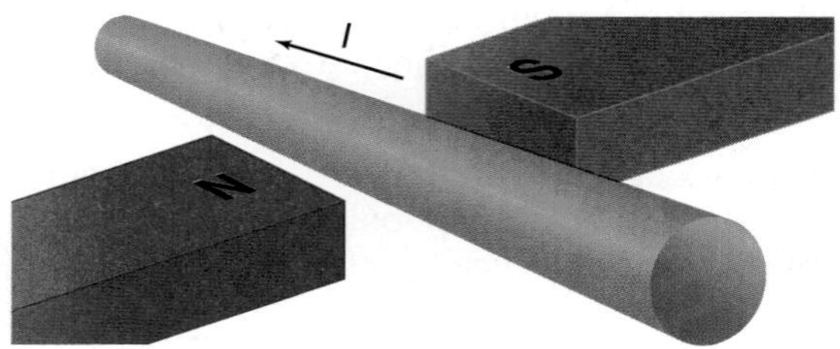

Figure 24-33

73. A wire that is 0.50 m long and carrying a current of 8.0 A is at right angles to a uniform magnetic field. The force on the wire is 0.40 N. What is the strength of the magnetic field?

74. The current through a wire that is 0.80 m long is 5.0 A. The wire is perpendicular to a 0.60-T magnetic field. What is the magnitude of the force on the wire?

75. A wire that is 25 cm long is at right angles to a 0.30-T uniform magnetic field. The current through the wire is 6.0 A. What is the magnitude of the force on the wire?

76. A wire that is 35 cm long is parallel to a 0.53-T uniform magnetic field. The current through the wire is 4.5 A. What force acts on the wire?

77. A wire that is 625 m long is perpendicular to a 0.40-T magnetic field. A 1.8-N force acts on the wire. What current is in the wire?

78. The force on a 0.80-m wire that is perpendicular to Earth's magnetic field is 0.12 N. What is the current in the wire? Use 5.0×10^{-5}T for Earth's magnetic field.

79. The force acting on a wire that is at right angles to a 0.80-T magnetic field is 3.6 N. The current in the wire is 7.5 A. How long is the wire?

 For more problems, go to Additional Problems, Appendix B.

80. A power line carries a 225-A current from east to west, parallel to the surface of Earth.
- **a.** What is the magnitude of the force resulting from Earth's magnetic field acting on each meter of the wire? Use $B_{\text{Earth}}=5.0\times10^{-5}$T.
- **b.** What is the direction of the force?
- **c.** In your judgment, would this force be important in designing towers to hold this power line? Explain.

81. Galvanometer A galvanometer deflects full-scale for a 50.0-μA current.
- **a.** What must be the total resistance of the series resistor and the galvanometer to make a voltmeter with 10.0-V full-scale deflection?
- **b.** If the galvanometer has a resistance of 1.0 kΩ, what should be the resistance of the series (multiplier) resistor?

82. The galvanometer in problem 81 is used to make an ammeter that deflects full-scale for 10 mA.
- **a.** What is the potential difference across the galvanometer (1.0 kΩ resistance) when a current of 50 μA passes through it?
- **b.** What is the equivalent resistance of parallel resistors having the potential difference calculated in a circuit with a total current of 10 mA?
- **c.** What resistor should be placed parallel with the galvanometer to make the resistance calculated in part b?

83. A beam of electrons moves at right angles to a magnetic field of 6.0×10^{-2} T. The electrons have a velocity of 2.5×10^{6} m/s. What is the magnitude of the force on each electron?

84. Subatomic Particle A muon (a particle with the same charge as an electron) is traveling at 4.21×10^{7} m/s at right angles to a magnetic field. The muon experiences a force of 5.00×10^{-12} N.
- **a.** How strong is the magnetic field?
- **b.** What acceleration does the muon experience if its mass is 1.88×10^{-28} kg?

85. A singly ionized particle experiences a force of 4.1×10^{-13} N when it travels at right angles through a 0.61-T magnetic field. What is the velocity of the particle?

86. A room contains a strong, uniform magnetic field. A loop of fine wire in the room has current flowing through it. Assume that you rotate the loop until there is no tendency for it to rotate as a result of the field. What is the direction of the magnetic field relative to the plane of the coil?

87. A force of 5.78×10^{-16} N acts on an unknown particle traveling at a 90° angle through a magnetic field. If the velocity of the particle is 5.65×10^{4} m/s and the field is 3.20×10^{-2} T, how many elementary charges does the particle carry?

Mixed Review

88. A copper wire of insignificant resistance is placed in the center of an air gap between two magnetic poles, as shown in **Figure 24-34.** The field is confined to the gap and has a strength of 1.9 T.
- **a.** Determine the force on the wire (direction and magnitude) when the switch is open.
- **b.** Determine the force on the wire (direction and magnitude) when the switch is closed.
- **c.** Determine the force on the wire (direction and magnitude) when the switch is closed and the battery is reversed.
- **d.** Determine the force on the wire (direction and magnitude) when the switch is closed and the wire is replaced with a different piece having a resistance of 5.5 Ω.

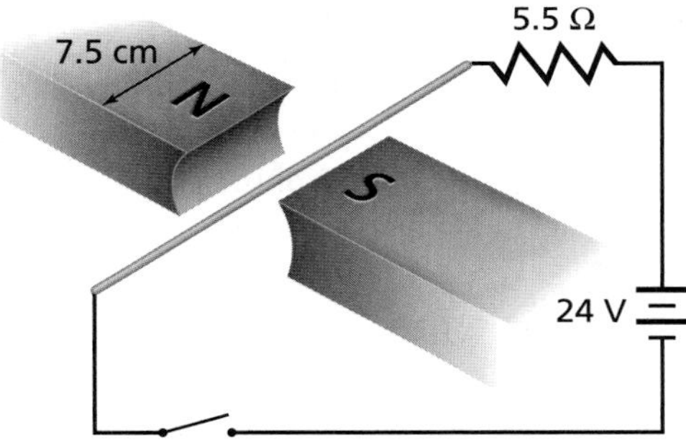

Figure 24-34

89. Two galvanometers are available. One has 50.0-μA full-scale sensitivity and the other has 500.0-μA full-scale sensitivity. Both have the same coil resistance of 855 Ω. Your challenge is to convert them to measure a current of 100.0 mA, full-scale.
- **a.** Determine the shunt resistor for the 50.0-μA meter.
- **b.** Determine the shunt resistor for the 500.0-μA meter.
- **c.** Determine which of the two is better for actual use. Explain.

90. Subatomic Particle A beta particle (high-speed electron) is traveling at right angles to a 0.60-T magnetic field. It has a speed of 2.5×10^{7} m/s. What size force acts on the particle?

91. The mass of an electron is 9.11×10^{-31} kg. What is the magnitude of the acceleration of the beta particle described in problem 90?

92. A magnetic field of 16 T acts in a direction due west. An electron is traveling due south at 8.1×10^5 m/s. What are the magnitude and the direction of the force acting on the electron?

93. Loudspeaker The magnetic field in a loudspeaker is 0.15 T. The wire consists of 250 turns wound on a 2.5-cm-diameter cylindrical form. The resistance of the wire is 8.0 Ω. Find the force exerted on the wire when 15 V is placed across the wire.

94. A wire carrying 15 A of current has a length of 25 cm in a magnetic field of 0.85 T. The force on a current-carrying wire in a uniform magnetic field can be found using the equation $F = ILB \sin\theta$. Find the force on the wire when it makes the following angles with the magnetic field lines of

a. 90° **b.** 45° **c.** 0°

95. An electron is accelerated from rest through a potential difference of 20,000 V, which exists between plates P_1 and P_2, shown in **Figure 24-35.** The electron then passes through a small opening into a magnetic field of uniform field strength, *B*. As indicated, the magnetic field is directed into the page.

a. State the direction of the electric field between the plates as either P_1 to P_2 or P_2 to P_1.

b. In terms of the information given, calculate the electron's speed at plate P_2.

c. Describe the motion of the electron through the magnetic field.

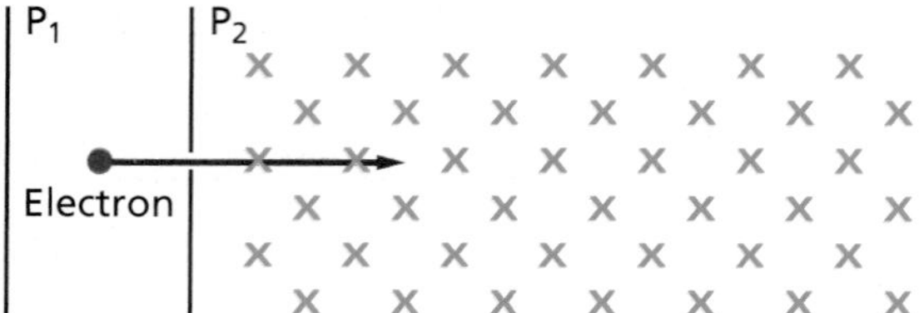

Figure 24-35

Thinking Critically

96. Apply Concepts A current is sent through a vertical spring, as shown in **Figure 24-36.** The end of the spring is in a cup filled with mercury. What will happen? Why?

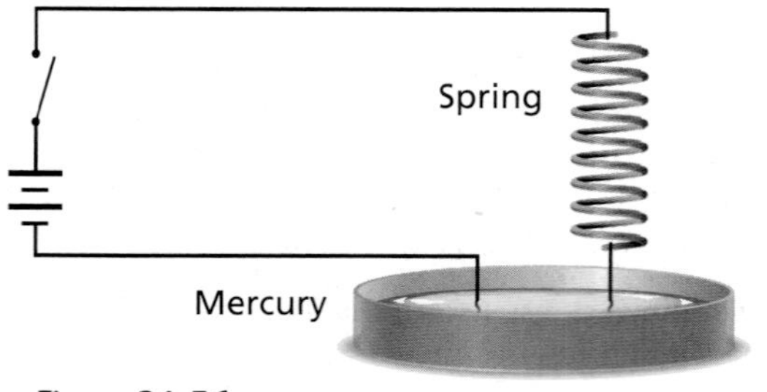

Figure 24-36

97. Apply Concepts The magnetic field produced by a long, current-carrying wire is represented by $B = (2 \times 10^{-7}\ \text{T·m/A})(I/d)$, where *B* is the field strength in teslas, *I* is the current in amps, and *d* is the distance from the wire in meters. Use this equation to estimate some magnetic fields that you encounter in everyday life.

a. The wiring in your home seldom carries more than 10 A. How does the magnetic field that is 0.5 m from such a wire compare to Earth's magnetic field?

b. High-voltage power transmission lines often carry 200 A at voltages as high as 765 kV. Estimate the magnetic field on the ground under such a line, assuming that it is about 20 m high. How does this field compare with a magnetic field in your home?

c. Some consumer groups have recommended that pregnant women not use electric blankets in case the magnetic fields cause health problems. Estimate the distance that a fetus might be from such a wire, clearly stating your assumptions. If such a blanket carries 1 A, find the magnetic field at the location of the fetus. Compare this with Earth's magnetic field.

98. Add Vectors In almost all cases described in problem 97, a second wire carries the same current in the opposite direction. Find the net magnetic field that is a distance of 0.10 m from each wire carrying 10 A. The wires are 0.01 m apart. Make a scale drawing of the situation. Calculate the magnitude of the field from each wire and use a right-hand rule to draw vectors showing the directions of the fields. Finally, find the vector sum of the two fields. State its magnitude and direction.

Writing In Physics

99. Research superconducting magnets and write a one-page summary of proposed future uses for such magnets. Be sure to describe any hurdles that stand in the way of the practical application of these magnets.

Cumulative Review

100. How much work is required to move a charge of 6.40×10^{-3} C through a potential difference of 2500 V? (Chapter 21)

101. The current flow in a 120-V circuit increases from 1.3 A to 2.3 A. Calculate the change in power. (Chapter 22)

102. Determine the total resistance of three, 55-Ω resistors connected in parallel and then series-connected to two 55-Ω resistors connected in series. (Chapter 23)

 For more problems, go to Additional Problems, Appendix B.

Multiple Choice

1. A straight wire carrying a current of 7.2 A has a field of 8.9×10^{-3} T perpendicular to it. What length of wire in the field will experience a force of 2.1 N?

   (A) 2.6×10^{-3} m (C) 1.3×10^{-1} m
   (B) 3.1×10^{-2} m (D) 3.3×10^{1} m

2. Assume that a 19-cm length of wire is carrying a current perpendicular to a 4.1-T magnetic field and experiences a force of 7.6 mN. What is the current in the wire?

   (A) 3.4×10^{-7} A (C) 1.0×10^{-2} A
   (B) 9.8×10^{-3} A (D) 9.8 A

3. A 7.12-μC charge is moving at the speed of light in a magnetic field of 4.02 mT. What is the force on the charge?

   (A) 8.59 N (C) 8.59×10^{12} N
   (B) 2.90×10^{1} N (D) 1.00×10^{16} N

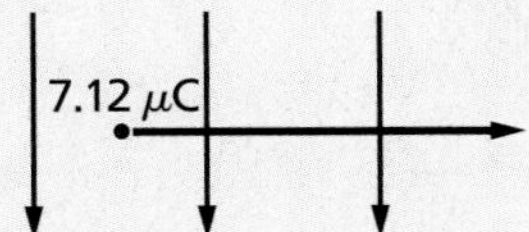

4. An electron is moving at 7.4×10^{5} m/s perpendicular to a magnetic field. It experiences a force of 18 N. What is the strength of the magnetic field?

   (A) 6.5×10^{-15} T (C) 1.3×10^{7} T
   (B) 2.4×10^{-5} T (D) 1.5×10^{14} T

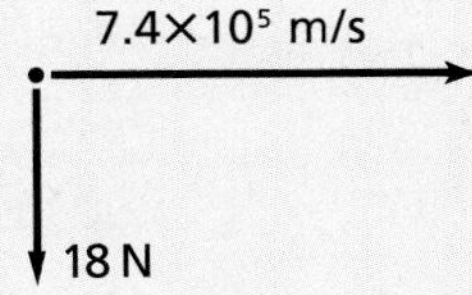

5. Which factor will not affect the strength of a solenoid?

   (A) number of wraps
   (B) strength of current
   (C) thickness of wire
   (D) core type

6. Which statement about magnetic monopoles is false?

   (A) A monopole is a hypothetical separate north pole.
   (B) Research scientists use them for internal medical testing applications.
   (C) A monopole is a hypothetical separate south pole.
   (D) They don't exist.

7. A uniform magnetic field of 0.25 T points vertically downward. A proton enters the field with a horizontal velocity of 4.0×10^{6} m/s. What are the magnitude and direction of the instantaneous force exerted on the proton as it enters the magnetic field?

   (A) 1.6×10^{-13} N to the left
   (B) 1.6×10^{-13} N downward
   (C) 1.0×10^{6} N upward
   (D) 1.0×10^{6} N to the right

Extended Answer

8. Derive the units of teslas in kilograms, meters, seconds, and coulombs using dimensional analysis and the formulas $F = qvB$ and $F = ILB$.

9. A wire attached to a 5.8-V battery is in a circuit with 18 Ω. 14 cm of the wire is in a magnetic field of 0.85 T and the force on the wire is 22 mN. What is the angle of the wire in the field given that the formula for angled wires in fields is $F = ILB \sin \theta$?

✓ Test-Taking TIP

Read the Directions

No matter how many times you've taken a particular test or practiced for an exam, it's always a good idea to read through the directions provided at the beginning of each section. It only takes a moment and could prevent you from making a simple mistake throughout the test that could cause you to do poorly.

Chapter 25 Electromagnetic Induction

What You'll Learn

- You will describe how changing magnetic fields can generate electric potential differences.
- You will apply this phenomenon to the construction of generators and transformers.

Why It's Important

The relationship between magnetic fields and current makes possible the three cornerstones of electrical technology: motors, generators, and transformers.

Hydroelectric Generators Dams commonly are built on rivers to provide a source of power for nearby communities. Within the dam, the potential and kinetic energy of water is turned into electric energy.

Think About This ▶ How do the generators located inside the dam convert the kinetic and potential energy of the water into electric energy?

physicspp.com

LAUNCH Lab

What happens in a changing magnetic field?

Question

How does a changing magnetic field affect a coil of wire passed through it?

Procedure

1. Place two bar magnets about 8 cm apart.
2. Attach a sensitive galvanometer to either end of a piece of coiled copper wire.
3. Slowly move the wire between the magnets. Note the galvanometer reading.
4. Vary the angle of the movement of the copper wire and the velocity of the wire. Note your results.

Analysis

What causes the galvanometer to move?

What situation makes the galvanometer deflect the most?

Critical Thinking When the wire is moved between the magnets, what is happening to the wire?

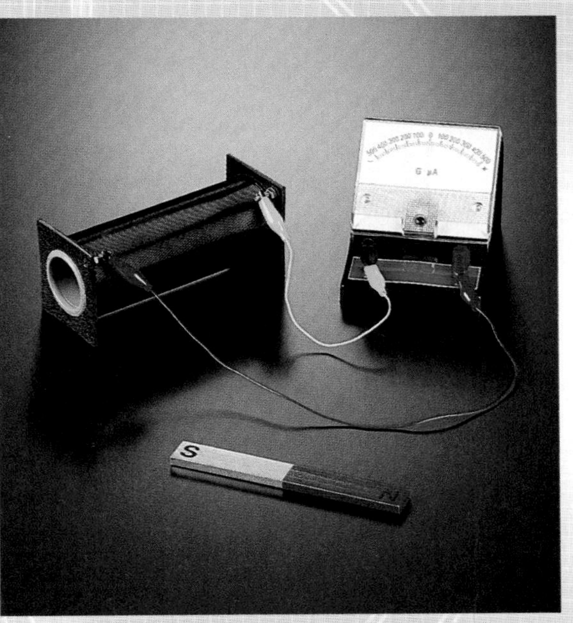

25.1 Electric Current from Changing Magnetic Fields

In Chapter 24, you learned how Hans Christian Oersted discovered that an electric current produces a magnetic field. Michael Faraday thought that the reverse must also be true: that a magnetic field produces an electric current. In 1822, Michael Faraday wrote a goal in his notebook: "Convert magnetism into electricity." Faraday tried many combinations of magnetic fields and wires without success. After nearly ten years of unsuccessful experiments, Faraday found that he could induce electric current by moving a wire through a magnetic field. In the same year, Joseph Henry, an American high-school teacher, also showed that a changing magnetic field could produce electric current. Henry took an idea developed by another scientist and broadened the application to other educational demonstration devices to make them more sensitive or powerful. Henry's versions of these devices were not new discoveries, but he made the devices more dramatic and effective as educational aids. However, Henry, unlike Faraday, chose not to publish his discoveries.

► **Objectives**

- **Explain** how a changing magnetic field produces an electric current.
- **Define** electromotive force.
- **Solve** problems involving wires moving in magnetic fields.

► **Vocabulary**

electromagnetic induction
fourth right-hand rule
electromotive force
electric generator
average power

Figure 25-1 When a wire is moved in a magnetic field, there is an electric current in the wire, but only while the wire is moving. The direction of the current depends on the direction in which the wire is moving through the field. The arrows indicate the direction of conventional current.

Concepts In Motion

Interactive Figure To seee an animation on electromagnetic induction, visit physicspp.com.

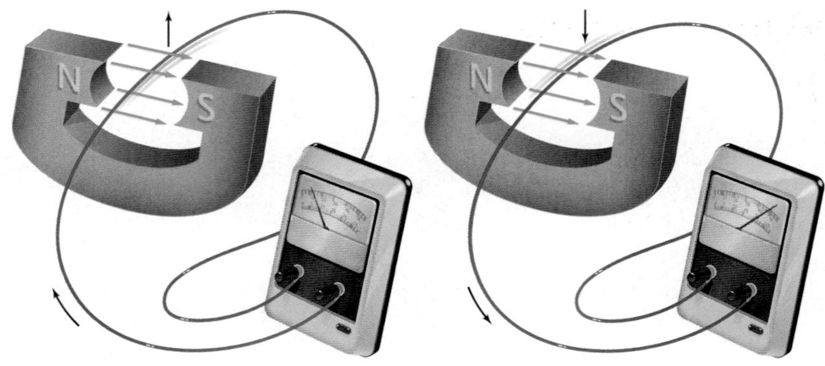

Electromagnetic Induction

Figure 25-1 shows one of Faraday's experiments, in which a wire loop that is part of a closed circuit is placed in a magnetic field. When the wire is held stationary or is moved parallel to the magnetic field, there is no current, but when the wire moves up through the field, the current is in one direction. When the wire moves down through the field, the current is in the opposite direction. An electric current is generated in a wire only when the wire cuts magnetic field lines.

Faraday found that to generate current, either the conductor can move through a magnetic field or a magnetic field can move past the conductor. It is the relative motion between the wire and the magnetic field that produces the current. The process of generating a current through a circuit in this way is called **electromagnetic induction.**

How can you tell the direction of the current? To find the force on the charges in the wire, use the **fourth right-hand rule** to hold your right hand so that your thumb points in the direction in which the wire is moving and your fingers point in the direction of the magnetic field. The palm of your hand will point in the direction of the conventional (positive) current, as illustrated in **Figure 25-2.**

Figure 25-2 The fourth right-hand rule can be used to find the direction of the forces on the charges in a conductor that is moving in a magnetic field.

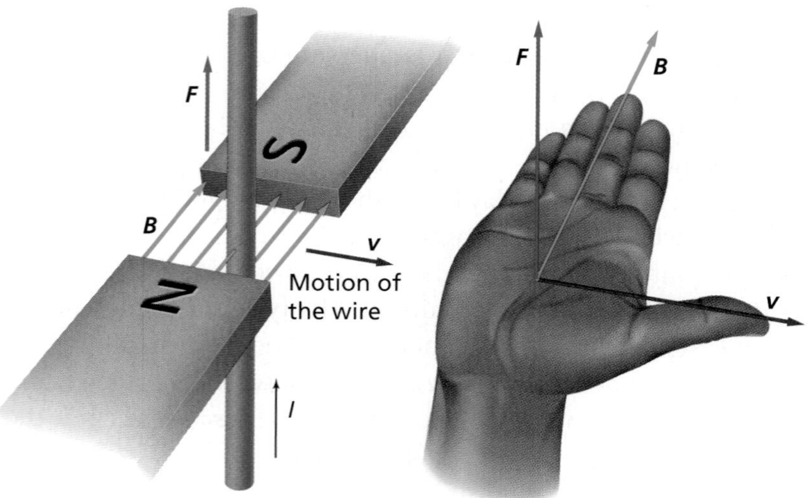

Electromotive Force

When you studied electric circuits, you learned that a source of electrical energy, such as a battery, is needed to produce a continuous current. The potential difference, or voltage, given to the charges by a battery is called the **electromotive force,** or *EMF*. Electromotive force, however, is not actually a force; instead, it is a potential difference and is measured in volts. Thus, the term electromotive force is misleading. Like many other historical terms still in use, it originated before the related principles—in this case, those of electricity—were well understood. The *EMF* is the influence that makes current flow from lower to higher potential, like a water pump in a water fountain.

What created the potential difference that caused an induced current in Faraday's experiment? When you move a wire through a magnetic field, you exert a force on the charges and they move in the direction of the force. Work is done on the charges. Their electrical potential energy, and thus their potential, is increased. The difference in potential is called the induced *EMF*. *EMF* depends on the magnetic field, B, the length of the wire in the magnetic field, L, and the velocity of the wire in the field that is perpendicular to the field, $v(\sin\theta)$.

Electromotive Force $EMF = BLv(\sin\theta)$

Electromotive force is equal to the magnitude of the magnetic field, times the length of the wire times the component of the velocity of the wire in the field perpendicular to the field.

If a wire moves through a magnetic field at an angle to the field, only the component of the wire's velocity that is perpendicular to the direction of the magnetic field generates *EMF*. If the wire moves through the field with a velocity that is exactly perpendicular to the field, then the above equation reduces to $EMF = BLv$, because $\sin 90° = 1$. Checking the units of the *EMF* equation will help you work algebra correctly in related problems. The unit for measuring *EMF* is the volt, V. In Chapter 24, B was defined as F/IL; therefore, the units for B are N/A·m. Velocity is measured in m/s. Using dimensional analysis, $(\text{N/A·m})(\text{m})(\text{m/s}) = \text{N·m/A·s} = \text{J/C} = \text{V}$. Recall from previous chapters that $\text{J} = \text{N·m}$, $\text{A} = \text{C/s}$, and $\text{V} = \text{J/C}$.

Application of induced *EMF* A microphone is a simple application that depends on an induced *EMF*. A dynamic microphone is similar in construction to a loudspeaker. The microphone shown in **Figure 25-3** has a diaphragm attached to a coil of wire that is free to move in a magnetic field. Sound waves vibrate the diaphragm, which moves the coil in the magnetic field. The motion of the coil, in turn, induces an *EMF* across the ends of the coil. The induced *EMF* varies as the frequency of the sound varies. In this way, the sound wave is converted to an electrical signal. The voltage generated is small, typically 10^{-3} V, but it can be increased, or amplified, by electronic devices.

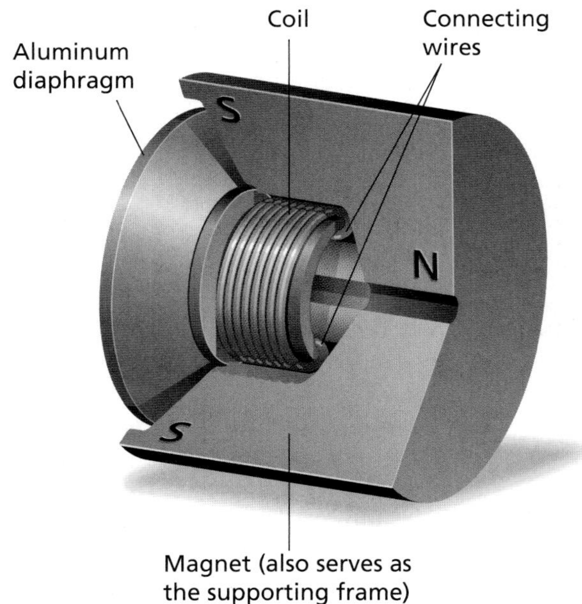

Figure 25-3 In this drawing of a moving coil microphone, the aluminum diaphragm is connected to a coil in a magnetic field. When sound waves vibrate the diaphragm, the coil moves in the magnetic field and generates a current that is proportional to the sound wave.

▶ EXAMPLE **Problem 1**

Induced *EMF* **A straight wire, 0.20 m long, moves at a constant speed of 7.0 m/s perpendicular to a magnetic field of strength 8.0×10^{-2} T.**

a. **What *EMF* is induced in the wire?**

b. **The wire is part of a circuit that has a resistance of 0.50 Ω. What is the current through the wire?**

c. **If a different metal was used for the wire, which has a resistance of 0.78 Ω, what would the new current be?**

1 Analyze and Sketch the Problem

- Establish a coordinate system.
- Draw a straight wire of length *L*. Connect an ammeter to the wire to represent a measurement of current.
- Choose a direction for the magnetic field that is perpendicular to the length of wire.
- Choose a direction for the velocity that is perpendicular to both the length of the wire and the magnetic field.

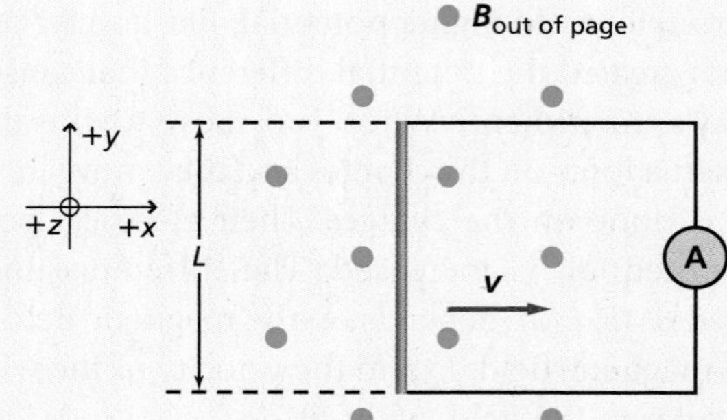

Known:	Unknown:
$v = 7.0$ m/s	$EMF = ?$
$L = 0.20$ m	$I = ?$
$B = 8.0\times10^{-2}$ T	
$R_1 = 0.50\ \Omega$	
$R_2 = 0.78\ \Omega$	

Physics Online
Personal Tutor For an online tutorial on induced EMF, visit physicspp.com.

2 Solve for the Unknown

a. $EMF = BLv$

$= (8.0\times10^{-2}\ \text{T})(0.20\ \text{m})(7.0\ \text{m/s})$ — Substitute $B = 8.0\times10^{-2}$, $L = 0.20$ m, $v = 7.0$ m/s

$= 0.11\ \text{T}\cdot\text{m}^2/\text{s}$

$= 0.11$ V

b. $I = \frac{V}{R}$

$= \frac{EMF}{R}$ — Substitute $V = EMF$

$= \frac{0.11\ \text{V}}{0.50\ \Omega}$ — Substitute $EMF = 0.11$ V, $R_1 = 0.50\ \Omega$

$= 0.22$ A

Using the fourth right-hand rule, the direction of the current is counterclockwise.

c. $I = \frac{EMF}{R}$

$= \frac{0.11\ \text{V}}{0.78\ \Omega}$ — Substitute $EMF = 0.11$ V, $R_2 = 0.78\ \Omega$

$= 0.14$ A

The current is counterclockwise.

3 Evaluate the Answer

- **Are the units correct?** Volt is the correct unit for *EMF*. Current is measured in amperes.
- **Does the direction make sense?** The direction obeys the fourth right-hand rule: *v* is the direction of the thumb, *B* is the same direction as the fingers, and *F* is the direction that the palm faces. Current is in the same direction as the force.
- **Is the magnitude realistic?** The answers are near 10^{-1}. This agrees with the quantities given and the algebra performed.

PRACTICE Problems

- Additional Problems, Appendix B
- Solutions to Selected Problems, Appendix C

1. A straight wire, 0.5 m long, is moved straight up at a speed of 20 m/s through a 0.4-T magnetic field pointed in the horizontal direction.
   a. What *EMF* is induced in the wire?
   b. The wire is part of a circuit of total resistance of 6.0 Ω. What is the current in the circuit?
2. A straight wire, 25 m long, is mounted on an airplane flying at 125 m/s. The wire moves in a perpendicular direction through Earth's magnetic field ($B = 5.0\times10^{-5}$ T). What *EMF* is induced in the wire?
3. A straight wire, 30.0 m long, moves at 2.0 m/s in a perpendicular direction through a 1.0-T magnetic field.
   a. What *EMF* is induced in the wire?
   b. The total resistance of the circuit of which the wire is a part is 15.0 Ω. What is the current?
4. A permanent horseshoe magnet is mounted so that the magnetic field lines are vertical. If a student passes a straight wire between the poles and pulls it toward herself, the current flow through the wire is from right to left. Which is the north pole of the magnet?

Electric Generators

The **electric generator,** invented by Michael Faraday, converts mechanical energy to electrical energy. An electric generator consists of a number of wire loops placed in a strong magnetic field. The wire is wound around an iron core to increase the strength of the magnetic field. The iron and wires are called the armature, which is similar to that of an electric motor.

The armature is mounted so that it can rotate freely in the magnetic field. As the armature turns, the wire loops cut through the magnetic field lines and induce an *EMF*. Commonly called the voltage, the *EMF* developed by the generator depends on the length of wire rotating in the field. Increasing the number of loops in the armature increases the wire length, thereby increasing the induced *EMF*. Note that you could have a length of wire with only part of it in the magnetic field. Only the portion within the magnetic field induces an *EMF*.

Current from a generator When a generator is connected in a closed circuit, the induced *EMF* produces an electric current. **Figure 25-4** shows a single-loop generator without an iron core. The direction of the induced current can be found from the third right-hand rule. As the loop rotates, the strength and the direction of the current change.

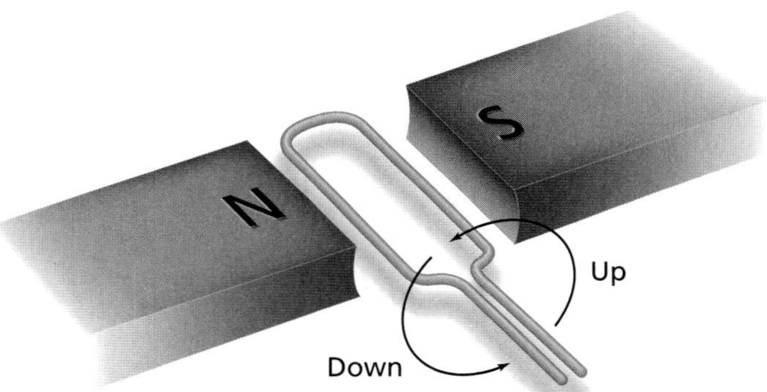

Figure 25-4 An electric current is generated in a wire loop as the loop rotates.

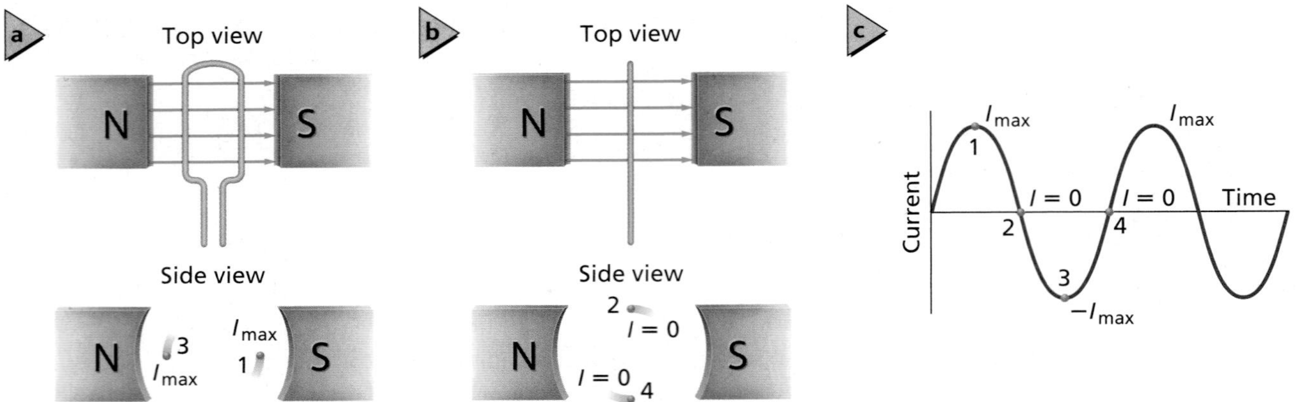

■ **Figure 25-5** The cross-sectional view of a rotating wire loop shows the position of the loop when maximum current is generated **(a).** When the loop is vertical, the current is zero **(b).** The current varies with time as the loop rotates **(c).** The variation of *EMF* with time can be shown with a similar graph.

The current is greatest when the motion of the loop is perpendicular to the magnetic field; that is, when the loop is in the horizontal position, as shown in **Figure 25-5a.** In this position, the component of the loop's velocity perpendicular to the magnetic field is greatest. As the loop rotates from the horizontal to the vertical position, as shown in **Figure 25-5b,** it moves through the magnetic field lines at an ever-increasing angle. Thus, it cuts through fewer magnetic field lines per unit of time, and the current decreases. When the loop is in the vertical position, the wire segments move parallel to the field and the current is zero. As the loop continues to turn, the segment that was moving up begins to move down and reverses the direction of the current in the loop. This change in direction takes place each time the loop turns through 180°. The current changes smoothly from zero to some maximum value and back to zero during each half-turn of the loop. Then it reverses direction. A graph of current versus time is shown in **Figure 25-5c.**

Does the entire loop contribute to the induced *EMF*? Look at **Figure 25-6,** where all four sides of the loop are depicted in the magnetic field. If the fourth right-hand rule is applied to segment *ab,* the direction of the induced current is toward the side of the wire. The same applies to segment *cd*. Thus, no current is induced along the length of the wire in *ab* or *cd*. But in segment *bc*, the direction of the induced current is from *b* to *c*, and in segment *ad*, the current is from *d* to *a*.

Because the conducting loop is rotating in a circular motion, the relative angle between a point on the loop and the magnetic field constantly changes. The electromotive force can be calculated by the electromotive force equation given earlier, $EMF = BLv(\sin\,\theta)$, except that L is now the length of segment *bc*. The maximum voltage is induced when a conductor is moving perpendicular to the magnetic field and thus $\theta = 90°$.

■ **Figure 25-6** Only the segments *bc* and *ad* have current induced through them. This can be shown using the fourth right-hand rule.

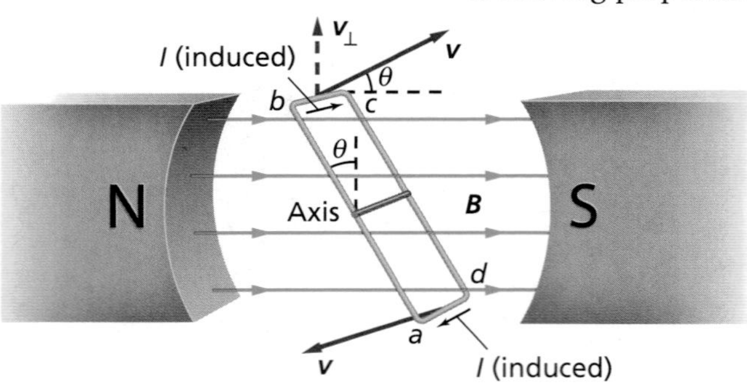

Generators, such as those in the chapter opening image, work in a similar fashion. Potential energy from water stored behind a dam is converted to kinetic energy, which spins the turbines. The turbines, in turn, turn coils of conductors in a magnetic field, thereby inducing an *EMF*. Generators and motors are almost identical in construction, but

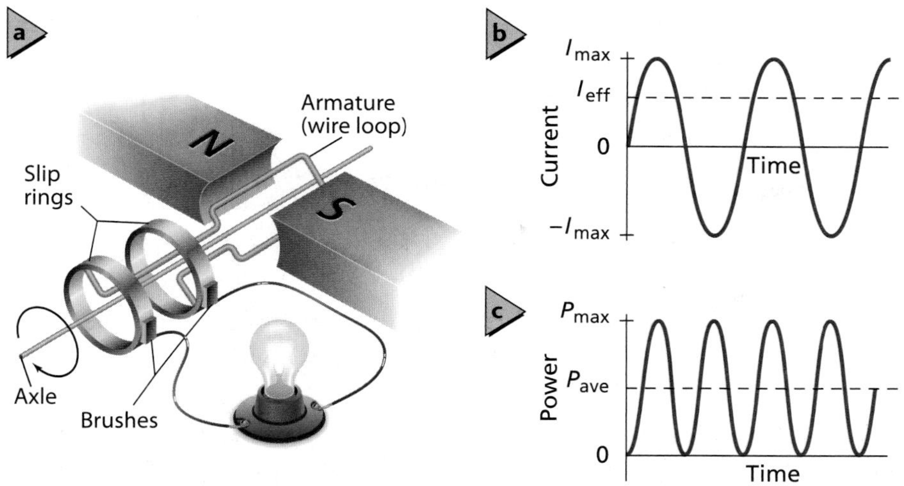

Figure 25-7 An alternating-current generator transmits current to an external circuit by way of a brush-slip-ring arrangement **(a).** The alternating current produced varies with time **(b).** The resulting power is always positive and also is sinusoidal **(c).**

they convert energy in opposite directions. A generator converts mechanical energy to electrical energy, while a motor converts electrical energy to mechanical energy.

Alternating-Current Generators

An energy source turns the armature of a generator in a magnetic field at a fixed number of revolutions per second. In the United States, electric utilities use a 60-Hz frequency, in which the current goes from one direction to the other and back to the first 60 times per second. **Figure 25-7a** shows how an alternating current, AC, in an armature is transmitted to the rest of the circuit. The brush-slip-ring arrangement permits the armature to turn freely while still allowing the current to pass into the external circuit. As the armature turns, the alternating current varies between some maximum value and zero, as shown in the graph in **Figure 25-7b.**

Average power The power produced by a generator is the product of the current and the voltage. Because both current and voltage vary, the power associated with an alternating current varies. **Figure 25-7c** shows a graph of the power produced by an AC generator. Note that power is always positive because I and V are either both positive or both negative. **Average power,** P_{AC}, is half the maximum power; thus, $P_{AC} = \frac{1}{2}P_{AC\ max}$.

Effective voltage and current It is common to describe alternating current and voltage in terms of effective current and voltage, rather than referring to their maximum values. Recall from Chapter 22 that $P = I^2R$. Thus, you can express effective current, I_{eff}, in terms of the average AC power as $P_{AC} = I_{eff}^2R$. To determine I_{eff} in terms of maximum current, I_{max}, start with the power relationship, $P_{AC} = \frac{1}{2}P_{AC\ max}$, and substitute in I^2R. Then solve for I_{eff}.

Effective Current $I_{eff} = \frac{\sqrt{2}}{2} I_{max} = 0.707\ I_{max}$

Effective current is equal to $\frac{\sqrt{2}}{2}$ times the maximum current.

Similarly, the following equation can be used to express effective voltage.

Effective Voltage $V_{eff} = \left(\frac{\sqrt{2}}{2}\right)V_{max} = 0.707\ V_{max}$

Effective voltage is equal to $\frac{\sqrt{2}}{2}$ times the maximum voltage.

Effective voltage, also is commonly referred to as RMS (root mean square) voltage. In the United States, the voltage generally available at wall outlets is described as 120 V, where 120 V is the magnitude of the effective voltage, not the maximum voltage. The frequency and effective voltage that are used vary in different countries.

PRACTICE Problems

• Additional Problems, Appendix B
• Solutions to Selected Problems, Appendix C

5. A generator develops a maximum voltage of 170 V.
   a. What is the effective voltage?
   b. A 60-W lightbulb is placed across the generator with an I_{max} of 0.70 A. What is the effective current through the bulb?
   c. What is the resistance of the lightbulb when it is working?
6. The RMS voltage of an AC household outlet is 117 V. What is the maximum voltage across a lamp connected to the outlet? If the RMS current through the lamp is 5.5 A, what is the maximum current in the lamp?
7. An AC generator delivers a peak voltage of 425 V.
   a. What is the V_{eff} in a circuit placed across the generator?
   b. The resistance is $5.0\times10^2\ \Omega$. What is the effective current?
8. If the average power dissipated by an electric light is 75 W, what is the peak power?

In this section you have explored how moving wires in magnetic fields can induce current. However, as Faraday discovered, changing magnetic fields around a conductor also can induce current in the conductor. In the next section, you will explore changing magnetic fields and the applications of induction by changing magnetic fields.

25.1 Section Review

9. **Generator** Could you make a generator by mounting permanent magnets on a rotating shaft and keeping the coil stationary? Explain.
10. **Bike Generator** A bike generator lights the headlamp. What is the source of the energy for the bulb when the rider travels along a flat road?
11. **Microphone** Consider the microphone shown in Figure 25-3. When the diaphragm is pushed in, what is the direction of the current in the coil?
12. **Frequency** What changes to the generator are required to increase the frequency?
13. **Output Voltage** Explain why the output voltage of an electric generator increases when the magnetic field is made stronger. What else is affected by strengthening the magnetic field?
14. **Generator** Explain the fundamental operating principle of an electric generator.
15. **Critical Thinking** A student asks, "Why does AC dissipate any power? The energy going into the lamp when the current is positive is removed when the current is negative. The net is zero." Explain why this reasoning is wrong.

Physics Online physicspp.com/self_check_quiz

25.2 Changing Magnetic Fields Induce *EMF*

In a generator, current is produced when the armature turns through a magnetic field. The act of generating current produces a force on the wires in the armature. In what direction is the force on the wires of an armature?

► **Objectives**

- **Apply** Lenz's law.
- **Explain** back-*EMF* and how it affects the operation of motors and generators.
- **Explain** self-inductance and how it affects circuits.
- **Solve** transformer problems involving voltage, current, and turn ratios.

► **Vocabulary**

Lenz's law
eddy current
self-inductance
transformer
primary coil
secondary coil
mutual inductance
step-up transformer
step-down transformer

Lenz's Law

Consider a section of one loop that moves through a magnetic field, as shown in **Figure 25-8a.** An *EMF*, equal to BLv, will be induced in the wire. If the magnetic field is out of the page and velocity is to the right, then the fourth right-hand rule shows a downward *EMF*, as illustrated in **Figure 25-8b,** and consequently a downward current is produced. In Chapter 24, you learned that a wire carrying a current through a magnetic field will experience a force acting on it. This force results from the interaction between the existing magnetic field and the magnetic field generated around all currents. To determine the direction of this force, use the third right-hand rule: if current, I, is down and the magnetic field, B, is out, then the resulting force is to the left, as shown in **Figure 25-8c.** This means that the direction of the force on the wire opposes the original motion of the wire, v. That is, the force acts to slow down the rotation of the armature. The method of determining the direction of a force was first demonstrated in 1834 by H.F.E. Lenz and is therefore called Lenz's law.

Lenz's law states that the direction of the induced current is such that the magnetic field resulting from the induced current opposes the change in the field that caused the induced current. Note that it is the change in the field and not the field itself that is opposed by the induced magnetic effects.

Opposing change **Figure 25-9,** on the next page, is an example of how Lenz's law works. The north pole of a magnet is moved toward the left end of a coil of wire. To create a force that will oppose the approach of the north pole, the left end of the coil also must become a north pole. In other words, the magnetic field lines must emerge from the left end of the coil. Using the second right-hand rule, which you learned in Chapter 24, you will see that if Lenz's law is correct, the induced current must be in a counterclockwise direction as viewed from the end of the coil where the magnet is inserted. Experiments have shown that this is so. If the magnet is turned so that a south pole approaches the coil, the induced current will flow in a clockwise direction.

Figure 25-8 A wire, length L, moving through a magnetic field, B, induces an electromotive force. If the wire is part of a circuit, then there will be a current, I. This current will interact with the magnetic field and produce a force, F. Notice that the resulting force opposes the motion, v, of the wire.

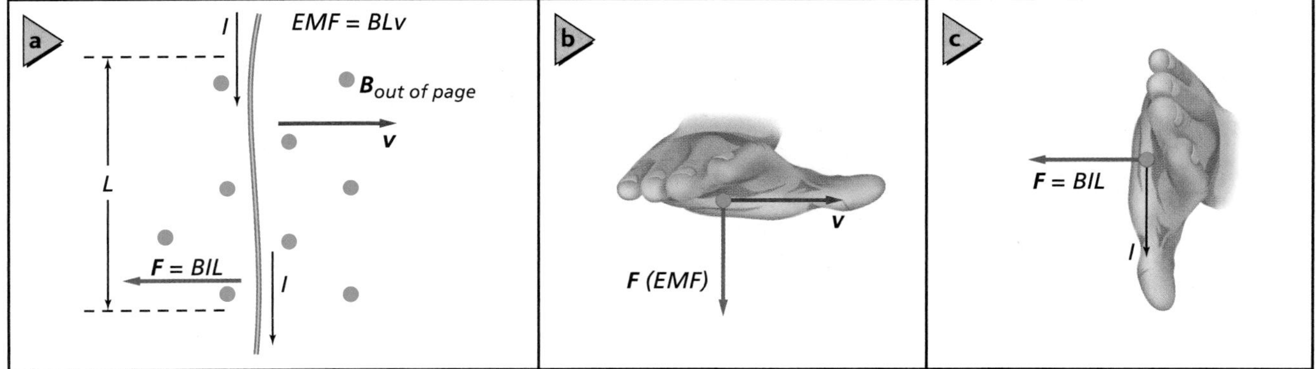

■ **Figure 25-9** The magnet approaching the coil causes an induced current to flow. Lenz's law predicts the direction of flow shown.

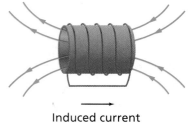

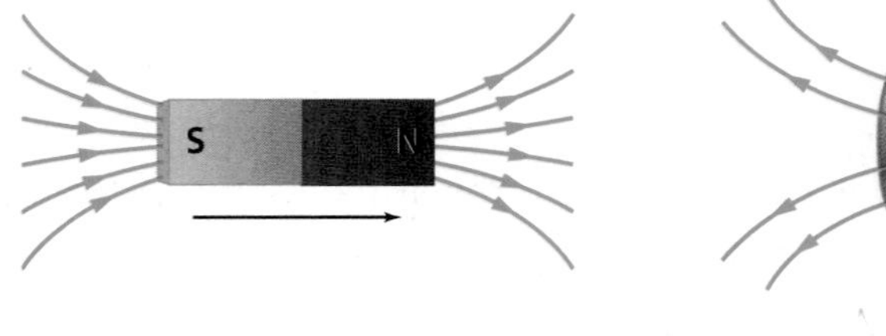

If a generator produces only a small current, then the opposing force on the armature will be small, and the armature will be easy to turn. If the generator produces a larger current, the force on the larger current will be greater, and the armature will be more difficult to turn. A generator supplying a large current is producing a large amount of electric energy. The opposing force on the armature means that mechanical energy must be supplied to the generator to produce the electric energy, consistent with the law of conservation of energy.

Motors and Lenz's law Lenz's law also applies to motors. When a current-carrying wire moves in a magnetic field, an *EMF* is generated. This *EMF*, called the back-*EMF*, is in a direction that opposes the current. When a motor is first turned on, there is a large current because of the low resistance of the motor. As the motor begins to turn, the motion of the wires across the magnetic field induces a back-*EMF* that opposes the current. Therefore, the net current through the motor is reduced. If a mechanical load is placed on the motor, as in a situation in which work is being done to lift a weight, the rotation of the motor will slow. This slowing down will decrease the back-*EMF*, which will allow more current through the motor. Note that this is consistent with the law of conservation of energy: if current increases, so does the rate at which electric power is being sent to the motor. This power is delivered in mechanical form to the load. If the mechanical load stops the motor, current can be so high that wires overheat.

As current draw varies with the changing speed of an electric motor, the voltage drop across the resistance of the wires supplying the motor also varies. Another device, such as a lightbulb, that is in parallel with the motor, also would experience the drop in voltage. This is why you may have noticed some lights in a house dimming when a large motorized appliance, such as an air conditioner or a table saw, starts operating.

■ **Figure 25-10** Sensitive balances use eddy-current damping to control oscillations of the balance beam **(a).** As the metal plate on the end of the beam moves through the magnetic field, a current is generated in the metal. This current, in turn, produces a magnetic field that opposes the motion that caused it, and the motion of the beam is dampened **(b).**

Concepts In Motion
Interactive Figure To see an animation on the application of Lenz's Law, visit physicspp.com.

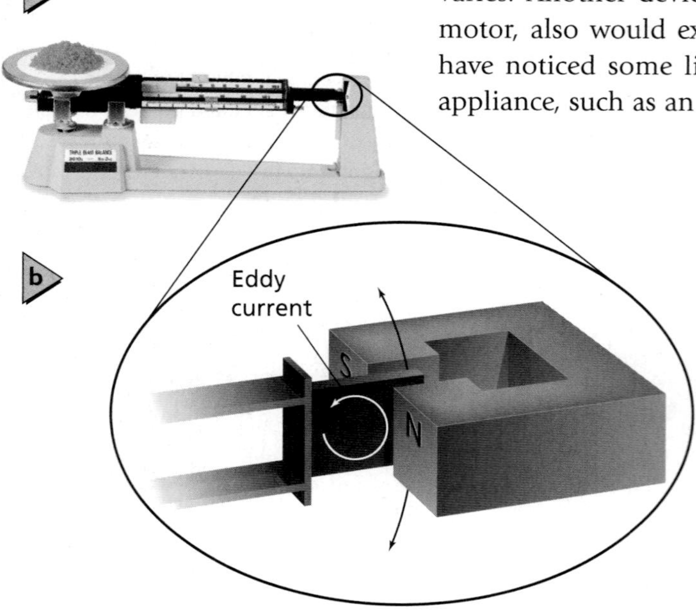

When the current to the motor is interrupted by a switch in the circuit being turned off or by the motor's plug being pulled from a wall outlet, the sudden change in the magnetic field generates a back-*EMF*. This reverse voltage can be large enough to cause a spark across the switch or between the plug and the wall outlet.

Application of Lenz's law A sensitive balance, such as the kind used in laboratories, uses Lenz's law to stop its oscillation when an object is placed on the pan. As shown in **Figure 25-10,** a piece of metal attached to the balance arm is located between the poles of a horseshoe magnet.

When the balance arm swings, the metal moves through the magnetic field. Currents called **eddy currents** are generated in the metal and produce a magnetic field that acts to oppose the motion that caused the currents. Thus, the metal piece is slowed down. The force opposes the motion of the metal in either direction but does not act if the metal is still. Thus, the force does not change the mass read by the balance. This effect is called eddy-current damping. A practical motor or transformer core is constructed from thin laminations, or layers, each one insulated from the other, to reduce the circulation of eddy currents.

Eddy currents are generated when a piece of metal moves through a magnetic field. The reverse is also true: a current is generated when a metal loop is placed in a changing magnetic field. According to Lenz's law, the current generated will oppose the changing magnetic field. The current generates a magnetic field of its own in the opposite direction that causes the uncut, aluminum ring in **Figure 25-11** to float. An AC current is in the coil, so a constantly changing magnetic field is generated. This changing magnetic field induces an *EMF* in the rings. If these rings were constructed from a nonconducting material such as nylon or brass, an *EMF* could not be induced. For the uncut ring, the *EMF* causes a current that produces a magnetic field that will oppose the change in the generating magnetic field. The interaction of these two magnetic fields causes the ring to push away from the coil, similar to the way in which the north poles of two magnets push away from each other. For the lower ring, which has been sawed through, an *EMF* is generated, but no current can result because of an incomplete path. Hence, no opposing magnetic field is produced by the ring.

Figure 25-11 Current is induced in the continuous metal ring, while there is no current in the cut ring.

Self-Inductance

Back-*EMF* can be explained in another way. As Faraday showed, *EMF* is induced whenever a wire cuts the lines of a magnetic field. The current through the wire shown in **Figure 25-12** increases from **Figure 25-12a** to **Figure 25-12c.** The current generates a magnetic field, shown by magnetic field lines. As the current and magnetic field increase, new lines are created. As more lines are added, they cut through the coil wires and generate an *EMF* to oppose the current changes. The *EMF* will make the potential of the top of the coil more negative than the bottom. This induction of *EMF* in a wire carrying changing current is called **self-inductance.**

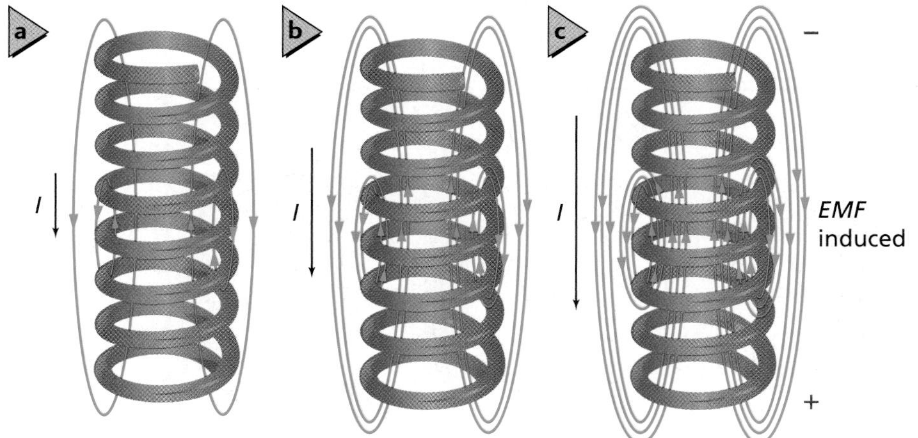

Figure 25-12 As the current in the coil increases from **(a)** on the left to **(c)** on the right, the magnetic field generated by the current also increases. This increase in the magnetic field produces an *EMF* that opposes the current direction.

MINI LAB

Motor and Generator

Motors and generators differ, mainly in the way they convert energy—electric to mechanical compared to mechanical to electric.

1. Make a series circuit with an efficient DC motor, a miniature lamp, and an ammeter.

2. Rotate the handle, or motor shaft, to try to light the lamp.

Analyze and Conclude

3. What happens if you vary the speed at which you rotate the handle?

4. Predict what will happen if you connect your motor to a second motor.

The size of the induced *EMF* is proportional to the rate at which field lines cut through the wires. The faster the current is changed, the larger the opposing *EMF*. If the current reaches a steady value, the magnetic field is constant, and the *EMF* is zero. When the current is decreased, an *EMF* is generated that tends to prevent the reduction in the magnetic field and current. Because of self-inductance, work has to be done to increase the current flowing through the coil. Energy is stored in the magnetic field. This is similar to the way in which a charged capacitor stores energy in the electric field between its plates.

Transformers

Transformers are used to increase or decrease AC voltages. Usage of transformers is common because they change voltages with relatively little loss of energy. In fact, many of the devices in your home, such as game systems, printers, and stereos, have transformers inside their casings or as part of their cords.

How transformers work Self-inductance produces an *EMF* when current changes in a single coil. A transformer has two coils, electrically insulated from each other, but wound around the same iron core. One coil is called the **primary coil.** The other coil is called the **secondary coil.** When the primary coil is connected to a source of AC voltage, the changing current creates a changing magnetic field, which is carried through the core to the secondary coil. In the secondary coil, the changing field induces a varying *EMF*. This effect is called **mutual inductance.**

The *EMF* induced in the secondary coil, called the secondary voltage, is proportional to the primary voltage. The secondary voltage also depends on the ratio of the number of turns on the secondary coil to the number of turns on the primary coil, as shown by the following expressions.

$$\frac{\text{secondary voltage}}{\text{primary voltage}} = \frac{\text{number of turns on secondary coil}}{\text{number of turns on primary coil}}$$

$$\frac{V_s}{V_p} = \frac{N_s}{N_p}$$

If the secondary voltage is larger than the primary voltage, the transformer is called a **step-up transformer,** as shown in **Figure 25-13a.** If the voltage coming out of the transformer is smaller than the voltage put in, then it is called a **step-down transformer,** as shown in **Figure 25-13b.**

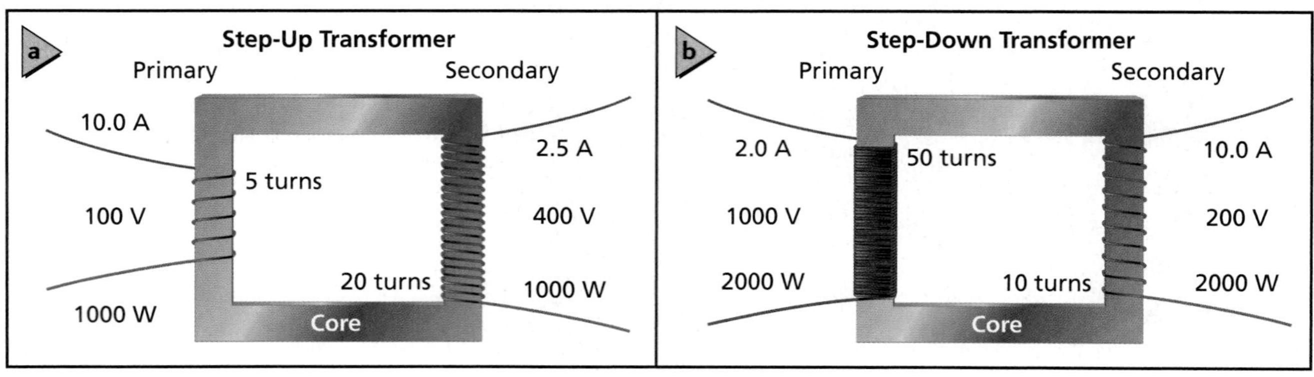

Figure 25-13 In a transformer, the ratio of input voltage to output voltage depends upon the ratio of the number of turns on the primary coil to the number of turns on the secondary coil. The output voltage can be the same as the input, greater than the input **(a),** or less than the input **(b)**.

In an ideal transformer, the electric power delivered to the secondary circuit equals the power supplied to the primary circuit. An ideal transformer dissipates no power itself, and can be represented by the following equations:

$$P_p = P_s$$

$$V_pI_p = V_sI_s$$

Rearranging the equation to find the ratio V_p/V_s shows that the current in the primary circuit depends on how much current is required by the secondary circuit. This relationship can be combined with the relationship shown earlier between voltage and the number of turns to result in the following.

Transformer Equation $\frac{I_s}{I_p} = \frac{V_p}{V_s} = \frac{N_p}{N_s}$

The ratio of the current in the secondary coil to the current in the primary coil is equal to the ratio of the voltage in the primary coil to the secondary coil, which is also equal to the ratio of the number of turns on the voltage in the primary coil to the number of turns on the secondary coil.

Figure 25-14 If the input voltage is connected to the coils on the left, where there is a larger number of turns, the transformer functions as a step-down transformer. If the input voltage is connected at the right, the transformer functions as a step-up transformer.

As mentioned previously, a step-up transformer increases voltage. Because transformers cannot increase the power output, there must be a corresponding decrease in current through the secondary circuit. Similarly, in a step-down transformer, the current is greater in the secondary circuit than it is in the primary circuit. A voltage decrease corresponds to a current increase, as shown in the Connecting Math to Physics. Another way to understand this is to consider a transformer as 100 percent efficient, as is typically assumed in industry. Therefore, in most cases, it may be assumed that the input power and the output power are the same. Figure 25-13 illustrates the principles of step-up and step-down transformers. As shown in **Figure 25-14,** some transformers can function either as step-up transformers or step-down transformers, depending on how they are hooked up.

APPLYING **PHYSICS**

Common Units Transformers typically are rated in volt-amps reactive (VA, kilo-VA, mega-VA). Technically, only pure resistive-type loads have their power expressed in watts and reactive loads in volt-amps.

Connecting **Math to Physics**

Inequalities Study the following expressions to help you understand the relationships among voltage, current, and the number of coils in step-up and step-down transformers.

Step-Up Transformer	Step-Down Transformer
$V_p < V_s$	$V_p > V_s$
$I_p > I_s$	$I_p < I_s$
$N_p < N_s$	$N_p > N_s$

▶ EXAMPLE Problem 2

Step-Up Transformers A step-up transformer has a primary coil consisting of 200 turns and a secondary coil consisting of 3000 turns. The primary coil is supplied with an effective AC voltage of 90.0 V.

a. What is the voltage in the secondary circuit?

b. The current in the secondary circuit is 2.0 A. What is the current in the primary circuit?

1 Analyze and Sketch the Problem

- Draw an iron core with turns of wire.
- Label the variables I, V, and N.

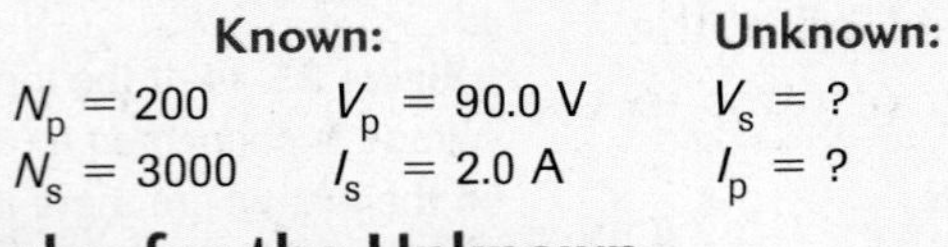

Known:		Unknown:
$N_p = 200$	$V_p = 90.0$ V	$V_s = ?$
$N_s = 3000$	$I_s = 2.0$ A	$I_p = ?$

2 Solve for the Unknown

a. Solve for V_s.

$$\frac{V_s}{V_p} = \frac{N_s}{N_p}$$

$$V_s = \frac{N_s V_p}{N_p}$$

$$= \frac{(3000)(90.0\text{ V})}{200}$$ **Substitute $N_s = 3000$, $V_p = 90.0$ V, $N_p = 200$**

$$= 1350\text{ V}$$

b. The power in the primary and secondary circuits are equal assuming 100 percent efficiency.

$$P_p = P_s$$

$$V_p I_p = V_s I_s$$ **Substitute $P_p = V_p I_p$, $P_s = V_s I_s$**

Solve for I_p.

$$I_p = \frac{V_s I_s}{V_p}$$

$$= \frac{(1350\text{ V})(2.0\text{ A})}{90.0\text{ V}}$$ **Substitute $V_s = 1350$ V, $I_s = 2.0$ A, $V_p = 90.0$ V**

$$= 3.0\times10^{1}\text{ A}$$

Math Handbook
Significant Digits
pages 833–834

3 Evaluate the Answer

- **Are the units correct?** Voltage should be in volts and current in amps.
- **Is the magnitude realistic?** A large step-up ratio of turns results in a large secondary voltage yet a smaller secondary current. The answers agree with this.

▶ PRACTICE Problems

• Additional Problems, Appendix B
• Solutions to Selected Problems, Appendix C

For the following problems, effective currents and voltages are indicated.

16. A step-down transformer has 7500 turns on its primary coil and 125 turns on its secondary coil. The voltage across the primary circuit is 7.2 kV. What voltage is being applied across the secondary circuit? If the current in the secondary circuit is 36 A, what is the current in the primary circuit?

17. A step-up transformer has 300 turns on its primary coil and 90,000 turns on its secondary coil. The *EMF* of the generator to which the primary circuit is attached is 60.0 V. What is the *EMF* in the secondary circuit? The current in the secondary circuit is 0.50 A. What current is in the primary circuit?

CHALLENGE PROBLEM

A distribution transformer (T_1) has its primary coil connected to a 3.0-kV AC source. The secondary coil is connected to the primary coil of a second transformer (T_2) by copper conductors. Finally, the secondary coil of transformer T_2 connects to a load that uses 10.0 kW of power. Transformer T_1 has a turn ratio of 5:1, and T_2 has a load voltage of 120 V. The transformer efficiencies are 100.0 percent and 97.0 percent, respectively.

1. Calculate the load current.
2. How much power is being dissipated by transformer T_2?
3. What is the secondary current of transformer T_1?
4. How much current is the AC source supplying to T_1?

Everyday uses of transformers As you learned in Chapter 22, long-distance transmission of electrical energy is economical only if low currents and very high voltages are used. Step-up transformers are used at power sources to develop voltages as high as 480,000 V. High voltages reduce the current required in the transmission lines, keeping the energy lost to resistance low. When the energy reaches the consumer, step-down transformers, such as those shown in **Figure 25-15,** provide appropriately low voltages for consumer use.

Transformers in home appliances further adjust voltages to useable levels. If you have ever had to charge a toy or operate a personal electronic device, you probably had to plug a large "block" into the wall outlet. A transformer of the type discussed in this chapter is contained inside of that block. In this case, it is probably reducing the household voltage of about 120 V to something in the 3-V to 26-V range.

Not all transformers are step-up or step-down. Transformers can be used to isolate one circuit from another. This is possible because the wire of the primary coil never makes direct contact with the wire of the secondary coil. This type of transformer would most likely be found in some small electronic devices.

Figure 25-15 Step-down transformers are used to reduce the high voltages in transmission lines to levels appropriate for consumers at the points of use.

25.2 Section Review

18. **Coiled Wire and Magnets** You hang a coil of wire with its ends joined so that it can swing easily. If you now plunge a magnet into the coil, the coil will swing. Which way will it swing relative to the magnet and why?

19. **Motors** If you unplugged a running vacuum cleaner from a wall outlet, you would be much more likely to see a spark than you would be if you unplugged a lighted lamp from the wall. Why?

20. **Transformers and Current** Explain why a transformer may only be operated on alternating current.

21. **Transformers** Frequently, transformer coils that have only a few turns are made of very thick (low-resistance) wire, while those with many turns are made of thin wire. Why?

22. **Step-Up Transformers** Refer to the step-up transformer shown in Figure 25-13. Explain what will happen to the primary current if the secondary coil is short-circuited.

23. **Critical Thinking** Would permanent magnets make good transformer cores? Explain.

PHYSICS LAB

Induction and Transformers

A transformer is an electric device without any moving components. It is made of two electric circuits interlinked by a magnetic field. A transformer is used to increase or decrease an AC potential difference, which often is called *voltage.* Transformers can be found everywhere. Every electronic device that plugs into your household electric circuits incorporates a transformer, usually to lower the voltage going to the device. Televisions that have standard cathode-ray picture tubes incorporate high-voltage transformers, which raise the standard household voltage to tens of thousands of volts. This accelerates electrons from the rear of the tube to the screen. In this experiment, you will use two coils with a removable iron core. One coil is called the primary coil, the other the secondary coil. When an AC voltage is applied to the primary coil, the changing magnetic field induces a current, and thus, a voltage in the secondary coil. This induced voltage is expressed by $V_s/V_p = N_s/N_p$, where N refers to the number of turns in the coils.

QUESTION

What is the relationship between voltages in the two coils of a transformer?

Objectives

- **Describe** how a transformer works.
- **Observe** the effect of DC voltage on a transformer.
- **Observe** the effect of AC voltage on a transformer.

Safety Precautions

Materials

primary and secondary coil apparatus
small AC power supply
AC voltmeter
DC power supply (0-6V, 0-5A)
connecting wires with alligator clips
small lightbulb with wires

Procedure

1. Estimate the number of coils of wire on the primary and secondary coils. Do this by counting the number of coils in 1 cm and multiplying by the coil's length in centimeters. The primary coil has one layer. The secondary coil has two layers of wire, so double the value for it. Record your results in the data table.
2. Place a small lightbulb across the contacts of the secondary coil. Carefully place the secondary coil into the primary coil. Slowly insert the iron core into the center of the secondary coil.
3. Attach two wires to the output of the DC power supply. Attach the positive wire from the power supply to one of the primary connections. Turn the power supply to nearly its maximum output setting. Holding the free end of the wire attached to the negative connection, gently tap its end to the other primary coil connection. Observe the area where you touch the wire to the connection. Record your observations in the data table.

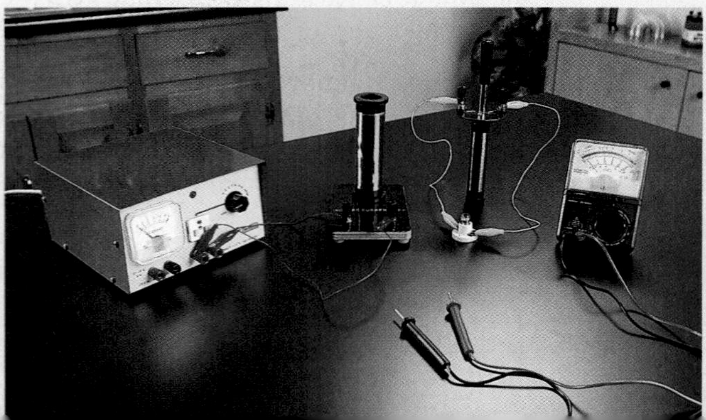

Data Table	
Number of primary coils	
Number of secondary coils	
Step 3 observation	
Step 4 observation	
Step 5 observation	
Step 6 observation	
Step 7 coil volts (V)	
Step 8 observation	
Step 9 iron core	

4. Observe the lightbulb while you are gently tapping the connection. What happens as the wire makes contact and then breaks the electric contact? Record your observation in the data table.
5. Hold the negative wire to the primary coil connection for 5 s and observe the lightbulb. Record your observation in the data table.
6. Disconnect the DC power supply and put it away while leaving the small lightbulb attached to the secondary coil. Attach the AC power supply to the two primary coil connections. Plug in the AC power supply and observe the lightbulb. Record your observations in the data table.
7. Select the AC scale for your voltmeter. Insert the probes into the voltmeter and carefully touch them to the primary coil and measure the applied voltage. Move the probe from the primary coil and measure the secondary coil voltage. Record both readings in the data table.
8. Repeat step 7, but slowly remove the iron core from the secondary coil. What happens to the lightbulb? Measure both primary and secondary coil voltages while the core is being removed. Record your observations in the data table.
9. Carefully feel the iron core. What is your observation? Record it in the data table.

Analyze

1. Calculate the ratio of N_s/N_p from your data.
2. Calculate the ratio of V_s/V_p from your data.
3. **Interpret Data** How do the ratios N_s/N_p and V_s/V_p compare?
4. **Recognize Cause and Effect** Based on the data for step 7, is this transformer a step-up or a step-down transformer? What evidence do you have to support this conclusion?

Conclude and Apply

1. **Infer** How can you explain your observation of the lightbulb in step 4?
2. **Infer** How can you explain the phenomena you observed at the negative connection of the primary coil in step 3?
3. **Infer** How can you explain your observations of the primary and the secondary coil voltages as you removed the iron core in step 8?
4. **Explain** Explain the temperature of the iron core you observed in step 9.

Going Further

Why does the transformer work only with alternating and not direct current?

Real-World Physics

Discuss the use of transformers to assist in the delivery of electricity from the power plant to your home.

To find out more about induction and transformers, visit the Web site: physicspp.com

How it Works How a Credit-Card Reader Works

Credit cards have revolutionized the world's economies by enabling money to be transferred quickly and easily. The credit-card reader, which captures data from a magnetic strip on the back of a card, is one of the most important links in the electronic transfer of money.

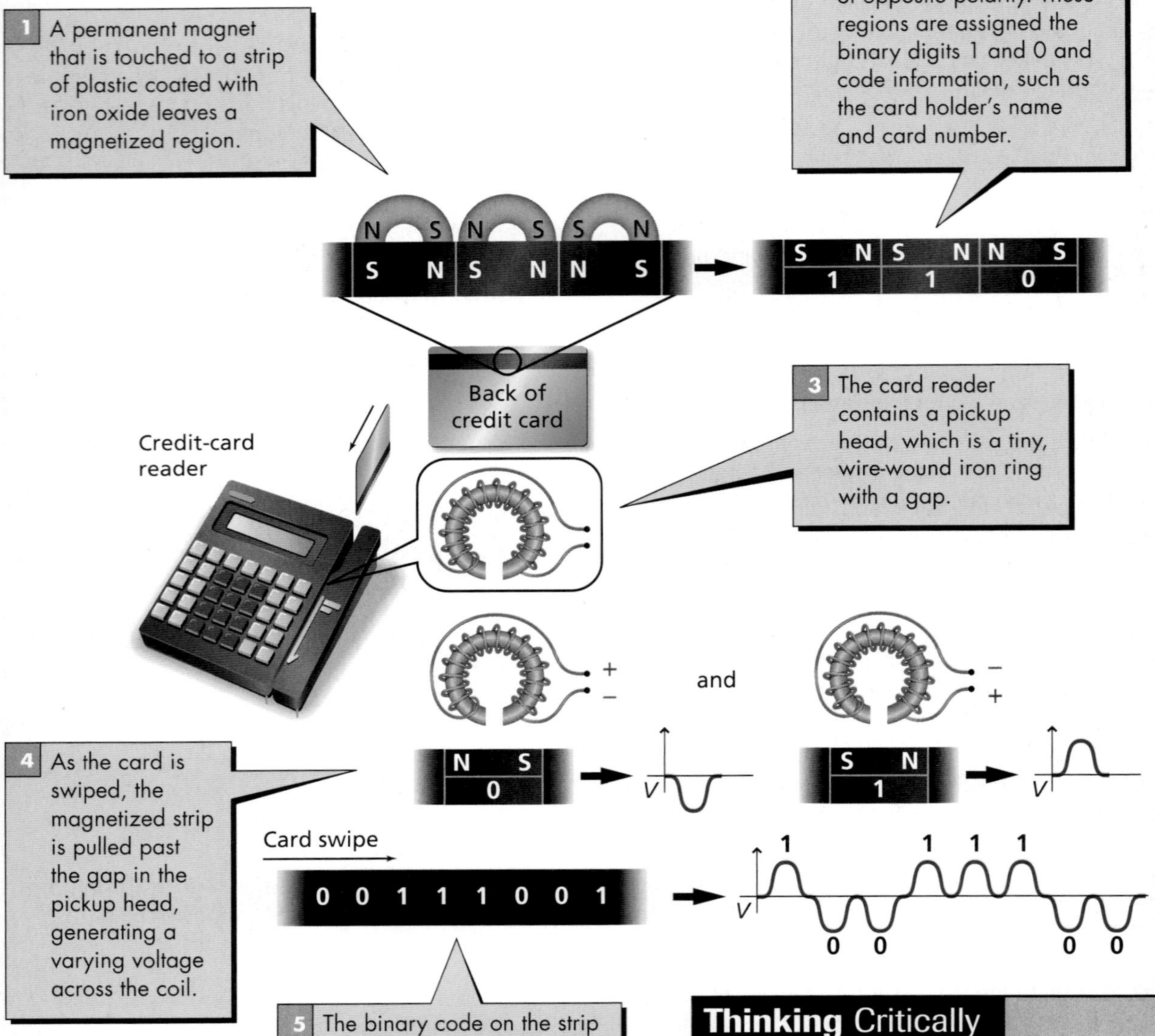

1 A permanent magnet that is touched to a strip of plastic coated with iron oxide leaves a magnetized region.
2 The magnet can be turned around to produce regions of opposite polarity. These regions are assigned the binary digits 1 and 0 and code information, such as the card holder's name and card number.
N S N S S N
S N S N N S
S N S N N S
1 1 0
Back of credit card
Credit-card reader
3 The card reader contains a pickup head, which is a tiny, wire-wound iron ring with a gap.
+ −
and
− +
N S
0
V
S N
1
V
4 As the card is swiped, the magnetized strip is pulled past the gap in the pickup head, generating a varying voltage across the coil.
Card swipe
0 0 1 1 1 0 0 1
1 1 1 1
V
0 0 0 0
5 The binary code on the strip is translated into a voltage waveform. The resulting waveform is stored in a computer's memory and then transmitted to the bank's verification office.

Thinking Critically

1. **Observe** Why is the order of the binar numbers in step 5 reversed from the magnetic strip to the voltage waveform
2. **Analyze** What would happen if the tiny iron ring did not have a gap in it?

Chapter

25 Study Guide

25.1 Electric Current from Changing Magnetic Fields

Vocabulary

- electromagnetic induction (p. 672)
- fourth right-hand rule (p. 672)
- electromotive force (p. 673)
- electric generator (p. 675)
- average power (p. 677)

Key Concepts

- Michael Faraday discovered that if a wire moves through a magnetic field, an electric current can flow.
- The current produced depends upon the angle between the velocity of the wire and the magnetic field. Maximum current occurs when the wire is moving at right angles to the field.
- Electromotive force, *EMF*, is the potential difference created across the moving wire. *EMF* is measured in volts.
- The *EMF* in a straight length of wire moving through a uniform magnetic field is the product of the magnetic field, B, the length of the wire, L, and the component of the velocity of the moving wire through the field that is perpendicular to the field, $v(\sin \theta)$.

$$EMF = BLv \sin \theta$$

- Effective current and voltage can be used to describe alternating current and voltage.

$$I_{eff} = 0.707\, I_{max}$$

$$V_{eff} = 0.707\, V_{max}$$

- A generator and a motor are similar devices. A generator converts mechanical energy to electric energy, whereas a motor converts electric energy to mechanical energy.

25.2 Changing Magnetic Fields Induce *EMF*

Vocabulary

- Lenz's law (p. 679)
- eddy current (p. 681)
- self-inductance (p. 681)
- transformer (p. 682)
- primary coil (p. 682)
- secondary coil (p. 682)
- mutual inductance (p. 682)
- step-up transformer (p. 682)
- step-down transformer (p. 682)

Key Concepts

- Lenz's law states that an induced current is always produced in a direction such that the magnetic field resulting from the induced current opposes the change in the magnetic field that is causing the induced current.
- Back-*EMF* is created by a current-carrying wire moving in a magnetic field. Back-*EMF* opposes the current.
- Self-inductance is a property of a wire carrying a changing current. The faster the current is changing, the greater the induced *EMF* that opposes that change.
- A transformer has two coils wound about the same core. An AC current through the primary coil induces an alternating *EMF* in the secondary coil. The voltages in alternating-current circuits may be increased or decreased by transformers.

$$\frac{I_s}{I_p} = \frac{V_p}{V_s} = \frac{N_p}{N_s}$$

Chapter 25 Assessment

Concept Mapping

24. Complete the following concept map using the following terms: *generator, back-EMF, Lenz's law.*

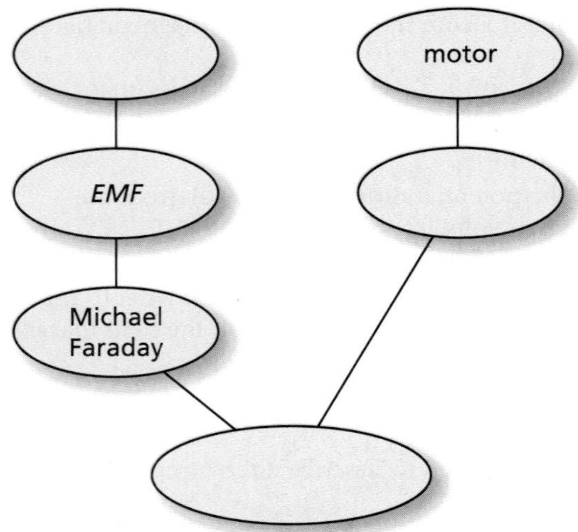

Mastering Concepts

25. What is the armature of an electric generator? (25.1)

26. Why is iron used in an armature? (25.1)

For problems 27–29, refer to ***Figure 25-16.***

27. A single conductor moves through a magnetic field and generates a voltage. In what direction should the wire be moved, relative to the magnetic field to generate the minimum voltage? (25.1)

28. What is the polarity of the voltage induced in the wire when it passes the south pole of the magnetic field? (25.1)

29. What is the effect of increasing the net conductor length in an electric generator? (25.1)

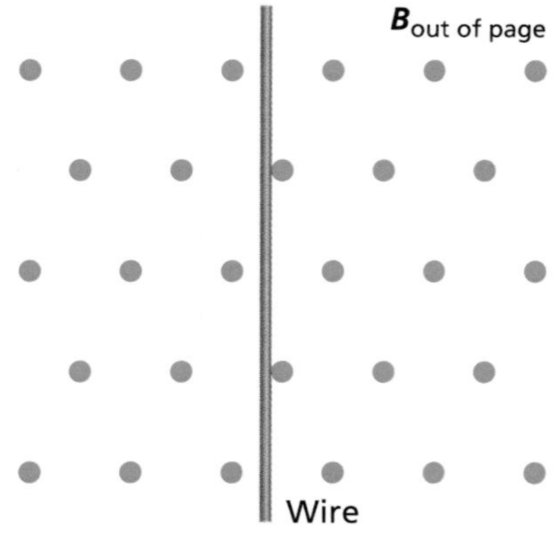

Figure 25-16

30. How were Oersted's and Faraday's results similar? How were they different? (25.1)

31. You have a coil of wire and a bar magnet. Describe how you could use them to generate an electric current. (25.1)

32. What does *EMF* stand for? Why is the name inaccurate? (25.1)

33. What is the difference between a generator and a motor? (25.1)

34. List the major parts of an AC generator. (25.1)

35. Why is the effective value of an AC current less than its maximum value? (25.1)

36. Hydroelectricity Water trapped behind a dam turns turbines that rotate generators. List all the forms of energy that take part in the cycle that includes the stored water and the electricity produced. (25.1)

37. State Lenz's law. (25.2)

38. What causes back-*EMF* in an electric motor? (25.2)

39. Why is there no spark when you close a switch and put current through an inductor, but there is a spark when you open the switch? (25.2)

40. Why is the self-inductance of a coil a major factor when the coil is in an AC circuit but a minor factor when the coil is in a DC circuit? (25.2)

41. Explain why the word *change* appears so often in this chapter. (25.2)

42. Upon what does the ratio of the *EMF* in the primary circuit of a transformer to the *EMF* in the secondary circuit of the transformer depend? (25.2)

Applying Concepts

43. Use unit substitution to show that the units of BLv are volts.

44. When a wire is moved through a magnetic field, does the resistance of the closed circuit affect current only, *EMF* only, both, or neither?

45. Biking As Logan slows his bike, what happens to the *EMF* produced by his bike's generator? Use the term *armature* in your explanation.

46. The direction of AC voltage changes 120 times each second. Does this mean that a device connected to an AC voltage alternately delivers and accepts energy?

 For more problems, go to Additional Problems, Appendix B.

47. A wire is moved horizontally between the poles of a magnet, as shown in **Figure 25-17.** What is the direction of the induced current?

Figure 25-17

48. You make an electromagnet by winding wire around a large nail, as shown in **Figure 25-18.** If you connect the magnet to a battery, is the current larger just after you make the connection or several tenths of a second after the connection is made? Or, is it always the same? Explain.

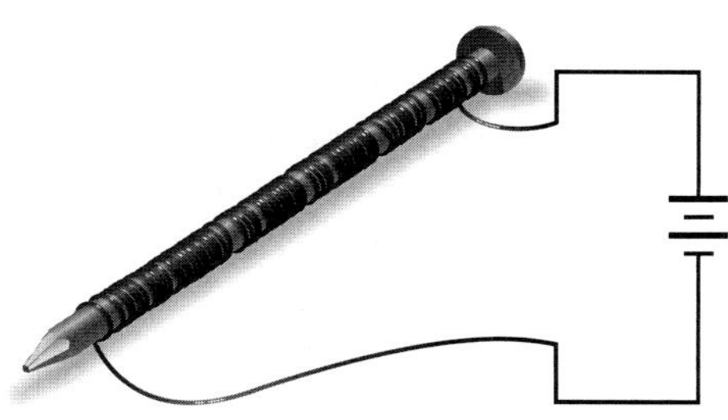

Figure 25-18

49. A segment of a wire loop is moving downward through the poles of a magnet, as shown in **Figure 25-19.** What is the direction of the induced current?

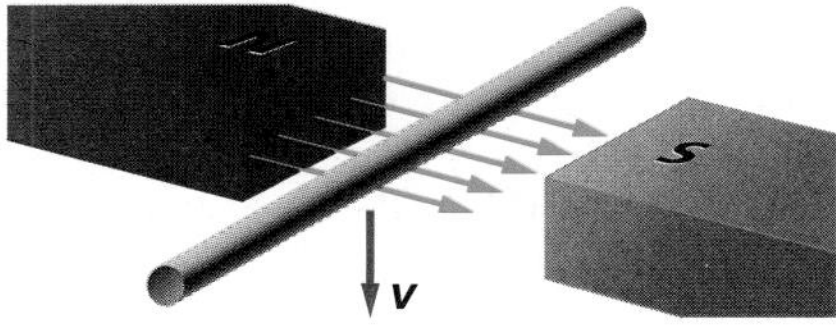

Figure 25-19

50. A transformer is connected to a battery through a switch. The secondary circuit contains a lightbulb, as shown in **Figure 25-20.** Will the lamp be lighted as long as the switch is closed, only at the moment the switch is closed, or only at the moment the switch is opened? Explain.

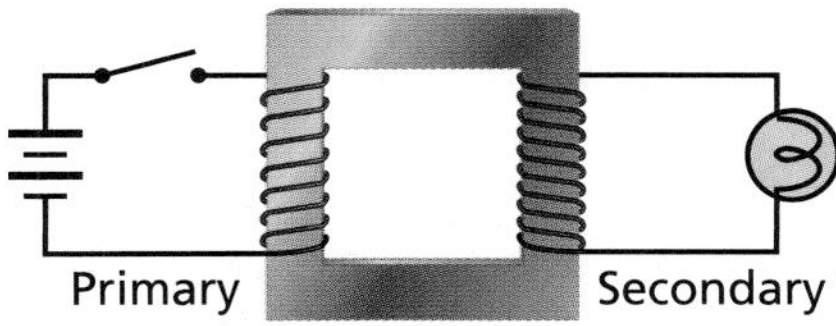

Figure 25-20

51. Earth's Magnetic Field The direction of Earth's magnetic field in the northern hemisphere is downward and to the north as shown in **Figure 25-21.** If an east-west wire moves from north to south, in which direction is the current?

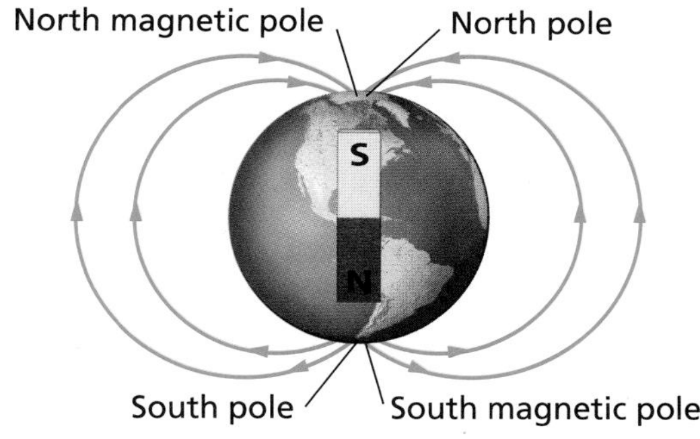

Figure 25-21

52. You move a length of copper wire down through a magnetic field, B, as shown in Figure 25-19.

a. Will the induced current move to the right or left in the wire segment in the diagram?

b. As soon as the wire is moved in the field, a current appears in it. Thus, the wire segment is a current-carrying wire located in a magnetic field. A force must act on the wire. What will be the direction of the force acting on the wire as a result of the induced current?

53. A physics instructor drops a magnet through a copper pipe, as illustrated in **Figure 25-22.** The magnet falls very slowly, and the students in the class conclude that there must be some force opposing gravity.

a. What is the direction of the current induced in the pipe by the falling magnet if the south pole is toward the bottom?

b. The induced current produces a magnetic field. What is the direction of the field?

c. How does this field reduce the acceleration of the falling magnet?

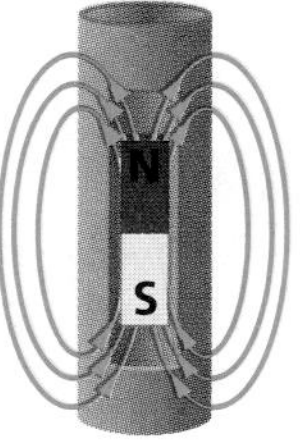

Figure 25-22

54. Generators Why is a generator more difficult to rotate when it is connected to a circuit and supplying current than it is when it is standing alone?

55. Explain why the initial start-up current is so high in an electric motor. Also explain how Lenz's law applies at the instant $t > 0$.

56. Using Figure 25-10 in conjunction with Lenz's law, explain why all practical transformer cores incorporate a laminated core.

57. A practical transformer is constructed with a laminated core that is not a superconductor. Because the eddy currents cannot be completely eliminated, there is always a small core loss. This results, in part, in a net loss of power within the transformer. What fundamental law makes it impossible to bring this loss to zero?

58. Explain the process of mutual induction within a transformer.

59. Shawn drops a magnet, north pole down, through a vertical copper pipe.
   a. What is the direction of the induced current in the copper pipe as the bottom of the magnet passes?
   b. The induced current produces a magnetic field. What is the direction of the induced magnetic field?

Mastering Problems

25.1 Electric Current from Changing Magnetic Fields

60. A wire, 20.0-m long, moves at 4.0 m/s perpendicularly through a magnetic field. An *EMF* of 40 V is induced in the wire. What is the strength of the magnetic field?

61. **Airplanes** An airplane traveling at 9.50×10^2 km/h passes over a region where Earth's magnetic field is 4.5×10^{-5} T and is nearly vertical. What voltage is induced between the plane's wing tips, which are 75 m apart?

62. A straight wire, 0.75-m long, moves upward through a horizontal 0.30-T magnetic field, as shown in **Figure 25-23,** at a speed of 16 m/s.
   a. What *EMF* is induced in the wire?
   b. The wire is part of a circuit with a total resistance of 11 Ω. What is the current?

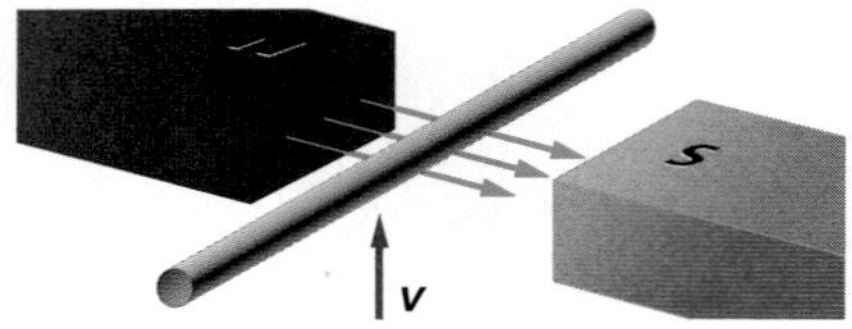

Figure 25-23

63. At what speed would a 0.20-m length of wire have to move across a 2.5-T magnetic field to induce an *EMF* of 10 V?

64. An AC generator develops a maximum *EMF* of 565 V. What effective *EMF* does the generator deliver to an external circuit?

65. An AC generator develops a maximum voltage of 150 V. It delivers a maximum current of 30.0 A to an external circuit.
   a. What is the effective voltage of the generator?
   b. What effective current does the generator deliver to the external circuit?
   c. What is the effective power dissipated in the circuit?

66. **Electric Stove** An electric stove is connected to an AC source with an effective voltage of 240 V.
   a. Find the maximum voltage across one of the stove's elements when it is operating.
   b. The resistance of the operating element is 11 Ω. What is the effective current?

67. You wish to generate an *EMF* of 4.5 V by moving a wire at 4.0 m/s through a 0.050-T magnetic field. How long must the wire be, and what should be the angle between the field and direction of motion to use the shortest wire?

68. A 40.0-cm wire is moved perpendicularly through a magnetic field of 0.32 T with a velocity of 1.3 m/s. If this wire is connected into a circuit of 10.0-Ω resistance, what is the current?

69. You connect both ends of a copper wire with a total resistance of 0.10 Ω to the terminals of a galvanometer. The galvanometer has a resistance of 875 Ω. You then move a 10.0-cm segment of the wire upward at 1.0 m/s through a 2.0×10^{-2}-T magnetic field. What current will the galvanometer indicate?

70. The direction of a 0.045-T magnetic field is 60.0° above the horizontal. A wire, 2.5-m long, moves horizontally at 2.4 m/s.
   a. What is the vertical component of the magnetic field?
   b. What *EMF* is induced in the wire?

71. **Dams** A generator at a dam can supply 375 MW (375×10^6 W) of electrical power. Assume that the turbine and generator are 85 percent efficient.
   a. Find the rate at which falling water must supply energy to the turbine.
   b. The energy of the water comes from a change in potential energy, $PE = mgh$. What is the change in *PE* needed each second?
   c. If the water falls 22 m, what is the mass of the water that must pass through the turbine each second to supply this power?

 For more problems, go to Additional Problems, Appendix B.

72. A conductor rotating in a magnetic field has a length of 20 cm. If the magnetic-flux density is 4.0 T, determine the induced voltage when the conductor is moving perpendicular to the line of force. Assume that the conductor travels at a constant velocity of 1 m/s.

73. Refer to Example Problem 1 and **Figure 25-24** to determine the following.
   a. induced voltage in the conductor
   b. current (I)
   c. direction of flux rotation around the conductor
   d. polarity of point A relative to point B

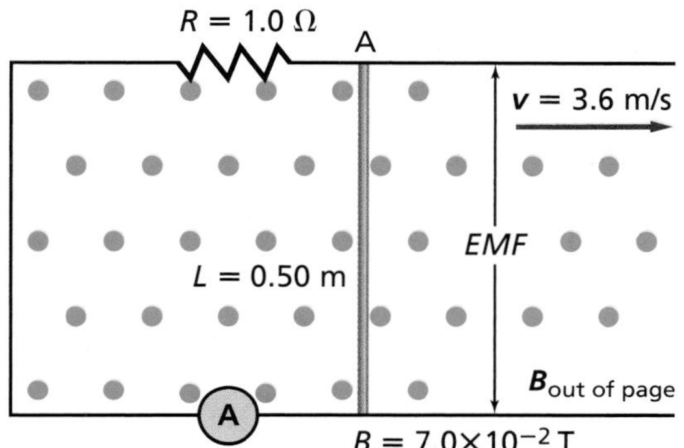

Figure 25-24

25.2 Changing Magnetic Fields Induce *EMF*

74. The primary coil of a transformer has 150 turns. It is connected to a 120-V source. Calculate the number of turns on the secondary coil needed to supply the following voltages.
   a. 625 V
   b. 35 V
   c. 6.0 V

75. A step-up transformer has 80 turns on its primary coil and 1200 turns on its secondary coil. The primary circuit is supplied with an alternating current at 120 V.
   a. What voltage is being applied across the secondary circuit?
   b. The current in the secondary circuit is 2.0 A. What current is in the primary circuit?
   c. What are the power input and output of the transformer?

76. **Laptop Computers** The power supply in a laptop computer requires an effective voltage of 9.0 V from a 120-V line.
   a. If the primary coil has 475 turns, how many does the secondary coil have?
   b. A 125-mA current is in the computer. What current is in the primary circuit?

77. **Hair Dryers** A hair dryer manufactured for use in the United States uses 10 A at 120 V. It is used with a transformer in England, where the line voltage is 240 V.
   a. What should be the ratio of the turns of the transformer?
   b. What current will the hair dryer now draw?

78. A 150-W transformer has an input voltage of 9.0 V and an output current of 5.0 A.
   a. Is this a step-up or step-down transformer?
   b. What is the ratio of V_{output} to V_{input}?

79. Scott connects a transformer to a 24-V source and measures 8.0 V at the secondary circuit. If the primary and secondary circuits were reversed, what would the new output voltage be?

Mixed Review

80. A step-up transformer's primary coil has 500 turns. Its secondary coil has 15,000 turns. The primary circuit is connected to an AC generator having an *EMF* of 120 V.
   a. Calculate the *EMF* of the secondary circuit.
   b. Find the current in the primary circuit if the current in the secondary circuit is 3.0 A.
   c. What power is drawn by the primary circuit? What power is supplied by the secondary circuit?

81. With what speed must a 0.20-m-long wire cut across a magnetic field for which B is 2.5 T if it is to have an *EMF* of 10 V induced in it?

82. At what speed must a wire conductor 50-cm long be moved at right angles to a magnetic field of induction 0.20 T to induce an *EMF* of 1.0 V in it?

83. A house lighting circuit is rated at 120-V effective voltage. What is the peak voltage that can be expected in this circuit?

84. **Toaster** A toaster draws 2.5 A of alternating current. What is the peak current through this toaster?

85. The insulation of a capacitor will break down if the instantaneous voltage exceeds 575 V. What is the largest effective alternating voltage that may be applied to the capacitor?

86. **Circuit Breaker** A magnetic circuit breaker will open its circuit if the instantaneous current reaches 21.25 A. What is the largest effective current the circuit will carry?

87. The electricity received at an electrical substation has a potential difference of 240,000 V. What should the ratio of the turns of the step-down transformer be to have an output of 440 V?

88. An alternating-current electric generator supplies a 45-kW industrial electric heater. If the system voltage is 660 V_{rms}, what is the peak current supplied?

89. A certain step-down transformer has 100 turns on the primary coil and 10 turns on the secondary coil. If a 2.0-kW resistive load is connected to the transformer, what is the effective primary current that flows? Assume that the secondary voltage is 60.0 V_{pk}.

90. A transformer rated at 100 kVA has an efficiency of 98 percent.
a. If the connected load consumes 98 kW of power, what is the input power to the transformer?
b. What is the maximum primary current with the transformer consuming its rated reactive power? Assume that V_P = 600 V.

91. A wire, 0.40-m long, cuts perpendicularly across a magnetic field for which *B* is 2.0 T at a velocity of 8.0 m/s.
a. What *EMF* is induced in the wire?
b. If the wire is in a circuit with a resistance of 6.4 Ω, what is the size of the current in the wire?

92. A coil of wire, which has a total length of 7.50 m, is moved perpendicularly to Earth's magnetic field at 5.50 m/s. What is the size of the current in the wire if the total resistance of the wire is 5.0×10^{-2} Ω? Assume Earth's magnetic field is 5×10^{-5} T.

93. The peak value of the alternating voltage applied to a 144-Ω resistor is 1.00×10^2 V. What power must the resistor be able to handle?

94. Television The CRT in a television uses a step-up transformer to change 120 V to 48,000 V. The secondary side of the transformer has 20,000 turns and an output of 1.0 mA.
a. How many turns does the primary side have?
b. What is the input current?

Thinking Critically

95. Apply Concepts Suppose that an "anti-Lenz's law" existed that meant a force was exerted to increase the change in a magnetic field. Thus, when more energy was demanded, the force needed to turn the generator would be reduced. What conservation law would be violated by this new "law"? Explain.

96. Analyze Real transformers are not 100 percent efficient. Write an expression for transformer efficiency in percent using power. A step-down transformer that has an efficiency of 92.5 percent is used to obtain 28.0 V from a 125-V household voltage. The current in the secondary circuit is 25.0 A. What is the current in the primary circuit?

97. Analyze and Conclude A transformer that supplies eight homes has an efficiency of 95 percent. All eight homes have operating electric ovens that each draw 35 A from 240-V lines. How much power is supplied to the ovens in the eight homes? How much power is dissipated as heat in the transformer?

Writing in Physics

98. Common tools, such as an electric drill, are typically constructed using a universal motor. Using your local library, and other sources, explain how this type of motor may operate on either AC or DC current.

Cumulative Review

99. Light is emitted by a distant star at a frequency of 4.56×10^{14} Hz. If the star is moving toward Earth at a speed of 2750 km/s, what frequency light will be detected by observers on Earth? (Chapter 16)

100. A distant galaxy emits light at a frequency of 7.29×10^{14} Hz. Observers on Earth receive the light at a frequency of 6.14×10^{14} Hz. How fast is the galaxy moving, and in what direction? (Chapter 16)

101. How much charge is on a 22-μF capacitor with 48 V applied to it? (Chapter 21)

102. Find the voltage across a 22-Ω, 5.0-W resistor operating at half of its rating. (Chapter 22)

103. Determine the total resistance of three, 85-Ω resistors connected in parallel and then series-connected to two 85-Ω resistors connected in parallel, as shown in **Figure 25-25.** (Chapter 23)

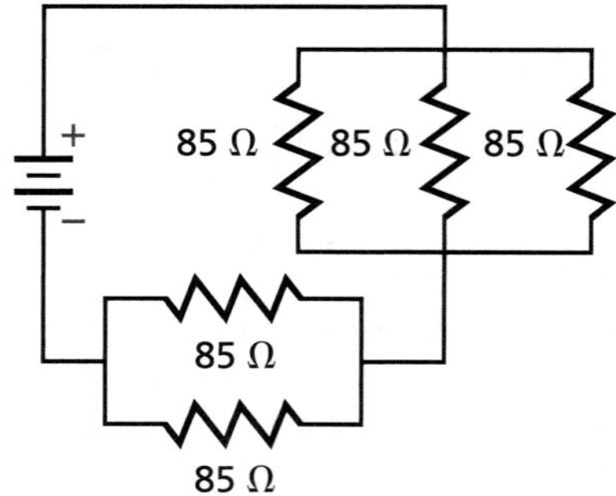

Figure 25-25

104. An electron with a velocity of 2.1×10^6 m/s is at right angles to a 0.81-T magnetic field. What is the force on the electron produced by the magnetic field? What is the electron's acceleration? The mass of an electron is 9.11×10^{-31} kg. (Chapter 24)

 For more problems, go to Additional Problems, Appendix B.

Multiple Choice

1. Which dimensional analysis is correct for the calculation of *EMF*?

   (A) (N·A·m)(J)
   (B) (N/A·m)(m)(m/s)
   (C) J·C
   (D) (N·m·A/s)(1/m)(m/s)

2. An electromotive force of 4.20×10^{-2} V is induced on a 427-mm-long piece of wire that is moving at a rate of 18.6 cm/s. What is the magnetic field that induced the *EMF*?

   (A) 5.29 T
   (B) 1.89 T
   (C) 3.34×10^{-3} T
   (D) 5.29×10^{-1} T

3. Which of the following will fail to induce an electric current in the wire?

   (A)

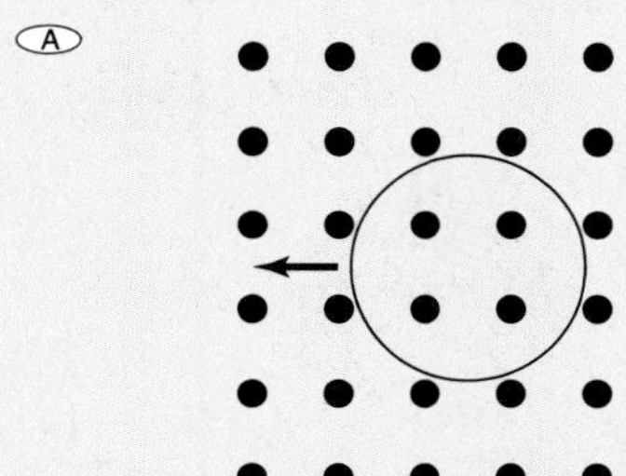

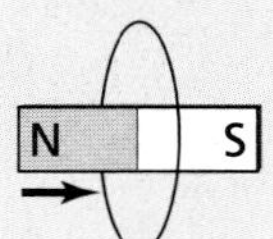

   (C)

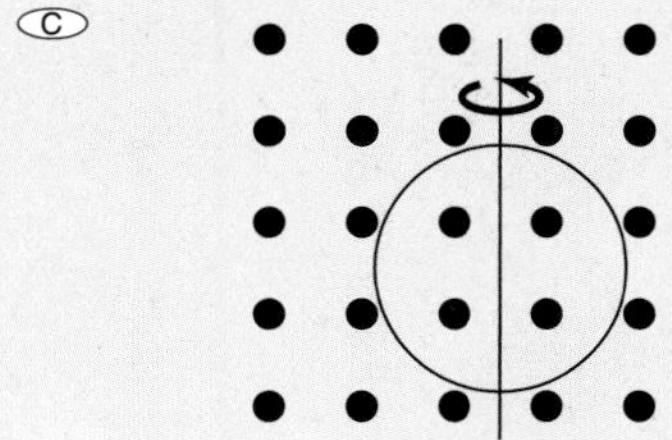

   (D)

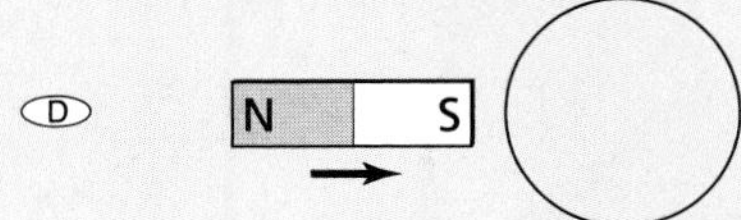

4. 15 cm of wire is moving at the rate of 0.12 m/s through a perpendicular magnetic field of strength 1.4 T. Calculate the *EMF* induced in the wire.

   (A) 0 V
   (B) 0.018 V
   (C) 0.025 V
   (D) 2.5 V

5. A transformer uses a 91-V supply to operate a 13-V device. If the transformer has 130 turns on the primary coil and the device uses 1.9 A of current from the transformer, what is the current supplied to the primary coil?

   (A) 0.27 A
   (B) 0.70 A
   (C) 4.8 A
   (D) 13.3 A

6. An AC generator that delivers a peak voltage of 202 V connects to an electric heater with a resistance of 4.80×10^{2} Ω. What is the effective current in the heater?

   (A) 0.298 A
   (B) 1.68 A
   (C) 2.38 A
   (D) 3.37 A

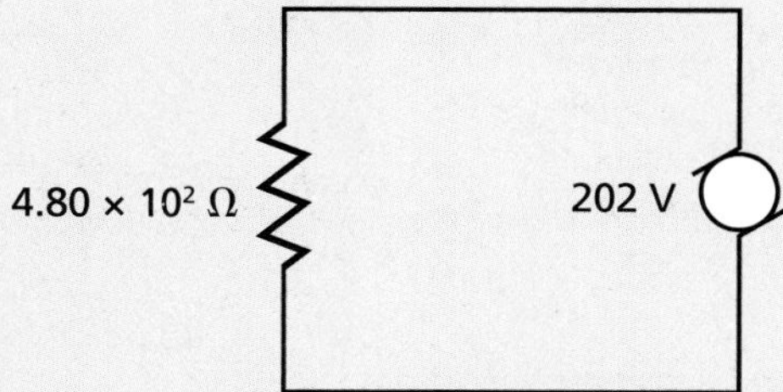

Extended Answer

7. Compare the power lost in transmission for an 800-W line at 160 V to the same power on a line at 960 V. Assume the resistance of the line is 2 Ω. What conclusion can you draw?

✓ Test-Taking TIP

Investigate

Ask your instructor what kinds of questions to expect on the test. Also ask for practice tests so that you can become familiar with the test-taking materials.

20 Electromagnetism

What You'll Learn

- You will learn how combined electric and magnetic fields can be used to determine the masses of electrons, atoms, and molecules.
- You will explain how electromagnetic waves are created, travel through space, and are detected.

Why It's Important

Many electromagnetic waves—from radio and television waves, to visible light, microwaves, and X rays—play vital roles in our lives.

Parabolic Receivers
This parabolic dish antenna is designed to receive radio waves from satellites orbiting hundreds of kilometers above Earth's surface and objects well beyond the solar system.

Think About This ▶

A parabolic dish gets its name from the shape of its reflecting surface—a parabola. Why are parabolic dish antennas well suited for receiving weak television signals?

physicspp.com

From where do radio stations broadcast?

Question

Radio signals are electromagnetic waves. How far away are the transmitters that are used to broadcast the radio station signals that you can tune in on the AM band?

Procedure

1. The AM radio band ranges from 540 kHz to 1690 kHz. Make a data table with columns for *Frequency (kHz), Station Call Sign, Signal Strength, Location, and Distance (km).*
2. Turn on the radio, adjust it to 540 kHz, and set the volume to a moderate level.
3. **Collect and Organize Data** Slowly adjust the frequency upward until you hear a radio station whose broadcast you can clearly understand. Listen to the broadcast for a short time to hear if the station identifies its call sign. Record the station's frequency, signal strength (strong, medium, or weak), and call sign in your data table.
4. Repeat step 3 until you have reached the upper end of the AM radio band, 1690 kHz.
5. Determine where each station broadcasts its signal from. Record the city each station broadcasts from in your data table.
6. **Measure in SI** Using maps, locate the city from which each radio station broadcasts. Determine the distance to each transmitter and record the value in your data table.

Analysis

How far away was the farthest radio station broadcast? Does the distance from the transmitter affect the station's signal strength?

Critical Thinking Changing the position of the antenna often affects a station's signal strength. What does this imply about the nature of radio waves?

26.1 Interactions of Electric and Magnetic Fields and Matter

Although you may not know what the terms stand for, you have probably heard of shortwave radio, microwaves, and VHF and UHF television signals. Each of these terms is used to describe one of the many types of electromagnetic waves that are broadcast through the air to provide you with radio, television, and other forms of communication. All of these waves consist of electric and magnetic fields propagating through space.

The key to understanding how these waves behave is understanding the nature of the electron. Why? Because electromagnetic waves are produced by accelerating electrons—the electrons' charge produces electric fields and the electrons' motion produces magnetic fields. Furthermore, the waves are broadcast and received by antennas, devices made of matter that also contains electrons. Thus, the logical first step in understanding how electromagnetic waves are produced, propagated, received, and used for so many devices is to learn about the properties of the electron.

▸ **Objectives**

- **Describe** the operation of a cathode-ray tube.
- **Solve** problems involving the interaction of charged particles with the electric and magnetic fields in cathode-ray tubes and mass spectrometers.
- **Explain** how a mass spectrometer separates ions of different masses.

▸ **Vocabulary**

isotope
mass spectrometer

Mass of an Electron

How do you determine the mass of something that cannot be seen with the unaided eye and whose mass is so small that it cannot be measured even by the most sensitive scale? Such was the challenge—that of determining the mass of an electron—facing physicists in the late 1800s. The solution required a series of discoveries. The first piece of the puzzle came from Robert Millikan. As described in Chapter 21, Millikan balanced charged oil droplets in an electric field and was able to determine the charge, q, of an electron (1.602×10^{-19} C). Next, British physicist J. J. Thomson was able to determine the charge-to-mass ratio, q/m, of an electron. Knowing both the charge-to-mass ratio, q/m, and the charge of an electron, q, Thomson was able to calculate the mass of an electron.

Thomson's experiments with electrons In 1897, J. J. Thomson performed the first experimental measurement of the charge-to-mass ratio of an electron. Thomson used a cathode-ray tube, a device that generates a stream of electrons. **Figure 26-1** shows the setup used in the experiment. In order to minimize collisions between the electron beam and the molecules in the air, Thomson removed virtually all of the air from within the glass tube.

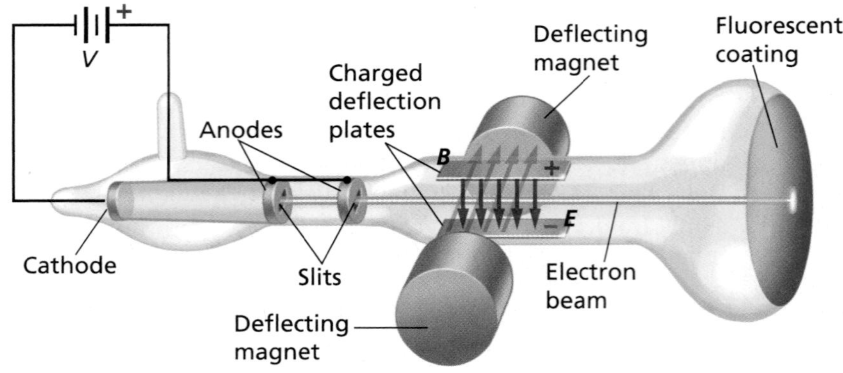

Figure 26-1 The charge-to-mass ratio of an electron was first measured with the Thomson adaptation of a cathode-ray tube. The electromagnets and charged deflection plates both are used to alter the path of the electron beam.

Concepts In Motion
Interactive Figure To see an animation on Thomson's experiments with electrons, visit physicspp.com.

Inside the cathode-ray tube, an electric field is produced by a large potential difference between the cathode and anode. Electrons are emitted from the cathode and are accelerated toward the anodes by the electric field. Some of the electrons pass through slits in the anodes, forming a narrow beam. When the electrons reach the end of the tube they strike a fluorescent coating, causing it to glow.

Thomson used electric and magnetic fields to exert force on and deflect the electron beam as it passed through the tube. The electric field, which was produced by charged parallel plates, was oriented perpendicular to the beam. The electric field (of strength E) produced a force (equal to qE) that acted on the electrons and deflected them upward, toward the positive plate. The magnetic field was produced by two electromagnets and was oriented at right angles to both the beam and the electric field. Recall from Chapter 24 that the force exerted by a magnetic field is perpendicular to the field and to the direction of motion of the electrons. Thus, the magnetic field (of strength B) produced a force (equal to Bqv, where v is the electron velocity) that acted on the electrons and deflected them downward.

The electric and magnetic fields could be adjusted until the beam of electrons followed a straight, or undeflected, path. When this occurred, the forces due to the two fields were equal in magnitude and opposite in direction. Mathematically, this can be represented as follows:

$$Bqv = Eq$$

Solving this equation for v yields the following expression:

$$v = \frac{Eq}{Bq} = \frac{E}{B}$$

This equation shows that the forces are balanced only for electrons that have a specific velocity, v. If the electric field is turned off, only the force due to the magnetic field remains. The magnetic force is perpendicular to the direction of motion of the electrons, causing them to undergo centripetal (center-directed) acceleration. The accelerating electrons follow a circular path of radius r. Using Newton's second law of motion, the following equation can be written to describe the electron's path:

$$Bqv = \frac{mv^2}{r}$$

Solving for q/m results in the following equation.

Charge-to-Mass Ratio in a Thomson Tube $\frac{q}{m} = \frac{v}{Br}$

In a Thomson tube, the ratio of an electron's charge to its mass is equal to the ratio of the electron's velocity divided by the product of the magnetic field strength and the radius of the electron's circular path.

Thomson calculated the straight trajectory velocity, v, using measured values of E and B. Next, he measured the distance between the spot formed by the undeflected beam and the spot formed when the magnetic field acted on the beam. Using this distance, he calculated the radius of the electron's circular path, r. Knowing the value of r, Thomson was able to calculate q/m. By averaging many experimental trials, he determined that $q/m = 1.759\times10^{11}$ C/kg. Using this value for q/m and the known value of q, the mass of the electron (m) was calculated.

$$m = \frac{q}{q/m} = \frac{1.602\times10^{-19}\text{ C}}{1.759\times10^{11}\text{ C/kg}} = 9.107\times10^{-31}\text{ kg}$$

$$m \cong 9.11\times10^{-31}\text{ kg}$$

Thomson's experiments with protons Thomson also used his cathode-ray test apparatus to determine the charge-to-mass ratio for positive ions. He took advantage of the fact that positively charged particles undergo the opposite deflection experienced by electrons moving through an electric or magnetic field. The differing deflection of electrons and positive ions can be seen in **Figure 26-2.**

To accelerate positively charged particles into the deflection region, Thomson reversed the direction of the electric field between the cathode and anodes. He also added a small amount of hydrogen gas to the tube. The electric field pulled electrons off the hydrogen atoms, changing the atoms into positive ions. These positive hydrogen ions, or protons, were then accelerated through a tiny slit in the anode. The resulting proton beam passed through electric and magnetic fields on its way toward the end of the tube.

Figure 26-2 This photograph shows the circular tracks of electrons (e^-) and positrons (e^+) moving through the magnetic field in a bubble chamber, a type of particle detector used in the early years of high-energy physics. Electrons and positrons curve in opposite directions.

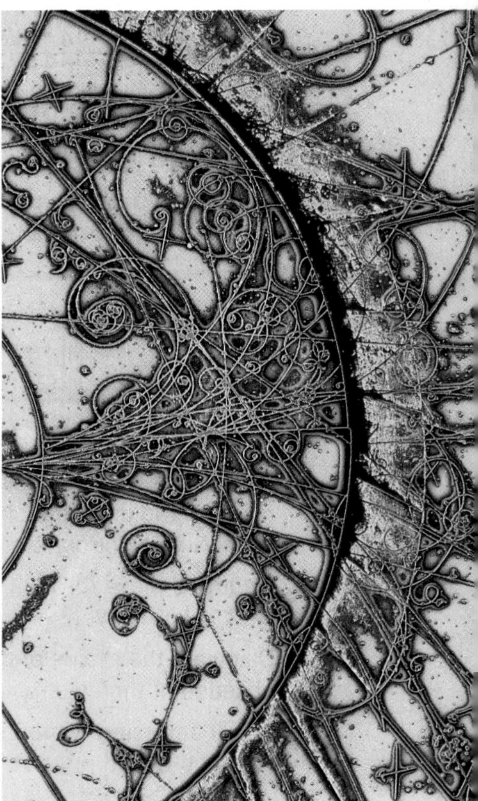

Using this technique, the mass of a proton was determined in the same manner as was the mass of the electron. The mass of a proton was found to be 1.67×10^{-27} kg. Thomson went on to use this technique to determine the masses of heavier ions produced when electrons were stripped from gases, such as helium, neon, and argon.

▶ EXAMPLE Problem 1

Path Radius An electron with a mass of 9.11×10^{-31} kg moves through a cathode-ray tube at 2.0×10^{5} m/s perpendicular to a magnetic field of 3.5×10^{-2} T. The electric field is turned off. What is the radius of the circular path that is followed by the electron?

1 Analyze and Sketch the Problem

- Draw the path of the electron and label the velocity, v.
- Sketch the magnetic field perpendicular to the velocity.
- Diagram the force acting on the electron. Add the radius of the electron's path to your sketch.

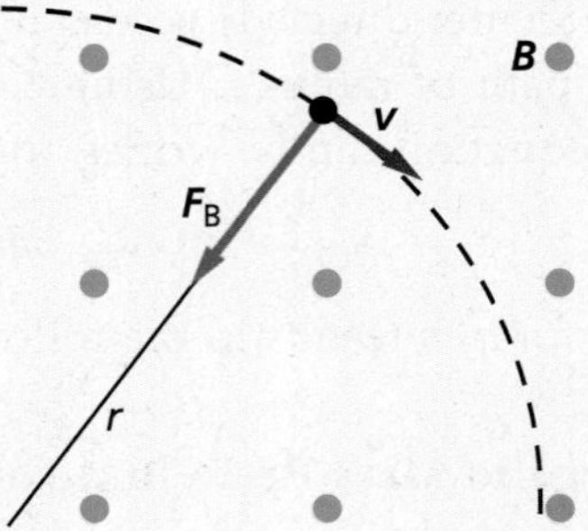

Known:

$v = 2.0\times10^{5}$ m/s
$B = 3.5\times10^{-2}$ T
$m = 9.11\times10^{-31}$ kg
$q = 1.602\times10^{-19}$ C

Unknown:

$r = ?$

2 Solve for the Unknown

Use Newton's second law of motion to describe an electron in a cathode-ray tube subjected to a magnetic field.

Math Handbook
Operations with Scientific Notation
pages 842–843

$$Bqv = \frac{mv^2}{r}$$

$$r = \frac{mv}{Bq}$$

$$= \frac{(9.11\times10^{-31}\text{ kg})(2.0\times10^{5}\text{ m/s})}{(3.5\times10^{-2}\text{ T})(1.602\times10^{-19}\text{ C})}$$

Substitute $m = 9.11\times10^{-31}$ kg, $v = 2.0\times10^{5}$ m/s, $B = 3.5\times10^{-2}$ T, $q = 1.602\times10^{-19}$ C

$$= 3.3\times10^{-5}\text{ m}$$

3 Evaluate the Answer

- **Are the units correct?** The radius of the circular path is a length measurement, given in units of meters.

▶ PRACTICE Problems

- Additional Problems, Appendix B
- Solutions to Selected Problems, Appendix C

Assume that all charged particles move perpendicular to a uniform magnetic field.

1. A proton moves at a speed of 7.5×10^{3} m/s as it passes through a magnetic field of 0.60 T. Find the radius of the circular path. Note that the charge carried by the proton is equal to that of the electron, but is positive.

2. Electrons move through a magnetic field of 6.0×10^{-2} T balanced by an electric field of 3.0×10^{3} N/C. What is the speed of the electrons?

3. Calculate the radius of the circular path that the electrons in problem 2 follow in the absence of the electric field.

4. Protons passing without deflection through a magnetic field of 0.60 T are balanced by an electric field of 4.5×10^{3} N/C. What is the speed of the moving protons?

The Mass Spectrometer

An interesting thing happened when Thomson put neon gas into the cathode-ray tube—he observed two glowing dots on the screen instead of one. Each dot corresponded to a unique charge-to-mass ratio; that is, he was able to calculate two different values for q/m. Thomson concluded that different atoms of the same element could have identical chemical properties, but have different masses. Each of the differing forms of the same atom, with the same chemical properties but different masses, is an **isotope.**

A device similar to Thomson's cathode-ray tube that is commonly used to study isotopes is the **mass spectrometer.** The mass spectrometer is able to precisely measure the charge-to-mass ratios of positive ions. From the charge-to-mass ratio, the mass of each isotope can be calculated. The material under investigation is called the ion source, as it is used to produce the positive ions. The ion source must either be a gas or a material that can be heated to form a vapor. The positive ions are formed when accelerated electrons strike the gas or vapor atoms. The collisions knock electrons off the atoms, forming positive ions. A potential difference, V, between the electrodes produces an electric field that is used to accelerate the ions. **Figure 26-3** shows one type of mass spectrometer.

Figure 26-3 The mass spectrometer is used to analyze isotopes of an element. Inside the spectrometer, a magnet causes the positive ions in a vacuum chamber to be deflected according to their mass **(a).** In the vacuum chamber, the process is recorded on a photographic plate or a solid-state detector **(b).**

Concepts In Motion
Interactive Figure To see an animation on a mass spectrometer, visit physicspp.com.

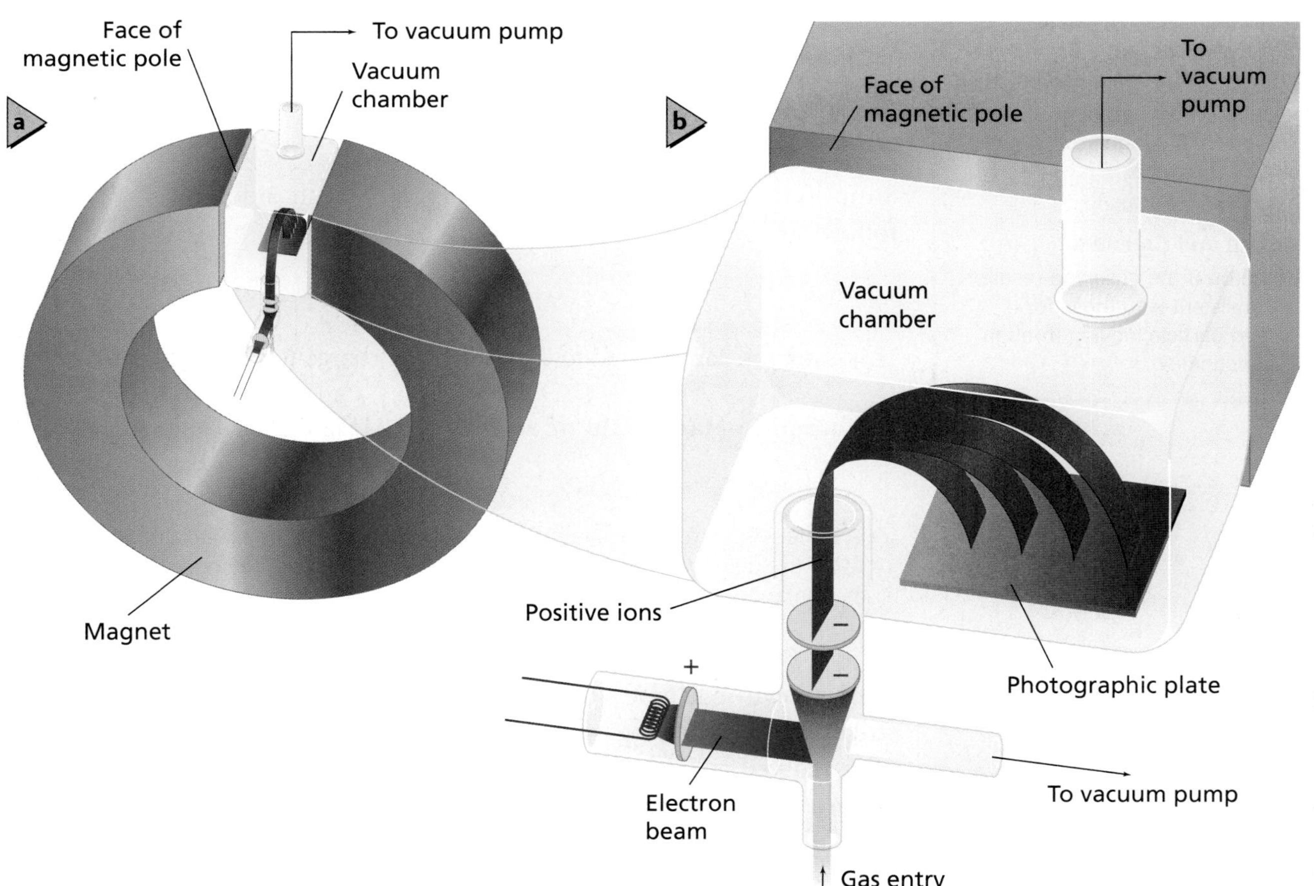

To select ions with a specific velocity, the ions first are made to pass through electric and magnetic deflecting fields, as in Thomson's cathode-ray tube. Ions that pass undeflected through these two fields then move into a region that is subject only to a uniform magnetic field. There, the ions follow a circular path. The radii of the paths can be used to determine the charge-to-mass ratios of the ions. The path radius, r, for an ion can be calculated from Newton's second law of motion.

$$Bqv = \frac{mv^2}{r}$$

Solving for r yields the following equation:

$$r = \frac{mv}{qB}$$

The velocity of an undeflected ion can be calculated from the equation for the kinetic energy of ions accelerated from rest, through a known potential difference, V.

$$KE = \frac{1}{2}mv^2 = qV$$

$$v = \sqrt{\frac{2qV}{m}}$$

Substituting this expression for v in the equation $r = mv/qB$ gives the radius of the circular path.

$$r = \frac{mv}{qB}$$

$$= \frac{m}{qB}\sqrt{\frac{2qV}{m}}$$

$$= \frac{1}{B}\sqrt{\frac{2Vm}{q}}$$

Simplifying this equation by multiplying both sides by B yields the following:

$$Br = \sqrt{\frac{2mV}{q}}$$

This equation can be used to determine the charge-to-mass ratio of an ion.

Charge-to-Mass Ratio of an Ion in a Mass Spectrometer

$$\frac{q}{m} = \frac{2V}{B^2r^2}$$

In a mass spectrometer, the ratio of an ion's charge to its mass is equal to twice the potential difference divided by the product of the square of the magnetic field strength and the square of the radius of the ion's circular path.

As shown in Figure 26-3, in one type of mass spectrometer, the ions strike a plate of photographic film, where they leave a mark. The diameter of the curved path traveled by the ions in the vacuum chamber can be easily measured because it is the distance between the mark made on the film and the slit in the electrode. Therefore, the radius of the path, r, is half of this measured distance.

MINI LAB

Modeling a Mass Spectrometer

Make a ramp by placing a small ball of clay under one end of a grooved ruler. Place a 6-mm-diameter steel ball about halfway up the ramp and release it.

1. Observe the ball as it rolls down the ramp and along the table top.

2. Experiment with the location of a strong magnet near the path of the ball on the table top. Place the magnet close to the path so that it causes the ball's path to curve, but not so close that it causes the ball to collide with the magnet. Repeat step 1 as needed.

3. Predict what will happen to the ball's path when the ball is released from a higher or lower point on the ramp.

4. Test your prediction.

Analyze and Conclude

5. Explain if the observed results are consistent with those for a charged particle moving through a magnetic field.

▶ EXAMPLE Problem 2

Mass of a Neon Atom **The operator of a mass spectrometer produces a beam of doubly ionized (2+) neon atoms. They first are accelerated by a potential difference of 34 V. Then, as the ions pass through a magnetic field of 0.050 T, the radius of their path is 53 mm. Determine the mass of the neon atom to the closest whole number of proton masses.**

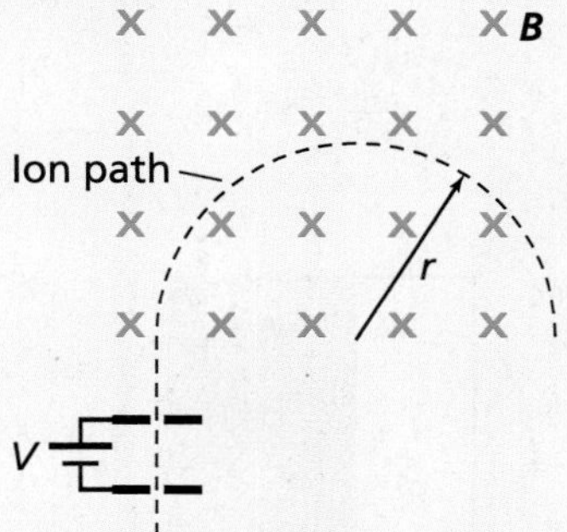

1 Analyze and Sketch the Problem

- Draw the circular path of the ions. Label the radius.
- Draw and label the potential difference between the electrodes.

	Known:	Unknown:
$V = 34$ V	$m_{proton} = 1.67\times10^{-27}$ kg	$m_{neon} = ?$
$B = 0.050$ T	$q = 2(1.60\times10^{-19}$ C)	$N_{proton} = ?$
$r = 0.053$ m	$= 3.20\times10^{-19}$ C	

2 Solve for the Unknown

Use the equation for the charge-to-mass ratio of an ion in a mass spectrometer.

$$\frac{q}{m_{neon}} = \frac{2V}{B^2r^2}$$

$$m_{neon} = \frac{qB^2r^2}{2V}$$

$$= \frac{(3.20\times10^{-19}\text{ C})(0.050\text{ T})^2(0.053\text{ m})^2}{2(34\text{ V})}$$

Substitute $q = 3.20\times10^{-19}$ C, $B = 0.050$ T, $r = 0.053$ m, and $V = 34$ V

$$= 3.3\times10^{-26}\text{ kg}$$

Physics Online

Personal Tutor For an online tutorial on the mass of an atom, visit physicspp.com.

Divide the mass of neon by the mass of a proton to find the number of proton masses.

$$N_{proton} = \frac{m_{neon}}{m_{proton}} = \frac{3.3\times10^{-26}\text{ kg}}{1.67\times10^{-27}\text{ kg/proton}}$$

$$\cong 20\text{ protons}$$

3 Evaluate the Answer

- **Are the units correct?** Mass should be measured in grams or kilograms. The number of protons should not be represented by any units.
- **Is the magnitude realistic?** Neon has two isotopes, with masses of approximately 20 and 22 proton masses.

▶ PRACTICE Problems

- Additional Problems, Appendix B
- Solutions to Selected Problems, Appendix C

5. A beam of singly ionized (1+) oxygen atoms is sent through a mass spectrometer. The values are $B = 7.2\times10^{-2}$ T, $q = 1.60\times10^{-19}$ C, $r = 0.085$ m, and $V = 110$ V. Find the mass of an oxygen atom.
6. A mass spectrometer analyzes and gives data for a beam of doubly ionized (2+) argon atoms. The values are $q = 2(1.60\times10^{-19}$ C), $B = 5.0\times10^{-2}$ T, $r = 0.106$ m, and $V = 66.0$ V. Find the mass of an argon atom.
7. A stream of singly ionized (1+) lithium atoms is not deflected as it passes through a magnetic field of 1.5×10^{-3} T that is perpendicular to an electric field of 6.0×10^2 N/C. What is the speed of the lithium atoms as they pass through the two fields?
8. In Example Problem 2, the mass of a neon isotope is determined. Another neon isotope is found to have a mass of 22 proton masses. How far apart on the photographic film would these two isotopes land?

Chemistry Connection

Isotopic analysis The approximate spacing between marks on the film for an ionized chromium (Cr) sample is shown in **Figure 26-4.** The four distinct red marks indicate that a naturally occurring sample of chromium is composed of four isotopes. The width of the mark corresponds to the abundance of the isotope. Note that the isotope with a mass number of 52 is the most abundant isotope, and that the sum of the percentages for the four isotopes equals 100 percent. As you may recall from chemistry, the mass of each element listed in the periodic table is actually a weighted average of the masses of all of the stable isotopes of that element.

Note that all of the chromium ions that hit the film have the same charge. Their charge depends on how many electrons were removed from the neutral chromium atoms used as the ion source. Recall that the ions are formed when accelerated electrons are used to knock electrons off neutral atoms. After the first electron is removed, producing a singly ionized (1+) atom, more energy is required to remove the second electron and produce a double ionized (2+) atom. This additional energy can be provided by electrons that undergo a greater acceleration because they are subjected to a greater electric field. Thus, higher-energy accelerated electrons can produce both singly and doubly charged ions. In this way, the operator of the mass spectrometer can choose the charge on the ion to be studied.

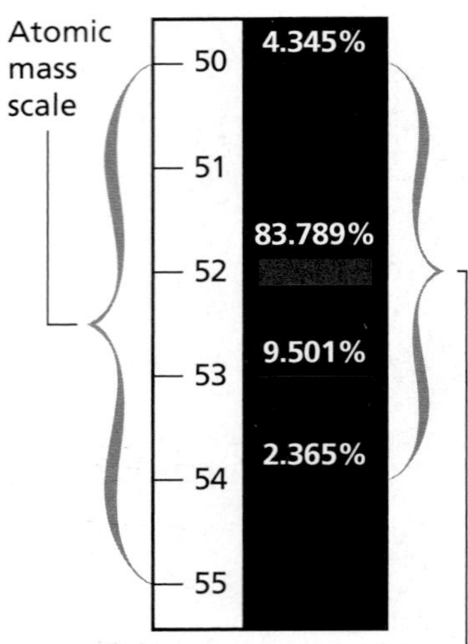

Figure 26-4 Mass spectrometers are widely used to determine the isotopic composition of an element. The above illustration shows the results of analyzing the marks left on a film plate by chromium isotopes.

Other applications Mass spectrometers have numerous applications. For example, a mass spectrometer can be used to "purify" a sample of uranium into its component isotopes. Rather than striking a detector to measure relative abundance, the separated isotopes are collected. The different isotopes are, in turn, used in varying applications. Mass spectrometers also are used to detect and identify trace amounts of molecules in a sample, an application extensively used in the environmental and forensic sciences. The device is so sensitive that researchers are able to separate ions with mass differences as small as one ten-thousandth of one percent and are able to identify the presence of a single molecule within a 10 billion-molecule sample.

26.1 Section Review

9. **Cathode-Ray Tube** Describe how a cathode-ray tube forms an electron beam.

10. **Magnetic Field** The radius of the circular path of an ion in a mass spectrometer is given by $r = (1/B)\sqrt{2Vm/q}$. Use this equation to explain how a mass spectrometer is able to separate ions of different masses.

11. **Magnetic Field** A modern mass spectrometer can analyze molecules having masses of hundreds of proton masses. If the singly charged ions of these molecules are produced using the same accelerating voltage, how would the mass spectrometer's magnetic field have to be changed for the ions to hit the film?

12. **Path Radius** A proton moves at a speed of 4.2×10^4 m/s as it passes through a magnetic field of 1.20 T. Find the radius of the circular path.

13. **Mass** A beam of doubly ionized (2+) oxygen atoms is accelerated by a potential difference of 232 V. The oxygen then enters a magnetic field of 75 mT and follows a curved path with a radius of 8.3 cm. What is the mass of the oxygen atom?

14. **Critical Thinking** Regardless of the energy of the electrons used to produce ions, J. J. Thomson never could remove more than one electron from a hydrogen atom. What could he have concluded about the positive charge of a hydrogen atom?

Physics Online physicspp.com/self_check_quiz

26.2 Electric and Magnetic Fields in Space

Although you probably do not realize it, you rely on electromagnetic waves every day. Signals broadcast from television and radio stations, orbiting satellites, and even those emanating from distant galaxies are all electromagnetic waves. Electromagnetic waves also are used in common consumer products, such as microwaves ovens, remote-control garage door openers, and cellular phones, to name a few. In this section, you will learn about the fields that make up electromagnetic waves, and how the waves are produced and received.

▶ **Objectives**

- **Describe** how electromagnetic waves propagate through space.
- **Solve** problems involving electromagnetic wave properties.
- **Describe** the factors affecting an antenna's ability to receive an electromagnetic wave of a specific wavelength.
- **Solve** problems involving electromagnetic wave propagation through dielectrics.

▶ **Vocabulary**

electromagnetic wave
dielectrics
antenna
electromagnetic spectrum
electromagnetic radiation
piezoelectricity
receiver

Electromagnetic Waves

Great advancements in the understanding of electromagnetic waves were made during the nineteenth century. These advancements led to the development of new devices and technologies that had a huge impact on modern society.

A series of breakthroughs In 1821, while performing a demonstration for his students, Danish physicist Hans Christian Oersted noticed that an electric current caused the needle in a nearby compass to deflect. Oersted realized that his observation displayed a fundamental connection between electricity and magnetism. He concluded that an electric current in a conductor produces a magnetic field, and that a changing electric current produces a changing magnetic field. Oersted's discovery created excitement in the scientific community and led to a flood of new research.

Eleven years after Oersted, Englishman Michael Faraday, and American high school physics teacher Joseph Henry independently discovered induction. Induction is the production of an electric field due to a moving magnetic field. Interestingly, induced electric fields exist even if there is not a wire present, as shown in **Figure 26-5a.** Thus, a changing magnetic field produces a corresponding changing electric field. Notice that the field lines of the induced electric field, shown in Figure 26-5a, are closed loops. This is because, unlike an electrostatic field, there are no charges on which the field lines begin or end.

■ **Figure 26-5** These diagrams represent an induced electric field **(a),** a magnetic field **(b),** and both electric and magnetic fields **(c).**

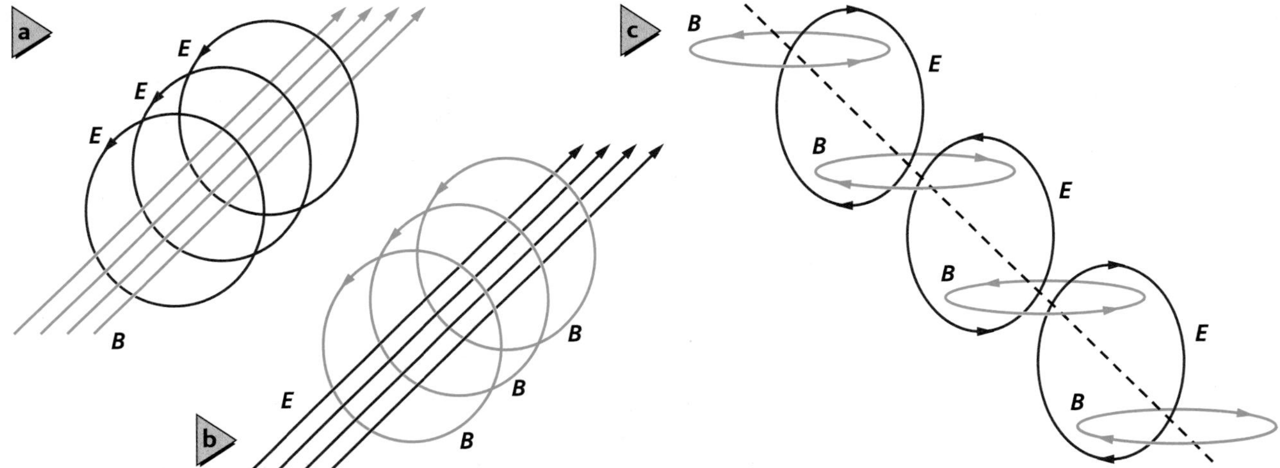

In 1860, Scottish physicist James Maxwell postulated that the opposite of induction is also true; that is, that a changing electric field produces a changing magnetic field. This is shown in **Figure 26-5b** on the previous page. Maxwell also suggested that charges were not necessary—a changing electric field alone would produce the magnetic field. He then predicted that both accelerating charges and changing magnetic fields would produce electric and magnetic fields that move through space.

A combined electric and magnetic field that travels through space is an **electromagnetic wave,** or EM wave. The orientations of the fields making up an electromagnetic wave are shown in **Figure 26-5c** on the previous page. In 1887, Heinrich Hertz, a German physicist, experimentally confirmed that Maxwell's theory was correct. Maxwell's theory led to a complete description of electricity and magnetism.

Electromagnetic wave properties The speed of an electromagnetic wave later was found to be approximately 3.00×10^8 m/s, now denoted as c, the speed of light. Light, a type of electromagnetic wave, and all other forms of electromagnetic waves, travel through space at c. The wavelength of an electromagnetic wave, its frequency, and the speed of light all are related.

Wavelength-Frequency Relationship for a Wave $\lambda = \frac{v}{f}$

The wavelength of a wave is equal to its speed divided by its frequency.

In this equation, the wavelength, λ, is measured in m; the speed, v, is measured in m/s; and the frequency, f, is measured in Hz. Note that for an electromagnetic wave traveling in air or a vacuum, the speed, v, is equal to c, the speed of light. Thus, for an electromagnetic wave, the equation becomes the following:

$$\lambda = \frac{c}{f}$$

In the equation, $c = 3.00\times10^8$ m/s.

Note that in the wavelength-frequency equation, the product of frequency and wavelength is constant—equal to c—for any electromagnetic wave. Thus, as wavelength increases, frequency decreases, and vice-versa. In other words, an electromagnetic wave with a long wavelength has a low frequency, whereas an electromagnetic wave with a short wavelength has a high frequency.

PRACTICE Problems

• Additional Problems, Appendix B
• Solutions to Selected Problems, Appendix C

15. What is the speed in air of an electromagnetic wave having a frequency of 3.2×10^{19} Hz?

16. What is the wavelength of green light having a frequency of 5.70×10^{14} Hz?

17. An electromagnetic wave has a frequency of 8.2×10^{14} Hz. What is the wavelength of the wave?

18. What is the frequency of an electromagnetic wave having a wavelength of 2.2×10^{-2} m?

Electromagnetic wave propagation through matter Electromagnetic waves also can travel through matter. Sunlight shining through a glass of water is an example of light waves traveling through three different forms of matter: air, glass, and water. Air, glass, and water are nonconducting materials known as **dielectrics.** The velocity of an electromagnetic wave through a dielectric is always less than the speed of the wave in a vacuum, and it can be calculated using the following equation:

$$v = \frac{c}{\sqrt{K}}$$

In this equation, the wave velocity, v, is measured in m/s; the speed of light, c, has a value of 3.00×10^8 m/s; and the relative dielectric constant, K, is a dimensionless quantity. In a vacuum, $K = 1.00000$, and the wave velocity is equal to c. In air, $K = 1.00054$, and electromagnetic waves move just slightly slower than c.

PRACTICE Problems

- Additional Problems, Appendix B
- Solutions to Selected Problems, Appendix C

19. What is the speed of an electromagnetic wave traveling through the air? Use $c = 299{,}792{,}458$ m/s in your calculation.

20. For light traveling through water, the dielectric constant is 1.77. What is the speed of light traveling through water?

21. The speed of light traveling through a material is 2.43×10^8 m/s. What is the dielectric constant of the material?

Electromagnetic wave propagation through space The formation of an electromagnetic wave is shown in **Figure 26-6.** An **antenna,** which is a wire designed to transmit or receive electromagnetic waves, is connected to an alternating current (AC) source. The AC source produces a varying potential difference in the antenna that alternates at the frequency of the AC source. This varying potential difference generates a corresponding varying electric field that propagates away from the antenna. The changing electric field also generates a varying magnetic field perpendicular to the page. Although the magnetic field is not shown in Figure 26-6, it also propagates away from the antenna. The combined electric and magnetic fields are electromagnetic waves that spread out into space, moving at the speed of light.

Figure 26-6 An alternating current source connected to the antenna produces a varying potential difference in the antenna. This varying potential difference generates a varying electric field **(a).** The varying electric field produces a changing magnetic field (not shown), and this magnetic field, in turn, generates an electric field. This process continues and the electromagnetic wave propagates away from the antenna, **(b)** and **(c).**

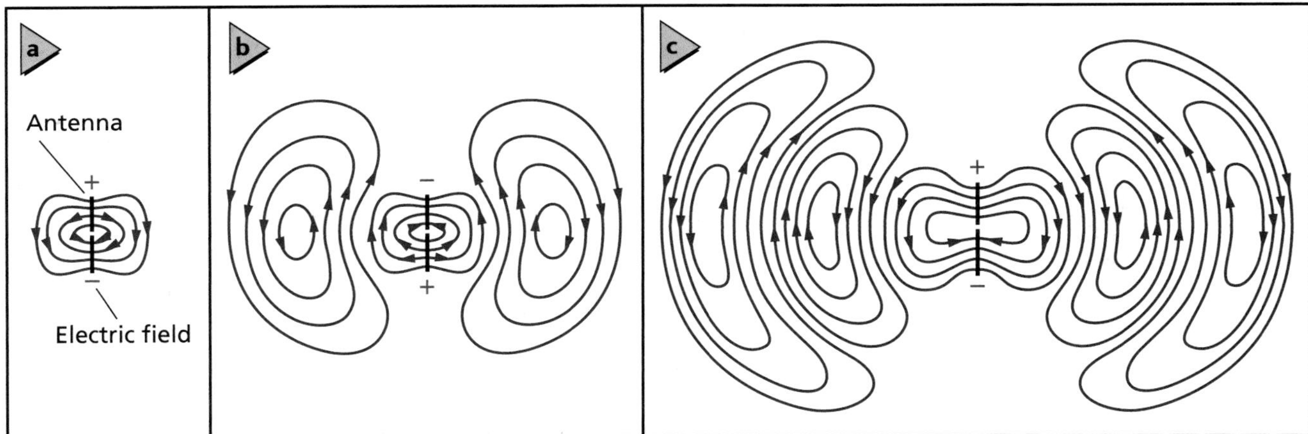

Figure 26-7 Portions of the electric and magnetic fields generated by an antenna might look like this at an instant in time **(a).** Note how the electric and magnetic fields are perpendicular to each other and to the direction of the wave velocity, ***v* (b).**

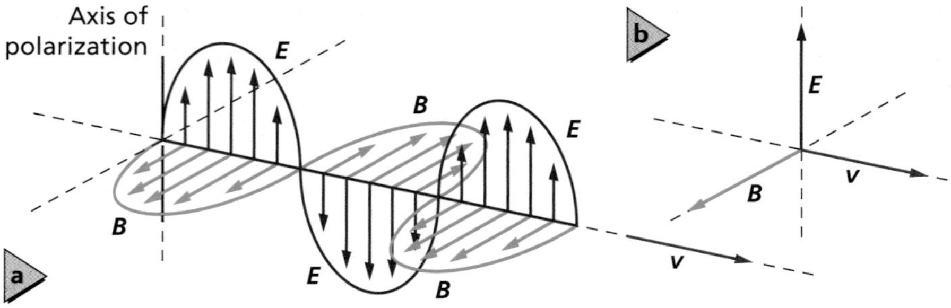

If it were possible to see invisible electromagnetic waves approaching, the changing fields would appear as in **Figure 26-7.** The electric field oscillates up and down, while the magnetic field oscillates at right angles to the electric field. Both of the fields are at right angles to the wave direction. Note that an electromagnetic wave produced by an antenna is polarized; that is, its electric field is parallel to the antenna's conductor.

Figure 26-8 The illustration below provides examples of various types of electromagnetic radiation and their wavelengths.

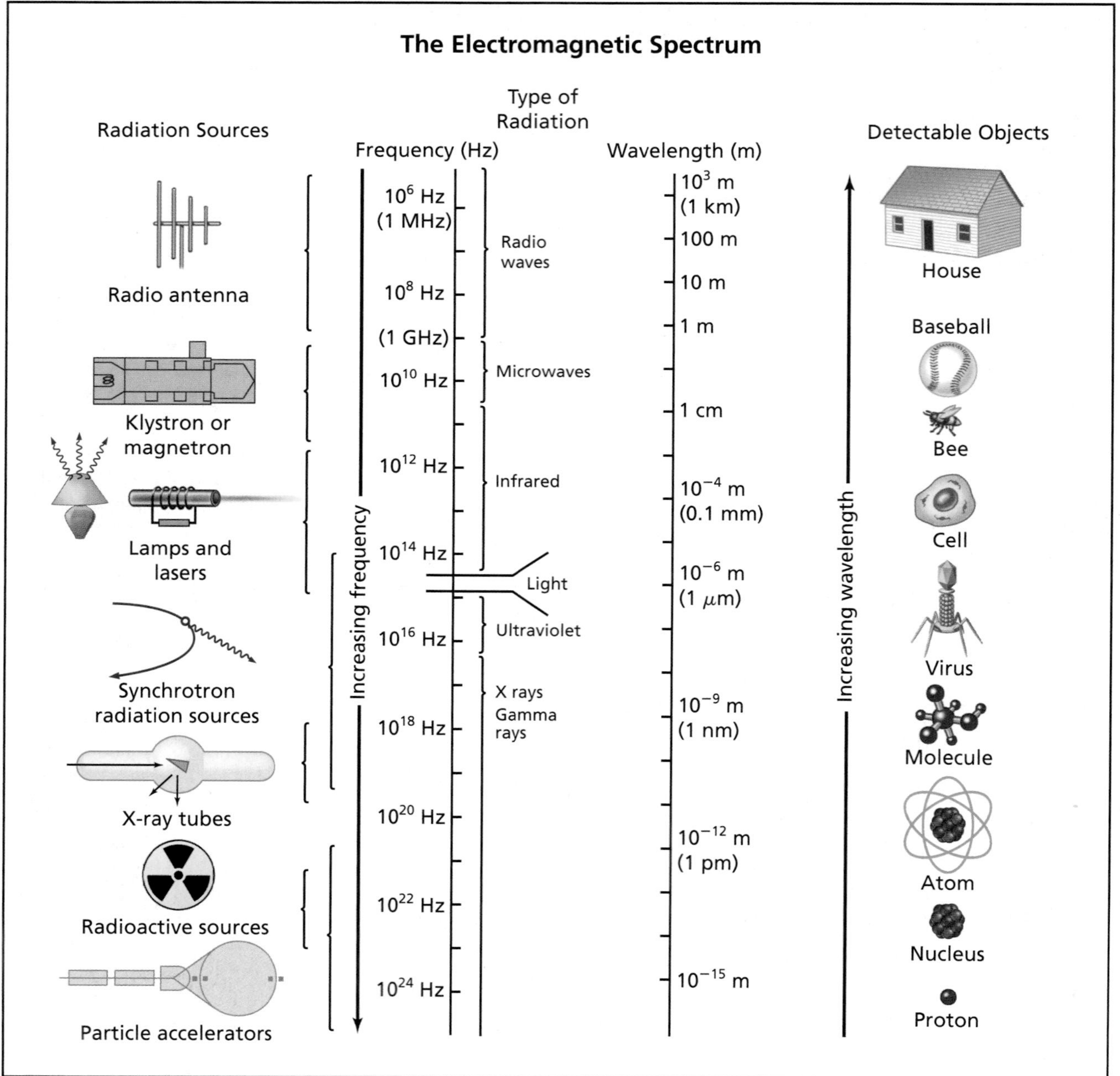

CHALLENGE PROBLEM

Visible light makes up only a very small portion of the entire electromagnetic spectrum. The wavelengths for some of the colors of visible light are shown in **Table 26-1.**

1. Which color of light has the longest wavelength?
2. Which color travels the fastest in a vacuum?
3. Waves with longer wavelengths diffract around objects in their path more than waves with shorter wavelengths. Which color will diffract the most? The least?
4. Calculate the frequency range for each color of light given in Table 26-1.

Table 26-1
Wavelengths of Visible Light

Color	Wavelength (nm)
Violet–Indigo	390 to 455
Blue	455 to 492
Green	492 to 577
Yellow	577 to 597
Orange	597 to 622
Red	622 to 700

Producing Electromagnetic Waves

Waves from an AC source As you just learned, an AC source connected to an antenna can transmit electromagnetic waves. The wave frequency is equal to the frequency of the rotating AC generator and is limited to about 1 kHz. The range of frequencies and wavelengths that make up all forms of electromagnetic radiation is shown in **Figure 26-8** and is called the **electromagnetic spectrum.**

Waves from a coil and a capacitor A common method of generating high-frequency electromagnetic waves is to use a coil and a capacitor connected in a series circuit. If the capacitor is charged by a battery, the potential difference across the capacitor creates an electric field. When the battery is removed, the capacitor discharges as the stored electrons flow through the coil, creating a magnetic field. When the capacitor is discharged, the coil's magnetic field collapses. A back-*EMF* then develops and recharges the capacitor in the opposite direction, and the process is repeated. When an antenna is connected across the capacitor, the fields of the capacitor are transmitted into space. One complete oscillation cycle is shown in **Figure 26-9.**

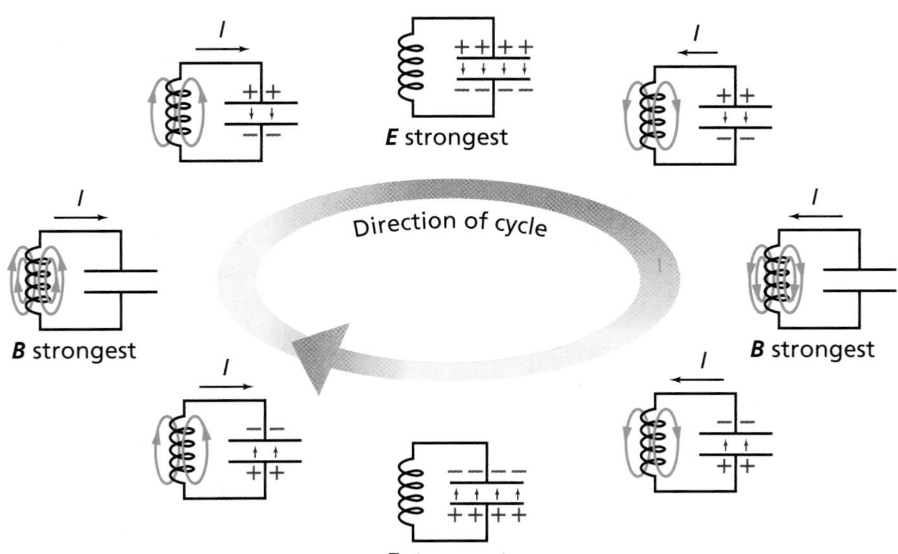

Figure 26-9 One complete oscillation cycle for a coil and capacitor circuit is shown. The capacitor and coil sizes determine the oscillations per second of the circuit, which equals the frequency of the waves produced.

APPLYING PHYSICS

Frequencies The Federal Communications Commission (FCC) assigns each radio and TV station a carrier wave with a specific frequency. A station broadcasts by varying its carrier wave. When the wave is received by your radio or TV set, the carrier wave is stripped away and the information from the wave is processed so that you can see or hear it.

The process occurring in the coil-and-capacitor circuit can be compared with the cyclic oscillations of a swinging pendulum, as shown in **Figure 26-10.** Assume that the electrons in the coil and capacitor are represented by the pendulum's bob. The moving bob has the greatest speed at the bottom of its swing, a position at which kinetic energy, *KE*, is maximized, and potential energy, *PE*, due to gravity is zero. This point in the pendulum's motion, shown in **Figure 26-10a,** is similar to the peak electric current flow in the coil when the charge on the capacitor is zero. When the bob reaches the peak of its swing, its vertical displacement and *PE* are maximized, whereas its *KE* is zero because the bob's velocity is zero. As shown in **Figure 26-10b,** this point in the motion is similar to when the capacitor holds the maximum charge and the current through the coil is zero.

Energy in the coil-and-capacitor circuit As you just learned, the *PE* of the pendulum is largest when its vertical displacement is greatest, and the *KE* is largest when the velocity is greatest. The sum of the *PE* and *KE*—the total energy—is constant throughout the motion of the pendulum. In the coil-and-capacitor circuit, both the magnetic field produced by the coil and the electric field in the capacitor contain energy. When the current is largest, the energy stored in the magnetic field is greatest. When the current is zero, the electric field of the capacitor is largest, and all the energy is contained in the electric field. The total energy of the circuit (the sum of the magnetic field energy, the electric field energy, the thermal losses, and the energy carried away by the generated electromagnetic waves) is constant. Energy that is carried, or radiated, in the form of electromagnetic waves is frequently called **electromagnetic radiation.**

Just as a pendulum eventually stops swinging if it is left alone, the oscillations in a coil and capacitor die out over time due to resistance in the circuit. The oscillations of both systems can be made to continue by adding energy. Gentle pushes, applied at the correct times, will keep a pendulum swinging. The largest amplitude swings occur when the frequency of pushes matches the frequency of swinging motion. This is the condition of resonance, which was discussed in Chapter 14. Similarly, voltage pulses applied to the coil-and-capacitor circuit at the right frequency keep the oscillations in the circuit going. One way of doing this is to add a second coil to the circuit, to form a transformer. A transformer is shown in **Figure 26-11.** The alternating current induced in the secondary coil is increased by an amplifier and added back to the coil and capacitor. This type of circuit can produce frequencies up to approximately 400 MHz.

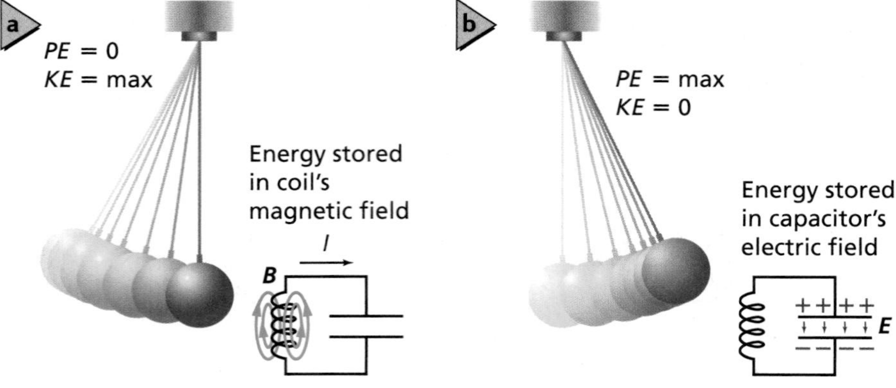

Figure 26-10 The motion of a pendulum is analogous to the action of electrons in a coil-and-capacitor circuit. Motion of the pendulum's bob is analogous to the current flow in the circuit **(a).** The point where the pendulum's motion comes to a stop is analogous to zero current in the circuit **(b).**

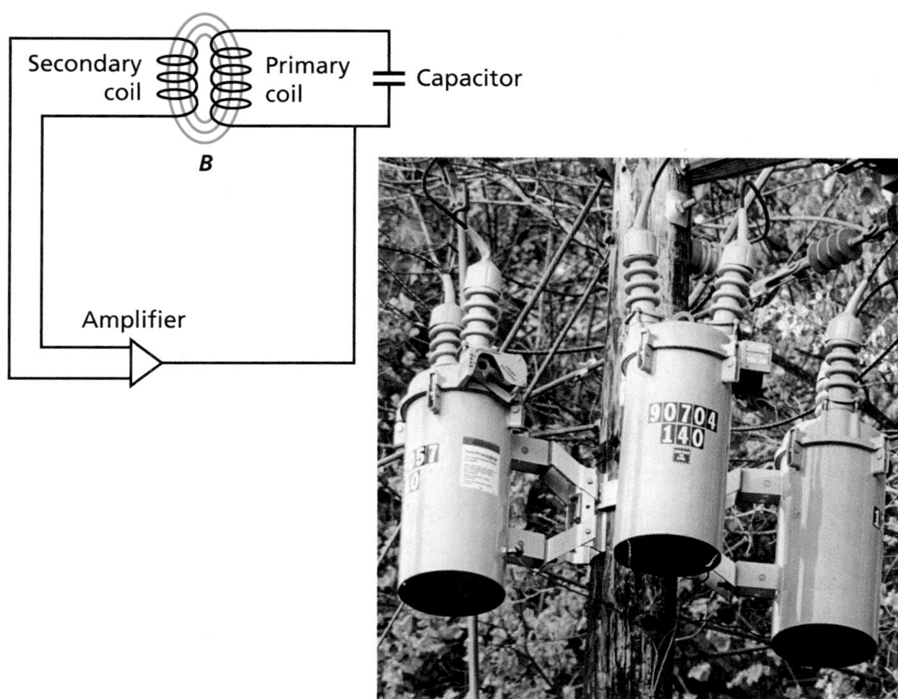

Figure 26-11 In a transformer, the amplified oscillation from the secondary coil is in resonance with the coil-and-capacitor circuit and keeps the oscillations going.

Waves from a resonant cavity The oscillation frequency produced by a coil-and-capacitor circuit can be increased by decreasing the size of the coil and capacitor used. However, above frequencies of 1 GHz, individual coils and capacitors can no longer be used. High frequency microwaves, with frequencies from 1 GHz to 100 GHz, are produced using a resonant cavity. The resonant cavity is a rectangular box that acts as both a coil and a capacitor. The size of the box determines the frequency of oscillation. Microwave ovens have resonant cavities that produce the microwaves used to cook food.

To produce even higher frequency infrared waves, the size of the resonant cavity would have to be reduced to molecular size. The oscillating electrons that produce infrared waves are, in fact, within the molecules. Visible and ultraviolet waves are generated by electrons within atoms. X rays and gamma rays are the result of accelerating charges in the nuclei of atoms. All electromagnetic waves arise from accelerated charges, and all travel at the speed of light.

Waves from piezoelectricity Coils and capacitors are not the only method of generating oscillation voltages. Quartz crystals deform when a voltage is applied across them, a property known as **piezoelectricity**. The application of an AC voltage to a cut section of quartz crystal results in sustained oscillations. An inverse linear relationship exists between crystal thickness and oscillation frequency. Just as a piece of metal vibrates at a specific frequency when it is bent and released, so does a quartz crystal. A crystal can be cut so that it vibrates at a specific desired frequency. An applied voltage deforms the crystal and starts the vibrations. The piezoelectric property also generates an *EMF* when the crystal is deformed. Because this *EMF* is produced at the vibrating frequency of the crystal, it can be amplified and returned to the crystal to keep it vibrating. Because of their nearly constant frequencies of vibration, quartz crystals commonly are used in watches.

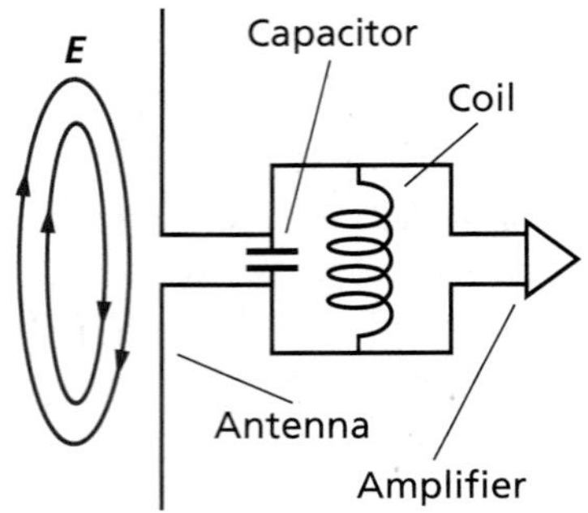

Figure 26-12 The changing electric fields from a radio station signal cause electrons in the antenna to accelerate. The information carried by the signal is then decoded and amplified and used to drive a loudspeaker.

Reception of Electromagnetic Waves

Now that you know how electromagnetic waves are produced and transmitted, how do you suppose the waves are detected? As you may have guessed, reception involves an antenna. As shown in **Figure 26-12,** the wave's electric fields accelerate the electrons of the material making up the antenna. The acceleration is largest when the antenna is positioned in the same direction as the wave polarization; that is, when it is parallel to the direction of the wave's electric fields. A potential difference across the terminals of the antenna oscillates at the frequency of the electromagnetic wave. This voltage is largest when the length of the antenna is one-half the wavelength of the wave it is to detect. Thus, an antenna's length is designed to be one-half of the wavelength of the wave it is supposed to receive. For this reason, an antenna designed to receive radio and television waves is much longer than one designed to receive microwaves.

While a simple wire antenna can detect electromagnetic waves, several wires are more effective. A television antenna often consists of two or more wires spaced about one-quarter wavelength apart. Electric fields that are generated in the individual wires form constructive interference patterns that increase the strength of the signal.

It is important to realize that all electromagnetic waves, not just visible light waves, undergo reflection, refraction, and diffraction. Thus, it should not be a surprise to learn that dish antennas, like the one shown at the beginning of this chapter, reflect very short wavelength electromagnetic signals, just as parabolic mirrors reflect visible light waves. A dish antenna's large surface area for collecting and focusing waves makes it well-suited to receive weak radio signals. A parabolic dish antenna works by reflecting and focusing the received signals off its surface and into a device called the horn. The horn, which is supported by a tripod structure over the main dish, contains a short dipole antenna. The horn channels the signals to a **receiver,** a device consisting of an antenna, a coil-and-capacitor circuit, a detector to decode the signal, and an amplifier.

Selection of waves As you know, many different radio and television stations transmit electromagnetic waves at the same time. If the information being broadcast is to be understood, the waves of a particular station must be selected. To select waves of a particular frequency (and reject the others) a tuner uses a coil-and-capacitor circuit connected to an antenna. The capacitance is adjusted until the oscillation frequency of the circuit equals the frequency of the desired wave. When this is done, only waves of the desired frequency can cause significant oscillations of the electrons in the circuit.

Energy from waves Waves carry energy as well as information. At microwave and infrared frequencies, waves accelerate electrons in molecules. The energy of the waves is converted to thermal energy in the molecules. This is how microwave ovens cook food.

Light waves also can transfer energy to electrons. Photographic film makes use of this fact by using the energy in light waves to drive a chemical reaction within the film. The result is a permanent record of the light from the subject that strikes the film. At higher frequencies, ultraviolet (UV) radiation causes many chemical reactions to occur, including those in living cells that produce sunburn and tanning.

X Rays

In 1895, German physicist Wilhelm Roentgen sent electrons through an evacuated glass tube, similar to the one shown in **Figure 26-13.** Roentgen used a very high voltage across the tube to give the electrons large kinetic energies. When the electrons struck the metal anode target within the tube, Roentgen noticed a glow on a phosphorescent screen a short distance away. The glow continued even when a piece of wood was placed between the tube and the screen. He concluded that some kind of highly penetrating rays were coming from the tube.

Because Roentgen did not know what these strange rays were, he called them X rays. A few weeks later, Roentgen found that photographic plates were darkened by X rays. He also discovered that soft body tissue was transparent to the rays, but that bone blocked them. He produced an X-ray picture of his wife's hand. Within months, doctors recognized the valuable medical uses of this phenomenon.

It now is known that an X ray is a high-frequency electromagnetic wave. In an X-ray tube, electrons first are accelerated to high speeds by means of potential differences of 20,000 V or more. When the electrons crash into matter, their kinetic energies are converted into the very high-frequency electromagnetic waves called X rays.

Electrons are accelerated to these speeds in cathode-ray tubes, such as the picture tube in a television. When the electrons hit the inside surface of a television screen's face plate, they come to a sudden stop and cause the colored phosphors to glow. This sudden stopping of the electrons also can produce harmful X rays. Thus, the face-plate glass in a television screen contains lead to stop the X rays and protect the viewers.

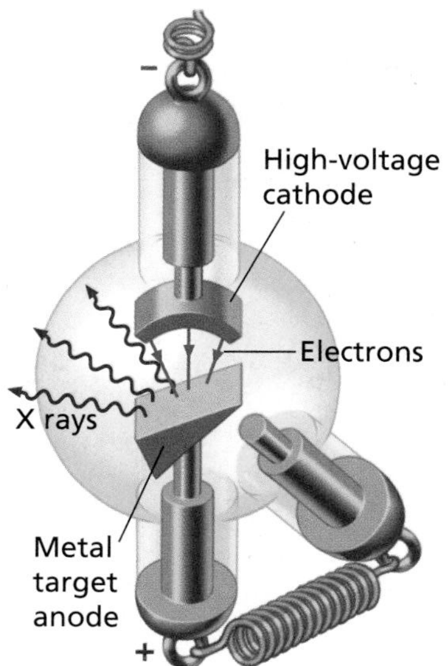

Figure 26-13 X rays are emitted when high energy electrons strike the metal target inside the X-ray tube. The target can be changed to produce X rays of different wavelengths.

26.2 Section Review

22. Wave Propagation Explain how electromagnetic waves are able to propagate through space.

23. Electromagnetic Waves What are some of the primary characteristics of electromagnetic waves? Do electromagnetic waves behave differently from the way that other waves, such as sound waves, behave? Explain.

24. Frequency An electromagnetic wave is found to have a wavelength of 1.5×10^{-5} m. What is the frequency of the wave?

25. TV Signals Television antennas normally have metal rod elements that are oriented horizontally. From this information, what can you deduce about the directions of the electric fields in television signals?

26. Parabolic Receivers Why is it important for a parabolic dish's receiving antenna to be properly aligned with the transmitter?

27. Antenna Design Television channels 2 through 6 have frequencies just below the FM radio band, while channels 7 through 13 have much higher frequencies. Which signals would require a longer antenna: those of channel 7 or those of channel 6? Provide a reason for your answer.

28. Dielectric Constant The speed of light traveling through an unknown material is 1.98×10^{8} m/s. Given that the speed of light in a vacuum is 3.00×10^{8} m/s, what is the dielectric constant of the unknown material?

29. Critical Thinking Most of the UV radiation from the Sun is blocked by the ozone layer in Earth's atmosphere. In recent years, scientists have discovered that the ozone layer over both Antarctica and the Arctic Ocean is thinning. Use what you have learned about electromagnetic waves and energy to explain why some scientists are very concerned about the thinning ozone layer.

Physics Online physicspp.com/self_check_quiz

PHYSICS LAB •

Electromagnetic Wave Shielding

Alternate CBL instructions can be found on the Web site.
physicspp.com

The electromagnetic spectrum consists of many types of electromagnetic radiation, each of which can be classified by its frequency or wavelength. Gamma rays have the highest frequency and energy, and have wavelengths that are a fraction of a nanometer. Following gamma rays in order of increasing wavelength (decreasing frequency and energy) are the following forms of electromagnetic radiation: X rays, ultraviolet, visible, infrared, microwaves, and radio waves. Only wavelengths that fall within the range of visible light can be detected by the human eye—all other forms of radiation are invisible.

Electromagnetic receivers, such as those used in radios and televisions, detect waves with an antenna. Because every electrical device with a changing or alternating current radiates electromagnetic waves, waves from these sources can interfere with the reception of a desired signal. Some materials are effective at blocking or shielding radio waves. In this lab you will investigate the radio wave shielding effectiveness of various materials.

QUESTION

What types of materials shield electromagnetic waves?

Objectives

- **Experiment** with various materials to determine if they are effective at shielding electromagnetic waves.
- **Observe and infer** about types of materials that shield radio waves.
- **Collect and organize** data on types of shielding.

Safety Precautions

- **Always wear safety goggles and a lab apron.**
- **Wear gloves when bending or handling the wire screen.**
- **Use caution when working with staples to avoid puncturing skin.**

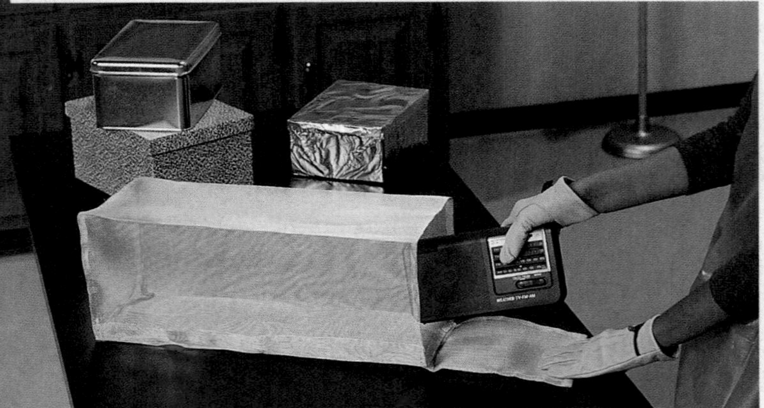

Materials

small battery-operated AM/FM radio
two small cardboard boxes
metal box or can with lid
aluminum foil
static shielding bag (type used to protect computer parts)
metal screen
masking tape
leather gloves
stapler

Procedure

1. Prepare the aluminum covered box. Cover the outside of one box and its lid with aluminum foil. Cover the lid separately from the rest of the box so the lid can be removed and replaced.
2. Prepare the wire screen box. Fold a piece of wire screen so that it forms a four-sided box shape with open ends. Use staples to hold connecting edges of the screen together. Make sure the wire box is big enough for the radio to fit inside. Next, cut a piece of wire screen to fit over each open end. Staple one of these pieces in place over one of the ends, making sure to leave no openings. Then, staple one edge of the remaining piece of wire screen to the other end of the wire box. This piece of screen will act as a door that can be opened and closed.

Data Table							
Band	**Frequency (Hz)**	**Enclosure**	**Observations**	**Band**	**Frequency (Hz)**	**Enclosure**	**Observations**
AM		A person's arms		FM		A person's arms	
AM		Cardboard box		FM		Cardboard box	
AM		Cardboard box covered with aluminum foil		FM		Cardboard box covered with aluminum foil	
AM		Wire screen box		FM		Wire screen box	
AM		Metal box		FM		Metal box	
AM		Static shielding bag		FM		Static shielding bag	

3. Turn on the radio and tune it to a strong signal from a station in the AM band. Record the frequency of the station. The frequency can be determined from the radio dial or display, or the station's broadcast frequency may be announced while you are listening.
4. Hold the radio next to your body and cover it with your arms. Ignoring the fact than the sound is muffled because you are covering the speaker, how has the reception of the signal been affected? Record your observations.
5. Place the radio inside a cardboard box and put the lid on the box. Listen to the radio's reception and record your observations.
6. Repeat step 5 four more times using the aluminum foil covered box, the wire screen box (with the door closed), the metal box (with lid on), and the static shielding bag, respectively.
7. Change the radio to the FM band and tune in a strong station. Record the frequency of the station and then repeat steps 4–6.

Analyze

1. **Summarize** Which materials were effective at shielding radio waves?
2. **Use Numbers** Calculate the wavelengths for each of the radio frequencies you used. Recall that $c = f\lambda$, where the velocity, c, of the electromagnetic waves is 3.00×10^8 m/s.
3. **Compare and Contrast** How does the wavelength of the tuned-in radio signal compare to the size of holes or openings in the materials used to shield the radio?
4. **Interpret Data** What did the materials that effectively shielded the radio from receiving signals have in common?

Conclude and Apply

1. **Explain** Offer an explanation of what might be happening to the electric and magnetic fields of the radio wave fields that are blocked from reaching the radio by the shielding materials.
2. **Infer** Why was covering the radio with your arms not effective at blocking the radio waves?
3. **Use Scientific Explanations** Ocean water absorbs radio waves, limiting their penetration below the surface to a depth equal to approximately one wavelength. Because of this, very low frequency (40–80 Hz) radio waves are used to contact submerged submarines. Why might the site for a high-powered radio transmitter used for submarine communication be located in a remote area far from the ocean? *(Hint: Estimate the length if a one-half wavelength antenna were used.)*

Going Further

How does the size of the small holes in the metal screen of a microwave oven door compare to the wavelength of a 2.4-GHz microwave?

Real-World Physics

Suppose you want to mail some photos or an audio recording stored on a magnetic computer diskette to a friend. What should you do to protect the diskette from electromagnetic waves during shipping?

To find out more about electromagnetic waves, visit the Web site: **physicspp.com**

Technology and Society

Cellular Phones

Do you own a cellular phone? Once rare and expensive, cell phones are now commonplace and relatively affordable. Approximately 60 percent of American teenagers have one.

Cellular Networks The cell phone gets its name from the way cell-phone companies divide a city into small regions called *cells.* Each cell is a hexagon-shaped zone within a larger hexagonal grid. Cells are typically about 26 square kilometers in area, though they vary in size depending on terrain and the number of cell phone users. Located within each cell is a base station, consisting of a tall tower-like structure and boxes or buildings that contain radio equipment. When you make a call, the signal from your phone is transmitted to the base station located within your cell. This signal then is transmitted from the local base station to the base station where the person you are calling is located.

How then, do cell phones communicate with base stations? Cell phones use radio waves to transmit information to and receive information from base stations. A cell phone is essentially a two-way radio containing both a radio transmitter and a radio receiver. The cell-phone transmitter takes the sound of your voice, encodes it onto a radio-frequency wave and then transmits the radio wave to a nearby base station. The base station then relays the signal to the destination base station by broadcasting radio waves through the air. On the receiving end of the call, the cell phone picks up the radio signal, decodes it, and reconverts it into audible sounds you can understand. By making use of two different frequencies (one frequency for talking and one frequency for listening) both people connected by the call can talk at the same time.

Network of Base Stations

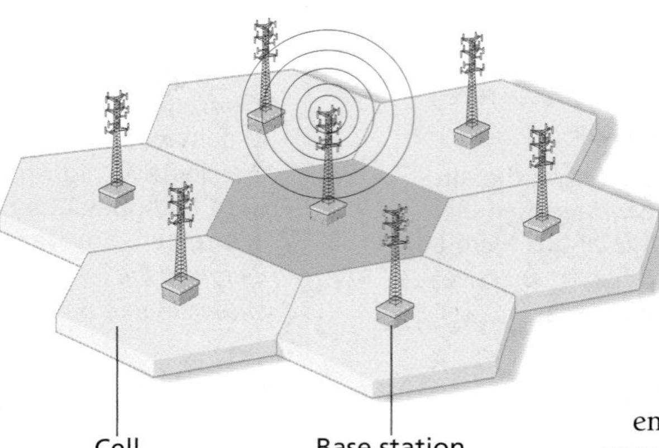

A cellular phone company's system of base stations can relay your call all the way across the country, even when both people involved in the call are moving. When you move from cell to cell, the signal is automatically relayed to the correct base station in the system.

Risks of Cell Phone Use Using a cell phone is not a risk-free activity. For example, driving while talking on a cell phone is dangerous. One study found that people talking on their cell phones while driving were four times more likely to be involved in an accident than non-cell phone users. Some service stations post warning signs prohibiting cell phone use. Static electricity generated by the cell phone might cause the gasoline fumes to ignite.

Another potential risk is more controversial. Because cell phones broadcast radio waves, they emit electromagnetic energy known as radio-frequency (RF) energy. There is some evidence that cell phones emit enough radiation to cause severe health problems, such as brain cancer and Alzheimer's disease. There is evidence supporting both sides of the debate. Currently, no one knows for certain what the long-term health effects of cell phone use are, if any.

Going Further

1. **Use Scientific Explanations** How did cellular phones get their name?
2. **Compare and Contrast** How are an AM/FM radio and a cellular phone similar? How are they different?
3. **Critical Thinking** Explain why the low-power transmitters used by cell phones are important in keeping the phones lightweight.

Chapter 26

Study Guide

26.1 Interactions of Electric and Magnetic Fields and Matter

Vocabulary

- isotope (p. 701)
- mass spectrometer (p. 701)

Key Concepts

- The ratio of charge to mass of an electron was measured by J. J. Thomson using balanced electric and magnetic fields in a cathode-ray tube.

$$\frac{q}{m} = \frac{v}{Br}$$

- An electron's mass can be found by combining Thomson's result with Millikan's measurement of the electron's charge.
- Atoms of the same element can have different masses.
- The mass spectrometer uses both electric and magnetic fields to measure the masses of ionized atoms and molecules.
- The mass spectrometer can be used to determine the charge-to-mass ratio of an ion.

$$\frac{q}{m} = \frac{2V}{B^2r^2}$$

26.2 Electric and Magnetic Fields in Space

Vocabulary

- electromagnetic wave (p. 706)
- dielectrics (p. 707)
- antenna (p. 707)
- electromagnetic spectrum (p. 709)
- electromagnetic radiation (p. 710)
- piezoelectricity (p. 711)
- receiver (p. 712)

Key Concepts

- Electromagnetic waves are coupled, changing electric and magnetic fields that move through space.
- The wavelength of a wave is equal to its speed divided by its frequency.

$$\lambda = \frac{v}{f}$$

  For an electromagnetic wave traveling in a vacuum, the speed in the above equation, v, is equal to the speed of light, c.
- The velocity of an electromagnetic wave through a dielectric is less than the speed of light in a vacuum.
- A changing current in a transmitting antenna is used to generate electromagnetic waves.
- Electromagnetic radiation can transmit energy and information through a medium or a vacuum.
- Piezoelectricity is the property of a crystal causing it to bend or deform and produce electrical vibrations when a voltage is applied across it.
- Receiving antennas convert electromagnetic waves to varying electric fields in conductors.
- Electromagnetic waves can be detected by the *EMF* that they produce in an antenna. Particular frequencies of electromagnetic waves can be selected by using a resonating coil-and-capacitor circuit, known as a tuner.
- A receiver obtains transmitted information from electromagnetic waves.
- The length of the most efficient antenna is one-half the wavelength of the wave to be detected.
- Microwave and infrared waves can accelerate electrons in molecules, thereby producing thermal energy.
- X rays are high-frequency electromagnetic waves emitted by rapidly accelerated electrons.

Chapter 26 Assessment

Concept Mapping

30. Complete the following concept map using the following term and symbols: *E*, *c*, *magnetic field*.

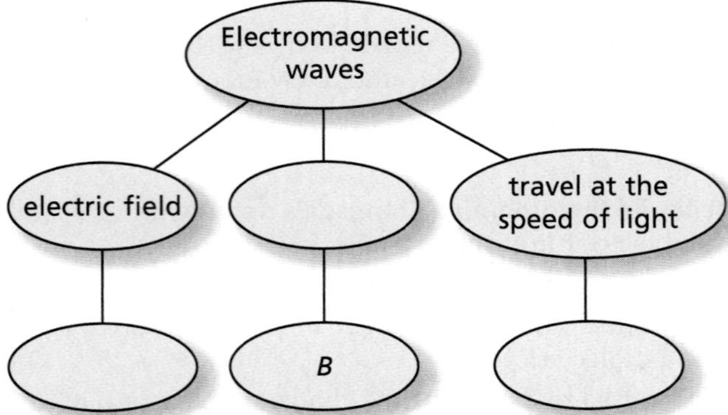

Mastering Concepts

31. What are the mass and charge of an electron? (26.1)

32. What are isotopes? (26.1)

33. The direction of an induced magnetic field is always at what angle to the changing electric field? (26.2)

34. Why must an AC generator be used to propagate electromagnetic waves? If a DC generator were used, when would it create electromagnetic waves? (26.2)

35. A vertical antenna wire transmits radio waves. Sketch the antenna and the electric and magnetic fields that it creates. (26.2)

36. What happens to a quartz crystal when a voltage is applied across it? (26.2)

37. How does an antenna's receiving circuit select electromagnetic radio waves of a certain frequency and reject all others? (26.2)

Applying Concepts

38. The electrons in a Thomson tube travel from left to right, as shown in **Figure 26-14.** Which deflection plate should be charged positively to bend the electron beam upward?

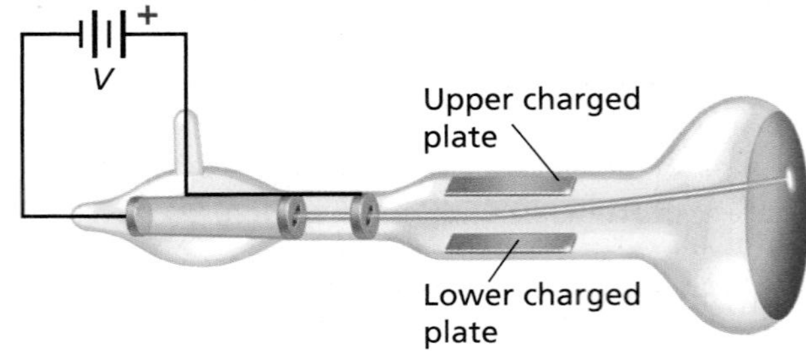

Figure 26-14

39. The Thomson tube in question 38 uses a magnetic field to deflect the electron beam. What would the direction of the magnetic field need to be to bend the beam downward?

40. Show that the units of E/B are the same as the units for velocity.

41. The vacuum chamber of a mass spectrometer is shown in **Figure 26-15.** If a sample of ionized neon is being tested in the mass spectrometer, in what direction must the magnetic field be directed to bend the ions into a clockwise semicircle?

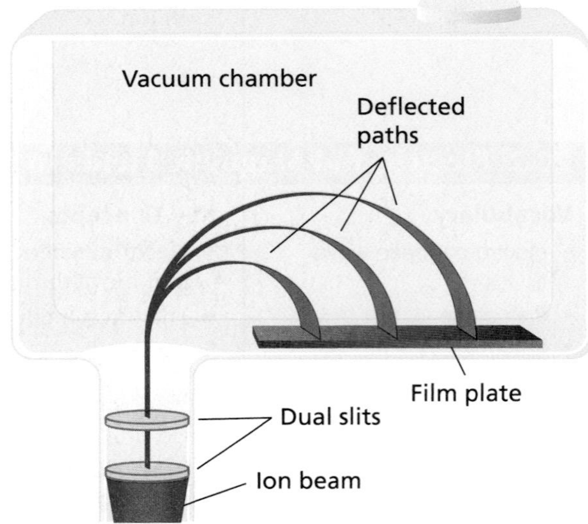

Figure 26-15

42. If the sign of the charge on the particles in question 41 is changed from positive to negative, do the directions of either or both of the fields have to be changed to keep the particles undeflected? Explain.

43. For each of the following properties, identify whether radio waves, light waves, or X rays have the largest value.
a. wavelength
b. frequency
c. velocity

44. TV Waves The frequency of television waves broadcast on channel 2 is about 58 MHz. The waves broadcast on channel 7 are about 180 MHz. Which channel requires a longer antenna?

45. Suppose the eyes of an alien being are sensitive to microwaves. Would you expect such a being to have larger or smaller eyes than yours? Why?

Mastering Problems

26.1 Interactions of Electric and Magnetic Fields and Matter

46. Electrons moving at 3.6×10^4 m/s pass through an electric field with an intensity of 5.8×10^3 N/C. How large a magnetic field must the electrons also experience for their path to be undeflected?

47. A proton moves across a 0.36-T magnetic field, as shown in **Figure 26-16.** If the proton moves in a circular path with a radius of 0.20 m, what is the speed of the proton?

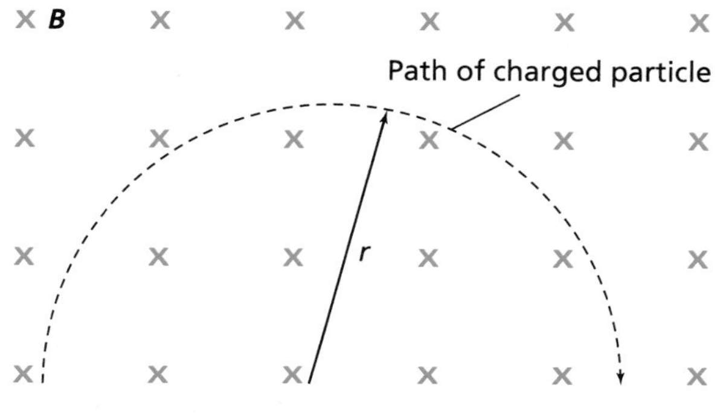

Figure 26-16

48. A proton enters a 6.0×10^{-2}-T magnetic field with a speed of 5.4×10^4 m/s. What is the radius of the circular path it follows?

49. An electron is accelerated by a 4.5-kV potential difference. How strong a magnetic field must be experienced by the electron if its path is a circle of radius 5.0 cm?

50. A mass spectrometer yields the following data for a beam of doubly ionized (2+) sodium atoms: $B = 8.0\times10^{-2}$ T, $q = 2(1.60\times10^{-19}$ C), $r = 0.077$ m, and $V = 156$ V. Calculate the mass of a sodium atom.

51. An alpha particle has a mass of approximately 6.6×10^{-27} kg and has a charge of 2+. Such a particle is observed to move through a 2.0-T magnetic field along a path of radius 0.15 m.

a. What speed does the particle have?

b. What is its kinetic energy?

c. What potential difference would be required to give it this kinetic energy?

52. A mass spectrometer analyzes carbon-containing molecules with a mass of 175×10^3 proton masses. What percent differentiation is needed to produce a sample of molecules in which only carbon isotopes of mass 12, and none of mass 13, are present?

53. Silicon Isotopes In a mass spectrometer, ionized silicon atoms have curvatures, as shown in **Figure 26-17.** If the smaller radius corresponds to a mass of 28 proton masses, what is the mass of the other silicon isotope?

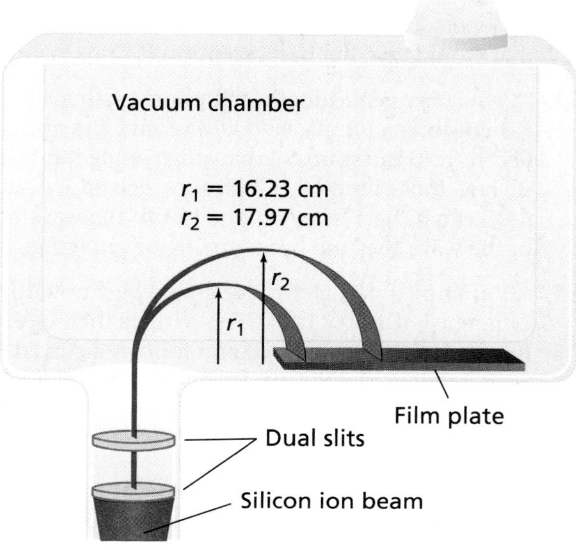

Figure 26-17

26.2 Electric and Magnetic Fields in Space

54. Radio Waves The radio waves reflected by a parabolic dish are 2.0 cm long. How long should the antenna be that detects the waves?

55. TV A television signal is transmitted on a carrier frequency of 66 MHz. If the wires on a receiving antenna are placed $\frac{1}{4}\lambda$ apart, determine the physical distance between the receiving antenna wires.

56. Bar-Code Scanner A bar-code scanner uses a laser light source with a wavelength of about 650 nm. Determine the frequency of the laser light source.

57. What is the optimum length of a receiving antenna that is to receive a 101.3-MHz radio signal?

58. An EM wave with a frequency of 100-MHz is transmitted through a coaxial cable having a dielectric constant of 2.30. What is the velocity of the wave's propagation?

59. Cell Phone A certain cellular telephone transmitter operates on a carrier frequency of 8.00×10^8 Hz. What is the optimal length of a cell phone antenna designed to receive this signal? Note that single-ended antennas, such as those used by cell phones, generate peak *EMF* when their length is one-fourth the wavelength of the wave.

Mixed Review

60. The mass of a doubly ionized (2+) oxygen atom is found to be 2.7×10^{-26} kg. If the mass of an atomic mass unit (amu) is equal to 1.67×10^{-27} kg, how many atomic mass units are in the oxygen atom?

61. **Radio** An FM radio station broadcasts on a frequency of 94.5 MHz. What is the antenna length that would give the best reception for this station?

62. At what frequency does a cell phone with an 8.3-cm-long antenna send and receive signals? Recall from question 59 that single-ended antennas, such as those used by cell phones, generate peak *EMF* when their length is one-fourth the wavelength of the wave they are broadcasting or receiving.

63. An unknown particle is accelerated by a potential difference of 1.50×10^{2} V. The particle then enters a magnetic field of 50.0 mT, and follows a curved path with a radius of 9.80 cm. What is the ratio of q/m?

Thinking Critically

64. **Apply Concepts** Many police departments use radar guns to catch speeding drivers. A radar gun is a device that uses a high-frequency electromagnetic signal to measure the speed of a moving object. The frequency of the radar gun's transmitted signal is known. This transmitted signal reflects off of the moving object and returns to the receiver on the radar gun. Because the object is moving relative to the radar gun, the frequency of the returned signal is different from that of the originally transmitted signal. This phenomenon is known as the Doppler shift. When the object is moving toward the radar gun, the frequency of the returned signal is greater than the frequency of the original signal. If the initial transmitted signal has a frequency of 10.525 GHz and the returned signal shows a Doppler shift of 1850 Hz, what is the speed of the moving object? Use the following equation:

$$v_{\text{target}} = \frac{cf_{\text{Doppler}}}{2f_{\text{transmitted}}}$$

Where,

v_{target} = velocity of target (m/s)

c = speed of light (m/s)

f_{Doppler} = Doppler shift frequency (Hz)

$f_{\text{transmitted}}$ = frequency of transmitted wave (Hz)

65. **Apply Concepts** H. G. Wells wrote a science-fiction novel called *The Invisible Man,* in which a man drinks a potion and becomes invisible, although he retains all of his other faculties. Explain why an invisible person would not be able to see.

66. **Design an Experiment** You are designing a mass spectrometer using the principles discussed in this chapter, but with an electronic detector replacing the photographic film. You want to distinguish singly ionized (1+) molecules of 175 proton masses from those with 176 proton masses, but the spacing between adjacent cells in your detector is 0.10 mm. The molecules must have been accelerated by a potential difference of at least 500.0 V to be detected. What are some of the values of V, B, and r that your apparatus should have?

Writing in Physics

67. Compose a 1–2 page report in which you outline the operation of a typical television, DVD, or VCR infrared remote-control unit. Explain why the simultaneous use of multiple remote-control units typically does not cause the units to interfere with each other. Your report should include block diagrams and sketches.

Cumulative Review

68. A He–Ne laser (λ = 633 nm) is used to illuminate a slit of unknown width, forming a pattern on a screen that is located 0.95 m behind the slit. If the first dark band is 8.5 mm from the center of the central bright band, how wide is the slit? (Chapter 19)

69. The force between two identical metal spheres with the charges shown in **Figure 26-18** is F. If the spheres are touched together and returned to their original positions, what is the new force between them? (Chapter 20)

Figure 26-18

70. What is the electric field strength between two parallel plates spaced 1.2 cm apart if a potential difference of 45 V is applied to them? (Chapter 21)

71. Calculate the daily cost of operating an air compressor that runs one-fourth of the time and draws 12.0 A from a 245-V circuit if the cost is $0.0950 per kWh. (Chapter 22)

72. A 440-cm length of wire carrying 7.7 A is at right angles to a magnetic field. The force on the wire is 0.55 N. What is the strength of the field? (Chapter 24)

73. A north-south wire is moved toward the east through a magnetic field that is pointing down, into Earth. What is the direction of the induced current? (Chapter 25)

Multiple Choice

1. For a charged particle moving in a circular trajectory, ______.

   Ⓐ the magnetic force is parallel to the velocity and is directed toward the center of the circular path
   Ⓑ the magnetic force may be perpendicular to the velocity and is directed away from the center of the circular path
   Ⓒ the magnetic force always remains parallel to the velocity and is directed away from the center of the circular path
   Ⓓ the magnetic force always remains perpendicular to the velocity and is directed toward the center of the circular path

2. The radius of the circular path that a proton travels while in a constant 0.10-T magnetic field is 6.6 cm. What is the velocity of the proton?

   Ⓐ 6.3×10^{5} m/s
   Ⓑ 2.0×10^{6} m/s
   Ⓒ 6.3×10^{7} m/s
   Ⓓ 2.0×10^{12} m/s

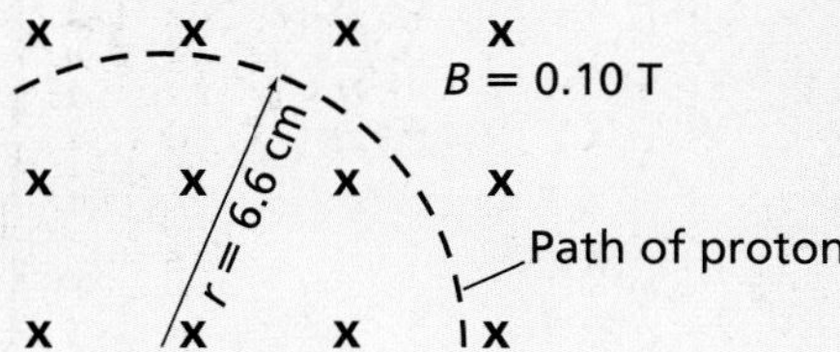

3. The dielectric constant of ruby mica is 5.4. What is the speed of light as it passes through ruby mica?

   Ⓐ 3.2×10^{3} m/s
   Ⓑ 9.4×10^{4} m/s
   Ⓒ 5.6×10^{7} m/s
   Ⓓ 1.3×10^{8} m/s

4. A certain radio station broadcasts with waves that are 2.87 m long. What is the frequency of the radio waves?

   Ⓐ 9.57×10^{-9} Hz
   Ⓑ 3.48×10^{-1} Hz
   Ⓒ 1.04×10^{8} Hz
   Ⓓ 3.00×10^{8} Hz

5. Which one of the following situations does not create an electromagnetic wave?

   Ⓐ Direct current (DC) voltage is applied to a piezoelectric quartz crystal.
   Ⓑ Current passes through a wire contained inside a plastic pipe.
   Ⓒ Current passes through a coil-and-capacitor circuit with a molecular-size resonant cavity.
   Ⓓ High energy electrons strike a metal target in an X-ray tube.

6. A proton beam has a radius of 0.52 m as it moves perpendicular to a magnetic field of 0.45 T. If the mass of an individual proton is 1.67×10^{-27} kg, what is the speed of the protons making up the beam?

   Ⓐ 1.2 m/s
   Ⓑ 4.7×10^{3} m/s
   Ⓒ 2.2×10^{7} m/s
   Ⓓ 5.8×10^{8} m/s

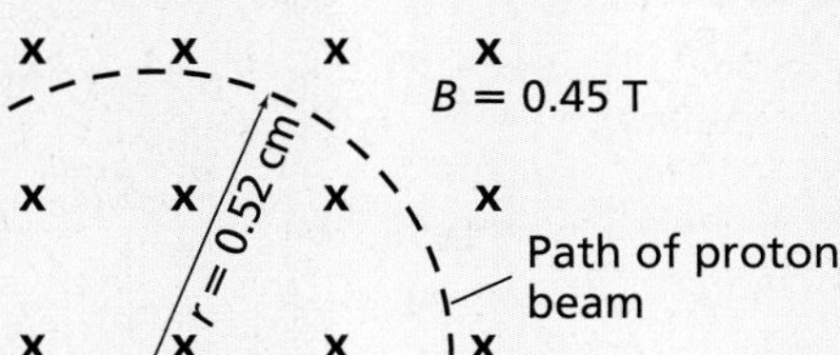

Extended Answer

7. A deuteron (the nucleus of deuterium) has a mass of 3.34×10^{-27} kg and a charge of $+e$. It travels in a magnetic field of 1.50 T in a circular path, with a radius of 0.0400 m. What is the velocity of the particle?

✓ Test-Taking TIP

Watch the Little Words

Underline words such as *never, always, least, not,* and *except* when you see them in test questions. These small words dramatically impact the meaning of a question.

Chapter 27

Quantum Theory

What You'll Learn

- You will understand that light behaves like particles having momentum and energy.
- You will learn that small particles of matter behave like waves and are subject to diffraction and interference.

Why It's Important

Quantum theory provides the basis for an amazing device called a scanning tunneling microscope (STM). The STM is vital to researchers studying DNA and chemical reaction mechanisms. It also is used in the development of smaller and faster computers.

Atomic-Scale Images Two types of silicon atoms, appearing in red and blue, are seen in this STM image of silicon.

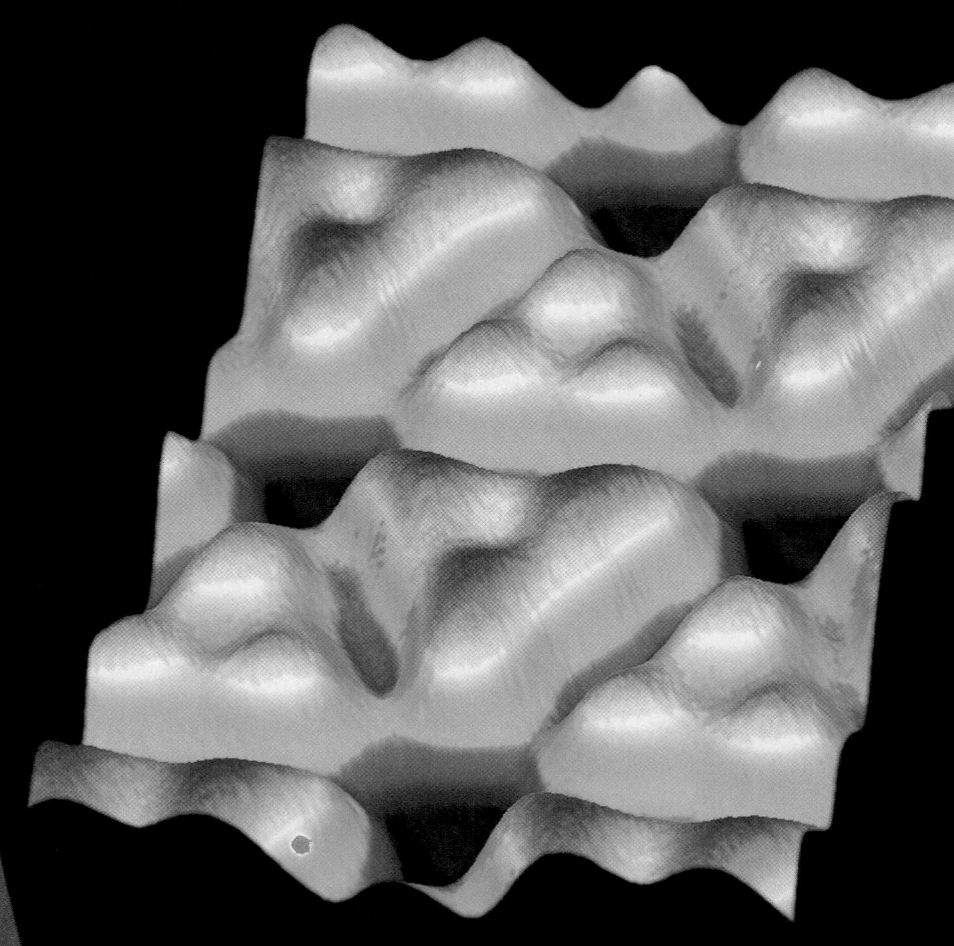

Think About This ▶

An STM was used by The Colorado School of Mines to produce this image of the surface of silicon. The STM uses the ability of electrons to jump across a barrier. How does this jump, which is impossible according to the law of conservation of energy, occur?

physicspp.com

What does the spectrum of a glowing lightbulb look like?

Question
What colors of visible light are emitted by a glowing, incandescent lightbulb?

Procedure

1. Screw a clear, incandescent lightbulb into a lamp socket.
2. Plug the lamp into an electrical outlet that is controlled by a dimmer switch. Turn the lamp on and dim it so that it glows weakly. ***CAUTION: Do not touch the glowing bulb, as it is very hot and can cause burns.***
3. Dim or turn off the other lights in the classroom.
4. Standing about 1-2 m away from the lightbulb, hold a holographic diffraction grating close to your eye. Observe the lightbulb through the diffraction grating. ***CAUTION: Do not directly view the glowing lightbulb without using the diffraction grating, as damage to your vision may result.***
5. **Make and Use Scientific Illustrations** Use colored pencils to sketch a diagram of what you observe.
6. Turn up the dimmer control to increase the lightbulb to its maximum brightness.
7. **Make and Use Scientific Illustrations** Use colored pencils to sketch a diagram of what you observe.

Analysis
Describe the spectrum emitted by the lightbulb. Is it continuous or a series of distinct colored lines? Describe how the observed spectrum changed when the lightbulb glowed brighter.

Critical Thinking
What is the source of the light emitted by the bulb? What happens to the temperature of the lightbulb's filament when the bulb glows brighter?

27.1 A Particle Model of Waves

James Maxwell's electromagnetic wave theory, which you learned about in the previous chapter, was proven to be correct by the experiments of Heinrich Hertz in 1889. Light was then firmly established as an electromagnetic wave. All of optics, including phenomena such as interference, diffraction, and polarization, seemed to be explainable in terms of the electromagnetic wave theory.

Problems remained for physicists, however, because Maxwell's notion of light as a purely electromagnetic wave could not explain several other important phenomena. These problems generally involved the absorption or emission of electromagnetic radiation. Two such problems were the emission spectrum given off by a hot body (hot object) and the discharge of electrically charged particles from a metal surface when ultraviolet radiation was incident upon it. As you will learn, these phenomena could be explained once it was understood that electromagnetic waves have particle-like properties in addition to wavelike properties.

▸ Objectives
- **Describe** the spectrum emitted by a hot body.
- **Explain** the photoelectric and Compton effects.
- **Solve** problems involving the photoelectric effect.

▸ Vocabulary
emission spectrum
quantized
photoelectric effect
threshold frequency
photon
work function
Compton effect

Radiation from Incandescent Bodies

Why was the radiation emitted from a hot body puzzling to physicists? The problem had to do with the intensity and frequency of the emitted radiation at different temperatures. Maxwell's electromagnetic wave theory could not account for the observed radiation emissions of hot bodies. What then, is the nature of the radiation emitted by hot bodies?

The lightbulb that you observed in the Launch Lab at the start of this chapter is an example of a hot body. As predicted by electromagnetic theory, light and infrared radiation are emitted by the vibrating charged particles within the filament. The filament glows because it is hot, and it is said to be incandescent; hence, the name incandescent lightbulb. The colors that you see depend upon the relative intensities of the emitted electromagnetic waves of various frequencies, and the sensitivity of your eyes to those waves.

When the dimmer control is used to increase the voltage to the bulb, the temperature of the glowing filament increases. As a result, the color changes from deep red to orange to yellow and finally, to white. This color change occurs because the higher-temperature filament emits higher-frequency radiation. The higher-frequency radiation comes from the higher-frequency end of the visible spectrum (the violet end) and results in the filament appearing to be whiter.

What would you expect to see if you viewed the glowing filament through a diffraction grating? When viewed in this way, all of the colors of the rainbow would be visible. The bulb also emits infrared radiation that you would not see. A plot of the intensity of the light emitted from a hot body over a range of frequencies is known as an **emission spectrum.** Emission spectra of the incandescent body at temperatures of 4000 K, 5800 K, and 8000 K are shown in **Figure 27-1.** Note that at each temperature, there is a frequency at which the maximum amount of energy is emitted. If you compare the location of each curve's maximum, you will see that as the temperature increases, the frequency at which the maximum amount of energy is emitted also increases.

MINI LAB

Glows in the Dark

Close the shades and turn off the lights in the room. Shine a flashlight at a beaker that contains fluorescein. Now place a red filter over the flashlight so that only red light hits the beaker.

1. **Describe** the results.
2. **Predict** how using a green filter instead of the red filter will affect the results.
3. **Test** your prediction.
4. **Explain** the results.
5. **Predict** whether the fluorescein will glow when a blue filter is used and provide an explanation for your prediction.
6. **Test** your prediction.

Analyze and Conclude

7. **Write** a brief explanation summarizing and explaining your observations.

Figure 27-1 This graph shows the emission spectra of an incandescent body at three different temperatures.

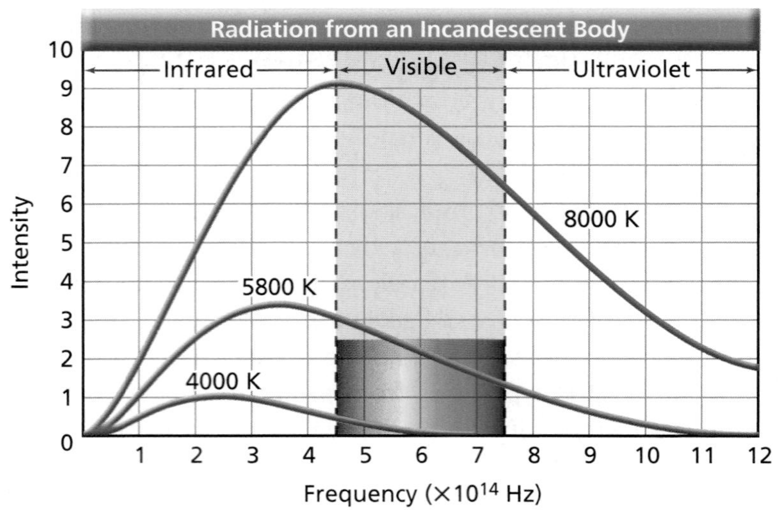

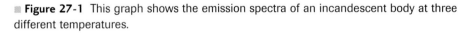

The total power emitted by a hot body also increases with temperature. The power (the energy emitted per second) of an electromagnetic wave is proportional to the hot body's kelvin temperature raised to the fourth power, $\propto T^4$. Thus, hotter bodies radiate considerably more power than do cooler bodies. Probably the most common example of a hot body radiating a great amount of power is the Sun, a dense ball of gases heated to incandescence by the energy produced within it. The Sun's surface temperature is 5800 K, and it radiates 4×10^{26} W of power, an enormous quantity. On average, each square meter of Earth's surface receives about 1000 J of energy per second (1000 W), enough to power ten 100-W lightbulbs.

Astronomy Connection

The problem with Maxwell's electromagnetic theory was that it is unable to explain the shape of the spectrum shown in Figure 27-1. Between 1887 and 1900, many physicists used existing classical physics theories to try to explain the shape. They all failed. In 1900, German physicist Max Planck found that he could calculate the spectrum if he introduced a revolutionary hypothesis: that atoms are not able to continuously change their energy. Planck assumed that the vibrational energy of the atoms in a solid could have only specific frequencies, as shown by the following equation.

Energy of Vibration $E = nhf$

The energy of a vibrating atom is equal to the product of an integer, Planck's constant, and the frequency of the vibration.

In the equation, f is the frequency of vibration of the atom, h is a constant, called Planck's constant, with a value of 6.626×10^{-34} J/Hz, and n is an integer such as 0, 1, 2, 3 . . .

$$n = 0:\ E = (0)hf = 0$$
$$n = 1:\ E = (1)hf = hf$$
$$n = 2:\ E = (2)hf = 2hf$$
$$n = 3:\ E = (3)hf = 3hf$$
and so on

Thus, the energy, E, can have the values hf, $2hf$, $3hf$, and so on, but never values such as $\frac{2}{3}hf$ or $\frac{3}{4}hf$. In other words, energy is **quantized**—it exists only in bundles of specific amounts. h is usually rounded to 6.63×10^{-34} J/Hz for calculations.

Planck also proposed that atoms do not always radiate electromagnetic waves when they are vibrating, as predicted by Maxwell. Instead, Planck proposed that atoms emit radiation only when their vibrational energy changes. For example, if the energy of an atom changes from $3hf$ to $2hf$, the atom emits radiation. The energy radiated is equal to the change in energy of the atom, in this case hf.

Planck found that the constant, h, has an extremely small value. This means that the energy-changing steps are too small to be noticeable in ordinary bodies. Still, the introduction of quantized energy was extremely troubling to physicists, especially to Planck himself. It was the first hint that the classical physics of Newton and Maxwell might be valid only under certain conditions. Planck was honored for his groundbreaking theory of quantized energy with a Nobel prize in 1918 and, more recently, with the postage stamp shown in **Figure 27-2.**

APPLYING PHYSICS

Temperature of the Universe The universe is filled with the radiation that was emitted when it was a very hot object. Currently, the emission spectrum of the universe matches that of a body with a temperature of 2.7 K. This is very cold. As you may recall, 0 K is the lowest possible temperature on the Kelvin temperature scale and is known as absolute zero.

Figure 27-2 This stamp honors Max Planck's work and refers to the constant that bears his name, Planck's constant, h. Planck's constant, which has a value of 6.626×10^{-34} J/Hz, is used in many quantum-related equations.

The Photoelectric Effect

Physicists of the early 1900s also were challenged by another troubling experimental result that could not be explained by Maxwell's wave theory. When ultraviolet radiation was incident on a negatively charged zinc plate, the plate discharged. When ordinary visible light was incident on the same charged plate, the plate did not discharge. This result was contrary to electromagnetic theory. Both ultraviolet radiation and visible light are forms of electromagnetic radiation, so why would the zinc plate be discharged by one but not by the other? And why would a positively charged zinc plate not be similarly discharged? Further study showed that the negatively charged zinc plate was discharging by emitting or ejecting electrons. The emission of electrons when electromagnetic radiation falls on an object is called the **photoelectric effect.**

The photoelectric effect can be studied in a photocell, such as the one shown in **Figure 27-3.** The cell contains two metal electrodes sealed in a tube from which the air has been removed. The evacuated tube keeps the metal surfaces from oxidizing and keeps the electrons from being slowed or stopped by particles in the air. The larger electrode, the cathode, usually is coated with cesium or another alkali metal. The smaller electrode, the anode, is made of a thin wire so that it blocks only a very small amount of radiation. The tube often is made of quartz so as to allow ultraviolet wavelengths to pass through it. A potential difference placed across the electrodes attracts electrons to the anode.

When no radiation falls on the cathode, there is no current in the circuit. When radiation falls on the cathode, a current is produced, which is measured by the ammeter, as shown in Figure 27-3. The current is produced because the photoelectric effect causes the ejection of electrons, also called photoelectrons, from the cathode. The flow of electrons is the current in the circuit. The electrons travel to the anode, the positive electrode.

Threshold frequency Not all radiation falling on the cathode results in a current. Electrons are ejected from the cathode only if the frequency of the radiation is greater than a certain minimum value, called the **threshold frequency,** f_0. The threshold frequency varies widely, depending on the type of metal. For example, all wavelengths of visible light except red will eject electrons from cesium, but no wavelength of visible light will eject electrons from zinc. Higher-frequency ultraviolet radiation is needed to produce the photoelectric effect in zinc.

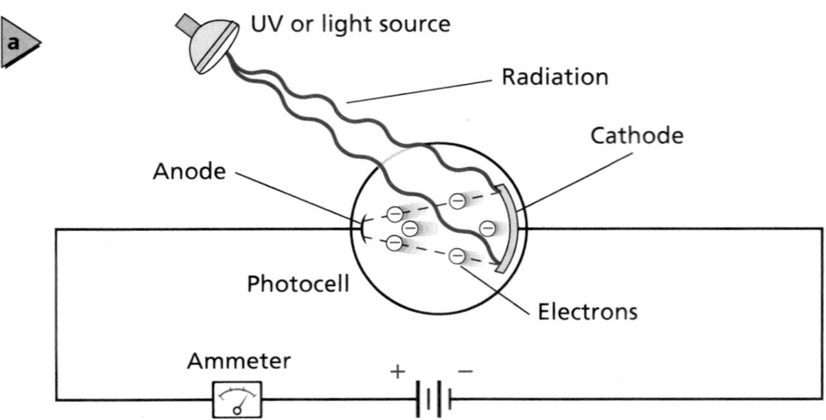

■ **Figure 27-3** In the photocell shown, electrons ejected from the cathode flow to the anode, completing the circuit and generating an electric current **(a).** This handheld light meter works because of the photoelectric effect and is used by photographers to measure light levels **(b).**

Concepts In Motion
Interactive Figure To see an animation on the photoelectric effect, visit physicspp.com.

No matter how intense, radiation with a frequency below f_0 will not cause the ejection of electrons from metal. Conversely, even very low-intensity radiation with a frequency at or above the threshold frequency causes the immediate ejection of electrons. When the incident radiation's frequency is equal to or greater than the threshold frequency, increasing the intensity of the radiation causes an increase in the flow of photoelectrons.

How does the electromagnetic wave theory explain the photoelectric effect? It can't. According to electromagnetic wave theory, the electric field accelerates and ejects the electrons from the metal, and the strength of the electric field is related to the intensity of the radiation (not to the radiation's frequency). Thus, it follows that electrons in the metal would need to absorb energy from a dim light source for a very long time before they gained enough energy to be ejected. As you just learned, however, this is not the case. Observations show that electrons are ejected immediately when even low-intensity radiation at or above the threshold frequency is incident on the metal.

Photons and quantized energy In 1905, Albert Einstein published a revolutionary theory that explained the photoelectric effect. According to Einstein, light and other forms of electromagnetic radiation consist of discrete, quantized bundles of energy, each of which was later called a **photon.** The energy of a photon depends on its frequency.

Energy of a Photon $E = hf$

The energy of a photon is equal to the product of Planck's constant and the frequency of the photon.

In the above equation, f is frequency in Hz, and h is Planck's constant. Because the unit Hz = 1/s or s^{-1}, the J/Hz unit of Planck's constant is also equivalent to J·s. Because the joule is too large a unit of energy to use with atomic-sized systems, the more convenient energy unit of the electron volt (eV) is usually used. One electron volt is the energy of an electron accelerated across a potential difference of 1 V.

$$1 \text{ eV} = (1.60\times10^{-19} \text{ C})(1 \text{ V})$$

$$= 1.60\times10^{-19} \text{ C}\cdot\text{V}$$

$$= 1.60\times10^{-19} \text{ J}$$

Using the definition of an electron volt allows the photon energy equation to be rewritten in a simplified form, as shown below.

Energy of a Photon $E = \frac{hc}{\lambda} = \frac{1240 \text{ eV}\cdot\text{nm}}{\lambda}$

The energy of a photon is equal to the constant 1240 eV·nm divided by the wavelength of the photon.

An explanation of the derivation of this equation and how to use it is given in the Problem-Solving Strategies on the next page.

Personal Tutor For an online tutorial on units of *hc* and photon energy, visit physicspp.com.

PROBLEM-SOLVING Strategies

Units of *hc* and Photon Energy

Connecting **Math to Physics**

Converting the quantity hc to the unit eV·nm results in a simplified equation that can be used to solve problems involving photon wavelength.

1. The energy of a photon of wavelength λ is given by the equation $E = hf$.
2. Because $f = c/\lambda$, this equation can be written as $E = hc/\lambda$.
3. When using the equation $E = hc/\lambda$, if the value of hc in eV·nm is divided by λ in nm, you will obtain the energy in eV. Thus, it is useful to know the value of hc in eV·nm.
4. The conversion of hc to the unit eV·nm is as follows:

$$hc = (6.626\times10^{-34}\ \text{J/Hz})(2.998\times10^{8}\ \text{m/s})\left(\frac{1\ \text{eV}}{1.602\times10^{-19}\ \text{J}}\right)\left(\frac{10^{9}\ \text{nm}}{1\ \text{m}}\right)$$

$$= 1240\ \text{eV}$$

5. Substituting $hc = 1240$ eV·nm into the equation for the energy of a photon yields the following, where λ is in nm and E is in eV:

$$E = \frac{hc}{\lambda} = \frac{1240\ \text{eV}\cdot\text{nm}}{\lambda}$$

6. Use the above equation to solve photon energy problems when energy in eV is desired.

It is important to note that Einstein's theory of the photon goes further than Planck's theory of radiation from hot bodies. While Planck had proposed that vibrating atoms emit electromagnetic radiation with energy equal to *nhf*, he did not suggest that light and other forms of electromagnetic radiation act like particles. Einstein's theory of the photon reinterpreted and extended Planck's theory of radiation from hot bodies.

Einstein's photoelectric-effect theory is able to explain the existence of a threshold frequency. A photon with a minimum frequency and energy, hf_0, is needed to eject an electron from metal. If the photon has a frequency below f_0, the photon will not have the energy needed to eject an electron. Because one photon interacts with one electron, an electron cannot simply accumulate subthreshold energy photons until it has enough energy to be ejected. On the other hand, radiation with a frequency greater than f_0 has more than enough energy to eject an electron. In fact, the excess energy, $hf - hf_0$, becomes the kinetic energy of the ejected electron.

Kinetic Energy of an Electron Ejected Due to the Photoelectric Effect $KE = hf - hf_0$

The kinetic energy of an ejected electron is equal to the difference between the incoming photon energy, hf, and the energy needed to free the electron from the metal, hf_0.

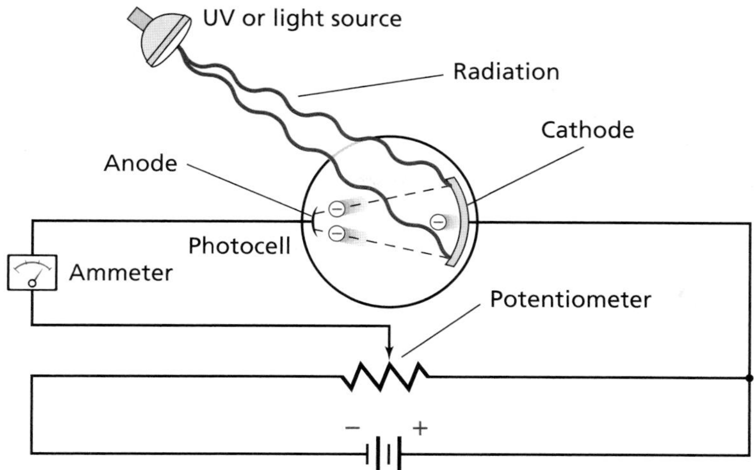

■ **Figure 27-4** The maximum kinetic energy of electrons ejected from the cathode can be measured using this apparatus. The ammeter measures the current through the circuit. By adjusting the potentiometer, the experimenter can determine the potential that results in zero current. At the zero current threshold, the maximum possible kinetic energy of the ejected electron can be calculated.

Note that hf_0 is the minimum energy needed to free the most loosely held electron in an atom. Because not all electrons in an atom have the same energy, some will require more than this minimum energy in order to escape. As a result, the ejected electrons have differing kinetic energies. Thus, it is important to realize that the phrase "kinetic energy of the ejected electrons" refers to the maximum kinetic energy that an ejected electron could have. Some of the ejected electrons will have less.

Testing the photoelectric theory How can Einstein's theory be tested? The kinetic energy of the ejected electrons can be measured indirectly by a device like the one illustrated in **Figure 27-4.** A variable electric potential difference is used to adjust the voltage across the tube. When the potential difference is adjusted to make the anode negative, the ejected electrons must expend energy to reach the anode. Only electrons ejected from the cathode with sufficient kinetic energy will be able to reach the anode.

As shown in Figure 27-4, light of the chosen frequency illuminates the cathode. Gradually, the experimenter increases the opposing potential difference, thereby making the anode more negative. As the opposing potential difference increases, more kinetic energy is needed for the electrons to reach the anode, and fewer electrons arrive there to complete the circuit. At a certain voltage, called the stopping potential, there are no electrons with enough kinetic energy to reach the anode, and the current stops.

At the stopping potential, the kinetic energy of the electrons at the cathode equals the work done by the electric field to stop them. This is represented in equation form as $KE = -qV_0$. In this equation, V_0 is the magnitude of the stopping potential in volts (J/C), and q is the charge of the electron (-1.60×10^{-19} C). Note that the negative sign in the equation along with the negative value of q yield a positive value for KE.

Applications The photoelectric effect is used in a variety of everyday applications. Solar panels, shown in **Figure 27-5,** use the photoelectric effect to convert the Sun's light into electricity. Garage-door openers have safety beams of infrared light that create current in the receiver through the photoelectric effect. If the beam of light is interrupted by an object as the garage door is closing, the current in the receiver stops and triggers the opener to open the door. The photoelectric effect also is used in nightlights and photo eyes that turn lights on and off automatically, depending on whether it is day or night.

■ **Figure 27-5** The solar panels on this building use the photoelectric effect to convert the Sun's light to electricity.

▶ EXAMPLE Problem 1

Photoelectron Kinetic Energy The stopping potential of a certain photocell is 4.0 V. What is the kinetic energy given to the electrons by the incident light? Give your answer in both joules and electron volts.

1 Analyze and Sketch the Problem

- Draw the cathode and anode, the incident radiation, and the direction of the ejected electron. Note that the stopping potential prevents electrons from flowing across the photocell.

Cathode Anode

Incident light λ

\+ −

Known:	Unknown:
$V_0 = 4.0$ V	KE (in J and eV) = ?
$q = -1.60\times10^{-19}$ C	

2 Solve for the Unknown

The electric field does work on the electrons. When the work done, W, equals the negative of the initial kinetic energy, KE, electrons no longer flow across the photocell.

$$KE + W = 0 \text{ J}$$

Solve for KE.

$$KE = -W$$
$$= -qV_0 \quad \text{Substitute } W = qV_0$$
$$= -(-1.60\times10^{-19}\text{ C})(4.0\text{ V}) \quad \text{Substitute } q = -1.60\times10^{-19}\text{ C},\ V_0 = 4.0\text{ V}$$
$$= +6.4\times10^{-19}\text{ J}$$

Convert KE from joules to electron volts.

$$KE = (6.4\times10^{-19}\text{ J})\left(\frac{1\text{ eV}}{1.60\times10^{-19}\text{ J}}\right)$$
$$= 4.0\text{ eV}$$

Math Handbook
Operations with Scientific Notation
pages 842–843

3 Evaluate the Answer

- **Are the units correct?** Joules and electron volts are both units of energy.
- **Does the sign make sense?** Kinetic energy is always positive.
- **Is the magnitude realistic?** The energy in electron volts is equal in magnitude to the stopping potential difference in volts.

▶ PRACTICE Problems

• Additional Problems, Appendix B
• Solutions to Selected Problems, Appendix C

1. An electron has an energy of 2.3 eV. What is the energy of the electron in joules?
2. What is the energy in eV of an electron with a velocity of 6.2×10^6 m/s?
3. What is the velocity of the electron in problem 1?
4. The stopping potential for a photoelectric cell is 5.7 V. Calculate the maximum kinetic energy of the emitted photoelectrons in eV.
5. The stopping potential required to prevent current through a photocell is 3.2 V. Calculate the maximum kinetic energy in joules of the photoelectrons as they are emitted.

CHALLENGE PROBLEM

Suppose a nickel with a mass of 5.0 g vibrates up and down while it is connected to a spring. The maximum velocity of the nickel during the oscillations is 1.0 cm/s. Assume that the vibrating nickel models the quantum vibrations of the electrons within an atom, where the energy of the vibrations is given by the equation $E = nhf$.

1. Find the maximum kinetic energy of the vibrating object.
2. The vibrating object emits energy in the form of light with a frequency of 5.0×10^{14} Hz. If the energy is emitted in a single step, find the energy lost by the object.
3. Determine the number of equally sized energy-step reductions that the object would have to make in order to lose all of its energy.

Mass = 5.0 g
Maximum velocity = 1.0 cm/s

A graph of the kinetic energies of the electrons ejected from a metal versus the frequencies of the incident photons is a straight line, as shown in **Figure 27-6.** All metals have similar graphs with the same slope. This slope equals the rise/run ratio of the line, which is equal to Planck's constant, h.

$$\text{slope} = \frac{\text{rise}}{\text{run}} = \frac{\text{change in maximum kinetic energy of ejected electrons}}{\text{change in frequency of incident radiation}}$$

$$= \frac{\Delta KE}{f} = h$$

The graphs of various metals differ only in the threshold frequency needed to free the electrons. In Figure 27-6, the threshold frequency, f_0, is the point at which $KE = 0$. In this case, f_0, located at the intersection of the line with the x-axis, is approximately 4.4×10^{14} Hz. The threshold frequency is related to the work function of the metal. The **work function** of a metal is the energy needed to free the most weakly bound electron from the metal. The magnitude of the work function is equal to hf_0. When a photon of frequency f_0 is incident on a metal, the energy of the photon is sufficient to release the electron, but not sufficient to provide the electron with any kinetic energy.

Between 1905 and 1916, American physicist Robert Millikan performed a brilliant set of experiments in which he attempted to disprove Einstein's photoelectric theory. While his results confirmed Einstein's equation, he did not accept Einstein's "radical" notion of the photon. Millikan's experiments made it possible for Einstein to receive a Nobel prize for his photoelectric theory in 1921. Two years later, in 1923, Millikan was awarded a Nobel prize for determining the charge of an electron and for his investigations into the photoelectric effect.

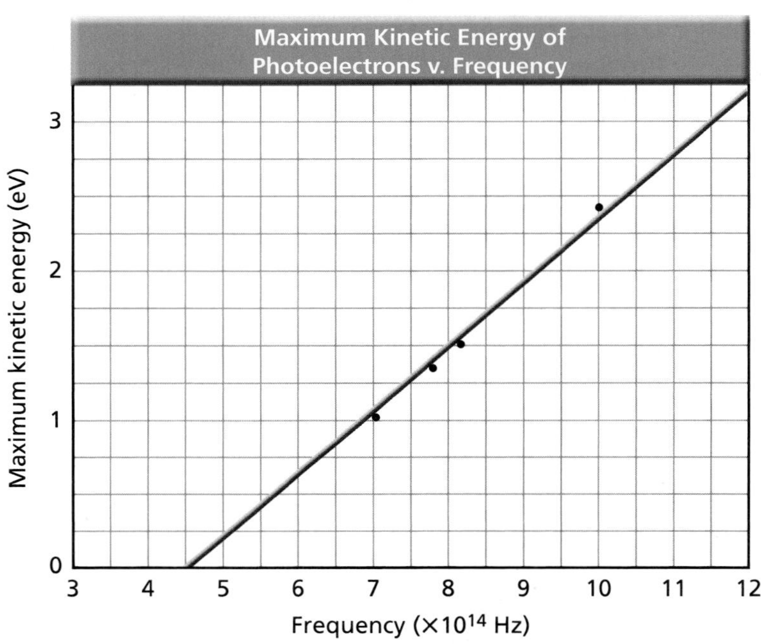

Figure 27-6 This graph shows that as the frequency of the incident radiation increases, the kinetic energy of the ejected electrons increases proportionally.

▶ EXAMPLE Problem 2

Work Function and Energy A photocell uses a sodium cathode. The sodium cathode has a threshold wavelength of 536 nm.

a. Find the work function of sodium in eV.

b. If ultraviolet radiation with a wavelength of 348 nm falls on sodium, what is the energy of the ejected electrons in eV?

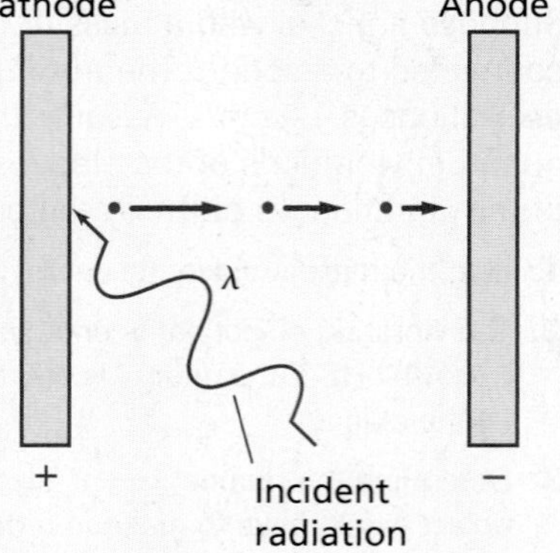

1 Analyze and Sketch the Problem

- Draw the cathode and anode, the incident radiation, and the direction of the ejected electron.

Known:		Unknown:	
$\lambda_0 = 536$ nm	$\lambda = 348$ nm	$W = ?$	$KE = ?$
$hc = 1240$ eV·nm			

2 Solve for the Unknown

a. Find the work function using Planck's constant and the threshold wavelength.

$$W = hf_0 = \frac{hc}{\lambda_0}$$

$$= \frac{1240 \text{ eV·nm}}{536 \text{ nm}}$$ **Substitute $hc = 1240$ eV·nm, $\lambda_0 = 536$ nm**

$$= 2.31 \text{ eV}$$

b. Use Einstein's photoelectric-effect equation to determine the energy of the incident radiation.

$$E = \frac{1240 \text{ eV·nm}}{\lambda}$$

$$= \frac{1240 \text{ eV·nm}}{348 \text{ nm}}$$ **Substitute $\lambda = 348$ nm**

$$= 3.56 \text{ eV}$$

Math Handbook
Operations with Significant Digits
pages 835–836

To calculate the energy of the ejected electron, subtract the work function from the energy of the incident radiation.

$$KE = hf - hf_0 = \frac{hc}{\lambda} - \frac{hc}{\lambda_0}$$

$$= E - W$$ **Substitute $\frac{hc}{\lambda} = E$, $\frac{hc}{\lambda_0} = W$**

$$= 3.56 \text{ eV} - 2.31 \text{ eV}$$ **Substitute $E = 3.56$ eV, $W = 2.31$ eV**

$$= 1.25 \text{ eV}$$

3 Evaluate the Answer

- **Are the units correct?** Performing dimensional analysis on the units verifies that eV is the proper unit for *KE*.
- **Does the sign make sense?** *KE* is always positive.
- **Are the magnitudes realistic?** Energies should be a few electron volts.

▶ PRACTICE Problems

- Additional Problems, Appendix B
- Solutions to Selected Problems, Appendix C

6. The threshold wavelength of zinc is 310 nm. Find the threshold frequency, in Hz, and the work function, in eV, of zinc.

7. The work function for cesium is 1.96 eV. What is the kinetic energy, in eV, of photoelectrons ejected when 425-nm violet light falls on the cesium?

8. When a metal is illuminated with 193-nm ultraviolet radiation, electrons with energies of 3.5 eV are emitted. What is the work function of the metal?

9. A metal has a work function of 4.50 eV. What is the longest-wavelength radiation that will cause it to emit photoelectrons?

The Compton Effect

The photoelectric effect demonstrates that a photon, even though it has no mass, has kinetic energy just as a particle does. In 1916, Einstein predicted that the photon should have another particle property: momentum. He showed that the momentum of a photon should be equal to E/c. Because $E = hf$ and $f/c = 1/\lambda$, the photon's momentum is given by the following equation.

Photon Momentum $p = \frac{hf}{c} = \frac{h}{\lambda}$

The momentum of a photon is equal to Planck's constant divided by the photon's wavelength.

Experiments done by an American physicist, Arthur Holly Compton, in 1922 tested Einstein's theory. The results of Compton's experiments further supported the particle model of light. Compton directed X rays of a known wavelength at a graphite target, as shown in **Figure 27-7a,** and measured the wavelengths of the X rays scattered by the target. He observed that some of the X rays were scattered without change in wavelength, whereas others had a longer wavelength than that of the original radiation. These results are shown in **Figure 27-7b.** Note that the peak wavelength for the unscattered X rays corresponds to the wavelength of the original incident X rays, whereas the peak wavelength for the scattered X rays is greater than that of the original incident X rays.

Recall that the equation for the energy of a photon, $E = hf$, also can be written as $E = hc/\lambda$. This second equation shows that the energy of a photon is inversely proportional to its wavelength. The increase in wavelength that Compton observed meant that the X-ray photons had lost both energy and momentum. The shift in the energy of scattered photons is called the **Compton effect.** This shift in energy is very small, only about 10^{-3} nm, and is a measurable effect only when X rays having wavelengths of 10^{-2} nm or less are used.

Figure 27-7 Compton used an apparatus similar to this one to study the nature of photons **(a).** The increased wavelength of the scattered photons is evidence that the X-ray photons have lost energy **(b).**

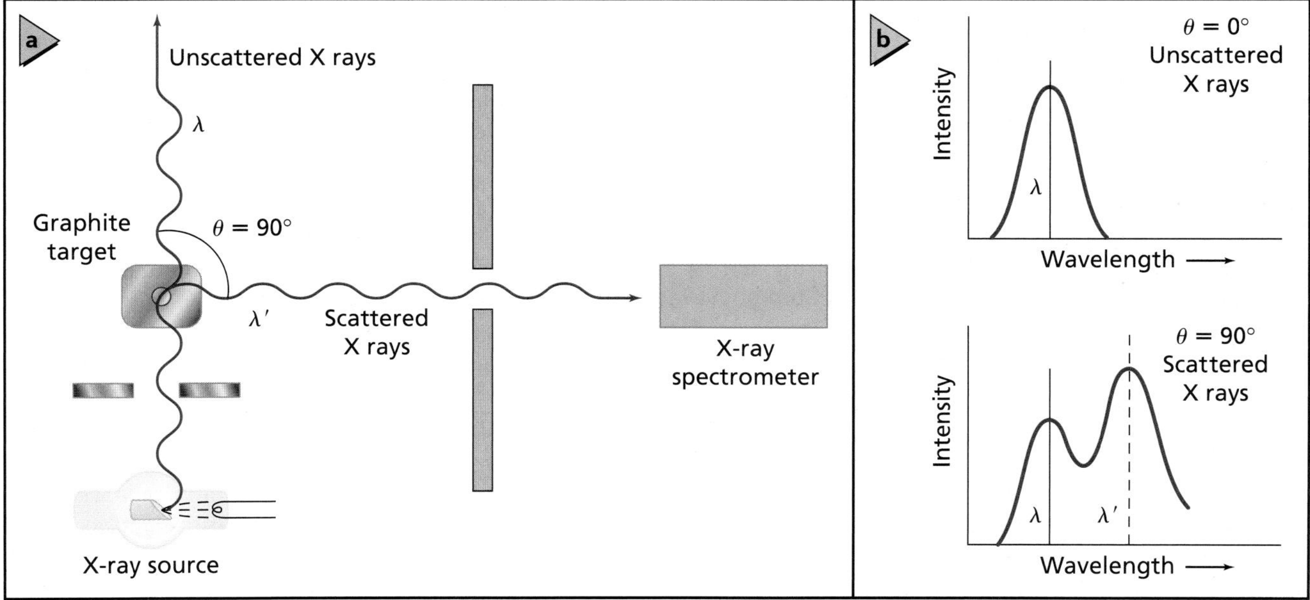

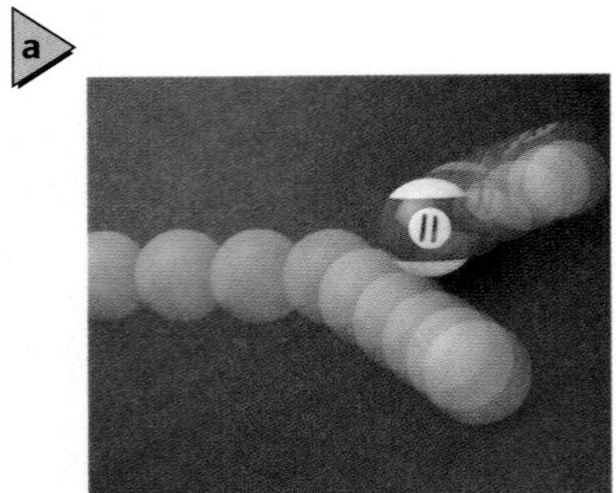

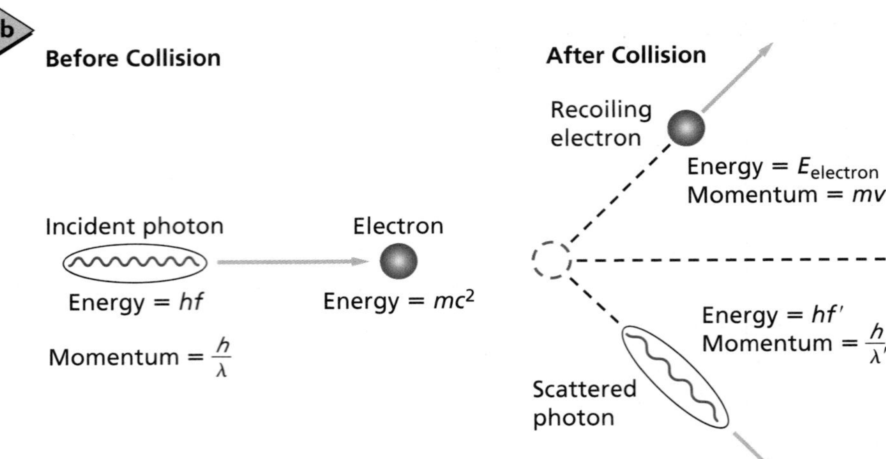

Figure 27-8 Much like the collision between two billiard balls **(a),** when a photon strikes an electron, the energy and momentum gained by the electron equal the energy and momentum lost by the photon **(b).**

In later experiments, Compton observed that electrons were ejected from the graphite block during the experiment. He suggested that the X-ray photons collided with electrons in the graphite target and transferred energy and momentum to them. Compton thought that these photon-electron collisions were similar to the elastic collisions experienced by billiard balls, as shown in **Figure 27-8.** He tested this idea by measuring the energy of the ejected electrons. Compton found that the energy and momentum gained by the electrons equaled the energy and momentum lost by the photons. Thus, photons obey the laws of conservation of momentum and energy when they are involved in collisions with other particles.

27.1 Section Review

10. Photoelectric Effect Why is high-intensity, low-frequency light unable to eject electrons from a metal, whereas low-intensity, high-frequency light can? Explain.

11. Frequency and Energy of Hot-Body Radiation As the temperature of a body is increased, how does the frequency of peak intensity change? How does the total amount of radiated energy change?

12. Photoelectric and Compton Effects An experimenter sends an X ray into a target. An electron, but no other radiation, emerges from the target. Explain whether this event is a result of the photoelectric effect or the Compton effect.

13. Photoelectric and Compton Effects Distinguish the photoelectric effect from the Compton effect.

14. Photoelectric Effect Green light ($\lambda = 532$ nm) strikes an unknown metal, causing electrons to be ejected. The ejected electrons can be stopped by a potential of 1.44 V. What is the work function, in eV, of the metal?

15. Energy of a Photon What is the energy, in eV, of the photons produced by a laser pointer having a 650-nm wavelength?

16. Photoelectric Effect An X ray is absorbed in a bone and releases an electron. If the X ray has a wavelength of approximately 0.02 nm, estimate the energy, in eV, of the electron.

17. Compton Effect An X ray strikes a bone, collides with an electron, and is scattered. How does the wavelength of the scattered X ray compare to the wavelength of the incoming X ray?

18. Critical Thinking Imagine that the collision of two billiard balls models the interaction of a photon and an electron during the Compton effect. Suppose the electron is replaced by a much more massive proton. Would this proton gain as much energy from the collision as the electron does? Would the photon lose as much energy as it does when it collides with the electron?

Physics Online physicspp.com/self_check_quiz

27.2 Matter Waves

The photoelectric effect and Compton scattering showed that a massless electromagnetic wave has momentum and energy, like a particle. If an electromagnetic wave has particlelike properties, could a particle exhibit interference and diffraction, as a wave does? In other words, does a particle have wavelike properties? In 1923, French physicist Louis de Broglie proposed just this, that material particles have wave properties. This proposal was so extraordinary that it was ignored by other scientists until Einstein read de Broglie's papers and supported his ideas.

▸ **Objectives**
- **Describe** evidence of the wave nature of matter.
- **Solve** problems involving the de Broglie wavelength of particles.
- **Describe** the dual nature of waves and particles, and the importance of the Heisenberg uncertainty principle.

▸ **Vocabulary**

de Broglie wavelength
Heisenberg uncertainty principle

De Broglie Waves

Recall that the momentum of an object is equal to its mass times its velocity, $p = mv$. By analogy with the momentum of a photon, $p = h/\lambda$, de Broglie proposed that the momentum of a particle is represented by the following equation:

$$p = mv = \frac{h}{\lambda}$$

The wavelength in the above relationship represents that of the moving particle and is known as the **de Broglie wavelength.** The following equation solves directly for the de Broglie wavelength.

De Broglie Wavelength $\lambda = \frac{h}{p} = \frac{h}{mv}$

The de Broglie wavelength of a moving particle is equal to Planck's constant divided by the particle's momentum.

According to de Broglie, particles such as electrons and protons should show wavelike properties. Effects such as diffraction and interference had never been observed for particles, so de Broglie's work was greeted with considerable doubt. In 1927, however, the results of two independent experiments proved that electrons are diffracted just as light is. In one experiment, English physicist George Thomson aimed a beam of electrons at a very thin crystal. Atoms in crystals are arrayed in a regular pattern that acts as a diffraction grating. Electrons diffracted from the crystal formed the same patterns that X rays of a similar wavelength formed. **Figure 27-9** shows the pattern made by diffracting electrons. In the United States, Clinton Davisson and Lester Germer performed a similar experiment using electrons reflected and diffracted from thick crystals. The two experiments proved that material particles have wave properties.

The wave nature of objects you see and handle every day is not observable because their wavelengths are extremely short. For example, consider the de Broglie wavelength of a 0.145-kg baseball when it leaves a bat with a speed of 38 m/s.

$$\lambda = \frac{h}{mv} = \frac{6.63\times10^{-34}\ \text{J·s}}{(0.145\ \text{kg})(38\ \text{m/s})} = 1.2\times10^{-34}\ \text{m}$$

This wavelength is far too small to have observable effects. As you will see in the following Example Problem, however, an extremely small particle, such as an electron, has a wavelength that can be observed and measured.

Figure 27-9 Electron diffraction patterns, such as this one for a cubic zirconium crystal, demonstrate the wave properties of particles.

▶ EXAMPLE Problem 3

De Broglie Wavelength An electron is accelerated by a potential difference of 75 V. What is its de Broglie wavelength?

1 Analyze and Sketch the Problem

- Include the positive and negative plates in your drawing.

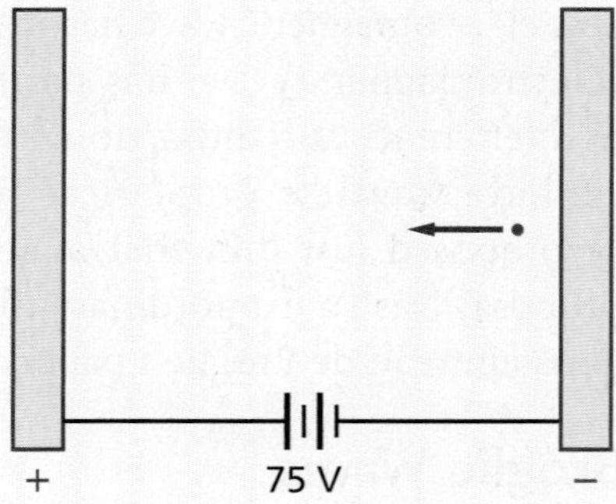

Known:		Unknown:
$V = 75$ V	$m = 9.11\times10^{-31}$ kg	$\lambda = ?$
$q = -1.60\times10^{-19}$ C	$h = 6.63\times10^{-34}$ J·s	

2 Solve for the Unknown

Write relationships for the kinetic energy of the electron based on potential difference and motion and use them to calculate the electron's velocity.

$KE = -qV$, and $KE = \frac{1}{2}mv^2$

$\frac{1}{2}mv^2 = -qV$ Equate both forms of *KE*.

Math Handbook
Isolating a Variable
page 845

Solve for v.

$$v = \sqrt{\frac{-2qV}{m}}$$

$$= \sqrt{\frac{-2(-1.60\times10^{-19}\text{ C})(75\text{ V})}{9.11\times10^{-31}\text{ kg}}}$$ Substitute $q = -1.60\times10^{-19}$ C, $V = 75$ V, $m = 9.11\times10^{-31}$ kg

$$= 5.1\times10^{6}\text{ m/s}$$

Solve for momentum, p.

$$p = mv$$

$$= (9.11\times10^{-31}\text{ kg})(5.1\times10^{6}\text{ m/s})$$ Substitute $m = 9.11\times10^{-31}$ kg, $v = 5.1\times10^{6}$ m/s

$$= 4.6\times10^{-24}\text{ kg·m/s}$$

Solve for the de Broglie wavelength, λ.

$$\lambda = \frac{h}{p}$$

$$= \frac{6.63\times10^{-34}\text{ J·s}}{4.6\times10^{-24}\text{ kg·m/s}}$$ Substitute $h = 6.63\times10^{-34}$ J·s, $p = 4.6\times10^{-24}$ kg·m/s

$= 1.4\times10^{-10}$ m, which is equivalent to 0.14 nm

3 Evaluate the Answer

- **Are the units correct?** Dimensional analysis on the units verifies m/s for v and nm for λ.
- **Do the signs make sense?** Positive values are expected for both v and λ.
- **Are the magnitudes realistic?** The wavelength is close to 0.1 nm, which is in the X-ray region of the electromagnetic spectrum.

▶ PRACTICE Problems

- Additional Problems, Appendix B
- Solutions to Selected Problems, Appendix C

19. A 7.0-kg bowling ball rolls with a velocity of 8.5 m/s.

a. What is the de Broglie wavelength of the bowling ball?

b. Why does the bowling ball exhibit no observable wave behavior?

20. What is the de Broglie wavelength and speed of an electron accelerated by a potential difference of 250 V?

21. What voltage is needed to accelerate an electron so it has a 0.125-nm wavelength?

22. The electron in Example Problem 3 has a de Broglie wavelength of 0.14 nm. What is the kinetic energy, in eV, of a proton ($m = 1.67\times10^{-27}$ kg) with the same wavelength?

Particles and Waves

Is light a particle or a wave? Evidence suggests that both particle and wave models are needed to explain the behavior of light. As you are about to discover, quantum theory and the dual nature of electromagnetic radiation led to fascinating scientific principles and applications. One such application, the scanning tunneling microscope (STM), is discussed in the How it Works feature on page 740.

Determining location and momentum It is logical to think that to accurately define the properties of an object, you would need to devise an experiment that directly measures the desired properties. For example, you cannot simply state that a particle is at a certain location moving with a specific speed. Rather, an experiment must be performed that locates the particle and measures its speed.

How can you detect the location of a particle? You must touch it or reflect light from it. If light is used, then the reflected light must be collected by an instrument or the human eye. Because of diffraction effects, the light used to detect the particle spreads out and makes it impossible to locate the particle exactly. The use of shorter-wavelength light or radiation decreases diffraction and allows the location of a particle to be more precisely measured.

Heisenberg uncertainty principle As a result of the Compton effect, however, when short-wavelength, high-energy radiation strikes a particle, the particle's momentum is changed, as shown in **Figure 27-10.** Therefore, the act of precisely measuring the location of a particle has the effect of changing the particle's momentum. The more precise the determination of a particle's location, the greater the uncertainty is in its momentum. In the same way, if the momentum of the particle is measured, the position of the particle changes and becomes less certain. This situation is summarized by the **Heisenberg uncertainty principle,** which states that it is impossible to measure precisely both the position and momentum of a particle at the same time. This principle, named for German physicist Werner Heisenberg, is the result of the dual wave and particle properties of light and matter. The Heisenberg uncertainty principle tells us there is a limit to how accurately position and momentum can be measured.

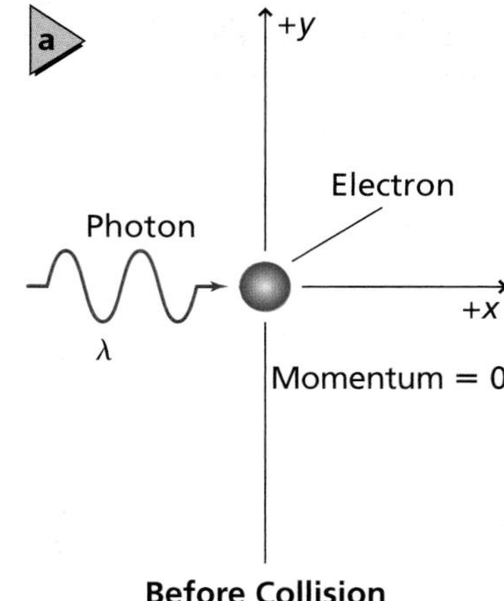

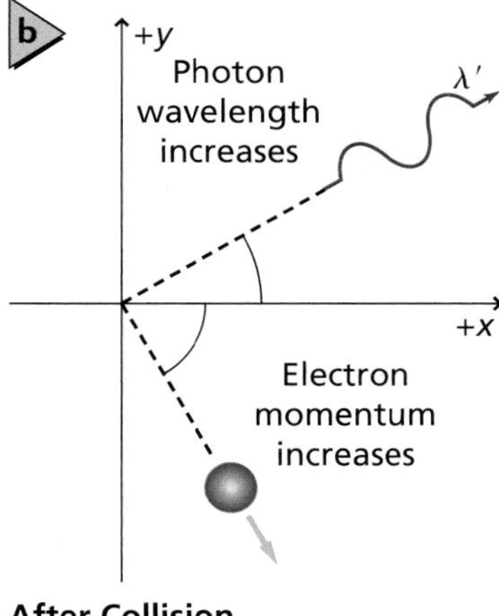

Figure 27-10 A particle can be seen only when light is scattered from it. Thus, the electron remains undetected **(a)** until an incident photon strikes it **(b).** The collision scatters the photon and electron and changes their momenta.

27.2 Section Review

23. **Wavelike Properties** Describe the experiment that confirmed that particles have wavelike properties.

24. **Wave Nature** Explain why the wave nature of matter is not obvious.

25. **De Broglie Wavelength** What is the de Broglie wavelength of an electron accelerated through a potential difference of 125 V?

26. **Wavelengths of Matter and Radiation** When an electron collides with a massive particle, the electron's velocity and wavelength decrease. How is it possible to increase the wavelength of a photon?

27. **Heisenberg Uncertainty Principle** When light or a beam of atoms passes through a double slit, an interference pattern forms. Both results occur even when atoms or photons pass through the slits one at a time. How does the Heisenberg uncertainty principle explain this?

28. **Critical Thinking** Physicists recently made a diffraction grating of standing waves of light. Atoms passing through the grating produce an interference pattern. If the spacing of the slits in the grating were $\frac{1}{2}\lambda$ (about 250 nm), what was the approximate de Broglie wavelength of the atoms?

PHYSICS LAB

Modeling the Photoelectric Effect

The emission of electrons from an object when electromagnetic radiation is incident upon it is known as the photoelectric effect. Electrons are ejected from the object only when the frequency of the radiation is greater than a certain minimum value, called the threshold frequency. In this investigation you will model the photoelectric effect using steel balls. You will examine why only certain types of electromagnetic radiation result in the emission of photoelectrons.

QUESTION

How can steel balls be used to model the photoelectric effect?

Objectives

- **Formulate a model** to investigate the photoelectric effect.
- **Describe** how the energy of a photon is related to its frequency.
- **Use scientific explanations** to explain why macroscopic phenomena cannot explain the quantum behavior of the atom.

Safety Precautions

- **Keep isopropyl alcohol away from open flame.**
- **Do not swallow isopropyl alcohol.**
- **Isopropyl alcohol can dry out skin.**

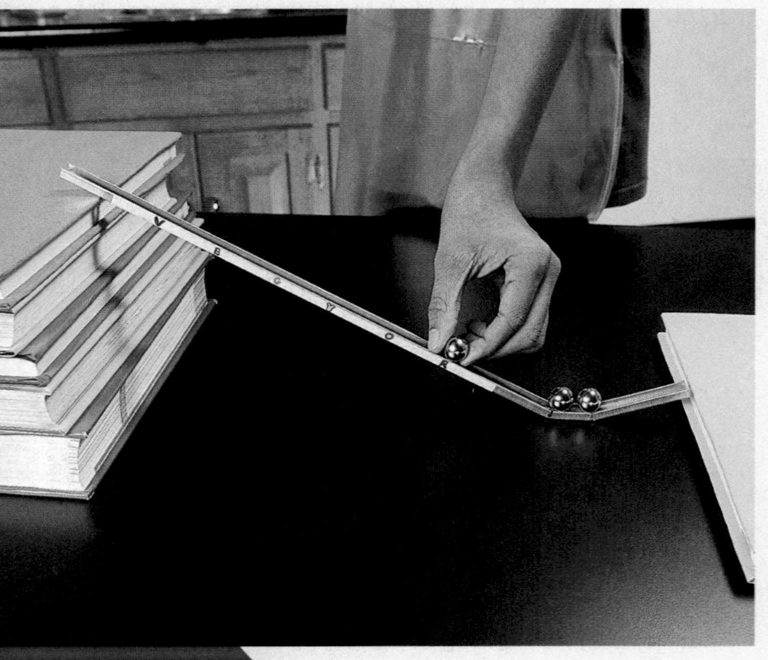

Materials

steel balls (3)
grooved channel (U-channel or shelf bracket)
books
red, orange, yellow, green, blue, and violet marking pens (or colored stickers)
metric ruler
isopropyl alcohol

Procedure

1. Shape the grooved channel, as shown in the photo, and use several books to support the channel, as shown. Make sure the books do not block the ends of the channel.
2. Mark a capital letter *R* with a red marking pen on the channel 4 cm above the table, as shown. The *R* represents red.
3. Mark a capital letter *V* with a violet marking pen on the channel at a distance 14 cm above the table as shown. The *V* represents violet. Use the other colored pens to place marks for blue, *B*, green, *G*, yellow, *Y*, and orange, *O*, uniformly between the marks for *R* and *V*.
4. Place two steel balls at the lowest point on the channel. These steel balls represent the atom's valence electrons.
5. Hold a steel ball in place on the channel at the *R* position. This steel ball represents an incident photon of red light. Note that the red photon has the lowest energy of the six colors of light being modeled.
6. Release the steel ball (photon) and see if it has enough energy to remove a valence electron from the atom; that is, observe if either of the two steel ball electrons escapes from the channel. Record your observations in the data table.

Data Table	
Color or Energy of Photon	**Observations**
Red	
Orange	
Yellow	
Green	
Blue	
Violet	
Less than red	
Greater than violet	

7. Remove the steel ball that represents the incident photon from the lower part of the channel. Replace the two steel balls used to represent the valence electrons at the lowest point on the channel.
8. Repeat steps 5–7 for each of the colors you marked on the channel. Be sure always to start with the two steel balls at the lowest point on the channel. Note that the violet photon has the greatest energy of the six colors of light being modeled. Record your observations in the data table.
9. Repeat steps 5–7, but release the steel ball representing the incident photon from a point slightly lower than the *R* position. Record your observations in the data table.
10. Repeat steps 5–7, but release the steel ball representing the incident photon from a point slightly higher than the *V* position. Record your observations in the data table.
11. Answer question 1 in the Conclude and Apply section and then test your prediction.
12. When you have finished the lab, return all materials to the locations specified by your instructor. Clean off the ink markings on the channel with isopropyl alcohol (or remove the colored stickers placed on the channel).

Analyze

1. **Interpret Data** Which color(s) of photons was able to remove at least one electron in your model?
2. **Interpret Data** Were any of the photons energetic enough to remove more than one electron? If so, identify the photon's color.
3. **Use Models** In step 9, what type of photon does the steel ball represent?
4. **Use Models** In step 10, what type of photon does the steel ball represent?
5. **Explain** Should photons of visible light be the only photons considered when investigating the photoelectric effect? Why or why not?
6. **Summarize** Summarize your observations in terms of the energies of photons.

Conclude and Apply

1. **Infer** What would happen if two red photons hit the two valence electrons at the same time? Test your prediction.
2. **Think Critically** Some materials hold on to their valence electrons more tightly than others. How could the model be modified to show this?
3. **Draw Conclusions** In this model, what happens to the photon's energy when it collides with an electron but does not remove the electron from the atom?

Going Further

Using the formula $E = hf$, where h is Planck's constant and f is the frequency of the electromagnetic radiation, calculate the energy of a red photon compared to the energy of a blue photon.

Real-World Physics

Photographers often have red lights in their darkrooms. Why do they use red light, but not blue light?

To find out more about the photoelectric effect and quantum theory, visit the Web site: **physicspp.com**

Scanning Tunneling Microscope

The scanning tunneling microscope (STM) was invented in 1981 by Gerd Binnig and Heinrich Rohrer. Five years later, they were awarded the Nobel Prize in physics. The atomic-level resolution of an STM allows scientists to form images of atoms, such as the image of silicon atoms shown on the monitor below. How does an STM work?

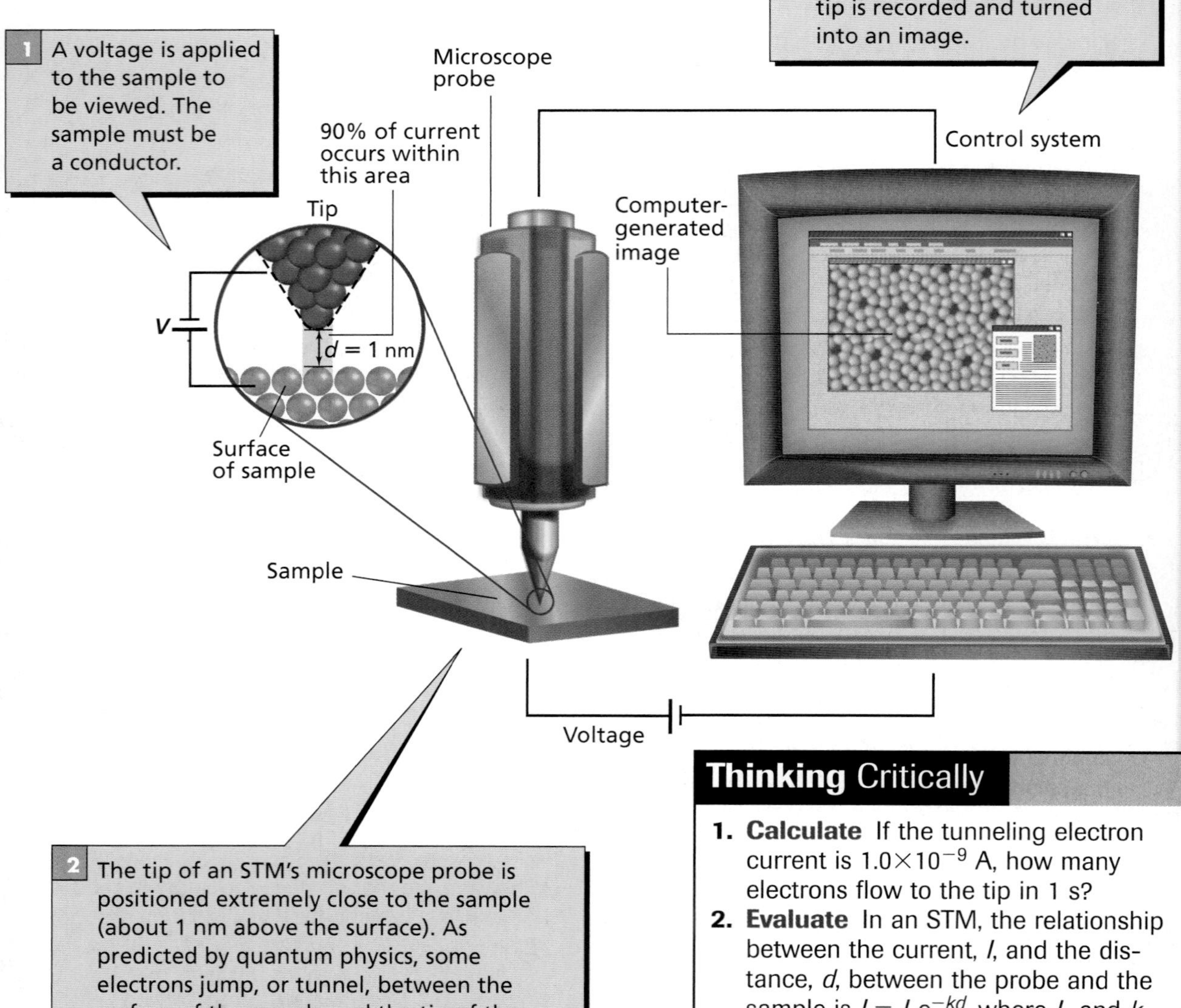

1 A voltage is applied to the sample to be viewed. The sample must be a conductor.

2 The tip of an STM's microscope probe is positioned extremely close to the sample (about 1 nm above the surface). As predicted by quantum physics, some electrons jump, or tunnel, between the surface of the sample and the tip of the probe. The moving electrons produce a current (measured in nanoamperes).

3 The control system scans the probe back and forth and up and down above the surface of the sample. The distance between the surface and the tip is kept constant, producing a constant current. The up-and-down movement of the tip is recorded and turned into an image.

Thinking Critically

1. **Calculate** If the tunneling electron current is 1.0×10^{-9} A, how many electrons flow to the tip in 1 s?
2. **Evaluate** In an STM, the relationship between the current, I, and the distance, d, between the probe and the sample is $I = I_0e^{-kd}$, where I_0 and k are constants. Use sample values to verify that the current decreases when the distance increases.
3. **Design an Experiment** What would you do if you wanted to use an STM to study a nonconductive sample?

Chapter 27

Study Guide

27.1 A Particle Model of Waves

Vocabulary

- emission spectrum (p. 724)
- quantized (p. 725)
- photoelectric effect (p. 726)
- threshold frequency (p. 726)
- photon (p. 727)
- work function (p. 731)
- Compton effect (p. 733)

Key Concepts

- Objects that are hot enough to be incandescent emit light because of the vibrations of the charged particles inside their atoms.
- The spectrum of incandescent objects covers a broad range of wavelengths. The spectrum depends upon the temperature of the incandescent objects.
- Planck explained the spectrum of an incandescent object by supposing that a particle can have only certain energies that are multiples of a constant, now called Planck's constant.

$$E = nhf$$

- Einstein explained the photoelectric effect by postulating that light exists in bundles of energy called photons.

$$E = hf = \frac{hc}{\lambda} = \frac{1240 \text{ eV·nm}}{\lambda}$$

- The photoelectric effect is the emission of electrons by certain metals when they are exposed to electromagnetic radiation.

$$KE = hf - hf_0$$

- The photoelectric effect allows the measurement of Planck's constant, h.
- The work function, which is equivalent to the binding energy of the electron, is measured by the threshold frequency in the photoelectric effect.
- The Compton effect demonstrates that photons have momentum, as predicted by Einstein.

$$p = \frac{hf}{c} = \frac{h}{\lambda}$$

- Even though photons, which travel at the speed of light, have zero mass, they do have energy and momentum.

27.2 Matter Waves

Vocabulary

- de Broglie wavelength (p. 735)
- Heisenberg uncertainty principle (p. 737)

Key Concepts

- The wave nature of material particles was suggested by de Broglie and verified experimentally by the diffraction of electrons through crystals. All moving particles have a wavelength, known as the de Broglie wavelength.

$$\lambda = \frac{h}{p} = \frac{h}{mv}$$

- The particle and wave aspects are complementary parts of the complete nature of both matter and light.
- The Heisenberg uncertainty principle states that it is not possible to simultaneously measure the precise position and momentum of any particle of light or matter.

Chapter 27 Assessment

Concept Mapping

29. Complete the following concept map using these terms: *dual nature, mass, wave properties, momentum, diffraction.*

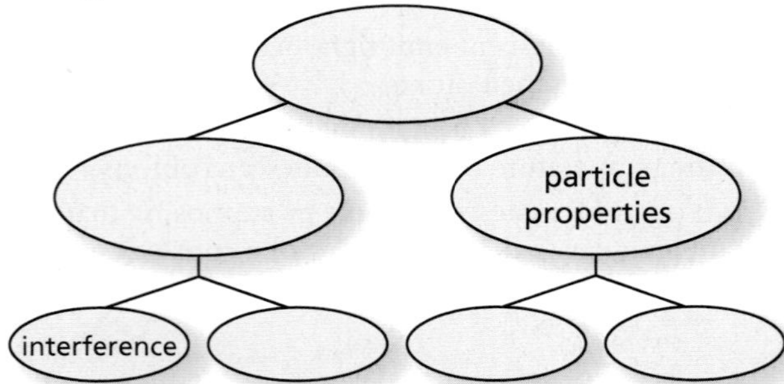

Mastering Concepts

30. Incandescent Light An incandescent lightbulb is controlled by a dimmer. What happens to the color of the light given off by the bulb as the dimmer control is turned down? (27.1)

31. Explain the concept of quantized energy. (27.1)

32. What is quantized in Max Planck's interpretation of the radiation of incandescent bodies? (27.1)

33. What is a quantum of light called? (27.1)

34. Light above the threshold frequency shines on the metal cathode in a photocell. How does Einstein's photoelectric effect theory explain the fact that as the light intensity increases, the current of photoelectrons increases? (27.1)

35. Explain how Einstein's theory accounts for the fact that light below the threshold frequency of a metal produces no photoelectrons, regardless of the intensity of the light. (27.1)

36. Photographic Film Because certain types of black-and-white film are not sensitive to red light, they can be developed in a darkroom that is illuminated by red light. Explain this on the basis of the photon theory of light. (27.1)

37. How does the Compton effect demonstrate that photons have momentum as well as energy? (27.1)

38. The momentum, p, of a particle of matter is given by $p = mv$. Can you calculate the momentum of a photon using the same equation? Explain. (27.2)

39. Explain how each of the following electron properties could be measured. (27.2)
a. charge **b.** mass **c.** wavelength

40. Explain how each of the following photon properties could be measured. (27.2)
a. energy **b.** momentum **c.** wavelength

Applying Concepts

41. Use the emission spectrum of an incandescent body at three different temperatures shown in Figure 27-1 on page 724 to answer the following questions.
a. At what frequency does the peak emission intensity occur for each of the three temperatures?
b. What can you conclude about the relationship between the frequency of peak radiation emission intensity and temperature for an incandescent body?
c. By what factor does the intensity of the red light given off change as the body's temperature increases from 4000 K to 8000 K?

42. Two iron bars are held in a fire. One glows dark red, while the other glows bright orange.
a. Which bar is hotter?
b. Which bar is radiating more energy?

43. Will high-frequency light eject a greater number of electrons from a photosensitive surface than low-frequency light, assuming that both frequencies are above the threshold frequency?

44. Potassium emits photoelectrons when struck by blue light, whereas tungsten emits photoelectrons when struck by ultraviolet radiation.
a. Which metal has a higher threshold frequency?
b. Which metal has a larger work function?

45. Compare the de Broglie wavelength of the baseball shown in **Figure 27-11** with the diameter of the baseball.

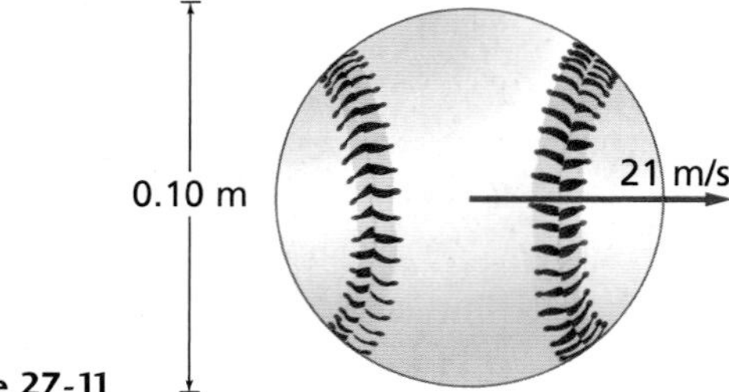

Figure 27-11

Mastering Problems

27.1 A Particle Model of Waves

46. According to Planck's theory, how does the frequency of vibration of an atom change if it gives off 5.44×10^{-19} J while changing its value of n by 1?

47. What potential difference is needed to stop electrons with a maximum kinetic energy of 4.8×10^{-19} J?

48. What is the momentum of a photon of violet light that has a wavelength of 4.0×10^{2} nm?

 For more problems, go to Additional Problems, Appendix B.

49. The stopping potential of a certain metal is shown in **Figure 27-12.** What is the maximum kinetic energy of the photoelectrons in the following units?
a. electron volts **b.** joules

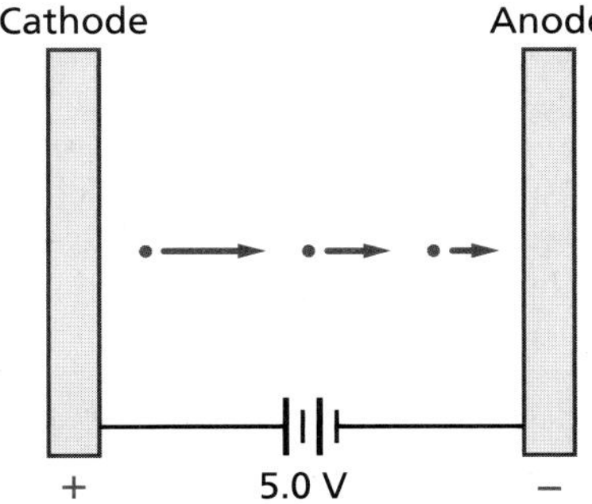

Figure 27-12

50. The threshold frequency of a certain metal is 3.00×10^{14} Hz. What is the maximum kinetic energy of an ejected photoelectron if the metal is illuminated by light with a wavelength of 6.50×10^{2} nm?

51. The threshold frequency of sodium is 4.4×10^{14} Hz. How much work must be done to free an electron from the surface of sodium?

52. If light with a frequency of 1.00×10^{15} Hz falls on the sodium in the previous problem, what is the maximum kinetic energy of the photoelectrons?

53. **Light Meter** A photographer's light meter uses a photocell to measure the light falling on the subject to be photographed. What should be the work function of the cathode if the photocell is to be sensitive to red light (λ = 680 nm) as well as to the other colors of light?

54. **Solar Energy** A home uses about 4×10^{11} J of energy each year. In many parts of the United States, there are about 3000 h of sunlight each year.
a. How much energy from the Sun falls on one square meter each year?
b. If this solar energy can be converted to useful energy with an efficiency of 20 percent, how large an area of converters would produce the energy needed by the home?

27.2 Matter Waves

55. What is the de Broglie wavelength of an electron moving at 3.0×10^{6} m/s?

56. What velocity would an electron need to have a de Broglie wavelength of 3.0×10^{-10} m?

57. A cathode-ray tube accelerates an electron from rest across a potential difference of 5.0×10^{3} V.
a. What is the velocity of the electron?
b. What is the wavelength associated with the electron?

58. A neutron is held in a trap with a kinetic energy of only 0.025 eV.
a. What is the velocity of the neutron?
b. Find the de Broglie wavelength of the neutron.

59. The kinetic energy of a hydrogen atom's electron is 13.65 eV.
a. Find the velocity of the electron.
b. Calculate the electron's de Broglie wavelength.
c. Given that a hydrogen atom's radius is 0.519 nm, calculate the circumference of a hydrogen atom and compare it with the de Broglie wavelength for the atom's electron.

60. An electron has a de Broglie wavelength of 0.18 nm.
a. How large a potential difference did it experience if it started from rest?
b. If a proton has a de Broglie wavelength of 0.18 nm, how large is the potential difference that it experienced if it started from rest?

Mixed Review

61. What is the maximum kinetic energy of photoelectrons ejected from a metal that has a stopping potential of 3.8 V?

62. The threshold frequency of a certain metal is 8.0×10^{14} Hz. What is the work function of the metal?

63. If light with a frequency of 1.6×10^{15} Hz falls on the metal in the previous problem, what is the maximum kinetic energy of the photoelectrons?

64. Find the de Broglie wavelength of a deuteron (nucleus of ^{2}H isotope) of mass 3.3×10^{-27} kg that moves with a speed of 2.5×10^{4} m/s.

65. The work function of iron is 4.7 eV.
a. What is the threshold wavelength of iron?
b. Iron is exposed to radiation of wavelength 150 nm. What is the maximum kinetic energy of the ejected electrons in eV?

66. Barium has a work function of 2.48 eV. What is the longest wavelength of light that will cause electrons to be emitted from barium?

67. An electron has a de Broglie wavelength of 400.0 nm, the shortest wavelength of visible light.
a. Find the velocity of the electron.
b. Calculate the energy of the electron in eV.

68. **Electron Microscope** An electron microscope is useful because the de Broglie wavelengths of electrons can be made smaller than the wavelength of visible light. What energy in eV has to be given to an electron for it to have a de Broglie wavelength of 20.0 nm?

69. Incident radiation falls on tin, as shown in **Figure 27-13.** The threshold frequency of tin is 1.2×10^{15} Hz.
  a. What is the threshold wavelength of tin?
  b. What is the work function of tin?
  c. The incident electromagnetic radiation has a wavelength of 167 nm. What is the kinetic energy of the ejected electrons in eV?

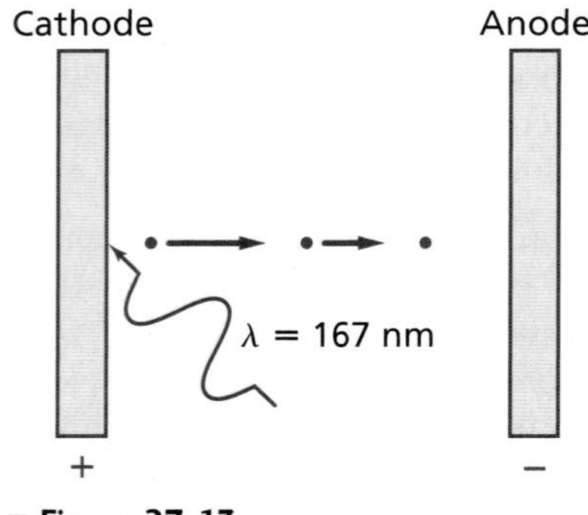

Figure 27-13

Thinking Critically

70. **Apply Concepts** A helium-neon laser emits photons with a wavelength of 632.8 nm.
  a. Find the energy, in joules, of each photon emitted by the laser.
  b. A typical small laser has a power of 0.5 mW (equivalent to 5×10^{-4} J/s). How many photons are emitted each second by the laser?

71. **Apply Concepts** Just barely visible light with an intensity of 1.5×10^{-11} W/m^2 enters a person's eye, as shown in **Figure 27-14.**
  a. If this light shines into the person's eye and passes through the person's pupil, what is the power, in watts, that enters the person's eye?
  b. Use the given wavelength of the incident light and information provided in Figure 27-14 to calculate the number of photons per second entering the eye.

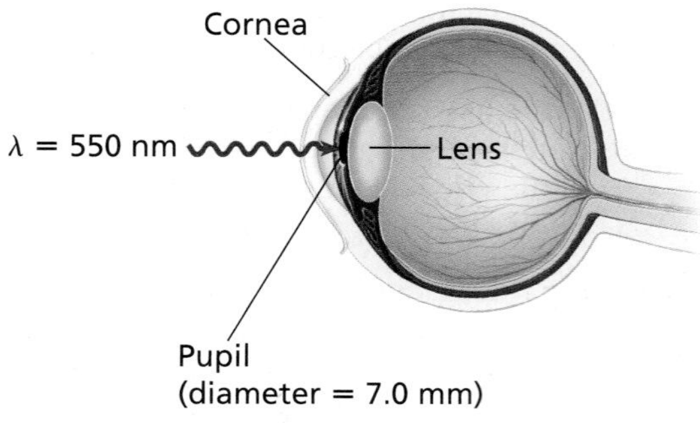

Figure 27-14

72. **Make and Use Graphs** A student completed a photoelectric-effect experiment and recorded the stopping potential as a function of wavelength, as shown in **Table 27-1.** The photocell had a sodium cathode. Plot the data (stopping potential versus frequency) and use your calculator to draw the best-fit straight line (regression line). From the slope and intercept of the line, find the work function, the threshold wavelength, and the value of h/q from this experiment. Compare the value of h/q to the accepted value.

Table 27-1
Stopping Potential v. Wavelength

λ (nm)	V_0 (eV)
200	4.20
300	2.06
400	1.05
500	0.41
600	0.03

Writing in Physics

73. Research the most massive particle for which interference effects have been seen. Describe the experiment and how the interference was created.

Cumulative Review

74. The spring in a pogo stick is compressed 15 cm when a child who weighs 400.0 N stands on it. What is the spring constant of the spring? (Chapter 14)

75. A marching band sounds flat as it plays on a very cold day. Why? (Chapter 15)

76. A charge of 8.0×10^{-7} C experiences a force of 9.0 N when placed 0.02 m from a second charge. What is the magnitude of the second charge? (Chapter 20)

77. A homeowner buys a dozen identical 120-V light sets. Each light set has 24 bulbs connected in series, and the resistance of each bulb is 6.0 Ω. Calculate the total load in amperes if the homeowner operates all the sets from a single exterior outlet. (Chapter 23)

78. The force on a 1.2-m wire is 1.1×10^{-3} N. The wire is perpendicular to Earth's magnetic field. How much current is in the wire? (Chapter 24)

 For more problems, go to Additional Problems, Appendix B.

Multiple Choice

1. The energy level of an atom changes as it absorbs and emits energy. Which of the following is NOT a possible energy level of an atom?

   (A) $\frac{3}{4}hf$
   (B) hf
   (C) $3hf$
   (D) $4hf$

2. How is the threshold frequency related to the photoelectric effect?

   (A) It is the minimum frequency of incident radiation needed to cause the ejection of atoms from the anode of a photocell.
   (B) It is the maximum frequency of incident radiation needed to cause the ejection of atoms from the anode of a photocell.
   (C) It is the frequency of incident radiation below which electrons will be ejected from an atom.
   (D) It is the minimum frequency of incident radiation needed to cause the ejection of electrons from an atom.

3. A photon has a frequency of 1.14×10^{15} Hz. What is the energy of the photon?

   (A) 5.82×10^{-49} J
   (B) 7.55×10^{-19} J
   (C) 8.77×10^{-16} J
   (D) 1.09×10^{-12} J

4. Radiation with an energy of 5.17 eV strikes a photocell as shown below. If the work function of the photocell is 2.31 eV, what is the energy of the ejected photoelectron?

   (A) 0.00 eV
   (B) 2.23 eV
   (C) 2.86 eV
   (D) 7.48 eV

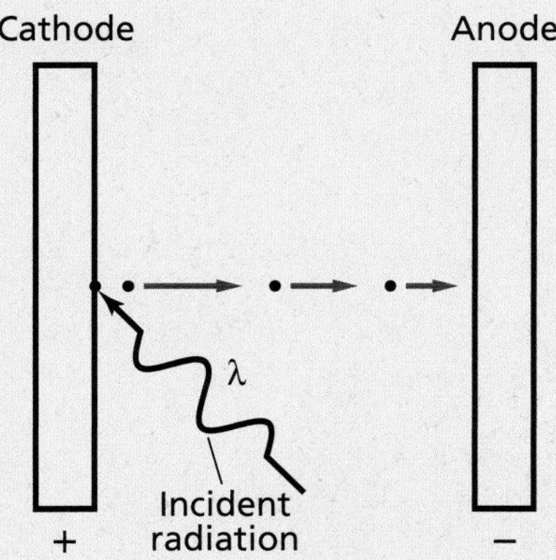

5. As shown in the diagram below, an electron is accelerated by a potential difference of 95.0 V. What is the de Broglie wavelength of the electron?

   (A) 5.02×10^{-22} m
   (B) 1.26×10^{-10} m
   (C) 2.52×10^{-10} m
   (D) 5.10×10^{6} m

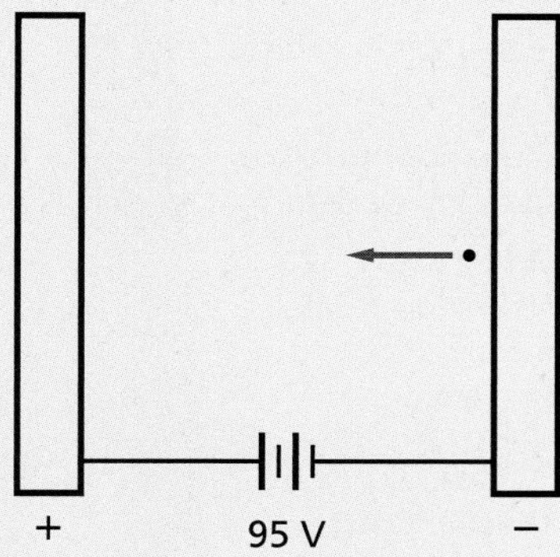

6. What is the de Broglie wavelength of an electron moving at 391 km/s? The mass of an electron is 9.11×10^{-31} kg.

   (A) 3.5×10^{-25} m
   (B) 4.79×10^{-15} m
   (C) 4.8×10^{-15} m
   (D) 1.86×10^{-9} m

7. What is the work function of a metal?

   (A) a measure of how much work an electron emitted from the metal can do
   (B) equal to the threshold frequency
   (C) the energy needed to free the metal atom's innermost electron
   (D) the energy needed to free the most weakly bound electron

Extended Answer

8. An object has a de Broglie wavelength of 2.3×10^{-34} m when its velocity is 45 m/s. What is the mass, in kg, of the object?

✓ Test-Taking TIP

Wear A Watch

If you are taking a timed test, make sure to pace yourself. Do not spend too much time on any one question. Skip over difficult questions and return to them after answering the easier questions.

Chapter 28

The Atom

What You'll Learn

- You will learn about the discovery of the atom's composition.
- You will determine energies of the hydrogen atom.
- You will learn how quantum theory led to the modern atomic model.
- You will learn how lasers work and what their applications are.

Why It's Important

The quantum model of the atom and the transition of electrons between energy levels explain much of the observed behavior of all matter.

Emission Spectra Each of these gas-filled tubes emits a unique spectrum of colors. The bright light is emitted when electrons in the gas make transitions to lower energy states.

Think About This ▶

Why are the colors of the lights different, and how could you identify what gases are used in each tube?

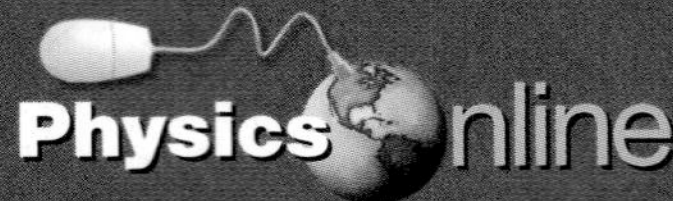

physicspp.com

How can identifying different spinning coins model types of atoms?

Question
When a quarter, a nickel, a penny, and a dime are spun on a tabletop, what characteristics allow you to identify the type of spinning coin?

Procedure
1. Hold a quarter up on its edge and flick it with your index finger to set it spinning. Note the appearance and sound of the spinning coin until it comes to a stop on the table.
2. Repeat step 1 three more times using a dime, a nickel, and a penny, respectively.
3. Have a classmate spin the coins, one at a time, in a random order. Observe each coin only after it is already spinning and then try to identify the type of coin that it is.
4. Repeat step 3, except this time, keep your eyes closed while trying to identify each of the spinning coins.

Analysis
How successful were you at identifying the individual coins when you were limited to listening to the sounds they made? What are the characteristics of a spinning coin that can be used to identify its type? What instruments might make the identification of the spinning coins easier?

Critical Thinking Excited atoms of an element in a high-voltage gas-discharge tube dissipate energy by emitting light. How might the emitted light help you identify the type of atom in the discharge tube? What instruments might help you do this?

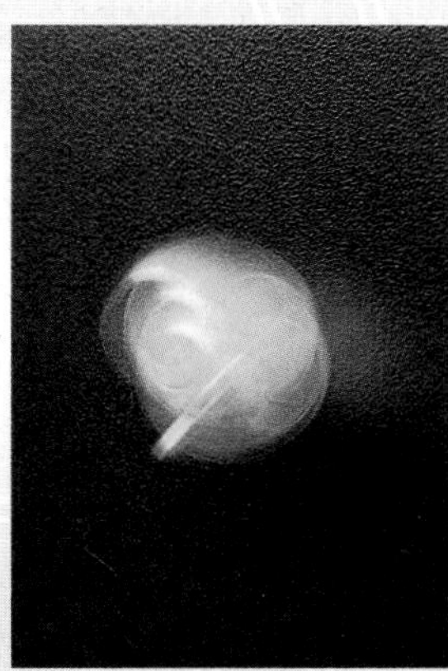

28.1 The Bohr Model of the Atom

► **Objectives**
- **Describe** the structure of the nuclear atom.
- **Compare and contrast** continuous spectra and line-emission spectra.
- **Solve** problems using orbital-radius and energy-level equations.

► **Vocabulary**
alpha particles
nucleus
absorption spectrum
energy level
ground state
excited state
principal quantum number

By the end of the nineteenth century, most scientists agreed on the existence of atoms. J. J. Thomson's discovery of the electron provided convincing evidence that the atom was made up of even smaller, subatomic particles. Every atom tested by Thomson contained negatively charged electrons, and these electrons possessed very little mass. Because atoms were known to be much more massive than the mass accounted for by the electrons they contain, scientists began looking for the missing mass that must be part of each atom. What was the nature of this yet-to-be-discovered massive part of the atom? How was this mass distributed within the atom?

Moreover, atoms were known to be electrically neutral, yet, only negatively charged electrons had been identified within the atom. How were the negatively charged electrons arranged in the atom? What was the source of the atom's neutrality? Were positively charged particles also present in the atom? Knowing that their understanding of the atom was far from complete, scientists began searching for answers to numerous and challenging questions.

The Nuclear Model

Many questions faced the researchers investigating the nature of the atom. What caused the emission of light from atoms? How were the electrons distributed in the atom? Physicists and chemists from many countries searched for the solutions to this puzzle. The results not only provided knowledge about the structure of the atom, but also a totally new approach to both physics and chemistry. The history of the research into the nature of the atom is one of the most exciting stories of the twentieth century.

J. J. Thomson believed that a massive, positively charged substance filled the atom. He pictured the negatively charged electrons as being distributed throughout this positively charged substance like raisins in a muffin. Ernest Rutherford, along with laboratory collaborators Hans Geiger and Ernest Marsden, however, performed a series of experiments that showed the atom had a very different structure.

Rutherford's experiments made use of radioactive compounds that emitted penetrating rays. Some of these emissions had been found to be massive, positively charged particles that moved at high speeds. These particles, which were later named **alpha particles,** are represented by the symbol α. The α-particles in Rutherford's experiments could be detected by the small flashes of light that were emitted when the particles collided with a zinc-sulfide-coated screen.

As shown in **Figure 28-1,** Rutherford directed a beam of α-particles at an extremely thin sheet of gold foil. Rutherford was aware of Thomson's model of the atom, and he expected only minor deflections of the α-particles as they passed through the thin gold foil. He thought that the paths of the massive, high-speed α-particles would be only slightly altered as they passed through the evenly distributed positive charge making up each gold atom. The test results amazed him. While most of the α-particles passed through the gold foil either undeflected or only slightly deflected, a few of the particles were scattered through very large angles. Some were even deflected through angles larger than 90°. A diagram of these results is shown in **Figure 28-2.** Rutherford compared his amazement to that of firing a 15-inch cannon shell at tissue paper and then having the shell bounce back and hit him.

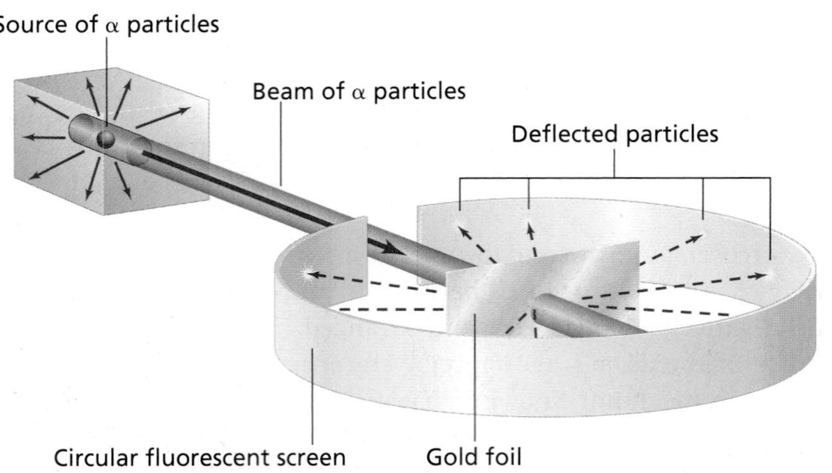

Figure 28-1 After bombarding metal foil with alpha particles, Rutherford's team concluded that most of the mass of the atom was concentrated in the nucleus.

Concepts In Motion
Interactive Figure To see an animation on Rutherford's experiment, visit physicspp.com.

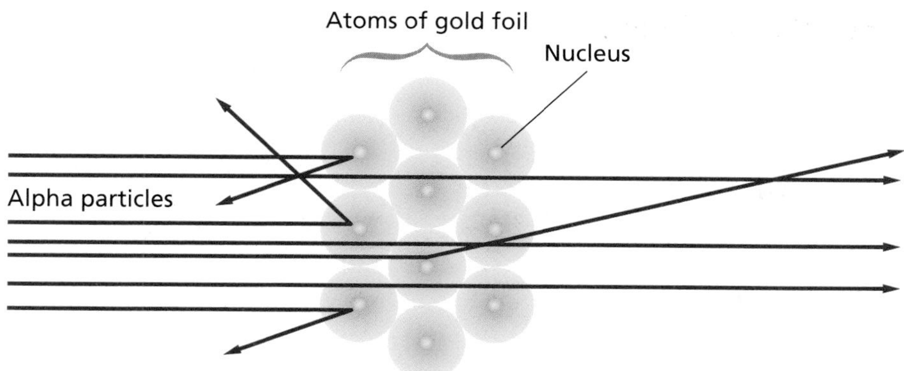

Figure 28-2 Most of the alpha particles directed at the thin sheet of gold foil passed through it without deflection. One in 20,000, however, were deflected at large angles.

Using Coulomb's force law and Newton's laws of motion, Rutherford concluded that the results could be explained only if all of the atom's positive charge were concentrated in a tiny, massive central core, now called the **nucleus.** Therefore, Rutherford's model of the atom is called the nuclear model. Researchers have since determined that all the positive charge and more than 99.9 percent of the mass of the atom are contained in its nucleus. The electrons, which do not contribute a significant amount of mass to the atom, are distributed outside of and far away from the nucleus. Thus, the space occupied by the electrons defines the overall size, or diameter, of the atom. Because the diameter of the atom is about 10,000 times larger than the diameter of the nucleus, the atom mostly is made up of empty space.

Emission spectra How are the electrons arranged around the nucleus of the atom? One of the clues that scientists used to answer this question came from the study of the light emitted by atoms. Recall from the previous chapter that the set of electromagnetic wavelengths emitted by an atom is called the atom's emission spectrum.

As shown in **Figure 28-3,** atoms of a gaseous sample can be made to emit light in a gas-discharge tube. You probably are familiar with the colorful neon signs used by some businesses. These signs work on the same principles as gas-discharge tubes do. A gas-discharge tube consists of a low-pressure gas contained within a glass tube that has metal electrodes attached to each end. The gas glows when high voltage is applied across the tube. What interested scientists the most about this phenomenon was the fact that each different gas glowed with a different, unique color. The characteristic glows emitted by several gases are shown in Figure 28-3.

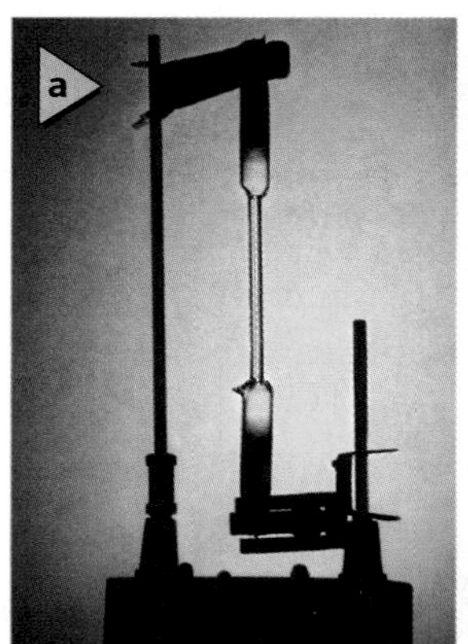

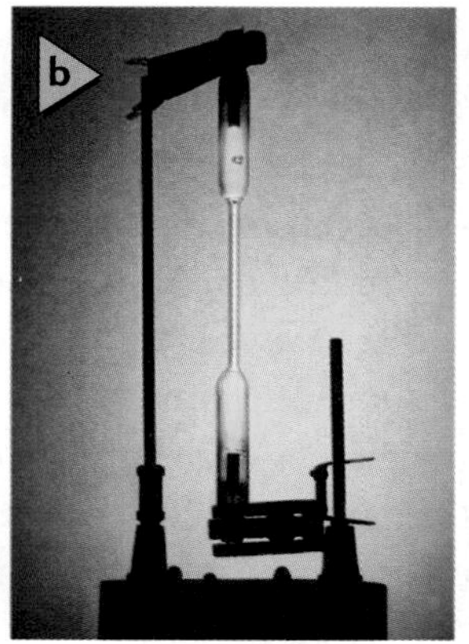

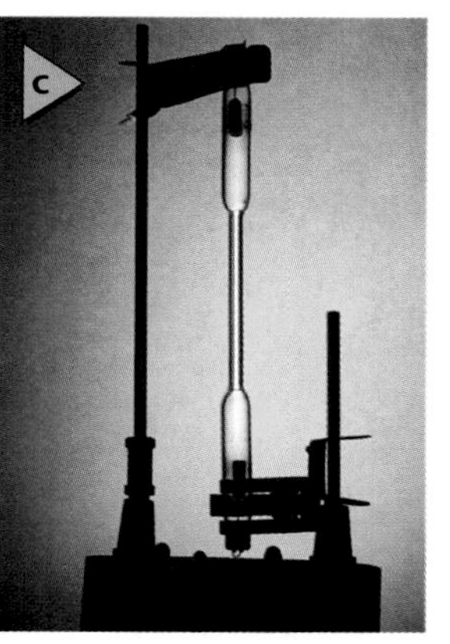

Figure 28-3 When high voltage is applied to a gas, the gas emits light, producing a unique glow. Hydrogen gas glows magenta **(a),** mercury glows bright blue **(b),** and nitrogen glows rose-orange **(c).**

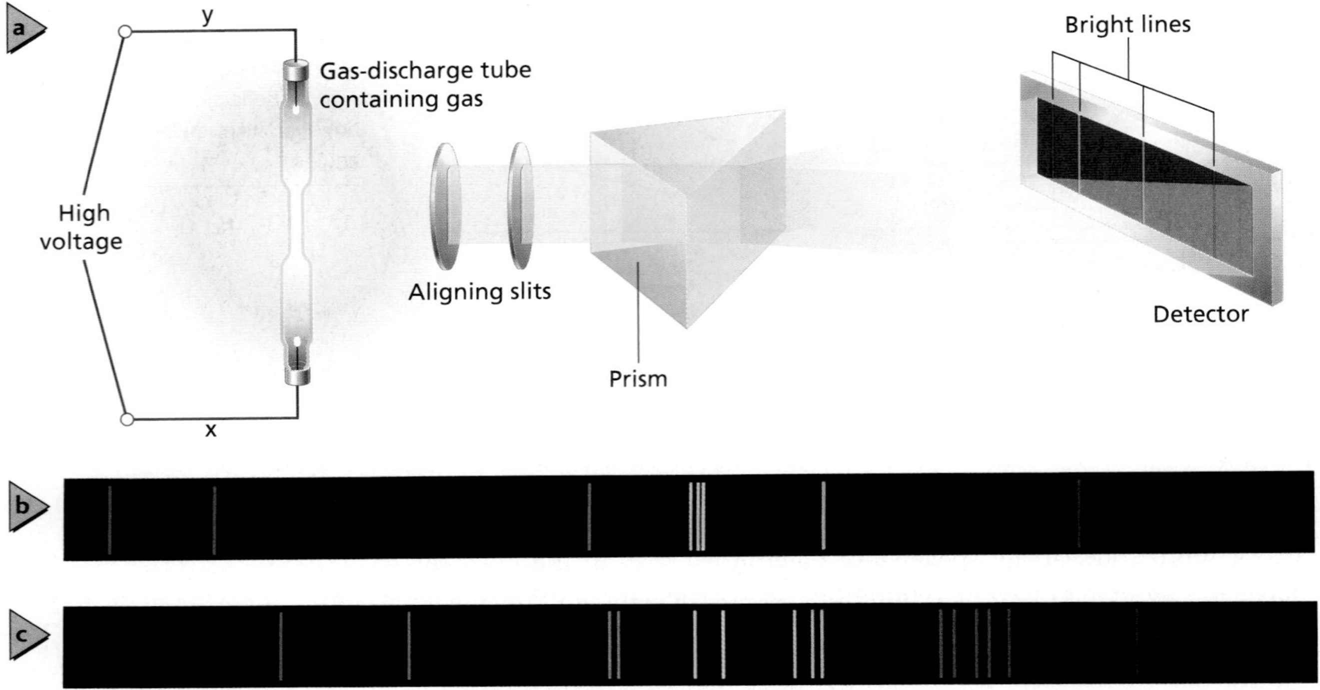

Figure 28-4 A prism spectroscope can be used to observe emission spectra **(a).** The emission spectra of mercury **(b)** and barium **(c)** show characteristic lines.

When the light emitted by the gas is passed through a prism or a diffraction grating, the emission spectrum of an atom is obtained. An emission spectrum can be studied in greater detail using an instrument called a spectroscope. As shown in **Figure 28-4a,** light in a prism spectroscope passes through a slit and is then dispersed as it travels through a prism. A lens system (not shown in the diagram) focuses the dispersed light so that it can be viewed, or recorded on a photographic plate or an electronic detector. The spectroscope forms an image of the slit at a different position for each wavelength.

The spectrum of a hot body, or incandescent solid, such as the filament in a lightbulb, is a continuous band of colors from red through violet. The spectrum of a gas, however, is a series of distinct lines of different colors. The bright-line emission spectra for mercury gas and barium gas are shown in **Figure 28-4b** and **Figure 28-4c,** respectively. Each colored line corresponds to a particular wavelength of light emitted by the atoms of that gas.

An emission spectrum is also a useful analytic tool, as it can be used to identify an unknown sample of gas. When the unknown gas is placed in a gas-discharge tube, it can be made to emit light. The emitted light consists of wavelengths that are uniquely characteristic of the atoms of that gas. Thus, the unknown gas can be identified by comparing its wavelengths with the wavelengths present in the spectra of known samples.

An emission spectrum also can be used to analyze a mixture of gases. When the emission spectrum of a combination of elements is photographed, an analysis of the lines on the photograph can indicate the identities and the relative concentrations of the elements present. If the material being examined contains a large amount of any particular element, the lines for that element are more intense on the photograph than those of the other elements of lesser quantities. Through a comparison of the line intensities, the percentage composition of the material can be determined.

■ **Figure 28-5** Fraunhofer lines appear in the absorption spectrum of the Sun. There are many lines, some of which are faint and others that are very dark, depending on abundances of the elements in the Sun.

Absorption spectra In 1814, Josef von Fraunhofer observed the presence of several dark lines in the spectrum of sunlight. These dark lines, now called Fraunhofer lines, are seen in **Figure 28-5.** He reasoned that as sunlight passed through the gaseous atmosphere surrounding the Sun, the gases absorbed certain characteristic wavelengths. These absorbed wavelengths produced the dark lines in the observed spectrum. The set of wavelengths absorbed by a gas is the **absorption spectrum** of the gas. The composition of the Sun's atmosphere was determined by comparing the missing lines in the observed spectrum with the known emission spectra of various elements. The compositions of many other stars have been determined using this technique.

Astronomy Connection

You can observe an absorption spectrum by passing white light through a gas sample and a spectroscope, as shown in **Figure 28-6a.** Because the gas absorbs specific wavelengths, the normally continuous spectrum of the white light has dark lines in it after passing through the gas. For a gas, the bright lines of the emission spectrum and the dark lines of the absorption spectrum often occur at the same wavelengths, as shown in **Figure 28-6b** and **Figure 28-6c,** respectively. Thus, cool, gaseous elements absorb the same wavelengths that they emit when excited. As you might expect, the composition of a gas can be determined from the wavelengths of the dark lines of the gas's absorption spectrum.

■ **Figure 28-6** This apparatus is used to produce the absorption spectrum of sodium **(a).** The emission spectrum of sodium consists of several distinct lines **(b),** whereas the absorption spectrum of sodium is nearly continuous **(c).**

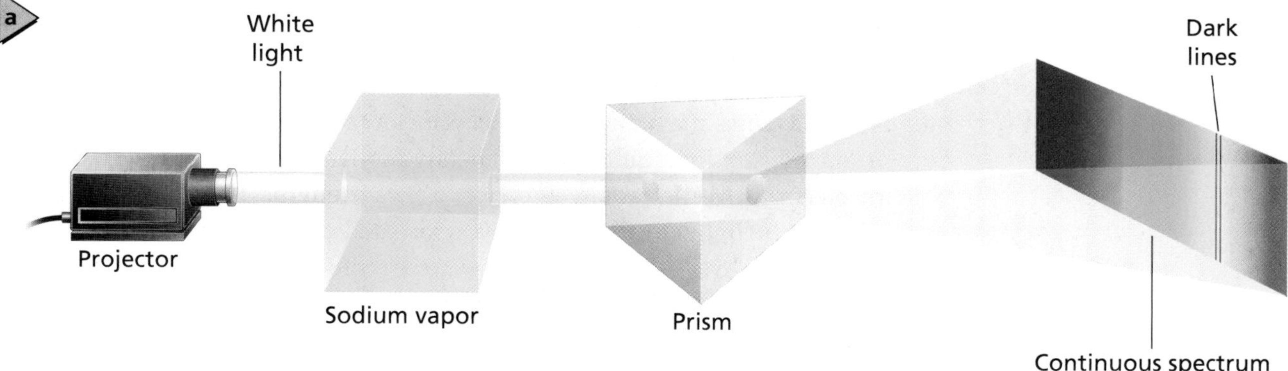

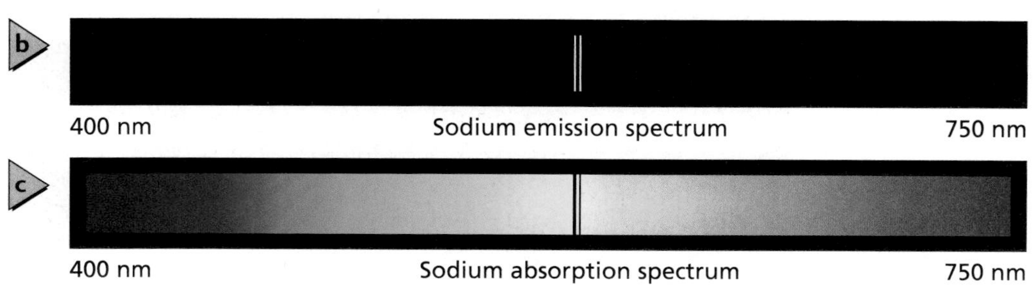

Spectroscopy Both emission and absorption spectra are valuable scientific tools. As a result of the elements' characteristic spectra, scientists are able to analyze, identify, and quantify unknown materials by observing the spectra that they emit or absorb. The emission and absorption spectra of elements are important in industry as well as in scientific research. For example, steel mills reprocess large quantities of scrap iron of varying compositions. The exact composition of a sample of scrap iron can be determined in minutes by spectrographic analysis. The composition of the steel can then be adjusted to suit commercial specifications. Aluminum, zinc, and other metal-processing plants employ the same method.

The study of spectra is a branch of science known as spectroscopy. Spectroscopists are employed throughout research and industrial communities. Spectroscopy has proven to be an effective tool for analyzing materials on Earth, and it is the only currently available tool for studying the composition of stars over the vast expanse of space.

The Bohr Model of the Atom

In the nineteenth century, many physicists tried to use atomic spectra to determine the structure of the atom. Hydrogen was studied extensively because it is the lightest element and has the simplest spectrum. The visible spectrum of hydrogen consists of four lines: red, green, blue, and violet, as shown in **Figure 28-7.** Any theory that explained the structure of the atom would have to account for these wavelengths and support the nuclear model. However, the nuclear model as proposed by Rutherford was not without its problems. Rutherford had suggested that electrons orbit the nucleus much like the planets orbit the Sun. There was, however, a serious flaw in this planetary model.

Problems with the planetary model An electron in an orbit constantly is accelerated toward the nucleus. As you learned in Chapter 26, accelerating electrons radiate energy by emitting electromagnetic waves. At the rate that an orbiting electron would lose energy, it should spiral into the nucleus within 10^{-9} s. This, however, must not be happening because atoms are known to be stable. Thus, the planetary model was not consistent with the laws of electromagnetism. In addition, the planetary model predicted that the accelerating electrons would radiate energy at all wavelengths. However, as you just learned, the light emitted by atoms is radiated only at specific wavelengths.

Danish physicist Niels Bohr went to England in 1911 and joined Rutherford's group to work on determining the structure of the atom. He tried to unite the nuclear model with Planck's quantized energy levels and Einstein's theory of light. This was a courageous idea because as of 1911, neither of these revolutionary ideas was widely understood or accepted.

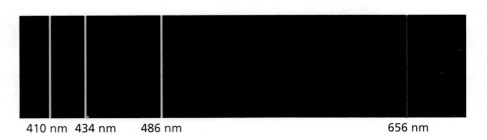

Figure 28-7 The emission spectrum of hydrogen in the visible range has four lines.

Quantized Energy

Bohr began with the planetary arrangement of electrons, as diagrammed in **Figure 28-8,** but then made the bold hypothesis that the laws of electromagnetism do not apply inside the atom. He postulated that an electron in a stable orbit does not radiate energy, even though it is accelerating. Bohr referred to this stable condition as a stationary state. He went on to assume that only stationary states with specific amounts of energy are allowed. In other words, Bohr considered the energy levels in an atom to be quantized.

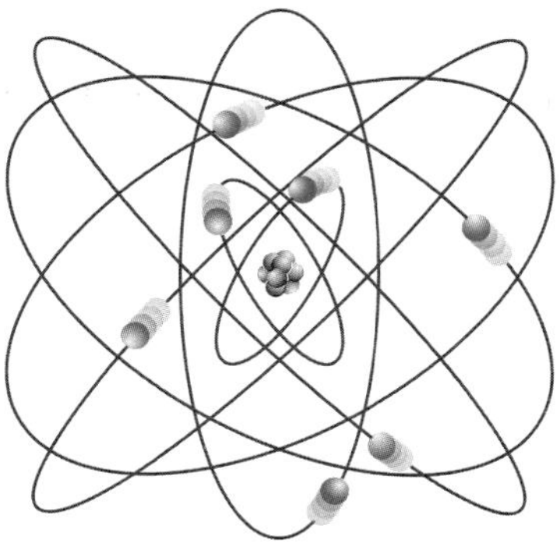

As shown in **Figure 28-9,** the quantization of energy in atoms can be likened to a flight of stairs with decreasing-height steps. To go up the stairs you must move from one step to the next—it is impossible to stop at a midpoint between steps. Instead of steps, atoms have quantized amounts of energy, each of which is called an **energy level.** Just as you cannot occupy a position between steps, an atom's energy cannot have a value between allowed energy levels. An atom with the smallest allowable amount of energy is said to be in the **ground state.** When an atom absorbs energy, it moves, or makes a transition to, a higher energy level. Any energy level above the ground state is called an **excited state.**

Figure 28-8 Bohr's planetary model of the atom was based on the postulation that electrons move in fixed orbits around the nucleus.

Energy of an Atom What determines the amount of energy an atom has? An atom's energy equals the sum of the kinetic energy of the electrons and the potential energy from the attractive force between the electrons and the nucleus. The energy of an atom with electrons in a nearby orbit is less than that of an atom with electrons in a faraway orbit because work must be done to move the electrons away from the nucleus. Thus, atoms in excited, higher-energy states have electrons in larger, or more distant, orbits. Because energy is quantized and energy is related to the size of the orbit, the size of the orbit also is quantized. The model of an atom just described, that of a central nucleus with orbiting electrons having specific quantized energy levels, is known as the Bohr model of the atom.

If Bohr was correct in hypothesizing that stable atoms do not radiate energy, then what is responsible for an atom's characteristic emission spectrum? To answer this question, Bohr suggested that electromagnetic energy is emitted when the atom changes from one stationary state to another. Incorporating Einstein's photoelectric theory, Bohr knew that the energy of every photon is given by the equation, $E_{photon} = hf$. He then postulated that when an atom absorbs a photon, the atom's energy increases by an amount equal to that of the photon. This excited atom then makes a transition to a lower energy level by emitting a photon.

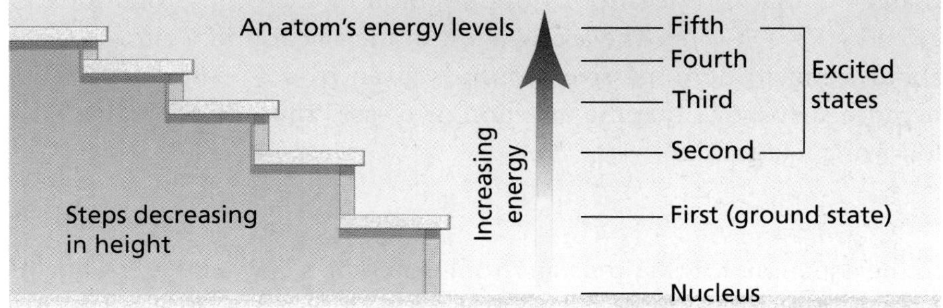

Figure 28-9 These decreasing-height steps are analogous to the allowed energy levels in an atom. Note how the difference in energy between adjacent energy levels decreases as the energy level increases.

$E_{photon\ 1} = E_3 - E_1$
$E_{photon\ 2} = E_2 - E_1$
$E_{photon\ 3} = E_3 - E_2$

$E_{photon\ 1} > E_{photon\ 2} > E_{photon\ 3}$
$\lambda_1 < \lambda_2 < \lambda_3$

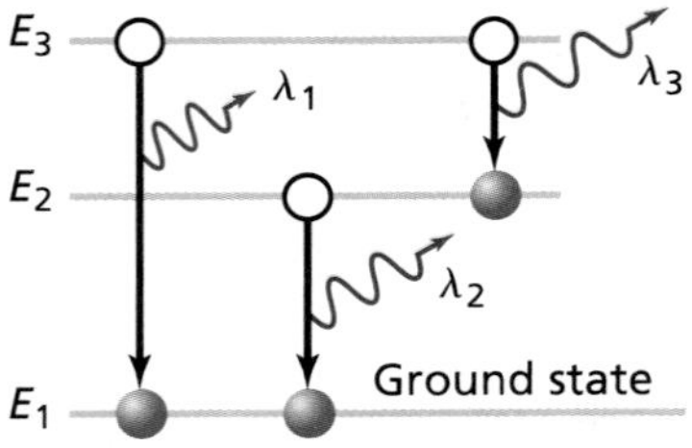

Figure 28-10 The energy of the emitted photon is equal to the difference in energy between the initial and final energy levels of the atom.

Concepts In Motion
Interactive Figure To see an animation on energy levels, visit physicspp.com.

When the atom makes the transition from its initial energy level, E_i, to its final energy level, E_f, the change in energy, ΔE_{atom}, is given by the following equation.

$$\Delta E_{atom} = E_f - E_i$$

As shown in **Figure 28-10,** the change in energy of the atom equals the energy of the emitted photon.

$$E_{photon} = \Delta E_{atom}$$

or

$$E_{photon} = E_f - E_i$$

The following equations summarize the relationships between the change in energy states of an atom and the energy of the photon emitted.

Energy of an Emitted Photon $E_{photon} = hf$, or $E_{photon} = \Delta E_{atom}$

The energy of an emitted photon is equal to the product of Planck's constant and the emitted photon's frequency. The energy of an emitted photon also is equal to the loss in the atom's energy.

Predictions of the Bohr Model

A scientific theory must do more than present postulates; it must allow predictions to be made that can be checked against experimental data. A good theory also can be applied to many different problems, and it ultimately provides a simple, unified explanation of some part of the physical world.

Bohr used his theory to calculate the wavelengths of light emitted by a hydrogen atom. The calculations were in excellent agreement with the values measured by other scientists. As a result, Bohr's model was widely accepted. Unfortunately, the model only worked for the element hydrogen; it could not predict the spectrum of helium, the next-simplest element. In addition, there was not a good explanation as to why the laws of electromagnetism should work everywhere but inside the atom. Not even Bohr believed that his model was a complete theory of the structure of the atom. Despite its shortcomings, however, the Bohr model describes the energy levels and wavelengths of light emitted and absorbed by hydrogen atoms remarkably well.

Development of Bohr's model Bohr developed his model by applying Newton's second law of motion, $F_{net} = ma$, to the electron. The net force is described by Coulomb's law for the interaction between an electron of charge $-q$ that is a distance r from a proton of charge $+q$. That force is given by $F = -Kq^2/r^2$. The acceleration of the electron in a circular orbit about a much more massive proton is given by $a = -v^2/r$, where the negative sign shows that the direction is inward. Thus, Bohr obtained the following relationship:

$$\frac{Kq^2}{r^2} = \frac{mv^2}{r}$$

In the equation, K is the constant from Coulomb's law and has a value of $9.0\times10^9\ N\cdot m^2/C^2$.

Next, Bohr considered the angular momentum of the orbiting electron, which is equal to the product of an electron's momentum and the radius of its circular orbit. The angular momentum of the electron is thus given by mvr. Bohr postulated that angular momentum also is quantized; that is, the angular momentum of an electron can have only certain values. He claimed that the allowed values were multiples of $h/2\pi$, where h is Planck's constant. Using n to represent an integer, Bohr proposed that $mvr = nh/2\pi$. Using $Kq^2/r^2 = mv^2/r$ and rearranging the angular momentum equation, $v = nh/2\pi mr$, Bohr found that the orbital radii of the electrons in a hydrogen atom are given by the following equation.

Electron Orbital Radius in Hydrogen $r_n = \frac{h^2n^2}{4\pi^2Kmq^2}$

The radius of an electron in orbit n is equal to the product of the square of Planck's constant and the square of the integer n divided by the quantity four times the square of π, times the constant K, times the mass of an electron, times the square of the charge of an electron.

You can calculate the radius of the innermost orbit of a hydrogen atom, also known as the Bohr radius, by substituting known values and $n = 1$ into the above equation.

$$r_1 = \frac{(6.626\times10^{-34}\ \text{J·s})^2(1)^2}{4\pi^2(9.0\times10^9\ \text{N·m}^2/\text{C}^2)(9.11\times10^{-31}\ \text{kg})(1.60\times10^{-19}\ \text{C})^2}$$

$$= 5.3\times10^{-11}\ \text{J}^2\text{·s}^2/\text{N·m}^2\text{·kg}$$

$$= 5.3\times10^{-11}\ \text{m, or } 0.053\ \text{nm}$$

By performing a little more algebra you can show that the total energy of the atom, which is the sum of the kinetic energy of the electron and the potential energy, and is given by $-Kq^2/2r$, is represented by the following equation:

$$E_n = \frac{-2\pi K^2mq^4}{h^2} \times \frac{1}{n^2}$$

By substituting numerical values for the constants, you can calculate the total energy of the atom in joules, which yields the following equation:

$$E_n = -2.17\times10^{-18}\ \text{J} \times \frac{1}{n^2}$$

Converting the relationship to units of electron volts yields the following equation.

Energy of a Hydrogen Atom $E_n = -13.6\ \text{eV} \times \frac{1}{n^2}$

The total energy of an atom with principal quantum number n is equal to the product of -13.6 eV and the inverse of n^2.

Both the electron's orbital radius and the energy of the atom are quantized. The integer, n, that appears in these equations is called the **principal quantum number.** It is the principal quantum number that determines the quantized values of r and E. In summary, the radius, r, increases as the square of n, whereas the energy, E, depends on $1/n^2$.

Physics Online

Personal Tutor For an online tutorial on the energy of the hydrogen atom, visit physicspp.com.

MINI LAB

Bright-Line Spectra

Turn on a gas-discharge tube power supply attached to a gas tube so that the tube glows. ***CAUTION: Handle gas tube carefully to avoid breaking. Do not touch any exposed metal when the power supply is turned on. Dangerous voltages are present. Always turn off the power supply before changing gas tubes.***

Turn off the room lights.

1. **Describe** the color that you observe.
2. **Observe** the gas-discharge tube through a diffraction grating.
3. **Sketch** the results of viewing the gas-discharge tube through the diffraction grating.
4. **Predict** whether the observed spectrum will change when a different gas-discharge tube is viewed through the diffraction grating.
5. **Test** your prediction.

Analyze and Conclude

6. **Sketch** the results of viewing the new gas-discharge tube through the diffraction grating.
7. **Explain** why there are differences in the two spectra.

Energy and electron transitions You may be wondering why the energy of an atom in the Bohr model has a negative value. Recall from Chapter 11 that only energy differences have meaning. The zero energy level can be chosen at will. In this case, zero energy is defined as the energy of the atom when the electron is infinitely far from the nucleus and has no kinetic energy. This condition exists when the atom has been ionized; that is, when the electron has been removed from the atom. Because work has to be done to ionize an atom, the energy of an atom with an orbiting electron is less than zero. The energy of an atom has a negative value. When an atom makes a transition from a lower to a higher energy level, the total energy becomes less negative, but the overall energy change is positive.

Some of hydrogen's energy levels and the possible energy level transitions that it can undergo are shown in **Figure 28-11.** Note that an excited hydrogen atom can emit electromagnetic energy in the infrared, visible, or ultraviolet range depending on the transition that occurs. Ultraviolet light is emitted when the atom drops into its ground state from any excited state. The four visible lines in the hydrogen spectrum are produced when the atom drops from the $n = 3$ or higher energy state into the $n = 2$ energy state.

Figure 28-11 The distinct set of color lines that make up a hydrogen atom's visible spectrum are known as the Balmer series. This visible light is the result of the photons emitted when electrons make transitions to the second energy level, $n = 2$. Other electron transitions in a hydrogen atom result in the emission of ultraviolet (Lyman series) and infrared (Paschen series) electromagnetic energy.

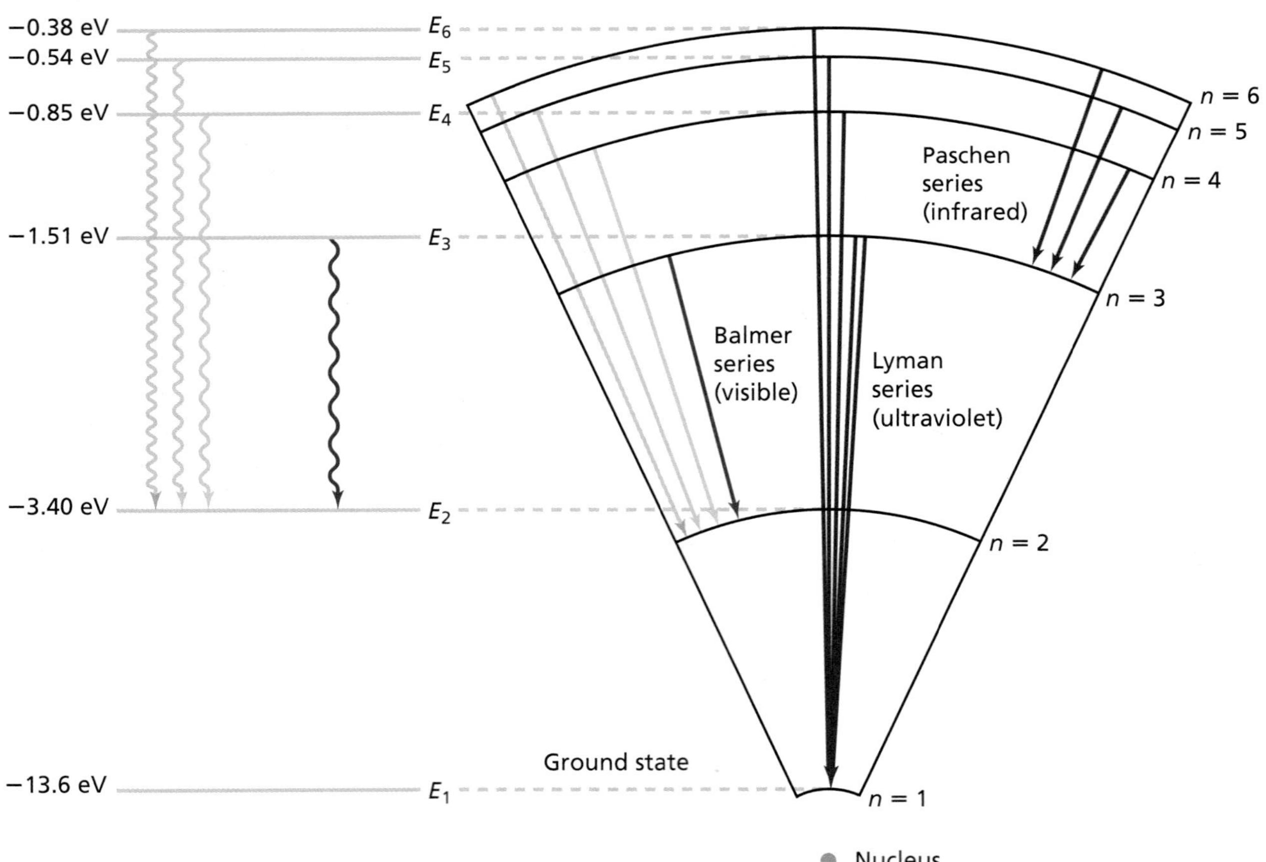

▶ EXAMPLE Problem 1

Energy Levels **A hydrogen atom absorbs energy, causing its electron to move from the innermost energy level ($n = 1$), to the second energy level ($n = 2$). Determine the energy of the first and second energy levels, and the energy absorbed by the atom.**

1 Analyze and Sketch the Problem

- Diagram the energy levels E_1 and E_2.
- Indicate the direction of increasing energy on the diagram.

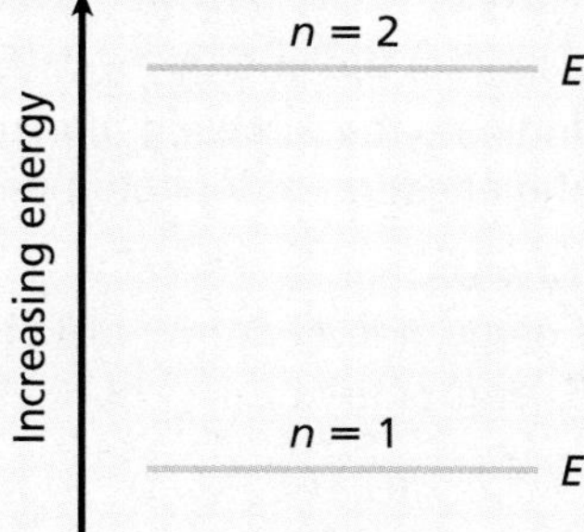

Known:	**Unknown:**
Quantum number of innermost energy level, $n = 1$	Energy of level E_1 = ?
Quantum number of second energy level, $n = 2$	Energy of level E_2 = ?
	Energy difference, ΔE = ?

2 Solve for the Unknown

Use the equation for the energy of an electron in its orbit to calculate the energy of each level.

$$E_n = -13.6 \text{ eV} \times \frac{1}{n^2}$$

$$E_1 = -13.6 \text{ eV} \times \frac{1}{(1)^2} \quad \text{Substitute } n = 1$$

$$= -13.6 \text{ eV}$$

$$E_2 = -13.6 \text{ eV} \times \frac{1}{(2)^2} \quad \text{Substitute } n = 2$$

$$= -3.40 \text{ eV}$$

Math Handbook
Operations with Significant Digits
pages 835–836

The energy absorbed by the atom, ΔE, is equal to the energy difference between the final energy level of the atom, E_f, and the initial energy level of the atom, E_i.

$$\Delta E = E_f - E_i$$

$$= E_2 - E_1 \quad \text{Substitute } E_f = E_2, E_i = E_1$$

$$= -3.40 \text{ eV} - (-13.6 \text{ eV}) \quad \text{Substitute } E_2 = -3.40 \text{ eV}, E_1 = -13.6 \text{ eV}$$

$$= 10.2 \text{ eV} \quad \text{Energy is absorbed.}$$

3 Evaluate the Answer

- **Are the units correct?** Orbital energy values should be measured in electron volts.
- **Is the sign correct?** The energy difference is positive when electrons move from lower energy levels to higher energy levels.
- **Is the magnitude realistic?** The energy needed to move an electron from the first energy level to the second energy level should be approximately 10 eV, which it is.

▶ PRACTICE Problems

- Additional Problems, Appendix B
- Solutions to Selected Problems, Appendix C

1. Calculate the energies of the second, third, and fourth energy levels in the hydrogen atom.
2. Calculate the energy difference between E_3 and E_2 in the hydrogen atom.
3. Calculate the energy difference between E_4 and E_2 in the hydrogen atom.
4. The text shows the solution of the equation $r_n = \frac{h^2n^2}{4\pi^2Kmq^2}$ for $n = 1$, the innermost orbital radius of the hydrogen atom. Note that with the exception of n^2, all factors in the equation are constants. The value of r_1 is 5.3×10^{-11} m, or 0.053 nm. Use this information to calculate the radii of the second, third, and fourth allowable energy levels in the hydrogen atom.
5. The diameter of the hydrogen nucleus is 2.5×10^{-15} m, and the distance between the nucleus and the first electron is about 5×10^{-11} m. If you use a ball with a diameter of 7.5 cm to represent the nucleus, how far away will the electron be?

▶ EXAMPLE Problem 2

Frequency and Wavelength of Emitted Photons **An excited hydrogen atom drops from the second energy level ($n = 2$) to the first energy level ($n = 1$). Calculate the energy and the wavelength of the emitted photon. Use E_1 and E_2 values from Example Problem 1.**

1 Analyze and Sketch the Problem

- Diagram the energy levels E_1 and E_2.
- Indicate the direction of increasing energy and show the emission of a photon on the diagram.

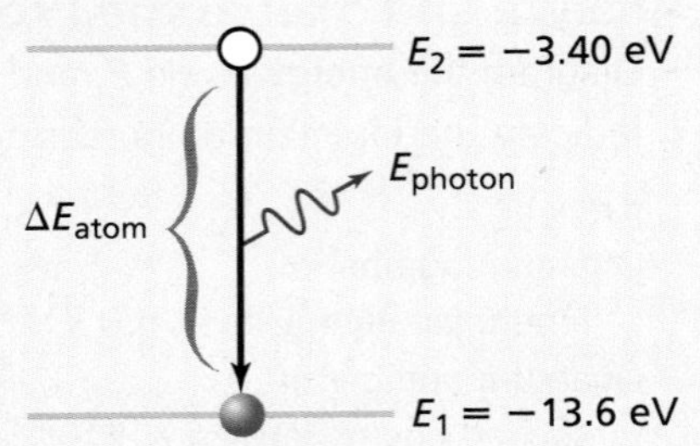

Known:	**Unknown:**
Energy of level $E_1 = -13.6$ eV	Frequency, $f = ?$
Energy of level $E_2 = -3.40$ eV	Wavelength, $\lambda = ?$
	Energy difference, $\Delta E = ?$

2 Solve for the Unknown

The energy of the emitted photon is equal to ΔE, the energy difference between the final energy level of the atom, E_f, and the initial energy level of the atom, E_i.

$\Delta E = E_f - E_i$

$= E_1 - E_2$ **Substitute $E_f = E_1$, $E_i = E_2$**

$= -13.6 \text{ eV} - (-3.40 \text{ eV})$ **Substitute $E_1 = -13.6$ eV, $E_2 = -3.40$ eV**

$= -10.2$ eV   Energy is emitted.

Math Handbook
Isolating a Variable
page 845

To determine the wavelength of the photon, use the following equations.

$\Delta E = hf$, so $f = \frac{\Delta E}{h}$ **Solve the photon energy equation for frequency.**

$c = \lambda f$, so $\lambda = \frac{c}{f}$ **Solve the wavelength-frequency equation for wavelength.**

$\lambda = \frac{c}{(\Delta E/h)}$ **Substitute $f = \frac{\Delta E}{h}$**

$= \frac{hc}{\Delta E}$

$= \frac{1240 \text{ eV·nm}}{10.2 \text{ eV}}$ **Substitute $hc = 1240$ eV·nm, $\Delta E = 10.2$ eV**

$= 122$ nm

3 Evaluate the Answer

- **Are the units correct?** Energy is measured in electron volts. The prefix *nano-* modifies the base SI unit, the meter, which is the correct unit for the wavelength.
- **Are the signs correct?** Energy is released when the atom emits a photon during the transition from the second energy level to the first; thus, the energy difference is negative.
- **Is the magnitude realistic?** Energy released in this transition produces light in the ultraviolet region below 400 nm.

▶ PRACTICE Problems

- Additional Problems, Appendix B
- Solutions to Selected Problems, Appendix C

6. Find the wavelength of the light emitted in Practice Problems 2 and 3. Which lines in Figure 28-7 correspond to each transition?

7. For a particular transition, the energy of a mercury atom drops from 8.82 eV to 6.67 eV.

a. What is the energy of the photon emitted by the mercury atom?

b. What is the wavelength of the photon emitted by the mercury atom?

8. The ground state of a helium ion is −54.4 eV. A transition to the ground state emits a 304-nm photon. What was the energy of the excited state?

CHALLENGE PROBLEM

Although the Bohr atomic model accurately explained the behavior of a hydrogen atom, it was unable to explain the behavior of any other atom. Verify the limitations of the Bohr model by analyzing an electron transition in a neon atom. Unlike a hydrogen atom, a neon atom has ten electrons. One of these electrons makes a transition between the $n = 5$ and the $n = 3$ energy states, emitting a photon in the process.

1. Assuming that the neon atom's electron can be treated as an electron in a hydrogen atom, what photon energy does the Bohr model predict?
2. Assuming that the neon atom's electron can be treated as an electron in a hydrogen atom, what photon wavelength does the Bohr model predict?
3. The actual wavelength of the photon emitted during the transition is 632.8 nm. What is the percent error of the Bohr model's prediction of photon wavelength?

E_5 $n = 5$
E_4 $n = 4$
$\lambda = 632.8$ nm
E_3 $n = 3$
E_2 $n = 2$
E_1 $n = 1$

The Bohr model was a major contribution to scientists' understanding of the structure of the atom. In addition to calculating the emission spectrum, Bohr and his students were able to calculate the ionization energy of a hydrogen atom. The ionization energy of an atom is the energy needed to completely free an electron from an atom. The calculated ionization value closely agreed with experimental data. The Bohr model further provided an explanation of some of the chemical properties of the elements. The idea that atoms have electron arrangements unique to each element is the foundation of much of our knowledge of chemical reactions and bonding. Niels Bohr, whose accomplishments are commemorated in the stamps shown in **Figure 28-12,** was awarded a Nobel prize in 1922.

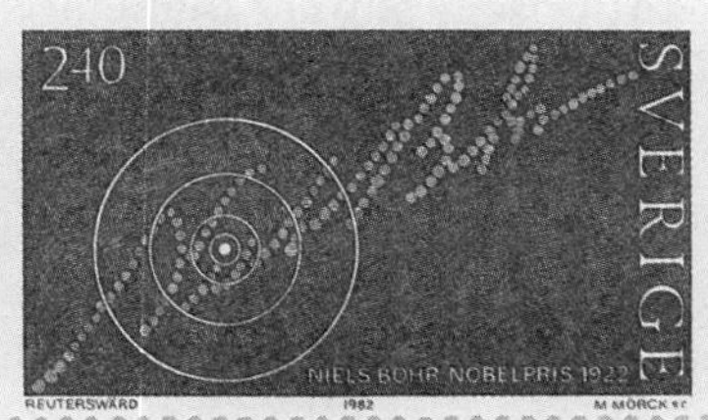

Figure 28-12 Honored in these postage stamps from Denmark and Sweden, Niels Bohr's great contribution to our understanding of the atom earned him worldwide recognition and a Nobel prize.

28.1 Section Review

9. **Rutherford's Nuclear Model** Summarize the structure of the atom according to Rutherford's nuclear model.
10. **Spectra** How do the emission spectra of incandescent solids and atomic gases differ? In what ways are they similar?
11. **Bohr Model** Explain how energy is conserved when an atom absorbs a photon of light.
12. **Orbit Radius** A helium ion behaves like a hydrogen atom. The radius of the ion's lowest energy level is 0.0265 nm. According to Bohr's model, what is the radius of the second energy level?
13. **Absorption Spectrum** Explain how the absorption spectrum of a gas can be determined. Describe the reasons for the spectrum's appearance.
14. **Bohr Model** Hydrogen has been detected transitioning from the 101st to the 100th energy levels. What is the radiation's wavelength? Where in the electromagnetic spectrum is this emission?
15. **Critical Thinking** The nucleus of a hydrogen atom has a radius of about 1.5×10^{-15} m. If you were to build a model of the hydrogen atom using a softball ($r = 5$ cm) to represent the nucleus, where would you locate an electron in the $n = 1$ Bohr orbit? Would it be in your classroom?

28.2 The Quantum Model of the Atom

▶ **Objectives**
- **Describe** the shortcomings of Bohr's atomic model.
- **Describe** the quantum model of the atom.
- **Explain** how a laser works.
- **Describe** the properties of laser light.

▶ **Vocabulary**
quantum model
electron cloud
quantum mechanics
coherent light
incoherent light
stimulated emission
laser

The postulates that Bohr made could not be explained on the basis of accepted physics principles of the time. For example, electromagnetic theory required that the accelerated particles radiate energy, causing the rapid collapse of the atom. In addition, the idea that electron orbits have well-defined radii was in conflict with the Heisenberg uncertainty principle. How could Bohr's work be put on a firm foundation?

From Orbits to an Electron Cloud

The first hint to the solution of these problems was provided by Louis de Broglie. Recall from Chapter 27 that de Broglie proposed that particles have wave properties, just as light has particle properties. The de Broglie wavelength of a particle with momentum mv is defined as $\lambda = h/mv$. The angular momentum of a particle can be defined as $mvr = hr/\lambda$. Thus, Bohr's required quantized angular-momentum condition, $mvr = nh/2\pi$, can be written in the following way:

$$\frac{hr}{\lambda} = \frac{nh}{2\pi} \quad \text{or} \quad n\lambda = 2\pi r$$

That is, the circumference of the Bohr orbit, $2\pi r$, is equal to a whole number multiple, n, of de Broglie wavelengths, λ. **Figure 28-13** illustrates this relationship.

In 1926, Austrian physicist Erwin Schröedinger used de Broglie's wave model to create a quantum theory of the atom based on waves. This theory did not propose a simple planetary model of an atom, as Bohr's model had. In particular, the radius of the electron orbit was not likened to the radius of the orbit of a planet about the Sun.

The Heisenberg uncertainty principle states that it is impossible to know both the position and momentum of an electron at the same time. Thus, the modern **quantum model** of the atom predicts only the probability that an electron is in a specific region. Interestingly, the quantum model predicts that the most probable distance between the electron and the nucleus for a hydrogen atom is the same as the radius predicted by the Bohr model.

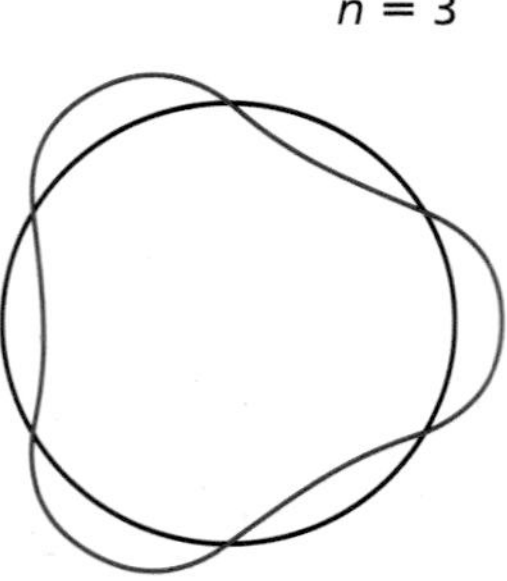

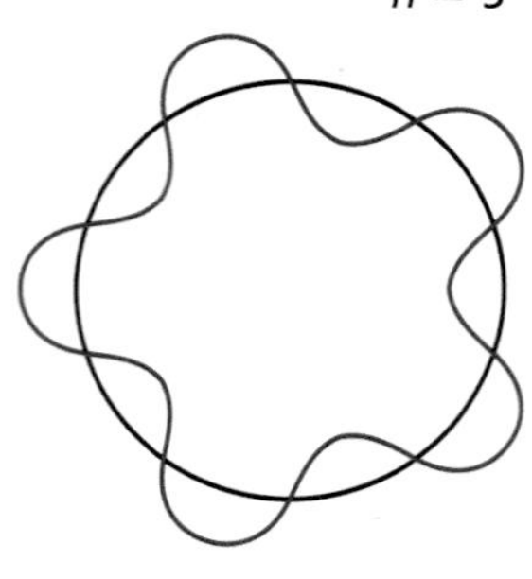

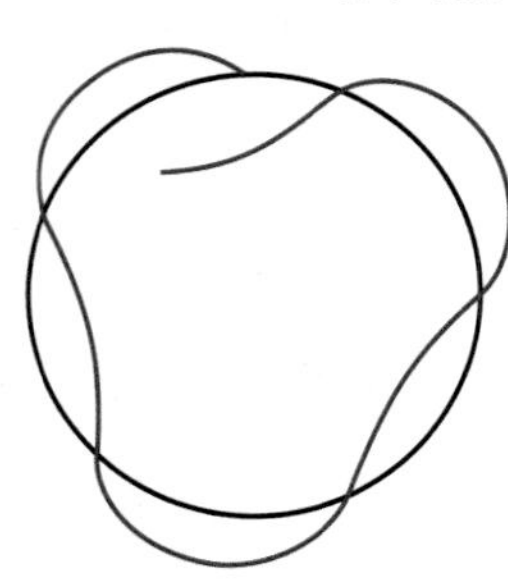

Figure 28-13 For an electron to have a stable orbit around the nucleus, the circumference of the orbit must be a whole-number multiple, n, of the de Broglie wavelength. Note that the whole-number multiples $n = 3$ and $n = 5$ are stable, whereas $n = 2.9$ is not.

The probability of the electron being at any specific radius can be calculated, and a three-dimensional plot can be constructed showing regions of equal probability. The region in which there is a high probability of finding the electron is called the **electron cloud.** A slice through the electron cloud for the two lowest energy states of a hydrogen atom is shown in **Figure 28-14.**

Even though the quantum model of the atom is difficult to visualize, **quantum mechanics,** the study of the properties of matter using its wave properties, makes use of this model and has been extremely successful in predicting many details of atomic structure. These details are very difficult to calculate precisely for all but the simplest atoms. Only very sophisticated computers can make highly accurate approximations for the heavier atoms. Quantum mechanics also enables the structures of many molecules to be calculated, allowing chemists to determine the arrangement of atoms in the molecules. Guided by quantum mechanics, chemists have been able to create new and useful molecules that do not occur in nature. Quantum mechanics also is used to analyze the details of the emission and absorption of light by atoms. As a result of quantum mechanical theory, a new source of light was developed.

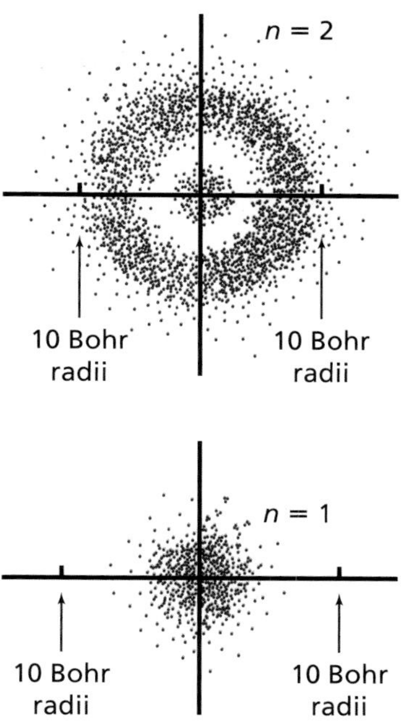

Figure 28-14 These plots show the probability of finding the electron in a hydrogen atom for distances up to approximately 10 Bohr radii from the nucleus for the first and second energy levels. The density of the distribution of dots corresponds to the probability of finding the electron. Note that the Bohr radius = 0.053 nm.

Lasers

As you know, light emitted by an incandescent source has a continuum of wavelengths, whereas light produced by an atomic gas consists of only a few distinct wavelengths. Light from both sources travels in all directions. Furthermore, light waves emitted by atoms at one end of a gas-discharge tube are not necessarily in step, or synchronized, with waves from the other end of the tube. That is, the waves are not necessarily all at the same point in their cycle at the same time. Recall from Chapter 19 that waves that are in step, with their minima and maxima coinciding, are said to be coherent. Light waves that are coherent are referred to as **coherent light.** Out-of-step light waves produce **incoherent light.** Both types of light waves are shown in **Figure 28-15.**

Light is emitted by atoms that have been excited. So far, you have learned about two ways in which atoms can be excited: thermal excitation and electron collision. Atoms also can be excited by collisions with photons of exactly the right energy.

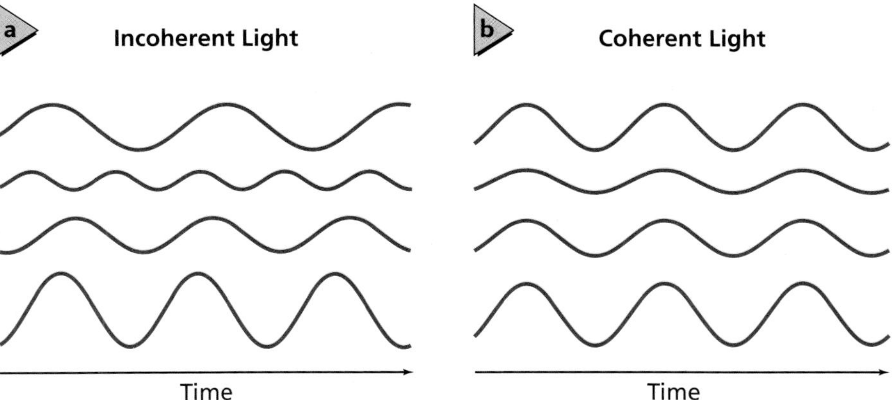

Figure 28-15 Waves of incoherent light **(a)** and coherent light **(b)** are shown.

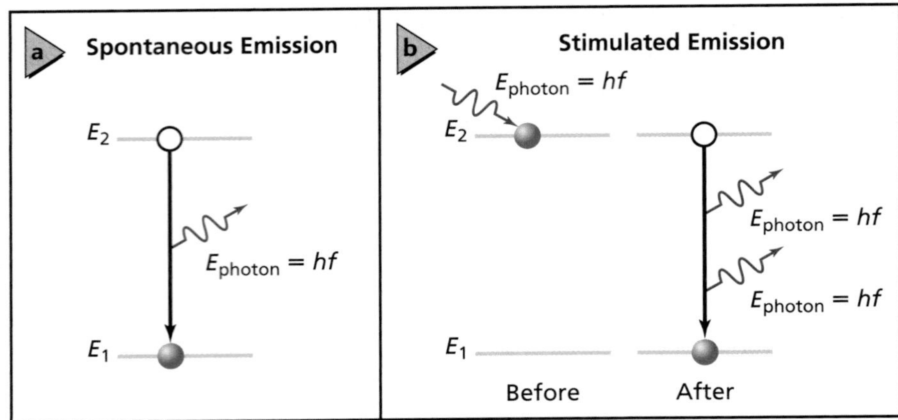

Figure 28-16 During spontaneous emission, an electron in an atom drops from the excited state, E_2, to the ground state, E_1, by spontaneously emitting a photon with energy, hf **(a).** During stimulated emission, an excited atom is struck by a photon with energy $E_2 - E_1$. The atom drops to the ground state and emits a photon. Both the incident and emitted photon have the same energy: $E_{photon} = E_2 - E_1$ **(b).**

Concepts In Motion
Interactive Figure To see an animation on spontaeous and stimulated emission, physicspp.com.

Spontaneous and stimulated emission What happens when an atom is in an excited state? After a very short time, it normally returns to the ground state, giving off a photon of the same energy that it absorbed, as shown in **Figure 28-16a.** This process is called spontaneous emission.

In 1917, Einstein considered what happens to an atom already in an excited state that is struck by a photon with an energy equal to the energy difference between the excited state and the ground state. He showed that the atom, by a process known as **stimulated emission,** returns to the ground state and emits a photon with an energy equal to the energy difference between the two states. The photon that caused, or stimulated, the emission is not affected. The two photons leaving the atom not only will have the same frequency, but they will be in step, or coherent, as shown in **Figure 28-16b.** Either of the two photons can now strike other excited atoms, thereby producing additional photons that are in step with the original photons. This process can continue and produce an avalanche of photons, all of the same wavelength and all having their maxima and minima at the same times.

For this process to occur, certain conditions must be met. First, there must be other atoms in the excited state. Second, the atoms must remain in the excited state long enough to be struck by a photon. Third, the photons must be contained so that they are able to strike other excited atoms. In 1959, a device called a **laser** was invented that fulfilled all the conditions needed to produce coherent light. The word *laser* is an acronym that stands for *l*ight *a*mplification by *s*timulated *e*mission of *r*adiation. An atom that emits light when it is stimulated in a laser is said to lase.

Atom excitation The atoms in a laser can be excited, or pumped, as outlined in **Figure 28-17.** An intense flash of light with a wavelength shorter than that of the laser can be used to pump the atoms. The shorter wavelength, higher energy photons produced by the flash collide with and excite the lasing atoms. When one of the excited atoms decays to a lower energy state by emitting a photon, the avalanche of photons begins. This results in the emission of a brief flash, or pulse, of laser light. Alternatively, the lasing atoms can be excited by collisions with other atoms. In the helium-neon lasers often seen in science classrooms, an electric discharge excites the helium atoms. These excited helium atoms collide with the neon atoms, pumping them to an excited state and causing them to lase. The laser light resulting from this process is continuous rather than pulsed.

Lasing The photons emitted by the lasing atoms are contained by confining the lasing atoms within a glass tube that has parallel mirrors at each end. One of the mirrors is more than 99.9% reflective and reflects nearly all of the light hitting it, whereas the other mirror is partially reflective and allows only about 1 percent of the light hitting it to pass through. Photons that are emitted in the direction of the ends of the tube will be reflected back into the gas by the mirrors. The reflected photons strike more atoms, releasing more photons with each pass between the mirrors. As the process continues, a high intensity of photons builds. The photons that exit the tube through the partially reflecting mirror produce the laser beam. **Figure 28-18** shows a laser being used in a laboratory.

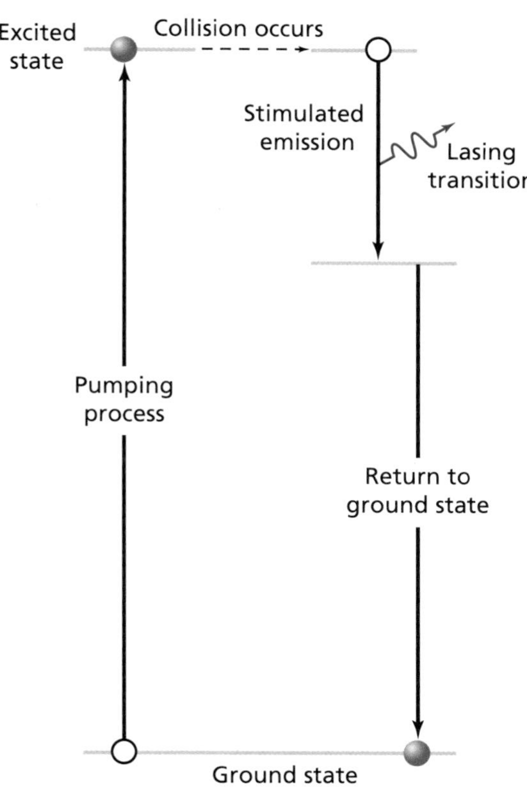

Figure 28-17 When a photon strikes an excited atom, it stimulates the atom to emit a second coherent photon and to make a transition to a lower state.

Because all the stimulated photons are emitted in step with the photons that struck the atoms, laser light is coherent light. The light is also all of one wavelength, or monochromatic, because the transition of electrons between only one pair of energy levels in one type of atom is involved. The parallel mirrors used in the laser result in the emitted laser light being highly directional. In other words, laser light does not diverge much as it travels. Because a typical laser beam is very small, often only about 2 mm in diameter, the light is very intense. Many solid, liquid, and gas substances can be made to lase. Most substances produce laser light at only one wavelength. The light from some lasers, however, can be tuned, or adjusted, over a range of wavelengths.

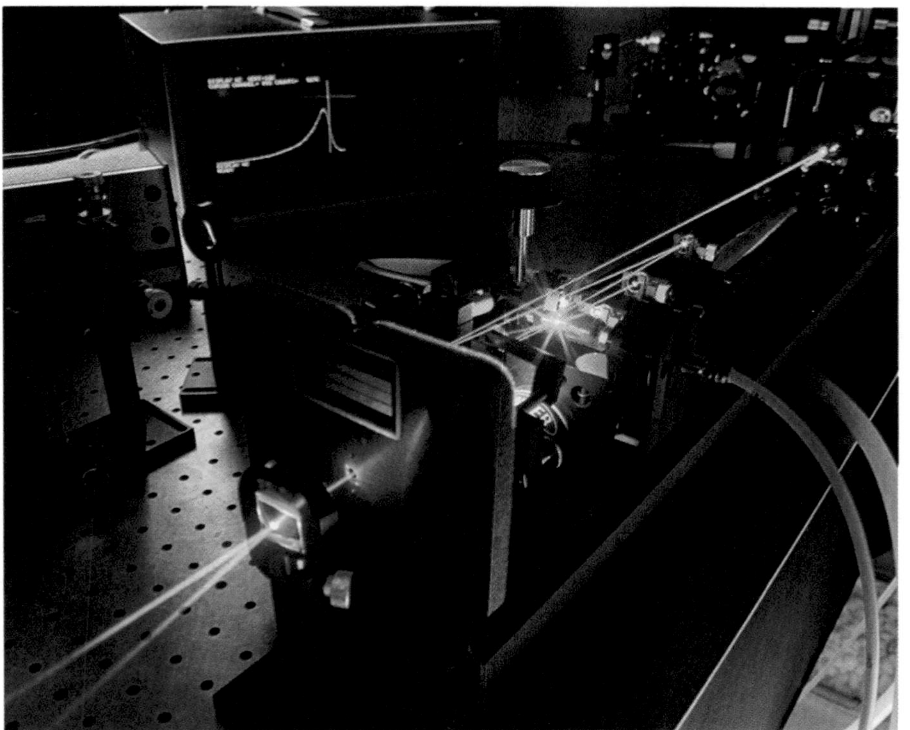

Figure 28-18 This argon laser produces a beam of coherent light.

Table 28-1		
Common Lasers		
Medium	**Wavelength (nm)**	**Type**
Krypton-fluoride excimer (KrF gas)	248 (ultraviolet)	Pulsed
Nitrogen (N_2 gas)	337 (ultraviolet)	Pulsed
Indium gallium nitride (InGaN crystal)	420	Continuous
Argon ion (Ar^+ gas)	476.5, 488.0, 514.5	Continuous
Neon (Ne gas)	632.8	Continuous
Gallium aluminum arsenide (GaAlAs crystal)	635, 680	Continuous
Gallium arsenide (GaAs crystal)	840–1350 (infrared)	Continuous
Neodymium (Nd: YAG crystal)	1064 (infrared)	Pulsed
Carbon dioxide (CO_2 gas)	10,600 (infrared)	Continuous

Laser Applications

If you have used a CD or a DVD player, then you have used a laser. These lasers, as well as the ones used in laser pointers, are made of semiconducting solids. The laser in a CD player is made of layers of gallium arsenide (GaAs) and gallium aluminum and arsenide (GaAlAs). The lasing layer is only 200-nm thick, and each side of the crystal is only 1–2 mm long. The atoms in the semiconducting solid are pumped by an electric current, and the resulting photons are amplified as they bounce between the polished ends of the crystal. **Table 28-1** shows the wavelength and type, pulsed or continuous, produced by some common lasers.

Most lasers are very inefficient. For example, no more than 1 percent of the electrical energy delivered to a gas laser is converted to light energy. Although crystal lasers have efficiencies near 20 percent, they often have much less power than gas lasers do. Despite their inefficiency, the unique properties of laser light have led to many applications. Laser beams are narrow and highly directional—they do not spread out over long distances. For this reason, surveyors use laser beams in applications such as checking the straightness of long tunnels and pipes. When astronauts visited the Moon, they left behind mirrors on the Moon's surface. Scientists on Earth have used the mirrors to reflect a laser beam transmitted from Earth. The distance between Earth and the Moon was thus accurately determined. By tracking the Moon's location from different parts of Earth, the movement of Earth's tectonic plates has been measured.

Laser light also is commonly used in fiber-optics communications. A fiber optic cable makes use of total internal reflection to transmit light over distances of many kilometers with little loss of signal energy. The laser, typically with a wavelength of 1300–1500 nm, is rapidly switched on and off, transmitting information as a series of pulses through the fiber. All over the world, optical fibers have replaced copper wires for the transmission of telephone calls, computer data, and even television pictures.

APPLYING **PHYSICS**

► **Laser Eye Surgery** Excimer lasers are used in laser eye surgery because the energy of the photons that they emit are able to destroy abnormal tissue without causing harm to surrounding healthy tissue. Thus, a skilled surgeon can use the laser to remove extremely thin layers of tissue in order to reshape a cornea. ◄

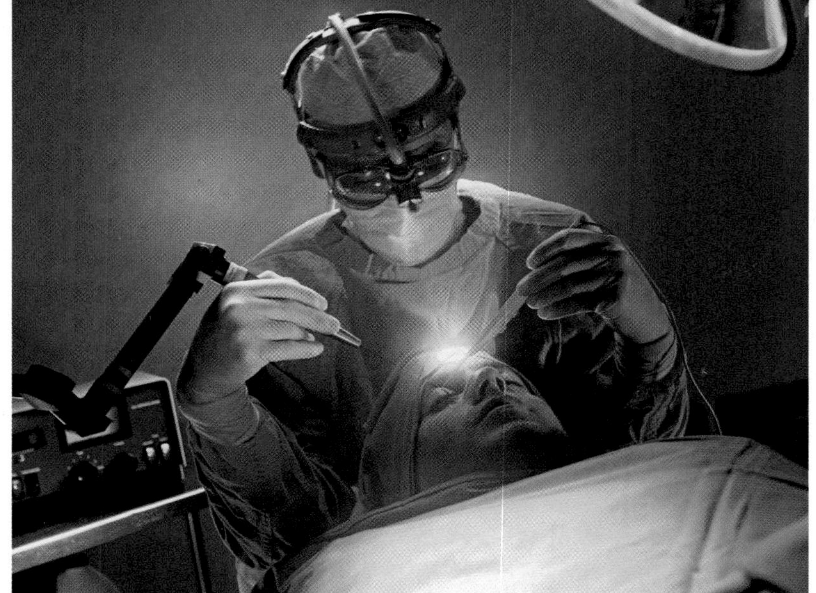

Figure 28-19 The ultraviolet photons emitted by this laser are able to strip electrons from atoms of the targeted tissue. The photons break chemical bonds and vaporize the tissue.

The single wavelength of light emitted by lasers makes them valuable in spectroscopy. Laser light is used to excite other atoms. The atoms then return to the ground state and emit characteristic spectra. Samples with extremely small numbers of atoms can be analyzed in this way. In fact, single atoms have been detected and held almost motionless by means of laser excitation.

The concentrated power of laser light is used in a variety of ways. In medicine, for example, lasers are used to reshape the corneas of eyes. Lasers also can be used in surgery, as shown in **Figure 28-19,** in place of a knife to cut flesh with little loss of blood. In industry, lasers are used to cut materials such as steel and to weld materials together. In the future, lasers may be able to produce nuclear fusion to create an almost inexhaustible energy source.

A hologram, shown in **Figure 28-20,** is a photographic recording of both the phase and the intensity of light. Holograms are made possible by the coherent nature of laser light. Holograms form realistic three-dimensional images and can be used, among other applications, in industry to study the vibrations of sensitive equipment and their components.

Figure 28-20 A hologram is formed when interference of two laser beams records both the intensity and phase of light from the object on film.

28.2 Section Review

16. Lasers Which of the lasers in Table 28-1 emits the reddest light (visible light with the longest wavelength)? Which of the lasers emit blue light? Which of the lasers emit beams that are not visible to the human eye?

17. Pumping Atoms Explain whether green light could be used to pump a red laser. Why could red light not be used to pump a green laser?

18. Bohr Model Limitations Although it was able to accurately predict the behavior of hydrogen, in what ways did Bohr's atomic model have serious shortcomings?

19. Quantum Model Explain why the Bohr model of the atom conflicts with the Heisenberg uncertainty principle, whereas the quantum model does not.

20. Lasers Explain how a laser makes use of stimulated emission to produce coherent light.

21. Laser Light What are the four characteristics of laser light that make it useful?

22. Critical Thinking Suppose that an electron cloud were to get so small that the atom was almost the size of the nucleus. Use the Heisenberg uncertainty principle to explain why this would take a tremendous amount of energy.

PHYSICS LAB

Finding the Size of an Atom

Ernest Rutherford used statistical analysis and probability to help analyze the results of his gold foil experiment. In this experiment, you will model the gold foil experiment using BBs and cups. You will then analyze your results in terms of probability to estimate the size of an object that cannot be seen.

QUESTION

How can probability be used to determine the size of an object that cannot be seen?

Objectives

- **Interpret data** to determine the probability of a BB striking an unseen object.
- **Calculate** the size of an unseen object based on probability.

Safety Precautions

- **Be sure to immediately pick up BBs that have fallen onto the floor.**

Materials

shoe box
three identical small paper cups
200 BBs
centimeter ruler
large towel or cloth

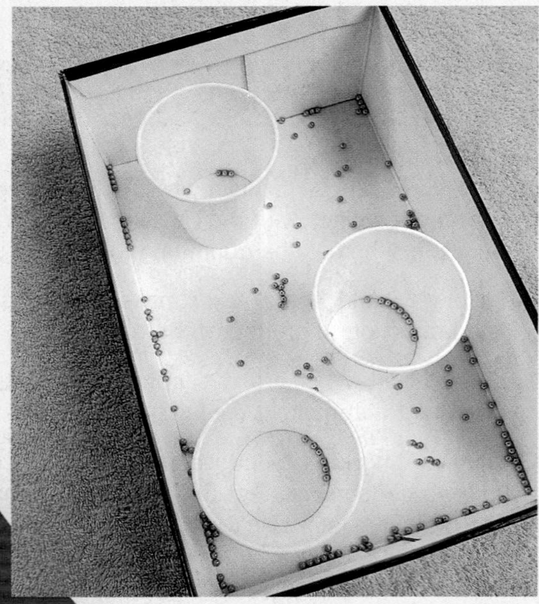

Procedure

1. Use the centimeter ruler to measure the length and width of the inside of the shoe box. Record the measurements in the data table.
2. Use the centimeter ruler to measure the diameter of the top of one of the cups. Record the measurement in the data table.
3. Place the shoe box in the center of a folded towel, such that the towel extends at least 30 cm beyond each side of the shoe box.
4. Randomly place the three paper cups in the bottom of the shoe box.
5. Have your lab partner randomly drop 200 BBs into the shoe box. Make sure he or she distributes the BBs evenly over the area of the shoe box. Note that some of the BBs may miss the shoe box and land on the towel.
6. Count the number of BBs in the cups and record the value in the data table.

Analyze

1. Calculate the area of the shoe box. The area of a rectangular shape is given by the equation Area $=$ Length $\times$ Width.
2. Calculate the area of a cup using the diameter you measured. The area of a circle is given by the equation Area $= \frac{\pi(\text{Diameter})^2}{4}$.
3. Calculate the total area of the cups by multiplying the area per cup by the total number of cups.
4. Calculate the percentage of shoe box that is occupied by the three cups by dividing the total area of the cups by the area of the shoe box and then multiplying by 100.
5. Calculate the percentage of BBs that landed in the cups by dividing the number of BBs in the cups by the number of BBs dropped, and then multiplying by 100.

Data Table

	Your Data	Data from Group 2	Data from Group 3	Data from Group 4	Data from Group 5	Class Average
Shoe box length (cm)						
Shoe box width (cm)						
Shoe box area (cm^2)						
Measured diameter of cup (cm)						
Calculated area of a cup (cm^2)						
Total number of cups	3	3	3	3	3	3
Total calculated area of cups (cm^2)						
Percentage of shoe box occupied by cups (%)						
Number of BBs dropped	200	200	200	200	200	200
Number of BBs in cups						
Percent of BBs in cups						
Percent of shoe box occupied by cups based on probability						
Total area of cups based on probability (cm^2)						
Number of cups	3	3	3	3	3	3
Area of one cup based on probability (cm^2)						

6. Determine the percentage of the shoe box occupied by the cups based on probability. Note that this percentage is (ideally) equal to the percentage of BBs that landed in cups.
7. Calculate the total area of the cups based on probability. To calculate this value, multiply the percentage of the shoe box occupied by the cups (based on probability) by the area of the shoe box.
8. Calculate the area of each cup based on probability by dividing the total area of the cups based on probability by three.
9. Record the experimental data from the other groups in the data table and then calculate classroom averages for all of the data.
10. **Error Analysis** Compare your calculated value for the area of the cup based on probability (experimental value) with the area of the cup calculated from the measured diameter (accepted value). What is the percent error in your value based upon probability? Calculate the percent error using the following equation:

$$\text{Percent error} = \frac{\left|\text{Accepted value} - \text{Experimental value}\right|}{\text{Accepted value}} \times 100$$

Conclude and Apply

1. Were you able to accurately determine the area of a cup based on probability? Explain in terms of the percent error.
2. List the error sources in this experiment and describe their effects on the results.

Going Further

If larger cups were used in the experiment, do you think you would need fewer, the same, or more BBs to achieve accurate results? Explain.

Real-World Physics

Your teacher polls the class about postponing a test. Does the accuracy of the poll depend on how many students are surveyed? Explain.

To find out more about the atom, visit the Web site: physicspp.com

Future Technology

Atom Laser

The recently developed atom laser is a technology with a promising future. Unlike traditional lasers that emit beams or pulses of coherent photons, atom lasers emit beams or pulses of coherent atoms. As explained below, coherent atoms are different from the incoherent atoms that make up ordinary matter.

History In 1923, Louis Victor de Broglie predicted that all particles, including atoms, have wave properties. Wavelength is inversely related to the particle's mass and velocity, and it is too small to be observed at room temperature. As the atom is cooled, its speed is reduced and its wavelength increases.

In the 1920s, Albert Einstein and Satyendra Nath Bose researched particles known as bosons. They predicted that if bosons could be cooled to their lowest possible energy state, then all of the particles would have the same phase and the same wavelength. In other words, the particles would have coherent properties. This unusual phase of matter is called a Bose-Einstein condensate.

The first Bose-Einstein condensates were formed in 1995 by Eric Cornell and Carl Wieman, and independently, by Wolfgang Ketterle. In further research, Ketterle brought two separate Bose-Einstein condensate samples near each other. Ketterle observed wavelike interference patterns from the atoms in the condensates. He went on to prove that all of the atoms in the condensate had the same wavelength and were in phase. The atoms in the condensate were coherent, just as Bose and Einstein predicted.

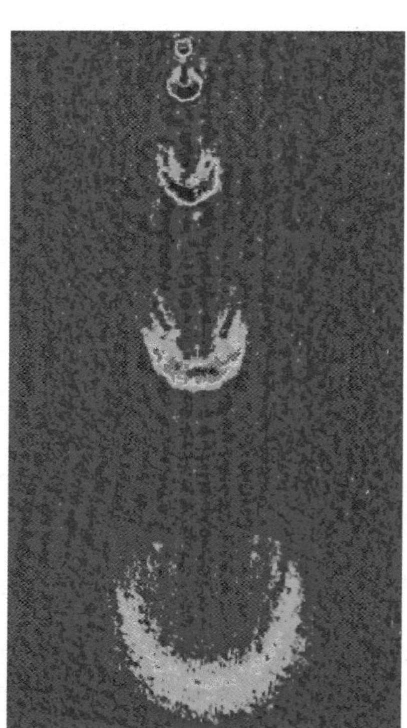

An atom laser emits pulses of coherent sodium atoms. Each pulse contains 10^5 to 10^6 atoms, and the pulses are accelerated downward due to gravity. The spreading of the pulses is a result of repulsive forces.

The First Atom Laser In 1997, Ketterle and his collaborators announced the first steps in the development of an atom laser. They developed a way to eject small pulses (between 100,000 and 1,000,000 atoms) of coherent atoms from a Bose-Einstein condensate into a beam.

In this first atom laser, pulses of coherent atoms could only travel in a single direction. The emitted atoms behaved like projectiles, following downward arcing paths due to the influence of gravity. As shown in the photo, the coherent atoms in each pulse tend to spread apart as the beam propagates. In 1999, William Phillips found a way to send pulses of coherent atoms in any direction and how to prevent the atoms from spreading out as the beam propagated. By making a series of very short pulses, Phillips was able to form a continuous beam of coherent atoms.

The Future Bose-Einstein condensates and atom lasers will be used to study the basic properties of quantum mechanics and matter waves. Scientists anticipate that atom lasers will be useful in making more-precise atomic clocks and in building small electronic circuits. The atom laser also may be used in atomic interferometry to precisely measure gravitational forces and to test relativity.

Going Further

1. **Research** Investigate what fermions are and if they can form Bose-Einstein condensates. (*Hint: See how the Pauli exclusion principle applies to fermions.*)
2. **Critical Thinking** Atom lasers operate in an ultrahigh vacuum environment. Why do you think this is so?

Chapter 28

Study Guide

28.1 The Bohr Model of the Atom

Vocabulary

- alpha particles (p. 748)
- nucleus (p. 749)
- absorption spectrum (p. 751)
- energy level (p. 753)
- ground state (p. 753)
- excited state (p. 753)
- principal quantum number (p. 755)

Key Concepts

- Ernest Rutherford directed positively charged, high-speed alpha particles at thin metal foils. By studying the paths of the reflected particles, he showed that an atom is mostly empty space with a tiny, massive, positively charged nucleus at its center.
- The spectrum produced by atoms of an element can be used to identify an unknown sample of that element.
- If white light passes through a gas, the gas absorbs the same wavelengths that it would emit if it were excited. If the light is then passed through a prism, the absorption spectrum of the gas is visible.
- Niels Bohr's model of the atom correctly showed that the energy of an atom can have only certain values; thus, it is quantized. He showed that the energy of a hydrogen atom in energy level n is equal to the product of -13.6 eV and the inverse of n^2.

$$E_n = -13.6 \text{ eV} \times \frac{1}{n^2}$$

- According to the Bohr model, atoms make transitions between allowable energy levels, absorbing or emitting energy in the form of photons (electromagnetic waves). The energy of the photon is equal to the difference between the initial and final states of the atom.

$$E_{\text{photon}} = E_{\text{f}} - E_{\text{i}}$$

- According to the Bohr model, the radius of an electron's orbit can have only certain (quantized) values. The radius of the electron in energy level n of a hydrogen atom is given by the following equation.

$$r_n = \frac{h^2 n^2}{4\pi^2 K m q^2}$$

28.2 The Quantum Model of the Atom

Vocabulary

- quantum model (p. 760)
- electron cloud (p. 761)
- quantum mechanics (p. 761)
- coherent light (p. 761)
- incoherent light (p. 761)
- stimulated emission (p. 762)
- laser (p. 762)

Key Concepts

- In the quantum-mechanical model of the atom, the atom's energy has only specific, quantized values.
- In the quantum-mechanical model of the atom, only the probability of finding the electron in a specific region can be determined. In the hydrogen atom, the most probable distance of the electron from the nucleus is the same as the electron's orbital radius in the Bohr model.
- Quantum mechanics is extremely successful in calculating the properties of atoms, molecules, and solids.
- Lasers produce light that is directional, powerful, monochromatic, and coherent. Each property gives the laser useful applications.

Concept Mapping

23. Complete the following concept map using these terms: *energy levels, fixed electron radii, Bohr model, photon emission and absorption, energy-level difference.*

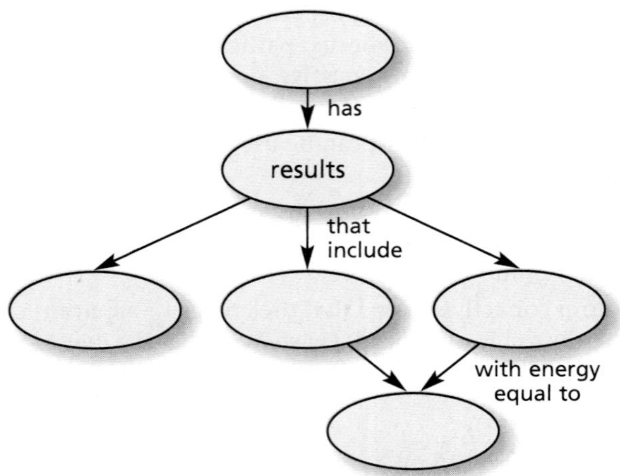

Mastering Concepts

24. Describe how Rutherford determined that the positive charge in an atom is concentrated in a tiny region, rather than spread throughout the atom. (28.1)

25. How does the Bohr model explain why the absorption spectrum of hydrogen contains exactly the same frequencies as its emission spectrum? (28.1)

26. Review the planetary model of the atom. What are some of the problems with a planetary model of the atom? (28.1)

27. Analyze and critique the Bohr model of the atom. What three assumptions did Bohr make in developing his model? (28.1)

28. **Gas-Discharge Tubes** Explain how line spectra from gas-discharge tubes are produced. (28.1)

29. How does the Bohr model account for the spectra emitted by atoms? (28.1)

30. Explain why line spectra produced by hydrogen gas-discharge tubes are different from those produced by helium gas-discharge tubes. (28.1)

31. **Lasers** A laboratory laser has a power of only 0.8 mW (8×10^{-4} W). Why does it seem more powerful than the light of a 100-W lamp? (28.2)

32. A device similar to a laser that emits microwave radiation is called a maser. What words likely make up this acronym? (28.2)

33. What properties of laser light led to its use in light shows? (28.2)

Applying Concepts

34. As the complexity of energy levels changes from atom to atom, what do you think happens to the spectra that they produce?

35. **Northern Lights** The northern lights are caused by high-energy particles from the Sun striking atoms high in Earth's atmosphere. If you looked at these lights through a spectrometer, would you see a continuous or line spectrum? Explain.

36. If white light were emitted from Earth's surface and observed by someone in space, would its spectrum appear to be continuous? Explain.

37. Is money a good example of quantization? Is water? Explain.

38. Refer to **Figure 28-21.** A photon with energy of 6.2 eV enters a mercury atom in the ground state. Will it be absorbed by the atom? Explain.

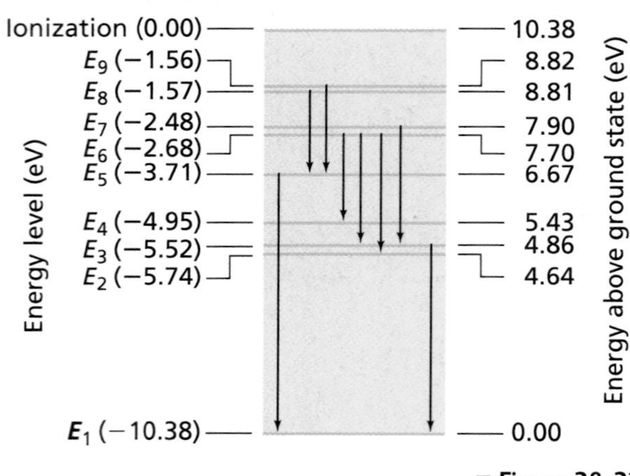

Figure 28-21

39. A certain atom has four energy levels, with E_4 being the highest and E_1 being the lowest. If the atom can make transitions between any two levels, how many spectral lines can the atom emit? Which transition produces the photon with the highest energy?

40. A photon is emitted when an electron in an excited hydrogen atom drops through energy levels. What is the maximum energy that the photon can have? If this same amount of energy were given to the atom in the ground state, what would happen?

41. Compare the quantum mechanical theory of the atom with the Bohr model.

42. Given a red, green, and blue laser, which produces photons with the highest energy?

 For more problems, go to Additional Problems, Appendix B.

Mastering Problems

28.1 The Bohr Model of the Atom

43. A calcium atom drops from 5.16 eV above the ground state to 2.93 eV above the ground state. What is the wavelength of the photon emitted?

44. A calcium atom in an excited state, E_2, has an energy level 2.93 eV above the ground state. A photon of energy 1.20 eV strikes the calcium atom and is absorbed by it. To what energy level is the calcium atom raised? Refer to **Figure 28-22.**

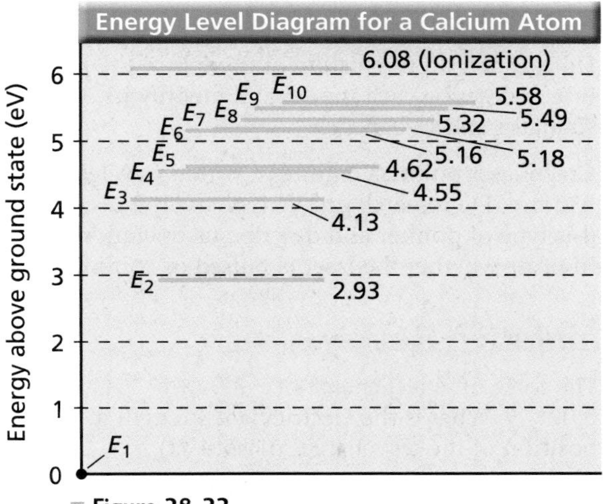

Figure 28-22

45. A calcium atom is in an excited state at the E_6 energy level. How much energy is released when the atom drops down to the E_2 energy level? Refer to Figure 28-22.

46. A photon of orange light with a wavelength of 6.00×10^2 nm enters a calcium atom in the E_6 excited state and ionizes the atom. What kinetic energy will the electron have as it is ejected from the atom?

47. Calculate the energy associated with the E_7 and the E_2 energy levels of the hydrogen atom.

48. Calculate the difference in energy levels in the previous problem.

Refer to Figure 28-21 for Problems 49 and 50.

49. A mercury atom is in an excited state at the E_6 energy level.
   a. How much energy would be needed to ionize the atom?
   b. How much energy would be released if the atom dropped down to the E_2 energy level instead?

50. A mercury atom in an excited state has an energy of -4.95 eV. It absorbs a photon that raises it to the next-higher energy level. What is the energy and the frequency of the photon?

51. What energies are associated with a hydrogen atom's energy levels of E_2, E_3, E_4, E_5, and E_6?

52. Using the values calculated in problem 51, calculate the following energy differences.
   a. $E_6 - E_5$
   b. $E_6 - E_3$
   c. $E_4 - E_2$
   d. $E_5 - E_2$
   e. $E_5 - E_3$

53. Use the values from problem 52 to determine the frequencies of the photons emitted when an electron in a hydrogen atom makes the energy level changes listed.

54. Determine the wavelengths of the photons having the frequencies that you calculated in problem 53.

55. A hydrogen atom emits a photon with a wavelength of 94.3 nm when its falls to the ground state. From what energy level did the electron fall?

56. For a hydrogen atom in the $n = 3$ Bohr orbital, find the following.
   a. the radius of the orbital
   b. the electric force acting between the proton and the electron
   c. the centripetal acceleration of the electron
   d. the orbital speed of the electron (Compare this speed with the speed of light.)

28.2 The Quantum Model of the Atom

57. **CD Players** Gallium arsenide lasers are commonly used in CD players. If such a laser emits at 840 nm, what is the difference in eV between the two lasing energy levels?

58. A GaInNi laser lases between energy levels that are separated by 2.90 eV.
   a. What wavelength of light does it emit?
   b. In what part of the spectrum is this light?

59. A carbon-dioxide laser emits very high-power infrared radiation. What is the energy difference in eV between the two lasing energy levels? Consult Table 28-1.

60. The power in a laser beam is equal to the energy of each photon times the number of photons per second that are emitted.
   a. If you want a laser at 840 nm to have the same power as one at 427 nm, how many times more photons per second are needed?
   b. Find the number of photons per second in a 5.0-mW 840-nm laser.

61. **HeNe Lasers** The HeNe lasers used in many classrooms can be made to lase at three wavelengths: 632.8 nm, 543.4 nm, and 1152.3 nm.
   a. Find the difference in energy between the two states involved in the generation of each wavelength.
   b. Identify the color of each wavelength.

Mixed Review

62. A photon with an energy of 14.0 eV enters a hydrogen atom in the ground state and ionizes it. With what kinetic energy will the electron be ejected from the atom?

63. Calculate the radius of the orbital associated with the energy levels E_5 and E_6 of the hydrogen atom.

64. A hydrogen atom is in the $n = 2$ level.

a. If a photon with a wavelength of 332 nm strikes the atom, show that the atom will be ionized.

b. When the atom is ionized, assume that the electron receives the excess energy from the ionization. What will be the kinetic energy of the electron in joules?

65. A beam of electrons is directed onto a sample of atomic hydrogen gas. What minimum energy of the electrons is needed for the hydrogen atoms to emit the red light produced when the atom goes from the $n = 3$ to the $n = 2$ state?

66. The most precise spectroscopy experiments use "two-photon" techniques. Two photons with identical wavelengths are directed at the target atoms from opposite directions. Each photon has half the energy needed to excite the atoms from the ground state to the desired energy level. What laser wavelength would be needed to make a precise study of the energy difference between $n = 1$ and $n = 2$ in hydrogen?

Thinking Critically

67. Apply Concepts The result of projecting the spectrum of a high-pressure mercury vapor lamp onto a wall in a dark room is shown in **Figure 28-23.** What are the differences in energy levels for each of the three visible lines?

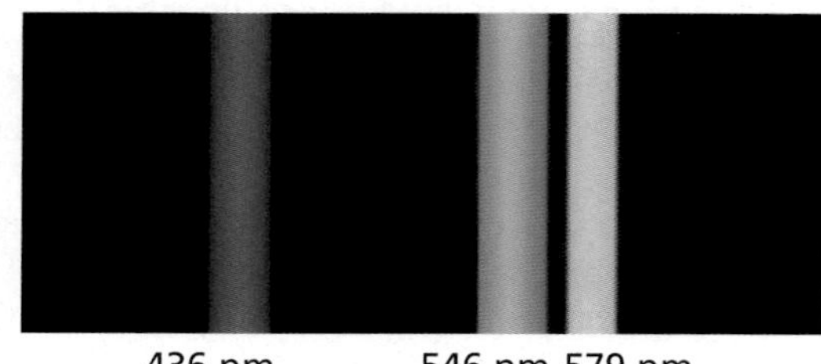

Figure 28-23

68. Interpret Scientific Illustrations After the emission of the visible photons described in problem 67, the mercury atom continues to emit photons until it reaches the ground state. From an inspection of Figure 28-21, determine whether or not any of these photons would be visible. Explain.

69. Anaylze and Conclude A positronium atom consists of an electron and its antimatter relative, the positron, bound together. Although the lifetime of this "atom" is very short—on the average it lives one-seventh of a microsecond—its energy levels can be measured. The Bohr model can be used to calculate energies with the mass of the electron replaced by one-half its mass. Describe how the radii of the orbits and the energy of each level would be affected. What would be the wavelength of the E_2 to E_1 transition?

Writing in Physics

70. Do research on the history of models of the atom. Briefly describe each model and identify its strengths and weaknesses.

71. Green laser pointers emit light with a wavelength of 532 nm. Do research on the type of laser used in this type of pointer and describe its operation. Indicate whether the laser is pulsed or continuous.

Cumulative Review

72. The force on a test charge of $+3.00\times10^{-7}$ C is 0.027 N. What is the electric field strength at the position of the test charge? (Chapter 21)

73. A technician needs a 4-Ω resistor but only has 1-Ω resistors of that value. Is there a way to combine what she has? Explain. (Chapter 23)

74. A 1.0-m-long wire is moved at right angles to Earth's magnetic field where the magnetic induction is 5.0×10^{-5} T at a speed of 4.0 m/s. What is the *EMF* induced in the wire? (Chapter 25)

75. The electrons in a beam move at 2.8×10^{8} m/s in an electric field of 1.4×10^{4} N/C. What value must the magnetic field have if the electrons pass through the crossed fields undeflected? (Chapter 26)

76. Consider the modifications that J. J. Thomson would need to make to his cathode-ray tube so that it could accelerate protons (rather than electrons), then answer the following questions. (Chapter 26)

a. To select particles of the same velocity, would the ratio E/B have to be changed? Explain.

b. For the deflection caused by the magnetic field alone to remain the same, would the B field have to be made smaller or larger? Explain.

77. The stopping potential needed to return all the electrons ejected from a metal is 7.3 V. What is the maximum kinetic energy of the electrons in joules? (Chapter 27)

 For more problems, go to Additional Problems, Appendix B.

Multiple Choice

1. Which model of the atom was based on the results of Rutherford's gold foil experiment?

   Ⓐ Bohr model
   Ⓑ nuclear model
   Ⓒ plum pudding model
   Ⓓ quantum mechanical model

2. A mercury atom emits light with a wavelength of 405 nm. What is the energy difference between the two energy levels involved with this emission?

   Ⓐ 0.22 eV Ⓒ 3.06 eV
   Ⓑ 2.14 eV Ⓓ 4.05 eV

3. The diagram below shows the energy levels for a mercury atom. What wavelength of light is emitted when a mercury atom makes a transition from E_7 to E_4?

   Ⓐ 167 nm Ⓒ 500 nm
   Ⓑ 251 nm Ⓓ 502 nm

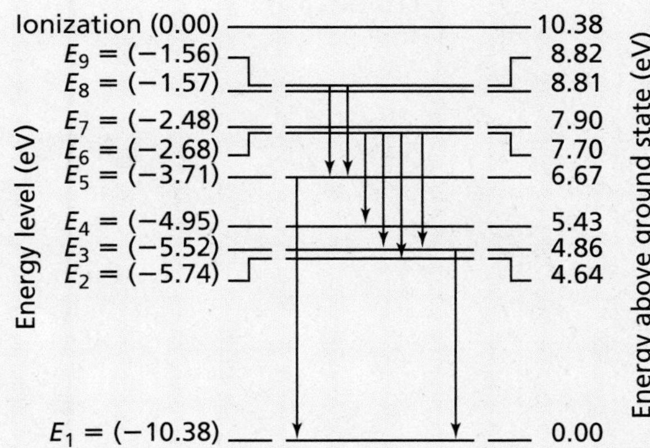

4. Which statement about the quantum model of the atom is false?

   Ⓐ The possible energy levels of the atom are quantized.
   Ⓑ The locations of the electrons around the nucleus are known precisely.
   Ⓒ The electron cloud defines the area where electrons are likely to be located.
   Ⓓ Stable electron orbits are related to the de Broglie wavelength.

Questions 5 and 6 refer to the diagram showing the Balmer series electron transitions in a hydrogen atom.

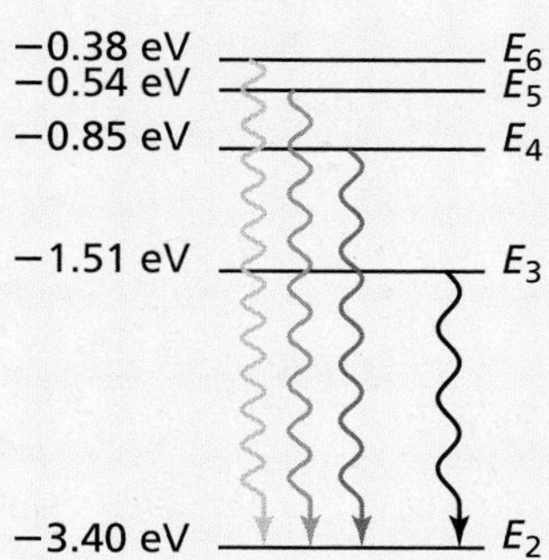

5. Which energy level transition is responsible for the emission of light with the greatest frequency?

   Ⓐ E_2 to E_5 Ⓒ E_3 to E_6
   Ⓑ E_3 to E_2 Ⓓ E_6 to E_2

6. What is the frequency of the Balmer series line related to the energy level transition from E_4 to E_2? (Note that 1 eV = 1.60×10^{-19} J.)

   Ⓐ 2.55×10^{14} Hz Ⓒ 6.15×10^{14} Hz
   Ⓑ 4.32×10^{14} Hz Ⓓ 1.08×10^{15} Hz

Extended Answer

7. Determine the wavelength of light emitted when a hydrogen atom makes a transition from the $n = 5$ energy level to the $n = 2$ energy level.

✓Test-Taking TIP

Stumbling Is Not Failing

Occasionally, you will encounter a question that you have no idea how to answer. Even after reading the question several times, it still may not make sense. If the question is multiple choice, focus on the part of the question that you know something about. Eliminate as many choices as you can. Then take your best guess and move on.

Chapter 29 Solid-State Electronics

What You'll Learn

- You will be able to distinguish among electric conductors, semiconductors, and insulators.
- You will examine how pure semiconductors are modified to produce desired electric properties.
- You will compare diodes and transistors.

Why It's Important

Semiconductors have electric properties that allow them to act as one-way conductors to amplify weak signals in many common electronic devices.

Fast Math Computers and electronic devices use the controlled movement of electrons and holes in semiconductors to do quick calculations and logical operations.

Think About This ▶

A silicon microchip might be small, but it may contain the equivalent of millions of resistors, diodes, and transistors. How can this level of complexity be produced in such a tiny structure?

physicspp.com

How can you show conduction in a diode?

Question
Which way does a two-color light-emitting diode (LED) conduct?

Procedure

1. Obtain a bi-color (red-green) LED and a 9- to 12-V AC power supply or transformer.
2. Wire a 100-Ω resistor and the LED in series with the AC source.
3. Be careful when plugging in the AC source so you are not shocked. Do not touch the resistor as it may become hot. Plug the AC source into a GFCI receptacle.
4. Record your observations of the LED.
5. Hold a stroboscopic disk in front of the LED and spin it. Record your observations of the LED as viewed through the disk.

Analysis
What color was the LED after you plugged in the power supply? What color was the LED as viewed through the stroboscopic disk?
Critical Thinking Suggest a possible explanation for your observations.

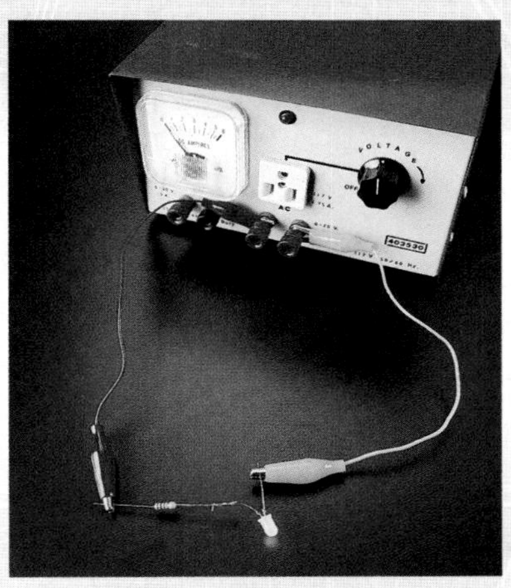

29.1 Conduction in Solids

► **Objectives**
- **Describe** electron motion in conductors and semiconductors.
- **Compare and contrast** *n*-type and *p*-type semiconductors.

► **Vocabulary**
semiconductors
band theory
intrinsic semiconductors
dopants
extrinsic semiconductors

Electronic devices depend not only on natural conductors and insulators, but also on materials that have been designed and produced by scientists and engineers working together. This brief investigation into electronics begins with a study of how materials conduct electricity.

All electronic devices owe their origins to the vacuum tubes of the early 1900s. In vacuum tubes, electron beams flow through space to amplify and control faint electric signals. Vacuum tubes are big, require lots of electric power, and generate considerable heat. They have heated filaments, which require the replacement of the tubes after one to five years.

In the late 1940s, solid-state devices were invented that could do the jobs of vacuum tubes. These devices are made of materials, such as silicon and germanium, known as **semiconductors.** The devices amplify and control very weak electric signals through the movement of electrons within a tiny crystalline space. Because very few electrons flow in them and they have no filaments, devices made from semiconductors operate with a low power input. They are very small, don't generate much heat, and are inexpensive to manufacture. The estimated useful life of these devices is 20 years or more.

Band Theory of Solids

You have learned about electric conductors and insulators. In conductors, electric charges can move easily, but not in insulators. When you examine these two types of materials at the atomic level, the difference in the way they are able to carry charges becomes apparent.

You learned in Chapter 13 that crystalline solids consist of atoms bound together in regular arrangements. You also know from Chapters 27 and 28 that an atom consists of a dense, positively charged nucleus surrounded by a cloud of negatively charged electrons. These electrons can occupy only certain allowed energy levels. Under most conditions, the electrons in an atom occupy the lowest possible energy levels. This condition is referred to as the ground state. Because the electrons can have only certain energies, any energy changes that occur are quantized; that is, the energy changes occur in specific amounts.

Energy bands Suppose you could construct a solid by assembling atoms together, one by one. You would start with an atom in the ground state. At large interatomic spacings (> 0.8 nm) with no very near neighbors, the graph in **Figure 29-1** shows two discrete energy levels for the atom. As the solid crystal forms by moving atoms closer to the atom, the electric fields of these other neighboring atoms affect the energy levels of its electrons. In the solid crystal, the result is that the ground state energy levels in each atom are split into multiple levels by the electric fields of all of its neighbors. There are so many of these levels, and they are so close together, that they no longer appear as distinct levels, but as the energy bands shown in Figure 29-1. The lower energy or valence bands are occupied by bonding electrons in the crystal, and the higher energy or conduction bands are available for electrons to move from atom to atom.

Notice in Figure 29-1 that the atomic separations for crystalline silicon and crystalline carbon (diamond) translate to valence bands and conduction bands that are separated by energy gaps. These gaps have no energy levels available for electrons. They are called forbidden energy regions. This description of valence and conduction bands, separated by forbidden energy gaps, is known as the **band theory** of solids and can be used to better understand electric conduction. For example, the band diagram in

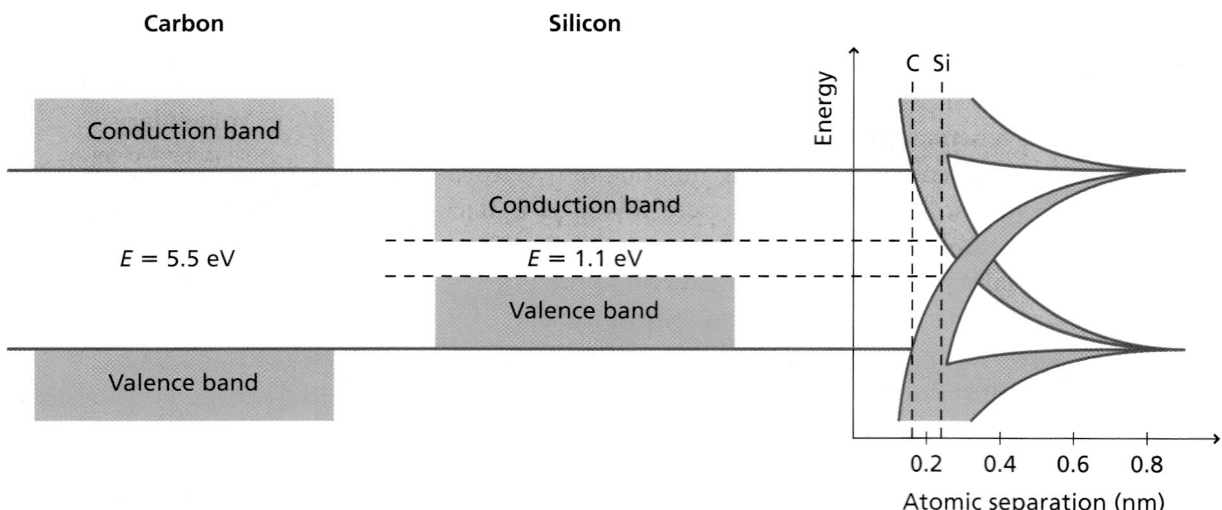

Figure 29-1 Energy levels of an atom are split apart when other atoms are brought closer, resulting in an energy gap between the valence and conduction bands.

Figure 29-1 suggests that a lot more energy will be required to move valence electrons from the valence band to the conduction band in the case of crystalline carbon (diamond structure) compared to silicon. Carbon in graphite form is a much better conductor because the structure of the atom arrangement in graphite gives it a smaller energy gap than that of diamond.

Crystalline silicon has a smaller energy gap than diamond does. At absolute zero, the valence band of silicon would be completely full and the conduction band would be completely empty. At room temperature, some number of valence electrons have enough thermal energy to jump the 1.1-eV gap to the conduction band and serve as charge carriers. As the temperature increases and more electrons gain enough energy to jump the gap, the conductivity of silicon will increase. Germanium has an energy gap of 0.7 eV, which is smaller than that of silicon. This means that germanium is a better conductor than silicon at any given temperature. However, it also means that germanium is too sensitive to heat for many electronic applications. Relatively small changes in temperature cause large changes in the conductivity of germanium, making control and stability of circuits troublesome.

Lead has an interatomic spacing of 0.27 nm. Figure 29-1 shows that this would translate to a band-gap diagram in which the conduction band overlaps the valence band. One would, therefore, expect lead to be a good conductor, and it is. Materials with overlapping, partially filled bands are conductors, as indicated in **Figure 29-2.**

Figure 29-2 In a material that is a good conductor, the conduction band is partially filled. The blue-shaded area shows energies occupied by electrons.

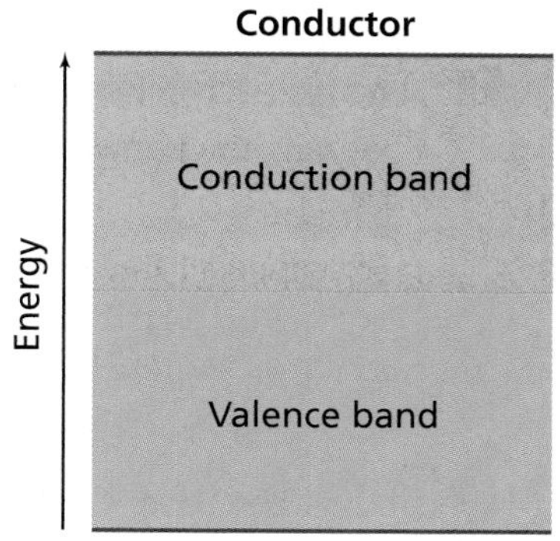

Conductor

Conduction band

Valence band

Energy

Conductors

When a potential difference is placed across a material, the resulting electric field exerts a force on the electrons. The electrons accelerate and gain energy and the field does work on them. If there are bands within the material that are only partially filled, then there are energy levels available that are only slightly higher than the electrons' ground state levels. As a result, the electrons that gain energy from the field can move from one atom to the next. Such movement of electrons from one atom to the next is an electric current, and the entire process is known as electric conduction. Materials with partially filled bands, such as the metals aluminum, lead, and copper, conduct electricity easily.

Random motion The free electrons in conductors move about rapidly in a random way, changing directions when they collide with the cores of the atoms. However, if an electric field is put across a length of wire, there will be a net force pushing the electrons in one direction. Although their motion is not greatly affected, they have a slow overall movement dictated by the electric field, as shown in **Figure 29-3.** Electrons continue to move rapidly with speeds of 10^6 m/s in random directions, and they drift very slowly at speeds of 10^{-5} m/s or slower toward the positive end of the wire. This model of conductors is called the electron-gas model. If the temperature is increased, the speeds of the electrons increase, and, consequently, they collide more frequently with atomic cores. Thus, as the temperature rises, the conductivity of metals is reduced. Conductivity is the reciprocal of resistivity. As conductivity is reduced, a material's resistance rises.

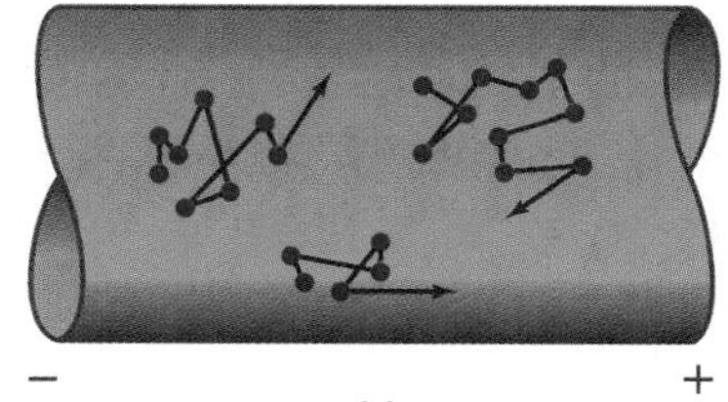

Figure 29-3 The electrons move rapidly and randomly in a conductor. If a field is applied across the wire, the electrons drift toward one end. Electron flow is opposite in direction to conventional current.

▶ EXAMPLE Problem 1

The Free-Electron Density of a Conductor How many free electrons exist in a cubic centimeter of copper? Each atom contributes one electron. The density, atomic mass, and number of atoms per mole of copper can be found in Appendix D.

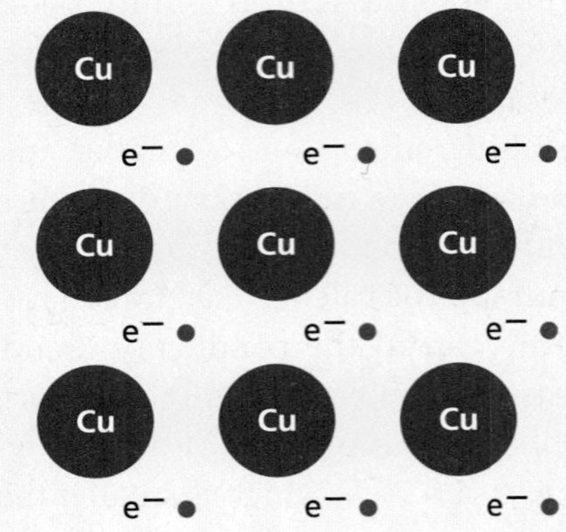

1 Analyze the Problem

- Identify the knowns using Appendix D.

Known:	Unknown:
For copper: 1 free e^- per atom	free e^-/cm^3 = ?
$\rho = 8.96$ g/cm^3	
$M = 63.54$ g/mol	
$N_A = 6.02\times10^{23}$ atoms/mol	

2 Solve for the Unknown

$$\frac{\text{free e}^-}{\text{cm}^3} = \frac{(\text{free e}^-)}{\text{atom}}(N_A)\left(\frac{1}{M}\right)(\rho)$$

$$= \left(\frac{1\text{ free e}^-}{1\text{ atom}}\right)\left(\frac{6.02\times10^{23}\text{ atoms}}{1\text{ mol}}\right)\left(\frac{1\text{ mol}}{63.54\text{ g}}\right)\left(\frac{8.96\text{ g}}{1\text{ cm}^3}\right)$$

Substitute free e^-/1 atom = 1 free e^-/1 atom, $N_A = 6.02\times10^{23}$ atoms/mol, $M = 63.54$ g/mol, $\rho = 8.96$ g/cm^3

$$= 8.49\times10^{22}\text{ free e}^-/\text{cm}^3 \text{ in copper}$$

3 Evaluate the Answer

- **Are the units correct?** Dimensional analysis on the units confirms the number of free electrons per cubic centimeter.
- **Is the magnitude realistic?** One would expect a large number of electrons in a cubic centimeter.

Math Handbook
Dimensional Calculations pages 846–847

▶ PRACTICE Problems

• Additional Problems, Appendix B
• Solutions to Selected Problems, Appendix C

1. Zinc, with a density of 7.13 g/cm^3 and an atomic mass of 65.37 g/mol, has two free electrons per atom. How many free electrons are there in each cubic centimeter of zinc?
2. Silver has 1 free electron per atom. Use Appendix D and determine the number of free electrons in 1 cm^3 of silver.
3. Gold has 1 free electron per atom. Use Appendix D and determine the number of free electrons in 1 cm^3 of gold.
4. Aluminum has 3 free electrons per atom. Use Appendix D and determine the number of free electrons in 1 cm^3 of aluminum.
5. The tip of the Washington Monument was made of 2835 g of aluminum because it was a rare and costly metal in the 1800s. Use problem 4 and determine the number of free electrons in the tip of the Washington Monument.

Insulators

In an insulating material, the valence band is filled to capacity and the conduction band is empty. As shown in **Figure 29-4,** an electron must gain a large amount of energy to go to the next energy level. In an insulator, the lowest energy level in the conduction band is 5–10 eV above the highest energy level in the valence band, as shown in **Figure 29-4a.** There is at least a 5-eV gap of energies that no electrons can possess.

Although electrons have some kinetic energy as a result of their thermal energy, the average kinetic energy of electrons at room temperature is not sufficient for them to jump the forbidden gap. If a small electric field is placed across an insulator, almost no electrons gain enough energy to reach the conduction band, so there is no current. Electrons in an insulator must be given a large amount of energy to be pulled into the conduction band. As a result, the electrons in an insulator tend to remain in place, and the material does not conduct electricity.

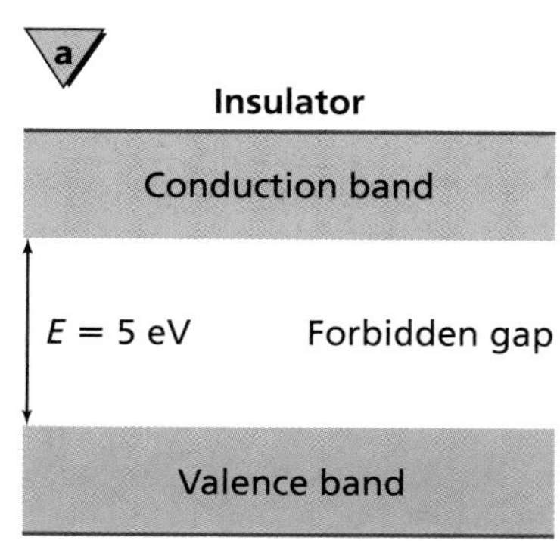

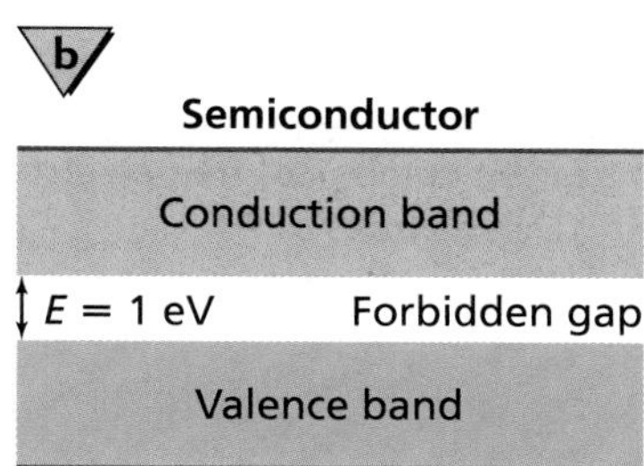

Figure 29-4 Compare the valence and conduction bands in an insulator **(a)** and in a semiconductor **(b).** Compare these diagrams with the one shown in Figure 29-2.

Semiconductors

Electrons can move more freely in semiconductors than in insulators, but not as easily as in conductors. As shown in **Figure 29-4b,** the energy gap between the valence band and the conduction band is approximately 1 eV. How does the structure of a semiconductor explain its electronic characteristics? Atoms of the most common semiconductors, silicon (Si) and germanium (Ge), each have four valence electrons. These four electrons are involved in binding the atoms together into the solid crystal. The valence electrons form a filled band, as in an insulator, but the forbidden gap between the valence and conduction bands is much smaller than in an insulator. Not much energy is needed to pull one of the electrons from a silicon atom and put it into the conduction band, as illustrated in **Figure 29-5a.** Indeed, the gap is so small that some electrons reach the conduction band as a result of their thermal kinetic energy alone. That is, the random motion of atoms and electrons gives some electrons enough energy to break free of their home atoms and wander around the silicon crystal.

If an electric field is applied to a semiconductor, electrons in the conduction band move through the solid according to the direction of the applied electric field. In contrast to the effect in a metal, the higher the temperature of a semiconductor, the more able electrons are to reach the conduction band, and the higher the conductivity.

An atom from which an electron has broken free is said to contain a hole. As shown in **Figure 29-5b,** a hole is an empty energy level in the valence band. The atom now has a net positive charge. An electron from the conduction band can jump into this hole and become bound to an atom once again. When a hole and a free electron recombine, their opposite charges neutralize each other.

Figure 29-5 Some electrons in semiconductors have enough thermal kinetic energy to break free and wander through the crystal, as shown in the crystal structure **(a)** and in the bands **(b).**

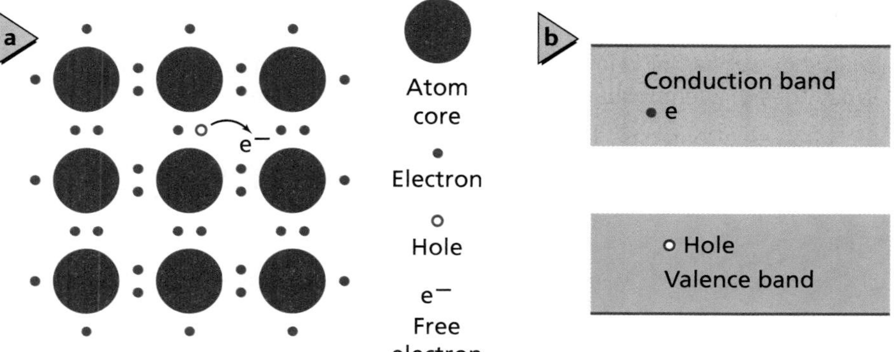

The electron, however, has left behind a hole at its previous location. Thus, as in a game of musical chairs, the negatively charged, free electrons move in one direction and the positively charged holes move in the opposite direction. Pure semiconductors that conduct as a result of thermally freed electrons and holes are called **intrinsic semiconductors.** Because so few electrons or holes are available to carry charge, conduction in intrinsic semiconductors is very low, making their resistances very high.

▶ EXAMPLE Problem 2

Fraction of Free Electrons in an Intrinsic Semiconductor **Because of the thermal kinetic energy of solid silicon at room temperature, there are 1.45×10^{10} free electrons/cm³. What is the number of free electrons per atom of silicon at room temperature?**

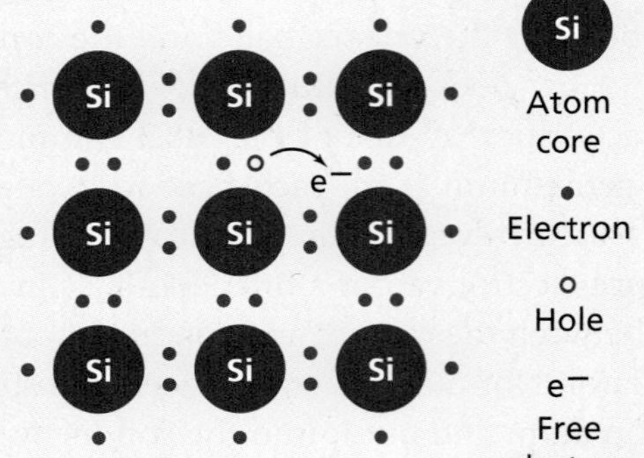

1 Analyze the Problem

- Identify the knowns and unknowns.

Known:

$\rho = 2.33\ \text{g/cm}^3$

$M = 28.09\ \text{g/mol}$

$N_A = 6.02\times10^{23}\ \text{atoms/mol}$

For Si: 1.45×10^{10} free e^-/cm^3

Unknown:

free e^-/atom of Si = ?

2 Solve for the Unknown

$$\frac{\text{free } e^-}{\text{atom}} = \left(\frac{1}{N_A}\right)(M)\left(\frac{1}{\rho}\right)(1.45\times10^{10}\text{ free } e^-/\text{cm}^3 \text{ for Si})$$

$$= \left(\frac{1\text{ mol}}{6.02\times10^{23}\text{ atoms}}\right)\left(\frac{28.09\text{ g}}{1\text{ mol}}\right)\left(\frac{1\text{ cm}^3}{2.33\text{ g}}\right)\left(\frac{1.45\times10^{10}\text{ free } e^-}{\text{cm}^3}\right)$$

Substitute $N_A = 6.02\times10^{23}$ atoms/mol, $M = 28.09$ g/mol, $\rho = 2.33$ g/cm³, free e^-/cm³ Si = 1.45×10^{10} free e^-/cm³

$$= 2.90\times10^{-13}\text{ free } e^-/\text{atom of Si}$$

or, roughly 1 out of 3 trillion Si atoms has a free electron

3 Evaluate the Answer

- **Are the units correct?** Using dimensional analysis confirms the correct units.
- **Is the magnitude realistic?** In an intrinsic semiconductor, such as silicon at room temperature, very few atoms have free electrons.

Math Handbook
Operations with Scientific Notation
pages 842–843

▶ PRACTICE Problems

• Additional Problems, Appendix B
• Solutions to Selected Problems, Appendix C

6. In pure germanium, which has a density of 5.23 g/cm³ and an atomic mass of 72.6 g/mol, there are 2.25×10^{13} free electrons/cm³ at room temperature. How many free electrons are there per atom?
7. At 200.0 K, silicon has 1.89×10^{5} free electrons/cm³. How many free electrons are there per atom at this temperature? What does this temperature represent on the Celsius scale?
8. At 100.0 K, silicon has 9.23×10^{-10} free electrons/cm³. How many free electrons are there per atom at this temperature? What does this temperature represent on the Celsius scale?
9. At 200.0 K, germanium has 1.16×10^{10} free electrons/cm³. How many free electrons are there per atom at this temperature?
10. At 100.0 K, germanium has 3.47 free electrons/cm³. How many free electrons are there per atom at this temperature?

Doped Semiconductors

The conductivity of intrinsic semiconductors must be increased greatly to make practical devices. **Dopants** are electron donor or acceptor atoms that can be added in low concentrations to intrinsic semiconductors. Dopants increase conductivity by making extra electrons or holes available. The doped semiconductors are known as **extrinsic semiconductors.**

***n*-type semiconductors** If an electron donor with five valence electrons, such as arsenic (As), is used as a dopant for silicon, the product is called an *n*-type semiconductor. **Figure 29-6a** shows a location in a silicon crystal where a dopant atom has replaced one of the silicon atoms. Four of the five As valence electrons bind to neighboring silicon. The fifth electron is called the donor electron. The energy of this donor electron is so close to the conduction band that thermal energy can easily move the electron from the dopant atom into the conduction band, as shown in **Figure 29-7a.** Conduction in *n*-type semiconductors is increased by the availability of these extra donor electrons to the conduction band.

***p*-type semiconductors** If an electron acceptor with three valence electrons, such as gallium (Ga), is used as a dopant for silicon, the product is called a *p*-type semiconductor. When a gallium atom replaces a silicon atom, one binding electron is missing, creating a hole in the silicon crystal, as shown in **Figure 29-6b.** Electrons in the conduction band can easily drop into these holes, creating new holes. Conduction in *p*-type semiconductors is enhanced by the availability of the extra holes provided by the acceptor dopant atoms, as shown in **Figure 29-7b.**

Both *p*-type and *n*-type semiconductors are electrically neutral. Adding dopant atoms of either type does not add any net charge to a semiconductor. Both types of semiconductor use electrons and holes in conduction. Only a few dopant atoms per million silicon atoms are needed to increase the conductivity of semiconductors by a factor of 1000 or more.

Silicon is doped by putting a pure silicon crystal in a vacuum with a sample of the dopant material. The dopant is heated until it is vaporized, and the atoms condense on the cold silicon. The dopant diffuses into the silicon on warming, and a thin layer of aluminum or gold is evaporated onto the doped crystal. A wire is welded to this metal layer, allowing the user to apply a potential difference across the doped silicon.

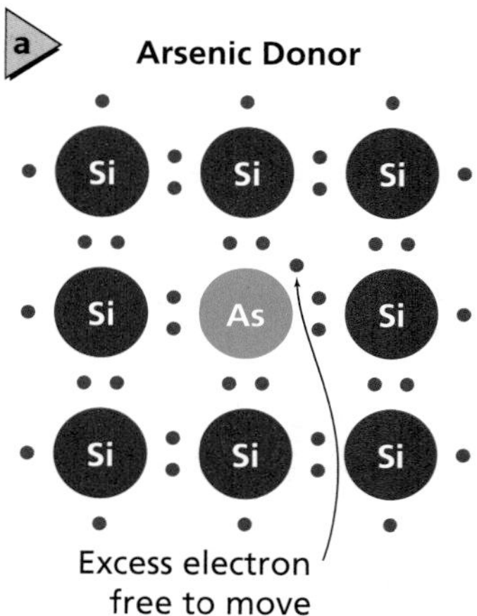

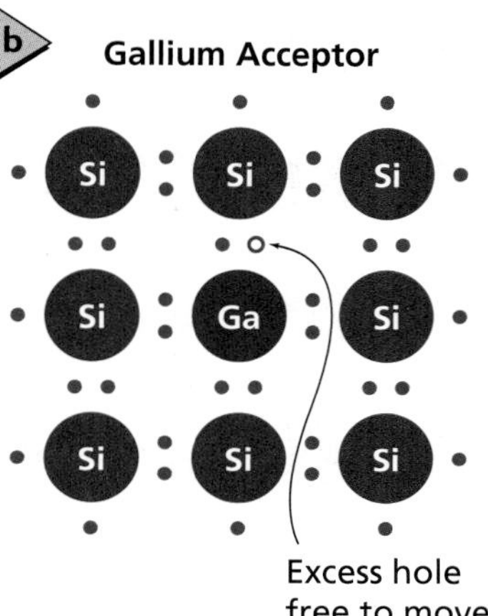

Figure 29-6 A donor atom of arsenic with five valence electrons replaces a silicon atom and provides an unbound electron in the silicon crystal **(a).** An acceptor atom of gallium with three valence electrons creates a hole in the crystal **(b).**

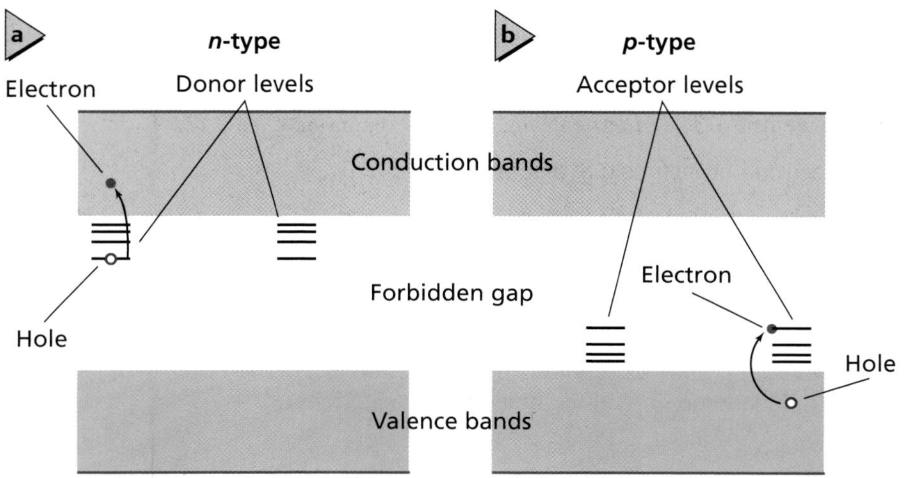

Figure 29-7 In an *n*-type semiconductor **(a),** donor energy levels place electrons in the conduction band. In a *p*-type semiconductor **(b),** acceptor energy levels result in holes in the valence band.

Thermistors The electric conductivity of intrinsic and extrinsic semiconductors is sensitive to both temperature and light. Unlike metals in which conductivity is reduced when the temperature rises, an increase in temperature of a semiconductor allows more electrons to reach the conduction band, and conductivity increases and resistance decreases. One semiconductor device, the thermistor, is designed so that its resistance depends very strongly on temperature. The thermistor can be used as a sensitive thermometer and to compensate for temperature variations of other components in an electric circuit. Thermistors also can be used to detect radio waves, infrared radiation, and other forms of radiation.

▶ EXAMPLE **Problem 3**

The Conductivity of Doped Silicon Silicon is doped with arsenic so that one in every million silicon atoms is replaced by an arsenic atom. Each arsenic atom donates one electron to the conduction band.

a. What is the density of free electrons?

b. By what ratio is this density greater than that of intrinsic silicon with 1.45×10^{10} free e^-/cm^3?

c. Is conduction mainly by the electrons of the silicon or the arsenic?

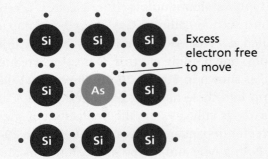

Arsenic Donor

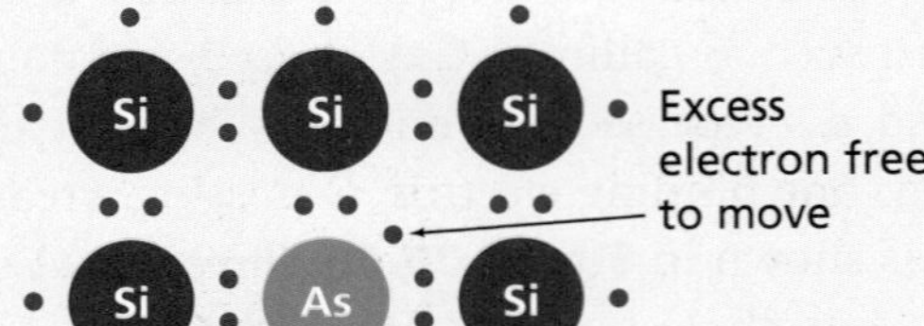

1 Analyze the Problem

- Identify the knowns and unknowns.

Known:	Unknown:
1 As atom/10^6 Si atoms	free e^-/cm^3 donated by As = ?
1 free e^-/As atom	ratio of As-donated free e^- to intrinsic free e^- = ?
4.99×10^{22} Si atoms/cm^3	
1.45×10^{10} free e^-/cm^3 in intrinsic Si	

2 Solve for the Unknown

a. $$\left(\frac{\text{free e}^-}{\text{cm}^3}\text{ from As}\right) = \left(\frac{\text{free e}^-}{\text{As atom}}\right)\left(\frac{\text{As atoms}}{\text{Si atoms}}\right)\left(\frac{\text{Si atoms}}{\text{cm}^3}\right)$$

$$\left(\frac{\text{free e}^-}{\text{cm}^3}\right) = \left(\frac{1\text{ free e}^-}{1\text{ As atom}}\right)\left(\frac{1\text{ As atom}}{1\times10^6\text{ Si atoms}}\right)\left(\frac{4.99\times10^{22}\text{ Si atoms}}{\text{cm}^3}\right)$$

Substitute free e^-/As atom = 1 free e^-/1 As atom, As atoms/Si atoms = 1 As atom/1×10^6 Si atoms, Si atoms/cm^3 = 4.99×10^{22} Si atoms/cm^3

$$= 4.99\times10^{16}\text{ free e}^-/\text{cm}^3\text{ from As donor in doped Si}$$

b. $$\text{Ratio} = \left(\frac{\text{free e}^-/\text{cm}^3\text{ in doped Si}}{\text{free e}^-/\text{cm}^3\text{ in intrinsic Si}}\right)$$

$$= \left(\frac{4.99\times10^{16}\text{ free e}^-/\text{cm}^3\text{ in doped Si}}{1.45\times10^{10}\text{ free e}^-/\text{cm}^3\text{ in intrinsic Si}}\right)$$

Substitute 4.99×10^{16} free e^-/cm^3 in doped Si, 1.45×10^{10} free e^-/cm^3 in intrinsic Si

$$= 3.44\times10^6\text{ As-donated electron per instrinsic Si electron}$$

c. Because there are over 3 million arsenic-donated electrons for every intrinsic electron, conduction is mainly by the arsenic-donated electrons.

Physics Online

Personal Tutor For an online tutorial on doped silicon, visit physicspp.com.

3 Evaluate the Answer

- **Are the units correct?** Using dimensional analysis confirms the correct units.
- **Is the magnitude realistic?** The ratio is large enough so that intrinsic electrons make almost no contribution to conductivity.

PRACTICE Problems

- Additional Problems, Appendix B
- Solutions to Selected Problems, Appendix C

11. If you wanted to have 1×10^4 as many electrons from arsenic doping as thermally free electrons in silicon at room temperature, how many arsenic atoms should there be per silicon atom?

12. If you wanted to have 5×10^3 as many electrons from arsenic doping as thermally free electrons in the germanium semiconductor described in problem 6, how many arsenic atoms should there be per germanium atom?

13. Germanium at 400.0 K, has 1.13×10^{15} thermally liberated carriers/cm^3. If it is doped with 1 As atom per 1 million Ge atoms, what is the ratio of doped carriers to thermal carriers?

14. Silicon at 400.0 K, has 4.54×10^{12} thermally liberated carriers/cm^3. If it is doped with 1 As atom per 1 million Si, what is the ratio of doped carriers to thermal carriers?

15. Based on problem 14, draw a conclusion about the behavior of germanium devices as compared to silicon devices at temperatures in excess of the boiling point of water.

Light meters Other useful applications of semiconductors depend on their light sensitivity. When light falls on a semiconductor, the light can excite electrons from the valence band to the conduction band in the same way that other energy sources excite atoms. Thus, the resistance decreases as the light intensity increases. Extrinsic semiconductors can be tailored to respond to specific wavelengths of light. These include the infrared and visible regions of the spectrum. Materials such as silicon and cadmium sulfide serve as light-dependent resistors in light meters used by lighting engineers to design the illumination of stores, offices, and homes; and by photographers to adjust their cameras to capture the best images, as shown in **Figure 29-8.**

Figure 29-8 Photographers use light meters to measure the intensity of incident light on an object.

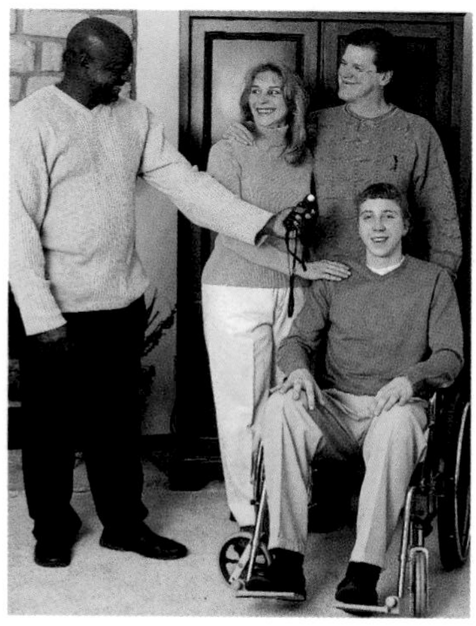

29.1 Section Review

16. Carrier Mobility In which type of material, a conductor, a semiconductor, or an insulator, are electrons most likely to remain with the same atom?

17. Semiconductors If the temperature increases, the number of free electrons in an intrinsic semiconductor increases. For example, raising the temperature by 8°C doubles the number of free electrons in silicon. Is it more likely that an intrinsic semiconductor or a doped semiconductor will have a conductivity that depends on temperature? Explain.

18. Insulator or Conductor? Silicon dioxide is widely used in the manufacture of solid-state devices. Its energy-band diagram shows a gap of 9 eV between the valence band and the conduction band. Is it more useful as an insulator or a conductor?

19. Conductor or Insulator? Magnesium oxide has a forbidden gap of 8 eV. Is this material a conductor, an insulator, or a semiconductor?

20. Intrinsic and Extrinsic Semiconductors You are designing an integrated circuit using a single crystal of silicon. You want to have a region with relatively good insulating properties. Should you dope this region or leave it as an intrinsic semiconductor?

21. Critical Thinking Silicon produces a doubling of thermally liberated carriers for every 8°C increase in temperature, and germanium produces a doubling of thermally liberated carriers for every 13°C increase. It would seem that germanium would be superior for high-temperature applications, but the opposite is true. Explain.

Physics Online physicspp.com/self_check_quiz

29.2 Electronic Devices

▸ **Objectives**

- **Describe** how diodes limit current to motion in only one direction.
- **Explain** how a transistor can amplify or increase voltage changes.

▸ **Vocabulary**

diode
depletion layer
transistor
microchip

Today's electronic instruments, such as radios, televisions, CD players, and microcomputers, rely on semiconductor devices that are combined on chips of silicon a few millimeters wide. In these devices, current and voltage vary in more complex ways than are described by Ohm's law. As a result, semiconductor devices can change current from AC to DC and amplify voltages.

Diodes

The simplest semiconductor device is the **diode.** A diode consists of a sandwich of *p*-type and *n*-type semiconductors. Rather than two separate pieces of doped silicon being joined, a single sample of intrinsic silicon is treated first with a *p*-dopant, then with an *n*-dopant. Metal contacts are coated on each region so that wires can be attached, as shown in **Figure 29-9a.** The boundary between the *p*-type and the *n*-type regions is called the *junction.* The resulting device, therefore, is called a *pn-junction diode.*

The free electrons on the *n*-side of the junction are attracted to the positive holes on the *p*-side. The electrons readily move into the *p*-side and recombine with the holes. Holes from the *p*-side similarly move into the *n*-side, where they recombine with electrons. As a result of this flow, the *n*-side has a net positive charge, and the *p*-side has a net negative charge. These charges produce forces in the opposite direction that stop further movement of charge carriers. The region around the junction is left with neither holes nor free electrons. This region, depleted of charge carriers, is called the **depletion layer.** Because it has no charge carriers, it is a poor conductor of electricity. Thus, a junction diode consists of relatively good conductors at the ends that surround a poor conductor.

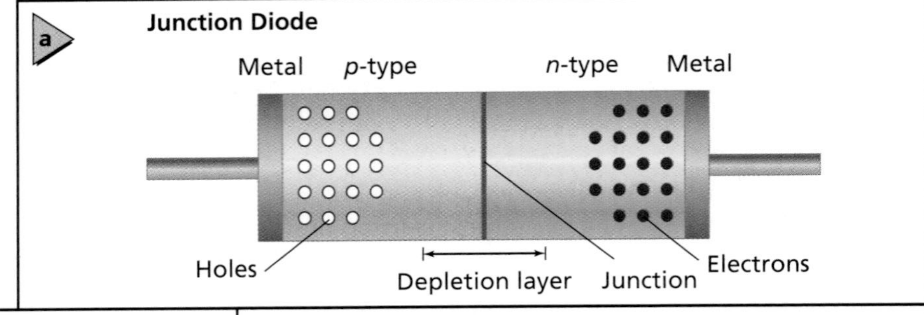

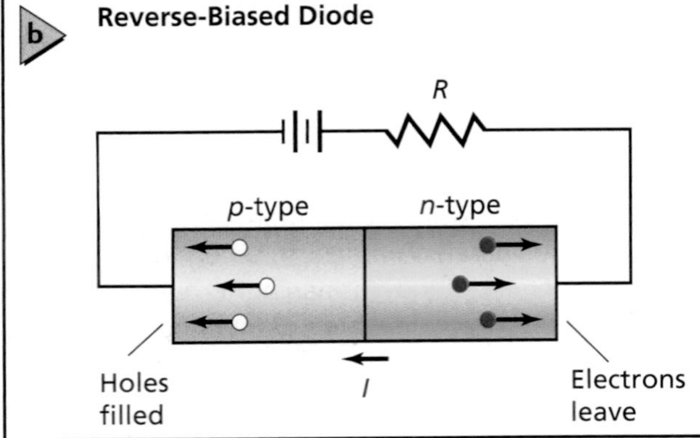

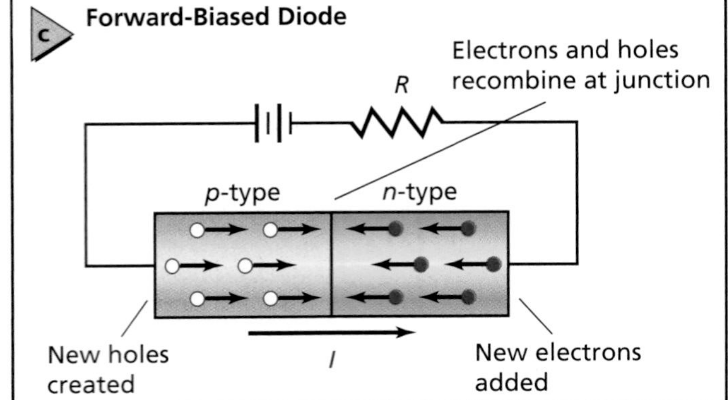

■ **Figure 29-9** A diagram of the *pn*-junction diode **(a)** shows the depletion layer, where there are no charge carriers. Compare the magnitude of current in a reverse-biased diode **(b)** and a forward-biased diode **(c).**

Concepts In Motion
Interactive Figure To see an animation on diodes, visit physicspp.com.

When a diode is connected into a circuit in the way shown in **Figure 29-9b,** both the free electrons in the *n*-type semiconductor and the holes in the *p*-type semiconductor are attracted toward the battery. The width of the depletion layer is increased, and no charge carriers meet. Almost no current passes through the diode: it acts like a very large resistor, almost an insulator. A diode oriented in this manner is a reverse-biased diode.

If the battery is connected in the opposite direction, as shown in **Figure 29-9c,** charge carriers are pushed toward the junction. If the voltage of the battery is large enough—0.6 V for a silicon diode—electrons reach the *p*-end and fill the holes. The depletion layer is eliminated, and a current passes through the diode. The battery continues to supply electrons for the *n*-end. It removes electrons from the *p*-end, which is the same as supplying holes. With further increases in voltage from the battery, the current increases. A diode in this kind of circuit is a forward-biased diode.

The graph shown in **Figure 29-10** shows the current through a silicon diode as a function of voltage across it. If the applied voltage is negative, the reverse-biased diode acts like a very high-value resistor and only a tiny current passes (about 10^{-11} A for a silicon diode). If the voltage is positive, the diode is forward-biased and acts like a low-value resistor, but not, however, one that obeys Ohm's law. One major use of a diode is to convert AC voltage to DC voltage with only one polarity. When a diode is used in a circuit that does this, it is called a rectifier. The arrow in the symbol for the diode, which you'll see in Example Problem 4, shows the direction of conventional current.

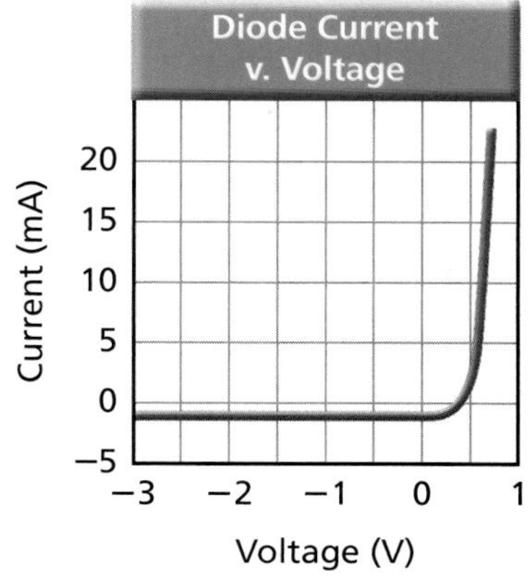

Figure 29-10 The graph indicates current-voltage characteristics for a silicon junction diode.

▶ EXAMPLE Problem 4

A Diode in a Simple Circuit **A silicon diode, with *I*/*V* characteristics like those shown in Figure 29-10, is connected to a power supply through a 470-Ω resistor. The power supply forward-biases the diode, and its voltage is adjusted until the diode current is 12 mA. What is the power supply voltage?**

1 Analyze and Sketch the Problem

- Draw a circuit diagram connecting a diode, a 470-Ω resistor, and a power supply. Indicate the direction of current.

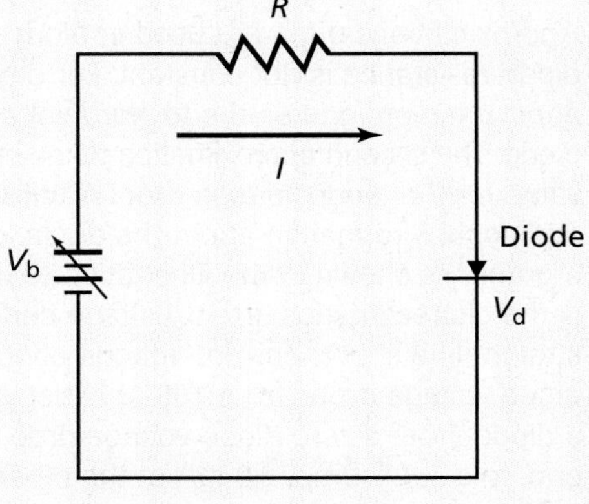

Known:	Unknown:
$I = 0.012$ A	$V_b = ?$
$V_d = 0.70$ V	
$R = 470\ \Omega$	

2 Solve for the Unknown

The voltage drop across the resistor is known from $V = IR$, and the power supply voltage is the sum of the resistor and the diode voltage drops.

$$V_b = IR + V_d$$

$$= (0.012\ \text{A})(470\ \Omega) + 0.70\ \text{V}$$ Substitute $I = 0.012$ A, $R = 470\ \Omega$, $V_d = 0.70$ V

$$= 6.3\ \text{V}$$

3 Evaluate the Answer

- **Are the units correct?** The power supply's potential difference is in volts.
- **Is the magnitude realistic?** It is in accord with the current and the resistance.

Math Handbook
Order of Operations page 843

PRACTICE Problems

- Additional Problems, Appendix B
- Solutions to Selected Problems, Appendix C

22. What battery voltage would be needed to produce a current of 2.5 mA in the diode in Example Problem 4?

23. What battery voltage would be needed to produce a current of 2.5 mA if another identical diode were added in series with the diode in Example Problem 4?

24. Describe how the diodes in the previous problem should be connected.

25. Describe what would happen in problem 23 if the diodes were connected in series but with improper polarity.

26. A germanium diode has a voltage drop of 0.40 V when 12 mA passes through it. If a 470-Ω resistor is used in series, what battery voltage is needed?

Figure 29-11 Diode lasers are used as both light emitters and detectors in bar-code scanners.

Light-emitting diodes Diodes made from combinations of gallium and aluminum with arsenic and phosphorus emit light when they are forward-biased. When electrons reach the holes in the junction, they recombine and release the excess energy at the wavelengths of light. These diodes are called light-emitting diodes, or LEDs. Some LEDs are configured to emit a narrow beam of coherent, monochromatic laser light. Such diode lasers are compact, powerful light sources. They are used in CD players, laser pointers, and supermarket bar-code scanners, as shown in **Figure 29-11.** Diodes can detect light as well as emit it. Light falling on the junction of a reverse-biased *pn*-junction diode creates electrons and holes, resulting in a current that depends on the light intensity.

CHALLENGE PROBLEM

Approximations often are used in diode circuits because diode resistance is not constant. For diode circuits, the first approximation ignores the forward voltage drop across the diode. The second approximation takes into account a typical value for the diode voltage drop. A third approximation uses additional information about the diode, often in the form of a graph, as shown in the illustration to the right. The curve is the characteristic current-voltage curve for the diode. The straight line shows current-voltage conditions for all possible diode voltage drops for a 180-Ω resistor, a 1.8-V battery, and a diode, from a zero diode voltage drop and 10.0 mA at one end, to a 1.8-V drop, 0.0 mA at the other end.

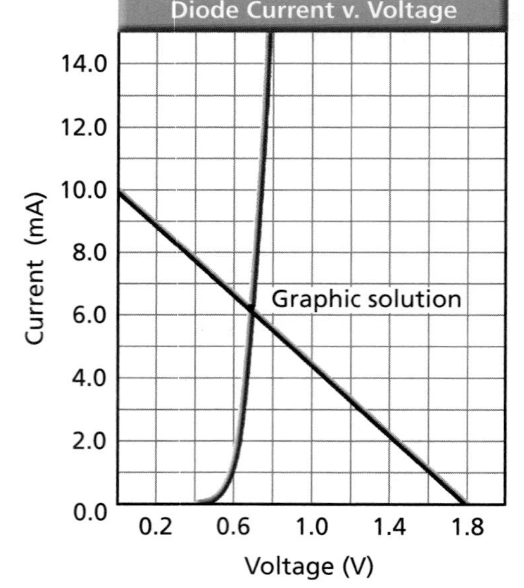

Use the diode circuit in Example Problem 4 with $V_b = 1.8$ V, but with $R = 180\ \Omega$:

1. Determine the diode current using the first approximation.
2. Determine the diode current using the second approximation and assuming a 0.70-V diode drop.
3. Determine the diode current using the third approximation by using the accompanying diode graph.
4. Estimate the error for all three approximations, ignoring the battery and resistor. Discuss the impact of greater battery voltages on the errors.

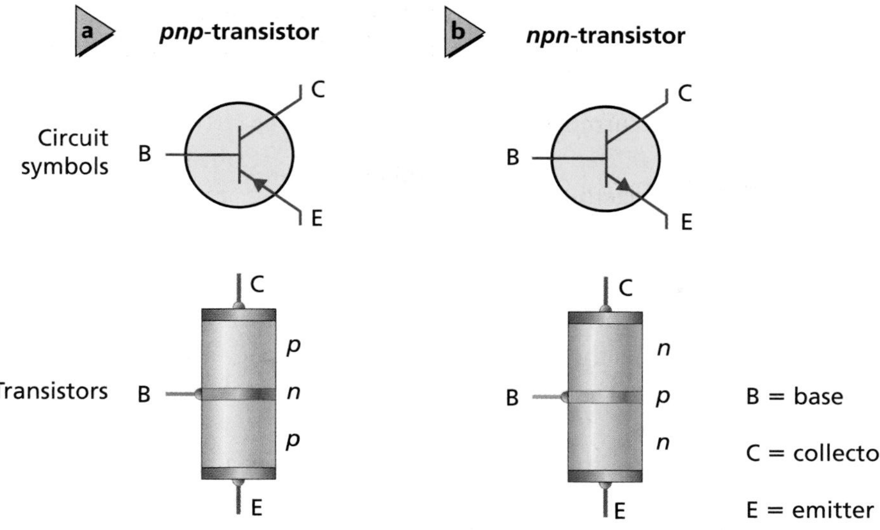

Figure 29-12 Compare the circuit symbols used to represent a *pnp*-transistor **(a)** and an *npn*-transistor **(b).**

Transistors and Integrated Circuits

A **transistor** is a simple device made of doped semiconductor material. An *npn*-transistor consists of layers of *n*-type semiconductor on either side of a thin *p*-type layer. The central layer is called the base and the regions on either side are the emitter and the collector. The schematic symbols for the two transistor types are shown in **Figure 29-12.** The arrow on the emitter shows the direction of conventional current.

The operation of an *npn*-transistor is illustrated in **Figure 29-13.** The two *pn*-junctions in the transistor can be thought of as initially forming two back-to-back diodes. The battery on the right, V_C, keeps the collector more positive than the emitter. The base-collector diode is reverse-biased, with a wide depletion layer, so there is no current from the collector to the base. When the battery on the left, V_B, is connected, the base is more positive than the emitter. That makes the base-emitter diode forward-biased, allowing current I_B from the base to the emitter.

The very thin base region is part of both diodes in the transistor. The charges injected by I_B reduce the reverse bias of the base-collector diode, permitting charge to flow from the collector to the emitter. A small change in I_B thus produces a large change in I_C.

The collector current causes a voltage drop across resistor R_C. Small changes in the voltage, V_B, applied to the base produce large changes in the collector current and thus changes in the voltage drop across R_C. As a result, the transistor amplifies small voltage changes into much larger changes. If instead the center layer is an *n*-type region, then the device is called a *pnp*-transistor. A *pnp*-transistor works the same way, except that the potentials of both batteries are reversed.

Current gain The current gain from the base circuit to the collector circuit is a useful indicator of the performance of a transistor. Although the base current is quite small, it is dependent on the base-emitter voltage that is controlling the collector current. For example, if V_B in Figure 29-13 is removed, the collector current will drop to zero. If V_B is increased, the base current, I_B, increases. The collector current, I_C, will also increase, but many times more (perhaps 100 times or so). The current gain from the base to the collector ranges from 50 to 300 for general-purpose transistors.

APPLYING PHYSICS

Diode Laser A typical diode laser emits light at 800 nm, which is the near infrared. The beam is output from a small spot on a GaAlAs chip, and when powered by 80 mA, the diode has a forward voltage drop of about 2 V. Diode lasers commonly are used in optical fiber transmissions.

Figure 29-13 A circuit using an *npn*-transistor demonstrates how voltage can be amplified.

Concepts In Motion

Interactive Figure To see an animation on transistors, visit physicspp.com.

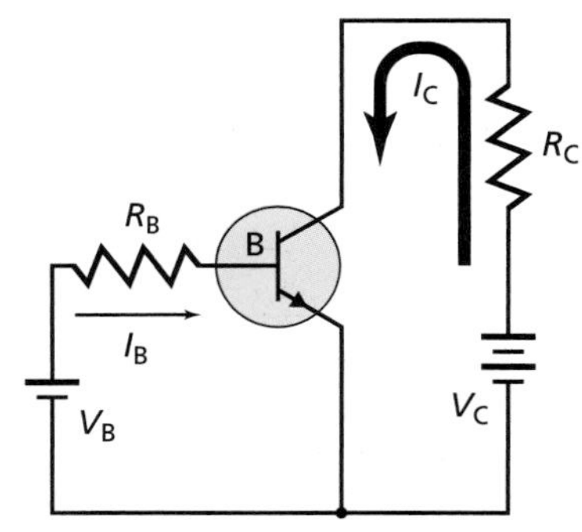

MINI LAB

Red Light 

Make a series circuit with a DC power supply, a 470-Ω resistor, and a red LED. Connect the short lead of the LED to the negative side of the power supply which is plugged into a GFCI-protected receptacle. Attach the other lead to the resistor. Connect the remaining resistor lead to the positive side of the power supply. Slowly increase the voltage until the LED glows. Note the voltage setting on the power supply.

1. Hypothesize what will happen if you reverse the direction of current.

2. Experiment by reversing the connections to the battery.

Analyze and Conclude

3. Explain your observations in terms of LED characteristics.

In a tape player, the small voltage variations from the voltage induced in a coil by magnetized regions on the tape are amplified to move the speaker coil. In computers, small currents in the base-emitter circuits can turn on or turn off large currents in the collector-emitter circuits. In addition, several transistors can be connected together to perform logic operations or to add numbers together. In these cases, they act as fast switches rather than as amplifiers.

Microchips An integrated circuit, called a **microchip,** consists of thousands of transistors, diodes, resistors, and conductors, each less than a micrometer across. All these components can be made by doping silicon with donor or acceptor atoms. A microchip begins as an extremely pure single crystal of silicon, 10–30 cm in diameter and 1–2 m long, as shown in **Figure 29-14.** The silicon is sliced by a diamond-coated saw into wafers less than 1-mm thick. The circuit is then built layer by layer on the surface of this wafer.

By a photographic process, most of the wafer's surface is covered by a protective layer, with a pattern of selected areas left uncovered so that they can be doped appropriately. The wafer is then placed in a vacuum chamber. Vapors of a dopant such as arsenic enter the machine, doping the wafer in the unprotected regions. By controlling the amount of exposure, the engineer can control the conductivity of the exposed regions of the chip. This process creates resistors, as well as one of the two layers of a diode or one of the three layers of a transistor. The protective layer is removed, and another one with a different pattern of exposed areas is applied. Then the wafer is exposed to another dopant, often gallium, producing *pn*-junctions. If a third layer is added, *npn*-transistors can be formed. The wafer also may be exposed to oxygen to produce areas of silicon dioxide insulation. A layer exposed to aluminum vapors can produce a pattern of thin conducting pathways among the resistors, diodes, and transistors.

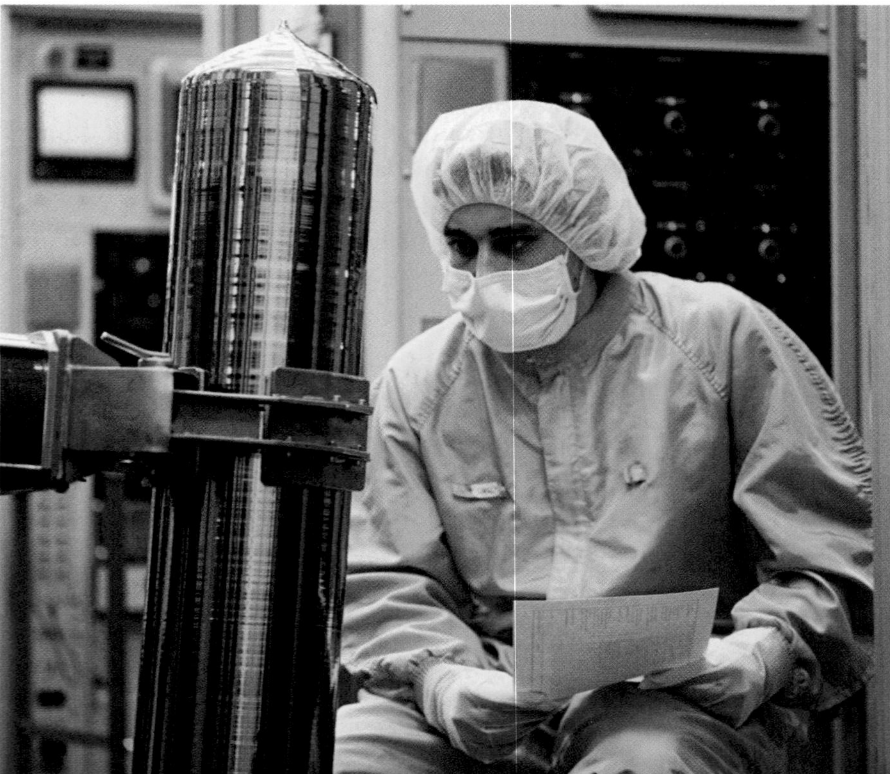

Figure 29-14 A technician prepares a large silicon crystal to be sliced into wafers for microchips.

Figure 29-15 Microchips form the heart of the central processing unit of computers. A penny is shown in the picture to represent scale.

Thousands of identical circuits, usually called *chips,* are produced at one time on a single wafer. The chips are then tested, sliced apart, and mounted in a carrier; wires are attached to the contacts; and the final assembly is then sealed into a protective plastic body. The tiny size of microchips, shown in **Figure 29-15,** allows the placement of complicated circuits in a small space. Because electronic signals need only travel tiny distances, this miniaturization has increased the speed of computers. Chips now are used in appliances and automobiles as well as in computers.

Semiconductor electronics requires that physicists, chemists, and engineers work together. Physicists contribute their understanding of the motion of electrons and holes in semiconductors. Physicists and chemists together add precisely controlled amounts of dopants to extremely pure silicon. Engineers develop the means of mass-producing chips containing thousands of miniaturized diodes and transistors. Together, their efforts have brought our world into this electronic age.

29.2 Section Review

27. Transistor Circuit The emitter current in a transistor circuit is always equal to the sum of the base current and the collector current: $I_E = I_B + I_C$. If the current gain from the base to the collector is 95, what is the ratio of emitter current to base current?

28. Diode Voltage Drop If the diode characterized in Figure 29-10 is forward-biased by a battery and a series resistor so that there is more than 10 mA of current, the voltage drop is always about 0.70 V. Assume that the battery voltage is increased by 1 V.

a. By how much does the voltage across the diode or the voltage across the resistor increase?

b. By how much does the current through the resistor increase?

29. Diode Resistance Compare the resistance of a *pn*-junction diode when it is forward-biased and when it is reverse-biased.

30. Diode Polarity In a light-emitting diode, which terminal should be connected to the *p*-end to make the diode light?

31. Current Gain The base current in a transistor circuit measures 55 μA and the collector current measures 6.6 mA. What is the current gain from base to collector?

32. Critical Thinking Could you replace an *npn*-transistor with two separate diodes connected by their *p*-terminals? Explain.

Physics Online physicspp.com/self_check_quiz

PHYSICS LAB

Diode Current and Voltage

Alternate CBL instructions can be found on the Web site.
physicspp.com

Semiconductor devices, such as diodes and transistors, are fabricated using a semiconductor that is made of partly *p*-type material and partly *n*-type material. A semiconductor doped with donor atoms is called an *n*-type semiconductor, while a semiconductor doped with an element leaving a vacancy or a hole in the lattice structure is referred to as a *p*-type semiconductor. A diode is made by doping adjacent regions of a semiconductor with donor and acceptor atoms, forming a *p-n* junction. In this lab, you will investigate the voltage and current characteristics of a diode that is placed in a direct current circuit and compare the response with your knowledge of resistors.

QUESTION

How do the current-voltage characteristics of a diode, an LED, and a resistor compare?

Objectives

- **Collect and organize data** of voltage drop and current for a diode and an LED.
- **Measure** the current passing through a diode and an LED as a function of voltage drop.
- **Compare and contrast** the current-voltage characteristics of a resistor with diodes.

Safety Precautions

- **Use caution with electric connections. Avoid contact with the resistor, which may become hot.**
- **Plug power supplies into only GFCI-protected receptacles to prevent shock hazard.**

Materials

DC power supply, variable, 0–12 VDC
100-Ω resistor, $\frac{1}{2}$- or 1-W
1N4002 diode
LED, red
ammeter, DC, 0–100 mA
voltmeter, 0–5 VDC
hook-up wire

Procedure

1. Prepare a data table similar to the one shown on page 791.
2. As indicated on the schematic diagram below, wire the negative terminal of the power supply to the negative side of the ammeter using the hook-up wire provided.

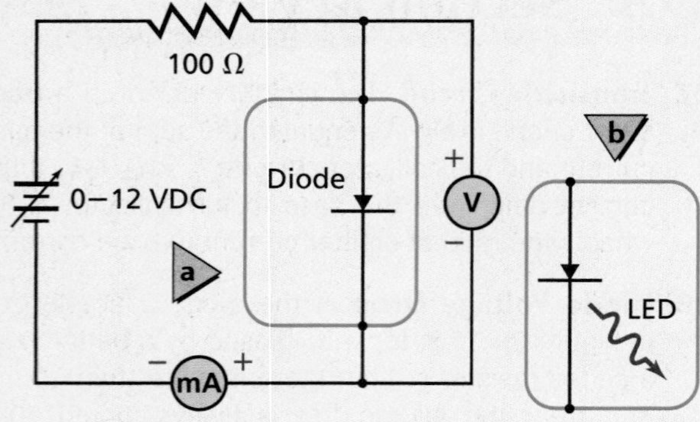

3. Locate the end of the diode with the silver band around it. Attach this end to the positive side of the ammeter.
4. Attach one end of the 100-Ω resistor to the free end of the diode.

Data Table

Voltage (V) Drop Across Diode	Diode Current (mA)	LED Current (mA)
0		
0.1		
0.2		
0.3		
1.7	———	
1.8	———	
1.9	———	
2.0	———	

5. Attach a wire from the free end of the 100-Ω resistor to the positive lead on the power supply.
6. As shown in the schematic, the voltmeter is in parallel with the diode. Attach a wire from the positive side of the voltmeter to the end of the diode attached to the resistor. Connect the negative side of the voltmeter to the end of the diode with the silver band, which is attached to the ammeter.
7. The diode circuit should look like part **a** of the schematic. Make sure the power supply is turned to zero and plug it in. Slowly turn up the power supply to increase the voltage drop across the diode from 0 up to 0.8 V, in 0.1-V increments. Record the corresponding current at each voltage. ***CAUTION: If your current goes higher than the capacity of your ammeter, do not increase the voltage any higher, and discontinue taking readings.*** Turn the power supply to zero and unplug it.
8. Observe the LED leads. One should be shorter than the other. Replace the 1N4002 diode with the LED so that it corresponds with part **b** of the schematic.

Physics Online

To find out more about solid-state electronics, visit the Web site: physicspp.com

9. Connect the shorter lead on the LED to the positive side of the ammeter (negative side of the voltmeter) where the silver banded end of the diode had been connected. Connect the longer lead of the LED to the resistor and to the positive side of the voltmeter.
10. Plug in the power supply. Slowly turn up the power supply to increase the voltage drop across the LED from 0 up to 2.0 V, in 0.1-V increments. Record the corresponding current at each voltage. Additionally, observe the LED and record your observations of it.

Analyze

1. **Make and Use Graphs** On one chart, sketch and label graphs of current versus voltage drop for both the diode and the LED. Place current on the *y*-axis and voltage on the *x*-axis. What are the shapes of these curves?
2. **Formulate Models** Using Ohm's law, compute and plot on the same graph the voltage-current relationship for a 100-Ω resistor from 0 to 2 V. Label this line 100 Ω. What is the form of this plot?

Conclude and Apply

1. **Compare and Contrast** How do the current-voltage curves for a diode, an LED, and a resistor compare?
2. Which of these devices follow Ohm's law?
3. **Analyze and Conclude** Diodes are described as having a turn-on voltage. What is the turn-on voltage for a silicon diode? For the LED you used?
4. **Explain** Why would the specifications for an LED give a light output at a specific current, such as 20 mA?

Going Further

What could be done to get better measurements of current for the diode?

Real-World Physics

Small incandescent lightbulbs typically draw 75–150 mA of current at a particular voltage. Why might manufacturers prefer using LEDs in a battery-powered CD or MP3 player?

Artificial Intelligence

The phrase *artificial intelligence* was first used in 1955. It is defined as "the scientific understanding of the mechanisms underlying thought and intelligent behavior and their embodiment in machines." Sometimes, a task needs artificial intelligence to be very logical. At other times, it may need artificial intelligence to think and behave with human biases. The goals in the field of artificial intelligence are to develop systems that can do both.

Artificial intelligence also is used to create expert systems in computers that are programmed with knowledge about specific topics. Humans can tell the computer the details of a specific situation, and the computer calculates the most logical course of action. In a medical environment, an expert system can be used to accurately diagnose disorders. Artificial intelligence weighs the facts of the situation and then infers which actions are most appropriate. However, artificial intelligence can operate only with facts that have been taught to the computer. Users must constantly be aware of this limitation of expert systems.

A prototype Mars rover decides how to navigate obstacles.

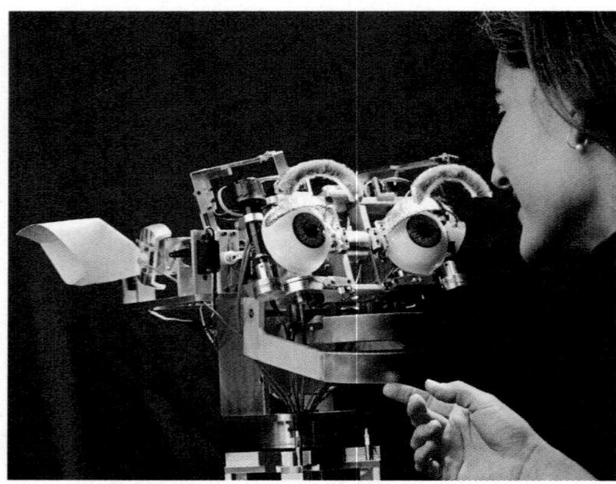

The robot, Kismet, displays human facial expressions.

Applications Artificial intelligence already is used in many areas, and it will do even more for us in the future. When a computer plays chess, it searches through hundreds of thousands of possible moves before selecting the best one. Research is being done to improve the efficiency of search algorithms.

Artificial intelligence currently is used for speech recognition to allow hands-free dialing of cell phones and for some interactive telephone transactions. It is not yet fully capable of understanding natural language, but that is a goal.

Three-dimensional computer vision is another future application. To mimic the sensory input and behaviors of humans, computers need to extract three-dimensional reality from two-dimensional images. Progress has been made, but humans are still much better than computers at this. With improved vision, artificial intelligence may control automobiles on Earth, or robots exploring another planet, with no human navigators needed.

Careers Studying mathematics, mathematical logic, and computer programming languages is important for developing systems that can make rational decisions. Knowledge of psychology assures that these decisions also can have a human character.

Going Further

1. **Debate the Issue** Are there ethical limits to the development of artificial intelligence?
2. **Recognize Cause and Effect** What problems might cause an expert system to make a poor decision?
3. **Critical Thinking** In what situations must artificial intelligence be absolutely rational, and in what situations should it include human biases?

Chapter 29

Study Guide

29.1 Conduction in Solids

Vocabulary

- semiconductors (p. 775)
- band theory (p. 776)
- intrinsic semiconductors (p. 780)
- dopants (p. 781)
- extrinsic semiconductors (p. 781)

Key Concepts

- Electric conduction may be explained by the band theory of solids.
- In solids, the allowed energy levels for outer electrons in an atom are spread into broad bands by the electric fields of electrons on neighboring atoms.
- The valence and conduction bands are separated by forbidden energy gaps; that is, by regions of energy levels that electrons may not possess.
- In conductors, electrons can move through the solid because the conduction band is partially filled.
- Electrons in metals have a fast random motion. A potential difference across the metal causes a slow drift of electrons, called an electric current.
- In insulators, more energy is needed to move electrons into the conduction band than is generally available.
- Conduction in semiconductors is enhanced by doping pure crystals with small amounts of other kinds of atoms, called dopants.
- *n*-type semiconductors are doped with electron donor atoms, and they conduct by the response of these donor electrons to applied potential differences.
- Arsenic, with five valence electrons, is an example of a donor atom.
- *p*-type semiconductors are doped with electron acceptor atoms, and they conduct by making holes available to electrons in the conduction band.
- Gallium, with three valence electrons, is an example of an acceptor atom.

29.2 Electronic Devices

Vocabulary

- diode (p. 784)
- depletion layer (p. 784)
- transistor (p. 787)
- microchip (p. 788)

Key Concepts

- A *pn*-junction diode consists of a layer of a *p*-type semiconductor joined with a layer of an *n*-type semiconductor.
- Diodes conduct charges in one direction only. They can be used in rectifier circuits to convert AC to DC.
- Electrons and holes near either side of the diode junction combine to produce a region without charge carriers known as the depletion layer.
- Applying a potential difference of the proper polarity across the diode makes the depletion layer even wider, no current is observed, and the diode is said to be reverse-biased.
- Reversing the polarity of the applied potential across the diode greatly reduces the depletion layer, current is observed, and the diode is said to be forward-biased.
- A transistor is a sandwich of three layers of semiconductor material, configured as either *npn*- or *pnp*-layers. The center base layer is very thin compared to the other layers, the emitter and collector.
- A transistor can act as an amplifier to convert a weak signal into a much stronger one.
- The ratio of the collector-emitter current to the base current is known as the current gain and is a useful measure of transistor amplification.
- Conductivity of semiconductors increases with increasing temperature or illumination, making them useful as thermometers or light meters. Diodes that emit light when a potential is applied are used in optical devices.

Chapter 29

Assessment

Concept Mapping

33. Complete the concept map using the following terms: *transistor, silicon diode, emits light, conducts both ways.*

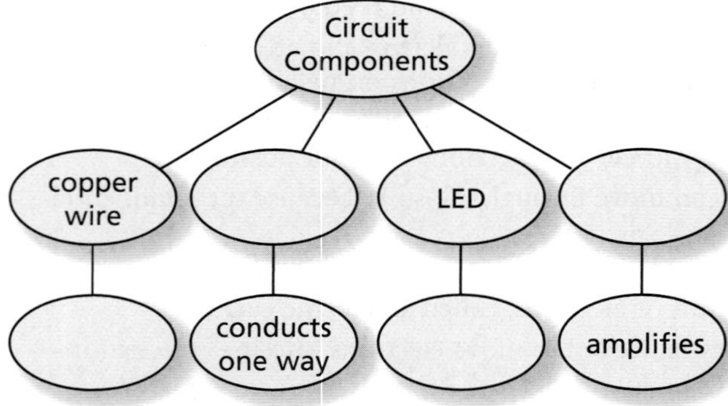

Mastering Concepts

34. How do the energy levels in a crystal of an element differ from the energy levels in a single atom of that element? (29.1)

35. Why does heating a semiconductor increase its conductivity? (29.1)

36. What is the main current carrier in a *p*-type semiconductor? (29.1)

37. An ohmmeter is an instrument that places a potential difference across a device to be tested, measures the current, and displays the resistance of the device. If you connect an ohmmeter across a diode, will the current you measure depend on which end of the diode was connected to the positive terminal of the ohmmeter? Explain. (29.2)

38. What is the significance of the arrowhead at the emitter in a transistor circuit symbol? (29.2)

39. Describe the structure of a forward-biased diode, and explain how it works. (29.2)

Applying Concepts

40. For the energy-band diagrams shown in **Figure 29-16,** which one represents a material with an extremely high resistance?

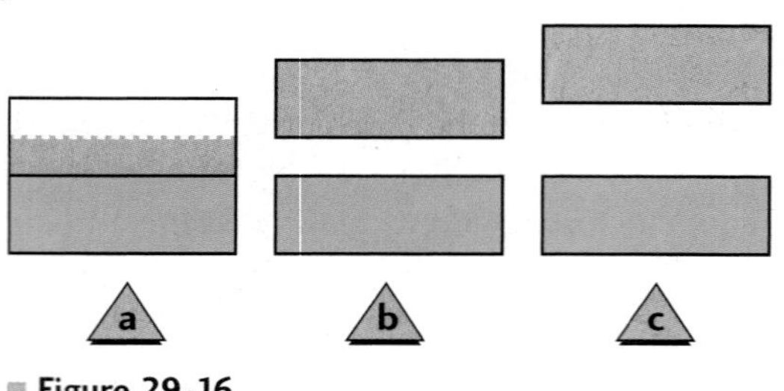

Figure 29-16

41. For the energy-band diagrams shown in Figure 29-16, which have half-full conduction bands?

42. For the energy-band diagrams shown in Figure 29-16, which ones represent semiconductors?

43. The resistance of graphite decreases as temperature rises. Does graphite conduct electricity more like copper or more like silicon does?

44. Which of the following materials would make a better insulator: one with a forbidden gap 8-eV wide, one with a forbidden gap 3-eV wide, or one with no forbidden gap?

45. Consider atoms of the three materials in problem 44. From which material would it be most difficult to remove an electron?

46. State whether the bulb in each of the circuits of **Figure 29-17 (a, b,** and **c)** is lighted.

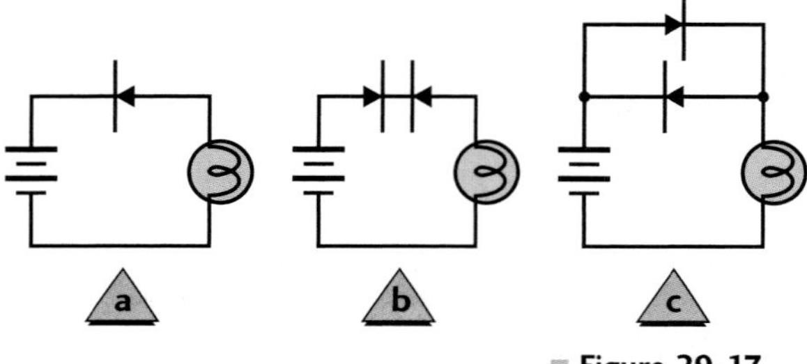

Figure 29-17

47. In the circuit shown in **Figure 29-18,** state whether lamp L_1, lamp L_2, both, or neither is lighted.

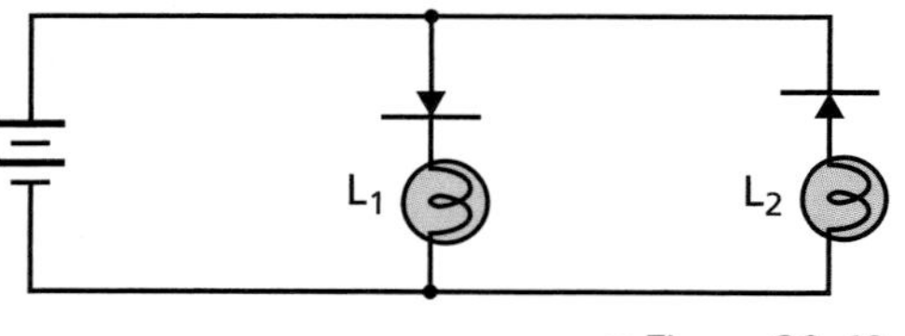

Figure 29-18

48. Use the periodic table to determine which of the following elements could be added to germanium to make a *p*-type semiconductor: B, C, N, P, Si, Al, Ge, Ga, As, In, Sn, or Sb.

49. Does an ohmmeter show a higher resistance when a *pn*-junction diode is forward-biased or reverse-biased?

50. If the ohmmeter in problem 49 shows the lower resistance, is the ohmmeter lead on the arrow side of the diode at a higher or lower potential than the lead connected to the other side?

51. If you dope pure germanium with gallium alone, do you produce a resistor, a diode, or a transistor?

 For more problems, go to Additional Problems, Appendix B.

52. Draw the time-versus-amplitude waveform for point A in **Figure 29-19a** assuming an input AC waveform, as shown in **Figure 29-19b.**

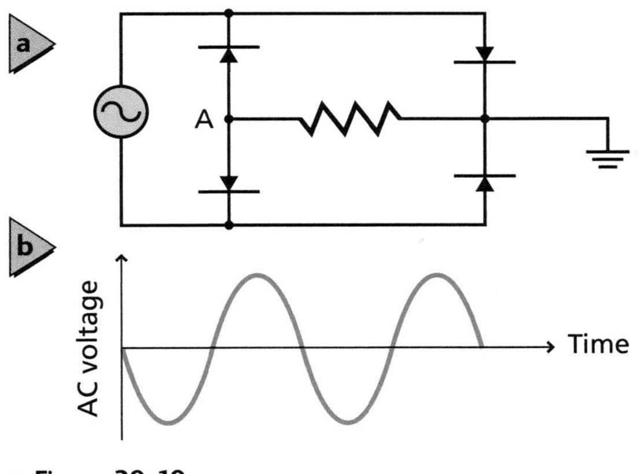

Figure 29-19

Mastering Problems

29.1 Conduction in Solids

53. How many free electrons exist in a cubic centimeter of sodium? Its density is 0.971 g/cm^3, its atomic mass is 22.99 g/mol, and there is 1 free electron per atom.

54. At a temperature of 0°C, thermal energy frees 1.55×10^9 e$^-$/cm^3 in pure silicon. The density of silicon is 2.33 g/cm^3, and the atomic mass of silicon is 28.09 g/mol. What is the fraction of atoms that have free electrons?

29.2 Electronic Devices

55. **LED** The potential drop across a glowing LED is about 1.2 V. In **Figure 29-20,** the potential drop across the resistor is the difference between the battery voltage and the LED's potential drop. What is the current through each of the following?
   a. the LED
   b. the resistor

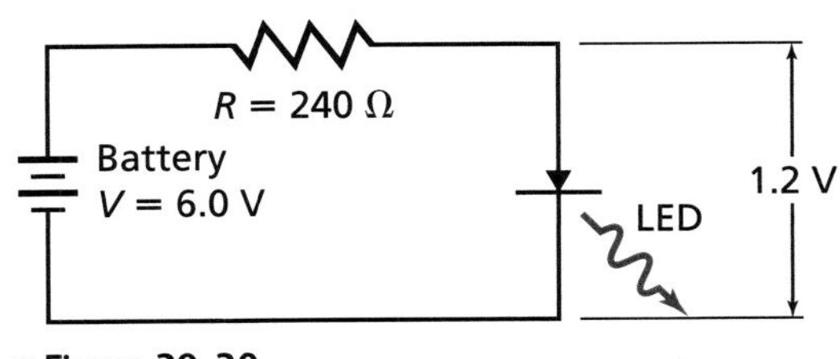

Figure 29-20

56. Jon wants to raise the current through the LED in problem 55 up to 3.0×10^1 mA so that it glows brighter. Assume that the potential drop across the LED is still 1.2 V. What resistor should be used?

57. **Diode** A silicon diode with I/V characteristics, as shown in Figure 29-10, is connected to a battery through a 270-Ω resistor. The battery forward-biases the diode, and the diode current is 15 mA. What is the battery voltage?

58. Assume that the switch shown in **Figure 29-21** is off.
   a. Determine the base current.
   b. Determine the collector current.
   c. Determine the voltmeter reading.

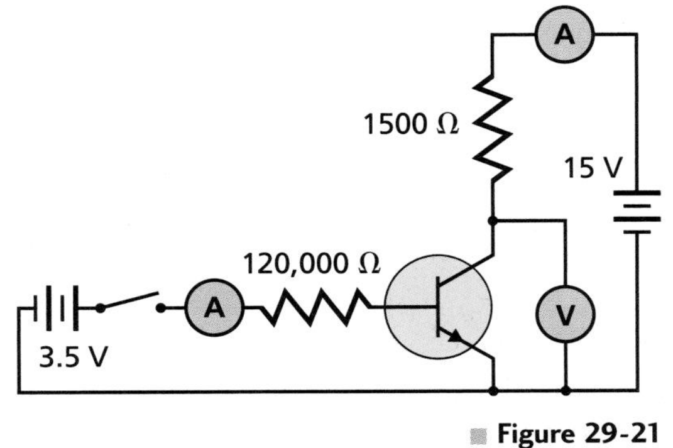

Figure 29-21

59. Assume that the switch shown in Figure 29-21 is on, and that there is a 0.70-V drop across the base-emitter junction and a current gain from base to collector of 220.
   a. Determine the base current.
   b. Determine the collector current.
   c. Determine the voltmeter reading.

Mixed Review

60. The forbidden gap in silicon is 1.1 eV. Electromagnetic waves striking the silicon cause electrons to move from the valence band to the conduction band. What is the longest wavelength of radiation that could excite an electron in this way? Recall that $E = 1240$ eV·nm/λ.

61. **Si Diode** A particular silicon diode at 0°C shows a current of 1.0 nA when it is reverse-biased. What current can be expected if the temperature increases to 104°C? Assume that the reverse-bias voltage remains constant. (The thermal carrier production of silicon doubles for every 8°C increase in temperature.)

62. **Ge Diode** A particular germanium diode at 0°C shows a current of 1.5 μA when it is reverse-biased. What current can be expected if the temperature increases to 104°C? Assume that the reverse-biasing voltage remains constant. (The thermal charge-carrier production of germanium doubles for every 13°C increase in temperature.)

63. LED A light-emitting diode (LED) produces green light with a wavelength of 550 nm when an electron moves from the conduction band to the valence band. Find the width of the forbidden gap in eV in this diode.

64. Refer to **Figure 29-22.**

a. Determine the voltmeter reading.

b. Determine the reading of A_1.

c. Determine the reading of A_2.

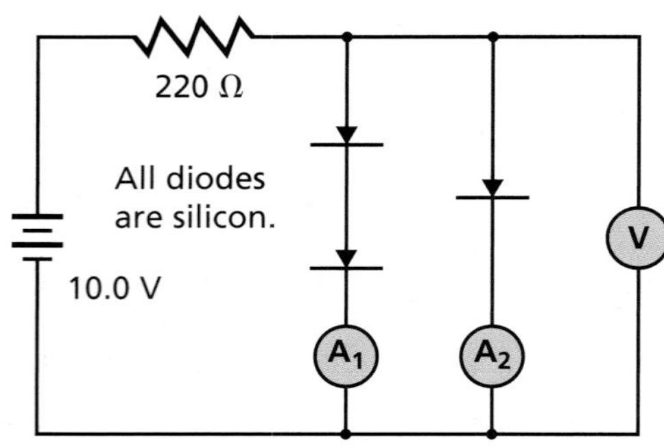

Figure 29-22

Thinking Critically

65. Apply Concepts A certain motor, in **Figure 29-23,** runs in one direction with a given polarity applied and reverses direction with the opposite polarity.

a. Which circuit (**a, b,** or **c**) will allow the motor to run in only one direction?

b. Which circuit will cause a fuse to blow if the incorrect polarity is applied?

c. Which circuit produces the correct direction of rotation regardless of the applied polarity?

d. Discuss the advantages and disadvantages of all three circuits.

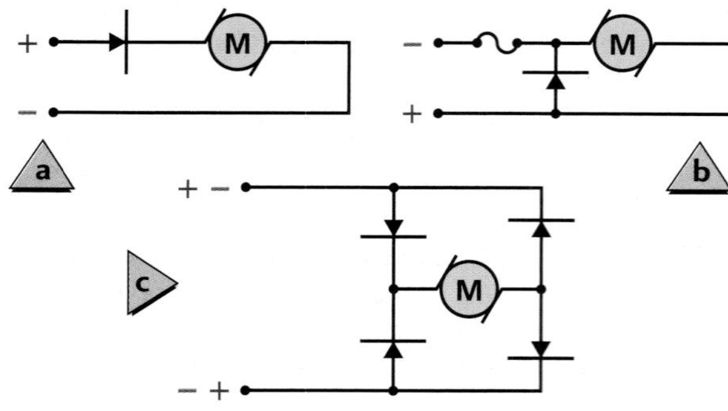

Figure 29-23

66. Apply Concepts The *I*/*V* characteristics of two LEDs that glow with different colors are shown in **Figure 29-24.** Each is to be connected through a resistor to a 9.0-V battery. If each is to be run at a current of 0.040 A, what resistors should be chosen for each?

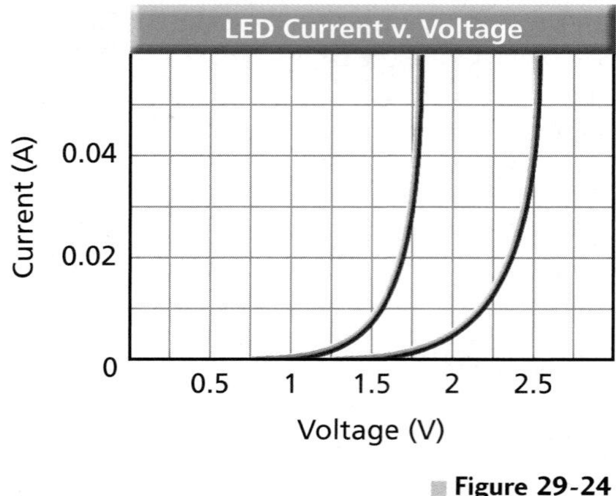

Figure 29-24

67. Apply Concepts Suppose that the two LEDs in problem 66 are now connected in series. If the same battery is to be used and a current of 0.035 A is desired, what resistor should be used?

Writing in Physics

68. Research the Pauli exclusion principle and the life of Wolfgang Pauli. Highlight his outstanding contributions to science. Describe the application of the exclusion principle to the band theory of conduction, especially in semiconductors.

69. Write a one-page paper discussing the Fermi energy level as it applies to energy-band diagrams for semiconductors. Include at least one drawing.

Cumulative Review

70. An alpha particle, a doubly ionized (2+) helium atom, has a mass of 6.7×10^{-27} kg and is accelerated by a voltage of 1.0 kV. If a uniform magnetic field of 6.5×10^{-2} T is maintained on the alpha particle, what will be the particle's radius of curvature? (Chapter 26)

71. What is the potential difference needed to stop photoelectrons that have a maximum kinetic energy of 8.0×10^{-19} J? (Chapter 27)

72. Calculate the radius of the orbital associated with the energy level E_4 of the hydrogen atom. (Chapter 28)

 For more problems, go to Additional Problems, Appendix B.

Multiple Choice

1. Which statement about diodes is false? Diodes can ___________.

   (A) amplify voltage (C) emit light
   (B) detect light (D) rectify AC

2. Cadmium has two free electrons per atom. How many free electrons are there per cm^3 of cadmium? The density of cadmium is 8650 kg/m^3.

   (A) 1.24×10^{21} (C) 9.26×10^{24}
   (B) 9.26×10^{22} (D) 1.17×10^{27}

3. The base current in a transistor circuit measures 45 μA and the collector current measures 8.5 mA. What is the current gain from base to collector?

   (A) 110 (C) 205
   (B) 190 (D) 240

4. In problem 3, if the base current is increased by 5 μA, how much is the collector current increased?

   (A) 5 μA (C) 10 mA
   (B) 1 mA (D) 190 μA

5. A transistor circuit shows a collector current of 4.75 mA, and the base to collector current gain is 250. What is the base current?

   (A) 1.19 μA (C) 4.75 mA
   (B) 18.9 μA (D) 1190 mA

6. Which line in the following table best describes both *n*- and *p*-type silicon semiconductors?

	n-type	*p*-type
(A)	Gallium-doped	Added electrons
(B)	Added electrons	Arsenic-doped
(C)	Arsenic-doped	Added holes
(D)	Added holes	Gallium-doped

7. Which line in the following table best describes the behavior of intrinsic silicon semiconductors to increasing temperature?

	Effect of Increasing Temperature on Intrinsic Silicon Semiconductors	
	Conductivity	**Resistance**
(A)	Increases	Increases
(B)	Increases	Decreases
(C)	Decreases	Increases
(D)	Decreases	Decreases

8. Thermal electron production in silicon doubles for every 8°C increase in temperature. A silicon diode at 0°C shows a current of 2.0 nA when reverse-biased. What will be the current at 112°C if the reverse-bias voltage is constant?

   (A) 11 μA (C) 44 μA
   (B) 33 μA (D) 66 μA

Extended Answer

9. A silicon diode is connected in the forward-biased direction to a power supply though a 485-Ω resistor, as shown below. If the diode voltage drop is 0.70 V, what is the power supply voltage when the diode current is 14 mA?

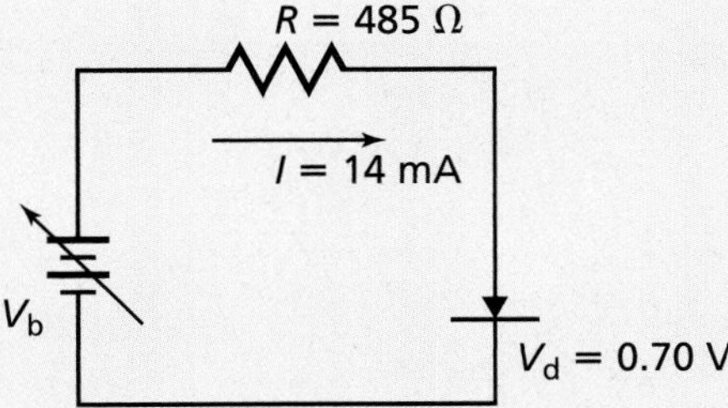

✓ Test-Taking TIP

Focus

If students near you are talking during a test, you should move. Respond only to the instructor when taking a test. Talking is a distraction and the instructor might think that you are cheating. Don't take the chance. Focus on the test.

Chapter 30

Nuclear Physics

What You'll Learn

- You will describe the components of a nucleus and how radioactive decay affects these components.
- You will calculate the energy released in nuclear reactions.
- You will examine how radioactive isotopes and nuclear energy are produced and used.
- You will understand the building blocks of matter.

Why It's Important

Nuclear physics has many applications, including medical research, energy production, and studying the structure of matter.

Medicine Radioactive isotopes are used to create images of the brain and other organs at work for medical diagnoses and research.

Think About This ▶

How does the radiation that is emitted by radioactive isotopes allow scientists and doctors to trace the operations of the human body?

physicspp.com

How can you model the nucleus?

Question
How is the force exerted by double-sided tape similar to the strong force?

Procedure

1. Cover the outer edges of 3–6 disk magnets with double-sided tape. Do the same thing to 3–6 disks of wood or aluminum of the same size. The magnets represent protons. The other disks represent neutrons.
2. Arrange the magnets with all the north poles facing up.
3. **Describe** the force exerted on one "proton" when another "proton" is brought closer, until they finally touch.
4. **Describe** the force exerted on a "neutron" by either another "neutron" or a "proton" when they are brought together, until they finally touch.

Analysis

The strong force drops to zero when the centers of two nucleons are farther apart than their diameters. How does that compare with the range of the force of the tape? The strong force is the same for neutrons and protons. Does this model behave this way?

Critical Thinking Stable nuclei usually have more neutrons than protons. Why would this model behave the same way?

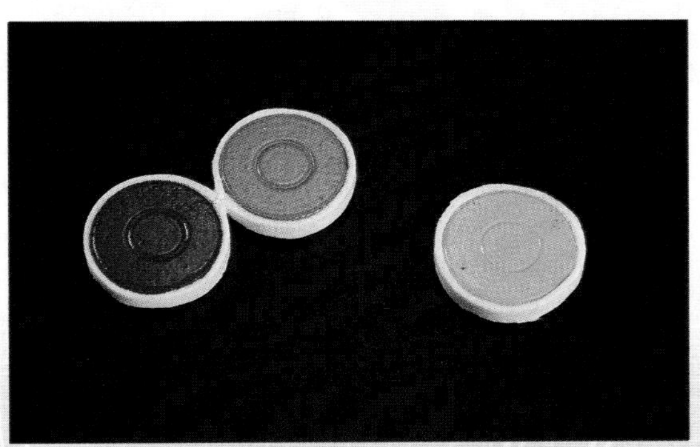

30.1 The Nucleus

> **▶ Objectives**
> - **Determine** the number of neutrons and protons in nuclides.
> - **Define** the binding energy of the nucleus.
> - **Relate** the energy released in a nuclear reaction to the change in binding energy during the reaction.
>
> **▶ Vocabulary**
> atomic number
> atomic mass unit
> mass number
> nuclide
> strong nuclear force
> nucleons
> binding energy
> mass defect

Ernest Rutherford not only established the existence of the nucleus, but he also conducted some of the early experiments to discover its structure. It is important to realize that neither Rutherford's experiments, nor the experiments of those who followed, offered the opportunity to observe the atom directly. Instead, inferences were drawn from the observations that researchers made. Recall from Chapter 28 that Rutherford's team made careful measurements of the deflection of alpha particles as they hit gold foil. These deflections could be explained if the atom was mostly empty space. The experiment showed that the atom had a small, dense, positively charged center surrounded by nearly massless electrons.

After the discovery of radioactivity by Becquerel in 1896, he and others looked at the effects produced when a nucleus breaks apart in natural radioactive decay. Marie and Pierre Curie discovered the new element radium and made it available to researchers all over the world, thereby increasing the study of radioactivity. Scientists discovered that through radioactivity one kind of atom could change into another kind, and thus, atoms must be made up of smaller parts. Ernest Rutherford and Fredrick Soddy used radioactivity to probe the center of the atom, the nucleus.

Description of the Nucleus

Are nuclei made up only of positively charged particles? At first, only the mass of the nucleus and the fact that the charge was positive were known. The magnitude of the charge of the nucleus was found as a result of X-ray scattering experiments done by Henry Moseley, a member of Rutherford's team. The results showed that the positively charged protons accounted for roughly half the mass of the nucleus. One hypothesis was that the extra mass was the result of protons, and that electrons in the nucleus reduced the charge to the observed value. This hypothesis, however, had some fundamental problems. In 1932, English physicist James Chadwick solved the problem when he discovered a neutral particle that had a mass approximately that of the proton. This particle, known as the neutron, accounted for the missing mass of the nucleus without increasing its charge.

Mass and charge of the nucleus The only charged particle in the nucleus is the proton. The **atomic number,** *Z*, of an atom is the number of protons. The total charge of the nucleus is the number of protons times the elementary charge, *e*.

$$\text{nuclear charge} = Ze$$

Both the proton and the neutron have a mass that is about 1800 times the mass of an electron. The proton and neutron masses are approximately equal to 1 u, where u is the **atomic mass unit,** 1.66×10^{-27} kg. To determine the approximate mass of the nucleus, multiply the number of neutrons and protons, or **mass number,** *A*, by u.

$$\text{nuclear mass} \cong A(\text{u})$$

Size of the nucleus Rutherford's experiments produced the first measurements of the size of the nucleus. He found that the nucleus has a diameter of about 10^{-14} m. A typical atom might have a radius 10,000 times larger than the size of the nucleus. Although the nucleus contains nearly all of the mass of an atom, proportionally the nucleus occupies less space in the atom than the Sun does in the solar system. The nucleus is incredibly dense—about 1.4×10^{18} kg/m^3. If a nucleus could be one cubic centimeter, it would have a mass of about one billion tons.

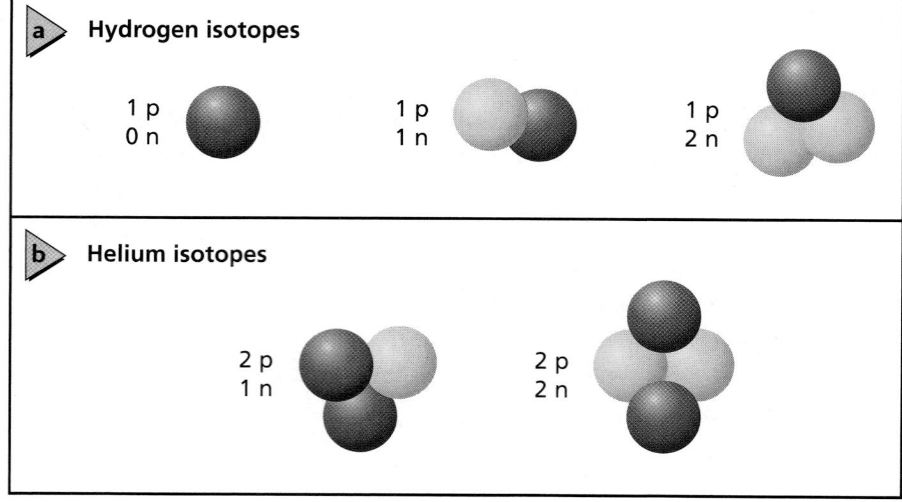

Figure 30-1 The nuclides of hydrogen **(a)** and helium **(b)** illustrate that all the nuclides of an element have the same numbers of protons, but have different numbers of neutrons. Protons are red and neutrons are gray in this drawing.

Do all elements have the same mass numbers?

Looking at the periodic table on page 916, you might notice that the first four elements all have an atomic mass, A, near a whole number. Boron, on the other hand, has a mass of 10.8 u. If the nucleus is made up of only protons and neutrons, each with a mass of approximately 1 u, shouldn't the total mass of any atom be near a whole number?

The puzzle of atomic masses that were not whole numbers was solved with the mass spectrometer. You learned in Chapter 26 how the mass spectrometer demonstrated that an element could have atoms with different masses. For example, in an analysis of a pure sample of neon, not one, but two spots appeared on the film of the spectrometer. The two spots were produced by neon atoms of different masses. One variety of the neon atom was found to have a mass of 20 u, while the second type had a mass of 22 u. All neutral neon atoms have 10 protons in the nucleus and 10 electrons in the atom. One kind of neon atom, however, has 10 neutrons in its nucleus, while the other has 12 neutrons. The two kinds of atoms are called *isotopes* of neon. The nucleus of an isotope is called a **nuclide.**

All nuclides of an element have the same number of protons, but have different numbers of neutrons, as illustrated by the hydrogen and helium nuclides shown in **Figure 30-1.** All isotopes of a neutral element have the same number of electrons around the nucleus and behave chemically in the same way.

Average mass The measured mass of neon gas is 20.183 u. This figure is now understood to be the average mass of the naturally occurring isotopes of neon. Thus, while the mass of an individual atom of neon is close to a whole number of mass units, the atomic mass determined from an average sample of neon atoms is not. Most elements have several isotopic forms that occur naturally. The mass of one isotope of carbon, carbon-12, is now used to define the mass unit. One u is defined to be $\frac{1}{12}$ the mass of the carbon-12 isotope.

A special method of notation is used to describe an isotope. A subscript representing the atomic number, or charge, Z, is written to the left of the symbol for the element. A superscript written to the left of the symbol is the mass number, A. This notation takes the form ${}^{A}_{Z}X$, where X is any element. For example, carbon-12 is ${}^{12}_{6}\mathrm{C}$, and the two isotopes of neon, with atomic number 10, are written as ${}^{20}_{10}\mathrm{Ne}$ and ${}^{22}_{10}\mathrm{Ne}$.

PRACTICE Problems — Additional Problems, Appendix B

1. Three isotopes of uranium have mass numbers of 234, 235, and 238. The atomic number of uranium is 92. How many neutrons are in the nuclei of each of these isotopes?
2. An isotope of oxygen has a mass number of 15. How many neutrons are in the nucleus of this isotope?
3. How many neutrons are in the mercury isotope ${}^{200}_{80}\mathrm{Hg}$?
4. Write the symbols for the three isotopes of hydrogen that have zero, one, and two neutrons in the nucleus.

What holds the nucleus together?

The negatively charged electrons that surround the positively charged nucleus of an atom are held in place by the attractive electromagnetic force. Because the nucleus consists of positively charged protons and neutral neutrons, the repulsive electromagnetic force among the protons might be expected to cause them to fly apart. Because this does not happen, an even stronger attractive force must exist within the nucleus.

The Strong Nuclear Force

APPLYING **PHYSICS**

Forces A positron is a positively charged electron. The electromagnetic attraction of an electron and a positron is 4.2×10^{42} times stronger than their gravitational attraction.

The **strong nuclear force,** also called the strong force, acts between protons and neutrons that are close together, as they are in a nucleus. This force is more than 100 times stronger than the electromagnetic force. The range of the strong force is short, only about the radius of a proton, 1.4×10^{-15} m. It is attractive and is of the same strength among protons and protons, protons and neutrons, and neutrons and neutrons.

Both neutrons and protons are called **nucleons.** The strong nuclear force holds the nucleons in the nucleus. If a nucleon were to be pulled out of a nucleus, work would have to be done to overcome the attractive force. Doing work adds energy to the system. Thus, the assembled nucleus has less energy than the separate protons and neutrons that make it up. The difference is the **binding energy** of the nucleus. Because the assembled nucleus has less energy, all binding energies are negative.

Binding Energy of the Nucleus

Einstein showed that mass and energy are equivalent, so the binding energy can be expressed in the form of an equivalent amount of mass, according to the following equation.

Energy Equivalent of Mass $E = mc^2$

The energy contained in matter is equal to the mass times the square of the speed of light in a vacuum.

Because energy has to be added to take a nucleus apart, the mass of the assembled nucleus is less than the sum of the masses of the nucleons that compose it.

For example, the helium nucleus, $^{4}_{2}\text{He}$, consists of two protons and two neutrons. The mass of a proton is 1.007276 u. The mass of a neutron is 1.008665 u. If the mass of the helium nucleus were equal to the sum of the masses of the two protons and the two neutrons, you would expect that the mass of the nucleus would be 4.031882 u. Careful measurement, however, shows that the mass of a helium nucleus is only 4.002603 u. The actual mass of the helium nucleus is less than the mass of its constituent parts by 0.029279 u. The difference between the sum of the masses of the individual nucleons and the actual mass is called the **mass defect.**

Mass spectrometers usually measure the masses of isotopes; that is, the nuclides with all their electrons. When calculating the mass defect of a nucleus, one must make sure that the mass of the electrons is accounted for properly. Thus, the mass of hydrogen (one proton and one electron) usually is given in mass-defect problems.

Masses normally are measured in atomic mass units. It will be useful, then, to determine the energy equivalent of 1 u (1.6605×10^{-27} kg). To determine the energy, you must multiply the mass by the square of the speed of light in a vacuum (2.9979×10^{8} m/s). This is expressed to five significant digits.

$$\begin{aligned} E &= mc^2 \\ &= (1.6605\times10^{-27}\ \text{kg})(2.9979\times10^{8}\ \text{m/s})^2 \\ &= 1.4924\times10^{-10}\ \text{kg}\cdot\text{m}^2/\text{s}^2 \\ &= 1.4924\times10^{-10}\ \text{J} \end{aligned}$$

Personal Tutor For an online tutorial on joule and electron volt, visit physicspp.com.

The most convenient unit of energy to use is the electron volt.

$$\begin{aligned} E &= (1.4924\times10^{-10}\ \text{J})(1\ \text{eV}/1.60217\times10^{-19}\ \text{J}) \\ &= 9.3149\times10^{8}\ \text{eV} \\ &= 931.49\ \text{MeV} \end{aligned}$$

Hence, 1 u of mass is equivalent to 931.49 MeV of energy. **Figure 30-2** shows how the binding energy per nucleon depends on the mass of the nucleus. Heavier nuclei are bound more strongly than lighter nuclei. Except for a few nuclei, the binding energy per nucleon becomes more negative as the mass number, *A*, increases to a value of 56, which is that of iron, Fe. The most tightly bound nucleus is $^{56}_{26}\text{Fe}$; thus, nuclei become more stable as their mass numbers approach that of iron. Nuclei whose mass numbers are larger than that of iron are less strongly bound, and are therefore less stable.

A nuclear reaction will occur naturally if energy is released by the reaction; that is, if the nucleus is transformed into one closer to the lowpoint of the graph at $A = 56$. At lower mass numbers, below $^{56}_{26}\text{Fe}$, a nuclear reaction will occur naturally if it increases the atomic mass. In the Sun and other stars, hydrogen is converted into helium, carbon, and other heavier elements in a reaction that releases energy, causing the electromagnetic radiation that you experience as visible light.

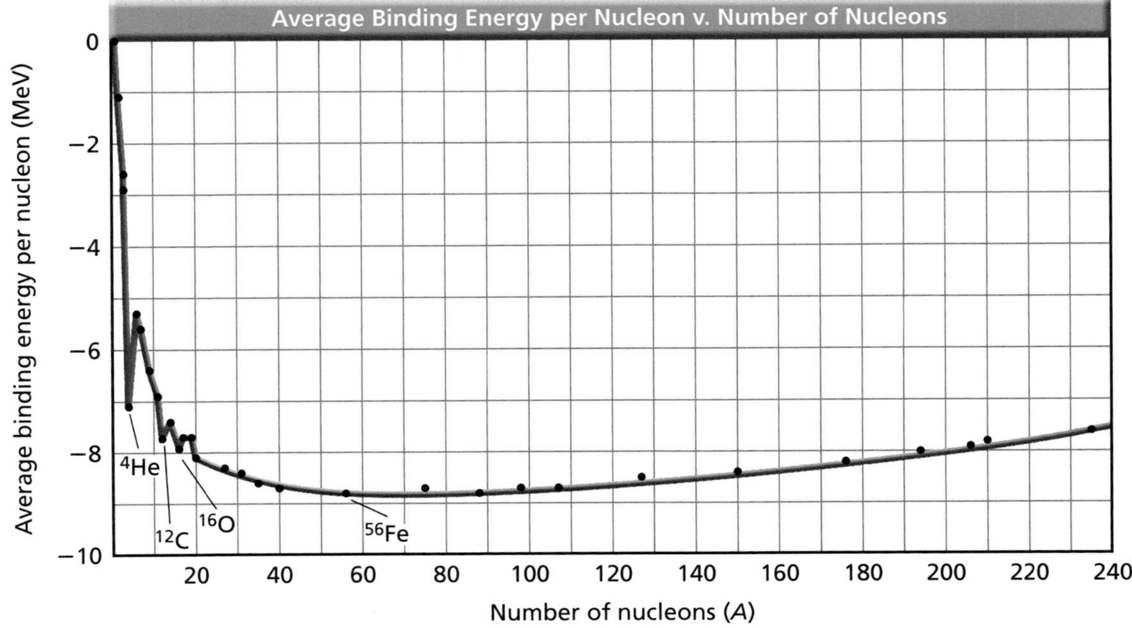

Figure 30-2 The binding energy per nucleon depends on the number of nucleons, *A*.

At mass numbers above 56, a nuclear reaction will occur naturally if it decreases the atomic mass. When uranium-238 decays to thorium-234, the resulting thorium nucleus is more stable than the uranium nucleus. Energy is released in the form of a radioactive particle with mass and kinetic energy. Thorium will not spontaneously transform into uranium because energy would have to be added to the nucleus. The heaviest nuclei in the periodic table are created in this way, by slamming together smaller nuclei in particle accelerators. In general, such heavy elements can exist only for fractions of a second before the nuclei decay into smaller, more stable nuclei. In contrast, when small nuclei gain nucleons, the binding energy of the larger nucleus is more negative, and thus, more stable than the sum of the binding energies of the smaller nuclei.

In the next section, calculations of binding energy will be used to understand nuclear reactions. The binding energy explains the energy released when small nuclei fuse, as in stars, and large nuclei split, as in the decay of radioactive elements.

▶ EXAMPLE Problem 1

Mass Defect and Nuclear Binding Energy **Find the mass defect and binding energy of tritium, $^{3}_{1}H$. The mass of the tritium isotope is 3.016049 u, the mass of a hydrogen atom is 1.007825 u, and the mass of a neutron is 1.008665 u.**

1 Analyze the Problem

Known:	**Unknown:**
mass of 1 hydrogen atom = 1.007825 u	total mass of nucleons and electron = ?
mass of 1 neutron = 1.008665 u	mass defect = ?
mass of tritium = 3.016049 u	binding energy of tritium = ?
binding energy of 1 u = 931.49 MeV	

2 Solve for the Unknown

Add the masses of the hydrogen atom (one proton and one electron) and two neutrons.

mass of 1 hydrogen atom:	1.007825 u
plus mass of 2 neutrons:	+ 2.017330 u
total mass of nucleons:	3.025155 u

Math Handbook
Operations with Significant Digits
pages 835–836

The mass defect is equal to the actual mass of tritium less the mass of the sum of its parts.

mass of tritium:	3.016049 u
less mass of nucleons:	− 3.025155 u
mass defect:	− 0.009106 u

The binding energy is the energy equivalent of the mass defect.

E = (mass defect in u)(binding energy of 1 u)

$E = (-0.009106\ \text{u})(931.49\ \text{MeV/u})$ Substitute mass defect = −0.009106 u, binding energy per u = 931.49 MeV

$= -8.4821\ \text{MeV}$

3 Evaluate the Answer

- **Are the units correct?** Mass is measured in u, and energy is measured in MeV.
- **Does the sign make sense?** Binding energy should be negative.
- **Is the magnitude realistic?** According to Figure 30-2, binding energies per nucleon in this range are between −2 MeV and −3 MeV, so the answer for three nucleons is reasonable.

▶ PRACTICE Problems

- Additional Problems, Appendix B
- Solutions to Selected Problems, Appendix C

Use these values to solve the following problems:
mass of hydrogen = 1.007825 u, mass of neutron = 1.008665 u,
1 u = 931.49 MeV.

5. The carbon isotope $^{12}_{6}C$ has a mass of 12.0000 u.

a. Calculate its mass defect.

b. Calculate its binding energy in MeV.

6. The isotope of hydrogen that contains one proton and one neutron is called deuterium. The mass of the atom is 2.014102 u.

a. What is its mass defect?

b. What is the binding energy of deuterium in MeV?

7. A nitrogen isotope, $^{15}_{7}N$, has seven protons and eight neutrons. It has a mass of 15.010109 u.

a. Calculate the mass defect of this nucleus.

b. Calculate the binding energy of the nucleus.

8. An oxygen isotope, $^{16}_{8}O$, has a nuclear mass of 15.994915 u.

a. What is the mass defect of this isotope?

b. What is the binding energy of its nucleus?

In few areas of physics has basic knowledge led to applications as quickly as in the field of nuclear physics. The medical use of the radioactive element radium began within 20 years of its discovery. Proton accelerators were used for medical applications less than one year after being invented. In the case of nuclear fission (splitting nuclei), the military application was under development before the basic physics was even known. Peaceful applications followed in less than ten years.

30.1 Section Review

9. Nuclei Consider these two pairs of nuclei: $^{12}_{6}C$ and $^{13}_{6}C$ and $^{11}_{5}B$ and $^{11}_{6}C$. In which way are the two alike? In which way are they different?

10. Binding Energy When tritium, $^{3}_{1}H$, decays, it emits a beta particle and becomes $^{3}_{2}He$. Which nucleus would you expect to have a more negative binding energy?

11. Strong Nuclear Force The range of the strong nuclear force is so short that only nucleons that are adjacent to each other are affected by the force. Use this fact to explain why, in large nuclei, the repulsive electromagnetic force can overcome the strong attractive force and make the nucleus unstable.

12. Mass Defect Which of the two nuclei in problem 10 has the larger mass defect?

13. Mass Defect and Binding Energy The radioactive carbon isotope $^{14}_{6}C$ has a mass of 14.003074 u.

a. What is the mass defect of this isotope?

b. What is the binding energy of its nucleus?

14. Critical Thinking In old stars, not only are helium and carbon produced by joining more tightly bound nuclei, but so are oxygen ($Z = 8$) and silicon ($Z = 14$). What is the atomic number of the heaviest nucleus that could be formed in this way? Explain.

30.2 Nuclear Decay and Reactions

▶ **Objectives**

- **Describe** three forms of radioactive decay.
- **Solve** nuclear equations.
- **Calculate** the amount remaining and activity of radioactive material after a given time.
- **Define** nuclear fission and fusion.
- **Describe** the operation of a nuclear reactor.

▶ **Vocabulary**

radioactive
alpha decay
beta decay
gamma decay
nuclear reaction
half-life
activity
fission
chain reaction
fusion

In 1896, Henri Becquerel was working with compounds containing the element uranium. To his surprise, he found that photographic plates that had been covered to keep out light became fogged, or partially exposed, when these uranium compounds were anywhere near the plates. This fogging suggested that some kind of ray from the uranium had passed through the plate coverings. Several materials other than uranium or its compounds also were found to emit these penetrating rays. Materials that emit this kind of radiation are now said to be **radioactive.** Because particles are emitted from the material, it is said to decay. Nuclei decay from a less stable form to a more stable form. Radioactivity is a natural process.

Radioactive Decay

In 1899, Ernest Rutherford and his colleagues discovered that the element radon spontaneously splits into a lighter nucleus and a light helium nucleus. The same year, Rutherford also discovered that uranium compounds produce three different kinds of radiation. He separated the types of radiation according to their penetrating ability and named them α (alpha), β (beta), and γ (gamma) radiation. Alpha radiation can be stopped by a thick sheet of paper, while 6 mm of aluminum is needed to stop most beta particles. Several centimeters of lead are required to stop gamma rays.

Alpha decay An alpha particle is the nucleus of a helium atom, ${}^{4}_{2}\text{He}$. The emission of an α particle from a nucleus is a process called **alpha decay.** The mass number of an α particle, ${}^{4}_{2}\text{He}$, is 4, and the atomic number is 2. When a nucleus emits an α particle, the mass number, A, of the decaying nucleus is reduced by 4, and the atomic number of the nucleus, Z, is reduced by 2. The element changes, or transmutes, into a different element. For example, ${}^{238}_{92}\text{U}$ transmutes into ${}^{234}_{90}\text{Th}$ through alpha decay.

Beta decay Beta particles are electrons emitted by the nucleus. The nucleus contains no electrons, however, so where do these electrons come from? **Beta decay** occurs when a neutron is changed to a proton within the nucleus. In all reactions, charge must be conserved. That is, the charge before the reaction must equal the charge after the reaction. In beta decay, when a neutron, charge 0, changes to a proton, charge $+1$, an electron, charge -1, also appears. In beta decay, a nucleus with N neutrons and Z protons ends up as a nucleus of $N-1$ neutrons and $Z+1$ protons. Another particle, an antineutrino, also is emitted in beta decay.

Gamma decay A redistribution of the energy within the nucleus results in **gamma decay.** The γ ray is a high-energy photon. Neither the mass number nor the atomic number is changed in gamma decay. Gamma radiation often accompanies alpha and beta decay. The three types of radiation are summarized in **Table 30-1.**

Radioactive elements often go through a series of successive decays until they form a stable nucleus. For example, ${}^{238}_{92}\text{U}$ undergoes 14 separate decays before the stable lead isotope ${}^{206}_{82}\text{Pb}$ is produced.

Table 30-1		
The Three Types of Radiation		
Alpha Particle	**Beta Particle**	**Gamma Ray**
Charge +2	Charge −1	No charge
Least penetration	Middle energy	Highest penetration
Transmutes nucleus: $A \rightarrow A - 4$ $Z \rightarrow Z - 2$ $N \rightarrow N - 2$	Transmutes nucleus: $A \rightarrow A$ $Z \rightarrow Z + 1$ $N \rightarrow N - 1$	Changes only energy: $A \rightarrow A$ $Z \rightarrow Z$ $N \rightarrow N$

Nuclear Reactions and Equations

A **nuclear reaction** occurs whenever the energy or number of neutrons or protons in a nucleus changes. Just as in chemical reactions, some nuclear reactions occur with a release of energy, while others occur only when energy is added to a nucleus.

One form of nuclear reaction is the emission of particles by radioactive nuclei. The reaction releases excess energy in the form of the kinetic energy of the emitted particles. Two such reactions are shown in **Figure 30-3.**

Nuclear reactions can be described by words, diagrams, or equations. The symbols used for the nuclei in nuclear equations make the calculation of atomic number and mass number in nuclear reactions simpler. For example, the equation for the reaction in **Figure 30-3a** is as follows.

$$^{238}_{92}\text{U} \rightarrow {}^{234}_{90}\text{Th} + {}^{4}_{2}\text{He}$$

The total number of nuclear particles stays the same during the reaction, so the sum of the superscripts on each side must be equal: 238 = 234 + 4. The total charge also is conserved, so the sum of the subscripts on each side must be equal: 92 = 90 + 2.

In beta decay, an electron, $^{0}_{-1}\text{e}$, and an antineutrino, $^{0}_{0}\bar{\nu}$, are produced. (The symbol for the antineutrino is the Greek letter *nu* with a bar over it; the bar denotes a particle of antimatter.) The transmutation of a thorium atom by the emission of a β particle is shown in **Figure 30-3b.**

$$^{234}_{90}\text{Th} \rightarrow {}^{234}_{91}\text{Pa} + {}^{0}_{-1}\text{e} + {}^{0}_{0}\bar{\nu}$$

Note that the sum of the superscripts on the left-hand side of the equation equals the sum of the superscripts on the right-hand side. Equality also exists between the subscripts on the two sides of the equation.

Figure 30-3 The emission of an alpha particle by uranium-238 results in the formation of thorium-234 **(a).** The emission of a beta particle by thorium-234 results in the formation of protactinium-234 **(b).**

Concepts In Motion
Interactive Figure To see an animation on alpha decay and beta decay, visit physicspp.com.

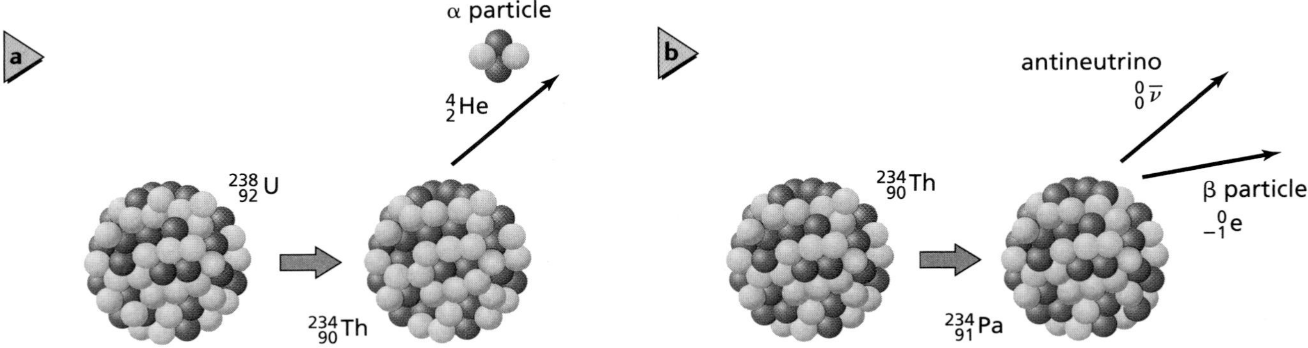

▶ EXAMPLE Problem 2

Alpha and Beta Decay **Write the nuclear equation for each radioactive process.**

a. **A radioactive radium isotope, $^{226}_{88}Ra$, emits an α particle and becomes the radon isotope $^{222}_{86}Rn$.**

b. **A radioactive lead isotope, $^{209}_{82}Pb$, decays into the bismuth isotope, $^{209}_{83}Bi$, by the emission of a β particle and an antineutrino.**

1 Analyze the Problem

Known:	Unknown:
a. $^{226}_{88}Ra \rightarrow \alpha$ particle $+ ^{222}_{86}Rn$	Is the decay possible?
α particle $= ^{4}_{2}He$	
b. $^{209}_{82}Pb \rightarrow ^{209}_{83}Bi + \beta$ particle + antineutrino	Is the decay possible?
β particle $= ^{0}_{-1}e$	
antineutrino $= ^{0}_{0}\bar{\nu}$	

2 Solve for the Unknown

a. $^{226}_{88}Ra \rightarrow ^{4}_{2}He + ^{222}_{86}Rn$ — Substitute $^{4}_{2}He$ for α particle

b. $^{209}_{82}Pb \rightarrow ^{209}_{83}Bi + ^{0}_{-1}e + ^{0}_{0}\bar{\nu}$ — Substitute $^{0}_{-1}e$ for β particle and $^{0}_{0}\bar{\nu}$ for antineutrino

3 Evaluate the Answer

- **Is the number of nucleons conserved?**
  - **a.** 226 = 222 + 4, so mass number is conserved.
  - **b.** 209 = 209 + 0 + 0, so mass number is conserved.
- **Is charge conserved?**
  - **a.** 88 = 86 + 2, so charge is conserved.
  - **b.** 82 = 83 − 1 + 0, so charge is conserved.

▶ PRACTICE Problems

• Additional Problems, Appendix B
• Solutions to Selected Problems, Appendix C

15. Write the nuclear equation for the transmutation of a radioactive uranium isotope, $^{234}_{92}U$, into a thorium isotope, $^{230}_{90}Th$, by the emission of an α particle.

16. Write the nuclear equation for the transmutation of a radioactive thorium isotope, $^{230}_{90}Th$, into a radioactive radium isotope, $^{226}_{88}Ra$, by the emission of an α particle.

17. Write the nuclear equation for the transmutation of a radioactive radium isotope, $^{226}_{88}Ra$, into a radon isotope, $^{222}_{86}Rn$, by α decay.

18. A radioactive lead isotope, $^{214}_{82}Pb$, can change to a radioactive bismuth isotope, $^{214}_{83}Bi$, by the emission of a β particle and an antineutrino. Write the nuclear equation.

19. A radioactive carbon isotope, $^{14}_{6}C$, undergoes β decay to become the nitrogen isotope $^{14}_{7}N$. Write the nuclear equation.

In alpha and beta decay, one nucleus, shown on the left of the equation, decays to another nucleus and one or more radioactive particles, shown on the right of the equation. Another example of transmutation occurs when a particle collides with the nucleus, often resulting in the emission of other particles, as in $^{12}_{6}C + ^{1}_{1}H \rightarrow ^{13}_{7}N$. Such reactions are illustrated in the next Example Problem, and in the discussion of fission later in the section.

▶ EXAMPLE Problem 3

Solve Nuclear Equations When nitrogen gas is bombarded with α particles, high-energy protons are emitted. What new isotope is created?

1 Analyze the Problem

Known:

nitrogen = ${}^{14}_{7}\text{N}$, $\alpha = {}^{4}_{2}\text{He}$

proton = ${}^{1}_{1}\text{H}$

Unknown:

What isotope belongs on the right side of the equation?

Math Handbook
Solving Equations
page 844

2 Solve for the Unknown

Write the equation for the nuclear reaction.

$${}^{4}_{2}\text{He} + {}^{14}_{7}\text{N} \rightarrow {}^{1}_{1}\text{H} + {}^{A}_{Z}X$$

Solve for Z and for A.

$Z = 2 + 7 - 1 = 8$ $\quad$ $A = 4 + 14 - 1 = 17$

Use the periodic table on page 916. The element with $Z = 8$ is oxygen.
The isotope must be ${}^{17}_{8}\text{O}$.

3 Evaluate the Answer

- **Does the equation balance?** The number of nucleons is conserved: $4 + 14 = 1 + 17$. The charge is conserved: $2 + 7 = 1 + 8$.

▶ PRACTICE Problems

- Additional Problems, Appendix B
- Solutions to Selected Problems, Appendix C

20. Use the periodic table on page 916 to complete the following.

a. ${}^{14}_{6}\text{C} \rightarrow ? + {}^{0}_{-1}\text{e} + {}^{0}_{0}\bar{\nu}$

b. ${}^{55}_{24}\text{Cr} \rightarrow ? + {}^{0}_{-1}\text{e} + {}^{0}_{0}\bar{\nu}$

21. Write the nuclear equation for the transmutation of a seaborgium isotope, ${}^{263}_{106}\text{Sg}$, into a rutherfordium isotope, ${}^{259}_{104}\text{Rf}$, by the emission of an alpha particle.

22. A proton collides with the nitrogen isotope ${}^{15}_{7}\text{N}$, forming a new isotope and an alpha particle. What is the isotope? Write the nuclear equation.

23. Write the nuclear equations for the beta decay of the following isotopes.

a. ${}^{210}_{82}\text{Pb}$ $\quad$ **b.** ${}^{210}_{83}\text{Bi}$ $\quad$ **c.** ${}^{234}_{90}\text{Th}$ $\quad$ **d.** ${}^{239}_{93}\text{Np}$

Half-Life

The time required for half of the atoms in any given quantity of a radioactive isotope to decay is the **half-life** of that element. After each half-life, the number of undecayed nuclei is cut in half, as shown in **Figure 30-4.** Each particular isotope has its own half-life.

For example, the half-life of the radium isotope ${}^{226}_{88}\text{Ra}$ is 1600 years. That is, in 1600 years, half of a given quantity of ${}^{226}_{88}\text{Ra}$ decays into another element, radon. In a second 1600 years, half of the remaining sample of radium will have decayed. In other words, one-fourth of the original amount still will remain after 3200 years. In contrast, a sample of polonium-210 will decay to one-fourth the original amount in just 276 days.

■ **Figure 30-4** Radioactive nuclei decay to more stable forms.

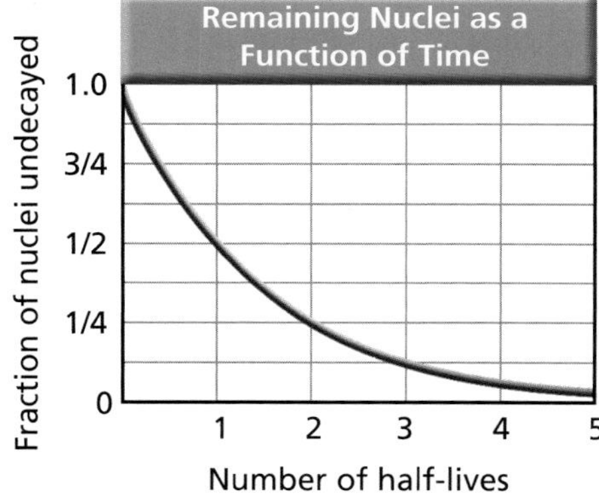

Table 30-2			
Half-Lives of Selected Isotopes			
Element	**Isotope**	**Half-Life**	**Radiation Produced**
Hydrogen	$^{3}_{1}H$	12.3 years	β
Carbon	$^{14}_{6}C$	5730 years	β
Cobalt	$^{60}_{27}Co$	30 years	β, γ
Iodine	$^{131}_{53}I$	8.07 days	β, γ
Lead	$^{212}_{82}Pb$	10.6 hours	β
Polonium	$^{194}_{84}Po$	0.7 seconds	α
Polonium	$^{210}_{84}Po$	138 days	α, γ
Uranium	$^{235}_{92}U$	7.1×10^{8} years	α, γ
Uranium	$^{238}_{92}U$	4.51×10^{9} years	α, γ
Plutonium	$^{236}_{94}Pu$	2.85 years	α
Plutonium	$^{242}_{94}Pu$	3.79×10^{5} years	α, γ

The half-lives of selected isotopes are shown in **Table 30-2.** If you know the original amount of a radioactive substance and its half-life, you can calculate the amount remaining after a given number of half-lives.

Half-Life $\text{remaining} = \text{original}\left(\frac{1}{2}\right)^{t}$

The amount of a radioactive isotope remaining in a sample equals the original amount times $\frac{1}{2}$ to the t, where t is the number of half-lives that have passed.

Half-lives of radioactive isotopes are used to date objects. The age of a sample of organic material can be found by measuring the amount of carbon-14 remaining. The age of Earth was calculated based on the decay of uranium into lead.

Personal Tutor For an online tutorial on half-life, visit physicspp.com.

The decay rate, or number of decays per second, of a radioactive substance is called its **activity.** Activity is proportional to the number of radioactive atoms present. Therefore, the activity of a particular sample also is reduced by one-half in one half-life. Consider $^{131}_{53}I$, with a half-life of 8.07 days. If the activity of a certain sample of iodine-131 is 8×10^{5} decays/s, then 8.07 days later, its activity will be 4×10^{5} decays/s. After another 8.07 days, its activity will be 2×10^{5} decays/s. The activity of a sample also is related to its half-life. The shorter the half-life, the higher the activity. Consequently, if you know the activity of a substance and the amount of that substance, you can determine its half-life. The SI unit for decays per second is a Becquerel (Bq).

PRACTICE Problems

• Additional Problems, Appendix B
• Solutions to Selected Problems, Appendix C

Refer to Figure 30-4 and Table 30-2 to solve the following problems.

24. A sample of 1.0 g of tritium, $^{3}_{1}H$, is produced. What will be the mass of the tritium remaining after 24.6 years?

25. The isotope $^{238}_{93}Np$ has a half-life of 2.0 days. If 4.0 g of neptunium is produced on Monday, what will be the mass of neptunium remaining on Tuesday of the next week?

26. A sample of polonium-210 is purchased for a physics class on September 1. Its activity is 2×10^{6} Bq. The sample is used in an experiment on June 1. What activity can be expected?

27. Tritium, $^{3}_{1}H$, once was used in some watches to produce a fluorescent glow so that the watches could be read in the dark. If the brightness of the glow is proportional to the activity of the tritium, what would be the brightness of such a watch, in comparison to its original brightness, when the watch is six years old?

Artificial Radioactivity

Radioactive isotopes can be formed from stable isotopes by bombardment with α particles, protons, neutrons, electrons, or gamma rays. The resulting unstable nuclei emit radiation until they are converted into stable isotopes. The radioactive nuclei may emit alpha, beta, and gamma radiation, as well as neutrinos, antineutrinos, and positrons. (A positron is a positively charged electron, ${}^{0}_{+1}\text{e}$.)

Artificially produced radioactive isotopes often are used in medicine and medical research. In many medical applications, patients are given radioactive isotopes of elements that are absorbed by specific parts of the body. A physician uses a radiation counter to monitor the activity in the region in question. A radioactive isotope also can be attached to a molecule that will be absorbed in the area of interest, as is done in positron emission tomography, better known as the PET scan. A PET scan of a brain is shown in **Figure 30-5.**

Radiation often is used to destroy cancer cells. These cells are more sensitive to the damaging effects of radiation because they divide more often than normal cells. Gamma rays from the isotope ${}^{60}_{27}\text{Co}$ are used to treat cancer patients. Radioactive iodine is injected to target thyroid cancer. In a third method, particles produced in a particle accelerator are beamed into tissue in such a way that they decay in the cancerous tissue and destroy the cells.

APPLYING PHYSICS

Radiation Treatment Gamma rays destroy both cancerous cells and healthy cells; thus, the beams of radiation must be directed only at cancerous cells.

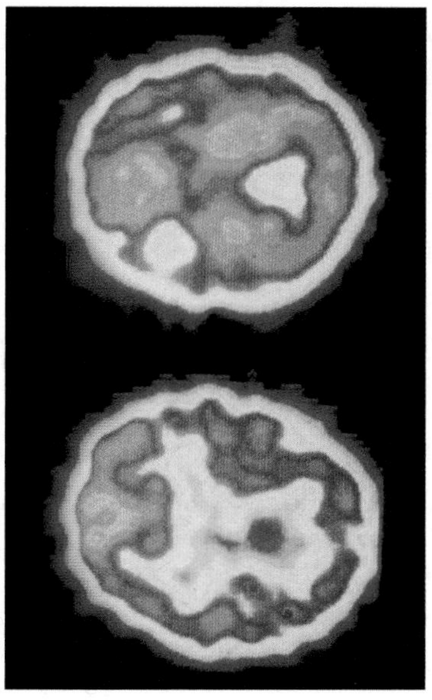

Figure 30-5 For a PET scan, physicians inject a solution in which a radioactive isotope, such as ${}^{18}_{9}\text{F}$, is attached to a molecule that will concentrate in the target tissues. When the ${}^{18}_{9}\text{F}$ decays, it produces positrons, which annihilate nearby electrons, producing gamma rays. The PET scanner detects the gamma rays. A computer then makes a three-dimensional map of the isotope distribution. Here, a normal brain (top), and the brain of a person suffering from dementia (bottom) are contrasted.

Nuclear Fission

The possibility of obtaining useful forms of energy from nuclear reactions was discussed in the 1930s. The most promising results came from bombarding substances with neutrons. In Italy, in 1934, Enrico Fermi and Emilio Segré produced many new radioactive isotopes by bombarding uranium with neutrons. German chemists Otto Hahn and Fritz Strassmann showed in 1939 that the resulting atoms acted chemically like barium. One week later, Lise Meitner and Otto Frisch proposed that the neutrons had caused a division of the uranium into two smaller nuclei, resulting in a large release of energy. Such a division of a nucleus into two or more fragments is called **fission.** The possibility that fission could be not only a source of energy, but also an explosive weapon, was immediately realized by many scientists.

Fission occurs when a nucleus is divided into two or more fragments, releasing neutrons and energy. The uranium isotope, ${}^{235}_{92}\text{U}$, undergoes fission when it is bombarded with neutrons. The elements barium and krypton are typical results of fission. The reaction is illustrated with the following equation.

$$ {}^{1}_{0}\text{n} + {}^{235}_{92}\text{U} \rightarrow {}^{92}_{36}\text{Kr} + {}^{141}_{56}\text{Ba} + 3{}^{1}_{0}\text{n} + 200\ \text{MeV} $$

The energy released by each fission can be found by calculating the masses of the atoms on each side of the equation. In the uranium-235 reaction, the total mass on the right side of the equation is 0.215 u smaller than that on the left. The energy equivalent of this mass is 3.21×10^{-11} J, or 2.00×10^{2} MeV. This energy appears as the kinetic energy of the products of the fission.

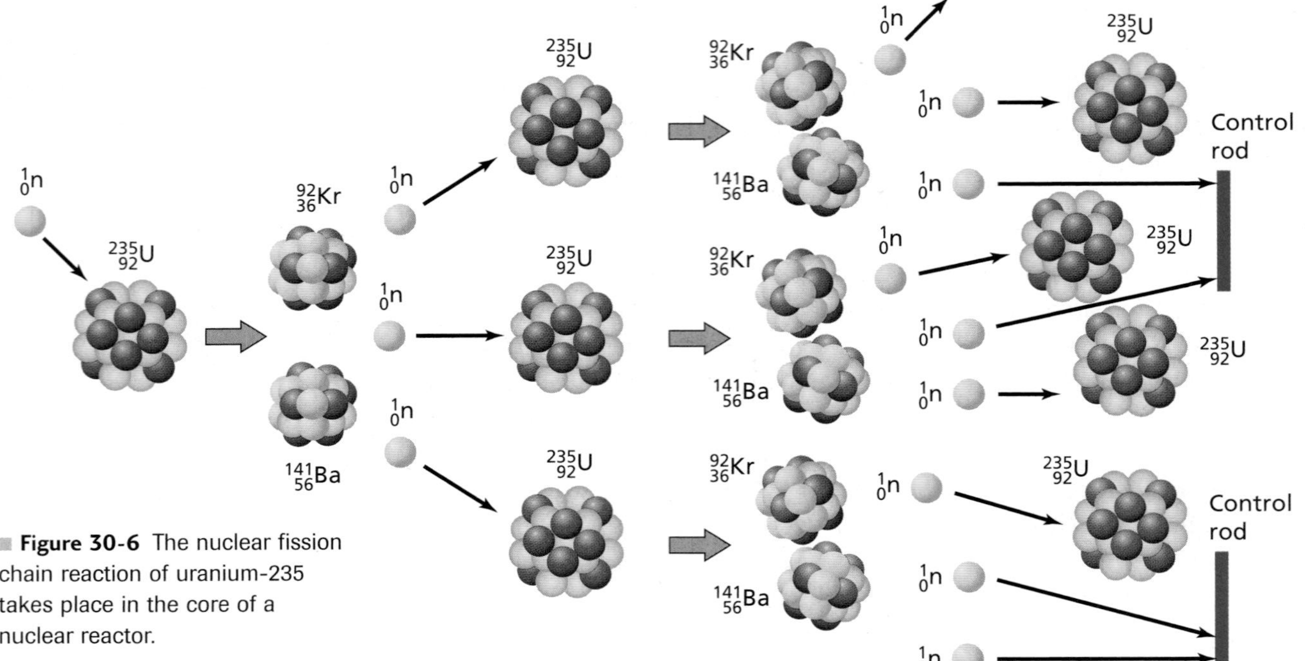

■ **Figure 30-6** The nuclear fission chain reaction of uranium-235 takes place in the core of a nuclear reactor.

Concepts In Motion

Interactive Figure To see an animation on chain reaction, visit physicspp.com.

The neutrons needed to cause the fission of additional $^{235}_{92}$U nuclei can be the neutrons produced once the fission process is started. If one or more of the neutrons cause a fission, that fission releases three more neutrons, each of which can cause more fission. This continual process of repeated fission reactions caused by the release of neutrons from previous fission reactions is called a **chain reaction.** The process is illustrated in **Figure 30-6.**

Nuclear Reactors

To create a controlled chain reaction and make use of the energy produced, the neutrons need to interact with the fissionable uranium at the right rate. Most of the neutrons released by the fission of $^{235}_{92}$U atoms are moving at high speeds. These are called fast neutrons. In addition, naturally occurring uranium consists of less than one percent $^{235}_{92}$U and more than 99 percent $^{238}_{92}$U. When a $^{238}_{92}$U nucleus absorbs a fast neutron, it does not undergo fission, but becomes a new isotope, $^{239}_{92}$U. The absorption of neutrons by $^{238}_{92}$U keeps most of the neutrons from reaching the fissionable $^{235}_{92}$U atoms. Thus, most neutrons released by the fission of $^{235}_{92}$U are unable to cause the fission of another $^{235}_{92}$U atom.

To control the reaction, the uranium is broken up into small pieces and placed in a *moderator,* a material that can slow down, or moderate, the fast neutrons. When a neutron collides with a light atom it transfers momentum and energy to the atom. In this way, the neutron loses energy. The moderator thus slows many fast neutrons to speeds at which they can be absorbed more easily by $^{235}_{92}$U than by $^{238}_{92}$U. The larger number of slow neutrons greatly increases the probability that a neutron released by the fission of a $^{235}_{92}$U nucleus will cause another $^{235}_{92}$U nucleus to fission. If there is enough $^{235}_{92}$U in the sample, a chain reaction can occur. To increase the amount of fissionable uranium, the uranium may be enriched by adding more $^{235}_{92}$U. Both types of uranium are used in nuclear reactors.

■ **Figure 30-7** The glow is due to the Cerenkov effect, which occurs when high speed particles entering the water exceed the speed of light in water. The electrons emit photons, which cause the water to glow when fuel rods are placed in it. This glow is not the result of radioactivity.

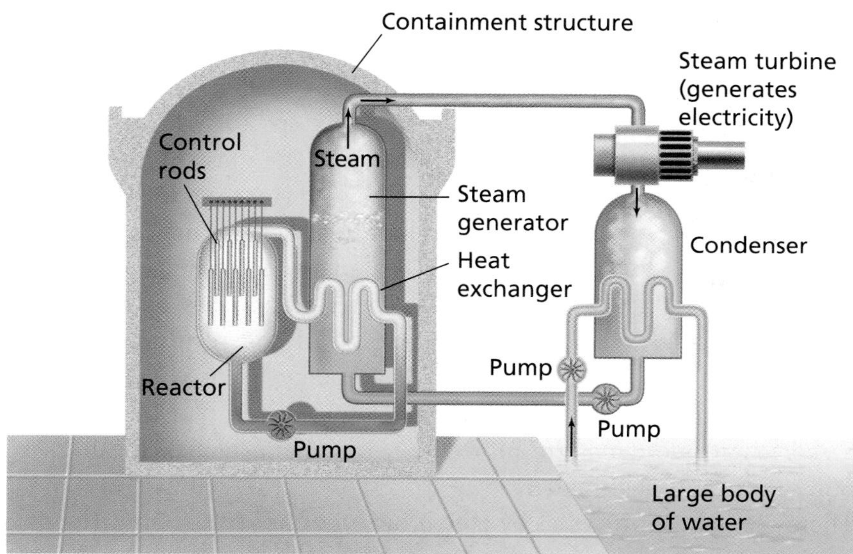

Figure 30-8 In a nuclear power plant, the thermal energy released in nuclear reactions is converted to electric energy.

The type of nuclear reactor used in the United States, the pressurized water reactor, contains about 200 metric tons of uranium sealed in hundreds of metal rods. The rods are immersed in water, as shown in **Figure 30-7.** Water not only is the moderator, but also transfers thermal energy away from the fission of uranium. Rods of cadmium metal are placed between the uranium rods. Cadmium absorbs neutrons easily and also acts as a moderator. The cadmium rods are moved in and out of the reactor to control the rate of the chain reaction. Thus, the rods are called control rods. When the control rods are inserted completely into the reactor, they absorb enough of the neutrons released by the fission reactions to prevent any further chain reaction. As the control rods are removed from the reactor, the rate of energy release increases, with more free neutrons available to continue the chain reaction.

Energy released by the fission heats the water surrounding the uranium rods. The water itself doesn't boil because it is under high pressure, which increases its boiling point. As shown in **Figure 30-8,** this water is pumped to a heat exchanger, where it causes other water to boil, producing steam that turns turbines. The turbines are connected to generators that produce electrical energy.

Fission of $^{235}_{92}U$ nuclei produces Kr, Ba, and other atoms in the fuel rods. Most of these atoms are radioactive. About once a year, some of the uranium fuel rods must be replaced. The old rods no longer can be used in the reactor, but they are still extremely radioactive and must be stored in a location that can be secured. Methods of permanently storing these radioactive waste products currently are being developed.

Nuclear Fusion

In nuclear **fusion,** nuclei with small masses combine to form a nucleus with a larger mass, as shown in **Figure 30-9** on the next page. In the process, energy is released. You learned earlier in this chapter that the larger nucleus is more tightly bound, so its mass is less than the sum of the masses of the smaller nuclei. This loss of mass corresponds to a release of energy.

MINI LAB

Modeling Radioactive Decay

You will need 50 pennies to represent 50 atoms of a radioactive isotope. In this simulation, heads indicates that a nucleus has not decayed.

1. Record 50 heads as the starting point.

2. Shake all the pennies in a large cup and pour them out. Remove all the tails and set them aside. Count and record the number of heads.

3. Repeat step 2 with the pennies that were heads in the last throw. Each throw simulates one half-life.

Analyze and Conclude

4. Graph the number of pennies as a function of the number of half-lives.

5. Collect the results from other students and use the totals to make a new graph.

6. Compare this graph with the individual ones. Which more closely matches the theoretical graph in Figure 30-4?

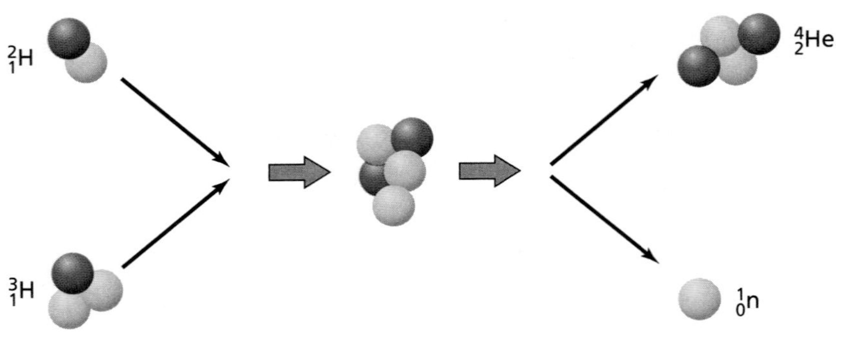

■ **Figure 30-9** The fusion of deuterium and tritium produces helium. Protons are red and neutrons are gray in the figure.

An example of fusion is the process that occurs in the Sun. Four hydrogen nuclei (protons) fuse in several steps to form one helium nucleus. The mass of the four protons is greater than the mass of the helium nucleus that is produced. The energy equivalent of this mass difference appears as the kinetic energy of the resultant particles. The energy released by the fusion that produces one helium-4 nucleus is 25 MeV. In comparison, the energy released when one dynamite molecule reacts chemically is about 20 eV, about 1 million times smaller.

There are several processes by which fusion occurs in the Sun. The most important process is the proton-proton chain.

$$^1_1H + ^1_1H \rightarrow ^2_1H + {}^0_{+1}e + ^0_0\nu$$
$$^1_1H + ^2_1H \rightarrow ^3_2He + \gamma$$
$$^3_2He + ^3_2He \rightarrow ^4_2He + 2^1_1H$$

The first two reactions must occur twice to produce the two ^3_2He particles needed for the final reaction. The net result (subtracting out the two protons produced in the final step) is that four protons produce one ^4_2He, two positrons, and two neutrinos.

The repulsive force between the charged nuclei requires the fusing nuclei to have high energies. Thus, fusion reactions take place only when the nuclei have large amounts of thermal energy. The proton-proton chain requires a temperature of about 2×10^7 K, such as that found in the center of the Sun. Fusion reactions also occur in a hydrogen, or thermonuclear, bomb. In this device, the high temperature necessary to produce the fusion reaction is produced by exploding a uranium fission, or atomic, bomb.

30.2 Section Review

28. Beta Decay How can an electron be expelled from a nucleus in beta decay if the nucleus has no electrons?

29. Nuclear Reactions The polonium isotope $^{210}_{84}Po$ undergoes alpha decay. Write the equation for the reaction.

30. Half-Life Use Figure 30-4 and Table 30-2 to estimate in how many days a sample of $^{131}_{53}I$ would have three-eighths its original activity.

31. Nuclear Reactor Lead often is used as a radiation shield. Why is it not a good choice for a moderator in a nuclear reactor?

32. Fusion One fusion reaction involves two deuterium nuclei, ^2_1H. A deuterium molecule contains two deuterium atoms. Why doesn't this molecule undergo fusion?

33. Energy Calculate the energy released in the first fusion reaction in the Sun, $^1_1H + ^1_1H \rightarrow ^2_1H + {}^0_{+1}e + ^0_0\nu$.

34. Critical Thinking Alpha emitters are used in smoke detectors. An emitter is mounted on one plate of a capacitor, and the α particles strike the other plate. As a result, there is a potential difference across the plates. Explain and predict which plate has the more positive potential.

Physics Online physicspp.com/self_check_quiz

30.3 The Building Blocks of Matter

The first physicists who studied the nucleus with high speed particles had to make do with α particles from radioactive sources. Other experimenters used cosmic rays, produced by not-yet-understood processes in stars and galaxies. In the early 1930s, the first laboratory devices that could accelerate protons and α particles to energies high enough to penetrate the nucleus were developed. Two devices, the linear accelerator and the synchrotron, are regularly used today.

► **Objectives**
- **Describe** the operation of particle accelerators and particle detectors.
- **Describe** the Standard Model of matter and explain the role of force carriers.

► **Vocabulary**

quarks
leptons
Standard Model
force carriers
pair production
weak nuclear force

Linear Accelerators

A linear accelerator can be used to accelerate protons or electrons. It consists of a series of hollow tubes within a long, evacuated chamber. The tubes are connected to a source of high-frequency alternating voltage, as illustrated in **Figure 30-10.** Protons are produced in an ion source similar to that described in Chapter 26. When the first tube has a negative potential, protons are accelerated into it. There is no electric field within the tube, so the protons move at a constant velocity. The length of the tube and the frequency of the voltage are adjusted so that when the protons have reached the far end of the tube, the potential of the second tube is negative in relation to that of the first. The resulting electric field in the gap between the tubes accelerates the protons into the second tube. This process continues, with the protons receiving an acceleration between each pair of tubes. The energy of the protons is increased by 10^5 eV with each acceleration. The protons ride along the crest of an electric field wave, much as a surfboard moves on the ocean. At the end of the accelerator, the protons can have energies of many millions or billions of electron volts. A similar method is used to accelerate electrons. Note that one can accelerate only charged particles.

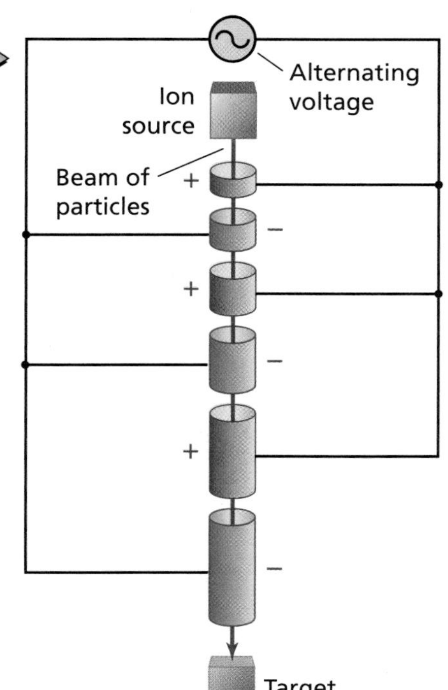

Figure 30-10 The linear accelerator at Stanford University is 3.3 km long and accelerates electrons to energies of 50 GeV **(a).** Protons in a linear accelerator are accelerated by changing the charge on the tubes as the protons move **(b).** (Not to scale)

Concepts In Motion
Interactive Figure To see an animation on a linear accelerator, visit physicspp.com.

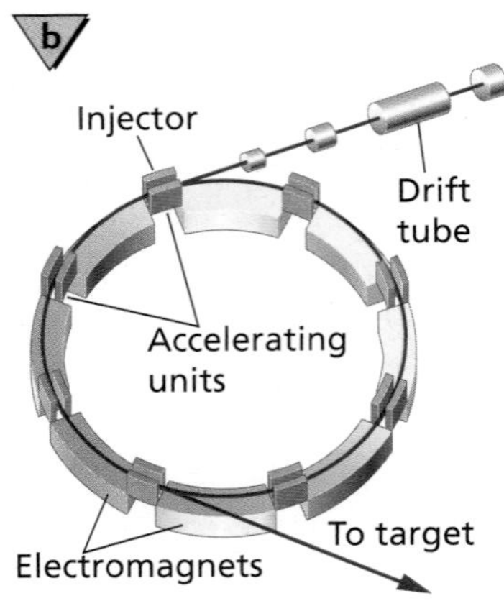

Figure 30-11 Fermi Laboratory's synchrotron has a diameter of 2 km **(a).** The synchrotron is a circular accelerator. Magnets are used to control the path and acceleration of the particles **(b).**

The Synchrotron

An accelerator may be made smaller by using a magnetic field to bend the path of the particles into a circle. In a synchrotron, the bending magnets are separated by accelerating regions, as shown in **Figure 30-11b.** In the straight regions, high-frequency alternating voltage accelerates the particles. The strength of the magnetic field and the length of the path are chosen so that the particles reach the location of the alternating electric field precisely when the field's polarity will accelerate them. One of the largest synchrotrons in operation is at the Fermi National Accelerator Laboratory near Chicago, shown in **Figure 30-11a.** Protons there reach energies of 1 TeV (10^{12} eV). A proton beam and an antiproton beam travel the circle in opposite directions. (An antiproton is the antimatter pair to a proton; it has the same mass and opposite charge.) The beams collide in several interaction regions and the results are studied.

Particle Detectors

Once particles are produced, the results of the collision need to be detected. In other words, they need to interact with matter in such a way that we can sense them with our relatively limited human senses. Your hand will stop an α particle, yet you will have no idea that the particle struck you. And as you read this sentence, billions of solar neutrinos pass through your body without so much as a twitch from you. Over the past century, scientists have devised tools to detect and distinguish the products of nuclear reactions.

In the last section, you learned that uranium samples fogged photographic plates. When α particles, β particles, or gamma rays strike photographic film, the film becomes fogged, or exposed. Thus, photographic film can be used to detect radiation. Many other devices are used to detect charged particles and gamma rays. Most of these devices make use of the fact that a collision with a high-speed particle will remove electrons from atoms. That is, the high-speed particles ionize the matter that they bombard. In addition, some substances fluoresce, or emit photons, when they are exposed to certain types of radiation. Thus, fluorescent substances also can be used to detect radiation. These three means of detecting radiation are illustrated in **Figure 30-12.**

Figure 30-12 Particles can be detected when they interact with matter, exposing film, charging the matter, or causing the matter to emit a photon.

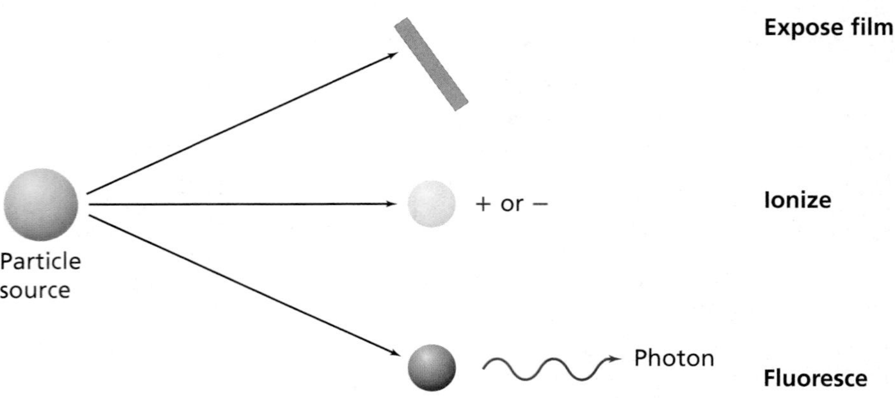

Geiger counter A Geiger-Mueller tube contains a copper cylinder with a negative charge. Down the center of this cylinder runs a rigid wire with a positive charge. The voltage across the wire and cylinder is kept just below the point at which a spontaneous discharge, or spark, occurs. When a charged particle or gamma ray enters the tube, it ionizes a gas atom between the copper cylinder and the wire. The positive ion that is produced is accelerated toward the copper cylinder by the potential difference, and the electron is accelerated toward the positive wire. As these charged particles move toward the electrodes, they create an avalanche of charged particles, and a pulse of current goes through the tube.

Condensation trails A device once used to detect particles was the Wilson cloud chamber. The chamber contained an area supersaturated with water vapor or ethanol vapor. When charged particles traveled through the chamber, leaving a trail of ions in their paths, vapor condensed into small droplets on the ions. In this way, visible trails of droplets, or fog, were formed. In a similar detector still used today, the bubble chamber, charged particles pass through a liquid held just above the boiling point. In this case, the trails of ions cause small vapor bubbles to form, marking the particles' trajectories, as shown in **Figure 30-13.**

Figure 30-13 This false-color bubble chamber photograph shows the tracks of charged particles.

Recent technology has produced detection chambers called wire chambers that are like giant Geiger-Mueller tubes. Huge plates are separated by a small gap filled with a low-pressure gas. A discharge is produced in the path of a particle passing through the chamber. A computer locates the discharge and records its position for later analysis.

Neutral particles do not leave trails because they do not produce discharges. The laws of conservation of energy and momentum in collisions can be used to tell if any neutral particles were produced. Other detectors measure the energy of the particles. The entire array of detectors used in high-energy accelerator experiments, such as the Collider Detector at Fermilab (CDF), can be up to three stories high, as shown in **Figure 30-14a.** The CDF is designed to monitor a quarter-million particle collisions each second, as though the detector functioned as a 5000-ton camera, creating a computer picture of the collision events, as shown in **Figure 30-14b.**

Figure 30-14 The Collider Detector at Fermilab (CDF) records the tracks from billions of collisions **(a).** A CDF computer image of a top quark event is shown **(b).**

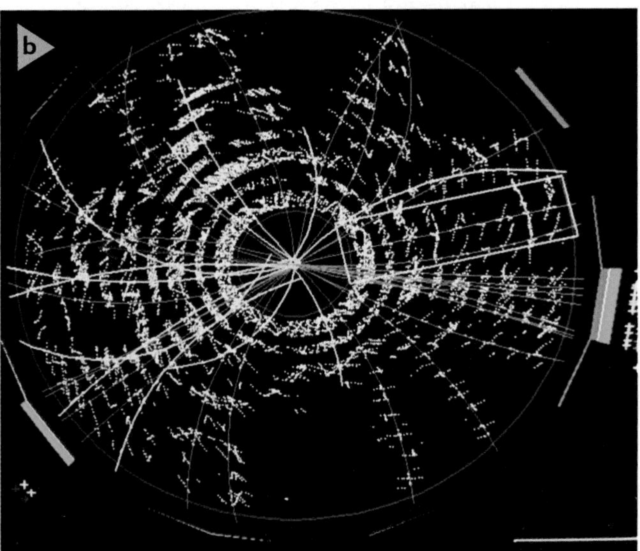

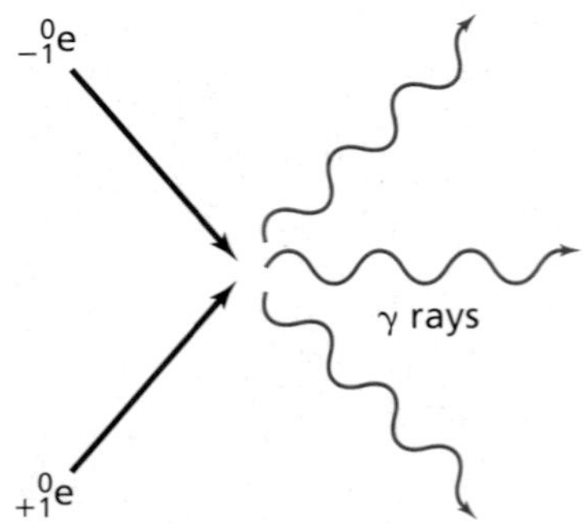

Figure 30-15 The collision of a positron and an electron results in gamma ray production.

Antimatter

In the late 1920s, Paul Dirac pradicted the existence of an antiparticle associated with each kind of particle. The positive electron, called a positron, is an example of an antiparticle, a particle of antimatter. The electron and positron have the same mass and charge magnitude; however, the signs of their charges are opposite. When a positron and an electron collide, the two can annihilate each other, resulting in energy in the form of gamma rays, as shown in **Figure 30-15.** The positron was not discovered until 1932.

Particles

The model of the atom in 1930 was fairly simple: protons and neutrons surrounded by electrons. More detailed studies of radioactive decay disturbed this simple picture. While the α particles and gamma rays emitted by radioactive nuclei have single energies that depend on the decaying nucleus, β particles are emitted with a wide range of energies. One might expect the energy of the β particles to be equal to the difference between the energy of the nucleus before decay and the energy of the nucleus produced by the decay. In fact, the wide range of energies of electrons emitted during beta decay suggested to Niels Bohr that another particle might be involved in nuclear reactions that carries away energy. Wolfgang Pauli in 1931 and Enrico Fermi in 1934 suggested that an unseen neutral particle was emitted with the β particle. Named the neutrino ("little neutral one" in Italian) by Fermi, the particle, which is actually an antineutrino, was not directly observed until 1956.

Other studies discovered more particles. The muon, which seemed to be a heavy electron, was discovered in 1937. In 1935, a remarkable hypothesis by Japanese physicist Hideki Yukawa spurred much research in the years to follow. Yukawa hypothesized the existence of a new particle that could carry the nuclear force through space, just as the photon carries the electromagnetic force. In 1947 a possible particle, the pion, was dicovered. While it was not the carrier of the strong force, it was a new type of matter.

Experiments with particle accelerators resulted in the identification of more and more particles, some with intermediate masses, others much more massive than the proton. They had positive and negative charges, or none at all. Some lifetimes were 10^{-23} s, while others had no detectable decays. At one point Enrico Fermi, asked to identify a particle track, replied "If I could remember the names of all these particles, I'd be a botanist!"

Figure 30-16 Even though quarks have fractional charges, all the particles they make have whole-number charges.

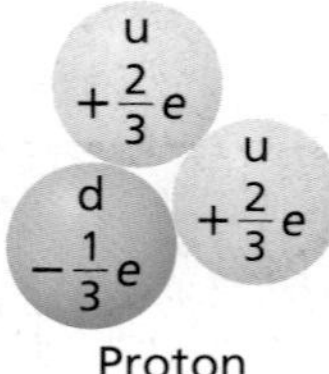

Proton

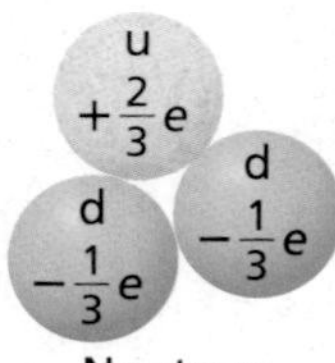

Neutron

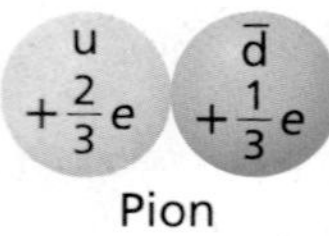

Pion

The Standard Model

By the late 1960s it became clear that neither protons, nor neutrons, nor pions are elementary particles. They are made up of a family of particles called **quarks,** as shown in **Figure 30-16.** Electrons and neutrinos belong to a different family, called **leptons.** Physicists now believe that there are three families of elementary particles: quarks, leptons, and **force carriers,** also called gauge bosons. This model of the building blocks of matter is called the **Standard Model.** Particles such as protons and neutrons, made of three quarks, are called baryons. A pair made of a

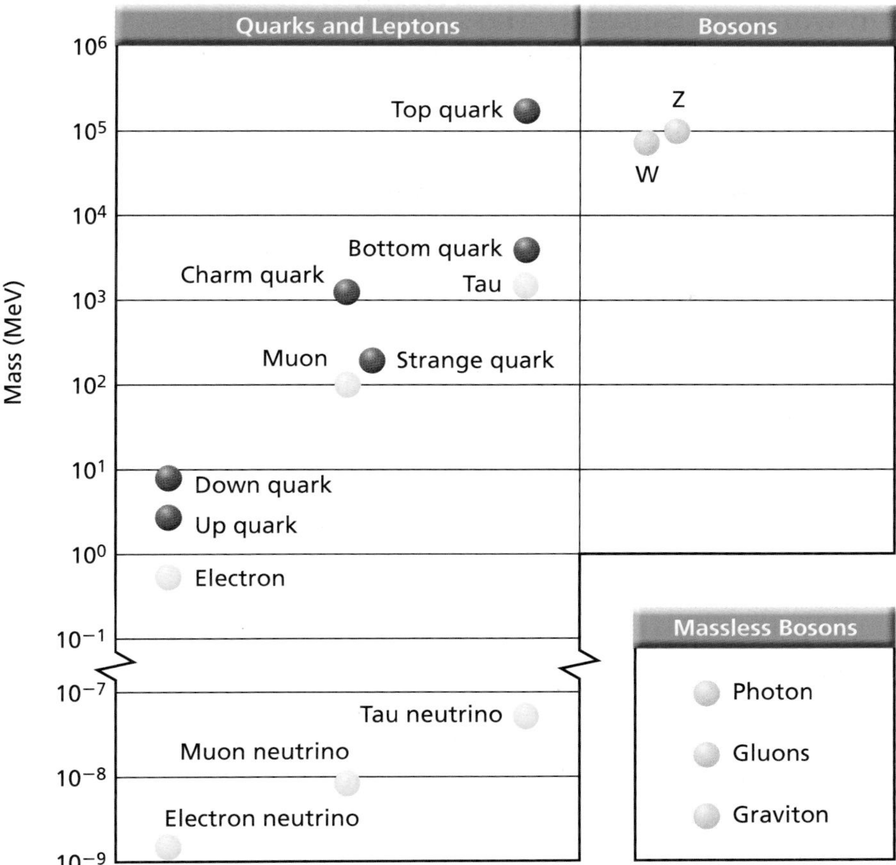

Figure 30-17 The known quarks and leptons are divided into three families. The everyday world is made from particles in the left-hand family (u, d, e). Particles in the middle group (c, s, μ) are found in cosmic rays and are routinely produced in particle accelerators. Particles in the right-hand family (b, t, τ) are believed to have existed briefly during the earliest moments of the Big Bang and are created in high-energy collisions. The gauge bosons carry the weak, electromagnetic, strong, and gravitational forces. Masses are stated as energy equivalents, given by Einstein's formula $E = mc^2$.

quark and an antiquark, like the pion, is called a meson. A new type of particle, made of four quarks and one antiquark, is called a pentaquark and might have been observed recently. There are six quarks and six leptons. Quarks and leptons form matter, while force carriers are particles that transmit forces. For example, photons carry the electromagnetic force. The eight gluons carry the strong nuclear force that binds quarks in baryons and mesons. Three weak gauge bosons are involved in β decay. The graviton is the name given to the yet-undetected carrier of gravitational force. The properties of elementary particles, which are the basis of the Standard Model, are summarized in **Figure 30-17.**

Protons and Neutrons

The quark model describes nucleons, the proton and the neutron, as an assembly of quarks. The nucleons each are made up of three quarks. The proton has two up quarks, u, (charge $+\frac{2}{3}e$) and one down quark, d, (charge $-\frac{1}{3}e$). A proton is described as p = uud. The charge on the proton is the sum of the charges of the three quarks, $(\frac{2}{3} + \frac{2}{3} + -\frac{1}{3})e = +e$. The neutron is made up of one up quark and two down quarks, n = udd. The charge of the neutron is zero, $(\frac{2}{3} + -\frac{1}{3} + -\frac{1}{3})e = 0$.

Individual free quarks cannot be observed, because the strong force that holds them together becomes larger as the quarks are pulled farther apart. In this case, the strong force acts like the force of a spring. It is unlike the electric force, which becomes weaker as charged particles are moved farther apart. In the quark model, the strong force is transmitted by gluons.

Conversions Between Mass and Energy

The amount of energy created in particle annihilation can be calculated using Einstein's equation for the energy equivalent of mass, $E = mc^2$. The mass of the electron is 9.11×10^{-31} kg. The mass of the positron is the same. Therefore, the energy equivalent of the positron and the electron together can be calculated as follows.

$$E = 2(9.11\times10^{-31}\text{ kg})(3.00\times10^{8}\text{ m/s})^2$$
$$E = (1.64\times10^{-13}\text{ J})(1\text{ eV} / 1.60\times10^{-19}\text{ J})$$
$$E = 1.02\times10^{6}\text{ eV or } 1.02\text{ MeV}$$

When a positron and an electron at rest annihilate each other, the sum of the energies of the gamma rays emitted is 1.02 MeV.

The inverse of annihilation also can occur. That is, energy can be converted directly into matter. If a gamma ray with at least 1.02 MeV of energy passes close by a nucleus, a positron and electron pair can be produced.

$$\gamma \rightarrow e^- + e^+$$

The conversion of energy into a matter-antimatter pair of particles is called **pair production.** Individual reactions, such as $\gamma \rightarrow e^-$ and $\gamma \rightarrow e^+$, however, cannot occur, because such events would violate the law of conservation of charge. Reactions such as $\gamma \rightarrow e^-$ + proton also do not occur; the pair must be a particle and its corresponding antiparticle.

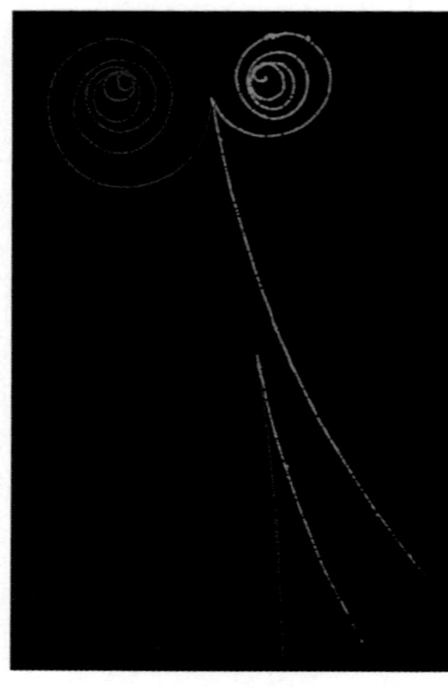

Figure 30-18 When a particle is produced, its corresponding antimatter particle is also produced. Here, a gamma ray decays into an electron-positron pair.

Matter and antimatter particles come in pairs The production of a positron-electron pair is shown in **Figure 30-18.** A magnetic field around the bubble chamber causes the oppositely charged particles' tracks to curve in opposite directions. The gamma ray produced no track. If the gamma ray's energy is larger than 1.02 MeV, the excess energy goes into the kinetic energy of the positron and electron. The positron soon collides with another electron and they are both annihilated, resulting in two or three gamma rays with a total energy of no less than 1.02 MeV.

Particle conservation Each quark and each lepton also has its antiparticle. The antiparticles are identical to the particles except that for charged particles, an antiparticle will have the opposite charge. For example, an up quark, u, has a charge of $+\frac{2}{3}$, while an anti-up quark, $\bar{u}$, has a charge of $-\frac{2}{3}$. A proton, uud, has a charge of +1 and the antiproton, $\bar{u}\bar{u}\bar{d}$, has a charge of $-\frac{2}{3} - \frac{2}{3} + \frac{1}{3} = -1$. When a particle and its antiparticle collide, they annihilate each other and are transformed into photons, or lighter particle-antiparticle pairs and energy. The total number of quarks and the total number of leptons in the universe is constant. That is, quarks and leptons are created or destroyed only in particle-antiparticle pairs. On the other hand, force carriers such as gravitons, photons, gluons, and weak bosons can be created or destroyed if there is enough energy.

Antiprotons also can be created. An antiproton has a mass equal to that of the proton but is negatively charged. Protons have 1836 times as much mass as electrons. Thus, the energy needed to create proton-antiproton pairs is comparably larger. The first proton-antiproton pair was produced and observed at Berkeley, California in 1955. Neutrons also have an antiparticle, called an antineutron.

PRACTICE Problems

• Additional Problems, Appendix B
• Solutions to Selected Problems, Appendix C

35. The mass of a proton is 1.67×10^{-27} kg.

a. Find the energy equivalent of the proton's mass in joules.

b. Convert this value to eV.

c. Find the smallest total γ-ray energy that could result in a proton-antiproton pair.

36. A positron and an electron can annihilate and form three gammas. Two gammas are detected. One has an energy of 225 keV, the other 357 keV. What is the energy of the third gamma?

37. The mass of a neutron is 1.008665 u.

a. Find the energy equivalent of the neutron's mass in MeV.

b. Find the smallest total γ-ray energy that could result in the production of a neutron-antineutron pair.

38. The mass of a muon is 0.1135 u. It decays into an electron and two neutrinos. What is the energy released in this decay?

Beta Decay and the Weak Interaction

The high energy electrons emitted in the beta decay of a radioactive nucleus do not exist in the nucleus. From where do they come? In the decay process a neutron is transformed into a proton. While in a stable nucleus, the neutron does not decay. A free neutron, or one in an unstable nucleus, however, can decay into a proton by emitting a β particle. Sharing the outgoing energy with the proton and β particle is an antineutrino, ${}^{0}_{0}\bar{\nu}$. The antineutrino has a very small mass and is uncharged, but like the photon, it carries momentum and energy. The neutron decay equation is written as follows.

$${}^{1}_{0}\text{n} \rightarrow {}^{1}_{1}\text{p} + {}^{0}_{-1}\text{e} + {}^{0}_{0}\bar{\nu}$$

When an isotope decays by emission of a positron, or antielectron, a process like beta decay occurs. While the decay of a free proton has never been observed, a proton within the nucleus can change into a neutron with the emission of a positron, ${}^{0}_{+1}\text{e}$, and a neutrino, ${}^{0}_{0}\nu$.

$${}^{1}_{1}\text{p} \rightarrow {}^{1}_{0}\text{n} + {}^{0}_{+1}\text{e} + {}^{0}_{0}\nu$$

The decay of neutrons into protons and protons into neutrons cannot be explained by the strong force. The existence of beta decay indicates that there must be another interaction, the **weak nuclear force,** acting in the nucleus. This force is much weaker than the strong nuclear force.

CHALLENGE PROBLEM

${}^{238}_{92}\text{U}$ decays by α emission and two successive β emissions back into uranium again.

1. Show the three nuclear decay equations.

2. Predict the atomic mass number of the uranium formed.

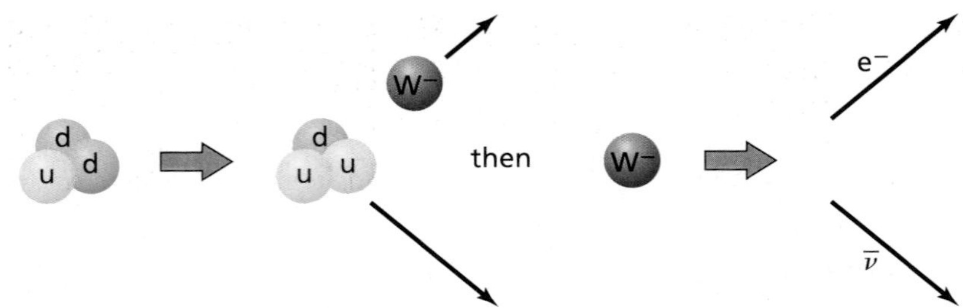

Figure 30-19 Beta decay of a neutron to a proton can be represented with the quark model: $d \rightarrow u + W^-$, then $W^- \rightarrow e^- + \bar{\nu}$.

Quark model of beta decay The difference between a proton, uud, and a neutron, udd, is only one quark. Beta decay in the quark model occurs in two steps, as shown in **Figure 30-19.** First, one d quark in a neutron changes to a u quark with the emission of a W^- boson. The W^- boson is one of the three carriers of the weak force. In the second step, the W^- boson decays into an electron and an antineutrino. Similarly, in the decay of a proton within a nucleus, a neutron and a W^+ boson would be emitted. The W^+ boson then decays into a positron and a neutrino.

The emission of the third weak force carrier, the Z^0 boson, is not accompanied by a change from one quark to another. The Z^0 boson produces an interaction between the nucleons and the electrons in atoms that is similar to, but much weaker than, the electromagnetic force holding the atom together. The interaction first was detected in 1979. The W^+, W^-, and Z^0 bosons first were observed directly in 1983.

While neutrinos and antineutrinos have long been thought to be massless, recent experiments that detect neutrinos from the Sun and from distant accelerators show that neutrinos do have mass, although these masses are much smaller than those of any other known particles.

Testing the Standard Model

You can see in Figure 30-16 that the quarks and leptons are separated into three families. The everyday world is made from particles in the left-hand family: protons, neutrons, and electrons. Particles in the middle group are found in cosmic rays and are routinely produced in particle accelerators. Particles in the right-hand family are believed to have existed briefly during the earliest moments of the Big Bang and are created in high-energy collisions.

What determines the masses of the quarks and leptons? The Higgs boson, which has been hypothesized as the particle that determines the masses of the leptons and quarks, has not yet been discovered. The Standard Model is not a theory; it does not explain the masses of particles, nor why there are three families of quarks and leptons.

Why are there four forces? The differences among the four fundamental interactions are evident: the forces may act on different quantities such as charge or mass, they may have different dependencies on distance, and the force carriers have different properties. There are, however, some similarities among the interactions. For instance, the force between charged particles, the electromagnetic interaction, is carried by photons in much the same way as weak bosons carry the weak interaction. The electric force acts over a long range because the photon has zero mass, while the weak force acts over short distances because the W and Z bosons are relatively massive. The mathematical structures of the theories of the weak interaction and electromagnetic interaction, however, are similar.

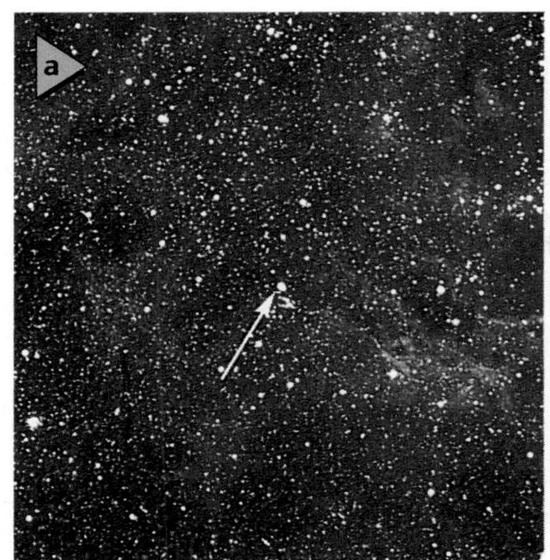

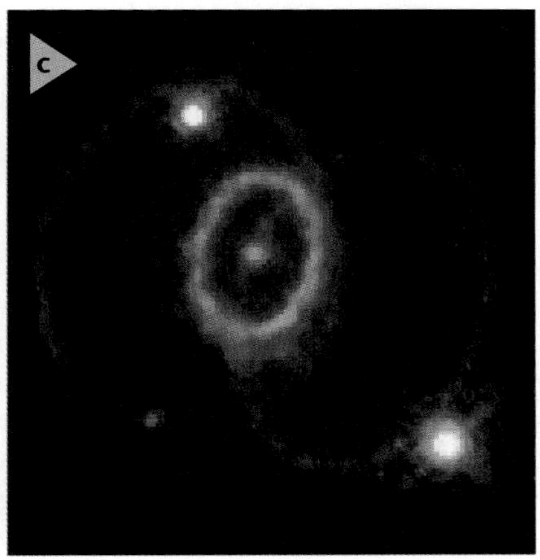

Figure 30-20 In a supernova, the electromagnetic and weak forces are indistinguishable. The increased light and neutrinos from supernova 1987A, shown here, reached Earth at the same time, demonstrating that neutrinos travel near the speed of light and are produced in supernovas, as predicted. Supernova 1987A is shown before exploding **(a),** during explosion **(b),** and in a close-up from *Hubble* **(c).**

Astrophysical theories of supernovae indicate that during massive stellar explosions, such as the one shown in **Figure 30-20,** the two interactions are identical. Present theories of the origin of the universe suggest that the two forces were identical during the early moments of the cosmos as well. For this reason, the electromagnetic and weak forces are said to be unified into a single force, called the electroweak force.

In the same way that the electromagnetic and weak forces were unified into the electroweak force during the 1970s, physicists presently are developing theories that include the strong force as well. Work is still incomplete. Theories are being improved, and experiments to test these theories are being planned. A fully unified theory that includes gravitation will require even more work.

Even more perplexing are the results of studies of the galaxies that suggest that matter described by the Standard Model makes up only a small fraction of the mass in the universe. A much larger amount is formed of dark matter, so called because it doesn't interact with photons or ordinary matter, except through the gravitational force. In addition, there appears to be dark energy, an unknown force that accelerates the expansion of the universe.

Thus the studies of the tiniest particles that make up nuclei are connected directly to investigations of the largest systems, the galaxies that make up the universe. Elementary particle physicists and cosmologists used to be at opposite ends of the length scale; now they question together "What IS the world made of?" Perhaps you will be able to answer that question.

30.3 Section Review

39. Nucleus Bombardment Why would a proton require more energy than a neutron when used to bombard a nucleus?

40. Particle Accelerator Protons in the Fermi Laboratory accelerator, Figure 30-11, move counterclockwise. In what direction is the magnetic field of the bending magnets?

41. Pair Production Figure 30-18 shows the production of two electron/positron pairs. Why does the bottom set of tracks curve less than the top pair of tracks?

42. The Standard Model Research the limitations of the Standard Model and possible replacements.

43. Critical Thinking Consider the following equations.

$$u \rightarrow d + W^{+} \quad \text{and} \quad W^{+} \rightarrow e^{+} + \nu$$

How could they be used to explain the radioactive decay of a nucleon that results in the emission of a positron and a neutrino? Write the equation involving nucleons rather than quarks.

PHYSICS LAB • Design Your Own

Exploring Radiation

Alternate CBL instructions can be found on the Web site.
physicspp.com

Radiation detectors use various means to detect radiation. One common type of detector in use is a Geiger-Mueller tube. It consists of a metal tube filled with gas at a low pressure and a long wire along the tube's axis. The wire is at a high potential difference, such as 400–800 V, relative to the metal tube. At one end of the tube is a thin, fragile window. When a high-energy photon or charged particle enters the tube through the window, some of the gas becomes ionized. The ionized electrons are attracted to the wire and speed up. Then they ionize additional atoms creating a pulse of charge striking the wire. This charge pulse is converted to a voltage pulse, amplified, and counted or sent to a speaker.

Previously, you learned that light and other electromagnetic radiation travels in all directions in straight lines from a source, such as the Sun. In this lab you will explore the relationship between distance from a radioactive gamma and beta source, and the measured radiation intensity.

QUESTION

What is the relationship between distance and radiation intensity from a gamma and a beta source?

Objectives

- **Measure** radiation.
- **Use variables, constants, and controls** to design your experiment.
- **Collect and organize data** from gamma and beta radiation activity compared to the distance from the source.
- **Compare and contrast** beta and gamma radiation activity.

Safety Precautions

- **If a Geiger counter is used, keep hands, pencils, etc. away from the end of the Geiger tube as the tube window is very thin and fragile.**
- **Plug equipment into only GFCI-protected receptacles to prevent shock hazard.**
- **Do not eat, drink, or apply makeup when working with radioactive materials.**
- **Be careful not to crack open the protective plastic case over the radioactive material. Inform your teacher immediately if this exposure happens.**

Materials

gamma and beta sources
radiation counter or student radiation monitor
meterstick
masking tape
stopwatch

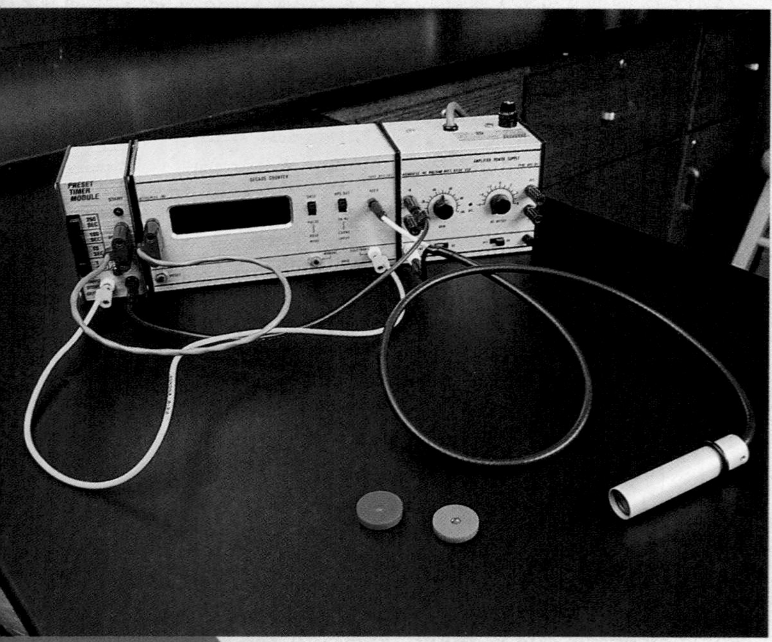

Data Table

Background Radiation (cpm = counts per minute)	cpm

Distance (cm)	Beta-Measured Count Rate (cpm)	Beta-Corrected Count Rate (cpm)	Gamma-Measured Count Rate (cpm)	Gamma-Corrected Count Rate (cpm)
2				
4				
6				
8				
10				
12				
14				

Procedure

1. The type of radiation counter or Geiger-Mueller tube and counter that is available in schools varies dramatically. Your procedure should take into consideration how to assemble and handle the type of equipment that is available for your use—both the detector and the radioactive material.
2. With the detector at least 1 m away from the radioactive materials, turn on the detector and measure the radiation. This is called background radiation. Record your data in the data table.
3. Measure the beta and gamma radiation from your sources at various distances.
4. Subtract the background count rate from the count rate recorded to obtain the corrected activity.
5. Be sure to check with your teacher and have your design approved before you proceed with the lab.

Analyze

1. **Observe and Infer** What is the background radiation source in this experiment?
2. **Make and Use Graphs** Make a plot of gamma count rate versus distance, placing distance on the horizontal axis and corrected sample count rate on the vertical axis. If the count rates are similar, plot the beta count rate on the same graph, and label the graph for each set of data.
3. **Make and Use Graphs** Make a plot of the corrected sample count rate versus $1/d^2$ for the beta and gamma data.

Conclude and Apply

1. **Explain** how the two graphs compare. What relationship exists between distance and count rate?
2. **Explain** how the background count rate would compare if you were at sea level, such as along the coast, compared to the level in the Rocky Mountains.
3. **Describe** what happens to the beta count rate when the Geiger-Mueller tube is moved back to three times the initial distance; for example, 18 cm as compared to 6 cm.

Going Further

What other physics phenomena follow similar patterns?

Real-World Physics

Explain how closeness to radioactive materials is a potential hazard for you or others.

To find out more about radiation, visit the Web site: physicspp.com

Future Technology

Thermonuclear Fusion

For several decades, physicists have been seeking to create and sustain a fusion reaction that will generate more energy than it consumes. A thermonuclear reactor would generate great heat from small amounts of deuterium, $^{2}_{1}H$, and tritium, $^{3}_{1}H$, which can be extracted from seawater.

To initiate a fusion reaction, a mixture of deuterium and tritium must be heated and compressed under conditions typical of those in the Sun. The required temperature would destroy the sort of containers used in fission plants. Confining the plasma is one of the chief design problems for fusion reactors.

Magnetic Confinement In a magnetic confinement reactor, a strong current is passed through a container of deuterium and tritium gas so that the plasma is compressed within the arc. Additional magnetic fields shape the plasma stream, as shown in the diagram, confining it away from the container walls. One promising configuration keeps the plasma in a toroid, or donut shape, which has the great advantage of having no ends to seal.

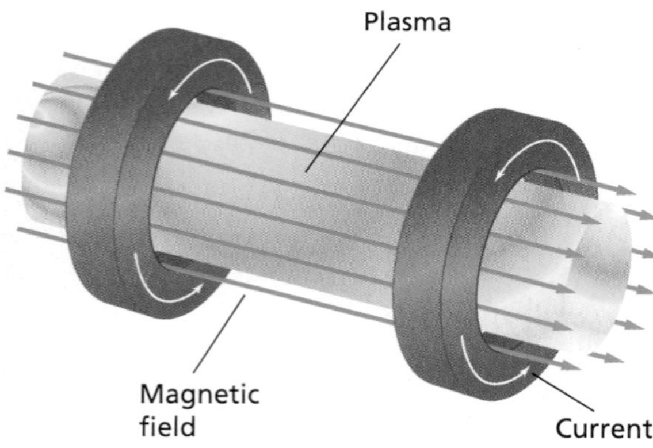

Magnetic confinement: the hot plasma is compressed and confined by the magnetic field.

Inertial Confinement If you look at the rapid, string-like motion of a continuous electric arc, you'll see that it's very difficult to form plasma into a stable shape. In the inertial confinement reactor, a microscopic pellet of frozen deuterium-tritium is illuminated on all sides by powerful laser beams. These lasers heat the outer layer of the pellet so quickly that it explodes. Simultaneously the remainder of the pellet is compressed and heated so greatly that a fusion reaction starts. The energy from the fusion of the pellet exceeds the energy used to heat the pellet. A stream of pellets is fused one after another to obtain a sustained reaction, and the obtained heat is captured to create steam for turbines.

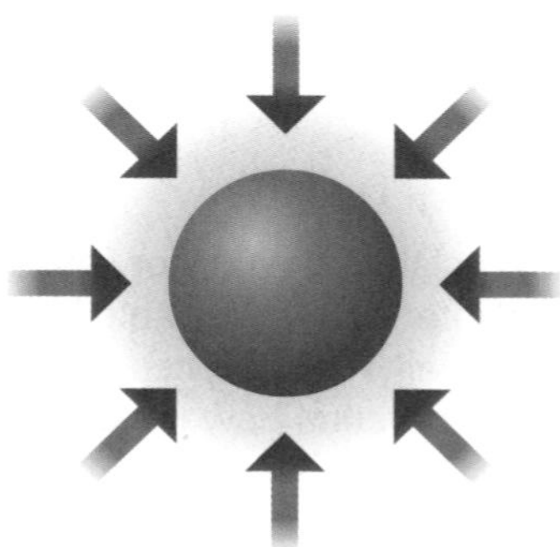

In inertial confinement, beams of light or X rays from a laser rapidly heat the surface of the pellet, forming a surrounding plasma envelope. The rest of the fuel is compressed by the blowoff of the hot surface material.

The Future While thermonuclear fusion has been sustained in both types of reactors, researchers have had trouble achieving a breakeven reaction (one in which the energy produced in the reaction exceeds the energy needed to sustain the reaction). Progress toward a practical thermonuclear reactor has been expensive and slow, but the promise is great. A fusion reactor is not completely free of radiation hazards because neutrons are produced in fusion reactions. But because the fuel itself is not radioactive, the amount of nuclear waste would be negligible.

Going Further

1. **Analyze** Why does the thermonuclear reactor appear to be such an attractive source of energy?
2. **Compare** You have seen three types of 'thermal' electric power plants. What features do they all have in common?

Chapter 30

Study Guide

30.1 The Nucleus

Vocabulary

- atomic number (p. 800)
- atomic mass unit (p. 800)
- mass number (p. 800)
- nuclide (p. 801)
- strong nuclear force (p. 802)
- nucleon (p. 802)
- binding energy (p. 802)
- mass defect (p. 802)

Key Concepts

- The number of protons in a nucleus is given by the atomic number, Z.
- The sum of the numbers of protons and neutrons in a nucleus is equal to the mass number, A.
- Atoms having nuclei with the same number of protons but different numbers of neutrons are called isotopes.
- The strong nuclear force binds the nucleus together.
- The energy released in a nuclear reaction can be calculated by finding the mass defect, the difference in mass of the particles before and after the reaction.

$$E = mc^2$$

- The binding energy is the energy equivalent of the mass defect.

30.2 Nuclear Decay and Reactions

Vocabulary

- radioactive (p. 806)
- alpha decay (p. 806)
- beta decay (p. 806)
- gamma decay (p. 806)
- nuclear reaction (p. 807)
- half-life (p. 809)
- activity (p. 810)
- fission (p. 811)
- chain reaction (p. 812)
- fusion (p. 813)

Key Concepts

- An unstable nucleus decays, transmuting into another element.
- Radioactive decay produces three kinds of particles. Alpha, α, particles are helium nuclei, beta, β, particles are high-speed electrons, and gamma, γ, rays are high-energy photons.
- In nuclear reactions, the sums of the mass number, A, and the total charge, Z, are not changed.
- The half-life of a radioactive isotope is the time required for half of the nuclei to decay. After t half-lives:

$$\text{remaining} = \text{original}\left(\frac{1}{2}\right)^t$$

- The number of decays of a radioactive sample per second is the activity.
- In nuclear fission, the uranium nucleus is split into two smaller nuclei with a release of neutrons and energy.
- Nuclear reactors use the energy released in fission to generate electrical energy.
- The fusion of hydrogen nuclei into a helium nucleus releases the energy that causes stars to shine.

30.3 The Building Blocks of Matter

Vocabulary

- quarks (p. 818)
- leptons (p. 818)
- Standard Model (p. 818)
- force carriers (p. 818)
- pair production (p. 820)
- weak nuclear force (p. 821)

Key Concepts

- Linear accelerators and synchrotrons produce high-energy particles.
- The Geiger-Mueller counter, cloud chamber, and other particle detectors use the ionization caused by charged particles passing through matter.
- All matter appears to be made up of quarks and leptons.
- Matter interacts with other matter through particles called force carriers.
- The Standard Model includes the quarks, leptons, and force carriers.
- When corresponding antimatter and matter particles combine, their mass and energy are converted into energy or lighter matter-antimatter particle pairs.
- By pair production, energy is transformed into a matter-antimatter pair.

Chapter 30 Assessment

Concept Mapping

44. Organize the following terms into the concept map: *Standard Model, quarks, gamma rays, force carriers, protons, neutrons, leptons, W bosons, neutrinos, electrons, gluons.*

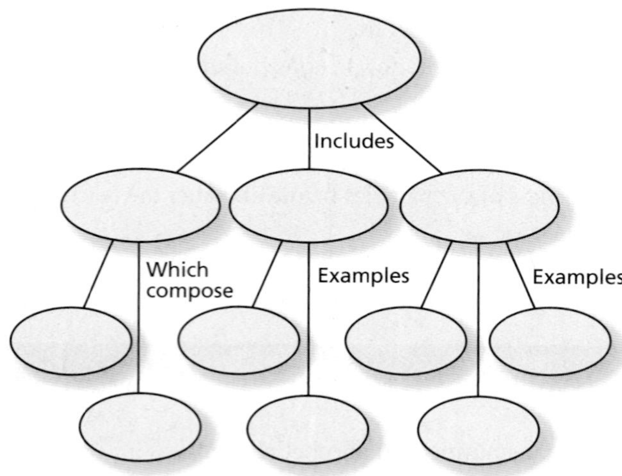

Mastering Concepts

45. What force inside a nucleus acts to push the nucleus apart? What force inside the nucleus acts to hold the nucleus together? (30.1)

46. Define the mass defect of a nucleus. To what is it related? (30.1)

47. Which are generally more unstable, small or large nuclei? (30.1)

48. Which isotope has the greater number of protons, uranium-235 or uranium-238? (30.1)

49. Define the term *transmutation* as used in nuclear physics and give an example. (30.2)

50. **Radiation** What are the common names for an α particle, β particle, and γ radiation? (30.2)

51. What two quantities must always be conserved in any nuclear equation? (30.2)

52. **Nuclear Power** What sequence of events must occur for a chain reaction to take place? (30.2)

53. **Nuclear Power** What role does a moderator play in a fission reactor? (30.2)

54. Fission and fusion are opposite processes. How can each release energy? (30.2)

55. **High-Energy Physics** Why would a linear accelerator not work with a neutron? (30.3)

56. **Forces** In which of the four interactions (strong, weak, electromagnetic, and gravitational) do the following particles take part? (30.3)
   a. electron
   b. proton
   c. neutrino

57. What happens to the atomic number and mass number of a nucleus that emits a positron? (30.3)

58. **Antimatter** What would happen if a meteorite made of antiprotons, antineutrons, and positrons landed on Earth? (30.3)

Applying Concepts

59. **Fission** A Web site claims that scientists have been able to cause iron nuclei to undergo fission. Is the claim likely to be true? Explain.

60. Use the graph of binding energy per nucleon in Figure 30-2 to determine whether the reaction $^{2}_{1}\text{H} + ^{1}_{1}\text{H} \rightarrow ^{3}_{2}\text{He}$ is energetically possible.

61. **Isotopes** Explain the difference between naturally and artificially produced radioactive isotopes.

62. **Nuclear Reactor** In a nuclear reactor, water that passes through the core of the reactor flows through one loop, while the water that produces steam for the turbines flows through a second loop. Why are there two loops?

63. The fission of a uranium nucleus and the fusion of four hydrogen nuclei to produce a helium nucleus both produce energy.
   a. Which produces more energy?
   b. Does the fission of a kilogram of uranium nuclei or the fusion of a kilogram of hydrogen produce more energy?
   c. Why are your answers to parts **a** and **b** different?

Mastering Problems

30.1 The Nucleus

64. What particles, and how many of each, make up an atom of $^{109}_{47}\text{Ag}$?

65. What is the isotopic symbol (the one used in nuclear equations) of a zinc atom composed of 30 protons and 34 neutrons?

66. The sulfur isotope $^{32}_{16}\text{S}$ has a nuclear mass of 31.97207 u.
   a. What is the mass defect of this isotope?
   b. What is the binding energy of its nucleus?
   c. What is the binding energy per nucleon?

 For more problems, go to Additional Problems, Appendix B.

67. A nitrogen isotope, $^{12}_{7}N$, has a nuclear mass of 12.0188 u.

a. What is the binding energy per nucleon?
b. Does it require more energy to separate a nucleon from a $^{14}_{7}N$ nucleus or from a $^{12}_{7}N$ nucleus? $^{14}_{7}N$ has a mass of 14.00307 u.

68. The two positively charged protons in a helium nucleus are separated by about 2.0×10^{-15} m. Use Coulomb's law to find the electric force of repulsion between the two protons. The result will give you an indication of the strength of the strong nuclear force.

69. The binding energy for $^{4}_{2}He$ is -28.3 MeV. Calculate the mass of a helium isotope in atomic mass units.

30.2 Nuclear Decay and Reactions

70. Write the complete nuclear equation for the alpha decay of $^{222}_{86}Rn$.

71. Write the complete nuclear equation for the beta decay of $^{89}_{36}Kr$.

72. Complete each nuclear reaction.

a. $^{225}_{89}Ac \rightarrow {}^{4}_{2}He + $ ____
b. $^{227}_{88}Ra \rightarrow {}_{-1}^{0}e + $ ____ $+$ ____
c. $^{65}_{29}Cu + {}^{1}_{0}n \rightarrow$ ____ $\rightarrow {}^{1}_{1}p + $ ____
d. $^{235}_{92}U + {}^{1}_{0}n \rightarrow {}^{96}_{40}Zr + 3({}^{1}_{0}n) + $ ____

73. An isotope has a half-life of 3.0 days. What percent of the original material will be left after

a. 6.0 days?
b. 9.0 days?
c. 12 days?

74. In an accident in a research laboratory, a radioactive isotope with a half-life of three days is spilled. As a result, the radiation is eight times the maximum permissible amount. How long must workers wait before they can enter the room?

75. When a boron isotope, $^{11}_{5}B$, is bombarded with protons, it absorbs a proton and emits a neutron.

a. What element is formed?
b. Write the nuclear equation for this reaction.
c. The isotope formed is radioactive and decays by emitting a positron. Write the complete nuclear equation for this reaction.

76. The first atomic bomb released an energy equivalent of 2.0×10^{1} kilotons of TNT. One kiloton of TNT is equivalent to 5.0×10^{12} J. Uranium-235 releases 3.21×10^{-11} J/atom. What was the mass of the uranium-235 that underwent fission to produce the energy of the bomb?

77. During a fusion reaction, two deuterons, $^{2}_{1}H$, combine to form a helium isotope, $^{3}_{2}He$. What other particle is produced?

78. $^{209}_{84}Po$ has a half-life of 103 years. How long would it take for a 100-g sample to decay so that only 3.1 g of Po-209 was left?

30.3 The Building Blocks of Matter

79. What would be the charge of a particle composed of three up quarks?

80. The charge of an antiquark is opposite that of a quark. A pion is composed of an up quark and an anti-down quark, $u\bar{d}$. What would be the charge of this pion?

81. Pions are composed of a quark and an antiquark. Find the charge of a pion made up of the following.

a. $u\bar{u}$
b. $d\bar{u}$
c. $d\bar{d}$

82. Baryons are particles that are made of three quarks. Find the charge on each of the following baryons.

a. neutron: ddu
b. antiproton: $\bar{u}\bar{u}\bar{d}$

83. The synchrotron at the Fermi Laboratory has a diameter of 2.0 km. Protons circling in it move at approximately the speed of light in a vacuum.

a. How long does it take a proton to complete one revolution?
b. The protons enter the ring at an energy of 8.0 GeV. They gain 2.5 MeV each revolution. How many revolutions must they travel before they reach 400.0 GeV of energy?
c. How long does it take the protons to be accelerated to 400.0 GeV?
d. How far do the protons travel during this acceleration?

84. **Figure 30-21** shows tracks in a bubble chamber. What are some reasons one track might curve more than another?

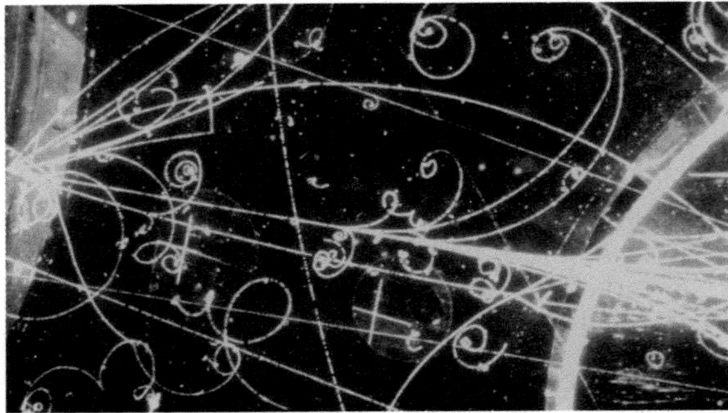

Figure 30-21

Mixed Review

85. Each of the following nuclei can absorb an α particle. Assume that no secondary particles are emitted by the nucleus. Complete each equation.

a. ${}^{14}_{7}\text{N} + {}^{4}_{2}\text{He} \rightarrow$ ____

b. ${}^{27}_{13}\text{Al} + {}^{4}_{2}\text{He} \rightarrow$ ____

86. ${}^{211}_{86}\text{Rn}$ has a half-life of 15 h. What fraction of a sample would be left after 60 h?

87. One of the simplest fusion reactions involves the production of deuterium, ${}^{2}_{1}\text{H}(2.014102\text{ u})$, from a neutron and a proton. Write the complete fusion reaction and find the amount of energy released.

88. A ${}^{232}_{92}\text{U}$ nucleus, mass = 232.0372 u, decays to ${}^{228}_{90}\text{Th}$, mass = 228.0287 u, by emitting an α particle, mass = 4.0026 u, with a kinetic energy of 5.3 MeV. What must be the kinetic energy of the recoiling thorium nucleus?

Thinking Critically

89. Infer Gamma rays carry momentum. The momentum of a gamma ray of energy E is equal to E/c, where c is the speed of light. When an electron-positron pair decays into two gamma rays, both momentum and energy must be conserved. The sum of the energies of the gamma rays is 1.02 MeV. If the positron and electron are initially at rest, what must be the magnitude and direction of the momentum of the two gamma rays?

90. Infer An electron-positron pair, initially at rest, also can decay into three gamma rays. If all three gamma rays have equal energies, what must be their relative directions? Make a sketch.

91. Estimate One fusion reaction in the Sun releases about 25 MeV of energy. Estimate the number of such reactions that occur each second from the luminosity of the Sun, which is the rate at which it releases energy, 4×10^{26} W.

92. Interpret Data An isotope undergoing radioactive decay is monitored by a radiation detector. The number of counts in each five-minute interval is recorded. The results are shown in **Table 30-3.** The sample is then removed and the radiation detector records 20 counts resulting from cosmic rays in 5 min. Find the half-life of the isotope. Note that you should first subtract the 20-count background reading from each result. Then plot the counts as a function of time. From your graph, determine the half-life.

Table 30-3
Radioactive Decay Measurements

Time (min)	Counts (per 5 min)
0	987
5	375
10	150
15	70
20	40
25	25
30	18

Writing in Physics

93. Research the present understanding of dark matter in the universe. Why is it needed by cosmologists? Of what might it be made?

94. Research the hunt for the top quark. Why did physicists hypothesize its existence?

Cumulative Review

95. An electron with a velocity of 1.7×10^{6} m/s is at right angles to a 0.91-T magnetic field. What is the force on the electron produced by the magnetic field? (Chapter 24)

96. An *EMF* of 2.0 mV is induced in a wire that is 0.10 m long when it is moving perpendicularly across a uniform magnetic field at a velocity of 4.0 m/s. What is the magnetic induction of the field? (Chapter 25)

97. An electron has a de Broglie wavelength of 400.0 nm, the shortest wavelength of visible light. (Chapter 27)

a. Find the velocity of the electron.

b. Calculate the energy of the electron in eV.

98. A photon with an energy of 14.0 eV enters a hydrogen atom in the ground state and ionizes it. With what kinetic energy will the electron be ejected from the atom? (Chapter 28)

99. A silicon diode (V = 0.70 V)that is conducting 137 mA is in series with a resistor and a 6.67-V power source. (Chapter 29)

a. What is the voltage drop across the resistor?

b. What is the value of the resistor?

 For more problems, go to Additional Problems, Appendix B.

Multiple Choice

1. How many protons, neutrons, and electrons are in the isotope nickel-60, $^{60}_{28}Ni$?

	Protons	Neutrons	Electrons
A	28	32	28
B	28	28	32
C	32	32	28
D	32	28	28

2. What has occurred in the following reaction?
$^{212}_{82}Pb \rightarrow {}^{212}_{83}Bi + e^- + \bar{\nu}$

   A. alpha decay
   B. beta decay
   C. gamma decay
   D. loss of a proton

3. What is the product when pollonium-210, $^{210}_{84}Po$, undergoes alpha decay?

   A. $^{206}_{82}Pb$
   B. $^{208}_{82}Pb$
   C. $^{210}_{85}Pb$
   D. $^{210}_{80}Pb$

4. A sample of radioactive iodine-131 emits beta particles at the rate of 2.5×10^8 Bq. The half-life is 8 days. What will be the activity after 16 days?

   A. 1.6×10^7 Bq
   B. 6.3×10^7 Bq
   C. 1.3×10^8 Bq
   D. 2.5×10^8 Bq

5. Identify the unknown isotope in this reaction: neutron + $^{14}_{7}N \rightarrow {}^{14}_{6}C$ + ?

   A. $^{1}_{1}H$
   B. $^{2}_{1}H$
   C. $^{3}_{1}H$
   D. $^{4}_{2}He$

6. Which type of decay does not change the number of protons or neutrons in the nucleus?

   A. positron
   B. alpha
   C. beta
   D. gamma

7. Polonium-210 has a half-life of 138 days. How much of a 2.34-kg sample will remain after four years?

   A. 0.644 mg
   B. 1.50 mg
   C. 1.51 g
   D. 10.6 g

8. An electron and a positron collide, annihilate one another, and release their energy as a gamma ray. What is the minimum energy of the gamma ray? (The energy equivalent of the mass of an electron is 0.51 MeV.)

   A. 0.51 MeV
   B. 1.02 MeV
   C. 931.49 MeV
   D. 1863 MeV

9. The illustration below shows the tracks in a bubble chamber produced when a gamma ray decays into a positron and an electron. Why doesn't the gamma ray leave a track?

   A. Gamma rays move too quickly for their tracks to be detected.
   B. Only pairs of particles can leave tracks in a bubble chamber.
   C. A particle must have mass to interact with the liquid and leave a track, and the gamma ray is virtually massless.
   D. The gamma ray is electrically neutral, so it does not ionize the liquid in the bubble chamber.

Extended Answer

10. The fission of a uranium-235 nucleus releases about 3.2×10^{-11} J. One ton of TNT releases about 4×10^9 J. How many uranium-235 nuclei are in a nuclear fission weapon that releases energy equivalent to 20,000 tons of TNT?

✓ Test-Taking TIP

Do Some Reconnaissance

Find out what the conditions will be for taking the test. Is it timed or untimed? Can you use a calculator or other tools? Will those tools be provided? Will mathematical constants be given? Know these things in advance so that you can practice taking tests under the same conditions.

Appendices Contents

I. Symbols

Δ	change in quantity	$a \times b$	a multiplied by b
$\pm$	plus or minus a quantity	ab	
$\propto$	is proportional to	$a(b)$	
$=$	is equal to	$a \div b$	a divided by b
$\approx$	is approximately equal to	a/b	
$\cong$	is approximately equal to	$\frac{a}{b}$	
$\leq$	is less than or equal to	$\sqrt{a}$	square root of a
$\geq$	is greater than or equal to	$\lvert a \rvert$	absolute value of a
$<<$	is much less than	$\log_b x$	log to the base, b, of x
$\equiv$	is defined as		

II. Measurements and Significant Digits

Connecting Math to Physics Math is the language of physics. Using math, physicists are able to describe relationships among the measurements that they make using equations. Each measurement is associated with a symbol that is used in physics equations. The symbols are called variables.

Significant Digits

All measured quantities are approximated and have significant digits. The number of significant digits indicates the precision of the measurement. Precision is a measure of exactness. The number of significant digits in a measurement depends on the smallest unit on the measuring tool. The digit farthest to the right in a measurement is estimated.

Example: What is the estimated digit for each of the measuring sticks in the figure below used to measure the length of the rod?

Using the lower measuring tool, the length is between 9 and 10 cm. The measurement would be estimated to the nearest tenth of a centimeter. If the length was exactly on the 9-cm or 10-cm mark, record it as 9.0 cm or 10.0 cm.

Using the upper measuring tool, the length is between 9.5 and 9.6 cm. The measurement would be estimated to the nearest hundredth of a centimeter. If the length was exactly on the 9.5-cm or 9.6-cm mark, record it as 9.50 cm or 9.60 cm.

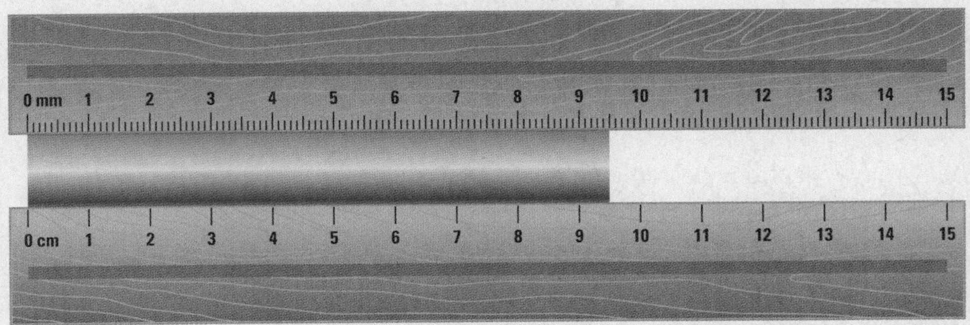

All nonzero digits in a measurement are significant digits. Some zeros are significant and some are not. All digits between and including the first nonzero digit from the left through the last digit on the right are significant. Use the following rules when determining the number of significant digits.

1. Nonzero digits are significant.
2. Final zeros after a decimal point are significant.
3. Zeros between two significant digits are significant.
4. Zeros used only as placeholders are not significant.

Example: State the number of significant digits in each measurement.

5.0 g has two significant digits.	**Using rules 1 and 2**
14.90 g has four significant digits.	**Using rules 1 and 2**
0.0 has one significant digit.	**Using rules 2 and 4**
300.00 mm has five significant digits.	**Using rules 1, 2, and 3**
5.06 s has three significant digits.	**Using rules 1 and 3**
304 s has three significant digits.	**Using rules 1 and 3**
0.0060 mm has two significant digits (6 and the last 0).	**Using rules 1, 2, and 4**
140 mm has two significant digits (just 1 and 4).	**Using rules 1 and 4**

▶ PRACTICE **Problems**

1. State the number of significant digits in each measurement.

a. 1405 m

b. 2.50 km

c. 0.0034 m

d. 12.007 kg

e. 5.8×10^{6} kg

f. 3.03×10^{-5} mL

There are two cases in which numbers are considered exact, and thus, have an infinite number of significant digits.

1. Counting numbers have an infinite number of significant digits.
2. Conversion factors have an infinite number of significant digits.

Examples:

The factor "2" in 2*mg* has an infinite number.

The number 2 is a counting number. It is an exact integer.

The number "4" in 4 electrons has an infinite number.

Because you cannot have a partial electron, the number 4, a counting number, is considered to have an infinite number of significant digits.

60 s/1 min has an infinite number.

There are exactly 60 seconds in 1 minute, thus there are an infinite number of significant digits in the conversion factor.

Rounding

A number can be rounded to a specific place value (like hundreds or tenths) or to a specific number of significant digits. To do this, determine the place being rounded, and then use the following rules.

1. When the leftmost digit to be dropped is less than 5, that digit and any digits that follow are dropped. Then the last digit in the rounded number remains unchanged.
2. When the leftmost digit to be dropped is greater than 5, that digit and any digits that follow are dropped, and the last digit in the rounded number is increased by one.
3. When the leftmost digit to be dropped is 5 followed by a nonzero number, that digit and any digits that follow are dropped. The last digit in the rounded number increases by one.
4. If the digit to the right of the last significant digit is equal to 5, and 5 is followed by a zero or no other digits, look at the last significant digit. If it is odd, increase it by one; if it is even, do not round up.

Examples: Round the following numbers to the stated number of significant digits.

8.7645 rounded to 3 significant digits is 8.76.	**Using rule 1**
8.7676 rounded to 3 significant digits is 8.77.	**Using rule 2**
8.7519 rounded to 2 significant digits is 8.8.	**Using rule 3**
92.350 rounded to 3 significant digits is 92.4.	**Using rule 4**
92.25 rounded to 3 significant digits is 92.2.	**Using rule 4**

▶ PRACTICE **Problems**

2. Round each number to the number of significant digits shown in parentheses.

a. 1405 m (2)

b. 2.50 km (2)

c. 0.0034 m (1)

d. 12.007 kg (3)

Operations with Significant Digits

When using a calculator, do all of the operations with as much precision as the calculator allows, and then round the result to the correct number of significant digits. The correct number of significant digits in the result depends on the measurements and on the operation.

Addition and subtraction Look at the digits to the right of the decimal point. Round the result to the least precise value among the measurements—the smallest number of digits to the right of the decimal points.

Example: Add 1.456 m, 4.1 m, and 20.3 m.

The least precise values are 4.1 m and 20.3 m because they have only one digit to the right of the decimal points.

$$\begin{array}{r} 1.456\ \text{m} \\ 4.1\ \text{m} \\ +\ 20.3\ \text{m} \\ \hline 25.856\ \text{m} \end{array}$$

Add the numbers

The sum is only as precise as the least precise number being added.

25.9 m **Round the result to the estimated place with the largest place value**

Multiplication and division Look at the number of significant digits in each measurement. Perform the calculation. Round the result so that it has the same number of significant digits as the measurement with the least number of significant digits.

Example: Multiply 20.1 m by 3.6 m.

$(20.1 \text{ m})(3.6 \text{ m}) = 72.36 \text{ m}^2$

The least precise value is 3.6 m with two significant digits. The product can only have as many digits as the least precise of the multiplied numbers.

72 m **Round the result to two significant digits**

▶ PRACTICE Problems

3. Simplify the following expressions using the correct number of significant digits.

a. 5.012 km + 3.4 km + 2.33 km

b. 45 g − 8.3 g

c. 3.40 cm × 7.125 cm

d. 54 m ÷ 6.5 s

Combination When doing a calculation that requires a combination of addition/subtraction and multiplication/division, use the multiplication/division rule.

Examples:

$$d = 19 \text{ m} + (25.0 \text{ m/s})(2.50 \text{ s}) + \frac{1}{2}(-10.0 \text{ m/s}^2)(2.50 \text{ s})^2$$
$$= 5.0\times10^1 \text{ m}$$

19 m only has two significant digits, so the answer should only have two significant digits

$$\text{slope} = \frac{70.0 \text{ m} - 10.0 \text{ m}}{29 \text{ s} - 11 \text{ s}}$$
$$= 3.3 \text{ m/s}$$

26 s and 11 s only have two significant digits each, so the answer should only have two significant digits

Multistep calculations Do not round to significant digits in the middle of a multistep calculation. Instead, round to a reasonable number of decimal places that will not cause you to lose significance in your answer. When you get to your final step where you are solving for the answer asked for in the question, you should then round to the correct number of significant digits.

Example:

$$F = \sqrt{(24 \text{ N})^2 + (36 \text{ N})^2}$$
$$= \sqrt{576 \text{ N}^2 + 1296 \text{ N}^2}$$ **Do not round to 580 N^2 and 1300 N^2**
$$= \sqrt{1872 \text{ N}^2}$$ **Do not round to 1800 N^2**
$$= 43 \text{ N}$$ **Final answer, so it should be rounded to two significant digits**

III. Fractions, Ratios, Rates, and Proportions

Fractions

A fraction names a part of a whole or a part of a group. A fraction also can express a ratio (see page 838). It consists of a numerator, a division bar, and a denominator.

$$\frac{\text{numerator}}{\text{denominator}} = \frac{\text{number of parts chosen}}{\text{total number of parts}}$$

Simplification Sometimes, it is easier to simplify an expression before substituting the known values of the variables. Variables often cancel out of the expression.

Example: Simplify $\frac{pn}{pw}$.

$\frac{pn}{pw} = \left(\frac{p}{p}\right)\left(\frac{n}{w}\right)$ **Factor out the *p* in the numerator and the denominator, and break the fraction into the product of two fractions**

$= (1)\left(\frac{n}{w}\right)$ **Substitute $(p/p) = 1$**

$= \frac{n}{w}$

Multiplication and division To multiply fractions, multiply the numerators and multiply the denominators.

Example: Multiply the fractions $\frac{s}{a}$ and $\frac{t}{b}$.

$\left(\frac{s}{a}\right)\left(\frac{t}{b}\right) = \frac{st}{ab}$ **Multiply the numerators and the denominators**

To divide fractions, multiply the first fraction by the reciprocal of the second fraction. To find the reciprocal of a fraction, invert it–switch the numerator and the denominator.

Example: Divide the fraction $\frac{s}{a}$ by $\frac{t}{b}$.

$\frac{s}{a} \div \frac{t}{b} = \left(\frac{s}{a}\right)\left(\frac{b}{t}\right)$ **Multiply the first fraction by the reciprocal of the second fraction**

$= \frac{sb}{at}$ **Multiply the numerators and the denominators**

Addition and subtraction To add or subtract two fractions, first write them as fractions with a common denominator; that is, with the same denominator. To find a common denominator, multiply the denominators of both fractions. Then, add or subtract the numerators and use the common denominator.

Example: Add the fractions $\frac{1}{a}$ and $\frac{2}{b}$.

$\frac{1}{a} + \frac{2}{b} = \left(\frac{1}{a}\right)\left(\frac{b}{b}\right) + \left(\frac{2}{b}\right)\left(\frac{a}{a}\right)$ **Multiply each fraction by a fraction equal to 1**

$= \frac{b}{ab} + \frac{2a}{ab}$ **Multiply the numerators and the denominators**

$= \frac{b + 2a}{ab}$ **Write a single fraction with the common denominator**

▶ PRACTICE Problems

4. Perform the indicated operation. Write the answer in simplest form.

a. $\frac{1}{x} + \frac{y}{3}$

b. $\frac{a}{2b} - \frac{3}{b}$

c. $\left(\frac{3}{x}\right)\left(\frac{1}{y}\right)$

d. $\frac{2a}{5} \div \frac{1}{2}$

Ratios

A ratio is a comparison between two numbers by division. Ratios can be written in several different ways. The ratio of 2 and 3 can be written in four different ways:

2 to 3 2 out of 3 2:3 $\frac{2}{3}$

Rates

A rate is a ratio that compares two quantities with different measurement units. A unit rate is a rate that has been simplified so that the denominator is 1.

Example: Write 98 km in 2.0 hours as a unit rate.

98 km in 2.0 hours is a ratio of $\frac{98\text{ km}}{2.0\text{ hours}}$

$\frac{98\text{ km}}{2.0\text{ hours}} = \left(\frac{98}{2.0}\right)\left(\frac{\text{km}}{\text{hour}}\right)$ **Split the fraction into the product of a number fraction and a unit fraction**

$= (49)\left(\frac{\text{km}}{\text{hour}}\right)$ **Simplify the number fraction**

$= 49$ km per hour or 49 km/h

Example: Write 16 Swedish crowns in 2 U.S. dollars as a unit rate.

16 Swedish crowns in \$2.00 American currency is a ratio of $\frac{16\text{ Swedish crowns}}{2\text{ U.S. dollars}}$

$\frac{16\text{ Swedish crowns}}{2\text{ U.S. dollars}} = \left(\frac{16}{2}\right)\left(\frac{\text{Swedish crowns}}{\text{U.S. dollars}}\right)$

$= (8)\left(\frac{\text{Swedish crowns}}{\text{U.S. dollars}}\right)$

$= 8$ Swedish crowns per U.S. dollar or 8 Swedish crowns/U.S. dollar

Proportions

A proportion is an equation that states that two ratios are equal:

$\frac{a}{b} = \frac{c}{d}$, where b and d are not zero.

Proportions used to solve ratio problems often include three numbers and one variable. You can solve the proportion to find the value of the variable. To solve a proportion, use cross multiplication.

Example: Solve the proportion $\frac{a}{b} = \frac{c}{d}$ for a.

Cross multiply: $\frac{a}{b} = \frac{c}{d}$

$ad = bc$ **Write the equation resulting from cross multiplying**

$a = \frac{bc}{d}$ **Solve for a**

▶ PRACTICE Problems

5. Solve the following proportions.

a. $\frac{4}{x} = \frac{2}{3}$

b. $\frac{13}{15} = \frac{n}{75}$

c. $\frac{36}{12} = \frac{s}{16}$

d. $\frac{2.5}{5.0} = \frac{7.5}{w}$

IV. Exponents, Powers, Roots, and Absolute Value

Exponents

An exponent is a number that tells how many times a number, *a*, is used as a factor. An exponent is written as a superscript. In the term a^n, *a* is the *base* and *n* is the exponent. a^n is called the *n*th power of *a* or *a* raised to the *n*th power.

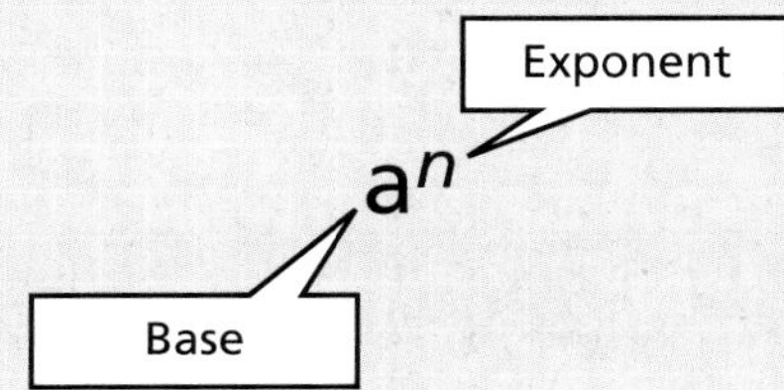

Connecting Math to Physics A subscript is not an exponent. In physics, a subscript is used to further describe the variable. For example, v_0 can be used to represent the velocity at time 0. A subscript is a part of the variable.

Positive exponent For any nonzero number, *a*, and any integer, *n*,

$$a^n = (a_1)(a_2)(a_3)\ldots(a_n)$$

Example: Simplify the following exponent terms.

$10^4 = (10)(10)(10)(10) = 10{,}000$

$2^3 = (2)(2)(2) = 8$

Zero exponent For any nonzero number, *a*,

$$a^0 = 1$$

Examples: Simplify the following zero exponent terms.

$2^0 = 1$

$13^0 = 1$

Negative exponent For any nonzero number, *a*, and any integer, *n*,

$$a^{-n} = \frac{1}{a^n}$$

Examples: Write the following negative exponent terms as fractions.

$$2^{-1} = \frac{1}{2^1} = \frac{1}{2} \qquad 2^{-2} = \frac{1}{2^2} = \frac{1}{4}$$

Square and Cube Roots

A square root of a number is one of its two equal factors. A radical sign, $\sqrt{}$, indicates a square root. A square root can be shown as exponent $\frac{1}{2}$, as in $\sqrt{b} = b^{\frac{1}{2}}$. You can use a calculator to find square roots.

Examples: Simplify the following square root terms.

$\sqrt{a^2} = \sqrt{(a)(a)} = a$

$\sqrt{9} = \sqrt{(3)(3)} = 3$

$\sqrt{64} = \sqrt{(8.0)(8.0)} = 8.0$ — **The answer has a zero to the right of the decimal to keep two significant digits**

$\sqrt{38.44} = 6.200$ — **Place two zeros to the right of the calculator answer to keep four significant digits**

$\sqrt{39} = 6.244997\ldots = 6.2$ — **Round the calculator answer to keep two significant digits**

Math Handbook

A cube root of a number is one of its three equal factors. A radical sign with the number 3, $\sqrt[3]{\ }$, indicates a cube root. A cube root also can be shown as exponent $\frac{1}{3}$, as in $\sqrt[3]{b} = b^{\frac{1}{3}}$.

Example: Simplify the following cube root terms.

$$\sqrt[3]{125} = \sqrt[3]{(5.00)(5.00)(5.00)} = 5.00$$

$$\sqrt[3]{39.304} = 3.4000$$

▶ PRACTICE Problems

6. Find each root. Round the answer to the nearest hundredth.

a. $\sqrt{22}$

b. $\sqrt[3]{729}$

c. $\sqrt{676}$

d. $\sqrt[3]{46.656}$

7. Simplify by writing without a radical sign.

a. $\sqrt{16a^2b^4}$

b. $\sqrt{9t^6}$

8. Write using exponents.

a. $\sqrt{n^3}$

b. $\frac{1}{\sqrt{a}}$

Operations with Exponents

In the following operations with exponents, a and b can be numbers or variables.

Product of powers To multiply terms with the same base, add the exponents, as in $(a^m)(a^n) = a^{m+n}$.

Quotient of powers To divide terms with the same base, subtract the bottom exponent from the top exponent, as in $a^m/a^n = a^{m-n}$.

Power of a power To calculate the power of a power, use the same base and multiply the exponents, as in $(a^m)^n = a^{mn}$.

***n*th root of a power** To calculate the root of a power, use the same base and divide the power exponent by the root exponent, as in $\sqrt[n]{a^m} = a^{\frac{m}{n}}$.

Power of a product To calculate the power of a product of a and b, raise both to the power and find their product, as in $(ab)^n = a^nb^n$.

▶ PRACTICE Problems

9. Write an equivalent form using the properties of exponents.

a. $\frac{x^2t}{x^3}$

b. $\sqrt{t^3}$

c. $(d^2n)^2$

d. $x^2\sqrt{x}$

10. Simplify $\frac{m}{q}\sqrt{\frac{2qv}{m}}$

Absolute Value

The absolute value of a number, n, is its magnitude, regardless of its sign. The absolute value of n is written as $|n|$. Because magnitudes cannot be less than zero, absolute values always are greater than or equal to zero.

Examples:

$|3| = 3$

$|-3| = 3$

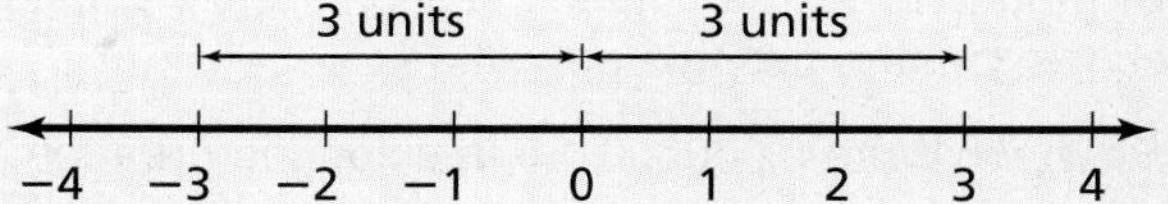

V. Scientific Notation

A number of the form $a \times 10^n$ is written in scientific notation, where $1 \leq a \leq 10$, and n is an integer. The base, 10, is raised to a power, n. The term a must be less than 10.

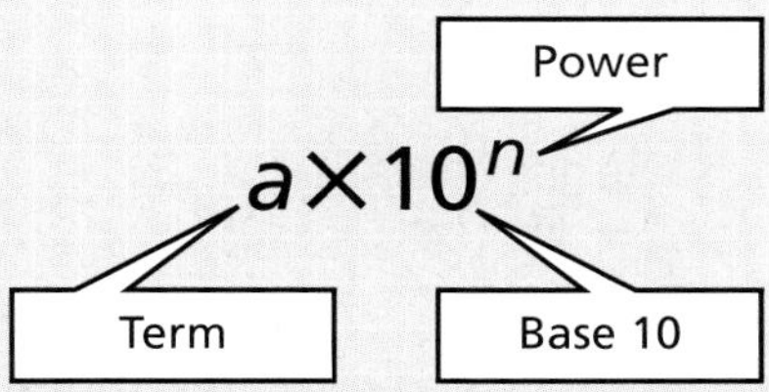

Connecting Math to Physics Physicists use scientific notation to express, compare, and calculate with measurements that are greater than 10 or less than 1. For example, the mass of a proton is written as 6.73×10^{-28} kg. The density of water is written as 1.000×10^3 kg/m^3. This shows, using significant digit rules, that this measurement is exactly 1000 to four significant digits. However, writing the density of water as 1000 kg/m^3 would imply that it has only one significant digit, which is incorrect. Scientific notation helps physicists keep accurate track of significant digits.

Large Numbers—Using Positive Exponents

Multiplying by a power of 10 is like moving the decimal point that same number of places to the left (if the power is negative) or right (if the power is positive). To express a large number in scientific notation, first determine the value for a, $1 \leq a < 10$. Count the number of decimal places from the decimal point in a to the decimal point in the number. Use that count as the power of 10. A calculator shows scientific notation with e for exponent, as in $2.4\text{e}+11 = 2.4 \times 10^{11}$. Some calculators use an E to show the exponent, or there is often a place on the display where the calculator can show smaller-sized digits representing the exponent.

Example: Write 7,530,000 in scientific notation.

The value for a is 7.53. (The decimal point is to the right of the first nonzero digit.) So the form will be 7.53×10^n.

$7{,}530{,}000 = 7.53 \times 10^6$ **There are six decimal places, so the power is 6**

To write the standard form of a number expressed in scientific notation, write the value of a, and place extra zeros to the right of the number. Use the power and move the decimal point in a that many places to the right.

Example: Write the following number in standard form.

$2.389 \times 10^5 = 2.38900 \times 10^5 = 238{,}900$

Small Numbers—Using Negative Exponents

To express a small number in scientific notation, first determine the value for a, $1 \leq a < 10$. Then count the number of decimal places from the decimal point in a to the decimal point in the number. Use that number as the power of 10. Multiplying by a number with a negative power is the same as dividing by that number with the corresponding positive power.

Example: Write 0.000000285 in scientific notation.

The value for a is 2.85. (The decimal point is to the right of the first nonzero digit.) So the form will be 2.85×10^n.

$0.000000285 = 2.85 \times 10^{-7}$ **There are seven decimal places, so the power is −7**

To express a small number in standard form, write the value for a and place extra zeros to the left of a. Use the power and move the decimal point in a that many places to the left.

Example:

$1.6 \times 10^{-4} = 00001.6 \times 10^{-4} = 0.00016$

▶ PRACTICE Problems

11. Express each number in scientific notation.

a. 456,000,000 **b.** 0.000020

12. Express each number in standard notation.

a. 3.03×10^{-7} **b.** 9.7×10^{10}

Operations with Scientific Notation

Calculating with numbers written in scientific notation uses the properties of exponents (see page 840).

Multiplication Multiply the terms and add the powers of 10.

Example: Simplify.

$(4.0 \times 10^{-8})(1.2 \times 10^{5}) = (4.0 \times 1.2)(10^{-8} \times 10^{5})$ **Group terms and bases of 10**

$= (4.8)(10^{-8+5})$ **Multiply terms**

$= (4.8)(10^{-3})$ **Add powers of 10**

$= 4.8 \times 10^{-3}$ **Recombine in scientific notation**

Division Divide the base numbers and subtract the exponents of 10.

Example: Simplify.

$\frac{9.60 \times 10^{7}}{1.60 \times 10^{3}} = \left(\frac{9.60}{1.60}\right) \times \left(\frac{10^{7}}{10^{3}}\right)$ **Group terms and bases of 10**

$= 6.00 \times 10^{7-3}$ **Divide terms and subtract powers of 10**

$= 6.00 \times 10^{4}$

Addition and subtraction Adding and subtracting numbers in scientific notation is more challenging, because the powers of 10 must be the same in order to add or subtract the numbers. This means that one of the numbers may need to be rewritten with a different power of 10. If the powers of 10 are equal, then use the distributive property.

Example: Simplify.

$$(3.2\times10^5) + (4.8\times10^5) = (3.2 + 4.8)\times10^5$$ **Group terms**

$$= 8.0\times10^5$$ **Add terms**

Example: Simplify.

$$(3.2\times10^5) + (4.8\times10^4) = (3.2\times10^5) + (0.48\times10^5)$$ **Rewrite 4.8×10^4 as 0.48×10^5**

$$= (3.2 + 0.48)\times10^5$$ **Group terms**

$$= 3.68\times10^5$$ **Add terms**

$$= 3.7\times10^5$$ **Round using addition/subtraction rule for significant digits**

▶ PRACTICE Problems

13. Evaluate each expression. Express the result in scientific notation.

a. $(5.2\times10^{-4})(4.0\times10^8)$ **b.** $(2.4\times10^3) + (8.0\times10^4)$

VI. Equations

Order of Operations

Scientists and mathematicians have agreed on a set of steps or rules, called the order of operations, so that everyone interprets mathematical symbols in the same way. Follow these steps in order when you evaluate an expression or use a formula.

1. Simplify the expressions inside grouping symbols, like parentheses ( ), brackets [ ], braces { }, and fraction bars.
2. Evaluate all powers and roots.
3. Do all multiplications and/or divisions from left to right.
4. Do all additions and/or subtractions from left to right.

Example: Simplify the following expression.

$$4 + 3(4 - 1) - 2^3 = 4 + 3(3) - 2^3$$ **Order of operations step 1**

$$= 4 + 3(3) - 8$$ **Order of operations step 2**

$$= 4 + 9 - 8$$ **Order of operations step 3**

$$= 8$$ **Order of operations step 4**

Connecting Math to Physics The previous example was shown step-by-step to demonstrate order of operations. When solving a physics problem, do not round to the correct number of siginificant digits until after the final calculation. In calculations involving an expression in a numerator and an expression in a denominator, the numerator and the denominator are separate groups and are to be calculated before dividing the numerator by the denominator. So, the multiplication/division rule is used to determine the final number of significant digits.

Solving Equations

To solve an equation means to find the value of the variable that makes the equation a true statement. To solve equations, apply the distributive property and the properties of equality. Any properties of equalities that you apply on one side of an equation, you also must apply on the other side.

Distributive property For any numbers *a, b,* and *c,*

$$a(b + c) = ab + ac \qquad a(b - c) = ab - ac$$

Example: Use the distributive property to expand the following expression.

$$\begin{aligned} 3(x + 2) &= 3x + (3)(2) \\ &= 3x + 6 \end{aligned}$$

Addition and subtraction properties of equality If two quantities are equal and the same number is added to (or subtracted from) each, then the resulting quantities also are equal.

$$a + c = b + c \qquad a - c = b - c$$

Example: Solve $x - 3 = 7$ using the addition property.

$$\begin{aligned} x - 3 &= 7 \\ x - 3 + 3 &= 7 + 3 \\ x &= 10 \end{aligned}$$

Example: Solve $t + 2 = -5$ using the subtraction property.

$$\begin{aligned} t + 2 &= -5 \\ t + 2 - 2 &= -5 - 2 \\ t &= -7 \end{aligned}$$

Multiplication and division properties of equality If two equal quantities each are multiplied by (or divided by) the same number, then the resulting quantities also are equal.

$$ac = bc$$

$$\frac{a}{c} = \frac{b}{c}, \text{ for } c \neq 0$$

Example: Solve $\frac{1}{4}a = 3$ using the multiplication property.

$$\begin{aligned} \frac{1}{4}a &= 3 \\ \left(\frac{1}{4}a\right)(4) &= 3(4) \\ a &= 12 \end{aligned}$$

Example: Solve $6n = 8$ using the division property.

$$\begin{aligned} 6n &= 8 \\ \frac{6n}{6} &= \frac{18}{6} \\ n &= 3 \end{aligned}$$

Example: Solve $2t + 8 = 5t - 4$ for *t*.

$$\begin{aligned} 2t + 8 &= 5t - 4 \\ 8 + 4 &= 5t - 2t \\ 12 &= 3t \\ 4 &= t \end{aligned}$$

Isolating a Variable

Suppose an equation has more than one variable. To isolate a variable—that is, to solve the equation for a variable—write an equivalent equation so that one side contains only that variable with a coefficient of 1.

Connecting Math to Physics Isolate the variable P (pressure) in the ideal gas law equation.

$$PV = nRT$$

$$\frac{PV}{V} = \frac{nRT}{V} \quad \text{Divide both sides by } V$$

$$P\left(\frac{V}{V}\right) = \frac{nRT}{V} \quad \text{Group } \frac{V}{V}$$

$$P = \frac{nRT}{V} \quad \text{Substitute } \frac{V}{V} = 1$$

▶ PRACTICE Problems

14. Solve for x.

a. $2 + 3x = 17$

b. $x - 4 = 2 - 3x$

c. $t - 1 = \frac{x + 4}{3}$

d. $a = \frac{b + x}{c}$

e. $\frac{2x + 3}{x} = 6$

f. $ax + bx + c = d$

Square Root Property

If a and n are real numbers, $n > 0$, and $a^2 = n$, then $a = \pm\sqrt{n}$.

Connecting Math to Physics Solve for v in Newton's second law for a satellite orbiting Earth.

$$\frac{mv^2}{r} = \frac{Gm_Em}{r^2}$$

$$\frac{rmv^2}{r} = \frac{rGm_Em}{r^2} \quad \text{Multiply both sides by } r$$

$$mv^2 = \frac{Gm_Em}{r} \quad \text{Substitute } \frac{r}{r} = 1$$

$$\frac{mv^2}{m} = \frac{Gm_Em}{rm} \quad \text{Divide both sides by } m$$

$$v^2 = \frac{Gm_E}{r} \quad \text{Substitute } \frac{m}{m} = 1$$

$$\sqrt{v^2} = \pm\sqrt{\frac{Gm_E}{r}} \quad \text{Take the square root}$$

$$v = \sqrt{\frac{Gm_E}{r}} \quad \text{Use the positive value for speed}$$

When using the square root property, it is important to consider for what you are solving. Because we solved for speed in the above example, it did not make sense to use the negative value of the square root. Also, you need to consider if the negative or positive value gives you a realistic solution. For example, when using the square root property to solve for time, a negative value may give you a time before the situation even started.

Math Handbook

Quadratic Equations

A quadratic equation has the form $ax^2 + bx + c = 0$, where $a \neq 0$. A quadratic equation has one variable with a power (exponent) of 2. It also may include that same variable to the first power. The solutions of a quadratic equation can be estimated by graphing on a graphing calculator. If $b = 0$, then there is no x-term in the quadratic equation. The equation can be solved by isolating the squared variable and finding the square root of each side of the equation using the square root property.

Quadratic Formula

The solutions of any quadratic equation can be calculated by using the quadratic formula. The solutions of $ax^2 + bx + c = 0$, where $a \neq 0$, are given by

$$x = \frac{-b \pm \sqrt{b^2 - 4ac}}{2a}$$

Like in the square root property, it is important to consider if the solutions to the quadratic formula give you a realistic answer to the problem you are solving. Usually, you can throw out one of the solutions because it is unrealistic. Projectile motion often requires the use of the quadratic formula when solving equations, so keep the realism of the solution in mind when solving.

▶ PRACTICE **Problems**

15. Solve for x.

a. $4x^2 - 19 = 17$

b. $12 - 3x^2 = -9$

c. $x^2 - 2x - 24 = 0$

d. $24x^2 - 14x - 6 = 0$

Dimensional Calculations

When doing calculations, you must include the units of each measurement that is written in the calculation. All operations that are performed on the number also are performed on its units.

Connecting Math to Physics The acceleration due to gravity, a, is given by the equation $a = \frac{2\Delta x}{\Delta t^2}$. A free-falling object near the Moon drops 20.5 m in 5.00 s. Find the acceleration, a. Acceleration is measured in meters per second squared.

$$a = \frac{2\Delta x}{\Delta t^2}$$

$$a = \frac{2(20.5\ \text{m})}{(5.00\ \text{s})^2}$$ **The number two is an exact number, so it does not affect the determination of significant digits**

$$a = \frac{1.64\ \text{m}}{\text{s}^2} \text{ or } 1.64\ \text{m/s}^2$$ **Calculate and round to three significant digits**

Unit conversion Use a conversion factor to convert from one measurement unit to another of the same type, such as from minutes to seconds. This is equivalent to multiplying by one.

Connecting Math to Physics Find Δx when $v_0 = 67$ m/s and $\Delta t = 5.0$ min. Use the equation, $\Delta x = v_0 \Delta t$.

$$\frac{60 \text{ seconds}}{1 \text{ minute}} = 1$$

$$\Delta x = v_0 \Delta t$$

$$\Delta x = \frac{67 \text{ m}}{\text{s}}\left(\frac{5.0 \text{ min}}{1}\right)\left(\frac{60 \text{ s}}{1 \text{ min}}\right)$$

Multiply by the conversion factor $\frac{60 \text{ s}}{1 \text{ min}} = 1$

$$\Delta x = 20100 \text{ m} = 2.0 \times 10^4 \text{ m}$$

Calculate and round to two significant digits. The numbers 60 s and 1 min are exact numbers, so they do not affect the determination of significant digits

▶ PRACTICE Problems

16. Simplify $\Delta t = \frac{4.0 \times 10^2 \text{ m}}{16 \text{ m/s}}$.

17. Find the velocity of a dropped brick after 5.0 s using $v = a\Delta t$ and $a = -9.80$ m/s^2.

18. Calculate the product: $\left(\frac{32 \text{ cm}}{1 \text{ s}}\right)\left(\frac{60 \text{ s}}{1 \text{ min}}\right)\left(\frac{60 \text{ min}}{1 \text{ h}}\right)\left(\frac{1 \text{ m}}{100 \text{ cm}}\right)$

19. An Olympic record for 100.0 m is 9.87 s. What is the speed in kilometers per hour?

Dimensional Analysis

Dimensional analysis is a method of doing algebra with the units. It often is used to check the validity of the units of a final result and the equation being used, without completely redoing the calculation.

Physics Example Verify that the final answer of $d_f = d_i + v_i t + \frac{1}{2}at^2$ will have the units m.

d_i is measured in m.
t is measured in s.
v_i is measured in m/s.
a is measured in m/s^2.

$$d_f = \text{m} + \left(\frac{\text{m}}{\text{s}}\right)(\text{s}) + \frac{1}{2}\left(\frac{\text{m}}{\text{s}^2}\right)(\text{s})^2$$

Substitute the units for each variable

$$= \text{m} + (\text{m})\left(\frac{\text{s}}{\text{s}}\right) + \frac{1}{2}(\text{m})\left(\frac{\text{s}^2}{\text{s}^2}\right)$$

Simplify the fractions using the distributive property

$$= \text{m} + (\text{m})(1) + \frac{1}{2}(\text{m})(1)$$

Substitute s/s = 1, s^2/s^2 = 1

$$= \text{m} + \text{m} + \frac{1}{2}\text{m}$$

Everything simplifies to m, thus d_f is in m

The factor of $\frac{1}{2}$ in the above does not apply to the units. It applies only to any number values that would be inserted for the variables in the equation. It is easiest to remove number factors like the $\frac{1}{2}$ when first setting up the dimensional analysis.

VII. Graphs of Relations

The Coordinate Plane

Points are located in reference to two perpendicular number lines, called axes. The horizontal number line is called the *x*-axis; the vertical number line is called the *y*-axis. The *x*-axis represents the independent variable. The *y*-axis represents the dependent variable. A point is represented by two coordinates (*x, y*), which also is called an ordered pair. The value of the independent variable, *x*, always is listed first in the ordered pair. The ordered pair (0, 0) represents the origin, which is the point where the two axes intersect.

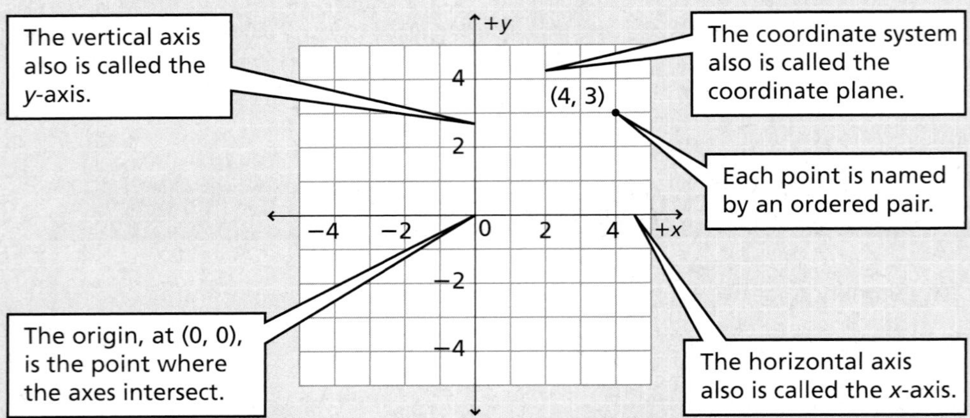
The vertical axis also is called the *y*-axis.
+y
4
2
−2
−4
−4
−2
0
2
4
+x
(4, 3)
The coordinate system also is called the coordinate plane.
Each point is named by an ordered pair.
The origin, at (0, 0), is the point where the axes intersect.
The horizontal axis also is called the *x*-axis.

Graphing Data to Determine a Relationship

Use the following steps to graph data.

1. Draw two perpendicular axes.
2. Identify the independent and the dependent variables. Label each axis using the variable names.
3. Determine the range of data for each variable. Use the ranges to decide on a convenient scale for each axis. Mark and number the scales.
4. Plot each data point.
5. When the points seem to lie approximately in a line, draw a "best-fit" line through the points. When the points do not lie in a line, draw a smooth curve through as many data points as possible. When there does not appear to be a trend or a pattern to the dots, do not draw any line or curve.
6. Write a title that clearly describes what the graph represents.

Expense	Pounds	Dollars
Hotel	40	58
Meals	30	43
Entertainment	15	22
Transportation	6	9

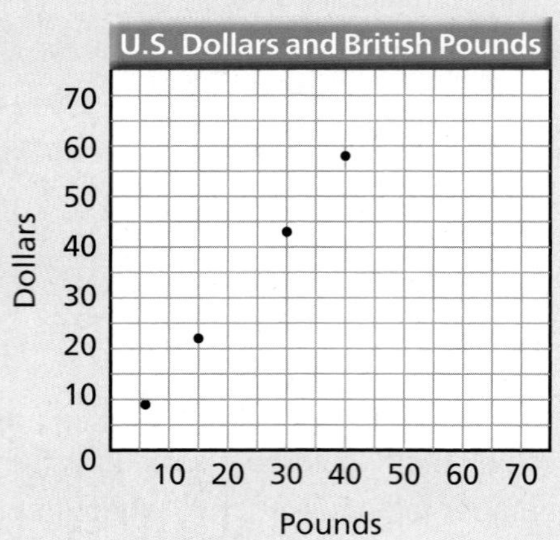
U.S. Dollars and British Pounds
Dollars
70
60
50
40
30
20
10
0
10 20 30 40 50 60 70
Pounds

Interpolating and Extrapolating

Interpolation is a process used to estimate a value for a relation that lies between two known values. Extrapolation is a process used to estimate a value for a relation that lies beyond the known values. The equation of a line through the points (the known values) helps you interpolate and extrapolate.

Example: Using the data and the graph, estimate the value, in dollars, of 20 pounds. Use interpolation.

Locate two points on either side of 20 pounds—15 pounds and 30 pounds. Draw a line (or place a straight edge) through the graphed points. Draw a line segment from 20 on the *x*-axis to that line. Draw a line segment from that intersection point to the *y*-axis. Read the scale on the *y*-axis. 20 pounds is about 28 or 29 dollars.

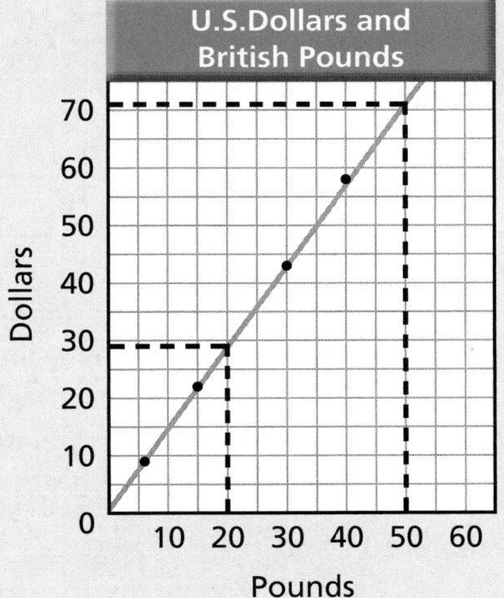

Example: Use extrapolation to estimate the value of 50 pounds.

Draw a line segment from 50 on the *x*-axis to the line through the points. Extend the line if necessary. Read the corresponding value on the *y*-axis. Extend the axis scale if necessary. 50 pounds is a little more than 70 dollars.

Interpreting Line Graphs

A line graph shows the linear relationship between two variables. Two types of line graphs that describe motion are used frequently in physics.

Connecting Math to Physics The following line graph shows a changing relationship between the two graphed variables.

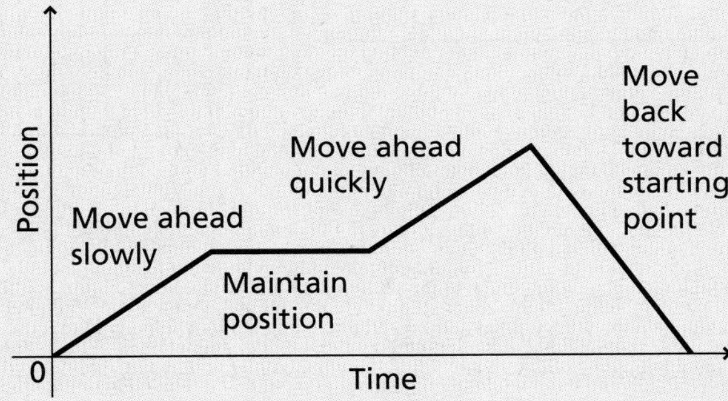

This line graph shows a constant relationship between the two graphed variables.

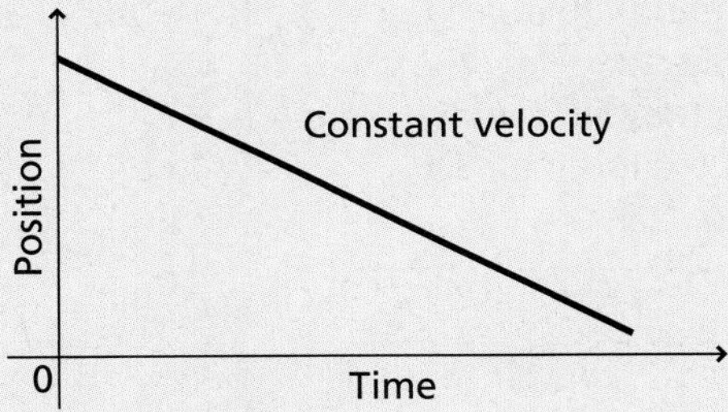

Math Handbook

Linear Equations

A linear equation can be written as a relation (or a function), $y = mx + b$, where m and b are real numbers, m represents the slope of the line, and b represents the y-intercept, the point at which the line crosses the y-axis.

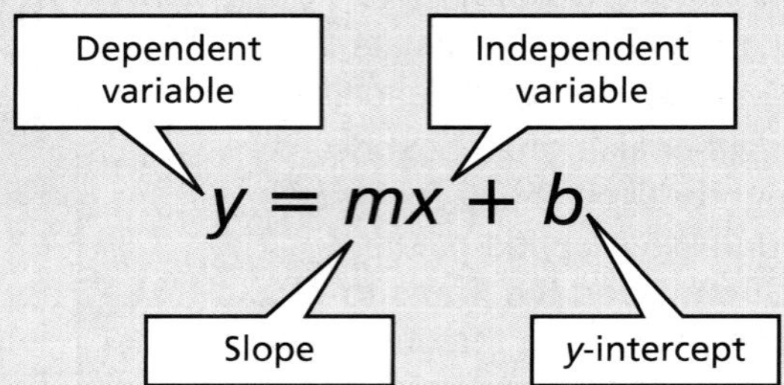

The graph of a linear equation is a line. The line represents all of the solutions of the linear equation. To graph a linear equation, choose three values for the independent variable. (Only two points are needed, but the third point serves as a check.) Calculate the corresponding values for the dependent variable. Plot each ordered pair (x, y) as a point. Draw a best-fit line through the points.

Example: Graph $y = -\frac{1}{2}x+3$

Calculate three ordered pairs to obtain points to plot.

Ordered Pairs	
x	**y**
0	3
2	2
6	0

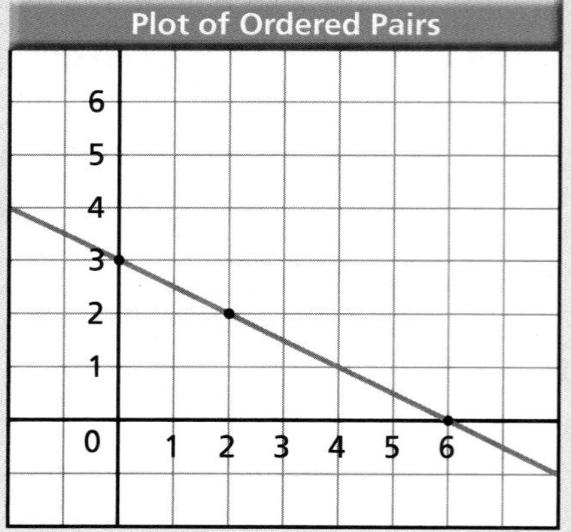

Slope

The slope of a line is the ratio of the change in y-coordinates to the change in x-coordinates. It is the ratio of the vertical change (rise) to the horizontal change (run). This number tells how steep the line is. It can be a positive or a negative number.

To find the slope of a line, select two points, (x_1, y_1) and (x_2, y_2). Calculate the run, which is the difference (change) between the two x-coordinates, $x_2 - x_1 = \Delta x$. Calculate the rise, which is the difference (change) between the two y-coordinates, $y_2 - y_1 = \Delta y$. Form the ratio.

$$\text{Slope } m = \frac{rise}{run} = \frac{y_2 - y_1}{x_2 - x_1} = \frac{\Delta y}{\Delta x},$$

where $x_1 \neq x_2$

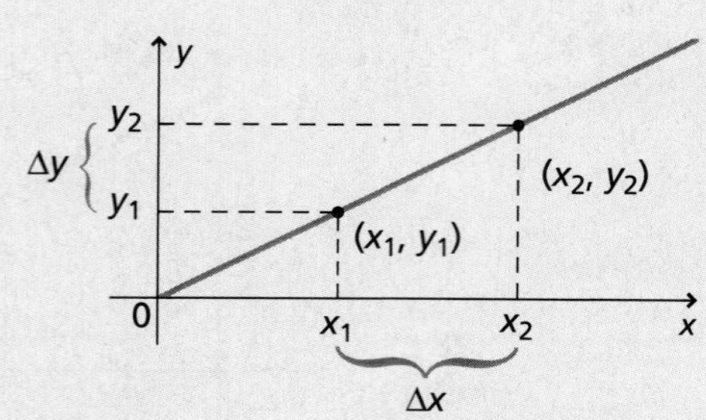

Direct Variation

If there is some nonzero constant, m, such that $y = mx$, then y varies directly as x. That means as the independent variable x increases, the dependent variable y increases. The variables x and y also are said to be proportional. This is a linear equation of the form $y = mx + b$ in which the value of b is zero. The graph passes through the origin, (0, 0).

Connecting Math to Physics In the force equation for an ideal spring, $F = kx$, where F is the spring force, k is the spring constant, and x is the spring displacement, spring force varies directly as (is proportional to) spring displacement. That is, the spring force increases as the spring displacement increases.

Inverse Variation

If there is some nonzero constant, m, such that $y = \frac{m}{x}$, then y varies inversely as x. That means as the independent variable x increases, the dependent variable y decreases. The variables x and y also are said to be inversely proportional. This is not a linear equation because it contains the product of two variables. The graph of an inverse relationship is a hyperbola. This relationship can be written as

$$y = \frac{m}{x}$$

$$y = m\frac{1}{x}$$

$$xy = m$$

Example: Graph the equation $xy = 90$.

Ordered Pairs	
x	**y**
−10	−9
−6	−15
−3	−30
−2	−45
2	45
3	30
6	15
10	9

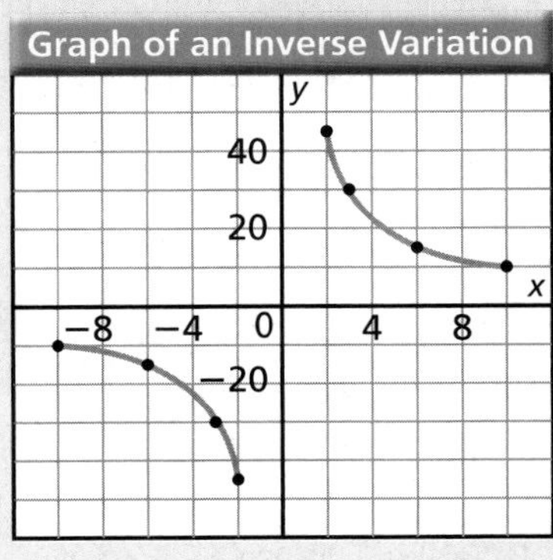

Connecting Math to Physics In the equation for the speed of a wave, $\lambda = \frac{v}{f}$, where λ is wavelength, f is frequency, and v is wave speed, wavelength varies inversely as (is inversely proportional to) frequency. That is, as the frequency of a wave increases, the wavelength decreases. v is constant.

Quadratic Graphs

A quadratic relationship is a relationship of the form

$y = ax^2 + bx + c$, where $a \neq 0$.

A quadratic relationship includes the square of the independent variable, x. The graph of a quadratic relationship is a parabola. Whether the parabola opens upward or downward depends on whether the value of the coefficient of the squared term, a, is positive or negative.

Example: Graph the equation $y = -x^2 + 4x - 1$.

Ordered Pairs	
x	y
−1	−6
0	−1
1	2
2	3
3	2
4	−1
5	−6

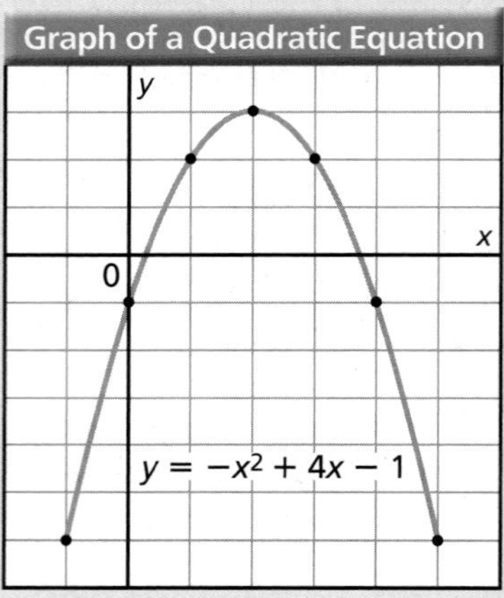

Connecting Math to Physics A position-time graph in the shape of a quadratic relation means that the object is moving at a constant acceleration.

Ordered Pairs	
Time (s)	**Position (m)**
1	3
2	6
3	11
4	18

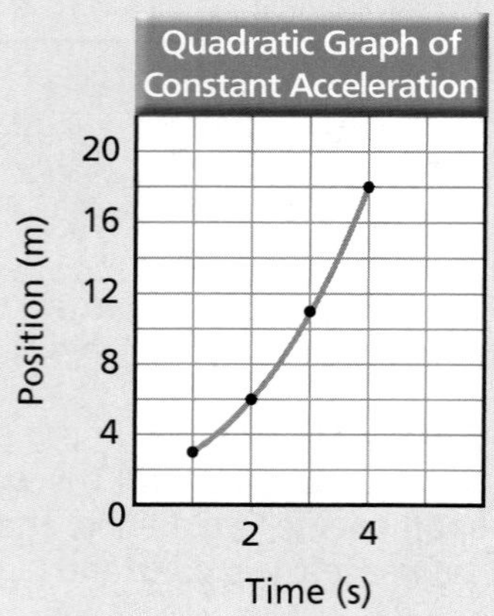

VIII. Geometry and Trigonometry

Perimeter, Area, and Volume

	Perimeter, Circumference Linear units	**Area** Squared units	**Surface Area** Squared units	**Volume** Cubic units
Square, side a	$P = 4a$	$A = a^2$		
Rectangle, length l width w	$P = 2l + 2w$	$A = lw$		
Triangle base b height h		$A = \frac{1}{2}bh$		
Cube side a			$SA = 6a^2$	$V = a^3$
Circle radius r	$C = 2\pi r$	$A = \pi r^2$		
Cylinder radius r height h			$SA = 2\pi rh + 2\pi r^2$	$V = \pi r^2 h$
Sphere radius r			$SA = 4\pi r^2$	$V = \frac{4}{3}\pi r^3$

Connecting Math to Physics Look for geometric shapes in your physics problems. They could be in the form of objects or spaces. For example, two-dimensional shapes could be formed by velocity vectors, as well as position vectors.

Area Under a Graph

To calculate the approximate area under a graph, cut the area into smaller pieces and find the area of each piece using the formulas shown above. To approximate the area under a line, cut the area into a rectangle and a triangle, as shown in **(a).**

To approximate the area under a curve, draw several rectangles from the x-axis to the curve, as in **(b).** Using more rectangles with a smaller base will provide a closer approximation of the area.

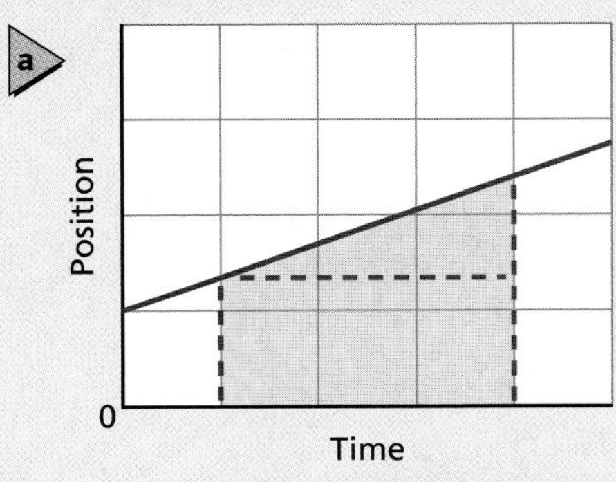

Total area = Area of the rectangle + Area of the triangle

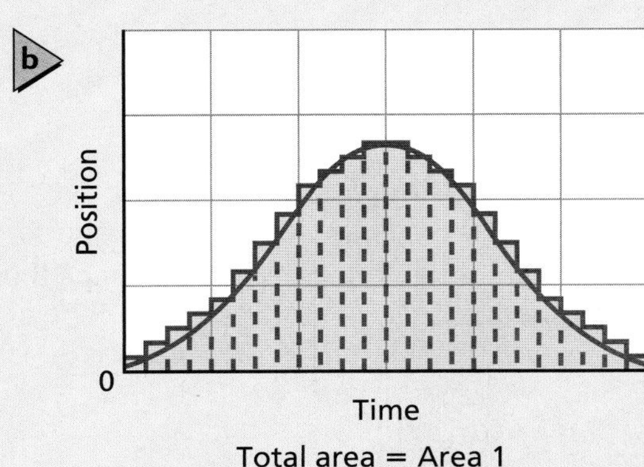

Total area = Area 1 + Area 2 + Area 3...

Right Triangles

The Pythagorean theorem states that if a and b are the measures of the legs of a right triangle and c is the measure of the hypotenuse, then

$$c^2 = a^2 + b^2.$$

To determine the length of the hypotenuse, use the square root property. Because distance is positive, the negative value does not have meaning.

$$c = \sqrt{a^2 + b^2}$$

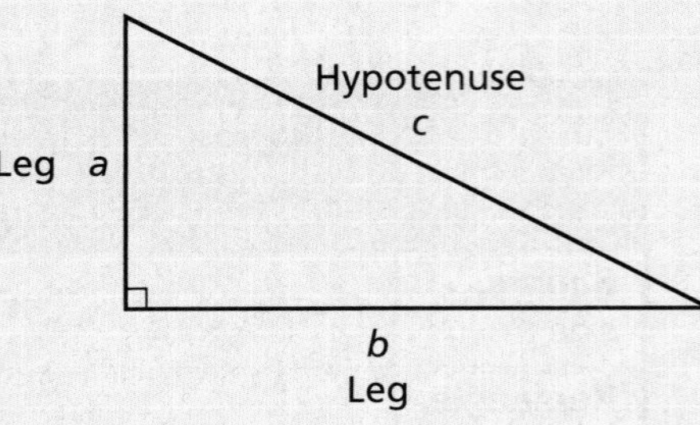

Example: In the triangle, $a = 4$ cm and $b = 3$ cm. Find c.

$$\begin{aligned} c &= \sqrt{a^2 + b^2} \\ &= \sqrt{(4\text{ cm})^2 + (3\text{ cm})^2} \\ &= \sqrt{16\text{ cm}^2 + 9\text{ cm}^2} \\ &= \sqrt{25\text{ cm}^2} \\ &= 5\text{ cm} \end{aligned}$$

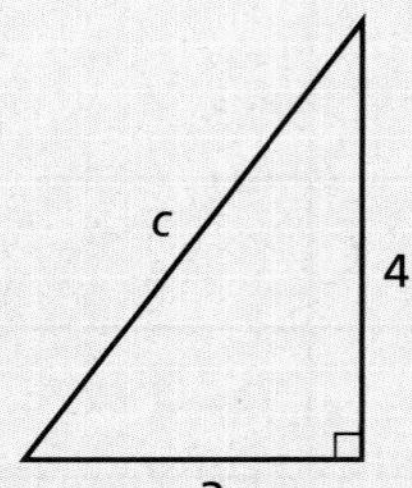

45°-45°-90° triangles The length of the hypotenuse is $\sqrt{2}$ times the length of a leg.

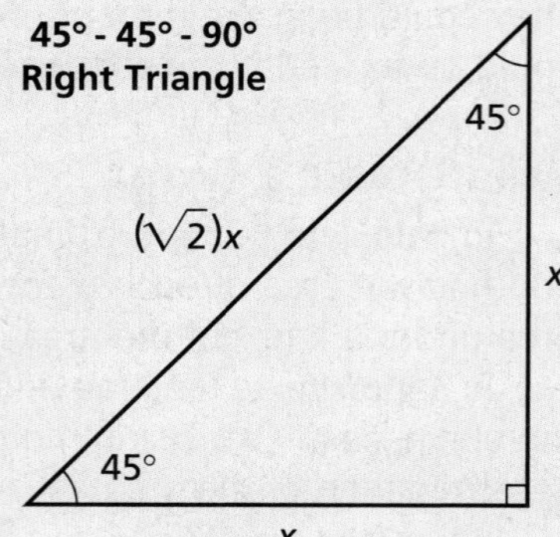

30°-60°-90° triangles The length of the hypotenuse is twice the length of the shorter leg. The length of the longer leg is $\sqrt{3}$ times the length of the shorter leg.

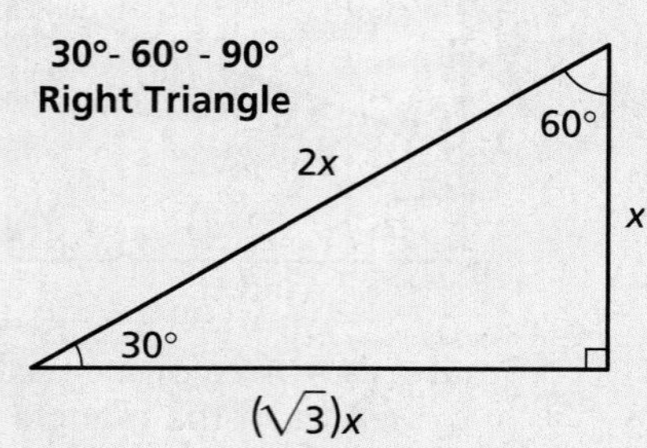

Trigonometric Ratios

A trigonometric ratio is a ratio of the lengths of sides of a right triangle. The most common trigonometric ratios are sine, cosine, and tangent.To memorize these ratios, learn the acronym SOH-CAH-TOA. SOH stands for Sine, Opposite, Hypotenuse. CAH stands for Cosine, Adjacent, Hypotenuse. TOA stands for Tangent, Opposite, Adjacent.

Words	Memory Aid	Symbols
The sine is the ratio of the length of the side opposite to the angle over the length of the hypotenuse.	SOH $\sin\theta = \frac{\text{opposite}}{\text{hypotenuse}}$	$\sin\theta = \frac{a}{c}$
The cosine is the ratio of the length of the side adjacent to the angle over the length of the hypotenuse.	CAH $\cos\theta = \frac{\text{adjacent}}{\text{hypotenuse}}$	$\cos\theta = \frac{b}{c}$
The tangent is the ratio of the length of the side opposite to the angle over the length of the side adjacent to the angle.	TOA $\tan\theta = \frac{\text{opposite}}{\text{adjacent}}$	$\tan\theta = \frac{a}{b}$

Example: In right triangle *ABC*, if $a = 3$ cm, $b = 4$ cm, and $c = 5$ cm, find $\sin\theta$ and $\cos\theta$.

$$\sin\theta = \frac{3\text{ cm}}{5\text{ cm}} = 0.6$$

$$\cos\theta = \frac{4\text{ cm}}{5\text{ cm}} = 0.8$$

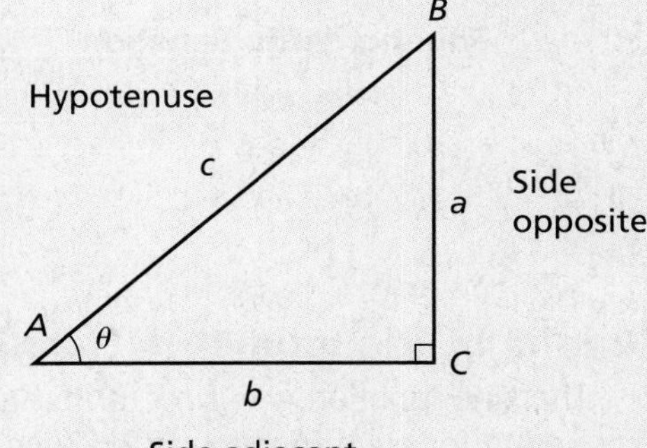

Example: In right triangle *ABC*, if $\theta = 30.0°$ and $c = 20.0$ cm, find *a* and *b*.

$$\sin 30.0° = \frac{a}{20.0\text{ cm}} \qquad \cos 30.0° = \frac{b}{20.0\text{ cm}}$$

$$a = (20.0\text{ cm})(\sin 30.0°) = 10.0\text{ cm}$$

$$b = (20.0\text{ cm})(\cos 30.0°) = 17.3\text{ cm}$$

Law of Cosines and Law of Sines

The laws of cosines and sines let you calculate sides and angles in any triangle.

Law of cosines The law of cosines looks like the Pythagorean theorem, except for the last term. θ is the angle opposite side *c*. If the angle θ is 90°, the $\cos\theta = 0$ and the last term equals zero. If θ is greater than 90°, its cosine is a negative number.

$$c^2 = a^2 + b^2 - 2ab\cos\theta$$

Example: Find the length of the third side of a triangle with $a = 10.0$ cm, $b = 12.0$ cm, $\theta = 110.0°$.

$$c^2 = a^2 + b^2 - 2ab\cos\theta$$

$$c = \sqrt{a^2 + b^2 - 2ab\cos\theta}$$

$$= \sqrt{(10.0\text{ cm})^2 + (12.0\text{ cm})^2 - 2(10.0\text{ cm})(12.0\text{ cm})(\cos 110.0°)}$$

$$= \sqrt{1.00\times10^2\text{ cm}^2 + 144\text{ cm}^2 - (60.0\text{ cm}^2)(\cos 110.0°)}$$

$$= 16.3\text{ cm}$$

Law of sines The law of sines is an equation of three ratios, where *a, b,* and *c* are the sides opposite angles *A, B,* and *C,* respectively. Use the law of sines when you know the measures of two angles and any side of a triangle.

$$\frac{\sin A}{a} = \frac{\sin B}{b} = \frac{\sin C}{c}$$

Example: In a triangle, $C = 60.0°$, $a = 4.0$ cm, $c = 4.6$ cm. Find the measure of angle *A*.

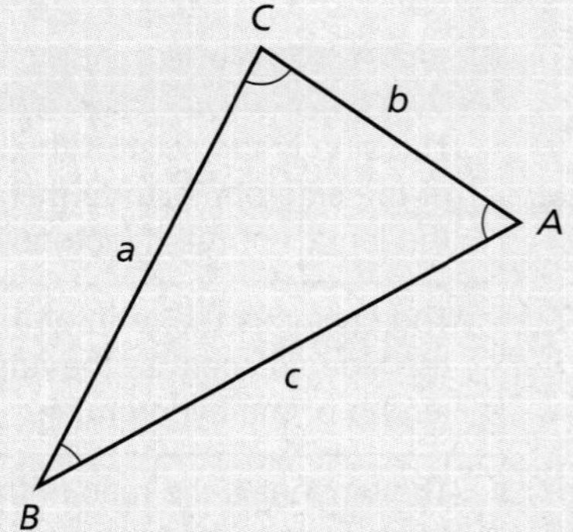

$$\frac{\sin A}{a} = \frac{\sin C}{c}$$

$$\sin A = \frac{a \sin C}{c}$$

$$= \frac{(4.0 \text{ cm})(\sin 60.0°)}{4.6 \text{ cm}}$$

$$= 49°$$

Inverses of Sine, Cosine, and Tangent

The inverses of sine, cosine, and tangent allow you to do the reverse of the sine, cosine, and tangent functions and find the angle. The trigonometric functions and their inverses are as follows:

Trigonometric Function	Inverse
$y = \sin x$	$x = \sin^{-1} y$ or $x = \text{arc} \sin y$
$y = \cos x$	$x = \cos^{-1} y$ or $x = \text{arc} \cos y$
$y = \tan x$	$x = \tan^{-1} x$ or $x = \text{arc} \tan y$

Graphs of Trigonometric Functions

The sine function, $y = \sin x$, and the cosine function, $y = \cos x$ are periodic functions. The period for each function is 2π. *x* can be any real number. *y* is any real numbers between -1 and 1, inclusive.

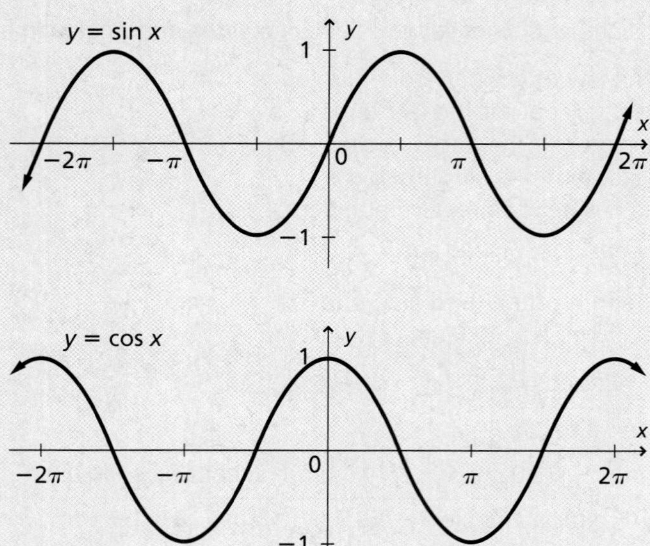

IX. Logarithms

Logarithms with Base b

Let b and x be positive integers, $b \neq 1$. The logarithm of x with base b, written $\log_b x$, equals y, where y is the exponent that makes the equation $b^y = x$ true. The log to the base b of x is the number to which you can raise b to get x.

$$\log_b x = y \text{ if and only if } b^y = x$$

Memory aid: "the log is the exponent"

Examples: Calculate the following logarithms.

$\log_2 \frac{1}{16} = -4$ **Because $2^{-4} = \frac{1}{16}$**

$\log_{10} 1000 = 3$ **Because $10^3 = 1000$**

When you want to find the log of a number, you also can use a calculator.

Connecting Math to Physics Physicists use logarithms to work with measurements that extend over several orders of magnitude, or powers of 10. Geophysicists use the Richter scale, a logarithmic scale that allows them to rate earthquakes from 0 to 7, or larger, though the power of earthquakes differ by 7 or more powers of 10.

Common Logarithms

Base 10 logarithms are called common logarithms. They often are written without the subscript 10.

$$\log_{10} x = \log x \quad x > 0$$

Antilogarithms or Inverse Logarithms

An antilogarithm is the inverse of a logarithm. An antilogarithm is the number which has a given logarithm.

Example: Solve $\log x = 4$ for x.

$\log x = 4$

$x = 10^4$ **10^4 is the antilogarithm of 4**

Connecting Math to Physics An equation for loudness, L, in decibels, is $L = 10 \log_{10} R$, where R is the relative intensity of the sound. Calculate R for a fireworks display with a loudness of 130 decibels.

$130 = 10 \log_{10} R$ **Divide by 10**

$13 = \log_{10} R$ **Use logarithm rule**

$R = 10^{13}$

When you know the logarithm of a number and want to know the number itself, use a calculator to find the antilogarithm.

▶ PRACTICE **Problems**

20. Write in exponential form. $\log_3 81 = 4$

21. Write in logarithmic form. $10^{-3} = 0.001$

22. Find x. $\log x = 3.125$

Appendix B: Additional Problems

Chapter 1

1. The density, ρ, of an object is given by the ratio of the object's mass, m, and volume, V, according to the equation $\rho = m/V$. What is the density of a cube that is 1.2 cm on each side and has a mass of 25.6 g?

2. An object that is moving in a straight line with speed v covers a distance, $d = vt$, in time t. Rewrite the equation to find t in terms of d and v. How long does it take a plane that is traveling at 350 km/h to travel 1750 km?

3. Convert 523 kg to milligrams.

4. The liquid measure milliliter, mL, is the same as 1 cm^3. How many milliliters of liquid can be held in a 2.5-m^3 container?

5. Part of the label from a vitamin container is shown below. The abbreviation "mcg" stands for micrograms. Convert the values to milligrams.

Each Tablet Contains	%DV
Folic Acid 400 mcg	100%
Vitamin B12 6 mcg	100%
Biotin 30 mcg	10%

6. What type of relationship is shown in the scatter plot? Write an equation to model the data.

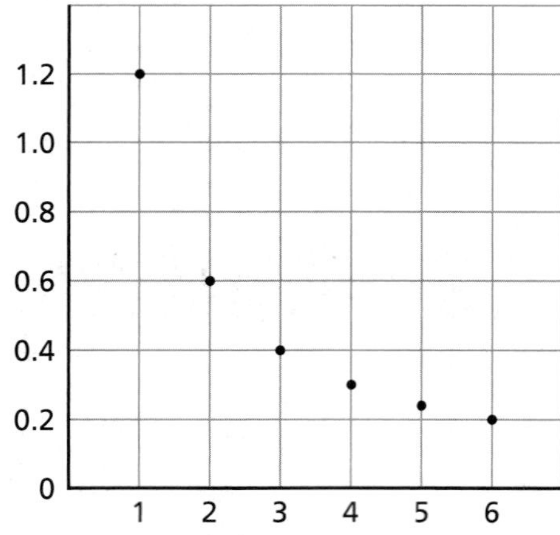

7. How many significant digits are there in each of the following measurements?

a. 100 m

b. 0.0023 m/s

c. 100.1 m

d. 2.0023

8. The buoyant (upward) force exerted by water on a submerged object is given by the formula $F = \rho V g$, where ρ is the density of water (1.0×10^3 kg/m^3), V is the volume of the object in m^3, and g is the acceleration due to gravity (9.80 m/s^2). The force is in newtons. Rewrite the equation to find V in terms of F. Use this to find the volume of a submerged barrel if the buoyant force on it is 9200 N.

9. What is 2.3 kg + 0.23 g?

10. Solve the following problems.

a. 15.5 cm $\times$ 12.1 cm

b. $\frac{14.678 \text{ m}}{3.2 \text{ m/s}}$

11. An experiment was performed to determine the period of a pendulum as a function of the length of its string. The data in the table below were measured.

a. Plot period, T, versus length, l.

b. Are the data linear?

c. Plot the period versus the square root of the length.

d. What is the relationship between the period and the square root of the length?

Length (m)	Period (s)
0.1	0.6
0.2	0.9
0.4	1.3
0.6	1.6
0.8	1.8
1.0	2.0

12. Based on the previous problem, what should be the period of a pendulum whose length is 0.7 m?

Chapter 2

1. A position-time graph for a bicycle is shown in the figure below.
   a. What is the position of the bicycle at 1.00 min?
   b. What is the position of the bicycle at 3.50 min?
   c. What is the displacement of the bicycle between the times 1.00 min and 5.00 min?
   d. Describe the motion of the bicycle.

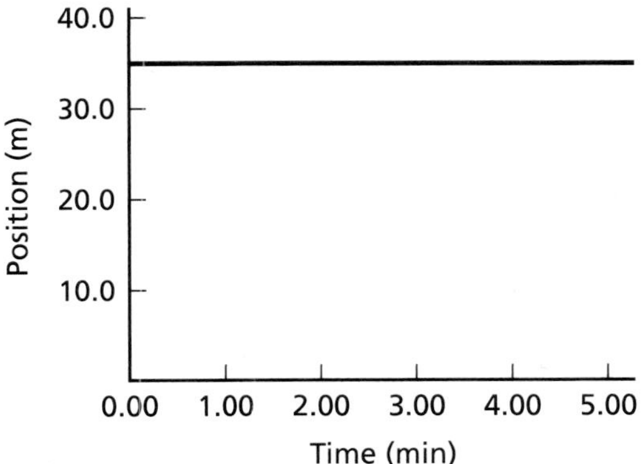

2. The position of an automobile is plotted as a function of time in the accompanying figure.
   a. What is the position of the car at 0.00 s?
   b. What is the position of the automobile after 2.00 s has elapsed?
   c. How far did the automobile travel between the times 1.00 s and 3.00 s?

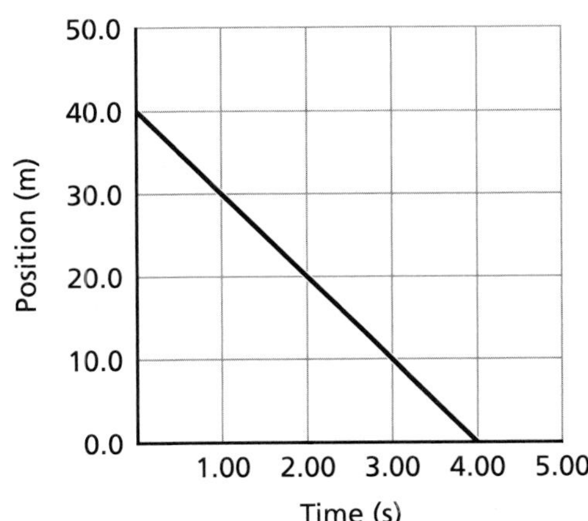

3. A jogger runs at a constant rate of 10.0 m every 2.0 s. The jogger starts at the origin, and runs in the positive direction for 3600.0 s. The figure below is a position-time graph showing the position of the jogger from time $t = 0.0$ s to time $t = 20.0$ s. Where is the runner at time $t = 5.0$ s? $t = 15.0$ s?

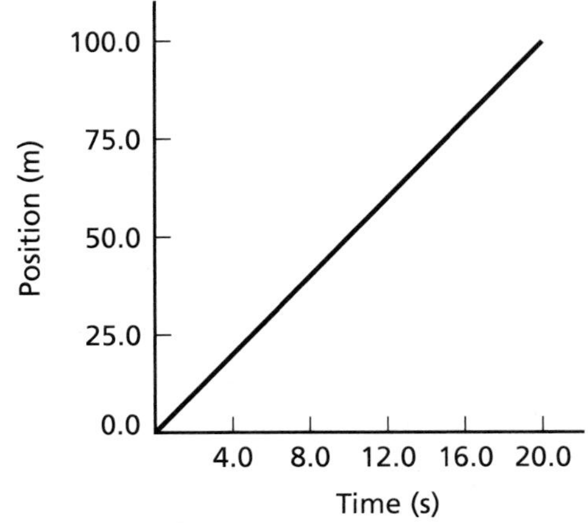

4. Two trains simultaneously leave the same train station at noon. One train travels north and the other travels south. The position-time graph for both trains is shown in the accompanying figure.
   a. What is the position of the train traveling north at 6.0 h?
   b. What is the position of the train traveling south at 6.0 h?
   c. What is the distance between the trains at 6.0 h? What is the distance at 10.0 h?
   d. At what time are the trains 600.0 km apart?
   e. Which train is moving more quickly?

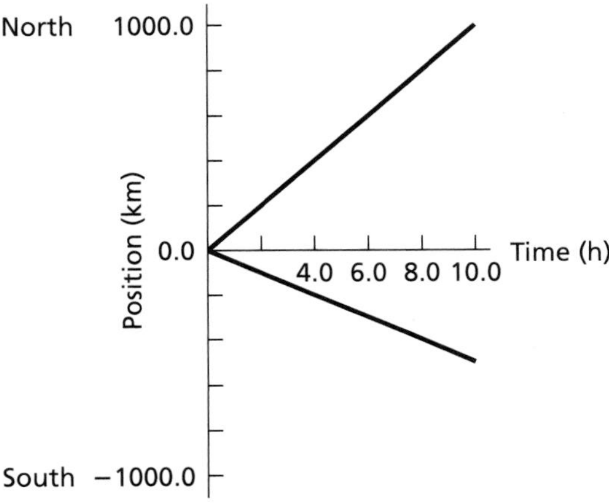

5. Two cars head out in the same direction. Car A starts 1.0 min before car B. The position-time graphs for both cars are shown in the accompanying figure.
   a. How far apart are the two cars when car B starts out at t = 1.0 min?
   b. At what time do the cars meet?
   c. How far apart are the cars at time t = 3.0 min?

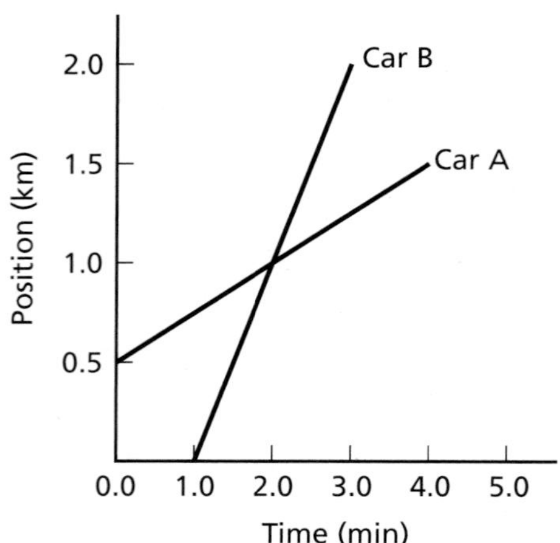

6. The position-time graph for two joggers, A and B, is shown in the accompanying figure.
   a. How far apart are the two runners at 10.0 min?
   b. At what time are they 1.00 km apart?
   c. How far apart are they at 50.0 min?
   d. At what time do they meet?
   e. What distance does jogger B cover between 30.0 min and 50.0 min?
   f. What distance does jogger A cover between 30.0 min and 50.0 min?

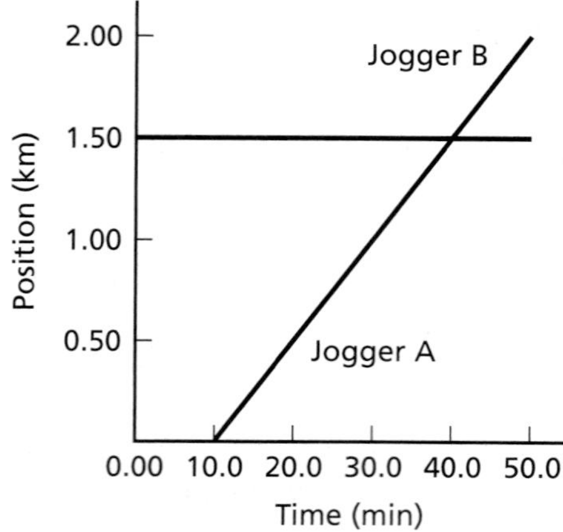

7. A child's toy train moves at a constant speed of 2.0 cm/s.
   a. Draw the position-time graph showing the position of the toy for 1.0 min.
   b. What is the slope of the line representing the motion of the toy?

8. The position of an airplane as a function of time is shown in the figure below.
   a. What is the average velocity of the airplane?
   b. What is the average speed of the airplane?

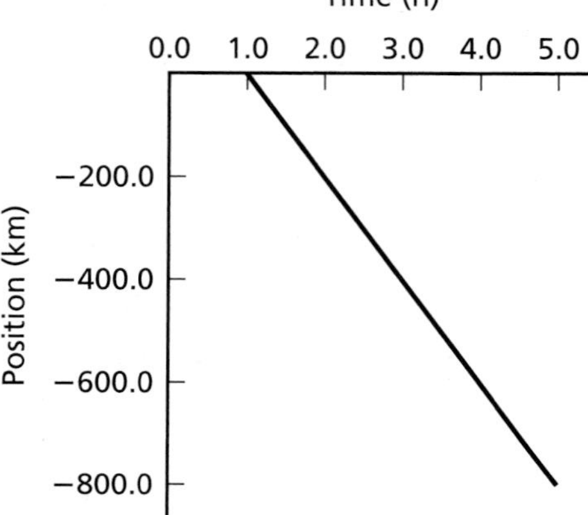

9. The position-time graph for a hot-air balloon that is in flight is shown in the accompanying figure.
   a. What is the average velocity of the balloon?
   b. What is the average speed of the balloon?

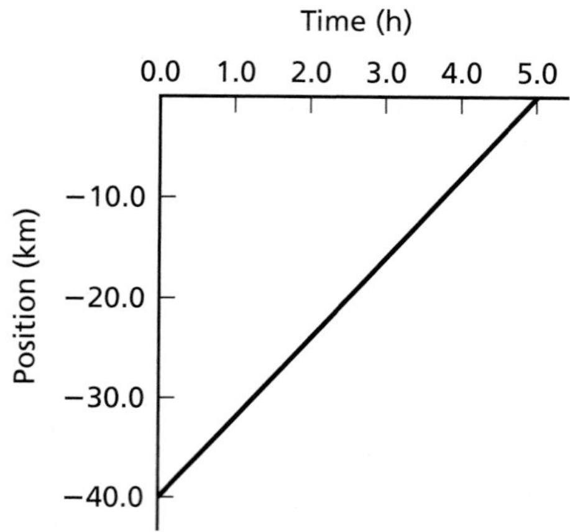

Chapter 3

1. Jason and his sister, Tara, are riding bicycles. Jason tries to catch up to Tara, who has a 10.0-s head start.
   a. What is Jason's acceleration?
   b. What is Tara's acceleration?
   c. At what time do they have the same velocity?

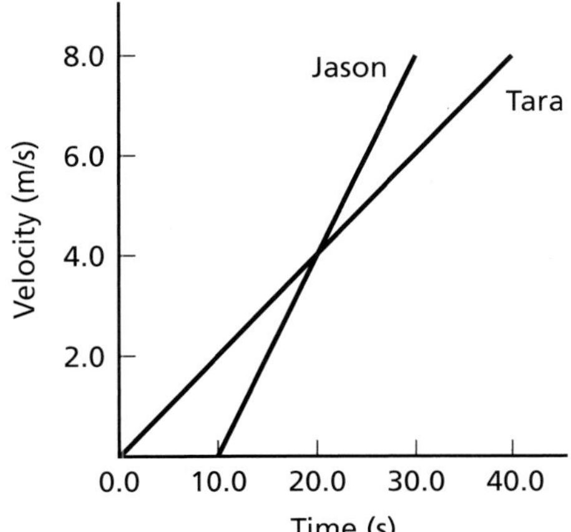

2. A dragster starts from rest and accelerates for 4.0 s at a rate of 5.0 m/s^2. It then travels at a constant speed for 2.5 s. A parachute opens, stopping the vehicle at a constant rate in 2.0 s. Plot the *v-t* graph representing the entire motion of the dragster.
3. A car traveling at 21 m/s misses the turnoff on the road and collides into the safety guard rail. The car comes to a complete stop in 0.55 s.
   a. What is the average acceleration of the car?
   b. If the safety rail consisted of a section of rigid rail, the car would stop in 0.15 s. What would be the acceleration in this case?
4. On the way to school, Jamal realizes that he left his physics homework at home. His car was initially heading north at 24.0 m/s. It takes him 35.5 s to turn his car around and head south at 15.0 m/s. If north is designated to be the positive direction, what is the average acceleration of the car during this 35.5-s interval?
5. A cheetah can reach a top speed of 27.8 m/s in 5.2 s. What is the cheetah's average acceleration?
6. After being launched, a rocket attains a speed of 122 m/s before the fuel in the motor is completely used. If you assume that the acceleration of the rocket is constant at 32.2 m/s^2, how much time does it take for the fuel to be completely consumed?
7. An object in free fall has an acceleration of 9.80 m/s^2 assuming that there is no air resistance. What is the speed of an object dropped from the top of a tall cliff 3.50 s after it has been released, if you assume the effect of air resistance against the object is negligible?
8. A train moving with a velocity of 51 m/s east undergoes an acceleration of -2.3 m/s^2 as it approaches a town. What is the velocity of the train 5.2 s after it has begun to decelerate?
9. The *v-t* graph of a runner is shown in the accompanying figure.
   a. What is the displacement of the runner between $t = 0.00$ s and $t = 20.0$ s?
   b. What is the displacement of the runner between $t = 20.0$ s and $t = 50.0$ s?
   c. What is the displacement of the runner between $t = 50.0$ s and $t = 60.0$ s?

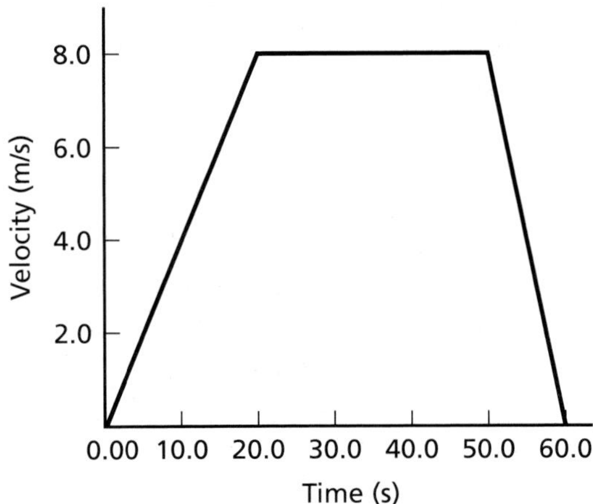

10. Draw the *v-t* graph of an automobile that accelerates uniformly from rest at $t = 0.00$ s and covers a distance of 180.0 m in 12.0 s.

11. The v-t graph of a car is shown in the accompanying figure. What is the displacement of the car from t = 0.00 s to t = 15.0 s?

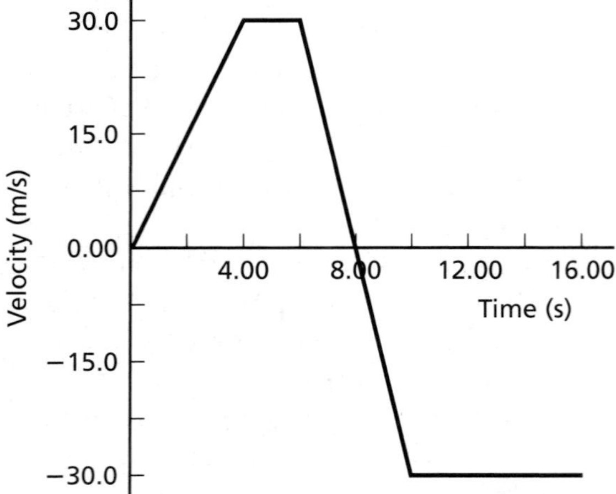

12. Suppose a car rolls down a 52.0-m-long inclined parking lot and is stopped by a fence. If it took the car 11.25 s to roll down the hill, what was the acceleration of the car before striking the fence?
13. A sky diver in free fall reaches a speed of 65.2 m/s when she opens her parachute. The parachute quickly slows her down to 7.30 m/s at a constant rate of 29.4 m/s^2. During this period of acceleration, how far does she fall?
14. A child rolls a ball up a hill at 3.24 m/s. If the ball experiences an acceleration of 2.32 m/s^2, how long will it take for the ball to have a velocity of 1.23 m/s down the hill?
15. A cheetah can accelerate from rest to a speed of 27.8 m/s in 5.20 s. The cheetah can maintain this speed for 9.70 s before it quickly runs out of energy and stops. What distance does the cheetah cover during this 14.9-s run?
16. A cab driver in a hurry is sitting at a red light. When the light turns green she rapidly accelerates for 3.50 s at 6.80 m/s^2. The next light is still red. She then slams on the brakes, accelerating at a rate of −9.60 m/s^2 before coming to rest at the stop light. What was her total distance for this trip?
17. A cyclist rides at a constant speed of 12.0 m/s for 1.20 min and then coasts to a stop with uniform acceleration 21.2 s later. If the total distance traveled is 1321 m, then what is the acceleration while the bike coasts to a stop?
18. A hiker tossed a water bottle to a friend down in a canyon. The friend caught it 4.78 s later. What is the height of the canyon?
19. A rock is thrown upward with a speed of 26 m/s. How long after it is thrown will the rock have a velocity of 48 m/s toward the ground?
20. The high-dive board at most pools is 3.00 m above the water. A diving instructor dives off the board and strikes the water 1.18 s later.
    - **a.** What was the initial velocity of the diver?
    - **b.** How high above the board did the diver rise?

Chapter 4

1. Draw a free-body diagram for the space shuttle just after it leaves the ground. Identify the forces acting on the shuttle. Make sure that you do not neglect air resistance. Also be sure that you indicate the direction of the acceleration, as well as the net force.
2. Draw a free-body diagram for a goldfish that is motionless in the middle of a fishbowl. Identify the forces acting on the fish. Indicate the direction of the net force on the fish and the direction of the acceleration of the fish.

3. Draw a free-body diagram for a submerged beach ball as it rises toward the surface just after being released. Identify the forces acting on the beach ball and indicate the direction of the net force and the acceleration.

4. Muturi is rearranging some furniture. He pushes the dresser with a force of 143 N, and there is opposing frictional force of 112 N. What is the net force?

5. One of the floats in a Thanksgiving Day parade requires four people pulling on ropes to maintain a constant speed of 3.0 km/h for the float. Two people pull with a force of 210 N each, and the other two pull with a force of 140 N each.
   a. Draw a free-body diagram.
   b. What is the force of friction between the float and the ground?

6. Five people are playing tug-of-war. Anders and Alyson pull to the right with 45 N and 35 N, respectively. Calid and Marisol pull to the left with 53 N and 38 N, respectively. With what force and in what direction does Benito pull if the game is tied?

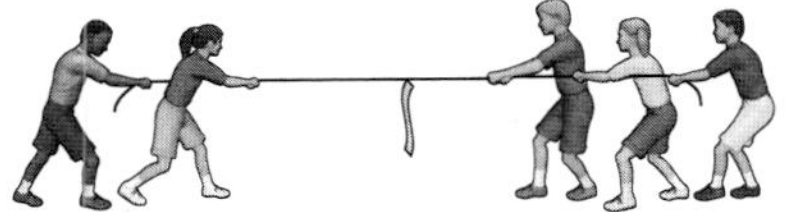

7. Two dogs fight over a bone. The larger of the two pulls on the bone to the right with a force of 42 N. The smaller one pulls to the left with a force of 35 N.
   a. Draw the free-body diagram for the bone.
   b. What is the net force acting on the bone?
   c. If the bone has a mass of 2.5 kg, what is its acceleration?

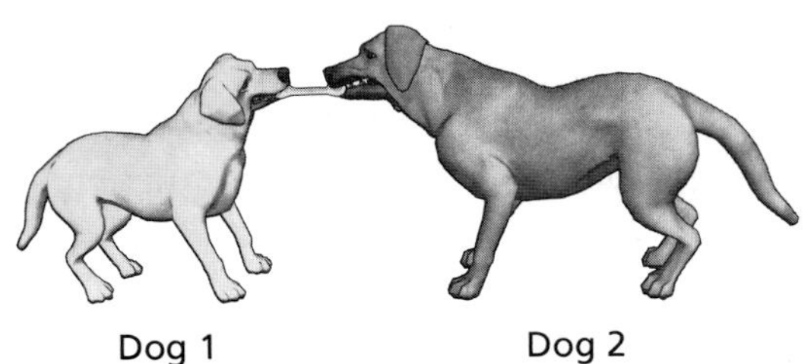

8. A large model rocket engine can produce a thrust of 12.0 N upon ignition. This engine is part of a rocket with a total mass of 0.288 kg when launched.
   a. Draw a free-body diagram of the rocket just after launch.
   b. What is the net force that is acting on the model rocket just after it leaves the ground?
   c. What is the initial acceleration of the rocket?

9. Erika is on an elevator and presses the button to go down. When the elevator first starts moving, it has an acceleration of 2.5 m/s^2 downward. Erika and the elevator have a combined mass of 1250 kg.
   a. Draw a free-body diagram for the elevator.
   b. What is the tension in the cable that provides the upward force on the elevator car?

10. Ngan has a weight of 314.5 N on Mars and a weight of 833.0 N on Earth.
   a. What is Ngan's mass on Mars?
   b. What is the acceleration due to gravity on Mars, g_{Mars}?

11. Alex is on the wrestling team and has a mass of 85.3 kg. Being a whiz at physics, he realizes that he is over the 830.0-N cut-off for his weight class. If he can convince the trainers to measure his weight in the elevator, what must be the acceleration of the elevator so that he just makes his weight class?

12. During a space launch, an astronaut typically undergoes an acceleration of 3 *g*s, which means he experiences an acceleration that is three times that of gravity alone. What would be the apparent weight of a 205-kg astronaut that experiences a 3-*g* liftoff?

13. Alfonso and Sarah like to go sky diving together. Alfonso has a mass of 88 kg, and Sarah has a mass of 66 kg. While in free fall together, Alfonso pushes Sarah horizontally with a force of 12.3 N.
   a. What is Alfonso's horizontal acceleration?
   b. What is Sarah's horizontal acceleration?

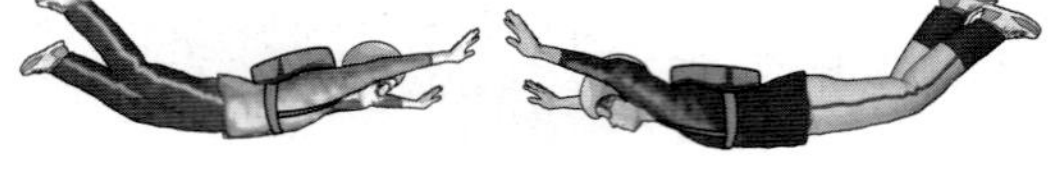

14. A 7.25-g bullet is fired from a gun. The muzzle velocity of the bullet is 223 m/s. Assume that the bullet accelerates at a constant rate along the barrel of the gun before it emerges with constant speed. The barrel of the gun is 0.203 m long. What average force does the bullet exert on the gun?

15. A 15.2-kg police battering ram exerts an average force of 125 N on a 10.0-kg door.
   a. What is the average acceleration of the door?
   b. What is the average acceleration of the battering ram?

16. As a demonstration, a physics teacher attaches a 7.5-kg object to the ceiling by a nearly massless string. This object then supports a 2.5-kg object below it by another piece of string. Finally, another piece of string hangs off the bottom of the lower object to be pulled with ever increasing force until the string breaks somewhere. The string will break when the tension reaches 156 N.
   a. Which length of string will break first?
   b. What is the maximum downward force the physics teacher can apply before the string breaks?

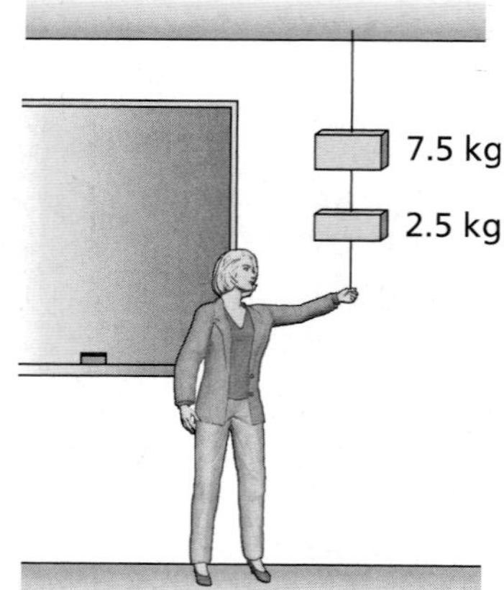

17. A 10.0-kg object is held up by a string that will break when the tension exceeds 1.00×10^2 N. At what upward acceleration will the string break?

18. A large sculpture is lowered into place by a crane. The sculpture has a mass of 2225 kg. When the sculpture makes contact with the ground, the crane slowly releases the tension in the cable as workers make final adjustments to the sculpture's position on the ground.
    a. Draw a free-body diagram of the sculpture when it is in contact with the ground, and there is still tension in the cable while the workers make the final adjustments.
    b. What is the normal force on the sculpture when the tension in the cable is 19,250 N?

Chapter 5

1. A soccer ball is kicked from a 22-m-tall platform. It lands 15 m from the base of the platform. What is the net displacement of the ball?
2. If the net displacement is 32 m for the same situation, as described in problem 1, how far from the platform base must the ball land?
3. For any single force vector, there is only one angle for which its x- and y-components are equal in size.
    a. What is that angle?
    b. How many times bigger is a vector at this particular angle than either of its components?
4. A cue ball on a billiards table travels at 1.0 m/s for 2.0 s. After striking another ball, it travels at 0.80 m/s for 2.5 s at an angle of 60.0° from its original path.
    a. How far does the cue ball travel before and after it strikes the other ball?
    b. What is the net displacement of the cue ball for the entire 4.5-s time interval?

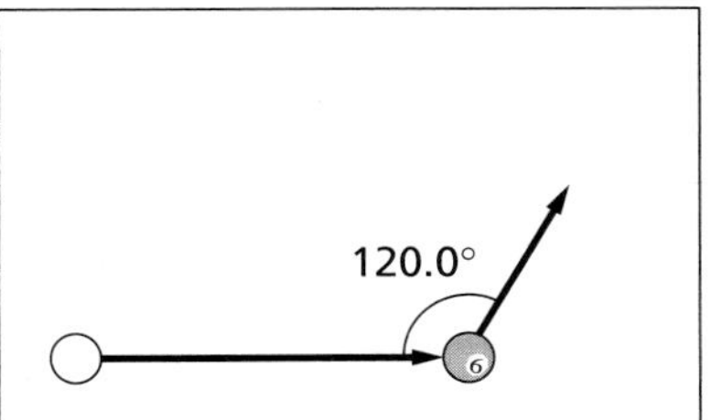

5. The table below represents a set of force vectors. These vectors begin at the origin of a coordinate system, and end at the coordinates given in the table.

Vector #	x-value (N)	y-value (N)
A	0.0	6.0
B	5.0	0.0
C	0.0	−10.0

    a. What is the magnitude of the resultant of the sum of these three vectors?
    b. What is the size of the angle, θ, that the resultant makes with the horizontal?
    c. Into which quadrant does the resultant point?

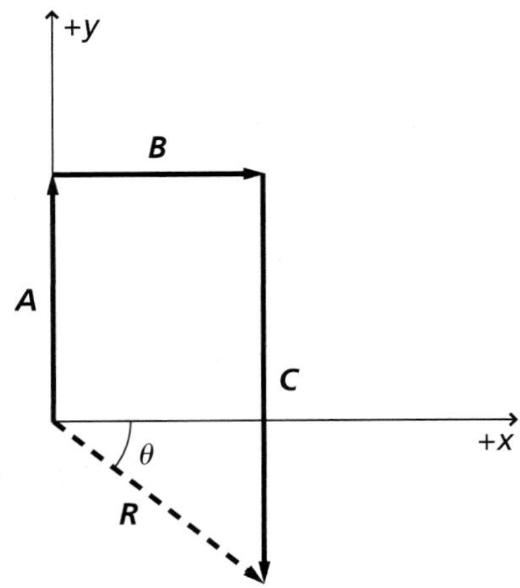

6. A 9.0-kg crate sits on a level, rough floor. A 61-N force is needed just to start it moving. What is the size of the coefficient of maximum static friction?

7. Given the graph below answer the following questions.
   a. What is the value of μ_k for this system?
   b. If the frictional force is 1.5 N, what is F_N?
   c. Does tripling F_N triple $F_{applied}$?
   d. Do $F_{applied}$ and F_N act in the same direction? Explain why or why not.

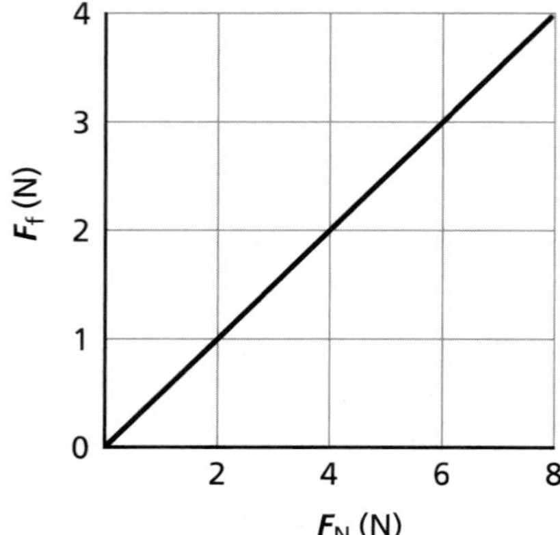

8. A wooden block sits on a level lab table. A string draped over a pulley connects to a bucket that can be filled with lead pellets. Maggie wants to measure how much applied mass (pellets + bucket) is needed to move the block along the table at a constant speed.
   a. If the applied mass of the bucket and the lead pellets is 0.255 kg, and the block has a weight of 12 N, what is the value of μ_k?
   b. If some extra pellets are added, describe the behavior of the block.

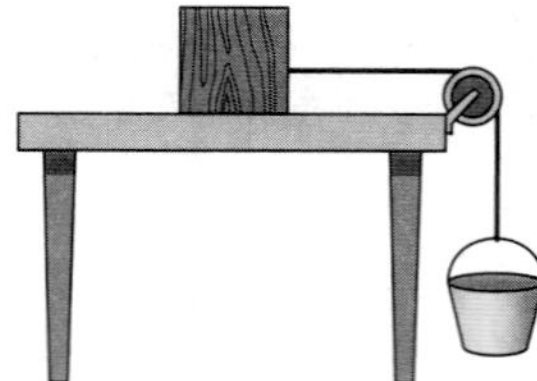

9. A boat travels 75 km southeast, then 56 km due east, then 25 km 30.0° north of east.
   a. Sketch the vector set on a N-E-S-W grid.
   b. Find its net E-W component of displacement and net N-S component of displacement.
   c. Find its net displacement.
   d. Find its net angle relative to an E-W axis.

10. A 1.25-kg box is being pulled across a level surface where μ_k is 0.80. If the string is suddenly cut, causing the applied force to immediately go to 0.0 N, what is the rate of acceleration of the block?

11. If the velocity of the box in problem 10 was 5.0 m/s at time zero, what will be its speed after 0.50 s? How far will it travel in that time interval?

12. A 0.17-kg hockey puck leaves the stick on a slap shot traveling 21 m/s. If no other forces act on the puck, friction eventually will bring it to rest 62 m away.
    a. Based on these data, determine the value of μ_k for this hockey puck on ice.
    b. Will the answer change if a puck of different mass, but same material and shape is used? Explain why or why not.

13. A 105-N suitcase sits on a rubber ramp at an airport carousel at a 25° angle. The weight vector can be broken into two perpendicular components.
    a. What is the magnitude of the component parallel to the ramp surface?
    b. What is the magnitude of the component at a right angle to the ramp surface?
    c. Which of those components is the normal force?

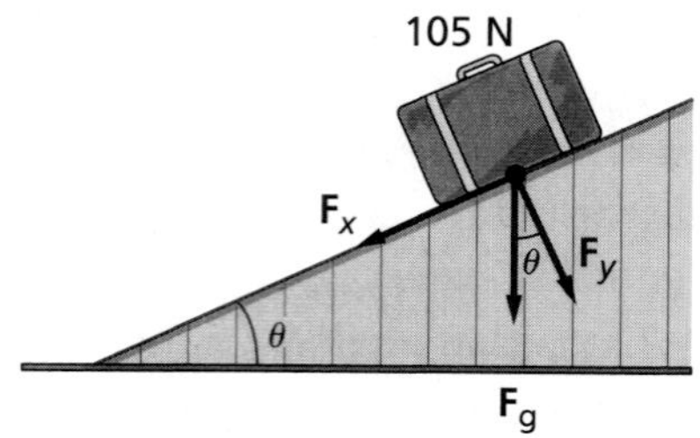

14. In problem 13, which force offsets the kinetic friction force? If F_x exceeds F_f, will the suitcase accelerate down the ramp?
15. Dalila decides to determine the coefficient of maximum static friction for wood against wood by conducting an experiment. First she places a block of wood on a small plank; next, she slowly lifts one end of the plank upward. She notes the angle at which the block just begins to slide, and claims that $\mu_s = \tan\,\theta$. She is correct. Show why.

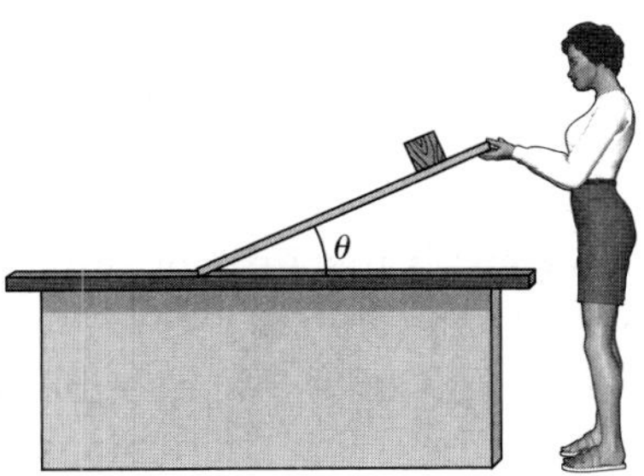

16. Jonathan, with a mass of 81 kg, starts at rest from the top of a water slide angled at 42° from the ground. He exits at the bottom 3.0 s later going 15 m/s. What is μ_k?
17. A 64-N box is pulled by a rope at a constant speed across a rough horizontal surface. If the coefficient of kinetic friction is 0.81, what is the magnitude of the applied force if that force is directed parallel to the floor?
18. Suppose that in problem 17 the applied force is directed by the rope at an angle, θ, to the floor.
    a. The normal force is no longer simply F_g. Show how to compute net vertical force.
    b. The frictional force still opposes the horizontal motion of the box. Show how to compute net horizontal force.

Chapter 6

1. A football player kicks a field goal from a distance of 45 m from the goalpost. The football is launched at a 35° angle above the horizontal. What initial velocity is required so that the football just clears the goalpost crossbar that is 3.1 m above the ground? Ignore air resistance and the dimensions of the football.
2. In a certain cathode-ray tube, a beam of electrons, moving at a constant velocity, enters a region of constant electric force midway between two parallel plates that are 10.0 cm long and 1.0 cm apart. In this region, the electrons experience an acceleration, a, toward the upper plate. If the electrons enter this region at a velocity of 3.0×10^6 m/s, what is the acceleration that needs to be applied so that the electrons just miss the upper deflection plate?

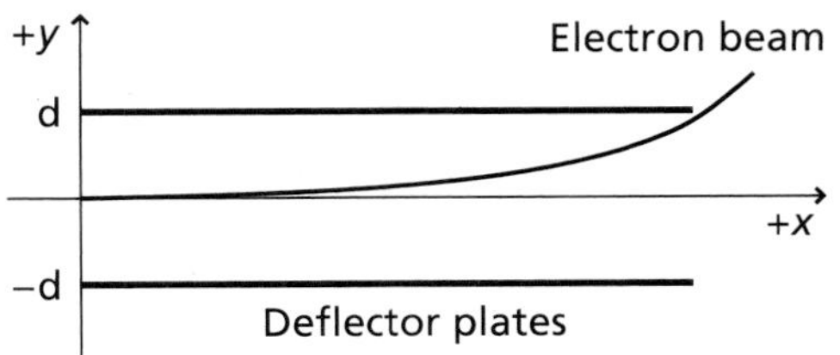

3. A skateboard track has a horizontal segment followed by a ramp that declines at a 45° angle, as shown.
    a. How long would the ramp need to be to provide a landing for a skateboarder who launches from the horizontal segment at a velocity of 5.0 m/s?
    b. If the initial velocity is doubled, what happens to the required length of the ramp?

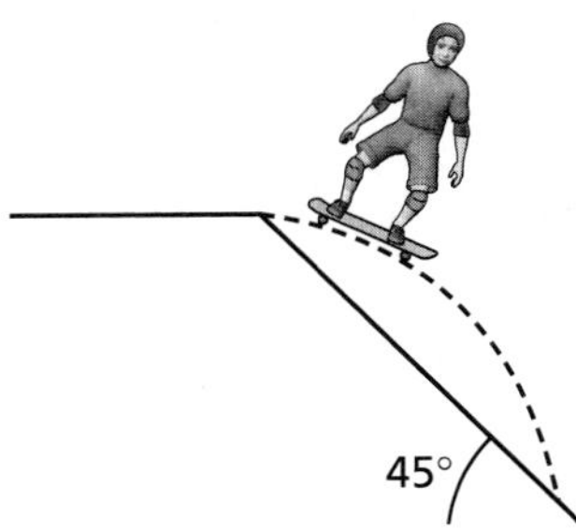

4. In an attempt to make a 3-point shot from a distance of 6.00 m, a basketball player lofts the ball at an angle of 68° above the horizontal. The ball has an initial velocity of 10.0 m/s and an initial distance from the floor of 1.50 m.
    a. What is the maximum height reached by the ball?
    b. If the basket rim height is 3.05 m, how far above the rim is the ball?

5. How far does a baseball that is thrown horizontally at 42.5 m/s drop over a horizontal distance of 18.4 m?
6. A projectile is launched from zero height with an initial angle, θ, above the horizontal and with an initial velocity, v_i.
   a. Show that the range of the projectile—the distance from the launch point at which the height is again zero—is given by: $R = \frac{v_i^2}{g}\sin 2\theta$
   b. What launch angle, θ, results in the largest range?
7. If a ring were constructed as part of a space station, how fast must a 50.0-m-radius ring rotate to simulate Earth's gravity?
8. A turntable for vinyl records works by constraining the needle to track inside a groove in a very close approximation of uniform circular motion. If the turntable rotational speed is $33\frac{1}{3}$ rpm, what is the needle's centripetal acceleration when it is 14.6 cm from the center?
9. A park ride is designed so that the rider is on the edge of a revolving platform, 3.5 m from the platform's center. This platform is mounted 8.0 m from the center of a larger revolving platform. The smaller platform makes one revolution in 6.0 s and completes 2.0 rev for each revolution of the larger platform. All rotations are counterclockwise. At the instant shown, what are the magnitude and the direction of the rider's velocity with respect to the ground?

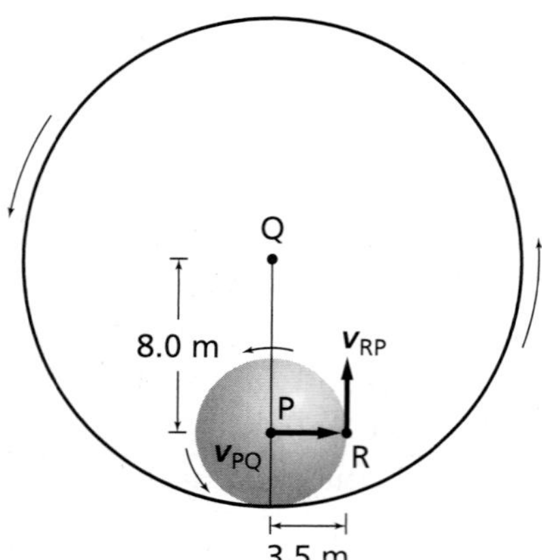

10. Two objects are placed on a flat turntable at 10.0 cm and 20.0 cm from the center, respectively. The coefficient of static friction with the turntable is 0.50. The turntable's rotational speed is gradually increased.
    a. Which of the two objects will begin to slide first? Why?
    b. At what rotational speed does the inner object begin to slide?
11. An airplane's airspeed is 2.0×10^2 km/h due east. Because of a wind blowing to the north, it is approaching its destination 15° north of east.
    a. What is the wind speed?
    b. How fast is the airplane approaching its destination?
12. A river is flowing 4.0 m/s to the east. A boater on the south shore plans to reach a dock on the north shore 30.0° downriver by heading directly across the river.
    a. What should be the boat's speed relative to the water?
    b. What is the boat's speed relative to the dock?

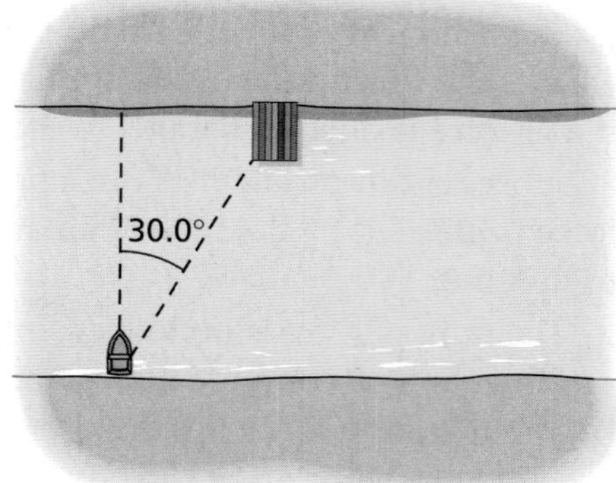

Chapter 7

1. A year is defined as the time it takes for a planet to travel one full revolution around the Sun. One Earth year is 365 days. Using the data in Table 7-1, calculate the number of Earth days in one Neptune year. If Neptune takes about 16 h to complete one of its days, how many Neptunian days long is Neptune's year?

2. Suppose a new planet was discovered to orbit the Sun with a period 5 times that of Pluto. Using the data in Table 7-1, calculate the average distance from the Sun for this new planet.
3. If a meteorite hit the Earth and moved it 2.41×10^{10} m closer to Venus, how many days would there be in an Earth year? Use the data in Table 7-1.
4. The Moon is a satellite of the Earth. The Moon travels one full revolution around the Earth in 27.3 days. Given that the mass of the Earth is 5.97×10^{24} kg, what is the average distance from the Earth to the Moon?
5. A satellite travels 7.18×10^{7} m from the center of one of the planets in our solar system at a speed of 4.20×10^{4} m/s. Using the data in Table 7-1, identify the planet.
6. Earth's atmosphere is divided into four layers: the troposphere (0–10 km), the stratosphere (10–50 km), the mesosphere (50–80 km), and the thermosphere (80–500 km). What is the minimum velocity an object must have to enter the thermosphere?

Chapter 8

1. The rotational velocity of a merry-go-round is increased at a constant rate from 1.5 rad/s to 3.5 rad/s in a time of 9.5 s. What is the rotational acceleration of the merry-go-round?
2. A record player's needle is 6.5 cm from the center of a 45-rpm record. What is the velocity of the needle?
3. Suppose a baseball rolls 3.2 m across the floor. If the ball's angular displacement is 82 rad, what is the circumference of the ball?
4. A painter uses a 25.8-cm long screwdriver to pry the lid off of a can of paint. If a force of 85 N is applied to move the screwdriver 60.0° from the perpendicular, calculate the torque.
5. A force of 25 N is applied vertically at the end of a wrench handle that is 45 cm long to tighten a bolt in the clockwise direction. What torque is needed by the bolt to keep the wrench from turning?
6. A 92-kg man uses a 3.05-m board to attempt to move a boulder, as shown in the diagram below. He pulls the end of the board with a force equal to his weight and is able to move it to 45° from the perpendicular. Calculate the torque applied.

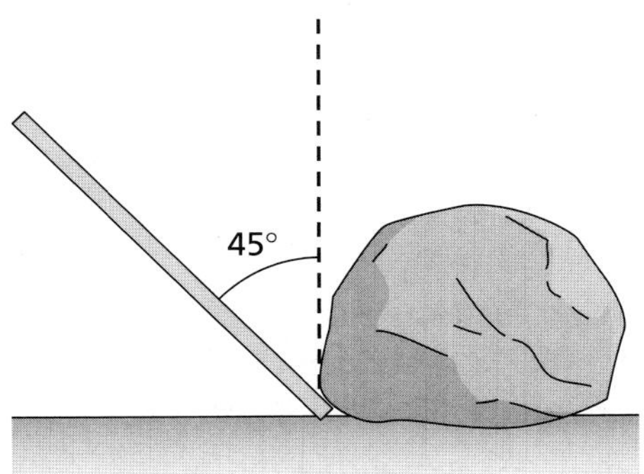

7. If a 25-kg child tries to apply the same torque as in the previous question, using only his or her weight for the applied force, what would the length of the lever arm need to be?
8. Logan, whose mass is 18 kg, sits 1.20 m from the center of a seesaw. If Shiro must sit 0.80 m from the center to balance Logan, what is Shiro's mass?
9. Two forces— 55-N clockwise and 35-N counter-clockwise —are applied to a merry-go-round with a diameter of 4.5 m. What is the net torque?

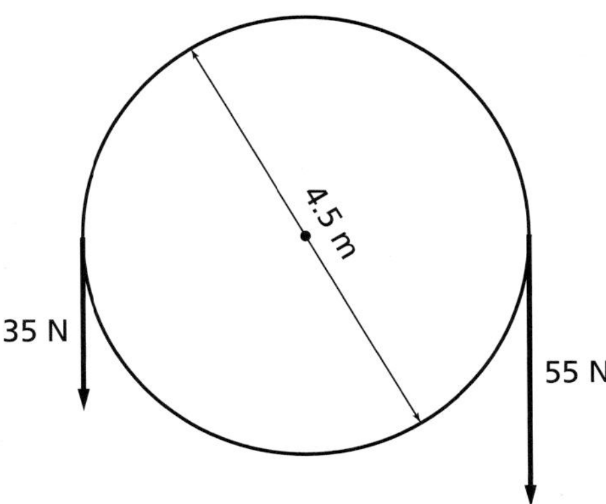

10. A student sits on a stool holding a 5.0-kg dumbbell in each hand. He extends his arms such that each dumbbell is 0.60 m from the axis of rotation. The student's moment of inertia is 5.0 kg·m^2. What is the moment of inertia of the student and the dumbbells?

11. A basketball player spins a basketball with a radius of 15 cm on his finger. The mass of the ball is 0.75 kg. What is the moment of inertia about the basketball?
12. A merry-go-round in the park has a radius of 2.6 m and a moment of inertia of 1773 kg·m^2. What is the mass of the merry-go-round?
13. The merry-go-round described in the previous problem is pushed with a constant force of 53 N. What is the angular acceleration?
14. What is the angular velocity of the merry-go-round described in problems 12 and 13 after 85 s, if it started from rest?
15. An ice-skater with a moment of inertia of 1.1 kg·m^2 begins to spin with her arms extended. After 25 s, she has an angular velocity of 15 rev/s. What is the net torque acting on the skater?
16. A board that is 1.5 m long is supported in two places. If the force exerted by the first support is 25 N and the forced exerted by the second is 62 N, what is the mass of the board?
17. A child begins to build a house of cards by laying an 8.5-cm-long playing card with a mass of 0.75 g across two other playing cards: support card A and support card B. If support card A is 2.0 cm from the end and exerts a force of 1.5×10^{-3} N, how far from the end is support card B located? Let the axis of rotation be at the point support card A comes in contact with the top card.
18. If support card A in the previous problem was moved so that it now is 2.5 cm from the end, how far from the other end does support card B need to be to reestablish equilibrium?

Chapter 9

1. A ball with an initial momentum of 6.00 kg·m/s bounces off a wall and travels in the opposite direction with a momentum of 4.00 kg·m/s. What is the magnitude of the impulse acting on the ball?
2. If the ball in the previous problem interacts with the wall for a time interval of 0.22 s, what is the average force exerted on the wall?
3. A 42.0-kg skateboarder traveling at 1.50 m/s hits a wall and bounces off of it. If the magnitude of the impulse is 150.0 kg·m/s, calculate the final velocity of the skateboarder.
4. A 50.0-g toy car traveling with a velocity of 3.00 m/s due north collides head-on with an 180.0-g fire truck traveling with a velocity of 0.50 m/s due south. The toys stick together after the collision. What are the magnitude and direction of their velocity after the collision?
5. A 0.040-kg bullet is fired into a 3.50-kg block of wood, which was initially at rest. The bullet remains embedded within the block of wood after the collision. The bullet and the block of wood move at a velocity of 7.40 m/s. What was the original velocity of the bullet?
6. Ball A, with a mass of 0.20 kg, strikes ball B, with a mass of 0.30 kg. The initial velocity of ball A is 0.95 m/s. Ball B is initially at rest. What are the final speed and direction of ball A and B after the collision if they stick together?
7. An ice-skater with a mass of 75.0 kg pushes off against a second skater with a mass of 42.0 kg. Both skaters are initially at rest. After the push, the larger skater moves off with a speed of 0.75 m/s eastward. What is the velocity (magnitude and direction) of the smaller skater after the push?
8. Suppose a 55.0-kg ice-skater, who was initially at rest, fires a 2.50-kg gun. The 0.045-kg bullet leaves the gun at a velocity of 565.0 m/s. What is the velocity of the ice-skater after she fires the gun?
9. A 1200-kg cannon is placed at rest on an ice rink. A 95.0-kg cannonball is shot from the cannon. If the cannon recoils at a speed of 6.80 m/s, what is the speed of the cannonball?

10. An 82-kg receiver, moving 0.75 m/s north, is tackled by a 110.0-kg defensive lineman moving 0.15 m/s east. The football players hit the ground together. Calculate their final velocity (magnitude and direction).
11. A 985-kg car traveling south at 29.0 m/s hits a truck traveling 18.0 m/s west, as shown in the figure below. After the collision, the vehicles stick together and travel with a final momentum of 4.0×10^4 kg·m/s at an angle of 45°. What is the mass of the truck?

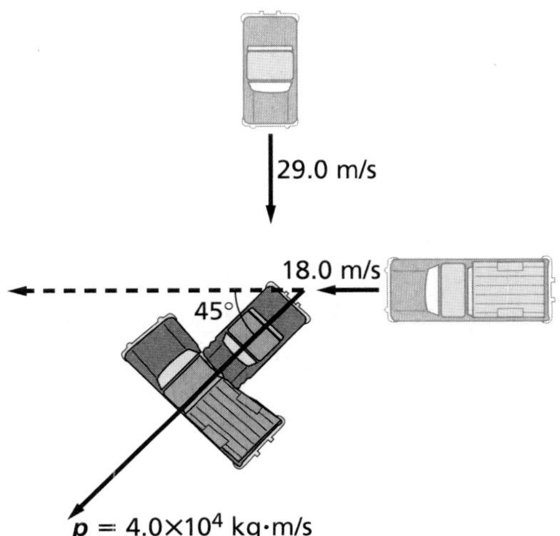

12. A 77.0-kg woman is walking 0.10 m/s east in the gym. A man throws a 15.0-kg ball south and accidentally hits the woman. The woman and the ball move together with a velocity of 0.085 m/s. Calculate the direction the woman and the ball move.

Chapter 10

1. A toy truck is pushed across a table 0.80 m north, and pulled back across the table 0.80 m south. If a constant horizontal force of 15 N was applied in both directions, what is the net work?
2. A 15-kg child experiences an acceleration of 0.25 m/s^2 as she is pulled 1.7 m horizontally across the floor by her sister. Calculate the change in the child's kinetic energy.
3. A man pushes a couch a distance of 0.75 m. If 113 J of work is done, what is the magnitude of the horizontal force applied?
4. Two blocks are tied together by a horizontal string and pulled a distance of 2.7 m across an air hockey table with a constant force of 35 N. The force is directed at an upward angle of 35° from the 9.0-kg block, as shown in the figure. What is the change in kinetic energy in the two-block system?

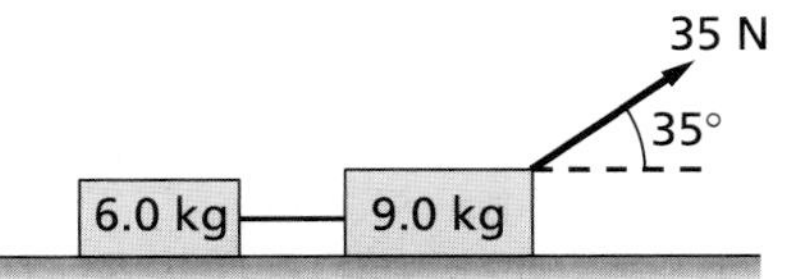

5. If the two-block system described in the previous problem was initially at rest, what is the final velocity?
6. A toy car with a mass of 0.75 kg is pulled 3.2 m across the floor with a constant force of 110 N. If 67 J of work is done, what is the upward angle of the force?
7. If a 75-W lightbulb is left on for 2.0 h, how much work is done?
8. A 6.50-horsepower (hp) self-propelled lawn mower is able to go from 0.00 m/s to 0.56 m/s in 0.050 s. If the mass of the lawn mower is 48.0 kg, what distance does the lawn mower travel in this time? (Use 1 hp = 746 W.)
9. A winch that's powered by a 156-W motor lifts a crate 9.8 m in 11 s. What is the mass of the crate?
10. A man exerts a force of 310 N on a lever to raise a crate with a mass of 910 kg. If the efficiency of the lever is 78 percent, what is the lever's *IMA*?
11. A worker uses a pulley to lift a 45-kg object. If the mechanical advantage of the pulley is 5.2, what is the effort force exerted by the worker?
12. When the chain on a bicycle is pulled 0.95 cm, the rear wheel rim moves a distance of 14 cm. If the gear has a radius of 3.5 cm, what is the radius of the rear wheel?

Chapter 11

1. A crate with a mass of 210 kg is horizontally accelerated by a force of 95 N. The force is directed at an upward angle so that the vertical part of the force is 47.5 N and the horizontal part of the force is 82.3 N (see the diagram below). If the crate is pulled 5.5 m across the floor, what is the change in kinetic energy of the crate?

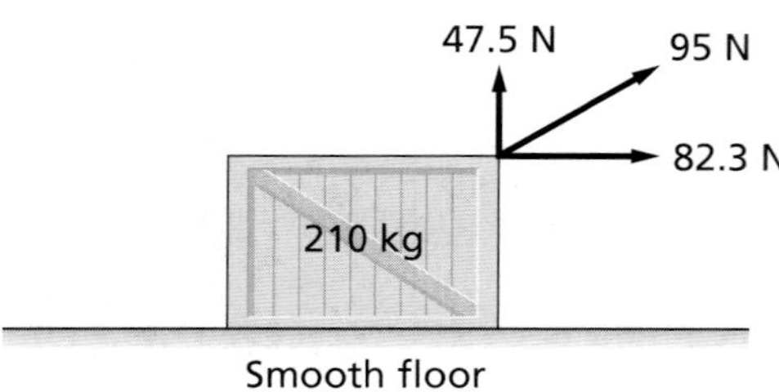

2. Assuming that the crate described in problem 1 was initially at rest, what is the final velocity of the crate?
3. If the crate described in problem 1 experienced a frictional force of 15 N, what is the final kinetic energy of the crate?
4. A 150-kg roller-coaster car climbs to the top of a 91-m-high hill. How much work is done against gravity as the car is lifted to the top of the hill?
5. A pendulum bob with a mass of 0.50 kg swings to a maximum height of 1.0 m. What is the kinetic energy when the pendulum bob is at a height of 0.40 m?
6. A sled and its rider, with a total mass of 95 kg, are perched on top of a 25-m-tall hill. A second hill is 12 m tall (see the diagram below). The rider is given an initial push providing 3674 J of kinetic energy. Neglecting friction, what is the velocity at the top of the top of the 12-m-tall hill?

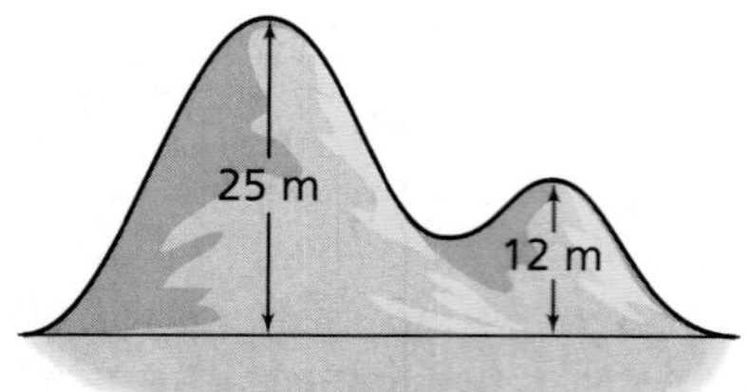

7. A 35-kg child is riding on a swing that rises to a maximum height of 0.80 m. Neglecting friction, what is the child's gravitational potential energy at the top of the swing? What is the child's kinetic energy at the top of the swing?
8. A sled and its rider are perched at the top of a hill that is 35 m tall. If the gravitational potential energy is 3.0×10^4 J, what is the weight of the sled and rider?
9. The big hill on a roller-coaster ride is 91 m tall. If the mass of the roller-coaster car and its two riders is 314 kg and the maximum velocity reached by this roller-coaster ride is 28 m/s, how much energy was lost to friction?
10. A dartboard with a mass of 2.20 kg is suspended from the ceiling such that it is initially at rest. A 0.030-kg dart is thrown at the dartboard with a velocity of 1.3 m/s. After the dart hits the dartboard, the dart and the board initially move with a velocity of 0.025 m/s. How much kinetic energy is lost in the system?
11. In a physics laboratory, students crash carts together on a frictionless track. According to the following data, was kinetic energy conserved?

	Mass (kg)	v_i (m/s)	v_f (m/s)
Cart A	0.25	0.18	−0.21
Cart B	0.36	−0.20	0.11

12. A 0.150-kg ball that is thrown at a velocity of 30.0 m/s hits a wall and bounces back in the opposite direction with a speed of 25.0 m/s. How much work was done by the ball?

Chapter 12

1. Convert the following Celsius temperatures to Kelvin temperatures.
   - **a.** −196°C
   - **b.** 32°C
   - **c.** 212°C
   - **d.** −273°C
   - **e.** 273°C
   - **f.** 27°C

2. Find the Celsius and the Kelvin temperatures for the following objects.
   a. average body temperature
   b. hot coffee
   c. iced tea
   d. boiling water
3. Convert the following Kelvin temperatures to Celsius temperatures.
   a. 4 K
   b. 25 K
   c. 272 K
   d. 373 K
   e. 298 K
   f. 316 K
4. A 9.8-g lead bullet with a muzzle velocity of 3.90×10^2 m/s is stopped by a wooden block. What is the change in temperature of the bullet if one-fourth of its original kinetic energy goes into heating the bullet?

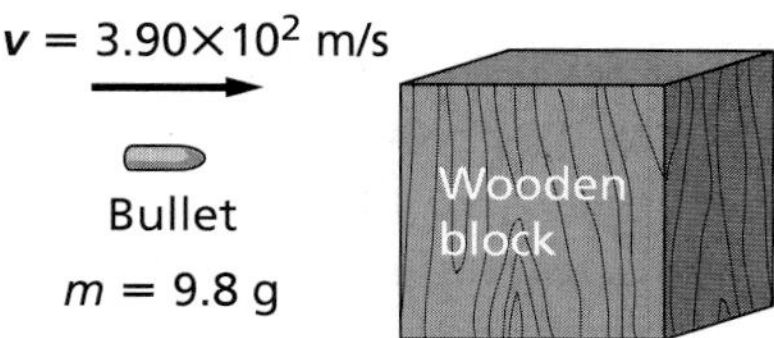

5. What is the change in temperature of 2.2 kg of the following substances if 8.5×10^3 J of thermal energy is added to each of the substances?
   a. ice
   b. water
   c. steam
   d. aluminum
   e. silver
   f. copper
6. A 2350-kg granite tombstone absorbs 2.8×10^7 J of energy from the Sun to change its temperature from 5.0°C at night to 20.0°C during an autumn day. Determine the specific heat of granite from this information.
7. A 2.00×10^3-g sample of water at 100.0°C is mixed with a 4.00×10^3-g sample of water at 0.0°C in a calorimeter. What is the equilibrium temperature of the mixture?
8. A 220-g iron horseshoe is heated to 825°C and then plunged into a bucket filled with 20.0 kg of water at 25.0°C. Assuming that no energy is transferred to the surroundings, what is the final temperature of the water at equilibrium?

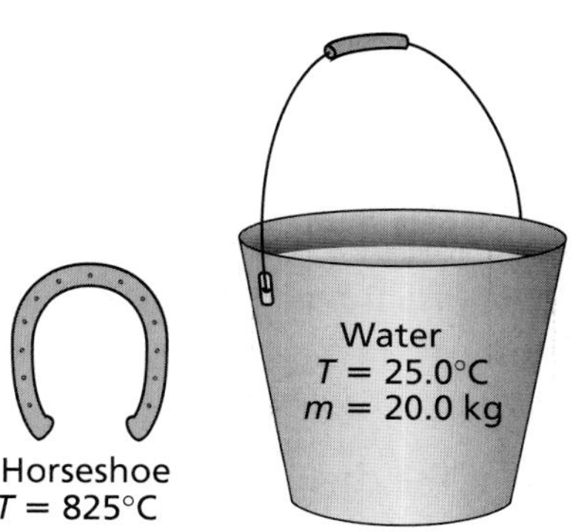

9. A cook adds 1.20 kg of soup bones to 12.5 kg of hot broth for flavoring. The temperatures of the bones and the broth are 20.0°C and 90.0°C, respectively. Assume that the specific heat of the broth is the same as that of water. Also assume that no heat is lost to the environment. If the equilibrium temperature is 87.2°C, what is the specific heat of the bones?

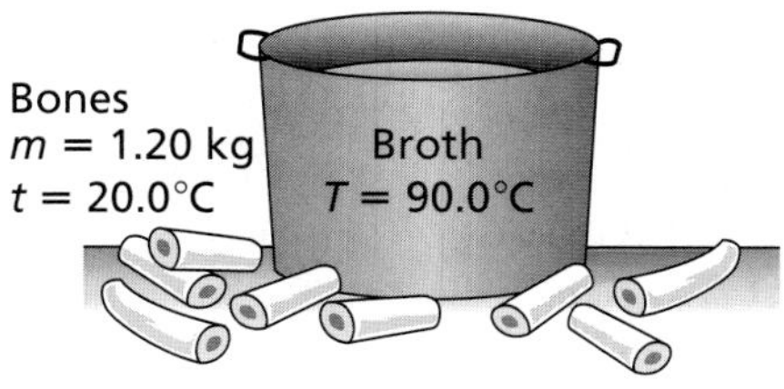

10. A 150.0-g copper cylinder at 425°C is placed on a large block of ice at 0.00°C. Assume that no energy is transferred to the surroundings. What is the mass of the ice that will melt?
11. How much energy is needed to melt one troy ounce, 31.1 g, of gold at its melting point?

12. Trying to make an interesting pattern, an artist slowly pours 1.00 cm^3 of liquid gold onto a large block of ice at 0.00°C. While being poured, the liquid gold is at its melting point of 1064°C. The density of gold is 19.3 g/cm^3, and its specific heat is 128 J/kg·K. What mass of ice will melt after all the gold cools to 0.00°C?

13. A cylinder containing 1.00 g of water at the boiling point is heated until all the water turns into steam. The expanding steam pushes a piston 0.365 m. There is a 215-N frictional force acting against the piston. What is the change in the thermal energy of the water?

14. What is the change in temperature of water after falling over a 50.0-m-tall waterfall. Assume that the water is at rest just before falling off and just after it reaches the bottom of the falls.

15. Two 6.35-kg lead bricks are rubbed against each other until their temperature rises by 1.50 K. How much work was done on the bricks?

Chapter 13

1. Use Table 13.1 to estimate the pressure in atmospheres (atm) on a climber standing atop Mt. Everest. Is this more or less than half standard atmospheric pressure (1.0 atm)?

2. A woman wearing high heels briefly supports all her weight on the heels. If her mass is 45 kg and each heel has an area of 1.2 cm^2, what pressure does she exert on the floor?

3. A typical brick has dimensions of 20.0 cm × 10.0 cm × 5.0 cm and weighs 20.0 N. How does the pressure exerted by a typical brick when it is resting on its smallest side compare to the pressure exerted when it is resting on its largest side?

4. As shown in the figure below, a bubble of gas with a volume of 1.20 cm^3 is released under water. As it rises to the surface, the temperature of the bubble increases from 27°C to 54°C, and the pressure is cut to one-third of its initial value. What is the volume of the bubble at the surface?

5. A sample of ethane gas (molar mass = 30.1 g/mol) occupies 1.2×10^{-2} m^3 at 46°C and 2.4×10^5 Pa. How many moles of ethane are present in the sample? What is the mass of the sample?

6. The constant R equals 8.31 Pa·m^3/mol·K as it is presented in the ideal gas law in this chapter. One mole of an ideal gas occupies 22.4 L at standard temperature and pressure, STP, which is defined as 0.00°C and 1.00 atm. Given this information, deduce the value of R in L·atm/mol·K.

7. Suppose that two linked pistons are both cylindrical in shape. Show that the ratio of forces generated is directly proportional to the square of the radii of the two cross-sectional circular areas.

8. The cross-sectional areas of the pistons in the system shown below have a ratio of 25 to 1. If the maximum force that can be applied to the small piston is 12 N, what is the maximum weight that can be lifted?

9. A car of mass 1.35×10^{3} kg sits on a large piston that has a surface area of 1.23 m^2. The large piston is linked to a small piston of area 144 cm^2. What is the weight of the car? What force must a mechanic exert on the small piston to raise the car?
10. At what depth in freshwater does the water exert a pressure of 1.00 atm (1 atm = 1.013×10^{5} Pa) on a scuba diver?
11. An iceberg floats in seawater, partly under water and partly exposed. Show that the value of $V_{submerged}/V_{total}$ equals $\rho_{ice}/\rho_{seawater}$. What percentage of the iceberg is exposed? Use 1.03×10^{3} kg/m^3 for the density of seawater, and 0.92×10^{3} kg/m^3 for the density of ice.
12. A concrete block ($\beta = 36\times10^{-6}$ °C^{-1}) of volume 0.035 m^3 at 30.0°C is cooled to −10.0°C. What is the change in volume?
13. A glass mirror used in a mountaintop telescope is subject to temperatures ranging from −15°C to 45°C. At the lowest temperature, it has a diameter of 5.1 m. If the coefficient of linear expansion for this glass is 3.0×10^{-6} °C^{-1}, what is the maximum change in diameter the mirror undergoes due to thermal expansion?
14. As shown in the figure below, a bimetallic strip is made from a piece of copper ($\alpha = 16\times10^{-6}$ °C^{-1}) and a piece of steel ($\alpha = 8\times10^{-6}$ °C^{-1}). The two pieces are equal in length at room temperature.
    a. As the strip is heated from room temperature, what is the ratio of the change in the length of the copper to that of the steel?
    b. How will the strip bend when it is heated above room temperature? When it is cooled below room temperature?

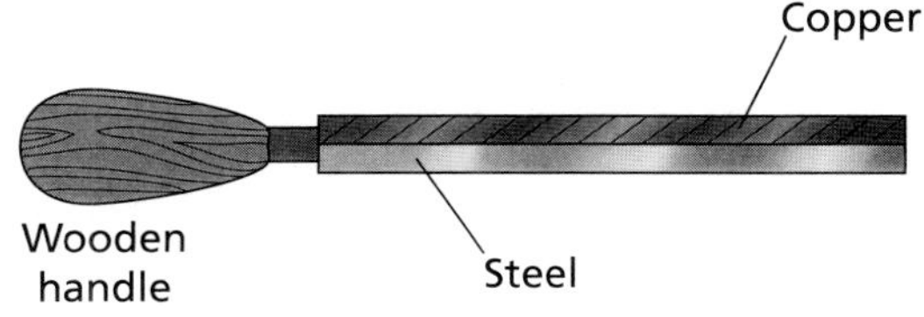

15. A fishing line's lead sinker has a volume of 1.40×10^{-5} m^3. The density of lead is 1.2×10^{4} kg/m^3. What is the apparent weight of the sinker when immersed in freshwater? Seawater is slightly denser than freshwater. Is the sinker's apparent weight bigger or smaller in seawater? Explain.

Chapter 14

1. What is the mass of the watermelon shown below, if the spring constant is 128 N/m?

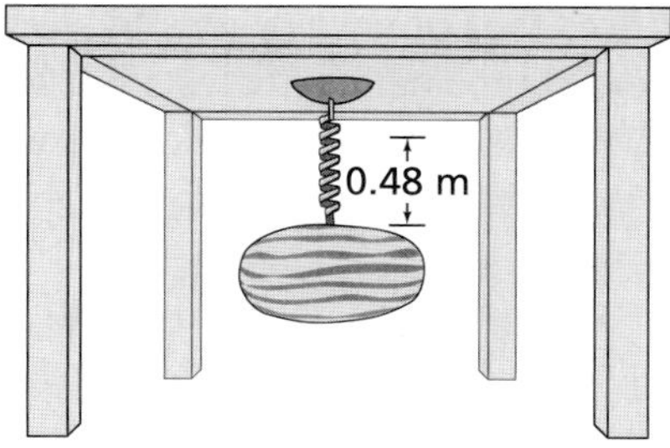

2. How many centimeters will a spring stretch when a 2.6-kg block is hung vertically from a spring with a spring constant of 89 N/m?
3. How much elastic potential energy does a spring with a constant of 54 N/m have when it is stretched 18 cm?
4. What is the period of the pendulum of the clock below?

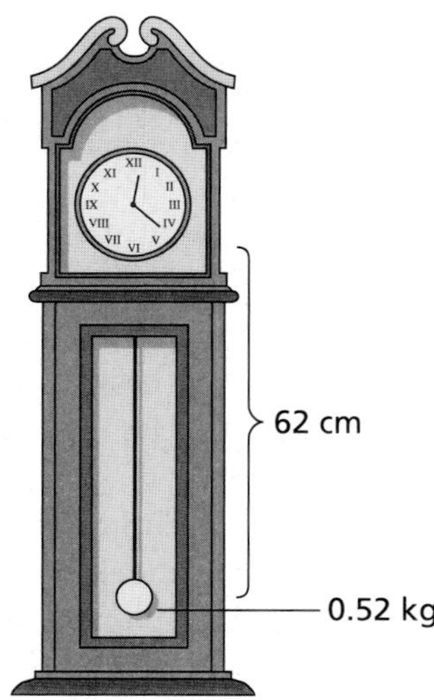

5. A clock pendulum has a period of 0.95 s. How much longer will it have to be to have a period of 1.0 s?

6. What is the length of a pendulum with a period of 89.4 ms?
7. Al is chopping wood across a clearing from Su. Su sees the axe come down and hears the sound of the impact 1.5 s later. How wide is the clearing?
8. When an orchestra is tuning up, the first violinist plays a note at 256 Hz.
   a. What is the wavelength of that sound wave if the speed of sound in the concert hall is 340 m/s?
   b. What is the period of the wave?
9. Geo is standing on a breakwater and he notices that one wave goes by every 4.2 s. The distance between crests is 12.3 m.
   a. What is the frequency of the wave?
   b. What is the speed of the wave?

Chapter 15

1. Sound waves are being used to determine the depth of a freshwater lake, as shown in the figure below. If the water is 25°C and it takes 1.2 s for the echo to return to the sensor, how deep is the lake?

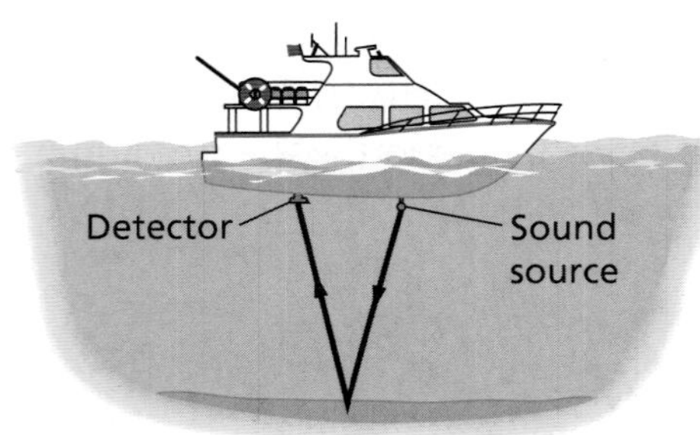

2. Find the wavelength of an 8300-Hz wave in copper.
3. Janet is standing 58.2 m from Ninovan. If Ninovan shouts, how long will it take Janet to hear her? Use 343 m/s for the speed of sound.
4. The engine on a motorcycle hums at 85 Hz. If the motorcycle travels toward a resting observer at a velocity of 29.6 m/s, what frequency does the observer hear? Use 343 m/s for the speed of sound.
5. If the speed of a wave on a 78-cm-long guitar string is known to be 370 m/s, what is the fundamental frequency?
6. Boat A is traveling at 4.6 m/s. Boat B is moving away from boat A at 9.2 m/s, as shown in the figure below. The captain of boat B blows an air horn with a frequency of 550 Hz. What frequency does boat A hear? Use 343 m/s for the speed of sound.

7. A submarine is traveling toward a stationary detector. If the submarine emits a 260-Hz sound that is received by the detector as a 262-Hz sound, how fast is the submarine traveling? Use 1533 m/s for the speed of sound in seawater.
8. The end of a pipe is inserted into water. A tuning fork is held over the pipe. If the pipe resonates at lengths of 15 cm and 35 cm, what is the frequency of the tuning fork? Use 343 m/s for the speed of sound.
9. A 350-Hz tuning fork is held over the end of a pipe that is inserted into water. What is the spacing between the resonance lengths of the pipe if the speed of sound is 348 m/s?

Chapter 16

1. What is the distance, *r*, between the lightbulb and the table in the figure below?

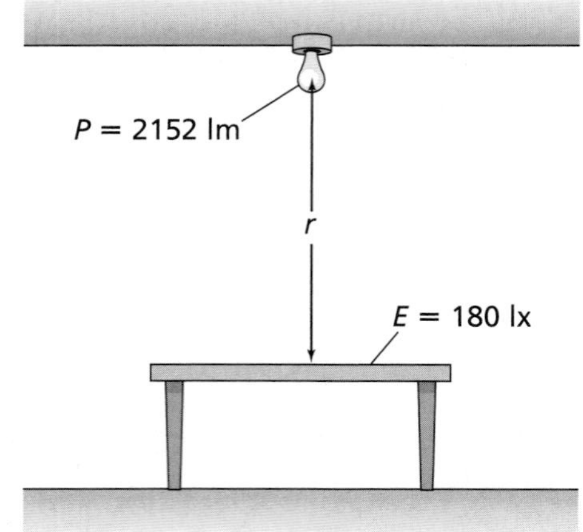

2. What is the luminous flux of a flashlight that provides an illuminance of 145 lx to the surface of water when held 0.50 m above the water?
3. An overhead light fixture holds three lightbulbs. Each lightbulb has a luminous flux of 1892 lm. The light fixture is 1.8 m above the floor. What is the illuminance on the floor?
4. What is the wavelength in air of light that has a frequency of 4.6×10^{14} Hz?
5. A helium atom is in a galaxy traveling at a speed of 4.89×10^{6} m/s away from the Earth. An astronomer on Earth observes a frequency from the helium atom of 6.52×10^{14} Hz. What frequency of light is emitted from the helium atom?
6. An astronomer observes that a molecule in a galaxy traveling toward Earth emits light with a wavelength of 514 nm. The astronomer identifies the molecule as one that actually emits light with a wavelength of 525 nm. At what velocity is the galaxy moving?

Chapter 17

1. A light ray is reflected off a plane mirror at an angle of 25° from the normal. A light ray from another source is reflected 54° from the normal. What is the difference in the angles of incidence for the two light sources?
2. A light ray is reflected off of a plane mirror, as shown. What is the angle of reflection?

3. The angle between an incident and a reflected ray is 70.0°. What is the angle of reflection?
4. An image is produced by a concave mirror, as shown. What is the object's position?

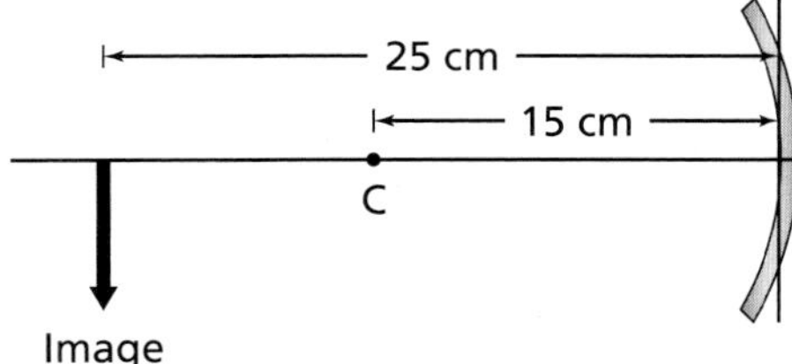

5. What is the magnification of the object in problem 4?
6. If the object in problem 4 is 3.5 cm tall, how tall is the image?
7. A 6.2-m-tall object is 2.3 m from a convex mirror with a −0.8-m focal length. What is the magnification?
8. How tall is the image in problem 7?
9. A ball is 6.5 m from a convex mirror with a magnification of 0.75. If the image is 0.25 m in diameter, what is the diameter of the actual ball?

Chapter 18

1. A piece of flint glass is lying on top of a container of water (see figure below). When a red beam of light in air is incident upon the flint glass at an angle of 28°, what is the angle of refraction in the flint glass?

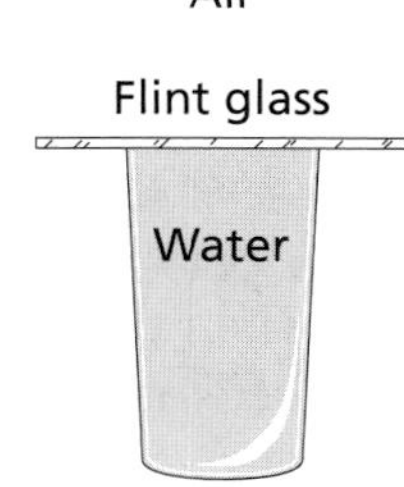

2. When the angle of refraction in the flint glass of problem 1 is 22°, what is the angle of refraction in the water?
3. When the beam of light in problem 1 enters the flint glass from the water, what is the maximum angle of incidence in the water such that light will transmit into the air above the flint glass? *Hint: Use an angle of refraction of the light beam in air that is almost 90°.*
4. An object that is 24 cm from a convex lens produces a real image that is 13 cm from the lens. What is the focal length of the lens?
5. A 5.0-cm-tall object is placed 16 cm from a convex lens with a focal length of 8.4 cm. What are the image height and orientation?

6. An object is 185 cm from a convex lens with a focal length of 25 cm. When an inverted image is 12 cm tall, how tall is the associated object?
7. What are the image height and orientation produced by the setup in the following figure?

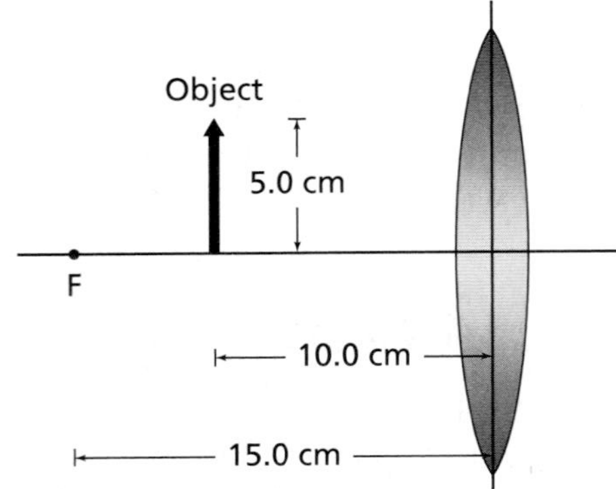

8. A convex lens can be used as a magnifying glass. When an object that is 15.0 cm from the lens has an image that is exactly 55 times the size of the object, what is the focal length of the lens?
9. A concave lens with a focal length of −220 cm produces a virtual image that is 36 cm tall. When the object is placed 128 cm from the lens, what is the magnification?

Chapter 19

1. A physics student performs a double-slit diffraction experiment on an optical bench, as shown in the figure below. The light from a helium-neon laser has a wavelength of 632.8 nm. The laser light passes through two slits separated by 0.020 mm. What is the distance between the centers of the central band and the first-order band?

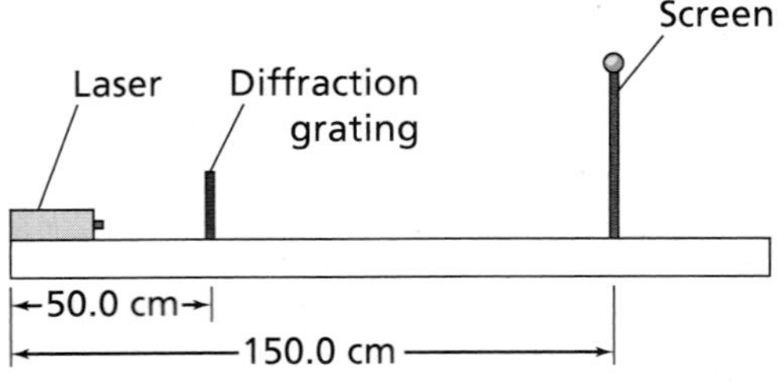

2. A laser of unknown wavelength replaces the helium-neon laser in the experiment described in problem 1. To obtain the best diffraction pattern, the screen had to be moved to 104.0 cm. The distance between the centers of the central band and the first-order band is 1.42 cm. What wavelength of light is produced by the laser?
3. Light with a wavelength of 454.5 nm passes through two slits that are 95.2 cm from a screen. The distance between the centers of the central band and the first-order band is 15.2 mm. What is the slit separation?
4. What color of light would be reflected from the soapy water film shown in the figure below?

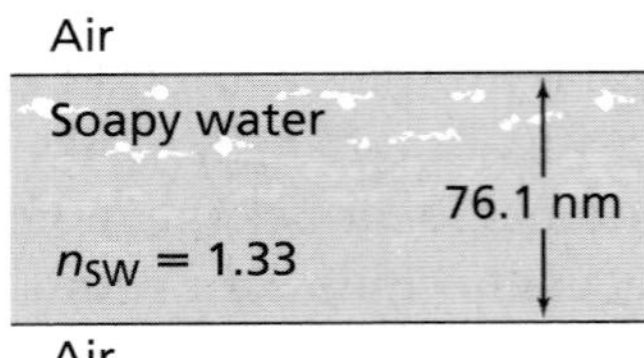

5. A 95.7-nm film of an unknown substance is able to prevent 555-nm light from being reflected when surrounded by air. What is the index of refraction of the substance?
6. An oil film (n = 1.45) on the surface of a puddle of water on the street is 118 nm thick. What frequency of light would be reflected?
7. Violet light (λ = 415 nm) falls on a slit that is 0.040 mm wide. The distance between the centers of the central bright band and the third-order dark band is 18.7 cm. What is the distance from the slit to the screen?
8. Red light (λ = 685 nm) falls on a slit that is 0.025 mm wide. The distance between the centers of the central bright band and the second-order dark band is 6.3 cm. What is the width of the central band?
9. The width of the central bright band of a diffraction pattern is 2.9 cm. Laser light of unknown wavelength passes through a single slit that is 0.042 mm wide and onto a screen that is 1.5 m from the slit. What is the wavelength of the light?

10. A diffraction grating has 13,400 lines per inch. What is the slit separation distance? (Use 1 in = 2.54 cm)
11. Light from a helium-neon laser (λ = 632.8 nm) passes through the diffraction grating described in problem 10. What is the angle between the central bright line and the first-order bright line?
12. A diffraction grating with slits separated by 3.40×10^{-6} m is illuminated by light with a wavelength of 589 nm. The separation between lines in the diffraction pattern is 0.25 m. What is the distance between the diffraction grating and the screen?

Chapter 20

1. The magnitude of a charge, q, is to be determined by transferring the charge equally to two pith balls. Each of the pith balls has a mass of m, and is suspended by an insulating thread of length l. When the charge is transferred, the pith balls separate to form an equilibrium in which each thread forms an angle, θ, with the vertical.
   a. Draw a force diagram showing the forces that are acting on the rightmost pith ball.
   b. Derive an expression for q as a function of θ, m, and l.
   c. Use your derived expression to determine the value of q when $\theta = 5.00°$, $m = 2.00$ g, and $l = 10.0$ cm.

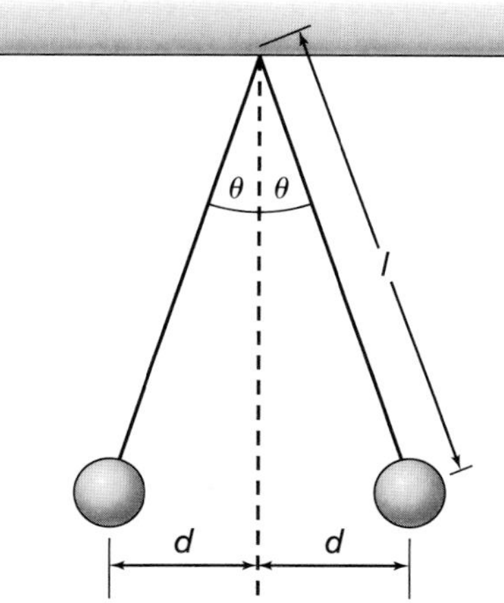

2. As shown in the figure below, four charges, each with charge q, are distributed symmetrically around the origin, O(0, 0), at A(1.000, 0), B(0, 1.000), C(−1.000, 0), and D(0, −1.000). Find the force on a fifth charge, q_T, located at T(5.000, 0).

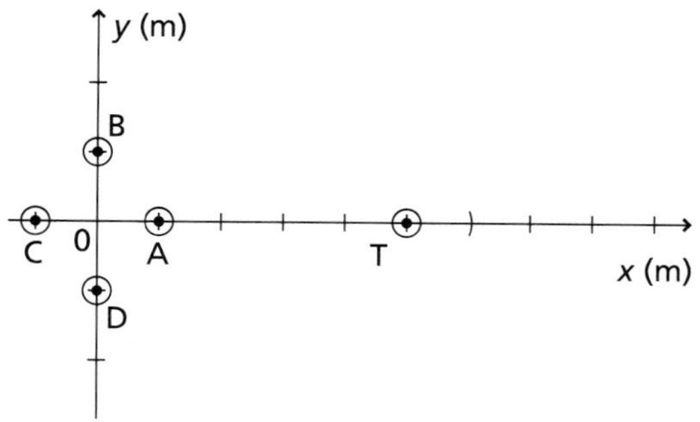

3. Consider that the four charges in the previous problem are now combined into a single charge, $4q$, located at the origin. What is the force on charge q_T?

Chapter 21

1. What are the magnitude and the sign of a point charge that experiences a force of 0.48 N east when placed in an electric field of 1.6×10^{5} N/C west?
2. A test charge of -1.0×10^{-6} C, located at the point T(0, 1) m, experiences a force of 0.19 N directed toward the origin, along the y-axis, due to two identical point charges located at A(−1, 0) m and B(1, 0) m.
   a. What is the sign of the charges at A and B?
   b. What are the magnitude and direction of the electric field at T?

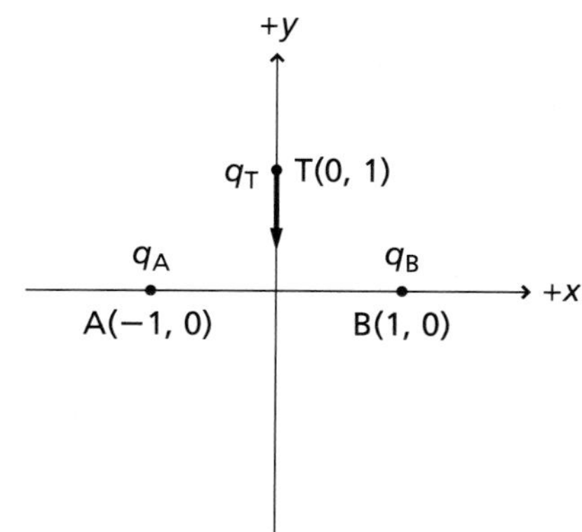

Additional Problems

3. A test charge of -0.5×10^{-7} C is placed in an electric field of 6.2×10^{4} N/C, directed 15° north of east. What is the force experienced by the charge?
4. By what percent must the distance from a point charge increase in order to have a reduction in the electric field strength by 40 percent?
5. A particle of mass $m = 2.0\times10^{-6}$ kg is in a circular orbit about a point charge. The charge, q, is 3.0×10^{-5} C and is at a distance of $r = 20.0$ cm.
   a. What is the electric field strength at all points on the orbit around the point charge?
   b. Considering only electrostatic forces, what charge, q, on the particle is required to sustain an orbital period of 3.0×10^{-3} s?

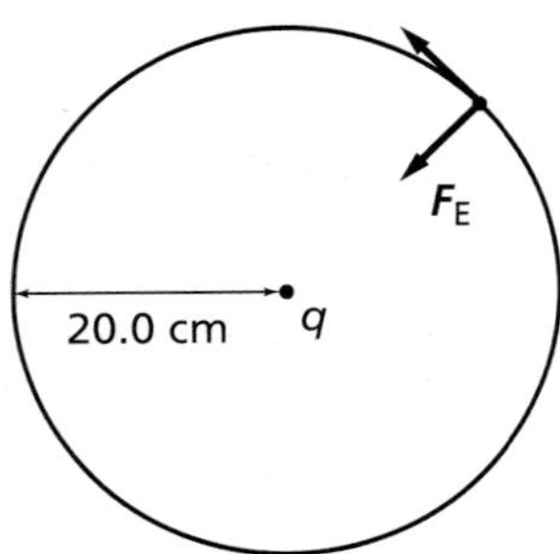

6. Two charges of equal magnitude and opposite sign are placed 0.50 m apart. The electric field strength midway between them is 4.8×10^{4} N/C toward the negative charge. What is the magnitude of each charge?

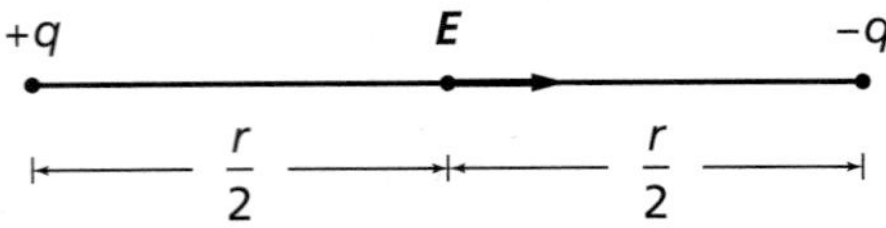

7. A pith ball weighing 3.0×10^{-2} N carries a charge of -1.0×10^{-6} C. The pith ball is placed between two large, parallel, metal plates, that are separated by 0.050 m. What is the potential difference, ΔV, that must be applied in order to suspend the pith ball between the plates?
8. The electric field strength, defined as a force per unit charge, has the units newtons per coulomb, N/C. The formula for electric potential difference, $\Delta V = Ed$, however, suggests that E also can be expressed as volts per meter, V/m.
   a. By analysis of the units, show that these two expressions for the units of electric field strength are equivalent.
   b. Suggest a reason why V/m is often a more useful way to express electric field strength.
9. Suppose you have two parallel, metal plates that have an electric field between them of strength 3.0×10^{4} N/C, and are 0.050 m apart. Consider a point, P, located 0.030 m from plate A, the negatively charged plate, when answering the following questions.
   a. What is the electric potential at P relative to plate A?
   b. What is the electric potential at P relative to plate B, the positively charged plate?
10. A certain 1.5-V size-AA battery has a storage capacity of 2500 C. How much work can this battery perform?
11. In a vacuum tube, electrons accelerate from the cathode element to the plate element, which is maintained at a positive potential with respect to the cathode. If the plate voltage is +240 V, how much kinetic energy has an electron acquired when it reaches the plate?
12. An oil drop with five excess electrons is suspended in an electric field of 2.0×10^{3} N/C. What is the mass of the oil drop?
13. An oil drop weighing 7.5×10^{-15} N carries three excess electrons.
   a. What potential difference is required to suspend the drop between parallel plates separated by 2.3 cm?
   b. If the oil drop picks up another electron, by how much must the potential difference between the plates be reduced to maintain the oil drop in suspension?

14. A charge of 2.00×10^{-6} C is moved against a constant electric field. If 4.50×10^{-4} J of work is done on the charge, what is the potential difference between the initial and the final locations of the charge?

15. Capacitors $C_1 = 220\ \mu F$ and $C_2 = 470\ \mu F$ are connected across a 48.0-V electric potential difference.
   a. What are the charges, q_1 and q_2, on each of the capacitors?
   b. What is the total charge, q_T, on both capacitors?
   c. Repeat steps a and b for a new potential difference, $\Delta v' = 96.0$ V.
   d. Now consider the two capacitors as a system. What would be a single equivalent capacitor, C_{eq}, that could replace C_1 and C_2, and be capable of yielding the same results?
   e. Based on your responses to the above, make a conjecture concerning the equivalent capacitance of a system of capacitors—all of which are connected across the same potential difference.

16. A new 90.0-V battery with a storage capacity of 2.5×10^4 C charges a 6800-μF capacitor with the switch in position A. Then the switch is thrown to position B to discharge the capacitor.
   a. How many times can this cycle be repeated before the battery is completely discharged?
   b. If the capacitor discharge occurs in 120 ms, what is the average power dissipated in the discharge circuit during one cycle?

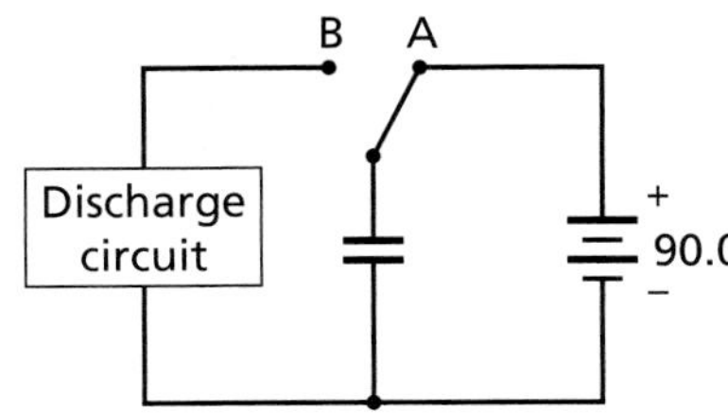

17. A 0.68-μF capacitor carries a charge on one plate of 1.36×10^{-5} C. What is the potential difference across the leads of this capacitor?

Chapter 22

1. A decorative lightbulb rated at 7.50 W draws 60.0 mA when lit. What is the voltage drop across the bulb?

2. A 1.2-V nickel-cadmium battery has a rated storage capacity of 4.0×10^3 mAh (milliamp-hours).
   a. What is the battery charge capacity in coulombs? *Hint: 1 C = 1 A·s.*
   b. How long can this battery supply a current of 125 mA?

3. A heating element of an electric furnace consumes 5.0×10^3 W when connected across a 240-V source. What current flows through the element?

4. The cold filament resistance of a lightbulb is 20.0 Ω. The bulb consumes 75 W when it is operating from a 120-V source. By what factor does the start-up current in the bulb exceed the operating current?

5. In the circuit shown below, a potentiometer is used to vary the current to the lamp. If the only resistance is due to the lamp, what is the current in the circuit?

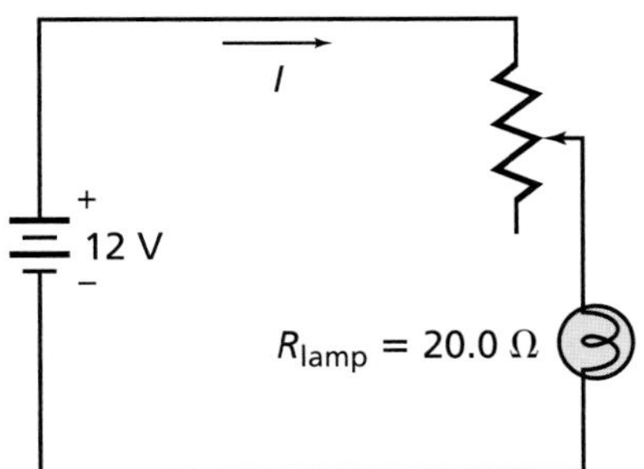

6. An electric toaster consumes 1875 W in operation. If it is plugged into a 125-V source, what is its resistance?

7. Draw a circuit diagram to include a 90.0-V battery, a 220-Ω resistor, and a 680-Ω resistor in series. Show the direction of conventional current.

8. Modify the diagram of the previous problem to include an ammeter and a voltmeter to measure the voltage drop across the 680-Ω resistor.

9. If the total resistance in the circuit in problem 8 is 9.0×10^2 Ω, what would the ammeter and the voltmeter indicate?

10. An electric motor with a load delivers 5.2 hp to its shaft (1 hp = 746 W). Under these conditions, it operates at 82.8 percent efficiency. (Efficiency is defined as the ratio of power output to power input.)
    - **a.** How much current does the motor draw from a 240-V source?
    - **b.** What happens to the remaining 17.2 percent of the input power?
11. An industrial heating process uses a current of 380 A supplied at 440-V potential.
    - **a.** What is the effective resistance of the heating element?
    - **b.** How much energy is used by this process during an 8-h shift?
12. An 8.0-Ω electric heater operates from a 120-V source.
    - **a.** How much current does the heater require?
    - **b.** How much time does the heater need to generate 2.0×10^4 J of thermal energy?
13. A manufacturer of lightbulbs advertises that its 55-W bulb, which produces 800.0 lm of light output, provides almost the same light as a 60.0-W bulb with an energy savings. The 60.0-W bulb produces 840.0 lm.
    - **a.** Which bulb most efficiently converts electric energy to light?
    - **b.** Assume a bulb has a lifetime of 1.0×10^3 h and an electric energy cost of $0.12/kWh. How much less does the 55-W bulb cost to operate over its lifetime?
    - **c.** Is most of the cost savings due to higher bulb efficiency, or due to the consumer's willingness to accept lower light output?
14. By what factor would the I^2R loss in transmission wires be reduced if the transmission voltage were boosted from 220 V to 22 kV? Assume that the rate of energy delivered is unchanged.
15. The electric-utility invoice for a household shows a usage of 1245 kWh during a certain 30-day period. What is the average power consumption during this period?

Chapter 23

1. A series circuit contains a 47-Ω resistor, an 82-Ω resistor, and a 90.0-V battery. What resistance, R_3, must be added in series to reduce the current to 350 mA?
2. What is the minimum number of 100.0-Ω resistors that must be connected in series together with a 12.0-V battery to ensure that the current does not exceed 10.0 mA?
3. A 120.0-V generator is connected in series with a 100.0-Ω resistor, a 400.0-Ω resistor, and a 700.0-Ω resistor.
    - **a.** How much current is flowing in the circuit?
    - **b.** What is the voltage drop in the 400-Ω resistor?
4. Show that the total power dissipated in a circuit of series-connected resistors is $P = I^2R$, where R is the equivalent resistance.
5. In an experiment, three identical lightbulbs are connected in series across a 120.0-V source, as shown. When the switch is closed, all of the bulbs are illuminated. However, when the experiment is repeated the next day, bulbs A and B are illuminated brighter than normal, and bulb C is dark. A voltmeter is used to measure the voltages in the circuit as shown. The voltmeter readings are:

$$V_1 = 120.0\text{ V}$$
$$V_2 = 60.0\text{ V}$$
$$V_3 = 0.0\text{ V}$$

    - **a.** What has happened in this circuit?
    - **b.** Explain why bulbs A and B are brighter than normal.
    - **c.** Is there more or less current flowing now in the circuit?

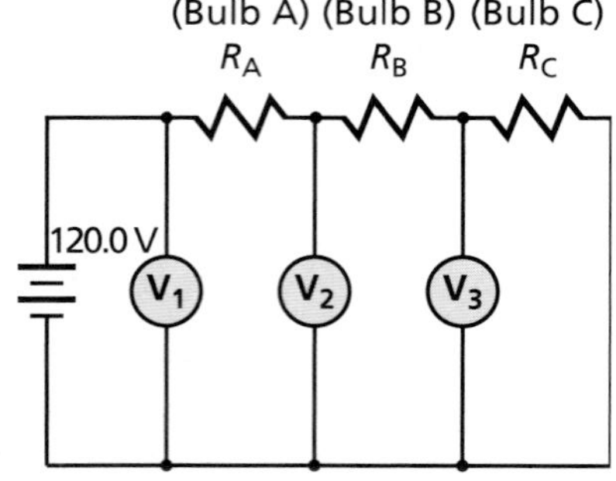

6. A string of holiday lights has 25 identical bulbs connected in series. Each bulb dissipates 1.00 W when the string is connected to a 125-V outlet.
   a. How much power must be supplied by the 125-V source?
   b. What is the equivalent resistance of this circuit?
   c. What is the resistance of each bulb?
   d. What is the voltage drop across each bulb?
7. Two resistors are connected in series across a 12.0-V battery. The voltage drop across one of the resistors is 5.5 V.
   a. What is the voltage drop across the other resistor?
   b. If the current in the circuit is 5.0 mA, what are the two resistor values?
8. In problem 7, the specifications for the resistors state that their resistance may vary from the listed nominal value. If the possible ranges of actual resistance values are as follows, $1050\ \Omega \leq R_1 \leq 1160\ \Omega$, and $1240\ \Omega \leq R_2 \leq 1370\ \Omega$, what is the possible minimum and maximum value of the nominal 5.5-V voltage drop?
9. A voltage-divider circuit is constructed with a potentiometer, as shown below. There are two fixed resistances of 113 kΩ and 294 kΩ. The resistance of the potentiometer can range from 0.0 to 100.0 kΩ.
   a. What is V_{out} when the potentiometer is at its minimum setting of 0.0 Ω?
   b. What is V_{out} when the potentiometer is at its maximum setting of 100.0 kΩ?
   c. What potentiometer setting is required to adjust V_{out} to exactly 65.0 V?

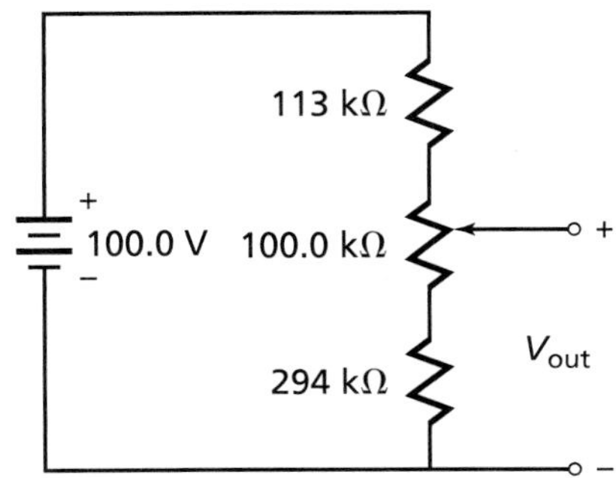

10. Two lightbulbs, one rated at 25.0 W and one rated at 75.0 W, are connected in parallel across a 125-V source.
    a. Which bulb is the brightest?
    b. What is the operating resistance of each of the bulbs?
    c. The bulbs are rewired in series. What is the dissipated power in each bulb? Which bulb is the brightest?
11. A 10.0-V battery has an internal resistance of 0.10 Ω. The internal resistance can be modeled as a series resistor, as shown.
    a. Derive an expression for the battery terminal voltage, V, as a function of the current, I.
    b. Create a graph of voltage versus current for a current range of 0.0 to 1.0 A.
    c. What value of load resistor R_L needs to be placed across the battery terminals to give a current of 1.0 A?
    d. How does the V-I function differ from that of an ideal voltage source?

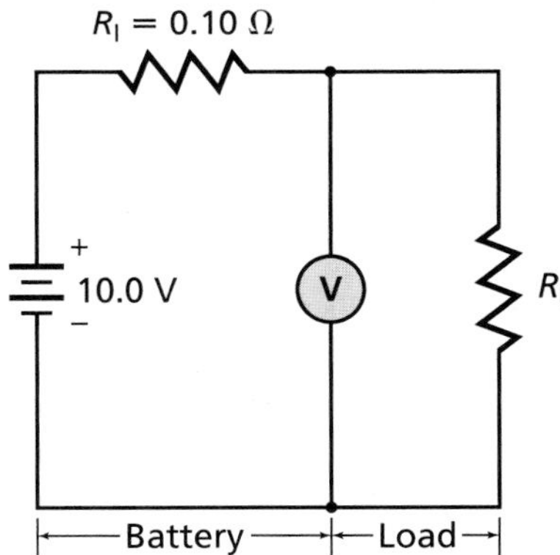

12. Show that the total power dissipated in a circuit of parallel-connected resistors is $P = \frac{V^2}{R}$, where R is the equivalent resistance.
13. A holiday light string of ten bulbs is equipped with shunts that short out the bulbs when the voltage drop increases to line voltage, which happens when a bulb burns out. Each bulb has a resistance of 200.0 Ω. The string is connected to a household circuit at 120.0 V. If the string is protected by a 250.0-mA fuse, how many bulbs can fail without blowing the fuse?

14. A 60.0-W lightbulb and a 75.0-W lightbulb are connected in series with a 120-V source.
    a. What is the equivalent resistance of the circuit?
    b. Suppose an 1875-W hair dryer is now plugged into the parallel circuit with the lightbulbs. What is the new equivalent resistance of the circuit?
15. What is the equivalent resistance of the following resistor network?

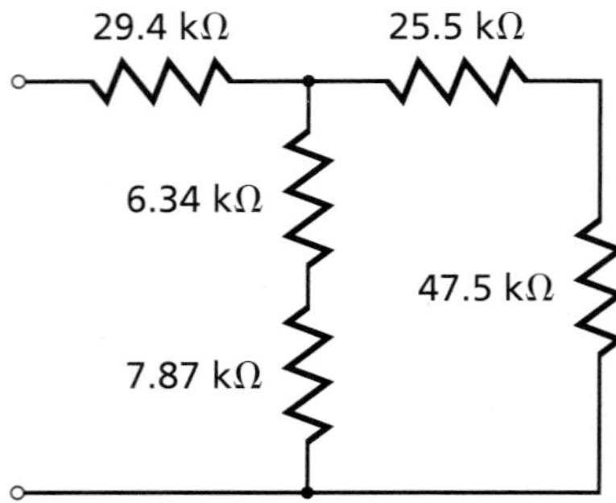

Chapter 24

1. The magnetic field of Earth resembles the field of a bar magnet. The north pole of a compass needle, used for navigation, generally points toward the geographic north pole. Which magnetic pole of the Earth is the compass needle pointing toward?
2. A magnet is used to collect some spilled paper clips. What is the magnetic pole at the end of the paper clip that is indicated in the figure?

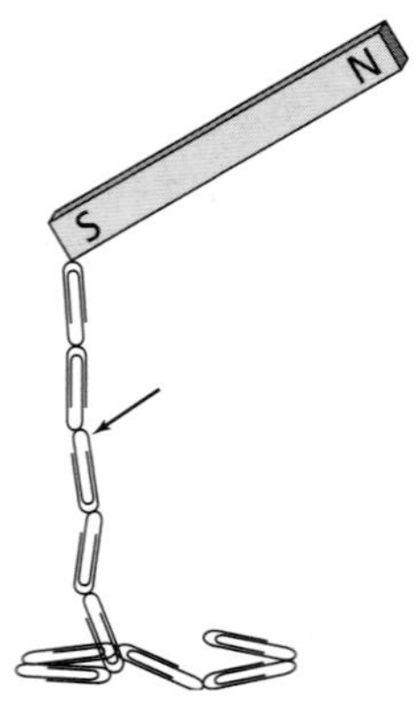

3. Jingdan turned a screwdriver into a magnet by rubbing it with a strong bar magnet to pick up a screw that has fallen into an inaccessible spot. How could he demagnetize his screwdriver after picking up the screw?
4. A long, straight, current-carrying wire carries a current from west to east. A compass is held above the wire.
    a. Which direction does the north pole of the compass point?
    b. When the compass is moved underneath the wire, in which direction does the north pole point?
5. Consider the sketch of the electromagnet in the figure below. Which components could you change to increase or decrease the strength of the electromagnet? Explain your answer.

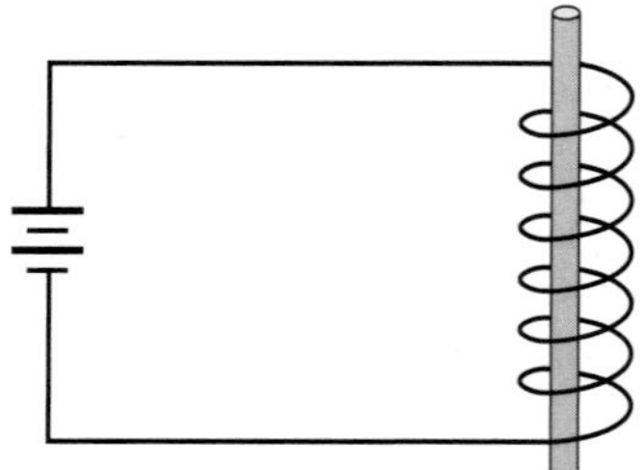

6. Sketch a graph showing the relationship between the magnetic field around a straight, current-carrying wire, and the distance from the wire.
7. How long is a wire in a 0.86-T field that carries a current of 1.4 A and experiences a force of 13 N?
8. A 6.0-T magnetic field barely prevents a 0.32-m length of copper wire with a current of 1.8 A from dropping to the ground. What is the mass of the wire?
9. How much current will be needed to produce a force of 1.1 N on a 21-cm-long piece of wire at right angles to a 0.56-T field?
10. Alpha particles (particles containing two protons, two neutrons, but no electrons) are traveling at right angles to a 47-μT field with a speed of 36 cm/s. What is the force on each particle?

11. A force of 7.1×10^{-12} N is exerted on some Al^{3+} ions (an atom missing three electrons) that are traveling at 430 km/s perpendicular to a magnetic field. What is the magnetic field?
12. Electrons traveling at right angles to a magnetic field experience a force of 8.3×10^{-13} N when they are in a magnetic field of 6.2×10^{-1} T. How fast are the electrons moving?

Chapter 25

1. A 72-cm wire is moved at an angle of 72° through a magnetic field of 1.7×10^{-2} T and experiences an *EMF* of 1.2 mV. How fast is the wire moving?
2. A 14.2-m wire moves 3.12 m/s perpendicular to a 4.21-T field.
   a. What *EMF* is induced in the wire?
   b. Assume that the resistance in the wire is 0.89 Ω. What is the amount of current in the wire?
3. A 3.1-m length of straight wire has a resistance of 3.1 Ω. The wire moves at 26 cm/s at an angle of 29° through a magnetic field of 4.1 T. What is the induced current in the wire?
4. A generator delivers an effective current of 75.2 A to a wire that has a resistance of 0.86 Ω.
   a. What is the effective voltage?
   b. What is the peak voltage of the generator?
5. What is the RMS voltage of a household outlet if the peak voltage is 165 V?
6. An outlet has a peak voltage of 170 V.
   a. What is the effective voltage?
   b. What effective current is delivered to an 11-Ω toaster?
7. A step-up transformer has a primary coil consisting of 152 turns and a secondary coil with 3040 turns. The primary coil receives a peak voltage of 98 V.
   a. What is the effective voltage in the primary coil?
   b. What is the effective voltage in the secondary coil?
8. A step-down transformer has 9000 turns on the primary coil and 150 turns on the secondary coil. The *EMF* in the primary coil is 16 V. What is the voltage being applied to the secondary coil?
9. A transformer has 124 turns on the primary coil and 18,600 turns on the secondary coil.
   a. Is this a step-down or a step-up transformer?
   b. If the effective voltage in the secondary coil is 3.2 kV, what is the peak voltage being delivered to the primary coil?

Chapter 26

1. A stream of singly ionized (1−) fluorine atoms passes undeflected through a magnetic field of 2.5×10^{-3} T that is balanced by an electric field of 3.5×10^{3} V/m. The mass of the fluorine atoms is 19 times that of a proton.
   a. What is the speed of the fluorine ions?
   b. If the electric field is switched off, what is the radius of the circular path followed by the ions?
2. An electron moves perpendicular to Earth's magnetic field with a speed of 1.78×10^{6} m/s. If the strength of Earth's magnetic field is about 5.00×10^{-5} T, what is the radius of the electron's circular path?
3. A proton with a velocity of 3.98×10^{4} m/s perpendicular to the direction of a magnetic field follows a circular path with a diameter of 4.12 cm. If the mass of a proton is 1.67×10^{-27} kg, what is the strength of the magnetic field?
4. A beam of doubly ionized (2+) calcium atoms is analyzed by a mass spectrometer. If $B = 4.5\times10^{-3}$ T, $r = 0.125$ m, and the mass of the calcium ions is 6.68×10^{-26} kg, what is the voltage of the mass spectrometer?
5. The speed of light in crown glass is 1.97×10^{8} m/s. What is the dielectric constant of crown glass?
6. The dielectric constant of diamond is 6.00. What is the speed of light in diamond?

7. By curving different isotopes through paths with different radii, a mass spectrometer can be used to purify a sample of mixed uranium-235 and uranium-238 isotopes. Assume that $B = 5.00\times10^{-3}$ T, $V = 55.0$ V, and that each uranium isotope has a 5+ ionization state. Uranium-235 has a mass that is 235 times that of a proton, while uranium-238 has a mass that is 238 times that of a proton. By what distance will the two isotopes be separated by the mass spectrometer?

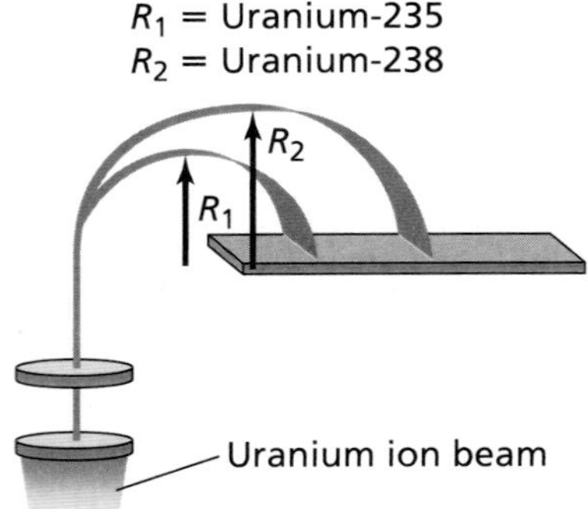

8. A mass spectrometer often is used in carbon dating to determine the ratio of C-14 isotopes to C-12 isotopes in a biological sample. This ratio then is used to estimate how long ago the once-living organism died. Because a mass spectrometer is sensitive to the charge-to-mass ratio, it is possible for a contaminant particle to alter the value measured for the C-14/C-12 ratio, and thus, yield erroneous results. When ionized, C-14 forms an ion with a +4 charge, and the mass of C-14 is 14 times that of a proton. Consider a contaminant lithium particle. If the most common lithium isotope has a mass that is seven times that of a proton, what must be the charge of the lithium ion needed to contaminate a carbon-14 experiment?
9. What is the wavelength of a radio wave with a frequency of 90.7 MHz?
10. What is the frequency of a microwave with a wavelength of 3.27 mm?
11. What is the frequency of an X ray with a wavelength of 1.00×10^{-10} m?
12. In recent years, physicists have slowed the speed of light passing through a material to about 1.20 mm/s. What is the dielectric constant of this material?

Chapter 27

1. If the maximum kinetic energy of emitted photoelectrons is 1.4×10^{-18} J, what is the stopping potential of a certain photocell?
2. The stopping potential of a photocell is 2.3 V. What is the initial velocity of an emitted photoelectron that is brought to a stop by the photocell?
3. If a photoelectron traveling at 8.7×10^{5} m/s is stopped by a photocell, what is the photocell's stopping potential?
4. Light with a frequency of 7.5×10^{14} Hz is able to eject electrons from the metal surface of a photocell that has a threshold frequency of 5.2×10^{14} Hz. What stopping potential is needed to stop the emitted photoelectrons?
5. A metal has a work function of 4.80 eV. Will ultraviolet radiation with a wavelength of 385 nm be able to eject a photoelectron from the metal?
6. When a metal is illuminated with radiation with a wavelength of 152 nm, photoelectrons are ejected with a velocity of 7.9×10^{5} m/s. What is the work function, in eV, of the metal?

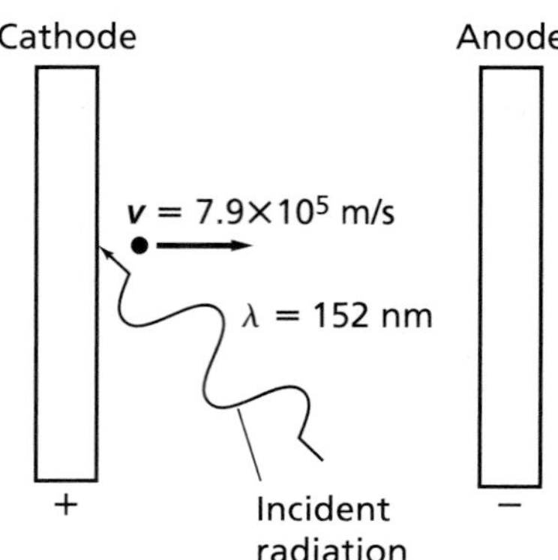

7. The de Broglie wavelength for an electron traveling at 9.6×10^{5} m/s is 7.6×10^{-10} m. What is the mass of the electron?
8. What is the de Broglie wavelength of a 68-kg man moving with a kinetic energy of 8.5 J?
9. An electron has a de Broglie wavelength of 5.2×10^{-10} m. What potential difference is responsible for this wavelength?

Chapter 28

1. An electron in a hydrogen atom makes a transition from E_3 to E_1. How much energy does the atom lose?
2. An electron in the hydrogen atom loses 3.02 eV as it falls to energy level E_2. From which energy level did the atom fall?
3. Which energy level in a hydrogen atom has a radius of 7.63×10^{-9} m?
4. When an electron falls from E_4 to E_1, what is the frequency of the emitted photon?
5. If an electron moves from E_3 to E_5, what is the wavelength of the photon absorbed by the atom?
6. If a hydrogen atom in its ground state absorbs a photon with a wavelength of 93 nm, it jumps to an excited state. What is the value of the energy in that excited state?

Chapter 29

1. Indium has 3 free electrons per atom. Use Appendix D and determine the number of free electrons in 1.0 kg of indium.
2. Cadmium has 2 free electrons per atom. Use Appendix D and determine the number of free electrons in 1.0 dm^3 of Cd.
3. Copper has 1 free electron per atom. What length of 1.00-mm diameter copper wire contains 7.81×10^{24} free electrons? Use Appendix D for physical constants.
4. At 400.0 K, germanium has 1.13×10^{15} free electrons/cm^3. How many free electrons per Ge atom are there at this temperature?
5. At 400.0 K, silicon has 4.54×10^{12} free electrons/cm^3. How many free electrons per Si atom are there at this temperature?
6. At 200.0 K, silicon has 3.79×10^{-18} free electrons per atom. How many free electrons/cm^3 are there in silicon at this temperature?
7. Silicon has 1.45×10^{10} free electrons/cm^3 at room temperature. If you wanted to have 3×10^{6} as many electrons from arsenic doping as thermal free electrons from silicon at room temperature, how many arsenic atoms should there be for each silicon atom? Each arsenic atom provides 1 free electron. Use Appendix D for physical constants.
8. At 200.0 K, germanium has 1.16×10^{10} thermally liberated charge carriers/cm^3. If it is doped with 1 As atom to 525,000 Ge atoms, what is the ratio of doped carriers to thermal carriers at this temperature? See Appendix D for physical constants.
9. At 200.0 K, silicon has 1.89×10^{5} thermally liberated charge carriers/cm^3. If it is doped with 1 As atom to 3.75 million Si atoms, what is the ratio of doped carriers to thermal carriers at this temperature? See Appendix D for physical constants.
10. The diode shown below has a voltage drop, V_d, of 0.45 V when I = 11 mA. If a 680-Ω resistor, R, is connected in series, what power supply voltage, V_b is needed?

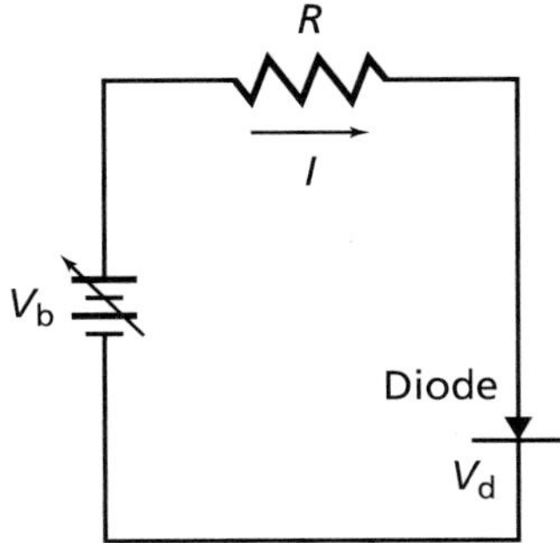

11. A diode in a circuit similar to the one in Figure 29-26 has a voltage drop, V_d, of 0.95 V when I = 18 mA. If a 390-Ω resistor, R, is connected in series, what power supply voltage, V_b is needed?

12. What power supply voltage would be needed to produce a current of 27 mA in the circuit in problem 10? Assume the diode voltage is unchanged.

Chapter 30

1. Carbon-14, or ^{14}C, is an isotope of the common $^{12}_{6}C$, and is used in dating ancient artifacts. What is the composition of its nucleus?
2. An isotope of iodine ($Z = 53$) is used to treat thyroid conditions. Its mass number is 131. How many neutrons are in its nucleus?
3. The only nonradioactive isotope of fluorine has nine protons and ten neutrons.
   a. What is its mass number?
   b. The atomic mass unit, u, is equal to 1.66×10^{-27} kg. What is fluorine-19's approximate mass in kilograms?
   c. Write the full symbol of this atom.
4. The magnesium isotope $^{25}_{12}Mg$ has a mass of 24.985840 u.
   a. Calculate its mass defect.
   b. Calculate its binding energy in MeV.
5. The isotope $^{10}_{5}B$ has a mass of 10.012939 u.
   a. Calculate the mass defect.
   b. Calculate its binding energy in MeV.
   c. Calculate its binding energy per nucleon.
6. The most stable isotope of all is $^{56}_{26}Fe$. Its binding energy per nucleon is −8.75 MeV/nucleon.
   a. What is the binding energy of this isotope?
   b. What is the mass defect of this isotope?
7. The isotope $^{239}_{94}Pu$ can be transmuted to an isotope of uranium, $^{235}_{92}U$.
   a. Write the nuclear equation for this transmutation.
   b. Identify the particle that is ejected.
8. The radioisotope $^{222}_{84}Po$ undergoes alpha decay to form an isotope of lead (lead has atomic number 82). Determine what the mass number of that isotope must be by writing a nuclear equation.
9. The graph below shows a sequence of alpha and beta decays, labeled *1, 2, 3,* and *4*. Consult Table 30-1 as needed.
   a. Which represent alpha decays, and which represent beta decays ?
   b. What is the overall change in mass in the sequence? In the number of neutrons?

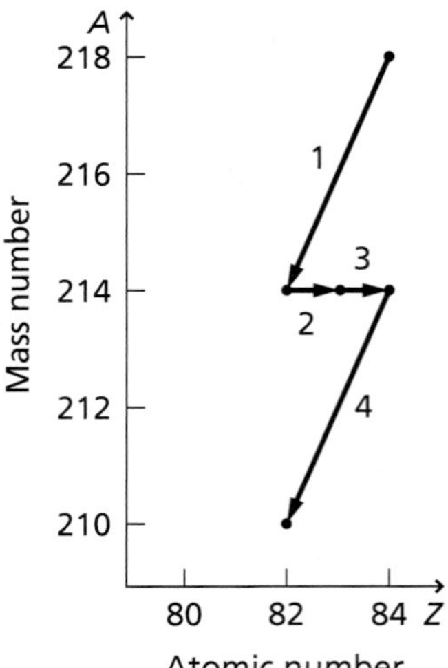

10. Use Appendix D to complete the two nuclear equations. Include correct subscripts and superscripts for each of the particles.
    a. $^{32}_{15}P \rightarrow ? + \beta +$ antineutrino
    b. $^{235}_{92}U \rightarrow ? + \alpha$
11. Write the complete nuclear equation for the beta decay of $^{34}_{16}S$.
12. A positron is identical to a beta particle, except that its charge is +1 instead of −1. $^{37}_{19}K$ undergoes spontaneous positron decay to form an isotope of argon. Identify the isotope of argon by writing a nuclear equation.
13. Iodine-131 has a half-life of 8.0 days. If there are 60.0 mg of this isotope at time zero, how much remains 24 days later?
14. Refer to Table 30-2. Given a sample of cobalt-60,
    a. how long a time is needed for it to go through four half-lives?
    b. what fraction remains at the end of that time?

15. The graph shows the activity of a certain radioisotope over time. Deduce its half-life from this data.

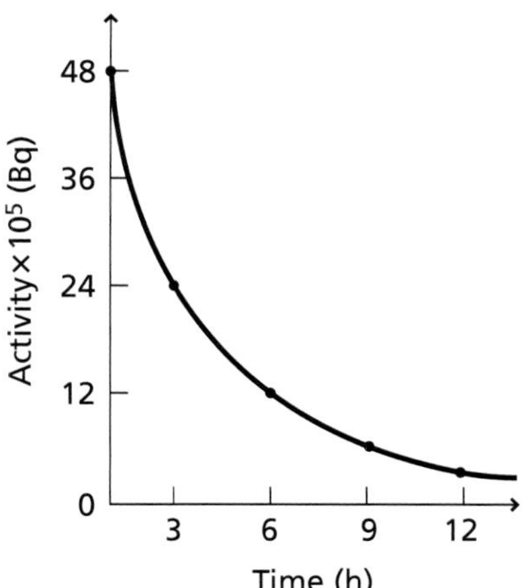

16. The mass of a proton and of an antiproton is 1.00728 u. Recall that the conversion of exactly 1 u into energy yields 931.5 MeV.
  a. Calculate the mass used up when a proton and an antiproton annihilate one another.
  b. Calculate the energy released here.

17. One source of a star's energy is the fusion of two deuterons to form an alpha particle plus a gamma ray.

Particle	Mass in u
${}^{2}_{1}H$	2.0136
${}^{4}_{2}He$	4.0026
${}^{0}_{0}\gamma$	0.0000

  a. Write the nuclear equation for this fusion.
  b. Calculate the mass "lost" in this process.
  c. Calculate the energy released in MeV.

18. When 1 mol (235 g) of uranium-235 undergoes fission, about 2.0×10^{10} kJ of energy are released. When 4.0 g of hydrogen undergoes fusion, about 2.0×10^{9} kJ are released.
  a. For each, calculate the energy yield per gram of fuel.
  b. Which process produces more energy per gram of fuel?

Appendix C: Solutions for Practice Problems

Chapter 1

1. $I = \frac{V}{R} = \frac{9.0\ \text{V}}{50.0\ \Omega} = 0.18\ \text{A}$

3. $t = \frac{v}{a} = \frac{4.00\ \text{m/s}}{0.400\ \text{m/s}^2} = 10.0\ \text{s}$

5. $750\ \cancel{\text{kHz}}\left(\frac{1000\ \cancel{\text{Hz}}}{1\ \cancel{\text{kHz}}}\right)\left(\frac{1\ \text{MHz}}{1{,}000{,}000\ \cancel{\text{Hz}}}\right) = 0.75\ \text{MHz}$

7. $366\ \cancel{\text{days}}\left(\frac{24\ \cancel{\text{h}}}{1\ \cancel{\text{day}}}\right)\left(\frac{60\ \cancel{\text{min}}}{1\ \cancel{\text{h}}}\right)\left(\frac{60\ \text{s}}{1\ \cancel{\text{min}}}\right) = 31{,}622{,}400\ \text{s}$

9. a.

6.201 cm
7.4 cm
0.68 cm
12.0 cm
26.281 cm
= 26.3 cm

b.

1.6 km =	1600 m
1.62 m =	1.62 m
1200 m =	12 m
	1613.62 m

= 1600 m or 1.6 km rounded to two significant digits

11. a. 320 cm^2 or 3.2×10^2 cm^2

b. 13.6 km^2

Chapter 2

9. The car begins at a position of 125.0 m and moves toward the origin, arriving at the origin 5.0 s after it begins moving. The car continues beyond the origin.

11. a. 4.0 s **b.** 100.0 m

13. a. 19 s **b.** 58 s

c.

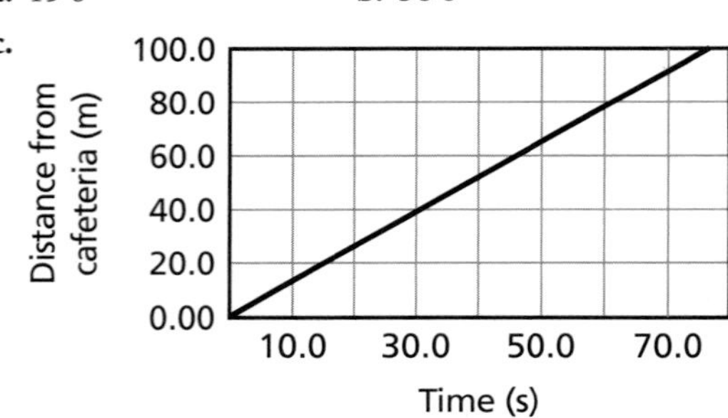

15. runner B

17. approximately 30 m

25. a.
$$\bar{v} = \left|\frac{\Delta d}{\Delta t}\right| = \left|\frac{d_2 - d_1}{t_2 - t_1}\right| = \left|\frac{-1\ \text{m} - 0\ \text{m}}{3\ \text{s} - 0\ \text{s}}\right| = |-0.3\ \text{m/s}| = 0.3\ \text{m/s}$$

b. The average velocity is the slope of the line, including the sign, so it is −0.33 m/s or 0.33 m/s north.

27. Because the bicycle is moving in the positive direction, the average speed and average velocity are the same. Using the points (0.00 min, 0.0 km) and (15.0 min, 10.0 km),

$$\bar{v} = \left|\frac{\Delta d}{\Delta t}\right| = \left|\frac{d_2 - d_1}{t_2 - t_1}\right| = \left|\frac{10.0\ \text{km} - 0.0\ \text{km}}{15.0\ \text{min} - 0.0\ \text{min}}\right| = 0.67\ \text{km/min}$$

The bicycle is moving in the positive direction at a speed of 0.67 km/min.

Chapter 3

1.

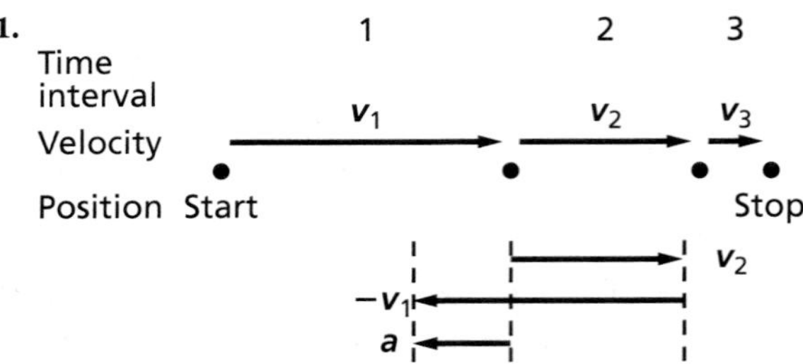

3. a. 5.0 to 15.0 s **b.** 0.0 to 5.0 s

c. 15.0 to 20.0 s

5.

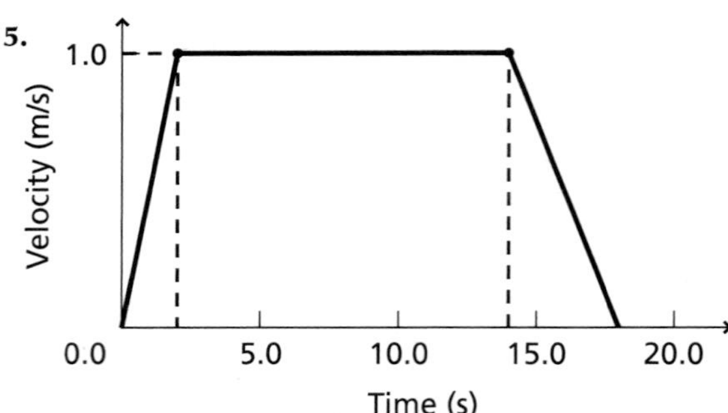

7. $\bar{a} = \frac{\Delta v}{\Delta t} = \frac{15\ \text{m/s} - 36\ \text{m/s}}{3.0\ \text{s}} = -7.0\ \text{m/s}^2$

9. a. $\bar{a} = \frac{\Delta v}{\Delta t} = \frac{0.0\ \text{m/s} - 25\ \text{m/s}}{3.0\ \text{s}} = -8.3\ \text{m/s}^2$

b. Half as great ($-4.2\ \text{m/s}^2$)

11. $\bar{a} = \frac{\Delta v}{\Delta t} = \frac{0.5\ \text{cm/y} - 1.0\ \text{cm/y}}{1.0\ \text{y}} = -0.5\ \text{cm/y}^2$

19.
$$v_f = v_i + at$$
$$= 30.0\ \text{km/h} + (3.5\ \text{m/s}^2)(6.8\ \text{s})\left(\frac{1\ \text{km}}{1000\ \text{m}}\right)\left(\frac{3600\ \text{s}}{1\ \text{h}}\right)$$
$$= 120\ \text{km/h}$$

21.
$$v_f = v_i + at$$
so $t = \frac{v_f - v_i}{a}$
$$= \frac{3.0\ \text{m/s} - 22\ \text{m/s}}{-2.1\ \text{m/s}^2}$$
$$= 9.0\ \text{s}$$

23. $\left(\frac{75\ \text{m}}{1\ \text{s}}\right)\left(\frac{3600\ \text{s}}{1\ \text{h}}\right)\left(\frac{1\ \text{km}}{1000\ \text{m}}\right) = 2.7 \times 10^2\ \text{km/h}$

25. a.

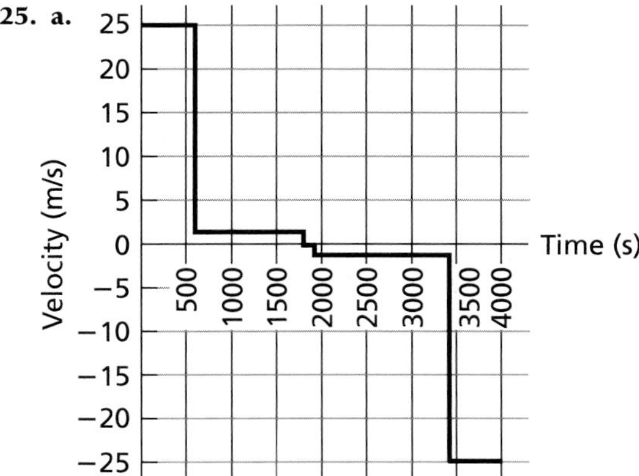

Distance the driver walked to the gas station:

$d = vt$

$= (1.5\ \text{m/s})(20.0\ \text{min})\left(\frac{60\ \text{s}}{1\ \text{min}}\right) = 1800\ \text{m}$

Time to walk back to the car:

$t = \frac{d}{v} = \frac{1800\ \text{m}}{1.2\ \text{m/s}} = 1500\ \text{s} = 25\ \text{min}$

b.

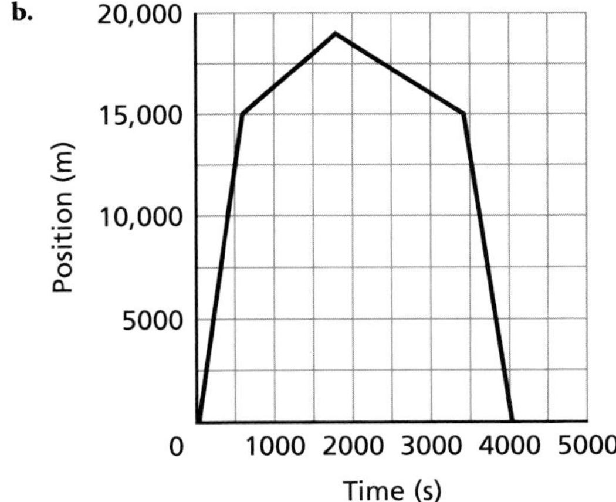

27. $v_f = v_i + at$

$a = \frac{v_f - v_i}{t} = \frac{22\ \text{m/s} - 44\ \text{m/s}}{11\ \text{s}} = -2.0\ m/s$

$d_f = d_i + v_i t_f + 1/2 a t_f^2$

$d_f = 0 + (44\ \text{m/s})(11\ \text{s}) + 1/2(-2.0\ \text{m/s}^2)(11\ \text{s})^2$

$d_f = 484\ \text{m} + (-121\ \text{m}) = 363\ \text{m}$

29. $v_i = v_f - a t_f$

$d_f = d_i + (v_f + a t_f)t_f + 1/2 a t_f^2$

$19\ \text{m} = 0 + [(7.5\ \text{m/s})(4.5\ \text{s}) - a(4.5\ \text{s})^2] + 1/2a(4.5\ \text{s})^2$

$\text{a} = 1.5\ \text{m/s}^2$

$v_f = v_i + a t_f$

$v_i = v_f - a t_f$

$= 7.5\ \text{m/s} - (1.5\ \text{m/s}^2)(4.5\ \text{s}) = 0.75\ \text{m/s}$

31.

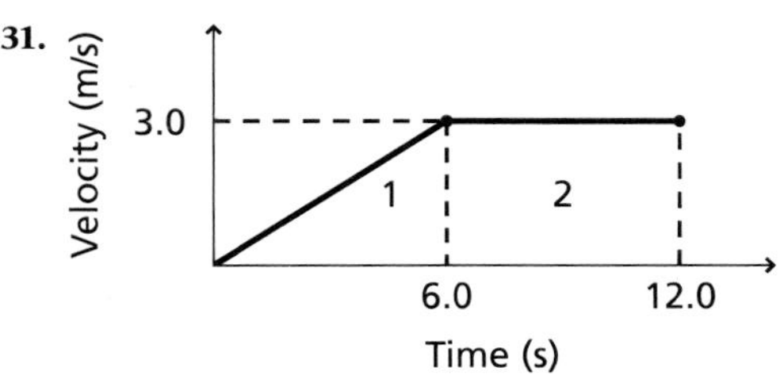

Part 1: Constant acceleration:

$d_1 = \frac{1}{2}(3.0\ \text{m/s})(6.0\ \text{s})$

$= 9.0\ \text{m}$

Part 2: Constant velocity:

$d_2 = (3.0\ \text{m/s})(12.0\ \text{s} - 6.0\ \text{s})$

$= 18\ \text{m}$

thus, $d = d_1 + d_2 = 9.0\ \text{m} + 18\ \text{m} = 27\ \text{m}$

33. Part 1: Constant velocity:

$d = vt = (4.3\ \text{m/s})(19\ \text{min})\left(\frac{60\ \text{s}}{1\ \text{min}}\right) = 4902\ \text{m}$

Part 2: Constant acceleration:

$d_f = d_i + v_i t + \frac{1}{2}at^2$

$a = \frac{2(d_f - d_i - v_i t)}{t^2}$

$= \frac{2(5.0\times10^3\ \text{m} - 4902\ \text{m} - (4.3\ \text{m/s})(19.4\ \text{s}))}{(19.4\ \text{s})^2}$

$= 0.077\ \text{m/s}^2$

43. a. Now the positive direction is downward.

$v_f = v_i + at$, where $a = g = 9.80\ \text{m/s}^2$

$v_f = 0.0\ \text{m/s} + (9.80\ \text{m/s}^2)(4.0\ \text{s})$

$= +39\ \text{m/s}$ when the downward direction is positive

b. $d = v_i t + \frac{1}{2}at^2$

$= (0.0\ \text{m/s})(4.0\ \text{s}) + \frac{1}{2}(9.80\ \text{m/s}^2)(4.0\ \text{s})^2$

$= +78\ \text{m}$

The brick still falls 78 m.

45. a. $a = -g$, and at the maximum height, $v_f = 0$

$v_f^2 = v_i^2 + 2ad$ becomes

$v_i^2 = 2gd$

$d = \frac{v_i^2}{2gd} = \frac{(22.5\ \text{m/s})^2}{2(9.80\ \text{m/s}^2)} = 25.8\ \text{m}$

b. Calculate time to rise using $v_f = v_i + at$, with $a = -g$ and $v_f = 0$.

$t = \frac{v_i}{g} = \frac{22.5\ \text{m/s}}{9.80\ \text{m/s}^2} = 2.30\ \text{s}$

The time to fall equals the time to rise, so the time to remain in the air is

$t_{air} = 2t_{rise} = (2)(2.30\ \text{s}) = 4.60\ \text{s}$

Chapter 4

1.

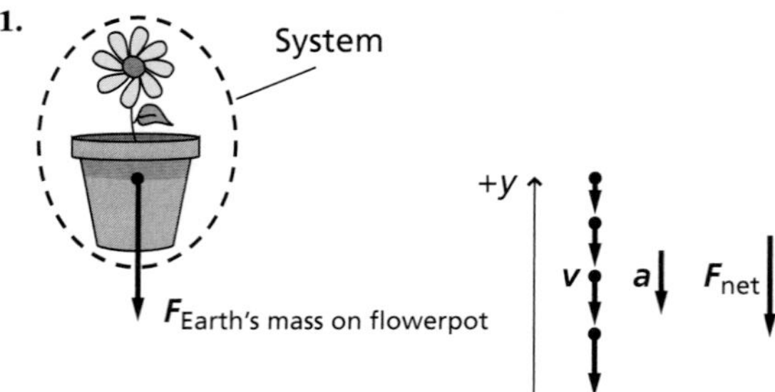

3.

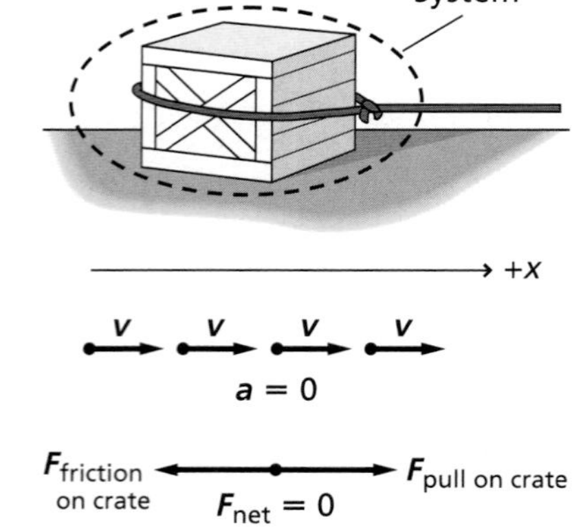

5.

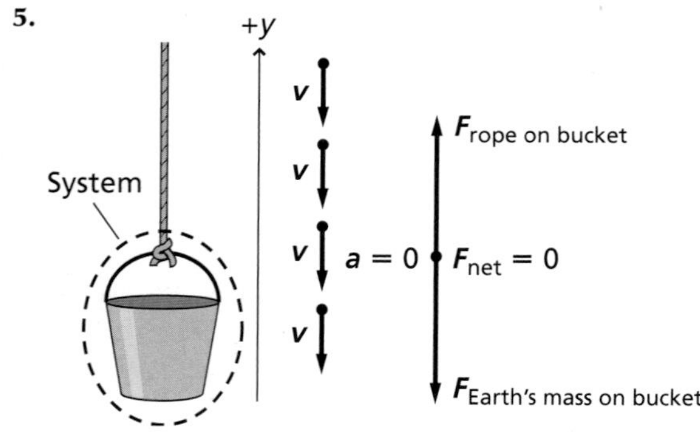

7. $F_{net} = 225\ N - 165\ N = 6.0\times10^{1}\ N$ in the direction of the larger force

15. The scale reads the weight of the watermelon:

$F_g = mg = (4.0\ kg)(9.80\ m/s^2) = 39\ N$

17. Identify Reiko's direction as positive and the rope as the system.

$F_{net} = F_{Reiko\ on\ rope} - F_{Taru\ on\ rope} = ma$

$F_{Reiko\ on\ rope} = ma + F_{Taru\ on\ rope}$

$= (0.75\ kg)(1.25\ m/s^2) + 16.0\ N$

$= 17\ N$

19. a. The scale reads 585 N. Since there is no acceleration, your weight equals the downward force of gravity:

$F_g = mg$

so $m = \frac{F_g}{g} = \frac{585\ N}{9.80\ m/s^2} = 59.7\ kg$

b. On the Moon, g changes:

$F_g = mg_{Moon}$

$= (59.7\ kg)(1.60\ m/s^2)$

$= 95.5\ N$

29. The only force acting on the brick is the gravitational attraction of Earth's mass. The brick exerts an equal and opposite force on Earth.

31.

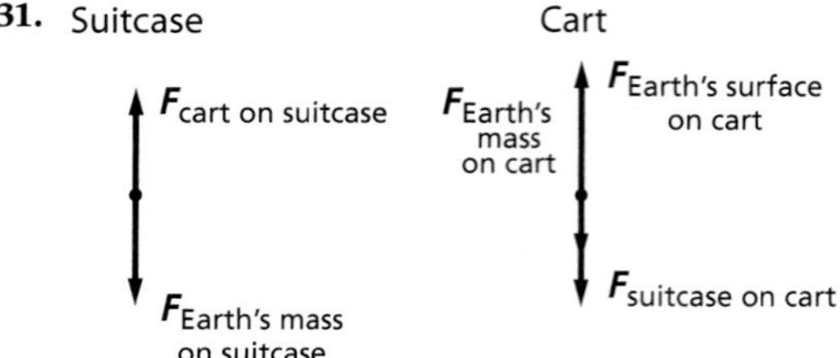

33. Identify the tire as the system and the direction of pulling as positive.

$F_{net} = F_{wheel\ on\ tire} - F_{Mika\ on\ tire} - F_{Diego\ on\ tire}$

$= ma = 0$

$F_{wheel\ on\ tire} = F_{Mika\ on\ tire} + F_{Diego\ on\ tire}$

$= 23\ N + 31\ N$

$= 54\ N$

Chapter 5

1.

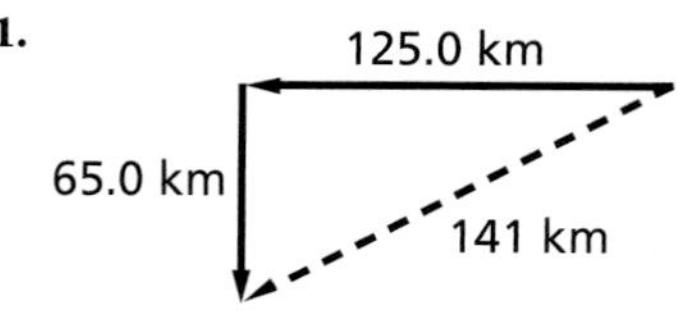

$R^2 = A^2 + B^2$

$R = \sqrt{A^2 + B^2}$

$= \sqrt{(65.0\ km)^2 + (125.0\ km)^2}$

$= 141\ km$

3. $R^2 = A^2 + B^2 - 2AB\cos\theta$

$R = \sqrt{A^2 + B^2 - 2AB\cos\theta}$

$= \sqrt{(4.5\ km)^2 + (6.4\ km)^2 - 2(4.5\ km)(6.4\ km)(\cos 135°)}$

$= 1.0\times10^{1}\ km$

5. Identify north and west as the positive directions.

$d_{1W} = d_1(\sin\theta) = (0.40\text{ km})(\sin 60.0°) = 0.35\text{ km}$

$d_{1N} = d_1(\cos\theta) = (0.40\text{ km})(\cos 60.0°) = 0.20\text{ km}$

$d_{2W} = 0.50\text{ km} \qquad d_{2N} = 0.00\text{ km}$

$R_W = d_{1W} + d_{2W} = 0.35\text{ km} + 0.50\text{ km} = 0.85\text{ km}$

$R_N = d_{1N} + d_{2N} = 0.20\text{ km} + 0.00\text{ km} = 0.20\text{ km}$

$$\begin{aligned} R &= \sqrt{{R_W}^2 + {R_N}^2} \\ &= \sqrt{(0.85\text{ km})^2 + (0.20\text{ km})^2} \\ &= 0.87\text{ km} \end{aligned}$$

$$\begin{aligned} \theta &= \tan^{-1}\left(\frac{R_W}{R_N}\right) = \tan^{-1}\left(\frac{0.85\text{ km}}{0.20\text{ km}}\right) \\ &= 77° \end{aligned}$$

0.87 km at 77° west of north

7. The resultant is 10.0 km. Using the Pythagorean theorem, the distance east is

$R^2 = A^2 + B^2$, so

$$\begin{aligned} B &= \sqrt{R^2 - A^2} \\ &= \sqrt{(10.0\text{ km})^2 - (8.0\text{ km})^2} \\ &= 6.0\text{ km} \end{aligned}$$

9. It could never be shorter than one of its components, but if it lies along either the *x*- or *y*-axis, then one of its components equals its length.

17. $F_N = mg = 52\text{ N}$

Since the speed is constant, the friction force equals the force exerted by the girl, 36 N.

$F_f = \mu_k F_N$

so $\mu_k = \dfrac{F_f}{F_N} = \dfrac{36\text{ N}}{52\text{ N}} = 0.69$

19.
$$\begin{aligned} F_{\text{Ames on box}} &= F_{\text{friction}} \\ &= \mu_s F_N \\ &= \mu_s mg \\ &= (0.55)(134\text{ N}) \\ &= 74\text{ N} \end{aligned}$$

21. $F_{\text{f, before}} = \mu_{\text{k, before}} F_N$

so $F_N = \dfrac{F_{\text{f, before}}}{\mu_{\text{k, before}}} = \dfrac{5.8\text{ N}}{0.58} = 1.0\times10^1\text{ N}$

$$\begin{aligned} F_{\text{f, after}} &= \mu_{\text{k, after}} F_N \\ &= (0.06)(1.0\times10^1\text{ N}) = 0.6\text{ N} \end{aligned}$$

23. $F_{\text{net}} = F - \mu_k F_N = F - \mu_k mg = ma$

$$\begin{aligned} \mu_k &= \frac{F - ma}{mg} \\ &= \frac{65\text{ N} - (41\text{ kg})(0.12\text{ m/s}^2)}{(41\text{ kg})(9.80\text{ m/s}^2)} \\ &= 0.15 \end{aligned}$$

25. The initial velocity is 1.0 m/s, the final velocity is 2.0 m/s, and the acceleration is 2.0 m/s², so

$a = \dfrac{v_f - v_i}{t_f - t_i}$; let $t_i = 0$ and solve for t_f.

$t_f = \dfrac{v_f - v_i}{a} = \dfrac{2.0\text{ m/s} - 1.0\text{ m/s}}{2.0\text{ m/s}^2} = 0.50\text{ s}$

33.

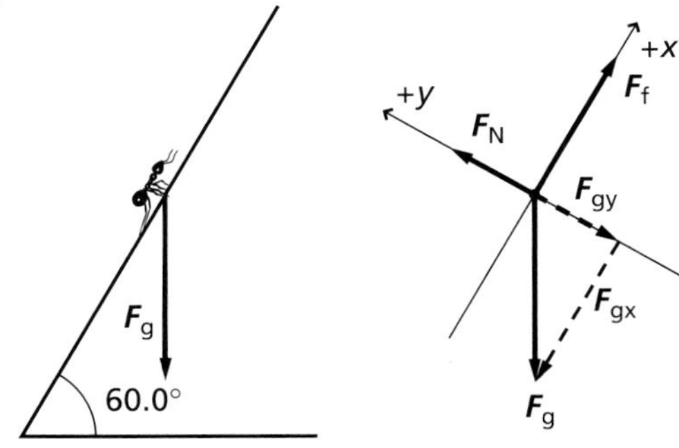

35. $F_{\text{g, perpendicular}} = F_g \cos\theta = mg\cos\theta$

$$\begin{aligned} \theta &= \cos^{-1}\frac{F_{\text{g, perpendicular}}}{mg} \\ &= \cos^{-1}\frac{449\text{ N}}{(50.0\text{ kg})(9.80\text{ m/s}^2)} \\ &= 23.6° \end{aligned}$$

37. $F_{\text{g, parallel}} = F_g(\sin\theta)$,
when the angle is with respect to the horizontal.
$F_{\text{g, perpendicular}} = F_g(\cos\theta)$, when the angle is with respect to the horizontal.

$F_{\text{g, perpendicular}} = 2F_{\text{g, parallel}}$

$2 = \dfrac{F_{\text{g, perpendicular}}}{F_{\text{g, parallel}}} = \dfrac{F_g\cos\theta}{F_g\sin\theta} = \dfrac{1}{\tan\theta}$

$\theta = \tan^{-1}\left(\frac{1}{2}\right) = 26.6°$ relative to the horizontal,

or 63.4° relative to the vertical.

39. Since $a = g(\sin\theta - \mu_k\cos\theta)$,

$$\begin{aligned} a &= (9.80\text{ m/s}^2)(\sin 31° - (0.15)(\cos 31°)) \\ &= 3.8\text{ m/s}^2 \end{aligned}$$

41. $a = g(\sin\theta - \mu_k\cos\theta)$

$a = g(\sin\theta) - g\mu_k(\cos\theta)$

If $a = 0$,

$0 = g(\sin\theta) - g\mu_k(\cos\theta)$

$\mu_k(\cos\theta) = \sin\theta$

$$\begin{aligned} \mu_k &= \frac{\sin\theta}{\cos\theta} \\ &= \frac{\sin 37°}{\cos 37°} \\ &= 0.75 \end{aligned}$$

Chapter 6

1. a. Since $v_y = 0$, $y - v_y t = -\frac{1}{2}gt^2$ becomes

$$y = -\frac{1}{2}gt^2$$

$$\text{or } t^2 = -\frac{2y}{g}$$

$$t = \sqrt{-\frac{2y}{g}}$$

$$= \sqrt{\frac{-2(-78.4\text{ m})}{9.80\text{ m/s}^2}}$$

$$= 4.00\text{ s}$$

b. $x = v_x t$

$$= (5.0\text{ m/s})(4.00\text{ s})$$

$$= 2.0\times10^1\text{ m}$$

c. $v_x = 5.0$ m/s. This is the same as the initial horizontal speed because the acceleration due to gravity influences only the vertical motion. For the vertical component, use $v = v_i + gt$ with $v = v_y$ and v_i, the initial vertical component of velocity, zero.

At $t = 4.00$ s

$$v_y = gt = (9.80\text{ m/s}^2)(4.00\text{ s}) = 39.2\text{ m/s}$$

3. $x = v_x t$;

$$t = \frac{x}{v_x}$$

$$y = -\frac{1}{2}gt^2$$

$$= -\frac{1}{2}g\left(\frac{x}{v_x}\right)^2$$

$$= -\frac{1}{2}(9.80\text{ m/s}^2)\left(\frac{0.070\text{ m}}{2.0\text{ m/s}}\right)^2$$

$$= 6.0\times10^{-3}\text{ m or } 0.60\text{ cm}$$

5. Following the method of Practice Problem 4,

Hangtime:

$$t = \frac{2v_i \sin\theta}{g} = \frac{2(27.0\text{ m/s})(\sin 60.0°)}{9.80\text{ m/s}^2} = 4.77\text{ s}$$

Distance:

$$x = (v_i \cos\theta)t$$

$$= (27.0\text{ m/s})(\cos 60.0°)(4.77\text{ s})$$

$$= 64.4\text{ m}$$

Maximum height:

at $t = \frac{1}{2}(4.77\text{ s}) = 2.38$ s

$$y = (v_i \sin\theta)t - \frac{1}{2}gt^2$$

$$= (27.0\text{ m/s})(\sin 60.0°)(2.38\text{ s}) - \frac{1}{2}(+9.80\text{ m/s}^2)(2.38\text{ s})^2$$

$$= 27.9\text{ m}$$

13. $a_c = \frac{v^2}{r} = \frac{(22\text{ m/s})^2}{56\text{ m}} = 8.6\text{ m/s}^2$

Recall that $F_f = \mu F_N$. The friction force must supply the centripetal force, so $F_f = ma_c$. The magnitude of the normal force is $F_N = mg$. The coefficient of friction must be at least

$$\mu = \frac{F_f}{F_N} = \frac{ma_c}{mg} = \frac{a_c}{g} = \frac{8.6\text{ m/s}^2}{9.80\text{ m/s}^2} = 0.88$$

15. $F_f = F_c$

$$= \frac{mv^2}{r}$$

$$= \frac{(45\text{ kg})(4.1\text{ m/s})^2}{6.3\text{ m}}$$

$$= 120\text{ N}$$

23. $v_{c/g} = v_{w/g} + v_{c/w}$

$$= 0.75\text{ m/s} - 0.020\text{ m/s}$$

$$= 0.73\text{ m/s}$$

25. $v = \sqrt{v_p{}^2 + v_w{}^2}$

$$= \sqrt{(150\text{ km/h})^2 + (75\text{ km/h})^2}$$

$$= 1.7\times10^2\text{ km/h}$$

Chapter 7

1. $\left(\frac{T_G}{T_I}\right)^2 = \left(\frac{r_G}{r_I}\right)^3$

$$r_G = \sqrt[3]{(4.2\text{ units})^3\left(\frac{7.15\text{ days}}{1.8\text{ days}}\right)^2}$$

$$= \sqrt[3]{1.17\times10^3\text{ units}^3}$$

$$= 11\text{ units}$$

3. $\left(\frac{T_M}{T_E}\right)^2 = \left(\frac{r_M}{r_E}\right)^3$ with $r_M = 1.52r_E$

$$\text{Thus, } T_M = \sqrt{\left(\frac{r_M}{r_E}\right)^3 T_E{}^2}$$

$$= \sqrt{\left(\frac{1.52r_E}{r_E}\right)^3(365\text{ days})^2}$$

$$= \sqrt{4.68\times10^5\text{ days}^2}$$

$$= 684\text{ days}$$

5. $\left(\frac{T_s}{T_M}\right)^2 = \left(\frac{r_s}{r_M}\right)^3$

$$\text{Thus, } r_s = \sqrt[3]{r_M{}^3\left(\frac{T_s}{T_M}\right)^2}$$

$$= \sqrt[3]{(3.90\times10^5\text{ km})^3\left(\frac{1.00\text{ days}}{27.3\text{ days}}\right)^2}$$

$$= \sqrt[3]{7.96\times10^{13}\text{ km}^3} = 4.30\times10^4\text{ km}$$

13. a. $v = \sqrt{\frac{Gm_E}{r_E + h}}$

$= \sqrt{\frac{(6.67\times10^{-11}\text{ N·m}^2/\text{kg}^2)(5.97\times10^{24}\text{ kg})}{(6.38\times10^6\text{ m} + 1.5\times10^5\text{ m})}}$

$= 7.9\times10^3$ m/s

b. $T = 2\pi\sqrt{\frac{(r_E + h)^3}{Gm_E}}$

$= 2\pi\sqrt{\frac{(6.38\times10^6\text{ m} + 1.5\times10^5\text{ m})^3}{(6.67\times10^{-11}\text{ N·m}^2/\text{kg}^2)(5.97\times10^{24}\text{ kg})}}$

$= 5.3\times10^3$ s

≈ 88 min

Chapter 8

1. a. $\Delta\theta = (60)(-2\pi\text{ rad}) = -120\pi$ rad or -377 rad

b. $\Delta\theta = -2\pi$ rad or -6.28 rad

c. $\Delta\theta = \left(\frac{1}{12}\right)(-2\pi\text{ rad}) = \frac{-\pi}{6}$ rad or -0.524 rad

3. a. The changes in velocity are the same, so the linear accelerations are the same.

b. Because the radius of the wheel is reduced from 35.4 cm to 24 cm, the angular acceleration will be increased.

$\alpha_1 = 5.23\text{ rad/s}^2$

$\alpha_2 = \frac{a_2}{r}$

$= \frac{1.85\text{ m/s}^2}{0.24\text{ m}}$

$= 7.7\text{ rad/s}^2$

11. $\tau = Fr\sin\theta$

so $F = \frac{\tau}{r\sin\theta}$

$= \frac{35\text{ N·m}}{(0.25\text{ m})(\sin 90.0°)}$

$= 1.4\times10^2$ N

13. $\tau = Fr\sin\theta$

so $\theta = \sin^{-1}\left(\frac{\tau}{Fr}\right)$

$= \sin^{-1}\left(\frac{32.4\text{ N·m}}{(232\text{ N})(0.234\text{ m})}\right)$

$= 36.6°$

15. $\tau = Fr\sin\theta$

Horizontal: $\theta = 90.0°$

$\tau = Fr\sin\theta$

$= mgr\sin\theta$

$= (65\text{ kg})(9.80\text{ m/s}^2)(0.18\text{ m})(\sin 90.0°)$

$= 1.1\times10^2$ N·m

Vertical: $\theta = 0.0°$

$\tau = Fr\sin\theta$

$= mgr\sin\theta$

$= (65\text{ kg})(9.80\text{ m/s}^2)(0.18\text{ m})(\sin 0.0°)$

$= 0.0$ N·m

17. $\tau_{chain} = F_g r = (-35.0\text{ N})(0.0770\text{ m}) = -2.70$ N·m
Thus, a torque of +2.70 N·m must be exerted to balance this torque.

19. $m_1 = \frac{m_2 r_2}{r_1}$

$m_1 = \frac{(0.23\text{ kg})(1.1\text{ cm})}{6.0\text{ cm}} = 0.042$ kg

21. For the mass of the two children,
$I = mr^2 + mr^2 = 2mr^2$
When r is doubled, I is multiplied by a factor of 4.

23. The moments of inertia are different. If the spacing between spheres is r and each sphere has mass m, then rotation about sphere A is
$I = mr^2 + m(2r)^2 = 5mr^2$

Rotation about sphere C is
$I = mr^2 + mr^2 = 2mr^2$

The moment of inertia is greater when rotating around sphere A.

25. Torque is now twice as great. The angular acceleration is also twice as great, so the change in angular velocity is twice as great. Thus, the final angular velocity is 32π rad/s, or 16 rev/s.

27. The torque on the wheel comes from either the chain or the string.

$\tau_{chain} = I_{wheel}\alpha_{wheel}$

$\tau_{wheel} = I_{wheel}\alpha_{wheel}$

Thus, $\tau_{chain} = \tau_{wheel}$

$F_{chain} r_{gear} = F_{string} r_{wheel}$

$F_{string} = \frac{F_{chain} r_{gear}}{r_{wheel}}$

$= \frac{(15\text{ N})(0.14\text{ m})}{0.38\text{ m}}$

$= 5.5$ N

29. $\alpha = \frac{\tau}{I} = \frac{Fr}{I}$

so $F = \frac{I\alpha}{r}$

$= \frac{(I_{small} + I_{large})\alpha}{r}$

$= \frac{\left(\frac{1}{2}m_{small}r_{small}^2 + I_{large}\right)\alpha}{r}$

$= \frac{\left(\left(\frac{1}{2}\right)(2.5\text{ kg})(0.090\text{ m})^2 + 0.26\text{ kg·m}^2\right)(2.57\text{ rad/s}^2)}{0.090\ m}$

$= 7.7$ N

Solutions for Practice Problems

37. **a.** clockwise: $\tau_A = F_A r_A$

$= -F_A(0.96\text{ m} - 0.30\text{ m})$

$= -(0.66\text{ m})F_A$

counterclockwise: $\tau_B = F_B r_B$

$= F_B(0.96\text{ m} - 0.45\text{ m})$

$= (0.51\text{ m})F_B$

b. $\tau_{net} = \tau_A + \tau_B = 0$

so $\tau_B = -\tau_A$

$(0.51\text{ m})F_B = (0.66\text{ m})F_A$

c. $F_g = F_A + F_B$

thus, $F_A = F_g - F_B$

$= F_g - \frac{(0.66\text{ m})F_A}{0.51\text{ m}}$

or $F_A = \frac{F_g}{1 + \frac{0.66\text{ m}}{0.51\text{ m}}}$

$= \frac{mg}{1 + \frac{0.66\text{ m}}{0.51\text{ m}}}$

$= \frac{(7.3\text{ kg})(9.80\text{ m/s}^2)}{1 + \frac{0.66\text{ m}}{0.51\text{ m}}}$

$= 31\text{ N}$

d. F_A would become greater, and F_B would be less.

39. Choose the center of mass of the board as the pivot. The force of Earth's gravity on the board is exerted totally on the support under the center of mass.

$\tau_{end} = -\tau_{diver}$

$F_{end}r_{end} = -F_{diver}r_{diver}$

Thus, $F_{end} = \frac{-F_{diver}r_{diver}}{r_{end}}$

$= \frac{-m_{diver}g r_{diver}}{r_{end}}$

$= \frac{-(85\text{ kg})(9.80\text{ m/s}^2)(1.75\text{ m})}{1.75\text{ m}}$

$= -8.3\times10^2\text{ N}$

To find the force on the center support, notice that because the board is not moving.

$F_{end} + F_{center} = F_{diver} + F_g$

Thus, $F_{center} = F_{diver} + F_g - F_{end}$

$= 2F_{diver} + F_g$

$= 2m_{diver}g + m_{board}g$

$= g(2m_{diver} + m_{board})$

$= (9.80\text{ m/s}^2)(2(85\text{ kg}) + 14\text{ kg})$

$= 1.8\times10^3\text{ N}$

Chapter 9

1.

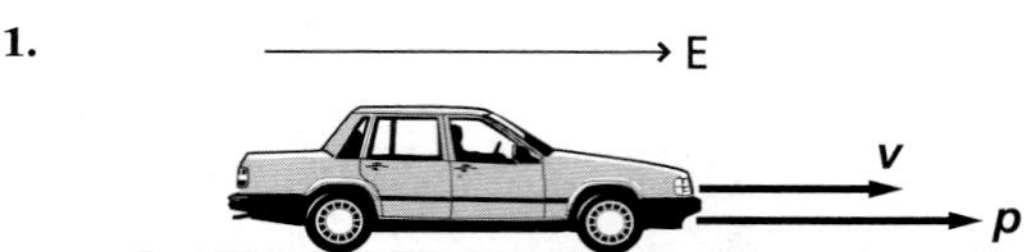

a. $p = mv$

$= (725\text{ kg})(115\text{ km/h})\left(\frac{1000\text{ m}}{1\text{ km}}\right)\left(\frac{1\text{ h}}{3600\text{ s}}\right)$

$= 2.32\times10^4\text{ kg·m/s}$ eastward

b. $v = \frac{p}{m}$

$= \frac{(2.32\times10^4\text{ kg·m/s})\left(\frac{3600\text{ s}}{1\text{ h}}\right)\left(\frac{1\text{ km}}{1000\text{ m}}\right)}{2175\text{ kg}}$

$= 38.4\text{ km/h}$ eastward

3. **a.** $F\Delta t = p_f - p_i = mv_f - mv_i$

$v_f = \frac{F\Delta t + mv_i}{m}$

$= \frac{(5.0\text{ N})(1.0\text{ s}) + (7.0\text{ kg})(2.0\text{ m/s})}{7.0\text{ kg}}$

$= 2.7\text{ m/s}$ in the same direction as the original velocity

b. $v_f = \frac{F\Delta t + mv_i}{m}$

$= \frac{(-5.0\text{ N})(1.0\text{ s}) + (7.0\text{ kg})(2.0\text{ m/s})}{7.0\text{ kg}}$

$= 1.3\text{ m/s}$ in the same direction as the original velocity

5.

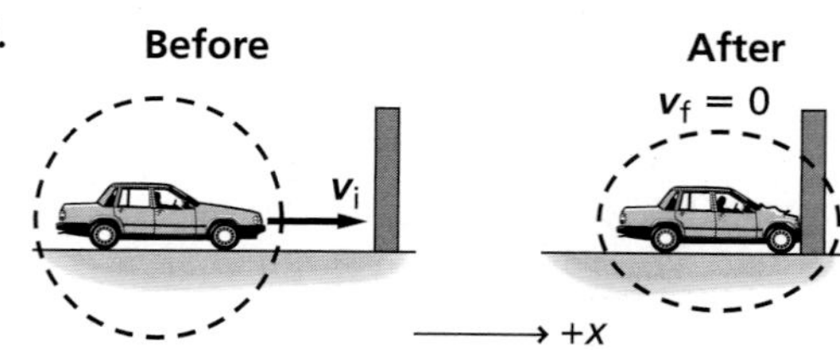

a. $F\Delta t = \Delta p = p_f - p_i$

$F = \frac{p_f - p_i}{\Delta t}$

$F = \frac{p_f - mv_i}{\Delta t}$

$= \frac{(0.0\text{ kg·m/s}) - (60.0\text{ kg})(94\text{ km/h})\left(\frac{1000\text{ m}}{1\text{ km}}\right)\left(\frac{1\text{ h}}{3600\text{ s}}\right)}{0.20\text{ s}}$

$= 7.8\times10^3\text{ N}$ opposite to the direction of motion

b. $F_g = mg$

$m = \frac{F_g}{g} = \frac{7.8\times10^3\text{ N}}{9.80\text{ m/s}^2} = 8.0\times10^2\text{ kg}$

Such a mass is too heavy to lift. You cannot safely stop yourself with your arms.

13. $p_i = p_f$

$mv_{Ai} + mv_{Bi} = 2mv_f$

$v_f = \frac{mv_{Ai} + mv_{Bi}}{2}$

$= \frac{2.2 \text{ m/s} + 0.0 \text{ m/s}}{2}$

$= 1.1 \text{ m/s}$

15. $m_b v_{bi} + m_w v_{wi} = (m_b + m_w)v_f$

where v_f is the common final speed of the bullet and piece of lumber.

Because $v_{wi} = 0.0$ m/s,

$v_{bi} = \frac{(m_b + m_w)v_f}{m_b}$

$= \frac{(0.0350 \text{ kg} + 5.0 \text{ kg})(8.6 \text{ m/s})}{0.0350 \text{ kg}}$

$= 1.2\times10^3 \text{ m/s}$

17. The system is the bullet and the ball.

$m_{bullet}v_{bullet,\,i} + m_{ball}v_{ball,\,i} = m_{bullet}v_{bullet,\,f} + m_{ball}v_{ball,\,f}$

$v_{ball,\,i} = 0.0$ m/s and $v_{bullet,\,f} = -5.0$ m/s

so $v_{ball,\,f} = \frac{m_{bullet}(v_{bullet,\,i} - v_{bullet,\,f})}{m_{ball}}$

$= \frac{(0.0350 \text{ kg})(475 \text{ m/s} - (-5.0 \text{ m/s}))}{2.5 \text{ kg}}$

$= 6.7 \text{ m/s}$

19. $p_{ri} + p_{fuel,\,i} = p_{rf} + p_{fuel,\,f}$

where $p_{ri} + p_{fuel,\,i} = 0.0$ kg·m/s

If the initial mass of the rocket (including fuel) is $m_r = 4.00$ kg, then the final mass of the rocket is

$m_{rf} = 4.00 \text{ kg} - 0.0500 \text{ kg} = 3.95 \text{ kg}$

$0.0 \text{ kg·m/s} = m_{rf}v_{rf} + m_{fuel}v_{fuel,\,f}$

$v_{rf} = \frac{-m_{fuel}v_{fuel,\,f}}{m_{rf}}$

$= \frac{-(0.0500 \text{ kg})(-625 \text{ m/s})}{3.95 \text{ kg}}$

$= 7.91 \text{ m/s}$

21. $p_{Ci} + p_{Ji} = p_{Cf} + p_{Jf}$

where $p_{Ci} = p_{Ji} = 0.0$ kg·m/s

$m_C v_{Cf} = -m_J v_{Jf}$

so $v_{Jf} = \frac{-m_C v_{Cf}}{m_J}$

$= \frac{-(80.0 \text{ kg})(4.0 \text{ m/s})}{115 \text{ kg}}$

$= 2.8$ m/s in the opposite direction

23. Car 1 is moving south. Car 2 is moving east.

Before:

$p_{i,\,x} = p_{1,\,x} + p_{2,\,x} = 0 + m_2 v_{2i}$

$p_{i,\,y} = p_{1,\,y} + p_{2,\,y} = m_1 v_{1i} + 0$

$p_f = p_i$

$= \sqrt{(p_{i,\,x})^2 + (p_{i,\,y})^2}$

$= \sqrt{(m_1 v_{2i})^2 + (m_1 v_{1i})^2}$

$v_f = \frac{p_f}{m_1 + m_2}$

$= \frac{\sqrt{(m_1 v_{2i})^2 + (m_1 v_{1i})^2}}{m_1 + m_2}$

$= \frac{\sqrt{((1732 \text{ kg})(31.3 \text{ m/s}))^2 + ((1383 \text{ kg})(-11.2 \text{ m/s}))^2}}{1383 \text{ kg} + 1732 \text{ kg}}$

$= 18.1 \text{ m/s}$

$\theta = \tan^{-1}\left(\frac{p_{i,\,y}}{p_{i,\,x}}\right)$

$= \tan^{-1}\left(\frac{m_1 v_{1i}}{m_2 v_{2i}}\right)$

$= \tan^{-1}\frac{(1383 \text{ kg})(-11.2 \text{ m/s})}{(1732 \text{ kg})(31.3 \text{ m/s})}$

$= 15.9°$ south of east

25. Before:

$p_{i,\,x} = m_1 v_{1i}$

$= (1345 \text{ kg})(15.7 \text{ m/s})$

$= 2.11\times10^4 \text{ kg·m/s}$

$p_f = p_i = (m_1 + m_2)v_f$

$= (1345 \text{ kg} + 1923 \text{ kg})(14.5 \text{ m/s})$

$= 4.74\times10^4 \text{ kg·m/s}$

$p_{f,\,y} = p_f \sin\theta = (4.74\times10^4 \text{ kg·m/s})(\sin 63.5°)$

$= 4.24\times10^4 \text{ kg·m/s}$

$p_{f,\,y} = p_{i,\,y} = m_2 v_{2i}$

$v_{2i} = \frac{p_{f,\,y}}{m_2} = \frac{4.24\times10^4 \text{ kg·m/s}}{1923 \text{ kg}}$

$= 22.1 \text{ m/s}$

Yes, it was exceeding the speed limit.

Chapter 10

1. a. Because $W = Fd$ and $\Delta KE = W$, doubling the force would double the work, which would double the change in kinetic energy to 1.35 J.

b. Because $W = Fd$, halving the distance would cut in half the work, which also would cut the change in kinetic energy in half, to 0.68 J.

3. a. $W = mgd$

$= (7.5 \text{ kg})(9.80 \text{ m/s}^2)(8.2 \text{ m})$

$= 6.0\times10^2 \text{ J}$

b. $W = Fd + 6.0\times10^2 \text{ J}$

$= (645 \text{ N})(8.2 \text{ m}) + 6.0\times10^2 \text{ J}$

$= 5.9\times10^3 \text{ J}$

c. $W = \Delta KE$

$\Delta KE = 5.9 \times 10^3$ J

5. $W = Fd\cos\theta$

$= 2(255\text{ N})(15\text{ m})(\cos 15°)$

$= 6.5\times10^3$ J

7. $W = Fd\cos\theta$

$= (628\text{ N})(15.0\text{ m})(\cos 46.0°)$

$= 6.54\times10^3$ J

9. $P = \frac{W}{t}$

$= \frac{(575\text{ N})(20.0\text{ m})}{10.0\text{ s}}$

$= 1.15\times10^3$ W

$= 1.15$ kW

11. $P = \frac{W}{t} = \frac{mgd}{t}$

$\frac{m}{t} = (35\text{ L/min})(1.00\text{ kg/L})$

$= 35$ kg/min

Thus,

$P = \left(\frac{m}{t}\right)gd$

$= (35\text{ L/min})(1\text{ min}/60\text{ s})(9.80\text{ m/s}^2)(110\text{ m})$

$= 0.63$ kW

13. $P = \frac{Fd}{t}$

$t = \frac{Fd}{P} = \frac{(6.8\times10^3\text{ N})(15\text{ m})}{0.30\times10^3\text{ W}}$

$= 340$ s

$= 5.7$ min

25. a. $IMA = \frac{d_e}{d_r} = \frac{(0.20\text{ m})}{(0.050\text{ m})} = 4.0$

b. $MA = \frac{F_r}{F_e} = \frac{(1.7\times10^4\text{ N})}{(1.1\times10^4\text{ N})} = 1.5$

c. efficiency $= \frac{MA}{IMA}\times100 = \frac{1.5}{4.0}\times100 = 38\%$

27. efficiency $= \frac{W_o}{W_i}\times100$

$= \frac{F_r d_r}{F_e d_e}\times100$

So $d_e = \frac{F_r d_r(100)}{F_e(\text{efficiency})}$

$= \frac{(1.25\times10^3\text{ N})(0.13\text{ m})(100)}{(225\text{ N})(88.7)}$

$= 0.81$ m

Chapter 11

1. To bring the skater to a stop:

$W = KE_f - KE_i$

$= \frac{1}{2}mv_f^2 - \frac{1}{2}mv_i^2$

$= \frac{1}{2}(52.0\text{ kg})(0.00\text{ m/s})^2 - \frac{1}{2}(52.0\text{ kg})(2.5\text{ m/s})^2$

$= -160$ J

To speed up again:

$W = KE_f - KE_i$

$= \frac{1}{2}mv_f^2 - \frac{1}{2}mv_i^2$

$= \frac{1}{2}(52.0\text{ kg})(2.5\text{ m/s})^2 - \frac{1}{2}(52.0\text{ kg})(0.00\text{ m/s})^2$

$= +160$ J

3. $KE = \frac{1}{2}mv^2$

$= \frac{1}{2}(7.85\times10^{11}\text{ kg})(2.50\times10^4\text{ m/s})^2$

$= 2.45\times10^{20}$ J

$\frac{KE_{comet}}{KE_{bomb}} = \frac{2.45\times10^{20}\text{ J}}{4.2\times10^{15}\text{ J}}$

$= 5.8\times10^4$ bombs would be required to produce the same amount of energy used by Earth in stopping the comet.

5. $W = Fd$

$= mg(h_f - h_i)$

$= (20.0\text{ kg})(9.80\text{ m/s}^2)(0.00\text{ m} - 1.20\text{ m})$

$= -2.35\times10^2$ J

7. Choose the ground as the reference level.

$\Delta PE = mg(h_f - h_i)$

$= (1.8\text{ kg})(9.80\text{ m/s}^2)(0.00\text{ m} - 6.7\text{ m})$

$= -1.2\times10^2$ J

15. The system is the bike + the rider + Earth. There are no external forces, so total energy is conserved.

$KE = \frac{1}{2}mv^2$

$= \frac{1}{2}(85.0\text{ kg})(8.5\text{ m/s})^2$

$= 3.1\times10^3$ J

$KE_i + PE_i = KE_f + PE_f$

$\frac{1}{2}mv^2 + 0 = 0 + mgh$

$h = \frac{v^2}{2g} = \frac{(8.5\text{ m/s})^2}{(2)(9.80\text{ m/s}^2)} = 3.7$ m

17. Bottom of the valley:

$KE_i + PE_i = KE_f + PE_f$

$0 + mgh = \frac{1}{2}mv^2 + 0$

$v = \sqrt{2gh}$

$= \sqrt{2(9.80\ \text{m/s}^2)(45.0\ \text{m})}$

$= 29.7\ \text{m/s}$

Top of the next hill:

$KE_i + PE_i = KE_f + PE_f$

$0 + mgh_i = \frac{1}{2}mv^2 + mgh_f$

$v = \sqrt{2g(h_i - h_f)}$

$= \sqrt{2(9.80\ \text{m/s}^2)(45.0\ \text{m} - 40.0\ \text{m})}$

$= 9.90\ \text{m/s}$

No, the angles of the hills do not affect the answers.

19. Conservation of momentum:

$mv = (m + M)V$, or

$v = \frac{(m + M)V}{m}$

$= \frac{(0.00800\ \text{kg} + 9.00\ \text{kg})(0.100\ \text{m/s})}{0.00800\ \text{kg}}$

$= 1.13\times10^2\ \text{m/s}$

Chapter 12

1. a. $T_C = T_K - 273 = 115 - 273 = -158°C$

b. $T_C = T_K - 273 = 172 - 273 = -101°C$

c. $T_C = T_K - 273 = 125 - 273 = -148°C$

d. $T_C = T_K - 273 = 402 - 273 = 129°C$

e. $T_C = T_K - 273 = 425 - 273 = 152°C$

f. $T_C = T_K - 273 = 212 - 273 = -61°C$

3. $Q = mC\Delta T$

$= (2.3\ \text{kg})(385\ \text{J/kg·K})(80.0°C - 20.0°C)$

$= 5.3\times10^4\ \text{J}$

5. $Q = mC\Delta T$

$= (75\ \text{kg})(4180\ \text{J/kg·K})(43°C - 15°C)$

$= 8.8\times10^6\ \text{J}$

$\frac{8.8\times10^6\ \text{J}}{3.6\times10^6\ \text{J/kWh}} = 2.4\ \text{kWh}$

(2.4 kWh)($0.15 per kWh) = $0.36

7. $m_A C_A(T_f - T_{Ai}) + m_W C_W(T_f - T_{Wi}) = 0$

Since in this particular case, $m_A = m_W$, the masses cancel and

$T_f = \frac{C_A T_{Ai} + C_W T_{Wi}}{C_A + C_W}$

$= \frac{(2450\ \text{J/kg·K})(16.0°C) + (4180\ \text{J/kg·K})(85.0°C)}{2450\ \text{J/kg·K} + 4180\ \text{J/kg·K}}$

$= 59.5°C$

9. $m_W C_W(T_f - T_{Wi}) + m_A C_A(T_f - T_{Ai}) = 0$

$C_A = \frac{m_W C_W(T_f - T_{Wi})}{m_A(T_{Ai} - T_f)}$

$= \frac{(0.100\ \text{kg})(4180\ \text{J/kg·°C})(25.0°C - 10.0°C)}{(0.100\ \text{kg})(100.0°C - 25.0°C)}$

$= 8.36\times10^2\ \text{J/kg·°C}$

19. $Q = mC\Delta T + mH_f$

$= (0.100\ \text{kg})(2060\ \text{J/kg·°C})(20.0°C)$

$+ (0.100\ \text{kg})(3.34\times10^5\ \text{J/kg})$

$= 3.75\times10^4\ \text{J}$

21. $Q = mC_{ice}\Delta T + mH_f + mC_{water}\Delta T + mH_v + mC_{steam}\Delta T$

$= (0.300\ \text{kg})(2060\ \text{J/kg·°C})(0.0°C - (-30.0°C))$

$+ (0.300\ \text{kg})(3.34\times10^5\ \text{J/kg}) +$

$(0.300\ \text{kg})(4180\ \text{J/kg·°C})(100.0°C - 0.0°C) +$

$(0.300\ \text{kg})(2.26\times10^6\ \text{J/kg}) + (0.300\ \text{kg})$

$(2020\ \text{J/kg·°C})(130.0°C - 100.0°C)$

$= 9.40\times10^2\ \text{kJ}$

23. $\Delta U = Q - W_{block}$; since $W_{drill} = -W_{block}$ and assume no heat added to drill:

$\Delta U = 0 + W_{drill} = mC\Delta T$

$W_{drill} = (0.40\ \text{kg})(897\ \text{J/kg·°C})(5.0°C)$

$= 1.8\times10^3\ \text{J}$

25. $\Delta U = mC\Delta T$

$= (0.15\ \text{kg})(4180\ \text{J/kg·°C})(2.0°C)$

$= 1.3\times10^3\ \text{J}$

The number of stirs is

$\frac{1.3\times10^3\ \text{J}}{0.050\ \text{J}} = 2.6\times10^4$

Chapter 13

1. $F = PA$

$= Plw$

$= (1.0\times10^5\ \text{Pa})(1.52\ \text{m})(0.76\ \text{m})$

$= 1.2\times10^5\ \text{N}$

3. $P = \frac{F}{A} = \frac{F_{g\,brick}}{A} = \frac{m_{brick}g}{lw} = \frac{\rho lwhg}{lw} = \rho hg$

$= (11.8\ \text{g/cm}^3)\left(\frac{1\times10^6\ \text{cm}^3}{1\ \text{m}^3}\right)\left(\frac{1\ \text{kg}}{1000\ \text{g}}\right)(0.200\ \text{m})(9.80\ \text{m/s}^2)$

$= 23.1\ \text{kPa}$

5. The maximum pressure is $P = \frac{F_g}{A} = \frac{mg}{A}$.

Therefore

$$A = \frac{mg}{P}$$
$$= \frac{(454\ \text{kg})(9.80\ \text{m/s}^2)}{5.0\times 10^4\ \text{Pa}}$$
$$= 8.9\times 10^{-2}\ \text{m}^2$$

7. $PV = nRT$; so

$$n = \frac{PV}{RT}$$
$$m = nM$$
$$m = \frac{PV}{RT}M$$
$$= \frac{(15.5\times 10^6\ \text{Pa})(0.020\ \text{m}^3)}{(8.31\ \text{Pa·m}^3/\text{mol·K})(293\ \text{K})}(4.00\ \text{g/mol})$$
$$= 5.1\times 10^2\ \text{g}$$

9. $PV = nRT$; so

$$V = \frac{nRT}{P}$$

where $n = \frac{m}{M} = \frac{1.0\times 10^3\ \text{g}}{29\ \text{g/mol}}$

$T = 20.0°\text{C} + 273 = 293\text{K}$

$$V = \frac{\left(\frac{1.0\times 10^3\ \text{g}}{29\ \text{g/mol}}\right)(8.31\ \text{Pa·m}^3/\text{mol·k})(293\ \text{K})}{(1.013\times 10^5\ \text{Pa})}$$
$$= 0.83\ \text{m}^3$$

23. $F_2 = \frac{F_1A_2}{A_1}$

$$= \frac{(1600\ \text{N})(72\ \text{cm}^2)}{1440\ \text{cm}^2}$$
$$= 8.0\times 10^1\ \text{N}$$

25. $F_2 = \frac{F_1A_2}{A_1}$. Thus,

$$\frac{A_2}{A_1} = \frac{F_2}{F_1} = \frac{400\ \text{N}}{1100\ \text{N}} = 0.4.$$

The adult stands on the larger piston.

27. $F_{\text{apparent}} = F_g - F_{\text{buoyant}}$

$$= \rho_{\text{brick}}Vg - \rho_{\text{water}}Vg$$
$$= (\rho_{\text{brick}} - \rho_{\text{water}})Vg$$
$$= (1.8\times 10^3\ \text{kg/m}^3 - 1.00\times 10^3\ \text{kg/m}^3)(0.20\ \text{m}^3)(9.80\ \text{m/s}^2)$$
$$= 1.6\times 10^3\ \text{N}$$

29. To hold the camera in place, the tension in the wire must equal the apparent weight of the camera.

$$T = F_{\text{apparent}}$$
$$= F_g - F_{\text{buoyant}}$$
$$= F_g - \rho_{\text{water}}Vg$$
$$= 1250\ \text{N} - (1.00\times 10^3\ \text{kg/m}^3)(16.5\times 10^{-3}\ \text{m}^3)(9.80\ \text{m/s}^2)$$
$$= 1.09\times 10^3\ \text{N}$$

31. The buoyant force on the foam must equal 480 N. We are assuming the canoe is made of dense material.

$$F_{\text{buoyant}} = \rho_{\text{water}}Vg$$

therefore, $V = \frac{F_{\text{buoyant}}}{\rho_{\text{water}}g}$

$$= \frac{480\ \text{N}}{(1.00\times 10^3\ \text{kg/m}^3)(9.80\ \text{m/s}^2)}$$
$$= 4.9\times 10^{-2}\ \text{m}^3$$

39. $L_2 = L_1 + \alpha L_1(T_2 - T_1)$; so

$$\Delta L = \alpha L_1(T_2 - T_1)$$
$$= (25\times 10^{-6}\ °\text{C}^{-1})(3.66\ \text{m})(39°\text{C} - (-28°\text{C}))$$
$$= 6.1\times 10^{-3}\ \text{m}$$
$$= 6.1\ \text{mm}$$

41. At the beginning, 400 mL of 4.4°C water is in the beaker. Find the change in volume at 30.0°C.

$$\Delta V = \beta V\Delta T$$
$$= (210\times 10^{-6}\ °\text{C}^{-1})(400\times 10^{-6}\ \text{m}^3)(30.0°\text{C} - 4.4°\text{C})$$
$$= 2\times 10^{-6}\ \text{m}^3$$
$$= 2\ \text{mL}$$

43. The aluminum shrinks more than the steel. Let L be the diameter of the rod.

$$\Delta L_{\text{aluminum}} = \alpha L\Delta T$$
$$= (25\times 10^{-6}\ °\text{C}^{-1})(0.85\ \text{cm})(0.0°\text{C} - 30.0°\text{C})$$
$$= -6.4\times 10^{-4}\ \text{cm}$$

The diameter of the steel hole shrinks by

$$\Delta L_{\text{steel}} = \alpha L\Delta T$$
$$= (12\times 10^{-6}\ °\text{C}^{-1})(0.85\ \text{cm})(0.0°\text{C} - 30.0°\text{C})$$
$$= -3.1\times 10^{-4}\ \text{cm}$$

The spacing between the rod and the hole will be $\frac{1}{2}(6.38\times 10^{-4}\ \text{cm} - 3.06\times 10^{-4}\ \text{cm}) = 1.7\times 10^{-4}\ \text{cm}$

Chapter 14

1. $F = kx$

$$= (95\ \text{N/m})(0.25\ \text{m})$$
$$= 24\ \text{N}$$

3. $F = kx$

$$k = \frac{F}{x}$$
$$= \frac{24\ \text{N}}{0.12\ \text{m}}$$
$$= 2.0\times 10^2\ \text{N/m}$$

5. $PE_{sp} = \frac{1}{2}kx^2$

$$x = \sqrt{\frac{2PE_{sp}}{k}}$$
$$= \sqrt{\frac{2(48\ \text{J})}{256\ \text{N/m}}}$$
$$= 0.61\ \text{m}$$

Appendix C

7. $T = 2\pi\sqrt{\frac{l}{g}}$

$l = g\left(\frac{T}{2\pi}\right)^2$

$= (1.6\ \text{m/s}^2)\left(\frac{2.0\ \text{s}}{2\pi}\right)^2$

$= 0.16\ \text{m}$

15. a. $v = \frac{d}{t} = \frac{515\ \text{m}}{1.50\ \text{s}} = 343\ \text{m/s}$

b. $T = \frac{1}{f} = \frac{1\ \text{m}}{436\ \text{Hz}} = 0.00229\ \text{s}$

c. $\lambda = \frac{v}{f} = \frac{343\ \text{m/s}}{436\ \text{Hz}} = 0.787\ \text{m}$

17. Lower

19. $v = \lambda f$, so $\lambda = \frac{v}{f} = \frac{15.0\ \text{m/s}}{6.00\ \text{Hz}} = 2.50\ \text{m}$

21. $v = \lambda f = (0.600\ \text{m})(20.0\ \text{Hz}) = 12.0\ \text{m/s}$

Chapter 15

1. $\lambda = \frac{v}{f} = \frac{343\ \text{m/s}}{18\ \text{Hz}} = 19\ \text{m}$

3. $f = \frac{v}{\lambda} = \frac{5130\ \text{m/s}}{1.25\ \text{m}} = 4.10\times10^3\ \text{Hz}$

5. $\lambda = \frac{v}{f}$,

so $v = \lambda f = (0.655\ \text{m})(2280\ \text{Hz}) = 1490\ \text{m/s}$

This speed corresponds to water at 25°C.

7. $v = 343\ \text{m/s}$, $f_s = 365\ \text{Hz}$,

$v_s = 0\ \text{m/s}$, $v_d = -25.0\ \text{m/s}$

$f_d = f_s\left(\frac{v - v_d}{v - v_s}\right)$

$= 365\ \text{Hz}\left(\frac{343\ \text{m/s} + 25.0\ \text{m/s}}{343\ \text{m/s}}\right)$

$= 392\ \text{Hz}$

9. $v = 1482\ \text{m/s}$, $f_s = 3.50\ \text{MHz}$, $v_s = 9.20\ \text{m/s}$,

$v_d = 0\ \text{m/s}$

$f_d = f_s\left(\frac{v - v_d}{v - v_s}\right)$

$= 3.50\ \text{MHz}\left(\frac{1482\ \text{m/s}}{1482\ \text{m/s} - 9.20\ \text{m/s}}\right)$

$= 3.52\ \text{MHz}$

19. Resonance spacing $= \frac{\lambda}{2} = 1.1\ \text{m}$, so $\lambda = 2.2\ \text{m}$

$v = \lambda f = (2.2\ \text{m})(440\ \text{Hz}) = 970\ \text{m/s}$

21. a. $\lambda_1 = 2L = 2(2.65\ \text{m}) = 5.30\ \text{m}$

The lowest frequency is

$f_1 = \frac{v}{\lambda_1} = \frac{343\ \text{m/s}}{5.30\ \text{m}} = 64.7\ \text{Hz}$

b. $f_2 = \frac{v}{\lambda_2} = \frac{v}{L} = \frac{343\ \text{m/s}}{2.65\ \text{m}} = 129\ \text{Hz}$

$f_3 = \frac{v}{\lambda_3} = \frac{3v}{2L} = \frac{3(343\ \text{m/s})}{2(2.65\ \text{m})} = 194\ \text{Hz}$

Chapter 16

1. $\frac{E_{after}}{E_{before}} = \frac{\left(\frac{P}{4\pi d_{after}^2}\right)}{\left(\frac{P}{4\pi d_{before}^2}\right)}$

$= \frac{d_{before}^2}{d_{after}^2}$

$= \frac{(30\ \text{cm})^2}{(90\ \text{cm})^2}$

$= \frac{1}{9}$

Therefore, after the lamp is moved the illumination is one-ninth of the original illumination.

3. Illuminance of a 150-W bulb

$P = 2275$, $d = 0.50, 0.75, \ldots, 5.0$

$E(d) = \frac{P}{4\pi d^2}$

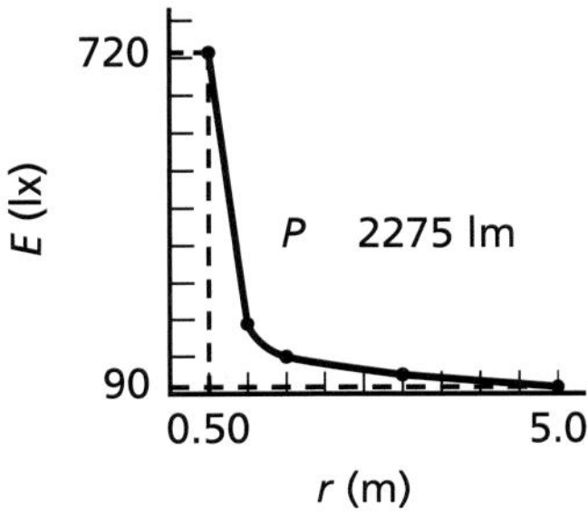

5. $E = \frac{P}{4\pi d^2}$

$P = 4\pi E d^2$

$= 4\pi(160\ \text{lm/m}^2)(2.0\ \text{m})^2$

$= 8.0\times10^3\ \text{lm}$

15. The relative speed along the axis is much less than the speed of light. Thus, you can use the observed light frequency equation.

Because the astronomer and the galaxy are moving away from each other, use the negative form of the observed light frequency equation.

$f_{obs} = f\left(1 - \frac{v}{c}\right)$

$= (6.16\times10^{14}\ \text{Hz})\left(1 - \frac{6.55\times10^6\ \text{m/s}}{3.00\times10^8\ \text{m/s}}\right)$

$= 6.03\times10^{14}\ \text{Hz}$

17. Assume that the relative speed along the axis is much less than the speed of light. Thus, you can use the Doppler shift equation.

$(\lambda_{obs} - \lambda) = \pm\frac{v}{c}\lambda$

The observed (apparent) wavelength appears to be longer than the known (actual) wavelength of the oxygen spectral line. This means that the astronomer and the galaxy are moving away from each other. So use the positive form of the Doppler shift equation.

$(\lambda_{obs} - \lambda) = +\frac{v}{c}\lambda$

Solve for the unknown variable.

$$v = c\frac{(\lambda_{obs} - \lambda)}{\lambda}$$

$$= (3.00\times10^8 \text{ m/s})\frac{(525 \text{ nm} - 513 \text{ nm})}{513 \text{ nm}}$$

$$= 7.02\times10^6 \text{ m/s}$$

Chapter 17

1. Polishing makes the surface smoother.

3. $\theta_i = \theta_r = 35°$

5. $\theta_{r1} = \theta_{i1} = 30°$

$\theta_{i2} = 90° - \theta_{r1} = 90° - 30° = 60°$

13. $\frac{1}{f} = \frac{1}{d_o} + \frac{1}{d_i}$

$$d_i = \frac{d_o f}{d_o - f}$$

$$= \frac{(36.0 \text{ cm})(16.0 \text{ cm})}{36.0 \text{ cm} - 16.0 \text{ cm}}$$

$$= 28.8 \text{ cm}$$

15. $\frac{1}{f} = \frac{1}{d_o} + \frac{1}{d_i}$

$$d_i = \frac{d_o f}{d_o - f}$$

$$= \frac{(16.0 \text{ cm})(7.0 \text{ cm})}{16.0 \text{ cm} - 7.0 \text{ cm}}$$

$$= 12.4 \text{ cm}$$

$$m = \frac{h_i}{h_o} = \frac{-d_i}{d_o}$$

$$h_i = \frac{-d_i h_o}{d_o}$$

$$= \frac{-(12.4 \text{ cm})(2.4 \text{ cm})}{16.0 \text{ cm}}$$

$$= -1.9 \text{ cm}$$

17.

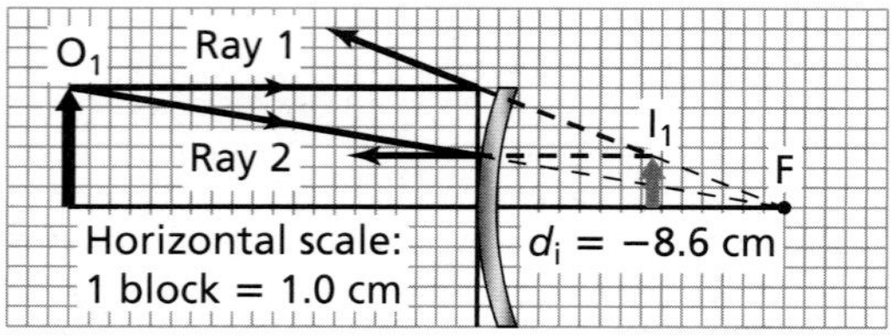

$$\frac{1}{d_o} + \frac{1}{d_i} = \frac{1}{f}$$

$$d_i = \frac{d_o f}{d_o - f}$$

$$= \frac{(20.0 \text{ cm})(-15.0 \text{ cm})}{20.0 \text{ cm} - (-15.0 \text{ cm})} = -8.57 \text{ cm}$$

19. $\frac{1}{f} = \frac{1}{d_i} + \frac{1}{d_o}$

so $f = \frac{d_o d_i}{d_o + d_i}$

and $m = \frac{-d_i}{d_o}$, so $d_o = \frac{-d_i}{m}$

$d_i = -24$ cm and $m = 0.75$, so

$$d_o = \frac{-(-24 \text{ cm})}{0.75} = 32 \text{ cm}$$

and

$$f = \frac{(32 \text{ cm})(-24 \text{ cm})}{32 \text{ cm} + (-24 \text{ cm})} = -96 \text{ cm}$$

21. $m = \frac{h_i}{h_o} = \frac{-d_i}{d_o}$

$$d_i = \frac{-d_o h_i}{h_o}$$

$$= \frac{-(2.4 \text{ m})(0.36 \text{ m})}{1.8 \text{ m}}$$

$$= -0.48 \text{ m}$$

$$\frac{1}{f} = \frac{1}{d_o} + \frac{1}{d_i}$$

$$f = \frac{d_i d_o}{d_i + d_o}$$

$$= \frac{(-0.48 \text{ m})(2.4 \text{ m})}{-0.48 \text{ m} + 2.4 \text{ m}}$$

$$= -0.60 \text{ m}$$

Chapter 18

1. $n_1 \sin\theta_1 = n_2 \sin\theta_2$

$$\theta_2 = \sin^{-1}\left(\frac{n_1 \sin\theta_1}{n_2}\right)$$

$$= \sin^{-1}\left(\frac{(1.00)(\sin 37.0°)}{1.36}\right)$$

$$= 26.3°$$

3. $n_1 \sin\theta_1 = n_2 \sin\theta_2$

$$\theta_2 = \sin^{-1}\left(\frac{n_1 \sin\theta_1}{n_2}\right)$$

$$= \sin^{-1}\left(\frac{(1.00)(\sin 30.0°)}{1.33}\right)$$

$$= 22.1°$$

5. $n_1 \sin\theta_1 = n_2 \sin\theta_2$

$$n_2 = \frac{n_1 \sin\theta_1}{\sin\theta_2}$$

$$= \frac{(1.33)(\sin 31°)}{\sin 27°}$$

$$= 1.5$$

15. $\frac{1}{f} = \frac{1}{d_o} + \frac{1}{d_i}$

$$d_i = \frac{d_o f}{d_o - f}$$

$$= \frac{(8.5 \text{ cm})(5.5 \text{ cm})}{8.5 \text{ cm} - 5.5 \text{ cm}}$$

$= 15.6$ cm, or 16 cm to two significant digits

$$m = \frac{h_i}{h_o} = \frac{-d_i}{d_o}$$

$$h_i = \frac{-d_i h_o}{d_o}$$

$$= \frac{-(15.6 \text{ cm})(2.25 \text{ cm})}{8.5 \text{ cm}}$$

$$= -4.1 \text{ cm}$$

17. $\frac{1}{f} = \frac{1}{d_i} + \frac{1}{d_o}$

with $d_o = d_i$ because

$m = \frac{-d_i}{d_o}$ and $m = -1$

Therefore,

$$\frac{1}{f} = \frac{2}{d_i}$$

$$d_i = 2f = 2(25 \text{ mm}) = 5.0\times10^1 \text{ mm}$$

$$d_o = d_i = 5.0\times10^1 \text{ mm}$$

19. $\frac{1}{f} = \frac{1}{d_o} + \frac{1}{d_i}$

$$d_i = \frac{d_o f}{d_o - f}$$

$$= \frac{(25 \text{ cm})(5.0 \text{ cm})}{25 \text{ cm} - 5.0 \text{ cm}}$$

$$= 6.2 \text{ cm}$$

$$m = \frac{h_i}{h_o} = \frac{-d_i}{d_o}$$

$$h_i = \frac{-d_i h_o}{d_o}$$

$$= \frac{-(6.2 \text{ cm})(2.0 \text{ cm})}{25 \text{ cm}}$$

$= -0.50$ cm (inverted image)

21. $\frac{1}{f} = \frac{1}{d_i} + \frac{1}{d_o}$

so $d_i = \frac{d_o f}{d_o - f}$

$$= \frac{(3.4 \text{ cm})(12.0 \text{ cm})}{3.4 \text{ cm} - 12.0 \text{ cm}}$$

$$= -4.7 \text{ cm}$$

$$h_i = \frac{-h_o d_i}{d_o}$$

$$= \frac{-(2.0 \text{ cm})(-4.7 \text{ cm})}{3.4 \text{ cm}}$$

$$= 2.8 \text{ cm}$$

23. $m = \frac{-d_i}{d_o}$

so $d_i = -md_o = -(4.0)(3.5 \text{ cm}) = -14 \text{ cm}$

$$\frac{1}{f} = \frac{1}{d_i} + \frac{1}{d_o}$$

so $f = \frac{d_o d_i}{d_o + d_i}$

$$= \frac{(3.5 \text{ cm})(-14 \text{ cm})}{3.5 \text{ cm} + (-14 \text{ cm})}$$

$$= 4.7 \text{ cm}$$

Chapter 19

1. $\lambda = \frac{xd}{L}$

$$= \frac{(13.2\times10^{-3} \text{ m})(1.90\times10^{-5} \text{ m})}{0.600 \text{ m}}$$

$$= 418 \text{ nm}$$

3. $\lambda = \frac{xd}{L}$

$$d = \frac{\lambda L}{x}$$

$$= \frac{(632.8\times10^{-9} \text{ m})(1.000 \text{ m})}{65.5\times10^{-3} \text{ m}}$$

$$= 9.66\times10^{-6} \text{ m}$$

$$= 9.66 \ \mu\text{m}$$

5. $2t = (m + \frac{1}{2})\frac{\lambda}{n_{oil}}$

For the thinnest film, $m = 0$.

$$t = \frac{\lambda}{4n_{oil}} = \frac{635 \text{ nm}}{(4)(1.45)} = 109 \text{ nm}$$

7. Because $n_{film} > n_{air}$, there is a phase inversion on the first reflection. Because $n_{silicon} > n_{film}$, there is a phase inversion on the second reflection.

For destructive interference to keep yellow-green from being reflected:

$$2t = \left(m + \frac{1}{2}\right)\frac{\lambda}{n_{film}}$$

For the thinnest film, $m = 0$.

$$t = \frac{\lambda}{4n_{film}} = \frac{555 \text{ nm}}{(4)(1.45)} = 95.7 \text{ nm}$$

9. For constructive interference

$$2t = \left(m + \frac{1}{2}\right)\frac{\lambda}{n_{film}}$$

For the thinnest film, $m = 0$.

$$t = \frac{\lambda}{4n_{film}} = \frac{521 \text{ nm}}{(4)(1.33)} = 97.9 \text{ nm}$$

17. $2x_1 = \frac{2\lambda L}{w}$

$$L = \frac{(2x_1)w}{2\lambda}$$

$$= \frac{(2.60\times10^{-2} \text{ m})(0.110\times10^{-3} \text{ m})}{2(589\times10^{-9} \text{ m})}$$

$$= 2.43 \text{ m}$$

19. $2x_1 = \frac{2\lambda L}{w}$

$$\lambda = \frac{(2x_1)w}{2L}$$

$$= \frac{(2.40\times10^{-3} \text{ m})(0.0295\times10^{-3} \text{ m})}{2(60.0\times10^{-2} \text{ m})}$$

$$= 5.90\times10^2 \text{ nm}$$

21. A full spectrum of color is seen. Because of the variety of wavelength, dark fringes of one wavelength would be filled by bright fringes of another color.

Solutions for Practice Problems

23. $\lambda = d \sin \theta$

$\sin \theta = \frac{\lambda}{d}$

$\theta = \sin^{-1}\left(\frac{\lambda}{d}\right)$

$\tan \theta = \frac{x}{L}$

$x = L \tan \theta$

$= L \tan\left(\sin^{-1}\left(\frac{\lambda}{d}\right)\right)$

$= (0.800 \text{ m})\left(\tan\left(\sin^{-1}\left(\frac{421\times10^{-9} \text{ m}}{8.60\times10^{-2} \text{ m}}\right)\right)\right)$

$= 0.449 \text{ m}$

25. $\lambda = d \sin \theta$

There is one slit per distance, d, so $\frac{1}{d}$ gives slits per centimeter.

$d = \frac{\lambda}{\sin \theta} = \frac{\lambda}{\sin\left(\tan^{-1}\left(\frac{x}{L}\right)\right)}$

$= \frac{632\times10^{-9}\text{m}}{\sin\left(\tan^{-1}\left(\frac{0.056 \text{ m}}{0.55 \text{ m}}\right)\right)}$

$= 6.2\times10^{-6} \text{ m} = 6.2\times10^{-4} \text{ cm}$

$\frac{1 \text{ slit}}{6.2\times10^{-4}\text{cm}}$

1.6×10^{3} slits/cm

Chapter 20

9. $F = \frac{Kq_Aq_B}{{d_{AB}}^2}$

$= \frac{(9.0\times10^{9} \text{ N·m}^2/\text{C}^2)(2.0\times10^{-4} \text{ C})(8.0\times10^{-4} \text{ C})}{(0.30 \text{ m})^2}$

$= 1.6\times10^{4} \text{ N}$

The force is attractive.

11.

Magnitudes of all forces remain the same. The direction changes to 42° above the $-x$ axis, or 138°.

13. $F_{\text{A on B}} = K\frac{q_Aq_B}{{d_{AB}}^2}$

$F_{\text{C on B}} = K\frac{q_Bq_C}{{d_{BC}}^2}$

$F_{\text{net}} = F_{\text{C on B}} - F_{\text{A on B}}$

$= K\frac{q_Bq_C}{{d_{BC}}^2} - K\frac{q_Aq_B}{{d_{AB}}^2}$

$= (9.0\times10^{9} \text{ N·m}^2/\text{C}^2)\frac{(3.6\times10^{-6} \text{ C})(4.0\times10^{-6} \text{ C})}{(0.20 \text{ m})^2}$

$- (9.0\times10^{9} \text{ N·m}^2/\text{C}^2)\frac{(2.0\times10^{-6} \text{ C})(3.6\times10^{-6} \text{ C})}{(0.60 \text{ m})^2}$

$= 3.1$ N toward the right

Chapter 21

1. $E = \frac{F}{q} = \frac{2.0\times10^{-4} \text{ N}}{5.0\times10^{-6} \text{ C}} = 4.0\times10^{1} \text{ N/C}$

3. $E = \frac{F}{q}$ so $F = Eq$

$F = (27 \text{ N/C})(3.0\times10^{-7} \text{ C}) = 8.1\times10^{-6}$ N south

5. a. No. The force on the 2.0-μC charge would be twice that on the 1.0-μC charge.

b. Yes. You would divide the force by the strength of the test charge, so the results would be the same.

7. Because the field strength varies as the square of the distance from the point charge, the new field strength will be one-fourth of the old field strength, or 6.5×10^{3} N/C.

9. $E = \frac{F}{q'} = K\frac{q}{d^2}$

so $q = \frac{Ed^2}{K} = \frac{(450 \text{ N/C})(0.25 \text{ m})^2}{9.0\times10^{9} \text{ N·m}^2/\text{C}^2} = -3.1\times10^{-9} \text{ C}$

The charge is negative, because the field is directed toward it.

17. $\Delta V = Ed$

$E = \frac{\Delta V}{d} = \frac{400 \text{ V}}{0.020 \text{ m}} = 2\times10^{4} \text{ N/C}$

19. $\Delta V = Ed$, so $d = \frac{\Delta V}{E} = \frac{125 \text{ V}}{4.25\times10^{3} \text{ N/C}} = 2.94\times10^{-2} \text{ m}$

21. $W = q\Delta V = (3.0 \text{ C})(1.5 \text{ V}) = 4.5 \text{ J}$

23. $W = q\Delta V$

$= (1.60\times10^{-19} \text{ C})(1.8\times10^{4} \text{ V})$

$= 2.9\times10^{-15} \text{ J}$

25. $W = q\Delta V$

$= qEd$

$= (1.60\times10^{-19} \text{ C})(4.5\times10^{5} \text{ N/C})(0.25 \text{ m})$

$= 1.8\times10^{-14} \text{ J}$

27. $F_g = Eq$

$q = \frac{F_g}{E} = \frac{1.9\times10^{-15} \text{ N}}{6.0\times10^{3} \text{ N/C}} = 3.2\times10^{-19} \text{ C}$

electrons $= \frac{q}{q_e}$

$= \frac{3.2\times10^{-19} \text{ C}}{1.60\times10^{-19} \text{ C}}$

$= 2$

29. $E = \frac{\Delta V}{d} = \frac{240\ V}{6.4\times10^{-3}\ m} = 3.8\times10^{4}\ N/C$

$E = \frac{F}{q}$, so $q = \frac{F}{E} = \frac{1.2\times10^{-14}\ N}{3.8\times10^{4}\ N/C} = 3.2\times10^{-19}\ C$

$\#\ \text{electrons} = \frac{q}{q_e}$

$= \frac{3.2\times10^{-19}\ C}{1.60\times10^{-19}\ C}$

$= 2$

31. $q = C\Delta V$, so the larger capacitor has a greater charge.

$q = (6.8\times10^{-6}\ F)(24\ V) = 1.6\times10^{-4}\ C$

33. $q = C\Delta V$

so $\Delta q = C(\Delta V_2 - \Delta V_1)$

$= (2.2\times10^{-6}\ F)(15.0\ V - 6.0\ V)$

$= 2.0\times10^{-5}\ C$

Chapter 22

1. $P = IV = (0.50\ A)(125\ V) = 63\ J/s = 63\ W$

3. $P = IV$

$I = \frac{P}{V} = \frac{75\ W}{125\ V} = 0.60\ A$

5. $P = IV$

$I = \frac{P}{V} = \frac{0.90\ W}{3.0\ V} = 0.30\ A$

7. $V = IR = (3.8\ A)(32\ \Omega) = 1.2\times10^{2}\ V$

9. a. $R = \frac{V}{I} = \frac{120\ V}{0.50\ A} = 2.4\times10^{2}\ \Omega$

b. $P = IV = (0.50\ A)(120\ V) = 6.0\times10^{1}\ W$

11. a. The new value of the current is

$\frac{0.60\ A}{2} = 0.30\ A$

So $V = IR = (0.30\ A)(2.1\times10^{2}\ \Omega) = 6.3\times10^{1}\ V$

b. The total resistance of the circuit is now

$R_{total} = \frac{V}{I} = \frac{125\ V}{0.30\ A} = 4.2\times10^{2}\ \Omega$

Therefore,

$R_{res} = R_{total} - R_{lamp}$

$= 4.2\times10^{2}\ \Omega - 2.1\times10^{2}\ \Omega$

$= 2.1\times10^{2}\ \Omega$

c. $P = IV = (0.30\ A)(6.3\times10^{1}\ V) = 19\ W$

13.

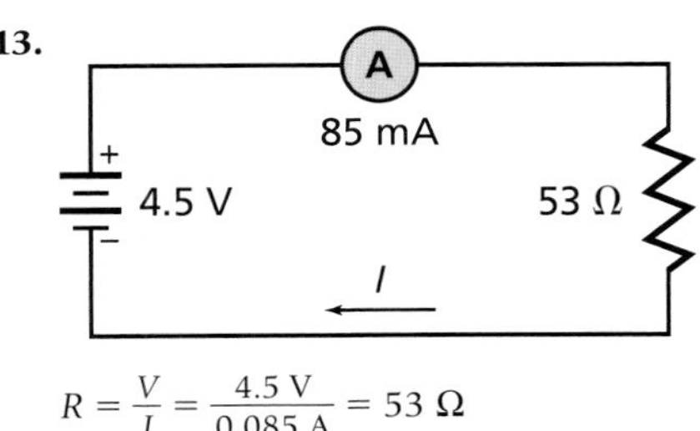

$R = \frac{V}{I} = \frac{4.5\ V}{0.085\ A} = 53\ \Omega$

15.

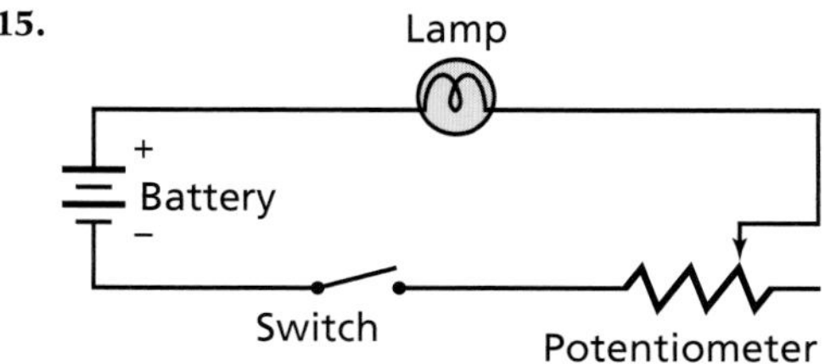

23. a. $I = \frac{V}{R} = \frac{120\ V}{15\ \Omega} = 8.0\ A$

b. $E = I^2Rt = (8.0\ A)^2(15\ \Omega)(30.0\ s) = 2.9\times10^{4}\ J$

c. 2.9×10^{4} J, because all electrical energy is converted to thermal energy.

25. a. $E = Pt = (0.22)(100.0\ J/s)(1.0\ min)\left(\frac{60\ s}{1\ min}\right) = 1.3\times10^{3}\ J$

b. $E = Pt = (0.78)(100.0\ J/s)(1.0\ min)\left(\frac{60\ s}{1\ min}\right) = 4.7\times10^{3}\ J$

27. $E = IVt = I(2V)\left(\frac{t}{2}\right)$

For a given amount of energy, doubling the voltage will divide the time by 2.

$t = \frac{2.2\ h}{2} = 1.1\ h$

29. a. $I = \frac{V}{R} = \frac{115\ V}{12{,}000\ \Omega} = 9.6\times10^{-3}\ A$

b. $P = VI = (115\ V)(9.6\times10^{-3}\ A) = 1.1\ W$

c. $\text{Cost} = (1.1\times10^{-3}\ kW)(\$0.12/kWh)$

$(30\ days)(24\ h/day)$

$= \$0.10$

31. $E_{charge} = (1.3)IVt = (1.3)(55\ A)(12\ V)(1.0\ h) = 858\ Wh$

$t = \frac{E}{IV} = \frac{858\ Wh}{(7.5\ A)(14\ V)} = 8.2\ h$

Chapter 23

1. $R = R_1 + R_2 + R_3$

$= 20\ \Omega + 20\ \Omega + 20\ \Omega$

$= 60\ \Omega$

$I = \frac{V}{R} = \frac{120\ V}{60\ \Omega} = 2\ A$

3. a. It will increase.

b. $I = \frac{V}{R}$, so it will decrease.

c. No. It does not depend on the resistance.

5. $V_1 = IR_1 = (3\ A)(10\ \Omega) = 30\ V$

$V_2 = IR_2 = (3\ A)(15\ \Omega) = 45\ V$

$V_3 = IR_3 = (3\ A)(5\ \Omega) = 15\ V$

$V_1 + V_2 + V_3 = 30\ V + 45\ V + 15\ V = 90\ V$

= voltage of battery

7. a. $I = \frac{V}{R} = \frac{17.0\ V}{255.0\ \Omega} = 66.7\ mA$

b. First, find the total resistance, then solve for voltage.

$R = R_A + R_B = 255\ \Omega + 292\ \Omega = 547\ \Omega$

$V = IR = (66.7\ \text{mA})(547\ \Omega) = 36.5\ \text{V}$

c. $P = IV = (66.7\ \text{mA})(36.5\ \text{V}) = 2.43\ \text{W}$

$P_A = I^2R_A = (66.7\ \text{mA})^2(255\ \Omega) = 1.13\ \text{W}$

$P_B = I^2R_B = (66.7\ \text{mA})^2(292\ \Omega) = 1.30\ \text{W}$

d. Yes. The law of conservation of energy states that energy cannot be created or destroyed; therefore, the rate at which energy is converted, or power is dissipated, will equal the sum of all parts.

9. The resistor with the lower resistance will dissipate less power; thus, it will be cooler.

11. a. $R = R_1 + R_2$

$= 22\ \Omega + 33\ \Omega$

$= 55\ \Omega$

b. $I = \frac{V}{R} = \frac{120\ \text{V}}{55\ \Omega} = 2.2\ \text{A}$

c. $V = IR$

Across a 22-Ω resistor,
$V_1 = IR_1 = \left(\frac{V}{R}\right)R_1 = \left(\frac{120\ \text{V}}{55\ \Omega}\right)(22\ \Omega) = 48\ \text{V}$

Across a 33-Ω resistor,
$V_2 = IR_2 = \left(\frac{V}{R}\right)R_2 = \left(\frac{120\ \text{V}}{55\ \Omega}\right)(33\ \Omega) = 72\ \text{V}$

d. $V = 48\ \text{V} + 72\ \text{V} = 1.20\times10^2\ \text{V}$

13. $V_B = \frac{VR_B}{R_A + R_B}$

$= \frac{(45\ \text{V})(235\ \text{k}\Omega)}{475\ \text{k}\Omega + 235\ \text{k}\Omega}$

$= 15\ \text{V}$

15. a. $\frac{1}{R} = \frac{1}{R_1} + \frac{1}{R_2} + \frac{1}{R_3}$

$= \frac{1}{15.0\ \Omega} + \frac{1}{15.0\ \Omega} + \frac{1}{15.0\ \Omega}$

$= \frac{3}{15.0\ \Omega}$

$R = 5.00\ \Omega$

b. $I = \frac{V}{R} = \frac{30.0\ \text{V}}{5.00\ \Omega} = 6.00\ \text{A}$

c. $I = \frac{V}{R_1} = \frac{30.0\ \text{V}}{15.0\ \Omega} = 2.00\ \text{A}$

17. a. Yes, it gets smaller.

b. Yes, it gets larger.

c. No, it remains the same. Currents are independent.

19. Neither. They both reach maximum dissipation at the same voltage.

$P = \frac{V^2}{R}$

$V = \sqrt{PR}$

The voltage is equal across parallel resistors, so:

$V = \sqrt{P_1R_1} = \sqrt{P_2R_2}$

$= \sqrt{(2\ \text{W})(12\ \Omega)} = \sqrt{(4\ \text{W})(6.0\ \Omega)}$

$= 5\ \text{V maximum}$

25. By conservation of energy (and power):

$P_T = P_1 + P_2 + P_3$

$= 2.0\ \text{W} + 3.0\ \text{W} + 1.5\ \text{W}$

$= 6.5\ \text{W}$

$P_T = IV$

$I = \frac{P_T}{V} = \frac{6.5\ \text{W}}{12\ \text{V}} = 0.54\ \text{A}$

27. Then, all of the working lights are in series. The 12 working lights will burn with equal intensity.

Chapter 24

1. a. repulsive

b. attractive

3. the bottom (the point)

5. a. from south to north

b. west

7. the pointed end

9. Yes. Connect the potentiometer in series with the power supply and the coil. Adjusting the potentiometer for more resistance will decrease the current flow and the strength of the field.

17. $F = ILB$

$= (0.50\ \text{m})(8.0\ \text{A})(0.40\ \text{N/A·m})$

$= 1.6\ \text{N}$

19. $F = ILB$, F = weight of wire

$B = \frac{F}{IL} = \frac{0.35\ \text{N}}{(6.0\ \text{A})(0.400\ \text{m})} = 0.15\ \text{T}$

21. Opposite to the direction of the electron motion

23. $F = qvB$

$= (2)(1.60\times10^{-19}\ \text{C})(3.0\times10^{4}\ \text{m/s})(9.0\times10^{-2}\ \text{T})$

$= 8.6\times10^{-16}\ \text{N}$

25. $F = qvB$

$= (2)(1.60\times10^{-19}\ \text{C})(4.0\times10^{4}\ \text{m/s})(5.0\times10^{-2}\ \text{T})$

$= 6.4\times10^{-16}\ \text{N}$

Chapter 25

1. a. $EMF = BLv$

$= (0.4\ \text{T})(0.5\ \text{m})(20\ \text{m/s})$

$= 4\ \text{V}$

b. $I = \frac{EMF}{R}$

$= \frac{4\ \text{V}}{6.0\ \Omega}$

$= 0.7\ \text{A}$

3. a. $EMF = BLv$

$= (1.0\ \text{T})(30.0\ \text{m})(2.0\ \text{m/s})$

$= 6.0\times10^{1}\text{V}$

b. $I = \frac{EMF}{R} = \frac{6.0\times10^{1}\ \text{V}}{15.0\ \Omega} = 4.0\ \text{A}$

5. a. $V_{eff} = (0.707)V_{max}$

$= (0.707)(170\ \text{V})$

$= 1.2\times10^{2}\ \text{V}$

b. $I_{eff} = (0.707)I_{max}$

$= (0.707)(0.70\ \text{A})$

$= 0.49\ \text{A}$

c. $R = \frac{V_{eff}}{I_{eff}} = \frac{\frac{V_{max}}{\sqrt{2}}}{\frac{I_{max}}{\sqrt{2}}} = \frac{V_{max}}{I_{max}}$

$= \frac{170\ \text{V}}{0.70\ \text{A}}$

$= 2.4\times10^{2}\ \Omega$

7. a. $V_{eff} = \frac{V_{max}}{\sqrt{2}}$

$= \frac{425\ \text{V}}{\sqrt{2}}$

$= 3.01\times10^{2}\ \text{V}$

b. $I_{eff} = \frac{V_{eff}}{R}$

$= \frac{3.01\times10^{2}\ \text{V}}{5.0\times10^{2}\ \Omega}$

$= 0.60\ \text{A}$

17. $\frac{V_P}{V_S} = \frac{N_P}{N_S}$

$V_S = \frac{V_P N_S}{N_P}$

$= \frac{(60.0\ \text{V})(90{,}000)}{300}$

$= 1.80\times10^{4}\ \text{V}$

$V_P I_P = V_S I_S$

$I_P = \frac{V_S I_S}{V_P}$

$= \frac{(1.80\times10^{4}\ \text{V})(0.50\ \text{A})}{60.0\ \text{V}}$

$= 1.5\times10^{2}\ \text{V}$

Chapter 26

1. $Bqv = \frac{mv^2}{r}$

$r = \frac{mv}{Bq}$

$= \frac{(1.67\times10^{-27}\ \text{kg})(7.5\times10^{3}\ \text{m/s})}{(0.60\ \text{T})(1.60\times10^{-19}\ \text{C})}$

$= 1.3\times10^{-4}\ \text{m}$

3. $Bqv = \frac{mv^2}{r}$

$r = \frac{mv}{Bq}$

$= \frac{(9.11\times10^{-31}\ \text{kg})(5.0\times10^{4}\ \text{m/s})}{(6.0\times10^{-2}\ \text{T})(1.60\times10^{-19}\ \text{C})}$

$= 4.7\times10^{-6}\ \text{m}$

5. $m = \frac{B^2r^2q}{2V}$

$= \frac{(7.2\times10^{-2}\ \text{T})^2(0.085\ \text{m})^2(1.60\times10^{-19}\ \text{C})}{2(110\ \text{V})}$

$= 2.7\times10^{-26}\ \text{kg}$

7. $Bqv = Eq$

$v = \frac{E}{B}$

$= \frac{6.0\times10^{2}\ \text{N/C}}{1.5\times10^{-3}\ \text{T}}$

$= 4.0\times10^{5}\ \text{m/s}$

15. All electromagnetic waves travel through air or a vacuum at c, 3.00×10^{8} m/s.

17. $\lambda = \frac{c}{f}$

$= \frac{3.00\times10^{8}\ \text{m/s}}{8.2\times10^{14}\ \text{Hz}}$

$= 3.7\times10^{-7}\ \text{m}$

19. $v = \frac{c}{\sqrt{K}}$

$= \frac{299{,}792{,}458\ \text{m/s}}{\sqrt{1.00054}}$

$= 2.99712\times10^{8}\ \text{m/s}$

21. $v = \frac{c}{\sqrt{K}}$

so $K = \left(\frac{c}{v}\right)^2$

$= \left(\frac{3.00\times10^{8}\ \text{m/s}}{2.43\times10^{8}\ \text{m/s}}\right)^2$

$= 1.52$

Chapter 27

1. $(2.3\ \text{eV})\left(\frac{1.60\times10^{-19}\ \text{J}}{1\ \text{eV}}\right) = 3.7\times10^{-19}\ \text{J}$

3. $m = 9.11\times10^{-31}$ kg, $KE = \frac{1}{2}mv^2$

so $v = \sqrt{\frac{2KE}{m}}$

$= \sqrt{\frac{2(3.7\times10^{-19}\ \text{J})}{9.11\times10^{-31}\ \text{kg}}}$

$= 9.0\times10^5$ m/s

5. $KE = -qV_0$

$= -(-1.60\times10^{-19}\ \text{C})(3.2\ \text{J/C})$

$= 5.1\times10^{-19}$ J

7. $KE_{max} = \frac{1240\ \text{eV·nm}}{\lambda} - hf_0$

$= \frac{1240\ \text{eV·nm}}{425\ \text{nm}} - 1.96\ \text{eV}$

$= 0.960$ eV

9. $hf_0 = 4.50$ eV, so $\frac{hc}{\lambda_0} = 4.50$ eV

Thus, $\lambda_0 = \frac{1240\ \text{eV·nm}}{4.50\ \text{eV}}$

$= 276$ nm

19. a. $\lambda = \frac{h}{mv}$

$= \frac{6.63\times10^{-34}\ \text{J·s}}{(7.0\ \text{kg})(8.5\ \text{m/s})}$

$= 1.1\times10^{-35}$ m

b. The wavelength is too small to show observable effects.

21. $\lambda = \frac{h}{p}$

so $p = \frac{h}{\lambda}$

$KE = \frac{1}{2}mv^2$

$= \frac{p^2}{2m}$

$= \frac{\left(\frac{h}{\lambda}\right)^2}{2m}$

$= \frac{\left(\frac{6.63\times10^{-34}\ \text{J·s}}{0.125\times10^{-9}\ \text{m}}\right)^2}{2(9.11\times10^{-31}\ \text{kg})}$

$= (1.544\times10^{-17}\ \text{J})\left(\frac{1\ \text{eV}}{1.60\times10^{-19}\ \text{J}}\right)$

$= 96.5$ eV, so it would have to be accelerated through 96.5 V.

Chapter 28

1. $E_n = \frac{-13.6\ \text{eV}}{n^2}$

$E_2 = \frac{-13.6\ \text{eV}}{(2)^2} = -3.40$ eV

$E_3 = \frac{-13.6\ \text{eV}}{(3)^2} = -1.51$ eV

$E_4 = \frac{-13.6\ \text{eV}}{(4)^2} = -0.850$ eV

3. $\Delta E = E_4 - E_2$

$= (-13.6\ \text{eV})\left(\frac{1}{4^2} - \frac{1}{2^2}\right)$

$= (-13.6\ \text{eV})\left(\frac{1}{16} - \frac{1}{4}\right)$

$= 2.55$ eV

5. $\frac{x}{0.075\ \text{m}} = \frac{5\times10^{-11}\ \text{m}}{2.5\times10^{-15}\ \text{m}}$

$x = 2\times10^3$ m

7. a. $\Delta E = 8.82\ \text{eV} - 6.67\ \text{eV}$

$= 2.15$ eV

b. $\lambda = \frac{hc}{\Delta E} = \frac{1240\ \text{eV.nm}}{\Delta E}$

$= \frac{1240\ \text{eV.nm}}{2.15\ \text{eV}}$

$= 577$ nm

Chapter 29

1. free e^-/cm^3

$= (2\ e^-/\text{atom})(6.02\times10^{23}\ \text{atoms/mol})\left(\frac{1\ \text{mol}}{65.37\ \text{g}}\right)(7.13\ \text{g/cm}^3)$

$= 1.31\times10^{23}$

3. free e^-/cm^3

$= (1\ e^-/\text{atom})(6.02\times10^{23}\ \text{atoms/mol})\left(\frac{1\ \text{mol}}{196.97\ \text{g}}\right)(19.31\ \text{g/cm}^3)$

$= 5.90\times10^{22}$

5. free $e^- = (1.81\times10^{23}\ \text{free}\ e^-/\text{cm}^3)\left(\frac{2835\ \text{g}}{2.70\ \text{g/cm}^3}\right)$

$= 1.90\times10^{26}$ free e^- in the tip

7. free e^-/atom

$= \left(\frac{1\ \text{mol}}{6.02\times10^{23}\ \text{atoms}}\right)\left(\frac{28.09\ \text{g}}{1\ \text{mol}}\right)\left(\frac{1\ \text{cm}^3}{2.33\ \text{g}}\right)\left(\frac{1.89\times10^5\ \text{free}\ e^-}{\text{cm}^3}\right)$

$= 3.78\times10^{-18}$ free e^-/atom

$T_K = T_C + 273$

$T_C = T_K - 273$

$= 200.0 - 273$

$= -73°\text{C}$

9. free e^-/atom

$$= \left(\frac{1 \text{ mol}}{6.02\times10^{23} \text{ atoms}}\right)\left(\frac{72.6 \text{ g}}{\text{mol}}\right)\left(\frac{\text{cm}^3}{5.23 \text{ g}}\right)\left(\frac{1.16\times10^{10} \text{ free e}^-}{\text{cm}^3}\right)$$

$= 2.67\times10^{-13}$

11. ratio $= \frac{\text{free e}^-/\text{cm}^3 \text{ in doped Si}}{\text{free e}^-/\text{cm}^3 \text{ in Si}}$

free e^-/cm^3 in doped Si = (ratio)(free e^-/cm^3 in Si)

$$\left(\frac{\text{As atoms}}{\text{Si atoms}}\right)\left(\frac{4.99\times10^{22} \text{ Si atoms}}{\text{cm}^3}\right) = (\text{ratio})(\text{free e}^-/\text{cm}^3 \text{ in Si})$$

$$\frac{\text{As atoms}}{\text{Si atoms}} = \frac{(1\times10^4)(1.45\times10^{10} \text{ free e}^-/\text{cm}^3)}{4.99\times10^{22} \text{ Si atoms/cm}^3}$$

$= 2.91\times10^{-9}$

13. ratio =

$$\frac{\left(\frac{1 \text{ free e}^-}{1 \text{ As atom}}\right)\left(\frac{1 \text{ As atom}}{1\times10^6 \text{ electrons}}\right)\left(\frac{4.34\times10^{22} \text{ electrons}}{\text{cm}^3}\right)}{1.13\times10^{15} \text{ thermal carriers}}$$

$= 38.4$

15. Germanium devices do not work well at such temperatures because the ratio of doped carriers to thermal carriers is small enough that temperature has too much influence on conductivity. Silicon is much better.

23. $V_b = IR + V_d + V_d$
$= (0.0025 \text{ A})(470 \ \Omega) + 0.50 \text{ V} + 0.50 \text{ V}$
$= 2.2 \text{ V}$

25. It would be impossible to obtain 2.5 mA of current with any reasonable power supply voltage because one of the diodes would be reverse-biased.

Chapter 30

1. $A - Z$ = neutrons
$234 - 92 = 142$ neutrons
$235 - 92 = 143$ neutrons
$238 - 92 = 146$ neutrons

3. $A - Z = 200 - 80 = 120$ neutrons

5. **a.** $12.000000 - 6(1.007825) - 6(1.008665)$
$= -0.098940$ u

b. $E = (-0.098940 \text{ u})(931.49 \text{ MeV/u})$
$= -92.161$ MeV

7. **a.** $15.010109 - 7(1.007825) - 8(1.008665)$
$= -0.113986$ u

b. $E = (-0.113986 \text{ u})(931.49 \text{ MeV/u})$
$= -106.18$ MeV

15. ${}^{234}_{92}\text{U} \rightarrow {}^{230}_{90}\text{Th} + {}^{4}_{2}\text{He}$

17. ${}^{226}_{88}\text{Ra} \rightarrow {}^{222}_{86}\text{Rn} + {}^{4}_{2}\text{He}$

19. ${}^{14}_{6}\text{C} \rightarrow {}^{14}_{7}\text{N} + {}^{0}_{-1}\text{e} + {}^{0}_{0}\bar{\nu}$

21. ${}^{263}_{106}\text{Sg} \rightarrow {}^{259}_{104}\text{Rf} + {}^{4}_{2}\text{He}$

23. **a.** ${}^{210}_{82}\text{Pb} \rightarrow {}^{210}_{83}\text{Bi} + {}^{0}_{-1}\text{e} + {}^{0}_{0}\nu$

b. ${}^{210}_{83}\text{Bi} \rightarrow {}^{210}_{84}\text{Po} + {}^{0}_{-1}\text{e} + {}^{0}_{0}\nu$

c. ${}^{234}_{90}\text{Th} \rightarrow {}^{234}_{91}\text{Pa} + {}^{0}_{-1}\text{e} + {}^{0}_{0}\nu$

d. ${}^{239}_{93}\text{Np} \rightarrow {}^{239}_{94}\text{Pu} + {}^{0}_{-1}\text{e} + {}^{0}_{0}\nu$

25. 8.0 days = 4(2.0 days), which is 4 half-lives

remaining = original $\left(\frac{1}{2}\right)^t$
$= (4.0 \text{ g})\left(\frac{1}{2}\right)^4$
$= 0.25$ g

27. Six years is about half of tritium's half-life of 12.3 years, thus, the brightness is $\left(\frac{1}{2}\right)^{\frac{1}{2}}$, or about $\frac{7}{10}$ the original brightness.

35. **a.** $E = mc^2$
$= (1.67\times10^{-27} \text{ kg})(3.00\times10^8 \text{ m/s})^2$
$= 1.50\times10^{-10}$ J

b. $E = \frac{1.50\times10^{-10} \text{ J}}{1.60217\times10^{-19} \text{ J/eV}}$
$= 9.36\times10^8$ eV
$= 936$ MeV

c. The minimum energy is $(2)(9.36\times10^8 \text{ eV}) = 1.87\times10^9$ eV

37. **a.** $E = (1.008655 \text{ u})(931.49 \text{ MeV})$
$= 939.56$ MeV

b. The total γ-ray energy is twice the energy equivalent of the neutron's mass, or 1879.1 MeV.

Appendix D: Tables

Color Conventions			
Displacement vectors (***d***)		Negative charges	−
Velocity vectors (***v***)		Positive charges	+
Acceleration vectors (***a***)		Current direction	
Force vectors (***F***)		Electron	
Momentum vectors (***p***)		Proton	
Light rays		Neutron	
Object		Coordinate axes	
Image			
Electric field (***E***)			
Magnetic field lines (***B***)			

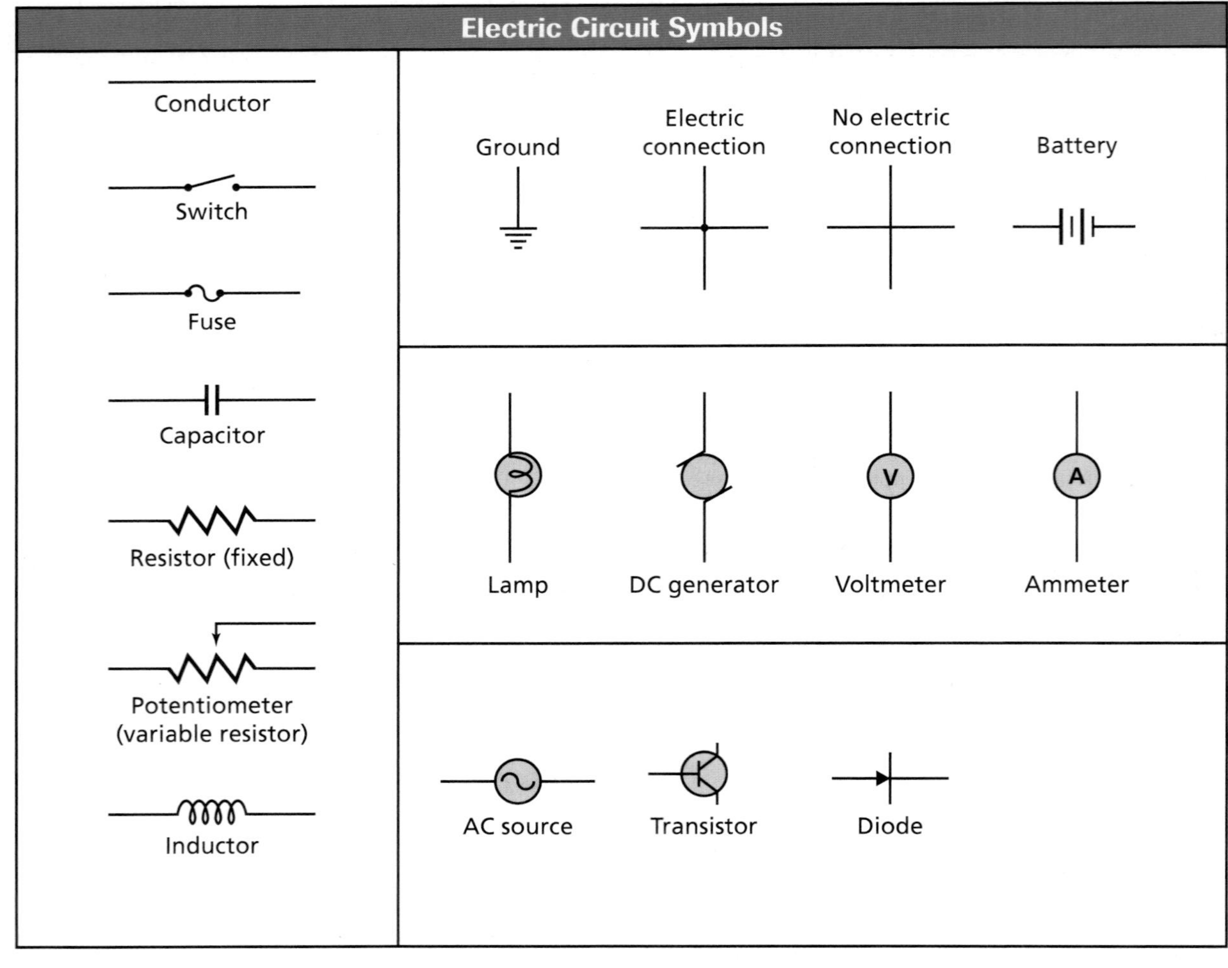

SI Base Units		
Quantity	**Name**	**Symbol**
Length	meter	m
Mass	kilogram	kg
Time	second	s
Temperature	Kelvin	K
Amount of a substance	mole	mol
Electric current	ampere	A
Luminous intensity	candela	cd

SI Derived Units				
Measurement	**Unit**	**Symbol**	**Expressed in Base Units**	**Expressed in Other SI Units**
Acceleration		m/s^2	m/s^2	
Area		m^2	m^2	
Capacitance	farad	F	$A^2 \cdot s^4/kg \cdot m^2$	
Density		kg/m^3	kg/m^3	
Electric charge	coulomb	C	$A \cdot s$	
Electric field		N/C	$kg \cdot m/C \cdot s^2$	
Electric resistance	ohm	Ω	$kg \cdot m^2/A^2 \cdot s^3$	V/A
EMF	volt	V	$kg \cdot m^2/A \cdot s^3$	
Energy, work	joule	J	$kg \cdot m^2/s^2$	$N \cdot m$
Force	newton	N	$kg \cdot m/s^2$	
Frequency	hertz	Hz	s^{-1}	
Illuminance	lux	lx	cd/m^2	
Magnetic field	tesla	T	$kg/A \cdot s^2$	$N \cdot s/C \cdot m$
Potential difference	volt	V	$kg \cdot m^2/A \cdot s^3$	W/A or J/C
Power	watt	W	$kg \cdot m^2/s^3$	J/s
Pressure	pascal	Pa	$kg/m \cdot s^2$	N/m^2
Velocity		m/s	m/s	
Volume		m^3	m^3	

Useful Conversions		
1 in = 2.54 cm	1 kg = 6.02×10^{26} u	1 atm = 101 kPa
1 mi = 1.61 km	1 oz ↔ 28.4 g	1 cal = 4.184 J
1 mi^2 = 640 acres	1 kg ↔ 2.21 lb	1 eV = 1.60×10^{-19} J
1 gal = 3.79 L	1 lb = 4.45 N	1 kWh = 3.60 MJ
1 m^3 = 264 gal	1 atm = 14.7 lb/in^2	1 hp = 746 W
1 knot = 1.15 mi/h	1 atm = 1.01×10^5 N/m^2	1 mol = 6.022×10^{23} items

Physical Constants			
Quantity	**Symbol**	**Value**	**Approximate Value**
Atomic mass unit	u	$1.66053886 \times 10^{-27}$ kg	1.66×10^{-27} kg
Avogadro's number	N_A	6.0221415×10^{23} mol^{-1}	6.022×10^{23} mol^{-1}
Boltzmann's constant	k	$1.3806505 \times 10^{-23}$ Pa·m^3/K	1.38×10^{-23} Pa·m^3/K
Constant in Coulomb's law	K	$8.987551788 \times 10^{9}$ N·m^2/C^2	9.0×10^{9} N·m^2/C^2
Elementary charge	e	$1.60217653 \times 10^{-19}$ C	1.602×10^{-19} C
Gas constant	R	8.314472 Pa·m^3/mol·K	8.31 Pa·m^3/mol·K
Gravitational constant	G	6.6742×10^{-11} N·m^2/kg^2	6.67×10^{-11} N·m^2/kg^2
Mass of an electron	m_e	$9.1093826 \times 10^{-31}$ kg	9.11×10^{-31} kg
Mass of a proton	m_p	$1.67262171 \times 10^{-27}$ kg	1.67×10^{-27} kg
Mass of a neutron	m_n	$1.67492728 \times 10^{-27}$ kg	1.67×10^{-27} kg
Planck's constant	h	$6.6260693 \times 10^{-34}$ J·s	6.63×10^{-34} J·s
Speed of light in a vacuum	c	2.99792458×10^{8} m/s	3.00×10^{8} m/s

SI Prefixes		
Prefix	**Symbol**	**Scientific Notation**
femto	f	10^{-15}
pico	p	10^{-12}
nano	n	10^{-9}
micro	μ	10^{-6}
milli	m	10^{-3}
centi	c	10^{-2}
deci	d	10^{-1}
deka	da	10^{1}
hecto	h	10^{2}
kilo	k	10^{3}
mega	M	10^{6}
giga	G	10^{9}
tera	T	10^{12}
peta	P	10^{15}

Moments of Inertia for Various Objects			
Object	**Location of Axis**	**Diagram**	**Moment of Inertia**
Thin hoop of radius r	Through central diameter	Axis, r	mr^2
Solid, uniform cylinder of radius r	Through center	Axis, r	$\frac{1}{2}mr^2$
Uniform sphere of radius r	Through center	Axis, r	$\frac{2}{5}mr^2$
Long, uniform rod of length l	Through center	Axis, l	$\frac{1}{12}ml^2$
Long, uniform rod of length l	Through end	Axis, l	$\frac{1}{3}ml^2$
Thin, rectangular plate of length l and width w	Through center	Axis, l, w	$\frac{1}{12}m(l^2 + w^2)$

Appendix D

Densities of Some Common Substances	
Substance	**Density (g/cm^3)**
Aluminum	2.702
Cadmium	8.642
Copper	8.92
Germanium	5.35
Gold	19.31
Hydrogen	8.99×10^{-5}
Indium	7.30
Iron	7.86
Lead	11.34
Mercury	13.546
Oxygen	1.429×10^{-3}
Silicon	2.33
Silver	10.5
Water (4°C)	1.000
Zinc	7.14

Melting and Boiling Points of Some Substances		
Substance	**Melting point (°C)**	**Boiling point (°C)**
Aluminum	660.37	2467
Copper	1083	2567
Germanium	937.4	2830
Gold	1064.43	2808
Indium	156.61	2080
Iron	1535	2750
Lead	327.5	1740
Silicon	1410	2355
Silver	961.93	2212
Water	0.000	100.000
Zinc	419.58	907

Specific Heats of Some Common Substances			
Material	**Specific Heat (J/kg·K)**	**Material**	**Specific Heat (J/kg·K)**
Aluminum	897	Lead	130
Brass	376	Methanol	2450
Carbon	710	Silver	235
Copper	385	Steam	2020
Glass	840	Water	4180
Ice	2060	Zinc	388
Iron	450		

Heats of Fusion and Vaporization of Some Common Substances		
Material	**Heat of Fusion, H_f (J/kg)**	**Heat of Vaporization, H_v (J/kg)**
Copper	2.05×10^5	5.07×10^6
Gold	6.30×10^4	1.64×10^6
Iron	2.66×10^5	6.29×10^6
Lead	2.04×10^4	8.64×10^5
Mercury	1.15×10^4	2.72×10^5
Methanol	1.09×10^5	8.78×10^5
Silver	1.04×10^5	2.36×10^6
Water (ice)	3.34×10^5	2.26×10^6

Coefficients of Thermal Expansion at 20°C		
Material	**Coefficient of Linear Expansion, α (°C)$^{-1}$**	**Coefficient of Volume Expansion, β (°C)$^{-1}$**
Solids		
Aluminum	25×10^{-6}	75×10^{-6}
Brass	19×10^{-6}	56×10^{-6}
Concrete	12×10^{-6}	36×10^{-6}
Copper	17×10^{-6}	48×10^{-6}
Glass (soft)	9×10^{-6}	27×10^{-6}
Glass (ovenproof)	3×10^{-6}	9×10^{-6}
Iron, steel	12×10^{-6}	35×10^{-6}
Platinum	9×10^{-6}	27×10^{-6}
Liquids		
Gasoline		950×10^{-6}
Mercury		180×10^{-6}
Methanol		1100×10^{-6}
Water		210×10^{-6}
Gases		
Air (and most other gases)		3400×10^{-6}

Speed of Sound in Various Media	
Medium	**m/s**
Air (0°)	331
Air (20°)	343
Helium (0°)	972
Hydrogen (0°)	1286
Water (25°)	1493
Seawater (0°)	1533
Rubber	1600
Copper (25°)	3560
Iron (25°)	5130
Ovenproof glass	5640
Diamond	12,000

Wavelengths of Visible Light	
Color	**Wavelength in Nanometers (nm)**
Violet light	380–430 nm
Indigo light	430–450 nm
Blue light	450–500 nm
Cyan light	500–520 nm
Green light	520–565 nm
Yellow light	565–590 nm
Orange light	590–625 nm
Red light	625–740 nm

Dielectric Constants, K (20°C)	
Vacuum	1.0000
Air (1 atm)	1.00059
Neon (1 atm)	1.00013
Glass	4–7
Quartz	4.3
Fused quartz	3.75
Water	80

Solar System Data									
	Mercury	**Venus**	**Earth**	**Mars**	**Jupiter**	**Saturn**	**Uranus**	**Neptune**	**Pluto**
Mass ($kg \times 10^{24}$)	0.3302	4.8685	5.9736	0.64185	1898.6	568.46	86.832	102.43	0.0125
Equatorial radius (km)	2439.7	6051.8	6378.1	3397	71,492	60,268	25,559	24,764	1195
Mean density (kg/m^3)	5427	5234	5515	3933	1326	687	1270	1638	1750
Albedo	0.119	0.750	0.306	0.250	0.343	0.342	0.300	0.290	0.4–0.6
Semimajor axis ($km \times 10^6$)	57.91	108.21	149.60	227.92	778.57	1433.53	2872.46	4495.06	5869.66
Orbital period (Earth days)	87.969	224.701	365.256	686.980	4332.589	10,759.22	30,685.4	60,189	90,465
Orbital inclination (degrees)	7.00	3.39	0.000	1.850	1.304	2.485	0.772	1.769	17.16
Orbital eccentricity	0.2056	0.0067	0.0167	0.0935	0.0489	0.0565	0.0457	0.0113	0.2444
Rotational period (hours)	1407.6	5832.5^R	23.9345	24.6229	9.9250	10.656	17.24^R	16.11	153.2928^R
Axial tilt (degrees)	0.01	177.36	23.45	25.19	3.13	26.73	97.77	28.32	122.53
Average surface temperature (K)	440	737	288	210	112	84	53	55	50

R indicates retrograde motion.

The Moon	
Mass	0.07349×10^{24} kg
Equatorial radius	1738.1 km
Mean density	3350 kg/m^3
Albedo	0.11
Semimajor axis	0.3844 $km \times 10^6$
Revolution period	27.3217 Earth days
Synodic (lunar) period	29.53 Earth days
Orbital inclination	18.28°–28.58°
Orbital eccentricity	0.0549
Rotational period	655.728 h

The Sun	
Mass	$1,989,100 \times 10^{24}$ kg
Equatorial radius	696000 km
Mean density	1408 kg/m^3
Absolute magnitude	+4.83
Luminosity	384.6×10^{24} J/s
Spectral type	G2 V
Rotational period	609.12 h
Mean energy production	0.1937×10^{-3} J/kg
Average temperature	5778 K

Appendix D: Tables

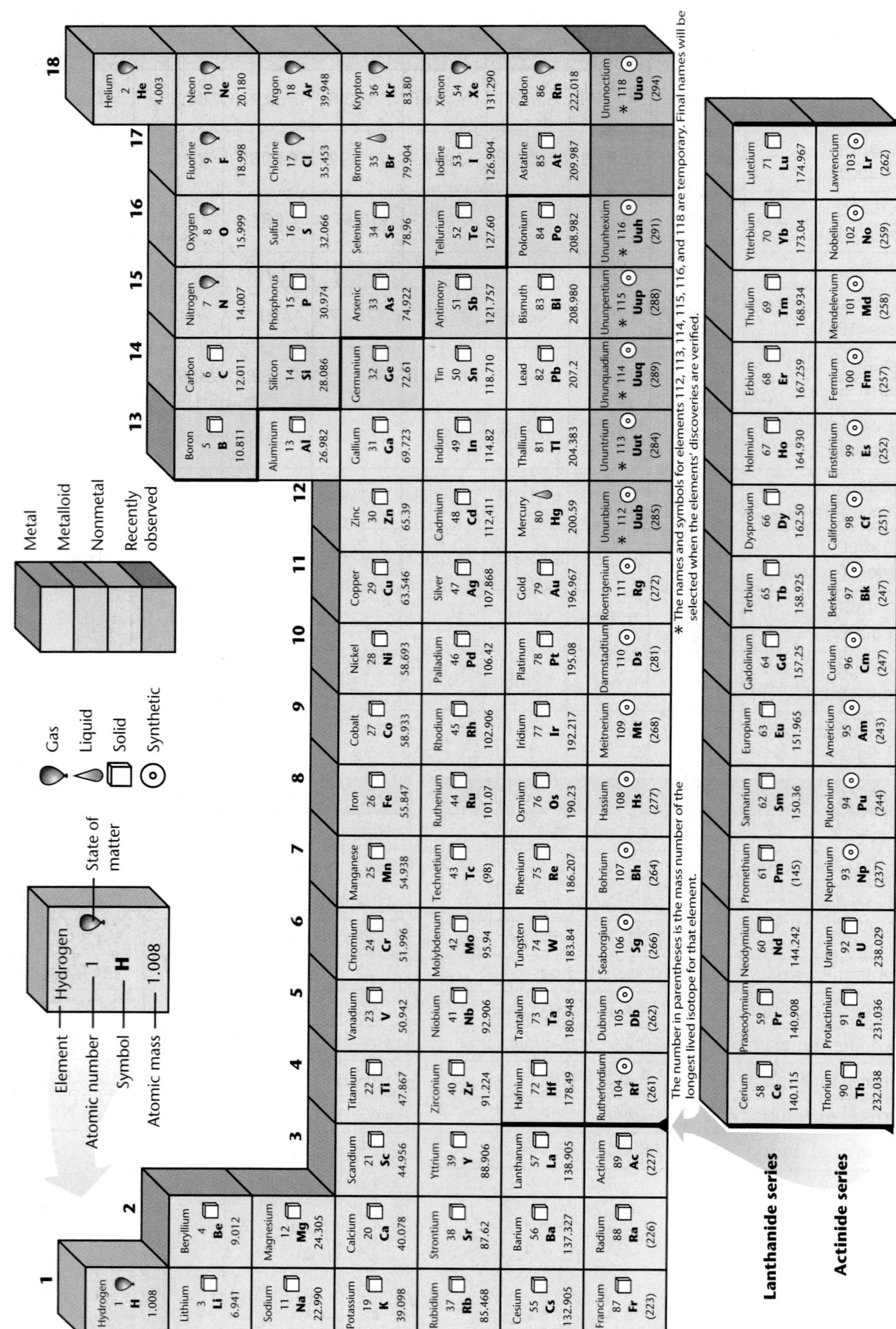

PERIODIC TABLE OF THE ELEMENTS

Period	1	2	3	4	5	6	7	8	9	10	11	12	13	14	15	16	17	18
1	Hydrogen 1 H 1.008																	Helium 2 He 4.003
2	Lithium 3 Li 6.941	Beryllium 4 Be 9.012											Boron 5 B 10.811	Carbon 6 C 12.011	Nitrogen 7 N 14.007	Oxygen 8 O 15.999	Fluorine 9 F 18.998	Neon 10 Ne 20.180
3	Sodium 11 Na 22.990	Magnesium 12 Mg 24.305											Aluminum 13 Al 26.982	Silicon 14 Si 28.086	Phosphorus 15 P 30.974	Sulfur 16 S 32.066	Chlorine 17 Cl 35.453	Argon 18 Ar 39.948
4	Potassium 19 K 39.098	Calcium 20 Ca 40.078	Scandium 21 Sc 44.956	Titanium 22 Ti 47.867	Vanadium 23 V 50.942	Chromium 24 Cr 51.996	Manganese 25 Mn 54.938	Iron 26 Fe 55.847	Cobalt 27 Co 58.933	Nickel 28 Ni 58.693	Copper 29 Cu 63.546	Zinc 30 Zn 65.39	Gallium 31 Ga 69.723	Germanium 32 Ge 72.61	Arsenic 33 As 74.922	Selenium 34 Se 78.96	Bromine 35 Br 79.904	Krypton 36 Kr 83.80
5	Rubidium 37 Rb 85.468	Strontium 38 Sr 87.62	Yttrium 39 Y 88.906	Zirconium 40 Zr 91.224	Niobium 41 Nb 92.906	Molybdenum 42 Mo 95.94	Technetium 43 Tc (98)	Ruthenium 44 Ru 101.07	Rhodium 45 Rh 102.906	Palladium 46 Pd 106.42	Silver 47 Ag 107.868	Cadmium 48 Cd 112.411	Indium 49 In 114.82	Tin 50 Sn 118.710	Antimony 51 Sb 121.757	Tellurium 52 Te 127.60	Iodine 53 I 126.904	Xenon 54 Xe 131.290
6	Cesium 55 Cs 132.905	Barium 56 Ba 137.327	Lanthanum 57 La 138.905	Hafnium 72 Hf 178.49	Tantalum 73 Ta 180.948	Tungsten 74 W 183.84	Rhenium 75 Re 186.207	Osmium 76 Os 190.23	Iridium 77 Ir 192.217	Platinum 78 Pt 195.08	Gold 79 Au 196.967	Mercury 80 Hg 200.59	Thallium 81 Tl 204.383	Lead 82 Pb 207.2	Bismuth 83 Bi 208.980	Polonium 84 Po 208.982	Astatine 85 At 209.987	Radon 86 Rn 222.018
7	Francium 87 Fr (223)	Radium 88 Ra (226)	Actinium 89 Ac (227)	Rutherfordium 104 Rf (261)	Dubnium 105 Db (262)	Seaborgium 106 Sg (266)	Bohrium 107 Bh (264)	Hassium 108 Hs (277)	Meitnerium 109 Mt (268)	Darmstadtium 110 Ds (281)	Roentgenium 111 Rg (272)	Ununbium * 112 Uub (285)	Ununtrium * 113 Uut (284)	Ununquadium * 114 Uuq (289)	Ununpentium * 115 Uup (288)	Ununhexium * 116 Uuh (291)		Ununoctium * 118 Uuo (294)

The number in parentheses is the mass number of the longest lived isotope for that element.

Lanthanide series	Cerium 58 Ce 140.115	Praseodymium 59 Pr 140.908	Neodymium 60 Nd 144.242	Promethium 61 Pm (145)	Samarium 62 Sm 150.36	Europium 63 Eu 151.965	Gadolinium 64 Gd 157.25	Terbium 65 Tb 158.925	Dysprosium 66 Dy 162.50	Holmium 67 Ho 164.930	Erbium 68 Er 167.259	Thulium 69 Tm 168.934	Ytterbium 70 Yb 173.04	Lutetium 71 Lu 174.967
Actinide series	Thorium 90 Th 232.038	Protactinium 91 Pa 231.036	Uranium 92 U 238.029	Neptunium 93 Np (237)	Plutonium 94 Pu (244)	Americium 95 Am (243)	Curium 96 Cm (247)	Berkelium 97 Bk (247)	Californium 98 Cf (251)	Einsteinium 99 Es (252)	Fermium 100 Fm (257)	Mendelevium 101 Md (258)	Nobelium 102 No (259)	Lawrencium 103 Lr (262)

* The names and symbols for elements 112, 113, 114, 115, 116, and 118 are temporary. Final names will be selected when the elements' discoveries are verified.

The Elements							
Element	**Symbol**	**Atomic Number**	**Atomic Mass**	**Element**	**Symbol**	**Atomic Number**	**Atomic Mass**
Actinium	Ac	89	(227)	Mercury	Hg	80	200.59
Aluminum	Al	13	26.982	Molybdenum	Mo	42	95.94
Americium	Am	95	(243)	Neodymium	Nd	60	144.24
Antimony	Sb	51	121.760	Neon	Ne	10	20.180
Argon	Ar	18	39.948	Neptunium	Np	93	(237)
Arsenic	As	33	74.922	Nickel	Ni	28	58.693
Astatine	At	85	(210)	Niobium	Nb	41	92.906
Barium	Ba	56	137.327	Nitrogen	N	7	14.007
Berkelium	Bk	97	(247)	Nobelium	No	102	(259)
Beryllium	Be	4	9.012	Osmium	Os	76	190.23
Bismuth	Bi	83	208.980	Oxygen	O	8	15.999
Bohrium	Bh	107	(264)	Palladium	Pd	46	106.42
Boron	B	5	10.811	Phosphorus	P	15	30.974
Bromine	Br	35	79.904	Platinum	Pt	78	195.078
Cadmium	Cd	48	112.411	Plutonium	Pu	94	(244)
Calcium	Ca	20	40.078	Polonium	Po	84	(209)
Californium	Cf	98	(251)	Potassium	K	19	39.098
Carbon	C	6	12.011	Praseodymium	Pr	59	140.908
Cerium	Ce	58	140.116	Promethium	Pm	61	(145)
Cesium	Cs	55	132.905	Protactinium	Pa	91	231.036
Chlorine	Cl	17	35.453	Radium	Ra	88	(226)
Chromium	Cr	24	51.996	Radon	Rn	86	(222)
Cobalt	Co	27	58.933	Rhenium	Re	75	186.207
Copper	Cu	29	63.546	Rhodium	Rh	45	102.906
Curium	Cm	96	(247)	Rubidium	Rb	37	85.468
Darmstadtium	Ds	110	(281)	Ruthenium	Ru	44	101.07
Dubnium	Db	105	(262)	Rutherfordium	Rf	104	(261)
Dysprosium	Dy	66	162.500	Samarium	Sm	62	150.36
Einsteinium	Es	99	(252)	Scandium	Sc	21	44.956
Erbium	Er	68	167.259	Seaborgium	Sg	106	(266)
Europium	Eu	63	151.964	Selenium	Se	34	78.96
Fermium	Fm	100	(257)	Silicon	Si	14	28.086
Fluorine	F	9	18.998	Silver	Ag	47	107.868
Francium	Fr	87	(223)	Sodium	Na	11	22.990
Gadolinium	Gd	64	157.25	Strontium	Sr	38	87.62
Gallium	Ga	31	69.723	Sulfur	S	16	32.065
Germanium	Ge	32	72.64	Tantalum	Ta	73	180.948
Gold	Au	79	196.967	Technetium	Tc	43	(98)
Hafnium	Hf	72	178.49	Tellurium	Te	52	127.60
Hassium	Hs	108	(277)	Terbium	Tb	65	158.925
Helium	He	2	4.003	Thallium	Tl	81	204.383
Holmium	Ho	67	164.930	Thorium	Th	90	232.038
Hydrogen	H	1	1.008	Thulium	Tm	69	168.934
Indium	In	49	114.818	Tin	Sn	50	118.710
Iodine	I	53	126.904	Titanium	Ti	22	47.867
Iridium	Ir	77	192.217	Tungsten	W	74	183.84
Iron	Fe	26	55.845	Uranium	U	92	238.029
Krypton	Kr	36	83.798	Vanadium	V	23	50.942
Lanthanum	La	57	138.906	Xenon	Xe	54	131.293
Lawrencium	Lr	103	(262)	Ytterbium	Yb	70	173.04
Lead	Pb	82	207.2	Yttrium	Y	39	88.906
Lithium	Li	3	6.941	Zinc	Zn	30	65.409
Lutetium	Lu	71	174.967	Zirconium	Zr	40	91.224
Magnesium	Mg	12	24.305	Element 111[*]	Uuu	111	(272)
Manganese	Mn	25	54.938	Element 112[*]	Uub	112	(285)
Meitnerium	Mt	109	(268)	Element 114[*]	Uuq	114	(289)
Mendelevium	Md	101	(258)				

* Names have not yet been approved by IUPAC.

Tables

SAFETY SYMBOLS	HAZARD	EXAMPLES	PRECAUTION	REMEDY
DISPOSAL	Special disposal procedures need to be followed.	Certain chemicals, living organisms	Do not dispose of these materials in the sink or trash can.	Dispose of wastes as directed by your teacher.
BIOLOGICAL	Organisms or other biological materials that might be harmful to humans	Bacteria, fungi, blood, unpreserved tissues, plant materials	Avoid skin contact with these materials. Wear mask or gloves.	Notify your teacher if you suspect contact with material. Wash hands thoroughly.
EXTREME TEMPERATURE	Objects that can burn skin by being too cold or too hot	Boiling liquids, hot plates, dry ice, liquid nitrogen	Use proper protection when handling.	Go to your teacher for first aid.
SHARP OBJECT	Use of tools or glassware that can easily puncture or slice skin	Razor blades, pins, scalpels, pointed tools, dissecting probes, broken glass	Practice common-sense behavior and follow guidelines for use of the tool.	Go to your teacher for first aid.
FUME	Possible danger to respiratory tract from fumes	Ammonia, acetone, nail polish remover, heated sulfur, moth balls	Make sure there is good ventilation. Never smell fumes directly. Wear a mask.	Leave foul area and notify your teacher immediately.
ELECTRICAL	Possible danger from electrical shock or burn	Improper grounding, liquid spills, short circuits, exposed wires	Double-check setup with teacher. Check condition of wires and apparatus.	Do not attempt to fix electrical problems. Notify your teacher immediately.
IRRITANT	Substances that can irritate the skin or mucous membranes of the respiratory tract	Pollen, moth balls, steel wool, fiberglass, potassium permanganate	Wear dust mask and gloves. Practice extra care when handling these materials.	Go to your teacher for first aid.
CHEMICAL	Chemicals that can react with and destroy tissue and other materials	Bleaches such as hydrogen peroxide; acids such as sulfuric acid, hydrochloric acid; bases such as ammonia, sodium hydroxide	Wear goggles, gloves, and an apron.	Immediately flush the affected area with water and notify your teacher.
TOXIC	Substance may be poisonous if touched, inhaled, or swallowed.	Mercury, many metal compounds, iodine, poinsettia plant parts	Follow your teacher's instructions.	Always wash hands thoroughly after use. Go to your teacher for first aid.
FLAMMABLE	Flammable chemicals may be ignited by open flame, spark, or exposed heat.	Alcohol, kerosene, potassium permanganate	Avoid open flames and heat when using flammable chemicals.	Notify your teacher immediately. Use fire safety equipment if applicable.
OPEN FLAME	Open flame in use, may cause fire.	Hair, clothing, paper, synthetic materials	Tie back hair and loose clothing. Follow teacher's instruction on lighting and extinguishing flames.	Notify your teacher immediately. Use fire safety equipment if applicable.

Eye Safety
Proper eye protection should be worn at all times by anyone performing or observing science activities.

Clothing Protection
This symbol appears when substances could stain or burn clothing.

Radioactivity
This symbol appears when radioactive materials are used.

Handwashing
After the lab, wash hands with soap and water before removing goggles.

Physics Online A science multilingual glossary is available at physicspp.com. The glossary includes the following languages: Arabic, Bengali, Chinese, English, Haitian Creole, Hmong, Korean, Portuguese, Russian, Spanish, Tagalog, Urdu, and Vietnamese.

Glossary

absorption spectrum (751): The characteristic set of wavelengths absorbed by a gas, which can be used to identify that gas.

acceleration (59): The rate at which the velocity of an object changes.

acceleration due to gravity (72): The acceleration of an object in free fall, resulting from the influence of Earth's gravity; acceleration due to gravity on Earth, g, is 9.80 m/s^2 toward the center of Earth.

accuracy (13): A characteristic of a measured value that describes how well the results of a measurement agree with the "real" value, which is the accepted value, as measured by competent experimenters.

achromatic lens (499): A combination of two or more lenses with different indices of refraction (such as a concave lens with a convex lens) that is used to minimize a chromatic aberration.

activity (810): The number of decays per second of a radioactive substance.

adhesive forces (350): The electromagnetic forces of attraction that particles of different substances exert on one another; responsible for capillary action.

alpha decay (806): The radioactive decay process in which an alpha (α) particle is emitted from a nucleus.

alpha particles (748): Massive, positively charged atomic particles that move at high speed; are represented by the symbol α.

ammeter (631): A low-resistance device connected in series that is used to measure the electric current in any branch or part of a circuit.

amorphous solid (359): A substance having a definite shape and volume, but lacking a regular crystal structure.

ampere (593): A flow of electric charge, or electric current, equal to one coulomb per second (1 C/s).

amplitude (375): In any periodic motion, the maximum distance an object moves from equilibrium.

angular acceleration (199): The change in angular velocity divided by the time needed to make the change; is measured in rad/s^2.

angular displacement (198): The change in the angle as an object rotates.

angular impulse–angular momentum theorem (234): States that the angular impulse on an object is equal to the change in the object's angular momentum.

angular momentum (233): The product of a rotating object's moment of inertia and its angular velocity; is measured in $kg{\cdot}m^2/s$.

angular velocity (198): The angular displacement of an object divided by the time needed to make the displacement.

antenna (707): A wire designed to transmit or receive electromagnetic waves.

antinode (389): The point with the largest displacement when two wave pulses meet.

apparent weight (98): The force experienced by an object, resulting from all the forces acting on it, giving the object an acceleration.

Archimedes' principle (354): States that an object immersed in a fluid has an upward force on it that equals the weight of the fluid displaced by the object.

armature (656): The wire coil of an electric motor, made up of many loops mounted on an axle or shaft; torque on an armature, and the motor's resultant speed, is controlled by varying the current through the motor.

atomic mass unit (800): A unit of mass, u, where u is equal to 1.66×10^{-27} kg.

atomic number (800): The number of protons in an atom's nucleus.

average acceleration (59): The change in an object's velocity during a measurable time interval, divided by that specific time interval; is measured in m/s^2.

average power (677): Half the maximum power associated with an alternating current.

average speed (44): How fast an object is moving; is the absolute value of the slope of an object's position-time graph.

average velocity (44): The change in position, divided by the time during which the change occurred; is the slope of an object's position-time graph.

band theory (776): The theory that electric conduction in solids can be better understood in terms of valance and conduction bands separated by forbidden energy gaps.

battery (592): A device made up of several galvanic cells connected together that converts chemical energy to electric energy.

beat (418): The oscillation of wave amplitude that results from the superposition of two sound waves with almost identical frequencies.

Bernoulli's principle (357): States that as the velocity of a fluid increases, the pressure exerted by that fluid decreases.

beta decay (806): The radioactive decay process in which a neutron is changed to a proton within the nucleus and a beta (β) particle and antineutrino are emitted.

binding energy (802): The energy equivalent of the mass defect; is always negative.

buoyant force (354): The upward force exerted on an object immersed in a fluid.

capacitance (577): The ratio of an object's stored charge to its electric potential difference.

capacitor (577): An electric device used to store charge that is made up of two conductors separated by an insulator.

center of mass (211): The point on the object that moves in the same way that a point particle would move.

centrifugal "force" (216): The apparent force that seems to pull on a moving object, but does not exert a physical outward push on it, and is observed only in rotating frames of reference.

centripetal acceleration (153): The center-seeking acceleration of an object moving in a circle at a constant speed.

centripetal force (154): The net force exerted toward the center of the circle that causes an object to have a centripetal acceleration.

chain reaction (812): Continual process of repeated fission reactions caused by the release of neutrons from previous fission reactions.

charging by conduction (547): The process of charging a neutral object by touching it with a charged object.

charging by induction (548): The process of charging an object without touching it, which can be accomplished by bringing a charged object close to a neutral object, causing a separation of charges, then separating the object to be charged, trapping opposite but equal charges.

chromatic aberration (499): A spherical lens defect in which light passing through a lens is focused at different points, causing an object viewed through a lens to seem to be ringed with color.

circuit breaker (627): An automatic switch that acts as a safety device in an electric circuit by opening and stopping the current flow when too much current flows through a circuit.

closed system (236): A system that does not gain or lose mass.

closed-pipe resonator (412): A resonating tube with one end closed to air; its resonant frequencies are odd-numbered multiples of the fundamental.

coefficient of kinetic friction (127): The slope of a line, μ_k, between two surfaces, relating frictional force to normal force.

coefficient of linear expansion (361): The change in length divided by the original length and the change in temperature.

coefficient of static friction (127): A dimensionless constant depending on the two surfaces in contact. It is used to calculate the maximum static frictional force that needs to be overcome before motion begins.

coefficient of volume expansion (361): The change in volume divided by the original volume and the change in temperature; is about three times the coefficient of linear expansion because solids expand in three directions.

coherent light (516): Light from two or more sources, whose additive superposition produces smooth wave fronts. (761): Light whose waves are in step, with coinciding maxima and minima.

cohesive forces (349): The electromagnetic forces of attraction that like particles exert on one another; responsible for surface tension and viscosity.

combination series-parallel circuit (629): A complex electric circuit that includes both series and parallel branches.

combined gas law (345): For a fixed amount of an ideal gas, the pressure times the volume, divided by the Kelvin temperature equals a constant; reduces to Boyle's law if temperature is constant and to Charles's law if pressure is constant.

complementary color (441): A color of light, which when combined with another color of light, produces white light.

components (122): Projections of the component vectors.

compound machine (269): A machine consisting of two or more simple machines that are connected so that the resistance force of one machine becomes the effort force of the second machine.

Compton effect (733): The shift in the energy of scattered photons.

concave lens (493): A diverging lens, thinner at its middle than at its edges, that spreads out light rays passing through it when surrounded by material with a lower index of refraction; produces a smaller, virtual, upright image.

concave mirror (464): A mirror that reflects light from its inwardly curving surface and can produce either an upright, virtual image or an inverted, real image.

conduction (315): The process by which kinetic energy is transferred when particles collide.

conductor (544): A material, such as copper, through which a charge will move easily.

consonance (418): A pleasant set of pitches.

convection (317): A type of thermal energy transfer that occurs from the motion of fluid in liquid or gas that is caused by differences in temperature.

conventional current (592): A flow of positive charges that move from higher potential to lower potential.

convex lens (493): A converging lens, thicker at its center than at its edges, that refracts parallel light rays so the rays meet at a point when surrounded by material with a lower index of refraction; can produce a smaller, inverted, real image, or a larger, upright, virtual image.

convex mirror (471): A mirror that reflects light from its outwardly curving surface and produces an upright, reduced, virtual image.

coordinate system (34): A system used to describe motion that gives the zero point location of the variable being studied and the direction in which the values of the variable increase.

Coriolis "force" (217): The apparent force that seems to deflect a moving object from its path and is observed only in rotating frames of reference.

coulomb (549): The SI standard unit of charge; one coulomb, C, is the magnitude of the charge of 6.24×10^{18} electrons or protons.

Coulomb's law (549): States that the force between two charges varies directly with the product of their charge and inversely with the square of the distance between them.

crest (383): The high point of a wave.

critical angle (489): The certain angle of incidence in which the refracted light ray lies along the boundary between two media.

crystal lattice (359): A fixed, regular pattern formed when the temperature of a liquid is lowered, the average kinetic energy of its particles decreases and, for many solids, the particles become frozen but do not stop moving and instead, vibrate around their fixed positions.

de Broglie wavelength (735): The wavelength associated with a moving particle.

decibel (406): The unit of measurement for sound level; also can describe the power and intensity of sound waves.

dependent variable (15): The factor in an experiment that depends on the independent variable.

depletion layer (784): The region around a *pn*-junction diode where there are no charge carriers and electricity is poorly conducted.

dielectrics (707): Nonconducting materials—glass, air, and water—through which electromagnetic waves move at less than the speed of light in a vacuum.

diffraction (439): The bending of light around a barrier.

diffraction grating (527): A device consisting of large numbers of single slits that are quite close together, diffract light, and form a diffraction pattern that is an overlap of single-slit diffraction patterns; can be used to precisely measure light wavelength or to separate light of different wavelengths.

diffraction pattern (524): A pattern on a screen of constructive and destructive interference of Huygens' wavelets.

diffuse reflection (459): A scattered, fuzzy reflection produced by a rough surface.

dimensional analysis (6): A method of treating units as algebraic quantities, which can be cancelled; can be used to check that an answer will be in the correct units.

diode (784): The simplest semiconductor device; conducts charges in one direction only and consists of a sandwich of *p*-type and *n*-type semiconductors.

dispersion (491): The separation of white light into a spectrum of colors by such means as a glass prism or water droplets in the atmosphere.

displacement (36): A change in position having both magnitude and direction; is equal to the final position minus the initial position.

dissonance (418): An unpleasant, jarring set of pitches.

distance (34): A scalar quantity that describes how far an object is from the origin.

domain (650): A very small group, usually 10–1000 μ, that is formed when the magnetic fields of the electrons in a group of neighboring atoms are aligned in the same direction.

dopants (781): Electron donor or acceptor atoms that can be added in low concentration to intrinsic semiconductors, increasing their conductivity by making either extra electrons or holes available.

Doppler effect (407): The change in the frequency of sound caused by the movement of either the source, the detector, or both the detector and the source.

Doppler shift (446): The difference between the observed wavelength of light and actual wavelength of light based on the relative speed of the observer and the source of the light.

drag force (100): The force exerted by a fluid on an object moving through the fluid; depends on the object's motion and properties and the fluid's properties.

eddy current (681): A current generated in a piece of metal that is moving through a changing magnetic field, producing a magnetic field that opposes the motion that caused the current.
efficiency (268): The ratio of output work to input work.
effort force (266): The force a person exerts on a machine.
elastic collision (298): A type of collision in which the kinetic energy before and after the collision remains the same.
elastic potential energy (291): The potential energy that may be stored in an object, such as a rubber band, as a result of its change in shape.
electric circuit (592): A closed loop or pathway that allows electric charges to flow.
electric current (592): A flow of charged particles.
electric field (563): The field that exists around any charged object; produces forces that can do work, transferring energy from the field to another charged object.
electric field lines (567): Lines that provide a picture of an electric field, indicate the field's strength by the spacing between the lines, never cross, and are directed toward negative charges and away from positive charges.
electric generator (675): A device that converts mechanical energy to electric energy and consists of a number of wire loops placed in a strong magnetic field.
electric motor (656): An apparatus that converts electric energy into rotational kinetic energy.
electric potential difference (569): The change in potential energy per unit charge in an electric field.
electromagnet (649): A magnet created when current flows through a wire coil.
electromagnetic induction (672): The process of generating a current through a circuit due to the relative motion between a wire and a magnetic field when the wire is moved through the magnetic field or the magnetic field moves past the wire.
electromagnetic radiation (710): Energy that is carried, or radiated, in the form of electromagnetic waves.
electromagnetic spectrum (709): The entire range of frequencies and wavelengths that make up all forms of electromagnetic radiation, including radio waves, microwaves, visible light, and X rays.
electromagnetic wave (706): Coupled, changing electric and magnetic field that travels through space.
electromotive force (673): The potential difference, measured in volts, given to the charges by a battery; is abbreviated *EMF*.
electron cloud (761) The region in which there is a high probability of finding an electron.
electroscope (546): A device that is used to detect electric charges and consists of a metal knob connected by a metal stem to two thin metal leaves.
electrostatics (541): The study of electric charges that can be collected and held in one place.
elementary charge (550): The magnitude of the charge of an electron.
emission spectrum (724): A plot of the intensity of light emitted from a hot body over a range of frequencies.
energy (258): The ability of an object to produce a change in itself or in the world around it.
energy level (753): The quantized amount of energy that an atom may have in each level.
entropy (328): A measure of the disorder in a system.
equilibrant (131): A force that places an object in equilibrium; is the same magnitude as the resultant, but opposite in direction.
equilibrium (95): The condition in which the net force on an object is zero.
equipotential (570): The electric potential difference of zero between two or more positions in an electric field.
equivalent resistance (619): For resistors in a series, is the sum of all the individual resistances.
excited state (753): Any energy level of an atom that is higher than its ground state.
extrinsic semiconductors (781): Semiconductors with greatly enhanced conductivity resulting from the addition of dopants.

farsightedness (501): A vision defect in which a person cannot see close objects clearly because images are focused behind the retina; can be corrected with a convex lens.
first law of thermodynamics (326): States that the change in thermal energy of an object is equal to the heat that is added to the object, minus the work done by the object.
first right-hand rule (648): A method used to determine the direction of a magnetic field relative to the direction of conventional current.
fission (811): The process in which a nucleus is divided into two or more fragments, and neutrons and energy are released.
fluids (342): Materials (liquids and gases) that flow and have no definite shape of their own.
focal length (464): The position of the focal point with respect to the mirror along the principal axis.
focal point (464): The point where incident light rays that are parallel to the principal axis converge after reflecting from the mirror.
force (88): A push or pull exerted on an object that causes a change in motion; has both direction and magnitude and may be a contact or a field force.
force carriers (818): Particles that transmit, or carry forces between matter.
fourth right-hand rule (672): A method used to determine the direction of the forces on the charges in a conductor that is moving in a magnetic field.
free-body diagram (89): A physical model that represents the forces acting on a system.
free fall (72): The motion of a body when air resistance is negligible and the motion can be considered due to the force of gravity alone.
frequency (384): The number of complete oscillations that a wave makes each second; is measured in hertz, Hz.
fundamental (417): For a musical instrument, the lowest frequency of sound that will resonate.
fuse (627): A short piece of metal that acts as a safety device in an electric circuit by melting and stopping the current from flowing if a dangerously high current passes through the circuit.
fusion (813): The process in which nuclei with small masses combine to form a nucleus with a larger mass and energy is released.

galvanometer (655): A device that is used to measure very small currents; can be used as a voltmeter or an ammeter.
gamma decay (806): The radioactive decay process in which there is a redistribution of energy within the nucleus, but no change in atomic mass or charge.
gravitational field (183): The field that surrounds any object with mass; equals the universal gravitational constant, times the mass of the object, divided by the square of the distance from the object's center.

gravitational force (175): The attractive force between two objects that is directly proportional to the mass of the objects.

gravitational mass (184): The size of the gravitational force between two objects.

gravitational potential energy (288): The stored energy in a system resulting from the gravitational force between Earth and the object.

ground-fault interrupter (627): A device that contains an electronic circuit that detects small current differences caused by an extra current path; it opens the circuit, prevents electrocution, and often is required as a safety measure for bathroom, kitchen, and exterior outlets.

grounding (548): The process of removing excess charge by touching an object to Earth.

ground state (753): State of an atom with the smallest allowable amount of energy.

half-life (809): The time required for half the atoms in a given quantity of a radioactive isotope to decay.

harmonics (417): Higher frequencies, which are odd-numbered multiples of the fundamental frequency; give certain musical instruments their own unique timbre.

heat (317): Energy transferred between two objects in contact with one another and always flows from the hotter object to the cooler object.

heat engine (326): A device that continuously converts thermal energy to mechanical energy; requires a high-temperature thermal energy source, a low-temperature receptacle (a sink), and a way to convert the thermal energy into work.

heat of fusion (324): The amount of heat required to change 1 kg of a substance from a solid state to a liquid state at its melting point.

heat of vaporization (324): The amount of heat required to change 1 kg of a substance from a liquid state to a gaseous state at its boiling point.

Heisenberg uncertainty principle (737): States that it is impossible to simultaneously measure the exact position and momentum of a particle of light or matter.

Hooke's law (376): States that the force acting on a spring is directly proportional to the amount that the spring is stretched.

hypothesis (8): An educated, testable guess about how variables are related.

ideal gas law (345): For an ideal gas, the pressure times the volume is equal to the number of moles, times the constant, *R*, and the Kelvin temperature; predicts the behavior of gases remarkably well unless under high-pressure or low-temperature conditions.

ideal mechanical advantage (267): For an ideal machine, is equal to the displacement of the effort force, divided by displacement of the load.

illuminance (433): The rate at which light strikes a surface, or falls on a unit area; is measured in lumens per square meter, lm/m^2, or lux, lx.

illuminated source (432): An object, such as the Moon, that becomes visible as a result of the light reflecting off it.

image (461): The combination of image points in a plane mirror from which the reflected object seems to originate.

impulse (230): The product of the average net force on an object and the time interval over which the force acts.

impulse-momentum theorem (230): States that the impulse on an object equals the object's final momentum minus the object's initial momentum.

incident wave (387): A wave that strikes a boundary between two media.

incoherent light (515): Light with unsynchronized wave fronts that illuminates objects with an even, white light. (761): Light whose waves are out of step, with their maxima and minima not coinciding.

independent variable (15): The factor that is changed or manipulated during an experiment.

index of refraction (486): For a medium, is the ratio of the speed of light in a vacuum to the speed of light in that medium.

inelastic collision (298): A type of collision in which the kinetic energy after the collision is less than the kinetic energy before the collision.

inertia (95): The tendency of an object to resist change.

inertial mass (183): A measure of an object's resistance to any type of force.

instantaneous acceleration (59): The change in an object's velocity at a specific instant of time.

instantaneous position (40): The position of an object at any particular instant in time.

instantaneous velocity (46): A measure of motion that tells the speed and direction of an object at a specific instant in time.

insulator (544): A material, such as glass, through which a charge will not move easily.

interaction pair (102): A pair of forces that are equal in strength, but opposite in direction.

interference (388): Results from the superposition of two or more waves; can be constructive (wave displacements in the same direction) or destructive (waves with equal but opposite amplitudes).

interference fringes (516): A pattern of light and dark bands on a screen, resulting from the constructive and destructive interference of light waves passing through two narrow, closely spaced slits in a barrier.

intrinsic semiconductors (780): Pure semiconductors that conduct charge as a result of thermally freed electrons and holes.

inverse relationship (18): A hyperbolic relationship that exists when one variable depends on the inverse of the other variable.

isolated system (237): A closed system on which the net external force is zero.

isotope (701): Each of the differing forms of the same atom that have different masses, but have the same chemical properties.

joule (259): Unit of energy, J; 1 J of work is done when a force of 1 N acts on an object over a displacement of 1 m.

Kepler's first law (172): States that the planets move in elliptical paths, with the Sun at one focus.

Kepler's second law (173): States that an imaginary line from the Sun to a planet sweeps out equal areas in equal time intervals.

Kepler's third law (173): States that the square of the ratio of the periods of any two planets is equal to the cube of the ratio of their average distances from the Sun.

kilowatt-hour (604): An energy unit used by electric companies to measure energy sales; 1 kWh is equal to 1000 watts, W, delivered continuously for 3600 s (1 h).

kinetic energy (258): The energy of an object, resulting from its motion.

kinetic friction (126): The force exerted on one surface by a second surface when the two surfaces rub against one another because one or both of the surfaces are moving.

laser (762): A device that produces powerful, coherent, directional, monochromatic light that can be used to excite other atoms; the acronym stands for *l*ight *a*mplification by *s*timulated *e*mission of *r*adiation.

law of conservation of angular momentum (243): States that if there are no net external torques on an object, then its angular momentum is conserved.

law of conservation of energy (293): States that in a closed, isolated system, energy is not created or destroyed, but rather, is conserved.

law of conservation of momentum (237): States that the momentum of any closed, isolated system does not change.

law of reflection (391): States that the angle of incidence is equal to the angle of reflection. (458): States that the angle a reflected ray makes, as measured from the normal to a reflective surface, equals the angle the incident ray makes, as measured from the same normal.

law of universal gravitation (175): States that gravitational force between any two objects is directly proportional to the product of their masses and inversely proportional to the square of the distance between their centers.

lens (493): A piece of transparent material, such as glass or plastic, that is used to focus light and form an image.

Lenz's law (679): States that an induced current always is produced in a direction such that the magnetic field resulting from the induced current opposes the change in the magnetic field that is causing the induced current.

leptons (818): Any of a family of particles like electrons and neutrinos.

lever arm (201): The perpendicular distance from the axis of rotation to the point where force is exerted.

linear relationship (16): A type of relationship that exists between two variables whose graphed data points lie on a straight line.

line of best fit (15): A line that best passes through or near graphed data; used to describe data and predict where new data will appear on the graph.

longitudinal wave (381): A mechanical wave in which the disturbance is in the same direction, or parallel to, the direction of wave motion.

loudness (406): Sound intensity as sensed by the ear and interpreted by the brain; depends mainly on the pressure wave's amplitude.

luminous flux (433): The rate at which light energy is emitted from a luminous source; is measured in lumens, lm.

luminous source (432): An object, such as the Sun or an incandescent lamp, that emits light.

machine (266): A tool that makes work easier (but does not change the amount of work) by changing the magnitude or the direction of the force exerted to do work.

magnetic field (645): The area around a magnet, or around any current-carrying wire or coil of wire, where a magnetic force exists.

magnetic flux (646): The number of magnetic field lines that pass through a surface.

magnification (468): The amount that an image is enlarged or reduced in size, relative to the object.

magnitude (35): A measure of size.

Malus's law (444): States that the intensity of light coming out of a second polarizing filter equals the intensity of polarized light coming out of a first polarizing filter, times the cosine, squared, of the angle between the polarizing axes of the two filters.

mass defect (802): The difference between the sum of the masses of individual nucleons and the actual mass.

mass number (800): The sum of the numbers of neutrons and protons in an atom's nucleus.

mass spectrometer (701): Device that uses both electric and magnetic fields to measure the masses of ionized atoms and molecules and can determine the charge-to-mass ratio of an ion.

measurement (11): A comparison between an unknown quantity and a standard.

mechanical advantage (266): The ratio of resistance force to effort force.

mechanical energy (293): The sum of kinetic and gravitational potential energy of a system.

microchip (788): An integrated circuit consisting of thousands of diodes, transistors, resistors, and conductors.

mirror equation (467): Relates the focal length, the object position, and the image position of a spherical mirror.

moment of inertia (205): The resistance to rotation.

momentum (230): The product of the object's mass and the object's velocity; is measured in kg·m/s.

monochromatic light (516): Light having only one wavelength.

motion diagram (33): A series of images showing the positions of a moving object taken at regular (equal) time intervals.

mutual inductance (682): Effect in which a changing current through a transformer's primary coil creates a changing magnetic field, which is carried through the core to the transformer's secondary coil, where the changing field induces a varying *EMF*.

nearsightedness (501): A vision defect in which a person cannot see distant objects clearly because images are focused in front of the retina; can be corrected with a concave lens.

net force (92): The vector sum of all the forces on an object.

neutral (543): An atom whose positively charged nucleus exactly balances the negative charge of the surrounding electrons.

Newton's first law (94): States that an object at rest will remain at rest, and a moving object will continue moving in a straight line with constant speed, if and only if the net force acting on that object is zero.

Newton's second law (93): States that the acceleration of an object is proportional to the net force and inversely proportional to the mass of the object being accelerated.

Newton's second law for rotational motion (208): States that the angular acceleration of an object is directly proportional to the net torque and inversely proportional to the moment of inertia.

Newton's third law (102): States that all forces come in pairs and that the two forces in a pair act on different objects and are equal in strength and opposite in direction.

node (389): The stationary point where two equal wave pulses meet and are in the same location, having a displacement of zero.

normal (391): The line in a ray diagram that shows the direction of the barrier and is drawn at a right angle, or perpendicular, to the barrier.

normal force (107): The perpendicular contact force exerted by a surface on another object.

nuclear reaction (807): Occurs when the number of neutrons or protons in a nucleus changes; can be produced by bombardment of nuclei by gamma rays, other nuclei, protons, neutrons, alpha particles, or electrons.

nucleons (802): Protons and neutrons.

nucleus (749): The tiny, massive, positively charged central core of an atom.

nuclide (801): The nucleus of an isotope.

object (461): A luminous or illuminated source of light rays that are to be reflected by a mirrored surface.

observed light frequency (446): The frequency of light as seen by an observer.

opaque (433): A medium that absorbs light and reflects some light rather than transmitting it, preventing objects from being seen through it.

open-pipe resonator (412): A resonating tube with both ends open that also will resonate with a sound source; its resonant frequencies are whole-number multiples of the fundamental.

origin (34): The point at which both variables in a coordinate system have the value zero.

pair production (820): The conversion of energy into a matter-antimatter pair of particles.

parallel circuit (623): A type of electric circuit in which there are several current paths; its total current is equal to the sum of the currents in the individual branches, and if any branch is opened, the current in the other branches remains unchanged.

parallel connection (600): A type of connection in which the circuit component and the voltmeter are aligned parallel to one another in the circuit, the potential difference across the voltmeter equals the potential difference across the circuit element, and there are two or more current paths to follow.

particle model (33): A simplified version of a motion diagram in which the moving object is replaced by a series of single points.

pascal (342): The SI unit of pressure.

Pascal's principle (352): States that any change in pressure applied at any point on a confined fluid is transmitted undiminished throughout the fluid.

pendulum (378): A device that can demonstrate simple harmonic motion when its bob (a massive ball or weight), suspended by a string or light rod, is pulled to one side and released, causing it to swing back and forth.

period (375): In any periodic motion, the amount of time required for an object to repeat one complete cycle of motion.

periodic motion (375): Any motion that repeats in a regular cycle.

periodic wave (381): A mechanical wave that moves up and down at the same rate.

photoelectric effect (726): The emission of electrons by certain metals that is produced when they are exposed to electromagnetic radiation.

photon (727): A discrete, quantized bundle of radiation that travels at the speed of light, has zero mass, and has energy and momentum.

physics (3): The study of matter and energy and their relationships.

piezoelectricity (711): The property of a crystal that causes it to bend or deform, producing electric vibrations, when a voltage is applied across it.

pitch (406): The highness or lowness of a sound wave, which depends on the frequency of vibration.

plane mirror (461): A flat, smooth surface from which light is reflected by regular reflection, producing a virtual image that is the same size as the object, has the same orientation, and is the same distance from the mirror as the object.

plasma (348): A gaslike, fluid state of matter made up of negatively charged electrons and positively charged ions that can conduct electricity; makes up most of the matter in the universe, such as stars.

polarization (443): Light whose waves oscillate only in a single plane. (644): For a magnet, describes the property of having two distinct, opposite ends, one of which is a north-seeking pole and the other of which is a south-seeking pole.

position (34): The separation between an object and the origin; it can be either positive or negative.

position-time graph (38): A graph that can be used to determine an object's velocity and position, as well as where and when two objects meet, by plotting the time data on a horizontal axis and the position data on a vertical axis.

power (263): The work done, divided by the time needed to do the work.

precision (12): A characteristic of a measured value describing the degree of exactness of a measurement.

pressure (342): The force on a surface, divided by the surface's area.

primary coil (682): An electrically insulated transformer coil, which, when connected to an AC voltage source, induces an alternating *EMF* in the secondary coil.

primary color (440): Red, green, and blue, which can be combined to form white light and mixed in pairs to produce the secondary colors: yellow, cyan, and magenta.

primary pigment (441): Cyan, magenta, and yellow, each of which absorbs one primary color from white light and reflects two primary colors; can be mixed in pairs to produce the secondary pigments: red, green, and blue.

principal axis (464): A straight line perpendicular to the surface of a mirror that divides the mirror in half.

principle of superposition (388): States that the displacement of a medium caused by two or more waves is the algebraic sum of the displacements of the individual waves.

principal quantum number (755): The integer, n, that determines the quantized values of r and E for an electron's orbital radius in hydrogen and the energy of a hydrogen atom—the radius, r, increases as the square of n, whereas the energy, E, depends on $1/n^2$.

projectile (147): An object shot through the air, such as a football, that has independent vertical and horizontal motions and, after receiving an initial thrust, travels through the air only under the force of gravity.

quadratic relationship (17): A parabolic relationship that results when one variable depends on the square of another variable.

quantized (725): Energy exists only in bundles of specific amounts.

quantum mechanics (761): The study of the properties of matter, using its wave properties.

quantum model (760): A model that predicts only the probability that an electron is in a particular region.

quarks (818): Tiny particles that make up protons, neutrons, and pions.

radian (197): $\frac{1}{2}\pi$ of a revolution; abbreviated "rad."

radiation (317): The thermal transfer of energy by electromagnetic waves through the vacuum of space.

radioactive (806): Materials that undergo radioactive decay and emit penetrating rays.

ray (390): A line that can show the direction a wave is traveling and is drawn at a right angle to a wave's crest.

ray model of light (432): A model that represents light as a ray that travels in a straight path, whose direction can be changed only by putting an obstruction in the path.

Rayleigh criterion (530): States that if the central bright spot of one image falls on the first dark ring of the second image, the images are at the limit of resolution.

real image (465): An inverted optical image that is smaller than the object and is formed by the converging of light rays.

receiver (712): A device used to obtain information from electromagnetic waves; consists of an antenna, a coil-and-capacitor circuit, a detector to decode the signal, and an amplifier.

reference level (288): The position where gravitational potential energy is defined as zero.

reflected wave (387): An erect or inverted returning wave that results from some of the energy of the incident wave's pulse being reflected backward.

refraction (391): The change in direction of waves at the boundary between two different media.

resistance (595): A property that determines how much current will flow; is equal to voltage divided by current.

resistance force (266): The force exerted by a machine.

resistor (596): A device with a specific resistance; may be made of long, thin wires; graphite; or semiconductors and often is used to control the current in circuits or parts of circuits.

resonance (380): A special form of simple harmonic motion that occurs when small forces are applied at regular intervals to an oscillating or vibrating object and the amplitude of the vibration increases.

resultant (35): A vector that results from the sum of two other vectors; it always points from the first vector's tail to the last vector's tip.

rotational kinetic energy (287): The kinetic energy of an object, proportional to the object's moment of inertia and the square of its angular velocity.

scalars (35): Quantities, such as temperature or distance, that are just numbers without any direction.

scientific law (9): A well-established rule about the natural world that sums up, but doesn't explain, a pattern in nature.

scientific method (8): A systematic method of observing, experimenting, and analyzing to answer questions about the natural world.

scientific theory (10): An explanation based on numerous observations, supported by experimental results, that may explain why things work the way they do.

secondary coil (682): An electrically insulated transformer coil in which an alternating *EMF* is induced by an AC current through the primary coil.

secondary color (440): Yellow, cyan, and magenta, each of which is produced by combining two primary colors.

secondary pigment (441): Red, green, and blue, each of which absorbs two primary colors from white light and reflects one primary color; can be produced by mixing pairs of cyan, magenta, and yellow pigments.

second law of thermodynamics (330): States that natural processes go in a direction that maintains or increases the total entropy of the universe.

second right-hand rule (649): A method used to determine the direction of the field produced by an electromagnet, relative to the flow of conventional current.

self-inductance (682): The inductance of *EMF* in a wire carrying a changing current.

semiconductors (775): Conductive materials, such as silicon or germanium, which, when made into solid-state devices, can amplify and control weak electric signals through electron movement within a tiny crystalline space.

series circuit (618): A type of electric circuit in which all current travels through each device and is the same everywhere; its current is equal to the potential difference divided by the equivalent resistance.

series connection (600): A type of connection in which there is only a single current path.

short circuit (627): Occurs when a very low resistance circuit is formed, causing a very large current that could easily start a fire from overheated wires.

significant digits (7): All the valid digits in a measurement, the number of which indicates the measurement's precision.

simple harmonic motion (375): A motion that occurs when the restoring force on an object is directly proportional to the object's displacement from equilibrium.

Snell's law of refraction (486): States that the product of the index of refraction of a medium and the sine of the angle of incidence equals the product of the index of refraction of a second medium and the sine of the angle of refraction.

solenoid (649): A long coil of wire with many loops; fields from each loop add to the fields of the other loops, creating a greater total field strength.

sound level (406): A logarithmic scale that measures amplitudes; depends on the ratio of the pressure variation of a particular sound wave to the pressure variation in the most faintly heard sound; unit of measurement is the decibel, dB.

sound wave (404): A pressure variation transmitted through matter as a longitudinal wave; it reflects and interferes and has frequency, wavelength, speed, and amplitude.

specific heat (318): The amount of energy that must be added to a material to raise the temperature of a unit mass by one temperature unit; is measured in J/kg·K.

specular reflection (459): A reflection produced by a smooth surface in which parallel light rays are reflected in parallel.

spherical aberration (467): The image defect of a spherical mirror that does not allow parallel light rays far from the principal axis to converge at the focal point, and produces an image that is fuzzy, not sharp.

Standard Model (818): A model of the building blocks of matter in which all particles can be grouped into three families: quarks, leptons, and force carriers.

standing wave (389): A wave that appears to be standing still, produced by the interference of two traveling waves moving in opposite directions.

static friction (126): The force exerted on one surface by a second surface when there is no motion between the two surfaces.

step-down transformer (682): A type of transformer in which the voltage coming out of the transformer is smaller than the voltage put into the transformer.

step-up transformer (682): A type of transformer in which the secondary voltage coming out of the transformer is larger than the primary voltage going into the transformer.

stimulated emission (762): The process that occurs when an excited atom is struck by a photon having energy equal to the energy difference between the excited state and the ground state—the atom drops to the ground state and emits a photon with energy equal to the energy difference between the two states.

streamlines (358): Lines representing the flow of fluids around objects.

strong nuclear force (802): An attractive force that binds the nucleus together; is of the same strength between protons and protons, protons and neutrons, and neutrons and neutrons.

superconductor (603): A material with zero resistance that can conduct electricity without a loss of energy.

surface wave (382): A mechanical wave in which the particles move both parallel and perpendicular to the direction of wave motion.

tension (105): The specific name for the force exerted by a rope or a string.

terminal velocity (101): The constant velocity of an object that is reached when the drag force equals the force of gravity.

thermal energy (295): A measure of the internal motion of an object's particles.

thermal equilibrium (315): The state in which the rate of energy flow between two objects is equal and the objects are at the same temperature.

thermal expansion (347): A property of all forms of matter that causes the matter to expand, becoming less dense, when heated.

thin-film interference (520): A phenomenon in which a spectrum of colors is produced due to the constructive and destructive interference of light waves reflected in a thin film.

thin lens equation (493): States that the inverse of the focal length of a sperical lens equals the sum of the inverses of the image position and the object position.

third right-hand rule (652): A method that can be used to determine the direction of the force on a current-carrying wire in a magnetic field.

threshold frequency (726): The certain minimum value at or above which the frequency of radiation causes the ejection of electrons from a metal.

time interval (36): The difference between two times.

torque (202): A measure of how effectively a force causes rotation; is equal to the force times the lever arm.

total internal reflection (489): Occurs when light traveling through an area with a higher index of refraction to an area with a lower index of refraction hits a boundary at an angle that exceeds the critical angle and all light reflects back into the area with the higher index of refraction.

trajectory (147): The path of a projectile through space.

transformer (682): A device that can decrease or increase the voltages in AC circuits with relatively little energy loss.

transistor (787): A simple device made of doped semiconducting material that can act as an amplifier, converting a weak signal to a much stronger one.

translucent (433): A medium that transmits light and also can reflect a fraction of the light, but does not allow objects to be seen clearly through it.

transparent (433): A medium that transmits light and also can reflect a fraction of the light, allowing objects to be seen clearly through it.

transverse wave (381): A mechanical wave that vibrates perpendicular to the direction of the wave's motion.

trough (383): The low point of a wave.

uniform circular motion (153): The movement of an object or particle trajectory at a constant speed around a circle with a fixed radius.

vector resolution (122): The process of breaking a vector into its components.

vectors (35): Quantities, such as position, that have both magnitude and direction.

velocity-time graph (58): A graph that can be used to plot the velocity of an object versus time and to determine the sign of an object's acceleration.

virtual image (461): The image formed of diverging light rays; is always on the opposite side of the mirror from the object.

volt (569): The unit equal to one joule per coulomb, 1 J/C.

voltage divider (620): A series circuit that is used to produce a voltage source of desired magnitude from a higher-voltage battery; often is used with sensors such as photoresistors.

voltmeter (631): A high-resistance device used to measure the voltage drop across any portion or a combination of portions of a circuit and is connected in parallel with the part of the circuit being measured.

watt (263): Unit of power, W; 1 J of energy transferred in 1 s.

wave (381): A disturbance that carries energy through matter or space; transfers energy without transferring matter.

wave front (390): A line representing the crest of a wave in two dimensions that can show the wavelength, but not the amplitude, of the wave when drawn to scale.

wavelength (383): The shortest distance between points where the wave pattern repeats itself, such as from crest to crest or from trough to trough.

wave pulse (381): A single disturbance or pulse that travels through a medium.

weak nuclear force (821): A weak force acting in the nucleus in beta, β, decay.

weightlessness (98): An object's apparent weight of zero that results when there are no contact forces pushing up on the object.

work (258): The transfer of energy by mechanical means; is done when a constant force is exerted on an object in the direction of motion, times the object's displacement.

work-energy theorem (258): States that when work is done on an object, a change in kinetic energy occurs.

work function (731): The energy required to free the most weakly bound electron from a metal; measured by the threshold frequency in the photoelectric effect.

X ray (713): High-frequency electromagnetic wave emitted by rapidly accelerated electrons.

Index

Index

Index

Index

Index

COVER (t) Michael Kevin Daly/CORBIS, **(bl)**Don Farrell/Getty Images, **(br)**Volker Mohrke/CORBIS, **(inset)**Myron Jay Dorf/CORBIS; **2** NASA; **3** Horizons Companies; **4** FOXTROT ©1998 Bill Amend. Reprinted with permission of UNIVERSAL PRESS SYNDICATE. All rights reserved.; **5** Mark E. Gibson; **6** National Institute of Standards and Technology; **7** Horizons Companies, **8** Laura Sifferlin; **9 (t)**Tom Pantages, **(bc)**NASA; **(others)**Photo Researchers; **11** PhotoEdit; **12 13** Horizons Companies; **14** Bill Crouse; **19** Magma Photo News; **20** Benjamin Coifman; **22 (1)** Bettman/CORBIS, **(r)**Charles O' Rear/CORBIS **25** Laura Sifferlin; **26** Horizons Companies; **28** Princeton Plasma Physics Laboratory; **30** AFP/Corbis; **31** Horizons Companies; **32 (t)**Getty Images, **(others)**Hutchings Photography; **33 43** Hutchings Photography; **48** Horizons Companies; **50** National Institute of Standards and Technology; **56** Rob Tringali/SportsChrome; **57** Horizons Companies; **72** Richard Megna/Fundamental Photographs; **76** Horizons Companies; **82** Skip Peticolas/Fundamental Photographs; **83** Joel Bennett/Peter Arnold, Inc.; **86** Joe McBride/CORBIS; **87** Horizons Companies; **90** Matt Meadows; **92 96** Aaron Haupt; **105** Tim Fuller; **108** Laura Sifferlin; **113** Aaron Haupt; **118** CORBIS; **119 136** Horizons Companies; **138** Courtesy of Six Flags Amusement Park; **146** Gibson Stock Photography; **147** Horizons Companies; **154** Denis Charlet/AFP/Getty Images; **160** Horizons Companies; **168** Bill Aron/PhotoEdit; **170** Wally Pacholka/ Astropics.com; **172** Newberry Library/Stock Montage; **177** Courtesy of PASCO Scientific; **178** Photodisc/ Artbase; **180** Russ Underwood, Lockheed Martin Space Systems/NASA; **182** NASA; **183** Courtesy of PASCO Scientific; **184 (t)**Horizons Companies, **(b)**Ted Kinsman/Photo Researchers; **186** Horizons Companies; **188** NASA; **190** Barry Runk/Grant Heilman; **196** Paul L. Ruben; **197** Horizons Companies; **211 (t)**Richard Megna/Fundamental Photographs, **(others)**Hutchings Photography; **218** Horizons Companies; **223** Mark D. Phillips/AFP/CORBIS; **228** fotobankyokohama/firstlight.ca; **229** Horizons Companies; **231** Rick Fischer/Masterfile; **234** Tim Fuller; **235** Rick Stewart/Getty Images; **238** Laura Sifferlin; **239** NASA; **243** F. Scott Grant/IMAGE Communications; **245** file photo; **246** Horizons Companies; **256** Colorstock/Getty Images; **257** Horizons Companies; **259** Hutchings Photography; **263** Laura Sifferlin; **264** Warn Industries Inc.; **266** Hutchings Photography; **270** Laura Sifferlin; **274** Horizons Companies; **278** Hutchings Photography; **284** David Madison Sports Images; **285** Horizons Companies; **287** Ken Redmond/Ken Redmond Photography; **288** Hutchings Photography; **291** Luis Romero/AP Wide World Photos, **(r)**Getty Images; **292** Bob Daemmrich/The Image Works; **302 304** Horizons Companies, **312** CORBIS; **313** Horizons Companies; **315** Tom Pantages; **316 (r)**FPG/Getty Images, **(cl)**Getty Images, **(others)** CORBIS; **322 (l)**John Cancalosi/Peter Arnold, Inc., **(r)**Jenny Hager/The Image Works; **330 (t)**Doug Martin, **(others)**Richard Hutchings/CORBIS; **332** Horizons Companies; **340** Steve Kaufman/Peter Arnold, Inc.; **341** Horizons Companies; **342 (t)**Gerard Photography, **(b)**NASA's Goddard Space Flight Center; **344** Timothy O'Keefe/Bruce Coleman; **347** Gary Settles/ Pennsylvania State University; **348** Norman Tomalin/Bruce Coleman; **349** Frank Cezus; **350 (t)**Runk/Schoenberger/Grant Heilman Photography, **(others)**Matt Meadows; **351** Orville Andrews; **353** Ralph White/CORBIS; **357** Horizons Companies; **358** Andy Sacks/Getty Images; **359 (l)**Craig Kramer, **(r)**James L. Amos/Peter Arnold, Inc.; **360 (t)**file photo, **(b)**courtesy U.S. Department of Transportation; **363** Doug Martin; **364** Horizons Companies; **366** Courtesy JILA BEC Group/University of Colorado; **368 (l)**Richard Megna/Fundamental Photographs, **(r)**Tim Courlas; **374** Michael T. Sedam/CORBIS, **(inset)**Bettman/CORBIS; **375** Horizons Companies; **382 (t)**CORBIS, **(others)**Tom Pantages; **387 388** Tom Pantages; **389** Richard Megna/Fundamental Photographs; **390 (t)**Fundamental Photographs, **(b)**Runk/Shoenberger from Grant Heilman; **391** Tom Pantages; **392** Horizons Companies; **398** Bruce Leighty/Image Finders; **400** Darrell Gulin/CORBIS; **402** Tim Fuller; **403** Horizons Companies; **407 (t)**Willie Maldonado/Getty Images, **(b)**Ztek Co./University of Nebraska Lincoln; **410** Stephen Dalton/Animals Animals; **414** David Mechlin/Index Stock Images; **417 (l)**Geoff Butler, **(c)**HIRB/Index Stock Images, **(r)**Photodisc/Artbase; **420** Horizons Companies; **424 (t)**Tim Fuller, **(b)**Carolina Biological/Visuals Unlimited; **425** Getty Images; **427** Horizons Companies; **430** Getty Images; **431** Horizons Companies; **433 (l)**Andrew McKim/Masterfile, **(others)**Laura Sifferlin; **440 (t)**Kodak, **(b)**Matt Meadows; **441** Tom Pantages; **442 (t)**Laura Sifferlin, **(b)**file photo; **443** Tom Pantages; **444** Laura Sifferlin; **446** Maarten Schmidt; **448** Horizons Companies; **450 (tl)**Jerry Driendl/Getty Images, **(tr)**Getty Images, **(bl)**David Duran/Fundamental Photographs, **(br)**Burazin/Masterfile; **456** George Matchneer; **457** Horizons Companies; **459** Henry Leap/James Lehman; **461** Laura Sifferlin; **466** Lick Observatory; **470** Horizons Companies; **471** Stewart Weir/Eye Ubiquitous/CORBIS; **474** Horizons Companies; **478** Richard Megna/Fundamental Photographs; **479** Hutchings Photography; **481** CORBIS; **484** Ric Frazier/Masterfile; **485 486 488** Horizons Companies; **490** John M. Dunay IV/Fundamental Photographs; **491** David Parker/Photo Researchers; **492** Gloria H. Chomica/Masterfile; **493** Horizons Companies; **494** David Young-Wolff/PhotoEdit; **504** Horizons Companies; **506** NASA; **514** Adrienne Hart-Davis/Science Photo Library/Photo Researchers; **515** Horizons Companies; **516 520** Tom Pantages; **523** CORBIS; **524** Tom Pantages; **527 (t)**Central Scientific Company, **(bl)**Ron Tanaka/Artbase, **(br)**Mark Thayer; **528** Kodansha; **532** Horizons Companies; **540** Kent Wood/Photo Researchers; **541** Horizons Companies;

542 StudiOhio; **543** Tom Pantages; **544** Horizons Companies; **549** Jose Pessoa-Divisao de Documentacao fotografica do Instituto Portugues de Museus; **553** Mike McClure/Maxximages; **554** Horizons Companies; **556** NASA; **562 563** Horizons Companies; **567 (r)**Horizons Companies, **(l)**courtesy PASCO Scientific; **568 571** Kodansha; **576** Tim Courlas; **577 580 584, 586** Horizons Companies; **590** Lester Lefkowitz/CORBIS; **591** Horizons Companies; **601** Hutchings Photography; **603** Hans-Jurgen Burkard/Peter Arnold, Inc.; **604 (t)**Bischel Studios, **(b)**Hutchings Photography; **606** Horizons Companies; **616** Catherine Karnow/CORBIS, **(inset)**Horizons Companies; **617** Horizons Companies; **618** file photo; **620** Laura Sifferlin; **628 632 639** Horizons Companies; **642** Fermilab/Photo Researchers; **643 644** Horizons Companies; **645** Matt Meadows; **646** Horizons Companies; **647 (b)**Doug Martin, **(others)**Horizons Companies; **648** Kodansha; **660** Horizons Companies; **670** Tom Pantages, **(inset)**Robert Cameron/Getty Images; **671** Horizons Companies; **680** Matt Meadows; **681** Horizons Companies; **683** Horizons Companies; **685** Matt Meadows; **686** Horizons Companies; **696** Rick England/Getty Images; **697** Horizons Companies; **699** CERN, P. Loiez/Science Photo Library/Photo Researchers; **711** Bert Krages/Visuals Unlimited; **714** Horizons Companies; **722** Colorado School of Mines; **723 725 726** Horizons Companies; **729** Ruth Dixon; **734** Richard Megna/Fundamental Photographs; **735** Ztek Co./University of Nebraska Lincoln; **738** Horizons Companies; **746** Carl & Ann Purcell/CORBIS; **747** Horizons Companies; **749** Richard Megna/ Fundamental Photographs; **750 (t)**Wabash Instrument Corp./Fundamental Photographs, **(b)**Wabash Instrument Corp./Fundamental Photographs; **751** Images were created by Dr. Donald Mickey, University of Hawaii, Institute for Astronomy, from National Solar Observatory spectra data.; **759** Horizons Companies; **763** ImageState/Firstlight; **765 (t)**Getty Images, **(b)**Roger Ressmeyer/CORBIS; **766** Horizons Companies; **768** Wolfgang Ketterle/Massachusetts Institute of Technology; **774** Larry Hamill, **(inset)**Getty Images; **775** Horizons Companies; **783** Laura Sifferlin; **786** Getty Images; **788** Science Photo Library/Photo Researchers; **789 790** Horizons Companies; **792 (l)**NASA, **(r)**Sam Ogden/Photo Researchers; **798** Kay Chernush; **799** Horizons Companies; **811** CNRI/Photo Researchers; **812** Science VU/Visuals Unlimited; **815** Stanford Linear Accelerator Center/Photo Researchers; **816** Courtesy of Fermilab; **817 (t)**Science VU/Visuals Unlimited, **(bl)**David Parker/Photo Researchers, **(br)**Fermilab/Photo Researchers; **820** Science Photo Library/Photo Researchers; **823 (r)**NASA, **(others)**David Malin/Anglo-Australian Observatory; **824** Horizons Companies; **829** Fermilab/Peter Arnold, Inc.

PERIODIC TABLE OF THE ELEMENTS

Element — Hydrogen; Atomic number — 1; Symbol — **H**; Atomic mass — 1.008; State of matter

Gas, Liquid, Solid, Synthetic

Metal, Metalloid, Nonmetal, Recently observed

	1	2	3	4	5	6	7	8	9	10	11	12	13	14	15	16	17	18
1	Hydrogen 1 **H** 1.008																	Helium 2 **He** 4.003
2	Lithium 3 **Li** 6.941	Beryllium 4 **Be** 9.012											Boron 5 **B** 10.811	Carbon 6 **C** 12.011	Nitrogen 7 **N** 14.007	Oxygen 8 **O** 15.999	Fluorine 9 **F** 18.998	Neon 10 **Ne** 20.180
3	Sodium 11 **Na** 22.990	Magnesium 12 **Mg** 24.305											Aluminum 13 **Al** 26.982	Silicon 14 **Si** 28.086	Phosphorus 15 **P** 30.974	Sulfur 16 **S** 32.066	Chlorine 17 **Cl** 35.453	Argon 18 **Ar** 39.948
4	Potassium 19 **K** 39.098	Calcium 20 **Ca** 40.078	Scandium 21 **Sc** 44.956	Titanium 22 **Ti** 47.867	Vanadium 23 **V** 50.942	Chromium 24 **Cr** 51.996	Manganese 25 **Mn** 54.938	Iron 26 **Fe** 55.847	Cobalt 27 **Co** 58.933	Nickel 28 **Ni** 58.693	Copper 29 **Cu** 63.546	Zinc 30 **Zn** 65.39	Gallium 31 **Ga** 69.723	Germanium 32 **Ge** 72.61	Arsenic 33 **As** 74.922	Selenium 34 **Se** 78.96	Bromine 35 **Br** 79.904	Krypton 36 **Kr** 83.80
5	Rubidium 37 **Rb** 85.468	Strontium 38 **Sr** 87.62	Yttrium 39 **Y** 88.906	Zirconium 40 **Zr** 91.224	Niobium 41 **Nb** 92.906	Molybdenum 42 **Mo** 95.94	Technetium 43 **Tc** (98)	Ruthenium 44 **Ru** 101.07	Rhodium 45 **Rh** 102.906	Palladium 46 **Pd** 106.42	Silver 47 **Ag** 107.868	Cadmium 48 **Cd** 112.411	Indium 49 **In** 114.82	Tin 50 **Sn** 118.710	Antimony 51 **Sb** 121.757	Tellurium 52 **Te** 127.60	Iodine 53 **I** 126.904	Xenon 54 **Xe** 131.290
6	Cesium 55 **Cs** 132.905	Barium 56 **Ba** 137.327	Lanthanum 57 **La** 138.905	Hafnium 72 **Hf** 178.49	Tantalum 73 **Ta** 180.948	Tungsten 74 **W** 183.84	Rhenium 75 **Re** 186.207	Osmium 76 **Os** 190.23	Iridium 77 **Ir** 192.217	Platinum 78 **Pt** 195.08	Gold 79 **Au** 196.967	Mercury 80 **Hg** 200.59	Thallium 81 **Tl** 204.383	Lead 82 **Pb** 207.2	Bismuth 83 **Bi** 208.980	Polonium 84 **Po** 208.982	Astatine 85 **At** 209.987	Radon 86 **Rn** 222.018
7	Francium 87 **Fr** (223)	Radium 88 **Ra** (226)	Actinium 89 **Ac** (227)	Rutherfordium 104 **Rf** (261)	Dubnium 105 **Db** (262)	Seaborgium 106 **Sg** (266)	Bohrium 107 **Bh** (264)	Hassium 108 **Hs** (277)	Meitnerium 109 **Mt** (268)	Darmstadtium 110 **Ds** (281)	Roentgenium 111 **Rg** (272)	Ununbium * 112 **Uub** (285)	Ununtrium * 113 **Uut** (284)	Ununquadium * 114 **Uuq** (289)	Ununpentium * 115 **Uup** (288)	Ununhexium * 116 **Uuh** (291)		Ununoctium * 118 **Uuo** (294)

The number in parentheses is the mass number of the longest lived isotope for that element.

* The names and symbols for elements 112, 113, 114, 115, 116, and 118 are temporary. Final names will be selected when the elements' discoveries are verified.

Lanthanide series	Cerium 58 **Ce** 140.115	Praseodymium 59 **Pr** 140.908	Neodymium 60 **Nd** 144.242	Promethium 61 **Pm** (145)	Samarium 62 **Sm** 150.36	Europium 63 **Eu** 151.965	Gadolinium 64 **Gd** 157.25	Terbium 65 **Tb** 158.925	Dysprosium 66 **Dy** 162.50	Holmium 67 **Ho** 164.930	Erbium 68 **Er** 167.259	Thulium 69 **Tm** 168.934	Ytterbium 70 **Yb** 173.04	Lutetium 71 **Lu** 174.967
Actinide series	Thorium 90 **Th** 232.038	Protactinium 91 **Pa** 231.036	Uranium 92 **U** 238.029	Neptunium 93 **Np** (237)	Plutonium 94 **Pu** (244)	Americium 95 **Am** (243)	Curium 96 **Cm** (247)	Berkelium 97 **Bk** (247)	Californium 98 **Cf** (251)	Einsteinium 99 **Es** (252)	Fermium 100 **Fm** (257)	Mendelevium 101 **Md** (258)	Nobelium 102 **No** (259)	Lawrencium 103 **Lr** (262)

Physical Constants			
Quantity	**Symbol**	**Value**	**Approximate Value**
Atomic mass unit	u	$1.66053886 \times 10^{-27}$ kg	1.66×10^{-27} kg
Avogadro's number	N_A	6.0221415×10^{23} particles/mol	6.022×10^{23} particles/mol
Boltzmann's constant	k	$1.3806505 \times 10^{-23}$ Pa·m³/K	1.38×10^{-23} Pa·m³/K
Constant in Coulomb's law	K	$8.987551788 \times 10^{9}$ N·m²/C²	9.0×10^{9} N·m²/C²
Elementary charge	e	$1.60217653 \times 10^{-19}$ C	1.602×10^{-19} C
Gas constant	R	8.314472 Pa·m³/mol·K	8.31 Pa·m³/mol·K
Gravitational constant	G	6.6742×10^{-11} N·m²/kg²	6.67×10^{-11} N·m²/kg²
Mass of an electron	m_e	$9.1093826 \times 10^{-31}$ kg	9.11×10^{-31} kg
Mass of a proton	m_p	$1.67262171 \times 10^{-27}$ kg	1.67×10^{-27} kg
Mass of a neutron	m_n	$1.67492728 \times 10^{-27}$ kg	1.67×10^{-27} kg
Planck's constant	h	$6.6260693 \times 10^{-34}$ J·s	6.63×10^{-34} J·s
Speed of light in a vacuum	c	2.99792458×10^{8} m/s	3.0×10^{8} m/s

SI Derived Units				
Measurement	**Unit**	**Symbol**	**Expressed in Base Units**	**Expressed in Other SI Units**
Acceleration		m/s^2	m/s^2	
Area		m^2	m^2	
Capacitance	farad	F	$A^2 \cdot s^4/kg \cdot m^2$	
Density		kg/m^3	kg/m^3	
Electric charge	coulomb	C	A·s	
Electric field		N/C	$kg \cdot m/C \cdot s^2$	
Electric resistance	ohm	Ω	$kg \cdot m^2/A^2 \cdot s^3$	V/A
EMF	volt	V	$kg \cdot m^2/A \cdot s^3$	
Energy, work	joule	J	$kg \cdot m^2/s^2$	N·m
Force	newton	N	$kg \cdot m/s^2$	
Frequency	hertz	Hz	s^{-1}	
Illuminance	lux	lx	cd/m^2	
Magnetic field	tesla	T	$kg/A \cdot s^2$	N·s/C·m
Potential difference	volt	V	$kg \cdot m^2/A \cdot s^3$	W/A or J/C
Power	watt	W	$kg \cdot m^2/s^3$	J/s
Pressure	pascal	Pa	$kg/m \cdot s^2$	N/m^2
Velocity		m/s	m/s	
Volume		m^3	m^3	